Alcamo's FUNDAMENTALS OF Microbiology

EIGHTH EDITION

Jeffrey C. Pommerville
Professor of Biology and Microbiology
Glendale Community College
Glendale, Arizona

JONES AND BARTLETT PUBLISHERS
Sudbury, Massachusetts
BOSTON TORONTO LONDON SINGAPORE

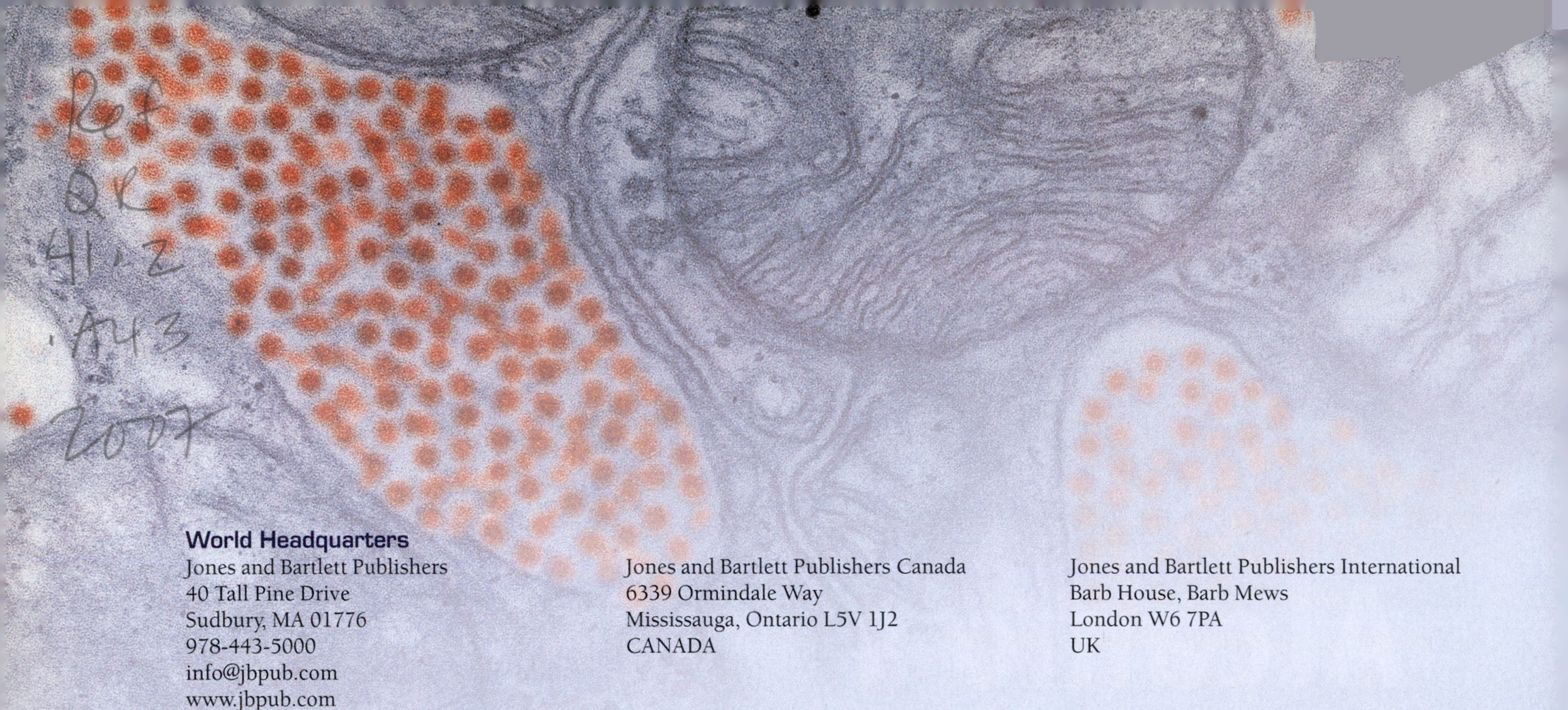

World Headquarters
Jones and Bartlett Publishers
40 Tall Pine Drive
Sudbury, MA 01776
978-443-5000
info@jbpub.com
www.jbpub.com

Jones and Bartlett Publishers Canada
6339 Ormindale Way
Mississauga, Ontario L5V 1J2
CANADA

Jones and Bartlett Publishers International
Barb House, Barb Mews
London W6 7PA
UK

Jones and Bartlett's books and products are available through most bookstores and online booksellers. To contact Jones and Bartlett Publishers directly, call 800-832-0034, fax 978-443-8000, or visit our website www.jbpub.com.
Substantial discounts on bulk quantities of Jones and Bartlett's publications are available to corporations, professional associations, and other qualified organizations. For details and specific discount information, contact the special sales department at Jones and Bartlett via the above contact information or send an email to specialsales@jbpub.com.

Production Credits

Chief Executive Officer: Clayton Jones
Chief Operating Officer: Don W. Jones, Jr.
President, Higher Education and Professional Publishing: Robert W. Holland, Jr.
V.P., Design and Production: Anne Spencer
V.P., Sales and Marketing: William Kane
V.P., Manufacturing and Inventory Control: Therese Connell
Acquisitions Editor, Science: Cathleen Sether
Managing Editor, Science: Dean W. DeChambeau
Editorial Assitant: Molly Steinbach
Senior Production Editor: Louis C. Bruno, Jr.
Marketing Manager: Andrea DeFronzo
Text and Cover Design: Anne Spencer
Illustrations: Elizabeth Morales
Associate Photo Researcher and Photographer: Christine McKeen
Photo Research Manager and Photographer: Kimberly L. Potvin
Composition: Shepherd, Inc.
Printing and Binding: Courier Kendallville
Cover Printing: Courier Kendallville
Cover Photo: © Dr. Dennis Kunkel/Visuals Unlimited

About the cover: A false-color scanning electron microscope view of a plastic cutting board surface. Rod-shaped and spherical bacterial cells are visible along with fungal hyphae with textured spores.

Library of Congress Cataloging-in-Publication Data

Pommerville, Jeffrey C.
Alcamo's fundamentals of microbiology / Jeffrey C. Pommerville. — 8th ed.
p. ; cm.
Includes bibliographical references and index.
ISBN-13: 978-0-7637-3762-7 (alk. paper)
ISBN-10: 0-7637-3762-3
1. Microbiology. 2. Medical microbiology. I. Alcamo, I. Edward. II. Title. III. Title: Fundamentals of microbiology.
[DNLM: 1. Microbiology. QW 4 P787a 2006]
QR41.2.A43 2006
579—dc22 2006010788
6048

Printed in the United States of America
10 09 08 07 06 10 9 8 7 6 5 4 3 2 1

Brief Contents

Contents

Courtesy of the CDC.

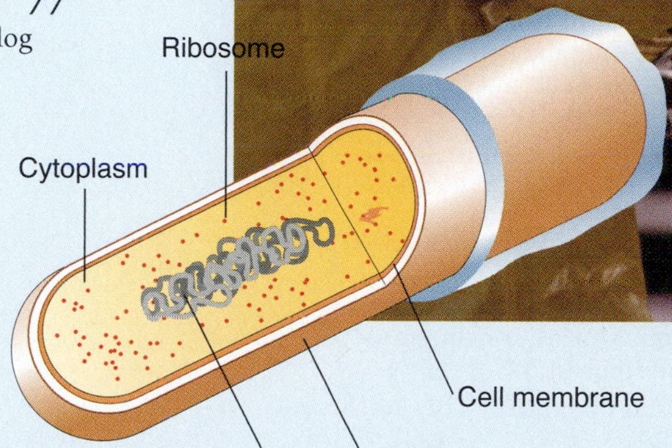

Ribosome
Cytoplasm
Cell membrane
Cell wall
DNA (chromosome)

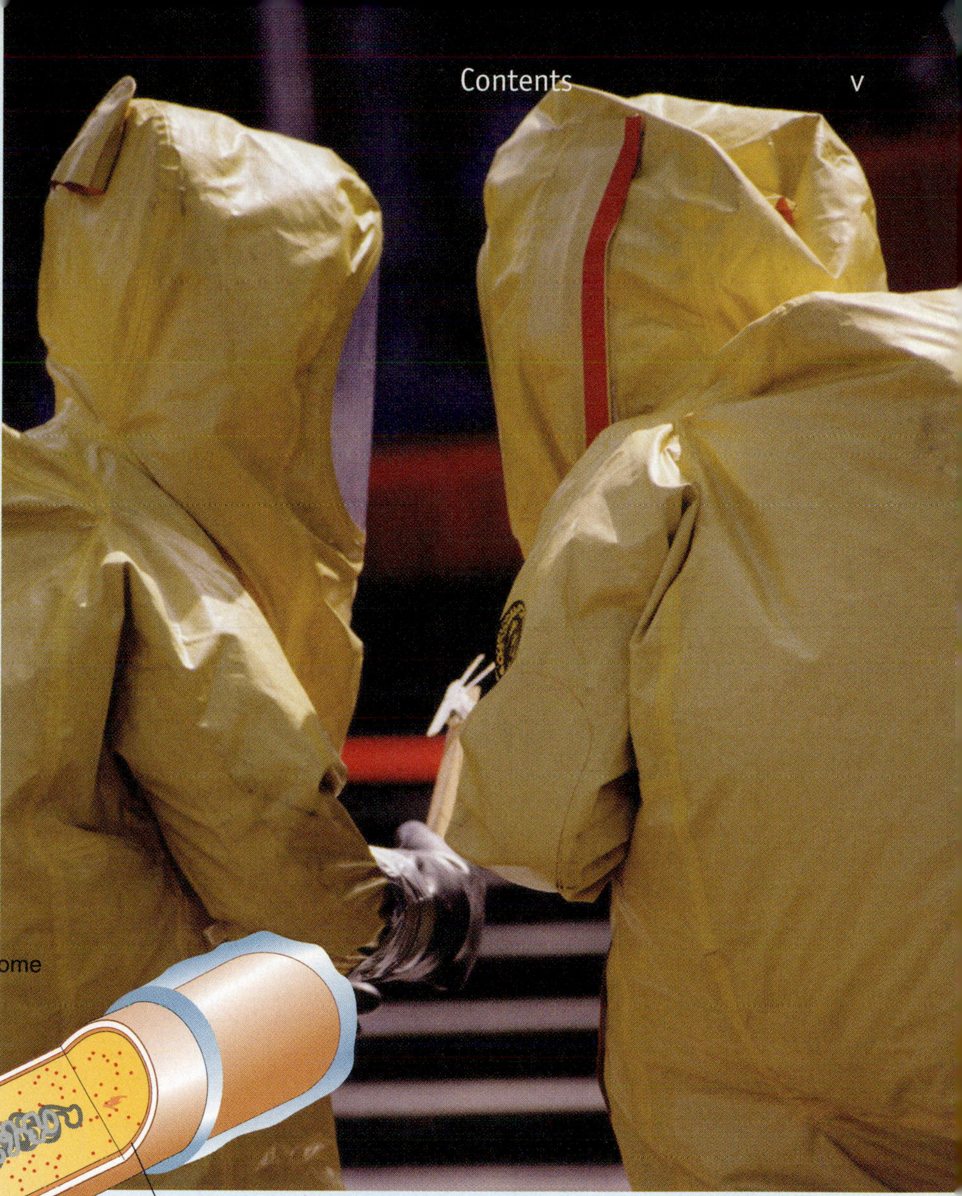

Courtesy of Roger Burks (University of California at Riverside), Mark Schneegurt (Wichita State University), and Cyanosite (www-cyanosite.bio.purdue.edu).

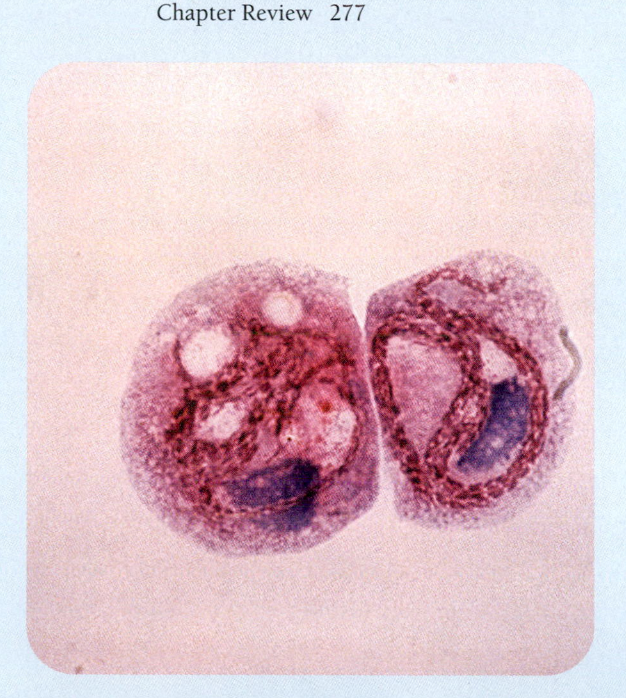

Courtesy of Don Howard/CDC.

Courtesy of Greg Knobloch/CDC.

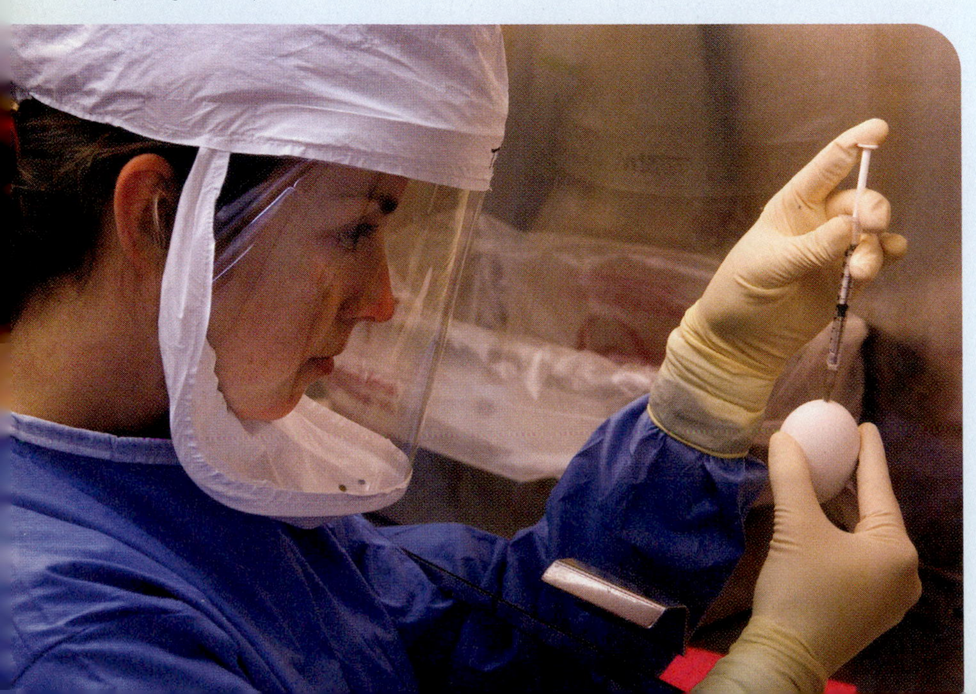

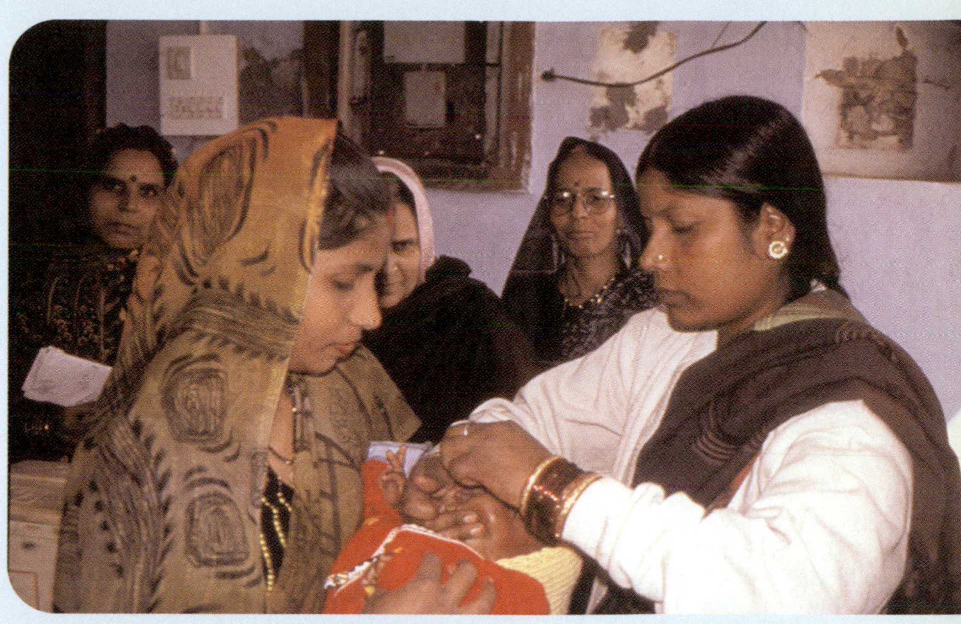

Courtesy of Chris Zahniser/CDC.

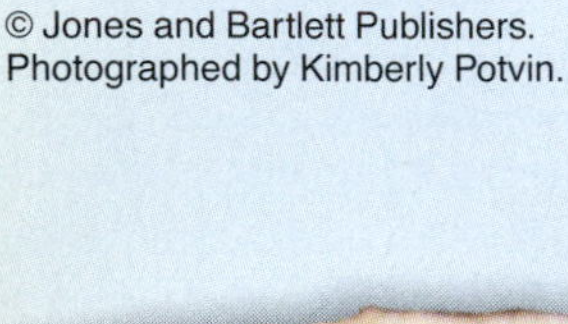

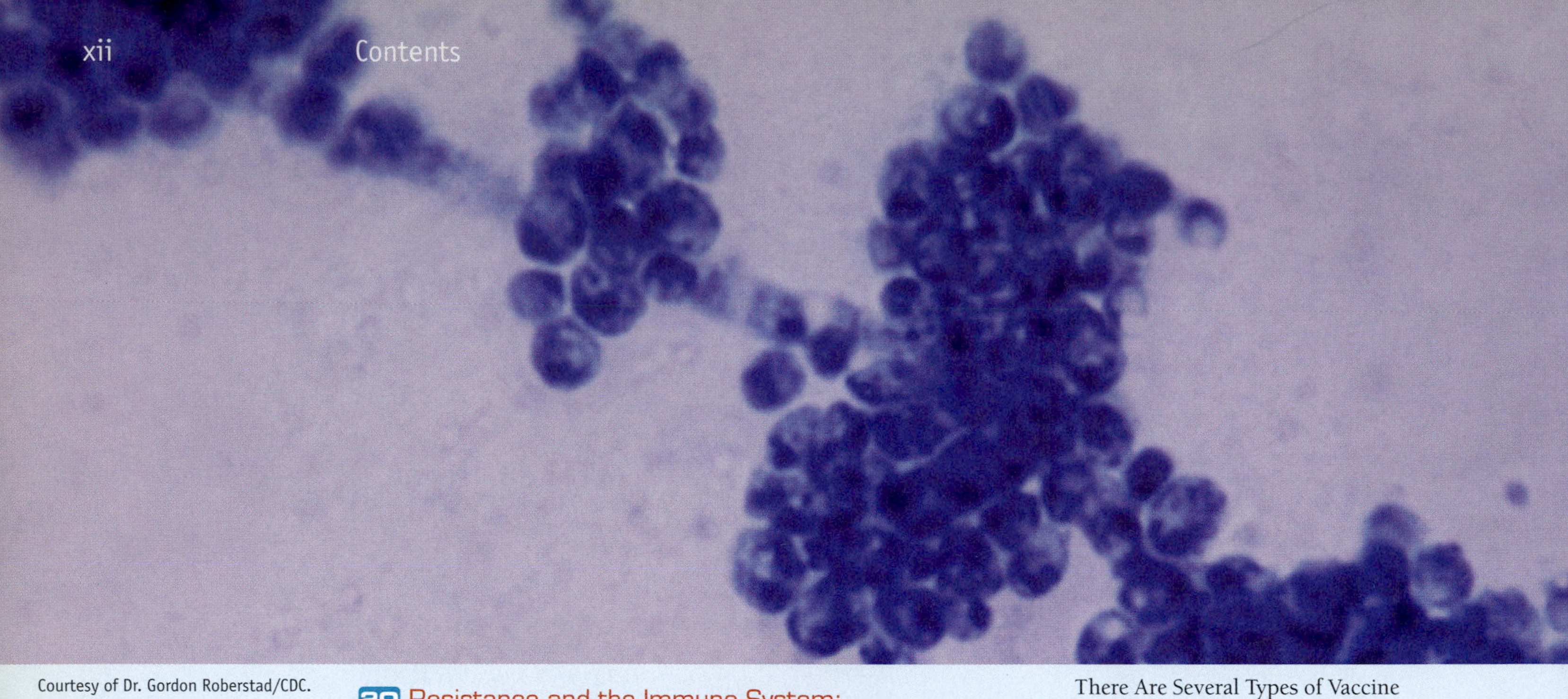

Courtesy of Dr. Gordon Roberstad/CDC.

Preface

THE IMPORTANCE OF THIS NEW EDITION

As the author of *Alcamo's Fundamentals of Microbiology*, this edition marks my second revision since the passing of Ed Alcamo in 2002. Most postsecondary textbooks, including those in the sciences, on average have a new edition published every three years. For this and several other reasons, the Government Accountability Office (GAO) published in 2004 a report, which, among other matters, evaluated the need for many textbooks to be revised so frequently. Therefore, is this revision necessary? The GAO study recommended textbook revisions be delayed until there are substantial changes in content or teaching methods to merit revision. If one were inclined to follow the GAO assessment—and more importantly the opinions and suggestions of many college and university instructors—the time is right for *Alcamo's Fundamentals of Microbiology, Eighth Edition*. Changes in content in this edition reflect the dynamic nature of microbiology, and the integration of new teaching pedagogies signal new approaches to facilitating learning and understanding in today's students.

In terms of content, microbiology is one of the most rapidly advancing scientific disciplines. Even though many of these discoveries may be beyond the technical or practical needs of the students for which this text is designed, many are directly applicable. Anxiety over a possible human pandemic evolving from the avian flu, alarm over the mumps outbreak in the Midwest in the spring of 2006, and health concerns over the 2006 outbreak of *E. coli* in packaged spinach illustrate a few reasons to understand microbiology and its practical applications. As in the previous editions, *Alcamo's Fundamentals of Microbiology, Eighth Edition*, retains the conversational style Ed developed over many editions and continues to provide you, the student, with up-to-date content so you can read—and evaluate—newspaper or magazine articles about microorganisms with understanding, or, as a future health giver, talk to patients with confidence about infectious diseases.

To perform well in your microbiology course requires a mastery of a substantial amount of new information. To facilitate this understanding and coordinate it with class material, I and the publishing group at Jones and Bartlett have developed a new "learning design" format for the textbook (described below) to make reading easier and more efficient. Most importantly, the design allows you to evaluate better your learning and provides you with the needed tools to probe your understanding; that is, chapter learning aids and assessment drills to evaluate your progress. Realize, a prepared student knows her or his mastery *before* an exam—not *as a result* of the exam!

AUDIENCE

Alcamo's Fundamentals of Microbiology, Eighth Edition, is written for introductory microbiology courses having an emphasis on the biology of human disease. It is geared toward students in health and allied health science curricula such as nursing, dental hygiene, medical assistance, sanitary science, and medical laboratory technology. It also will be an asset to students studying food science, agriculture, environmental science, and health administration. In addition, the text provides a firm foundation for advanced programs in biological sciences, as well as medicine, dentistry, and other health professions.

ORGANIZATION

Alcamo's Fundamentals of Microbiology, Eighth Edition, is divided into seven major areas of concentration. These areas use basic principles as frameworks to provide the unity and diversity of microbiology. Among the principles explored are the variations in structure and growth of microorganisms, the basis for infectious disease and resistance, and the beneficial effects microorganisms have on our lives.

Part 1 deals with the foundations of microbiology. It includes chapters on the origins of microbiology and the universal concepts that underpin the science. Part 2 then concentrates on the *Bacteria*, the microorganisms most of us think of when we say "germs." The discussions carry over to Part 3, where the spectrum of bacterial diseases is surveyed. Part 4 looks at the significance of other microorganisms, including viruses, fungi, and the protozoal and multicellular parasites.

In Part 5 of the text, the emphasis turns to infectious disease and the body's resistance through the immune system. Here, we study the reasons for disease and the means for surviving it. A logical segue is in Part 6, which presents various mechanisms for controlling microorganisms using physical and chemical agents such as heat, radiation, disinfectants, and antibiotics. Part 7 closes the text with brief discussions of how public health measures interrupt epidemics. Some key insights also are given on the positive effects microorganisms exert through biotechnology.

WHAT'S NEW

As mentioned, this edition represents a major make over, a new "learning design" for *Alcamo's Fundamentals of Microbiology*. In addition, you will see an added emphasis on a global perspective to infection and disease and more emphasis on taxonomy. Besides a detailed update to microbial structure and function, disease information and statistics, and the immune system, a few chapters have been reorganized; however, the sequence and number of chapters remains intact.

■ Rems (Roentgen Equivalent Man): A measure of radiation dose related to biological effect.

Marginal Definition

Chapter Organization

Chapter 4 on bacterial structure has been expanded with new information on topics such as the bacterial cytoskeleton. This has necessitated moving bacterial growth into its own chapter (Chapter 5). In Part 4, the chapters on protozoa and multicellular parasites have been combined into one chapter entitled "The Eukaryotic Parasites" (Chapter 17). Part 5 includes the basic material on the immune system, which has been separated into two chapters. Chapter 19 investigates innate immunity while a separate chapter (Chapter 20) now is devoted solely to acquired immunity. The immune system is a complex concept to understand, so two separate chapters better partitions this essential information. Finally, in Part 6, what were separate chapters for the physical and chemical control of microorganisms has been consolidated into one integrated chapter (Chapter 23).

The "Learning Design" Concept

The text format has been improved for reading and learning, and includes activities designed to encourage student interaction and assessment. These design elements form an integrated study and learning package (learning tools) for student understanding and assessment.

- **Chapter Introductions** provide a stimulating thought or historical perspective to set the tone for the chapter.
- **Key Concepts** present statements identifying the important concepts in the upcoming section and alert you to the significance of that written material.
- **Boldface Terms** highlight important terms and ideas in the text.
- **Marginal Definitions** present succinct definitions of notable terms as they enter the discussion.
- **Marginal Drawings** provide visual images of bacterial shapes and cell arrangements (Chapters 9–12) and eukaryotic cells (Chapters 16 and 17).
- **Marginal Chemical Structures** present structural formulas for many of the antimicrobial drugs described in Chapter 24.
- **Concept and Reasoning Checks** allow you to pause and either summarize the information presented in the previous section or critically reason through a question pertaining to the previous section.
- **MicroFocus Boxes** explore interesting topics concerning microbiology or microorganisms.
- **MicroInquiry Boxes** allow you to investigate (usually interactively) some important aspect of the chapter being studied.

- **Textbook Cases** are embedded in the disease chapters to help you understand pathogens by presenting contemporary disease outbreaks originally reported by the Centers for Disease Control and Prevention.
- **Figure Questions** (Chapters 2–27) further reinforce your understanding of microbiology concepts described in the text.
- **Summary Tables** pull together the similarities and differences of topics discussed in the chapter.
- **Pronouncing Microorganism Names** (inside front and back covers) helps you correctly pronounce those sometimes tongue twisting microorganism names.
- **Summaries of Key Concepts** condense the major ideas discussed in the chapter. The "learning design" package also includes many useful and important end-of-chapter student assessments.
- **Learning Objectives** outline the important concepts in the chapters through Bloom's Taxonomy, a classification of levels of intellectual skills important in learning.
- **Self-Test** questions are multiple-choice questions focusing on concrete "facts" learned in the chapter. Let's face it, there is information that needs to be memorized in order to reason critically.
- **Questions for Thought and Discussion** encourage students to use the text to resolve thought-provoking problems with contemporary relevance.
- **Applications** are questions requiring students to reason critically through a problem of practical significance.
- **Chapter Review** contains questions of a somewhat unconventional type to assist review of the chapter contents.

MICROFOCUS 5.2: Being Skeptical

Germination of 25 Million-Year-Old Endospores?

Endospores have been recovered and germinated from various archaeological sites and environments. Living spores have been recovered and germinated from the intestines of Egyptian mummies several thousand years old. In 1983, archaeologists found viable spores in sediment lining Minnesota's Elk Lake. The sediment was over 7,500 years old.

All these reports though pale in comparison to the controversial discovery reported in 1996 by researcher Raul Cano of California Polytechnic State University, San Louis Obispo. Cano found bacterial spores in the stomach of a fossilized bee trapped in amber—a hardened resin—produced from a tree in the Dominican Republic. The fossilized bee was about 25 million-years-old. When the amber was cracked open and the material from the abdomen of the bee extracted and placed in nutrient medium, the equally ancient spores germinated. With microscopy, the cells from a colony were very similar to *Bacillus sphaericus,* which is found today in bees in the Dominican Republic. Is it possible for an endospore to survive for 25 million years—even if it is encased in amber?

Critics were quick to claim the bacterial species may represent a modern-day species that contaminated the amber sample being examined. However, Professor Cano had carried out appropriate and rigorous decontamination procedures and sterilized the amber sample before cracking it open. He also carried out all the procedures in a class II laminar flow hood, which prevents outside contamination from entering the working area. In addition, the hood had never been used for any other bacterial extraction processes. Several other precautions were added to eliminate any chance that the spores were modern-day contaminants from an outside source. Still, many scientists question whether all contamination sources had been identified.

The major question that remains is whether DNA can remain intact and functional after so long a period of dormancy. Does it really have a capability of replication and producing new vegetative growth? Granted, the DNA presumably was protected in a resistant spore, but could DNA remain intact for 25 million years?

Research on bacterial DNA suggests the maximum survival time is about 400,000 to 1.5 million years. If true, then the 25 million-year-old spores could not be viable. But that is based on current predictions and they may be subject to change as more research is carried out with ancient DNA.

The verdict? It seems unlikely that such ancient endospores could germinate after 25 million years. Perhaps new evidence will change that perception.

MicroFocus Box

Being Skeptical

One of the seven types of essay boxes new to this edition is titled "Being Skeptical." A good scientist is a skeptic and skepticism is an important part of science. Skepticism, unlike cynicism, is not unwilling to accept a claim or observation. Skepticism simply says "Prove it!" Science applies scientific reasoning as the method for proof. Thus, a scientist, such as a microbiologist, must see the evidence and it must be compelling before the observation or statement is provisionally accepted. The claim is still open to further examination and experimentation.

The "Being Skeptical" essays scattered through the textbook present an often-fantastic statement or claim. The essay then examines the claim using reasoning skills and the scientific process, which is sometimes called the scientific method.

WHY PATHOGENS?

Microorganisms perform many useful services for humans when they produce food products, manufacture organic materials in industrial plants, and recycle such elements as carbon and nitrogen. The emphasis of this book, however, is on the tiny, but significant percentage of microorganisms causing human disease, the so-called pathogens. Why do we emphasize pathogens? Here are several reasons:

- Pathogens have regularly altered the course of human history.
- Pathogens are familiar to audiences of microbiology.
- Pathogens add drama to an invisible world of microorganisms.
- Pathogens illustrate ecological relationships between humans and microorganisms.
- Pathogens point up the diversity of microorganisms.

Moreover, the study of pathogens makes basic science relevant and shows how microbiology interfaces with other disciplines such as sociology, economics, history, politics, and geography. Finally, the study of pathogens helps us to understand contemporary newspaper articles, magazine headlines, and stories on the news. And in the end, that makes us better citizens. Indeed, the famous essayist Thomas Mann once wrote, "All interest in disease is only another expression of interest in life."

SUPPLEMENTS TO THE TEXT

Jones and Bartlett offers an array of ancillaries to assist instructors and students in teaching and mastering the concepts in this text. Additional information and review copies of any of the following items are available through your Jones and Bartlett sales representative or by going to www.bioscience.jbpub.com

For the Student

The web site we developed exclusively for the eighth edition of this text, http://microbiology.jbpub.com, offers a variety of resources to enhance understanding of microbiology. The site contains eLearning, a free on-line study guide with chapter outlines, chapter essay questions, key term reviews, and short study quizzes.

The *Study Guide* to accompany this textbook contains important information to help you study, take effective class notes, prepare properly for exams, and even to manage your time effectively. The latter is the single most common reason for poor performance in college courses. The *Study Guide* also contains over 3000 practice exercises and study questions of various types to help you learn and retain the information in the text.

Laboratory Fundamentals of Microbiology, Eighth Edition, is a series of over 30 multipart laboratory exercises providing basic training in the handling of microorganisms and reinforcing ideas and concepts described in the textbook.

Guide to Infectious Diseases by Body Systems is an excellent tool for learning about microbial diseases. Each of the fifteen body systems units presents a brief introduction to the anatomical system and the bacterial, viral, fungal, or parasitic organism infecting the system.

An anthology called *Encounters with Microbiology* brings together "Vital Signs" articles from *Discover Magazine* in which health professionals use their knowledge of microbiology in their medical cases.

For the Instructor

Compatible with Windows and Macintosh platforms, the **Instructor's ToolKit—CD-ROM** provides instructors with the following traditional ancillaries:

- The *Instructor's Manual,* provided as a text file, includes chapter summaries and complete chapter lecture outlines and answers to all the end-of-chapter assessments.
- *Chapter Assessments Answers* provide short answers to figure questions, Concept and Reasoning Checks, and all end-of-chapter materials.
- The *Test Bank* is available as straight text files.
- *The PowerPoint™ Image Bank* provides the illustrations, photographs, and tables (to which Jones and Bartlett Publishers holds the copyright or has permission to reproduce digitally) inserted into PowerPoint slides. You can quickly and easily copy individual images or tables into your existing lecture slides.
- The *PowerPoint Lecture Outline Slides* presentation package provides lecture notes, and images for each chapter of *Alcamo's Fundamentals of Microbiology.* Instructors with the Microsoft PowerPoint software can customize the outlines, art, and order of presentation.

Acknowledgments

My name may be on the cover of this textbook, but putting together a complete edition requires an experienced and capable publication team. At Jones and Bartlett Publishers, such experience has been instrumental. Cathleen Sether has been a most able acquisitions editor. Dean DeChambeau guided the project with his excellent management skills, including occasionally soothing an author's tantrums; Lou Bruno again supervised the production process with great skill; Anne Spencer brought the new design format to reality; Molly Steinbach deftly assisted everyone; Christine McKeen tracked down many of the great photos that embellish these pages; Deborah Patton read every page and created the index; Shellie Newell was again the "eagle-eye" copy editor; and Elizabeth Morales continued her expert art in translating my scribbling into first-class artwork.

The book benefited from the expertise of several fellow microbiologists and biologists. I wish to thank Michael McKinley and Stephen Williams for their input during the writing of this edition, and Teri Shors for providing many student feedback evaluations. Robert Bowker and Ryan Sawby provided photographic assistance. I especially want to thank Michelle Beach, a GCC microbiology student who, as her special project, read every word of the seventh edition, making note of typographic errors, syntax and grammatical errors, and all unclear statements found in the text, including all the excessive usage of the word *that!*

After more than 20 years of university and college instruction, I must thank all my former students who kept me on my toes in the classroom and required me to be always prepared. Your suggestions and evaluations have encouraged me to continually assess my instruction so it can be easily understood. I salute you, and I hope those of you who read this text will let me know what works and what still needs improvement to make your learning efficient and still enjoyable.

J. Pommerville
Scottsdale, AZ
Fall, 2006

About the Author

Today I am a microbiologist, researcher, and science educator. My plans did not start with that intent. While in high school in Santa Barbara, California, I wanted to play professional baseball, study the stars, and own a '66 Corvette. None of these desires would come true—my batting average was miserable (but I was a good defensive fielder), I hated the astronomy correspondence course I took, and I never bought that Corvette.

I found an interest in biology at Santa Barbara City College. After squeaking through college calculus, I transferred to the University of California at Santa Barbara (UCSB) where I received a B.S. in Biology and stayed on to pursue a Ph.D. degree in the lab of Ian Ross studying cell communication and sexual pheromones in a water mold. After receiving my doctorate in Cell and Organismal Biology, my graduation was written up in the local newspaper as a native son who was a fungal sex biologist—an image that was not lost on my three older brothers!

While in graduate school at UCSB, I rescued a secretary in distress from being licked to death by a German shepherd. Within a year, we were married (the secretary and I). When I finished my doctoral thesis, I spent several years as a postdoctoral fellow at the University of Georgia. Worried that I was involved in too many research projects, a faculty member told me something I will never forget. He said, "Jeff, it's when you can't think of a project or what to do that you need to worry." Well, I have never had to worry!

I then moved on to Texas A&M University, where I spent eight years in teaching and research—and telling Aggie jokes. Toward the end of this time, after publishing over 30 peer-reviewed papers in national and international research journals, I realized I had a real interest in teaching and education. Leaving the sex biologist nomen behind, I headed farther west to Arizona to join the biology faculty at Glendale Community College where I continue to teach introductory biology and microbiology.

I have been lucky to be part of several educational research projects and have been honored with two of my colleagues with a Team Innovation of the Year Award by the League of Innovation in the Community Colleges. In 2000, I became project director and lead principal investigator for a National Science Foundation grant to improve student outcomes in science through changes in curriculum and pedagogy. I had a fascinating three years coordinating more than 60 science faculty members (who at times were harder to manage than students) in designing and field testing 18 interdisciplinary science units. This culminated with me being honored in 2003 with the Gustav Ohaus Award (College Division) for Innovations in Science Teaching from the National Science Teachers Association.

I have been on the editorial board for *Microbiology Education*, the education research journal of the American Society for Microbiology

(ASM) and in 2004 was co-chair for the ASM Conference for Undergraduate Educators. Currently (2006–2007), I am the chair of Undergraduate Education Division of ASM. In 2006, I was selected as one of four outstanding instructors at Glendale Community College.

I mention all this not to impress but to show how the road of life sometimes offers opportunities in unexpected and unplanned ways. The key though is keeping your eyes open and your mind thinking; then unlimited opportunities will come your way. With the untimely passing of my friend and professional colleague Ed Alcamo, I was privileged in 2003 to be offered the opportunity to take over the authorship of *Fundamentals of Microbiology*. It is an undertaking I continue to relish as I (along with the wonderful folks at Jones and Bartlett) try to evolve a new breed of microbiology textbook reflecting the pedagogy change occurring in science classrooms today. And, hey, who knows—maybe that '66 Corvette could be in my garage yet.

DEDICATION

I dedicate this edition of the book to my wife Yvonne. The long days and nights of revisions that at times seemed almost insurmountable could not have been completed without her support, understanding, enthusiasm—and patience. She has been not only my loving wife but also my best friend and motivator, keeping me on track during wavering moments.

To the Student—Study Smart

Your success in microbiology and any college or university course will depend on your ability to study effectively and efficiently. Therefore, this textbook was designed with you, the student, in mind. The text's organization will help you improve your learning and understanding and, ultimately, your grades. The learning design concept described in the Preface and illustrated below reflects this organization. Study it carefully, and, if you adopt the flow of study shown, you should be a big step ahead in your preparation and understanding of microbiology—and for that matter any subject you are taking.

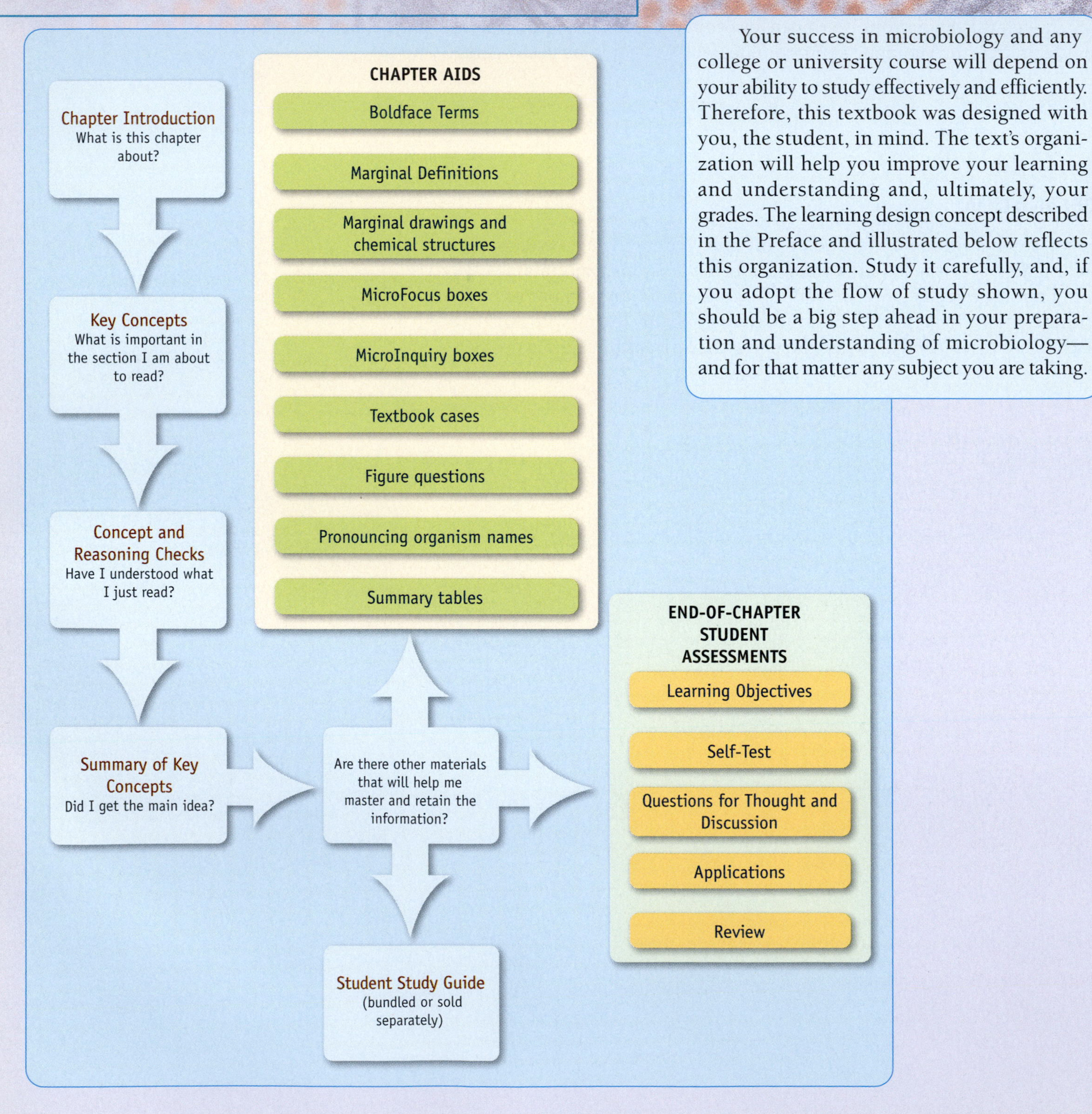

When I was an undergraduate student, I hardly ever read the "To the Student" section (if indeed one existed) in my textbooks because the section rarely contained any information of importance. This one does, so please read on.

In college, I was a mediocre student until my junior year. Why? Mainly because I did not know how to study properly, and, important here, I did not know how to read a textbook effectively. My textbooks were filled with underlined sentences (highlighters hadn't been invented yet!) without any plan on how I would use this "emphasized" information. In fact, most textbooks *assume* you know how to read a textbook properly. I didn't and you might not, either.

Reading a textbook is difficult if you are not properly prepared. So you can take advantage of what I learned as a student and have learned from instructing thousands of students; I have worked hard to make this text user friendly with a reading style that is not threatening or complicated. Still, there is a substantial amount of information to learn and understand, so having the appropriate reading and comprehension skills is critical. Therefore, I encourage you to spend 30 minutes reading this section, as I am going to give you several tips and suggestions for acquiring those skills. Let me show you how to be an active reader. Note: the student *Study Guide* also contains similar information on how to take notes from the text, how to study, how to take class (lecture) notes, how to prepare for and take exams, and perhaps most important for you, how to manage your time effectively. It all is part of this "learning design," my wish to make you a better student.

BE A PREPARED READER

Before you jump into reading a section of a chapter in this text, prepare yourself by finding the place and time and having the tools for study.

Place. Where are you right now as you read these lines? Are you in a quiet library or at home? If at home, are there any distractions, such as loud music, a blaring television, or screaming kids? Is the lighting adequate to read? Are you sitting at a desk or lounging on the living room sofa? Get where I am going? When you read for an educational purpose—that is, to learn and understand something—you need to maximize the environment for reading. Yes, it should be comfortable but not to the point that you will doze off.

Time. All of us have different times during the day when we perform some skill, be it exercising or reading, the best. The last thing you want to do is read when you are tired or simply not "in tune" for the job that needs to be done. You cannot learn and understand the information if you fall asleep or lack a positive attitude. I have kept the chapters in this text to about the same length so you can estimate the time necessary for each and plan your reading accordingly. If you have done your preliminary survey of the chapter or chapter section, you can determine about how much time you will need. If 40 minutes is needed to read—and comprehend (see below)—a section of a chapter, find the place and time that will give you 40 minutes of uninterrupted study. Brain research suggests that most people's brains cannot spend more than 45 minutes in concentrated, technical reading. Therefore, I have avoided lengthy presentations and instead have focused on smaller sections, each with its own heading. These should accommodate shorter reading periods.

Reading Tools. Lastly, as you read this, what study tools do you have at your side? Do you have a highlighter or pen for emphasizing or underlining important words or phrases? Notice, the text has wide margins, which allow you to make notes or to indicate something that needs further clarification. Do you have a pencil or pen handy to make these notes? Or, if you do not want to "deface" the text, make your notes in a notebook. Lastly, some students find having a ruler is useful to prevent your eyes from wandering on the page and to read each line without distraction.

BE AN EXPLORER BEFORE YOU READ

When you sit down to read a section of a chapter, do some preliminary exploring. Look at the section head and subheadings to get an idea of what is discussed. Preview any diagrams, photographs, tables, graphs, or other visuals used. They give you a better idea of what is going to occur. We have used a good deal of space in the text for these features, so use them to your advantage. They will help you learn the written information and comprehend its meaning. Do not try to understand all the visuals, but try to generate a mental "big picture" of what is to come. Familiarize yourself with any symbols or technical jargon that might be used in the visuals.

The end of each chapter contains a **Summary of Key Concepts** for that chapter. It is a good idea to read the summary before delving into the chapter. That way you will have a framework for the chapter before filling in the nitty-gritty information.

BE A DETECTIVE AS YOU READ

Reading a section of a textbook is not the same as reading a novel. With a textbook, you need to uncover the important information (the terms and concepts) from the forest of words on the page. So, the first thing to do is read the complete paragraph. When you have determined the main ideas, highlight or underline them. However, I have seen students' highlighting the entire paragraph in yellow, including every *a*, *the*, and *and*. This is an example of highlighting before knowing what is important. So, I have helped you out somewhat. Important terms and concepts are in **bold face** followed by the definition (or the definition might be in the margin). So only highlight or underline with a pen essential ideas and key phrases—not complete sentences, if possible. By the way, the important microbiological terms and major concepts also are in the **Glossary** at the back of the text.

What if a paragraph or section has no boldfaced words? How do you find what is important here? From an English course, you may know that often the most important information is mentioned first in the paragraph. If it is followed by one or more examples, then you can backtrack and know what was important in the paragraph. In addition, I have added section "speed bumps" (called **Concept and Reasoning Checks**) to let you test your learning and understanding before getting too far ahead in the material. These checks also are clues to what was important in the section you just read.

BE A REPETITIOUS STUDENT

Brain research has shown that each individual can only hold so much information in short-term memory. If you try to hold more, then something else needs to be removed—sort of like a full computer disk. So that you do not lose any of this important information, you need to transfer it to long-term memory—to the hard drive if you will. In reading and studying, this means retaining the term or concept; so, write it out in your notebook *using your own words*. Memorizing a term does not mean you have learned the term or understood the concept. By actively writing it out in your own words, you are forced to think and actively interact with the information. This repetition reinforces your learning.

BE A PATIENT STUDENT

In textbooks, you cannot read at the speed that you read your e-mail or a magazine story. There are unfamiliar details to be learned and understood—and this requires being a patient, slower reader. Actually, if you are not a fast reader to begin with, as I am, it may be an advantage in your learning process. Identifying the important information from a textbook chapter requires you to *slow down* your reading speed. Speed-reading is of no value here.

KNOW THE WHAT, WHY, AND HOW

Have you ever read something only to say, "I have no idea what I read!" As I've already mentioned, reading a microbiology text is not the same as reading *Sports Illustrated* or *People* magazine. In these entertainment magazines, you read passively for leisure or perhaps amusement. In *Alcamo's Fundamentals of Microbiology*, you must read actively for learning and understanding—that is, for *comprehension*. This can quickly lead to boredom unless you engage your brain as you read—that is, be an active reader. Do this by knowing the *what*, *why*, and *how* of your reading.

- *What* is the general topic or idea being discussed? This often is easy to determine because the section heading might tell you. If not, then it will appear in the first sentence or beginning part of the paragraph.
- *Why* is this information important? If I have done my job, the text section will tell you why it is important or the examples provided will drive the importance home. These sur-

rounding clues further explain why the main idea was important.

- *How* do I "mine" the information presented? This was discussed under being a detective.

A MARKED UP READING EXAMPLE

So let's put words into action. Below is a passage from the text. I have marked up the passage as if I were a student reading it for the first time. It uses many of the hints and suggestions I have provided. Remember, it is important to read the passage slowly, and concentrate on the main idea (concept) and the special terms that apply.

HAVE A DEBRIEFING STRATEGY

After reading the material, be ready to debrief. Verbally summarize what you have learned. This will start moving the short-term information into the long-term memory storage—that is, *retention*. Any notes you made concerning confusing material should be discussed as soon as possible with your instructor. For microbiology, allow time to draw out diagrams. Again, repetition makes for easier learning and better retention.

In many professions, such as sports or the theater, the name of the game is practice, practice, practice. The hints and suggestions I have given you form a skill that requires practice to perfect and use efficiently. Be patient, things will

Many Foodborne and Waterborne Diseases Have a Bacterial Cause

KEY CONCEPT

- Bacterial gastrointestinal diseases may arise from intoxications or infections.

We categorize the foodborne and waterborne bacterial diseases as either intoxications or infections. **Intoxications** are illnesses in which bacterial toxins are ingested in food or water. Examples are the toxins causing botulism, staphylococcal food poisoning, and clostridial food poisoning. By contrast, **infections** refer to illnesses in which live bacterial pathogens in food and water are ingested and subsequently grow in the body. Salmonellosis, shigellosis, and cholera are examples. Toxins may be produced, but they are the result of infection.

Etiology: The study of the causes (origins) of disease.

Determining the etiology of a bacterial disease depends on several factors.

Incubation period. If an individual ingests and swallows a contaminated food or beverage, there is a delay, called the **incubation period**, before the symptoms appear. This period can range from hours to days, depending on the bacterial species and on the infectious dose. During the incubation period, the toxins or microbes pass through the stomach into the intestine where they may directly affect gastrointestinal function or be absorbed into the bloodstream.

Infectious dose: The number of organisms consumed to give rise to symptoms of an illness.

Clinical symptoms. The symptoms produced by an intoxication or infection depend

not happen overnight; perseverance and willingness though will pay off with practice. You might also check with your college or university academic (or learning) resource center. These folks will have more ways to help you to read a textbook better and to study well overall.

CONCEPT MAPS

In science as well as in other subjects you take at the college or university, there often are concepts that appear abstract or simply so complex they are difficult to understand. A **concept map** is one tool to help you enhance your abilities to think and learn. Critical reasoning and the ability to make connections between complex, nonlinear information are essential to your studies and career.

Concept maps are a learning tool designed to represent complex or abstract information visually. Neurobiologists and psychologists tell us that the brain's primary function is to take incoming information and interpret it in a meaningful or practical way. They also have found that the brain has an easier time making sense of information when it is presented in a visual format. Importantly, concept maps not only present the information in "visual sentences" but also take paragraphs of material and present it in an "at-a-glance" format. Therefore, you can use concept maps to

- Communicate and organize complex ideas in a meaningful way
- Aid your learning by seeing connections within or between concepts and knowledge
- Assess your understanding or diagnose misunderstanding

There are many different types of concept maps. The two most used in this textbook are the *process map* or *flow chart* and the *hierarchical map*. The hierarchical map starts with a general concept (the most inclusive word or phrase) at the top of the map and descends downward using more specific, less general words or terms. In several chapters in this textbook process or hierarchical maps are drawn—and you have the opportunity to construct your own hierarchical maps as well.

Concept mapping is the strategy used to produce a concept map. So, let's see how one makes a hierarchical map.

How to Construct a Concept Map

1. Print the central idea (concept or question to be mapped) in a box at the top center of a blank, unlined piece of paper. Use uppercase letters to identify the central idea.
2. Once the concept has been selected, identify the key terms (words or short phrases) that apply to or stem from the concept. Often these may be given to you as a list. If you have read a section of a text, you can extract the terms from that material, as the words are usually boldfaced or italicized.
3. Now, from this list, try to create a hierarchy for the terms you have identified; that is, list them from the most general, most inclusive to the least general, most specific. This ranking may only be approximate and subject to change as you begin mapping.
4. Construct a preliminary concept map. This can be done by writing all of the terms on Post-its®, which can be moved around easily on a large piece of paper. This is necessary as one begins to struggle with the process of building a good hierarchical organization.
5. The concept map connects terms associated with a concept in the following way:
   - The relationship between the concept and the first term(s), and between terms, is connected by an arrow pointing in the direction of the relationship (usually downward or horizontal if connecting related terms).
   - Each arrow should have a label, a very short phrase that explains the relationship with the next term. In the end, each link with a label reads like a sentence.
6. Once you have your map completed, redraw it in a more permanent form. Box in all terms that were on the sticky notes. Remember there may be more than one way to draw a good concept map, and don't be scared off if at first you have some problems mapping; mapping will become more apparent to you after you have practiced this technique a few times using the opportunities given to you in the early chapters of the textbook.
7. Now look at the map and see if it answers the following. Does it:
   - Define clearly the central idea by positioning it in the center of the page?
   - Place all the terms in a logical hierarchy

and indicate clearly the relative importance of each term?

- Allow you to figure out the relationships among the key ideas more easily?
- Permit you to see all the information visually on one page?
- Allow you to visualize complex relationships more easily?
- Make recall and review more efficient?

Example

After reading the section in Chapter 7 on "Protein Synthesis," a student makes a list of the terms used and maps the concept. Using the steps outlined above, the student produces the following hierarchical map. Does it satisfy all the questions asked in (7)?

Practical Uses for Mapping

- **Summarizing textbook readings.** Use mapping to summarize a chapter section or a whole chapter in a textbook. This purpose for mapping is used many times in this text.
- **Summarizing lectures.** Although producing a concept map during the classroom period may not be the best use of the time, making a concept map or maps from the material after class will help you remember the important points and encourage high-level, critical reasoning, which is so important in university and college studies.
- **Reviewing for an exam.** Having concept maps made ahead of time can be a very useful and productive way to study for an exam, particularly if the emphasis of the course is on understanding and applying abstract, theoretical material, rather than on simply reproducing memorized information.
- **Working on an essay.** Mapping also is a powerful tool to use during the early stages of writing a course essay or term paper. Making a concept map before you write the first rough draft can help you see and ensure you have the important points and information you will want to make.

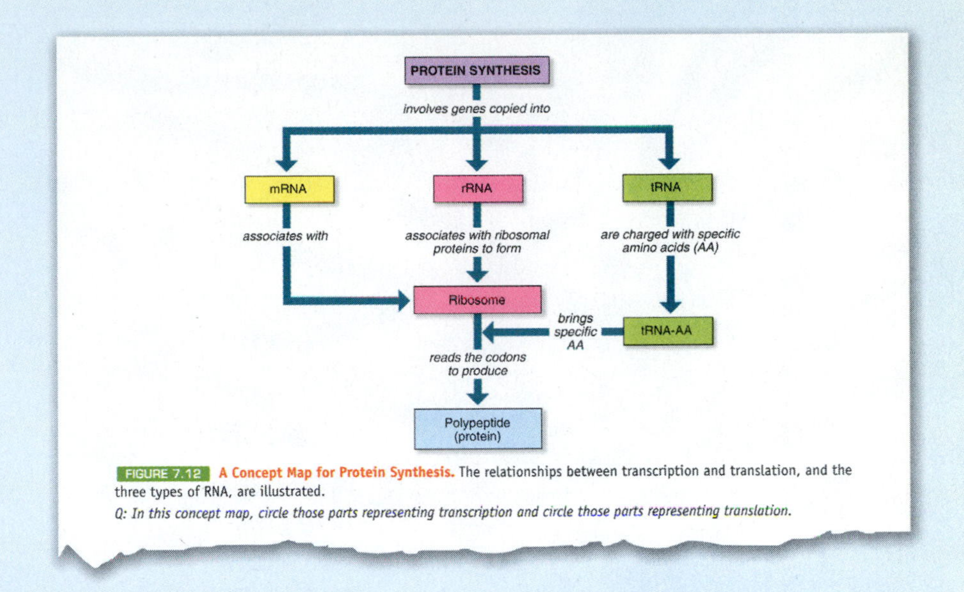

FIGURE 7.12 **A Concept Map for Protein Synthesis.** The relationships between transcription and translation, and the three types of RNA, are illustrated.

Q: In this concept map, circle those parts representing transcription and circle those parts representing translation.

SEND ME A NOTE

In closing, I would like to invite you to write me and let me know what is good about this textbook so I can build on it and what may need improvement so I can revise it. Also, I would be pleased to hear about any news of microbiology in your community, and I'd be happy to help you locate any information not covered in the text. I can be reached at the Department of Biology, Glendale Community College, 6000 W. Olive Avenue, Glendale, Arizona 85302. Feel free to e-mail me at: jeffrey.pommerville@gcmail.maricopa.edu.

I wish you great success in your microbiology course. Welcome! Let's now plunge into the wonderful and sometimes awesome world of microorganisms.

—Dr. P.

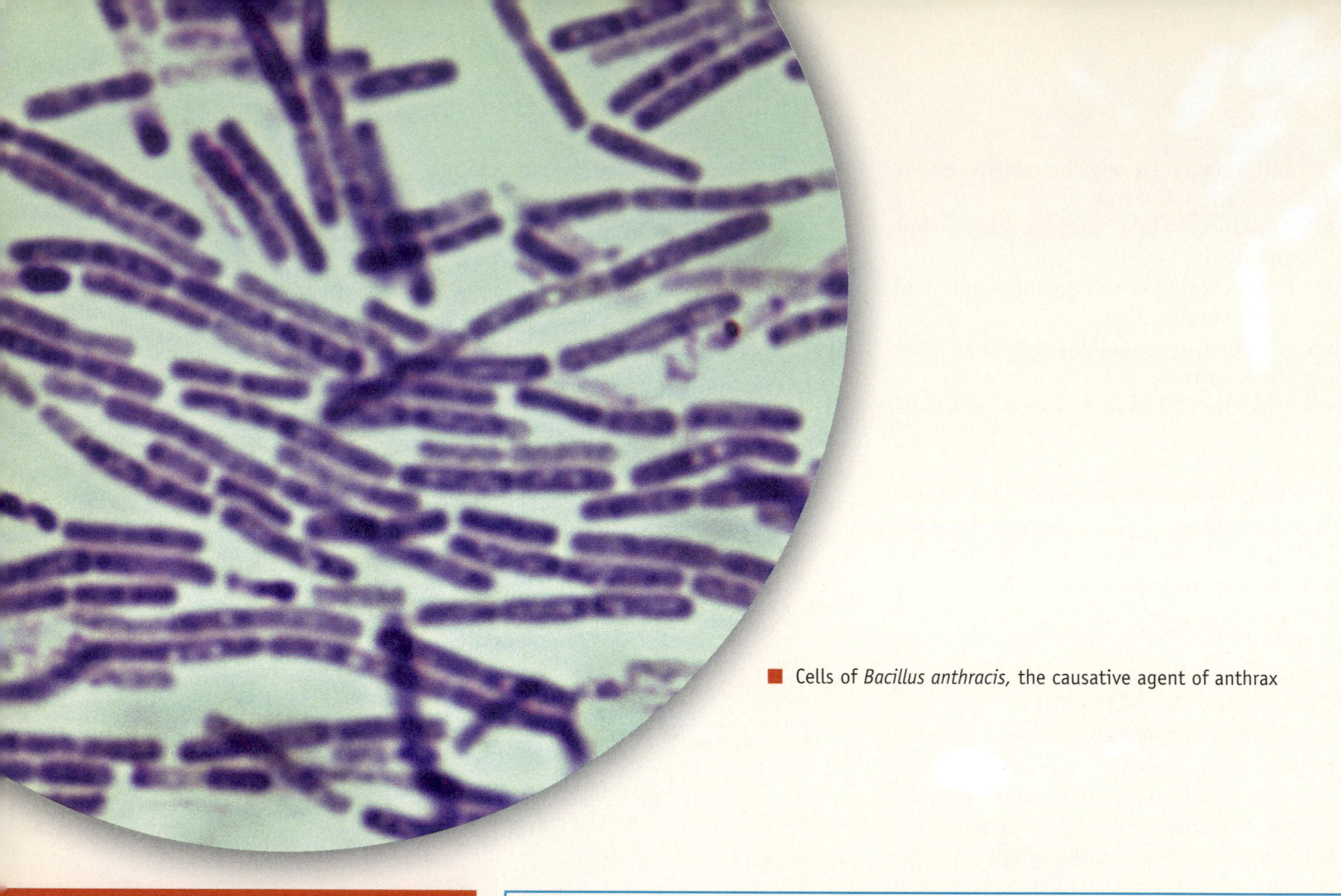

Cells of *Bacillus anthracis,* the causative agent of anthrax

PART 1 Foundations of Microbiology

In 1676, a century before the Declaration of Independence, a Dutch merchant named Anton van Leeuwenhoek sent a noteworthy letter to the Royal Society of London. Writing in the vernacular of his home in the United Netherlands, Leeuwenhoek described how he used a simple microscope to observe vast populations of minute, living creatures. His reports opened a chapter of science that would evolve into the study of microscopic organisms and the discipline of microbiology. At that time, few people, including Leeuwenhoek, attached any practical significance to the microorganisms, but during the next three centuries, scientists would discover how profoundly these organisms influence the quality of our lives.

We begin our study of the microorganisms by exploring the grassroot developments that led to the establishment of microbiology as a science. These developments are surveyed in Chapter 1, where we focus on the individuals who stood at the forefront of discovery. Today we are in the midst of a third Golden Age of microbiology and our understanding of microorganisms continues to grow even as you read this book. Chapter 1, therefore, is an important introduction to microbiology then and now.

Part 1 also contains a chapter on basic chemistry, inasmuch as microorganisms are chemical machines. Moreover, their activities are all related to chemistry. The third chapter in Part 1 will set down some basic concepts and tools for studying microorganisms, much as the alphabet applies to word development. In succeeding chapters, we will formulate words into sentences and sentences into ideas as we survey the different groups of microorganisms and concentrate on their importance to public health and human welfare.

MICROBIOLOGY PATHWAYS

Being a Scientist

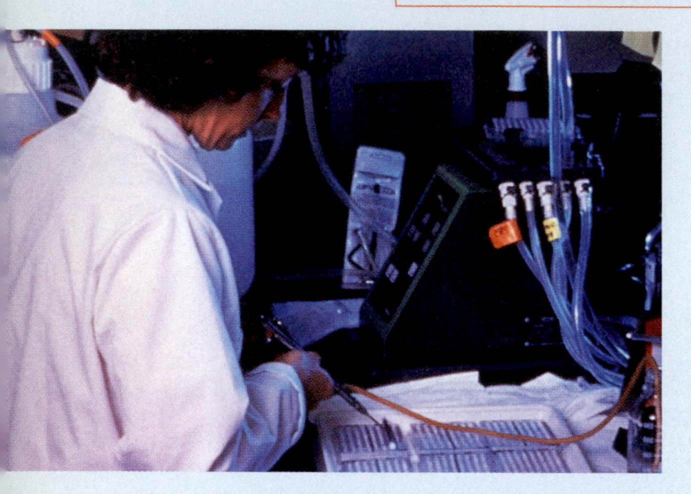

Science may not seem like the most glamorous profession. So, as you read many of the chapters in this text, you might wonder why many scientists have the good fortune to make key discoveries. At times it might seem like it is the luck of the draw, but actually many scientists have a set of characteristics that put them on the trail to success.

Robert S. Root-Bernstein, a physiology professor at Michigan State University, points out that many prominent scientists like to goof around, play games, and surround themselves with a type of chaos aimed at revealing the unexpected. Their labs may appear to be in disorder, but they know exactly where every tube or bottle belongs. Scientists also identify intimately with the organisms or creatures they study (it is said that Louis Pasteur actually dreamed about microorganisms), and this identification brings on an intuition—a "feeling for the organism." In addition, there is the ability to recognize patterns that might bring a breakthrough. (Pasteur had studied art as a teenager and, therefore, he had an appreciation of patterns.)

The geneticist and Nobel laureate Barbara McClintock once remarked, *"I was just so interested in what I was doing I could hardly wait to get up in the morning and get at it. One of my friends, a geneticist, said I was a child, because only children can't wait to get up in the morning to get at what they want to do."* Clearly, another characteristic of a scientist is having a child-like curiosity for the unknown.

Professor Alcamo once received a letter from a student, asking why he became a microbiologist. *"It was because I enjoyed my undergraduate microbiology course"* he said, *"and when I needed to select a graduate major, microbiology seemed like a good idea. I also think I had some of the characteristics described by Root-Bernstein: I loved to try out different projects; my corner of the world qualified as a disaster area; still I was a nut on organization, insisting that all the square pegs fit into the square holes."*

For this author, science has been an extraordinary opportunity to discover and understand something never before known. Science is fun, yet challenging—and at times arduous, tedious, and frustrating. As with most of us, we will not make the headlines for a breakthrough discovery or find a cure for a disease. However, as scientists we all hope our research will contribute to a better understanding of a biological (or microbiological) phenomenon and will push back the frontiers of knowledge. For me, the opportunity of doing something I love far outweighed the difficulties along the way.

Like any profession, being a scientist is not for everyone. Besides having a bachelor's or higher degree in biology or microbiology, you should be well read in the sciences and capable of working as part of an interdisciplinary team. Of course, you should have good quantitative and communication skills, have an inquisitive mind, and be goal oriented. If all this sounds interesting, then maybe you fit the mold of a scientist. Why not consider pursuing a career in microbiology? Some possibilities are listed in this book, but you should also visit with your instructor. Simply stop by the student union, buy two cups of coffee, and you are on your way.

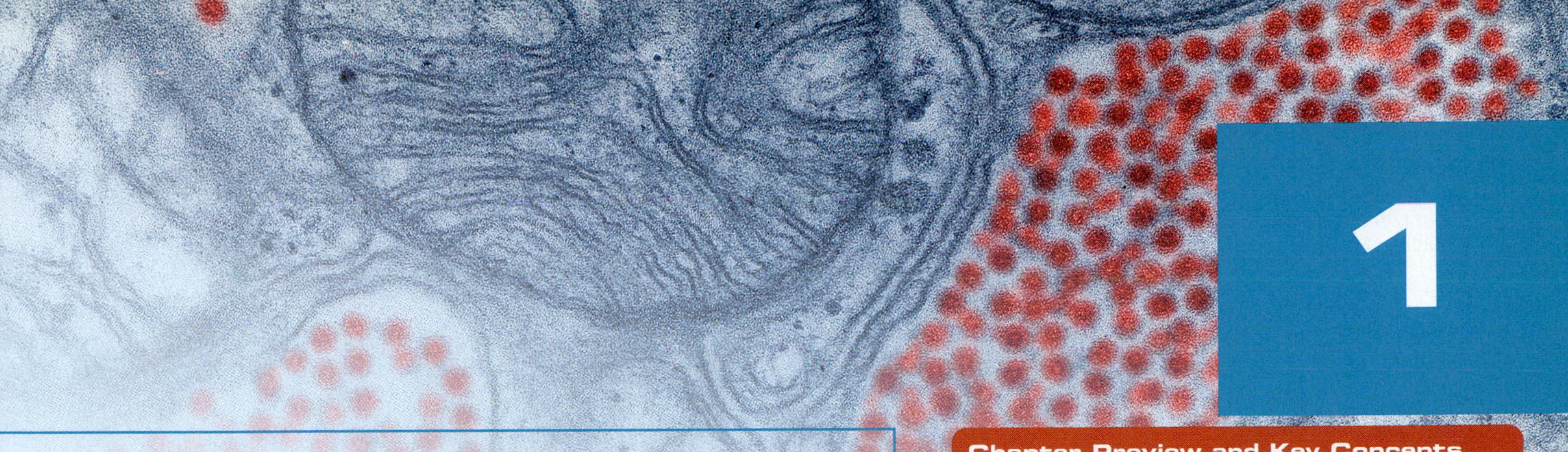

Microbiology: Then and Now

In the field of observation, chance favors only the prepared mind.
—Louis Pasteur (1822–1895)

Space. The final frontier! Really? *The* final frontier? There are an estimated 50 billion galaxies and 10^{21} stars in the visible universe. However, the microbial universe consists of more than 10^{31} microorganisms scattered among an estimated 2 to 3 billion species. So, could understanding these organisms on Earth be as important as studying galaxies in space?

In 1990, microbiologist Stephen Giovannoni of Oregon State University identified in the Sargasso Sea off the southeast United States what is perhaps the most abundant and successful organism on the planet. Called SAR11 (SAR for Sargasso), this bacterial organism, which now goes by the scientific name *Pelagibacter ubique,* has been identified across the oceans of the world. What makes it significant is its population size, which Giovannoni estimates to be 2.4×10^{28} cells. (Compare this to the 6 billion (6×10^{9}) humans and the 50 billion (5×10^{10}) known galaxies and you see how big a number this is!) SAR11 alone accounts for 20 percent of all oceanic bacterial species—and 50% of the bacterial species in the surface waters of temperate oceans in the summer!

SAR11's success story suggests the organism must have a significant impact on the planet. Although such roles remain to be identified and understood, Giovannoni believes SAR11 is responsible for up to 10 percent of all nutrient recycling on the planet, which influences the cycling of carbon and even affects global warming.

Also sailing the Sargasso Sea near Bermuda is Craig Venter and his team at the J. Craig Venter Institute. Fresh from his success with the

Chapter Preview and Key Concepts

1.1 The Beginnings of Microbiology
- The discovery of microorganisms was dependent on observations made with the microscope.
- The emergence of experimental science provided a means to test long-held beliefs and resolve controversies.

MicroInquiry 1: Experimentation and Scientific Inquiry

1.2 Microorganisms and Disease Transmission
- Early epidemiology studies suggested how diseases could be spread and be controlled.
- Resistance to a disease can come from exposure to and recovery from a mild form of (or a very similar) disease.

1.3 The Classical Golden Age of Microbiology (1854–1914)
- The germ theory was based on the observations that different microorganisms have distinctive and specific roles in nature.
- Antisepsis and identification of the cause of animal diseases reinforced the germ theory.
- Koch's postulates provided a way to identify a specific microorganism as causing a specific infectious disease.
- Laboratory science and teamwork stimulated the discovery of additional infectious disease agents.
- Viruses also can cause disease.
- Many beneficial bacterial species recycle nutrients in the environment.

1.4 Studying Microorganisms
- Microorganisms represent more than agents of disease.
- The organisms and agents studied in microbiology represent a diverse group.

1.5 The Second Golden Age of Microbiology (1943–1970)
- Microorganisms and viruses can be used as model systems to study phenomena common to all life.
- All microorganisms have a characteristic cell structure.
- Antimicrobial chemicals can be effective in treating infectious diseases.

1.6 The Third Golden Age of Microbiology—Now
- Infectious disease (natural and intentional) preoccupies much of microbiology.
- Microbial ecology and evolution are dominant themes in modern microbiology.

■ Genome: The complete set of genetic information in a human cell.

private sector effort to sequence the human genome, Venter's team in 2004 reported the discovery of over 1,800 new microbial species in Sargasso seawater and from them isolated 1.2 million new gene sequences.

Now Venter has begun a voyage—á la Charles Darwin—to sample seawater from around the world and determine the impact humans have on the microorganisms in the oceanic environment. For the emerging field of marine molecular microbiology, Venter believes the sequencing of marine microorganisms will provide examples of novel metabolic pathways, identify species that use alternative energy sources, and perhaps help solve critical environmental problems, including global warming.

Giovannoni and Venter are just two of many microbiologists trying to understand the role of microorganisms in the ocean's ecosystems and their dominant role on this planet. But most of all, as "Being a Scientist" identified, Giovannoni's and Venter's primary goal is a voyage of discovery. Since less than 1 percent of the marine microorganisms have been identified, the microbial universe does represent an inner final frontier!

The science of **microbiology** embraces a biologically diverse group of usually small life forms, encompassing primarily **microorganisms** (bacteria, fungi, algae, and protozoa) and viruses.

Microorganisms (or **microbes** for short) are present in vast numbers in nearly every environment and habitat on Earth, not just the Sargasso Sea. They survive in Antarctica, on top of the tallest mountains, near thermal vents in the deepest parts of the oceans, in the deepest, darkest caves, and even miles down within the crust of the earth. In all, by weight microbes make up about two thirds of Earth's living material. It is probably safe to say they have colonized every habitat or environment on and in the Earth.

The rich deversity of microorganisms illustrates their profound influence on all aspects of life. For example, they are essential to the recycling of nutrients that form the bodies of all organisms and sustaining all higher forms of life. They help control our climate and, as a group, produce about 50 percent of the oxygen gas we breathe and many other organisms use. They have influenced the evolution of life on Earth and actually have outpaced that of the more familiar plants and animals.

Microorganisms survive in, or are purposely put in, many of the foods we eat. Microorganisms and viruses also are in the air we breathe and, at times, in the water we drink. Even closer to home, microorganisms normally colonize our skin and grow in our mouth, ears, nose, throat, and digestive tract. Fortunately, the majority of these microbes are benign or may actually be beneficial in keeping dangerous microbes out.

When most of us hear the word *bacterium* or *virus* though, we think infection or disease. Although such **pathogens** (disease-causing agents) are rare, they periodically have carved out great swatches of humanity as epidemics passed over the land. Some diseases—such as plague, cholera, and smallpox—have become known as "slate-wipers," a reference to the barren towns they left in their wake (MicroFocus 1.1). Even today, with antibiotics and vaccines to cure and prevent many infectious diseases, they still bring concern and sometimes panic. Just think about the scares that AIDS, severe acquired respiratory syndrome (SARS), and most recently avian influenza have caused worldwide (FIGURE 1.1A).

Microbiology also provides the "tools" for answering questions about life. Many microorganisms and viruses are used as "experimental models" to understand better the chemical and physical activities, the genetic interactions, and the evolutionary relationships common to all life (FIGURE 1.1B). The science also has developed the methods to analyze and detect disease and has provided the means by which many can be prevented.

A major focus of this introductory chapter is to give you an introspective "first look" at microbiology—then and now. We will see how microbes were first discovered and how those that cause infectious disease preoccupied the minds and efforts of so many. Along the way, we continue to see how curiosity and scientific inquiry stimulated the quest for understanding.

Although the study of microorganisms began in earnest with the work of Pasteur and Koch, they were not the first to report microorganisms. To begin our story, we reach back to the 1600s, where we encounter some equally inquisitive individuals.

MICROFOCUS 1.1: History

The Tragedy of Eyam

On the last Sunday in August (Plague Sunday), English pilgrims gather in the English countryside outside the village of Eyam, to pay homage to the townsfolk who in 1665–1666 gave their lives so that others might live. The pilgrims pause, bow their heads, and remember. In 1665, bubonic plague was raging in London. In late August, a traveling tailor arrived in the village of Eyam, about 140 miles north of London. Unknown to him, cloth arriving from London was infected with plague-carrying fleas.

A dance in the graveyard to ward off the plague.

Within a few days, plague began to spread throughout Eyam and villagers debated whether they should flee north. The village rector realized that if the villagers left, they could spread the plague to other towns and villages. So, he made a passionate plea that they stay. After some deep soul-searching, most townsfolk resolved to remain, even though they knew that meant many would die.

The villagers marked off a circle of stones outside the village limits, and people from the adjacent towns brought food and supplies to the barrier, leaving them there for the self-quarantined villagers. Finally, in late 1666, the rector recorded, *"Now, blessed be God, all our fears are over for none have died of the plague since the eleventh of October and the pest-houses have long been empty.* In the end, 260 of the town's 350 residents succumbed to the plague. Some have suggested this self-sacrificing incident is commemorated in a familiar children's nursery rhyme, one version of which is:

A ring-a-ring of rosies
A pocketful of posies
A tishoo! A tishoo!
We all fall down.

The ring of rosies refers to the rose-shaped splotches on the chest and armpits of plague victims. Posies were tiny flowers the people hoped would sweeten the air and ward off the foul smell associated with the disease. "A tishoo!" refers to the fits of sneezing that accompanied the disease. The last line, the saddest of all, suggests the deaths that befell so many.

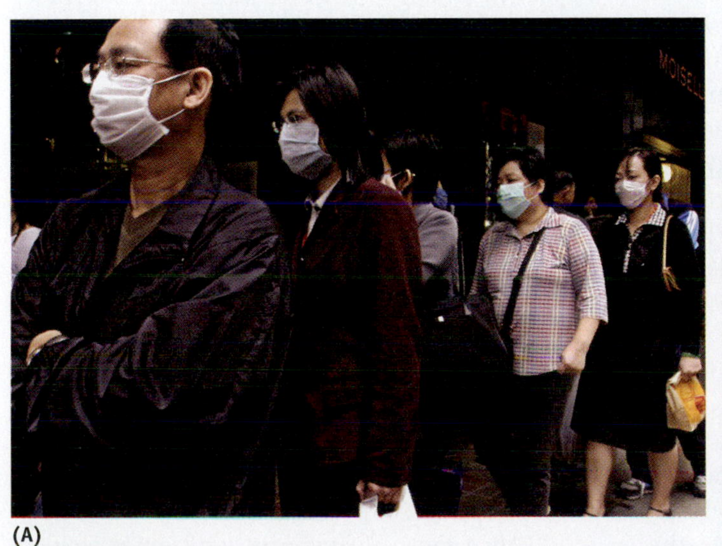

(A)

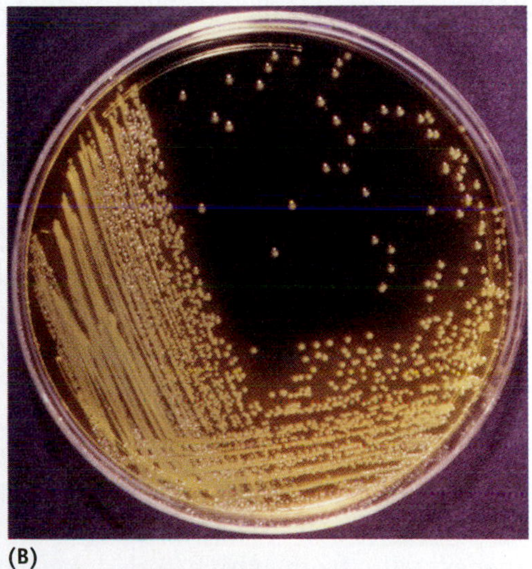

(B)

FIGURE 1.1 Infectious Disease and Model Organisms. (**A**) During the 2003 pandemic of SARS, people in affected areas such as Asia wore masks in an attempt to avoid being infected with the SARS virus. (**B**) Many microorganisms, such as this culture plate of the bacterium *Staphylococcus*, can be grown easily in culture, making them "experimental models" for study.

1.1 The Beginnings of Microbiology

■ Convex: Referring to a surface that curves outward.

Near the end of the 16th century, an observational revolution was about to begin: Dutch spectacle maker, Zacharias Janssen, was one of several individuals who discovered that if two convex lenses were put together, small objects would appear larger. Although the ability to magnify objects minimally with single pieces of glass (hand lenses) had been known for centuries, it was not until around 1600 that an instrument combined two lenses in a tube to increase magnification.

As the 17th century arrived, many individuals in Holland, England, and Italy further developed the instrument. In fact, it was in 1625 that the Italian Francesco Stelluti or Giovanni Faber used the term *microscopio* or microscope to refer to this new invention, which Galileo had suggested be called, "the small glass for spying things up close." This combination of lenses, or "compound microscope," would be the forerunner of the modern microscope.

Microscopy—Discovery of the Very Small

KEY CONCEPT

- The discovery of microorganisms was dependent on observations made with the microscope.

Robert Hooke, an English natural philosopher (the term *scientist* was not coined until 1833), was one of the most inventive and ingenious minds in the history of science. As the Curator of Experiments for the Royal Society of London, Hooke was the first to take advantage of the magnification abilities of the compound microscope. From these studies, the Royal Society in 1665 published *Micrographia* (FIGURE 1.2A). This book contained Hooke's descriptions of microscopes and was filled with stunning hand-drawn illustrations, including the first microorganism (a common bread mold) made from the objects he saw with his microscope.

Although these microscopes only magnified about 25 times (25×), Hooke's observations of thin slices of cork showed that these slices consisted of "a great many little boxes" (FIGURE 1.2B). He called the empty, enclosed spaces *cella*–from which today we have the word *cell*. However, another 200 years would pass before a formal theory of the cell was fully developed.

Micrographia represents one of the most important books in science history because Hooke's observations and meticulous illustrations awakened the learned and general population of Europe to the world of the very small, revolutionized the art of scientific investigation, and showed that the microscope was an important tool for unlocking the secrets of nature.

Anton van Leeuwenhoek, a contemporary of Hooke, was a successful tradesman and draper in Delft, Holland. As a cloth merchant, he used hand lenses to inspect the quality of cloth. After seeing Hooke's *Micrographia*, and without much education, Leeuwenhoek became skilled at grinding single pieces of glass into fine lenses, which he placed between two silver or brass plates riveted together (FIGURE 1.3A, B). Using only a single lens, no larger than the head of a pin, his "simple microscope" could magnify objects more than 200×.

The process of "observation" is an important skill for all scientists, including microbiologists—and Leeuwenhoek was an obsessed observer with the endless curiosity typical of today's scientists. He looked at the same sample with his microscopes repeatedly before he would accept his observations as accurate. He believed only sound observation and experimentation could be trusted—a requirement that remains a cornerstone of all science inquiry today.

Leeuwenhoek chose to communicate his observations through letters to the Royal Society. In 1674, one letter described a sample of cloudy surface water from a marshy lake. Placing the sample before his lens, he described hundreds of what he thought were tiny, living animals (probably protozoa), which he called **animalcules.** His curiosity aroused, Leeuwenhoek soon located animalcules in rainwater and in material from his own teeth and feces.

In 1683, he sent his 17th letter to the Royal Society in which he described and illustrated for the first time what almost certainly were bacterial cells from dental plaque (FIGURE 1.3C). Leeuwenhoek wrote:

> *"I then most always saw, with great wonder, that in the said matter there were many very little living animalcules, very prettily a-moving. The biggest sort . . . had a very*

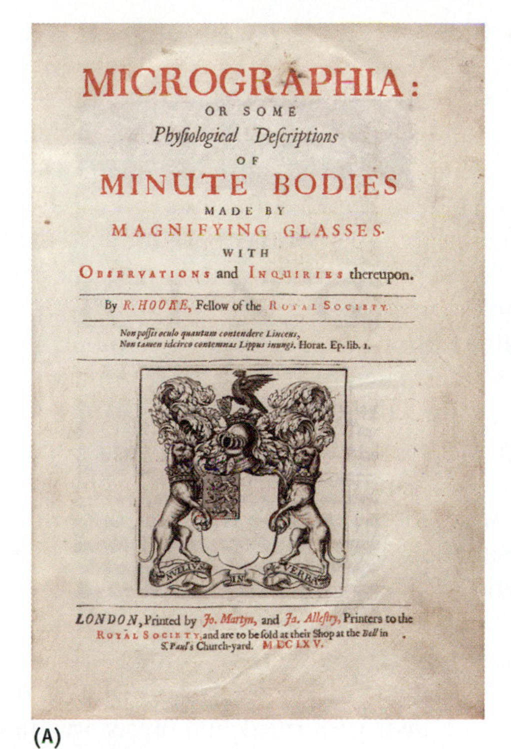

MICROGRAPHIA:
OR SOME
Physiological Descriptions
OF
MINUTE BODIES
MADE BY
MAGNIFYING GLASSES.
WITH
OBSERVATIONS and INQUIRIES thereupon.

By *R. HOOKE*, Fellow of the ROYAL SOCIETY.

Non possis oculo quantum contendere Linceus,
Non tamen idcirco contemnas Lippus inungi. Horat. Ep. lib. 1.

LONDON, Printed by *Jo. Martyn*, and *Ja. Allestry*, Printers to the ROYAL SOCIETY, and are to be sold at their Shop at the *Bell* in *S. Paul's* Church-yard. M DC LX V.

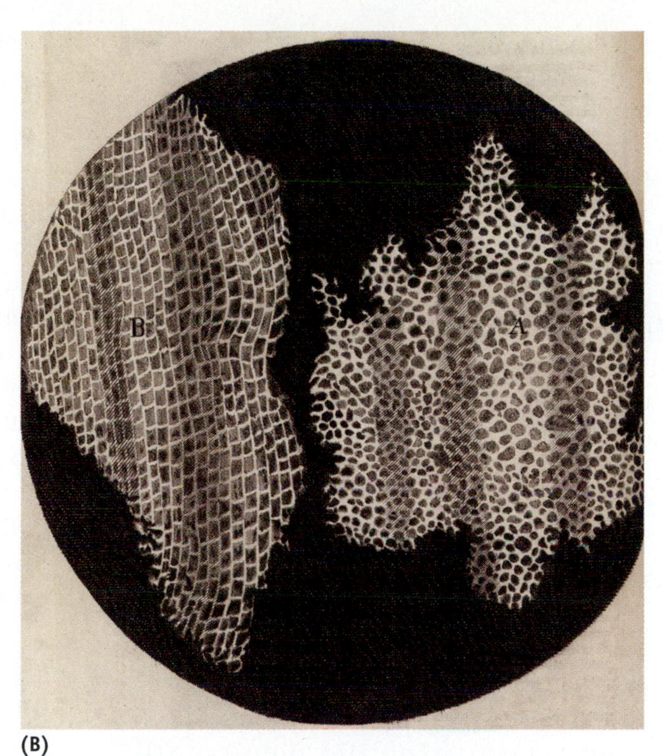

(A) (B)

FIGURE 1.2 **Hooke's *Micrographia* and Cork Cells.** In his *Micrographia* (**A**), Robert Hooke included a drawing of thin shavings of cork that he saw with his microscope (**B**). The empty compartments were called "cella" from which the term *cell* was derived.

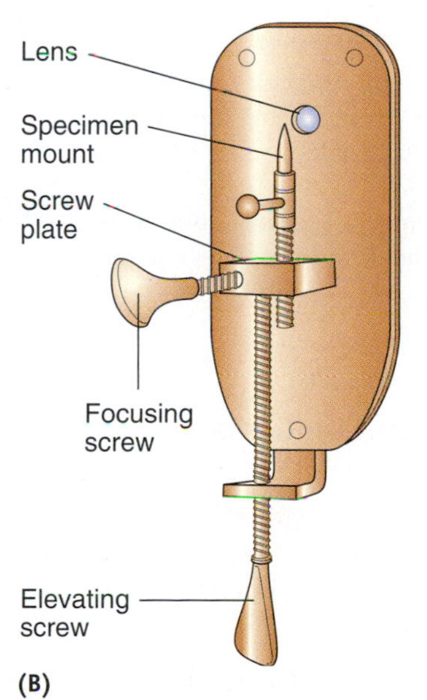

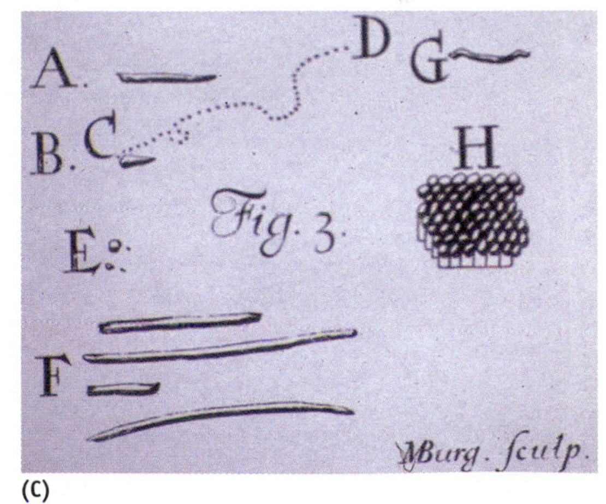

(A) (B) (C)

FIGURE 1.3 **Leeuwenhoek and His Animalcules.** (**A**) Leeuwenhoek looking through one of his simple microscopes. (**B**) For viewing, he placed a specimen on the tip of the specimen mount, which was attached to a screw plate. An elevating screw moved the specimen up and down while the focusing screw pushed against the metal plate, moving the specimen toward or away from the lens. (**C**) Leeuwenhoek's drawings of animalcules (bacterial cells) were included in a letter sent to the Royal Society in 1683. He found many of these organisms between his teeth and those of others.

strong and swift motion, and shot through the water (or spittle) like a pike does through the water. The second sort . . . ofttimes spun round like a top . . . and these were far more in number."

Leeuwenhoek's sketches were elegant in detail and clarity. Among the 165 letters sent to the Royal Society, he outlined structural details of protozoa and yeast, and described threadlike fungi and microscopic algae. Before his death in 1723 at age 90, Leeuwenhoek made numerous observations of cells and organisms, including red blood cells, spermatozoa, and plant cells.

Unfortunately, Leeuwenhoek was a very suspicious and secretive person. He invited no one to work with him, nor did he show anyone how he ground his lenses. Without these lenses, naturalists could not repeat his observations or verify his results, which are key components of scientific inquiry. Still, Leeuwenhoek's observations on the presence and diversity of his "marvelous beasties" and Hooke's *Micrographia* opened the door to a completely new world: the world of the microbe.

CONCEPT AND REASONING CHECKS

1.1. If you were alive in Leeuwenhoek's time, how would you explain the origin for the animalcules he found in materials such as lake water or dental plaque?

1.2. Regarding science inquiry, identify some pros and cons in Leeuwenhoek's studies with simple microscopes.

Experimentation—Can Life Generate Itself Spontaneously?

KEY CONCEPT

- The emergence of experimental science provided a means to test long-held beliefs and resolve controversies.

In the early 1600s, most naturalists were "vitalists," individuals who thought life depended on a mysterious "vital force" that pervaded all organisms. This force provided the basis for the doctrine of **spontaneous generation**, which suggested that organisms could arise from where there was putrefaction and decay. The renowned Flemish physician Jan Baptista van Helmont suggested that mice could spontaneously generate from decaying wheat bran and dirty shirts. Common people also embraced the idea, for they too witnessed what appeared to be slime that produced toads and decomposing wheat grains that generated wormlike maggots.

Regarding the latter, Leeuwenhoek suggested that maggots did not arise from wheat grains, but rather from tiny eggs laid in the grain that he could see with his microscope. Such divergent observations concerning spontaneous generation required a new form of investigation—"experimentation"—and a new generation of experimental naturalists arose.

Spontaneous generation had been accepted without any rigorous proof. Noting Leeuwenhoek's descriptions, the Italian naturalist Francesco Redi performed one of history's first biological experiments to see if maggots could arise from rotting meat. In 1688, he covered some jars of rotting meat with gauze, thereby preventing the entry of flies, while leaving other jars uncovered. If flies were prevented from landing on pieces of exposed meat, Redi predicted they could not lay their invisible eggs and no maggots would hatch (FIGURE 1.4). Indeed, that is exactly what Redi observed and the idea that spontaneous generation could produce larger living creatures soon subsided. However, what about the mysterious and minute animalcules that appeared to straddle the boundary between the nonliving and living world?

In 1748, a British clergyman and naturalist, John Needham, suggested that the spontaneous generation of animalcules resulted from the decay of more complex organisms; that is, as the organisms decayed, their molecules reorganized into animalcules. To prove this, Needham boiled several tubes of mutton gravy and sealed the tubes with corks. After several days, Needham proclaimed that the "*gravy swarm'd with life, with microscopical animals of most dimensions.*" He was convinced that putrefaction could generate the vital force needed for spontaneous generation.

(A)

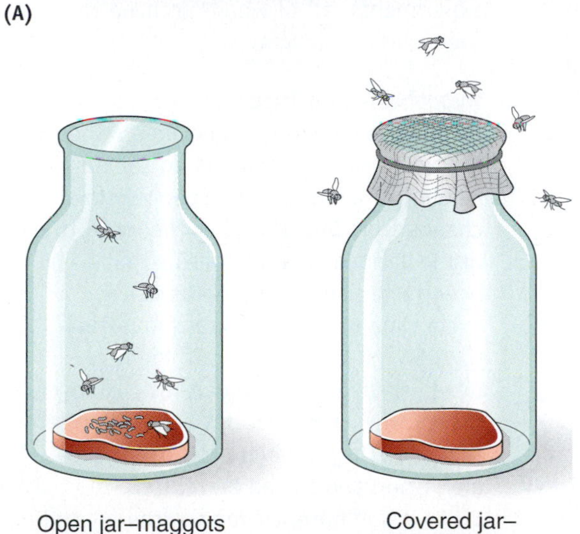

(B)

FIGURE 1.4 Redi's Experiments Refute Spontaneous Generation. In 1688, Francesco Redi (**A**) attempted to disprove the belief that maggots (fly larvae) arise spontaneously from decaying meat. (**B**) Redi carried out one of the first biological experiments by placing a piece of meat in an open jar and another in a jar covered with gauze. Maggots arose only in the open jar because flies had access to the meat where they laid their eggs. His experiment refuted the theory of spontaneous generation of larger organisms.

Since experiments almost always are subject to varying interpretations, the Italian cleric and naturalist Lazzaro Spallanzani challenged Needham's conclusions and suggested that the duration of heating might not have been long enough. He repeated Needham's experiments by placing meat and vegetable broth in a number of tubes, sealing the necks by melting the glass, and then boiling them for longer periods. As control experiments, he left some tubes open to the air and stoppered others loosely with corks. After two days, the open tubes were swarming with animalcules, but the stoppered ones had many fewer—and the sealed flasks contained none. Spallanzani proclaimed, *"the number of animalcula developed is proportional to the communication with the external air."*

Needham and others countered that Spallanzani's experiments had destroyed the vital force of life with excessive heating and excluded the air necessary for life. The controversy over spontaneous generation of animalcules continued into the mid-1800s. To solve the problem, a new and unique experimental strategy was needed.

To get at a resolution, the French Academy of Sciences sponsored a contest for the best experiment to prove or disprove spontaneous generation. In 1859, Louis Pasteur took up the challenge and, through an elegant series of experiments that were a variation of the methods of Needham and Spallanzani, discredited the idea. MicroInquiry 1 outlines the process of scientific inquiry and Pasteur's winning experiments.

Although Pasteur's experiments generated considerable debate for several years, his exacting and carefully designed experiments marked the end of the long and tenacious clashes over spontaneous generation that had begun two centuries earlier.

However, today there is another form of "spontaneous generation" occurring in the laboratory (MicroFocus 1.2).

CONCEPT AND REASONING CHECKS

1.3 Evaluate the role of experimentation as an important skill to the eventual rejection of spontaneous generation.

MICROINQUIRY 1

Experimentation and Scientific Inquiry

Science certainly is a body of knowledge as you can see from the thickness of this textbook! However, science also is a process—a way of learning. Often we accept and integrate into our understanding new information because it appears consistent with what we believe is true. But, are we confident our beliefs are always in line with what is actually true? To test or challenge current beliefs, scientists must present logical arguments supported by well-designed and carefully executed **experiments**.

The Components of Scientific Inquiry

There are many ways of finding out the answer to a problem. In science, **scientific inquiry**—or what has been called the "scientific method"—is the way problems are investigated. Let's understand how scientific inquiry works by following the logic of the experiments Louis Pasteur published in 1862 to refute the idea of spontaneous generation.

When studying a problem, the inquiry process usually begins with **observations**. For spontaneous generation, Pasteur's earlier observations suggested that organisms do not appear from nonliving matter (see text discussion of the early observations supporting spontaneous generation).

Next comes the **question**, which can be asked in many ways but usually as a "what," "why," or "how" question. For example, "What accounts for the generation of microorganisms in the beef broth?"

From the question, various **hypotheses** are proposed that might answer the question. A hypothesis is a provisional but testable explanation for an observed phenomenon. In almost any scientific question, several hypotheses can be proposed to account for the same observation. However, previous work or observations usually bias which hypothesis looks most promising, and scientists then put their "pet hypothesis" to the test first.

Pasteur's previous work suggested that the purported examples of life arising spontaneously in mutton gravy or other meat or vegetable broths were simply cases of airborne microorganisms landing on a suitable substance and then multiplying in such profusion that they could be seen as a cloudy liquid.

Pasteur's Experiments

Pasteur set up a series of experiments to test the hypothesis that "Life only arises from other life" (see facing page).

Experiment 1A and 1B: Pasteur sterilized a meat broth in glass flasks by heating. He then either left the neck open to the air (A) or sealed the glass neck (B). Organisms only appeared (turned the broth cloudy) in the open flask.

Experiment 2A and 2B: Pasteur sterilized a meat broth in necked glass flasks by heating. The glass neck was either heated (A) or left unheated in swan-neck flasks (B), so named because their S-shaped necks resembled a swan's neck. No organisms appeared in either case, even after several days.

Analysis of Pasteur's Experiments

Let's analyze the experiments. Pasteur had a preconceived notion of the truth and designed experiments to test his hypothesis. In his experiments, only one **variable** (an adjustable condition) changed. In experiment 1, the neck was open or closed; in experiment 2, the neck was heated or it wasn't heated. Pasteur kept all other factors the same; that is, the broth was the same in each experiment; it was heated the same length of time; and similar flasks were used. Thus, the experiments had rigorous **controls** (the comparative condition): in experiment 1, the control was the flask left open; in experiment 2, the control was the unheated swan-neck. Such controls are pivotal when explaining an experimental result. Pasteur's finding that no life appeared in the swan-necked flask (experiment 2B) is interesting, but tells us very little by itself. We only learn something by comparing this finding to the result in the control condition: that life did grow when the flask neck was straight (experiment 1A) or in an extension of the experiment when the neck was cut off.

Also note that the idea of spontaneous generation could not be dismissed by just one experiment (see "His critics" on facing page). Pasteur's experiments required the accumulation of many experiments, all of which pointed to the same conclusion.

Hypothesis and Theory

When does a hypothesis become a theory? The answer is that there is no set time or amount of evidence that specifies the change from hypothesis to theory. A **theory** is defined as a hypothesis that has been tested and shown to be correct every time by many separate investigators. So, at some point, sufficient evidence exists to say a hypothesis is now a theory. However, theories are not written in stone. They are open to further experimentation and so can be refuted.

As a side note, a theory often is used incorrectly in everyday speech and in the news media. In these cases, a theory is equated incorrectly with a hunch or belief—whether or not there is evidence to support it. In science, a theory is a general set of principles supported by large amounts of experimental evidence.

Discussion Point

Based on Pasteur's experiments, one could still argue that spontaneous generation could occur. Discuss how this could be so.

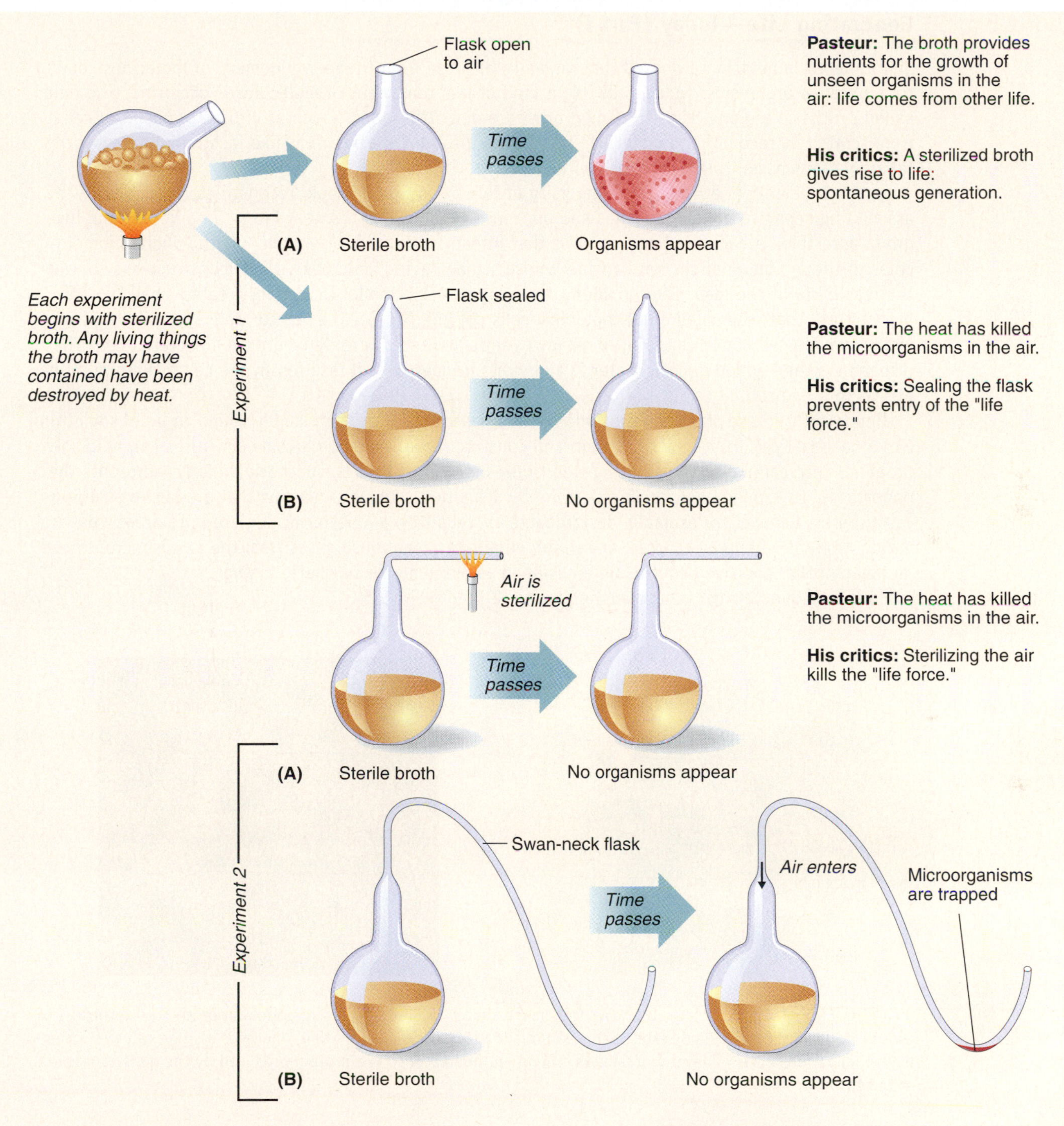

Pasteur and the Spontaneous Generation Controversy. (**1A**) When a flask of sterilized broth is left open to the air, organisms appear. (**1B**) When a flask of sterilized broth is boiled and sealed, no living things appear. (**2A**) When air entering a flask of sterilized broth is heated with a flame, no living things appear. (**2B**) Broth sterilized in a swan-neck flask is left open to the air. The curvature of the neck traps dust particles and microorganisms, preventing them from reaching the broth.

MICROFOCUS 1.2: Biotechnology

Generating Life—Today (Part I)

Spontaneous generation proposed that animalcules arose from the rearrangement of molecules coming from decayed organisms. Today, a different kind of rearrangement of molecules is occurring. The field, called synthetic biology, aims to rebuild or create new "life forms" (such as viruses or bacterial cells) from scratch by recombining molecules taken from different species. It is like fashioning a new car by taking various parts from a Ford, Chevy, and Toyota.

In 2002, scientists at the State University of New York, Stony Brook, reconstructed a poliovirus by assembling separate poliovirus genes and proteins (FIGURE A). A year later, Craig Venter and his group assembled a bacteriophage—a virus that infects bacterial cells—from "off-the-shelf" biomolecules. Although many might not consider viruses to be "living" microbes, these constructions showed the feasibility of the idea. Then in 2004, researchers at Rockefeller University created small "vesicle bioreactors" that resembled crude biological cells (FIGURE B). The vesicle walls were made of egg white and the cell contents, stripped of any genetic material, were derived from a bacterial cell. The researchers then added genetic material and viral enzymes, which resulted in the cell making proteins, just as in a live cell.

Importantly, these steps toward synthetic life have more uses than simply trying to build something like a bacterial cell from scratch. Design and construction of novel organisms or viruses can help solve problems that cannot be solved using traditional organisms; that is, synthetic biology represents the opportunity to expand evolution's repertoire by designing cells or organisms that are better at doing certain jobs. Can we, for example, design bacterial cells that are better at degrading toxic wastes, providing alternative energy sources, or helping eliminate greenhouse gases from the atmosphere? These and many other positive benefits are envisioned as outcomes of synthetic biology.

Part II of Generating Life appears in Chapter 2 (p. 63).

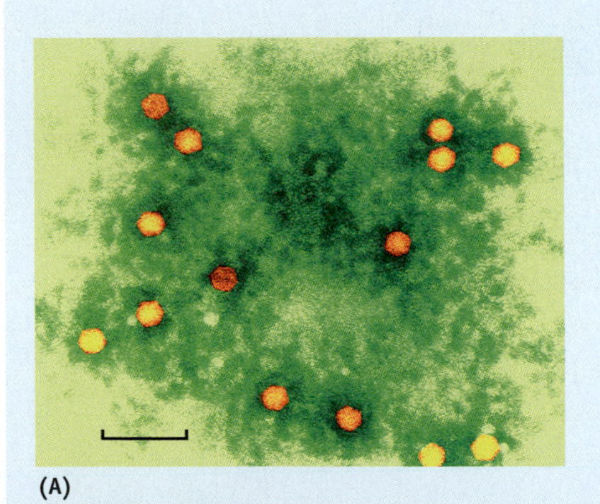
(A)

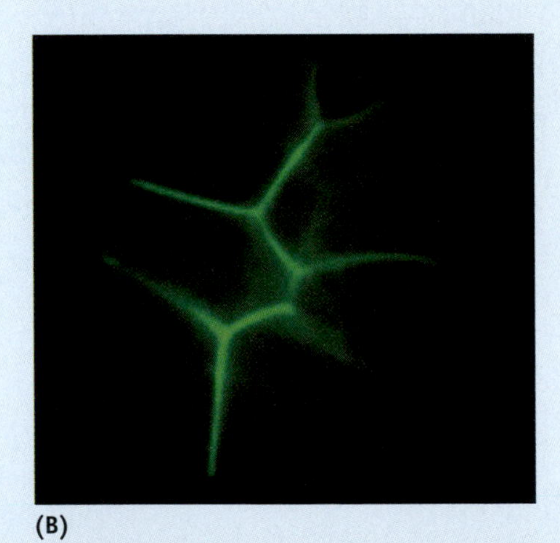
(B)

FIGURES A & B **Making New Life?** (**A**) This image shows naturally-occurring polioviruses, similar to those assembled from the individual parts. (Bar = 50 nm.) (**B**) A "vesicle reactor" that simulates a crude cell was assembled from various parts of several organisms. The green fluorescence is a protein produced by the genetic material added to the vesicle.

1.2 Microorganisms and Disease Transmission

In the 13th century, people knew diseases could be contagious, so quarantines were used to combat disease spread. In 1546, the Italian poet and naturalist Girolamo Fracostoro suggested that transmission could occur by direct human contact, from lifeless objects like clothing and eating utensils, or through the air.

By the mid-1700s, the prevalent belief among naturalists and common people was that disease resulted from an altered chemical quality of the atmosphere or from tiny poisonous particles in the air, an entity called **miasma** (the word malaria comes from *mala aria*, meaning "bad air"). To protect oneself from the black plague in Europe, for example, plague doctors often wore an elaborate costume they thought would protect them from the plague miasma (FIGURE 1.5A).

As the 19th century unfolded, more scientists relied on keen observations and experimentation as a way of knowing and explaining divergent observations, including contagion and disease.

Epidemiology—Understanding Disease Transmission

KEY CONCEPT

- Early epidemiology studies suggested how diseases could be spread and be controlled.

Epidemiology, as applied to infectious diseases, is the scientific study from which the source, cause, and mode of transmission of disease can be identified. The first scientific epidemiological studies, carried out by Ignaz Semmelweis and John Snow, were instrumental in suggesting how diseases were transmitted—and how simple measures could interrupt transmission.

Ignaz Semmelweis was a Hungarian obstetrician who was shocked by the numbers of pregnant women in his hospital who were dying of puerperal fever (a type of blood poisoning also called childbed fever) during labor. He determined the disease was more prevalent in the ward handled by medical students (29% deaths) than in the ward run by midwifery students (3% deaths). This comparative study suggested to Semmelweis that the mode of transmission must involve his medical students. He deduced that the source of contagion must be from cadavers on which the medical students previously had been performing autopsies because midwifery students did not work on cadavers. So, in 1847, Semmelweis directed his staff to wash their hands in chlorine water before entering the maternity ward (FIGURE 1.5B). Deaths from childbed fever dropped, showing that disease spread could be interrupted. Unfortunately, few physicians initially heeded Semmelweis' recommendations.

In 1854, a cholera epidemic hit London, including the Soho district. With residents dying, English surgeon John Snow set out to discover the reason for cholera's spread. He carried out one of the first thorough epidemiological studies by interviewing sick and healthy Londoners and plotting the location of each cholera case on a district map (FIGURE 1.6). The results indicated most cholera cases clustered to a sewage-contaminated street pump from which local residents obtained their drinking water. Snow then instituted the first known example of a public health measure to interrupt disease transmission—he requested the parish Board of Guardians to remove the street pump

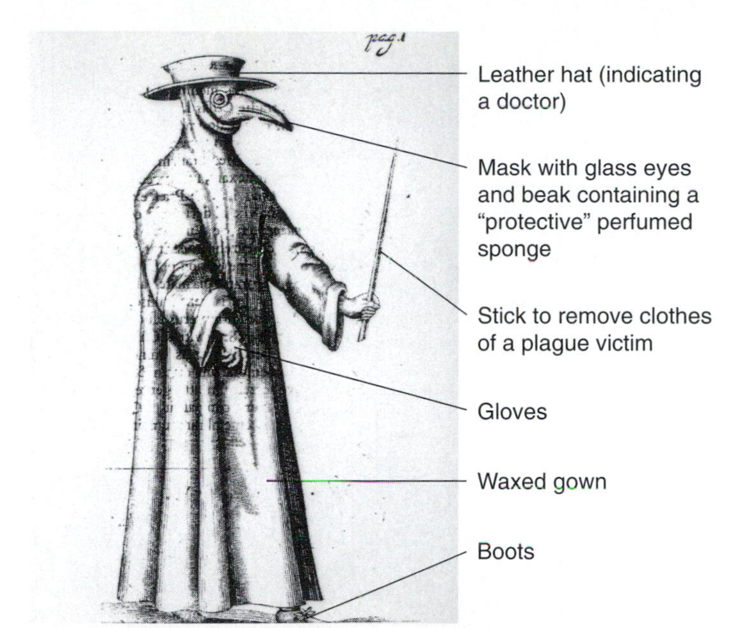

(A)

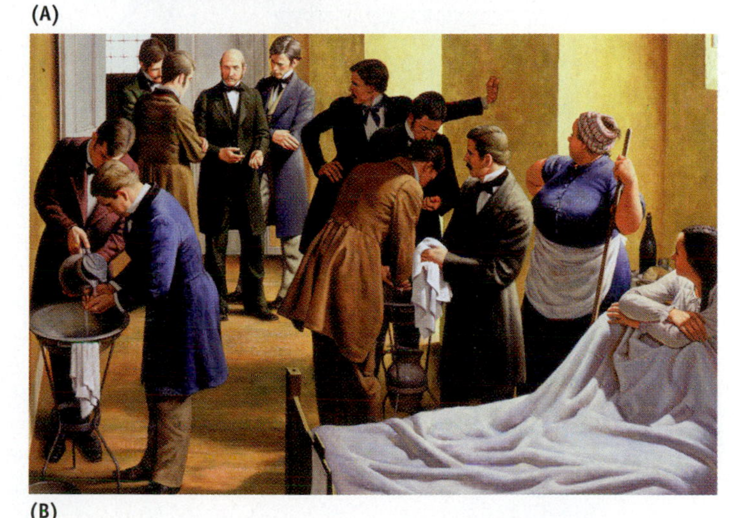

(B)

FIGURE 1.5 **Keeping Disease at Bay.** (**A**) This dress was thought to protect a plague doctor from the air (miasma) that caused the plague. (**B**) Semmelweis believed if hospital staff washed their hands, cases of puerperal fever would be reduced by preventing its spread from medical students to patient.

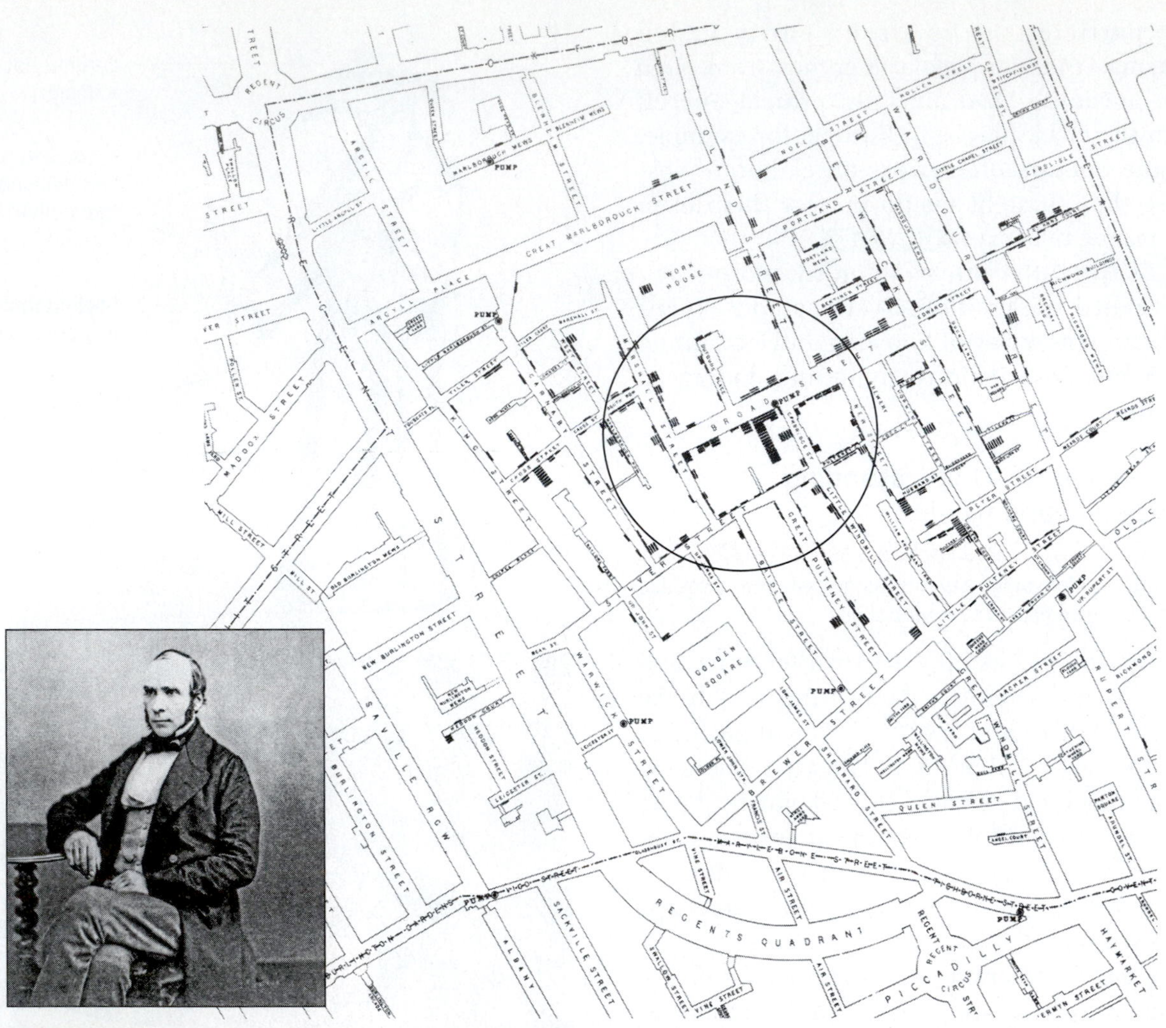

FIGURE 1.6 **John Snow and Cholera.** John Snow (inset) produced a map plotting all the cholera cases in the London Soho district and observed a cluster near to the Broad Street pump (circle).

handle! Again, disease spread was broken by a simple procedure.

Snow went on to propose that cholera was not spread by a miasma but rather was waterborne. In fact, he asserted that a specific germ caused cholera—an educated guess that proved to be correct even though the causative agent would not be identified for another 29 years.

■ **Germ:** Any microorganism capable of causing disease.

It is important to realize that although the miasma premise was incorrect, the fact that disease was associated with bad air and filth led to new hygiene measures, such as cleaning streets, laying new sewer lines in cities, and improving working conditions. These changes helped usher in the Sanitary Movement and create the infrastructure for the public health systems we have today (MicroFocus 1.3).

CONCEPT AND REASONING CHECKS

1.4 Contrast the importance of the observations and studies by Semmelweis and Snow toward providing a better understanding of disease transmission.

Variolation and Vaccination—Prevention of Infectious Disease

KEY CONCEPT

- Resistance to a disease can come from exposure to and recovery from a mild form of (or a very similar) disease.

Besides the controversies over the mechanism of disease transmission, ways to prevent disease from occurring were being considered. In the 1700s, smallpox was prevalent throughout Europe. In England, for example, smallpox epidemics were so severe that one third of the

MICROFOCUS 1.3: Public Health
Epidemiology Today

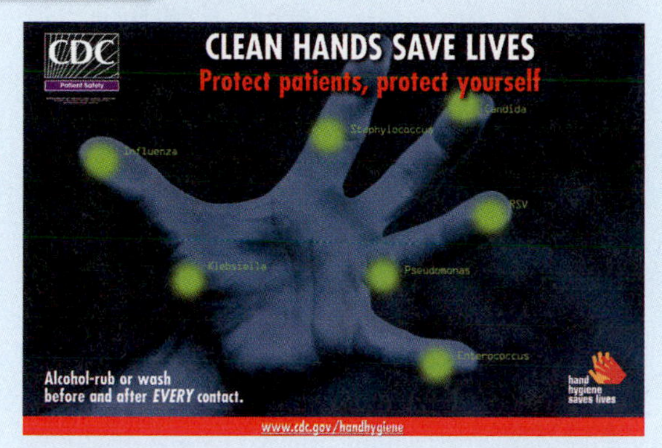

Today, we have a good grasp of disease transmission mechanisms, as we will discuss in Chapter 18. However, even with the advances in sanitation and public health, cholera remains a public health threat in parts of the developing world. In addition, almost 160 years after Semmelweis' suggestions, a lack of hand washing by hospital staff, even in developed nations, remains a major mechanism for disease transmission. The simple process of washing one's hands still could reduce substantially disease transmission among the public and in hospitals.

Two of the most important epidemiological organizations today are the Centers for Disease Control and Protection (CDC) in Atlanta, Georgia and, on a global perspective, the World Health Organization (WHO) in Geneva, Switzerland. Both employ numerous epidemiologists, popularly called "disease detectives," who, like Snow (but with more expertise), systematically gather information about disease outbreaks in an effort to discover how the disease agent is introduced, how it is spread in a community or population, and how the spread can be stopped. For example, in 1993 more than 200 people in Washington State developed similar gastrointestinal symptoms, which CDC investigators traced to bacterial contamination in hamburger meat from a fast-food chain. Warnings to cook beef until it was well done halted the outbreak and prevented further transmission. And to think, it all started with the seminal work of Semmelweis and Snow.

children died before reaching the age of three. Many victims who recovered were blinded from corneal infections and most were left pockmarked.

Significantly, survivors were protected from suffering the disease a second time. These observations suggested that if one contracted a weakened or mild form of the disease, perhaps such individuals would have lifelong resistance.

In the 14th century, the Chinese knew that smallpox survivors would not get reinfected. Spreading from China to India and Africa, the practice of **variolation** developed, which involved blowing a ground smallpox powder into the individual's nose. Europeans followed by inoculating dried smallpox scabs under the skin. Although some individuals did get smallpox, most contracted a mild form of the disease and, upon recovery, were resistant to future smallpox infections. Awareness of such successes often lead to "smallpox parties" at which individuals were inoculated with smallpox material.

As an English country surgeon, Edward Jenner learned that milkmaids who occasionally contracted cowpox, a disease of the udders of cows, would come down with a mild, smallpox-like disease. With recovery, they were protected from smallpox. Jenner wondered if intentionally giving cowpox to people would protect them against smallpox and be an effective alternative to variolation. In 1796, he put the matter to the test.

A dairy maid named Sarah Nelmes came to his office, the lesions of cowpox evident on her hand. Jenner took material from the lesions and scratched it into the skin of a boy named James Phipps (FIGURE 1.7). The boy developed a slight fever, but recovered. Six weeks later Jenner inoculated young Phipps with material from a smallpox lesion. Within

FIGURE 1.7 **The First Vaccination against Smallpox.** Edward Jenner performed the first vaccination against smallpox. On May 14, 1796, material from a cowpox lesion was scratched into the arm of eight-year-old James Phipps. The vaccination protected him from smallpox.

days, the boy developed a reaction at the site but failed to show any sign of smallpox.

Jenner repeated his experiments with other children, including his own son. His therapeutic technique of **vaccination** (*vacca* = "cow") worked in all cases and eliminated the risks associated with variolation. In 1798, he published a pamphlet on his work that generated considerable interest. Prominent physicians confirmed his findings, and within a few years, Jenner's method of vaccination spread through Europe and abroad. By 1801, some 100,000 people in England were vaccinated even though no one knew the biological mechanism generating disease resistance. President Thomas Jefferson wrote to Jenner, *"You have erased from the calendar of human afflictions one of its greatest. Yours is the comfortable reflection that mankind can never forget that you have lived."*

A hundred years would pass before scientists discovered the milder cowpox virus was setting up a defensive mechanism by the body's immune system against the deadlier smallpox virus. It is remarkable that without any knowledge of viruses or disease causation, Jenner accomplished what he did. Again, hallmarks of a scientist—keen observational skills and insight—led to a therapeutic intervention against disease.

CONCEPT AND REASONING CHECKS

1.5 Evaluate the effectiveness of variolation and vaccination as ways to produce disease resistance.

The Stage Is Set

During the early years of the 1800s, several events occurred that helped set the stage for the coming "germ revolution." In the 1830s, advances were made in microscope optics that allowed better resolution of objects. This resulted in improved and more widespread observations of tiny living organisms, many of which resembled short sticks. In fact, in 1838 the German biologist Christian Ehrenberg suggested these "rod-like" looking organisms be called **bacteria** (*bakterion* = "little rod"). A note in passing: the 1830s were quite noteworthy for biology. In 1831, Darwin set sail on his voyage of discovery aboard the HMS Beagle, and by the end of the decade the "cell theory"—that all organisms are made of cells—was being revealed by Schleiden and Schwann.

Advances also were being made in acknowledging that "particles" were involved in the transmission of disease. Although the miasma idea had suggested these were nonliving entities, the Swiss physician Jacob Henle suggested in 1840 that living organisms could cause disease. This was strengthened in 1854 by Filippo Pacini's discovery of rod-shaped cholera bacteria in stool samples from cholera patients.

Still, scientists debated whether bacterial organisms could cause disease because such living organisms also were found in healthy people. Therefore, how could these bacterial cells possibly cause disease?

To understand clearly the nature of infectious disease, a new conception of disease had to emerge. In doing so, it would be necessary to demonstrate that a specific bacterial species was associated with a specific infectious disease. This would require some very insightful work, guided by Louis Pasteur in France and Robert Koch in Germany.

TABLE 1.1 is a summary of the early accomplishments regarding microorganisms, spontaneous generation, and disease transmission and prevention.

TABLE 1.1 Some Early Accomplishments in Microbiology

Investigator	Time Frame	Accomplishment
Fracostoro	Mid-1500s	"Contagion" passes among individuals, objects, and air
Hooke	Late-1600s	The compound microscope is used for magnifying small objects; reproductive structures of a mold observed and described
Leeuwenhoek	Late 1600s	Animalcules are present in many environments
Fabricius	Early 1700s	Fungi cause diseases in plants
Jablot	Early 1700s	Various forms of protozoa observed
Needham	Mid-1700s	Animalcules in broth arise by spontaneous generation
Spallanzani	Mid-1700s	Heat destroys animalcules in broth
Jenner	Late 1700s	Vaccination against smallpox is successful
Ehrenberg	Early 1800s	Many of the microscopic animalcules are called bacteria
Henle	Mid-1800s	Living organisms could cause disease
Semmelweis	Mid-1800s	Chlorine hand washing prevents disease spread
Snow	Mid-1800s	Water is involved in disease transmission
Pasteur	Mid-1800s	Spontaneous generation does not occur

1.3 The Classical Golden Age of Microbiology (1854–1914)

Beginning around 1854, microbiology blossomed and continued until the advent of World War I. During these 60 years, many branches of microbiology were established, and the foundations were laid for the maturing process that has led to modern microbiology. We refer to this period as the first, or classical, Golden Age of microbiology because a case will be made for three such epochs.

Louis Pasteur Proposes That Germs Cause Infectious Disease

KEY CONCEPT

- The germ theory was based on the observations that different microorganisms have distinctive and specific roles in nature.

Born in 1822 in Dôle, France, Louis Pasteur studied chemistry at the École Normale Supérieure in Paris and, in 1854, was appointed Professor of Chemistry at the University of Lille in northern France (FIGURE 1.8A). Pasteur was among the first scientists who believed that problems in science could be solved in the laboratory with the results having practical applications. Always one to tackle big problems, in 1857 he set out to prove yeasts were the living organisms responsible for the chemical process of wine fermentation.

■ Fermentation: A splitting of sugar molecules into simpler products, including alcohol, acid, and gas (CO_2).

The prevailing theory held that wine fermentation resulted from the chemical breakdown of grape juice to alcohol. No living agent seemed to be involved. However, Pasteur's microscope observations consistently revealed large numbers of tiny yeast cells in wine that were overlooked by other scientists. When he mixed yeast in a sugar-water solution, the yeast grew and the quantity of yeast increased. Yeast must be living organisms.

Pasteur also demonstrated that wines, beers, and vinegar each contained different and specific types of microorganisms (FIGURE 1.8B). For example, in studying a local problem of wine souring, he observed that only soured wines contained populations of bacterial cells. These cells must have contaminated a batch of yeast and produced the acids that caused the souring.

Pasteur recommended a practical solution for the "wine disease" problem: heat the grape juice to destroy all the evidence of life, after which yeasts should be added to begin the fermentation. Alternatively, heat the wine after fermentation but before aging. His controlled

(A)

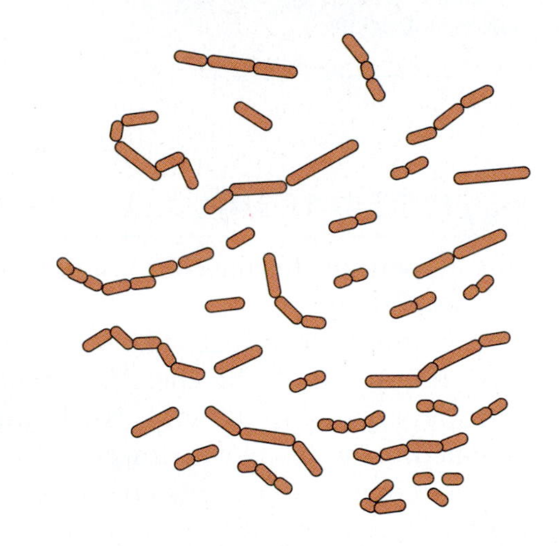

(B)

FIGURE 1.8 **Louis Pasteur and Fermentation Bacteria.** (**A**) Louis Pasteur as a 32 year-old professor of chemistry at the Universtiy of Lille. (**B**) The following is part of a description of the living bacterial cells he observed. *"A most beautiful object: vibrios all in motion, advancing or undulating. They have grown considerably in bulk and length since the 11th; many of them are joined together in long sinuous chains . . . "* Pasteur concluded these bacterial cells can live without air or free oxygen; in fact, *"the presence of gaseous oxygen operates prejudicially against the movements and activity of those vibrios."*

heating technique to kill pathogens, known as **pasteurization**, soon was applied to other products, especially milk.

Pasteur's experiments demonstrated that yeast and bacterial cells are tiny, living factories in which important chemical changes take place. Therefore, if microorganisms represented agents of change, perhaps human infections could be caused by those microorganisms that cause disease—**germs**.

In 1857, Pasteur published a short paper on wine souring by bacterial cells in which he implied that germs also could be related to human illness. Five years later, he formulated the **germ theory** of disease, which holds that some microorganisms are responsible for infectious disease.

Pasteur's discoveries would be tempered with sadness, however. In 1859, his daughter Jeanne died of typhoid fever.

CONCEPT AND REASONING CHECKS

1.6 Describe how wine fermentation and souring led Pasteur to propose the germ theory.

Pasteur's Work Stimulates Disease Control and Reinforces Disease Causation

KEY CONCEPT

- Antisepsis and identification of the cause of animal diseases reinforced the germ theory.

Pasteur had reasoned that if microorganisms were acquired from the environment, their spread could be controlled and the chain of disease transmission broken.

Joseph Lister was Professor of Surgery at Glasgow Royal Infirmary in Scotland, where more than half his amputation patients died—not from the surgery—but rather from postoperative infections. Hearing Pasteur's germ theory, Lister argued that these surgical infections resulted from living organisms in the air. Knowing that carbolic acid had been effective on sewage control, in 1865 he used a carbolic acid spray in surgery and on surgical wounds (FIGURE 1.9). The result was spectacular—the wounds healed without infection. His technique would soon not only revolutionize medicine and the practice of surgery, but also lead to the practice of **antisepsis**, the use of chemical methods for disinfection of external living surfaces, such as the skin.

Pasteur's belief in the germ theory was strengthened only once again to be shattered when his two-year-old daughter Camille developed a tumor and died of blood poisoning. Pasteur realized he was no closer to solving the riddle of disease.

In 1865, cholera engulfed Paris, killing 200 people a day. Pasteur tried to capture the

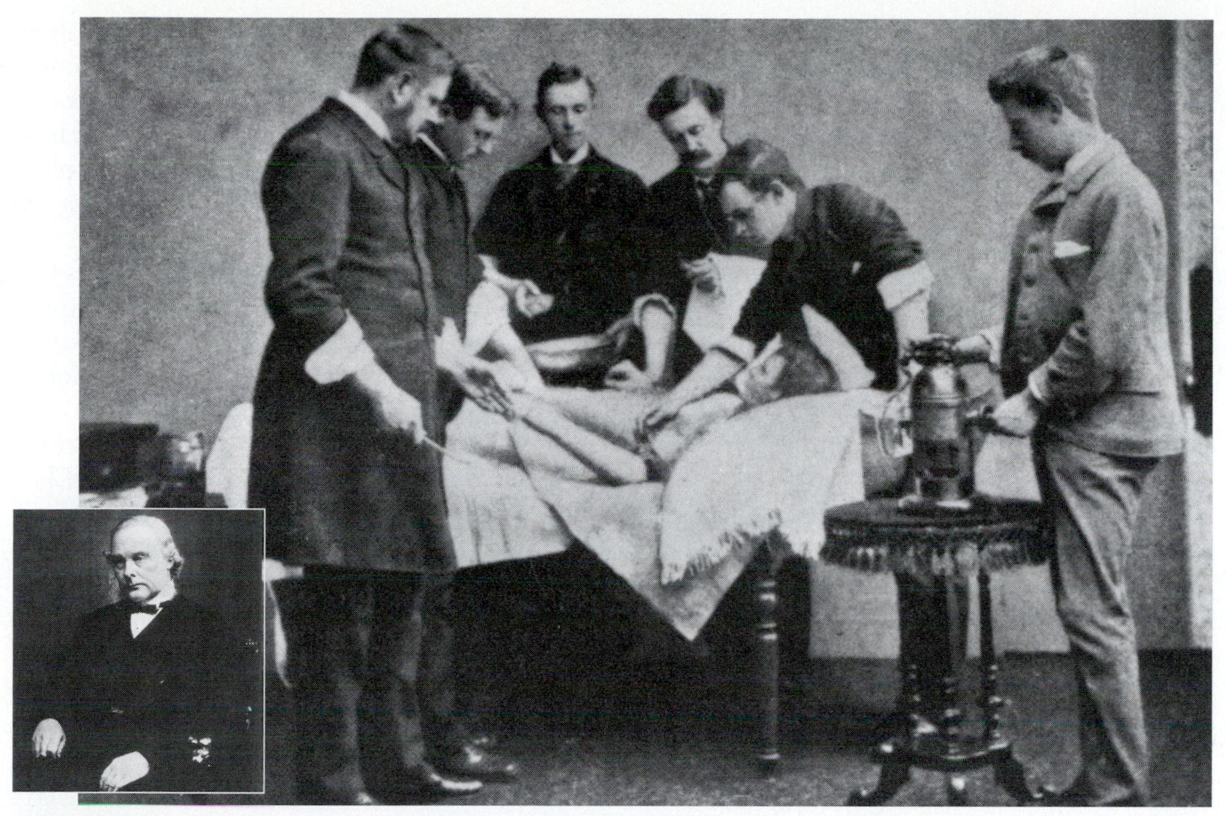

FIGURE 1.9 **Lister and Antisepsis.** By 1870, Joseph Lister (inset) was using a carbolic acid spray in surgery and on surgical wounds to prevent post-operative infections.

responsible pathogen by filtering the hospital air and trapping the bacterial cells in cotton. Unfortunately, Pasteur could not grow or separate one bacterial species apart from the others because his broth cultures allowed the organisms to mix freely. Although Pasteur demonstrated that bacterial inoculations made animals ill, he could not pinpoint an exact cause.

In an effort to help French industry again, Pasteur turned his attention to pébrine, the disease of silkworms. Late in 1865, he identified a protozoan infesting the silkworms and the mulberry leaves fed to the worms. By separating the healthy silkworms from the diseased silkworms and their food, he managed to quell the spread of disease. The identification of the protozoan was crucial to supporting the germ theory and Pasteur would never again doubt the ability of microorganisms to cause infectious disease. For Pasteur, however, it was another time of grief. Cecille, his third daughter, succumbed at the age of 12 to typhoid fever. Now infectious disease would be his only interest.

Although Pasteur failed to relate a specific microorganism to a specific human disease, his work stimulated others to investigate the nature of microorganisms and to ponder their association with disease. For example, Gerhard Hansen, a Norwegian physician, identified bacterial cells in the tissues of leprosy patients in 1871, and Otto Obermeier of Germany described bacterial organisms in the blood of relapsing fever patients in 1873. Another German bacteriologist, Ferdinand Cohn, discovered that bacterial cells multiply by dividing, suggesting that infecting bacterial cells could grow and multiply in number.

To completely prove the germ theory, what was missing was the ability to isolate a specific bacterial species from a diseased individual and demonstrate the isolated organism caused the same disease.

■ Broth:
A liquid containing nutrients for microbial growth.

CONCEPT AND REASONING CHECKS

1.7 Examine how Lister's antisepsis procedures and Pasteur's work on pébrine each supported the germ theory.

Robert Koch Formalizes Standards to Identify Germs with Infectious Disease

KEY CONCEPT

- Koch's postulates provided a way to identify a specific microorganism as causing a specific infectious disease.

Robert Koch (FIGURE 1.10A) was a German country doctor who was well aware of anthrax, a deadly disease that periodically ravaged cattle and sheep, and could cause disease in humans.

In 1875, Koch injected mice with the blood from such diseased sheep and cattle. He then performed meticulous autopsies and noted the same symptoms in the mice that had appeared in the sheep and cattle. Next, he isolated from the blood a few rod-shaped bacterial cells (called **bacilli**) and grew them in the aqueous humor of an ox's eye. Koch watched for hours as the bacilli multiplied, formed tangled threads, and finally reverted to highly resistant spores. He then took several spores on a sliver of wood and injected them into healthy mice. The symptoms of anthrax appeared within hours. Koch autopsied the animals and found their blood swarming with bacilli. He reisolated the bacilli in fresh aqueous humor. The cycle was now complete. The bacilli definitely were the causative agent of anthrax.

When Koch presented his work, scientists were astonished. Here was the verification of the germ theory that had eluded Pasteur. Koch's procedures became known as **Koch's postulates** and were quickly adopted as the formalized standards for relating a specific organism to a specific disease (FIGURE 1.10B).

Koch Develops Pure Culture Techniques

In 1877, Koch developed methods for staining bacterial cells and preparing permanent visual records. Then, in 1880, Koch accepted an appointment to the Imperial Health Office, and while there, he observed a slice of potato on

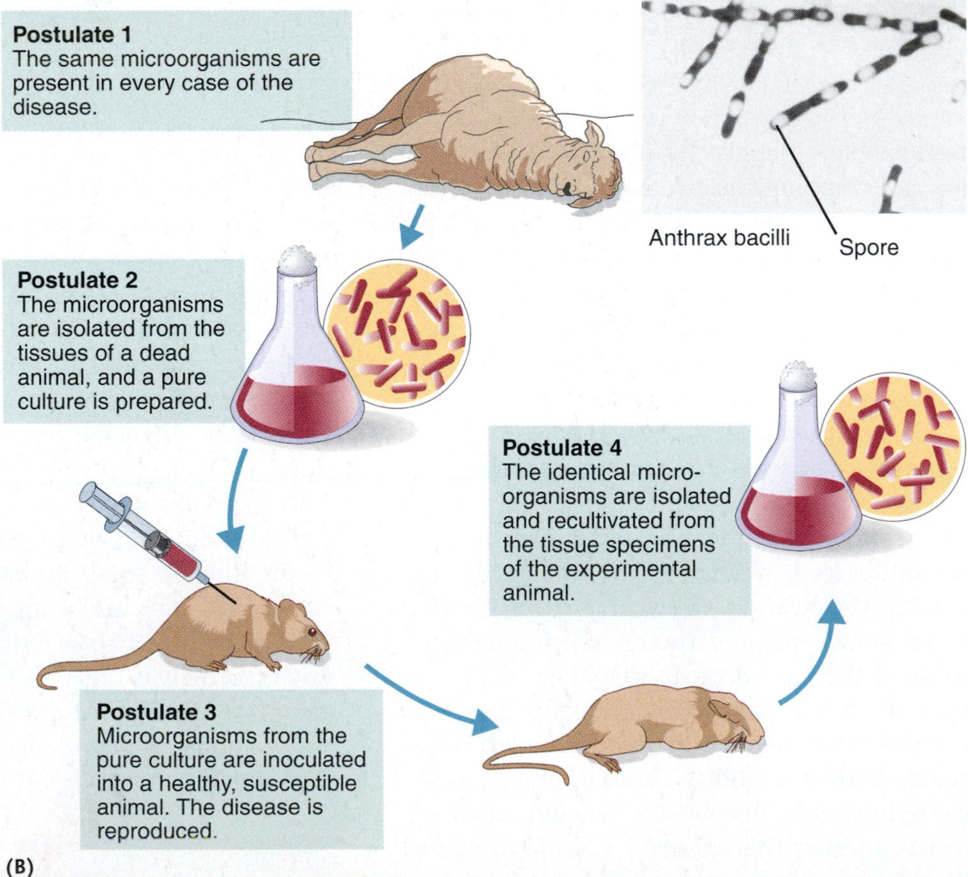

FIGURE 1.10 **A Demonstration of Koch's Postulates.** Robert Koch (**A**) developed what became known as Koch's postulates (**B**) that were used to relate a single microorganism to a single disease. The insert (in the upper right) is a photo of the rod-shaped anthrax bacteria. Many rods are swollen with spores (white ovals).

which small masses of bacterial cells, which he termed **colonies**, were growing and multiplying. So, Koch tried adding gelatin to his broth to prepare a solid culture surface in a culture (Petri) dish. He then inoculated bacterial cells on the surface and set the dish aside to incubate. Within 24 hours, visible colonies were present on the surface (MicroFocus 1.4 details a further advance in cultivation techniques).

Koch now could inoculate laboratory animals with a pure culture of bacterial cells and be certain that only one bacterial species was involved. In 1881, he outlined his pure culture techniques to the International Medical Congress in London. Pasteur and several of his coworkers were present. Several days later, Koch received a personal letter of congratulations from Pasteur.

CONCEPT AND REASONING CHECKS

1.8 Why was pure culture crucial to Koch's postulates?

Competition Fuels the Study of Infectious Disease

KEY CONCEPT

- Laboratory science and teamwork stimulated the discovery of additional infectious disease agents.

The years following the 1870 Franco-Prussian war were accompanied by fierce national pride. Both France and Germany were undergoing unification, and heroes, including scientists, played an important role in the spirit of nationalism. A competition arose that would last into the next century.

Research studies conducted in a laboratory were becoming the normal method of work. Pasteur's lab and coworkers were primarily interested in the mechanism of infection and immunity, and the practical applications that could be derived, while Koch's lab focused on procedural methods such as isolation, cultivation, and identification of specific pathogens.

MICROFOCUS 1.4: History

Jams, Jellies, and Microorganisms

One of the major developments in microbiology was Robert Koch's use of a solid culture surface on which bacterial colonies would grow. He accomplished this by solidifying beef broth with gelatin. When inoculated onto the surface of the nutritious medium, bacterial cells grew vigorously at room temperature and produced discrete, visible colonies.

On occasion, however, Koch was dismayed to find that the gelatin turned to liquid. It appeared that certain bacterial species were producing a chemical substance to digest the gelatin. Moreover, gelatin liquefied at the warm incubator temperatures commonly used to cultivate certain bacterial species.

Walther Hesse, an associate of Koch's, mentioned the problem to his wife and laboratory assistant, Fanny Eilshemius (Hesse). She had a possible solution. For years, she had been using a seaweed-derived powder called agar (pronounced ah'-gar) to solidify her jams and jellies. The formula had been passed to her by her mother, who learned it from Dutch friends living in Java. Agar was valuable because it mixed easily with most liquids and once gelled, it did not liquefy, even at the warm incubator temperatures.

Fanny Hesse

Hesse was sufficiently impressed to recommend agar to Koch. Soon Koch was using it routinely to grow bacterial species, and in 1884 he first mentioned agar in his paper on the isolation of the tubercle bacillus. It is noteworthy that Fanny Eilshemius may have been among the first Americans (she was originally from New Jersey) to make a significant contribution to microbiology.

Another point of interest: The common Petri dish also was invented about this time (1887) by Julius Petri, another of Koch's assistants.

Pasteur continued work with anthrax and found that anthrax bacilli were temperature sensitive. Chickens did not acquire anthrax at their normal body temperature of 42°C, but did so when the animals were cooled down to 37°C. Pasteur also recovered anthrax spores from the soil and suggested that disease transmission could be stopped if dead animals were burned or buried deeply in soil unfit for grazing.

One of Pasteur's more remarkable discoveries was made in 1880. For months, he had been working on ways to attenuate the bacterial cells of chicken cholera using heat, different growth conditions, successive inoculations in animals, and virtually anything that might damage the cells. Finally, he developed a weak strain by suspending the bacterial cells in a mildly acidic medium and allowing the culture to remain undisturbed for a long period. When the bacterial cells were inoculated into chickens and later followed by a dose of lethal bacterial cells, the animals did not develop cholera. This principle is the basis for many vaccines today. Pasteur applied the principle to anthrax in 1881 and found he could protect sheep against the disease (FIGURE 1.11).

■ **Attenuate:** To reduce or weaken.

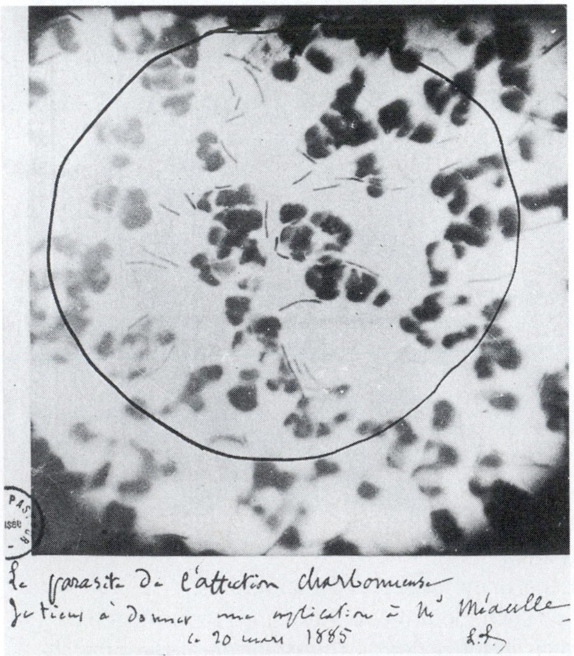

FIGURE 1.11 The Anthrax Bacillus. A photomicrograph of the anthrax bacillus taken by Louis Pasteur in 1885. Pasteur circled the bacilli (the tiny rods) in tissue and annotated the photograph, "the parasite of Charbonneuse." ("Charbonneuse" is the French equivalent of anthrax.)

Both labs also were making several discoveries regarding diphtheria. In Pasteur's lab, Émile Roux and Alexandre Yersin linked diphtheria to a toxin produced by the bacterial cells infecting the body. Two years later, Koch's coworker, Emil von Behring, successfully treated diphtheria by injecting an **antitoxin,** a preparation of antibodies obtained from animals immunized against diphtheria.

There also was an international flavor in the French and German laboratories. Shibasaburo Kitasato of Japan studied with Koch and successfully cultivated the tetanus bacillus. One of Pasteur's associates was Elie Metchnikoff, a native of Ukraine. In 1884, Metchnikoff published an account of **phagocytosis,** an immunological defensive process in which certain groups of the body's white blood cells engulf and destroy microorganisms.

In 1885, Pasteur reached the zenith of his career when he successfully immunized a young boy against the dreaded disease rabies. Although he never saw the causative agent of rabies, Pasteur cultivated it in the brain of animals and injected the boy with bits of the tissue (Microfocus 1.5). The experiment was a triumph because it fulfilled his dream of applying the principles of science to practical problems. Such successes helped establish the Pasteur Institute in Paris, one of the world's foremost scientific establishments. Pasteur presided over the Institute until his death in 1895.

Koch also reached the height of his influence in the 1880s. In 1883, he interrupted his work on tuberculosis to lead groups studying cholera in Egypt and India. In both countries, Koch isolated a comma-shaped bacillus and confirmed the suspicion first raised by John Snow 30 years earlier that water is the key to transmission. In 1891, Koch became director of Berlin's Institute for Infectious Diseases. At various times, he studied malaria, plague, and sleeping sickness, but his work with tuberculosis ultimately gained him the 1905 Nobel Prize in Physiology or Medicine. He died of a stroke in 1910 at the age of 66.

The germ theory set a new course for studying and treating infectious disease. The studies carried out by Pasteur and Koch, and their colleagues, put the discipline of **bacteriology,** the study of bacterial organisms, on the map. By

MICROFOCUS 1.5: History

The Private Pasteur

The notebooks of Louis Pasteur had been an enduring mystery of science ever since the scientist himself requested his family not to show them to anyone. But in 1964, Pasteur's last surviving grandson donated the notebooks to the National Library in Paris, and after soul-searching for a decade, the directors made them available to a select group of scholars. Among the group was Gerald Geison of Princeton University. What Geison found stripped away part of the veneration conferred on Pasteur and showed another side to his work.

In 1881, Pasteur conducted a trial of his new anthrax vaccine by inoculating half a flock of animals with the vaccine, then exposing the entire flock to the disease. When the vaccinated half survived, Pasteur was showered with accolades. However, Pasteur's notebooks, according to Geison, reveal that he had prepared the vaccine not by his own method, but by a competitor's. (Coincidentally, the competitor suffered a nervous breakdown and died a month after the experiment ended.)

Pasteur also apparently sidestepped established protocols when he inoculated two boys with a rabies vaccine before it was tested on animals. Fortunately, the two boys survived, possibly because they were not actually infected or because the vaccine was, indeed, safe and effective. Nevertheless, the untested treatment should not have been used, says Geison. His book, *The Private Science of Louis Pasteur* (Princeton University Press, 1995) places the scientist in a more realistic light and shows that today's pressures to succeed in research are little different than they were more than a century ago.

1880, bacteriology was a well-respected field that promised other disease agents also would be quickly identified. In fact, a new generation of international scientists stepped in to expand the work on infectious disease.

CONCEPT AND REASONING CHECKS

1.9 Assess the importance of the science laboratory and teamwork to the increasing identification of pathogenic bacteria.

Other Global Pioneers Contribute to New Disciplines in Microbiology

KEY CONCEPTS

- Viruses also can cause disease.
- Many beneficial bacterial species recycle nutrients in the environment.

Although the list of identified microbes was growing, the agents responsible for diseases such as measles, mumps, smallpox, and yellow fever continued to elude identification. In 1892, a Russian scientist, Dimitri Ivanowsky, used a filter developed by Pasteur's group to trap what he thought were bacterial cells responsible for tobacco mosaic disease, which produces mottled and stunted tobacco leaves. Surprisingly, Ivanowsky discovered that when he applied the liquid that passed through the filter to healthy tobacco plants, the leaves became mottled and stunted. Ivanowsky assumed bacterial cells somehow had slipped through the filter.

Unaware of Ivanowsky's work, Martinus Beijerinck, a Dutch investigator, did similar experiments in 1899 and suggested tobacco mosaic disease was a "contagious, living liquid" that acted like a poison or virus (*virus* = "poison"). In 1898, the first "filterable virus" responsible for an animal disease—hoof-and-mouth disease—was discovered, and in 1901 American Walter Reed concluded that the agent responsible for yellow fever in humans also was a filterable agent. With theses discoveries, the discipline of **virology**, the study of viruses, was launched.

Contagious: Capable of being transmitted from one person to the next.

While many scientists were advancing medical microbiology, others devoted their research to the environmental importance of microorganisms. The Russian scientist Sergei Winogradsky discovered bacterial cells that metabolized sulfur and developed the concept of **nitrogen fixation**, where bacterial cells convert nitrogen gas (N_2) into ammonia (NH_3). Beijerinck was the first to pure culture microorganisms from the soil and showed that trapped ammonia could be made available to plants for growth. Together with Winogradsky, he developed many of the laboratory materials essential to the study of environmental

TABLE 1.2 Some Notable Figures and Their Accomplishments during the Classical Golden Age of Microbiology, 1854–1914

Investigator (Year)	Country	Accomplishment
Joseph Lister (1865)	Great Britain	Developed the principles of aseptic surgery
Otto Obermeier (1868)	Germany	Observed bacterial cells in relapsing fever patients
Ferdinand Cohn (1872)	Germany	Established bacteriology as a science; produced the first bacterial taxonomy scheme
Gerhard Hansen (1873)	Norway	Observed bacterial cells in leprosy patients
Ernst Karl Abbé (1878)	Germany	Developed the oil-immersion lens and Abbé condenser for the compound microscope
Friedrich Löeffler (1883)	Germany	Isolated diphtheria bacillus
Georg Gaffky (1884)	Germany	Cultivated the typhoid bacillus
Hans Christian Gram (1884)	Denmark	Introduced staining system to identify bacterial cells
Elie Metchnikoff (1884)	Ukraine	Described phagocytosis
Paul Ehrlich (1885)	Germany	Suggested some dyes might control bacterial infections
Daniel E. Salmon (1886)	United States	Studied swine plague
Emile Roux and Alexandre Yersin (1888)	France	Identified the diphtheria toxin
Shibasaburo Kitasato (1889)	Japan	Isolated the tetanus bacillus
Emil von Behring (1890)	Germany	Developed the diphtheria antitoxin
Sergius Winogradsky (1891)	Russia	Studied the biochemistry of soil bacteria
Dimitri Ivanowsky (1892)	Russia	Studied tobacco mosaic disease from which he isolated a filterable agent
Richard Pfeiffer (1892)	Germany	Identified a cause of meningitis
William Welch (1892)	United States	Isolated the gas gangrene bacillus
Theobald Smith (1893)	United States	Proved that ticks transmit Texas fever
Masaki Ogata (1897)	Japan	Discovered that rat fleas transmit plague
Ronald Ross (1898)	Great Britain	Showed mosquitoes can transmit malaria
Kiyoshi Shiga (1898)	Japan	Isolated a cause of bacterial dysentery
Martinus Beijerinck (1899)	Netherlands	Developed the discipline of environmental microbiology and provided some of the first clues for viruses as infectious agents
Walter Reed (1901)	United States	Studied mosquito transmission of yellow fever in Cuba
David Bruce (1903)	Great Britain	Proved that tsetse flies transmit sleeping sickness
Almroth Wright (1903)	Great Britain	Described opsonins to assist phagocytosis
Jules Bordet (1906)	France	Isolated the pertussis bacillus
Albert Calmette (1906)	France	Developed immunization process for tuberculosis
Howard Ricketts (1906)	United States	Showed that ticks transmit Rocky Mountain spotted fever
Charles Nicolle (1909)	France	Proved that lice transmit typhus fever

microbiology, while adding to the understanding of the essential roles microorganisms play in the environment.

The advent of World War I brought a pause to microbiology research. Nevertheless, the discoveries made about infectious disease would set the stage for their accurate diagnoses, prevention, and cure. These accomplishments are listed in TABLE 1.2.

Today, along with Giovannoni and Venter, many microbiologists continue to search for and understand the roles of microorganisms.

In fact, it has been suggested that less than 2 percent of all microorganisms on Earth have been identified and many fewer have been cultured or studied thoroughly. There is still a lot to be discovered in the microbial world!

CONCEPT AND REASONING CHECKS

1.10 Describe how viruses were discovered as disease-causing agents.

1.11 Judge the significance of the work pioneered by Winogradsky and Beijerinck.

1.4 Studying Microorganisms

Besides bacteriology and virology, other disciplines also were developing at the beginning of the 20th century. This included **mycology**, the study of fungi; **protozoology**, the study of the protozoa; and **phycology**, the study of algae (FIGURE 1.12).

The applications of microbiological knowledge also were important to the development of epidemiology, infection control, and **immunology**, which is the study of bodily defenses against microorganisms and other agents.

Why Study Microorganisms and Viruses Today?

KEY CONCEPT

- Microorganisms represent more than agents of disease.

Today, it is almost impossible to pick up a newspaper, read a magazine, or hear a news broadcast without there being a story about microorganisms or some "killer microbe." For example, a story might describe an outbreak of an illness caused by eating or drinking a contaminated food product or describe how human infectious diseases are becoming resistant to antibiotics (FIGURE 1.13). Sometimes media information distorts the truth or is simply incorrect. Therefore, a strong foundation in microbiology is the only way to evaluate the substance of such "stories."

So, one simple answer to the question is: there still is so much to learn and understand. A good portion of the classical Golden Age was devoted to identifying the agents responsible for human infectious disease. Such knowledge was critical in developing disease prevention strategies. Today, many classical diseases, such as tuberculosis, malaria, and cholera remain and many new ones are being discovered as they infect the human population in some part of the world (e.g., SARS, Ebola hemorrhagic fever, and AIDS). Therefore, epidemiology and infection control remain critical factors of public health systems worldwide and represent topics that medical and nursing students as well as others in the health care professions, need to understand.

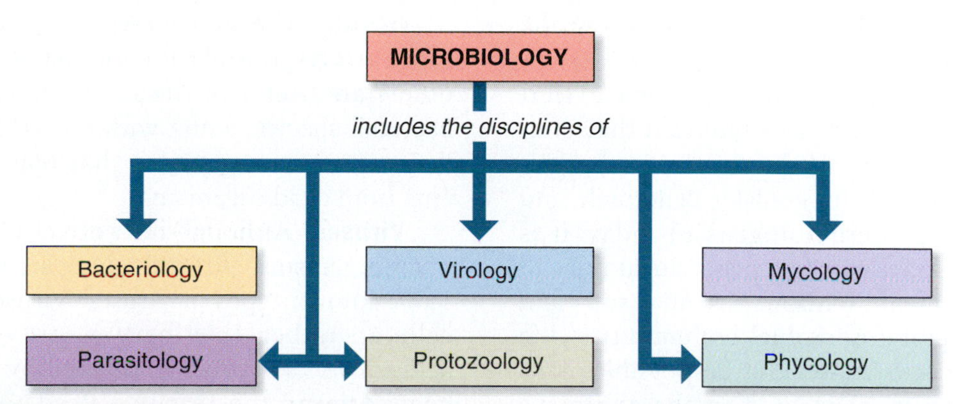

FIGURE 1.12 **Microbiology Disciplines by Organism or Agent Studied.** This simple concept map shows the relationship between microbiology and the organisms or agents that make up the various disciplines. Parasitology is the study of animal parasites. Some of these parasites cause disease in humans, which is why parasitology is included with the other disciplines of microbiology.

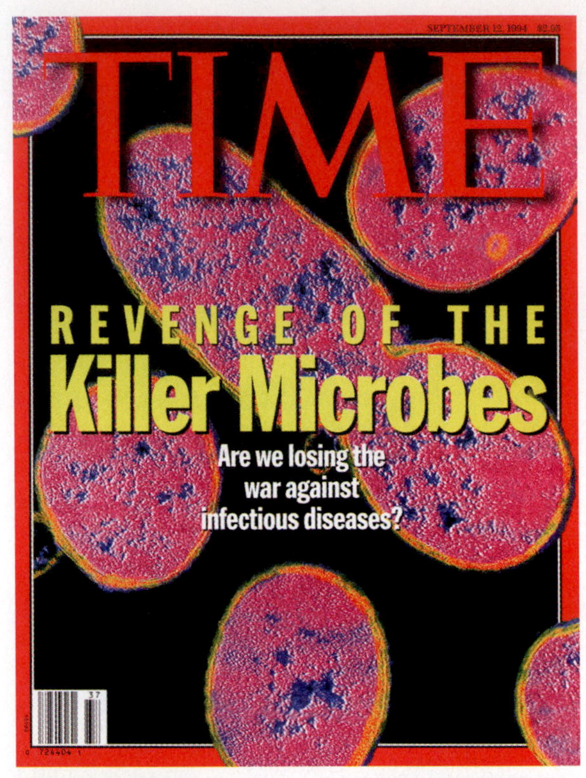

FIGURE 1.13 **Microbes in the News.** The September 12, 1994 cover of *Time* magazine.

The study of microorganisms also provides the opportunity to better understand processes that are common to all life, be it bacterial, fungal, or human. In the next section, we see how microbes have been used as experimental models to better understand the cellular processes involved with aspects of molecular genetics that are common to all living organisms. Many bacterial species grow very rapidly, which facilitates their use in the research laboratory.

You also should now be aware that microorganisms are more significant than simply acting as agents of disease. From the pioneering work by Winogradsky, Beijerinck, and others to the microbiologists of today, it is clear that microorganisms play critical roles in the environment. Without essential soil- and aquatic-dwelling microbial communities, life on planet Earth would not be possible. This appreciation is illustrated in the explosive interest in microbial discovery by Giovannoni, Venter, and thousands of other microbiologists. So, today we need to continue to discover and identify microorganisms and then study their roles and functions in nature.

CONCEPT AND REASONING CHECKS

1.12 As a student, why should you have an understanding of microorganisms and viruses?

The Spectrum of Microorganisms Is Diverse

KEY CONCEPT

- The organisms and agents studied in microbiology represent a diverse group.

By the end of the classical Golden Age of microbiology, the diversity of microbes included more than just bacterial species. Let's briefly survey what we know about these groups today.

Bacteria. Today, it is estimated that there may be more than 10 million bacterial species. Most are very small, single-celled organisms (although some form filaments, and many associate in a bacterial mass called a "biofilm"). The cells may be spherical, spiral, or rod-shaped (FIGURE 1.14A), and they lack the cell nucleus and most of the typical cellular compartments typical of other microbes and multicellular organisms.

Based on recent biochemical and molecular studies, these bacterial species have been divided into two domains, called the ***Bacteria*** and the ***Archaea***. Both groups are metabolically more diverse than the other microbes. Most bacterial and archaeal species absorb their food from the environment, but some bacterial species, like the **cyanobacteria**, carry out photosynthesis (FIGURE 1.14B).

Besides the disease-causing members, some are responsible for food spoilage while others are useful in the food industry. Many bacterial species, along with several fungi, are **decomposers**, organisms that recycle nutrients from dead organisms.

Viruses. Although not correctly labeled as microorganisms, currently there are more than 3,600 known types of viruses. Viruses are not cellular; rather, they have a core of nucleic acid (DNA or RNA) surrounded by a protein coat. Among the features used to identify viruses are morphology (size, shape), genetic material (RNA, DNA), and biological properties (organism or tissue infected).

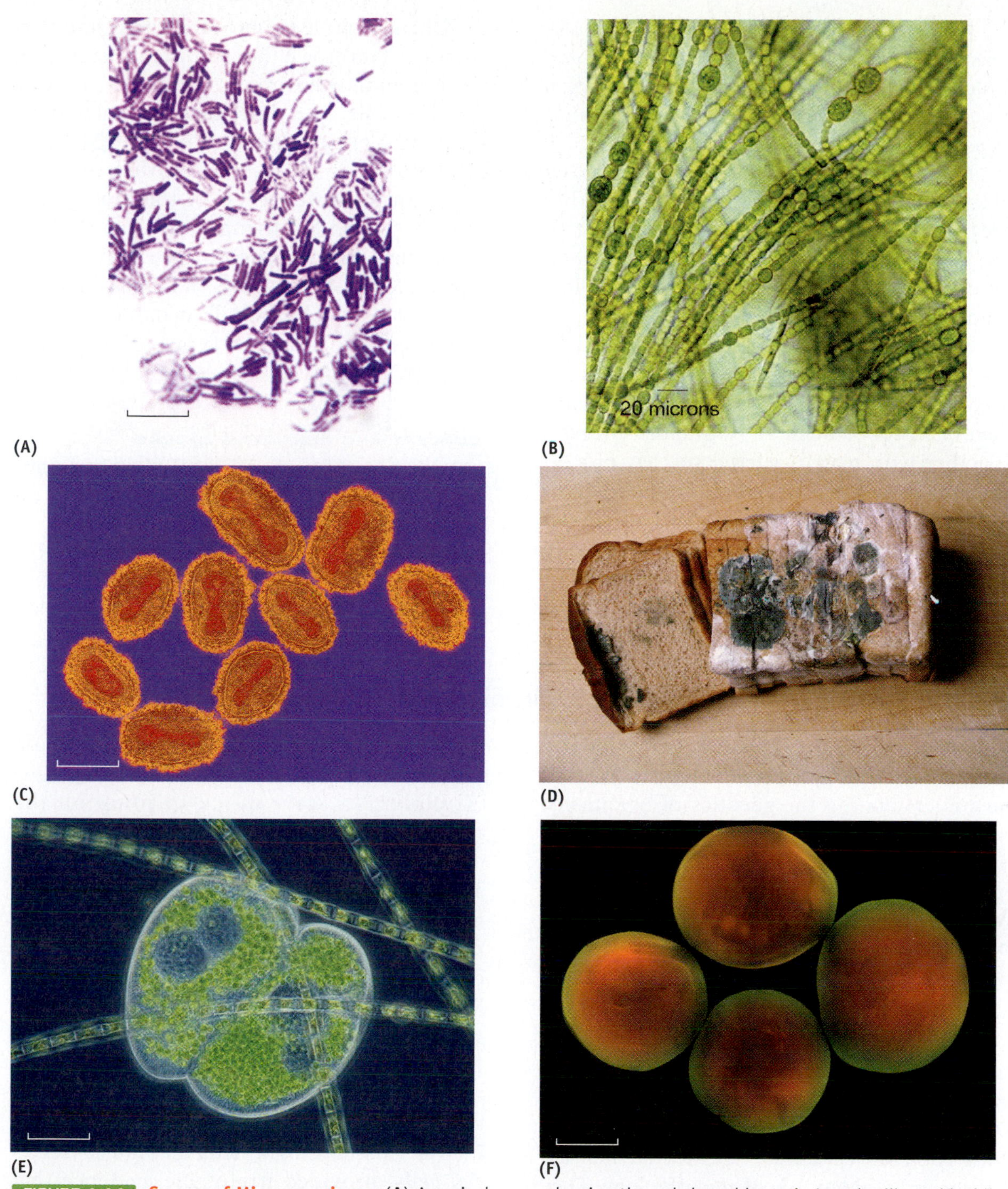

(A) (B) (C) (D) (E) (F)

FIGURE 1.14 **Groups of Microorganisms.** (**A**) A vaginal smear showing the rod shaped bacteria *Lactobacillus acidophilus* (stained purple), a common inhabitant of the vagina. (Bar = 10 µm.) (**B**) Filamentous strands of *Nostoc*, a cyanobacterium that carries out photosynthesis. (Bar = 20 µm.) (**C**) Smallpox viruses. (Bar = 100 nm.) (**D**) A bread mold growing on a loaf of bread. (**E**) Assorted green algae. (Bar = 10 µm.) (**F**) Cells of the protozoan *Cryptosporidium*, a diarrheal disease agent. (Bar = 2.5 µm.)

Viruses infect organisms for one reason only—to replicate. Viruses in the air or water, for example, cannot replicate because they need the metabolic machinery inside a cell. Of the known viruses, only a small percentage causes disease in humans. Polio, the flu, measles, AIDS, and smallpox are examples (FIGURE 1.14C).

The other groups of microbes all have a cell nucleus and a variety of internal cellular compartments.

Fungi. The fungi include the unicellular yeasts and the multicellular mushrooms and molds (FIGURE 1.14D). About 70,000 species of fungi have been described; however, there may be as many as 1.5 million species in nature.

Most fungi grow best in warm, moist places and secrete digestive enzymes that break down nutrients into smaller bits that can be absorbed easily. Fungi thus live in their own food supply. If that food supply is a human, disease may result.

Some fungi provide useful products including antibiotics, such as penicillin. Others are used in the food industry to impart distinctive flavors in foods such as Roquefort cheeses. Together with many bacterial species, numerous fungi play a major role as decomposers.

Protista. The protista consist of single-celled protozoa and algae. Some are free living while others live in association with plants or animals. Locomotion may be achieved by flagella or cilia, or by a crawling movement.

Different protista obtain nutrients in different ways. Protozoa either absorb nutrients from the surrounding environment or ingest algae or bacterial cells. The unicellular or filamentous algae carry out photosynthesis (FIGURE 1.14E). Most protozoa are helpful in that they are important in lower levels of the food chain, providing food for living organisms such as snails, clams, and sponges. Some protozoa are capable of causing diseases in animals, including humans; these include malaria, sleeping sickness, and several types of diarrhea (FIGURE 1.14F).

CONCEPT AND REASONING CHECKS

1.13 Construct a concept map for the **Microbial Agents** in this section.

Terms for map:

Algae	Bacteria	Protozoa
Archaea	Fungi	Viruses
Cyanobacteria	Microorganisms	
Decomposers	Protista	

1.5 The Second Golden Age of Microbiology (1943–1970)

The 1940s brought the birth of molecular genetics to biology. Many biologists focused on understanding the genetics of organisms, including the nature of the genetic material and its regulation.

Molecular Biology Relies on Microorganisms

KEY CONCEPT

- Microorganisms and viruses can be used as model systems to study phenomena common to all life.

In 1943, the Italian-born microbiologist Salvador Luria and the German physicist Max Dulbrück carried out a series of experiments with bacterial cells and viruses that marked the second Golden Age of microbiology. They used a common gut-inhabiting bacterium, *Escherichia coli*, to address a basic question regarding evolutionary biology: do mutations occur spontaneously or does the environment induce them? Luria and Dulbrück showed that bacterial cells could develop spontaneous mutations that generate resistance to viral infection. Besides the significance of their findings to microbial genetics, the use of *E. coli* as a microbial model system showed to other researchers that microorganisms could be used to study general principles of biology.

■ Mutation: A permanent alteration in a DNA base sequence.

Biologists were quick to jump on the "microbial bandwagon." Experiments carried out by Americans George Beadle and Edward Tatum in the 1940s ushered in the field of molecular biology by using the fungus *Neurospora* to show that one gene codes for one enzyme. Oswald Avery, Colin MacLeod, and Maclyn McCarty, working with the bacterial species *Streptococcus pneumoniae*, suggested in 1944 that deoxyribonucleic acid (DNA) is the genetic material in cells. In 1953, American biochemist Alfred Hershey and geneticist Martha Chase, using a virus that infects bacterial cells, provided irrefutable evidence that DNA is the substance of the genetic material. These experiments and discoveries, which will be discussed in more detail in Chapter 7, placed microbiology in the middle of the molecular biology revolution.

Two Types of Cellular Organization Are Realized

KEY CONCEPT

- All microorganisms have a characteristic cell structure.

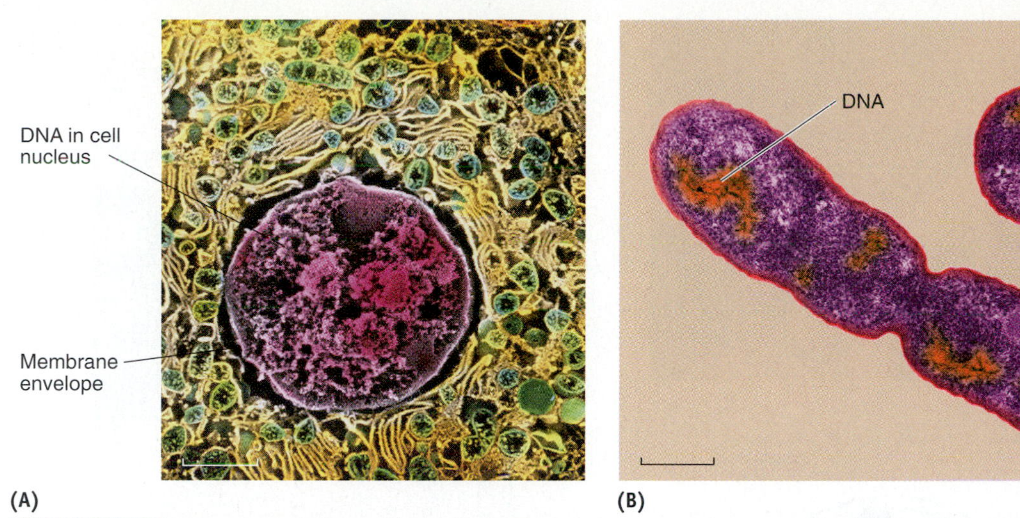

FIGURE 1.15 **False Color Images of Eukaryotic and Prokaryotic Cells.** (**A**) A scanning electron micrograph of a eukaryotic cell. All eukaryotes, including the protozoa, algae, and fungi, have their DNA (pink) enclosed in a cell nucleus with a membrane envelope. (Bar = 3 μm.) (**B**) A transmission electron micrograph of a dividing *Escherichia coli* cell. The DNA (orange) is not surrounded by a membrane. (Bar = 0.5 μm.)

The small size of bacterial cells hindered scientists' abilities to confirm that these cells were "cellular" in organization. In the 1940s and 1950s, a new type of microscope—the electron microscope—was being developed that could magnify objects and cells thousands of times better than typical light microscopes. With the electron microscope, for the first time bacterial cells were seen as being cellular like all other microbes, plants, and animals. However, studies showed that they were organized in a fundamentally different way from other organisms.

It was known that animal and plant cells contained a cell nucleus that houses the genetic instructions in the form of chromosomes and was separated physically from other cell structures by a membrane envelope (FIGURE 1.15A). This type of cellular organization is called **eukaryotic** (*eu* = "true"; *karyon* = "nucleus"). Microscope observations of the protista and fungi had revealed that these organisms also had a eukaryotic organization. Thus, not only are all plants and animals eukaryotes, so are the microorganisms that comprise the fungi and protists.

Studies with the electron microscope revealed that bacterial (and archaeal) cells had few of the cellular compartments typical of eukaryotic cells. They lacked a cell nucleus, indicating the bacterial chromosome (DNA) was not surrounded by a membrane envelope (FIGURE 1.15B). Therefore, members of the *Bacteria* and *Archaea* have a **prokaryotic** (*pro* = "before") type of cellular organization and represent prokaryotes. (By the way, because viruses lack a cellular organization, they are neither prokaryotes nor eukaryotes.)

CONCEPT AND REASONING CHECKS

1.14 Distinguish between prokaryotic and eukaryotic cells.

Antibiotics Are Used to Cure Infectious Disease

KEY CONCEPT

- Antimicrobial chemicals can be effective in treating infectious diseases.

In 1910, another coworker of Koch's, Paul Ehrlich, synthesized the first "magic bullet"—a chemical that could kill pathogens. Called Salvarsan, Ehrlich showed that this arsenic-containing compound cured syphilis, a sexually-transmitted disease prevalent at the time. Antibacterial **chemotherapy**, the use of antimicrobial chemicals to kill microbes, was born.

In 1929, Alexander Fleming, a Scottish scientist, went on vacation leaving several culture plates of bacterial colonies on the lab bench. When he returned, he found a mold growing in one of the bacterial cultures (FIGURE 1.16). Upon further inspection, Fleming observed that the mold, a species of *Penicillium*, killed the bacterial cells and colonies that were near

(A)

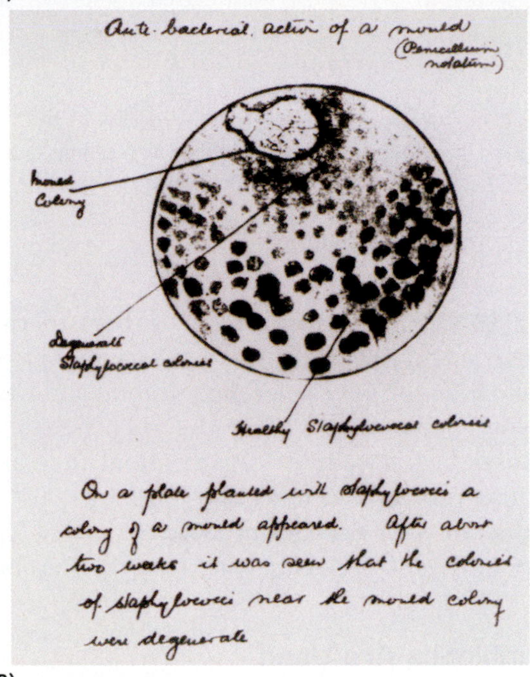

(B)

FIGURE 1.16 Fleming and the Discovery of Penicillin. (**A**) Fleming in his laboratory. (**B**) Fleming's notes on the inhibition of bacterial growth by the fungus *Penicillium*.

FIGURE 1.17 Penicillin—One of the Magic Bullets. A World War II poster touting the benefits of penicillin and illustrating the great enthusiasm in the United States for treating infectious diseases in war casualties.

it. He named the antimicrobial substance penicillin and developed an assay for its production. In 1940, British biochemists Howard Florey and Ernst Chain purified penicillin and carried out clinical trials that showed the antimicrobial potential of the natural drug (MicroFocus 1.6).

Additional "wonder drugs" also were being discovered. The German chemist Gerhard Domagk discovered a synthetic chemical dye, called Prontosil, which was effective in treating *Streptococcus* infections. Examination of soil bacteria led Selman Waksman to the discovery of actinomycin and streptomycin, the latter being the first effective agent against tuberculosis. He coined the term **antibiotic** to refer to those antimicrobial substances naturally produced by mold and bacterial species that inhibit growth or kill other microorganisms.

The push to market effective antibiotics was stimulated by a need to treat deadly infections in casualties of World War II (FIGURE 1.17). By the 1950s, penicillin and several additional antibiotics were established treatments in medical practice. In fact, the growing arsenal of antibiotics convinced many that the age of infectious disease was waning. In 1969, then U.S. Surgeon General William Stewart confidently declared to Congress that it was time to "close the books on infectious diseases."

Partly due to the perceived benefits of antibiotics, interest in microbes was waning by the end of the 1960s. Research funding became harder to obtain and the knowledge gained from bacterial studies was being applied to eukaryotic organisms, especially animals. What was ignored was the mounting evidence that bacterial species were becoming resistant to antibiotics.

CONCEPT AND REASONING CHECKS

1.15 Contrast Ehrlich's Salvarsan and Domagk's Prontosil from those drugs developed by Fleming, Florey and Chain, and Waksman.

MICROFOCUS 1.6: History

Hiding a Treasure

Their timing could not have been worse. Howard Florey, Ernest Chain, Norman Heatley, and others of the team had rediscovered penicillin, refined it, and proved it useful in infected patients. But it was 1939, and German bombs were falling on London. This was a dangerous time to be doing research into new drugs and medicines. What would they do if there was a German invasion of England? If the enemy were to learn the secret of penicillin, the team would have to destroy all their work. So, how could they preserve the vital fungus yet keep it from falling into enemy hands?

Heatley made a suggestion. Each team member would rub the mold on the inside lining of his coat. The *Penicillium* mold spores would cling to the rough coat surface where the spores could survive for years (if necessary) in a dormant form. If an invasion did occur, hopefully at least one team member would make it to safety along with his "moldy coat." Then, in a safe country the spores would be used to start new cultures and the research could continue. Of course, a German invasion of England did not occur, but the plan was an ingenious way to hide the organism.

The whole penicillin story is well told in *The Mold in Dr. Flory's Coat* by Eric Lax (Henry Holt Publishers, 2004)

1.6 The Third Golden Age of Microbiology—Now

Microbiology finds itself on the world stage again, in part from the biotechnology advances made in the latter part of the 20th century. **Biotechnology**, which frequently uses the natural and genetically-engineered abilities of microbial agents to carry out biological processes for industrial/commercial/medical applications, has revolutionized the way microorganisms are genetically manipulated to act as tiny factories producing human proteins, such as insulin, or new synthetic vaccines, such as the hepatitis B vaccine. In the latest Golden Age, microbiology again is making important contributions to the life sciences.

Microbiology Continues to Face Many Challenges

KEY CONCEPT

- Infectious disease (natural and intentional) preoccupies much of microbiology.

The third Golden Age of microbiology is addressing several challenges, many of which still concern the infectious diseases that kill 15 million people globally each year.

A New Infectious Disease Paradigm. Infectious disease remains a major concern worldwide. The fact that more than 11 million people die each year from tuberculosis (TB), malaria, lower respiratory infections, diarrheal diseases, and AIDS has led to the United States Leadership Against Global HIV/AIDS, Tuberculosis, and Malaria Act of 2003. This act requires U.S. agencies to work with foreign governments and international organizations to expand programs and improve coordination in the fight against HIV/AIDS, TB, and malaria in Africa and the Caribbean.

Today, our view of infectious diseases also has changed. In Pasteur and Koch's time, it was mainly a problem of finding the germ that caused a specific disease. Today, a new paradigm is emerging. Pathogens are being discovered that were never known to be associated with infectious disease and some of these agents actually cause more than one disease. In addition, there are **polymicrobial diseases**; that is, diseases caused by more than one infectious agent. Equating a microbe with a disease is made even more challenging because of the inability of culture many of these pathogens. Even some noninfectious diseases, such as heart disease, may have a microbial component that triggers the illness.

Increased Antibiotic Resistance. Another challenge concerns our increasing inability to fight infectious disease because pathogens have become or are becoming resistant to antibiotics and other antimicrobial drugs. Ever since it was recognized that pathogens could mutate into "supermicrobes" that are resistant to many drugs, a crusade has been waged to restrain the inappropriate use of these drugs by doctors and to educate patients not to demand them in uncalled-for situations.

The challenge facing microbiologists and drug companies is to find new and effective antibiotics to which pathogens will not quickly develop resistance before the current arsenal is completely useless. One benefit from sequencing the genome of pathogens is to discover a molecular "Achilles heel" or sensitive spot where the organisms would be vulnerable to antimicrobial drugs or to which effective vaccines could be generated.

Emerging and Reemerging Infectious Diseases. From time to time, infectious diseases, such as SARS or West Nile fever, seem to pop up from nowhere and threaten the health of populations and even whole nations. With the globalization of today's societies and unprecedented human mobility, it is important to quickly identify a disease outbreak and control it before it can spread to epidemic or pandemic proportions.

Public health systems are particularly concerned with two groups of infectious diseases. **Emerging infectious diseases** are those that have recently surfaced in a population. Among the more newsworthy have been AIDS, hantavirus pulmonary syndrome, Lyme disease, mad cow disease, and most recently, SARS. **Reemerging infectious diseases** are ones that have existed in the past but are now showing a resurgence in incidence or a spread in geographic range. Often the cause for the reemergence is antibiotic resistance or an increase in susceptible individuals. Climate change also has been implicated in the upsurge in disease. Among the more prominent reemerging diseases are cholera, tuberculosis, dengue fever, and, for the first time in the Western Hemisphere, West Nile fever (FIGURE 1.18A). Again, these diseases are of concern to public health officials because lack of identification or control measures could bring on a serious health crisis.

Bioterrorism. Perhaps it is the potential misuse of microbiology that has brought microbiology to the attention of the life science community and the public. **Bioterrorism** involves the intentional or threatened use of biological agents to cause fear in or actually inflict death or disease upon a large population. Most of the recognized biological agents are microorganisms, viruses, or microbial toxins that are bringing diseases like anthrax, smallpox, and plague back into the human psyche (FIGURE 1.18B). To minimize the use of

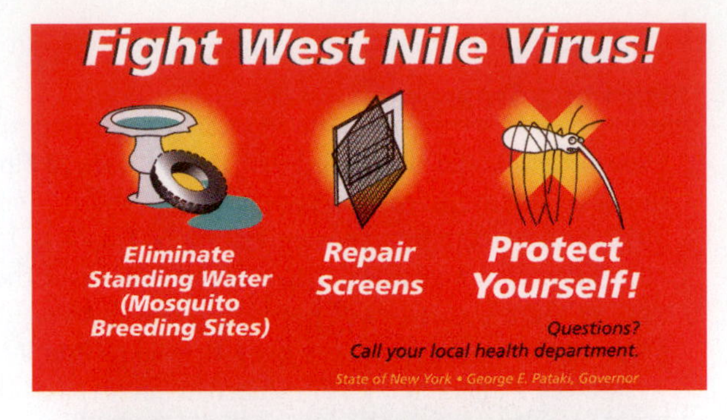

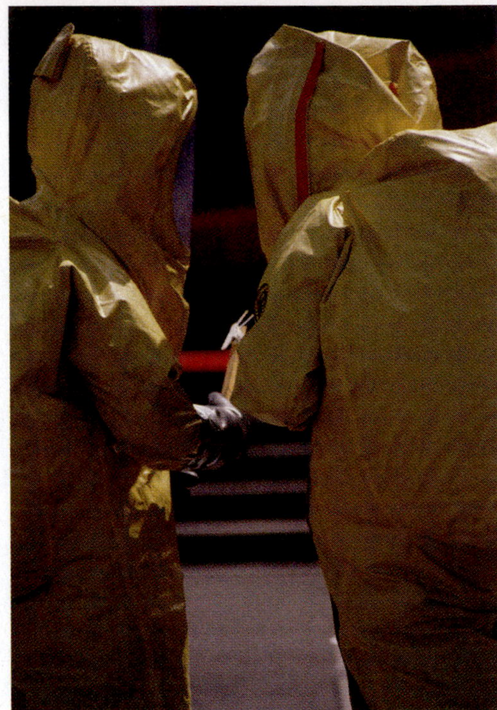

FIGURE 1.18 **Emerging Disease Threats: Natural and Intentional.** (**A**) There have been and will continue to be natural disease outbreaks. West Nile virus (WNV) is just one of several agents responsible for emerging or reemerging diseases. Methods have been designed that individuals can use to protect themselves from mosquitoes that spread the WNV. (**B**) Combating the threat of bioterrorism often requires special equipment and protection because many agents seen as possible bioweapons could be spread through the air.

these agents to inflict mass casualties, the challenge to the scientific community and microbiologists is to improve the ways that bioterror agents are detected, discover effective measures to protect the public, and develop new and effective treatments for individuals or whole populations. If there is anything good to come out of such challenges, it is that we will be better prepared for potential natural emerging infectious disease outbreaks, which initially might be difficult to tell apart from a bioterrorist attack.

CONCEPT AND REASONING CHECKS

1.16 Describe the natural and intentional disease threats challenging microbiology.

Microbial Ecology and Evolution Are Helping to Drive the New Golden Age

KEY CONCEPT

- Microbial ecology and evolution are dominant themes in modern microbiology.

Since the time of Pasteur, microbiologists have wanted to know how a microbe interacts, survives, and thrives in the environment. Today, microbiology is less concerned with a specific microbe and more concerned with the process and mechanisms that link microbial agents.

Microbial Ecology. Traditional methods of microbial ecology require organisms from an environment be cultivated in the laboratory so that they can be characterized and identified. However, up to 99 percent of microorganisms do not grow well in the lab (if at all) and therefore could not be studied. Today, many microbiologists, armed with genetic, molecular, and biotechnological tools, can study and characterize these unculturable microbes. Such investigations are producing a new understanding of microbial communities and their influence on the ecology of all organisms. SAR11 and the plans of Craig Venter, mentioned in the chapter's opening piece, are but two examples.

Today we are learning that many microbes do not act as individual entities; rather, in nature they survive in complex communities called a **biofilm** (FIGURE 1.19A). Microbes in biofilms act very differently than individual cells and can be difficult to treat when biofilms cause infectious disease. If you or someone you know has had a middle ear infection, the cause was a bacterial biofilm.

(A)

(B)

FIGURE 1.19 **Microbial Ecology—Biofilms and Bioremediation.** (**A**) This 3-month-old lab grown biofilm contains at least two microbial communities. The algae (top) carry out photosynthesis but depend on "waste products" produced by the bacteria (bottom). (**B**) Microbes can be used to clean up toxic spills. A shoreline coated with oil from an oil spill can be sprayed with microorganisms that, along with other measures, help degrade oil.

The discovered versatility of many bacterial and archaeal species is being applied to problems that have the potential to benefit humankind. Bioremediation is one example where the understanding of microbial ecology has produced a useful outcome (FIGURE 1.19B). Other microbes hold potential to solve ecological impacts caused by toxic wastes, fertilizers, and pesticides released into the environment.

■ Bioremediation: The use of microorganisms to remove or decontaminate toxic materials in the environment.

Microbial Evolution. Evolution represents the foundation for all biology and medicine. Biologists have estimated that there are millions

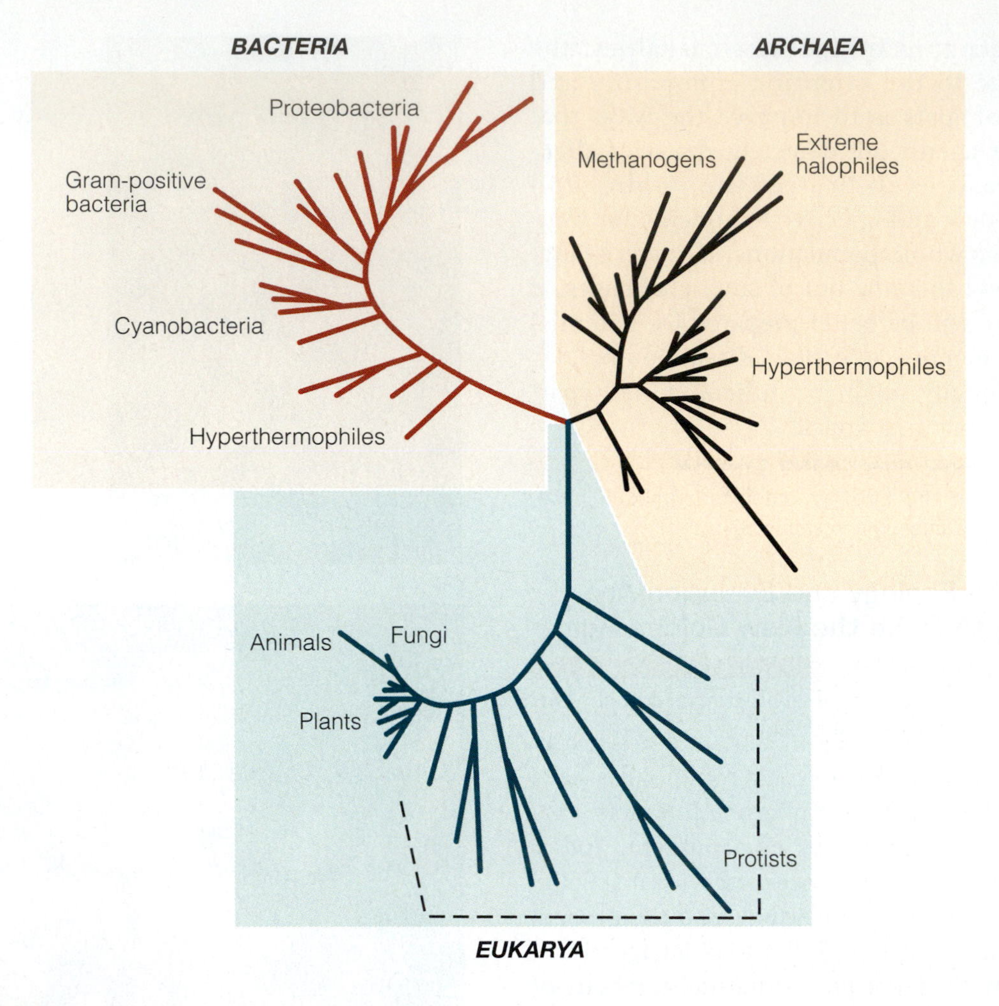

FIGURE 1.20 **A Phylogenetic Tree of Life.** A phylogenetic tree based on sequencing the ribosomal RNA of organisms. The tree shows the relationships and separation of *Bacteria* and *Archaea*. The *Eukarya* consist of the protists, plants, fungi, and animals. Reproduced from Norman Pace 2001. *Proc. Natl. Acad. Sci. USA*. 98:805–808. Copyright 2006 National Academy of Science, U.S.A.

of species of organisms living on Earth today. The study of **phylogeny** involves the identification of evolutionary relationships between these organisms and has resulted in a complex genealogical tree, called the "phylogenetic tree of life." Using ribosomal RNA (rRNA) base sequences from different organisms, one can determine evolutionary relationships. Such studies have identified three distinct phylogenetic lineages (FIGURE 1.20). As mentioned earlier, the prokaryotes, or classical "bacteria," actually sort out into two separate lineages and so have been split into two separate domains, the *Bacteria* and *Archaea*. All the eukaryotic organisms remain in one domain, called the ***Eukarya***. This phylogenetic tree of life provides a working framework for the life sciences and represents the model for the organization of biological and microbiological knowledge in this text.

■ Ribosomal RNA: The RNA types that are major components of the ribosome.

CONCEPT AND REASONING CHECKS

1.17 As microbiologists continue to explore the microbial universe, it is becoming more apparent that microbes are "invisible emperors" that rule the world. Now that you have completed Chapter 1, provide examples to support the statement: Microbes Rule!

SUMMARY OF KEY CONCEPTS

1.1 The Beginnings of Microbiology

- The observations with the microscope made by Hooke and especially Leeuwenhoek, who reported the existence of animalcules (microorganisms), sparked interest in an unknown world of microscopic life.
- The controversy over spontaneous generation initiated the need for accurate scientific experimentation, which then provided the means to refute the concept.

1.2 Microorganisms and Disease Transmission

- Semmelweis and Snow believed that infectious disease could be caused by something transmitted from the environment and that the transmission could be interrupted.
- Edward Jenner determined that disease (smallpox) could be prevented through vaccination with a similar disease-causing agent.

1.3 The Classical Golden Age of Microbiology (1854–1914)

- Pasteur's fermentation experiments indicated that microorganisms could induce chemical changes. He proposed the germ theory of disease, which stated that human disease could be due to chemical changes brought about by microorganisms in the body.
- Lister's use of antisepsis techniques and Pasteur's studies of pébrine supported the germ theory and showed how diseases can be controlled.
- Koch's work with anthrax allowed him to formalize the methods (Koch's postulates) for relating a single microorganism to a single disease. These postulates were only valid after he discovered how to make pure cultures of bacterial species on a solid surface.
- Laboratory science arose from the intense rivalry that developed between the labs of Pasteur and Koch as they hunted down the microorganisms of infectious disease. Pasteur's lab studied the mechanisms for infection and developed vaccines for chicken cholera and human rabies. Koch's lab focused on isolation, cultivation, and identification of pathogens responsible for tuberculosis, typhoid fever, and diphtheria.
- Ivanowsky and Beijerinck provided the first evidence for viruses as infectious agents.
- Winogradsky and Beijerinck examined the beneficial roles of noninfectious microorganisms in the environment.

1.4 Studying Microorganisms

- Studying microbes involves not only those causing infectious diseases, but also the majority, which have beneficial roles.
- Microbes include the "bacteria" (*Bacteria* and *Archaea*), viruses, fungi (yeasts and molds), and protista (protozoa and algae).

1.5 The Second Golden Age of Microbiology (1943–1970)

- Many of the advances toward understanding molecular biology and general principles in biology were based on experiments using microbial model systems.
- With the advent of the electron microscope, microbiologists realized that there were two basic types of cellular organization: eukaryotic and prokaryotic.
- Following from the initial work by Ehrlich, antibiotics were developed as "magic bullets" to cure many infectious diseases.

1.6 The Third Golden Age of Microbiology—Now

- In the 21st century, fighting infectious disease, combating increasing antibiotic resistance, identifying emerging and reemerging infectious diseases, and countering the bioterrorism threat are challenges facing microbiology, health care systems, and society.
- Microbial ecology is providing new clues to the roles of microorganisms in the environment. The understanding of microbial evolution has advanced with the use of genomic technologies and has provided a new "phylogenetic tree of life."

LEARNING OBJECTIVES

After understanding the textbook reading, you should be capable of writing a paragraph that includes the appropriate terms and pertinent information to answer the objective.

1. Identify the significant contributions made by Hooke and Leeuwenhoek that foreshadowed the beginnings of microbiology.
2. Discuss spontaneous generation and compare the experiments that led to its downfall.
3. Identify the components of scientific inquiry.
4. Assess the importance of the work carried out by Semmelweis and by Snow that went against the miasma idea and established the field of epidemiology.
5. Explain how Jenner's work led to a way of preventing infectious disease.
6. Judge the importance of (a) the germ theory of disease and (b) Koch's postulates to the identification of microbes as agents of infectious disease.
7. Describe how viruses were discovered.
8. Identify several nondisease-causing roles that microorganisms play in the environment.
9. Explain why *Bacteria* and *Archaea* are domains of prokaryotic cells and all other organisms in the *Eukarya* domain are composed of eukaryotic cells.
10. Outline the major challenges facing microbiology today.
11. Draw a "tree of life" and identify the types of organisms and characteristics found in the three major branches (domains).

SELF-TEST

Answer each of the following questions by selecting the **one** answer that best fits the question or statement. Answers to the even-numbered questions can be found in **Appendix C**.

1. The first person to see bacterial cells with the microscope was
A. Pasteur.
B. Koch.
C. Leeuwenhoek.
D. Stelluti.
E. Hooke.

2. Who among the following was *not* involved with proving or disproving spontaneous generation?
A. Semmelweis
B. Needham
C. Redi
D. Pasteur
E. Spallanzani

3. A hypothesis is
A. a conclusion resulting from the results of an experiment.
B. a testable explanation for an observed phenomenon.
C. the same thing as a theory.
D. an idea based on many experiments.
E. a controlled experiment.

4. The observations and work of Semmelweis and Snow formed the basis for a new field of microbiology called
A. immunology.
B. bacteriology.
C. virology.
D. parasitology.
E. epidemiology.

5. The process of ______ involved the inoculation of dried smallpox scabs under the skin.
A. vaccination
B. fermentation
C. antisepsis
D. variolation
E. immunization

6. The process of controlled heating that was used to keep wine from spoiling is called
A. curdling.
B. fermentation.
C. pasteurization.
D. variolation.
E. immunization.

7. The first person to employ antisepsis in surgery was
A. Lister.
B. Semmelweis.
C. Koch.
D. Pasteur.
E. Jenner.

8. All of the following are part of Koch's postulates *except*:
A. the microorganism must be isolated from a dead animal and pure cultured.
B. a specific organism is related to a specific disease.
C. the pure cultured organism is inoculated into a healthy, susceptible animal.
D. the same microorganism must be present in every case of the disease.
E. All the above statements (**A**–**D**) are true of Koch's postulates.

9. The first microbiologists to study the role of nonpathogenic microbes in the environment were
A. Ivanowsky and Beijerinck.
B. Pasteur and Koch.
C. Winogradsky and Beijerinck.
D. Giovannoni and Venter.
E. Metchnikoff and Kitasato.

10. What group of microorganisms has a variety of internal cell compartments and acts as decomposers?
A. *Bacteria*
B. Protista
C. Viruses
D. *Archaea*
E. Fungi

11. What group of microbial agents lacks the cellular structures characteristic of prokaryotic and eukaryotic organisms?
A. *Bacteria*
B. Protozoa
C. Viruses
D. Algae
E. Fungi

12. Salvarsan was a chemical derived from _____ and used to treat _____.
A. mercury; cholera
B. silver; gonorrhea
C. arsenic; syphilis
D. bacteria; cholera
E. fungi; anthrax

13. The term antibiotic was coined by _____ to refer to antimicrobial substances naturally derived from _____.
A. Waksman; bacteria and fungi
B. Fleming; fungi
C. Domagk; other living organisms
D. Fleming; fungi and bacteria
E. Ehrlich; bacteria

14. Which one of the following is *not* an emerging infectious disease?
A. SARS
B. Polio
C. Hantavirus pulmonary disease
D. Lyme disease
E. AIDS

15. The domain *Eukarya* contains all the following groups *except* the
A. protozoa.
B. fungi.
C. viruses.
D. algae.
E. animals.

QUESTIONS FOR THOUGHT AND DISCUSSION

Answers to the even-numbered questions can be found in **Appendix C.**

1. Many people are fond of pinpointing events that alter the course of history. In your mind, which single event described in this chapter had the greatest influence on the development of microbiology? What event would be in second place?
2. One of the foundations of scientific inquiry is proper experimental design involving the use of controls. What is the role of a control in an experiment? For each of the experiments described in the section on spontaneous generation, identify the control(s) and explain how the interpretation of the experimental results would change without such controls.
3. In 1878, after studying bacteriology in Germany, William Welch returned to the United States, eager to apply the principles he had learned to his medical practice. Welch asked administrators at New York's Bellevue Hospital Medical College for space and money to conduct research, and he was awarded a paltry $25 for three kitchen tables to serve as laboratory benches. Why do you believe Welch was rebuffed?
4. One reason for the rapid advance in knowledge concerning molecular biology during the second Golden Age of microbiology was because many researchers used microorganisms as model systems. Why would something like a bacterial cell be more advantageous to use for research than, say, rats or guinea pigs?
5. When you tell a friend that you are taking microbiology this semester, she asks, *"Exactly what is microbiology?"* How do you answer her?
6. On the front page of this chapter there is a quote from Louis Pasteur. How does this quote apply to the work done by (a) Semmelweis, (b) Snow, and (c) Fleming?
7. Who would you select as the "father of microbiology?" (a) Leeuwenhoek or (b) Pasteur and Koch. Support your decision.

APPLICATIONS

Answers to the even-numbered questions can be found in **Appendix C.**

1. As a microbiologist in the 1940s, you are interested in discovering new antibiotics that will kill bacterial pathogens. You have been given a liquid sample of a chemical substance to test in order to determine if it kills bacterial cells. Drawing on the culture techniques of Robert Koch, design an experiment that would allow you to determine the killing properties of the sample substance.
2. As an environmental microbiologist, you discover a new species of microbe. How could you determine if it is a prokaryote or eukaryote? Suppose it is a prokaryote. What technique would you use to determine if it is a member of the domain *Bacteria* or *Archaea*?

REVIEW

On completing your study of these pages, test your understanding of their contents by deciding whether the following statements are true (T) or false (F). If the statement is false, substitute a word or phrase for the underlined word or phrase to make the statement true. The answers to even-numbered statements can be found in **Appendix C.**

1. _____ Leeuwenhoek was a vitalist who believed mice could spontaneously generate from wheat bran and dirty shirts.
2. _____ Pasteur proposed that "wine disease" was a souring of wine caused by yeast cells.
3. _____ Antisepsis is the use of chemical methods for disinfecting living surfaces.
4. _____ Separate bacterial colonies can be observed in a broth culture.
5. _____ Semmelweis proposed that cholera was a waterborne disease.
6. _____ Some bacterial species can convert nitrogen gas (N_2) into ammonia (NH_3).
7. _____ Fungi are eukaryotic microorganisms.
8. _____ Robert Koch was a French country doctor.
9. _____ "Smallpox parties" involved inoculating individuals with smallpox scabs.
10. _____ Mycology is the scientific study of viruses.

HTTP://MICROBIOLOGY.JBPUB.COM/

The site features learning, an on-line review area that provides quizzes and other tools to help you study for your class. You can also follow useful links for in-depth information, read more MicroFocus stories, or just find out the latest microbiology news.

Chapter Preview and Key Concepts

2.1 The Elements of Life
- Atoms are composed of charged and uncharged particles.
- Isotopes are atoms of an element with varying mass numbers.
- Ions are atoms with an electrical charge.
- The chemical properties of atoms are strongly governed by the number of electrons in their outermost electron shell.

2.2 Chemical Bonding
- Unstable atoms can be linked through ionic interactions.
- Unstable atoms can interact through the sharing of outer shell electron pairs.
- Hydrogen bonding is the electrostatic attraction between molecules.
- Chemical reactions convert reactants into products.

2.3 Water, pH, and Buffers
- Water is the solvent of life.
- Cell chemistry is sensitive to pH changes.
- Buffers prevent pH shifts.

2.4 Major Organic Compounds of Living Organisms
- Functional groups represent the set of atoms involved in chemical reactions.
- Carbohydrates provide energy and structural materials.
- Lipids store energy and are components of membranes.
- Nucleic acids store, transport, and control hereditary information.
- Protein function depends on its three-dimensional shape.

MicroInquiry 2: Isotopes as Tracers in Drug Discovery

The Chemical Building Blocks of Life

The significant chemicals in living tissue are rickety and unstable, which is exactly what is needed for life.
—Isaac Asimov (1920–1992)

They are found around the world; some of the most intriguing being in the arid regions of the American Southwest. These are ancient petroglyphs, symbols often in the form of zigzags and spirals, or human and animal figures, which were carved into rocks and cliff faces by the ancestors of many Native American cultures. When you first see the petroglyphs, some perhaps over 3,000 years old, you wonder what they were meant to represent. Unfortunately, deciphering and understanding their meaning has been difficult.

What has been discovered is the chemical process making petroglyphs possible. At first glance, the rocks into which the petroglyphs were etched appear as though the ancients had first painted the rocks a shade of orange, red, dark brown, or gray. In reality, the rock color represents a thin coating (patina), commonly referred to as *rock varnish* or *desert varnish*. By carving and scraping through the dark patination into the lighter rock beneath, the petroglyphs appear as light colored images against a dark background.

Until 2006, evidence suggested the rock varnish was a product of the bacterial cells living on the rock. Although only a hundredth of a millimeter (0.01 mm) in thickness, this patina was thought to represent a bacterial biofilm, coloring the rocks and cliff faces.

It now appears that the rock coatings, which can vary from place to place, are formed by nonliving components present in the environment. The mechanism involves the mobilization and redistribution of silica. During desert varnish formation, organic products on the rock surface can be preserved by interaction with condensing silicic acid. Since microbes are rarely visible on or in varnish surfaces, these organisms do not appear to be the cause for desert varnish formation. However, entombed *Bacteria*, *Archaea*, and *Eukarya*, along with their organic compounds, have been identified in varnishes, suggesting microorganisms may strongly influence varnish formation.

Importantly, DNA and amino acids from a diverse group of bacterial, archaeal, and fungal cells have been isolated from the varnish. Thus, by analyzing the rock coatings, past environments on Earth can be studied—and possibly other planets. In fact, similar silica coatings may exist on rocks observed on other planets, such as Mars. If these are rock varnishes, the coatings might also hold traces of microbial life that may have existed on the Red Planet.

In Chapter 1, you learned that microorganisms are found in most, if not all, habitats on Earth and perhaps their "signatures" will be found on Mars! How can microbes survive, for example, in the high temperatures of a hot spring, the interior of a rock, or the acidic runoff from a mine (FIGURE 2.1)? The answer in all cases involves **cellular chemistry**, the chemical reactions between atoms and molecules that provides for the unique metabolism in cells.

The basic principles of chemistry also permit microbiologists to understand how pathogens make a living in the human body and how the body responds. For example, studying microbial chemistry means determining:

- How microbes cause disease.
- How the immune system attempts to combat infections.
- How antibiotics and vaccines can eliminate or protect against infections.

This chapter serves as a primer or review of the fundamental concepts of chemistry that form a foundation for the chapters ahead. We will identify the elements making up all known substances and show how these elements combine to form the major groups of organic compounds found in virtually all forms of life. Realize the time invested now to understand or refresh your memory about chemistry will make subsequent chapters easier and prepare you for a rewarding learning experience as you continue your study of microbiology.

(A)

(B)

(C)

FIGURE 2.1 **Cellular Chemistry Allows Microbes to Colonize and Survive in Earth's Environments.** (**A**) Yellowstone's Grand Prismatic Spring. The gentle flow of heated water spreads out in terraces, with green and red algae thriving in the warm, shallow water toward the edge. (**B**) A vertical profile of microbes inside calcite rock. Buried below the surface lies a black layer of lichen. (**C**) Acid-loving bacterial species thrive in this stream receiving acid drainage from a Missouri coal mine.

Q: What other "exotic" environments can you identify where microbes might survive?

2.1 The Elements of Life

■ Matter:
Anything that occupies space and has mass.

■ Mass:
The quantity of matter in a sample.

As far as scientists know, all matter in the physical universe—be it a rock, a tree, or a microbe—is built of substances called chemical elements. **Chemical elements** are the most basic forms of matter and they cannot be broken down into other substances by ordinary chemical means.

Ninety-two naturally occurring elements have been discovered, while additional elements have been made in the laboratory or nuclear reactor. One or two letters, many standing for its English, Latin, or Greek name, designate each element. For example, H is the symbol for hydrogen, O for oxygen, Cl for chlorine, and Mg for magnesium. Some Latin abbreviations include Na (*natrium* = "sodium"), K (*kalium* = "potassium"), and Fe (*ferrum* = "iron").

Only about 25 of the 92 naturally occurring elements are essential to the survival of living organisms. Many of these are major elements needed in relatively large amounts (TABLE 2.1). Note that just six of these elements—carbon, hydrogen, nitrogen, oxygen, phosphorus, and sulfur—make up 98 percent of the weight in both human and bacterial cells. (The acronym CHNOPS is helpful in remembering these six important elements.)

In addition, there are a number of elements needed in much smaller amounts. These elements vary from organism to organism, but often include such elements as sodium (Na), calcium (Ca), manganese (Mn), iron (Fe), copper (Cu), and zinc (Zn).

Matter Is Composed of Atoms

KEY CONCEPT

- Atoms are composed of charged and uncharged particles.

An **atom** is the smallest unit of an element having the properties of that element; it cannot be broken down further without losing the quality of the element. Simply stated, carbon consists of carbon atoms, oxygen of oxygen atoms, and so forth. If you split a carbon atom into simpler parts, it no longer has the properties of carbon. Because atoms are so small, they cannot be seen with ordinary biological microscopes. Still, scientists have carried out experiments revealing the structure of atoms.

An atom consists of a positively charged core, the **atomic nucleus** (FIGURE 2.2A). The atomic nucleus contains most of the atom's mass and two kinds of tightly packed particles called **protons** and **neutrons**. Although protons and neutrons have about the same mass, protons bear a positive electrical charge (value = +1), while neutrons have no charge.

The number of protons in an atom defines each element. For example, carbon atoms always have six protons. If there are seven protons, it is no longer carbon but rather the element nitrogen. The number of protons also represents the **atomic number** of the atom. As shown in Table 2.1, carbon with six protons has an atomic number of 6. The **mass number** is the number of protons and neutrons combined. Because carbon atoms have six protons and usually six neutrons in the atomic nucleus, the mass number of carbon is 12.

Surrounding the atomic nucleus is a negatively charged cloud of **electrons** (value = –1). In any atom, the number of electrons is equal to the number of protons; that is, an atom has no net electrical charge. Because it is impossible to predict at any moment where a particular electron might be located, the electrons often are said to exist within a cloud. However, we can identify the spaces within the atom where electrons are usually located. These spaces are called **electron shells**, each shell representing a different energy level. FIGURE 2.2B provides a simple diagram of the structures, atomic numbers, and mass numbers of four atoms essential to life.

CONCEPT AND REASONING CHECKS

2.1 Draw a concept map for **Atomic Structure** using the boldfaced terms in this section.

Atoms Can Vary in the Number of Neutrons or Electrons

KEY CONCEPTS

- Isotopes are atoms of an element with varying mass numbers.
- Ions are atoms with an electrical charge.

Although the number of protons is the same for all atoms in an element, the number of neutrons in an element may vary, altering its mass number. Most carbon atoms, for example, have a mass number of 12, but some carbon atoms have eight neutrons, rather than six, in the atomic nucleus and, hence, a mass number of 14.

TABLE 2.1 Some of the Major Elements of Humans

Element	Symbol	Percentage by Weight in Body	Atomic Number	Mass Number
Oxygen	O	65	8	16
Carbon	C	18	6	12
Hydrogen	H	10	1	1
Nitrogen	N	3	7	14
Phosphorus	P	1	15	31
Sulfur	S	0.9	16	32
Sodium	Na	0.2	11	23
Magnesium	Mg	0.1	12	24
Chlorine	Cl	0.2	17	36
Potassium	P	0.4	19	39

Some of the Major Elements of a Prokaryotic Cell

Element	Symbol	Percentage by Weight
Oxygen	O	72
Carbon	C	12
Hydrogen	H	10
Nitrogen	N	3
Phosphorus	P	0.6
Sulfur	S	0.3

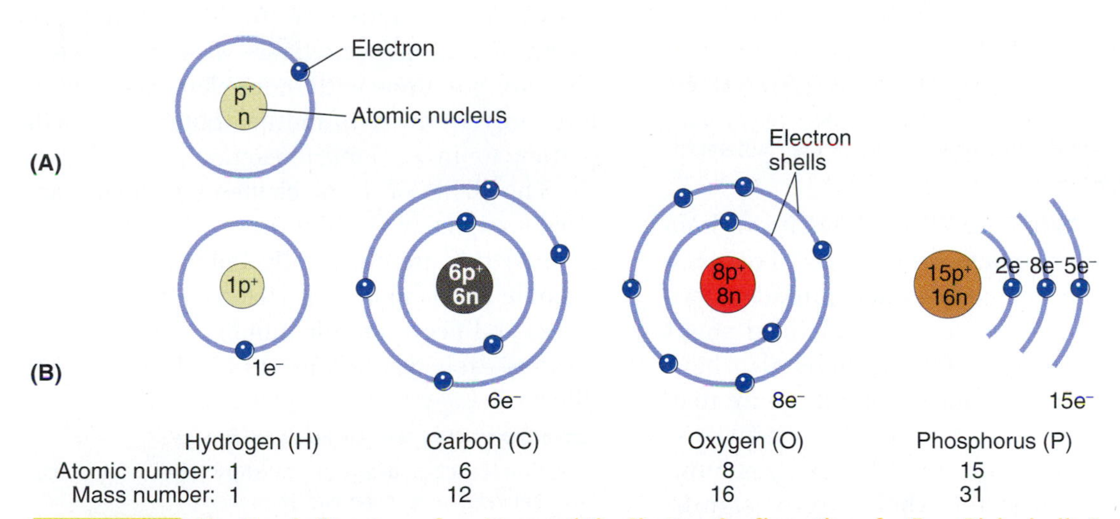

FIGURE 2.2 **The Atomic Structure of an Atom and the Electron Configurations for Four Biologically Important Elements.** (**A**)The atom is composed of protons and neutrons in the atomic nucleus, and electrons that move about the nucleus in electron shells. (**B**) The atomic structure of four biologically essential elements illustrates that the number of protons equals the number of electrons (though not necessarily equal to the number of neutrons).

Q: Knowing the mass number for an element, what does that tell you about the mass of an electron?

Atoms of the same element that have different numbers of neutrons are called **isotopes**. Therefore, carbon-12 and carbon-14 (symbolized as ^{12}C and ^{14}C) are isotopes of carbon.

Some isotopes are unstable and give off energy in the form of radiation. Such **radioisotopes** are useful in research and medicine. ^{14}C can be incorporated into an organic substance and used as a radioactive tracer to follow the fate of the substance, as MicroInquiry 2 demonstrates (see page 64).

Atoms are uncharged when they contain equal numbers of electrons and protons. Atoms normally do not gain or lose protons, but a gain or loss of electrons is possible. If this takes place, the atom achieves an electrostatic charge and is called an **ion** (FIGURE 2.3). The addition of one or more electrons to an atom means there is/are more negatively-charged electrons than positively-charged protons. Such a negatively-charged ion is called an **anion**. By contrast, the loss of one or more electrons leaves the atom with extra protons and yields a positively-charged ion, called a **cation**. As we will see, ion formation is important to some forms of chemical bonding.

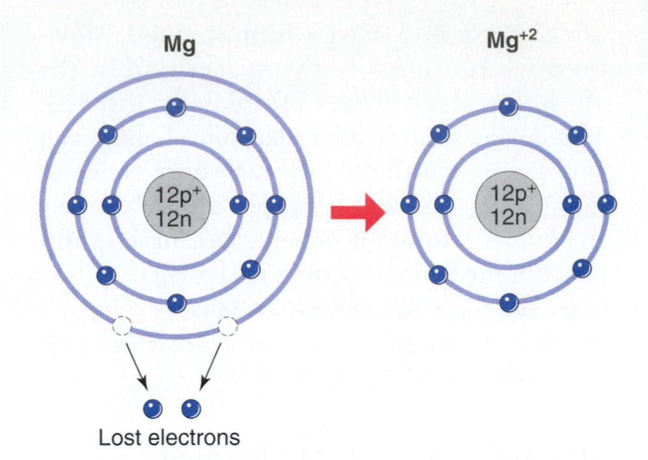

FIGURE 2.3 **Formation of an Ion.** Ions can be formed by the loss or gain of one or more electrons. Here, a magnesium (Mg) atom has lost two electrons to become a magnesium ion (Mg^{+2}).

Q: For Mg, what does the superscript denote?

Electron Placement Determines Chemical Reactivity

KEY CONCEPT

- The chemical properties of atoms are strongly governed by the number of electrons in their outermost electron shell.

As shown in Figure 2.2B, each shell can hold a maximum number of electrons. The shell closest to the nucleus can accommodate two electrons, while the second and third shells each can hold eight. Other shells also have maximum numbers but usually no more than 18 are present in those outer shells. Since the 25 essential elements are of lower mass number, only the first few shells are of significance to life. Inner shells are filled first and, if there are not enough electrons to completely fill the shell, the outermost shell is left incompletely filled.

Atoms with an unfilled outer electron shell are unstable but can become stable by interacting with another unstable atom. A carbon atom, with six electrons, has two electrons in its first shell and only four in the second (see Figure 2.2B). For this reason, carbon is extremely reactive in "finding" four more electrons and, as we will see, forms innumerable combinations with other elements. Therefore, only atoms with unfilled outer shells will participate in a chemical reaction.

The shells of a few elements normally are filled completely. Each of these elements, called an inert gas, are chemically stable and exist as separate atoms in nature. Helium (atomic number 2) and neon (atomic number 10) are examples, as each has its outermost electron shell filled.

CONCEPT AND REASONING CHECKS

2.2 Construct a diagram to show the difference between an isotope and an ion.

2.3 Looking at Figure 2.2b, do these atoms have filled outer electron shells? Explain.

2.2 Chemical Bonding

Isaac Asimov's opening quote that chemicals are "rickety and unstable" applies to how atoms interact. When the electron shells of two unstable atoms come close, the electron shells overlap, an energy exchange takes place, and each of the participating atoms assumes an electron configuration more stable than its original unstable configuration. When two or more atoms are linked together, the force holding them is called a **chemical bond**. Chemical bonds are the result of these rickety, unstable atoms filling their outer electron shells.

The rearrangement of atoms through chemical bonding can occur in one of two major ways: atoms, as ions, can interact electrostatically; or, each uncharged atom can share electrons with one or more other atoms. In both cases, the result is atoms having full electron shells.

Ionic Bonds Form between Oppositely Charged Ions

KEY CONCEPT

- Unstable atoms can be linked through ionic interactions.

In the formation of an **ionic bond**, one atom gives up its outermost electrons to another. The reaction between sodium and chlorine is illustrative (FIGURE 2.4A). Sodium has an atomic number of 11, with electrons arranged in shells containing of 2, 8, and 1. Chlorine, with an atomic number of 17, has its electrons in shells containing 2, 8, and 7. To achieve stability, sodium atoms only need to lose one electron and chlorine atoms only need to gain one electron. Thus, when atoms of the two elements are brought together, the sodium atoms donate one electron to the chlorine atoms. Both atoms now have their outermost shell filled.

The transfer of electrons leads to ion formation. By acquiring one electron, chlorine atoms become chloride ions (Cl^-). By contrast, sodium atoms now have an extra proton and therefore are sodium ions (Na^+). Because opposite electrical charges attract each other, the chloride ions and sodium ions come together to form stable sodium chloride (NaCl).

Salts are typically formed through ionic bonding. Besides sodium and chloride, important salts are formed from other ions, including calcium (Ca^{+2}), potassium (K^{+2}), magnesium (Mg^{+2}), and iron (Fe^{+2} or Fe^{+3}). Although ionic bonds are relatively weak, they play important roles in protein structure and the reactions between antigens and antibodies in the immune response (Chapter 20).

When two or more different elements interact with one another to achieve stability, they form a **compound**. Each compound, like each element, has a definite formula and set of properties that distinguish it from its components. For example, sodium (Na) is an explosive metal and chlorine (Cl) is a poisonous gas, but the compound they form is edible table salt (NaCl).

CONCEPT AND REASONING CHECKS

2.4 Construct a diagram to show how the salt calcium chloride ($CaCl_2$) is formed.

Covalent Bonds Share Electrons

KEY CONCEPT

- Unstable atoms can interact through the sharing of outer shell electron pairs.

Atoms also can achieve stability by sharing electrons between the atoms, the sharing producing a **covalent bond**. Such bonds are very important in biology because the CHNOPS elements of life usually enter into covalent bonds with themselves or one another.

Covalent bonding occurs frequently in carbon because this element has four electrons in its outer shell. The carbon atom is not strong enough to acquire four additional electrons, but it is sufficiently strong to retain the four it has. It therefore enters into a variety of covalent bonds with four other atoms or groups of atoms. The vast array of carbon compounds that can be formed is responsible for the chemistry of life.

Many of the microbes residing in the ruminant stomach of a cow produce methane or natural gas (CH_4) as a by-product of cellulose digestion. This gas is a good example to illustrate covalent bonding between carbon and hydrogen (FIGURE 2.4B). A carbon atom shares each of its four outer shell electrons

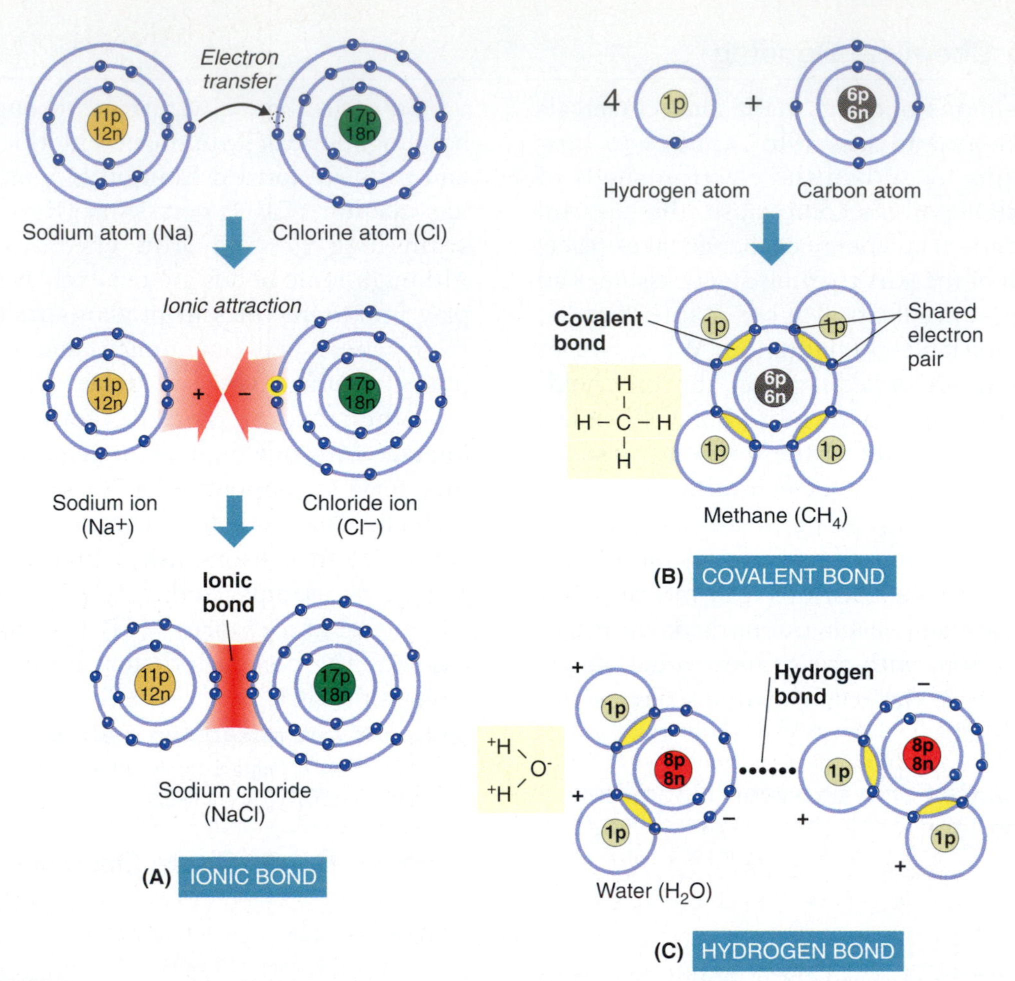

FIGURE 2.4 **Chemical Bonding.** (**A**) The transfer of an electron from an atom of sodium to an atom of chlorine generates oppositely charged ions. The electrical attraction between these ions creates an ionic bond. The resulting compound is sodium chloride. (**B**) A covalent bond involves the sharing of electron pairs between atoms, the example shown here being the simple organic compound methane. (**C**) Hydrogen bonds form when unequal sharing of electrons creates polarized "ends" within a molecule. Hydrogen atoms are typically interact with nitrogen or oxygen atoms, as is the case with the hydrogen bonding that occurs between water molecules.

Q: How does an ionic bond differ from a covalent bond?

with the electron of a hydrogen atom, forming four single covalent bonds.

Scientists often draw chemical structures as **structural formulas**; that is, chemical diagrams showing the order and arrangement of atoms. In Figure 2.4B, each line between carbon and hydrogen (C—H) represents a single covalent bond between a pair of shared electrons. Other molecules, such as carbon dioxide (CO_2), share two pairs of electrons and therefore two lines are used to indicate the double covalent bond: O=C=O. However, in all cases, the atoms now are stable because the outer electron shell of each atom is filled through this sharing.

A **molecule** is two or more atoms held together by covalent bonds. Molecules may be composed of only one kind of atom, as in oxygen gas (O_2), or they may consist of different kinds of atoms in substances such as water (H_2O), carbon dioxide (CO_2), and the simple sugar glucose ($C_6H_{12}O_6$). As shown by these examples, the kinds and amounts of atoms (the subscript) in a molecule is called the **molecular formula**. (Note that the presence of one atom is represented without a subscript 1).

The simplest derivatives of carbon are the **hydrocarbons**, molecules consisting solely of hydrogen and carbon. Methane is the most fundamental hydrocarbon (MicroFocus 2.1).

MICROFOCUS 2.1: Evolution

Earth's Early Microbial Chemistry

In 2005, NASA's Cassini spacecraft made several flybys of Saturn's largest moon, Titan (see figure). Although the presence of methane gas (CH_4) in the upper atmosphere was no surprise, the detection of several types of complex organic materials were surprising. In many ways, this is similar to the conditions that might have existed on Earth early in its history when life was first getting started.

Many scientists believe that long before oxygen gas (O_2) dominated the atmosphere, methane was a major component, giving the atmosphere a pinkish-orange color similar to Titan's today. If so, then the first organisms to evolve on Earth might have been oxygen-intolerant methane-producers called *methanogens*. These microbes sustained the atmosphere for perhaps a billion years before the oxygen-producing microbes, the cyanobacteria, took hold some 2.7 billion years ago.

In the previous billion years, methanogens thrived in many of the very warm environments where hydrogen gas (H_2) dominated and could used H_2 with CO_2 for energy production. Methane gas would be a by-product. The continued accumulation of methane would warm up the planet. But with the increasing concentration of methane, sunlight would link methane molecules together and produce hydrocarbons.

Hydrocarbons would condense as a haze of atmospheric particles. Importantly, the haze would have a cooling effect on the atmosphere and shift life to those methanogens that preferred cooler temperatures. The changing chemistry also may have given oxygen-producing microbes a foothold and along with the hydrocarbon haze, lead to the first ice age about 2.3 billion years ago.

Eventually, methanogens either died out or "retreated" to oxygen-free environments where methane still dominated. Today, methanogens make up about half of all the species in the domain *Archaea*, further supporting their ancestors as being among the first life to evolve.

On Earth, we may never be able to verify any hypothesis concerning the pre-biological chemistry or origins of life; perhaps we can by exploring other worlds, such as Mars or Saturn's moon Titan.

An artist's rendering of Huygens, the probe carried by Cassini and sent through Titan's atmosphere to land on the moon's surface, which occurred on January 14, 2005.

Other hydrocarbons consist of chains of carbon atoms and, in some cases, the chains may be closed to form a ring. Structural formula examples are presented on the next page with their molecular formulas in parentheses.

When atoms bond together to form a molecule, they establish a geometric relationship determined largely by the electron configuration. Notice in the hydrocarbons drawn on the next page that the covalent bonds are distributed equally around each carbon atom. Each of these examples of the equal sharing of electron pairs represents a **nonpolar molecule**—there are no electrical charges (poles) and the bonds are called nonpolar covalent bonds.

Methane (CH_4)

Propane (C_3H_8)

Benzene (C_6H_6)

Not all molecules are nonpolar. Indeed, one of the most important molecules to life, water, is a **polar molecule**—it has electrically charged poles. In a water molecule, the two hydrogen atoms are attached to one side of the oxygen atom (FIGURE 2.3C). As a result, the protons gather at this side and give the molecule a slightly positive charge. By contrast, the other side of the molecule carries a slightly negative charge owing to the accumulation of electrons. The water molecule therefore consists of polar covalent bonds.

CONCEPT AND REASONING CHECKS

2.5 Why is NaCl considered a compound but H_2O a molecule?

Hydrogen Bonds Form between Polar Groups or Molecules

KEY CONCEPT

- Hydrogen bonding is the electrostatic attraction between molecules.

A **hydrogen bond** involves the attraction of a partially positive hydrogen atom that is covalently bonded to one polar molecule toward another polar molecule having either a partially negative oxygen atom ($H^+–O^-$) or nitrogen atom ($H^+–N^-$). Although hydrogen bonds are much weaker than covalent bonds, hydrogen bonds provide the "glue" to hold water molecules together (FIGURE 2.5). These bonds also are important to the structure of proteins and nucleic acids, two of the major components of living cells.

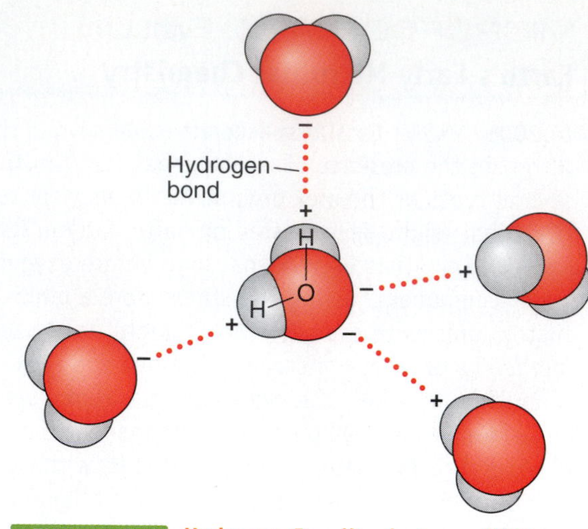

FIGURE 2.5 **Hydrogen Bonding between Water Molecules.**

Q: What is the origin of the (+) and (–) charges on hydrogen and oxygen in a water molecule?

TABLE 2.2 summarizes the different types of chemical bonds we have discussed.

CONCEPT AND REASONING CHECKS

2.6 Construct a diagram to show the hydrogen bonding in liquid ammonia (NH_3).

Chemical Reactions Change Bonding Partners

KEY CONCEPT

- Chemical reactions convert reactants into products.

A **chemical reaction** is a process in which atoms or molecules interact to form new

TABLE 2.2 Three Types of Chemical Bonds in Organic Compounds

Type	Chemical Basis	Strength	Example
Ionic	Attraction between oppositely-charged ions	Weak	Sodium chloride; salts
Covalent	Sharing of electron pairs between atoms	Strong	Glucose
Hydrogen	Attraction of a hydrogen nucleus (a proton) to negatively charged oxygen or nitrogen atoms in the same or neighboring molecules	Weak	Water molecules

bonds. Different combinations of atoms or molecules result from the reaction; that is, bonding partners change. However, the total number of interacting atoms remains constant. For chemical reactions, an arrow is used to indicate in which direction the reaction will proceed. By convention, the atoms or molecules drawn to the left of the arrow are the **reactants** and those to the right are the **products**.

In biology, many chemical reactions are based on the assembly of larger compounds or the tearing down of larger compounds into smaller ones. In a synthesis reaction, smaller reactants are put together into larger products. If water is involved as a product, often it is called a **dehydration synthesis reaction**:

$C_6H_{12}O_6$	+	$C_6H_{12}O_6$	→	$C_{12}H_{22}O_{11}$	+	H_2O
Glucose		Glucose		Maltose		Water

The reverse is a *decomposition* reaction, where a larger reactant is broken into smaller products. Often in biology, water is one of the reactants used to break a molecule, so it is referred to as a **hydrolysis reaction** (*hydro* = "water"; *lysis* = "break"):

$C_{12}H_{22}O_{11}$	+	H_2O	→	$C_6H_{12}O_6$	+	$C_6H_{12}O_6$
Maltose		Water		Glucose		Glucose

Most importantly, the new products formed have the same number and types of atoms that were present in the reactants. In forming new products, chemical reactions only involve a change in the bonding partners. No atoms have been gained or lost from any of these reactions.

■ Synthesis: The formation of larger compounds through the union of smaller ones.

CONCEPT AND REASONING CHECKS

2.7 Identify other synthesis and decomposition reactions with which you are familiar from your daily life. Realize these will occur in microbes as well.

2.3 Water, pH, and Buffers

All organisms are composed primarily of water. In fact, most eukaryotic organisms are about 90 percent water while prokaryotic ones are about 70 percent water. No organism, not even the prokaryotes, can develop and grow without water.

Water Has Several Unique Properties

KEY CONCEPT

- Water is the solvent of life.

Liquid water is the medium in which all cellular chemical reactions occur. Being polar, water molecules are attracted to other polar molecules and act as the universal solvent in cells. Take for example what happens when you put a solute like salt in water (FIGURE 2.6). The solute dissolves into separate sodium and chloride ions because water molecules break the weak ionic bonds and surround each ion in a sphere of water molecules. An **aqueous solution**, which consists of solutes in water, is essential for chemical reactions to occur (MicroFocus 2.2).

Water molecules also are reactants in many chemical reactions. The example of the hydrolysis reaction shown at the top of this page involved water in splitting maltose into separate molecules of glucose and glucose.

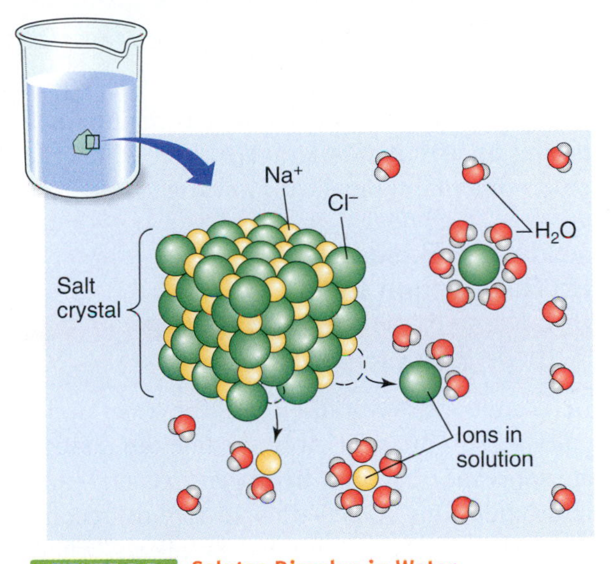

FIGURE 2.6 Solutes Dissolve in Water.

Q: When dissolving, why do the H's of water surround Cl^- while the O's surround the Na^+?

■ Solvent: The liquid doing the dissolving to form a solution.

■ Solute: The substance dissolved in the solvent.

As you have learned, the polar nature of water molecules leads to hydrogen bonding. By forming a large number of hydrogen bonds between water molecules, it takes a large amount of heat energy to increase the temperature of water. Likewise, a large amount of heat must be lost before water decreases temperature. So, by being 70 to 90 percent water,

MICROFOCUS 2.2: Tools

The Relationship between Mass Number and Molecular Weight

Often we need to know how much a particular molecule or solute weighs, which is referred to as the **molecular weight**. The calculation of the molecular weight simply consists of adding together the mass number of all the individual atoms in a molecule, such as water, carbon dioxide, or glucose. Thus, the molecular weight of a water molecule is 18 daltons, while the molecular weight of a glucose molecule is 180 daltons. Other molecules can reach astonishing proportions—antibodies of the immune system may have a molecular weight of 150,000 daltons and the bacterial toxin causing botulism is of over 900,000 daltons.

■ Dalton:
A unit to measure the weight of atomic particles or molecules; equivalent to the atomic mass unit used in chemistry (one-twelfth the weight of an atom of ^{12}C).

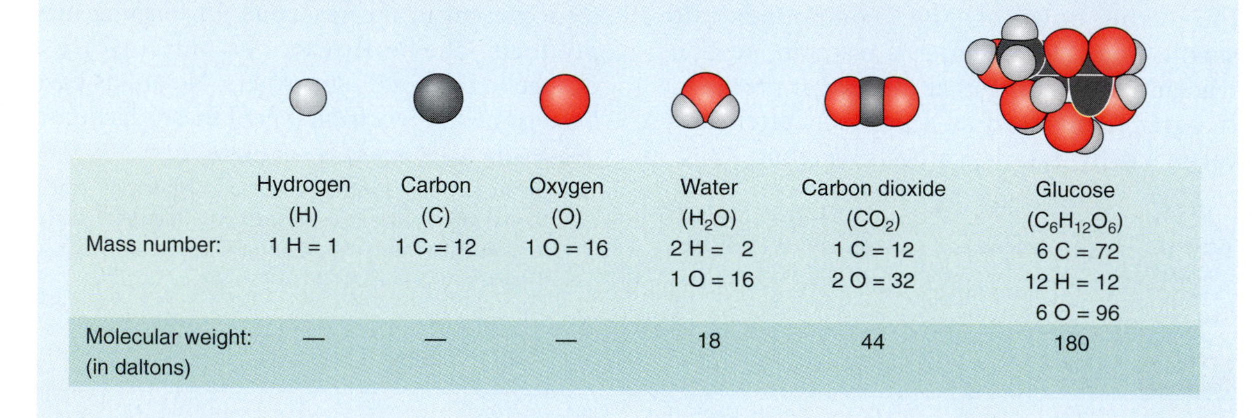

	Hydrogen (H)	Carbon (C)	Oxygen (O)	Water (H_2O)	Carbon dioxide (CO_2)	Glucose ($C_6H_{12}O_6$)
Mass number:	1 H = 1	1 C = 12	1 O = 16	2 H = 2 1 O = 16	1 C = 12 2 O = 32	6 C = 72 12 H = 12 6 O = 96
Molecular weight: (in daltons)	—	—	—	18	44	180

cells are bathed in a solvent that maintains a more consistent temperature even when the environmental temperatures change.

Acids and Bases Must Be Balanced in Cells

KEY CONCEPT

- Cell chemistry is sensitive to pH changes.

In an aqueous solution, most of the water molecules remain intact. However, some can dissociate spontaneously into hydrogen ions (H^+) and hydroxide ions (OH^-) only to rapidly recombine. This can be represented as follows, where the double arrow indicates a reversible reaction:

$$H_2O \leftrightarrow H^+ + OH^-$$

For the chemical reactions in all cells to work properly, there must be a correct balance of H^+ and OH^-. Besides water, other compounds in cells can release H^+ or OH^- and affect cell function. For our purposes, an **acid** is a chemical substance that donates H^+ to water or another solution. By contrast, a **base** (or alkali) is a substance that accepts hydrogen ions in solution, often by combining them with the hydroxyls of the base.

Acids are distinguished by their sour taste. Some common examples are acetic acid in vinegar, citric acid in citrus fruits, and lactic acid in sour milk products. Strong acids can donate large numbers of hydrogen ions to a solution. Hydrochloric acid (HCl), sulfuric acid (H_2SO_4), and nitric acid (HNO_3) are examples. Weak acids, typified by carbonic acid (H_2CO_3), donate a smaller number of hydrogen ions.

Bases have a bitter taste. Strong bases take up numerous hydrogen ions from a solution leaving the solution with an excess of hydroxyl ions. Potassium hydroxide (KOH), a material used to make soap, is among them.

Acids and bases frequently react with each other because of their opposing chemical characteristics. An exchange reaction involving hydrochloric acid (HCl) and sodium hydroxide (NaOH) is one example:

$$HCl + NaOH \rightarrow NaCl + H_2O$$

To indicate the degree to which a solution is acidic, the Danish chemist Søren P. L. Sørensen introduced the symbol **pH** (potential hydrogen) and the **pH scale**. The scale extends

MICROFOCUS 2.3: Environmental Microbiology

Just South of Chicago

All you need is a map, some pH paper, and a few collection vials. When in Chicago, use your map to find the Lake Calumet region just southeast of Chicago. When you arrive, pull out your pH paper and sample some of the ground water in the region near the Calumet River. You will be shocked to discover the pH is greater than 12—almost as alkaline as oven cleaner! In fact, this might be one of the most extreme pH environments on Earth.

How did the water get this alkaline and could anything possibly live in the groundwater?

The groundwater in the area near Lake Calumet became strongly alkaline as a result of the steel slag that has been dumped into the area for more than 100 years. Used to fill the wetlands and lakes, water and air chemically react with the slag to produce lime [calcium hydroxide, $Ca(OH)_2$]. It is the estimated that 10 trillion cubic feet of slag and generated lime has pushed the pH to such a high value.

Now use your collection vials to collect some samples of the water. Back in the lab you will be surprised to find there are bacterial communities present in the water. Hydrogeologists who have collected such samples have discovered some bacterial species that until then had only been found in Greenland and deep gold mines of South Africa. Other identified species appear to use the hydrogen resulting from the corrosion of the iron for energy.

How did these bacterial organisms get there? The hydrogeologists propose that the bacterial species have been there and have simply adapted to the environment over the last 100 years when slag has been dumped. Otherwise, the microbes must have been imported in some way.

So, once again, provide a specific environment and they will come (or evolve)—the microbes that is.

from 0 (extremely acidic; high H^+) to 14 (extremely basic; low H^+) and is based on actual calculations of the number of hydrogen ions present when a substance mixes with water. Realize that every time the pH changes by one unit, the hydrogen ion concentration changes 10 times. A substance with a pH of 7, such as pure water, is said to be neutral; it is neither acidic nor basic because it has equal numbers of H^+ and OH^-. FIGURE 2.7 summarizes the pH values of several common substances.

The greatest diversity of microorganisms occurs in environments where the pH is near neutral, although there are some spectacular exceptions (MicroFocus 2.3). In fact, the acidic soils of the Amazon rain forest contain less than half as many bacterial species as do the neutral soils of the deserts in the American Southwest.

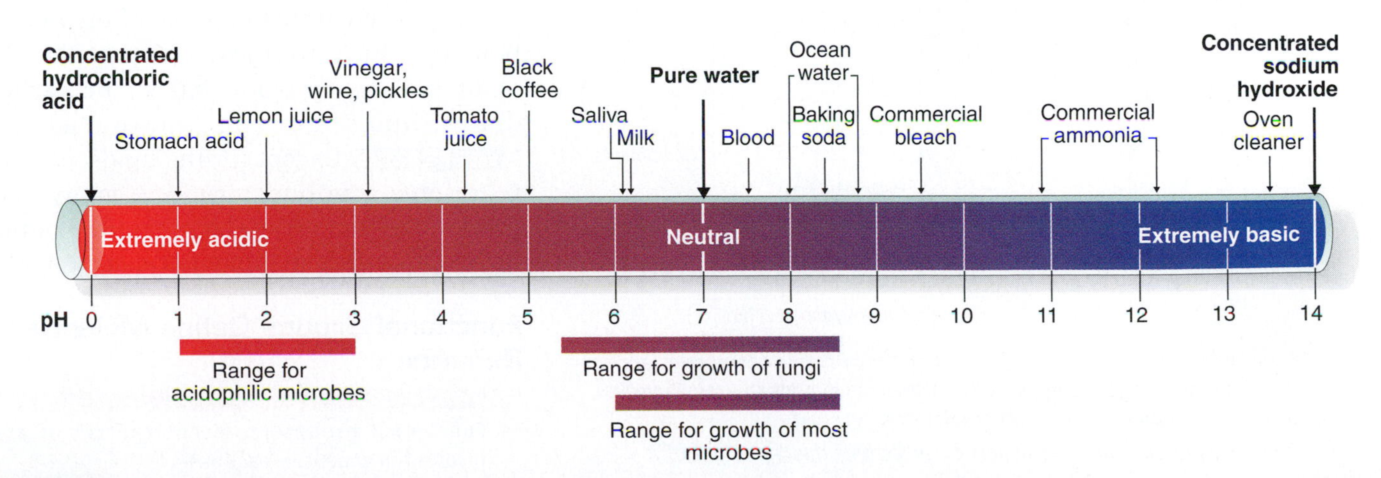

FIGURE 2.7 A Sample of pH Values for Some Common Substances.

Q: On the pH scale, notice that many of the beverages we drink (e.g., wine, tomato juice, coffee) are fairly acidic. However, we would never drink an alkaline solution (e.g., commercial bleach, ammonia). Propose an explanation for these observations.

Buffers Are a Combination of a Weak Acid and Base

KEY CONCEPT

- Buffers prevent pH shifts.

As microorganisms—and all organisms—take up or ingest nutrients and undergo metabolism, chemical reactions occur that use up or produce H^+. It is important for all organisms to balance the acids and bases in their cells because chemical reactions and organic compounds are very sensitive to pH shifts. Proteins are especially vulnerable, as we will soon see. If the internal cellular pH is not maintained, these proteins may be destroyed.

Likewise, when most microbes grow in a microbiological nutrient medium, the waste products produced may lower the pH of the medium, which could kill the organisms.

To prevent pH shifts, cells and the growth media contain **buffers**, which are compounds that maintain a specific pH. The buffer does not necessarily maintain a neutral pH, but rather whatever pH is required for that environment.

Most biological buffers consist of a weak acid and a weak base (FIGURE 2.8). If an excessive number of H^+ are produced (potential pH drop), the base can absorb them. Alternatively, if there is a decrease in the hydrogen ion concentration (potential pH increase), the weak acid can dissociate, replacing the lost hydrogen ions.

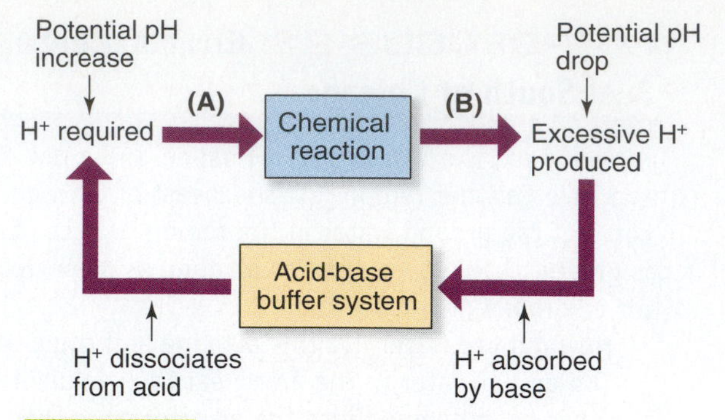

FIGURE 2.8 **A Hypothetical Example for pH Shifts.** An acid/base buffer system can prevent pH shifts from occurring as a result of a chemical reaction. If the reaction is using up H^+ (**A**), the acid component prevents a pH rise by donating H^+ to offset those used. If the reaction is producing excess H^+ (**B**), the base can prevent a pH drop by "absorbing" them.

Q: Propose what would happen if a chemical reaction continued to release excessive H^+ for a prolonged period of time.

CONCEPT AND REASONING CHECKS

2.8 If the pH of an aqueous solution changes from pH 7 to a pH 5, how many times has the H^+ concentration changed?

2.9 If the pH does drop in a cell, what does that tell you about the buffer system?

2.4 Major Organic Compounds of Living Organisms

As mentioned in the last section, a typical prokaryotic cell is about 70 percent water. If all the water is evaporated, the predominant *dry weight* remaining consists of **organic compounds**, which are those compounds related to or having a carbon basis: the carbohydrates, lipids, proteins, and nucleic acids (FIGURE 2.9). Except for the lipids, each class represents a **polymer** (*poly* = many; *mer* = part) built from a very large number of building blocks called **monomers** (*mono* = one).

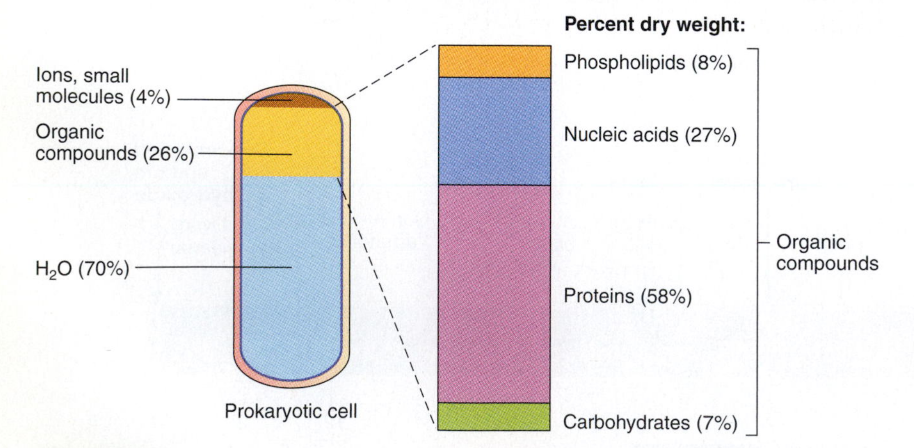

FIGURE 2.9 **Organic Compounds in Prokaryotic Cells.** Organic compounds are abundant in cells. The approximate composition of these compounds in a prokaryotic cell is similar to the percentages found in other microbes.

Q: Propose a reason why proteins make up almost 60 percent of the dry weight of a prokaryotic cell.

Functional Groups Define Molecular Behavior

KEY CONCEPT

- Functional groups represent the set of atoms involved in chemical reactions.

Before we look at the major classes of organic compounds, we need to address one question. The monomers building carbohydrates, nucleic acids, and proteins are essentially stable molecules because their outer shells are filled through covalent bonding. Why then should these molecules take part in chemical reactions to build polymers?

The answer is that these monomers are not completely stable. Projecting from the carbon skeletons or other atoms on these biological molecules are groups of atoms called functional groups. **Functional groups** represent points where further chemical reactions can occur if facilitated by a specific enzyme. The reactions will not happen spontaneously.

There is a small number of functional groups but their differences and placement on compounds makes possible a large variety of chemical reactions. The important functional groups in living organisms, include the: hydroxyl group (–OH); carbonyl group (–CHO); carboxyl group (–COOH); amino group ($-NH_2$); sulfhydryl group (–SH), and phosphate group ($-OPO_3^{-2}$).

Functional groups on monomers can interact to form larger molecules or polymers through dehydration synthesis reactions. In addition, functional groups can be critical for the decomposition of larger polymers into monomers through hydrolysis reactions.

If you are unsure about the role of these functional groups, do not worry. We will see how specific functional groups interact through dehydration synthesis reactions as we now visit each of the four classes of organic compounds.

CONCEPT AND REASONING CHECKS

2.10 Why must dehydration synthesis and hydrolysis reactions be controlled by enzymes?

Carbohydrates Consist of Sugars and Sugar Polymers

KEY CONCEPT

- Carbohydrates provide energy and structural materials.

Carbohydrates are organic compounds composed of carbon, hydrogen, and oxygen atoms that build sugars and starches. In simple sugars, like glucose ($C_6H_{12}O_6$), the ratio of hydrogen to oxygen is 2 to 1, the same as in water. The carbohydrates therefore are considered *hydrated carbon*. However, the atoms are not present as water molecules bound to carbon but rather carbon covalently bonded to hydrogen and hydroxyl groups (H–C–OH).

Carbohydrates function as energy sources in cells. They also function as structural molecules in cell walls and nucleic acids. Often the carbohydrates are termed *saccharides* (*sacchar* = sugar) and can divided into three groups.

Monosaccharides are simple sugars; they represent the monomers for disaccharides and polysaccharides. Monosaccharides have three to seven carbon atoms. Glucose and fructose are both six-carbon sugars and are among the most widely encountered monosaccharide isomers (FIGURE 2.10A).

Glucose serves as the basic supply for cellular energy in the world. Estimates vary, but many scientists estimate half the world's carbon exists as glucose. Such sugars are synthesized from water and carbon dioxide through the process of photosynthesis. Algae and cyanobacteria have the chemical machinery for this process, which is described in Chapter 6.

Disaccharides (*di* = two) are double sugars. They are composed of two monosaccharides held together by a covalent bond. Sucrose (table sugar) is an example (Figure 2.10A). This disaccharide is constructed from a glucose and fructose molecule through a dehydration synthesis reaction. Sucrose is a starting point in wine fermentations, and it is often involved in tooth decay (MicroFocus 2.4). Maltose is another disaccharide but is composed of two glucose monomers (FIGURE 2.10B). Maltose occurs in cereal grains, such as barley, and is fermented by yeasts for energy. An important by-product of the fermentation is the formation of alcohol in beer. Lactose is a third common disaccharide. It is composed of the monosaccharides glucose and galactose. Lactose is known as milk sugar because it is the principal carbohydrate in milk. Under controlled industrial conditions, microorganisms digest the lactose for energy; in the process, they produce the acid in yogurt, sour cream, and other sour dairy products.

In most bacterial cells, the cell wall is composed of carbohydrate and protein. The carbohydrate is a disaccharide linked in long

- Enzyme: A protein that facilitates a specific chemical reaction.
- Isomers: Molecules with the same molecular formula but having different structural and functional properties.

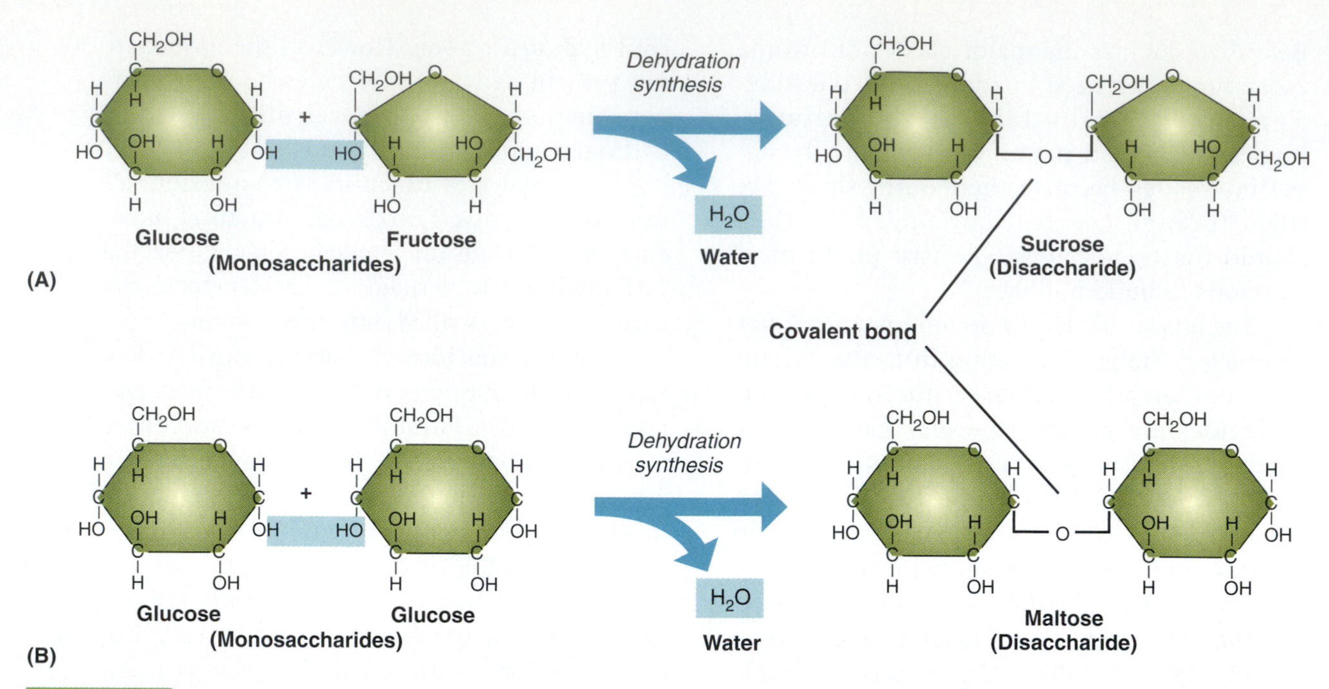

FIGURE 2.10 **Structural Formulas for Monosaccharides and Disaccharides.** Note that in the synthesis of the disaccharides, specific hydroxyl functional groups are the site of the dehydration synthesis reaction.

Q: In these two diagrams of dehydration synthesis reactions, what key component for the reactions is not shown?

MICROFOCUS 2.4: Public Health

Sugars, Acid, and Dental Cavities

Each of us at some time probably has feared the dentist's drill after a cavity has been detected. Dental cavities usually result from eating too much sugar or sweets. These sugars contribute to cavity formation only in an indirect way. The real culprits are oral bacteria.

Many species of microorganisms normally inhabit the mouth (see figure). Some of the bacterial species, along with saliva and food debris, form a gummy layer called dental plaque, a type of biofilm mentioned in Chapter 1. If not removed, plaque accumulates on the grooved chewing surfaces of back molars and at the gum line.

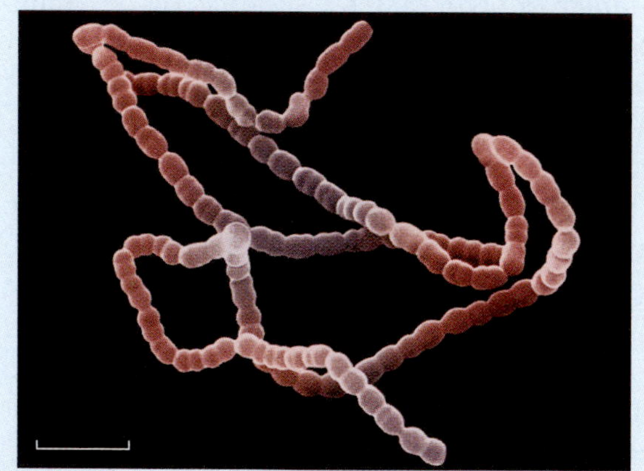

Cells of *Streptococcus mutans*, one of the major agents in dental plaque that produces cavity-causing acid. (Bar = 10 μm.)

Plaque starts to accumulate within 20 minutes after eating. As the bacteria multiply, they digest sucrose (table sugar) in sweets for energy. The metabolism of sucrose has two consequences. Some plaque bacteria produce dextran, an adhesive polysaccharide that increases the thickness of plaque. They also produce lactic acid as a by-product of sugar metabolism. Being trapped under the plaque, the acid is not neutralized by the saliva. When the pH drops to 5.5 or lower, the hydrogen ions start to dissolve or demineralize the dental enamel. Over time, a depression or cavity forms. When the soft dental tissues underneath the enamel are reached, toothache pain results from the exposure of the sensitive nerve endings in the soft tissues.

Good oral hygiene, including flossing, brushing, and regular professional dental cleaning, can keep plaque to a minimum. At home, watch what you eat. Consuming sugary foods with a meal or for dessert is less likely to cause cavities because the increased saliva produced while eating helps wash food debris off the tooth surface and neutralize any acids produced. However, snacking on sugary foods that are sticky, like caramel, toffee, dried fruit, or candies, allows the food debris to cling to teeth for a longer time, causing more plaque and providing continuous acid attack on your teeth. No wonder cavities are one of the most prevalent infectious diseases, second only to the common cold. Additional information on preventing tooth decay is provided in MicroFocus 12.5.

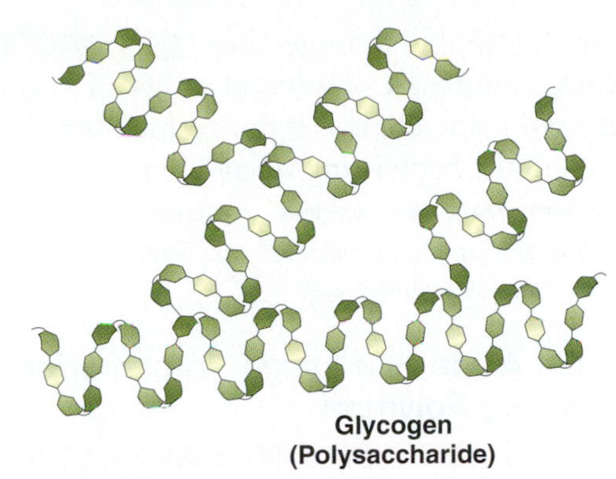

FIGURE 2.11 Structural Formula for a Polysaccharide. The polysaccharide glycogen is a complex arrangement of covalently-bonded glucose molecules.

Q: What type of reaction forms polysaccharides?

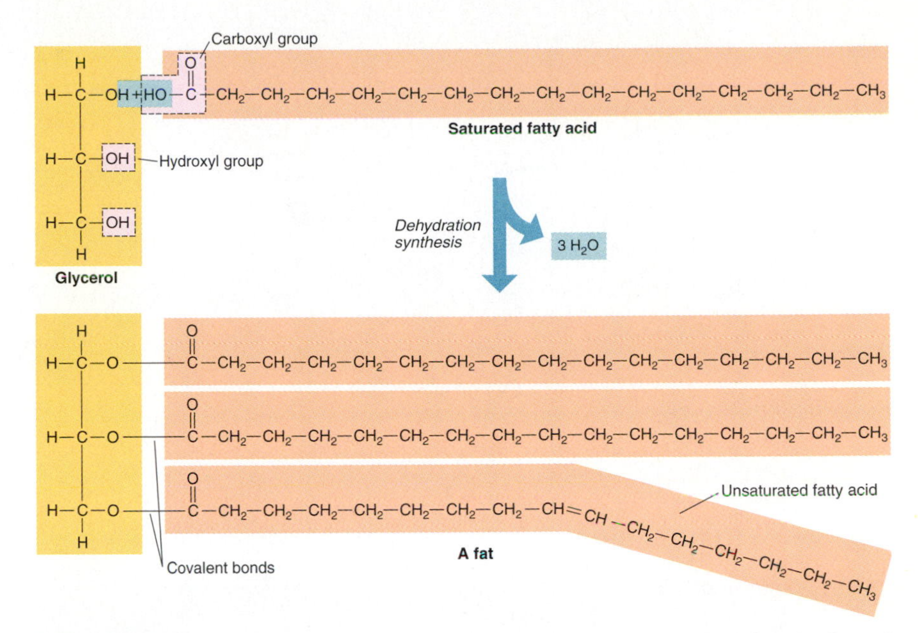

FIGURE 2.12 The Molecular Components and Synthesis of a Fat. A fat consists of fatty acids and glycerol. Fat formation occurs by dehydration synthesis, as the components of water are removed from the reactants and covalent bonds are formed. Each fatty acid contains a long hydrocarbon chain. Some chains may contain saturated fatty acids while others may have unsaturated fatty acids.

Q: How could this unsaturated fat become a saturated fat?

chains. In Chapter 4, we will examine the cell wall in more detail.

Polysaccharides (*poly* = many) are complex carbohydrates, representing large polymers formed by joining together hundreds of thousands of glucose monomers. Covalent bonds resulting from the reactions link the units together.

Starch, a storage polysaccharide, can be broken down through hydrolysis reactions into glucose and used by some eukaryotic microbes as an energy source. Glycogen, the common storage polysaccharide in humans, also functions as a stored energy source in some prokaryotes (FIGURE 2.11). The hydrolysis of these polysaccharides into glucose monomers requires the enzyme amylase, which will facilitate the hydrolysis reaction. Cellulose, a structural polysaccharide, is a component of the cell walls of plants and many algae. Some bacterial, archaeal, and protozoal cells have enzymes to digest cellulose.

CONCEPT AND REASONING CHECKS

2.11 The microbial community in a termite's gut contains the enzyme cellulase. How does this benefit the termite and the termite's microbial community?

Lipids Are Water-Insoluble Compounds

KEY CONCEPT

- Lipids store energy and are components of membranes.

The **lipids** are a broad group of nonpolar organic compounds that are **hydrophobic**; they do not dissolve in water. Like carbohydrates, lipids are composed of carbon, hydrogen, and oxygen, but the proportion of oxygen is much lower.

The best-known lipids are the **fats**. They serve many organisms, but not bacterial species, as important stored energy sources. Fats consist of a three-carbon glycerol molecule and up to three long-chain fatty acids (FIGURE 2.12). Each fatty acid is a long nonpolar hydrocarbon chain containing between 16 and 18 carbon atoms. Bonding of each fatty acid to the glycerol molecule occurs by a dehydration synthesis reaction between the hydroxyl and carboxyl functional groups.

There are two major types of fats. **Saturated fats** contain three fatty acids with the maximum number of hydrogen atoms extending from the carbon backbone. **Unsaturated fats** contain one or more fatty acid tails with less than the maximum hydrogen atoms because of double covalent bonds between a few carbon atoms.

Another type of lipid found in cell membranes is the **phospholipids**. Although similar to fats, phospholipids have only two fatty acid tails attached to glycerol (FIGURE 2.13A). In

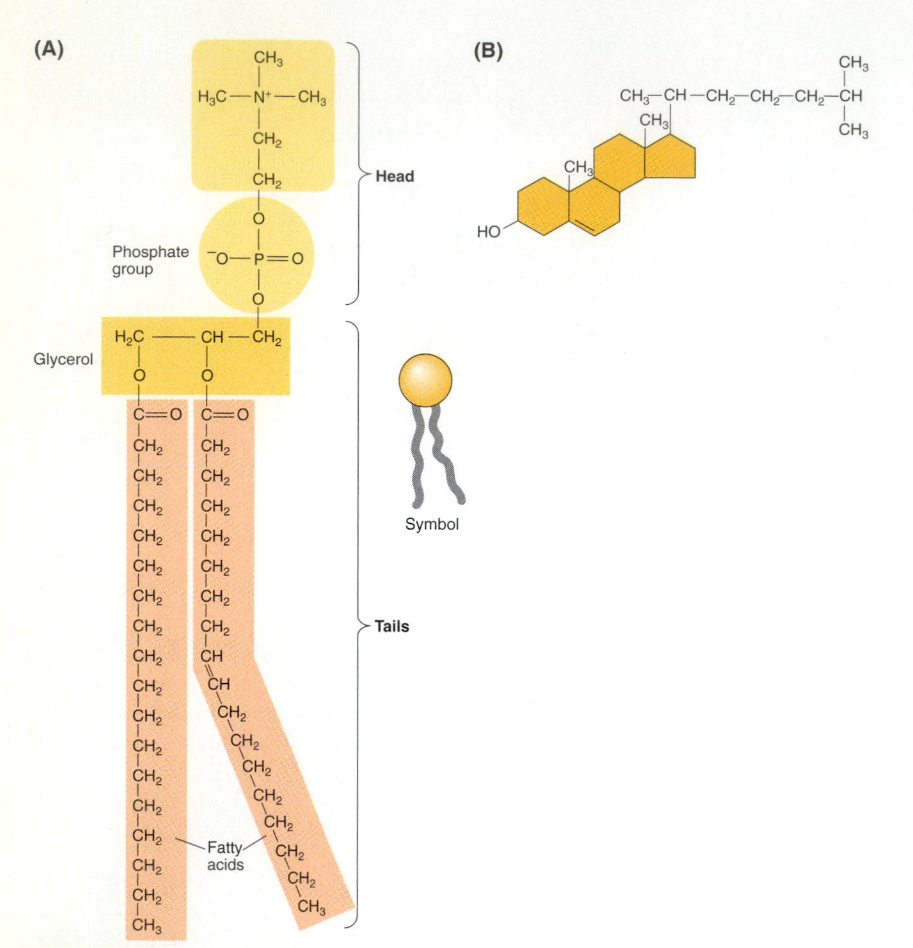

FIGURE 2.13 **Some Other Lipids of Importance.** (**A**) A phospholipid molecule. The polar head group is attached to fatty acid tails. The tails can consist of saturated or unsaturated fatty acids. Inset: The symbol for the structure of a phospholipid. (**B**) The steroid cholesterol. Steroids have four carbon rings (yellow). In this diagram the carbons making up the rings and the attached hydrogen atoms are omitted.

Q: Steroids can integrate into the fatty acid tails of phospholipids. What property gives them this ability?

place of the third fatty acid there is a phosphate group, representing a functional group that is polar and can actively interact with other polar molecules. We will have more to say about phospholipids and membranes in the next chapter. Some bacterial toxins are a combination of polysaccharide and lipid (Textbook Case 2).

Other types of lipids include the waxes and sterols. Waxes are composed of long chains of fatty acids and form part of the cell wall in *Mycobacterium tuberculosis*, the bacterium causing tuberculosis. **Sterols**, such as cholesterol, are very different from lipids and are included with lipids solely because they too are hydrophobic molecules (FIGURE 2.13B). Sterols, composed of several rings of carbon atoms with side chains, stabilize membranes of fungi and the bacterium *Mycoplasma*.

CONCEPT AND REASONING CHECKS

2.12 Why are fats not considered polymers in the sense that polysaccharides are?

Nucleic Acids Are Large, Information-Containing Polymers

KEY CONCEPT

- Nucleic acids store, transport, and control hereditary information.

The **nucleic acids**, among the largest organic compounds found in organisms, are composed of carbon, hydrogen, oxygen, nitrogen, and phosphorus atoms. Two types function in all living things: **deoxyribonucleic acid (DNA)** and **ribonucleic acid (RNA)**.

Both DNA and RNA are composed of repeating monomers called **nucleotides** (FIGURE 2.14A). Each nucleotide has three components: a sugar molecule, a phosphate group, and a nitrogenous base. The sugar in DNA is deoxyribose, while in RNA it is ribose. The **nitrogenous bases** are nitrogen-containing compounds. In DNA, the purine bases are adenine (A) and guanine (G), while the pyrimidine bases are cytosine (C) and thymine (T). In RNA, adenine, guanine, and cytosine also are present, but uracil (U) is found instead of thymine. MicroInquiry 2 (see page 64) uses these base differences to investigate the mode of action of a potential antibiotic.

Nucleotides are covalently joined through dehydration synthesis reactions between the sugar of one nucleotide and the phosphate of the adjacent nucleotide to eventually form a **polynucleotide** (FIGURE 2.14B).

DNA. In 1953, James Watson, Francis Crick, Rosalind Franklin, and Maurice Wilkins published papers describing how a complete DNA molecule consists of two polynucleotide strands opposed to each other in a ladder-like arrangement (FIGURE 2.14C). Guanine and cytosine line up opposite one another, and thymine and adenine oppose each other in the two strands. The complementary base pairs in the double-stranded DNA molecule are held together by hydrogen bonds. The double

Textbook CASE 2

An Outbreak of *Salmonella* Food Poisoning

1 During the morning of October 17, 1991, a restaurant employee prepared a Caesar salad dressing, cracking fresh eggs into a large bowl containing olive oil.

2 Anchovies, garlic, and warm water then were mixed into the eggs and oil.

3 The warm water raised the temperature of the mixture slightly before the dressing was placed in the refrigerator.

4 Later that day, the Caesar dressing was placed at the salad bar in a cooled compartment having a temperature of about 60°F. The dressing remained at the salad bar until the restaurant closed, a period of 8 to 10 hours. During that time, many patrons helped themselves to the Caesar salad.

5 Within three days, fifteen restaurant patrons experienced gastrointestinal illness. Symptoms included diarrhea, fever, abdominal cramps, nausea, and chills. Thirteen sought medical care, and eight (all elderly over 65 years of age) required intravenous rehydration.

6 From the stool samples of all 13 patrons who sought medical attention, bacteria were cultured (see figure) and laboratory tests identified *Salmonella enterica* as the causative agent.

7 *S. enteritidis* produces a lipopolysaccharide toxin that causes the symptoms experienced by all the affected patrons.

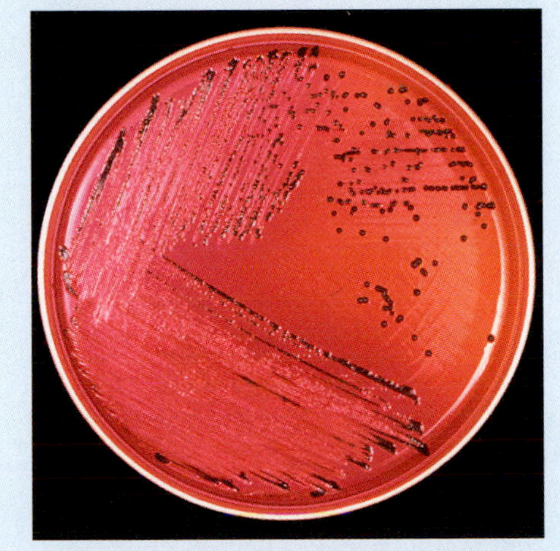

A culture plate of *Salmonella*.

Questions

A. *What might have been the origin of the bacterial contamination?*

B. *What conditions would have encouraged bacterial growth?*

C. *How could the outbreak have been prevented?*

D. *What types of organic compounds form the bacterial toxin?*

E. *Why did so many of the elderly patrons develop a serious illness?*

For additional information see http://www.cdc.gov/ncidod/dbmd/diseaseinfo/salment_g.htm.

strand then twists to form a spiral arrangement called the DNA **double helix**.

DNA is the genetic material in all organisms in the domains of life. The genetic information exists in discrete units called **genes**, which are sequences of nucleotides that encode information to regulate and synthesize proteins (Chapter 7). All microbes can be packed with thousands of genes. In bacterial and archaeal cells, these genes are found on a single circular chromosome, while in most eukaryotic microbes, the genes are located on several linear chromosomes. Genes only carry the information to regulate or synthesize proteins.

■ Chromosome: A DNA molecule containing the hereditary information in the form of genes.

RNA. Besides having uracil as a base and ribose as the sugar, RNA molecules in cells are single-stranded polynucleotides. Biologists once viewed RNAs as the intermediaries,

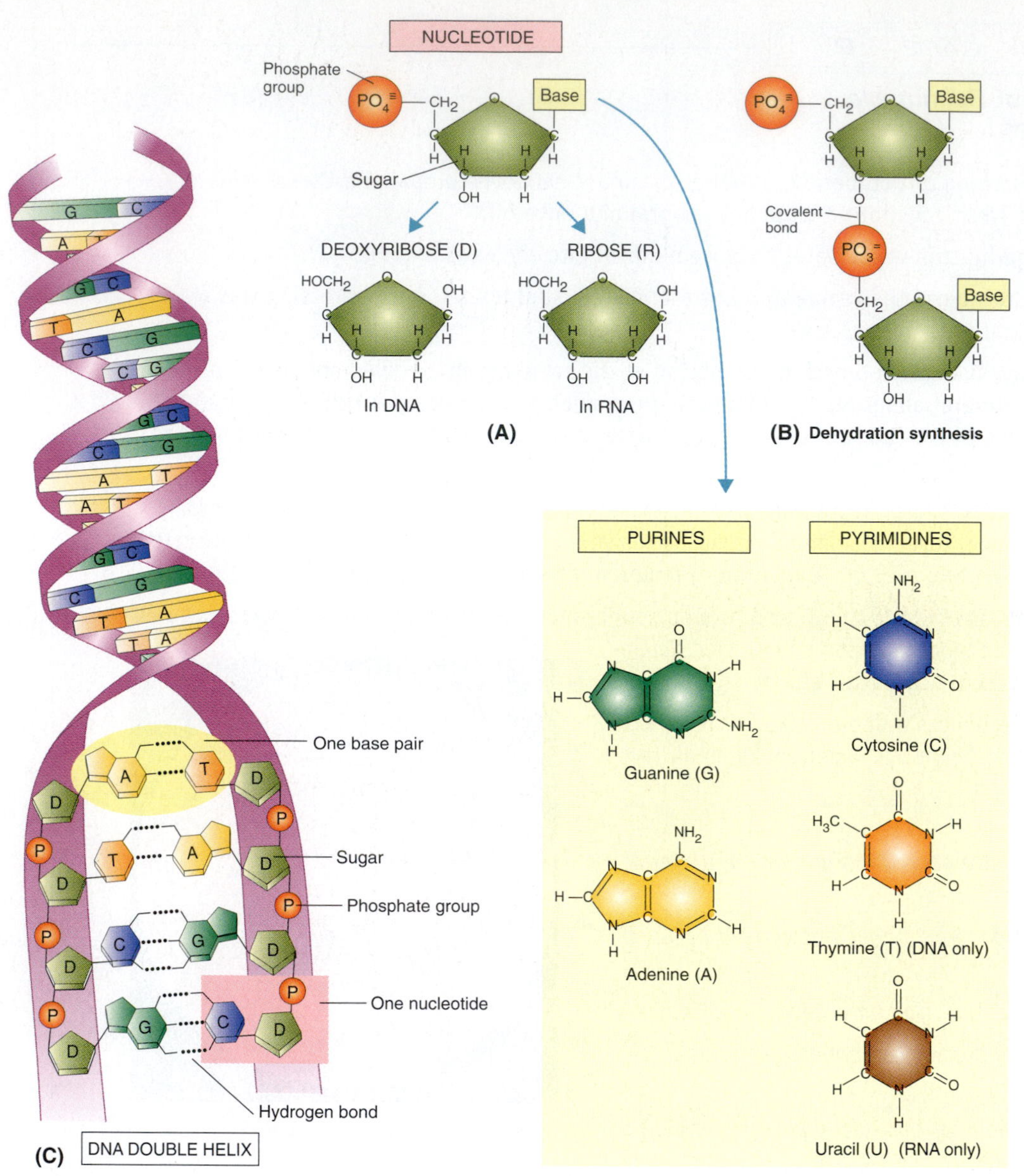

FIGURE 2.14 **The Molecular Structures of Nucleotide Components and the Construction of DNA.** (**A**) The sugars in nucleotides are ribose and deoxyribose, which are identical except for one additional oxygen atom in RNA. The nitrogenous bases include adenine and guanine, which are large purine molecules, and thymine, cytosine, and uracil, which are smaller pyrimidine molecules. Note the similarities in the structures of these bases and the differences in the side groups. (**B**) Nucleotides are bonded together by dehydration synthesis reactions. (**C**) The two polynucleotides of DNA are bonded together by hydrogen bonds between adenine (A) and thymine (T) or guanine (G) and cytosine (C) to form a double helix.

Q: If a segment of one strand of DNA has the bases TTAGGCACG, what would be the sequence of bases in the complementary strand?

involved in copying the gene information and using the information to construct proteins. This certainly is a major role for RNA but not the only role.

In viruses like the herpesviruses and smallpox virus, DNA is the genetic information. However, many other viruses have RNA as the genetic information. For example, the influenza virus and measles virus never have DNA present. Other RNA molecules play key roles in regulating gene activity, while several small RNAs control various cellular processes in microbial cells.

The nucleic acids cannot be altered without injuring the organism or killing it. Ultraviolet light damages DNA, and thus it can be used to control microbes on an environmental surface. Chemicals, such as formaldehyde, alter the nucleic acids of viruses and can be used in the preparation of vaccines. Certain antibiotics interfere with DNA or RNA function and thereby kill bacteria.

Chapters 7 and 8 are devoted to the role of nucleic acids in the genetics of microbes and how the genetic information can be manipulated for medical or industrial applications.

It also is important to point out that nucleotides have other roles in cells beside being part of DNA or RNA. Adenine nucleotides with three attached phosphate groups form **adenosine triphosphate (ATP)**, which is the cellular energy currency in all cells. Other nucleotides can be part of the structure of some enzymes, while independent, modified nucleotides called cyclic adenosine monophosphate (cAMP) act as chemical signals in many microbes.

CONCEPT AND REASONING CHECKS

2.13 How does the structure of DNA differ from that of RNA?

Proteins Are the Workhorse Polymers in Cells

KEY CONCEPT

- Protein function depends on its three-dimensional shape.

Proteins are the most abundant organic compounds in microorganisms and all living organisms, making up about 58 percent of the cell's dry weight. The high percentage of protein indicates their essential and diverse roles. Many proteins function as structural components of cells and cell walls, and as transport agents in membranes. A large number of proteins serve as enzymes. Proteins are composed of carbon, hydrogen, oxygen, nitrogen, and usually sulfur atoms.

Proteins are built from nitrogen-containing monomers called **amino acids** (MicroFocus 2.5). At the center of each amino acid is a carbon atom attached to two functional groups: an amino group ($-NH_2$) and a carboxyl group ($-COOH$) (FIGURE 2.15A). Also attached to the carbon is a side chain, called the **R group**. Each of the 20 amino acids differs only by the atoms composing the R group (FIGURE 2.15B). These side chains, many being functional groups, are essential in determining the final shape, and therefore function, of the protein.

In protein formation, two amino acids (sometimes called peptides) are joined together by a covalent bond when the amino group of one amino acid is linked to the carboxyl of another amino acid through a dehydration synthesis reaction (FIGURE 2.16). Repeating the reaction hundreds of times produces a long chain of amino acids called a **polypeptide** and the covalent bond therefore is called a **peptide bond**. How the amino acids

MICROFOCUS 2.5: Environmental Microbiology

Triple Play—Bacteria to Plants to Humans

About 80 percent of the atmosphere is nitrogen gas (N_2). Nitrogen gas, as you have discovered, contains a triple covalent bond that is very hard to break. Yet, one of the essential elements in nucleic acids and proteins in all organisms is nitrogen. So, how can the gaseous form of nitrogen be converted into a form that can be used to make essential biological compounds?

The most important biological process to break the triple covalent bond in N_2 is accomplished by a few bacterial species commonly found in root nodules of pea and bean plants (legumes) or in the soil close to the plant roots. These bacterial organisms contain an enzyme, called nitrogenase, which converts N_2 into ammonia that then can be further converted by microbial action into forms used by legumes and other plants. The process is called **nitrogen fixation:**

$$N_2 \longrightarrow NH_3 \text{ (ammonia)}$$

In contact with water, the gaseous ammonia is converted to ammonium ions (NH_4^+), which serve as a source of nitrogen for nucleic acid and amino acid synthesis by bacterial cells and plants. We then get our nitrogen for amino acids and nucleotides from eating plants or through exchange reactions of carbohydrate metabolism that convert sugars into nucleotides or amino acids.

The important point to remember in all this chemistry is that the initial fixation of nitrogen is dependent on bacterial chemistry. In fact, without nitrogen fixation, life as we know it would not exist.

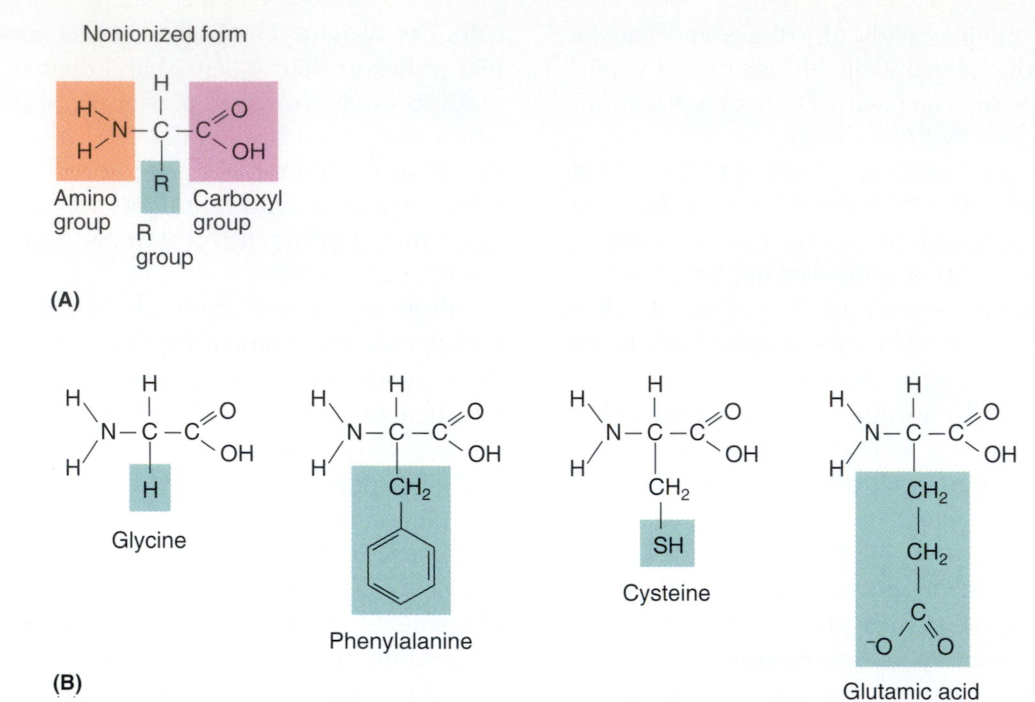

FIGURE 2.15 **Examples of Amino Acids.** (**A**) The generalized structure for an amino acid. (**B**) Four amino acids that differ by the molecular nature of the R group. Cysteine contains a sulfhydryl functional group and glutamic acid has a carboxyl functional group as the R group.

Q: What is common and what is unique to all 20 amino acids?

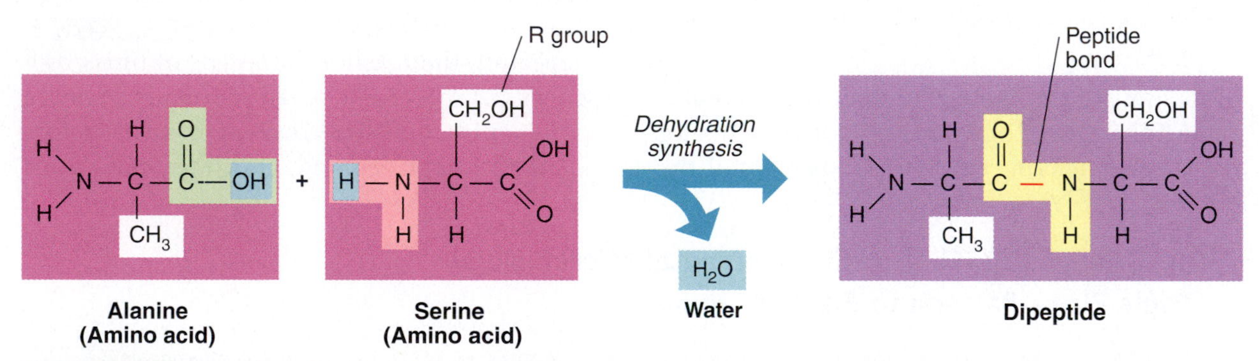

FIGURE 2.16 **Formation of a Dipeptide by Dehydration Synthesis.** The amino acids alanine and serine are shown, with the differences in white. The –OH from the carboxyl group of alanine combines with the –H from the amino group of serine to form water. The open bonds then link together, yielding a peptide bond. The process is repeated hundreds of times, adding additional amino acids to form a polypeptide.

Q: What are the functional groups that interact to form a peptide bond?

are slotted into position to build a polypeptide is a complex process discussed in Chapter 7. However, the sequence of amino acids is critical because a single amino acid improperly positioned may change the three-dimensional shape and function of the protein.

Because proteins have tremendously diverse roles, they come in many sizes and shapes. The final shape depends on several factors associated with the amino acids. The sequence of amino acids in the polypeptide represents the **primary structure** (FIGURE 2.17A). Each protein that has a different function will have a different primary structure. However, the sequence of amino acids alone is not sufficient to confer function.

Many polypeptides have regions folded into a corkscrew shape or **alpha helix**. These regions represent part of the protein's **secondary structure** (FIGURE 2.17B). Hydrogen bonds between

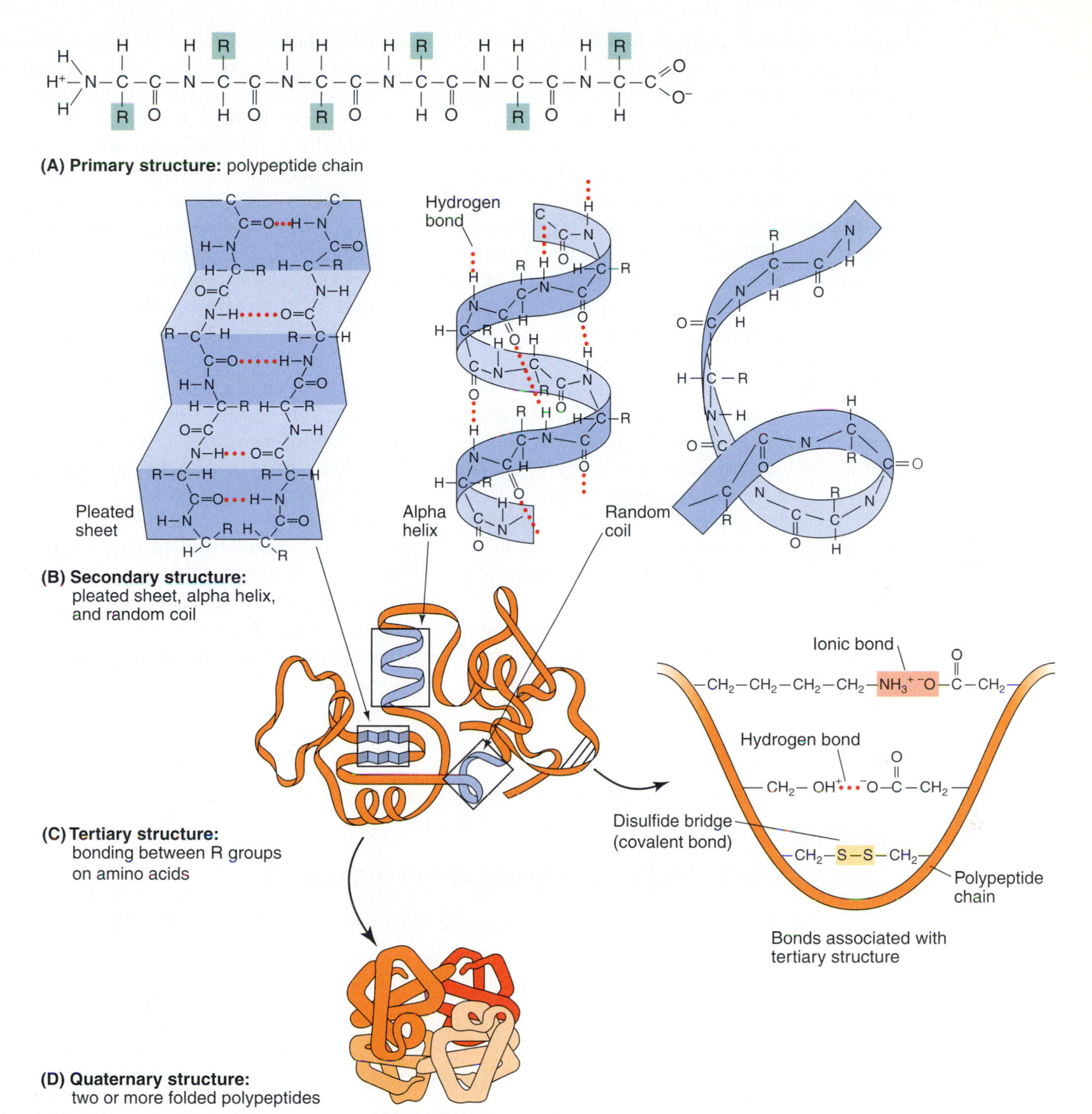

FIGURE 2.17 **Protein Folding.** The function of a protein is dependent on how it folds. Folding depends on the primary structure (**A**), which dictates all higher levels of folding. Secondary structure (**B**) is dependent on the interactions between amino and carbonyl groups on amino acids, while tertiary structure (**C**) depends on bonding interactions between R groups. Some proteins consist of more than one polypeptide (**D**), and this quaternary structure also is determined by the bonding interactions between each polypeptide.

Q: Explain why each and every protein must have 3 to 4 levels of protein folding.

amino groups (–NH) and carbonyl groups (–C=O) on nearby amino acids maintain this structure. A secondary structure also may form when the hydrogen bonds cause portions of the polypeptide chain to line up alongside one another, forming a pleated sheet. Other regions may not interact and remain in a random coil.

Many polypeptides also have a **tertiary structure** (FIGURE 2.17C). Such a polypeptide is folded back on itself much like a spiral telephone cord. Ionic and hydrogen bonds between R groups on amino acids in close proximity help form and maintain the polypeptide in its tertiary structure. In addition, covalent bonds, called **disulfide bridges**, between sulfur atoms in R groups are important in stabilizing tertiary structure.

Many proteins are single polypeptides. However, other proteins contain two or more polypeptides to form the complete and functional protein; this is called the **quaternary structure** (FIGURE 2.17D). Each polypeptide chain is folded into its tertiary structure and the unique association between separate polypeptides produces the quaternary structure. The same types of chemical bonds are involved as in tertiary structure.

The ionic and hydrogen bonds helping hold a protein in its functional shape are relatively weak associations. As such, these interactions in a protein are influenced by environmental conditions. When subjected to heat, pH changes, or certain chemicals, these bonds may break, leaving only the primary structure intact. This loss of three-dimensional shape is referred to as **denaturation**. For example, the white of a boiled egg is denatured egg protein (albumen) and cottage cheese is denatured milk protein. Should enzymes be denatured, the important chemical reactions they facilitate will be interrupted and death of the organism may result. Viruses also can be destroyed by denaturing the proteins found in the viral protein coat.

Now you should understand the importance of buffers in cells; by preventing pH shifts, they prevent protein denaturation and maintain protein function.

The four major classes of organic compounds are summarized in TABLE 2.3. MicroFocus 2.6 looks at the origins of the monomers and polymers discussed in this chapter, while MicroInquiry 2 uses several attributes of the chemical elements to discover how an antibacterial chemical works.

CONCEPT AND REASONING CHECKS

2.14 Why does a denatured protein no longer have a functional role?

TABLE
2.3 Major Organic Compounds of Microorganisms

Organic Compound	Monomer	Some Major Functions	Examples
Carbohydrate			
Monosaccharides	–	Energy source	Glucose, fructose
Disaccharides	Monosaccharides	Energy source; physical structure	Lactose, sucrose; bacterial walls
Polysaccharides	Monosaccharides	Energy storage; physical structure	Starch, glycogen; cellulose, cell walls
Lipid			
Fats	Fatty acids and glycerol	Energy source	Fat; oil
Phospholipids	Fatty acids, glycerol, phosphate	Foundation for membranes	Cell membranes
Waxes	Fatty acids	Physical structures	*Mycobacterium* cell walls
Sterols	–	Membrane stability	Cholesterol
Nucleic acid	Nucleotides	Inheritance; control of protein synthesis	DNA, RNA
Protein	Amino acids	Enzymes; toxins; physical structures	Antibodies; viral surface; flagella; membranes

MICROFOCUS 2.6: Evolution

Generating Life—Today (Part II)

In MicroFocus 1.2, we discussed the idea of creating life artificially—or what is called synthetic biology. The idea concerns the ability to create "new" life in a test tube. As you discovered, some scientists already have made initial strides in this direction by reconstructing a virus from scratch. Others are trying to design "new organisms" by piecing together specific cellular and genetic parts taken from various microbes. So far, these "bioreactors" do not represent cellular life. However, the potential to build synthetic life also opens up the possibility of assembling in the lab a "primordial organism" similar to what might have started life on Earth.

Chemical precursors of life may have involved gases like formaldehyde, methane, water, hydrogen cyanide, and ammonia. As the Earth formed, these "molecular seeds" were brought together and concentrated at the Earth's surface. Through a process of chemical evolution, chemical reactions between these precursors might have produced larger and more complicated compounds. If the products could be stabilized in some way, they could form into simple sugars, nucleotides, and amino acids that are the building blocks of carbohydrates, nucleic acids, and proteins.

In fact, in the 1950s Stanley Miller and Harold Urey carried out experiments demonstrating that such chemical evolution could occur. Mixing the primeval gases with an energy source produced within days a few amino acids and nitrogenous bases. Although scientists still argue about the validity of the Miller-Urey experiments, they do show that complex compounds can be generated.

Most scientists today believe that RNA was the original "genetic information" in primitive cells and could also act as an enzyme (see MicroFocus 6.2, page 167). Such ribozymes may have arose from sugars, such as ribose, combining with the nitrogenous bases (A, G, C, and U). Along with phosphate, which existed in volcanic areas and other deposits on Earth, the bases and sugars could be linked together into nucleotides, which then would form into long chains of RNA. So, RNA-based life forms may have spread and evolved on Earth for millions of years.

Generating or recreating such life today may be the way to verify the hypothesis. Scientists, such as John Szostak and his team at Harvard Medical School, are mixing RNA nucleotides with clay and then adding fatty acids. The spontaneous result is a primordial cell—a lipid (membrane) bubble containing short RNA polynucleotides. A key experiment is to demonstrate that the RNAs can carry out simple chemical reactions such as hydrolysis or dehydration synthesis reactions by acting as a ribozyme.

The researchers have not created synthetic life—yet. However, the experiments do indicate their ideas have promise and may go a long way toward telling scientists how life originated on Earth.

MICROINQUIRY 2

Isotopes as Tracers in Drug Discovery

Several biologically important elements have isotopes that are radioactive. The table below lists several of those commonly used in research.

Some Radioactive Elements

ELEMENT	COMMON FORM	RADIOACTIVE FORM
Hydrogen	^{1}H	^{3}H (tritium)
Carbon	^{12}C	^{14}C
Sulfur	^{32}S	^{35}S

Let's examine the use of these isotopes in drug development. Note that today there are more sophisticated ways to accomplish the same thing.

Scenario: As a researcher for a drug company, you are given a chemical that seems to kill the bacterium *Escherichia coli*. However, your job is to discover how the potential antibacterial chemical works; that is, what organic compound (DNA, RNA, or protein) is affected that brings about the death of the bacterial cells?

2a. What bases are unique to RNA and DNA?

2b. How could isotopes be used to label radioactively these nucleic acids so they can be identified separately?

2c. Which of the three elements in the table is specific to proteins?

Here is one way we could approach the problem. We can purchase the following items from science supply houses: ^{3}H-uridine (i.e., radioactive uracil, the base found only in RNA); ^{14}C-thymine (the base found only in DNA); and ^{35}S-methionine (one of the 20 amino acids that contains sulfur in its structure). So, we have specific "labels" that would be incorporated into RNA, DNA, and protein.

The experiment will involve growing the bacterium *E. coli* in each of three broth cultures. To these actively growing cultures, we will add ^{3}H-uridine to one culture, ^{14}C-thymine to another, and ^{35}S-methionine to the third. If our potential antimicrobial chemical is not added to the cultures, the following occurs: as DNA is copied, the molecules should become radioactive with ^{14}C-thymine; as RNA is synthesized, the RNAs should become radioactive with ^{3}H-uridine; and as proteins are synthesized, they will become radioactive with ^{35}S-methionine. We can assume we have instruments to measure the radioactivity.

To these actively growing cultures containing the isotope, we add our chemical (antimicrobial drug) to be tested. With time, we remove samples from each of the cultures and determine the level of radioactivity in the DNA, RNA, and proteins. The results of the experiment are shown in the graph below.

2d. What do these isotope experiments indicate about the target for the chemical? Which of the three organic compounds is affected?

Answers can be found in Appendix D.

The effect of an antimicrobial drug on DNA, RNA, and protein synthesis.

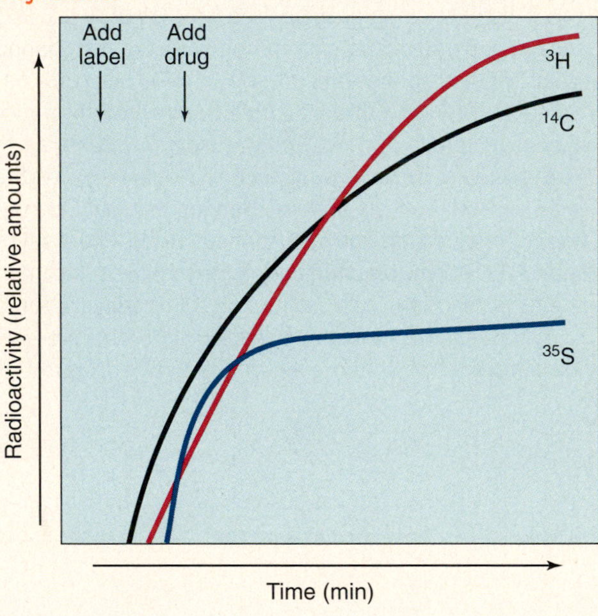

SUMMARY OF KEY CONCEPTS

2.1 The Elements of Life

- Atoms consist of an atomic nucleus (with neutrons and positively-charged protons) surrounded by a cloud of negatively-charged electrons.
- Isotopes of an element have different numbers of neutrons. Some unstable ones, called radioisotopes, are useful in research and medicine.
- If an atom gains or loses electrons, it becomes an electrically charged ion. Many ions are important in microbial metabolism.
- Each electron shell holds a maximum number of electrons. Interactions occur between atoms to fill the outer shells with electrons.

2.2 Chemical Bonding

- Ionic bonds result from the attraction of oppositely charged ions. Compounds called salts result.
- Most atoms achieve stability through a sharing of electrons, forming covalent bonds. The equal sharing of electrons produces molecules that are nonpolar (no electrical charge). Atomic interactions between hydrogen and oxygen or nitrogen produce unequal sharing of electrons, which generate polar molecules (have electrical charges).
- Separate polar molecules, like water, are electrically attracted to one another and form hydrogen bonds, involving positively charged hydrogen atoms and negatively charged oxygen or nitrogen atoms.
- In a chemical reaction, the atoms in the reactant change bonding partners in forming one or more products. Two common chemical reactions in cells are dehydration synthesis reactions and hydrolysis reactions. In these reactions, the number of atoms is the same in the reactants and products.

2.3 Water, pH, and Buffers

- All chemical reactions in organisms occur in liquid water. Being polar, water has unique properties. These include its role as a solvent, as a chemical reactant, and as a factor to maintain a fairly constant temperature.
- Acids donate hydrogen ions to a solution while bases remove hydrogen ions from a solution. The pH scale indicates the number of hydrogen ions in a solution and denotes the relative acidity of a solution.
- Buffers are a mixture of a weak acid and a weak base that maintain acid/base balance in cells. Excess hydrogen ions can be absorbed by the base and too few hydrogen ions can be provided by the acid.

2.4 Major Organic Compounds of Living Organisms

- The building of large organic compounds depends on the functional groups found on the building blocks, called monomers. Functional groups on monomers interact through dehydration synthesis reactions to form a covalent bond between monomers.
- Carbohydrates are used primarily as energy sources for life processes, and they include monosaccharides such as glucose; disaccharides such as sucrose; and polysaccharides such as glycogen.
- Lipids serve as energy sources, but their major role is as phospholipids in the construction of cell membranes. Other lipids include the sterols.
- The genetic instructions for living organisms are composed of two types of nucleic acids: deoxyribonucleic acid (DNA), which stores and encodes the hereditary information; and ribonucleic acid (RNA), which transmits the information to make proteins, controls genes, and helps regulate genetic activity.
- Proteins are chains of amino acids connected by peptide bonds, a type of covalent bond. Proteins are used as enzymes and as structural components of cells. Primary, secondary, and tertiary structures form the functional shape of many proteins, while quaternary structure is needed for others.

LEARNING OBJECTIVES

After understanding the textbook reading, you should be capable of writing a paragraph that includes the appropriate terms and pertinent information to answer the objective.

1. Identify (by name and chemical symbol) the chemical elements making up 98% of life.
2. Contrast the properties of protons, neutrons, and electrons. Assess the importance of these atomic particles to atomic structure.
3. Summarize how elements differ from one another.
4. Distinguish between ionic bonds and covalent bonds. Contrast these to hydrogen bonds.
5. Justify why all life is considered to be carbon-based.
6. Identify the role of functional groups on organic molecules.
7. Explain how organic polymers are built. Identify the four types of organic compounds.
8. Draw the structural formula for glucose.
9. Identify a disaccharide. Name three disaccharides.
10. List the polysaccharides found in cells or organisms. Explain their role in microorganisms.
11. Explain how dehydration synthesis forms a triglyceride. Contrast between saturated and unsaturated fats.
12. Explain how phospholipids differ from triglycerides in structure and function.
13. Explain how steroids are different from triglyceride.
14. Identify a nucleotide and explain how a polynucleotide is formed.
15. Summarize how DNA and RNA differ in structure and function.
16. List some functions of proteins in cells and organisms.
17. Identify the monomers making proteins. Explain how these monomers differ from one another.
18. Show how amino acids link together and name the specific type of bond formed between these amino acids.
19. Compare the four levels of protein structure. Identify the types of bonds involved in each level of structure.
20. Explain why protein shape and function depend on the sequence of amino acids in the protein.
21. Describe what happens to proteins when they are exposed to a change in pH.

SELF-TEST

Answer each of the following questions by selecting the ***one*** answer that best fits the question or statement. Answers to even-numbered questions can be found in **Appendix C**.

1. An element with a mass number of 14 has how many electrons?
   A. 7
   B. 14
   C. 21
   D. 28
   E. None of the above (**A–D**) are correct.
2. When full, an atom with two electron shells will have _____ electrons in the inner shell and _____ electrons in the outer shell.
   A. 2; 2
   B. 2; 4
   C. 2; 8
   D. 4; 8
   E. 4; 4
3. Atoms of the same element that have different numbers of neutrons are called
   A. isotopes.
   B. ions.
   C. functional groups.
   D. isomers.
   E. inert elements.
4. For ______ bonding, one or more electrons are transferred between atoms.
   A. hydrogen
   B. ionic
   C. peptide
   D. polar covalent
   E. nonpolar covalent
5. The covalent bonding of atoms forms a/an
   A. molecule.
   B. ion.
   C. element.
   D. isomer.
   E. buffer.
6. The ______ bond is a weak bond that exists between poles of adjacent molecules.
   A. hydrogen
   B. ionic
   C. polar covalent
   D. peptide
   E. nonpolar covalent
7. Which one of the following is *not* a molecular formula?
   A. CO_2
   B. Na^+
   C. $C_6H_{12}O_6$
   D. CH_4
   E. O_2
8. In what type of chemical reaction are the products of water removed during the formation of covalent bonds?
   A. Hydrolysis
   B. Ionization
   C. Dehydration synthesis
   D. Decomposition
   E. None of the above (**A–D**) are correct.
9. An acid is a chemical substance that donates ______ ions to a solution such as water.
   A. calcium
   B. magnesium
   C. hydrogen
   D. calcium
   E. hydroxyl
10. The pH scale relates the measure of ______ of a chemical substance.
    A. isotopes
    B. ionization
    C. denaturation
    D. acidity
    E. buffering
11. A functional group designated —CHO is known as a/an
    A. carboxyl.
    B. carbonyl.
    C. amino.
    D. hydrocarbon.
    E. hydroxyl.
12. Which one of the following is *not* a carbohydrate?
    A. Sucrose
    B. Glycogen
    C. Glucose
    D. Lipid
    E. Maltose
13. Examples of disaccharides include lactose, sucrose, and
    A. cellulose.
    B. glucose.
    C. fructose.
    D. maltose.
    E. galactose.
14. Both DNA and RNA are composed of ______.
    A. polynucleotides
    B. genes
    C. polysaccharides
    D. polyglycerides
    E. polypeptides
15. The two strands of the DNA double helix are held together by ______ bonds.
    A. hydrogen
    B. peptide
    C. nonpolar covalent
    D. polar covalent
    E. ionic
16. Genetic information is encoded in the
    A. sequence of DNA nucleotides.
    B. phosphate-sugar backbone.
    C. protein primary structure.
    D. protein quaternary structure.
    E. None of the above (**A–D**) are correct.
17. The ______ structure of a protein is the sequence of amino acids.
    A. primary
    B. secondary
    C. polynucleotide
    D. tertiary
    E. quaternary
18. Which one of the following is *not* a polymer?
    A. Cellulose
    B. Protein
    C. Glucose
    D. Polypeptide
    E. Glycogen

QUESTIONS FOR THOUGHT AND DISCUSSION

Answers to even-numbered questions can be found in **Appendix C**.

1. Calcium atoms have two electrons in their outer shell; magnesium atoms also have two. Would you expect these elements to interact through bonding when their atoms are brought together? Explain.
2. Propose a reason why organic molecules tend to be so large.
3. Bacterial cells do not grow on bars of soap even though the soap is wet and covered with bacterial organisms after one has washed. Explain this observation.
4. Suppose you had the choice of destroying one class of organic compounds in bacterial cells to prevent their spread. Which class would you choose? Why?
5. Milk production typically has the bacterium *Lactobacillus* added to the milk before it is delivered to market. This organism produces lactic acid. (a) Why would this organism be added to the milk and (b) why was it chosen?
6. The toxin associated with the foodborne disease botulism is a protein. To avoid botulism, home canners are advised to heat preserved foods to boiling for at least 12 minutes. How does the heat help?
7. Proteins are made up of chains of amino acids, yet the proteins themselves are not acidic. Why do you think this is so?
8. Justify Isaac Asimov's quote, "The significant chemicals in living tissue are rickety and unstable, which is exactly what is needed for life," to the atoms, molecules, and compounds described in this chapter.

APPLICATIONS

Answers to even-numbered questions can be found in **Appendix C**.

1. You want to grow a bacterial species that is acid-loving; that is, it grows best in very acid environments. Would you want to grow it in a culture that has a pH of 2.0, 6.8, or 11.5? Explain.
2. You are given two beakers of a broth growth medium. However, only one of the beakers of broth is buffered. How could you determine which beaker contains the buffered broth solution? Hint: You are provided with a bottle of concentrated HCl and pH papers that indicate a solution's pH.

HTTP://MICROBIOLOGY.JBPUB.COM/

The site features learning, an on-line review area that provides quizzes and other tools to help you study for your class. You can also follow useful links for in-depth information, read more MicroFocus stories, or just find out the latest microbiology news.

REVIEW

Answers to even-numbered questions can be found in **Appendix C**.

1. Use the following list to identify the structure (i–v) drawn below.
   A. Amino acid
   B. Monosaccharide
   C. Nucleotide
   D. Fat
   E. Disaccharide
   F. Polysaccharide
   G. Steroid
2. Identify any and all functional groups on each structure (i–v).

(i)

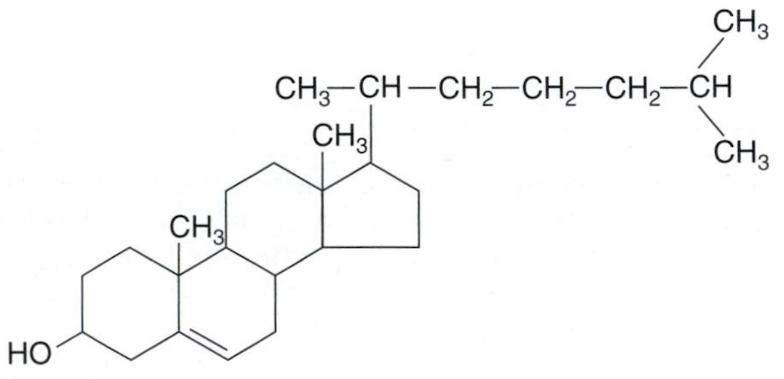

CH_3

$CH_3-CH-CH_2-CH_2-CH_2-CH$

CH_3

CH_3

CH_3

HO

(ii)

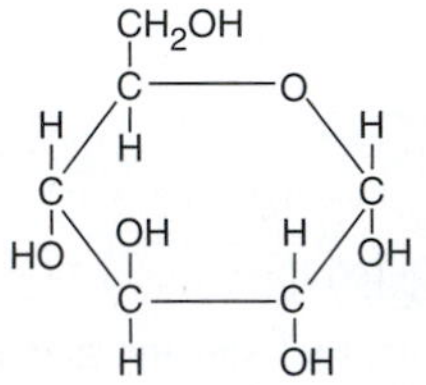

CH_2OH

C O

H H H

C C

HO OH H OH

C C

H OH

(iii)

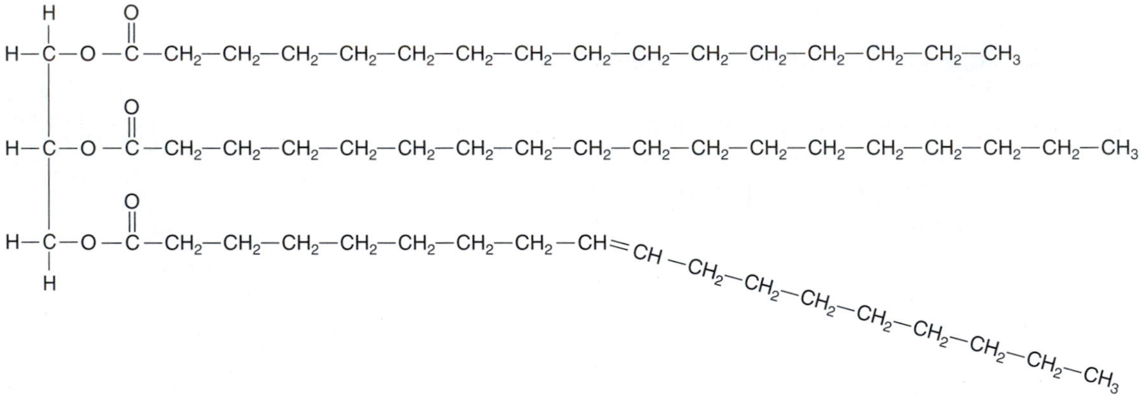

H O

$H-C-O-C-CH_2-CH_2-CH_2-CH_2-CH_2-CH_2-CH_2-CH_2-CH_2-CH_2-CH_2-CH_2-CH_2-CH_2-CH_3$

O

$H-C-O-C-CH_2-CH_2-CH_2-CH_2-CH_2-CH_2-CH_2-CH_2-CH_2-CH_2-CH_2-CH_2-CH_2-CH_2-CH_2-CH_2-CH_3$

O

$H-C-O-C-CH_2-CH_2-CH_2-CH_2-CH_2-CH_2-CH_2-CH=CH-CH_2-CH_2-CH_2-CH_2-CH_2-CH_2-CH_2-CH_3$

H

(iv)

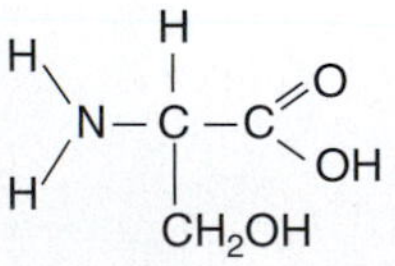

H

H

$N-C-C$ O

H OH

CH_2OH

(v)

CH_2OH

C O CH_2OH O

H H H H

C C C C

HO OH H O H HO CH_2OH

C C C C

H OH OH H

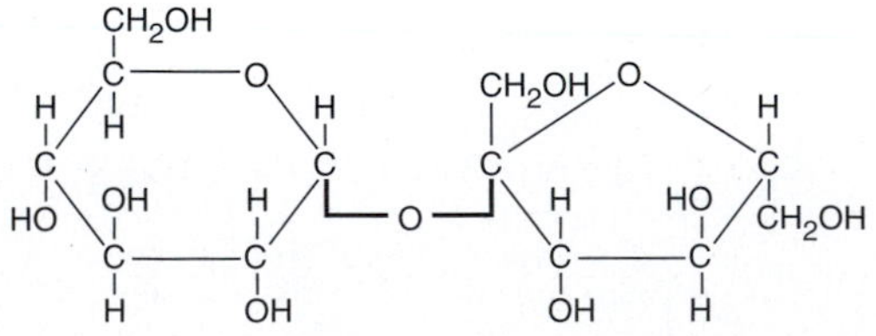

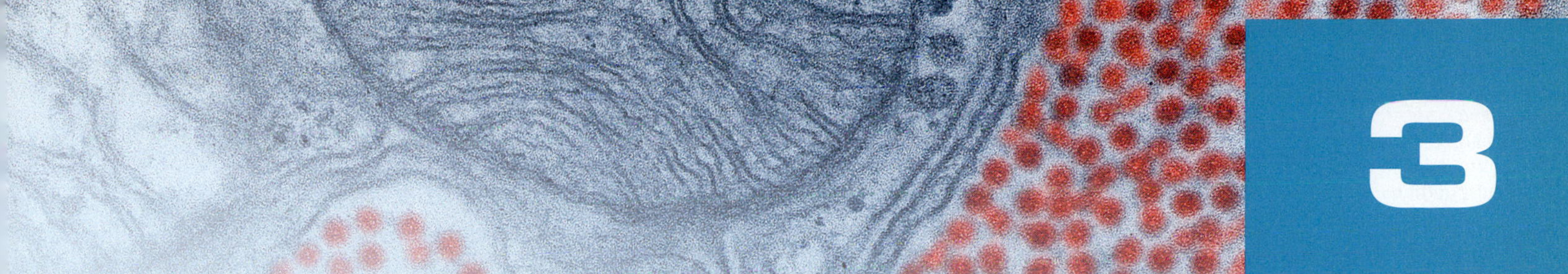

3

Concepts and Tools for Studying Microorganisms

It's as if [he] lifted a whole submerged continent out of the ocean.
—A prominent biologist speaking of Carl Woese, whose work led to the three-domain concept for living organisms

Chapter Preview and Key Concepts

3.1 The Prokaryotic/Eukaryotic Paradigm
- Prokaryotes undergo biological processes as complex as in eukaryotes.
- There are organizational patterns common to all prokaryotes and eukaryotes.
- Prokaryotes and eukaryotes have distinct subcellular compartments.

MicroInquiry 3: The Evolution of Eukaryotic Cells

3.2 Cataloging Microorganisms
- Organisms historically were grouped by shared characteristics.
- The binomial system identifies each organism by a universally-accepted scientific name.
- Species can be organized into higher, more inclusive groups.
- Historically, most organisms were assigned to one of several kingdoms.
- The three domain system uses nucleotide sequence analysis to catalog organisms.
- Taxonomy now includes many criteria to distinguish one prokaryote from another.

3.3 Microscopy
- Metric system units are the standard for measurement.
- Light microscopy uses visible light to magnify and resolve specimens.
- Specimens stained with a dye are contrasted against the microscope field.
- Different optical configurations provide detailed views of cells.
- Electron microscopy uses a beam of electrons to magnify and resolve specimens.

The oceans of the world are a teeming but invisible forest of microorganisms and viruses. As we discovered in Chapter 1, these microscopic agents are not only found around the world but, as Steven Giovannoni had pointed out for SAR11, being in such numbers implies they must be doing something important.

A substantial portion of the marine microbes represent the **phytoplankton** (*phyto* = "plant"; *plankto* = "wandering"), which are floating communities of prokaryotic cyanobacteria and eukaryotic algae. Besides forming the foundation for the marine food web, the phytoplankton account for 50 percent of the photosynthesis on earth and, in so doing, supply about half the oxygen gas we and other organisms breathe.

While sampling ocean water, scientists from MIT's Woods Hole Oceanographic Institution discovered that many of their samples were full of a marine cyanobacterium, which they eventually named *Prochlorococcus*. Inhabiting tropical and subtropical oceans, a typical sample often contained more than 200,000 (2×10^5) cells in one drop of seawater.

Studies with *Prochlorococcus* suggest the organism is responsible for more than 50 percent of the photosynthesis among the phytoplankton (FIGURE 3.1). This makes *Prochlorococcus* the smallest and most abundant marine photosynthetic organism yet discovered.

■ Ecotype: A subgroup of species having special characteristics to survive in its ecological surroundings.

The success of *Prochlorococcus* is due, in part, to the presence of different ecotypes inhabiting different ocean depths. For example, the high sunlight ecotype occurs in the top 100 meters while the low-light type is found between 100 and 200 meters. This latter ecotype compensates for the decreased light by increasing the amount of cellular chlorophyll that can capture the available light.

In terms of nitrogen sources, the high-light ecotype only uses ammonium ions (NH_4^+) (see MicroFocus 2.5). At increasing depth, NH_4^+ is less abundant so the low-light ecotype compensates by using a wider variety of nitrogen sources.

These and other attributes of *Prochlorococcus* illustrate how microbes survive through change. They are of global importance to the functioning of the biosphere as well as affecting our lives on Earth.

■ Biosphere: That part of the earth—including the air, soil, and water—where life occurs.

Once again, we encounter an interdisciplinary group of scientists studying how microorganisms influence our lives and life on this planet. Microbial ecologists study how the phytoplankton communities help in the natural recycling and use of chemical elements such as nitrogen. Evolutionary microbiologists look at these microorganisms to learn more about their taxonomic relationships, while microscopists, biochemists, and geneticists study how *Prochlorococcus* cells compensate for a changing environment of sunlight and nutrients.

This chapter focuses on many of the aspects described above. We examine how prokaryotes maintain a stable internal state and how they can exist in multicellular, complex communities. Throughout the chapter we are concerned with the relationships between prokaryotic and eukaryotic organisms and the many attributes they share. Then, we explore the methods used to name and catalog microorganisms. Finally, we discuss the tools and techniques used to observe microorganisms.

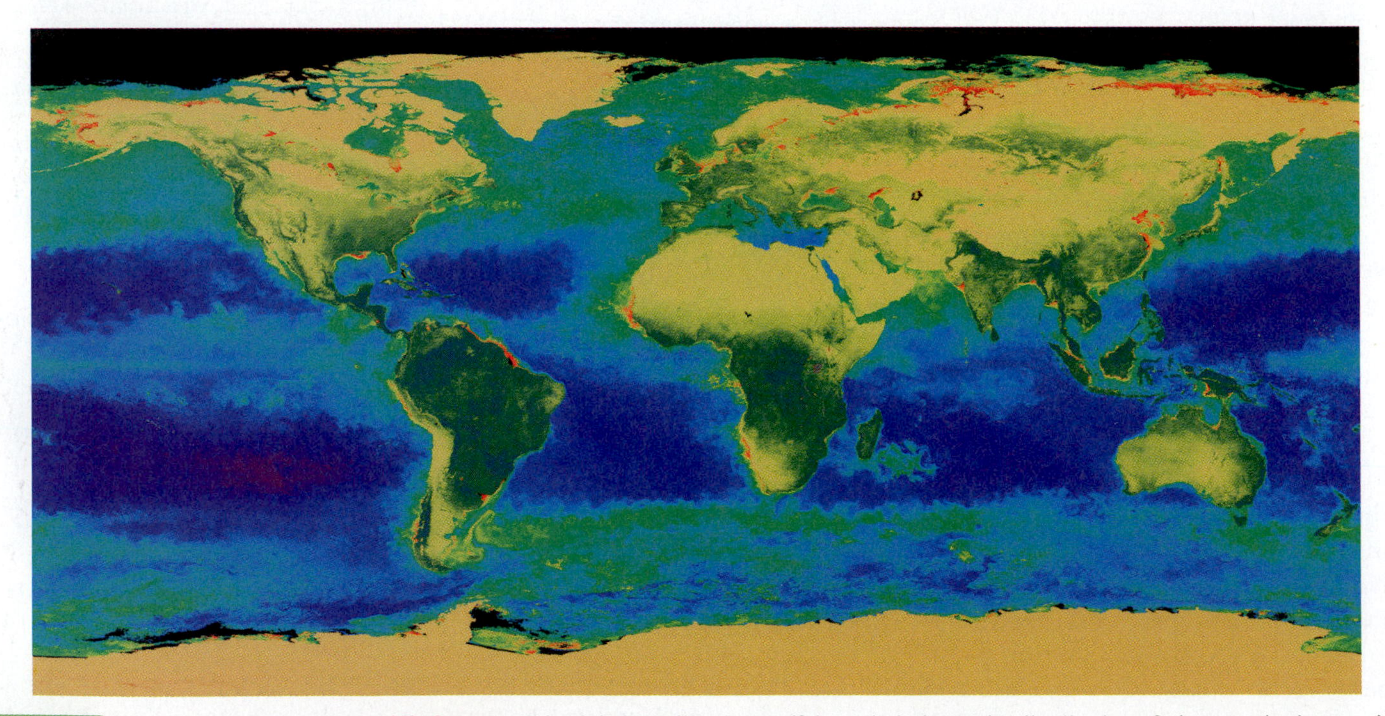

FIGURE 3.1 **Photosynthesis in the World's Oceans.** This global satellite image (false color) shows the distribution of photosynthetic organisms on the planet. In the aquatic environments, red colors indicate high levels of chlorophyll and productivity, yellow and green are moderate levels, and blue and purple areas are the "marine deserts."

Q: How do the landmasses where photosynthesis is most productive (green) compare in size to photosynthesis in the oceans?

3.1 The Prokaryotic/Eukaryotic Paradigm

In Chapter 1 we introduced the concept of prokaryotes and eukaryotes. In the news media or even in scientific magazines and textbooks, bacterial and archaeal cells—the prokaryotes—often are described as "simple organisms" compared to the "complex organisms" representing multicellular plants and animals. This view represents a mistaken perception. Despite their microscopic size, the prokaryotes exhibit every complex feature, or emerging property, common to all living organisms. These include:

- DNA as the hereditary material and complex gene control.
- Complex biochemical patterns of growth and energy conversions.
- Complex responses to stimuli.
- Reproduction.
- Adaptation.
- Complex organization.

Why then have the prokaryotes inherited the misnomer of "simple organisms?" To answer that question, we need to examine the similarities between prokaryotes and eukaryotes. They share many universal processes even though in some cases the structures to carry out these processes may be different. Having done this, we can then better distinguish the significant differences that set prokaryotes apart. In the next chapter, we look more closely at the differences between *Bacteria* and *Archaea*.

Prokaryote/Eukaryote Similarities

KEY CONCEPT

- Prokaryotes undergo biological processes as complex as in eukaryotes.

All organisms continually battle their external environment, where factors such as temperature, sunlight, or toxic chemicals can have serious consequences. Organisms strive to maintain a stable internal state by making appropriate metabolic or structural adjustments. This ability to adjust yet maintain a relatively steady internal state is called **homeostasis** (*homeo* = "similar"; *stasis* = "state"). Two examples illustrate the concept.

The low-light *Prochlorococcus* ecotype mentioned in the chapter introduction lives at depths of 80 to 200 meters. At these varying depths, transmitted sunlight decreases and any one nitrogen source is less accessible. The ecotype compensates for the light reduction and nitrogen limitation by (1) increasing the amount of cellular chlorophyll to capture light and (2) using a wider variety of available nitrogen sources. These adjustments maintain a relatively stable internal state.

Chlorophyll: The major pigment used to convert sunlight into sugar.

Suppose a patient is given an antibiotic to combat a bacterial infection (FIGURE 3.2). In response, the infecting bacterium may compensate for the change by pumping the drug out of the cell. The adjustment maintains homeostasis in the bacterial cell.

In both these examples, the internal environment is maintained despite a changing environment. Such, often complex, homeostatic controls are as critical to prokaryotes as they are to eukaryotes (MicroFocus 3.1).

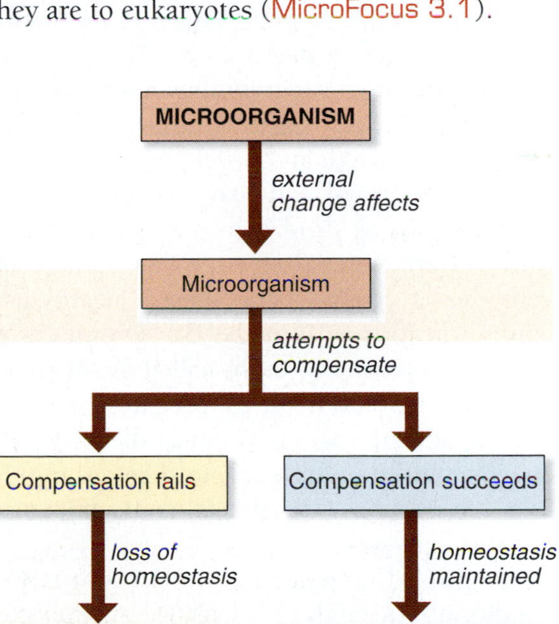

FIGURE 3.2 **Homeostasis.** A concept map illustrating the ability of a bacterial pathogen to compensate for the presence of an antibiotic. Survival is dependent on its homeostatic abilities.

Q: Redraw the concept map for the low-light response by Prochlorococcus.

CONCEPT AND REASONING CHECKS

3.1 How does homeostasis relate to the listed emerging properties of living organisms?

Another misunderstanding about prokaryotes is that they represent "unicellular" organisms. And why not? When they are studied, often they appear in the microscope as single cells. The early studies of disease causation done by Koch (see Chapter 1) certainly required pure cultures to associate a specific disease with one specific microbe. However, today it is necessary to abolish the impression that bacteria are self-contained, independent organisms. In nature few species live such a pure and solitary life. In fact, it has been estimated that as many as 99 percent of prokaryotic species live in communal associations called **biofilms**; that is, in a "multicellular state" where survival requires chemical communication and cooperation between cells.

One example of prokaryotic multicellularity within a species involves a group of soil-dwelling bacteria called the **myxobacteria**, which exhibit the most complex behavior among the known prokaryotes. The cells represent a social community dependent on cell-to-cell interactions and communication throughout growth and development.

Myxobacterial cells can obtain nutrients by decomposing dead plant or animal matter. More often, when live prey (e.g., bacterial cells, yeasts, or algae) are "sensed" nearby, cell communication within the flat colonies triggers (1) a swarming or so-called "wolf pack" behavior characterized by predatory feeding and (2) an increasing secretion of hydrolytic enzymes by the wolf pack to digest the prey. Like a lone wolf, a single cell could not carry out this behavior.

At the other extreme, under starvation conditions and high cell densities, myxobacterial cells chemically communicate to form a mound of 100,000 cells and cooperate to build a three-dimensional structure called a fruiting body (FIGURE 3.3). Within this structure, a cell-secreted slime forms the stalk on top of which the remaining viable cells develop into resting structures called myxospores.

FIGURE 3.3 **A Myxobacterial Fruiting Body.** The stalk (lower left) is made of slime secreted by the cells that eventually form the resting spores in the head. (Bar = 10 μm.)

Q: How does the myxobacterial fruiting body represent a multicellular structure?

Here then is one example where the term *unicellular* does not apply. MicroFocus 3.2 further examines the concept of prokaryotic multicellularity.

The final misconception to be addressed concerns the reference to prokaryotes as "simple cells" or "simpler organisms" than their eukaryotic counterparts. Historically, when one looks at bacterial cells even with an electron microscope, often there is little to see (FIGURE 3.4). "Cell structure," representing the cell's physical appearance or its components and the "pattern of organization," referring to the configuration of those structures and their relationships to one another, does give the impression of simpler cells.

But what has been overlooked is the "cellular process," the activities all cells carry out for the continued survival of the cell (and organism). At this level, the complexity is just as intricate as in any eukaryotic cell. So, in reality, prokaryotes carry out many of the same cellular processes as eukaryotes—only

MICROFOCUS 3.1: Being Skeptical

Are Cyanobacterial Food Supplements Better for What's Ailing You?

The food supplement business is always looking for new products to boost human health. Take the phytoplankton. The chapter introduction highlighted the importance of these microbial communities to the marine food web. Some individuals therefore have perceived this as indicating these microorganisms must be very nutritious.

Among the items found on today's health food shelves and the Internet are capsules, pills, and powders consisting of dried cyanobacteria, primarily from the fresh-water species *Aphanizomenon flos-aquae* (AFA) and the genus *Spirulina* (see figure). These products are promoted as a treatment or cure for a wide variety of human ailments, including asthma, allergies, chronic viral hepatitis, anxiety, depression, fatigue, hypoglycemia, digestive problems, and attention deficit disorder. The microbial products purport to help individuals lose weight, improve memory and mental ability, boost immune system function, and restore overall cellular balance.

Are these claims reasonable? What potentially unique nutrients must these cyanobacterial supplements have to provide relief from all these ailments? The answer—none that we know. The organisms contain large amounts of chlorophyll, but humans cannot carry out photosynthesis so why eat chlorophyll? Cyanobacteria do contain small amounts of protein, beta carotene, and a few vitamins and minerals. However, you would have to eat shovels full of the cyanobacterial powders to get a sufficient level of these nutrients—there are more sensible and cheaper ways to get these same nutrients in easily consumable amounts.

Importantly, having come from natural lakes, the products may be contaminated with toxic substances, such as heavy metals that polluted the lake water.

The verdict? Yes, microbes do have many useful and important roles in the human body. However, medical organizations agree there is not enough evidence to support their use for treatment of any medical condition. Whenever a product proclaims such a wide-range of curative powers, be skeptical. As the old adage says, "If it is too good to be true," it probably is.

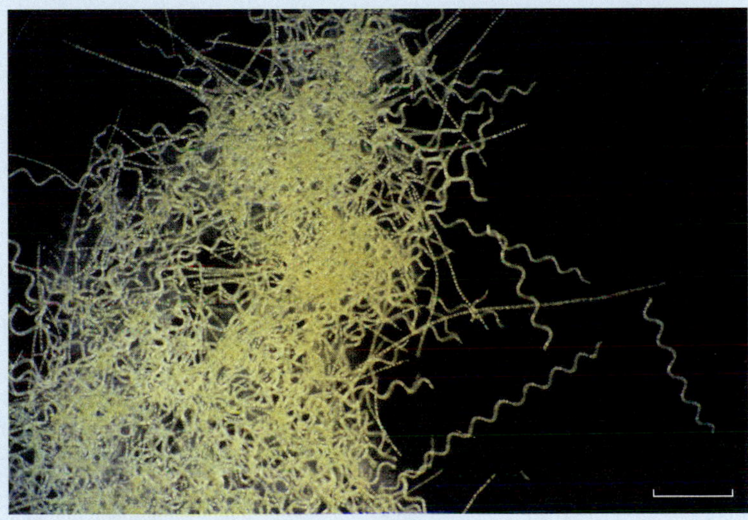

A mass of *Spirulina* cells. (Bar = 5 μm.)

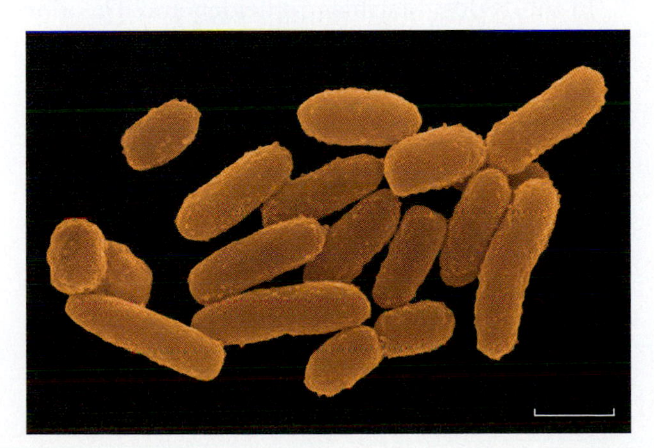

FIGURE 3.4 **Simpler Organisms?** This false-color photo of the bacterial species *Yersinia pestis,* the causative agent of plague, was taken with an electron microscope. (Bar =1 μm.)

Q: Why does this view suggest to the observer that prokaryotes are simple cells?

MICROFOCUS 3.2: Environmental Microbiology
Multicellular Prokaryotes

As microbiologists continue to study prokaryotic organisms, it is becoming quite obvious that these microbes are social creatures. Whether it is studying myxobacteria, mixed bacterial populations in a biofilm, or other communities of microorganisms, there is more to bacterial life than just existing in a multicellular community.

Quorum Sensing

Several bacterial behaviors are dependent on population density. When population numbers reach a critical level, chemical signals are produced that alter gene activity. Such **quorum sensing (QS)** can control spore formation as described for the myxobacteria or enhance the ability of a pathogen to cause disease.

Chemical communication can occur between different bacterial species. This QS cross talk allows the cells of one species to detect the presence of other species, which could be a good thing or a bad thing. In biofilms, QS is necessary for the development of the complex structures within the biofilm. However, *Pseudomonas aeruginosa,* a pathogen of the human gut, can sense an immune system response to infection. Through QS, *P. aeruginosa* counters immune activation by strengthening or enhancing its ability to cause disease by disrupting the function of epithelial cells in the gut. These chemical signals would not be produced by solitary, non-communicative cells.

Programmed Cell Death

The multicellular nature of microbial communities also demonstrates cell suicide; that is, genetically programmed cell death (PCD). Such a phenomenon (called apoptosis) was discovered in eukaryotes many years ago, but a death program in bacterial cells is a newly discovered phenomenon that is based on multicellular behavioral responses.

One example of PCD is the myxobacteria, where up to 80 percent of the cells die in the process of fruiting body formation, the remaining 20 percent becoming the myxospores. Also, during biofilm development, overcrowding can be prevented by cells dying within the biofilm, which hollows out the structure and makes room for new cell reproduction. "Sensed" by the multicellular aggregate, specific toxic proteins trigger this form of PCD.

Perhaps most interesting (and controversial) is research looking at the death of defective or injured cells resulting from antibiotics. In the face of a chemical attack, cells damaged by the antibiotic undergo PCD so they do not become a burden to the remaining cells. Yet other damaged cells in the community apparently turn off the PCD program and become persistor cells—they survive by growing so slowly, they are not susceptible to the action of antibiotics.

Interestingly, many of the multicellular behaviors observed only occur in wild strains in nature. Evidently, the "good life" in a culture plate has made such cell-cell behaviors unnecessary—the cells have become unicellular "couch bacteria."

without the need for an elaborate, visible pattern of organization.

Let's now look at the cell structure relationships in terms of organized patterns and processes.

CONCEPT AND REASONING CHECKS

3.2 Respond to the statement that prokaryotes are simple, unicellular organisms.

Prokaryotes and Eukaryotes: The Similarities in Organizational Patterns

KEY CONCEPT

- There are organizational patterns common to all prokaryotes and eukaryotes.

In the 1830s, Matthias Schleiden and Theodor Schwann developed part of the **cell theory** by demonstrating all organisms are composed of one or more cells, making the cell the fundamental unit of life. (Note: about 20 years later, Rudolph Virchow added that all cells arise from pre-existing cells.) Although the concept of a microorganism was just in its infancy at the time, the doctrine suggests that there are certain organizational patterns common to prokaryotes and eukaryotes.

Genetic organization. All organisms have a similar genetic organization whereby the hereditary material is communicated or

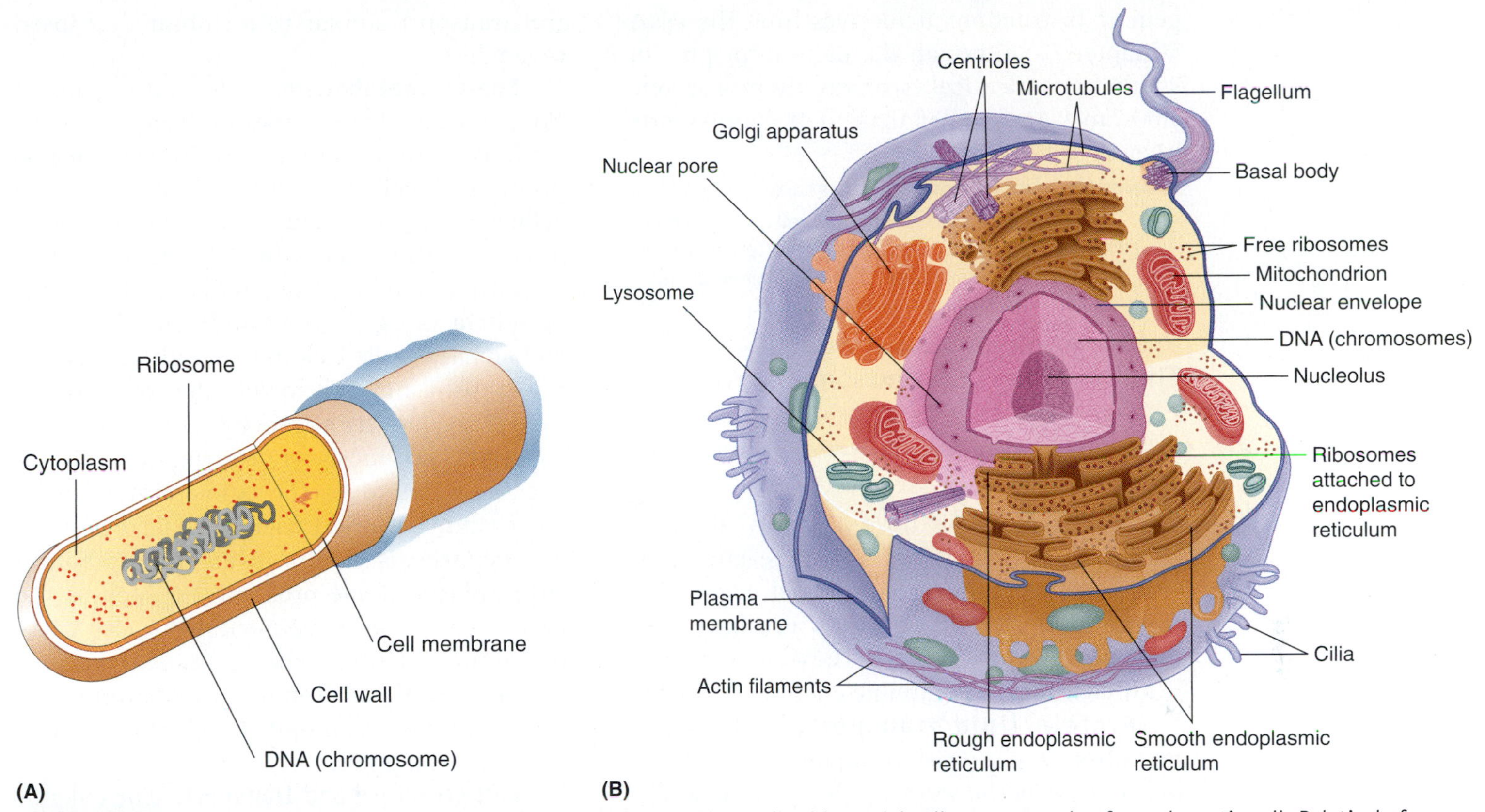

FIGURE 3.5 **A Comparison of Prokaryotic and Eukaryotic Cells.** (**A**) A stylized bacterial cell as an example of a prokaryotic cell. Relatively few visual compartments are present. (**B**) A protozoan cell as a typical eukaryotic cell. Note the variety of cellular subcompartments, many of which are discussed in the text.

Q: List the ways you could microscopically distinguishing a eukaryotic microbial cell from a prokaryotic cell.

expressed (Chapter 7). The organizational pattern for the hereditary material is in the form of one or more chromosomes. Structurally, most prokaryotic cells have a single, circular DNA molecule without an enclosing membrane (FIGURE 3.5.) Eukaryotic cells, however, have multiple, linear chromosomes enclosed by the membrane envelope of the cell nucleus.

Compartmentation. All prokaryotes and eukaryotes have an organizational pattern separating the internal compartments from the surrounding environment but allowing for the exchange of solutes and wastes. The pattern for compartmentation is represented by the cell. All cells are surrounded by a **cell membrane** (known as the **plasma membrane** in eukaryotes), where the phospholipids form the impermeable boundary to solutes while proteins regulate the exchange of solutes and wastes across the membrane. We have more to say about membranes in the next chapter.

Metabolic organization. The process of metabolism is a consequence of compartmentation. By being enclosed by a membrane, all cells have an internal environment in which chemical reactions occur. This space, called the **cytoplasm**, represents everything surrounded by the membrane and, in eukaryotic cells, exterior to the cell nucleus. If the cell structures are removed from the cytoplasm, what remains is the **cytosol**, which consists of water, salts, ions, and organic compounds as described in Chapter 2.

■ Metabolism: All the chemical reactions occurring in an organism or cell.

Protein synthesis. All organisms must make proteins, which we learned in Chapter 2 are the workhorses of cells and organisms. The structure common to all prokaryotes and eukaryotes is the **ribosome**, an RNA-protein machine that cranks out proteins based on the genetic instructions it receives from the DNA (Chapter 7). Although the pattern for protein synthesis is identical, structurally prokaryotic ribosomes are smaller than their counterparts in eukaryotic cells.

CONCEPT AND REASONING CHECKS

3.3 The cell theory states that the cell is the fundamental unit of life. Summarize those processes all cells have that contribute to this fundamental unit.

Prokaryotes and Eukaryotes: The Structural Distinctions

KEY CONCEPT

- Prokaryotes and eukaryotes have distinct subcellular compartments.

Eukaryotic microbes have a variety of structurally discrete, often membrane-enclosed, subcellular compartments called **organelles** (Figure 3.5). Prokaryotic cells also have subcellular compartments—they just are not readily visible or membrane enclosed.

Protein/lipid transport. Eukaryotic microbes have a series of membrane-enclosed organelles in the cytosol that compose the cell's **endomembrane system**, which is designed to transport protein and lipid cargo through and out of the cell. This system includes the **endoplasmic reticulum (ER)**, which consists of flat membranes to which ribosomes are attached (rough ER) and tube-like membranes without ribosomes (smooth ER). These portions of the ER are involved in protein and lipid synthesis and transport, respectively.

The **Golgi apparatus** is a group of independent stacks of flattened membranes and vesicles where the proteins and lipids coming from the ER are processed, sorted, and packaged for transport. **Lysosomes**, somewhat circular, membrane-enclosed sacs containing digestive (hydrolytic) enzymes, are derived from the Golgi apparatus and are vital to the killing of many pathogens by white blood cells.

Prokaryotes lack an endomembrane system, yet they are capable of manufacturing and modifying proteins and lipids just as their eukaryotic relatives do. However, many bacterial cells contain so-called **microcompartments** surrounded by a protein shell. These microcompartments represent a type of organelle since the shell proteins can control transport similar to membrane-enclosed organelles.

Energy metabolism. Cells and organisms carry out one or two types of energy transformations. Through a process called **cellular respiration**, all cells convert chemical energy into cellular energy for cellular work. In eukaryotic microbes, this occurs in the cytosol and in membrane-enclosed organelles called **mitochondria** (sing., mitochondrion). Bacterial and archaeal cells lack mitochondria; they use the cytosol and cell membrane to complete the energy converting process.

A second energy transformation, **photosynthesis**, involves the conversion of light energy into chemical energy. In algal protists, photosynthesis occurs in membrane-bound **chloroplasts**. Some prokaryotes, such as the cyanobacteria we have mentioned, also carry out almost identical energy transformations. Again, the cell membrane or elaborations of the membrane represent the chemical workbench for the process.

Cell structure and transport. The eukaryotic **cytoskeleton** is organized into an interconnected system of fibers, threads, and interwoven molecules that give structure to the cell and assist in the transport of materials throughout the cell. The main components of the cytoskeleton are microtubules and actin filaments, each assembled from different protein subunits. Prokaryotes to date have no similar physical cytoskeleton, although proteins related to those that construct microtubules and actin filaments aid in determining the shape in some bacterial cells as we will see in Chapter 4.

Cell motility. Many microbial organisms live in watery or damp environments and use the process of cell motility to move from one place to another. Many algae and protozoa have long, thin protein projections called **flagella** (sing., flagellum) that, covered by the plasma membrane, extend from the cell. By beating back and forth, the flagella provide a mechanical force for motility. Many prokaryotic cells also exhibit motility; however, the flagella are structurally different and without a cell membrane covering. The pattern of motility also is different, providing a rotational propeller-like force for motility (Chapter 4).

TABLE 3.1 A Comparison of Prokaryotes and Eukaryotes

Process	Microbial Cell Structure or Compartment	
	Prokaryotic	Eukaryotic
Genetic organization	Circular DNA chromosome	Linear DNA chromosomes
Compartmentation	Cell membrane	Plasma membrane
Metabolic organization	Cytoplasm	Cytoplasm
Protein synthesis	Ribosomes	Ribosomes
Protein/lipid transport	Cytoplasm	Endomembrane system
Energy metabolism	Cytoplasm and cell membrane	Mitochondria and chloroplasts
Cell structure and transport	Proteins in cytoplasm	Protein filaments in cytoplasm
Cell motility	Prokaryotic flagella	Eukaryotic flagella or cilia
Water regulation	Cell wall	Cell wall

Some protozoa also have other membrane-enveloped appendages called **cilia** (sing., cilium) that are shorter and more numerous than flagella. In some motile protozoa, they wave in synchrony and propel the cell forward. No prokaryotes have cilia.

Water balance. The aqueous environment in which many microorganisms live presents a situation where the process of diffusion occurs, specifically the movement of water, called **osmosis**, into the cell. Continuing unabated, the cell would eventually swell and burst (cell lysis) because the cell or plasma membrane does not provide the integrity to prevent lysis.

Many prokaryotic and eukaryotic cells contain a **cell wall** exterior to the cell or plasma membrane. Although the structure and organization of the wall differs between groups, all cell walls provide support for the cells, give them shape, and help them resist the pressure exerted by the internal water pressure.

A summary of the prokaryote and eukaryote processes and structures is presented in TABLE 3.1. MicroInquiry 3 examines a scenario for the evolution of the eukaryotic cell.

■ Diffusion: The movement of a substance from where it is in a higher concentration to where it is in a lower concentration.

CONCEPT AND REASONING CHECKS

3.4 Explain how variation in cell structure between prokaryotes and eukaryotes can be compatible with a similarity in cellular processes between prokaryotes and eukaryotes.

3.2 Cataloging Microorganisms

If you open any catalog, items are separated by types, styles, or functions. For example, in a fashion catalog, watches are separated from shoes and, within the shoes, men's, women's, and children's styles are separated from one another. Even the brands of shoes or their use (e.g., dress, casual, athletic) may be separated.

The human drive to catalog objects extends to the sciences, especially biology where historically the process was not much different from cataloging watches and shoes; it was based on shared characteristics. In this section, we shall explore the principles on which microorganisms are cataloged.

Classification Attempts to Catalog Organisms

KEY CONCEPT

- Organisms historically were grouped by shared characteristics.

The science of classification, called **taxonomy**, involves the systematized arrangement or cataloging of related organisms into logical categories. Taxonomy is essential for understanding the relationships among living organisms, discovering unifying concepts among organisms, and providing a universal "language" for communication among biologists.

MICROINQUIRY 3

The Evolution of Eukaryotic Cells

Biologists and geologists have speculated for decades about the chemical evolution that led to the origins of the first prokaryotic cells on Earth (see MicroFocus 2.1 and 2.6). Whatever the origin, the first ancestral prokaryotes arose about 3.8 billion years ago.

Scientists also have proposed various scenarios to account for the origins of the first eukaryotic cells. A key concern here is figuring out how different membrane compartments arose to evolve into what are found in the eukaryotic cells today. Debate on this long intractable problem continues, so here we present some of the ideas that have fueled such discussions.

At some point around 2 billion years ago, the increasing number of metabolic reactions occurring in some prokaryotes started to interfere with one another. Complexity would necessitate more extensive compartmentation.

The Endomembrane System May Have Evolved through Invagination

Similar to today's prokaryotic cells, the cell membrane of an ancestral prokaryote may have had specialized regions involved in protein synthesis, lipid synthesis, and nutrient hydrolysis. If the invagination of these regions occurred, the result could have been the internalization of these processes as independent internal membrane systems. For example, the membranes of the endoplasmic reticulum may have originated by multiple invagination events of the plasma membrane (see **Figure A1**).

Biologists have suggested that the elaboration of the evolving ER created the nuclear envelope. Surrounded and protected by a double membrane, greater genetic complexity could occur as the primitive eukaryotic cell continued to evolve in size and function.

Chloroplasts and Mitochondria Arose from a Symbiotic Union of Engulfed Prokaryotes

Mitochondria and chloroplasts are not part of the extensive endomembrane system. Therefore, these energy-converting organelles probably originated in a different way.

The structure of modern-day chloroplasts and mitochondria is very similar to the description of a bacterial cell. In fact, mitochondria, chloroplasts, and bacteria share a large number of similarities (see **Table**). In addition, there are bacterial cells alive today that carry out cellular respiration similarly to mitochondria and other bacterial cells that can carry out photosynthesis similarly to chloroplasts.

These similar functional patterns, along with other chemical and molecular similarities, suggested to Lynn Margulis that present-day chloroplasts and mitochondria represent modern representatives of what were once, many eons ago, free-living prokaryotes. Margulis, therefore, proposed the **endosymbiotic theory** for the origin of mitochondria and chloroplasts. The hypothesis suggests, in part, that mitochondria evolved from a prokaryote that carried out cellular respiration and which was "swallowed" (engulfed) by a primitive eukaryotic cell. The bacterial partner then lived within (*endo*) the eukaryotic cell in a mutually beneficial association (*symbiosis*) (see **Figure A2**).

Likewise, a photosynthetic prokaryote, perhaps a primitive cyanobacterium, was engulfed and evolved into the chloroplasts present in plants and algae today (see **Figure A3**). The theory also would explain why both organelles have two membranes. One was the cell membrane of the engulfed bacterial cell and the other was the plasma membrane resulting from the engulfment process. By engulfing these prokaryotes and not destroying them, the evolving eukaryotic cell gained energy-conversion abilities, while the symbiotic bacterial cells gained a protected home.

Obviously, laboratory studies can only guess at mechanisms to explain how cells evolved and can only suggest—not prove—what might have happened billions of years ago. The description here is a very simplistic view of how the first eukaryotic cells might have evolved. Short of inventing a time machine, we may never know the exact details for the origin of eukaryotic cells and organelles.

Discussion Point

Determine which endosymbiotic event must have come first: the engulfment of the bacterial progenitor of the chloroplast or the engulfment of the bacterial progenitor of the mitochondrion.

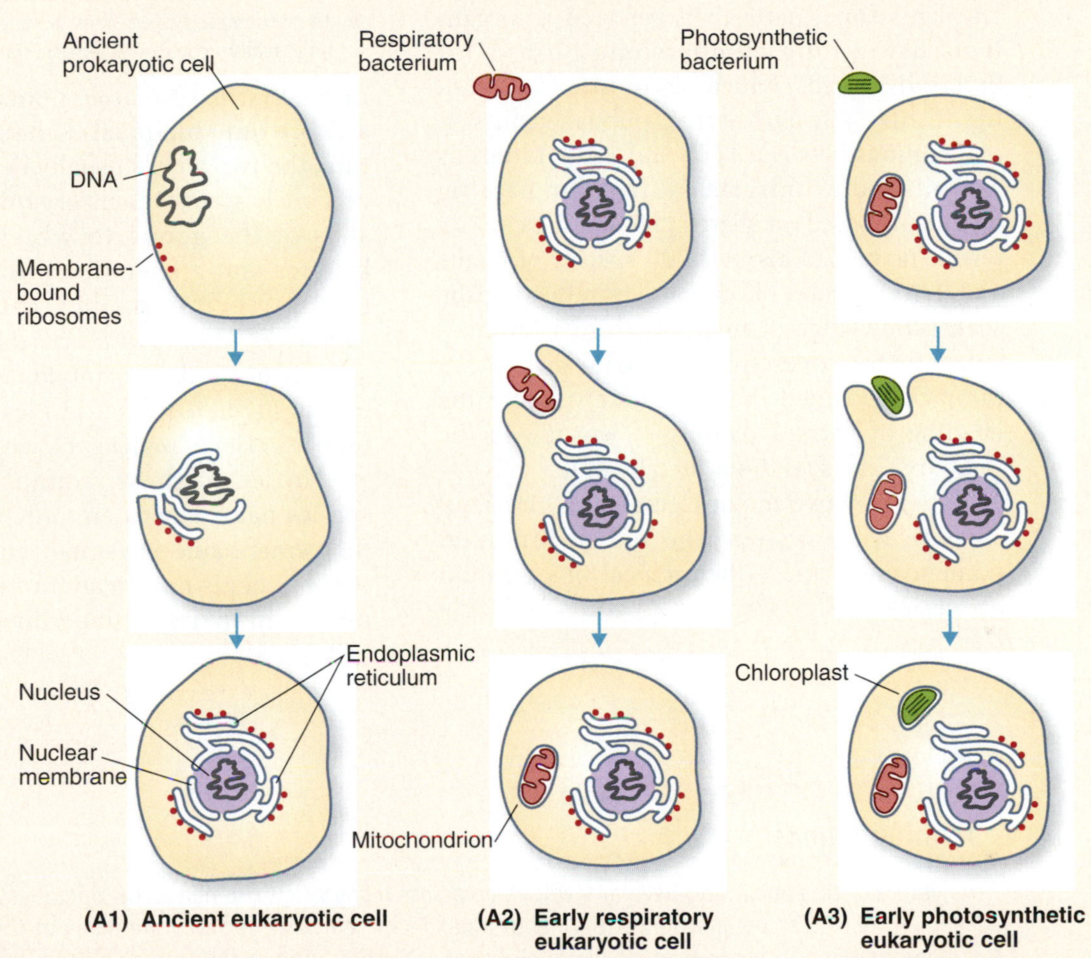

FIGURE A **Possible Origins of Eukaryotic Cell Compartments.** (**A1**) Invagination of the cell membrane from an ancient prokaryotic cell may have lead to the development of the cell nucleus as well as to the membranes of the endomembrane system, including the endoplasmic reticulum. (**A2**) The mitochondrion may have resulted from the uptake and survival of a bacterial cell that carried out cellular respiration. (**A3**) A similar process, involving a bacterial cell that carried out photosynthesis, could have accounted for the origin of the chloroplast.

TABLE

Similarities between Mitochondria, Chloroplasts, Prokaryotes, and Microbial Eukaryotes

Characteristic	Mitochondria	Chloroplasts	Prokaryotes	Microbial Eukaryotes
Average size	1–5 μm	1–5 μm	1–5 μm	10–20 μm
Nuclear envelope present	No	No	No	Yes
DNA molecule shape	Circular	Circular	Circular	Linear
Ribosomes	Yes; bacterial-like	Yes; bacterial-like	Yes	Yes; eukaryotic-like
Protein synthesis	Make some of their proteins	Make some of their proteins	Make all of their proteins	Make all of their proteins
Reproduction	Binary fission	Binary fission	Binary fission	Mitosis and cytokinesis

The Swedish botanist Carl von Linné, better known to history as Carolus Linnaeus, laid down the foundation of modern taxonomy. Until his time, naturalists referred to organisms using long cumbersome phrases or descriptive terms, which often varied by country. In his *Systema Naturae,* published in several editions between 1735 and 1759, Linnaeus established a uniform system for naming organisms based on shared characteristics. As a result, he named about 7,700 species of plants and 4,400 species of animals. In reflecting the scant knowledge of microorganisms (animalcules) at the time and his general disinterest, Linnaeus grouped them separately under the heading of Vermes (*vermis* = "worm") in the category Chaos (*chaos* = "an abyss").

There are two major elements to Linnaeus' classification system: the use of binomial nomenclature and a hierarchical organization of species.

Nomenclature Gives Scientific Names to Organisms

KEY CONCEPT

- The binomial system identifies each organism by a universally-accepted scientific name.

In his *Systema Naturae,* Linnaeus popularized a two word (binomial) scheme of nomenclature, the two words usually derived from Latin or Greek stems. Each organism's name consists of the **genus** to which the organism belongs and a **specific epithet**, a descriptor that further describes the genus name. Together these two words make up the **species** name. For example, the common bacterium *Escherichia coli* resides in the gut of all humans (*Homo sapiens*) (MicroFocus 3.3).

Notice in these examples that when a species name is written, only the first letter of the genus name is capitalized, while the specific epithet is not. In addition, both words are printed in italics or underlined. After the first

MICROFOCUS 3.3: Tools

Naming Names

As you read this book, you have and will come across many scientific names for microbes, where a species name is a combination of the genus and specific epithet. Not only are many of these names tongue twisting to pronounce (they are all listed with their pronunciation inside the front and back covers), but how in the world did the organisms get those names? Here are a few examples.

Genera Named after Individuals

Escherichia coli: named after Theodore Escherich who isolated the bacterial cells from infant feces in 1885. Being in feces, it commonly is found in the colon.

Neisseria gonorrhoeae: named after Albert Neisser who discovered the bacterial organism in 1879. As the specific epithet points out, the disease it causes is gonorrhea.

Genera Named for a Microbe's Shape

Vibrio cholerae: vibrio means "comma-shaped," which describes the shape of the bacterial cells that cause cholera.

Staphylococcus epidermidis: staphylo means "cluster" and *coccus* means "spheres." So, these bacterial cells form clusters of spheres that are found on the skin surface (epidermis).

Genera Named after an Attribute of the Microbe

Saccharomyces cerevisiae: in 1837, Theodor Schwann observed yeast cells and called them *Saccharomyces* (*saccharo* = "sugar;" *myce* = "fungus") because the yeast converted grape juice (sugar) into alcohol; *cerevisiae* (from *cervisia* = "beer") refers to the use of yeast since ancient times to make beer.

Myxococcus xanthus: myxo means "slime," so these are slime-producing spheres that grow as yellow (*xantho* = "yellow") colonies on agar.

Thiomargarita namibiensis: see MicroFocus 3.5.

time a species name has been spelled out, biologists usually abbreviate the genus name using only its initial genus letter or some accepted substitution, together with the full specific epithet; that is, *E. coli* or *H. sapiens*. A cautionary note: often in magazines and newspapers, proper nomenclature is not followed, so our gut bacterium would be written as Escherichia coli.

CONCEPT AND REASONING CHECKS

3.5 Which one of the following is a correctly written scientific name for the bacterium that causes anthrax? (a) *bacillus Anthracis;* (b) *Bacillus Anthracis;* or (c) *Bacillus anthracis*.

Classification Uses a Hierarchical System

KEY CONCEPT

- Species can be organized into higher, more inclusive groups.

Linnaeus' cataloging of plants and animals used shared and common characteristics. Such similar organisms that could interbreed were related as a species, which formed the least inclusive level of the hierarchical system. Part of Linnaeus' innovation was the grouping of species into higher taxa that also were based on shared, but more inclusive, similarities.

Today several similar species are grouped together into a genus (pl. genera). A collection of similar genera makes up a family and families with similar characteristics make up an order. Different orders may be placed together in a class and classes are assembled together into a phylum (pl. phyla) or division (in bacteriology and botany). All phyla or divisions would be placed together in a kingdom and/or domain, the most inclusive level of classification. TABLE 3.2 outlines the taxonomic hierarchy for three organisms.

In prokaryotes, an organism may belong to a rank below the species level to indicate a special characteristic exists within a subgroup of the species. Such ranks have practical usefulness in helping to identify an organism. For example, two biotypes of the cholera bacterium, *Vibrio cholerae*, are known: *Vibrio cholerae* classic and *Vibrio cholerae* El Tor. Other designations of ranks include subspecies, serotype, strain, morphotype, and variety.

■ Biotype: A population or group of individuals having the same genetic constitution (genotype).

CONCEPT AND REASONING CHECKS

3.6 How would you describe an order in the taxonomic classification?

Kingdoms and Domains: Trying to Make Sense of Taxonomic Relationships

KEY CONCEPT

- Historically, most organisms were assigned to one of several kingdoms.

As more microorganisms were described after Linnaeus' time, some were considered plants (bacterial, algal, and fungal organisms) while the protozoa were categorized as animals.

■ Taxa (sing. Taxon): Subdivisions used to classify organisms.

In 1866, the German naturalist Ernst H. Haeckel disturbed the tidiness of Linnaeus' plant and animal kingdoms. How could mushrooms be plants when they do not carry out photosynthesis? Haeckel therefore coined the term "protist" for a microorganism, and

TABLE 3.2 Taxonomic Classification of Humans, Brewer's Yeast, and a Common Bacterium

	Human Being	Brewer's Yeast	*Escherichia coli*
Domain	*Eukarya*	*Eukarya*	*Bacteria*
Kingdom	Animalia	Fungi	
Phylum (Division)	Chordata	Ascomycota	Proteobacteria
Class	Mammalia	Saccharomycotina	Gamma-Proteobacteria
Order	Primates	Saccharomycetales	Enterobacteriales
Family	Hominidae	Saccharomycetaceae	Enterobacteriaceae
Genus	*Homo*	*Saccharomyces*	*Escherichia*
Species	*H. sapiens*	*S. cerevisiae*	*E.coli*

he placed all bacterial, protozoal, algal, and fungal species in this third kingdom. Although not everyone liked his system, it did provoke discussions about the nature of microorganisms.

During the twentieth century, advances in cell biology and interest in evolutionary biology led scientists to question the two- or three-kingdom classification schemes. In 1969, Robert H. Whittaker of Cornell University proposed a system that quickly gained wide acceptance in the scientific community. Further expanded in succeeding years by Lynn Margulis of the University of Massachusetts, the **five kingdom system** of organisms was established (FIGURE 3.6).

In the five-kingdom system, bacterial organisms were so different from other organisms they must be in their own kingdom. Protista were limited to the protozoa and the unicellular algae. Fungi included non-green, non-photosynthetic eukaryotic organisms that, along with other characteristics, had cell walls that were chemically different from those in bacterial, algal, and plant cells.

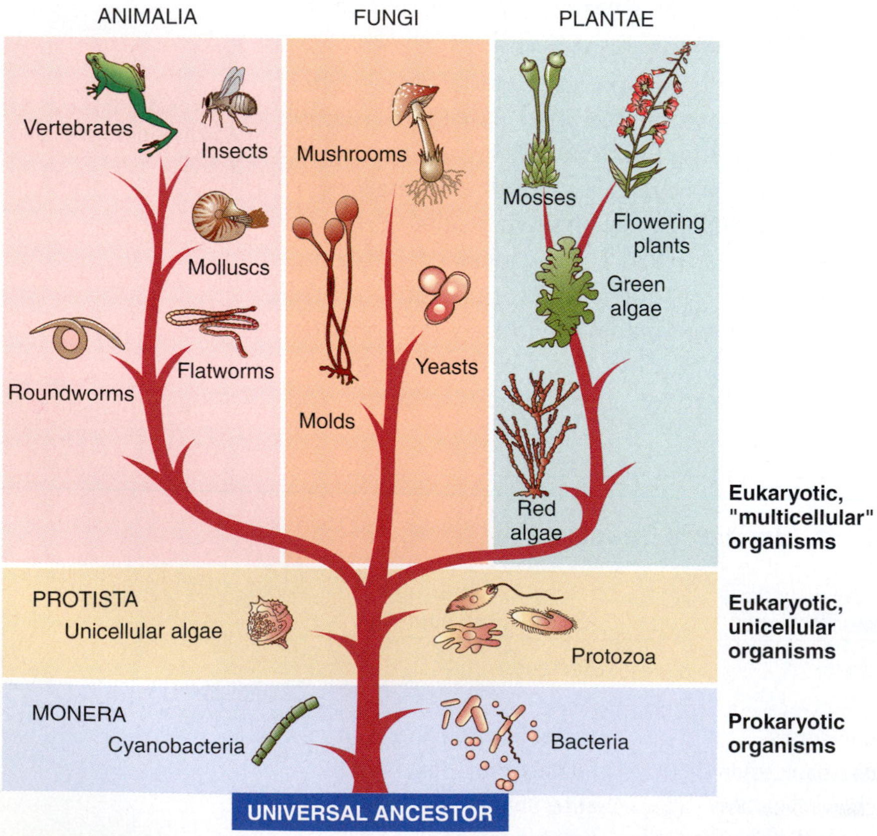

FIGURE 3.6 **Five-Kingdom System.** This system implies an evolutionary lineage, beginning with the Monera and extending to the Protista. In this scheme, certain Protista were believed to be ancestors of the Plantae, Fungi, and Animalia.

Q: How many of the kingdoms consist of microorganisms?

CONCEPT AND REASONING CHECKS

3.7 Could an organism be assigned to a kingdom based on one characteristic? Two characteristics? How many characteristics?

The Three-Domain System Places the Prokaryotes in Separate Lineages

KEY CONCEPT

- The three domain system uses nucleotide sequence analysis to catalog organisms.

Often it is difficult to make sense of taxonomic relationships because new information that is more detailed keeps being discovered about organisms. This then motivates taxonomists to figure out how the new information fits into the known classification schemes—or how the schemes need to be modified to fit the new information. This is no clearer than the most recent taxonomic revolution that, as the opening quote states, "It's as if [he] lifted a whole submerged continent out of the ocean."

"He" is Carl Woese, who along with his coworkers at the University of Illinois proposed a new classification scheme with a new most inclusive taxon, the **domain**. The new scheme primarily came from work that compared the nucleotide base sequences of the RNA in ribosomes, those protein manufacturing machines needed by all cells. Woese's results in the late 1970s were especially relevant when comparing those sequences from a group of bacterial organisms formerly called the archaebacteria (*archae* = "ancient"). Many of these bacterial forms are known for their ability to live under extremely harsh environments. Woese discovered that the nucleotide sequences in these archaebacteria were different from those in other bacterial species and in eukaryotes. After finding other differences, including cell wall composition, membrane lipids, and sensitivity to certain antibiotics, the evidence pointed to there being three taxonomic lines to the tree of life.

In Woese's **three domain system**, one domain includes the former archaebacteria and is called the domain *Archaea* (FIGURE 3.7). The second encompasses all the remaining true bacteria and is called the domain *Bacteria*. The third domain, the *Eukarya*, includes the

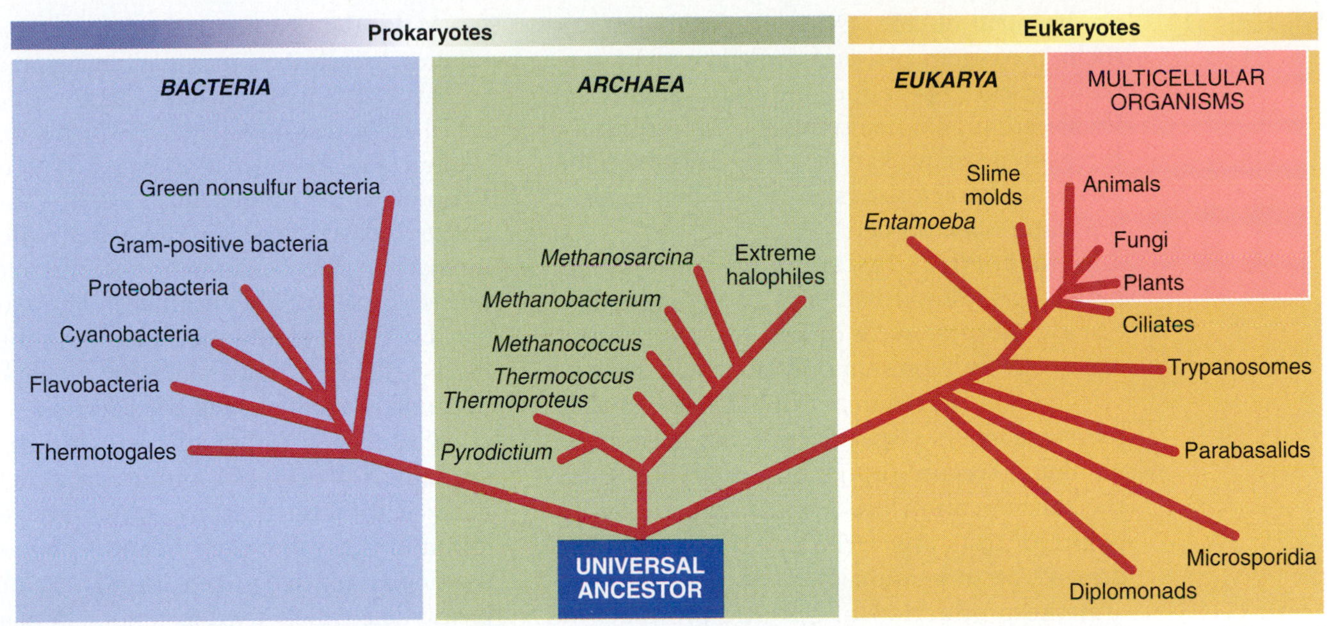

FIGURE 3.7 **The Three-Domain System.** Fundamental differences in genetic endowments are the basis for the three domains of all organisms on Earth. The line length between any two groups is proportional to their genetic differences.

Q: What cellular characteristic was the major factor stimulating the development of the three-domain system?

four remaining kingdoms (i.e., Protista, Plantae, Fungi, and Animalia).

Scientists initially were reluctant to accept the three-domain system of classification, and many deemed it a threat to the tenet that all living things are either prokaryotes or eukaryotes. Then, in 1996, Craig Venter and his coworkers deciphered the DNA base sequence of the archaean *Methanococcus jannaschii* and showed that almost two thirds of its genes are different from those of the *Bacteria*. They also found that proteins replicating the DNA and involved in RNA synthesis have no counterpart in the *Bacteria*. It appears that the three-domain system now is on firm ground.

CONCEPT AND REASONING CHECKS

3.8 What is the "whole submerged continent" that Woese lifted out of the ocean and why is the term "lifted" used?

Distinguishing between Prokaryotes

KEY CONCEPT

- Taxonomy now includes many criteria to distinguish one prokaryote from another.

We shall consider the taxonomy of most groups of microorganisms in their respective chapters. Prokaryotic taxonomy (referring to the *Bacteria* and *Archaea*), however, merits special attention because these organisms occupy an important position in microbiology and have had a complex system of taxonomy. David Hendricks Bergey devised one of the first systems of classification for the prokaryotes in 1923. His book, *Bergey's Manual of Determinative Bacteriology,* was updated and greatly expanded in the decades that followed. Now in its 9th edition, *Bergey's Manual* is available for use as a reference guide to identifying unknown bacteria that previously have been isolated in pure culture.

The five-volume *Bergey's Manual of Systematic Bacteriology* is intended as a classification guide and the official listing for all recognized *Bacteria* and *Archaea*. Since the first edition in 1984, bacterial taxonomy has gone through tremendous changes. For example, in Volume 1 of the second edition (2001), more than 2,200 new species and 390 new genera have been added.

There are several traditional and more modern criteria that microbiologists can use to identify (and classify) prokaryotes. Let's briefly review these methods.

Physical characteristics. These include staining reactions to help determine the organism's shape (morphology), and the size and arrangement of cells. Other characteristics can include oxygen, pH, and growth temperature requirements. Spore-forming ability and motility are additional determinants. Unfortunately, there are many prokaryotes that have the same physical characteristics, so other distinguishing features are needed.

Biochemical tests. As microbiologists better understood bacterial physiology, they discovered there were certain metabolic properties that were present only in certain groups.

Today, a large number of biochemical tests exist and often a specific test can be used to eliminate certain groups of prokaryotes from the identification process. Among the more common tests are: fermentation of carbohydrates, the utilization of a specific substrate, and the production of specific products or waste products. But, as with the physical characteristics, often several biochemical tests are needed to differentiate between species.

■ Substrate: The reactants on which an enzyme acts to produce one or more products.

These identification tests are important clinically, as they can be part of the arsenal available to the clinical lab that is trying to identify a pathogen. Many of these tests can be rapidly identified using modern procedures and automated systems (FIGURE 3.8).

Serological tests. Microorganisms are antigenic, meaning they are capable of triggering the production of antibodies. Solutions of such collected antibodies, called **antisera**, are commercially available for many medically-important pathogens. For example, mixing a *Salmonella* antiserum with *Salmonella* cells will cause the cells to clump together or agglutinate. If a foodborne illness occurs, the antiserum may be useful in identifying if *Salmonella* is the pathogen. More information about serological testing will be presented in Chapter 21.

■ Antibodies: Proteins produced by the immune system in response to a specific chemical configuration (antigen).

Nucleic acid analysis. In 1984, the editors of *Bergey's Manual* noted that there is no "official" classification of bacterial species and that the closest approximation to an official classification is the one most widely accepted by the community of microbiologists. The editors stated that a comprehensive classification might one day be possible. Today, the fields of molecular genetics and genomics have advanced the analysis and sequencing of nucleic acids. This has given rise to a new era of molecular taxonomy.

Molecular taxonomy is based on the universal presence of ribosomes in all living organisms. In particular, it is the RNAs in the ribosome, called ribosomal RNA (rRNA), which are of most interest and the primary basis of Woese's construction of the three-domain system. Many scientists today believe the rRNA molecule is the ultimate "molecular chronometer" that allows for precise bacterial classification in all taxonomic classes. Other techniques, including the polymerase chain reaction and nucleic acid hybridization, will be mentioned in later chapters.

The vast number of tests and analyses available for bacterial and archaeal cells can make it difficult to know which are relevant taxonomically for classification purposes or medically for pathogen identification purposes. One widely used technique in many disciplines is the **dichotomous key**. There are various forms of dichotomous keys, but one very useful construction is a flow chart where a series of positive or negative test procedures are listed down the page. Based on the dichotomous nature of the test (always a positive or negative result), the flow chart immediately leads to the next test result. The result is the identification of a specific organism. A simplified example is shown in MicroFocus 3.4.

CONCEPT AND REASONING CHECKS

3.9 Why are so many tests often needed to identify a specific prokaryotic species?

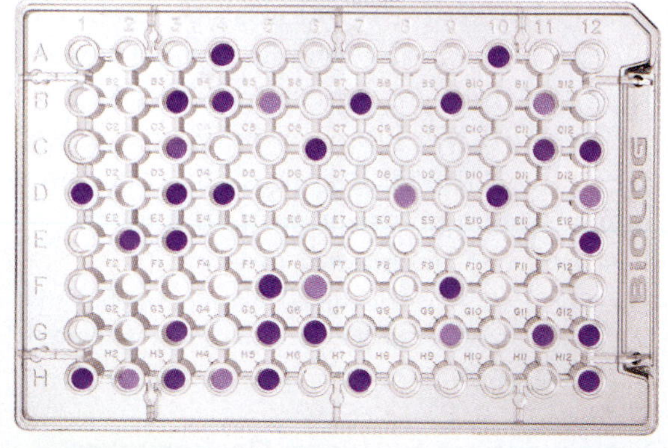

FIGURE 3.8 **A Biolog MicroPlate®.** New microbiological system that tests bacterial respiration simultaneously with 95 different substrates and then identifies species and types within species by their metabolic fingerprint. The 96th well is a negative control with no substrate.

Q: Of the methods described on the this page, which is/are most likely to be used in this more automated system? Explain.

MICROFOCUS 3.4: Tools
Dichotomous Key Flow Chart

A medical version of a taxonomic key (in the form of a dichotomous flow chart) can be used to identify very similar bacterial species based on physical and biochemical characteristics.

In this simplified scenario, an unknown bacterium has been cultured and several tests run. The test results are shown in the box at the bottom. Using the test results and the flow chart, identify the bacterial species that has been cultured.

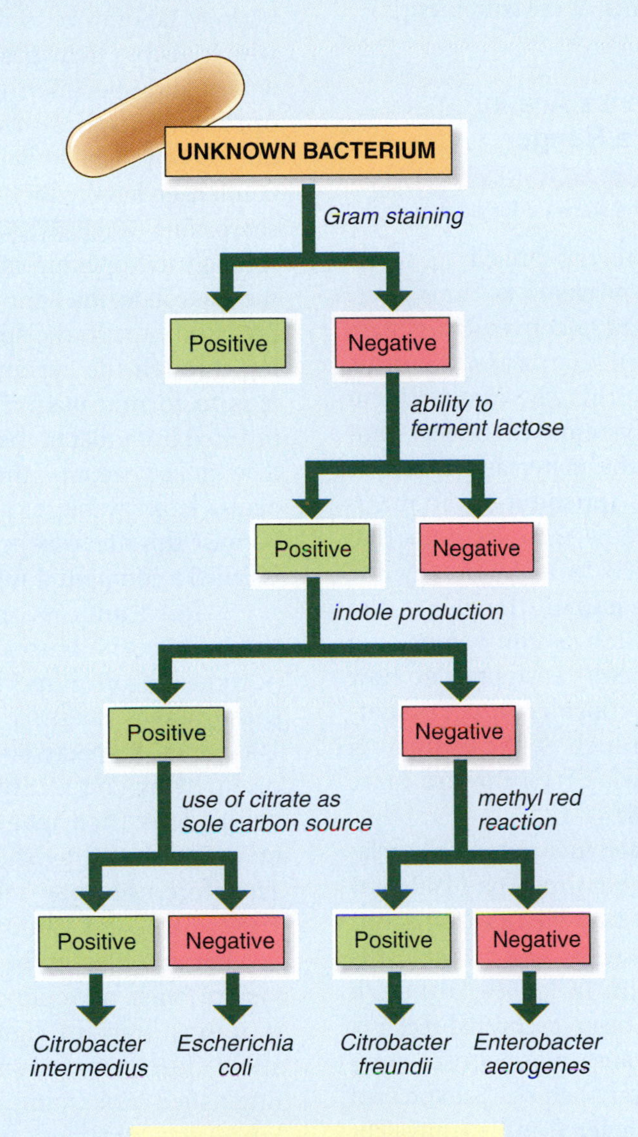

Microbiology Test Results

- Gram stain: gram-negative rods

Biochemical tests:
- Citrate test: negative
- Lactose fermentation: positive
- Indole test: positive
- Methyl red test: positive

3.3 Microscopy

The ability to see small objects all started with the microscopes used by Robert Hooke and Anton van Leeuwenhoek. By now, you should be aware that microorganisms usually are very small. Before we examine the instruments used to "see" these tiny creatures, we need to be familiar with the units of measurement.

Most Microbial Agents Are In the Micrometer Size Range

KEY CONCEPT

- Metric system units are the standard for measurement.

One physical characteristic used to study microorganisms and viruses is their size. Because they are so small, a convenient system of measurement is used that is the scientific standard around the world. The measurement system is the metric system, where the standard unit of length is the meter and is a little longer than a yard (see **Appendix A**). To measure microorganisms, we need to use units that are a fraction of a meter. In microbiology, the common unit for measuring length is the **micrometer (μm)**, which is equivalent to a millionth (10^{-6}) of a meter. To appreciate how small a micrometer is, consider this: Comparing a micrometer to an inch is like comparing a housefly to New York City's Empire State Building, 1,472 feet high.

Microbial agents range in size from the relatively large, almost visible protozoa (100 μm) down to the incredibly tiny viruses (0.02 μm) (FIGURE 3.9). Most bacterial and archael cells are about 1 μm to 5 μm in length, although notable exceptions have been discovered recently (MicroFocus 3.5). Because most viruses are a fraction of one micrometer, their size is expressed in nanometers. A **nanometer (nm)** is equivalent to a billionth (10^{-9}) of a meter; that is, 1/1,000 of a μm. Using nanometers, the size of the poliovirus, among the smaller viruses, measures 20 nm (0.02 μm) in diameter.

CONCEPT AND REASONING CHECKS

3.10 If a bacterial cell is 0.75 μm in length, what is its length in nanometers?

Light Microscopy Is Used to Observe Most Microorganisms

KEY CONCEPT

- Light microscopy uses visible light to magnify and resolve specimens.

The basic microscope system used in the microbiology laboratory is the **light microscope**, in which visible light passes directly through the lenses and specimen (FIGURE 3.10A). Such an optical configuration is called **bright-field microscopy**. Visible light is projected through a condenser lens, which focuses the light into a sharp cone (FIGURE 3.10B). The light then passes through the opening in the stage. When hitting the glass slide, the light is reflected or refracted as it passes through the specimen. Next, light passing through the specimen enters the objective lens to form a magnified intermediate image inverted from that of the specimen. This intermediate image becomes the object magnified by the ocular lens (eyepiece) and seen by the observer. Because this microscope has several lenses, it also is called a **compound microscope**.

A light microscope usually has at least three objective lenses: the low-power, high-power, and oil-immersion lenses. In general, these lenses magnify an object 10, 40, and 100 times, respectively. (Magnification is represented by the multiplication sign, ×.) The ocular lens then magnifies the intermediate image produced by the objective lens by 10×. Therefore, the total magnification achieved is 100×, 400×, and 1,000×, respectively.

■ **Total magnification:** The magnification of the ocular multiplied by the magnification of the objective lens being used.

For an object to be seen distinctly, the lens system must have good **resolving power**; that is, it must transmit light without variation and allow closely spaced objects to be clearly distinguished. For example, a car seen in the distance at night may appear to have a single headlight because at that distance the unaided eye lacks resolving power. However, using binoculars, the two headlights can be seen clearly as the resolving power of the eye increases.

When switching from the low-power (10×) or high-power (40×) lens to the oil-

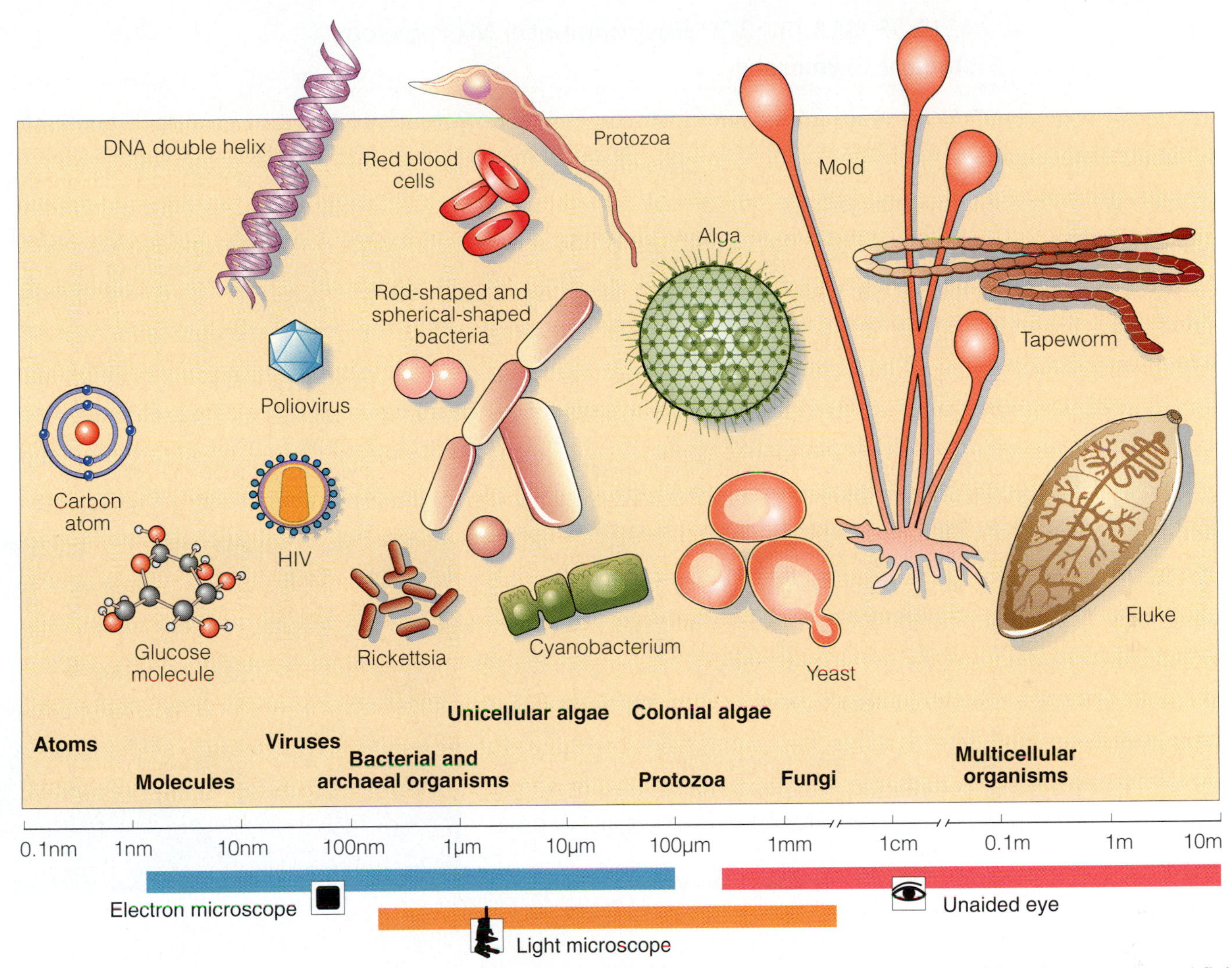

FIGURE 3.9 **Size Comparison among Various Atoms, Molecules, and Microorganisms (not drawn to scale).** Although tapeworms and flukes usually are macroscopic, the diseases these parasites cause are studied by microbiologists.

Q: Which domain on average has the smallest organisms and which has the largest?

immersion lens (100×), one quickly finds that the image has become fuzzy. The object lacks resolution, and the resolving power of the lens system appears to be poor. The poor resolution results from the refraction of light.

Both low-power and high-power objectives are wide enough to capture sufficient light for viewing. The oil-immersion objective, on the other hand, is so narrow that most light bends away and would miss the objective lens FIGURE 3.10C. The **index of refraction** (or refractive index) is a measure of the light-bending ability of a medium. Immersion oil has an index of refraction of 1.5, which is almost identical to the index of refraction of glass. Therefore, by immersing the 100× lens in oil, the light does not bend away from the lens as it passes from the glass slide and the specimen.

The oil thus provides a homogeneous pathway for light from the slide to the objective, and

MICROFOCUS 3.5: Environmental Microbiology
Biological Oxymorons

An oxymoron is a pair of words that seem to refer to opposites, such as jumbo shrimp, holy war, old news, and sweet sorrow. One of the characteristics we used for microorganisms is that most are invisible to the naked eye; you need a microscope to see them. Always true? So how about the oxymoron: macroscopic microorganism?

In 1993, researchers at Indiana University discovered near an Australian reef macroscopic bacterial cells in the gut of surgeonfish. Each cell was so large that a microscope was not needed to see it. The spectacular giant, measuring over 0.6 mm in length (that's 600 μm compared to 2 μm for *Escherichia coli*) even dwarfs the protozoan *Paramecium*.

While on an expedition off the coast of Namibia (western coast of southern Africa), Heide Schultz and teammates from the Max Planck Institute for Marine Microbiology in Bremen, Germany, found another bacterial monster in sediment samples from the sea floor (see figure). These bacterial cells were spherical being about 0.1 to 0.3 mm in diameter—but some as large as 0.75 mm—about the diameter of the period in this sentence. Their volume is about 3 million times greater than that of *E. coli*. The cells, shining white with enclosed sulfur granules, were held together in chains by a mucus sheath looking like a string of pearls. Thus, the bacterial species was named *Thiomargarita namibiensis* (meaning "sulfur pearl of Namibia").

How does a prokaryotic cell survive in so large a size? The trick is to keep the cytoplasm as a thin layer plastered against the edge of the cell so materials do not need to travel (diffuse) far to get into or out of the cell. The rest of the cell is a giant "bubble," called a vacuole, in which nitrate and sulfur are stored as potential energy sources. Thus, the actual cytoplasmic layer is microscopic and as close to the surface as possible.

Yes, the vast majority of microorganisms are microscopic, but exceptions have been found in some exotic places.

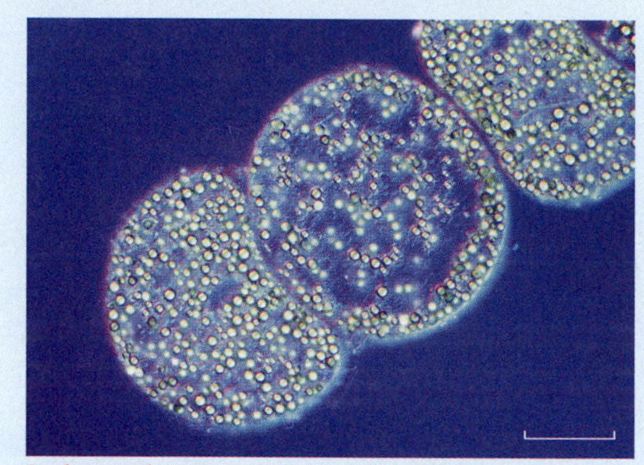

A phase microscopy micrograph showing a chain of *Thiomargarita namibiensis* cells. (Bar = 250 μm.)

the resolution of the object increases. With the oil-immersion lens, the highest resolution possible with the light microscope is attained, which is near 0.2 μm (200 nm) (MicroFocus 3.6).

CONCEPT AND REASONING CHECKS

3.11 What are the two most important properties of the light microscope?

Staining Techniques Provide Contrast

KEY CONCEPT

- Specimens stained with a dye are contrasted against the microscope field.

Microbiologists commonly stain bacterial cells before viewing them because the cytoplasm of bacterial cells lacks color, making it hard to see the cells on a bright background. Several staining techniques have been developed to provide **contrast** for bright-field microscopy.

To perform the **simple stain technique**, bacterial cells in a droplet of water or broth are smeared on a glass slide and the slide air-dried. Next, the slide is passed briefly through a flame in a process called **heat fixation**, which bonds the cells to the slide, kills any organisms still alive, and increases stain absorption. Now the slide is flooded with a **basic** (cationic) **dye** such as methylene blue (FIGURE 3.11A). Because cationic dyes have a positive charge, the dye is attracted to the cytoplasm and cell wall, which primarily have negative charges. By contrasting the blue cells against the bright background, the staining procedure allows the observer to measure cell size and determine cell shape. It also can provide information about how cells are arranged with respect to one another (Chapter 4).

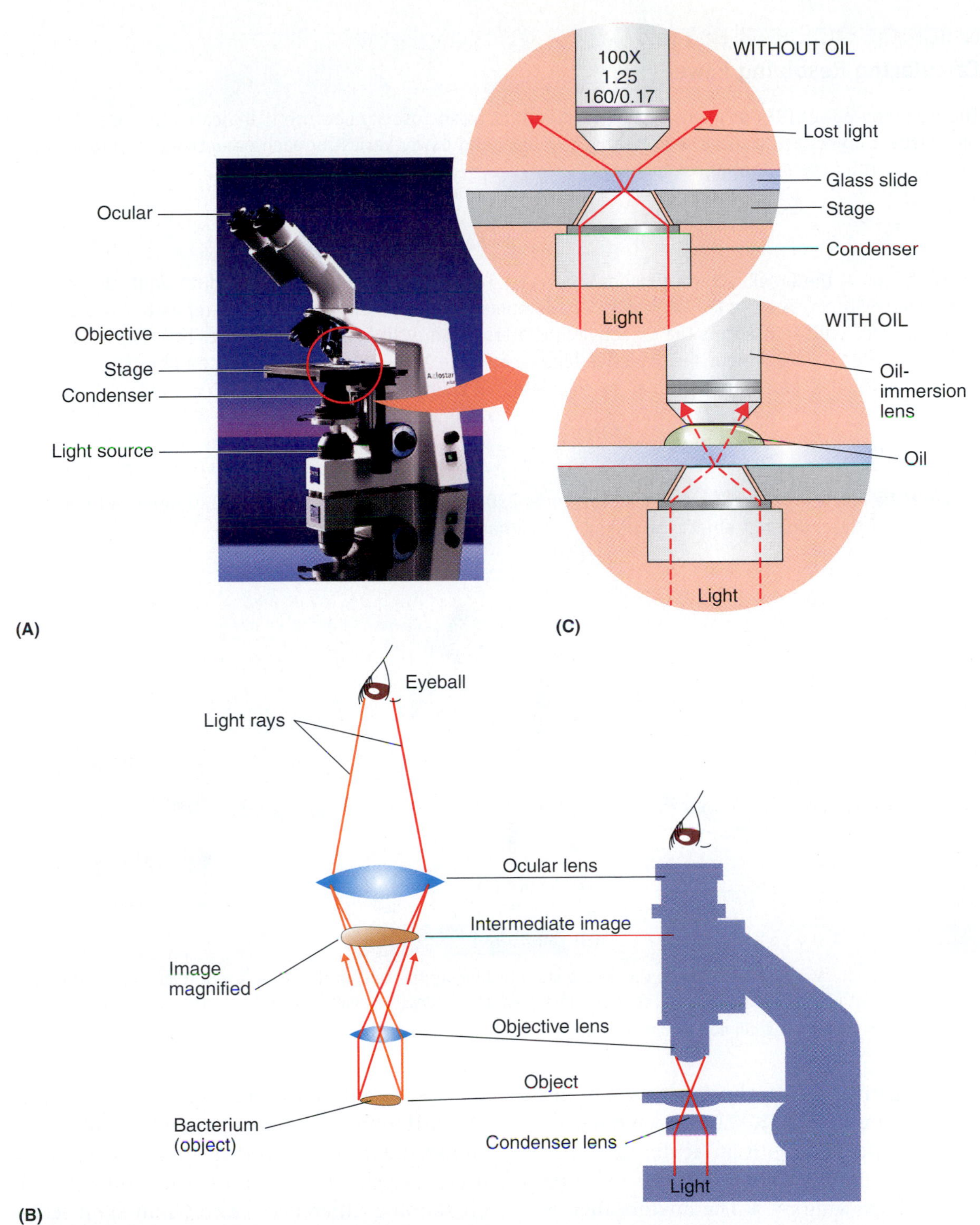

FIGURE 3.10 **The Light Microscope.** (**A**) The light microscope is used in many instructional and clinical laboratories. Note the important features of the microscope that contribute to the visualization of the object. (**B**) Image formation in the light microscope requires light to pass through the objective lens, forming an intermediate image. This image serves as an object for the ocular lens, which magnifies the image and forms the final image the eye perceives. (**C**) When using the oil immersion lens (100×), oil must be placed between and continuous with the slide and objective lens.

Q: Why must oil be used with the 100× oil-immersion lens?

MICROFOCUS 3.6: Tools

Calculating Resolving Power

The resolving power (RP) of a lens system is important in microscopy because it indicates the size of the smallest object that can be seen clearly. The resolving power varies for each objective lens and is calculated using the following formula:

$$RP = \frac{\lambda}{2 \times NA}$$

In this formula, the Greek letter λ (lambda) represents the wavelength of light; for white light, it averages about 550 nm. The symbol NA stands for the numerical aperture of the lens and refers to the size of the cone of light that enters the objective lens after passing through the specimen. This number generally is printed on the side of the objective lens (see Figure 3.10C). For a oil-immersion objective with an NA of 1.25, the resolving power may be calculated as follows:

$$RP = \frac{550\,\text{nm}}{2 \times 01.25} = \frac{550}{2.5} = 220 \text{ nm or } 0.22\ \mu\text{m}$$

Because the resolving power for this lens system is 220 nm, any object smaller than 220 nm could not be seen as a clear, distinct object. An object larger than 220 nm would be resolved.

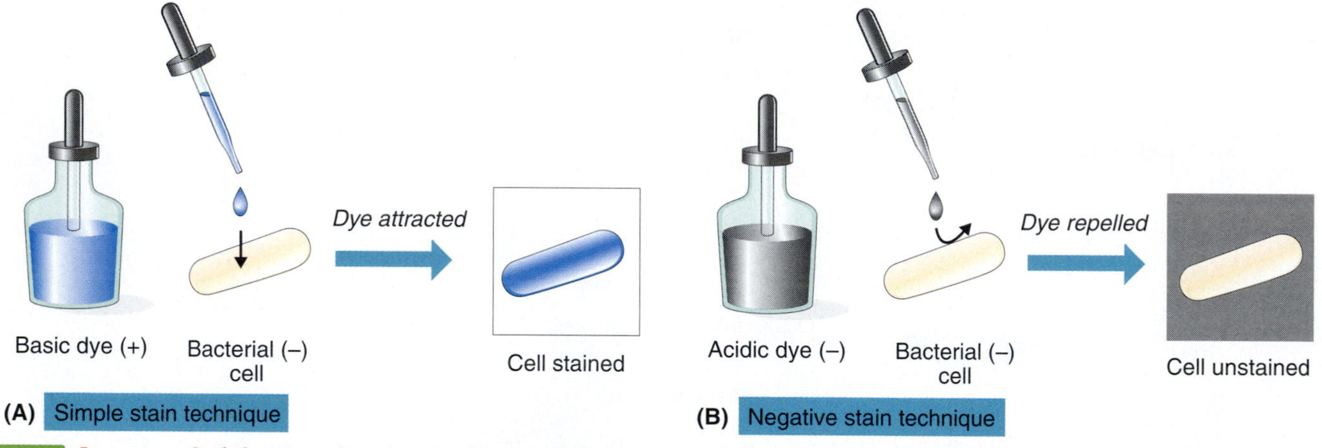

FIGURE 3.11 **Important Staining Reactions in Microbiology.** (**A**) In the simple stain technique, the cells in the smear are stained and contrasted against the light background. (**B**) With the negative stain technique, the cells are unstained and contrasted against a dark background.

Q: Explain how the simple and negative staining procedures stain and do not stain cells, respectively.

The **negative stain technique** works in the opposite manner (FIGURE 3.11B). Bacterial cells are mixed on a slide with an **acidic** (anionic) **dye** such as nigrosin (a black stain) or India ink (a black drawing ink). The mixture then is pushed across the face of the slide and allowed to air-dry. Because the anionic dye carries a negative charge, it is repelled from the cell wall and cytoplasm. The stain does not enter the cells and the observer sees clear or white cells on a black or gray background. Because this technique avoids chemical reactions and heat fixation, the cells appear less shriveled and less distorted than in a simple stain. They are closer to their natural condition.

The **Gram stain technique** is an example of a **differential staining procedure**; that is, it allows the observer to differentiate (separate) bacterial cells visually into two groups based on staining differences. The Gram stain technique is named for Christian Gram, the Danish physician who first perfected the technique in 1884.

The first two steps of the technique are straightforward (FIGURE 3.12A). Air-dried, heat-fixed smears are (1) stained with crystal violet, rinsed, and then (2) a special Gram's iodine solution is added. All bacterial cells would appear blue-purple if the procedure was stopped and the sample viewed with the light

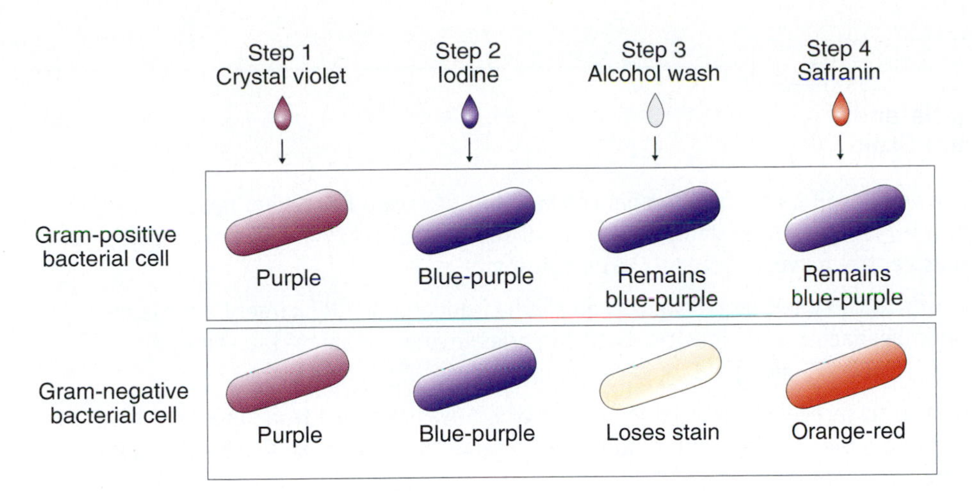

(A) Gram stain technique

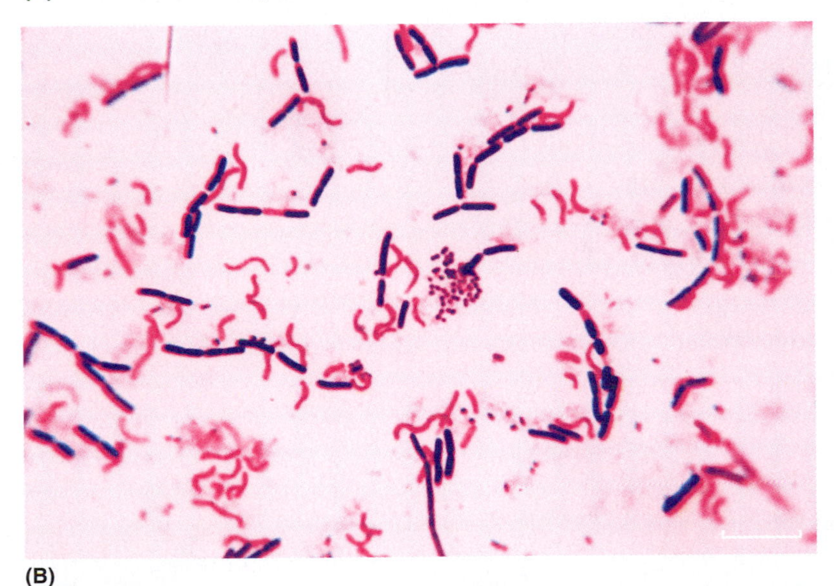

(B)

FIGURE 3.12 **Important Staining Reactions in Microbiology.** The Gram stain technique is a differential staining procedure. (**A**) All bacterial cells stain with the crystal violet and iodine, but only gram-negative cells lose the color when alcohol is applied. Subsequently, these bacterial cells stain with the safranin dye. Gram-positive cells remain blue purple. (**B**) This light micrograph demonstrates the staining results of a Gram stain for differentiating between gram-positive and gram-negative cells. (Bar = 10 µm.)

Q: Besides identifying the Gram reaction, what other characteristics can be determined using the Gram stain procedure?

microscope. Next, the smear is (3) rinsed with a decolorizer, such as 95 percent alcohol or an alcohol-acetone mixture. Observed at this point, certain bacterial cells may lose their color and become transparent. These are the **gram-negative** bacterial cells. Others retain the crystal violet and represent the **gram-positive** bacterial cells. The last step (4) uses safranin, a red cationic dye, to counterstain the gram-negative organisms; that is, give them a orange-red color. So, at the technique's conclusion, gram-positive cells are blue-purple while gram-negative cells are orange-red (FIGURE 3.12B). Similar to the simple staining, gram staining also allows the observer to determine size, shape, and arrangement of cells.

Knowing whether a bacterial cell is gram-positive or gram-negative is important for microbiologists and clinical technicians who use the results from the Gram stain technique to classify it in *Bergey's Manual* or aid in the identification of an unknown bacterial species (Textbook Case 3).

Gram-positive and gram-negative bacterial cells also differ in their susceptibility to chemical substances such as antibiotics (gram-positive cells are more susceptible to penicillin, gram-negative cells to tetracycline). Also, gram-negative cells have more complex cell walls, as

Textbook CASE 3

Bacterial Meningitis and a Misleading Gram Stain

1. A woman comes to the hospital emergency room complaining of severe headache, nausea, vomiting, and pain in her legs. On examination, cerebral spinal fluid (CFS) was observed leaking from a previous central nervous system (CNS) surgical site.
2. The patient indicates that 6 weeks and 8 weeks ago she had undergone CNS surgery after complaining of migraine headaches and sinusitis. Both surgeries involved a spinal tap. Analysis of cultures prepared from the CFS indicated no bacterial growth.
3. The patient was taken to surgery where a large amount of CFS was removed from underneath the old incision site. The pinkish, hazy fluid indicated bacterial meningitis, so among the laboratory tests ordered was a Gram stain.
4. The patient was placed on antibiotic therapy, consisting of vancomycin and cefotaxime.
5. Laboratory findings from the gram-stained CFS smear showed a few gram-positive, spherical bacterial cells that often appeared in pairs. The results suggested a *Streptococcus pneumoniae* infection.
6. However, upon reexamination of the smear, a few gram-negative spheres were observed.
7. When transferred to a blood agar plate, growth occurred and a prepared smear showed many gram-negative spheres (see figure). Further research indicated that several genera of gram-negative bacteria, including *Acinetobacter,* can appear gram-positive due to under-decolorization.
8. Although complicated by the under-decolorization outcome, the final diagnosis was bacterial meningitis due to *Acinetobacter baumanii*.

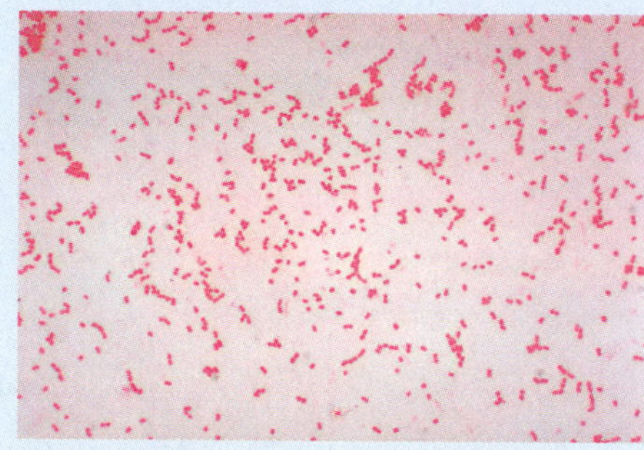

A gram-stained preparation from the blood agar plate.

Questions

A. *From the gram-stained CSF smear, what color were the bacterial spheres?*

B. *After reexamination of the CFS smear, assess the reliability of the gram-stained smear.*

C. *What reagent is used for the decolorization step in the Gram stain?*

Adapted from: Harrington, B. J. and Plenzler, M., 2004. Misleading gram stain findings on a smear from a cerebrospinal fluid specimen. *Lab. Med.* 35(8): 475–478.
For additional information see http://www.cdc.gov/ncidod/hip/aresist/acin_general.htm.

described in Chapter 4, and gram-positive and gram-negative bacterial species can produce different types of toxins.

■ Toxins: Chemical substances that are poisonous.

One other differential staining procedure, the **acid-fast technique**, deserves mention. This technique is used to identify members of the genus *Mycobacterium*, one species of which causes tuberculosis. These bacterial organisms are normally difficult to stain with the Gram stain because the cells have very waxy walls. However, the cell will stain red when treated with carbol-fuchsin (red dye) and heat (or a lipid solubilizer) (FIGURE 3.13). The cells then retain their color when washed with a dilute acid-alcohol solution. Other stained genera lose the red color easily during the acid-alcohol wash. The *Mycobacterium* species, therefore, is called acid resistant or *acid fast*. Because they stain red and break sharply when they reproduce, *Mycobacterium* species often are referred to as "red snappers."

CONCEPT AND REASONING CHECKS

3.12 What would happen if a student omitted the alcohol wash step when doing the Gram stain procedure?

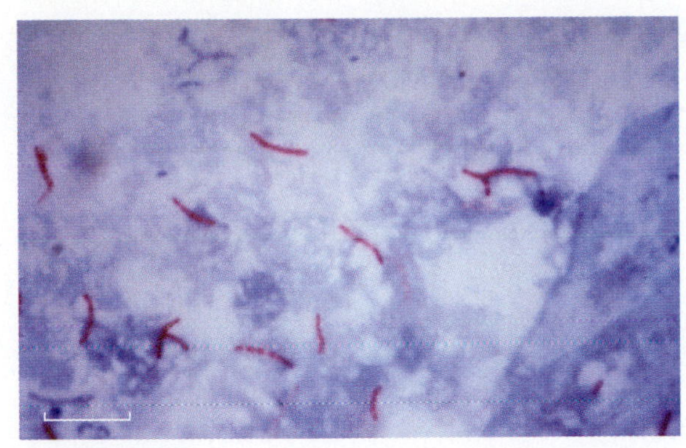

FIGURE 3.13 *Mycobacterium tuberculosis.* The acid-fast technique is used to identify species of *Mycobacterium*. The cells retain the red dye after an acid-alcohol wash. (Bar = 10 µm.)

Q: Why are cells of Mycobacterium *resistant to Gram staining?*

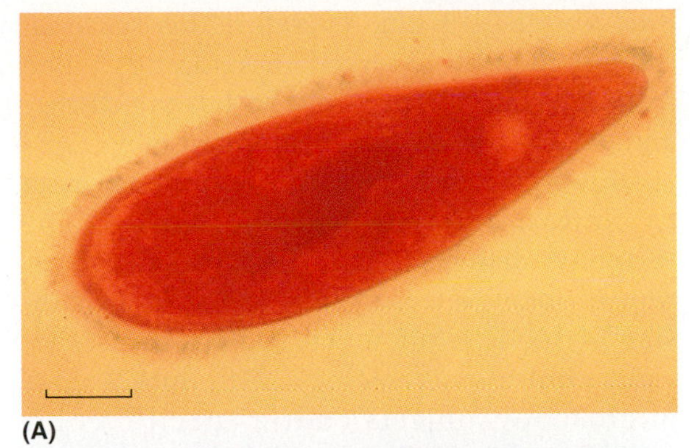

(A)

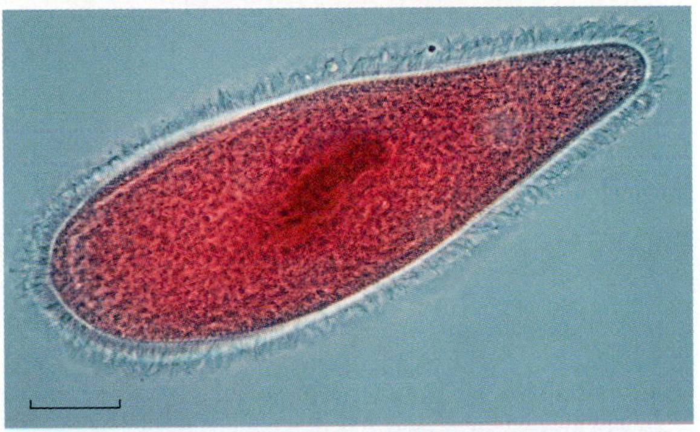

(B)

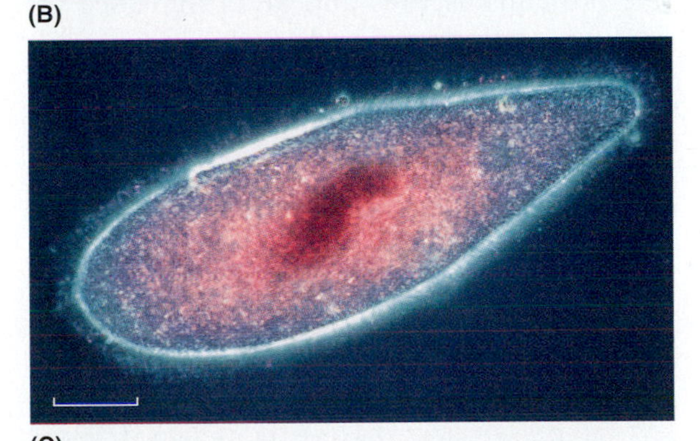

(C)

FIGURE 3.14 **Variations in Light Microscopy.** The same *Paramecium* specimen seen with three different optical configurations: (**A**) bright-field, (**B**) phase-contrast, and (**C**) dark-field. (Bar = 25 µm.)

Q: What advantage is gained by each of the three microscopy techniques?

Light Microscopy Has Other Optical Configurations

KEY CONCEPT

- Different optical configurations provide detailed views of cells.

Bright-field microscopy provides little contrast (FIGURE 3.14A). However, a light microscope can be outfitted with other optical systems to improve contrast of prokaryotic and eukaryotic cells without staining. Three systems commonly employed are mentioned here.

Phase-contrast microscopy uses a special condenser and objective lenses. This condenser lens on the light microscope splits a light beam and throws the light rays slightly out of phase. The separated beams of light then pass through and around the specimen, and small differences in the refractive index within the specimen show up as different degrees of brightness and contrast. With phase-contrast microscopy, microbiologists can see organisms alive and unstained (FIGURE 3.14B). The structure of yeasts, molds, and protozoa is studied with this optical configuration.

Dark-field microscopy also uses a special condenser lens mounted under the stage. The condenser scatters the light and causes it to hit the specimen from the side. Only light bouncing off the specimen and into the objective lens makes the specimen visible, as the surrounding area appears dark because it lacks background light (FIGURE 3.14C). Dark-field microscopy provides good resolution and often illuminates parts of a specimen not seen with bright-field optics. Dark-field microscopy also is the preferred way to study motility of live cells.

Dark-field microscopy helps in the diagnosis of diseases caused by organisms near the limit of resolution of the light microscope. For example, syphilis, caused by the spiral bacterium *Treponema pallidum*, has a diameter of only about 0.15 μm. Therefore,this bacterial species may be observed in scrapings taken from a lesion of a person who has the disease and observed with dark-field microscopy.

Fluorescence microscopy is a major asset to clinical and research laboratories. The technique has been applied to the identification of many microorganisms and is a mainstay of modern microbial ecology and especially clinical microbiology. With fluorescence microscopy, objects emit a specific color (wavelength) of light after absorbing a shorter wavelength radiation.

Microorganisms are coated with a fluorescent dye, such as fluorescein, and then illuminated with ultraviolet (UV) light. The energy in UV light excites electrons in fluorescein, and they move to higher energy levels. However, the electrons quickly drop back to their original energy levels and give off the excess energy as visible light. The coated microorganisms thus appear to fluoresce; in the case of fluorescein, they glow a greenish yellow. Other dyes produce other colors (FIGURE 3.15).

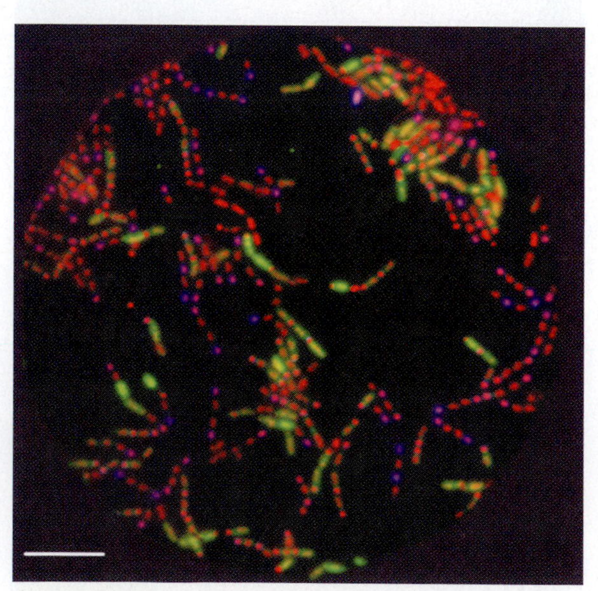

FIGURE 3.15 **Fluorescence Microscopy.** Fluorescence microscopy of sporulating cells of *Bacillus subtilis*. DNA has been stained with a dye that fluoresces red and a sporulating protein with fluorescein (green). RNA synthesis activity is indicated by a dye that fluoresces blue. (Bar = 15 μm.)

Q: What advantage is gained by using fluorescence optics over the other light microscope optical configurations?

An important application of fluorescence microscopy is the **fluorescent antibody technique** used to identify an unknown organism. In one variation of this procedure, fluorescein is chemically attached to antibodies, the protein molecules produced by the body's immune system. These "tagged" antibodies are mixed with a sample of the unknown organism. If the antibodies are specific for that organism, they will bind to it and coat the cells with the dye. When subjected to UV light, the organisms will fluoresce. If the organisms fail to fluoresce, the antibodies were not specific to that organism and a different tagged antibody is tried.

CONCEPT AND REASONING CHECKS

3.13 What optical systems can improve specimen contrast over bright-field microscopy?

Electron Microscopy Provides Detailed Images of Cells, Cell Parts, and Viruses

KEY CONCEPT

- Electron microscopy uses a beam of electrons to magnify and resolve specimens.

The **electron microscope** grew out of an engineering design made in 1933 by the German physicist Ernst Ruska (winner of the 1986 Nobel Prize in Physics). Ruska showed that electrons will flow in a sealed tube if a vacuum is maintained to prevent electron scattering. Magnets, rather than glass lenses, pinpoint the flow onto an object, where the electrons are absorbed, deflected, or transmitted depending on the density of structures within the object (FIGURE 3.16). When projected onto a screen underneath, the electrons form a final image that outlines the structures. As mentioned in Chapter 1, the early days of electron microscopy produced electron micrographs that showed bacterial cells indeed were cellular but their structure was different from eukaryotic cells. This led to the development of the prokaryotic and eukaryotic groups of organisms.

■ Electron micrographs: Images recorded on electron-sensitive film.

The power of electron microscopy is the extraordinarily short wavelength of the beam of electrons. Measured at 0.005 nm (compared to 550 nm for visible light), the short wavelength dramatically increases the resolving power of the system and makes possible the visualization of viruses and detailed cellular structures, often called the **ultrastructure** of cells. The practical limit of resolution of biological samples with

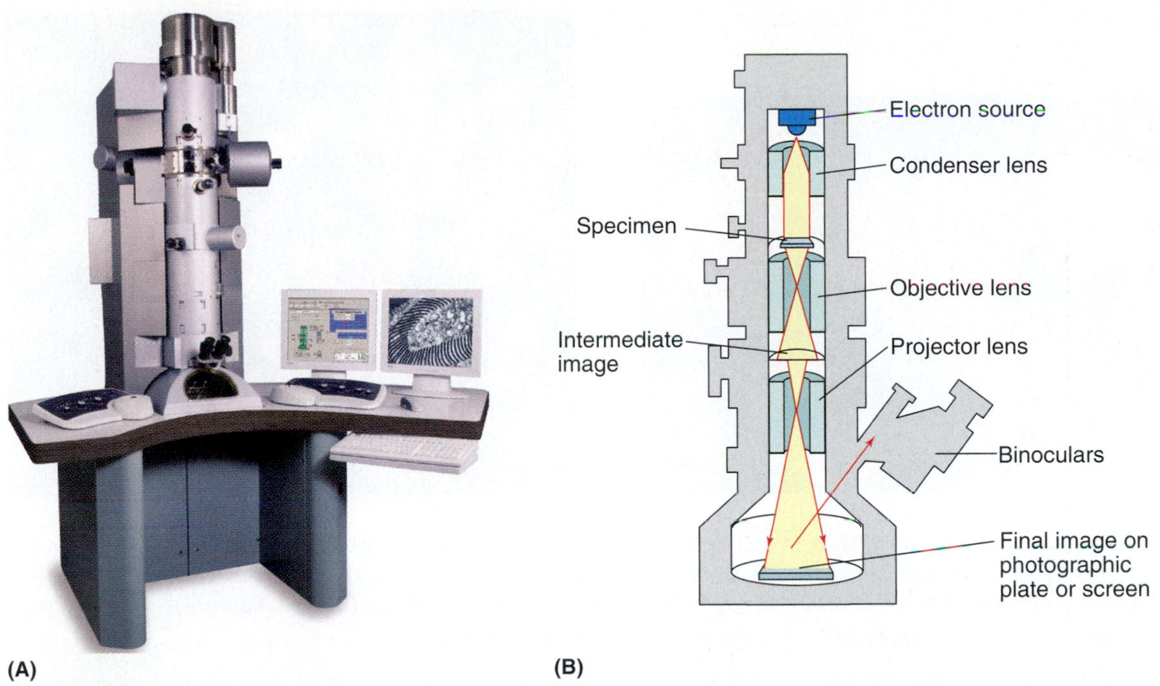

FIGURE 3.16 **The Electron Microscope.** (**A**) A transmission electron microscope (TEM). (**B**) A schematic of a TEM. A beam of electrons is emitted from the electron source and electromagnets are used to focus the beam on the specimen. The image is magnified by objective and projector lenses. The final image is projected on a screen, television monitor, or photographic film.

Q: How does the path of the image for the transmission electron microscope compare with that of the light microscope (Figure 3.10)?

the electron microscope is about 2 nm, which is 100× better than the resolving power of the light microscope. The drawback of the electron microscope is that the method needed to prepare a specimen kills the cells or organisms.

Two types of electron microscopes are currently in use. The **transmission electron microscope** (**TEM**) is used to view and record detailed structures within cells (FIGURE 3.17A). Ultrathin sections of the prepared specimen must be cut because the electron beam can penetrate matter only a very short distance. After embedding the specimen in a suitable plastic mounting medium or freezing it, scientists cut the specimen into sections with a diamond knife. In this manner, a single bacterial cell can be sliced, like a loaf of bread, into hundreds of thin sections.

Several of the sections are placed on a small grid and stained with heavy metals such as lead and osmium to provide contrast. The microscopist then inserts the grid into the vacuum chamber of the microscope and focuses a 100,000-volt electron beam on one portion of a section at a time. An image forms on the screen below or can be recorded on film. The electron micrograph may be enlarged with enough resolution to achieve a final magnification of over 20 million ×.

The **scanning electron microscope** (**SEM**) was developed in the late 1960s to enable researchers to see the surfaces of objects in the natural state and without sectioning. The specimen is placed in the vacuum chamber and covered with a thin coat of gold. The electron beam then scans across the specimen and knocks loose showers of electrons that are captured by a detector. An image builds line by line, as in a television receiver. Electrons that strike a sloping surface yield fewer electrons, thereby producing a darker contrasting spot and a sense of three dimensions. The resolving power of the conventional SEM is about 7 nm and magnifications with the SEM are limited to about 50,000×. However, the instrument provides vivid and undistorted views of an organism's surface details (FIGURE 3.17B).

The electron microscope has added immeasurably to our understanding of the structure and function of microorganisms by letting us penetrate their innermost secrets. In the chapters ahead, we will encounter many of the

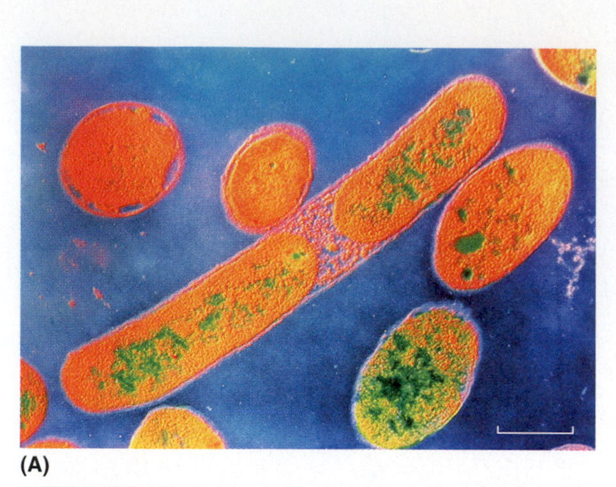

(A)

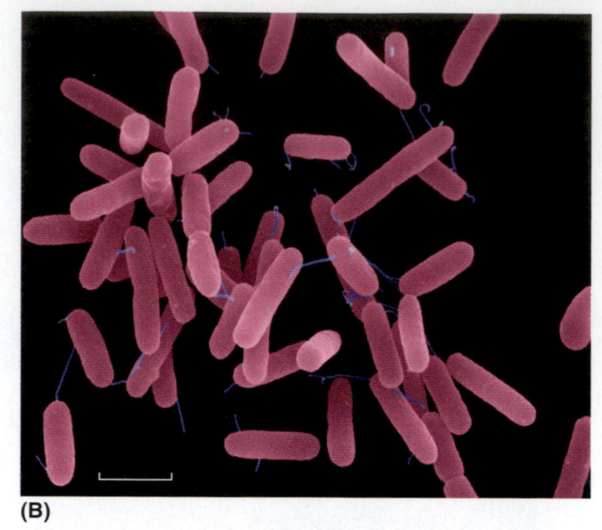

(B)

FIGURE 3.17 **Transmission and Scanning Electron Microscopy Compared.** The bacterium *Pseudomonas aeruginosa* as seen with two types of electron microscopy. (**A**) A view of sectioned cells seen with the transmission electron microscope. (Bar = 1.0 μm.) (**B**) A view of whole cells seen with the scanning electron microscope. (Bar = 3.0 μm.)

Q: What types of information can be gathered from each of these electron micrographs?

TABLE

3.3 Comparison of Various Types of Microscopy

Type of Microscopy	Special Feature	Appearance of Object	Magnification Range	Objects Observed
Light				
Bright-field	Visible light illuminates object	Stained microorganisms on clear background	100× – 1,000×	Arrangement, shape, and size of killed microorganisms (except viruses)
Phase-contrast	Special condenser throws light rays "out of phase"	Unstained microorganisms with contrasted structures	100× – 1,000×	Internal structures of live, unstained eukaryotic microorganisms
Dark-field	Special condenser scatters light	Unstained microorganisms on dark background	100× – 1,000×	Live, unstained microorganisms (e.g., spirochetes)
Fluorescence	UV light illuminates fluorescent-coated objects	Fluorescing microorganisms on dark background	100× – 1,000×	Outline of microorganisms coated with fluorescent-tagged antibodies
Electron				
Transmission	Short-wavelength electron beam penetrates sections	Alternating light and dark areas contrasting internal cell structures	100× – 200,000×	Ultrathin slices of microorganisms and internal components of eukaryotic and prokaryotic cells
Scanning	Short-wavelength electron beam knocks loose electron showers	Microbial surfaces	10× – 50,000×	Surfaces and textures of microorganisms and cell components

structures displayed by electron microscopy, and we will better appreciate microbial physiology as it is defined by microbial structures.

The various types of light and electron microscopy are compared in TABLE 3.3.

CONCEPT AND REASONING CHECKS

3.14 What type of electron microscope would be used to examine (a) the surface structures on a *Paramecium* cell and (b) the organelles in an algal cell?

SUMMARY OF KEY CONCEPTS

3.1 The Prokaryotic/Eukaryotic Paradigm

- All living organisms, both prokaryotic and eukaryotic, share the common emergent properties of life.
- All organisms and cells must attempt to maintain a stable internal state called homeostasis.
- Most prokaryotic species interact through a multicellular association involving chemical communication and cooperation between cells.
- Although prokaryotic and eukaryotic cells may differ in the structures they contain and the pattern of organization of those structures, both groups carry out similar and complex cellular processes.
- Prokaryotic and eukaryotic cells share certain organizational processes, including genetic organization, compartmentation, metabolic organization, and protein synthesis.
- Although prokaryotic and eukaryotic cells carry out many similar processes, eukaryotic cells often contain membrane-enclosed compartments (organelles) to accomplish the processes.

3.2 Cataloging Microorganisms

- Many systems of classification have been devised to catalog organisms. The work of Linnaeus and Haeckel represents early efforts based on shared characteristics.
- Part of an organism's binomial name is the genus name; the remaining part is the specific epithet that describes the genus name. Thus, a species name consists of the genus and specific epithet.
- Organisms are properly classified using a standardized hierarchical system from species (the least inclusive) to domain (the most inclusive).
- In the 1960s, Whittaker established the five kingdom system. In this classification, bacterial organisms are placed in the kingdom Monera, protozoa and algae in the kingdom Protista, and fungi in the kingdom Fungi.
- Based on several molecular and biochemical differences within the kingdom Monera, Woese proposed a three domain system. The Kingdom Monera is separated into two domains, the *Bacteria* and *Archaea*. The remaining kingdoms (Protista, Fungi, Plantae, and Animalia) are placed in the domain *Eukarya*.
- *Bergey's Manual* is the standard reference to identify and classify bacterial species. Criteria have included traditional characteristics, but modern molecular methods have led to a reconstruction of evolutionary events and organism relationships.

3.3 Microscopy

- Another criterion of a microorganism is its size, a characteristic that varies among members of different groups. The micrometer (μm) is used to measure the dimensions of bacterial, protozoal, and fungal cells. The nanometer (nm) is commonly used to express viral sizes.
- The instrument most widely used to observe microorganisms is the light microscope. Light passes through several lens systems that magnify and resolve the object being observed. Although magnification is important, resolution is key. The light microscope can magnify up to 1,000× and resolve objects as small as 0.2 μm.
- For bacterial cells, staining generally precedes observation. The simple, negative, Gram, acid-fast, and other staining techniques can be used to impart contrast and determine structural or physiological properties.
- Microscopes employing phase-contrast, dark field, or fluorescence optics have specialized uses in microbiology to contrast cells without staining.
- To increase resolution and achieve extremely high magnification, the electron microscope employs a beam of electrons to magnify and resolve specimens. To observe internal details (ultrastructure), the transmission electron microscope is most often used; to see whole objects or surfaces, the scanning electron microscope is useful.

LEARNING OBJECTIVES

After understanding the textbook reading, you should be capable of writing a paragraph that includes the appropriate terms and pertinent information to answer the objective.

1. Identify the six emerging properties common to prokaryotes and eukaryotes.
2. Assess the importance of homeostasis to cell (organismal) survival.
3. Contrast prokaryotes as unicellular and multicellular organisms.
4. Apply the concepts of cell structure, the pattern of organization, and cellular process to prokaryotes and eukaryotes.
5. Describe the four organizational patterns common to all organisms.
6. Identify the structural distinctions between prokaryotic and eukaryotic cells.
7. Explain how prokaryotic cells carry out cellular processes common to eukaryotic cells without needing membrane-bound organelles.
8. Distinguish between the processes of invagination and endosymbiosis for the origin of eukaryotic cells.
9. Evaluate Linnaeus' contributions to modern taxonomy.
10. Write scientific names of organisms using the binomial system.
11. Identify the taxa used to classify organisms from least to most inclusive taxa.
12. Discuss the assigning of organisms according to the five kingdom system of classification.
13. Explain the assignment of organisms to the three domain system of classification.
14. Contrast the procedures used to identify and classify bacterial and archaeal species.
15. Use a dichotomous key.
16. Measure sizes of microbial agents using metric system units.
17. Assess the importance of magnification and resolving power to microscopy.
18. Compare and contrast the procedures used to stain bacterial cells.
19. Summarize the Gram stain procedure.
20. Identify the optical configurations that provide contrast with light microscopy.
21. Compare the uses of the transmission and scanning electron microscopes.

SELF-TEST

Answer each of the following questions by selecting the ***one*** answer that best fits the question or statement. Answers to even-numbered questions can be found in **Appendix C**.

1. ______ describes the ability of organisms to maintain a stable internal state.
A. Metabolism
B. Homeostasis
C. Biosphere
D. Ecotype
E. None of the above (**A–D**) is correct.

2. Which one of the following phrases would *not* apply to prokaryotes?
A. Programmed cell death
B. Multicellular communities
C. Cell-cell communication
D. Cell cooperation
E. Nucleated cells

3. Proteins are made by the
A. mitochondria.
B. lysosomes.
C. Golgi apparatus.
D. ribosomes.
E. cytoskeleton.

4. Cell walls are necessary for
A. nutrient transport regulation.
B. DNA compartmentation.
C. protein transport.
D. energy metabolism.
E. water balance.

5. Which one of the following is *not* found in prokaryotic cells?
A. Cell membrane
B. Ribosomes
C. DNA
D. Mitochondria
E. Cytoplasm

6. The two functions of the endoplasmic reticulum are
A. protein and lipid transport.
B. cell respiration and photosynthesis.
C. osmotic regulation and genetic control.
D. cell respiration and protein synthesis.
E. sorting and packaging of proteins.

7. Mitochondria differ from chloroplasts in that only mitochondria
A. carry out photosynthesis.
B. are membrane-bound.
C. are found in the *Eukarya*.
D. convert chemical energy to cellular energy.
E. transform sunlight into chemical energy.

8. Who is considered to be the father of modern taxonomy?
A. Woese
B. Whittaker
C. Aristotle
D. Haeckel
E. Linnaeus

9. Which one of the following is the correct genus name for the bacterial organism that causes syphilis?
A. treponema
B. pallidum
C. *Treponema*
D. *pallidum*
E. *T. pallidum*

10. Several classes of organisms would be classified into one
A. family.
B. genus.
C. species.
D. order.
E. phylum (division).

11. The domain *Eukarya* includes all the following *except*
A. fungi.
B. protozoa.
C. archaeal cells.
D. algae.
E. animals.

12. ______ was first used to catalog organisms into one of three domains.
A. Photosynthesis
B. Ribosomes
C. Ribosomal RNA
D. Nuclear DNA
E. Mitochondrial DNA

13. Resolving power is the ability of a microscope to
A. estimate cell size.
B. magnify an image.
C. see two close objects as separate.
D. keep objects in focus.
E. Both **B** and **D** are correct.

14. Calculation of total magnification involves which microscope lens or lenses?
A. Ocular and condenser
B. Objective only
C. Ocular only
D. Objective and ocular
E. Objective and condenser

15. Before bacterial cells are simple stained and observed with the light microscope, they must be
A. smeared on a slide.
B. heat fixed.
C. killed.
D. air dried.
E. All the above (**A–D**) are correct.

16. The counterstain used in the Gram stain procedure is
A. iodine.
B. alcohol.
C. carbol-fuchsin.
D. safranin.
E. malachite green.

17. The practical limit of resolution with the transmission electron microscope is ______.
A. 10 μm
B. 1 μm
C. 50 nm
D. 10 nm
E. 2 nm

18. For transmission electron microscopy, contrast is provided by
A. heavy metals.
B. a coat of gold.
C. crystal violet.
D. plastic.
E. copper ions.

19. If you wanted to study the surface of a bacterial cell, you would use
A. a transmission electron microscope.
B. a light microscope with phase-contrast optics.
C. a scanning electron microscope.
D. a light microscope with dark-field optics.
E. a light microscope with bright-field optics.

20. If you wanted to study bacterial motility you would most likely use
A. a transmission electron microscope.
B. a light microscope with phase-contrast optics.
C. a scanning electron microscope.
D. a light microscope with dark-field optics.
E. a light microscope with bright-field optics.

QUESTIONS FOR THOUGHT AND DISCUSSION

Answers to even-numbered questions can be found in **Appendix C.**

1. A local newspaper once contained an article about "the famous bacteria E. coli." How many things can you find wrong in this phrase? Rewrite the phrase correctly.
2. Microorganisms have been described as the most chemically diverse, the most adaptable, and the most ubiquitous organisms on Earth. Although your knowledge of microorganisms still may be limited at this point, try to add to this list of "mosts."
3. Prokaryotes lack the cytoplasmic organelles commonly found in the eukaryotes. Provide a reason for this structural difference.
4. A new bacteriology laboratory is opening in your community. What is one of the first books that the laboratory director will want to purchase? Why is it important to have this book?
5. In 1987, in a respected science journal, an author wrote, "Linnaeus gave each life form two Latin names, the first denoting its genus and the second its species." A few lines later, the author wrote, "Man was given his own genus and species *Homo sapiens*." What is conceptually and technically wrong with both statements?
6. A student of general biology observes a microbiology student using immersion oil and asks why the oil is used. "To increase the magnification of the microscope" is the reply. Do you agree? Why?
7. Every state has an official animal, flower, or tree, but only Oregon has a bacterial species named in its honor: *Methanohalophilus oregonese*. The specific epithet *oregonese* is obvious, but can you decipher the meaning of the genus name?

APPLICATIONS

Answers to even-numbered questions can be found in **Appendix C.**

1. A student is performing the Gram stain technique on a mixed culture of gram-positive and gram-negative bacterial cells. In reaching for the counterstain in step 4, he inadvertently takes the methylene blue bottle and proceeds with the technique. What will be the colors of gram-positive and gram-negative bacteria at the conclusion of the technique?
2. Would the best resolution with a light microscope be obtained using red light (λ = 680 nm), green light (λ = 520 nm), or blue light (λ = 500 nm)? Explain your answer.
3. The electron micrograph below shows several bacterial cells. The micrograph has been magnified 12,000×. At this magnification, the cells are about 25 mm in length. Calculate the actual length of the bacterial cells in micrometers (μm)?

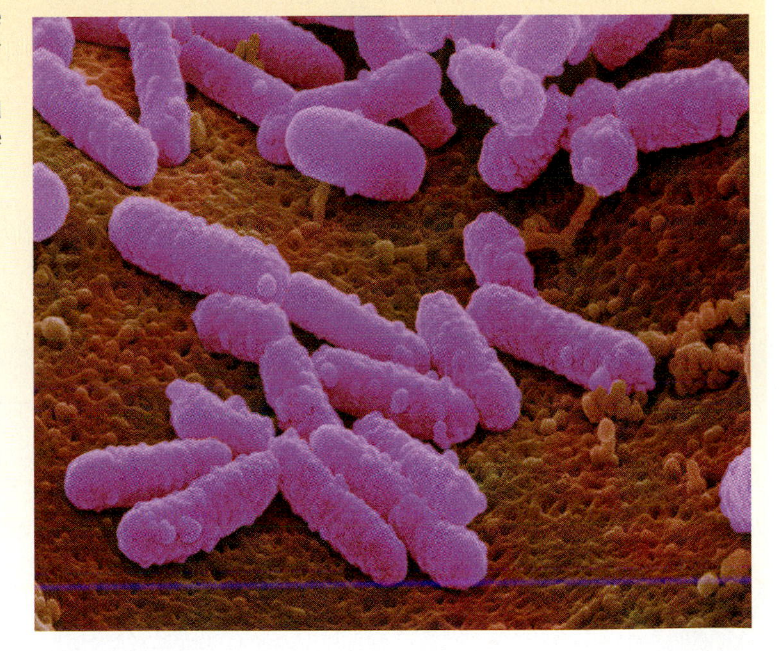

REVIEW

Match the statement on the left to the term on the right by placing the letter of the term in the available space. **Appendix C** contains the correct answers to the even-numbered statements.

Statement

1. _____ System of nomenclature used for microorganisms and other living things.
2. _____ Unit of measurement used for viruses and equal to a billionth of a meter.
3. _____ Major group of organisms whose cells have no nucleus or organelles in the cytoplasm.
4. _____ Bacterial organisms capable of photosynthesis.
5. _____ Devised the three domain system of classification in which two domains contain prokaryotes.
6. _____ Type of microscope that uses a special condenser to split the light beam.
7. _____ Type of electron microscope for which cell sectioning is not required.
8. _____ The ability of an organism to maintain a stable internal state.
9. _____ These structures carry out protein synthesis in all cells.
10. _____ The organelle, absent in prokaryotes, which carries out the conversion of chemical energy to cellular energy in eukaryotes.
11. _____ Domain in which many prokaryotes are classified.
12. _____ Staining technique that differentiates bacterial cells into two groups.
13. _____ Author of an early system of classification for bacterial organisms.
14. _____ Category into which two or more genera are grouped.
15. _____ Coined the name protista for microorganisms.
16. _____ The staining technique employing a single cationic dye.
17. _____ The convention for writing the binomial name of microorganisms.
18. _____ Unit of measurement for bacterial cells and equal to a millionth of a meter.
19. _____ Type of microscopy using UV light to excite a dye-coated specimen.
20. _____ Staining technique in which the background is colored and the cells are clear.

Term

A. Archaea
B. Bergey
C. Binomial
D. Boldface
E. Chloroplast
F. Cyanobacteria
G. Dark-field
H. Eukarya
I. Family
J. Fluorescence
K. Fungi
L. Gram
M. Haeckel
N. Homeostasis
O. Italics
P. Micrometer
Q. Mitochondrion
R. Nanometer
S. Negative
T. Prokaryotes
U. Ribosomes
V. Scanning
W. Simple
X. Taxonomy
Y. Transmission
Z. Woese

HTTP://MICROBIOLOGY.JBPUB.COM/

The site features learning, an on-line review area that provides quizzes and other tools to help you study for your class. You can also follow useful links for in-depth information, read more MicroFocus stories, or just find out the latest microbiology news.

REVIEW—CELL STRUCTURE

Identify the cell structures indicated in drawings (A) and (B) below. Answers to the even-numbered structures can be found in **Appendix C.**

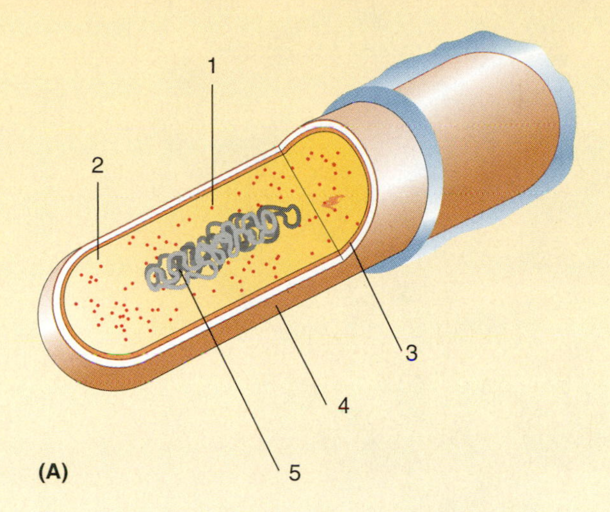

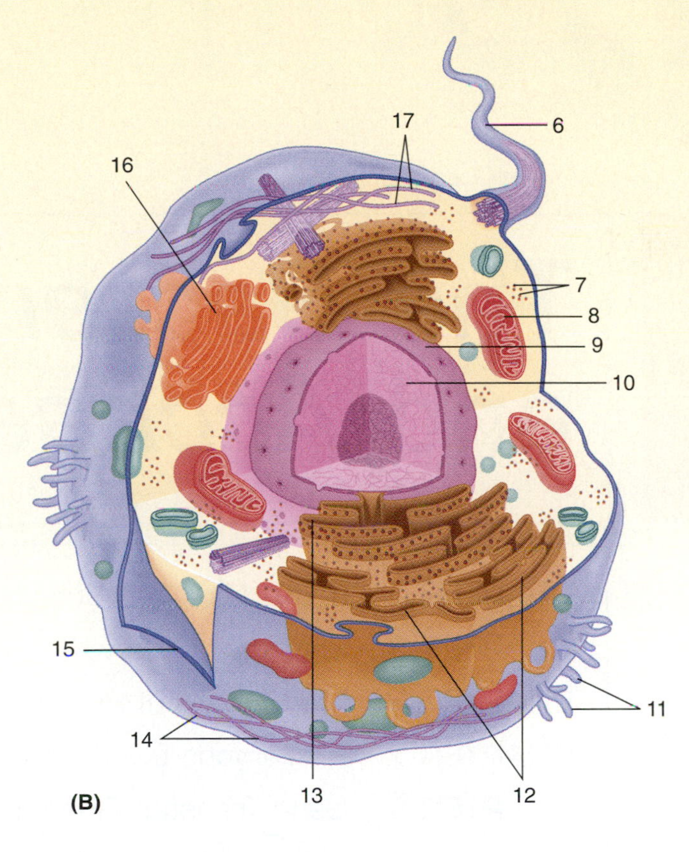

False-color electron micrograph of *Escherichia coli* cells

PART 2

The *Bacteria* and *Archaea*

We live at the center of a microbial universe. On all sides, microscopic organisms surround us and make their presence felt—for good or ill. The useful species outnumber the harmful ones by thousands and are so valuable we could not live without them. Some of the remaining species can be agents of disease and death.

Since the mid-1800s, scientists have linked microorganisms to events of human importance. Before then, disease processes now attributed to microorganisms seemed to happen almost spontaneously and without apparent cause.

In Part 2 of this text, we focus on two domains of organisms—the *Bacteria* and *Archaea*—that make up a major part of the tree of life. The domain *Bacteria* has traditionally occupied an important niche in microbiology because scientists, until recently, probably knew more about them than any other organisms. Some species have been involved in the great plagues of history, and for centuries their effects have captured the imagination of scientists and writers. Many species can be studied easily in the laboratory, and their chemical activities have been charted and well documented. Also, many helpful ones play key roles in industrial processes. The *Archaea* share some similarities in structure and metabolism with the *Bacteria*, while other characteristics set them apart and place them in their own domain.

Small as they are, the prokaryotes are endowed with the ability to perform certain acts characteristic of all living things. They take in food, grow, excrete waste products, reproduce, and die. In addition, they are sensitive to external agents and stimuli, and respond in a characteristic manner.

Since almost all the prokaryotic human pathogens known are in the domain *Bacteria,* we will concentrate on these species. In Chapter 4, we survey their structural frameworks and, in Chapter 5, examine their growth patterns and nutritional requirements. Chapter 6 describes the metabolism of prokaryotic cells, including those chemical reactions that produce energy and use energy. Chapter 7 is devoted to the basics of prokaryotic genetics, while Chapter 8 explores the modern fields of genetic engineering and bacterial genomics. The discussions in Part 2 have broad significance not only in medicine and industry but also in our daily lives.

MICROBIOLOGY PATHWAYS

Biotechnology

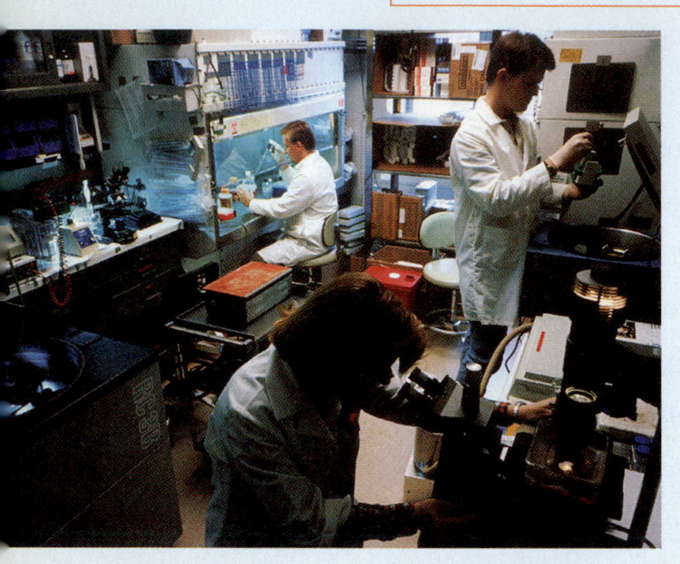

During the 1980s, the editors of *Time Magazine* referred to DNA technology as "the most awesome skill acquired by man since the splitting of the atom." Indeed, the work with DNA, begun in the 1950s and continuing today, has opened vistas previously unimagined. Scientists can now remove bits of DNA from organisms, snip and rearrange the genes, and insert them into different species, where the genes will express themselves. Practical results of these experiments have led to the mass production of hormones, blood-clotting factors, and other pharmaceutical products. They also have given us diagnostic methods based on DNA fingerprinting; advances in gene therapy; a revolution in agricultural research; barnyard animals producing human hemoglobin; and a colossal project that has mapped the entire human genome.

Industrial microbiology or microbial biotechnology (the terms "industrial microbiology" and "biotechnology" are often one and the same) applies scientific and engineering principles to the processing of materials by microorganisms and viruses, or plant and animal cells, to create useful products or processes. Because biotechnology essentially uses the basic ingredients of life to make new products, it is both a cutting-edge technology and an applied science.

If you would like to be part of what analysts predict will be one of the most important applied sciences of this century, then microbiology is the place to start. You would be well advised to take a course in biochemistry as well as one in genetics. Courses in physiology and cell biology are also helpful. Employers will be looking for individuals with good laboratory skills, so be sure to take as many lab courses as you can. And don't be afraid to become a "lab-rat" (the scientific equivalent of basketball's "gym-rat").

You may enter the biotechnology field with an associate's, bachelor's, master's, or doctoral degree. This is because there are so many levels at which individuals are hired. Most professional levels of employment require a college degree (BS) in biology, microbiology, or biotechnology with minors in one or more of the complementary sciences. Persons who have project responsibilities often have one or more advanced degrees (MS and/or PhD) in biology, microbiology, or some other allied field such as molecular biology, biochemistry, biotechnology, chemical engineering, or genetics.

An employer also will be looking for work experience, which you can obtain by assisting a senior scientist, doing an internship, or working summers in a biotech firm (usually for slave wages). The campus research lab is another good place to obtain work experience. It also might be a good idea to sharpen your writing skills, because you will be preparing numerous reports.

As Chapter 8 explains, the novel and imaginative research that established biotechnology was founded in microbiology, and it continues to call on microbiology for its continuing growth.

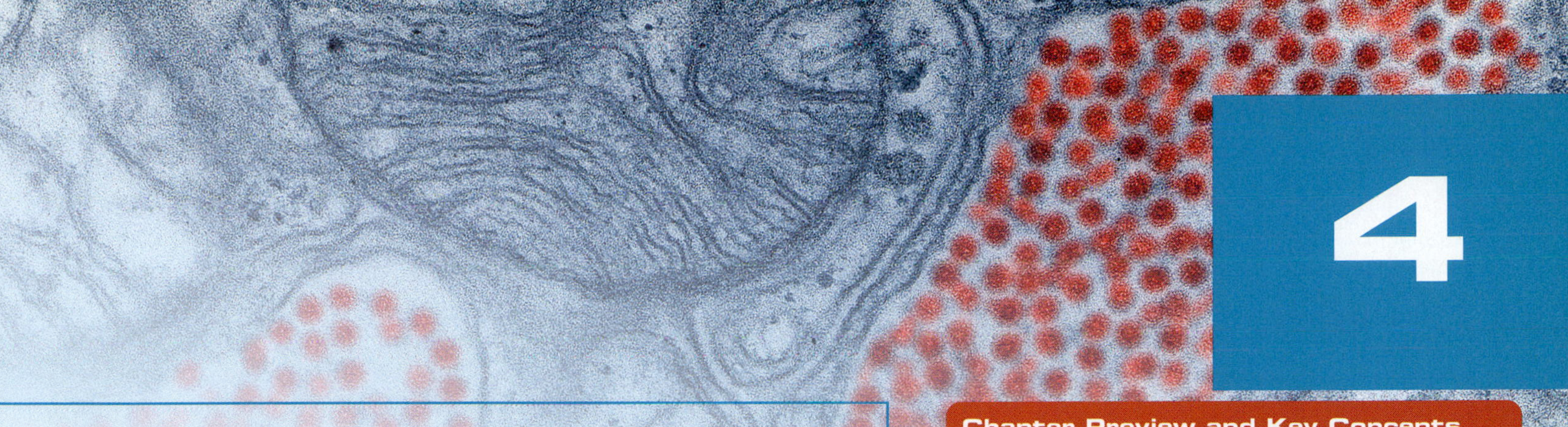

Prokaryotic Cell Structure and Function

Our planet has always been in the "Age of Bacteria," ever since the first fossils—bacteria of course—were entombed in rocks more than 3 billion years ago. On any possible, reasonable criterion, bacteria are—and always have been—the dominant forms of life on Earth.
—Paleontologist Stephen J. Gould (1941–2002)

Chapter Preview and Key Concepts

"Double, double toil and *trouble; Fire burn, and cauldron bubble*" is the refrain repeated several times by the chanting witches in Shakespeare's *Macbeth* (Act IV, Scene 1). This image of a hot, boiling cauldron actually describes the environment in which many prokaryotes happily grow! For example, many prokaryotic species can be isolated from the hot, acidic mouth of volcanic vents.

When the eminent evolutionary biologist and geologist Stephen J. Gould wrote the opening quote of this chapter, he as well as most microbiologists had no idea that embedded in these "bacteria" was another whole domain of organisms. Thanks to the pioneering studies of Carl Woese and his colleagues, it now is quite evident there are two distinctly different groups of prokaryotes—the *Bacteria* and the *Archaea*. Many of the organisms Woese and others studied are organisms that would live a happy life in a witch's cauldron because they can grow at high temperatures, produce methane gas, or survive in extremely acidic and hot environments—a real cauldron! Termed **extremophiles**, these *Archaea* have a unique genetic makeup and possess structures enabling them to tolerate extreme environmental conditions.

In fact, Gould's "first fossils" may have been archaeal species. Many microbiologists believe the ancestors of today's archaeal species might represent a type of organism that first inhabited planet Earth when it was a young hot place (MicroFocus 2.1). These unique characteristics

■ Histones: Proteins that bind to DNA and package the molecule in a chromosome.

led Woese to propose these organisms be lumped together and called the Archaebacteria (*archae* = "ancient").

Since then, the domain name has been changed to *Archaea* because (1) not all members are extremophiles or related to these possible ancient ancestors and (2) they are not *Bacteria*—they are *Archaea*. Some might also debate using the term prokaryotes when referring to both domains, but we will continue the practice and refer to both domain *Bacteria* and *Archaea* as the prokaryotes.

As more microbes have had their complete genomes sequenced, it now is clear that there are unique as well as shared characteristics between species in the domains *Bacteria*, *Archaea*, and *Eukarya* (TABLE 4.1). Overall, these characteristics strongly support the three-domain tree of life. Some characteristics, such as the presence of a peptidoglycan cell wall may be almost universal within a domain, while other characteristics such as chlorophyll-based photosynthesis are specific only to certain organisms within a domain. Yet other characteristics, such as chromosome form, are similar between the *Bacteria* and *Archaea* while the presence of histones is similar between the *Archaea* and *Eukarya*.

In this chapter, we examine briefly some of the organisms in the two prokaryotic domains *Bacteria* and *Archaea*. However, because almost all known prokaryotic pathogens of humans are in the domain *Bacteria*, we emphasize structure within this domain. As we see in this chapter, a study of the structural features of bacterial cells provides a window to their activities and illustrates how the *Bacteria* relate to other living organisms.

As we examine prokaryotic structure, we can assess the dogmatic statement that prokaryotic cells are characterized by a lack of a cell nucleus and internal membrane-bound organelles. Before you finish this chapter, you will be equipped to revise this view.

4.1 Prokaryotic Diversity

In this section, we discuss prokaryotic diversity using the current classification scheme, which is based in large part on nucleotide sequence data. There are some 7,000 known species of prokaryotes and a suspected 10 million. In this section, we will highlight a few phyla and groups using the phylogentic tree in FIGURE 4.1.

The Domain *Bacteria* Contains Some of the Best Studied Prokaryotes

KEY CONCEPT

- The *Bacteria* are classified into several major phyla.

There are about 18 phyla of *Bacteria* identified from culturing or nucleotide sequencing. It should come as no shock to you by now to read

TABLE 4.1 Some Major Differences between *Bacteria*, *Archaea*, and *Eukarya*

Characteristic	*Bacteria*	*Archaea*	*Eukarya*
Cell nucleus	No	No	Yes
Chromosome form	Single, circular	Single, circular	Multiple, linear
Histone proteins present	No	Yes	Yes
Peptidoglycan cell wall	Yes	No	No
Membrane lipids	Ester-linked	Ether-linked	Ester-linked
Ribosome sedimentation value	70S	70S	80S
Ribosome sensitivity to diphtheria toxin	No	Yes	Yes
First amino acid in a protein	Formylmethionine	Methionine	Methionine
Chlorophyll-based photosynthesis	Yes (cyanobacteria)	No	Yes (algae)
Growth above 80°C	Yes	Yes	No
Growth above 100°C	No	Yes	No

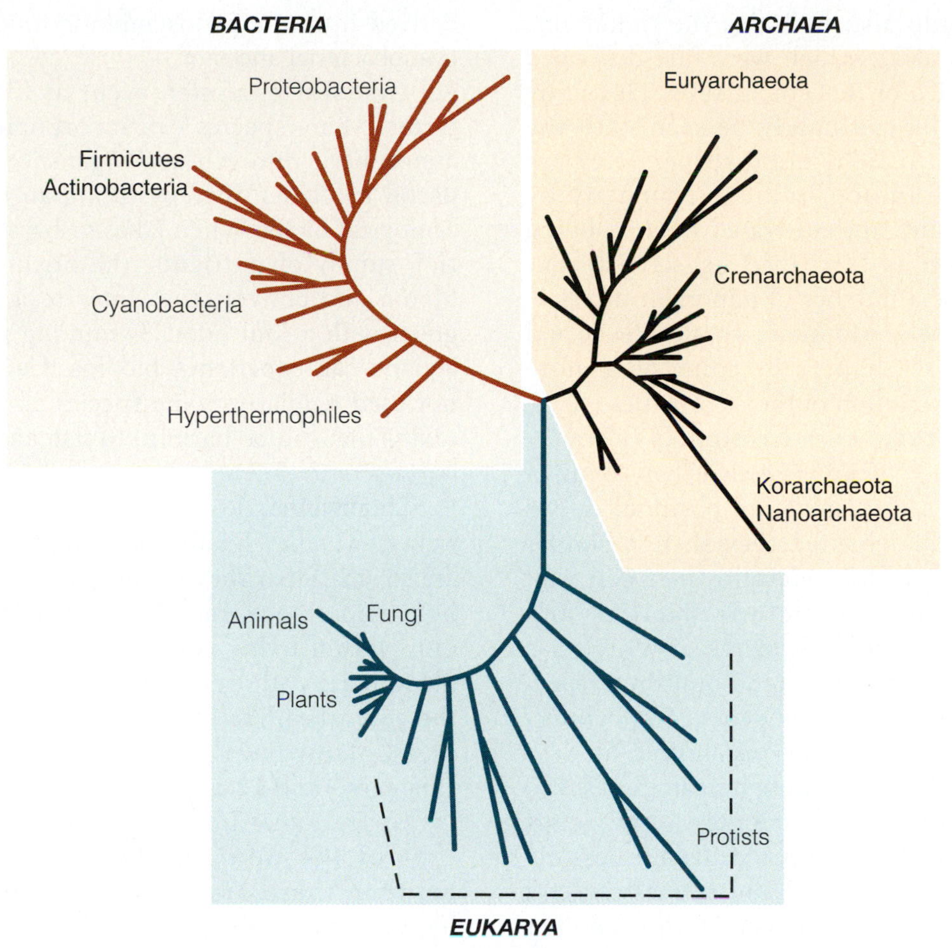

FIGURE 4.1 **The Phylogenetic Tree of Life.** The tree shows the evolutionary relationships between groups of organisms based on ribosomal RNA analysis.

Q: Which branches include the prokaryotes and eukaryotes?

that the vast majority of these phyla play a positive role in nature. Although not unique to just the bacterial phyla, they digest sewage into simple chemicals; they extract nitrogen from the air and make it available to plants for protein production; they break down the remains of all that die and recycle the carbon and other elements; and they produce foods for human consumption and products for industrial technology.

Of course, we know from Chapter 1 and personal experience that some bacterial organisms are harmful—most all known human pathogens are members of the domain *Bacteria*. Certain species multiply within the human body, where they disrupt tissues or produce toxins that result in disease.

The *Bacteria* have adapted to the diverse environments on Earth, inhabiting the air, soil, and water, and they exist in enormous numbers on the surfaces of virtually all plants and animals. They can be isolated from Arctic ice, thermal hot springs, the fringes of space, and the tissues of animals. Bacterial species have so completely colonized every part of the Earth that their mass is estimated to outweigh the mass of all plants and animals combined. Let's look briefly at some of the major phyla and other groups.

Proteobacteria. The proteobacteria (*proteo* = "first") contain the largest and most diverse group of species. The phylum includes many familiar gram-negative genera, such as *Escherichia*, and some of the most recognized human pathogens, including species of *Shigella*, *Salmonella*, *Neisseria* (responsible for gonorrhea), *Yersinia* (responsible for plague), and *Vibrio* (responsible for cholera). It is likely that the mitochondria of the *Eukarya* were derived through endosymbiosis from an ancestor of the proteobacteria (see MicroInquiry 3).

The group also includes the rickettsiae (sing., rickettsia), which were first described by Howard Taylor Ricketts in 1909. These tiny bacterial cells can barely be seen with the most powerful light microscope. They are transmitted among humans primarily by arthropods, and are cultivated only in living tissues such as fertilized eggs. Different species cause a number of important diseases, including Rocky Mountain spotted fever and typhus fever. Chapter 11 contains a more thorough description of their properties.

- **Arthropods:** Animals having jointed appendages and segmented body (e.g., ticks, lice, fleas, mosquitoes).
- **Algaecide:** A chemical that kills algae.

Firmicutes. The firmicutes (*firm* = "strong"; *cuti* = "skin") is a clade consisting of many species that are gram-positive. As we will see in this chapter, they share a similar thick "skin," which refers to their cell wall structure. Genera include *Bacillus* and *Clostridium*, specific species of which are responsible for anthrax and botulism, respectively. Species within the genera *Staphylococcus* and *Streptococcus* are responsible for several mild to life-threatening human diseases.

Also within the Firmicutes is the genus *Mycoplasma*, which lacks a cell wall. Possibly the smallest bacterial cell, one species causes a form of pneumonia (Chapter 9) while another mycoplasmal illness represents a sexually-transmitted disease (Chapter 12).

Actinobacteria. Within this phylum are the actinomycetes. These bacterial organisms form a system of branched filaments that somewhat resemble the growth form of fungi. The genus *Streptomyces* is the source for important antibiotics. Another medically-important genus is *Mycobacterium*, one species of which is responsible for tuberculosis.

Cyanobacteria. In Chapter 3 we discussed the cyanobacteria. Once known as blue-green algae because of their pigmentation, pigments also may be black, yellow, green, or red. The periodic redness of the Red Sea, for example, is due to blooms of cyanobacteria whose members contain large amounts of red pigment. Cyanobacteria are unique among prokaryotes because they carry out photosynthesis similar to unicellular algae (Chapter 6) using the light-trapping pigment chlorophyll. Their evolution on earth was responsible for the "oxygen revolution" that transformed life on the young planet. In addition, chloroplasts probably are derived from the endosymbiotic union with a cyanobacterial ancestor.

- **Blooms:** Sudden increases in the numbers of cells of an organism in an environment.

Cyanobacteria often occur as filamentous forms. Many species can incorporate ("fix") atmospheric nitrogen into organic compounds useful to plants, thereby filling an important ecological niche. When lakes or bays contain a rich supply of nutrients, the organisms may bloom and convert the water to a pea-soup green with a foul odor. Swimming pools and aquaria can experience blooms if algaecide is not used regularly. Some species also produce toxins that can be harmful to fish and humans (MicroFocus 4.1).

Chlamydiae. Roughly half the size of the rickettsiae, the chlamydiae (sing., chlamydia) are so small that they cannot be seen with the light microscope and they can be cultivated only within living cells. Most species in the phylum are pathogens and one species causes the gonorrhea-like disease known as chlamydia. Chlamydial diseases are described in Chapters 9 and 12.

Several other lineages branch off near the root of the domain. The common link between these organisms is that they are **hyperthermophiles**; they grow at high temperatures. Examples include *Aquifex* and *Thermotoga*, which typically are found in earthly cauldrons such as hot springs.

CONCEPT AND REASONING CHECKS

4.1 Construct a simple concept map for the domain *Bacteria* using the boldfaced terms in this section.

The Domain *Archaea* Contains Many Extremophiles

KEY CONCEPT

- The *Archaea* are currently classified into two major phyla.

Classification within the domain *Archaea* has been more difficult than within the domain *Bacteria,* in large part because they have not been studied as long as their bacterial counterparts.

Many genera are extremophiles, meaning they grow best at environmental extremes, such as very high temperatures, high salt concentrations, or extremes of pH. Many live deep in the oceans although there are archaeal genera that thrive under more modest conditions. The archaeal genera can be placed into one of two major phyla.

MICROFOCUS 4.1: Public Health/Environmental Microbiology

Toxic Cyanobacteria

In the past chapter, we discovered the important role cyanobacteria play as part of the phytoplankton in marine environments. In addition, some cyanobacteria produce toxins dangerous to fish and humans. However, in April 2005, a group of international scientists reported most species of cyanobacteria they examined were capable of producing another toxin that can cause neurological disease in humans.

The toxin in question is a non-protein amino acid called BMAA. BMAA is also found in the seeds of cycad plants, the BMAA produced by a symbiotic cyanobacterium found in the cycad roots. The Chamorro people of Guam eat the cycad seeds and, interestingly, biopsied brain tissue of those individuals suffering neurological diseases similar to Parkinson's was associated with BMAA. BMAA also has been found in some Alzheimer's patients.

Thick mats of cyanobacteria in the Baltic Sea often arise in nutrient-rich or polluted waters.

When the international group tested marine and freshwater cyanobacteria (free or in symbiotic association with plants) for BMAA, 90 percent of the 41 strains examined were positive. Often the toxin was associated with "water blooms," many of which can cover thousands of square kilometers. The scientists emphasize that if there is a correlation between BMAA and neurological diseases, cyanobacteria blooms and the amounts of toxin released could be of great health concern.

Euryarchaeota. The Euryarchaeota contain organisms with varying physiologies. Some groups, such as the **methanogens** (*methano* = "methane"; *gen* = "produce") are killed by oxygen gas and therefore are found in environments devoid of oxygen gas. The production of methane (natural) gas is important in their energy metabolism. In fact, these archaeal species release more than 2 billion tons of methane gas into the atmosphere every year. About a third comes from the archaeal cells living in the stomach (rumen) of cows (see Chapter 2).

Another group is the **extreme halophiles** (*halo* = "salt"; *phil* = "loving"). They are distinct from the methanogens in that they require oxygen gas for energy metabolism and need high concentrations of salt (NaCl) to grow and reproduce. The fact that they often contain pink pigments makes their identification easy (FIGURE 4.2A). In addition, some extreme halophiles have been found in lakes where the pH is greater than 11.

A third group is perhaps the most interesting. The **thermoacidophiles** grow optimally at high temperatures but extremely low pH.

Crenarchaeota. The second phylum, the Crenarchaeota, tend to grow in hot, acidic environments. Hot sulfur springs are one environment where these archaeal species thrive. The temperature is around 75°C but the springs are extremely acidic (pH of 2–3). Volcanic vents are another place where these organisms can survive quite happily (FIGURE 4.2B). The other two tentative phyla, the Korarchaeota and Nanoarchaeota, also consist of hyperthermophiles.

Realize in this section we have discussed archaeal species that tend to inhabit relatively extreme environments. However, many others live in more normal environments with regard to temperature, salt, and pH.

CONCEPT AND REASONING CHECKS

4.2 Compared to the more moderate environments in which some archaea species grow, why have others adapted to such extreme environments?

(A)

(B)

FIGURE 4.2 **Extremophiles.** Many archaeal species live in extreme environments. (**A**) The extreme halophiles often inhabit commercial seawater evaporation ponds such as these in the east bay of San Francisco. The color is due to the presence of pigments in the microbes. (**B**) Archaeal species (mats) also can be found in volcanic vents, which represent hot, very acidic environments.

Q: Explain how these archaeal species can survive when all other microorganisms would be killed by these extreme conditions.

4.2 The Shapes and Arrangements of Prokaryotic Cells

Prokaryotes vary greatly in size, shape, and arrangement of cells. As described in Chapter 3 for the simple and Gram stains, these three characteristics are studied best by viewing stained cells with the light microscope. Such studies show that most, including the clinically significant ones, appear in one of three different shapes: the rod, the sphere, or the spiral.

Prokaryotes Vary in Cell Shape and Cell Arrangement

KEY CONCEPT

- Most prokaryotic cells have a rod, spherical, or spiral shape and are organized into a specific cellular arrangement.

A prokaryotic cell with a rod shape is called a **bacillus** (pl., bacilli). In various species of rod-shaped bacteria, the cylindrical cell may be as long as 20 µm or as short as 0.5 µm. Certain bacilli are slender, such as those of *Salmonella typhi* that cause typhoid fever; others, such as the agent of anthrax *(Bacillus anthracis)*, are rectangular with squared ends; still others, such as the diphtheria bacilli *(Corynebacterium diphtheriae)*, are club shaped. Most rods occur singly, but some are arranged into a long chain called **streptobacillus** (*strepto* = "chains") (FIGURE 4.3A). Realize there are two ways to use the word "bacillus": to denote a rod-shaped bacterial cell, and as a genus name.

A spherically shaped bacterial cell is known as a **coccus** (pl., cocci; *kokkos* = "berry"). Cocci tend to be quite small, being only 0.5 µm to 1.0 µm in diameter. Although they are usually round, they also may be oval, elongated, or indented on one side. Many bacterial species that are cocci stay together after division and take on cellular arrangements characteristic of the species (FIGURE 4.3B). Cocci remaining in a pair after reproducing represent a **diplococcus**. The organism that causes gonorrhea, *Neisseria gonorrhoeae,* and one type of bacterial meningitis *(N. meningitidis)* are examples. Cocci that remain in a chain are called **streptococcus**. Certain species of streptococci are involved in strep throat *(Streptococcus pyogenes)* and tooth decay *(S. mutans)*, while other species are harmless enough to be used for producing dairy prod-

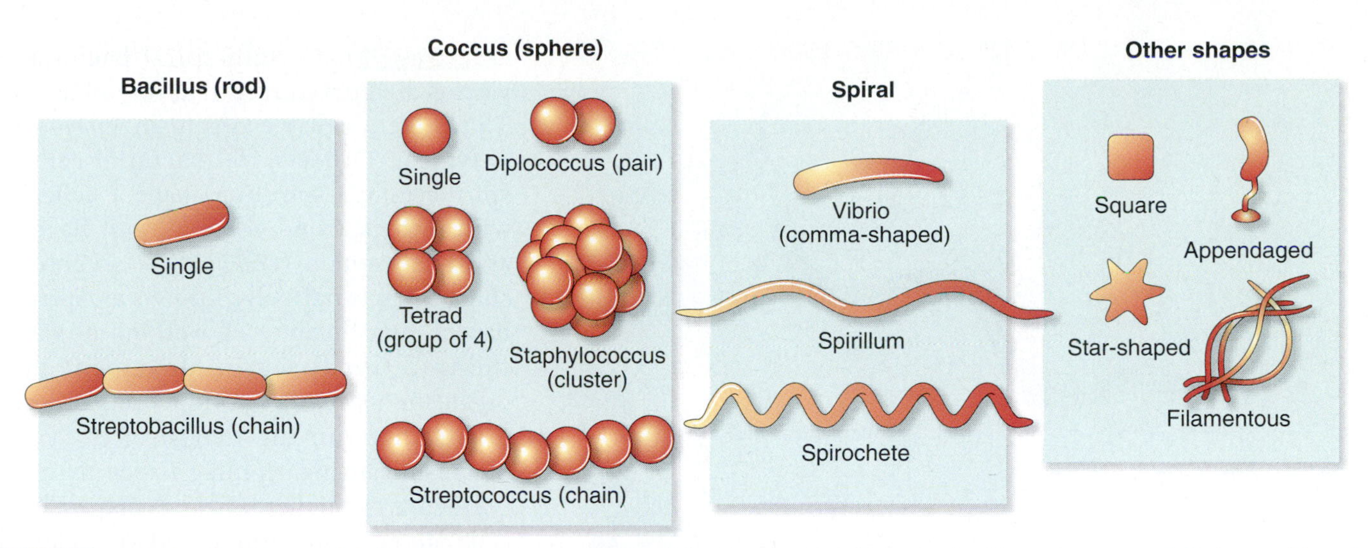

FIGURE 4.3 **Variation in Shape and Cell Arrangements.** Most bacterial and archaeal cells have a bacillus, coccus, or spiral shape, although other shapes can exist in specific species. Most spiral-shaped cells are not organized into a specific arrangement.

Q: Propose why the spiral-shaped cells are not organized into a specific arrangement of cells.

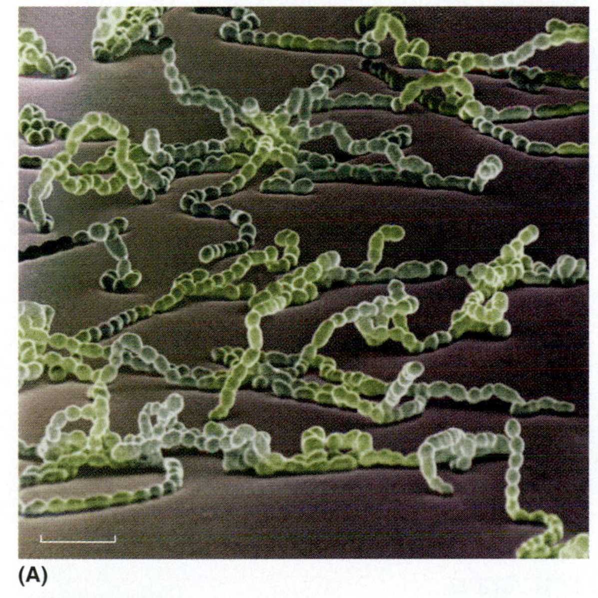
(A)

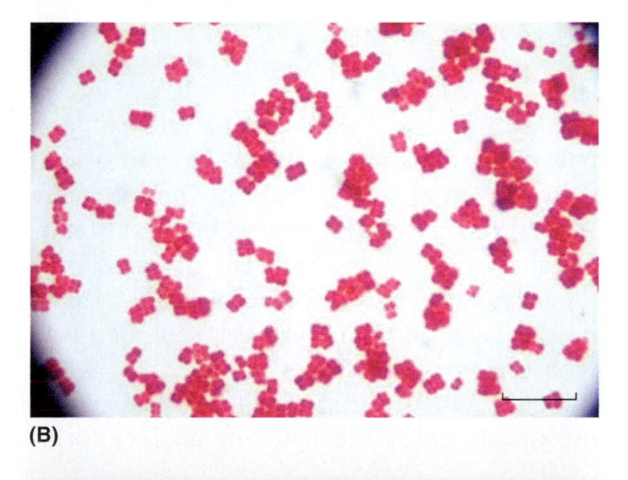
(B)

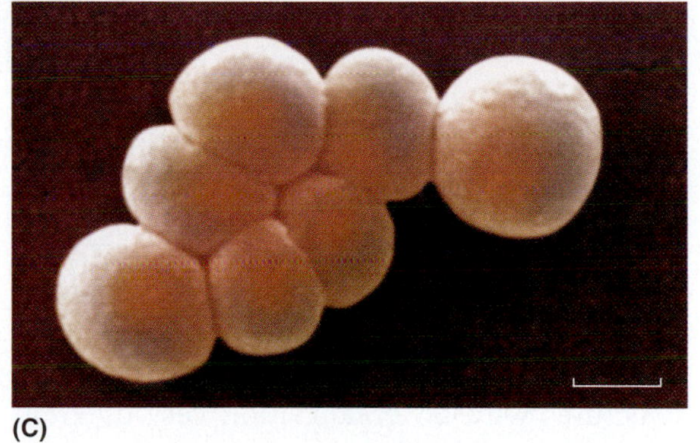
(C)

FIGURE 4.4 **Arrangements of Bacterial Cocci.** (**A**) This false-color scanning electron micrograph shows a streptococcus chain with its characteristic string-of-beads appearance. (Bar = 5 μm.) (**B**) Clusters of sarcinae in cube-like packets of eight or more. (Bar = 10 μm.) (**C**) Several cocci form a cluster typical of *Staphylococcus aureus* in this false-color scanning electron micrograph. (Bar = 0.5 μm.)

Q: In what plane must the cells divide to produce these arrangements?

ucts such as yogurt (*S. lactis*) (FIGURE 4.4A). Another arrangement of cocci is the **tetrad**, consisting of four cocci forming a square. A cube-like packet of eight cocci is called a **sarcina** (*sarcina* = "bundle"). *Micrococcus luteus*, a common inhabitant of the skin, is one example (FIGURE 4.4B). Other cocci may divide randomly and form an irregular grape-like cluster of cells called a **staphylococcus** (*staphylo* = "cluster"). A well-known example,

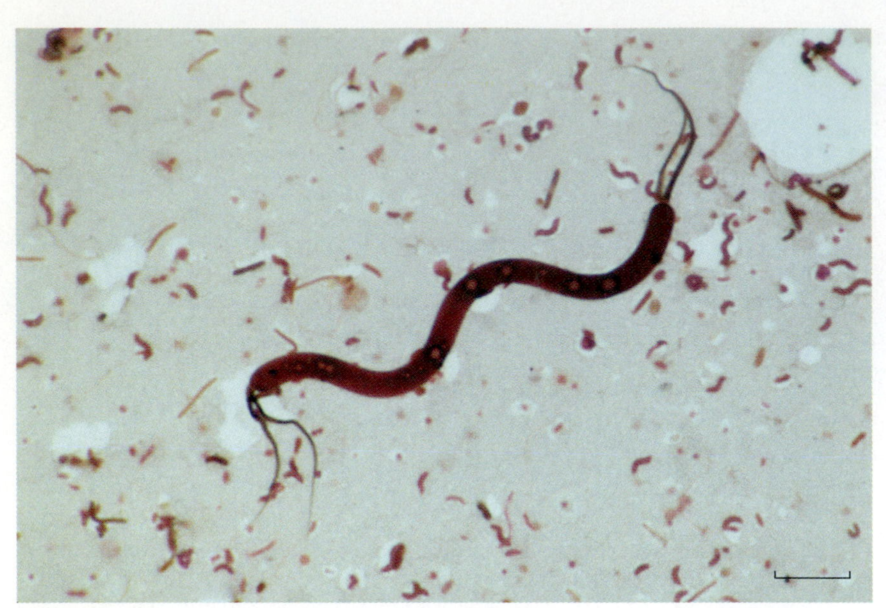

FIGURE 4.5 **Light Micrograph of a Spiral-Shaped Bacterial Cell.** The spiral shape is seen clearly in this cell, and flagella are visible at the poles of the cell. (Bar = 1 µm.)

Q: To which form of spiral bacteria does this species belong?

Staphylococcus aureus, is often a cause of food poisoning, toxic shock syndrome, and several skin infections (FIGURE 4.4C). The latter are known in the modern vernacular as "staph" infections. Notice again that the words streptococcus and staphylococcus can be used to describe cell shape and arrangement, or a bacterial genus.

The third major shape of bacterial cells is the **spiral**, which can take one of three forms (FIGURE 4.3C). Certain spiral bacteria called **vibrios** are curved rods that resemble commas. The cholera-causing organism *Vibrio cholerae* is typical. Other spiral bacterial cells called **spirilla** (sing., spirillum) have a helical shape with a thick, rigid cell wall and flagella that assist movement (FIGURE 4.5). Those spiral-shaped bacterial cells known as **spirochetes** have a thin, flexible cell wall but no flagella in the traditional sense. Movement in these organisms occurs by contractions of axial filaments that run the length of the cell. The organism causing syphilis, *Treponema pallidum,* typifies a spirochete. Spiral-shaped bacterial cells can be from 1 µm to 100 µm in length.

In addition to the bacillus, coccus, and spiral shapes, other variations exist (FIGURE 4.3D). The genus *Caulobacter* has appendaged bacterial cells; members of the genus *Nocardia* consist of branching filaments; and some archaeal species have square and star shapes.

Stained cells or live cells viewed with the light microscope do not show striking internal structure. When the electron microscope is used, however, resolution increases down to 2 nm, allowing scientists to observe a level of bacterial detail not possible with the light microscope.

CONCEPT AND REASONING CHECKS

4.3 Propose why different species of *Bacteria* have different shapes.

4.3 An Overview to Prokaryotic Cell Structure

In the last chapter we discovered that bacterial and archaeal cells appear to have little visible structure when observed with a light microscope. This, along with their small size, gave the impression they are homogeneous, static structures with an organization very different from eukaryotic cells.

However, the point was made that bacterial and archaeal species still have all the complex processes typical of eukaryotic cells. It is simply a matter that, in most cases, the structure and sometimes pattern to accomplish these processes is different from the membranous organelles of the eukaryotic cells.

Recent advances in understanding prokaryotic cell biology indicate these organisms exhibit a highly ordered intracellular organization. This organization is centered on three specific processes that need to be carried out (FIGURE 4.6). These are:

- **Sensing and responding to the exterior environment.** Because most prokaryotic cells are surrounded by a cell wall, some pattern of "external structures" is necessary to sense their environment and respond to it or other cells.
- **Compartmentation of metabolism.** As described in Chapter 3, cell metabolism

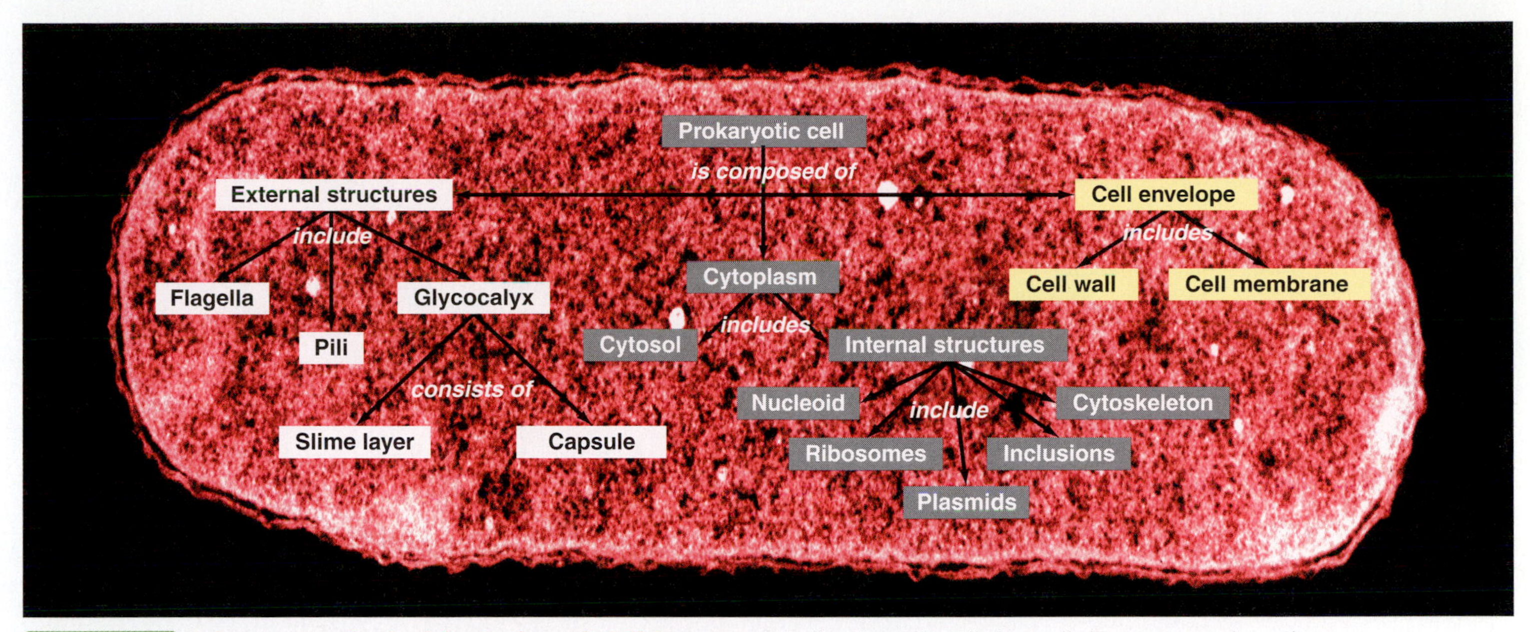

FIGURE 4.6 **A Concept Map for Studying Prokaryotic Cell Structure.** Not all prokaryotic cells have all the structures shown here.
Q: Why can't we see in the TEM image of a bacterial cell all the structures outlined in the concept map?

must be segregated from the exterior environment and yet be able to transport materials to and from that environment. In addition, protection from osmotic pressure due to water movement into cells must be in place. The "cell envelope" fulfills those roles.

- **Growth and reproduction.** Survival of prokaryotic cells demands a complex metabolism that occurs within the aqueous "cytoplasm." These processes and reproduction exist as internal structures or subcompartments localized to specific areas within the cytoplasm.

Our understanding of prokaryotic cell biology is still an emerging field of study. However, it has become very clear that there is more to the prokaryotic cells than previously thought—smallness does not equate with simplicity.

Although this chapter is primarily looking at prokaryotic cells at the cellular level, it also is important to realize that prokaryotic cells, like their eukaryotic counterparts, are organized on the molecular level as well. Specific cellular proteins can be localized to specific regions of the cell. For example, as the name suggests, *Streptococcus pyogenes* has spherical cells. Yet many of the proteins that confer its pathogenic nature in causing diseases like strep throat are secreted from a specific area of the surface. *Yersinia pestis,* which is the agent responsible for plague, contains a specialized secretion apparatus through which proteins are released. This apparatus only exists on the bacterial surface that is in contact with the target human cells.

So, the advances being made in prokaryotic cell biology and cell structure are not only important in their own right in understanding the prokaryotic cell, these studies also may have important significance to clinical microbiology and the fight against infectious disease. As more is discovered about prokaryotic cells and how they truly differ from eukaryotic cells, the better equipped we will be to develop new antimicrobial agents that will target the subcellular organization of pathogens. In an era when we have fewer effective antibiotics to fight infections, the application of the understanding of prokaryotic structure and function may be very important.

On the following pages, we examine some of the common prokaryotic structures found in an idealized bacterial cell, as no single species contains all the structures (FIGURE 4.7). Our journey starts by examining the structures on or extending from the surface of the cell. Then, we examine the cell envelope and spend some time discussing the cell membrane. Our journey then plunges into the cell cytoplasm. All

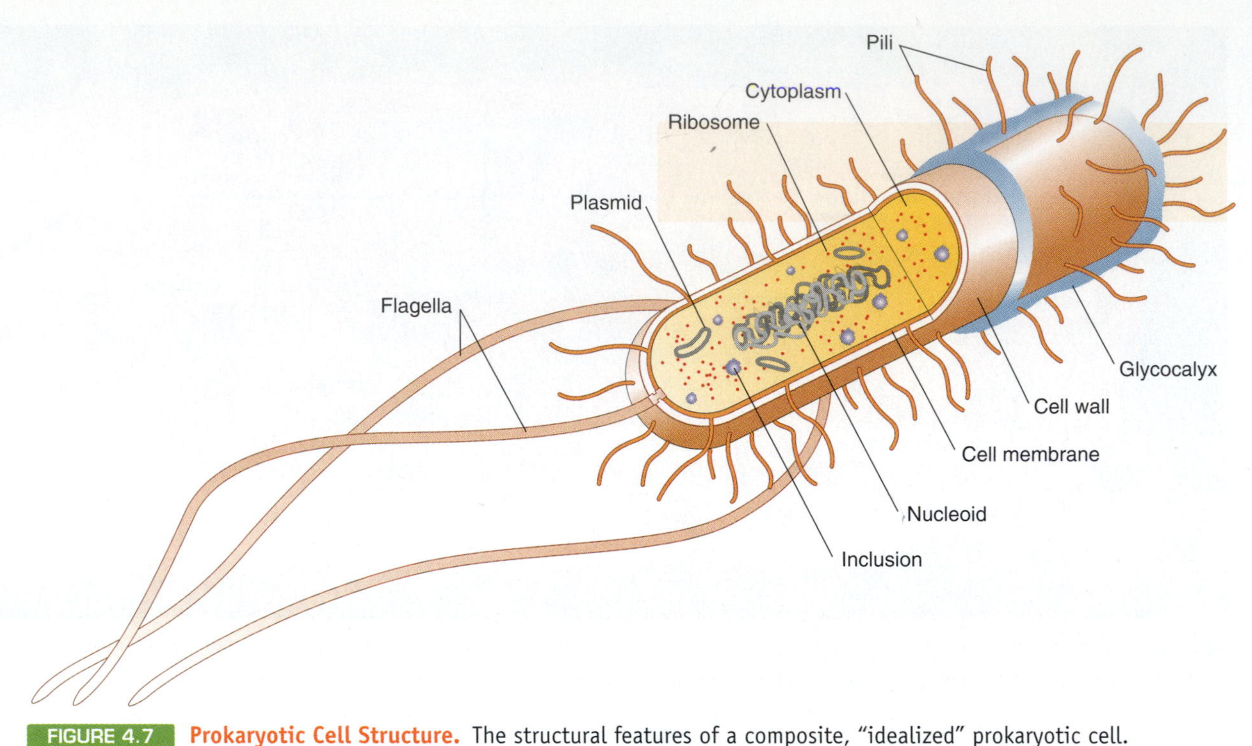

FIGURE 4.7 **Prokaryotic Cell Structure.** The structural features of a composite, "idealized" prokaryotic cell.
Q: Which structures represent (A) external structures, (B) the cell envelope, and (C) cytoplasmic structures?

cells must control and coordinate many metabolic process that need to be separated from one another. We will discover the cytoplasmic subcellular compartmentation that provides this function.

CONCEPT AND REASONING CHECKS

4.4 What is gained by prokaryotic cells having three general sets of structures—external, envelope, and cytoplasmic?

4.4 External Prokaryotic Cell Structures

Prokaryotic cells need to respond to and monitor their external environment. This is made difficult by having a cell wall that sort of "blindfolds" the cell. Many prokaryotes have solved this sensing problem by possessing structures that extend from the cell surface into the environment.

Pili Are Protein Fibers Extending from the Surface of Many Prokaryotes

KEY CONCEPT

- Pili allow cells to attach to surfaces or other cells.

Numerous short, thin, proteinaceous fibers, called **pili** (sing., pilus; *pilus* = "hair"), protrude from the surface of most gram-negative bacteria (FIGURE 4.8). The fibers, composed of protein, act as scaffolding onto which specific adhesive molecules, called **adhesins**, are attached. Therefore, the function of pili is to attach cells to surfaces to form biofilms or, in the case of human pathogens, to form microcolonies on human cell and tissue surfaces. This requires that the pili on different bacterial species have specialized adhesins to "sense" the appropriate cell. For example, the adhesins on *Neisseria gonorrhoeae* cells specifically anchor the cells to surfaces of the urogenital tract whereas the adhesins on *Bordetella pertussis* (causative agent of whooping cough) adhere to cells of the upper respiratory tract. In this way, the short fibers enhance attachment to host cells, facilitating tissue colonization and possibly disease development. Without the chemical mooring line lashing the bacterial cells to host cells, it is less likely the cells could infect host tissue (MicroFocus 4.2).

MICROFOCUS 4.2: Public Health

Diarrhea Doozies

They gathered at the clinical research center at Stanford University to do their part for the advancement of science (and earn a few dollars as well). They were the "sensational sixty"—sixty young men and women who would spend three days and nights and earn $300 to help determine whether hair-like structures called pili have a significant place in disease.

A number of nurses and doctors were on hand to help them through their ordeal. The students would drink a fruit-flavored cocktail containing a special diarrhea-causing strain of *Escherichia coli*. Thirty cocktails had *E. coli* with normal pili, while thirty had *E. coli* with pili mutated beyond repair. The hypothesis was that the bacterial cells with normal pili would latch onto intestinal tissue and cause diarrhea, while those with mutated pili would be unable to attach and would be swept away by the rush of intestinal movements and not cause intestinal distress. At least that's what the sensational sixty would either verify or prove false.

On that fateful day in 1997, the experiment began. Neither the students nor the health professionals knew who was drinking the diarrhea cocktail and who was getting the "free pass"; it was a so-called double-blind experiment. Then came the waiting. Some experienced no symptoms, but others felt the bacterial onslaught and clutched at their last remaining vestiges of dignity. For some it was three days of hell, with nausea, abdominal cramps, and numerous bathroom trips; for others, luck was on their side, and investing in a lottery ticket seemed like a good idea.

When it was all over, the numbers appeared to bear out the hypothesis: The great majority of volunteers who drank the mutated bacterial cells experienced no diarrhea, while the great majority of those who drank the normal bacterial cells had attacks of diarrhea, in some cases real doozies.

All appeared to profit from the experience: The scientists had some real-life evidence that pili contribute to infection; the students made their sacrifice to science and pocketed $300 each; and the local supermarket had a surge of profits from unexpected sales of toilet paper, Pepto-Bismol, and Imodium.

Besides these attachment pili, some bacterial species produce **conjugation pili** that establish contact between appropriate cells, facilitating the transfer of genetic material from donor to recipient through a process called conjugation (Chapter 8). Conjugation pili are longer than attachment pili and only one or a few are produced on a cell.

Until recently, attachment pili were thought to be specific to only certain species of gram-negative bacteria. However, research now indicates that pili are present on some gram-positive bacteria, including the pathogenic species *Corynebacterium diphtheriae* and group B *Streptococcus*. However, very little is known about their function, although they probably play a very similar role to the pili on gram-negative cells.

It should be noted that microbiologists use the term *pili* interchangeably with *fimbriae* (sing., fimbria; *fimbria* = "fibers").

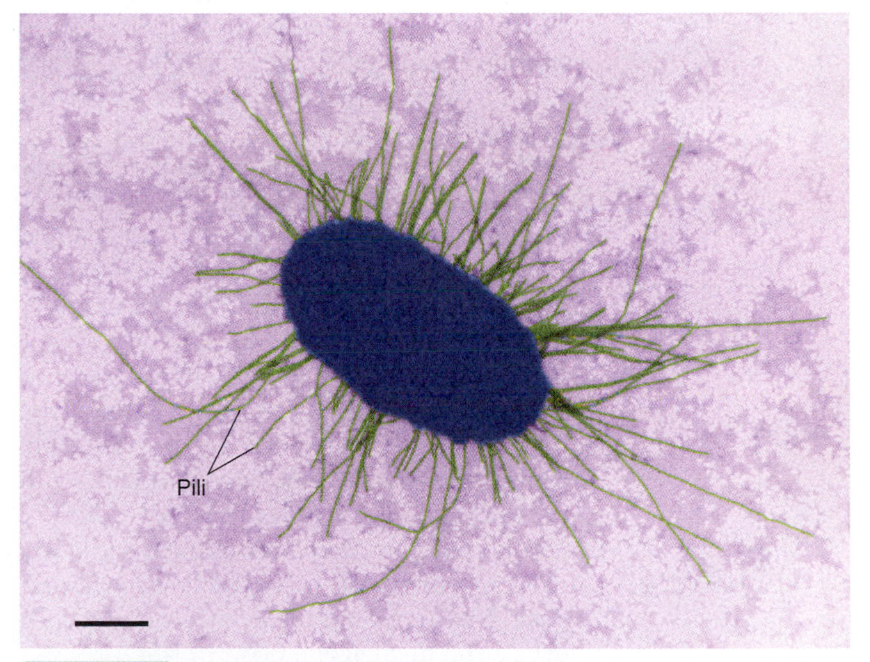

FIGURE 4.8 **Bacterial Pili.** False-color transmission electron micrograph of an *E. coli* cell (blue) with many pili (green). (Bar = 0.5 μm.)

Q: What function do pili play?

CONCEPT AND REASONING CHECKS

4.5 What would happen if pili lacked adhesins?

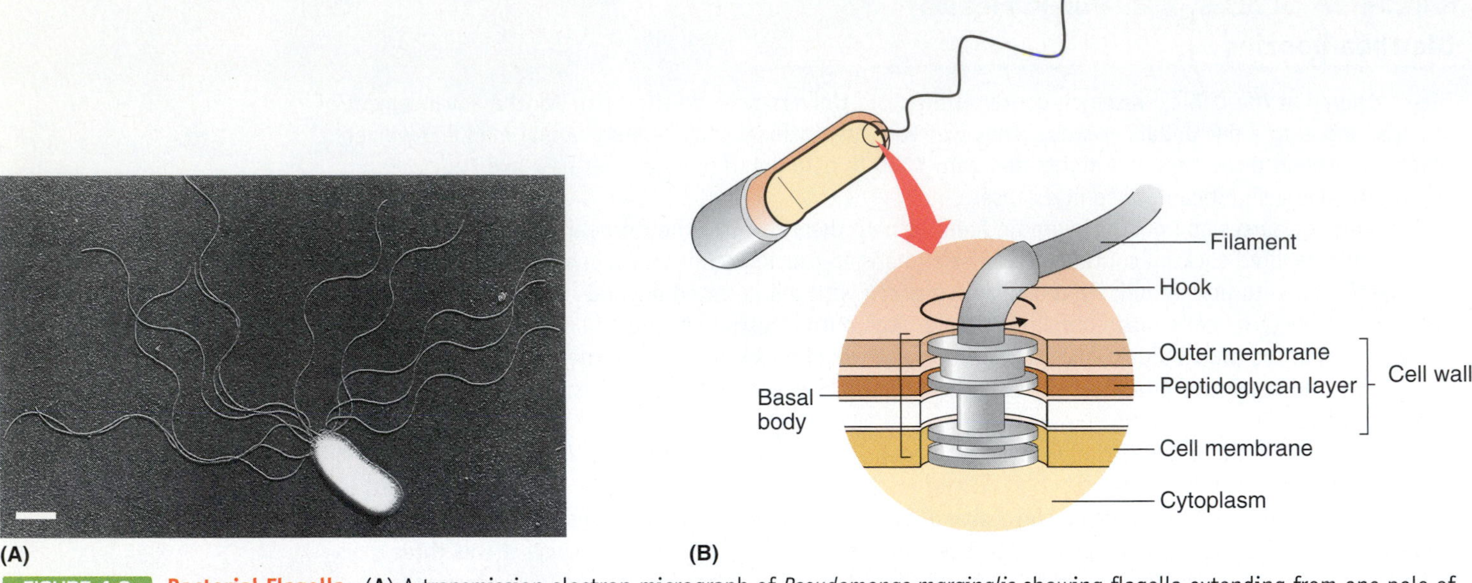

FIGURE 4.9 **Bacterial Flagella.** (**A**) A transmission electron micrograph of *Pseudomonas marginalis* showing flagella extending from one pole of the cell. (Bar = 1 μm.) Note that the flagella are many times the width of the cell. (**B**) The flagellum on a gram-negative bacterial cell is attached to the cell wall and membrane by a pair of protein rings in the basal body.

Q: Why is the flagellum referred to as a "nanomachine?"

Prokaryotic Flagella Are Long Appendages Extending from the Cell Surface

KEY CONCEPT

- Flagella provide motility for many prokaryotes.

Numerous species of *Bacteria* and *Archaea* are capable of some type of locomotion. This can be in the form of flagellar motility or gliding motility.

Flagellar motility. Many prokaryotes are motile by using a remarkable "nanomachine" called the **flagellum** (pl. flagella). Depending on the species, one or more flagella may be attached to one or both ends of the cell, or at positions distributed over the cell surface.

Flagella range in length from 10 μm to 20 μm and are many times longer than the diameter of the cell. Because they are only about 20 nm thick, they cannot be seen with the light microscope unless stained. However, their existence can be inferred by using dark-field microscopy to watch the live cells dart about. The flagella can be seen clearly with the electron microscope (FIGURE 4.9A).

In the domain *Bacteria*, each flagellum is composed of a helical filament, hook, and basal body (FIGURE 4.9B). The hollow filament is composed of long, rigid strands of protein while the hook attaches the filament to a basal body anchored in the cell membrane and cell wall.

The basal body is an assembly of more than 20 different proteins that form a central rod and set of enclosing rings. Gram-positive bacteria have one ring embedded in the cell membrane and one in the cell wall, while gram-negative bacteria have a pair of rings embedded in the cell membrane and another pair in the cell wall.

In the domain *Archaea*, flagellar protein composition and structure differs from that of the *Bacteria*; motility appears similar though.

The basal body represents a powerful biological motor or rotary engine that generates a propeller-type rotation of the flagellum. The energy for rotation comes from by the diffusion of protons (hydrogen ions; H^+) into the cell through proteins associated with the basal body. This energy is sufficient to produce up to 1,500 rpm by the filament, driving the cell forward.

What advantage is gained by cells having flagella? In nature, there are many chemical nutrients that cells need to survive. Cells will be attracted to such **attractants** by using their flagella to move up the concentration gradient; that is, toward the attractant. The process is called **chemotaxis**.

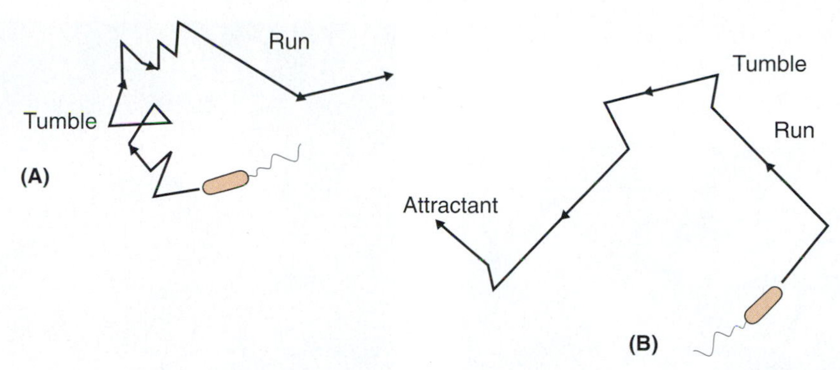

FIGURE 4.10 **Chemotaxis.** Chemotaxis represents a behavioral response to chemicals. (**A**) Rotation of the flagellum counterclockwise causes the bacterial cell to "run," while rotation of the flagellum clockwise causes the bacterial cell to "tumble," as shown. (**B**) During chemotaxis to an attractant, such as sugar, flagellum behavior leads to longer runs and fewer tumbles, which will result in biased movement toward the attractant.

Q: Predict the behavior of a bacterial cell if it sensed a repellant; that is a potential harmful or lethal chemical.

Being so small, prokaryotes sense their chemical surroundings using a temporal sensing system. In the absence of a gradient, the flagellar motor causes the flagella to rotate counterclockwise and the cell moves straight ahead in short bursts called "runs" (FIGURE 4.10A). These runs can last a few seconds and the cells can move up to 10 body lengths per second (the fastest human can run about 5–6 body lengths per second). When the flagellar motor changes direction (rotates clockwise), the cell "tumbles" randomly for a second. Then, the motor again reverses direction and another run occurs in a new direction.

■ Temporal sensing: One that compares the chemical environment and concentration from one second to the next.

If an attractant gradient is present, cell behavior changes; cells moving up the gradient now experience longer periods when the motor turns counterclockwise (lengthened runs) and shorter periods when it turns clockwise (shortened tumbles) (FIGURE 4.10B). The combined result is a net movement toward the attractant; that is, up the concentration gradient.

Similar types of motile behavior are seen in photosynthetic prokaryotes moving toward light (phototaxis) or other prokaryotes moving toward oxygen gas (aerotaxis).

One additional type of flagellar organization is found in the spirochetes, a group of gram-negative, coiled bacterial species. The cells are motile by flagella that extend from one or both poles of the cell but fold back along the cell body (FIGURE 4.11). Such **endoflagella** and the cell body are surrounded by an outer sheath membrane. Motility results from the torsion generated on the cell by the normal rotation of

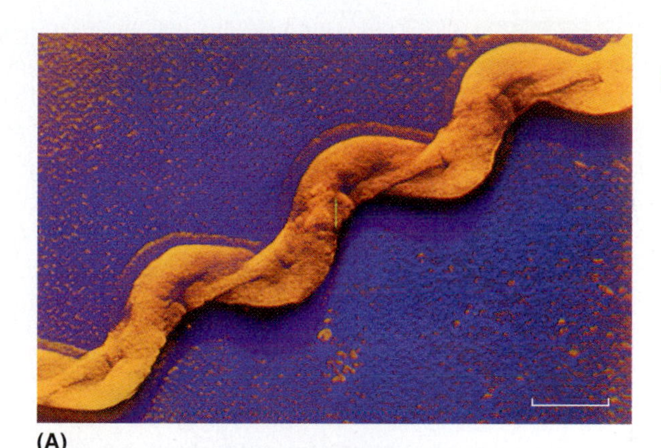

(A)

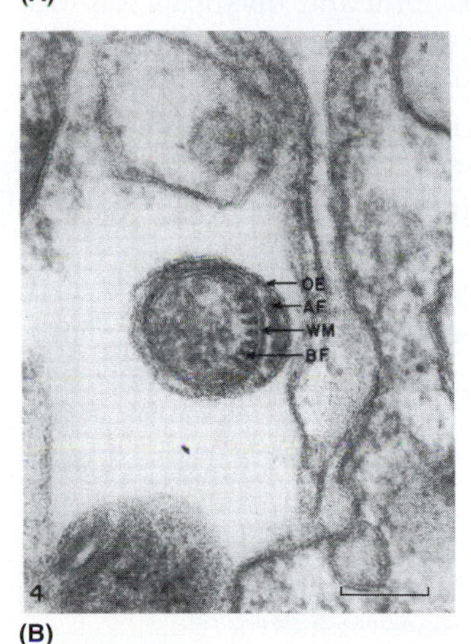

(B)

FIGURE 4.11 **The Spirochete Endoflagella.** Spirochetes contain endoflagella (axial filaments). (**A**) A false-color electron micrograph shows the endoflagella rod running down the length of the corkscrew-shaped cell. (Bar = 1 µm.) (**B**) A cross section through the endoflagellum can be seen in this transmission electron micrograph. (Bar = 0.6 µm.)

Q: How are endoflagella different from true bacterial flagella?

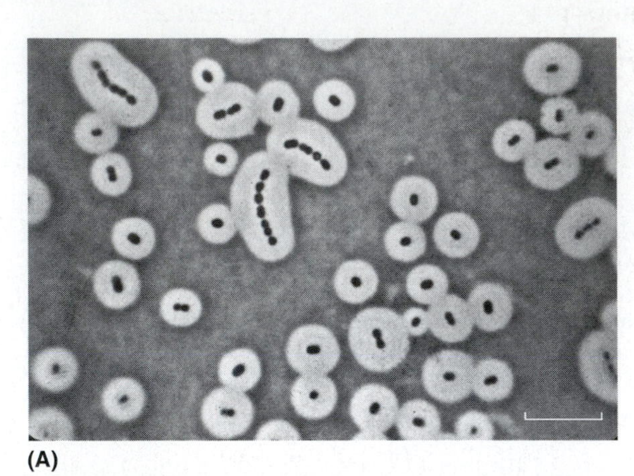
(A)

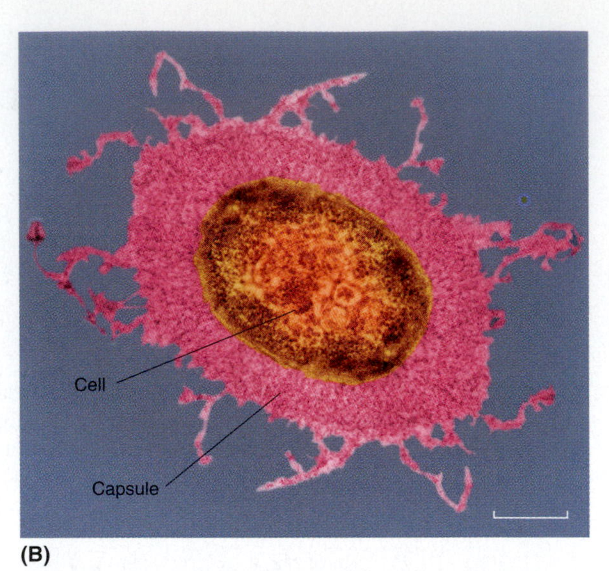

(B)

FIGURE 4.12 **The Bacterial Glycocalyx.** (**A**) Demonstration of the presence of a capsule in an *Acinetobacter* species by negative staining and observed by phase-contrast microscopy. (Bar = 10 μm.) (**B**) A false-color transmission electron micrograph of *Escherichia coli*. The cell is surrounded by a thick capsule (pink). (Bar = 0.5 μm.)

Q: How does the capsule provide protection for a prokaryotic cell?

the flagella. The resulting motility is less regular and more jerky than with flagellar motility.

Gliding motility. Some prokaryotes can move about without flagella by gliding across a solid surface. The motility occurs along the long axis of bacillus- or filamentous-shaped cells and usually is slower than flagellar motility. The cyanobacteria and myxobacteria (see Chapter 3) are two examples or organisms with gliding motility.

How the cells actually move is not completely understood. The cyanobacteria are thought to excrete a slime on which the cell glides. In other bacterial species, surface proteins may undergo a ratcheting motion, which pulls the cell forward. As with flagellar motility, the movement allows cells to explore new environments or, as in the myxobacteria, interact with other cells in their social community (Chapter 3).

MicroFocus 4.3 investigates how flagella may have evolved.

CONCEPT AND REASONING CHECKS

4.6 Explain how flagella move cells during a "run."

■ Virulence factor: A pathogen-produced molecule or structure that allows the cell to invade or evade the immune system and possibly cause disease.

■ Encapsulated: A cell having a capsule.

■ Phagocytosis: A process whereby certain white blood cells (phagocytes) engulf foreign matter and often destroy microorganisms.

The Glycocalyx Is an Outer Layer External to the Cell Wall

KEY CONCEPT

- A glycocalyx protects against desiccation, attaches cells to surfaces, and helps pathogens evade the immune system.

Most prokaryotic cells secrete an adhering layer of polysaccharides, or polysaccharides and small proteins, called the **glycocalyx**. (*glyco* = "sweet"; *calyx* = "coat"). The layer can be thick and covalently bound to the cell, in which case it is known as a **capsule**. Thinner, diffuse polysaccharides are referred to as a **slime layer**. Colonies containing cells with a glycocalyx appear moist and glistening. The actual capsule can be seen by light microscopy when observing a negative stain preparation or by transmission electron microscopy (FIGURE 4.12).

The glycocalyx serves as a buffer between the cell and the external environment. Because of its high water content, the glycocalyx can protect cells from desiccation. Another major role of the glycocalyx is to allow the cells to attach to surfaces. The glycocalyx of *V. cholerae*, for example, permits the cells to attach to the intestinal wall of the host. The glycocalyx of pathogens therefore represents a virulence factor.

Other encapsulated pathogens, such as *Streptococcus pneumoniae* (a principal cause of bacterial pneumonia) and *Bacillus anthracis*, evade the immune system because they cannot be easily engulfed by white blood cells during phagocytosis. Scientists believe the repulsion between bacterial cell and phagocyte is due to strong negative charges on the capsule and phagocyte surface.

MICROFOCUS 4.3: Evolution

The Origin of the Bacterial Flagellum

Flagella are found on many prokaryotic (bacterial and archaeal) and eukaryotic cells. Although they are all structurally different from one another, they have the same function—to propel the cell through its moist environment. Does the presence of these structures argue for a common "ancestor flagellum"? Probably not.

In the 1970s, Lynn Margulis argued that the eukaryotic flagellum, which she called the undulapodium, evolved from a symbiotic association with a spirochete. By attaching to an evolving eukaryotic cell, the spirillum provided a source for motility. In fact, there are some eukaryotic organisms alive today that live in the termite gut that have commandeered spirochetes as motility organelles. However, the hypothesis has never been widely accepted because the eukaryotic flagellum does not resemble any bacterium, including the spirochetes, and prokaryotes have their own version of the eukaryotic flagellar protein (see section 4.6) and would not have needed to form an association with a spirochete.

So, what about the origin of the bacterial flagellum then? Are there any other structures found in bacterial cells that resemble a flagellum? The answer is yes! Several bacterial species, including *Yersinia pestis*, the agent of bubonic plague, contain structures designed to inject toxins into an appropriate eukaryotic host cell. These bacterial cells have a hollow tube or needle to accomplish this process, just as the prokaryotic flagellum and filament are hollow (see diagram below). In addition, many of the flagellar proteins are similar to part of the injection proteins. In 2004, investigations discovered that *Y. pestis* cells contain all the genes needed for a flagellum—but the cells have lost the ability to use these genes. *Y. pestis* is nonmotile and it appears that the bacterial cells have a subset of flagellar proteins that are used to build the injection device.

One scenario then is that ancient prokaryotes evolved a structure that was the progenitor of the injection and flagellar systems. Perhaps the filaments that evolved into an injection system diverged and were refined into a flagellum.

The fascinating result of these investigations and proposals is it demonstrates that structures can evolve from other structures with a different function. It is not necessary that evolution "design" a structure from scratch but rather it can modify existing structures for other functions.

Individuals have proposed that the complexity of structures like the bacterial flagellum are just too complex to arise gradually through a step-by-step process. Some outside force must be involved. However, the investigation being conducted illustrates that a step-by-step evolution of a specific structure is not required. Rather, there can be cooperation, where one structure is modified to have other functions. The bacterial flagellum almost certainly falls into that category.

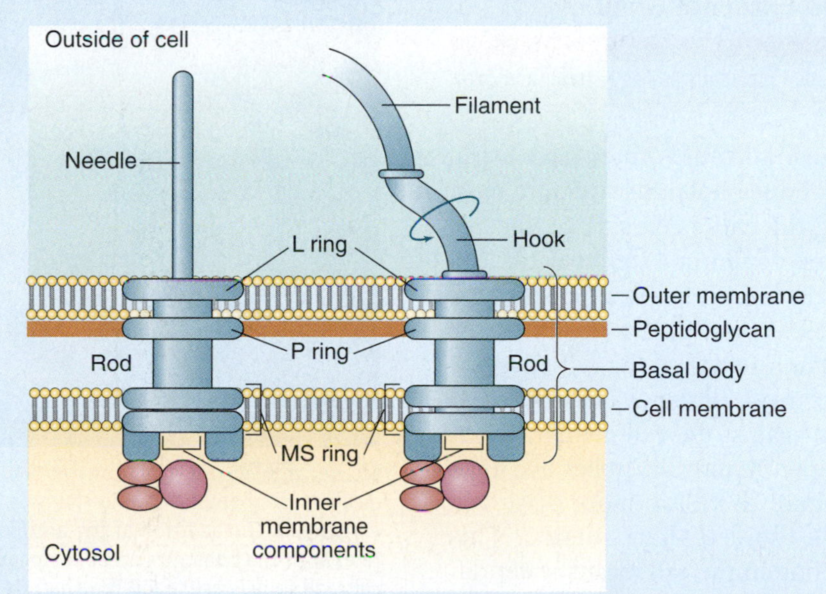

A bacterial injection device (left) compared to a bacterial flagellum (right).

A slime layer usually contains a mass of tangled fibers of a polysaccharide called dextran. The fibers attach the bacterial cell to tissue surfaces. A case in point is *Streptococcus mutans*, an important cause of tooth decay previously discussed in MicroFocus 2.4.

A glycocalyx allows *S. mutans* and other bacterial species to form dental plaque, which represents a type of biofilm on the tooth surface. Introduced in Chapter 1, biofilms are embedded microcolonies of bacteria that can attach to each other as well as to surfaces such as industrial pipelines, sewage-treatment systems, medical instruments, or body tissues. In biofilms, a carbohydrate matrix binds the microcolonies together, and surrounding water channels deliver nutrients and remove waste. In the example for dental plaque, a slime layer about 10 μm thick lies at the base of the biofilm and attaches the bacterial cells to the surface. Living within such biofilms effectively shields the cells from the body's immune defenses as well as from antibiotics and other therapies. Textbook Case 4 details a medical consequence of a biofilm.

In food products, the slime-producing bacteria may cause an unsightly and distasteful experience. For instance, the glue-like slime produced by *Alcaligenes viscolactis* accumulates in milk, causing it to become thick and stringy. The result is "ropy milk." Bread may also become ropy if contaminated with slime-producing *Bacillus subtilis*.

CONCEPT AND REASONING CHECKS

4.7 Under what circumstances might it be advantageous to a prokaryotic cell to have a capsule rather than a slime layer?

4.5 The Cell Envelope

The **cell envelope** is a complex structure that forms the two "wrappers"—the **cell wall** and the **cell membrane**—surrounding the cell cytoplasm. The cell wall is relatively porous to the movement of substances whereas the cell membrane regulates transport of nutrients and metabolic products.

The Prokaryotic Cell Wall Is a Tough and Protective External Shell

KEY CONCEPT

- Bacterial cell walls help maintain cell shape and prevent cell rupture.

The fact that most all prokaryotes have a cell wall suggests the critical role this structure must play. By covering the entire cell surface, the cell wall acts as an exoskeleton to protect the cell from injury and damage. It helps, along with the cytoskeleton (see Section 4.6), to maintain the shape of the cell and reinforce the cell envelope against the high intracellular water (osmotic) pressure pushing against the cell membrane. As described in Chapter 3, most microbes live in an environment where there are more dissolved materials inside the cell than outside. This hypertonic condition in the cell means water diffuses inward, accounting for the increased osmotic pressure. Without a cell wall, the cell would rupture or undergo **lysis** (FIGURE 4.13).

■ Hypertonic: A solution with more dissolved material (solutes) than the surrounding solution.

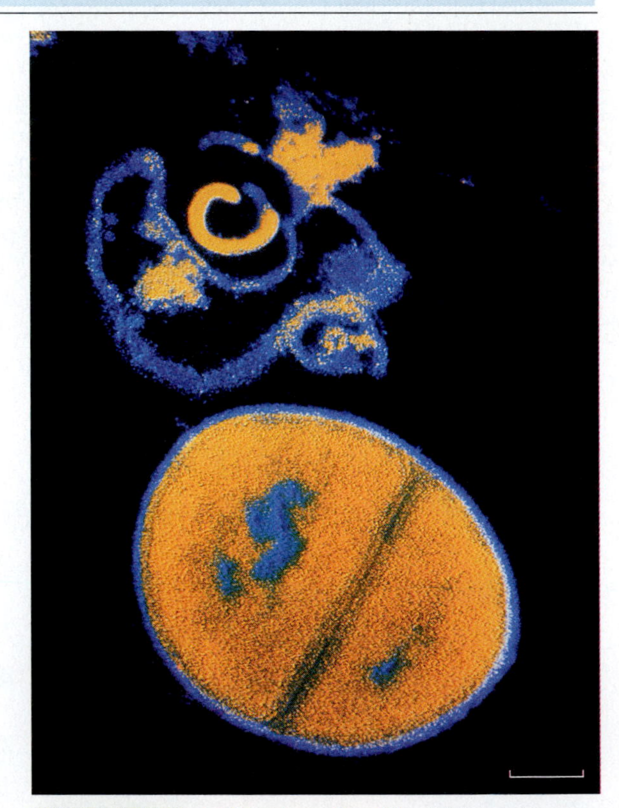

FIGURE 4.13 **Cell Rupture (Lysis).** A false-color electron micrograph showing the lysis of a *Staphylococcus aureus* cell. The addition of the antibiotic penicillin interferes with the construction of the peptidoglycan of the cell wall in new cells, and they quickly burst (top cell). (Bar = 0.25 μm.)

Q: Where is the concentration of dissolved substances (solutes) higher, inside the cell and outside? Explain.

Textbook CASE 4

An Outbreak of *Enterobacter cloacae* Associated with a Biofilm

Hemodialysis is a treatment for people with severe chronic kidney disease (kidney failure). The treatment filters the patient's blood to remove wastes and excess water. Before a patient begins hemodialysis, an access site is created on the lower part of one arm. Similar to an intravenous (IV) site, a tiny tube runs from the arm to the dialysis machine. The patient's blood is pumped through the dialysis machine, passed through a filter or artificial kidney called a *dialyzer,* and the cleaned blood returned to the patient's body at the access site. The complete process can take 3 to 4 hours.

1 During September 1995, a patient at an ambulatory hemodialysis center in Montreal, Canada received treatment on a hemodialysis machine to help relieve the effects of kidney disease. The treatment was performed without any unusual incident.

2 The next day, a second patient received treatment on the same hemodialysis machine. His treatment also went normally, and he returned to his usual activities after the session was completed.

3 In the following days, both patients experienced bloodstream infections (BSIs). They had high fever, muscular aches and pains, sore throat, and impaired blood circulation. Because the symptoms were severe, the patients were hospitalized. Both patients had infections of *Enterobacter cloacae,* a gram-negative rod.

4 In the following months, an epidemiological investigation reviewed other hemodialysis patients at that center. In all, seven additional adult patients were identified who had used the same hemodialysis machine. They discovered all seven had similar BSIs.

5 Inspection of the hemodialysis machine used by these nine patients indicated the presence of biofilms containing *Enterobacter cloacae,* which was identical to those samples taken from the patients' bloodstreams (see figure).

6 Further study indicated that the dialysis machine was contaminated with *E. cloacae,* specifically where fluid flows.

7 It was discovered that hospital personnel were disinfecting the machines correctly. The problem was that the valves in the drain line were malfunctioning, allowing a backflow of contaminated material.

8 Health officials began a hospital education program to ensure that further outbreaks of infection were curtailed.

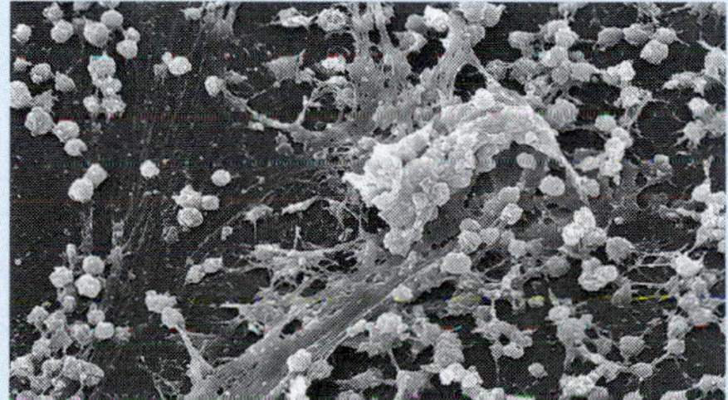

Similar to the description in this textbook case, biofilms consisting of *Staphylococcus* cells can contaminate hemodialysis machines.

Questions

A. *Suggest how the hemodialysis machine originally became contaminated.*

B. *Why weren't the other five cases of BSI correlated with the hemodialysis machine until the epidemiological investigation was begun?*

C. *How could future outbreaks of infection be prevented?*

For additional information see http://www.cdc.gov/mmwr/preview/mmwrhtml/00051244.htm.

It is similar to blowing so much air into a balloon that the air pressure bursts the balloon.

The bacterial cell wall differs markedly from the walls of archaeal cells and cells of other microorganisms in containing **peptidoglycan**, which consists of chains of disaccharides (glycan strands) cross-linked by peptide bridges (FIGURE 4.14A). Each disaccharide in this very large molecule, also called murein, is composed of two monosaccharides, N-acetylglucosamine (NAG) and N-acetylmuramic acid (NAM). The carbohydrate backbone can occur in multiple layers connected by side chains of four amino acids and peptide cross-bridges. There can be more to a bacterial cell wall than just peptidoglycan, so two types of bacterial cell walls are recognized, which also provide the means to identify stained bacteria as gram-positive (blue-purple) or gram-negative (orange-red).

■ Endotoxin: A poison that can activate inflammatory responses, leading to high fever, shock, and organ failure.

Gram-positive bacteria. The gram-positive bacterial cells have a very thick peptidoglycan cell wall (FIGURE 4.14B). The abundance and thickness (25 nm) of this material may be one reason why they retain the crystal violet in the Gram stain (see Chapter 3). The multiple layers of glycan strands are cross-linked to one another both in the same layer as well as between layers.

The gram-positive cell wall also contains an anionic polysaccharide derivative called **teichoic acid**. Wall teichoic acids, which are bound to NAM in the glycan chains, are essential for cell viability—if the genes for teichoic acid synthesis are deleted, cell death occurs. Still, the function of the teichoic acids remains unclear. They may help maintain a surface charge on the cell wall, control the activity of autolytic enzymes acting on the peptidoglycan, and/or maintain permeability of the cell wall layer.

■ Autolytic enzymes: Enzymes that break bonds in the peptidoglycan, thereby causing lysis of the cell.

Gram-negative bacteria. The cell wall of gram-negative bacterial cells is structurally quite different from that of the gram-positive wall (FIGURE 4.14C). The peptidoglycan layer is two-dimensional; the glycan strands compose but a single layer with cross-linking only between strands in the single layer. This is one reason why it loses the crystal violet dye during the Gram stain. Also, there is no teichoic acid present.

The unique feature of the gram-negative cell wall is the presence of an **outer membrane**, which is separated by a gap, called the **periplasmic space**, from the cell membrane. This gel-like area contains digestive enzymes and transport proteins to speed entry of nutrients into the cell. The peptidoglycan layer is located in the periplasmic space.

The inner half of the outer membrane contains phospholipids similar to the cell membrane. However, the outer half is composed of **lipopolysaccharide (LPS)**, which consists of an O-polysaccharide attached to a unique lipid molecule known as **lipid A**. The O-polysaccharide is used to identify variants of a species (e.g., strain O157:H7 of *E. coli*). On cell death, lipid A is released and represents an endotoxin that can be toxic if ingested (Chapter 18).

The outer membrane also contains unique proteins called **porins**. These proteins form pores in the outer membrane through which small molecules pass into the periplasmic space. Larger molecules cannot pass, partly accounting for the resistance of gram-negative cells to antimicrobial agents, dyes, disinfectants, and lysozyme.

TABLE 4.2 summarizes the major differences between the two types of bacterial cell walls.

CONCEPT AND REASONING CHECKS

4.8 Penicillin and lysozyme primarily affect peptidoglycan synthesis in gram-positive bacterial cells. Why are these agents less effective against gram-negative bacterial cells?

The Archaeal Cell Wall Also Provides Mechanical Strength

KEY CONCEPT

- Archaeal cell walls do not contain peptidoglycan.

Archaeal species vary in the type of wall they possess. None have the peptidoglycan wall structure, so the cells would be resistant to lysozyme and penicillin action. However, some species have a **pseudopeptidoglycan** where the NAM is replaced by N-acetyltalosamine uronic acid (NAT). Other archaeal cells have walls made of polysaccharide, protein, or both.

The most common cell wall among archaeal species is a surface layer called the **S-layer**. It consists of hexagonal patterns of protein or

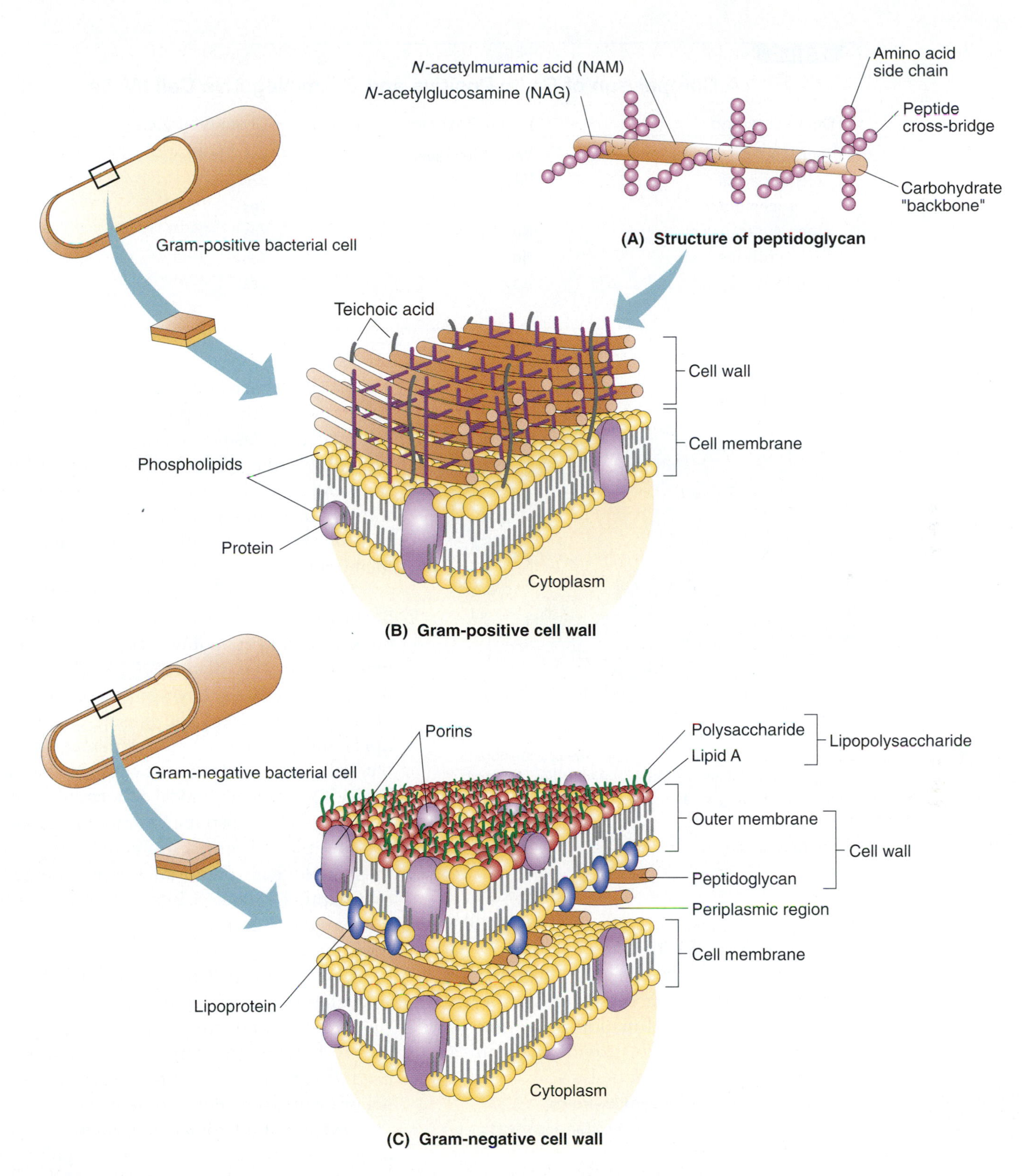

FIGURE 4.14 **A Comparison of the Cell Walls of Gram-Positive and Gram-Negative Bacterial cells.** (**A**) The structure of peptidoglycan is shown as units of NAG and NAM joined laterally by amino acid cross-bridges and vertically by side chains of four amino acids. (**B**) The cell wall of a gram-positive bacterial cell is composed of peptidoglycan layers combined with teichoic acid molecules. (**C**) In the gram-negative cell wall, the peptidoglycan layer is much thinner, and there is no teichoic acid. Moreover, an outer membrane overlies the peptidoglycan layer such that both comprise the cell wall. Note the structure of the outer membrane in this figure. The outer half is unique in containing lipopolysaccharide and porin proteins.

Q: Simply based on cell wall structure, assess the potential of gram-positive and gram-negative cells as pathogens.

TABLE **4.2** A Comparison of Gram-Positive and Gram-Negative Cell Walls

Characteristic	Gram Positive	Gram Negative
Peptidoglycan	Yes, thick layer	Yes, thin layer
Teichoic acids	Yes	No
Outer membrane	No	Yes
Lipopolysaccharides	No	Yes
Porin proteins	No	Yes
Periplasmic space	No	Yes

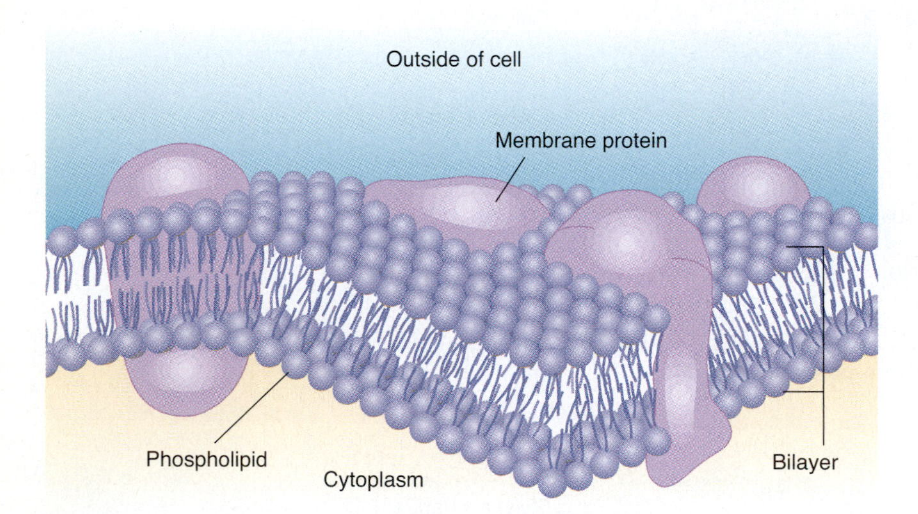

FIGURE 4.15 **The Structure of the Bacterial Cell Membrane.** The cell membrane of a bacterial cell consists of a phospholipid bilayer in which are embedded integral membrane proteins. Other proteins and ions may be associated with the integral proteins or the phospholipid heads.

Q: Why is the cell membrane referred to as a fluid mosaic structure?

glycoprotein that self-assemble into a crystalline lattice 5 to 25 nm thick.

Although the walls may be structurally different and the molecules form a different structural pattern, the function is the same as in bacterial species—to provide mechanical support and prevent osmotic lysis.

- **Hydrophobic:** Pertaining to molecules or parts of molecules that are not soluble in water.
- **Hydrophilic:** Pertaining to molecules or parts of molecules that are soluble in water.

CONCEPT AND REASONING CHECKS

4.9 Distinguish between peptidoglycan and pseudopeptidoglycan cell walls.

The Cell Membrane Represents the Interface between the Cell Environment and the Cell Cytoplasm

KEY CONCEPT

- Molecules and ions cross the cell membrane by facilitated diffusion or active transport.

A cell (or plasma) membrane is a universal structure that separates external from internal (cytoplasmic) environments, preventing soluble materials from simply diffusing into and out of the cell. One exception is water, which due to its small size and overall lack of charge can diffuse slowly across the membrane.

The bacterial cell membrane, which is about 5 nm thick, is 40 percent phospholipid and 60 percent protein. In illustrations, the cell membrane appears very rigid (FIGURE 4.15). In reality, it is quite fluid, having the consistency of olive oil. This means the mosaic of phospholipids and proteins are not cemented in place, but rather they can move laterally in the membrane. This dynamic model of membrane structure therefore is called the **fluid mosaic model**.

The phospholipid molecules, typical of most biological membranes, are arranged in two parallel layers (a bilayer) and represent the barrier function of the membrane. The phospholipids contain a charged phosphate head group attached to two hydrophobic fatty acid chains (see Chapter 2). The fatty acid "tails" are the portion that forms the permeability barrier. In contrast, the hydrophilic head groups are exposed to the aqueous external or cytoplasmic environments.

Several antimicrobial substances act on the membrane bilayer. The antibiotic polymyxin pokes holes in the bilayer, while some detergents and alcohols dissolve the bilayer. Such action allows the cytoplasmic contents to leak out of bacterial cells, resulting in death through cell lysis.

A diverse population of membrane proteins populates the phospholipid bilayer. These membrane proteins often have stretches of hydrophobic amino acids that interact with the hydrophobic fatty acid chains in the membrane. These proteins span the width of the bilayer and are referred to as integral membrane proteins. Other proteins, called peripheral membrane proteins, are associated with the polar heads of the bilayer.

The membrane proteins carry out numerous important functions. Some represent enzymes needed for cell wall synthesis or for energy metabolism. As mentioned, prokaryotes lack mitochondria and part of that organelle's function is carried out by the cell membrane. Other membrane proteins help anchor the DNA to the membrane during replication or act as receptors of chemical information, sensing changes in environment conditions and triggering appropriate responses.

Perhaps the largest group of integral membrane proteins is involved as transporters of charged solutes, such as amino acids, simple sugars, nitrogenous bases, and ions across the lipid bilayer. The transport systems are highly specific though, only transporting a single molecular species or a very similar class of molecules. Therefore, there must be many different transport proteins to regulate the diverse molecular traffic that must flow into or out of a cell.

The transport process can be passive or active. In **facilitated diffusion**, integral membrane proteins facilitate the movement of materials down their concentration gradient; that is, from an area of higher concentration to one of lower concentration (FIGURE 4.16). By acting as a conduit for diffusion or as a transporter through the nonpolar bilayer, hydrophilic solutes can enter or leave without the need for cellular energy.

Unlike facilitated diffusion, **active transport** allows different concentrations of solutes to be established outside or inside of the cell against the concentration gradient. These membrane proteins act as "pumps" and, as such, demand an energy input from the cell. Besides certain solutes, large molecules such as proteins, which cannot freely diffuse across membranes, are transported across the membrane by active transport.

CONCEPT AND REASONING CHECKS

4.10 Justify the necessity for phospholipids and proteins in the cell membrane.

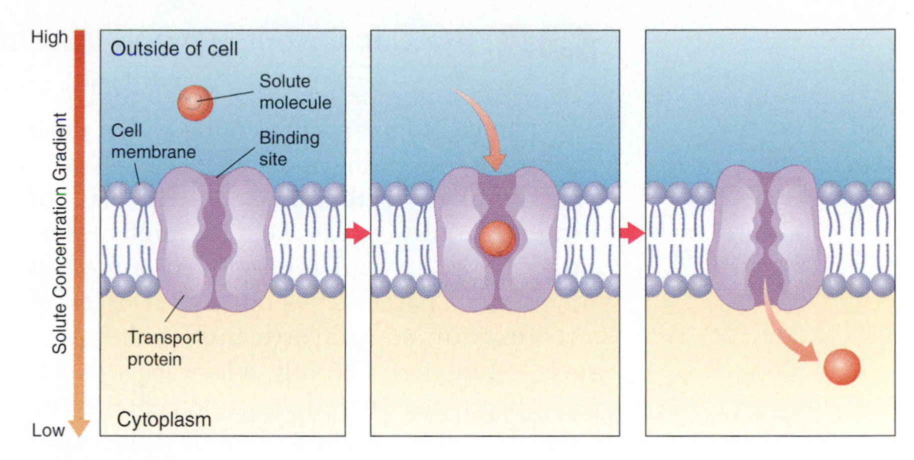

FIGURE 4.16 Facilitated Transport through a Membrane Protein. Many transport proteins facilitate the diffusion of nutrients across the lipid bilayer. The transport protein forms a hydrophilic channel through which a specific solute can diffuse.

Q: Why would a solute move through a membrane protein rather than simply across the lipid bilayer?

The Archaeal Cell Membrane Differs from Bacterial and Eukaryal Membranes

KEY CONCEPT

- Archaeal phospholipids are structurally unique.

Besides the differences in ribosomal RNA sequences in the domain *Archaea*, another major difference used to separate the archaeal organisms into their own domain is the chemical nature of the cell membrane.

The manner in which the hydrophobic lipid tails are attached to the glycerol is different in the *Archaea*. The tails are bound to the glycerol by ether linkages rather than the ester linkages found in the domains *Bacteria* and *Eukarya*.

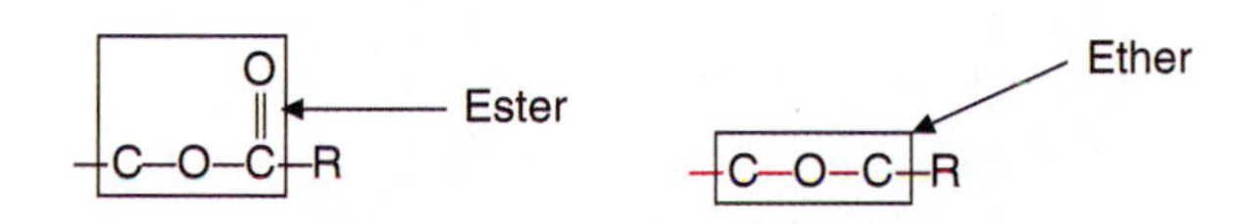

Also, typical fatty acid tails are absent from the membranes; instead, repeating five-carbon units form lipid tails longer than the fatty acid tails. In addition, the tails of adjacent sides of the cell membrane are bonded together forming a lipid monolayer rather than a bilayer. This provides an advantage to the hyperthermophiles by preventing a peeling in two of the membrane, which would occur with a typical phospholipid bilayer structure.

CONCEPT AND REASONING CHECKS

4.11 What is unique about archaeal membrane structure?

4.6 The Cell Cytoplasm and Internal Structures

The cell membrane encloses the **cytoplasm**, which is the compartment within which most growth and metabolism occur. The cytoplasm consists of the **cytosol**, a semifluid mass of proteins, amino acids, sugars, nucleotides, salts, vitamins, and ions—all dissolved in water (see Chapter 2)—and several bacterial structures or subcompartments, each with a specific function. Recently, *microcompartments* have been discovered in prokaryotes. The exact function of these protein-lined subcompartments is being investigated.

■ Haploid: Having a single set of genetic information.

The Nucleoid Represents a Subcompartment Containing the Chromosome

KEY CONCEPT

- The nucleoid contains the cell's essential genetic information.

The chromosome region in a prokaryotic cell appears as a diffuse mass termed the **nucleoid** (FIGURE 4.17). The nucleoid does not contain a covering or membrane; rather, it represents a central subcompartment in the cytoplasm where the DNA aggregates and ribosomes are absent. Usually there is a single chromosome per cell and, with few exceptions, exists as a closed loop of DNA and protein.

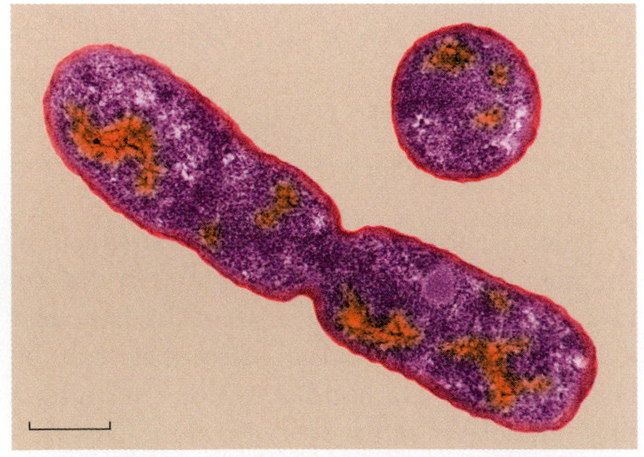

FIGURE 4.17 **The Prokaryotic Nucleoid.** In this false-color transmission electron micrograph of *Escherichia coli*, nucleoids (orange) occupy a large area in a bacterial cell. Both longitudinal (center cells) and cross sections (cell upper right) of *E. coli* are visible. (Bar = 0.5 µm.)

Q: How does a nucleoid differ from the eukaryotic cell nucleus described in the previous chapter?

The DNA contains the hereditary information or **genes** of the cell. Since most cells only have one copy of each gene, the cells are genetically haploid. Unlike eukaryotic microorganisms and other eukaryotes, the nucleoid and chromosome do not undergo mitosis and having but the one set of genetic information cannot undergo meiosis.

The complete set of genes in an organism, called the **genome**, varies by species. For example, the genome of *E. coli*, a typical bacterial species in the mid-size range, contains about 4300 genes. In all cases, these genes determine what proteins and enzymes the cell can make; that is, what metabolic reactions and activities can be carried out. For *E. coli*, this equates to some 2,000 different proteins. Extensive coverage of bacterial DNA is presented in Chapter 7.

Plasmids Are Found in Many Prokaryotic Cells

KEY CONCEPT

- Plasmids contain nonessential genetic information.

Besides a nucleoid, many prokaryotic cells also contain smaller molecules of DNA called **plasmids**. About a tenth the size of a prokaryotic chromosome, these stable, extrachromosomal DNA molecules exist as closed loops containing 5 to 100 genes. There can be one or more plasmids in a cell and these may contain similar or different genes. Plasmids replicate independently of the chromosome and can be transferred between cells during recombination. They also represent important "vectors" in industrial technologies that use genetic engineering. Both topics are covered in Chapter 8.

■ Vectors: Genetic elements capable of incorporating and transferring genetic information.

Although plasmids may not be essential for cellular growth, they provide a level of genetic flexibility. For example, some plasmids possess genes for disease-causing toxins and many carry genes for chemical or antibiotic resistance. For this latter reason, these genetic elements often are called **R plasmids** (R for resistance).

CONCEPT AND REASONING CHECKS

4.12 What properties distinguish the bacterial chromosome from a plasmid?

Other Subcompartments Exist in the Prokaryotic Cytoplasm

KEY CONCEPT

- Ribosomes and inclusion bodies carry out specific intracellular functions.

One of the universal cell structures mentioned in Chapter 3 was the **ribosome**. There are hundreds of thousands of these nearly spherical particles in the cell cytoplasm, which gives it a granular appearance when viewed with the electron microscope. Their relative size is determined by how fast they settle when spun in a centrifuge. Measured in Svedberg units (S), prokaryotic ribosomes represent 70S particles.

Prokaryotic ribosomes are built from RNA and protein and are composed of a small subunit (30S) and a large subunit (50S). For proteins to be synthesized, the two subunits come together to form a 70S functional ribosome (Chapter 7). Some antibiotics, such as streptomycin and tetracycline, prevent prokaryotic ribosomes from carrying out protein synthesis.

Cytoplasmic structures, called **inclusion bodies**, are found in many prokaryotes. Many of these bodies store nutrients or the monomers for cellular structures. For example, some inclusions consist of aggregates or granules of polysaccharides (glycogen), globules of elemental sulfur, or lipid. Other inclusion bodies can serve as important identification characters for bacteria pathogens. One example is the diphtheria bacilli that contain **metachromatic granules**, or **volutin**, which are deposits of polyphosphate (long chains of inorganic phosphate) along with calcium and other ions. These granules stain with dyes such as methylene blue.

Some aquatic and marine prokaryotes float on the water surface. Floatation is made possible by the presence of **gas vesicles**, cytoplasmic compartments built from a water-tight protein shell. These vesicles decrease the density of the cell, which generates and regulates their buoyancy.

The **magnetosome**, another type of subcompartment, is described in MicroFocus 4.4. These inclusions are invaginations of the cell membrane, which are coordinated and positioned by cytoskeletal filaments similar to eukaryotic actin filaments.

CONCEPT AND REASONING CHECKS

4.13 Provide the roles for the cytoplasmic inclusions in bacterial and archaeal cells.

Prokaryotic Cells Have a "Cytoskeleton"

KEY CONCEPT

- Prokaryotic cytoskeletal proteins regulate cell division and help determine cell shape.

Until recently, it appeared prokaryotic cells lacked a cytoskeleton, which is a common feature in eukaryotic cells (see Chapter 3). However, it now appears cytoskeletal proteins homologous to those in the eukaryotic cytoskeleton are present in prokaryotes. Let's briefly look at these proteins.

The first prokaryotic protein discovered was a homolog of the eukaryotic protein tubulin, which forms filaments that assemble into microtubules. The prokaryotic homolog forms filaments similar to those in microtubules but the filaments do not assemble into microtubules. The prokaryotic proteins have been found in all bacterial and archaeal cells examined and appear to function in the regulation of cell division. During the process of cell division, the protein localizes around the neck of dividing cell. Its central presence is essential for recruiting other proteins essential for the deposition of a new cell wall between the dividing cells (MicroFocus 4.5).

Protein homologs remarkably similar in three-dimensional structure to eukaryotic actin filaments assemble into filaments that help determine cell shape in *E. coli* and *Bacillus subtilis*. The prokaryotic homologs have been found in most non-spherical cells where they form a helical network beneath the cell membrane to guide the proteins involved in cell wall formation (FIGURE 4.18A). The homologs also are involved with chromosome segregation during cell division and magnetosome formation.

Intermediate filaments (IF), another component of the eukaryotic cytoskeleton in some metazoans, have a prokaryotic homolog as well. The protein, called crescentin, helps determine the characteristic crescent shape of *Caulobacter crescentus* cells. In older cells that become filamentous, crescentin maintains the helical shape of the cells by aligning with the inner cell curvature beneath the cytoplasmic membrane (FIGURE 4.18B).

Even though the evolutionary relationships are quite distant between prokaryotic

- **Centrifuge:** An instrument that spins particles suspended in lipid at high speed.
- **Homolog:** An entity with similar attributes.
- **Metazoans:** Members of the vertebrates, nematodes, and mollusks.

MICROFOCUS 4.4: Environmental Microbiology

A "Not So Fatal" Attraction

To get from place to place, humans often require the assistance of maps, compasses, or gas station attendants. In the microbial world, life is generally more simple, and traveling is no exception.

In the early 1980s, Richard P. Blakemore and his colleagues at the University of New Hampshire observed mud-dwelling bacterial cells gathering at the north end of water droplets. On further study, they discovered each cell had a chain of magnetic particles acting like a compass directing the organism's movements (magnetotaxis). Additional studies have shown that the magnetotactic bacteria swim toward the north in the Northern Hemisphere and toward the south in the Southern Hemisphere.

Additional interdisciplinary investigations by microbiologists and physicists have shown the magnetotactic bacteria contain a linear array of membrane-bound vesicles along the cell's long axis (see figure). Each vesicle, called a **magnetosome**, is an invagination of the cell membrane and contains the protein machinery to nucleate and grow a crystal of magnetite (Fe_3O_4) or greigite (Fe_3S_4).

To date, all magnetotactic bacterial cells are motile, gram-negative cells common in aquatic and marine habitats, including sediments where oxygen is absent. This last observation is particularly noteworthy because it explains why these organisms have magnetosomes.

It originally was thought that magnetotaxis was used to guide cells in a parallel (Northern Hemisphere) or anti-parallel (Southern Hemisphere) manner along the geomagnetic field lines to those regions of the habitat with no oxygen; in other words, they travel downward toward the sediment. More recent studies have shown that some magnetotactic bacteria actually prefer low concentrations of oxygen. So the opinion now is that both magnetotaxis and aerotaxis work together to allow cells to "find" the optimal point within an oxygen gradient. This "not so fatal" attraction permits the bacteria to reach a sort of biological nirvana and settle in for a life of environmental bliss.

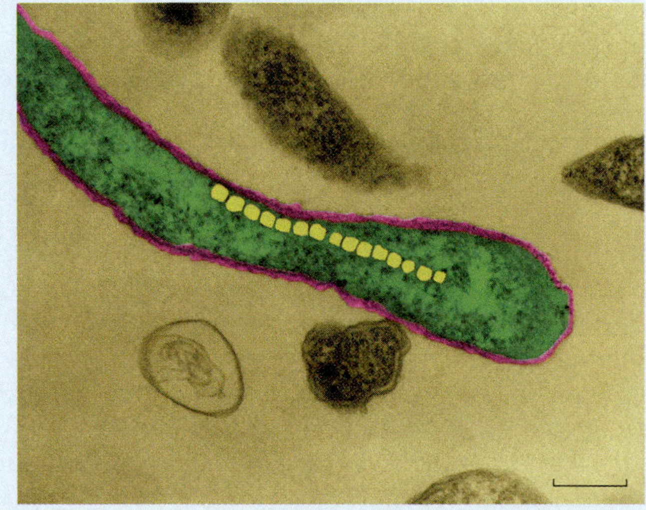

Bacterial magnetosomes (yellow) are seen in this false-color transmission electron micrograph of a magnetotactic marine spirillum. (Bar = 1 µm.)

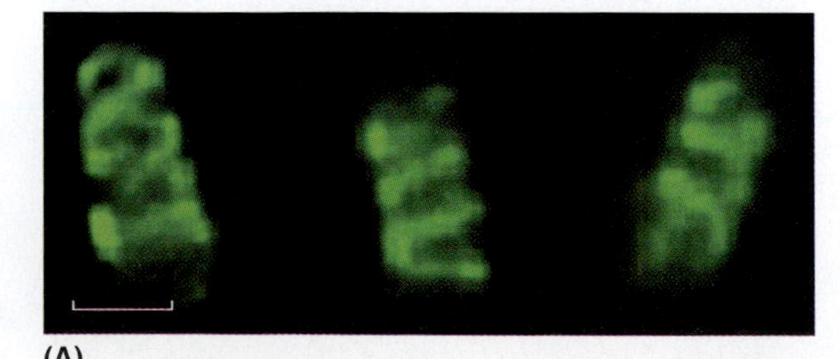

(A)

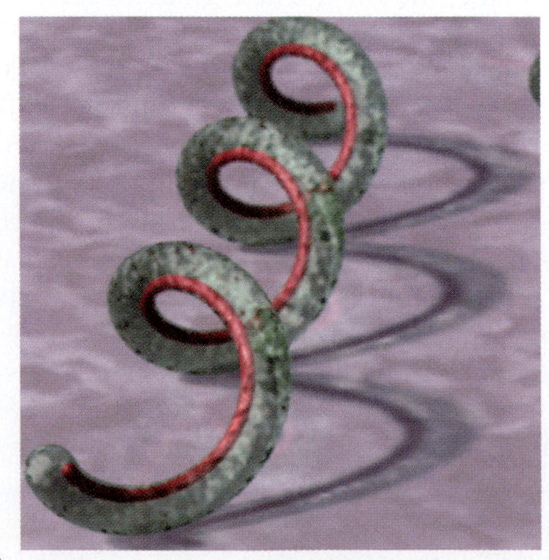

(B)

FIGURE 4.18 A Prokaryotic Cytoskeleton. Prokaryotes have proteins similar to those that form the eukaryotic cytoskeleton. (**A**) Actin-like proteins form helical filaments that curve around the edges of these cells of *Bacillus subtilis*. (Bar = 1.5 µm.) (**B**) Three-dimensional model of a helical *C. crescentus* cell (green) with a helical cytoskeletal filament of crescentin (pink).

Q: What would be the shape of these cells without the cytoskeletal proteins?

MICROFOCUS 4.5: Public Health

The Wall-less Cytoskeleton

Sometimes the lack of something can speak loudly. Take for example the mycoplasmas such as *Mycoplasma pneumoniae* that causes primary atypical pneumonia (walking pneumonia). This, as well as other *Mycoplasma* species, lack a cell wall. How then can they maintain a defined cell shape (see figure)?

Transmission electron microscopy has revealed that mycoplasmal cells contain a very complex cytoskeleton and further investigations indicate the cytoskeletal proteins are very different from the typical cytoskeletal homologs found in other groups of the Firmicutes. For example, *Spiroplasma citri,* which causes infections in other animals, has a fibril protein cytoskeleton that is laid down as a helical ribbon. This fibril protein has not been found in any other organisms, prokaryotic or eukaryotic. Because the cells are spiral shaped, the ribbon probably is laid down in such a way to determine cell shape. So, cell shape is not totally dependant on a cell wall.

In *Mycoplasma genitalium,* which is closely related to *M. pneumoniae* and causes human urethral infections, a prokaryotic tubulin homolog has been identified, but none of the other proteins have been identified that it recruits at the division neck for cell division. Surprising? Not really. Important? Immensely! Since mycoplasmas do not have a cell wall, why would they require those proteins that lay down a peptidoglycan cross wall between cells? So the lack of something (wall-forming proteins) tells us what those proteins must do in their gram-positive relatives that do have walls.

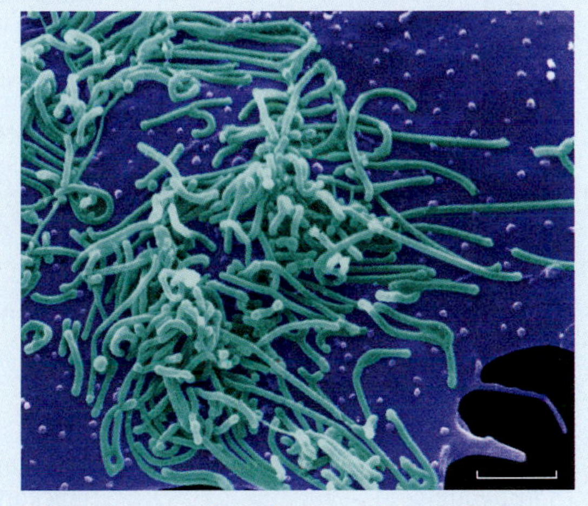

False-color scanning electron micrograph of *Mycoplasma pneumoniae* cells. (Bar = 2.5 μm.)

and eukaryotic cytoskeletal proteins based on protein sequence data, the similarity of their three-dimensional structure and function is strong evidence supporting homologous eukaryotic and prokaryotic cytoskeletons.

CONCEPT AND REASONING CHECKS

4.14 Evaluate the relationship between the eukaryotic cytoskeleton and the cytoskeletal protein homologs in prokaryotes.

4.7 The Prokaryotic/Eukaryotic Cell—Revisited

TABLE 4.3 summarizes the structural features of prokaryotic cells. One of the take-home lessons from the table and discussions of prokaryotic cell structure and function explored in this chapter is the ability of prokaryotes to carry out the "complex" metabolic and biochemical processes typically associated with eukaryotic cells—usually without the need for elaborate membrane-enclosed subcompartments—although a few such organelles are now being discovered (MicroInquiry 4). Clearly, prokaryotic cells are not simply bags of water and chemicals surrounded by a membrane and rigid cell wall. After all, in the 1920s and 30s, biologists were mistaken in believing eukaryotic cells were just bags of enzymes and water.

Earlier in this chapter, the intricate subcellular compartmentation in prokaryotes was discussed for several cell structures. What about other major cellular processes such as making proteins? This requires two processes, that of transcription and translation (Chapter 7). In eukaryotic cells, these processes are spatially separated into the cell nucleus (transcription) and the cytoplasm (translation).

TABLE

4.3 A Summary of the Structural Features of Prokaryotic Cells

Structure	Chemical Composition	Function	Comment
External Structures			
Pili	Protein	Attachment to surfaces Genetic transfer	Found primarily in gram-negative bacteria
Flagella	Protein	Motility	Present in many rods and spirilla; few cocci; vary in number and placement
Glycocalyx	Polysaccharides and small proteins	Buffer to environment Attachment to surfaces	Capsule and slime layer Contributes to disease Found in plaque bacteria and biofilms
Cell Envelope			
Cell wall		Cell protection Shape determination Prevents cell lysis	
Bacterial	Gram-positives: thick peptidoglycan + teichoic acid Gram-negatives: little peptidoglycan and an outer membrane		Site of activity of penicillin and lysozyme Gram-negative bacteria release endotoxins
Archaeal	Pseudopeptidoglycan Protein		S-layer
Cell membrane	Protein Phospholipid	Cell boundary Transport into/out of cell Site of enzymatic reactions	Conforms to fluid mosaic model
Internal Structures			
Nucleoid	DNA	Site of essential genes	Exists as single, closed loop chromosome
Plasmids	DNA	Site of nonessential genes	R plasmids
Ribosomes	RNA and protein	Protein synthesis	Inhibited by certain antibiotics
Inclusion bodies	Glycogen, sulfur, lipid	Nutrient storage	Used as nutrients during starvation periods
Metachromatic granules	Polyphosphate	Storage for ATP and nucleic acid synthesis	Found in diphtheria bacilli
Gas vesicles	Protein shells	Buoyancy	Helps cells float
Magnetosome	Magnetite/greigite	Cell orientation	Helps locate preferred habitat
Acidocalcisomes	Phosphate, magnesium, potassium, calcium	Storage compartment?	Formerly identified as volutin inclusions
Cytoskeleton	Proteins	Cell division, chromosomal segregation, cell shape	Functionally similar to eukaryotic cytoskeletal proteins

In prokaryotes, there also can be spatial separation (FIGURE 4.19). Transcription appears to be separated in space from translation. The RNA polymerase molecules needed for transcription are localized to a region separate from the ribosomes and other proteins that perform translation. So, even without a nuclear membrane, prokaryotic cells can separate the process involved in making cellular proteins, in a manner similar to eukaryotic cells.

As we continue into the chapters immediately ahead, we will continue to see other examples where the old prokaryotic/eukaryotic paradigm appears to be eroding away.

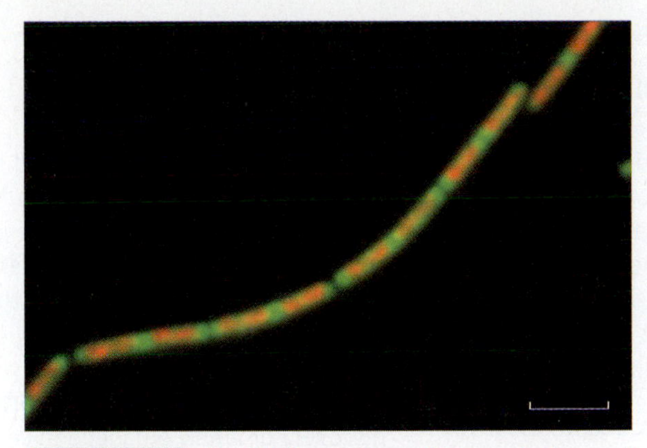

FIGURE 4.19 **Spatial Separation of Transcription and Translation.** A nuclear membrane is not required to separate the processes of transcription and translation in prokaryotic cells. In these cells of *Bacillus subtilis,* fluorescence microscopy was used to identify RNA polymerase (transcription) using a red fluorescent protein and ribosomes (translation) using a green fluorescent protein. Separate subcompartments are very evident. (Bar = 3 µm.)

MICROINQUIRY 4

Inquiring into Membrane Organelles in Prokaryotes

One of the guiding principles of cell biology has been that the way to tell prokaryotic and eukaryotic cells apart is to look for a cell nucleus and other membrane-enclosed organelles. If membrane-bound organelles are absent, the cell must be a prokaryote.

Question
Do membrane-bound organelles exist in prokaryotic cells?

Background
Cytoplasmic "membranous" structures have been identified in several bacterial species. *Rhodospirillum rubrum* and *Rhodobacter capsulatus* are two species of proteobacteria that contain abundant membrane vesicles called chromatophores on which photosynthesis occurs. However, rather than being independent subcompartments, the vesicles are invaginations of the cell membrane. Cyanobacteria possess sac-like membrane systems (thylakoids) with the same function. Both photosynthetic membrane systems are without enzymatic function.

Another group of intracellular "membranous" structures are the inclusion bodies discussed in this chapter. These and other microcompartments would be considered equivalent to organelles found in the eukaryotes if they (1) had enzymatic function and (2) the protein shell regulated transport into and out of the microcompartment.

In 1994, scientists at the University of Illinois, Urbana-Champagne reported the discovery of a new organelle in several eukaryotic microbial parasites, including *Trypanosoma,* one species of which causes sleeping sickness. This independent, membrane-enclosed subcompartment contained enzymes that pumped phosphate and hydrogen ions into the vesicle, making it an acidic compartment. It also contained high concentrations of polyphosphate (long chains of inorganic phosphate), calcium and other ions. Thus, these structures were called **acidocalcisomes**.

Then in 2003, the same group announced they found acidocalcisomes in bacterial cells. But prokaryotes do not have independent membrane-bound organelles with enzymatic function. Right?

Hypothesis
Prokaryotic cells contain acidocalcisomes. If correct, then the newly discovered organelles must be enclosed by a true phospholipid bilayer membrane.

Investigation
Using electron microscopy, the University of Illinois scientists demonstrated the bacterial structure identified as acidocalcisomes had a true membrane surrounding the organelle (see **Figure**). The organelles were randomly distributed in the cell although in some cells they were close to one end (pole) of the cell.

Hypothesis
Prokaryotic cells contain acidocalcisomes. If so, then the organelles should contain enzymes and substances typical of the eukaryotic acidocalcisomes.

Investigation
Further biochemical analyses of the "bacterial acidocalcisomes" indicated they contain large amounts of phosphorus, magnesium, potassium, and calcium—just like their eukaryotic cousins. This indicates that one function for acidocalcisomes would be as storage subcompartments for these ions and polyphosphate.

SUMMARY OF KEY CONCEPTS

4.1 Prokaryotic Diversity

- The phylogenetic tree of life contains many bacterial phyla and groups, including the Proteobacteria, Firmicutes, Cyanobacteria, and Hyperthermophiles.
- Many organisms in the domain *Archaea* live in extreme environments. The Euryarchaeota (methanogens, extreme halophiles, and the thermoacidophiles) and the Crenarchaeota are two major groups.

4.2 The Shapes and Arrangements of Prokaryotic Cells

- Bacilli have a cylindrical shape and can remain as single cells or be arranged into chains (streptobacilli). Cocci are spherical and form a variety of arrangements, including the diplococcus, streptococcus, and staphylococcus. The spiral-shaped bacteria can be curved rods (vibrios) or spirals (spirochetes and spirilla). Spirals generally appear as single cells. Other bacterial cells have a stalked or filamentous shape.

4.3 An Overview to Prokaryotic Cell Structure

- Cell organization is centered on three specific processes: sensing and responding to environmental changes, compartmentalizing metabolism, and growing and reproducing.

4.4 External Prokaryotic Cell Structures

- Pili are short hair-like appendages found on many gram-negative bacteria to facilitate attachment to a surface. Conjugation pili are used for genetic transfer of DNA.
- One or more flagella, found on many rods and spirals, provide for cell motility. Each flagellum consists of a basal body attached to the flagellar filament. In nature, flagella propel bacterial cells toward nutrient sources (chemotaxis). Spirochetes have endoflagella, while other prokaryotes undergo gliding motility.
- The glycocalyx is a sticky layer of polysaccharides that protects the cell against desiccation, attaches it to surfaces, and helps evade immune cell attack. The glycocalyx can be thick and tightly bound to the cell (capsule) or thinner and loosely bound (slime layer).

4.5 The Cell Envelope

- The cell wall provides structure and protects against cell lysis. Gram-positive walls have a thick wall of peptidoglycan strengthened with teichoic acids. Gram-negative cell walls have a single layer of peptidoglycan and an outer membrane

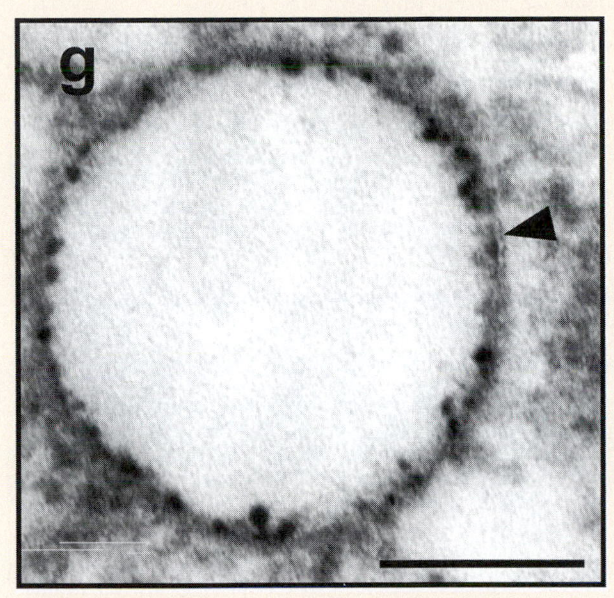

An acidocalcisome in the cytoplasm of *Agrobacterium tumefaciens*. The arrowhead indicates a portion of the bilayer membrane. (Bar = 0.1 μm.)

Observations

In fact, volutin or metachromatic granules are subcellular inclusions found in many prokaryotic cells that play this storage role. These inclusions supposedly lack a surrounding membrane though. Structures recognized as volutin granules also have been found in algae, where the granules have a surrounding membrane, an acidic pH, and high concentrations of calcium, magnesium, and polyphosphate. In fact, these eukaryotic volutin granules actually are acidocalcisomes.

Question

Could the volutin granules of bacterial cells actually be acidocalcisomes?

Hypothesis

Volutin inclusions are acidocalcisomes. If so, then volutin inclusions should contain similar storage materials.

Investigation

Isolation and biochemical characterization of volutin granules from the bacteria *Agrobacterium tumefaciens* and *Rhodospirillum rubrum* indicate that the volutin granules, as with true acidocalcisomes, are surrounded by a membrane, and are acidic subcompartments that are rich in polyphosphate and able to accumulate calcium and other elements.

Conclusion

The structural and biochemical resemblance of volutin granules of *A. tumefaciens* with eukaryotic acidocalcisomes is striking. Therefore, acidocalcisomes are the first organelle common to both prokaryotes and eukaryotes. The results further suggest a common origin for acidocalcisomes in an ancient ancestor that existed before the prokaryotic and eukaryotic lineages on the tree of life diverged.

Discussion Point

The prokaryotic/eukaryotic dividing line is becoming more fuzzy. Membrane organelles (acidocalcisomes, magnetosomes) exist and cytoskeletal proteins structurally and functionally resembling their counterparts in eukaryotic cells are present.

Discuss the prokaryotic/eukaryotic cell paradigm taking into account compartmentation and cytoskeletal organization.

---

containing lipopolysaccharide and porin proteins. Archaeal cell walls lack peptidoglycan but may have either a pseudopeptidoglycan or S-layer.

- The cell membrane represents a permeability barrier and the site of transfer for nutrients and metabolites into and out of the cell. The cell membrane reflects the fluid mosaic model for membrane structure in that the lipids are fluid and the proteins are a mosaic that can move laterally in the bilayer. The archaeal cell membrane links lipids through an ether linkage and the lipid tails are bonded together into a single monolayer.

4.6 The Cell Cytoplasm and Internal Structures

- The cytoplasm consists of the cytosol (solutes and water) and several subcompartments.
- The DNA (bacterial chromosome), located in the nucleoid, is the essential genetic information and represents the organism's genome.
- Prokaryotic cells may contain one or more plasmids, circular pieces of nonessential DNA that replicate independently of the chromosome.
- Ribosomes carry out protein synthesis while inclusion bodies store nutrients or structural building blocks.
- The prokaryotic cytoskeleton contains protein homologs to the cytoskeletal proteins in eukaryotic cells. The prokaryotic cytoskeleton proteins help determine cell shape, regulate cell division, or control chromosomal segregation during cell division.

4.7 The Prokaryotic/Eukaryotic Cell—Revisited

- Investigations of the cell biology of the prokaryotic cell are showing that compartmentation can occur; it simply does not require the diverse membranous organelles typical of eukaryotic cells.

LEARNING OBJECTIVES

After understanding the textbook reading, you should be capable of writing a paragraph that includes the appropriate terms and pertinent information to answer the objective.

1. Identify the major bacterial phyla and characteristics of each group.
2. Explain why many archaeal organisms are considered extremophiles.
3. Compare the various shapes and arrangements of prokaryotic cells.
4. Summarize how the prokaryotic processes of sensing and responding to the environment, compartmentation of metabolism, and growth and metabolism are linked to prokaryotic cell structure.
5. Assess the role of pili to bacterial colonization and infection.
6. Describe the structure of prokaryotic flagella and discuss their role to growth and survival.
7. Differentiate between a capsule and slime layer. Identify their roles in prokaryotic survival.
8. Compare and contrast the structure of a gram-positive cell wall with a gram-negative cell wall.
9. Summarize the differences between bacterial and archaeal cell walls.
10. Justify the need for a cell membrane surrounding all prokaryotic cells.
11. Explain how the structure of archaeal cell membranes differs from bacterial cell membranes.
12. Describe the structure of the nucleoid.
13. Judge the usefulness of plasmids to prokaryotic metabolism and survival.
14. List the typical inclusion bodies found in prokaryotic cells and identify their contents.
15. Describe three roles that the prokaryotic cytoskeleton plays.
16. Justify the statement, "Prokaryotic cells are as highly organized subcellularly as eukaryotic cells."

SELF-TEST

Answer each of the following questions by selecting the ***one*** answer that best fits the question or statement. Answers to even-numbered questions can be found in **Appendix C**.

1. Proteobacteria is a phylum within the domain
   A. *Bacteria.*
   B. *Eukarya.*
   C. *Eukaryotae.*
   D. *Prokaryotae.*
   E. *Archaea.*
2. Which one of the following is *not* a group within the Firmicutes?
   A. Gram-positive bacteria
   B. *Streptococcus*
   C. Methanogens
   D. *Mycoplasma*
   E. *Bacillus* and *Clostridium*
3. The _____ are prokaryotes that can be grown only in living cells.
   A. Cyanobacteria
   B. Mycoplasmas
   C. Rickettsiae
   D. Chlamydiae
   E. Actinomycetes
4. The domain *Archaea* includes all the following groups *except* the:
   A. mycoplasmas.
   B. extreme halophiles.
   C. thermoacidophiles.
   D. Crenarchaeota.
   E. Euryarchaeota.
5. Spherical bacterial cells in chains would be a referred to as a _____ arrangement.
   A. bacillus
   B. vibrio
   C. streptococcus
   D. staphylococcus
   E. tetrad
6. Which one of the following statements does *not* apply to pili?
   A. Pili are made of protein.
   B. Pili allow for attachment.
   C. Pili facilitate tissue colonization.
   D. Pili provide for motility.
   E. Pili contain adhesins.
7. Flagella are
   A. made of carbohydrate and lipid.
   B. found on all prokaryotic cells.
   C. shorter than pili.
   D. important for chemotaxis.
   E. All the above (**A–D**) are true.
8. A capsule is similar to the pili because both
   A. can represent virulence factors.
   B. are made of protein.
   C. contain dextran fibers.
   D. permit attachment to surfaces.
   E. Both **A** and **D** are true.
9. Which one of the following (**A–D**) is *not* part of the bacterial cell envelope? If all are, select **E.**
   A. Gram-negative cell wall
   B. Cell membrane
   C. Gram-positive cell wall
   D. Pili and capsule
   E. All the above (**A–D**) are part of the cell envelope.
10. Gram-negative bacteria would stain _____ with the Gram stain and have _____ in the wall.
    A. orange-red; teichoic acid
    B. orange-red; lipopolysaccharide
    C. purple; peptidoglycan
    D. purple; teichoic acid
    E. orange-red; pseudopeptidoglycan

11. A periplasmic space is found in _____ bacteria and the space contains _____.
 A. gram-negative; peptidoglycan
 B. gram-positive; digestive enzymes
 C. gram-positive; porin proteins
 D. gram-negative; outer membrane
 E. gram-positive; LPS material

12. When comparing bacterial and archaeal cell membranes, only bacterial cell membranes
 A. have three layers of phospholipids
 B. have a phospholipid bilayer.
 C. contain membrane proteins.
 D. are fluid.
 E. have ether linkages

13. Which one of the following statements about the nucleoid is *not* true?
 A. It contains a DNA chromosome.
 B. It represents a nonmembranous subcompartment.
 C. It represents an area devoid of ribosomes.
 D. It is usually a closed loop of DNA.
 E. It contains nonessential genetic information.

14. Plasmids
 A. replicate with the bacterial chromosome.
 B. contain essential growth information.
 C. may contain antibiotic resistance genes.
 D. are as large as the bacterial chromosome.
 E. Both **A** and **C** are true.

15. The prokaryotic cytoskeleton
 A. regulates cell division.
 B. helps determine cell shape.
 C. contains homologs to eukaryotic cytoskeletal proteins.
 D. Helps in the segregation of chromosomes.
 E. All the above (**A–D**) are correct.

QUESTIONS FOR THOUGHT AND DISCUSSION

Answers to even-numbered questions can be found in **Appendix C**.

1. In reading a story about a bacterium that causes a human disease, the word "bacillus" is used. How would you know if the article is referring to a bacterial shape or a bacterial genus?

2. Suppose this chapter on the structure and growth of bacteria had been written in 1940, before the electron microscope became available. Which parts of the chapter would probably be missing?

3. Why has it taken so long for microbiologists to discover membranous organelles and a cytoskeleton in prokaryotic cells?

APPLICATIONS

Answers to even-numbered questions can be found in **Appendix C**.

1. A bacterium has been isolated from a patient and identified as a gram-positive rod. Knowing that it is a human pathogen, what structures would it most likely have? Explain your reasons for each choice.

2. Another patient has a blood infection caused by a gram-negative bacterium. Why might it be dangerous to prescribe an antibiotic to treat the infection?

3. In the research lab, the gene for the cytoskeletal protein similar to eukaryotic tubulin is transferred into the DNA chromosome of a coccus-shaped bacterium. When this cell undergoes cell division, predict what shape the daughter cells will exhibit. Explain your answer.

IDENTIFICATION

Identify and label the structure on the accompanying prokaryotic cell from each of the following descriptions. Some separate descriptions may apply to the same structure. Answers to even-numbered descriptions can be found in **Appendix C**.

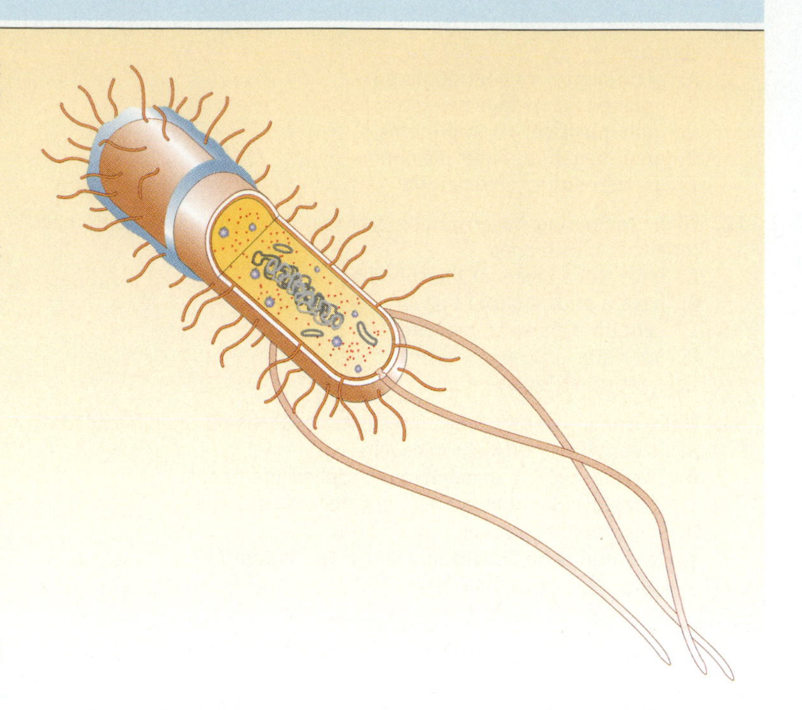

Descriptions

1. An essential structure for chemotaxis, aerotaxis, or phototaxis.
2. Contains nonessential genetic information that provides genetic variability.
3. The structure that synthesizes proteins.
4. The protein structures used for attachment to surfaces.
5. Contains essential genes for metabolism and growth.
6. Prevents cell desiccation.
7. A 70S particle.
8. Contains peptidoglycan.
9. Stores granules of starch.
10. Composed of phospholipids and protein.
11. The compartment within which most growth and metabolism occurs.
12. Regulates the passage of substances into and out of the cell.
13. Extrachromosomal loops of DNA.
14. Represents a capsule or slime layer.
15. The semifluid mass of proteins, amino acids, sugars, salts, and ions dissolved in water.

HTTP://MICROBIOLOGY.JBPUB.COM/

The site features learning, an on-line review area that provides quizzes and other tools to help you study for your class. You can also follow useful links for in-depth information, read more MicroFocus stories, or just find out the latest microbiology news.

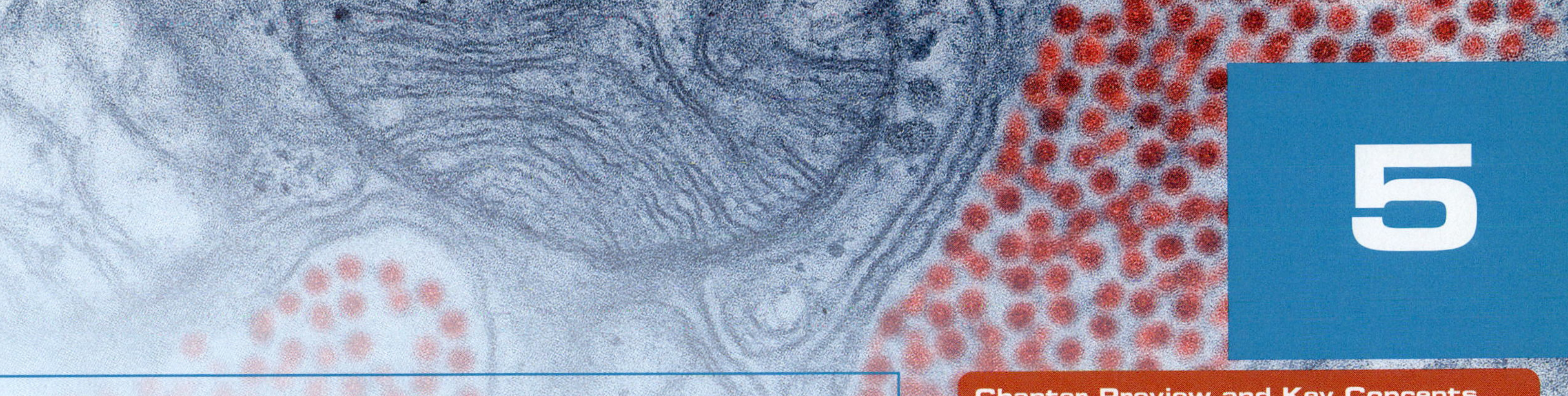

5

Prokaryotic Growth and Nutrition

But who shall dwell in these worlds if they be inhabited? . . . Are we or they Lords of the World? . . .
—Johannes Kepler (quoted in *The Anatomy of Melancholy*)

Chapter Preview and Key Concepts

5.1 Prokaryotic Reproduction
- Binary fission produces genetically-identical daughter cells.
- Prokaryotes vary in their generation times.

5.2 Prokaryotic Growth
- Bacterial population growth goes through four phases.
- Endospores are dormant structures to endure times of nutrient stress.
- Growth of prokaryotic populations is sensitive to temperature, oxygen gas, and pH.

5.3 Culture Media and Growth Measurements
- Culture media contain the nutrients needed for optimal prokaryotic growth.
- Special chemical formulations can be devised to isolate and identify some prokaryotes.
  MicroInquiry 5: Identification of Bacterial Species
- Two standard methods are available to produce pure cultures.
- Prokaryotic growth can be measured by direct and indirect methods.

Books have been written about it; movies have been made; even a radio play in 1938 about it frightened thousands of Americans. What is it? Martian life. In 1877 the Italian astronomer, Giovanni Schiaparelli, saw lines on Mars, which he and others assumed were canals built by intelligent beings. It wasn't until well into the 20th century that this notion was disproved. Still, when we gaze at the red planet, we wonder: Did life ever exist there?

We are not the only ones wondering. Astronomers, geologists, and many other scientists have asked the same question. Today microbiologists have joined their comrades, wondering if microbial life once existed on the Red Planet or, for that matter, elsewhere in our Solar System. In 1996, NASA scientists reported finding what looked like fossils of microbes inside a meteorite thought to have come from Mars. Although most now believe these "fossils" are not microbial, it only fueled the debate.

Could microbes, as we know them here on Earth, survive on Mars where the temperatures are far below 0°C and—as far as we know—there is little, if any, water? Researchers, using a device to simulate the Martian environment, placed in it microbes known to survive extremely cold environments here on Earth. Their results indicated that members of the *Archaea*, specifically the methanogens, could grow at the cold temperatures and low pressures known to exist on Mars. They concluded that life could have existed on the Red Planet in the past or "dwell in these worlds [today] if they be inhabited."

TABLE 5.1 Some Microbial Record Holders

Record
Hottest environment (Volcano Island, Italy)—235°F (113°C) *Pyrolobus fumarii* (*Archaea*)
Coldest environment (Antarctica)—5°F (−15°C) Cryptoendoliths (*Bacteria* and lichens)
Highest radiation survival—5MRad, or 5000X what kills humans *Deinococcus radiodurans* (*Bacteria*)
Deepest—3.2 km underground Many *Bacterial* and *Archaeal* species
Most acid environment—pH 0.0 (most life is at least factor of 100,000 less acidic) *Ferroplasma acidarmanus* (*Archaea*)
Most alkaline environment—pH 12.8 (most life is at least factor of 1000 less basic) Proteobacteria (*Bacteria*)
Longest in space: 6 years (NASA satellite) *Bacillus subtilis* (*Bacteria*)
High pressure environment—1200 times atmospheric pressure (Mariana Trench) *Moritella, Shewanella* and others (*Bacteria*)
Saltiest environment—47% salt, (15 times human blood saltiness) several *Bacterial* and *Archaeal Species*

Source: http://www.astrobio.net/news/.

So, microbiologists have joined the search for extraterrestrial life. This seems a valid pursuit since the extremophiles found here on Earth survive, and even require, living in extreme environments (TABLE 5.1)—some not so different from Mars (FIGURE 5.1). If life did or does exist on Mars, it almost certainly was or is microbial—most likely prokaryotic.

In 2004, NASA sent two spacecrafts to Mars to look for indirect signs of past life. Scientists here on Earth monitored instruments on the Mars rovers, *Spirit* and *Opportunity*, designed to search for signs suggesting water once existed on the planet. Some findings suggest there are areas where salty seas once washed over the plains of Mars, creating a life-friendly environment.

Opportunity found evidence for ancient shores of a large body of surface water that contained currents, which left their marks in rocks at the bottom of what once was a sea. The rover also found a distinct chemical makeup in the rocks and unique layering patterns suggestive of slow-moving water in an evaporating sea.

Did or does life exist on Mars? Perhaps one day when human explorers or more sophisticated spacecraft reach Mars, we will know.

FIGURE 5.1 **The Martian Surface?** This barren-looking landscape is not Mars but the Atacama Desert in Chile. It looks similar to photos taken by the Mars rovers *Spirit* and *Opportunity*.

Q: Does this area look like a habitable place for life, even microbial life?

Whether microorganisms are here on Earth in a moderate or extreme environment, or on Mars, there are certain physical and chemical requirements they must possess to survive, reproduce, and grow. In this chapter, we explore the process of cell reproduction in prokaryotic cells as compared to that in eukaryotic microbial cells. We also examine the physical and chemical conditions required for growth of bacterial and archaeal cells, and discover the ways that prokaryotic growth can be measured.

As we have been emphasizing in this text, the domains of organisms may have different structures and patterns, yet carry out the same process. This again is illustrated clearly by cell division and growth processes.

5.1 Prokaryotic Reproduction

Growth in the microbial world usually refers to an increase in the numbers of individuals; that is, an increase in the population size with each cell carrying the identical genetic instructions of the parent cell. **Asexual reproduction** is a process to maintain genetic constancy while increasing cell numbers. In eukaryotic microbes, an elaborate interaction of microtubules and proteins with chromosomes allows for the precise events of mitosis and cytokinesis. Prokaryotes can accomplish the same thing without the microtubular involvement.

Most Prokaryotes Reproduce by Binary Fission

KEY CONCEPT

- Binary fission produces genetically-identical daughter cells.

Most prokaryotes reproduce by an asexual process called **binary fission**, which usually occurs after a period of growth in which the cell doubles in mass. At this time, the chromosome (DNA) replicates and the two DNA molecules separate (FIGURE 5.2). Chromosome segregation in prokaryotes is not well understood. Unlike eukaryotes, prokaryotes lack a mitotic spindle to separate replicated chromosomes. The segregation process in prokaryotes involves specialized chromosomal-associated proteins but there is no clear picture describing how most of these proteins work to ensure accurate chromosome segregation.

In any event, cell fission at midcell involves **cytokinesis**, an inward pinching of the cell envelope (cell membrane and cell wall) to separate the mother cell into two genetically identical daughter cells. The tubulin homolog found in prokaryotic cells (see Chapter 4) is part of the fission ring apparatus that organizes invagination of the cell envelope.

Cytokinesis occurs in two different ways in the *Bacteria*. In gram-negative cells, like *Escherichia coli*, division occurs by the constriction of the cell envelope, followed by cell separation. In gram-positive cells, such as *Bacillus licheniformis*, constriction allows a newly synthesized wall to form a septum between daughter cells. Cell separation then

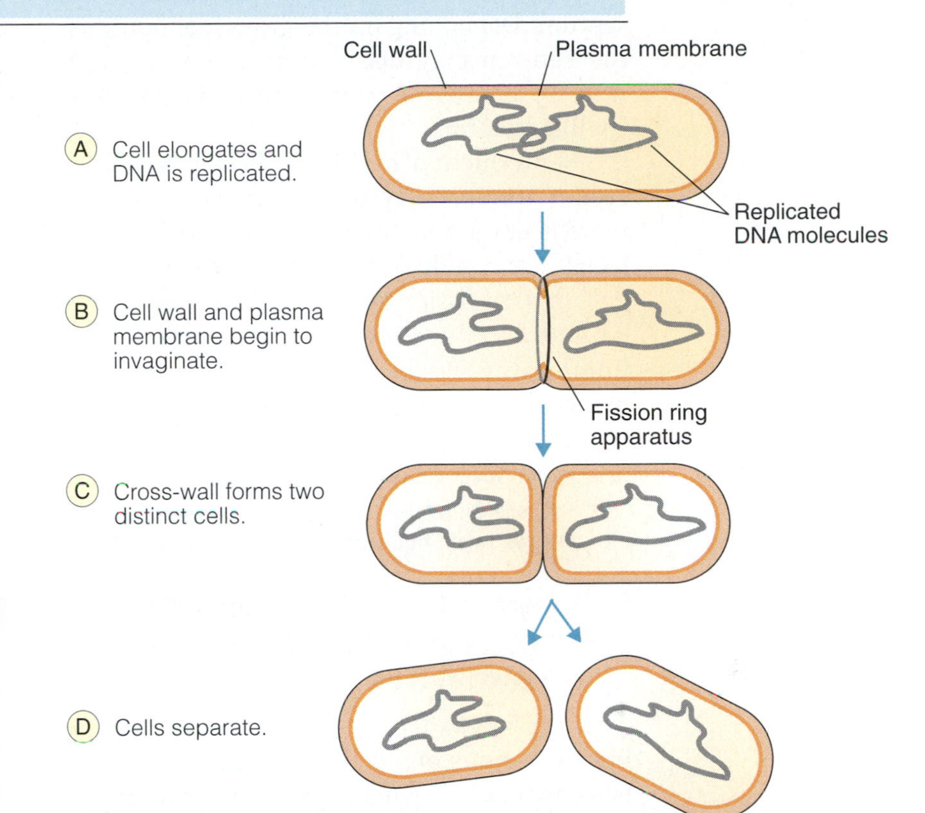

(A)

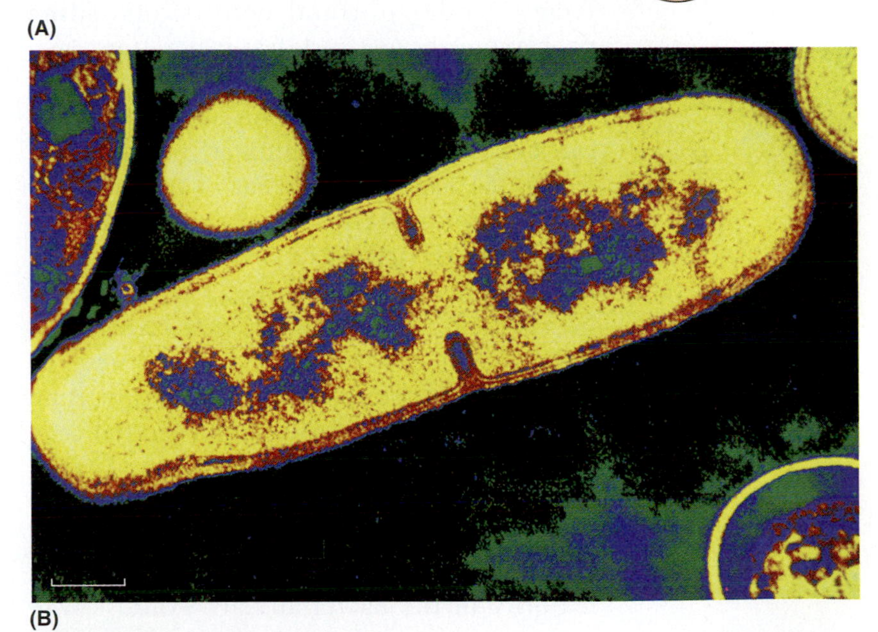

(B)

FIGURE 5.2 **The Process of Binary Fission.** (**A**) Binary fission in a rod-shaped cell begins with DNA replication and segregation. Cytokinesis occurs at the midcell fission ring apparatus as a constriction of the cell envelope where the mother cell separates into two genetically-identical daughter cells. (**B**) A false-color transmission electron micrograph of a cell of *Bacillus licheniformis* undergoing binary fission. The invagination of the cell wall and membrane is evident at midcell. (Bar = 0.25 μm.)

Q: How would binary fission differ for a prokaryotic organism having cells arranged in chains and another that forms single cells?

occurs by dissolution of the material in the septum. Depending on the growth conditions, the septum may dissolve at a slow enough rate for chains of connected cells (streptobacilli) to form.

Reproduction by binary fission seems to confer immortality to prokaryotes because there is never a moment at which the first bacterial cell has died. Each mother cell undergoes binary fission to become the two young daughter cells. However, the perception of immortality has been challenged by experiments suggesting prokaryotes do age (MicroFocus 5.1).

CONCEPT AND REASONING CHECKS

5.1 Propose an explanation of how a bacterial cell "knows" when to divide.

Prokaryotes Reproduce Asexually

KEY CONCEPT

- Prokaryotes vary in their generation times.

The interval of time between successive binary fissions of a cell or population of cells is known as the **generation time** (or doubling time). Under optimal conditions, some prokaryotes have a very fast generation time; for others, it is much slower. For example, the optimal generation time for *Staphylococcus aureus* is about 30 minutes; for *Mycobacterium tuberculosis*, the agent of tuberculosis, it is approximately 15 hours; and for the syphilis spirochete, *Treponema pallidum*, it is a long 33 hours.

One enterprising mathematician calculated that if *E. coli* binary fissions were to continue at their optimal generation time (15 minutes) for 36 hours, the bacterial cells would cover the surface of the Earth! Thankfully, this will not occur because of the limitation of nutrients and the loss of ideal physical factors required for growth. The majority of the bacterial cells would starve to death or die in their own waste.

The generation time is useful in determining the amount of time that passes before disease symptoms appear in an infected individual; faster division times often mean a shorter incubation period for a disease. Suppose you eat an undercooked hamburger contaminated with the pathogen *E. coli* O157:H7, which has one of the shortest generation times—just 20 minutes under optimal conditions (FIGURE 5.3). If you ingested one cell (more likely several hundred at least) at 8:00 PM this evening, two would be present by 8:20, four by 8:40, and eight by 9:00. You would have over 4,000 by midnight. By 3:00 AM, there would be over 2 million. Depending on the response of the immune system, it is quite likely that sometime during the night you would know you have food poisoning.

■ **Incubation period:** The time from entry of a pathogen into the body until the first symptoms appear.

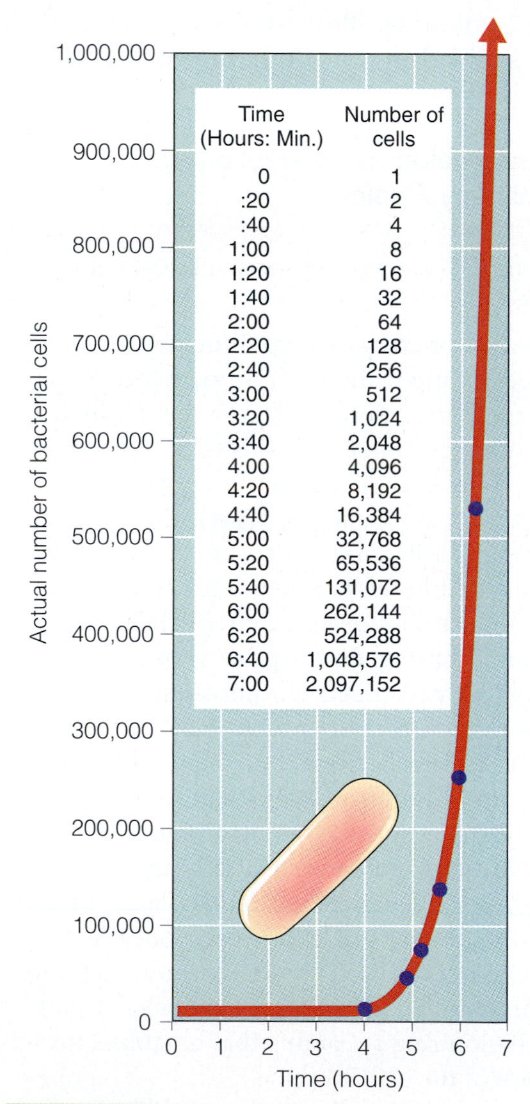

Time (Hours: Min.)	Number of cells
0	1
:20	2
:40	4
1:00	8
1:20	16
1:40	32
2:00	64
2:20	128
2:40	256
3:00	512
3:20	1,024
3:40	2,048
4:00	4,096
4:20	8,192
4:40	16,384
5:00	32,768
5:20	65,536
5:40	131,072
6:00	262,144
6:20	524,288
6:40	1,048,576
7:00	2,097,152

FIGURE 5.3 A Skyrocketing Bacterial Population. The number of *E. coli* cells progresses from 1 cell to 2 million cells in a mere 7 hours. The J-shaped growth curve gets steeper and steeper as the hours pass. Only a depletion of food, buildup of waste, or some other limitation will halt the progress of the curve.

Q: What is the generation time for the bacterial species in this figure?

MICROFOCUS 5.1: Evolution/Environmental Microbiology

A Microbe's Life

It used to be thought that prokaryotes do not age—they are immortal. This might seem obvious considering a mother cell divides at the mid-point by binary fission to become the two genetically equal daughter cells. However, new research suggests that although the DNA may be identical, after several generations of binary fission, the population consists of cells of different ages and the oldest ones have the longest generation time.

Eric Stewart and his collaborators at INSERM, the National Institute of Health and Medical Research in Paris, filmed *Escherichia coli* cells as they divided into daughter cells on a specially designed microscope slide. A record of every daughter cell (total of 35,000 individual cells) was recorded for nine generations over a period of six hours. Then, a custom-designed computer system analyzed the micrographs.

The group's results suggest that although the cells may divide symmetrically, daughter cells are not morphologically or physiologically symmetrical. Each contains cellular poles of different ages.

When a mother cell divides, each daughter cell inherits one end or pole of the mother cell. The region where the cells split develops into the other pole (see figure). For example, in the first division, the mother cell splits with a new wall (red). When the daughter cells grow in size they contain an old pole (brown) and a new one (red). When each of these cells divides, two have the oldest pole (brown) and youngest pole (green) while the two other cells have a younger pole (red) and a youngest pole (green). So after just two divisions, there are two populations of daughter cells: two have oldest and youngest poles while two have younger poles and youngest poles. According to Stewart's group, the two cells with the oldest/youngest poles grew 2.2 percent slower than the cells with younger/youngest poles.

As more and more binary fissions occur, the difference in age between daughter cells will continue to increase. The bottom line is that cells inheriting older and older poles experience longer generation times, reduced rates of offspring formed, and increased risk of dying compared to cells with younger, newer poles. This loss of fitness is called **senescence**. Note: Stewart's group could not follow any cells to actual death because the cell populations eventually had so many cells, even their computer program could not keep them all independently recoded.

Exactly why the older cells senesce is not understood. However, if the results from Stewart's group are verified by others, it would at least appear that bacterial cells cannot escape the aging process. Even a microbe's life is limited.

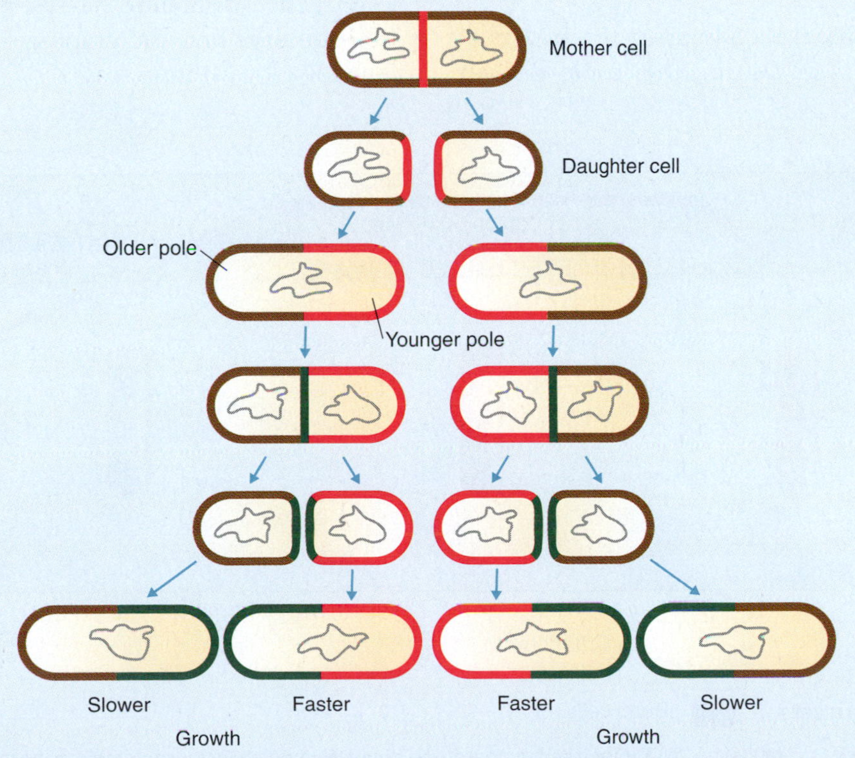

Two successive binary fissions produce daughter cells with various aged poles (brown = oldest; red = younger; green = youngest).

Prokaryotes are subject to the same controls on growth as all other organisms on Earth. Let's examine the most important growth factors conferring optimal generation times.

CONCEPT AND REASONING CHECKS

5.2 Identify several factors that would slow the generation time of a bacterial species like *E. coli*.

5.2 Prokaryotic Growth

In the previous section, we discovered how fast prokaryotic cells can grow under ideal circumstances. Let's look at the growth of bacterial populations in a little more detail.

A Bacterial Growth Curve Illustrates the Dynamics of Growth

KEY CONCEPT

- Bacterial population growth goes through four phases.

A typical **bacterial growth curve** for a population illustrates the events occurring over time (FIGURE 5.4). Whether several bacterial cells infect the human respiratory tract or are transferred to a tube of fresh growth medium in the laboratory, four distinct phases of growth occur: the lag phase; the logarithmic phase; the stationary phase; and the decline phase.

The **lag phase** encompasses the first portion of the curve. During this time, no cell divisions occur. Rather, bacterial cells are adapting to their new environment. In the respiratory tract, scavenging white blood cells may engulf and destroy some of the cells; in growth media, some cells may die from the shock of transfer or the inability to adapt to the new environment. The actual length of the lag phase depends on the metabolic activity in the remaining cells. They must grow in size, store nutrients, and synthesize enzymes—all in preparation for binary fission.

The population then enters an active stage of growth called the **logarithmic phase** (or **log phase**). This is the exponential growth described above for *E. coli*. In the log phase, all cells are undergoing binary fission and the generation time is dependent on the species and environmental conditions present. As each generation time passes, the number of cells doubles and the graph rises in a straight line on a logarithmic scale.

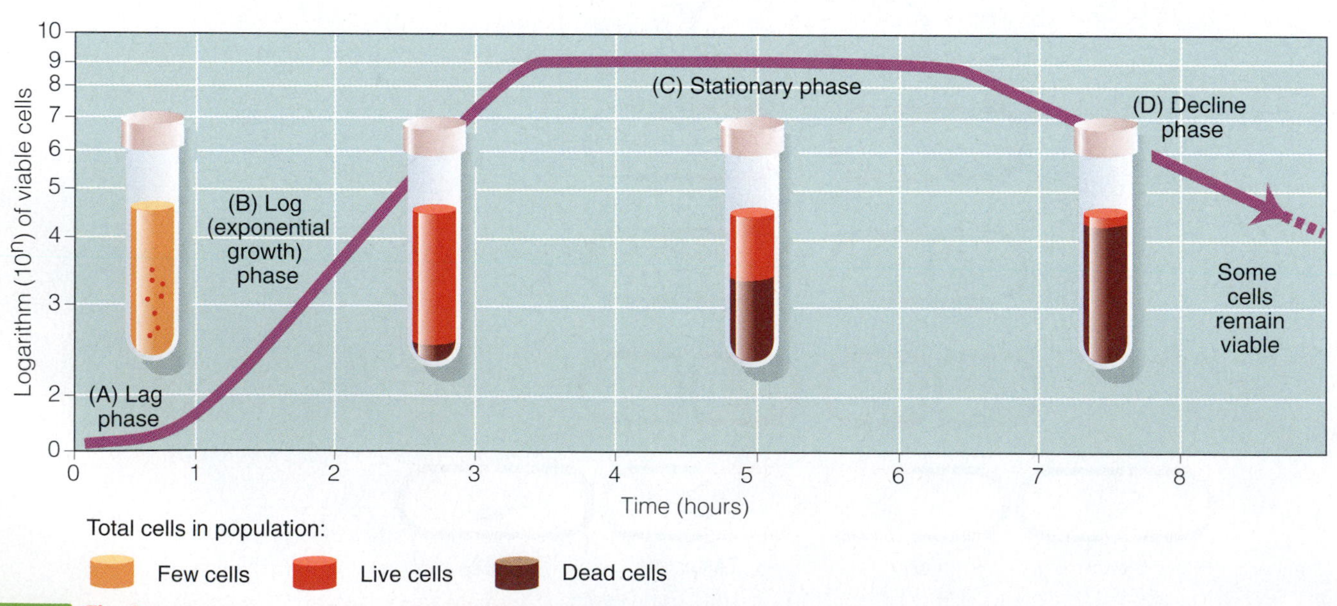

FIGURE 5.4 **The Growth Curve for a Bacterial Population.** (**A**) During the lag phase, the population numbers remain stable as bacterial cells prepare for division. (**B**) During the logarithmic (exponential growth) phase, the numbers double with each generation time. Environmental factors later lead to cell death, and (**C**) the stationary phase shows a stabilizing population. (**D**) The decline phase is the period during which cell death becomes substantial.

Q: Why would antibiotics work best on cells in the log phase?

In humans, disease symptoms usually develop during the log phase because the bacterial cells cause tissue damage. Coughing or fever may occur, and fluid may enter the lungs if the air sacs are damaged. If the bacterial cells produce toxins, tissue destruction may become apparent. During the log phase in our broth tube, the medium becomes cloudy (turbid) due to increasing cell numbers. If plated on solid medium, bacterial growth will be so vigorous that visible colonies appear and each colony may consist of millions of cells (FIGURE 5.5). Vulnerability to antibiotics is also highest at this active stage of growth because many antibiotics affect actively metabolizing cells.

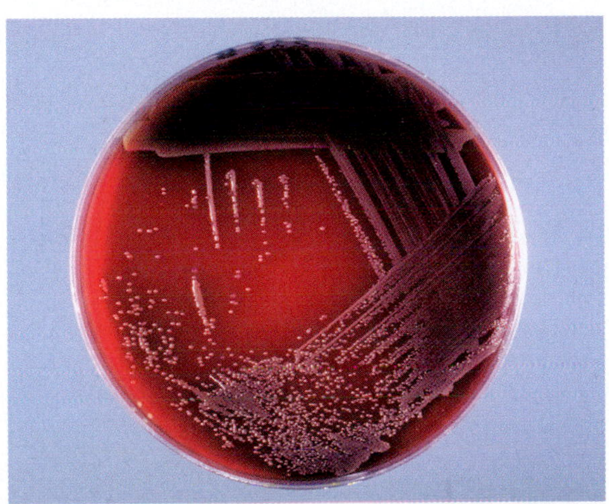

(A)

(B)

FIGURE 5.5 **Two Views of Bacterial Colonies.**
(**A**) Bacterial colonies cultured on blood agar in a culture dish. Blood agar is a mixture of nutrient agar and blood cells. It is widely used for growing bacterial colonies.
(**B**) Close-up of typhoid bacteria *(Salmonella typhi)* colonies being cultured on a growth medium.

Q: How did each colony in (A) or (B) start?

After some days (in an infection) or hours (in a culture tube), the vigor of the population changes and, as the reproductive and death rates equalize, the population enters a plateau, called the **stationary phase**. In the respiratory tract, antibodies from the immune system are attacking the bacterial cells, and phagocytosis by white blood cells adds to their destruction. In the culture tube, available nutrients become scarce and waste products accumulate. Factors such as oxygen also may be in short supply. This limitation of nutrients and buildup of waste materials leads to the death of many cells.

If nutrients in the external environment remain limited or the quantities become exceeding low, the population enters a **decline phase** (or exponential death phase). Now the number of dying cells far exceeds the number of new cells formed. A bacterial glycocalyx may forestall death by acting as a buffer to the environment, and flagella may enable organisms to move to a new location. For many species, though, the history of the population ends with the death of the last cell. When we discuss the progression of human diseases in Chapter 18, we will see a similar curve for the stages of a disease.

For some bacterial species, especially soil bacteria, they can escape cell death by forming endospores. Let's examine these amazing dormancy structures next.

CONCEPT AND REASONING CHECKS

5.3 Suppose the bacterial growth curve in Figure 5.4 was produced for a bacterium growing at an optimal temperature of 37°C. Construct a growth curve if the same bacterium was grown at a suboptimal temperature of 23°C.

Endospores Are a Response to Nutrient Limitation

KEY CONCEPT

- Endospores are dormant structures to endure times of nutrient stress.

A few gram-positive bacterial species, especially soil bacteria belonging to the genera *Bacillus* and *Clostridium*, produce highly resistant structures called **endospores** or, simply, spores (FIGURE 5.6). As described in the previous section, bacterial cells normally grow,

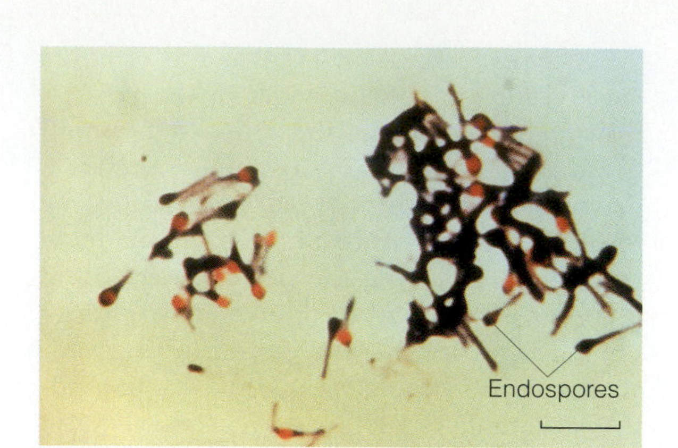

(A)

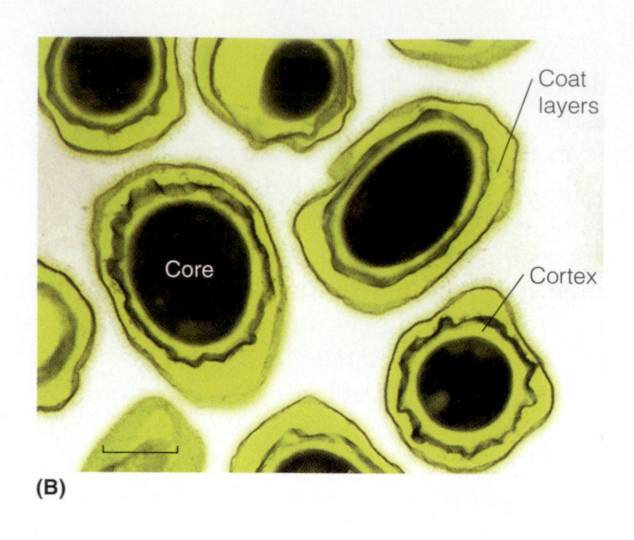

(B)

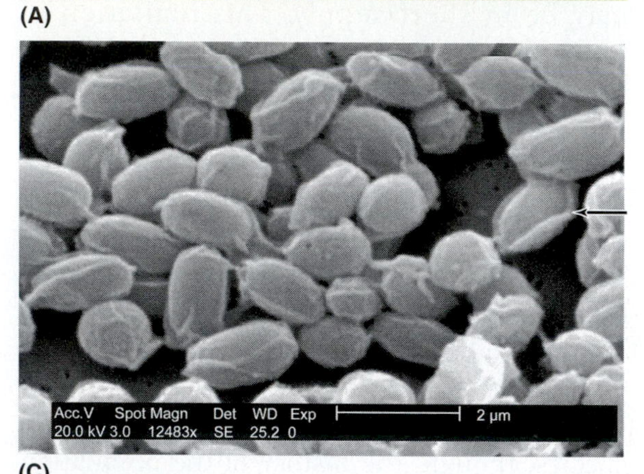

(C)

FIGURE 5.6 Three Different Views of Bacterial Spores. (**A**) A view of *Clostridium* with the light microscope, showing terminal spore formation. Note the characteristic drumstick appearance of the spores. (Bar = 5.0 µm.) (**B**) The fine structure of a *Bacillus anthracis* spore seen using the transmission electron microscope. The visible spore structures include the core, cortex and coat layers. (Bar = 0.5 µm.) (**C**) A scanning electron microscope view of a germinating spore (arrow). Note that the spore coat divides equatorially along the long axis, and as it separates, the vegetative cell emerges. (Bar = 2.0 µm.)

Q: If an endospore is resistant to so many environmental conditions, how does a spore "know" conditions are favorable for germination?

Vegetative: Referring to cells actively metabolizing and obtaining nutrients.

mature, and reproduce as vegetative cells. However, when nutrients such as carbon or nitrogen are limiting and the population density reaches a critical mass, species of *Bacillus* and *Clostridium* enter stationary phase and begin spore formation or **sporulation**.

Sporulation begins when the bacterial chromosome replicates and binary fission is characterized by an asymmetric cell division (FIGURE 5.7). The smaller cell, the prespore, will become the mature endospore, while the larger mother cell will commit itself to maturation of the endospore before undergoing lysis.

Depending on the exact asymmetry of cell division, the endospore may develop at the end of the cell, near the end, or at the center of the cell (the position is useful for species identification purposes).

The prespore cell contains cytoplasm and DNA, and a large amount of **dipicolinic acid**, a unique organic substance that helps stabilize the proteins and DNA. After the cell is engulfed by the mother cell, thick layers of peptidoglycan form the cortex, followed by a series of protein coats that protect the contents further. The mother cell then disintegrates and the spore is freed. It should be stressed that sporulation is not a reproductive process. Rather, the endospore represents a dormant stage in the life of the bacterial species.

Endospores are probably the most resistant living things known. Desiccation has little

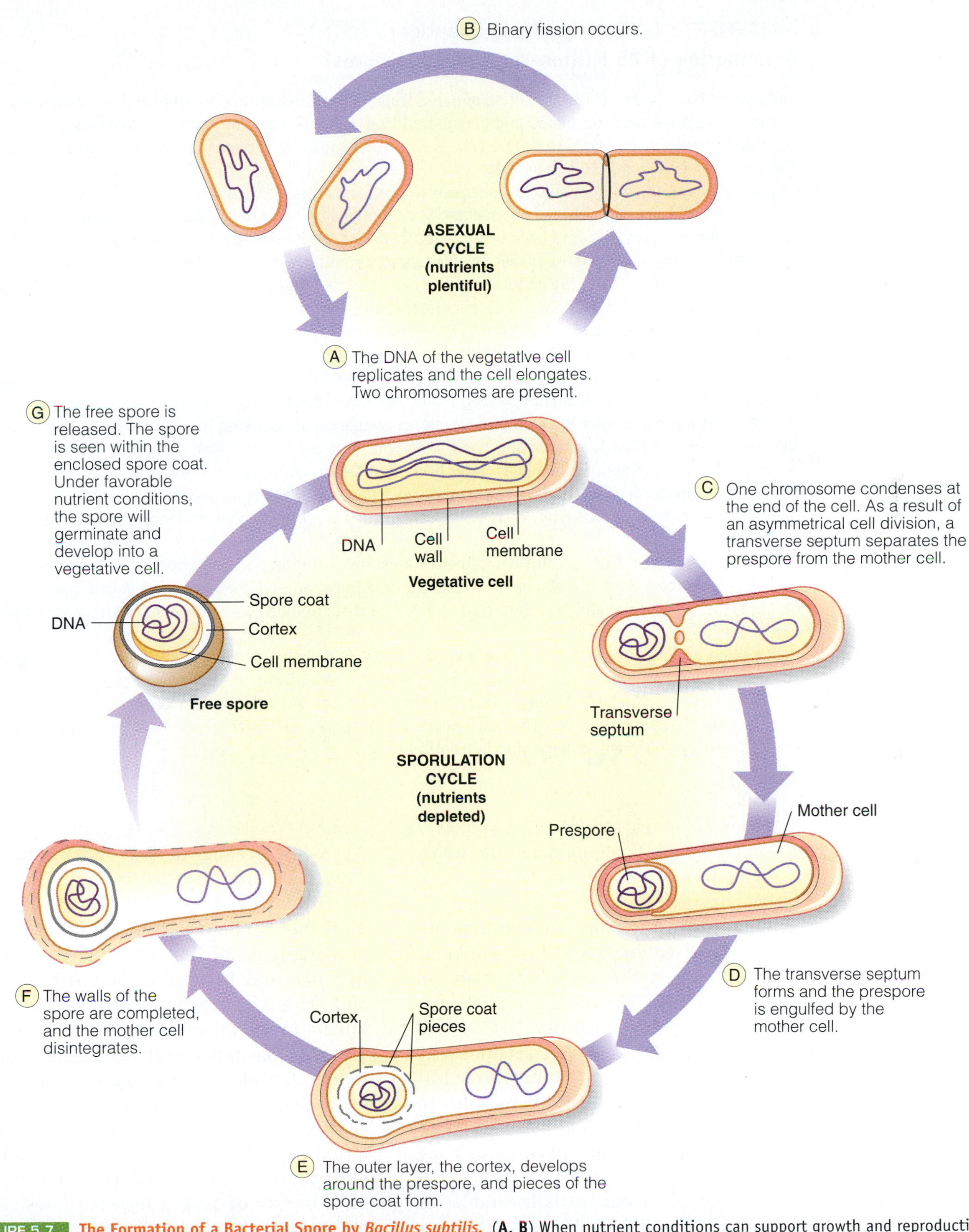

FIGURE 5.7 **The Formation of a Bacterial Spore by *Bacillus subtilis*.** (**A, B**) When nutrient conditions can support growth and reproduction, vegetative cells continue through asexual cycles of binary fission. (**C–G**) When nutrient conditions become limiting (e.g., carbon, nitrogen), endospore-formers, such as *B. subtilis,* enter the sporulation cycle shown here.

Q: Hypothesize how a vegetative cell "knows" nutrient conditions are limiting.

MICROFOCUS 5.2: Being Skeptical
Germination of 25 Million-Year-Old Endospores?

Endospores have been recovered and germinated from various archaeological sites and environments. Living spores have been recovered and germinated from the intestines of Egyptian mummies several thousand years old. In 1983, archaeologists found viable spores in sediment lining Minnesota's Elk Lake. The sediment was over 7,500 years old.

All these reports though pale in comparison to the controversial discovery reported in 1996 by researcher Raul Cano of California Polytechnic State University, San Louis Obispo. Cano found bacterial spores in the stomach of a fossilized bee trapped in amber—a hardened resin—produced from a tree in the Dominican Republic. The fossilized bee was about 25 million-years-old. When the amber was cracked open and the material from the abdomen of the bee extracted and placed in nutrient medium, the equally ancient spores germinated. With microscopy, the cells from a colony were very similar to *Bacillus sphaericus,* which is found today in bees in the Dominican Republic. Is it possible for an endospore to survive for 25 million years—even if it is encased in amber?

Critics were quick to claim the bacterial species may represent a modern-day species that contaminated the amber sample being examined. However, Professor Cano had carried out appropriate and rigorous decontamination procedures and sterilized the amber sample before cracking it open. He also carried out all the procedures in a class II laminar flow hood, which prevents outside contamination from entering the working area. In addition, the hood had never been used for any other bacterial extraction processes. Several other precautions were added to eliminate any chance that the spores were modern-day contaminants from an outside source. Still, many scientists question whether all contamination sources had been identified.

The major question that remains is whether DNA can remain intact and functional after so long a period of dormancy. Does it really have a capability of replication and producing new vegetative growth? Granted, the DNA presumably was protected in a resistant spore, but could DNA remain intact for 25 million years?

Research on bacterial DNA suggests the maximum survival time is about 400,000 to 1.5 million years. If true, then the 25 million-year-old spores could not be viable. But that is based on current predictions and they may be subject to change as more research is carried out with ancient DNA.

The verdict? It seems unlikely that such ancient endospores could germinate after 25 million years. Perhaps new evidence will change that perception.

effect on the spore. By containing little water, endospores also are heat resistant and undergo very few chemical reactions. These properties make them difficult to eliminate from contaminated medical materials and food products. For example, endospores can remain viable in boiling water (100°C) for 2 hours. When placed in 70 percent ethyl alcohol, endospores have survived for 20 years. Humans can barely withstand 500 rems of radiation, but endospores can survive one million rems. In this dormant condition, endospores can "survive" for extremely long periods of time (MicroFocus 5.2).

Rems (Roentgen Equivalent Man): A measure of radiation dose related to biological effect.

When the environment is favorable for cell growth, the protective layers break down and each endospore germinates into a vegetative cell.

A few serious diseases in humans are caused by spore formers. The most newsworthy has been anthrax, the agent of the 2001 bioterror attack through the mail. This potentially deadly disease, originally studied by Koch and Pasteur, is caused by *Bacillus anthracis* (Chapter 11). Inhaled spores germinate in the lower respiratory tract and the resulting vegetative cells secrete three deadly toxins. Botulism, gas gangrene, and tetanus are diseases caused by different species of *Clostridium.* Clostridial endospores often are found in soil, as well as in human and animal intestines. However, the environment must be free of oxygen for the spores to germinate to vegetative cells. Dead tissue in a wound provides such an environment for the development of tetanus and gas gangrene

(Chapter 11), and a vacuum-sealed can of food is suitable for the development of botulism (Chapter 10).

Killing endospores can be a tough task. Heating them for many hours under high pressure will do the trick. If they contaminate machinery, such as they did in mail sorting equipment in the 2001 anthrax attacks, there are potent but highly dangerous chemical methods to kill the spores (Chapter 23). Postal workers who were exposed to the spores were effectively treated with antibiotics that can kill any newly-germinated endospores before the vegetative cells can produce and secrete the deadly toxins.

CONCEPT AND REASONING CHECKS

5.4 Hypothesize why gram-negative and most gram-positive bacterial species cannot produce endospores.

Optimal Prokaryotic Growth Is Dependent on Several Physical Factors

KEY CONCEPT

- Growth of prokaryotic populations is sensitive to temperature, oxygen gas, and pH.

Now that we have examined the reproduction and growth of prokaryotes, let's examine the essential physical and chemical factors influencing prokaryotic cell growth.

Temperature. Temperature is one of the most important factors governing growth. Each prokaryotic species has an optimal growth temperature and an approximate 30° range, from minimum to maximum, over which the cells will grow but with a slower generation time (FIGURE 5.8). In general, prokaryotes can be assigned to one of four groups based on their optimal growth temperature.

Prokaryotes that have their optimal growth rates below 15°C but can still grow at 0°C to 20°C are called **psychrophiles** (*psychro* = "cold"). Since about 70 percent of the Earth is covered by oceans having deep water temperatures below 5°C, psychrophiles represent a group of bacterial and archaeal extremophiles that make up the largest portion of the global prokaryotic community. In fact, many psychrophiles can grow as fast at 4°C as *E. coli* does at 37°C. On the other hand, at these low temperatures, psychrophiles could not be human pathogens because they cannot grow at the warmer 37°C body temperature.

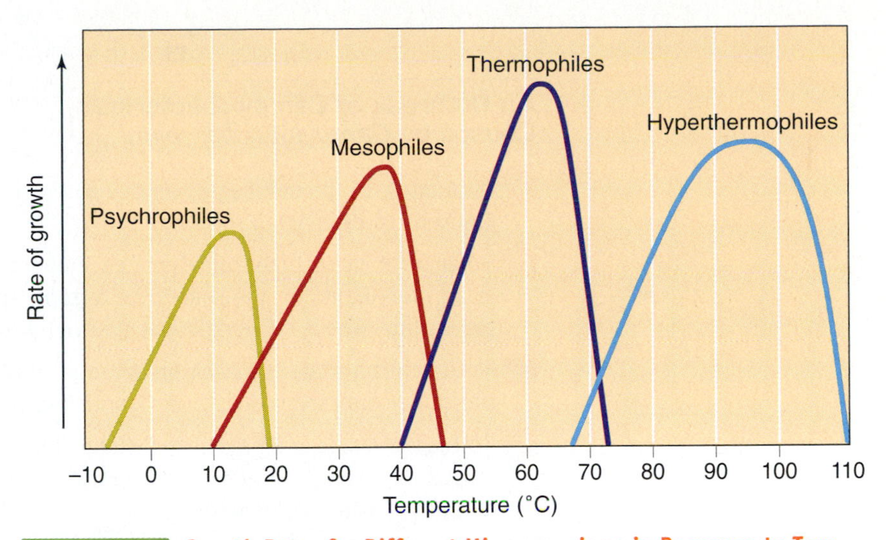

FIGURE 5.8 Growth Rates for Different Microorganisms in Response to Temperature. Temperature optima and ranges define the growth rates for different types of microorganisms. Notice that the growth rates decline quite rapidly to either side of the optimal growth temperature.

Q: Propose what adaptations are needed for prokaryotes to survive at the psychrophilic or thermophilic extremes.

At the opposite extreme are the **thermophiles** (*thermo* = "heat") that multiply best at temperatures around 60°C but still multiply from 40°C to 70°C. Thermophiles are present in compost heaps and hot springs, and are important contaminants in dairy products because they survive pasteurization temperatures. However, thermophiles pose little threat to human health because they do not grow well at the cooler temperature of the body. Opposite to the psychrophiles, thermophiles have highly saturated fatty acids in their cell membranes to stabilize these structures. They also contain heat-stable proteins and enzymes.

There also are many *Archaea* that grow optimally above 80°C. These **hyperthermophiles** have been isolated from seawater brought up from hot-water vents along rifts on the floor of the Pacific Ocean. Because the high pressure keeps the water from boiling, some of these prokaryotes can grow at an astonishing 113°C (see Table 5.1).

Most of the best-characterized prokaryotes are **mesophiles** (*meso* = "middle"), which thrive at the middle temperature range of 10° to 45°C. This includes the pathogens able to grow in the human body. Mesophiles often can grow at temperatures substantially below

Textbook CASE 5

An Outbreak of Campylobacteriosis Caused by *Campylobacter jejuni*

1. On August 15, a cook began his day by cutting up raw chickens to be roasted for dinner.
2. He also cut up lettuce, tomatoes, cucumbers, and other salad ingredients on the same countertop. The countertop surface where he worked was unusually small.
3. For lunch that day, the cook prepared sandwiches on the same countertop. Most were garnished with lettuce.
4. Restaurant patrons enjoyed sandwiches for lunch and roasted chicken for dinner. Many patrons also had a portion of salad with their meal.
5. During the next three days, 14 people experienced stomach cramps, nausea, and vomiting.
6. Public health officials learned that all the affected patrons had eaten salad with lunch or dinner. *Campylobacter,* a bacterial pathogen of the intestines, was located in their stools.
7. On inspection, microbiologists concluded that the chicken was probably contaminated with *Campylobacter jejuni* (see figure). However, the microbiologists concluded that the cooked chicken was not the cause of the illness.
8. Microbiologists concluded that *C. jejuni* from the raw chicken was the source.

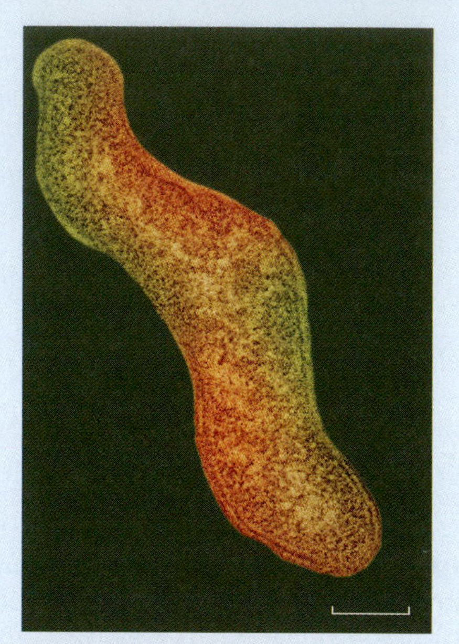

False-color transmission electron micrograph of *C. jejuni.* (Bar = 0.5 μm.)

Questions

A. *Why would the chicken not be the source for the illness?*

B. *Why was the raw chicken identified as the source?*

C. *How, in fact, did the patrons become ill?*

For additional information see http://www.cdc.gov/mmwr/preview/mmwrhtml/00051427.htm.

their normal range. For example, refrigerated foods can harbor mesophiles that will grow very slowly and cause food spoilage. *Staphylococcus aureus* can contaminate improperly handled or prepared cold cuts, salads, or various leftovers. The slow growth of these organisms at refrigeration temperature (5°C) can result in the deposit of toxins in the food products. When such foods are consumed without heating, the toxins may cause food poisoning. Other examples of mesophiles growing in the cold are *Campylobacter* species, which are the most frequently identified cause of infective diarrhea (Textbook Case 5). Because these organisms are not truly psychrophilic, some microbiologists prefer to describe them as psychrotrophic or psychrotolerant; they will survive at 0°C but prefer to grow at typical mesophile temperatures.

Oxygen. The growth of many prokaryotes depends on a plentiful supply of oxygen, and in this respect, such **obligate aerobes** are similar to eukaryotic organisms—they must use oxygen gas as a final electron acceptor to

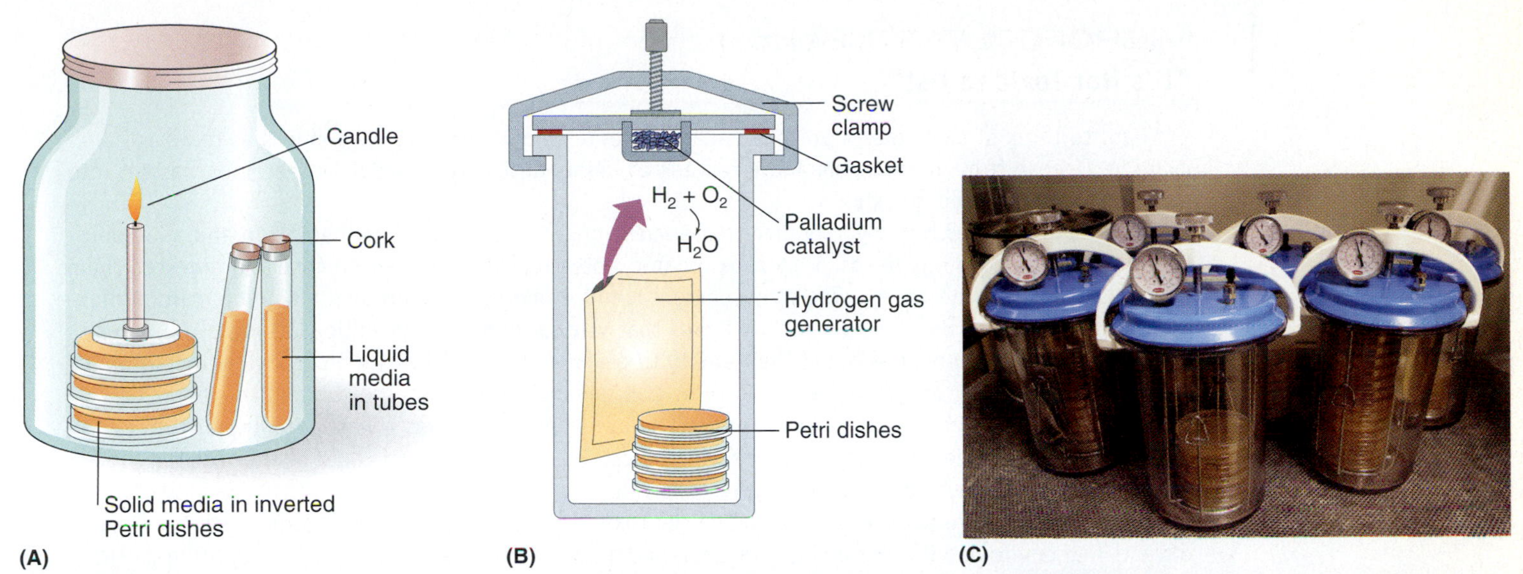

FIGURE 5.9 **Bacterial Cultivation in Different Gas Environments.** Two types of cultivation methods are shown for bacterial species that grow poorly in an oxygen-rich environment. (**A**) A candle jar, in which microaerophilic bacterial cells grow in an atmosphere where the oxygen is reduced by the burning candle. (**B, C**) An anaerobic jar, in which hydrogen is released from a generator and then combines with oxygen through a palladium catalyst to form water and create an anaerobic environment.

Q: In which jar would a facultative aerobe grow?

make cellular energy. Other species, such as *Treponema pallidum*, the agent of syphilis, are termed **microaerophiles** because they survive in environments where the concentration of oxygen is relatively low. In the body, certain microaerophiles cause disease of the oral cavity, urinary tract, and gastrointestinal tract. Conditions can be established in the laboratory to study these microbes (FIGURE 5.9A).

The **anaerobes**, by contrast, are prokaryotes that do not or cannot use oxygen. Many are **aerotolerant**, meaning they are insensitive to oxygen. Others are **obligate anaerobes**, which are inhibited or killed if oxygen is present. This means they need other ways to make cell energy. Some anaerobic prokaryotes, such as *Thiomargarita namibiensis* discussed in MicroFocus 3.5, use sulfur in their metabolic activities instead of oxygen, and therefore they produce hydrogen sulfide (H_2S) rather than water (H_2O) as a waste product of their metabolism. Others we have already encountered, such as the ruminant archaeal organisms that produce methane as the by-product of the energy conversions. In fact, life originated on Earth in an anaerobic environment consisting of methane and other gases (MicroFocus 5.3).

Some species of anaerobic bacteria cause disease in humans. For example, the *Clostridium* species that cause tetanus and gas gangrene multiply in the dead, anaerobic tissue of a wound and produce toxins causing tissue damage. Another species of *Clostridium* multiplies in the oxygen-free environment of a vacuum-sealed can of food, where it produces the lethal toxin of botulism.

Among the most widely used methods to establish anaerobic conditions in the laboratory is the GasPak system, in which hydrogen reacts with oxygen in the presence of a catalyst to form water, thereby creating an oxygen-free atmosphere (FIGURE 5.9B).

Many prokaryotes are neither aerobic nor anaerobic, but "facultative." **Facultative** prokaryotes grow in either the presence or a reduced concentration of oxygen. This group includes many staphylococci and streptococci as well as members of the genus *Bacillus* and a variety of intestinal rods, among them *E. coli*. A facultative aerobe prefers anaerobic conditions (but also grows aerobically), while a facultative anaerobe prefers oxygen-rich conditions (but also grows anaerobically).

A common way to test an organism's oxygen sensitivity is to use a **thioglycollate broth**, which binds free oxygen so that only fresh oxygen entering at the top of the tube would be available (FIGURE 5.10).

MICROFOCUS 5.3: Evolution

"It's Not Toxic to Us!"

It's hard to think of oxygen as a poisonous gas, but billions of years ago, oxygen was as toxic as cyanide. One whiff by an organism and a cascade of highly destructive oxidation reactions was set into motion. Death followed quickly.

Difficult to believe? Not if you realize that ancient organisms relied on fermentation and anaerobic chemistry for their energy needs. They took organic materials from the environment and digested them to release the available energy. The atmosphere was full of methane, hydrogen, ammonia, carbon dioxide, and other gases. But no oxygen. And it was that way for hundreds of millions of years.

Then came the cyanobacteria and their ability to perform photosynthesis. Chlorophyll and chlorophyll-like pigments evolved, and organisms could now trap radiant energy from the sun and convert it to chemical energy in carbohydrates. But there was a downside: Oxygen was a waste product of the process—and it was deadly because the oxygen radicals (O_2^-, OH·) produced could disrupt cellular metabolism by "tearing away" electrons from other molecules.

As millions of microbial species died off in the toxic oceans and atmosphere, others "escaped" to oxygen-free environments that are still in existence today. A few species survived by adapting to the new oxygen environment. They survived because they evolved the enzymes to safely tuck away oxygen atoms in a nontoxic form. That form was water. They used oxygen as a final electron acceptor in an electron transport system to tap foods for large amounts of energy. And so the Krebs cycle and oxidative phosphorylation came into existence.

Also coming into existence were millions of new species, some merely surviving and others thriving in the oxygen-rich environment. The face of planet Earth was changing as anaerobic and fermenting species declined and aerobic species proliferated. A couple of billion years would pass before one particularly well-known species of oxygen-breathing creature evolved: *Homo sapiens*.

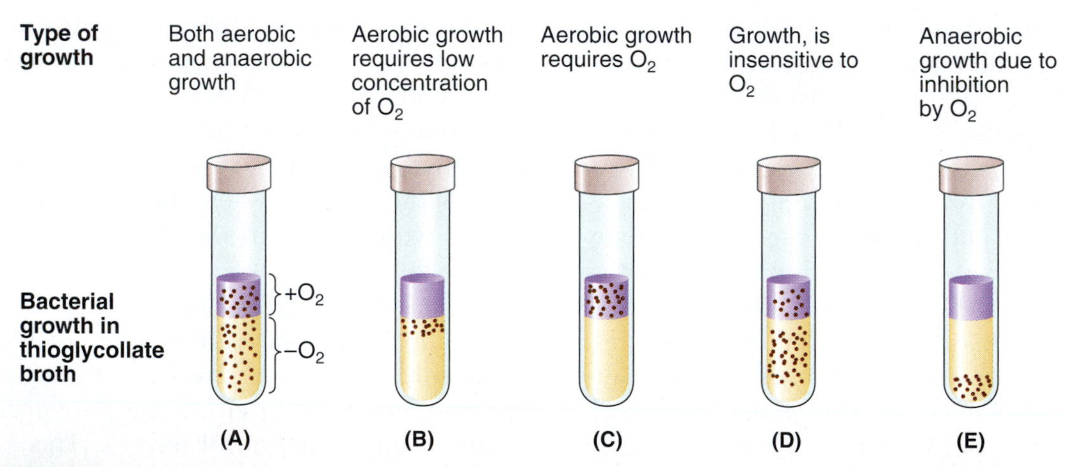

FIGURE 5.10 The Effect of Oxygen on Prokaryotic Growth. Each tube contains a thioglycollate broth into which was inoculated a different bacterial species.

Q: Identify the O_2 requirement in each thioglycollate tube based on the growth density [example: (A) represents facultative].

Finally, there are bacterial species said to be **capnophilic** (*capno* = "smoke"); they require an atmosphere low in oxygen but rich in carbon dioxide. Members of the genera *Neisseria* and *Streptococcus* are capnophiles.

pH. The cytoplasm of most prokaryotes has a pH near 7.0. This means that the majority of species grow optimally at neutral pH (see Chapter 2). Human blood and tissues, with a pH of approximately 7.2 to 7.4, provide a suitable environment for the proliferation of many pathogens. Most bacterial species have a pH range under which they will grow more slowly and this minimum to maximum range usually covers three pH units. However, some pH-hearty prokaryotes, such as *Vibrio cholerae,* can tolerate acidic conditions as low as pH 2.0 and alkaline conditions as high as pH 9.5.

Acid-tolerant bacteria called **acidophiles** are valuable in the food and dairy industries. For example, certain species of *Lactobacillus* and *Streptococcus* produce the acid that converts milk to buttermilk and cream to sour cream. These species pose no threat to good health even when consumed in large amounts. The "active cultures" in a cup of yogurt are actually acidophilic bacterial species. Extreme acidophiles are found among the *Archaea* as we saw in the MicroFocus 2.3.

The majority of known bacterial species, however, do not grow well under acidic conditions. Thus, the acidic environment of the stomach helps deter disease, while providing a natural barrier to the organs beyond. In addition, you may have noted certain acidic foods such as lemons, oranges, and other citrus fruits as well as tomatoes and many vegetables are hardly ever contaminated by bacterial growth.

Hydrostatic and Osmotic Pressure. Further environmental factors can influence the growth of prokaryotic cells. Psychrophiles in deep ocean waters and sediments are under extremely high hydrostatic pressure. In some deep marine trenches the hydrostatic pressure is tremendous—as high as 16,000 pounds per square inch (psi). Prokaryotes may be the only organisms able to withstand the pressure. Such **barophiles** in fact will die quite quickly at normal atmospheric pressures (14.7 psi).

We have discussed osmotic pressure previously in regard to the pressure water exerts on cells and the necessity for cells to have cell walls to prevent rupture (see Chapter 4). In a reverse scenario, should the environment have more dissolved materials, water would leave the cells and the cells would plasmolyze. This is the principle behind salting meats and other food products, and using sugar as a preservative in jams and jellies. A high salt or sugar concentration will prevent growth and may even kill the cells (Chapter 23).

There are prokaryotes though that are salt-loving. These **halophiles** require relatively high levels of salt (sodium chloride) to survive. Again marine prokaryotes represent halophiles surviving well in 3.5 percent salt. Several mesophilic species also are salt-tolerant. These include the species of *Staphylococcus*. The extreme halophiles represent groups of the *Archaea* that tolerate salt concentrations of 20 to 30 percent.

The ability of microbes to withstand some very extreme conditions suggests they could live on other worlds (MicroFocus 5.4).

CONCEPT AND REASONING CHECKS

5.5 Construct a concept map for the physical factors influencing prokaryotic growth.

■ Hydrostatic pressure: The pressure exerted by the weight of water.

5.3 Culture Media and Growth Measurements

In this chapter, we have been discussing prokaryotic growth and the physical factors that control growth. To complete our analysis of growth and nutrition, we need to identify the chemical media used to grow and separate specific prokaryotes, and consider the measurements used to evaluate growth.

A critical development in the design of culture media and the analysis of cell growth was the introduction of agar by Robert Koch (see Chapter 1). **Agar** is a polysaccharide derived from marine red algae. It contains no essential nutrients and is a unique colloid that remains liquid until cooled to below approximately 36°C. The solidified medium can be used to cultivate prokaryotes, isolate pure cultures, or accomplish other tasks, such as a medium for measuring population growth.

■ Colloid: Aggregates of molecules in a finely divided state dispersed in a solid medium.

Culture Media Are of Two Basic Types

KEY CONCEPT

- Culture media contain the nutrients needed for optimal prokaryotic growth.

MICROFOCUS 5.4: Environmental Biology

War of the Worlds—On Mars

In Steven Spielberg's 2005 film *War of the Worlds,* adapted from H. G. Wells' 1898 novel, what were thought to be falling stars or meteorites turn out to be Martian spaceships fleeing a dying world. When the curious come to examine the crash sites in the countryside, they discover the alien spacecrafts are filled with tentacled Martian invaders and their robotic war machines. Metallic appendages emerge from the crash craters and begin to destroy everything in their path. The war between Mars and Earth has begun. Although this is science fiction, in reality the scenario could happen—only on Mars.

The United States has sent several spacecraft to Mars since the first Viking landers in 1976. Recently, an international team of scientists carried out studies suggesting terrestrial microbes could hitch a ride to Mars on such a craft—and even survive the journey. The team believes most spacecraft that have touched down on Mars were not thoroughly sterilized by heat or radioactivity, so they could be carrying living microbes from Earth. NASA scientists have assumed Mars' thin atmosphere, which allows intense ultraviolet (UV) radiation to reach the planet's surface—triple Earth's intensity—would kill any life inadvertently carried on the spacecraft. In laboratory tests, Martian-level doses of UV radiation destroyed most microbes in just seconds.

The reason the international team has raised the microbe alarm is from the tests they carried out. They tested the endurance of a particularly hardy cyanobacterium that thrives in the dry deserts of Antarctica. The extremophile, called *Chroococcidiopsis,* inhabits porous rocks near the rock surface where temperature and humidity are very low. The team found that most dormant spores of *Chroococcidiopsis* were killed after five minutes of a Martian UV dose. However, a few spores remained alive if they were buried by just 1 mm of soil. So, microbes might survive—and potentially grow—if protected from UV radiation and present in an environment with water and nutrients.

Until now, American spacecraft have not landed in areas known to have such "habitable" conditions. That is not to say there are not such places though. NASA's next Mars lander, the Phoenix mission, will land in the northern arctic region in 2008. It will dig into the subsurface to detect water ice and probe for habitats of present day life—areas where earthly microbial aliens could establish a foothold from a contaminated spacecraft.

If true, and Martian life also was present in these regions, is it possible that earthly tentacled (piliated) bacterial or radiation-resistant archaeal invaders might start a war of the worlds—on Mars?

Since the time of Pasteur and Koch, microbiologists have tried to grow prokaryotes in laboratory cultures; that is, in ways to mimic the natural environment. Today, many of the media used in the medical diagnostic bacteriology laboratory have their origins in the first Golden Age of Microbiology (see Chapter 1). These early media often contain blood or serum to mimic the environment in the human body.

For the isolation and identification of prokaryotes, two types of culture media are commonly used. **Nutrient broth** and **nutrient agar** media are examples of a chemically undefined medium, or **complex medium**. It is called complex because the exact components or their quantity is not known for certain (TABLE 5.2). For example, it is not known precisely what carbon and energy sources or other growth factors are present. Complex media are commonly used in the teaching laboratory because the purpose is simply to grow prokaryotes and not be concerned about what specific nutrients are needed to accomplish this action.

The other type of medium is a chemically defined or **synthetic medium**. In this medium, the chemical composition and amount of all components are known (Table 5.2). This medium is used when trying to determine an organism's specific growth requirements.

Culture Media Can Be Devised to Select for or Differentiate between Prokaryotic Species

KEY CONCEPT

- Special chemical formulations can be devised to isolate and identify some prokaryotes.

Most culturable prokaryotes grow well in common complex or synthetic media. Since we want to focus in human pathogens, the basic ingredients of the growth media can be modified in one of three ways to provide fast and critical information about the organism causing an infection or disease (TABLE 5.3).

A **selective medium** contains ingredients to inhibit the growth of certain prokaryotes in a mixture while allowing the growth of others. The basic growth medium may contain extra salt (NaCl) or an antibiotic to inhibit the growth of some organisms but permits the growth of those prokaryotes or pathogens one wants to isolate.

Another modification to a basic growth medium is the addition of one or more substances that allow one to differentiate between very similar species based on specific biochemical or physiological properties. This **differential medium** contains in the culture plate specific chemicals to indicate which species possess and which lack a particular biochemical process. Such indicators make it easy to distinguish visually colonies of one organism from colonies of other similar organisms on the same culture plate. MicroInquiry 5 looks closer at these two approaches to identify or separate bacterial species.

Although most common prokaryotes grow well in nutrient broth and nutrient agar, certain so-called fastidious organisms may require an **enriched medium** containing special nutrients (MicroFocus 5.5).

Other prokaryotes are simply impossible to cultivate in any laboratory culture medium yet devised. In fact, less than 1% of the microorganisms in natural water and soil samples can be cultured. So, it is impossible to estimate accurately microbial diversity in an environment based solely on culturability. Such prokaryotes are said to be in a **VBNC** (viable but non-culturable) state. Procedures for identifying VBNC organisms included

■ Fastitious: Having complex nutritional requirements.

TABLE 5.2 Composition of a Complex and a Chemically Defined Growth Medium

Ingredient	Nutrient Supplied	Amount
Complex Agar Medium		
Peptone	Amino acids, peptides	5.0 g
Beef extract	Vitamins, minerals, other nutrients	3.0 g
Sodium chloride (NaCl)	Sodium and chloride ions	8.0 g
Agar		15.0 g
Water		1.0 liter
Synthetic Broth Medium		
Glucose	Simple sugar	5.0 g
Ammonium phosphate ($(NH_4)_2HPO_4$)	Nitrogen, phosphate	1.0 g
Sodium chloride (NaCl)	Sodium and chloride ions	5.0 g
Magnesium sulfate ($MgSO_4 \cdot 7H_2O$)	Magnesium ions, sulphur	0.2 g
Potassium phosphate (K_2HPO_4)	Potassium ions, phosphate	1.0 g
Water		1.0 liter

TABLE 5.3 A Comparison of Special Culture Media

Name	Components	Uses	Examples
Selective medium	Growth stimulants Growth inhibitors	Selecting certain prokaryotes out of mixture	Mannitol salt agar for staphylococci
Differential medium	Dyes Growth stimulants Growth inhibitors	Distinguishing different prokaryotes in a mixture	MacConkey agar for gram-negative bacteria
Enriched medium	Growth stimulants	Cultivating fastidious prokaryotes	Blood agar for streptococci; chocolate agar for *Neisseria* species

MICROINQUIRY 5

Identification of Bacterial Species

It often is necessary to identify a bacterial species or be able to tell the difference between similar-looking species in a mixture. In microbial ecology, it might be necessary to isolate certain naturally-growing species from others in a mixture. In the clinical and public health setting, microbes might be pathogens associated with disease or poor sanitation. In addition, some may be resistant to standard antibiotics normally used to treat an infection. In all these cases, identification can be accomplished by modifying the composition of a complex or synthetic growth medium. Let's go through several scenarios.

- Suppose you are an undergraduate student in a marine microbiology course. On a field trip, you collect some seawater samples and, now back in the lab, you want to grow only photosynthetic microbes.

  How would you select for photosynthetic microbes? First, you know the photosynthetic organisms manufacture their own food, so their energy source will be sunlight and not the organic compounds typically found in culture media (see Table 5.2). So, you would need to use a synthetic medium but leave out the glucose. Also, knowing the salts typically in ocean waters, you would want to add them to the medium. You would then inoculate a sample of the collected material into a broth tube, place the tube in the light, and incubate for one week at a temperature typical of where the organisms were collected.

5a. What would you expect to find in the broth tube after one week's incubation?

What you have used in this scenario is a **selective medium**; that is, one that will encourage the growth of photosynthetic microbes (light and sea salts) and suppress the growth of non-photosynthetic microorganisms (no carbon = no energy source).

- As an infection disease officer in a local hospital, you routinely swab critical care areas to determine if there are any antibiotic resistant bacteria present. You are especially concerned about methicillin-resistant *Staphylococcus aureus* (MRSA) as it frequently can cause disease outbreaks in a hospital setting. One swab you put in a broth tube showed turbidity after 48 hours.

5b. Knowing that *Staphylococcus* species are halotolerant, how could you devise an agar medium to visually determine if any of the growth is due to *Staphylococcus?*

Again, a selective medium would be used. It would be prepared by adding 7.5% salt to a complex agar medium. A sample from the broth tube would be streaked on the plate and incubated at 37°C for 48 hours.

5c. What would you expect to find on the agar plate after 48 hours?

Your selective medium contained 10 discrete colonies. You do a Gram stain and discover that all the colonies contain clusters of purple spheres; they are gram-positive. However, there are other species of *Staphylococcus* that do not cause disease. One is *S. epidermidis,* a common skin bacterium. A Gram stain therefore is of no use to differentiate *S. aureus* from *S. epidermidis*.

5d. Knowing that only *S. aureus* will produce acid in the presence of the sugar mannitol, how could you design a differential broth medium to determine if any of the colonies are *S. aureus?* (Hint: phenol red is a pH indicator that is red at neutral pH and yellow at acid pH).

You can identify each bacterial species by taking a complex broth medium, such as nutrient broth, and adding salt and mannitol (mannitol salts broth) and phenol red. Next, you inoculate a sample of each colony into a separate tube. You inoculate the 10 tubes and incubate them for 48 hours at 37°C.

5e. The broth tubes are shown below. What do the results signify? Which tubes contain which species of *Staphylococcus?*

This method is an example of a **differential medium** because it allowed you to visually differentiate or distinguish between two very similar bacterial species.

Knowing which colonies on the original selective medium plate are *S. aureus,* you need to determine which, if any, are resistant to the antibiotic methicillin.

5f. How could you design an agar medium to identify any MRSA colonies?

5g. If the plates are devoid of growth, what can you conclude?

Again, you have used a selective medium; the addition of methicillin will permit the growth of any MRSA bacteria and suppress the growth of *S. epidermidis* (sensitive to methicillin).

Answers can be found in **Appendix D**.

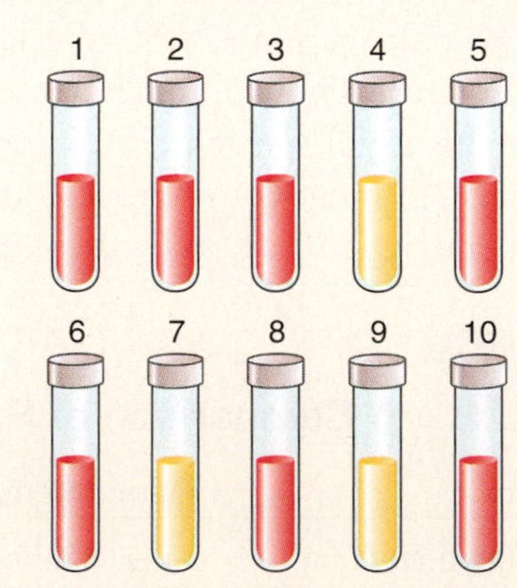

Results from differential broth tubes.

MICROFOCUS 5.5: Public Health

"Enriching" Koch's Postulates

On July 21–23, 1976, some 5,000 Legionnaires attended the Bicentennial Convention of the American Legion in Philadelphia, PA. About 600 of the Legionnaires stayed at the Bellevue Stratford Hotel. As the meeting was ending, several Legionnaires who stayed at the hotel complained of flu-like symptoms. Four days after the convention ended, an Air Force veteran who had stayed at the hotel died. He would be the first of 34 Legionnaires over several weeks to succumb to a lethal pneumonia, which became known as Legionnaires' disease or legionellosis.

As with any new disease, epidemiological studies look for the source of the disease. The Centers for Disease Control and Prevention (CDC) had an easy time tracing the source back to the Bellevue Stratford Hotel. Epidemiological studies also try to identify the causative agent. Using Koch's postulates, CDC staff collected tissues from lung biopsies and sputum samples. However, no microbes could be detected on slides of stained material. By December 1976, they were no closer to identifying the infectious agent.

How can you verify Koch's postulates if you have no infectious agent? It was almost like being back in the times of Pasteur and Koch. Why was this bacterial species so difficult to culture on bacteriological media? Perhaps it was a virus.

After trying 17 different culture media formulations, the agent was finally cultured. It turns out it was a bacterial species, named *Legionella pneumophila,* but one with fastidious growth requirements. The initial agar medium contained a beef infusion, amino acids, and starch. When this medium was enriched with 1% hemoglobin and 1% isovitalex, small, barely visible colonies were seen after five days of incubation at 37°C. Investigators then realized the hemoglobin was supplying iron to the bacterium and the isovitalex was a source of the amino acid cysteine. Using these two chemicals in pure form, along with charcoal to absorb bacterial waste, a pH of 6.9, and an atmosphere of 2.5 percent CO_2, the growth of *L. pneumophila* was significantly enhanced. From these cultures, a gram-negative rod was confirmed (see figure).

With an enriched medium to pure culture the organism, susceptible animals (guinea pigs) could be injected as required by Koch's postulates. *L. pneumophila* then was recovered from infected guinea pigs, verifying the organism as the causative agent of Legionnaires' disease.

Today, we know *L. pneumophila* is found in many aquatic environments, both natural and artificial. At the Bellevue Stratford Hotel, epidemiological studies indicated guests were exposed to *L. pneumophila* as a fine aerosol emanating from the air-conditioning system. Through some type of leak, the organism gained access to the system from the water cooling towers.

Koch's postulates are still useful—it's just hard sometimes to satisfy the postulates without an isolated pathogen.

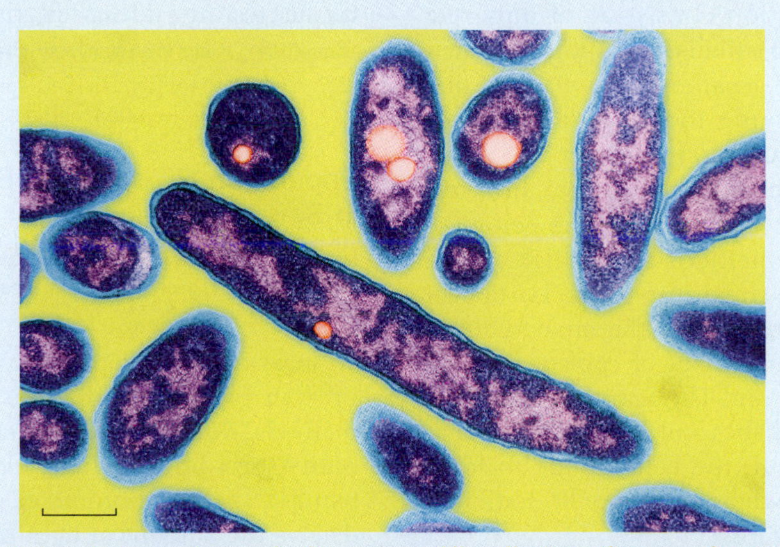

False-color transmission electron micrograph of *L. pneumophila* cells. Note the expansive nucleoid region (pink). (Bar = 0.5 μm.)

direct microscopic examination and, most commonly, amplification of diagnostic gene sequences or 16S rRNA sequences as mentioned in the introduction to Chapter 1.

Why are these organisms non-culturable? Some organisms have metabolic injuries, such as damage to the cell membrane or ribosomes. More often the reason is simply a lack of knowledge about the nutritional requirements of prokaryotes. For example, most species of the rickettsiae and chlamydiae can only be grown in mammalian cell cultures or animals as they fail to grow even in enriched nutrient media. Studies on VBNC prokaryotes present a vast and as yet unexplored field, which is important not only for detection of human pathogens, but also to reveal the diversity of *Bacteria* and *Archaea*.

CONCEPT AND REASONING CHECKS

5.6 List reasons why many prokaryotes cannot be cultured in existing complex or synthetic growth media.

Population Measurements Are Made Using Pure Cultures

KEY CONCEPT

- Two standard methods are available to produce pure cultures.

Prokaryotes rarely occur in nature as a single species. Rather, they are mixed with other species, a so-called mixed culture. Therefore, to study a species, microbiologists and laboratory technologists must use a **pure culture**—that is, a population consisting of only one species. This is particularly important when trying to identify a pathogen, as Pasteur discovered when trying to discover the agent responsible for cholera (see Chapter 1).

■ Subculturing: The process of transferring bacteria from one tube or plate to another.

If one has a mixed broth culture, how can the prokaryotes be isolated as pure colonies? Two established methods are available. The first method is the **pour-plate isolation method**. Here, a sample of the mixed culture is diluted in several tubes of cooled, but still molten, agar medium. The agar then is poured into sterile Petri dishes and allowed to harden. During incubation, the cells divide to form discrete colonies where they have been diluted the most (FIGURE 5.11).

The second method, called the **streak-plate isolation method**, uses a single plate of nutrient agar (FIGURE 5.12). An inoculum from a mixed culture is removed with a sterile loop or needle, and a series of streaks is made on the surface of one area of the plate. The loop is flamed, touched to the first area, and a second series is made in a second area. Similarly, streaks are made in the third and fourth areas, thereby spreading out the individual cells. On incubation, each cell will grow exponentially to form discrete colonies. In a sense, the cells are being "diluted" to where there are single cells on the agar.

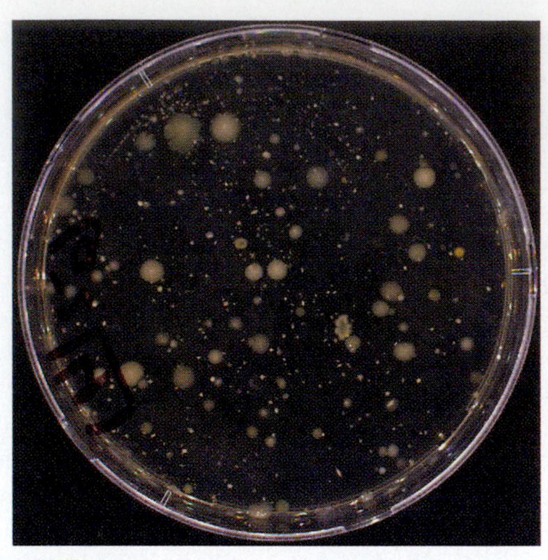

FIGURE 5.11 **A Pour Plate.** The dispersed bacterial cells grow as individual, discrete colonies.

Q: By looking at this plate, how would you know the original broth culture was a mixture of bacterial species?

In both methods, the researcher or technologist can select samples of the colonies for further testing and subculturing.

CONCEPT AND REASONING CHECKS

5.7 Explain the difference between the pour-plate and streak-plate isolation methods.

Population Growth Can Be Measured in Several Ways

KEY CONCEPT

- Prokaryotic growth can be measured by direct and indirect methods.

To measure the amount (mass) of prokaryotic growth in a medium, there are numerous methods. For example, the cloudiness, or **turbidity**, of a broth culture may be determined using a spectrophotometer. This instrument detects the amount of light scattered by a suspension of cells. The amount of light scatter (optical density, OD) is a function of the cell number; that is, the more cells present, the more light is scattered and the higher the OD

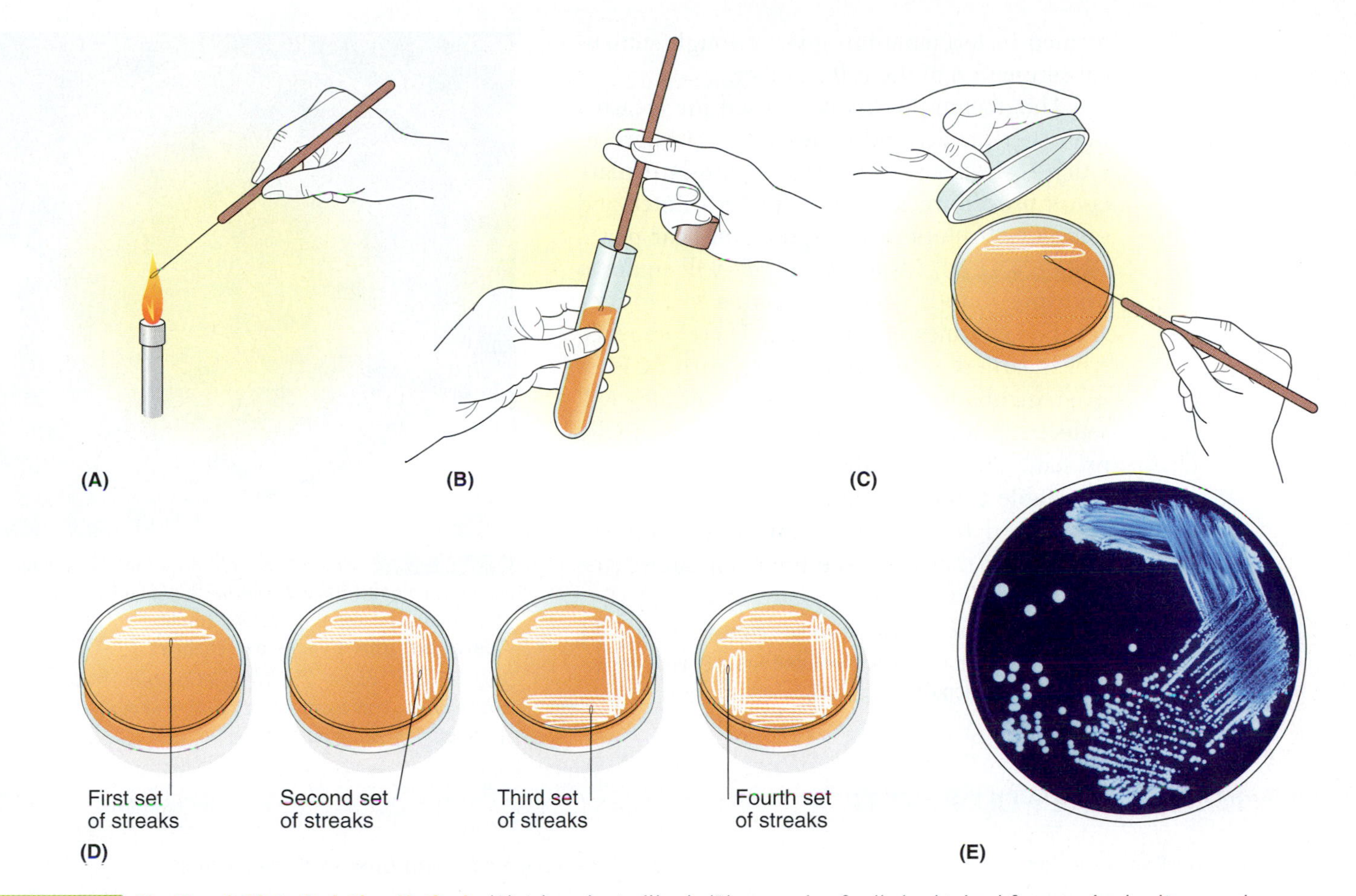

FIGURE 5.12 **The Streak-Plate Isolation Method.** (**A**) A loop is sterilized, (**B**) a sample of cells is obtained from a mixed culture, and (**C**) streaked near one edge of the plate of medium. (**D**) Successive streaks are performed, and the plate is incubated. (**E**) Well-isolated and defined colonies illustrate a successful isolation.

Q: Justify the need to streak a mixed sample over four areas on a culture plate.

reading on the spectrophotometer. A standard curve can be generated to serve as a measure of cell numbers in other situations.

There are a number of ways to directly measure cell numbers. Scientists may wish to perform a **direct microscopic count** using a known sample of the culture on a specially designed slide (FIGURE 5.13). However, this procedure will count both live and dead cells. Other indirect methods include measuring the dry weight of the prokaryotes which gives an indication of the cell mass. Oxygen uptake in metabolism also can be measured as an indication of metabolic activity and therefore cell number.

Cell estimations also use indirect methods. Two common methods are used. In the **most probable number test**, samples of prokaryotes are added to numerous lactose broth tubes and the presence or absence of gas

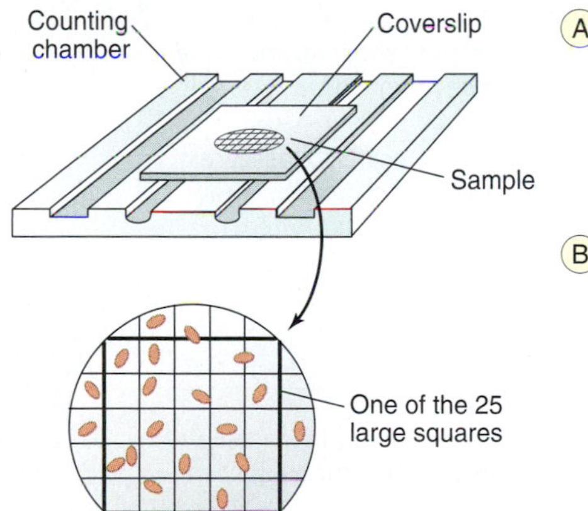

(A) The counting chamber is a specially marked slide containing a grid of 25 large squares of known area. The total volume of liquid held is 0.00002 ml (2×10^{-5} ml).

(B) The counting chamber is placed on the stage of a light microscope. The number of cells are counted in several of the large squares to determine the average number.

FIGURE 5.13 **Direct Microscopic Counting Procedure Using the Petroff-Hausser Counting Chamber.** This procedure can be used to estimate the number of live and dead cells in a culture sample.

Q: Suppose the average number of cells per square was 14. Calculate the number of cells in a 10 ml sample.

formed in fermentation gives a rough statistical estimation of the cell number.

This technique has been used for measuring water quality and is described in MicroInquiry 26 in Chapter 26. In the **standard plate count procedure**, a broth culture is diluted and samples of dilutions are spread on agar plates (FIGURE 5.14). Ideally, each cell will undergo multiple rounds of cell divisions to produce separate colonies on the plate. Therefore, each cell is called a **colony-forming unit** (CFU). After incubation, the number of colonies will reflect the number of cells (CFUs) originally present. This test is desirable because it gives the **viable count** of cells (the living cells only), compared to a microscopic count or dry weight test that gives the total cell count (the living as well as dead).

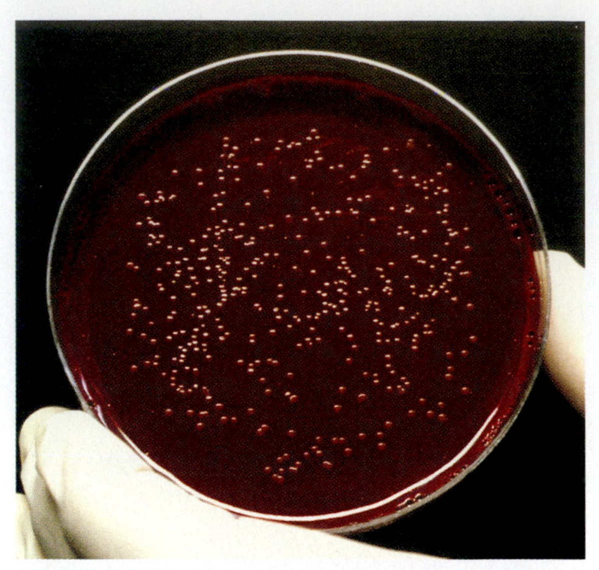

FIGURE 5.14 **The Standard Plate Count.** Individual bacterial colonies have grown on this blood agar plate. Each colony represents a colony-forming unit (CFU) since it developed from a single bacterial cell.

Q: If a 0.1 ml sample of a 10^4 dilution contained 250 colonies, how many bacterial cells were in 10 ml of the original broth culture?

CONCEPT AND REASONING CHECKS

5.8 Distinguish between direct and indirect methods to measure population growth.

SUMMARY OF KEY CONCEPTS

5.1 Prokaryotic Reproduction

- Prokaryotic reproduction involves DNA replication and binary fission to produce genetically identical daughter cells.
- Binary fissions occur at intervals called the generation time, which for prokaryotes may be as short as 20 minutes.

5.2 Prokaryotic Growth

- The dynamics of the bacterial growth curve show how a population grows exponentially, reaches a certain peak and levels off, and then may decline.
- Sporulation is a dormancy response to nutrient limitation and high population density. The endospores formed are resistant to many harsh environmental conditions.
- Temperature, oxygen, pH, and hydrostatic/osmotic pressure are physical factors that influence prokaryotic growth. Away from the optimal condition, growth slows or is inhibited.

5.3 Culture Media and Growth Measurements

- Complex and synthetic media contain the nutrients to grow prokaryotes.
- Complex or synthetic media can be modified to select for a desired prokaryotic organism, to differentiate between two similar prokaryotic species, or to enrich for species requiring special nutrients.
- Pure cultures can be produced from a mixed culture by the pour-plate isolation method or the streak-plate isolation method. In both cases, discrete colonies can be identified that represent only one prokaryotic species.
- Prokaryotic growth can be measured by direct microscopic count, dry weight, or oxygen uptake. Indirect methods include the most probable number test and the standard plate count procedure.

LEARNING OBJECTIVES

After understanding the textbook reading, you should be capable of writing a paragraph that includes the appropriate terms and pertinent information to answer the objective.

1. Distinguish between the phases of binary fission.
2. Summarize the uses for knowing a prokaryote's generation time.
3. Compare the events of each phase of a bacterial growth curve.
4. Contrast the stages of sporulation and assess the importance of the process.
5. Identify the 4 groups of microorganisms based on temperature requirements.
6. Differentiate between obligate aerobes, obligate anaerobes, and facultative organisms.
7. Summarize the role of pH to prokaryotic growth.
8. Assess the role of salt in controlling prokaryotic growth.
9. Contrast the chemical composition of complex and synthetic media.
10. Explain how selective and differential media are each constructed.
11. Explain the procedures used in the pour-plate and streak-plate isolation methods.
12. Judge the usefulness of direct and indirect methods to measure prokaryotic growth.

SELF-TEST

Answer each of the following questions by selecting the ***one*** answer that best fits the question or statement. Answers to even-numbered questions can be found in **Appendix C**.

1. Which one of the following does *not* apply to prokaryotic reproduction?
A. Presence of a fission ring apparatus.
B. Invagination of cell wall and cell membrane.
C. A spindle apparatus.
D. Cytokinesis follows DNA replication.
E. A symmetrical division occurs.

2. If a bacterial cell in a broth tube has a generation time of 40 minutes, how many cells will there be after 5 hours of optimal exponential growth?
A. 64
B. 128
C. 200
D. 280
E. 320

3. A prokaryote's generation time would be determined during the _____ phase.
A. decline
B. death
C. lag
D. log
E. stationary

4. All the following are events of the sporulation cycle *except*:
A. symmetrical cell divisions
B. transverse septum formation.
C. mother cell disintegration.
D. DNA replication.
E. Prespore engulfment by the mother cell.

5. Endospore formers include species of
A. gram-negative bacilli.
B. *Escherichia*.
C. cyanobacteria.
D. *Clostridium*.
E. All of the above (**A–D**).

6. A _____ has an optimal growth temperature at human body temperature.
A. hyperthermophile
B. mesophile
C. thermophile
D. extremophile
E. psychrophile

7. A prokaryote that uses high concentrations of oxygen gas would be a/an _____ organism.
A. obligately aerobic
B. facultative
C. microaerophilic
D. obligately anaerobic
E. Both **A** and **B** are correct.

8. The most likely microbes to grow on fruits or vegetables would be
A. capnophiles.
B. thermophiles.
C. acidophiles.
D. barophiles.
E. halophiles.

9. The use of agar as a growth medium was introduced by
A. Louis Pasteur.
B. Christian Gram.
C. Edward Jenner.
D. Robert Fleming.
E. Robert Koch.

10. If the carbon source in a growth medium is beef extract, the medium must be an example of a _____ medium.
A. chemically-defined
B. complex
C. chemically-undefined
D. synthetic
E. Both **B** and **C** are correct.

11. A _____ medium would involve the addition of the antibiotic methicillin to identify methicillin-resistant bacteria.
A. differential
B. selective
C. thioglycollate
D. VBNC
E. enriched

12. Bacteria in a VBNC state could be due to
A. cell membrane damage.
B. unknown nutritional requirements.
C. metabolic injuries.
D. damaged ribosomes.
E. All of the above (**A–D**).

13. Which one of the following (**A–D**) is *not* part of the streak-plate method? If all are, select **E.**
A. Making four sets of streaks on a plate.
B. Diluting a mixed culture in molten agar.
C. Using a mixed culture.
D. Using a sterilized loop.
E. All the above (**A–D**).

14. Indirect methods to measure bacterial growth would *not* include
A. total bacterial count.
B. microscopic count.
C. viable count.
D. most probable number.
E. standard plate count.

QUESTIONS FOR THOUGHT AND DISCUSSION

Answers to even-numbered questions can be found in **Appendix C**.

1. To prevent decay by bacterial species and to display the mummified remains of ancient peoples, museum officials place the mummies in glass cases where oxygen has been replaced with nitrogen gas. Why do you think nitrogen is used?
2. Extremophiles are of interest to industrial corporations, who see the prokaryotes as important sources of enzymes that function at temperatures of 100°C and pH levels of 10 (the enzymes have been dubbed "extremozymes"). What practical uses can you foresee for these enzymes?
3. During the filming of the movie *Titanic,* researchers discovered at least 20 different species of prokaryotes literally consuming the ship, especially a rather large piece of the midsection. What type of prokaryotes would you expect were at work on the ship?
4. Although thermophilic prokaryotes are presumably harmless because they do not grow at body temperatures, they may still present a hazard to good health. Can you think of a situation in which this might occur?

APPLICATIONS

Answers to even-numbered questions can be found in **Appendix C**.

1. Consumers are advised to avoid stuffing a turkey the night before cooking, even though the turkey is refrigerated. A homemaker questions this advice and points out that the bacterial species of human disease grow mainly at warm temperatures, not in the refrigerator. What explanation might you offer to counter this argument?
2. Public health officials found that the water in a Midwestern town was contaminated with sewage bacteria. The officials suggested that homeowners boil their water for a couple of minutes before drinking it. (a) Would this treatment sterilize the water? Why? (b) Is it important that the water be sterile? Explain.

REVIEW

On completing your study of these pages, test your understanding of their contents by deciding whether the following statements are true (T) or false (F). If the statement is false, substitute a word or phrase for the <u>underlined</u> word or phrase to make the statement true. Answers to even-numbered statements are listed in **Appendix C**.

1. _____ Endospores are produced by some <u>gram-negative</u> bacterial species.
2. _____ Obligate aerobes use <u>oxygen gas</u> as a final electron acceptor in energy production.
3. _____ The most common growth medium used in the teaching laboratory is a <u>complex medium</u>.
4. _____ The <u>majority</u> of prokaryotes that have been discovered can be cultured in growth media.
5. _____ A standard plate count procedure is an example of a <u>direct method</u> to estimate population growth.
6. _____ In attempting to culture a fastidious prokaryote, a <u>differential medium</u> would be used.
7. _____ Acidophiles grow best a pHs <u>greater than 9</u>.
8. _____ <u>Mesophiles</u> have their optimal growth near 37°C.
9. _____ Prokaryotes lack a <u>mitotic spindle</u> to separate chromosomes.
10. _____ The fastest doubling time would be found in the <u>lag phase</u> of a bacterial growth curve.

HTTP://MICROBIOLOGY.JBPUB.COM/

The site features learning, an on-line review area that provides quizzes and other tools to help you study for your class. You can also follow useful links for in-depth information, read more MicroFocus stories, or just find out the latest microbiology news.

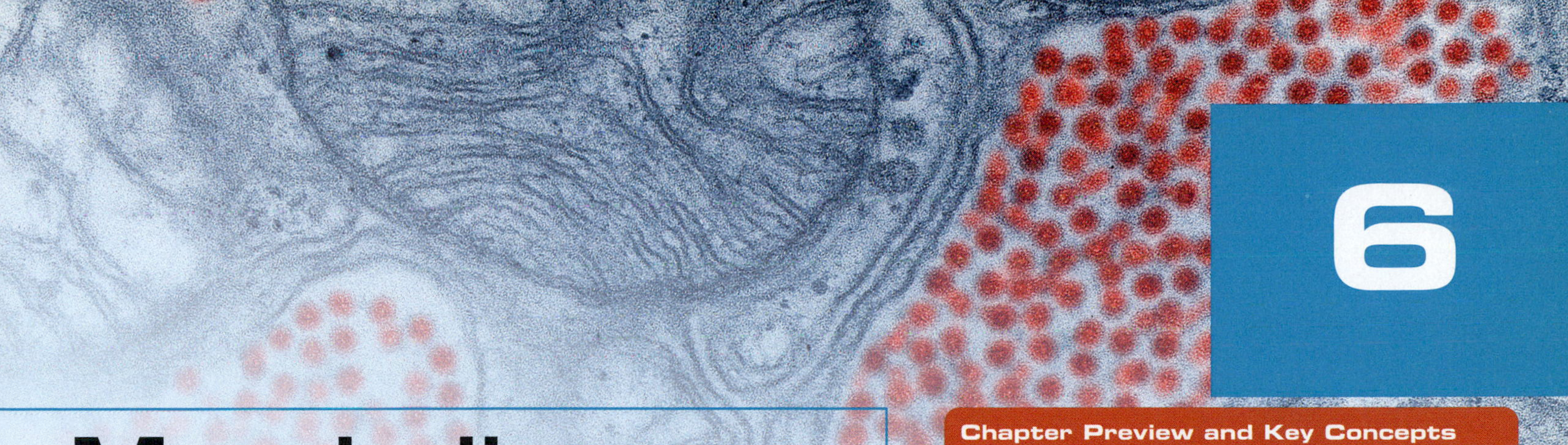

6

Metabolism of Prokaryotic Cells

Life is like a fire; it begins in smoke and ends in ashes.
—Ancient Arab proverb connecting energy to life

Charlie Swaart had been a social drinker for years. A few beers or drinks with his pals, but no lasting alcoholic consequences. Then, in 1945, he began a nightmare that would make medical history.

One October day, while stationed in Tokyo after World War II, Swaart suddenly became drunk for no apparent reason. He had not had any alcohol for days, but all of a sudden he would feel like he had been partying all night. After sleeping it off, he would be fine the next day.

Unfortunately, this "behavior" returned time and time again. For years thereafter, the episodes continued—bouts of drunkenness and monumental hangovers without drinking so much as a beer! Doctors were puzzled as they could detect alcohol on his breath or in his blood. Was this some type of internal metabolism gone haywire? Was it the result of a bacterial infection? It didn't seem likely. They warned him though not to drink any additional alcohol for fear of damaging his liver. Swaart followed their advice to the letter; still, he experienced periods of drunkenness.

Twenty years passed before Swaart learned of a similar case in Japan. A Japanese businessman had endured years of social and professional disgrace before doctors discovered a yeast-like fungus in his intestine. Studying this eukaryotic microbe showed that the fungal cells were fermenting carbohydrates to alcohol right there in his intestine. The fungus was identified as *Candida albicans* (FIGURE 6.1). Now having *C. albicans* in one's intestine is not uncommon; but, finding fermenting *C. albicans* was historic.

Chapter Preview and Key Concepts

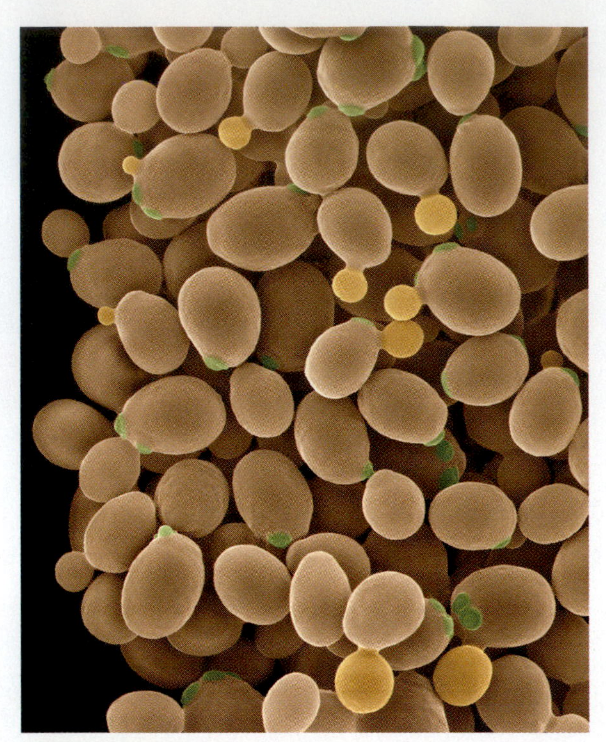

FIGURE 6.1 **False-Color Scanning Electron Micrograph of *Candida albicans* Cells.** *C. albicans* is different from the yeasts normally associated with the fermentation process in beer and wine.

Q: To what domain of organisms does C. albicans *belong?*

Swaart learned that an antibiotic had worked to kill the yeast cells. With this knowledge, he approached his doctor. Sure enough, lab tests showed massive colonies of *C. albicans* in Swaart's intestine too. The sugar in a cup of coffee or any carbohydrate in pasta, cake, or candy could bring on drunkenness. However, to cure his illness Swaart had to travel back to Japan to get the effective antibiotic.

Researchers believed the atomic blasts of Hiroshima and Nagasaki in 1945 may have caused a normal *C. albicans* to mutate to a fermenting form, which somehow found its way into Swaart's digestive system. One can only wonder if there are many other individuals who have become living fermentation vats for the fungus. For Charlie Swaart, though, the nightmare was finally over.

TABLE 6.1 A Comparison of Two Key Aspects of Cellular Metabolism

Anabolism	Catabolism
Buildup of small molecules	Breakdown of large molecules
Products are large molecules	Products are small molecules
Photosynthesis	Glycolysis, Krebs cycle
Mediated by enzymes	Mediated by enzymes
Energy generally is required (endergonic)	Energy generally is released (exergonic)

The process of fermentation described here was in a eukaryotic microbe. However, other types of fermentation processes occur in prokaryotes and they are but one aspect of the broad topic of prokaryotic metabolism.

Metabolism refers to all the biochemical processes taking place in living cells or what today is called **metabolomics**, the measurement and analysis of metabolites in an organism. These processes are divided into two general categories: **Anabolism**—the synthesis of large organic compounds; and **catabolism**—the hydrolysis of such compounds (see Chapter 2).

From an energy perspective, anabolic reactions form bonds, which require energy. Such energy-requiring processes are **endergonic** (*end* = "inner"; *ergon* = "work") **reactions**. In contrast, catabolic reactions break bonds, releasing energy. These processes are called **exergonic** (*ex* = "outside of") **reactions**. TABLE 6.1 compares anabolism and catabolism.

In this chapter, we examine the types of metabolism exhibited by prokaryotes. Much of the chapter discusses the catabolic reactions involved in energy conversions forming adenosine triphosphate (ATP). Since the chapter emphasizes the role of carbohydrates in the energy conversions, it might be worthwhile to revisit Chapter 2 and refresh yourself with these types of organic compounds.

6.1 Enzymes and Energy in Metabolism

The growth we examined in the previous chapter depends on metabolic processes that occur in the cell cytosol, on the cell membrane, in the periplasmic space (gram-negative bacteria), and outside the cell. To carry out these reactions, cells need a large variety of enzymes. Therefore, we begin our study of prokaryotic metabolism with a discussion of these essential proteins, which have been known only since the early 1900s (MicroFocus 6.1).

Enzymes Catalyze All Cellular Reactions

KEY CONCEPT

- Cellular chemical reactions do not occur spontaneously; they are controlled by enzymes.

Enzymes are a group of organic molecules that increase the probability of chemical reactions while themselves remaining unchanged. They accomplish in fractions of a second what otherwise might take hours, days, or longer to happen spontaneously under normal biological conditions. For example, even though organic molecules like amino acids have functional groups, it is highly unlikely they would randomly bump into one another in the precise way needed for a chemical reaction (dehydration synthesis) to occur and for a new peptide bond to be formed (see Chapter 2). Thus, the reaction rate would be very slow were it not for the activity of enzymes.

Enzymes are reusable. Once a chemical reaction has occurred, the enzyme is released to participate in another identical reaction. Enzyme activity also is highly specific; an enzyme that functions in one chemical reaction usually will not participate in another type of reaction. That means there must be thousands of different enzymes to catalyze the thousands of different chemical reactions of metabolism occurring in a prokaryotic cell.

The substance acted upon by the enzyme is called the **substrate** and the product(s)

MICROFOCUS 6.1: History

"Hans, Du Wirst Das Nicht Glauben!"

Louis Pasteur's discovery of the role yeast cells play in fermentation heralded the beginnings of microbiology because it showed that tiny organisms could bring about important chemical changes. However, it also opened debate on how yeasts accomplished fermentation. Lively controversies ensued among scientists; some suggesting sugars from grape juice entered yeast cells to be fermented; others believing fermentation occurred outside the cells. The question would not be resolved until a fortunate accident happened in the late 1890s.

In 1897, two German chemists, Eduard and Hans Buchner, were preparing yeast as a nutritional supplement for medicinal purposes. They ground yeast cells with sand and collected the cell-free "juice." To preserve the juice, they added a large quantity of sugar (as was commonly done at that time) and set the mixture aside. Several days later Eduard noticed an unusual alcoholic aroma coming from the mixture. Excitedly, he called to his brother, "Hans, Du Wirst Das Nicht Glauben!" ("Hans, you'll never believe this!") One taste confirmed their suspicion: The sugar had fermented to alcohol.

The discovery by the Buchner brothers was momentous because it demonstrated that a chemical substance inside yeast cells brings about fermentation, and that fermentation can occur without living cells. The chemical substance came to be known as an *enzyme,* meaning "in yeast."

In 1905, the English chemist Arthur Haden expanded the Buchner study by showing that "enzyme" is really a multitude of chemical compounds and should better be termed "enzymes." Thus, he added to the belief that fermentation is a chemical process. Soon, many chemists became biochemists, and biochemistry gradually emerged as a new scientific discipline.

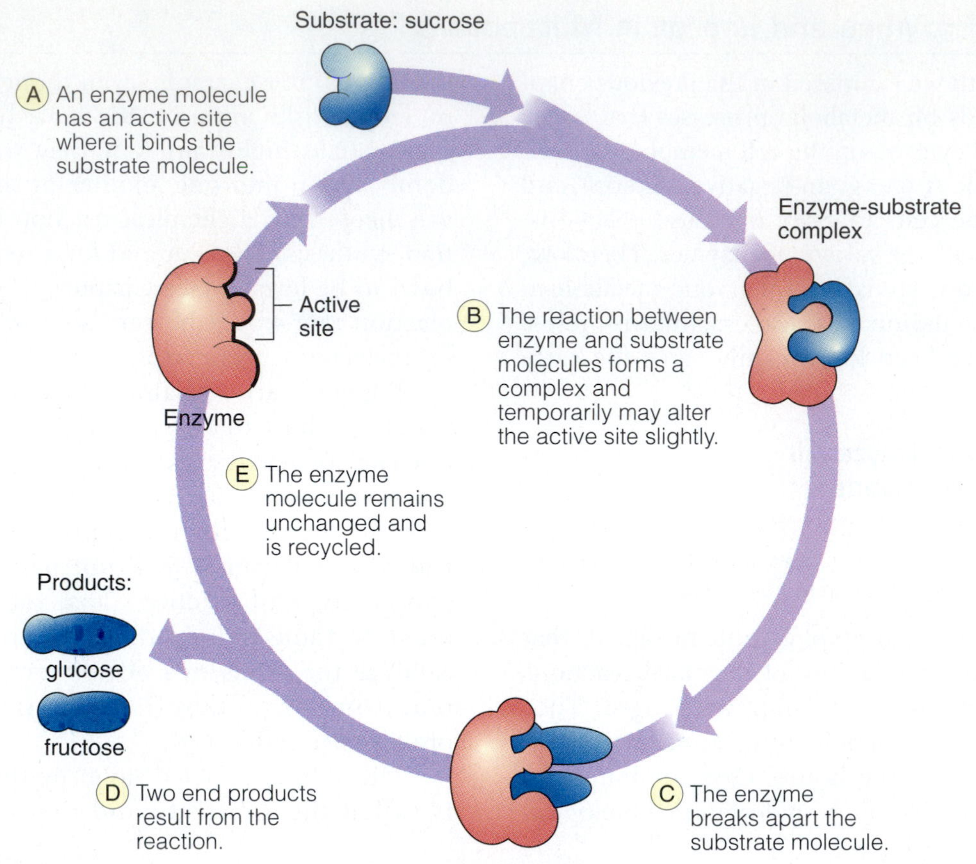

FIGURE 6.2 **The Mechanism of Enzyme Action.** Although this example shows an enzyme hydrolyzing a substrate (sucrose), enzymes also catalyze dehydration synthesis reactions, which, in this case, would combine glucose and fructose into sucrose.

Q: How do enzymes recognize specific substrates?

formed are appropriately termed **products**. Because the reactions are usually reversible, enzymes may bring about synthesis as well as hydrolysis. This factor is important in metabolism because anabolism often occurs by a reversal of many steps in catabolism.

Many enzymes can be identified by their names, which often end in "-ase." Some enzymes are named for the substrate on which they act. For example, sucrase is the enzyme that breaks down sucrose and ribonuclease digests ribonucleic acid. Hydrolases catalyze hydrolysis reactions and include lipase, which breaks down lipids, and peptidase, which breaks the peptide bond between amino acids.

CONCEPT AND REASONING CHECKS

6.1 List the characteristics of enzymes.

Enzymes Act through Enzyme-Substrate Complexes

KEY CONCEPT

- Cellular chemical reactions occur at the enzyme's active site.

Enzymes function by aligning substrate molecules in such a way that a reaction is highly favorable. In the hydrolysis reaction shown in FIGURE 6.2, the three-dimensional shape of the enzyme molecule recognizes and holds the substrate in an **enzyme-substrate complex**. While in the complex, chemical bonds in the substrate are stretched or weakened by the enzyme, causing the bond to break. In a synthesis reaction, by contrast, the electron shells of the substrates in the enzyme-substrate complex are forced to overlap in the spot where the chemical bond will form.

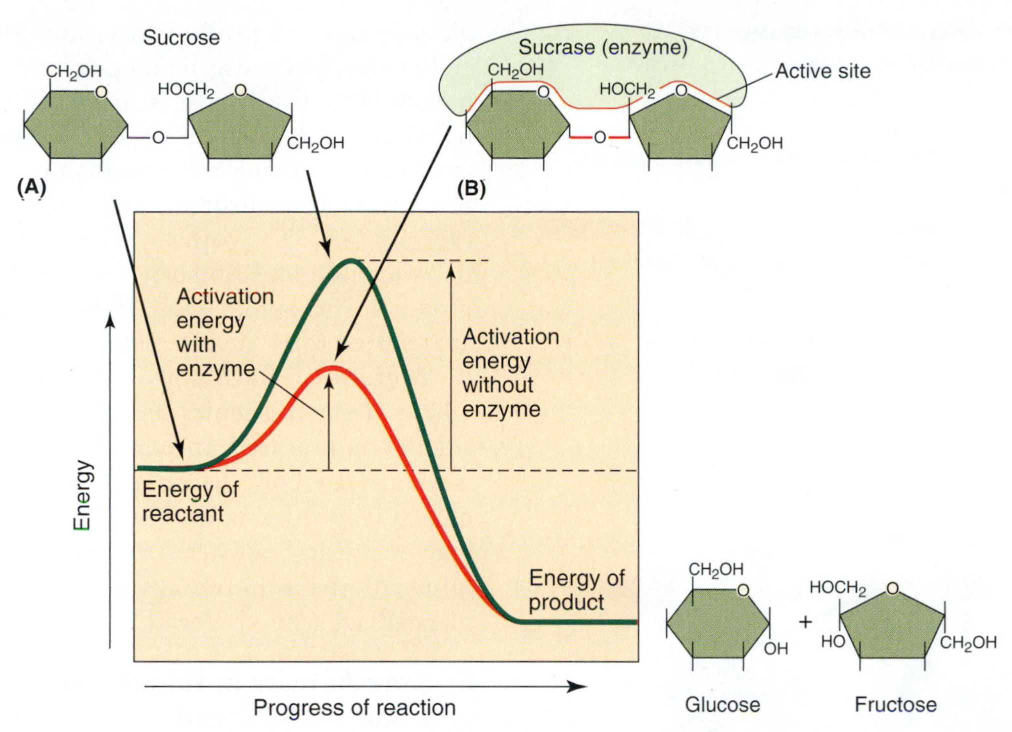

FIGURE 6.3 **Enzymes and Activation Energy.** Enzymes lower the activation energy barrier required for chemical reactions of metabolism. (**A**) The hydrolysis of sucrose is unlikely because of the high activation energy barrier. (**B**) When sucrase is present, the enzyme effectively lowers the activation energy barrier, making the hydrolysis reaction highly likely.

Q: How does the enzyme lower the activation energy of a stable substrate?

Thus, in a hydrolysis or synthesis reaction, recognition of the substrate(s) is a precisely controlled, nonrandom event. Each enzyme has a critical area on the surface, called the **active site**, where the substrate(s) is positioned and oriented to react. In Figure 6.2, the active site has a shape that closely matches the parts of the substrate that will interact in this hydrolysis reaction. The shape of an active site is determined by the levels of protein folding (see Chapter 2).

Looking at sucrose again, the bonds holding glucose and fructose together will not break spontaneously. The reason is the bond between the monosaccharides is stable and there is a substantial energy barrier preventing a reaction (FIGURE 6.3A). The job of sucrase is to bind the substrate and lower the energy barrier so that it is much more likely that the reaction will occur. In other words, the bond holding glucose to fructose needs to be destabilized (i.e., stretched, weakened) by the enzyme (FIGURE 6.3B). The energy required to do this is called the **activation energy**. Enzymes, then, play a key role in metabolism because they lower the amount of activation energy required for a reaction to take place. They assist in the destabilization of chemical bonds and the formation of new ones by separating or joining atoms in a carefully orchestrated fashion.

Some enzymes are made up entirely of protein. An example is lysozyme, the enzyme in human tears and saliva that digests the cell walls of gram-positive bacterial cells. Other enzymes, however, may contain small, nonprotein substances that participate in the catalytic reaction. If the complementing substance is a metal ion, such as magnesium (Mg^{2+}), iron (Fe^{2+}), or zinc (Zn^{2+}), it is called a **cofactor**. When the nonprotein participant is a small organic molecule, it is referred to as a **coenzyme**. Examples of two important coenzymes are nicotinamide adenine dinucleotide (NAD^+) and flavin adenine dinucleotide (FAD). These coenzymes play a significant role as electron carriers in metabolism, and we shall encounter them in our ensuing study of prokaryotic metabolism.

CONCEPT AND REASONING CHECKS

6.2 Assess the role of the active site in "stimulating" a chemical reaction.

Enzymes Often Team Up in Metabolic Pathways

KEY CONCEPT

- Metabolism often involves a series of chemical reactions controlled by separate enzymes.

There are many examples, such as the sucrose example, where an enzymatic reaction is a single substrate to product reaction. However, cells more often use metabolic pathways. A **metabolic pathway** is a sequence of chemical reactions, each reaction catalyzed by a different enzyme, in which the product of one reaction serves as a substrate for the next reaction (FIGURE 6.4A). The pathway starts with the initial substrate and finishes with the final end product. The products of "in-between" stages are referred to as intermediates.

Metabolic pathways can be anabolic, where larger molecules are synthesized from smaller monomers. In contrast, other pathways are catabolic because they break larger molecules into smaller ones. Such pathways may be linear, branched, or cyclic. We will see many of these pathways in the microbial metabolism sections ahead.

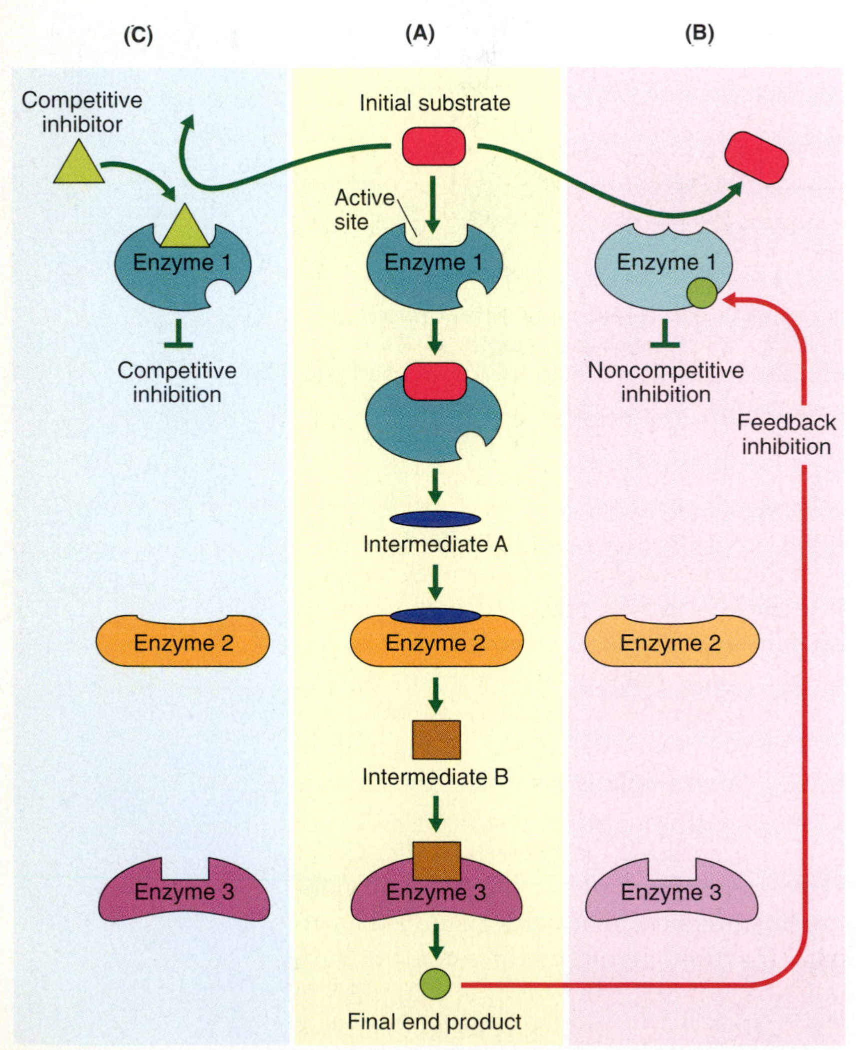

FIGURE 6.4 **Metabolic Pathways and Enzyme Inhibition.** (**A**) In a metabolic pathway, a series of enzymes transforms an initiate substrate into a final end product. (**B**) If excess final end product accumulates, it "feeds back" on the first enzyme in the pathway and inhibits the enzyme by binding at another site on the enzyme. (**C**) In competitive inhibition, a substrate that resembles the normal substrate competes with the substrate for the enzyme's active site. Competitive inhibition would reduce the productivity of the metabolic pathway by slowing down or stopping the pathway. In both (**B**) and (**C**), the whole pathway can become temporarily inoperative.

Q: Hypothesize why most substrates cannot be converted into a final end product in one enzymatic step.

Enzyme Activity Is Regulated and Can Be Inhibited

KEY CONCEPT

- Metabolism can control and be controlled by enzymes.

The same chemical reaction does not occur in a cell all the time, even if the substrate is present. Rather, cells regulate the enzymes so that they are present or active only at the appropriate time during metabolism.

One of the most common ways of regulating enzymes is for the final end product of a metabolic pathway to inhibit an enzyme in that pathway (FIGURE 6.4B). If the first enzyme in the pathway is inhibited, then no product is available to "feed" the rest of the pathway. Such **feedback inhibition** is typical of many metabolic pathways in cells. In general, when the final end product or any molecule binds to a non-active site on the enzyme, the shape of the active site changes and can no longer bind substrate. This type of inhibition is referred to as **noncompetitive inhibition**.

Another way of inhibiting an enzyme is by blocking its active site so the normal substrate cannot bind. Such **competitive inhibition** occurs in the following way (FIGURE 6.4C). If a molecule resembles the normal substrate, it will compete with the normal substrate for the active site. Sitting in the active site, this competitive inhibitor does not allow the normal substrate to bind.

Sulfonamide drugs operate in this way so that the enzyme cannot carry out its normal function (Chapter 24).

Because most enzymes are protein molecules, they are sensitive to any physical or chemical agents that damage proteins. Heat can be used to kill bacteria because heat alters the tertiary structure of enzymes (see Chapter 2). Chemicals, such as alcohol and phenol, precipitate enzyme proteins as well as other proteins, and therefore act as disinfectants. Any antibiotic that interferes with protein synthesis automatically interferes with enzyme production.

The dogma in biology used to say that all enzymes were proteins. Although this statement usually is true, there are cases where ribonucleic acid (RNA) can have catalytic effects (MicroFocus 6.2).

CONCEPT AND REASONING CHECKS

6.3 Distinguish between competitive and noncompetitive inhibition.

Energy in the Form of ATP Is Required for Metabolism

KEY CONCEPT

- ATP is the universal energy currency in cells.

In many metabolic reactions, energy is needed, along with enzymes, for the reactions to occur. The cellular "energy currency" is a compound called **adenosine triphosphate (ATP)**

■ **Sulfonamide drugs:** (Sulfa drugs) Antimicrobial chemicals that interfere with folic acid synthesis in susceptible bacterial cells.

MICROFOCUS 6.2: History/Biotechnology

Ribozymes—Telling Us about Our Past and Helping with Our Future

Until the 1980s, one of the bedrock principles of biology held that nucleic acids (DNA and RNA) were the informational molecules responsible for directing the metabolic reactions in the cell. Proteins, specifically the enzymes, were the workhorses responsible for catalyzing the thousands of chemical reactions taking place in the cell. The dogma was, "All enzymes are proteins."

In 1981, new research evidence suggested that RNA molecules could act as catalysts in certain circumstances. Today, scientists believe RNA acting by itself can trigger specific chemical reactions.

The seminal research on RNA was performed independently by Thomas R. Cech of the University of Colorado and Sidney Altman of Yale University. Altman had found an unusual enzyme in some bacterial cells, an enzyme composed of RNA and protein. Initially, he thought the RNA was a contaminant, but when he separated the RNA from the protein, the bacterial enzyme could not function. After several years, Altman and his colleagues showed that RNA was the enzyme's key component because it could act alone. At about the same time, Cech discovered that RNA molecules from *Tetrahymena,* a protozoan, could catalyze certain reactions under laboratory conditions. He showed that a molecule of RNA could cut internal segments out of itself and splice together the remaining segments.

Many biologists responded to the findings of Cech and Altman with disbelief. The implication of the research was that proteins and nucleic acids are not necessarily interdependent, as had been assumed. The research also opened the possibility that RNA could have evolved on Earth without protein. In fact, a number of scientists have proposed that life may have started in a primeval "RNA world." This world would have been swarming with self-catalyzing forms of RNA having the ability to reproduce and carry genetic information. In essence, there arose a whole new way of imagining how life might have begun. The Nobel Prize committee was equally impressed. In 1989, it awarded the Nobel Prize in Chemistry to Cech and Altman.

By 1990, these self-reproducing molecules of RNA had a name—ribozymes. They share many similarities with their protein counterparts, including the presence of binding pockets that, like active sites on enzymes, recognize specific molecular shapes. Biochemists at Massachusetts General Hospital showed that one type of ribozyme could join together separate short nucleotide segments. The research was a step toward designing a completely self-copying RNA molecule.

Today, the understanding of catalytic ribozymes goes beyond the research laboratory. Several companies are using new molecular techniques to construct new catalytic ribozymes in what is termed "directed evolution." Development of these ribozymes may have uses in clinical diagnostics and as therapeutic agents. For example, in diagnostic applications ribozymes are being developed to identify potential new drugs. Other companies are using ribozymes as biosensors to detect viral contaminants in blood. These catalytic molecules also may be useful in fighting infectious diseases by inactivating RNA molecules in viruses or other pathogens.

So, ribozymes have much to offer in understanding our very distant past as well as providing for a more healthy future.

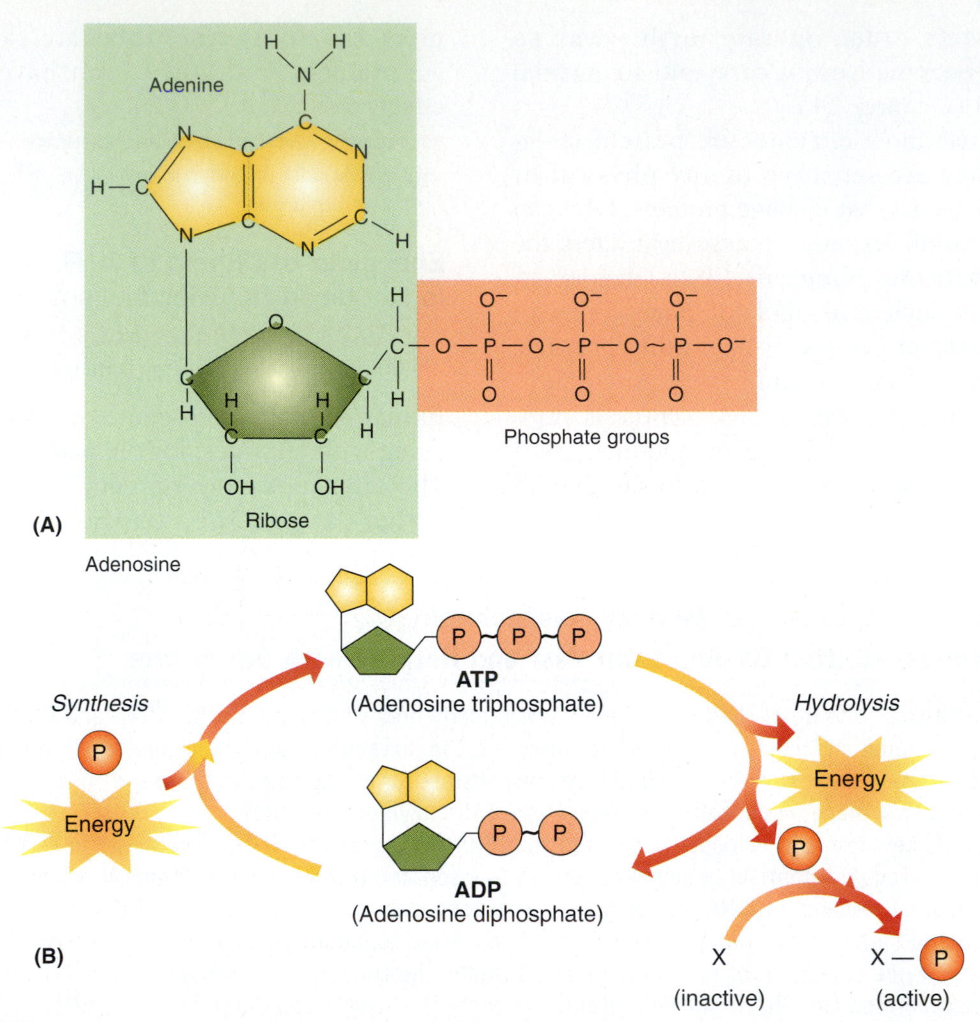

FIGURE 6.5 **Adenosine Triphosphate and the ATP/ADP Cycle.** Adenosine triphosphate (ATP) is a key immediate energy source for bacterial cells and other living things. (**A**) The ATP molecule is composed of adenine and ribose bonded to one another and to three phosphate groups. (**B**) When the ATP molecule breaks down, it releases a phosphate group and energy and becomes adenosine diphosphate (ADP). The freed phosphate can activate another chemical reaction through phosphorylation. For the synthesis of ATP, energy and a phosphate group must be supplied to an ADP molecule. *Q: What genetic molecule closely resembles ATP?*

(FIGURE 6.5A). In prokaryotes, the ATP is formed on the cell membrane, while in eukaryotes the reactions occur primarily in the mitochondria. An ATP molecule acts like a portable battery. It provides the needed energy for activities such as binary fission, flagellar motion, and spore formation. On a more chemical level, it fuels protein synthesis and carbohydrate breakdown. It is safe to say that a major share of prokaryotic functions depends on a continual supply of ATP. Should the supply be cut off, the cell dies very quickly, as ATP cannot be stored.

ATP molecules are relatively unstable. In Figure 6.5a, notice that the three phosphate groups all have negative charges on an oxygen atom. Like charges repel, so the phosphate groups in ATP, being tightly packed together, are very unstable. Breaking the so-called "high-energy bond" holding the last phosphate group on the molecule produces a more stable **adenosine diphosphate** (**ADP**) molecule and a free phosphate group (FIGURE 6.5B). ATP hydrolysis is analogous to a spring compacted in a box. Open the box (hydrolyze the phosphate group) and you have a more stable spring (a more stable ADP molecule). The release of the spring (the freeing of a phosphate group) provides the means by which work can be done. Thus, the hydrolysis of the unstable phosphate groups in ATP molecules to a more stable condition is what drives other energy-requiring reactions through the transfer of phosphate groups (Figure 6.5b). The addition of a phosphate group to another molecule is called **phosphorylation**.

Because ATP molecules are unstable, they cannot be stored. Therefore, prokaryotic cells synthesize large organic compounds like glycogen or lipids for energy storage. As needed, the chemical energy in these molecules can be released in catabolic reactions and used to reform ATP from ADP and phosphate (Figure 6.5b). This ATP/ADP cycle occurs continuously in cells. It has been estimated that a typical prokaryotic cell must reform about 3 million ATP molecules per second from ADP and phosphate to supply its energy needs.

It might be a good idea to review what we have covered to this point, which is summarized in FIGURE 6.6. Enzymes regulate metabolic reactions by binding an appropriate substrate at its active site. Often a series of metabolic steps (metabolic pathway) are required to form the final product. Some steps in the pathway may require energy (endergonic); this energy is supplied by ATP as it is hydrolyzed to ADP. Other reactions release energy (exergonic), which may be used to reform ATP from ADP.

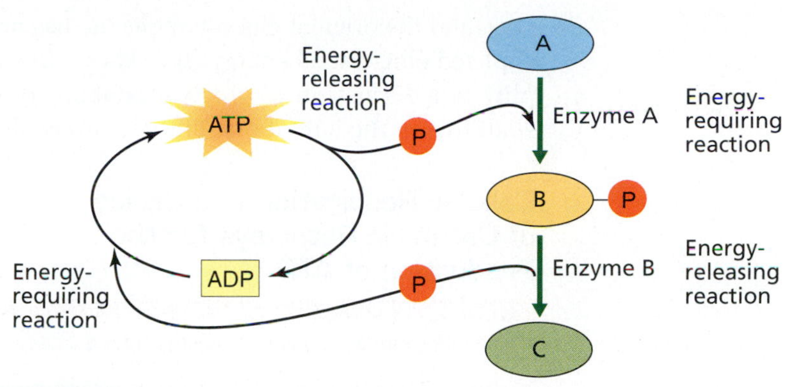

FIGURE 6.6 **A Metabolic Pathway Coupled to the ATP/ADP Cycle.** In this metabolic pathway, enzyme A catalyzes an energy-requiring reaction where the energy comes from ATP hydrolysis. Enzyme B converts the phosphorylated substrate to the end product. Being an energy-releasing reaction, the free phosphate can be coupled to the reformation of ATP.

Q: What are the terms energy-requiring and energy-releasing reactions used?

CONCEPT AND REASONING CHECKS

6.4 Judge the importance of the ATP cycle to prokaryotic metabolism.

6.2 The Catabolism of Glucose

Since the early part of the twentieth century, the chemistry of glucose catabolism has been the subject of intense investigation by biochemists because glucose is a key source of energy for ATP production. Moreover, the process of glucose catabolism is very similar in all organisms, making this "metabolic interlock" one feature that unites all life.

Glucose Contains Stored Energy That Can Be Extracted

KEY CONCEPT

- Glucose is a primary source for generating ATP.

A mole of glucose (180 g) contains about 686,000 calories of energy. This fact can be demonstrated in the laboratory by setting fire to a mole of glucose and measuring the energy released. In a prokaryotic cell, however, not all the energy is set free from glucose, nor can the cell trap all that is released. The process accounts for the transfer of about 40 percent of the glucose energy to ATP energy; that is, chemical energy to cellular energy.

The catabolism of a glucose molecule does not take place in one chemical reaction, nor do ATP molecules form all at once. Rather, the energy in glucose is extracted (converted) slowly to ATP through metabolic pathways. It is similar to the proverb quoted at the beginning of this chapter, "*Life is like a fire; it begins in smoke and ends in ashes.*" The catabolism of glucose starts with a little energy being converted to ATP (the smoke), which builds to a point where large amounts of energy are converted to ATP (the

- **Mole:** The molecular weight of a substance expressed in grams.
- **Calories:** Units of energy defined in the amount of heat required to raise one kilogram of water 1°C.

fire), and the original glucose molecule has been depleted of its useful energy (the ashes). To simplify our discussion of glucose catabolism, we shall follow the fate of one glucose molecule.

Cellular Respiration Is a Series of Catabolic Pathways for the Production of ATP

KEY CONCEPT

- Cellular respiration harvests energy from glucose.

Virtually all cells make ATP by harvesting energy from exergonic metabolic pathways. Such a process is called **cellular respiration**. If cells consume oxygen in making ATP, the catabolic process is called **aerobic respiration**. In other instances, cells can make almost equally substantial amounts of ATP without using oxygen, in which case it is called **anaerobic respiration**. To begin our study of cellular respiration, we shall follow the process of aerobic respiration as it occurs in prokaryotic cells. The process is represented by the following summary equation:

$$\underset{\text{Glucose}}{C_6H_{12}O_6} + \underset{\text{Oxygen}}{6\ O_2} + 38\ ADP + 38\ P$$
$$\downarrow$$
$$\underset{\text{Carbon dioxide}}{6\ O_2} + \underset{\text{Water}}{6\ H_2O} + 38\ ATP$$

The events summarized in the equation are conveniently divided into three processes: glycolysis, the Krebs cycle, and oxidative phosphorylation. Let's examine each of these in sequence.

Glycolysis Is the First Stage of Energy Extraction

KEY CONCEPT

- Glycolysis is a metabolic pathway yielding ATP and NADH.

The splitting of glucose, called **glycolysis** (*glyco* = "sweet"), occurs in the cytosol of prokaryotes and involves a metabolic pathway that converts an initial 6-carbon substrate, **glucose**, into two 3-carbon molecules called **pyruvate**. Between glucose and pyruvate, there are eight intermediates formed, each catalyzed by a specific enzyme. FIGURE 6.7 illustrates the process. For easy referral, numbers in circles identify each reaction, and it would be helpful to refer to the figure as the discussion proceeds.

The first part of glycolysis is endergonic; one molecule of ATP is hydrolyzed (consumed) in reaction (1) and a second in reaction (3). In both cases, the phosphate group from ATP attaches to the product. Thus, reaction (1) produces glucose-6-phosphate, and reaction (3) yields fructose-1,6-bisphosphate (bis means "two separate;" that is, two separate phosphate molecules).

After the splitting of fructose-1,6-bisphosphate into two 3-carbon molecules, each passes through an additional series of conversions that ultimately form pyruvate. During reactions (6) and (9), ATP is generated. In both steps, enough energy is released to synthesize an ATP molecule from ADP and phosphate, resulting in a total of four ATP molecules. Considering two ATP molecules were consumed in reactions (1) and (3), the net gain from glycolysis is two molecules of ATP.

Before we proceed, take note of reaction (5). This enzymatic reaction releases two high-energy electrons and two protons (H^+), which are picked up by the coenzyme NAD^+, reducing each to NADH. This and similar events will have great significance shortly as an additional source to generate ATP.

■ Reducing (reduction): The process of a substance gaining electron pairs.

CONCEPT AND REASONING CHECKS

6.5 Besides the final end product, what are the other products of glycolysis?

The Krebs Cycle Extracts More Energy from Pyruvate

KEY CONCEPT

- The Krebs cycle yields additional ATP and NADH, as well as $FADH_2$.

The **Krebs cycle** (also called the **citric acid cycle**) is a series of chemical reactions named for Hans A. Krebs, who won the 1953 Nobel Prize for the discovery of several participating substances in the cycle. This metabolic pathway is referred to as a cycle because the end product formed is used as one substrate to initiate the pathway. All of the reactions are catalyzed by enzymes, and all take place along the cell membrane of prokaryotic cells. In eukaryotic microbes, including the protozoa, algae, and fungi, the Krebs cycle reactions occur in the mitochondria.

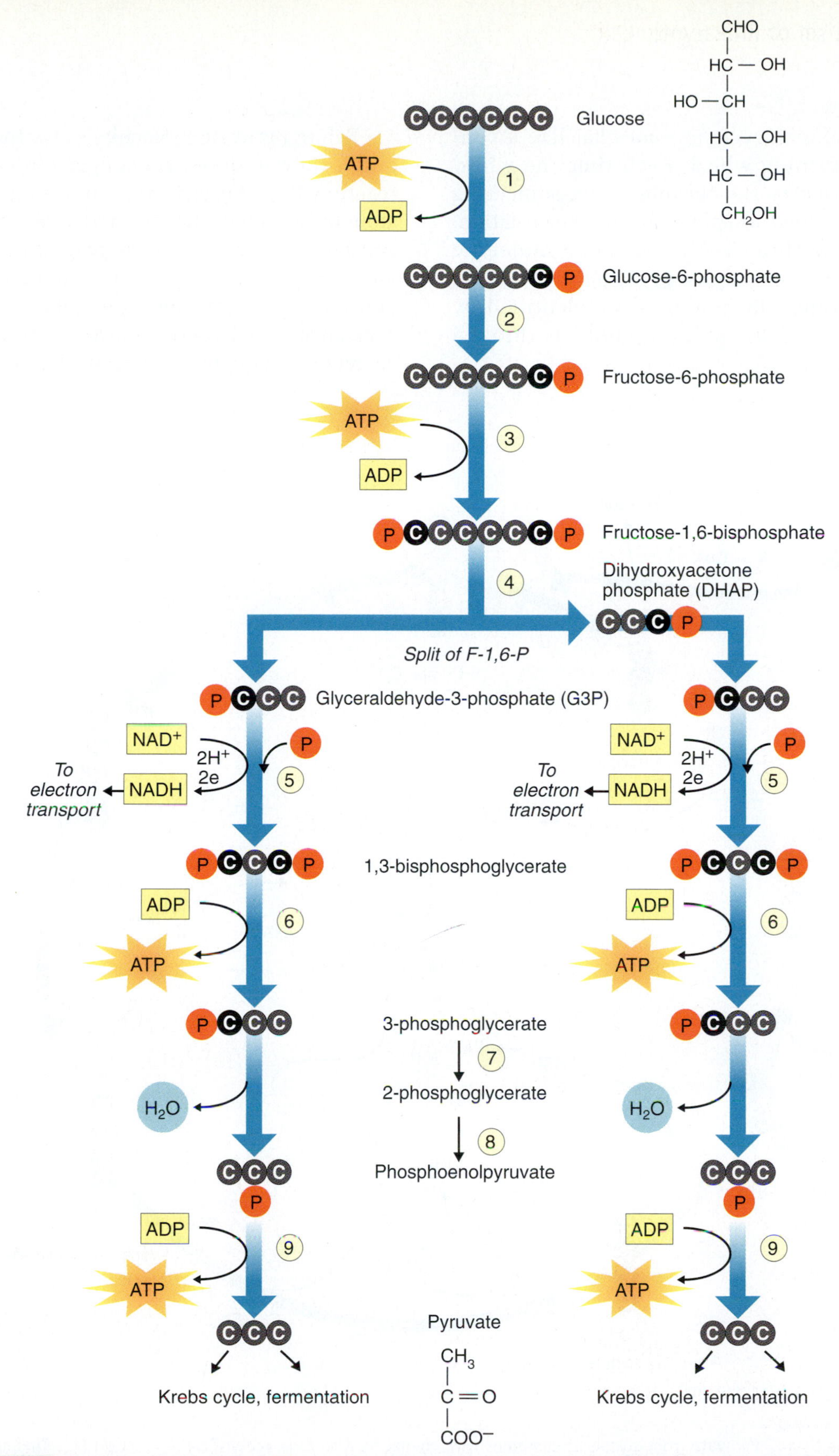

FIGURE 6.7 **The Steps of Glycolysis.** Carbon atoms are represented by circles. The dark circles represent carbon atoms bonded to phosphate groups. ATP is supplied to glucose in reaction **(1)** and to fructose-6-phosphate in reaction **(3)**. Two ATP molecules are generated in reaction **(6)** and two more in reaction **(9)**. The total ATP produced is four molecules but since two are used up in glycolysis, the net gain is two molecules of ATP. Also, in reaction **(5)**, high-energy electrons and protons (H^+) are captured by NAD^+ molecules, forming NADH. Note that the original six carbon atoms of glucose exist as two pyruvate molecules.

Q: At the end of glycolysis, where is the energy that was originally in glucose?

The Krebs cycle is somewhat like a constantly turning wheel. Each time the wheel comes back to the starting point, something must be added to spin it for another rotation. That something is the pyruvate molecule derived from glycolysis. FIGURE 6.8 shows the Krebs cycle. The reactions are identified by capital letters in circles to guide us through the cycle.

Before pyruvate molecules enter the Krebs cycle, they undergo a change, indicated in reaction (A). An enzyme removes a carbon atom from each of the two pyruvate molecules and releases the carbons as two carbon dioxide molecules ($2CO_2$). The remaining two carbon atoms of pyruvate are combined with a substance called **coenzyme A (CoA)** to form **acetyl-CoA**. Equally important, this is another

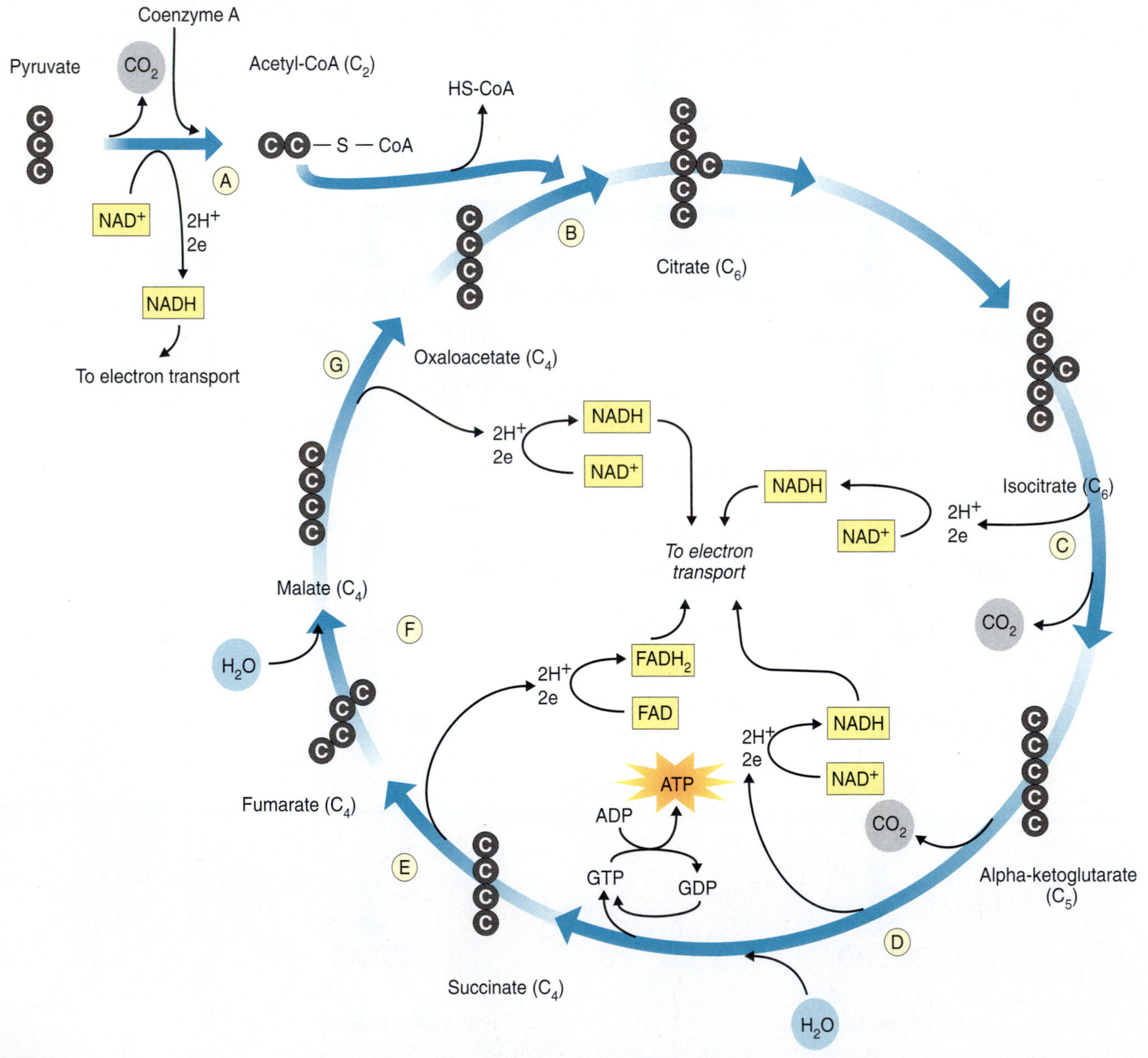

FIGURE 6.8 **The Steps of the Krebs Cycle.** Pyruvate from glycolysis combines with coenzyme A to form acetyl-CoA in reaction (**A**). This molecule then joins with oxaloacetate to form citrate (**B**). In reactions (**C**) through (**F**), citrate enters the pathways, where intermediates form and regenerate oxaloacetate in the last reaction (**G**). During the process, the three carbons of pyruvate are liberated as three molecules of carbon dioxide. ATP is formed in reaction (**D**) via GTP, and high-energy electrons and protons are liberated during reactions (**A**), (**C**), (**D**), (**E**), and (**G**). These are captured by NAD^+ or FAD molecules.

Q: Why are so many reactions required to extract energy out of pyruvate?

reaction associated with electron and proton transfer to NAD^+ to form NADH.

The two remaining carbons from pyruvate are now ready to enter the Krebs cycle. In reaction (**B**), each acetyl-CoA unites with a 4-carbon oxaloacetate to form citrate, a 6-carbon molecule. (Citrate, or citric acid, may be familiar to you as a component of soft drinks.) Note that isocitrate changes in reaction (**C**) to a substance having only five carbon atoms. The sixth atom has been released as CO_2. In the next step, (**D**), another carbon atom is released as CO_2. The 4-carbon molecule then undergoes a series of modifications (reactions **E**, **F**, and **G**) reforming oxaloacetate. The cycle is now complete, and oxaloacetate is ready to unite with another molecule of acetyl-CoA.

Several features of the Krebs cycle merit closer scrutiny. First, we shall follow the carbon. Pyruvate, with three carbon atoms, emerges from glycolysis, but after one turn of the Krebs cycle, its carbon atoms exist in three molecules of CO_2. There are two molecules of pyruvate from glycolysis, so when the second molecule enters the Krebs cycle, its carbon atoms also will form three CO_2 molecules. Remember that we began with a 6-carbon glucose molecule; six CO_2 molecules now have been produced. This fulfills part of the equation for aerobic respiration:

$$\underline{C_6}H_{12}O_6 + 6\ O_2 + 38\ ADP + 38\ P$$
$$\downarrow$$
$$\underline{6\ C}O_2 + 6\ H_2O + 38\ ATP$$

The second feature of the Krebs cycle is reaction (**D**). Here a molecule of guanosine triphosphate (GTP) forms and is used immediately to form ATP from ADP. Because we have two pyruvate molecules entering the cycle (per molecule of glucose), a second ATP molecule will form from GTP when the second pyruvate passes through the cycle.

Last, and perhaps of most importance, are reactions **C**, **D**, **E**, and **G**. Reactions **C**, **D**, and **G**, like reaction 5 in glycolysis and in the conversion of pyruvate to acetyl-CoA, are associated with NAD^+ and again are reduced to NADH. Two NADH molecules are produced in each step for a total of six. In addition, reaction **E** accomplishes much of the same result except it is associated with another coenzyme, FAD. It too receives two electrons and two protons from the reaction, being reduced to $FADH_2$. For the two pyruvate molecules starting the process, two $FADH_2$ molecules are formed.

In summary, glycolysis and the Krebs cycle have extracted as much energy as possible from glucose and pyruvate (FIGURE 6.9). This has amounted to a small gain of ATP molecules formed from one glucose molecule. However, the 10 NADH and 2 $FADH_2$ molecules formed are most significant. Let's see how.

CONCEPT AND REASONING CHECKS

6.6 Identify the initial substrates and final end products of the Krebs cycle.

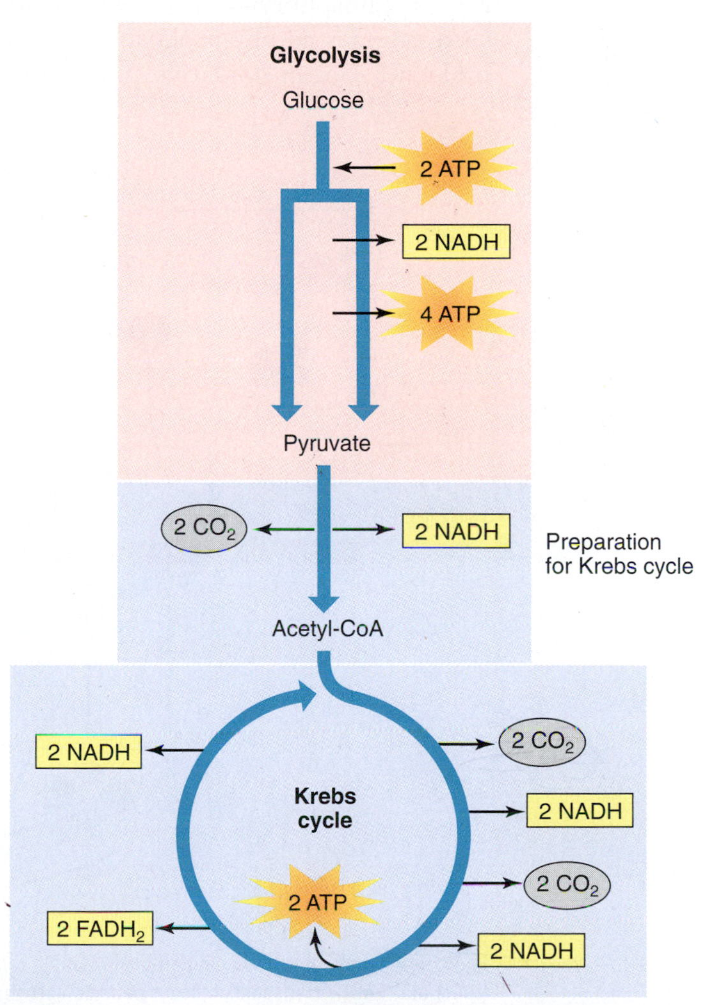

FIGURE 6.9 **Summary of Glycolysis and the Krebs Cycle.** Glycolysis and the Krebs cycle are metabolic pathways to extract chemical energy from glucose to generate cellular energy (ATP).

Q: Every time one glucose molecule is broken down to CO_2 and water, how many ATP, NADH, and $FADH_2$ molecules are gained?

Oxidative Phosphorylation Is the Process by Which Most ATP Molecules Form

KEY CONCEPT

- NADH and $FADH_2$ provide the source for oxidative phosphorylation.

Oxidative phosphorylation refers to a sequence of reactions in which two events happen: Pairs of electrons are passed from one chemical substance to another (electron transport), and the energy released during their passage is used to combine phosphate with ADP to form ATP (ATP synthesis). The adjective oxidative is derived from the term **oxidation**, which refers to the loss of electron pairs from molecules. Its counterpart is **reduction**, which refers to a gain of electron pairs by molecules. Phosphorylation, as we already have seen, implies adding a phosphate to another molecule. So, in oxidative phosphorylation, the loss and transport of electrons will enable ADP to be phosphorylated to ATP. Like the Krebs cycle, oxidative phosphorylation takes place at the cell membrane in prokaryotes and in the mitochondria of eukaryotic microbes.

Oxidative phosphorylation is responsible for producing 34 molecules of ATP per glucose. The overall sequence involves the NAD^+ and FAD coenzymes that underwent reduction to NADH and $FADH_2$ during glycolysis and the Krebs cycle; remember, they gained two electrons in those metabolic pathways. In oxidative phosphorylation, the coenzymes will be reoxidized by passing off those two electrons to a series of electron carriers (FIGURE 6.10). These carriers, called **cytochromes** (*cyto* = "cell"; *chrome* = "color"), are a set of protein pigments containing iron ions that accept and release electron pairs. Together, the cytochromes form an **electron transport chain**. The last link in the chain is oxygen gas.

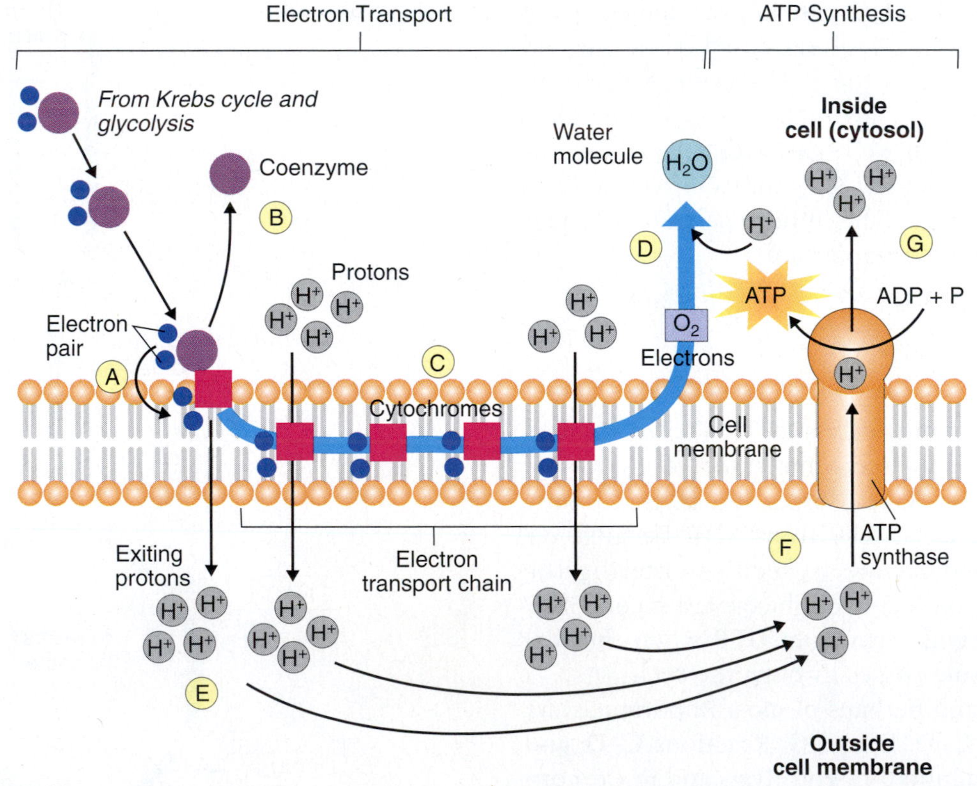

FIGURE 6.10 Oxidative Phosphorylation in Prokaryotes. (**A**) Originating in glycolysis or the Krebs cycle, coenzymes NADH and $FADH_2$ transport electron pairs to cytochromes in the cell membrane. (**B**) The NAD^+ or FAD coenzyme becomes reoxidized and is reused. (**C**) Cytochromes transport the electron pairs, which combine with oxygen and protons to form water molecules (**D**). As the electrons are transported, they release energy, which fuels the transport of protons (H^+) across the cell membrane at three points (**E**). (**F**) Protons then reenter the cytoplasm of the cell through a protein channel in the ATP synthase enzyme. (**G**) ADP molecules join with phosphates as protons move through the channel, producing the ATP molecules.

Q: What would happen to the oxidative phosphorylation process if this cell were deprived of oxygen?

In the oxidative phosphorylation process shown in Figure 6.10, the two electrons with each NADH and $FADH_2$ are passed to the first cytochrome in the chain (**A**). The reoxidized coenzymes, NAD^+ and FAD, return to the cytosol (**B**) to be used again in glycolysis or the Krebs cycle. Like walking along stepping stones, each electron pair is passed from one cytochrome to the next down the chain (**C**) until the electron pair is accepted by oxygen. Oxygen also acquires two protons (2 H^+) from the cytosol (**D**) and becomes water (H_2O). Oxygen's role is of great significance because if oxygen were not present, there would be no way for cytochromes to unload their electrons and the entire system would soon back up like a jammed conveyer belt and come to a halt. Oxygen's role also is reflected in the equation for aerobic respiration:

$$C_6H_{12}O_6 + \underline{6}\ \underline{O_2} + 38\ ADP + 38\ P$$
$$\downarrow$$
$$6\ CO_2 + \underline{6}\ \underline{H_2O} + 38\ ATP$$

So, what is the importance of the electron transport chain since no ATP has been made? The actual mechanism for ATP synthesis comes from the pumping of protons by a process called **chemiosmosis** (*osmos* = "push"). First proposed by Nobel Prize winner Peter Mitchell, chemiosmosis uses the power of proton movement across a membrane to conserve energy for ATP synthesis.

What happens in chemiosmosis also is shown in Figure 6.10. As the electrons pass from cytochrome to cytochrome, the electrons gradually lose energy. The energy, however, is not lost in the sense that it is gone forever. Instead, the energy is used at three transition points to "pump" protons (H^+) across the membrane from the cytosol to the area outside of the cell membrane (**E**). Soon a large number of protons have built up outside the membrane, and because they cannot easily reenter the cell, they represent a large concentration of potential energy (much like a boulder at the top of a hill). The protons are positively charged, so there also is a buildup of charges outside the membrane.

Suddenly, a series of channels opens and the proton flow reverses (**F**). Each "channel" is contained within a large enzyme complex called **ATP synthase**, which has binding sites for ADP and phosphate. As the protons rush through the channel, they release their energy, and the energy is used to synthesize ATP molecules from ADP and phosphate ions (**G**), as MicroInquiry 6 explains. Three molecules of ATP can be synthesized for each pair of electrons originating from NADH; two molecules of ATP are produced for each pair of electrons from $FADH_2$ because the coenzyme interacts further down the chain. MicroFocus 6.3 highlights a novel way of using the bacterial respiratory process to generate electricity.

Chemiosmosis occurs only in structurally intact membranes. If the cell membrane is damaged so proton movement cannot take place, the synthesis of ATP ceases even though electron transport through the cytochrome system continues. Without ATP production, the organism rapidly dies. This is one reason why damage to the prokaryotic cell membrane, such as with antibiotics or detergent disinfectants, is so harmful.

The ATP yield from aerobic respiration is summarized in FIGURE 6.11. The grand total

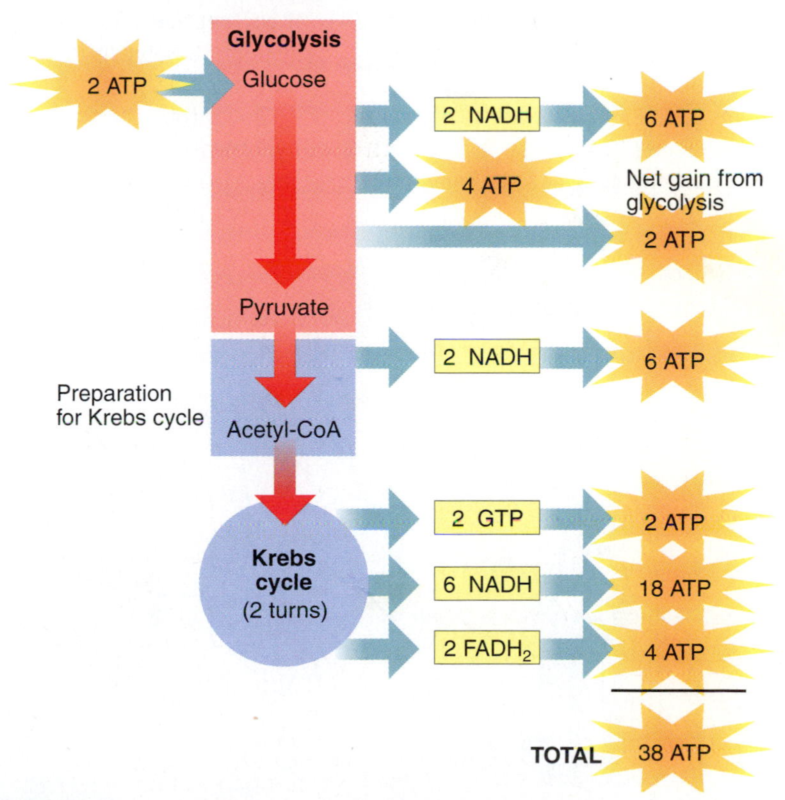

FIGURE 6.11 **The ATP Yield from Aerobic Respiration.** In a prokaryotic cell, 38 molecules of ATP can result from the metabolism of a molecule of glucose. Each NADH molecule accounts for the formation of three molecules of ATP; each molecule of $FADH_2$ accounts for two ATP molecules.

Q: From this diagram, what is the single most important product for ATP synthesis?

MICROINQUIRY 6

The Machine That Makes ATP

Every day, an adult human weighing 160 pounds uses up about 80 pounds of ATP (about half of his or her weight). The ATP is changed to its two breakdown products, ADP and phosphate, and enormous amounts of energy are made available to do metabolic work.

However, the body's weight does not go down, nor does it change perceptibly because the cells are constantly regenerating ATP from the breakdown products. Discovering how this is accomplished and how the recycling works were the seminal achievements of 1997's Nobel Prize winners in Chemistry.

The Chemiosmotic Basis for ATP Synthesis

One of the three 1997 winners was Paul D. Boyer at the University of California at Los Angeles. Boyer's work expanded the pioneering work of Peter Mitchell, who developed the concept of chemiosmosis. Chemiosmosis proposes that electron transport between cytochromes provides the energy to "pump" protons (H^+) across the membrane; in the case of prokaryotes, this is from the cytosol to the environment. As explained in the text, this proton gradient provides the force or potential to drive the protons back into the cell through an enzyme called ATP synthase. This flow of rapidly streaming H^+ brings together ADP and phosphate to form ATP.

How Does Proton Flow Cause ATP Synthesis?

The groundbreaking research as to how the ATP synthase works came from studies with *Escherichia coli* cells. Researchers knew that ATP synthase consisted of three functional regions (see **Figure A**). Embedded in the cell membrane was an F_0 complex consisting of nine polypeptides of three different types. Extending from the cytoplasmic end of F_0 is a stalk made of three different proteins. The stalk connects the F_0 complex to the F_1 complex, which consists of six polypeptides, three called α polypeptides and three called β polypeptides. So, an ATP synthase consists of 18 polypeptides—a veritable nanomachine.

Boyer took the three complexes and hypothesized how they could manufacture ATP. His ideas plus newer findings have been merged into the current model (see **Figure B**):

1. The flow of protons through F_0 causes F_0 and the stalk to spin as protons stream by (somewhat reminiscent of a turning water wheel).
2. One of the polypeptides (γ) in the stalk extends into the F_1 complex and makes contact with each of the b-subunits as it rotates.

Because there are three β-subunits, three ATP molecules are produced each time the stalk makes a complete rotation. Each β-subunit is like an active site in an enzyme. It recognizes and binds substrate [ADP and P], forms an enzyme-substrate complex, and releases the product (ATP).

Discussion Point

Ribosomes, flagella, and ATP synthase all represent "nanomachines" to carry out specific functions in cells. Discuss the concept of a prokaryotic cell as being an assemblage of nanomachines.

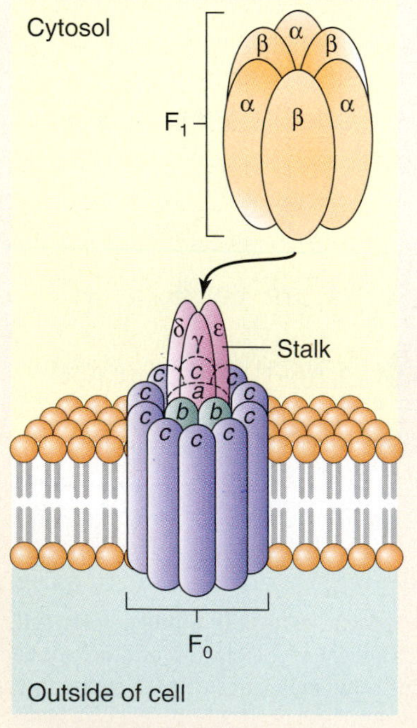

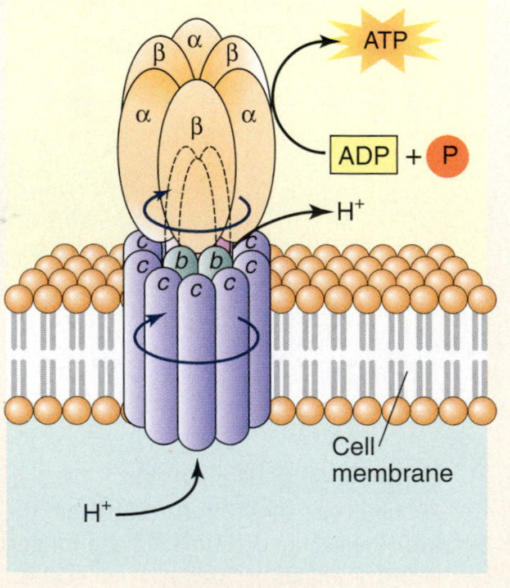

FIGURE A ATP synthase enzyme consists of 18 polypeptides in three complexes, the F_0 and stalk that are embedded in the cell membrane and the F_1 that projects into the cytosol.

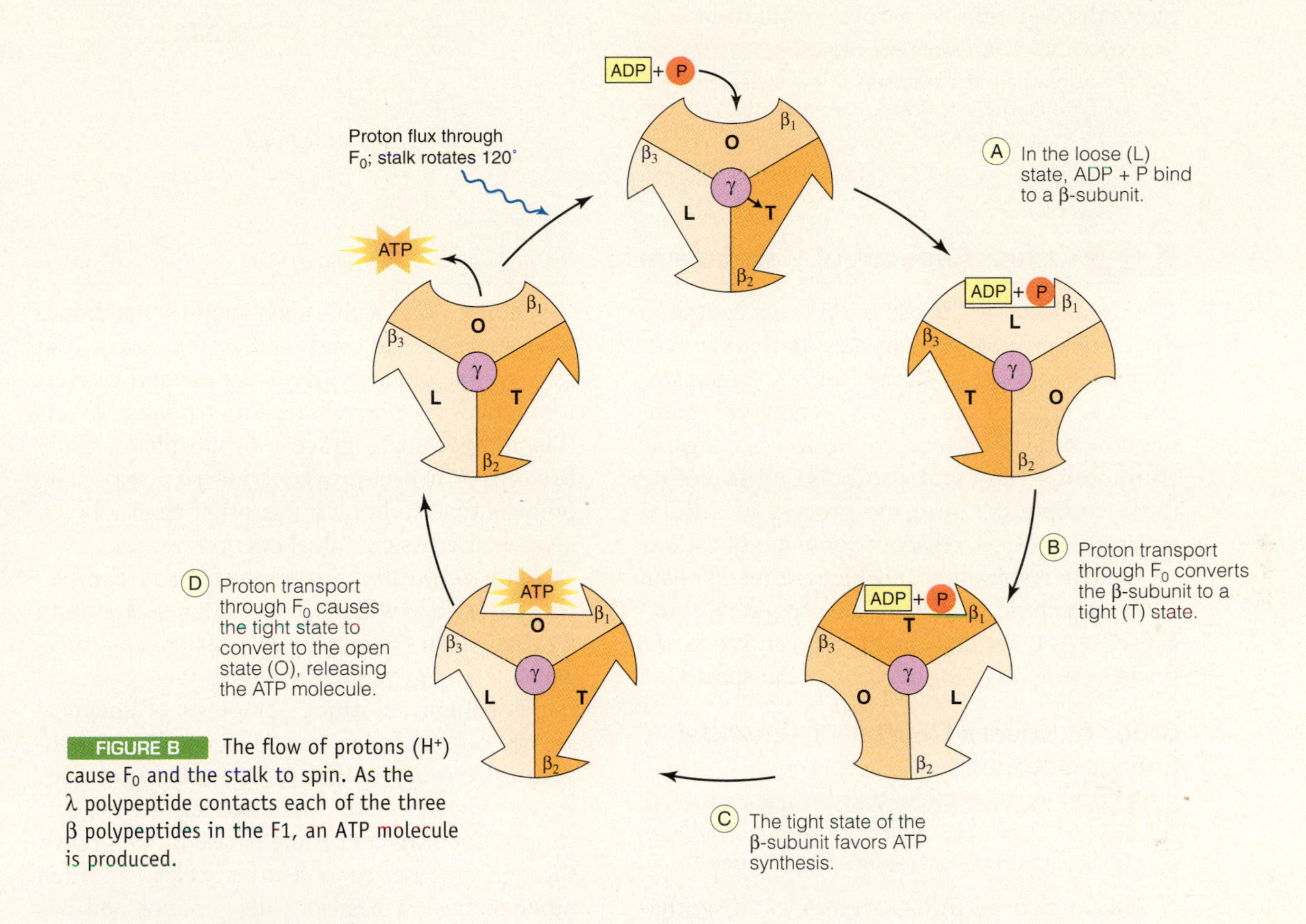

FIGURE B The flow of protons (H^+) cause F_0 and the stalk to spin. As the λ polypeptide contacts each of the three β polypeptides in the F1, an ATP molecule is produced.

MICROFOCUS 6.3: Biotechnology

Bacteria Not Included

How many toys (child or adult) or electronic devices do you purchase each year where batteries are needed to run the device? And often batteries are not included. Today, a new type of battery is being developed—one that converts sugar not into ATP but rather into electricity. The battery is one packed with bacterial cells.

Realize that cellular respiration involves minute electrical currents being generated. During cellular respiration, electrons are transferred to cofactors like NAD^+ and they are passed along a chain of cytochromes during oxidative phosphorylation. Swades Chaudhuri and Derek Lovely of the University of Massachusetts of Amherst have taken this idea and applied it to developing a new type of fuel cell or battery.

The scientists mixed the bacterial species *Rhodoferax ferrireducens*, which they found in aquifer sediments in Virginia, with any of a variety of common sugars. When placed in a chamber with a graphite electrode, *R. ferrireducens* metabolized the sugar, stripped off the electrons, and transferred them directly to the electrode. The result: a current was produced. In addition, the bacterial cells continued to grow, so a stable current could be produced with high efficiency.

It is still a long way from producing a reliable, long-lasting bacterial battery. However, the researchers believe much of the agricultural or industrial waste produced today could be the "sugar" used in making these bacterial batteries. So, as Sarah Graham reported for *Scientific American.com*, "Perhaps one day electronics will be sold with the caveat 'bacteria not included.' "

of ATP molecules from the energy extracted from one glucose molecule is 38. It also completes the equation for aerobic respiration:

$$C_6H_{12}O_6 + 6\ O_2 + \underline{38\ ADP} + \underline{38\ P} \rightarrow 6\ CO_2 + 6\ H_2O + \underline{38\ ATP}$$

CONCEPT AND REASONING CHECKS

6.7 Identify where the reactions of glycolysis and aerobic respiration occur in (a) a prokaryotic cell and (b) a eukaryotic cell.

6.3 Other Aspects of Catabolism

The catabolism of glucose is a process central to the metabolism of prokaryotes and eukaryotes alike as it provides a glimpse of how organisms obtain energy for life. In this section, we examine how cells obtain energy from other organic compounds (fats and proteins) by directing those compounds into the process of cellular respiration. We also discover how modifications to cellular respiration and glucose metabolism allow anaerobic organisms to use glucose and generate ATP without having oxygen gas as the final electron acceptor in electron transport.

Other Nutrients Represent Potential Energy Sources

KEY CONCEPT

- Other carbohydrates as well as fats and proteins also can supply chemical energy for ATP production.

A wide variety of monosaccharides, disaccharides, and polysaccharides serve as useful energy sources for prokaryotes. All must go through a series of preparatory conversions before they are processed in glycolysis, the Krebs cycle, and oxidative phosphorylation.

In preparation for entry into the scheme of metabolism, different carbohydrates use different pathways (FIGURE 6.12). Sucrose, for example, is firs
into its consti
fructose. The g
colysis pathwa
cule first is co
The latter the
and a molecu
scheme as D
another disacc
enzyme lactase
tose undergoes
ready to enter g
6-phosphate [r

Stored polysaccharides, such as starch and glycogen, are metabolized by enzymes that remove one glucose unit at a time and convert it to glucose-1-phosphate. An enzyme converts this compound to glucose-6-phosphate, ready for entry into the glycolysis pathway. The point is that carbohydrates other than glucose also are used as chemical energy sources.

The economy of metabolism is demonstrated further when we consider protein and fat catabolism (Figure 6.12). Fats are extremely valuable energy sources because their chemical bonds contain enormous amounts of chemical energy. Although proteins are generally not considered energy sources, cells use them for energy when carbohydrates and fats are in short supply. Both fats and proteins are broken down through glucose catabolism as well as through other pathways. Basically, the proteins and fats undergo a series of enzyme-catalyzed conversions and form components normally occurring in carbohydrate metabolism. These components then continue along the metabolic pathways as if they originated from carbohydrates.

Proteins are broken down to amino acids (see Chapter 2). Enzymes then convert many amino acids to pathway components by
the amino group and substituting a
group. This process is called **deami-**
or example, alanine is converted to
and aspartic acid is converted to
ate. For certain amino acids, the
more complex, but the result is the
amino acids become pathway inter-
of cellular respiration.
onsist of three fatty acids bonded to a
olecule (see Chapter 2). To be useful
purposes, the fatty acids are sepa-
the glycerol by the enzyme lipase.
has taken place, the glycerol portion

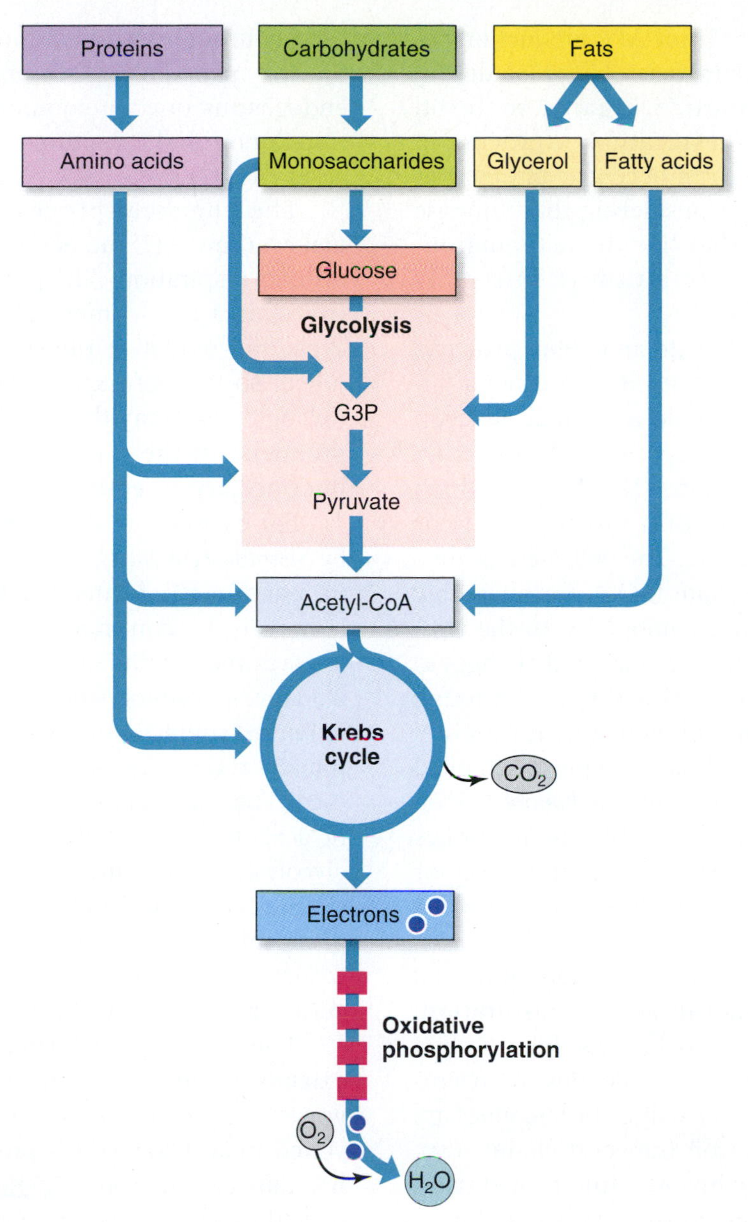

FIGURE 6.12 Carbohydrate, Protein, and Fat Metabolism. Besides glucose, other carbohydrates as well as proteins and fats can be sources of energy by providing electrons and protons for cellular respiration. The intermediates enter the pathway at various points.

Q: Why would more ATP be produced from the products of one fatty acid entering the pathway at acetyl-CoA than from one amino acid entering at acetyl-CoA?

is converted to DHAP. For fatty acids, there is a complex series of conversions called **beta oxidation**, in which each long-chain fatty acid is broken by enzymes into 2-carbon units. Other enzymes then convert each unit to a molecule of acetyl-CoA ready for the Krebs cycle. We previously noted that for each turn of the Krebs cycle, 16 molecules of ATP are derived. A quick calculation should illustrate the substantial energy output from a 16-carbon fatty acid (eight 2-carbon units).

CONCEPT AND REASONING CHECKS

6.8 Describe how lipids and proteins are prepared for entry into the cellular respiration pathway.

Anaerobic Respiration Produces ATP Using Other Final Electron Acceptors

KEY CONCEPT

- ATP can be produced through chemiosmosis without oxygen gas.

Because the solubility of O_2 in water is quite low, many prokaryotes rely on terminal electron acceptors other than O_2 for ATP production. To accomplish this, certain anaerobic or facultative prokaryotes metabolize organic material through **anaerobic respiration**, a process in which the final electron acceptor in electron transport is not O_2. Considering the immense number of species that live in such environments, anaerobic respiration is extremely important ecologically.

Facultative or obligate anaerobic prokaryotes carry out anaerobic respiration using a different inorganic molecule as a final electron acceptor. The facultative species *Escherichia coli*, for example, uses nitrate (NO_3^-) with which electrons combine to form nitrite (NO_2^-) or another nitrogen product. The obligate anaerobe *Desulfovibrio* uses sulfate ($SO_4^=$) for anaerobic respiration. The sulfate combines with the electrons from the cytochrome chain and changes to hydrogen sulfide (H_2S). This gas gives a rotten egg smell to the environment (as in a tightly compacted landfill). A final example is exhibited by the archaeal methanogens, *Methanobacterium* and *Methanococcus*. These obligate anaerobes use carbonate ($CO_3^=$) as a final electron acceptor and, with hydrogen nuclei, form large amounts of methane gas (CH_4).

In anaerobic respiration, the amount of ATP produced is less than in aerobic respiration. There are several reasons for this. First, only a portion of the Krebs cycle functions in anaerobic respiration, so fewer reduced coenzymes are available to the electron transport chain. Also, not all of the cytochromes function during anaerobic respiration, so the ATP yield will be less. The exact amount of ATP produced therefore will depend on the organism and where in the respiratory pathway intermediates enter.

CONCEPT AND REASONING CHECKS

6.9 Why do obligate anaerobes tend to grow slower than obligate aerobes?

Fermentation Produces ATP Using an Organic Final Electron Acceptor

KEY CONCEPT

- Fermentation generates ATP in the absence of exogenous electron acceptors.

In environments that are anoxic and without the alternative electron acceptors needed by anaerobes, much of the organic material will be catabolized through fermentation. **Fermentation** is the enzymatic process for producing ATP using endogenous organic compounds as both electron donors and acceptors—exogenous electron acceptors (O_2, NO_3^-, $SO_4^=$, $CO_3^=$) are absent.

■ Anoxic: Without oxygen gas (O_2).

The chemical process of fermentation makes a few ATP molecules in the absence of cellular respiration. The products of glycolysis (pyruvate) are formed. However, the Krebs cycle and oxidative phosphorylation are shut down, so the pyruvate molecules are shuttled through a pathway that produces fermentation products. In these pathways, pyruvate is the intermediary accepting the electrons.

For example, in the fermentation of glucose by *Streptococcus lactis*, the intermediary molecule in reaction (5) forms NADH (FIGURE 6.13A). However, the continued production of NADH requires the continued input of NAD^+. Without oxidative phosphorylation running, NAD^+ exists in limited supply in the cytosol and must be continually regenerated so that glycolysis may proceed. The conversion of pyruvate to lactate (lactic acid) is a way to reform NAD^+ coenzymes so glycolysis can still make two ATP molecules for every glucose molecule consumed. In a dairy plant, this process of **lactic acid fermentation** by *S. lactis* is carefully controlled so the acid will curdle fresh milk to make buttermilk.

The diversity of fermentation chemistry extends to some eukaryotic microbes as well. In yeasts like *Saccharomyces*, pyruvate first is converted to acetaldehyde, a process in which carbon dioxide is released (FIGURE 6.13B). Acetaldehyde then serves as the acceptor for the electrons and protons of NADH. As the acetaldehyde is converted to ethyl alcohol (ethanol), NAD^+ is reformed. The liquor industry uses the ethyl alcohol produced in **alcoholic fermentation** to make alcoholic beverages such as beer and wine (Chapter 27). Fermentation of carbohydrates to alcohol by *C. albicans* also may take place in the human body, such as that of Charlie Swaart's, as explained in the opener of this chapter.

The energy benefits to fermentative organisms are far less than in cellular respiration. In fermentation, each glucose passing through glycolysis yields two ATP molecules and the production of fermentation end products. This is in sharp contrast to the 38 molecules evolving in cellular respiration. It is

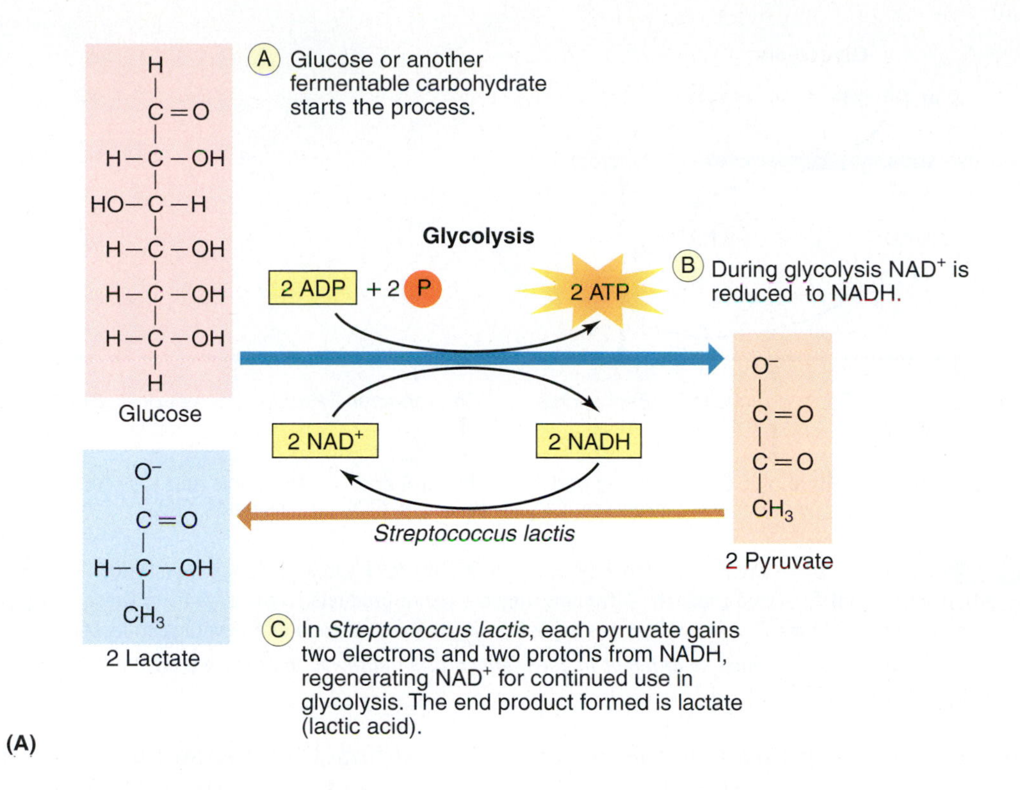

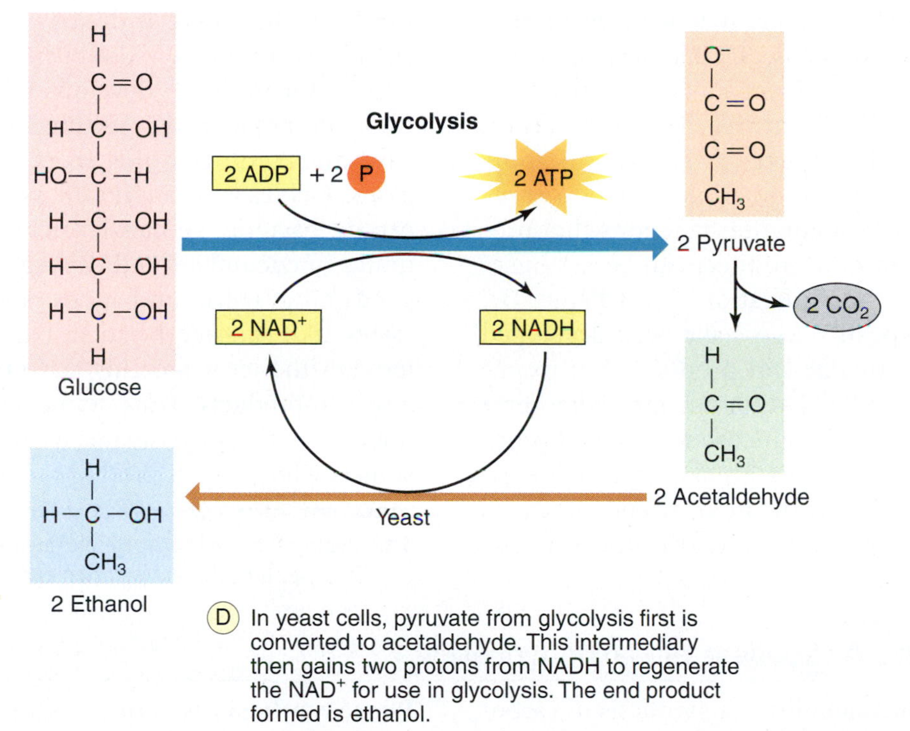

FIGURE 6.13 **Microbial Fermentation.** Fermentation is an anaerobic process that reoxidizes NADH to NAD^+ by converting organic materials into fermentation products. (**A**) Lactic acid fermentation; (**B**) alcoholic fermentation.

Q: How are lactic acid and alcoholic fermentation identical in purpose?

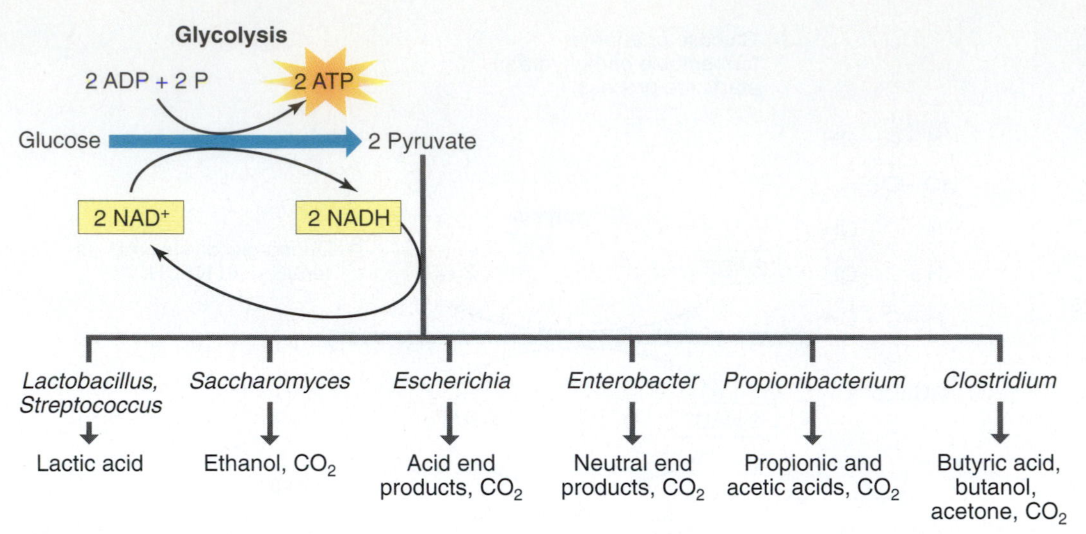

FIGURE 6.14 **Variations in Fermentation End Products.** Different prokaryotes (or fungal yeasts) can carry out fermentation and produce different final end products. Although they are waste products to the organism, these and many other end products may be useful in medical identifications as well as in commercial food and beverage industries.

Q: What specific metabolic catalyst must be different in each fermentation pathway to produce a different end product from the same substrate (pyruvate)?

clear that cellular respiration is the better choice for energy conservation, but under anoxic conditions, there may be little alternative if life for *S. lactis*, *Saccharomyces*, or any fermentative microorganism is to continue.

FIGURE 6.14 identifies the fermentation pathways and products of several microorganisms. The ability of prokaryotes to carry out different fermentation reactions that produce different end products can be very useful in species identification (see Chapter 3). Therefore, specific tests have been developed to detect particular end products. For example, the methyl red test indicates whether a species can ferment glucose to acid end products, while the Voges-Proskauer test identifies species that form neutral end products from the acids produced through glucose fermentation.

Although fermentation end products are waste products of the microorganisms, the food and beverage industries see many of these products in a different light. Swiss cheese, for instance, develops its flavor partly from propionic acid resulting from fermentation and gets its holes from fermentation gases. Pickles and sauerkraut are sour because certain bacterial species ferment the carbohydrates in cucumbers and cabbage, respectively, producing acetic and lactic acids. Sausage tastes like sausage because bacterial species ferment the meat proteins and produce mixed acid end products. Thus, fermentation is useful not only to the microorganisms, but also to consumers who enjoy its products (Chapter 27).

CONCEPT AND REASONING CHECKS

6.10 Justify the need for some prokaryotes (and eukaryotes) to produce fermentation end products.

6.4 The Anabolism of Carbohydrates

Although the anabolism or synthesis of carbohydrates takes place through various mechanisms in prokaryotes, the unifying feature is the requirement for energy.

Photosynthesis Is a Process to Acquire Chemical Energy

KEY CONCEPT

- Photosynthesis converts light energy into chemical energy usually in the form of carbohydrates.

Photosynthesis is a process by which light energy is converted to chemical energy that is then stored as carbohydrate or other organic compounds. In prokaryotes, the process takes place in the cell membrane, which contains chlorophyll or chlorophyll-like pigments. ATP is a key intermediary compound in the process, and glucose is a major end product. Among prokaryotes, photosynthesis occurs in the green sulfur bacteria, the purple sulfur bac-

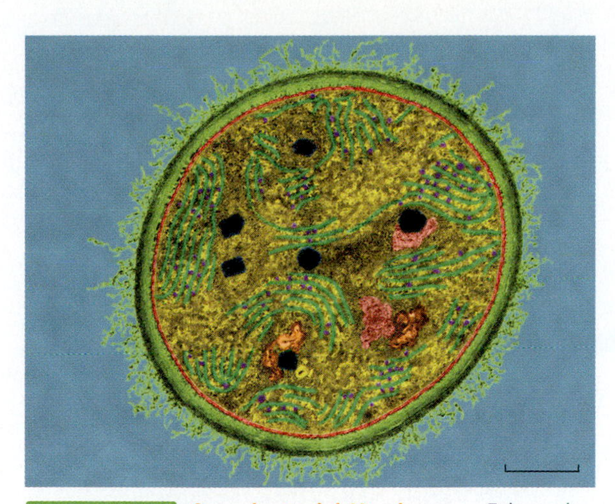

FIGURE 6.15 **Cyanobacterial Membranes.** False-color transmission electron micrograph of a cyanobacterium displaying the membranes (green) along which photosynthetic pigments are located. (Bar = 2 μm.)

Q: What membranes in algae and plant chloroplasts are analagous to the cyanobacterial membranes?

teria, and the cyanobacteria (FIGURE 6.15). Among eukaryotes, photosynthesis occurs in the chloroplasts of such organisms as diatoms, dinoflagellates, and other algae. Our discussion focuses on the prokaryotes, specifically the cyanobacteria, which carry out photosynthesis in much the same manner as algae.

The phases of photosynthesis are shown in FIGURE 6.16, where the sequence of stages is labeled by number. Light energy is absorbed by the green pigment **chlorophyll *a***, a magnesium-containing, lipid-soluble compound (Figure 6.16A). Chlorophylls and accessory pigments make up light-receiving systems called **photosystems**. The light excites pigment molecules in photosystem II, and each molecule loses one electron (**1**). Dislodged electrons are accepted by the first of a series of electron carriers (**2**). The electrons are passed along the series of cytochromes, and eventually the electrons are taken up by other chlorophyll pigments that form photosystem I.

As the electrons move between cytochromes, energy is made available for proton pumping across the cell membrane of the cyanobacterium, followed by chemiosmosis. As described for oxidative phosphorylation, ATP is formed when protons pass back across the membrane and release their energy. Because light was involved in the formation of ATP, this process is called **photophosporylation**.

The electrons in photosystem I again are excited by light energy (**3**) and are boosted out of the pigment molecules to the first of another set of electron carriers, and finally to a coenzyme called nicotinamide adenine dinucleotide phosphate ($NADP^+$). The coenzyme functions much like NAD^+ in that $NADP^+$ receives pairs of electrons and protons from water molecules to form NADPH (**4**).

Because two major energy products, ATP and NADPH, result from this phase of photosynthesis, the phase is termed the **energy-fixing reactions** because light energy is trapped and converted to (or "fixed" as) chemical energy. It is important to note that electrons are replaced in the chlorophyll molecules by electrons from water molecules (which also supply hydrogen ions, as noted above). The residual portions of the water molecules recombine with one another and yield oxygen gas that fills the atmosphere and is used by aerobic organisms for cellular respiration.

In the second phase of photosynthesis, another cyclic metabolic pathway forms carbohydrates (Figure 6.16B). The process is known as the **carbon-fixing reactions** because the carbon in carbon dioxide is trapped. An enzyme bonds carbon dioxide to a 5-carbon organic substance called ribulose 1,5-bisphosphate (RuBP)(**1**). (The enzyme is called ribulose bisphosphate carboxylase.) The resulting 6-carbon molecule then splits to form two molecules of 3-phosphoglycerate (3PG). In the next step, the products of the energy-fixing reactions, ATP and NADPH, drive the conversion of 3PG to glyceraldehyde-3-phosphate (G3P)(**2**). Two molecules of G3P then condense with each other to form a molecule of glucose (**3**). Thus, the overall formula for photosynthesis may be expressed as:

$$6\ CO_2 + 6\ H_2O + ATP \\ \downarrow \\ C_6H_{12}O_6 + 6\ O_2 + ADP + P$$

Notice that this equation is the reverse of the equation for aerobic respiration. The fundamental difference is that aerobic respiration is an energy-yielding process, while photosynthesis is an energy-trapping process.

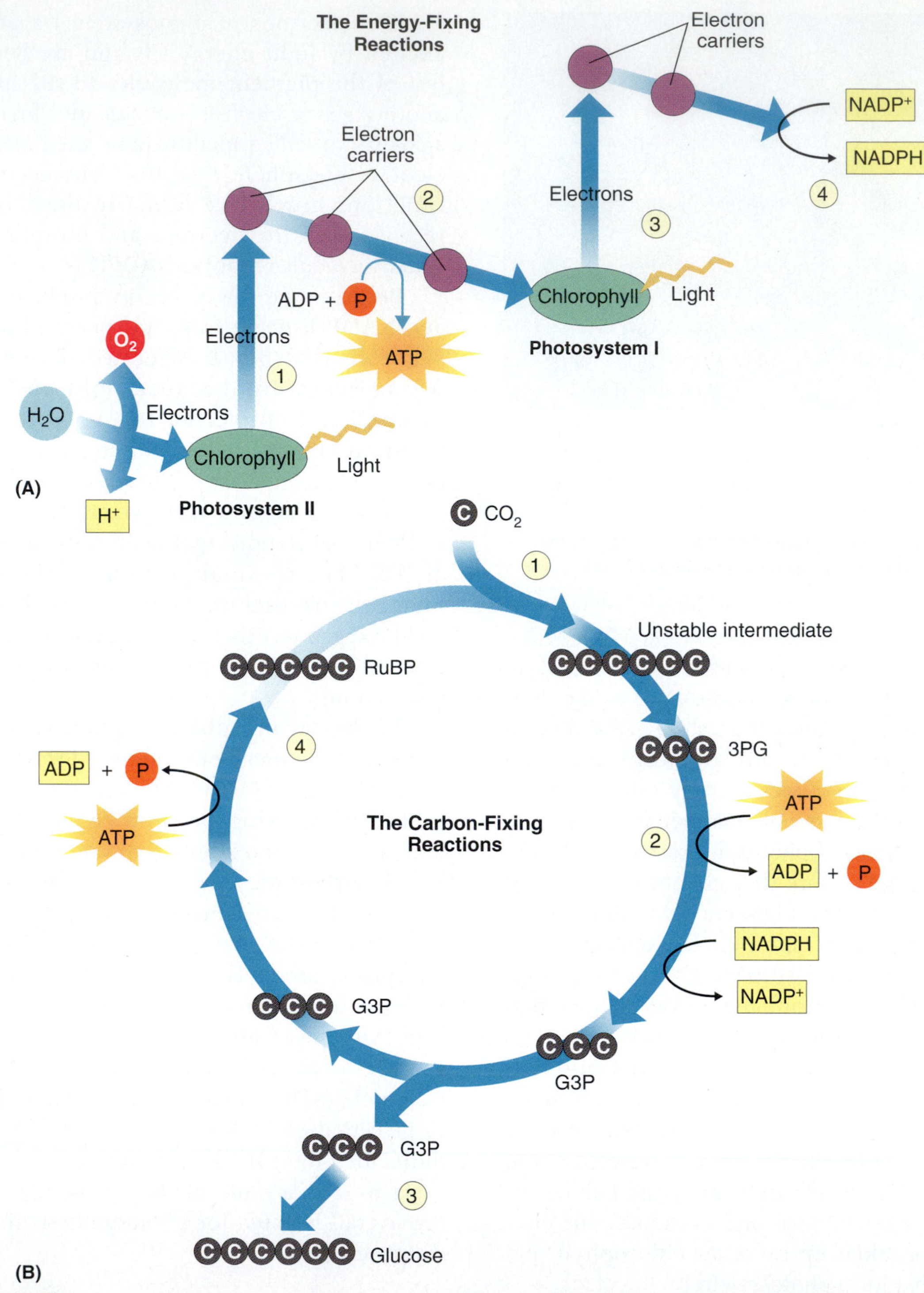

FIGURE 6.16 **Photosynthesis in Microorganisms.** (**A**) The energy-fixing reactions occurring along the cell membrane of a cyanobacterium. (**1**) Electrons in chlorophyll receive a boost in energy from light and (**2**) ATP is synthesized as the electrons pass among electron carriers. (**3**) The electrons receive a second boost, and (**4**) the energy is used to form high-energy NADPH. The ATP and NADPH are used in the carbon-fixing reactions. (**B**) The carbon-fixing reactions. (**1**) Carbon dioxide unites with ribulose bisphosphate (RuBP) to form an unstable 6-carbon molecule. (**2**) The latter splits to form two molecules of 3-phosphoglycerate (3PG) then glyceraldehyde-3-phosphate (G3P). ATP and NADPH from the light-fixing reaction are used in the latter conversion. (**3**) Condensations of two 3-carbon G3P molecules yields glucose, and (**4**) the remainder is used to form RuBP to continue the process. ATP is used in the latter reaction.

Q: How are the carbon-fixing reactions of photosynthesis dependent on the energy-fixing reactions?

To finish off the cycle, most G3P molecules also undergo a complex series of enzyme-catalyzed reactions that require ATP to reform RuBP (4). However, some G3P exits the cycle and combines in pairs to form glucose. The sugar then can be used for cell respiration, stored as glycogen, or used for other cellular purposes.

In addition to the cyanobacteria, several other groups of prokaryotes trap energy by photosynthesis. Two such groups are the green sulfur bacteria and purple sulfur bacteria, so named because of the colors imparted by their pigments. These bacterial organisms have chlorophyll-like pigments known as **bacteriochlorophylls** to distinguish them from other chlorophylls. In the energy-fixing reactions, the organisms do not use water as a source of hydrogen ions. Consequently, no oxygen is liberated. Instead of water, a series of inorganic or organic substances, such as fatty acids, are used as a source of hydrogen ions. Certain species of green sulfur bacteria use hydrogen sulfide (H_2S) as a hydrogen ion source. Thus, the green and purple sulfur bacteria commonly live under anaerobic conditions in environments such as sulfur springs and stagnant ponds.

Another variation of bacterial photosynthesis occurs in some members of the *Archaea*. Instead of the usual chlorophylls, the extreme halophiles of this group contain a pigment called **bacteriorhodopsin** (which is similar to the rhodopsin of the human eye). In the presence of oxygen, the extreme halophiles can synthesize ATP with the aid of this pigment.

CONCEPT AND REASONING CHECKS

6.11 Compare and contrast the processes of glucose catabolism (aerobic respiration) with glucose anabolism (photosynthesis).

6.5 Patterns of Metabolism

Bacteria must meet certain nutritional requirements for growth. Besides water, which is an absolute necessity, prokaryotes need nutrients that can serve as energy sources and raw materials for the synthesis of cell components. These generally include proteins for structural compounds and enzymes, carbohydrates for energy, and a series of vitamins, minerals, and inorganic salts.

Autotrophs and Heterotrophs Get Their Energy and Carbon in Different Ways

KEY CONCEPT

- Autotrophs and heterotrophs vary in their energy and carbon sources.

Two different patterns exist for satisfying an organism's metabolic needs. These patterns are called **autotrophy** and **heterotrophy**. They are primarily based on the source of carbon used for making cell components.

Autotrophs. Organisms that synthesize their own foods from simple carbon sources such as CO_2 are referred to as **autotrophs** (*auto* = "self"; *troph* = "nourish"). Those that use light as the energy source, such as the cyanobacteria, are **photoautotrophs**. Another group of autotrophs do not use light as an energy source. Instead, they use inorganic compounds and are referred to as **chemoautotrophs**. For example, species of *Nitrosomonas* convert ammonium ions (NH_4^+) into nitrite ions (NO_2^-) under aerobic conditions, thereby obtaining ATP. The genus *Nitrobacter* then converts the nitrite ions into nitrate ions (NO_3^-), also as an ATP-generating mechanism. In addition to providing energy to both bacterial species, these reactions have great significance in the environment as a critical part of the nitrogen cycle (Chapter 26). By preserving nitrogen in the soil in the form of nitrate or ammonia, it can be used by green plants to form amino acids.

■ Nitrogen cycle: A biogeochemical cycle that cycles nitrogen gas into nitrogenous compounds and back again.

Heterotrophs. The majority of microorganisms are **heterotrophs** (hetero = "other"). Such heterotrophic organisms obtain their energy and carbon in one of two ways. The **photoheterotrophs** use light as their energy source and preformed organic compounds such as fatty acids and alcohols as sources of carbon. Photoheterotrophs include certain green nonsulfur and purple nonsulfur bacteria.

TABLE 6.2 A Nutritional Classification of Microorganisms

Nutritional Type	Energy Source	Carbon Source	Examples
Autotrophs			
Photoautotroph	Light	Carbon dioxide (CO_2)	Photosynthetic bacteria (green sulfur and purple sulfur bacteria), cyanobacteria, algae
Chemoautotroph	Inorganic compounds	Carbon dioxide (CO_2)	*Nitrosomonas, Nitrobacter*
Heterotrophs			
Photoheterotroph	Light	Organic compounds	Purple nonsulfur and green nonsulfur bacteria
Chemoheterotroph	Organic compounds	Organic compounds	Most prokaryotes; all fungi and protozoa

The **chemoheterotrophs** use preformed organic compounds for both their energy and carbon sources. Glucose would be one example. Those chemoheterotrophic prokaryotes that feed exclusively on dead organic matter are commonly called **saprobes**. In contrast, chemoheterotrophs that feed on living organic matter, such as human tissues, are commonly known as **parasites**. The term *pathogen* is used if the parasite causes disease in its host organism. We will certainly see many examples of this in upcoming chapters.

TABLE 6.2 summarizes this nutritional classification.

CONCEPT AND REASONING CHECKS

6.12 Produce a concept map for the nutritional classification of microorganisms.

SUMMARY OF KEY CONCEPTS

6.1 Enzymes and Energy in Metabolism

- The two major themes of prokaryotic metabolism are catabolism (the breakdown of organic molecules) and anabolism (the synthesis of organic molecules).
- Prokaryotes and all living things use enzymes, protein molecules that speed up a chemical change, to control cellular reactions.
- Enzymes bind to substrates at their active site (enzyme-substrate complex) where functional groups are destabilized, lowering the activation energy.
- Many metabolic processes occur in metabolic pathways, where a sequence of chemical reactions is catalyzed by different enzymes.
- Enzymes are regulated through feedback inhibition. The final end product often inhibits the first enzyme in the pathway.
- Enzymes also can be inhibited by heavy metals and some antibiotics. Physical and chemical agents can denature enzymes and inhibit their action in cells.
- Many metabolic reactions in cells require energy in the form of adenosine triphosphate (ATP). The breaking of the terminal phosphate produces enough energy to supply an endergonic reaction, and often involves the addition of the phosphate to another molecule (phosphorylation).

6.2 The Catabolism of Glucose

- Cellular respiration is a series of metabolic pathways in which chemical energy is converted to cellular energy (ATP). It may require oxygen gas (aerobic respiration) or another inorganic final electron acceptor (anaerobic respiration).
- Glycolysis, the catabolism of glucose to pyruvate, extracts some energy from which two ATP and 2 NADH molecules result.
- The catabolism of pyruvate into carbon dioxide and water in the Krebs cycle extracts more energy as ATP, NADH, and $FADH_2$. Carbon dioxide gas is released.
- The process of oxidative phosphorylation involves the oxidation of NADH and $FADH_2$, the transport of freed electrons along a cytochrome chain, the pumping of protons across the cell membrane, and the synthesis of ATP from a reversed flow of protons. These last two steps are referred to as chemiosmosis.

6.3 Other Aspects of Catabolism

- Other carbohydrates, such as sucrose, lactose, and polysaccharides, represent energy sources that can be metabolized through cellular respiration.
- Besides carbohydrates, proteins and, especially, fats can be metabolized through the cellular respiratory pathways to produce ATP.
- The anaerobic respiration of glucose uses different final electron acceptors in oxidative phosphorylation. Glycolysis and the Krebs cycle still function and ATP synthesis occurs.
- In fermentation, the catabolism of glucose can continue without a functional Krebs cycle or oxidative phosphorylation process. To maintain a steady supply of NAD^+ for glycolysis and ATP synthesis, pyruvate is redirected into other pathways that reoxidize NADH to NAD^+. End products include lactate or ethanol. Only the two ATP molecules of glycolysis are synthesized in fermentation from each molecule of glucose.

6.4 The Anabolism of Carbohydrates

- The anabolism of carbohydrates can occur by photosynthesis, the process whereby light energy is used to synthesize ATP, and the latter is then used to fix atmospheric carbon dioxide into carbohydrate molecules. Other bacterial species carry out photosynthesis using other pigments and without water as a source of hydrogen ions.

6.5 Patterns of Metabolism

- Autotrophs synthesize their own food from carbon dioxide and light energy (photoautotrophs) or carbon dioxide and inorganic compounds (chemoautotrophs). Heterotrophs obtain their carbon from organic compounds and energy from light (photoheterotrophs) or from organic compounds (chemoheterotrophs).

LEARNING OBJECTIVES

After understanding the textbook reading, you should be capable of writing a paragraph that includes the appropriate terms and pertinent information to answer the objective.

1. Contrast anabolism and catabolism as biochemical reactions and as energy processes.
2. Describe the properties of enzymes.
3. State the role of an enzyme-substrate complex to regulating metabolism.
4. Judge the importance of metabolic pathways in prokaryotic cells.
5. Compare the mechanisms of noncompetitive and competitive inhibition.
6. Assess the role of ATP and the ATP/ADP cycle in cell metabolism.
7. Explain the importance of glucose to energy metabolism.
8. Summarize the important steps of glycolysis.
9. Identify the importance of the Krebs cycle to aerobic and anaerobic respiration.
10. Construct an electron transport pathway, indicating the important steps in the synthesis of ATP.
11. Identify where each reactant is used and where each product is produced in the cellular respiration summary equation.
12. Measure the yield of ATP molecules in each phase of aerobic respiration.
13. Identify what other compounds can be used to supply chemical energy for ATP production.
14. Compare and contrast aerobic and anaerobic respiration.
15. Summarize the steps in fermentation and identify the reason why pyruvate is converted into a final end product.
16. Summarize the importance of (a) the energy-fixing reactions and (b) the carbon-fixing reaction of photosynthesis.
17. Distinguish the energy and carbon sources for the four nutritional classes of microorganisms.

SELF-TEST

Answer each of the following questions by selecting the ***one*** answer that best fits the question or statement. Answers to even-numbered questions can be found in **Appendix C**.

1. An exergonic reaction
A. releases energy.
B. is usually a catabolic reaction.
C. is part of metabolism.
D. could be coupled to an anabolic reaction.
E. All the above (**A–D**) are correct.

2. Enzymes are
A. inorganic compounds.
B. destroyed in a reaction.
C. proteins.
D. biological catalysts.
E. Both **C** and **D** are correct.

3. Enzymes combine with a _____ at the _____ site to lower the activation energy.
A. substrate; active
B. product; noncompetitive
C. product; active
D. coenzyme; competitive
E. substrate; coenzyme

4. Which one of the following is *not* a metabolic pathway?
A. Krebs cycle.
B. Fermentation.
C. The carbon-fixing reactions.
D. Glycolysis.
E. Sucrose → glucose + fructose.

5. If an enzyme's active site becomes deformed, _____ inhibition was likely responsible.
A. noncompetitive
B. pathway
C. competitive
D. metabolic
E. cellular

6. Which one of the following is *not* part of an ATP molecule?
A. Phosphate groups
B. Cofactor
C. High-energy bonds
D. Ribose
E. Adenine

7. The addition of a phosphate group to a molecule is called
A. feedback.
B. reduction.
C. phosphorylation.
D. oxidation.
E. Both **B** and **D** are true.

8. All of the following are examples of part of the cellular respiration process *except:*
A. fermentation.
B. glycolysis.
C. oxidative phosphorylation.
D. Krebs cycle.
E. electron transport.

9. Which one of the following (**A–D**) is *not* produced during glycolysis?
A. ATP
B. NADH
C. Pyruvate
D. Glucose
E. All the above (**A–D**) are produced.

10. All the following are produced during the Krebs cycle *except:*
A. CO_2.
B. O_2.
C. ATP.
D. $FADH_2$.
E. NADH.

11. The electron transport chain is directly involved with
A. H^+ pumping.
B. CO_2 production.
C. ATP synthesis.
D. reducing NAD^+.
E. generating oxygen gas.

12. For the methanogens, anaerobic respiration uses _____ as a final electron acceptor.
A. O_2
B. alcohol
C. $CO_3^=$
D. lactate
E. pyruvate

13. In fermentation, the conversion of pyruvate into a final end product is critical for the production of
A. CO_2.
B. glucose.
C. enzymes.
D. O_2.
E. NAD^+.

14. Which one of the following is the correct sequence for the flow of electrons in the energy-fixing reactions of photosynthesis?
A. Water—photosystem I—photosystem II—NADPH
B. Photosystem I—NADPH—water—photosystem II
C. Water—photosystem II—photosystem I—NADPH
D. NADPH—photosystem II—photosystem I—water
E. Photosystem II—water—NADPH—photosystem I

15. Microorganisms that use organic compounds as energy and carbon sources are
A. photoautotrophs.
B. chemoautotrophs.
C. chemoheterotrophs.
D. autotrophs.
E. photoheterotrophs.

QUESTIONS FOR THOUGHT AND DISCUSSION

Answers to even-numbered questions can be found in **Appendix C**.

1. A student goes on a college field trip and misses the microbiology exam covering prokaryotic metabolism. Having made prior arrangements with the instructor for a make-up exam, he finds one question on the exam: "Discuss the interrelationships between anabolism and catabolism." How might you have answered this question?
2. If ATP is such an important energy source for prokaryotes, why do you think it is not added routinely to the growth medium for these organisms?
3. One of the most important steps in the evolution of life on Earth was the appearance of certain organisms in which photosynthesis takes place. Why was this critical?
4. A population of a *Bacillus* species is growing in a soil sample. Suppose glycolysis came to a halt in these bacterial cells. Would this mean that the Krebs cycle would also stop? Why?

APPLICATIONS

Answers to even-numbered questions can be found in **Appendix C**.

1. You have two flasks with broth media. One contains a species of cyanobacteria. The other flask contains *E. coli*. Both flasks are sealed and incubated under optimal growth conditions for two days. Assuming the cell volume and metabolic rate of the bacterial cells is identical in each flask, why would the carbon dioxide concentration be higher in the *E. coli* flask than in the cyanobacteria flask after the two-day incubation?
2. A stagnant pond usually has a putrid odor because hydrogen sulfide has accumulated in the water. A microbiologist recommends that tons of green sulfur bacteria be added to remove the smell. What chemical process does the microbiologist have in mind? Do you think it will work?
3. Citrase is the enzyme that converts citrate to α-ketoglutarate in the Krebs cycle. A chemical company has located a mutant microorganism that cannot produce this enzyme and proposes to use the microorganism to manufacture a particular product. What do you suppose the product is? How might this product be useful?

REVIEW

For each choice, circle the word or term that best completes each of the following statements. The answers to even-numbered statements are listed in **Appendix C**.

1. The sum total of all an organism's biochemical reactions is known as (catabolism, metabolism); it includes all the (synthesis, digestion) reactions called anabolism and all the breakdown reactions known as (inactivation, catabolism).
2. Enzymes are a group of (carbohydrate, protein) molecules that generally (slow down, speed up) a chemical reaction by converting the (substrate, active site) to end products.
3. The aerobic respiration of glucose begins with the process of (oxidative phosphorylation, glycolysis) and requires that (amino acids, energy) be supplied by (ATP, NADH) molecules.
4. The process of (fermentation, the Krebs cycle) takes place in the absence of (oxygen, carbon dioxide) and begins with a molecule of (glucose, protein) and ends with molecules of (amino acid, an organic end product).
5. In oxidative phosphorylation, pairs of (protons, electrons) are passed among a series of (chromosomes, cytochromes) with the result that (oxygen, energy) is released for (NAD^+, ATP) synthesis.
6. In the Krebs cycle, (glucose, pyruvate) undergoes a series of changes and releases its (carbon, nitrogen) as (carbon dioxide, nitrous oxide) and its electrons to (NAD^+, ATP).
7. For use as energy compounds, proteins are first digested to (uric, amino) acids, which then lose their (carboxyl, amino) groups in the process of (fermentation, deamination) and become intermediates of cellular respiration.
8. Ribulose 1,5-bisphosphate bonds with (carbon monoxide, carbon dioxide) molecules during (fermentation, photosynthesis), a process that ultimately results in molecules of (pyruvate, glucose).
9. Chemoautotrophs use energy from (light, inorganic compounds) to synthesize (carbohydrates, oxygen gas) and are typified by species of *(Staphylococcus, Nitrosomonas)*.
10. Fats are broken down to (fatty acids, coenzymes), which are converted through (beta oxidation, deamination) reactions to (glucose, two-carbon units) and eventually enter (cellular respiration, photosynthesis).

HTTP://MICROBIOLOGY.JBPUB.COM/

The site features learning, an on-line review area that provides quizzes and other tools to help you study for your class. You can also follow useful links for in-depth information, read more MicroFocus stories, or just find out the latest microbiology news.

7

Chapter Preview and Key Concepts

Prokaryotic Genetics

We wish to suggest a structure for the salt of deoxyribose nucleic acid (DNA). This structure has novel features which are of considerable biological interest.
—In the first 1953 paper by Watson and Crick describing the structure of DNA

In our fast-paced world, we often measure time in minutes and seconds, so our minds find it difficult to imagine the colossal 4.5 billion years that the Earth has been in existence. It may help, however, to think of Earth's history as a single year.

In the months of January and February, Earth was a hot, volcanic, lifeless ball of rock bombarded by material left over from the formation of the solar system. As the earth cooled during March, water vapor condensed into oceans and seas, providing conditions more amenable for the origin of life. Around April, prokaryotes, or something akin to prokaryotes, first appeared. As they evolved, they thrived and diversified in both environments without oxygen gas and, by mid-June, environments with oxygen gas (see MicroFocus 2.1 and 5.3). Prokaryotes were the only organisms on Earth until early August, when single-celled eukaryotes, such as the algae, emerged. These organisms flourished and represent ancestors of present-day species. About mid-September, multicellular eukaryotes arose, whose descendants would evolve into diverse plants, fungi, and animals. Not until mid-November did the first of the plants, fungi, and animals move out of the sea onto the land. The dinosaurs were in existence from December 19 to December 25, and by December 27, the Earth bore a resemblance to modern Earth. Finally, on December 31, close to midnight, humans appeared.

We take this trek through geologic time to help us appreciate why prokaryotes have prospered genetically and in evolutionary terms. They have been successful primarily because they have been around the longest and have adapted well (FIGURE 7.1). In fact, *Bacteria* and

Archaea have been on Earth about 3.7 billion years (versus about 200,000 years for humans), as FIGURE 7.2 shows. During this time, gene changes have been occurring regularly and nature has used the prokaryotes to test its newest genetic traits. The detrimental traits have been eliminated (together with the organisms unlucky enough to have them), while the beneficial traits have thrived and have been passed on to the next generation—and onto the present day.

FIGURE 7.1 **Fossil Microbes.** This photograph is looking down on an ancient sea floor in Western Australia's Pilbara region. The wavy markings and cone-shaped formations may be evidence for a microbial reef made of cyanobacteria that existed 3.4 billion years ago.

Q: What would it mean for the environment once cyanobacteria predominated?

Modern prokaryotes, therefore, enjoy the fruits of genetic change. Because of their diverse genes, they can thrive in the varied environments on Earth, whether it is the snows of the Arctic or the boiling hot volcanic vents of the ocean depths. No other organisms can compare to prokaryotes in sheer numbers.

Finally, consider a prokaryote's multiplication rate—a new generation every half hour—and it is easy to see how a useful genetic change (such as drug resistance) can be propagated quickly in a stressful environment (one containing a drug).

Any one of these factors—time on Earth, sheer numbers, multiplication rate—would be sufficient to explain how prokaryotes have evolved to their current form. However, when taken together, the factors help us appreciate why they have done very well in the evolutionary lottery—very well, indeed.

In this chapter, we examine mutation, one of the two processes that have brought ancient prokaryotes to the myriad forms we observe on Earth today. However, to understand the material in these topics, we first must look at DNA replication and how the information in DNA is processed and regulated in making proteins—something alluded to by the remarkable discovery made in 1953 by James Watson and Francis Crick (see chapter opening quotation).

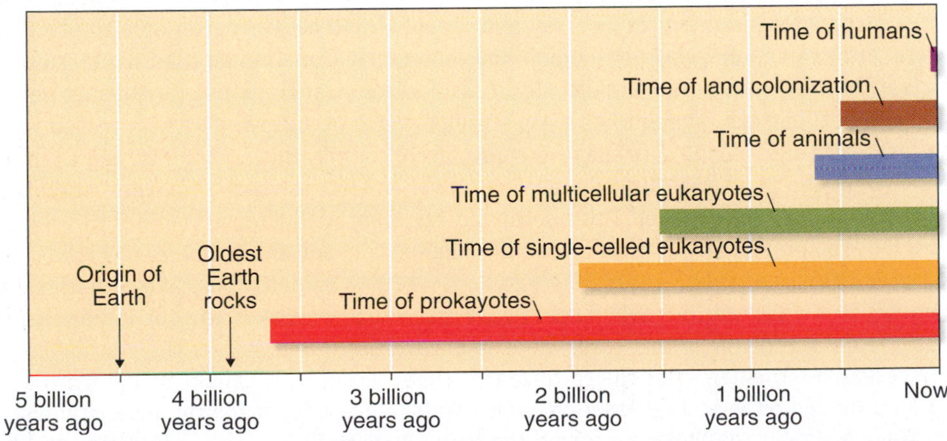

FIGURE 7.2 **The Appearance of Life on Earth.** This time line shows the relative amounts of time that various groups of organisms have existed on Earth. The prokaryotes have been in existence for a notably longer period than any other group, particularly humans. They have adapted well to Earth simply because they have had the longest opportunity to evolve through adaptation.

Q: Why did it take so long for eukaryotes to appear on Earth?

7.1 Prokaryotic DNA

In 1953, James Watson and Francis Crick worked out DNA's double helix structure based, in part, on the X-ray studies of Rosalind Franklin (MicroFocus 7.1). This discovery would set the stage for all we know today about DNA replication, protein synthesis, and gene control. In Chapter 2, we described the structure of the DNA molecule, so it might be helpful to review that material before proceeding too far in this chapter.

MICROFOCUS 7.1: History

The Tortoise and the Hare

We all remember the children's story of the tortoise and the hare. The moral of the story was those who plod along slowly and methodically (the tortoise) will win the race over those who are speedy and impetuous (the hare). The race to discover the structure of DNA is a story of collaboration and competition—a science tortoise and the hare.

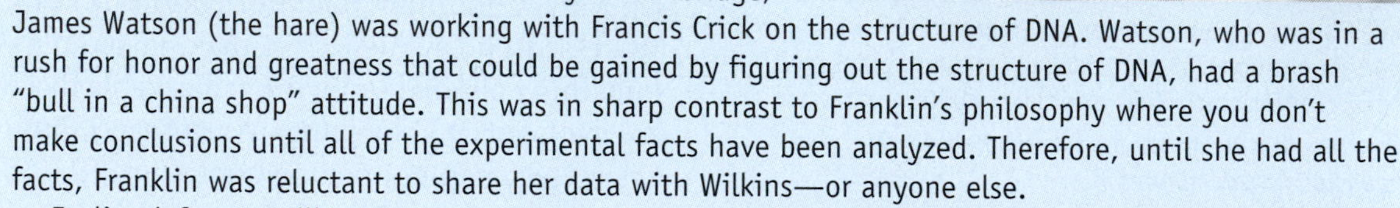

Rosalind Franklin (the tortoise) was 31 when she arrived at King's College in London in 1951 to work in J. T. Randell's lab. Having received a Ph.D. in physical chemistry from Cambridge University, she moved to Paris where she learned the art of X-ray crystallography. At King's College, Franklin was part of Maurice Wilkins's group and she was assigned the job of using X-ray crystallography to work out the structure of DNA fibers. Her training and constant pursuit of excellence allowed her to produce excellent, high-resolution X-ray photographs of DNA.

Meanwhile, at the Cavendish Laboratory in Cambridge, James Watson (the hare) was working with Francis Crick on the structure of DNA. Watson, who was in a rush for honor and greatness that could be gained by figuring out the structure of DNA, had a brash "bull in a china shop" attitude. This was in sharp contrast to Franklin's philosophy where you don't make conclusions until all of the experimental facts have been analyzed. Therefore, until she had all the facts, Franklin was reluctant to share her data with Wilkins—or anyone else.

Feeling left out, Wilkins was more than willing to help Watson and Crick. Because Watson thought Franklin was "incompetent in interpreting X-ray photographs" and he was better able to use the data, Wilkins shared with Watson an X-ray photograph and report that Franklin had filed. From these materials, it was clear that DNA was a helical molecule. It also seems clear Franklin knew this as well but, perhaps being a physical chemist, she did not grasp its importance because she was concerned with getting all the facts first and making sure they were absolutely correct. But, looking through the report that Wilkins shared, the proverbial "light bulb" went on when Crick saw what Franklin had missed; that the two DNA strands were antiparallel. This knowledge, together with Watson's ability to work out the base pairing, led Watson and Crick to their "leap of imagination" and the structure of DNA.

In her book entitled, *Rosalind Franklin: The Dark Lady of DNA* (HarperCollins, 2002), author Brenda Maddox suggests it is uncertain if Franklin could have made that leap as it was not in her character to jump beyond the data in hand. In this case, the leap of intuition won out over the methodical, data collecting in research—the hare beat the tortoise this time. However, it cannot be denied that Franklin's data provided an important key from which Watson and Crick made the historical discovery.

In 1962, Watson, Crick, and Wilkins received the Nobel Prize in Physiology or Medicine for their work on the structure of DNA. Should Franklin have been included? The Nobel Prize committee does not make awards posthumously and Franklin had died four years earlier from ovarian cancer. So, if she had lived, did Rosalind Franklin deserve to be included in the award?

Prokaryotic DNA Is Organized within the Nucleoid

KEY CONCEPT

- The DNA in most prokaryotic cells exists as a single, circular chromosome.

Most of the genetic information in prokaryotic cells is contained within the **chromosome**, the cell's intracellular source of genetic information. Usually, this is a single, circular molecule of DNA that is haploid, although a few species may have multiple chromosomes.

The chromosome exists as thread-like fibers associated with some protein and is localized in the cytosol within a space called the **nucleoid** (see Chapter 4). In prokaryotic cells, remember that one of the unique features of the nucleoid area is the absence of a surrounding membrane envelope typical of the cell nucleus in eukaryotic cells.

The circular chromosome of *Escherichia coli* probably has been studied more thoroughly than that of any other prokaryote. The genome of *E. coli* has about 4,300 genes coding for growth and metabolic activities. Some viruses, by contrast, have as few as seven genes, while the human genome has some 35,000 genes. Characteristics of prokaryotic and eukaryotic chromosomes are compared in TABLE 7.1.

Genome studies of prokaryotes have shown that many genes involved with the same process are clustered. For example, in *E. coli*, many genes involved with biosynthetic or degradative processes are grouped together. Such clustering makes genetic control of these processes more efficient since their activity can be coordinated within one cluster.

TABLE 7.1 Characteristics of Prokaryotic and Eukaryotic Chromosomes

Prokaryotic Chromosome	Eukaryotic Chromosome
Organized in the nucleoid	Organized in the nucleus with a membrane envelope
Chromosome usually circular	Chromosomes are linear
About 1.5 mm in length	Tens or hundreds of millimeters in length
Single molecule of DNA per genetic trait (haploid)	Two molecules of DNA per genetic trait (diploid); some organisms haploid
4,300 genes in *E. coli* genome	35,000 genes in human genome
Little protein present	Histone protein present
No dominance or recessiveness in genes	Genes may be dominant or recessive, or have other inheritance patterns
Introns very rare	Introns present
Mutations occur in DNA	Mutations occur in DNA
Replicates just prior to binary fission	Replicates just prior to mitosis
Single or multiple replication origins	Multiple replication origins

DNA within a Chromosome Is Highly Compacted

KEY CONCEPT

- Prokaryotic DNA must be tightly packed to fit in the cell.

In all prokaryotic cells, including *E. coli*, the DNA occupies about one third of the total volume of the cell, and when extended its full length, it is about 1.5 millimeters (mm) long. This is approximately 500 times the length of the bacterial cell. So, how can a 1.5-mm-long circular chromosome fit into a 1.0- to 2.0-μm *E. coli* cell?

The answer is **supercoiling**, a twisting and tight packing caused by a number of abundant nucleoid-associated proteins. Thus, the DNA double helix twists on itself like a wound-up rubber band. The coils are folded further into loops of 10,000 bases, each forming a **supercoiled domain** (FIGURE 7.3A) and there are about 400 such domains in an *E. coli* chromosome, giving the molecule an overall "flower" structure called the **looped domain structure**. The high level of compaction is evident when the cell envelope is broken, releasing the DNA in a looped form (FIGURE 7.3B). How the loops are anchored in the nucleoid is not understood.

■ Haploid: Having a single set of genetic information.

CONCEPT AND REASONING CHECKS

7.1 Justify the necessity for DNA supercoiling and looped domains.

Many Prokaryotic Cells also Contain Plasmids

KEY CONCEPT

- Plasmids carry nonessential, but often useful information.

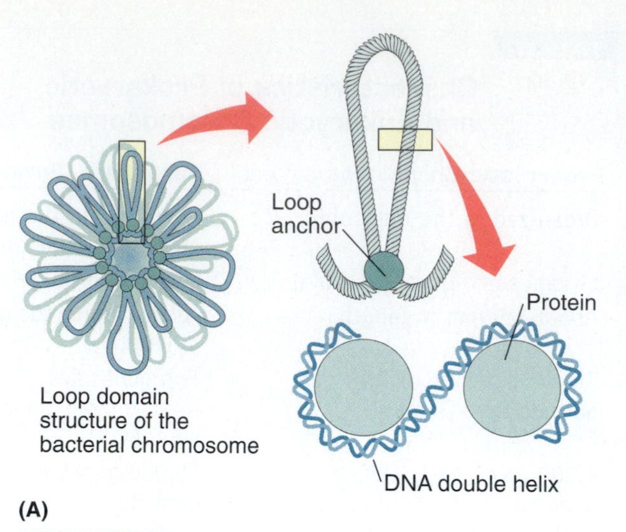

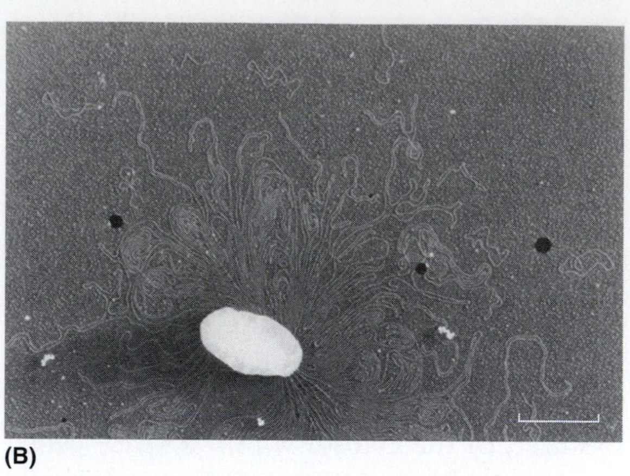

FIGURE 7.3 **Prokaryotic DNA Packing.** (**A**) The loop domain structure of the chromosome, as seen head-on. The loops in DNA help account for the compacting of a large amount of DNA in a relatively small cell. (**B**) An electron micrograph of an *E. coli* cell immediately after cell lysis. The uncoiled DNA fiber exists in loops attached to the disrupted cell envelope. A plasmid also is seen at the center, far right. (Bar = 1 μm.)

Q: How does plasmid structure compare to that of the prokaryotic chromosome?

Many prokaryotes contain **plasmids**, which are stable extrachromosomal DNA elements that do not carry genetic information essential for normal structure, growth, or metabolism. This means a plasmid could be removed from a cell without affecting its viability, assuming the cell is in a nutrient-rich environment free of toxic materials.

Most plasmids are circular, one or more may be found in most prokaryotic cells, and they are easily transferred between cells (Chapter 8). Plasmids exist and replicate as independent genetic elements in the cytosol (Figure 7.3b) where they typically contain about 2 percent of the total genetic information of the cell. Exceptions include some plasmids that can be quite large because they can integrate into a chromosome and excise from it many additional chromosomal genes, some of which may be essential for cell growth.

In most cases though, plasmids are not essential to the normal survival of the prokaryotic cell but they can confer selective advantages and provide genetic flexibility for those organisms possessing them. For example, some plasmids, called **F plasmids**, allow for the transfer of genetic material from donor to recipient through a recombination process (Chapter 8). Other plasmids may play protective roles. **R plasmids** ("resistance" factors), for example, carrying genes for antibiotic resistance. Others contain genes for resistance to potentially toxic heavy metals (e.g., silver, mercury).

Some plasmids provide offensive abilities. Some species of *Streptomyces* carry plasmids for the production of antibiotics while plasmids in other prokaryotes contain genes for the production of **bacteriocins**, a group of proteins that inhibit or kill other bacterial species.

Finally, there are plasmids containing genes coding for toxins affecting human cells and disease processes. The genes encoding the toxins responsible for anthrax are carried on a plasmid. We shall have much to say about these extrachromosomal units when we discuss recombination and genetic engineering in the next chapter.

CONCEPT AND REASONING CHECKS

7.2 What does it mean to say plasmids carry nonessential genetic information?

7.2 DNA Replication

Watson and Crick's 1953 paper on the structure of DNA provided a glimpse of how DNA might be copied. They concluded, "*It has not escaped our notice that the specific pairing we have postulated immediately suggests a possible copying mechanism for the genetic material.*" In fact, the copying of the genetic material, called **DNA replication**, occurs with such precision that the two daughter cells from binary fission are genetically identical to the parent cell.

DNA Replication Is a Highly Regulated Process

KEY CONCEPT

- DNA replication is a three-phase event requiring an array of proteins working in sequence.

Prokaryotic chromosome replication requires the products of more than 20 genes and, although it occurs in a smooth process, we can separate it into three stages (FIGURE 7.4): **initiation** when the DNA unwinds and the strands separate; **elongation** when enzymes synthesize a new polynucleotide strand of DNA for each of the two old (parental) strands; and **termination**, when each of the two DNA helices separate from one another. This combination of a new and old strand was first observed in *E. coli* in 1958 by Matthew J. Meselson and Franklin W. Stahl. It is called **semiconservative replication** because each old strand of the replicated DNA is conserved in the new DNA molecule and one strand is newly synthesized. Let's look at each stage in more detail using Figure 7.4 to guide us.

Initiation. DNA replication starts at a fixed point in the *Bacteria* and most *Archaea* or at multiple points (in some *Archaea*). This point is called the **origin of replication** (*ori*C), which is a sequence of about 250 base pairs. A group of initiator proteins binds at the origin and, using the energy from ATP, starts to unwind and separate the two polynucleotide strands. Since replication is bidirectional, two V-shaped **replication forks** form and move in opposite directions away from the origin.

Elongation. Synthesis of DNA then occurs on each old strand, which represents a template for the synthesis of a new complementary strand. Many proteins are involved in DNA synthesis. Some stabilize the single-stranded templates while **DNA polymerase III** moves along each strand, catalyzing the insertion of new complementary nucleotides to each template strand.

DNA polymerase can "read the DNA" only in the 3′ to 5′ direction (see inset in Figure 7.4). Therefore, at each replication fork, one strand has new nucleotides added in a continuous fashion, beginning at the origin. This is called the **leading strand**. However, since DNA polymerase reads in the 3′ to 5′ direction, the other template strand in each fork will produce discontinuous strands. Here, the new polynucleotide strand is synthesized in a series of segments called the **lagging strand**, which later are joined into a single strand with the help of an enzyme called **DNA ligase**. These segments, which are about 1000 nucleotides long, are called **Okazaki fragments**, after Reiji Okazaki, who discovered them in 1968.

In *E. coli*, DNA synthesis takes about 40 minutes, which means at each replication fork DNA polymerase III is adding new complementary bases at the rate of about 1,000 per second! At this pace, errors occur where an incorrect base is added. Such mutations could be potentially lethal, so there must be a mechanism to correct any errors. DNA polymerases III and I detect any mismatched nucleotides, remove the incorrect nucleotide in the pair, and add the correct nucleotide. Such proofreading reduces replication errors to about 1 in every 10 billion bases added. We will have more to say about mutations later in this chapter.

Termination. In about 40 minutes, the two replication forks meet 180° from *ori*C. At the termination point (*ter*C), there are additional terminator proteins that block the replication forks from advancing any further. Then, the two intertwined DNA molecules (chromosomes) are separated by other enzymes, guaranteeing that each daughter cell will inherit one complete chromosome.

■ Mutations: Changes in the base sequence in the DNA.

■ Base pairs: The complementary pairing of A—T and G—C on the two opposite polynucleotide strands.

CONCEPT AND REASONING CHECKS

7.3 Describe the roles for DNA proteins in the replication process.

7.3 Protein Synthesis

The discovery of the structure of DNA also provided a glimpse into understanding how a cell makes proteins. **Protein synthesis** is a process in which amino acids are precisely bound together in a three-dimensional structure determined by the hereditary information, the

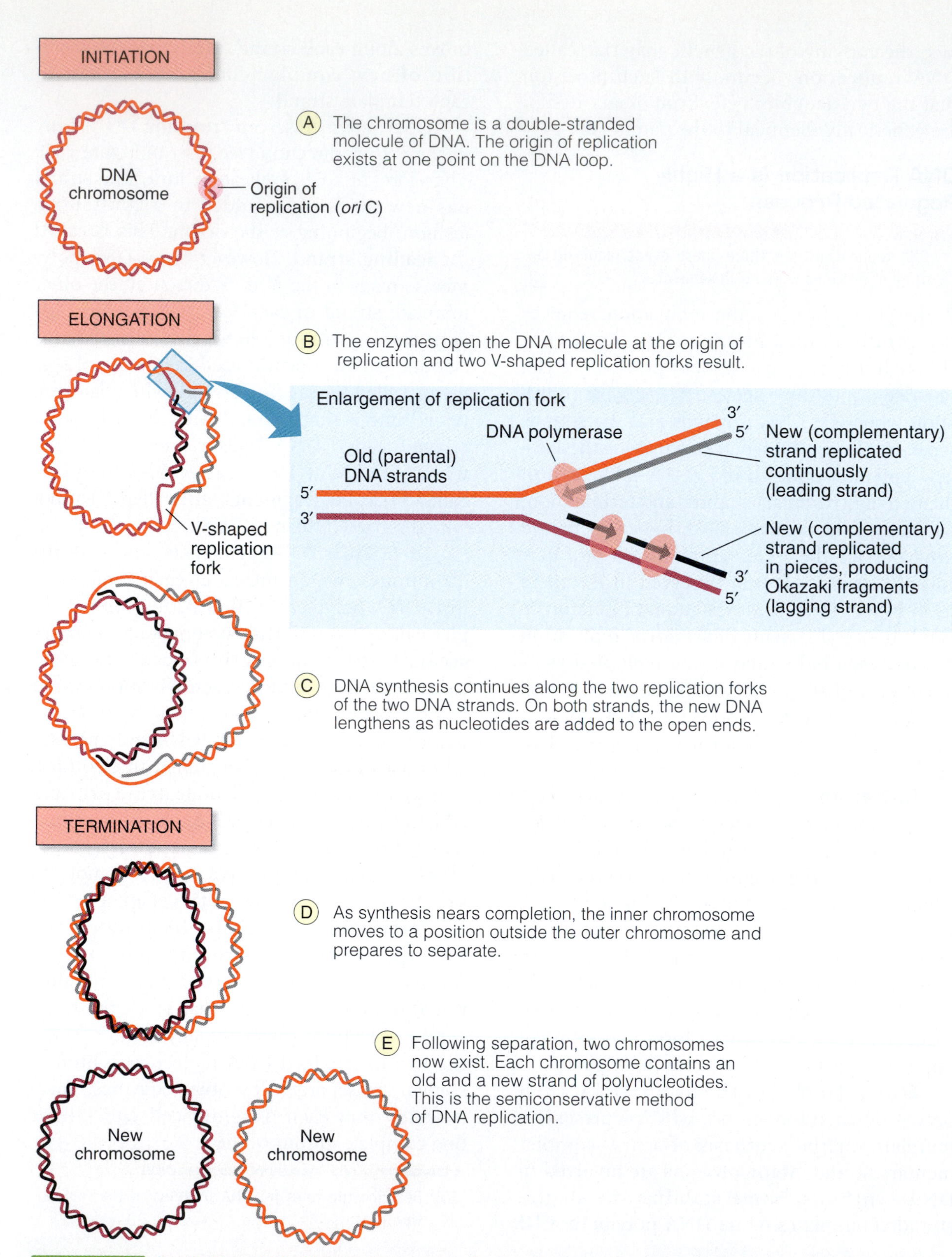

FIGURE 7.4 Replication of the Circular Chromosome in *E. coli*. DNA replication occurs in three continuous stages. Initiation occurs at the origin of replication and generates two replication forks moving clockwise and counterclockwise around the circular chromosome. Elongation involves the addition of complementary nucleotides by DNA polymerase III to the template strands. Based on 5′ to 3′ reading, leading and lagging stands are formed at each replication fork. When replication forks reach the opposite side of the DNA, DNA synthesis stops and the two DNA molecules (chromosomes) separate.

Q: Distinguish between leading and lagging strands during DNA synthesis.

TABLE 7.2 A Comparison of DNA and RNA

DNA (Deoxyribonucleic Acid)	RNA (Ribonucleic Acid)
In prokaryotes, found in the nucleoid and plasmids; in eukaryotes, found in the nucleus, mitochondria, and chloroplasts	In prokaryotes and eukaryotes, found in the cytosol and in ribosomes; in eukaryotes, found in the nucleolus
Always associated with chromosome (genes); each chromosome has a fixed amount of DNA	Found mainly in combinations with proteins in ribosomes (ribosomal RNA) in the cytosol, as messenger RNA, and as transfer RNA
Contains a 5-carbon sugar called deoxyribose	Contains a 5-carbon sugar called ribose
Contains bases adenine, guanine, cytosine, thymine	Contains bases adenine, guanine, cytosine, uracil
Contains phosphorus (in phosphate groups) that connects deoxyribose sugars with one another	Contains phosphorus (in phosphate groups) that connects ribose sugars with one another
Functions as the molecule of inheritance	Functions in protein synthesis and gene regulation
Double stranded	Usually single stranded
Larger size	Smaller size

genes, in the cell. The process requires not only DNA, but also ribonucleic acid (RNA). Both DNA and RNA were described in Chapter 2. A review of their structure is recommended so the discussion to follow can be fully comprehended (TABLE 7.2).

The Central Dogma Identifies the Flow of Genetic Information

KEY CONCEPT

- The hereditary information in DNA is the template to make or control the synthesis of proteins.

One of the central truths in biology states that the genetic information in DNA first is expressed as RNA by a process called **transcription**. One type of RNA then functions as a messenger by carrying the genetic code to areas of the cytosol where the ribosomes are located. There, amino acids are fitted together in a precise sequence to form the protein. This sequencing process, called **translation**, reflects the genetic information in the DNA. This **central "dogma"** (*dogma* = "opinion") of biology can be summarized as follows:

$$\text{DNA} \xrightarrow{\text{transcription}} \text{RNA} \xrightarrow{\text{translation}} \text{Protein}$$

As we will see in Chapter 13, a few viruses modify this rule.

Transcription Copies Genetic Information into RNA

KEY CONCEPT

- Different DNA segments are transcribed into one of three types of RNA.

In transcription, various types of RNA are produced according to the code of nitrogenous bases in the DNA molecule. The DNA thus serves as a template for new RNA molecules. The process initiates with the DNA double helix of a gene unwinding, followed by an uncoupling of the two DNA strands as the hydrogen bonds between opposing bases break (FIGURE 7.5).

RNA polymerase is the enzyme that synthesizes an RNA polynucleotide from a DNA template. In the *Bacteria* and *Archaea*, there is but one RNA polymerase with the archaeal enzyme being more similar to the eukaryal polymerases.

Only one of the two DNA strands within a gene is transcribed. The RNA polymerase recognizes this DNA template strand by a sequence of bases called the **promoter** located on one of the two strands. The polymerase binds to the promoter, unwinds the helix, and separates the two strands within the gene. As the enzyme moves along the DNA template strand, complementary pairing brings RNA

■ Genetic code: The sequence of bases in the DNA or codons in the RNA that specify a specific polypeptide.

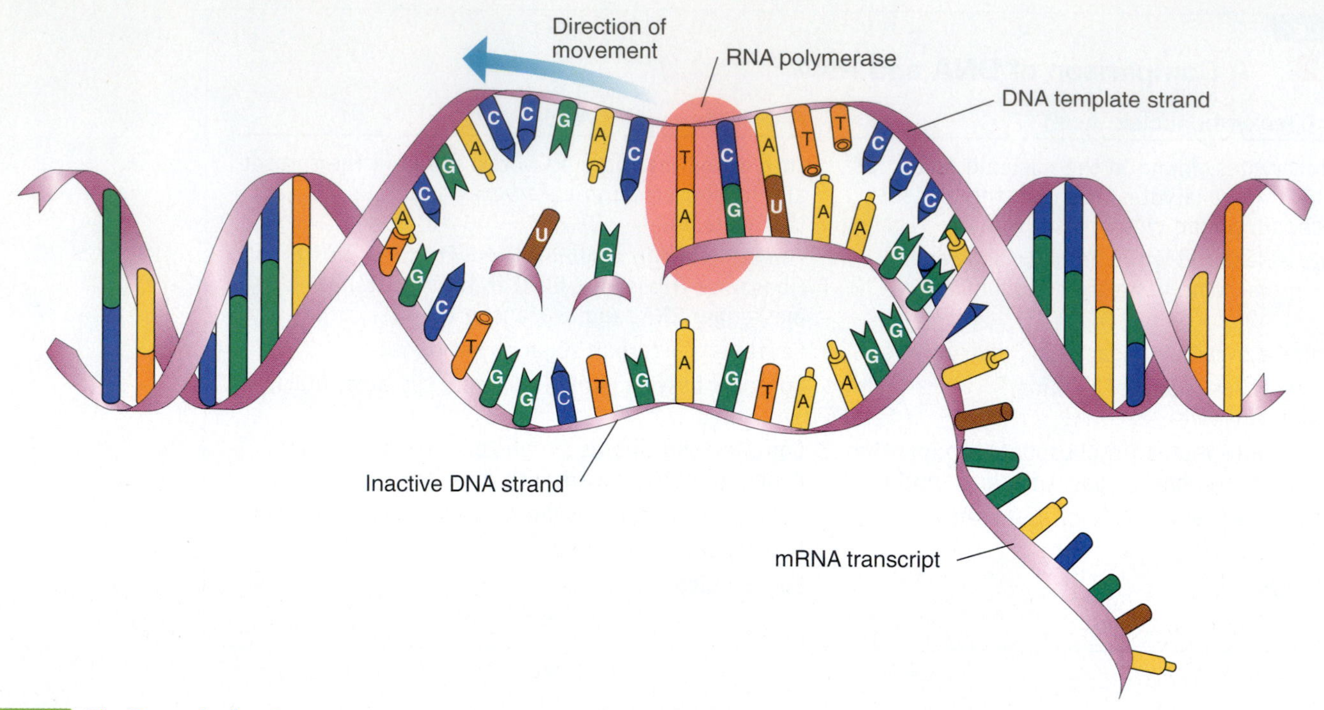

FIGURE 7.5 **The Transcription Process.** For transcription to occur, a gene must unwind and the base pairs separate. The enzyme RNA polymerase does this as it moves along the template strand of the DNA and adds complementary RNA nucleotides. Note that the other DNA strand of the gene is not transcribed.

Q: Justify the need for a promoter sequence in a gene.

nucleotides to the template strand—guanine (G) and cytosine (C) pair with one another and thymine (T) in the DNA template pairs with adenine (A) in the RNA. However, an adenine base on the DNA template pairs with a uracil (U) base in the RNA because RNA nucleotides contain no thymine bases (see Chapter 2).

Like initiation, termination of transcription occurs at specific base sequences on the DNA template strand. The RNA transcription product released represents a complementary image to the sequence of bases in the DNA template strand.

Transcription produces three types of RNA, all of which are needed for translation.

Messenger RNA (mRNA). This RNA carries the genetic information as to what protein or polypeptide will be synthesized. Each mRNA transcribed from a different gene carries a different message; that is, different information as to what protein will be synthesized. The message is encoded in a series of three-base codes or **codons** found along the length of the mRNA. Each codon specifies an individual amino acid to be slotted into position during translation.

Ribosomal RNA (rRNA). Three rRNAs are transcribed from specific regions of the DNA. Together with protein, these RNAs serve a structural role as the framework of the ribosomes, which are the sites at which amino acids assemble into proteins. They also serve a functional role in the translation process.

Transfer RNA (tRNA). The conventional drawing for a tRNA is in a shape roughly like a cloverleaf (FIGURE 7.6). One point presents a sequence of three nitrogenous bases, which functions as an **anticodon**; that is, a sequence that complementary binds to an mRNA codon. The tRNAs have a structural role in delivering amino acids to the ribosome for assembly into proteins. Each tRNA has a specific amino acid attached through an enzymatic reaction involving ATP. For example, the amino acid alanine binds only to the tRNA specialized to transport alanine; glycine is transported by a different tRNA.

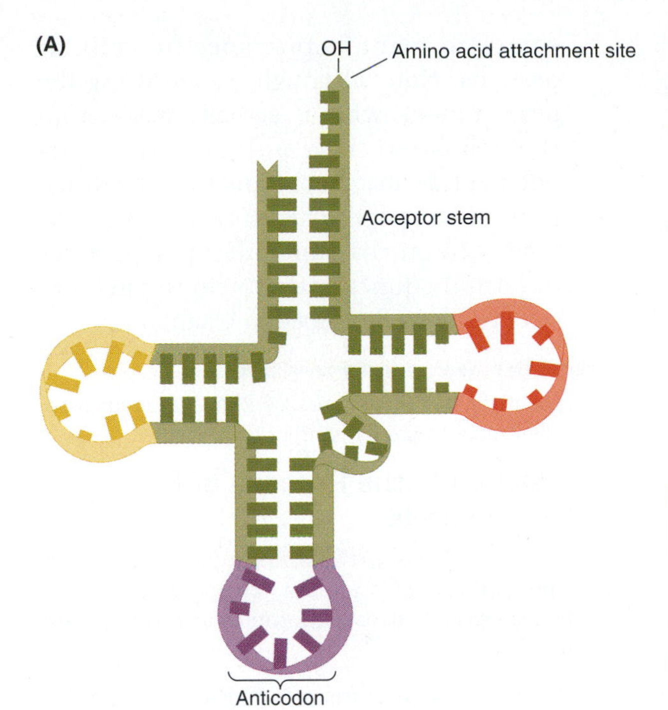

FIGURE 7.6 **Structure of a Transfer RNA (tRNA).** (**A**) The traditional "cloverleaf" configuration for tRNA. The anticodon will pair up with a complementary codon in the mRNA. The appropriate amino acid attaches to the end of the acceptor arm. (**B**) A schematic diagram of the more correct three-dimensional structure of a tRNA.

Q: What part of the tRNA is critical for the complementary binding to a codon in a mRNA?

There is one important difference between prokaryotic and eukaryotic RNAs. In most prokaryotes, all of the bases in a gene are transcribed and used to specify a particular protein. In 1977, Philip Sharp and his associates at the Massachusetts Institute of Technology found that certain portions of eukaryotic DNA are not part of the final RNA and are removed from the RNA before the molecule can function. These intervening DNA segments removed after transcription are called **introns**, while the functioning, expressed segments are called **exons**.

CONCEPT AND REASONING CHECKS

7.4 Identify the three types of RNA and summarize their roles in protein synthesis.

The Genetic Code Is Degenerate

KEY CONCEPT

- More than one codon often specifies a specific amino acid.

One of the startling discoveries of biochemistry is that the **genetic code**, the sequence of bases in the DNA or codons in the RNA to specify a specific polypeptide, is redundant. In fact, in most cases there is more than one codon for each amino acid. Because there are four nitrogenous bases, mathematics tells us 64 possible combinations can be made of the four bases, using three at a time. But there are only 20 amino acids for which a code must be supplied. How do scientists account for the remaining 44 codes?

It is now known that 61 of the 64 codons specify an amino acid and most of those amino acids have multiple codons (as shown in TABLE 7.3). For example, GCU, GCC, GCA, and GCG all code for alanine (ala). This lack of a one-to-one relationship between codon and amino acid is said to be degenerate. In Table 7.3, notice one of the 64 codons, AUG, represents the **start codon** for making a protein. This codon specifies the amino acid methionine (met). Three additional codons, which do not code for an amino acid (UGA, UAG, UAA), are called **stop codons** because they terminate the addition of amino acids to a growing polypeptide chain. The genetic code of bases is nearly universal for all species, be they prokaryotic or eukaryotic. Thus, the GCU codon for alanine in prokaryotes is also the codon for alanine in human cells.

CONCEPT AND REASONING CHECKS

7.5 What is meant by the genetic code being degenerate? Give two examples.

Before we proceed to the last stage, translation, let's summarize the protein synthesis process to this point (FIGURE 7.7).

1. Each gene of the DNA contains information to manufacture a specific form of RNA.
2. The information can be transcribed into:
   - mRNAs, which are produced from genes carrying the information as to what protein will be made during translation;
   - rRNAs, which form part of the structure of the ribosomes and help in the translation of the mRNA; and
   - tRNAs, each of which carries a specific amino acid needed for the translation process.
3. With the tRNAs and mRNAs present in the cytosol, they can combine within ribosomes to manufacture specific cellular proteins. Note: although we are using the term protein, what is actually made from the ribosome is a polypeptide. This polypeptide may represent the functional protein (tertiary structure) or first combine with one or more other polypeptides to form the functional protein (quaternary structure), as described in Chapter 2.

CONCEPT AND REASONING CHECKS

7.6 Identify the three types of RNA and summarize their roles in protein synthesis.

Translation Is the Process of Making the Polypeptide

KEY CONCEPT

- The synthesis of a protein (polypeptide) occurs through chain initiation, elongation, and termination/release.

In the process of translation, the language of the genetic code (nucleotides) is translated

TABLE 7.3 The Genetic Code Decoder

The genetic code embedded in an mRNA is decoded by knowing which codon specifies which amino acid. On the far left column, find the first letter of the codon; then find the second letter from the top row; finally read up or down from the right-most column to find the third letter. The three-letter abbreviations for the amino acids are in parentheses. Note: In the *Bacteria*, AUG codes for formylmethionine.

FIRST LETTER	Second letter: U	Second letter: C	Second letter: A	Second letter: G	THIRD LETTER
U	UUU UUC Phenylalanine (Phe) UUA UUG Leucine (Leu)	UCU UCC UCA UCG Serine (Ser)	UAU UAC Tyrosine (Tyr) UAA Stop codon UAG Stop codon	UGU UGC Cysteine (Cys) UGA Stop codon UGG Tryptophan (Trp)	U C A G
C	CUU CUC CUA CUG Leucine (Leu)	CCU CCC CCA CCG Proline (Pro)	CAU CAC Histidine (His) CAA CAG Glutamine (Gln)	CGU CGC CGA CGG Arginine (Arg)	U C A G
A	AUU AUC AUA Isoleucine (Ile) AUG Formylmethionine (Fmet) or Methionine (Met); start codon	ACU ACC ACA ACG Threonine (Thr)	AAU AAC Asparagine (Asn) AAA AAG Lysine (Lys)	AGU AGC Serine (Ser) AGA AGG Arginine (Arg)	U C A G
G	GUU GUC GUA GUG Valine (Val)	GCU GCC GCA GCG Alanine (Ala)	GAU GAC Aspartic acid (Asp) GAA GAG Glutamic acid (Glu)	GGU GGC GGA GGG Glycine (Gly)	U C A G

into the language of proteins (amino acids). The process takes place at the ribosome, where an mRNA molecule meets tRNA molecules bound to their appropriate amino acids. As the translation process is described, it would be helpful for you to follow along using FIGURE 7.8. In the figure, several of the early events in the process have already occurred. A refresher on the structure of proteins and the peptide bonds holding them together will be of value (see Chapter 2).

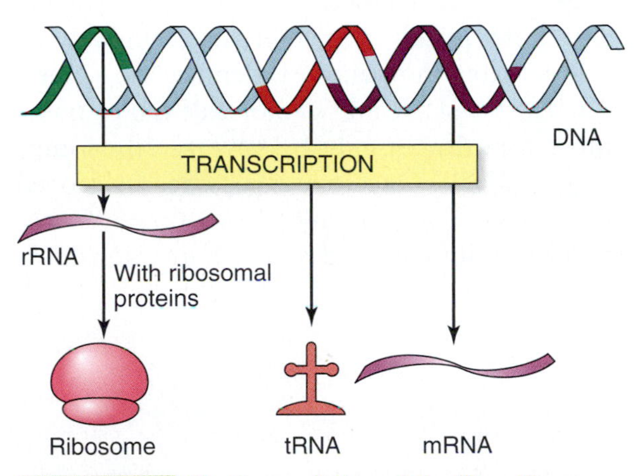

FIGURE 7.7 **The Transcription of the Three Types of RNA.** Genes in the DNA contain the information to produce three types of RNA: mRNA, rRNA, and tRNA.

Q: What is each type of RNA used for in the prokaryotic cell?

Translation takes place as the ribosome moves along the mRNA, with the codons in mRNA exposed to tRNA binding sites within the ribosome.

Chain initiation. Translation began with the addition of the tRNA whose anticodon recognized the start codon AUG binding to the mRNA on the 30S subunit of the ribosome (see Chapter 4). Then, a 50S ribosomal subunit was added to form the functional 70S ribosome. In *Bacteria*, the first amino acid is formylmethionine (fmet) while in the *Archaea* and *Eukarya*, it is methionine (met).

■ Formylmethionine: The presence of the formyl group (H-CO—) attached to methionine.

Chain elongation. The second tRNA inserted into the ribosome was one that recognized the UCC codon in the mRNA. From the genetic code decoder (Table 7.3), the amino acid attached to the tRNA must have been serine (ser). Hydrogen bonds between the codon and anticodon bases temporarily held the tRNA in position, while an enzyme attached met to ser by a peptide bond. ATP and guanosine triphosphate (GTP) supplied the energy for the reaction.

The first tRNA then was released, leaving its met molecule on the amino acid chain. Moving right one codon, the ribosome exposes the next codon (GCC), and the appropriate tRNA with the amino acid alanine (ala) attached. Again, an

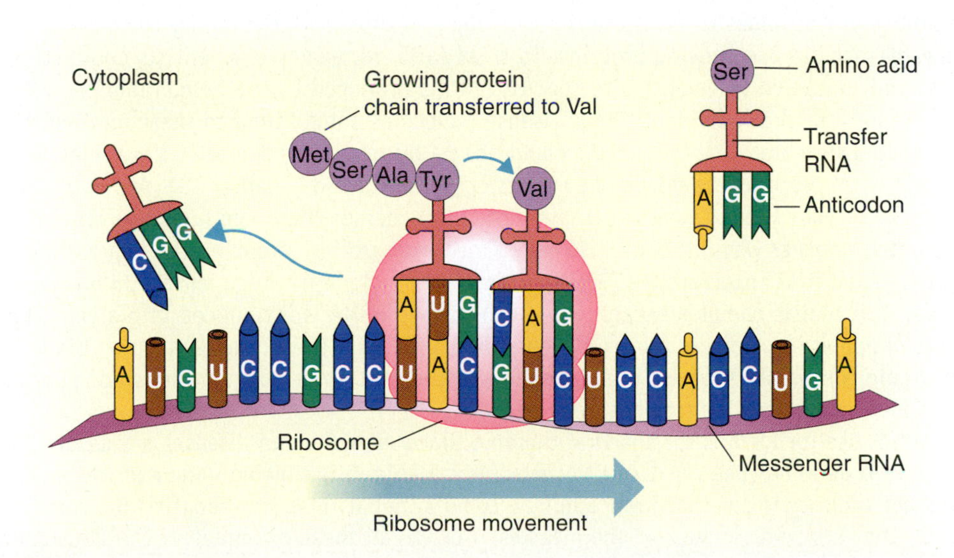

FIGURE 7.8 **The Translation Process in Protein Synthesis.** The messenger RNA moves to the ribosome, where it is met by transfer RNA molecules bonded to different amino acids. Within the ribosome, the tRNA molecules align themselves opposite the mRNA codon and bring the amino acids into position. A peptide bond forms between adjacent amino acids on the growing protein chain, after which the amino acid leaves the tRNA. The tRNA returns to the cytoplasm to bond with another molecule of the same amino acid. Note that this exceptionally short mRNA has a start and a stop codon.

Q: After the tRNA carrying ser is positioned in the ribosome, what amino acid will the next two tRNAs be carrying?

enzyme transferred the dipeptide met-ser to alanine. The tRNA that carried serine exited the ribosome and the process of chain elongation continued as the ribosome moved to expose the next codon. In the step actually shown in Figure 7.8, the tRNA recognizing the GUC codon arrives. This tRNA carries the amino acid valine (val). Again, the enzyme will transfer the growing polypeptide chain (met-ser-ala-tyr) to val. The translation process then repeats itself with the next codon and on down the mRNA.

Chain termination/release. The process of adding tRNAs and transferring the elongating polypeptide to the entering amino acid/tRNA continues until the ribosome reaches a stop codon (UGA in this example). There is no tRNA to recognize any of these codons. Rather, proteins called **releasing factors** bind where the tRNA would normally attach. This triggers the release of the polypeptide and a disassembly of the ribosome subunits, which can be reassembled for translation of another mRNA.

During synthesis, the polypeptide already may start to twist into its secondary and tertiary structure. For many polypeptides, groups of cytoplasmic proteins called **chaperones** ensure the folding process occurs correctly.

MicroFocus 7.2 describes how the understanding of protein synthesis has been used to block "harmful" proteins from being made.

Cells typically make hundreds if not thousands of copies of each protein. Producing such large amounts of a protein can be done efficiently and quite quickly. Remember each prokaryotic cell contains thousands of identical ribosomes. Therefore, a single mRNA molecule can be translated simultaneously by several ribosomes (FIGURE 7.9). Once one ribosome has moved far enough along the mRNA, another 30S subunit can "jump on" and initiate translation. Such

MICROFOCUS 7.2: Biotechnology
Antisense and Interference Makes Sense

With the recent outbreak of fatal encephalitis caused by the West Nile virus and atypical pneumonia caused by the severe acute respiratory syndrome (SARS) coronavirus, scientists have been trying to find ways to treat and cure these and other viral diseases since they are not affected by antibiotics.

One of the potential approaches being considered is the use of antisense molecules as therapeutic agents. **Antisense molecules** are RNA fragments that are the complement of an mRNA that carries a specific genetic message for protein synthesis. By binding to the mRNA, antisense molecules should block the ability of ribosomes to translate the message and thus have the ability to shut off the production of unwanted or disease-causing proteins. To treat AIDS, for example, scientists could create an antisense strand that is complementary to specific mRNAs produced by the human immunodeficiency virus (HIV). In an infected individual, the antisense molecules should bind to these viral mRNAs and, as double-stranded RNA molecules, the mRNAs could not be translated by the cell's ribosomes. Without these essential viral proteins, no new HIV particles could be formed (Chapter 15). Although such strategies make sense on paper, they have yet to produce the successes that were hoped for in clinical trials.

More recently, another way has been discovered for turning off or silencing the expression of specific genes. This is called **RNA interference** (**RNAi**). This is a technique in which extracellular, double-stranded (ds) RNA that is complementary to a known target mRNA is introduced into a cell. The dsRNA in the cell is chopped into smaller pieces by cellular enzymes and these fragments then bind to the target mRNA. Again these dsRNA pieces are degraded and the protein or polypeptide is not produced. Indirectly, the gene for that polypeptide has been silenced.

One potential use for RNAi is for antiviral therapy. Since many human diseases are caused by viruses that have an RNA genome (Chapter 13), RNAi may be valuable in inhibiting gene expression. For example, RNAi could silence viruses that induce human tumors, hepatitis A and hepatitis B viruses, influenza viruses, and other RNA viruses such as the measles virus. In all these examples, if the virus cannot replicate, new viruses cannot be produced—and disease development would be prevented.

The potential value of RNAi has recently been recognized. In 2006, Andrew Z. Fire (Stanford University School of Medicine) and Craig C. Mello (University of Massachusetts Medical School) were awarded the Nobel Prize in Physiology or Medicine "for their discovery of RNA interference—gene silencing by double-stranded RNA."

a cluster of ribosomes all translating the same mRNA is called a **polysome**.

CONCEPT AND REASONING CHECKS

7.7 Explain why the ribosome can be portrayed as a "cellular translator."

Antibiotics Interfere with Protein Synthesis

KEY CONCEPT

- Many antibiotics can inhibit transcription or translation.

Many antibiotics affect protein synthesis in prokaryotic cells and therefore are clinically useful in treating human infections and disease. Several antibiotics interfere with transcription. Rifampin binds to the RNA polymerase so that transcription cannot initiate whereas actinomycin binds to the DNA template so the RNA polymerase cannot transcribe the complete DNA template strand.

A very large number of antibiotics inhibit translation by binding to the prokaryotic ribosome. For example, streptomycin prevents chain initiation, while drugs like chloramphenicol and tetracycline inhibit the chain elongation step. We will have much more to learn about antibiotics in Chapter 24.

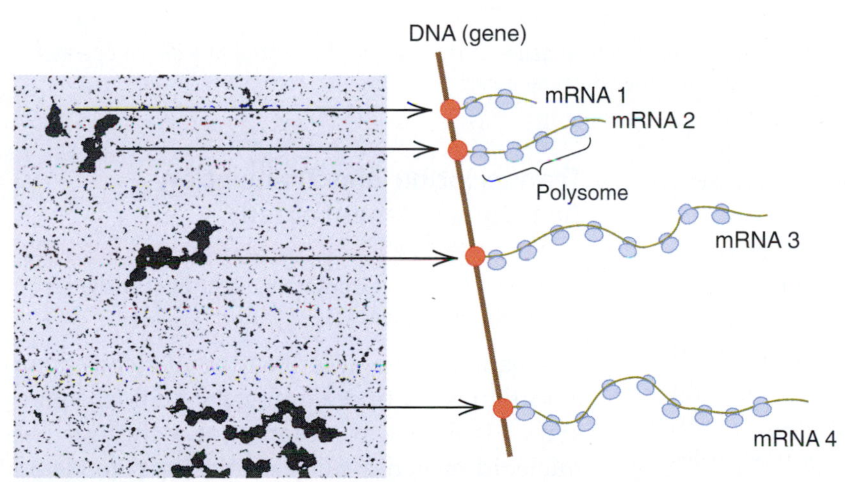

FIGURE 7.9 **Coupled Transcription and Translation in Prokaryotes.** The electron micrograph shows transcription of a gene in *E. coli* and translation of the mRNA. The dark spots are ribosomes, which coat the mRNA. An interpretation of the electron micrograph is at the right. Each mRNA has ribosomes attached along its length. The large red dots are the RNA polymerase molecules; they are too small to be seen in the electron micrograph. The length of each mRNA is equal to the distance that each RNA polymerase has progressed from the transcription-initiation site. For clarity, the polypeptides elongating from the ribosomes are not shown.

Q: From the interpretation of the micrograph, (a) how many times has this gene been transcribed and (b) how many identical polypeptides are being translated?

CONCEPT AND REASONING CHECKS

7.8 Propose a hypothesis to explain why so many antibiotics specifically affect protein synthesis.

Protein Synthesis Can Be Controlled in Several Ways

KEY CONCEPT

- Many genes are controlled by operons.

In Chapter 6, we described how negative feedback can control enzyme activity. Another control mechanism simply is to not make the enzyme (or any other protein in general) when it is not needed. Since transcription is the first step leading to protein manufacture in cells, another way to control what proteins and enzymes are present is to regulate the mechanisms that induce ("turn on") or repress ("turn off") transcription of a gene or set of genes.

In 1961, two Pasteur Institute scientists, Françoise Jacob and Jacques Monod, proposed such a mechanism for controlling protein synthesis. They suggested segments of prokaryotic DNA are organized into functional units called **operons** (FIGURE 7.10). Their pioneering research along with more recent studies indicates each operon consists of a cluster of **structural genes** that provides genetic codes for proteins having metabolically related functions. Adjacent to the structural genes is the **operator**, which is a sequence of bases controlling the expression (transcription) of the structural genes. Next to the operator is a **promoter**, which represents the sequence of bases to which the RNA polymerase binds to initiate transcription of the structural genes. Also important, but not part of the operon is a distant **regulatory gene** that codes for a **repressor protein**.

In the operon model, the repressor protein binds to the operator. Binding prevents the RNA polymerase from moving down the

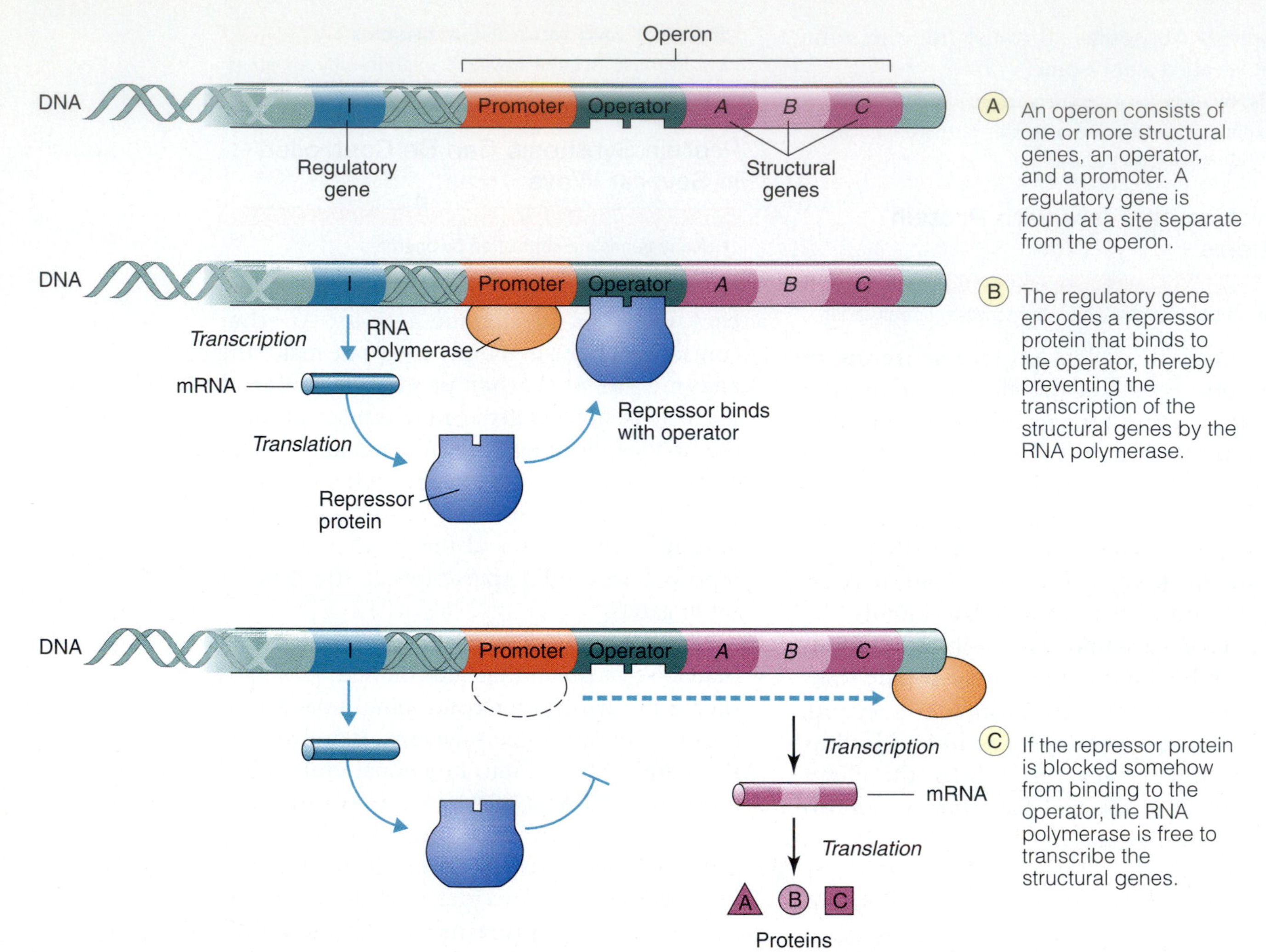

FIGURE 7.10 **The Operon and Negative Control.** An operon consists of a group of structural genes that are under the control of a single operator. Negative control exists if the operator prevents the RNA polymerase from transcribing the structural genes.

Q: How does the operator prevent structural gene transcription in negative control?

operon and thus cannot transcribe the structural genes. This is called **negative control** of protein synthesis because the repressor protein inhibits or "turns off" gene transcription within the operon. When the repressor in some way is prevented from binding to the operator, the RNA polymerase has clear sailing and transcribes the structural genes, which then are translated into the final polypeptides.

MicroInquiry 7 presents two contrasting examples of how an operon works to induce or repress gene transcription. Following the series of observations and explanations given in the MicroInquiry, you should have a firm understanding of how prokaryotic cells can control protein synthesis through transcription.

CONCEPT AND REASONING CHECKS

7.9 Why is transcription described in this section referred to as negative control?

Transcription and Translation Are Compartmentalized

KEY CONCEPT

- Transcription and translation occur in spatially separated compartments.

In Chapter 4, we described the nucleoid as an amorphous area containing the cell's chromosome. It also was noted that although the nucleoid lacked a nuclear envelope, nucleoid and cytoplasmic activities were segregated much as they are in eukaryotic cells.

Research studies have shown that, at least in *Bacillus subtilis*, RNA polymerases are concentrated within the nucleoid core in the central portion of the cell (FIGURE 7.11). If correct, then most of the cell's transcription presumably occurs in the same region. In fact, RNA polymerase was often localized to specific regions of the nucleoid, somewhat similar

MICROINQUIRY 7

The Operon Theory and the Control of Protein Synthesis

The best way to visualize and understand the operon model for control of protein synthesis is by working through a couple of examples.

The Lactose *(lac)* Operon

Here is a piece of experimental data. The disaccharide lactose represents a potential energy source for *E. coli* cells if it can be broken into its monomers of glucose and galactose. One of the enzymes involved in the metabolism of lactose is β-galactosidase. If *E. coli* cells are grown in the absence of lactose, β-galactosidase activity cannot be detected as shown in the graph (**Figure A**).

However, when lactose is added to the nutrient broth, very quickly enzyme activity is detected. How can this change from inhibition to expression be explained in the operon model?

Based on the operon theory, we would propose that when lactose is absent from the growth medium, the repressor protein for the lac operon binds to the operator and blocks passage of the RNA polymerase that is attached to the adjacent promoter (**Figure Bi**). Being unable to move past the operator, the polymerase cannot transcribe the structural genes, one of which codes for β-galactosidase.

When lactose is added to the growth medium, lactose will be transported into the bacterial cell, where the disaccharide binds to the repressor protein and inactivates it (**Figure Bii**). With the repressor protein inactive, it no longer can recognize and bind to the operator. The RNA polymerase now is not blocked and can translocate down the operon and transcribe the structural genes. Lactose is called an inducer because its presence has induced, or "turned on," structural gene transcription in the lac operon. It explains why β-galactosidase activity increases when lactose was present.

Now let's see if you can figure out this scenario.

Tryptophan *(trp)* Operon

E. coli cells have a cluster of structural genes that code for five enzymes in the metabolic pathway for the synthesis of the amino acid tryptophan (trp). Therefore, if *E. coli* cells are grown in a broth culture lacking trp, they continue to grow normally by synthesizing their own tryptophan, as shown in the graph (**Figure A**).

However, as the graph shows, when trp is added to the growth medium, new enzyme synthesis is repressed or "turned off" and cells use the trp supplied in the growth medium.

How can enzyme repression be explained by the operon model? The solution is provided in **Appendix D**.

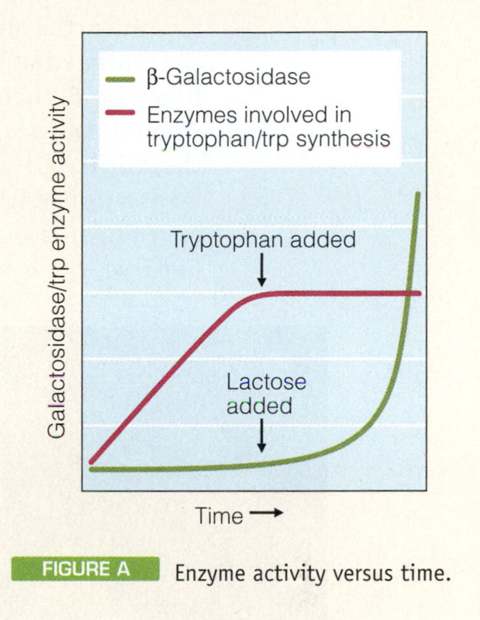

FIGURE A Enzyme activity versus time.

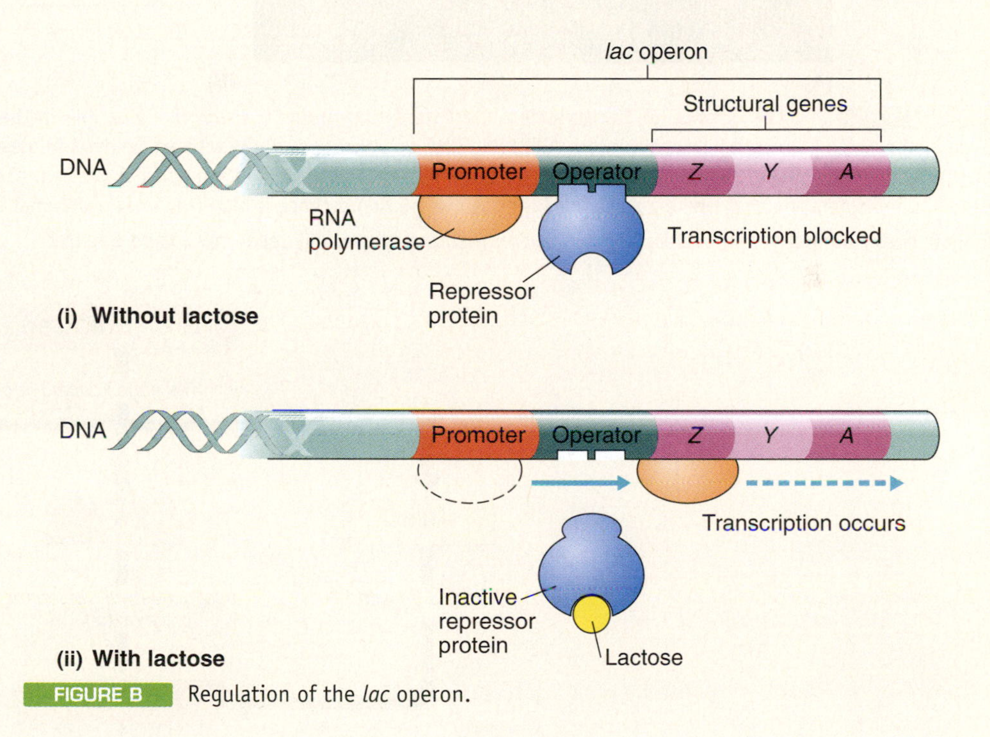

FIGURE B Regulation of the *lac* operon.

to its localization in eukaryotic nucleoli carrying out rRNA synthesis. This is in contrast to the localization of ribosomes, which are absent in the nucleoid region and primarily concentrated at the cell poles. If the observation with *B. subtilis* holds true for other *Bacteria* and *Archaea*, then prokaryotic cells have the ability to segregate transcription and translation without the need for a nuclear envelope in a manner analogous to that of the *Eukarya*.

Other research suggests that individual genes within the chromosome also have specific positions within the nucleoid and are not randomly positioned. For example, the *oriC* region of the chromosome lies at one end of the nucleoid and the *terC* region at the opposite end. The precise mechanisms responsible for nucleoid gene organization and the establishment of core and peripheral zones remain to be elucidated.

FIGURE 7.12 summarizes the protein synthesis process.

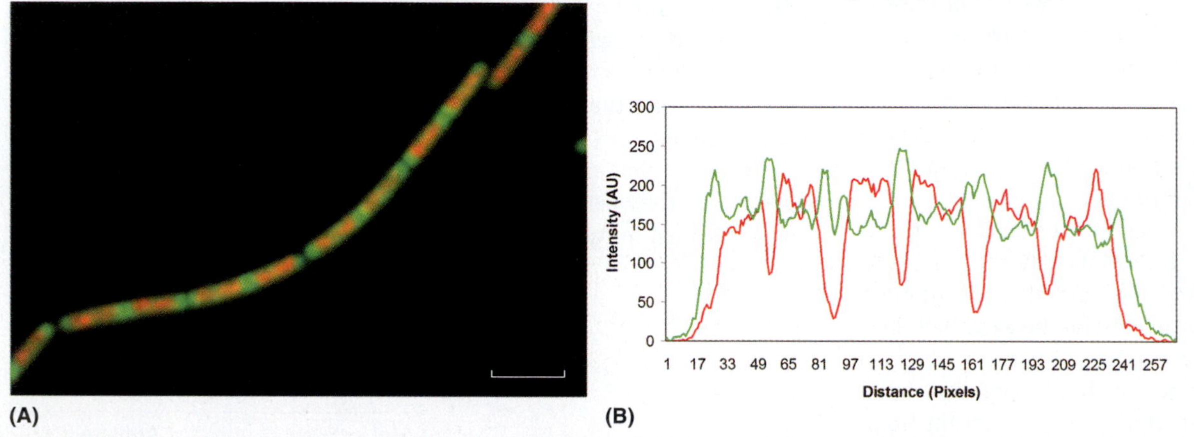

FIGURE 7.11 **The Localization of Transcription and Translation in *Bacillus subtilis* Cells.** (**A**) In these dividing *B. subtilis* cells, ribosomal subunits have been labeled with a green fluorescent protein (GFP) and RNA polymerase subunits with a label that fluoresces red. The RNA polymerase (transcription) is found mainly in the nucleoid core while the ribosomes (translation) are concentrated at the poles of the cell. (Bar = 3 μm.) (**B**) This linescan through the cells confirms the interpretation that where polymerase RNA polymerase fluorescence is high (red line), ribosomal fluorescence is low (green line) and vice-versa.

Q: How does fluorescence microscopy aid in the identification of spatially separated compartments?

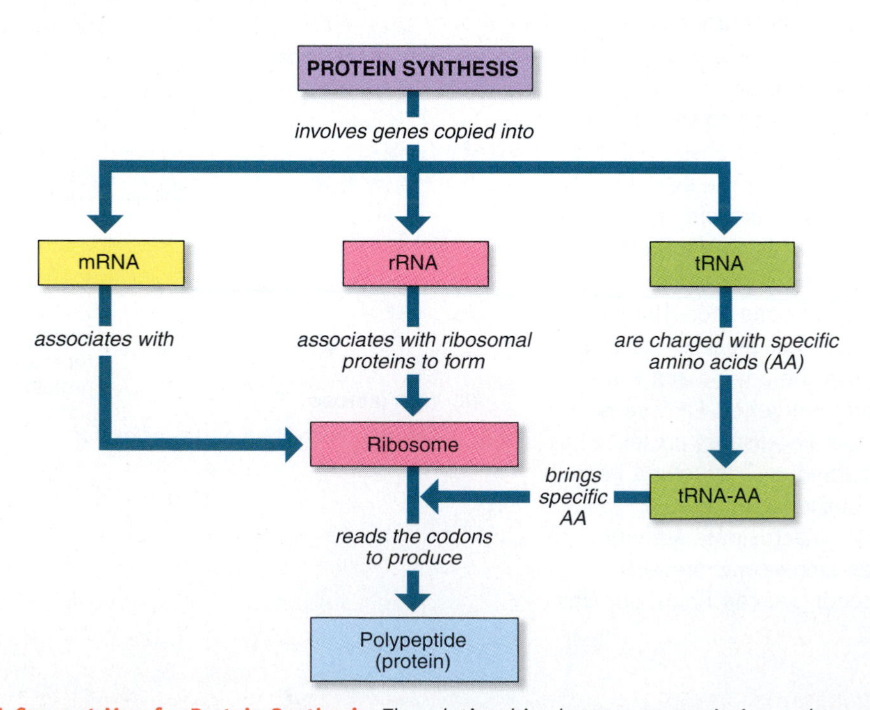

FIGURE 7.12 **A Concept Map for Protein Synthesis.** The relationships between transcription and translation, and the three types of RNA, are illustrated.

Q: In this concept map, circle those parts representing transcription and circle those parts representing translation.

7.4 Mutations

The information in a prokaryotic chromosome may be altered through a permanent change in the DNA called a **mutation**. In most cases, a mutation involves a disruption of the nitrogenous base sequence in the DNA molecule. This can result in the production of a miscoded mRNA, resulting in the insertion of one or more incorrect amino acids into the polypeptide during translation. Because proteins govern numerous cellular activities, mutations may alter some aspect of these activities—for better or worse (MicroFocus 7.3).

Mutations Are the Result of Natural Processes or Induced

KEY CONCEPT

- Mutations can be spontaneous or induced.

Spontaneous mutations are heritable changes to the base sequence in the DNA that result from natural phenomena. These changes could be from everyday radiation penetrating the atmosphere or errors made and not corrected by DNA polymerase III during replication. It has been estimated that one such mutation may occur for every 10^6 to 10^{10} divisions of a prokaryotic cell.

A mutant cell arising from a spontaneous mutation usually is masked by the normal wild type cells in the population. However, should some agent be present from which the mutant survives, it may multiply and emerge as the predominant form. For many decades, for example, doctors used penicillin to treat gonorrhea. Then, in 1976, a penicillin-resistant strain of *Neisseria gonorrhoeae* emerged in human populations. Many investigators believe the resistant strain had been present for perhaps centuries, but only now with heavy use of penicillin could it arise and fill the niche once held by the penicillin-sensitive forms.

Most of our understanding of mutations has come from experiments in which scientists

■ Wild type: The common or native form of a gene or organism.

■ Niche: The functioning of a species in relation to other species and its physical environment.

MICROFOCUS 7.3: Evolution

Evolution of An Infectious Disease

Could the Black Death of the fourteenth century and the 25 million Europeans that succumbed to plague have been the result of a few genetic changes to a bacterial cell? Could the entire course of Western civilization have turned based on these changes?

Possibly so, maintain researchers from the federal Rocky Mountain Laboratory in Montana. In 1996, a research group led by Joseph Hinnebusch reported that three genes missing in the plague bacillus *Yersinia pestis* are present in a fairly harmless form of the organism *(Y. pseudotuberculosis)* that causes mild food poisoning. Thus, it is possible that the entire story of plague's pathogenicity revolves around a small number of gene changes.

Bubonic, septicemic, and pneumonic plague are caused by *Y. pestis*, a rod-shaped bacterium transmitted by the rat flea (Chapter 11). In an infected flea, the bacterial cells eventually amass in its foregut and obstruct its gastrointestinal tract. Soon the flea is starving, and it starts biting victims (humans and rodents) uncontrollably and feeding on their blood. During the bite, the flea regurgitates some 24,000 plague bacilli into the bloodstream of the unfortunate victim.

At least three genes appear important in the evolution of plague. It appears that nonpathogenic plague bacilli have these genes, which encourage the bacilli to remain harmlessly in the midgut of the flea. Pathogenic plague bacilli, by contrast, do not have the genes. Free of control, the bacteria migrate from the midgut to the foregut and form a plug of packed bacilli that are passed on to the victim.

In 2002, Hinnebusch and colleagues published evidence that another gene, carried on a plasmid, codes for an enzyme that is required for the initial survival of *Y. pestis* bacilli in the flea midgut. By acquiring this gene from another unrelated organism, *Y. pestis* made a crucial jump in its host range. It now could survive in fleas and became adapted to relying on its blood-feeding host for transmission. So, a few genetic changes may have been a key force leading to the evolution and emergence of plague. This is just another example of the flexibility that many microbes have to repackage themselves constantly into new and, sometimes, more dangerous agents of infectious disease.

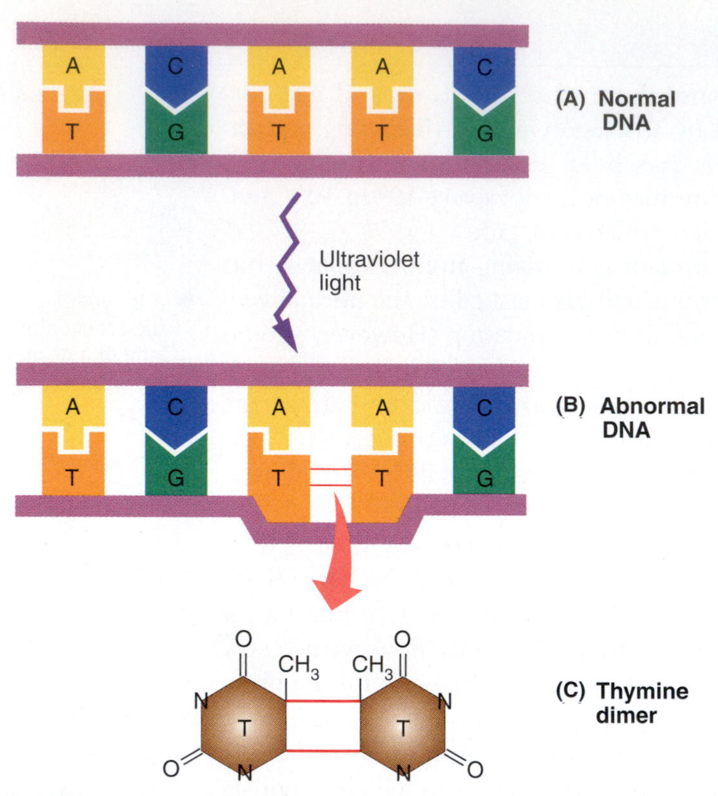

FIGURE 7.13 **The Formation of Spontaneous and Induced Mutations.** When cells are irradiated with ultraviolet (UV) light either naturally or through experiment, the radiations may affect the cell's DNA. UV light can cause adjacent thymine molecules to pair within the DNA strand to form a thymine dimer.

Q: How might a thymine dimer block the movement of RNA polymerase?

purposely generate mutations in prokaryotic cells. Such **induced mutations** are produced by chemical or physical agents called **mutagens**.

Physical mutagens. **Ultraviolet (UV) light** is a physical mutagen whose energy induces adjacent thymine (or cytosine) bases in the DNA to covalently link together forming dimers (FIGURE 7.13). If these dimers occur in a protein-coding gene, the RNA polymerase cannot insert the correct bases (A—A) in mRNA molecules where the dimers are located.

Chemical mutagens. **Nitrous acid** is an example of a chemical mutagen that converts DNA's adenine bases to hypoxanthine bases (FIGURE 7.14A). Adenine would normally base pair with thymine, but the presence of hypoxanthine causes a base pairing with cytosine after replication. Later, should replication occur from the gene with the cytosine mutation, the mRNA will contain a guanine rather than an adenine if no mutation had occurred.

Mutations also are induced by **base analogs**, such as **5-bromouracil**, which bears a close chemical resemblance to thymine (FIGURE 7.14B). During replication, the base analog could pair with adenine when thymine should be present.

Other base analogs resemble other DNA bases and are useful as antiviral agents in the treatment of diseases caused by DNA viruses, such as the herpesviruses (Chapter 14). **Acyclovir**, for example, is a base analog that can substitute for guanine during viral replication. The presence of acyclovir blocks viral replication, so new virus particles cannot be produced. As such, treatment with acyclovir can be effective in decreasing the frequency and severity of genital herpes.

CONCEPT AND REASONING CHECKS

7.10 How do chemical mutagens interfere with DNA replication or protein synthesis?

Point Mutations Are a Result of Spontaneous or Induced Mutations

KEY CONCEPT

- Point mutations affect one base pair in a DNA sequence.

Regardless of the cause of the mutation, one of the most common results is a **point mutation**, which affects just one point (base pair) in a gene. Such mutations may be a change to or substitution of a different base pair, or a deletion or addition of a base pair.

Base-pair substitutions. If a point mutation causes a base-pair substitution, then the transcription of that gene will have one incorrect base in the mRNA sequence of codons. Perhaps one way to see the effects of such changes is using an English sentence made up of three-letter words (representing codons) where one letter has been changed. As three-letter words, the letter substitution still reads correctly, but the sentence makes less sense.

Normal sequence: THE FAT CAT ATE THE RAT

Substitution: THE FAT CA<u>R</u> ATE THE RAT

As shown in FIGURE 7.15A, depending on the placement of the substituted base, when the mRNA is translated this may cause no change (silent mutation), lead to the insertion of the wrong amino acid (missense mutation), or generate a stop codon (nonsense mutation), prematurely terminating the polypeptide.

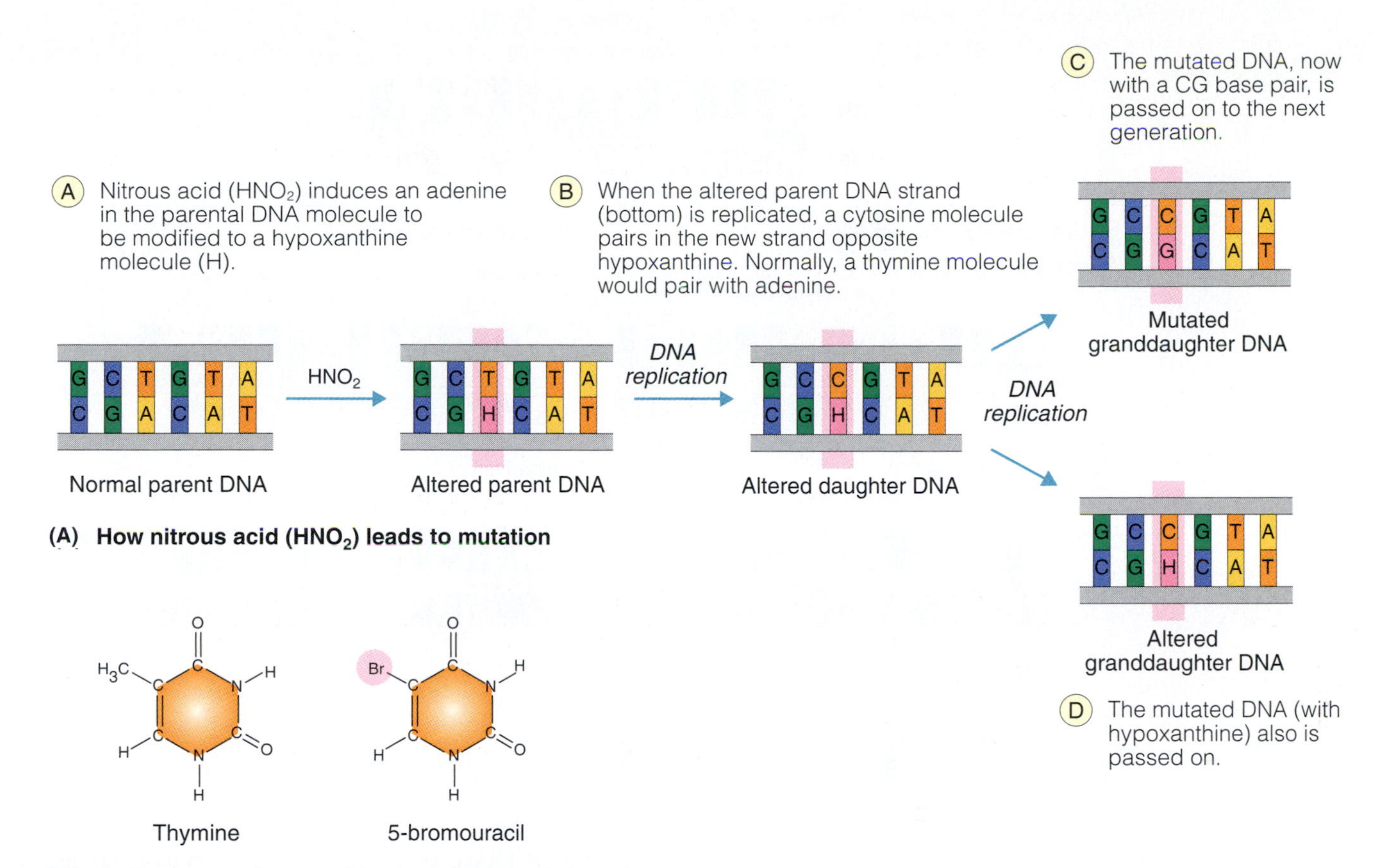

FIGURE 7.14 **The Effect of Chemical Mutagens.** (**A**) Nitrous acid induced an adenine to hypoxanthine change. After replication of the hypoxanthine-containing strand, the granddaughter DNA has a mutated C—G base pair. (**B**) Base analogs induce mutations by substituting for nitrogenous bases in the synthesis of DNA. Note the similarity in chemical structure between thymine and the base analog 5-bromouracil.

Q: How does nitrous acid differ from 5-bromouracil in inducing mutations?

Base-pair Deletion or Insertion. Point mutations also can cause the loss or addition of a base in a gene, resulting in an inappropriate number of bases. Again, using our English sentence, we can see how a deletion or insertion of one letter affects the reading of the three-letter word sentence.

Normal sequence: THE FAT CAT ATE THE RAT

Deletion: THE F_TC ATA TET HER AT

Insertion: THE FAT ACA TAT ETH ERA T

As you can see, the "sentence mutations" are nonsense when reading the sentence as three-letter words. The same is true in a cell. Ribosomes always read three letters (one codon) at one time, generating potentially extensive mistakes in the amino acid sequence (FIGURE 7.15B). Thus, like our English sentence, the deletion or addition of a base will cause a reading frameshift because the ribosome always reads the genetic code in groups of three bases. Therefore, loss or addition of a base shifts the reading of the code by one base. The result is serious sequence errors in the amino acids, which will probably produce an abnormal protein (nonsense) unable to carry out its role in metabolism.

CONCEPT AND REASONING CHECKS

7.11 Justify the statement: "A frameshift mutation potentially is more dangerous to an organism's viability than a base-pair substitution."

Repair Mechanisms Attempt to Correct Mistakes or Damage in the DNA

KEY CONCEPT

- Cells have the ability to repair damaged DNA.

During the life of a prokaryotic cell (indeed, of every prokaryotic and eukaryotic cell), cellular DNA endures thousands of damaging events

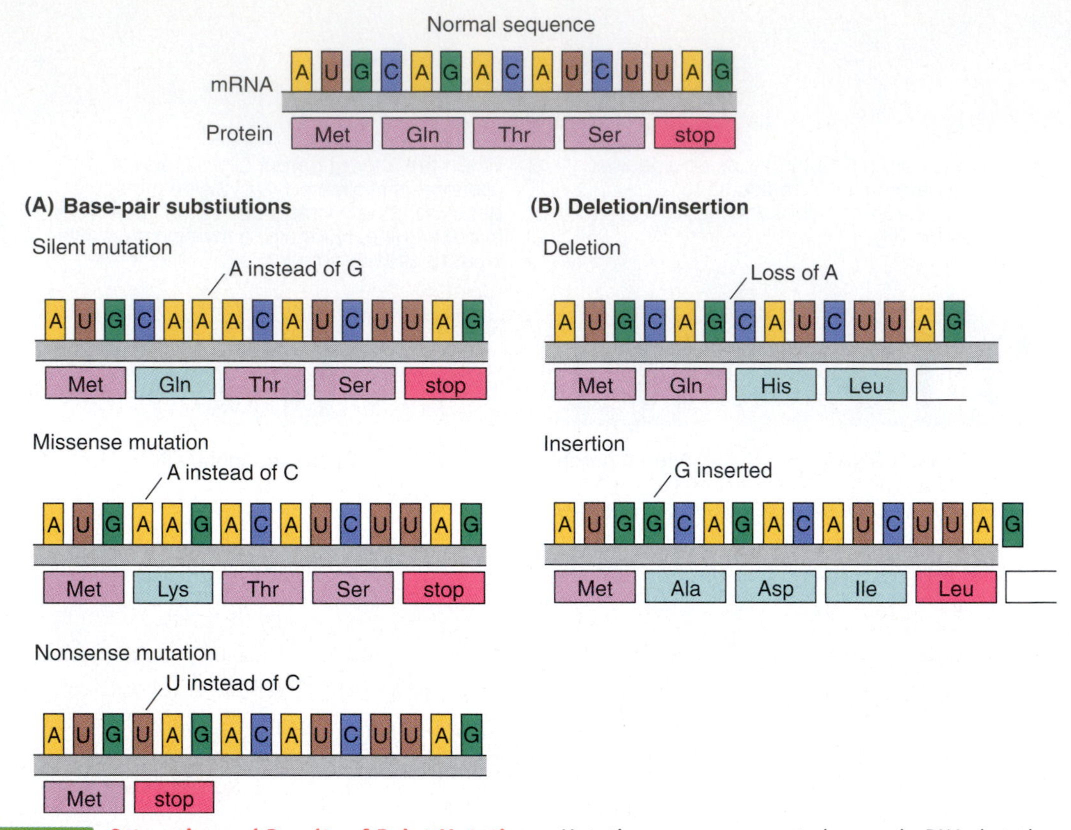

FIGURE 7.15 **Categories and Results of Point Mutations.** Mutations are permanent changes in DNA, but they are represented here as they are reflected in mRNA and its protein product. (**A**) Base-pair substitutions can produce silent, missense, or nonsense mutations. (**B**) Deletions or insertions shift the reading frame of the ribosome.

Q: Determine the normal sequence of bases in the template strand of the gene and the base change that gave rise to each of the "mutated" mRNAs.

resulting from DNA replication errors and other base changes caused by mutagens. Because these errors could disrupt metabolism, cells attempt to correct such DNA damage by using a variety of DNA repair mechanisms.

One type of repair mechanism is called **mismatch repair**. As described earlier in this chapter, as the DNA polymerase adds new complementary bases to the DNA template strand during replication, it makes mistakes. Therefore, as it adds bases, it also "proofreads" its work and removes mismatched nucleotides (FIGURE 7.16). So, the repair process is somewhat like driving with a mechanic in the back seat.

The fact that DNA is double-stranded is not a fluke. By being double-stranded, one strand can act as a template to correct mismatches. It is estimated that about 1 in 10,000 bases is mismatched during DNA replication. *E. coli* has about 4.6 million bases in its chromosome, so mathematics says over 460 mismatches will occur and must be repaired every replication. Considering the enzyme is catalyzing the addition of 50,000 bases every minute, an initial one percent error rate is very efficient.

Mutations caused by physical mutagens also can be corrected. Almost 100 different nuclease enzymes are known to exist in *E. coli* cells. When DNA is damaged by a physical mutagen, such as UV light, several of these nucleases execute **excision repair** (FIGURE 7.17). First, nucleases cut out (excise) the damaged DNA. Then, a different DNA polymerase from the one used in replication replaces the missing nucleotides with the correct ones. Finally, DNA ligase seals the new strand into the rest of the polynucleotide.

CONCEPT AND REASONING CHECKS

7.12 Explain why cells need at least two repair mechanisms (mismatch and excision).

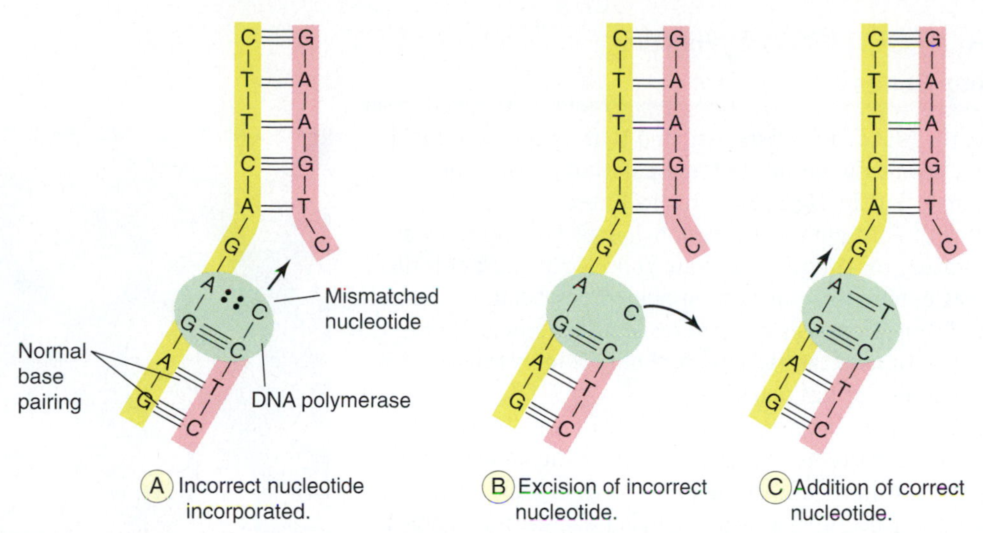

FIGURE 7.16 **Mismatch Repair Mechanisms.** Mismatch repair during DNA replication. As a result of an incorrectly paired nucleotide, DNA polymerase removes (excises) the nucleotide and adds the correct complementary nucleotide.

Q: Propose a hypothesis to explain how the DNA polymerase "knows" there is a mismatch during replication.

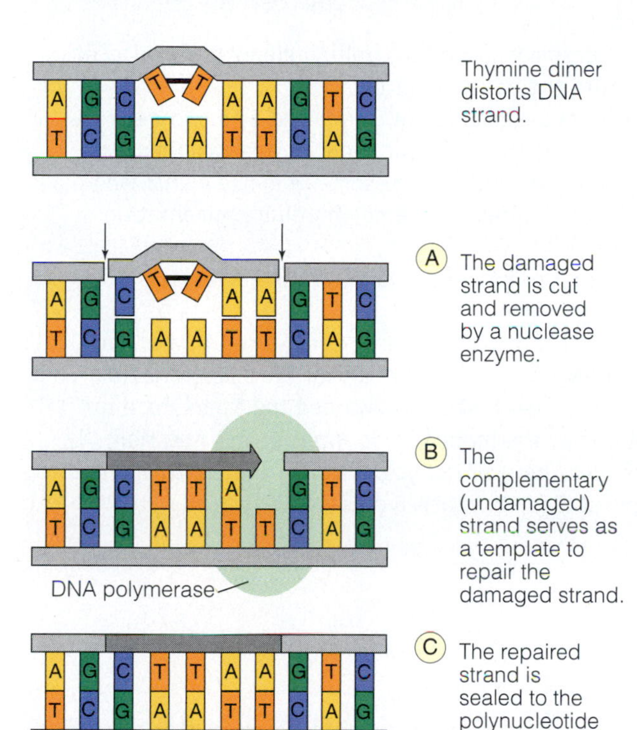

FIGURE 7.17 **Excision Repair Mechanism.** Thymine dimer distortion triggers nuclease repair enzymes that excise the damaged DNA and permit resynthesis of the correct nucleotides.

Q: How might excision nucleases recognize an error in a DNA fragment?

Transposable Genetic Elements Can Cause Mutations

KEY CONCEPT

- Insertion sequences and transposons move from one DNA location to another.

Mutations of a different nature may be caused by fragments of DNA called **transposable genetic elements**. Two types are known in prokaryotes: insertion sequences and transposons. **Insertion sequences (IS)** are small segments of DNA with about 1,000 base pairs. They are found at one or more sites on a chromosome or plasmid. IS have no genetic information other than for the ability to insert into a chromosome. IS form copies of themselves, and the copies move into other areas of the chromosome. These events are rare, but when they occur they can interrupt the coding sequence in a gene, thereby inducing the wrong protein or, more likely, no protein to form. IS may be a prime force behind spontaneous mutations.

A second type of transposable genetic element is the **transposon**. These are the so-called "jumping genes" for which Barbara McClintock won the 1983 Nobel Prize in Physiology or Medicine (MicroFocus 7.4). Transposons are larger than IS and carry information, such as antibiotic resistance, which can be conferred to the prokaryotic cell. Like IS, they can interrupt the genetic code of a gene.

The movement of transposons appears to be nonreciprocal, meaning an element moves ("jumps") away from its location and nothing takes its place. (This contrasts with insertion sequences, where copies move.) Transposons can move from plasmid to plasmid, from plasmid to

MICROFOCUS 7.4: History

Jumping Genes

In the early 1950s, scientists assumed that genes were fixed elements, always found in the same position on the same chromosome. But in 1951, Barbara McClintock unveiled her research with corn plants at a symposium at Cold Spring Harbor Laboratory on Long Island, New York. McClintock described genes that apparently moved from one chromosome to another. The audience listened in respectful silence. There were no questions after her talk, and only three people requested copies of her paper.

McClintock grew Indian corn, or maize. In the 1940s, she noticed curious patterns of pigmentation on the kernels. Other scientists might have missed the patterns as random variations of nature, but McClintock's record keeping and careful analysis revealed a method to nature's madness. The pigment genes causing the splotches of color appeared to be switched on or off in particular generations. Still more remarkable, the "switches" seemed to occur at different places along the same chromosome. Some switches even showed up in different chromosomes. Such "controlling elements," as McClintock called them, were available whenever needed to turn the genes on or off.

In the modern lexicon of molecular genetics, McClintock's elements are recognized as a two-gene system. One is an activator gene, the other a dissociation gene. The activator gene, for reasons unknown, can direct a dissociation gene to "jump" along the arm of the ninth chromosome in maize plants where color is regulated. When the jumping gene reinserts itself, it turns off the neighboring pigmentation genes, thereby altering the color of the kernel.

The jumping gene is identical to the transposon found in bacteria and certainly serves as a driving force in evolution.

For Barbara McClintock, recognition came 30 years after that symposium at Cold Spring Harbor. In 1981 (at the age of 79), she received eight awards, among them a $60,000-a-year lifetime grant from the MacArthur Foundation and the $15,000 Lasker prize. In 1983, she was awarded the Nobel Prize in Physiology or Medicine. When informed of the Nobel award, she replied to an interviewer's question, "It seemed unfair to reward a person for having so much pleasure over the years, asking the maize plants to solve specific problems and then watching their response." Dr. McClintock died in 1992.

chromosome, or from chromosome to plasmid. The presence of inverted repetitive base sequences at the ends of the element appears to be important in establishing the ability to move (FIGURE 7.18).

Of particular significance is the finding that many transposons contain genes for antibiotic resistance. For example, if a plasmid containing a transposon moves from one bacterial cell to another, as plasmids are known to do, the transposon will move along with it, thus spreading the genes for antibiotic resistance. Moreover, the movement of transposons among plasmids helps explain how a single plasmid acquires numerous genes for resistance to different antibiotics.

CONCEPT AND REASONING CHECKS

7.13 What role do insertion sequences and transposons play?

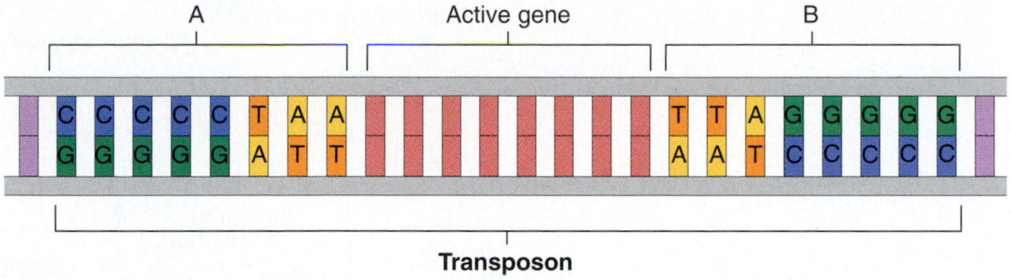

A
Active gene
B
C C C C C T A A ... T T A G G G G G
G G G G G A T T ... A A T C C C C C
Transposon

FIGURE 7.18 **Transposon Structure.** A transposon contains one or more active genes bordered by inverted repetitive sequences. Note that the base sequence in A is the reverse and complement of the base sequence in B. Also note the presence of inverted repetitive base sequences (C-G and G-C) at the ends of the transposon.

Q: What is the significance of the inverted repetitive base sequences?

7.5 Identifying Mutants

A prokaryote (or any organism) carrying a mutation is called a **mutant**, while the normal strain isolated from nature is the **wild type**. Some mutants are easy to identify because the **phenotype** or physical appearance, of the organism or the colony has changed from the wild type. For example, some bacterial colonies appear red because they produce a red pigment. Treat the colonies with a mutagen and, after plating on nutrient agar, mutants form colorless colonies. However, not all mutants can be identified solely by their "looks."

Plating Techniques Select for Specific Mutants or Characteristics

KEY CONCEPT

- Mutant identification can involve negative or positive selection techniques.

Selection is a very useful technique to identify and isolate a single mutant from among thousands of possible cells or colonies. Let's look at two selection techniques that make this search possible.

First, in both techniques, the chemical composition of the transfer plate is key to visual identification of the colonies being hunted. The use of a replica plating device makes the identification possible. The device consists of a sterile velveteen cloth or filter paper mounted on a solid support. When an agar plate (master plate) with bacterial colonies is gently pressed against the surface of the velveteen, some cells from each colony stick to the velveteen. If another agar plate then is pressed against this velveteen cloth, some cells will be transferred (replicated) in the same pattern as on the master plate.

Now, suppose you want to find a nutritional mutant unable to grow without the amino acid histidine. This mutant (written his$^-$) has lost the ability that the wild type strain (his$^+$) has to make its own histidine. Such a mutant having a nutritional requirement for growth is called an **auxotroph** (*auxo* = "grow"; *troph* = "nourishment"), while the wild type is a **prototroph** (*proto* = "original"). Phenotypically, there is no difference between the two strains. However, you can visually identify the auxotroph using a **negative selection** plating technique (FIGURE 7.19). Any colonies missing on the minimal medium plate must be *his*$^-$.

As another example, suppose you want to see if there are any bacteria in a hospital ward that are resistant to the antibiotic tetracycline. Again, phenotypically there is no difference between those strains sensitive to tetracycline and those resistant to the antibiotic. However, a **positive selection** plating technique permits visual identification of such tetracycline resistant mutants (FIGURE 7.20).

CONCEPT AND REASONING CHECKS

7.14 How does negative selection differ from positive selection?

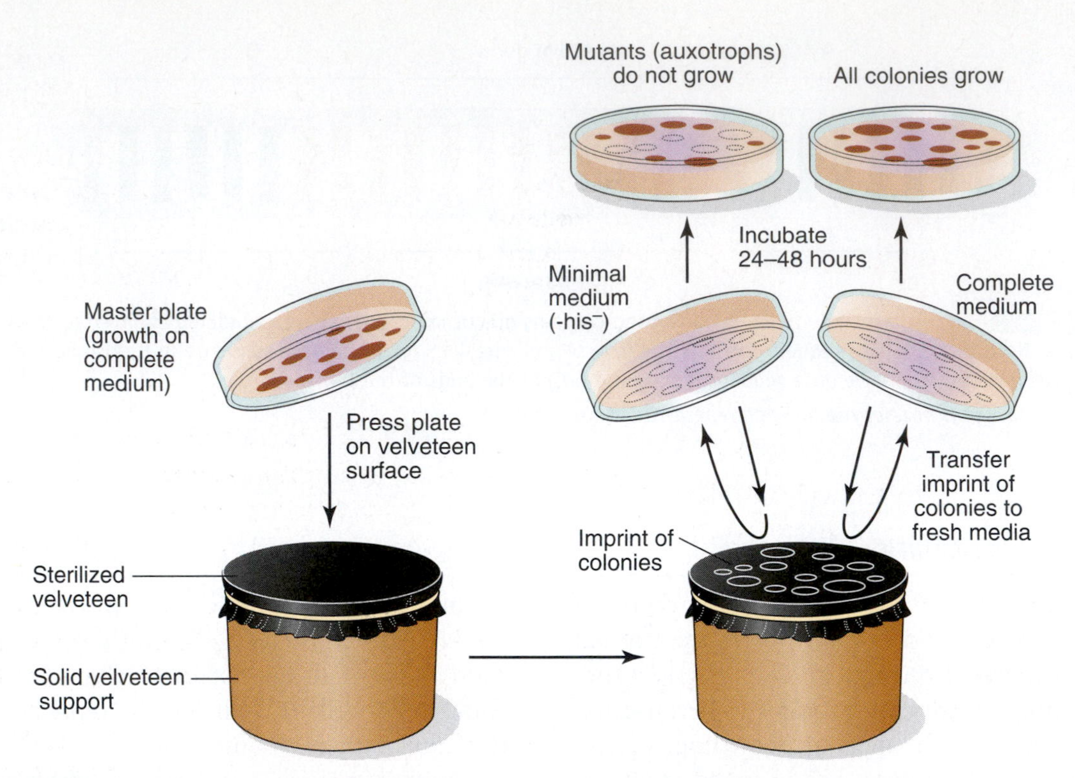

FIGURE 7.19 **Negative Selection Identifies Auxotrophs.** Negative selection plating techniques can be used to detect nutritional mutants (auxotrophs) that fail to grow when replica plated on minimal medium (in this example, a growth medium lacking histidine). Comparison to replica plating on complete medium visually identifies the auxotrophic mutants.

Q: In this example, which colonies transferred to the complete medium represent the auxotrophs (his⁻)?

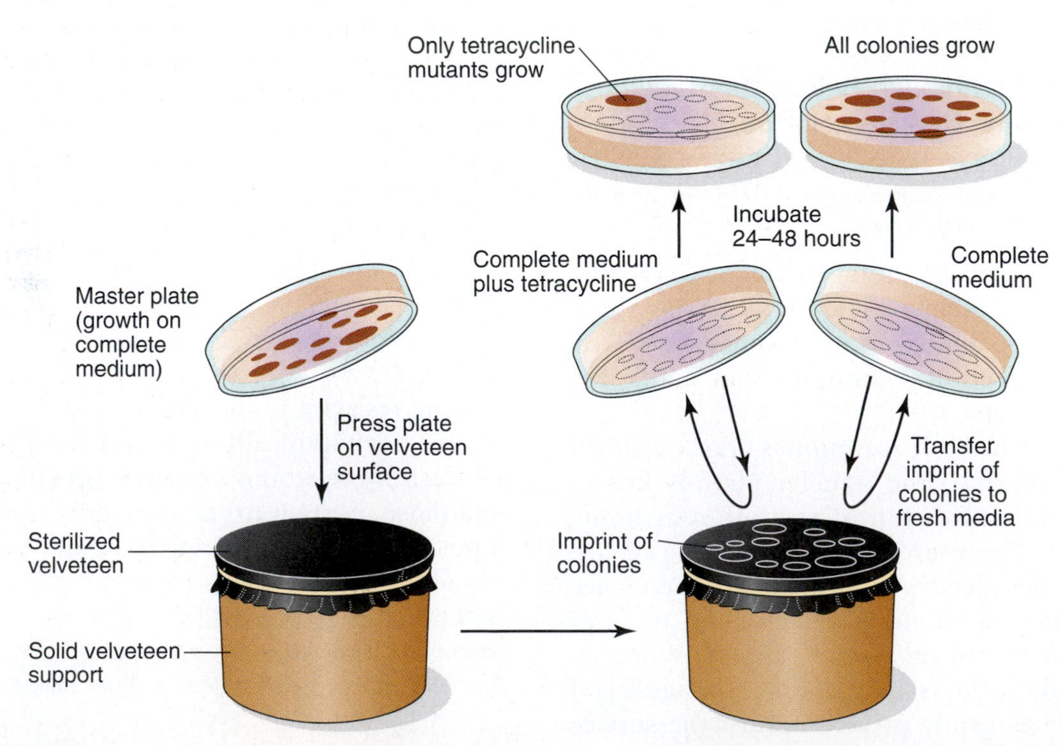

FIGURE 7.20 **Positive Selection of Mutants.** Positive selection plating techniques can be used to identify antibiotic resistant mutants.

Q: What do the vacant spots on the complete medium plus tetracycline represent?

The Ames Test Can Identify Potential Mutagens

KEY CONCEPT

- Ames test revertants suggest a chemical is a potential carcinogen in humans.

Some years ago, scientists observed that about 90 percent of human **carcinogens**, agents causing tumors in humans, also induce mutations in bacterial cells. Working on this premise, Bruce Ames of the University of California developed a procedure to help identify potential human carcinogens by determining whether the agent can mutate bacterial auxotrophs. The procedure, called the **Ames test**, is a widely used, relatively inexpensive, rapid, and accurate screening test. For the Ames test, an auxotrophic, histidine-requiring strain (his^-) of *Salmonella enterica* serotype Typhimurium is used. If inoculated onto a plate of nutrient medium lacking histidine, no colonies will appear because in this auxotrophic strain the gene inducing histidine synthesis is mutated and hence not active.

In preparation for the Ames test, the potential carcinogen is mixed with a liver enzyme preparation. The reason for doing this is because often chemicals only become tumor causing and mutagenic in humans after they have been modified by liver enzymes.

To perform the Ames test, the his^- strain is inoculated onto an agar plate lacking histidine (FIGURE 7.21). A well is cut in the middle of

■ Screening test: A process for detecting mutants by examining numerous colonies.

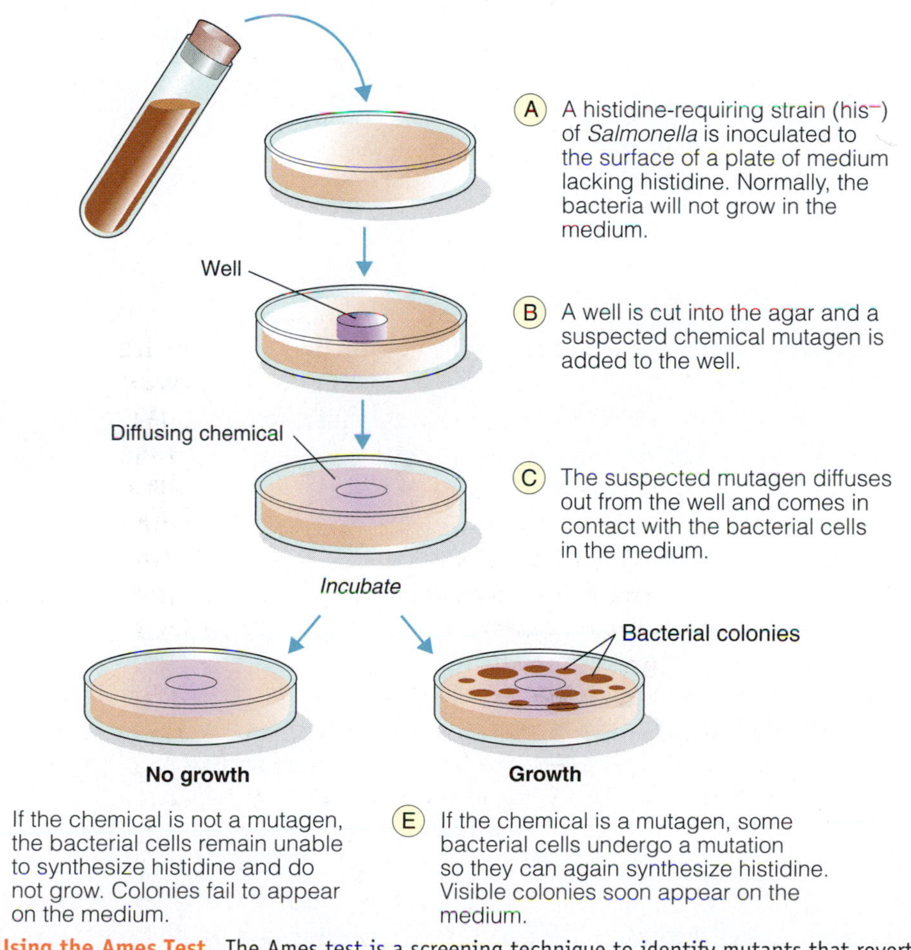

FIGURE 7.21 **Using the Ames Test.** The Ames test is a screening technique to identify mutants that reverted back to the wild type because of the presence of a mutagen.

Q: As a control, a plate with Salmonella *(his^-) but without mutagen is incubated. After 24–48 hours, a few colonies are seen on the plate. Explain this observation.*

the agar, and the potential liver-modified carcinogen is added to the well (or a filter paper disk with the chemical is placed on the agar surface). The chemical diffuses into the agar. The plate is incubated for 24 to 48 hours. If bacterial colonies appear, one may conclude the agent mutated the bacterial his$^-$ gene back to the wild type (his$^+$); that is, **revertants** were generated that could again encode the enzyme needed for histidine synthesis. Because the agent is a mutagen, it is therefore a possible carcinogen in humans. If bacterial colonies fail to appear, one assumes that no mutation took place. However, it is possible the mutation did occur, but was repaired by a DNA repair enzyme. This possibility has been overcome by using bacterial strains known to be inefficient at repairing errors.

CONCEPT AND REASONING CHECKS

7.15 Just because a potential carcinogen generates revertants in *Salmonella*, does that always mean it is cancer causing in humans? Explain.

SUMMARY OF KEY CONCEPTS

7.1 Prokaryotic DNA

- The DNA of a prokaryote's chromosome is supercoiled and folded into a series of looped domains. Each loop consists of 10,000 bases.
- Prokaryotic cells may have one or more plasmids. These small closed loops of DNA carry information that can confer selective advantages (e.g., antibiotic resistance, protein toxins) to the cells.

7.2 DNA Replication

- DNA replicates by a semiconservative mechanism where each strand of the original DNA molecule acts as a template to synthesize a new strand.
- DNA replication starts with initiator proteins binding at the origin of replication, forming two replication forks moving in opposite directions around the chromosome. DNA polymerase III moves along the strands inserting the correct DNA nucleotide to complementary bind with the template strand. At each replication fork, one of the two strands is synthesized in a continuous fashion while the other strand is formed in a discontinuous fashion, forming Okazaki fragments, which are joined by a DNA ligase.

7.3 Protein Synthesis

- The synthesis of a protein is a complex mechanism in which the genetic information in DNA is first transcribed to a genetic message in RNA and then translated to a sequence of amino acids in the protein.
- Transcription occurs when the RNA polymerase binds to a promoter sequence on the DNA template strand. Various forms of RNA, including mRNA, tRNA, and ribosomal RNA, are transcribed from the DNA and are important in the translation process.
- Translation occurs on ribosomes, which bring together mRNA and tRNAs. The ribosome "reads" the mRNA codons and inserts the correct tRNA to match codon and anticodon. Translation continues until the ribosome encounters a stop codon when the protein translation machinery disassembles and the protein is released. Chaperone proteins help the elongating polypeptide to fold properly during translation. A polyribosome is a string of ribosomes all translating the same mRNA.
- Different control factors influence protein synthesis. The best understood is the operon model where binding of a repressor protein to the operon represses transcription.

7.4 Mutations

- Mutations are permanent changes in the cellular DNA. This can occur spontaneously in nature resulting from a replication error or the effects of natural radiation.
- In the laboratory, physical and chemical mutagens can induce mutations. Deletions or insertions involving the chromosome as well as transposons and insertion sequences also may be mutagenic.
- Base pairs in the DNA can change in one of two ways. A base-pair substitution does not change the reading frame of an mRNA but can result in a silent, missense, or nonsense mutation. A point mutation also can occur from the loss or gain of a base pair. Such mutations change the reading frame and often lead to loss of protein function.
- Replication errors or other damage done to the DNA often can be repaired. Mismatch repair replaces an incorrectly matched base pair with the correct pair. Excision repair removes a section of damaged DNA and replaces it with the correctly paired bases.
- Transposable genetic elements exist in many prokaryotes. Insertion sequences only carry information to copy the sequences and insert them into another location in the DNA. Transposons are genetic elements that "jump" from one location in the DNA.

7.5 Identifying Mutants

- Auxotrophic mutants can be identified by negative selection plating techniques. Positive selection can be used to identify mutants having certain attributes, such as antibiotic resistance.
- The Ames test is a method of using an auxotrophic bacterial species to identify mutagens that may be carcinogens in humans. The test is based on the ability of a potential mutagen to revert an auxotrophic mutant to its prototrophic form.

LEARNING OBJECTIVES

After understanding the textbook reading, you should be capable of writing a paragraph that includes the appropriate terms and pertinent information to answer the objective.

1. Summarize the organization of the DNA and chromosome within the nucleoid.
2. Assess the role of plasmids in prokaryotic cells.
3. Identify and explain the events of the three phases of DNA replication.
4. State the central dogma of genetics.
5. Describe the role of RNA polymerase in the transcription process.
6. Read the genetic code and identify start and stop codons.
7. Identify and describe the events occurring in the three stages of translation.
8. Evaluate the effect of antibiotics on the protein synthesis process.
9. Label the sequences composing an operon and compare the functions of each sequence to the transcription process.
10. Identify the spatial relationships between transcription and translation with the nucleoid and cytoplasm.
11. Compare and contrast spontaneous and induced mutations.
12. Differentiate between physical and chemical mutagens.
13. Distinguish between a base-pair substitution and a deletion or insertion mutation.
14. Explain how mismatch repair works.
15. Describe how cells use excision repair to correct UV-induced mutations.
16. Contrast how insertion sequences and transposons can cause mutations.
17. Compare and contrast negative and positive selection plating techniques for identifying mutants.
18. Evaluate the use of the Ames test to identify chemicals that are potential carcinogens in humans.

SELF-TEST

Answer each of the following questions by selecting the ***one*** answer that best fits the question or statement. Answers to even-numbered questions can be found in **Appendix C**.

1. Which one of the following is not *true* of the prokaryotic chromosome?
A. It is located in the nucleoid.
B. It usually is a single, circular molecule.
C. Some genes are dominant to others.
D. Many genes may be clustered together.
E. It usually is haploid.

2. DNA compaction involves
A. nuclear proteins.
B. a twisting and packing of the DNA.
C. supercoiling.
D. the formation of looped domains.
E. All the above **(A–D)** are correct.

3. Plasmids are
A. another name for transposons.
B. accessory genetic information.
C. domains within a chromosome.
D. formed during part of the replication process.
E. daughter chromosomes.

4. The enzyme _____ adds complementary bases to the DNA template strand during replication.
A. ligase
B. helicase
C. DNA polymerase III
D. protease
E. RNA polymerase

5. At a chromosome replication fork, the lagging strand consists of _____ that are joined by _____.
A. plasmid fragments; DNA polymerase
B. RNA sequences; ligase
C. Okazaki fragments; RNA polymerase
D. RNA sequences; ribosomes
E. Okazaki fragments; ligase

6. Another name for transcription would be
A. RNA synthesis.
B. negative control.
C. DNA synthesis.
D. excision repair.
E. polysome formation.

7. RNA polymerase initiates transcription by binding to the
A. operator base sequence.
B. promoter base sequence.
C. *ori*C region.
D. mRNA transcript.
E. Both **A** and **C** are true.

8. The information as to what protein will be synthesized in a cell is carried by the
A. tRNA.
B. promoter.
C. ribosome.
D. mRNA.
E. rRNA.

9. Which one of the following codons would terminate translation?
A. AUG
B. *ter*C
C. UUU
D. UAA
E. *ori*C

10. The translation of a mRNA by multiple ribosomes is called _____ formation.
A. Okazaki
B. polysome
C. plasmid
D. domain
E. transposon

11. Which one of the following is *not* part of an operon? If the answer is all, then select **E.**
A. Regulatory gene
B. Operator
C. Promoter
D. Structural genes
E. All the above **(A–D)** are part.

12. All of the following are examples of chemical mutagens *except:*
A. nitrous acid.
B. base analogs.
C. UV light.
D. 5-bromouracil.
E. acyclovir.

13. UV light causes dimers to form between _____ bases.
A. adenine
B. guanine
C. uracil
D. analog
E. thymine

14. Which one of the following **(A–D)** could *not* cause a change in the mRNA reading frame? If all of them cause a change, then select **E**.
A. Insertion sequence
B. Base-pair substitution
C. Base addition
D. Base deletion
E. All the above **(A–D)** are correct.

15. Which one of the following statements is incorrect with regard to transposable genetic elements?
A. They can be insertion sequences.
B. They were first discovered by Watson and Crick.
C. They may be found in plasmids.
D. Transposons are an example.
E. Some may have information for antibiotic resistance.

QUESTIONS FOR THOUGHT AND DISCUSSION

Answers to even-numbered questions can be found in **Appendix C**.

1. The author of a general biology textbook writes in reference to the development of antibiotic resistance, "The speed at which bacteria reproduce ensures that sooner or later a mutant bacterium will appear that is able to resist the poison." How might this mutant bacterial cell appear? Do you agree with the statement? Does this bode ill for the future use of antibiotics?

2. Many viruses have double-stranded DNA as their genetic information while many others have single-stranded RNA as the genetic material. Which group of viruses do you believe is more likely to efficiently repair its genetic material? Explain.

3. Some scientists suggest that mutation is the single most important event in evolution. Do you agree? Why or why not?

APPLICATIONS

Answers to even-numbered questions can be found in **Appendix C**.

1. You are working for a bioremediation company that wants to develop a bacterial strain that will degrade toxic benzene found in many hazardous waste sites. How would you go about visually identifying bacterial colonies that have this characteristic?

2. A chemical is tested with the Ames test to see if the chemical is mutagenic and therefore possibly a cancer-causing chemical in humans. On the test plate containing the chemical, no his^+ colonies are seen near the central well. However, many colonies are growing some distance from the well. If these colonies truly represent his^+ colonies, why are there no colonies closer to the central well?

3. Bioremediation is a process that uses prokaryotes to degrade environmental pollutants. You want to use one of these organisms to clean up a toxic waste site (biochemical refinery) that contains benzene in the soil around the refinery. Benzene (molecular formula = C_6H_6) is a major contaminant found around many of these chemical refineries. You have a culture of bacterial cells growing on a culture plate that were derived from a soil sample from the refinery area. You also have a supply of benzene. Explain how you could visually identify chemoheterotrophic bacterial colonies on agar that use benzene as their sole carbon and energy source for metabolism.

4. Suppose you now have such benzene colonies growing on agar. However, it also has been discovered that material containing radioactive phosphorus (^{32}P) is in the soil around the refinery and this radioactive material can kill bacterial organisms. Since you want to identify colonies that might be sensitive to ^{32}P, you obtain a sample of the material containing ^{32}P. Explain how you could visually determine if any of your colonies are sensitive to radioactive phosphorus (^{32}P).

REVIEW

Answer the following questions that pertain to (1) transcription and translation, and (2) mutations. Use the genetic code (Table 7.3) on page 200 as needed. Answers to even-numbered questions can be found in **Appendix C**.

1. The following base sequence is a complete polynucleotide made in a bacterial cell.

   AUGGCGAUAGUUAAACCCGGAGGGUGA

   With this sequence, answer the following questions.
   - **A.** Provide the sequence of nucleotide bases found in the inactive DNA strand of the gene.
   - **B.** How many codons will be transcribed in the mRNA made from the template DNA strand?
   - **C.** How many amino acids are coded by the mRNA made and what are the specific amino acids?
   - **D.** Why isn't the number of codons in the template DNA the same as the number of amino acids in the polypeptide?

2. Use the sequence of bases in the box to answer the following questions about mutations in prokaryotes.

   TACACGATGGTTTTGAAGTTACGTATT

   - **A.** Is the sequence in the box a single strand of DNA or RNA? Why?
   - **B.** Using the sequence in the box, show the translation result if a mutation results in a C replacing the T at base 12 from the left end of the sequence. Is this an example of a silent, missense, or nonsense mutation?
   - **C.** Using the sequence in the box, show the translation result if a mutation results in an A inserted between the T (base 12) and the T (base 13) from the left end of the sequence. Is this an example of a silent, missense, or nonsense mutation?

HTTP://MICROBIOLOGY.JBPUB.COM/

The site features learning, an on-line review area that provides quizzes and other tools to help you study for your class. You can also follow useful links for in-depth information, read more MicroFocus stories, or just find out the latest microbiology news.

8

Chapter Preview and Key Concepts

8.1 Genetic Recombination in Prokaryotes
- Prokaryotic genes can be transferred between generations and between organisms of the same generation.
- Transformation involves the transfer of exposed DNA fragments from donor to recipient.
- Conjugation uses plasmids for DNA transfer from donor to recipient.
- To transfer chromosomal DNA, conjugation requires chromosomally-integrated plasmids.
- On occasion, independently replicating bacterial viruses can transfer chromosomal DNA from donor to recipient.

8.2 Genetic Engineering and Biotechnology
- Plasmids can be spliced open and a gene of interest inserted to form a recombinant DNA molecule.
- Prokaryotes can be transformed by the acquisition of a human gene.

MicroInquiry 8: Molecular Cloning of a Human Gene into Bacterial Cells

8.3 Microbial Genomics
- Genome sequences are rapidly expanding for many prokaryotes and eukaryotes.
- Some human genes and human DNA sequences may have prokaryotic origins.
- Understanding gene organization and function provides for new microbial applications.
- Comparative genomics compares the DNA sequences of related or unrelated species.

Gene Transfer, Genetic Engineering, and Genomics

Genetic engineering is the most powerful and awesome skill acquired by man since the splitting of the atom.
—The editors of *Time* magazine describing the potential for genetic engineering

Medicinal microbe's genome sequenced. How often have we read in the newspaper or heard from the news media about an organism's genes (genome) being sequenced or its DNA being mapped? It must be significant, right? After all it made the news! But what is the underlying significance of such sequencing? Here is a good example.

In late spring of 2003, a group of British scientists announced they had mapped the genome of a very important bacterial species, *Streptomyces coelicolor.* This is a gram-positive organism commonly found in the soil. The mapping project began in 1997 and took six years to complete partly because the organism's genome is one of the largest ever sequenced. It has 8.6 million base pairs and some 7,825 genes. On completion of the sequencing, one of the scientists on the project said, "*It is a fabulous resource for scientists.*" Why?

Well, here is where microbial genomics shows its power. *S. coelicolor* belongs to a family of *Bacteria*, the streptomycetes, which are responsible for producing over 65 percent of the naturally-known antibiotics used today (see Chapter 4). This includes tetracycline and erythromycin. By analyzing the genome of *S. coelicolor* and other *Streptomyces* species, additional metabolic pathways may be discovered for the production of other, yet unidentified and perhaps novel antibiotics. In fact, the researchers have identified 18 gene clusters they suspect are involved with the production of antibiotics. If correct, knowing the genome and its organization might allow scientists to transform the organism into an "antibiotic factory" and add to the dwindling armada of usable antibiotics to which prokaryotes are not yet resistant (FIGURE 8.1). Using genetic engineering techniques, they could rearrange gene clusters and perhaps produce even more useful and potent antibiotics than presently available.

One example illustrating the need for newer antibiotics concerns "staph infections", most commonly caused by **methicillin-resistant *Staphylococcus aureus*** (MRSA). These pathogens are resistant to methicillin and other common antibiotics, such as oxacillin, penicillin, and amoxicillin. MRSA infections occur most frequently among persons in hospitals and healthcare facilities who have weakened immune systems (TABLE 8.1). According to MRSAInfection.org, "a patient with a hospital acquired [MRSA] infection is about 7 times more likely to die than an uninfected patient." So, new and unique antibiotics might be able to attack and inhibit or kill the drug resistant *S. aureus*.

FIGURE 8.1 ***Streptomyces coelicolor* Colonies.** In this photograph of *S. coelicolor*, colonies growing on agar are secreting droplets of liquid containing antibiotics.

Q: What is the advantage to the bacterial cells to secrete a chemical with antibiotic properties?

But there is more. *S. coelicolor* is a close relative of the tuberculosis, leprosy, and diphtheria bacilli. By comparing genomes, scientists hope to learn why. *S. coelicolor* are not pathogenic, while the other three are pathogens. What is different about their genomes might be important in understanding the pathogenicity of their relatives and perhaps even designing new antibiotics through genetic engineering to attack these pathogens.

Genetic engineering involves the manipulating of genes in organisms or between organisms in order to introduce new characteristics into the recipient to either produce a useful product or to actually generate **genetically modified organisms** (GMOs). **Genomics** is the study of an organism's genome; its study has the potential of offering new therapeutic methods for the treatment of several human diseases. So, this chapter addresses more than simply research techniques; rather, we discuss how their applications can have far-reaching consequences for all of us.

Before we can explore these topics, we need to understand the process of **genetic recombination**, a natural mechanism for DNA transfer from one microorganism to another. Its understanding provides a unique perspective on prokaryotic evolution, microbial ecology, and molecular biology while providing insights for the techniques of genetic engineering and the field of microbial genomics.

TABLE 8.1 Hospital-Acquired Infection Statistics, Pennsylvania—2004*

Number of hospital-acquired infections	11,668
Number of deaths associated with hospital-acquired infections	1,510
Extra number of hospital days associated with these infections	205,000
Additional hospital charges	$2 billion

*Data are from: MRSAInfection.org

8.1 Genetic Recombination in Prokaryotes

Traditionally, when one thinks about the inheritance of genetic information, one envisions genes passed from parent to offspring. However, imagine being able to transfer genes between members of your own family, or between you and one of your classmates. Prokaryotes are accomplished at doing both types of information transfer.

Genetic Information in Prokaryotes Can Be Transferred Vertically and Horizontally

KEY CONCEPT

- Prokaryotic genes can be transferred between generations and between organisms of the same generation.

In Chapter 7, we discussed mutations, which were one of the ways by which the genetic material in a cell can be permanently altered. Because the permanent change occurred in the parent cell, all future generations derived by binary fission from the parent also will have the mutation. This form of genetic transfer is referred to as **vertical gene transfer** (FIGURE 8.2A).

Prokaryotes lack sexual reproduction as a mechanism for genetic diversity. However, they still possess a process by which genetic alterations and diversity can arise. This is through the process of **horizontal gene transfer** (HGT), which involves the intercellular transfer of DNA from a donor cell to a recipient cell (FIGURE 8.2B). If, for example, the recipient cell receives from the donor a chromosomal DNA fragment, a plasmid, a transposon (see Chapter 7), or some combination of these elements containing a gene for antibiotic resistance, the new DNA pairs with a complementary region of recipient DNA and replaces it. In this case, there is no change in quantity of the recipient's chromosomal DNA, but there is a substantial change in its quality and cell physiology—an antibiotic sensitive cell has become resistant. In fact, the increasing global resistance

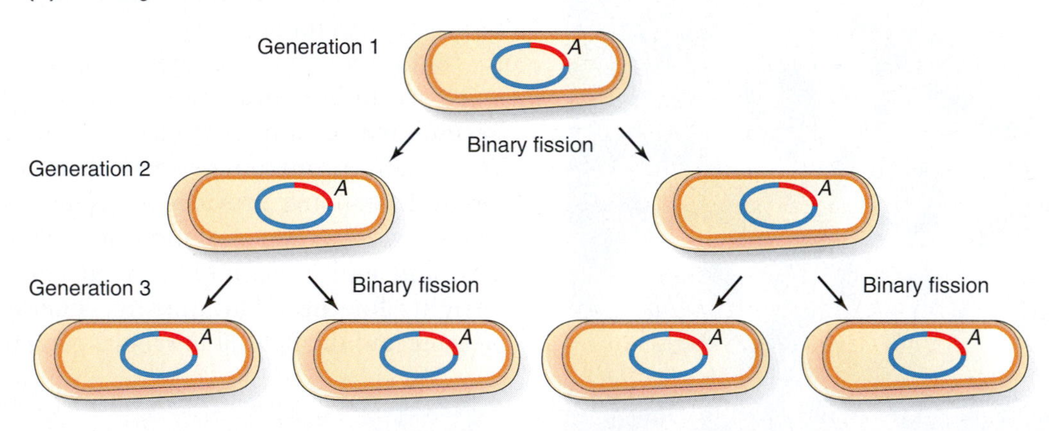

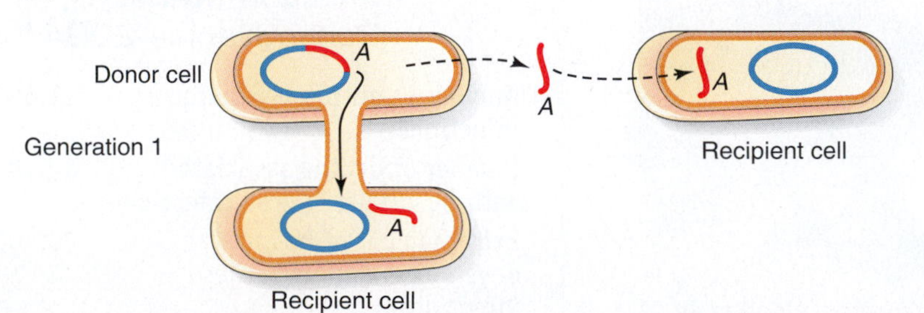

FIGURE 8.2 **Gene Transfer Mechanisms.** Genes can be transferred between prokaryotes in two ways. (**A**) In vertical transfer, a cell undergoes binary fission and the daughter cells of the next generation contain the identical genes found in the parent. (**B**) In horizontal transfer, genes are transferred by various mechanisms to other cells of the same generation.

Q: Determine which transfer mechanism (vertical or horizontal) provides the potential for the most genetic diversity?

MICROFOCUS 8.1: Environmental Microbiology

Gene Swapping in the World's Oceans

Many of us are familiar with the accounts of microorganisms in and around us, but we are less familiar with the massive numbers of microbes in the world's oceans. For example, microbial ecologists estimate there are an estimated 10^{29} prokaryotes in the world's oceans. Also, there are some 10^{30} viruses called bacteriophages ("bacteria eaters") in the oceans that infect these oceanic prokaryotes.

In the infection process, sometimes bacteriophages by mistake carry pieces of the prokaryotic chromosome (rather than viral DNA) from the infected cell to another recipient cell. In the recipient cell, the new DNA fragment can be swapped for an existing part of the recipient's chromosome. It is a fairly rare event, occurring only once in every 100 million (10^8) virus infections. That doesn't seem very significant until you now consider the number of bacteriophages and susceptible prokaryotes existing in the oceans. Working with these numbers and the potential number of virus infections, scientists suggest that if only one in every 100 million infections brings a fragment of prokaryotic DNA to a recipient cell, there are about 10 million billion (that's 10,000,000,000,000,000 or 10^{16}) such gene transfers per second in the world's oceans. That is about 10^{21} infections per day!

We do not understand what all this recombination means. What we can conclude is there's an awful lot of gene swapping going on!

to antibiotics by human pathogens, such as *Streptococcus pneumoniae, Staphylococcus aureus,* and *Pseudomonas aeruginosa,* is one example of the prevalence of HGT. MicroFocus 8.1 provides another spectacular example for a very extensive rate for genetic recombination through HGT in the world's oceans.

Three distinctive mechanisms mediate the horizontal transfer of DNA between cells: transformation, conjugation, and transduction. All three processes involve a similar four steps. The donor DNA must be:

1. Readied for transfer.
2. Transferred to the recipient cell.
3. Taken up successfully by the recipient cell.
4. Incorporated in a stable state in the recipient.

Let's now look at each of the three recombination mechanisms.

CONCEPT AND REASONING CHECKS

8.1 How does HGT differ from vertical gene transfer?

Transformation Is the Uptake and Expression of DNA in a Recipient Cell

KEY CONCEPT

- Transformation involves the transfer of exposed DNA fragments from donor to recipient.

Transformation is the uptake of a free DNA fragment from the surrounding environment and the expression of the genetic information in the recipient cell; that is, by integration of the DNA fragment, the recipient (transformant) has gained some ability it previously lacked. Today, transformation represents a gene transfer process building genetic diversity in a prokaryotic population.

This form of HGT was the first to be described in prokaryotes. In 1928, the English bacteriologist Frederick Griffith published the results of an interesting set of experiments with *S. pneumoniae*. This bacterial species, referred to as a pneumococcus (pl., pneumococci), is a major cause of pneumonia (Chapter 9). Pneumococci occur in two different strains: a wild-type encapsulated strain, designated S, because the organisms grow in smooth colonies and cause pneumonia; and an unencapsulated mutant strain, designated R, because the colonies appear rough and are harmless. Griffith showed that mice injected with living S strain die, while those mice injected with living R strain or heat-killed S strain survive (FIGURE 8.3).

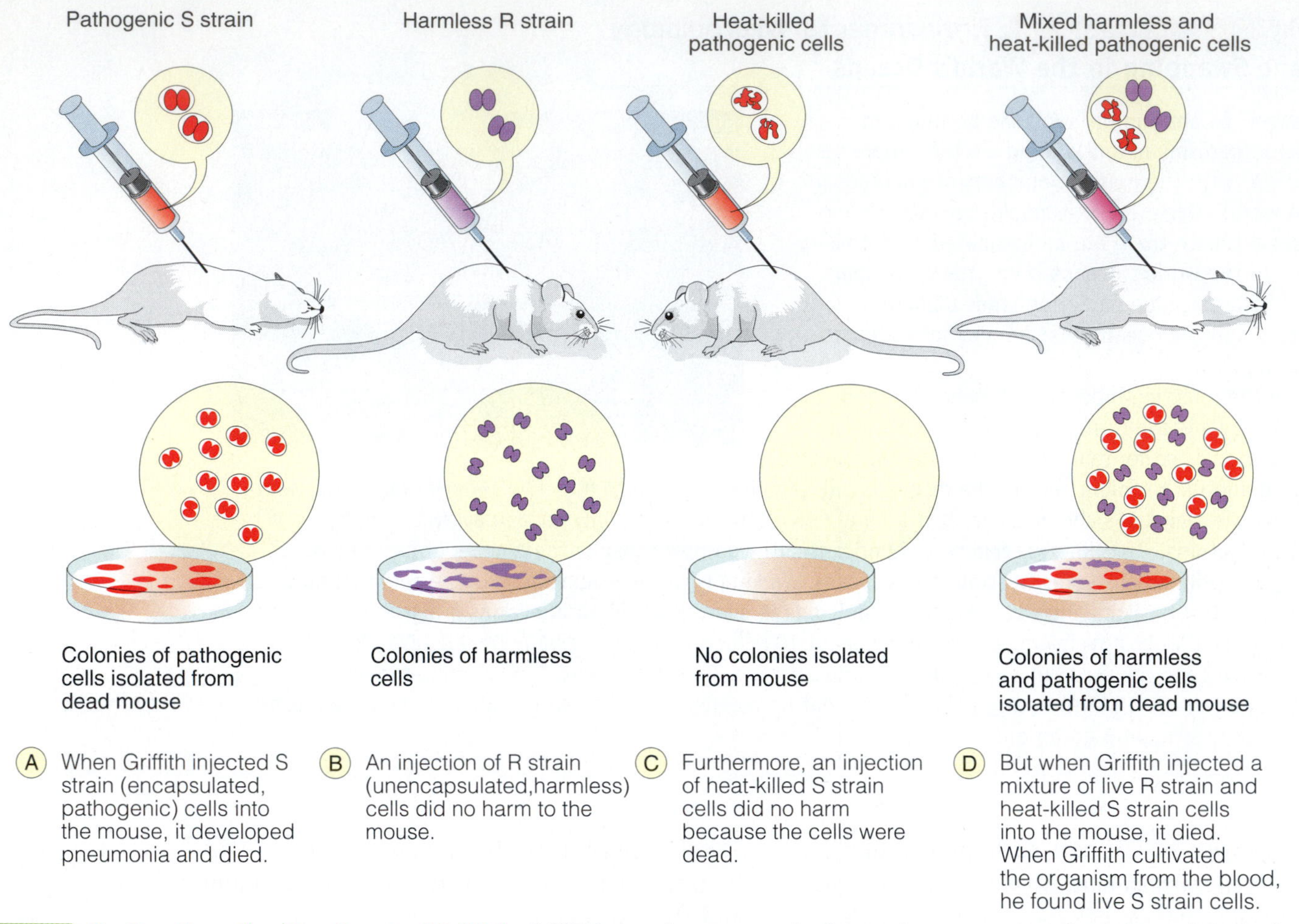

FIGURE 8.3 **The Transformation Experiments of Griffith.** Griffith's experiments were the first to demonstrate transformation and the horizontal transfer of genetic information.

Q: What is unique about the S strain pneumococci that is responsible, in part, for making them pathogenic?

What happened next surprised Griffith. He mixed heat-killed, S strain cells with live R strain cells (both of which were non-lethal) and let the mixture incubate; then he injected the mixture into mice. The mice died! Griffith wondered how a mixture of live harmless R strain cells and debris from the dead pathogenic S strain cells could kill the mice. His answer came when he autopsied the animals: microscopic examination of their blood showed the presence of live S strain pneumococci. Knowing that spontaneous generation does not occur, Griffith reasoned that somehow the live R strain cells had been transformed into live S strain cells. Though he could not explain how this happened, he published his results but never tried to understand the nature of the transforming agent.

However, in 1944 Oswald T. Avery and his associates Colin M. MacLeod and Maclyn N. McCarty of the Rockefeller Institute, purified and identified the transforming substance; it was DNA.

Modern scientists regard transformation as an important genetic recombination method even though it takes place in less than 1 percent of a prokaryotic cell population. Transformation occurs regularly when prokaryotes exist in crowded conditions, such as in rich soil or the human intestinal tract. Under natural conditions, about 40 known prokaryotic species are highly transformable when their DNA is very similar to the DNA being received.

The ability of a cell to be transformed depends on its **competence**, which refers to the ability of a recipient cell to take up DNA from the environment. Competence is an intriguing but variable property among prokaryotes. For example, growing cells of *S. pneumoniae* secrete a competence factor to induce the competence state, while *Haemophilus influenzae* cells become competent when the culture is switched from a rich to a minimal growth medium. In both cases, several genes encode proteins for binding and uptake of DNA fragments. In fact, in *Bacillus* cells more than 16 different proteins function in DNA fragment uptake.

The transformation process occurs as follows. In natural environments, when prokaryotic cells die and lyse, the chromosome typically breaks apart into fragments of DNA composed of about 10 to 20 genes (FIGURE 8.4). Uptake of DNA fragments appears to occur only at the cell poles, as this is where the competence factors are found and actual DNA uptake has been observed.

A competent cell usually incorporates only one or at most a few DNA fragments. Internalization of DNA is an ATP-dependent process and requires DNA-binding proteins, cell wall degradation proteins, and cell membrane transport proteins. Bacterial species such as *S. pneumoniae* and *Bacillus subtilis*, degrade one strand of the DNA fragment as it is being taken into the cell. Such single-stranded DNA, if stably associated with a similar region of the recipient's chromosome, will replace a similar chromosome sequence. In *H. influenzae*, a double-stranded DNA fragment is internalized, but then one strand is digested by a nuclease before incorporation into the bacterial chromosome.

One potential effect of transformation is to increase an organism's

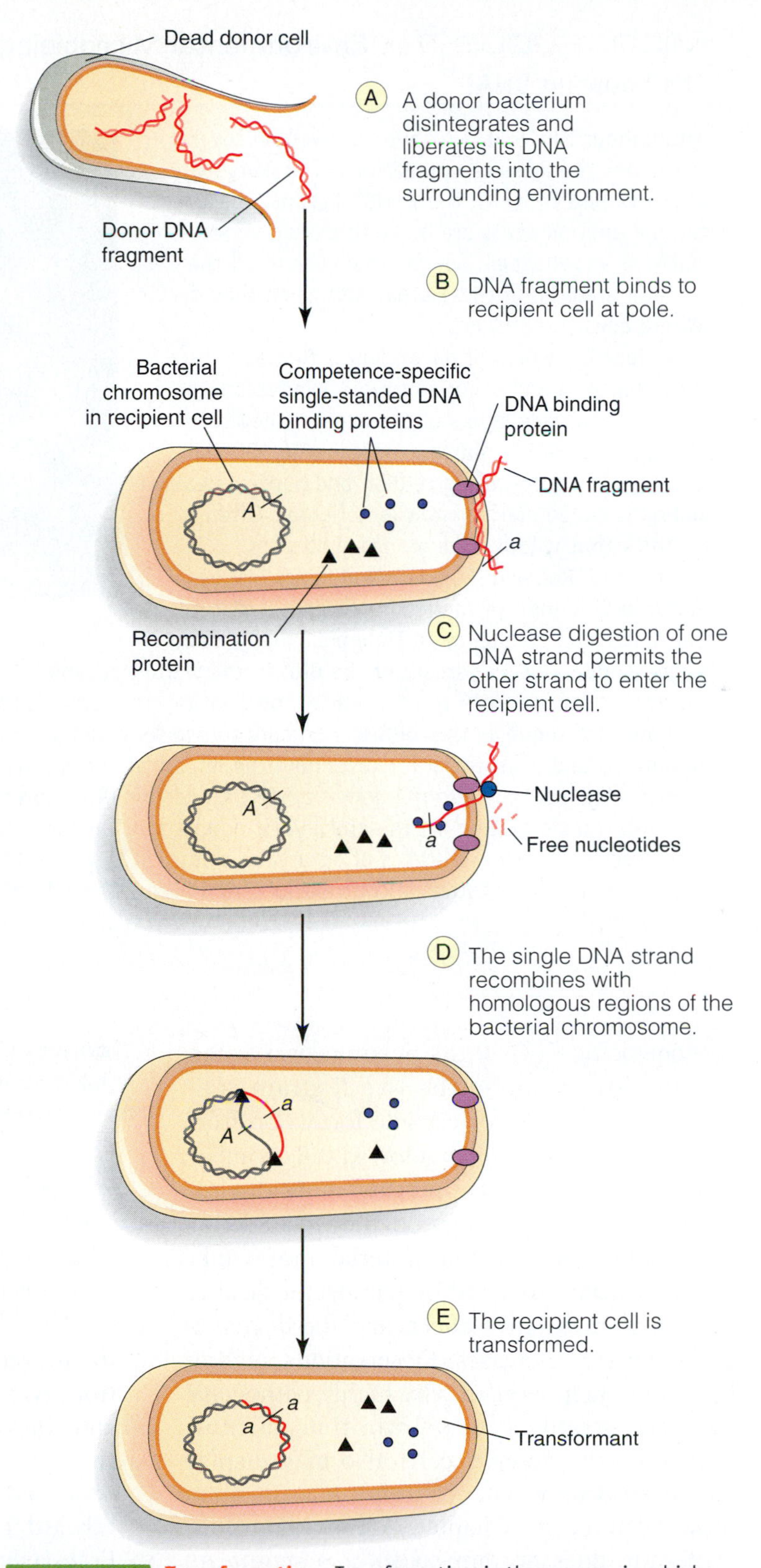

FIGURE 8.4 **Transformation.** Transformation is the process in which DNA fragments from the environment bind to a competent recipient cell, pass into the recipient, and incorporate into the recipient's chromosome.

Q: Using this diagram, illustrate how Griffith's experiment with mixed heat-killed S strain and live R stain pneumococci produced live S strain encapsulated cells.

MICROFOCUS 8.2: Environmental Microbiology

It's Snowing DNA!

Throughout this text, we have observed on several occasions the massive populations of prokaryotes that live and thrive in the world's oceans. These marine environments are home to a huge variety of *Bacteria* and *Archaea*. So what happens to all the genetic material in these organisms when they die? Where does the DNA go?

A decade's worth of data suggest that as cyanobacteria and other eukaryotic phytoplankton die in the surface layers of the oceans, they sink through the water column at rates up to 100 meters per day. This free-falling cellular and particulate debris is called marine snow (see figure). Often it is so thick that it looks like an ocean blizzard!

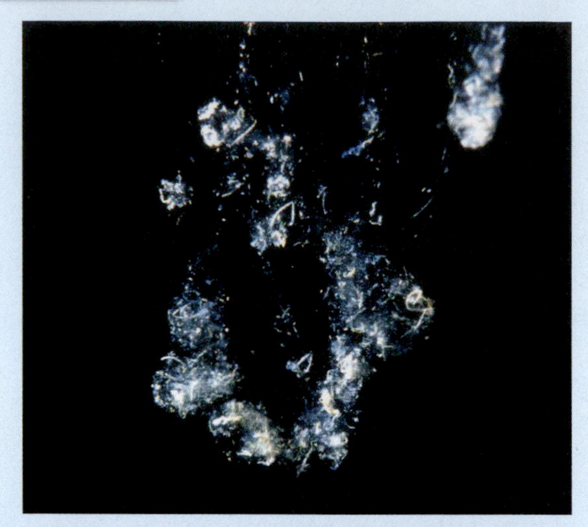

Marine snow represents microbial cellular debris.

In 2005, Roberto Danovaro and his colleague Antonio Dell'Anno of the Marine Science Institute of the University of Ancona in Italy published a paper suggesting about 65 percent of the DNA in the world's oceans is found in the sea-bed sediments; 90 percent of that DNA is extracellular, originating from organisms in the marine snow.

Danovaro suggests this DNA is a primary source for carbon, nitrogen, and phosphorus for sea-bed organisms and is essential for sustaining the microbial communities on and in the sediments.

But what about horizontal gene transfer? Could this DNA represent a source of genetic diversity through transformation for the prokaryotic populations on the ocean bottom? Danovaro hypothesizes that most of the snowy DNA is of eukaryotic origin. Still, with all the marine prokaryotes, why couldn't they be making a substantial contribution? And if so, is transformation a significant phenomenon on the sea beds?

■ Pathogenicity: The ability of a pathogen to cause disease.

pathogenicity. In Griffith's pneumococci experiments, for example, the live R strain cells acquired the genes for capsule formation from the dead S strains, which allowed the organism to avoid body defenses and thus cause disease. Microbiologists also have demonstrated that when mildly pathogenic bacterial strains take up DNA from other mildly pathogenic strains, there is a cumulative effect, and the degree of pathogenicity increases. Observations such as these may help explain why highly pathogenic bacterial strains appear from time to time. Transformed bacterial cells also may display enhanced drug resistance from the acquisition of R plasmids (see Chapter 7). However, transformation does not appear to be a significant contributor to the dispersal of antibiotic-resistance genes.

Extracellular DNA can be quite prevalent. MicroFocus 8.2 identifies a massive source of such "dead" DNA.

CONCEPT AND REASONING CHECKS

8.2 Why is competence key to the transformation process?

Conjugation Involves Cell-to-Cell Contact for Horizontal Gene Transfer

KEY CONCEPT

- Conjugation uses plasmids for DNA transfer from donor to recipient.

In the recombination process called **conjugation**, two live prokaryotic cells come together and the donor cell directly transfers DNA to the recipient cell. This process was first observed in 1946 by Joshua Lederberg and Edward Tatum in a series of experiments with *Escherichia coli*. Lederberg and Tatum mixed two different strains of *E. coli* and found genetic traits could be transferred if contact occurred.

The process of conjugation requires a special conjugation apparatus called the

conjugation pilus that was first described in Chapter 4 (FIGURE 8.5). For cell-to-cell contact, the donor cell, designated F^+, produces the conjugation pilus that contacts the recipient cell, known as an F^- cell. The donor cell is called F^+ because it contains an **F factor**, which is a plasmid that contains about 100 genes, most of which are associated with plasmid DNA replication and production of the conjugation pilus (See Chapter 7). The F^- cell lacks the plasmid. Once contact is made, the pilus shortens to bring the two cells close together.

Following pilus formation (FIGURE 8.6A), the F factor DNA replicates by a **rolling-circle mechanism**; one strand of plasmid DNA remains in a closed loop, while an enzyme nicks the other strand at a point called the **origin of transfer** (*ori*T). This single-stranded DNA then "rolls off" the loop and passes through the channel to the recipient cell. As the horizontal transfer occurs, DNA synthesis in the donor cell produces a new complementary strand to replace the transferred strand. Once DNA transfer is complete, the two cells separate.

In the recipient cell, the new single-stranded DNA serves as a template for synthesis of a complementary polynucleotide strand, which then circularizes to reform an F factor. This completes the conversion of the recipient from F^- to F^+ and this cell now represents a donor cell (F^+) capable of conjugating with another F^- recipient. Transfer of the F factor does not involve the chromosome; therefore, the recipient does not acquire new genes other than those on the F factor.

The high efficiency of DNA transfer by conjugation shows that conjugative plasmids can spread rapidly, converting a whole population into plasmid-containing cells. Indeed, conjugation appears to be the major mechanism for antibiotic resistance transfer. In laboratory experiments, for example, prokaryotic strains carrying plasmid antibiotic-resistance genes were introduced into mice. HGT through conjugation of these R factors rapidly occurred. In nature, conjugation readily occurs between prokaryotes in soil and in water.

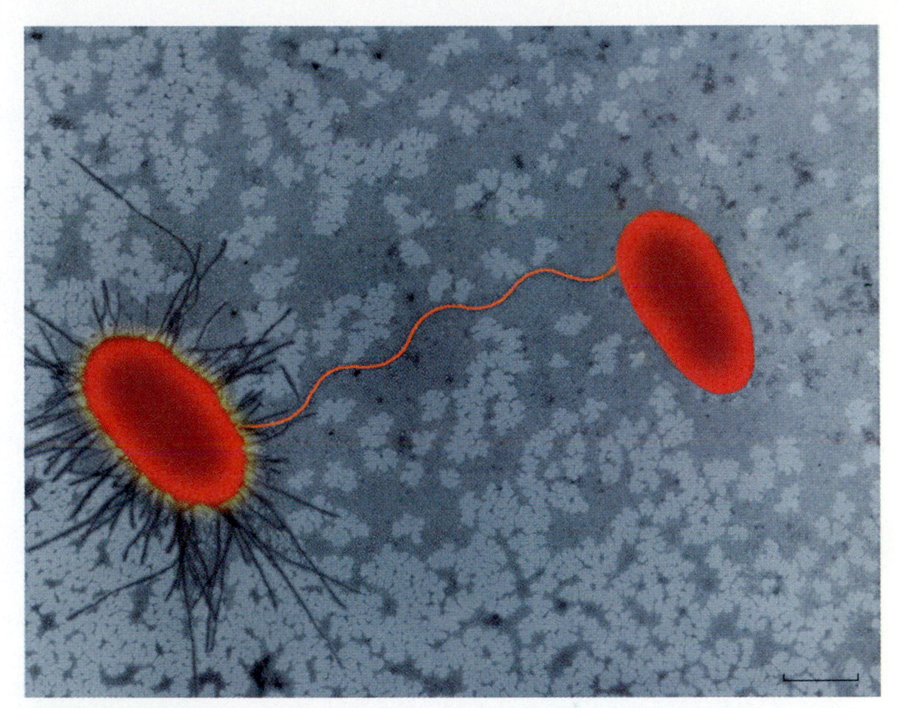

FIGURE 8.5 **Bacterial Conjugation in *E. coli*.** The direct transfer of DNA between live donor (F^+; left cell) and recipient (F^-) cells requires a conjugation pilus. In this false-color transmission electron micrograph, the F^+ cell has produced a conjugation pilus that has contacted the F^- cell. (Bar = 0.5 μm.)

Q: What are the other structures projecting from the F^+ cell?

CONCEPT AND REASONING CHECKS

8.3 What genes must be transferred to an F^- cell to convert it to F^+?

Conjugation also Can Transfer Chromosomal DNA

KEY CONCEPT

- To transfer chromosomal DNA, conjugation requires chromosomally-integrated plasmids.

Prokaryotes also can undergo a type of conjugation that accounts for the horizontal transfer of chromosomal material from donor to recipient cell. Prokaryotes exhibiting the ability to donate chromosomal genes are called **high frequency of recombination** (**Hfr**) strains.

In Hfr strains, the F factor attaches to the chromosome (FIGURE 8.6B). This attachment is a rare event requiring an insertion sequence to recognize the F factor. Once incorporated into the chromosomal DNA, the F factor no longer controls its own replication. However, the Hfr cell triggers conjugation just like an F^+ cell. When a recipient cell is present, a conjugation pilus forms and attaches to the F^- cell. The two cells are brought together. One strand of the donor chromosome is nicked at *ori*T

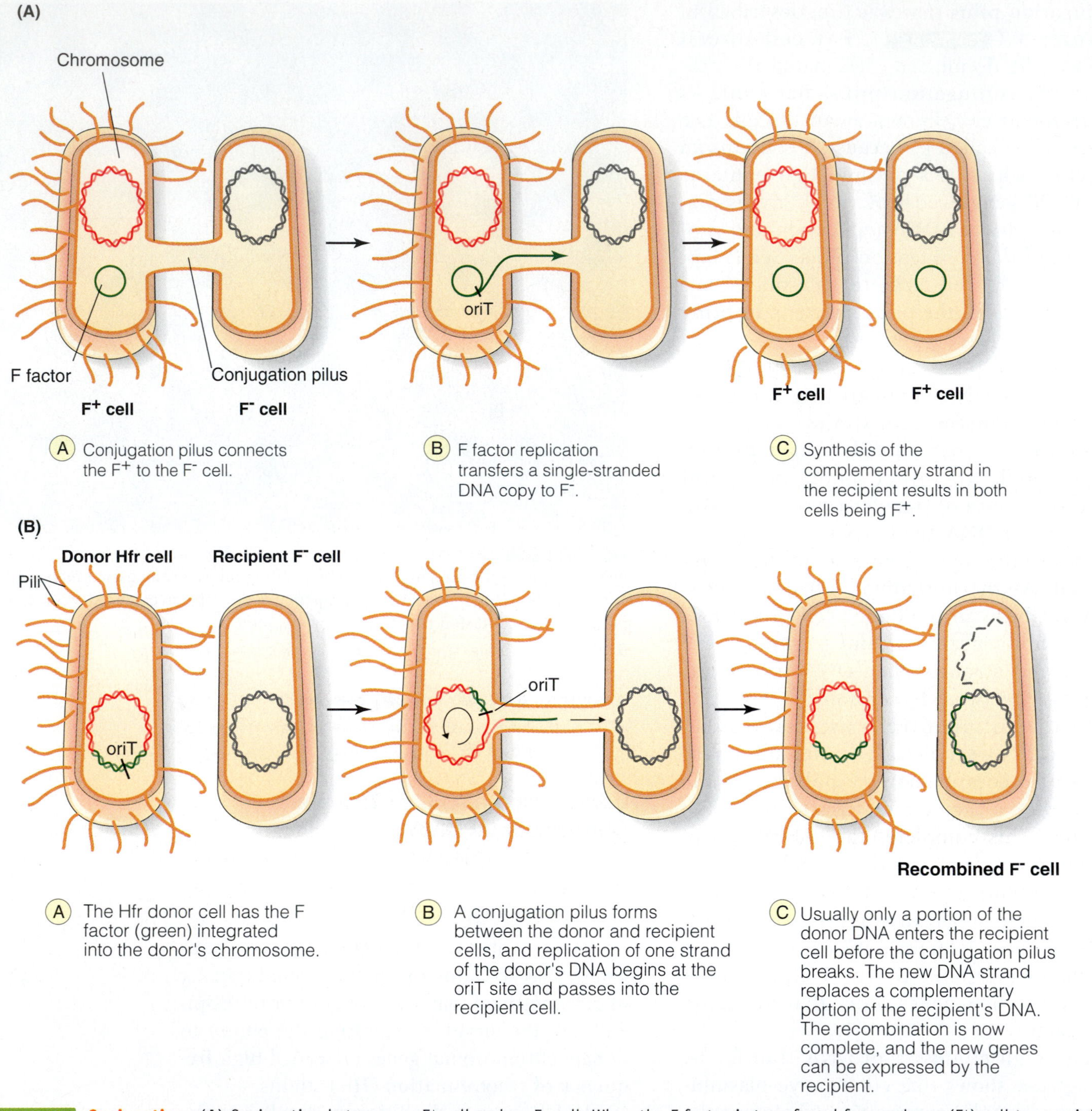

FIGURE 8.6 **Conjugation.** (**A**) Conjugation between an F$^+$ cell and an F$^-$ cell. When the F factor is transferred from a donor (F$^+$) cell to a recipient (F$^-$) cell, the F$^-$ cell becomes an F$^+$ cell as a result of the presence of an F factor. (**B**) Conjugation between an Hfr and an F$^-$ cell allows for the transfer of some chromosomal DNA from donor to recipient cell.

Q: Propose a hypothesis to explain why only a single-stranded DNA molecule is transferred across the conjugation pilus.

and a portion of the single-stranded chromosomal/plasmid DNA then passes into the recipient cell.

The *ori*T site in the chromosome is in the middle of the F plasmid genes. Therefore, in the recipient cell, the first genes to enter are only a part of the F factor and these genes are not the ones to make the cell an F$^+$ donor. Rather, the last genes transferred to the recipient cell would control the donor state. However, these rarely enter the recipient because conjugation usually is interrupted by movements that break the bridge between cells before complete transfer is accomplished. An estimated 100 minutes is required for the transfer of a complete *E. coli* chromosome with plasmid genes—something rarely occurring in nature. Thus, the F^- cell usually remains a recipient, although it now has some recombined chromosomal genes, and is referred to as a **recombinant F^-** cell.

Should the entire chromosome be transferred to the recipient, the F factor usually detaches from the chromosome, and enzymes synthesize a strand of complementary DNA. The F factor now forms a loop to assume an existence as a plasmid, and the recipient becomes a donor (F^+) cell.

Occasionally, in an Hfr cell, the integrated F plasmid breaks free from the chromosome and, in the process, takes along a fragment of chromosomal DNA. The plasmid with its extra DNA is now called an **F' plasmid** (pronounced "F-prime"). When the F' plasmid is transferred during a subsequent conjugation, the recipient acquires those chromosomal genes excised from the donor. This process results in a recipient having its own genes for a particular process as well as additional genes from the plasmid DNA for the same process. In the genetic sense, the recipient is a partially diploid organism because there are two genes for a given function.

Conjugation has been demonstrated to occur between cells of various genera of prokaryotes. For example, conjugation occurs between such gram-negative genera as *Escherichia* and *Shigella, Salmonella* and *Serratia,* and *Escherichia* and *Salmonella.* HGT has great significance because of the possible transfer of antibiotic-resistance genes carried on plasmids. Moreover, when the genes are attached to transposons, the transposons may "jump" from ordinary plasmids to F factors, after which transfer by conjugation may occur. (MicroFocus 8.3 describes one case with serious medical overtones.)

Although conjugative pili are found only on some gram-negative bacteria, gram-positive bacteria also appear capable of conjugation. Microbiologists have experimented extensively with *Streptococcus mutans,* a common cause of dental caries. In this organism, conjugation appears to involve only plasmids, particularly those carrying genes for antibiotic resistance. Moreover, the conjugation does not involve pili. Rather, the recipient cell apparently secretes substances encouraging the donor cell to produce clumping factors composed of protein. The factors bring together (clump) the donor and recipient cell, and pores form between the cells to permit plasmid transfer. Chromosomal transfer has not been demonstrated.

CONCEPT AND REASONING CHECKS

8.4 Unlike conjugation between an F^+ and F^-, where the receipient cell becomes F^+, why doesn't an Hfr and F^- conjugation result in the recipient cell being an Hfr?

Transduction Involves Viruses as Agents for Horizontal Transfer of DNA

KEY CONCEPT

- On occasion, independently replicating bacterial viruses can transfer chromosomal DNA from donor to recipient.

Another form of genetic recombination was reported in 1952 by Joshua Lederberg and Norton Zinder. While working with mutant cells of *Salmonella*, Lederberg and Zinder observed recombination, but ruled out conjugation and transformation because the cells were separated by a thin membrane and DNA was absent in the extracellular fluid. Eventually, they discovered a virus in the fluid and uncovered the details of recombination.

Transduction is the third form of HGT and it requires a virus to carry a chromosomal DNA fragment from donor to recipient cell. The virus participating in transduction is called a **bacteriophage** (literally "bacteria eater") or simply **phage**. As with all viruses, the bacteriophages have a core of DNA or RNA surrounded by a coat of protein (Chapter 13).

In the replication cycle of a bacteriophage, different phages can interact with prokaryotes in one of two different ways (FIGURE 8.7). In a **lytic cycle**, the phage DNA penetrates the cell, destroys the host chromosome, replicates itself within the cell, and then destroys (lyses) the cell as new phages are released. Because the phages killed the cell, they are called **virulent phages**.

MICROFOCUS 8.3: Public Health

Staph Detectives and Jumping Genes

It was crazy—local and CDC health officials seemed to be everywhere. They were testing people and rushing off with the samples. What was happening?

In the summer of 2002, the inevitable happened. A diabetes patient, while in a Detroit dialysis center, complained of foot ulcers. When the staff took a culture from the ulcer and had it analyzed, a new strain of *Staphylococcus aureus* was identified. Why the concern? The bacterial strain was resistant to vancomycin, one of the last resort antibiotics in the medical arsenal that was effective in killing staph. Luckily, none of the more than 300 individuals who had come in contact with the dialysis patient were positive for the vancomycin-resistant *S. aureus* (VRSA) strain.

Although *S. aureus* occurs commonly in the nose and skin of healthy people, it seldom causes a problem. In hospital patients though, whose immune systems are weak, it can be life-threatening. And if the pathogen is resistant to vancomycin, not much is left to help VRSA-infected patients.

So how did VRSA arise in the dialysis patient? Additional samples taken from the patient showed that two *S. aureus* strains were present. One was resistant to most antibiotics in common use while the other, from the foot ulcer, was resistant to all those antibiotics and to vancomycin. Somehow the first strain must have "picked up" the gene for vancomycin resistance, accounting for the second strain found in the skin ulcer of the patient.

Further detective work isolated vancomycin-resistant *Enterococcus faecalis,* another species common to hospitals, from the patient's ulcer. Doing genetic analysis of all three organisms showed only *E. faecalis* and the VRSA stain had a plasmid with a gene for vancomycin resistance. The conclusion: the plasmid jumped species from *E. faecalis* to *S. aureus.*

The most interesting detective work concerned how it jumped species. The plasmid of *E. faecalis* included a transposon (see Chapter 7), which contained the vancomycin resistant gene. But the VRSA strain had the transposon with resistance gene on a plasmid different from that in *E. faecalis.* Explanation? In the patient's foot ulcer, a conjugation process occurred whereby *E. faecalis* transferred the vancomycin resistance gene as a transposon-incorporated plasmid to *S. aureus.* After transfer, the transposon with the resistance gene must have "jumped" out of the plasmid and integrated itself in an *S. aureus* plasmid before the *S. aureus* cell destroyed the *E. faecalis* plasmid.

How the *E. faecalis* cell became resistant in the first place was not identified.

Other phages interact with prokaryotic cells in a slightly different way, called a **lysogenic cycle**. These phages also invade the host but without directly causing cell lysis. Instead, the phage DNA integrates into the host chromosome as a **prophage** and the phages participating in this cycle are known as **temperate phages**. The host cell survives and as it undergoes DNA replication and binary fission, the prophage is copied and vertically transferred to daughter cells as well. However, eventually the prophage will excise itself and go through a lytic cycle (Chapter 13).

Let's examine how these two types of phage can transfer fragments of donor cell DNA to a recipient cell.

Generalized transduction is carried out by virulent phages, such as the P1 phage that infects *E. coli* (FIGURE 8.8). After injecting the DNA into the cell, the host cell's chromosome is digested into small fragments. When the new phage DNA is produced, the DNA normally is packaged into new phage particles. However, on rare occasions (1 in 100,000,000) a random (general) fragment of host cell DNA may accidentally be captured in the packaging process and end up in a phage head rather than phage DNA. These phages are fully formed though and they can infect another cell. However, they are defective because they carry no phage genes and cannot replicate themselves after infection. This is the type of genetic recombination described in MicroFocus 8.1.

Following release from the lysed host cell, the transducing phage attaches to a new (recipient) cell and injects its chromosomal DNA into the host cell. Once in the recipient, new

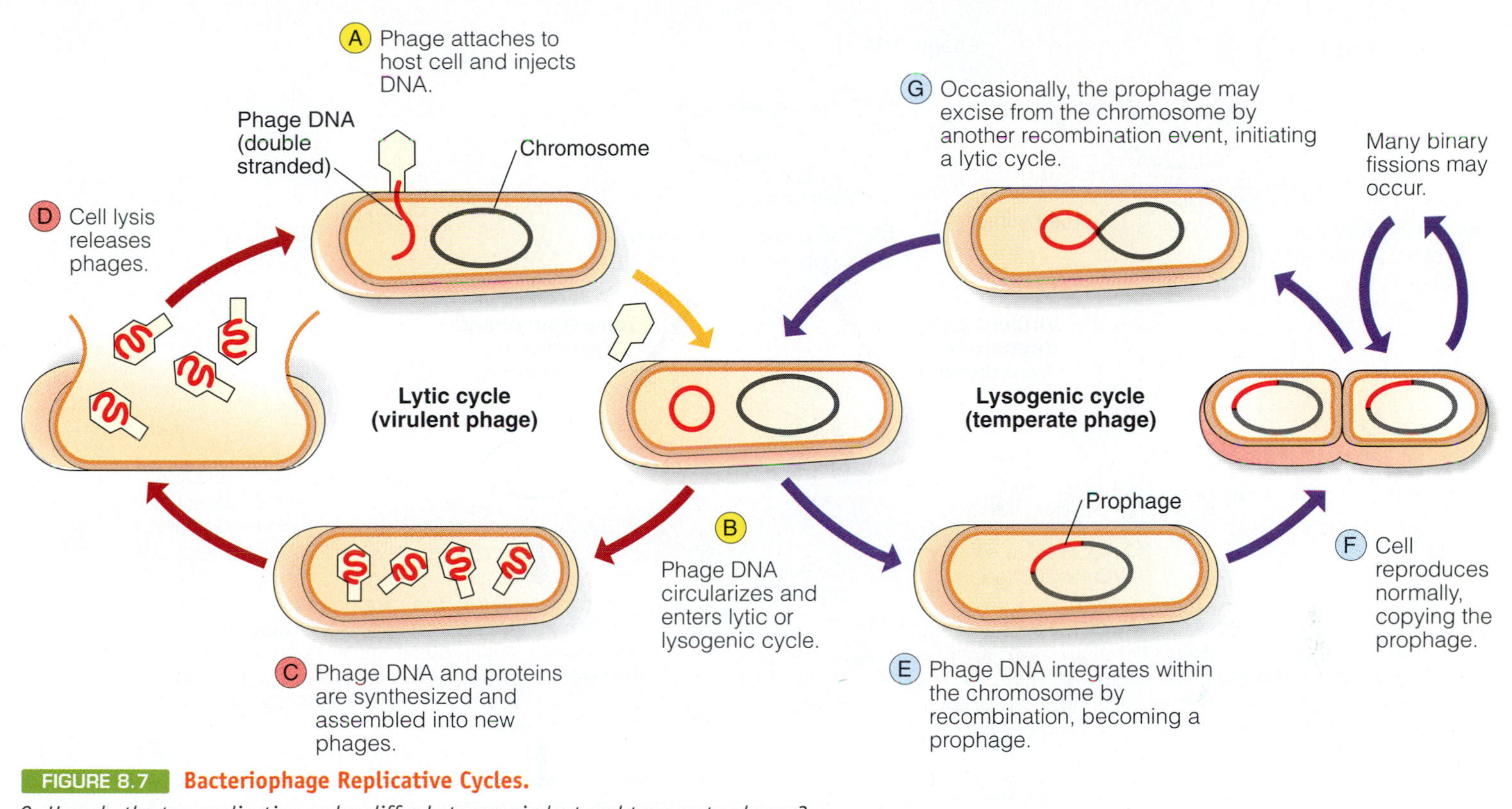

FIGURE 8.7 Bacteriophage Replicative Cycles.

Q: How do the two replicative cycles differ between virulent and temperate phages?

genes can pair with a section of the recipient's DNA and replace the section in a fashion similar to conjugation. The recipient has now been transduced (changed) using genes from the donor prokaryotic cell.

Specialized transduction occurs as a result of a lysogenic cycle and unlike generalized transduction, results in the transfer of specific genes. One of the most studied temperate viruses is phage lambda, which also infects *E. coli*. Being a temperate phage, the lambda phage DNA is integrated as a prophage into the chromosomal DNA (Figure 8.8).

At some time in the future, the prophage undergoes excision from the chromosome and enters the lytic cycle. Most of the time, the excision occurs precisely and the intact phage DNA is released. Sometimes, however, an imprecise excision occurs, and the excised prophage takes along a few flanking *E. coli* genes while leaving behind a few phage genes. At the conclusion of phage replication, multiple copies of the phage, each with a donor gene, are produced. Again, these phages would be defective because they are missing a few phage genes needed for replication. Such a transducing phage can infect another cell and transfer its genes, which cannot encode a replicative cycle. Instead, the genes integrate into the recipient chromosome, carrying the donor's genes with them. As before, the recipient cell acquired genes from the original donor cell and the recipient is now considered transduced.

Specialized transduction is an extremely rare event in comparison to the generalized form. However, there are exceptions. For example, the diphtheria bacillus, *Corynebacterium diphtheriae,* harbors proviral DNA providing the genetic code for a toxin causing diphtheria. Other toxins encoded by proviral DNA include staphylococcal enterotoxins in food poisoning, clostridial toxins in some forms of botulism, and streptococcal toxins in scarlet fever.

TABLE 8.2 compares the three forms of genetic recombination and horizontal gene transfer.

CONCEPT AND REASONING CHECKS

8.5 What is the major difference between transformation and transduction?

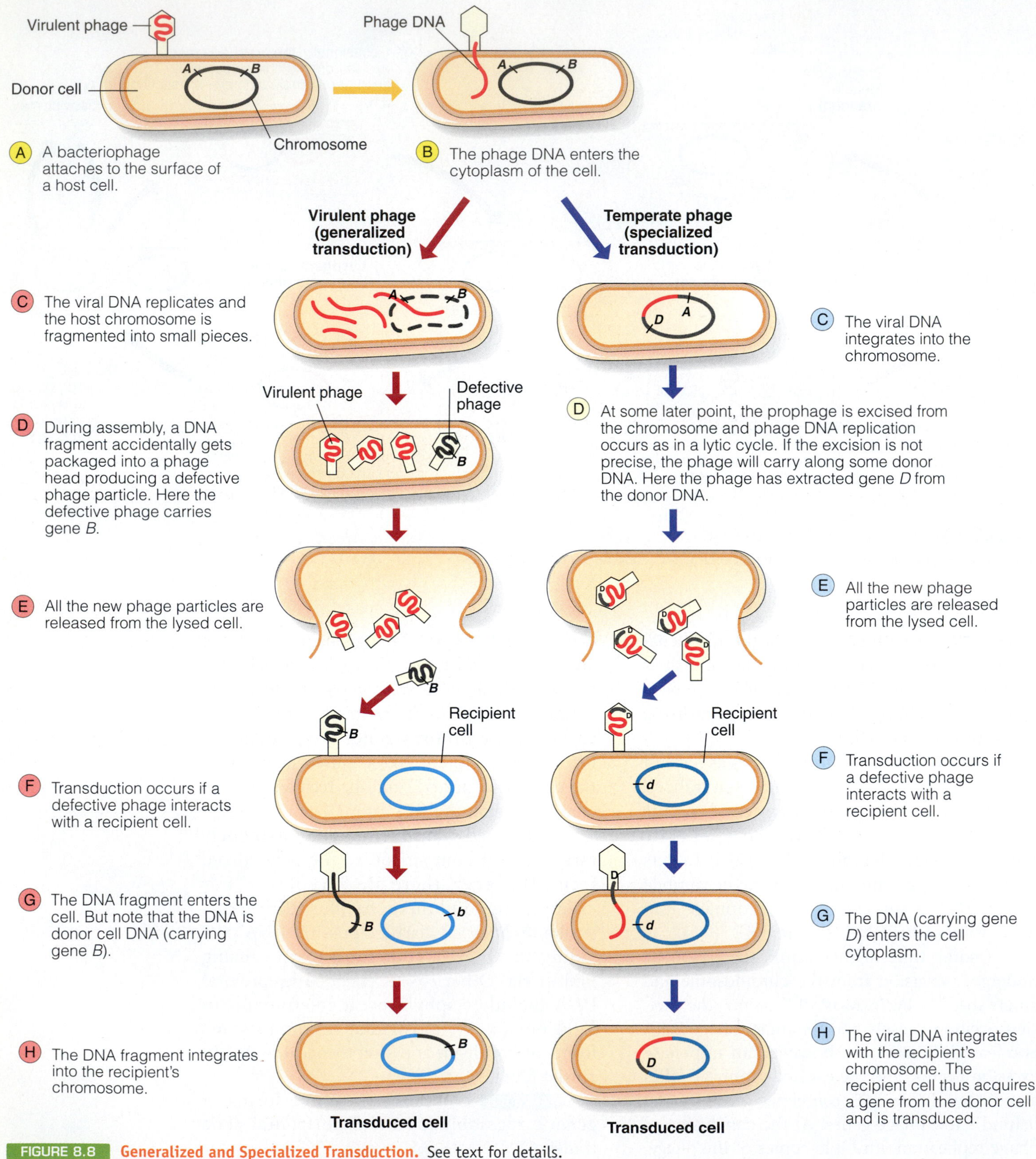

FIGURE 8.8 **Generalized and Specialized Transduction.** See text for details.

Q: How does the outcome of transduction differ between generalized and specialized?

TABLE 8.2 A Comparison of Transformation, Conjugation, and Transduction

Characteristic	Transformation	Conjugation	Transduction
Method of DNA transfer	Naked DNA moves across wall and membrane of recipient	Through cell-to-cell contact	By an intermediary virus
Amount of DNA transferred	Few genes	Variable	Few genes
Plasmid transfer possible	Yes	Yes	Not likely
Entire cell chromosome transferred	No	Sometimes	No
Virus required	No	No	Yes
Extracellular DNA fragments required	Yes	No	No
Used to acquire antibiotic resistance	Yes	Yes	Not likely

8.2 Genetic Engineering and Biotechnology

Prior to the 1970s, *Bacteria* having special or unique metabolic properties were detected through mutant analysis or by screening for cells with certain metabolic talents (MicroFocus 8.4). Experiments in genetic recombination entered a new dimension in the late 1970s, when it became possible to insert genes into prokaryotic DNA and thereby establish a genetically identical population that would produce proteins from the inserted genes. The use of microbial genetics, including the isolation, manipulation, and control of gene expression had far-reaching ramifications, leading to the entirely new field of genetic engineering.

Many of the products derived from genetic engineering have advanced the field of medicine and industrial production. **Biotechnology** is the name given to the commercial and industrial applications derived from genetic engineering.

Genetic Engineering Was Born from Genetic Recombination

KEY CONCEPT

- Plasmids can be spliced open and a gene of interest inserted to form a recombinant DNA molecule.

The science of **genetic engineering** involves an alteration to the genetic material in an organism to change its traits or to allow the organism to produce a biological product, usually a protein, that the organism was previously incapable of producing. The field surfaced in the early 1970s when the techniques became available to manipulate DNA. Among the first scientists to attempt genetic manipulation was Paul Berg of Stanford University. In 1971, Berg and his coworkers opened the circular DNA molecule from simian virus-40 (SV40) and spliced it into a bacterial chromosome. In doing so, they constructed the first **recombinant DNA molecule**—a DNA molecule containing DNA segments spliced together from two or more organisms. This human-manipulated genetic recombiration process was extremely tedious though because the cut bacterial and viral DNAs had blunt ends, making sealing of the two DNAs difficult. Berg therefore had to use exhaustive enzyme chemistry to form staggered ends that would combine easily through complementary base pairing. While Berg was performing his experiments, an important development came from Herbert Boyer and his group at the University of California. Boyer isolated a **restriction endonuclease** enzyme that recognizes and cuts specific short stretches of nucleotides. Importantly, the enzyme leaves the DNA with mortise-like staggered ends. The bits of single-stranded DNA extending out from the chromosome easily attached to the complementary ends protruding from another fragment of DNA in recombinant experiments. Scientists quickly dubbed the single-stranded extensions "sticky ends."

Today, there is a vast array of restriction enzymes from *Bacteria* and *Archaea*, each recognizing a specific nucleotide sequence (TABLE 8.3). Enzyme designations are derived from the species from which they were

MICROFOCUS 8.4: History/Biotechnology

Clostridium Acetobutylicum and the Jewish State

In 1999, scientists completed sequencing the genome of *Clostridium acetobutylicum*, a nonpathogenic bacterial species. Because some other species of *Clostridium* are major pathogens (one produces the food toxin that causes botulism, and another is responsible for tetanus), the scientists hope their sequencing work will yield insights into what enables some species to become pathogens while others remain harmless. However, the organism's ability to convert starch into the organic solvents acetone and butanol is what has a prominent place in history.

In 1900, an outstanding chemist named Chaim Weizmann, a Russian-born Jew, completed his doctorate at the University of Geneva in Switzerland. He also was an active Zionist and advocated the creation of a Jewish homeland in Palestine. In 1904, Weizmann moved to Manchester, England, where he became a research fellow and senior lecturer at Manchester University. During this time, he was elected to the General Zionist Council.

Weizmann began working in the laboratory of Professor William Perkin, where he attempted to use microbial fermentation to produce industrially useful substances. He discovered that *C. acetobutylicum* converted starch to a mixture of ethanol, acetone, and butanol, the latter an important ingredient in rubber manufacture. The fermentation process seemed no other commercial value—until World War I broke out in 1914.

At that time, the favored propellant for rifle bullets and artillery projectiles was a material called cordite. To produce it, a mixture of cellulose nitrate and nitroglycerine was combined into a paste using acetone and petroleum jelly. Before 1914, acetone was obtained through the destructive distillation of wood. However, the supply was inadequate for wartime needs, and by 1915, there was a serious shell shortage, mainly due to the lack of acetone for making cordite.

After his inquiries to serve the British government were not returned, a friend of Weizmann's went to Lloyd George, who headed the Ministry of Munitions. Lloyd George was told about Weizmann's work and how he could synthesize acetone in a new way. The conversation resulted in a London meeting between Weizmann, Lloyd George, and Winston Churchill. After explaining the capabilities of *C. acetobutylicum,* Weizmann became director of the British admiralty laboratories where he instituted the full-scale production of acetone from corn. Additional distilleries soon were added in Canada and India. The shell shortage ended.

After the war ended, now British Prime Minister Lloyd George wished to honor Weizmann for his contributions to the war effort. Weizmann declined any honors but asked for support of a Jewish homeland in Palestine. Discussions with Foreign Minister Earl Balfour led to the Balfour Declaration of 1917, which committed Britain to help establish the Jewish homeland. Weizmann went on to make significant contributions to science—he suggested that other organisms be examined for their ability to produce industrial products and is considered the father of industrial fermentation. Weizmann also laid the foundations for what would become the Weizmann Institute of Science, one of Israel's leading scientific research centers. His political career also moved upward—he was elected the first President of Israel in 1948. Chaim Weizman died in 1952.

isolated. For example, the restriction enzyme *Eco*RI stands for *Escherichia coli* **r**estriction enzyme **I**.

Each enzyme cuts both strands of the DNA because the recognition sequences are a palindrome. Each strand has the same complementary set of nucleotide bases. Thus, restriction enzymes are "molecular scissors" used by genetic engineers to open a prokaryotic chromosome or plasmid at specific locations and insert a DNA segment from another organism.

■ Palindrome:
A series of letters reading the same left to right and right to left.

To seal the recombinant DNA segments, **DNA ligase** was used. This enzyme normally functions during the DNA replication and repair to seal together DNA fragments (see Chapter 7).

Meanwhile, Stanley Cohen, also at Stanford University, was accumulating data on the plasmids of *E. coli.* Cohen found he could isolate plasmids from the bacterial cells and insert them into fresh bacterial cells by suspending the organisms in calcium chloride and heating them suddenly. This made the *E. coli* cells

TABLE **8.3** Examples of Restriction Endonuclease Recognition Sequences

Organism	Restriction Enzyme	Recognition Sequence*
Escherichia coli	*Eco*RI	G ↓ AATTC CTTAA ↓ G
Streptomyces albus	*Sal*I	G ↓ TCGAC CAGCT ↓ G
Haemophilus influenzae	*Hind*III	A ↓ AGCTT TTCGA ↓ A
Bacillus amyloliquefaciens	*Bam*HI	G ↓ GATC^{m}C CC^{m}TAG ↓ G
Providencia stuartii	*Pst*I	CTGCA ↓ G G ↓ ACGTC

*Arrows indicate where the restriction enzyme cuts the two strands of the recognition sequence; C^{m} = methylcytosine.

competent to take up the plasmid via the transformation process. Once inside the cells, the plasmids multiply independently and produce **clones**; that is, copies of themselves.

Working together, Boyer and Cohen isolated plasmids from *E. coli* and opened them with restriction enzymes (FIGURE 8.9). Next, they inserted a segment of foreign DNA into the plasmids and sealed the segment using DNA ligase. Mimicking natural genetic recombination, they then inserted the plasmids (recombinant DNA molecules) into fresh *E. coli* cells. By 1973, their technique had successfully spliced genes across genera, from *S. aureus* into *E. coli*. These genetic engineering experiments intrigued the scientific community because for the first time they could manipulate genes from a wide variety of species and splice them together.

CONCEPT AND REASONING CHECKS

8.6 List the steps required to form a recombinant DNA molecule.

Genetic Engineering Has Many Commercial and Practical Applications

KEY CONCEPT

- Prokaryotes can be transformed by the acquisition of a human gene.

Today, over 18 million people in the United States have **diabetes**, a group of diseases resulting from abnormally high blood glucose levels. One cause is from the inability of the body to produce sufficient levels of insulin (referred to as juvenile or insulin-dependent diabetes or type I diabetes) to control the blood glucose level. This means that diabetics must receive daily injections of the protein to survive. Before 1982, diabetics received purified insulin extracted and purified from the pancreas of cattle and pig cadavers. This can pose a problem because animal insulin could trigger allergenic reactions and possibly contain unknown viruses that had infected the animal. The solution was to produce insulin using genetic engineering techniques. Eli Lilly marketed the first such synthetic insulin, called Humulin, in 1982. Since then, other genetically-engineered insulin products have been developed. Such commercial successes of biotechnology were a sign of things to come.

The best way to understand how genetic engineering operates is to follow an actual procedure for the production of insulin. MicroInquiry 8 describes one of the early methods—by cloning the human gene for insulin into bacterial cells. This involves isolating the piece of DNA containing the human insulin gene, precisely cutting the gene out, and splicing the insulin gene into a bacterial plasmid. The recombinant DNA then is placed in bacterial cells like *E. coli*, forming clones. The cells transcribe the mRNA, translate it into the protein, and then secrete the human insulin.

Besides insulin, a number of other proteins of important pharmaceutical value to humans have been produced by genetically-engineered microorganisms. Many of these

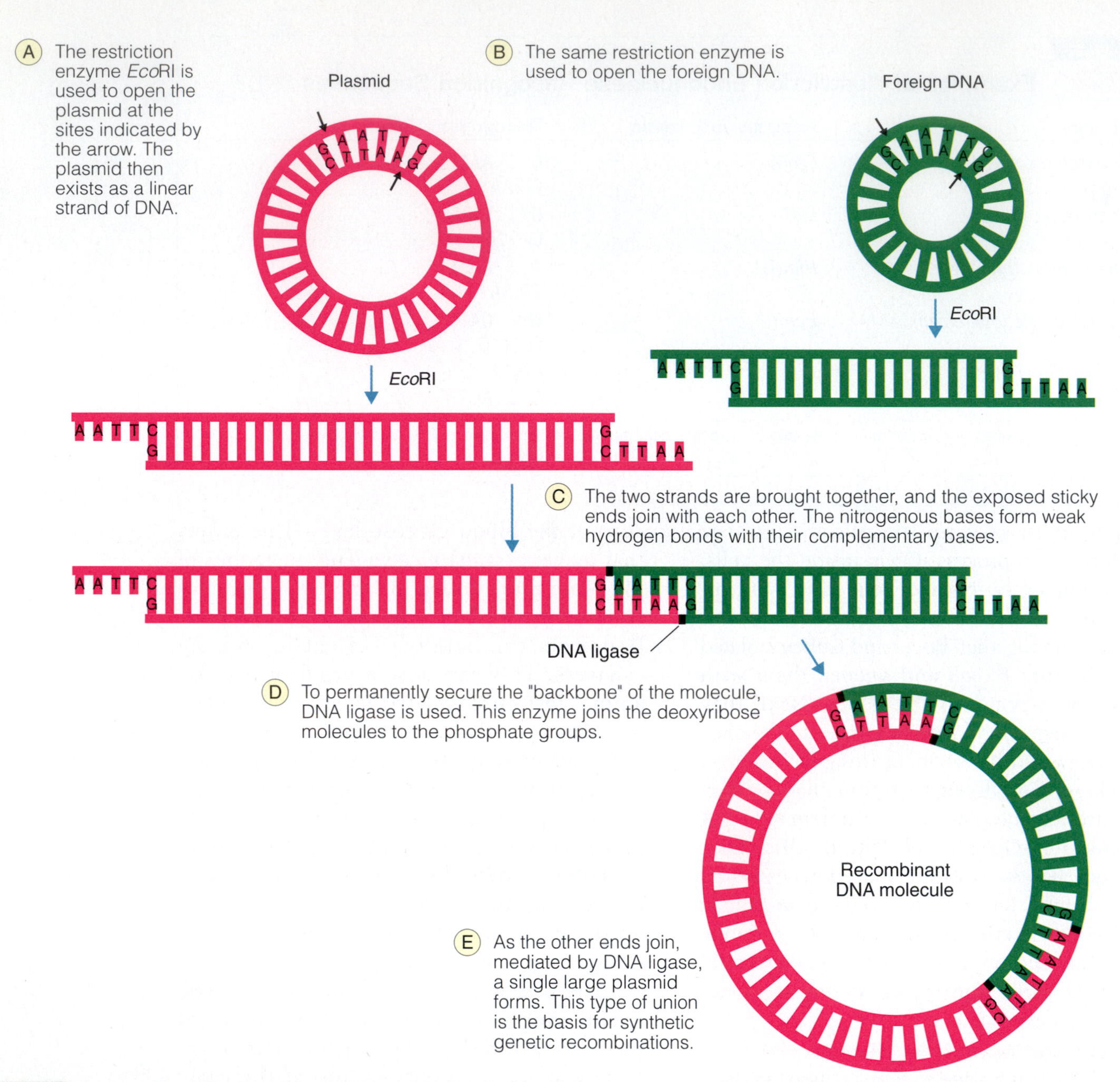

FIGURE 8.9 **Construction of a Recombinant DNA Molecule.** In this construction, two unrelated plasmids (loops of DNA) are united to form a single plasmid representing a recombinat DNA molecule.

Q: Why is the product of genetic engineering called a recombinant DNA molecule?

■ Interferons: Naturally-produced human proteins that interfere with viral replication.

proteins, such as the interferons, are produced in relatively low amounts in the body, making purification extremely costly. Therefore, the only economical solution to obtain significant amounts of the product is through genetic engineering.

In 2005, thousands of biotechnology companies worldwide were working on the commercial and practical applications of genetic engineering. Some are research companies with special units for these studies, while others were established solely to pursue and develop new products by gene engineering techniques. Let's look at a few examples.

Environmental biology. We already have mentioned in this chapter and previous ones the usefulness of prokaryotes as a source of genes. Prokaryotes represent a huge, mostly untapped gene pool representing metabolically diverse processes. Examples such as bioremediation have been discussed where genetically engineered or genetically recombined prokaryotic cells are provided with specific genes whose products will break down toxic pollutants, clean up waste materials, or degrade oil spills. We have barely scratched the surface to take advantage of the metabolic diversity present in prokaryotes.

Antibiotic production. The presence or threat of infectious disease represents a high demand for antibiotics. Therefore, many antibiotics today are produced commercially using microorganisms. Although antibiotics are produced in nature, the bacterial or fungal cell often does not produce these compounds in high yield. Furthermore, as no new class of antibiotics has been identified since the early 1990s, existing compounds need to be redesigned to reach their targets more efficiently. This means the microbes must be genetically engineered to produce larger quantities of antibiotics and/or to produce modified antibiotics to which infectious microbes have yet to show resistance.

Pharmaceutical applications. The pharmaceutical products of DNA are numerous and diverse (TABLE 8.4). Many of the genetically-engineered products are either the organism itself or proteins expressed by recombinant DNA in clones (FIGURE 8.10).

Agricultural applications. Genetic engineering has extended into many realms of science. In agriculture, for example, genes for herbicide resistance have been transplanted from bacterial cells into tobacco plants, demonstrating that these transgenic plants better tolerate the herbicides used for weed control. For tomato growers, a notable advance was made when researchers at Washington University spliced genes from a pathogenic virus into tomato plant cells and demonstrated the cells would produce viral proteins at their surface. The viral proteins blocked viral infection, providing resistance to the transgenic tomato plants.

For gene transfer experiments in plants, the vector DNA often used for transfer is a plasmid from the bacterium *Agrobacterium tumefaciens*. This organism causes a plant tumor called crown gall, which develops when DNA from the bacterial cells inserts itself into the plant cell's chromosomes (FIGURE 8.11). Researchers remove the tumor-inducing gene from the plasmid and then splice the desired gene into the plasmid and allow the bacterial cells to infect the plant.

TABLE 8.4 Examples of Therapeutic Products Derived from Genetic Engineering

Product	Treatment/Use
Factors VIII, IX	Replace clotting factors missing in hemophiliacs
Tissue plasminogen activator (TPA)	Prevents blood clotting after heart attacks or strokes
Epidermal growth factor (EGF)	Promotes wound healing
Insulin	Used to stimulate glucose uptake from the blood in diabetics
Somatotropin	A growth hormone used to replace the missing hormone in people of short stature
Colony stimulating factor (CSF)	Used to stimulate white blood cell production in cancer and AIDS patients
α-Interferon	Used with other antiviral agents to fight viral infections
Vaccine proteins	Used to treat and prevent infectious diseases

The dairy industry was the first to feel the dramatic effect of the new DNA technology. In the 1980s, researchers at Cornell University injected dairy cows with bovine growth hormone (BGH) produced from bacterial cells engineered with the *BGH* gene. They reported a 41 percent increase in milk from the experimental cows. Also being developed is a pig with more meat and less fat, a product of genetically engineered porcine growth hormone. Scientists at Auburn University have endowed young carp with extra copies of activated growth hormone genes, hoping to enable the fish to grow more efficiently in aquacultural surroundings.

■ Transgenic: Referring to an organism containing a stable gene from another organism.

Detection and diagnosis. The genes of an organism contain the essential information responsible for its behavior and characteristics. Bacterial pathogens, for example, contain specific sequences of nucleotides that can confer on the pathogen the ability to infect and cause disease in the host. Because these nucleotide sequences are distinctive and often unique, if detectable, they can be used as a definitive diagnostic determinant.

MICROINQUIRY 8

Molecular Cloning of a Human Gene into Bacterial Cells

Genetic engineering has been used to produce pharmaceuticals of human benefit. One example concerns the need for insulin injections in people suffering from diabetes (an inability to produce the protein insulin to regulate blood glucose level). Prior to the 1980s, the only source for insulin was through a complicated and expensive extraction procedure from cattle or pig pancreases. But, what if you could isolate the human insulin gene and, through transformation, place it in bacterial cells? These cells would act as factories churning out large amounts of the pure protein that diabetics could inject.

To do molecular cloning of a gene, besides the bacterial cells, we need three ingredients: a cloning vector, the human gene of interest, and restriction enzymes. Plasmids are the **cloning vector**, a genetic element used to introduce the gene of interest into the bacterial cells. Human DNA containing the insulin gene must be obtained. We will not go into detail as to how the insulin gene can be "found" from among 35,000 human genes. Suffice it to say that there are standard procedures to isolate known genes. Restriction enzymes will cut open the plasmids and cut the gene fragments that contain the insulin gene, generating complementary sticky ends.

The following description represents one procedure to genetically engineer the human insulin gene into cells of *Escherichia coli*. (There are other procedures, some more direct. However, this method reinforces many of the concepts we have learned in the last few chapters).

1. Plasmids often carry genes, such as antibiotic resistance. We are going to use the cloning vector shown in **Figure A** because it contains genes for resistance to tetracycline (tet^R) and ampicillin (amp^R) antibiotics. This will be important for identification of clones that have been transformed. In addition, this vector has a single restriction sequence for the restriction enzyme SalI (see Table 8.3). Importantly, this cut site is within the tet^R gene. Also, the plasmid will replicate independently in *E. coli* cells and can be placed in the cells by transformation.
2. The vector and human DNA are cut with SalI to produce complementary sticky ends on both the opened vector and the insulin gene (**Figure B**). Vector and the insulin gene then are mixed together in a solution with DNA ligase, which will covalently link the sticky ends. Some vectors will be recombinant plasmids; that is, plasmids containing the insulin gene. Other plasmids will close back up without incorporating the gene.
3. The plasmids are placed in *E. coli* cells by transformation. The plasmids replicate independently in the bacterial cells, but as the bacterial cells multiply, so do the plasmids. By allowing the plasmids to replicate, we have cloned the plasmids, including any that contain the insulin gene.
4. Because we do not know which plasmids contain the insulin gene, we need to screen the clones to identify the recombinant plasmids. This is why we selected a plasmid with tet^R and amp^R resistance genes.

Because the SalI cut site is within *tetR*, any recombinant plasmids will have a defective tetracycline resistance gene and the bacterial cells carrying those plasmids will not grow on a medium containing tetracycline. Plasmids without the insulin gene have an intact tet^R gene and can grow on the medium. In addition, there will be bacterial cells that did not take up a plasmid and will be sensitive to tetracycline and ampicillin. Thus, using the positive selection technique described in Chapter 7 and a selective medium as described in Chapter 5, we can identify which bacterial cells contain the recombinant plasmids.

Therefore, we plate all our bacterial cells onto a master plate—a nutrient medium without tetracycline. All bacterial cells should grow and form colonies. Using the replica plating technique mentioned in Chapter 7, we now replica plate all the colonies onto (a) a growth medium containing tetracycline and (b) onto a growth medium containing ampicillin.

Question 8. Explain which clones should grow on which plates. The answer can be found in **Appendix D**.

5. These colonies can now be isolated and grown in larger batches of liquid medium. These batch cultures then are inoculated into large "production vats," called **bioreactors**, in which the cells grow to massive numbers while secreting large quantities of insulin into the liquid.

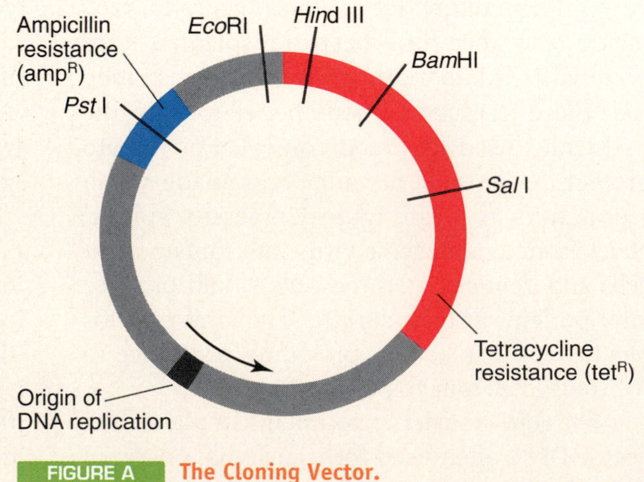

FIGURE A The Cloning Vector.

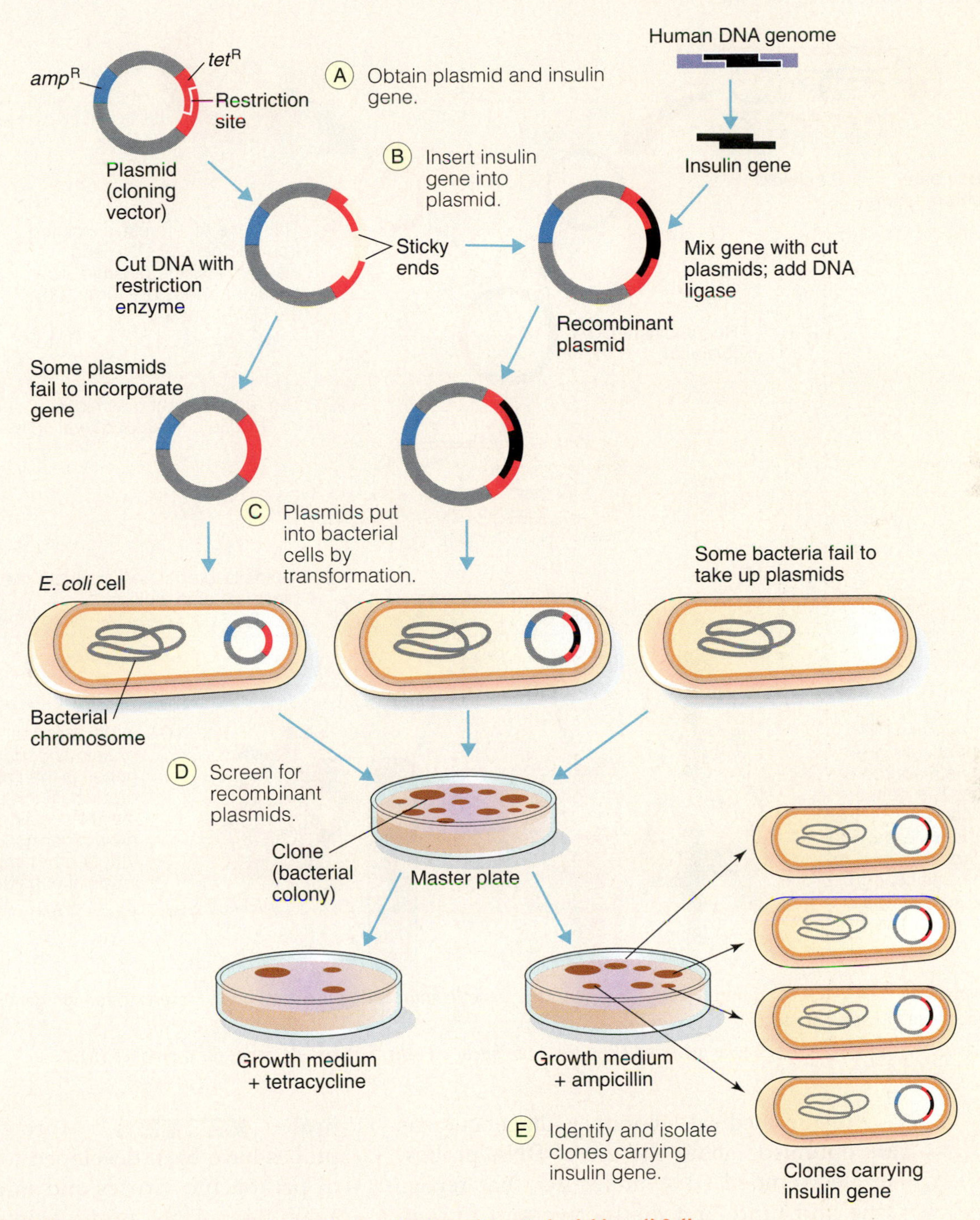

FIGURE B **The Sequence of Steps to Engineer the Insulin Gene into *Escherichia coli* Cells.**

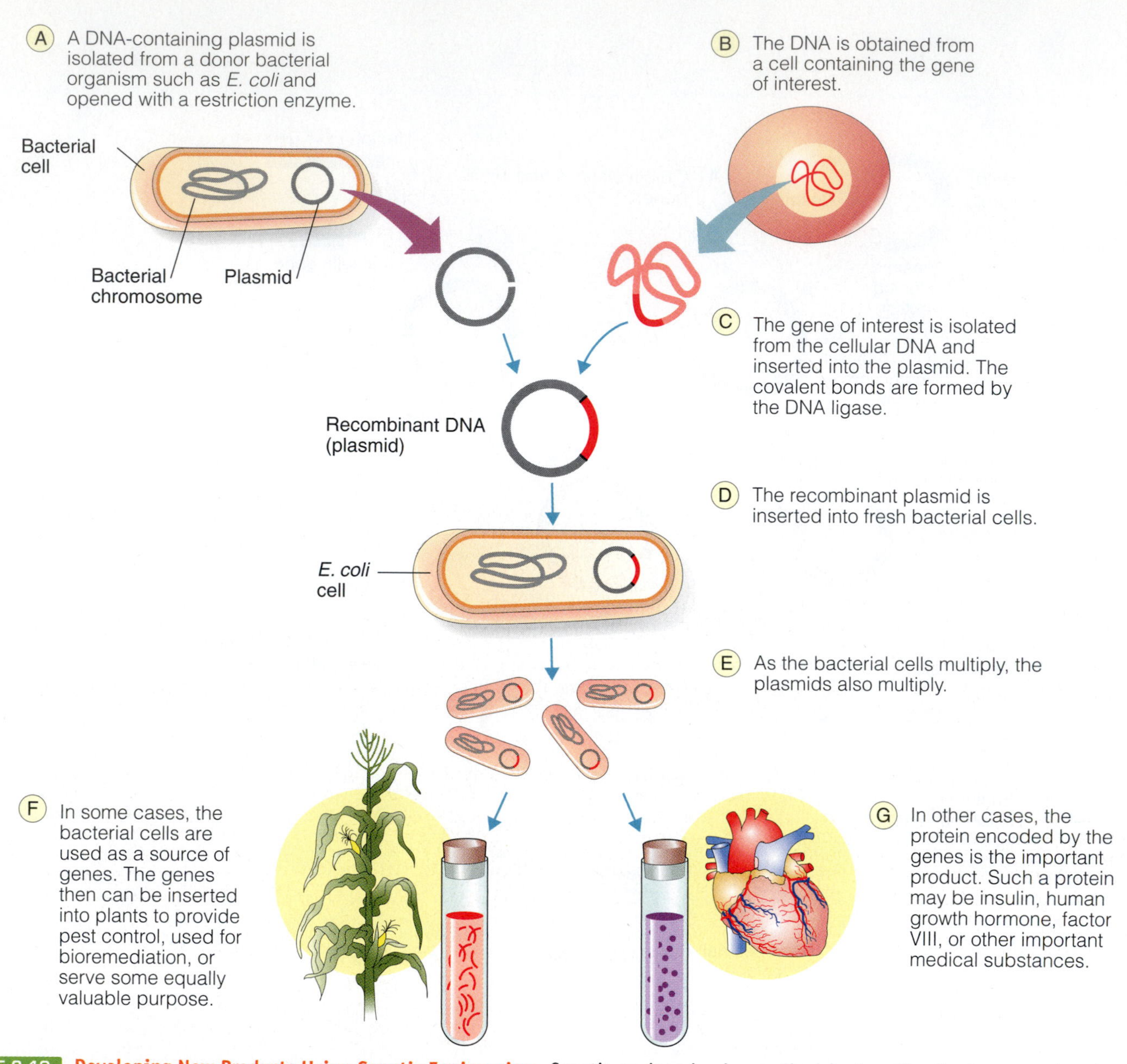

FIGURE 8.10 **Developing New Products Using Genetic Engineering.** Genetic engineering is a method for inserting foreign genes into bacterial cells and obtaining chemically useful products.

Q: How do bacterial cells that have been genetically-modified (F) differ from bacterial cells that encode genes for a product (G)?

In the medical laboratory, diagnosticians are optimistic about the use of **DNA probes**, single-stranded DNA molecules that recognize and bind to a distinctive and unique nucleotide sequence of a pathogen. The DNA probe binds (hybridizes) to its complementary nucleotide sequence from the pathogen, much like strips of Velcro stick together. To make a probe, scientists first identify the DNA segment (or gene) in the pathogen that will be the target of a probe. Using this segment, they construct the single-stranded DNA probe (FIGURE 8.12). More than 100 DNA probes have been developed for the detection of pathogenic viruses and microbes. As one example, a DNA probe exists for the early detection of malarial infections caused by the protozoan *Plasmodium falciparum*. DNA is isolated from the tissues of the suspected patient and fragmented into single-stranded DNA segments. These segments are attached to a solid support.

The labeled DNA probe is added. If the DNA sample from the patient contains the

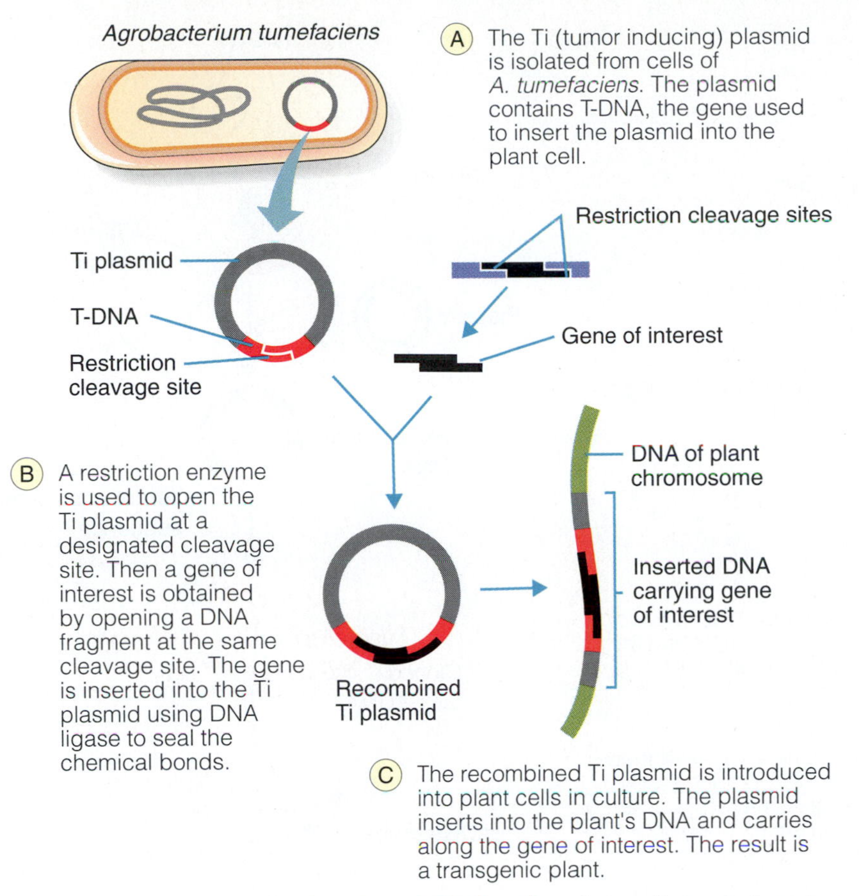

FIGURE 8.11 **The Ti Plasmid as a Vector in Plant Genetic Engineering.** *A. tumefaciens* induces tumors in plants and causes a disease called crown gall. The catalyst for infection is a tumor-inducing (Ti) plasmid. This plasmid, without the tumor-causing gene, is used to carry a gene of interest into plant cells.

Q: Why must the tumor-inducing gene in the Ti plasmid be removed before the plasmid is used in genetic engineering procedures?

target *P. falciparum* DNA segment, the probe binds to the complementary nucleotide sequence. In doing so, it concentrates the label (radioactivity, fluorescence, or a colored chemical) at that site and indicates a match has been made. Clearly, the use of DNA probes represents a reliable and rapid method for detecting and diagnosing many human infectious diseases.

Vaccine production. Microbial vaccines traditionally have been either killed or attenuated (inactivated) preparations. Sometimes attenuated microbial preparations are not completely inactivated and a slight chance exists that the patient could develop the disease for which they have been vaccinated. Today, genetic engineering has developed subunit vaccines, which are risk free for people being vaccinated (TABLE 8.5). The hepatitis B vaccine, for example, is perfectly safe because the vaccine is made from only the protein coat subunits of the virus. The viral genes for the coat are cloned in yeast cells, and the viral protein harvested as the vaccine. Active hepatitis B viruses cannot arise from such vaccine material.

Recombinant vaccines also are attractive because the precise genetic makeup of the vaccine is known, often making the administration of higher doses possible without the development of serious side effects. Such vaccines can be produced faster than traditional vaccines, although they are not always cheaper.

Also being developed are **DNA vaccines** where the genetic material (a gene) is the vaccine. Such gene vaccines often produce immunity in the individual in which the gene has been taken up. Again, they should be safe and cheaper to produce. Chapter 21 will describe more fully the nature of vaccines.

■ Subunit vaccines: Vaccines containing only a protein fragment from the pathogen.

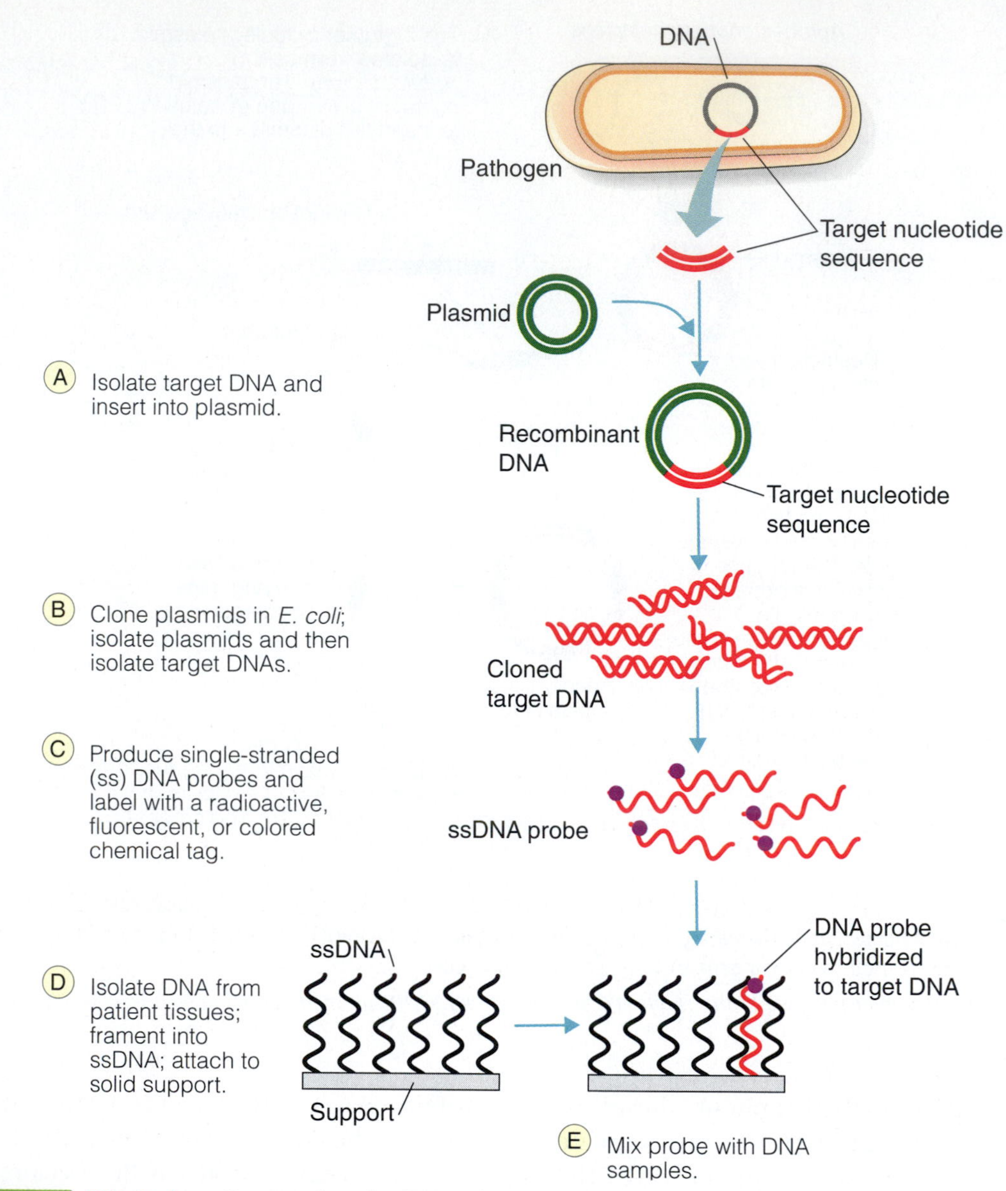

FIGURE 8.12 **DNA Probes.** Construction of a DNA probe and its use in disease detection and diagnosis.

Q: Why must the DNA probes be single stranded?

TABLE 8.5 Selected Genetically-Engineered Vaccines

Disease	Virus/Bacterial Species	Vaccine Function
Viral diseases		
Cervical cancer	Papilloma viruses 16/18	Prevention of cervical cancer
Hepatitis B	Hepatitis B virus	Prevention of hepatitis B
Measles	Measles virus	Prevention of measles
Rabies	Rabies virus	Prevention of rabies
Bacterial diseases		
Diphtheria	*Corynebacterium diphtheriae*	Prevention of diphtheria
Tetanus	*Clostridium tetani*	Prevention of tetanus

Lastly, it should be mentioned that genetically engineered products are not always a "no brainer" in terms of their development. This is no clearer than in the attempts to develop a vaccine for AIDS. Since 1987, scientists and genetic engineers have tried to identify viral subunits that can be used to develop an AIDS vaccine. However, it is not so much that genetic engineering can't be done as it is the virus just seems to find ways to circumvent a vaccine and the immunity developing in the patient. Still, scientists hope a safe and effective genetically-engineered vaccine can be developed. We will have much more to say about AIDS and vaccines in Chapter 15.

CONCEPT AND REASONING CHECKS

8.7 Why are genetically-engineered vaccines safer than many traditional vaccines?

8.3 Microbial Genomics

In April 2003, exactly 50 years to the month after Watson and Crick announced the structure of DNA (see Chapter 2), a publicly financed, $3 billion international consortium of biologists, industrial scientists, computer experts, engineers, and ethicists completed perhaps the most ambitious project in the history of biology. The Human Genome Project, as it was called, had succeeded in mapping the **human genome**—that is, the 3 billion nitrogenous bases in a human cell were identified and strung together in the correct order (sequenced). The completion of the project represents a scientific milestone with unimaginable health benefits.

Many Microbial Genomes Have Been Sequenced

KEY CONCEPT

- Genome sequences are rapidly expanding for many prokaryotes and eukaryotes.

If the human genome was represented by a rope two inches in diameter, it would be 32,000 miles long. The genome of a bacterial species like *E. coli* at this scale would be only 1,600 miles long or about 1/20 the length of human DNA. So, being substantially smaller, microbial genomes are easier and much faster to sequence.

In May 1995, the first complete genome of a free-living organism was sequenced: the 1.8 million base pairs (1.8 Mb) in the genome of the bacterial species *Haemophilus influenzae* (MicroFocus 8.5). In a few short months, the genome for a second organism, *Mycoplasma genitalium* was reported. This reproductive tract pathogen has one of the smallest known bacterial genomes, consisting of only 580,000 base pairs and 250 genes. In 1996, the genome for the yeast *Saccharomyces cerevisiae* was sequenced. In a field already filled with milestones, the sequencing of *S. cerevisiae* marked the first glimpse into the eukaryotic genome. Sixteen chromosomes were analyzed, 12 million bases were sequenced, and 6,000 genes were identified. The sequencing revealed many genes wholly new to biology.

Since then, hundreds more microbial genomes have been sequenced, including a variety of pathogens (TABLE 8.6). Sequences by themselves, although the result of very impressive work, do not tell us much. So, what do these sequences tell us and what practical use can be derived from this information?

Segments of the Human Genome May Have "Microbial Ancestors"

KEY CONCEPT

- Some human genes and human DNA sequences may have prokaryotic origins.

With the sequencing of the human genome, one interesting development was to compare the human nucleotide gene sequences to prokaryotic sequences. Are there any similarities in the genes each contains? The conclusion? Maybe.

Some comparisons indicate as many as 200 of our 35,000 genes are essentially identical to those found in members of the *Bacteria*. However, these human genes were not acquired directly from bacterial species, but rather were genes picked up by animals representing early ancestors of humans. So important were these genes, they have been preserved and passed along from organism to organism throughout evolution. For example, several researchers suggest some genes coding

MICROFOCUS 8.5: Biotechnology

Putting Humpty Back Together Again

Humpty Dumpty sat on a wall.
Humpty Dumpty had a great fall.
All the king's horses,
And all the king's men,
Couldn't put Humpty together again.

A battery of DNA sequencing machines.

We all remember this nursery rhyme, but it can be an analogy for the efforts required in sequencing the genome of an organism. To sequence a whole genome, you have to take thousands of small DNA fragments and, after sequencing them, try to "put the fragments together again." The good news in genomics is that you can "put Humpty together again."

Small fragments must be used when sequencing a whole genome because current methods will not work with the tremendously long stretches of DNA, even those shorter ones found in bacterial genomes. Therefore, one strategy is to break the genome into small fragments. These fragments then are sequenced using sequencing machines (see figure) and the fragments reassembled into the full genome. This technique is called the "whole-genome shotgun method." It can be extremely fast, but there are so many little pieces that it can be very difficult to put the whole genome together again.

The "shotgun" strategy first was used in 1995 by Craig Venter, Hamilton Smith, and their colleagues to sequence the genome of a self-replicating, free-living organism—*Haemophilus influenzae*. This gram-negative bacterial rod causes ear and respiratory infections and bacterial meningitis in children. With 1.8 million base pairs, the size of its genome is average for prokaryotes, but is about ten times larger than any virus that had been sequenced.

To sequence *H. influenzae,* Venter and Smith cut copies of the DNA into pieces of 1,600 to 2,000 base pairs. The segments then were partly sequenced at both ends, using automated sequencing machines. These base-pair sequences—with their many overlaps—became the sequence information that was entered into the computer. Using innovative computer software, the 24,304 DNA fragments generated from the *H. influenzae* genome were compared, clustered, and matched for assembling the genome.

Assembling the *H. influenzae* genome was a considerable achievement—and to some observers a surprise. The genome contains 1,830,137 base pairs, in which 1,749 genes are embedded. Once assembled, the genes could be located, compared to known genes, and a detailed map developed. Sequencing *H. influenzae* took about a year, but demonstrated that "the king's horses" (supercomputers and shotgun sequencing) and "the king's men" (the large group of collaborators) could "put Humpty together again"—and with speed and accuracy. Note: Since 1995, great strides have been made in sequencing technology. If *H. influenzae* were to be sequenced today, it would take about five days, rather than an entire year.

for brain signaling chemicals and for communication between cells did not evolve gradually in human ancestors; rather, these ancestors acquired the genes directly from prokaryotes. This provocative claim remains highly controversial though and more work is needed to better analyze this possibility.

Another discovery from the human genome project indicates that only about 5 to 10 percent of our DNA appears to code for proteins and regulatory RNAs. A number of scientists believe some of the non-gene DNA may be "genetic debris" from viruses and bacterial cells that infected other cells during cell evolution hundreds of millions of years ago. So, microbial genomes will have much to tell us about our past as comparisons continue.

CONCEPT AND REASONING CHECKS

8.8 How do microbial genomes compare in size and composition to the human genome?

TABLE **8.6** **Microbial Genomes Completely Sequenced and Representative Examples***

	Genomes Sequenced	Genome Size (Mb†)	Comments
Viruses	>1,100		
HIV		0.009	Causative agent of AIDS
Poliovirus		0.007	Causative agent of polio
Rabies virus		0.012	Causative agent of rabies
SARS coronavirus		0.029	Causes an atypical pneumonia (SARS)
Variola virus		0.185	Causative agent of smallpox
West Nile virus		0.011	Causes a fatal encephalitis (inflammation of the brain)
Domain *Archaea*	24		
Methanococcus jannaschii		1.66	First archaean sequenced
Nanoarchaeum equitans		0.50	Smallest known archaeal genome
Domain *Bacteria*	291		
Bacillus anthracis		5.23	Causative agent of anthrax
Carsonella ruddii		0.16	Smallest known cellular genome
Clostridium tetani		2.80	Causative agent of tetanus
Escherichia coli		4.64	Common intestinal bacterium in humans
E. coli 0157:H7		5.59	Causes bloody diarrhea; hemolytic uremic syndrome
Mycobacterium tuberculosis		4.42	Causative agent of tuberculosis
Mycoplasma genitalium		0.58	Causes reproductive tract infections
Staphylococcus aureus		2.82	Causes skin lesions, food poisoning,toxic shock syndrome
Treponema pallidum		1.14	Causative agent of syphilis
Vibrio cholerae		4.00	Causative agent of cholera
Yersinia pestis		4.65	Causative agent of plague
Domain *Eukarya*	31		
Saccharomyces cerevisiae		12.07	Baker's yeast; used in fermentation
Plasmodium falciparum		22.90	Causative agent of malaria

*Data published in the Genomes OnLine Database. Available at: http://wit.integratedgenomics.com/ERGO_supplement/genomes.html/. Accessed October 14, 2006.

†Mb = 1,000,000 bases

Microbial Genomics Will Advance Our Understanding of the Microbial World

KEY CONCEPT

- Understanding gene organization and function provides for new microbial applications.

Microorganisms have existed on Earth for more than 3.7 billion years, although we have known about them for little more than 300 years. Over this long period of evolution, they have become established in almost every environment on Earth and, although they are the smallest organisms on the planet, they influence—if not control—some of the largest events. Yet, with few exceptions, we do not know a great deal about any of these microbes, and we have been able to culture and study in the laboratory less than 1 percent of all microorganism species.

However, our limited knowledge is changing. With the advent of microbial genomics, we have begun the third Golden Age of microbiology, a time when remarkable scientific discoveries will be made toward understanding the workings and interactions of the microbial world. Some potential consequences from the understanding of microbial genomes are outlined below and in the chapters ahead.

Safer food production. Several bacterial species cause food-borne illnesses (Chapter 10) and the genomes of many of these species have been sequenced. This knowledge will provide the tools to develop rapid and accurate detection methods for such pathogens.

Identification of unculturable prokaryotes. Since the vast majority of prokaryotes cannot be cultured (see Chapter 5), genomics offers a way to identify these organisms. Craig Venter is just one of many scientists studying the gene sequences of prokaryotes. Venter and others are attempting to sequence and identify the collective genomes, called the **metagenome**, of all prokaryotes in a microbial community, such as the Sargasso Sea (see Chapter 1). This ability to identify unculturable prokaryotes is opening up the new discipline of **environmental genomics**. Such genomic information can allow microbiologists to better understand how microbial communities function and how the organisms interact with one another.

Genomic information also is being used to discover if there are unculturable microbial representatives that could be used as alternative energy sources to solve critical environmental problems, including global warming.

Cleaner environments. Growing out of environmental genomics are applications to environmental cleanup or **bioremediation**. For example, one environmental concern is radioactive waste sites remaining from atomic weapons programs and the civilian nuclear power industry. These sites contain not only dangerous radioactive isotopes (see Chapter 2), but also high concentrations of mercury, toluene, and other toxic compounds. Although a few microbes, such as *E. coli* and *Pseudomonas*, can degrade many of these toxic materials into less harmful products, the microbes are sensitive to the high levels of radiation and will develop lethal mutations.

Scientists have discovered a few bacterial species that can withstand such radiations. Perhaps the toughest is *Deinococcus radiodurans* (FIGURE 8.13). This bacterium can withstand 1,500 times the radiation exposure that would kill other prokaryotes or humans! Its genome was sequenced in 1999. Through genetic analysis and genomics, scientists have discovered the organism contains a highly efficient DNA repair process capable of quickly correcting the damage caused by radiation exposure (see Chapter 7).

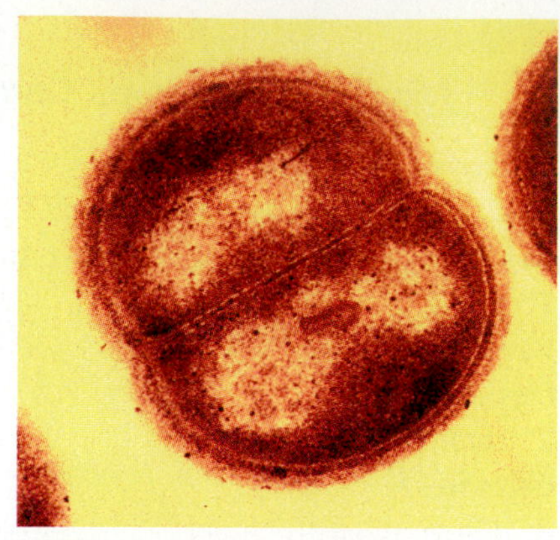

FIGURE 8.13 ***Deinococcus radiodurans.*** This false-color transmission electron micrograph of *D. radiodurans* shows a cell undergoing binary fission. This bacterial species is very resistant to genetic damage caused by radiation.

Q: How does this bacterium resist damage by radiation?

The proposal is to engineer the genes for mercury and toluene degradation into the DNA of *D. radiodurans*. The organism then could be sprayed over a toxic site and the bacterial cells could clean up the toxic chemicals. Using today's genomics information and genetic engineering capabilities, it should be possible to develop genetically modified organisms (GMOs) with the appropriate ensemble of engineered genes capable of degrading whatever compounds are present.

Improved biosensing. The advent of the anthrax bioterrorism events of 2001 and the continued threat of bioterrorism has led many researchers to look for ways to more efficiently and more rapidly detect the presence of such bioweapons (see Chapter 1). Many of the diseases caused by these potential biological agents cause no symptoms for at least several days after infection, and when symptoms appear, initially they are flu-like. The genomes of several microbes representing potential bioweapons have been sequenced and the knowledge gained will be useful in designing rapid detection systems.

Beyond bioterrorism, biosensors could be developed to detect warfare agents. For example, prokaryotes could be used to detect the location of land mines. GMOs, engineered with the ability to detect trinitrotoluene (TNT), the explosive that leaks from land mines, could be sprayed over a suspect area. When monitored from the air, TNT leaks (land mines) could be detected by a fluorescent glow engineered into the cells and genetically programmed to be produced when in contact with TNT. Similar scenarios could be devised for other pollutants and pathogens typically found in soil, food, or water.

CONCEPT AND REASONING CHECKS

8.9 How is environmental genomics contributing to a better world?

Comparative Genomics Brings a New Perspective to Defining Infectious Diseases and Studying Evolution

KEY CONCEPT

- Comparative genomics compares the DNA sequences of related or unrelated species.

Sequencing the DNA bases of a microorganism (or any other organism) has and still does provide important information concerning the number of bases and genes comprising the organism. However, such sequences provide little understanding of how prokaryotic genes work together to run the metabolism of an organism. One needs to understand how the prokaryote uses its genome to form the functioning organism. Sequencing is only the first part of a deeper understanding.

Having sequenced a microbial genome, the next step is to discover the functions for the genes. Sequences need to be analyzed (annotated) to identify the location of the genes and the function of their RNA or protein products. For example, in most of the microbial genomes sequenced to date, nearly 50 percent of the identified genes encoding proteins have not yet been connected with a cellular function. About 30 percent of these proteins are unique to each species. The challenging discipline of **functional genomics** attempts to discover what these proteins do and how those genes interact with others and the environment to maintain and allow the microbe to grow and reproduce.

One of the most important areas beyond DNA sequencing is the field of **comparative genomics**, which compares the DNA sequence from one microbe with the DNA sequence of another similar or dissimilar organism. Comparing sequences of similar genes indicates how genomes have evolved over time and provides clues to the relationships between microbes on the phylogenetic tree of life (see Chapter 1).

Comparisons indicate some strains of a bacterial species may contain **genomic islands**, sequences of up to 25 genes that are absent from other strains of the same prokaryotic species. Many of these islands can be identified as having come from an altogether different prokaryotic species, suggesting some form of HGT, such as conjugation. It is believed the nonpathogenic bacterial species *Thermotoga maritima* has acquired about 25 percent of its genome from HGT. In addition, sequence analysis indicates its genome is a mixture of bacterial and archaeal genes. This suggests *T. maritima* evolved before the split of *Bacteria* and *Archaea* domains.

One of the most interesting aspects of comparative genomics relates to infectious disease. By comparing the genomes of pathogenic and nonpathogenic bacterial species, or between pathogens with different host ranges, microbiologists are learning a lot about pathogen evolution. Here are a few examples.

There are three bacterial species of *Bordetella* (Chapter 9). *B. pertussis* causes whooping cough in humans, *B. parapertussis* causes whooping cough in infants, but also infects sheep, and *B. bronchiseptica* produces respiratory infections in other animals. Comparative genomic analysis of these three species reveals that *B. pertussis* and *B. parapertussis* are missing large segments of DNA (1719 genes), which are present in *B. bronchiseptica*. This analysis suggests (1) *B. pertussis* and *B. parapertussis* evolved from a *B. bronchiseptica*-like ancestor; and (2) the adaptation of *B. pertussis* and *B. parapertussis* to their more restrictive hosts is due to the loss of the 1719 genes. In fact, only *B. bronchiseptica* is capable of surviving outside the host. So, in this genome comparison between similar species, survival of *B. pertussis* and *B. parapertussis* requires they infect organisms supplying them with the materials they no longer can make; that is, pathogenicity has evolved from the loss of gene function.

At the opposite extreme is the evolution of pathogenicity through the acquisition of new genes. *Corynebacterium diphtheriae* is the causative agent for diphtheria (Chapter 9). Genome analysis indicates this species in the not too distant past has acquired through HGT 13 genetic regions, each representing a genomic island. These islands are called **pathogenicity islands** because they encode many of the pathogenic characteristics of the bacterial species (e.g., pili formation and iron uptake).

E. coli O157:H7 has recently become a dangerous threat to human health worldwide, causing severe gastrointestinal ailments (Chapter 10). One of the most recent outbreaks involved the contamination of bagged spinach. Some 200 Americans became ill and at least two died. When the genome of *E. coli* 0157:H7 was compared to the nucleotide sequence of a non-pathogenic strain, another example for the presence of pathogenicity islands was discovered. Both strains have a large genome and have evolved from a common ancestor. Both have genomic islands acquired through HGT. However, the genomic islands in *E. coli* O157:H7 code for the known pathogenicity genes (e.g., pili and toxins) and therefore represent pathogenicity islands. The genomic islands in the non-pathogenic strain lack these pathogenicity genes. What is not clear is if the pathogenicity islands were acquired only by the O157:H7 strain or the non-pathogenic strain lost the pathogenicity islands.

These few examples represent examples of the power of comparative genomics to resolve differences between species and shed light on the evolution of bacterial pathogens—something we will explore in detail in the next section of the text.

CONCEPT AND REASONING CHECKS

8.10 Explain how comparative genomics helps explain some aspects of pathogenicity.

SUMMARY OF KEY CONCEPTS

8.1 Genetic Recombination in Prokaryotes

- Recombination implies a horizontal transfer of DNA fragments between prokaryotes and an acquisition of genes by the recipient cell. All three forms of recombination are characterized by the introduction of new genes to a recipient cell by horizontal gene transfer.
- In transformation, a competent prokaryotic cell takes up DNA fragments from the local environment. The new DNA fragment displaces a segment of equivalent DNA in the recipient cell, and new genetic characteristics may be expressed.
- In one form of conjugation, a live donor (F^+) cell transfers an F factor (plasmid) to a recipient cell (F^-), which then becomes F^+.
- In another form of conjugation, Hfr strains contribute a portion of the donor's chromosomal genes to the recipient cell.
- Transduction involves a virus entering a prokaryote and later replicating within it. In generalized transduction, a prokaryotic DNA fragment is mistakenly incorporated into an assembling phage. In specialized transduction, the virus first incorporates itself into, then detaches from, the chromosome, taking a segment of chromosomal DNA with it. In both forms, the phage transports the DNA to a new recipient cell (transduced cell).

8.2 Genetic Engineering and Biotechnology

- Genetic engineering is an outgrowth of studies in prokaryotic genetic recombination. The ability to construct recombinant DNA molecules was based on the ability of restriction endonucleases to form sticky ends on DNA fragments.
- Plasmids are isolated from a prokaryotic cell, spliced with foreign genes, then inserted into fresh bacterial cells where the foreign genes are expressed as protein. Prokaryotes are used as the biochemical factories for the synthesis of such proteins as insulin, interferon, and human growth hormone.
- Genetic engineering is only one branch of modern biotechnology. This technology applies DNA-based genetic engineering techniques to environmental biology, antibiotic production, pharmaceutical and agricultural industries, pathogen detection and identifications, and vaccine production improvements.

8.3 Microbial Genomics

- Since 1995, increasingly more microbial genomes have been sequenced, that is, the linear sequence of bases has been annotated.
- A comparison of prokaryotic genomes with the human genome has shown there may be some 200 genes in common between these organisms. Comparisons between microbial genomes indicate almost 50 percent of the identified genes have yet to be associated with a protein or function in the cell.
- With the understanding of the relationships between sequenced microbial DNA molecules comes the potential for safer food production, the identification of unculturable prokaryotes, a cleaner environment, and improved monitoring of the environment.
- Sequencing is only the first step in understanding the behaviors and capabilities of microorganisms. Functional genomics attempts to determine the functions of the sequenced genes and how those genes interact with one another and with the environment. Comparative genomics compares the similarities and differences between microbial genome sequences. Such information provides an understanding of the evolutionary past and how pathogens might have arose through the gain or loss of pathogenicity islands.

LEARNING OBJECTIVES

After understanding the textbook reading, you should be capable of writing a paragraph that includes the appropriate terms and pertinent information to answer the objective.

1. Contrast vertical and horizontal gene transfer mechanisms.
2. Describe and assess the role of transformation as a genetic recombination mechanism.
3. Distinguish between F^+- and Hfr-induced conjugation mechanisms.
4. Summarize the steps involved in (a) generalize and (b) specialized transduction.
5. Differentiate between genetic engineering and biotechnology.
6. Explain how a recombinant DNA molecule is similar to natural genetic recombination, especially transformation.
7. Identify the role of plasmids and restriction endonucleases in the genetic engineering process.
8. Construct a diagram illustrating how a human gene can be spliced into plasmids and then expressed in bacterial cells.
9. List and explain how genetic engineering and biotechnology have been important to environmental biology, antibiotic production, pharmaceutical applications, detection and diagnosis of pathogens, and vaccine production.
10. Assess the importance of prokaryotic genomics in understanding the human genome.
11. Summarize some potential consequences arising from studies in microbial genomics.
12. Justify the need for comparative genomics as related to pathogen evolution.

SELF-TEST

Answer each of the following questions by selecting the ***one*** answer that best fits the question or statement. Answers to even-numbered questions can be found in **Appendix C**.

1. Which one of the following (**A–D**) is *not* an example of genetic recombination? If all are examples, select **E**.
A. F^+ and F^- conjugation
B. Binary fission
C. Transduction
D. Transformation
E. All the above (**A–D**) are examples.

2. Transformation refers to
A. using a virus to transfer DNA fragments.
B. DNA fragments transfered between live donor and recipient cells.
C. the formation of a transduced cell.
D. the formation of an F^- recombinant cell.
E. the transfer of naked fragments of DNA.

3. Competence is the ability of
A. donor cells to transfer DNA fragments.
B. phage to carry chromosomal DNA fragments.
C. recipient cells to take up DNA fragments.
D. F^+ and F^- cells to conjugate.
E. plasmids to be spliced with a foreign DNA.

4. An F^- cell is unable to initiate conjugation because it lacks
A. double-stranded DNA.
B. an F factor.
C. a prophage.
D. DNA polymerase.
E. DNA ligase.

5. An Hfr cell
A. has a free F factor in the cytoplasm.
B. lacks pili.
C. has a chromosomally-integrated F factor.
D. contains a prophage for conjugation.
E. cannot conjugate with a F^- recombinant.

6. A _____ is *not* associated with specialized transduction.
A. virulent phage
B. lysogenic cycle
C. prophage
D. recipient cell
E. lysed host cell

7. Which complementary sequence would *not* be recognized by a restriction endonuclease?
A. GAATTC
CTTAAG
B. AAGCTT
TTCGAA
C. GTCGAC
CAGCTG
D. AATTCC
TTAAGG
E. CTGCAG
GACGTC

8. Plasmids are
A. closed loops of DNA.
B. accessory genetic information.
C. used for genetic engineering.
D. opened by restriction endonucleases.
E. All the above (**A–D**) are correct.

9. A _____ seals sticky ends of recombinant DNA segments.
A. DNA ligase
B. helicase
C. restriction endonuclease
D. protease
E. RNA polymerase

10. The first genetically-engineered human protein was
A. interferon.
B. insulin.
C. blood clotting factor IX.
D. TPA.
E. somatotropin.

11. Gene transfer experiments in plants often use a plasmid from
A. *E. coli*.
B. fungi.
C. *Agrobacterium*.
D. *Bacillus*.
E. protozoa.

12. _____ are single-stranded DNA molecules that can recognize and bind to a distinctive nucleotide sequence of a pathogen.
A. Prophages
B. Plasmids
C. Nucleases
D. Cloning vectors
E. DNA probes

13. The first completely sequenced genome from a free-living organism was from
A. humans.
B. *E. coli*.
C. *Saccharomyces*.
D. *Haemophilus*.
E. *Bordetella*.

14. A metagenome refers to
A. a large genome in an organism.
B. the collective genomes of many organisms.
C. the genome of a metazoan.
D. the genome of an archaeal species.
E. two identical genomes in different species.

15. Comparative genomics compares the genomes of
A. two similar species.
B. two different species within a domain.
C. two similar strains of a species.
D. two species in different domains.
E. All the above (**A–D**) are correct.

16. Genomic islands are
A. gene sequences not part of the chromosomal genes.
B. adjacent gene sequences unique to one or a few strains in a species.
C. acquired by HGT.
D. gene sequences that move from one location to another within a chromosome.
E. Both **B** and **C** are correct.

QUESTIONS FOR THOUGHT AND DISCUSSION

Answers to even-numbered questions can be found in **Appendix C**.

1. Which of the recombination processes (transformation, conjugation, or transduction) would be most likely to occur in the natural environment? What factors would encourage or discourage your choice from taking place?
2. Some prokaryotes can take up DNA via the transformation process. From an evolutionary perspective, what might have been the original advantage for a prokaryote taking up naked DNA fragments from the extracellular environment?
3. Since the 1950s, the world has been plagued by a broad series of influenza viruses that differ genetically from one another. For example, we have heard of swine flu, Hong Kong flu, Bangkok flu, and avian flu. How might the process of transduction help explain this variability?
4. It is not uncommon for students of microbiology to confuse the terms reproduction and recombination. How do the terms differ?
5. While studying for the microbiology exam covering the material in this chapter, a friend and biology major asks you why genomics, and especially prokaryotic genomics, was emphasized. How would you answer this question?

APPLICATIONS

Answers to even-numbered questions can be found in **Appendix C**.

1. In 1976, an outbreak of pulmonary infections among participants at an American Legion convention in Philadelphia led to the identification of a new disease, Legionnaires' disease. The bacterial organism responsible for the disease had never before been known to be pathogenic. From your knowledge of bacterial genetics, can you postulate how it might have acquired the ability to cause disease?
2. You are going to do a genetic engineering experiment, but the labels have fallen off the bottles containing the restriction endonucleases. One loose label says *Eco*RI and the other says *Pvu*I. How could you use the plasmid shown in MicroInquiry 8 to determine which bottle contains the *Pvu*I restriction enzyme?
3. As a research member of a genomics company, you are asked to take the lead on sequencing the genome of *Legionella pneumophila*. (a) Why is your company interested in sequencing this bacterial species, and (b) what possible applications are possible from knowing its DNA sequence?

REVIEW

Use the following syllables to compose the term that answers each of the clues below. The number of letters in each term is indicated by the blank lines, and the number of syllables is shown by the number in parentheses. Each syllable is used only once, and the answers to even-numbered statements are in **Appendix C**.

ASE BAC CLE COC COM CON CUS DO DROME EN FER FITH GA GASE GE GE GRIF HOR I I IN IN JU LENT LI LI LY MIDS MO NOME NU NY O ON PAL PE PHAGE PI PLAS PNEU SO TAL TENCE TER TER TION U VIR ZON

1. Closed loops of DNA (2) ___ ___ ___ ___ ___ ___ ___ ___
2. Restriction recognition sequence (3) ___ ___ ___ ___ ___ ___ ___ ___ ___ ___
3. Transforming property (3) ___ ___ ___ ___ ___ ___ ___ ___ ___ ___
4. Transduction virus (5) ___ ___ ___ ___ ___ ___ ___ ___ ___ ___ ___ ___ ___
5. Recombinant DNA enzyme (5) ___ ___ ___ ___ ___ ___ ___ ___ ___ ___ ___ ___
6. Transformed bacterium (4) ___ ___ ___ ___ ___ ___ ___ ___ ___ ___ ___ ___
7. Conjugation structures (2) ___ ___ ___ ___
8. Viral non-replication (4) ___ ___ ___ ___ ___ ___ ___ ___
9. DNA linking enzyme (2) ___ ___ ___ ___ ___ ___
10. Gene engineered drug (4) ___ ___ ___ ___ ___ ___ ___ ___ ___ ___
11. Discovered transformation (2) ___ ___ ___ ___ ___ ___ ___ ___
12. Type of HGT (4) ___ ___ ___ ___ ___ ___ ___ ___ ___ ___ ___
13. Phage that causes lysis (3) ___ ___ ___ ___ ___ ___ ___ ___
14. Complete set of genes (2) ___ ___ ___ ___ ___ ___
15. Type of gene transfer (4) ___ ___ ___ ___ ___ ___ ___ ___ ___ ___

HTTP://MICROBIOLOGY.JBPUB.COM/

The site features learning, an on-line review area that provides quizzes and other tools to help you study for your class. You can also follow useful links for in-depth information, read more MicroFocus stories, or just find out the latest microbiology news.

False-color scanning electron micrograph of *Staphylococcus aureus* cells

PART 3

Bacterial Diseases of Humans

Throughout history, bacterial diseases have posed a formidable challenge to humans and often swept through populations virtually unchecked. In the eighteenth century, the first European visitors to the South Pacific found the islanders robust, happy, and well adapted to their environment. But the explorers introduced syphilis, tuberculosis, and pertussis (whooping cough) to a susceptible population. Soon these diseases spread like wildfire. For example, the Hawaiian population was about 300,000 when Captain Cook landed in 1778; by 1860, disease had reduced the population to fewer than 37,000.

With equally devastating results, the Great Plague came to Europe from Asia, and cholera spread westward from India. Together with tuberculosis, diphtheria, and dysentery, these bacterial diseases ravaged European populations for centuries and insidiously wove themselves into the pattern of life. Infant mortality was partic-ularly shocking: England's Queen Anne, who reigned in the early 1700s, lost 16 of her 17 babies to disease; and until the mid-1800s, only half the children born in the United States reached their fifth year.

Today, humans can cope better with bacterial diseases. Though credit often is given to antimicrobial drugs, the major health gains have resulted from understanding disease and the body's resistance mechanisms, coupled with modern sanitary methods to prevent microorganisms from reaching their targets. Immunization also has played a key role in preventing disease. Indeed, very few people in our society die of the bacterial diseases that once accounted for the majority of all deaths.

In Part 3 of this text, we study the bacterial diseases of humans. The diseases have been grouped according to their major mode of transmission. Airborne diseases are discussed in Chapter 9; foodborne and waterborne diseases in Chapter 10; soilborne and arthropodborne diseases in Chapter 11; and sexually transmitted, contact, and miscellaneous diseases in Chapter 12. Many of these diseases are of historical interest and are currently under control. However, the human body is continually confronted with newly emerging or resurgent infectious diseases. In this regard, disease has not changed. Only the pattern of disease has changed.

MICROBIOLOGY PATHWAYS

Clinical Microbiology

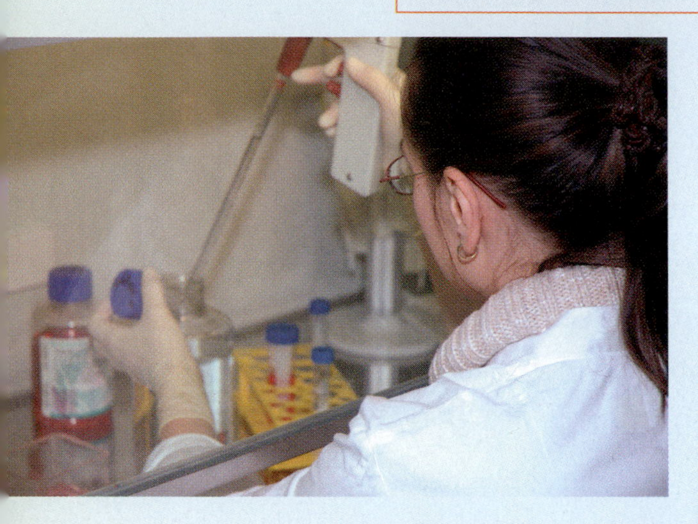

One of the most famous books of the twentieth century was *Microbe Hunters* by Paul de Kruif. First published in the 1920s, de Kruif's book describes the joys and frustrations of Pasteur, Koch, Ehrlich, von Behring, and many of the original microbe hunters. The exploits of these scientists make for fascinating reading and help us understand how the concepts of microbiology were formulated. I would urge you to leaf through the book at your leisure.

Microbe hunters did not come to an end with Pasteur, Koch, and their contemporaries, nor did the stories of microbe hunters end with the publication of de Kruif's book. Approximately 25 percent of all deaths worldwide and 60 percent of all deaths in children under four years of age are due to infectious agents. Today, clinical microbiology is concerned with the microbiology of infectious diseases, and the men and women working in hospital, public, and private laboratories are today's diseases detectives. These individuals search for the pathogens of disease. Many travel to far corners of the world studying organisms, and many more remain close to home, identifying the pathogens in samples sent by physicians, identifying their interactions with the immune system, and working out the diagnosis and epidemiology of these diseases.

In fact, a well-developed knowledge of clinical microbiology is critical for the physician and medical staff who are faced with the concepts of disease and antimicrobial therapy. Microbiologists even work in dental clinical labs, since many bacterial species are involved in tooth decay and periodontal disease. Microbiology is one of the few courses where much of the fundamentals of microbiology are used regularly. This includes the clinical aspects of infectious diseases: manifestations (signs and symptoms), diagnosis, treatment, and prevention.

A career in clinical microbiology usually requires a master of science in clinical microbiology. With such a degree, jobs include supervisory positions in medical centers or private reference laboratories, infection control positions in clinical settings, public health, marketing and sales in the pharmaceutical and biotechnology industries, teaching at community colleges or technical colleges, or research in academic, government or industry (pharmaceutical and biotechnology) settings.

Clinical microbiology also offers an outlet for the talents of those who prefer to tinker with machinery. New instruments and laboratory procedures are constantly being designed and developed in an effort to shorten the time between detection and identification of microorganisms. Many tests used in the clinical laboratory reflect human ingenuity. For example, there is a test that detects bacterial species by their interference with the passage of light and their ability to scatter light at peculiar angles. Such modern devices as laser beams are used in this kind of instrumentation.

The microbe hunters have not changed materially in the past 100 years. The objectives of the search may be different, but the fundamental principles of the detective work remain the same. The clinical microbiologist is today's version of the great masters of a bygone era.

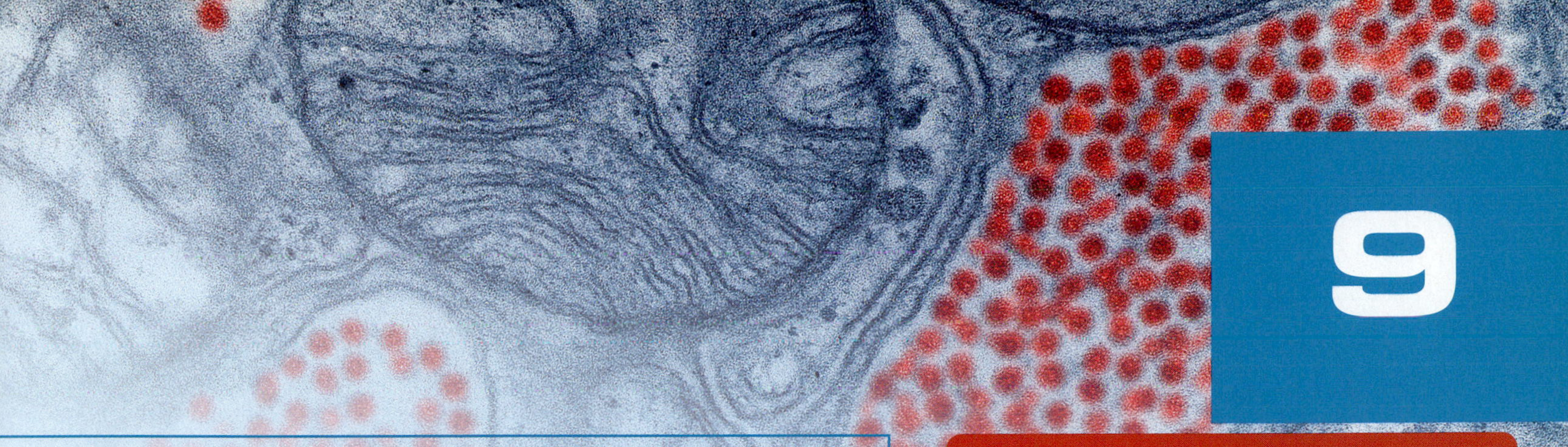

9

Airborne Bacterial Diseases

Pertussis is the only vaccine-preventable childhood illness that has continued to rise since the 1980s with an increasing proportion of cases in adolescents and adults.
—Centers for Disease Control and Prevention

Chapter Preview and Key Concepts

9.1 Diseases of the Upper Respiratory Tract
- Group A streptococci attach to cells and secrete toxins.
- *Corynebacterium diphtheriae* secretes a toxin that inhibits protein synthesis in epithelial cells.
- *Bordetella pertussis* secretes a toxin that destroys cells of the ciliated epithelium.
- Several bacterial species infect the nasopharynx and spread rapidly to the meninges.

9.2 Diseases of the Lower Respiratory Tract
- *Mycobacterium tuberculosis* can cause a lifelong infection.
- "Typical" bacterial pneumonia can be community or hospital acquired.
- "Atypical" bacterial pneumonia has an insidious onset.
- Some rickettsiae and chlamydiae are transmitted by dust particles or animal droppings.

MicroInquiry 9: Infectious Disease Identification

On August 14, 2002, a 39-year-old male oil refinery worker in Crawford County, Illinois, visited the refinery's health unit complaining of a two-week cough. Later that day, the worker's 50-year-old supervisor also visited the unit with a spastic cough, which had started three days earlier. Both patients were advised to see their own health care provider where blood samples indicated a recent infection with *Bordetella pertussis*. The Crawford County Health Department and Illinois Department of Public Health were contacted because a possible outbreak could be brewing.

In the early parts of the 20th century, one of the most common childhood diseases and causes of death in the United States was pertussis, commonly called whooping cough. Before the introduction of a pertussis vaccine in 1940, *B. pertussis* was responsible for infection and disease in 150 out of every 100,000 people. By 1980, the **incidence**, or frequency with which the disease occurs, had dropped to one in every 100,000 individuals. The vaccine had almost eliminated the pathogen.

At the oil refinery, active surveillance and case investigations were initiated by the health officials. Those workers with a persistent and spastic cough were sent to the local hospital for evaluation and interviews. Health department officials needed to know the time of illness onset, where workers worked in the refinery, work schedule, and individuals with whom they had close contacts. Local school officials and health care providers were alerted and given guidelines on ways

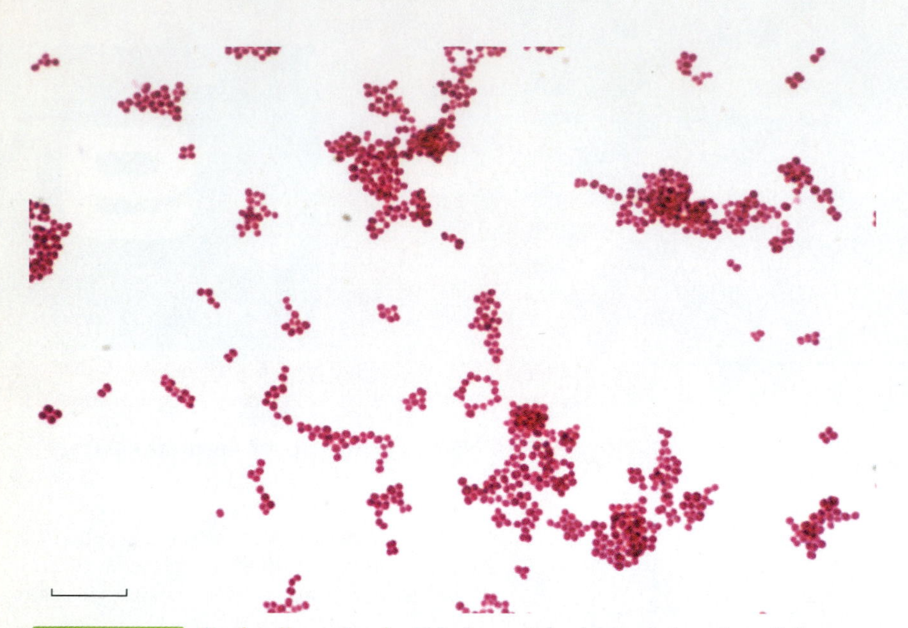

FIGURE 9.1 ***Bordetella pertussis.*** This Gram stain shows chains of small *B. pertussis* cells. (Bar = 10 μm.)

Q: What is the Gram reaction of these stained cells?

to recognize pertussis and preventive measures to prevent its spread.

In the course of the epidemiological investigation, 17 cases of pertussis were identified at the refinery, 15 having had close contact with the supervisor originally diagnosed; 7 cases occurred among the community and had no apparent relation to the refinery. In all, 21 of the cases occurred in adults 20 years of age or older. Patients received an antibiotic effective against the pathogen and all recovered.

How the disease was passed from the supervisor remains unclear. *B. pertussis* is spread by airborne droplets (FIGURE 9.1). Other than an indoor, 5-minute morning meeting each day, work assignments were all outdoors, although workers often congregated in an indoor dining area at lunch.

Every 3 to 4 years, a pertussis outbreak occurs in the United States—and, as indicated above, many of these cases occur in adults. Although nearly all youngsters growing up receive the pertussis vaccine, vaccine-induced protection does not last a lifetime; therefore, adolescents and adults can become susceptible to disease when vaccine-induced immunity wanes, approximately 5 to 10 years after vaccination. As a result, college students and adults (like the refinery workers) may be vulnerable (FIGURE 9.2).

Pertussis is but one of a group of airborne diseases spread by respiratory droplets. We will divide the airborne bacterial diseases into two general categories. The first category will include diseases of the upper respiratory tract, such as strep throat, scarlet fever, diphtheria, pertussis, and several forms of meningitis. The second category will include diseases of the lower respiratory tract: tuberculosis, pneumococcal pneumonia, primary atypical pneumonia, Legionnaires' disease, and others. As we proceed, note how the initial focus of infection is followed by spread to other organs. Also note that antibiotics are available for treating these diseases and immunizations are used for protecting the community at large.

9.1 Diseases of the Upper Respiratory Tract

The bacterial diseases of the upper respiratory tract (URT) can be severe, as several diseases in this section illustrate. One reason is because the respiratory tract is a portal of entry to the blood, and from there, the bacterial pathogen can spread to other sensitive internal organs.

■ Portal of entry: Refers to the site at which a pathogen enters the host.

Streptococcal Diseases Can Be Mild to Severe

KEY CONCEPT

- Group A streptococci attach to cells and secrete toxins.

Streptococci are a large and diverse group of encapsulated, nonmotile, facultatively anaerobic, gram-positive cocci. The bacterial cells divide in one plane and cling together to form pairs or chains of various lengths (FIGURE 9.3A).

Microbiologists classify the streptococci by two widely accepted systems. The first system, developed by J. H. Brown in 1919, divides streptococci into hemolytic (*hemo* = "blood") groups, depending on how they affect sheep red blood cells in blood agar. The **α-hemolytic** streptococci turn blood agar an olive-green

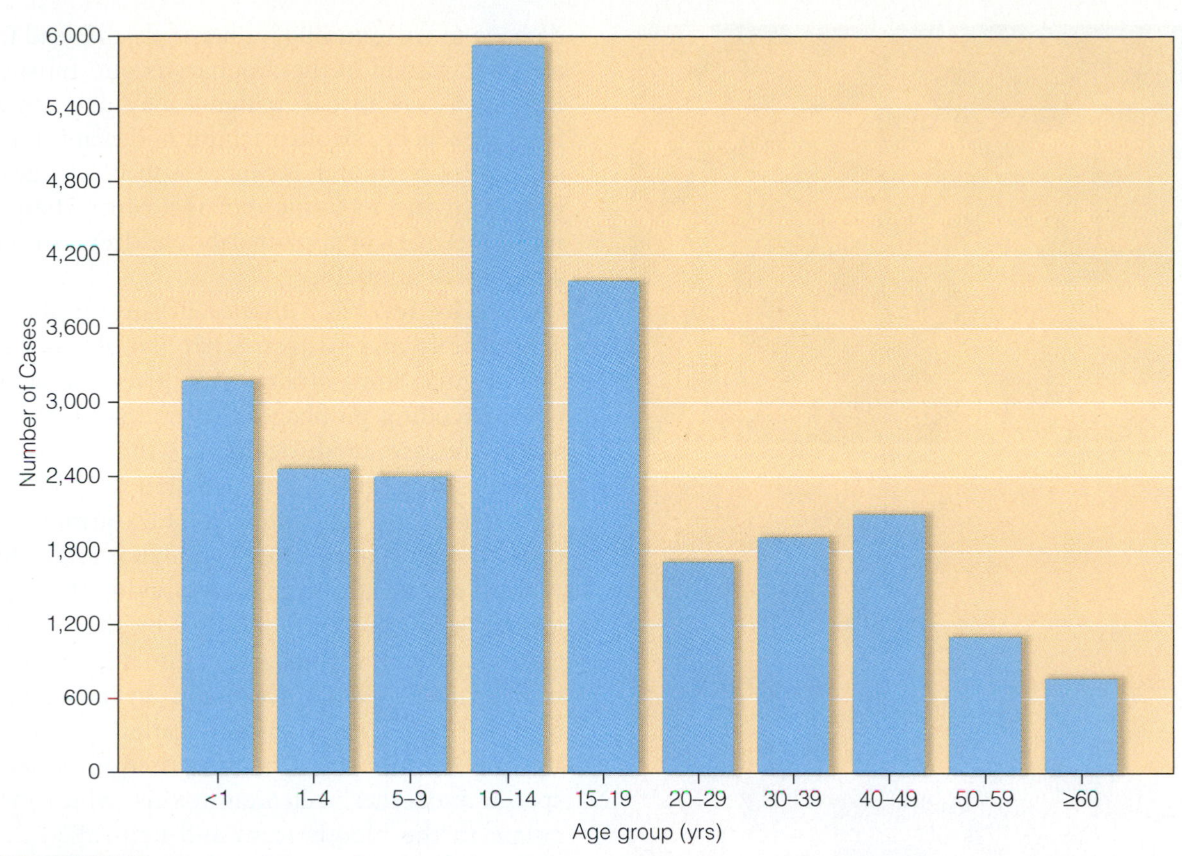

FIGURE 9.2 **Number of Cases of Pertussis by Age Group—United States 2004.** The actual number of pertussis cases (especially among adolescents [10–19 years] and adults) is substantially underreported because the illness resembles other conditions, so infected individuals might not seek medical care. Since 1990, the percentage of cases in adolescents and adults has increased from 20 percent to 67 percent. (*Source*: http://www.cdc.gov/mmwr/pdf/wk/mm5353.pdf, page 59.)

Q: Which age group (infants [<1], children [1–9 years], adolescents, or adults) accounts for the majority of reported cases?

color as a secreted toxin partially destroys red blood cells in the medium; colonies of **β-hemolytic** streptococci produce clear, colorless zones around the colonies due to the complete destruction of red blood cells (FIGURE 9.3B); and **non-hemolytic** (gamma) streptococci have no effect on red blood cells and thus cause no change in blood agar.

The second classification system, developed by American bacteriologist Rebecca Lancefield in 1933, is based on variants of a carbohydrate located in the cell walls of streptococci. Groups A and B are the most important to human disease. In fact, most streptococcal diseases of the upper respiratory tract are caused by members of group A; *Streptococcus pyogenes* is the most common species. This β-hemolytic organism is generally implied when physicians refer to **group A streptococci** (GAS).

The pathogenicity of *S. pyogenes* is enhanced by the presence of a capsule, pili, and a substance called **M protein**. This protein is anchored in the cell membrane, traverses the cell wall, and appears as fibrils protruding from the cell surface (FIGURE 9.3C). The protein helps the cells adhere to pharyngeal tissue and retards phagocytosis. Over 100 serotypes of M protein have been identified.

The GAS are among the most common human bacterial pathogens. They can be carried asymptomatically but also can be responsible for a variety of diseases; most are relatively mild, such as **streptococcal pharyngitis**, popularly known as **strep throat**. The *S. pyogenes* cells reach the upper respiratory tract within respiratory droplets expelled by infected persons during coughing and sneezing. If the cells grow and secrete toxins, these substances cause

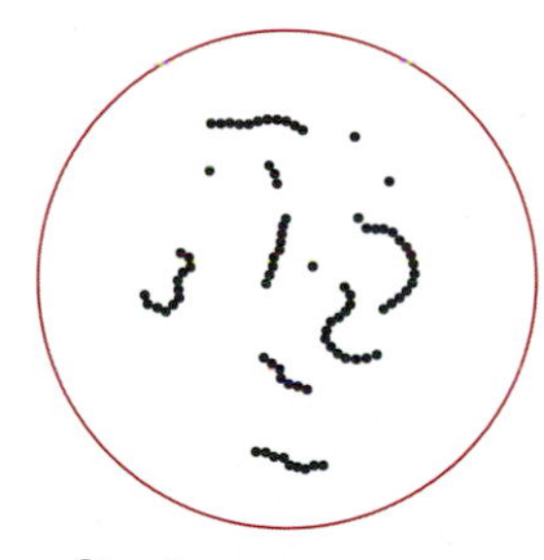

Streptococcus pyogenes

- **Serotypes:** Closely related groups of microorganisms or structures distinguished by a specific set of antigens.
- **Respiratory droplets:** Relatively large mucus particles that travel less than one meter.

■ Inflammation:
A defensive response to injury.

■ Exotoxins:
Toxic proteins released by certain bacterial species into the environment.

■ Symptoms:
Effects of an infection or disease experienced by the patient.

damage to surrounding human cells and lead to an inflammation of the oropharynx and tonsils. Besides a sore throat, patients may develop a fever, headache, swollen lymph nodes and tonsils, and a beefy red appearance to pharyngeal tissues owing to tissue damage. More than a million Americans, primarily children, suffer strep throat annually.

Scarlet fever is a disease arising in about 10 percent of children with streptococcal pharyngitis. Some strains of *S. pyogenes* carry toxin-encoding prophages. In the throat, these strains release **erythrogenic** (*erythro* = "red") exotoxins that cause a pink-red skin rash on the neck, chest, and soft-skin areas of the arms. The rash, which usually occurs in children under 18 years of age, results from blood leaking through the walls of capillaries damaged by the toxins. Other symptoms include a sore throat, fever, and swollen neck glands. Normally, an individual experiences only one case of scarlet fever in a lifetime because the immune system produces special antibodies, called **antitoxins**, which circulate in the bloodstream and neutralize the toxins during succeeding episodes.

Individuals with strep throat or scarlet fever usually get better within two weeks without treatment (MicroFocus 9.1). Treatment with antibiotics, such as penicillin or clarithromycin, can shorten the duration of symptoms and prevent serious complications.

A serious complication is **rheumatic fever**, which is most common in young school-age children. This condition, which is not an infection but rather an inflammation in response to the throat infection, primarily affects the joints and heart. It is characterized by fever and joint pain. The most significant long-range effect is permanent scarring and distortion of the heart valves, a condition called **rheumatic heart disease**. The damage arises from a response of the body's antibodies to streptococcal M proteins cross reacting with similar proteins on heart tissue. Rheumatic fever cases have been declining in the United States due to antibiotic treatment. In 1994, the last year the Centers for Disease Control and Prevention (CDC) required reporting of rheumatic fever cases, 112 cases were reported versus 10,000 reported cases in 1961. However, in developing nations, rheumatic fever remains a serious problem.

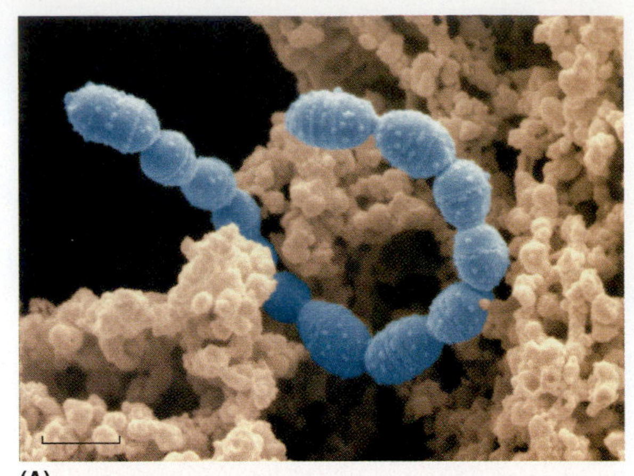

(A)

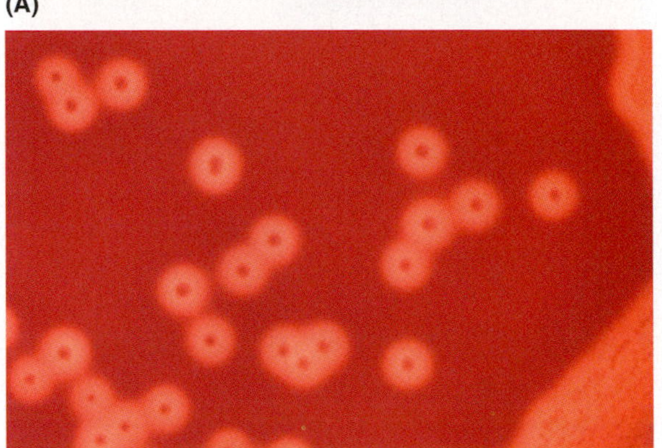

(B)

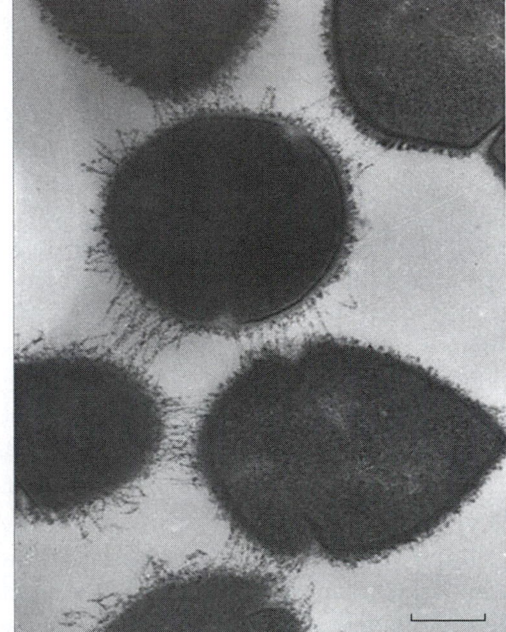

(C)

FIGURE 9.3 **Streptococci.** (**A**) A false-color scanning electron micrograph of streptococci growing as long chains. (Bar = 2 μm.) (**B**) An example of β-hemolysis caused by a toxin released from *Streptococcus pyogenes* cells. (**C**) A close-up view of *S. pyogenes* cells showing strands of M protein protruding through the capsule. (Bar = 0.5 μm.)

Q: What function does M protein play in pathogenesis?

MICROFOCUS 9.1: Public Health
A Wakeup Call

In life, sometimes the best way to move past a road block is to simply face it head on—or attack the defensive team by running right at them. Often in the microbial world, bacterial species may have the same idea—but with a twist.

One of the most common host responses to a group A streptococci (GAS) infection in humans is to try to contain the infection by forming a blood blot around the infected area (see figure)—in other words, the host attempts to entomb the bacterial cells in the clot. Unfortunately, one of the abilities GAS possess is to simply break through the defensive wall set up to contain the infection. How do they do this?

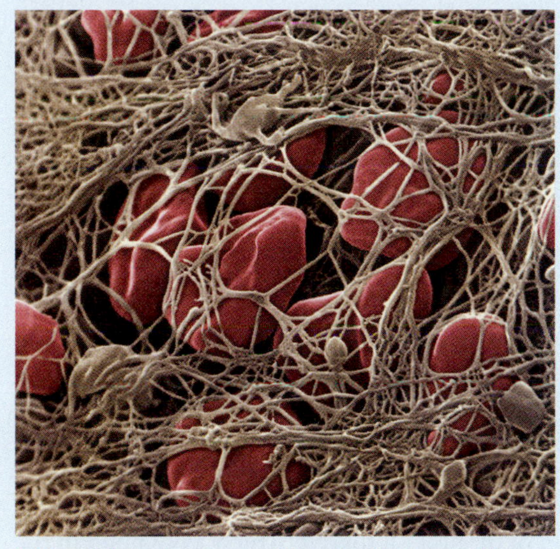
A blood clot, showing red blood cells trapped in a fibrin mesh.

Normally a blood clot stays as a clot because the protein plasmin that would dissolve the clot stays in an inactive form called plasminogen. What *Streptococcus pyogenes* does is "wake up" or activate plasminogen.

Trapped within a clot, the *S. pyogenes* cells secrete an enzyme called streptokinase. What streptokinase does is to catalyze the conversion of inactive plasminogen into active plasmin—in other words, it wakes the protein up. As plasmin, the protein triggers a series of reactions leading to the dissolving of the clot. Now the bacteria can escape and perhaps cause a serious infection elsewhere in the body.

Another complication arising from streptococcal pharyngitis is **acute glomerulonephritis**, which is a rare inflammatory response to specific types of M proteins. It is most common in young patients, who usually have an uneventful recovery. Progressive, irreversible renal damage may occur in adults.

GAS diseases also can occur when streptococci infect the skin, muscle, or blood (Chapter 12). Infections of the lower respiratory tract, causing streptococcal pneumonia are described later in this chapter.

CONCEPT AND REASONING CHECKS

9.1 What makes *S. pyogenes* such a potentially dangerous pathogen in the upper respiratory tract?

Diphtheria Is a Life-Threatening Illness

KEY CONCEPT

- *Corynebacterium diphtheriae* secretes a toxin that inhibits protein synthesis in epithelial cells.

Diphtheria is a local infection of the throat. The disease was first recognized in 1826 by French pathologist Pierre F. Bretonneau. He named the disease la diphtherite, from the Greek *diphthera* for "membrane," a reference to the exudate appearing in the throats of patients. In 1883, Edwin Klebs observed bacterial cells in material from a patient's throat, and the next year, Friederich Löeffler successfully cultivated the organism, which appeared as club-shaped cells. The organism was named *Corynebacterium diphtheriae* (*coryne* = "club").

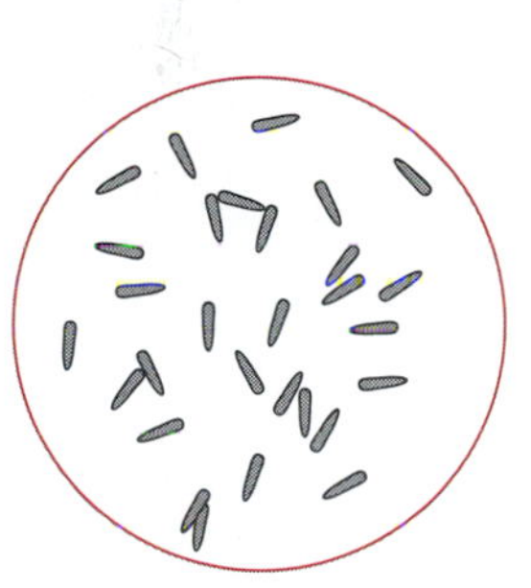
Corynebacterium diphtheriae

The cells of *C. diphtheriae* are noncapsulated, nonmotile, aerobic, gram-positive bacilli. When stained with methylene blue, numerous blue cytoplasmic dots are seen that represent **metachromatic granules** (see Chapter 4). The bacterial cells remain in clumps after multiplying and form a picket fence-like arrangement called a palisade layer.

Diphtheria is acquired by inhaling respiratory droplets from an infected person. Initial symptoms include a sore throat and low-grade fever. In epithelial cells, the bacterial cells secrete a potent exotoxin, which is encoded by a corynebacterium-containing prophage gene. The exotoxin inhibits the translation process by ribosomes. The result

■ Epithelial cells: Cube-like cells lining the skin and body cavities such as the respiratory tract.

is the accumulation of dead tissue, mucus, white blood cells, and fibrous material, called a **pseudomembrane** ("pseudo" because it does not fit the definition of a true membrane) on the tonsils or pharynx. Mild cases fade after a week while more severe cases can persist for two to six weeks.

Complications can arise if the thickened pseudomembrane results in respiratory blockage. If the exotoxin spreads to the bloodstream, heart damage and destruction of the fatty sheath surrounding nerves can lead to cardiac arrhythmia and coma. Left untreated, 5 to 10 percent of respiratory cases result in death.

■ Arrhythmia: An irregular heart beat.

Treatment requires antibiotics (penicillin or erythromycin) to eradicate the pathogen and antitoxins to neutralize the exotoxins. Immunization against diphtheria may be rendered by an injection of diphtheria **toxoid**, which, contained in the diphtheria-tetanus-acellular pertussis (DTaP) vaccine, consists of toxin molecules treated with formaldehyde or heat to destroy their toxic qualities. The toxoid induces the immune system to produce antitoxins (antibodies) that circulate in the bloodstream throughout the person's life.

Due to immunization starting in early childhood, the number of cases of diphtheria in the United States is less than a dozen annually (one confirmed case in 2003). However, the disease remains a health problem in many regions of the world and booster doses are required. From 1990 to 1997, for example, a major diphtheria outbreak occurred in the Newly Independent and Baltic States of the former Soviet Union due to inadequate immunization practices; it affected almost 150,000 people, with about 5,000 deaths.

CONCEPT AND REASONING CHECKS

9.2 In 17th century Spain, diphtheria was called "el garatillo" = the strangler. Why was it given this name?

Pertussis (Whooping Cough) Is Highly Contagious

KEY CONCEPT

- *Bordetella pertussis* secretes a toxin that destroys cells of the ciliated epithelium.

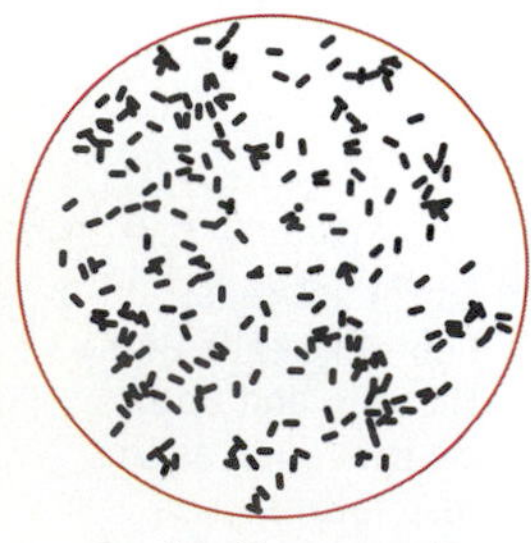

Bordetella pertussis

Pertussis, also known as **whooping cough**, is caused by *Bordetella pertussis*, a small, nonmotile, aerobic, gram-negative rod first isolated by Jules Bordet and Octave Gengou in 1906. The bacilli are spread by respiratory droplets that adhere to and aggregate on the cilia of epithelial cells in the upper respiratory tract (FIGURE 9.4). Cytotoxin production paralyzes the ciliated cells and impairs mucus movement, potentially causing pneumonia. Although the bacterial cells produce several toxins, no one toxin is considered the prime factor in the disease.

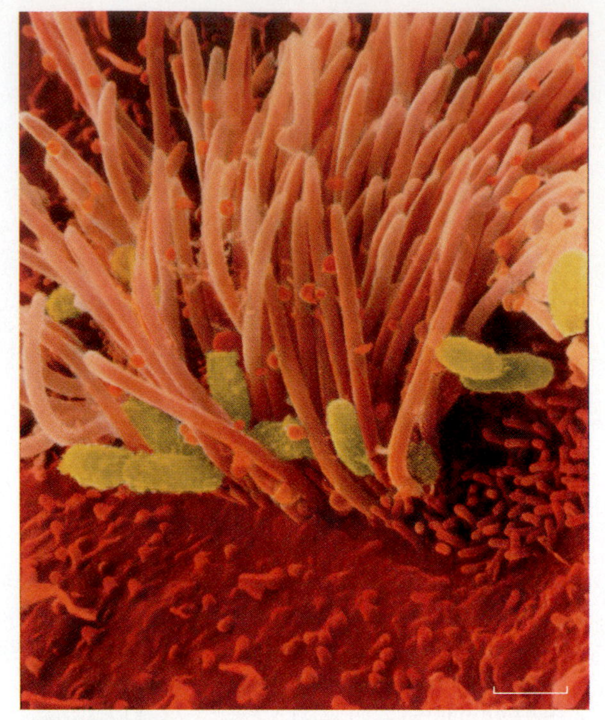

FIGURE 9.4 ***Bordetella pertussis* Associated with the Ciliated Epithelium.** False-color scanning electron micrograph of human tracheal epithelium. The pertussis cells (green) cause a dramatic loss of cilia function, which is to protect the respiratory tract from dust and particles. (Bar = 1 μm.)

Q: Propose a hypothesis to explain why cilia loss would lead to fits of coughing.

Pertussis is one of the more dangerous and highly contagious diseases of childhood years. Typical cases of pertussis occur in two phases. The initial stage is marked by general malaise, low-grade fever, and increasingly severe cough. During the next phase, disintegrating cells and mucus accumulate in the airways and cause labored breathing. Patients experience multiple **paroxysms**, which consist of rapid-fire staccato coughs all in one exhalation, followed by a forced inhalation over a partially closed glottis. The rapid inhalation results in the characteristic "whoop" (hence, the name whooping cough). Ten to fifteen paroxysms may occur daily, and exhaustion usually follows each. Sporadic coughing continues during several weeks of convalescence, even after the pathogen has vanished. (Doctors call it the "100-day cough."). Convalescence depends on the speed at which the ciliated epithelium regenerates.

Eradication of the bacterial cells is generally successful when erythromycin is administered before the respiratory passageways become blocked. However, antibiotic treatment only reduces the duration and severity of the illness.

As with diphtheria, the low incidence of pertussis stems partly from use of a pertussis vaccine. The older vaccine (diphtheria-pertussis-tetanus, or DPT) contained merthiolate (Thimerosal)-killed *B. pertussis* cells and was considered risky because about 1 in 300,000 vaccinees suffered high fevers and seizures. As of 1993, public health officials recommended the newer acellular pertussis (aP) vaccine prepared from *B. pertussis* chemical extracts. Combined with diphtheria and tetanus toxoids, the triple vaccine has the acronym DTaP; commercially, it is known as Tripedia.

Although the incidence of pertussis has declined substantially since introduction of an effective vaccine in 1949, the number of cases in the United States has been rising since 1981. In fact, the CDC recorded 25,827 cases in 2004, the largest number of pertussis cases since 1959. The majority of reported cases occur in adolescents (nearly 6,000 in 2004) because adolescents (and adults) lose vaccine-induced immunity 5 to 10 years after vaccination (see chapter opener). In 2005, the US Food and Drug Administration (FDA) licensed two new single-dose booster vaccines (Tdap) to provide adolescents and adults with protection against tetanus, diphtheria, and pertussis.

CONCEPT AND REASONING CHECKS

9.3 Why have reported cases of pertussis been increasing in the United States?

Bacterial Meningitis Can Be Life Threatening

KEY CONCEPT

- Several bacterial species infect the nasopharynx and spread rapidly to the meninges.

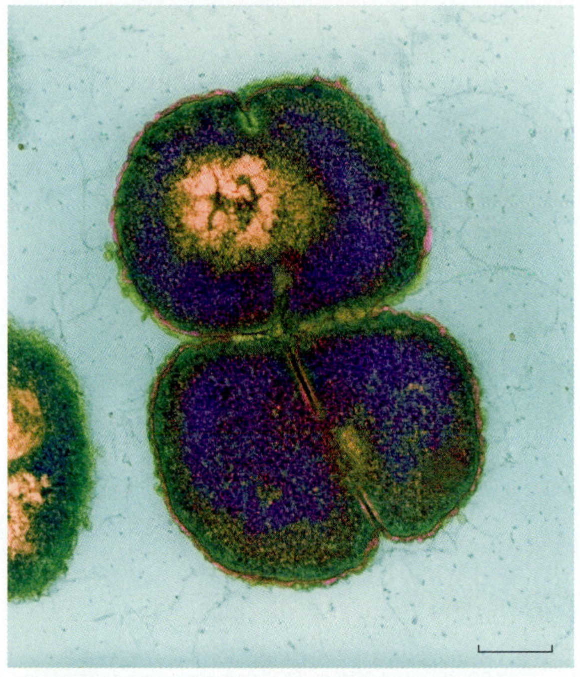

FIGURE 9.5 ***Neisseria meningitidis.*** A false-color transmission electron micrograph of *N. meningitidis* cells. The cytoplasm (blue) and cell membrane are surrounded by a capsule (green). (Bar = 0.5 µm.)

Q: What role does the capsule play in the disease process?

Merthiolate: A mercury derivative compound formerly used in vaccines as a disinfectant and preservative.

The term **meningitis** (*meningo* = "membrane") refers to several diseases of the **meninges**, the three membranous coverings of the brain and spinal cord. Although not a highly contagious disease, meningitis can be caused by viruses or one of several bacterial species. Cases of bacterial meningitis arising from the URT are primarily caused by *Neisseria meningitidis, Streptococcus pneumoniae,* or *Haemophilus influenzae* type b.

Neisseria meningitidis. A particularly dangerous form of meningitis is meningococcal meningitis caused by *N. meningitidis* (FIGURE 9.5). The pathogen is a small, nonmotile, encapsulated, aerobic, gram-negative diplococcus that attaches to the nasopharyngeal mucosa by pili. It also forms a capsule of which there are more than 14 serogroups. Most infections in the United States are due to serogroups B, C, Y, and W-135, with B the major cause of disease and mortality. According to the World Health Organization (WHO), bacterial meningitis is responsible for more than 700,000 cases globally each year. Ninety percent of cases occurred in the 18 countries

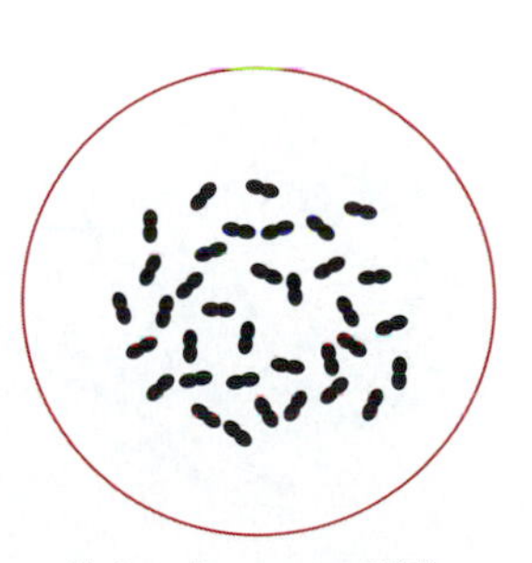

Neisseria meningitidis

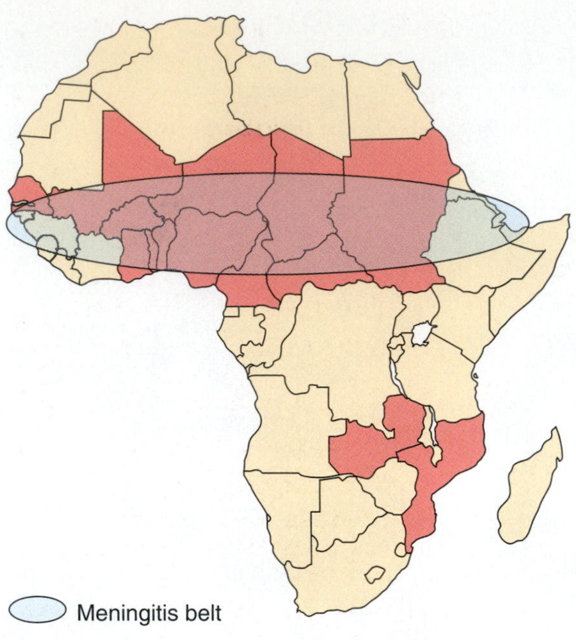

FIGURE 9.6 **The African Meningitis Belt.** This map of Africa shows the so-called "meningitis belt" where deadly epidemics caused by *N. meningitidis* occur almost every year. *Source:* http://www.medilinkz.org.

Q: How is N. meningitis *spread between individuals?*

forming sub-Saharan Africa's so-called "meningitis belt" (FIGURE 9.6).

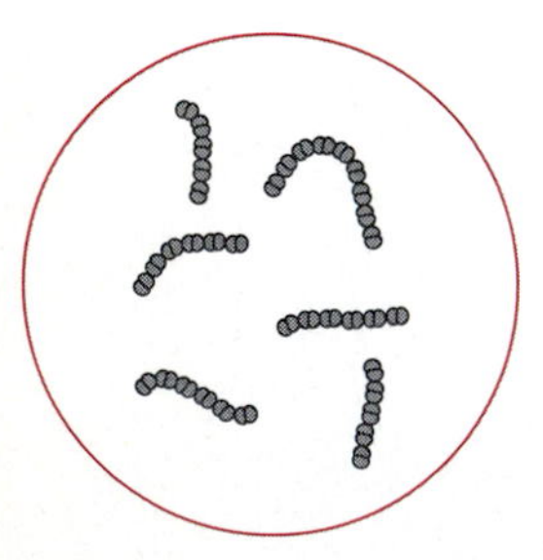

Streptococcus pneumoniae

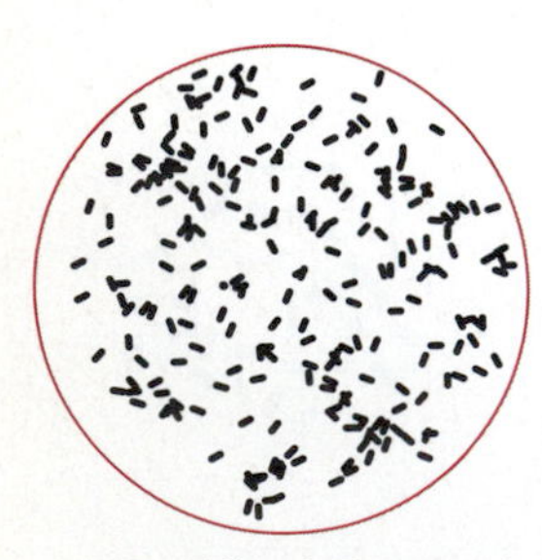

Haemophilus influenzae b

N. meningitidis is a fragile organism that does not survive easily in the environment and must be maintained in nature by person-to-person transfer of large droplet respiratory secretions. Meningococcal meningitis is therefore prevalent where people are in close proximity for long periods of time. Grade-school classrooms, military camps, college dorms, and prisons are examples. Though most people suffer nothing worse than a respiratory disease, the CDC reported 1,361 cases of meningococcal meningitis in 2004. Globally, it is the agent responsible for about 50 percent of meningitis cases in 2- to 18-year-olds.

Cases of meningococcal meningitis in very young children are sometimes complicated by the formation of lesions in the adrenal glands and accompanying hormone imbalances. This condition, called the **Waterhouse-Friderichsen syndrome**, results from the release of a bacterial endotoxin into the blood. Death can occur within 10 to 12 hours.

Streptococcus pneumoniae. Referred to as the pneumococci, *S. pneumoniae* has the same cell characteristics as *S. pyogenes*. The pneumococcus is not only responsible for many cases of pneumonia (see below) but also is the agent responsible for about 30 percent of meningitis cases. **Pneumococcal meningitis** has a high mortality rate (20–30 percent).

***Haemophilus influenzae* type b.** In 1892, Richard Pfeiffer isolated a small, nonmotile, encapsulated, gram-negative rod he thought was the cause of influenza. However, during the great influenza epidemic of 1918 to 1919, Pfeiffer's bacillus was identified as a secondary cause of the disease, and influenza was attributed to a virus rather than a bacterium (Chapter 14).

Haemophilus influenzae type b (Hib) once was the most prevalent bacterial species causing meningitis (***Haemophilus* meningitis**) in children between the ages of two months and five years. In 1986, about 18,000 cases of *Haemophilus* meningitis occurred in the United States annually. At that time, however, a vaccine had been licensed by the FDA and the epidemic had peaked. As of 1993, the vaccine was combined with the DTaP vaccine for distribution to children as Tetramune, and in 2004, the number of reported cases among children under 5 years of age in the United States was down to 19.

All three bacterial species discussed above enter the body by respiratory droplets from prolonged contact, such as coughing, sneezing, or kissing. They then colonize the nasopharynx and sinus cavities (FIGURE 9.7). In the case of *N. meningitidis,* the disease consists of an influenza-like upper respiratory infection called **meningococcal pharyngitis**. However, should the organism invade into the nonciliated epithelium and spread to the blood, a condition called **meningococcemia** occurs and the bacterial cells multiply rapidly.

Once in the blood, all three pathogens are capable of crossing the blood-brain barrier. The meninges then become inflamed, causing pressure on the spinal cord and brain. Patients normally experience a fever and stiff

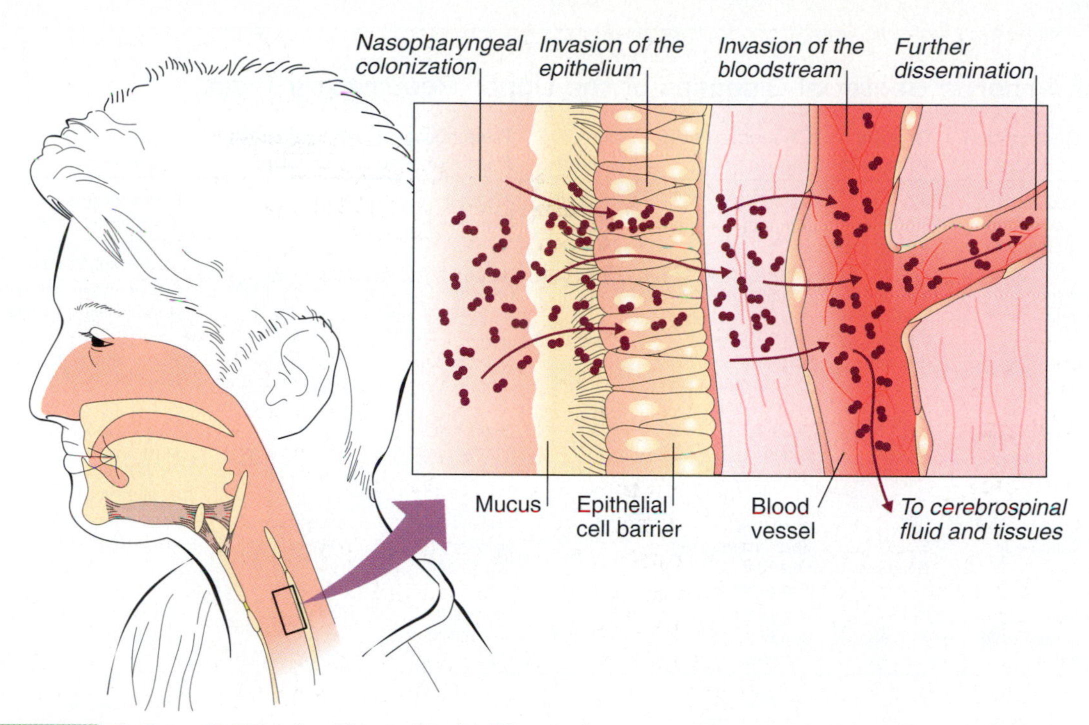

FIGURE 9.7 **Pathogenic Steps Leading to Meningitis.** The bacterial species capable of causing meningitis (*N. meningitidis, S. pneumoniae,* and *H. influenzae*) first colonize the nasopharynx, then invade the epithelium causing respiratory distress. They then pass into the bloodstream. Finally, they disseminate to tissues near the spinal cord, causing inflammation and meningitis.

Q: What bacterial structures would facilitate (a) attachment to the nasopharynx and (b) survival in the bloodstream?

neck; symptoms rapidly evolve into a pounding headache, nausea and vomiting, and often sensitivity to bright light. With meningococcal meningitis, a rash also appears on the skin, beginning as bright-red patches, which progress to blue-black spots. Left untreated, within a few hours, fifty percent of cases of meningococcal meningitis result in coma and death. The disease also can produce lasting disabilities, such as deafness, blindness, and paralysis.

Bacterial meningitis, especially, meningococcal meningitis, represents a medical emergency. Early diagnosis and treatment are crucial to prevent disabilities or death. A principal criterion for diagnosis is the observation and/or cultivation of the *N. meningitidis* cells in samples of spinal fluid obtained by a spinal tap. Treatment with antibiotics, such as penicillin, cefotaxime, or ceftriaxone, usually is recommended, often in large intravenous doses.

No single vaccine provides immunity to all causes of meningitis. A polysaccharide vaccine (Menomune) has been available for 25 years and is safe for all age groups. A new vaccine (Menactra) containing capsular serogroups (A, C, Y, and W–135) was licensed in 2005 for immunization against and more durable protection from meningococcal meningitis in persons 11 to 55 years. The CDC strongly recommends all college freshman be immunized before taking up residence in a college dorm. Vaccines against pneumococcal and *Haemophilus* meningitis also are available.

The airborne bacterial diseases of the upper respiratory tract are summarized in TABLE 9.1.

CONCEPT AND REASONING CHECKS

9.4 Draw a concept map for bacterial meningitis using the bacteria and bold-faced terms in this section.

TABLE 9.1 A Summary of Airborne Bacterial Diseases of the Upper Respiratory Tract

Disease	Causative Agent	Description of Agent	Organs Affected	Characteristic Signs	Toxin Involved	Treatment Administered	Immunization Available	Comment
Strep Throat Scarlet fever	*Streptococcus pyogenes*	Gram-positive encapsulated streptococcus	URT Blood Skin	Sore throat Skin rash	Erythrogenic toxin	Penicillin Erythromycin	None	Rheumatic fever or glomerulonephritis possible Beta-hemolytic strains
Diphtheria	*Corynebacterium diphtheriae*	Gram-positive rod	URT Heart, nerve fibers	Pseudomembrane Sore throat, moderate fever	Yes	Penicillin Antitoxin	Toxoid in DTaP	Bacteriophage involved Metachromatic granules
Pertussis (Whooping cough)	*Bordetella pertussis*	Gram-negative rod	URT	Mucous plugs Paroxysms of cough with "whoop"	Yes	Erythromycin	Acellular vaccine available as DTaP	Mucus movement impaired
Meningococcal Meningitis	*Neisseria meningitidis*	Gram-negative encapsulated diplococcus	URT Blood Meninges	Toxemia Paralysis Skin spots	Yes	Cefotaxime Ceftriaxone	Capsule polysaccharide vaccine	High mortality if untreated Adrenal gland involvement
Pneumococcal Meningitis	*Streptococcus pneunoniae*	Gram-positive diplococcus	URT Blood Meninges	Headache Sever nausea and vomiting Photophobia	Yes	Penicillin Cefotaxime Ceftriaxone	Pneumococcal vaccine	Responsible for 30% of meningitis cases
Haemophilus Meningitis	*Haemophilus influenzae* b	Gram-negative encapsulated rod	URT Meninges	Respiratory symptoms Paralysis	Not established	Cefotaxime Ceftriaxone	Conjugated vaccine to type b only	Dramatic decline in young children

Abbreviation: URT, upper respiratory tract.

9.2 Diseases of the Lower Respiratory Tract

In the lower respiratory tract (LRT), a number of bacterial diseases affect the lung tissues. As injury occurs, fluid builds up in the lung cavity, and the space for obtaining oxygen and eliminating carbon dioxide is reduced. This is the basis for a possibly fatal pneumonia.

Tuberculosis Is a Major Cause of Death Worldwide

KEY CONCEPT

- *Mycobacterium tuberculosis* can cause a lifelong infection.

At the turn of the 20th century, **tuberculosis (TB)** was the world's leading cause of death from all causes, accounting for one fatality in every seven cases. Today's statistics, though improved, are still very high. In the United States, the CDC reported 14,517 cases for 2004, over half of which occurred in foreign-born persons. MicroFocus 9.2 recounts its existence in the Americas. In developing nations however, health officials report more deaths from TB than from any other infectious disease.

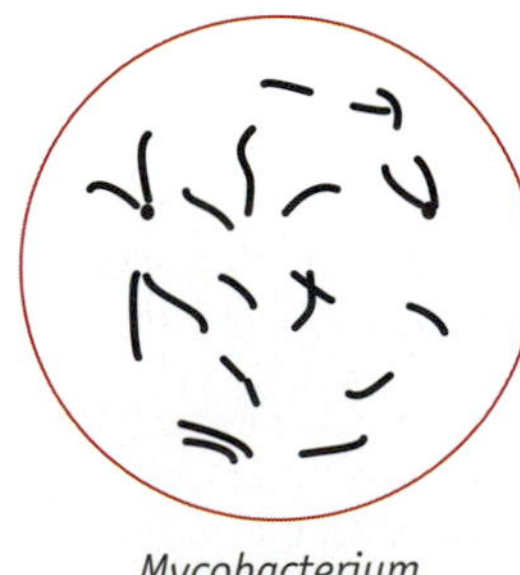

Mycobacterium tuberculosis

The WHO estimates that globally 2 million die of TB every year. The organization also believes one billion people globally will become infected and 36 million will die of tuberculosis by 2020, unless control measures are strengthened.

Tuberculosis is caused by *Mycobacterium tuberculosis*, the "tubercle" bacillus first isolated by Robert Koch in 1882. It is a small, aerobic, nonmotile rod whose cell wall contains a layer of waxy material that greatly enhances resistance to environmental pressures. In the laboratory, a sputum sample for staining must be accompanied by heat to penetrate this barrier, or a lipid-dissolving material must be used. Once stained, however, the organisms resist decolorization, even when subjected to a 5 percent acid-alcohol solution. Thus, the bacilli are said to be acid resistant, or **acid fast** (see Chapter 4).

M. tuberculosis enters the respiratory tract in small aerosolized droplets (multiple exposures are generally necessary). Crowded conditions and poor ventilation contribute to disease spread. Thus, people who live in overcrowded, urban ghettoes often contract TB. Malnutrition and a generally poor quality of life contribute to the establishment of disease.

Unlike many of the other infectious diseases where the individual becomes ill within a week or two of infection, TB takes much

MICROFOCUS 9.2: History

Not Guilty!

Poor Christopher Columbus! Historians have accused you of bringing smallpox to the New World—and measles, whooping cough, tuberculosis, and almost every other conceivable disease. One can almost imagine that the stately Santa Maria was a hospital ship!

Well, rest easy, Chris, as scientists have cleared you of bringing at least one disease—tuberculosis. Your defense is based on 1995 research by Arthur Aufderheide from the University of Minnesota. Some years before, Aufderheide was studying the remains of a mummified woman from Peru when he noticed in her lung tissues several lumps reminiscent of tuberculosis. He enlisted the help of a molecular biologist to extract DNA from the lumps and amplify it so there was enough to identify. The DNA turned out to be identical to that of *Mycobacterium tuberculosis*.

Why was that important? Well, Chris, the mummy was a thousand years old—that's right, one thousand years. Apparently, tuberculosis was already here hundreds of years before you arrived. At least you're not guilty on that count.

longer. After being inhaled, the bacterial cells enter the alveoli where host-pathogen interactions occur (FIGURE 9.8).

If the person has had no previous exposure with the pathogen, 70 percent of the individuals do not become infected. The other 30 percent develop **primary tuberculosis**. In about 90 percent of cases, the infection is arrested, lung lesions heal, and the individuals never are aware they are infected, although they may have a positive tuberculin test (see below). This is referred to as **latent tuberculosis** and is carried by 2 billion people worldwide. Of these, 90 percent will never develop an active infection.

However, 10 percent of people who have primary or latent TB will contract clinical TB and develop active disease that can be transmitted to others. Individuals will become ill within three months, experience chronic cough, chest pain, and high fever, and they expel **sputum**, the thick matter accumulating in the lower respiratory tract. (Often the sputum is rust colored, indicating that blood has entered the lung cavity.).

In these cases, the body responds to the disease by forming a wall of macrophages. As these materials accumulate in the lung, a hard nodule called a **tubercle** arises (hence the name tuberculosis). This tubercle may be visible in a chest X ray. Surrounding tissue may be damaged, although the disease may become dormant. The tubercle undergoes fibrosis and calcification.

Unfortunately, the bacilli in the tubercle are not killed, and the tubercle may expand as the lung tissue progressively deteriorates. In many instances, the tubercle breaks apart and bacterial cells spread through the blood and lymph to other organs such as the liver, kidney, meninges, and bone. If active tubercles develop throughout the body, the disease is called **miliary tuberculosis** (*milium* = "seed;" in reference to the tiny lesions resembling the millet seeds in bird food). Tubercle bacilli produce no discernible toxins, but growth is so unrelenting the tissues are literally consumed, a factor that gave tuberculosis its alternate name, **consumption**.

Early detection of tuberculosis is aided by the **tuberculin reaction**, a test that begins with the application of a purified protein derivative (PPD) of *M. tuberculosis* to the skin. One method of application uses an injection of PPD intradermally into the forearm (the **Mantoux test**). Depending on the patient's risk of exposure, the skin becomes thick, and a raised, red welt of a defined diameter develops. A positive test does not necessarily reflect the presence of active disease, but may indicate a recent immunization, previous tuberculin test, or past exposure to *M. tuberculosis*. It suggests a need for further tests.

Tuberculosis is an extremely stubborn disease. Physicians once recommended fresh air, but now they treat it with such drugs as isoniazid (INH), pyrazinamide, and rifampin.

■ **Host:** An organism on or in which another organism resides.

■ **Macrophages:** Large white blood cells that remove waste products, microorganisms, and foreign material from the bloodstream.

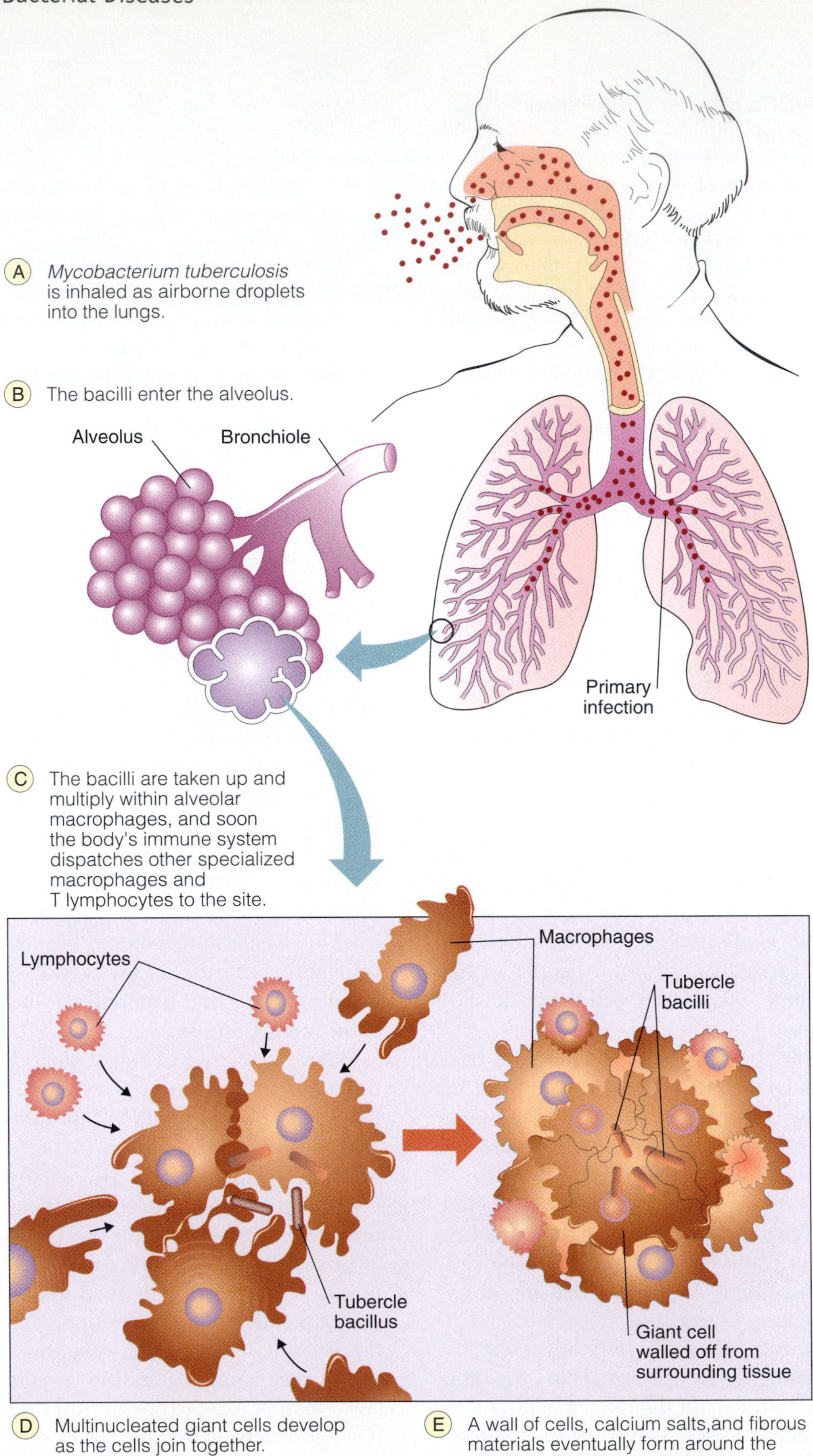

FIGURE 9.8 **The Progress of Tuberculosis.** Following invasion of the alveoli, the tubercle bacilli are taken up by macrophages and "walled off."

Q: What is the immune system attempting to do by forming tubercles?

The recent appearance of **multidrug-resistant *Mycobacterium tuberculosis*** (MDR-TB) has necessitated the use of ethambutol and streptomycin to help delay the emergence of resistant strains. In addition, antimicrobial drug therapy is intensive and must be extended over a period of six to nine months or more, partly because the organism multiplies at a very slow rate (its generation time is about 18 hours). Early relief, boredom, and forgetfulness often cause the patient to stop taking the medication, and the disease flares anew. In 1998, the FDA approved the drug rifapentine, which is taken only once a week. MicroFocus 9.3 describes the WHO treatment strategy.

MICROFOCUS 9.3: Public Health

Tragic Endings but Hopeful Futures

Tuberculosis (TB) is a contagious disease caused by *Mycobacterium tuberculosis*. When an infected person coughs, sneezes, or spits, they project TB bacilli into the air that can be inhaled by a vulnerable contact. In 2005, someone in the world was being newly infected with the bacillus every second. Overall, 33 percent of the world's population currently is infected and TB is killing more people every year; in fact, in 2005 someone was dying of TB every 15 seconds. Among the reasons for the rise in TB cases is that infected individuals are not completing the full course of antibiotics once they start to feel better. Not only does this behavior fail to cure the disease, it also helps generate multidrug-resistant TB (MDR-TB).

To address these issues, the World Health Organization (WHO) has developed the most widely accepted TB treatment program available today. DOTS (Direct Observation Treatment System) is a program to detect and cure TB. Once a patient with infectious TB is identified by a sputum smear, their treatment follows the DOTS strategy. A physician, community worker, or trained volunteer observes and records the patient swallowing four basic medications over a six- to eight-month period. A sputum smear is repeated after two months to check progress and again at the end of treatment.

DOTS appears to be very effective when it is used. Take for example the following case. In the early 1990s, a powerful strain of drug-resistant TB emerged in New York City. Affecting hundreds of people in hospitals and prisons, the outbreak killed 80 percent of the infected patients. The city responded with DOTS, which seemed to work because the outbreak subsided.

However, in 1997 it was back, this time in South Carolina. A New York patient (not on DOTS) moved to South Carolina. His TB had lingered and he infected three family members in his new community. Soon, another six members of the community were sick with TB. However, these six individuals had not had contact with the family; in fact, they did not even know the family. Investigators from the Centers for Disease Control and Prevention (CDC) were called in to investigate. They learned that one family member had been in the hospital, where he was examined with a bronchoscope (a lighted tube extended into the air passageways). Unfortunately, the bronchoscope was not disinfected properly after the examination and was used to examine these other individuals.

Many stories have happy endings, but this is not one of them. Of the six patients infected with the drug-resistant strain of *M. tuberculosis*, two died from TB and three died from other causes while battling TB. Only one recovered.

Thus, one sees the importance and effectiveness of DOTS. The WHO says that DOTS produces a 95 percent cure rate and thus prevents new infections. DOTS also prevents the development of MDR-TB by making sure TB patients take the full course of treatment.

Since DOTS was introduced in 1995, 10 million infectious patients have been treated successfully. In China, there has been a 96 percent cure rate and in Peru a 91 percent cure rate for new cases of TB. Overall, WHO has set a goal of detecting 70 percent of new infectious TB cases and curing 85 percent of the detected cases. Meeting the goal in all countries remains a formidable task, but one that offers a hopeful future.

Insidious: Refers to a disease that progresses gradually and often asymptomatically.

Tuberculosis is a particularly insidious problem to those who have AIDS. In these patients, the **T lymphocytes**, immune system cells that normally mount a response to *M. tuberculosis*, are being destroyed, and the patient cannot respond to the bacterial infection. HIV-infected patients face a mortality rate from TB of 70 to 90 percent, usually within one to four months of developing symptoms. Unlike most other TB patients, those with HIV usually develop tuberculosis in the lymph nodes, bones, liver, and numerous other organs. Ironically, AIDS patients often test negative for the tuberculin skin test because without T lymphocytes, they cannot produce the telltale red welt signaling exposure. Tuberculosis often is the first disease to occur in the AIDS patient, even before any of the other opportunistic illnesses appear, and it is generally more intractable than in non-AIDS patients. The WHO estimates that worldwide, about 4.4 million people are coinfected with HIV and *M. tuberculosis*. AIDS is discussed in more detail in Chapter 15.

Attenuate: To weaken by chemical or culture processes.

Immunization to tuberculosis may be rendered by injections of an attenuated strain of *Mycobacterium bovis*. This species causes tuberculosis in cows as well as humans. The attenuated strain is called bacillé Calmette Guérin, or **BCG**, after Albert Calmette and Camille Guérin, the two French investigators who developed it in the 1920s (Chapter 21). Though the vaccine is used in parts of the world where the disease causes significant mortality and morbidity, many health officials oppose its use in the United States because they point to the success of early detection and treatment, and to the vaccine's occasional side effects. New vaccines consisting of subunits, molecules of DNA, and attenuated strains of mycobacteria are currently being developed.

Several other species of *Mycobacterium* deserve a brief mention. The first, *M. chelonae,* is an acid-fast rod frequently found in soil and water. During the 1980s, microbiologists first recognized this fast-growing bacillus as a cause of lung diseases, wound infections, arthritis, and skin abscesses. *M. haemophilum* surfaced as a pathogen in 1991 when 13 cases occurred in immunocompromised individuals in New York City hospitals. Cutaneous ulcerating lesions and respiratory symptoms were observed in the patients. *M. kansasii* causes infections that are indistinguishable from tuberculosis. In the United States, infections are most common in the central states and rare in the southeast. The group known as *M. avium* complex (MAC) tends to cause infection only in compromised individuals. Thus, in the United States, MAC represents an opportunistic infection that is responsible for most cases of miliary TB in AIDS patients. For all species mentioned here, there is no evidence for spread between individuals; rather, infection comes from contacting soil, or ingesting food or water contaminated with the organism.

CONCEPT AND REASONING CHECKS

9.5 Construct a concept map for tuberculosis starting with primary TB and showing both the dormant and active disease paths.

"Typical" Pneumonia Can Be Caused by Several Bacteria

KEY CONCEPT

- "Typical" bacterial pneumonia can be community or hospital acquired.

The term **pneumonia** refers to microbial disease of the bronchial tubes and lungs. A wide spectrum of organisms, including viruses, fungi, and bacterial species, may cause pneumonia. "Typical" pneumonia refers to patients complaining of a cough, fever, and chest pain. Over 80 percent of bacterial cases of typical pneumonia are due to *S. pneumoniae;* however, several other species also can cause the lung infection.

Streptococcus pneumoniae. Besides being the second leading cause of bacterial meningitis, *S. pneumoniae* also causes **pneumococcal pneumonia**.

As already described, *S. pneumoniae* is a gram-positive, encapsulated chain of diplococci (FIGURE 9.9). Pneumococcal pneumonia is community acquired. Although it exists in all age groups, the mortality rate is highest among infants, the elderly, and those with underlying medical conditions. More than 500,000 cases are reported each year in the United States, resulting in approximately 40,000 deaths.

S. pneumoniae is usually acquired by aerosolized droplets or contact, and the pneumococci exist in the upper respiratory tract of the majority of healthy Americans. However, the

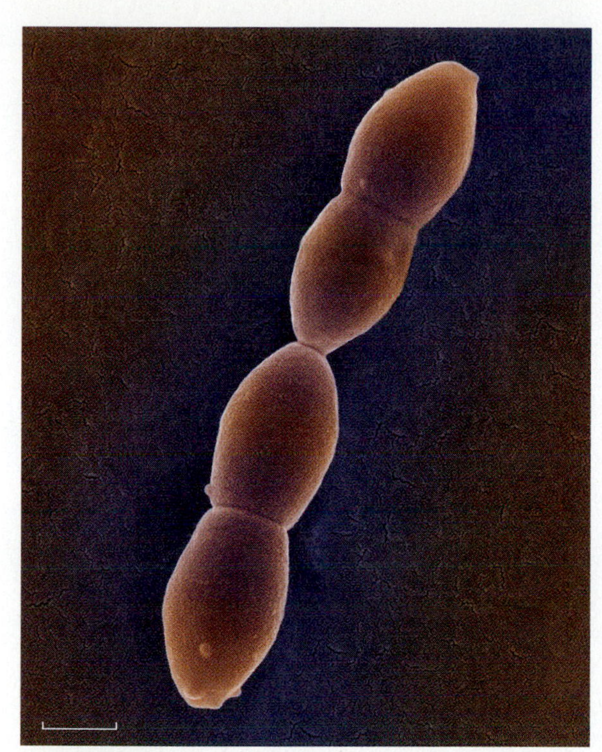

FIGURE 9.9 ***Streptococcus pneumoniae.*** False-color transmission electron micrograph of *S. pneumoniae,* the cause of pneumococcal pneumonia. (Bar = 0.5 µm.)

Q: How would you describe the arrangement of S. pneumoniae *cells?*

natural resistance of the body is high, and disease usually does not develop until the defenses are compromised. Malnutrition, smoking, viral infections, and treatment with immune-suppressing drugs most often predispose one to *S. pneumoniae* infections.

Patients with pneumococcal pneumonia experience high fever, sharp chest pains, difficulty breathing, and rust-colored sputum. The color results from blood seeping into the alveolar sacs of the lung as bacterial cells multiply and cause the tissues to deteriorate. The involvement of an entire lobe of the lung is called **lobar pneumonia**. If both left and right lungs are involved, the condition is called **double pneumonia**. Scattered patches of infection in the respiratory passageways are referred to as **bronchopneumonia**.

The antibiotic for pneumococcal pneumonia has been penicillin. However, increasing penicillin resistance has shifted antibiotic drug choice to cefotaxime or ceftriaxone.

Unfortunately, recovery from one serotype does not confer immunity to another serotype (over 90 capsular serotypes are known). A polyvalent antipneumococcal capsular polysaccharide vaccine immunizes against 23 serotypes, which are responsible for almost 90 percent of pneumococcal pneumonia cases.

Haemophilus influenzae. Some 10 percent of "typical" pneumonia cases, especially in the elderly and compromised individuals, are caused by inhaling respiratory droplets containing unencapsulated *H. influenzae* strains. The infection can become systemic and cause **otitis media** (middle ear infection) and **sinusitis** (inflammation of the sinuses). These URT infections are treated with trimethoprim-sulfamethoxazole.

Staphylococcus aureus. One of the most common causes of hospital-acquired pneumonia results from an infection by *Staphylococcus aureus*, a facultatively anaerobic, gram-positive coccus. If bacterial cells infect the lungs, a severe, necrotizing pneumonia may occur.

■ Necrotizing: Refers to cell or tissue death.

Klebsiella pneumoniae. In 1882, Carl Friedländer isolated *Klebsiella pneumoniae*, an important cause of bacterial pneumonia. In the years thereafter, Friedländer's bacillus was related to about 5 percent of cases.

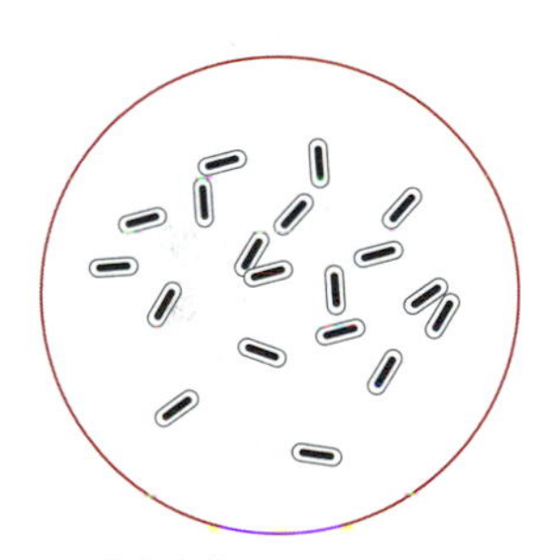

Klebsiella pneumoniae

K. pneumoniae is a nonmotile, gram-negative rod with a prominent capsule. The bacillus is acquired by droplets, and often it occurs naturally in the respiratory tracts of humans. The *K. pneumoniae* may be a primary disease or a secondary disease in alcoholics or people with impaired pulmonary function. As a primary lobar pneumonia, it is characterized by sudden onset and gelatinous reddish-brown sputum. The organisms grow over the lung surface and rapidly destroy the tissue, often causing death. In its secondary form, *K. pneumoniae* occurs in already ill individuals and is a hospital-acquired disease spread by such routes as clothing, intravenous solutions, foods, and the hands of health-care workers.

■ Predispose: Referring to an individual susceptible to a condition.

"Atypical" Pneumonia Can Be Caused by a Diverse Group of Bacterial Species

KEY CONCEPT

- "Atypical" bacterial pneumonia has an insidious onset.

"Atypical" pneumonia is more insidious than "typical" pneumonia. Patient complaints

■ Myalgia:
Pain in the muscles or muscle groups.

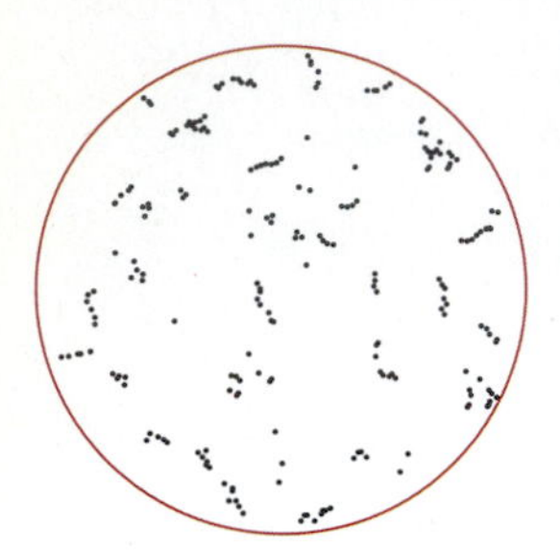

Mycoplasma pneumoniae

include fever, cough, headache, and "myalgia." Several bacterial species can cause this form of pneumonia.

Mycoplasma pneumoniae. During the 1940s, a sudden increase in the number of cases of pneumonia prompted a search for a new infectious agent. At Harvard University, Monroe Eaton isolated a tiny agent from the respiratory tracts of patients and cultivated it on media supplemented with blood (FIGURE 9.10A.) The organism was subsequently named *Mycoplasma pneumoniae*. The disease came to be known as **primary atypical pneumonia**—"primary" because it occurs in previously healthy individuals (pneumococcal pneumonia is usually a secondary disease); "atypical" because the organism differs from the typical pneumococcus and symptoms are unlike those in lobar pneumonia. Today, this community-acquired disease causes about 20 percent of pneumonias.

M. pneumoniae is recognized as one of the smallest bacterial species causing human disease. Mycoplasmas measure about 0.2 μm in size and are pleomorphic; that is, they assume a variety of shapes (FIGURE 9.10B.) Because they have no cell wall, they have no Gram reaction or sensitivity to penicillin. *M. pneumoniae* cells are very fragile and do not survive for long outside the human or animal host. Therefore, they must be maintained in nature by passage in droplets from host to host.

Most patients, who are usually between 6 and 20 years old, experience gradual symptoms of headache, fever, fatigue, and a characteristic dry, hacking cough. Research indicates the organisms attach to and destroy ciliated bronchial epithelial cells. Blood invasion does not occur, and the disease is rarely fatal. Often it is called **walking pneumonia** (even though the term has no clinical significance). Epidemics are common where crowded conditions exist, such as in college dormitories, military bases, and urban ghettoes. Erythromycin and tetracycline are commonly used as treatments.

Research in the 1940s established that antibodies produced against *M. pneumoniae* agglutinate type O human red blood cells at 4°C but not at 37°C. This observation was used to develop the **cold agglutinin screening test (CAST)**: A patient's serum is combined with red blood cells at cold temperatures, and the red cells are observed for clumping. Diagnosis is assisted also by isolation of the organism on blood agar and observation of a distinctive "fried-egg" colony appearance (Figure 9.10A).

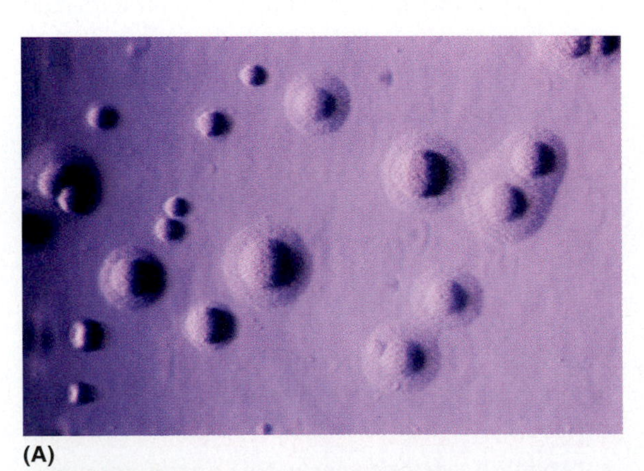

(A)

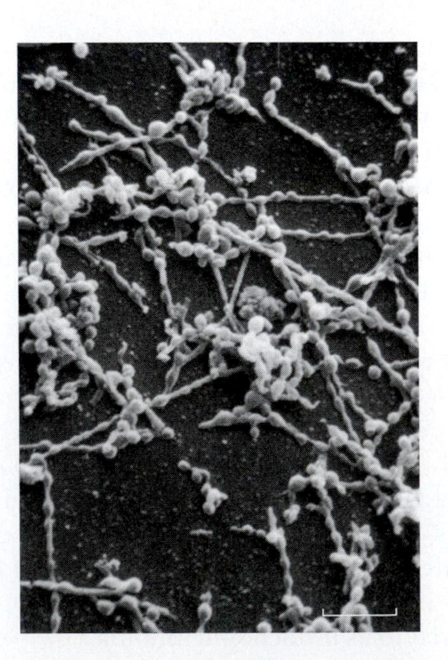

(B)

FIGURE 9.10 ***Mycoplasma pneumoniae.*** Two views of *M. pneumoniae*, the agent of primary atypical pneumonia. (**A**) Colony morphology on agar shows the typical "fried egg" appearance. (**B**) A scanning electron micrograph, demonstrating the pleomorphic shape exhibited by mycoplasmas. (Bar = 2 μm.)

Q: What structural feature is missing from the mycoplasma cells that allows for their pleomorphic shape?

MicroFocus 9.4 describes an interesting application for another bacterial species known to cause respiratory infections.

Legionella pneumophila. From July 21 to July 24, 1976, the Bellevue-Stratford Hotel in Philadelphia was the site of the 58th annual convention of Pennsylvania's chapter of the American Legion. Toward the end of the convention, 140 conventioneers and 72 other people in or near the hotel became ill with headaches, fever, coughing, and pneumonia. Eventually, 34 individuals died of the disease and its complications. In January 1977, investigators from the Centers for Disease Control and Prevention (CDC) announced the isolation of the infecting bacterial species from the lung tissue of one of the patients. The organism appeared responsible not only for the Legionnaires' disease (as the disease had come to be known), but also for a number of other unresolved pneumonia-like diseases.

Legionnaires' disease (also called **legionellosis**) is caused by a motile, unencapsulated, gram-negative rod named *Legionella pneumophila* (FIGURE 9.11A.) The bacillus exists where water collects, and apparently it becomes airborne in wind gusts and breezes. Cooling towers, industrial air-conditioning units, lakes, stagnant pools, and puddles of water have been identified as sources of the pathogen (Textbook Case 9). Older adults are most susceptible. After breathing the contaminated aerosolized droplets into the respiratory tract, the disease develops within a week. Human-to-human transmission does not occur.

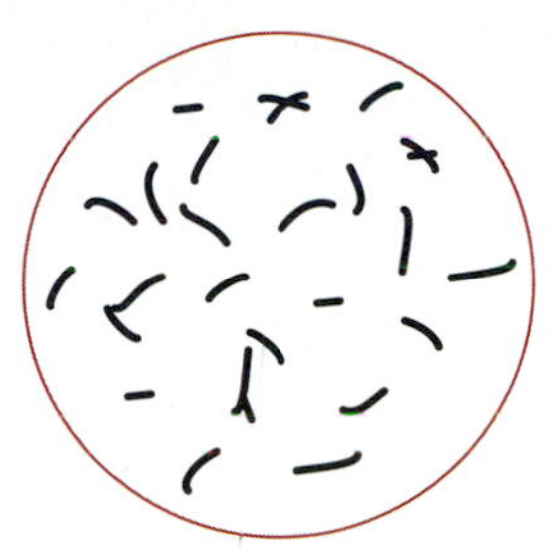
Legionella pneumophila

The symptoms of Legionnaires' disease include headache, fever, a dry cough with little sputum, and some diarrhea and vomiting. In addition, chest X rays show a characteristic pattern of lung involvement, and necrotizing pneumonia is the most dangerous effect of the disease. Erythromycin is effective for treatment.

After *L. pneumophila* was isolated in early 1977, microbiologists realized the organism was responsible for another milder infection called **Pontiac fever**. This is an influenza-like illness that lasts 2 to 5 days but does not cause pneumonia. Symptoms disappear without treatment.

Since the discovery of *Legionella* in water, microbiologists have been perplexed how

MICROFOCUS 9.4: History
"Keep It Short, Please!"

Defining, developing, and proving the germ theory of disease was one of the great triumphs of scientists in the late 1800s. Applying the theory to practical problems was another matter, however, because people were reluctant to change their ways. It would take some rather persuasive evidence to move them.

For many decades, microbiologists considered *Serratia marcescens* a nonpathogenic bacillus. Today, this motile, gram-negative rod is viewed as a cause of respiratory disease in "compromised" patients.

In the summer of 1904, influenza struck with terrible force among members of Britain's House of Commons. Soon the members began wondering aloud whether they should ventilate their crowded chamber. They decided to hire British bacteriologist Mervyn Henry Gordon to determine whether "germs" were being transferred through the air and whether ventilation would help the situation.

Gordon devised an ingenious experiment. He selected as his test organism *Serratia marcescens* because the bacterium forms bright-red visible colonies in Petri dishes of nutrient agar. Gordon prepared a liquid suspension of the bacterial cells and gargled with it!

Gordon then stood in the chamber and delivered a two-hour oration consisting of selections from Shakespeare's *Julius Caesar* and *Henry V*. His audience was hundreds of open Petri dishes. The theory was simple: If bacteria were transferred during Gordon's long-winded speeches, then they would land on the agar plates and form red colonies.

And land they did. After several days, red colonies appeared on plates placed right in front of Gordon, as well as in distant reaches of the chamber. The members were impressed. They proposed a more constant flow of fresh air to the chamber, as well as shorter speeches. No one was about to object to either solution, especially the latter.

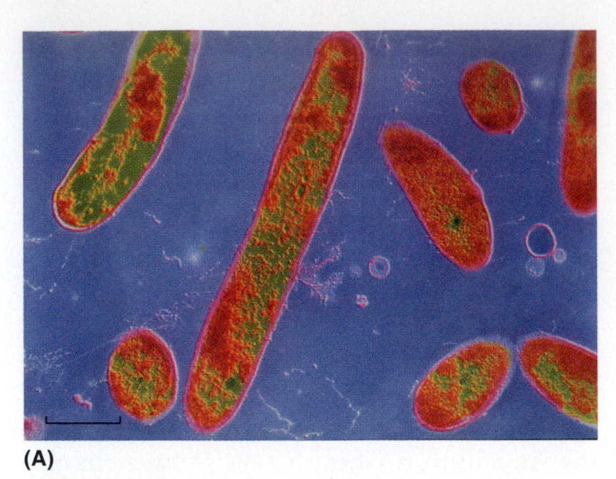

(A)

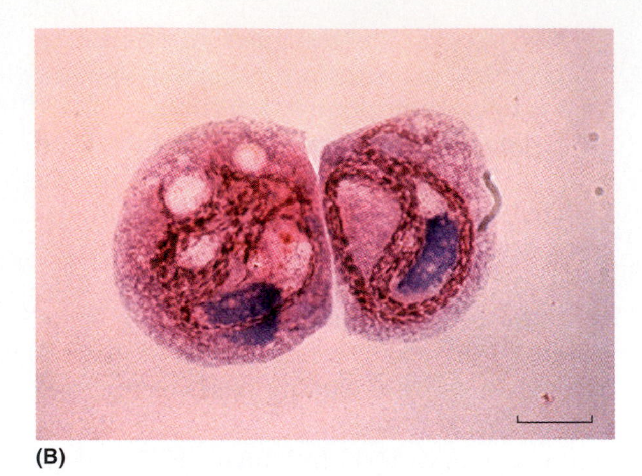

(B)

FIGURE 9.11 ***Legionella pneumophila.*** Two views of *L. pneumophila*, the agent of Legionnaires' disease. (**A**) A false-color transmission electron micrograph of *L. pneumophila* cells. (Bar = 1 µm.) (**B**) A false-color transmission electron micrograph of the protozoan *Tetrahymena pyriformis* infected with chains of *L. pneumophila* cells (dark red). (Bar = 10 µm.)

Q: How does infecting a protozoan benefit the bacterium when in its natural environment?

such fastidious bacilli survive in often hostile, aquatic environments. An answer was suggested by studies that showed the bacilli could live and grow within the protective confines of waterborne protozoa (FIGURE 9.11B). In a South Dakota outbreak, amoebas were found in abundance in the hospital's water supply. The water was treated by heating and adding chlorine to help quell the spread.

CONCEPT AND REASONING CHECKS

9.6 Hypothesize why some pneumonia-causing species are community acquired and others are hospital acquired?

Some Rickettsiae and Chlamydiae Also Cause Pneumonia

KEY CONCEPT

- Some rickettsiae and chlamydiae are transmitted by dust particles or animal droppings.

Pneumonia also can be caused by some of the smallest bacterial organisms, the rickettsiae and chlamydiae (see Chapter 4). They are obligate, intracellular parasites, meaning they only grow inside host cells.

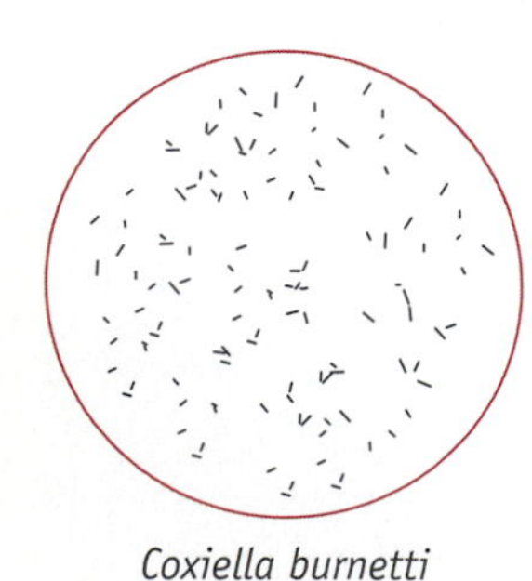

Coxiella burnetti

Coxiella burnetii. Q fever (the "Q" is derived from "query," originally referring to the unknown cause of the disease) is caused by the rickettsia named *Coxiella burnetii* (FIGURE 9.12). It was named after H. R. Cox, who isolated it in Montana in 1935, and Frank Macfarlane Burnet, who studied its properties in the late 1930s.

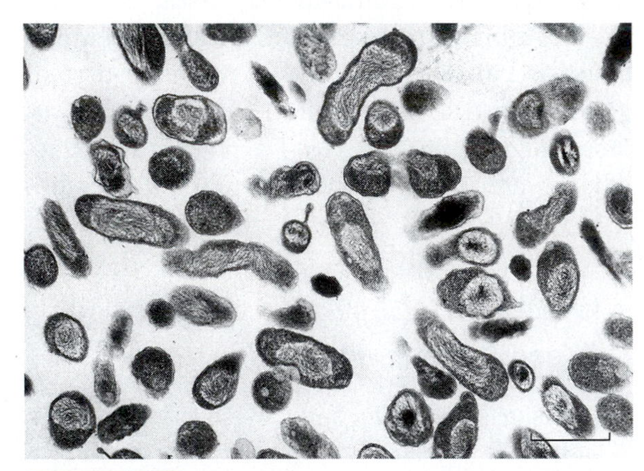

FIGURE 9.12 ***Coxiella burnetii.*** An electron micrograph of *C. burnetii*, the agent of Q fever. Note the oval-shaped rods of the organism. (Bar = 1 µm.)

Q: What are the unusual features of these bacteria?

Q fever is prevalent worldwide among livestock, especially in dairy cows, goats, and sheep, and therefore, outbreaks may occur wherever these animals are raised, housed, or transported. In 2004, more than 13 confirmed cases of Q fever were reported in Colorado. Transmission among diseased livestock and to humans occurs primarily by inhaling the organisms in dust particles or handling infected animals. In addition, humans may acquire the disease by consuming raw milk

Textbook CASE  9

Legionnaires' Disease Outbreak—Bogalusa, Louisiana

1 On October 31, 1989, two physicians in Bogalusa, Louisiana reported an outbreak of more than 50 cases of acute pneumonia to the state department of health. Most cases occurred within a 3-week interval in mid- to late October; six persons had died. All cases had occurred in older adults and 76 percent of the cases were female. Lab analysis confirmed 33 cases were caused by *Legionella pneumophila*, the agent of Legionnaires' disease.

2 When investigators from the Centers for Disease Control and Prevention (CDC) arrived, it was imperative to determine quickly the source and mode of transmission of *L. pneumophila*.

3 Most cases were among residents of Bogalusa. A total of 28 patients and 56 controls were interviewed. Patients and controls were asked about exposures to cooling towers and nearby buildings.

4 Of the 28 patients, three reported visiting Hospital B with a cooling tower; of the 56 controls, seven reported visiting Hospital B. Similarly, 7 of the patients and 12 of the controls reported having visited the nearby post office.

5 Microbiological analysis was unable to confirm any contamination in the hospital B cooling tower. In addition, visiting the post office was eliminated as a possible source.

6 Further interviews identified two other potential sources, butcher shop A and grocery store B. Butcher shop A was visited by 12 patients and 19 controls, while 25 patients and 28 controls had visited grocery store B.

7 A detailed microbiological investigation of grocery store B was undertaken. Several days later, the CDC investigators had the answer. *L. pneumophila* had been isolated from an ultrasonic misting machine close to where shoppers selected produce in the vegetable section (see figure). The machine's aerosol action had sprayed bacterial cells into the air to which shoppers were exposed. No cases were reported among employees.

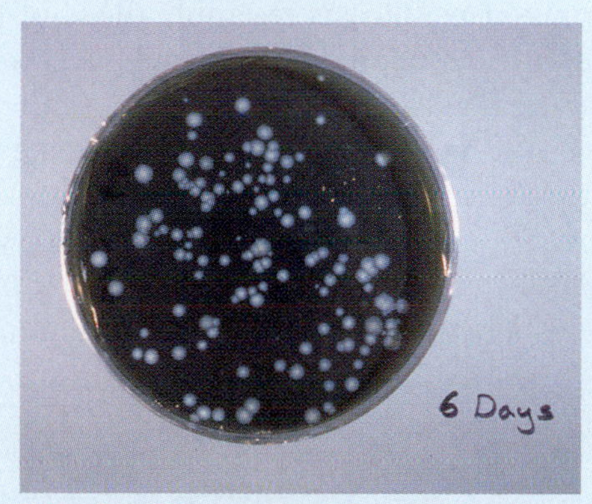

Colonies of *L. pneumophila* growing on an enriched agar medium.

Questions

A. *Why was it important to know that most cases were in the city of Bogalusa? Who are the controls?*

B. *Why was a cooling tower first suspected as the source of the outbreak?*

C. *Why was the post office and butcher shop A eliminated as possible sources?*

D. *Who needs to know about these findings? How would you go about reporting the findings?*

E. *Explain the significance of (i) 76 percent of the cases being female and (ii) no cases occurring among grocery store employees.*

For additional information see: http://www.cdc.gov/mmwr/preview/mmwrhtml/00001563.htm.

infected with *C. burnetii* or milk that has been improperly pasteurized. Patients experience severe headache, high fever, a dry cough, and occasionally, lesions on the lung surface. The mortality rate is low, and treatment with doxycycline is effective for chronic infections. A vaccine is available for workers in high-risk occupations.

Zoonotic: Refers to a disease transmitted from animals to humans.

Chlamydia psittaci. Psittacosis is a zoonotic disease and is transmitted to humans by infected parrots, parakeets, canaries, and other members of the psittacine family of birds (*psittakos* = "parrot"). The disease also occurs in pigeons, chickens, turkeys, and seagulls, and some microbiologists prefer to call it **ornithosis** (*ornith* = "bird") to reflect the more widespread occurrence in bird species. Humans acquire *C. psittaci* by inhaling airborne dust or dried droppings of infected birds. Sometimes the disease is transmitted by a bite from a bird or via the respiratory droplets from another human. The symptoms of psittacosis resemble those of primary atypical pneumonia. Fever is accompanied by headaches, dry cough, and scattered patches of lung infection. Doxycycline is commonly used in therapy.

In 2004, the CDC reported 12 cases based on reports from less than 40 states. The incidence of psittacosis in the United States remains low, partly because federal law requires a 30-day quarantine for imported psittacine birds. In addition, birds are given water treated with chlortetracycline hydrochloride (CTC) and CTC-impregnated feed.

Chlamydia pneumoniae. **Chlamydial pneumonia** is caused by *Chlamydia pneumoniae*. The organism is transmitted by respiratory droplets and causes a mild walking pneumonia, principally in young adults and college students. The disease is clinically similar to psittacosis and primary atypical pneumonia, and is characterized by fever, headache, and nonproductive cough. Treatment with doxycycline or erythromycin hastens recovery from the infection. The organism was first observed in Seattle, Washington, in 1983 and now is believed to infect many thousands annually in the United States. Its relationship to cardiovascular disease is explored in MicroFocus 9.5.

The airborne bacterial diseases of the lower respiratory tract are summarized in TABLE 9.2. MicroInquiry 9 presents several case studies concerning some of the bacterial diseases discussed in this chapter.

CONCEPT AND REASONING CHECKS

9.7 Summarize (1) the unique characteristics and (2) the mode of transmission of the rickettsiae and chlamydiae.

MICROFOCUS 9.5: Being Skeptical

Can Heart Disease Be Caused by an Infection?

Here's one to startle the senses: the "germ theory of cardiovascular diseases." Skeptical? Let's look deeper.

Coronary artery disease (CAD) remains a leading cause of death even with current medications and laser, angioplasty, or other innovative devices that are available. Perhaps antimicrobial drugs would help.

But, antibiotics for what? Why not start, scientists say, with *Chlamydia pneumoniae?* In several studies, *C. pneumoniae* has been associated with heart attack patients and in numerous males with CAD. Researchers suggest the chlamydiae could injure the blood vessels, triggering an inflammatory response where immune system cells attack the vessel walls and induce the formation of large, fibrous lesions, or plaques. When pieces of plaque break free, they start blood clots that clog the arteries and cause heart attacks (the condition is known as atherosclerosis). Also, inoculation of mice and rabbits with *C. pneumoniae* accelerates atherosclerosis, even more so if the animals are fed cholesterol-enriched diets.

A second suspect is cytomegalovirus (CMV), an animal virus of the herpes family. In fact, more than 70 percent of the CAD population is CMV-positive. Scientists have known that people infected with CMV respond very poorly to arterial cleaning, the technique of angioplasty, and the arteries quickly close up. CMV also accelerates atherosclerosis in mice and rats.

A weaker association with CAD is seen with *Streptococcus sanguis,* an agent of periodontal disease. Some microbiologists suggest that poor oral hygiene and bleeding gums give the bacterial species

TABLE 9.2 A Summary of Major Airborne Bacterial Diseases of the Lower Respiratory Tract

Disease	Causative Agent	Description of Agent	Organs Affected	Characteristic Signs	Treatment Administered	Immunization Available	Comment
Tuberculosis	*Mycobacterium tuberculosis*	Acid-fast rod	Lungs, bones, other organs	Tubercle	Isoniazid Rifampin Pyrazinamide	Bacille Calmette Guérin (BCG)	Extended treatment necessary Diagnosis by tuberculin test Related to AIDS
Pneumococcal pneumonia	*Streptococcus pneumoniae*	Gram-positive encapsulated diplococcus in chains	Lungs	Rust-colored sputum	Penicillin Cefotaxime Ceftriaxone	Polysaccharide vaccine to 23 strains	Over 90 strains identified Natural resistance high Deterioration of alveoli
Klebsiella pneumonia	*Klebsiella pneumoniae*	Gram-negative encapsulated rod	Lungs	Pneumonia	Various antibiotics	None	Common in people with impaired pulmonary function
Primary atypical pneumonia	*Mycoplasma pneumoniae*	No cell wall Pleomorphic	Lungs	Dry cough	Erythromycin Tetracycline	None	Called walking pneumonia Diagnosis by CAST
Legionnaires' disease (Legionellosis)	*Legionella pneumophila*	Gram-negative rod	Lungs	Pneumonia	Erythromycin	None	Associated with airborne water droplets
Q fever	*Coxiella burnetii*	Rickettsia	Lungs	Influenza-like symptoms	Doxycycline	Vaccine for high-risk workers	Reservoirs include dairy cows, goats, sheep Associated with raw milk
Psittacosis	*Chlamydia psittaci*	Chlamydia	Lungs	Influenza-like symptoms	Doxycycline	None	Occurs in parrots and parrot-like birds
Chlamydial pneumonia	*Chlamydia pneumoniae*	Chlamydia	Lungs	Influenza-like symptoms	Doxycycline Erythromycin	None	Young adults affected

access to the blood, where it produces blood-clotting proteins. It's no coincidence, they maintain, that people with unhealthy teeth and gums tend to have more heart trouble. Unfortunately, such groups also have other lifestyle factors that confound the association.

The final suspect is *Helicobacter pylori,* the cause of most peptic ulcers. Italian scientists have linked a virulent strain of *H. pylori* with increased incidence of heart disease. However, other trials do not find the organism as significant in predicting CAD as the other agents.

To account for these observations and research findings, many medical researchers and microbiologists are proposing the concept of a total pathogen burden. First proposed in 2000, the concept suggests that while a single infectious agent may only minimally increase the risk of atherosclerosis, the burden of several agents could greatly increase the risk. In fact, several recent studies suggest that exposure to several microbial suspects does correlate with increased risk for CAD and, in established CAD cases, incident death. The studies propose that *C. pneumoniae* and CMV probably play a direct role in atherosclerosis while other agents, like *S. sanguis* and *H. pylori,* contribute indirectly to inflammation. The concept does not prove causality, but suggests avenues for further study, including the possible use of antibiotics, antivirals, or vaccines to prevent or cure CAD.

Still skeptical? Many scientists are too but they keep an open mind to new evidence. So, before you run out to buy antibiotics, stick with the more likely anti-inflammatory factors of diet, exercise, and not smoking.

MICROINQUIRY 9

Infectious Disease Identification

Below are several descriptions of infectious diseases based on material presented in this chapter. Read the case history and then answer the questions posed. Answers can be found in **Appendix D.**

Case 1

The patient, a 33-year-old male, arrives at a local health clinic complaining that he has felt "out of sorts," has a fever, and has lost over 10 percent of his body weight in the last month. He also has a cough that produced rust-colored sputum. The patient is referred to a local hospital where a chest X ray and sputum sample are taken. Upon further questioning, the patient admits to having tested HIV-positive about one year ago. A tuberculin test also is ordered. Additional questioning of the patient reveals he had been living with two roommates for two years. Before that, he had lived for eight years with another roommate who had tested positive for tuberculosis about 6 months before the onset of the patient's symptoms. The sputum samples are negative for the two roommates, but both have a positive tuberculin test result. Both test negative for HIV. Following diagnosis, the patient was placed on isoniazid (INH) for 12 months.

9.1a. Why was a chest X ray ordered?
9.1b. Why was a sputum sample taken?
9.1c. What should a positive tuberculin skin test look like?
9.1d. What does such a test result indicate?
9.1e. Based on the symptoms and laboratory results, what infectious disease does the patient suffer? What is the agent?
9.1f. How did the patient contract the disease?
9.1g. Why is the infectious agent more virulent in HIV-infected patients?
9.1h. Most treatment procedures call for a 6 to 8 month program. Why was the patient placed in INH for an extended period?

Case 2

The parents bring their 2-year-old daughter to the hospital emergency room. She appears to have an upper respiratory infection that her parents think started about one week previously. They say that their daughter had lost her appetite and appeared especially sleepy about four days ago. She complained of a sore throat. Examination indicates that she has a moderate fever but no chest congestion. Throat and blood cultures are taken and their daughter is put on penicillin.

On returning to the hospital three days later, her throat culture shows gram-positive rods. The blood culture is negative. Her parents remark that this morning their daughter started breathing harder. Examination of her pharynx indicates the presence of a leathery membrane. On questioning the parents, it is discovered that the child has had no immunizations. The child is admitted to the hospital and treatment immediately begun.

9.2a. What infectious agent does the child have?
9.2b. The medical staff is concerned about the seriousness of the disease. Why does the presence of gram-positive rods in the throat cause such concern?
9.2c. How could this disease have been prevented?
9.2d. What is the prescribed treatment protocol?

Case 3

A 63-year-old retired steel worker who is a heavy smoker and alcoholic comes to the emergency room complaining of having a fever and shortness of breath for the last two days. This morning he has developed a cough with rust-colored sputum. A chest X ray is taken and shows involvement in the left lower lobe of the lungs. A sputum sample and blood sample are taken for Gram staining, and the patient is checked into the hospital where he is placed on penicillin. Two days later the patient is feeling much improved. His physician tells him that both sputum and blood cultures indicate the presence of gram-positive diplococci. The patient is released from the hospital and recovers completely after finishing antibiotic therapy.

9.3a. What organism is responsible for the patient's infection?
9.3b. Why is this patient at high risk for becoming infected with the bacterial organism?
9.3c. How could the patient likely have prevented contracting the disease?
9.3d. What bacterial factors are responsible for virulence?
9.3e. If plated on blood agar, what type of hemolytic reaction should be seen?

SUMMARY OF KEY CONCEPTS

9.1 Diseases of the Upper Respiratory Tract

- Streptococcal diseases are caused by *Streptococcus pyogenes.*
  - Symptoms of streptococcal pharyngitis (strep throat): Sore throat, fever, cough, and swollen lymph nodes and tonsils. Scarlet fever: Fever, cough, swollen lymph nodes and tonsils, and a skin rash.
  - Transmission: Airborne droplets.
  - Treatment: Penicillin or clarithromycin.
  - Complications: Rheumatic fever and acute glomerulonephritis.
- Diphtheria is caused by a prophage-harboring strain of *Corynebacterium diphtheriae.* As dead tissue accumulates, a pseudomembrane forms in the throat or nasopharynx.
  - Symptoms: Sore throat, low-grade fever.
  - Transmission: Respiratory droplets.
  - Treatment: Antitoxins and penicillin or erythromycin.
  - Prevention: DTaP or Tdap vaccine.
- Pertussis (whopping cough) is caused by *Bordetella pertussis.*
  - Symptoms: Fever, increasingly severe cough, leading to the characteristic "whoop."
  - Transmission: Respiratory droplets.
  - Treatment: Erythromycin (does not shorten the illness)
  - Prevention: DTaP or Tdap vaccine.
- Bacterial meningitis is an inflammation of the membranes surrounding the brain and spinal cord. The infection can be caused by *Neisseria meningitidis, Streptococcus pneumoniae,* or *Haemophilus influenzae.*
  - Symptoms: Headache, rigid neck; vomiting and sensitivity to bright light.
  - Transmission: Respiratory droplets.
  - Treatment: Early diagnosis and penicillin, cefotaxime, or ceftriaxone antibiotics.
  - Prevention: Vaccines available for each.

9.2 Diseases of the Lower Respiratory Tract

- Tuberculosis (TB) is caused by the acid-fast bacilli of *Mycobacterium tuberculosis.*
  - Symptoms: Chronic cough, chest pain, and high fever.
  - Transmission: Inhalation of bacilli.
  - Treatment: Isoniazid, pyrazinamide, and rifampin requires six to nine months.
  - Prevention: BCG vaccine available in endemic areas.
- "Typical" bacterial pneumonia is primarily caused by *Streptococcus pneumoniae* (pneumococcal pneumonia). *Haemophilus influenzae, Staphylococcus aureus,* and *Klebsiella pneumoniae* are other species.
  - Symptoms: Fever, sharp chest pains, cough and rust-colored sputum.
  - Transmission: Aerosolized droplets.
  - Treatment: Penicillin or erythromycin.
  - Prevention: Pneumococcal vaccine.
- "Atypical" bacterial pneumonia can be caused by *Mycoplasma pneumoniae* and *Legionella pneumophila,* the latter being responsible for Legionnaires' disease (legionellosis).
  - Symptoms: Fever, dry cough, and diarrhea and vomiting.
  - Transmission: Aerosolized droplets.
  - Treatment: Erythromycin.
- Rickettsiae and chlamydiae also can cause airborne infections. Q fever is caused by *Coxiella burnetii.*
  - Symptoms: Severe headache, high fever, and dry cough.
  - Transmission: Inhalation of contaminated dust particles or consuming unpasteurized, or improperly pasteurized, milk.
  - Treatment: Doxycycline.
- Psittacosis is caused by *Chlamydia psittaci.*
  - Symptoms: Fever, headache, and dry cough.
  - Transmission: Inhaling contaminated dust or dried droppings from infected pet birds.
  - Treatment: Doxycycline.
- Chlamydial pneumonia is caused by *Chlamydia pneumoniae.*
  - Symptoms: Fever, headache, and dry cough.
  - Transmission: Respiratory droplets.
  - Treatment: Doxycycline or erythromycin.

LEARNING OBJECTIVES

After understanding the textbook reading, you should be capable of writing a paragraph that includes the appropriate terms and pertinent information to answer the objective.

1. Summarize the clinical aspects of group A streptococci (GAS) and the complications arising from streptococcal pharyngitis.
2. Identify the bacterial species responsible for and describe the clinical aspects, treatment, and prevention of diphtheria.
3. Justify why pertussis is viewed as one of the more dangerous contagious diseases.
4. Distinguish between the bacterial species responsible for meningitis.
5. Summarize the clinical aspects of *Mycobacterium tuberculosis* and the problems concerning antibiotic resistance.
6. Distinguish between the bacterial species responsible for "typical" pneumonia.
7. Compare and contrast the bacterial species causing "atypical" pneumonia.
8. Summarize the mode of transmission and types of pneumonia caused by rickettsiae and chlamydiae.

SELF-TEST

Answer each of the following questions by selecting the ***one*** answer that best fits the question or statement. Answers to even-numbered questions can be found in **Appendix C**.

1. β hemolytic streptococci produce _____ zones around colonies on blood agar.
A. green
B. red
C. clear
D. yellow
E. blue

2. Which one of the following is a complication of streptococcus pharyngitis?
A. Rheumatic fever
B. Pseudomembrane blockage
C. Tubercle formation.
D. Meningococcemia
E. Waterhouse-Friderichsen syndrome

3. Methylene blue staining of metachromatic granules is diagnostic for
A. *Mycobacterium tuberculosis.*
B. *Corynebacterium diphtheriae.*
C. *Chlamydia pneumoniae.*
D. *Bordetella pertussis.*
E. None of the above (**A–D**) is correct.

4. The presence of a pseudomembrane is a result of
A. strep throat.
B. meningitis.
C. scarlet fever.
D. diphtheria.
E. psittacosis.

5. A toxoid is a/an
A. heat- or chemically-inactivated toxin.
B. diphtheria exotoxin.
C. antibody of the immune system.
D. mercury derivative.
E. Both **A** and **B** are correct.

6. A paroxysm is a _____ and is characteristic of _____.
A. fever; meningitis
B. rapid-fire cough; meningitis
C. fever; pneumonia
D. headache; pertussis
E. rapid-fire cough; pertussis

7. Complaints of headache, fever, and photophobia are characteristic of
A. diphtheria.
B. pneumonia.
C. whooping cough.
D. meningitis.
E. strep throat.

8. Bacterial meningitis
A. is a disease crossing the blood-brain barrier.
B. is a disease of the membranes covering the brain.
C. can be caused by *S. pneumoniae.*
D. often starts as a nasopharynx infection.
E. All the above (**A–D**) are correct.

9. Acid-fast staining is characteristic of
A. *Mycoplasma.*
B. *Streptococcus.*
C. *Klebsiella.*
D. *Coxiella.*
E. *Mycobacterium.*

10. The formation of giant cells are found in
A. brain cells.
B. tubercles.
C. meningitis.
D. *Haemophilus.*
E. Q fever.

11. The Mantoux test is used to identify
A. TB exposure.
B. Q fever.
C. ornithosis
D. pertussis.
E. meningitis.

12. This bacterial species is the most common cause of "typical" pneumonia.
A. *H. influenzae*
B. *S. pneumoniae*
C. *M. pneumoniae*
D. *L. pneumophila*
E. *B. pertussis*

13. A bacterial species commonly causing hospital-acquired pneumonia is
A. *H. influenzae.*
B. *K. pneumoniae.*
C. *L. pneumophila.*
D. *C. pneumoniae.*
E. Both **A** and **B** are correct.

14. Necrotizing pneumonia is a serious consequence of this disease.
A. Q fever
B. Psittacosis
C. Legionnaires' disease
D. Tuberculosis
E. Ornithosis

15. Humans can acquire _____ from the droppings of infected birds.
A. Q fever
B. Legionella
C. Chlamydial pneumonia
D. Tuberculosis
E. Ornithosis

QUESTIONS FOR THOUGHT AND DISCUSSION

Answers to even-numbered questions can be found in **Appendix C**.

1. Between 1986 and 1996 *Haemophilus* meningitis was virtually eliminated in the United States. Indeed, at the beginning of the period there were 18,000 cases annually, but in 1996, only 254 cases were reported. What factors probably contributed to the decline of the disease?
2. A bacteriophage is responsible for the ability of the diphtheria bacillus to produce the toxin that leads to disease. Do you believe that having the virus is advantageous to the infecting bacillus? Why or why not?
3. At present there is no licensed vaccine for the prevention of any streptococcal diseases, even though these are among the most commonly experienced bacterial diseases in the United States. Can you postulate why a vaccine, especially one composed of killed streptococci, might pose a threat to health?
4. In 1998, Primary Children's Hospital in Salt Lake City reported a dramatic increase in the number of rheumatic fever cases. Doctors were alerted to start monitoring sore throats more carefully. Why do you suppose this prevention method was recommended?

APPLICATIONS

Answers to even-numbered questions can be found in **Appendix C**.

1. A patient is admitted to the hospital with high fever and a respiratory infection. Pneumococci and streptococci were eliminated as causes. Penicillin was ineffective. The most unusual sign was a continually dropping count of red blood cells. Can you guess the final diagnosis?
2. One of the major world health stories of 1995 was the outbreak of diphtheria in the New Independent and Baltic States of the former Soviet Union. If you were in charge of this international public health emergency, what would be your plan to help quell the spread of *Corynebacterium diphtheriae?*
3. Meningococcal meningitis outbreaks sweep through the meningitis belt (see Figure 9.6) annually. Usually the epidemics arrive with the exceptionally dry and dusty season and subside with the rainy season. Explain why meningococcal meningitis would occur in this season.
4. The CDC reports that an estimated 40,000 people in the United States die annually from pneumococcal pneumonia. Despite this high figure, only 30 percent of older adults who could benefit from the pneumococcal vaccine are vaccinated (compared to over 50 percent who receive an influenza vaccine yearly). As an epidemiologist in charge of bringing the pneumonia vaccine to a greater percentage of older Americans, what would you do to convince older adults to be vaccinated?

REVIEW

Answer each of the following by filling in the blank with the correct word or phrase. Answers to even-numbered statements can be found in **Appendix C**.

1. Scarlet fever is caused by a species of ________________.
2. ________________ is caused by a species of *Mycobacterium*.
3. The bacterial species ________________ causes psittacosis and pneumonia.
4. ________________ is caused by transmission in droplets of airborne water.
5. ________________ is a gram-negative rod that causes whooping cough.
6. The development of active tuberculosis throughout the body is called ________________ tuberculosis.
7. ________________ is a disease of parrots, parakeets, and canaries as well as humans.
8. Another name of *Streptococcus pneumoniae* is ________________.
9. A bacterial form of ________________ may be caused by *Neisseria* or *Haemophilus*.
10. *Mycoplasma* species have no ________________, which often gives them a pleomorphic shape.
11. *Coxiella burnetti* is the causative agent of ________________.
12. A spinal ________________ may be necessary to locate meningitis organisms.
13. The genus ________________ consists of acid-fast rods.
14. The agent of pneumococcal pneumonia is a gram-________________ organism.
15. Species of ________________ may be alpha, beta, or nonhemolytic.

HTTP://MICROBIOLOGY.JBPUB.COM/

The site features learning, an on-line review area that provides quizzes and other tools to help you study for your class. You can also follow useful links for in-depth information, read more MicroFocus stories, or just find out the latest microbiology news.

10

Foodborne and Waterborne Bacterial Diseases

Chapter Preview and Key Concepts

In the aftermath of the Rwanda crisis in 1994, outbreaks of cholera caused at least 48,000 cases and 23,800 deaths within one month in the refugee camps in Goma, the Congo.
—Statement by the World Health Organization

On August 31, 1854, London was experiencing another epidemic of cholera. More than 500 people had died in just 10 days from a disease characterized by stomachache, vomiting, and diarrhea so profuse victims would die from dehydration. Previous epidemics had killed thousands of people on the European continent and in England. Through his pioneering epidemiological studies (see Chapter 1), John Snow, a London physician, pinpointed a water pump on Broad Street as the source of transmission for the current epidemic (FIGURE 10.1). Snow believed the only way to stop the spread was to remove the pump handle so no one could get water from the pump. On September 8, 1854, city officials took Snow's advice and had the pump handle removed. The action was successful, supporting Snow's belief that cholera was a waterborne, contagious disease. Few at the time believed Snow's theory for a waterborne disease, and it would not be until 1883 that the cholera bacterium, *Vibrio cholerae,* was isolated.

The year 2004 marked the 150th anniversary of John Snow's landmark epidemiological studies. Unfortunately, cholera still exists today. Before 2000, cholera epidemics had rampaged through Zambia on the African continent, causing 13,154 cases in 1991, 11,659 cases in 1992, and 11,327 in 1999. Instituting in-home chlorination of drinking water hopefully would bring an end to these epidemics—and no epidemics occurred between 2000 and 2002. Then, in November 2003, cases of cholera were

FIGURE 10.1 **Cholera and the Broad Street Pump.** John Snow was able to correlate the spread of cholera with a contaminated water pump on Broad Street in London in 1866. The outbreak was controlled by removing the pump handle.

Q: What is the significance of the "death" caricature in this piece of historical artwork from the time?

confirmed in Lusaka, the capital city of Zambia with a population of 2 million people. Between November 28, 2003 and January 4, 2004, there were an estimated 2,529 cases of cholera and 128 deaths from the disease in the capital. The Centers for Disease Control and Prevention (CDC) was invited in by the Zambian government to help deal with the epidemic, which by March 1 had added another 2,101 cases and 25 deaths.

Had the water purification system broken down? Or was there another source for the outbreak? CDC officials undertook an extensive epidemiological field investigation by comparing many daily activities involving control (uninfected) individuals and case patients.

The CDC analysis indicated consumption of untreated drinking water and chlorine-treated water was essentially identical between the two groups, so a waterborne source was not the cause of this epidemic. Rather, analysis eventually pointed to raw vegetables as being associated with cholera. Somewhere during the transport, delivery, or use in the home, vegetables were contaminated with the *V. cholerae*

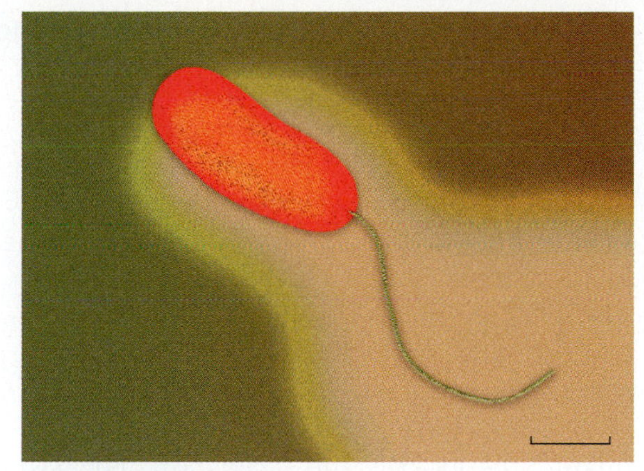

FIGURE 10.2 ***Vibrio cholerae.*** This false-color transmission electron micrograph shows a cell of *V. cholerae*. Its flagellum can be seen. (Bar = 1 µm.)

Q: What is the shape of these cells?

bacterium (FIGURE 10.2). So, in this cholera epidemic, the mode of disease transmission was foodborne.

In addition, hand soap (indicative of hand washing) was found much more frequently in the homes of control-group individuals, suggesting hand washing reduced the risk for transmission of the disease. By instituting new food control methods and urging hand washing, the number of cholera cases in Lusaka declined dramatically.

This outbreak is remarkable because it reaffirms the 140-year old work of Snow and Semmelweis, who implicated contaminated water and poor hand hygiene in disease transmission. So, sometimes the more things change, the more they stay the same. Field epidemiology and simple prevention strategies, like hand washing, remain critical to meeting the public health challenges we face in today's modern world—as cholera epidemics continue. Between 1 January and 30 August 2006, some 25,000 cases and more than 700 deaths due to cholera had been reported throughout the southern part of Sudan in Africa.

Cholera is just one of many illnesses transmitted through food and water. In this chapter, we will first go through a brief introduction to the intoxications and infections caused by the foodborne and waterborne bacteria. Then, we will discuss disease intoxications and the bacterial infections. As we proceed, take note of the illnesses that have similar symptoms.

■ Symptoms: Effects of an infection or disease experienced by a patient.

10.1 Introduction to Foodborne and Waterborne Bacterial Diseases

There is a substantial number of known diseases caused by bacterial pathogens, viruses, or parasites—or their toxins—that are spread through food or water. These diseases may be mild or life-threatening for millions of people in the United States and around the world. The CDC estimates 76 million people in the United States suffer foodborne illnesses each year, accounting for 325,000 hospitalizations and more than 5,000 deaths—primarily infants, the elderly, and immunocompromised individuals. According to the World Health Organization (WHO), waterborne diseases account for an estimated 1.7 million deaths worldwide each year. Most of these deaths are from diarrheal diseases, especially in children in developing nations.

■ Immunocompromised: Refers to the lack of an adequate immune response resulting from disease, exposure to radiation, or treatment with immunosuppressive drugs.

Many Foodborne and Waterborne Diseases Have a Bacterial Cause

KEY CONCEPT

- Bacterial gastrointestinal diseases may arise from intoxications or infections.

We categorize the foodborne and waterborne bacterial diseases as either intoxications or infections. **Intoxications** are illnesses in which bacterial toxins are ingested in food or water. Examples are the toxins causing botulism, staphylococcal food poisoning, and clostridial food poisoning. By contrast, **infections** refer to illnesses in which live bacterial pathogens in food and water are ingested and subsequently grow in the body. Salmonellosis, shigellosis, and cholera are examples. Toxins may be produced, but they are the result of infection.

Determining the etiology of a bacterial disease depends on several factors.

■ Etiology: The study of the causes (origins) of disease.

Incubation period. If an individual ingests and swallows a contaminated food or beverage, there is a delay, called the **incubation period**, before the symptoms appear. This period can range from hours to days, depending on the bacterial species and on the infectious dose. During the incubation period, the toxins or microbes pass through the stomach into the intestine where they may directly affect gastrointestinal function or be absorbed into the bloodstream.

■ Infectious dose: The number of organisms consumed to give rise to symptoms of an illness.

Clinical symptoms. The symptoms produced by an intoxication or infection depend on the specific toxin or microbe, and the number of toxins or cells ingested (**toxic** or **infectious dose**). Although the intoxications and infections often have different symptoms, nausea, abdominal cramps, vomiting, and diarrhea often are common. Because these symptoms are so universal, it can be difficult to identify the microbe causing a disease unless the disease is part of a recognized outbreak or laboratory tests are done to identify the causative agent.

Duration of illness. Intoxications and infections can have very different lengths of time during which the symptoms persist. Some may be very insidious and disappear quite quickly, while others may be longer causative agent.

Demographics. Certain individuals within a population may be more prone to infections or the effects of a toxin. Often infants, whose immune system is still maturing or the elderly, whose immune system is waning, are more susceptible. Individuals who are immunocompromised due to another illness or chemotherapy may be more vulnerable. Also, populations living in unsanitary and overcrowded conditions where public health measures are lacking will be more likely to become ill from contaminated food or water.

There Are Several Ways Foods or Water Become Contaminated

KEY CONCEPT

- Unsanitary processing procedures and fecal material often contaminate foods and water.

We live in a microbial world, and there are many opportunities for food to become contaminated as it is produced and prepared, or during its distribution. Many foodborne microbes are present in healthy animals (usually in their intestines) raised for food. The carcasses of cattle and poultry can become contaminated during slaughter if they are exposed to small amounts of intestinal contents. Fresh fruits and vegetables can be contaminated if they are washed or irrigated with water contaminated with animal manure or human sewage (Chapter 26).

Other foodborne microbes can be introduced through the fecal-oral route; from infected humans who handle the food, or by cross-contamination from some other raw agricultural product. For example, *Shigella* bacteria can contaminate foods from the unwashed hands of food handlers who are infected. In the home kitchen, microbes can be transferred from one food to another food by using the same knife, cutting board, or other utensil to prepare both without washing the surface or utensil between uses. A food fully cooked can become recontaminated if it touches raw foods or drippings containing pathogens.

Water can become contaminated in several ways. A common example is when an ill child or adult swimmer with diarrhea has an "accident" in a public pool. If the pool is not sufficiently chlorinated, such recreational water diseases can be passed to other individuals if they swallow the feces-contaminated water. Disease pathogens also can be spread by surface or groundwater contaminated with untreated or poorly treated sewage. In either case, individuals can become sick.

Finally, some bacterial pathogens cause foodborne or waterborne bacterial diseases only when they are in large numbers. With warm, moist conditions and plenty of nutrients, lightly contaminated food left out overnight can be highly contaminated by the next day. *Escherichia coli*, for instance, dividing every 30 minutes can produce 17 million progeny in 12 hours. If the food were refrigerated promptly, the bacterial cells would not multiply at all. However, two foodborne bacterial species we will discuss, *Listeria monocytogenes* and *Yersinia enterocolitica*, can grow at refrigerator temperatures.

Several bacterial species are commonly found on many raw foods. Therefore, illness can be prevented by (1) controlling the initial number of bacteria present, (2) preventing the small number from growing, (3) destroying the bacteria by proper cooking, and (4) avoiding re-contamination. FIGURE 10.3 identifies where many foodborne illnesses occur.

CONCEPT AND REASONING CHECKS

10.1 What types of etiologic information may help in identifying the cause of an intoxication or infection?

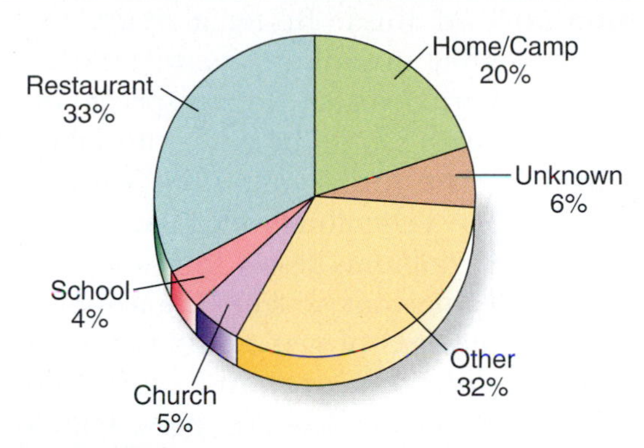

FIGURE 10.3 Occurrence of Foodborne Illnesses. The mishandling of food can lead to many cases of foodborne illness. Restaurants are the single most common source. (Data from CDC, *Morbidity and Mortality Weekly Report*, October 25, 1996, Atlanta: Centers for Disease Control and Prevention.)

Q: What might be some of the "other" sources that make up 32 percent of cases?

10.2 Foodborne Intoxications

Intoxications generally involve a brief incubation period and the situation also resolves in a relatively brief period. We shall observe this pattern in the diseases below.

Bacterial Food Poisoning Can Result from an Intoxication

KEY CONCEPT

- Exotoxins produced by a few bacterial species can be responsible for foodborne disease.

Three bacterial species are noted for their ability to secrete exotoxins, which may cause life-threatening intoxications. These species include *Clostridium botulinum, Staphylococcus aureus*, and *Clostridium perfringens*.

Clostridium botulinum. Of all the foodborne intoxications in humans, none is more dangerous than **botulism**. It is a rare but serious illness caused by *Clostridium botulinum*, a spore-forming, obligately anaerobic, gram-positive bacillus. The endospores exist in the intestines of humans as well as fish, birds, and barnyard animals. They reach the soil in manure, organic fertilizers, and sewage, and often, they cling to harvested products. When spores enter the anaerobic environment of cans or jars, they

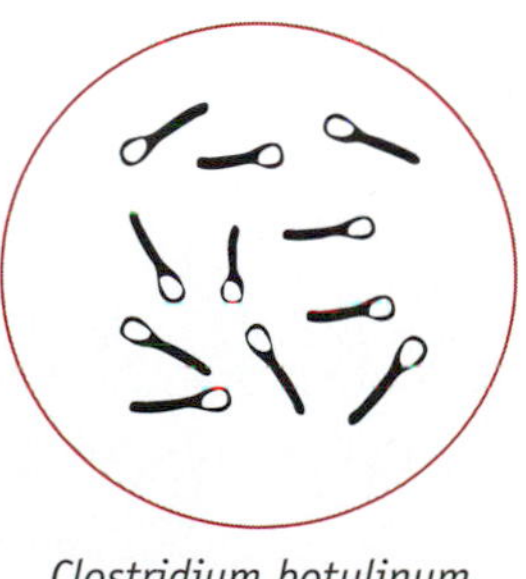

Clostridium botulinum

germinate to vegetative bacilli, and the bacilli produce the toxin.

The protein toxin is called an **exotoxin** because the live bacilli release the toxin into the food environment as it is synthesized. The bacterial cells themselves are of little consequence, but the botulinum toxin is so powerful that one ounce in an aerosolized form could kill all the people in the United States. Scientists have identified seven types of *C. botulinum,* depending on the variant of toxin produced. Types A, B, and E cause most human disease, with type E associated with most cases of foodborne transmission.

The symptoms of botulism usually develop within 18 to 36 hours after ingesting the toxin-contaminated food. Being a neurotoxin, patients suffer neurologic manifestations, including blurred vision, slurred speech, difficulty swallowing and chewing, and labored breathing. The limbs lose their tone and become flabby, a condition called **flaccid paralysis**. These symptoms result from the toxin's effects on the nervous system. The neurotoxin penetrates the ends of nerve cells and inhibits the release of the neurotransmitter acetylcholine. Without acetylcholine, nerve impulses cannot pass into the muscles, and the muscles do not contract. Failure of the diaphragm and rib muscles to function leads to respiratory paralysis and death within a day or two.

■ Neurotoxin: A substance that damages, destroys, or impairs the functioning on nerve tissue.

Since botulism is a type of foodborne intoxication, antibiotics are of no value as a treatment against the toxin. Instead, if treated early, large doses of specific antibodies called **antitoxins** can be administered to neutralize the unbound toxins. Therefore, knowing which type of botulinum toxin caused the disease is important because antitoxin therapy must be type specific. Life-support systems such as mechanical ventilators also are used. For example, in September, 2006, four cases of foodborne botulism (three in the United States and one in Canada) were reported in individuals who consumed improperly refrigerated carrot juice. The lives of these patients depended on botulinum antitoxin and assistance of hospital ventilators.

■ Syndrome: A collection of symptoms.

Although the annual number of cases in the United States is low (average of 110 cases per year), the number of deaths is about 8 percent. Complete recovery can take up to one year.

Botulism can be avoided by heating foods before eating them, because the toxin is destroyed on exposure to temperatures of 90°C for 10 minutes. However, experience shows that most outbreaks are related to home-canned foods having a low acid content, such as asparagus, green beans, beets, and corn, and from foods eaten cold. Other foods linked to botulism include mushrooms, olives, salami, and sausage. In fact, the word botulism is derived from the Latin *botulus,* for "sausage."

Although foodborne botulism is very dangerous, other forms of botulism exist. **Wound botulism** is caused by toxins produced in the anaerobic tissue of a wound infected with *C. botulinum.* Penicillin is an effective treatment. **Infant botulism** is the most common form of botulism in the United States, accounting for over 70 percent of the total annual cases reported. Unlike foodborne botulism where the botulinum toxin is ingested, infant botulism results from the ingestion of food containing *C. botulinum* endospores. Thus, parents should not give their infant honey, which is the most common food triggering infant botulism. Spores that are in about 10 percent of honey germinate and grow in the intestinal tract where the botulinum cells release toxin. This form of botulism typically affects infants 3 to 24 months old because they have not established the normal balance of bowel microbes. Since the toxin produces lethargy and poor muscle tone, infant botulism often is referred to as floppy baby syndrome. Although hospitalization may be necessary, intoxication normally does not require antitoxin treatment. In addition, some cases of sudden infant death syndrome (SIDS) have been associated with infant botulism.

In recent years, one of the botulinum toxins has been put to practical use. In minute doses, **Botox®** (botulism toxin type A) can relieve temporarily a number of movement disorders (so-called dystonias) caused by involuntary sustained muscle contractions. For example, the toxin is used to treat strabismus, or misalignment of the eyes, commonly known as cross-eye; it also is used against blepharospasm, or involuntarily clenched eyelids. The toxin may be valuable in relieving stuttering, uncontrolled blinking, and musician's cramp (the bane of the violinist). Botox® also has been approved for use in the temporary relief of facial wrinkles and frown lines. In 2004, it was approved for temporary relief of hyperhidrosis (excessive body sweating).

Staphylococcus aureus. **Staphylococcal food poisoning** is caused by *Staphylococcus aureus,* a nonmotile, facultatively anaerobic, gram-positive rod that tends to grow in clusters of cells (FIGURE 10.4A). Today, staphylococcal food poisoning ranks as the second most reported of all types of foodborne disease (*Campylobacter*-related illnesses are first). Because most staphylococcal outbreaks probably go unreported, staphylococcal food poisoning could actually be the most common type.

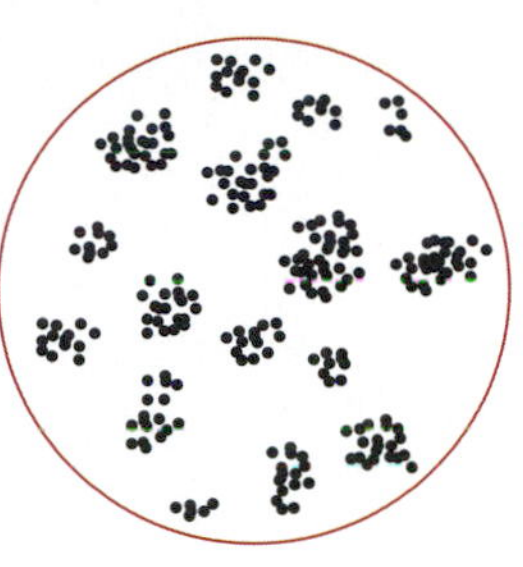

Staphylococcus aureus

Like botulism, staphylococcal food poisoning is caused by exotoxin-contaminated foods. The incubation period is a brief 1 to 6 hours, so the individual usually can think back and pinpoint the source, which often is a protein rich food, such as meats or fish. Contaminated dairy products, cream-filled pastries, or salads such as egg salad also can be a cause. Ham is particularly susceptible because staphylococci tolerate salt.

Because the symptoms are restricted to the intestinal tract, the toxin is called an **enterotoxin** (*entero* = "intestine"). Patients experience **gastroenteritis**, which consists of abdominal cramps, nausea, vomiting, prostration, and diarrhea as the toxin encourages the release of water. (The word diarrhea is derived from the Greek stems *dia* = "through" and *rhein* = "to flow"; hence, water "flows through" the intestines). The symptoms last for several hours, and recovery is usually rapid and complete in 1 to 2 days.

A key reservoir of *S. aureus* in humans is the nose. Thus, an errant sneeze by a food handler may be the source of staphylococci in foods. Studies indicate, however, the most common mode of transmission is from boils or abscesses (Chapter 12) on the skin that shed staphylococci into the food product.

■ Reservoir: The natural host or habitat of a pathogen.

Foods containing *S. aureus* lack an unusual taste, odor, or appearance, and the only clues to possible contamination are factors such as moisture content, low acidity, and improper heating previous to arrival at the table; or by improper refrigeration. The staphylococci grow over a broad temperature range of 8°C to 45°C, and

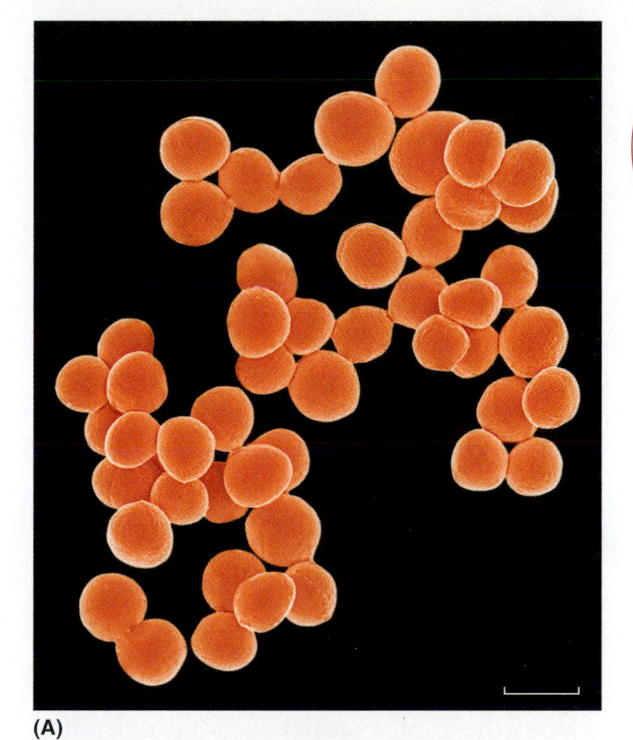

(A)

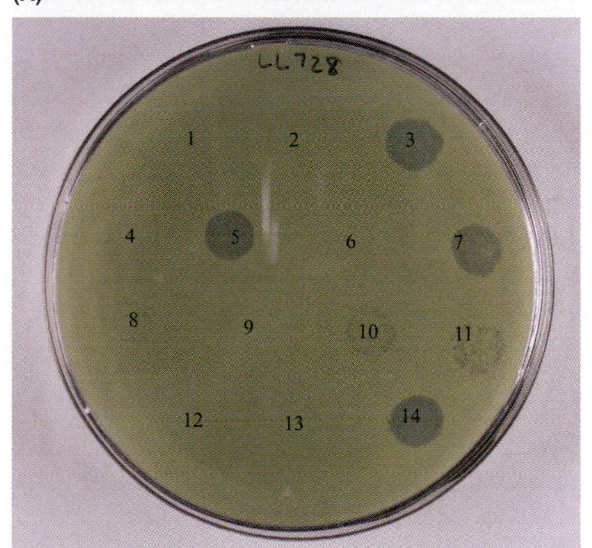

(B)

FIGURE 10.4 ***Staphylococcus aureus.*** (**A**) A false-color scanning electron micrograph of *S. aureus,* a common cause of food poisoning arising from the ingestion of foods contaminated with an exotoxin. The typical grape-like cluster formation of the cocci can be seen. (Bar = 1 μm.) (**B**) Typing of *S. aureus* strains with bacteriophages. The plate of nutrient medium was seeded with the unknown strain of *Staphylococcus*, and numbered bacteriophages were then placed into different areas. The clear areas indicate which of the phages interacted specifically with the bacteria and killed then, leaving a clear circle. In this case, the strain of *S. aureus* is one that interacts with phages 3, 5, 7, and 14.

Q: Why is it important to type S. aureus *strains?*

because refrigerator temperatures are generally set at about 5°C, refrigeration is not an absolute safeguard against contamination. The staphylococcal enterotoxin is among the most heat resistant of all exotoxins. Heating at 100°C for 30 minutes will not denature the protein toxin.

S. aureus normally does not grow in human intestines because of competition by other organisms. Therefore, public health investigators usually are unable to locate the organisms in stool samples. Moreover, the contaminated food often has been consumed completely. Thus, case reports often are based on symptoms, patterns of outbreak, and type of food eaten. When investigators locate staphylococci, they can identify the organisms by growth on mannitol salt agar, Gram staining, and testing with bacteriophages to learn the strain involved, as FIGURE 10.4B illustrates.

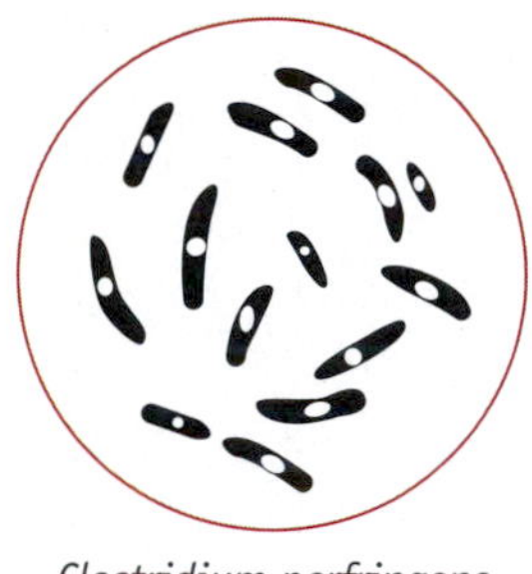

Clostridium perfringens

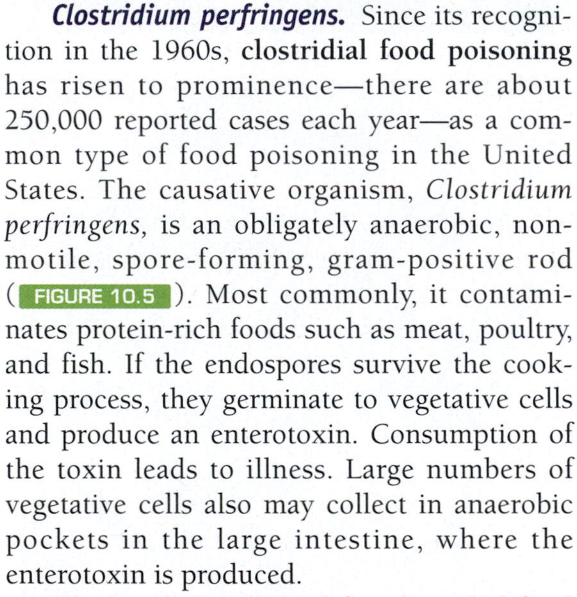

Clostridium perfringens. Since its recognition in the 1960s, **clostridial food poisoning** has risen to prominence—there are about 250,000 reported cases each year—as a common type of food poisoning in the United States. The causative organism, *Clostridium perfringens,* is an obligately anaerobic, nonmotile, spore-forming, gram-positive rod (FIGURE 10.5). Most commonly, it contaminates protein-rich foods such as meat, poultry, and fish. If the endospores survive the cooking process, they germinate to vegetative cells and produce an enterotoxin. Consumption of the toxin leads to illness. Large numbers of vegetative cells also may collect in anaerobic pockets in the large intestine, where the enterotoxin is produced.

The incubation period for clostridial food poisoning is 8 to 24 hours, a factor that distinguishes it from staphylococcal food poisoning. Clinical symptoms require a high infectious dose (10^8 or greater). Moderate to severe abdominal cramping and watery diarrhea are common symptoms, as the enterotoxin encourages the outward movement of water from epithelial cells lining the intestinal tract. Recovery is rapid, often within 1 to 2 days, and antibiotic therapy is generally unnecessary.

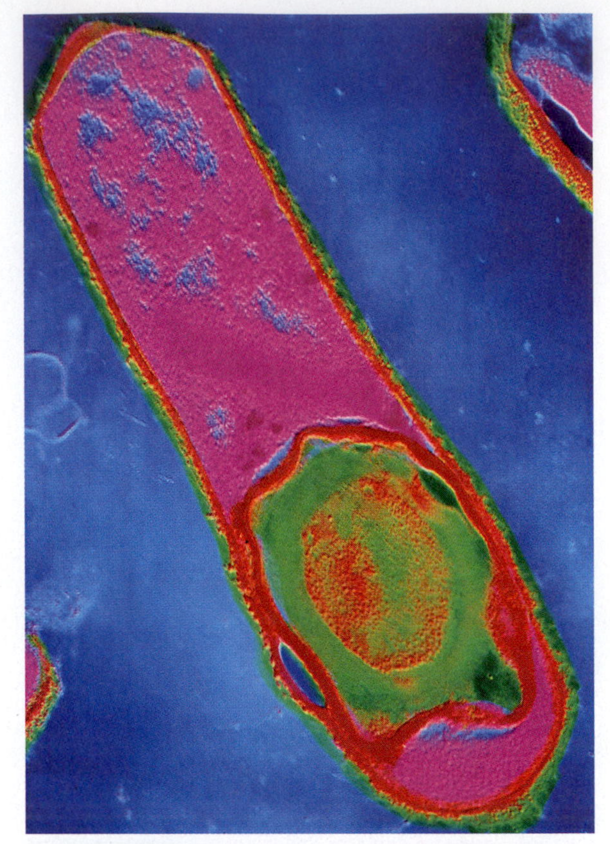

FIGURE 10.5 ***Clostridium perfringens.*** False-color transmission electron micrograph of a *C. perfringens* cell. It produces a heat-resistant enterotoxin that can contaminate protein-rich foods.

Q: What is the "green" structure in the bottom portion of the cell?

CONCEPT AND REASONING CHECKS

10.2 Identify how staphylococcal and clostridial food poisoning could be told apart.

10.3 Foodborne and Waterborne Infections

Foodborne and waterborne infections have a longer incubation period than intoxications because bacterial cells must establish themselves in the body before symptoms develop. Since 1996, there has been a substantial reduction in the incidences of several of these illnesses in the United States.

Typhoid Fever Involves a Blood Infection

KEY CONCEPT

- *Salmonella enterica* serotype Typhi can cause a systemic, life-threatening illness.

Typhoid fever is among the classical diseases (the "slate-wipers") that have ravaged human

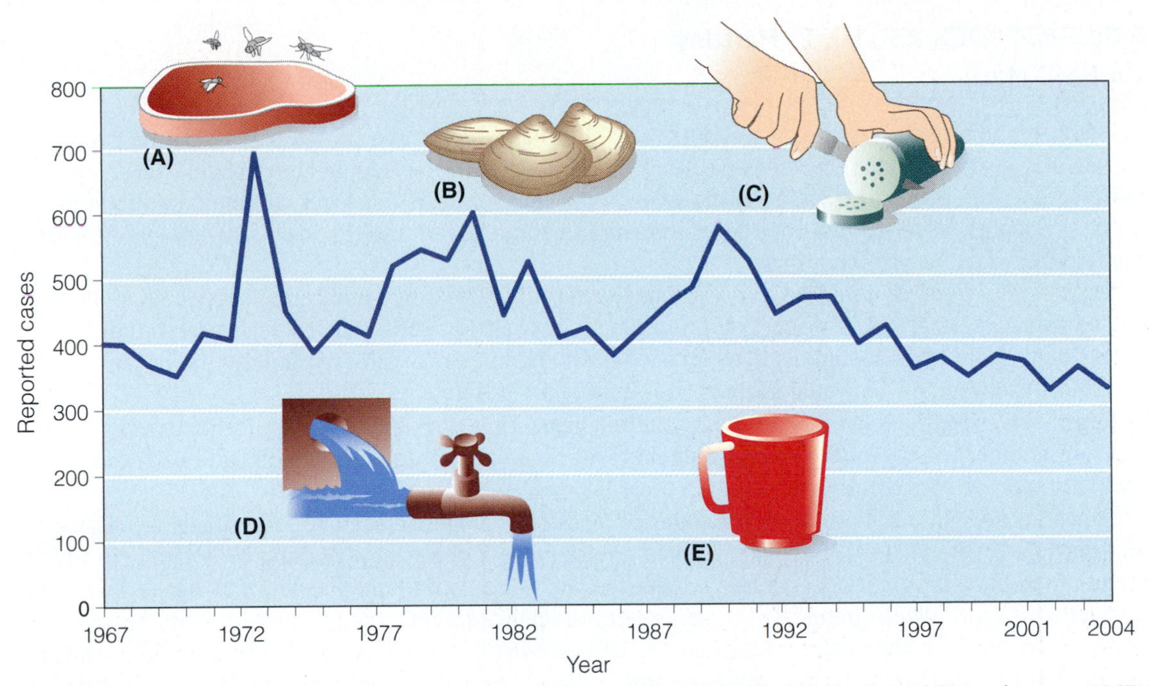

FIGURE 10.6 **The Incidence of Typhoid Fever.** Reported cases of typhoid fever in the United States by year, 1967 to 2004. For 2004, 322 cases were reported. (Data from CDC, 2004, *Summary of Notifiable Diseases,* Atlanta: Centers for Disease Control and Prevention.)

Q: What do the five images (A–E) portray? Name them.

populations for generations. The disease was first studied by Karl Eberth and Georg Gaffky, both coworkers of Robert Koch. Eberth observed the causative organism in 1880 and Gaffky isolated it in pure culture four years later.

There are some 2500 unique serotypes of *Salmonella*. Serotypes are used for *Salmonella* instead of species because of the uncertain relationships existing among the organisms. Typhoid fever is caused by *S. enterica* serotype Typhi, a motile, nonspore-forming, facultatively anaerobic, gram-negative rod. For convention sake, we will refer to it as *S.* Typhi.

■ Serotypes: Closely related groups of microorganisms or structures distinguished by a specific set of antigens.

S. Typhi displays high resistance to environmental conditions outside the body, which enhances its ability to remain alive for long periods in water, sewage, and certain foods exposed to contaminated water. It is transmitted by the five Fs: flies, food, fingers, feces, and fomites (FIGURE 10.6). Humans are the only host for *S.* Typhi.

The organism is acid resistant, and with the buffering effect of food and beverages, it survives passage through the stomach. During the 5- to 21-day incubation period, it invades the small intestine, causing deep ulcers, bloody stools, and abdominal pain. Blood invasion leads to a systemic illness and the patient experiences mounting fever, lethargy, and delirium. (The word typhoid is derived from the Greek *typhos* = "smoke," a reference to the delirium.) In about 30 percent of cases, the abdomen becomes covered with a faint rash (**rose spots**), indicating blood hemorrhage in the skin. Symptoms last for 3 to 4 weeks. If untreated, about 15 percent of individuals die.

Less than 400 cases of typhoid fever are reported annually to the CDC (Figure 10.6). However, about 85 percent of infections are acquired during international travel to endemic areas. Treatment is generally successful with the antibiotic ceftriaxone. About 5 percent of recoverers are carriers and continue to harbor and shed the organisms for a year or more. Because food handlers can be carriers of disease, the public health department usually monitors the activities of carriers. The experiences of one of history's most famous carriers, Typhoid Mary, are recounted in MicroFocus 10.1.

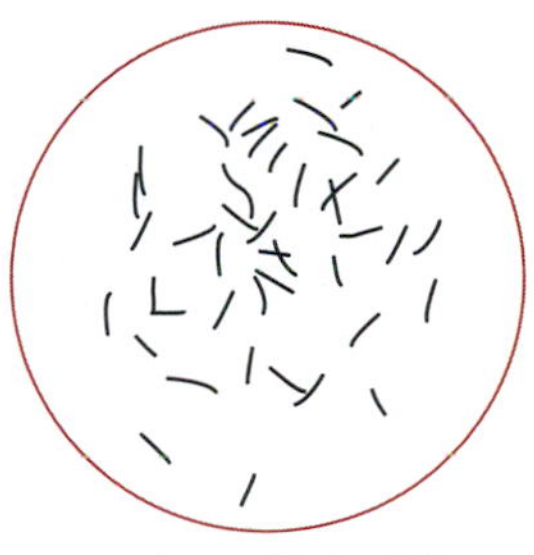

Salmonella Typhi

Traditional vaccines for typhoid fever have consisted of dead *S.* Typhi cells, but adverse reactions have introduced an element of risk. A

MICROFOCUS 10.1: History

Typhoid Mary

By 1906, typhoid fever was claiming about 25,000 lives annually in the United States. During the summer of that year, a puzzling outbreak occurred in the town of Oyster Bay on Long Island, New York. One girl died and five others contracted typhoid fever, but local officials ruled out contaminated food or water as sources. Eager to find the cause, they hired George Soper, a well-known sanitary engineer from the New York City Health Department.

Soper's suspicions centered on Mary Mallon, the seemingly healthy family cook. But she had disappeared three weeks after the disease surfaced. Soper was familiar with Robert Koch's theory that infections like typhoid fever could be spread by people who harbor the organisms. Quietly, he began to search for the woman who would become known as Typhoid Mary.

Soper's investigations led him back over the ten years' time during which Mary Mallon cooked for several households. Twenty-eight cases of typhoid fever occurred in those households, and each time, the cook left soon after the outbreak.

Soper tracked Mary Mallon through a series of leads from domestic agencies and finally came face-to-face with her in March 1907. She had assumed a false name and was now working for a family in which typhoid had broken out. Soper told her he believed she was a carrier and pleaded with her to be tested for typhoid bacilli. When she refused to cooperate, the police forcibly brought her to a city hospital on an island in the East River off the Bronx shore. Tests showed her stools teemed with typhoid organisms, but fearing her life was in danger, she adamantly refused the gall bladder operation that would eliminate them. As news of her imprisonment spread, Mary became a celebrity. Soon public sentiment led to a health department policy deploring the isolation of carriers. She was released in 1910.

But Mary's saga had not ended. In 1915, she turned up again at New York City's Sloane Hospital working as a cook under a new name. Eight people had recently died of typhoid fever, most of them doctors and nurses. Mary was taken back to the island, this time in handcuffs. Still, she refused the operation and vowed never to change her profession. Doctors placed her in isolation in a hospital room while trying to decide what to do. The weeks wore on.

Eventually, Mary became less incorrigible and assumed a permanent residence in a cottage on the island. She gradually accepted her fate and began to help out with routine hospital work. However, she was forced to eat in solitude and was allowed few visitors. Mary Mallon died in 1938 at the age of 70 from the effects of a stroke. She was buried without fanfare in a local cemetery.

newer oral Ty21a vaccine is composed of a weakened (attenuated) strain of *S.* Typhi. Another injectable vaccine (ViCPS) consists of capsular polysaccharides from *S.* Typhi. Its supporters point to a stronger response than that rendered by the Ty21a vaccine. Both vaccines lose effectiveness after a few years, so booster shots are needed for international travelers. Microfocus 10.2 describes an interesting relationship between *S.* Typhi and cystic fibrosis.

CONCEPT AND REASONING CHECKS

10.3 Why is typhoid fever often called enteric fever?

Salmonellosis Can Be Contracted from a Variety of Foods

KEY CONCEPT

- Several *Salmonella* serotypes are responsible for gastroenteritis.

There are about 1.3 million cases of **salmonellosis** each year in the United States, with more than 500 deaths. The most common serotypes involved with illness are *S. enterica* serotype Enteritidis and *S. enterica* serotype Typhimurium.

A relatively large infectious dose is required to initiate an illness. After an incubation period of 6 to 48 hours, the patient with salmonellosis experiences gastroenteritis (fever, nausea, vomiting, diarrhea, and abdominal cramps). Intestinal ulceration is usually less severe than in typhoid fever, and blood invasion is uncommon. The symptoms typically last 4 to 7 days. Dehydration may occur in some patients, necessitating fluid replacement. Diagnosis usually consists of isolating the *Salmonella* serotype from stool specimens or rectal swabs, using differential media.

MICROFOCUS 10.2: Public Health

The Downside, the Upside

Cystic fibrosis is a gruesome disease. Abnormal amounts of thick, sticky mucus build up in the respiratory tracts of children and clog the air passageways. Parents have to slap children on the back repeatedly to help clear the clogging. Life-threatening respiratory infections often accompany the disease.

Scientists now know that cystic fibrosis is a genetic disease that develops when two mutant genes are inherited from the parents. With the mutation in place, a regulatory protein is not produced, and the sticky mucus accumulates. About 30,000 individuals in the United States suffer the misery of cystic fibrosis each year.

But there is a strange twist to this story. Harvard researchers have discovered that the regulatory protein is an attachment site for the typhoid bacillus *Salmonella* Typhi. Apparently, the regulatory protein protrudes from cells and binds to *S.* Typhi at the start of the infectious disease process. Now the bacilli are free to invade and destroy the tissues. However, if there is no protein, then there is no binding, and no disease.

Occasionally, Mother Nature does things that seem to defy explanation. Evidence suggests that the cystic fibrosis patient is immune to typhoid fever. Before the age of antibiotics, that would have been a worthwhile trade-off, since typhoid fever was a major killer. Indeed, in developing countries, typhoid fever is still a serious health problem. Is it possible that cystic fibrosis is another of those "diseases of civilization"? What do you think?

With increased awareness and modern methods of detection, salmonellosis has been linked to a broad variety of foods, including unpasteurized milk and poultry products. *Salmonella* serotypes commonly infect chickens and turkeys when the normal bacterial species of the gut are absent (FIGURE 10.7). In 2004–2005, *Salmonella* outbreaks in the United States included 561 cases associated with eating raw roma tomatoes, 304 cases arising from undercooked turkey, and 15 cases from ice cream contaminated after pasteurization. Also in 2005, more than 1,000 people in Spain came down with salmonellosis after consuming contaminated precooked chicken. This was the largest outbreak of the disease in Spain's history. In 2004, 5 million pounds of raw almonds had to be recalled in the United States because of *Salmonella* contamination, and in 2006 more than 55 people fell ill in the United Kingdom from eating *Salmonella*-contaminated chocolate bars.

Eggs are another source of salmonellosis when used in foods such as custard pies, cream cakes, eggnog, ice cream, and mayonnaise (Chapter 26). Cases have been associated with Caesar salad dressing made with raw eggs. Salmonellosis also has been associated with cracked or contaminated egg shells, but researchers now believe that *Salmonella* serotypes infect the ovary of the hen and pass into the egg before the shell forms. If this is so, then even the highest-grade eggs may be contaminated. Consumers should store eggs in the main compartment of the refrigerator, refrigerate leftover egg dishes quickly in small containers (to accelerate cooling), and avoid "runny" or undercooked eggs.

Live animals also are known to transmit salmonellosis, and many states now prohibit the sale of Easter chicks and ducklings for this reason. Moreover, in 1975, the FDA prohibited the distribution of pet turtles less than 4 inches long because more than 100,000 cases of salmonellosis were attributed to handling these animals. In 2005, an estimated 850,000 to 1.7 million households have reptiles as pets. Thus, the sale of reptiles such as iguanas is restricted. Still, three children were admitted to hospitals in 2002 with bloody diarrhea due to contact with infected iguanas.

CONCEPT AND REASONING CHECKS

10.4 What etiologic characteristic(s) would separate staphylococcal food poisoning from salmonellosis?

Shigellosis (Bacterial Dysentery) Occurs Where Sanitary Conditions Are Lacking

KEY CONCEPT

- *Shigella* produces an exotoxin that leads to illness.

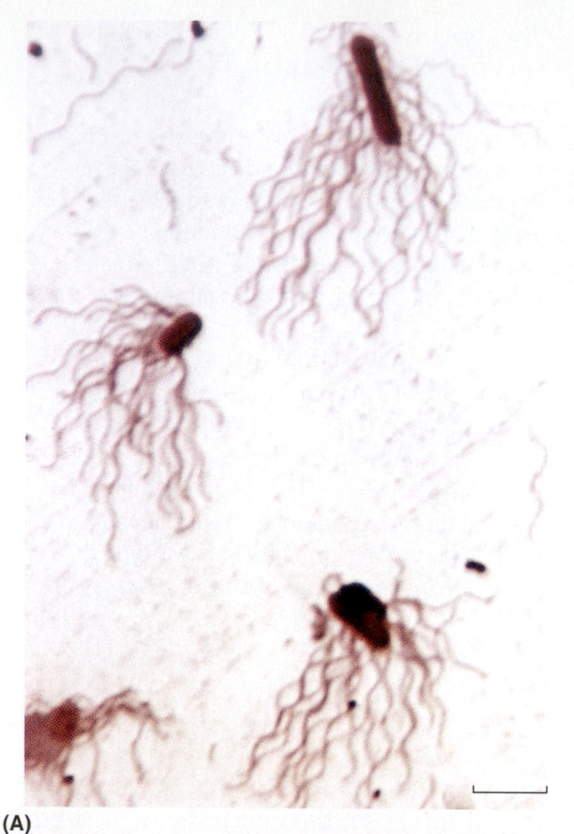

(A)

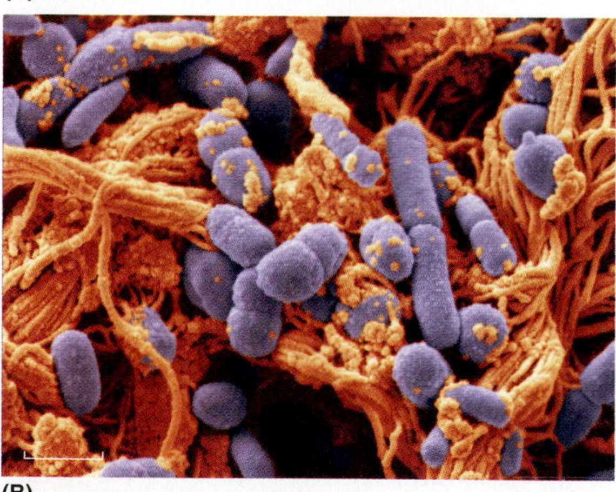

(B)

FIGURE 10.7 ***Salmonella enterica.*** Two views of *S. enterica.* (**A**) A light microscopy photo of *S. enterica* cells. Note the long length of the flagella relative to the cell. (Bar = 2 μm.) (**B**) *S. enterica* serotype Typhimurium observed on the collagen fibers of muscle tissue from an infected chicken. (Bar = 3 μm.)

Q: How might Salmonella *cells attach to the collagen fibers?*

■ Dysentery:
A syndrome manifested by waves of intense abdominal cramps and frequent passage of small-volume, bloody mucoid stools.

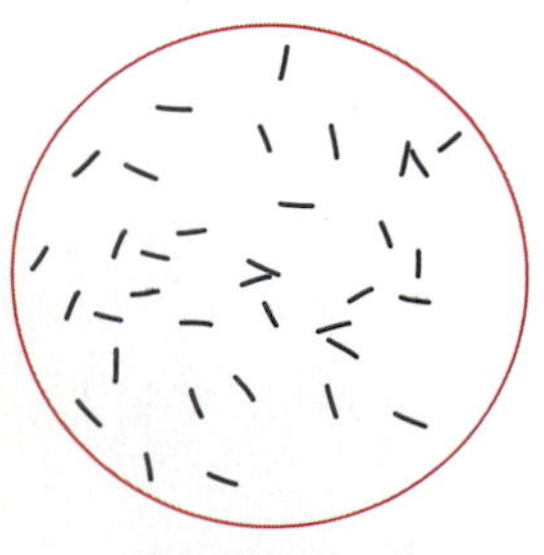

Shigella sonnei

Members of the genus *Shigella* were first described by the Japanese investigator Kiyoshi Shiga in 1898, and two years later by the European microbiologist Simon Flexner.

Shigellosis in the United States is caused primarily by *Shigella sonnei,* which continues to cause outbreaks in day care centers and accounts for the majority of shigellosis cases. About 14,000 cases of gastroenteritis are reported annually in the United States, which represents about 3 percent of the estimated infections. Another species, *S. dysenteriae,* causes deadly epidemics in the developing world.

Shigella species are small unencapsulated, nonmotile, gram-negative rods found mainly in humans and other primates. Humans ingest the organisms in contaminated water as well as in many contaminated foods, especially eggs, vegetables, shellfish, and dairy products.

An infectious dose requires fewer than 200 organisms, which penetrate the epithelial cells lining the intestine and, after 2 to 3 days, produce sufficient exotoxins (**Shiga toxin**) to trigger gastroenteritis. Infection of the large intestine results in bloody mucoid stools that sometimes produce a fatal dysentery. In the Battle of Crécy in 1346, the English army was racked with dysentery. When the French attacked, they literally caught the English with their pants down. For many years, *Shigella* diseases were known as **bacterial dysentery**. However, symptoms of gastroenteritis without blood or mucus are more common than dysentery, and the name shigellosis is preferred.

Shigella species can be isolated from stools, and most cases of shigellosis subside within a week. Usually there are few complications, but patients who lose excessive fluids must be given salt tablets, oral solutions, or intravenous injections of salt solutions for rehydration. Careful hygiene is most important although antibiotics are sometimes effective in reducing the duration of illness and the number of bacilli shed, but many strains of *Shigella* are becoming more resistant to antibiotics. Recoverers generally are carriers for a month or more and continue to shed the bacilli in their feces. Vaccines are not available.

CONCEPT AND REASONING CHECKS

10.5 What control measure (hygiene practice) would most likely decrease transmission rates of shigellosis?

Cholera Can Involve Enormous Fluid Loss

KEY CONCEPT

- *Vibrio cholerae* is transmitted by contaminated water and food.

No diarrheal disease can compare with the extensive diarrhea associated with **cholera** (see chapter opener). The WHO estimates there are more than 100,000 cases and over 1,900 deaths annually. The organism, *Vibrio cholerae,* was first isolated by Robert Koch in 1883.

V. cholerae is a motile, facultatively anaerobic, gram-negative, curved rod (see Figure 10.1). The bacilli enter the intestinal tract in contaminated food, such as raw oysters, or in water (FIGURE 10.8A). *V. cholerae* is extremely susceptible to stomach acid. However, if high numbers are ingested, enough survive to colonize the intestines. There the bacilli do not invade the mucosa but rather move along the intestinal epithelium, secreting an enterotoxin (**cholera toxin**) that stimulates the unrelenting loss of fluid and electrolytes.

In the most severe cases, a patient may lose up to one liter of fluid every hour for several hours. The fluid, referred to as **rice-water stools**, is colorless and watery, reflecting the conversion of the intestinal contents to a thin liquid-like barley soup. The patient's eyes become gray and sink into their orbits. The skin is wrinkled, dry, and cold, and muscular cramps occur in the arms and legs. Despite continuous thirst, sufferers cannot hold fluids. The blood thickens, urine production ceases, and the sluggish blood flow to the brain leads to shock and coma. In untreated cases, the mortality rate may reach 70 percent. However, it is easily treated and prevented. Although prevalent in the United States in the 1800s, good sanitation and water treatment have eliminated the pathogen as a health threat.

Antibiotics such as tetracycline may be used to kill the bacterial cells, but the key treatment is restoration of the body's water and electrolyte balance. Often, this entails intravenous injections of salt solutions. More commonly, patients can be treated with **oral rehydration solution** (ORS), a solution of electrolytes and glucose designed to restore the normal balances in the body.

Cholera has been observed for centuries in human populations, and seven pandemics have been documented since 1817. The first six were caused by *V. cholerae* biotype Classic. The current seventh pandemic, caused by *V. cholerae* biotype El Tor, began in 1961 in Indonesia and now involves about 35 countries (FIGURE 10.8B). A major outbreak occurred in Peru and Ecuador early in 1991 and from there spread throughout the region, accounting for 731,000 cases by the

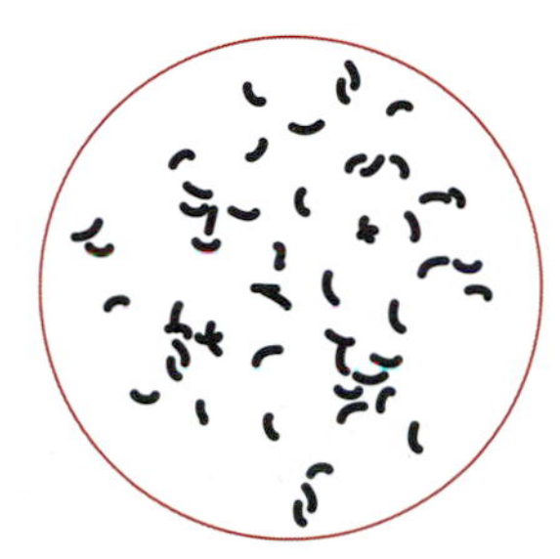

Vibrio cholerae

■ Electrolytes: Any ions in cells, blood, or other organic material.

(A)

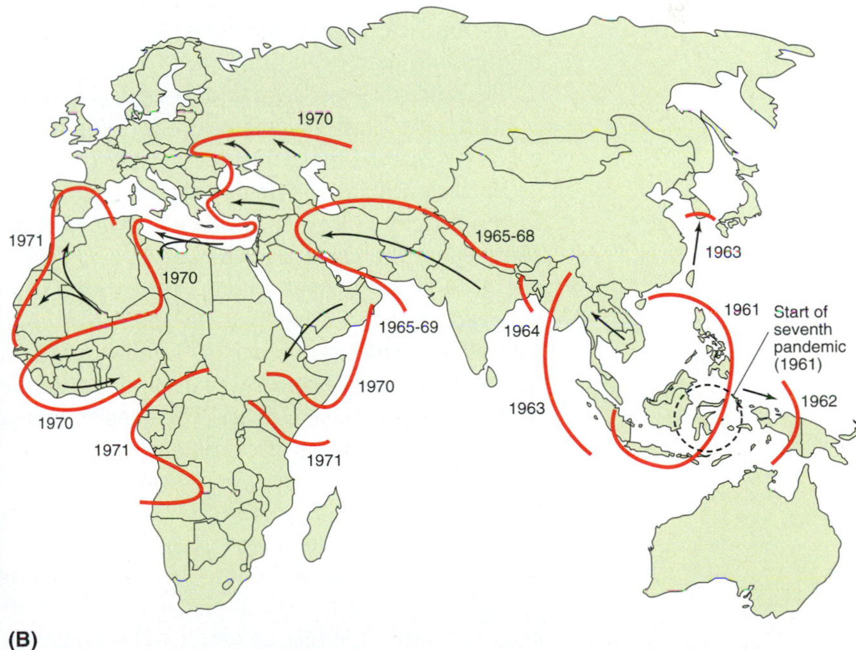

(B)

FIGURE 10.8 **Cholera Disease Spread.** (**A**) The spread of cholera is associated with poor sanitation and water supplies contaminated by feces. The Ganges River delta in India is a central point of endemic cholera. Cholera bacilli frequently inhabit the river and pass easily among unsuspecting bathers. (**B**) The spread of *V. cholerae* biotype El Tor started in 1961 and spread eastward and westward.

Q: How would V. cholerae *be spread?*

end of 1992 and 6,300 deaths in 21 countries in the Western Hemisphere. Microfocus 10.3 describes how the responsible biotype may have emerged. An eighth pandemic emerged in India in 1992 and has spread across Asia. This strain, *V. cholerae* serotype O139, likely is derived from the El Tor biotype since comparative genomics (see Chapter 8) indicates the two serotypes share 99 percent of their genes.

At present, travelers to cholera regions of the world are immunized with preparations of dead *V. cholerae*, thereby obtaining protection for about six months. This approach may change, however, because public health officials are anticipating a genetically-engineered cholera vaccine that will give longer-lasting protection without the risk of using dead pathogenic cells. In 1992, researchers at the University of Maryland identified the genes for enterotoxin production and successfully removed them from experimental *V. cholerae* cells. So treated, the vibrios became nonpathogenic but still alive (i.e., attenuated), and they could be used in a vaccine to stimulate an antibody response. Field trials for the new vaccine are ongoing at this writing.

For generations, scientists believed the cholera bacillus existed only within a human host. That idea was refuted by University of Maryland researchers led by Rita Colwell. Colwell and her colleagues found *V. cholerae* in waters from the Chesapeake Bay, even though no cholera outbreaks were remotely close to the site. They postulated the organisms persist in a spore-like dormant state that cannot be cultivated. (Highly sensitive antibody tests detected the organism when more traditional tests failed.) Apparently, in the cold, nutrient-poor water, the organism's metabolic rate diminishes, and it stops reproducing. This state, says Colwell, allows the cholera bacillus to survive in habitats and environments ranging from seawater to the human intestinal tract. In the sea, moreover, the bacillus

MICROFOCUS 10.3: Biotechnology

How the Sheep Got the Wolf's Clothes

The seventh pandemic of cholera began in Sulawesi, Indonesia, in 1961. Soon it spread to India, the Soviet Union, and the Middle East. By 1991, it had reached Latin America and affected all countries of South America, except Argentina. By 1995, over 5,000 people had died of cholera, and many hundreds of thousands had been terribly sick.

The pandemic was due to a toxin-producing serotype of *Vibrio cholerae* known as O1. The cholera toxin binds to intestinal cells, setting off a cascade of reactions, and water pours out of the cells—up to 5 gallons of water per day. Where did the toxin come from? Microbiologists from Harvard think they have the answer: a virus.

Matthew Waldor and John Mekalanos had been studying cholera for many years. They were impressed by the toxicity of the O1 serotype and wondered why other serotypes were far less lethal. Their interest centered on the cholera toxin gene in *V. cholerae,* and they speculated the gene might have been delivered by a bacteriophage through the process of transduction (see Chapter 8). To test their theory, they removed the entire toxin gene by sophisticated genetic engineering techniques, and then replaced it with an antibiotic-resistance gene. Now they cultured the new antibiotic-resistant cholera cells along with normal cholera cells susceptible to antibiotics. Bingo! The susceptible cells soon became antibiotic resistant. Something (a phage?) seemed to be leaving the antibiotic-resistant cells and ferrying the resistance gene to the susceptible cells.

But, maybe the bacteria were conjugating and exchanging their genes directly. To test this theory, Waldor and Mekalanos passed the genetically engineered (antibiotic-resistant) cells through a filter that would trap everything except phages. They took the clear, cell-free liquid and added it to a fresh batch of normal cells. Double bingo! The cells became resistant to antibiotics. The phage theory strengthened.

Still another test: They treated the clear, cell-free fluid with enzymes that destroy free-floating nucleic acids but have no effect on viruses. (This would eliminate any molecular DNA or RNA that might transform cells.) Then they combined the fluid with normal cells. Once again, the cells became antibiotic resistant. And the coup de grace: Electron micrographs revealed long, stringy phages in the cell-free fluid.

To be sure, the cholera bacterium is not the first to have its toxicity associated with a phage (the diphtheria and botulism organisms are others), but the finding helps explain how an organism can suddenly become lethal. The sheep had acquired the wolf's clothing.

appears to be a regular gut inhabitant of a tiny crustacean known as a copepod. The findings are novel, and they may signal a new outlook for cholera in the decades ahead.

CONCEPT AND REASONING CHECKS

10.6 How is the mechanism of action of cholera toxin similar to that of other enterotoxins mentioned in this chapter?

E. coli Diarrheas Cause Various Forms of Gastroenteritis

KEY CONCEPT

- *E. coli* can be pathogenic within the gastrointestinal tract.

Escherichia coli is normally found in the human colon. However, other serotypes can be pathogenic. In fact, they are one of the major causes of infantile diarrhea.

E. coli is the motile, facultatively anaerobic, gram-negative rod (FIGURE 10.9). Transmission follows the fecal-oral route where contaminated food or water represents the vehicle for transmission. The pathogenic serotypes may induce watery diarrhea in several ways. Three are described below.

Enterotoxigenic *E. coli* (ETEC). ETEC penetrate the intestinal epithelium and produce a toxin that causes gastroenteritis. This illness, typically referred to as **traveler's diarrhea**, affects 20 to 50 percent of travelers within two weeks of traveling to tropical and subtropical destinations. (A number of organisms including several bacterial species, viruses, and protozoa may cause traveler's diarrhea, but recent studies point to *E. coli* as the principal agent.) The volume of fluid lost is usually low, but occasionally it may be considerable and dysentery may occur. The diarrhea lasts 3 to 7 days. The possibility of traveler's diarrhea may be reduced by careful hygiene and attention to the food and water consumed during visits to other countries.

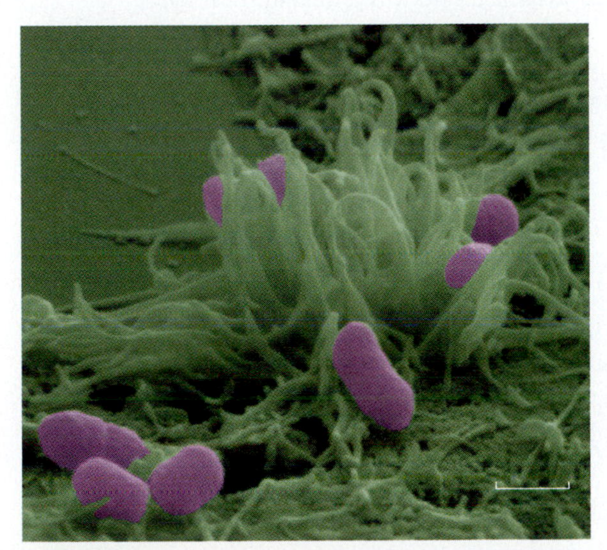

FIGURE 10.9 ***Escherichia coli.*** False-color scanning electron micrograph of *E. coli* cells inhabiting the surface of the intestines. *E. coli* are part of the normal intestinal microbiota of both humans and animals. When other strains of *E. coli* enter the intestines through contaminated food or water, infections such as gastroenteritis can occur. (Bar = 2 μm.)

Q: Why doesn't the normal intestinal E. coli *cause gastroenteritis?*

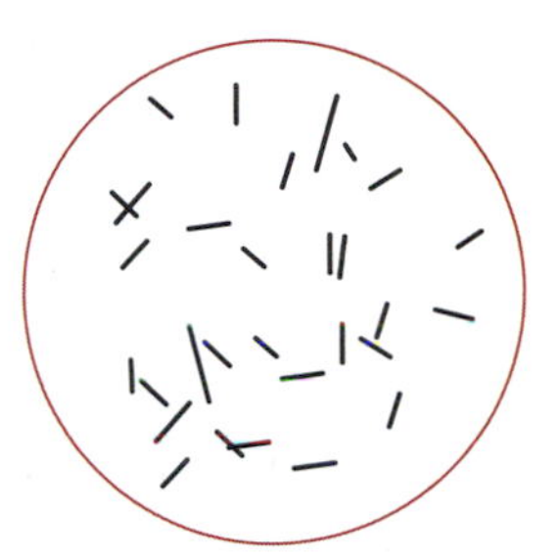

Escherichia coli

Enteropathogenic *E. coli* (EPEC). EPEC is an important cause of diarrhea in infants, especially where sanitation conditions are poor. The infection occurs during birth and the *E. coli* cells attach to the intestinal mucosa and produce a Shiga-like toxin. Watery diarrhea results.

Enterohemorrhagic *E. coli* (EHEC). The most common and publicized EHEC serotype is O157:H7, which was first recognized in 1982. A major source of transmission is from undercooked ground beef, although other sources including consumption of unpasteurized milk and juice, sprouts, lettuce, and salami, and contact with cattle, have been documented. In fact, in 2004–2005, three outbreaks in North Carolina, Florida, and Arizona sickened 173 people attending agricultural fairs and petting zoos. Textbook Case 10 presents an example. Waterborne transmission occurs through swimming in contaminated lakes and pools or drinking water inadequately chlorinated.

The organism is particularly pathogenic because less than 1000 bacilli can establish infection, it is acid-tolerant, it produces toxins at an unusually high rate, and, because it colonizes in the intestines, it can deliver toxins to this area efficiently. After an incubation period of 1 to 8 days, O157:H7 produces sufficient shiga-like exotoxin to cause **hemorrhagic colitis**, a severe diarrhea with grossly bloody diarrhea. In uncomplicated cases, the symptoms resolve within 5 to 10 days.

Complications involving the kidneys and leading to kidney failure is called **hemolytic uremic syndrome** (HUS); seizures, coma, colonic perforation, liver disorder, and heart muscle infections have been associated with HUS. Children less than 5 years old and the elderly are more likely to develop these serious complications.

Textbook CASE 10

Outbreak of *E. coli* 0157:H7 Infection

1 On October 6, 1996, family A from a large city in Connecticut decided to take a drive in the country. Along the way they stopped at a general store for a bite of lunch. The father and two daughters had apple cider; the mother had a soda instead.

2 Three days later, the father and children began to experience serious abdominal pains and vomiting. Moreover, there was blood in their stools. The mother had no symptoms.

3 One of the daughters of family A became worse and had to be admitted to the hospital. The presence of bloody diarrhea was noted by the doctor.

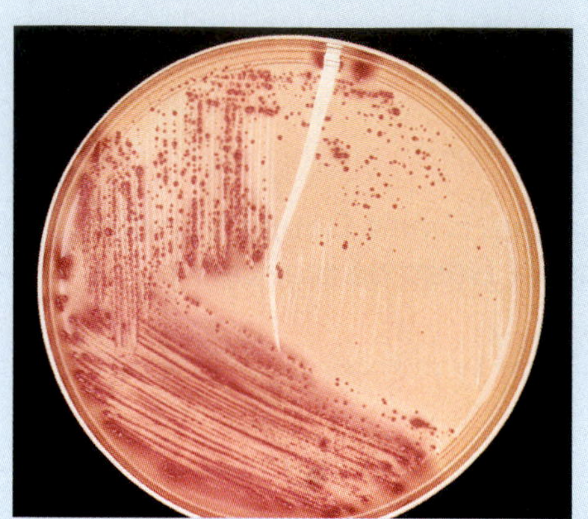

E. coli 0157:H7 growing on a culture plate.

4 On October 11, the Connecticut Department of Public Health (DPH) was notified of the three illnesses plus five more with disease onset during the same period. A case definition was defined and a stool sample from the daughter was sent to the clinical lab for identification.

5 Meanwhile, the kidneys of family A's daughter were suffering and the doctor advised dialysis to assist the kidney function.

6 The laboratory results identified and confirmed *E. coli* 0157:H7 as the cause of infection (see figure).

7 Health officials were notified, and they began a telephone survey to find out if anyone else was similarly infected. Over two dozen cases were found. All were asked about food consumption during the seven days preceding the illness.

8 Based on the interviews, increased risk of illness was associated with drinking fresh apple cider from cider plant A.

9 Inspectors visited the cider plant A. They were told the apples were taken from a pasture where cattle, sheep, and wild deer grazed. The apples were picked directly from the trees. What most interested investigators was hearing many apples also were picked up from the ground.

10 Appropriate control measures were instituted immediately to prevent further cases.

Questions

A. *What would be the case definition defined by the DPH?*

B. *Hearing that the cause of infection was E. coli 0157:H7, what types of food might the DPH investigators be most interested in from the phone survey?*

C. *What is the infection complication exhibited by family A's daughter?*

D. *Why were the DPH investigators most interested in the "drop" apples collected from the soil surface of the pasture?*

E. *What control measures were instituted to prevent further outbreak cases?*

For additional information, see: http://www.cdc.gov/mmwr/preview/mmwrhtml/00045558.htm.

Epidemiologists at the CDC estimated that in 2004 there were 3,168 reported cases of EHEC infections of which 2,544 were due to O157:H7. Most states require *E. coli* O157:H7 isolates be reported to the state health department, and physicians are alerted to watch for cases of bloody diarrhea. Such a situation arose in 2006 when *E. coli* O157:H7 contaminated bagged spinach and sickened almost 200 Americans across the nation (FIGURE 10.10). One death was reported from HUS.

FIGURE 10.10 **Fresh Bagged Spinach.** The 2006 outbreak of *E. coli* 0157:H7 contaminated spinach forced the recall of all bagged products.

Q: Propose a mechanism by which the spinach became contaminated with the E. coli *strain.*

The prevailing wisdom is that *E. coli* O157:H7 exists in the intestines of healthy cattle but causes no disease in these animals. Slaughtering brings *E. coli* to beef products, and excretion to the soil accounts for transfer to plants and fruits. As of this writing, the source of the spinach contamination remains unknown but probably involves an animal or water source. *E. coli* O157:H7 ferments the alcoholic carbohydrate sorbitol very slowly, and this factor is useful in a diagnostic test: In MacConkey agar, the lactose is replaced by sorbitol, and *E. coli* O157:H7 produces white colonies, while other *E. coli* strains produce red or pink colonies.

In recent years, other serotypes have been identified as causes of intestinal illness. All told, over 100 serotypes of *E. coli* have been implicated in HUS and hemorrhagic colitis. Treatment regimens are not established and research continues on a vaccine for all *E. coli* diarrheas.

CONCEPT AND REASONING CHECKS

10.7 Why are children and infants more susceptible to EHEC and EPEC, respectively?

Gastric Ulcer Disease Can Be Spread Person to Person

KEY CONCEPT

- *Helicobacter pylori* causes gastritis and duodenal and gastric ulcers.

Approximately 25 million Americans suffer from gastric ulcers during their lifetime. For decades, scientists believed all ulcers resulted from "excess acid" due to factors such as nervous stress, smoking, alcohol consumption, diet, and physiological dysfunction. However, the work of two Australian gastroenterologists, Barry Marshall and J. Robin Warren, made it clear that the bacterial species *Helicobacter pylori* is involved (Microfocus 10.4).

This motile, microaerophilic, gram-negative curved rod infects half the world's population, yet only 2 percent are afflicted with duodenal or gastric ulcers. One percent has stomach cancer, which also is associated with an *H. pylori* infection.

It is uncertain how *H. pylori* is transmitted. Most likely, it is spread person to person through contaminated food or water. How *H. pylori* manages to survive in the intense acidity of the stomach is interesting. When *H. pylori* penetrates the stomach mucous layer, it attaches to the stomach wall. There it secretes the enzyme urease. Urease digests urea in the area and produces ammonia as an end product (FIGURE 10.11). The ammonia neutralizes acid in the vicinity of the infection. The ammonia and an *H. pylori* cytotoxin cause destruction of the mucous-secreting cells, exposing the underlying connective tissue to the stomach acid. In the stomach lining, a sore 0.6 to 12 cm in diameter appears (although some ulcers may be up to 30 cm in diameter). The pain is severe and is not relieved by food or an antacid (as is a duodenal ulcer).

In the past, physician biopsies of a patient's stomach tissue were used to detect *H. pylori*, but in 1996, a new and relatively simple, noninvasive urea breath test was approved by the United States Food and Drug Administration (FDA). The patient drinks a urea solution fortified with harmless carbon-13 isotopes (see Chapter 2). Because *H. pylori* breaks down urea rapidly, the carbon-13 is quickly expelled as CO_2 in the patient's breath, and it can be detected easily if the bacterial cells are present. If no isotope is detected, the organism is probably not present. The test also can be used to document eradication of the organism after therapy. Thus, doctors have revolutionized the treatment of ulcers by prescribing antibiotics such

MICROFOCUS 10.4: Being Skeptical

Are Ulcers an Infectious Disease?

Could the old claim, "You're giving me an ulcer!" actually be true? Could ulcers be a transmissible disease?

In 1982, two Australian gastroenterologists, Barry J. Marshall and J. Robin Warren, identified bacterial cells living in the stomach lining of over 100 patients who had ulcers. The initial discovery was serendipitous because Marshall and Warren could not cultivate the organism under normal conditions. Only when they were swamped with work did they leave their culture plates in the incubator too long. And only then did bacterial colonies appear. Marshall (see **Figure A**) and Warren identified the organism as the gram-negative curved rod *Campylobacter pyloridis*. Since then, the organism's name has been changed to *Helicobacter pylori* (see **Figure B**).

Marshall and Warren speculated that inflammation in the stomach (gastritis) and ulceration of the stomach or duodenum (peptic ulcer disease) was the result of an infection of the stomach caused by *H. pylori*. This speculation was met with widespread skepticism. Even additional research did not change many minds because the conventional wisdom said that stress, diet, or other factors trigger excess acid secretion and ulcer formation. Indeed, two of the top-selling drugs at the time in the United States were antacids cimetidine (Tagamet) and ranitidine (Zantac), both used to control acid secretion. So, most microbiologists and physicians doubted the duo's claims.

This so frustrated Marshall that he decided to do the ultimate experiment—become a guinea pig. So, in July, 1984, he made a dilute solution of the *H. pylori* bacterial cells and drank it. "I drank it down very quickly, like a tequila shot," he said. If his theory was correct, he should get an ulcer. And guess what? His big gulp paid off. Fourteen days later he had the telltale signs of stomach inflammation. Luckily, his immune system then knocked out the infection.

In 1993, a study published in the *New England Journal of Medicine* indicated that 48 of 52 peptic ulcer patients could be cured of their ulcers in six weeks if treated with two antibiotics over a 12-day period. Another 52 patients received a placebo and 39 seemed to be cured, but a year later, the ulcers had returned in all 39 patients. By comparison, only 4 of the patients receiving antibiotics experienced a recurrence.

In all, it took ten years for the medical community to agree with Marshall's assertion. But the ultimate reward came in 2005 when he and Warren won the 2005 Nobel Prize in Physiology or Medicine for their discovery. As the Nobel Assembly announced, "Thanks to the pioneering discovery by Marshall and Warren, peptic ulcer disease is no longer a chronic, frequently disabling condition, but a disease that can be cured by a short regimen of antibiotics and acid secretion inhibitors." Three cheers (not a glassful of *H. pylori* though) to Marshall and Warren!

FIGURE A **Doctor Barry J. Marshall.**

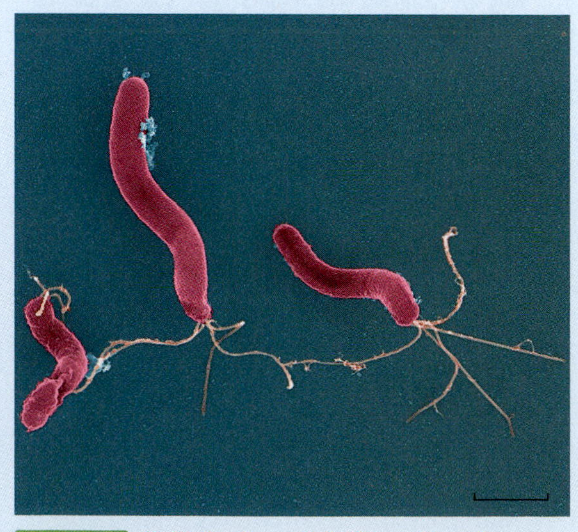

FIGURE B ***Helicobacter pylori*. (Bar = 1 μm.)**

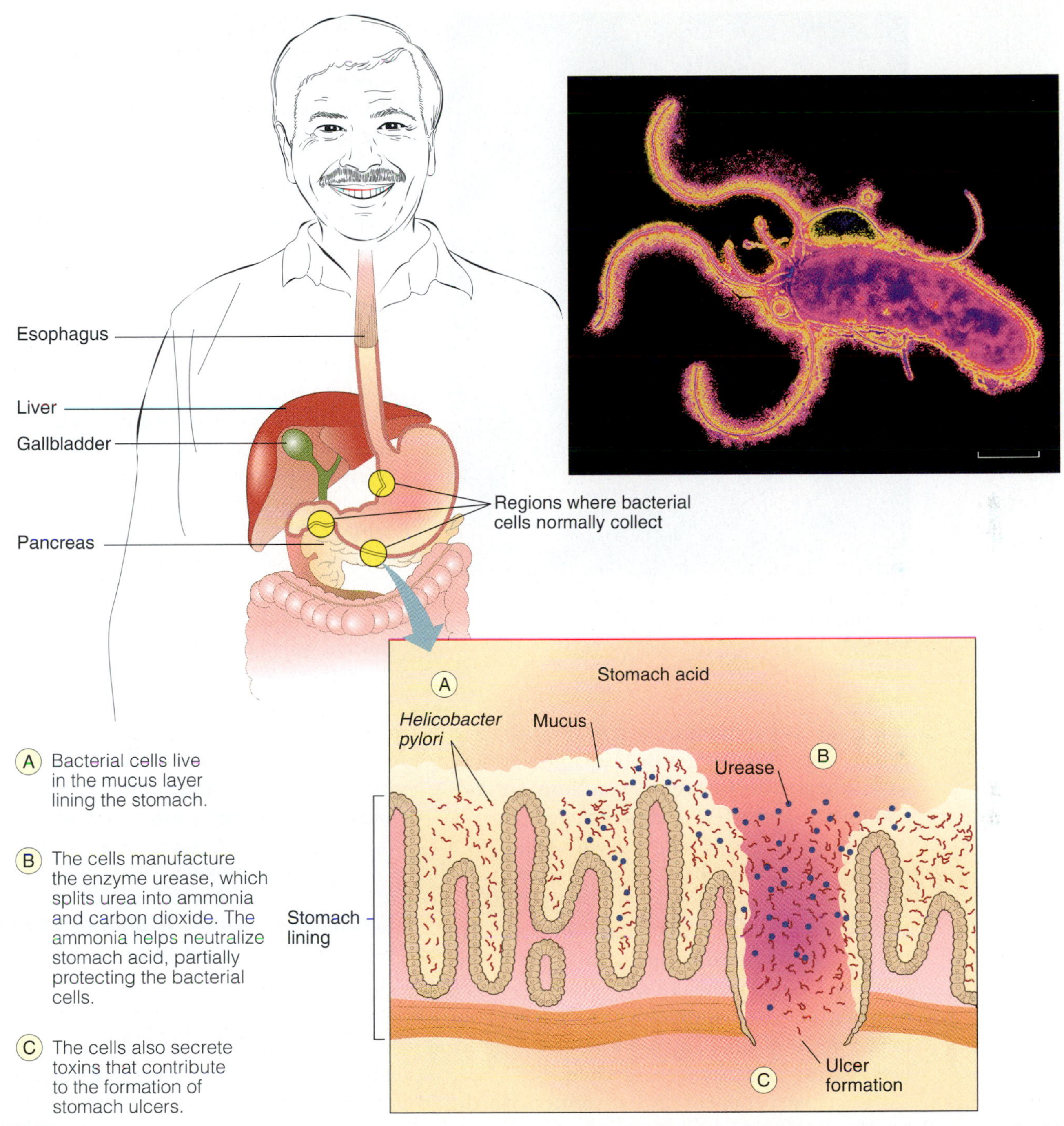

FIGURE 10.11 **The Progression of Gastric Ulcers.** The majority of gastric ulcers are caused by *Helicobacter pylori* [inset] (Bar = 1 µm). This figure illustrates how they cause an ulcer.

Q: How might gastric ulcer disease lead to stomach cancer?

as amoxicillin, tetracycline, or clarithromycin (Biaxin), along with omeprazole (Prilosec) for acid suppression. They have achieved cure rates of up to 94 percent; relapses are uncommon.

CONCEPT AND REASONING CHECKS

10.8 Summarize how an *H. pylori* infection causes stomach ulceration.

Campylobacteriosis Results from Consumption of Contaminated Poultry or Dairy Products

KEY CONCEPT

- *Campylobacter jejuni* is transmitted to humans through the fecal-oral route.

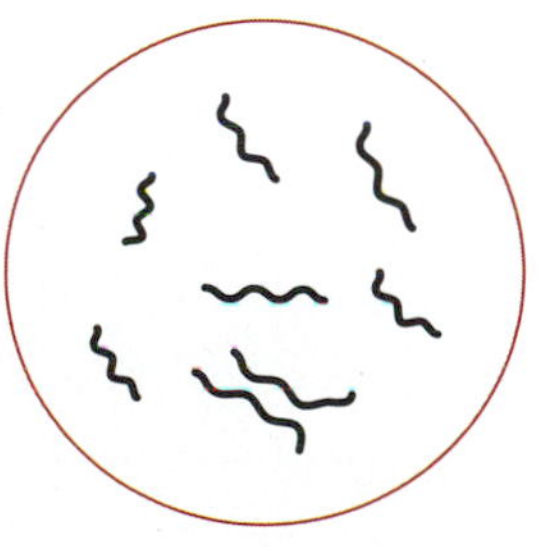

Campylobacter jejuni

Since the early 1970s, **campylobacteriosis** has emerged from an obscure disease in animals to

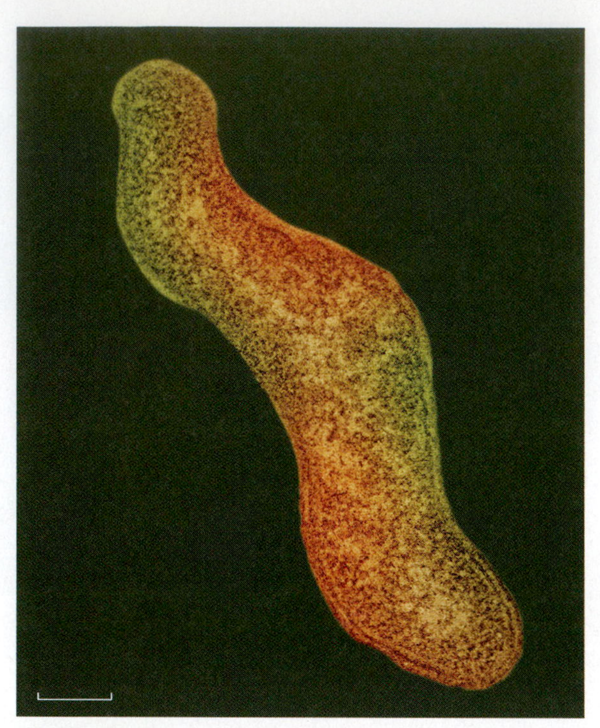

FIGURE 10.12 ***Campylobacter jejuni.*** A false-color transmission electron micrograph of a *C. jejuni* cell. (Bar = 0.5 μm.)

Q: What shape are these cells?

being the most commonly reported bacterial cause of gastroenteritis in the United States. The illness affects over 2.5 million persons and causes about 100 deaths each year.

The pathogen, *Campylobacter jejuni,* is a motile, microaerophilic, curved (*campylo* = "bending"), gram-negative rod (FIGURE 10.12). Reservoirs for the organism are the intestinal tracts of many animals, including dairy cattle, chickens, and turkeys. In fact, chickens raised commercially are colonized with *C. jejuni* by the fourth week of life. Unpasteurized dairy products also can be a source of infection.

C. jejuni primarily is transmitted via the fecal/oral route through contact or exposure to contaminated foods or water. During an incubation period of 2 to 7 days, the bacterial cells colonize the small or large intestine. Invasion of the mucosa leads to inflammation and occasional mild ulceration. However, the signs and symptoms of campylobacteriosis are not unique as they range from mild diarrhea to severe gastroenteritis with bloody stools. Most patients recover in less than a week without treatment, but some have high fevers and bloody stools for prolonged periods. Erythromycin therapy hastens recovery.

Some people may develop a rare nervous system disease several weeks after the diarrheal illness. This disease, called **Guillain-Barré syndrome**, results from the immune system attacking the body's own nerves. The resulting nerve damage can cause paralysis lasting several weeks and usually requires intensive care. Approximately 1 in every 1,000 reported cases of campylobacteriosis leads to Guillain-Barré syndrome and up to 40 percent of Guillain-Barré syndrome cases in the United States might be caused by campylobacteriosis.

Raw *Campylobacter*-contaminated poultry often is the source of infection. A common route of infection is to cut chicken on a cutting board and then use the unwashed cutting board to prepare other raw foods or vegetables. Other foods also can be contaminated. In 1998, 79 persons at a summer camp were infected with *C. jejuni* through ingestion of contaminated tuna salad. In 2003, an investigative study identified salad vegetables and bottled water as newly identified risk factors.

Other species, including *C. coli* and *C. fetus,* also have been identified as human pathogens, and as methods of detection improve, the number of reported cases of campylobacteriosis probably will continue to rise.

CONCEPT AND REASONING CHECKS

10.9 Explain how *C. jejuni* could be responsible for Traveler's diarrhea.

Listeriosis Usually Manifests Itself as Meningoencephalitis or Septicemia

KEY CONCEPT

- *Listeria monocytogenes* species can be transmitted by a variety of contaminated foods.

The CDC estimates that there are 2,500 cases of **listeriosis** in the United States annually and 20 percent of those cases result in death. The illness does not appear to be transmissible from person to person.

Listeriosis is caused by *Listeria monocytogenes,* a small, nonspore-forming, facultatively anaerobic, gram-positive rod that is motile at room temperature. The bacillus is commonly found in the soil and in the intestines of many animals, including birds, fish, barnyard animals, dairy cattle, and household pets. It is transmitted to humans by food contaminated with fecal matter, as well as by the consumption of contaminated animal foods. Delicatessen cold cuts, as well as soft cheeses (e.g., Brie, Camembert, feta, and blue-veined cheeses) have been associated with a significant number of cases. The pathogen is considered psychrotrophic, meaning it is able to grow at refrigerator temperatures (see Chapter 5). Therefore, refrigerated foods contaminated with *L. monocytogenes* will not suppress growth of the bacterial cells.

The incubation period can be prolonged, anywhere from 2 to 6 weeks. Healthy adults experience few symptoms of *L. monocytogenes* ingestion. The disease primarily affects newborns, pregnant women, the elderly, and immunocompromised individuals. In these cases, the illness can become systemic with symptoms of fever, malaise, arthritis, and jaundice.

Listeriosis occurs in many forms. One form, called **meningoencephalitis**, is characterized by headaches, stiff neck, delirium, and coma. Another form is **septicemia**, a blood disease accompanied by high numbers of infected white blood cells called monocytes (hence the organism's name) (FIGURE 10.13). The intracellular bacterial cells secrete a toxin that damages the membrane enveloping the internalized bacterial cells so they cannot be destroyed by lysosomes (see Chapter 3).

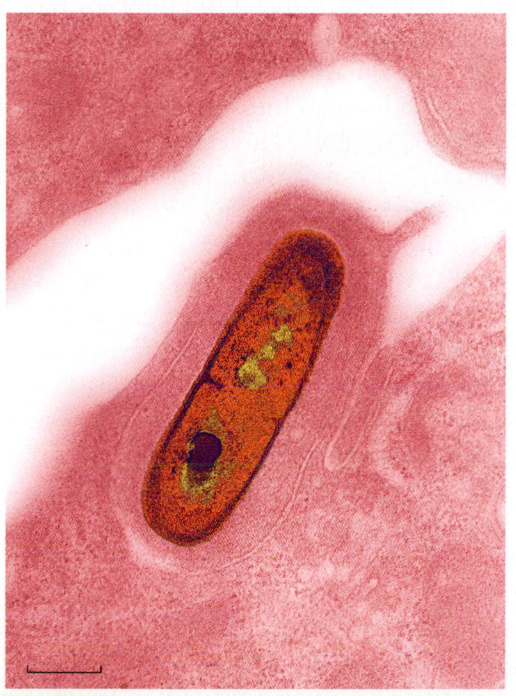

FIGURE 10.13 ***Listeria monocytogenes.*** A transmission electron micrograph of *L. monocytogenes* entering a monocyte. (Bar = 1 μm.)

Q: How does L. monocytogenes *evade destruction by the monocyte's lysosomes?*

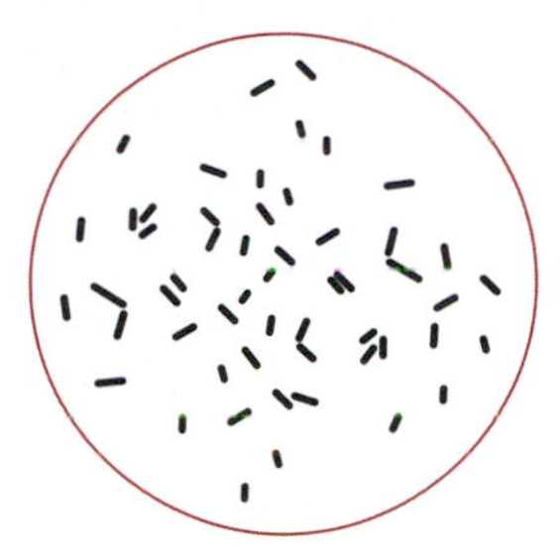

Listeria monocytogenes

A third form, occurring in females is characterized by infection of the uterus, with vague flu-like symptoms. If contracted during pregnancy, the disease may result in miscarriage of the fetus or mental damage in the newborn (newborn meningitis). Other individuals suffer respiratory distress, diarrhea, back pain, skin itching, and other nonspecific symptoms, which makes diagnosing the disease difficult.

Although ampicillin is an effective treatment, relapses are common. A notable outbreak of listeriosis occurred in late 1998 and early 1999. Close to 100 cases of illness were reported in 22 states, all linked to commercial brands of hot dogs and deli meats. Health officials suspected that dust from construction work on air-conditioning units at the production plant may have been involved. Fourteen adults died during the outbreak, and six pregnant women suffered miscarriages.

■ Jaundice: A condition in which there is yellowing of the whites of the eyes, skin, and mucous membranes, caused by bile pigments in the blood.

CONCEPT AND REASONING CHECKS

10.10 What are the three forms of listeriosis?

Several Other Bacterial Species Can Be Transmitted through Food or Water

KEY CONCEPT

- Several known and emerging pathogens can cause gastrointestinal illness.

A number of other bacterial organisms transmitted by food or water merit brief consideration in this chapter.

***Brucella* species.** A bacterial illness called **brucellosis** is an occupational hazard of farmers, veterinarians, dairy and meat plant workers, and others who work with large ruminant animals. Among the species responsible for the most common zoonotic infection worldwide are *Brucella abortus* (cattle), *B. suis* (swine), *B. melitensis* (goats and sheep), and *B. canis* (dogs). All species are small, nonmotile, gram-negative rods causing disease in humans.

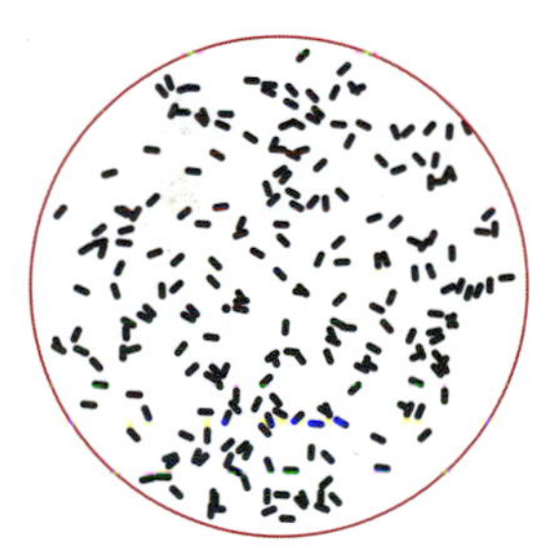

Brucella abortus

■ Zoonotic: Refers to a disease transmitted from animals to humans.

Bacterial transmission can occur by splashing contaminated milk into the eye, by the accidental passage of contaminated fluids through skin abrasions, by contact with infected animals, and by the consumption of contaminated milk or other dairy products. In one example of food transmission, the CDC reported 29 cases in Mexican emigrants living in Houston, Texas. All had eaten goat cheese made from unpasteurized goat's milk. Human-to-human transmission is unknown.

B. abortus has been effectively eliminated from American livestock and most human cases of brucellosis are in travelers or immigrants from an endemic region. The consumption of unpasteurized milk from outside the United States remains a source of infection from *B. abortus* and *B. melitensis*.

■ Cellulitis: A diffuse inflammation of the connective tissues of the skin.

On infection, the bacterial cells are transported to the blood-rich organs, such as the spleen and lymph glands. Patients experience flu-like weakness, as well as backache, joint pain, and a high fever (with drenching sweats) in the daytime and low fever (with chills) in the evening. This fever pattern gives the disease its alternate name, **undulant fever** (*undulat* = "wavy"). The disease seldom causes death in humans, and a combination of doxycycline and gentamicin speeds recovery.

■ Emetic: Refers to vomiting.

The pasteurization of milk and livestock immunization have reduced the incidence of human disease in the United States from 6,300 cases in 1947 to 114 cases in 2004. However, in 2003/2004 an outbreak of brucellosis was discovered in two livestock herds in Wyoming. The disease probably was transmitted to them from an infected elk herd. No human cases were reported.

***Vibrio* species.** A major cause of foodborne infections in Japan and other areas of the world where seafood is the main staple of the diet is *Vibrio parahaemolyticus*, a curved, gram-negative rod. After about a 24-hour incubation period, the bacilli invade the mucosa, causing acute abdominal pain, vomiting, diarrhea, and watery stools. In 2004, 54 people in Alaska became ill from eating raw oysters, and in 2006, 177 cases were reported in New York, Oregon, and Washington from eating raw shellfish.

In 1996, the CDC reported an outbreak of intestinal illness due to another *Vibrio* species, *V. vulnificus*. This species occurs naturally in brackish and seawaters, where oysters and clams live. People who consume these mollusks raw are at risk, especially immunocompromised individuals and those who suffer from liver disease or low stomach acid. (Indeed, taking an antacid after a meal of any contaminated food may neutralize stomach acid and facilitate the passage of the pathogen to the bloodstream.) The gastrointestinal infection involves fever, nausea, and severe abdominal cramps. The infection can become systemic and cause necrotic skin lesions and cellulitis. The mortality rate for the disease can be as high as 50 percent. There are some 100 reported cases in the United States each year, usually limited to the Gulf Coast. However, in 2006, *V. vulnificus* was identified for the first time in the ocean waters of Rhode Island.

***Bacillus cereus*.** The motile, spore-forming, gram-positive bacillus, *Bacillus cereus*, causes food poisoning in two distinct forms, both due to enterotoxins. A 2 to 6 hour incubation period can produce either a diarrheal illness, accompanied by abdominal pain, or an emetic form, characterized by substantial vomiting, frequently experienced after consuming contaminated cooked grains. Neither form involves fever, and most patients recover within two days without treatment. In the 1980s, investigators traced a notable outbreak to a college cafeteria where a macaroni and cheese dish harbored a million bacilli per gram. Heat-resistant spores had apparently survived the cooking process.

***Yersinia enterocolitica*.** Another emerging cause of foodborne illness is *Yersinia enterocolitica* (FIGURE 10.14). This motile (at room temperature), gram-negative rod is widely distributed in animals and in river and lake waters. Infections occur from consuming food that came in contact with infected domestic animals or raw pork. Following a 1 to 2 day incubation, affected individuals experience fever, diarrhea, and abdominal pain. The symptoms last 1 to 3 weeks. Unless the illness becomes systemic, antibiotic therapy is unnecessary.

The largest episode of disease thus far recorded in the United States occurred in 1982 and involved 172 patients in Arkansas, Tennessee, and Mississippi. Most required hospitalization. Investigators believed contaminated milk was the source. A 1990 outbreak in Georgia involved raw chitterlings (pork intestines). In this case, *Y. enterocolitica* was apparently transferred to a group of children by hand-to-hand contact with the individual preparing the food.

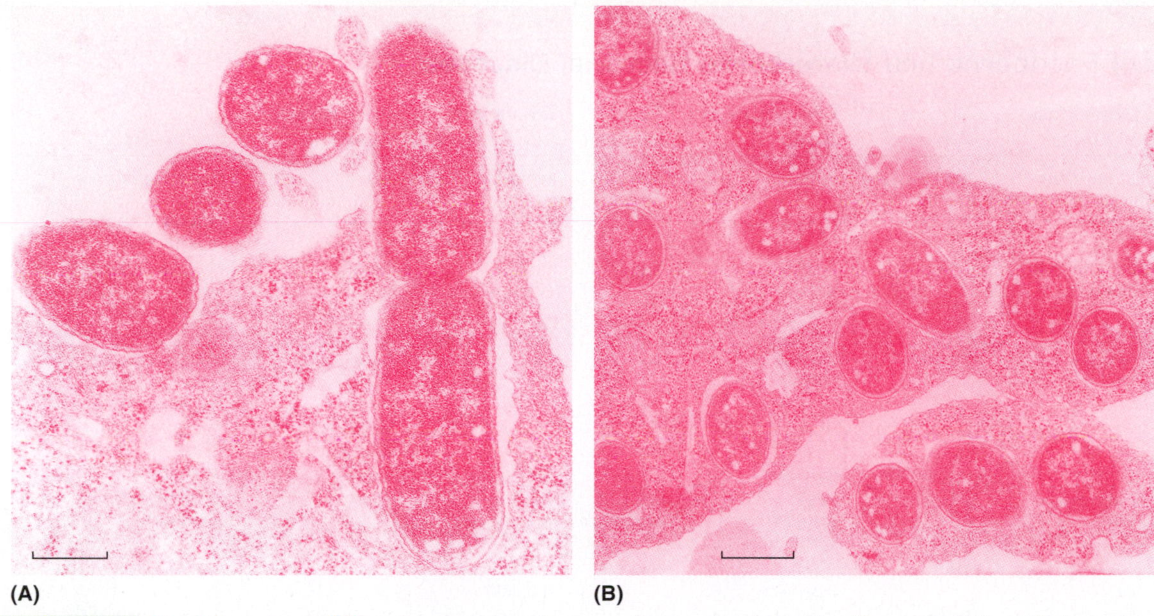

(A) (B)

FIGURE 10.14 ***Yersinia enterocolitica.*** Transmission electron micrographs of invasive *Y. enterocolitica.* (**A**) A number of bacterial cells attached to the plasma membrane of a host cell. One bacterial cell appears to be undergoing cell division and entering the cell at the same time. (Bar = 1 µm.) (**B**) Several *Y. enterocolitica* cells observed within the cytoplasm of an infected cell. The bacterial cells are in membrane-enclosed vacuoles. Researchers postulate that a single gene is responsible for the invasive capacity displayed by this organism. (Bar = 2 µm.)

Q: What types of foods or beverages might be contaminated by Y. enterocolitica*?*

Other bacterial species. Another cause of intestinal illness is *Plesiomonas shigelloides,* a nonmotile, facultatively anaerobic, gram-negative rod. *P. shigelloides* normally is limited to the tropical and subtropical climates of Asia and Africa, and is commonly found in the gut of tropical fish. It can cause intestinal illness in people who eat raw seafood or who travel in the tropics and consume contaminated water or eat contaminated food. A 1990 case, however, appeared to fit neither description. The case occurred in a year-old child who had apparently been infected by water in the bathroom tub. Her parents previously had poured water from their aquarium into the tub. The aquarium was home to a collection of piranhas, a type of tropical fish the family kept as pets.

Clinical reports in the 1980s documented that *Aeromonas* species, especially *A. hydrophila,* cause human gastrointestinal disease. The organisms are motile, gram-negative rods commonly found in soil and water. They appear to be transmitted by food. Both cholera-like and dysentery-like diarrheas, ranging from mild to severe, have been reported in patients. An enterotoxin may be responsible for the symptoms.

The foodborne and waterborne bacterial diseases discussed in this chapter are summarized in TABLE 10.1. Several of the diseases mentioned in this chapter could be agents for bioterrorism, as Microfocus 10.5 describes. Also, MicroInquiry 10 provides you with an opportunity to identify some of the diseases and bacterial pathogens discussed in this chapter.

CONCEPT AND REASONING CHECKS

10.11 How do these other bacterial species cause gastrointestinal illness?

TABLE

10.1 A Summary of Foodborne and Waterborne Bacterial Diseases

Disease	Causative Agent	Description of Agent	Organs Affected	Characteristic Signs	Toxin Involved	Treatment Administered	Immunization Available	Comment
Botulism	*Clostridium botulinum*	Gram-positive spore-forming rod	Neuromuscular junction	Paralysis	Yes	Antitoxin	None	Most powerful toxin known Infant and wound botulism possible
Staphylococcal food poisoning	*Staphylococcus aureus*	Gram-positive sphere	Intestine	Diarrhea Nausea Vomiting	Yes	None	None	Affects millions of Americans annually Due to an enterotoxin Brief incubation period
Clostridial food poisoning	*Clostridium perfringens*	Gram-positive spore-forming rod	Intestine	Diarrhea Cramping	Yes	None	None	Common in protein-rich foods Spores survive cooking
Typhoid fever	*Salmonella* Typhi	Gram-negative rod	Intestine Blood Gall bladder	Ulcers Fever Rose spots	Not established	Chloramphenicol	Vaccine of dead bacterial cells	Spread from carriers
Salmonellosis	*Salmonella* serotypes	Gram-negative rods	Intestine	Fever Diarrhea Vomiting	Not established	Not recommended	None	Associated with poultry products
Shigellosis	*Shigella* serotypes	Gram-negative rod	Intestine	Diarrhea Dysentery	Yes	Rehydration Antibiotics	None	May be accompanied by dysentery
Cholera	*Vibrio cholerae*	Gram-negative curved rod	Intestine	Rice-water stools Extreme diarrhea Shock	Yes	Rehydration	Vaccine of dead bacterial cells	Danger from dehydration Rare in US Raw seafood involved
E. coli diarrheas	*Escherichia coli*	Gram-negative rod	Intestine	Diarrhea	Yes	Antibiotics Rehydration	None	Symptoms due to enterotoxin Enteroinvasive strains observed Possible hemolytic uremic syndrome
Campylo-bacteriosis	*Campylobacter jejuni*	Gram-negative curved rod	Intestine	Diarrhea Fever	Not established	Erythromycin	None	Associated with raw milk Symptoms mild to serious
Listeriosis	*Listeria monocytogenes*	Gram-positive small rod	Intestine Meninges Monocytes	Diarrhea Meningoencephalitis	Not established	Antibiotics	None	Dangerous in pregnancy Associated with animals
Brucellosis	*Brucella* species	Gram-negative rod	Spleen Lymph glands	Undulating fever Joint pain	Not established	Doxycycline Rifampin	None	Induces abortion in barnyard animals Hazardous to animal workers
Other foodborne and waterborne diseases	*Vibrio parahaemolyticus*	Gram-negative rod	Intestine	Diarrhea	Not established	Various antibiotics	None	Associated with seafood
	Vibrio vulnificus	Gram-negative rod	Intestine	Diarrhea	Not established	Various antibiotics	None	Associated with shellfish
	Bacillus cereus	Gram-positive spore-forming rod	Intestine	Diarrhea	Yes	None	None	Spores survive in cooked foods
	Yersinia enterocolitica	Gram-negative rod	Intestine	Diarrhea	Not established	Various antibiotics	None	Widely found in animals
	Pleisomonas shigelloides	Gram-negative rod	Intestine	Diarrhea	Not established	Various antibiotics	None	Occurs in tropical climates
	Aeromonas hydrophila	Gram-negative rod	Intestine	Diarrhea	Yes	Various antibiotics	None	Common in soil and water

MICROFOCUS 10.5

Foodborne and Waterborne Bacterial Diseases as Biological Weapons

There are several bacterial species and bacterial toxins the Centers for Disease Control and Prevention (CDC) and the United States government have identified as potential biological weapons that could be used through foodborne or waterborne routes. Category A agents are of highest risk to national security because they could (1) be easily disseminated or transmitted person to person, (2) cause high mortality with a major public health impact, and (3) cause public panic and social disruption. The toxins of botulism fall into this category.

Category B agents are moderately easy to disseminate and cause moderate morbidity and low mortality. Several foodborne and waterborne pathogens are in this category, including *Clostridium perfringens, Staphylococcus aureus, Salmonella* species, *Shigella dysenteriae, Escherichia coli* 0157:H7, *Vibrio cholerae,* and *Brucella* species.

Some of these agents actually have been used to commit **biocrimes**, the intentional introduction of a biological agent into food or water to sicken small groups of people. The first known biocrime in the United States occurred in The Dalles, Oregon, in 1984 when the Rajneeshee religious cult, in an effort to influence and win seats in the local election, intentionally contaminated salad bars and a city water tank with *Salmonella enterica* serotype Typhimurium. The result was unsuccessful, although the crime did sicken over 750 citizens and hospitalized 40. Another example occurred in 1996 at a Texas Medical Center. A disgruntled employee deliberately contaminated doughnuts and muffins in a hospital workroom with *Shigella dysenteriae.* All 12 employees who ate the food became ill and 4 were hospitalized.

Bioterrorism has similar aims but on a much larger scale, because the use of such biological agents would cause fear in or actually inflict death or disease upon a large population. Botulinum toxin is one such biological agent that disseminated in food or water, could be used as a bioterror weapon. The muscle-paralyzing botulinum toxins are among the most powerful toxins produced by living organisms.

If large amounts of the toxin could be produced and disseminated in food or water, the net result would be symptoms similar to typical foodborne botulism. A bioterror act should be relatively easy to identify as an act of bioterrorism rather than a natural outbreak because fewer than 150 cases of botulism are reported each year in the United States. Provided medical and health authorities act quickly, antitoxin treatment given before the onset of symptoms would be very effective against a botulism incident. Also, an investigational vaccine, available for high-risk exposures, is being tested.

The United States has more than 57,000 food processors and 1.2 million food retailers amounting to a $200 billion annual business. In May 2003, the Food and Drug Administration (FDA) proposed new regulations requiring food companies to keep better records for tracking foods involved in any future emergencies or terrorism-related contamination. The agency also plans to require advance information of food import shipments to intercept any contaminated products. In July 2003, the FDA began evaluating ways to prevent or reduce the risk of deliberate contamination of the nation's food supply. This might include chemical treatments, temperature controls, and technology intervention.

Still, for most of the agents identified at the top of this box, a large outbreak should signal bioterrorism because these agents normally do not produce large-scale outbreaks in the United States.

MICROINQUIRY 10

Foodborne and Waterborne Disease Identification

Below are several descriptions of foodborne and waterborne diseases based on material presented in this chapter. Read the case history and then answer the questions posed. Answers can be found in **Appendix D**.

Case 1

An 8-year-old male having diarrhea and abdominal distress is brought to a medical clinic by his mother. He had just returned from a weekend camping trip with friends where the abdominal pain began two days before his return home. Examination of the patient indicates he has a fever and fecal examination presents a bloody stool. A Gram stain reveals curved, gram-negative rods.

10.1a. Based on the examination and laboratory results, what organism most likely is responsible for the patient's symptoms?

10.1b. Propose several ways that the patient could have been exposed to this organism.

10.1c. Describe how foods or water can become contaminated with this organism.

10.1d. What complications can result from the diarrheal illness? What is the reason for these complications?

10.1e. In severe cases of the disease, what antibiotic could be given to shorten the duration of the symptoms?

Case 2

A 23-year-old woman comes to the emergency room complaining of lethargy and appears disoriented. The patient admits to abdominal pain and has a fever. The patient indicates that she recently returned from an Amazon River cruise where, on the last day, she ate with the local people and drank local water. She had not been vaccinated for any infectious diseases before her trip. The patient exhibits a bloody stool and a blood smear indicates the presence of gram-negative rods.

10.2a. Given her travel history, identify three bacterial agents with which she has an increased likelihood of being infected.

10.2b. Of the bacterial diseases presented in this chapter, which ones can be ruled out?

10.2c. Based on her symptoms and laboratory results, what bacterial disease is mostly likely responsible for her infection?

10.2d. Give two possible ways she could have contracted the disease.

10.2e. How might she have prevented getting the disease?

Case 3

A 27-year-old man arrives at the hospital emergency room complaining of headache, stiff neck, and irritability. He indicates that he had a kidney transplant about a year ago that was necessitated by an automobile accident. A blood culture is taken. A blood smear shows the presence of gram-positive rods and an assay for cell motility is positive. Asked about his recent eating habits, the patient says that he had purchased some feta cheese recently from a local food market.

10.3a. Given the patient's symptoms and recent eating habits, what bacterial agent is most likely the cause of the disease?

10.3b. What is the significance of the headache, stiff neck, and irritability symptoms?

10.3c. What antibiotics should be given to the patient?

10.3d. What should immunocompromised individuals do to avoid this scenario and why doesn't it apply to healthy adults?

10.3e. Why is it significant that this agent is psychrotrophic?

Case 4

A previously healthy 12-year-old male comes to the emergency with his father. The son has exhibited bloody diarrhea and abdominal pain for the past 24 hours. The diarrhea is now worse, triggering the visit to the emergency room. His physical examination is unremarkable except for dehydration. Questioning the father, neither has been around anyone else with diarrhea and they have not traveled out of the country. His father does admit to cooking his son a hamburger using meat that had been sitting on the kitchen counter for "some time."

10.4a. Based on the symptoms and family activities, what organism and strain is the cause of the disease?

10.4b. What two clues lead you to this conclusion?

10.4c. Explain what other bacterial agents, although similar, would be ruled out.

10.4d. What is unique about this disease-causing strain?

10.4e. How can this strain of the bacterial species be culturally identified from other strains of the species?

SUMMARY OF KEY CONCEPTS

10.1 Introduction to Foodborne and Waterborne Bacterial Diseases

- Intoxications involve illnesses caused by bacterial toxins while infections refer to diseases arising from bacterial growth in the body. An incubation period occurs before the symptoms appear. Symptoms depend on the specific toxin or microbe and the numbers ingested.
- Cattle and poultry carcasses can become contaminated during slaughter. Fresh fruits and vegetables can become contaminated when washed or irrigated with water that is contaminated with animal manure or human sewage. Infected humans can contaminate food they handle (fecal/oral route), or there can be cross-contamination from other raw agricultural products. Some bacterial pathogens cause foodborne or waterborne bacterial diseases only when they are in large numbers.

10.2 Foodborne Intoxications

- Botulism is a severe form of food poisoning caused by nerve toxins produced by *Clostridium botulinum*. Symptoms include blurred vision, slurred speech, and difficulty swallowing and breathing. Botulism is treated by administration of an antitoxin. Patients often need to be placed on a ventilator.
- *Staphylococcus aureus* is a pathogen causing food poisoning through ingestion of pre-formed enterotoxins. Abdominal cramps, nausea, vomiting, and diarrhea are common symptoms.
- *Clostridium perfringens* causes food poisoning through the ingestion of the enterotoxin. Symptoms include watery diarrhea and abdominal pain.

10.3 Foodborne and Waterborne Infections

- *Salmonella* Typhi is an acid-resistance bacillus that invades the blood. The infection is characterized by fever, abdominal pain, rose spots, lethargy, and delirium. Typhoid vaccines are available and chloramphenicol is the antibiotic of choice.
- Salmonellosis is an infection caused by ingestion of a *Salmonella* serotype. Symptoms include fever, nausea, vomiting, diarrhea, and abdominal cramps.
- Ingestion of *Shigella* and the production of enterotoxins cause shigellosis. Symptoms include fever, abdominal cramps, and sometimes bloody diarrhea.
- *Vibrio cholerae* infects the small intestines, producing an enterotoxin causing profuse watery diarrhea and vomiting. Oral rehydration therapy restores electrolyte and glucose balance in the body.
- Several strains of *E. coli* can cause various forms of gastroenteritis. Watery diarrhea, caused by enterotoxin production, is typical of traveler's diarrhea. *E. coli* 0157:H7 is a more serious form of diarrhea that can lead to complications called hemorrhagic colitis. Often this involves the kidneys, causing hemolytic uremic syndrome (HUS).
- The presence of *Helicobacter pylori* is responsible for most cases of gastric ulcers. The exact transmission mechanism is not understood, although contaminated food or water is a likely candidate. High cure rates have been attained with antibiotics and acid suppressors.
- *Campylobacter jejuni* is the causative agent for this foodborne and waterborne infection. Symptoms include fever, abdominal pains, and bloody stools. If treatment is needed, erythromycin is the antibiotic used.
- Fecal-contaminated foods often contain *Listeria monocytogenes*. In individuals with a weakened immune system, meningoencephalitis can occur. Infection in pregnant women can lead to miscarriage or mental damage to the newborn.
- Several other bacteria can cause foodborne and waterborne diseases. Four species of *Brucella* can cause brucellosis. Other gastrointestinal illnesses are caused by *Vibrio parahaemolyticus, Vibrio vulnificus, Bacillus cereus,* and *Yersinia enterocolitica, Plesiomonas shigelloides,* and *Aeromonas hydrophila.*

LEARNING OBJECTIVES

After understanding the textbook reading, you should be capable of writing a paragraph that includes the appropriate terms and pertinent information to answer the objective.

1. Differentiate between a bacterial intoxication and a bacterial infection.
2. Identify ways that foods and water become contaminated with bacterial pathogens.
3. Distinguish between the bacterial species causing intoxications.
4. Summarize the clinical significance of *Salmonella* Typhi infections.
5. Distinguish between the symptoms of salmonellosis and shigellosis.
6. Summarize the clinical significance of *Vibrio cholerae* and assess the impact of cholera pandemics.
7. Compare and contrast the various forms of *E. coli* infections.
8. Diagram the steps involved in gastric ulcer disease.
9. Explain why campylobacteriosis has become the most commonly reported form of bacterial gastroenteritis.
10. Explain why an infection by *Listeria monocytogenes* is most dangerous to newborns, the elderly, and pregnant women.
11. List several other bacterial pathogens responsible for foodborne or waterborne illnesses.

SELF-TEST

Answer each of the following questions by selecting the ***one*** answer that best fits the question or statement. Answers to even-numbered questions can be found in **Appendix C**.

1. A major symptom in patients experiencing botulism is _____ of the limbs and respiratory muscles.
A. infection
B. paralysis
C. involuntary movements
D. necrosis
E. contraction

2. To treat patients who have ingested botulism toxin, large doses of _____ must be administered.
A. antibiotic
B. aspirin
C. Pepto-Bismol
D. antitoxin
E. electrolytes

3. _____ food poisoning has a relatively short incubation period and is caused by a toxin deposited in food during aerobic bacterial growth.
A. Staphylococcal
B. *E. coli*
C. Clostridial
D. *Listeria*
E. *Yersinia*

4. The organism *Clostridium perfringens* multiplies in foods only under _____ conditions.
A. aerobic
B. freezing
C. anaerobic
D. saline
E. permissive

5. A small percentage of those who recover from _____ remain carriers and can spread the organism in their feces.
A. botulism
B. undulant fever
C. brucellosis
D. listeriosis
E. typhoid fever

6. Disease associated with *Shigella* species is usually accompanied by a syndrome of cramps and bloody stools called
A. hemorrhagic colitis.
B. HUS.
C. rose spots.
D. dysentery.
E. rice-water stools.

7. One of the most excessive diarrheas observed in an intestinal disease is associated with
A. staphylococcal food poisoning.
B. typhoid fever.
C. cholera.
D. gastric ulcer disease.
E. brucellosis.

8. *Escherichia coli* is a common gram-_____ that can be a cause of _____.
A. negative coccus; botulism
B. positive rod; hemorrhagic colitis
C. negative rod; traveler's diarrhea
D. positive coccus; typhoid fever
E. negative rod; cholera

9. The most commonly reported cause of bacterial gastroenteritis is due to _____.
A. *Campylobacter*
B. *Listeria*
C. *Escherichia*
D. *Shigella*
E. *Clostridium*

10. The fever pattern in brucellosis gives the disease its alternate name of _____ fever.
A. typhoid
B. undulant
C. *Yersinia*
D. yellow
E. None of the above (**A–D**) are correct.

11. Those who work with _____ may be exposed to the bacterial pathogen that causes brucellosis.
A. chickens
B. reptiles
C. agriculture products
D. shellfish
E. large ruminant animals

12. Consumers of seafood are particularly vulnerable to intestinal infection due to a species of _____.
A. *Vibrio*
B. *Salmonella*
C. *Helicobacter*
D. *Listeria*
E. *Campylobacter*

13. A spore-forming, motile rod that can cause foodborne illness in a diarrheal or emetic form is *Bacillus*
A. *anthracis.*
B. *abortus.*
C. *enterocolitica.*
D. *cereus.*
E. *botulinum.*

QUESTIONS FOR THOUGHT AND DISCUSSION

Answers to even-numbered questions can be found in **Appendix C**.

1. In 1997, researchers in Boston reported that *Helicobacter pylori* accumulates in the gut of houseflies after the flies feed on food containing the pathogen. What are the implications of this research?
2. Some years ago, the CDC noticed a puzzling trend: Reported cases of salmonellosis seemed to soar in the summer months and then drop radically in September. Can you venture a guess as to why this is so?
3. Most physicians agree that the illness called "stomach flu" is not influenza at all. They say the cramps, diarrhea, and vomiting can be due to a variety of bacterial species. Which organisms in this chapter might be good candidates?
4. A frozen-food manufacturer recalls thousands of packages of jumbo stuffed shells and cheese lasagna after a local outbreak of salmonellosis. Which parts of the pasta products would attract the attention of inspectors as possible sources of salmonellosis? Why?

APPLICATIONS

Answers to even-numbered questions can be found in **Appendix C**.

1. You are doing the supermarket shopping for the upcoming class barbecue. What are some precautions you can take to ensure that the event is remembered for all the right reasons?
2. You read in the newspaper that botulism was diagnosed in 11 patrons of a local restaurant. The disease was subsequently traced to mushrooms bottled and preserved in the restaurant. What special cultivation practice enhances the possibility that mushrooms will be infected with the spores *Clostridium botulinum?*
3. In preparation for a summer barbecue, your roommate cuts up chickens on a wooden carving board. After running the board under water for a few seconds, he uses it to cut up tomatoes, lettuce, peppers, and other salad ingredients. What sort of trouble may occur?
4. The state department of health received reports of illness in 18 workers at a local pork processing plant. All the affected employees worked on the plant's "kill floor." All had gram-negative rods in their blood. Their symptoms included fever, chills, fatigue, sweats, and weight loss. Which disease was pinpointed in the workers?
5. A classmate plans to travel to a tropical country for spring break. To prevent traveler's diarrhea, she was told to take 2 ounces or 2 tablets of Pepto-Bismol four times a day for 3 weeks before travel begins. Short of turning pink, what better measures can you suggest she use to prevent traveler's diarrhea?

REVIEW

When you have completed your study of foodborne and waterborne bacterial diseases, test your knowledge of the important terms by circling the choices that best complete each of the following statements. The answers to even-numbered statements are listed in **Appendix C**.

1. To treat patients who have botulism, large doses of (antitoxin, antibiotic) must be administered.
2. Disease associated with *Shigella* species can produce (diarrhea, dysentery), which is identified by the presence of cramps and bloody stools.
3. (Neurotoxins, Enterotoxins), such as those found with botulism, can cause flaccid paralysis.
4. Many foodborne and waterborne bacterial diseases have ill-defined (symptoms, syndromes), making pathogen identification difficult.
5. Only a small percentage of those who recover from typhoid fever remain (carriers, free) of the bacterial cells.
6. Many individuals who work for the CDC are concerned with the causes or origins of a disease, the study of which is called (serology, etiology).
7. If an infection leads to (cellulitis, encephalitis), the individual develops a diffuse inflammation in the connective tissues of the skin.
8. The genus *Salmonella* is made up of many (genotypes, serotypes); that is, there are many closely related groups that are identified by a specific set of antigens.
9. An infection of the blood, which can occur with listeriosis, is referred to as (septicemia, gastroenteritis.)
10. If one has ingested botulism exotoxins, the individual is considered to be (infected, intoxicated).

HTTP://MICROBIOLOGY.JBPUB.COM/

The site features learning, an on-line review area that provides quizzes and other tools to help you study for your class. You can also follow useful links for in-depth information, read more MicroFocus stories, or just find out the latest microbiology news.

11

Chapter Preview and Key Concepts

11.1 Soilborne Bacterial Diseases
- *Bacillus anthracis* produces a capsule and three exotoxins.
- *Clostridium tetani* produces a powerful neurotoxin.
- *Clostridium perfringens* produces a group of toxins and hydrolytic enzymes.
- *Leptospira interrogans* infection produces a diverse set of symptoms.

11.2 Arthropodborne Bacterial Diseases
- *Yersinia pestis* is carried in rodents and their fleas.
- *Francisella tularensis* infections can involve an extremely small infectious dose.
- *Borrelia burgdorferi* is transmitted by the bite of a small tick.
- Other *Borrelia* species cause a relapsing, febrile illness.

11.3 Rickettsial and Ehrlichial Arthropodborne Diseases
- *Rickettsia* infections often involve a characteristic rash.
- *Ehrlichia* species infect different groups of leukocytes.

MicroInquiry 11: Soilborne and Arthropodborne Disease Identification

Soilborne and Arthropodborne Bacterial Diseases

Father abandoned child, wife husband, one brother another . . . And I, Agnolo di Tura . . . buried my five children with my own hands . . . . So many died that all believed that it was the end of the world.
—Agnolo di Tura, describing bubonic plague in his chronicle (*The Plague of Siena*) in 1348

As the above quote attests, bubonic plague, commonly known as "the Black Death," was probably the greatest catastrophe ever to strike Europe. It swept back and forth across the continent for almost a decade, each year increasing in ferocity. By 1348, two-thirds of the European population was stricken and half of the sick had died. Houses were empty, towns were abandoned, and a dreadful solitude hung over the land. The sick died too quickly for the living to bury them, so victims often were buried in "plague pits" (FIGURE 11.1). At one point, the Rhône River was consecrated as a graveyard for plague victims. Contemporary historians wrote that posterity would not believe such things could happen, because those who saw them were themselves appalled. The horror was almost impossible to imagine; to many people, including Agnolo di Tura, "it was the end of the world."

Before the century concluded, the Black Death visited Europe at least five more times in periodic reigns of terror. During one epidemic in Paris, an estimated 800 people died each day; in Siena, Italy, the population dropped from 42,000 to 15,000; and in Florence, almost 75 percent of the citizenry perished. Flight was the chief recourse for people who could afford it, but ironically, the escaping travelers spread the disease. Those who remained in the cities were locked in their homes until they succumbed or recovered.

The immediate effect of the plague was a general paralysis in Europe. Trade ceased and wars stopped. Bewildered peasants who survived encountered unexpected prosperity because landowners had to pay higher wages to obtain help. Land values declined and class relationships were upset, as the system of feudalism gradually crumbled. However, medical practices became increasingly sophisticated, with new standards of sanitation and a 40-day period of detention (quarantine) imposed on vessels docking at ports.

The graveyard of plague left fertile ground for the renewal of Europe during the Renaissance. To many historians, the Black Death remains a major turning point in Western civilization.

Plague is a disease caused by *Yersinia pestis*, an organism found in rodents and their fleas in many areas around the world. Although there are sporadic outbreaks of plague today around the world, there is another reason to be concerned about the potential horrors of the disease. Many microbiologists and government officials see *Y. pestis* as a possible bioterror agent. Used in an aerosol attack, the pathogen could cause cases of pneumonic plague, an infection of the lungs. One to six days after becoming infected with the bacilli, people would develop fever, weakness, and rapidly developing pneumonia with shortness of breath, chest pain, cough, and sometimes bloody or watery sputum. Nausea, vomiting, and abdominal pain may also occur. Without early treatment, pneumonic plague usually leads to respiratory failure, shock, and rapid death.

Because of the delay between being exposed to *Y. pestis* and becoming sick, people could travel over a large area before becoming contagious and possibly infecting others. Once people have the disease, the bacterial cells can spread to others who have close contact with them. Several types of antibiotics are effective for curing the disease provided they are given within 24 hours of the first symptoms.

A bioweapon carrying *Y. pestis* is possible because the pathogen occurs in nature and could be isolated and grown in an appropriately equipped laboratory, although weaponizing *Y. pestis*—that is, making it easily transmissible through the air—would require advanced knowledge and technology.

FIGURE 11.1 **The Plague Pit.** This painting shows the unloading of dead bodies during the plague of 1665. These pits were no more than mass graves used to bury the plague victims that had been gathered up from the streets on the dead carts.

Q: Why would these plague victims be buried in mass graves rather than by traditional funerals—and at night?

Besides plague, other diseases like typhus and relapsing fever are transmitted by arthropods, and both can be interrupted by arthropod control. Neither is a major problem in our society, but a substantial number of cases of other arthropodborne diseases, such as tularemia, Rocky Mountain spotted fever, and Lyme disease are reported each year. We shall study each of these diseases in this chapter.

■ **Arthropods:** Animals having jointed appendages and segmented bodies (e.g., ticks, lice, fleas, mosquitoes).

First we will examine a number of soilborne diseases where organisms enter the body through a cut, wound, or abrasion, or by inhalation. Among these are anthrax, another feared disease in bioterrorism, and tetanus, a concern to anyone who has stepped on a nail or piece of glass. We also will study other diseases receiving wider recognition as detection methods improve. The soilborne diseases, as well as the arthropodborne diseases, are primarily problems of the blood.

11.1 Soilborne Bacterial Diseases

Soilborne bacterial diseases are those whose agents are transferred from the soil to the unsuspecting individual. To remain alive in the soil, the bacterial cells must resist environmental extremes, and often the cells form endospores, as the first three diseases illustrate.

Anthrax Is an Enzootic Disease

■ Enzootic: Refers to a disease endemic to a population of animals.

KEY CONCEPT

- *Bacillus anthracis* produces a capsule and three exotoxins.

Bacillus anthracis was the first bacterial species shown by Koch to be the causative agent of an infectious disease (see Chapter 1). **Anthrax** is primarily an endemic disease of large, domestic herbivores, such as cattle, sheep, and goats. Animals ingest the spores from the soil during grazing, and soon they are overwhelmed with vegetative bacterial cells as their organs fill with bloody black fluid (*anthrac* = "coal"; the disease name is thus a reference to the blackening of the blood). About 80 percent of untreated animals die.

Bacillus anthracis

Anthrax is caused by *Bacillus anthracis*, a spore-forming, aerobic, gram-positive rod (FIGURE 11.2). Endospores germinate rapidly on contact with human tissues to produce vegetative cells (see Chapter 5). The thick capsule of the cells impedes phagocytosis and the organisms produce three exotoxins that work together to cause disease. Capsule and toxins are coded by genes carried on two plasmids.

Humans acquire anthrax from infected animal products or contaminated dust. This can happen in one of three ways.

Inhalation anthrax. Workers who tan hides, shear sheep, or process wool may inhale the spores and contract inhalational (pulmonary) anthrax as a form of pneumonia called **woolsorter's disease**. It initially resembles a common cold (fever, chills, cough, chest pain, headache, and malaise). After several days, the symptoms may progress to severe breathing problems and shock. Inhalation anthrax is usually fatal without early treatment.

Intestinal anthrax. Consumption of contaminated and undercooked meat may lead to gastrointestinal anthrax. It is characterized by an acute inflammation of the intestinal tract. Initial signs include nausea, loss of appetite, vomiting, and fever. This is followed by abdominal pain, vomiting of blood, and severe diarrhea. Intestinal anthrax results in death in 25 to 60 percent of untreated cases.

Cutaneous anthrax. Skin abrasions with spore-contaminated animal products, including

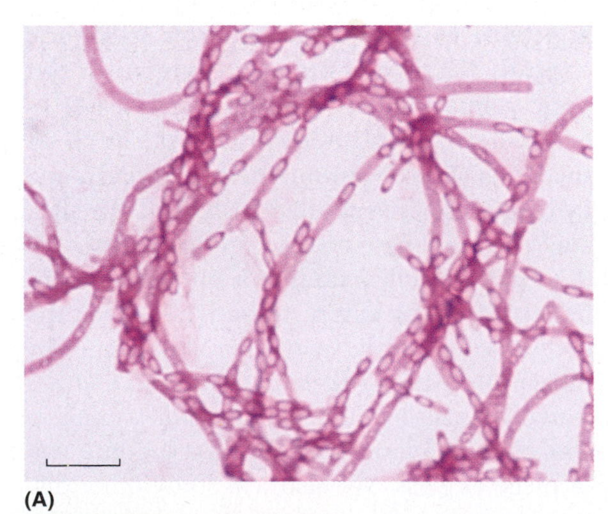

(A)

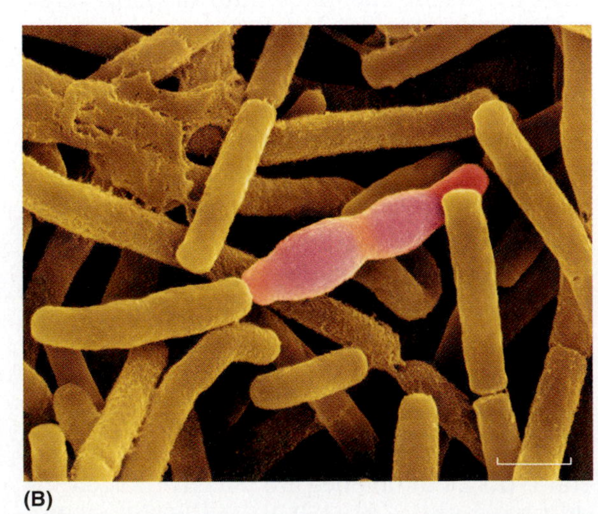

(B)

FIGURE 11.2 ***Bacillus anthracis.*** *B. anthracis* is the cause of anthrax. (**A**) Spores (white ovals) in vegetative cells can be seen in this photomicrograph. (Bar = 10 μm.) (**B**) A false-color scanning electron micrograph of vegetative cells. It is from such cells that the exotoxins are produced. (Bar = 2 μm.)

Q: What advantage is provided to the organism by producing endospores?

violin bows, shaving bristles, goatskin drums, and leather jackets, can lead to infection. Skin infection begins as a papule, but within one to two days it develops into a pustule of black, necrotic (dying) tissue that eventually crusts over (FIGURE 11.3). Lymph glands in the adjacent area may be invaded and swell. Cutaneous anthrax accounts for more than 95 percent of all anthrax infections. About 20 percent of untreated cases will result in death.

B. anthracis infections can be treated with penicillin or ciprofloxacin. In 2000 to 2002, the total number of anthrax cases in the United States was two per year. At present, there is no vaccine for civilian use, although there is a cell-free vaccine for veterinarians and others who work with livestock.

Anthrax also is considered a threat in bioterrorism and in biological warfare (MicroFocus 11.1). The seriousness of using biological agents as a means for bioterrorism was underscored in October 2001 when *B. anthracis* spores were distributed intentionally through the United States mail. In all, 22 cases of anthrax (11 inhalation and 11 cutaneous) were identified, making the case-fatality rate among patients with inhalation anthrax 45 percent

■ Papule:
A raised itchy bump that resembles an insect bite.

■ Pustule:
A papule containing pus.

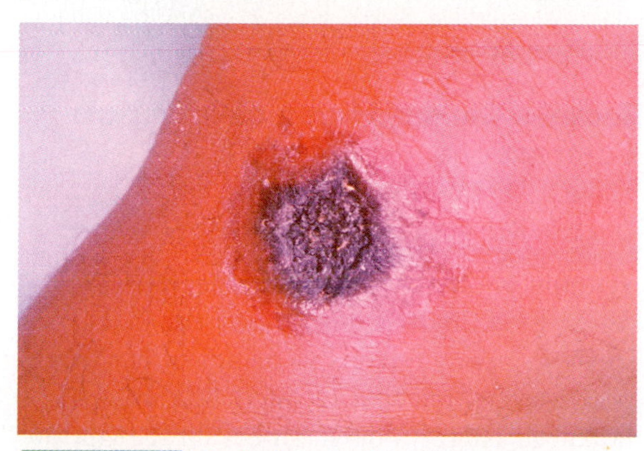

FIGURE 11.3 **An Anthrax Lesion.** This cutaneous lesion is a result of infection with anthrax bacilli. Lesions like this one develop when anthrax spores contact the skin, germinate to vegetative cells, and multiply.

Q: Why was a skin ulceration like this given the name "anthrax?"

MICROFOCUS 11.1: History

The Legacy of Gruinard Island

In 1941, the specter of airborne biological warfare hung over Europe. Fearing that the Germans might launch an attack against civilian populations, British authorities performed a series of experiments to test their own biological weapons. Anthrax spores were seen as an agent that could be aerosolized and released unobtrusively. Drifting over a large city, they would be undetectable and could infect thousands of individuals.

To test this possibility, investigators placed 60 sheep on Gruinard Island, a mile-long patch of land off the coast of Scotland. An anthrax spore-containing bomb was exploded over the island. Within days, all the sheep were dead.

Warfare with biological weapons never came to reality in World War II, but the contamination of Gruinard Island remained. A series of tests in 1971 discovered anthrax spores still viable at and below the upper crust of the soil. Fearing they could be spread by earthworms, British officials posted signs warning people not to set foot on the island (see figure) but did little else.

Then a strange protest occurred in 1981. Activists demanded that the British government decontaminate the island. They backed their demands with packages of soil taken from the island. Notes led government officials to two 10-pound packages of spore-laden soil, and the writers threatened that 280 pounds were hidden elsewhere.

Partly because of the protests, the British government instituted a decontamination of the island in 1986. Technicians used a powerful brushwood killer, combined with burning and treatment with formalin in seawater. Finally, they managed to rid the soil of anthrax spores. By April 1987, sheep were once again grazing on the island. However, people were somewhat reluctant to return. Gruinard Island remains a monument of sorts to the effects of biological warfare testing.

(5/11). The six other individuals with inhalation anthrax and all the individuals with cutaneous anthrax recovered. Had it not been for antibiotic therapy, many more might have been stricken.

CONCEPT AND REASONING CHECKS

11.1 What cellular factors make *B. anthracis* a dangerous pathogen?

Tetanus Causes Hyperactive Muscle Contractions

KEY CONCEPT

- *Clostridium tetani* produces a powerful neurotoxin.

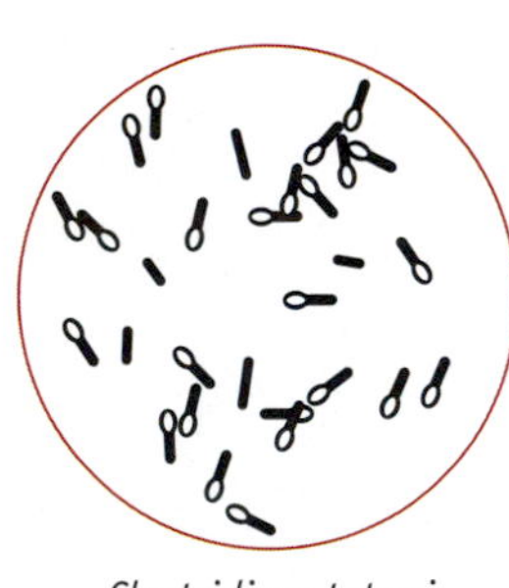

Clostridium tetani

Tetanus is one of the most dangerous human diseases. *Clostridium tetani*, the bacterial species causing tetanus, is an anaerobic, gram-positive bacillus first isolated in 1889 by the Japanese bacteriologist Shibasaburo Kitasato (see Chapter 1). *C. tetani* forms endospores typically found in barnyard and garden soils containing animal manure.

Spores in very small numbers enter the body through a wound. The spore-containing wounds may result from a fracture, gunshot, animal bite, or puncture by a piece of glass, a thorn, or a rusty nail. Even illicit drugs can contain spores (MicroFocus 11.2). In dead oxygen-free tissue of the wound, spores germinate into vegetative bacilli that produce several toxins. The most important of these toxins is the exotoxin **tetanospasmin**, the second most powerful toxin known to science (after the botulism toxin).

Once inside the tissue, the spores germinate to vegetative cells. At the neuromuscular junction, tetanospasmin prevents the release of glycine (an amino acid) and other neurotransmitters needed to inhibit muscle contraction. Without any inhibiting influence, volleys of spontaneous impulses arise in the motor neurons, causing the muscles to contract continuously and without control.

Symptoms of tetanus intoxication develop rapidly, often within hours of exposure. A patient experiences generalized muscle stiffness, especially in the facial and swallowing muscles. Spasms of the jaw muscles cause the teeth to clench and bring on a condition called **trismus**, or **lockjaw**. Severe cases are characterized by a "fixed smile" (risus sardonicus) and muscle spasms cause an arching of the back (**opisthotonus**). Spasmodic inhalation and seizures in the diaphragm and rib muscles leads to reduced ventilation, and patients often experience violent deaths.

Patients are treated with sedatives and muscle relaxants and are placed in quiet, dark rooms as noise and bright light can trigger muscle spasms. Physicians prescribe penicillin to destroy the organisms and tetanus antitoxin to neutralize the toxin.

Immunization for children involves injections of tetanus toxoid in the diphtheria-tetanus-acellular pertussis (DTaP) vaccine (see Chapter 9). The toxoid is prepared by treating the toxin with formaldehyde to eliminate its toxic quality. Children usually receive injections at 2, 4, 6, and 18 months and 5 years of age. Booster injections of tetanus toxoid in the **Td vaccine** (a "tetanus shot") are recommended every 10 years to keep the level of immunity high.

The United States has had a steady decline in the incidence of tetanus, with 34 cases confirmed in 2004. Two individuals died. Older Americans are primarily affected because they either have not been immunized or kept up their booster immunizations. Forty-six percent of cases in 2004 occurred in persons over 60 years of age.

In other parts of the world, tetanus remains a major health problem. A common source for illness occurs at birth when the umbilical stump becomes infected from non-sterile instruments or dressings.

CONCEPT AND REASONING CHECKS

11.2 What are the definitive symptoms of tetanus?

Gas Gangrene Causes Massive Tissue Damage

KEY CONCEPT

- *Clostridium perfringens* produces a group of toxins and hydrolytic enzymes.

Gangrene (*gangren* = "a sore") develops when the blood flow ceases to a part of the body, usually as a result of blockage by dead tissue. The body part, generally an extremity, becomes dry and shrunken, and the skin color changes to purplish or black. The gangrene may spread as enzymes from broken cells destroy other cells, and the tissue may have to be debrided or the body part amputated.

Debrided: Referring to the removal of dead, damaged, or infected tissue.

Gas gangrene, or **myonecrosis** (*myo* = "muscle"; *necros* = "death") is caused by *Clostridium perfringens*, a nonmotile, encapsu-

MICROFOCUS 11.2: Public Health

Tetanus Outbreak among Injecting Drug Users

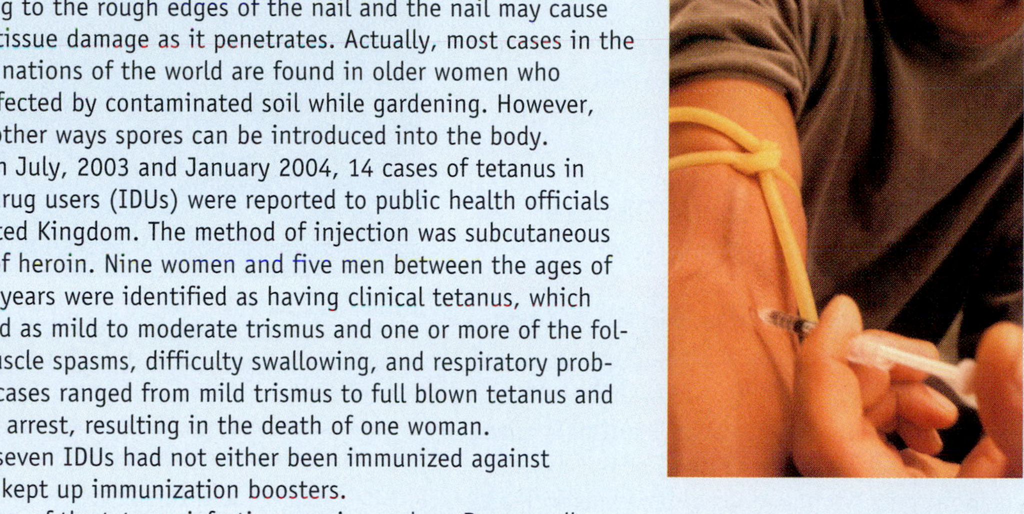

When one thinks about tetanus, the typical image coming into mind is stepping on a rusty nail. Such nails do pose a threat because spores cling to the rough edges of the nail and the nail may cause extensive tissue damage as it penetrates. Actually, most cases in the developed nations of the world are found in older women who become infected by contaminated soil while gardening. However, there are other ways spores can be introduced into the body.

Between July, 2003 and January 2004, 14 cases of tetanus in injecting drug users (IDUs) were reported to public health officials in the United Kingdom. The method of injection was subcutaneous injection of heroin. Nine women and five men between the ages of 20 and 53 years were identified as having clinical tetanus, which was defined as mild to moderate trismus and one or more of the following: muscle spasms, difficulty swallowing, and respiratory problems. The cases ranged from mild trismus to full blown tetanus and respiratory arrest, resulting in the death of one woman.

At last seven IDUs had not either been immunized against tetanus or kept up immunization boosters.

The source of the tetanus infection remains unclear. Because all cases were clustered in a short period, the most likely source is the drug or the adulterant (a substance added to the drug to make it less pure).

Public health officials as well as intensive care facilities throughout England, Scotland, and Wales were notified of the outbreak as were drug action teams and hospitals.

lated, anaerobic, spore-forming, gram-positive rod. After endospores in contaminated dirt are introduced through a severe, open wound, the spores germinate and the vegetative cells multiply rapidly in the anaerobic environment. As they grow, they ferment muscle carbohydrates and decompose the muscle proteins (thus the term "myonecrosis"). Large amounts of gas may result from this metabolism, causing a crackling sound as the gas accumulates under the skin. The gas also presses against blood vessels, thereby blocking the flow and forcing cells away from their blood supply. In the infection process, the organisms secrete at least 12 exotoxins. The most important is α-toxin, which damages and lyses blood cells. Degradative enzymes, such as DNase and hyaluronidase, are produced that disrupt cell tissues, facilitating the passage of bacterial cells and spread of infection.

The symptoms of gas gangrene include a foul odor and intense pain and swelling at the wound site (FIGURE 11.4). Initially the site turns dull red, then green, and finally blue black. Anemia is common, and bacterial toxins may damage the

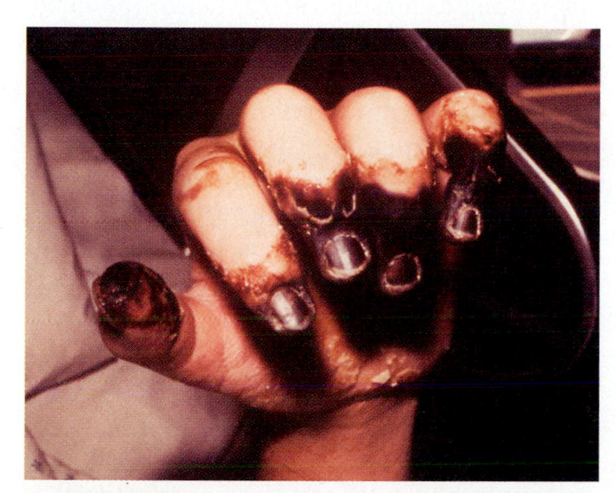

FIGURE 11.4 Gas Gangrene of the Hand. A severely infected hand showing gangrene (blackened tissue necrosis). This gangrene developed from an infection from an accident while the patient was scaling fish. Antibiotic drugs may prevent the infection leading to gangrene, but in this advanced stage, amputation of the hand may be necessary.

Q: What properties of Clostridium perfringens *would cause the tissue necrosis seen here?*

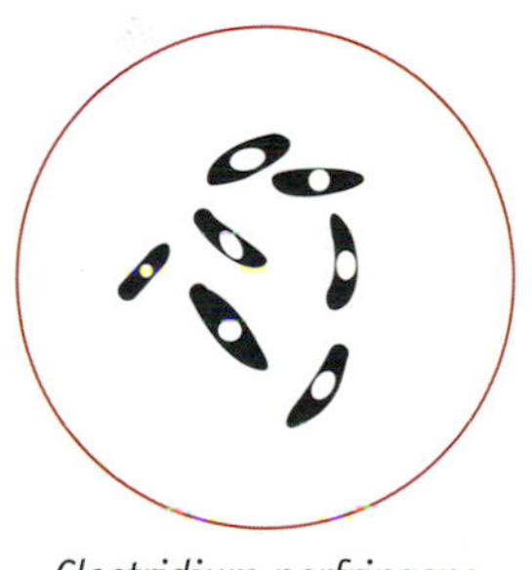

Clostridium perfringens

heart and nervous system. Treatment consists of antibiotic therapy as well as debridement, amputation, or exposure in a hyperbaric oxygen chamber. However, without treatment, disease spreads rapidly, and death frequently occurs within days of gangrene initiation.

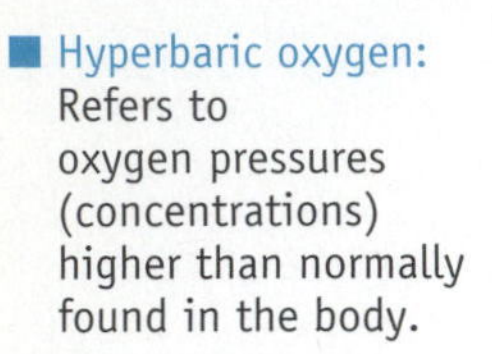

■ **Hyperbaric oxygen:** Refers to oxygen pressures (concentrations) higher than normally found in the body.

CONCEPT AND REASONING CHECKS

11.3 What characteristic of *C. perfringens* is being "attacked" by placing a gas gangrene patient in a hyperbaric chamber?

Leptospirosis Is a Zoonotic Disease Found Worldwide

KEY CONCEPT

- *Leptospira interrogans* infection produces a diverse set of symptoms.

Leptospirosis (*lepto* = "thin"; *spir* = "spiral") is a typical **zoonosis**—a disease of animals that can spread to humans. The disease affects household pets such as dogs and cats as well as barnyard and wild animals. Humans acquire it by contact with these animals or from soil, food, or water contaminated with their urine.

The agent of leptospirosis is *Leptospira interrogans,* a thin, aerobic, coiled, gram-negative spirochete usually with a hook at one end resembling a question mark, hence the name interrogans (*roga* = "ask") (FIGURE 11.5). The undulating movements of these organisms result from contractions of submicroscopic fibers called endoflagella (see Chapter 4). In infected animals, the spirochetes colonize the kidney tubules and are excreted in the urine to the soil.

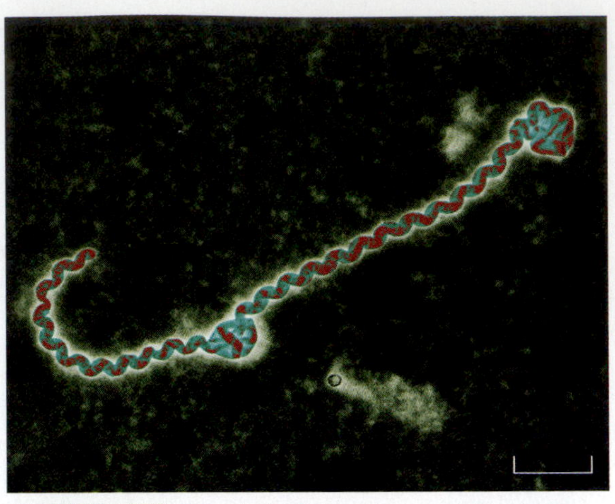

FIGURE 11.5 ***Leptospira interrogans.*** A false-color transmission electron micrograph of *L. interrogans,* the agent of leptospirosis. Note the tightly coiling spirals. (Bar = 2 μm.)

Q: Besides the coil shape, what are the other notable features of this bacterial species?

As a result of a person swimming or wading in contaminated water, *L. interrogans* enters the human body through the mucus membranes of the eyes, nose, and mouth, or through the skin, especially through abrasions and the soft parts of the feet. The bacterial cells multiply rapidly and most infected individuals experience vague flu-like symptoms. However, for 5 to 10 percent of these individuals, the illness progresses to a systemic form, which can be lethal. In the first phase, there is an acute onset of headache, muscle aches, vomiting and nausea; the eyes become very red indicating conjunctivitis. Episodes of fever and chills occur for 4 to 9 days, but then disappear.

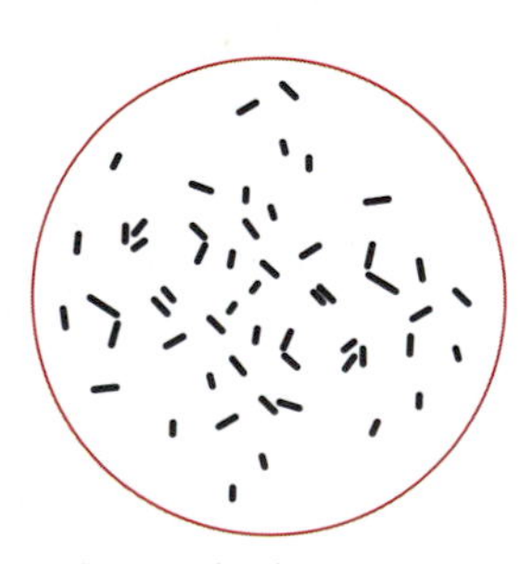

Leptospira interrogans

■ **Conjunctivitis:** An inflammation of the conjunctiva of the eye.

As the spirochetes directly invade and infect various organs, a second phase occurs as the immune system reacts to the infection. A fever returns and meningitis is common, which can lead to stupor and coma. Inflammation also may occur in the liver and lungs. In addition, kidney damage and jaundice (Weil disease) may be present. In other even more severe cases, the patient may vomit blood from gastric hemorrhages and have liver and kidney dysfunction. Despite the numerous tissues involved, the mortality rate from leptospirosis is low (about 10 percent), and doxycycline is used with success.

Leptospirosis incidence in the United States is low, with 40 to 100 cases identified annually (50 percent of cases occur in Hawaii). However, leptospirosis is emerging as a globally important infectious organism, especially in warm, subtropical regions. The disease is of particular concern to adventure travelers to distant, exotic locales (MicroFocus 11.3). Closer to home, in 1997 the largest ever outbreak of leptospirosis in the United States broke out at a triathlete competition in Wisconsin where 98 of 1,000 athletes became ill. Twenty-eight were hospitalized. The source of infection was from swimming in a lake used for the competition. However, health officials have been unable to identify how the lake became a breeding ground for *L. interrogans.*

The soilborne diseases are summarized in TABLE 11.1.

CONCEPT AND REASONING CHECKS

11.4 What bacterial and host factors make *L. interrogans* a potent pathogen?

MICROFOCUS 11.3: Public Health/Environmental Microbiology

A Real Eco-Challenge!

Space photo of river deltas in Borneo.

It started as a headache on the plane back from Borneo to the States. Within three days, Steve went to a hospital emergency room in Los Angeles complaining of fever and chills, muscle aches, vomiting, and nausea.

The Eco-Challenge 2000 in Sabah, Borneo was the site for the annual adventure race. Some 304 participants composing 76 teams from 26 countries competed in the 10-day endurance event, which was designed to push participants and their teams beyond their athletic limits. During Eco-Challenge 2000, teams would kayak on the open ocean, mountain bike into the rainforest, spelunk in hot caves, and swim in local rivers (see figure). Of the 76 teams starting the challenge, 44 teams finished.

About the time Steve arrived at the emergency room, the Centers for Disease Control and Prevention (CDC) in Atlanta received calls from the Idaho Department of Health, the Los Angeles County Department of Health Services, and the GeoSentinel Network (a network of international travel clinics) reporting cases of a febrile illness similar to Steve's.

The CDC quickly carried out a phone questionnaire to 158 participants in the Eco-Challenge. Many reported symptoms similar to Steve's, including chills, fever, headache, diarrhea, and conjunctivitis. Twenty-five respondents had been hospitalized. Within a few days, antibiotic therapy had Steve recovering. In fact, all 135 affected participants completely recovered.

The similar symptoms suggested leptospirosis and laboratory tests either confirmed the presence of *Leptospira* antibodies or positive culture of the bacterial genus from serum samples collected from ill participants.

To identify the source and the exposure risk, information was gathered from participants about various portions of the race course. Analysis identified swimming in and kayaking on the Sagama River as the probable source.

Several participants who did not become ill had taken doxycycline as a preventative for malaria and leptospirosis, as race organizers had advised. Unfortunately, Steve had not heeded those words and suffered a real "eco-challenge." Asked if he would participate in the 2001 Eco-Challenge in New Zealand, he said, "Heck yes! It just adds to pushing the limits."

TABLE 11.1 A Summary of Soilborne Bacterial Diseases

Disease	Causative Agent	Description of Agent	Organs Affected	Characteristic Signs	Toxin Involved	Treatment Administered	Immunization Available	Comment
Anthrax	*Bacillus anthracis*	Gram-positive spore-forming aerobic rod	Blood Lungs Skin	Hemorrhaged blood Boil-like lesions	Yes	Penicillin Ciprofloxacin	For animals	High mortality rate Affects many organs Rare disease in humans
Tetanus	*Clostridium tetani*	Gram-positive spore-forming anaerobic rod	Nerves at synapse	Spasms Tetanus	Yes	Metronidazole Antitoxin	Toxoid in DTaP	Second most powerful toxin
Gas gangrene	*Clostridium perfringens*	Gram-positive spore-forming anaerobic rod	Muscles Nerves Blood cells	Gangrene Swollen tissue	Yes	Penicillin	None	Gas blocks flow of blood Called clostridial myonecrosis
Leptospirosis	*Leptospira interrogans*	Spirochete	Kidney Liver Spleen	Jaundice Vomiting	Not established	Penicillin Doxycycline	For animals	Common in animals Typical zoonosis

11.2 Arthropodborne Bacterial Diseases

A living organism such as an arthropod that transmits disease agents is called a **vector** (*vect* = "carried"). Arthropods transmit diseases to humans usually by taking a blood meal from another animal and themselves becoming infected. Then they pass the organisms to another individual during the next blood meal. Arthropodborne diseases occur primarily in the bloodstream, and they often are characterized by a high fever and a body rash.

Plague Can Be a Highly Fatal Disease

KEY CONCEPT

- *Yersinia pestis* is carried in rodents and their fleas.

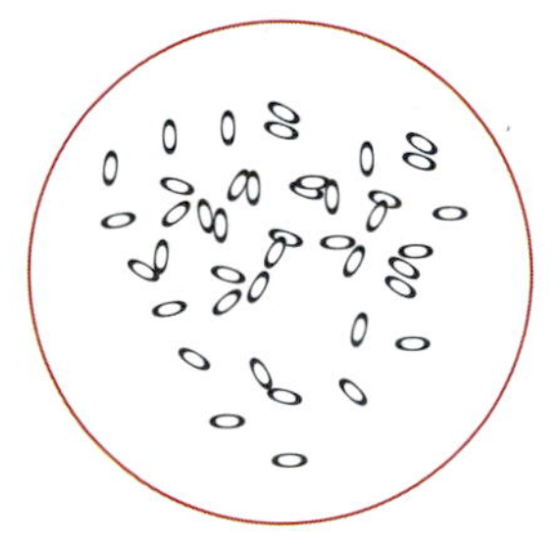

Yersinia pestis

Few diseases have had the rich and terrifying history of bubonic plague, nor can any match the array of social, economic, and religious changes wrought by this disease (see chapter introduction).

The first documented pandemic of plague probably began in Africa during the reign of the Roman emperor Justinian in AD 542. It lasted 60 years, killed millions, and contributed to the downfall of Rome. The second pandemic was known as the Black Death because of the purplish-black splotches on victims and the terror it evoked in the 1300s (Microfocus 11.4). The Black Death killed an estimated 40 million people in Europe, almost one third of the population on the continent. A deadly epidemic also occurred in London in 1665, where 70,000 people succumbed to the disease.

■ Septicemic: Referring to growth of bacterial cells in the blood.

The third pandemic occurred in the late 1800s, when Asian warfare facilitated the spread of a Burmese focus of plague, and migrations brought infected individuals to China and Hong Kong. During an epidemic in 1894, the causative organism was isolated by Alexandre Yersin and, independently, by Shibasaburo Kitasato (see Chapter 1). Plague first appeared in the United States in San Francisco in 1900, carried by rats on ships from Asia. The disease spread to ground squirrels, prairie dogs, and other wild rodents, and it is now endemic in the southwestern states, where it is commonly called **sylvatic plague** (*sylva* = "forest").

Plague is caused by *Yersinia pestis* (*pestis* = "plague") (FIGURE 11.6A). This nonmotile, gram-negative rod stains heavily at the poles of the cell, giving it a safety-pin appearance and a characteristic called **bipolar staining** when direct smears from infected specimens are observed. The bacillus is transmitted by the oriental rat flea *Xenopsylla cheopis*. The bacterial cells in an infected flea often clot in the digestive system, starving the flea (FIGURE 11.6B). This causes the flea to become even more voracious in finding a blood meal. Normally, the fleas infest only rats, but as septicemic rats die, the fleas may jump to another animal, such as humans, in an attempt to feed in the skin. In this process, bacterial cells are regurgitated into the human bloodstream.

Bubonic plague is a blood disease. The bacterial cells multiply in the bloodstream and local-

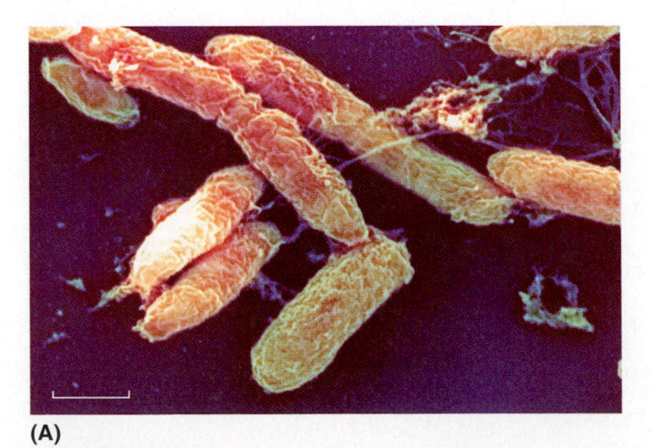

(A)

(B)

FIGURE 11.6 ***Yersinia Pestis* and the Flea Vector.** (**A**) False-color scanning electron micrograph of *Yersinia pestis*, the causative agent of bubonic plague. (Bar = 1 µm.) (**B**) *Xenopsylla cheopis* (oriental rat flea) with clotted *Y. pestis* mass (red foregut).

Q: How does clotting in the foregut lead to human infections?

MICROFOCUS 11.4: History

Catapulting Terror

Bodies came flying across the walls. People in panic either fled in horror or tried to remove the dead, decaying bodies. Caffa was in chaos.

One of the most horrendous emerging infectious diseases was starting to spread to Europe, North Africa, and the Near East in the mid-fourteenth century. It was the Black Death—bubonic plague—which historians believe moved out of the lands north of the Caspian and Black Seas.

Caffa (today Feodosija, Ukraine) was a port city situated on the north shore of the Black Sea. Through an agreement with the local Tartars (Mongols) who controlled the area, Caffa was placed under control of Genoa, Italy, and Christian merchants were allowed to trade goods with the Far East. In 1343, a group of Italian merchants from Genoa found themselves trapped behind the walls of Caffa after a brawl between the Italians and Tartars. The dreaded Tartars laid siege to the city and over the next five years, Genoa lost and regained control of the city several times.

During the siege of 1346, the Tartars were unable to drive the Italians and other Christians from the city. Then, plague broke out. Large numbers of Tartars started dying. Losing interest in the siege, the Tartars had the bodies of their dead plague victims placed in catapults and lobbed over the walls of Caffa into the city. The Tartars hoped the stench would kill everyone in Caffa; and, in fact, soon plague was sweeping through Caffa. The townspeople were terrified: Either the plague would kill them inside the walls, or the Tartars would kill them outside the walls. But the Tartars were equally terrified of the plague, and they were withdrawing.

Sensing an opportunity to escape, the merchants ran for their ships and sailed off to Genoa, Venice, and other homeports in the Mediterranean. Unfortunately, their voyage home would be a voyage of death. Many died of the plague onboard, and the survivors spread the disease wherever they stopped to replenish their supply of food and water.

Could such a tale be true? Could the dead diseased bodies catapulted into Caffa transmit plague? Almost certainly they could. City defenders would have carried away the dead, mangled bodies, which would spread the disease by contact. Poor sanitation and health of its citizens in Caffa would make transmission even easier and more widespread, especially if pneumonic plague broke out.

The attempted siege of Caffa in 1346 represents the most spectacular early episode of biological warfare, with the Black Death as its consequence. It demonstrates the very essence of terrorism as defined today—the intentional or threatened use of biological agents to cause fear in or actually inflict death or disease upon a large population for political, religious, or ideological reasons. The siege of Caffa shows us the horrifying consequences that can come from the use of infectious disease as a weapon.

ize in the lymph nodes, especially those of the armpits, neck, and groin. Hemorrhaging in the lymph nodes causes painful and substantial swellings called **buboes** (*bubon* = "the groin"). Dark, purplish splotches from hemorrhages also can be seen through the skin (the "rosies" in "Ring-a-ring of rosies"; see Chapter 1). Blood pressure drops and, without treatment, mortality reaches 60 percent. From the buboes, the bacilli may spread to the bloodstream, where they cause **septicemic plague** or lead to **plague meningitis**. Nearly 100 percent of untreated cases are fatal.

Human-to-human transmission of plague during epidemics is spread by respiratory droplets because septicemic cases can progress to the lungs. In this form, the disease is called **pneumonic plague** and is highly contagious. Lung symptoms are similar to pneumonia, with extensive coughing and sneezing. Hemorrhaging and fluid accumulation are common. Many suffer cardiovascular collapse with death common within 48 hours of the onset of symptoms. Mortality rates for pneumonic plague approach 100 percent.

When detected early, plague can be treated with streptomycin or tetracycline, reducing mortality to about 18 percent. Diagnosis consists of the laboratory isolation of *Y. pestis*, together with tests for plague antibodies and typing with bacteriophages. A vaccine consisting of dead *Y. pestis* cells is available for high-risk groups.

The disease occurs sporadically in Native American populations and in travelers through the Southwest. Small-game hunters, taxidermists, veterinarians, zoologists, and others who handle small rodents must be aware of possibly contracting plague. About 12 cases are reported to the Centers for Disease control and Prevention (CDC) annually. In 2004, three cases were reported, all acquired from infected fleas or contact with infected rabbits in Colorado and Wyoming. One case was fatal. Overall, about 14 percent of all plague cases in the United States are fatal. MicroFocus 11.5 recounts a case associated with wildlife.

CONCEPT AND REASONING CHECKS

11.5 Explain how bubonic plague can develop into communicable pneumonic plague.

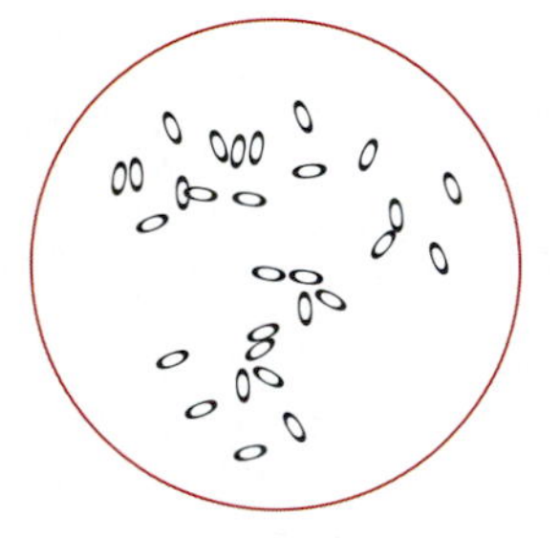

Francisella tularensis

Tularemia Has More Than One Disease Presentation

KEY CONCEPT

- *Francisella tularensis* infections can involve an extremely small infectious dose.

Tularemia is one of several microbial diseases first recorded in the United States (others include St. Louis encephalitis, Rocky Mountain spotted fever, Lyme disease, and Legionnaires' disease). In 1911, George W. McCoy, a Public Health Service investigator, reported a plague-like disease in ground squirrels from Tulare County, California, and within a year he isolated the responsible bacillus. By 1920, researchers identified tularemia in humans, and Edward Francis assumed a detailed study of the disease. Over the next 25 years, Francis amassed data on over 10,000 cases, including his own. In the 1960s, when a name was coined for the causative organism, *Francisella tularensis* was selected to honor him and identify the California county of first observation.

F. tularensis is a small, aerobic, encapsulated, gram-negative rod. It is extremely virulent because as few as 10 to 50 CFUs (see Chapter 5) can cause disease. It occurs in a broad variety of wild animals, especially rodents, and it is particularly prevalent in rabbits (in which case it is known as **rabbit fever**). Cats and dogs may acquire the bacillus during romps in the woods, and humans are infected by arthropods from the fur of animals. Ticks are important vectors in this regard, as evidenced by an outbreak of 20 tickborne cases in South Dakota in the 1980s. Other methods of transmission include consumption of infected rabbit meat, splashing bacilli into the eye, inhaling bacilli, or contact with an infected animal.

MICROFOCUS 11.5: Public Health
"What a Tragedy!"

"Just an incredible day," he must have been thinking, as he trudged along. It was great to be out in the desert. Great to get away from the books for a while. Final exams were only a few weeks away, then summer, then college, then who knows? Maybe a career in physical education? Maybe a coaching job? Three miles completed, five miles to go. He had to keep moving to make it back home for dinner.

He didn't mind the solitude of the Arizona desert. It was cool up here in Flagstaff, just right for hiking. And there were plenty of animals to see and interesting plants to watch for. He would stop now and then for some water or to watch the horizon or feed a colony of prairie dogs as they bustled about the terrain. Then he continued on to home.

He enjoyed wrestling as well, and he was the team captain. Two days after the hike, he sustained a groin injury while wrestling. When the ache remained, he went to the doctor. The doctor noted the groin swelling and wondered whether it had anything to do with the fever the young man was experiencing. Was it just a coincidence—the ache, the swelling, the fever? This was no ordinary groin injury. "We'll watch it for a couple of days," the doctor said as he gave the young man a pain reliever.

Tragedy struck the next morning. All his mother remembered was a loud thud from the bathroom. She didn't remember her terrified shriek or dialing the emergency number. The emergency medical technicians were there in a flash, but it was too late. He was dead.

The investigation that followed took public health officials down many dead-ends. It wasn't the wrestling injury, they concluded. Still, the groin swelling made them suspicious. "That hiking trail," they asked his mother, "where is it?"

They set out in search of an elusive answer. About three and a half miles out, they came upon a colony of prairie dogs. The animals didn't look well. In fact, some were dead nearby. Carefully, they trapped a sick animal and carried it back to the lab. Two days later, the lab report was ready—the animal was sick with plague. Then the young man's tissues were tested—again, plague. The investigators shook their heads. "What a tragedy!"

Various forms of tularemia exist, depending on where the bacilli enter the body. An arthropod bite, for example, may lead to swollen lymph nodes and a skin ulcer at the bite site (FIGURE 11.7). Individuals typically experience flu-like symptoms. Inhalation of *F. tularensis* cells may lead to respiratory disease and produce swollen lymph nodes, a dry cough, and pain under the breast bone.

Tularemia usually resolves on treatment with streptomycin, and few people die of the disease. Epidemics are unknown, and evidence suggests that tularemia may not be communicable among humans despite the many modes of entry to the body. Physicians reported 134 cases to the CDC in 2004.

CONCEPT AND REASONING CHECKS

11.6 Propose a hypothesis to explain why the *F. tularensis* infectious dose is so small.

Lyme Disease Can Be Divided into Three Stages

KEY CONCEPT

- *Borrelia burgdorferi* is transmitted by the bite of a small tick.

One of the major emerging infectious diseases is **Lyme disease**, currently the most commonly reported arthropodborne illness in the United States. Although 95 percent of cases occur in the northeastern and north-central states, cases also have been reported in the Pacific coast states and the Southeast. In 2004 it accounted for 19,804 U.S. cases of infectious disease.

Lyme disease is named for Old Lyme, Connecticut, the suburban community where a cluster of cases was observed in 1975. The disease is caused by the spirochete *Borrelia burgdorferi* (FIGURE 11.8A). It was named for Willy Burgdorfer, the microbiologist who studied the spirochete in the gut of infected ticks.

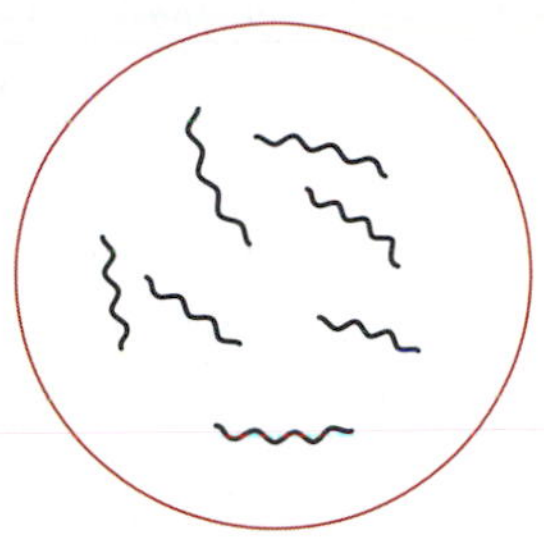

Borrelia burgdorferi

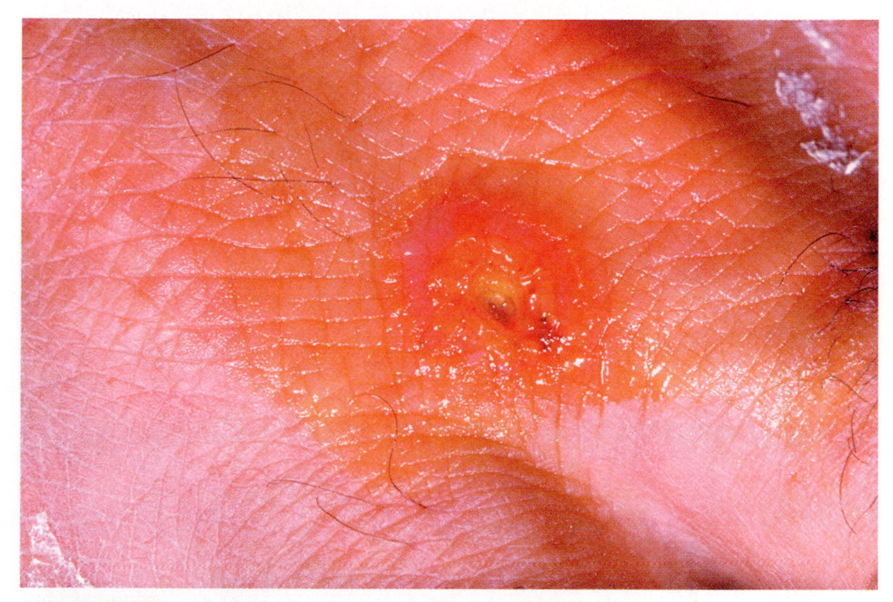

FIGURE 11.7 **A Tularemia Lesion.** The lesions of tularemia occur where the bacilli enter the body. The disease can be acquired by handling infected rabbit meat or from the bite of an infected arthropod.

Q: What are the symptoms associated with lesion formation?

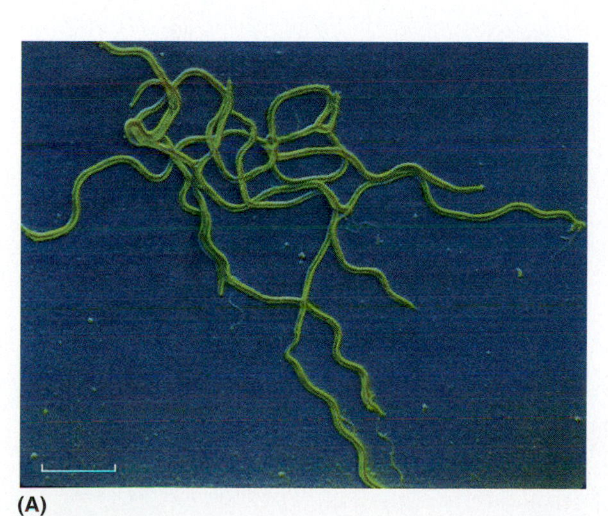

(A)

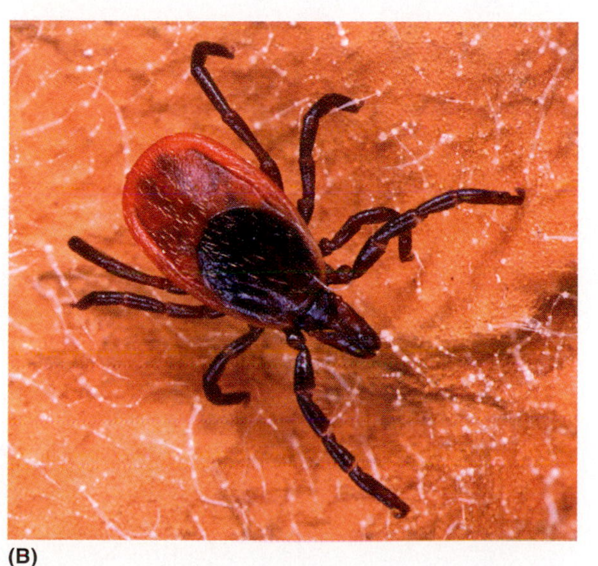

(B)

FIGURE 11.8 ***Borrelia burgdorferi* and Its Tick Vector.** (**A**) A false-color transmission electron micrograph showing an abundance of *B. burgdorferi* spirochetes, the agents of Lyme disease. (Bar = 1 μm.) (**B**) A photograph of *Ixodes scapularis,* a tick species that transmits Lyme disease.

Q: Although not evident in the electron micrograph, what structure typical of spirochetes contributes to their motility?

The tick that transmits most cases of Lyme disease in the Northeast and Midwest is the deer tick *Ixodes scapularis* (formerly *I. dammini*) (FIGURE 11.8B); in the West, the major vector is the western black-legged tick *I. pacificus*. It is smaller than the American dog tick, and it lives and mates in the fur of white-tailed deer. Eventually it falls into the tall grass, where it waits for an unsuspecting dog, rodent, or human to pass by. The tick then attaches to its new host and penetrates into the skin. During the next 24 to 36 hours, it takes a blood meal and swells to the size of a small pea. While sucking the blood, it also defecates into the wound, and, if the tick is infected, spirochetes are transmitted (FIGURE 11.9). If the tick is observed on the skin, it should be removed with forceps or tweezers, and the area should be thoroughly cleansed with soap and water and an antiseptic applied.

Lyme disease has a variable incubation period of 3 to 31 days and, untreated, typically has three stages. The **early localized stage** involves a slowly expanding red rash at the site of the tick bite. The rash is called **erythema** (red) **migrans** (expanding), or **EM**. Beginning as a small flat or raised lesion, the rash increases in diameter in a circular pattern over a period of weeks, sometimes reaching a diameter of 10 to 15 inches (FIGURE 11.10). It has an intense red border and central clearing, termed the **bull's-eye rash**. It can vary in shape and is usually hot to the touch, but it need not be present in all cases of disease. Indeed, about 20 percent of patients do not develop EM. The tick bite can be distinguished from a mosquito bite because the latter itches, while a tick bite does not. Fever, aches and pains, and flu-like symptoms usually accompany the rash. Effec-

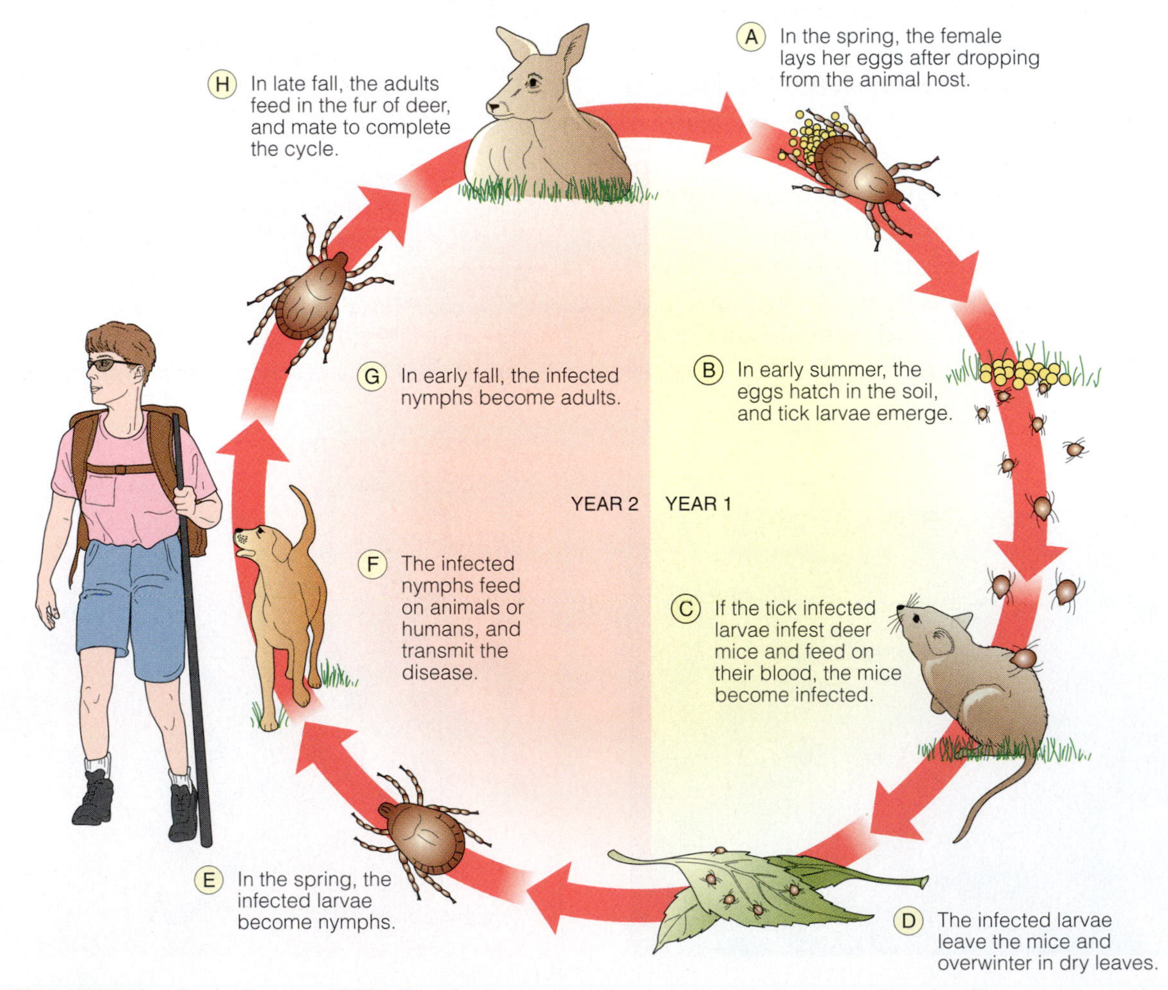

FIGURE 11.9 **Life Cycle of Ixodid Ticks.** The life cycle of *Ixodes scapularis* and *I. pacificus,* the ticks that transmit Lyme disease.

Q: Why is Lyme disease considered a zoonotic disease?

tive treatment can be rendered with amoxicillin or doxycycline.

Left untreated, an **early disseminated stage** begins weeks to months later with the spread of *B. burgdorferi* to the skin, heart, nervous system, and joints. On the skin, multiple smaller EMs develop while invasion of the nervous system can lead to meningitis, facial palsy, and peripheral nerve disorders. Cardiac abnormalities are the most common, such as brief, irregular heartbeats. Joint and muscle pain also occur.

If still left untreated, a **late stage** occurs months to years later. About 10 percent of patients develop chronic arthritis with swelling in the large joints, such as the knee. Although Lyme disease is not known to have a high mortality rate, the overall damage to the body can be substantial.

A diagnosis of Lyme disease is usually based on symptoms and the patient's recent activities are noted (e.g., hiking in the woods, living in a tick-infested area, camping). Blood tests may not be accurate so tests designed to detect spirochetal DNA (polymerase chain reaction; PCR) have been developed (Chapter 22).

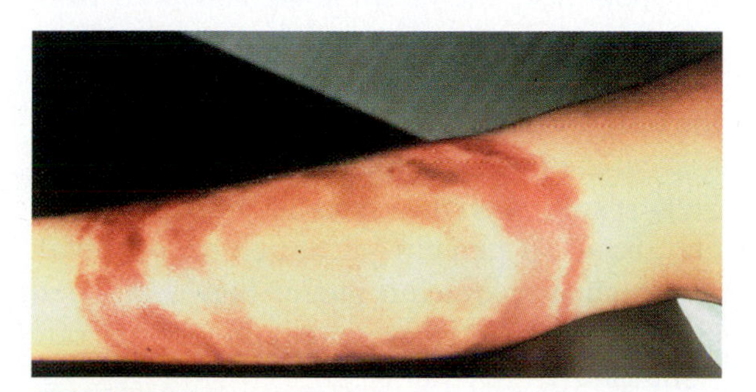

FIGURE 11.10 **Erythema Migrans.** Erythema migrans (EM) is the rash accompanying 80 percent of cases of Lyme disease. The rash consists of a large patch with an intense red border. It is usually hot to the touch, and it expands with time.

Q: What is the common name for the rash?

A vaccine for dogs (LymeVax) has been licensed and is in routine use. A vaccine for humans (LYMErix®) was removed from the market in February 2002 apparently because of poor sales, a complicated immunization schedule, and side effects.

CONCEPT AND REASONING CHECKS

11.7 Describe the three clinical stages of untreated Lyme disease.

Relapsing Fever Is Carried by Ticks and Lice

KEY CONCEPT

- Other *Borrelia* species cause a relapsing, febrile illness.

Relapsing fever is caused by a long spirochete responsible for two forms of the disease. It is called relapsing fever because infected individuals go through periods of fever and chills when many spirochetes are present in the blood. As the spirochetes decline in number, the individual recovers for several days before a recurrence of the symptoms (FIGURE 11.11).

Endemic relapsing fever often is caused by *Borrelia hermsii* and *B. turicatae*. It is transmitted by the bite of the tick *Ornithodoros* from rodent hosts to humans. Ticks normally inhabit the rodent burrows and nests, where the natural infection cycle proceeds without apparent disease in the rodents. Humans are incidental hosts, often bitten briefly (5–20 minutes) at night and without notice by the infected ticks. With tickborne relapsing fever, up to 13 relapses can occur and, untreated, mortality can be 2 to 5 percent.

Cabins in wilderness areas of the northwest and southwest United States are favorable

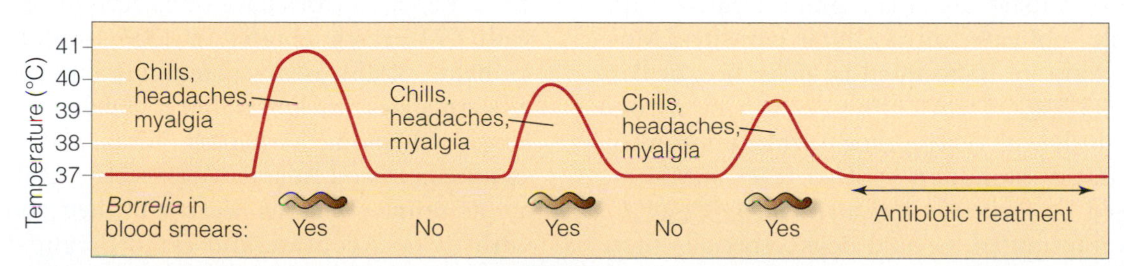

FIGURE 11.11 **The Cycles of Relapsing Fever.** In relapsing fever, chills, headache, and fever peak with high numbers of *Borrelia* spirochetes in the blood.

Q: During which part of each cycle would specific antibodies against Borrelia *be produced?*

nesting sites for infected rodents and their ticks, especially rustic cabins where rodents have access (Textbook Case 11).

Epidemic relapsing fever is caused by *Borrelia recurrentis,* which is carried by the body louse, *Pediculus humanus.* The disease is spread between humans by the lice, and infection occurs when the louse is crushed into the bite wound. No cases of epidemic relapsing fever have been recorded in the United States since 1906. It mainly is found in Africa, China, and parts of South America in overcrowded, poverty-stricken regions where the public health systems have failed. It also is a disease of war, such as in World War II when some 10 million people were infected.

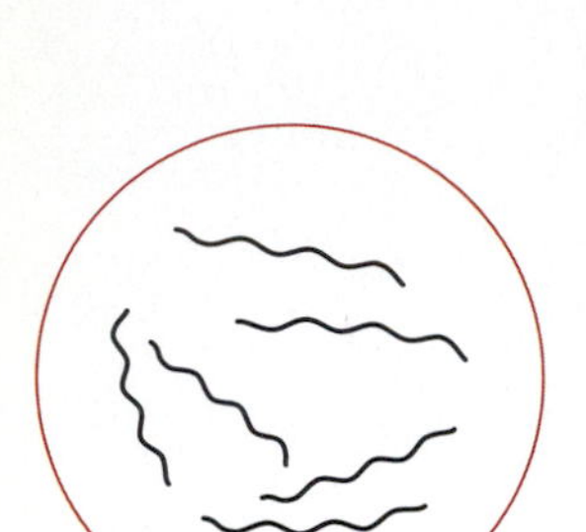

Borrelia recurrentis

Cases of louseborne relapsing fever are characterized by substantial fever, shaking, chills, headache, prostration, and drenching sweats. The symptoms last for a couple of days, disappear for about eight days, then reappear up to 5 times during the following weeks. Left untreated, mortality rates can be as high as 40 percent.

Both types of relapsing fever can be effectively treated with doxycycline or erythromycin to hasten recovery.

CONCEPT AND REASONING CHECKS

11.8 Explain the reason that infected individuals experience relapsing fevers.

11.3 Rickettsial and Ehrlichial Arthropodborne Diseases

In 1909, Howard Taylor Ricketts, a University of Chicago pathologist, described a new organism in the blood of patients with Rocky Mountain spotted fever and showed that ticks transmit the disease. A year later, he located a similar organism in the blood of animals infected with Mexican typhus, and discovered that fleas were the important vectors in this disease. Unfortunately, in the course of his work, Ricketts fell victim to the disease and died. When later research indicated that Ricketts had described a unique group of microorganisms, the name rickettsiae was coined to honor him.

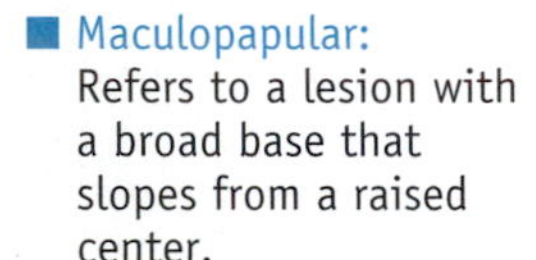

■ Maculopapular: Refers to a lesion with a broad base that slopes from a raised center.

Rickettsial Infections Are Transmitted by Arthropods

KEY CONCEPT

- *Rickettsia* infections often involve a characteristic rash.

The rickettsiae are small, gram-negative, nonmotile, obligate, intracellular parasites. Most infections are transmitted by ticks, lice, or fleas (FIGURE 11.12). All illnesses can be treated effectively with doxycycline or chloramphenicol.

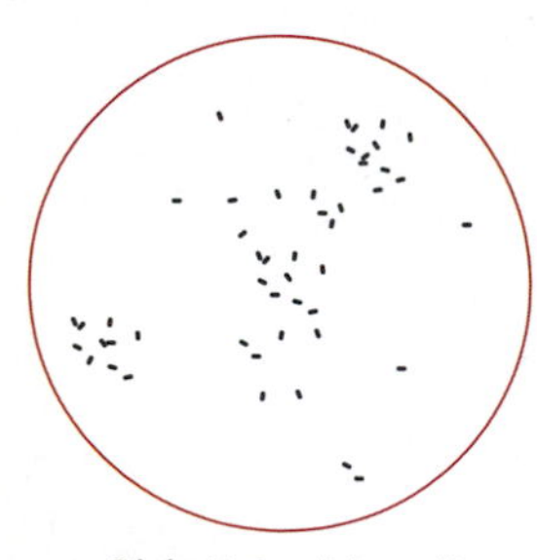

Rickettsia rickettsii

Rocky Mountain spotted fever (RMSF) is caused by *Rickettsia rickettsii* (FIGURE 11.13A). It is transmitted by hard ticks, especially those of the genera *Amblyomma* and *Dermacentor* (Figure 11.12A). Following a tick bite, the hallmarks of RMSF are a high fever lasting for many days, severe headaches, and a skin rash reflecting damage to the small blood vessels. The rash begins as pink spots (macules), and progresses to pink-red papules (FIGURE 11.13B). Where the spots fuse, they form a maculopapular rash, which becomes dark red and then fades without evidence of scarring. As the disease progresses, the rash appears on the palms of the hands and soles of the feet and progressively spreads to the body trunk. Mortality rates of untreated cases are about 30 percent, although an outbreak in Montana recorded a rate as high as 75 percent. Antibiotic treatment reduces this rate significantly.

RMSF was first observed in early settlers to the American Northwest. About a thousand cases were reported annually in the United States in the early 1980s, but public education about the disease, along with improved methods of diagnosis and treatment, caused a drop until 1998. Reported cases have since been increasing with 1,713 cases reported in 2004. Contrary to its name, RMSF is not commonly reported in western states any longer, but it remains a problem in Oklahoma, Arkansas, Texas, and many southeastern and Atlantic Coast states. Children are its primary victims because of their contact with ticks. Accurate and rapid diagnosis is essential in treating RMSF.

Epidemic typhus (also called **typhus fever**) is one of the most notorious of all bacterial diseases. It is considered a prolific killer of

Textbook CASE 11

Tickborne Relapsing Fever Outbreak

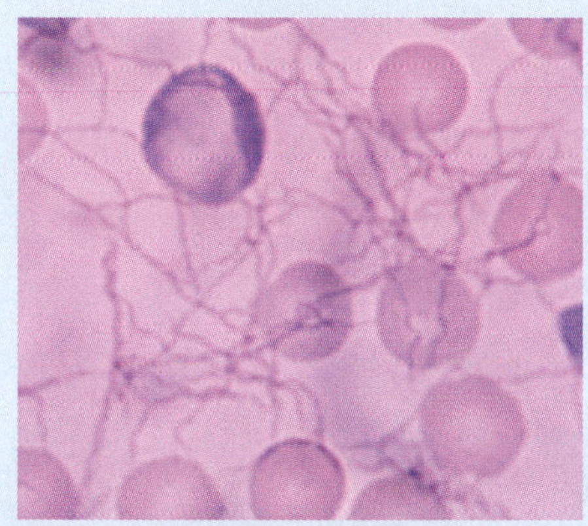
Tickborne relapsing fever spirochetes.

1 In late July 2002, a family reunion was planned at a remote, previously uninhabited cabin in the mountains of northern New Mexico. Three days before the reunion, three family members arrived to clean the cabin. After the reunion, about half of the 39 family members slept in the cabin overnight.

2 Four days after the reunion, one of the family members who cleaned the cabin arrived at a local hospital complaining of fever, chills, muscle aches, and a rash on the forearms. The symptoms had started two days previously.

3 A hospital lab technician took a blood sample from the patient and identified spirochetes on a blood smear (see figure). The immediate diagnosis was tickborne relapsing fever (TBRF). On August 2, 2002, the New Mexico Department of Health and the Indian Health Services were immediately notified.

4 By August 7, 2002, 39 family members had sought medical care or were visited by a public health nurse. A study was begun to determine risk factors for infection, better understand the outbreak, and provide prevention measures.

5 A case of TBRF was defined. Based on the definition, 14 family members suffering the illness were identified.

6 Blood samples were taken from all 14 symptomatic patients. Spirochetes were identified in the blood smears of nine patients.

7 Samples from 13 patients were sent to the Centers for Disease Control and Prevention (CDC) in Atlanta. Eleven samples, including two samples negative for spirochetes by blood smear, were cultured and all grew *Borrelia hermsii*.

8 All 14 patients received antibiotic therapy and recovered.

9 Risk analysis revealed that all three family members who cleaned the cabin and eight of the other 36 were more likely to be TBRF patients.

10 Inspection of the cabin by Indian Health Service workers identified abundant rodent nesting materials and droppings within the walls of the cabin.

Questions

A. *Why was such a prompt diagnosis and notification of health services necessary?*

B. *How would you define a case of TBRF?*

C. *Why were spirochetes only found in 9 of the 14 patients?*

D. *Why did all three who cleaned the cabin become ill?*

E. *Of the other eight patients, what is most likely the reason for them contracting the disease?*

For additional information see http://www.cdc.gov/mmwr/preview/mmwrhtml/mm5234a1.htm.

(A)

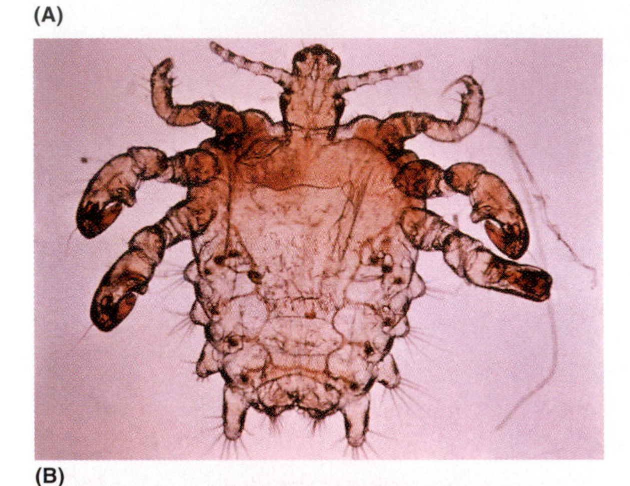

(B)

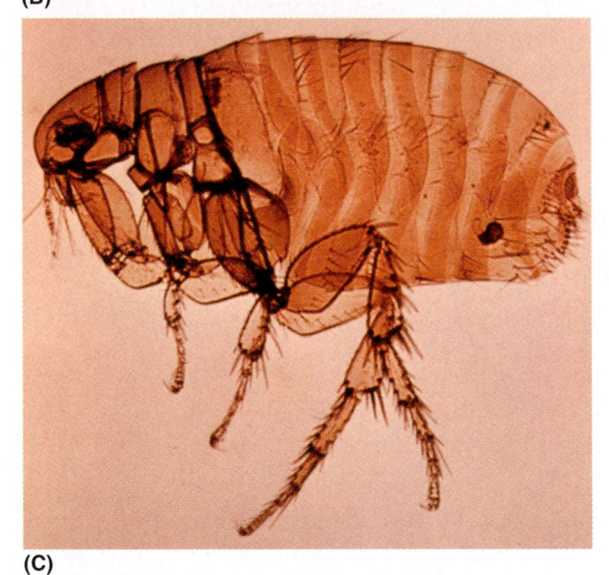

(C)

FIGURE 11.12 **Three Arthropods that Transmit Rickettsial Diseases.** (**A**) Rocky Mountain woodtick *(Dermacentor)*. (**B**) The body louse *(Pediculus)*. (**C**) The flea *(Xenopsylla)*.

Q: Why are arthropods such excellent vectors for the transmission of rickettsial diseases?

humans, and on several occasions it has altered the course of history, as when it helped decimate the Aztec population in the 1500s. Historians report that Napoleon marched into Russia in 1812 with over 200,000 French soldiers but his forces were hit hard by typhus. Hans Zinsser's classic book *Rats, Lice, and History* describes events such as these and provides an engaging look at the effects of several infectious diseases on civilization (MicroFocus 11.6).

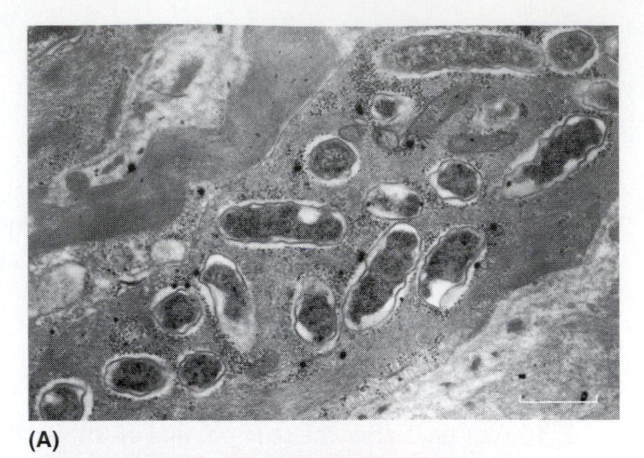

(A)

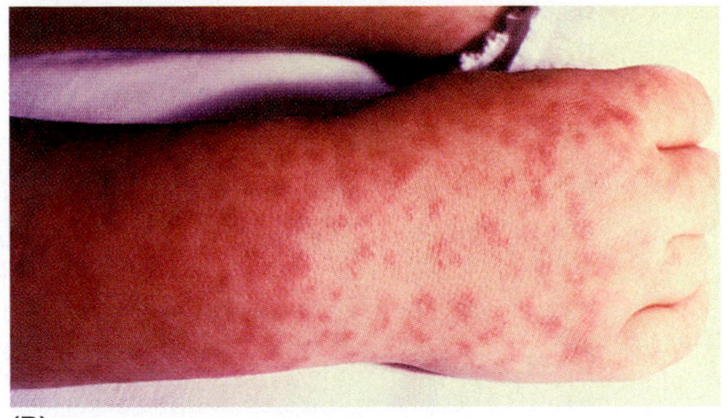

(B)

FIGURE 11.13 ***Rickettsia rickettsii* and Symptoms of Rocky Mountain Spotted Fever.** (**A**) *R. rickettsii* in the cytoplasm of a kidney cell. (Bar = 0.5 μm.) (**B**) Child's right hand and wrist displaying the characteristic spotted rash of Rocky Mountain spotted fever.

Q: What is happening in the host to produce a spotted rash?

Epidemic typhus is caused by *Rickettsia prowazekii*. The rickettsiae are transmitted to humans by body lice of the genus *Pediculus* (Figure 11.12B). Lice are natural parasites of humans and flourish where sanitation measures are lacking and hygiene is poor. Often these conditions are associated with war, famine, poverty, and a generally poor quality of life.

The rickettsiae are excreted in feces from the lice, so scratching the bite will facilitate infection into the wound. The characteristic fever and rash of rickettsial disease are particularly evident in epidemic typhus. There is a maculopapular rash, but unlike the rash of RMSF, it appears first on the body trunk and progresses to the extremities. Intense fever, sometimes reaching 40°C, remains for many days as the patient hallucinates and becomes delirious. Some patients suffer permanent damage to the blood vessels and heart, and over 75 percent of sufferers die in epidemics. Again, antibiotic therapy reduces this percentage substantially, and the disease is rare in the United States.

Endemic typhus is a second but milder form of typhus. This disease, called **Mexican typhus** in the Southwest, occurs sporadically in human populations because the rat flea that transmits it, *Xenopsylla cheopis,* is not a natural parasite of humans (Figure 11.12C). However, the disease is prevalent in rodent populations where fleas abound (such as rats and squirrels), and thus it is called **murine typhus** (*murin* = "of the mouse"). Cats and their fleas are involved also, and lice may harbor the bacilli.

The agent of endemic typhus is *Rickettsia typhi*. When an infected flea feeds in the human skin, it deposits the organism into the wound. Endemic typhus is usually characterized by a mild fever, persistent headache, and a maculopapular rash that spreads from trunk to extremities. Often the recovery is spontaneous, without the need of drug therapy. However, lice may transport the rickettsiae to other individuals and initiate an epidemic. Most recent cases have occurred in Southern California and south Texas. In 2004, 27 people in Nueces County, Texas, contracted murine typhus. All recovered.

Scrub typhus occurs in Asia and the Southwest Pacific. It is so named because *Trombicula,* the mite transmitting the disease, lives in scrubland where the soil is sandy or marshy and vegetation is poor. Scrub typhus also is called **tsutsugamushi fever**, from the Japanese words tsutsuga for "disease" and mushi for "mite." The causative agent, *Rickettsia tsutsugamushi,* enters the skin during mite infestations and soon causes fever and a rash, together with other typhus-like symptoms. Outbreaks may be significant, as evidenced by the 7,000 U.S. servicemen affected in the Pacific during World War II.

Rickettsialpox was first recognized in 1946 in an apartment complex in New York City. Investigators traced the disease to mites in the fur of local house mice, and named the disease rickettsialpox because the skin rash was similar

MICROFOCUS 11.6: **History**

Students and Typhus Fever

Hans Zinsser once remarked, ". . .*[the students] force a teacher continually to renew the fundamental principles of the sciences from which his specialty takes off. So while we are, technically speaking, professors, we are actually older colleagues of our students, from whom we often learn as much as we teach them.*"

Hans Zinsser was born in New York City in 1878. He achieved fame for isolating the bacterial agent of epidemic typhus and was a pioneer in the study of autoimmunity and how microbes could induce disease, as in rheumatic fever. Zinsser was equally well known for his writing and wrote some of the most uniquely personal, wise, and witty prose of the twentieth century.

Zinsser also was a dedicated teacher. Writing at a time when most medical students were male, he remarked, ". . . *as we grow wiser we learn that the relatively small fractions of our time which we spend with well-trained, intelligent young men are more of a privilege than an obligation.*"

Dr. Zinsser.

Perhaps the culmination of his writing was his most famous book entitled (with tongue in cheek) *Rats, Lice, and History: Being a Study in Biography, Which, After Twelve Preliminary Chapters Indispensable for the Preparation of the Lay Reader, Deals with the Life History of Typhus Fever.* It was an international best seller.

Here is a sample of Zinsser's writing from his 1935 book: "*Soldiers have rarely won wars. They often mop up after the barrage of epidemics. And typhus, with its brothers and sisters—plague, cholera, typhoid, dysentery—has decided more campaigns than Caesar, Hannibal, Napoleon, and all the other generals of history. The epidemics get the blame for defeat, the generals the credit for victory. It ought to be the other way 'round. . . .*" Zinsser died in 1940 from lymphatic cancer at the age of 61.

If you think you'd enjoy learning more about disease and its effect on civilization, Zinsser's book is still in print. It would be a worthwhile investment of your time.

to a chickenpox rash. Rickettsialpox is now considered a benign disease. It is caused by *Rickettsia akari* (*acari* = "mite"). Fever and rash are typical symptoms and fatalities are rare.

Brill-Zinsser disease was first described in textbooks from the early 1900s as a mild form of typhus. The high fever was a distinguishing characteristic, and it was recognized that if a patient was infested with lice, an epidemic of typhus might ensue. Today microbiologists believe Brill-Zinsser disease is a relapse of an earlier case of typhus in which *R. prowazekii* lay dormant in the patient for many years. However, symptoms are usually milder.

Trench fever was the most widespread disease encountered during World War I (except for the great influenza pandemic of 1918 to 1919). An estimated 1 million soldiers are thought to have been infected. The disease is caused by *Bartonella (Rochalimaea) quintana*, one of the few rickettsial-like organisms cultivated outside of living cells. It is a gram-negative rod.

Transmission takes place by body lice. Symptoms include a maculopapular rash and a fever occurring at irregular intervals. Trench fever is prevalent where lice abound, and trench warfare provided a near-ideal setting. It also has been implicated in heart muscle inflammation in homeless individuals.

Diagnosis for all rickettsial diseases takes into account evidence of an arthropod bite, progress of any rash, and recent activities of the patient (e.g., camping, backpacking, and other outdoor activities where contact with arthropods may have occurred). Antibody-detection tests are advised to make a correct diagnosis.

CONCEPT AND REASONING CHECKS

11.9 What distinguishes the different rickettsial diseases from one another?

Ehrlichial Infections Are Emerging Diseases in the United States

KEY CONCEPT

- *Ehrlichia* species infect different groups of leukocytes.

Ehrlichiosis, the final disease we shall consider, was first described in humans in 1986. Formerly believed to be confined to dogs, ehrlichiosis has been recognized in two forms in the United States: **human monocytic ehrlichiosis** (HME), which is caused by *Ehrlichia chaffeensis* (because the first case was observed at Fort Chaffee, Arkansas); and **human granulocytic ehrlichiosis** (HGE) caused by *Ehrlichia phagocytophila* (FIGURE 11.14).

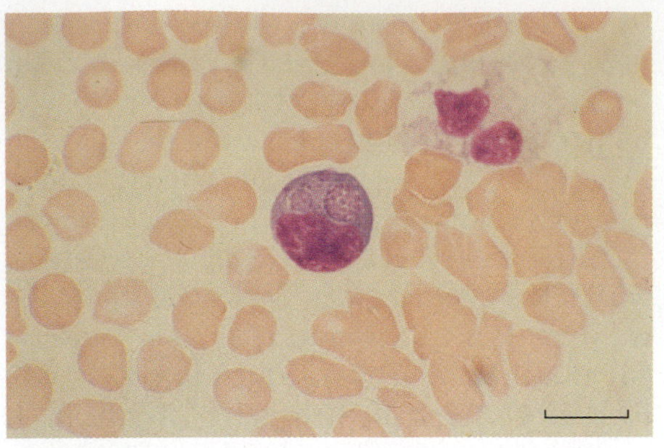

FIGURE 11.14 ***Ehrlichia phagocytophila* in a Blood Smear.** *E. phagocytophila* causes human granulocytic ehrlichiosis (HGE). This tickborne organism multiplies in human white blood cells (stained purple) called granulocytes. The granulocytes are neutrophils. (Bar = 10 µm.)

Q: What are the predominant light pink cells in this blood smear?

HME is transmitted by the Lone Star tick (prevalent in the South), while HGE is transmitted by the dog tick and the deer tick, the same one transmitting Lyme disease (prevalent in the Northeast). Patients with HME or HGE suffer from headache, malaise, and fever, with some liver disease and, infrequently, a maculopapular rash. Indeed, both HGE and HME are quite similar to Lyme disease, except the symptoms come on faster in HGE and HME, they clear more quickly, and the rash is infrequent. Because HME affects the body's monocytes (hence "monocytic") and HGE neutrophils (a type of granulocyte, hence "granulocytic") both bring about a lowering of the white blood cell count (**leukopenia**).

The CDC reported 537 cases of HGE and 338 cases of HMG in 2004. However, this may be an underestimate because identification and reporting of human ehrlichioses are incomplete at the state level. Still, HGE cases predominate in the upper Midwest and Northeast while HME cases predominate in New York, North Carolina, and the central Midwest.

The arthropodborne diseases covered in this chapter are summarized in TABLE 11.2. MicroInquiry 11 presents four cases for study involving soilborne and arthropodborne diseases.

CONCEPT AND REASONING CHECKS

11.10 Draw a concept map for the rickettsial diseases indicating how the diseases differ from one another.

TABLE 11.2 A Summary of Arthropodborne Bacterial Diseases

Disease	Causative Agent	Description of Agent	Organs Affected	Characteristic Signs	Vector	Treatment Administered	Comment
Bubonic plague	*Yersinia pestis*	Gram-negative bipolar staining rod	Lymph nodes Blood Lungs	Buboes Pneumonia Septicemia	Rat flea	Tetracycline Streptomycin	Septicemic and pneumonic forms Rodents infected
Tularemia	*Francisella tularensis*	Gram-negative bipolar staining rod	Eyes Skin Blood	Eye lesion Skin ulcer Pneumonia	Flea Tick	Tetracycline Streptomycin	Resembles many other diseases Many modes of transmission No epidemics
Lyme disease	*Borrelia burgdorferi*	Spirochete	Skin Joints Heart	Erythema chronicum migrans Arthritis	Tick	Penicillin Doxycycline	Most prevalent tickborne disease Three possible stages
Relapsing fever	*Borrelia recurrentis*	Spirochete	Blood Liver	Fever Jaundice	Louse Tick	Doxycycline	Relapses common Rose spots develop on skin
Rocky Mountain spotted fever	*Rickettsia rickettsii*	Rickettsia	Blood Skin	Fever Rash	Tick	Doxycycline	Most common rickettsial disease Rash first on extremities
Epidemic typhus	*Rickettsia prowazekii*	Rickettsia	Blood Skin	Fever Rash	Louse	Doxycycline Chloramphenicol	High mortality rate Rash first on body trunk
Endemic typhus	*Rickettsia typhi*	Rickettsia	Blood Skin	Fever Rash	Flea	Doxycycline Chloramphenicol	Mild typhus-like disease Prevalent in rodents
Scrub typhus	*Rickettsia tsutsugamushi*	Rickettsia	Blood Skin	Fever Rash	Mite	Doxycycline Chloramphenicol	Occurs in scrubland Prevalent in Asia
Rickettsialpox	*Rickettsia akari*	Rickettsia	Blood Skin	Fever Rash	Mite	Doxycycline Chloramphenicol	Rare disease Resembles chickenpox
Trench fever	*Bartonella quintana*	Rickettsia	Blood Skin	Fever Rash	Louse	Doxycycline Chloramphenicol	Common in World War I Cultivation in artificial medium
Ehrlichiosis (HME) (HGE)	 *Ehrlichia chaffeensis* *E. phagocytophila*	Rickettsia	Blood	Fever	Tick	Doxycycline Chloramphenicol	Recognized since 1986

MICROINQUIRY 11

Soilborne and Arthropodborne Disease Identification

Below are several descriptions of soilborne and arthropodborne bacterial diseases based on material presented in this chapter. Read the case history and then answer the questions posed. Answers can be found in **Appendix D**.

Case 1

An 18-year-old girl who had been in good health goes to her physician in Seattle, Washington, complaining of a headache and flu-like symptoms. Further questioning indicates she had developed a red rash on her thigh that enlarged but then disappeared after two weeks. During this time, she also had a fever. The physician discovers the patient had been hiking in the hills east of Seattle ten days prior to developing the rash. The patient is placed on doxycycline and recovers.

11.1a. What important clues toward identifying the disease are indicated from the patient's case?

11.1b. What disease does she have and what organism is responsible for the disease?

11.1c. How was the organism responsible for the disease transmitted to the patient?

11.1d. What complications could occur if the patient had not visited her physician?

11.1e. How could this disease have been prevented?

Case 2

A 33-year-old indigent man comes to the emergency room of the county hospital. The emergency room nurse immediately notices that the man cannot open his mouth because of facial muscle spasms. The physician on duty detects right-sided face pain and trismus. She is able to ascertain from the patient that he has not been able to eat for two days because of jaw pain. Examination of his body shows necrotic, blackened areas on the bottom of his left foot. Based on the signs and physical exam, treatment is started.

11.2a. Based on these signs and the physical exam results, what organism has infected this man and what disease does he have?

11.2b. What treatment should be provided?

11.2c. Should treatment have waited until the diagnosis was confirmed by laboratory results? Explain.

11.2d. What is the significance of the necrotic areas of the patient's left foot?

11.2e. How might this indigent man have become infected?

Case 3

A 10-year-old boy is brought to the emergency room in Charleston, South Carolina, by his mother and father. He had been in good health when the family went for a July Fourth holiday camping trip nine days ago in the Appalachian Mountains. The day after returning from the trip, a tick was discovered in his scalp and was removed. In the emergency room, the boy complains of a headache. He has a fever and the physician observes a pink rash on the palms of his hands and soles of his feet. His white blood cell count is slightly elevated. The physician starts the patient on erythromycin therapy and the boy eventually recovers completely.

11.3a. Identify two organisms that could produce the finding described in this case study.

11.3b. What is the agent responsible for the patient's disease? What were the clues for the diagnosis?

11.3c. What infectious bacterial diseases are spread by ticks?

11.3d. What test(s) might be used to confirm the diagnosis?

11.3e. How would organisms like *Ehrlichia chaffeensis* be eliminated as the agent responsible for the patient's disease?

Case 4

A 34-year-old woman arrives in the emergency room complaining of extreme pain in the right shin with limited mobility in the leg. Her breathing is normal. Examination of the lower leg indicates trauma and the skin is discolored a greenish blue. An additional finding was a crackling sound in her lower leg. The patient tells the physician that one week ago she had a severe mountain biking accident and had several deep cuts to her leg. She is given antibiotics and taken to the operating room where necrotic muscle is discovered. A biopsy Gram stain from the tissue shows gram-positive rods.

11.4a. Based on the emergency room findings, what disease does the patient have and what bacterium is responsible for the infection?

11.4b. What physical conditions of growth should be used when incubating the blood culture?

11.4c. In the operating room, what should the surgeons do once they know the identity of the disease?

11.4d. To ensure a systemic infection does not develop, what physical treatment can be used to slow down or stop the growth of the infecting bacterial cells? Why would this treatment work?

SUMMARY OF KEY CONCEPTS

11.1 Soilborne Bacterial Diseases

- Anthrax is an acute infectious disease caused by *Bacillus anthracis.* Human contact can be by inhalation, consumption, or skin contact with spores. Inhalation produces symptoms of respiratory distress and causes a blood infection. Consumption and skin contacts lead to boil-like lesions. Ciprofloxacin has been the antibiotic of choice.
- *Clostridium tetani* is the causative agent of tetanus. Symptoms of generalized muscle stiffness and trismus lead to convulsive contractions with an unnatural fixed smile. Antitoxin and antibiotics can be used to neutralize the toxin and kill the bacterial cells. A vaccine is available for prevention.
- Gas gangrene is caused by the anaerobic endospore-forming *Clostridium perfringens.* Symptoms include intense pain, swelling, and a foul odor at the wound site. Penicillin and removal of dead tissue help recovery.
- Leptospirosis is a disease spread from animals to humans (zoonosis) by *Leptospira interrogans.* Infected individuals have flu-like symptoms. Up to 10 percent of patients experience a systemic form of the disease. Penicillin or doxycycline is the antibiotic of choice.

11.2 Arthropodborne Bacterial Diseases

- *Yersinia pestis* is the causative agent of plague. This highly fatal infectious disease is transmitted to humans by the bites of infected fleas. Bubonic plague is characterized by the formation of buboes. Spreading of the bacilli to the blood leads to septicemic plague. Localization in the lungs is characteristic of pneumonic plague, which can be spread person to person. Without treatment, septicemic and pneumonic plague are nearly 100 percent fatal.
- Tularemia is caused by *Francisella tularensis,* which is highly infectious at low doses. Various forms of the disease occur depending on where the bacilli enter the body. Skin ulcers, eye lesions, and pulmonary symptoms can result. Treatment is effective using tetracycline or streptomycin.
- Lyme disease results from an infection by the spirochete *Borrelia burgdorferi.* It involves three stages: the erythema migrans (EM) rash; neurological and cardiac disorders of the central nervous system; and migrating arthritis. Penicillin or doxycycline is effective.
- Relapsing fever results in recurring attacks of high fever caused by ticks carrying *Borrelia hermsii* (endemic relapsing fever) or body lice carrying *Borrelia recurrentis* (epidemic relapsing fever). Symptoms include substantial fever, shaking chills, headache, and drenching sweats. Treatment with doxycycline or erythromycin speeds recovery.

11.3 Rickettsial and Ehrlichial Arthropodborne Diseases

- Rocky Mountain spotted fever is caused by *Rickettsia rickettsii,* an intracellular parasite. Carried by ticks, the symptoms include a high fever for several days, severe headache, and a maculopapular skin rash. Effective treatment requires doxycycline or chloramphenicol.
- Epidemic typhus is a potentially fatal disease occurring in unsanitary conditions and in overcrowded living conditions. It is caused by *Rickettsia prowazekii* that is carried by body lice. A maculopapular rash progresses to the extremities and an intense fever, hallucinations, and delirium are characteristic symptoms of the disease. In epidemics, mortality can be as high as 75 percent. Effective treatment requires doxycycline or chloramphenicol.
- Endemic typhus is transmitted by fleas. The causative agent, *Rickettsia typhi,* causes a mild fever, headache, and maculopapular rash. Recovery often is spontaneous, although doxycycline or chloramphenicol can be helpful.
- Other diseases caused by rickettsiae include scrub typhus *(Rickettsia tsutsugamushi),* rickettsialpox *(Rickettsia akari),* Brill-Zinsser disease (recurring *Rickettsia prowazekii*), and trench fever *(Bartonella quintana).*
- Ehrlichiosis has been recognized in two forms: human monocytic ehrlichiosis *(Ehrlichia chaffeensis)* and human granulocytic ehrlichiosis *(Ehrlichia phagocytophila).*

LEARNING OBJECTIVES

After understanding the textbook reading, you should be capable of writing a paragraph that includes the appropriate terms and pertinent information to answer the objective.

1. Describe zoonotic bacterial diseases.
2. Summarize the clinical significance of the three forms of anthrax.
3. Identify how *Clostridium tetani* causes tetanus and list the symptoms of the disease.
4. Explain why gas gangrene is referred to as myonecrosis and identify the toxins and enzymes involved with the disease.
5. Distinguish between the more mild and systemic forms of leptospirosis.
6. Describe a vector in relation to infectious disease.
7. Contrast between bubonic, septicemic, and pneumonic plague.
8. Summarize the clinical significance of glandular and inhalation tularemia.
9. Distinguish between the three stages of untreated Lyme disease.
10. Compare and contrast the two forms of relapsing fever.
11. Summarize the characteristics of the rickettsiae.
12. Identify the hallmarks of Rocky Mountain spotted fever.
13. Differentiate between epidemic and endemic typhus.
14. Discuss the characteristics of the two forms of ehrlichiosis.

SELF-TEST

Answer each of the following questions by selecting the ***one*** answer that best fits the question or statement. Answers to even-numbered questions can be found in **Appendix C**.

1. The soilborne bacterial species described in this chapter survive in the soil by producing
A. toxins.
B. endospores.
C. antitoxins.
D. toxoids.
E. pili.

2. Woolsorter disease applies to the _____ form of _____.
A. inhalation; tularemia
B. toxic; myonecrosis
C. intestinal; anthrax
D. inhalation; anthrax
E. ingested; leptospirosis

3. A raised, itchy bump typical of anthrax that resembles an insect bite is called a
A. pustule.
B. welt.
C. papule.
D. lesion.
E. None of the above (**A–D**) is correct.

4. Which one of the following describes the mode of action of tetanospasmin?
A. It inhibits muscle contraction.
B. It damages and lyses red blood cells.
C. It disrupts cell tissues,
D. It inhibits muscle relaxation.
E. It lyses white blood cells.

5. Opisthotonus refers to _____ caused by _____.
A. arching of the back; tetanus
B. clenched teeth; tularemia
C. a rash; Lyme disease
D. a papule; anthrax
E. an inflammation; tetanus

6. A crackling sound associated with myonecrosis is due to
A. respiratory distress due to plague.
B. nerve contractions due to tetanus.
C. a toxin produced by *B. anthracis.*
D. lymph node swelling due to plague.
E. gas produced by *C. perfringens.*

7. Leptospirosis is caused by a _____ bacterial cell.
A. spiral-shaped
B. gram-positive, rod-shaped
C. gram-negative, coccus-shaped
D. rickettsial
E. gram-positive, coccus-shaped

8. *Leptospira interrogans* has all the following characteristics *except*:
A. axial filaments.
B. aerobic metabolism.
C. exotoxin production.
D. a hook at one end of the cell.
E. rapid multiplication.

9. A characteristic sign of systemic leptospirosis is
A. a bull's eye rash.
B. conjunctivitis.
C. gangrene.
D. arthritis.
E. foul odor at the wound site.

10. Endemic plague in the American Southwest is called _____ plague.
A. systemic
B. sylvatic
C. murine
D. pneumonic
E. bubonic

11. A characteristic of cell staining of *Y. pestis* is a
A. gram-negative staining.
B. bipolar staining.
C. gram-positive staining.
D. gram-variable staining.
E. Both **A** and **B** are correct.

12. Hemorrhaging and swelling of lymph nodes in the groin is typical of
A. anthrax.
B. relapsing fever.
C. leptospirosis.
D. bubonic plague.
E. Lyme disease.

13. What disease is referred to as rabbit fever?
A. Relapsing fever
B. Tularemia
C. Endemic typhus
D. Leptospirosis
E. Epidemic typhus

14. Skin ulcers are a common lesion resulting from
A. swimming in water contaminated with *L. interrogans.*
B. being bitten by a tick infected with *B. burgdorferi.*
C. being bitten by fleas infected with *Y. pestis.*
D. handling rabbit meat infected with *F. tularensis.*
E. a puncture wound infected with *C. tetani.*

15. Lyme disease is spread by
A. ticks.
B. fleas.
C. mites.
D. lice.
E. Both **B** and **D** are correct.

16. Erythema migrans is typical of the _____ stage of Lyme disease.
A. asymptomatic
B. early localized
C. early disseminated
D. late
E. recurrent

17. A brief tick bite and a small number of recurring periods of fever and chills is typical of
A. louseborne relapsing fever.
B. ehrlichiosis.
C. Rocky Mountain spotted fever.
D. epidemic typhus.
E. rickettsialpox.

18. Rocky Mountain spotted fever is most common in the
A. southeastern United States.
B. Rocky Mountains.
C. Pacific northwest.
D. New England.
E. upper Midwest.

19. *Rickettsia prowazekii* is the agent causing
A. epidemic typhus.
B. Rocky Mountain spotted fever.
C. scrub typhus.
D. trench fever.
E. relapsing fever.

20. Infection of leukocytes is characteristic of _____ species.
A. *Rickettsia*
B. *Borrelia*
C. *Ehrlichia*
D. *Francisella*
E. *Leptospira*

QUESTIONS FOR THOUGHT AND DISCUSSION

Answers to even-numbered questions can be found in **Appendix C.**

1. Although the tetanus toxin is second in potency to the toxin of botulism, many physicians consider tetanus to be a more serious threat than botulism. Would you agree? Why?
2. Endemic typhus was observed in five members of a Texas household. On investigation, epidemiologists learned that family members had heard rodents in the attic, and two weeks previously they had used rat poison on the premises. Investigators concluded that both the rodents and the rat poison were related to the outbreak. Why?
3. Centuries ago, the habit of shaving one's head and wearing a wig probably originated in part as an attempt to reduce lice infestations in the hair. Why would this practice also reduce the possibilities of certain diseases? Which diseases?
4. Even before October 2001, people from various government and civilian agencies were concerned about a terrorist attack using anthrax spores. In various scenarios, try to paint a picture of how such an attack might happen. Then, using your knowledge of microbiology, present your vision of how agencies might deal with such an attack.
5. At various times, local governments are inclined to curtail deer hunting. How might this lead to an increase in the incidence of Lyme disease?
6. In autumn, it is customary for homeowners in certain communities to pile leaves at the curbside for pickup. How might this practice increase the incidence of tularemia, Lyme disease, and Rocky Mountain spotted fever in the community?

APPLICATIONS

Answers to even-numbered questions can be found in **Appendix C.**

1. In February 1980, a patient was admitted to a Texas hospital complaining of fever, headache, and chills. He also had greatly enlarged lymph nodes in the left armpit. A sample of blood was taken and Gram stained, whereupon gram-positive diplococci were observed. The patient was treated with cefoxitin, a drug for gram-positive organisms, but soon thereafter he died. On autopsy, *Yersinia pestis* was found in his blood and tissues. Why was this organism mistakenly thought to be diplococci, and what error was made in the laboratory? Why were the symptoms of plague missed?
2. Some estimates place epidemic typhus among the all-time killers of humans; one listing even has it in third place behind malaria and plague. In 1997, during a civil war, an outbreak of epidemic typhus occurred in the African country of Burundi. The outbreak was estimated to be the worst since World War II. What conditions may have led to this epidemic?
3. Leptospirosis has been contracted by individuals working in such diverse locales as subway tunnels, gold mines, rice paddies, and sewage-treatment plants. As an epidemiologist, what precautions would you suggest these workers take to protect themselves against the disease?
4. A young woman was hospitalized with excruciating headache, fever, chills, nausea, muscle pains in her back and legs, and a sore throat. Laboratory tests ruled out meningitis, pneumonia, mononucleosis, toxic shock syndrome, and other diseases. On the third day of her hospital stay, a faint pink rash appeared on her arms and ankles. By the next day, the rash had become darker red and began moving from her hands and feet to her arms and legs. Can you guess the eventual diagnosis?
5. In Chapter 9 of the Bible, in the Book of Exodus, the sixth plague of Egypt is described in this way: "Then the Lord said to Moses and Aaron, 'Take a double handful of soot from a furnace, and in the presence of Pharaoh, let Moses scatter it toward the sky. It will then turn into a fine dust over the whole land of Egypt and cause festering boils on man and cattle throughout the land.'" Which disease in this chapter is probably being described?

REVIEW

The bacterial diseases transmitted by soil and arthropods are the main focus of this chapter. To test your understanding of the chapter contents, match the statement on the left with the disease on the right by placing the correct letter in the available space. A letter may be used once, more than once, or not at all. **Appendix C** contains the answers to even-numbered statements.

Statement

____ **1.** Accompanied by erythema migrans.

____ **2.** Transmitted by lice; caused by *Rickettsii prowazekii.*

____ **3.** Affects neutrophils in the body; transmitted by ticks.

____ **4.** Caused by a spore-forming rod that produces hyaluronidase and α-toxin.

____ **5.** Primarily endemic in large herbivores, such as cattle, sheep, and goats.

____ **6.** Treated with antitoxins; caused by an anaerobic spore former.

____ **7.** Caused by *Borrelia burgdorferi;* transmitted by a tick.

____ **8.** Bubonic, septicemic, and pneumonic stages.

____ **9.** Maculopapular rash beginning on extremities and progressing to body trunk.

____ **10.** Caused by a spirochete that infects kidney tissues in pets and humans.

____ **11.** Occurs in small game animals, especially rabbits.

____ **12.** Caused by a gram-negative rod with bipolar staining; transmitted by the rat flea.

____ **13.** Long-range complications include arthritis in large joints.

____ **14.** Up to 13 attacks of substantial fever, joint pains, and skin spots; *Borrelia* involved.

____ **15.** Immunization rendered by the DTaP vaccine.

____ **16.** Pulmonary, intestinal, and skin forms possible; due to a *Bacillus* species.

____ **17.** A typical zoonosis; caused by a spiral bacterium.

____ **18.** Epidemics where sanitation is poor; louseborne rickettsial disease.

____ **19.** Sustained and uncontrolled contractions of the body's muscles.

____ **20.** Most commonly reported tickborne disease in the United States.

Disease

A. Anthrax
B. Ehrlichiosis
C. Endemic typhus
D. Epidemic typhus
E. Gas gangrene
F. Leptospirosis
G. Lyme disease
H. Plague
I. Relapsing fever
J. Rickettsialpox
K. Rocky Mountain spotted fever
L. Tetanus
M. Tularemia

HTTP://MICROBIOLOGY.JBPUB.COM/

The site features learning, an on-line review area that provides quizzes and other tools to help you study for your class. You can also follow useful links for in-depth information, read more MicroFocus stories, or just find out the latest microbiology news.

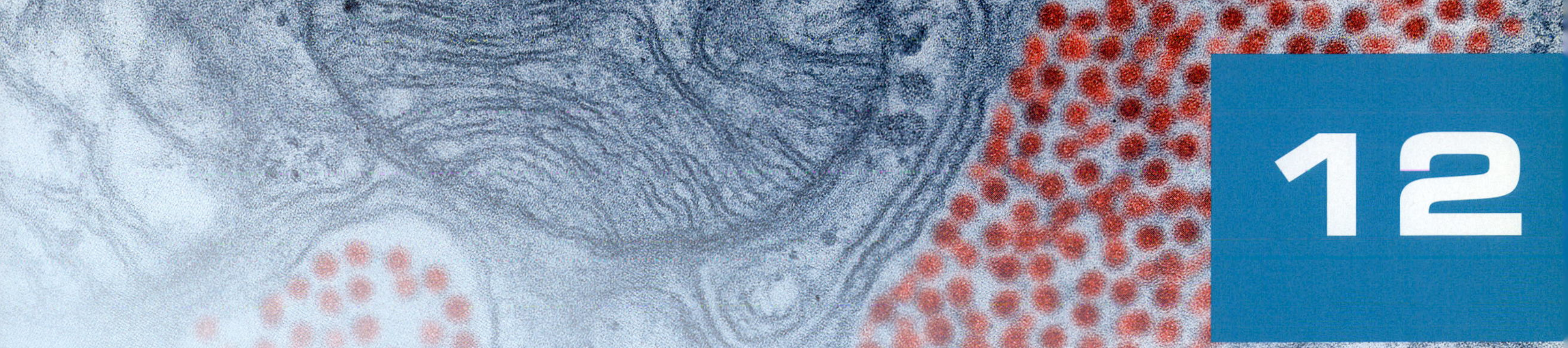

12

Sexually Transmitted, Contact, and Miscellaneous Bacterial Diseases

Chapter Preview and Key Concepts

12.1 Sexually Transmitted Diseases
- *Treponema pallidum* transmission almost always is by sexual contact or from mother to fetus.
- *Neisseria gonorrhoeae* infects the urogenital tract.
- *Chlamydia trachomatis* can damage the female reproductive organs.
- *Haemophilus ducreyi* infections produce a highly contagious but curable STD.
- *Ureaplasma urealyticum* causes a form of nongonococcal urethritis.
- *Chlamydia, Mycoplasma,* and other bacterial genera are responsible for STDs that are not life-threatening.

12.2 Contact Bacterial Diseases
- *Mycobacterium leprae* infects the skin and peripheral nerves.
- *Staphylococcus aureus* infections can create abscesses and/or produce exotoxins.
- *Chlamydia trachomatis* infections can lead to blindness.
- *Haemophilus aegyptius* can cause an inflammation of the conjunctiva.
- *Treponema pertenue* cause nonsexually-transmitted infections that can result in very destructive disease.

12.3 Miscellaneous Bacterial Diseases
- Endogenous bacterial diseases are caused by organisms normally inhabiting the body.
- Bacterial disease can arise from animal bites or scratches.
- Bacterial species in the human oral cavity can trigger tooth decay and gum disease.
- Bacterial disease can arise in any part of the urinary system, usually through fecal contamination.

  MicroInquiry 12: Sexually Transmitted, Contact, and Miscellaneous Bacterial Disease Identification
- Nosocomial infections often represent cross contamination from health care worker to patient.

I just froze. Then I closed the door, and went in my room and cried.
—A soft-spoken, 38-year-old woman recalling her reaction when she was visited by a health official and told she had syphilis

What do Abraham Lincoln, Adolf Hitler, Friedrich Nietzsche, Oscar Wilde, Ludwig van Beethoven, and Vincent van Gogh have in common? Very likely all suffered from syphilis if Deborah Hayden's research is correct—and it more than likely is. In 2003, she wrote a book entitle *Pox: Genius, Madness, and the Mysteries of Syphilis* (New York: Basic Books) in which she looks at 14 eminent figures from the 15th to 20th centuries whose behavior, careers, or personalities were more than likely shaped by this sexually transmitted disease.

Syphilis originally was called the Great Pox to separate it from smallpox and until the introduction of penicillin in 1943, was untreatable. It caused a chronic and relapsing disease that could disseminate itself throughout the body, only to reappear later as so-called tertiary syphilis. In this most dangerous and terminal form, the disease produces excruciating headaches, gastrointestinal pains, and eventually deafness, blindness, paralysis, and insanity.

Yet sometimes ecstasy and fierce creativity were part of the "symptoms." As Deborah Hayden says, "one of the 'warning signs' of tertiary syphilis is the sensation of being serenaded by angels." In fact, writer Karen Blixen (Isak Dinesen) once said that "Syphilis sold her soul to the devil for the ability to tell stories." Deborah Hayden believes it is just such emotions that provided much of the creative spark for many of the notable historical figures she describes in her book. Perhaps the most intriguing is the debated proposal that syphilis may have driven Hitler mad and that he was dying of syphilis when he committed suicide in his Berlin bunker in the final days of World War II.

If Deborah Hayden's arguments are true, it is amazing how a bacterial organism has affected the body and mind in different ways, shaping the thoughts of writers and philosophers, the creative genius of artists, composers, and scientists—and yes, the madness of dictators.

Sexually transmitted diseases (STDs) remain a health problem in the United States and around the world, the seriousness of which is underscored by the number of infections. The Centers for Disease Control and Prevention (CDC) estimate that for each day in 2004, over 52,000 Americans contracted a sexually transmitted disease; almost half among the 15 to 24 year age group. Importantly, more than one STD can be acquired at the same time. In addition, individuals infected with an STD are two to five times more likely to acquire HIV than uninfected individuals if they are exposed to the virus through sexual contact.

The STDs discussed in this chapter make up four of the top 10 infectious diseases in the United States (FIGURE 12.1). However, the increase in STDs is but one example of how changing social patterns can affect the incidence of other bacterial diseases discussed in this chapter. For example, the incidence of leprosy in the United States has risen because in the last decades immigrant groups have brought the disease with them. Toxic shock syndrome was first recognized widely in 1980 when a new brand of high-absorbency tampon appeared on the commercial market. Finally, the mortality rate from nocardiosis rose when this disease was found to complicate cases of acquired immunodeficiency syndrome (AIDS). These and other contact and miscellaneous diseases will be addressed in this chapter.

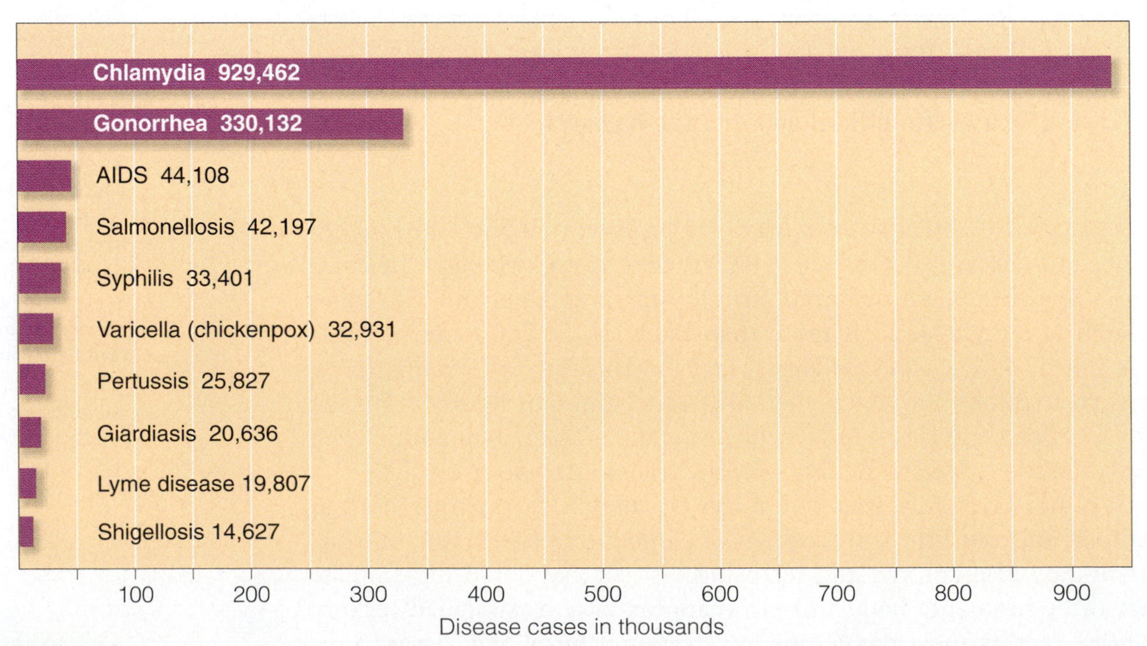

FIGURE 12.1 **Top Ten Reported Cases of Notifiable Infectious Diseases in the United States, 2004.** Four of the top five most reported microbial diseases in the United States are sexually transmitted.

Q: How many of the top ten diseases are of bacterial origin?

12.1 Sexually Transmitted Diseases

The sexually transmitted diseases of bacterial origin belong to a broad category of diseases transmitted by contact. The contact in this case is with the reproductive organs. This type of person-to-person transmission is necessary for bacterial survival because the bacterial cells usually cannot remain alive outside the body tissues. Viral, fungal, and protozoan STDs are discussed in Chapters 15, 16, and 17.

Syphilis Is a Chronic, Infectious Disease

KEY CONCEPT

- *Treponema pallidum* transmission almost always is by sexual contact or from mother to fetus.

Over the centuries, Europeans have had to contend with four pox diseases: chickenpox, cowpox, smallpox, and the Great Pox, a disease now known as syphilis. The first European epidemic was recorded in the late 1400s, shortly after the conquest of Naples by the French army. For decades the disease had various names, but by the 1700s, it had come to be called syphilis (MicroFocus 12.1).

Syphilis is currently ranked among the top five most reported microbial diseases in the United States (see Figure 12.1). Statistics indicate over 33,000 people are afflicted with the disease annually and the number has been rising since 2000. Twelve million cases are reported each year worldwide. Taken alone, these figures suggest the magnitude of the syphilis epidemic, but some public health microbiologists believe for every case reported, as many as nine cases go unreported.

Syphilis is caused by *Treponema pallidum*, (*trepo* = "turn"; *nema* = "thread"; *pallid* = "pale"; thus literally "pale turning thread). This spirochete moves by means of endoflagella. Humans are the only host for *T. pallidum*, so the organism must spread by human-to-human contact, usually during sexual intercourse. It penetrates the skin surface through the mucous membranes or via a wound, abrasion, or hair follicle. The variety of clinical symptoms accompanying the stages, and their similarity to other diseases, have led some physicians to call syphilis the "great imitator." The three stages of chronic, untreated disease are described next.

The incubation period for syphilis varies greatly, but it averages about three weeks. **Primary syphilis** is characterized by a lesion, called a **chancre**, which is a painless circular, purplish ulcer with a small, raised margin with hard edges (FIGURE 12.2A). The chancre develops at the site of entry of the spirochetes, often the genital organs. However, any area of the skin may be affected, including the pharynx, rectum, or lips. The chancre teems with spirochetes. It persists for three to eight weeks, and then it heals spontaneously. However, the infection has not been eliminated, as the spirochetes spread through the blood and lymph to other body organs.

Several weeks later, the untreated patient experiences **secondary syphilis**. Symptoms include fever and a flu-like illness as well as swollen lymph nodes reminiscent of infectious mononucleosis. The skin rash may be mistaken for measles, rubella, or chickenpox (FIGURE 12.2B). It usually involves the palms, soles, face, and scalp. Loss of the eyebrows often occurs, and a patchy loss of hair results in "moth-eaten" areas commonly seen on the head. Involvement of the liver may lead to jaundice and suspicion of hepatitis. In untreated patients, the symptoms last several weeks. Most patients recover, but they bear pitted scars from the healed lesions and remain "pockmarked." In 2004, there were 7,980 cases of primary and secondary syphilis reported to the CDC—the highest level since 1997.

These individuals now enter a chronic, latent stage that can last 3 to 30 years. Many patients will have relapses of secondary syphilis during which time they remain infectious. Within four years, the relapses cease and the disease is no longer infectious (except in pregnant females). Patients either remain asymptomatic or slowly progress to late-stage symptoms.

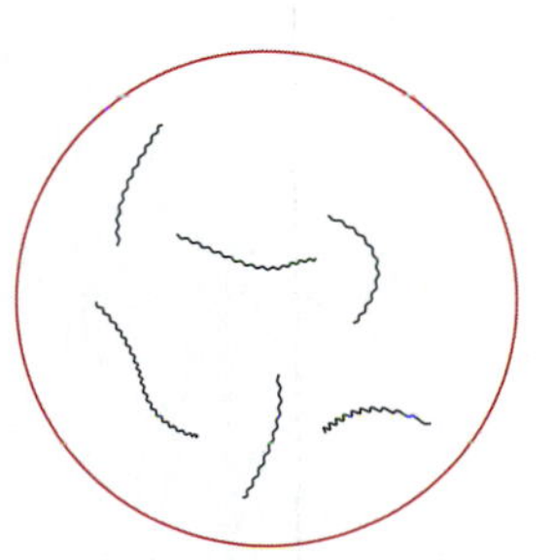

Treponema pallidum

About 40 percent of untreated patients eventually develop **tertiary syphilis**. This stage occurs in many forms, but most commonly it involves the skin, skeletal, or cardiovascular and nervous systems. The hallmark of tertiary syphilis is the **gumma**, a soft, painless, gummy noninfectious granular lesion (FIGURE 12.2C). In the cardiovascular system,

MICROFOCUS 12.1: History

The Origin of the Great Pox

Among the more intriguing questions in medical history are how and why syphilis suddenly emerged in Europe in the late 1400s. Writers of that period tell of an awesome new disease that swept over Europe and on to India, China, and Japan. But where did the disease come from?

One oft-told story is that syphilis existed in the New World, and that members of Columbus's crew acquired it during stopovers in the Caribbean islands. Columbus returned to Palos in northern Italy in 1493, and some of his crew reportedly joined the army of Charles VII of France. In 1494, Charles's army attacked Naples in Italy, but mounting losses from the strange new sickness forced him to withdraw. His army of 30,000 French, German, Swiss, English, Polish, Spanish, and Hungarian troops returned to their native lands, and apparently brought the disease home with them.

A second theory holds that the disease first came to Spain and Portugal with slaves imported from Africa in the mid-1400s. An African disease called yaws is very similar to syphilis in causative organism, transmission, and stages of development. Certain historians believe that some unknown factor caused yaws to flare up in the form of syphilis in the late 1400s. They speculate that the army of Charles VII provided a highly susceptible population of diverse men who spread the disease wherever they traveled. Indeed, recent studies of Native American burial grounds show no traces of syphilis before the arrival of Columbus. By contrast, remains of those dying after Columbus's arrival show signs that the disease was present in the community.

With its devastating effects, syphilis inspired a variety of epithets. The Italians called it the French disease (morbus Gallicus), while the French called it the Italian disease (la maladie Italienne); to the Japanese, it was the Chinese disease. The English impartially termed it the Great Pox. The name "syphilis" derives from the works of Girolamo Fracostoro, a sixteenth-century poet-scientist of Verona (see Chapter 1). In 1530, Fracostoro wrote a long poem about a shepherd named Syphilus who momentarily left his pastoral responsibilities to commit a sexual indiscretion. The angered gods punished him with the horrible sores of the disease. In a later work on disease transmission, Fracostoro suggested that the illness be called syphilis after the mythical shepherd.

Syphilis was as international in effect as in name, and proved to be no respector of rank. Henry VIII of England, Napoleon of France, and Peter the Great of Russia all contracted the disease. Poets such as Keats, musicians such as Beethoven, and artists such as Gauguin also succumbed (see chapter opener), as did millions of common people. For many generations, epidemics of syphilis swept back and forth across the world. As Lord Byron wrote in one of his poems:

The smallpox has gone out of late; Perhaps it may be follow'd by the Great.

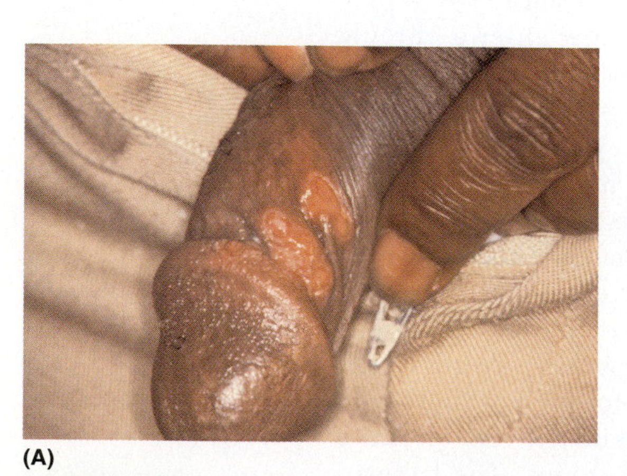
(A)

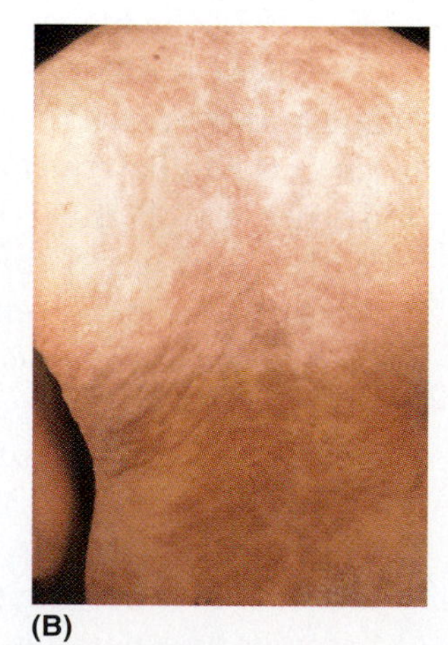
(B)

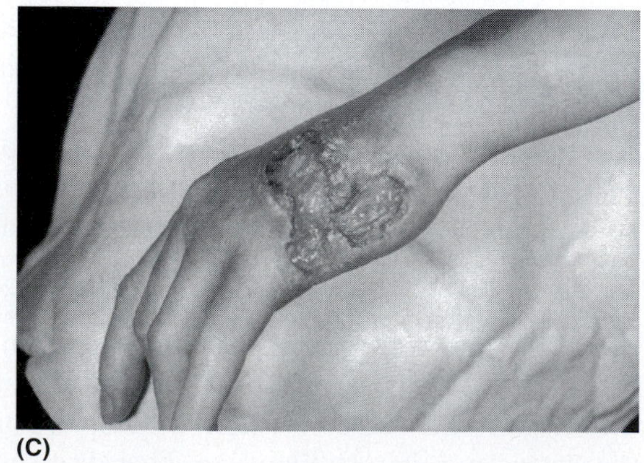
(C)

FIGURE 12.2 **The Stages of Syphilis.** (**A**) The chancre of primary syphilis as it occurs on the penis. The chancre is circular with raised margins and is usually painless. (**B**) A skin rash is characteristic of secondary syphilis. (**C**) The gumma that forms in tertiary syphilis is a granular, diffuse lesion compared with the primary chancre.

Q: From which of these stages would the spirochete be contagious?

gummas weaken the major blood vessels, causing them to bulge and burst. In the spinal cord and meninges, gummas lead to degeneration of the tissues and paralysis. In the brain, they alter the patient's personality and judgment and cause insanity so intense that for many generations, people with tertiary syphilis were confined to mental institutions. Damage can be so serious as to cause death.

Syphilis is a serious problem in pregnant women because the spirochetes penetrate the placental barrier after the third or fourth month of pregnancy, causing **congenital syphilis** in the fetus. Infection can lead to death (stillbirth). Surviving infants develop skin lesions and open sores. Affected children often suffer poor bone formation, meningitis, or **Hutchinson's triad**, a combination of deafness, impaired vision, and notched, peg-shaped teeth. In 2004, there were 353 congenital syphilis cases reported to the CDC, a number that has been declining since 1991.

The cornerstone of syphilis control is the identification and treatment of the sexual contacts of patients. Penicillin is the drug of choice and a single dose usually is curative for primary and secondary syphilis. Unfortunately, *T. pallidum* cannot be cultivated on laboratory media. Diagnosis in the primary stage therefore depends on the observation of spirochetes from the chancre using fluorescence or dark-field microscopy (FIGURE 12.3). As the disease progresses, a number of tests to detect syphilis antibodies becomes useful, including the rapid plasma reagin (RPR) test, the Venereal Disease Research Laboratory (VDRL) test, and others (Chapter 22).

CONCEPT AND REASONING CHECKS

12.1 Describe the three stages of syphilis.

Gonorrhea Can Be an Infection in Any Sexually Active Person

KEY CONCEPT

- *Neisseria gonorrhoeae* infects the urogenital tract.

Gonorrhea is the second most frequently reported notifiable microbial disease in the United States (Figure 12.1). During the 1960s, the incidence of gonorrhea rose dramatically; since 1975, several hundred thousand cases have been reported annually, the highest percentage being in persons under 24 years of age. Epidemiologists suggest that 3 to 4 million cases go undetected or unreported each year. Despite effective antibiotic therapy and historically low rates, gonorrhea remains an epidemic.

The agent of gonorrhea is *Neisseria gonorrhoeae*, a small, unencapsulated, nonmotile, gram-negative diplococcus named for Albert Neisser, who isolated it in 1879 (FIGURE 12.4). The organism, commonly known as the **gonococcus**, has a characteristic double-bean shape.

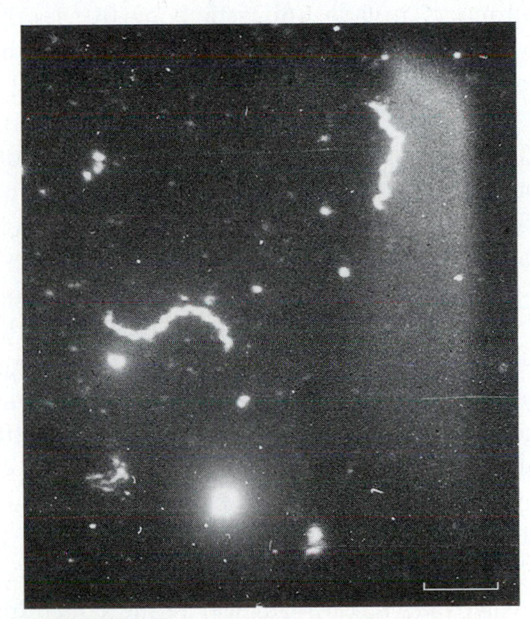

FIGURE 12.3 **The *Treponema pallidum* Spirochete.** A dark-field microscope view of the syphilis spirochete seen in a sample taken from the chancre of a patient. (Bar = 5 µm.)

Q: Why are all samples of T. pallidum *taken from patient samples and not from a culture?*

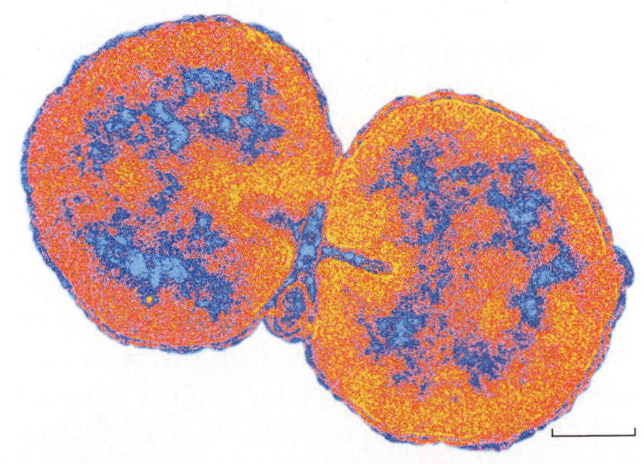

FIGURE 12.4 ***Neisseria gonorrhoeae.*** A false-color transmission electron micrograph of *N. gonorrhoeae* cells. (Bar = 0.25 µm.)

Q: What is the unique feature of these cells?

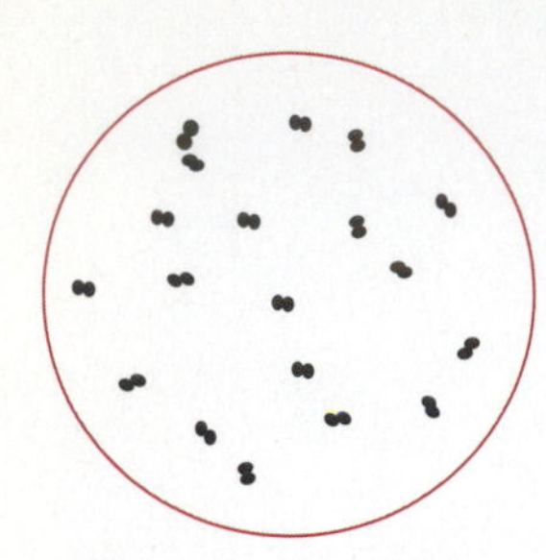

Neisseria gonorrhoeae

N. gonorrhoeae is a very fragile organism susceptible to most antiseptics and disinfectants. Being sensitive to dehydration, it survives only a brief period outside the body and is rarely contracted from a dry surface such as a toilet seat. The great majority of cases of gonorrhea therefore are transmitted during sexual intercourse. (Gonorrhea is sometimes called "the clap," from the French "clappoir" for "brothel.")

The incubation period for gonorrhea ranges from two to six days. In females, the gonococci invade and attach by pili to the epithelial surfaces of the cervix and the urethra. The cervix may be reddened, and a discharge may be expressed by pressure against the pubic area. Patients often report abdominal pain and a burning sensation on urination, and the normal menstrual cycle may be interrupted.

- **Cervix:** The neck of the womb, consisting of a narrow passage leading to the vagina.
- **Urethra:** The tube that passes from the bladder to the exterior that carries urine and, in males, semen during ejaculation.

In some females, gonorrhea also spreads to the fallopian tubes, which extend from the uterus to the ovaries. As these thin passageways become riddled with pouches and adhesions, the passage of egg cells becomes difficult. Complete blockage or inflammation of the fallopian tubes is called **salpingitis**. A condition of the pelvic organs such as this is called **pelvic inflammatory disease (PID)**. Sterility may result from scar tissue remaining after the disease has been treated, or a woman may experience an ectopic pregnancy. It should be noted that symptoms are not universally observed in females, and an estimated 50 percent of affected women exhibit no symptoms. Such asymptomatic women may spread the disease unknowingly.

- **Ectopic pregnancy:** Development of a fertilized egg outside of the womb, often in a fallopian tube.
- **Epididymis:** A coiled tube attached to the back and upper side of the testicle that stores sperm.
- **Keratitis:** An inflammation and swelling of the cornea.

Symptoms tend to be more acute in males than in females, and males thus tend to seek diagnosis and treatment more readily. In males, symptoms of a gonococcal infection are more acute and easier to identify. When gonococci infect the mucus membranes of the urethra, symptoms include a tingling sensation in the penis, followed in a few days by pain when urinating. There is also a thin, watery discharge at first, and later a more obvious yellow, thick fluid resembling semen. Frequent urination and an urge to urinate develop as the disease spreads further into the urethra. The lymph nodes of the groin may swell also, and sharp pain may be felt in the testicles. Unchecked infection of the epididymis may lead to sterility.

Gonorrhea does not restrict itself to the urogenital organs. **Gonococcal pharyngitis**, for example, may develop in the pharynx if bacterial cells are transmitted by oral-genital contact; patients complain of sore throat or difficulty in swallowing. Infection of the rectum, or **gonococcal proctitis**, also is observed, especially in individuals performing anal intercourse. Transmission to the eyes may occur by fingertips or towels, developing into keratitis.

Gonorrhea is particularly dangerous to infants born to infected women. The infant may contract gonococci during passage through the birth canal and develop an inflammation of the eyes called **gonococcal ophthalmia**. To preclude the blindness that may ensue, most states have laws requiring the eyes of newborns be treated with drops of 1 percent silver nitrate or preferably antibiotics such as erythromycin, which will also eliminate a potential chlamydial infection.

Gonorrhea can be treated effectively with doxycycline together with ceftriaxone. However, antibiotic resistance is an important challenge to controlling gonorrhea. Over 16 percent of isolates collected in 2003 were resistant to penicillin, tetracycline, or both due to plasmid- and chromosomally-mediated resistance genes.

Gonorrhea can be detected by observing gram-negative diplococci from cultivated swab samples as well as in white blood cells from the urogenital tract (FIGURE 12.5). For an immunological test, physicians take a swab sample and dip the swab into an antibody solution. An immunoassay reaction takes place (Chapter 22), and within a few hours, a color reaction indicates the presence or absence of gonococci. The test allows doctors to detect gonorrhea early so that treatment can start immediately. A test to detect the DNA of gonococci also is available.

CONCEPT AND REASONING CHECKS

12.2 Contrast the symptoms of gonorrhea in females and males.

Chlamydial Urethritis Can Be Asymptomatic

KEY CONCEPT

- *Chlamydia trachomatis* can damage the female reproductive organs.

Among the chlamydiaceae, three species cause human illness. *Chlamydia psittaci* and *C. pneumoniae* were discussed in Chapter 9. The disease "**chlamydia**" or **chlamydial urethritis** is caused by *C. trachomatis,* an exceptionally small (0.35 μm), round to ovoid-shaped organism with a cell membrane and outer membrane, but without any peptidoglycan (see Chapter 4). It is an obligate "energy" parasite, meaning it is unable to make its own ATP and must rely on the host cell to complete the process.

Chlamydia is the most frequently reported STD and notifiable disease in the United States (Figure 12.1). In 2004, almost 930,000 new cases of chlamydia were reported to the CDC. This was the highest rate since reports began in the mid-1980s and the highest since mandatory reporting in 1995. Seventy-five percent of these cases were in individuals under 25 years of age. Part of the reason for the increase is due to the increased number of screening programs available and the development of better diagnostic tests.

C. trachomatis has a biphasic and unique reproductive cycle (FIGURE 12.6). There is a non-replicating, extracellular, infectious **elementary body** (**EB**) and a replicating, intracellular, noninfectious **reticulate body**. Humans appear to be the only host for the organism.

Chlamydial urethritis represents one of several diseases collectively known as **nongonococcal urethritis**, or **NGU**. NGU is a general term for a condition in which people without gonorrhea have a demonstrable infection of the urethra usually characterized by inflammation, and often accompanied by a discharge. Over 50 percent of cases of NGU are due to chlamydial urethritis, 25 percent to ureaplasmal urethritis (see below), and the remaining 25 percent to unknown causes.

C. trachomatis causes a gonorrhea-like disease transmitted by vaginal, anal, or oral sex. Any sexually-active individual can be infected. The disease has an incubation period of about one to three weeks. The disease often is referred to as the "silent disease" because the organism does not cause extensive tissue injury directly. Thus, some 85 to 90 percent of infected individuals have no symptoms, often do not seek health care, and can unknowingly pass the disease on to others. If symptoms do occur, they are due to the inflammation caused by the immune system's response to limit the spread of the infection.

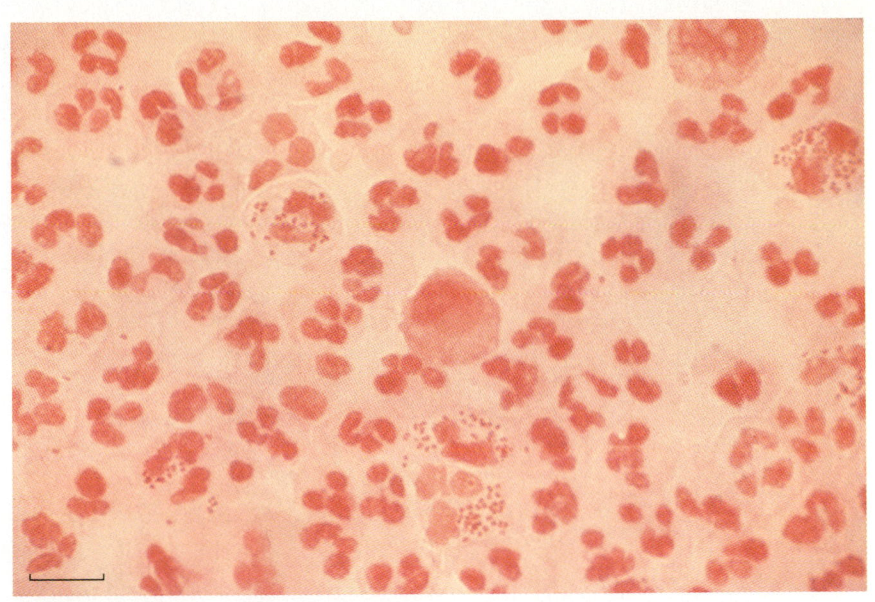

FIGURE 12.5 **Gonococcal Urethral Smear.** A gram-stained smear of discharge from the male urethra showing the gonococci (pink diplococci) in the cytoplasm of white blood cells. (Bar = 10 μm.)

Q: What is the Gram reaction for N. gonorrhoeae?

Females often note a slight vaginal discharge, as well as inflammation of the cervix. Burning pain also is experienced on urination, reflecting infection in the urethra. In complicated cases, the disease may spread to the fallopian tubes, causing salpingitis (FIGURE 12.7). About 40 percent of untreated infections progress to pelvic inflammatory disease (PID), which is more likely a consequence of chlamydia than of gonorrhea. PID from gonorrhea and chlamydia is believed to affect about 50,000 women in the United States annually. Often, however, there are few symptoms of disease before the salpingitis manifests itself, thus adding to the danger of infertility and ectopic pregnancy (Textbook Case 12). Chlamydial urethritis is the number one cause of first trimester pregnancy-related deaths in the United States.

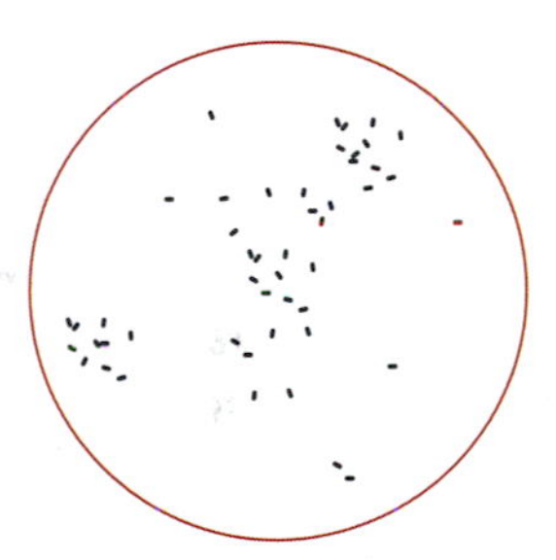

Chlamydia trachomatis

In males, chlamydia is characterized by painful urination and a discharge that is more watery and less copious than in gonorrhea. The discharge often is observed after urinating for the first time in the morning. Tingling sensations in the penis are generally evident. Inflammation of the epididymis may result in sterility, but this complication is uncommon.

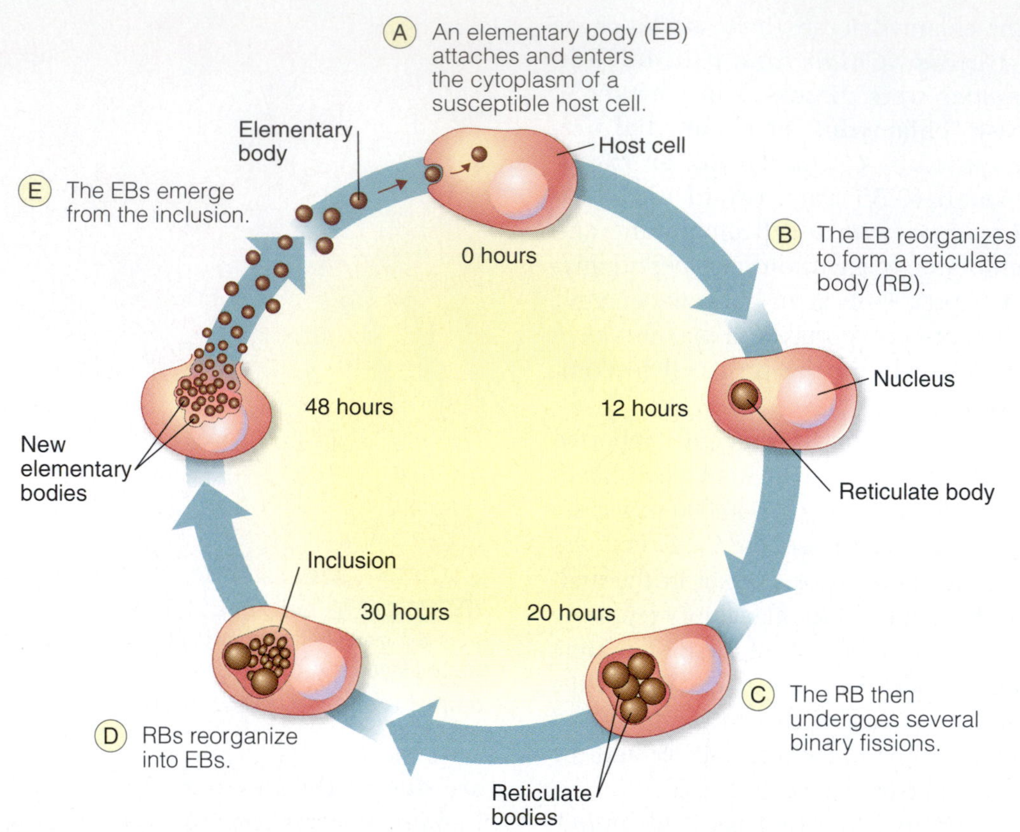

FIGURE 12.6 **The Chlamydial Life Cycle.** The reproductive cycle of the chlamydiae involves two types of cells. After infection, nonreplicating elementary bodies (EB) reorganize into reticulate bodies (RB), which divide to form additional RBs. Within 30 hours, the RBs begin to reorganized into EBs within an inclusion. The EBs are released from the inclusion body.

Q: Because C. trachomatis *only reproduces inside cells, what cellular products might the pathogen require from its host to survive and grow?*

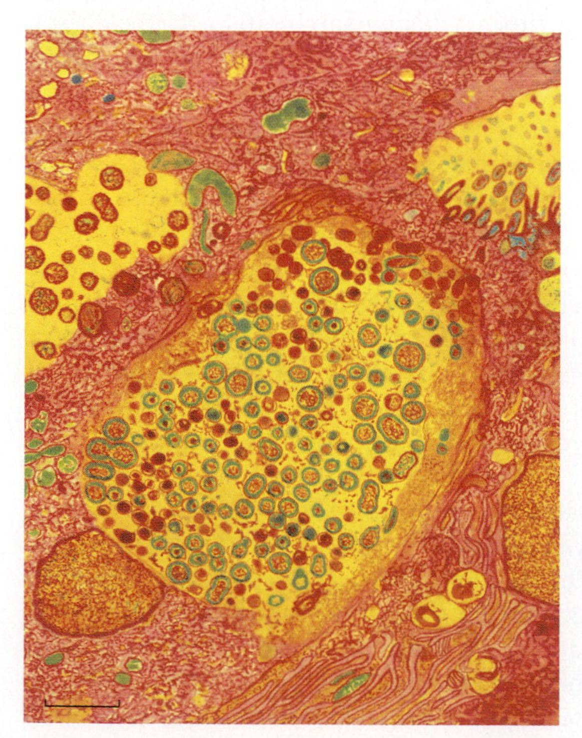

FIGURE 12.7 **Chlamydial Inclusion Body.** False-color transmission electron micrograph of cells sectioned from a woman's fallopian tube infected with *Chlamydia trachomatis*. Spherical cells (green/brown) are elementary bodies and reticulate bodies seen inside an inclusion body (yellow). Cell cytoplasm is pink; a portion of a cell nucleus (brown) is seen at the lower left. (Bar = 2 μm.)

Q: What complications can arise from an infection of the fallopian tubes?

Textbook CASE 12

A Case of Chlamydia Leading to Sterility

1 An educated, professional woman met a gentleman at a friend's house one evening. She was director of a New York law firm. The man was equally successful in his professional career.

2 The couple hit it off immediately. There were many evenings of quiet candlelit dinners, and soon, they became sexually intimate. Neither one used condoms or other means of protection.

3 Three days after having intercourse, the woman began experiencing fever, vomiting, and severe abdominal pains. She immediately made an appointment to see her doctor.

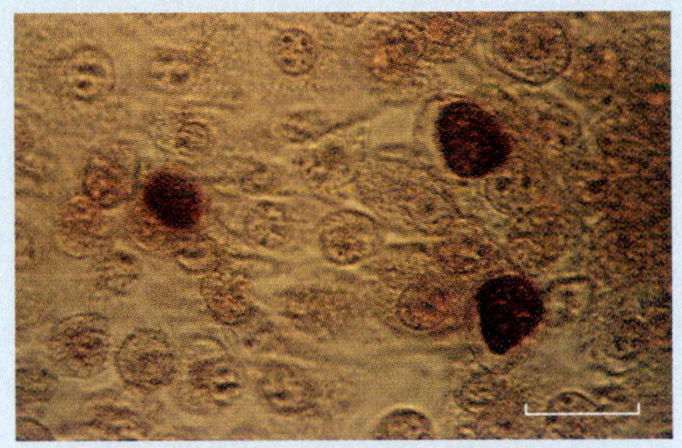

Dark inclusion bodies typical of a chlamydial infection. (Bar = 2 μm.)

4 Assuming the illness was an intestinal upset, the physician prescribed appropriate medication. The woman was actually suffering from chlamydia, but the fact that she was sexually active did not come up during the examination.

5 Feeling better, the woman continued her normal routine. But six months later, with no apparent warning, she collapsed on a New York City sidewalk.

6 The woman was rushed to a local hospital, where doctors diagnosed a chlamydial infection by observing the dark inclusion bodies typical of chlamydia (see figure). They performed emergency surgery: her uterus and fallopian tubes were badly scarred. She recovered; however, the scarring left her unable to have children.

Questions

A. *Would condom use block the transfer of C. trachomatis cells?*

B. *Does fever, vomiting, and abdominal pains indicate a chlamydial infection?*

C. *What "complications" were developing during the six months when the woman was in improved health?*

D. *What are inclusion bodies?*

E. *How would scarring lead to infertility?*

Chlamydial pharyngitis or proctitis is possible as a result of oral or anal intercourse. Cell research suggests the urogenital infections in males may influence male fertility and reduce sperm quality. If this proves true in actual individuals, then *C. trachomatis* infections in males could lead to infertility.

Newborns may contract *C. trachomatis* during delivery from an infected mother and develop a disease of the eyes known as **chlamydial ophthalmia**. The silver nitrate used to prevent gonococcal ophthalmia is not effective as a preventative, so erythromycin therapy is required. Studies in the 1980s also suggested chlamydial pneumonia may develop in newborns from an exposure to *C. trachomatis* during birth. Health officials estimate that each year, over 75,000 newborns suffer chlamydial ophthalmia and 30,000 newborns experience chlamydial pneumonia.

Chlamydial infections may be treated successfully (95 percent cure rate) with one dose of azithromycin or 7 days with doxycycline. If a woman is pregnant, erythromycin is substituted because doxycycline affects bone formation in newborns.

Timely treatment of a patient's sexual partners is critical to prevent reinfection. Sexual partners are evaluated, tested, and treated if they have had contact with the patient in the previous 60 days.

Two relatively fast and simple laboratory tests have been available to detect *C. trachomatis*. In the first test, a physician takes a swab sample from the penis or the cervix (as in a Pap smear). The sample is used for a fluorescent antibody test using monoclonal antibodies (Chapter 22). Within 30 minutes the results are available. An immunoassay test, also performed with a swab sample, can be completed in the doctor's office and is similar to the test for gonorrhea. A test to detect the DNA of *C. trachomatis* using polymerase chain reaction (PCR) technology also is available. It uses a urine sample and is said to detect as few as five cells in a sample.

■ Monoclonal antibodies: Antibodies experimentally produced against a single type of cell or substance.

In 1998, the genome of *C. trachomatis* was sequenced. It is a streamlined genome with only 894 protein-coding genes. One interesting discovery was a set of genes for synthesizing peptidoglycan—even though the bacterium apparently has no peptidoglycan in its cell wall. The organism also has about 20 genes it apparently "borrowed" from a host cell sometime during its evolution millennia ago.

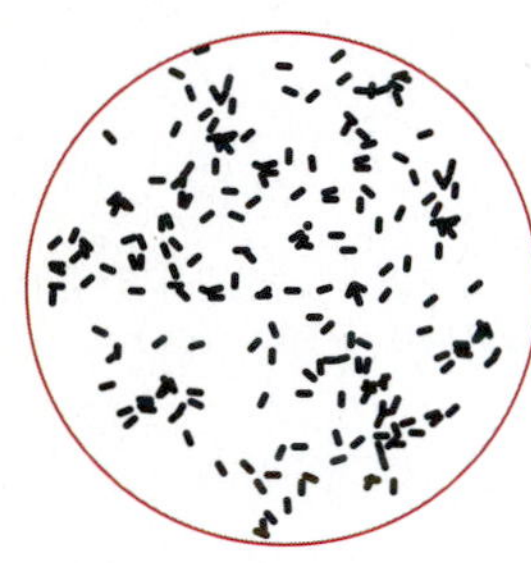

Haemophilus ducreyi

■ Erythema: A narrow zone of redness.

The sequences of both *C. trachomatis* and *C. pneumoniae* have been determined with the hope that comparative genomics (see Chapter 8) will significantly enhance the understanding of both pathogens. Identification of genes particular to one or the other species could indicate mutually exclusive biological, virulence, and pathogenesis capabilities, while genes the two have in common will help researchers better understand the metabolic capabilities necessary for living in a human host.

At least 186 genes have been identified in the *C. pneumoniae* genome that are not homologous to sequences in the *C. trachomatis* genome, and 70 genes on the *C. trachomatis* genome that are unrepresented on the *C. pneumoniae* genome. Also, these pathogenic chlamydiae have been compared to an endosymbiotic species living in amoebae. The symbiont also is missing many of the genes needed for essential metabolic pathways, suggesting a long evolutionary heritage of intracellular parasitism.

CONCEPT AND REASONING CHECKS

12.3 Why is chlamydia referred to as the "silent disease?"

Chancroid Causes Painful Genital Ulcers

KEY CONCEPT

- *Haemophilus ducreyi* infections produce a highly contagious but curable STD.

Chancroid is an STD believed to be more prevalent worldwide than gonorrhea or syphilis. The disease is endemic in many developing nations, and it is common in tropical climates and where public health standards are low. The disease has been decreasing in the United States since 1987 and there were 30 cases reported in 2004. However, it is difficult to culture and therefore the disease could be substantially underdiagnosed.

The causative agent of chancroid is *Haemophilus ducreyi*, a small gram-negative rod named for Augusto Ducrey, who observed it in skin lesions in 1889. After a three- to five-day incubation period, a tender papule surrounded by erythema forms at the entry site. The papule quickly becomes pus filled and then breaks down, leaving a shallow, saucer-shaped ulcer that bleeds easily and is painful in men. The ulcer has ragged edges and soft borders, a characteristic distinguishing it from the primary lesion of syphilis. For this reason, the disease is often called **soft chancre**.

The lesions in chancroid most often occur on the penis in males and the labia or clitoris in females. Substantial swelling in the lymph nodes of the groin may be observed. However, the disease generally goes no further; in fact, in women, it often goes unnoticed. The clinical picture in chancroid makes the disease recognizable, but definitive diagnosis depends on isolating *H. ducreyi* from the lesions.

The transmission of chancroid depends on contact with the lesion, although sexual contact is not required. Contact with open ulcers or their fluid can spread the disease. Azithromycin, erythromycin, or ceftriaxone drugs are useful for therapy, but the disease often disappears in 10 to 14 days without treatment. However, treatment is essential because the open lesion increases the risk of HIV transmission.

CONCEPT AND REASONING CHECKS

12.4 How could a *H. ducreyi* infection make one more at risk of an HIV infection?

Ureaplasmal Urethritis Produces Mild Symptoms

KEY CONCEPT

- *Ureaplasma urealyticum* causes a form of nongonococcal urethritis.

Ureaplasmal urethritis is another type of NGU. It is caused by *Ureaplasma urealyticum*, a mycoplasma (see Chapter 3), so named because of its ability to digest urea in culture media. The organism often is referred to as a **T-mycoplasma** because "tiny" colonies of the organisms develop on laboratory media. At about 0.15 µm in size, *U. urealyticum* is one of the smallest known bacterial cells to cause human disease. Transmission is generally by sexual contact.

The symptoms of ureaplasmal urethritis are similar to those of gonorrhea and chlamydial urethritis. A distinction can be made between the diseases because in ureaplasmal urethritis, the discharge is variable in quantity, and the urethral pain is usually aggravated during urination. Symptoms are often very mild. Penicillin cannot be used to treat ureaplasmal urethritis because *U. urealyticum* has no cell wall. Tetracycline is currently the drug of choice. Diagnosis often depends on eliminating gonorrhea or other types of NGU as possibilities, and for this reason, cases are not often recognized.

Infertility is one consequence of ureaplasmal urethritis because, like chlamydia. low sperm counts and poor movement of sperm cells have been observed in males. Salpingitis in females also has been described. Moreover, *U. urealyticum* is capable of colonizing the placenta during pregnancy, and reports have linked it to spontaneous abortions and premature births. As previously noted, 25 percent of NGU cases may represent ureaplasmal infections. NGU also can be caused by *Mycoplasma genitalium*.

CONCEPT AND REASONING CHECKS

12.5 Why is ureaplasmal urethritis considered an NGU?

Other Sexually Transmitted Diseases Also Exist

KEY CONCEPT

- *Chlamydia*, *Mycoplasma*, and other bacterial genera are responsible for STDs that are not life-threatening.

A few other STDs merit brief attention in this chapter.

Lymphogranuloma venereum (LGV) is a systemic STD caused by a different serotype of *C. trachomatis*. LGV is more common in males than females, and is accompanied by fever, malaise, and swelling and tenderness in the lymph nodes of the groin. Females may experience infection of the rectum (proctitis), if the chlamydiae pass from the genital opening to the nearby intestinal opening. LGV is prevalent in Southeast Asia and Central and South America. Sexually active individuals returning from these areas may show symptoms of the disease, but treatment with doxycycline leads to rapid resolution. It rarely occurs in the United States and other developed nations. However, in 2004 and 2005 outbreaks of LGV among men who have sex with men in Canada, the Netherlands, and other European countries were reported.

Granuloma inguinale is a rare STD in Europe and North America (typically in the Southeast), but it remains an endemic problem in tropical and subtropical areas of the world, such as Caribbean countries and Africa. It is caused by *Calymmatobacterium granulomatis*, a small, gram-negative, encapsulated bacillus. The disease begins with a primary lesion starting as a nodule and progressing to a granular ulcer that bleeds easily. In most cases, this ulcer forms in the external genital organs but it may spread to other regions by contaminated fingers. The lymph nodes in the groin may swell, but fever and other body symptoms are usually absent, a factor distinguishing the disease from LGV. Tissue samples reveal masses of bacterial cells called **Donovan bodies** within white blood cells in the lesion. Anal intercourse, rather than vaginal intercourse, is the most frequent source of infection. About 50% of infected men and women have lesions in the anal area. The disease responds well to doxycycline.

Vaginitis is a general term for various mild infections of the vagina and sometimes the vulva. **Bacterial vaginosis**, the most common form of vaginitis, occurs when healthy vaginal organisms are replaced by harmful ones. One is *Gardnerella vaginalis*, formerly called *Haemophilus vaginalis*. This small, gram-negative rod usually lives uneventfully in the vagina, but flare-ups of disease may take place and transmission by sexual contact may occur during these times. A foul-smelling discharge is the most prominent symptom, and clindamycin or metronidazole therapy generally provides relief.

Mycoplasmal urethritis is a disease similar to ureaplasmal urethritis. The organism, *Mycoplasma hominis*, causes an opportunistic infection by colonizing the placenta and causing spontaneous abortion or premature birth. *M. hominis* is distinguished from *Ureaplasma* by its inability to digest urea, its preference for anaerobic conditions during laboratory cultivation, and its larger colony growth on agar (FIGURE 12.8). Tetracycline is prescribed for active cases of mycoplasmal urethritis.

TABLE 12.1 summarizes the sexually transmitted bacterial diseases.

CONCEPT AND REASONING CHECKS

12.6 Propose an explanation as to why all the STD-causing walled bacteria are gram-negative.

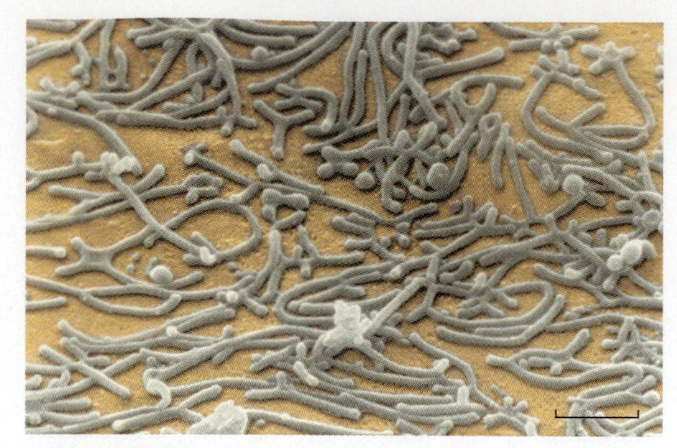

FIGURE 12.8 **A Species of *Mycoplasma*.** False-color scanning electron micrograph of a species of *Mycoplasma*. *M. hominis* causes mycoplasmal urethritis, an STD that is similar to chlamydia. (Bar = 2 μm.)

Q: What cell structure is missing from Mycoplasma *cells that contributes to their pleomorphic shape?*

12.2 Contact Bacterial Diseases

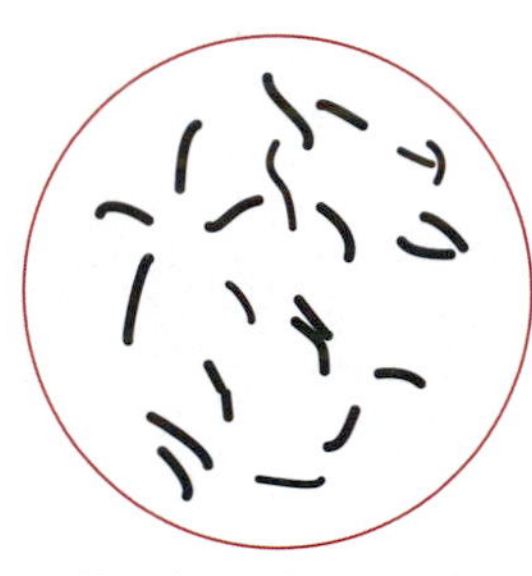

Mycobacterium leprae

Numerous bacterial diseases are transmitted by contact other than sexual. Usually, some form of skin contact is required, as these diseases illustrate.

Leprosy (Hansen Disease) Is a Chronic, Systemic Infection

KEY CONCEPT

- *Mycobacterium leprae* infects the skin and peripheral nerves.

For most of the past 2600 years, **leprosy** has been considered a curse of the damned. It did not kill, but neither did it seem to end. Instead, it lingered for years, causing the tissues to degenerate and deforming the body. In biblical times, the afflicted were ostracized from the community, though what was called leprosy in the Old Testament often was not that specific disease. Among the more heroic stories of medicine is the work of Father Damien de Veuster, the Belgian priest who in 1870 established a hospital for leprosy patients on Molokai, Hawaii. An equally heroic story was written more recently (MicroFocus 12.2).

The agent of leprosy is *Mycobacterium leprae*, an acid-fast rod related to *M. tuberculosis* (see Chapter 9). *M. leprae* was observed for the first time in 1874 by the Norwegian physician Gerhard Hansen. It is referred to as Hansen bacillus, and leprosy is commonly called **Hansen disease**. *M. leprae* is an obligate intracellular parasite, so it cannot be cultivated in artificial laboratory media. In 1960, researchers at the CDC succeeded in cultivating the bacillus in the footpads of mice, and in 1969, scientists discovered it would grow in the skin of nine-banded armadillos.

Leprosy is hard to transmit because about 95 percent of the world's population has a natural immunity to the disease. It is spread by nasal secretions, which are taken up through the upper respiratory tract. The disease has an unusually long incubation period of three to six years, a factor making diagnosis very difficult. Because the organisms are heat sensitive, the symptoms occur in the skin and peripheral nerves in the cooler parts of the body, such as the hands, feet, face, and earlobes. Severe cases also involve the eyes and the respiratory tract. Susceptibility is highest in childhood and decreases with age. More males appear to be infected than females.

Patients with leprosy experience disfiguring of the skin and bones, twisting of the limbs, and curling of the fingers to form the characteristic claw hand. The largest number of deformities develops from the loss of pain sensation

TABLE

12.1 A Summary of Sexually Transmitted Bacterial Diseases

Disease	Causative Agent	Description of Agent	Organs Affected	Characteristic Signs	Treatment Administered	Comment
Syphilis	*Treponema pallidum*	Spirochete	Skin Cardiovascular organs Nervous system	Chancre Skin lesions Gumma	Penicillin	Primary, secondary, and tertiary stages Congenital transmission
Gonorrhea	*Neisseria gonorrhoeae*	Gram-negative diplococcus	Urethra, cervix Fallopian tubes Epididymis Eyes, pharynx	Pain on urination Discharge Salpingitis	Tetracycline Ceftriaxone	The second most reported infectious disease Complicated by pelvic inflammatory disease Eye infection in newborns
Chlamydia (chlamydial urethritis)	*Chlamydia trachomatis*	Chlamydia	Urethra, cervix Fallopian tubes Epididymis Eyes, pharynx	Pain on urination Watery discharge Salpingitis	Doxycycline Erythromycin	The most reported infectious disease Leads to infertility Antibody test for diagnosis
Ureaplasma urethritis	*Ureaplasma urealyticum*	Mycoplasma	Urethra Fallopian tubes Epididymis	Pain on urination Variable discharge Salpingitis	Tetracycline	Leads to infertility Linked to spontaneous abortion
Chancroid (soft chancre)	*Haemophilus ducreyi*	Gram-negative rod	External genital organs Inguinal lymph nodes	Soft chancre Erythema Swollen inguinal lymph nodes	Azithromycin Erythromycin Ceftriaxone	Few complications Many immigrant cases
Lymphogranuloma Venereum	*Chlamydia trachomatis*	Chlamydia	Lymph nodes of groin Rectum	Swollen lymph nodes Proctitis	Doxycycline	Prevalent in Central and South America
Granuloma inguinale	*Calymmatobacterium granulomatis*	Gram-negative rod	External genital organs	Bleeding ulcer Swollen inguinal lymph nodes	Tetracycline	Donovan bodies seen in lesions
Vaginitis	*Gardnerella vaginalis*	Gram-negative rod	Vagina	Foul-smelling discharge	Tetracycline	Organism commonly found in the vagina
Mycoplasmal urethritis	*Mycoplasma hominis* *M. genitalium*	Mycoplasma	Urethra Fallopian tubes Epididymis	Pain on urination Variable discharge Salpingitis	Tetracycline	Similar to ureaplasmal urethritis

MICROFOCUS 12.2: History

The "Star" of Carville

On December 1, 1894, seven leprosy patients arrived at an old plantation on a crook in the Mississippi River. Soon thereafter, four nuns of the Order of St. Vincent de Paul joined them. Together this small band formed the nucleus of what was to become the National Hansen's Disease Center at Carville, Louisiana.

Change came slowly. In 1921, the United States Public Health Service acquired the institution, but it remained essentially a prison, patrolled by guards and surrounded by a chain link fence with barbed wire. Then, in 1931, a leprosy patient named Stanley Stein arrived. Stanley Stein was not his real name—he had forsaken that for fear of bringing shame to his family. Soon, Stein instituted a weekly paper to bring a sense of community to the patients. Originally named *The Sixty-Six Star* (Carville was United States Marine Hospital Number 66), the name was eventually shortened to *The Star*.

As the circulation of *The Star* increased, Stein and others launched a campaign for change. In 1936, the patients acquired a telephone so they could hear the voices of their families. Three years later, the swamps were drained to reduce the incidence of malaria. Soon there came a better infirmary, a new recreation hall, and removal of the barbed wire. In 1946, the State of Louisiana allowed the patients to vote in local and national elections.

Stanley Stein stands next to the printing press as copies of *The Star* are printed.

Through all these years, Stein's leprosy worsened. Originally he had tuberculoid leprosy, the form in which the nerves are damaged. Afterward, however, he developed lepromatous leprosy, which causes lesions to form on the face, ears, and eyes. Soon he was totally blind. Without feeling in his fingers, he could not even learn Braille.

But Stein was not finished. He and his newspaper tirelessly fought for a new post office and weekend passes for patients. In 1961, President Kennedy paid tribute to *The Star* on its thirtieth anniversary and singled out its indomitable editor for praise. Stanley Stein died in 1968. By that time, *The Star* had a circulation of 80,500 in all 50 states and 118 foreign countries.

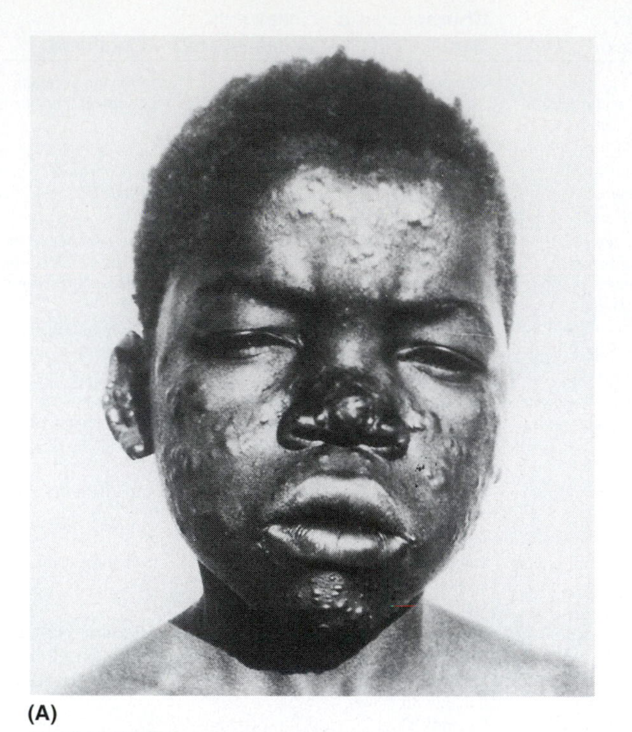

(A)

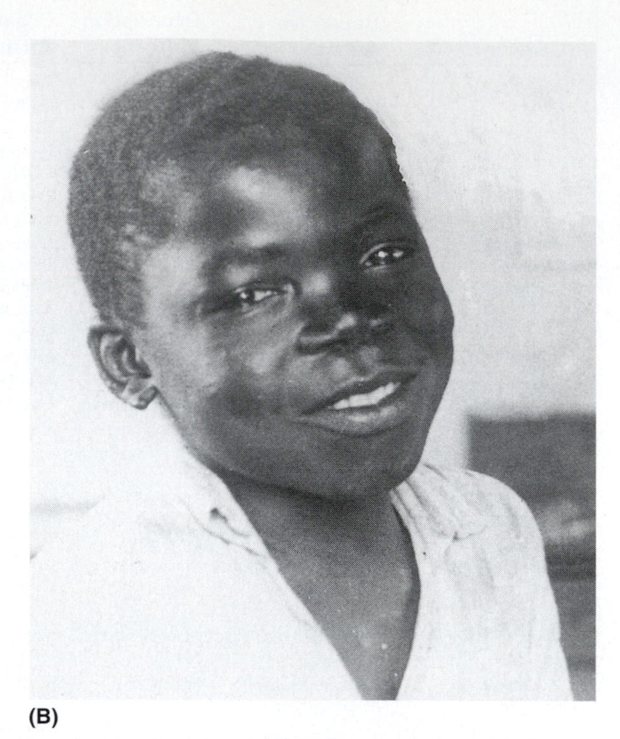

(B)

FIGURE 12.9 **Treating Leprosy.** The young boy with lepromatous leprosy is pictured (**A**) before treatment with dapsone and (**B**) some months later, after treatment. Note that the lesions of the ear and face and the swellings of the lips and nose have largely disappeared.

Q: Why is the disfiguring caused by Mycobacterium leprae *limited to the body extremities?*

due to nerve damage caused by lower numbers of bacilli. This form is called **paucibacillary** (pauci = "few") or **tuberculoid leprosy**. Inattentive patients, for example, might pick up a pot of boiling water without flinching.

Disease progression can lead to the loss of facial features accompanied by a thickening of the outer ear and collapse of the nose (FIGURE 12.9A). Many tumor-like growths called **lepromas** form on the skin and in the respiratory tract. This form is referred to as **multibacillary** or **lepromatous leprosy**. This is the most serious form of the disease because the immune system fails to react, meaning there can be millions of bacilli in the body.

For many years, the principal drug for the treatment and cure of leprosy was a sulfur compound known commercially as **dapsone**. In many cases, such as the one shown in FIGURE 12.9B, the results were dramatic. Today, treatment involves multidrug therapy with dapsone, rifampin, and clofazimine. In the United States, there are approximately 6,500 patients currently being treated (MicroFocus 12.3). In 2004, 105 new cases were reported to the CDC.

In 1992, the WHO began a campaign to "eliminate" leprosy by 2005. However, elimination meant "controlling" the number of registered cases to 1 registered case in every 10,000 people in a population. In 2005, all but nine countries in Asia, Africa, and South America have met this goal. Some attempts have been made to immunize populations with BCG, the vaccine used for tuberculosis (see Chapter 9).

The genomes of both *M. leprae* and *M. tuberculosis* have been sequenced and comparative genomics has revealed some interesting findings. The most interesting discovery so far is that more than half of the genes in *M. tuberculosis* are missing from *M. leprae*. This suggests at some time in the distant past *M. leprae* lost a substantial number of essential genes for metabolism, requiring the organism to depend on infection of host cells to provide its necessary metabolic and growth needs.

CONCEPT AND REASONING CHECKS

12.7 Distinguish between tuberculoid and lepromatous leprosy.

MICROFOCUS 12.3: History and Public Health
Migrations, Armadillos, and Graveyards

The WHO estimated there were almost 300,000 new cases of leprosy in 2005. Fortunately, this number is far below historical numbers where infections seemed to follow people wherever they migrated.

Microbiologist Stewart Cole of the Pasteur Institute in Paris and his international team of colleagues have attempted to pinpoint the source where leprosy originated. Sampling some 175 worldwide strains of *Mycobacterium leprae,* the team was able to catalog the isolates into four genetic strains based on rare DNA differences between strains.

A nine-banded armadillo.

Comparing these strains, Cole's team announced in 2005 that leprosy originated in either East Africa or Central Asia because the strains found here represent the rarest (and probably oldest). Human migrations from here then spread the pathogen.

Each year about 300 people come to the National Hansen's Disease Programs (NHDP) in Baton Rouge, Louisiana, for two to six weeks of initial diagnosis and treatment, followed by subsequent care at any of 14 specially designated outpatient clinics around the country.

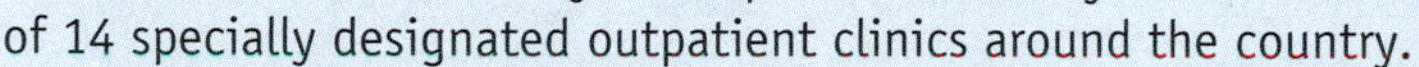

In the United States, more than 200 new leprosy cases are reported to the NHDP registry annually; about 160 are new cases diagnosed for the first time in individuals returning from abroad. Another 6 to 8 cases in Louisiana and about 30 cases in Texas are reported in residents living along the Gulf Coast. How these individuals contracted leprosy remains unknown, as they report they had not had contact with infected individuals.

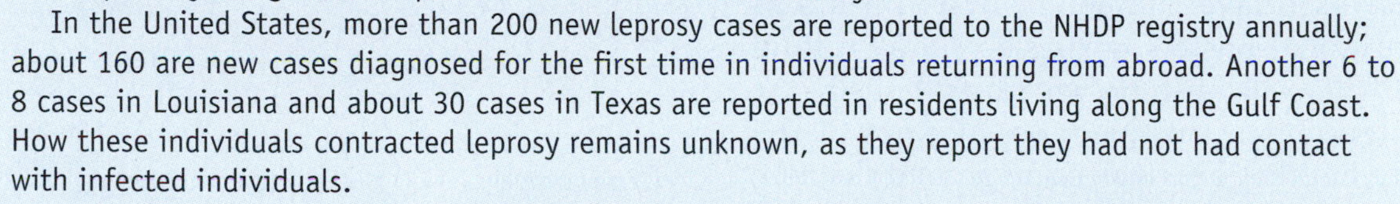

A clue though has been discovered. The Gulf Coast cases of leprosy involve a strain of *M. leprae* closely related to the strain normally found in nine-banded armadillos (see figure) in the area—and nowhere else in the world. Reports suggest up to 30 percent of the armadillos are infected. The nine-banded armadillo has a habit of digging and burrowing, including in graveyards where past generations may have come across fresh leprosy corpses.

So, perhaps that is how the armadillos became infected. But how about the Gulf Coast residents? Could there be an interspecies transfer occurring whereby some form of human contact with infected armadillos has transferred *M. leprae?*

Staphylococcal Contact Diseases Have Several Manifestations

KEY CONCEPT

- *Staphylococcus aureus* infections can create abscesses and/or produce exotoxins.

Staphylococci are normal inhabitants of the human skin, mouth, nose, and throat. Although they generally live in these areas without causing harm, they can initiate disease when they penetrate the skin barrier or the mucous membranes. Penetration is assisted by open wounds, damaged hair follicles, ear piercing, dental extractions, or irritation of the skin by scratching. *Staphylococcus aureus,* the grape-like cluster of gram-positive cocci, is the species usually involved in these contact diseases.

Localized skin infections. The hallmark of staphylococcal skin disease is the production of puss-filled pockets in the skin. **Folliculitis** is an infection at the base of a hair follicle (FIGURE 12.10A). Another infection involves the formation of an **abscess**, a circumscribed pus-filled lesion. A **furuncle** (boil) is a warm, painful abscess below the skin surface. Rupturing of the abscess can lead to infection of the surrounding tissue. Treatment requires a physician to open and drain the pus, and then rinse the abscess with sterile saline solution to

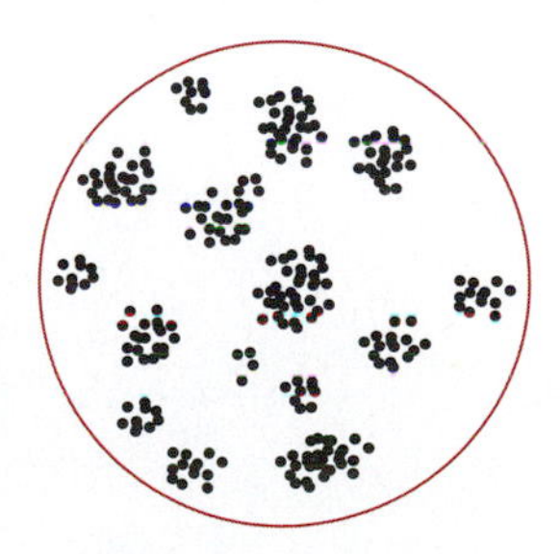

Staphylococcus aureus

remove any remaining bacterial cells. The disease needs to be treated with caution because staphylococci commonly invade the blood and penetrate to other organs. For example, septicemia may develop, as well as staphylococcal pneumonia, endocarditis, meningitis, or nephritis. A trivial skin boil is often the source.

Carbuncles are a group of connected, deeper abscesses (FIGURE 12.10B). Skin contact with other people spreads the disease. Infected individuals often are tired and may have a fever. Bacteremia can occur, requiring antibiotic treatment and debridement.

Debridement: The removal of dead, damaged, or infected tissue.

Food handlers should be aware that staphylococci from furuncles and carbuncles can be transmitted to food, where they can cause food poisoning (see Chapter 10).

A more widespread and highly contagious staphylococcal skin disease is **impetigo**. Here the infection, which is more common in children, is more superficial and involves localized patches of epidermis just below the outer skin layer. Impetigo first appears as thin-walled blisters oozing a yellowish fluid and form yellowish-brown crusts. Usually the blisters occur on the exposed parts of the body, but they also may occur around the nose and upper lip after a child has had a cold with a runny nose, because the constant irritation provides a mechanism for penetration by the staphylococci. *Streptococcus pyogenes* (see Chapter 9) also may cause impetigo.

Staphylococcal skin diseases and blood-borne infections are commonly treated with penicillin, but resistant strains of *S. aureus* are well known, and physicians may need to test a series of alternatives before an effective antibiotic is selected. This problem was brought to public awareness during the 1990s, when reports of antibiotic resistance in staphylococci became widespread. Resistances to numerous antibiotics were reported, and **multidrug-resistant *Staphylococcus aureus*** (**MRSA**) soon appeared in many hospitals. It has been estimated up to 100,000 people are hospitalized each year with MRSA infections. The last-resort antibiotic against MRSA is vancomycin, a very expensive and somewhat toxic drug. However, by 1997, **vancomycin-resistant *Staphylococcus aureus*** (**VRSA**) was detected in clinical settings in

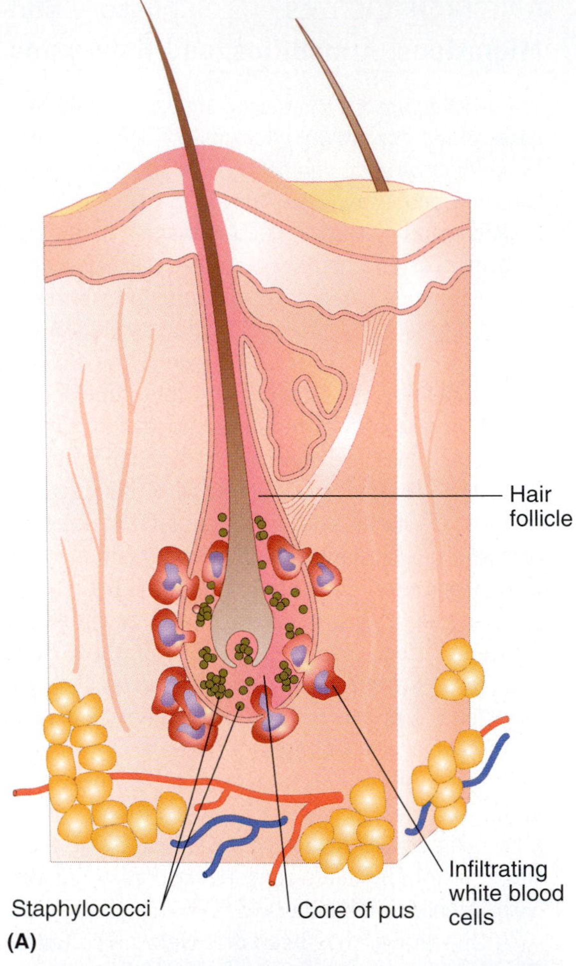

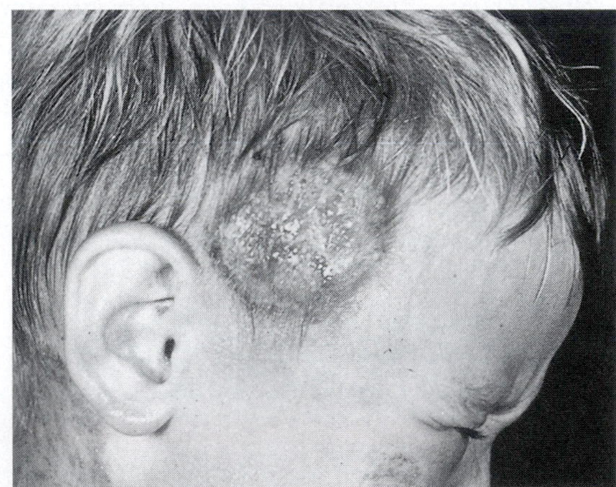

FIGURE 12.10 Staphylococci and Skin Abscesses.
(**A**) Staphylococci at the base of a hair follicle during the development of a skin lesion. White blood cells that engulf bacterial cells have begun to collect at the site as pus accumulates, and the skin has started to swell.
(**B**) A severe carbuncle on the head of a young boy. Abscesses often begin as trivial skin pimples and boils, but they can become serious as staphylococci penetrate to the deeper tissues.
Q: How would a severe carbuncle be treated?

Japan. In 2002, the first two cases of VRSA were identified in patients in the United States. New drugs and new treatment approaches are being made available to the medical community to help stem the tide of MRSA and VRSA. Chapter 24 discusses the issue of antibiotic resistance in more detail.

CONCEPT AND REASONING CHECKS

12.8 Compare and contrast furuncles and carbuncles.

Toxin-generated *S. aureus* contact diseases. Several *S. aureus* strains produce toxins. Some cause gastrointestinal illnesses (see Chapter 10) while other strains produce toxins that can initiate diseases coming from contact with the toxin.

Scalded skin syndrome is usually seen in infants, young children, and immunocompromised individuals. The skin becomes red, wrinkled, and tender to the touch, with a sandpaper appearance. It may then peel off. Exotoxins produced by staphylococci living at a point distant from the skin appear to be responsible for this condition. Mortality rates may be high in untreated cases due to bacterial invasion of the lungs or blood. Cefazolin is the drug of choice for treatment.

Toxic shock syndrome (TSS) is the term for a disease characterized by sudden fever and circulatory collapse. TSS remained in relative obscurity until the fall of 1980, when a major outbreak occurred in menstruating women who used a particular brand of highly absorbent tampons. News of TSS dominated the media for about six months and led to a recall of the tampons. With that episode, toxic shock syndrome assumed a position of significance in modern medicine. In 2004, 95 cases were reported to the CDC.

Toxic shock syndrome is caused by another toxin-producing strain of *S. aureus*. The earliest symptoms of disease include a rapidly rising fever, accompanied by vomiting and watery diarrhea. Patients then experience a sore throat, severe muscle aches, and a sunburn-like rash. Between the third and seventh days peeling of the skin, especially on the palms of the hands and soles of the feet, occurs (FIGURE 12.11). A sudden drop in blood pressure can possibly lead to shock and heart failure. Antibiotics may be used to control the growth of bacterial cells, but measures such as blood transfusions must be taken to control the shock.

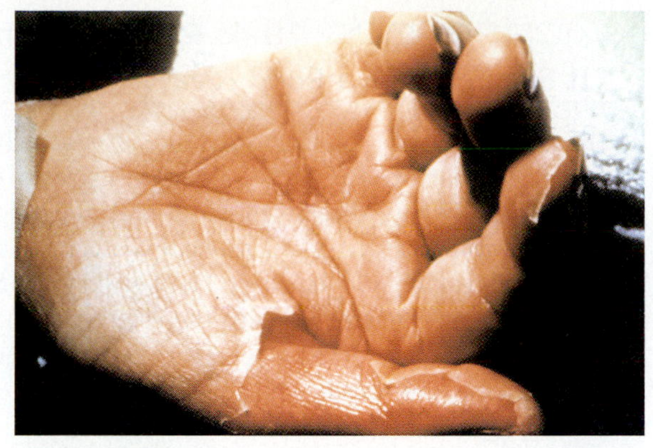

FIGURE 12.11 **Toxic Shock Syndrome.** The photograph shows the peeling skin often associated with TSS.

Q: What characteristic does Staphylococcus aureus *possess to cause toxic shock?*

Although the staphylococci involved in TSS exist in various places in the body, the ones inhabiting the vagina have received the most attention. During the 1980 outbreak, scientists speculated that lacerations or abrasions of the tissue by tampon inserters gave the staphylococci access to the tissues. Others suggested staphylococci grow in the warm, stagnant fluid during the long period the tampon is in place. It appears certain multiple factors play a role in TSS, because males, prepubertal girls, and postmenopausal women also have been stricken. Indeed, in 1994 one woman contracted TSS from a contaminated needle (MicroFocus 12.4).

CONCEPT AND REASONING CHECKS

12.9 Summarize the characteristics of toxin-generated *S. aureus* infections.

Trachoma Is Transmitted by Personal Contact

KEY CONCEPT

- *Chlamydia trachomatis* infections can lead to blindness.

Trachoma is a disease of the eyes and is the world's leading cause of preventable blindness. It occurs in hot, dry regions of the world and it is prevalent in Mediterranean countries, parts of Africa and Asia, and in the southwestern United States in Native American populations. There are 500 million infections, mostly children, worldwide and 7 to 9 million individuals have been blinded by trachoma (FIGURE 12.12).

MICROFOCUS 12.4: Public Health

"It Seemed Like a Good Idea!"

The idea of a tattoo seemed okay. All her friends had them, and a tattoo would add a sense of uniqueness to her personality. After all, she was already 22. It took some pushing from her friends, but she finally made it into the tattoo parlor that day in Fort Worth, Texas.

Two weeks later the pains started—first in her stomach, then all over. Her fever was high, and now a rash was breaking out; it looked like her skin was burned and was peeling away. There was one visit to the doctor, then immediately to the emergency room of the local hospital. The gynecologist guessed it was an inflammation of the pelvic organs (pelvic inflammatory disease, they called it), so he gave her an antibiotic and sent her home.

But it got worse—the fever, the rash, the peeling, the pains. Back she went to the emergency room. This time they would admit her to the hospital, give her intravenous blood transfusions and antibiotics, keep her for 11 days, and discover a severe blood infection due to *Staphylococcus aureus*. And there was an unusual diagnosis: toxic shock syndrome. Don't women get that from tampons? Most do, she was told, but a few get it from staphylococci entering a skin wound—a wound that can be made by a contaminated tattooing needle.

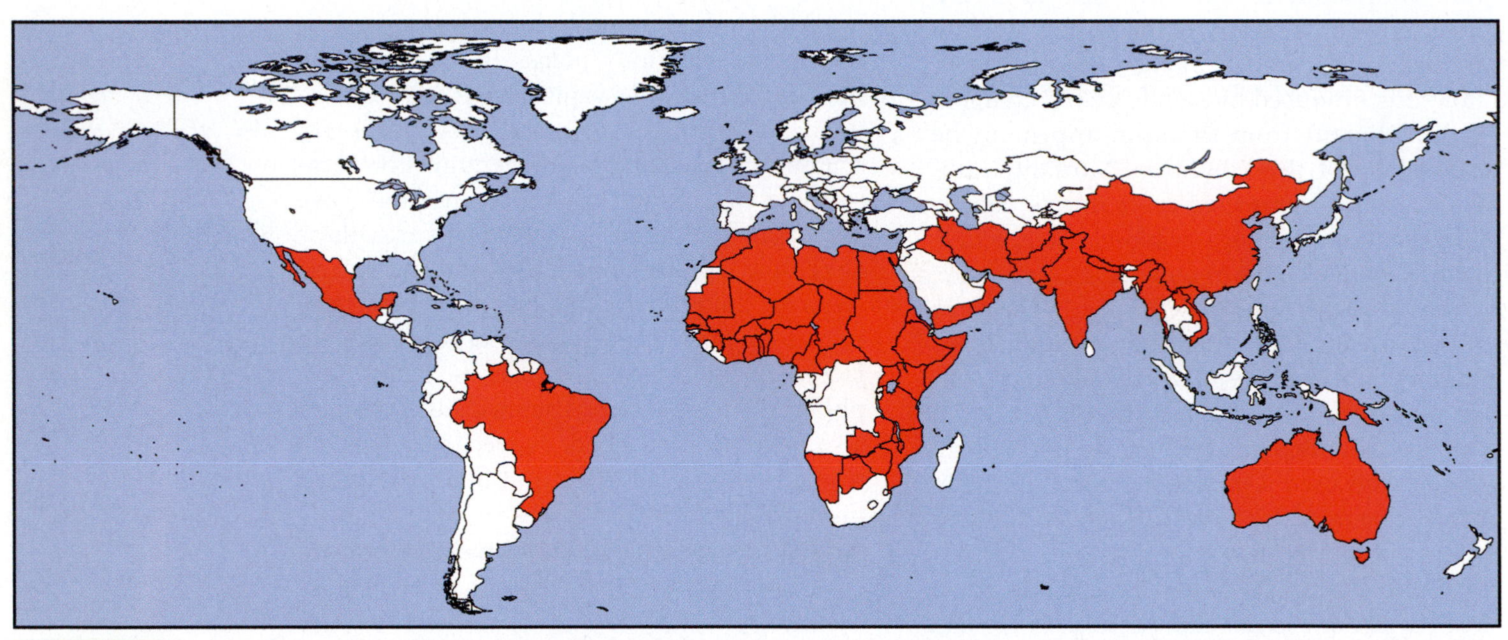

FIGURE 12.12 **Worldwide Trachoma.** Blinding or suspected blinding trachoma affects individuals worldwide, but especially in Africa, Southeast Asia, Mexico, and parts of South America. (Map by Silvio Mariotti/WHO.)

Q: How can trachoma be treated?

■ Conjunctiva: The thin membrane covering the cornea and forming the inner eyelid.

■ Secondary infections: Infections that can occur as a result of the primary infection.

Trachoma is caused by a serotype of *Chlamydia trachomatis* that is different from those responsible for chlamydia urethritis and lymphogranuloma venereum. It is not sexually transmitted but rather by personal contact with contaminated fingers, towels, and optical instruments. Face-to-face contact and flies also are modes of transmission.

The chlamydiae multiply in the conjunctiva of the eye. A series of tiny, pale nodules form on this membrane, giving it a rough appearance (trachoma come from *trachi* = "rough"). In serious cases, the upper eyelid turns in, causing abrasion of the cornea by the eyelashes. Blindness develops over 10 to 15 years from corneal abrasions and lesions. Inflammation also interferes with the flow of tears in the eye. Without sufficient tears, which contain antimicrobial chemicals, such as lysozyme (see Chapter 4), secondary infections can occur.

Tetracycline and erythromycin help reduce the symptoms of trachoma, but in many patients, the relief is only temporary because chlamydiae reinfect the tissues. In 1997, the WHO established the Alliance for Global Elimination of Trachoma by 2020 (GET 2020). Since then, ten national programs, making up 50 percent of the global trachoma burden, have reduced acute infections in children by 50 percent. This has involved using a **SAFE strategy**; that is, Surgery of the eyelids; Antibiotics for acute infections; Facial hygiene improvements; and Environmental access to safe water.

CONCEPT AND REASONING CHECKS

12.10 What are the challenges facing health officials trying to eliminate trachoma?

Bacterial Conjunctivitis (Pinkeye) Is Very Common

KEY CONCEPT

- *Haemophilus aegyptius* can cause an inflammation of the conjunctiva.

Several microorganisms cause conjunctivitis, among them the bacterial species *Haemophilus aegyptius;* also called the Koch-Weeks bacillus. This organism also is known in the literature as *Haemophilus influenzae* biotype III because of its close relationship to *H. influenzae* type b (see Chapter 9). The organism is a small, gram-negative rod that grows in chocolate agar, a rich medium containing disrupted red blood cells.

Conjunctivitis is a disease of the conjunctiva. When infected, the membrane becomes inflamed and the blood vessels dilate, a factor imparting a brilliant pink color to the white of the eye (hence the name **pinkeye**, which can be caused by other bacterial species and viruses as well). A copious discharge runs down the cheek in the waking hours and crusts the eyelids shut during sleep. The eyes are swollen and itch intensely, and vision in bright light is impaired (photophobia).

Conjunctivitis is extremely contagious, especially where people congregate. The inflammation may be transmitted in a number of ways, including face-to-face contact and via airborne droplets. Contaminated optometric instruments, microscopes, and towels also may transmit the bacilli. The disease normally runs its course in about two weeks, and therapy usually is not required. Neomycin may be administered to hasten recovery.

In recent years, a variant of *H. aegyptius* also has been isolated from the blood of patients suffering from a disease called **Brazilian purpuric fever**. This life-threatening disease is accompanied by nausea, vomiting, fever, and hemorrhagic skin lesions. Many patients display conjunctivitis before the onset of more serious symptoms.

CONCEPT AND REASONING CHECKS

12.11 How does bacterial conjunctivitis differ from trachoma?

Yaws Starts as Skin Sores

KEY CONCEPT

- *Treponema pertenue* cause nonsexually-transmitted infections that can result in very destructive disease.

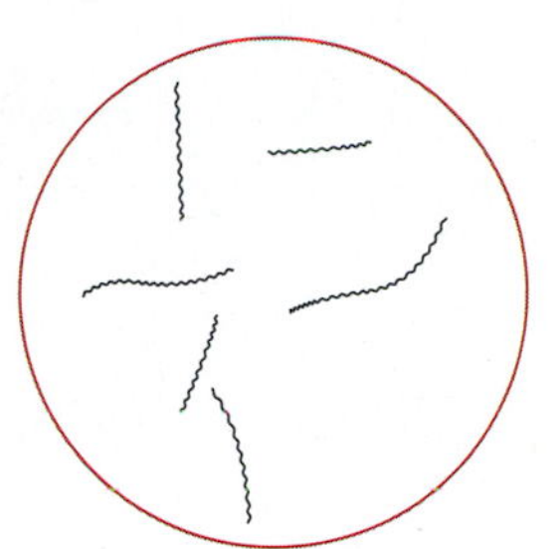

Treponema pertenue

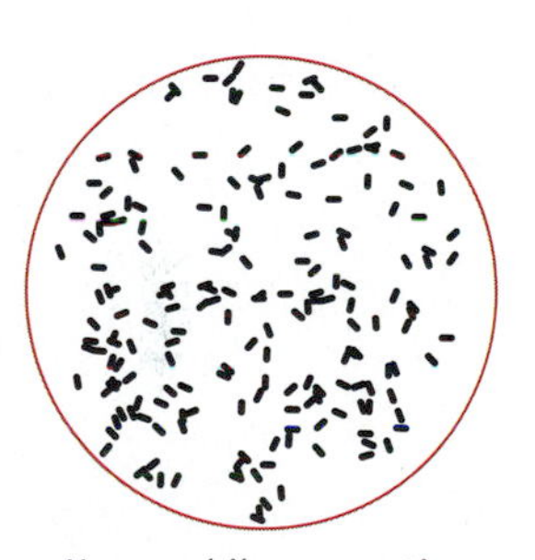

Haemophilus aegyptius

Yaws commonly occurs in tropical countries of Africa, South America, and Southeast Asia. It is caused by *Treponema pertenue*, a spirochete identical in appearance and similar in chemistry to the syphilis spirochete. Yaws is usually acquired by nonsexual skin contact, most often among children living under conditions of poor hygiene.

The disease starts several weeks after exposure as a red, raised lesion or skin sore at the site of entry, which is usually a leg. Blood associated with the lesion gives it the appearance of a raspberry, and the disease is sometimes called **frambesia** (framboise = "raspberry").

In time the lesion disappears, but months later the patient develops numerous soft granular nodules on the face, arms, and legs. Left untreated, these also disappear slowly. Later, the disease continues to destroy skin and bone, causing disfigurement (destruction of limb and face flesh) and disability. Yaws is seldom lethal but open sores are subject to secondary infections.

The similarities have led investigators to postulate that syphilis may have originated as yaws, or vice versa. **Bejel** and **pinta** are two other diseases very similar to yaws. The three are referred to as **treponematoses**.

TABLE 12.2 provides a summary of the contact bacterial diseases.

CONCEPT AND REASONING CHECKS

12.12 Although yaws and syphilis come from similar spirochetes, describe how yaws is different from syphilis.

TABLE

12.2 A Summary of Contact Bacterial Diseases

Disease	Causative Agent	Description of Agent	Organs Affected	Characteristic Signs	Toxin Involved	Treatment Administered	Comment
Leprosy (Hansen disease)	*Mycobacterium leprae*	Acid-fast rod	Skin, bones Peripheral nerves	Tumor-like growths Skin disfigurement "Claw hand"	None	Dapsone Rifampin Clofazimine	Lepromin skin test for diagnosis Long incubation period
Staphylococcal skin diseases	*Staphylococcus aureus*	Gram-positive cluster of cocci	Skin	Abscess, boil Scalded-skin syndrome Impetigo	Probable	Penicillin Vancomycin	Antibiotic resistance in staphylococci Food handlers involved
Toxic shock syndrome	*Staphylococcus aureus*	Gram-positive cluster of cocci	Blood	Fever Watery diarrhea Sore throat Sunburn-like rash	Yes	Pencillin Blood transfusions	Occurs in all groups, especially menstruating women Over 100 cases annually
Trachoma	*Chlamydia trachomatis*	Chlamydia	Eyes	Nodules on conjunctiva Scarring in eye	Not established	Tetracycline Erythromycin	Leading cause of preventable blindness
Bacterial conjunctivitis	*Haemophilus aegyptius*	Gram-negative rod	Eyes	Pinkeye Photophobia	Not established	Neomycin	Extremely contagious Copious discharge
Yaws	*Treponema pertenue*	Spirochete	Skin	Red, raised lesion	Not established	Penicillin	Found in tropical countries Related to bejel and pinta

12.3 Miscellaneous Bacterial Diseases

A few other bacterial diseases are related to animal bites, poor oral hygiene, and hospital stays. Also in this group are diseases related to organisms already in the body (not transferred person-to-person), the so-called **endogenous diseases**. We discuss these first.

There Are Several Endogenous Bacterial Diseases

KEY CONCEPT

- Endogenous bacterial diseases are caused by organisms normally inhabiting the body.

Natural host resistance generally prevents proliferation of endogenous bacteria (FIGURE 12.13). However, when the resistance is suppressed, disease may follow.

Bacteroides is the predominant anaerobic, gram-negative rod inhabiting the large intestine. *B. fragilis* is the most common pathogenic species of the group. When a person sustains an intestinal injury, these organisms may enter the bloodstream and cause peritonitis or form abdominal abscesses. Metronidazole is used to treat infections.

■ Peritonitis: An inflammation of the peritoneum, the membrane lining the abdomen.

One of the side effects of excessive antibiotic use is the elimination of many intestinal bacterial species that normally keep other species in check. Under these circumstances, an endogenous disease may develop from infection by *Clostridium difficile*, an anaerobic, spore-forming, gram-positive rod. Three percent of healthy individuals and 20 to 40 percent of hospitalized patients are colonized with this organism. In 2005, a new, dangerous strain was identified that produces two exotoxins at 20 times the concentration of the previous known strains. Old and new strains

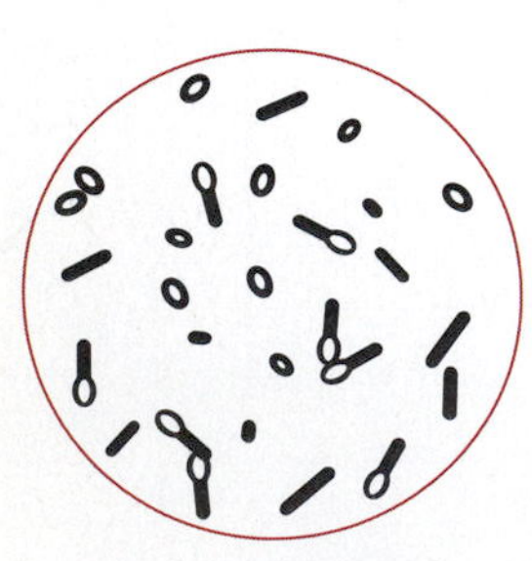
Clostridium difficile

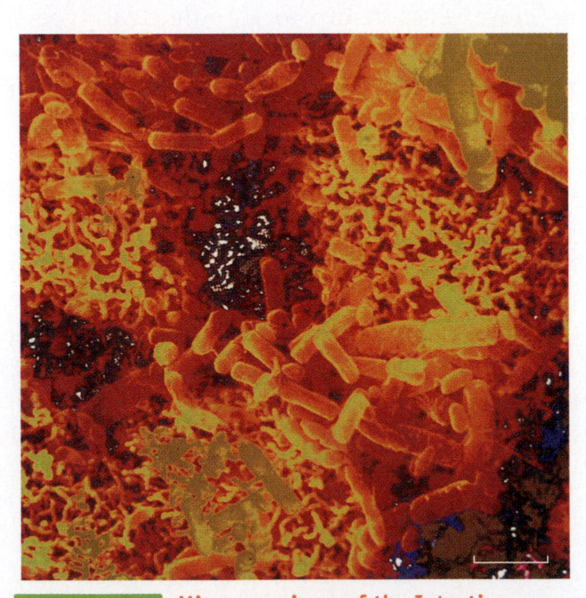
FIGURE 12.13 **Microorganisms of the Intestine.** A false-color scanning electron micrograph of the human large intestine. Endogenous disease may develop when bacterial cells like these invade the tissue during periods of suppressed host resistance. (Bar = 2 µm.)

Q: Why are many of the endogenous bacterial species of the intestine anaerobic?

induce a condition called **pseudomembranous colitis**. Yellow membranous lesions cover the intestinal lining, and patients experience diarrhea with watery stools. While the old strain primarily affects older or ill patients, the new strain can cause disease in healthy individuals.

Two bacterial diseases are caused by **actinomycetes**. These organisms are filamentous, slender, gram-positive rods resembling fungi in their morphology and growth.

Actinomycosis is caused by *Actinomyces israelii*, named for James A. Israel, who described the bacillus in 1878. It is an anaerobic species often found in the human gastrointestinal and respiratory tracts. Infection follows trauma to body tissues. For example, during a dental extraction, *A. israelii* can enter the gum tissues, multiply, and grow toward the facial surface, causing a red, lumpy and hard swelling. The condition, known as **lumpy jaw**, may develop into a skin problem with draining sinuses. Another form of actinomycosis involves draining sinuses of the chest wall, while a third form is characterized by abdominal sinuses, often as a complication of ulcers. Penicillin is used for treatment.

Nocardiosis is caused by species of the aerobic *Nocardia;* 50 percent of invasive infections are due to *N. asteroides*. When inhaled, the bacterial cells enter the lungs, causing a pulmonary infection (pneumonia). Fever, coughing, and chest pain develop. Spread to the brain can occur and be life-threatening. Up to 15 percent of nocardiosis patients also have an HIV infection. Sulfamethoxazole and trimethoprim are effective treatment drugs.

CONCEPT AND REASONING CHECKS

12.13 What are the characteristics of the endogenous bacterial diseases?

Animal Bite Diseases Occasionally Occur

KEY CONCEPT

- Bacterial disease can arise from animal bites or scratches.

Each year in the United States, about 3.5 million people are bitten by animals. Most of these wounds heal without complications, but in certain cases, bacterial disease may develop.

Pasteurellosis is caused by *Pasteurella multocida*, a gram-negative rod common in the pharynx in cats and dogs. The pathogen causes most wound infections resulting from a dog or cat bite. In humans, the symptoms of pasteurellosis develop rapidly, with local redness, warmth, swelling, and tenderness at the wound site. Abscesses frequently form, especially if the wound has been sutured. Some patients may experience arthritis. The disease responds slowly to penicillin therapy.

Cat-scratch disease (CSD) affects an estimated 20,000 Americans each year, primarily children. Although cats transmit few diseases to humans, CSD is transmitted by a scratch, bite, or lick from an infected cat. Most cases are associated with *Bartonella henselae*, a rickettsia (FIGURE 12.14A). Symptoms of CSD include red-crusted blisters at the site of entry, headache, malaise, and low-grade fever. Swollen lymph glands, generally on the side of the body near the bite, accompany the disease (FIGURE 12.14B).

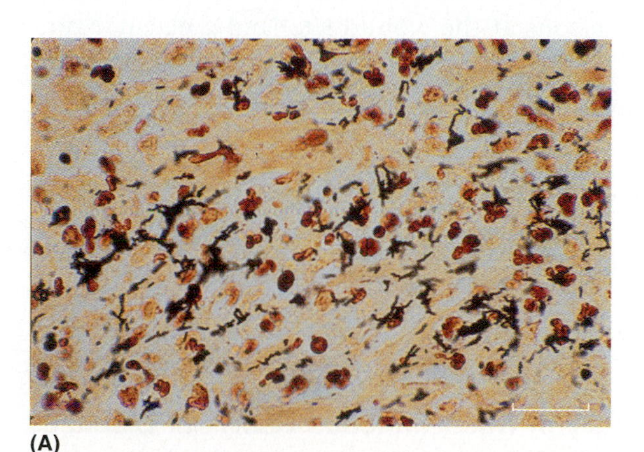

(A)

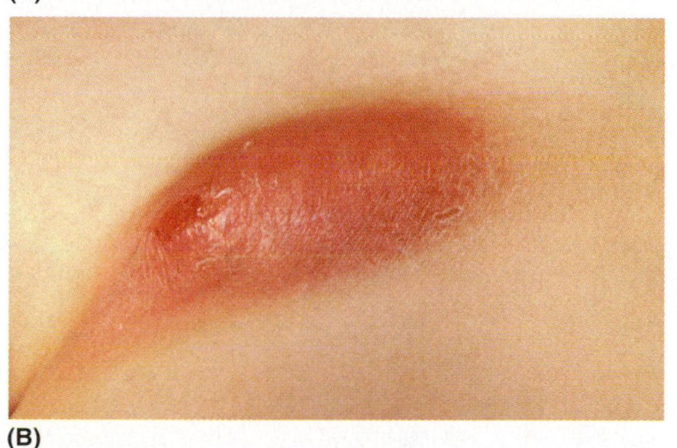

(B)

FIGURE 12.14 **Cat-Scratch Disease.** **(A)** A photomicrograph of *Bartonella henselae* (dark rods), the causative agent of cat-scratch disease. (Bar = 10 μm.) **(B)** A patient displaying in the groin the considerable lymph node swelling accompanying cat-scratch disease.

Q: Can B. henselae *be cultured in microbiological growth media? Explain.*

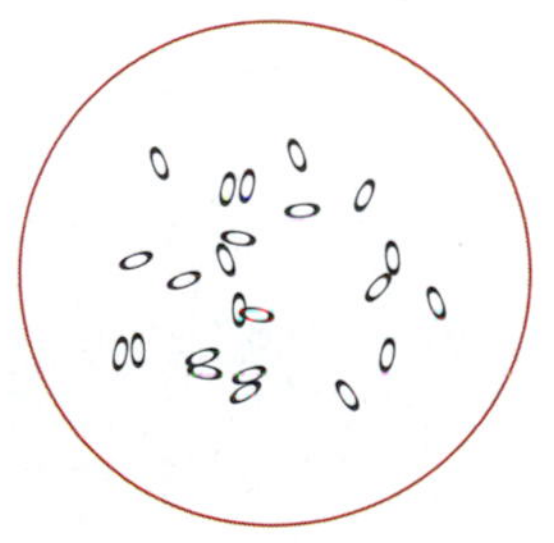

Pasteurella multocida

B. henselae and *B. quintana* also appear to cause **bacillary angiomatosis** (BA), a disease often found in individuals with lowered immune system function, such as AIDS. Infection from exposure to flea-infested cats (*B. henselae*) or the human body louse (*B. quintana*) produces tumors whose cells organize themselves to form blood vessels. While BA is treatable and curable with rifampin plus doxycycline, it can be life-threatening if untreated.

Rat-bite fever can be caused by one of two different bacterial species. In the United States, *Actinobacillus muris* (formerly *Streptobacillus moniliformis*), a gram-negative rod that occurs in long chains, is found in the pharynx of wild rats and other rodents feeding on infected rodents. Patients experience a lesion at the site of the bite or scratch, then a typical triad of fever, pain in the back and joints, and skin rash. In Asia, *Spirillum minus*, a rigid spiral cell with polar flagella, causes lesions at the wound site, and a maculopapular rash that spreads out from this point. In Japan and other parts of Asia, *Spirillum*-related rat-bite fever is known as **sodoku**. Antibiotic therapy with penicillin or erythromycin is recommended for either form of the disease.

Actinobacillus muris and *Spirillum minus*

TABLE 12.3 summarizes the endogenous and animal bite bacterial diseases.

12.14 Describe the symptoms of the diseases caused by animal bites or scratches.

Oral Diseases Cause Pain and Disability for Affected Individuals

KEY CONCEPT

- Bacterial species in the human oral cavity can trigger tooth decay and gum disease.

The oral cavity is a type of ecosystem, with complex interrelationships among the members of the resident population of microorganisms and the moist oral environment. The cavity has various ecological niches, each with a different physical property and nutrient supply dictating the number and type of microorganisms that can survive.

At least 600 bacterial species inhabit the human mouth, including a variety of streptococci, diphtheria-like bacilli, lactobacilli, spirochetes, and filamentous bacterial cells. Scientists estimate there are between 50 billion and 100 billion bacterial cells in the adult mouth at any one time.

The material accumulating on the tooth surface is known by many terms, the most common of which is plaque. **Plaque** is essentially a biofilm, a deposit of dense gelatinous material consisting of salivary proteins, food debris, and

TABLE 12.3 A Summary of Miscellaneous Bacterial Diseases

Disease	Causative Agent	Description of Agent	Organs Affected	Characteristic Signs	Treatment Administered	Comment
Endogenous diseases						
Bacteroides infection	*Bacteroides fragilis*	Gram-negative anaerobic rod	Intestine Blood	Blood clots	Clindamycin Chloramphenicol	May lead to gangrene Normal intestinal inhabitant
Pseudomembranous colitis	*Clostridium difficile*	Gram-positive spore-forming rod	Intestine	Intestinal lesions Diarrhea	Withdrawal of causative antibiotic	Accompanies excessive antibiotic use
Actinomycosis	*Actinomyces israelii*	Gram-positive anaerobic fungus-like rod	Gum tissues Chest wall Abdominal organs	Draining sinus	Penicillin Tetracycline	Lumpy jaw possible Associated with IUD use Normal lung inhabitant
Nocardiosis	*Nocardia asteroides*	Acid-fast fungus-like rod	Lungs	Abscesses	Co-trimoxazole Sulfonamides	Associated with AIDS Endogenous disease Soil transmission via wound
Animal bite diseases						
Pasteurellosis	*Pasteurella multocida*	Gram-negative rod	Skin	Abscess at site of bite Arthritis	Penicillin Doxycycline	Common in cats and dogs
Cat-scratch disease	*Bartonella henselae*	Rickettsia Gram-negative rod	Skin	Lesion at site of bite Swollen lymph glands	Symptomatic management Various antibiotics	Treatment not always necessary
Rat-bite fever	*Actinobacillus muris*	Gram-negative rod	Skin	Lesion at site of bite Rash, fever	Penicillin Tetracycline	Typical fever, arthritis, and rash
	Spirillum minus	Spirochete	Skin	Lesion at site of bite Rash, fever	Tetracycline Streptomycin	Common in Japan

an enormous mass of bacterial cells and their products (see Chapter 3). Many of these species have been cultivated in the laboratory, and two thirds are either anaerobic or facultative species.

Tooth loss remains a problem among children and older adults. In fact, 25 percent of adults over age 60 have lost all of their teeth, primarily because of tooth decay, which affects 95 percent of all adults, and advanced gum disease. The human oral diseases then can be separated into those on the tooth surface and those below the gum line.

Dental caries (*caries* = "rottenness"), or tooth decay, affects more than 20 percent of American children aged 2 to 4, 50 percent of those aged 6 to 8, and nearly 60 percent of those aged 15. Dental caries develops if three factors are present: a caries-susceptible tooth with a buildup of plaque; dietary carbohydrate, usually in the form of sucrose (sugar); and acidogenic (acid-producing) bacterial species (FIGURE 12.15A).

One of the primary bacterial causes of caries is *Streptococcus mutans*, a gram-positive organism with ovoid cells (FIGURE 12.15B). Its enzymes react with sucrose and convert it into a long-chain carbohydrate called dextran (see Chapter 2). These materials give *S. mutans* its special adherence qualities to the smooth surfaces, pits, and fissures of a tooth. The bacilli then ferment dietary carbohydrates to lactic acid, with smaller amounts of acetic acid, formic acid, and butyric acid. The acids break down the calcium phosphate salts in hydroxyapatite, the mineral constituent of bone and enamel. Protein-digesting enzymes break down any remaining organic materials. Other streptococcal species involved in caries include *S. sanguis*, *S. mitis*, and *S. salivarius*.

The prevention of dental caries depends on protecting the teeth by the ingestion and topical application of fluorides and modifying the diet by minimizing sucrose in foods (MicroFocus 12.5). Combating the acidogenic bacterial cells involves dextran synthesis inhibition by streptococci as well as stimulating antibody production. Some novel approaches to cavity prevention are explored in MicroFocus 12.6.

Caries is not the only form of dental disease. **Periodontal disease** (**PD**) is the result of an inflammation of the periodontal tissues, which surround the teeth and attach to the jaw to support essential tooth function. In individuals with good oral hygiene, subgingival bacterial species are common. However, they are aerobic, gram-positive organisms that are held in check by the environmental conditions and do not proliferate.

If left unchecked, PD can develop and lead to tooth loss. Poor oral hygiene leads to an increase in subgingival plaque. In these situations, the environment changes and undergoes a process of maturation into an anaerobic environment. The plaque becomes populated

(A)

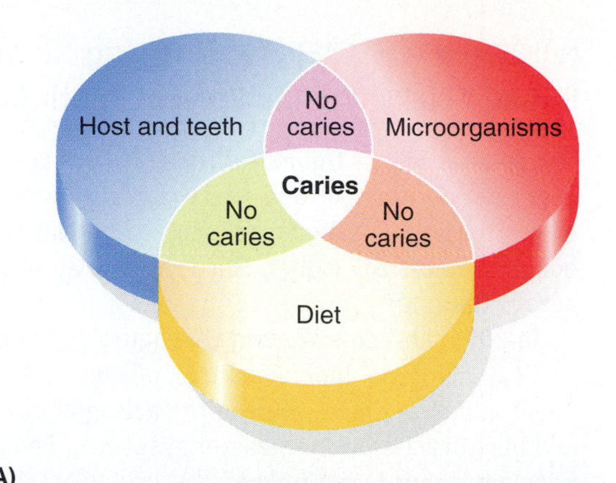

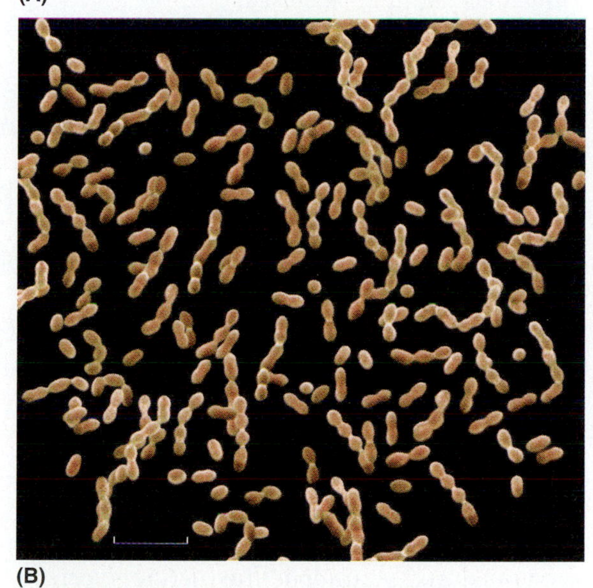

(B)

FIGURE 12.15 **Dental Caries.** (**A**) Overlapping circles depicting the interrelationships of the three factors that lead to caries activity. (**B**) False-color scanning electron microscope image of *Streptococcus mutans*, a major cause of dental caries. (Bar = 10 μm.)

Q: What role does S. mutans *play in the development of dental caries?*

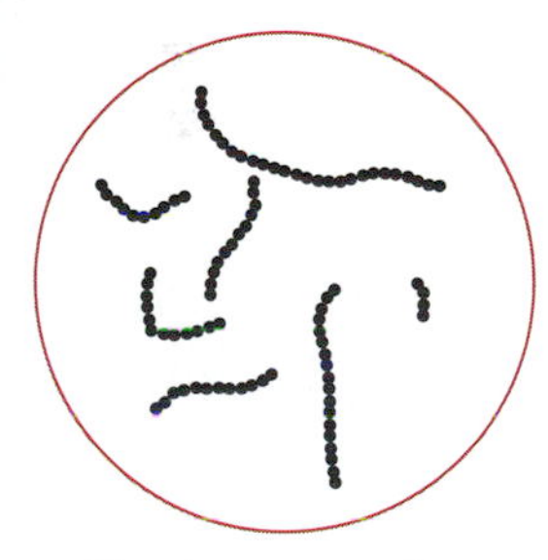

Streptococcus mutans

MICROFOCUS 12.5: Being Skeptical

Sweet, Sticky Raisins Prevent Tooth Decay

The general rule of thumb for tooth decay says frequent eating of sweet and sticky foods is likely to promote cavities. Even MicroFocus 2.4 in this textbook reiterates this belief. But is a general rule always the rule?

In 2005, Christine Wu and colleagues at the University of Chicago announced that raisins, certainly a sweet and sticky food, impede the action of cavity- and gum disease-causing bacterial species. Perhaps rules are meant to be broken?

So what did they discover? Wu and her colleagues found five chemicals in raisins that suppressed the growth of oral bacteria. Three of the chemicals inhibited the growth of *Streptococcus mutans,* which can cause cavities, and *Porphyromonas gingivalis,* which is instrumental in triggering periodontal disease.

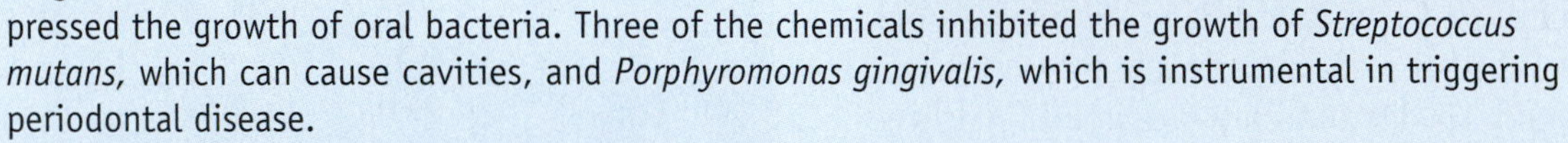

Analysis also indicated one of the raisin chemicals actually prevented *S. mutans* from attaching to the tooth surface, which is critical for this bacterium to initiate a cavity. Although raisins may be sweet and sticky, the sugars they contain are glucose and fructose; not the sucrose needed for dextran formation.

So, it appears that raisins may actually promote oral health.

with anaerobic, gram-negative rods and spirochetes (FIGURE 12.16). The primary agents include *Porphyromonas gingivalis, Bacteroides forsythus*, and *Actinobacillus actinomycetemcomitans*. This environment, consisting of perhaps 400 different species embedded in a polysaccharide matrix, represents a biofilm adhering to the teeth at and below the gum line. Unlike plaque on tooth surfaces, the subgingival biofilm contains no sugars, but rather the bacterial cells metabolize proteins in the crevices between teeth.

Importantly, the presence of the bacterial cells in subgingival plaque is not sufficient for the development of PD. Rather, an inflammatory response triggered by these cells is the primary reason for tissue destruction in PD. Other risk factors include genetic factors, smoking, and diabetes.

PD is characterized by ulcers appearing first along the gingival margin accompanied by severe pain and spontaneous bleeding. Immune cells, along with their chemical mediators and hydrolytic enzymes, degrade the periodontal ligaments and the surrounding bone. As the periodontal tissue decays, the

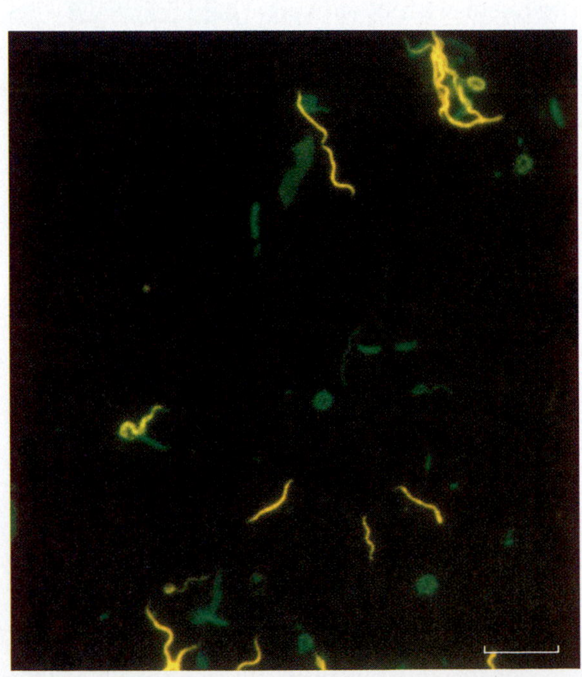

FIGURE 12.16 **Bacteria of the Oral Plaque.** A photomicrograph of treponemes found in the subgingival plaque of a patient with periodontal disease (PD). These spirochetes have been stained and observed with fluorescence microscopy. The organisms cannot be cultivated in laboratory media and must be studied directly as they come from the patient. (Bar = 10 μm.)

Q: Besides spirochetes, what other general group of bacterial species is associated with PD?

MICROFOCUS 12.6: Biotechnology

Watch out *S. Mutans*—Your Days May Be Numbered!

In the war against tooth decay, people have armed themselves with floss, toothpaste, toothbrushes, and many innovative variations of these. Still, the bacterial cells seem to win, especially *Streptococcus mutans*. Modifying the diet helps and adding fluoride to the water tips the balance still further. And newer innovations may be just around the corner.

British researcher Charles G. Kelly and his research team have produced an antibacterial coating for the teeth. The preparation binds to the tooth surface where *S. mutans* would attach and prevents bacterial attachment. In their studies, they smeared the synthetic preparation on the teeth of volunteers whose mouths were cleansed of bacterial cells. Control volunteers were treated with a placebo. Those receiving the synthetic preparation remained free of *S. mutans* for over three months while the bacterial cells appeared in the mouths of control volunteers after only three weeks.

Could a vaccine be the death knell for *S. mutans?* In London, Julian Ma and coworkers at Guy's Hospital are experimenting with a vaccine containing antibodies that latch onto *S. mutans,* preventing the bacterial cells from binding to the tooth surface. Because the vaccine, called CaroRx™, does not trigger an immune reaction, such treatments would need to be repeated every year or two. CaroRx™ is currently in phase two clinical trials.

In Boston, Martin Taubman, Daniel Smith, and colleagues at the Forsyth Institute have developed another vaccine against *S. mutans*. This vaccine blocks the enzyme responsible for synthesizing the long-chain bacterial carbohydrates that stick to the tooth surface. Taubman and Smith believe "immunizing" children 18 months to 3 years of age would give lifelong protection. So, these three innovations come from different research directions but all would accomplish the same result: no attachment means no bacterial cells, which means no acid, which means no tooth decay.

Jeffrey Hillman and his associates at the University of Florida and Oragenics, Inc. in Alachua, Florida, have come up with perhaps the most ingenious way to rid the mouth of *S. mutans*—chemically attack them. Hillman's goal is to replace the caries-causing *S. mutans* secreting lactic acid with "good" *S. mutans* unable to produce the acid. By collecting mouth samples from hundreds of patients, his group isolated a strain of *S. mutans* that produced a toxin that kills the other strains. The group then took this toxin-producing strain and genetically engineered it so that it would not secrete lactic acid. Using this "replacement therapy," the strain was squirted onto the teeth of rats. By producing the toxin, the genetically engineered bacterial species killed and replaced the resident *S. mutans* population and the rats exhibited much reduced levels of cavities. Similarly, three human volunteers had their mouths rinsed for five minutes with the engineered strain. At this writing, they either have no *S. mutans* in their mouth or only the engineered strain. Again, replacement therapy on this very limited group of volunteers worked. Clinical trials were set to begin in 2005 and Hillman hopes that a commercial product will be available to dentists in four or five years. If so, a five-minute mouth rinse would do the trick.

For the present, the dental wars go on. It is comforting to know, however, that scientists have imaginative solutions that extend beyond a new flavor of toothpaste. Dental caries is the most widespread infectious disease in today's world. Putting it to an end would be a considerable feather in the scientific cap. So, watch out *S. mutans,* your days may be numbered!

teeth may become loosened and eventually dislodged completely.

Diagnosis depends on the observation of gram-negative rods and spirochetes from the ulcers. The disease is sometimes called **fusospirochetal disease** because the rods have a long, thin fusiform shape and are mixed with spirochetes. Antibiotic washes may be used in therapy, and some physicians suggest painting the area with a traditional remedy of gentian violet.

CONCEPT AND REASONING CHECKS

12.15 Assess the differing roles of bacterial organisms in dental caries and periodontal disease.

Urinary Tract Infections Can Become Serious Health Problems

KEY CONCEPT

- Bacterial disease can arise in any part of the urinary system, usually through fecal contamination.

Over 4 million episodes of **urinary tract infections** (**UTIs**) occur annually in the

■ Fusiform: Refers to a spindle-shape with pointed ends.

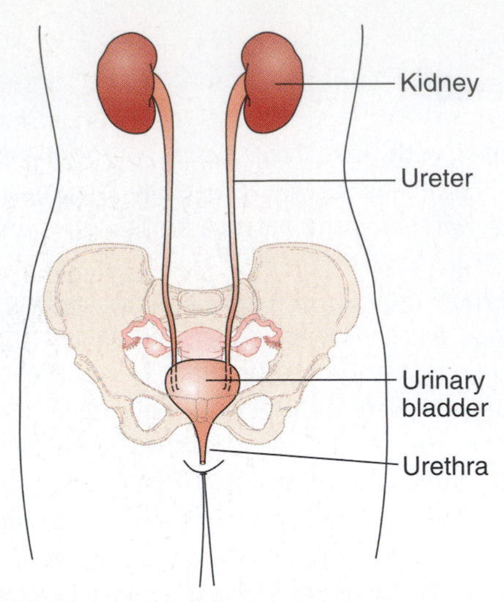

FIGURE 12.17 The Female Urinary Tract. The urinary system consist of the kidneys, which remove excess liquid and waste from the blood as urine. The ureters carry the urine to the bladder in the lower abdomen. Stored urine in the bladder is emptied through the urethra.

Q: Explain how an infection of the kidney would arise.

United States and represent the second most frequent cause of ambulatory visits to the doctor's office (after respiratory tract infections). A UTI can occur anywhere in the urinary tract (FIGURE 12.17). Most develop in the urethra and cause a disease called **urethritis**. The bacterial cells then can move to the bladder, causing a bladder infection (**cystitis**). Left untreated, the infection may move up the ureters to infect the kidneys (**pyelonephritis**).

UTIs usually occur when bacterial cells enter the urinary tract through the urethra and start to multiply in the bladder. *Escherichia coli* is the major cause of UTIs, although other genera such as *Chlamydia*, *Mycoplasma*, and *Proteus* can be involved (FIGURE 12.18A).

Women are infected more often than men, most likely because of anatomical differences: the female urethra is much shorter than that in men, so infecting organisms have a much shorter distance to travel. Also, the female urethra opens near the anus, permitting bacterial cells close access to the bladder. Sexual intercourse also can trigger an infection in women. About 20 percent of women who have a UTI will suffer from recurrent infections, each usually involving a different bacterial strain. In men, UTIs are rarer but when they occur, they are serious infections and are difficult to treat.

Sepsis: The presence of bacteria or toxins in the blood or tissues.

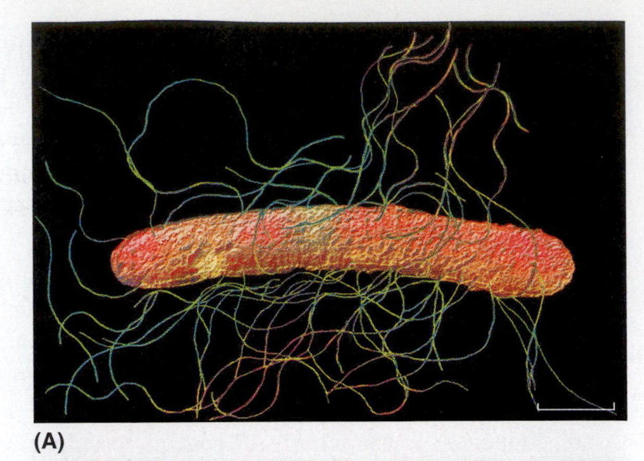

(A)

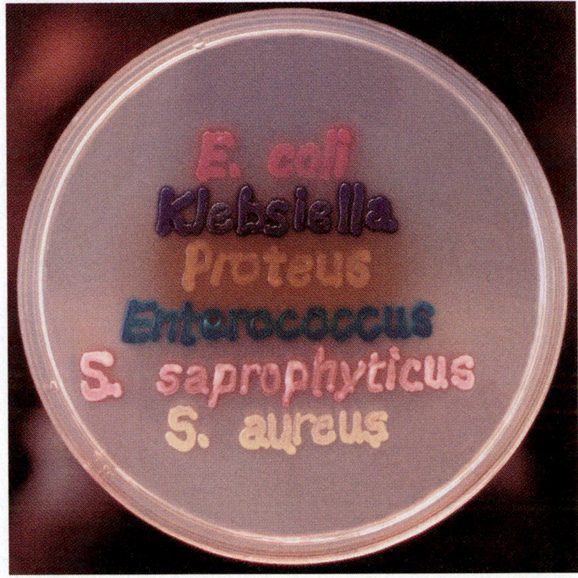

(B)

FIGURE 12.18 Pathogens of the Urinary Tract and Their Detection. (**A**) A false-color transmission electron micrograph of *Proteus mirabilis*. This large cell contains hundreds of flagella. (Bar = 5 μm.) (**B**) Various urinary tract pathogens are cultivated on a laboratory medium called CHROMagar. The different bacteria display different color reactions, which aid identification: *Escherichia coli*, *Klebsiella pneumoniae*, *Proteus mirabilis*, *Enterococcus faecium*, *Staphylococcus saprophyticus*, and *Staphylococcus aureus*.

Q: What advantage is it to P. mirabilis *to possess flagella?*

Sufferers report abdominal or back pain, burning pain on urination, and frequent urges to urinate. In the case of pyelonephritis, fever, sepsis, and decreased kidney function occur.

A UTI is diagnosed by microscopic examination of a sample of urine for blood cells and bacterial cells. The bacterial cells then are grown in culture, if possible. Various cultivation methods can be used in diagnosis, including one in which different bacterial species give different color reactions (FIGURE 12.18B).

A sensitivity test is performed to determine which antibiotic is best for treatment (Chapter 24). Trimethoprim with or without sulfamethoxazole are typical drugs for routine, uncomplicated UTIs.

Changing daily habits can help avoid UTIs. Drinking plenty of water will flush any infecting bacterial cells from the urinary system. Studies have shown that cranberry juice and vitamin C may inhibit bacterial growth by increasing the acidity of the urinary tract. Cranberry juice also inhibits adhesion of bacterial cells to the bladder wall. Avoiding tight-fitting clothes and urinating soon after sexual intercourse can reduce the possibility of infection.

Women who suffer from repeated UTIs often have lower antibody levels than those women who have normal levels of antibodies. Because antibodies help fight infection (Chapter 20), researchers are investigating the possibility of a vaccine against *E. coli* as a way of preventing urinary tract infections. Research groups are experimenting with vaccines composed of killed bacterial cells delivered as an injectable or oral vaccine, or applied directly to the vaginal tract by suppository. Local delivery of the vaccine would ensure a local immune response and avoid inflammation caused by injection at distant sites.

MicroInquiry 12 presents four case studies covering some of the diseases described in this chapter.

CONCEPT AND REASONING CHECKS

12.16 Why do more women suffer UTIs than men?

Nosocomial Infections Can Be Acquired in a Healthcare Setting

KEY CONCEPT

- Nosocomial infections often represent cross contamination from health care worker to patient.

Since the time of Ignaz Semmelweis in 1847, **nosocomial infections or hospital-acquired infections (HAIs)** have been a major health care issue. HAIs are acquired during hospitalization (nosocomial comes from *noso* = "disease"; *comi* = "care") or while in chronic care facilities, including nursing homes and outpatient facilities. The CDC has estimated that up to 10 percent of all hospital patients in the United States may develop a nosocomial disease during their stay, with surgical patients particularly susceptible. Certain types of operations such as amputations and intestinal surgery are accompanied by an infection rate approaching 30 percent. Over 2 million patients may be involved annually, and 20,000 die each year in the United States from HAIs.

These numbers make identifying high-risk patients, instituting new or better prevention strategies, and following standard precautions as critical interventions toward disrupting the spread of infectious diseases in any health care setting. All these settings represent high-density communities composed of unusually susceptible individuals. A variety of pathogenic bacterial species abound, and new ones are continually being introduced as new patients arrive. In addition, the extensive use of antimicrobial agents contributes to the development of resistant strains of microorganisms (such as MRSA), and staff members become carriers of these strains (FIGURE 12.19). Many of the patients have already experienced some interference with their normal immune defenses, such as a breach of the skin barrier in surgery, radiation therapy for cancer, immunosuppressive medication, or indwelling apparatuses such as catheters and intravenous tubes. When all these factors meet, HAIs break out.

Up to 80 percent of nosocomial diseases are caused by microbes brought with the patient at the time of admission. These organisms are generally **opportunistic**—they do not cause disease in healthy individuals, but they are dangerous in compromised persons. Ten to twenty percent of HAIs develop following infection with bacterial species inhabiting the health care environment.

UTIs account for about 50 percent of nosocomial infections, primarily from catheterization. Another 25 percent arise from post-operative infections and the rest primarily come from upper respiratory infections (see Chapter 9). Among the most common prevalent bacterial species are gram-negative rods such as *Escherichia coli, Pseudomonas aeruginosa, Serratia marcescens, Enterobacter aerogenes, Enterobacter cloacae, Klebsiella pneumoniae*, and *Proteus* species. Gram-positive *Staphylococcus aureus* and

MICROINQUIRY 12

Sexually Transmitted, Contact, and Miscellaneous Bacterial Disease Identification

Below are several descriptions of sexually transmitted, contact, and miscellaneous bacterial diseases based on material presented in this chapter. Read the case history and then answer the questions posed. Answers can be found in **Appendix D**.

Case 1

The patient is a 17-year-old woman who comes to the clinic indicating that several days ago she started feeling nauseous but had not experienced any vomiting. She tells the physician that the day before coming to the clinic she had a fever and chills; she also has been urinating more frequently and the urine has a foul smell. She is diagnosed as having a urinary tract infection.

12.1a. What types of bacterial species could be responsible for her illness?

12.1b. Why are these types of diseases more prevalent in women than they are in men?

12.1c. What types of urinary infections can occur?

12.1d. How could this patient attempt to avoid another UTI?

12.1e. What role do biofilms play in UTIs?

Case 2

A 19-year-old unwed mother arrives at the emergency room of the county hospital complaining of having cramps and abdominal pain for several days. She says she had never had a urinary tract infection and could not have gonorrhea, as she was treated and cured of that two years ago. She has not experienced nausea or vomiting. When questioned, she tells the emergency room nurse that she has a single male sexual partner and condoms always are used. Based on further examination, the patient is diagnosed with pelvic inflammatory disease (PID). An endocervical swab is used for preparing a tissue culture. Staining results indicate the presence of cell inclusions.

12.2a. What bacterial species can be associated with PID? What disease does she most likely have?

12.2b. Why was a tissue culture inoculum ordered? Describe the reproductive cycle of this organism.

12.2c. What other tests could be ordered for the patient's infection?

12.2d. Why was the emergency room concerned about her sexual activity?

12.2e. What misconception does the patient have about her past gonorrhea infection?

Case 3

A 71-year-old man visits the local hospital emergency room at 6:00 PM after noticing a red infection streak running up his left forearm. He tells the physician that he was playing with his cat this morning when it bit him on the left wrist. Thinking nothing of it and being an amateur photographer, he went about printing some photographs in his darkroom. At 3:00 PM, he finished and, in the daylight, noticed that his wrist was swollen and painful. The physician also notes that the patient experiences tenderness at the site. She then notices a small puncture wound on the wrist and a small abscess. A Gram stain smear from the abscess indicates the presence of gram-negative rods.

12.3a. What bacterial species is responsible for the patient's illness?

12.3b. What clues lead you to identify this specific organism?

12.3c. Where is this organism normally found in cats?

12.3d. What could the patient have done to make it less likely that an infection occurred?

Case 4

A 17-year-old man comes to a free neighborhood clinic. He says that he noticed some white pus-like discharge and a tingling sensation in his penis. Since yesterday, he has had pain on urinating. He tells the physician that he has been sexually active with several female partners over the past eight months, but no one has had any sexually transmitted disease. Examination determines that there is no swelling of the lymph nodes in the groin or pain in the testicles. A Gram stain indicates the presence of gram-negative diplococci. The patient is given antibiotics, instructed to tell his female partners they both should be medically examined, and then he is released.

12.4a. Based on the clinic findings, what disease does the patient have and what bacterial species is responsible for the infection?

12.4b. Why is it important for his sexual partners to be medically examined, even if they experience no symptoms? What complications could arise if they are infected?

12.4c. For which other organisms is this patient at increased risk? Why?

12.4d. What significance can be drawn from the fact that the patient does not have any swelling of the lymph nodes in the groin or pain in the testicles?

12.4e. What antibiotics most likely would be given to the patient?

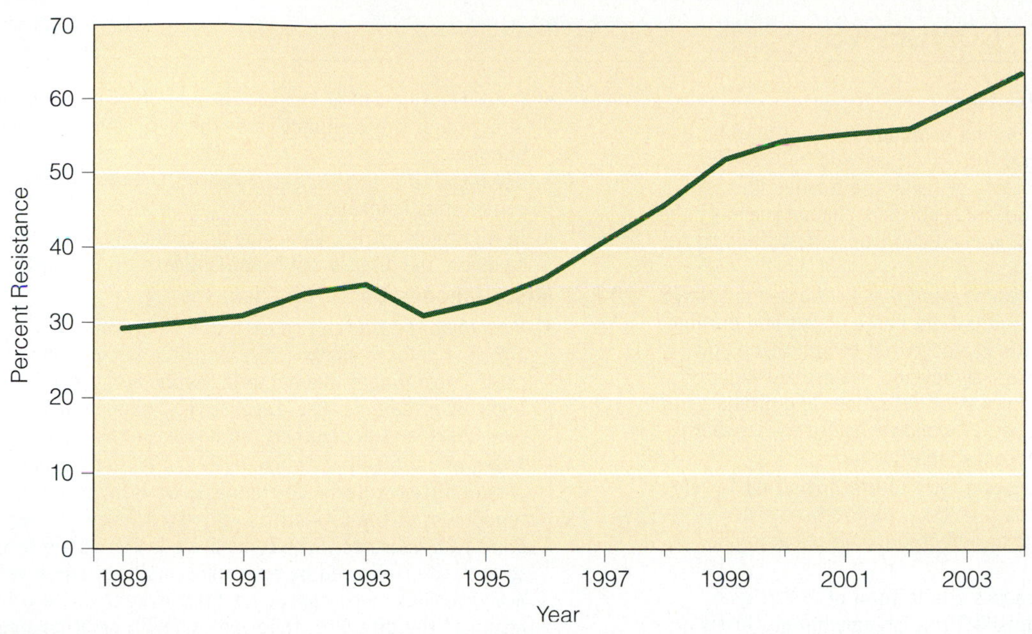

FIGURE 12.19 **Nosocomial Infections Caused by MRSA.** MRSA resistance in intensive care units (ICUs) has continued to increase, surpassing the 50 percent mark in 1999. *Source*: Data from the National Nosocomial Infections Surveillance System/CDC.

Q: Why is there special concern for MRSA resistance in the ICU?

Streptococcus pneumoniae are other important causes of nosocomial diseases.

Another major concern is the development of antibiotic resistance, especially methicillin resistance by *S. aureus* (MRSA). The presence of such organisms in the health care environment requires the use of more powerful antibiotics to treat infections and disease. Unfortunately, in many cases this has only accelerated the potential for multidrug resistant superbugs as a nosocomial threat.

Contact in the health care facility often comes from the hands of health care workers. Such medical personnel moving from patient to patient establish a **chain of transmission** by which infectious diseases can be easily transported from one patient to another via the hands of a hospital worker. Contaminated equipment or instruments, such as urinary catheters, colonoscopes, endoscopes, and arthroscopes, also can be spread disease because the invasive procedures allow pathogens to breach the skin barrier.

To deal with nosocomial diseases, hospitals designate a specialist, usually a **nurse epidemiologist**, whose primary responsibility is to locate problem areas and report them to an **infection control officer** or committee. The local committee consists of nurses, doctors, dieticians, engineers, and housekeeping and laboratory workers who monitor equipment and procedures to interrupt the disease cycle.

One of the most basic ways to disrupt the disease cycle is by hand washing between patient contacts. In ICUs, emergency rooms, urgent care clinics, and other areas such a simple procedure often is overlooked during the rush of crisis care.

Approximately one third of HAIs are preventable. By identifying at-risk patients, hand washing, and protecting equipment and instruments through sterilization and disinfection procedures, control or elimination of nosocomial diseases is possible.

■ Superbugs: Microbes resistant to a wide variety of antibiotics.

CONCEPT AND REASONING CHECKS

12.17 Assess the current state of HAIs in the United States and the prevention measures to limit disease spread.

SUMMARY OF KEY CONCEPTS

12.1 Sexually Transmitted Diseases

- Syphilis is caused by *Treponema pallidum*. Primary syphilis is characterized by a chancre. Untreated, secondary syphilis occurs and presents symptoms of fever and flu-like illness along with a skin rash. Tertiary syphilis is characterized by the gumma. Infected pregnant women can transmit syphilis to the fetus (congenital syphilis).
- Gonorrhea is caused by *Neisseria gonorrhoeae*. Women may have symptoms of vaginal and urethral discharge of pus and burning during urination. Salpingitis is a potential complication and pelvic inflammatory disease may develop. Males may experience urethral inflammation, a burning on urination, and discharge of pus. Infected women can pass *N. gonorrhoeae* to the newborn during birth, causing gonococcal ophthalmia.
- *Chlamydia trachomatis* causes a form of urethritis that is very similar to a mild gonorrhea infection. Newborns can be infected during birth from an infected mother, leading to chlamydial ophthalmia.
- *Ureaplasma urealyticum* causes a mild form of urethritis. Symptoms can be distinguished from chlamydial urethritis or gonorrhea by the presence of urethral pain during urination.
- *Haemophilus ducreyi* causes genital ulcers and swollen lymph nodes in the groin.
- Another serotype of *Chlamydia trachomatis* causes lymphogranuloma venereum. *Calymmatobacterium granulomatis* causes granuloma inguinale, while *Gardnerella vaginalis* is one bacterial species that can cause vaginitis. Mycoplasmal urethritis can be caused by *Mycoplasma hominis*.

12.2 Contact Bacterial Diseases

- Leprosy is a slowly progressive infection caused by *Mycobacterium leprae*. Patients suffer disfiguring of the skin and bones. Lepromas form on the skin.
- *Staphylococcus aureus* can cause furuncles and carbuncles. Other strains cause impetigo. Scalded skin syndrome and toxic shock syndrome are the result of toxin production by *S. aureus* serotypes.
- Other serotypes of *Chlamydia trachomatis* cause trachoma, an infection of the conjunctiva of the eye, which can lead to blindness.
- *Haemophilus aegyptius* causes an inflammation of the conjunctiva (pinkeye).
- In tropical countries, yaws may occur and is caused by *Treponema pertenue*. The disease can eventually lead to severe deformities.

12.3 Miscellaneous Bacterial Diseases

- *Bacteroides fragilis* can cause blood clots while *Clostridium difficile* produces exotoxins responsible for pseudomembranous colitis. Two fungal-like bacteria, *Actinomyces israelii* and *Nocardia asteroids* cause, respectively, lumpy jaw in the oral gum tissue and nocardiosis, an abscess of the lungs.
- A dog or cat bite can transmit *Pasteurella multocida*. The person bitten experiences redness, warmth, swelling, and tenderness at the bite site. Cat-scratch fever is caused by *Bartonella henselae*. Symptoms are a lesion at the bite or scratch site, followed by headache, malaise, and fever. *Actinobacillus muris* causes rat-bite fever characterized by a lesion at the bite site, followed by fever, arthritis-like pain in the joints, and a skin rash.
- Dental caries is most often caused by *Streptococcus mutans*. Poor hygiene can lead to the development of acute necrotizing ulcerative gingivitis caused by a variety of gram-negative bacterial species and spirochetes.
- Urinary tract infections develop in the urethra and cause urethritis. The bacterial cells then can move to the bladder, causing cystitis. Left untreated, the infection may move up the ureters and cause pyelonephritis. *Escherichia coli* is one of the most common causes of UTIs.
- The majority of nosocomial diseases are acquired in hospitals brought with the patient at the time of admission. Hand washing between patient contacts, recognizing at-risk patients, and protecting equipment and instruments from contamination are crucial in decreasing disease transmission to or between patients.

LEARNING OBJECTIVES

After understanding the textbook reading, you should be capable of writing a paragraph that includes the appropriate terms and pertinent information to answer the objective.

1. Summarize the current state of sexually transmitted disease prevalence.
2. Distinguish between the three possible stages of syphilis.
3. Describe (a) the possible complications resulting from gonorrhea in females and (b) explain the danger of gonorrhea in pregnant females.
4. Identify the infectious and noninfectious forms of *Chlamydia trachomatis.*
5. Distinguish between the signs and symptoms of chlamydial urethritis in males and females.
6. Describe the signs and symptoms of chancroid and ureaplasmal urethritis.
7. Summarize the key symptoms of LGV, granuloma inguinale, vaginitis, and mycoplasmal urethritis.
8. Summarize the clinical significance of leprosy (Hansen disease).
9. Assess the role of *Staphylococcus aureus* as an agent for contact diseases.
10. Explain how the *Chlamydia trachomatis* serotype causes trachoma.
11. Identify the characteristics of bacterial conjunctivitis (pinkeye).
12. Explain how *Treponema pertenue* can cause disfigurement.
13. Summarize how endogenous bacterial diseases initiate infection.
14. Identify and discuss the characteristics of the three animal bite diseases.
15. Assess the role of anaerobic bacterial species in dental caries and periodontal disease.
16. Summarize the ways the sterile urinary tract can become infected.
17. Assess the significance of nosocomial infections today and how they can be prevented.

SELF-TEST

Answer each of the following questions by selecting the ***one*** answer that best fits the question or statement. Answers to even-numbered questions can be found in **Appendix C**.

1. The American health organization that charts epidemics, including STDs, is the
A. WHO.
B. FDA.
C. HHS.
D. CDC.
E. NHDP.

2. According to the CDC, how many people contract an STD each day in the United States?
A. 150
B. 1500
C. 8,000
D. 15,000
E. 52,000

3. A chancre is typical of which stage of syphilis?
A. Primary
B. Secondary
C. Tertiary
D. Chronic, latent
E. Gumma

4. Salpingitis is associated with _____ and can lead to _____.
A. syphilis; gumma formation
B. gonorrhea; sterility
C. chlamydia; ophthalmia
D. chancroid; soft chancre
E. gonorrhea; pharyngitis

5. Which one of the following (**A–D**) is *not* correct concerning the chlamydial life cycle? If all are correct, then select **E**.
A. It occurs independently of a host.
B. Reticulate bodies are noninfectious.
C. Reticulate bodies reorganize into elementary bodies.
D. Elementary bodies infect host cells.
E. All the above (**A–D**) are correct.

6. Ophthalmia is associated with what two STDs?
A. Syphilis and chlamydia
B. Gonorrhea and chlamydia
C. Syphilis and chancroid
D. Syphilis and gonorrhea
E. Chlamydia and chancroid

7. Leprosy is spread by
A. coughing.
B. sneezing.
C. nasal secretions.
D. sexual contact.
E. Both **A** and **B** are correct.

8. Lepromas are
A. caused by animal bites.
B. inflammatory blisters.
C. produced by *Treponema pallidum* infections.
D. formed during urinary tract infections.
E. tumor-like growths.

9. Toxin-generated *S. aureus* diseases resulting from contact include
A. purpuric fever.
B. scalded skin syndrome.
C. toxic shock syndrome.
D. pseudomembranous colitis.
E. Both **B** and **C** are correct.

10. The SAFE strategy is used to reduce acute infections of
A. gonorrhea.
B. yaws.
C. chlamydial urethritis.
D. trachoma.
E. LGV.

11. The bacterial species responsible for pinkeye is
A. *Treponema pertenue.*
B. *Bacteroides fragilis.*
C. *Nocardia asteroides.*
D. *Haemophilus aegyptius.*
E. *Clostridium difficile.*

12. Endogenous actinomycete-caused diseases or conditions include
A. pseudomembranous colitis.
B. bacillary angiomatosis.
C. lumpy jaw.
D. pinkeye.
E. Both **A** and **D** are correct.

13. A pulmonary, pneumonia-like infection that can spread to the brain describes this disease.
A. Pasteurellosis
B. Nocardiosis
C. Rat-bite fever
D. Cat-scratch disease
E. Actinomycosis

14. *Bartonella henselae* infections, especially in children, can lead to this disease.
A. Rat-bite fever
B. Cat-scratch disease
C. Yaws
D. Lumpy jaw
E. None of the above (**A–D**) is correct

15. *Streptococcus mutans* is associated with
A. dental caries.
B. urinary tract infections.
C. nosocomial infections.
D. cystitis.
E. periodontal disease.

16. Pyelonephritis is a bacterial infection of the
A. bladder.
B. lungs.
C. kidneys.
D. liver.
E. urethra.

QUESTIONS FOR THOUGHT AND DISCUSSION

Answers to even-numbered questions can be found in **Appendix C**.

1. One of the major problems of the current worldwide epidemic of AIDS is the possibility of transferring the human immunodeficiency virus (HIV) among those who have a sexually transmitted disease. Which diseases in this chapter would make a person particularly susceptible to penetration of HIV into the bloodstream?
2. Studies indicate that most cases of *Staphylococcus*-related impetigo occur during the summer months. Why do you think this is the case?
3. In some African villages, blindness from trachoma is so common that ropes are strung to help people locate the village well, and bamboo poles are laid to guide farmers planting in the fields. What measures can be taken to relieve such widespread epidemics as this?
4. At a specified hospital in New York City, hundreds of patients pay a regular visit to the "neurology ward." Some sign in with numbers; others invent fictitious names. All receive treatment for leprosy. Why do you think this disease still carries such a stigma?
5. Several years ago the Rockefeller Foundation offered a $1 million prize to anyone who could successfully develop a simple and rapid test to detect chlamydia and/or gonorrhea. The test had to use urine as a test sample, and be performed and interpreted by someone with a high school education. No one ever claimed the prize. Can you guess why?
6. Researchers once suggested a "dip and brush" method for controlling oral diseases. The idea was to dip a toothbrush into an antiseptic several times while brushing to reduce the level of plaque bacteria. Do you think this method will reduce dental problems? Explain.

APPLICATIONS

Answers to even-numbered questions can be found in **Appendix C**.

1. Suppose a high incidence of leprosy existed in a particular part of the world. Why is it conceivable that there might be a correspondingly low level of tuberculosis?
2. An African patient reports to a local hospital with an upper lip swollen to about three times its normal size. Probing with a safety pin at facial points where major nerve endings terminate showed that the area to the left of the nose and above the lip was without feeling. When a biopsy of the tissue was examined, it revealed round reservoirs of immune system cells called granulomas within the nerves. On bacteriological analysis, acid-fast rods were observed in the tissue. What disease do all these data suggest?
3. During a field trip, an undergraduate biology student is bitten on the left index finger by a wild rat. Within 12 hours, her finger is swollen and throbbing. Soon thereafter she is hospitalized with swollen lymph nodes, a skin rash, fever, and exquisite sensitivity of the finger. Gram-negative, branching rods were found in the tissue. What infectious disease had she contracted?
4. Certain microscopes have the added feature of a small hollow tube that fits over the eyepieces or oculars. Viewers are encouraged to rest their eyes against the tube and thereby block out light from the room. Why is this feature hazardous to health?
5. After a young man suffers an abrasion on the right arm, his affectionate cat licks the wound. Several days later, a pustular lesion appears at the site and a low-grade fever develops. He also experiences "swollen glands" on the right side of his neck. What disease has he acquired?
6. A woman suffers two miscarriages, each after the fourth month of pregnancy. She then gives birth to a child, but impaired hearing and vision become apparent as it develops. Also, the baby's teeth are shaped like pegs and have notches. What medical problem existed in the mother?

REVIEW

Answer each of the following by filling in the blank with the correct word or phrase. Answers to even-numbered questions can be found in **Appendix C**.

1. Ureaplasmal urethritis is a type of __________.
2. __________ is the disease caused by *Haemophilus ducreyi*.
3. Lymphogranuloma venereum (LGV) prevalent in __________-Asia.
4. A __________ is absent in *Mycoplasma* species.
5. Granuloma inguinale is characterized by having bacterial masses called __________ bodies.
6. *Gardnerella* is the cause of __________.
7. *Mycoplasma* __________ is the agent responsible for mycoplasmal urethritis.
8. __________ are a group of connected, deep abscesses caused by *S. aureus.*
9. Yaws is thought to have originated from or given rise to __________.
10. __________ is the most common agent of urinary tract infections.
11. Sufferers of UTIs have a __________ on urination.
12. Transport of infectious disease from patient to patient by medical personnel is called a __________.

HTTP://MICROBIOLOGY.JBPUB.COM/

The site features learning, an on-line review area that provides quizzes and other tools to help you study for your class. You can also follow useful links for in-depth information, read more MicroFocus stories, or just find out the latest microbiology news.

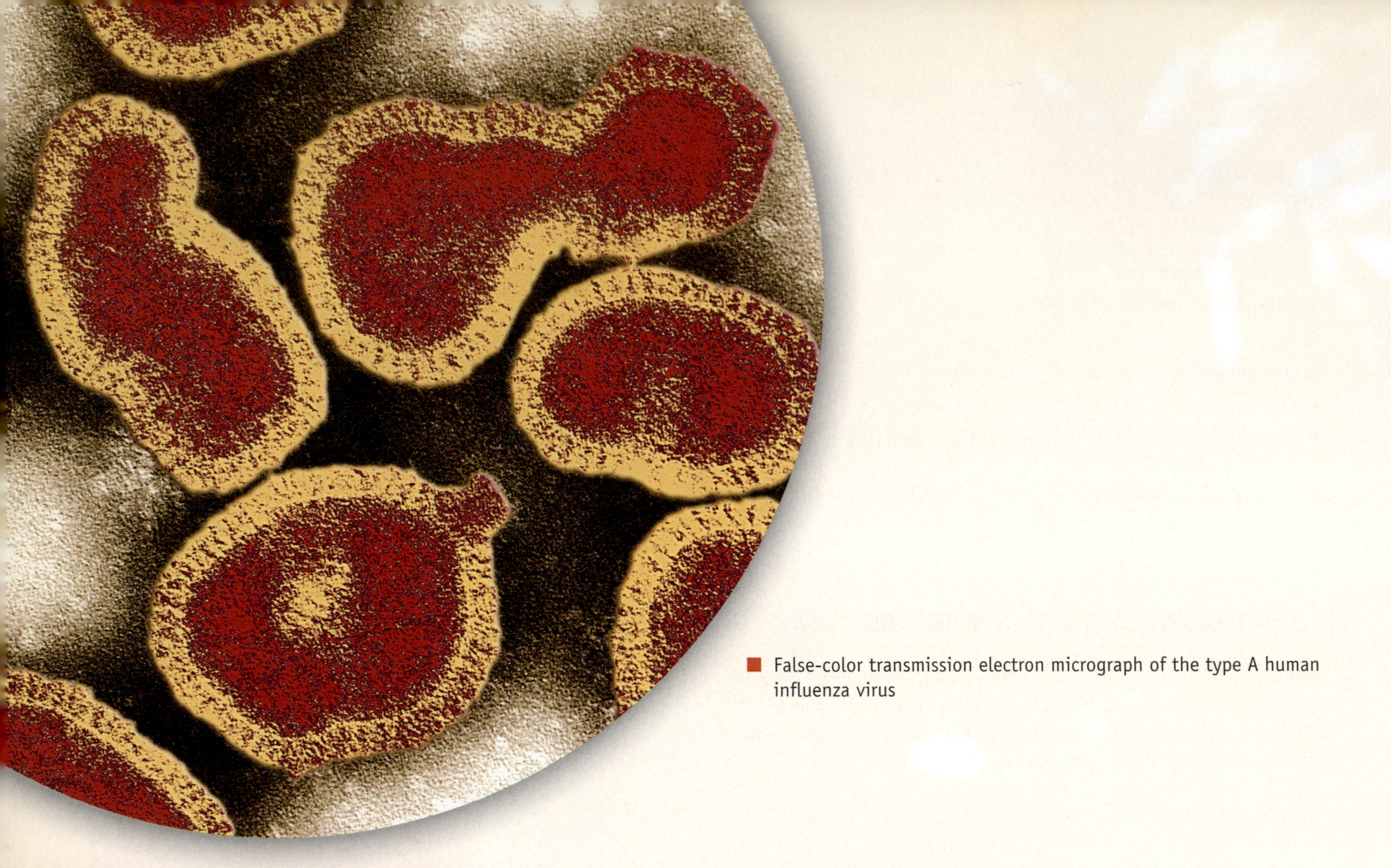

False-color transmission electron micrograph of the type A human influenza virus

PART 4

Viruses and Eukaryotic Microorganisms

The bacterial species we have examined in the previous chapters are but one of several groups of microbial agents interwoven with the lives of humans. Other prominent groups are the viruses, fungi, and parasites. Knowledge of these groups developed slowly during the early 1900s, partly because they were generally more difficult to isolate and cultivate than bacterial organisms. Also, the established methods for research into bacterial growth were more advanced than for other microorganisms, and investigators often chose to build on established knowledge rather than pursue uncharted courses of study. Moreover, the urgency to learn about the other groups was not as great because they did not appear to cause such great epidemics and pandemics.

This perception changed in the second half of the 1900s. Many bacterial diseases came under control with the advent of vaccines and antibiotics, and the increased funding for biological research allowed attention to shift to other infectious agents. The viruses finally were identified and cultivated, and microbiologists laid the foundations for their study. Fungi gained prominence as tools in biological research, and scientists soon recognized their significance in ecology and industrial product manufacturing. As remote parts of the world opened to trade and travel, public health microbiologists realized the global impact of parasitic diseases. Moreover, as concern for the health of the world's people increased, observers expressed revulsion at the thought that hundreds of millions of human beings were infected with these parasites.

In Part 4, we examine the viruses and eukaryotic microorganisms over the course of five chapters. Chapter 13 is devoted to a study of the viruses and viral-like agents, while Chapters 14 and 15 outline the multiple diseases caused by these infectious particles. In Chapter 16, the discussion moves to fungi, while in Chapter 17, the area of interest is the protozoa and the multicellular parasites. Throughout these chapters, the emphasis is on human disease. You will note some familiar diseases, such as hepatitis, chickenpox, and malaria, as well as some less familiar ones, such as dengue fever, toxoplasmosis, and schistosomiasis. The spectrum of diseases continues to unfold as scientists develop new methods for the detection, isolation, and cultivation of viruses and eukaryotic microorganisms.

MICROBIOLOGY PATHWAYS

Virology

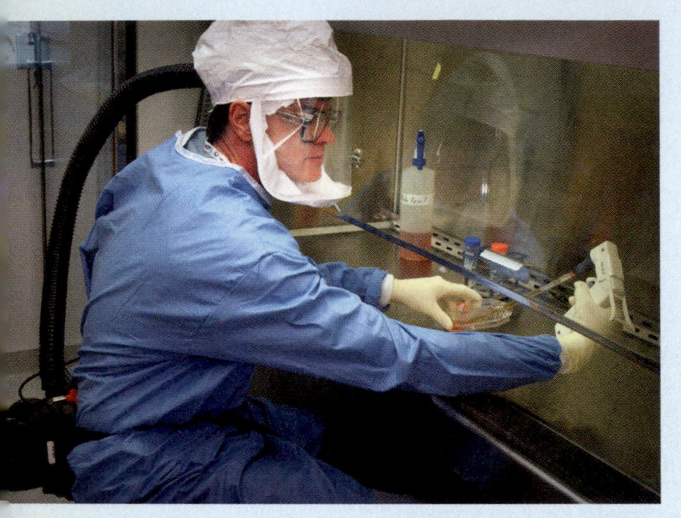

When Ed Alcamo was in college, he was part of a group of twelve biology majors. Each of them had a particular area of "expertise." John was going to be a surgeon, Jim was interested in marine biology, Walt was a budding dentist, and Ed was the local virologist. He was fascinated with viruses, the ultramicroscopic bits of matter, and at one time wrote a term paper summarizing arguments for the living or nonliving nature of viruses. (At the time, neither side was persuasive, and even his professor gracefully declined to take a stand.)

Ed never quite made it to being a virologist, but if your fascination with these infectious particles is as keen as his, you might like to consider a career in virology.

Virologists investigate dreaded diseases such as AIDS, polio, and rabies, while others (epidemiologists) investigate disease outbreaks. Virologists also concern themselves with many types of cancer, and others study the chemical interactions of viruses with various tissue culture systems and animal models.

Virologists also are working to replace agricultural pesticides with viruses able to destroy mosquitoes and other pests. Some virologists are inserting viral genes into plants and are hoping the plants will produce viral proteins to lend resistance to disease. One particularly innovative group is trying to insert genes from hepatitis B viruses into bananas. They hope that one day we can vaccinate ourselves against hepatitis B by having a banana for lunch.

If you wish to consider the study of viruses, an undergraduate major in biology would be a good choice. Because the biology of viruses is related to the biology of cells, courses in biochemistry and cell biology will be required.

Following completion of college, most virologists study for an M.D. or Ph.D. degree. M.D.s pursue virology research in the context of patients or disease and become investigators with an interest in infectious disease or epidemiology. Most Ph.D.s pursue more basic questions with academic institutions, or in industrial or governmental organizations.

A great way to find out if you have the "research bug" is to work in a college or university laboratory. Many colleges and universities with research programs employ undergraduate students (near minimal wage!) in the research laboratory. So, here is a good way to "get your feet wet." If you find it interesting, it also will enhance your chances to be accepted by a top-flight graduate school.

Not interested in the lab bench research? Virologists also can find careers in full time teaching. In addition, your knowledge can be used to pursue a career in communication, serving as a science writer or reporter. They also pursue careers in business administration or law, especially involving the pharmaceutical industry or patent law.

But no matter what avenue chosen, you should be a curious and hardworking student with a passion for science, like Ed Alcamo and your author. Do you have these interests?

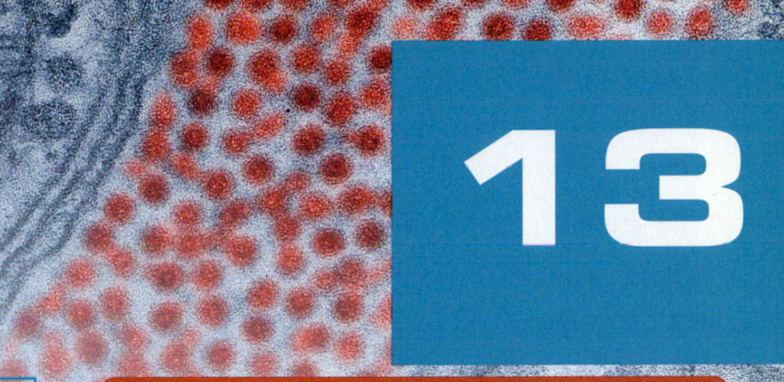

The Viruses and Virus-Like Agents

It's just a piece of bad news wrapped up in protein.
—Nobel laureate Peter Medawar (1915–1987) describing a virus

Ed Alcamo remembers the spring of 1954.

"I was a lad of thirteen growing up in the Bronx and looking forward to a carefree summer. But I could feel the tension in my parents' voices as they anticipated the months ahead, for summer was the dreaded polio season.

And sure enough, by early July the tension had turned to outright fear. I was told to avoid the public pool and the lusciously cool air-conditioned movie house. I had to report any cough or stiff neck promptly. 'Stick with your old friends,' my father told me. 'You've already got their germs.' Most of the time I was indoors, and the only baseball I got to play was in my imagination, as I listened to the Yankees every afternoon on my portable radio.

Our family was one of the lucky few to have a television, and each night we watched row upon row of iron lungs, and we saw the faces of the kids whose bodies were captured forever in their iron prisons (see FIGURE 13.1 next page). Iron lungs, I was told, help you breathe when paralysis affects the respiratory muscles. We heard and read about the daily toll from polio, where the victims lived, and how many kids had died.

But there was hope. My mom and her friends were out collecting dimes to fight polio (they called it the Mothers' March Against Polio), and the National Foundation of Infantile Paralysis said it had 75 million dimes to help fund the tests of a new vaccine—Dr. Salk's vaccine. Two million children would be getting shots. Maybe next year would be different.

Chapter Preview and Key Concepts

13.1 Foundations of Virology
- Viruses were first associated with diseases in plants.

13.2 What Are Viruses?
- Viruses are submicroscopic and have either a DNA or RNA genome.
- Viruses can have helical, icosahedral, or complex symmetry.
- Viruses infect specific organisms and tissues within multicellular organisms.

13.3 The Classification of Viruses
- Many criteria are used to classify the viruses.

13.4 Viral Replication and Its Control
- Bacteriophages undergo a lytic or lysogenic cycle of infection.
- Both naked and enveloped animal viruses share a similar series of infection and replication events.
- Some animal viruses can integrate their genome into a host cell chromosome.
- Antiviral agents interfere with specific phases of the viral replication cycle.
- Interferons are natural proteins capable of triggering proteins to block viral replication.

13.5 The Cultivation and Detection of Viruses
- Cytology can provide a rapid initial diagnosis to identify an unknown viral infection.
- Viruses can be "grown" in various types of tissue culture and detected by the formation of plaques.

MicroInquiry 13: The One-Step Growth Cycle

13.6 Tumors and Viruses
- Tumors are the result of uncontrolled cell divisions.
- A few animal viruses can represent agents of tumor development.
- Oncogenes represent cancer-causing genes.

13.7 Emerging Viruses and Virus Evolution
- Viral recombination and mutation can give rise to new viruses.
- Viruses may have preceded cellular life.

13.8 Virus-Like Agents
- Viroids lack a protein coat.
- Prions appear to lack nucleic acid.

FIGURE 13.1 **Iron Lung Ward—1953.** This iron lung ward in Boston is filled with rows of polio patients. The iron lung sealed the thoracic cavity in an air-tight chamber. The chamber created a negative pressure around the thoracic cavity, thereby causing air to rush into the lungs to equalize intrapulmonary pressure.

Q: What is the iron lung attempting to do for the patient?

Boy, next year sure was different! On April 12, 1955, at a televised news conference, Dr. Jonas Salk declared, 'The vaccine works!' The celebration was wild. Our school closed for the day. And the church bells rang, even though it was a Thursday. I could tell my mother was relieved—we had steak for dinner that night.

By summertime, things were back to normal. I was now fourteen and eager to show off my baseball skills to any girl who cared to watch. Down at the neighborhood pool I was learning how to dive (when no one was watching). And for a quarter, I got to cheer for the cavalry at the Saturday afternoon movie. Summer was back."

The polio virus is one of the smallest viruses, about the same diameter as a cell ribosome (FIGURE 13.2). At the upper end of the spectrum is the smallpox virus, which approximates the size of the smallest bacterial cells, such as the chlamydiae and mycoplasmas (see Chapter 4). You will note a simplicity in viruses that has led many microbiologists to question whether they are living organisms or fragments of genetic material leading an independent existence.

In this chapter, we study the properties of viruses, especially animal viruses, focusing on their unique mechanism for replication. We see how they are classified, how they are eliminated outside the body, and how the body deals with them during a period of disease. The chapter also discusses even more bizarre virus-like agents that can cause disease in plants and animals.

13.1 Foundations of Virology

The development of the germ theory recognized disease patterns associated with a specific bacterial species (see Chapter 1). However, some diseases resisted identification and many of these would turn out to be viral diseases. Our survey begins with a review of some important historical events that identified viruses as infective agents of disease.

Many Scientists Contributed to the Early Understanding of Viruses

KEY CONCEPT

- Viruses were first associated with diseases in plants.

■ **Tobacco mosaic disease:** A viral disease causing tobacco leaves to shrivel and assume a mosaic appearance.

In Chapter 1, we mentioned the work of the Russian pathologist Dimitri Ivanowsky who studied tobacco mosaic disease (FIGURE 13.3). Ivanowsky filtered the crushed leaves of a diseased plant and found that the clear sap dripping from the filter (rather than the crushed leaves on the filter) contained the infectious agent. Unable to see any microorganisms, Ivanowsky suspected that a filterable virus (*virus* = "a poison") was the agent of disease. Six years later, Martinus Beijerinck repeated Ivanowsky's work and demonstrated the virus was inactivated by boiling. Beijerinck concluded the disease agent was a contagious, living fluid.

In 1898, foot-and-mouth disease was suspected as being a filterable virus, implying that a virus could be transmitted among animals as well as plants. Three years later, Walter Reed and his group in Cuba provided

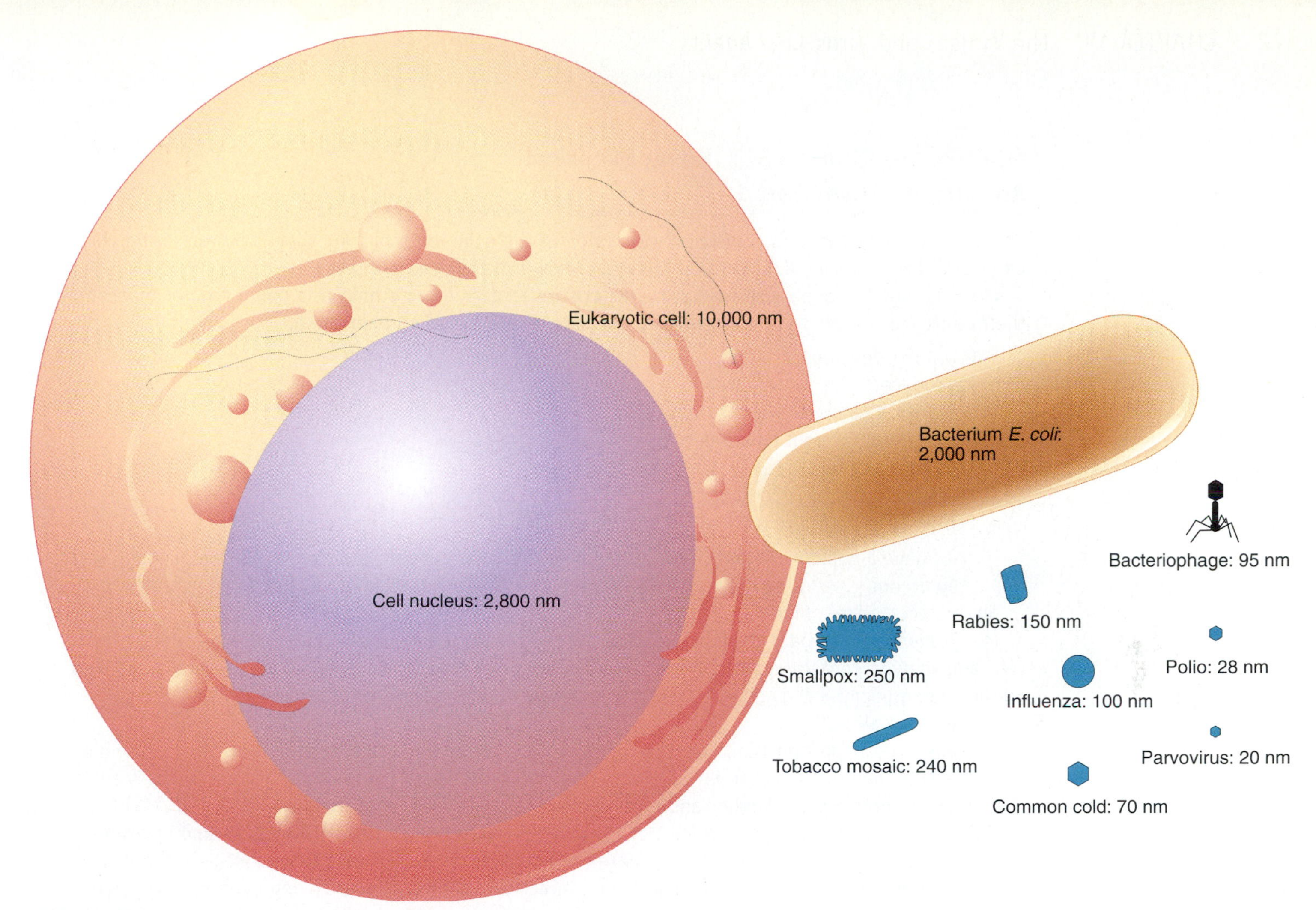

FIGURE 13.2 **Size Relationships Among Microorganisms and Viruses.** The sizes of various viruses relative to a eukaryotic cell, a cell nucleus, and the bacterial species *E. coli*. Viruses range from the very small poliovirus to the much larger smallpox virus.

Q: Propose a hypothesis to explain why viruses are so small.

(A)

(B)

FIGURE 13.3 **The Investigator, the Disease, and the Virus.** In 1892, the Russian pathologist Dimitri Ivanowsky (**A**) used an ultramicroscopic filter to separate the clear juice from crushed leaves of tobacco plants suffering from tobacco mosaic disease. He placed the juice on healthy leaves and reproduced the disease as the leaves (**B**) became shriveled with a mosaic appearance.

Q: Why was it impossible for Ivanowsky and his contemporaries to see viruses with the light microscope?

MICROFOCUS 13.1: Being Skeptical
Are Viruses Alive? Part 1

Viruses are on the edge of life. But on which side of the edge are they? Are they living or, being able to be crystallized, are they nonliving? If nonliving, how can they cause disease?

Most biology textbooks use several characteristics to define something as a living organism based on their ability to exhibit certain emergent properties. These include an ability to:

- Grow and develop.
- Reproduce.
- Establish a complex organization.
- Regulate its internal environment.
- Transform energy.
- Respond to the environment.
- Evolve by adapting to a changing environment.

Viruses have a complex organization, respond to the environment, and evolve. As independent entities, they do not grow or develop, reproduce, regulate their internal environment, or transform energy.

So, are they alive? Hmm.

Many scientists consider them living because they can reproduce if given a chance to infect a host cell, and they do contain genetic information like all living organisms. Other scientists would say they are not living because they do not satisfy all seven emergent properties of life.

So, are they alive? Hmm.

Perhaps more important than debating whether they are living is to determine how they cause disease and how they relate to the phylogenetic tree of life. On these two points, they are in the mainstream of microbiological thought and investigation.

These agents of disease also are part of the genomic sequence of all life. Biologists and microbiologists estimate that viral gene sequences make up 8 percent of the human genome. Most of these genetic sequences are remnants of ancient viral infections, the sequences having been passed down from generation to generation. However, some make us human. One ancient viral protein, for example, is critical in placental formation.

Are they alive or not? Take your pick, but make sure you read MicroFocus 13.2 first. In any case, they certainly are intriguing agents for study.

Yellow fever: A mosquito-borne viral disease of the human liver and blood.

evidence linking yellow fever with an unfilterable virus, and with this report, viruses were associated with human disease.

In 1915, English bacteriologist Frederick Twort discovered viruses that infected bacterial cells. Two years later, such viruses were identified by French-Canadian scientist Felix d'Herrelle. He called them **bacteriophages** (*phage* = "eat"), or simply **phages**, for their ability to destroy the bacterial cells. When a drop of phages was placed in a broth culture of cells, the cells disintegrated within minutes. Bacteriophages have since become important tools in genetic transduction (see Chapter 8) and in bacterial identification by phage typing (see Chapters 9 and 10).

By the early 1930s, it was generally assumed viruses were living microorganisms below the resolving power of available light microscopes. However, in 1935 the tobacco mosaic virus (TMV) was crystallized, suggesting viruses might be nonliving agents of disease (MicroFocus 13.1). Additional work with TMV revealed the virus was composed exclusively of nucleic acid and protein.

Because viruses will not grow on a nutrient agar plate the way bacterial cells do, some other form for virus cultivation was needed. In 1931, Alice M. Woodruff and Ernest W. Goodpasture described how fertilized chicken eggs could be used to cultivate some viruses. The shell of the egg was a natural culture dish containing nutrient medium, and viruses multiplied within the chick embryo tissues.

The invention of the electron microscope revealed the nature of the viruses (see Chapter 3). By 1941, virologists were beginning to visualize viruses, including TMV (FIGURE 13.4).

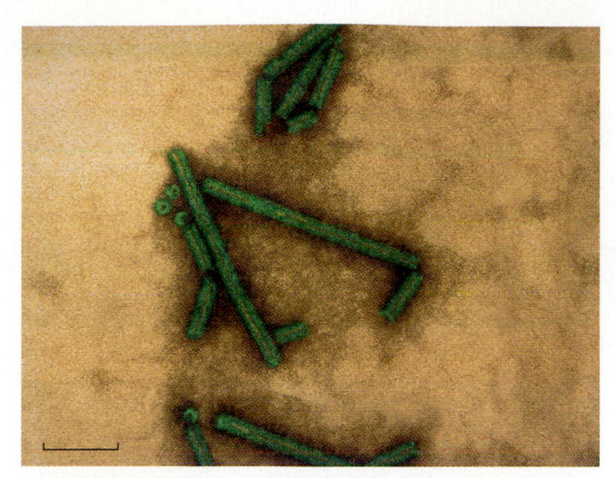

FIGURE 13.4 The Tobacco Mosaic Virus (TMV). This false-colored transmission electron micrograph of TMV shows the rod-shaped structure of the virus particles. (Bar = 80 nm.)

Q: From this micrograph, why would virologists call viruses "crystallizable" particles?

Another key development occurred in the 1940s as a result of the national polio epidemic. Attempts at vaccine production were stymied by the inability to cultivate polioviruses outside the body, but John Enders, Thomas Weller, and Frederick Robbins of Children's Hospital in Boston solved that problem. Meticulously, they developed a test tube medium of nutrients, salts, and pH buffers in which living animal cells would remain alive. In these living cells, polioviruses replicated to huge numbers, and by the late 1950s, Jonas Salk and Albert Sabin had adapted the technique to produce massive quantities of virus for use in polio vaccines (Chapter 15).

Viruses, as agents of disease, had now come into the national psyche and were agents of great interest to microbiologists and molecular biologists.

CONCEPT AND REASONING CHECKS

13.1 Describe the major events leading to the recognition of viruses as pathogens.

13.2 What Are Viruses?

Today more than 5,000 viruses have been identified. Amazingly, this is only a small proportion of the estimated 400,000 different viruses virologists believe may exist.

Viruses Are Tiny Infectious Agents

KEY CONCEPT

- Viruses are submicroscopic and have either a DNA or RNA genome.

Viruses are small, obligate, intracellular particles; that is, most can be seen only with the electron microscope and they must infect and take over a host cell in order to replicate. This is because they lack the chemical machinery for generating energy and synthesizing large molecules. Viruses, therefore, must find an appropriate eukaryotic or prokaryotic cell in which they can replicate—and, as a result, often cause disease.

Viruses have some unique features not seen with the living microorganisms. They have no organelles, no cytoplasm, and no cell nucleus or nucleoid (see Chapter 4). Instead, they are comprised of two basic components: a nucleic acid genome and a surrounding coat of protein; thus, as Peter Medawar remarked (chapter opening quote), a virus is "just a piece of bad news [meaning they cause disease] wrapped up in protein."

The **viral genome** contains either DNA or RNA, but not both, and the nucleic acid occurs in either a double-stranded or a single-stranded form. Usually the nucleic acid is a linear or circular molecule, although in some instances (as in influenza viruses) it exists as separate, nonidentical segments. The viral genome is folded or coiled, which allows the viruses to maintain their extremely small size.

The protein coat of a virus particle, called a **capsid**, gives shape or symmetry to the virus (FIGURE 13.5). Generally, the capsid is subdivided into individual protein subunits called **capsomeres** (the organization of capsomeres yields the viral symmetry). The capsid with its enclosed genome is referred to as a **nucleocapsid**.

The capsid also provides a protective covering for the viral genome because the construction of its amino acids resists temperature, pH, and other environmental fluctuations. In some viruses, special capsid proteins called **spikes** help attach the virus to the host cell and facilitate penetration of the cell. Viruses composed solely of a nucleocapsid are called **naked viruses** (Figure 13.5A).

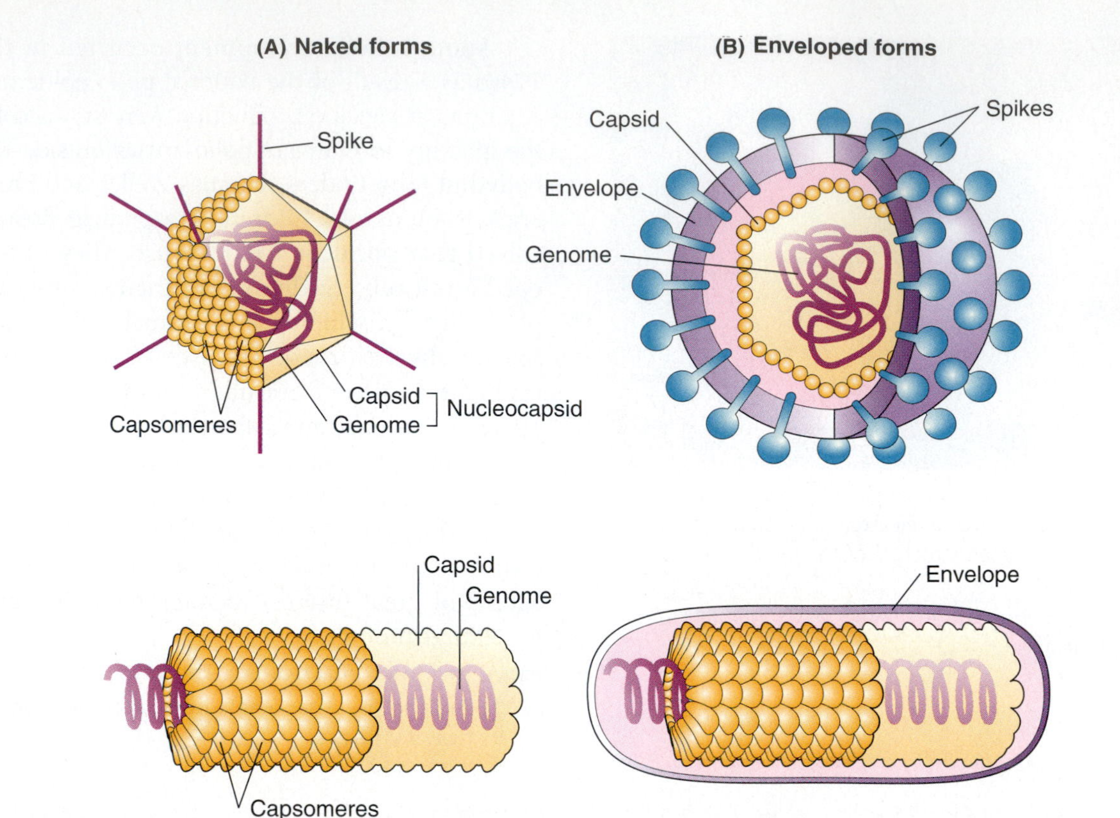

FIGURE 13.5 **The Components of Viruses.** (**A**) Naked viruses consist of a nucleic acid genome (either DNA or RNA) and a protein capsid. Capsomere units are shown on one face of the capsids. Spikes may be present on the capsid. (**B**) Enveloped viruses have an envelope that surrounds the nucleocapsid. Again spikes usually are present.

Q: What important role do spikes play in the infective behavior of viruses?

Many other viruses are surrounded by a flexible membrane known as an **envelope** and are referred to as **enveloped viruses** (Figure 13.5B). The envelope is composed of lipids and protein, similar to the host cell membrane; in fact, it is acquired from the host cell during replication and is unique to each type of virus. These viruses may lose their infectivity if the envelope is destroyed. Also, when the envelope is present, the symmetry of the capsid may not be apparent because the envelope is generally a loose-fitting structure over the nucleocapsid. Many enveloped viruses also contain spikes projecting from the envelope. These proteins also function for attachment and host cell penetration.

A completely assembled and infectious virus outside its host cell is known as a **virion**. MicroFocus 13.2 revisits the question if viruses are alive.

CONCEPT AND REASONING CHECKS

13.2 Identify the role of each structure found on (A) a naked and (B) an enveloped virus.

Viruses Are Grouped by Their Shape

KEY CONCEPT

- Viruses can have helical, icosahedral, or complex symmetry.

Viruses can be separated into groups based on their nucleocapsid symmetry; that is, their three-dimensional shapes (FIGURE 13.6). Certain viruses, such as rabies and tobacco mosaic viruses, exist in the form of a helix and are said to have **helical symmetry** (Figure 13.6A). The helix is a tightly wound coil resembling a corkscrew or spring.

Other viruses, such as herpes simplex and polioviruses, have the shape of an icosahedron and hence, **icosahedral** (*icos* = "twenty," *edros* = "side") **symmetry** (Figure 13.6B). The icosahedron then has 20 triangular faces and 12 corners.

A few viruses have a combination of helical and icosahedral symmetry, a construction described as **complex symmetry** (Figure 13.6C). Some bacteriophages, for example, have complex symmetry, with an icosahedral

MICROFOCUS 13.2: Environmental Microbiology

Are Viruses Alive? Part 2

The news headline states, "*Virus Discovered That Is As Large As Some Bacteria!*" But how can that be? Dogma says that viruses are so small you cannot see them with the light microscope. So, could the news media have it wrong?

In 2002, researchers at the CNRS in France were looking for *Legionella* bacteria (see Chapter 9) in water samples when they stumbled upon a virus in water-borne amoebae. The "virus" was first identified in England during a pneumonia outbreak in 1992, but the investigators thought it was a bacterial cell because of its large size. However, closer observation by the French group showed that the coccus-like particles in the amoebae had the structure and morphologic characteristics of a virus. The name mimivirus (*mimi*cking *virus*) was given to the virus and a separate family, the *Mimiviridae*, established.

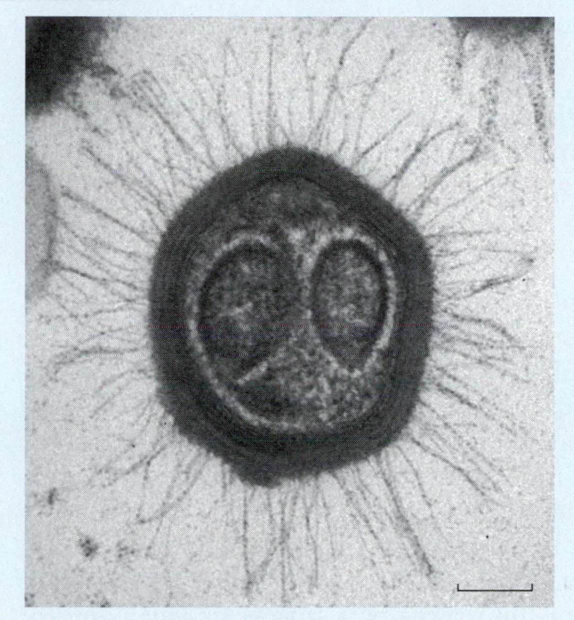

Transmission electron micrograph of the mimivirus. (Bar = 150 nm.)

The mimivirus is a naked, double-stranded DNA virus about 400 nm in diameter. Its genome was sequenced in 2004 and found to contain 1262 genes, which is more than three times the size of other virus genomes. Like other viruses though, it cannot convert energy or replicate on its own. But that's where the similarities end.

The mimiviruses contain both DNA and RNA, something viruses by definition are not supposed to contain. They also have seven genes shared with all three domains of life: *Bacteria, Archaea,* and *Eukarya*. However, it is not yet known if the virus uses these genes.

So, are viruses alive? These discoveries with the mimivirus certainly blur the lines between life and nonlife.

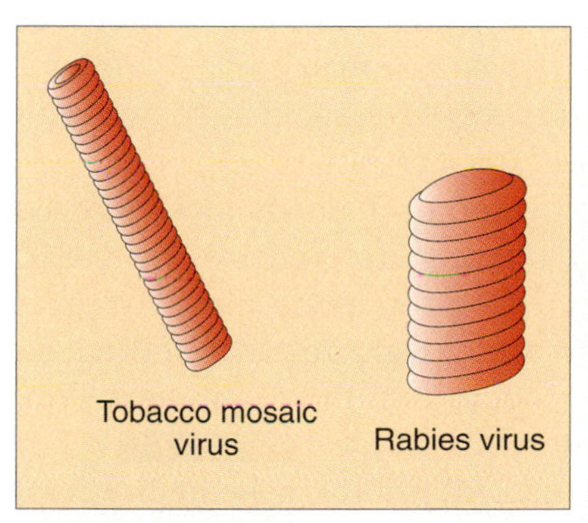

(A) Helical viruses

(B) Icosahedral viruses

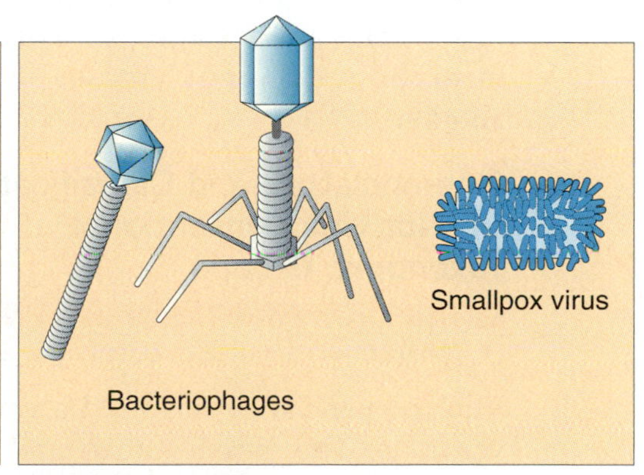

(C) Complex viruses

FIGURE 13.6 Various Viral Shapes. Viruses exhibit numerous variations in symmetry. The nucleocapsid may have helical symmetry (**A**) in the tobacco mosaic and rabies viruses or (**B**) icosahedral symmetry typical of the herpesviruses and polioviruses. In other viruses, (**C**) a complex symmetry exists. The smallpox virus has a series of rod-like filaments embedded within the membranous envelope at its surface.

Q: Although the bacteriophages are classified as complex, what two symmetries do these viruses exhibit?

head and a collar and tail assembly in the shape of a helical sheath. Poxviruses, by contrast, are brick shaped, with submicroscopic filaments occurring in a swirling pattern at the periphery of the virus.

Viruses Have a Host Range and Tissue Specificity

KEY CONCEPT

- Viruses infect specific organisms and tissues within multicellular organisms.

As a group, viruses can infect almost any cellular organism. There are specific viruses able to infect bacterial cells, while others infect protozoa, fungi, plants, or animals. A virus' **host range** refers to what organisms (hosts) the virus can infect and it is based on a virus' capsid structure. Most viruses have a very narrow host range. A specific bacteriophage, for example, only infects specific bacterial species, the smallpox virus only infects humans, and the poliovirus infects only humans and primates. A few viruses may have a broader host range, as the rabies viruses infect humans and most warm-blooded animals.

Even within its host range, many viruses only infect certain cell types or tissues within a multicellular plant or animal. This limitation is called **tissue tropism** (tissue attraction). For example, the host range for the human immunodeficiency virus (HIV) is a human. In humans, HIV primarily infects a specific group of white blood cells called T helper cells because the envelope has protein spikes for binding to receptor molecules on these cells. The virus does not infect cells in other tissues or organs such as the heart or liver. Rabies virus is best at infecting cells of the nervous system and brain because its envelope contains proteins recognizing receptors only on these tissues. Therefore, a virus' host range and tissue tropism are linked to infectivity. If a potential host cell lacks the appropriate receptor or the virus lacks the complementary protein, the virus usually cannot bind to or infect the host cell.

CONCEPT AND REASONING CHECKS

13.3 How does viral structure determine host range and tissue tropism?

13.3 The Classification of Viruses

A classification system for viruses that is similar to that for living organisms (see Chapter 3) has been slow in coming, in part because sufficient data are not always available to determine how different viruses relate to one another.

Nomenclature and Classification Do Not Use Conventional Taxonomic Groups

KEY CONCEPT

- Many criteria are used to classify the viruses.

Viral nomenclature has used a variety of conventions. The measles virus and poxviruses, for example, are named after the disease they cause; the Ebola and Marburg viruses after where they were originally isolated; and the Epstein-Barr virus after the researchers who studied it. Others are named after morphologic factors—the coronaviruses (*corona* = "crown") have a crown-like capsid and the picornaviruses (*pico* = "small"; *rna* = "ribonucleic acid") are very small viruses with an RNA genome.

One classification scheme is loosely based on the body tissues affected by the virus. Though inexact, this scheme places viruses into four convenient groups, depending on whether they replicate in the respiratory, skin, visceral, or nervous tissue (TABLE 13.1). For convenience, we will use this scheme in Chapters 14 and 15.

A more encompassing classification system is being devised by the International

TABLE 13.1 Classification of Human Viral Diseases by Tissue Affected

Group	Tissues Affected	Important Diseases
Pneumotropic	Respiratory system	Influenza, respiratory syncytial disease, rhinovirus infection, SARS
Dermotropic	Skin and subcutaneous tissues	Chickenpox, warts, measles, mumps, smallpox, rubella
Viscerotropic	Blood, lymphatic, and gastrointestinal organs	Yellow fever, dengue fever, infectious mononucleosis, viral fevers, viral gastroenteritis, hepatitis A, hepatitis B, AIDS
Neurotropic	Central nervous system	Rabies, polio, West Nile fever (encephalitis)

Committee on Taxonomy of Viruses (ICTV). At this writing, higher order taxa (phyla and classes) have not been developed. In 2005, three orders were recognized that comprised more than 80 families, each ending with *-viridae* (e.g., *Herpesviridae; Coronaviridae*). Viruses have been categorized into hundreds of genera; each genus name ends with the suffix *-virus* (e.g., *Herpesvirus; Coronavirus*). Names (binomial nomenclature) have not yet been agreed upon for species. In this text, we use the colloquial names (e.g., herpesviruses, coronaviruses). Note that due to the rapid changes taking place in viral taxonomy, some new names may be in use by the time you read this text.

Viruses from 24 different families cause disease in humans. A selection of viral families affecting humans, together with some of their characteristics, is presented in TABLE 13.2. These viruses have been split into two board classes based on their genome type and strand type.

DNA viruses contain either single-stranded or double-stranded DNA genomes, which are replicated by direct DNA-to-DNA copying. Most DNA viruses replicate in the host cell's nucleus. One exception to direct replication is the hepadnaviruses, which replicate their DNA through an RNA intermediate catalyzed by reverse transcriptase.

RNA viruses contain either single-stranded or double-stranded RNA genomes, which are replicated by direct RNA-to-RNA copying. Some of these viruses, such as the picornaviruses and coronaviruses, have their RNA genome in the form of messenger RNA (mRNA). These viruses are referred to as **positive-strand (+ strand) RNA**.

Other RNA viruses, such as the orthomyxoviruses and paramyxoviruses, have RNA genomes consisting of RNA strands that would be complementary to a mRNA; these genomes are referred to as **negative-strand (– strand) RNA**.

Although usually grouped with the RNA viruses, the retroviruses are replicated indirectly through a DNA intermediate (RNA-to-DNA-to-RNA). Each virion contains two copies of + strand RNA. During the infection process, a DNA intermediate will be formed using a reverse transcriptase enzyme carried with the virion.

A complete listing of all families based on the 2005 8th ICTV Report can be found on their Web site (http://www.ncbi.nlm.nih.gov/ICTVdb/Ictv/fr-fst-a.htm).

CONCEPT AND REASONING CHECKS

13.4 Draw a concept map for virus families using nucleic acid type, strand type (single or double; + or –), and naked or enveloped virions.

■ Reverse transcriptase: An enzyme that copies single-stranded RNA into double-stranded DNA.

TABLE
13.2 Major Families of Human Viruses and Their Characteristics

Family	Strand Type*	Capsid Symmetry	Envelope or Naked Virion	Diameter (nm)	Disease Examples
DNA Viruses					
Poxviridae	Double	Complex	Envelope	170–300	Smallpox, monkeypox
Herpesviridae	Double	Icosahedral	Envelope	150–200	Cold sores, genital herpes, chickenpox, shingles, infectious mononucleosis
Adenoviridae	Double	Icosahedral	Naked	70–90	Common cold, viral meningitis
Papovaviridae	Double	Icosahedral	Naked	45–55	Warts, genital warts, cervical cancer
Hepadnaviridae	Double (w/RNA intermediate)	Icosahedral	Envelope	42	Hepatitis B, liver cancer
Parvoviridae	Single	Icosahedral	Naked	18–26	Fifth disease
RNA Viruses					
Reoviridae	Double	Icosahedral	Naked	60–80	Gastroenteritis
Picornaviridae	Single (+)	Icosahedral	Naked	28–30	Polio, some colds, hepatitis A
Caliciviridae	Single (+)	Icosahedral	Naked	35–40	Gastroenteritis
Togaviridae	Single (+)	Icosahedral	Envelope	60–70	Rubella, encephalitis
Flaviviridae	Single (+)	Icosahedral	Envelope	40–50	Yellow fever, dengue fever, hepatitis C, West Nile fever (encephalitis)
Coronaviridae	Single (+)	Helical	Envelope	80–160	SARS
Filoviridae	Single (–)	Helical	Envelope	80–10,000	Ebola and Marburg hemorrhagic fevers
Bunyaviridae	Single (–)	Helical	Envelope	90–120	Hantavirus pulmonary syndrome
Orthomyxoviridae	Single (–)	Helical	Envelope	90–120	Influenza
Paramyxoviridae	Single (–)	Helical	Envelope	150–300	Mumps, measles
Rhabdoviridae	Single (–)	Helical	Envelope	70–380	Rabies
Arenaviridae	Single (–)	Helical	Envelope	50–300	Lassa fever
Retroviridae	Single (+) (w/DNA intermediate)	Icosahedral	Envelope	80–130	AIDS, human adult T-cell leukemia

*(+) = positive-stranded; (–) = negative-stranded

13.4 Viral Replication and Its Control

The process of viral replication is one of the most remarkable events in nature. A virion invades a living cell a thousand or more times its size, uses the metabolism of the cell to produce copies of itself, and often destroys the host cell when new virions are released.

Replication has been studied in a wide range of virions and their host cells. We examine the bacteriophages first and then discuss the animal viruses.

The Replication of Bacteriophages Is a Five-Step Process

KEY CONCEPT

- Bacteriophages undergo a lytic or lysogenic cycle of infection.

One of the best studied processes of replication is that carried on by bacteriophages of the T-even group (T for "type"). Bacteriophages T2, T4, and T6 are in this group. They are large,

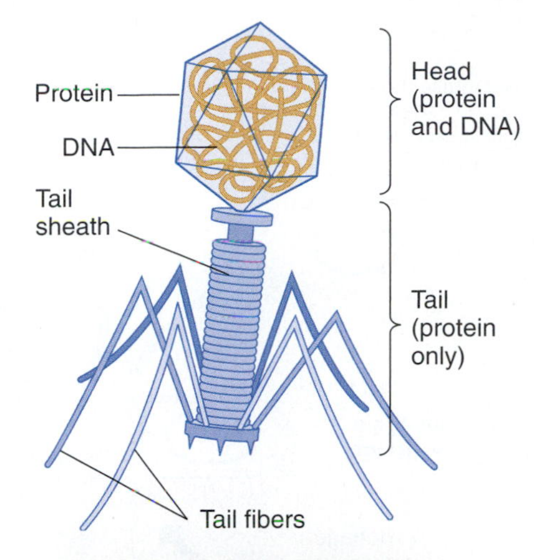

(A)

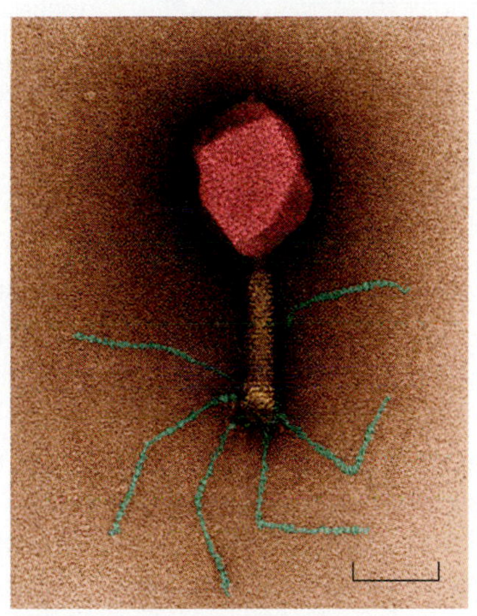

(B)

FIGURE 13.7 **Bacteriophage Structure.** (**A**) The structure of a bacteriophage consists of the head, inside of which is the nucleic acid, and the tail. The tail sheath is hollow to allow transfer of nucleic acid during infection. (**B**) A false-color transmission electron micrograph of a T-even bacteriophage (Bar = 70 nm.)

Q: What is the purpose of the tail fibers?

complex, naked DNA virions with the characteristic head and tail of bacteriophages (FIGURE 13.7). They contain tail fibers, which function similar to spikes on animal viruses and identify what bacterial species the phage will be able to infect. The T-even phages are **virulent viruses**, meaning they lyse the host cell while carrying out a **lytic cycle** of infection.

It is important to note that the nucleic acid in a phage contains only a few of the many genes needed for viral synthesis and replication. It contains, for example, genes for synthesizing viral structural components, such as capsid proteins, and for a few enzymes used in the synthesis; but it lacks the genes for many other key enzymes, such as those used during nucleic acid synthesis. Therefore, its dependence on the host cell is substantial.

We shall use phage replication in *E. coli* as a model for the viruses. An overview of the process is presented in FIGURE 13.8.

1. **Attachment** is the first phase in the replication cycle of a virulent phage. When phage and bacterial cell collide randomly, sites on the phage's tail fibers must match with a complementary receptor site on the cell wall of the bacterium. The actual attachment consists of a weak chemical union between virion and receptor site. In some cases, the bacterial flagellum or pilus contains the receptor site.

2. **Penetration**, the second stage, is accomplished when the tail of the phage releases the enzyme lysozyme to dissolve a portion of the bacterial cell wall. Then the tail sheath contracts and the tail core drives through the cell wall. As the tip of the core reaches the cell membrane below, the DNA passes through the hollow tail core and on through the cell membrane into the bacterial cytoplasm. For most bacteriophages, the capsid remains outside.

3. **Biosynthesis** involves the production of new phage genomes and capsid parts. Once inside the host cell, phage genes code for the disruption of the host chromosome. The phage DNA then uses the bacterial nucleotides and cell enzymes to synthesize multiple copies of its genome. Messenger RNA molecules transcribed from phage DNA appear in the cytoplasm, and the biosynthesis of phage enzymes and capsid proteins begins. Bacterial ribosomes, amino acids, and enzymes are all enlisted for biosynthesis. Because viral capsids are repeating units of capsomeres, a relatively simple genetic code can be used over and over.

Maturation is the assembly of viral parts into complete virus particles. The enzymes encoded by viral genes guide the assembly in step-by-step fashion. In one area, phage heads and tails are assembled from protein subunits; in another, the heads are packaged with DNA; and in a third, the tails are attached to the heads.

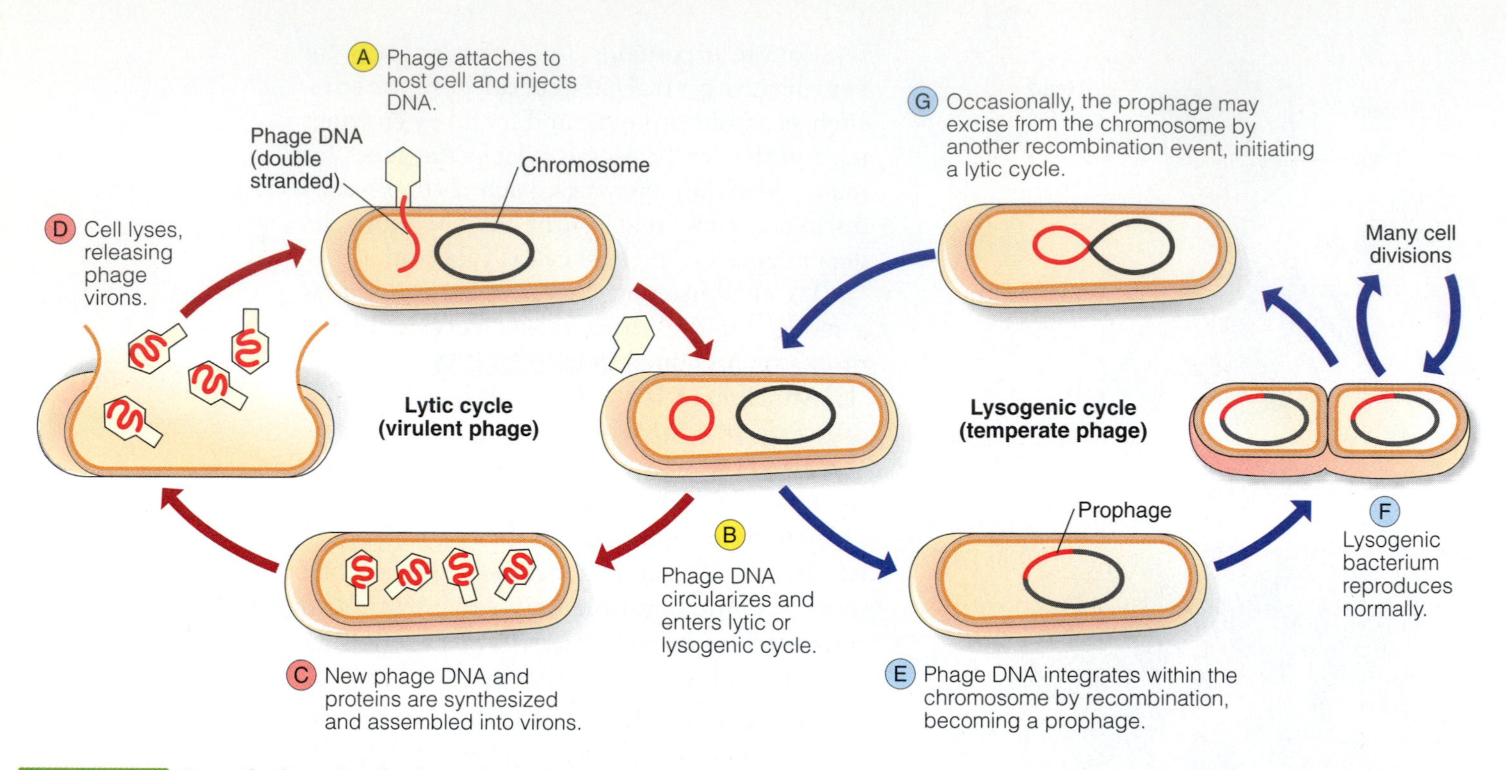

FIGURE 13.8 **Bacteriophage Replication.** The pattern of replication in bacteriophages (tail fibers not shown) can involve a lytic or lysogenic cycle. Replication serves as a model for the replication of animal viruses.

Q: How can bacteriophages facilitate the horizontal transfer of genetic material?

4. **Release** involves the exit of virions from their bacterial shell. For bacteriophages, this also is called the **lysis stage** because the cell ruptures. For some phages, the important enzyme in this process is lysozyme, encoded by the bacteriophage genes late in the sequence of events. The enzyme degrades the bacterial cell wall, and the newly released bacteriophages are set free to infect more bacterial cells.

MicroFocus 13.3 describes the use of phages as a way to combat bacterial diseases.

CONCEPT AND REASONING CHECKS

13.5 Summarize the steps in a lytic cycle of infection by a bacteriophage.

Other phages interact with prokaryotic cells in a slightly different way, called a **lysogenic cycle** (Figure 13.8). For example, lambda (λ) phage also invades *E. coli* but without immediately causing cell lysis. Instead, the phage DNA integrates into the bacterial chromosome as a **prophage** (Figure 13.8E). Bacteriophages participating in this cycle are known as **temperate phages**. The bacterial cell survives the infection and continues to grow and divide. As the bacterial cell undergoes DNA replication and binary fission, the prophage is copied and vertically transferred to daughter cells as part of the replicated bacterial chromosome (Figure 13.8F). Thus, as cells divide, each daughter cell is "infected;" that is, it contains the viral genome as a prophage.

Such binary fissions can continue for an undefined period of time. Usually at some point, the bacterial cells become stressed (e.g., lack of nutrients, presence of noxious chemicals). This triggers the prophage to excise itself from the bacterial chromosomes (Figure 13.8G) and go through a lytic cycle, lysing the bacterial cells as new λ phage are released.

CONCEPT AND REASONING CHECKS

13.6 Why would it be advantageous for a phage to carry out a lysogenic cycle rather than a lytic cycle?

Animal Virus Replication Has Similarities to Phage Replication

KEY CONCEPT

- Both naked and enveloped animal viruses share a similar series of infection and replication events.

Unlike bacteriophages, animal viruses attach to receptor sites on the host plasma membrane rather than the cell wall. However, animal

MICROFOCUS 13.3: History/Public Health/Biotechnology

Phage Therapy

When bacteriophages were identified in 1915, some scientists, such as their discoverer Felix d'Herrelle of the Pasteur Institute in Paris, started to promote phages as therapy for curing dreaded bacterial diseases such as cholera and bubonic plague. It even captured the imagination of writers like Sinclair Lewis who, in his 1925 novel *Arrowsmith,* had his idealistic young physician, Martin Arrowsmith, travel to the West Indies to treat bubonic plague—with phage therapy.

D'Herrelle and others believed that if phages could destroy bacterial cells in test tubes, why not try them in the human body? Unfortunately, the bacteriophages turned out to be highly specific viruses with a narrow host range and would attack only certain bacterial strains. Moreover, no one knew much about them at the time or had the means to work with them effectively. Phage therapy never was deeply embraced in the United States, partly due to the rise of antibiotics in the 1940s and 1950s. Who needed phage therapy! The same could not be said for scientists Eastern Europe and the former Soviet Union, where research with phages continued.

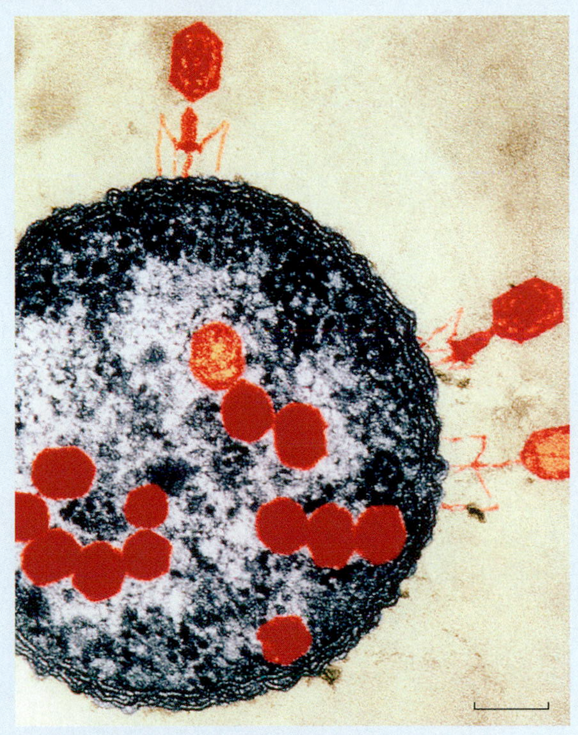

Phage infecting an *E. coli* cell. (Bar = 100 nm)

Fast-forward to the modern era. In the 1980s, the Polish microbiologist Stefan Slopek identified a bacterial species in the blood of an ill patient, then searched out and isolated a bacteriophage specific for that pathogen. He injected a solution of the phages into the patient. Over a period of days, the infection gradually resolved. Slopek tried again—this time with 137 patients. Each patient benefited from the treatment, and several patients seemed completely cured.

Today, much more is known about the genetics and biology of phages than in the 1940s—and biotechnology offers techniques to engineer the bacteriophages to attack specific bacterial targets.

However, there are hurdles to overcome. For example, the narrow host range meant having a huge stable of different bacteriophages or a phage "cocktail" composed of numerous phages with a broad host range to attack a variety of bacterial diseases. Today, it is possible to prepare pure preparations of phages with the appropriate host range and in sufficient quantity to be effective.

Another historical problem concerned the lytic potential of phage. When phages lyse gram-negative bacteria, a large amount of cell wall material representing a dangerous toxin can be released into the body, possibly causing circulatory shock. In 2004, researchers in Vienna, Austria engineered bacteriophage that would kill bacteria but not lyse the cells. More surprising was the discovery that these phages were ten times more lethal than the natural phage.

In 2006, the US Food and Drug Administration (FDA) approved a "cocktail" of six purified phages that can be safely sprayed on luncheon meats (cold cuts, hot dogs, sausages) to kill any of 170 strains of *Listeria monocytogenes*. As described in Chapter 10, listeria is a serious foodborne disease, primarily affecting newborns, pregnant women, and immunocompromised individuals. Plans are in the works to get FDA approval for another bacteriophages to kill *Escherichia coli* that can contaminate beef (see figure).

So, the phage therapy as envisioned by d'Herrelle is on the rebound and holds great promise for the treatment of acute and chronic infections, as well as in the food industry.

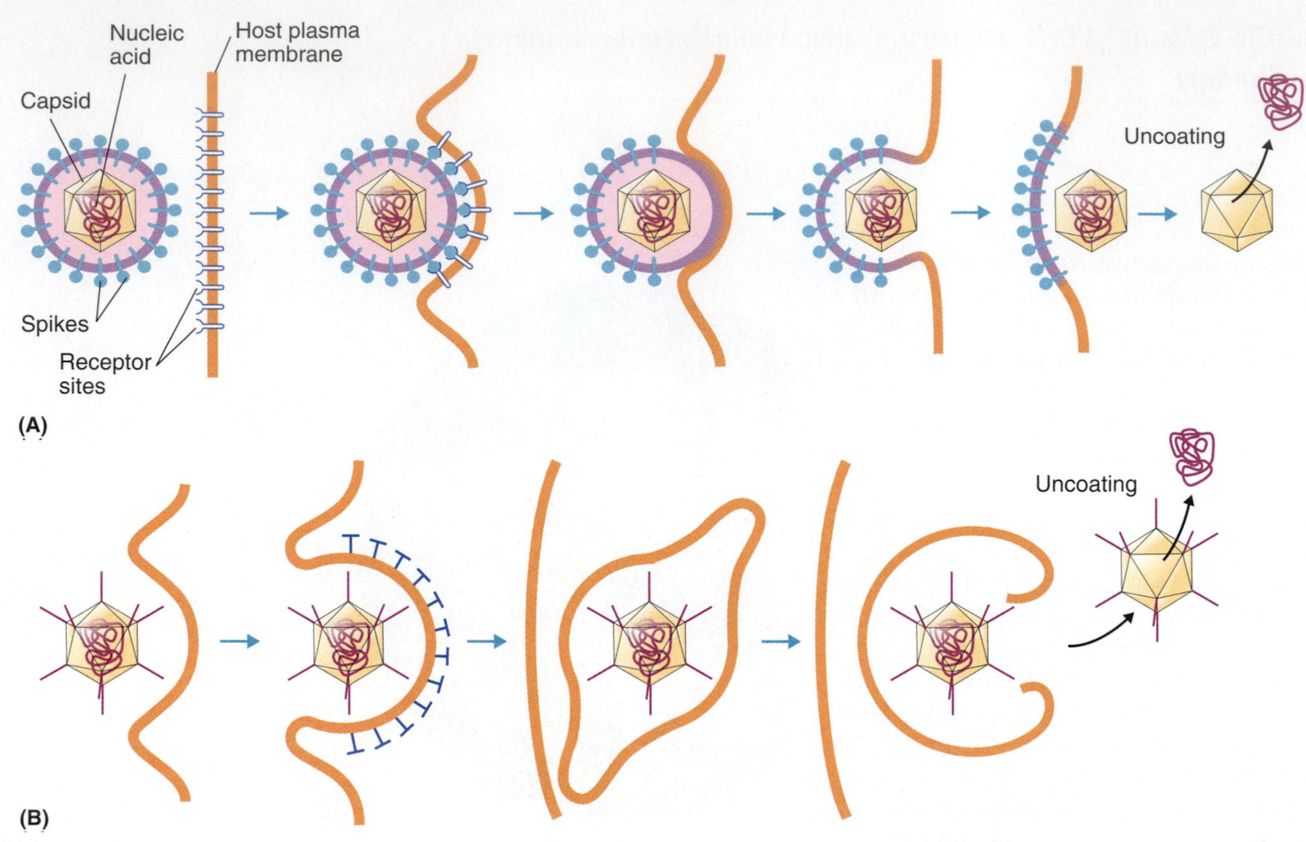

FIGURE 13.9 **The Entry of Animal Viruses into Their Host Cells.** Animal viruses enter their host cells by one of two pathways. (**A**) An enveloped virus, such as HIV, contacts the plasma membrane and the spikes interact with receptor sites on the membrane surface. The envelope fuses with the plasma membrane, and the nucleocapsid passes into the host cell's cytoplasm. Then, the nucleic acid is released (uncoated). (**B**) A second entry method also involves a specific interaction between spikes and receptor sites on the plasma membrane. However, for this adenovirus, it undergoes endocytosis into a vacuole forming around the virus. The vacuole pinches off into the cytoplasm. Loss of the vacuole membrane liberates the nucleocapsid into the cytoplasm and uncoating occurs.

Q: Propose a hypothesis to explain why two different entry mechanisms exist for virion entry into a host cell.

■ Endocytosis: A process by which insoluble materials are taken into a cell by invagination of the plasma.

viruses have no tail fibers, so the attachment sites often are the spikes distributed over the surface of the capsid (adenovirus) or envelope (HIV). The spikes are what determine the host range for the virus.

An understanding of the attachment phase can have practical consequences because the host's receptor sites are inherited characteristics; that is, the sites vary from person to person, accounting for the susceptibility of different individuals to a particular virus.

Penetration is also different. Phages inject their DNA into the host cell cytoplasm, but animal viruses usually are taken into the cytoplasm as intact nucleocapsids. For viruses like HIV, the viral envelope fuses with the plasma membrane and releases the nucleocapsid into the cytoplasm (FIGURE 13.9A).

In other cases, such as the adenoviruses or influenzavirus, the virion is taken into the cell by endocytosis. The cell then enfolds the virion within a vacuole and brings it into the cytoplasm (FIGURE 13.9B). Once in the cell, the vacuole membrane breaks down, releasing the nucleocapsid into the cytoplasm. In both examples, the capsid is separated from the genome in a process referred to as **uncoating**.

One way we split the virus families was based on whether they have DNA or RNA as their genetic information. The DNA of a DNA virus supplies the genetic codes for enzymes that synthesize viral parts from available building blocks. Although the poxviruses replicate entirely in the host cell cytoplasm, most of the DNA viruses employ a division of labor: DNA genomes are synthesized in the host cell

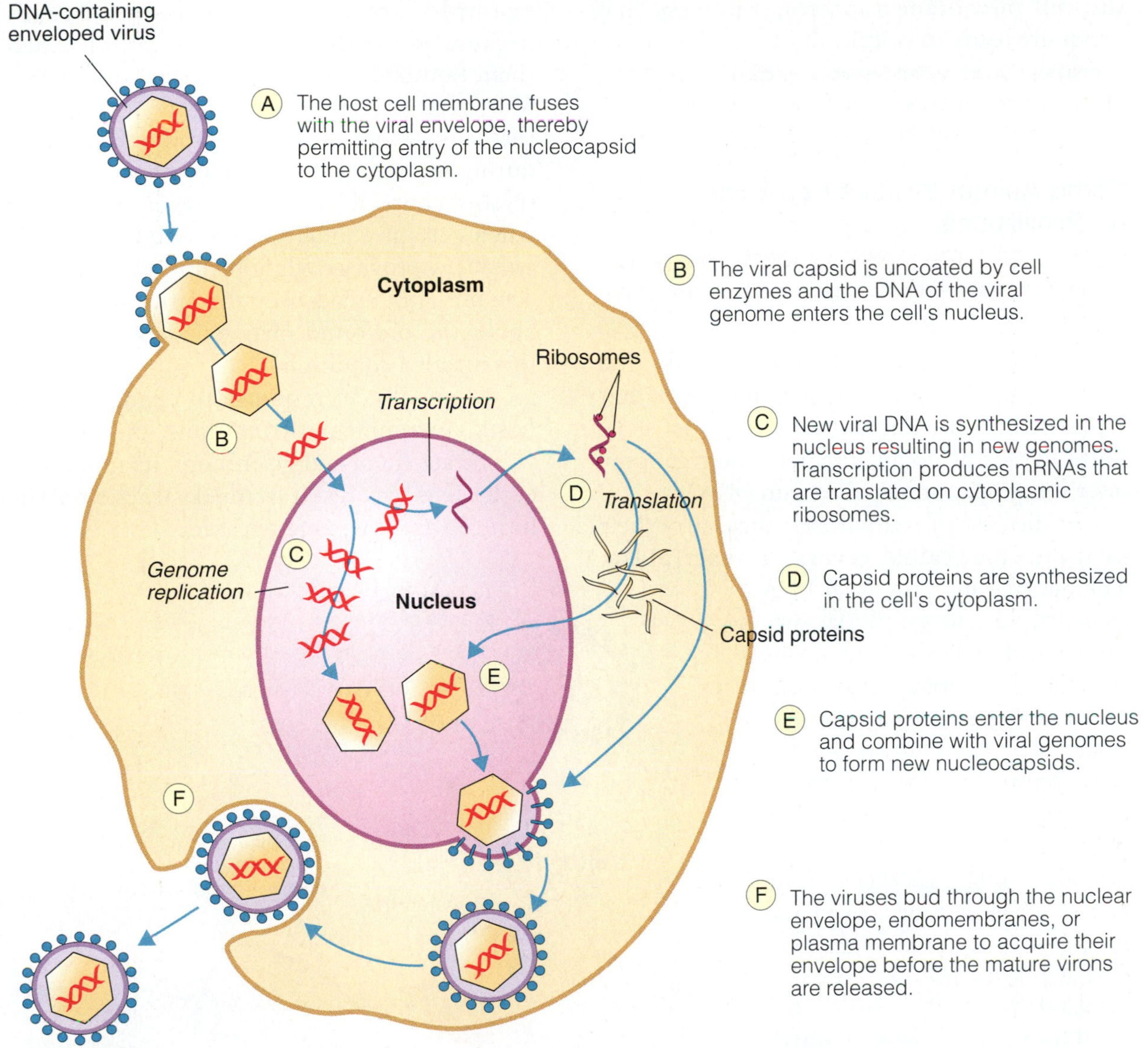

FIGURE 13.10 **Replication of a DNA Animal Virus.** The virus illustrated here is a herpesvirus (such as one that might cause chickenpox), and the host cell is from human skin.

Q: Why must the viral DNA enter the cell nucleus to replicate?

nucleus, and capsid proteins are produced in the cytoplasm (FIGURE 13.10). The proteins then migrate to the nucleus and join with the nucleic acid molecules for assembly. Adenoviruses and herpesviruses follow this pattern.

RNA viruses follow a slightly different pattern. The + strand RNA can act as a messenger RNA molecule (see Chapter 6) and immediately begin supplying the codes for protein synthesis. Other – strand RNA viruses, such as the orthomyxoviruses, use their RNA as a template to synthesize a complementary strand of RNA. Usually the enzyme RNA polymerase is present in the virus to synthesize the complementary strand. The synthesized RNA then is used as a messenger RNA molecule for protein synthesis.

The final steps of viral replication may include the acquisition of an envelope. In this step, envelope proteins are synthesized and incorporated into a nuclear or cytoplasmic membrane, or the plasma membrane. Then, the virus pushes through the membrane, forcing a portion of the membrane ahead of it and around it, resulting in an envelope. This process, called **budding**, need not necessarily kill the cell during the virus' exit. The same cannot be said for naked viruses, however. They leave the cell when the cell membrane ruptures, a process that generally leads to cell death.

CONCEPT AND REASONING CHECKS

13.7 Distinguish between the ways animal viruses can penetrate host animal cells.

Some Animal Viruses Can Exist as Proviruses

KEY CONCEPT

- Some animal viruses can integrate their genome into a host cell chromosome.

In a replication cycle, DNA viruses like the some of the herpesviruses may integrate their DNA into a chromosome of the host cell. Retroviruses also can integrate their genome after it has been transcribed into DNA.

In the case of retroviruses, they carry their own enzyme, called **reverse transcriptase**. The enzyme uses the viral RNA as a template to synthesize single-stranded DNA (the terms reverse transcriptase and retrovirus are derived from this reversal of the central dogma; see Chapter 7). Once formed, the DNA serves as a template to form a complementary DNA strand (FIGURE 13.11). The viral RNA is then destroyed and the DNA complex moves to the cell nucleus using the cell's cytoskeleton.

The herpesvirus or retrovirus DNA now enters the cell nucleus and integrates into one of the host cell's chromosomes, where the viral genome is known as a **provirus**. The process we have described here is very similar to prophage formation in bacterial cells. In both cases, the viral genome encodes a repressor protein preventing activation of the viral genes necessary for replication. Thus, it is said that the virus is in a state of **latency**.

Latency may have several implications. Proviruses, for example, are immune to body defenses because the body's antibodies cannot reach them (antibodies do not penetrate into cells). Moreover, the virus is propagated each time the cell's chromosome is reproduced, such as during mitosis in animal cells. As we see later in this chapter, some cancers may develop when certain animal viruses enter a cell and assume a provirus relationship with that cell. The proteins encoded by the virus often bring about the profound changes associated with this dreaded condition.

Eventually, in response to changes in the host's environmental conditions, the provirus will be activated and go through its replicative cycle. This may result in the destruction of the host cell as viruses are released.

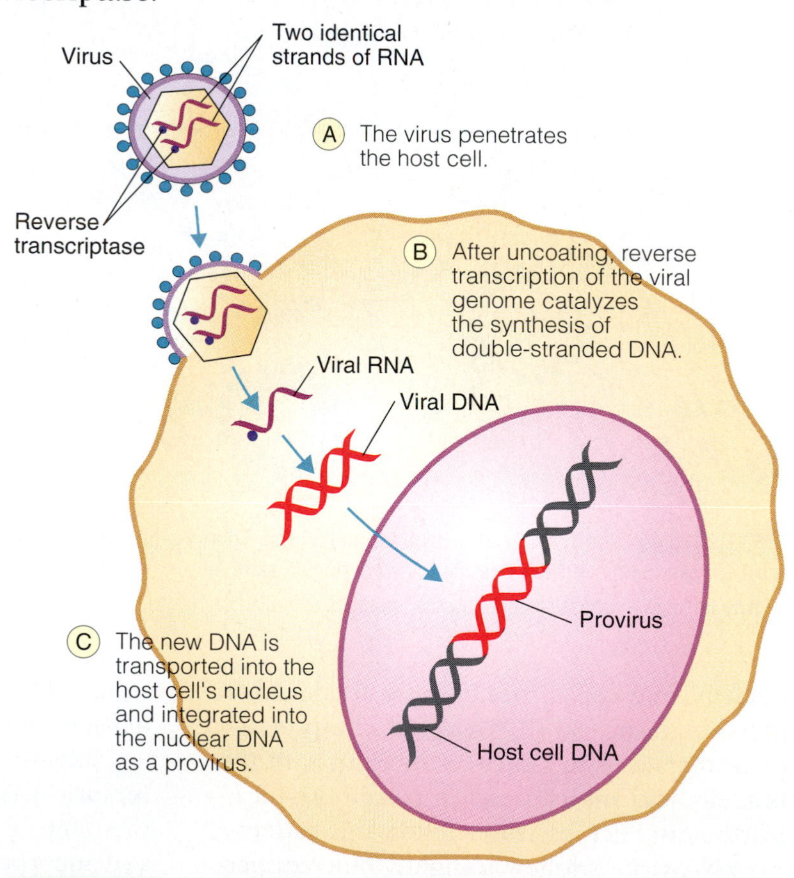

FIGURE 13.11 **The Formation of a Provirus by HIV.** Some DNA viruses and retroviruses can incorporate their genome into a host cell chromosome. Shown here is the human immunodeficiency virus (HIV), a retrovirus causing HIV disease and AIDS. The single-stranded RNA genome can be reverse transcribed into double-stranded DNA by a reverse transcriptase enzyme. The DNA then enters the cell nucleus and integrates into a chromosome.

Q: Propose a reason why genome integration is an advantage for the virus.

CONCEPT AND REASONING CHECKS

13.8 Why don't RNA viruses, like the coronaviruses or the orthomyxoviruses, undergo provirus formation?

Antiviral Drugs Can Be Used to Treat a Limited Number of Human Viral Diseases

KEY CONCEPT

- Antiviral agents interfere with specific phases of the viral replication cycle.

The body's normal immunological defenses can be summoned to actively fight disease either naturally as a result of being infected (Chapter 20) or artificially as a result of a vaccination (Chapter 21). However, antibiotics will not work against viral diseases because viruses lack the structures and metabolic machinery with which antibiotics interfere (Chapter 24). For example, penicillin is useless for inhibiting viruses because viruses have no cell wall.

A limited number of antiviral drugs are of value for treating diseases caused by picornaviruses, herpesviruses, hepatitis B and C viruses, HIV, and influenza viruses. These drugs affect viral penetration/uncoating, genome replication, or viral maturation/release (TABLE 13.3).

Some antivirals block the ability of viruses to penetrate or uncoat in the host cell. Antiviral drugs such as pleconaril integrate into the picornavirus capsid and prevent capsid uncoating. Amantadine and rimantadine prevent the influenza A viruses from escaping the endocytotic vacuole in which they are contained. HIV penetration is blocked by enfuvirtide, which blocks the fusion of the HIV envelope with the host cell plasma membrane.

Most of the antiviral drugs designed target the DNA polymerase of the herpesviruses or the reverse transcriptase of HIV and hepatitis B virus. Most of these drugs represent **base analogs** (see Chapter 7) and insert themselves into the replicating DNA strand. Insertion blocks the ability of the virus to continue replicating its genome. Azidothymidine (AZT) and several other drugs have been used to treat HIV infections and AIDS. Another drug called foscarnet is used in patients having retinal disease (retinitis) caused by the cytomegalovirus (CMV).

Reverse transcriptase inhibitors bind directly to reverse transcriptase and inhibit its activity, thereby preventing the synthesis of DNA in retroviruses. Other antiviral drugs work by blocking the maturation process or release of the virions from the host cells. **Protease inhibitors** react with the HIV protease, the enzyme that trims viral proteins down to working size for the construction of the capsid. In the type A influenzaviruses, neuraminidase is an enzyme in the spike that helps the viruses spread to other cells (Chapter 14). The **neuraminidase inhibitors** zanamivir (Relenza) and oseltamivir (Tamiflu) block the action of neuraminidase, preventing release of new virions—and thereby limiting disease spread in the body. The development of all these drugs is based on biochemical knowledge of virus function coupled with human ingenuity. Indeed, medical science one day may provide an antiviral drug to deal with obesity (MicroFocus 13.4).

CONCEPT AND REASONING CHECKS

13.9 Why have most of the antiviral agents synthesized been targeted against DNA polymerases or reverse transcriptases?

Interferon Puts Cells in an Antiviral State

KEY CONCEPT

- Interferons are natural proteins capable of triggering proteins to block viral replication.

Interferon (IFN) represents a group of naturally-produced proteins that "alert" cells to a viral infection. IFN-alpha and IFN-beta are produced by viral infection of almost any type, be it DNA or RNA. These IFNs trigger a nonspecific reaction designed to protect against the stimulating virus, as well as many other viruses. In addition, some IFNs have anticancer properties. However, human IFN is the only one that will work in humans as mouse, chicken, dog, or other animal IFNs are ineffective. IFN-gamma mobilizes natural killer cells, which attack tumor cells (Chapter 22).

TABLE **13.3** Examples of Antiviral Drugs, Their Mode of Action, and Targeted Viruses

Antiviral Drug	Mode of Action	Example
Blocking of Penetration/Uncoating		
Amantadine (Symmetrel)[1] Rimantadine (Flumadine)	Block viral uncoating	Influenza A viruses
Pleconaril	Blocks uncoating	Picornavirus infections
Enfuvirtide (Fuzeon)	Blocks penetration/uncoating	HIV disease/AIDS
Inhibition of Genome Replication		
Acyclovir (Zovirax)	Base analog (mimics guanine) inhibiting DNA replication	Genital herpes, chickenpox, shingles
Ganciclovir (Cytovene)	Base analog (mimics guanine) inhibiting DNA polymerase	Herpesvirus infections (retinitis)
Ribavirin (Virazole)	Base analog (mimics guanine) inhibiting viral replication	Broad range of viruses, including hepatitis B and C
Azidothymidine (AZT; Retrovir)	Base analog (mimics adenine) terminating DNA chain elongation	HIV and other retrovirus infections
Vidarabine (Vira-A)	Terminates DNA chain elongation	Shingles and herpesvirus-caused encephalitis
Idoxuridine (Stoxil) Trifluridine (Viroptic)	Base analogs (mimic thymidine) causing replication errors (mutations)	Keratitis caused by herpesvirus infection
Dideoxycytidine (Hivid) Dideoxyinosine (Videx)	Base analogs (mimic cytosine and adenine, respectively) terminating DNA chain elongation	HIV disease/AIDS
Foscarnet (Foscavir)	Inhibits DNA polymerases	Retinitis caused by herpesvirus infection
Nevirapine (Viramune) Delavirdine (Rescriptor)	Reverse transcriptase inhibitors	Used primarily against HIV
Inhibition of Virion Maturation/Release		
Indinavir (Crixivan) Ritonavir (Norvir) Saquinavir (Invirase)	Inhibit HIV protease	HIV disease/AIDS
Oseltamivir (Tamiflu) Zanamivir (Relenza)	Inhibit viral release	Influenza A viruses

[1]Trade name

IFN-alpha and IFN-beta do not interact directly with viruses, but rather with the cells they protect. The IFNs are produced when a virion releases its genome into the cell. High concentrations of double-strand RNA in the infected cells induce the cell to synthesize and secrete IFNs (FIGURE 13.12). These bind to specific receptor sites on the surfaces of adjacent cells and trigger the production of **antiviral proteins** within those cells. Such cells are now in an **antiviral state** and capable of inhibiting viral replication by inhibiting protein synthesis.

In 1980, Swiss and Japanese scientists deciphered the genetic code for interferon, and spliced the gene into *E. coli* plasmids (see Chapter 8). Experiments showed that IFN from bacterial factories would reduce hepatitis symptoms, diminish the spread of herpes zoster, and shrink certain cancers. In 1984, a Swiss biotechnology firm began marketing IFN-alpha using the trade name Intron.

MICROFOCUS 13.4: Being Skeptical

Obesity Is Infectious!?

In 1997, University of Wisconsin researchers announced that obesity could be linked to a viral infection. They proposed it might even be possible one day to eliminate the virus and slim down quickly by taking an antiviral drug or being vaccinated. Is there any truth here?

The road to this startling announcement began in 1990 with a veterinarian's chance remark that chickens infected with adenoviruses gain considerable weight. The conversation took place in India, where Nikhil V. Dhurandhar, a nutritionist, was on his way to Wisconsin to do research. The veterinarian's observation intrigued Dhurandhar.

At the medical school in Wisconsin, the fledgling researcher and his colleague Richard Atkinson obtained a flock of chickens and a culture of adenoviruses (respiratory viruses often involved in the common cold). They soon confirmed that when chickens, mice, and monkeys were infected with the viruses, the animals rapidly gained weight.

But was this true in humans as well? Dhurandhar solicited volunteers and got responses from 45 lean individuals and 154 obese people, each weighing over 250 pounds. Working with adenovirus 36 (AD-36), Dhurandhar and Atkinson believed the presence of antibodies against AD-36 could prove the virus was currently in the body or that it had once been there. They took blood samples and analyzed them: 10 percent of the lean individuals had AD-36 antibodies, but 30 percent of the obese people had the telltale antibodies.

The percentage of AD-36 positive volunteers was not as spectacular as one might hope, so Dhurandhar, now at Wayne State University in Detroit, measured antibody presence in two identical twins, one who was lean and the other who had gained weight after going to college. He found anti-AD-36 antibodies only in the heavier twin.

Richard Atkinson has recently started his own company, the Obetech Obesity Research Center, in Richmond, VA. He believes AD-36 infects fat cells, causing an eventual increase in fat cell numbers. Atkinson suggests there is a 70 to 100 percent chance of getting fat if one becomes infected with AD-36.

One very interesting discovery made is that obese volunteers have lower levels of serum cholesterol and triglycerides than thin volunteers. Normally, one associates weight gain with high cholesterol and triglycerides.

At this time, it is still hard to see how an adenovirus normally associated with short-term "cold infections" could be linked to such a long-term chronic infection of fat cells. Dhurandhar and Atkinson do but the evidence needs to be more convincing. Both researchers would agree though that proper diet, exercise, and good lifestyle habits are keys to staying fit and healthy.

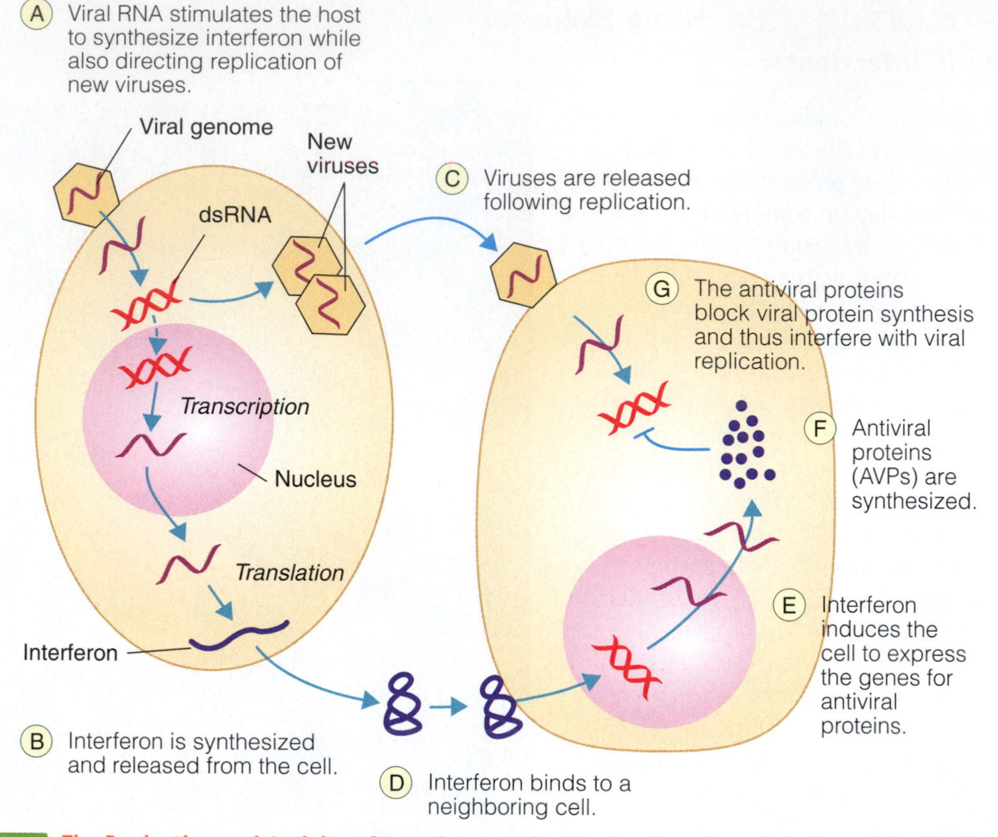

FIGURE 13.12 The Production and Activity of Interferon. Left: A host cell produces interferon following infection by an RNA virus. The virus replicates in the same cell. Right: The interferon reacts with receptors at the surface of a neighboring cell and induces the cell to produce antiviral proteins and the cell enters an activated, antiviral state. The proteins interfere with viral replication in the cell, possibly by binding to messenger RNA molecules.

Q: Why would interferon itself not be considered an antiviral compound?

In 1986, the US Food and Drug Administration approved the sale of IFN-alpha for use against a form of leukemia; in 1988, it approved its use against genital warts and in 1992, against chronic hepatitis B.

CONCEPT AND REASONING CHECKS

13.10 Distinguish between the role of interferon and antiviral proteins.

13.5 The Cultivation and Detection of Viruses

If an individual contracts a viral disease, there are various ways the viral agent can be cultivated and detected for identification and eventual diagnosis. A prompt identification often is necessary for selecting the appropriate antiviral therapy.

Detection of Viruses Often Is Critical to Disease Identification

KEY CONCEPT

- Cytology can provide a rapid initial diagnosis to identify an unknown viral infection.

The diagnosis of viral diseases like the flu and a cold are usually straightforward and do not require further laboratory confirmation. In some cases, viral infections leave their mark on the infected individuals. Measles is accompanied by **Koplik spots**, which are a series of bright red patches with white pimple-like centers on the lateral mouth surfaces. Swollen salivary glands and teardrop-like skin lesions are associated with mumps and chickenpox, respectively. Some viral diseases also can be detected by how they react with red blood cells, as MicroFocus 13.5 indicates.

MICROFOCUS 13.5: Tools

The Hemagglutination-Inhibition Test

An indirect method for detecting viruses is to search for viral antibodies in a patient's serum. This can be done by combining serum (the blood's fluid portion) with known viruses.

Certain viruses—such as those of influenza, measles, and mumps—have the ability to agglutinate (clump) red blood cells. This phenomenon, called **hemagglutination** (**HA**) will produce a thin layer of cells over the bottom of a plastic dish (**Figure A**). If the cells do not agglutinate (no hemagglutination), they fall to the bottom of the well as a small "button."

HA can be used for detection and identification purposes. It is possible to detect antibodies against certain viruses because antibodies react with viruses and tie up the reaction sites that otherwise would bind to red blood cells, thereby inhibiting hemagglutination. Thus, a laboratory test called the **hemagglutination-inhibition** (**HAI**) **test** can be performed. In the test, the patient's serum is combined with known viruses and then red blood cells are added (**Figure B**). Hemagglutination inhibition indicates antibodies are present in the serum. Such a finding implies the patient has been exposed to the virus.

Figure C presents a scenario for detecting exposure to the measles virus.

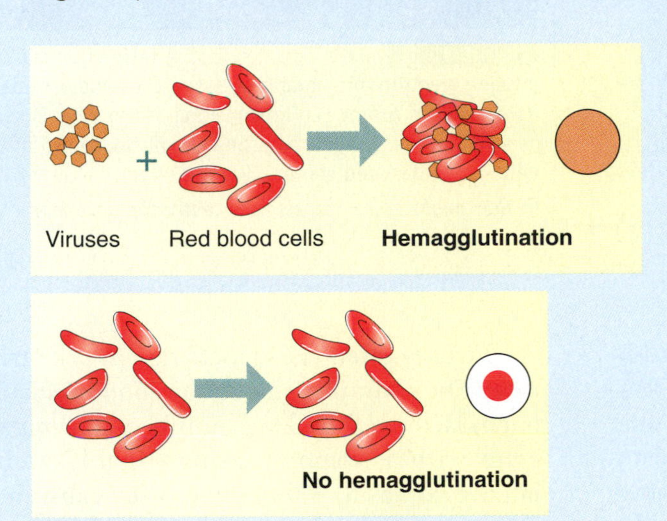

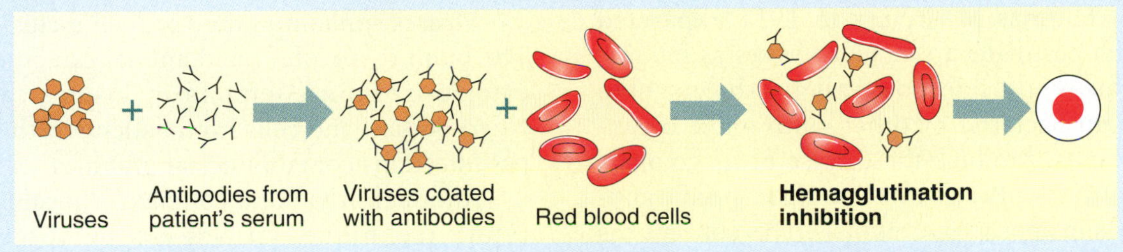

(B)

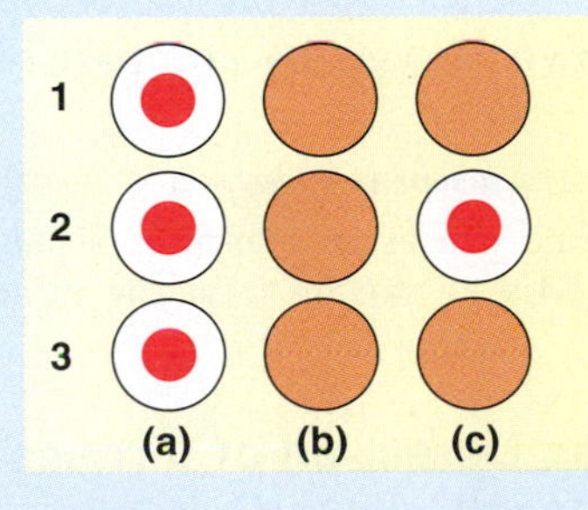

(C)

FIGURES A AND B Hemagglutination and the Hemagglutination-Inhibition Test. (**A**) Viruses of certain diseases, such as measles and mumps, can agglutinate (clump) red blood cells, forming a film of cells over the bottom of a plastic well. No agglutination produced a small pellet or "button" at the bottom of the cell. (**B**) In the hemagglutination-inhibition (HAI) test, known viruses are combined with serum from a patient. If the serum contains antibodies for that virus, the antibodies coat the virus, and when the viruses are then combined with red blood cells, no agglutination will take place.

FIGURE C A Hemagglutination-Inhibition Test. Three young children (1, 2, 3) who attended the same daycare center were admitted to the hospital with what appeared to be measles. All three were tested for the presence of measles antibodies. Which child or children likely had the measles?

(a) = positive control; (b) = negative control; (c) = patient's serum.

TABLE **13.4** Examples of Virus Cytopathic Effects

Virus	Cytopathic Effect
	Changes in cell structure
Picornaviridae	Shrinking of cell nucleus
Papovaviridae	Cytoplasmic vacuoles
Paramyxoviridae, Coronaviridae	Cell fusion syncytia
Herpesviridae	Chromosome breakage
Herpesviridae, Adenoviridae, Picornaviridae, Rhabdoviridae	Rounding and detachment of cells from culture
	Cell inclusions
Adenoviridae	Virions in nucleus
Rhabdoviridae	Virions in cytoplasm (Negri bodies)
Poxviridae	"Viral factories" in cytoplasm
Herpesviridae	Nuclear granules

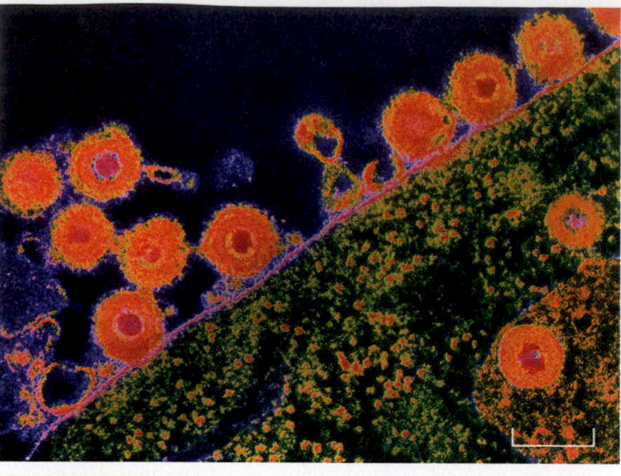

FIGURE 13.13 **Cells and Viruses.** A false-color transmission electron micrograph of tissue cells and associated herpesviruses (orange). The envelope of the virus is seen as a ring at the viral surface, and the nucleocapsid of the virus is the dark red center. (Bar = 200 nm.)

Q: How might these herpesviruses enter the host cell?

However, virus detection is not straightforward and the proper identification has a bearing on relating a particular virus to a particular disease. In the classical sense, Koch's postulates cannot be applied to a viral disease because, unlike bacterial cells, viruses cannot be cultivated in pure culture. To circumvent this problem, Thomas M. Rivers in 1937 expanded Koch's postulates to include viruses as follows: Filtrates of the infectious material shown not to contain bacterial or other cultivatable organisms must produce the disease or its counterpart; or, the filtrates must produce specific antibodies in appropriate animals. This concept has come to be known as **Rivers' postulates**.

The laboratory diagnosis of a viral disease can be carried out using microscopy. **Cytology** uses light microscopy to examine cells obtained from body tissue or fluids. Such cytological techniques may be used in the clinical laboratory where viruses are being identified from samples taken from patients. When viruses replicate in host cells, often a noticeable deterioration or structural change occurs. This is called a **cytopathic effect** (CPE). Characteristic CPEs are listed in (TABLE 13.4).

Viruses often cause a change in cell structure. For example, infectious mononucleosis is characterized by large numbers of lymphocytes with a "foamy-looking," highly vacuolated cytoplasm. Paramyxoviruses cause host cells to fuse together into multinucleate giant cells called **syncytia**.

Viruses sometimes produce cell inclusions. The brain tissue of a rabid animal can contain cytoplasmic nucleoprotein inclusions called **Negri bodies**, and cells from patients with herpes infections contain nuclear granules.

Although it is not a common clinical laboratory technique, viruses can be observed directly by electron microscopy (FIGURE 13.13). Virologists often use the electron microscope to identify unknown viruses by comparison to known viruses.

CONCEPT AND REASONING CHECKS

13.11 How can some viruses be identified simply by observing a tissue sample from the infected patient?

Cultivation and Detection of Viruses Most Often Uses Cells in Culture

KEY CONCEPT

- Viruses can be "grown" in various types of tissue culture and detected by the formation of plaques.

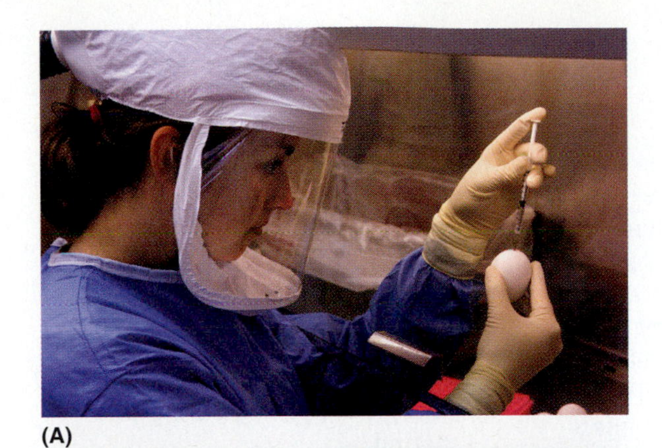
(A)

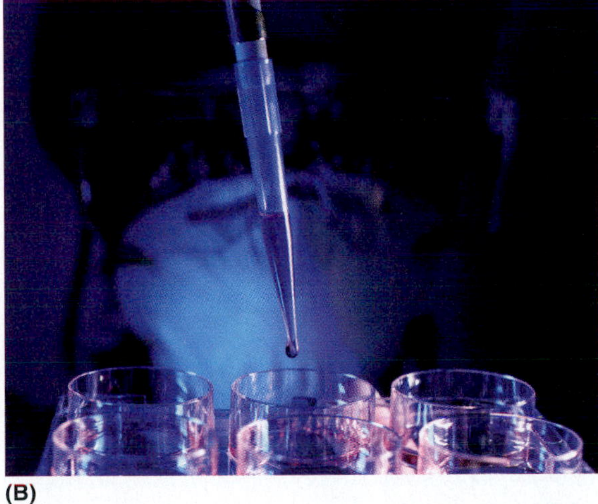
(B)

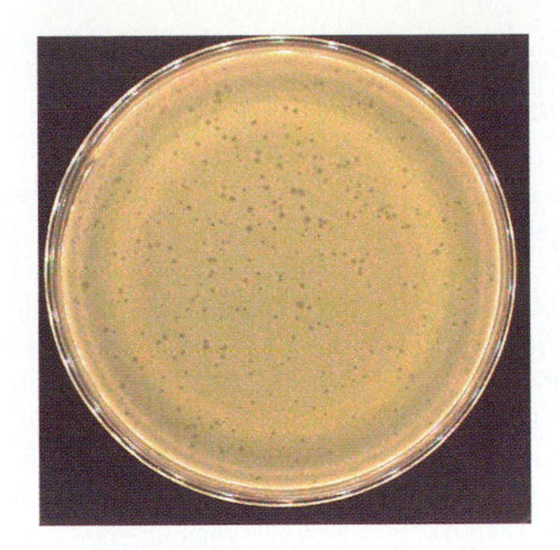
(C)

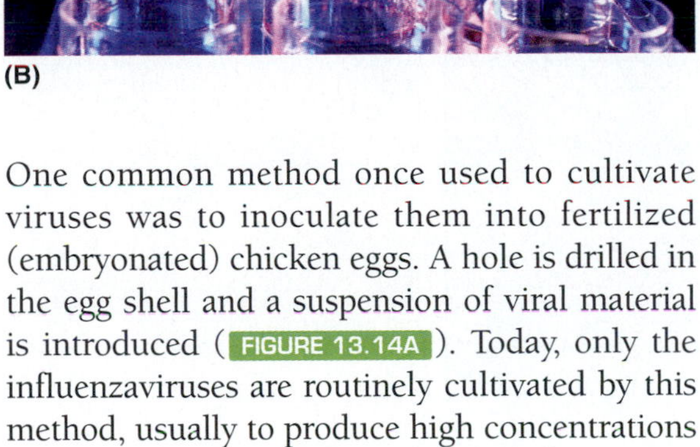

FIGURE 13.14 **Infection of Cells in Embryonated Eggs and Cell Culture.** (**A**) Inoculation of fertilized eggs. Techniques such as these are standard practice for growing the flu virus. (**B**) A masked researcher uses a pipette to infect a culture of human cells with a virus. The dishes contain a culture medium, which allows the cells to survive outside the body. The effect of the virus on the infected cells will then be studied. (**C**) Plaque formation in a cell culture. Plaques are evident as the clear areas in the dish.

Q: Why must a mask be worn when infecting embryonated eggs and cells in culture?

One common method once used to cultivate viruses was to inoculate them into fertilized (embryonated) chicken eggs. A hole is drilled in the egg shell and a suspension of viral material is introduced (FIGURE 13.14A). Today, only the influenzaviruses are routinely cultivated by this method, usually to produce high concentrations of viruses for vaccine production.

The most common method of cultivating and detecting viruses is to infect cell cultures. To prepare the culture, animal cells are separated from a tissue with enzymes and suspended in a solution of nutrients, growth factors, pH buffers, and salts. In such a **primary cell culture**, the cells adhere to the bottom of a plastic dish or well and reproduce to form a single layer, called a **monolayer**. The different cell types in a primary cell culture can be separated enzymatically and isolated as single cell type, called a **cell line**. The type of cell culture used will depend on the virus species to be cultivated in the monolayer. Viruses are then introduced into the culture (FIGURE 13.14B).

Viruses can be detected by the formation of plaques. A **plaque** is a clear zone within the cloudy "lawn" of bacterial cells or monolayer of animal cells (FIGURE 13.14C). The viruses infect and replicate in the cells, thereby destroying them and forming plaques. Epidemiologic surveys of several bacterial diseases such as staphylococcal food poisoning are aided by plague formation called **phage typing** (see Chapter 10). MicroInquiry 13 uses the plaque assay to monitor intracellular virus production and extracellular virus release during an infection cycle of animal cells in culture.

CONCEPT AND REASONING CHECKS

13.12 Explain how plaques would be formed on a lawn of bacterial cells.

MICROINQUIRY 13

The One-Step Growth Cycle

In the research laboratory, we can follow the replication of viruses by generating a one-step growth curve. Realize that we are not really looking at growth, but the replication and increase in number of virus particles. There are several periods associated with a virus growth cycle that you should remember because you will need to identify the periods in the growth curve. The **eclipse period** is the time when no virions can be detected inside cells and the **latent phase** is the time during which no extracellular virions can be detected. Also, the **burst size** is the number of virions released per infected cell.

To generate our growth curve, we start by inoculating our viruses onto a susceptible cell culture. There are 100,000 (10^5) cells in each of 10 cultures. We then add ten times the number of viruses (10^6) in a small volume of liquid to each culture to make sure all the cells will be infected rapidly. After 60 minutes of incubation, we wash each culture to remove any viruses that did not attach to the cells and add fresh cell growth medium. Then, at 0 hour and every four hours after infection, we remove one culture, pour off the growth medium, and lyse the cells. The number of viruses in the growth medium and in the lysed cells are determined. Because we cannot see viruses, we measure any viruses in the growth medium or in the lysed cells using a plaque assay. We assume that one **plaque-forming unit** (**PFU**) is equal to one virion.

The curve drawn plots the results from our growth experiment.

Answer the following questions based on the one-step growth curve. Answers can be found in **Appendix D**.

13.1a. **How long is the eclipse period for this viral infection? Explain what is happening during this period.**

13.1b. **How long is the latent period for this infection? Explain what is happening during this period.**

13.1c. **What is the burst size?**

13.1d. **Explain why the growth curve shows a decline in phages between 0 and 4 hours.**

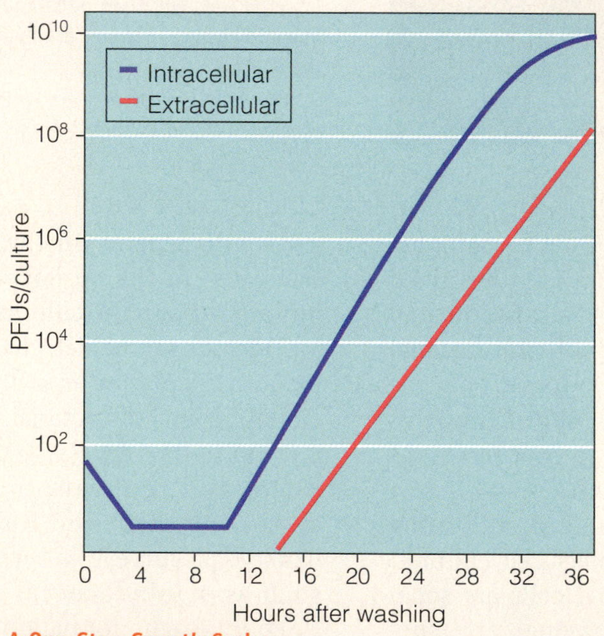

A One-Step Growth Cycle

13.6 Tumors and Viruses

Cancer is indiscriminate. It affects humans and animals, young and old, male and female, rich and poor. In the United States, over 557,000 people die of cancer annually, making the disease the leading cause of death after heart disease and stroke. Worldwide, over 7 million people die of cancer each year.

Cancer Is an Uncontrolled Growth and Spread of Cells

KEY CONCEPT

- Tumors are the result of uncontrolled cell divisions.

Cancer results, in part, from the uncontrolled reproduction (mitosis) of a single cell. The cell escapes controlling factors and as it continues to multiply faster than normal cells, a cluster of cells soon forms. Eventually, the cluster yields a clone of abnormal cells referred to as a **tumor**. Normally, the body will respond to a tumor by surrounding it with a capsule of connective tissue. Such a local tumor is designated **benign** because it usually is not life threatening.

Additional changes can occur to tumor cells that release them from their specific confines. Often they stick together less firmly than normal cells and fail to stop dividing when cells come in contact with one another. They may break out of the capsule and **metastasize**, a spreading of the cells to other tissues of the body. Such a tumor now is described as **malignant** and the individual now has **cancer** (*cancer* = "crab"; a reference to the radiating spread of cells, which resembles a crab). So, a series of changes must occur to a healthy cell before it can become a tumor or cancer cell.

How can such a mass of cells bring illness and misery to the body? By their sheer numbers, cancer cells invade and erode local tissues, interrupt normal functions, and choke organs to death by robbing them of vital nutrients. Thus, the cancer patient will commonly experience weight loss even while maintaining a normal diet.

CONCEPT AND REASONING CHECKS

13.13 How does a benign tumor differ from a malignant tumor?

Viruses Are Responsible for Up to 20 Percent of Human Tumors

KEY CONCEPT

- A few animal viruses can represent agents of tumor development.

The World Health Organization (WHO) estimates that 60 to 90 percent of all human cancers are associated with **carcinogens**, chemicals and physical agents that produce cellular changes leading to cancer (FIGURE 13.15). Among the known chemical carcinogens are the hydrocarbons found in cigarette smoke as well as asbestos, nickel, certain pesticides and dyes, and environmental pollutants in high amounts. Physical agents include ultraviolet (UV) light and X rays (see Chapter 7).

A few viruses also act as carcinogens. In 1911, Peyton Rous demonstrated that a virus

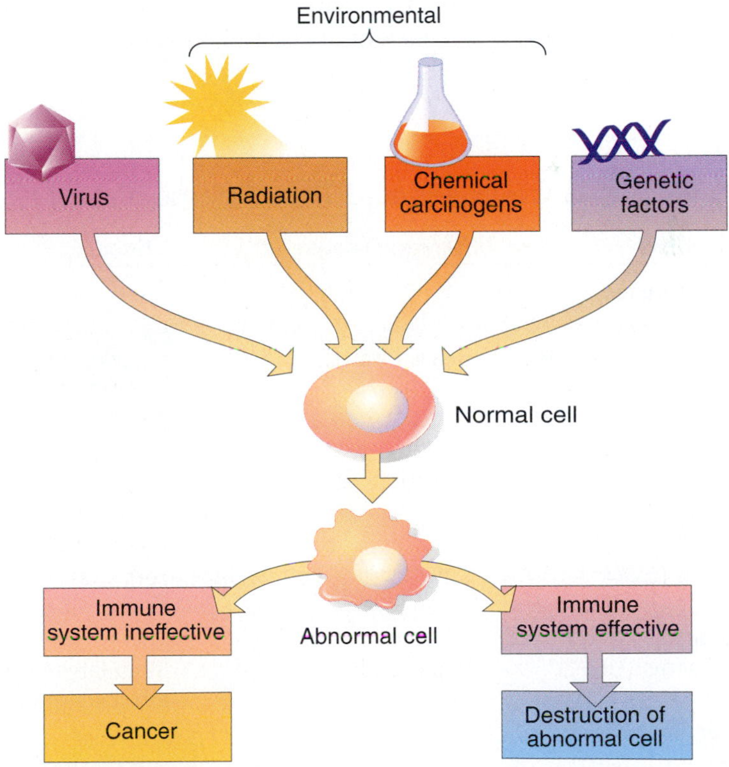

FIGURE 13.15 **The Onset of Cancer.** Viruses and other factors induce a normal cell to become abnormal. When the immune system is effective, it destroys the abnormal cell and no tumor develops. However, when the abnormal cell evades the immune system, it develops first into a tumor and then may become malignant, spreading to other tissues in the body.

Q: What factors are required for a tumor to become malignant?

■ Sarcoma: A malignant tumor that begins growing in connective tissue such as muscle, bone, fat, or cartilage.

■ Transformation: The conversion of a normal cell into a malignant cell brought about by the action of a carcinogen.

■ PAP smear: A test to detect cancerous or precancerous cells of the cervix, allowing for early diagnosis of cancer.

caused a sarcoma in chickens. Further experiments with animals also indicated other viruses can induce tumor formation.

A number of viruses now have been isolated from human cancers (TABLE 13.5). When these viruses are transferred to animals and cell cultures, an observable transformation of normal cells to tumor cells takes place. Examples of such **oncogenic viruses** (tumor-causing viruses) are the herpesviruses associated with tumors of the human cervix and the Epstein-Barr virus, which is linked to **Burkitt lymphoma**, a tumor of the jaw. Of special note is **cervical cancer**, the second most common cancer in women under age 35. The human papilloma virus (HPV) subtypes (family *Papovaviridae*) have a strong correlation with this cancer. In one study, 71 percent of the women developing cervical cancer had HPV present in their PAP smear. The good news is there is now a vaccine, called Gardasil, that provides almost 100 percent protection against the two most common HPV strains causing 70 percent of cervical cancers.

CONCEPT AND REASONING CHECKS

13.14 Name several viruses involved in the development of human cancers.

Oncogenic Viruses Transform Infected Cells

KEY CONCEPT

- Oncogenes represent cancer-causing genes.

The mechanism by which viruses transform normal cells into tumor cells remained obscure until the oncogene theory was developed in the 1970s. First postulated by Robert Huebner and George Todaro in 1969, the **oncogene theory** suggests that transforming genes, the so-called **oncogenes**, normally reside in the chromosomal DNA of a cell. In the late 1970s, researchers J. Michael Bishop and Harold Varmus, of the University of California at San Francisco, located oncogenes in a wide variety of creatures from fruit flies to humans. Bishop and Varmus also made the astonishing discovery that practically the same genes exist in certain viruses, and they hypothesized these

TABLE 13.5 Human Oncoviruses and Their Effects on Cell Growth

Oncogenic Virus	Benign Disease	Effect	Tumor
DNA Tumor Viruses			
Papovaviridae Human papilloma virus (HPV)	Benign warts	Encodes genes that inactivate cell growth regulatory proteins	Cervical and penile cancer
Herpesviridae Epstein-Barr virus (EBV)	Infectious mononucleosis	Stimulates cell growth and activates a host cell gene that prevents cell death	Burkitt lymphoma Hodgkin lymphoma
Herpesvirus 8 (HV8)	Growths in lymph nodes	Forms lesions in connective tissue	Kaposi's sarcoma
Hepadnaviridae Hepatitis B virus (HBV)	Hepatitis B	Stimulates overproduction of a transcriptional regulator	Liver cancer
RNA Tumor Viruses			
Flaviviridae Hepatitis C virus (HCV)	Hepatitis C	Being investigated	Liver cancer
Retroviridae Human T-cell leukemia virus (HTLV-1)	Weakness of the legs	Encodes a protein that activates growth-stimulating gene expression	Adult T-cell leukemia/lymphoma
XMRV	Unknown	Effect unclear	Prostate cancer (?)

genes could have been captured by the viruses from previous infections. It appeared that the oncogenes were not viral in origin but part of the genetic endowment of every living cell.

The discovery of oncogenes demonstrated some forms of cancer have a genetic basis. As research continued, oncology researchers extracted DNA from tumors and used it to turn healthy cells into tumorous ones in the test tube. Moreover, they surmised the transforming substance was in a small segment of the tumor cell DNA—probably a single gene. Finally, in 1981, three separate research groups isolated an oncogene residing in a human bladder cancer. At this writing, over 60 different oncogenes have been identified.

In recent years, the theory of oncogene activity has been revised slightly. Researchers now propose that normal genes, called **protooncogenes**, are the forerunners of oncogenes (FIGURE 13.16). Protooncogenes may have important functions as regulators of growth and mitosis. Because protooncogenes exist in diverse forms of life, they must be important in cell metabolism, perhaps as growth regulators. (As one researcher notes, "They would not have survived through evolution just to make tumors.") In fact, protooncogenes can be converted to oncogenes by radiation, by chemical carcinogens, by chromosomal breakage and rearrangement, or by viruses after which tumor formation begins.

But how do viruses trigger the transformation? Virologists have observed that when some viruses enter a cell, they may become a provirus. Whether it is a retrovirus or a DNA virus, the double-stranded DNA inserts into the cell chromosome. Sometimes the integration of the provirus will be adjacent to a protooncogene. When virus replication is triggered, the viral DNA (provirus) not only replicates its own viral DNA, but also a few neighboring host genes (FIGURE 13.17). Thus, when new viruses are produced, the viral DNA contains the protooncogene as part of its genome. Such protooncogenes "captured" in the viral genome are

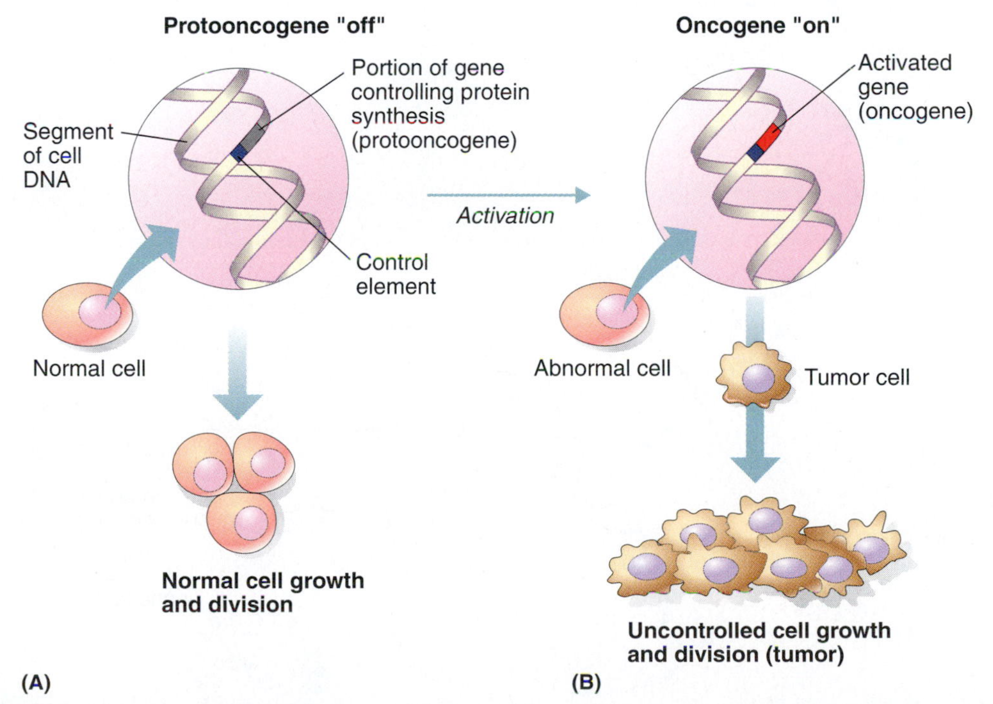

FIGURE 13.16 **The Oncogene Theory.** The oncogene theory helps explain the process leading to tumor development. (**A**) The normal cell grows and divides without complications. Within its DNA it contains protooncogenes that are "turned off." When the genes are activated by viruses or other factors, they revert to oncogenes, which "turn on." An abnormal cell (**B**) results. The oncogenes encode proteins that regulate the development of a tumor and may contribute to transformation from a normal cell to a cancer cell.

Q: What properties are common to tumor cells?

called **v-oncogenes**. The same scenario would hold for the retroviruses except the provirus produces RNA. Still, the viral RNA packaged into the virions will contain the viral information and the protooncogene information.

When these oncogenic viruses infect another cell and incorporate as a provirus, the v-oncogene is under control of the virus—not the host cell. So, expression of the v-oncogene can influence cellular growth and mitosis in several ways. V-oncogenes, for example, may provide the genetic sequences for growth factors stimulating uncontrolled cell development and reproduction. Or the oncogenes may become incapable of encoding substances that turn off cell growth, a function the protooncogenes once had.

Another mechanism of transformation has been studied with the virus of **Burkitt lymphoma**. In this cancer of the lymphoid connective tissues of the jaw, the viral genome inserts itself into a chromosome of **B lymphocytes**, a type of white blood cell important in immunity. The insertion triggers the protooncogene *c-myc*, which is involved in cell growth, to move from chromosome 8 to chromosome 14, far from the influence of its control genes. The protooncogene, now an oncogene, appears to produce elevated amounts of its protein product, leading to tumor formation.

Infection by an oncogenic virus is more common than the cancer that may develop. So, infection may cause a tumor, but additional genetic and/or environmental events are needed before the cell is truly a cancer cell. The reason the viruses are called oncogenic viruses is because the v-oncogene or activated oncogene is telling the infected cells, "Divide, divide, divide!" In a short period, a tumor forms. The excessive divisions provide an

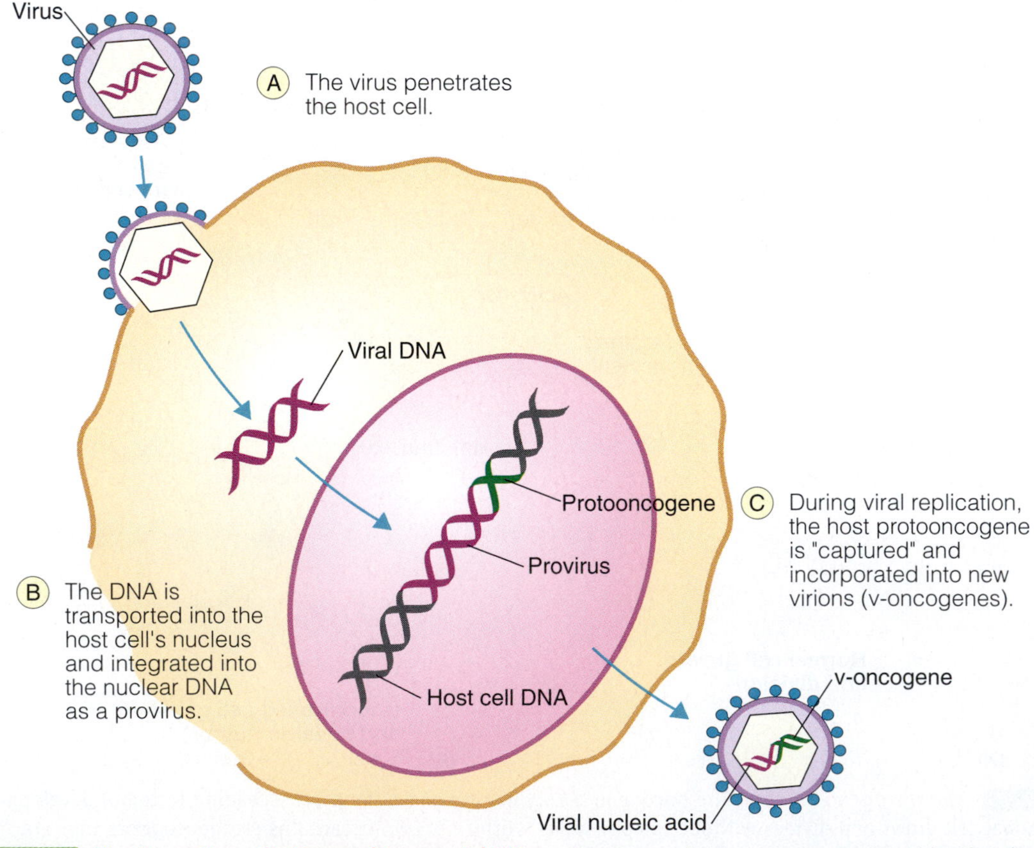

FIGURE 13.17 The Formation of a Tumor (Oncogenic) Virus. An oncogenic virus can be produced by the viral genome integrating into a host chromosome adjacent to a protooncogene. During replication, the provirus along with the protooncogene will be replicated and incorporated into new virions.

Q: How would the oncogenic virus with the v-oncogene lead to tumor formation?

opportunity for mutations to occur that can lead the infected cells down the path to malignancy and cancer formation.

This ability of viruses to get inside cells and deliver their viral information has been manipulated by medical research teams using genetic engineering techniques to deliver genes to cure genetic illnesses. MicroFocus 13.6 looks at the pros and cons of using viruses for this purpose.

CONCEPT AND REASONING CHECKS

13.15 Explain how the oncogene theory is related to virus infection.

MICROFOCUS 13.6: Biotechnology

The Power of the Virus

How would you like to be injected with an adenovirus (see figure) or herpesvirus? How about the virus used to make the smallpox vaccine or a retrovirus? This doesn't sound like a great idea, but in fact it might be a way to treat genetic diseases and many forms of cancer. **Virotherapy**, as it is called, may be useful in curing disease rather than causing it.

Ever since the power of genetic engineering made it possible to transfer genes between organisms, scientists and physicians have wondered how they could use viruses to cure disease.

One potential way is to use viruses against cancer. In 1997, a herpesvirus gene was mutated so the virus would replicate only in tumor cells, which possessed the missing enzyme. Because viruses make the ideal killers, perhaps they would kill the infected cancer cells. Used on a terminal patient with a form of brain cancer, the herpesvirus seems to have worked as the patient is still alive in 2006.

Additional cancer-killing viruses are being developed; some cause the cancer cells to commit cell suicide while others alert the immune system to the cancer danger.

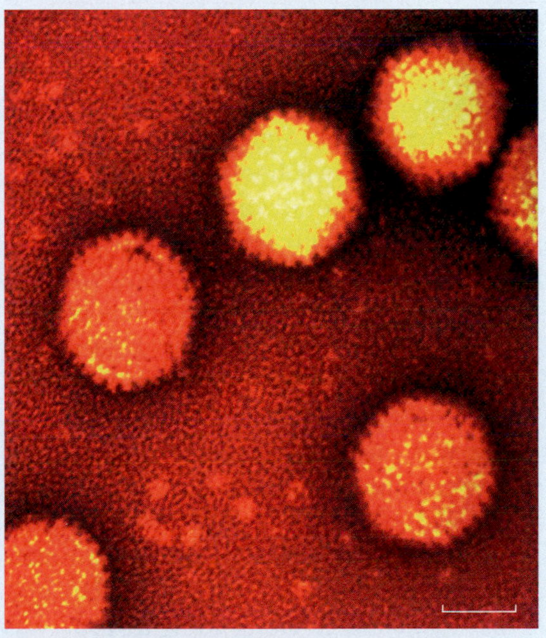

False-color transmission electron micrograph of adenoviruses. (Bar = 50 nm.)

Viral gene therapy has shown promise but it has had its misfires as well. In 1999, an 18-year-old patient, Jesse Gelsinger, died as a result of an immune response that developed in response to the adenovirus used to treat a rare metabolic disorder. Importantly, seventeen patients had been successfully treated using the same adenovirus prior to Gelsinger. In 2002, French scientists used virus-inserted genes to cure four young boys suffering with a nonfunctioning immune system. Again, the therapy worked, but the virus-inserted gene disrupted other cellular functions, leading to leukemia (cancer of the white blood cells) in two of the boys.

So, the idea of using virotherapy certainly works, but delivering the virus without complications and targeting the genes to the correct place can be a difficult assignment. With regard to cancer, the only adverse effects have been patients reporting flu-like symptoms, again due to the immune system's response to the detection of virus in the body.

In this regard, one problem of virotherapy is the immune system response. Like any infection, the immune system mounts an attack on the injected viruses, which, therefore, may be destroyed before the viruses reach their target. To overcome this problem, researchers have developed a version of the vaccinia virus and adenovirus coated with an "extracellular envelope" that is invisible to the immune system yet able to search out and find its target—the cancer cells.

Another approach is to load up cancer-targeted viruses with radioactive materials, a toxic drug, or a gene that codes for an anti-cancer proteins. These "stealth viruses" would deliver their cargo only to cancer cells and destroy the cells.

In 2006, virotherapy is in high gear with more than 15 early clinical trials in progress to examine the power of the virus to directly or indirectly kill cancer cells.

13.7 Emerging Viruses and Virus Evolution

Almost every year a newly emerging influenza virus descends upon the human population. Other viruses not even heard of a few decades ago, such as HIV, Ebolavirus, Hantavirus, West Nile virus, and SARS virus, are often in the news. Where are these viruses coming from?

Emerging Viruses Usually Arise Through Natural Phenomena

KEY CONCEPT

- Viral recombination and mutation can give rise to new viruses.

Many **emerging diseases** are the result of viruses appearing for the first time in a population or rapidly expanding their host range with a corresponding increase in detectable disease (TABLE 13.6). Most of these viruses are not new. Some, like Hantavirus, West Nile virus, and others are just new to certain populations or geographical areas (MicroFocus 13.7). Others, like HIV, have crossed host ranges and resulted in the development of new diseases.

One way "new" viruses arise is through **genetic recombination** (see Chapter 8). Take for example influenza. Genetic recombination allows two different influenza viruses to reassort genome segments (Chapter 14). The "bird flu" that broke out in Hong Kong in 1997 was the result of the reassortment of genome segments from a strain of avian flu virus with those of a human flu virus. This produced a virus with the capability of infecting humans.

Viruses also arise as a result of the second force driving evolution—**mutation** (see Chapter 7). For example, when a single nucleotide is altered (point mutation) in an RNA virus genome in the host cytoplasm, there is no way to "proofread" or correct the mistake during replication. Occasionally, one of these mutations may be advantageous. In the case of HIV, a beneficial mutation could generate a new virus strain resistant to an antiviral drug. With a rapid replication rate and burst size, it does not take long for a beneficial mutation to establish itself within a population.

Even if a new virus has emerged, it must encounter an appropriate host to replicate and spread. It is believed that smallpox and measles both evolved from cattle viruses, while flu probably originated in ducks and pigs. HIV almost certainly has evolved from a monkey (simian) immunodeficiency virus (Chapter 15). Consequently, at some time such viruses had to make a species jump. What could facilitate such a jump?

Our proximity to animals and pathogens makes such a jump possible. Today, population pressure is pushing the human population into new areas where potentially virulent viruses may be lurking. Evidence shows the Machupo and Junin viruses (*Arenaviridae*) jumped from rodents to humans as a result of increased agricultural practices that, for the

TABLE 13.6 Examples of Emerging Viruses

Virus	Family	Emergence Factor
Influenza	*Orthomyxoviridae*	Mixed pig and duck agriculture, mobile population
Dengue fever	*Flaviviridae*	Increased population density, environments that favor breeding mosquitoes
Sin Nombre (Hantavirus)	*Bunyaviridae*	Large deer mice population and contact with humans
Ebola/Marburg	*Filoviridae*	Human contact with fruit bats
HIV	*Retroviridae*	Increased host range, blood and needle contamination, sexual transmission, social factors
West Nile	*Flaviviridae*	Vector transported unknowingly to New York City
Machupo/Junin	*Arenaviridae*	Rodent contact via agricultural practices
Nipah/Hendra	*Paramyxoviridae*	Unknown
SARS-associated	*Coronaviridae*	Contact with horeshoe bat

first time, brought infected rodents into contact with humans.

An increase in the size of the animal host population carrying a viral disease also can "explode" as an emerging viral disease. Spring 1993 in the American Southwest was a wet season, providing ample food for deer mice. The expanding deer mice population brought them into closer contact with humans. Leaving behind mouse feces and dried urine containing the hantavirus made infection in humans likely. The deaths of 14 people with a mysterious respiratory illness in the Four Corners area that spring eventually were attributed to this newly recognized virus.

So, emerging viruses are not really new. They are simply evolving from existing viruses and, through human changes to the environment, are given the opportunity to spread or to increase their host range.

CONCEPT AND REASONING CHECKS

13.16 Distinguish the ways that emerging viral disease arise.

There Are Three Hypotheses for the Origin of Viruses

KEY CONCEPT

- Viruses may have preceded cellular life.

Although no one can know for certain how viruses originated, scientists and virologists have put forward three hypotheses.

Regressive evolution hypothesis. Viruses are degenerate life-forms; that is, they are derived from intracellular parasites that have lost many functions other organisms possess and have retained only those genes essential for their parasitic way of life.

Cellular origins hypothesis. Viruses are derived from subcellular components and functional assemblies of macromolecules that

MICROFOCUS 13.7 Environmental Microbiology

Escape from Wardang Island

European rabbits had been a scourge in Australia ever since English settlers released a dozen of them in 1840. Without natural predators, the rabbits thrived and, true to their reputation, bred prolifically and became a pest. By 1995, over 300 million rabbits dotted the Australian landscape, bringing destruction to farmland and agriculture. Something had to be done.

Australian scientists had a plan for rabbit biocontrol: They would quarantine several thousand rabbits on Wardang Island, about 7 miles off Australia's southern coast, and infect them with caliciviruses (*calic* = "cup"; the caliciviruses are a group of RNA viruses with cup-like projections on their surfaces). The virus quickly kills rabbits by causing blood clots in vital organs. If the infection was a success, then a plan for the mainland could be developed.

False-color transmission electron micrograph of caliciviruses. (Bar = 15 nm.)

To reduce the spread of either viruses or rabbits outside the enclosure on Wardang Island, the scientists maintained fence security and tested the blood of wild rabbits outside the pens for evidence of calicivirus escape. However, by September 1995, scientists were finding dead rabbits outside the pens—and then on the mainland! The dead rabbits had evidence of calcivirus disease. The virus had escaped! (Research conducted in 1997 would indicate the bush fly might be the carrier to the mainland.)

Now the scientists would have to go ahead with the general release. So, virologists and veterinarians acted swiftly. With government approval, they released infected rabbits at hundreds of sites on the mainland. Soon, dead rabbits were everywhere. Within two years, the rabbit population was reduced by 95 percent in some areas; and many plant and animal species, preyed on by the rabbits, were rebounding after remaining unseen for generations. At this writing, the native wildlife remains unaffected.

At first, scientists were discouraged at having to act in haste, but in this instance, the escape from Wardang Island apparently had a beneficial (and lucky) twist.

have escaped their origins inside cells by being able to replicate autonomously in host cells.

Independent entities hypothesis. Viruses coevolved with cellular organisms from the self-replicating molecules believed to have existed in the primitive prebiotic earth.

While each of these theories has its supporters, the topic generates strong disagreements among experts. In the end, it is not so much a matter of how viruses arose, but rather how we become infected with them.

CONCEPT AND REASONING CHECKS

13.17 Which came first—viruses or cells? Support your answer.

13.8 Virus-Like Agents

When viruses were discovered, scientists believed they were the ultimate infectious particles. It was difficult to conceive of anything smaller than viruses as agents of disease in plants, animals, and humans. However, the perception was revised as scientists discovered new disease agents—the subviral particles referred to as **virus-like agents**.

Viroids Are Infectious RNA Particles

KEY CONCEPT

- Viroids lack a protein coat.

Viroids are tiny fragments of nucleic acid known to cause diseases in crop plants. In the 1960s, Theodore O. Diener and colleagues at the US Department of Agriculture's Beltsville, Maryland, research center were investigating a suspected viral disease, potato spindle tuber (PST), which results in long, pointed potatoes shaped like spindles. Nothing would destroy the disease agent except an RNA-degrading enzyme, and in 1971, the group postulated that a fragment of single-stranded RNA was involved. Diener called the agent a viroid, meaning "virus-like."

Today, more than two dozen crop diseases have been related to viroids. The largest of these particles is about one-fifteenth of the size of the smallest virus (FIGURE 13.18). The RNA chain of the PST viroid has a known molecular sequence (359 nucleotides), but it contains so few genetic sequences that the replication cycle is not understood. Diener speculates the viroids originated as introns, the sections of RNA spliced out of messenger RNA molecules before the messengers are able to function (see Chapter 7). Because the viroid RNA encodes no proteins, another hypothesis suggests the viroid RNA interacts with host cell RNA, inactivating proteins that bring about disease through loss of cell function. A similar proposal suggests the viroid RNA "silences" host cell "target RNA," again bringing about disease through loss of cell regulation. Disease causation remains to be determined.

CONCEPT AND REASONING CHECKS

13.18 How do viroids differ from viruses?

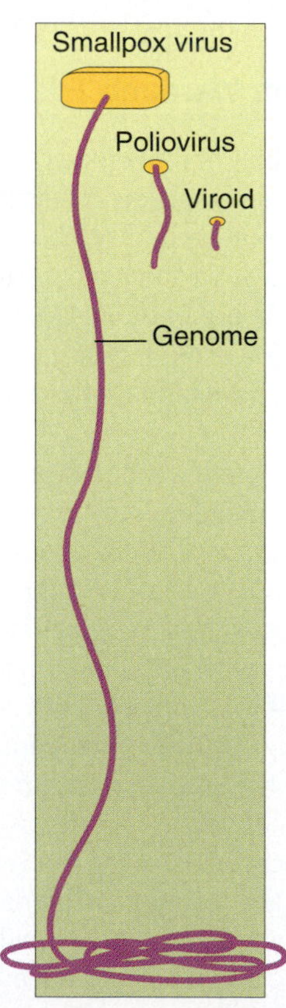

FIGURE 13.18 **Genome Relationships.** The size relationships of a smallpox virus, poliovirus, and viroid. The genome of the potato spindle tuber viroid has 359 nucleotides, while that of the smallpox virus has almost 200,000.

Q: Estimate whether viroid RNA is large enough to code for a variety of proteins.

Prions Are Infectious Proteins

KEY CONCEPT

- Prions appear to lack nucleic acid.

In 1986, cattle in Great Britain began dying from a mysterious illness. The cattle experienced weight loss, became aggressive, lacked coordination, and were unsteady on their hooves. These detrimental effects became known as **mad cow disease** and were responsible for the eventual death of these animals. A connection between mad cow disease and a similar human disease surfaced in Great Britain in the early 1990s when several young people died of a human brain disorder resembling mad cow disease. Symptoms included dementia, weakened muscles, and loss of balance. Health officials suggested the human disease was caused by eating beef that had been processed from cattle with mad cow disease. It appeared the disease agent was transmitted from cattle to humans.

Besides mad cow disease, similar neurologic degenerative diseases have been discovered and studied in other animals and humans. These include scrapie in sheep and goats, wasting disease in elk and deer, and Creutzfeldt-Jakob disease in humans. All are examples of a group of rare diseases called **transmissible spongiform encephalopathies** (**TSEs**) because, like mad cow disease, they can be transmitted to other animals of the same species and possibly to other animal species, including humans, and the disease causes the formation of "sponge-like" holes in brain tissue (FIGURE 13.19A).

Many scientists originally believed these agents were a new type of virus. However, in the early 1980s, Stanley Prusiner and colleagues isolated an unusual protein from scrapie-infected tissue, which they thought represented the infectious agent. Prusiner called the proteinaceous infectious particle a **prion** (pronounced pree-on). The sequencing of the protein led to the identification of the coding gene, called *PrP*. The *PrP* gene is primarily expressed in the brain.

■ Dementia: A loss of memory.

This led Prusiner and colleagues to propose the **protein-only hypothesis**, which predicts that prions are composed solely of protein and contain no nucleic acid. The protein-only hypothesis further proposes there are two types of prion proteins (FIGURE 13.19B, C). Normal cellular prions (PrP^{C}) are found on the surfaces of brain cells while abnormal prions, as found in scrapie (PrP^{SC}), have a different shape. These latter forms are the suspected infectious agents

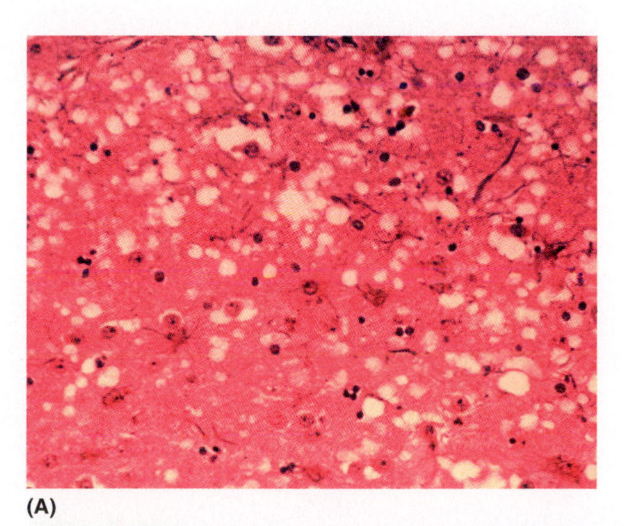
(A)

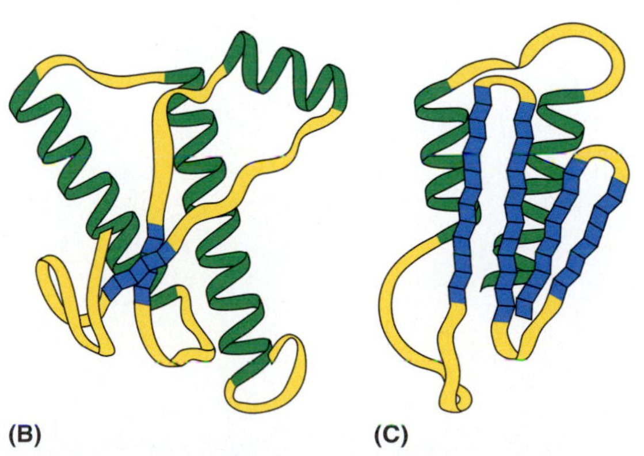
(B) (C)

FIGURE 13.19 **Prion "Infection" and Structure.** (**A**) A photomicrograph showing the vacuolar degeneration of gray matter characteristic of human and animal prion diseases. (**B**) This drawing shows the tertiary structure of the normal prion protein (Prp^{C}). The helical regions of secondary structure are green. (**C**) A misfolded prion (Prp^{SC}). This infectious form of the prion protein results from the helical regions in the normal protein unfolding and an extensive pleated sheet secondary structure (blue ribbons) forming within the protein. The misfolding allows the proteins to clump together and react in ways not completely understood that contribute to disease.

Q: How does a change in tertiary structure affect the functioning of a protein?

of diseases like mad cow and scrapie. More definitive proof came in 2004 when Prusiner's group demonstrated that purified prions can cause disease when injected into brains of genetically-engineered mice.

Many researchers believe prion diseases—that is, TSEs—are spread by the infectious PrP^{SC} binding to normal PrP^{C}, causing the latter to change shape. In a domino-like scenario, the newly converted PrP^{SC} proteins, in turn, would cause more PrP^{C} proteins to become abnormal. The PrP^{SC} proteins then form insoluble aggregates, which are responsible for the large, sponge-like holes left where groups of nerve cells have died. Importantly, PrP^{SC} does not trigger an immune response. Death of the animal occurs from the numerous nerve cell deaths that lead to loss of brain function.

The human form of TSE appears similar to classical Creutzfeldt-Jakob disease (CJD). The new form of CJD, called **variant CJD (vCJD)** is characterized clinically by neurologic abnormalities such as dementia. Neuropathology shows a marked spongiform appearance throughout the brain and death occurs within 3 to 12 months after symptoms appear.

As of July 2006, the number of vCJD definite or probable deaths in Great Britain was 156, although there have been more than 184,000 cases reported in farmed cattle in Great Britain through the end of 2005.

In January 2001, over 1,000 head of cattle in Texas were feared to be the first American cattle to suffer mad cow disease. Luckily, tests conducted by the US Food and Drug Administration (FDA) indicated there was no mad cow disease present. However, in 2003 the first mad cow was discovered in Alberta, Canada, and in December the first such cow was reported in the state of Washington. A second BSE-positive cow was identified in June 24, 2005. New protection measures include condemning those animals having signs of neurologic illness and holding any cows suspected of having BSE until test results are known. In addition, "downer cattle"—those unable to walk on their own—cannot be used for human food or animal feed.

A similar situation exists in other animals typically eaten by humans, such as, deer and elk (MicroFocus 13.8).

CONCEPT AND REASONING CHECKS

13.19 How are prions (a) similar to and (b) different from viruses?

MICROFOCUS 13.8 Environmental Microbiology/Public Health

Bambi—Stay Safe

In 2002, the Wisconsin Department of Natural Resources permitted the killing of some 18,000 deer in the state. The dead deer had their heads cut off and shipped in sealed plastic bags for testing. The carcasses were incinerated. What seemingly horrific act was going on here?

The kill was an attempt to keep an infectious disease called **chronic wasting disease** (**CWD**) from spreading through the state's deer population. Officials hoped the killing would thin out the deer population and slow the spread of the disease. The culling also would allow health examiners to determine how many of the killed animals had CWD.

CWD is a transmissible spongiform encephalopathy (TSE) of deer and elk. First recognized as a clinical "wasting" syndrome in 1967, the disease has been identified in wild and captive mule deer, white-tailed deer, North American elk, and in captive black-tailed deer in 12 states and two Canadian provinces.

Typical of TSEs, such as BSE in cattle, scrapie in sheep and goats, and variant Creutzfeldt-Jacob disease (vCJD) in humans, CWD produces a progressive, degenerative disease that affects infected animals. In deer and elk, clinical signs include chronic weight loss, behavioral changes, excessive salivation, and increased drinking and urination. CWD always is fatal in these animals. There is no known treatment or vaccine for CWD. However, CWD is many times more infectious than BSE or vCJD.

Being a TSE, the infectious agent that causes CWD is a prion. And that is cause for worry. Evidence suggests CWD is transmitted horizontally between animals. The route of infection remains unclear, although urine has been suggested. In an animal that ate urine-contaminated grass, prions could be transmitted. A 2006 study suggests CWD may be spread through animal saliva and blood. Whether these studies hold up to further investigations, hunters need to take appropriate precautions when handling killed animals. Although there is no known relationship between CWD and other TSEs of animals or people, in one experiment, CWD prions were mixed with normal prions from deer and humans. Less than 7 percent of the normal human prions (PrP^{C}) were induced to misfold into the abnormal form (PrP^{SC}). However, the percentage of misfolding is about the same as that induced when BSE prions are mixed with human prions.

The CDC has issued this statement: "It is generally prudent to avoid consuming food derived from any animal with evidence of a TSE. To date, there is no evidence that CWD has been transmitted or can be transmitted to humans under natural conditions. However, there is not yet strong evidence that such transmissions could not occur." The states with known cases of CWD in free-range animals are Colorado, Illinois, Nebraska, New Mexico, South Dakota, Utah, Wisconsin, and Wyoming. States with known CWD occurrences in farmed or captive animals are Arizona, Idaho, Iowa, Indiana, Kansas, Kentucky, Michigan, Minnesota, Missouri, Montana, Nevada, North Dakota, Oklahoma, and Texas.

SUMMARY OF KEY CONCEPTS

13.1 Foundations of Virology

- The concept of a virus as a filterable agent of disease was first suggested by Ivanowsky. Other similar agents were soon identified with human disease, while Twort and d'Herrelle identified viruses as infective agents in bacterial organisms. Studying viruses was extremely difficult until the 1940s, when the electron microscope enabled scientists to see viruses and innovative methods allowed researchers to cultivate them.

13.2 What Are Viruses?

- Viruses are small, obligate, intracellular particles. All viruses are composed of nucleic acid as either DNA or RNA and in either a single-stranded or double-stranded form. The genome is surrounded by a protein capsid, which is usually subdivided into capsomeres. Many viruses have an envelope surrounding the nucleocapsid. Spikes protruding from the capsid or envelope are used for attachment to host cells.
- Viruses have helical, icosahedral, or complex symmetry.
- The host range of a virus refers to what organisms (hosts) the virus can infect. Many viruses only infect certain cell types or tissues within the host, referred to as a tissue tropism (tissue attraction).

13.3 The Classification of Viruses

- Viruses have no agreed upon binomial names. Classification schemes have been based on physiological and biochemical characteristics or on the tissue where replication takes place. Two broad classes of viruses can be organized based on their genome and strand type: the single-stranded or double-stranded DNA viruses and the single-stranded or double-stranded RNA viruses.

13.4 Viral Replication and Its Control

- Virulent bacteriophages undergo a lytic replication cycle involving attachment, penetration, biosynthesis, maturation, and release phases.
- Temperate phages can enter a lysogenic cycle where the phage genome integrates into the bacterial chromosome as a prophage. A lytic cycle is triggered under environmentally stressful conditions.
- Animal viruses also progress through the same phases of replication. However, penetration can occur by membrane fusion or endocytosis, and the biochemistry of nucleic acid synthesis varies among DNA and RNA viruses.
- Many of the DNA viruses and the retroviruses can incorporate their viral DNA within a host chromosome as a provirus.
- Antiviral drugs are targeted at several points in the viral replicative cycle, including penetration, genome replication, and virion maturation and release.
- Interferon represents a group of antiviral substances naturally produced by cells when infected. Secretion of interferon stimulates uninfected neighboring cells to enter an antiviral state by producing antiviral proteins.

13.5 The Cultivation and Detection of Viruses

- Various detection methods for viruses are based on characteristic cytologic changes referred to as the cytopathic effect.
- The cultivation of animal viruses is maintained by using cell cultures. The detection of viruses can be determined by plaque formation.
- Virus-antibody interactions can be used to detect and identify viruses using a hemagglutination-inhibition test.

13.6 Tumors and Viruses

- Cancer is a complex condition in which cells multiply without control (tumors) and possibly develop into cancers.
- There are at least six carcinogenic viruses known to cause human tumors. Many of these tumors may develop into cancer.
- Viruses may bring about tumors by converting protooncogenes into tumor oncogenes. If such genes are carried in a virus, they are called viral oncogenes.

13.7 Emerging Viruses and Virus Evolution

- Viruses use genetic recombination and mutations as mechanisms to evolve.
- Human population expansion into new areas and increased agricultural practices in previously forested areas have exposed humans to existing animal viruses. Such viruses have to make the jump to humans and be able to replicate in the human host. If this occurs, new (emerging) viral diseases are a predictable outcome.
- Three hypotheses have been proposed for the origin of viruses. These include the regressive evolution hypothesis, the cellular origins hypothesis, and the independent entities hypothesis.

13.8 Virus-Like Agents

- Viroids are infectious particles made of RNA. They infect a few plant species and cause disease. Viroids lack protein and a capsid, but can replicate themselves inside the host.
- Prions are infectious particles made of protein. They are capable of causing a number of animal diseases, including mad cow disease in cattle and variant Creutzfeldt-Jakob disease (vCJD) in humans. Prions cause disease by folding improperly and in the misfolded shape, cause other prions to misfold. These and other similar diseases in animals are examples of a group of rare diseases called transmissible spongiform encephalopathies (TSEs).

LEARNING OBJECTIVES

After understanding the textbook reading, you should be capable of writing a paragraph that includes the appropriate terms and pertinent information to answer the objective.

1. Identify the important historical developments in the identification and role of viruses.
2. Distinguish between the structures common to all viruses and those that separate naked from enveloped viruses.
3. Identify the three shapes viruses may exhibit.
4. Contrast viral host range with tissue tropism.
5. Summarize the classification schemes used to group viruses.
6. Explain the difference between DNA and RNA viruses.
7. Identify the five phases of and explain the events occurring in a phage lytic viral replicative cycle.
8. State how a lysogenic cycle differs from a lytic cycle in bacteriophage.
9. Describe how animal viruses penetrate and replicate in a host cell.
10. Explain how some DNA viruses and retroviruses can exist in a provirus form.
11. Identify the three major targets of antiviral drugs and assess their effectiveness.
12. Diagram the pathway by which interferon interferes with viral replication.
13. Assess the role of the cytopathic effect for virus detection.
14. Explain how animal cell cultures are produced and how viruses can be visually detected in these infected cultures.
15. Define how benign and malignant tumors arise.
16. Identify three viruses that are involved in causing human tumors.
17. Explain how viruses can cause tumors.
18. Identify the two mechanisms by which emerging viruses arise and how an animal virus could jump to humans.
19. Contrast the three hypotheses for the origin viruses.
20. Summarize the properties of viroids.
21. Summarize the properties of and diseases resulting from prions.

SELF-TEST

Answer each of the following questions by selecting the ***one*** answer that best fits the question or statement. Answers to even-numbered questions can be found in **Appendix C**.

1. Which one of the following scientists was *not* involved with discovering viruses?
   A. Walter Reed
   B. Felix d'Herrelle
   C. Dimitri Ivanowsky
   D. Robert Fleming
   E. Martinus Beijerinck
2. A viral genome may consist of
   A. DNA only.
   B. RNA only.
   C. DNA and protein.
   D. DNA or RNA.
   E. DNA and RNA.
3. A nucleocapsid can have _____ symmetry.
   A. radial
   B. icosahedral
   C. helical
   D. bilateral
   E. Both **B** and **C** are correct.
4. An RNA virus genome in the form of messenger RNA is referred to as a
   A. + strand RNA.
   B. double-stranded RNA.
   C. – strand RNA.
   D. reverse strand RNA.
   E. complex strand RNA.
5. Retroviruses are able to
   A. convert protein to RNA.
   B. form double-stranded RNA from single-stranded RNA.
   C. convert RNA into DNA.
   D. form + and – strand RNA.
   E. Both **A** and **C** are correct.
6. A virulent bacteriophage can
   A. undergo a lysogenic cycle.
   B. carry out a lytic cycle.
   C. integrate its genome in the host cell.
   D. remain dormant in a bacterial cell.
   E. exist as a prophage.
7. The release of the viral genome from the capsid is called
   A. uncoating.
   B. endocytosis.
   C. penetration.
   D. biosynthesis.
   E. maturation.
8. Provirus formation is possible in members of the
   A. singe-stranded DNA viruses.
   B. single-stranded (+ strand) RNA viruses.
   C. retroviruses.
   D. double-stranded RNA viruses.
   E. single-stranded (– strand) RNA viruses.

9. Antiviral drugs that are base analogs affect the _____ phase of the viral replication cycle.
   A. attachment
   B. penetration
   C. release
   D. genome replication
   E. uncoating

10. Which one of the following statements about interferon is false?
   A. Interferon is produced in response to a viral infection.
   B. Interferon is a naturally produced protein.
   C. Interferon triggers the production of antiviral proteins in neighboring cells.
   D. Interferon puts cells in an antiviral state.
   E. Interferon is a protein that binds to RNA virus genomes.

11. Cytopathic effects would include all the following *except*
   A. changes in cell structure.
   B. plaque formation.
   C. presence of Negri bodies.
   D. vacuolated cytoplasm.
   E. syncytia formation.

12. A benign tumor
   A. will metastasize.
   B. represents cancer.
   C. is malignant.
   D. is a clone of dividing cells.
   E. All the above (**A–D**) are true.

13. Which of the following is *not* a carcinogen?
   A. Genetic factors
   B. UV light
   C. Certain chemicals
   D. X rays
   E. Radiation

14. The oncogene theory states that transforming genes
   A. normally occur in the host genome.
   B. can exist in viruses.
   C. are not of viral origin.
   D. originate from protooncogenes.
   E. All of the above (**A–D**) are correct.

15. Newly emerging viruses causing human disease can arise from
   A. species jumping.
   B. mutations.
   C. genetic recombination.
   D. infected animal populations increasing in size.
   E. All of the above (**A–D**) are correct.

16. Viroids contain
   A. RNA and DNA.
   B. only DNA.
   C. only RNA.
   D. DNA and a capsid.
   E. RNA and an envelope.

17. Which one of the following statements about prions is false?
   A. Prions are infectious proteins.
   B. Abnormal prions misfold.
   C. Prions have caused disease in Americans.
   D. Human disease is called variant CJD.
   E. Prions can be transmitted to humans from infected beef.

QUESTIONS FOR THOUGHT AND DISCUSSION

Answers to even-numbered questions can be found in **Appendix C**.

1. If you were to stop 1,000 people on the street and ask if they recognize the term "virus," all would probably respond in the affirmative. If you were then to ask the people to describe a virus, you might hear answers like, "It's very small" or "It's a germ," or a host of other colorful but not very descriptive terms. As a student of microbiology, how would you describe a virus?

2. Oncogenes have been described in the recent literature as "Jekyll and Hyde genes." What factors may have led to this label, and what does it imply? In your view, is the name justified?

3. Researchers studying bacterial species that live in the oceans have long been troubled by the question of why these microorganisms have not saturated the oceanic environments. What might be a reason?

4. When discussing the multiplication of viruses, virologists prefer to call the process replication, rather than reproduction. Why do you think this is so? Would you agree with virologists that replication is the better term?

5. Bacterial species can cause disease by using their toxins to interfere with important body processes; by overcoming body defenses, such as phagocytosis; by using their enzymes to digest tissue cells; or other similar mechanisms. Viruses, by contrast, have no toxins and produce no digestive enzymes. How, then, do viruses cause disease?

6. How have revelations from studies on viruses, viroids, and prions complicated some of the traditional views about the principles of biology?

APPLICATIONS

Answers to even-numbered questions can be found in **Appendix C.**

1. A friend has the flu and knows you are taking microbiology. So, he asks you, "What antibiotic would be best to fight the flu?" You tell him that antibiotic therapy is not appropriate. Why is this your answer?
2. A person appears to have died of rabies. As a coroner, you need to verify that the cause of death was rabies. What procedures could you use to confirm the presence of rabies virus in the brain tissue of the deceased?
3. A technician mixes influenzavirus with a person's serum (antibodies). Red blood cells are then added and hemagglutination occurs. Does this result indicate that the person had been exposed to the flu virus? Explain.

REVIEW

Use the following syllables to compose the term that answers each clue from virology. The number of letters in the term is indicated by the dashes, and the number of syllables in the term is shown by the number in parentheses. Each syllable is used only once. The answers are listed in Appendix D.

A A A BAC CAP CAP CLO CO COS CY DINE DRON EN FER GE GENE HE I I I IN LENT MAN MERES MOR NEG NOME O ON ON ON ONS OPE PHAGE PRI RI SID SO TA TER TER TU U VEL VIR VIR VIR

1. Viral protein coat (2) ___ ___ ___ ___ ___ ___
2. Bacterial virus (5) ___ ___ ___ ___ ___ ___ ___ ___ ___ ___ ___ ___ ___
3. Rabies granules (2) ___ ___ ___ ___ ___
4. Herpes drug (4) ___ ___ ___ ___ ___ ___ ___ ___ ___
5. Natural antiviral protein (4) ___ ___ ___ ___ ___ ___ ___ ___ ___ ___
6. Clone of abnormal cells (2) ___ ___ ___ ___ ___
7. Cancer gene (3) ___ ___ ___ ___ ___ ___ ___ ___
8. Phages that lyse bacterial cells (3) ___ ___ ___ ___ ___ ___ ___ ___
9. Shape of poliovirus (5) ___ ___ ___ ___ ___ ___ ___ ___ ___ ___ ___
10. Viral genetic information (2) ___ ___ ___ ___ ___ ___
11. Completely assembled virus (3) ___ ___ ___ ___ ___ ___
12. Drug for influenza (4) ___ ___ ___ ___ ___ ___ ___ ___ ___ ___
13. Surrounds the capsid (3) ___ ___ ___ ___ ___ ___ ___ ___
14. Infectious protein particles (2) ___ ___ ___ ___ ___ ___
15. Capsid subunits (3) ___ ___ ___ ___ ___ ___ ___ ___ ___ ___

HTTP://MICROBIOLOGY.JBPUB.COM/

The site features learning, an on-line review area that provides quizzes and other tools to help you study for your class. You can also follow useful links for in-depth information, read more MicroFocus stories, or just find out the latest microbiology news.

14

Chapter Preview and Key Concepts

14.1 Viral Infections of the Upper Respiratory Tract
- Influenza A and B viruses evolve through antigenic drift and antigenic shift.
  MicroInquiry 14: Drifting and Shifting—How Influenza Viruses Evolve
- Rhinoviruses are responsible for the common-cold syndrome.
- Adenovirus respiratory infections typically cause a severe sore throat (pharyngitis).

14.2 Viral Infections of the Lower Respiratory Tract
- The paramyxoviruses are a group of viruses causing similar symptoms.
- The SARS coronavirus causes a unique form of pneumonia.

14.3 Diseases of the Skin Caused by Herpesviruses
- Human herpes simplex viruses cause an array of viral diseases.
- The varicella-zoster virus causes chickenpox and shingles.
- Human herpesvirus 6 infection primarily causes a short-term fever and rash in infants.
- The Epstein-Barr virus and human herpesvirus 8 have oncogenic potential.

14.4 Other Viral Diseases of the Skin
- Two genera of paramyxoviruses cause measles and mumps.
- The rubella virus produces a pale-pink rash on the trunk and extremities.
- Human parvovirus B19 is most common among elementary school-age children.
- Human papillomaviruses can cause common or genital warts.
- The poxviruses produce contagious and dangerous diseases.

Viral Infections of the Respiratory Tract and Skin

Sometime soon we will face a biological attack that has nothing to do with terrorists and everything to do with viruses mutating in the chicken farms of southeast Asia.
—*New Scientist* magazine, January 10, 2004

Every year there are seasonal flu outbreaks that may become epidemic; that is, they spread quickly through a large population. However, once in a while there is a **pandemic**, a worldwide epidemic. The biggest influenza pandemic, called the "Spanish flu," occurred in 1918 to 1919 (FIGURE 14.1). A fifth of the world's population was infected by a virus thought to have originated in birds. Estimates suggest between 20 and 100 million people died, young adults being the worst hit. As we see in this chapter, such pandemics can occur when a new infectious virus appears for which the human population has no immunity. Thus, many of us get a flu shot every year to protect us from the current dominant strains of circulating flu viruses.

Influenza continues to be an ongoing problem in the twenty-first century. Health experts and microbiologists believe we are overdue for another serious and perhaps very deadly flu pandemic. In fact, one may be "brewing" in 2007. An avian flu virus has been spreading throughout Asia and into Europe and Africa. By mid-October, 2006, this influenza virus had killed at least 151 people of the 256 infected in Asia and Southeast Asia since 2003. At this time, it is still difficult for people to contract the disease because it is spread mainly from infected fowl to humans. Human-to-human transmission has not yet been documented. However, it might only take a mutation or two to transform the avian flu virus into a form that can become highly contagious, as the chapter opening quote suggests.

FIGURE 14.1 **U.S. Soldiers—1918.** This photo of American soldiers marching near the end of World War I shows them wearing gauze masks to protect against the flu. The masks were totally ineffective.

Q: Why would these masks be useless against spreading or contracting the flu virus?

There are several other viruses with tropisms to the respiratory tract or skin. Chickenpox is still among the most commonly reported diseases of childhood years. Genital herpes has become so rampant that, by some estimates, 10 to 20 million Americans are currently infected.

The news on viruses is not all bad. We now have vaccines to many viral diseases that once wrecked havoc. For example, two skin diseases, mumps and rubella, were part of the fabric of life only a generation ago, but the annual case reports have dropped from hundreds of thousands to mere hundreds (and some officials are bold enough even to whisper the word "eradication").

In this chapter we focus on several viral diseases, besides influenza, which affect the respiratory tract—the so-called pneumotropic diseases. We also examine viral diseases affecting the skin—the dermotropic viral diseases. Note: the division of the pneumotropic and dermotropic viral diseases is an artificial classification simply for grouping convenience (see Chapter 13). Therefore, the symptoms described may go beyond the respiratory tract or skin, respectively. As in Chapter 15, each disease will be presented as an independent essay, so you can establish an order that best suits your study needs.

■ Contagious: Capable of being transmitted from one person to the next.

14.1 Viral Infections of the Upper Respiratory Tract

The upper respiratory tract (URT) consists of the nose, pharynx (throat), and the middle ear and auditory tubes. Influenza (the flu) and the common cold are the most common viral infections of the URT, accounting for more than 1 billion infections each year in the United States alone.

Influenza Is a Highly Communicable Acute Respiratory Infection

KEY CONCEPT

- Influenza A and B viruses evolve through antigenic drift and antigenic shift.

Influenza is a highly contagious acute disease of the respiratory tract that is transmitted by airborne respiratory droplets. The disease is believed to take its name from the Italian word for "influence," a reference either to the influence of heavenly bodies, or to the *influenza di freddo*, "influence of the cold." Since the first recorded epidemic in 1510, scientists have described 31 pandemics. The most notable pandemic of the twentieth century was the "Spanish" flu in 1918–1919; others took place in 1957 (the "Asian" flu) and in 1968 (the "Hong Kong" flu). Today, there are about 35,000 excess deaths in the United States and 250,000 to 500,000 deaths worldwide annually related to influenza infections.

The enveloped influenza virion belongs to the *Orthomyxoviridae* family. It is composed of eight single-stranded (– strand) segments of RNA, each wound helically and associated with protein to form a nucleocapsid (FIGURE 14.2). An additional structural protein, the **matrix protein**, surrounds the core of RNA segments. An envelope covers the matrix protein.

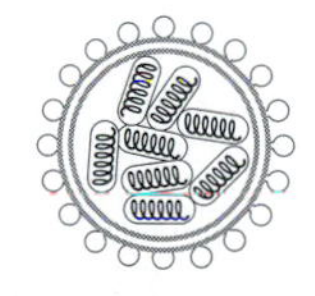

Influenza virus

■ Acute disease: One that develops rapidly, exhibits substantial symptoms, and then comes to a climax.

Projecting through the envelope are two types of spikes. One type contains the enzyme **hemagglutinin** (H), a substance facilitating the attachment and penetration of influenza viruses into host cells. Its shape determines the virus' host range and tropism. The second type contains another enzyme, **neuraminidase** (N), a

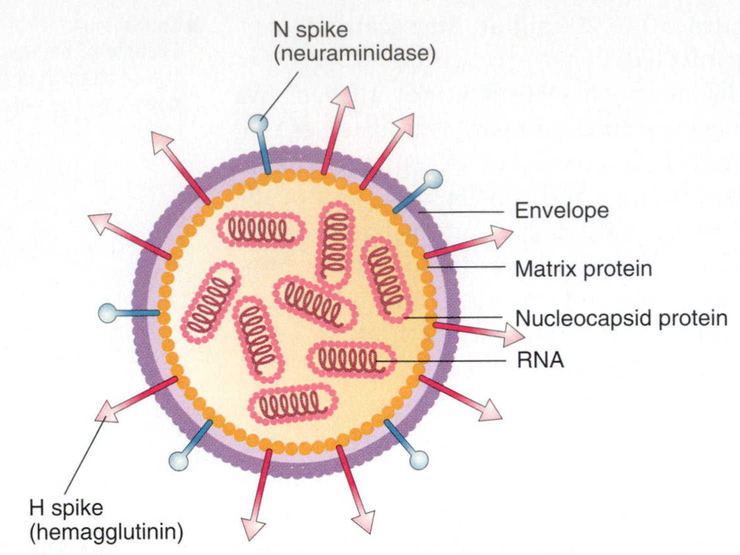

FIGURE 14.2 **The Influenzavirus.** This diagram of the influenzavirus shows its eight segments of RNA, matrix protein, and envelope with hemagglutinin and neuraminidase spikes protruding.

Q: What is the role of the hemagglutinin and neuraminidase spikes?

compound assisting in the release of the virions from the host cell when replication is complete.

Three types of influenza viruses are recognized.

Influenza A strikes every year and causes most epidemics. It circulates in many animals, including ducks, chickens, pigs, and humans. Type A is divided into subtypes based on the H and N surface proteins. There are 15 known H subtypes and 9 different N subtypes. The current subtypes in humans are A(H1N1) and A(H3N2).

Influenza B also strikes every year but is less widespread than type A. It only circulates between humans and is not divided into subtypes. Each year's flu vaccine is a mixture of the most prevalent A and B subtypes.

Influenza C causes a mild respiratory illness but not epidemics.

Each year a slightly different flu strain evolves, based, in part, on changes to H and/or N spike proteins; thus, there is a need for a new flu vaccine each year. Sometimes the new strain is quite mild, while in other years, a predominant strain will be more dangerous, such as the "Spanish" flu. How do new flu strains arise? MicroInquiry 14 examines their evolution.

The onset of influenza is abrupt after an incubation period of one to four days. The individual with an uncomplicated illness develops sudden chills, fatigue, headache, and pain most pronounced in the chest, back, and legs. Over a 24-hour period, body temperature can rise to 40°C, and a severe cough develops. Individuals may experience sore throat, nasal congestion, sneezing, and tight chest, the latter a probable reflection of viral invasion of tissues of the trachea and bronchi. Despite these severe symptoms, influenza is normally short-lived and has a favorable prognosis. The disease is self-limiting and usually resolves in one week to ten days.

Most of the annual deaths from influenza are due to pneumonia caused by the virus spreading into the lungs. Secondary complications in unvaccinated individuals, especially those over 65 years, infants, or with underlying medical conditions (immunocompromised), may lead to bacterial pneumonia if *Staphylococcus aureus* or *Haemophilus influenzae* invade the damaged respiratory tissue. MicroFocus 14.1 describes the role of preschoolers in driving flu outbreaks that can lead to pneumonia deaths in the community.

Influenza infection in rare cases is associated with two serious complications. **Guillain-Barré syndrome** (**GBS**) occurs when the body damages its own peripheral nerve cells, causing muscle weakness and sometimes paralysis. **Reye syndrome** usually makes its appearance in young people after they use aspirin to treat fever or pain associated with influenza. It begins with nausea and vomiting, but the progressive mental changes (such as confusion or delirium) may occur. Very few children develop Reye syndrome (less than 0.03–1 case per 100,000 persons younger than 18 years).

Because influenza-like symptoms can be caused by bacterial and other viral infections, a diagnosis of influenza, if necessary, is based on several factors. This includes the pattern of spread in the community, observation of disease symptoms, laboratory isolation of viruses, or hemagglutination of red blood cells (see Chapter 13).

The best prevention for the flu is an annual flu vaccination. Each year's batch of vaccine

MICROFOCUS 14.1: Public Health

Preschoolers Drive Flu Outbreaks

Every October to November in the Northern Hemisphere, we brace for another outbreak or epidemic spread of the flu. Many people go to their physician, clinic, or even grocery store for a flu shot. Others of us will "catch" the flu and suffer through several days of agony. For some 35,000, especially the elderly and immunocompromised, contracting the flu will lead to pneumonia and death. Wouldn't it be great if there was some way to predict pneumonia and flu deaths in a population? That now may be possible.

Researchers at the Children's Hospital Boston and Harvard Medical School reported biosurveillance data suggesting that otherwise healthy preschoolers (3- to 4-year-olds) drive flu epidemics. The researchers found that by late September, kids in this age group were the first to develop flu symptoms.

Why is this group most vulnerable? Current immunization policies suggest that infants 6 to 23 months old be vaccinated against the flu. Policy also suggests that older children, including those preschoolers, only be vaccinated if they have high medical-risk conditions. That means they will be vulnerable to the flu. And with many being in daycare and preschool, close contact makes spreading the flu effortless as those exposed bring the infection home.

The data also demonstrate that when preschoolers start sneezing, the unvaccinated elderly become ill. Thus, the research suggests vaccinating those individuals who are driving and transmitting flu to others—the preschoolers. They are the sentinels by which an ensuing flu outbreak or epidemic can be identified. Immunizing preschoolers will decrease flu transmission and limit adult mortality in the unvaccinated.

(see MicroFocus 21.1) is based on the previous year's predominant influenza A and B viruses and is about 75 percent effective. Because the viruses are grown in chicken embryos (see Chapter 13), people who are allergic to chickens or eggs should not be given the vaccine.

Treatment of uncomplicated cases of influenza requires bed rest, adequate fluid intake, and aspirin (or acetaminophen in children) for fever and muscle pain. Because antibiotics do not work against viral infections, more serious cases may respond to amantadine or rimantadine, drugs that interfere with uncoating in the replication cycle (see Chapter 13). Two antiviral drugs, zanamivir (Relenza) and, in the United States, oseltamivir (Tamiflu), are available by prescription. These drugs target the neuraminidase spikes projecting from the influenza virus envelope and block the release of new virions (see Chapter 13). If given to otherwise healthy adults or children early in disease onset, these drugs can reduce the duration of illness by one day and make complications less likely to occur. However, these drugs should not be taken in place of vaccination, which remains the best prevention strategy.

CONCEPT AND REASONING CHECKS

14.1 From MicroInquiry 14, distinguish between antigenic shift and antigenic drift.

MICROINQUIRY 14

Drifting and Shifting—How Influenza Viruses Evolve

Influenza viruses evolve in two different ways. Both involve **antigenic variation**, a process in which chemical and structural changes occur periodically in hemagglutinin (H) and neuraminidase (N) spike proteins (antigens), thereby yielding new virus strains.

Antigenic drift involves small changes to the virus. These changes involve minor point mutations resulting from RNA replication errors. The mutations will be expressed in the new virions produced. Although such mutations may be detrimental, on occasion one might confer an advantage to the virus, such as being resistant to a host's immune system. For example, a spike protein might have a subtle change in shape (that is, the structural shape has "drifted") so they are not recognized by the host's immune system. This is what happens prior to most flu seasons. The virus spikes are different enough from the previous season that the host's antibodies fail to recognize the new strain. Both influenza A and B viruses can undergo antigenic drift.

Antigenic shift is an abrupt, major change in structure to influenza A viruses. Antigenic shift may give rise to new strains with completely new hemagglutinin and neuraminidase spike proteins (that is, the spike structure has "shifted") to which everyone is totally defenseless, and from which a pandemic may ensue (see figure, next page).

Two mechanisms account for antigenic shift.

The "Spanish" flu was the introduction of a completely new flu strain (H1N1) into the human population from birds. The H1N1 strain jumped directly to humans and adapted quickly to replicate efficiently in humans (**Panel A**).

The second mechanism involves "gene swapping" or reassortment between different flu viruses (**Panel B**). The 1957 influenza virus ("Asian" influenza; H2N2), for example, was a reassortment, where the human H1N1 virus acquired new spike genetic segments (H2 and N2) from an avian species. The 1968 influenza virus ("Hong Kong" influenza; H3N2) was the result of another reassortment; the human H2N2 strain acquired a new hemagglutinin genetic segment (H3) from another avian species (**Panel C**).

Pigs usually are the reassortment "vessels" (or intermediate host) because they can be infected by both avian and human flu viruses.

Future pandemic strains could arise through either antigenic drift or shift (**Panel D**). The 2005 H5N1 avian flu strain could mutate further or recombine with H3N2 to produce a potentially deadly new strain in humans.

Discussion Point

The current avian (or bird) flu (H5N1 strain) is lethal to domestic fowl and can be transmitted to humans. As mentioned in the chapter introduction, as of late October, 2006, at least 151 of the 256 people infected have died (see map).

If a human pandemic should occur, health care providers would play a crucial role in minimizing the pandemic. Therefore, planning for pandemic influenza is crucial.

Discuss the specific steps that should be taken by individuals, colleges and universities, and communities in planning for a pandemic outbreak.

Once you have come up with a plan, check out what the CDC suggests (http://www.pandemicflu.gov/) and what your state is doing.

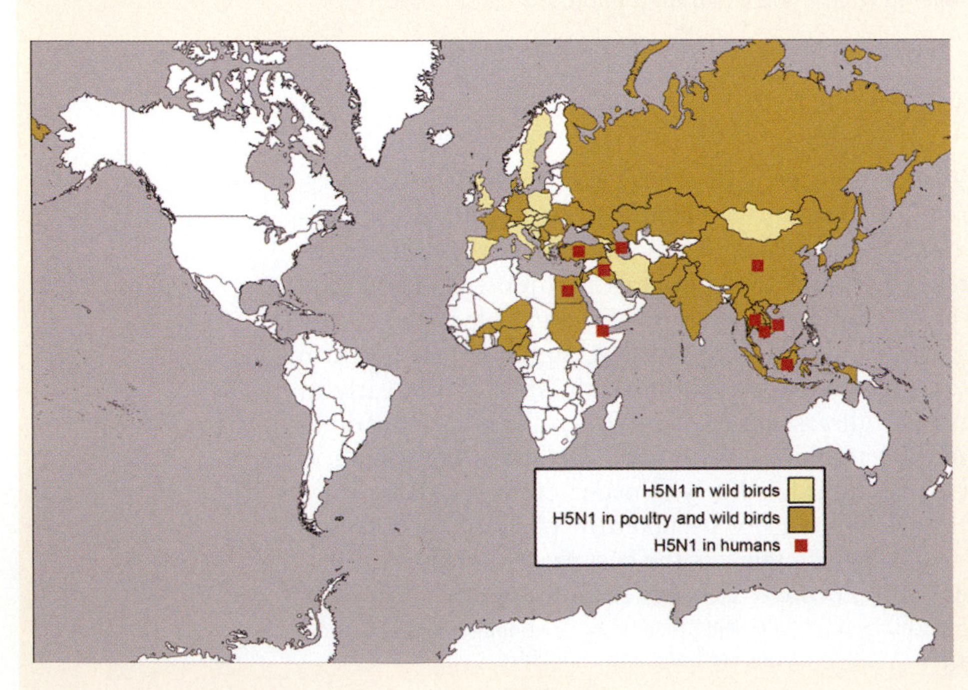

Nations with confirmed cases of H5N1 avian influenza—July 7, 2006. (Courtesy of PandemicFlu.gov.)

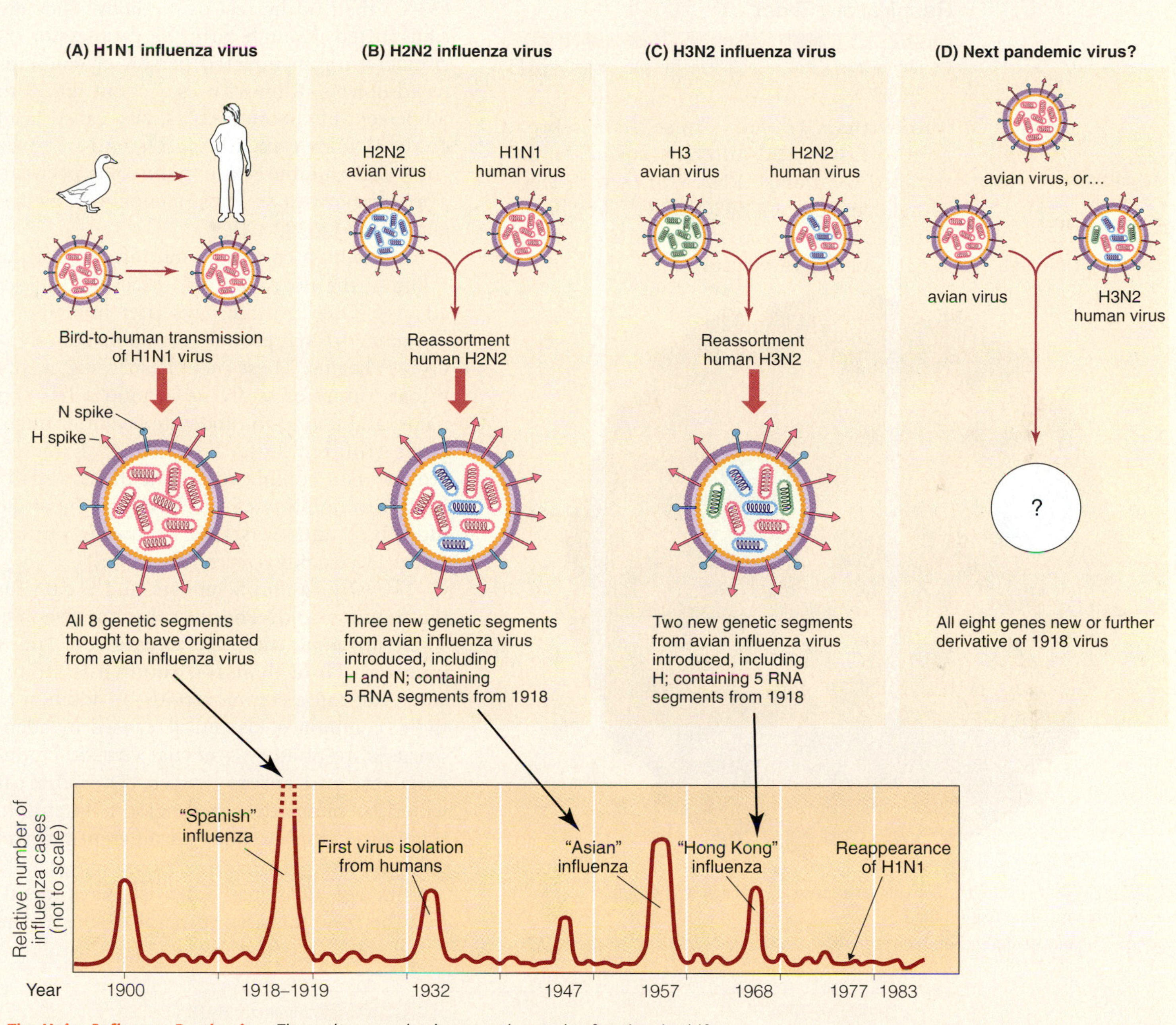

The Major Influenza Pandemics. These three pandemics were the result of antigenic shifts.

Rhinovirus Infections Produce Inflammation in the Upper Respiratory Tract

KEY CONCEPT

- Rhinoviruses are responsible for the common-cold syndrome.

Rhinoviruses (*rhino* = "nose") are a broad group of over 100 different naked, single-stranded (+ strand), RNA viruses with icosahedral symmetry (FIGURE 14.3). They belong to the family *Picornaviridae* (*pico* = "small"; hence, small-RNA-viruses).

■ Croup:
An inflammation of the larynx and trachea, causing a cough, hoarseness, and breathing difficulties.

FIGURE 14.3 A Rhinovirus. A computer-generated image of the surface (capsid) of human rhinovirus 16. (Bar = 3 nm.)

Q: What is unique about each rhinovirus that makes vaccine development impractical?

Rhinoviruses thrive in the human nose, where the temperature is a few degrees cooler (33°C) than in the rest of the body. They are transmitted through airborne droplets or by contact with an infected person or contaminated objects. Rhinoviruses account for 30 to 50 percent of **common colds**, also called head colds. Adults typically suffer two or three colds and children six to eight colds per year. The rhinoviruses are most common in the fall and spring (FIGURE 14.4).

A common cold is a viral infection of the lining of the nose, sinuses, throat, and upper airways. One to three days after infection, a sequence of symptoms (common-cold syndrome) begins. These include sneezing, a sore throat, runny or stuffy nose, mild aches and pains, and a mild-to-moderate hacking cough. Some children suffer from croup. Earaches also can be a complication. The illness usually is of short duration of 7 to 10 days. One of the old wives tales was that becoming chilled could cause colds (MicroFocus 14.2).

So, why cannot scientists find a cure for the common cold? The answer is quite simple. There are more than 120 viruses and strains involved. More than 100 rhinovirus strains alone can cause common colds. In addition to these, common colds can be caused by adenoviruses, respiratory syncytial viruses, coronaviruses, and enteroviruses. Therefore, it would be impractical to develop a vaccine to immunize people against all different types of cold viruses.

But not all is lost. Scientists have identified the receptor sites in nasal tissues where

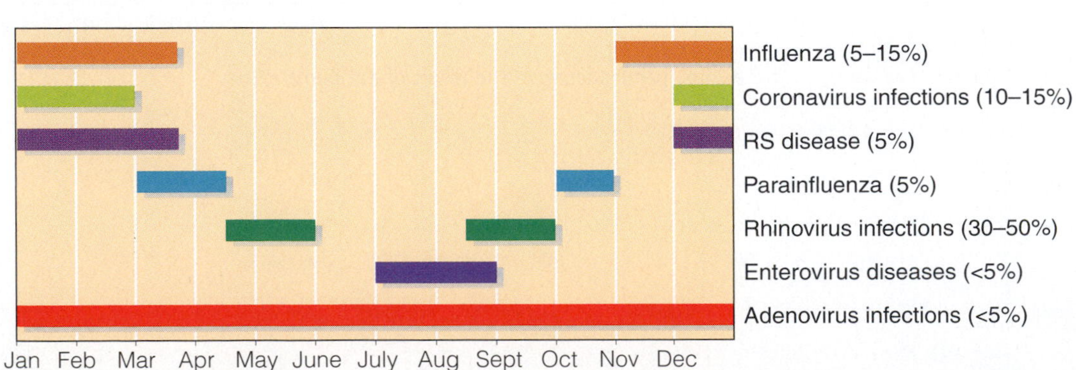

FIGURE 14.4 The Seasonal Variation and Estimated Annual Proportion of Viral Respiratory Diseases. This chart shows the seasons associated with various viral diseases of the respiratory tract (and their annual percentage). Enteroviruses (Chapter 15) cause diseases of the gastrointestinal tract as well as respiratory disorders.

Q: Hypothesize as to why different cold viruses cause diseases at different times of the year.

rhinoviruses attach. Furthermore, they have synthesized an antibody that binds to these sites and blocks viral attachment. Supposedly, the antibody could be used as an anti-cold drug. Another group of researchers have used copies of the receptor sites as a drug to bind to the virus blocking tissue tropism.

Antibiotics will not prevent or cure a cold. Antihistamines can sometimes be used to treat the symptoms of a cold; however, they do not shorten the length of the illness. For other remedies, such as vitamin C, zinc, and herbs like echinacea, there is no verified scientific evidence these substances limit or prevent colds.

Because rhinoviruses spread by respiratory droplets, washing hands and sneezing into a handkerchief are important to decrease transmission of the viruses. Interestingly, one can now buy anti-viral facial tissues that purportedly "kill" 99.9% of cold and flu viruses (rhinoviruses 1 and 2; influenza A and B) in the tissue within 15 minutes.

CONCEPT AND REASONING CHECKS

14.2 How could you determine by symptoms if an individual has a cold or the flu?

Adenovirus Infections Also Produce Symptoms Typical of a Common Cold

KEY CONCEPT

- Adenovirus respiratory infections typically cause a severe sore throat (pharyngitis).

Adenoviruses (family *Adenoviridae*) are a group of over 50 antigenic types (serotypes) of nonenveloped, icosahedral virions having double-stranded DNA (FIGURE 14.5). They multiply in the nuclei of several human host tissues and induce the formation of **inclusion bodies**, which are a series of bodies composed of numerous virions arranged in a crystalline

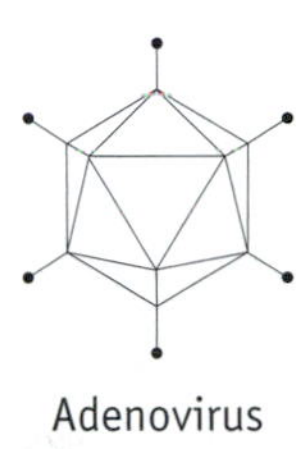
Adenovirus

MICROFOCUS 14.2: Being Skeptical

Catching a Chill: Can It Cause a Cold?

How many times can you remember your mom or a family member saying to you, "Bundle up or you will catch a cold!" Can you actually "catch" a cold from a body chill?

During the last 50 years of research, scientists have claimed that one cannot get a cold from a chill. Rather, colds result from more people being cooped up indoors during the winter months, making virus transmission from person-to-person very likely. At least, that was the scientific dogma until 2005.

In the November 2005 issue of *Family Practice*, British researchers Claire Johnson and Professor Ron Eccles at Cardiff University's Common Cold Center announced that a drop in body temperature can allow a cold to develop.

The researchers signed up 180 volunteers between October and March to participate in the study. Split into two groups, one group put their bare feet into bowls of ice cold water for 20 minutes. The other group put their bare feet in similar but empty bowls.

Over the next five days, 29 percent of individuals who had their feet chilled developed cold symptoms, while just 9 percent of the control group developed symptoms.

Professor Eccles suggests that colds may develop not because the volunteers actually "caught" a cold virus, but rather that they harbored the virus all along. Chilling lowers the person's immunity, providing the opportunity for the virus to produce more severe symptoms.

Chilling causes a constriction of blood vessels in the nose, limiting the warm blood flow supplying white blood cells to eliminate or control the cold viruses present. Professor Eccles states that, "A cold nose may be one of the major factors that causes common colds to be seasonal."

The verdict: Well, the results of one study do not make for a general consensus. More definitive studies may help verify or refute the "cold nose" claim.

■ **Febrile:** Relates to fever.

■ **Conjunctivitis:** An inflammation of the conjunctiva of the eye.

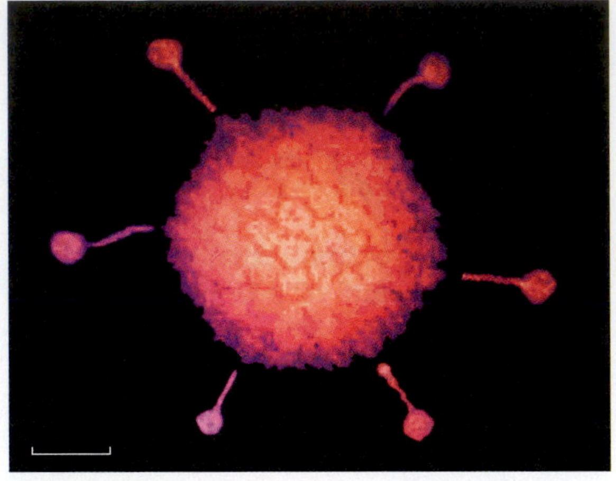

FIGURE 14.5 **An Adenovirus.** This false-color transmission electron micrograph of an adenovirus shows the icosahedral symmetry. The rough surface of the virus is due to the presence of capsomeres. (Bar = 25 nm.)

Q: What are the projections from the capsid surface?

■ **Adenoid tissue:** Refers to the pharyngeal tonsil located in the upper rear of the pharynx.

pattern (see Chapter 13). The viruses take their name from the adenoid tissue from which they were first isolated.

Adenoviruses can be highly infectious and are frequent causes of URT diseases often symptomatic of a common cold in infants and young children. Transmitted through respiratory droplets, the viruses most often cause distinctive symptoms because the fever is substantial, the throat is very sore (**acute** febrile **pharyngitis**), and the cough is usually severe. In addition, the lymph nodes of the neck swell and a whitish-gray material appears over the throat surface.

Some adenoviruses may produce a form of conjunctivitis called **pharyngoconjunctival fever**, which is most commonly contracted by swimming in virus-contaminated water. There have been more than 42,000 cases reported in Japan in 2006. New military recruits may suffer **acute respiratory disease** (**ARD**) as a result of adenovirus transmission in crowded conditions. Any of these conditions can progress to **viral pneumonia**.

No antiviral agents currently available treat adenoviral infections. An adenovirus vaccine is available for ARD in military recruits. Some adenoviruses are oncogenic (see Chapter 13), but none have been associated with malignancies in humans. In recent years, adenoviruses have developed a more positive image as vectors (carriers) for genes in viral gene therapy experiments (see MicroFocus 13.6).

CONCEPT AND REASONING CHECKS

14.3 Why isn't an adenovirus vaccine available to the general public?

14.2 Viral Infections of the Lower Respiratory Tract

■ **Bronchioles:** The narrow tubes inside of the lungs that branch off of the main air passages (bronchi).

The lower respiratory tract (LRT) in humans consists of the larynx (voice box), trachea (windpipe), bronchial tubes, and the alveoli.

Paramyxovirus Infections Affect the Lower Respiratory Tract

KEY CONCEPT

- The paramyxoviruses are a group of viruses causing similar symptoms.

A number of viruses, primarily in the *Paramyxoviridae*, are associated with the LRT (FIGURE 14.6). All these viruses are enveloped, single-stranded (–strand) RNA viruses. The mumps and measles viruses, which cause skin diseases, will be discussed in the next section.

Respiratory syncytial (RS) disease is caused by the respiratory syncytial virus (RSV). Since 1985, RS disease has been the most common lower respiratory tract disease affecting infants and children under one year of age. Infection takes place in the bronchioles and air sacs of the lungs, and the disease is often described as **viral pneumonia**. When the virus infects tissue cells, the latter tend to fuse together, forming giant multinucleate cells called **syncytia** (see Chapter 13).

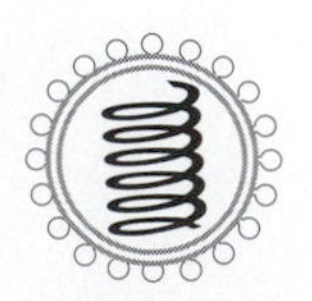

Respiratory syncytial virus

RS disease can occur in adults as an upper respiratory disease with influenza-like symptoms. Outbreaks occur yearly throughout the United States, but most cases are misdiagnosed or unreported. Some virologists believe that up to 95 percent of all children have been exposed to the disease by the age of five, and Centers for Disease Control and Prevention (CDC) epidemiologists estimate there are 51,000 to 82,000 hospitalizations and 4,500 deaths in infants and children each year in the United States as a result of RS disease.

Maternal antibodies passed from mother to child probably provide protection during

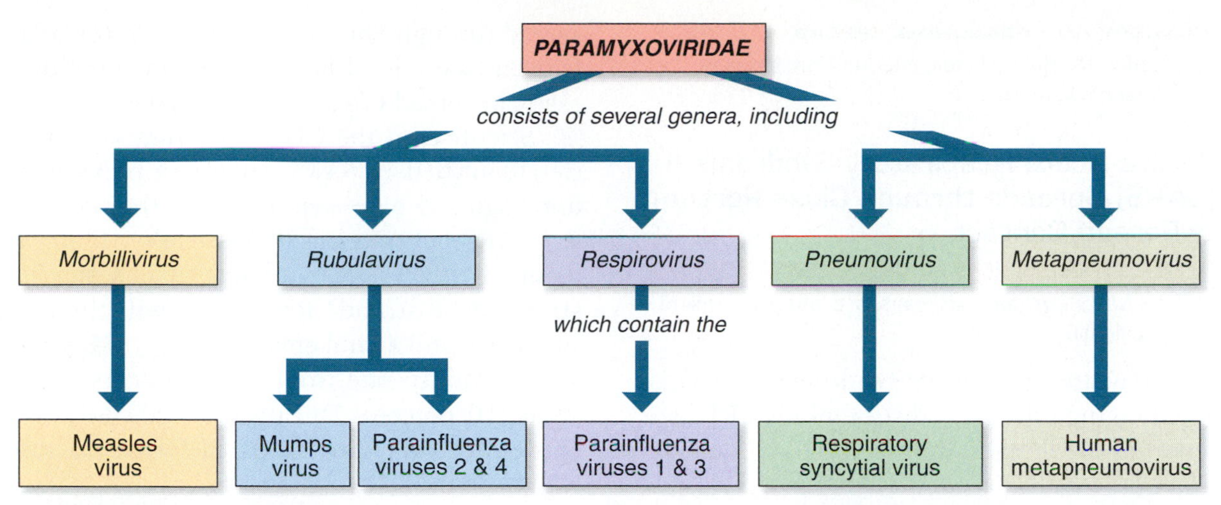

FIGURE 14.6 The Relationships between the Human Paramyxoviruses. This concept map illustrates the relationships between the paramyxoviruses that cause human respiratory and skin diseases. Note the relationship between measles and mumps (skin diseases).

Q: Which virus is most closely related to the respiratory syncytial virus?

the first few months of life, but the risk of infection increases as these antibodies disappear. Indeed, researchers have successfully used preparations of antibodies to lessen the severity of established cases of RS disease. Aerosolized ribavirin also has been used with success.

Parainfluenza (*para* = "near") infections are caused primarily by human parainfluenza viruses 1 and 3 (FIGURE 14.7). They account for 40 percent of acute respiratory infections in children. Although as widespread as influenza, parainfluenza is a much milder disease and is transmitted by direct contact or aerosolized droplets. It is characterized by minor upper respiratory illness, often referred to as a cold. "Bronchiolitis" and croup may accompany the disease, which is most often seen in children under the age of six. The disease predominates in the late fall and early spring, and is seasonal, as Figure 14.4 indicated. No specific therapy exists.

RSV-like illnesses, originally cataloged as illnesses of unknown origin, recently have been identified as caused by the **human metapneumovirus (hMPV)**. Just about every child in the world has been infected by the virus by age 5. Discovered in 2003, hMPV is responsible for 12 percent of LRT infections and 15 percent of common colds in children. Human MPV appears to be milder than RS disease. Like RS disease, ribavirin is active against hMPV.

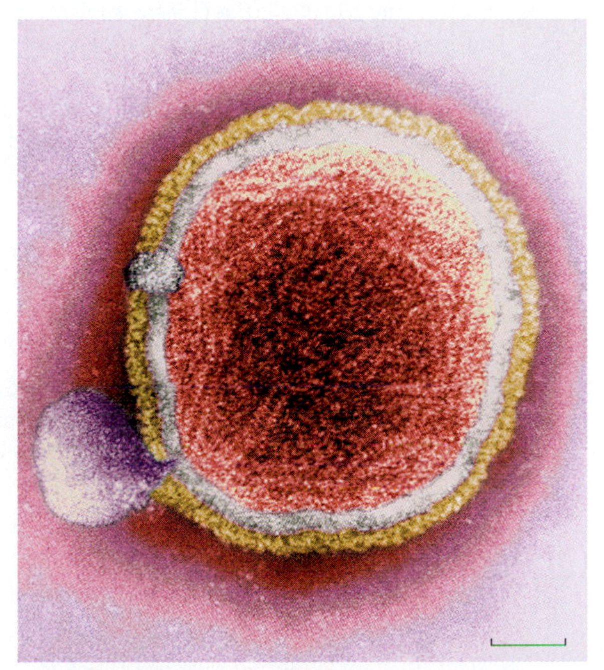

FIGURE 14.7 The Parainfluenza Virus. False-color transmission electron micrograph of a parainfluenza virus. The envelope of individual virions often gives the virus a pleomorphic shape, although they do have a helical capsid. (Bar = 50 nm.)

Q: Why is the virus referred to as parainfluenza?

■ Bronchiolitis: An inflammation of the bronchioles.

You might ask why it has taken so long to discover such a prevalent virus. The answer is that hMPV does not grow well in a cell culture and so was not detected in clinical specimens until better molecular methods became available.

CONCEPT AND REASONING CHECKS

14.4 Identify the common relationships between the paramyxoviruses.

Severe Acute Respiratory Syndrome (SARS) Spreads through Close Person-to-Person Contact

KEY CONCEPT

- The SARS coronavirus causes a unique form of pneumonia.

Severe acute respiratory syndrome (SARS) is an emerging infectious disease of the LRT first reported in China in spring 2003 and quickly spread through Southeast Asia and to Canada. It is an example of how fast an emerging disease can spread (Textbook Case 14).

Scientists at the CDC and other laboratories identified in SARS patients a previously unrecognized coronavirus, which they named the SARS coronavirus (SARS-CoV). Being a member of the *Coronaviridae*, it is a single-stranded (+ strand) RNA virus with helical symmetry and a viral envelope (FIGURE 14.8). The mortality rate from SARS appears to be about 10 percent. During the 2002 to 2003 epidemic, the World Health Organization

Textbook CASE 14

The Outbreak of SARS: 2002–2003

1. On February 11, 2003, the Chinese Ministry of Health notified the World Health Organization (WHO) of a mystery respiratory illness that had been occurring since November 2002 in Guangdong province in southern China. However, officials of China refused to allow WHO officials to investigate.
2. February 21, a 64-year-old doctor from Guangdong came to Hong Kong to attend a wedding. He stayed at a Hong Kong hotel, infecting 16 people who spread the disease to Hanoi, Vietnam, Singapore, and Toronto, Canada.
3. February 23, a Canadian woman tourist checked out of the same Hong Kong hotel and returned to Toronto, where her family greeted her. She died 10 days later as five family members were hospitalized.
4. February 28, a WHO doctor in Hanoi, Carlo Urbani, treated one of the people infected in Hong Kong and realized this is a new disease, which he named "severe acute respiratory syndrome" (SARS). He died of the disease 29 days later.
5. March 15, the WHO declared SARS a worldwide health threat. To block the chain of transmission, isolation and quarantine were instituted. Over half of those infected were health care providers.
6. April 16, the identification was made by 13 laboratories around the world that a new, previously unknown, coronavirus caused SARS (see figure).
7. June 30, the WHO announced there had been no new cases of SARS for two weeks. During the 114-day epidemic, more than 8,000 people from 29 countries were infected and 774 died.

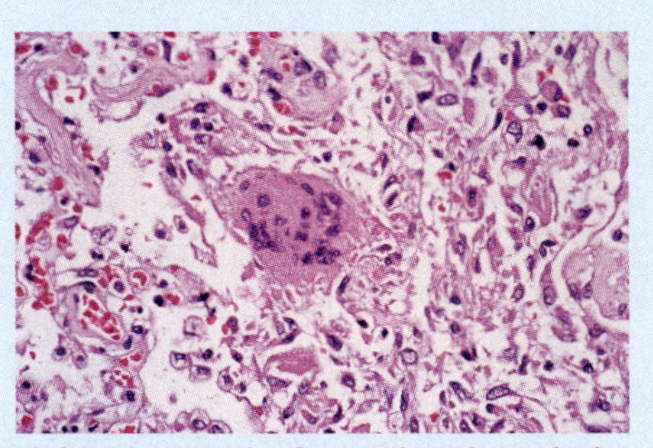

Light micrograph of a section through tissue from the lung of a patient with SARS. A giant cell is seen at center, which has numerous nuclei (dark purple). The white spaces are alveoli, tiny sacs in the lungs through whose walls gaseous exchange takes place. Several red blood cells (red) are seen among the tissue. Alveolar damage causes a cough and breathing difficulties.

Questions

A. *Why might Chinese officials be reluctant to allow a WHO investigation into the mystery illness?*

B. *Justify the use of quarantine and isolation to break the chain of SARS transmission.*

C. *Propose an explanation as to why there was such a disproportionately high number of infections in health care providers.*

D. *How could such a large number of infections in healthcare workers have been prevented?*

E. *Why is SARS a textbook case for an emerging infectious disease?*

For additional information see http://www.cdc.gov/mmwr/preview/mmwrhtml/mm5226a4.htm.

(WHO) identified 8,098 reported cases from 29 countries. There were 774 deaths, the majority being in China, Hong Kong, Taiwan, Vietnam, Singapore, and Canada.

SARS-CoV can be spread through close person-to-person contact by touching one's eyes, nose, or mouth after contact with the skin of someone with SARS. Spreading also comes from contact with objects contaminated through coughing or sneezing with infectious droplets by a SARS-infected individual. Whether SARS can spread through the air or in other ways remains to be discovered. Bats may be the natural host of SARS-CoV.

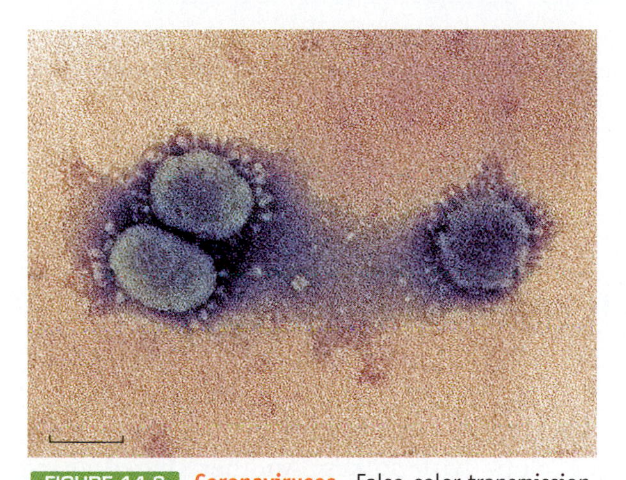

FIGURE 14.8 **Coronaviruses.** False-color transmission electron micrograph of three human coronaviruses. The spikes can be seen clearly extending from the viral envelope. Viruses similar to these are responsible for severe acute respiratory syndrome (SARS). (Bar = 100 nm.)

Q: By looking at the micrograph, explain why the virus is referred to as a "corona" virus.

Many people remain asymptomatic after contacting SARS-CoV. However, in affected individuals, moderate URT illness may occur and include fever (greater than 38°C), headache, an overall feeling of discomfort, and body aches. After two to seven days, SARS patients may develop a dry cough and have trouble breathing. In those patients progressing to a severe respiratory illness, pneumonia develops with insufficient oxygen reaching the blood. In 10 to 20 percent of cases, patients require mechanical ventilation.

Most of the cases of SARS in the United States in 2003 occurred among travelers returning from other parts of the world where SARS was present. Transmission of SARS to health care workers appears to have occurred after close contact with sick people before recommended infection control precautions were put into place.

Prevention requires limiting future outbreaks or epidemics. All suspected SARS patients should be isolated for at least 10 days. For known cases, isolation should be continued for 10 days after the SARS fever has broken.

Because this is a newly emerging disease, treatment options remain unclear. Antiviral therapy, including oseltamivir (Tamiflu) or ribavirin, has proved ineffective.

The pneumotropic viral diseases are summarized in TABLE 14.1.

CONCEPT AND REASONING CHECKS

14.5 How does SARS appear to be spread between individuals?

TABLE 14.1 A Summary of Viral Infections of the Respiratory Tract

Disease	Classification of Virus	Transmission	Organs Affected	Vaccine	Special Features	Complications
Influenza	*Orthomyxoviridae*	Droplets	Upper Respiratory tract	Available	Antigenic variation Strains A, B, C	Reye syndrome Guillain-Barré syndrome
Rhinovirus infections	*Picornaviridae*	Droplets Contact	Upper respiratory tract	Not available	Head colds	None
Adenovirus infections	*Adenoviridae*	Droplets Contact	Lungs, meninges Eyes	Not available	Common cold syndrome	ARD Viral pneumonia
Respiratory syncytial disease	*Paramyxoviridae*	Droplets	Respiratory tract	Not available	Syncytia of respiratory cells	Pneumonia
Parainfluenza	*Paramyxoviridae*	Droplets Contact	Lower respiratory tract	Not available	Common cold syndrome	None
SARS	*Coronaviridae*	Droplets Contact	Lower Respiratory tract	None available	Newly emerged disease	Pneumonia

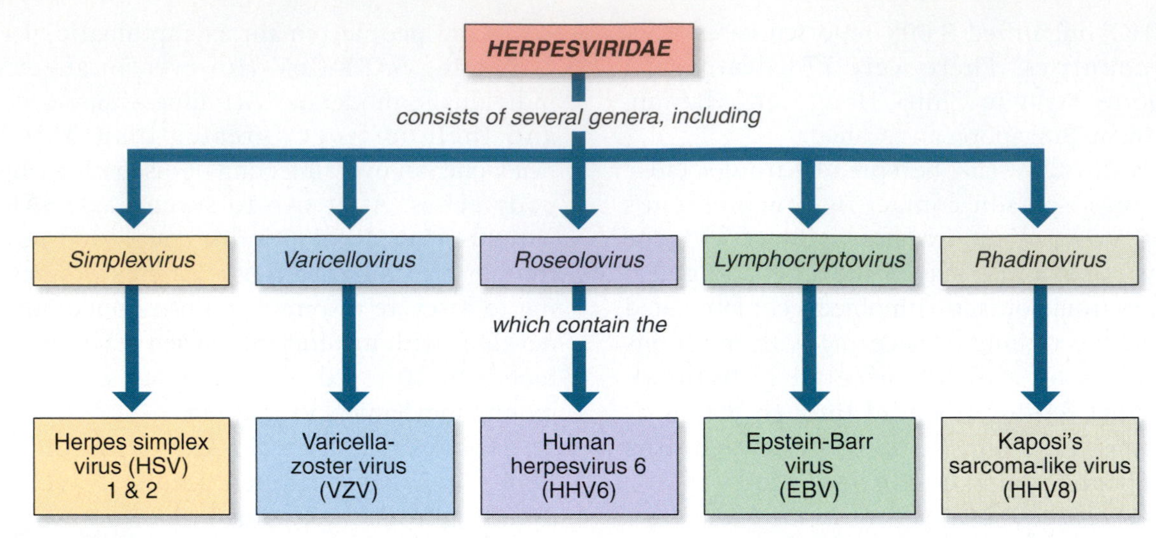

FIGURE 14.9 The Relationships between the Human Herpesviruses. This concept map illustrates the relationships between six of the eight viral species in the family *Herpesviridae* that cause human disease. Only EBV and HHV8 are known to be oncogenic.

Q: What does it mean for a virus to be oncogenic?

14.3 Diseases of the Skin Caused by Herpesviruses

Several viral diseases of the skin are caused by members of the *Herpesviridae*. Certain ones, such as herpes simplex, remain epidemic in contemporary times; others such as chickenpox are being brought under control through effective vaccination programs. However, there are relatively few antiviral drugs for, and prevention programs remain a major course of action in, dealing with these diseases.

Presently, there are eight known viral species in the family *Herpesviridae* that infect humans (FIGURE 14.9). However, the small number of viruses should not be an indication of their significance. For example, some virologists believe over 90 percent of Americans have been exposed to the herpes simplex virus (HSV) by age 18 and more than 90 percent of adults worldwide have been infected with the Epstein-Barr virus (EBV). In addition, genital herpes is one of the most common sexually-transmitted diseases (STDs) in the world. The word *herpes* is Greek for "creeping," referring to the spreading of herpes infections through the body after contact has been made.

Herpes simplex virus

All human herpesviruses are large virions with a double-stranded DNA genome. They have icosahedral symmetry and an envelope with spikes. Another shared characteristic is their ability to undergo **latency** (see Chapter 13) and then reactivate at some later period. The herpesviruses can infect several types of cells in epithelial and neural tissue, causing a variety of skin and related diseases. We examine several of these following the order presented in Figure 14.9.

Human Herpes Simplex Infections Are Widespread and Often Recurrent

KEY CONCEPT

- Human herpes simplex viruses cause an array of viral diseases.

Two of the most prevalent herpesviruses are herpes simplex viruses 1 and 2 (HSV-1, HSV-2; FIGURE 14.10A).

Cold sores are caused primarily by HSV-1 and they are contagious. Such sores and blisters of herpes infections have been known for centuries. In ancient Rome, an epidemic was

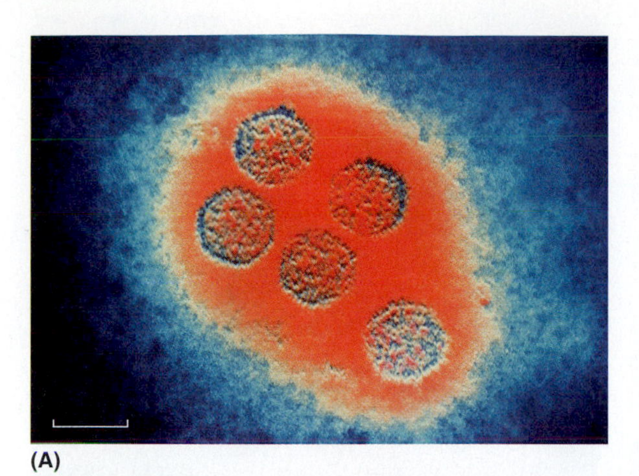

(A)

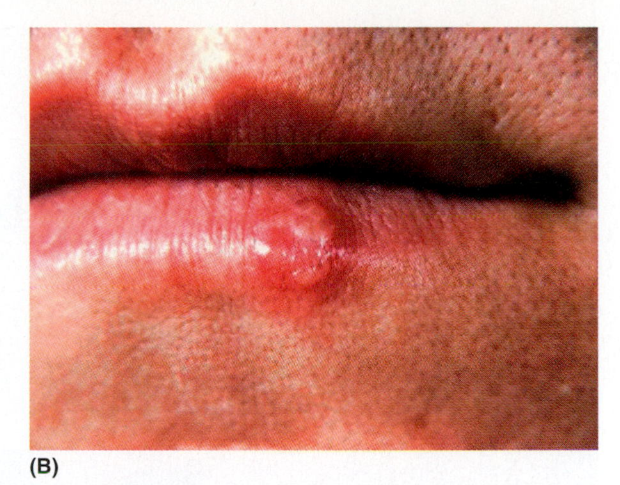

(B)

FIGURE 14.10 **Herpes Simplex Virus and Cold Sores.** (**A**) A false-color transmission electron micrograph of a cell infected with herpes simplex viruses. (Bar = 150 nm.) (**B**) Cold sores (fever blisters) of herpes simplex erupt as tender, itchy papules and progress to vesicles that burst, drain, and form scabs. Contact with the sores accounts for spread of the virus.

Q: How would the release phase of the HSV replicative cycle lead to blisters and vesicles on the lips?

so bad that the Emperor Tiberius banned kissing and Shakespeare, in Romeo and Juliet, writes of "blisters o'er ladies lips."

Cold sores, also called fever blisters, start as a tingling sensation and the presence of a small, hard spot on the lip. Within a couple of days red, fluid-filled blisters appear (FIGURE 14.10B). The blisters eventually break, releasing the fluid containing infectious virions. A yellow crust forms and then peels off without leaving a scar. Cold sores generally clear up without treatment in 7 to 10 days. A person is most likely to transmit the infection from the time the blisters appear until they have completely dried and crusted over.

After the primary infection, the viruses enter the sensory neurons and become latent in the nearby sensory ganglia. Viral reactivation and movement to the epithelia can trigger another round of cold sores. Reactivation of the dormant viruses often occurs after some form of trauma, often in response to stressful triggers, such as fever, menstruation, or emotional disturbance. Even environmental factors like sunburn (exposure to ultraviolet light) can trigger reactivation (FIGURE 14.11).

Preventing the transmission of cold sores means not kissing others while the blisters are present. Washing one's hands often and not touching other areas of the body also help limit virus spread. For example, the eyes can become infected through touching with a contaminated finger. Infection of the eye, called **herpes keratitis**, causes scarring of the cornea and is a leading cause of blindness in the United States. Some 400,000 Americans suffer from a form of ocular herpes.

Genital herpes is one of the most common STDs in the United States, accounting for more than 600,000 new cases diagnosed each year. It usually is caused by HSV-2, although the incidence of HSV-1 infections has been increasing. Genital herpes is spread primarily by sexual contact. It is highly unlikely a person could contract an infection through contact with toilets or other objects used by an infected person because the virus "dies" quickly outside the body.

Signs generally appear within a few days of sexual contact, often as itching or throbbing in the genital area. This is followed by reddening and swelling of a small area where painful, thin blisters erupt. The blisters crust over and the sores disappear, usually within about three weeks. The signs often are very mild and may go unnoticed. Fifty percent of infected individuals experience only one outbreak in their lifetime.

■ Ganglia:
Structures containing a dense cluster of nerve cells.

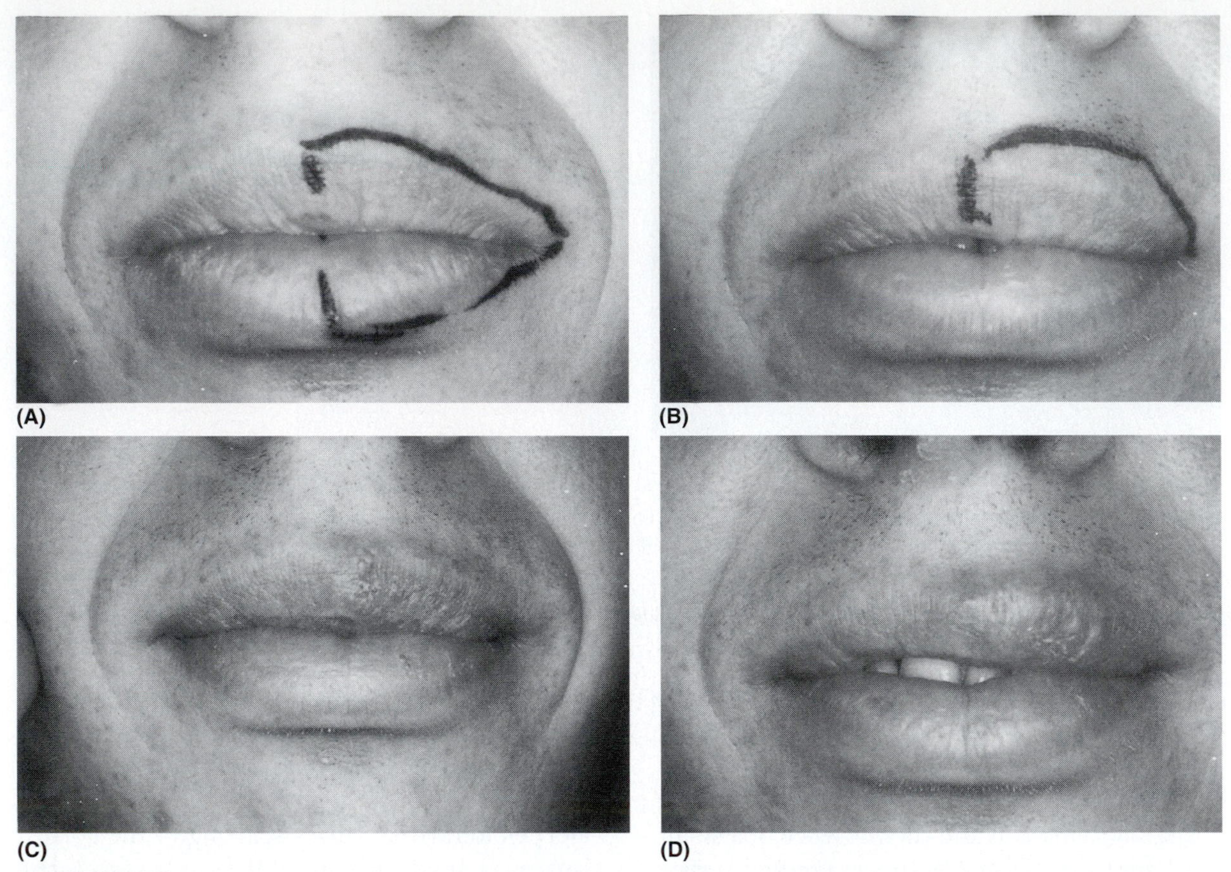

FIGURE 14.11 **Ultraviolet Light and Herpes.** An experiment showing the effects of ultraviolet (UV) light on the formation of herpes simplex sores of the lips. This patient usually experienced sores on the left upper lip. (**A**) The patient was exposed to UV light from a retail cosmetic sunlamp on the left upper and lower lips in the area designated by the line. The remainder of the face was protected by a sunscreen. (**B**) Sores formed on the left upper lip. (**C**) The patient was later exposed a second time to the sunlamp, but only on the left upper lip. (**D**) Sores formed and were larger than the previous ones.

Q: What would you conclude from these experiments regarding the reactivation of herpesviruses?

However, for many infected individuals, a latent period occurs during which time the virus remains in nerve cells near the blisters. In the majority of cases, the virus becomes reactivated again by stressful situations and symptoms reappear. The cycle of latency and recurrent infections can occur three to eight times a year. People with active herpes lesions are highly infectious and can pass the viruses to others during sexual contact.

In healthy individuals, genital herpes usually causes no serious complications. Importantly though, a person with the infection has an increase risk of transmitting or contracting other sexually transmitted diseases, including AIDS.

Prevention is similar to that for any STD—abstain from sexual activity or limit sexual contact to only one person who is infection-free. Although there is no cure for genital herpes, antiviral drugs such as acyclovir can help heal the sores sooner and reduce the frequency of recurrent infections.

Neonatal herpes is a devastating and life-threatening disease transmitted by infected mothers to newborns during childbirth. Although most pregnant women with genital herpes have healthy babies, a small number spread the infection to their newborns during labor and delivery. Even with adequate treatment for the newborn, mental development can be delayed, blindness can occur, and persistent seizures can result. If symptoms and a diagnosis of an active genital herpes infection are made prior to delivery, the obstetrician may recommend birth by cesarean section.

In recent years, the acronym **TORCH** has been coined to focus attention on diseases with congenital significance and induced by microbial teratogens: **T** for toxoplasmosis, **O** is for other diseases, such as syphilis; **R** for rubella, **C** for cytomegalovirus, and **H** for herpes simplex virus.

CONCEPT AND REASONING CHECKS

14.6 Explain how a primary infection and a latent infection differ for HSV.

Chickenpox Is No Longer a Prevalent Disease in the United States

KEY CONCEPT

- The varicella-zoster virus causes chickenpox and shingles.

In the centuries when pox diseases regularly swept across Europe, people had to contend with the Great Pox (syphilis), the smallpox, the cowpox, and the chickenpox. Prior to the availability of a vaccine for chickenpox in 1995, about 4 million children contracted chickenpox each year in the Unites States. In 2004, there were 32,931 cases reported.

Chickenpox (varicella) is a highly communicable disease caused by the varicella-zoster virus (VZV), another virus species in the family *Herpesviridae*. It is transmitted by respiratory droplets and skin contact, and it has an incubation period of two weeks. The disease begins in the respiratory tract, with fever, headache, and malaise. Viruses then pass into the bloodstream and localize in the peripheral nerves and skin. As it multiplies in the cutaneous tissues, VZV produces a red, itchy rash on the face, scalp, chest, and back, although it can spread across the entire body. The rash quickly turns into small, teardrop-shaped, fluid-filled vesicles (FIGURE 14.12A). The vesicles in chickenpox develop over three or four days in a succession of "crops." They itch intensely and eventually break open to yield highly infectious virus-laden fluid. Although many refer to the vesicles as pox, the latter term is more correctly reserved for the pitted scars of smallpox. In chickenpox, the vesicles form crusts that fall off. A person who has chickenpox can transmit the virus for up to 48 hours before the telltale rash appears and remains contagious until all spots crust over. Chickenpox usually lasts two weeks or less and rarely causes complications.

The most common complication of chickenpox is a bacterial infection of the skin. Chickenpox may also lead to pneumonia or an inflammation of the brain (encephalitis), both of which can be very serious if not treated. Reye syndrome may occur during the recovery period, so aspirin should not be used to reduce fever in both children and adults.

Acyclovir has been used in high risk groups to lessen the symptoms of chickenpox and hasten recovery. A varicella-zoster immune globulin (VZIG), which contains antibodies to the chickenpox virus, may also be used. The chickenpox vaccine (Varivax) is the safest, most effective way to prevent chickenpox and its possible complications. The vaccine consists of attenuated viruses administered subcutaneously and is recommended for all individuals over one year of age. One dose is given to children between ages 1 and 12, and two doses are given to adolescents and adults (pregnant women should not be immunized). The varicella vaccine is 85 percent effective in disease prevention.

Shingles (zoster) is an adult disease produced by the same virus causing chickenpox. Anyone who has had chickenpox as a child is at risk for the latent illness, although only about 10 percent of adults actually develop shingles. Still, this amounts to some one million new cases each year.

After an infection, some VZV may remain in nerve cells. Many years later, the virus can reactivate, travel down the nerves to the skin of the body trunk, and resurface as shingles (FIGURE 14.12B). Here they cause blisters with blotchy patches of red that appear to encircle the trunk (*zoster* = "girdle"). Many sufferers also experience a series of headaches as well as facial paralysis and sharp "ice-pick" pains described as among the most debilitating known. The condition can occur repeatedly and is linked to emotional and physical stress (such as radiation therapy) as well as to a suppressed immune system or aging.

■ Teratogens: Agents that interfere with the normal development of the fetus.

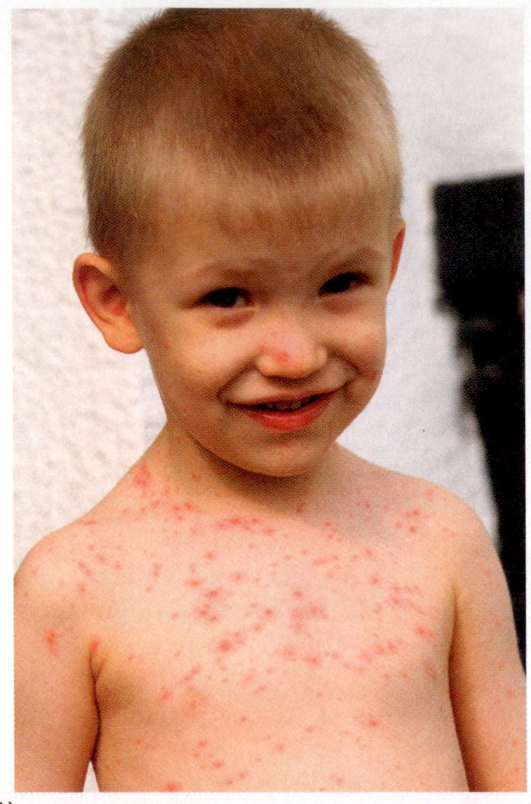

(A)

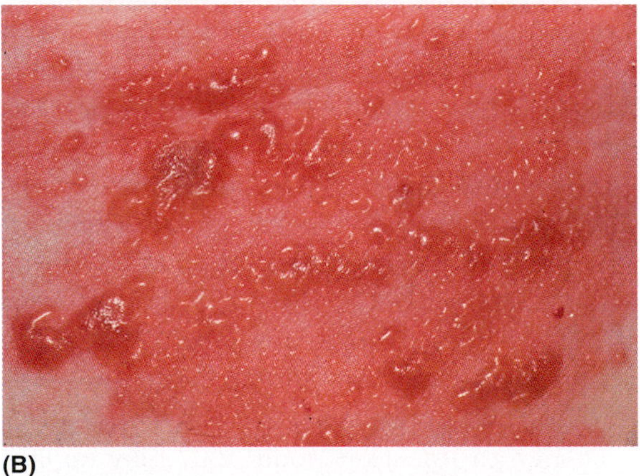

(B)

FIGURE 14.12 **The Lesions of Chickenpox and Shingles.** (**A**) A typical case of chickenpox. The lesions may be seen in various stages, with some in the early stage of development and others in the crust stage. (**B**) Dermal distribution of shingles lesions on the skin of the body trunk. The lesions contain less fluid than in chickenpox and occur in patches as red, raised blotches.

Q: Summarize how varicella latency could lead to shingles.

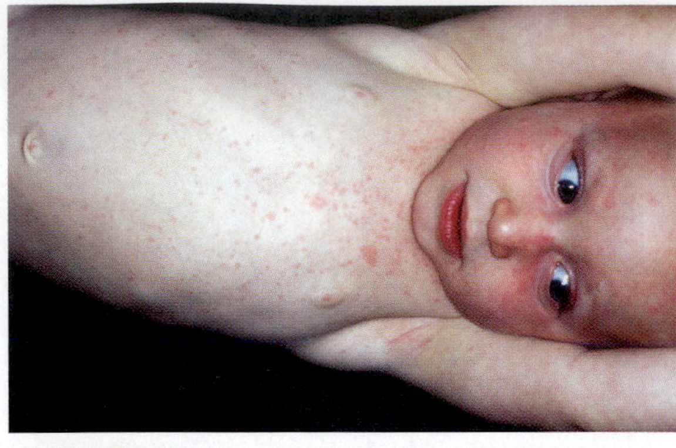

FIGURE 14.13 **Roseola.** Following a high fever, a red body rash may appear on the neck and trunk of infants or young children.

Q: Why should aspirin not be given to infants or young children suffering a viral infection such as roseola?

Shingles can lead to its own complication—a condition in which the pain of shingles persists for years after the blisters disappear. This complication, called **postherpetic** neuralgia, can be severe. Most cases occur in people over age 50, and a person with an active case of shingles can induce chickenpox (but not shingles) in another susceptible person. AIDS patients may be susceptible to the disease because of their compromised immune system.

■ Neuralgia: Severe pain in a part of the body through which a particular nerve runs.

In 2006, a live, attenuated vaccine, called zostavax, was approved to prevent shingles in people over 60 years of age.

For herpes zoster, acyclovir therapy lessens the symptoms and some unconventional methods also may help (MicroFocus 14.3).

CONCEPT AND REASONING CHECKS

14.7 Assess the consequences of someone having shingles being able to transmit the virus to a susceptible person.

Human Herpesvirus 6 Infections Primarily Occur in Infancy

KEY CONCEPT

- Human herpesvirus 6 infection primarily causes a short-term fever and rash in infants.

Human herpesvirus 6 (HHV-6) belongs to a different genus of herpesviruses from HSV-1 and HSV-2. HHV-6 primarily affects infants and young children six months to three years old. The virus causes **roseola infantum**, an acute, self-limiting condition marked by high fever. This often is followed by a red body rash (FIGURE 14.13). Roseola spreads from person to person through contact with an infected person's respiratory secretions or saliva and the infection usually lasts about one

MICROFOCUS 14.3: Being Skeptical

Can Chinese Tai Chi Prevent Shingles?

We all know that exercise is a good thing. It builds strength, endurance, and cardiovascular health. One Chinese form of physical and aerobic exercise for attaining good health is through Tai Chi. Tai Chi involves a series of 20 slow and deliberate body movements that purport to produce a calm and tranquil mind. In 2003, researchers announced that Tai Chi Chih, a low-impact form of Tai Chi, boosts shingles immunity in the elderly. Can Tai Chi actually do this?

The theory goes that when a person's immunity is compromised or wanes as one ages, the individual is susceptible to more infections. For example, it is true that individuals over 50 are more prone to recurrent shingles attacks as the varicella-zoster virus (VZV) reactivates because it may no longer be kept in check by the immune system. Stress and other trauma also can intensify the reactivation. Therefore, Michael R. Irwin and colleagues at UCLA's Neuropsychiatric Institute decided to examine if Ti Chi could reduce the levels of stress and thereby boost a person's immunity—and reduce the chances for shingles.

Thirty-six men and women over 60 were enrolled in the study. Half were started on a 15-week program of Tai Chi Chih. The other half were asked to postpone the program for 15 weeks.

After the 15-week period, analysis indicated the Tai Chi group had a 50 percent increase over the control group in immune memory function for the shingles virus. The researchers state the increase was sufficient to actually help prevent shingles from reoccurring. Because Tai Chi did not improve physical movement within the group, Irwin speculates that Tai Chi's calming influence reduces stress levels, which could boost immune system function—including keeping VZV in check.

Such studies are part of a growing field called **psychoneuroimmunology**, which looks at the relationships between the nervous system and the immune system. Professor Irwin believes if memory to the VZV can be maintained, memory to other infections might also be maintained as one ages.

The verdict: This was a small study involving only 36 volunteers. A more thorough study would be to follow groups of Tai Chi Chih volunteers and monitor how many actually report a flare-up of shingles compared to the general population. Still, this study does support the mind-body connection as a way to maintain good health.

week. Treatment requires bed rest, fluids, and fever-reducing (nonaspirin) medications.

The prevalence of the virus is seen by the fact that most children have been infected by the time they enter kindergarten and up to 80 percent of adults have antibodies against HHV-6. Like all herpesviruses, it can remain latent in saliva and monocytes. As such, 30 to 60 percent of bone marrow transplant recipients suffer a HHV-6 viremia during the first few weeks after transplantation. The recurrence can potentially lead to pneumonia or encephalitis. No drugs have been approved for HHV-6 infections.

Some researchers have suggested but not proved that HHV-6 may remain dormant in the body from the childhood years, and then resurface to be part of the chain of events leading to multiple sclerosis (Chapter 15).

CONCEPT AND REASONING CHECKS

14.8 Determine why a bone-marrow transplant recipient is at risk of contracting an HHV-6 infection.

- **Monocytes:** A type of white blood cell formed in the bone marrow.
- **Viremia:** The presence of viruses in the blood.

A Few Herpesvirus Infections Are Oncogenic

KEY CONCEPT

- The Epstein-Barr virus and human herpesvirus 8 have oncogenic potential.

Only two human herpesviruses, the Epstein-Barr virus (EBV) and human herpesvirus 8 (HHV-8), have been identified as oncogenic agents. EBV more commonly is associated with benign infections, such as infectious mononucleosis, and will be discussed in Chapter 15.

■ Angiogenic: Refers to the generation of many blood vessels.

Kaposi sarcoma (KS) is a highly angiogenic tumor of the blood vessel walls. It is most commonly seen in individuals with a weakened immune system, such as those with HIV/AIDS. The malignancy is caused by human herpesvirus 8 (HHV-8) or Kaposi sarcoma–associated herpesvirus. KS, which is marked by dark or purple skin lesions, has become one of the most common tumors in AIDS patients (FIGURE 14.14). The DNA of HHV-8 is present in most biopsies of tissue from KS patients and antibodies against the virus are invariably detected in those with the disease or at risk of developing it.

Treatment of smaller skin lesions involves the use of liquid nitrogen, low-dose radiation, or chemotherapy applied directly to the lesion.

TABLE 14.2 summarizes the characteristics of the herpesviruses that affect the skin.

CONCEPT AND REASONING CHECKS

14.9 Although all herpesviruses appear to have a latent state, why are only EBV and HHV-8 potentially carcinogenic?

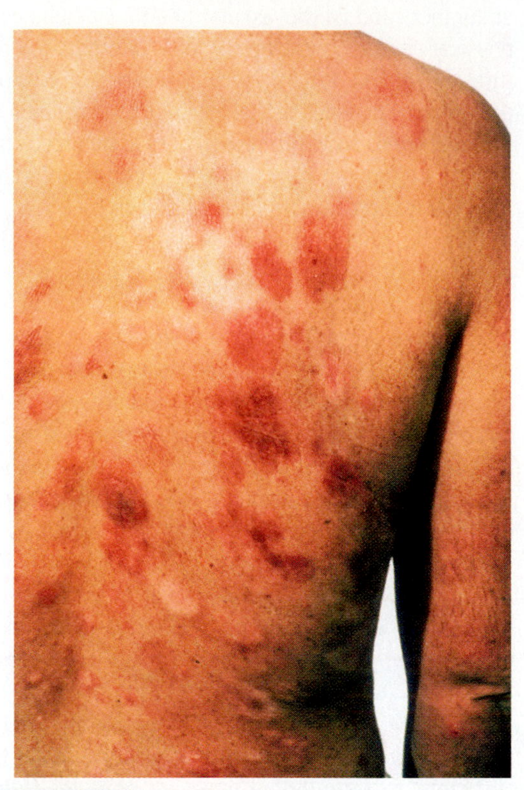

FIGURE 14.14 **Kaposi Sarcoma.** This man is suffering from AIDS. Kaposi sarcoma is an opportunistic disease, appearing as dark lesions on the skin.

Q: What is meant by an opportunistic disease?

TABLE 14.2 A Summary of Viral Diseases Caused by the Herpesviruses

Disease	Classification of Virus	Transmission	Organs Affected	Vaccine	Special Features	Complications
Herpes simplex (HSV-1 & HSV-2)	*Herpesviridae*	Contact	Skin Pharynx Genital organs	Not available	Characteristic lesions Acyclovir treatment	Encephalitis Congenital infections Neonatal herpes
Chickenpox (varicella)	*Herpesviridae*	Droplets Contact	Skin Nervous system	Attenuated viruses	Characteristic lesions Crops Acyclovir treatment	Herpes zoster (shingles) Reye syndrome Pneumonia
Shingles (zoster)	*Herpesviridae*		Nervous system	Not yet available	Facial paralysis "ice-pick" pains	Postherpetic neuralgia
Roseola	*Herpesviridae*	Contact	Skin	None	High Fever Red body rash	Pneumonia Encephalitis (transplant recipients)
Kaposi sarcoma	*Herpesviridae*	Contact (sexual and nonsexual)	Skin	None	Dark purple lesions	Reoccurance and spread

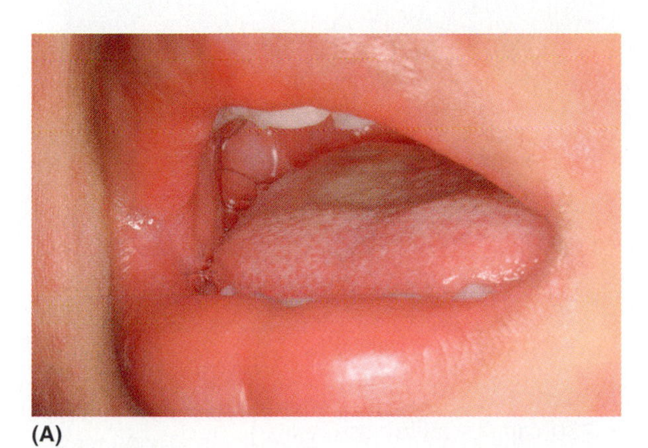

(A)

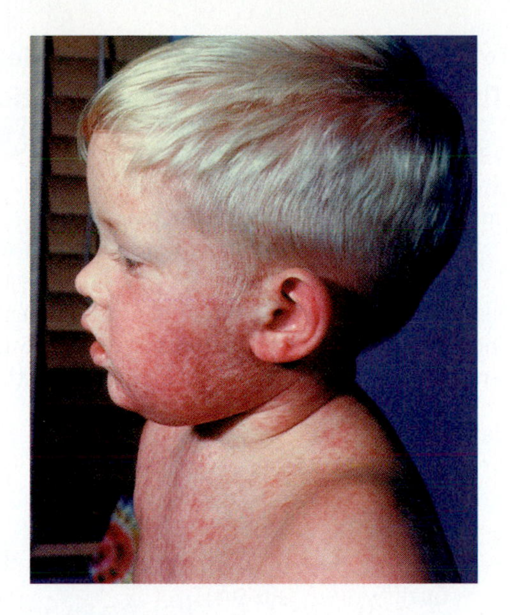

(B)

FIGURE 14.15 **Koplik Spots and the Measles Rash.** (**A**) Koplik spots in the mouth of a child suffering from measles. The red spots with white centers are a frequent symptom in this highly infectious viral disease. (**B**) A child with measles, showing the typical rash on face and torso that mainly affects children.

Q: Based on the rash, how would you distinguish measles from scarlet fever?

14.4 Other Viral Diseases of the Skin

There are a number of other viral infections and diseases associated with the skin. Some are common childhood diseases, such as mumps and measles, while another, smallpox, has been eradicated but was once a human scourge worldwide.

Paramyxovirus Infections Can Cause Typical Childhood Diseases

KEY CONCEPT

- Two genera of paramyxoviruses cause measles and mumps.

Measles is a highly contagious disease caused by a viral species of the *Paramyxoviridae* family (see Figure 14.6). Even though measles is a vaccine-preventable disease, there are some 30 to 40 million cases of measles worldwide each year, resulting in more than 750,000 deaths.

Transmission of the measles virus usually occurs by respiratory droplets during the early stages of the disease. Symptoms commonly include a hacking cough, sneezing, nasal discharge, eye redness, sensitivity to light, and a high fever. Red patches with white grain-like centers appear along the gum line in the mouth two to four days after the onset of symptoms. These diagnostic patches are called **Koplik spots** after Henry Koplik, the New York pediatrician who described them in 1896 (FIGURE 14.15A).

The characteristic red rash of measles appears about two days after the first evidence of Koplik spots. Beginning as pink-red pimple-like spots (maculopapules), the rash breaks out at the hairline, then covers the face and spreads to the trunk and extremities (FIGURE 14.15B). **Rubeola**, the alternative name for measles, is derived from the Latin *rube* for "red." Rashes resemble those in scarlet fever, but the severe sore throat of scarlet fever generally does not develop. Within a week, the rash turns brown and fades.

Measles usually is characterized by complete recovery. In some cases, however, bacterial disease may develop in the damaged respiratory tissue. Another possible problem is subacute sclerosing panencephalitis (SSPE), a rare brain disease characterized by a decrease in cognitive skills and loss of nervous function. Measles is a major cause of death in children in developing countries.

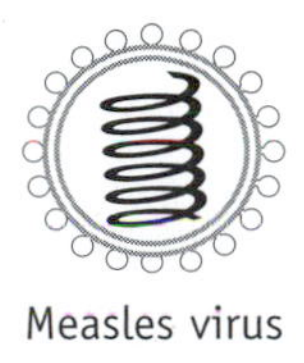

Measles virus

Prevention is accomplished with the measles vaccine, which usually is given as a measles-mumps-rubella (MMR) inoculation.

MICROFOCUS 14.4: Public Health

The Wanderings of a Measles Infection

Very few measles cases are reported in the United States anymore. However, imported cases could spread the disease to at-risk individuals.

The following example, which occurred in Arizona in January 2005, demonstrates how one infectious individual could spread the disease to a community including thousands of students at Arizona State University (ASU). Times and locations are estimates.

January 9

- 3:30 PM: McDonald's (Camp Verde)
- 4:00 PM: Target (Flagstaff)
- 5:00 PM: Burger King (Flagstaff)
- Evening: Beaver Street Brewery (Flagstaff)
- Quality Inn (Flagstaff)

January 10

- AM: Chapel of the Holy Cross (Sedona)
- ??: Subway/gas station (Camp Verde)
- Evening: Safeway (Tempe)

January 11

- ASU campus buildings (Tempe)

January 12

- ASU campus buildings (Tempe)
- 4:00 PM: Phoenix City Hall (Phoenix)
- 5:00 PM: Hotel San Carlos (Phoenix)

January 13

- ASU campus buildings (Tempe)

January 14–17

- ??

January 18

- 9:00 AM: Advanced Urgent Care (Phoenix)
- 1:00 PM: Emergency room, Tempe St. Luke's Hospital (Tempe)

No measles cases were reported, but the potential to infect unvaccinated individuals was of great concern.

Immunization of children entering grade school is now mandatory in all 50 states. In 1978, the US Public Health Service launched a campaign to eliminate measles in the United States. By 1983, the total number of reported cases was 1,463, a reduction of 99.7 percent from the prevaccine era. In 2004, there were 37 reported cases, 27 of which were internationally imported. Still, measles can be cause for alarm, as MicroFocus 14.4 points out.

Mumps is caused by another member of the *Paramyxoviridae*. Its name comes from the English "to mump," meaning to be sullen or to sulk. The characteristic sign of the disease is enlarged jaw tissues arising from swollen salivary glands, especially the parotid glands (FIGURE 14.16). **Infectious parotitis** is an alternate name for the disease.

■ Parotid glands: One of three pairs of salivary glands, located below and in front of the ears.

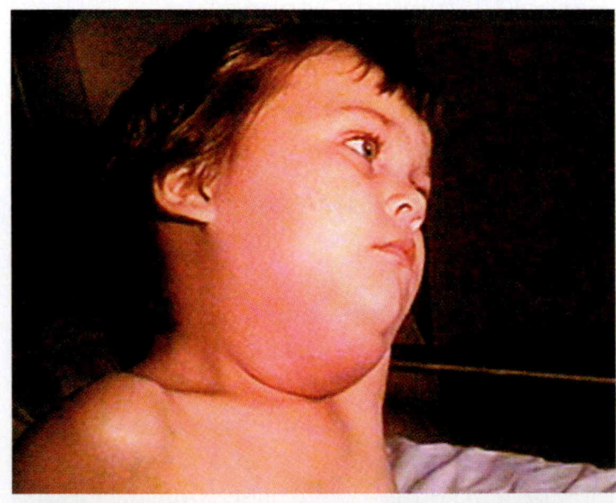

FIGURE 14.16 **Mumps.** Close-up of a young child with mumps (infectious parotitis).

Q: What causes the swelling in the jaw region?

Mumps is spread by respiratory droplets or contact with contaminated objects; it is considered less contagious than measles or chickenpox. The virus is found in human blood, urine, and cerebrospinal fluid, even though its effects are observed primarily in the parotid glands.

About one third of people infected with the mumps virus have no signs or symptoms. Obstruction of the ducts leading from the parotid glands retards the flow of saliva, which causes the characteristic swelling. The skin overlying the glands is usually taut and shiny, and patients experience pain when the glands are touched.

In male patients, the mumps virus may pose a threat to the reproductive organs. As long ago as 1790, the Scottish physician Robert Hamilton observed swelling and damage to the testes and named the condition **orchitis** (*orchi* = "the testicles"). The sperm count may be reduced, but sterility is not common. An estimated 25 percent of mumps cases in post-adolescent males develop into orchitis.

Prevention is achieved by the mumps vaccine, which consists of attenuated viruses and is usually combined with the measles and rubella vaccines (MMR). 258 cases occurred in 2004. However, in the spring of 2006, a mumps outbreak in the Midwest infected more than 2,600 people. Most cases were related to incomplete MMR vaccinations.

CONCEPT AND REASONING CHECKS

14.10 Compare measles and mumps as skin diseases.

Rubella (German Measles) Is an Acute, Mildly Infectious Disease

KEY CONCEPT

- The rubella virus produces a pale-pink rash on the trunk and extremities.

For generations, **rubella** (*rube* = "reddish"; *ella* = "small") or **German measles** (*germanus* = "similar") was thought to be a mild form of measles. In 1829, the German physician Rudolph Wagner noted the differences between the diseases. The disease occurs in several countries, although more than half now use a rubella vaccine.

Rubella is caused by a single-stranded (+ strand) RNA genome virus of the *Togaviridae* (*toga* = "cloak") family (FIGURE 14.17). The virion is icosahedral with a coat-like envelope and spikes. Viral transmission generally occurs by contact or respiratory droplets, and the disease is usually mild. Rubella—the **R** in the **TORCH** group of diseases—is accompanied by occasional fever with a variable, pale-pink maculopapular rash beginning on the face and spreading to the body trunk and extremities. The rash develops rapidly, often within a day, and fades after another two days. Recovery is usually prompt, but relapses appear to be more common than with other diseases, possibly because the viruses remain active within body cells.

Rubella is dangerous to the developing fetus in a pregnant woman. Called **congenital rubella**, transplacenta infection of the fetus can lead to destruction of the fetal capillaries, and blood insufficiency follows. The organs most often affected are the eyes, ears, and cardiovascular organs, and children may be born with cataracts, glaucoma, deafness, or heart defects.

■ Transplacental: Refers to movement from mother to fetus.

Since its introduction in 1969, the rubella vaccine, which is combined with the measles and mumps vaccines (MMR), has had a dramatic effect on the rate of incidence of the

FIGURE 14.17 **The Rubella Virus.** This false-color transmission electron micrograph shows a group of rubella viruses. (Bar = 100 nm.)

Q: What visual characteristic would assign the virus to the togaviruses?

Rubella virus

disease in the United States. In 1969, physicians reported 58,000 cases of rubella, but by 2004, the number was down to 10 reported cases and rubella is no longer considered an **endemic** disease in the United States. The vaccine consists of attenuated viruses cultivated in human tissue cultures. It is for subcutaneous inoculation of children. More than 95 percent of children entering grade school now provide evidence of rubella vaccination.

■ Endemic: Referring to a disease occurring within a specific area, region, or locale.

Rubella symptoms are so mild that no treatment is required.

CONCEPT AND REASONING CHECKS

14.11 Hypothesize why rubella has been called three-day measles.

Fifth Disease (Erythema Infectiosum) Produces a Mild Rash

KEY CONCEPT

- Human parvovirus B19 is most common among elementary school-age children.

In the late 1800s, numbers were assigned to diseases accompanied by skin rashes. Disease I was measles, II was scarlet fever, III was rubella, IV was Duke's disease (also known as roseola and now recognized as any rose-colored rash), and V was erythema infectiosum. This so-called fifth disease remained a mystery until the modern era.

Fifth disease virus

The agent of **fifth disease** is human parvovirus B19, which is a small, single-stranded DNA virion of the *Parvoviridae* (*parv* = "small") family, having icosahedral symmetry. Like all parvoviruses, B19 is dependent on the host cell or other viruses (adenoviruses, herpesviruses) to replicate.

Community outbreaks of fifth disease occur worldwide and transmission appears to be by respiratory droplets. Fifth disease primarily affects children although most infections are asymptomatic. If symptoms occur, the outstanding characteristic is a fiery red rash on the cheeks and ears (FIGURE 14.18). The rash may spread to the trunk and extremities, but it fades within several days, leaving a "lacy" rash on the skin. An acute infection can cause a reduction in red blood cell production, which can be very serious in children suffering from malaria who already are anemic from the parasitic infection.

Human papilloma virus

Parvovirus B19 only infects humans (dog and cat parvoviruses do not infect humans).

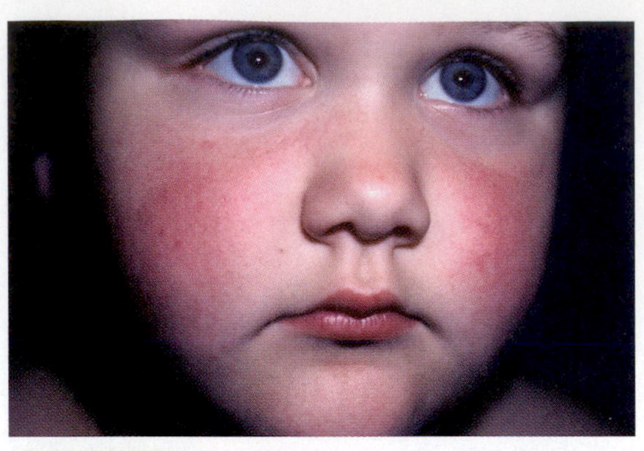

FIGURE 14.18 **Fifth Disease.** The fiery red rash of a child with fifth disease (erythema infectiosum). The confluent red rash is seen.

Q: Why is fifth disease sometimes called slapped-cheek disease?

However, fifth disease is not limited to children. Adults suffer from painful joints similar to the symptoms of rheumatoid arthritis, especially in the fingers, wrists, knees, and ankles. Infection of the bone marrow also may lead to anemia, and pregnant women may suffer miscarriage (but birth defects generally do not occur). Although antibody preparations (immune globulin) are available for treatment, the symptoms usually resolve spontaneously.

CONCEPT AND REASONING CHECKS

14.12 What is the unique feature of parvovirus B19?

Some Human Papillomavirus Infections Cause Warts

KEY CONCEPT

- Human papillomaviruses can cause common or genital warts.

Common warts are small, usually benign skin growths resulting from infection by a specific strain of the human papillomavirus (HPV). These viruses represent a collection of over two dozen types of icosahedral, naked, double-stranded DNA virions of the *Papovaviridae* family.

Common warts typically appear on the hands or fingers, and **plantar warts** occur on the soles of the feet. In most cases, these skin warts are white or pink and cause no pain. Although rare, common warts can be acquired through direct contact with HPV from another person or by direct contact with a towel or object used by someone who has the virus. Therefore, prevention requires maintaining proper cleanliness and not picking at them, which can lead to their spread.

MICROFOCUS 14.5: Being Skeptical
Skin Test Antigens Eliminate Warts

Many of us suffer from allergies and have had skin tests done to discover what substances (allergens) we are sensitive to. Researchers now suggest that injections of these same allergens can cure common or genital warts. So, what's the evidence?

Researchers at the University of Arkansas School for Medical Sciences reported in 2005 that injecting substances representing skin test antigens could eliminate warts—all of them! All that was required was to inject a single wart.

The researchers stated that 50 percent of volunteers injected with the antigens had complete eradication of all body warts, including genital warts. The volunteers were wart free!

How could this be? The research team suggests that the antigen injection into a single wart (see figure) stimulates an attack by the immune system. This attack is not only on the antigens injected, but also the human papillomaviruses in the mix of antigens.

The verdict: It is hard to argue against results that completely abolish the infection. And besides being effective, it is safe and relatively painless.

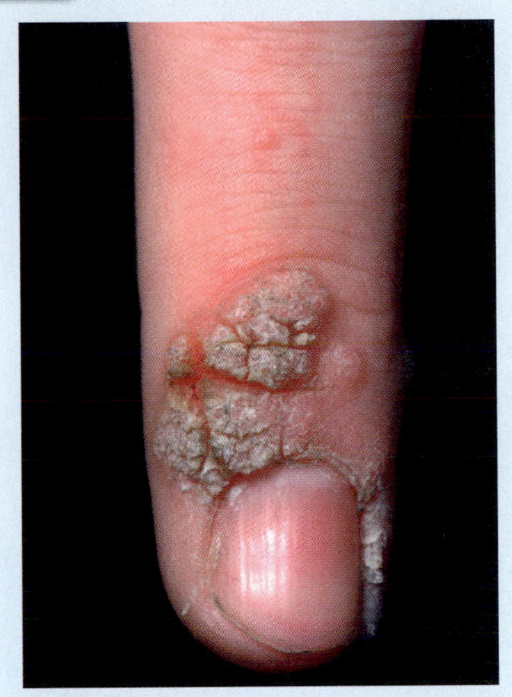

Skin warts on a finger.

Common warts can be difficult to treat and eradicate. A physician may try freezing or minor surgery, or prescribe a chemical treatment. Recently, a new and innovative treatment has shown great success in eliminating common warts (MicroFocus 14.5).

Genital warts are often transmitted in sexual contact. These warts are sometimes called **condylomata** (condylo = "knob"), a reference to the fig-like appearance of the warts. The HPV strains can be transferred during sexual intercourse, and some virologists suggest that this condition may be more prevalent than genital herpes. An estimated 5.5 million Americans are infected every year, and many harbor and transmit the virus without experiencing symptoms.

Evidence shows some strains of HPV are associated with cervical cancer as well as cancer of the vulva, anus, and penis (see Chapter 13). Moreover, a pregnant woman with genital warts may transmit the viruses during the birth process, and studies show that in the newborn the viruses may lodge in the larynx, trachea, or the lungs.

Prevention requires practicing safe sex. Treatment for genital warts can involve medications or surgical procedures. HPV never is totally eradicated and genital warts can reappear after treatment. However, a vaccine is now available that is almost 100 percent effective in protecting women from cervical cancer and genital warts (see Chapter 13).

CONCEPT AND REASONING CHECKS

14.13 Propose a mechanism by which papillomaviruses stimulate cell division to form common warts.

Poxvirus Infections Have Had Great Medical Impacts on Populations

KEY CONCEPT

- The poxviruses produce contagious and dangerous diseases.

Though most dermotropic viral diseases tend not to be life-threatening, a few, such as smallpox, have exacted heavy tolls of human misery.

Smallpox is a contagious and sometimes fatal disease that until recently had ravaged people around the world since pre-biblical times. Few people escaped the pitted scars accompanying the disease, and children were

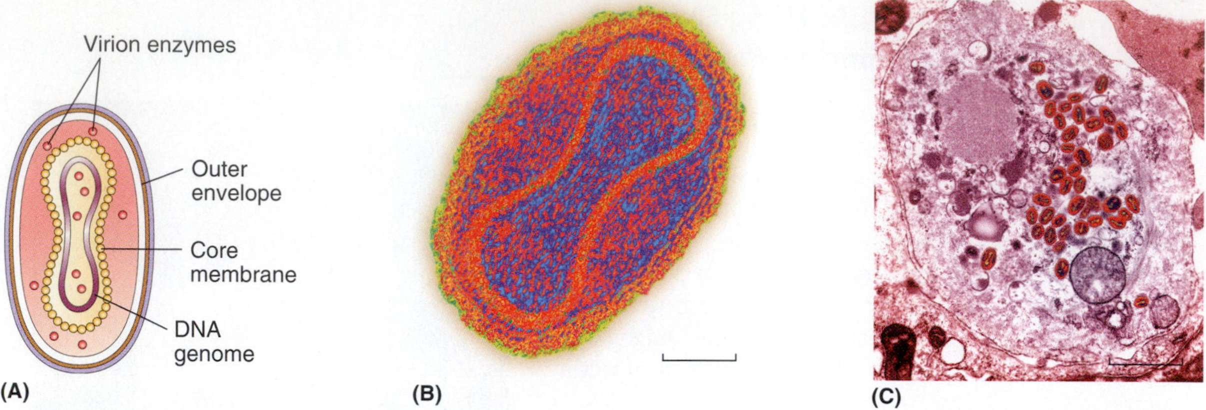

FIGURE 14.19 **The Smallpox Virus.** (**A**) A drawing of the smallpox virus, showing its complex features. (**B**) A false-color transmission electron micrograph of the smallpox virus cultivated in cell culture. (Bar = 50 nm.) (**C**) A false-color transmission electron micrograph of a cell infected with smallpox viruses. Rectangular mature virions (red) can be observed. (Bar = 800 nm.)

Q: Although variola virus has a DNA genome, it replicates totally in the cell cytoplasm. What enzyme must it carry to ensure proper DNA replication in the cytoplasm?

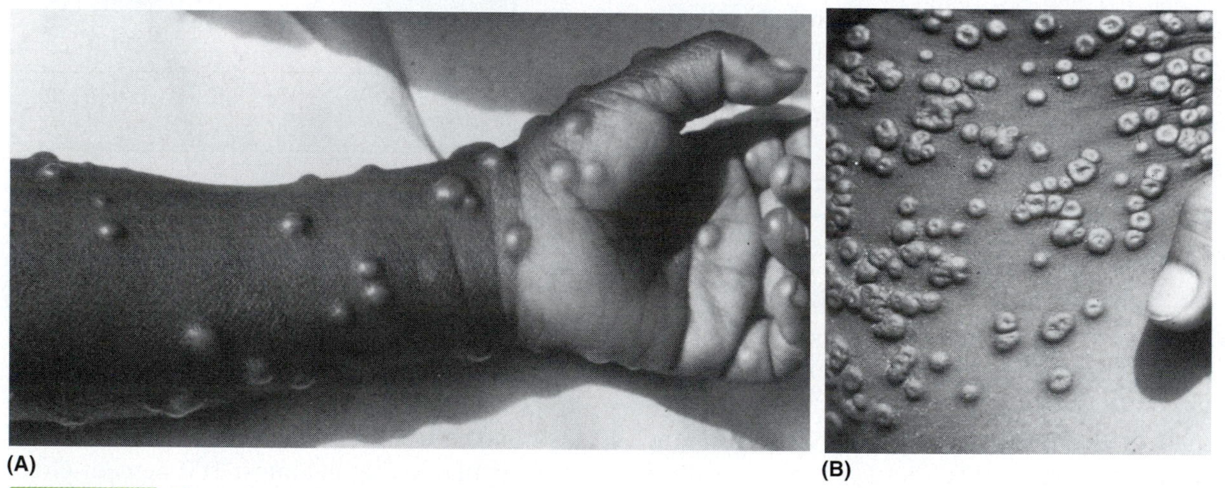

FIGURE 14.20 **The Lesions of Smallpox.** (**A**) The smallpox lesions are raised, fluid-filled vesicles similar to those in chickenpox. For this reason, cases of chickenpox sometimes are misdiagnosed as smallpox. Later, the lesions will become pustules (**B**), and then form pitted scars, the pocks.

Q: What is the difference between a vesicle and a pustule?

not considered part of the family until they had survived smallpox. Thanks to Edward Jenner, the first attempts at a vaccine were begun (see Chapter 1).

Smallpox is caused by a brick-shaped double-stranded DNA virus of the *Poxviridae* family (FIGURE 14.19). It is one of the largest virions, approximately the size of chlamydiae (see Chapter 12). The nucleocapsid is surrounded by a series of fiber-like rods with an envelope. Transmission is by contact.

The earliest signs of smallpox are high fever and general body weakness. Pink-red spots, called **macules**, soon follow, first on the face and then on the body trunk. The spots become pink pimples, called **papules** followed by fluid-filled vesicles so large and obvious that the disease is also called variola (*varus* = "vessel"; FIGURE 14.20). The vesicles become deep **pustules**, which break open and emit pus. If the person survives, the pustules leave pitted scars, or **pocks**. These are generally smaller than the lesions of syphilis (the Great Pox) or chickenpox. TABLE 14.3 summarizes the stages of smallpox.

Vaccination has been hailed as one of the greatest medical and social advances because it was the first attempt to control disease on a

TABLE **14.3** Stages of Smallpox

Stage	Explanation	Duration	Contagious?
Incubation period	Following exposure to the virus, people do not have any symptoms and usually feel fine during the incubation period.	7–17 days (Average: 12–14 days)	No
Initial symptoms	First symptoms include high fever (38°–40°C), malaise, head and body aches, and sometimes vomiting. Affected individuals are too sick to carry on their normal activities.	2–4 days	No
Early rash	**Days 1 & 2:** A rash emerges as small red spots on the tongue and in the mouth, and develop into sores that break open and spread large amounts of the virus into the mouth and throat. About the time the sores in the mouth break down, a rash appears on the skin, starting on the face, and spreads to the arms and legs and then to the hands and feet. Usually the rash spreads to all parts of the body within 24 hours. As the rash appears, the fever usually falls and the person may start to feel better. **Day 3:** The rash becomes raised bumps. **Day 4:** The bumps fill with thick, opaque fluid and often have a depression in the center that looks like a belly button. (This is a major distinguishing characteristic of smallpox.) Fever often occurs again until scabs form over the bumps.	4 days	Very contagious
Pustular rash	The bumps become pustules—sharply raised, usually round and firm to the touch, as if there is a small round object under the skin. People often said the bumps feel like BB pellets embedded in the skin.	5 days	Yes
Pustules and scabs	By the end of the second week after the rash appears, the pustules begin to form a crust and then scab.	5 days	Yes
Resolving scabs	Within three weeks after the rash appears, the scabs begin to fall off, leaving marks on the skin that become pitted scars.	6 days	Yes
Scabs resolved	Scabs have fallen off.		No

Source: Centers for Disease Control. Available at: http://www.cdc.gov. Accessed July 2003.

national scale. It was also the first effort to protect the community rather than the individual. In 1966, the WHO received funding to attempt the global eradication of smallpox. Surveillance containment methods were used to isolate every known smallpox victim, and all contacts were vaccinated. The eradication was aided by the fact that smallpox viruses do not exist anywhere in nature except in humans. On October 26, 1977, health care workers reported isolation of the last case. In 1979, the WHO announced worldwide smallpox eradication, the first such claim made for any disease.

There are two known stocks of smallpox virus, one at the CDC in Atlanta and the other at a similar facility in Russia. However, the former Soviet Union produced massive amounts of smallpox virus during the Cold War years, so there may be stocks that "walked away" to rogue nations or terrorist organizations after the fall of the Soviet Union in 1991. The destruction of the remaining smallpox stocks at the CDC and in Russia has been planned by the WHO. However, it has been postponed several times because of the controversy over the value of keeping smallpox stocks (MicroFocus 14.6).

Since vaccinations against smallpox stopped in the United States in 1972 and elsewhere soon after, a majority of people lack immunity to the disease. This makes smallpox one of the most dangerous weapons of bioterrorism, even though a 2003 study suggested

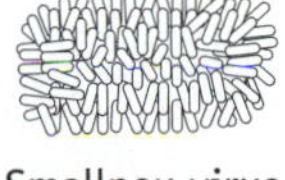

Smallpox virus

MICROFOCUS 14.6: Public Health

"Should We or Shouldn't We?"

One of the liveliest debates in microbiology is whether the last remaining stocks of smallpox viruses should be destroyed. Here are some of the arguments.

For Destruction

- People are no longer vaccinated, so if the virus should escape the laboratory, a deadly epidemic could ensue.
- The DNA of the virus has been sequenced and many cloned fragments are available for performing research experiments; therefore, the whole virus is no longer necessary.
- Eradicating the disease means eradicating the remaining stocks of laboratory virus, and the stocks must be destroyed to complete the project.
- If the United States and Russia destroy their smallpox stocks, it will send a message that biological warfare will not be tolerated.

Against Destruction

- Future studies of the virus are impossible without the whole virus. Indeed, certain sequences of the viral genome defy deciphering by current laboratory means.
- Studying the genome of the virus without the whole virus will not provide insights into how the virus causes disease.
- Mutated viruses could cause smallpox-like diseases, so continued research on smallpox is necessary in order to be prepared.
- Smallpox viruses may be secretly retained in other labs in the world for bioterrorism purposes, so destroying the stocks may create a vulnerability. Smallpox viruses also may remain active in buried corpses.
- Destroying the virus impairs the scientists' right to perform research, and the motivation for destruction is political, not scientific.

Now it's your turn. Can you add any insights to either list? Which argument do you prefer? P.S. In April 2002, the World Health Assembly of the World Health Organization recommended postponing the destruction of all remaining smallpox stocks until all research and drug development is concluded. This will allow time to prepare for a natural outbreak or potentially deliberate (bioterrorist) release of variola virus.

those who were vaccinated prior to 1972 still have some level of immunity. In addition, the United States has stated that it now has adequate stockpiles of smallpox vaccine to vaccinate every American if necessary.

Other diseases caused by poxviruses also are potentially dangerous (MicroFocus 14.7).

Molluscum Contagiosum is another viral disease that forms mildly contagious, wart-like skin lesions. The virus of molluscum contagiosum is an enveloped, double-stranded DNA virion of the *Poxviridae* family. Transmission is generally by sexual contact.

The lesions are firm, waxy, and elevated with a depressed center. When pressed, they yield a milky, curd-like substance. Although usually flesh toned, the lesions may appear white or pink. Possible areas involved include the facial skin and eyelids in children, and the external genitals in adults. The lesions may be removed by excising them (cutting them out) and pose no public health threat. A characteristic feature of the disease is the presence of large cytoplasmic bodies called **molluscum bodies** in infected cells from the base of the lesion.

TABLE 14.4 presents a summary of these other dermotropic viral diseases.

CONCEPT AND REASONING CHECKS

14.14 Describe how one could identify smallpox from chickenpox based on the rash formed.

MICROFOCUS 14.7: Public Health

The First American Case of Monkeypox

Monkeypox is a rare viral disease occurring mostly in central and western Africa. It is called "monkeypox" because it was first found in 1958 in laboratory primates. Blood tests of animals in Africa later found other types of animals also had monkeypox. In 1970, the first case of monkeypox was reported in humans.

The monkeypox virus is a double-stranded DNA virus in the same group of *Poxviridae* as the smallpox and cowpox viruses. People get monkeypox from an infected animal through bites or contact with the animal's blood, body fluids, or its rash. The disease also can spread from person to person through large respiratory droplets during long periods of face-to-face contact, or by touching body fluids of a sick person or objects such as bedding or clothing contaminated with the virus.

The first outbreak of monkeypox in the Western Hemisphere occurred in the Midwest in June 2003. On April 9, 2003, a Texas animal distributor received a shipment of some 800 small mammals from Accra, Ghana. This included rope squirrels, Gambian giant rats, and dormice.

Twelve days later, an Illinois distributor received Gambian rats and dormice from the Texas distributor. An infected Gambian rat was housed with prairie dogs that the distributor had before the prairie dogs were sold to distributors in six states. Additional prairie dogs were sold at animal swap meets.

Starting on May 15, doctor reports from the Midwest came in that people were exhibiting symptoms similar to but much milder and less contagious than smallpox. Individuals became ill with a fever, respiratory symptoms, and swollen lymph nodes. A rash developed, which progressed into raised bumps filled with fluid. The rash started on the face and spread across the body. Eventually the rash crusted over and the scabs fell off. The illness lasted for two to four weeks.

Over the next few weeks, outbreaks of monkeypox were identified in Wisconsin, Illinois, and Indiana where the Texas distributor had sent prairie dogs to pet dealers. On June 20, the last case was reported.

On June 30, 2003, a final report to the CDC identified 71 cases of monkeypox from Wisconsin, Illinois, Indiana, Missouri, Kansas, and Ohio. There were no deaths, although 18 were hospitalized. Of the 71 cases, 35 were laboratory confirmed and 36 were suspect and probable cases.

This outbreak of monkeypox underlined the need to closely screen and protect the public from exotic animals exported to the United States. This time the infection was fairly mild as it was caused by a weak monkeypox strain; the next time, it or another disease might be contagious and lethal.

TABLE

14.4 A Summary of Other Viral Diseases of the Skin

Disease	Classification of Virus	Transmission	Organs Affected	Vaccine	Special Features	Complications
Measles (rubeola)	*Paramyxoviridae*	Droplets Contact	Respiratory tract Skin Blood	Attenuated viruses	Koplik spots Progress of rash Hemagglutination inhibition	SSPE Pneumonia Encephalitis
Mumps	*Paramyxoviridae*	Droplets	Salivary glands Blood	Attenuated viruses	Swollen glands Hemagglutination inhibition	Orchitis Encephalitis Meningitis
Rubella (German measles)	*Togaviridae*	Droplets Contact	Skin Blood	Attenuated viruses	Skin rash Mild cold symptoms	Congenital rubella syndrome
Fifth disease (erythema infectiosum)	*Parvoviridae*	Droplets (?)	Skin Blood	Not available	"Slapped-cheek" appearance Lacy skin patterns	None established
Warts	*Papovaviridae*	Contact	Skin	Not available	Characteristic lesions	None
Smallpox (variola)	*Poxviridae*	Contact Droplets Fomites	Skin Blood	Cowpox viruses	Characteristic lesions Macules, papules, vesicles, pox	Permanent scarring
Molluscum contagiosum	*Poxviridae*	Contact	Skin	Not available	Characteristic lesions	None

SUMMARY OF KEY CONCEPTS

14.1 Viral Infections of the Upper Respiratory Tract

- Influenza is caused by three different orthomyxoviruses: types A, B, and C. The spike proteins hemagglutinin and neuraminidase are necessary for viral entry and exit during an infection. Symptoms include sudden chills, headache, fatigue, and chest pain. Influenza is best prevented with yearly vaccination, although new antiviral drugs can shorten the duration of symptoms. Antigenic drift and antigenic shift account for the yearly differences in flu strains and for flu pandemics.
- Over 100 rhinoviruses are members of the picornaviruses. They are transmitted through airborne droplets or by contaminated objects. Symptoms produce a typical head cold.
- Adenoviruses cause some types of colds. These DNA viruses also cause keratoconjunctivitis. They have been used in gene therapy as the vector to carry therapeutic genes.

14.2 Viral Infections of the Lower Respiratory Tract

- The RS virus is a paramyxovirus. After infecting cells, it causes the cells to fuse into syncytia. RS disease often causes a type of viral pneumonia in children and an upper respiratory disease in adults. Immune globulin and antivirals have been used to lessen the symptoms.
- Parainfluenza is another paramyxovirus that produces milder symptoms than influenza. No specific therapy exists.
- Other RSV-like illnesses have been caused by unknown viruses. Most of those illnesses are now thought to be caused by the human metapneumovirus, another member of the paramyxoviruses.
- SARS represents a newly emerging viral disease caused by a coronavirus. It is spread by person-to-person contact. Symptoms include fever, headache, feeling of discomfort, and body aches. A dry cough and difficulty breathing often occur. In severe illness, insufficient oxygen reaches the blood and mechanical ventilation is required.

14.3 Diseases of the Skin Caused by Herpesviruses

- Herpes simplex describes a wide spectrum of viral diseases commonly found in the environment. Among the herpesviruses are ones that cause cold sores (HSV-1) and genital herpes (primarily HSV-2). Several antiviral drugs have been developed to treat herpes simplex infections. Neonatal herpes is a possible life-threatening disease transmitted from a herpes simplex-infected mother during childbirth.
- Another member of the herpesviruses is varicella-zoster. This virus causes chickenpox, which is one of the most highly contagious diseases. The same virus causes shingles in adults, which can be a painfully debilitating disease. Acyclovir has been successful at lessening the symptoms.
- Human herpesvirus 6 causes roseola and has been implicated in multiple sclerosis.
- The Epstein-Barr virus can cause infectious mononucleosis in adolescents and young adults. It also represents an oncogenic virus and is the cause of Burkitt lymphoma. It also is associated with Hodgkin disease and multiple sclerosis.
- Human herpesvirus 8 causes Kaposi sarcoma, which is very prevalent in AIDS patients.

14.4 Other Viral Diseases of the Skin

- The measles virus is a member of the paramyxoviruses. Symptoms of measles include a hacking cough, sneezing, eye redness, sensitivity to light, and a high fever. Koplik spots are common along the gum line. The characteristic red rash appears a few days after the Koplik spots appear.
- Mumps is produced by a viral infection from another member of the paramyxoviruses. The disease produces swollen parotid (salivary) glands. The swelling of the testes (orchitis) may be a complication of the disease.
- Rubella is a member of the togaviruses, a group of single-stranded (+ strand) RNA viruses. Infection leads to a fever and a pale-pink rash beginning on the face and spreading over the body and extremities. Congenital rubella syndrome can develop if a pregnant woman has rubella.
- Fifth disease is caused by the B19 virus, which is a member of the parvoviruses. These single-stranded DNA viruses are spread by the respiratory route. Infection produces a red rash that resembles a slapped cheek. In adults, an infection can produce painful joints.
- Different strains of the human papillomaviruses (HPV) produce warts or cervical cancer. Many of the strains of HPV produce benign viral tumors, such as plantar warts, on the skin. Genital warts are caused by other HPV strains and are transmitted by sexual intercourse. Malignancies of the cervix, vagina, and penis have been associated with such HPV infections.
- The most dangerous virus in the groups described in this chapter is variola, a member of the poxviruses. Smallpox symptoms start with a fever and body weakness followed by the development of pustules on the face and body. Survival leaves pockmarks where pustules had formed and scabbed. Smallpox was the first disease purposely eradicated from planet Earth.
- Molluscum contagiosum is a disease caused by another poxvirus but not closely related to smallpox. Through sexual contact, the virus infection produces mild wart-like skin lesions that pose no public health threat.

LEARNING OBJECTIVES

After understanding the textbook reading, you should be capable of writing a paragraph that includes the appropriate terms and pertinent information to answer the objective.

1. Identify the major influenza viruses and the structures involved in generating subtypes.
2. Summarize the genetic mechanisms by which influenza A and B viruses evolve.
3. Explain why a vaccine against rhinoviruses is not feasible.
4. List the types of diseases associated with adenovirus infections.
5. Organize the paramyxoviruses into related species.
6. Describe RS disease and explain how syncytia are formed.
7. Summarize the infections caused by the human parainfluenza viruses.

8. Identify the significance of human metapneumovirus infections.
9. Distinguish how SARS differs from other respiratory tract infections.
10. Determine the general characteristics of the herpesviruses.
11. Describe the infections caused by herpes simplex virus-1 and herpes simplex virus-2.
12. Identify the diseases associated with the TORCH acronym.
13. Explain why the incidence of chickenpox has declined in the United States and how the varicella-zoster virus causes shingles.
14. Summarize the diseases caused by human herpesvirus-6.
15. Describe the relationship between human herpesvirus-8 and HIV/AIDS.
16. Summarize the characteristics of the paramyxovirus infections causing (1) measles and (2) mumps.
17. State the potential complication to a pregnant mother who has contracted rubella.
18. Identify the characteristics associated with fifth disease.
19. Distinguish between common and genital warts, including possible complications.
20. Summarize the clinical and social significance of smallpox.

SELF-TEST

Answer each of the following questions by selecting the ***one*** answer that best fits the question or statement. Answers to even-numbered questions can be found in **Appendix C**.

1. Which one of the following statements is *not* true of the influenza viruses?
   A. They have a segmented genome.
   B. The genome is double-stranded DNA.
   C. The viruses have an envelope.
   D. There are three types of flu viruses.
   E. Spikes project from the virus surface.

2. Aspirin should not be given to children for most viral diseases because the drug
   A. leads to viral resistance.
   B. stimulates virus replication.
   C. can cause an antigenic shift.
   D. may cause Reye syndrome.
   E. may cause a bacterial infection.

3. There are more than _____ different rhinoviruses, which belong to the _____ family.
   A. 10; *Herpesviridae*
   B. 50; *Orthomyxoviridae*
   C. 100; *Adenoviridae*
   D. 30; *Paramyxoviridae*
   E. 100; *Picornaviridae*

4. All of the following are diseases caused by the adenoviruses *except*
   A. viral pneumonia.
   B. acute respiratory disease.
   C. parainfluenza.
   D. pharyngoconjunctival fever.
   E. common cold.

5. Which one of the following is *not* a member of the *Paramyxoviridae*. If all are, select (**E**).
   A. RSV
   B. Human metapneumovirus
   C. SARS-CoV
   D. Parainfluenza virus
   E. All the above (**A–D**) are members.

6. The human metapneumovirus
   A. has infected almost every child around the world.
   B. causes life-threatening diseases.
   C. undergoes a latent stage.
   D. is a member of the *Coronaviridae*.
   E. Both **C** and **D** are correct.

7. SARS is
   A. an LRT infection.
   B. spread through close person-to-person contact.
   C. a mild, respiratory infection.
   D. most often seen in young children.
   E. Both **A** and **B** are correct.

8. Which one of the following is *not* true of the *Herpesviridae?*
   A. They have a single-stranded genome.
   B. Eight species infect humans.
   C. They have icosahedral symmetry.
   D. Many can undergo latency.
   E. They have an envelope.

9. Cold sores and genital herpes can be caused by
   A. EBV.
   B. HSV-1.
   C. HHV-6.
   D. VZV.
   E. HHV-8.

10. A teratogen is a/an
    A. cancer-causing agent.
    B. toxin produced by the TORCH viruses.
    C. protein found in some viruses.
    D. agent interfering with fetal development.
    E. complication of an influenza infection.

11. A red, itchy rash that forms small, teardrop-shaped, fluid-filled vesicles is typical of
    A. measles.
    B. rubella.
    C. chickenpox.
    D. mumps.
    E. roseola.

12. ______ are diagnostic for measles.
A. Koplik spots
B. Wart-like lesions
C. Swollen lymph nodes
D. Blisters on the body trunk
E. Swollen salivary glands

13. Mumps is an infection of the ______ and, in male patients, the virus may pose a threat to the ______.
A. respiratory tissue; brain
B. parotid glands; reproductive organs
C. skin; kidneys
D. blood; reproductive organs
E. parotid glands; lungs

14. The R in TORCH stands for
A. rhinovirus.
B. rubeola.
C. respiratory.
D. rubella.
E. roseola.

15. Some papillomaviruses are capable of causing
A. a "lacy" rash on the skin.
B. lymphomas.
C. lung cancer.
D. pneumonia.
E. cervical cancer.

16. Which one of the following statements applies to smallpox?
A. The disease is associated with macules, papules, and pustules.
B. The disease has been eradicated worldwide.
C. It can be sexually-transmitted.
D. The virus can lie dormant in host cells.
E. Both **A** and **B** are correct.

QUESTIONS FOR THOUGHT AND DISCUSSION

Answers to even-numbered questions can be found in **Appendix C**.

1. Thomas Sydenham was an English physician who, in 1661, differentiated measles from scarlet fever, smallpox, and other fevers, and set down the foundations for studying these diseases. How would you go about distinguishing the variety of look-alike skin diseases discussed in this chapter?
2. Most physicians agree that there would be great demand for a genital herpes vaccine. However, there is much opposition to marketing a vaccine containing attenuated HSV. Why is this so? What alternatives are there for a useful vaccine against genital herpes?
3. A Boeing 737 bound for Kodiak, Alaska, developed engine trouble and was forced to land. While the airline rounded up another aircraft, the passengers sat for 4 hours in the unventilated cabin. One passenger, it seemed, was in the early stages of influenza and was coughing heavily. By the week's end, 38 of the 54 passengers on the plane had developed influenza. What lessons about infectious disease does this incident teach?
4. In the United Kingdom, the approach to rubella control is to concentrate vaccination programs on young girls just before they enter the childbearing years. In the United States, the approach is to immunize all children at the age of 15 months. Which approach do you believe is preferable? Why?

APPLICATIONS

Answers to even-numbered questions can be found in **Appendix C**.

1. The CDC reports an outbreak of measles at an international gymnastics competition. A total of 700 athletes and numerous coaches and managers from 51 countries are involved. What steps would you take to avert a disastrous international epidemic?
2. A child experiences "red bumps" on her face, scalp, and back. Within 24 hours, they have turned to tiny blisters and become cloudy, some developing into sores. Finally, all become brown scabs. New "bumps" keep appearing for several days, and her fever reaches 39°C by the fourth day. Then the blisters stop coming and the fever drops. What disease has she had?
3. A man experiences an attack of shingles and you warn him to stay away from children as much as possible. Why did you give him this advice?

REVIEW

On completing your study of viral diseases of the respiratory tract and skin, test your comprehension of the chapter contents by circling the choices that best complete each of the following statements. The answers to even-numbered statements are listed in **Appendix C**.

1. Rhinoviruses are a collection of (RNA, DNA) viruses having (helical, icosahedral) symmetry and the ability to infect the (air sacs, nose), causing (mild, serious) respiratory symptoms.
2. Herpes simplex is a viral disease transmitted by (breathing contaminated air, contact) and is characterized by thin-walled (blisters, ulcers) that often appear during periods of (emotional stress, exercising).
3. In children, the skin lesions of chickenpox occur (all at once, in crops) and resemble (teardrops, pitted scars), but in adults the lesions are known as (shingles, erythemas).
4. The complications of influenza include (Reye, Koplik) syndrome; for mumps, the complication is a disease of the (testes, pancreas) called (colitis, orchitis).
5. One of the early signs of (smallpox, measles) is a series of (Koplik spots, inclusion bodies) occurring in the (lungs, mouth) and signaling a (red, purple) rash is forthcoming.
6. Antigenic variation among (mumps, influenza) viruses seriously hampers the development of a highly effective (vaccine, treatment), and a life-threatening situation can occur if secondary infection due to (fungi, bacteria) complicates the primary infection.
7. Respiratory syncytial disease is caused by a (DNA, RNA) virus infecting the (lungs, intestines) of (adults, children) and inducing cells to (fuse together, cluster) and form giant cells called (syncytia, tumors).
8. SARS is caused by a (coronavirus, orthomyxovirus), a/an (naked, enveloped) virus spread by (sexual, person-to-person) contact.
9. Adenoviruses include a collection of (DNA, RNA) viruses and are responsible for (yellow fever, common colds), as well as infections of the (eye, ear).
10. Genital herpes is caused by a (helical, icosahedral) virus most often (HSV-1, HSV-2) and causes blisters with (thick, thin) walls that disappear in about three (days, weeks), only to reappear when (stress, physical injury) occurs.
11. The TORCH diseases are a set of (infectious, physiological) diseases transmitted by (airborne droplets, transplacental passage), occurring in (the elderly, newborns), and including (rubeola, rubella) and (herpes simplex, humoral disease).

HTTP://MICROBIOLOGY.JBPUB.COM/

The site features learning, an on-line review area that provides quizzes and other tools to help you study for your class. You can also follow useful links for in-depth information, read more MicroFocus stories, or just find out the latest microbiology news.

15

Viral Infections of the Blood, Lymphatic, Gastrointestinal, and Nervous Systems

Chapter Preview and Key Concepts

I don't think we are losing the war, but we're certainly not finished with the war.
—Ronald Valdiserri, CDC Deputy Director, speaking on the upsurge of HIV cases in the United States

2006 marks 25 years into the epidemic of acquired immunodeficiency syndrome (AIDS), and there still are only somber numbers reported. In 2005, the Joint United Nations Programme on HIV/AIDS (UNAIDS) reported 5 million new AIDS cases globally, which represents the highest number of new cases since the beginning of the epidemic in 1981 (FIGURE 15.1). Twenty million have died from AIDS. Worldwide, in 2004 there were an estimated 40 million people infected with the human immunodeficiency virus (HIV); over 3 million died; and some 750,000 babies were born with HIV infection. UNAIDS has projected that 70 million people will die from AIDS in the next 20 years—if a cure is not found. Indeed, there still is no cure or vaccine, and antiretroviral therapies can be toxic and quickly become less useful as HIV develops resistance.

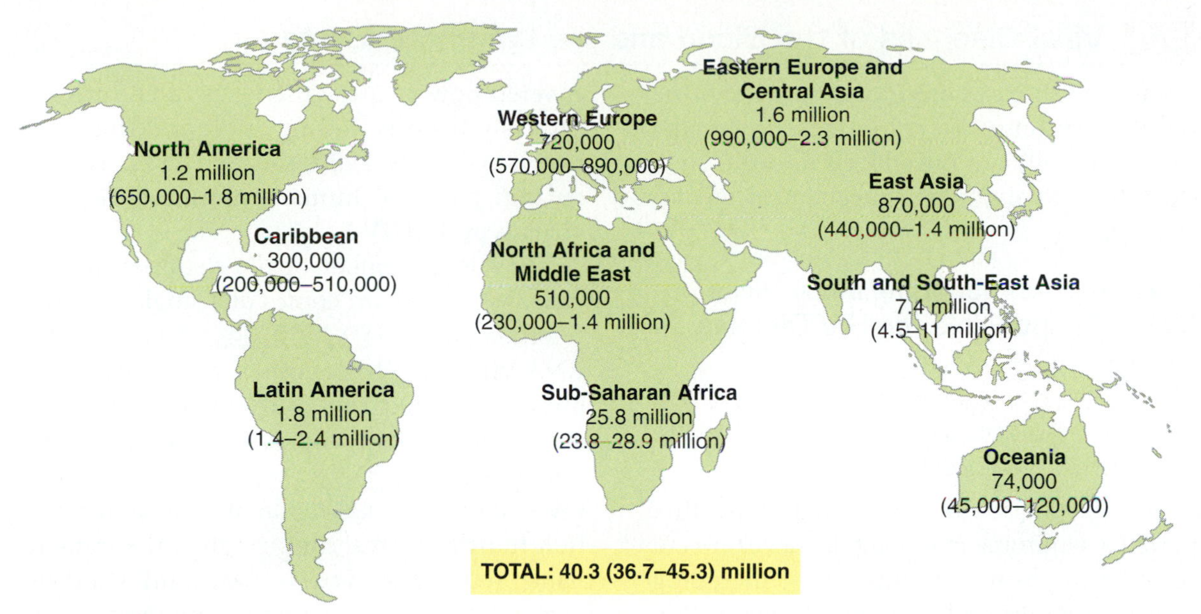

FIGURE 15.1 **Adults and Children Living with HIV Infection—2005.** The numbers of HIV infections is an estimate made by the Joint United Nations Program on HIV/AIDS. The total is 40.3 (36.7–45.3) million. Source: UNAIDS.org.
Q: Provide some reasons why these numbers represent only estimates.

The epidemic is worst in sub-Saharan Africa where almost 26 million people have HIV infection. In some African cities, a staggering 30 to 50 percent—almost half the population—is HIV-positive. Without a cure, by 2020 more than 25 percent of the workforce in some African cities will be lost because of AIDS. In July 2003, President Bush pledged $15 billion to fight AIDS in Africa and the Caribbean where almost half of the world's HIV infections are located. The plan to combat AIDS calls for antiviral treatment for 2 million HIV-infected people who cannot afford the costly cocktail of drugs that can prolong and improve their lives.

Africa is not the only worry. In South and Southeast Asia, there are an estimated 7 million people infected, 5 million in India alone. In some areas, UNAIDS reports the HIV infection rate among drug users is greater than 70 percent. In Eastern Europe, sexually transmitted diseases have been doubling every year since 1998, which means HIV infections also are increasing at alarming rates.

Compared to Africa, Asia, and Eastern Europe, the AIDS prevalence in most developed nations has remained flat, in part, because of the number of people on antiretroviral drugs. In 2003, there were more than 40,000 new AIDS infections in adults and children in the United States, bringing the total number to more than one million. Most alarming, 25 percent of those infected with HIV do not know it. There were 14,000 deaths in 2003, bringing the total to 530,000 since 1981.

In 2002, federal health authorities announced there was a 2.2 percent increase in HIV infections from the previous year, which represents the first rise since 1993. American health officials attributed this disturbing upswing to a growing complacency about the dangers of AIDS. In addition, younger people do not remember the devastation of the AIDS epidemic and perhaps feel safer with the advent of life-extending antiretroviral drugs now available.

A major focus of this chapter is on AIDS. However, it by no means represents the only viral infection, so we also examine several other diseases of concern. Overall, the diseases addressed in this chapter fall into three general categories. Some illnesses, such as AIDS, yellow fever, and mononucleosis are diseases of the blood, while others, including the noroviruses, affect the digestive system. Still others, such as rabies, polio, and West Nile encephalitis affect the nervous system. As in Chapter 14, each disease is presented as a separate essay, and you may select the order of study most suitable to your study needs.

15.1 Viral Diseases of the Blood and the Lymphatic Systems

Several viral diseases are of the blood and the lymphatic systems. To reach these areas, the viruses generally are introduced into the body tissues by sexual contact, mechanical means, or arthropods.

The Human Immunodeficiency Virus (HIV) Is Responsible for HIV Disease and AIDS

KEY CONCEPT

- Acquired immunodeficiency syndrome (AIDS) results from immune system dysfunction.

In 1981, physicians in the United States first reported a syndrome involving the development of certain opportunistic infections, including fungal pneumonia and an unusual type of skin cancer called Kaposi sarcoma. These illnesses, along with sudden weight loss, swollen lymph nodes, and immune system deficiencies represented the signs and symptoms for a disease that became know as **acquired immunodeficiency syndrome** (AIDS).

By 1983, the most plausible factor responsible for AIDS was a virus. In that year, Luc Montagnier and his French group isolated the infectious agent from a patient and Robert Gallo's group in the United States discovered how to grow the virus in culture and published convincing evidence that HIV causes AIDS. By 1986, the virus was given its current name of **human immunodeficiency virus type 1** (**HIV-1**).

HIV is a member of the *Retroviridae*. The virion contains two copies of a single-stranded (+ strand) RNA (FIGURE 15.2). Unique to the RNA viruses, the genome is packaged with enzymes called **reverse transcriptase** that are needed to copy single-stranded RNA into double-stranded DNA (see Chapter 13). This reversal of the usual mode of genetic information transfer (transcription) gives the virus its name, retrovirus (*retro* = "backward") and the enzyme its name, reverse transcriptase.

The HIV genome is surrounded by a cone-shaped icosahedral capsid. Between the capsid and envelope is a **matrix protein** that facilitates viral penetration. Like other enveloped viruses, the HIV envelope contains spikes for attachment and entry into the host cell.

In 1986, a second type of HIV, called HIV-2, was isolated from AIDS patients in West Africa. Comparative studies indicate both types have the same mode of transmission and are associated with similar infections. However, HIV-2 appears to develop more slowly so people infected with HIV-2 are less infectious early in the course of infection. There are few reported cases of HIV-2 in the United States. It is important to note that for both types, AIDS is the end result of an HIV infection.

MicroFocus 15.1 discusses the origins of HIV.

As mentioned, the viral replication process involves a reverse transcription process (FIGURE 15.3). When the RNA is released in the cytoplasm of a host cell, the reverse transcriptase synthesizes a molecule of DNA using the genetic message in the RNA as a template. The DNA molecule then integrates into the host's DNA as a provirus. Most of the time, the provirus initiates DNA transcription and translation, resulting in the biosynthesis and maturation of new virions. The new HIV virions then "bud" from the host cell to infect other cells. The infected individual now has a **HIV infection**.

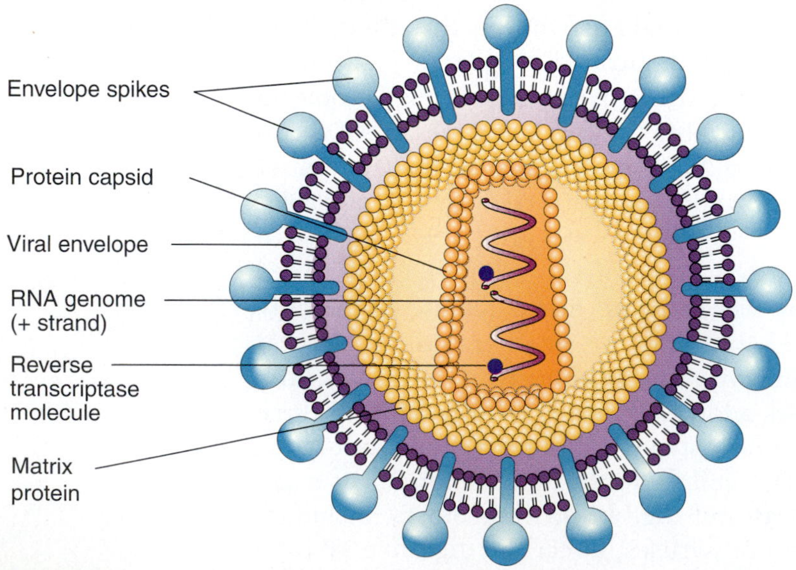

Envelope spikes
Protein capsid
Viral envelope
RNA genome (+ strand)
Reverse transcriptase molecule
Matrix protein

FIGURE 15.2 **A Diagram of the Human Immunodeficiency Virus (HIV).** The virus consists of two molecules of RNA and molecules of reverse transcriptase. A protein capsid surrounds the genome and an envelope with spikes of protein lies outside the capsid and matrix.

Q: What are the unique features about the structure of HIV?

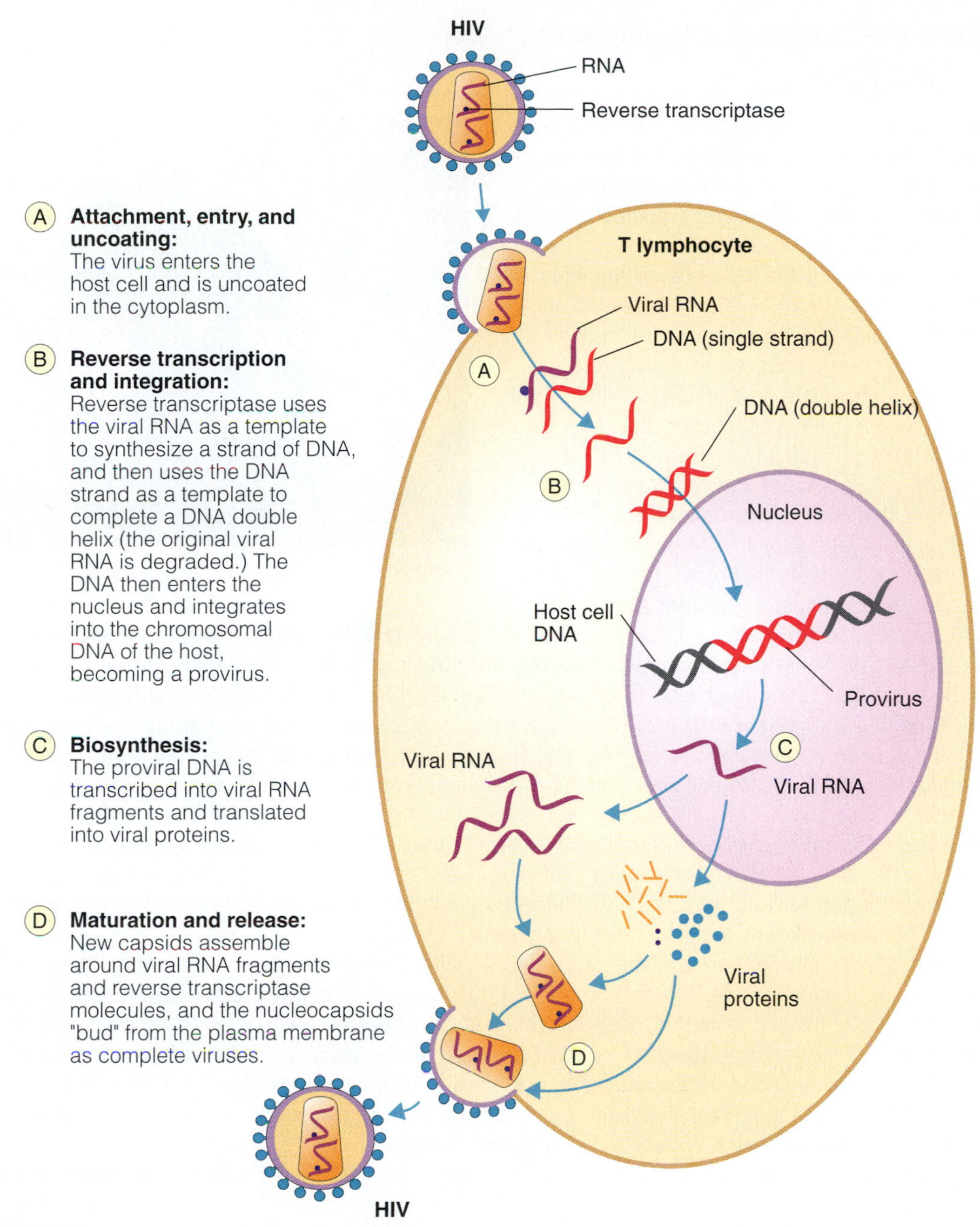

FIGURE 15.3 **The Replication Cycle of the Human Immunodeficiency Virus (HIV).** Replication is dependent on the presence and activity of the reverse transcriptase enzyme.

Q: What is the role of the reverse transcriptase enzyme?

The tissue tropism of HIV is determined by cellular receptors on specific host cells to which the virus attaches and penetrates the cell. HIV normally infects immune system cells of the blood and lymphatic system, including the **T lymphocytes**. These cells, which are the backbone of immune system defenses, also are commonly called **CD4$^+$ T cells** because they have a protein, CD4, on the cell surface to which HIV attaches. If these T cells cannot function properly, the immune system will not respond to other pathogens that now have the opportunity to infect the body. (These processes are explored in more detail in Chapter 20.)

At least 50 percent of the body's CD4$^+$ T cells are located in the mucosal lining of the

MICROFOCUS 15.1: Evolution

HIV's Family Tree

For years, scientists and other people have debated the origin of the human immunodeficiency virus (HIV). Did it come from contaminated polio vaccine? Was it a government secret project that went awry? Where did the virus arise?

HIV is a member of the genus *Lentivirus*, all of which produce slow, (*lent* = "slow") incessant infections of the immune system. These viruses have been found in several animals, including cats, sheep, horses and cattle, and monkeys, where the simian immunodeficiency virus (SIV) has been isolated. It is now generally accepted that HIV is a descendant of SIV because certain strains of SIV bear a very close resemblance to HIV-1 and HIV-2, the two types of HIV. HIV-2 corresponds to the SIV strain found in the sooty mangabey (or green monkey), which is indigenous to western Africa.

A young chimpanzee.

Until recently, the origin for the more virulent HIV-1 was more difficult to place. The closest counterpart was to SIV found in chimpanzees. However, this virus still had certain significant differences from HIV-1. In 2002, a group of researchers from the University of Alabama announced that they had found a type of SIV almost identical to HIV-1 in a frozen sample taken from a subgroup of chimpanzees once common in west central Africa (see figure). They concluded that wild chimps had been infected simultaneously with two different SIVs, which had recombined to form a third virus that could be passed on to other chimps and, more significantly, was capable of infecting humans and causing AIDS.

These two different viruses were traced back to an SIV that infected red-capped mangabeys and one found in greater spot-nosed monkeys. They believe that the hybridization took place inside chimps that had become infected with both strains of SIV after they hunted and killed the two smaller species of monkeys. (Yes, some chimps do hunt and kill animals!)

The most likely scenario is that in the 1930s, HIV-1 jumped to humans who were eating "bush meat" (the term used for monkey meat). In fact, transfer of retroviruses from primates to hunters can still be documented. HIV-2 jumped from the sooty mangabey in the 1960s.

The sudden arrival of HIV in the 1980s could have been due to increased international air travel. The use of blood transfusions and intravenous drug use, both of which increased in the 1970s, also could have contributed to the emergence HIV. Both have been a documented source of HIV transmission as well as for other diseases, such as hepatitis C, between humans.

It is doubtful we will ever know exactly how HIV evolved, but the current scientific evidence appears quite strong.

gastrointestinal tract. HIV specifically targets and infects those subsets of $CD4^+$ T cells primed to attack and destroy HIV. Within two to three weeks after infection, HIV destroys half of this HIV-fighting population of T cells.

HIV also infects and paralyzes the **B lymphocytes** (**B cells**), whose job is to mount an antibody response to infection. HIV infection of B cells causes these cells to behave abnormally and be unable to respond correctly to the infection. Without a functioning T and B cell population, collapse of the immune system places the host at risk for opportunistic pathogens and cancers, such as fungal pneumonia and Kaposi sarcoma.

Although the rate of clinical disease progression varies among individuals, let's look at the classical progression of stages leading

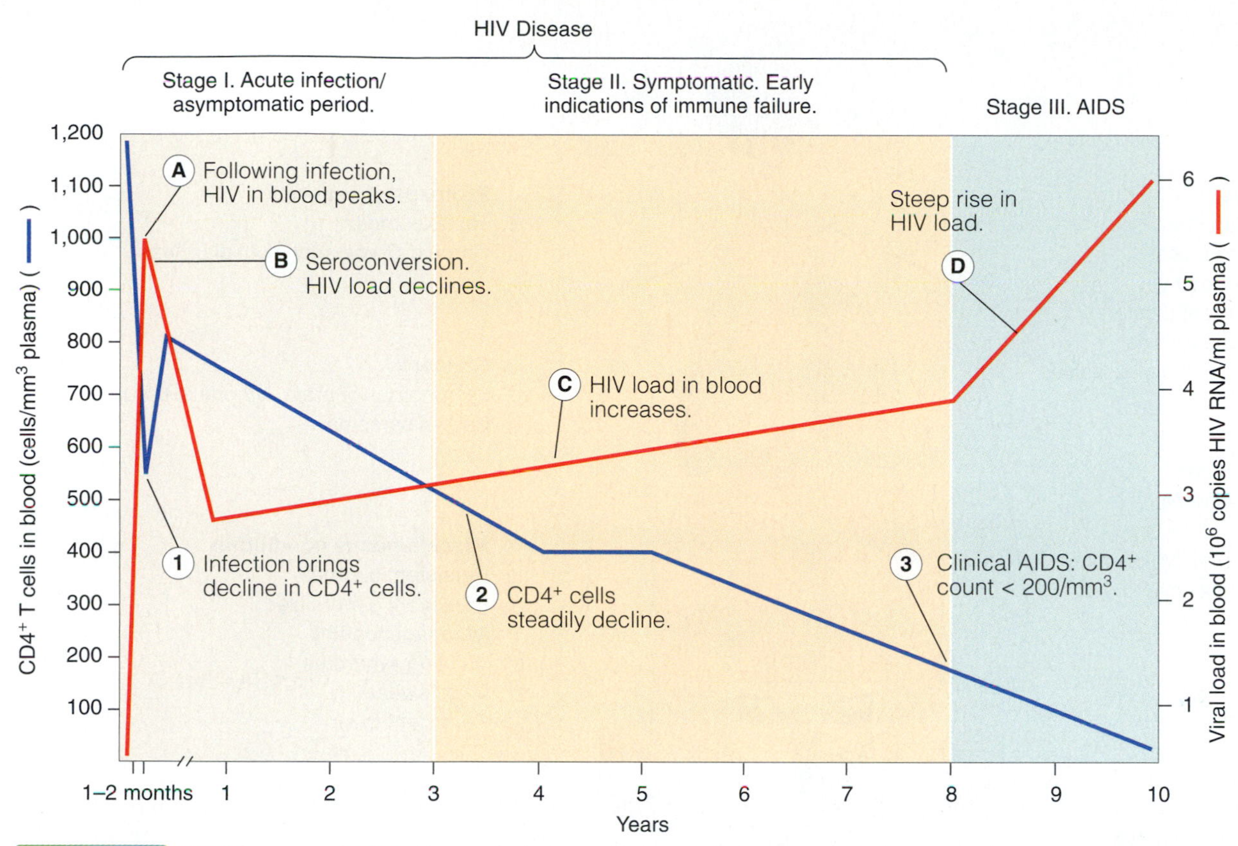

FIGURE 15.4 **HIV Disease and AIDS.** Once infected with HIV, an individual has HIV disease. An immune response brings about an abrupt drop in the blood HIV population, which rises as the immune system fails (A–D). Without antiretroviral therapy, the T cell count slowly drops (1–3). Once the T cell count is below 200/mm^3, the person is said to have AIDS.
Q: Explain why is takes so long for the T cell number to drop.

to AIDS. Stages I and II are referred to as **HIV disease**.

Stage I. Many people are asymptomatic when they first become infected with HIV. Others experience an acute HIV infection, which produces a flu-like illness within a month or two of HIV exposure. The symptoms of fever, headache, tiredness, and enlarged lymph nodes in the neck and groin usually disappear within a few weeks to several months. Importantly, during this period, the immune system is at war with HIV (FIGURE 15.4).

The immune system starts responding to the infection by producing antibodies and T cells directed against HIV. Such a process in the body is referred to as **seroconversion**, meaning HIV antibodies can be detected in the blood. This usually has occurred by three to four weeks post-infection.

Stage II. Without treatment, the infected individual often remains free of major diseases, although early signs of immune failure may occur as indicated by periods of swollen lymph nodes, upper respiratory infections, and some weight loss. Specific infections, such as *Pneumocystis* pneumonia, also may develop. Depending on the genetic makeup of the individual, this stage lasts for up to six to eight years, during which time the level of HIV in the blood slowly rises as the virus is actively multiplying, infecting, and killing CD4+ T cells.

Stage III. Over time, the immune system loses the fight against HIV. At full bore, HIV can produce more than 10 billion new virions per day. Thus, lymph nodes and tissues become damaged from the battles and the immune system can no longer replace T cells at the rate HIV destroys them. In addition, HIV may have mutated into a form that is more pathogenic and more aggressive in destroying T cells and B cells.

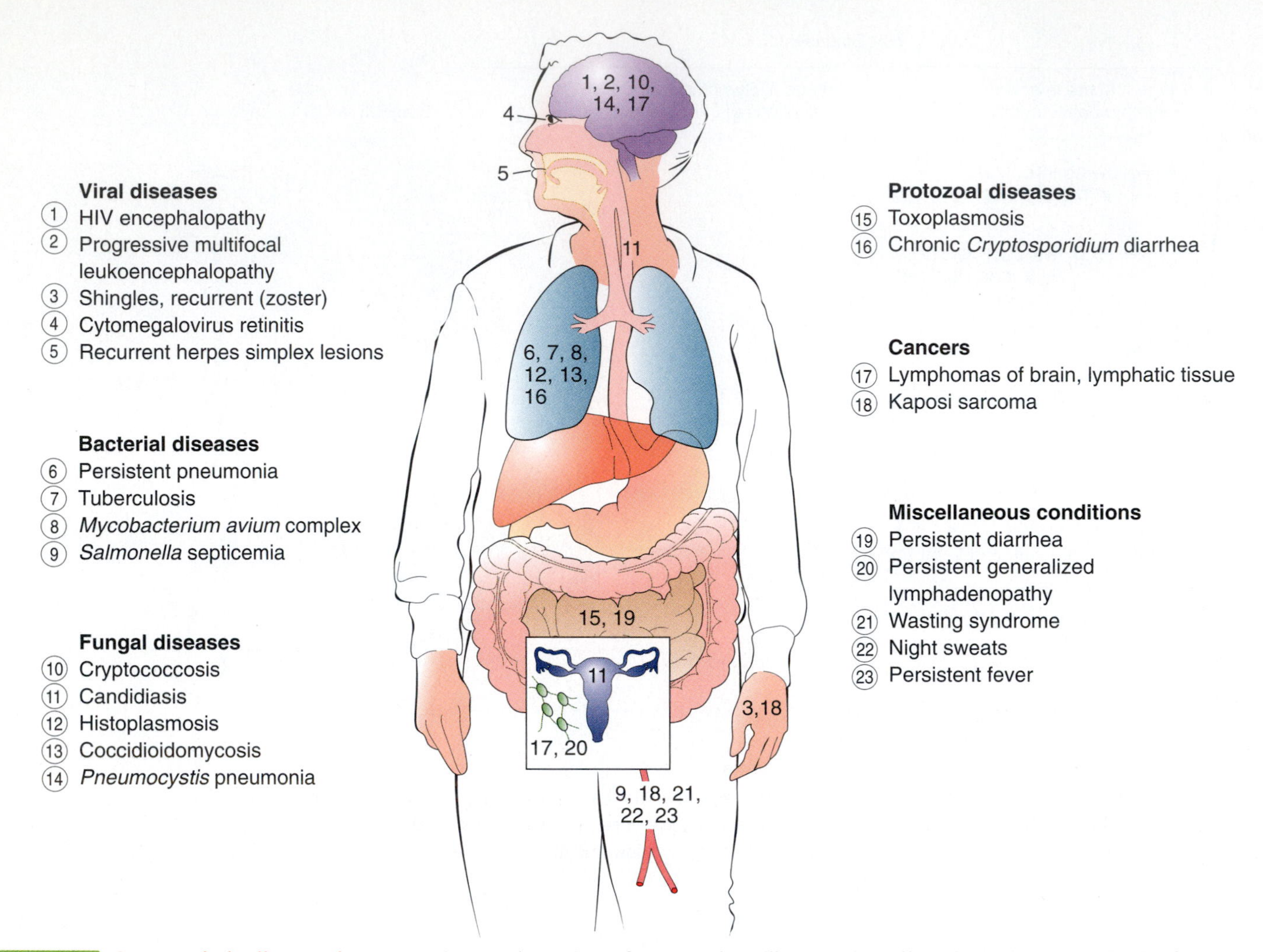

FIGURE 15.5 **Opportunistic Illnesses in AIDS Patients.** The variety of opportunistic illnesses that affect the body as a result of infection with HIV. Note the various systems that are affected and the numerous opportunistic pathogens involved.

Q: Why is there such a diverse group of potential pathogens that can cause opportunistic infections?

As the immune system deteriorates, the symptoms worsen and **opportunistic infections** develop. These infections can occur in many parts of the body, but common examples are listed in FIGURE 15.5.

Infected individuals do not have AIDS until they reach this most advanced stage of HIV disease. The Centers for Disease Control and Prevention (CDC) defines an HIV-infected individual as having AIDS if their $CD4^+$ T cell count is below 200 cells per cubic millimeter of blood. (Healthy adults usually have $CD4^+$ T cell counts of 1,000 or higher.)

Opportunistic infections now can be severe and eventually fatal because the immune system is so damaged by HIV that the body cannot fight off multiple pathogen infections. People with AIDS also are likely to develop various cancers, especially those caused by viruses, such as Kaposi sarcoma and cervical cancer, or cancers of the lymphatic system known as lymphomas.

Scientists have been studying those few individuals who were infected with HIV ten or more years ago and yet HIV is virtually undetectable. Scientists are trying to discover what factors allow these people to keep the viral load down. Is there something different about their immune systems; are they infected with a less aggressive HIV strain; or are their genes in some way protecting them from the effects of HIV? It is hoped that by understanding the body's natural method of control, ideas will surface for protective HIV vaccines to prevent the disease from progressing.

Transmission. HIV can be transmitted by "risky behaviors," such as sharing of blood-contaminated needles and having unprotected sexual contact, including vaginal intercourse or anal intercourse with an infected individual, or having unprotected intercourse with a person of unknown HIV status. Rectal tissues bleed and give access to the virus and through lesions, cuts, or abrasions of the vaginal tract. Importantly, an individual who has a sexually transmitted disease (STD), including gonorrhea, syphilis, chlamydia, or genital herpes, is more susceptible to getting HIV infection during sex with infected partners. The use of condoms has been shown to decrease the transmission of HIV significantly.

Packed red blood cells and blood factor concentrates may contain the virus, but extensive tests now are performed to preclude their presence. Health care workers are at risk of acquiring HIV during their professional activities, such as through an accidental needle stick. Individuals at high risk always should practice established infection-control procedures (standard precautions; Chapter 12) even though chances are rare that an accidental stick with a contaminated needle or other medical instrument would transmit HIV.

HIV can be transferred from an infected mother to her fetus (transplacental transfer) or baby at birth. An infected mother also can transmit HIV to her baby through breast milk.

Diagnosis. Because early HIV infection may be asymptomatic, many tests rely on the detection of antibodies in the blood to HIV. HIV antibodies are detected by the ELISA test (discussed in Chapter 21). In 2003, the FDA approved the OraQuick Rapid HIV-1 test for clinical and hospital use. It may soon be available commercially (MicroFocus 15.2). For HIV antibody tests, if a person is tested and the result is negative, it may mean the person was exposed to HIV within the past three months, which often is too early for the test to detect HIV antibodies. A repeat test at a later time is recommended. The Western blot analysis is used to confirm an antibody test result.

In 2006, the U.S. Food and Drug Administration (FDA) approved the APTIMA assay, which detects HIV RNA, as a diagnosis of primary HIV-1 infection. The **viral load test** detects the RNA of HIV and is available to monitor HIV-1 virus circulating in the blood of patients with established infections.

Treatment. The first drug used for treatment was azidothymidine, commonly known as AZT. AZT interferes with reverse transcriptase activity and acts as a chain terminator as it inhibits DNA synthesis. Other antiretroviral drugs also were discussed in Chapter 13. They included reverse transcriptase inhibitors and protease inhibitors. These drugs interfere with viral genome replication and the processing step of capsid production, respectively. Other antiretroviral drugs include the fusion inhibitors, which work by blocking receptor recognition to the surface of $CD4^+$ T cells. Such inhibition blocks viral entry into the CD4 cells.

HIV can become resistant to any of these drugs, so more effective treatment requires a combination (or "cocktail") of drugs. When three or more drugs are used together, the combination is referred to as **highly active antiretroviral therapy** (**HAART**). Although HAART is not a cure, it has been significant in reducing the risk of HIV transmission as well as the number of deaths. Overall, antiretroviral therapy has extended the life of a HIV/AIDS patient by about eight years. However, HIV can still hide out from drug attacks (MicroFocus 15.3).

Prevention. The only way to prevent HIV infection is to avoid behaviors putting a person at risk of infection, such as sharing needles or having unprotected sex.

A successful vaccine for HIV disease and AIDS has yet to be developed. Two types of vaccines are being considered. Preventive vaccines for HIV-negative individuals (health care providers) could be given to prevent infection. Therapeutic vaccines, for HIV-positive individuals, would help the individual's immune system control HIV either by blocking virus entry, provirus formation, or replication such that the disease could not progress or be transmitted to others (see Chapter 13).

There are two major reasons why a successful vaccine has been lacking. First, HIV continually mutates and recombines. Such behavior by the virus means that a vaccine has to protect individuals against a moving target. In addition, because HIV infects $CD4^+$ T cells,

■ Western blot:
A technique to identify protein constituents, such as from a virus.

MICROFOCUS 15.2: Tools

Checking for HIV Infection—At Home

If you thought you might be infected with HIV, would you be more likely to: (a) be too afraid to be tested; (b) go to a clinic or family physician to be tested; (c) go to the drug store and purchase an HIV testing kit that is sent to a lab for processing; or (d) go to your medicine cabinet and do the test right at home? The latter may soon be possible.

A quick HIV test, called the OraQuick Advance Rapid HIV-1/2 Antibody Test, currently is being used in clinics and hospitals with great success. It is 99.3% accurate in detecting antibodies to HIV, meaning the person tested has been exposed to HIV.

The test works as follows: The person uses a test stick (see figure) supplied in the kit to swab the upper and lower gum line. The oral fluid sample on the swab is placed in a special solution for 20 minutes. The indicator location on the device contains a strip of synthetic proteins that can detect HIV antibodies.

If a reddish band appears at the control (C) location, the test result is negative for HIV antibodies. If reddish bands appear at both the control (C) location and the test (T) location, the test is "reactive," meaning the result is a preliminary positive for HIV antibodies. Importantly, this is a "preliminary" result and does not definitely mean the person is HIV-positive. Rather, the individual would need to go to a clinic to have a more definitive test done to confirm the result. *The person is considered HIV-positive only if the confirmatory test result is positive.*

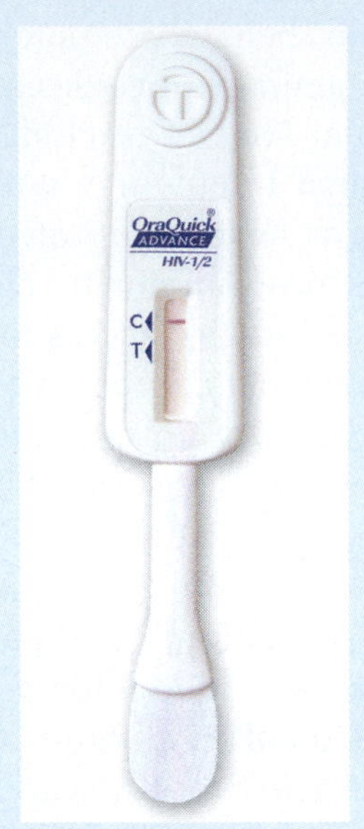

The OraQuick test stick.

As of early 2006, the OraQuick test was not available for purchase by the general public because of the controversy surrounding its use and a potential positive result. One side of the debate says people are very likely to become depressed knowing they are "possibly" infected with HIV. Such individuals may take drastic action, such as committing suicide.

The other side of the debate suggests the test may identify many of the 300,000 individuals in the United States who do not know they are infected. In fact, these individuals are responsible for 65 percent of all new HIV infections each year. In addition, individuals who know they are infected are 50 percent less likely to transmit HIV to another uninfected individual.

In late 2005, the CDC received reports from several cities experiencing clusters of false-positive test results using oral fluid samples. Further tests are ongoing.

the vaccine needs to activate the very cells infected by the virus. Although at times it appears that vaccine research is an uphill battle, scientists are optimistic that a safe, effective, affordable, and stable HIV vaccine can be produced. Importantly, progress in basic and clinical research is moving forward and scientists are inching closer to identifying products suitable for a successful HIV vaccine. More information about a possible AIDS vaccine is presented in Chapter 21.

CONCEPT AND REASONING CHECKS

15.1 Distinguish between the three stages of HIV disease and AIDS.

Two Herpesviruses Cause Blood Diseases

KEY CONCEPT

- Infectious mononucleosis and cytomegalovirus disease affect adolescents and produce congenital defects, respectively.

Although most members of the human herpesviruses primarily affect the skin (see Chapter 14), two species are associated with blood infections and disease.

Infectious mononucleosis is a blood disease, especially of antibody-producing B lymphocytes (a type of mononuclear white blood cell) in the lymph nodes and spleen. The

MICROFOCUS 15.3: Public Health

HIV: The Escape Artist

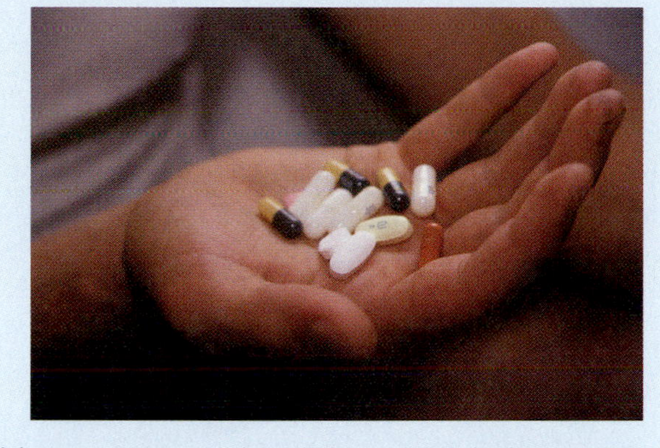

Even though the levels of HIV in an individual on HAART are extremely low, the person is not cured. The virus remains at low levels in locations where antiretroviral drugs appear unable to reach. These hiding places include: the T cells; the brain; semen; and, being a retrovirus, integrated into the DNA molecule itself.

Because of this ability of HIV to escape attack, some researchers, such as the HIV co-discoverer Robert Gallo, believe there never will be a cure for the disease. With HIV's ability to hide out in the body, patients would have to keep taking drugs for their entire life. If they stop, the virus comes back with a vengeance. Others believe it might be possible to seek out and destroy the viruses in their hideouts.

One place HIV hides is in memory T cells, a subgroup of T cells that spring into action in a subsequent infection. (This is why you only get diseases like the mumps once.) For some reason, HIV does not destroy these cells—and no drug currently available attacks them. Being integrated into the DNA as a provirus, these memory T cells represent a potent hideout. Researchers are working on ways to activate the cells and make them divide, so antiretroviral drugs will then destroy these cells.

Resting T cells also will not be destroyed by antiretroviral drugs because the drugs only are effective in actively dividing cells. In these cells, HIV replicates slowly but surely.

In addition, many immune cells have the ability to pump out toxins that have been imported. To these cells, antiretroviral drugs look like toxins and they are effectively pumped out of the cell.

Other sites also need to be attacked. One is the brain where HIV also takes refuge. Even individuals on HAART can develop HIV-associated dementia, a neurological problem due to HIV infection. Here, the blood-brain barrier prevents the entry of potentially effective drugs. HIV also hides out in the male semen, where immune cells appear to be inefficient at activating antiviral drugs.

If new drugs can be developed to seek out and destroy the HIV hideouts, future HAART medications may become much more effective. Still others, like Gallo, believe there are just too many hideouts and they all cannot be eliminated.

Time will tell.

name infectious mononucleosis (or "mono") is familiar to young adults because the disease is common in this age group. It is sometimes called the "kissing disease" because it is spread by contact with saliva. The disease strikes an estimated 100,000 people annually in the United States.

Infectious mononucleosis is caused by the Epstein-Barr virus (EBV) of the *Herpesviridae*. This is one of the most common human viruses; up to 95 percent of the American population between 35 and 40 years of age has been infected with EBV. Infants usually are susceptible to EBV as soon as their maternal antibodies (present at birth) disappear. Many children who become infected with EBV show no symptoms or the symptoms are indistinguishable from other typical childhood illnesses.

If a person is not infected as an infant or young child, an infection with EBV during adolescence or young adulthood runs a 35 to 50 percent chance of causing infectious mononucleosis. **EBV disease**, involving infection of B lymphocytes, is assumed to be a precursor to mononucleosis.

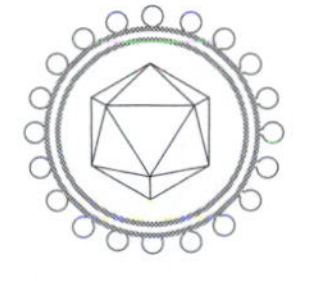

Epstein-Barr virus

EBV is spread person-to-person via saliva or by saliva-contaminated objects, such as table utensils and drinking glasses. Major symptoms are sore throat, enlargement of the lymph nodes ("swollen glands"), and fever, giving the disease another name—**glandular fever**. Mononucleosis usually runs its course in three to four weeks and is not highly contagious.

Among the most dangerous complications are defects of the heart, paralysis of the face, and rupture of the spleen. The liver may be involved and jaundice may occur, a condition some physicians refer to as hepatitis. Those who recover usually become carriers for several months and shed the viruses into their saliva.

■ Heterophile antibodies: Antibodies nonspecifically reacting with proteins or cells from unrelated animal species.

The diagnostic procedures for mononucleosis include detection of an elevated lymphocyte count and the observation of **Downey cells**, the damaged B cells with vacuolated and granulated cytoplasm. The **Monospot test** can be used to detect heterophile antibodies to EBV (FIGURE 15.6).

No vaccine is available for EBV infections and no drugs have proven effective in treating mononucleosis.

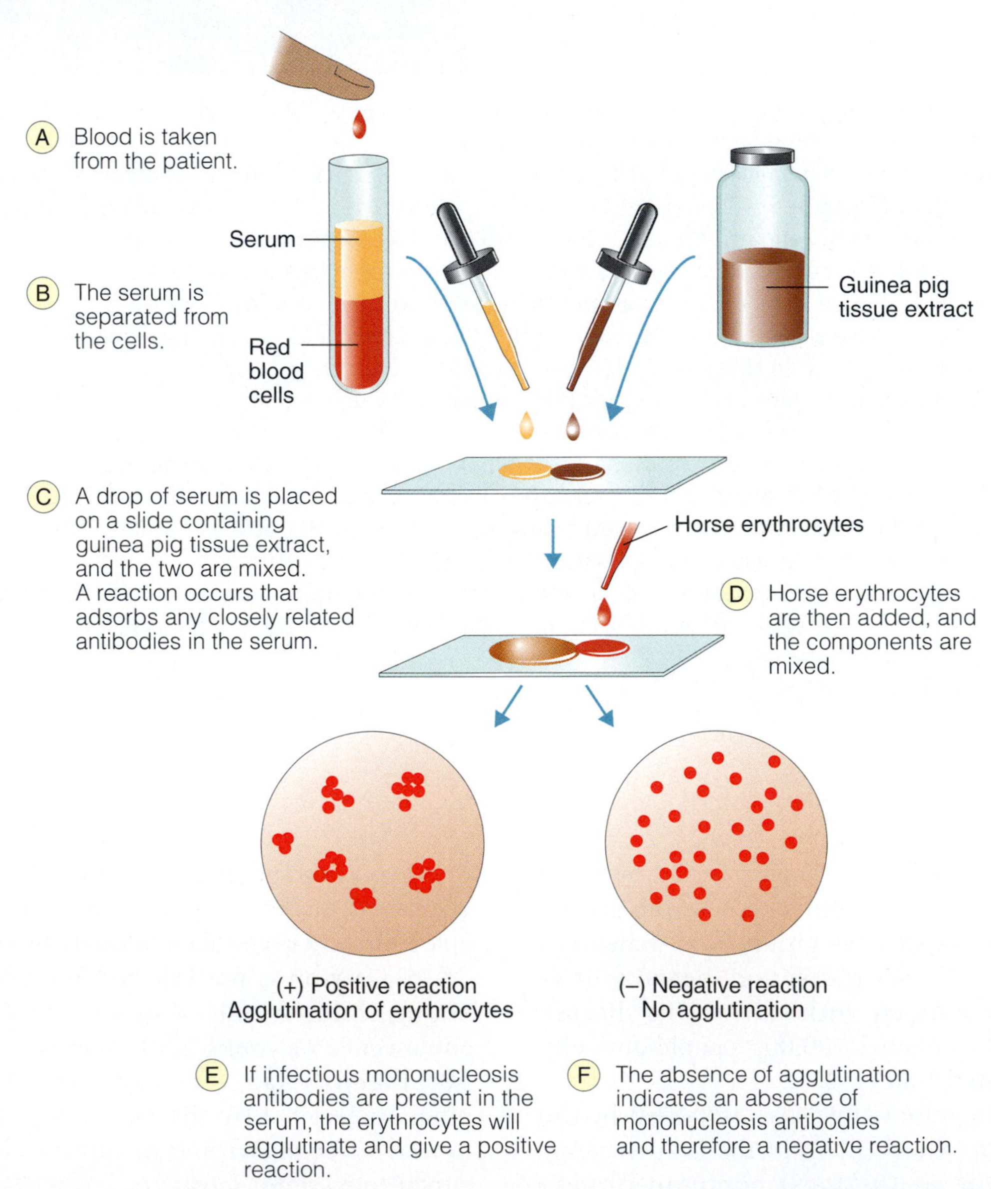

FIGURE 15.6 **The Monospot Slide Test for Infectious Mononucleosis.** About one week after the onset of infection by EBV, many patients develop heterophile antibodies. The antibodies peak at weeks two to five and may persist for several months to one year. The Monospot test is performed by first mixing samples of the patient's serum with a guinea pig tissue extract, which binds to and removes any closely related antibodies. Horse erythrocytes then are added and the red blood cells are observed for agglutination. Suitable controls (not shown) must also be included.

Q: What types of controls should be included?

EBV has been detected in patients who have **Burkitt lymphoma**, a tumor of the connective tissues of the jaw that is prevalent in areas of Africa. In fact, EBV was the first human virus associated with a malignancy (see Chapter 13). Contemporary virologists continue to search for reasons why EBV is associated with tumors on one continent and infectious mononucleosis on another. Some cancer specialists theorize the malaria parasite, common in central Africa, acts as an irritant of the lymph gland tissue, thereby stimulating tumor development. All Burkitt lymphoma cells have a chromosomal translocation that activates an oncogene (see Chapter 13). Therefore, EBV infection may stimulate or increase the frequency of the translocation.

EBV also has been associated with T-cell malignancies, including nasopharyngeal carcinoma. In immune-suppressed transplant patients, EBV can cause B cell lymphomas. In addition, there is growing evidence suggesting a possible link between EBV and **Hodgkin disease**, a lymphoma of the lymph nodes and spleen. Also, evidence is mounting for an association between EBV and an increased risk of developing **multiple sclerosis**, a muscle-weakening disease of the central nervous system.

Cytomegalovirus disease can produce serious birth defects. The cytomegalovirus (CMV) is the largest member of the *Herpesviridae*. The virus takes its name from the enlarged cells (*cyto* = "cell"; *megalo* = "large") found in infected tissues. Usually, these are cells of the salivary glands, epithelium, or liver. After the primary infection, CMV undergoes lifelong latency.

CMV disease in a healthy individual may be among the most common diseases in American communities. Infection may produce an infectious mononucleosis syndrome, involving fever and malaise. Most patients recover uneventfully.

However, if a CMV-infected woman is pregnant, a serious congenital disease may ensue in 5 to 10 percent of infants if the virus passes into the fetal bloodstream and damages the fetal tissues. Mental impairment is sometimes observed among the more than 17,000 congenital CMV-associated infected infants in the United States and Europe every year. The **C** in the **TORCH** group of diseases refers to congenital CMV-associated disease (see Chapter 13).

CMV infection and latency generates an immune response that keeps the virus in check. However, in immunocompromised individuals, CMV can reactivate. Prior to HAART, up to 25 percent of AIDS patients experienced CMV-induced retinitis. CMV also can accelerate the progression to AIDS and infect the lungs, liver, brain, and kidneys and cause death. Patients undergoing cancer therapy or receiving organ transplants may be susceptible to CMV disease. In the latter case, immunosuppressive drugs often are administered to "knock out" the immune system's ability to reject the transplant.

A vaccine for congenital CMV-associated infections is being researched. Several drugs, including acyclovir and valacyclovir provide effective treatment for transplant patients and as a pre-emptive therapy to CMV disease.

■ Retinitis: A serious infection of the retina that can lead to blindness.

■ Lymphomas: Malignant tumors originating in a lymph node.

CONCEPT AND REASONING CHECKS

15.2 Identify the serious complication that may result from (a) EBV and (b) CMV infections.

Several Hepatitis Viruses Are Bloodborne

KEY CONCEPT

- Hepatitis B and C can become chronic and cause liver disease.

Hepatitis B (formerly called serum hepatitis) is a global health problem, accounting for 1 million deaths every year. Two billion people, one third of the world's population, have been exposed to the hepatitis B virus (HBV) and some 350 million have chronic HBV infections. About 20 percent of these individuals are at risk of dying from HBV-related liver disease.

Hepatitis B virus

HBV is in the family *Hepadnaviridae* (*hepa* = "liver") and is the smallest known DNA virus. It has a partially double-stranded, circular DNA genome. Normal virions consist of a nucleocapsid surrounded by a **hepatitis B core antigen** (**HBcAg**) and envelope containing **hepatitis B surface antigen** (**HBsAg**).

Transmission of hepatitis B usually involves direct or indirect contact with an infected body fluid such as blood or semen. For example, transmission may occur by contact with blood-contaminated needles, such as hypodermic syringes or those used for tattooing,

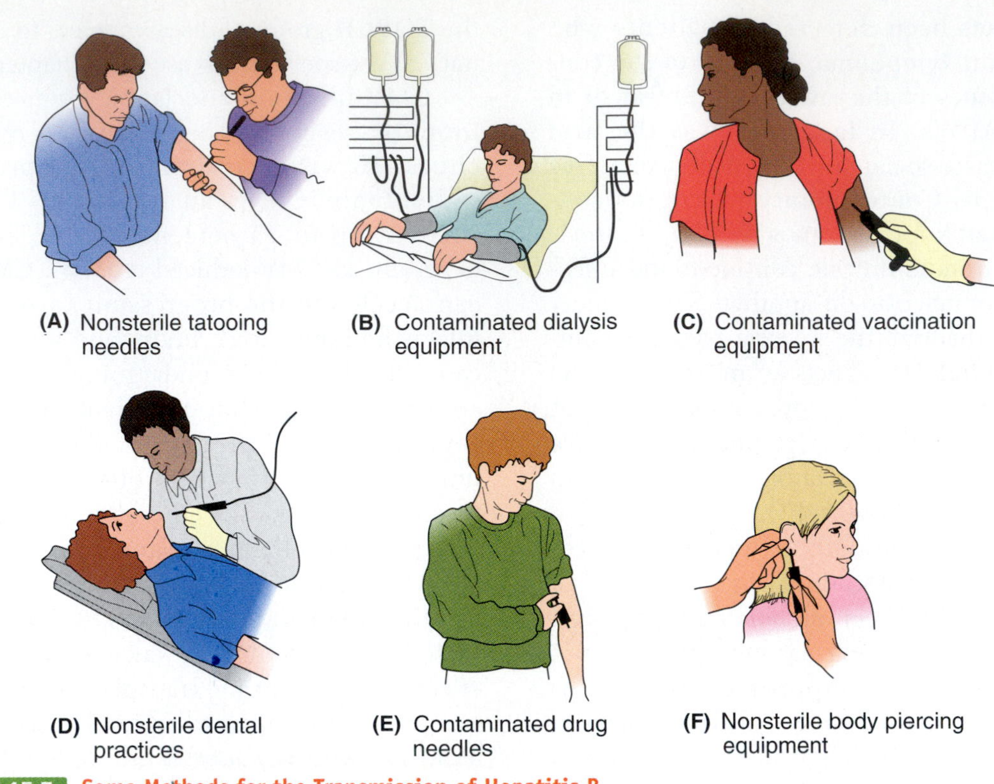

FIGURE 15.7 **Some Methods for the Transmission of Hepatitis B.**

Q: What is the common denominator in all these methods of transmission?

acupuncture, or ear piercing (FIGURE 15.7). Blood-contaminated objects such as endoscopes, saliva-contaminated (non-sterile) dental instruments, and renal dialysis tubing also are implicated.

■ **Endoscopes:** Instruments consisting of a fiber-like strand that are inserted through an incision for diagnostic or surgical procedures.

Hepatitis B is also an important sexually transmitted disease through vaginal, anal, or oral sex with an infected partner. HBV can enter through small tissue tears, allowing the virus to enter the bloodstream of the receptive partner from blood, saliva, semen, or vaginal secretions from the infected individual.

Hepatitis B has a long incubation period of four weeks to six months during which time HBV infects the liver but is not cytolytic. Among children and adults, the primary (acute) infection may be asymptomatic. Symptoms are more likely in adults and include fatigue, loss of appetite, nausea and vomiting, and dark urine. Patients experience jaundice weeks later. Abdominal pain and tenderness are felt in the upper-right quadrant of the abdomen (the liver is located on the upper right side of the abdomen). Recovery from an acute infection usually occurs about three to four months after the onset of jaundice. The virus is cleared from the blood and liver, and the individuals develop immunity to reinfection.

■ **Cytolytic:** Refers to the destruction (lysis) of cells.

■ **Jaundice:** A condition in which bile pigments seep into the circulatory system, causing the skin and whites of the eyes to have a dull yellow color.

About 5 percent of patients develop persistent infections and may or may not have symptoms. In rare cases, **cirrhosis**, an extensive hepatocellular injury, can occur due to immune system reactions to the infection. In addition, chronically-infected carriers run a 100-times higher chance of developing liver cancer or **hepatocellular carcinoma** (HCC) than non-carriers. The cellular and molecular reasons for carcinoma development are not completely understood.

The risk of developing a persistent, chronic state decreases with age. Newborns have a 90 percent risk, and 25 percent of infants and young children with persistent infections will eventually die from cirrhosis or HCC. The mortality rate drops to 15 percent if a persistent infection develops as adolescents or young adults.

Hepatitis B can be prevented by immunization with any of several hepatitis B vaccines. These vaccines consist of HBsAg produced by genetically-engineered yeast cells (see Chapter 8). Recommended for all age groups

MICROFOCUS 15.4: History

What's Worse—the Disease or Medical Intervention?

Egypt has a population of 62 million and contains the highest prevalence of hepatitis C in the world. The Egyptian Ministry of Health estimates a national prevalence rate of at least 12 percent (or 7.2 million people). Chronic hepatitis C is the main cause of liver cirrhosis and liver cancer in Egypt, and indeed, one of the top five leading causes of death. The highest concentration of the hepatitis C virus (HCV) appears in farming people living in the Nile delta and rural areas. So, why does Egypt have such a high prevalence rate of HCV?

Several recent studies suggest that HCV was transmitted through the contamination of reusable needles and syringes used in the treatment of schistosomiasis, a disease caused by a blood parasite (Chapter 17). Schistosomiasis is a common parasitic disease in Egypt and can cause urinary or liver damage over many years. Farmers and rural populations are at greatest risk of acquiring the disease through swimming or wading in contaminated irrigation channels or standing water.

Prior to 1984, the treatment for schistosomiasis was intravenous tartar emetic. Between the 1950s and the 1980s, hundreds of thousands of Egyptians received this standard treatment, called parenteral antischistosomal therapy (PAT). Today, drugs for schistosomiasis are administered in pill form.

Evidence suggests inadequately sterilized needles used in the PAT campaign contributed to the transmission of HCV and set the stage for the world's largest transmission of blood-borne pathogens resulting from medical intervention. The PAT treatment campaign was conducted with the best of intentions, using accepted sterilization techniques of the time. However, much of the PAT campaign was carried out before disposable syringes and needles were available. In addition, no one was aware of HCV prior to the 1980s or the dangers associated with blood exposure.

Further evidence for the correlation between the PAT campaign and hepatitis C was the drop in the hepatitis C rate when PAT injections were replaced with oral medications. Sadly, in part because of the high number of people who were infected, the risk of transmission remains high in the Egyptian population today.

■ Emetic: A chemical substance that induces vomiting.

(including infants), they are particularly valuable for health care workers who might be exposed to blood from patients. For infant use, it is combined with the Hib vaccine as Comvax. As a result of child and adolescent vaccinations, there has been a 68 percent decrease in reported cases of hepatitis B in the United States (6,212 cases in 2004) since 1990.

Injections of interferon alfa (Intron A) can influence the course of hepatitis B. Because the side-effects are not tolerated well by many patients, lamivudine (Epivir), adefovir (Hepsera), and entecavir (Baraclude) are base analogs offering effective therapy. Moreover, injections of hepatitis B immune globulin, which consists of antibodies concentrated from the serum of blood donors, can be used for persons without known immunity who have come in contact with HBV.

Hepatitis C causes liver disease in 123 million people worldwide, with the highest proportion in Asia and Africa. Almost 4 million Americans are infected with the hepatitis C virus and 2.7 million of those have chronic infections. Still, new infections have declined since 1980 to 720 in 2004.

The hepatitis C virus (HCV) is an enveloped, single-stranded (+ strand) RNA virus of the *Flaviviridae* family and is primarily transmitted by blood, injection drug use, or blood transfusions (in countries without a blood screening program). However, it also has been spread inadvertently through medical intervention, as MicroFocus 15.4 clearly demonstrates.

There are few symptoms associated with the primary infection and 20 to 50 percent of patients fully recover, although no permanent immunity is generated. Most cases (50 to 80 percent) develop a symptomless, insidious chronic infection, which takes the lives of almost 10,000 Americans who die from cirrhosis, HCC, or other complications every year. Indeed, damage from hepatitis C is the primary reason for liver transplants in the United States. Alcoholism and intravenous

TABLE
15.1 Viral Diseases Associated with the Blood and Lymphatic Systems

Disease	Classification of Virus	Transmission	Organs Affected	Vaccine	Special Features	Complications
Acquired immuno-deficiency syndrome	*Retroviridae*	Contact with body fluids and blood products	Blood Lymph nodes Brain	Not available	Immune deficiency	Opportunistic illnesses
Infectious mononucleosis	*Herpesviridae*	Saliva Contact Droplets	Blood Lymph nodes Spleen	Not available	Downey cells Monospot slide test	Splenic rupture Jaundice
Cytomegalo-virus disease	*Herpesviridae*	Contact Congenital transfer	Blood Lung	Not available	Enlarged cells	Fetal damage
Hepatitis B	*Hepadnaviridae*	Contact with body fluids	Liver	Synthetic proteins	Jaundice	Liver cancer
Hepatitis C, G	*Flaviviridae*	Contact with body fluids	Liver	Not available	Jaundice	Liver damage

drug use are among the major cofactors accelerating the incidence of chronic hepatitis C.

Interferon alfa and ribavirin, standard therapies for the disease, cures less than 50 percent of those infected. No vaccine is available.

Hepatitis D is caused by two viruses: HBV and the hepatitis D virus (HDV). The latter virus consists of a protein fragment called the delta antigen and a segment of RNA, and can only cause liver damage when HBV is present. HDV requires the outside coat of HBV to infect cells. Therefore, one cannot become infected with hepatitis D unless he or she already is infected with HBV. Chronic liver disease with cirrhosis is two to six times more likely in a co-infection.

Hepatitis G also is a chronic liver illness. The hepatitis G virus (HGV) is an enveloped, single-stranded (+ strand) RNA virus of the *Flaviviridae*. Like HBV and HCV, HGV is transmitted by blood, blood products, or sexual intercourse. It appears to cause persistent infections in 15 to 30 percent of infected adults.

TABLE 15.1 summarizes the viral diseases associated with the blood and the lymphatic system.

CONCEPT AND REASONING CHECKS

15.3 Summarize the similarities in symptoms between hepatitis B and C.

15.2 Viral Diseases Causing Hemorrhagic Fevers

In March 2005, the Angola Ministry of Health and the World Health Organization (WHO) reported 63 hemorrhagic deaths (mostly children along with three health care workers) at the Uige Provincial Hospital. By late July, there were a total of 374 cases of which 329 died from an outbreak of Marburg hemorrhagic fever. This is just one of several illnesses called **viral hemorrhagic fevers (VHFs)** caused by four families of RNA viruses. The illnesses, characterized by vascular system damage, occur sporadically and are rare in the United States.

Yellow fever virus

Flaviviruses Can Cause a Terrifying and Severe Illness

KEY CONCEPT

- Yellow fever and dengue fever can produce fever, bleeding, and/or circulatory failure.

The *Flaviviridae* are enveloped, icosahedral, (+ strand) RNA virions. They also are referred to as **arboviruses** because they are *ar*thropod-*bo*rne viruses. The two most virulent are the yellow fever and dengue fever viruses.

Yellow fever was the first human disease associated with a virus. As a result of the slave trade from Africa, the disease spread rapidly in large regions of the Caribbean and tropical Americas, and was common in the southern and eastern United States for many generations in the 18th and 19th centuries. In 1901, a group led by the American Walter Reed identified mosquitoes as the agents of transmission. With widespread mosquito control, the incidence of the disease gradually declined. Today, yellow fever is endemic in 33 countries in Africa and South America, and causes over 200,000 cases and 30,000 deaths annually.

Jungle yellow fever occurs in monkeys and other jungle animals, where the virus is taken up by blood-feeding mosquitoes. Epidemics occur when a person is bitten by an infected mosquito in the forest and then travels to urban areas where the virus is passed via a blood meal to another mosquito species, *Stegomyia* (formerly *Aedes*) *aegypti*. This species then transmits the virus, now causing an urban yellow fever among humans.

S. aegypti injects the virus into the human bloodstream, causing abrupt symptoms of headache, fever, and muscle pain lasting three to five days. Then, for 2 to 24 hours, the symptoms abate as the virus is cleared by antibodies and the immune system. Many patients recover at this stage.

In 15 to 25 percent of patients, the illness reappears in a much more terrifying form, causing severe nausea, uncontrollable hiccups, and a violent, black vomit. Liver damage produces jaundice (the disease often is called "yellow jack") and major hemorrhaging from the gums, mouth, and nose occur as the patient becomes delirious. Up to 50 percent of patients go into a coma and die from internal bleeding.

As a zoonosis, yellow fever cannot be eradicated. It can be prevented by immunization with either of two vaccines. Except for supportive therapy, no treatment exists for yellow fever.

Dengue fever takes its name from the Swahili word *dinga,* meaning "cramp-like attack," a reference to the symptoms. It is the most important arboviral disease of humans, as there are 50 million new infections and over 12,000 deaths annually. Dengue fever traditionally was confined to Southeast Asia. However, in 1963, the disease broke out in Central America and continues to occur sporadically in the Americas.

There are four types of dengue fever virus (FIGURE 15.8A). Transmission is by the *S. aegypti* mosquito and the tiger mosquito *S. albopicta*. While taking a blood meal, infected mosquitoes inject viruses into the blood where they multiply in white blood cells and platelets.

High fever and prostration are early signs of infection. These are followed by sharp pain in the muscles and joints. Patients often report sensations feeling like their bones are breaking; thus, the disease has been called **breakbone fever**. After about a week, the symptoms fade.

Serious complications or death are uncommon unless one of the other three types of dengue virus later enters the body of a recovered patient. Then, a condition called **dengue hemorrhagic fever** may occur. In this condition, the immune system reacts to the memory of the first dengue infection, allowing the new one to replicate unchallenged. A rash from skin hemorrhages appears on the face and extremities, and severe vomiting and shock ensue (**dengue shock syndrome**) as the blood pressure decreases dramatically.

Several vaccines are in clinical trials. No successful antiviral therapy has been identified.

CONCEPT AND REASONING CHECKS

15.4 Compare the similarities and differences between yellow fever and dengue fever.

■ **Endemic:** Referring to a disease occurring within a specific area, region, or locale.

■ **Prostration:** Drained of strength.

Bunyaviruses Are Spread by Infected Animals

KEY CONCEPT

- Hantavirus disease is associated with high mortality.

Several human hemorrhagic fevers are caused by viruses in the family *Bunyaviridae,* which are enveloped, single-stranded (– strand) RNA viruses with helical symmetry. The genome consists of three segments.

In the summer of 1993, a brief epidemic occurred among residents of the southwestern United States. By the end of 1993, 91 cases were confirmed in 20 states and 48 patients died. The disease was given the technical name **hantavirus pulmonary syndrome** (**HPS**). Between 1993 and 2005, there have

■ **Zoonosis:** A disease transmitted from animals to humans.

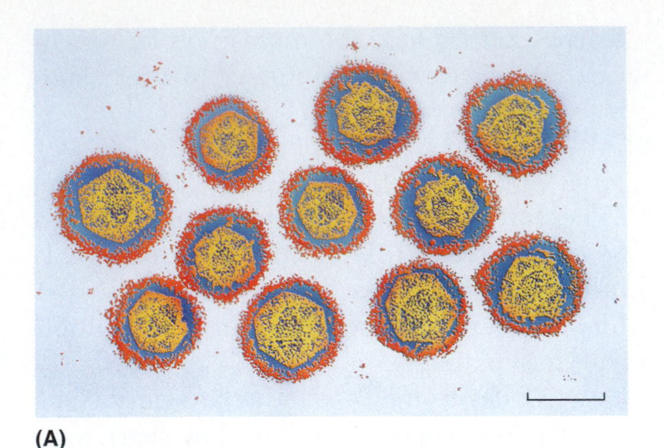
(A)

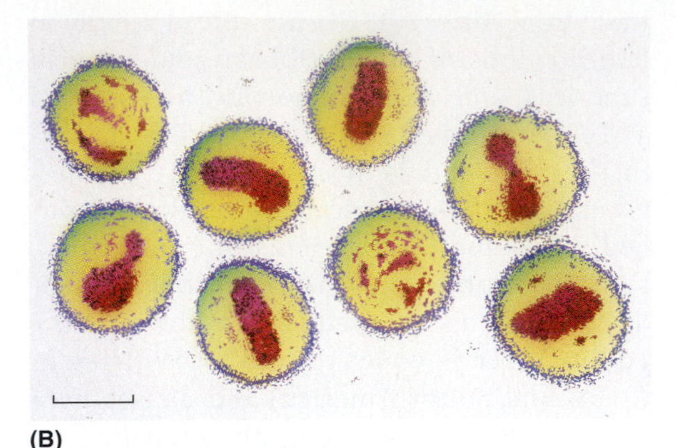
(B)

FIGURE 15.8 **The Dengue Fever Virus and Hantavirus.** False-color transmission electron micrographs of (**A**) dengue fever viruses (Bar = 50 nm) and (**B**) hantaviruses. Both are enveloped viruses that are members of the *Flaviviridae* and *Bunyaviridae* families, respectively. (Bar = 90 nm.)

Q: What do the yellow geometric shapes represent in (A) and the red-colored bars in (B)?

been 396 cases of HPS of which 146 have died in the United States.

Hantavirus disease is caused by a strain of the hantavirus that was named the Sin Nombre virus (Spanish for "no name") (FIGURE 15.8B). The deer mouse is the host for the Sin Nombre virus, and it sheds the virus in saliva, urine, and feces. Humans usually are infected by breathing the infectious aerosolized dried urine or feces.

In one to five weeks after exposure, early symptoms of infection include fatigue, fever, and muscle aches. About half of all HPS patients experience headaches, dizziness, difficulty breathing, and low blood pressure that can lead to respiratory failure as the lungs fill with fluid.

Prevention consists of eliminating rodent nests and minimizing contact with them. There is no vaccine for hantavirus infection.

HPS has now become an established disease in the lexicon of medicine not only in the United States but throughout much of the Americas (FIGURE 15.9). There have been 1,910 cases reported through 2004.

Sandfly fever is transmitted by sandflies of the genus *Phlebotomus*. It occasionally breaks out in Mediterranean regions, Southeast Asia, and parts of Central America. Patients suffer recurrent high fever and joint and bone pains resembling those in dengue fever.

Rift Valley fever is named for an immense earthquake-prone region in eastern Africa called the Rift Valley. In addition to affecting humans, the disease affects animals and causes extensive losses of sheep and cattle. Transmission is by several genera of mosquitoes, and dengue-like pain in the bones and joints accompanies the fever, which lasts for about a week.

CONCEPT AND REASONING CHECKS

15.5 Identify the serious complication that may result from hantavirus infection.

Members of the *Filoviridae* Produce Severe Hemorrhagic Lesions of the Tissues

KEY CONCEPT

- Ebola and Marburg hemorrhagic fevers are among the most deadly.

Marburg and Ebola hemorrhagic fevers are severe illnesses caused by viruses in the *Filoviridae* (*filo* = "thread") family. They consist of long thread-like, single-stranded (– strand) RNA viruses, (FIGURE 15.10).

Ebola hemorrhagic fever (EHF) captured headlines in 1976 and 1979 when the first outbreaks occurred in Sudan and Zaire. When it was over, 88 percent (280) infected people died. Through 2004, there have been 17 confirmed outbreaks of EHF in Africa. Over 1,800 cases have been reported. Mortality ranges from 50 to 90 percent.

EHF is caused by infection with the ebolavirus (Ebola is the river in the Democratic Republic of the Congo (formerly Zaire) where the illness was first recognized). Recent evidence suggests the virus is zoonotic and is

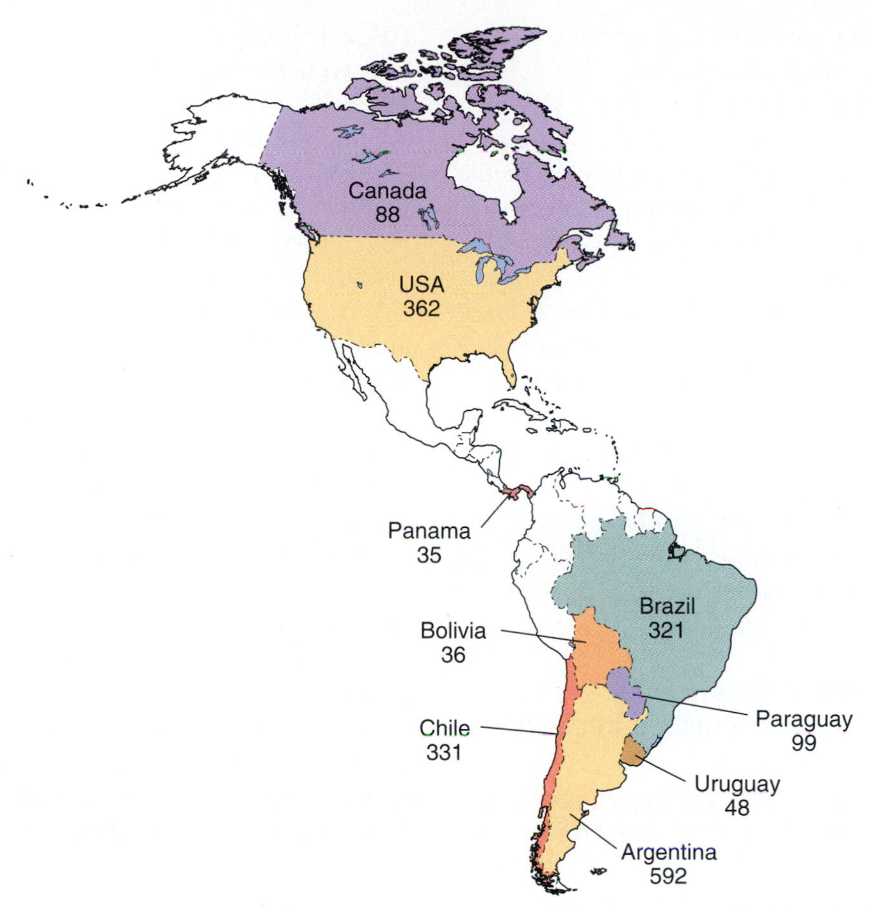

FIGURE 15.9 **Hantavirus Pulmonary Syndrome Cases, the Americas 1993–2004.** Cases of HPS have been reported by the Pan American Health Organization in North, Central, and South America. The number of reported cases are indicated. *Source:* Data from the Pan American Health Organization (PAHO), http://www.paho.org. Accessed October 21, 2006.

Q: Provide a hypothesis why the majority of HPS cases are in more temperate regions (United States and Argentina) of the Americas.

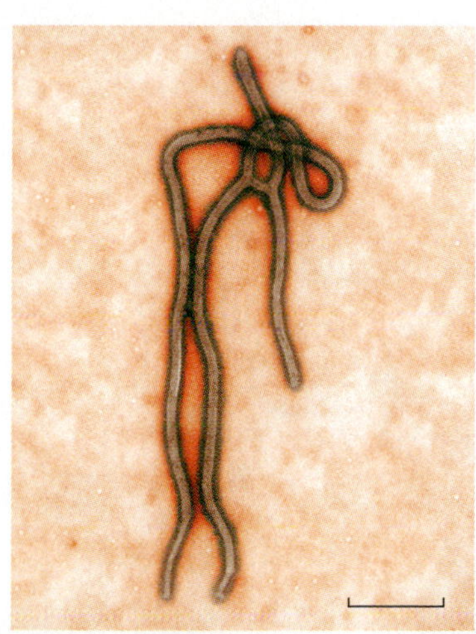

FIGURE 15.10 **The Ebolavirus.** False-color transmission electron micrograph of an ebolavirus. It is one of the group of filoviruses, so-called for its thin and long shape. Here the viral filament is seen looping back on itself. (Bar = 140 nm.)

Q: Why are these viruses placed in the Filoviridae *family?*

normally maintained in an animal host native to Africa. Fruit bats have been identified as a possible reservoir.

Uninfected individuals, including health care workers, can be exposed to the virus by direct contact with the blood and/or secretions of an infected person or through contact with contaminated objects, such as needles. Following an incubation period of a few days to two weeks, symptoms include fever, headache, joint and muscle aches, sore throat, and weakness. This is followed by diarrhea, vomiting, and stomach pain. A rash, red eyes, hiccups and internal and external bleeding may be seen in some patients.

The ebolavirus damages endothelial cells, causing massive internal bleeding and hemorrhaging. Infection also limits the immune response. Thus, the patient bleeds to death internally before a reasonable immunologic defense can be mounted.

■ **Reservoir:** The natural host or habitat of a pathogen.

■ **Endothelial:** Refers to the layer of cells liniing body cavities, such as the veins and arteries.

There are no guaranteed prevention measures and a vaccine is being tested.

Marburg hemorrhagic fever (MHF), named for Marburg, West Germany, where an outbreak occurred in 1967, was first identified in tissues of green monkeys imported from Africa. How transmission of the marburgvirus occurs is unclear.

After an incubation period of five to ten days, MHF develops symptoms similar to EHF but usually has a lower fatality rate. In 2005, the biggest outbreak of MHF occurred in Angola. The virus infected 374 people and killed 329.

Again, prevention measures have not been established and a vaccine is being tested.

CONCEPT AND REASONING CHECKS

15.6 Why is EHF such a deadly disease?

Members of the *Arenaviridae* Are Associated with Chronic Infections in Rodents

KEY CONCEPT

- Lassa fever can lead to hemorrhage and shock.

Lassa fever is so named because it was first reported in the town of Lassa, Nigeria, in 1969. The disease is caused by a zoonotic, single-stranded (– strand) RNA virus of the *Arenaviridae* family that has sandy-looking granules in the virion (*arena* = "sand"). Lassa fever is responsible for about 5,000 deaths per year in West Africa.

Transmission of the Lassa fever virus is shown in FIGURE 15.11. Infection leads to severe fever, exhaustion, and patchy blood-filled hemorrhagic lesions of the throat. The fever persists for weeks and profuse internal hemorrhaging is common. Overall, only about 1% of infections result in death.

Prevention involves avoiding contact with rodents and keeping homes clean. Ribavirin is effective in treating the disease.

Other viral hemorrhagic fevers, all caused by *Bunyaviridae* or *Arenaviridae* are Congo-Crimea hemorrhagic fever, which occurs worldwide; Oropouche fever, which affects regions of Brazil; and Junin and Machupo, the hemorrhagic fevers of Argentina and Bolivia,

FIGURE 15.11 Transmission of Lassa Fever Virus. Transmission of Lassa fever is primarily through aerosol or direct contact with excreta from infected rodents, or from contaminated food.

Q: Why might Lassa fever-infected rodents be a health hazard around homes or villages?

respectively. An arenavirus called the Sabia virus has caused hemorrhagic illnesses in Brazil, while the Guanarito virus is associated with Venezuelan hemorrhagic fever.

In the 1960s and 1970s, public health officials believed they had triumphed over infectious disease. Smallpox was on the way to extinction; polio was under control; and—thanks to antibiotics, sanitation, and pesticides—such maladies as tuberculosis, cholera, and malaria were disappearing. But then in one wave after another, nature counterattacked with Marburg disease, Lassa fever, Ebola fever, and hantavirus disease as well as Legionnaires' disease, hepatitis C, and AIDS. In all, at least 30 newly identified pathogens emerged and taught science a valuable lesson for the new century: don't ever let your guard down.

TABLE 15.2 summarizes the diseases caused by the hemorrhagic fever viruses.

CONCEPT AND REASONING CHECKS

15.7 How are humans most likely infected by the Lassa fever virus?

15.3 Viral Infections of the Gastrointestinal Tract

Several human viruses are responsible for illnesses and diseases of the digestive (gastrointestinal) system. These illnesses include hepatitis and viral gastroenteritis.

Hepatitis Viruses A and E Are Transmitted by the Gastrointestinal Tract

KEY CONCEPT

- Hepatitis A and E are spread by the fecal-oral route.

Hepatitis A is an acute inflammatory disease of the liver most commonly transmitted by food or water contaminated by the feces of an infected individual. Approximately 1.5 million cases of hepatitis A occur each year worldwide and it remains one of the most frequently reported vaccine-preventable diseases in the United States. In 2004, there were 5,683 cases reported to the CDC.

Hepatitis A is caused by a small, naked, icosahedral, single-stranded (+ strand), RNA virion belonging to the *Picornaviridae* family (FIGURE 15.12). The hepatitis A virus (HAV) is very resistant to chemical and physical agents, and several minutes of exposure to boiling water may be necessary to inactivate the virion.

Hepatitis A virus

Transmission of hepatitis A, sometimes referred to as infectious hepatitis, often involves an infected food handler and outbreaks have been traced also to day-care centers where workers contacted contaminated feces. Interestingly, children often serve as the principal reservoir because their infections

TABLE 15.2 Viral Diseases Causing Hemorrhagic Fevers

Disease	Classification of Virus	Transmission	Organs Affected	Vaccine	Special Features	Complications
Yellow fever	*Flaviviridae*	*Stegomyia aegypti* (mosquito)	Liver Blood	Inactivated viruses	Jaundice	Hemorrhaging
Dengue fever	*Flaviviridae*	*Stegomyia aegypti* (mosquito)	Blood Muscles	Not available	Breakbone fever	Hemorrhaging
Hantavirus disease	*Bungaviridae*	Contact Animals	Lungs	Not available	Fever Breathing difficulties	Respiratory failure
Ebola/Marburg Hemmorrhagic fever	*Filoviridae*	Contact with body fluids	Blood Multisystem dysfunction	Not available	Fever	Hemorrhaging
Lassa fever	*Arenaviridae*	Contact with animal fluids	Blood	Not available	Fever	Hemorrhaging

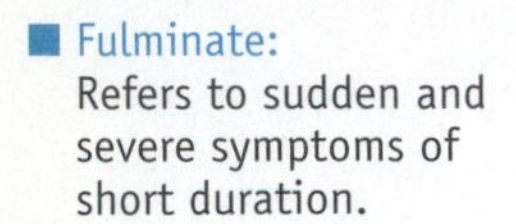

■ **Fulminate:** Refers to sudden and severe symptoms of short duration.

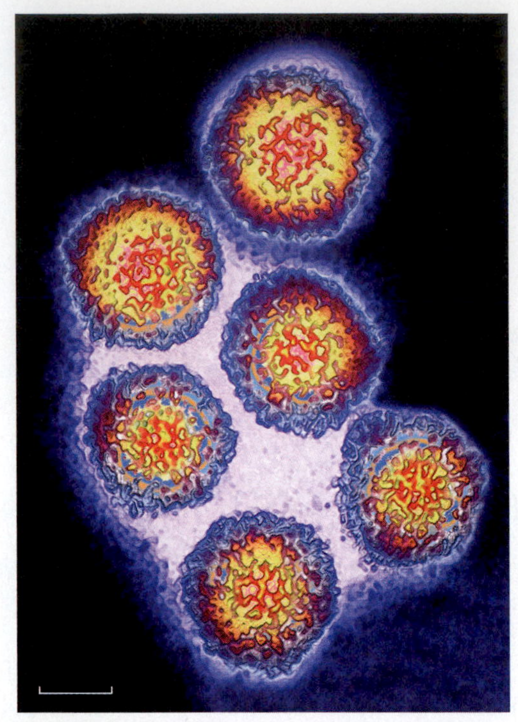

FIGURE 15.12 **Hepatitis A Viruses.** A false-color transmission electron micrograph of hepatitis A viruses. (Bar = 15 nm.)

Q: In what organ would one most likely find these viruses replicating?

usually are asymptomatic. In addition, the disease may be transmitted by raw shellfish such as clams and oysters, since these animals filter and concentrate the viruses from contaminated seawater.

The incubation period is usually between two and four weeks. Because the primary site of replication is the gastrointestinal tract, initial symptoms include anorexia, nausea, vomiting, and low-grade fever. The virus then is transported to the liver, its major site of replication. Discomfort in the upper-right quadrant of the abdomen follows as the liver enlarges. Considerable jaundice usually follows the onset of symptoms by one or two weeks (the urine darkens, as well), but many cases are without jaundice. Passage into the bile occurs and then new virions are shed through the intestines and into the feces.

The symptoms may last for several weeks, and relapses are common. A long period of convalescence generally is required, during which alcohol and other liver irritants are excluded from the diet. Recovery brings lifelong immunity. However, about 100 Americans die each year from fulminate hepatitis A infections.

Diagnostic procedures for hepatitis A are based on liver function tests, observation of characteristic symptoms, and the demonstration of hepatitis A antibodies in the serum. The largest ever outbreak of hepatitis A in the United States occurred in late 2003 (Textbook Case 15).

Maintaining high standards of personal and environmental hygiene, and removing the source of contamination are essential to preventing the spread of hepatitis A. Two vaccines, known commercially as Havrix and Vaqta, are available in the United States for people over two years of age. Also available for individuals over 17 years of age is Twinex, a combination vaccine for hepatitis A and B.

There is no treatment for hepatitis A except for prolonged rest and relieving symptoms. In those exposed to HAV, it is possible to prevent development of the disease by administering hepatitis A immune globulin within two weeks of infection. This preparation consists of antiviral antibodies obtained from blood donors (Chapter 21). Blood is routinely screened for hepatitis antibodies.

Hepatitis E is an opportunistic, emergent disease caused by a naked, single-stranded (+ strand) RNA virus of the *Caliciviridae* family. It shares many clinical characteristics and symptoms with hepatitis A. The hepatitis E virus is transmitted to humans via fecal-contaminated drinking water. Evidence also suggests it may be a zoonosis.

The disease affects primarily young adults and pregnant women seem to be particularly susceptible to illness, with mortality being as high as 30 percent. No evidence of chronic infection has been noted.

The impact of hepatitis E can be appreciated from two 2004 outbreaks coming from contaminated water. One outbreak was in the Greater Darfur region of Sudan, where almost 7,000 cases and 87 deaths were reported; the other was in neighboring Chad, where there have been some 1,500 cases and 46 deaths in refugee camps.

Textbook CASE 15

Hepatitis A Outbreak—Monaca, PA, 2003

1. On November 5, 2003, an outbreak of hepatitis A was confirmed through lab analysis reported by the Pennsylvania Health Department. At that time, there were 34 cases confirmed at a mall restaurant with 10 customers and 12 restaurant employees reporting symptoms of hepatitis A infection.
2. By Friday, November 7, 130 people had contracted hepatitis A (see figure) and the health department provided injections of immunoglobulin as a precaution for anyone who had eaten at the restaurant between October 22 and November 2.
3. At this time, state officials and arriving CDC investigators suspected the virus was spread by an infected worker who failed to wash his or her hands before handling food.
4. On November 8, the first fatality from the hepatitis A outbreak was reported. The person died from liver failure. The outbreak had risen to 240 confirmed cases. Officials still believed the problem centered on an infected food worker.
5. On November 12, health officials announced the confirmed case count had risen to 340. Mention also was made of a recent multi-state (Tennessee, Georgia, and North Carolina) outbreak of hepatitis A in late September and early October. These 250 cases resulted from eating contaminated green onions at a few local restaurants.
6. By November 13, 410 illnesses were reported. Transmission from an infected food worker was ruled out as all employees became ill after the outbreak started. Interviews of restaurant patrons began in order to identify what and how much they ate at the mall restaurant.
7. By November 15, two more people had died from the hepatitis A outbreak and more than 500 illnesses had been reported.
8. Patron interviews and further menu item investigations pointed to Mexican salsa containing green and white onions as the prime source of illness.
9. Ninety-eight percent of patrons who became ill reported eating salsa containing raw green onions. Illness was not associated with eating salsa containing raw white onions.
10. In all, more than 550 people had been stricken during the hepatitis A outbreak. Genetic analysis of the virus implicated raw green onions imported from three firms in Mexico.

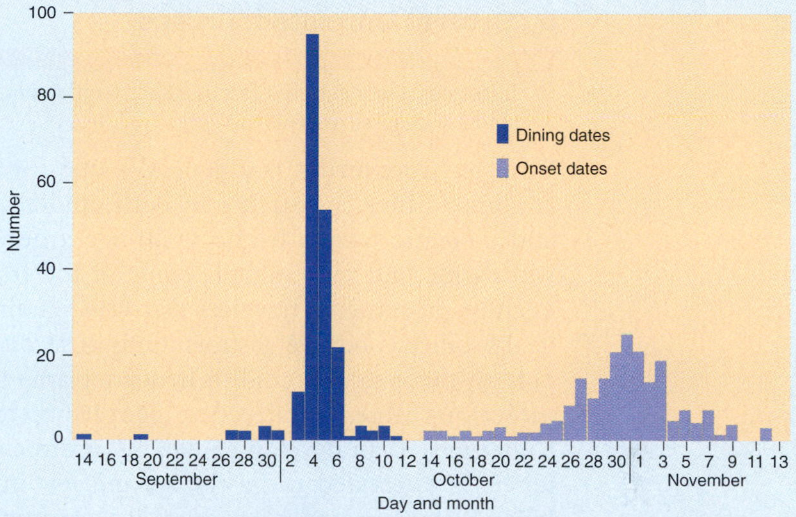

Number of hepatitis A cases by date of eating at resturant A and illness onset.

Questions

A. *Why would the infections of restaurant employees be significant?*

B. *Explain why immunoglobulin injections were recommended.*

C. *How common are deaths from hepatitis A?*

D. *Why was it important to discover what and how much food the affected restaurant customers ate?*

E. *Provide some ways that the green onions may have initially become contaminated with the hepatitis A virus.*

For additional information see http://www.cdc.gov/mmwr/preview/mmwrhtml/mm5247a5.htm.

CONCEPT AND REASONING CHECKS

15.8 Compare and contrast the three major types of hepatitis; that is, A, B, and C.

Viral Gastroenteritis Is Caused by Several Unrelated Viruses

KEY CONCEPT

- Infectious gastroenteritis can cause substantial morbidity in children and adults.

Viral gastroenteritis is a general name for a common illness occurring in both epidemic and endemic forms. It affects all age groups worldwide and may include some of the frequently encountered traveler's diarrheas. Public health officials believe gastroenteritis is second only to the common cold in frequency among infectious illnesses affecting people in the United States. In developing nations, gastroenteritis is estimated to be the second leading killer of children under the age of 5, accounting for 23 percent of all deaths in this age group.

Clinically the disease varies, but usually it has an explosive onset with varying combinations of diarrhea, nausea, vomiting, low-grade fever, cramping, headache, and malaise. It can be severe in infants, the elderly, and patients whose immune systems are compromised by other illnesses. Some people mistakenly call it "stomach flu."

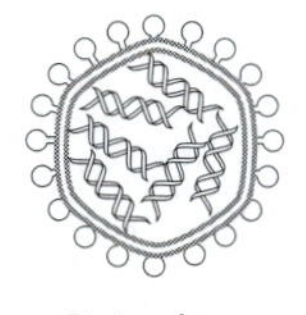

Rotavirus

Rotavirus infections represent one of the world's deadliest infections in children. The diarrhea-related illness is associated with 25 million clinic visits, 2 million hospitalizations, and more than 600,000 deaths worldwide among children younger than five years of age.

The rotavirus (*rota* = "wheel") is a naked, circular-shaped virion whose genome contains eleven segments of double-stranded RNA (FIGURE 15.13). It is a member of the *Reoviridae* family.

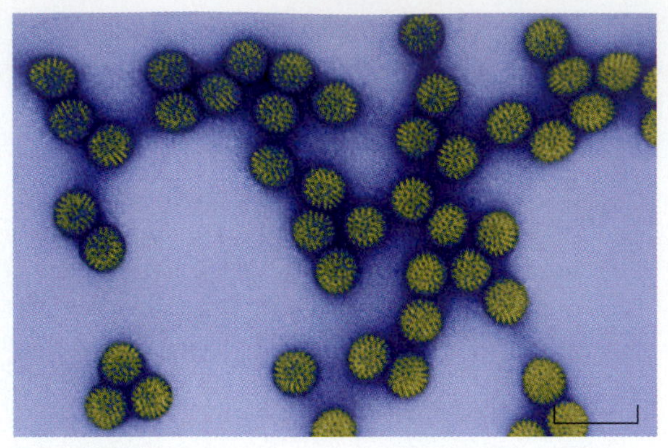

FIGURE 15.13 **Rotaviruses.** False-color transmission electron micrograph of rotaviruses. (Bar = 140 nm.)

Q: From this micrograph, why were these viruses called rotaviruses?

Transmission occurs by ingestion of contaminated food or water (fecal-oral route), or from contaminated surfaces. The rotaviruses make their way to the small intestine where they infect enterocytes, inducing diarrhea.

■ Enterocyte: Cells of the intestinal epithelium.

Recovery from the infection does not guarantee immunity, as many children have multiple rounds of reinfection. The CDC considers rotaviruses the single most important cause of diarrhea in infants and young children admitted to hospitals. Some 75,000 hospitalizations and 20 to 40 deaths occur each year in the United States.

These numbers, but especially the 600,000 childhood deaths worldwide make development of a safe and effective rotavirus vaccine a high priority objective. In 1998, RotaShield was the vaccine to protect against rotavirus infection. However, in July 1999, the FDA removed the RotaShield vaccine after some infants developed bowel obstructions following vaccination. Thankfully, in early 2006, a new vaccine, RotaTeq, was licensed in the United States while another, Rotarix, was licensed in 30 other countries.

Treatment, as for most gastrointestinal diseases, requires oral rehydration salt solutions.

Norovirus infections are the most likely cause of nonbacterial gastroenteritis in adults. Noroviruses (formerly called the Norwalk-like viruses) are transmitted primarily through the fecal-oral route, either by consumption of contaminated food or water, or by direct person-to-person spread. Contamination of surfaces also may act as a source of infection. The CDC estimates at least 50 percent of all foodborne outbreaks (23,000,000) of viral gastroenteritis are caused by norovirus infections. In 2006, more than 45 outbreaks of norovirus infections occurred on cruise ships in European waters, similar to those on U.S. ships between 2001 and 2004.

TABLE 15.3 Viral Infections of the Gastrointestinal Tract

Disease	Classification of Virus	Transmission	Organs Affected	Vaccine	Special Features	Complications
Hepatitis A	*Picornaviridae*	Food Water Contact	Liver	Inactivated virus	Jaundice	Liver damage
Rotavirus infections	*Reoviridae*	Food Water	Intestine	In clinical trials	Diarrhea	Dehydration
Norovirus infections	*Caliciviridae*	Food Water Contact	Intestine	Not available	Diarrhea	Dehydration
Enterovirus infections	*Picornaviridae*	Food Water	Intestine	Not available	Diarrhea	Dehydration Meningitis

The noroviruses belong to the family *Caliciviridae*. These viruses are naked, icosahedral virions with a single-stranded (+ strand) RNA genome. Currently, there are at least four norovirus groups. Noroviruses are highly contagious, and as few as 10 virions can cause illness in an individual.

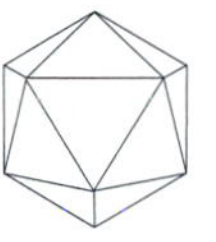

Norovirus

The incubation period for norovirus gastroenteritis is 15 to 48 hours. Typical gastrointestinal symptoms last 12 to 60 hours and recovery is complete. Dehydration is the most common complication. Washing hands and having safe food and water are the only preventions and the only treatment for norovirus gastroenteritis is fluid and electrolytes.

Enterovirus infections also can cause viral gastroenteritis. Enteroviruses are small, icosahedral single-stranded (+ strand) RNA virions of the *Picornaviridae* family.

One enterovirus is the Coxsackie virus, first isolated from a patient residing in Coxsackie, New York. The virus occurs in many strains, with B4 and B5 associated most commonly with gastroenteritis. Group B viruses also are implicated in pleurodynia (or Bornholm disease), a painful disease of the rib muscles; and myocarditis, a serious disease of the heart muscle and valves. In addition, group B Coxsackie viruses are among the most frequent causes of aseptic meningitis. Group A viruses have been isolated from cases of respiratory infections, conjunctivitis, and herpangina, a disease of children with abrupt fever onset and punched-out vesicles on the soft palate, tongue, tonsils, and hands ("hand, foot, and mouth disease"). Enterovirus 71 is involved. The Coxsackie viruses may be the so-called 24-hour viruses responsible for brief bouts of diarrhea.

The second enterovirus is the echovirus, which gets its name from *e*nteric (intestinal), *c*ytopathogenic (pathogenic to cells), *h*uman (human host), and *o*rphan (a virus without a famous disease). Echoviruses occur in many strains and cause gastroenteritis as well as aseptic meningitis. The meningitis, however, is usually less severe than bacterial meningitis. Echoviruses are also a cause of respiratory infections and rapidly developing maculopapular skin rashes called **exanthemas**.

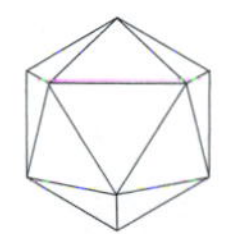

Coxsackie virus

TABLE 15.3 summarizes the characteristics of the virally caused gastrointestinal illnesses.

CONCEPT AND REASONING CHECKS

15.9 Explain why oral rehydration salt solutions are the treatment for most forms of gastroenteritis.

15.4 Viral Diseases of the Nervous System

Several viral diseases affect the human nervous system, which can suffer substantial damage when viruses replicate in the tissue. Rabies, polio, and West Nile fever are perhaps the most recognized diseases.

The Rabies Virus Is of Great Medical Importance Worldwide

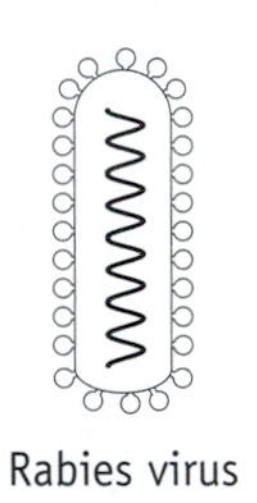

Rabies virus

KEY CONCEPT

- Rabies is a highly fatal disease once symptoms arise.

Rabies is notable for having the highest mortality rate of any human disease, once the symptoms have fully materialized; there are an estimated 55,000 deaths annually, mostly in rural areas of Africa and Asia. Few people in history have survived rabies, and in those who did, it is uncertain whether the symptoms were due to the disease or the therapy (MicroFocus 15.5).

The rabies virus is a single-stranded (– strand) RNA virion of the *Rhabdoviridae* family with a meager five genes in its genome. It is rounded on one end, flattened on the other, and looks like a bullet (FIGURE 15.14).

Animal rabies can occur in most warm-blooded animals, including dogs, cats, prairie dogs, and bats. In 2004, more than 6,345 wildlife cases were reported throughout the United

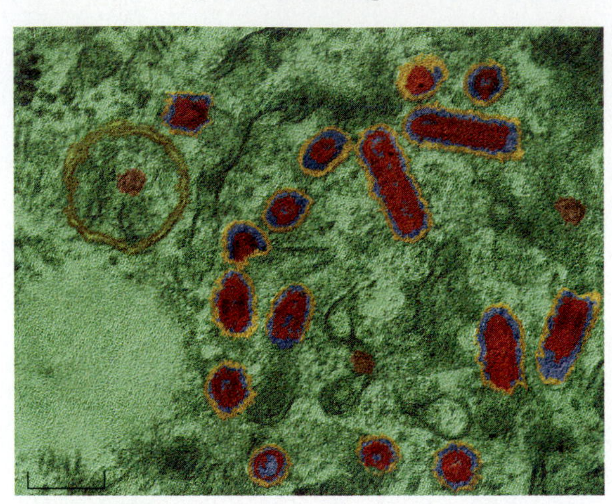

FIGURE 15.14 **The Rabies Virus.** An accumulation of rabies viruses in the salivary gland tissue of a canine. In this false-color transmission electron micrograph, the viruses are red. (Bar = 100 nm.)

Q: In this micrograph, identify the viruses that most closely have the shape characteristic of rhabdoviruses.

MICROFOCUS 15.5: Surviving Rabies

On September 4, 2004, 16-year-old Jeanne Giese of Fond du Lac, Wisconsin was bitten by a rabid bat outside her church. Jeanne did not seek immediate treatment and became gravely ill a month later.

Normally, once the symptoms of rabies set in, recovery is not an option and death occurs within a week. Jeanne was rushed to Children's Hospital in Milwaukee where doctors administered an unproven treatment. She was placed in an induced coma and given a combination of antiviral drugs (ketamine, midazolam, ribavirin, and amantadine) in their effort to save her life.

About a week later, Jeanne was brought out of the coma. She was paralyzed, unable to speak or walk, and without sensation, as she suffered from the effects of rabies on her nervous system. Physicians detected brain-wave activity, but were unsure what was ahead after the drugs would wear off. However, when they did wear off, Jeanne demonstrated some eye movements and reflexes—a real positive sign. Progress continued and after 76 days, Jeanne was released from the hospital in a wheelchair. She had a long road to recovery, if indeed recovery would be complete. She would need to regain her faculties, including her ability to speak and walk.

On Christmas of 2005, Jeanne celebrated her improbable survival at home with her family. The teenager is the world's only known unvaccinated human rabies survivor. Through physical and speech therapy, she is regaining her ability to walk and talk, and nerves damaged by rabies are recommunicating with muscles. Jeanne has returned to school and wants to rejoin the girls' volleyball team next year and, eventually, attend college and major in veterinary science.

Jeanne's lead physician, Dr. Rodney E. Willoughby, Jr. at Children's Hospital and the Medical College of Wisconsin, has been disappointed that his published technique allowing for Jeanne's recovery has not been tried elsewhere. In part, that might be due to the fact that it consumes so many intensive care resources. In developing nations, where most of the 55,000 human rabies cases occur, such resources are not available.

For Jeanne, she was glad the resources were there, which has allowed her to accomplish every teenager's dream—she has earned her temporary driver's license.

States, with raccoons and skunks accounting for the majority of the cases (FIGURE 15.15).

The virus enters the tissue through a skin wound contaminated with the saliva, urine, blood, or other fluid from a rabid animal. The air in a cave inhabited by diseased bats can transmit the virus also. Indeed, rabid bats are a primary source of human infection in the United States.

The incubation period for rabies varies according to the amount of virus entering the tissue and the wound's proximity to the central nervous system. As few as six days or as long as a year may elapse before symptoms appear. A bite from a rabid animal does not ensure transmission, however, because experience shows that only 5 to 15 percent of bitten individuals develop the disease.

Early signs of rabies are abnormal sensations such as tingling, burning, or coldness at the site of the bite. Fever, headache, and increased muscle tension develop, and the patient becomes alert and aggressive. Soon there is paralysis, especially in the swallowing muscles of the pharynx, and saliva drips from the mouth. Brain degeneration, together with an inability to swallow, increases the violent reaction to the sight, sound, or thought of water (*rabies* = "rage"). The disease therefore has been called **hydrophobia**—literally, the fear of water. Death usually comes within days from respiratory paralysis.

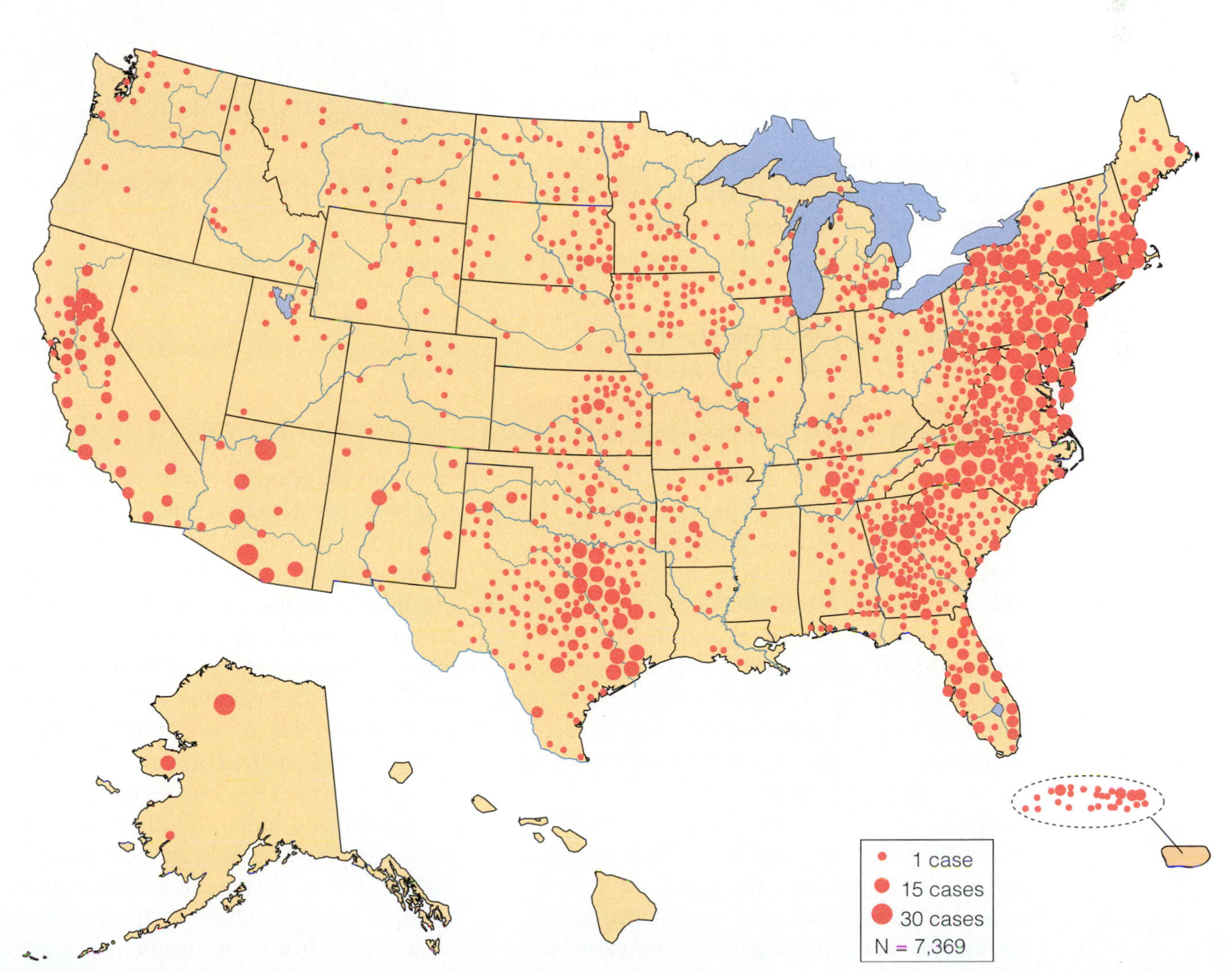

FIGURE 15.15 **Reported Cases of Animal Rabies, 2003.** This geographic map identifies the location of animal rabies reported in 49 states, the District of Columbia, and Puerto Rico in 2003.

Source: Centers for Disease Control. Available at http://www.cdc.gov/ncidod/dvrd/rabies/Epidemiology/Epidemiology.htm.

Q: Why are so many of the animal rabies cases found in the Northeast?

An estimated 10 million people worldwide receive **post-exposure immunization** each year after being exposed to rabies-suspect animals. This entails five injections on days 0, 3, 7, 14, and 28 in the shoulder muscle of the arm. These injections are preceded by thorough cleansing and one dose of rabies immune globulin to provide immediate antibodies at the site of the bite. This usually is accompanied by a tetanus booster, but the latter is omitted if exposure is not certain. For high-risk individuals such as veterinarians, trappers, and zoo workers, a preventive immunization of three injections may be given.

Rabies historically has been a major threat to animals. One form, called **furious rabies**, is accompanied by violent symptoms as the animal becomes wide eyed, drools, and attacks anything in sight. In the second form, **dumb rabies**, the animal is docile and lethargic, with few other symptoms. Health departments now are conducting a novel campaign to immunize wild animals using vaccine air-dropped within biscuit-sized baits of dog food and fish meal.

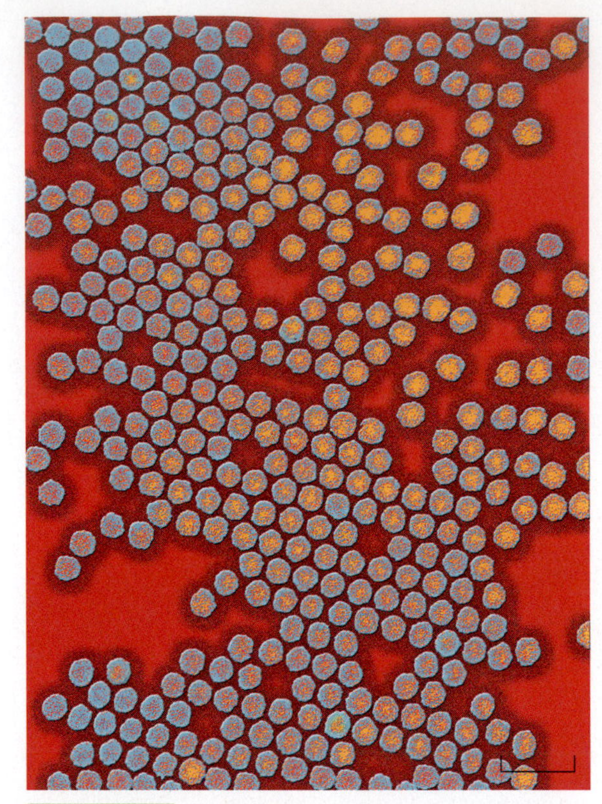

FIGURE 15.16 **Polioviruses.** A false-color transmission electron micrograph of the polioviruses. Although the particles appear circular, their symmetry actually is icosahedral. With a diameter of about 27 nm, these are among the smallest viruses that cause human disease. (Bar = 100 nm.)

Q: How can such a small virus cause such a paralytic disease?

CONCEPT AND REASONING CHECKS

15.10 Why is the incubation period so variable for the rabies virus?

The Polio Virus May Be the Next Infectious Disease Eradicated

KEY CONCEPT

- Polio historically has been a severe paralytic disease.

The name **polio** is a shortened form of **poliomyelitis** (*polio* = "gray"; *myelo* = for "spinal cord"), referring to the "gray matter," which is the nerve tissue of the spinal cord and brain in which the virus infects.

The polioviruses, being in the *Picornaviridae* family, are among the smallest virions, measuring 27 nm in diameter. They are composed of a naked icosahedron-shaped capsid containing a single-stranded (+ strand) RNA genome (FIGURE 15.16).

Poliovirus

Polioviruses usually enter the body by contaminated water and food. They multiply first in the tonsils and then in lymphoid tissues of the gastrointestinal tract, causing nausea, vomiting, and cramps. In many cases, this is the extent of the problem. Sometimes, however, the viruses pass through the bloodstream and localize on the meninges, where they cause **meningitis**. Paralysis of the arms, legs, and body trunk may result.

■ Meninges: The membranes surrounding and protecting the brain and spinal chord.

In the most severe form of polio, the viruses infect the medulla of the brain, causing **bulbar polio** (the medulla is bulb-like). Nerves serving the upper body torso are affected. Swallowing is difficult, and paralysis develops in the tongue, facial muscles, and neck. Paralysis of the diaphragm muscle causes labored breathing and may lead to death (see Chapter 13 opener).

Virologists have identified three types of poliovirus: type I, the Brunhilde strain, causes a major number of epidemics and is sometimes a cause of paralysis; type II, the Lansing strain, occurs sporadically but invariably causes paralysis; and type III, the Leon strain, usually remains in the intestinal tract. Once the method of laboratory cultivation of polioviruses was

MICROFOCUS 15.6: History/Public Health

Blitz

This was the National Immunization Day for India, one in a series of such events across the world. The campaigns were coordinated by the WHO, with help from UNICEF, Rotary International, and the CDC. On this day in December 1997, two million people set out to eradicate polio in India. Coming from every conceivable corner of a vast country, they arrived at 650,000 Indian villages, where they set up immunization posts. Children came to them by the thousands, then the hundreds of thousands, and then the millions (see figure). By the end of the day, 127 million children had received polio immunizations—just one of the success stories from around the world to eradicate polio.

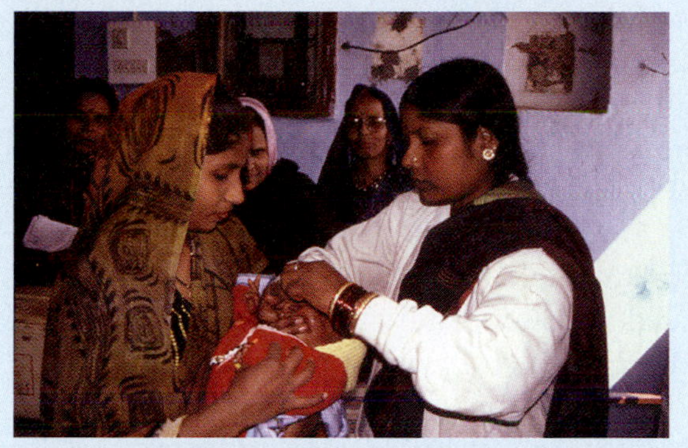

A child being immunized against polio in India.

There are also setbacks in the vaccination blitz. In 2004, polio cases broke out in Nigeria, which had been polio-free. Back in 2003, rumors of contaminated" polio vaccine caused an immediate cessation of immunization. Polio returned, paralyzing 491 Nigerian children. It was not until mid-2004 that the vaccine was deemed safe and the Nigerian government relaunched the immunization effort.

Wars and civil strife also make eradication difficult. There are the Days of Tranquility, where pauses in wars and civil strife allow children to be immunized. In Sri Lanka, for example, polio vaccine was passed across front lines during Days of Tranquility in 1995 and 1996.

In other countries it has not been so easy. In Somalia, a country plagued by ongoing conflicts, 131 polio cases were reported in January 2006. Somalia is the only country in the world with a geographically expanding polio outbreak.

But optimism remains high. With sufficient funding and additional immunization activities—and perseverance—the day is near when polio may be nothing more than a memory.

established, a team led by Jonas Salk grew large quantities of the viruses and inactivated them with formaldehyde to produce the first polio vaccine in 1955. Albert Sabin's group subsequently developed a vaccine containing attenuated (weakened) polioviruses. This vaccine was in widespread use by 1961 and could be taken orally (OPV) as compared with Salk's vaccine, which had to be injected (IPV). Both are referred to as **trivalent** vaccines because they contain all three strains of virus. One drawback of the Sabin vaccine is that being a live vaccine, a few cases of vaccine-caused polio have occurred (1 in 2.4 million).

The vaccines have contributed substantially to the reduction of polio. In 1988, the forty-first World Health Assembly, with funds raised by Rotary International, launched a global initiative to eradicate polio by the end of the year 2000. To meet this goal, progress was made in the 1980s to eliminate the poliovirus from the Americas, and to protect all children from the disease. In the 15 years since the Global Polio Eradication Initiative was launched, the number of cases of polio has fallen by over 99 percent, from more than 350,000 cases in 1988 to less than 700 reported cases in 2003. In the same period, the number of polio-infected countries was reduced from 125 to 7 (MicroFocus 15.6).

Widely endemic on five continents in 1988, polio now is limited to parts of Africa and south Asia. Obviously, the goal to eradicate polio by year 2000 has not been met. In fact, there has been a major outbreak in India in 2006 and several outbreaks in polio-free countries resulting from importation from countries with endemic virus.

A discouraging legacy of the polio epidemics is **postpolio syndrome (PPS)**. Apparently, many people who had polio decades ago now are experiencing the initial ailments they had with polio, including muscle weakness and atrophy, general fatigue and exhaustion, muscle and joint pain, and breathing or swallowing problems. The National Institute of Neurological Disorders and Stroke (NINDS) estimates there are 300,000 polio survivors in the United States and PPS will affect 25 to 50 percent or more of these individuals.

Several theories have been put forward to explain the cause of PPS, including a reactivated virus or an autoimmune reaction (Chapter 22), which over the years has caused the body's immune system to attack motor neurons as if they were foreign substances.

CONCEPT AND REASONING CHECKS

15.11 Explain why the American polio vaccination schedule has dropped the OPV as part of the vaccination.

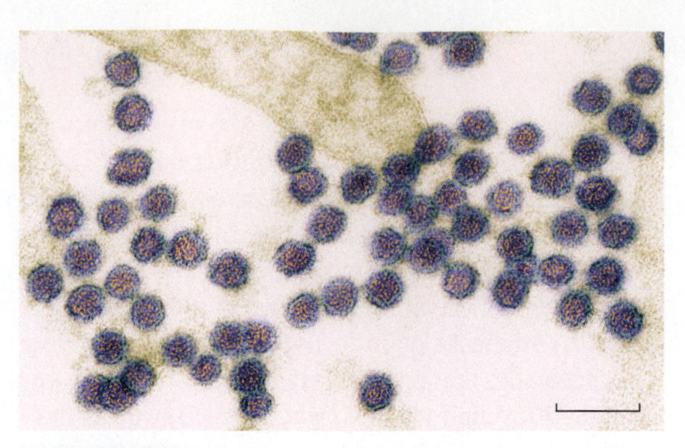

FIGURE 15.17 **West Nile Virus.** False-color transmission electron micrograph of the West Nile virus (WNV). WNV is transmitted by mosquitoes and is known to infect both humans and animals (such as birds). (Bar = 100 nm.)

Q: How would you descibe the shape of the West Nile virus?

Different Flaviviruses Can Be Carried by Blood-Sucking Arthropods

KEY CONCEPT

- West Nile fever can progress to encephalitis and meningitis.

Encephalitis is an acute inflammation of the brain. Used in the general sense, encephalitis may refer to any brain disorder, much as pneumonia refers to a lung disorder. In this section, we use encephalitis to mean a number of viral disorders that are arthropodborne (hence arboviral).

Arboviral encephalitis may be caused by a series of RNA viruses, usually of the *Togaviridae* or *Bunyaviridae* families. In humans, viral encephalitis is characterized by sudden, very high fever and a severe headache. Normally, the patient experiences pain and stiffness in the neck, with general disorientation. Patients become drowsy and stuporous, and may experience a number of convulsions before lapsing into a coma. Paralysis and mental disorders may afflict those who recover. Mortality rates are generally high.

There are many forms of arboviral encephalitis and various vectors of the disease. One form is **St. Louis encephalitis** (SLE), named for the city where it was first identified in 1933. This disease, transmitted by mosquitoes, resurfaced in 1975 with 1,300 cases nationwide. Other forms are California encephalitis, La Crosse encephalitis, and Japanese B encephalitis, all transmitted by mosquitoes. Russian encephalitis is transmitted by ticks.

Arboviral encephalitis is also a serious problem in horses, causing erratic behavior, loss of coordination, and fever. The disease can be transmissible from horses to humans by ticks, mosquitoes, and other arthropods. Important forms are **Eastern equine encephalitis** (EEE), **Western equine encephalitis** (WEE), and **Venezuelan Eastern equine encephalitis** (VEEE).

West Nile fever is an emerging disease in the Western Hemisphere. It is caused by the West Nile virus (WNV), another member of the *Flaviviridae* (FIGURE 15.17).

Before the outbreak in the United States in 1999, WNV was established in Africa, western Asia, and the Middle East. It is closely related to St. Louis encephalitis virus. The virus has a somewhat broad host range and can infect humans, birds, mosquitoes, horses, and some other mammals. Since the first outbreak in 1999 in New York City, each year WNV has moved farther west across the United States. By 2006, cases had been reported in every state except Alaska and Hawaii. Now, experts believe the virus is here to stay and will cause seasonal epidemics that flare up in the summer and continue into the fall. In 2005, there were 3,000 cases reported and 119 deaths.

WNV is spread by the bite of an infected mosquito, which itself becomes infected when it feeds on infected birds (FIGURE 15.18).

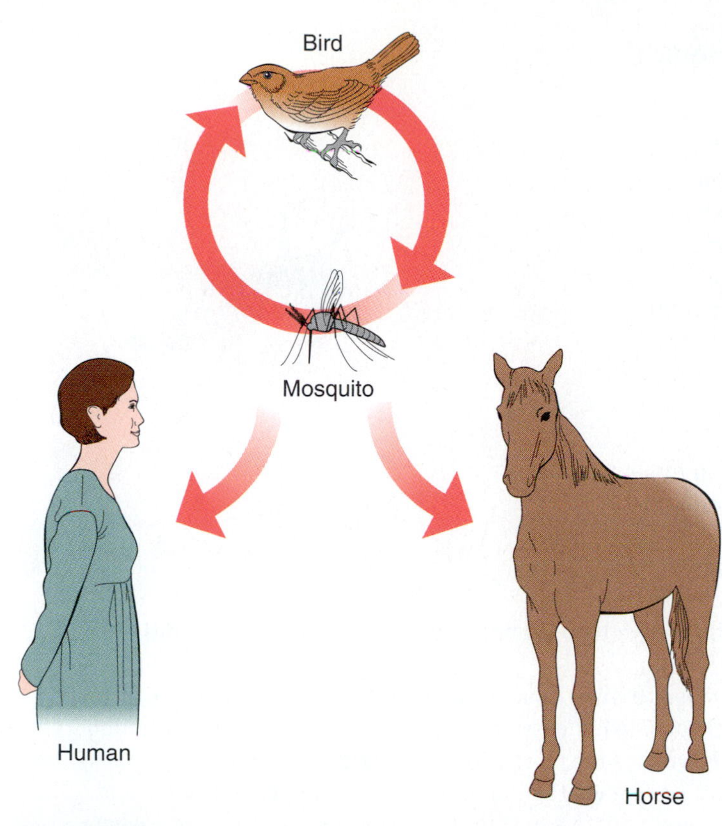

FIGURE 15.18 **The Transmission of West Nile Encephalitis.** The generalized pattern of viral encephalitis transmission among various animals, including humans. Epidemics of viral encephalitis, such as West Nile, are difficult to interrupt because numerous animals are involved.

Q: Although it may be difficult to interrupt viral encephalitis transmission, why would mosquito elimination be the best bet for interruption?

Infected mosquitoes then spread WNV to humans and other animals when they bite. In 2002, health officials reported a very small number of cases where WNV was transmitted through blood transfusions, organ transplants, breast-feeding, and even during pregnancy from mother to baby. The virus is not spread through casual contact such as touching or kissing a person who is infected.

WNV causes a potentially serious illness. People typically develop symptoms between 3 and 14 days after they are bitten by the infected mosquito. Approximately 80 percent of people who are infected remain asymptomatic. Most of the remaining infected individuals display West Nile fever, consisting of a mild fever, headache, body aches, nausea, vomiting, and sometimes swollen lymph glands. They also may develop a skin rash on the chest, stomach, and back. Symptoms typically last a few days. About one in 150 people infected with WNV will develop encephalitis or meningitis that affects the central nervous system. Symptoms can include high fever, headache, stiff neck, stupor, disorientation, coma, convulsions, muscle weakness, vision loss, numbness, and paralysis. These symptoms may last several weeks, and the neurologic effects may be permanent. Death can result.

There is no vaccine yet available nor is there any specific treatment for WNV infection. In cases of encephalitis and meningitis, people may need to be hospitalized so they can receive supportive treatment including intravenous fluids, help with breathing, and nursing care. People who spend a lot of time outdoors are more likely to be bitten by an infected mosquito. These individuals should take special care to avoid mosquito bites (MicroFocus 15.7). Also, people over the age of 50 are more likely to develop serious symptoms from a WNV infection if they do get sick, so they too should take special care to avoid mosquito bites.

Lymphocytic Choriomeningitis (LCM) is usually found in mice, hamsters, and other rodents, where it is transmitted by feces and dust-borne viruses from urine. Several outbreaks in humans have been recorded since 1960, most related to pet hamsters or hamster colonies in laboratories.

The lymphocytic choriomeningitis virus is a single-stranded (– strand) RNA virus of the *Arenaviridae* family. LCM produces a mild illness, often with influenza-like symptoms. Fever and malaise precede headache, drowsiness, and stupor, and the meninges of the brain are infiltrated with large numbers of lymphocytes (hence the disease's name). The symptoms subside within a week, and the mortality rate is very low. Aseptic meningitis may be the only symptom of the disease.

TABLE 15.4 summarizes the neurotropic viral diseases. MicroInquiry 15 presents several case studies concerning some of the viral diseases discussed in this chapter.

CONCEPT AND REASONING CHECKS

15.12 Assess the consequences of being infected with the West Nile virus.

MICROFOCUS 15.7: Public Health

Avoiding Mosquitoes

Have you ever felt that you were "picked on" by mosquitoes while others were essentially left alone? Well, your feelings are correct and avoiding the diseases carried by them can be difficult because mosquitoes "love" some of us more than others (see figure).

There are more than 170 species of mosquitoes in the United States. Luckily, not all of them like humans and, of those that do, only the females take a blood meal from their victim. So, what makes for a good victim?

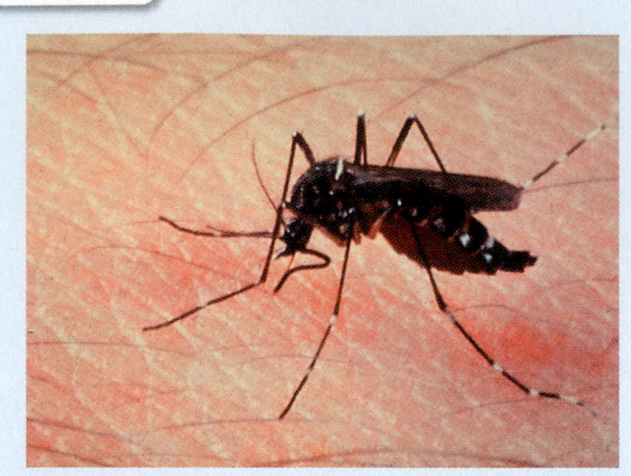

Mosquito feeding on human.

Compounds, such as lactic acid and carbon dioxide gas, as well as some perfumes attract mosquitoes. If you are hot and sweaty, you are a more likely target than someone whose body temperature is cooler and dry. Mosquitoes tend to be drawn to the face, ankles, and hands. Men are more attractive than women and adults more often are bitten than children.

One's daily physiology may influence a mosquito bite. Women, for example, at certain points in their menstrual period are more likely to be attacked. Walking outside in the early morning or late evening makes one more likely to be bitten as that is when many mosquito species are most active.

So, it is best to wear insect repellent, such as one containing at least 20 to 35 percent DEET, and cover the extremities as best as possible. Also, eliminate any standing water around your home where mosquitoes may breed. As the CDC says, "Tell mosquitoes to buzz off!"

TABLE 15.4 A Summary of Neurotropic Viral Diseases

Disease	Classification of Virus	Transmission	Organs Affected	Vaccine	Special Features	Complications
Rabies	*Rhabdoviridae*	Contact with body fluids	Brain Spinal cord	Inactivated viruses	Paralysis Hydrophobia	Death
Polio	*Picornaviridae*	Food Water Contact with human feces	Intestine Spinal cord Brain	Inactivated viruses Attenuated viruses	Infection of medulla	Permanent paralysis
Arboviral encephalitis	Many viruses	Arthropods	Brain	Not available	Encephalitis	Coma Seizures
West Nile fever, encephalitis, and meningitis	*Flaviviridae*	Mosquitoes	Brain	Not available	Neurological effects can be permanent	Death
Lymphocytic choriomeningitis	*Arenaviridae*	Dust Contact	Brain Meninges	Not available	Lymphocytes in brain	Paralysis

MICROINQUIRY 15

Viral Disease Identification

Below are descriptions of several viral diseases based on material presented in this chapter. Read the case history and then answer the questions posed. Answers can be found in **Appendix D**.

Case 1

A 27-year-old male with a history of intravenous and over-the-counter drug abuse comes to the emergency room complaining of nausea, vomiting, headache, and abdominal pain. He indicates he has had these symptoms for two days. His vital signs are normal and he presents with slight jaundice. Physical examination indicates discomfort in the upper right quadrant. Serologic tests for HBsAg and serum antibodies to HAV are negative.

15.1a. Based on these symptoms and clinical findings, what disease does the patient have and what virus is the cause?

15.1b. What factors lead you to this conclusion?

15.1c. What is the significance of discomfort in the upper right quadrant?

15.1d. What is the probable source of the infection?

15.1e. To what other diseases is the patient at risk?

Case 2

A 19-year-old female comes to a neighborhood clinic complaining of difficulty swallowing. She tells the physician she has had a sore throat and fever for about a week and has been extremely tired. Examination indicates she has enlarged tonsils and a high count of B lymphocytes. Her airways are clear and she has a clean chest X ray. Throat cultures for a bacterial infection are negative but a Monospot test is positive.

15.2a. What viral disease would be consistent with the patient's symptoms and clinical tests? What virus is the causative agent?

15.2b. What is the significance of the elevated B lymphocyte count?

15.2c. What dangerous complications can occur from this disease?

15.2d. In developing nations, what tumor often is associated with an infection by the same virus? How might the virus influence tumor formation?

Case 3

A 59-year-old man visits the local hospital emergency room complaining of nausea, abdominal discomfort with vomiting, and fever. On physical examination, he has a temperature of 38°C and appears jaundiced. He indicates the symptoms appeared abruptly two weeks ago after he had returned from a tour to Egypt. While there, the patient indicated he lived like the locals, ate in local restaurants, and often drank the local water. He did not have any vaccinations before leaving on his trip.

15.3a. Name two viral diseases discussed in this chapter that the patient might have.

15.3b. What is this patient's most likely disease? Why would you make that conclusion?

15.3c. Identify a possible transmission route for this virus.

15.3d. What serologic tests could be done to pinpoint the illness?

15.3e. If the patient had sought medical intervention earlier in the illness, what treatment could have been prescribed?

Case 4

A 27-year-old woman comes to a neighborhood clinic with a fever, backache, headache, and bone and joint pain. She indicates these symptoms appeared about two days ago. She also complains of eye pain. Questioning the woman reveals that she has just returned from a trip to Bangladesh where she was doing ecological research in the tropical forests. Skin examination shows remnants of several mosquito bites, which the woman corroborates. She indicates this was her first trip to Southeast Asia. She also reports that she had been taking some antibiotics that she had been given.

15.4a. Based on the woman's symptoms, what viral disease does she most likely have?

15.4b. What clues lead you to this diagnosis?

15.4c. Explain why the antibiotics did not help the woman's condition.

15.4d. Why might the woman want to seriously consider not returning to Southeast Asia to continue her ecological work?

SUMMARY OF KEY CONCEPTS

15.1 Viral Diseases of the Blood and the Lymphatic Systems

- AIDS (acquired immunodeficiency syndrome) is the final stage of the HIV (human immunodeficiency virus) infection. HIV infects and destroys $CD4^+$ T cells, eventually leading to an inability of the immune system to fend off opportunistic diseases. Transmission is through blood, blood products, contaminated needles, or unprotected sexual intercourse. Many antiretroviral drugs are available to slow the progression of disease.
- Mononucleosis (Epstein-Barr virus; EBV) produces fever, sore throat, and lymph node swelling. EBV also is responsible for Burkitt lymphoma typically found in developing nations.
- Cytomegalovirus causes a severe disease in newborns and birth defects.
- Hepatitis B (hepatitis B virus; HBV) is a serious inflammation of the liver. Blood transfusions, contaminated needles, and sexual intercourse are typical means of transmission.
- Hepatitis C (hepatitis C virus; HCV) produces a slow and damaging liver disease, causing cirrhosis and potential liver failure.
- Hepatitis D (delta hepatitis) causes liver infection in the presence of HBV. Hepatitis G is transmitted by contaminated blood products or sexual intercourse.

15.2 Viral Diseases Causing Hemorrhagic Fevers

- Yellow fever produces fever, jaundice, bleeding gums, a bloody stool, internal bleeding, and delirium. A vaccine is available for prevention.
- Dengue fever causes fever, and muscle and joint pain. Recovery usually is complete; however, infection by another strain of the virus can cause dengue hemorrhagic fever. Vascular shock can lead to death.
- Hantavirus infection in a minority of individuals causes hantavirus pulmonary syndrome. Blood hemorrhaging, and renal and respiratory failure can occur.
- Viral hemorrhagic fevers include a variety of viruses responsible for Colorado tick fever, sandfly fever, and Rift Valley fever.
- Viral hemorrhagic fevers also include Marburg and Ebola hemorrhagic fevers. Both have caused substantial numbers of deaths.
- Lassa fever is accompanied by severe fever, prostration, and patchy blood-filled hemorrhagic lesions of the throat.

15.3 Viral Infections of the Gastrointestinal Tract

- Hepatitis A (hepatitis A virus; HAV) follows the fecal-oral route. Symptoms of infection include anorexia, nausea, vomiting, and fever, followed by jaundice. Hepatitis E, also transmitted via the fecal-oral route produces no evidence of a chronic infection.
- Viral gastroenteritis can be caused by rotaviruses. Severe dehydration or death can occur in children. Norovirus symptoms include acute-onset vomiting (more common in children), watery non-bloody diarrhea with abdominal cramps, and nausea. The Coxsackie virus and echovirus also cause viral gastroenteritis.

15.4 Viral Diseases of the Nervous System

- Rabies infections lead to fever, headache, aggressive behavior, and sensitivity to light. Without treatment, infection is almost 100 percent fatal.
- Polio affects nervous tissue, often leading to permanent paralysis and muscle atrophy. A polio vaccine has eliminated the disease from most of the world.
- St. Louis encephalitis and several other types of equine encephalitis can be transmitted to humans from horses.
- West Nile virus (WNV) produces a mild infection causing fever, headache, body aches, nausea, and vomiting. About 1 in 150 people infected develop encephalitis or meningitis. These symptoms may last several weeks and death can result.
- Lymphocytic choriomeningitis comes from exposure to contaminated rodent feces. The virus produces a mild illness, and the symptoms of fever, headache, drowsiness, and stupor subside in about a week.

LEARNING OBJECTIVES

After understanding the textbook reading, you should be capable of writing a paragraph that includes the appropriate terms and pertinent information to answer the objective.

1. Diagram how the human immunodeficiency virus (HIV) infects a cell.
2. Summarize the stages of HIV disease leading to AIDS.
3. Identify the prevention and treatment methods used for HIV disease and AIDS.
4. Describe the symptoms of mononucleosis and the potential complications arising from the illness.
5. Assess the potential of a cytomegalovirus infection to a pregnant woman.
6. Compare the similarities and differences between the nature of the hepatitis B and C viruses and the illnesses they cause.
7. Recommend treatment methods for hepatitis B and C.
8. Define a viral hemorrhagic fever and explain the relationship to arboviruses.
9. Discuss the potential complications of a yellow fever illness.
10. Explain why dengue fever is considered the most important arboviral disease of humans.
11. Distinguish between hantavirus disease and hantavirus pulmonary syndrome.
12. Summarize the symptoms of Ebola and Marburg hemorrhagic fevers.
13. Describe how Lassa fever is transmitted.
14. Explain how the hepatitis A virus is spread and prevented.
15. Identify the symptoms of viral gastroenteritis.

16. Explain why rotavirus infections are so deadly in children.
17. Describe how noroviruses are transmitted.
18. Identify the specific viruses associated with enterovirus infections.
19. Assess the outcome to someone bit by a rabid animal and recommend treatment if symptoms have not yet appeared.
20. Explain how the polio virus causes disease and identify the two types of polio vaccines.
21. Judge the seriousness of an equine encephalitis to humans.
22. List the possible outcomes from an infection with the West Nile virus.

SELF-TEST

Answer each of the following questions by selecting the ***one*** answer that best fits the question or statement. Answers to even-numbered questions can be found in **Appendix C**.

1. The human immunodeficiency virus (HIV) is
   A. a member of the *Reoviridae*.
   B. capable of causing cancer.
   C. integrates into a host cell chromosome.
   D. capable of converting RNA to DNA.
   E. Both **C** and **D** are correct.
2. HIV primarily infects _____ cells.
   A. $CD4^+$ T
   B. red blood
   C. skin
   D. lung
   E. liver
3. An HIV patient with swollen lymph nodes, a mild upper respiratory infection, and $CD4^+$ T cell count of _____ would be in stage _____ of HIV disease.
   A. 1,000; stage II
   B. 700; stage I
   C. 400; stage II
   D. 400; stage III
   E. 150; stage I
4. Mononucleosis is an infection of _____ cells by the _____.
   A. T; cytomegalovirus
   B. liver; hepatitis virus
   C. B; Epstein-Barr virus
   D. lung; cytomegalovirus
   E. red blood; Epstein-Barr virus
5. Which of the following is *not* a transmission mechanism for hepatitis B?
   A. Contaminated dialysis equipment
   B. Sexual contact
   C. Non-sterile body piercing equipment
   D. Fecal-oral route
   E. Blood-contaminated needles
6. Cirrhosis could be a complication from an infection by
   A. the hepatitis A virus.
   B. the hepatitis B virus.
   C. the hepatitis C virus.
   D. the hepatitis G virus.
   E. Both **B** and **C** are correct.
7. Which one of the following is *not* a hemorrhagic fever virus?
   A. Dengue fever virus
   B. Ebola virus
   C. Marburg virus
   D. West Nile virus
   E. Yellow fever virus
8. Symptoms of headache, fever, and muscle pain lasting 3 to 5 days, followed by a 2 to 24 hour abating of symptoms is characteristic of
   A. Yellow fever.
   B. Hepatitis C.
   C. Dengue fever.
   D. Ebola hemorrhagic fever.
   E. West Nile fever.
9. There are _____ types of dengue fever virus, which infect _____.
   A. 1; liver cells
   B. 2; red blood cells
   C. 4; white blood cells
   D. 6; lung cells
   E. 8; kidney cells
10. Fatigue, fever, and muscle aches along with headaches, dizziness, difficulty breathing, and low blood pressure are symptoms of
    A. hantavirus pulmonary syndrome.
    B. Lassa fever.
    C. yellow fever.
    D. Marburg hemorrhagic fever.
    E. None of the above (**A–D**) are correct.
11. A long thread-like RNA virus is typical of the _____ viruses.
    A. hepatitis C
    B. Ebola
    C. polio
    D. West Nile
    E. Coxsackie
12. Which one of the following characteristics pertains to hepatitis A?
    A. Transmission is by the fecal-oral route.
    B. The incubation period is 2 to 4 weeks.
    C. It is an acute, inflammatory liver disease.
    D. About 100 Americans die each year from the "fulminate" form.
    E. All of the above (**A–D**) are correct.

13. _____ are the single most important cause of diarrhea in infants and young children admitted to American hospitals.
A. Noroviruses
B. Echoviruses
C. Hepatitis A viruses
D. Rotaviruses
E. Coxsackie viruses

14. Hydrophobia is a term applied to
A. rotavirus infections.
B. West Nile fever.
C. arboviral encephalitis.
D. rabies.
E. polio.

15. These viruses multiply first in the tonsils and then the lymphoid tissues of the gastrointestinal tract.
A. Rabies virus
B. Rotavirus
C. Polio virus
D. Hepatitis A virus
E. West Nile virus

16. West Nile virus can be transmitted by
A. blood transfusions.
B. organ transplants.
C. mosquitoes.
D. breast-feeding.
E. All the above (**A–D**) are correct.

QUESTIONS FOR THOUGHT AND DISCUSSION

Answers to even-numbered questions can be found in **Appendix C.**

1. Health authorities panicked when an outbreak of Ebola hemorrhagic disease occurred among imported macaques in a quarantine facility in Reston, Virginia, in 1989. What sparks such a dramatic response when a disease like Ebola fever breaks out?

2. In many diseases, the immune system overcomes the infectious agent, and the person recovers. In other diseases, the infectious agent overcomes the immune system, and death follows. Compare this broad overview of disease and resistance to what is taking place with AIDS, and explain why AIDS is probably unlike any other disease encountered in medicine.

3. A diagnostic test has been developed to detect hepatitis C in blood intended for transfusion purposes. Obviously, if the test is positive, the blood is not used. However, there is a lively controversy as to whether the blood donor should be informed of the positive result. What is your opinion? Why?

4. In the southwestern United States, abundant rain and a mild winter often bring conditions that encourage a burgeoning rodent population. Under these circumstances, what viral disease would health officials anticipate and what precautions should they give residents?

5. Disney World uses "sentinel chickens" strategically placed on the grounds to detect any signs of viral encephalitis. Why do you suppose they use chickens? Why is Disney World particularly susceptible to outbreaks of viral encephalitis? And what recommendations might be offered to tourists if the disease broke out?

APPLICATIONS

Answers to even-numbered questions can be found in **Appendix C.**

1. Written on some blood donor cards is the notation "CMV." What do you think the letters mean, and why are they placed there?

2. Sicilian barbers are renowned for their skill and dexterity with razors (and sometimes their singing voices). In 1995, French researchers studied a group of 37 Sicilian barbers and found that 14 had antibodies against hepatitis C, despite never having been sick with the disease. By comparison, when a random group of 50 blood donors was studied, none had the antibodies. As an epidemiologist, what might account for the high incidence of exposure to hepatitis C among the barbers?

3. As a state health inspector, you are suggesting all restaurant workers should be immunized with the hepatitis A vaccine. Why would restaurant owners agree or disagree with your idea?

4. An epidemiologist notes that India has a high rate of dengue fever but a very low rate of yellow fever. What might be the cause of this anomaly?

REVIEW

On completing your study of these pages, test your understanding of their contents by deciding whether the following statements are true or false. If the statement is true, write "T" in the space. If false, write "F" and substitute a word for the *underlined* word or phrase to make the statement true. The answers to even-numbered statements are listed in **Appendix C**.

1. _____ Both yellow fever and dengue fever are caused by a DNA virus transmitted by the mosquito.
2. _____ Eighty percent of people infected by West Nile virus experience flu-like symptoms.
3. _____ Downey cells are a characteristic sign of the viral disease infectious mononucleosis.
4. _____ The term "hydrophobia" means "fear of water," and it is commonly associated with patients who have encephalitis.
5. _____ The cell most often affected by HIV, the AIDS virus, is the human monocyte.
6. _____ Norovirus and rotavirus are both considered to be agents of viral encephalitis.
7. _____ The Epstein-Barr virus is the cause of infectious mononucleosis.
8. _____ Hepatitis is primarily a disease of the liver.
9. _____ A vaccine is available to prevent yellow fever but not to prevent AIDS.
10. _____ The Salk and Sabin vaccines are used for immunizations against hepatitis.
11. _____ HIV is a reovirus in which the RNA of the genome is used as a template to synthesize DNA.
12. _____ Hepatitis B is most commonly transmitted by contact with infected semen or infected blood.
13. _____ One of the most important causes of diarrhea in infants and young children admitted to hospitals is the Epstein-Barr virus.
14. _____ Filoviruses are long, thread-like viruses that cause hemorrhagic fevers and include the marburgvirus.
15. _____ Enlargement of the lymph nodes, sore throat, mild fever, and a high count of B-lymphocytes are characteristic symptoms in people who have polio.
16. _____ The C in the TORCH group of diseases stands for the cytomegalovirus, which can be transmitted from a pregnant woman to her unborn child.

HTTP://MICROBIOLOGY.JBPUB.COM/

The site features learning, an on-line review area that provides quizzes and other tools to help you study for your class. You can also follow useful links for in-depth information, read more MicroFocus stories, or just find out the latest microbiology news.

16

Eukaryotic Microorganisms: The Fungi

Chapter Preview and Key Concepts

16.1 Characteristics of Fungi
- Most fungi grow as a branching filamentous network.
- Fungal growth is dependent on moisture, adequate chemical nutrients, and physical factors.
- Fungal reproduction may take place by asexual as well as by sexual processes.

16.2 The Classification of Fungi
- Fungal classification usually is based on the form of sexual reproduction.

  MicroInquiry 16: Evolution of the Fungi
- Baker's yeast and other yeasts are important eukaryotic organisms in industry and research.

16.3 Fungal Diseases of the Skin
- Dermatophytes can cause athlete's foot and ringworm.
- *Candida albicans* most commonly causes vaginitis or thrush.
- *Sporothrix schenkii* forms lesions under the skin.

16.4 Fungal Diseases of the Lower Respiratory Tract
- *Cryptococcus neoformans* causes a rare, but dangerous respiratory disease.
- *Histoplasma capsulatum* most often causes a mild flu-like illness.
- *Blastomyces dermatitidis* can cause lung and skin infections.
- *Coccidioides immitis* infections can disseminate and be life-threatening.
- *Pneumocystis jiroveci* is a fungal pathogen in immunocompromised hosts.
- *Aspergillus* and *Claviceps* species can cause human illnesses.

That mold. It smells like death.
—Veronica Randazzo after returning to her house following hurricane Katrina

In August 2005, hurricane Katrina left unimaginable devastation everywhere along the south-central coast of the United States. In New Orleans, thousands of homes were flooded and left sitting in feet of water for weeks. Many health experts were concerned about outbreaks of infectious diseases like cholera, West Nile fever, and gastrointestinal illnesses. Homes sitting in stagnant water could become breeding grounds for microorganisms.

Thankfully, most of these infectious disease scenarios did not occur. However, what did break out in many of the parishes in New Orleans was mold. There was mold on walls, ceilings, cabinets, clothes, and just about anything that provided a source of moisture and nutrients (FIGURE 16.1). It formed carpets of spore-forming colonies everywhere.

If you see small spots of mold in your home, a bleach solution will do a great job to kill and eliminate the problem. But what about when an entire home and its contents are one giant culture dish? More than likely, most of these homes will have to be demolished and furniture and other home contents will have to be destroyed. Many home items, like beds, couches, or cabinets that were above the water level probably contain mold due to the prolonged humidity, and will have to be replaced. Most health officials told residents to follow the same slogan used for potentially spoiled food: "When in doubt, throw it out."

At Tulane Hospital, the first floor was covered with mold, which made cleanup and reopening of many wards a very difficult chore. In homes as well as offices and hospitals, molds were discovered growing in ventilation systems and the ventilation ducts. If the ventilation

(A)

(B)

FIGURE 16.1 A Wall of Mold. (A) Every circular spot seen on this wall and (B) on the ceiling and other objects represents a mold colony. Each colony started from a single spore; now each mature colony contains millions of spores.

Q: How do you think molds reproduce?

fans were turned on, literally hundreds of billions of spores would be blown and spread to new areas to germinate and grow. In fact, six weeks after Katrina, the mold spore count in the air at various sites in New Orleans was as high as 102,000 spores per cubic meter—twice the number considered normal for New Orleans.

What about illnesses or disease from breathing the mold spores? By November, many New Orleans residents who had returned were suffering from upper respiratory problems—the residents "affectionately" called it the "Katrina cough." In fact, residents with asthma, bronchitis, and allergies who had left New Orleans were asked not to return just yet. Exactly how dangerous the cough might be has yet to be determined.

This discussion indicates that molds grow in many natural and constructed environments. They often break down dead or decaying matter, such as the cases described in New Orleans. Many molds though are of great importance in the natural recycling in the biosphere. Still others can act as pathogens and cause some dangerous and debilitating diseases.

■ Biosphere: The whole area of Earth's surface, atmosphere, and sea that is inhabited by living things.

From the description of the situation in New Orleans, you can appreciate fungi as producers of massive numbers of **spores**, representing microscopic cells for disseminating the organisms to new territories and environments where they germinate, grow, and again reproduce.

Molds and yeasts are fungi, and they contain many species some of which we survey and study in this chapter. We will encounter many beneficial fungi such as those used to make antibiotics or in commercial and industrial processes. We also will identify and discuss several widespread human diseases caused by fungi. Many are of concern to immunocompromised individuals.

■ Immunocompromised: Refers to the lack of an adequate immune response resulting from disease, exposure to radiation, or treatment with immunosuppressive drugs.

Our study begins with a focus on the structures, growth patterns, and life cycles of fungi—something quite unique from the other groups of microorganisms.

16.1 Characteristics of Fungi

The **fungi** (sing., fungus) are a diverse group of eukaryotic microorganisms. Some 75,000 species have been described, although as many as 1.5 million may exist. For many decades, fungi were classified as plants, but laboratory studies have revealed at least four properties that distinguish fungi from plants:

- Fungi lack chlorophyll, while plants have this pigment.
- Fungal cell walls contain a carbohydrate called chitin; plant cell walls have cellulose.
- Most fungi are not truly multicellular like plants.
- Fungi are heterotrophic, while plants are autotrophic.

Mainly for these reasons, fungi are placed in their own kingdom Fungi, within the domain *Eukarya* of the "tree of life" (see Chapter 3). The study of fungi is called **mycology** (*myco* = "fungus") and a person who studies fungi is a mycologist.

Fungi Share a Combination of Characteristics

KEY CONCEPT

- Most fungi grow as a branching filamentous network.

Fungi generally have life cycles involving two phases: a growth (vegetative) phase and a reproductive phase. A major group of fungi, the **molds**, which you have come to appreciate already from the chapter opener, grow as long, tangled filaments of cells that give rise to visible colonies (FIGURE 16.2A). Another group, the **yeasts**, are unicellular organisms whose colonies on agar visually resemble bacterial colonies (FIGURE 16.2B). Yet other forms are **dimorphic**; usually at ambient temperature (25°C) they grow as filamentous molds, but at body temperature (37°C) they convert to unicellular pathogenic yeast forms.

With the notable exception of yeasts, fungi consist of masses of intertwined filaments called **hyphae** (sing., hypha). The hyphae are the morphological unit of a filamentous fungus and individual hyphae usually are visible only with the aid of a microscope (FIGURE 16.3A). Hyphae have a broad diversity of forms and can be highly branched. A thick mass of hyphae is called a **mycelium** (pl., mycelia). This mass is usually large enough to be seen with the unaided eye, and generally it has a rough, cottony texture. The mycelium along with any reproductive structures would represent the fungal organism.

(A)

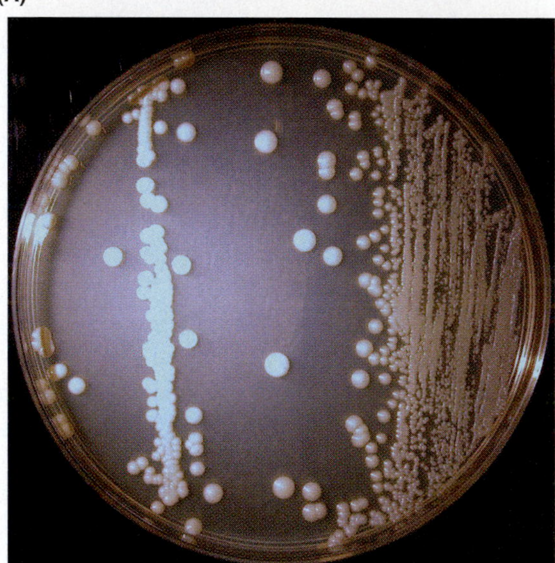

(B)

FIGURE 16.2 **Fungal Colonies.** (**A**) On growth media, molds, such as *Penicillium chrysogenum,* grow as fuzzy colonies visible to the naked eye. The spores are the darker blue-green regions of the colonies. (**B**) Petri dish culture showing colonies of the yeastlike fungus *Torulopsis glabrata*. A food spoilage organism, this species obtains its nutritional requirements from the breakdown of dead or decaying material.

Q: What substance would you predict is in the liquid droplets on the Penicillium *colony surface?*

Being eukaryotic organisms, fungi have one or more nuclei as well as a range of organelles including mitochondria, an endomembrane system, ribosomes, and a cytoskeleton. The cell wall is composed of large amounts of chitin. **Chitin** is a carbohydrate polymer of acetylglucosamine units; that is, glucose molecules containing amino and acetyl groups. The cell wall provides rigidity

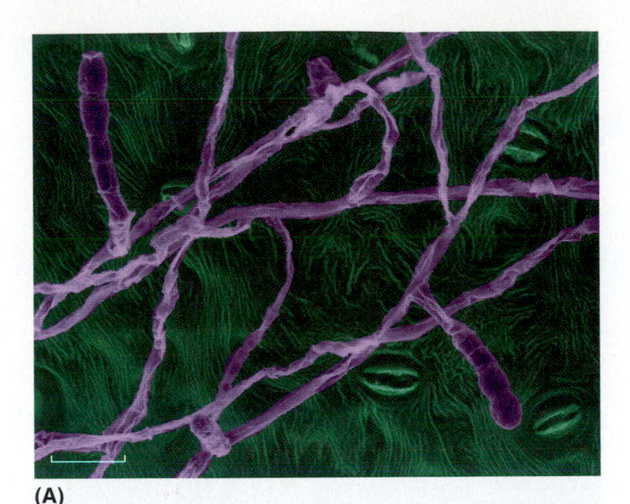
(A)

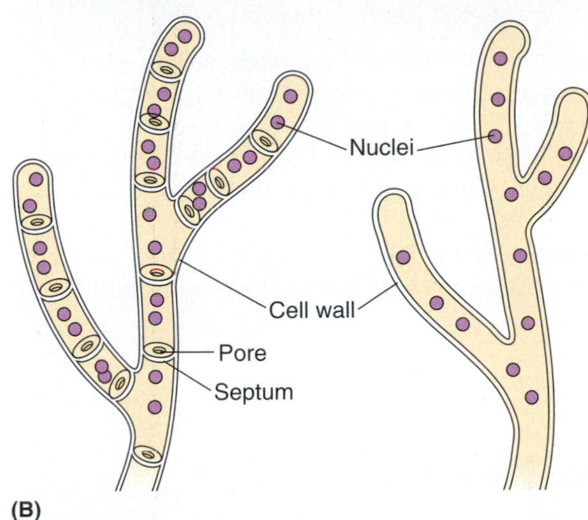

(B)

FIGURE 16.3 **Hypha Structure.** (**A**) A false-color scanning electron micrograph of fungal hyphae growing on a leaf surface. (Bar = 40 μm.) (**B**) Molds have hyphae that are either septate or nonseptate. Septa compartmentalize hyphae into separate cells, although the septa have a pore through which cytoplasm and nuclei can move. Other fungi have nonseptate hyphae.

Q: What unique structural features are presented in this figure?

and strength, which, like the cell wall of bacterial cells, allows the cells to resist bursting due to high internal water pressure.

In many species of fungi, hyphal cross walls, called **septa** (sing., septum), divide the cytoplasm into separate cells (FIGURE 16.3B). Such fungi are described as **septate**. In the blue-green mold *Penicillium chrysogenum*, the septa are incomplete, however, and pores in the cross walls allow adjacent cytoplasms to mix. In other fungal species, such as the common bread mold *Rhizopus*, the filaments are **nonseptate**. In both examples, such hyphae are considered **coenocytic**, meaning they contain many nuclei in a common cytoplasm.

Because fungi absorb preformed organic matter, they are described as **heterotrophic** organisms. Most are **saprobes**, feeding on dead or decaying organic matter. Together with many bacterial species, these fungi make up the **decomposers** (see Chapter 3), recycling vast quantities of organic matter (MicroFocus 16.1). In industrial settings, these decompositions can be profitable. The fungus *Trichoderma*, for example, produces enzymes to degrade cellulose and give jeans a "stone-washed" appearance.

Other fungi are pathogens, living on plants or animals, and often causing disease. Most such fungi are opportunistic and only cause disease when given the opportunity.

CONCEPT AND REASONING CHECKS

16.1 Assess the role of hyphae to fungal growth.

Fungal Growth Is Influenced by Several Factors

KEY CONCEPT

- Fungal growth is dependent on moisture, adequate chemical nutrients, and physical factors.

Fungi acquire their nutrients through absorption, either as saprobes or pathogens. Being mostly terrestrial organisms, the molds and yeasts secrete enzymes into the surrounding environment that break down (hydrolyze) complex organic compounds into simpler ones. As a result of this extracellular digestion, simpler compounds, like glucose and amino acids, can be absorbed.

The mycelium formed by a mold represents the "feeding network" for these fungi. In some cases the mycelium can form a tremendously large surface area for nutrient absorption (MicroFocus 16.2). For the yeasts, nutrients are absorbed across the cell surface, similar to the way bacterial cells obtain their nutrients.

Mycelial extension occurs by the elongation or growth of the hyphae. The nuclei divide by mitosis and growth of these mycelial fungi involves lengthening and branching of the hyphae at the hyphal tips. Yeast growth in number occurs by cell division as the cells undergo many rounds of mitosis and cytokinesis to form a large population of individual cells.

MICROFOCUS 16.1: Evolution

When Fungi Ruled the Earth

A mushroom species growing on a rotting log.

About 250 million years ago, at the close of the Permian period, a catastrophe of epic proportions visited the Earth. Apparently, over 90 percent of animal species in the seas vanished. The great Permian extinction, as it is called, also wreaked havoc on land animals and cleared the way for dinosaurs to inherit the planet.

But land plants managed to survive, and before the dinosaurs came, they spread and enveloped the world. At least, that is what paleobiologists traditionally believed.

Now, however, they are revising their outlook and finding a significant place for the fungi. In 1996, Dutch scientists from Utrecht University presented evidence that land plants were decimated by the extinction and that for a brief geologic span, dead wood covered the planet. During this period, they suggest the fungi emerged and wood-rotting species experienced a powerful spike in their populations (see figure). Support for their theory is offered by numerous findings of fossil fungi from the post-Permian period. The fossils are bountiful and they come from all corners of the globe. Significantly, they contain fungal hyphae, the active feeding forms rather than the dormant spores.

And so it was that fungi proliferated wildly and entered a period of feeding frenzy where they were the dominant form of life on Earth. It's something worth considering next time you kick over a mushroom growing in a lawn.

MICROFOCUS 16.2: Environmental Microbiology

A Humongous and Ancient Individual

Armillaria mushrooms growing on a tree trunk.

What's bigger than a blue whale and older than a California redwood? A fungus, of course! *Armillaria,* a plant pathogen, infects evergreen trees. The visible part of the fungus is the golden mushrooms it produces (see figure). However, underground in the soil lies an invisible mycelium invading the roots of the evergreen tree and absorbing its water and carbohydrates, which often spells death for the infected tree.

In 2001, scientists studying in Oregon's Malheur National Forest discovered a single *Armillaria* spreading through 2,200 acres of forest soil. It measures over nine square kilometers—about the area of 1,600 football fields! To be of this size, scientists believe the fungus germinated from a single spore somewhere between 2,000 and 8,500 years ago. Now that's old and humongous, considering the mycelium of most *Armillaria* species covers only 20 or 30 acres.

This discovery is not unique. Another species of *Armillaria* was discovered in a Michigan hardwood forest that measured about 37 acres in size and, in eastern Washington, an *Armillaria* mycelium measuring about 1,500 acres in size has been discovered.

The world's biggest fungus is challenging the concept of what constitutes an individual. Is the Oregon *Armillaria* a single individual organism? "If you could take away the soil and look at it, it's just one big heap of fungus with all of these filaments that go out under the surface," says Dr. Catherine Parks, who was one of its discoverers.

Fungal growth is influenced by many factors in the environment. Besides the availability of chemical nutrients, oxygen, temperature, and pH influence growth.

Oxygen. The majority of fungi are aerobic organisms, with the notable exception of the facultative yeasts, which can grow in either the presence of oxygen or under fermentation conditions (see Chapter 6).

Temperature. Most fungi grow best at about 23°C, a temperature close to normal room temperature. Notable exceptions are the pathogenic fungi, which grow optimally at 37°C, which is body temperature. As mentioned, dimorphic fungi grow as yeastlike cells at 37°C and a mycelium at 23°C. Psychrophilic fungi grow at still lower temperatures, such as the 5°C found in a normal refrigerator.

pH. Many fungi thrive under mildly acidic conditions at a pH between 5 and 6. Acidic soil therefore may favor fungal turf diseases, in which case lime (calcium carbonate) is added to neutralize the soil. Mold contamination also is common in acidic foods such as sour cream, citrus fruits, yogurt, and most vegetables. Moreover, the acidity in breads and cheese encourages fungal growth. Blue (Roquefort) cheese, for example, consists of milk curds in which the mold *Penicillium roqueforti* is growing (FIGURE 16.4).

FIGURE 16.4 **Roquefort Cheese.** Roquefort cheese is made from cow's milk and contains the mold *Penicillium roquefortii*.

Q: Why would a mold be added to the ripening process of cheeses such as this?

■ Psychrophilic: Refers to organisms that prefer to grow at cold temperatures.

Normally, high concentrations of sugar are conducive to growth, and laboratory media for fungi usually contain extra glucose; examples include Sabouraud dextrose (glucose) agar and potato dextrose agar.

Fungal growth in nature forms important links in ecological cycles because fungi, along with bacterial species, rapidly decompose animal and plant matter (Chapter 26). Working in immense numbers, fungi release carbon and minerals back to the environment, making them available for recycling.

Many fungi also live in a mutually beneficial relationship with other species in nature through a symbiotic association called mutualism (see Chapter 3). Fungi called **mycorrhizae** (*rhiza* = "root") live harmoniously with plants where the hyphae of these fungi invade or envelop the roots of plants (FIGURE 16.5A). These mycorrhizae consume some of the carbohydrates produced by the plants, but in return act as a second root system, contributing essential minerals and water to promote plant growth (FIGURE 16.5B). Mycorrhizae have been found in plants from salt marshes, deserts, and pine forests. In fact, these beneficial fungi live in and around the roots of 95% of examined plant species.

Besides the mycorrhizae, most plants examined also contain fungal **endophytes**, which are fungi living and growing entirely within plants, especially leaf tissue. They do not cause disease but, rather like mycorrhizae, they provide better or new growth opportunities for the plant. In the southwestern Rocky Mountains, for instance, a fungus thrives on the blades of a species of grass, producing a powerful poison that can put horses and other animals to sleep for about a week (the grass is called "sleepy grass" by the locals). Thus, the grass survives while other species are nibbled to the ground. MicroFocus 16.3 describes a few remarkable examples.

■ Mutualism: A symbiotic relationship between two organisms of different species that benefits both.

CONCEPT AND REASONING CHECKS

16.2 Estimate the value of mycorrhizae and fungal endophytes to plant survival.

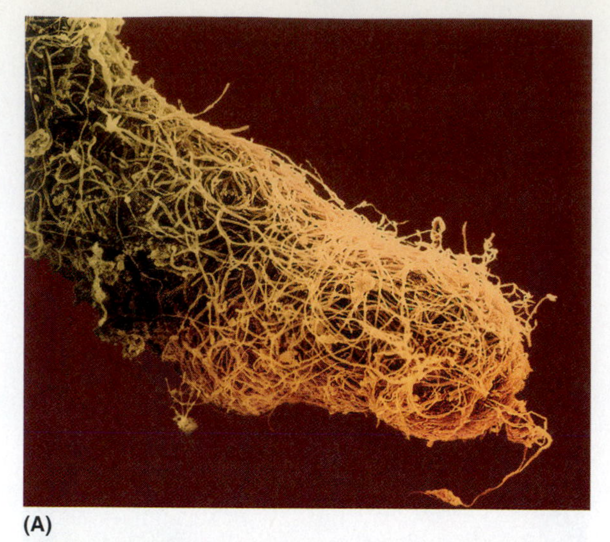
(A)

(B)

FIGURE 16.5 **Mycorrhizae and Their Affect on Plant Growth.** (**A**) Mycorrhizae surround the root of a Eucalyptus tree in this false-color scanning electron micrograph. These fungi are involved symbiotically with their plant host, such as aiding in mineral metabolism. (**B**) An experiment analyzing the mycorrhizal effects on plant growth.

Q: Which plant or plants (CK, GM, GE) represent(s) the control and experimental set ups?

MICROFOCUS 16.3: Environmental Microbiology

Fungi as Protectors

What would you think if you saw wheat growing in the deserts of California or Arizona? You probably would think it was a mirage or you were hallucinating from the lack of water. Well, in the near future, such an apparition might indeed be real.

It turns out that mycorrhizae are not the only fungi providing protection and growth benefits for hundreds of species of plants. The hyphae of endophytes grow within plant tissue and between cells of many healthy leaves where nutrients are exchanged between host and fungus. No damage is done to the leaves and only the reproductive structures make it to the surface to release spores into the wind.

In cacao leaves, the more mature (green) leaves gain more benefit from endophyte associations than young (red) leaves.

One example of endophyte benefit was discovered by Regina Redman and Russell Rodriguez of the U.S. Geological Survey in Seattle, Washington. They took perennial grasses normally growing in hot soils around geysers and grew them in the lab. Those plants exposed to hot soil but lacking endophytes died while those living with endophytes survived. Most interesting, the fungi by themselves also died in the hot soils. Apparently, there needs to be a give and take between host and fungus—representing a true mutualistic relationship.

Redman and Rodriguez then teamed up with Joan Henson at Montana State University to show that these same endophytes protect other plants. At least in the lab, when endophyte spores were placed on tomato, watermelon, or wheat seedlings, the tomato and watermelon seedlings with endophytes survived the stresses of high temperatures (50°C) or drought conditions. Although the wheat seedlings died, the plants survived about a week longer than those without endophytes.

Meanwhile in Panama, another group of researchers, lead by A. Elizabeth Arnold, now at the University of Arizona, discovered that chocolate-tree (cacao) leaves harboring endophytes are more resistant to pathogen attack, while fungus-free leaves become diseased. In their studies, when leaves were purposely inoculated with one of the major pathogens of cacao, leaves associated with several different endophytes were less likely to be invaded and would most likely survive. Although young leaves benefited from the symbiosis, the older leaves appeared to benefit the most from the endophyte association (see figure).

So, at least in the lab, endophytic fungi play various but important roles in plant survival. As more is discovered, perhaps the day will come when wheat will grow in the desert.

Reproduction in Fungi Involves Spore Formation

KEY CONCEPT

- Fungal reproduction may take place by asexual as well as by sexual processes.

Sporulation is the process of spore formation. It usually occurs in structures called **fruiting bodies**, which represent the part of a fungus in which spores are formed and from which they are released. These structures may be asexual and invisible to the naked eye, or sexual structures, such as the macroscopic mushrooms.

Asexual reproduction. Asexual reproductive structures develop at the ends of specialized hyphae. As a result of mitotic divisions, thousands of spores are produced, all genetically identical.

Many asexual spores develop within sacs or vessels called **sporangia** (sing., sporangium; *angio* = "vessel") (FIGURE 16.6A). Appropriately, the spores are called **sporangiospores**. Other fungi produce spores on supportive structures called **conidiophores** (FIGURE 16.6B). These unprotected, dust-like spores are known as **conidia** (sing., conidium; *conidio* = "dust"). Fungal spores are extremely light and are blown about in huge numbers by wind currents. In yet other fungi, spores may form simply by fragmentation of the hyphae yielding **arthrospores** (*arthro* = "joint"). The fungi that cause athlete's foot multiply in this manner.

Many yeasts reproduce asexually by **budding**. In this process, the cell becomes swollen at one edge, and a new cell called a **blastospore**, (*blasto* = "bud") develops (buds) from the parent cell (FIGURE 16.7). Eventually, the spore breaks free to live independently. The parent cell can continue to produce additional blastospores.

Once free of the fruiting body, spores landing in an appropriate environment have the capability of germinating to reproduce new unicellular yeast cells or a new hypha (FIGURE 16.8). Continued growth will eventually form a mycelium.

Sexual reproduction. Many fungi also produce spores by sexual reproduction. In this process, opposite mating types come together and fuse FIGURE 16.9). Because the nuclei are genetically different in each mating type, the fusion cell represents a **heterokaryon** (*hetero* = "different"; *karyo* = "nucleus"); that is, a cell with genetically dissimilar nuclei existing for some length of time in a common cytoplasm. Eventually the nuclei fuse and a diploid cell is formed. The chromosome number soon is halved by meiosis, returning the cell or organism to a haploid condition.

(A)

(B)

FIGURE 16.6 **Fungal Fruiting Bodies.** False-color electron micrographs of sporangia and conidia. (**A**) Sporangia of the common bread mold *Rhizopus*. Each round sporangium contains thousands of sporangiospores. (Bar = 20 μm.) (**B**) The conidiophores and conidia in the mold-like phase of *Penicillium roquefortii*. Many conidiophores are present within the mycelium. Conidiophores (orange) containing conidia (blue) are formed at the end of specialized hyphae (green). (Bar = 20 μm.)

Q: Why must fungal spores be elevated on the tips of hyphae?

■ Mating types: Separate mycelia of the same fungus or separate hyphae of the same mycelium.

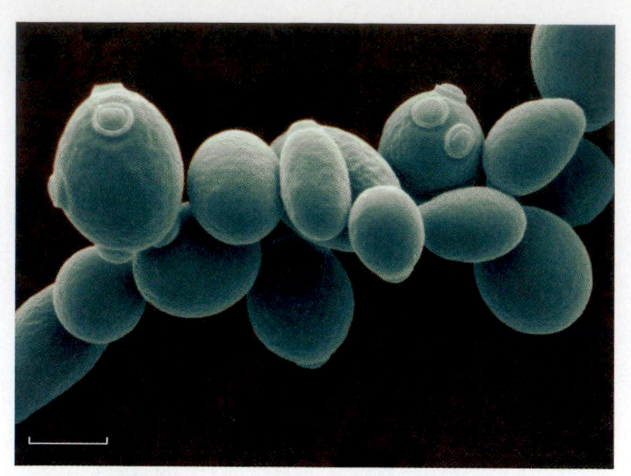

FIGURE 16.7 **Yeast Budding.** False-color scanning electron micrograph of the unicellular yeast *Saccharomyces cerevisiae*. (Bar = 3 μm.)

Q: Propose what the circular areas represent on the parent cell at the left and right center of the micrograph.

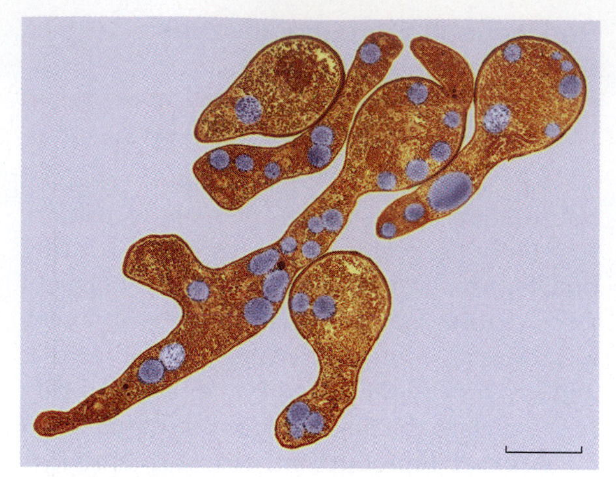

FIGURE 16.8 **Germinating Fungal Spores.** A false-color transmission electron micrograph of germinating fungal spores. (Bar = 3 μm.)

Q: What do the elongated structures protruding from the round spore represent?

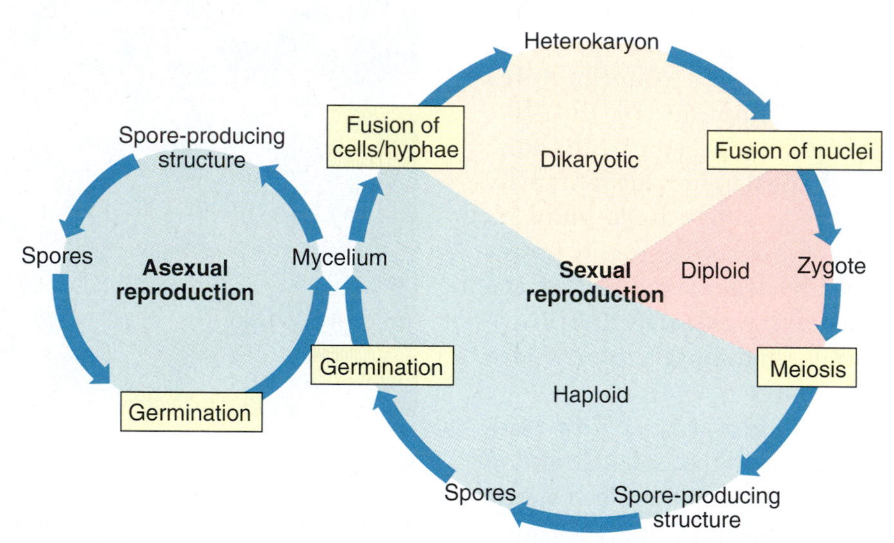

FIGURE 16.9 **A Typical Life Cycle of a Mold.** Many fungi have both an asexual and sexual reproduction characterized by spore formation. The unique phase in the life cycle is the presence of a dikaryotic phase in the sexual reproduction cycle. In this phase, nuclei from two different mating types remain as independent organelles in a common cytoplasm.

Q: How could an organism like a mold survive without a sexual cycle?

A visible fruiting body often results during sexual reproduction and it is the location of the spores. Perhaps the most recognized fruiting body from which spores are produced is the mushroom. Haploid spores develop from specialized cells.

Sexual reproduction is advantageous because it provides an opportunity for the evolution of new genetic forms better adapted to the environment than the parent forms. For example, a fungus may become resistant to fungicides as a result of chromosomal changes during sexual reproduction.

CONCEPT AND REASONING CHECKS

16.3 Assess the role of asexual and sexual reproduction in fungi.

16.2 The Classification of Fungi

Originally, fungi were considered part of the plant kingdom. Then in 1968, they received their own kingdom status with Whittaker's classification scheme (see Chapter 3). Currently, they are still considered a kingdom, but under the domain *Eukarya* in Woese's three-domain system.

Fungi Can Be Classified into Five Different Phyla

KEY CONCEPT

- Fungal classification usually is based on the form of sexual reproduction.

Historically, fungal distinctions were made on the basis of either structural differences, or physiological and biochemical patterns. However, DNA analyses and genome sequencing are becoming an important tool for drawing relationships among various fungi.

A fungus can be cataloged into one of five phyla, depending on its mode of sexual reproduction. These phyla are the Chytridiomycota, Glomeromycota, Zygomycota, Ascomycota, and the Basidiomycota. If the fungus lacks a recognized sexual cycle, it is placed into an informal group called the mitosporic fungi.

The Chytridiomycota and Glomeromycota. The oldest known fungi are related to certain members of the **Chytridiomycota**, commonly called **chytrids**. Members of the phylum give us clues about the origin of fungi. First, chytrids are predominantly aquatic, and not terrestrial, organisms. This means the fungi originated in the water along with plants and animals. Secondly, being aquatic, chytrids have flagellated reproductive cells. No other fungi have motile flagellate cells, suggesting the other fungi lost this trait at some point in their evolutionary history. Finally, like other fungi, chytrids have chitin strengthening their cell walls. Until recently, few chytrids had any noticeable impact—for good or bad (MicroFocus 16.4).

The **Glomeromycota** form what some consider the most extensive symbiosis on Earth. These fungi represent a group of mycorrhizae that exist within the roots of more than 80 percent of the world's land plants. They do not kill the plants but rather interact mutualistically by providing essential phosphate and other micronutrients (see Chapter 2) to the plant. In return, the fungi receive needed organic compounds from the plant. In fact, some mycologists believe that plant evolution onto land more than 400 million years ago depended on the symbiosis with the ancestral Glomeromycota, which were able to provide plants with needed nutrients from the soil.

CONCEPT AND REASONING CHECKS

16.4 What are the unique features of the Chytridiomycota and Glomeromycota?

The Zygomycota. The phylum **Zygomycota** consists of a group of fungi (zygomycetes) inhabiting terrestrial environments. Familiar representatives include fast-growing bread molds and other molds typically growing on spoiled fruits with high sugar content or on acidic vegetables (FIGURE 16.10). On these and similar materials, the heterotrophic fungi typically grow inside their food, dissolving the substrate with extracellular enzymes, and taking up nutrients by absorption.

Members of the phylum make up about one percent of the described species of fungi. The zygomycetes have chitinous cell walls and grow as mycelia with nonseptate hyphae. During sexual reproduction, sexually opposite mating types fuse, forming a unique, heterokaryotic, diploid **zygospore** (FIGURE 16.11). After a period of dormancy, the zygospore germinates and releases haploid sporangiospores from a sporangium. Elsewhere in the mycelium, thousands of asexually produced sporangiospores are produced within sporangia. Both sexually-produced and asexually-produced spores are dispersed on wind currents. Several members can cause infections and disease in humans.

CONCEPT AND REASONING CHECKS

16.5 Describe the properties of the Zygomycota.

The Ascomycota. Members of phylum **Ascomycota** (*asco* = "sac") or sac fungi, commonly are called the ascomycetes. They are very diverse, account for about 75 percent of all known fungi. The phyla contains many common and useful fungi, including *Saccharomyces cerevisiae* (Baker's yeast), *Morchella esculentum* (the edible morel), and *Penicillium*

MICROFOCUS 16.4: Environmental Microbiology

The Day the Frogs Died

A species of harlequin frog, one of many species being wiped out by a chytrid infection.

Many of us remember the days in high school biology lab when we had an opportunity to dissect a frog. Little did we realize that these imported African frogs may be responsible for a disease that is decimating the frog populations worldwide.

In the early 1990s, researchers in Australia and Panama started reporting massive declines in the number of amphibians in ecologically pristine areas. As the decade progressed, massive die-offs occurred in dozens of frog species and a few species even became extinct. Once filled with frog song, the forests were quiet. "They're just gone," said one researcher.

Hypotheses to explain these declines included, among others, habitat modification, introduction of new predators, increased ultraviolet radiation, pollution, adverse weather changes, and infectious disease.

By 1998, infectious disease was identified as one of the most likely reasons for the decline. More than 100 amphibian species on four continents, including North and South America, were infected with a chytrid called *Batrachochytrium dendrobatidis*. This chytrid, the only one known to infect vertebrates, uses the frog's keratinized skin as a nutrient source; epidermal sloughing of the frog's skin is one sign of the disease called chytridiomycosis. Roughly one third of the world's 6,000 amphibian species are now considered under threat of extinction (see figure). But, why the sudden die off in recent years?

Alan Pounds, a researcher at the Golden Toad Laboratory for Conservation in Costa Rica, suggests that global warming is creating optimal conditions for the fungus. His work, published in 2005, indicates climate change has led to cooler days and warmer nights on tropical mountainsides, creating the ideal growth—and infection—conditions for *B. dendrobatidis*.

Pounds also suggests the chytrid may have come from some of the African frogs exported around the world. His hypothesis is that some escaped frogs passed the fungus to hardier ones, like bullfrogs, which in turn infected more susceptible frog species.

Are frog deaths from chytridiomycosis a sign of a yet unseen shift in the ecosystem, much like a canary in a coal mine? Some believe this type of "pathogen pollution" may be as serious as chemical pollution.

"Disease is killing the frogs, but climate change is pulling the trigger," Dr. Pounds said.

FIGURE 16.10 Mold Growing on a Tomato. Zygomycete molds typically grow and reproduce on overripe fruits or vegetables, such as this tomato.

Q: Identify the black structures and the white fuzzy growth on the tomato.

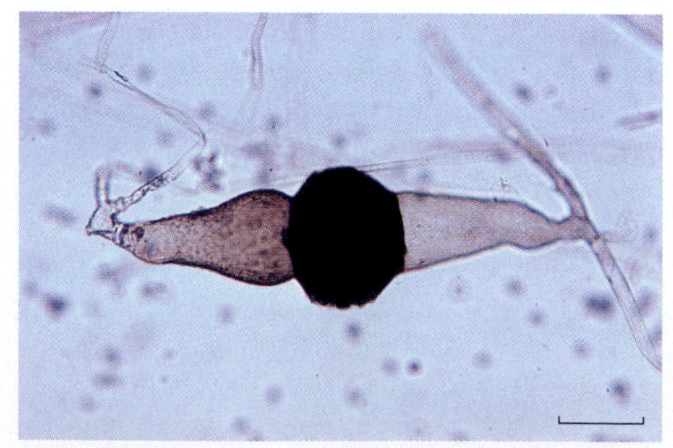

FIGURE 16.11 Sexual Reproduction in the Zygomycota. Sexual reproduction between hyphae of different mating types (+ and −) produces a darkly pigmented zygospore. (Bar = 30 μm.)

Q: When the zygospore germinates, what will be produced from the structure?

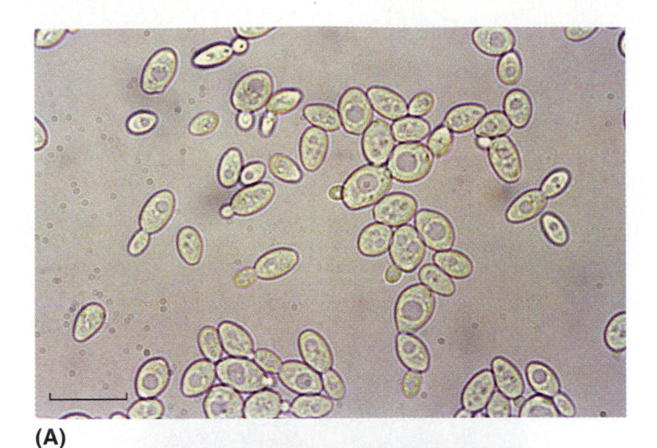

(A)

(B)

(C)

FIGURE 16.12 **Representative Ascomycetes.** (**A**) Light micrograph of yeast cells. Several of the cells are undergoing budding. (Bar = 10 µm.) (**B**) The edible morel, an ascomycete prized for its delicate taste. (**C**) A *Penicillium* species growing on an orange.

Q: What does the dark green part of the fungus on the orange represent?

chrysogenum (the mold that produces penicillin) (FIGURE 16.12). The phylum also has several members associated with illness and disease. *Aspergillus flavus*, produces **aflatoxin**, a fungal contaminant of nuts and stored grain that is both a toxin and the most potent known natural carcinogen; *Cryphonectria parasitica*, responsible for the death of 4 billion chestnut trees in the eastern United States; and *Candida albicans*, cause of thrush, diaper rash, and vaginitis. These and other human diseases will be discussed in the next section.

Penicillium and *Candida* species lack a sexual reproductive cycle. However, comparative genomics (see Chapter 8) and other nonsexual phenotypic characters have shown that these fungi actually are members of the Ascomycota.

The Ascomycota is a sister group to the Basidiomycota because the hyphae of both phyla are septate; both have cross-walls dividing the hyphae into segments, but with large pores allowing a continuous flow of cytoplasm. The hyphae, like other filamentous fungi, form a mycelium to obtain nutrients from dead or living organisms. In fact, their biggest ecological role is in decomposing and recycling plant material.

As a group, the ascomycetes have the ability to form conidia through asexual reproduction or **ascospores** through sexual reproduction. Ascospores are formed within a reproductive structure called an **ascus** (pl. asci), within which eight haploid ascospores form (FIGURE 16.13).

Some ascomycetes form symbioses with plant roots (mycorrhizae) or the leaves and stems of plants (endophytes). Ascomycetes also are the most frequent fungal partner in lichens. A **lichen** is a mutualistic association between a fungus and a photosynthetic organism such as a cyanobacterium or green alga (FIGURE 16.14A, B). Most of the visible body of a lichen is the fungus. Its hyphae penetrate the cells of the photosynthetic partner and receive carbohydrate nutrients. The photosynthetic organism receives fluid from the water-husbanding fungus.

Lichens often are grouped by appearance into leafy lichens (foliose), shrubbery lichens (fruticose), and crusty lichens (crustose), as FIGURE 16.14C shows. Together, the organisms

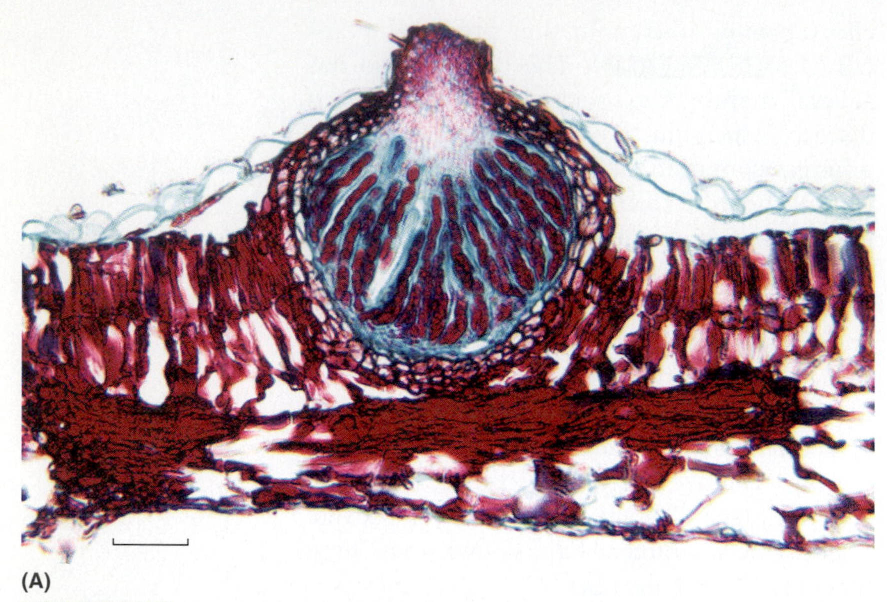

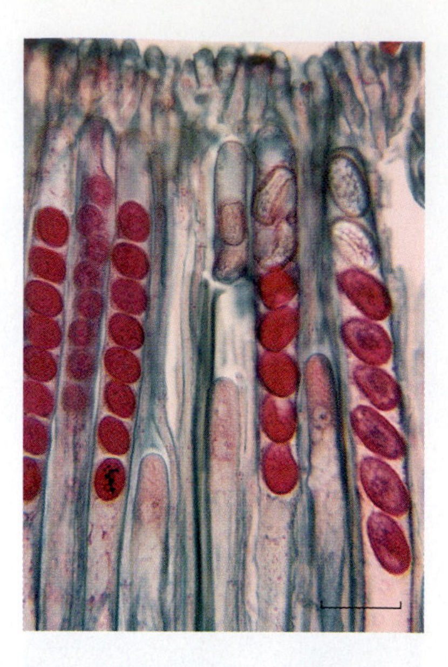

(A) (B)

FIGURE 16.13 **Ascomycetes and Spores.** (**A**) Cross section of an ascocarp on an apple leaf. (Bar = 20 μm.) (**B**) A higher magnification of several asci, each containing eight ascospores. (Bar = 5 μm.)

Q: Ascospores are representative of what stage of the ascomycete life cycle?

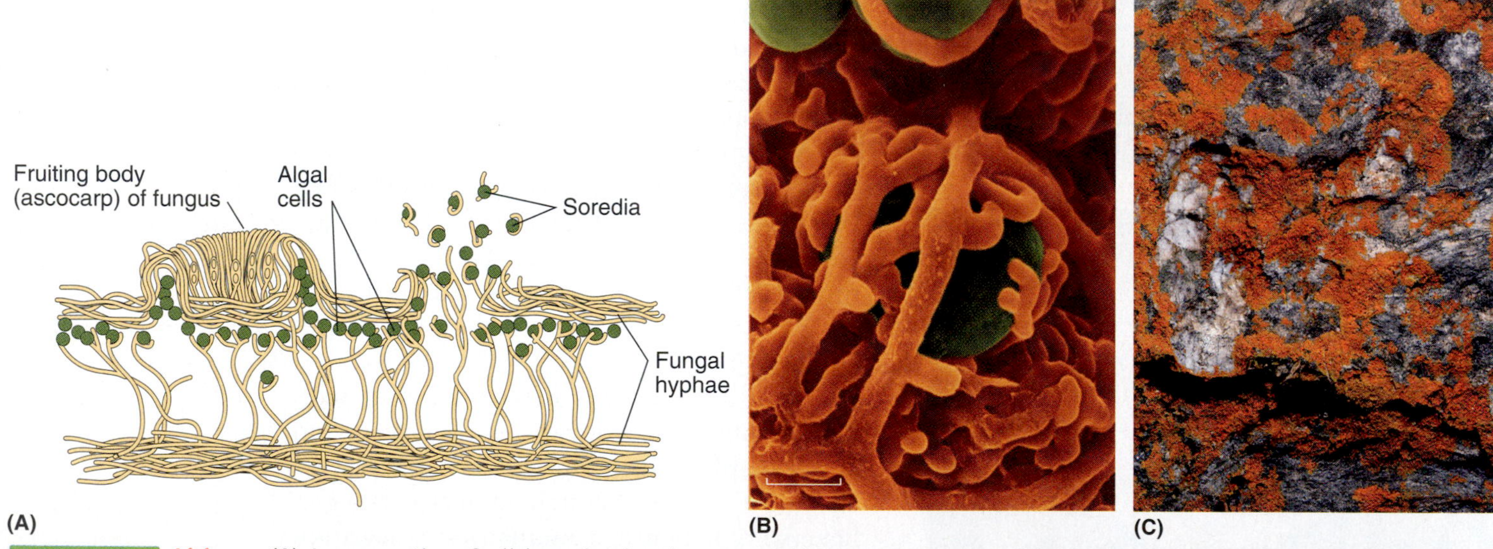

FIGURE 16.14 **Lichens.** (**A**) A cross section of a lichen, showing the upper and lower surfaces where tightly coiled fungal hyphae enclose photosynthetic algal cells. On the upper surface, a fruiting body, or ascocarp, has formed. Airborne clumps of algae and fungus called soredia are dispersed from the ascocarp to propagate the lichen. Loosely woven fungi at the center of the lichen permit the passage of nutrients, fluids, and gases. (**B**) A false-color scanning electron micrograph of the close, intimate contact between fungal hyphae (orange) and an alga cell (green). (Bar = 2 μm.) (**C**) A typical "crusty" lichen growing on the surface of a rock. Lichens are rugged organisms that can tolerate environments where there are few nutrients and extreme conditions.

Q: What attributes of the fungus and alga permit the lichen to withstand extreme environmental conditions?

form an association that readily grows in environments where neither organism could survive by itself (e.g., rock surfaces). Indeed, in some harsh environments, lichens support entire food chains. In the Arctic tundra, for example, reindeer graze on carpets of reindeer moss, which actually is a type of lichen.

CONCEPT AND REASONING CHECKS

16.6 Summarize the properties of the Ascomycota.

(A) (B) (C)

FIGURE 16.15 **The Basidiomycota.** The basidiomycetes are characterized by sexual reproductive structures that usually are macroscopic. These include (**A**) the mushrooms, such as this species of *Amanita muscaria,* and (**B**) the puffballs, which release spores in what appears like a puff of smoke (arrow). (**C**) The outward spreading of a fungal mycelium can be visually detected by the formation of a ring of mushrooms, often called a "fairy ring."

Q: Knowing how a mycelium grows, where on the mycelium (fairy ring) in (C) do the mushrooms form?

(A) (B) (C)

FIGURE 16.16 **Mushroom Gills and Basidiospores.** (**A**) A group of *Agaricus* mushroom caps, showing the gills on their underside. (**B**) A false-color scanning electron micrograph of the mushroom gills. The rough appearance represents forming basidia on which basidiospores will form. (**C**) Another scanning EM showing the basidiospores (dark brown) attached to basidia. (Bar = 20 μm.)

Q: Why is the specific epithet for this species of Agaricus *called* bisporis?

The Basidiomycota. Members of the phylum **Basidiomycota**, commonly known as basidiomycetes, are club fungi. The Basidiomycota contains about 30,000 identified species, representing 37 percent of the known species of true fungi. The Basidiomycota are unicellular or multicellular, sexual or asexual, and terrestrial or aquatic. The most recognized members are the mushrooms and puffballs (FIGURE 16.15A, B). Some Basidiomycota form mycorrhizae while others are important plant pathogens, such as the so-called "rusts" and "smuts" that infect cereals and other grasses. Rust fungi are so named because of the rusty, orange-red color of the infected plant, while smut fungi are characterized by sooty black masses of spores forming on leaves and other plant parts. Some Basidiomycota cause serious diseases in animals, including humans as we will soon discuss.

The name basidiomycete refers to the reproductive structure on which sexual spores are produced. In many mushrooms, the underside of the cap is lined with "gills" on which club-shaped **basidia** (sing. basidium; *basidium* = "small pedestal") are formed (FIGURE 16.16). Within these basidia, the haploid, sexual spores, called **basidiospores**, are produced.

In the soil, the basidiospores germinate and grow as a mycelium. When mycelia of different mating types come in contact, they fuse into a heterokaryon (FIGURE 16.17). A long-lived **dikaryon**, in which each cell contains

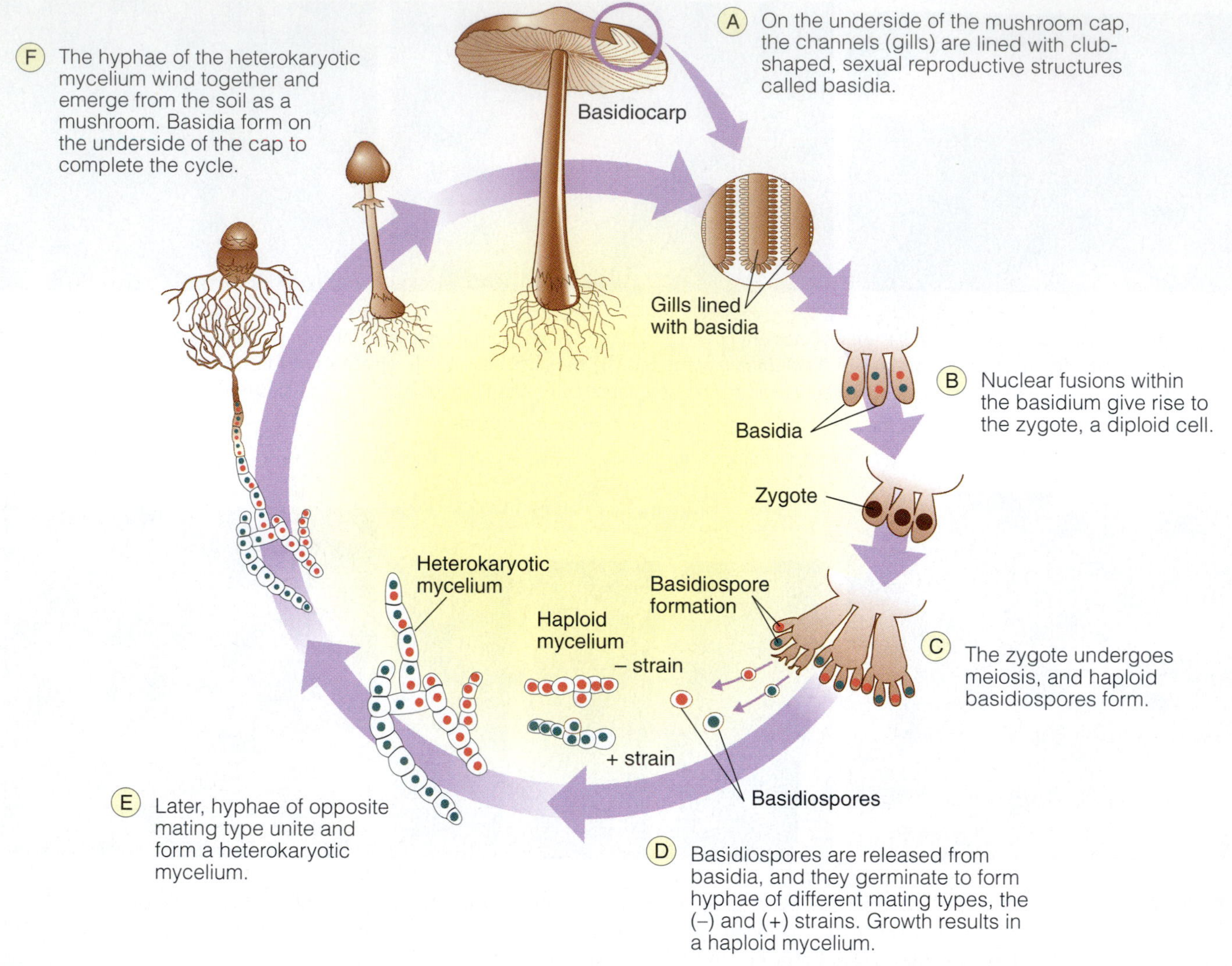

FIGURE 16.17 **The Sexual Reproductive Cycle of a Typical Basidiomycete.** The basidiomycets include the mushrooms from which basidiospores are produced.

Q: Identify the dikaryon phase of the basidiomycete sexual cycle in this figure.

two genetically different haploid nuclei resulting from the hyphal union, is another characteristic feature. Under appropriate environmental conditions, some of the hyphae become tightly compacted and force their way to the surface and grow into a fruiting body typically called a mushroom. Often a ring of mushrooms forms, which has been called a "fairy ring" because centuries ago people thought the mushrooms appeared where "fairies" had danced the night before (FIGURE 16.15C).

Edible mushrooms belong to the genus *Agaricus*, but some of the most deadly toxins are produced by another species, *Amanita phalloides*. Every year outbreaks of mushroom poisoning, most related to this genus, are reported to the Centers for Disease Control and Prevention (CDC).

MicroInquiry 16 looks at the relationship of the fungi to other eukaryotic kingdoms.

CONCEPT AND REASONING CHECKS

16.7 Identify the properties of the Basidiomycota.

The Mitosporic Fungi. Certain fungi lack a known sexual cycle of reproduction; consequently, they are labeled with the term **mitosporic fungi** because the asexual spores are the product of mitosis.

Many mitosporic fungi are reclassified when a sexual cycle is observed or compara-

TABLE 16.1 Comparisons of the Fungal Phyla Important to Human Disease

Phylum	Common Name	Cross Walls (Septa)	Sexual Structure	Sexual Spore	Asexual Spore	Representatives
Zygomycota	Zygomycetes	No	Zygospore	Zygospore	Sporangiospore	*Rhizopus*
Ascomycota	Ascomycetes, sac fungi	Yes	Ascus	Ascospore	Conidia	*Saccharomyces* *Aspergillus* *Penicillium*
Basidiomycota	Basidiomycetes, club fungi	Yes	Basidium	Basidiospore	Conidia fragments	*Agaricus* *Amanita*

tive genomics identifies a close relationship to a known phylum. We already mentioned this situation for *Penicillium* and *Candida*.

Another case in point is the fungus *Histoplasma capsulatum*, which causes a human disease we will discuss in the next section. When the organism was found to produce ascospores, it was reclassified with the ascomycetes and given the new name *Emmonsiella capsulata*. However, some traditions die slowly, and certain mycologists insisted on retaining the old name because it was familiar in clinical medicine. Thus, mycologists decided to use two names for the fungus: the new name, *Emmonsiella capsulata*, for the sexual stage; and the old name, *Histoplasma capsulatum*, for the asexual stage.

Among the mitosporic fungi a few are pathogenic for humans. These fungi usually reproduce by asexual spores, budding, or fragmentation, where the hyphal segments are commonly blown about with dust or deposited on environmental surfaces. For example, fragments of the athlete's foot fungus are sometimes left on towels and shower room floors.

The three phyla of fungi important to human health are compared in TABLE 16.1.

CONCEPT AND REASONING CHECKS

16.8 Distinguish the characteristics of the mitosporic fungi.

Yeasts Represent a Term for Any Single-Celled Stage of a Fungus

KEY CONCEPT

- Baker's yeast and other yeasts are important eukaryotic organisms in industry and research.

The word "yeast" refers to a large variety of unicellular fungi (as well as the single-cell stage of any fungus). Included in the group are non-spore-forming yeasts of the mitosporic fungi, as well as certain yeasts belonging to the Basidiomycota or Ascomycota. The yeasts we consider here are the species of *Saccharomyces* used extensively in industry and research. Pathogenic yeasts will be discussed with human diseases.

Saccharomyces (*saccharo* = "sugar") is a fungus with the ability to ferment sugars. The most commonly used species of *Saccharomyces* are *S. cerevisiae* and *S. ellipsoideus*, the former used for bread baking (Baker's yeast) and alcohol production, the latter for alcohol production. Yeast cells reproduce chiefly by budding, as described previously (see Figure 16.7), but a sexual cycle also exists in which cells fuse and form an enlarged structure (an ascus) containing spores (ascospores). *S. cerevisiae* therefore is an ascomycete.

The cytoplasm of *Saccharomyces* is rich in B vitamins, a factor making yeast tablets valuable nutritional supplements (ironized yeast) for people with iron-poor blood. The baking industry relies heavily upon *S. cerevisiae* to supply the texture in breads (MicroFocus 16.5). During the dough's rising period, yeast cells break down glucose and other carbohydrates, producing carbon dioxide gas (see Chapter 6). The carbon dioxide expands the dough, causing it to rise. Protein-digesting enzymes in yeast partially digest the gluten protein of the flour to give bread its spongy texture.

Yeasts are plentiful where there are orchards or fruits (the haze on an apple is a layer of

MICROINQUIRY 16

Evolution of the Fungi

The fossil record provides evidence that fungi and plants "hit the land" about the same time. There are fossilized fungi that are over 450 million years old, which is about the time that plants started to colonize the land. In fact, as described in this chapter, some fossilized plants, perhaps representing some of the first to colonize land, appear to have mycorrhizal associations.

Taxonomists believe that a single ancestral species gave rise to the five fungal phyla (Chytridiomycota, Glomeromycota, Zygomycota, Ascomycota, and Basidiomycota). Only the Chytridiomycota (chytrids) have flagella, suggesting they are the oldest line and that fungal ancestors were flagellated and aquatic (see figure). The chytrids may have evolved from a protistan ancestor and, as fungi colonized the land, they evolved different reproductive styles, which taxonomists have separated into the four nonflagellated phyla described in this chapter.

So, what evolutionary relationships do the fungi have with the eukaryotic kingdoms Plantae and Animalia? The **Table** below has some data that you need to analyze and from which you need to make a conclusion as to what relationships are best supported by the evidence. You do not need to understand the function or role for every characteristic listed. The answers can be found in **Appendix D**.

16.1. Based on the data presented, are fungi more closely related to animals or plants? Justify your answer.

16.2. Draw a taxonomic scheme showing the relationships between Plantae, Animalia, and Fungi kingdoms.

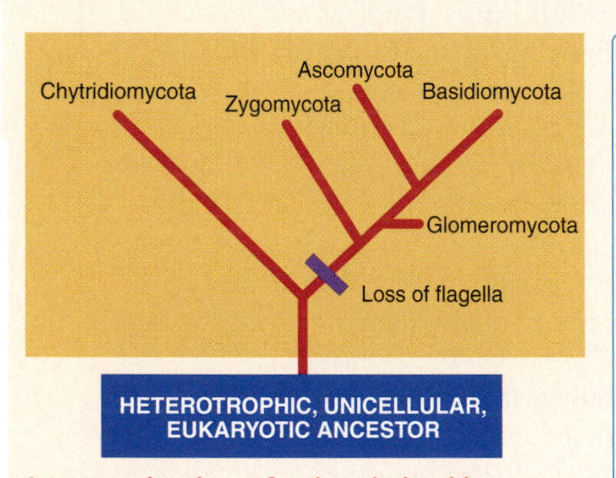

A taxonomic scheme for the relationships between fungal phyla.

TABLE

Comparisons of Organismal, Cellular, and Biochemical Characteristics

Characteristic	Fungi	Plants	Animals
Protein sequence similarities	✓		✓
Elongation factor 3 protein (translation) similarities	✓		✓
Types of polyunsaturated fatty acids			
Alpha linolenic		✓	
Gamma linolenic	✓		✓
Cytochrome system similarities	✓		✓
Mitochondrial UGA codes for tryptophan	✓		✓
Plate-like mitochondrial cristae	✓		✓
Presence of chitin	✓		some
Type of glycoprotein bonding			
O-linked	✓		✓
N-linked	✓		✓
Absorptive nutrition	✓		some
Glycogen reserves	✓		✓
Pathway for lysine biosynthesis			
Aminoadipic acid pathway	✓		
Diaminopimelic pathway		✓	
Type of sterol intermediate			
Lanosterol	✓		✓
Cycloartenol		✓	
Mitochondrial ribosome 5S RNA present		✓	
Presence of lysosomes	✓		✓

MICROFOCUS 16.5: History

The Drier the Better

ALL NATURAL YEAST • NO PRESERVATIVES
Fleischmann's ACTIVE DRY Yeast
NET WT. 1/4 OZ (7g)

Need a quick microbiology laboratory? Simply go to the grocery store and purchase a package of "active dry yeast." Open the package, pour the contents into a bit of warm water, and you're on your way—instant microorganisms! You can study the yeasts with the microscope, investigate their physiology, and if you're really technically proficient, rearrange their genes. All this from a package.

It wasn't always that way, though. Scientists could never figure out how to keep yeasts alive and in a dry state at the same time. If someone wanted to make bread, they had to get a starter from the "mother" dough where the yeast was growing; if the objective was wine fermentation, a trip to the culture supply or the old wine was necessary.

Then, during World War II, German prisoners were found in possession of a curious brown powder—it was the elusive dry yeast. They had the yeast for use as a nutritious food, or for bread when added to a bit of dough (or "Battlefield Red," if added to crushed grapes). Of course, the prisoners were not about to reveal the secret of the dried yeast, probably because they did not know what it was.

The postwar period was a different matter. In the euphoria of sharing, the Western world learned the secret of trehalose. Trehalose is a disaccharide. Added to yeast cells, it stabilizes the plasma membrane, prevents cell damage due to drying, and keeps the yeast alive and active. Now everyone knew the answer, including an entrepreneur named Arthur Fleischmann—millionaire founder and owner of Fleischmann's Active Dry Yeast.

yeasts). In natural alcohol fermentations, wild yeasts of various *Saccharomyces* species are crushed with the fruit; in controlled fermentations, *S. ellipsoideus* is added to the prepared fruit juice. The fruit juice bubbles profusely as carbon dioxide evolves. When the oxygen is depleted, the yeast metabolism shifts to fermentation and the pyruvate from glycolysis changes to consumable ethyl alcohol (see Chapter 6). The huge share of the American economy taken up by the wine and spirits industries is testament to the significance of the fermentation yeasts. A fuller discussion of fermentation processes is presented in Chapter 27.

S. cerevisiae probably is the most understood eukaryotic organism at the molecular and cellular levels. Its complete genome was sequenced in 1997, the first eukaryotic **model organism** to be completely sequenced. *S. cerevisiae* contains about 6000 genes. It might appear initially that *S. cerevisiae* would have little in common with human beings. However, both are eukaryotic organisms with a cell nucleus, chromosomes, and a similar mechanism for cell division (mitosis).

S. cerevisiae has been used to better understand not only cell function in animals, but also as a model for human disease. In fact, about 20 per cent of human disease genes have counterparts in yeast. For example, the chemical and signaling process by which a cell prepares itself for mitosis was either first discovered using yeast cells or major contributions to the understanding came from research with yeast cells. Research with yeast has identified the presence of prions, which have helped in the understanding of human prion diseases (see Chapter 13).

Potential drugs useful in disease treatment also can be screened using yeast cells. A yeast mutant, for example, with the equivalent of a human disease-causing gene, can be treated with potential therapeutic drugs to identify a compound able to restore normal function to the yeast cell gene. Such drugs, or modifications of them, might also be useful in humans.

Model organism: A relatively simple organism used to study general principles of biology.

CONCEPT AND REASONING CHECKS

16.9 Summarize the importance of yeasts to commercial interests and research.

16.3 Fungal Diseases of the Skin

In humans, fungal diseases, called **mycoses**, often affect many body regions. Some of these affect the skin or body surfaces. For example, several diseases, including ringworm and athlete's foot, involve the skin areas. One disease, candidiasis, may take place in the oral cavity, intestinal tract, skin, vaginal tract, and other body locations depending on the conditions stimulating its development.

Dermatophytosis Is an Infection of the Body Surface

KEY CONCEPT

- Dermatophytes can cause athlete's foot and ringworm.

Dermatophytosis (*dermato* = "skin"; phyto = "plant", referring to the days when fungi were grouped with plants) is a general name for a fungal disease of the hair, skin, or nails. The diseases are commonly known as **tinea infections** (*tinea* = "worm") because in ancient times, worms were thought to be the cause. Tinea infections include athlete's foot (tinea pedis); head ringworm (tinea capitis); body ringworm (tinea corporis); groin ringworm or "jock itch" (tinea cruris); and nail ringworm (tinea unguium).

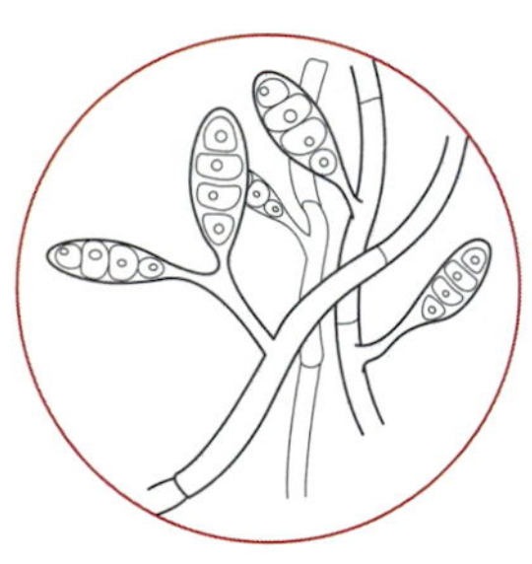

Trichophyton

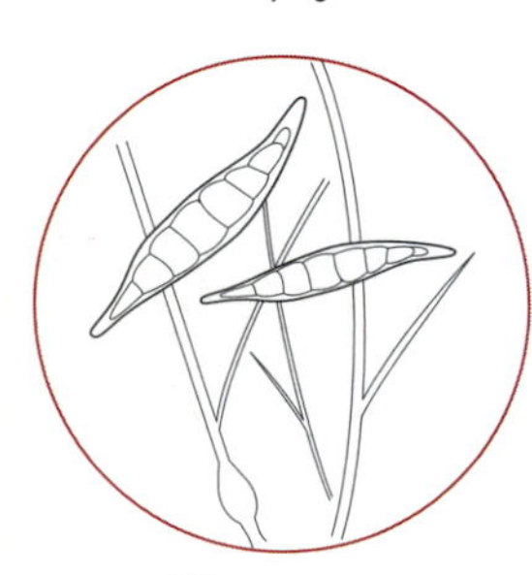

Microsporum

The causes of dermatophytosis are a group of fungi called **dermatophytes**. *Epidermophyton* currently is considered a mitosporic fungus, while species of *Trichophyton* (sexual stage *Arthroderma*) and *Microsporum* (sexual stage *Nannizzia*) are ascomycetes (FIGURE 16.18A).

If protected from dryness, the dermatophytes live for weeks on wooden floors of shower rooms or on mats. People transmit the fungi by contact and on towels, combs, hats, and numerous other types of **fomites** (inanimate objects). Because tinea diseases affect cats and dogs, they can transmit the fungi to humans (FIGURE 16.18B).

Dermatophytosis is commonly accompanied by blister-like lesions appearing along the nail plate, in the webs of the toes or fingers, or on the scalp or skin. Often a thin, fluid discharge exudes when the blisters are scratched or irritated. As the blisters dry, they leave a scaly ring. There also can be loss of hair, change of hair color, and local inflammatory reactions.

Treatment of dermatophytosis often is directed at changing the conditions of the skin environment. Commercial powders dry the diseased area, while ointments change the pH to make the area inhospitable for the fungus. Certain acids such as undecylenic acid (Desenex) and mixtures of acetic acid and benzoic acid (Whitfield's ointment) are active against the fungi. Also, tolnaftate (Tinactin) and miconazole (Micatin) are useful as topical agents for infections not involving the nails and hair. Griseofulvin, administered orally, is a highly effective chemotherapeutic agent for severe dermatophytosis.

CONCEPT AND REASONING CHECKS

16.10 Describe the characteristics of dermatophytosis.

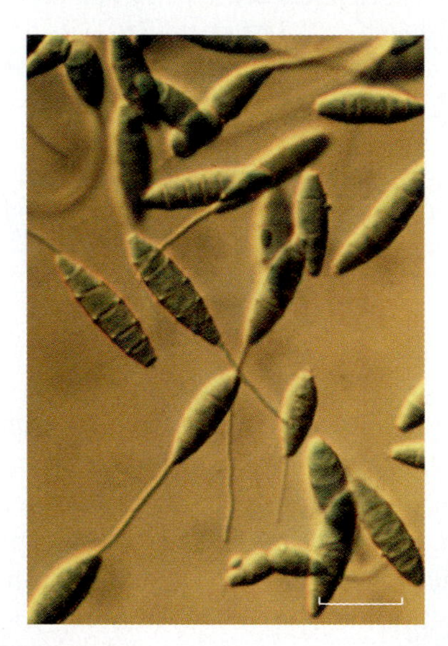

(A)

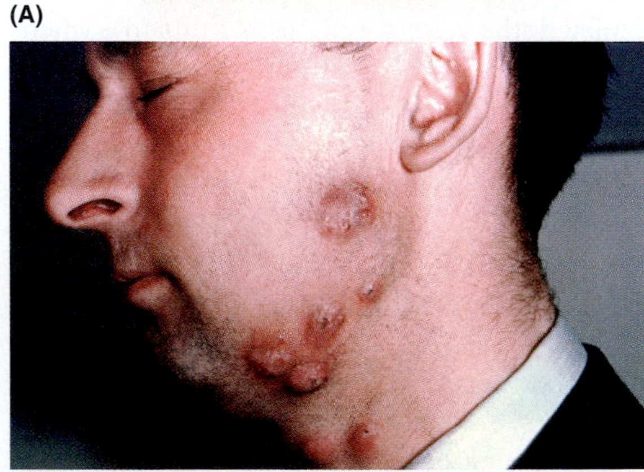

(B)

FIGURE 16.18 **Ringworm.** Ringworm of the skin or body (tinea corporis). (**A**) Light microscope photograph of *Microsporum,* one of the fungi causing ringworm on the scalp and body. (Bar = 15 μm.) (**B**) A case of body ringworm on the face and neck that was "caught" from a cat.

Q: Why is the shape of each ringworm skin lesion circular?

Candidiasis Often Is a Mild, Superficial Infection

KEY CONCEPT

- *Candida albicans* most commonly causes vaginitis or thrush.

Candida albicans often is present in the skin, mouth, vagina, and intestinal tract of healthy humans, where it lives without causing disease. The organism is a small mitosporic yeast that forms filaments called pseudohyphae when cultivated in laboratory media. When immune system defenses are compromised, or when changes occur in the normal microbial population in the body, *C. albicans* flourishes and causes numerous forms of **candidiasis**.

One form of candidiasis occurs in the vagina and is often referred to as **vulvovaginitis**, or a "yeast infection." There are some 20 million cases reported every year in the United States. Symptoms include itching sensations (pruritis), burning internal pain, and a white "cheesy" discharge. Reddening (erythema) and swelling of the vaginal tissues also occur. Diagnosis is performed by observing *C. albicans* in a sample of vaginal discharge or vaginal smear (FIGURE 16.19A), and by cultivating the organisms on laboratory media. Treatment is usually successful with nystatin (Mycostatin) applied as a topical ointment or suppository. Miconazole, clotrimazole, and ketoconazole are useful alternatives.

Vulvovaginitis is considered a sexually transmitted disease (but the disease is usually much milder in men than in women). In addition, studies have shown that excessive antibiotic use may encourage loss of the rod-shaped lactobacilli normally present in the vaginal environment. Without lactobacilli as competitors, *C. albicans* flourishes. Other predisposing factors are the contraceptive intrauterine device (IUD), corticosteroid treatment, pregnancy, diabetes, and tight-fitting garments, which increase the local temperature and humidity.

Oral candidiasis is known as **thrush**. This disease is accompanied by small, white flecks that appear on the mucous membranes of the oral cavity and then grow together to form soft, crumbly, milk-like curds (FIGURE 16.19B). When scraped off, a red, inflamed base is revealed. Oral suspensions of gentian violet and nystatin ("swish and spit") are effective for therapy. The disease is common in newborns, who acquire it during passage through the vagina (birth canal) of infected mothers. Children also may contract thrush from nursery utensils, toys, or the handles of shopping carts. Candidiasis may be related to a suppressed immune system. Indeed, thrush may be an early sign of AIDS in an adult patient.

Candidiasis in the intestinal tract is closely tied to the use of antibiotics. Certain drugs destroy the bacterial cells normally found there and allow *C. albicans* to flourish. In the 1950s, yogurt became popular as a way of replacing the bacterial cells. Today when intestinal surgery is anticipated, the physician often uses antifungal agents to curb *Candida* overgrowth.

■ Pseudohyphae: Cells formed by budding that are more elongated than typical oval cells.

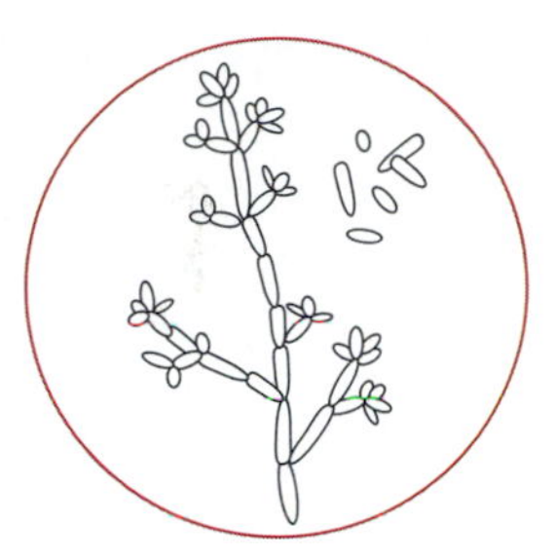

Candida albicans

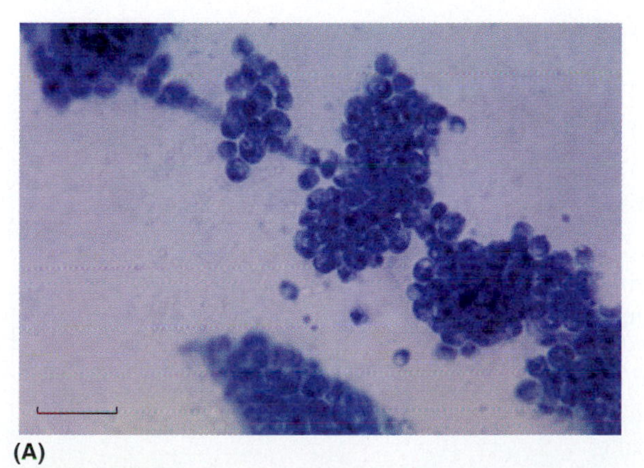

(A)

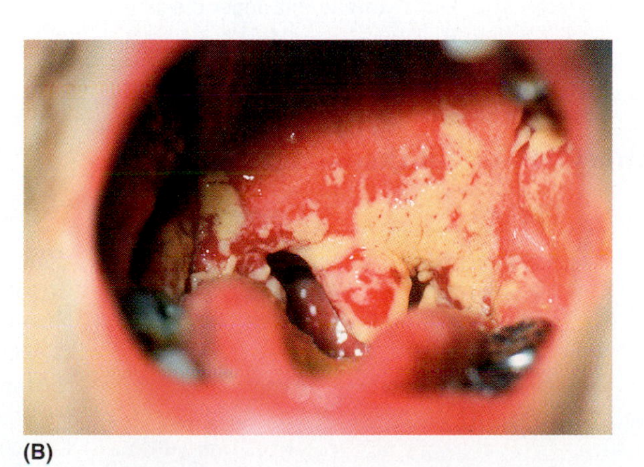

(B)

FIGURE 16.19 ***Candida albicans.*** **(A)** A light micrograph of stained *C. albicans* cells from a vaginal swab. (Bar = 60 μm.) **(B)** A severe case of oral candidiasis, showing a thick, creamy coating over the tongue.

Q: From the photo (A), how do you know this specimen of C. albicans *was from a human infection?*

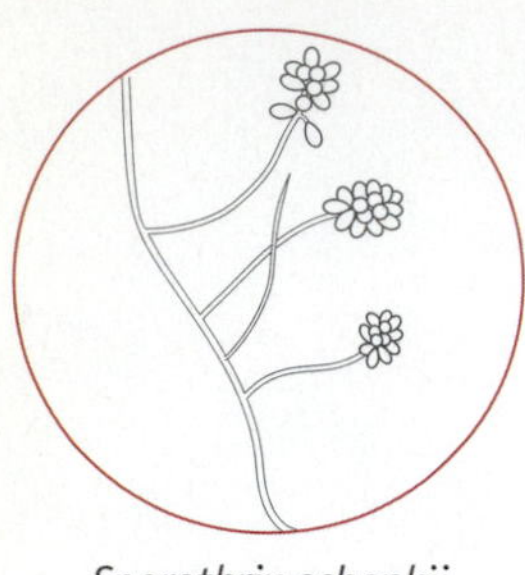

Sporothrix schenkii

Moreover, people whose hands are in constant contact with water may develop a hardening, browning, and distortion of the fingernails called **onychia**, also caused by *C. albicans*.

CONCEPT AND REASONING CHECKS

16.11 Describe the different forms of candidiasis based on body location.

Sporotrichosis Is an Occupational Hazard

KEY CONCEPT

- *Sporothrix schenkii* forms lesions under the skin.

People who work with wood, wood products, or the soil may contract **sporotrichosis**. Handling sphagnum (peat) moss used to pack tree seedlings or skin punctures with rose thorns (rose thorn disease) can lead to the disease as the result of infection by conidia from *Sporothrix schenkii* (FIGURE 16.20A). Pus-filled purplish lesions form at the site of entry, and "knots" may be felt under the skin (FIGURE 16.20B). Dissemination, though rare, may occur to the bloodstream, where blockages may cause swelling of the tissues (edema). In 1988, an outbreak of 84 cases of cutaneous sporotrichosis occurred in people who handled conifer seedlings packed with sphagnum moss from Pennsylvania. Cutaneous infections are controlled with potassium iodide, but systemic infections require amphotericin B therapy.

TABLE 16.2 summarizes the fungal diseases of the skin.

CONCEPT AND REASONING CHECKS

16.12 Describe the characteristics of sporotrichosis.

(A)

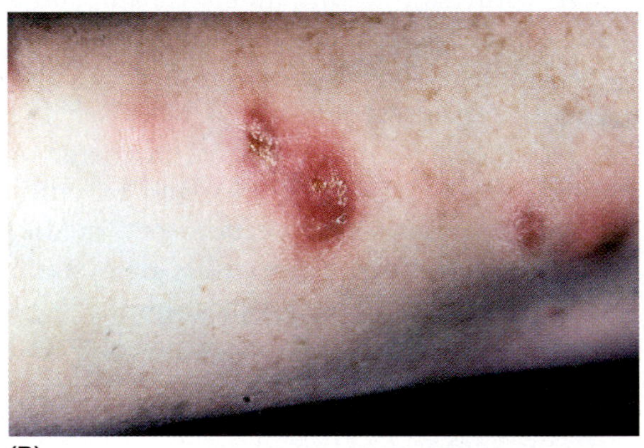

(B)

FIGURE 16.20 *Sporothrix schenkii.* (**A**) A false-color scanning electron micrograph of hyphae (orange) and conidia (purple) formed on conidiophores. (Bar = 8 μm.) (**B**) A patient showing the lesions of sporotrichosis on an infected arm. Characteristic "knots" can be felt under the skin.

Q: Explain how a rose thorn harboring S. schenkii *conidia can cause a skin disease.*

TABLE 16.2 A Summary of Fungal Diseases of the Skin

Organism	Phylum	Disease	Treatment	Comment
Trichophyton species *Microsporum species* *Epidermophyton species*	Ascomycota Ascomycota Mitosporic fungus	Tinea pedis Tinea capitis Tinea corporis	Undecylenic acid Griseofulvin Miconazole Itraconazole	Widely encountered in skin diseases
Candida albicans	Mitosporic fungus	Candidiasis Vaginitis Thrush, onychia	Nystatin Miconazole Ketoconazole	Normally in human intestine
Sporothrix schenkii	Ascomycota	Sporotrichosis	Amphotericin B Potassium iodide	Associated with rotten wood

16.4 Fungal Diseases of the Lower Respiratory Tract

Additional mycoses affect other body parts in humans, with a primary infection in the lungs that can spread to other body areas. In many of these fungal diseases, a weakened immune system contributes substantially to the occurrence of the infection.

Cryptococcosis Usually Occurs in Immunocompromised Individuals

KEY CONCEPT

- *Cryptococcus neoformans* causes a rare, but dangerous respiratory disease.

Cryptococcosis is among the most dangerous fungal diseases in humans. It affects the lungs and the meninges and is estimated to account for over 25 percent of all deaths from fungal disease.

■ Meninges: The membranes surrounding and protecting the brain and spinal chord.

Cryptococcosis is caused by an oval-shaped yeast known as *Cryptococcus neoformans* (sexual stage *Filobasidiella neoformans*) and is a member of the Basidiomycota. The organism is found in the soil of urban environments and grows actively in the droppings of pigeons, but not within the pigeon tissues. Cryptococci may become airborne with gusts of wind, and the organisms subsequently enter the respiratory passageways of humans. Air conditioner filters are hazardous because they trap large numbers of cryptococci.

C. neoformans cells, having a diameter of about 5 to 6 μm, are embedded in a gelatinous capsule that provides resistance to phagocytosis (FIGURE 16.21A). The cells penetrate to the air sacs of the lungs, but symptoms of infection are generally rare (FIGURE 16.21B). However, if the cryptococci pass into the bloodstream and localize in the meninges and brain, the patient experiences piercing headaches, stiffness in the neck, and paralysis. Diagnosis is aided by the observation of encapsulated yeasts in respiratory secretions or cerebrospinal fluid (CSF) obtained by a spinal tap.

Untreated cryptococcal meningitis may be fatal. However, intravenous treatment with the antifungal drug amphotericin B is usually successful, even in severe cases. Because this drug has toxic side effects, such as kidney damage and anemia, the patient should be monitored continually.

Resistance to cryptococcal meningitis appears to depend upon the proper functioning of the branch of the immune system governed by T lymphocytes (T cells). When these cells are absent in sufficient quantities, the immune system becomes severely compromised, and cryptococci can invade the tissues as opportunists. In AIDS patients cryptococcosis can be life-threatening.

CONCEPT AND REASONING CHECKS

16.13 Explain why cryptococcosis is such a dangerous fungal disease.

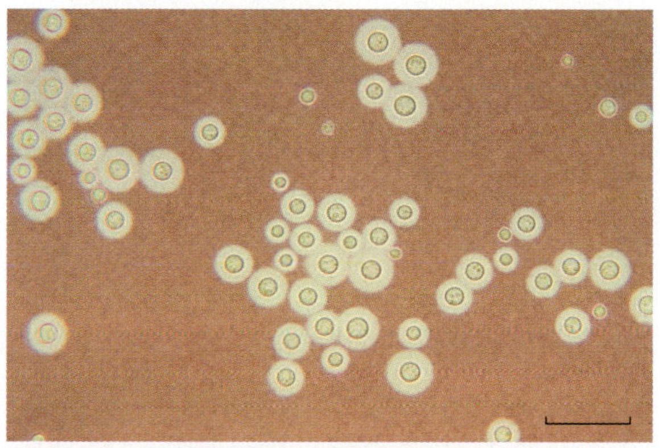

(A)

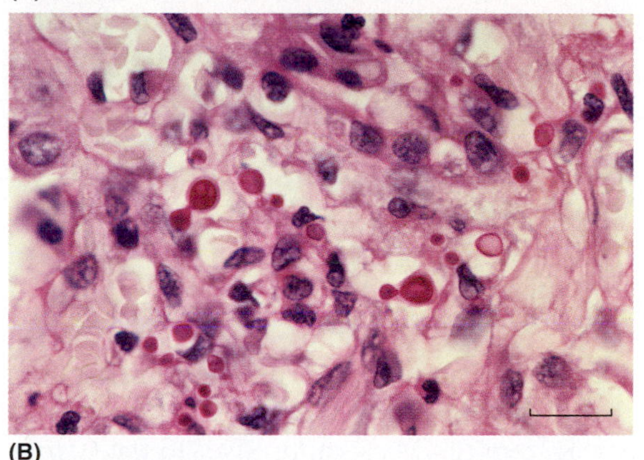

(B)

FIGURE 16.21 ***Cryptococcus neoformans.*** **(A)** A light microscope photomicrograph of *C. neoformans* cells. Internalization of the fungal cells is retarded by the capsules (white halo) they possess. (Bar = 20 μm.) **(B)** A stained photomicrograph of *C. neoformans* cells (red) from lung tissue of an AIDS patient. The capsule surrounding the cells provides resistance to phagocytosis and enhances the pathogenic tendency of the fungus. (Bar = 20 μm.)

Q: Why would C. neoformans *be a serious health threat to AIDS patients?*

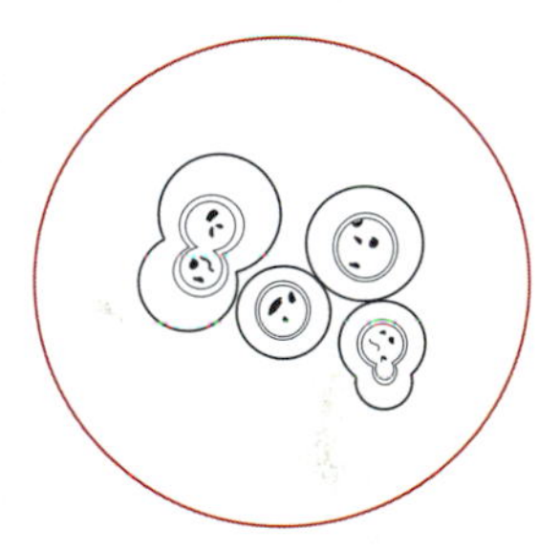

Cryptococcus neoformans

Histoplasmosis Can Produce a Systemic Disease

KEY CONCEPT

- *Histoplasma capsulatum* most often causes a mild flu-like illness.

In 1988, a group of 17 American university students crawled into a cave in Costa Rica to observe the numerous bats whose droppings covered the floor. Within three weeks, 15 students developed fever, headache, cough, and severe chest pains. Twelve patients tested positive for *Histoplasma capsulatum,* and all were treated for histoplasmosis.

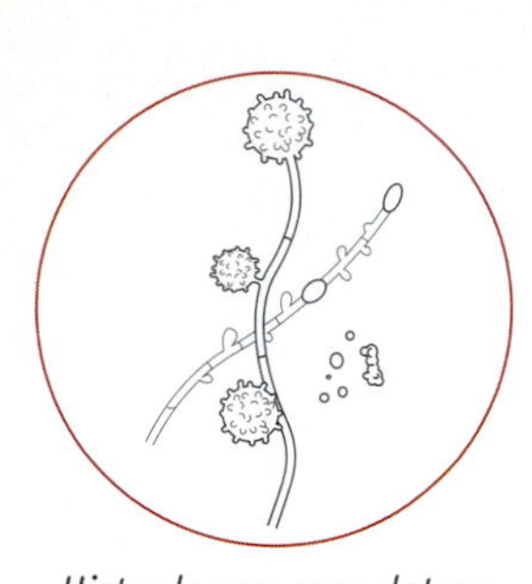

Histoplasma capsulatum

Histoplasmosis is a lung disease most prevalent in the Ohio and the Mississippi River valleys where it is often is called summer flu. The causative agent is *Histoplasma capsulatum.* Infection usually occurs from the inhalation of spores present in dry, dusty soil or found in the air of chicken coops and bat caves (FIGURE 16.22A). Prolonged exposure to the air therefore may be hazardous, as the Costa Rica outbreak illustrates. Being a dimorphic fungus, it grows as a yeast form at 37°C.

Most people experience mild influenza-like illness and recover without treatment. However, in immunocompromised people a disseminated form with tuberculosis-like lesions of the lungs and other internal organs may occur, making AIDS patients especially vulnerable. Amphotericin B or ketoconazole may be used in treatment.

Blastomycosis Usually Is Acquired Via the Respiratory Route

KEY CONCEPT

- *Blastomyces dermatitidis* can cause lung and skin infections.

Blastomycosis occurs principally in Canada, the Great Lakes region, and areas of the United States from the Mississippi River to the Carolinas. The pathogen is *Blastomyces dermatitidis* (sexual stage *Ajellomyces dermatitidis*). The fungus is dimorphic and produces conidia that are inhaled. Within the lungs, the conidia germinate as the yeast form.

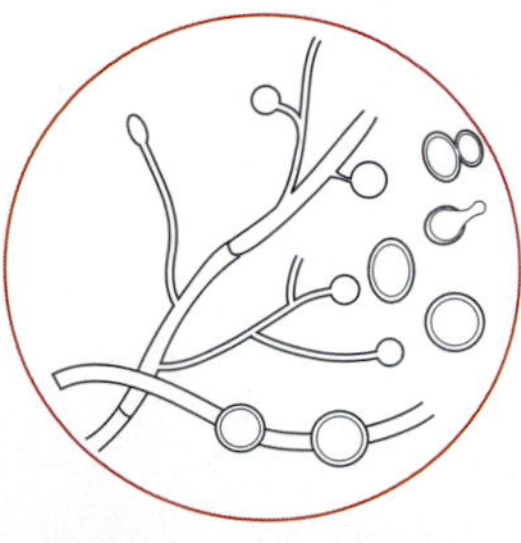

Blastomyces dermatitidis

Acute blastomycosis is associated with dusty soil and bird droppings, particularly in moist soils near barns and sheds. Inhalation leads to lung lesions with persistent cough and chest pains. Entry to the body also may occur through cuts and abrasions, and raised, wart-like lesions often are observed on the face, legs, or hands (FIGURE 16.22B). Healing is generally spontaneous (in 2 to 3 weeks).

Although blastomycosis is rare, it can affect immunocompromised patients, such as those with AIDS. Chronic pneumonia is the

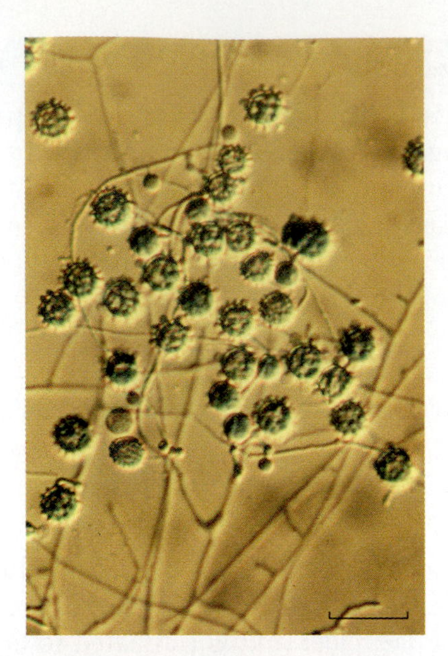

(A)

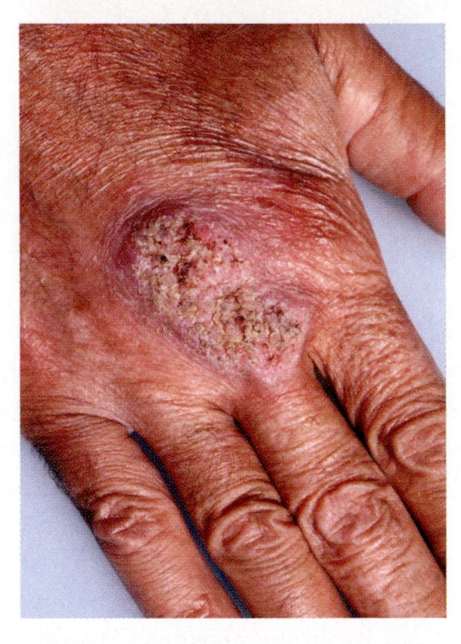

(B)

FIGURE 16.22 ***Histoplasma* Mycelium and *Blastomyces* Skin Infection.** (**A**) *H. capsulatum* grows as a mycelium in soil enriched by animal excrement. The spores are produced from the hyphal tips. (Bar = 20 μm.) (**B**) A skin infection of blastomycosis. The fungus usually affects the lungs after inhalation of fungal spores, but may become disseminated to the skin.

Q: Which dimorphic form of these fungi is associated with human infections?

most common manifestation. The disseminated form of blastomycosis may involve many internal organs (bones, liver, spleen, or nervous system) and may prove fatal. Amphotericin B is used in therapy.

CONCEPT AND REASONING CHECKS

16.14 Compare and contrast histoplasmosis and blastomycosis.

Coccidioidomycosis Can Become a Potentially Lethal Infection

KEY CONCEPT

- *Coccidioides immitis* infections can disseminate and be life-threatening.

Travelers to the San Joaquin Valley of California and dry regions of the southwestern United States may contract a fungal disease known as **coccidioidomycosis**, known more commonly as valley or desert fever. It is caused by *Coccidioides immitis*, an ascomycete fungus. The organism produces arthrospores by a unique process of endospore and spherule formation (FIGURE 16.23). Coccidioidomycosis is usually transmitted by dust particles laden with fungal spores. Cattle, sheep, and other animals deposit the spores in soil, and they become airborne with gusts of wind.

When inhaled into the human lungs, *C. immitis* induces an influenza-like disease, with a dry, hacking cough, chest pains, and high fever. During most of the 1980s, about 450 annual cases of coccidioidomycosis were reported to the CDC. In 1991, that number jumped to over 1,200 cases, and in 2004, the number of reports exceeded 6,400, the majority occurring in California and Arizona.

Although most cases are self-limiting, a small number of cases (1 out of 1,000) become disseminated and involve myriad internal organs and structures, including the meninges of the spinal cord. Recovery brings lifelong immunity. Amphotericin B is prescribed for severe cases and a vaccine is in the early stages of development.

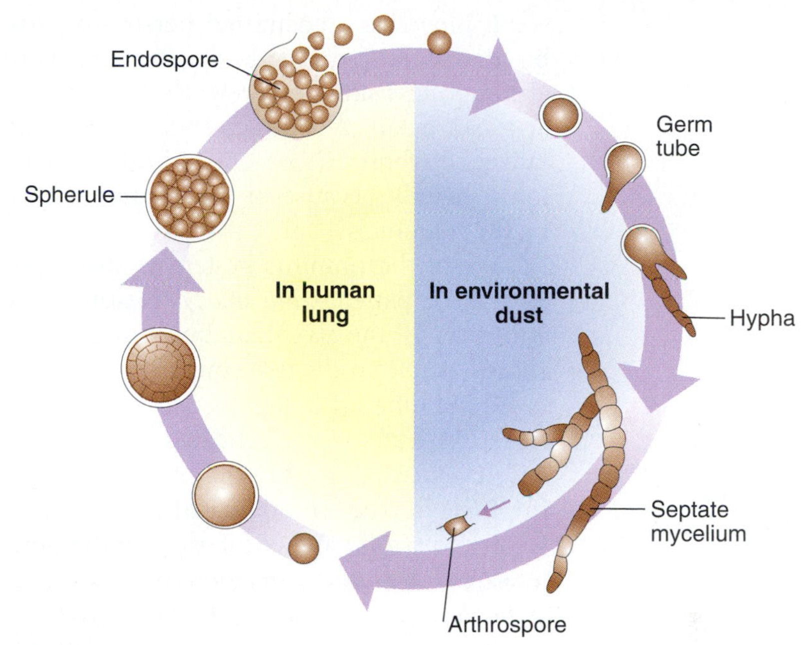

FIGURE 16.23 **The Life Cycle of *Coccidioides immitis*.** Outside the body, the organism exists as a septate mycelium. It fragments into airborne arthrospores, which are inhaled. In the respiratory tract, the arthrospores swell to yield a large body, the spherule, that segments and breaks down to release endospores. When released to the environment, the endospores form germ tubes and then the mycelium.

Q: Explain why C. immitis *is considered a dimorphic fungus.*

CONCEPT AND REASONING CHECKS

16.15 What is the unique feature of a coccidioidomycosis infection?

Pneumocystis Pneumonia Can Cause a Lethal Pneumonia

KEY CONCEPT

- *Pneumocystis jiroveci* is a fungal pathogen in immunocompromised hosts.

Pneumocystis pneumonia (**PCP**) currently is the most common cause of nonbacterial pneumonia in Americans with suppressed immune systems. The causative organism, *Pneumocystis jiroveci* (previously called *Pneumocystis carinii*) produces an atypical pneumonia that remained in relative obscurity until the 1980s, when it was recognized as the cause of death in over 50 percent of patients dying from the effects of AIDS. Although the organism's structure and behavior are more typical of a protozoan (Chapter 17), recent evidence supports the assignment of *P. jiroveci* as an ascomycete. Analysis of its ribosomal RNA, for example, shows a closer relationship to fungal RNA than to protozoal RNA.

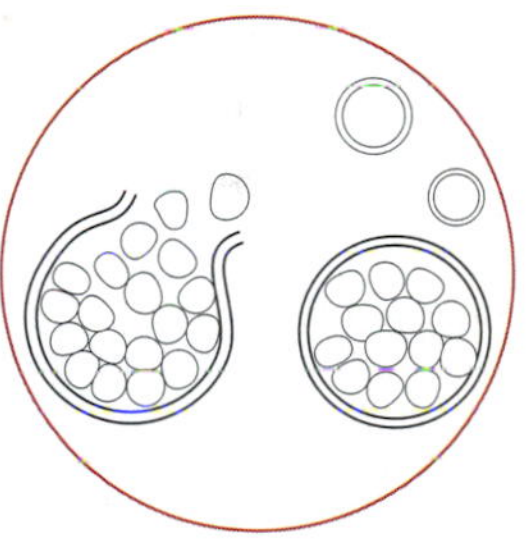

Coccidioides immitis

P. jiroveci has a complex life cycle taking place entirely in the alveoli of the lung. A feeding stage, called the **trophozoite**, swells to become a precyst stage, in which up to eight sporozoites develop in forming a cyst. When the cyst is mature, it opens and liberates the sporozoites, which enlarge and undergo further reproduction and maturation to trophozoites.

P. jiroveci is transmitted person-to-person by droplets from the respiratory tract, although transmission from the environment also can occur. A wide cross section of individuals harbors the organism without symptoms, mainly because of the control imposed by the immune system.

When the immune system is suppressed, as in AIDS patients, *Pneumocystis* trophozoites and cysts fill the alveoli and occupy all the air spaces. A nonproductive cough develops, with fever and difficult breathing. Progressive deterioration leads to consolidation of the lungs and, eventually, death. The current treatment for PCP is trimethoprim-sulfamethoxazole (co-trimoxazole) and corticosteroid therapy.

■ Consolidation: Formation of a firm dense mass in the alveoli.

CONCEPT AND REASONING CHECKS

16.16 What are the unique features of *P. jiroveci?*

Other Fungi Also Cause Mycoses

KEY CONCEPT

- *Aspergillus* and *Claviceps* species can cause human illnesses.

A few other ascomycete fungal diseases deserve brief mention because they are important in certain parts of the United States or they affect individuals in certain professions. Generally the diseases are mild, although complications may lead to serious tissue damage.

Invasive **aspergillosis** is a unique disease because the fungus enters the body as conidia and then grows as a mycelium. Disease usually occurs in an immunosuppressed host or where an overwhelming number of conidia has entered the tissue. The most common cause is *Aspergillus fumigatus*. An opportunistic infection of the lung may yield a round ball of mycelium called an **aspergilloma**, requiring surgery for removal. In addition, conidia in the earwax can lead to a painful ear disease known as **otomycosis** while disseminated *Aspergillus* causes blockage of blood vessels, inflammation of the inner lining of the heart, or clots in the heart vessels. Voriconazole therapy is usually necessary.

Another species of *Aspergillus* have great commercial value. Beano™ is an enzyme preparation from *A. niger* that breaks down carbohydrates typically found in beans, cabbage, broccoli, and other high fiber foods, which are foods tending to produce flatulence as a result of microbial action in the intestines. The *A. niger* enzyme preparation breaks down these carbohydrates before microbes can get to them and produce gas as an end product.

Fungal toxins are called **mycotoxins** and two closely related fungi, *A. flavus* and *A. parasiticus,* produce mycotoxins called **aflatoxins**. The mold is found primarily in warm, humid climates, where it contaminates agricultural products such as peanuts, grains, cereals, sweet potatoes, corn, rice, and animal feed. Aflatoxins are deposited in these foods and ingested by humans where they are thought to be carcinogenic, especially in the liver. Contaminated meat and dairy products are also sources of the toxins. Half of the cancers in sub-Saharan Africa are liver cancers and 40 percent of the analyzed foods contain aflatoxins, highlighting the threat.

Ergotism is caused by *Claviceps purpurea,* another ascomycete fungus producing a powerful toxin. *C. purpurea* grows as hyphae on kernels of rye, wheat, and barley. As hyphae penetrate the plant, the fungal cells gradually consume the substance of the grain, and the dense tissue hardens into a purple body called a **sclerotium**. A group of peptide derivatives called alkaloids are produced by the sclerotium and deposited in the grain as a substance called **ergot**. Products such as bread made from rye grain may cause ergot rye disease, or ergotism.

Symptoms may include numbness, hot and cold sensations, convulsions with epileptic-type seizures, and paralysis of the nerve endings. Lysergic acid diethylamide (LSD) is a derivative of an alkaloid in ergot. Commercial derivatives of these alkaloids are used to cause contractions of the smooth muscles, such as to induce labor or relieve migraine headaches. MicroFocus 16.6 relates another possible effect of ergotism.

TABLE 16.3 summarizes the fungal diseases of the respiratory tract.

CONCEPT AND REASONING CHECKS

16.17 Summarize the unique features of aspergillosis and ergotism.

MICROFOCUS 16.6: History

"The Work of the Devil"

As an undergraduate, Linnda Caporael was missing a critical history course for graduation. Little did she know that through this class she was about to provide a possible answer for one of the biggest mysteries of early American history: the cause of the Salem Witch Trials. These trials in 1692 led to the execution of 20 people who had been accused of being witches in Salem, Massachusetts (see figure).

A witch trial in Massachusetts, 1692.

Linnda Caporael, now a behavioral psychologist at New York's Rensselaer Polytechnic Institute, in preparation of a paper for her history course had read a book where the author could not explain the hallucinations among the people in Salem during the witchcraft trials. Caporael made a connection between the "Salem witches" and a French story of ergot poisoning in 1951. In Pont-Saint-Esprit, a small village in the south of France, more than 50 villagers went insane for a month after eating ergotized rye flour. Some people had fits of insomnia, others had hallucinogenic visions, and still others were reported to have fits of hysterical laughing or crying. In the end, three people died.

Caporael noticed a link between these bizarre symptoms, those of Salem witches, and the hallucinogenic effects of drugs like LSD, which is a derivative of ergot. Could ergot possibly have been the perpetrator?

During the Dark Ages, Europe's poor lived almost entirely on rye bread. Between 1250 and 1750, ergotism, then called "St. Anthony's fire," led to miscarriages, chronic illnesses in people who survived, and mental illness. Importantly, hallucinations were considered "work of the devil."

Toxicologists know that eating ergotized food can cause violent muscle spasms, delusions, hallucinations, crawling sensations on the skin, and a host of other symptoms—all of which, Linnda Caporael found in the records of the Salem witchcraft trials. Ergot thrives in warm, damp, rainy springs and summers, which were the exact conditions Caporael says existed in Salem in 1691. Add to this that parts of Salem village consisted of swampy meadows that would be the perfect environment for fungal growth and that rye was the staple grain of Salem—and it is not a stretch to suggest that the rye crop consumed in the winter of 1691–1692 could have been contaminated by large quantities of ergot.

Caporael concedes that ergot poisoning can't explain all of the events at Salem. Some of the behaviors exhibited by the villagers probably represent instances of mass hysteria. Still, as people reexamine events of history, it seems that just maybe ergot poisoning did play some role—and, hey, not bad for an undergraduate history paper!

TABLE 16.3 A Summary of Fungal Diseases of the Respiratory Tract

Organism	Phylum	Disease	Treatment	Comment
Cryptococcus neoformans	Basidiomycota	Cryptococcosis	Amphotericin B	Associated with pigeon droppings
Histoplasma capsulatum	Ascomycota	Histoplasmosis	Amphotericin B	Associated with birds and bats
Blastomyces dermatitidis	Ascomycota	Blastomycosis	Amphotericin B	Associated with bird droppings
Coccidioides immitis	Ascomycota	Coccidioidomycosis	Amphotericin B	Common in southwestern US
Pneumocystis jiroveci	Ascomycota?	*Pneumoncystis* pneumonia	Trimethoprim-sulfamethoxazole	Associated with AIDS
Aspergillus fumigatus	Ascomycota	Aspergillosis Otomycosis	Voriconazole	Hyphae grow in body

SUMMARY OF KEY CONCEPTS

16.1 Characteristics of Fungi

- Fungi are eukaryotic microorganisms with heterotrophic metabolism. Most fungi consist of masses of hyphae that form a mycelium. Cross-walls separate the cells of hyphae in many fungal species. Most fungi are coenocytic.
- Fungi secrete enzymes into the surrounding environment and absorb the breakdown products. Tremendous absorption can occur when there is a large mycelium surface. Most fungi are aerobic, grow best around 25°C, and prefer slightly acidic conditions.
- Reproductive structures generally occur at the tips of hyphae. Masses of asexually- or sexually-produced spores are formed within or at the tips of fruiting bodies.

16.2 The Classification of Fungi

- The phylum Chytridiomycota is characterized by motile cells, while the Glomeromycota represent fungi living symbiotically with land plants.
- In the phylum Zygomycota, the sexual phase is characterized by the formation of a zygospore, which releases haploid spores that germinate into a new mycelium.
- The phylum Ascomycota includes the unicellular yeasts and filamentous molds. Ascospores are produced that germinate to form a new haploid mycelium, while asexual reproduction is through the dissemination of conidia. Lichens are a mutualistic association between an ascomycete and either a cyanobacterium or a green alga.
- The phylum Basidiomycota includes the mushrooms. Within these fruiting bodies, basidiospores are produced. On germination, they produce a new haploid mycelium. Rusts and smuts that cause many plant diseases are additional members of the phylum.
- The mitosporic fungi lack a sexual phase. Many human fungal diseases involve fungi in this informal group.
- *Saccharomyces* is a notable unicellular ascomycete yeast involved in baking and brewing, and scientific research.

16.3 Fungal Diseases of the Skin

- Fungal dermatophytes cause skin diseases, including athlete's foot (tinea pedis) and various forms of ringworm.
- Candidiasis, caused by *Candida albicans,* is an opportunistic disease. Candidiasis can occur in numerous organs such as the vaginal tract (vulvovaginitis), and oral cavity (thrush). Various antifungal drugs are available for treatment.
- *Sporothrix schenkii* causes sporotrichosis, which is transmitted by skin punctures. Pus-filled lesions form, generating "knots" under the skin. Treatment involves potassium iodide.

16.4 Fungal Diseases of the Lower Respiratory Tract

- Cryptococcosis is caused by *Cryptococcus neoformans.* It often infects immunosuppressed individuals, such as AIDS patients, and therefore represents an opportunistic disease. Cryptococcosis affects the lungs and spinal cord. Amphotericin B is available to alleviate the symptoms of the disease.
- Histoplasmosis develops from the inhalation of spores of *Histoplasma capsulatum.* The infection usually does not require treatment. However, if the disease becomes disseminated, it produces tuberculosis-like lesions that can be fatal.
- Blastomycosis develops from breathing dusty soil or aerosols from bird droppings containing spores of *Blastomyces dermatitidis.* In humans, the disease can disseminate and produce wart-like lesions on the face, hands, and legs. Amphotericin B can be used for treatment.
- *Coccidioides immitis* causes coccidioidomycosis, a respiratory infection that develops from breathing the fungal spores. For most infected individuals, it produces flu-like symptoms. In a few people, the disease becomes disseminated and severe cases can be fatal. Treatment involves amphotericin B therapy.
- *Pneumocystis jiroveci* is an opportunistic fungus in individuals with an impaired immune system. The fungus infects the alveoli of the lungs leading to a nonproductive cough and fever. Progressive deterioration is called pneumocystis pneumonia (PCP). Without treatment, infection often leads to death.
- *Aspergillus fumigatus* causes aspergillosis, which is a lung infection. *A. flavus* produces a mycotoxin, called aflatoxin, which is carcinogenic. *Claviceps purpurea* produces alkaloid derivatives in grain called ergot. Eating ergotized breads can cause ergotism, which produces a variety of symptoms including convulsions with epileptic-type seizures and paralysis of nerve endings.

LEARNING OBJECTIVES

After understanding the textbook reading, you should be capable of writing a paragraph that includes the appropriate terms and pertinent information to answer the objective.

1. Justify why the fungi are in a separate kingdom from the plants.
2. Differentiate between molds, yeasts, and dimorphic fungi.
3. Summarize the structure and function of fungal hyphae.
4. Identify and describe the physical factors governing fungal growth.
5. Discuss examples of fungal symbioses, including the mycorrhizae and the fungal endophytes.
6. Distinguish between the forms of asexual and sexual spores produced by fungi.
7. Describe the generalized sexual life cycle of a mold.
8. Summarize the characteristics of the Chytridiomycota and their relevance to fungal taxonomy.
9. Identify the key characteristics of the Zygomycota, Ascomycota, and Basidiomycota.
10. Describe the organization of lichens.
11. Assess the usefulness of yeasts, such as *Saccharomyces*, to industry and scientific research.

12. Summarize the symptoms and treatment of dermatophytosis.
13. Describe the major types of *Candida* infections.
14. Identify common mechanisms for transmission of *Sporothrix schenkii.*
15. Summarize the symptoms and complications of cryptococcosis.
16. Discuss the similarities between histoplasmosis and blastomycosis.
17. Review the symptoms of and the complications arising from coccidioidomycosis.
18. Evaluate the potential seriousness of a pneumocystis pneumonia infection.
19. Summarize the possible affects of an *Aspergillus* species infection.
20. Assess the origins and consequences of ergotism.

SELF-TEST

Answer each of the following questions by selecting the ***one*** answer that best fits the question or statement. Answers to even-numbered questions can be found in **Appendix C**.

1. Which one of the following statements about fungi is *not* true?
   A. Some fungi are photosynthetic.
   B. Fungi have cell walls made of chitin.
   C. Fungi are heterotrophic organisms.
   D. Some fungi are unicellular.
   E. Fungi consist of the yeasts and molds.
2. Most dimorphic pathogens are _____ at _____ °C and _____ at _____ °C.
   A. asexual; 23; sexual; 37
   B. sexual; 23; asexual; 37
   C. filamentous; 37; yeast forms; 23
   D. yeasts; 10; molds; 23
   E. filamentous; 23; yeast forms; 37
3. A fungal mycelium may contain
   A. septate hyphae.
   B. a coenocytic cytoplasm.
   C. chitinous cell walls.
   D. mitochondria.
   E. All the above (**A–D**).
4. Which one of the following best describes the growth conditions for a typical fungus?
   A. pH 3; 23°C; no oxygen gas present
   B. pH 8; 30°C; no oxygen gas present
   C. pH 6; 30°C; oxygen gas present
   D. pH 3; 23°C; oxygen gas present
   E. None of the above (**A–D**) is correct.
5. Mycorrhizae and fungal endophytes
   A. are types of lichen associations.
   B. are plant pathogens.
   C. are human pathogens.
   D. form mutualistic associations with plants.
   E. cause diseases in animals.
6. All the following are examples of asexual spore formation *except:*
   A. arthrospores.
   B. conidia.
   C. sporangiospores.
   D. basidiospores.
   E. yeast budding.
7. A heterokaryon refers to
   A. different types of asexual spores.
   B. a cell with genetically-different nuclei.
   C. yeast and hyphal growth forms.
   D. different modes of disease transmission.
   E. different types of sexual spores.
8. This phylum makes up about 1 percent of fungal species, is nonmotile, and has nonseptate hyphae.
   A. Mitosporic fungi
   B. Chytridiomycota
   C. Zygomycota
   D. Basidiomycota
   E. Ascomycota
9. An organism without a known sexual stage would be classified in the
   A. Mitosporic fungi.
   B. Chytridiomycota.
   C. Zygomycota.
   D. Basidiomycota.
   E. Ascomycota.
10. This fungal disease can cause a blister-like lesion on the scalp.
    A. Candidiasis
    B. Dermatophytosis
    C. Cryptococcosis
    D. Histoplasmosis
    E. None of the above (**A–D**) is correct.
11. _____ causes more than 20 million cases each year and symptoms include an itching sensation and burning internal pain.
    A. Onychia
    B. Thrush
    C. "Jock itch"
    D. Vulvovaginitis
    E. Sporotrichosis
12. Which one of the following fungi would most likely be found in pigeon droppings?
    A. *Trichophyton*
    B. *Pneumocystis*
    C. *Cryptococcus*
    D. *Coccidioides*
    E. *Sporothrix*

13. Moving to the Ohio or Mississippi River valleys might make one susceptible to
- **A.** PCP.
- **B.** thrush.
- **C.** valley fever.
- **D.** dermatophytosis.
- **E.** histoplasmosis.

14. Amphotericin B would be used to treat all of the following *except:*
- **A.** blastomycosis.
- **B.** histoplasmosis.
- **C.** aspergillosis.
- **D.** valley fever.
- **E.** ergotism.

15. The formation of arthrospores and spherules is characteristic of
- **A.** coccidioidomycosis.
- **B.** histoplasmosis.
- **C.** candidiasis.
- **D.** aspergillosis.
- **E.** blastomycosis.

16. Aflatoxin is produced by _____ and is _____.
- **A.** *Sporothrix schenkii;* a hallucinogen
- **B.** *Aspergillus flavus;* carcinogenic
- **C.** *Claviceps purpurea;* a hallucinogen
- **D.** *Aspergillus niger;* a mycotoxin
- **E.** *Aspergillus fumigatus;* carcinogenic

QUESTIONS FOR THOUGHT AND DISCUSSION

Answers to even-numbered questions can be found in **Appendix C**.

1. In a suburban community, a group of residents obtained a court order preventing another resident from feeding the flocks of pigeons that regularly visited the area. Microbiologically, was this action justified? Why?
2. Mr. A and Mr. B live in an area of town where the soil is acidic. Oak trees are common, and azaleas and rhododendrons thrive in the soil. In the spring, Mr. A spreads lime on his lawn, but Mr. B prefers to save the money. Both use fertilizer, and both have magnificent lawns. Come June, however, Mr. B notices that mushrooms are popping up in his lawn and that brown spots are beginning to appear. By July, his lawn has virtually disappeared. What is happening in Mr. B's lawn and what can Mr. B learn from Mr. A?
3. Residents of a New York community, unhappy about the smells from a nearby composting facility and concerned about the health hazard posed by such a facility, had the air at a local school tested for the presence of fungal spores. Investigators from the testing laboratory found abnormally high levels of *Aspergillus* spores on many inside building surfaces. Is there any connection between the high spore count and the composting facility? Is there any health hazard involved?
4. On January 17, 1994 a serious earthquake struck the Northridge section of Los Angeles County in California. From that date through March 15, 170 cases of coccidioidomycosis were identified in adjacent Ventura County. This number was almost four times the previous year's number of cases. What is the connection between the two events?

APPLICATIONS

Answers to even-numbered questions can be found in **Appendix C**.

1. You decide to make bread. You let the dough rise overnight in a warm corner of the room. The next morning you notice a distinct beer-like aroma in the air. What did you smell, and where did the aroma come from?
2. In a Kentucky community, a crew of five workers demolished an abandoned building. Three weeks later, all five required treatment for acute respiratory illness, and three were hospitalized. Cells obtained from the patients by lung biopsy revealed oval bodies and epidemiologists found an accumulation of bat droppings at the demolition site. As the head epidemiologist, what disease did the workers contract?
3. A woman comes to you complaining of a continuing problem of ringworm, especially of the lower legs in the area around the shins. Questioning her reveals she has five very affectionate cats at home. What disease does she have and what would be your suggestion to her?

REVIEW

To test your knowledge of the important fungi, match the statement on the left to the organism on the right by placing the correct letter in the available space. A letter may be used once, more than once, or not at all. Answers to even-numbered statements are listed in **Appendix C**.

_____ **1.** Produces a widely used antibiotic.
_____ **2.** Used for bread baking.
_____ **3.** Causes "valley fever" in the southwestern US.
_____ **4.** Common white or gray bread mold.
_____ **5.** Sexual phase known as *Emmonsiella*.
_____ **6.** Agent of rose thorn disease.
_____ **7.** Edible mushroom.
_____ **8.** Has nonseptate hyphae.
_____ **9.** Associated with the droppings of pigeons.
_____ **10.** Agent of ergot disease in rye plants.
_____ **11.** Cause of vaginal yeast infections.
_____ **12.** One of the causes of dermatophytosis.
_____ **13.** Often found in chicken coops and bat caves.
_____ **14.** Produces a toxic aflatoxin.
_____ **15.** Reproduction includes a spherule.
_____ **16.** Involves a trophozoite stage.

A. *Agaricus* species
B. *Aspergillus flavus*
C. *Aspergillus* species
D. *Batrachochytrium dendrobatidis*
E. *Blastomyces dermatitidis*
F. *Candida albicans*
G. *Claviceps purpurea*
H. *Coccidioides immitis*
I. *Cryphonectria parasitica*
J. *Cryptococcus neoformans*
K. *Epidermophyton* species
L. *Histoplasma capsulatum*
M. *Penicillium notatum*
N. *Pneumocystis jiroveci*
O. *Rhizopus stolonifer*
P. *Saccharomyces cerevisiae*
Q. *Saccharomyces ellipsoideus*
R. *Sporothrix schenkii*
S. *Taxomyces andreanae*

HTTP://MICROBIOLOGY.JBPUB.COM/

The site features learning, an on-line review area that provides quizzes and other tools to help you study for your class. You can also follow useful links for in-depth information, read more MicroFocus stories, or just find out the latest microbiology news.

17

Eukaryotic Microorganisms: The Parasites

Chapter Preview and Key Concepts

17.1 Classification and Characteristics of Protists
- Protista include plant-, fungal-, and animal-like microorganisms.
- Protozoa represent several different branches on the "tree of life".
- Few protozoal but many helminthic parasites require two or more hosts.

17.2 Protozoal Diseases of the Skin, and the Digestive and Urinary Tracts
- Leishmaniasis is a vector-born disease.
- Amoebiasis, giardiasis, cryptosporidiosis, and cyclosporiasis result from ingestion of contaminated food or water.
- Trichomoniasis affects over 10 percent of sexually active individuals.

17.3 Protozoal Diseases of the Blood and Nervous System
- Malaria is caused by four different apicomplexan species.
- Trypanosomiasis embodies African sleeping sickness and Chagas disease.
- Babesiosis is a malaria-like disease.
- Toxoplasmosis is extremely contagious.
- Primary amoebic meningoencephalitis (PAM) is a rare but deadly infection of the brain.

17.4 The Multicellular Helminths and Helminthic Infections
- Parasitic helminths include the flatworms and the roundworms.
- Schistosomiasis, and human lung and liver fluke diseases, are due to trematode infections.
  MicroInquiry 17: Parasites as Manipulators
- Tapeworm diseases result from eating undercooked meat contaminated with cysts.
- Roundworm diseases affect the digestive system and muscles.
- Filariasis is a swelling of the lymphatic tissues.

It races through the bloodstream, hunkers down in the liver, then rampages through red blood cells before being sucked up by its flying, buzzing host to mate, mature, and ready itself for another wild ride through a two-legged motel.
—The editor of *Discover* magazine describing, in flowery terms, the life cycle of the malarial parasite

Approximately 2 million people die every year from malaria, an infection caused by a protozoan parasite and carried from person to person by mosquitoes. The disease is one of the most severe public health problems worldwide and is a leading cause of death and disease in many developing countries. Yet in 1957, a global program to eradicate the parasite commenced only to end in failure 21 years later. What happened?

Believe it or not, malaria once was an infectious killer in the United States. In the late 1880s, malaria was quite common in the American plains and southeast with epidemics reaching as far north as Montreal, Canada. Malaria was a major source of casualties in the American Civil War and until the 1930s, malaria remained endemic in the southern states.

To eliminate malaria, American officials established the National Malaria Eradication Program on July 1, 1947. This was a cooperative undertaking between the newly established Communicable Disease Center (the original CDC and a new component of the US Public Health Service) and state and local health agencies of the 13 malaria-affected Southeastern states. In 1947, 15,000 malaria cases were reported. However, by 1950, after more than 4,650,000 homes had been sprayed with the DDT (dichlorodiphenyltrichloroethane) pesticide to kill the mosquitoes, only 2,000 malaria cases were reported. By 1952, the United States was malaria free and the program ended.

Encouraged by the success of the American eradication effort, in 1957 the World Health Organization (WHO) began a similar effort to eradicate malaria worldwide. These efforts involved house spraying with insecticides, antimalarial drug treatment, and surveillance. Successes were made in some countries, but the emergence of drug resistance, widespread mosquito resistance to DDT, wars, and massive population movements made the eradication efforts unsustainable. The eradication campaign was eventually abandoned in 1978.

Now, almost 30 years later, the WHO estimates 40 percent of the world's population is at risk of malaria (FIGURE 17.1). The good news is another global campaign, initiated by the WHO, several United Nations agencies, and the World Bank, is underway. Named "Roll Back Malaria," the program calls for a 50 percent reduction in the burden of malaria by 2010. Let's hope there is great success this time around.

Malaria is just one of a number of human diseases caused by **parasites**, organisms that must live in or on a different species to get their nourishment. There are two different groups of parasites of concern to microbiology because of their ability to cause infectious disease.

One group contains single-celled **protists** (see Chapter 3). Some of the diseases they cause are familiar to us, such as malaria. Others affect the intestine, blood, or other organs of the body.

The second group are multicellular parasites, referred to as **helminths** (*helminth* = "worm"). These include the flatworms and roundworms, which together probably infect more people worldwide than any other group of organisms. In the strict sense, flatworms and roundworms are animals, but they are studied in microbiology because of their ability to cause disease. Together with the parasitic protists, they are the subject of study of the biological discipline known as **parasitology**.

Our study begins with a focus on the characteristics and classification of the parasitic protists, followed by a survey of human diseases they cause. Our review of the helminths is brief. We include descriptions of several parasites, their life cycles, and the types of organisms and tissues they infect.

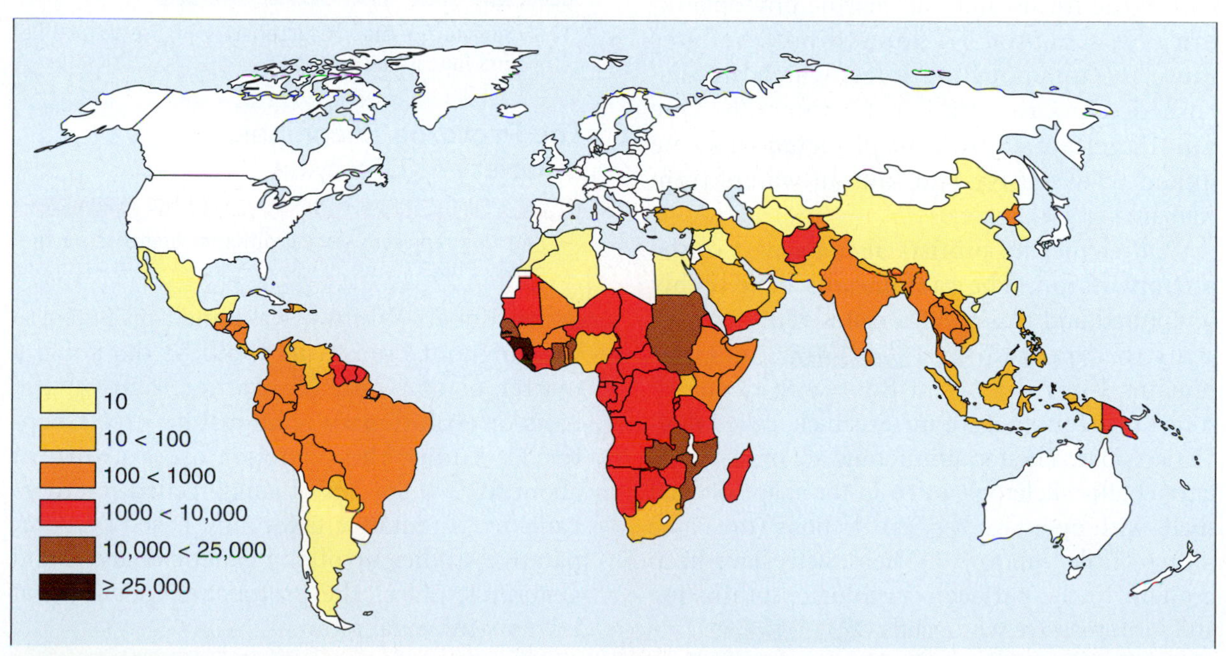

FIGURE 17.1 **Malaria Cases (per 100,000) by Country—2004.** The WHO has stated that 40 percent of the world's population is at risk of malaria. *Source*: Data from the WHO/Malaria Department.

Q: How can 40 percent of the world's population be at risk?

17.1 Classification and Characteristics of Protists

Biologists estimate there are some 200,000 named species of **protists**, those eukaryotic microorganisms that are not fungi, plants, or animals (see Chapter 3). As such, their taxonomic relationships are diverse and not always well understood.

The Protista Are a Perplexing Group of Microorganisms

KEY CONCEPT

- Protista include plant-, fungal-, and animal-like microorganisms.

The protists are an extremely diverse group. Most are unicellular; however, the functions of the single cell bear a resemblance to the functions of multicellular organisms rather than to those of an isolated cell from the organism. Most protists are free-living and thrive where there is water. They may be located in damp soil, mud, drainage ditches, and puddles. Some species remain attached to aquatic plants or rocks, while other species swim about.

Protists also are very diverse nutritionally. Some, including the unicellular **green algae**, contain chloroplasts and carry out photosynthesis similar to green plants (FIGURE 17.2A). Other protists, such as the **dinoflagellates**, are part of the freshwater and marine phytoplankton (see Chapter 3). Some dinoflagellates cause the infamous red tides, which are discussed in more detail in Chapter 26. One dinoflagellate, *Pfiesteria piscicida*, may be linked to extensive fish kills in waters from Alabama to Delaware.

■ Red tides: Brownish-red discoloration in seawater that is caused by an increased presence of dinoflagellates.

Other marine protists also are part of the phytoplankton. The **radiolarians** have highly sculptured and glassy silica plates with radiating arms to capture prey (FIGURE 17.2B). Their remains litter the ocean floor with deposits, sometimes hundreds of meters thick, called radiolarian ooze. The **foraminiferans** are protists that have chalky skeletons, often in the shape of snail shells with openings between sections (the name means "little window"). Their shells have been brought to the surface by geologic upthrusting and form massive white cliffs (FIGURE 17.2C).

Still other protists are completely heterotrophic and absorb extracellularly-digested food materials. Many of these are sometimes referred to as "fungal-like" protists primarily because they produce a mycelium-type of filamentous growth. Some are plant pathogens. *Phytophthora ramorum,* for example, infects the California coastal live oak and causes sudden oak death. *Phytophthora infestans* is the agent of late blight in potatoes. It was responsible for the Irish famine of the 19th century and its infection had great ecological impact on humans and society (MicroFocus 17.1).

The protists also include many motile, predatory, or parasitic species that absorb or ingest food. These protists traditionally have been called the **protozoa** (*proto* = "first; *zoo* = "animal"), referring to their animal-like properties that incorrectly suggested to biologists that protozoa were close evolutionary ancestors of the first animals. Though often studied by zoologists, protozoa also interest microbiologists because they are unicellular, most have a microscopic size, and several are responsible for infectious disease. Therefore, these are the microbial parasites we will emphasize next.

CONCEPT AND REASONING CHECKS

17.1 Summarize the characteristics of the plant-like and fungal-like protists.

The Protozoa Encompass a Variety of Lifestyles

KEY CONCEPT

- Protozoa represent several different branches on the "tree of life".

When Robert Whittaker assigned protozoa to the kingdom Protista in 1969, he did so as a matter of convenience, rather than on the basis of evolutionary relationships (see Chapter 3). Today, the protozoa are a group of about 65,000 species of single-celled microorganisms. A tentative taxonomy, based on comparative studies involving genetic analysis and genomics, places the protozoans in one of at least six informal groups.

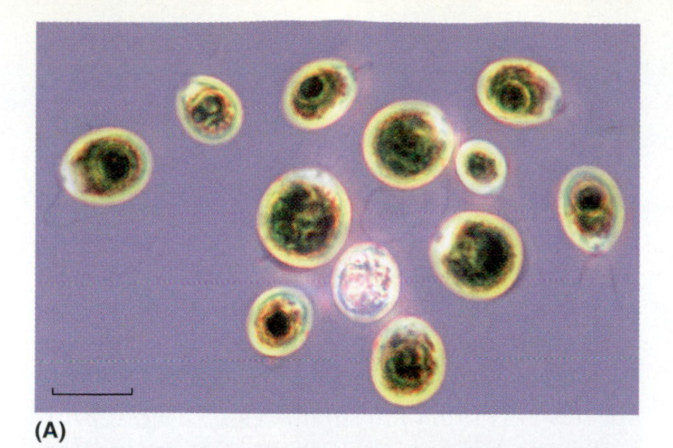

(A)

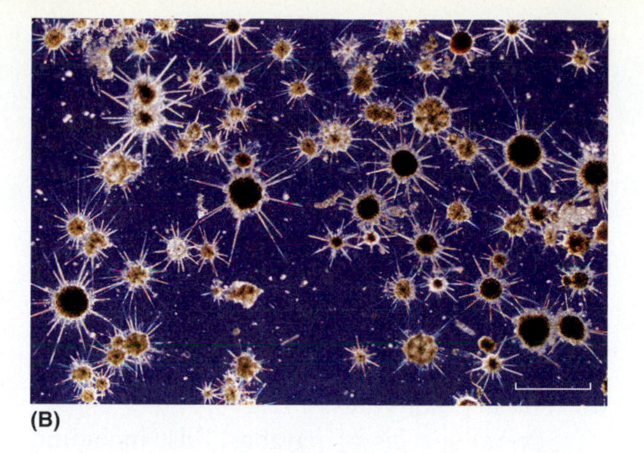

(B)

(C)

FIGURE 17.2 **Algae and Phytoplankton.** (**A**) A light micrograph of the green alga *Chlamydomonas*. (Bar = 10 μm.) (**B**) Light micrograph of an assortment of radiolarians. Radiolarians build hard skeletons made of silica around themselves as they float in seas with other plankton. (Bar = 120 μm.) (**C**) The white chalk cliffs at Beachy Head in Sussex, England consist of the ancient shells of foraminiferans.

Q: What characteristics are similar in the algae and phytoplankton?

MICROFOCUS 17.1: History

The Great Irish Potato Famine

Ireland of the 1840s was an impoverished country of 8 million people. Most were tenant farmers paying rent to landlords who were responsible, in turn, to the English property owners. The sole crop of Irish farmers was potatoes, grown season after season on small tracts of land. What little corn was available was usually fed to the cows and pigs.

Early in the 1840s, heavy rains and dampness portended calamity. Then, on August 23, 1845, The Gardener's Chronicle and Agricultural Gazette reported: "*A fatal malady has broken out amongst the potato crop. On all sides we hear of the destruction. In Belgium, the fields are said to have been completely desolated.*"

Potatoes covered with *Phytophthora infestans* (dark brown regions).

The potatoes had suffered before. There had been scab, drought, "curl," and too much rain, but nothing was quite like this new disease. It struck down the plants like frost in the summer. Beginning as black spots, it decayed the leaves and stems, and left the potatoes a rotten, mushy mass with a peculiar and offensive odor (see figure). Even the harvested potatoes rotted.

The winter of 1845 to 1846 was a disaster for Ireland. Farmers left the rotten potatoes in the fields, and the disease spread. The farmers first ate the animal feed and then the animals. They also devoured the seed potatoes, leaving nothing for spring planting. As starvation spread, the English government attempted to help by importing corn and establishing relief centers. In England, the potato disease had few repercussions because English agriculture included various grains. In Ireland, however, famine spread quickly.

After two years, the potato rot seemed to slacken, but in 1847 ("Black '47") it returned with a vengeance. Despite relief efforts by the English, over two million Irish people died from starvation. At least one million Irish people left the land and moved to cities or foreign countries. During the 1860s, great waves of Irish immigrants came to the United States—and in the next century, their Irish American descendants would influence American culture and politics. And to think—it all resulted from *Phytophthora infestans* that caused late blight of potatoes.

Parabasalids and Diplomonads. These protozoan groups are heterotrophic, have flagella for motility, and live in low oxygen or anaerobic environments. As such, they contain few, if any mitochondria. The organisms in these groups may represent organisms whose ancestors were the earliest forms of unicellular eukaryotes.

■ Bilateral symmetry: A form of symmetry where an imaginary plane divides an object into right and left halves, each side being a mirror of the other.

Several species of **parabasalids**, including *Trichonympha*, are found in the guts of termites where the protozoa participate in a mutualistic relationship (FIGURE 17.3A). The cells contain hundreds of flagella with the typical 9+2 arrangement of microtubules found in all eukaryotic flagella (see Chapter 3). Undulations sweep down the flagella to the tip, and the lashing motion forces water outward to provide locomotion. Other species, such as *Trichomonas vaginalis*, are parasitic in humans.

The **diplomonads** have two haploid nuclei and three pair of flagella at the anterior end and one pair at the posterior end, giving the cell bilateral symmetry. The most notable species is *Giardia intestinalis* (FIGURE 17.3B). It lacks several typical eukaryotic organelles, including a Golgi apparatus and lysosomes, and has primitive mitochondria. It can survive outside the anaerobic environment of the intestine by forming a **cyst**, which is a dormant, highly resistant stage that develops in some protozoa when the organism secretes a thick case around itself during times of environmental stress.

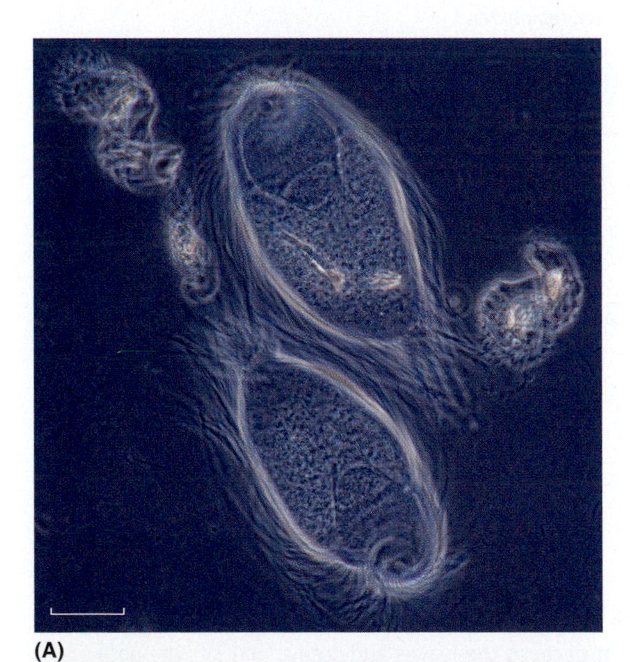

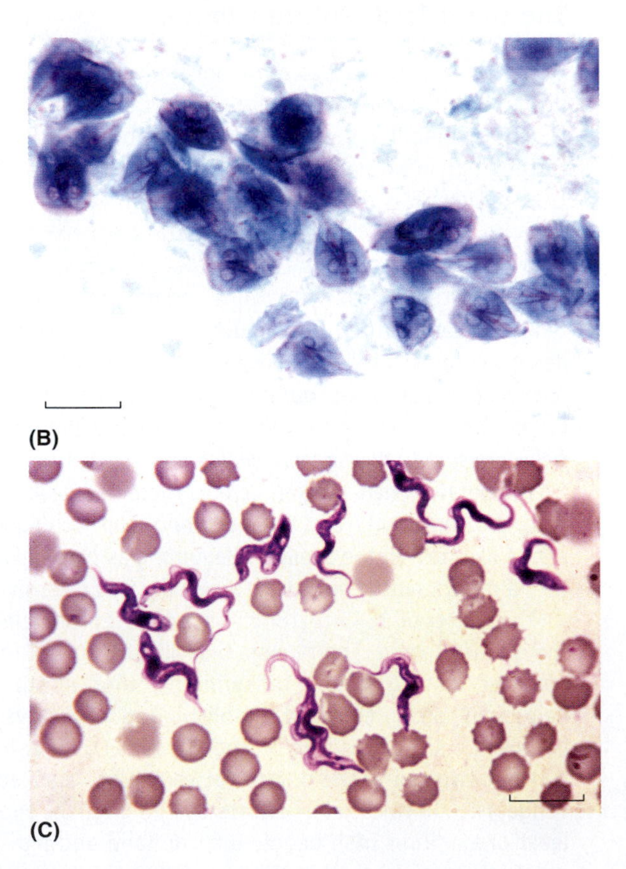

FIGURE 17.3 **Parabasalid, Diplomonad, and Kinetoplastid Parasites.** (**A**) A light micrograph of *Trichonympha,* a parabasalid parasite found in the gut of termites. Each thin, wispy line represents a flagellum used for motility. (Bar = 25 µm.) (**B**) Another light micrograph of stained *Giardia intestinalis* cells. The pear-shaped cell body of this diplomonad and flagella are typical features. (Bar = 5 µm.) (**C**) Light micrograph of *Trypanosoma* in a blood smear. The wavy cell appearance is due to the membrane covering the flagellum. (Bar = 10 µm.)

Q: What type of heterotrophic nutrition (absorptive or ingestive) can you infer from these micrographs?

Kinetoplastids. Another ancient lineage is members of the **kinetoplastids**, which are heterotrophic and have a single posterior flagellum (FIGURE 17.3C). A wavy, undulating membrane is attached to the flagellum. The protozoa have the typical array of eukaryotic organelles and the single mitochondrion contains a mass of DNA called the kinetoplast. All of the almost 800 species are parasitic and live in aerobic or anaerobic environments.

Some 60 percent of the species are trypanosomes (*trypano* = "hole"; *soma* = "body"), referring to the hole the organism bores to infect the host. Two *Trypanosoma* species cause forms of sleeping sickness in Africa and South America, while species of *Leishmania* can produce a skin disease or often fatal visceral infection.

CONCEPT AND REASONING CHECKS

17.2 What features can be used to separate the parabasilids, diplomonads, and kinetoplastids?

Amoebozoans. Another group of protozoa, the **amoebozoans** (*amoeba* = "change"), are mostly free-living and may be as large as 1 millimeter in diameter. They usually live in freshwater and reproduce by binary fission. The amoebozoans are soft bodied forms having the ability to change shape as their cytoplasm flows into temporary formless cytoplasmic projections called **pseudopods** (*pseudo* = "false"; *pod* = "a food"); thus, the motion is called **amoeboid motion.** Pseudopods also capture small algae and other protozoa through the ingestive process of phagocytosis (FIGURE 17.4A). The pseudopods enclose the particles to form an organelle called a **food vacuole**, which then joins with lysosomes. The lysosomes contain digestive enzymes to digest the material in the captured prey. Nutrients are absorbed from the vacuole, and any undigested residue is eliminated from the cell.

The genus *Entamoeba* can be far more serious, as all species are parasitic. In humans, amoebic dysentery or encephalitis may result from an infection.

Ciliates. The ciliated protozoans, or **ciliates**, are among the most complex cells on Earth and have been the subject of biological investigations for many decades. They are found readily in almost any pond water sample; they have a variety of shapes; they exist in several colors, including light blue and pink; and they exhibit elaborate and controlled behavior patterns.

Ciliates range in size from a microscopic 10 μm to a huge 3 mm. All ciliates are covered with hair-like cilia (sing., cilium) in longitudinal or spiral rows. Cilia beat in a synchronized and coordinated pattern, much like a field of wheat bending in the breeze or the teeth on a comb bending if you pass your thumb across the row. The organized "rowing" action moves the ciliate along in one direction. By contrast, flagellar motion tends to be jerky and much slower.

The cytoplasm contains the typical eukaryotic organelles and each cell has two types of cell nuclei. All ciliates have a complex series of alveolar sacs, called a **pellicle**, underlying the plasma membrane. The pellicle provides cell structure and stores calcium ions that control the rate of ciliary beating.

Ciliates, like the amoebozoans, are heterotrophic by ingestion through a primitive gullet, which sweeps in food particles for digestion. In addition, freshwater protozoa continually take in water by the process of osmosis and eliminate the excess water via organelles called **contractile vacuoles** (FIGURE 17.4B). These vacuoles expand with water drawn from the cytoplasm and then appear to "contract" as they release water through a temporary opening in the cell membrane.

Asexual reproduction occurs by binary fission. The complexity of ciliates however is illustrated by the nature of sexual reproduction (FIGURE 17.5). Ciliates have two types of nuclei; there is a single large **macronucleus** controlling cell metabolism and one or more **micronuclei** used for genetic recombination. During sexual recombination, called **conjugation**, two cells make contact, and a cytoplasmic bridge forms between them. A micronucleus from each cell undergoes two divisions to form four micronuclei, of which three disintegrate and only one remains to

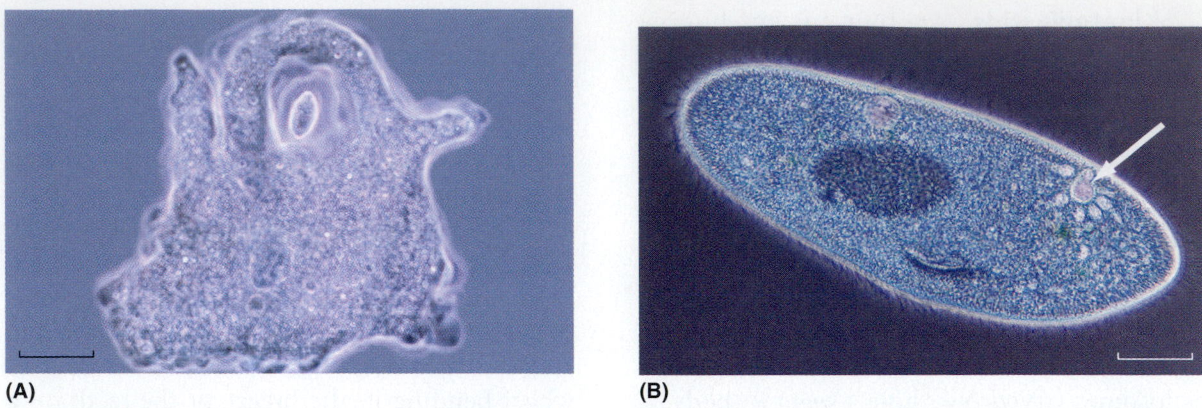

FIGURE 17.4 **An Amoeba and a Ciliate.** (**A**) A light micrograph of *Amoeba proteus*. The pseudopodia are extending around the prey, beginning the process of phagocytosis. (Bar = 100 μm.) (**B**) This light micrograph of the ciliate *Paramecium* shows the contractile vacuole (arrow). (Bar = 50 μm.)

Q: What is the function of the contractile vacuole?

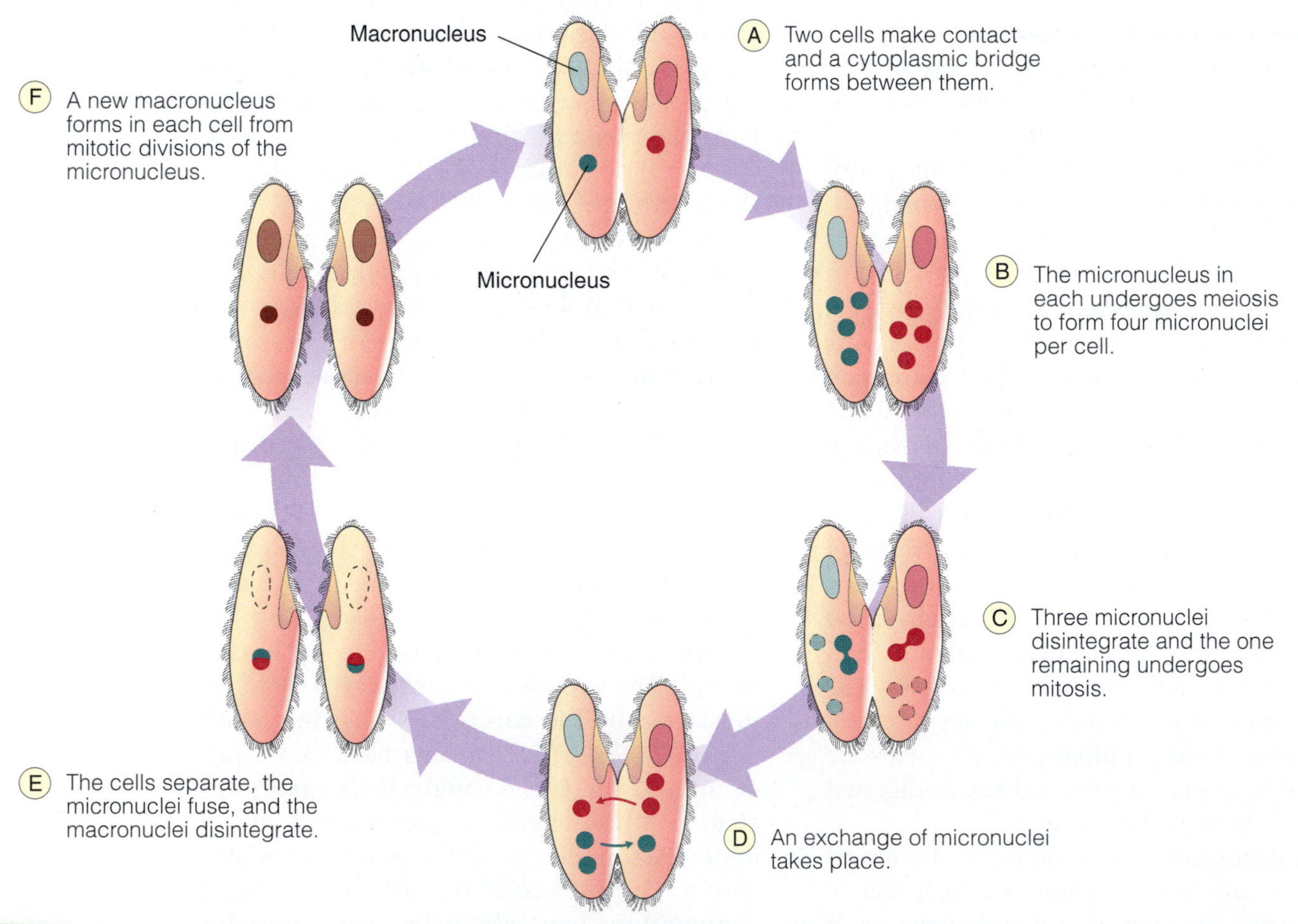

FIGURE 17.5 **Sexual Recombination in *Paramecium*.** Ciliates, such as *Paramecium*, reproduce sexually by the process of conjugation. In this process, an exchange of micronuclei gives rise to new macronuclei.

Q: Provide a hypothesis to explain why ciliates, unlike most other organisms, must have two types of nuclei.

TABLE

17.1 Comparison of the Protozoan Groups of the Protists

Group	Pellicle	Motility	Other Characteristics	Representatives
Parabasalids	No	Flagella	Undulating membrane; modified mitochondria	*Trichomonas*
Diplomonads	No	Flagella	Two nuclei; modified mitochondria	*Giardia*
Kinetoplastids	No	Flagella	Kinetoplast DNA	*Trypanosoma, Leishmania*
Amoebozoans	No	Amoeboid movement	Pseudopodia	*Entamoeba, Naegleria*
Ciliates	Yes	Cilia	Macro- and micronuclei	*Paramecium*
Apicomplexans	Yes	Flagella (gametes only)	Apical complex	*Plasmodium, Toxoplasma*

undergo mitosis. Now a "swapping" of micronuclei takes place, followed by a union to re-form the normal micronucleus.

This genetic recombination is somewhat analogous to what occurs in bacterial cells. It is observed during periods of environmental stress, a factor that suggests the formation of a genetically different and, perhaps, better-adapted organism.

CONCEPT AND REASONING CHECKS

17.3 How does conjugation in ciliates differ from that in the *Bacteria?*

Apicomplexans. The last group of protozoa we will mention is the **apicomplexans**, so named because the *api*cal tip of the cell contains a *complex* of organelles used for penetrating host cells. Adult apicomplexan cells have no cilia or flagella projecting through the flattened pellicle, but the gametes are flagellated.

These parasitic protozoa have a complex life cycle including alternating sexual and asexual reproductive phases. These phases often occur in different hosts. Two parasitic species, *Plasmodium* and *Toxoplasma,* are of special significance because the first causes one of the most prolific killers of humans—malaria—and the second is associated with AIDS.

TABLE 17.1 summarizes some of the attributes of the protozoan groups of protists.

CONCEPT AND REASONING CHECKS

17.4. From this brief description, identify why apicomplexans are closely related to the ciliates.

Parasite Life Cycles Have Some Unique Features

KEY CONCEPT

- Few protozoal but many helminthic parasites require two or more hosts.

Before we describe some of the diseases caused by parasitic protozoa and then helminths, it is worth mentioning a few details concerning the parasitic life cycles.

Most protozoa and some helminths require only one host for the completion of their life cycle. For example, most parasitic protozoa have a stage of their life cycle that is viable in the environment or can survive in the environment. Thus, the infective form, which is the form that enters the human body, differs from one parasitic species to another. An active feeding stage, called a **trophozoite**, is the infective form for some protozoa like *Trichomonas.* For others, including *Entamoeba,* the **cyst**, a dormant, highly resistant stage that can survive long periods in the environment, enters the human body.

Some protozoa and most helminths require two or more hosts to complete their life cycle. The host organism in which the sexual cycle occurs is called the **definitive host** while the host in which the asexual cycle (or larval development) occurs is the **intermediate host**.

For example, the protozoan *Plasmodium* produces infective **sporozoites**, cells formed from multiple fissions (asexual reproduction) of an encysted zygote, in mosquitoes (definitive host). These cells enter the human body. Later, within the human host (intermediate host), plasmodial gametes, called **gametocytes**, are produced. These are taken up by mosquitoes in a blood meal.

17.2 Protozoal Diseases of the Skin, and the Digestive and Urinary Tracts

Protozoal diseases occur in a variety of systems of the human body. For example, some diseases, such as amoebiasis and giardiasis, take place in the digestive system, while others, such as trichomoniasis, develop in the urogenital tract. We start with a cutaneous protozoal disease.

Leishmania Can Cause a Cutaneous or Visceral Infection

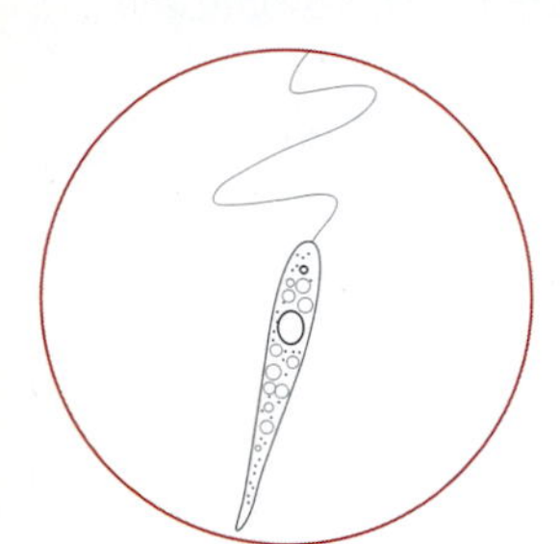

Leishmania donovani

KEY CONCEPT

- Leishmaniasis is a vector-born disease.

Leishmaniasis is a rare disease in the United States, but it occurs in 88 countries on four continents with an at-risk population of 350 million people. The responsible protozoa are in the kinetoplastid group and include several species of *Leishmania*, including *L. major* and *L. donovani* (FIGURE 17.6A). Transmission is by the bite of an infected female sandfly of the genus *Phlebotomus* (FIGURE 17.6B). The sandfly vector, which is only about one third the size of a mosquito, becomes infected by biting an infected animal, such as a rodent, dog, or another human.

■ Vector: An insect that transits pathogens or parasites from an infected animal to humans.

There are two main forms of leishmaniasis. *L. major* causes a disfiguring **cutaneous** (skin) **disease**. Within a few weeks after being bitten, a sore appears on the skin. The sore then expands and ulcerates to resemble a volcano with a raised edge and central crater (FIGURE 17.6C). Some sores may be painless and become covered by a scab. There are about 500,000 new cases of cutaneous leishmaniasis each year worldwide.

The other form of leishmaniasis is a **visceral** (body organ) **disease** called **kala azar**, meaning "black fever." It is caused by *L. donovani*. Symptoms do not appear until several months after being bitten by a sandfly. Infection of the white blood cells leads to irregular bouts of fever, swollen spleen and liver, progressive anemia, and emaciation. About 90 percent of cases are fatal, if not treated. There are about 1.5 million new cases globally each year, 90 percent occurring in India, Bangladesh, Nepal, Sudan, and Brazil. American soldiers have been infected during the Iraq conflicts with the cutaneous or visceral form (MicroFocus 17.2).

Control of the sandfly remains the most important method for preventing outbreaks of leishmaniasis. The antimony compound, stibogluconate, is used to treat established cases.

(A)

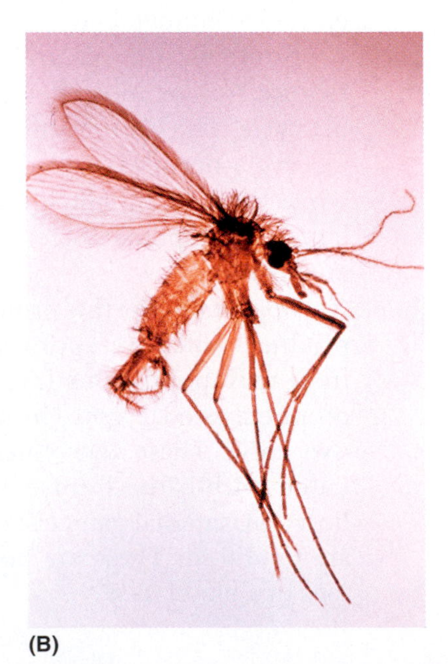

(B)

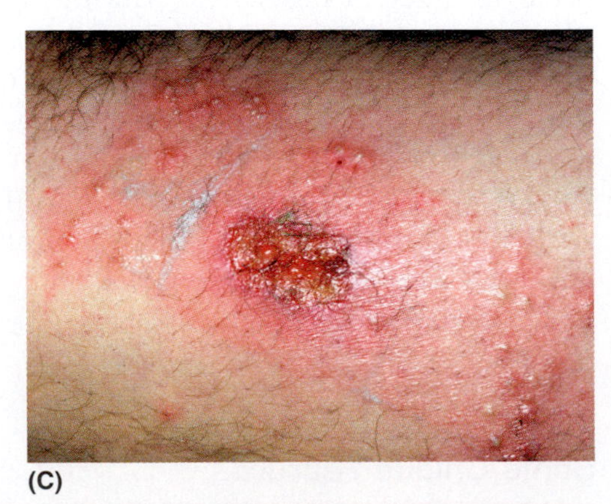

(C)

FIGURE 17.6 **Leishmaniasis.** (**A**) A light micrograph showing a cluster of *Leishmania* cells, which are long, thin, and flagellated. (Bar = 20 µm.) (**B**) The sandfly, *Phlebotomus,* which is the vector for transmission of leishmaniasis. (**C**) Skin lesion due to cutaneous leishmaniasis.

Q: Explain how the parasite can bring about the physical skin lesion.

CONCEPT AND REASONING CHECKS

17.5 Summarize the two types of leishmaniasis and their health consequences.

Several Protozoal Parasites Cause Diseases of the Digestive System

KEY CONCEPT

- Amoebiasis, giardiasis, cryptosporidiosis, and cyclosporiasis result from ingestion of contaminated food or water.

There are three groups of parasitic diseases caused by different groups of protozoa, including the amoebozoans, the apicomplexans, and the kinetoplastids.

Amoebiasis. The second leading cause of death from parasitic disease, only surpassed by malaria, is **amoebiasis**. The parasitic diseases occur worldwide and primarily affect children and adults who are undernourished and living in unsanitary conditions. Although an intestinal illness at first, it can spread to various organ systems. Some 40,000 to 100,000 people die each year from amoebiasis.

The causative agent of amoebiasis is *Entamoeba histolytica*. In nature, the protozoan exists in the cyst form, which enters the body by food or water contaminated with human or animal feces, or by direct contact with feces. The cysts pass through the stomach and emerge as amoebal trophozoites in the distant portion of the small intestine and in the large intestine (FIGURE 17.7).

MICROFOCUS 17.2: Public Health
The "Baghdad Boil"

Between August 2002 and February 2004, the United States Department of Defense (DoD) announced they had identified 522 cases of cutaneous leishmaniasis (CL) among military personnel serving in Afghanistan and Iraq. The disease was "affectionately" called the "Baghdad boil" (see figure).

Leishmania major, which is endemic in Southwest/Central Asia, was the parasitic species identified in the 176 cases analyzed. Patients were treated with sodium stibogluconate.

The DoD has implemented prevention measures to decrease the risk of CL. These procedures included improving hygiene conditions, instituting a CL awareness program among military personnel, using permethrin-treated clothing and bed nets to kill or repel sandflies, and applying insect repellent containing 30 percent DEET to exposed skin. However, despite these measures, there were more than 250 new cases in March and April of 2005.

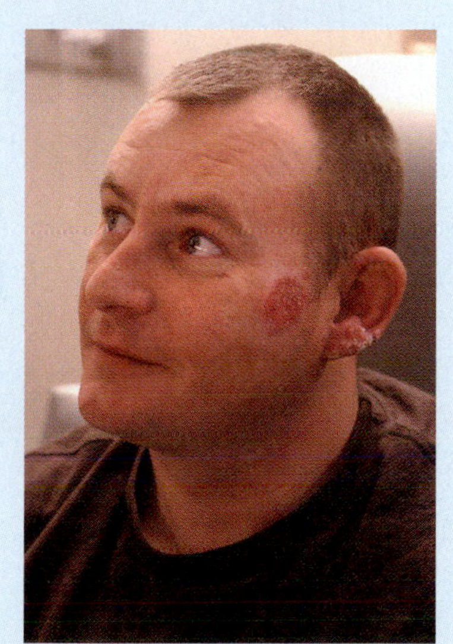

Cutaneous leishmaniasis

Over the same period in 2002–2004, only two cases of visceral leishmaniasis (VL) were reported in military personnel—both in Afghanistan. Both patients exhibited a persistent fever, enlargement of the spleen, and progressive anemia. Although both patients had used personal protective measures as outlined above, both remarked they had been bitten by insects many times. The seriousness of the diseases required aggressive treatment. Patient A had an incubation period of three months before symptoms became apparent. On hospitalization and identification of VL, he received antileishmanial therapy, which consisted of a lipid formulation of amphotericin B. After one week, the fever dissipated and the patient resumed his duties after one additional week.

Patient B had an unusual 14 month incubation period, during which time his deployment ended. His self-reported onset required the administration of lipid amphotericin B and symptoms briefly improved before relapsing. The patient was rehospitalized and received a 28-day treatment with sodium stibogluconate.

This latter case highlights the need for civilian health care workers in the United States to be aware of potential VL in persons who had been deployed to Southwest/Central Asia.

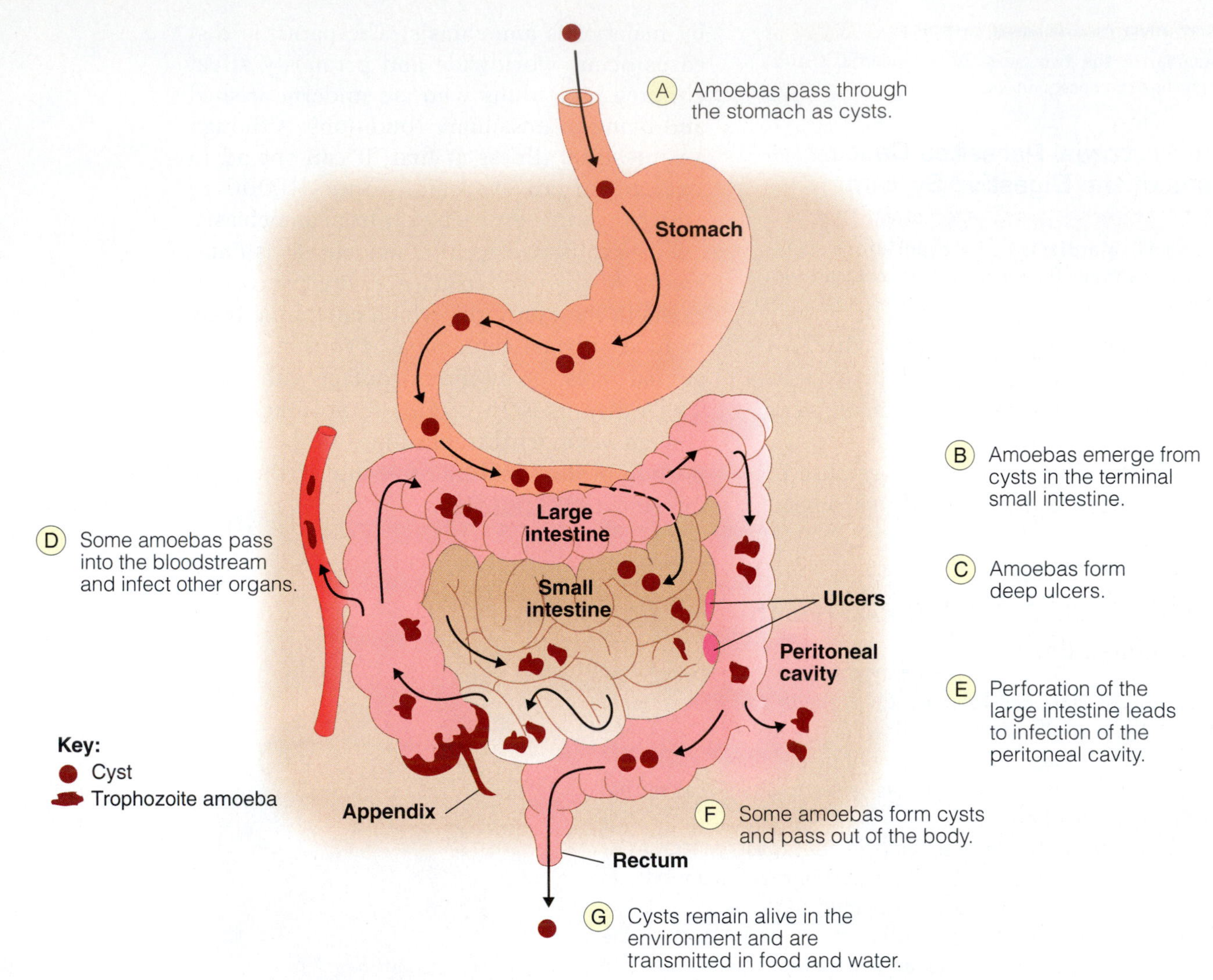

FIGURE 17.7 **The Course of Amoebiasis Due to *Entamoeba histolytica*.** The parasite enters the body as a cyst, which develops into an amoeboid form that causes deep ulcers.

Q: What is the advantage for the parasite to enter the body as a cyst?

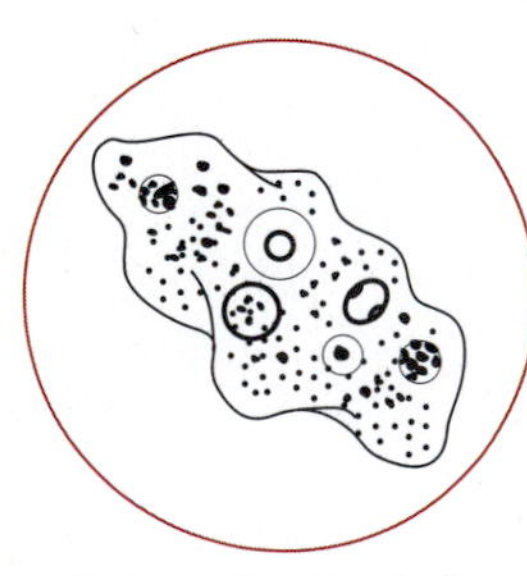

Entamoeba histolytica

On average, about one in 10 people who are infected with *E. histolytica* becomes sick from the infection. The symptoms often are quite mild and can include loose stools, stomach pain, and stomach cramping. The trophozoites multiply by binary fission and produce cysts, which are passed in the feces.

E. histolytica has the ability to destroy tissue (*histo* = "tissue"; *lyt* = "loosened"). Using their protein-digesting enzymes, the trophozoites can penetrate the wall of the large intestine, causing lesions and deep ulcers. Patients who experience stomach pain, bloody stools, and fever have a more severe form of the disease called **amoebic dysentery**. In rare cases, the parasites invade the blood and spread to the liver, lung, or brain, where fatal abscesses may develop.

Prevention is a matter of not eating potentially contaminated food or drinking unpasteurized milk or other diary products. Bottled water or boiled water should be consumed in countries where amoebiasis occurs. Metronidazole and paromomycin commonly are used to treat amebiasis, but the drugs do not affect the cysts, and repeated attacks of amoebiasis may occur for months or years. The patient often continues to shed cysts in the feces to infect other people.

MicroFocus 17.3 describes a parasitic amoebal infection of the eye.

Giardiasis. In 2004, the Centers for Disease Control and Prevention (CDC) received

MICROFOCUS 17.3: Public Health

What's Growing in Your Plumbing?

In 1986, the Centers for Disease Control and Prevention (CDC) received reports of 22 cases from 14 states of eye infections specifically in individuals who wore contact lenses. Testing identified the free-living amoebozoan *Acanthamoeba* as the infecting agent and the disease was *Acanthamoeba* keratitis (AK).

AK is a rare but very painful and potentially blinding infection of the cornea, the transparent covering at the front of the eye (see figure). *Acanthamoeba* has been found in virtually every environment, including soil, dust, fresh water, and seawater. The parasite sometimes resides in untreated swimming pools, hot tubs, and even in bottled water.

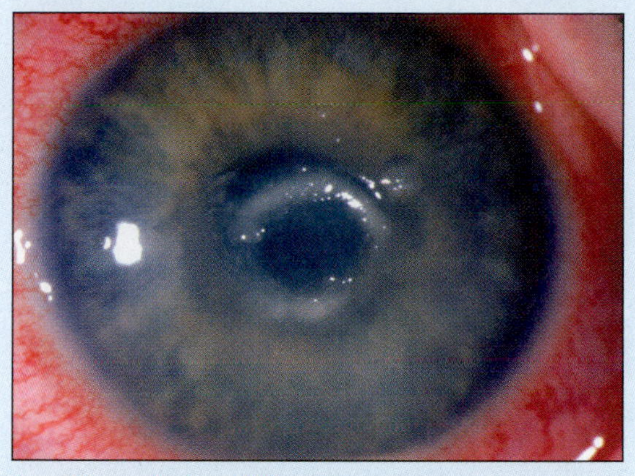

***Acanthamoeba* keratitis.**

The 24 cases reported to the CDC resulted from contact lenses becoming contaminated after improper cleaning and handling. In fact, people who make their own lens cleaning solution are more at risk because proper sterile conditions often are not followed.

In the United Kingdom (UK), an infection of AK occurs about once in every 30,000 contact lens wearers, which is a rate 15-times higher than in the United States. Why?

That is exactly what John Dart, an ophthalmologist at Moorefields Eye Hospital in London, England, wanted to know. Dart knew that until recently all homes in the UK had to have cold-water storage tanks. Could this be the breeding ground for *Acanthamoeba?*

Dart and his colleagues compared the *Acanthamoeba* mitochondrial DNA from eight patients with that from their home water supply. In six cases, the DNAs were identical; the organisms came from the home water supply. However, not all patients stored their contact lenses using the home tap water, so Dart believes the source of the eye infections comes either from the patients having washed their faces with water while wearing their contact lenses or handling the lenses with wet hands. A program for better hygiene practices has been introduced.

reports of 20,636 cases of **giardiasis**, making it the most commonly detected protozoal disease of the intestinal tract in the United States. Because not all cases are reported to the CDC, the disease is estimated to cause 100,000 to 2.5 million infections, primarily in the summer and early fall. The disease is sometimes mistaken for viral gastroenteritis and is considered a type of traveler's diarrhea.

The causative agent is the diplomonad *Giardia intestinalis* (also called *G. lamblia*). This organism is distinguished by four pairs of anterior flagella and two nuclei that stain darkly to give the appearance of eyes on a face. The protozoan can be divided equally along its longitudinal axis and is therefore said to display bilateral symmetry.

Giardiasis is commonly transmitted by food or water containing *Giardia* cysts stemming from cross-contamination of drinking water with sewage. The cysts pass through the stomach and the trophozoites emerge as flagellated cells in the duodenum. They multiply rapidly by binary fission and adhere to the intestinal lining using a sucking disk located on the lower cell surface (FIGURE 17.8).

■ Duodenum: The first short section of the small intestine immediately beyond the stomach.

Giardia intestinalis

Acute giardiasis develops after an incubation period of about seven days. The patient feels nauseous, experiences gastric cramps and flatulence, and emits a foul-smelling watery diarrhea sometimes lasting for one or three weeks. Infectious cysts are excreted in the feces.

Treatment of giardiasis may be administered with drugs such as metronidazole or tinidazole. However, these drugs have side effects and the physician may wish to avoid them by letting the disease run its course without treatment. Those who recover often become carriers and excrete the cysts for years.

CONCEPT AND REASONING CHECKS

17.6 Identify why most cases of giardiasis are reported in the summer to early fall.

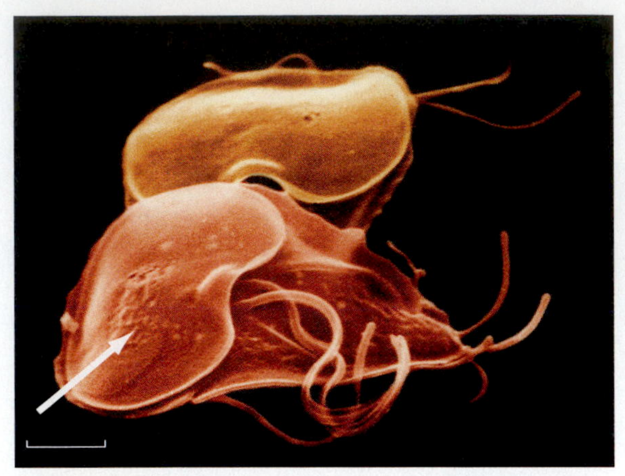

FIGURE 17.8 ***Giardia intestinalis.*** False-color scanning electron micrograph of *G. intestinalis* in the human small intestine. The pear-shaped cell body and sucker device (arrow) are evident, as are the flagella of this diplomonad. (Bar = 1 µm.)

Q: What is the function of the sucker device?

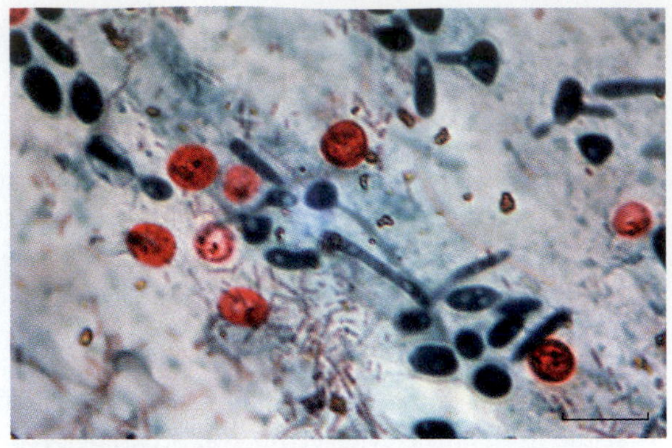

FIGURE 17.9 ***Cryptosporidium* Oocysts.** A light micrograph of a stained fecal smear. *Cryptosporidium* oocysts are red. (Bar = 2 µm.)

Q: What is an oocyst?

Cryptosporidiosis. Since 1976, outbreaks of **cryptosporidiosis** have been reported in several countries. The most remarkable outbreak occurred in 1993 in Milwaukee, Wisconsin, where more than 400,000 people were affected, making it the largest waterborne infection ever recorded in the United States. In 2004, there were 3,577 reported cases, primarily in the upper Midwest, and in 2005 there was an outbreak in a waterpark in Geneva, New York that sickened more than 3,000.

Cryptosporidiosis is caused by the apicomplexans *Cryptosporidium parvum* and *C. hominis.* The organisms have a complex life cycle involving trophozoite, sexual, and oocyst stages (FIGURE 17.9). Transmission occurs mainly through contact with contaminated water (such as drinking or recreational water containing oocysts). Physical contact also can transmit *Cryptosporidium* oocysts, making children in daycare centers at risk.

■ Oocyst: The thick-walled fertilized gamete of apicomplexans.

Cryptosporidiosis has an incubation period of about one week, which explains why few cases are diagnosed properly because most people relate their nausea and diarrhea to something they ate a day or two before. Patients with competent immune systems appear to suffer limited diarrhea lasting one or two weeks where newly-formed infectious oocysts are excreted in the feces.

In immunocompromised individuals, such as AIDS patients, *Cryptosporidium* is an opportunistic infection. Patients experience cholera-like profuse diarrhea that can be severe and irreversible. These patients undergo dehydration and emaciation, and often die of the disease. There has recently been a reduction in the number of infections due to HAART (see Chapter 15).

Cyclosporiasis. In the late 1990s, public health officials in the United States identified a series of clusters of intestinal disease related to the consumption of raspberries imported from Guatemala. In 2004 and 2005, outbreaks in Texas, Illinois, and Florida sickened over 400 people after eating raw basil. In all cases, the outbreaks were related to the apicomplexan *Cyclospora cayetanensis.*

Fresh produce and water can serve as vehicles for transmission of oocysts, which are ingested in contaminated food or water. The oocysts are similar to, but larger than, those of *Cryptosporidium parvum.* Differential diagnosis is important because *C. cayetanensis* responds to the drug combination of trimethoprim-sulfamethoxazole, whereas *Cryptosporidium* does not.

Cyclosporiasis has an incubation period of one week, again making disease identification difficult. Also, by that time affected individuals may have discarded the contaminated food, making source identification also difficult. Symptoms of the disease include watery diarrhea, nausea, abdominal cramping, bloating, and vomiting. Treatment is successful with the drugs noted above, but the symptoms often return. Moreover, the symptoms often remain for over one month during the first illness.

How fresh produce becomes contaminated during outbreaks is not clear. Tainted water used for washing the produce may be the source, the produce may be handled by someone whose hands are contaminated, or the produce may be contaminated during shipping. Regardless of the source, better control measures focusing on improved water quality and better sanitation methods on local farms are needed.

CONCEPT AND REASONING CHECKS

17.7 Compare and contrast cryptosporidiosis and cyclosporiasis.

A Protozoan Parasite Also Infects the Urinary Tract

KEY CONCEPT

- Trichomoniasis affects over 10 percent of sexually active individuals.

Trichomoniasis is among the most common pathogenic protozoan diseases in industrialized countries, including the United States, where an estimated 2.5 million people are affected annually. The disease is transmitted primarily by sexual contact and is considered a sexually transmitted disease (STD).

Trichomonas vaginalis, the causative agent, is a pear-shaped protozoan of the parabasalids (FIGURE 17.10). The organism has no cyst stage and its only host is humans, where it thrives and replicates by binary fission in the slightly acidic environment of the female vagina and the male urethra and prostate. Establishment may be encouraged by physical or chemical trauma, including poor hygiene, drug therapy, diabetes, or mechanical contraceptive devices such as the intrauterine device (IUD).

The incubation period is 5 to 28 days. In females, trichomoniasis is accompanied by intense itching (pruritis) and burning pain during urination. Usually, a creamy white, frothy discharge also is present. The symptoms are frequently worse during menstruation, and erosion of the cervix may occur. In males, the disease may be asymptomatic. Symptoms occur primarily in the urethra, with pain on urination and a thin, mucoid discharge. The disease can occur concurrently with gonorrhea.

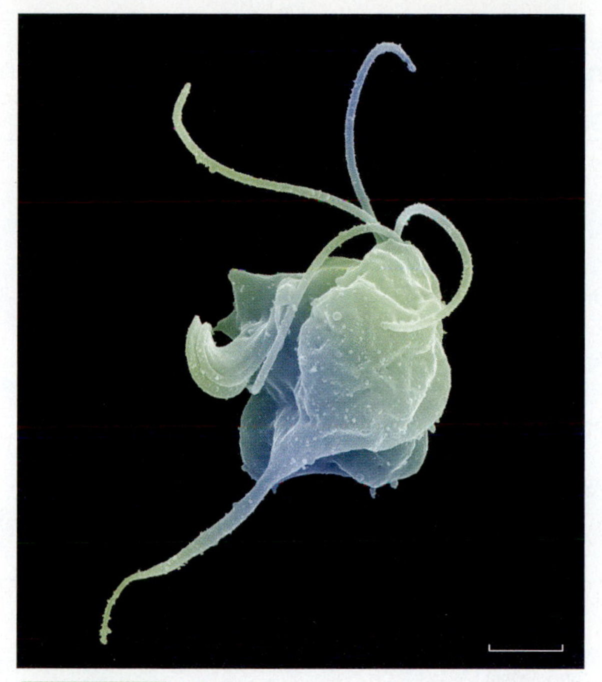

FIGURE 17.10 ***Trichomonas vaginalis.*** False-color scanning electron micrograph of a *T. vaginalis* trophozoite. (Bar = 5 μm.)

Q: What are the whip-like structures evident in the micrograph?

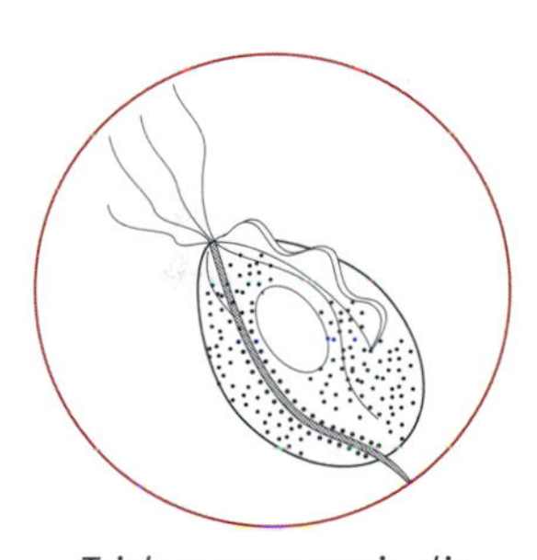

Trichomonas vaginalis

The drug of choice for treatment is orally administered metronidazole or tinidazole, and both patient and sexual partners should be treated concurrently to prevent transmission or reinfection. Some drug resistance has been reported.

TABLE 17.2 summarizes the protozoal diseases of the skin, and the digestive and urinary tracts.

CONCEPT AND REASONING CHECKS

17.8 Compare the symptoms of trichomoniasis to other STDs caused by bacterial species.

TABLE
17.2 A Summary of Protozoal Diseases of the Skin, and the Digestive and Urinary Tracts

Organism	Group	Disease	Transmission	Treatment	Comment
Leishmania donovani	Kinetoplastid	Leishmaniasis (Kala-azar)	Sandfly	Antimony	Ulcers yield skin disfiguration
Entamoeba histolytica	Amoebozoan	Amoebiasis	Water Food	Paromomycin Metronidazole	Deep intestinal ulcers
Giardia intestinalis	Diplomonad	Giardiasis	Water Contact	Quinacrine Tinidazole	Incidence increasing in US
Cryptosporidium parvum	Apicomplexan	Cryptosporidiosis	Water Food	None effective	Associated with AIDS
Cyclospora cayetanensis	Apicomplexan	Cyclosporiasis	Food	Trimethoprim Sulfamethoxazole	Long incubation period
Trichomonas vaginalis	Parabasalid	Trichomoniasis	Sexual contact	Metronidazole Tinidazole	May result in sterility

17.3 Protozoal Diseases of the Blood and Nervous System

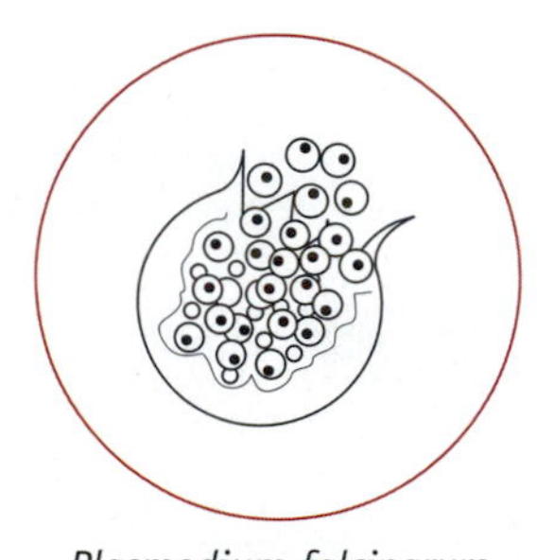
Plasmodium falciparum

We finish our discussion of the protozoal parasites by examining those protozoa that cause infections in the blood or in the nervous system. This includes two of the most prevalent diseases, malaria and sleeping sickness. Visceral leishmaniasis was discussed in the previous section with its cutaneous form.

The *Plasmodium* Parasite Infects the Blood

KEY CONCEPT

- Malaria is caused by four different apicomplexan species.

Malaria is a disease that has been known since at least 1000 BC. During the 1700s, Europeans suffered wave after wave of malaria and few regions were left untouched. Even American pioneers settling in the Mississippi and Ohio valleys suffered great losses from the disease.

Today, between 300 and 500 million of the world's population suffer from malaria, which exacts its greatest toll in Africa. The WHO estimates over 1 million children under the age of 5 die from malaria annually; that is equivalent to one child dying every 30 seconds! No infectious disease of contemporary times can claim such a dubious distinction. And it is worldwide. In 2006, more than 18,000 cases were reported in China's Anhui Province. This is a 90 percent increase from 2005. Even the United States is involved in the malaria pandemic—over 1,400 imported cases were reported in 2004, although a few apparently have been by local transmission (Textbook Case 17).

Malaria is caused by four species of the apicomplexan genus *Plasmodium: P. vivax, P. ovale, P. malariae,* and *P. falciparum.* All are transmitted by the female *Anopheles* mosquito, which consumes human blood to provide chemical components for her eggs. The life cycle of the parasites has three important stages: the sporozoite, the merozoite, and the gametocyte. Each is a factor in malaria. The most serious infections that can be life threatening are caused by *P. falciparum.*

The mosquito (definitive host) sucks blood from a person with malaria and acquires gametocytes, the form of the protozoan found in the blood (FIGURE 17.11). Within the insect sexual reproduction occurs and a transition to sporozoites takes place. The sporozoites then migrate to the salivary gland.

Sporozoite infection causes the female *Anopheles* mosquito to increase its biting frequency. When the mosquito bites another human, several hundred sporozoites enter the person's bloodstream and quickly migrate to the liver. After several hours, the transformation of one sporozoite to 25,000 **merozoites** is completed, and the merozoites emerge from the liver to invade the red blood cells (RBCs). While in the RBCs, the merozoites can synthesize any of about 150 proteins that attach to

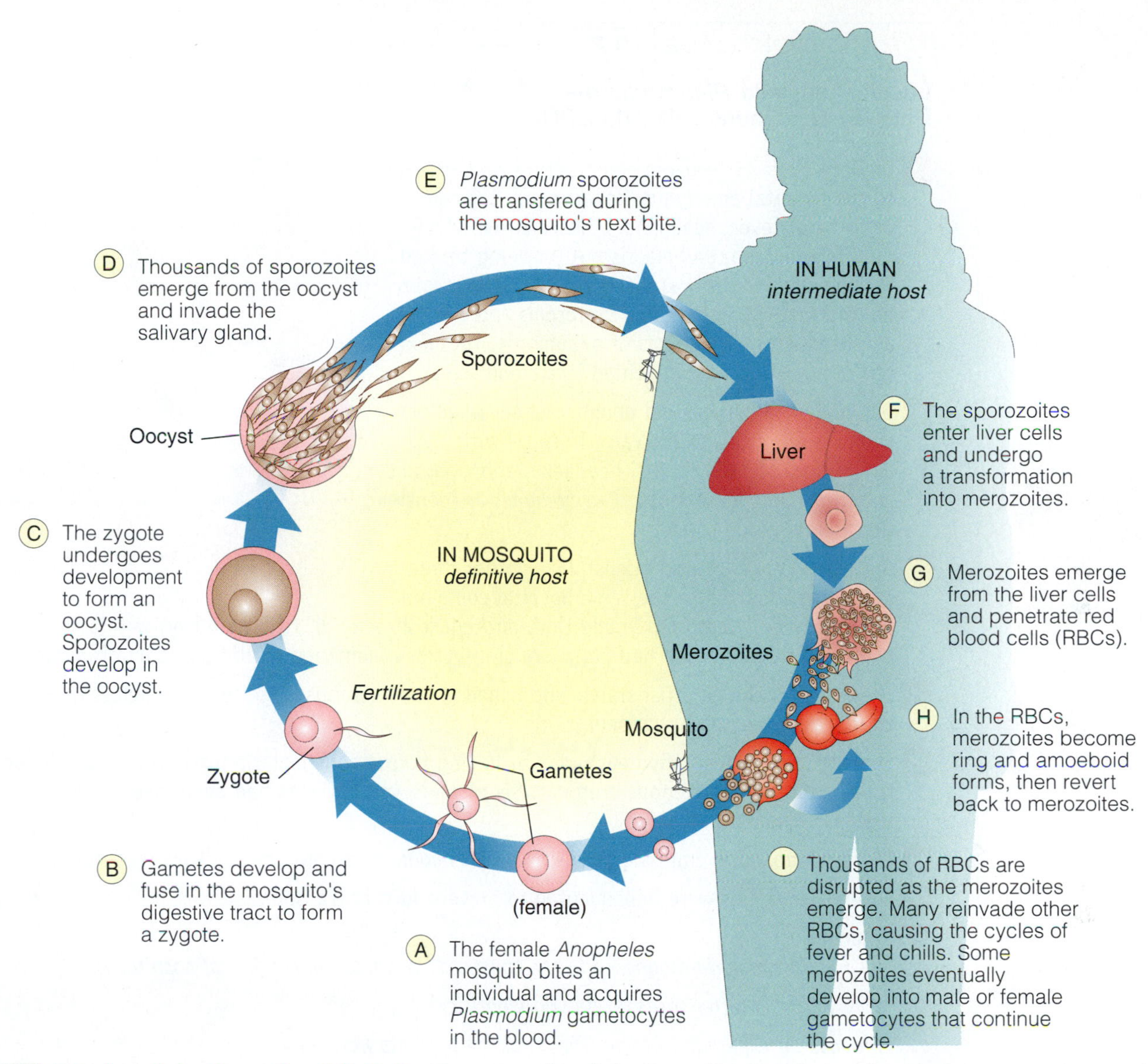

FIGURE 17.11 **Malaria and the *Plasmodium* Life Cycle.** The propagation of the *Plasmodium* parasite requires two hosts, mosquitoes and humans. *Q: Why are two hosts required to complete the* Plasmodium *life cycle?*

RBC membranes and cause the RBCs to cluster in the blood vessels. By constantly switching among 150 genes (for 150 proteins), the malarial parasite avoids detection by the body's immune system.

Within RBCs, the merozoites undergo another series of transformations resulting in several gametocytes and thousands of new merozoites being formed. In response to a biochemical signal, thousands of RBCs rupture simultaneously releasing the parasites and their toxins.

Now the excruciating malaria attack begins. First, there is intense cold, with shivers and chattering teeth. The temperature then rises rapidly to 40°C, and the sufferer develops intense fever, headache, and delirium. After two or three hours, massive perspiration ends the hot stage, and the patient often falls asleep, exhausted. During this quiet period, the merozoites enter a new set of red blood cells and repeat the cycle of transformations.

Death from malaria is due to a number of factors related to the loss of red blood cells. Substantial anemia develops, and the hemoglobin from ruptured blood cells darkens the urine; malaria is, therefore, sometimes called **black-water fever**. Cell fragments and RBC clustering accumulate in the small vessels of the brain, kidneys, heart, liver, and other vital organs and cause clots to form. Heart attacks, cerebral hemorrhages, and kidney failure are common.

Textbook CASE 17

Locally-Acquired *Plasmodium*—Palm Beach County, Florida 2003

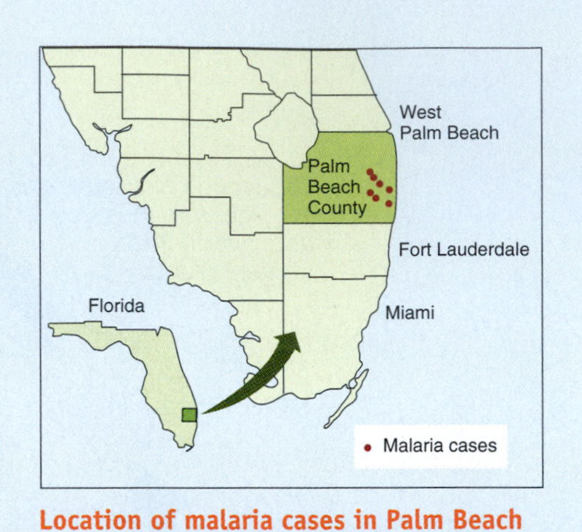

Location of malaria cases in Palm Beach County, Florida, from July to August 2003.

1. On July 22, a 46-year-old construction worker came to the hospital emergency department (ED) complaining of fever, headache, chills, nausea, vomiting, dehydration, and malaise. After being treated for dehydration, he was discharged. Two days later he returned with worsening symptoms and was admitted with a diagnosis of pneumonia. The next day *Plasmodium* was identified in a blood smear.
2. On July 24, a 37-year-old plumber who worked outside was admitted to the same hospital with a six-day history of fever, chills, headache, anorexia, and vomiting. On the next day, *Plasmodium* was identified from a blood smear.
3. Between August 15 and August 24, another three men were admitted to the same hospital complaining of similar symptoms. In addition, on August 25 and 26, two more men were admitted at two other hospitals. All had *Plasmodium* infections identified within 24 hours of admittance.
4. These men included a fisherman who fished in the evenings, a golfer, a homeless individual, an outdoorsman, and a carpenter.
5. Epidemiologists identified each man as having no previous malaria infection, recent blood transfusion, or intravenous drug use. Six of the seven patients had never been to a malarious region of the world.
6. Targeted mosquito trapping produced no *Plasmodium*-infected mosquitoes (see figure).
7. Several strategies were implemented to prevent further transmission.

Questions

A. *Why wasn't malaria diagnosed in the construction worker on the initial hospital visit?*

B. *Why were the latter cases of malaria identified so quickly?*

C. *Propose an explanation as to why all of the patients were men.*

D. *What could the epidemiologists infer from their interviews of the seven patients?*

E. *Suggest what strategies might be invoked to impede further transmission.*

For additional information see http://www.cdc.gov/mmwr/preview/mmwrhtml/mm5238a3.htm.

Since its discovery around 1640, quinine has been the mainstay for treating malaria. During World War II, American researchers developed the drug chloroquine, which remained an important mode of therapy until recent years, when drug resistance began emerging in *Plasmodium* species. Since 1989, an alternative drug called mefloquine has been recommended for individuals entering malaria regions of the world. However, serious medical side effects, including cognitive functioning problems, have been associated with some people taking the drug. Another drug, artemisinin, is effective in curing malaria especially when combined with other drugs to limit the development of drug resistance. Many other drugs are in clinical trials.

Experimental vaccines directed against the sporozoite or merozoite stage are being tested.

MICROFOCUS 17.4: Public Health

Roll Back Malaria

The World Health Organization (WHO) estimates there are over 300 million cases of malaria each year, resulting in more than two million deaths.

The United Nations General Assembly has designated 2001–2010 the decade to "Roll Back Malaria" in developing nations. Roll Back Malaria is a global partnership designed to cut by 50 percent the world's malaria burden by 2010. The program will enable countries to take effective and sustained action against malaria by providing their citizens with rapid and effective treatment, and attempting to prevent and control malaria in pregnant women.

Children in Guayoquil, Ecuador, receiving malaria tablets.

Because over 70 percent of all malarial deaths occur in children under five years of age, a child's most vulnerable period for contracting malaria starts at six months, when the mother's protective immunity wears off and before the infant has established its own fully functional immune system. During this window of vulnerability, a child's condition can deteriorate quickly and the child can die within 48 hours after the first symptoms appear. Therefore, a key part of Roll Back Malaria is aimed at children (see figure) and a series of interim goals have been put forward. They propose that:

- At least 60 percent of those suffering from malaria should have access to inexpensive and correct treatment within 24 hours of the onset of symptoms.
- At least 60 percent of those at risk of malaria, particularly pregnant women and children less than five years old, should have access to suitable personal and community protective measures, such as insecticide-treated mosquito nets.
- At least 60 percent of all pregnant women who are at risk of malaria, especially those in their first pregnancies, should receive intermittent preventive treatment.

Through worldwide partnerships, the Roll Back Malaria program will allow countries that experience a high malaria burden to take effective and sustainable action to halve the malaria burden by 2010.

Though progress has been made in the control of malaria, the mortality and morbidity figures remain appallingly high. MicroFocus 17.4 looks at a program to reduce this "malaria burden."

CONCEPT AND REASONING CHECKS

17.9 Summarize the life cycle of the *Plasmodium* parasite.

The *Trypanosoma* Parasites Can Cause Life-Threatening Systemic Diseases

KEY CONCEPT

- Trypanosomiasis embodies African sleeping sickness and Chagas disease.

Trypanosomiasis is a general name for two diseases caused by parasitic species of the kinetoplastid *Trypanosoma* (FIGURE 17.12A). The two diseases caused by trypanosomes are traditionally known as human African sleeping sickness and Chagas disease.

Human African sleeping sickness is endemic in 36 African countries and exerts a level of mortality greater than that of HIV disease/AIDS. The WHO estimates there are more than 500,000 new cases every year with some 65,000 deaths.

Trypanosoma cycles between humans and the tsetse fly (FIGURE 17.12B). The insect bites an infected patient or animal, and the trypanosomes localize in the insect's salivary gland. After a two-week development, transmission occurs during a bite. The point of

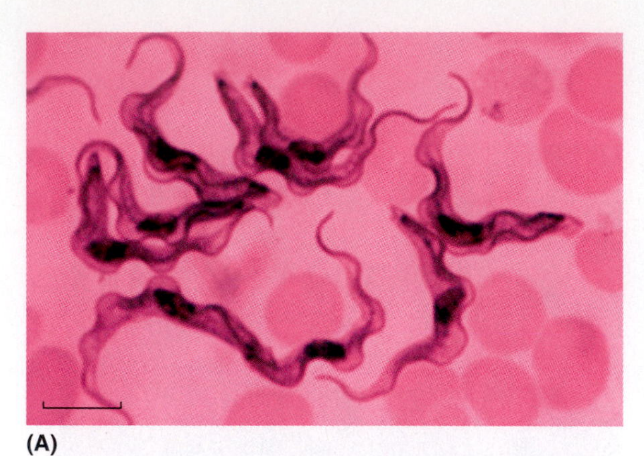

(A)

(B)

FIGURE 17.12 *Trypanosoma*. (A) A light micrograph of stained *Trypanosoma* among red blood cells. (Bar = 5 µm.) (B) The tsetse fly, a vector for *Trypanosoma*.

Q: What is unique about the Trypanosoma *cells?*

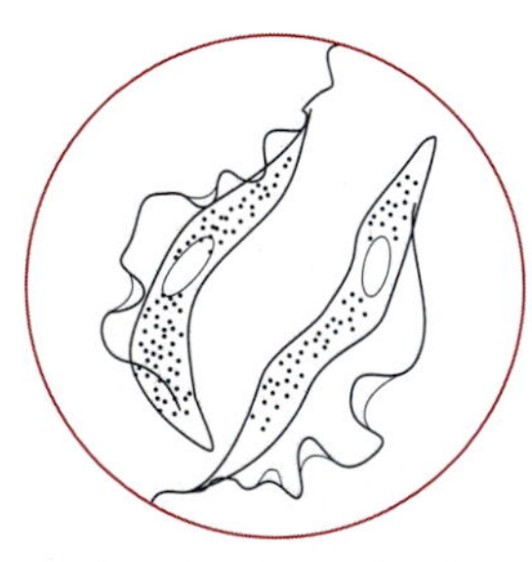

Trypanosoma brucei

Babesia microti

entry becomes painful and swollen in several days and a chancre similar to that in syphilis is observed. Invasion of, and multiplication in, the bloodstream follows (stage 1). It then spreads to the central nervous system (stage 2).

Two types of African sleeping sickness exist. A chronic form, common in central and western Africa, is caused by *Trypanosoma brucei* variety *gambiense*. It is accompanied by chronic bouts of fever, as well as severe headaches, changes in sleep patterns and behavior, and a general wasting away. As the trypanosomes invade the brain, the patient slips into a coma (hence the name, "sleeping sickness"). The second form, common in eastern and southern Africa, is due to *Trypanosoma brucei* variety *rhodesiense*. The disease is more acute with high fever and rapid coma preceding death.

Prevention involves clearing brushlands and treating areas where the tsetse flies breed. Patients are treated with either the drug pentamidine, melarsoprol, or eflornithine, depending on the form and stage of the disease.

American trypanosomiasis or **Chagas disease** is found in Mexico and 17 countries in Central and South America. A recent estimate put the number of cases in South and Central America at 18 million with approximately 200,000 new cases every year and some 50,000 deaths.

Chagas disease is caused by *Trypanosoma cruzi* and is transmitted by the triatomid (reduviid) bugs. The insects feed at night and bite where the skin is thin, such as on the forearms, face, and lips. For this reason, they are called "kissing bugs."

Once in the blood, the trypanosomes invade many cell types and undergo multiple rounds of binary fission. During this acute phase, the individual experiences high parasite numbers even though most infections are asymptomatic. Following the acute phase, the parasite number declines. In 20 to 30 percent of infected individuals, a chronic irreversible disease occurs that in 10 to 30 years can develop clinical symptoms, which vary by geographical region. Individuals may experience widespread tissue damage including intestinal tract abnormalities and extensive cardiac nerve destruction that is so thorough the victim experiences sudden heart failure. Benznidazole and nifurtimox have proved useful for acute disease; no effective drug is available for chronic infections.

CONCEPT AND REASONING CHECKS

17.10 What is the vector difference between African and American trypanosomiasis?

Babesia Is an Apicomplexan Parasite

KEY CONCEPT

- Babesiosis is a malaria-like disease.

Babesiosis, found in the northeastern United States, is a malaria-like disease caused by *Babesia microti*. The protozoa live in ticks of the genus *Ixodes*, and are transmitted when these arthropods feed in human skin. Areas of coastal Massachusetts, Connecticut, and Long Island, New York, have experienced outbreaks in recent years.

Babesia microti penetrates human red blood cells. As the cells disintegrate, a mild anemia develops. Piercing headaches accompany the disease and, occasionally, meningitis occurs. A suppressed immune system appears to favor establishment of the disease. However, babesiosis is rarely fatal and drug therapy is not recommended. Carrier conditions may develop in recoverers, and spread by blood transfusion is possible. Travelers returning from areas of high incidence therefore are advised to wait several weeks before donating blood to blood banks. Tick control is considered the best method of prevention.

Babesia has a significant place in the history of American microbiology. In the late 1800s, Theobald Smith located *B. bigemina* in the blood of cattle suffering from Texas fever. His report was one of the first linking protozoa to disease, and, in part, it necessitated that the then-prevalent "bacterial" theory of disease be modified to include eukaryotic microorganisms.

CONCEPT AND REASONING CHECKS

17.11 Why is babesiosis considered to be a malaria-like disease?

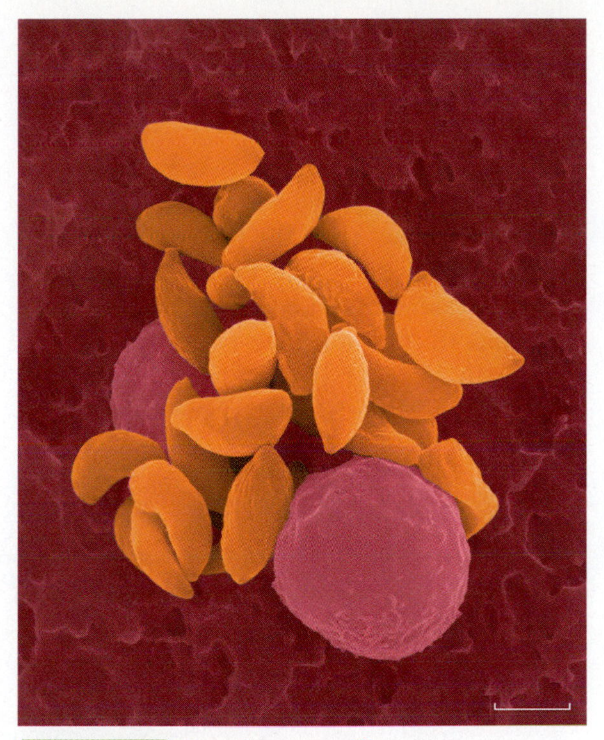

FIGURE 17.13 ***Toxoplasma gondii.*** A false-color scanning electron micrograph of numerous crescent-shaped *T. gondii* trophozoites (orange). (Bar = 5 µm.)

Q: What infective stage is represented by these cells?

Toxoplasma Causes a Relatively Common Blood Infection

KEY CONCEPT

- Toxoplasmosis is extremely contagious.

Toxoplasmosis affects up to 50 percent of the world's population, including 50 million Americans. Thus, the causative agent, *Toxoplasma gondii*, is regarded as a universal parasite. Some researchers believe it is the most common parasite of humans and other vertebrates. *T. gondii* exists in three forms: the trophozoite, the cyst, and the oocyst. Trophozoites are crescent-shaped or oval organisms without flagella (FIGURE 17.13). Located in tissue during the acute stage of disease, they force their way into all mammalian cells, with the notable exception of erythrocytes. To enter cells, the parasites form a ring-shaped structure on the host cell membrane and then pull the membrane over themselves, much like pulling a sock over the foot. Cysts develop from the trophozoites within host cells and may be the source of repeated infections. Muscle and nerve tissue are common sites of cysts. Oocysts are oval bodies that develop from the cysts by a complex series of asexual and sexual reproductive processes.

Toxoplasma gondii

Toxoplasma gondii exists in nature in the cyst and oocyst forms. Grazing animals acquire these forms from the soil and pass them to humans via contaminated beef, pork, or lamb (FIGURE 17.14). Rare hamburger meat is a possible source. Domestic cats acquire the cysts from the soil or from infected birds or rodents. Oocysts then form in the cat. Humans are exposed to the oocysts when they forget to wash their hands after contacting cat feces while changing the cat litter or working in the garden. Touching the cat also can bring oocysts to the hands, and contaminated utensils, towels, or clothing can contact the mouth and transfer oocysts.

Toxoplasmosis develops after trophozoites are released from the cysts or oocysts in the host's gastrointestinal tract. *T. gondii* rapidly invades the intestinal lining and spreads throughout the body via the blood. However, for most healthy individuals, the parasite causes no serious illness even though one remains infected for life. The highly contagious nature of the parasite stems from its need to get back into its definitive host, the cat, where oocysts are produced.

Pregnant women are at risk of developing a dangerous toxoplasmosis infection because the protozoa may cross the placenta and infect the fetal tissues (Figure 17.14). Neurologic damage, lesions of the fetal visceral organs, or spontaneous abortion may result. Congenital infection is least likely during the first trimester, but damage may be substantial when it occurs. By contrast, congenital infection is more common if the woman is infected in the third trimester, but fetal damage is less severe. Lesions of the retina are the most widely documented complication in congenital infections. The **T** in the **TORCH** group of diseases refers to toxoplasmosis (the others are rubella, cytomegalovirus, and herpes simplex; O is for other diseases, such as syphilis).

T. gondii also is known to cause severe disease in immunosuppressed individuals. For example, in patients with AIDS the parasite

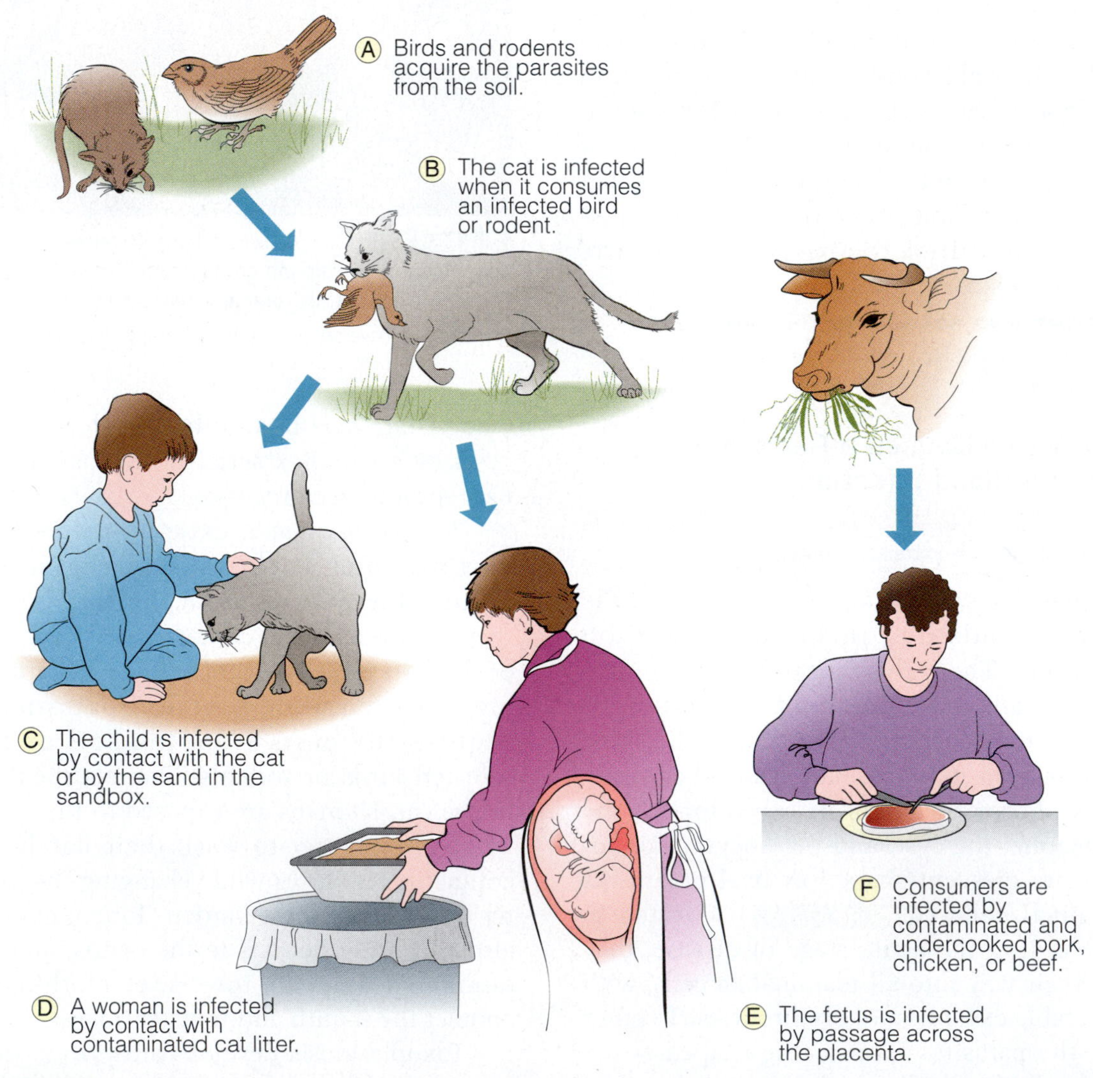

FIGURE 17.14 **The Cycle of Toxoplasmosis in Nature.** Humans become infected through contact with the feces of an infected cat or consumption of undercooked, contaminated beef.

Q: What form of the Toxoplasma *parasite is passed to humans?*

attacks the brain tissue, causing an inflammation and swelling that often results in cerebral lesions, seizures, and death.

CONCEPT AND REASONING CHECKS

17.12 Assess the impact of *Toxoplasma* to pregnant women and immunocompromised people.

Naegleria Can Infect the Central Nervous System

KEY CONCEPT

- Primary amoebic meningoencephalitis (PAM) is a rare but deadly infection of the brain.

Primary amoebic meningoencephalitis (PAM) is a rare disease with less than 200 cases reported worldwide since 1965. However, it is the most deadly disease of the central nervous system after rabies (see Chapter 15). Ninety-five percent of patients die within four to five days of infection.

PAM is caused by several species of thermophilic amoebozoans in the genus *Naegleria*, especially *N. fowleri* (FIGURE 17.15). It also can be caused by *Acanthamoeba* and *Hartmannella* species.

Naegleria is an opportunistic pathogen of humans, causing meningoencephalitis when inhaled, often by swimming in contaminated warm surface water. The free-living trophozoites appear to enter the body through the mucous membranes of the nose and then follow the olfactory tracts to the brain.

The symptoms resemble those in other forms of encephalitis and meningitis. Nasal congestion precedes piercing headaches, fever, delirium, neck rigidity, and occasional seizures if the victim is not treated with amphotericin B combined with miconazole and rifampin.

TABLE 17.3 summarizes the protozoal diseases of the blood and nervous system.

CONCEPT AND REASONING CHECKS

17.13 What similarities exist between *Naegleria* and *Entamoeba*?

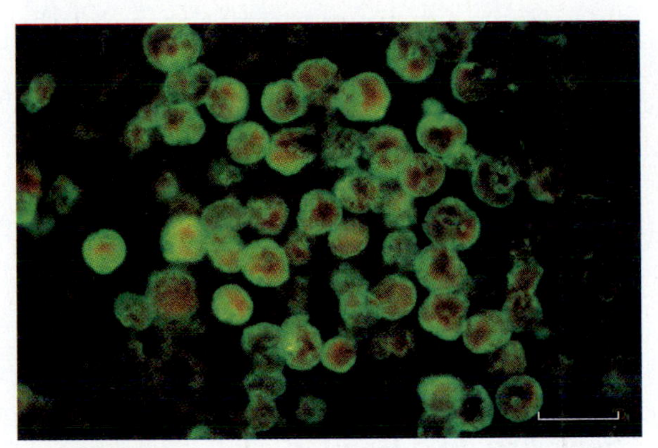

FIGURE 17.15 ***Naegleria fowleri.*** A fluorescent antibody stain of *N. fowleri* amoebae in tissue from a brain autopsy. (Bar = 40 µm.)

Q: What is meant by fluorescent antibody staining?

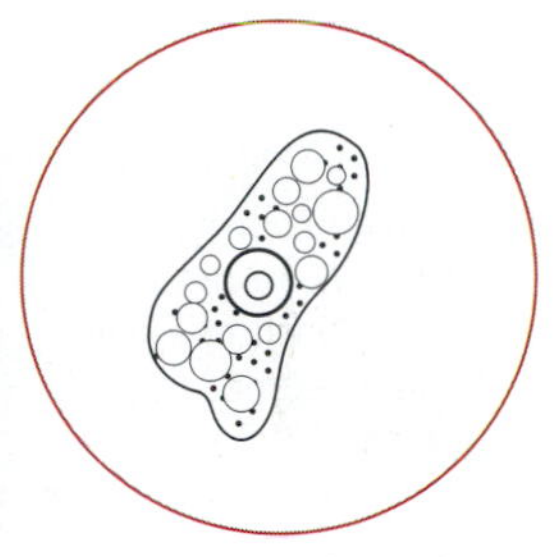

Naegleria fowleri

Meningoencephalitis: An inflammation of the brain and meninges.

TABLE 17.3 A Summary of Protozoal Diseases of the Blood and Nervous System

Organism	Group	Disease	Transmission	Treatment	Comment
Plasmodium species	Apicomplexan	Malaria	Mosquito (*Anopheles*)	Quinine Mefloquine	World's most urgent public health problem
Trypanosoma brucei	Kinetoplastid	Human African sleeping sickness	Tsetse fly	Various drugs	Two types, depending on region
Trypanosoma cruzi	Kinetoplastid	Chagas disease	Triatomid (kissing) bug	Various drugs	Common in South America
Babesia microti	Apicomplexan	Babesiosis	Tick (*Ixodes*)	None recommended	Carrier state possible
Toxoplasma gondii	Apicomplexan	Toxoplasmosis	Domestic cats Food	Sulfonamide drugs	Congenital damage possible Associated with AIDS
Naegleria fowleri	Amoebozoan	Primary amoebic meningoencephalitis	Water	None effective	Uncommon in US

17.4 The Multicellular Helminths and Helminthic Infections

The **helminths** are among the world's most common animal parasites. For example, 2 billion people—approximately 33 percent of the human population—are infected with soil-transmitted helminths! Therefore, in concluding this chapter on the parasites, we are concerned with medical helminthology and the diseases caused by the parasitic worms.

As mentioned in the introduction to this chapter, such parasites are of interest to microbiologists because of the parasites' ability to cause an enormous level of morbidity worldwide. However, different from many of the viral and bacterial pathogens we have discussed, most parasitic helminths are dependent on the host or hosts for sustenance, so it is to the helminth's benefit that the host stays alive. Therefore, the helminths tend to cause diseases of debilitation and chronic morbidity resulting from physical factors related to the helminthic load (number of worms present) or location in the body.

■ Hermaphroditic: Refers to an organism having both male and female reproductive organs.

There Are Two Groups of Parasitic Helminths

KEY CONCEPT

- Parasitic helminths include the flatworms and the roundworms.

The helminths of medical significance are the flatworms and the roundworms.

Flatworms. Animals in the phylum Platyhelminthes (*platy* = "flat"; "*helmin*" = "worm") are the **flatworms**. As multicellular animals, they have tissues functioning as organs in organ systems. However, they have no specialized respiratory or circulatory structures, and they lack a digestive tract. The gut (gastrovascular cavity) simply consists of a sac with a single opening, thus placing the worm in close contact with its surroundings. Complex reproductive systems are found in many species within the phylum, and a large number of species are hermaphroditic. Two groups of flatworms are of concern regarding human disease.

The **trematodes**, includes the **flukes**, which have flattened, broad bodies (FIGURE 17.16A). The animals exhibit bilateral symmetry. Trematodes have a complex life cycle that may include encysted egg stages and temporary larval forms. Sucker devices are commonly present to enable the parasite to attach to its host. In many cases, two hosts exist: an intermediate host, which harbors the larval form, and a definitive host, which harbors the mature adult form. In this chapter, we shall be mostly concerned with parasites whose definitive host is a human.

(A)

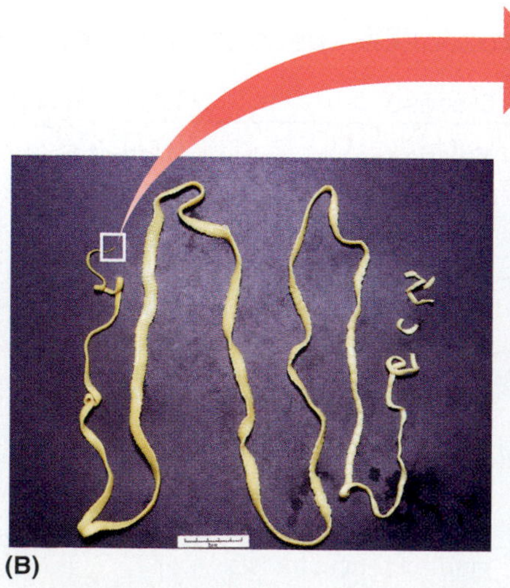
(B)

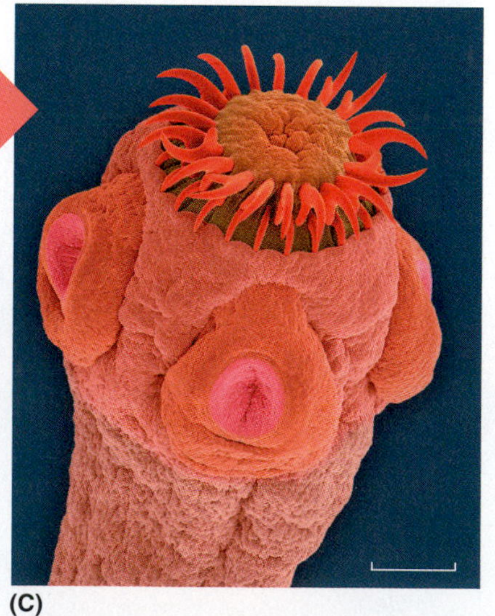
(C)

FIGURE 17.16 **Flukes and Tapeworms.** (**A**) Light micrograph of a liver fluke, *Fasciola hepatica,* a parasite infecting sheep, cattle, and humans. (Bar = 50 µm.) (**B**) Photograph of a coiled beef tapeworm, *Taenia saginata,* which has grown to several meters in length. (**C**) A false-color scanning electron micrograph of the head, scolex, of *T. saginata* showing the suckers and hooks. (Bar = 1 mm.)

Q: What does the dark branched structures form in the liver fluke (A) and what is the purpose of the suckers and hooks on the scolex head of the tapeworm (C)?

The life cycle of a fluke often contains several phases. In the human host, the parasite produces fertilized eggs generally released in the feces. When the eggs reach water, they hatch and develop into tiny ciliated larvae called **miracidia** (sing., miracidium). The miracidia penetrate snails (the intermediate host) and go through a series of asexual reproductive stages. The cyst makes its way back to humans.

The trematode life style requires the parasite to evade the host's immune system. It accomplishes this by having its surface resemble the surface of the host cells, so the immune system "sees" the worm as a "normal" cell, not an invader. This mimicry is quite effective as some flukes can remain on a human host for 40 years or more.

The other group of flatworms is the **cestodes**, which includes the **tapeworms**. These parasitic worms have a head region, called the **scolex**, and a ribbon-like body consisting of segments called **proglottids** (FIGURE 17.16B, C). The proglottids most distant from the scolex are filled with fertilized eggs. As the proglottids break free, they spread the eggs.

Tapeworms generally live in the intestines of a host organism. In this environment, they are constantly bathed by nutrient-rich fluid, from which they absorb food already digested by the host. Tapeworms have adapted to a parasitic existence and have lost their intestines, but they still retain well-developed muscular, excretory, and nervous systems.

Tapeworms are widespread parasites infecting practically all mammals, as well as many other vertebrates. Because they are more dependent on their hosts than flukes, tapeworms have precarious life cycles. Tapeworms have a limited range of hosts, and the chances for completing the cycle are often slim. With rare exceptions, tapeworms require at least two hosts. Humans often become infected by eating undercooked meat containing tapeworm cysts, which then develop into mature adult worms.

Roundworms. Among the most prevalent animals are the **roundworms** in the phylum Nematoda (*nema* = "thread"). These parasites have a thread-like body and occupy every imaginable habitat on Earth. They live in the sea, in freshwater, and in soil from polar regions to the tropics. Good topsoil, for example, may contain billions of nematodes per acre. They parasitize every conceivable type of plant and animal, causing both economic crop damage and serious disease in animals. Yet, they may even have a beneficial effect for humans (MicroFocus 17.5).

Roundworms have separate sexes. Following fertilization of the female by the male, the eggs hatch to larvae that resemble miniature adults. Growth then occurs by cellular enlargement and mitosis. Damage in hosts is generally caused by large worm burdens in the blood vessels, lymphatic vessels, or intestines (FIGURE 17.17). Also, the infestation may result in nutritional deficiency or damage to the muscles.

CONCEPT AND REASONING CHECKS

17.14 Compare the body plan of the flatworms with that of the roundworms.

Several Trematodes Can Cause Human Illness

KEY CONCEPT

- Schistosomiasis, and human lung and liver fluke diseases, are due to trematode infections.

Schistosomiasis. The WHO estimates that 200 million people in 74 countries suffer from schistosomiasis, which kills approximately 200,000 every year. There are even

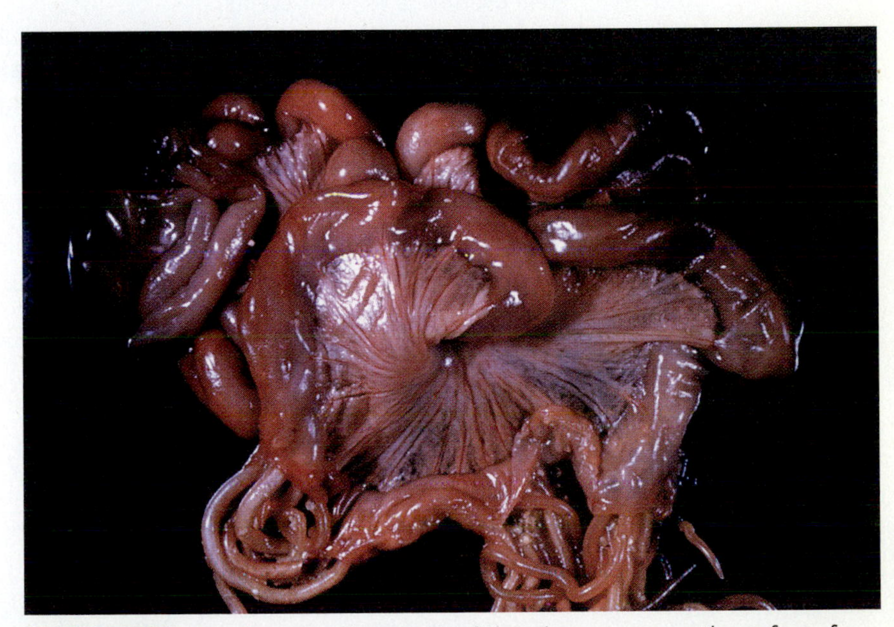

FIGURE 17.17 **Roundworms.** Photograph of threadworms seen on the surface of, and burrowing into, the gut wall of a pig.

Q: How do the nematodes differ from the flatworms?

MICROFOCUS 17.5: Being Skeptical

Eat Worms and Cure Your Ills—and Allergies!

Adult whipworms (male top; female bottom). (Bar = 1 cm.)

Doctors and researchers have shown that eating worms (actually drinking worm eggs) can fight disease. What! Are you nuts? Drink worm eggs!

Joel Weinstock, a gastroenterologist at the University of Iowa, discovered that as allergies and other diseases have increased in Western countries, infections by roundworm parasites have declined. This is not the case in other countries where allergies are rare and worm infections are quite common. Weinstock wondered if there was a correlation between allergy increase and parasite decline.

To test his hypothesis, Weinstock "brewed" a liquid concoction consisting of thousands of pig whipworm eggs (ova). The whipworms are called *Trichuris suis*, so his product with ova is called TSO. This roundworm (see figure) was chosen because once the ova hatch, they will not survive long in the human digestive system and will be passed out in the feces.

One woman in Iowa was suffering from incurable ulcerative colitis, which is caused by the immune system overreacting. Immune cells start attacking the person's own gut lining, making it bleed. The symptoms are severe cramps and acute, intense, diarrhea. In a trial run, Weinstock gave the woman a small glass full of his TSO. Every three weeks she downed another glass. Guess what? Her ulcerative colitis is in remission and she no longer suffers any disease symptoms.

Further trials in 2004 involved 100 people suffering the same disease and a further 100 suffering Crohn disease, which is another type of inflammatory bowel disease related to immune function. In this study, 50 percent of the volunteers suffering ulcerative colitis and 70 percent of those suffering Crohn disease went into remission, as identified by no symptoms of abdominal pain, ulcerative bleeding, and diarrhea.

Weinstock believes some parasites are so intimately adapted with the human gut that if they are eradicated, bad things may happen, such as the bowel disorders mentioned. He says the immune system has become so involved with defending against parasites that if you take them away, the immune system overreacts to other events.

Is there something to this? Alan Brown, an academic researcher in the United Kingdom (UK) picked up a hookworm infection while on a field trip outside the UK. Being that he was a well-nourished Westerner, the 300 hookworms in his gut caused no major problem. However—since being infected, his hayfever allergy has disappeared!

The result: I need more evidence before I would drink worm eggs for a gastrointestinal disease.

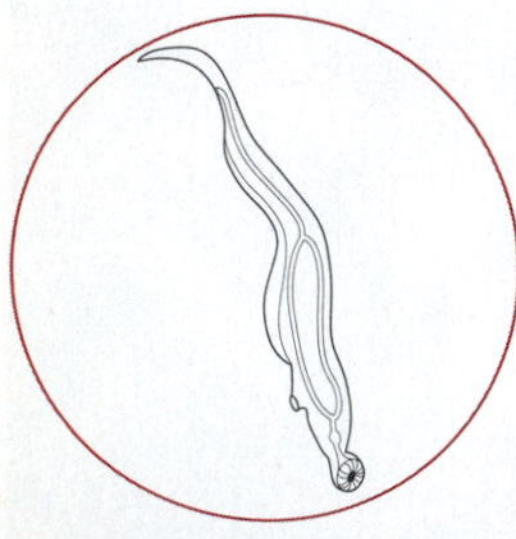

Schistosoma mansoni

about 400,000 individuals in the United States who suffer from a mild form of the disease.

Schistosomiasis is caused by several species of blood flukes, including *Schistosoma mansoni* (Africa and South America), *S. japonicum* (Asia), and *S. haematobium* (Africa and India). In some regions, the term **bilharziasis** is still used for the disease; it comes from the older name for the genus, *Bilharzia*.

Species of *Schistosoma* measure about 10 mm in length. The eggs hatch in freshwater to produce miracidia, which then make their way to snails. In the snails, miracidia convert to a second larval form called **cercariae** (FIGURE 17.18). The cercariae escape from the snails and attach themselves to the bare skin of humans. Cercariae infect the blood and mature into adult flukes, which cause fever and chills. However, the major effects of disease are due to eggs: carried by the bloodstream to the liver, they cause substantial liver damage; in the intestinal wall, ulceration, diarrhea, and abdominal pain occur; and in the bladder, egg infection causes bloody urine and pain on urination. Male and female species mate

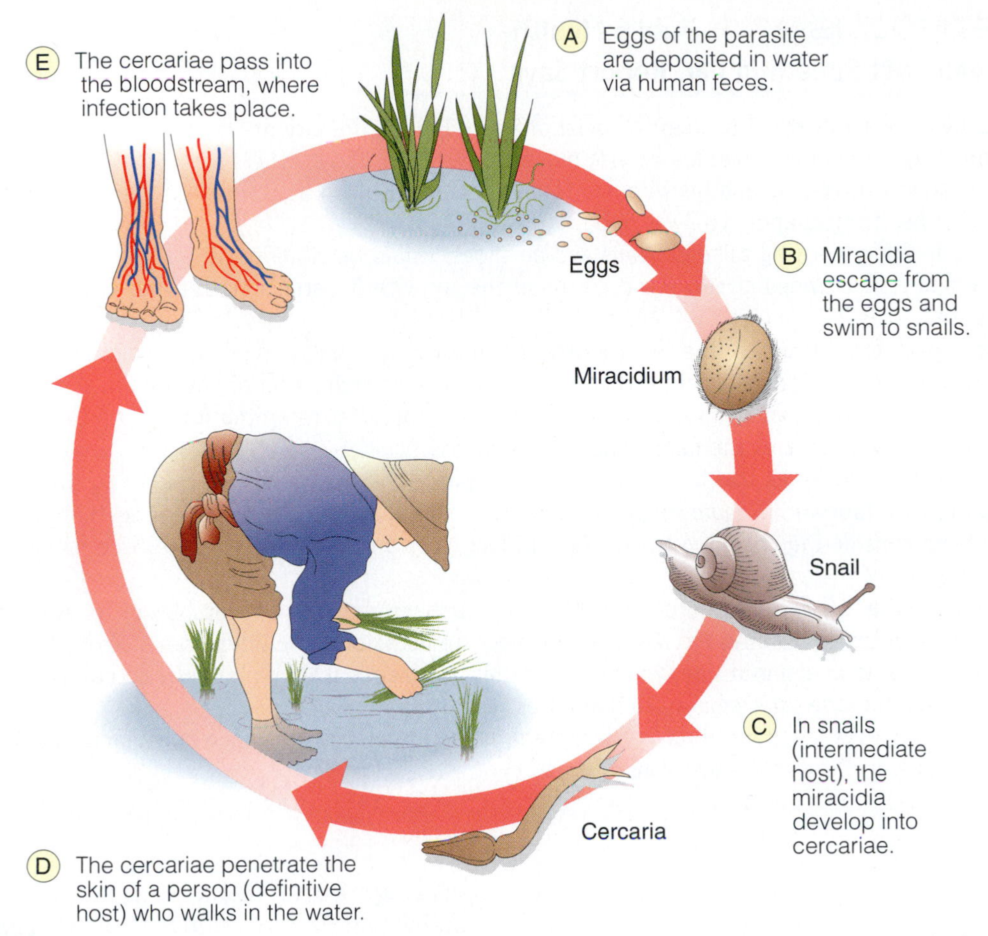

FIGURE 17.18 **The Life Cycle of the Blood Fluke *Schistosoma mansoni*.** The life cycle of the trematode *S. mansoni* involves an intermediate host, the snail, and a definitive host, a human.

Q: What defines a definitive and intermediate host?

in the human liver and produce eggs that are released in the feces.

The antihelminthic drug praziquantel is used for treatment. MicroFocus 14.6 discussed how treatment for schistosomiasis spread hepatitis C through much of Egypt.

Certain species of *Schistosoma* penetrate no farther than the skin because the definitive hosts are birds instead of humans. Often after swimming in schistosome-contaminated water, the cercariae penetrate the skin but are attacked and destroyed by the body's immune system. However, they release allergenic substances that cause an itching and body rash, commonly known as **swimmer's itch**. The condition, not a serious threat to health, is common in northern lakes in the United States.

CONCEPT AND REASONING CHECKS

17.15 Summarize how schistosomiasis causes such great morbidity worldwide.

Human Lung and Liver Fluke Diseases. *Paragonimus westermani* is a trematode common in Asia and the South Pacific. Cercariae develop in snails, and metacercariae later form in freshwater crayfish and crabs. Infection follows when people eat the poorly cooked shellfish, and the flukes pass from the intestine to the blood and then to the lungs. Eggs of *P. westermani* are coughed up from the lung, swallowed, and then excreted in the feces. Difficult breathing and chronic cough develop as the parasites accumulate in the lungs. Fatalities are possible.

The human liver fluke *Fasciola hepatica* is a flatworm common in cattle and especially in sheep. Eggs from the animal reach the soil in

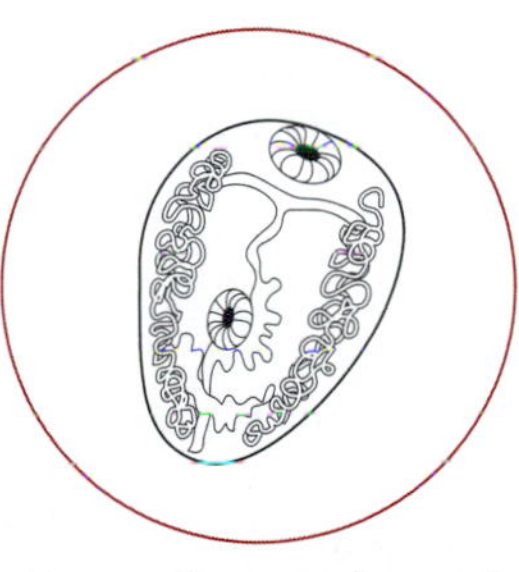

Paragonimus westermani

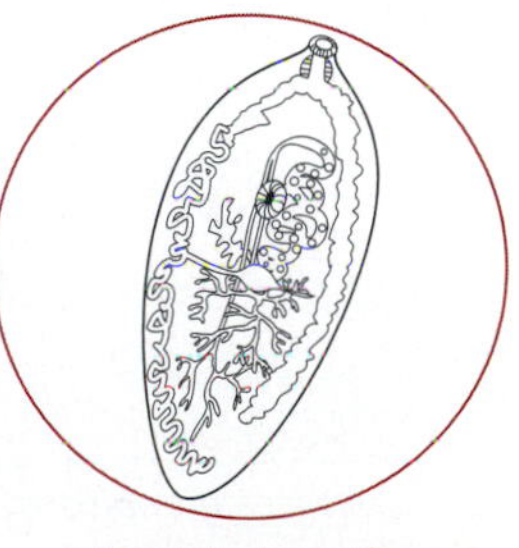

Fasciola hepatica

MICROFOCUS 17.6: Public Health

Man Leaps Off Speeding Car, Report Says

It was truly a bazaar death. The chief financial officer (CFO) for the city of Phoenix, Arizona was a very professional, quiet man who had a way with numbers. He cared about others and worked with the underprivileged. In fact, he and his wife had formed a foundation to help working women and children in need find health insurance. Life was good.

So, why in 2004, wearing tattered jeans and no shoes, would he climb out the window of his car going 55 mph on a crowded street, stand on top of the car with his arms outstretched—and then jump to his death?

A few days before his death, the finance officer had taken a sick day to see his doctor complaining of feeling "tired and worn out." But other than that, there was no indication of any abnormal behavior.

However, it happens that two years earlier, he had fallen ill after returning from a trip to Mexico. At that time, nobody outside of the family knew of his illness.

According to his wife, he had been suffering from cysticercosis, an infection caused by the larval stage of the pork tapeworm *Taenia solium*. He had been taking medication for the parasitic infection and had been considering changing the dosage. In fact, after his death it was discovered he had been having flare-ups of the parasite.

Apparently, the CFO had been infected in Mexico probably from eating vegetables or fruits contaminated with pig feces containing the parasite. On very rare occasions, the larvae can move to the brain and cause frontal lobe disinhibition, which means that they can make an individual do bazaar things, including impairing one's decision-making abilities.

So, it appears, like MicroInquiry 17 describes, parasites can cause strange behaviors. In this case, the result was a very sad and tragic consequence.

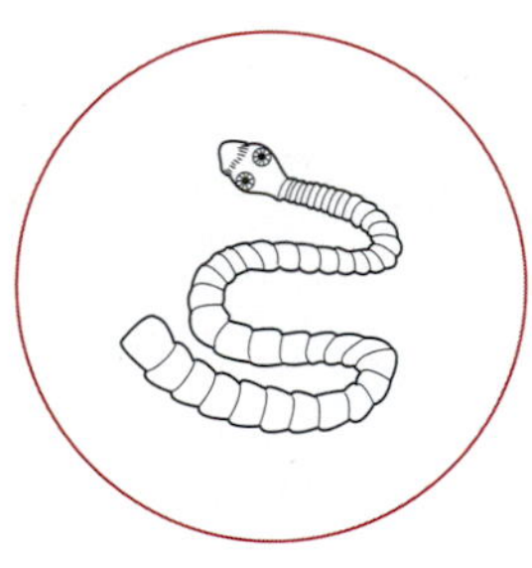

Taenia saginata

feces, and if snails are present, the conversions to cercariae and encysted metacercariae follow. Parasite cysts gather on vegetation, such as watercress and other water plants, and ingestion by humans follows. The parasites penetrate the intestinal wall and migrate to the liver, where tissue damage may be substantial, especially if fluke numbers are high. Treatment involves bithionol.

MicroInquiry 17 examines some amazing phenomena of how parasitic flukes can manipulate their host for their own benefit—and survival.

CONCEPT AND REASONING CHECKS

17.16 Why do all three fluke diseases described here have snails as a stage in their life cycles?

Tapeworms Survive in the Human Intestines

KEY CONCEPT

- Tapeworm diseases result from eating undercooked meat contaminated with cysts.

Beef and Pork Tapeworm Diseases. Approximately 50,000 people die each year from beef and pork tapeworm diseases. Humans are the definitive hosts for both the beef tapeworm *Taenia saginata* and the pork tapeworm *Taenia solium*. Humans acquire the tapeworm cysts by eating poorly cooked beef or pork. The beef tapeworm may reach 10 meters in length, while the pork tapeworm length is 2 to 8 meters.

Attachment via the scolex occurs in the small intestine and obstruction of this organ may result. In most cases, however, there are few symptoms other than mild diarrhea, and a mutual tolerance may develop between parasite and host. Each tapeworm may have up to 2,000 proglottids and infected individuals will expel numerous gravid proglottids daily. The proglottids accumulate in the soil and are consumed by cattle or pigs. Embryos from the eggs travel to the animal's muscle, where they form cysts.

■ Gravid: Refers to carrying eggs.

In rare instances, infection can lead to very bizarre behaviors (MicroFocus 17.6).

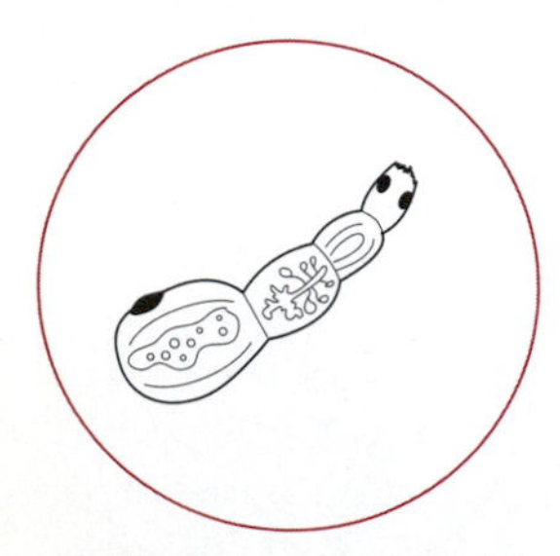

Echinococcus granulosus

Dog Tapeworm Disease. Dogs and other canines such as wolves, foxes, and coyotes are the definitive hosts for the dog tapeworm *Echinococcus granulosus* (FIGURE 17.19A). Eggs reach the soil in feces and spread to numerous intermediary hosts, one of which is humans.

MICROINQUIRY 17

Parasites as Manipulators

Over the last several chapters, we have been discussing viral, bacterial, and fungal pathogens, and protozoan and multicellular parasites. For the latter parasites, you might conclude they make their living "sponging off" their hosts with the ultimate aim to reproduce. In fact, for a long time scientists thought these organisms simply evolved so they could take advantage of what their hosts had to offer. In reality, the parasites may be quite a bit more sophisticated than previously thought. Two cases involving multicellular parasites are submitted for your inquiry.

Case 1

The lancet fluke, *Dicrocoelium dendriticum,* infects cows. Because the infected cow is the definitive host, the fluke needs to get out of the cow to form cysts in another animal. If the cysts can then be located on grass blades, another cow may eat those cyst-containing grass blades and another round of infection by *D. dendriticum* can occur.

17.1a. As a parasite, how do you ensure cysts will be on grass blades that cows eat?

Infected cows excrete dung containing fluke eggs. Snails (intermediate host #1) forage on the dung and in the process ingest the fluke eggs. The eggs hatch and bore their way from the snail gut to the digestive gland where fluke larvae are produced. In an attempt to fight off the infection, the snail smothers the parasites in balls of slime coughed up into the grass.

Ants (intermediate host #2) come along and swallow these nutrient-rich slime balls. Now, here is the really interesting part, because the lancet fluke has yet to get cysts onto grass blades so they will be eaten by more cows. In infected ants, the larvae travel to the ant's head and specifically the nerves that control the ant's mandibles. While most of the larvae then return to the abdomen where they form cysts, a few flukes remain in the head and control the ant's behavior. As evening comes and the temperature cools, the head-infected ants with a belly full of cysts leave their colony, climb to the tip of grass blades, and hold on tight with their mandibles (see figure). The lancet fluke has taken control of the ants and placed them in a position most favorable to be eaten by a grazing cow.

17.1b. What happens if no cow comes grazing that evening?

If a grazing cow does not come by, the next morning the flukes release their influence on the ants, which return to their "normal duties." But, the next evening, the flukes again take control and the ants march back up to the tips of grass blades. Should a cow eat the grass with the ants, the fluke cysts in the ant abdomen quickly hatch and another cycle of reproduction begins in the cow. Now, that is quite a behavioral driving force to control your destiny; in this case, to ensure fluke cysts are positioned so they will be eaten by the cow.

An ant clutching a grass blade by its feet and mandibles.

Case 2

(This case is based on research carried out in a coastal salt marsh by Kevin Lafferty's group at the University of California at Santa Barbara.) Another fluke, *Euhaplorchis californiensis,* uses shorebirds as its definitive host. Infected birds drop feces loaded with fluke eggs into the marsh. Horn snails (intermediate host) eat the droppings containing the eggs. The eggs hatch and castrate the snails. Larvae are produced in the water and latch onto the gills of their second intermediate host, the California killifish. In the killifish, the larvae move from the blood vessels to a nerve that carries the larvae to the brain where the larvae form a thin layer on top of the brain. There they stay, waiting for the fish to be eaten by a shorebird. Once in the bird's gut, the adults develop and produce another round of fertilized eggs. Here, we have two questions.

17.2a. What is the purpose for the larvae castrating the snails?

An experiment was set up to answer this question. Throughout the salt marsh, Lafferty set up cages with uninfected snails and other cages with infected ones. The results were as expected—the uninfected snails produced many more offspring than did the infected snails. Lafferty believes a marsh full of uninfected snails would reproduce so many offspring as to deplete the algae and increase the population of crabs that feed on the snails. By castrating the snails, the parasite actually is controlling the snail population and keeping the salt marsh ecosystem balanced—again for its benefit.

17.2b. Why do the larvae in the killifish take up residence on top of the fish's brain?

Through a series of experiments, Lafferty's group discovered that infected fish underwent a swimming behavior not observed in the uninfected fish. The infected killifish darted near the water surface, a very risky behavior that makes it more likely the fish would be caught by shorebirds. In fact, experiments showed infected fish were 30 times more likely to be caught by shorebirds than uninfected fish. Now, from the parasite's perspective, they want to be caught so another round of reproduction can start in the shorebird. So again, the fluke has manipulated the situation to its benefit.

Parasites may cause disease, but they also maximize their chances for ensuring infection. Pretty amazing!

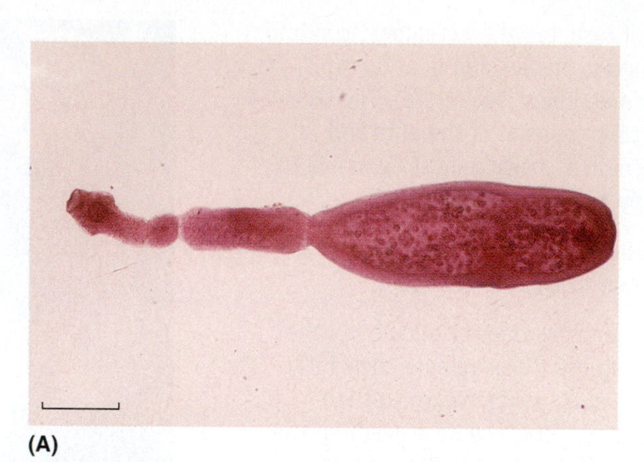
(A)

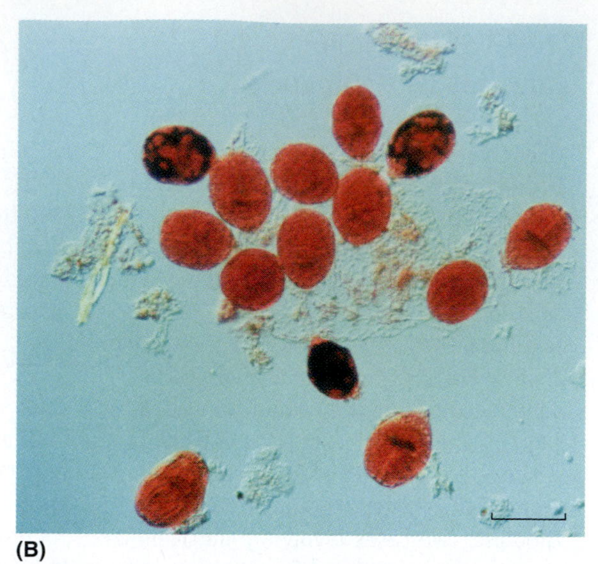
(B)

FIGURE 17.19 **The Dog Tapeworm.** (**A**) A light micrograph of the dog tapeworm, *Echinococcus*. A large section of the worm (right) contains numerous eggs. (Bar = 2 mm.) (**B**) In humans, the tapeworm migrates to the liver or lungs, where they form slow-growing hydatid cysts (red). (Bar = 200 μm.)

Q: Identify the scolex in the light micrograph of Echinococcus. *What was the identifying feature?*

Contact with a dog also may account for transmission. In humans, the parasites travel by the blood to the liver, where they form thick-walled **hydatid cysts** (FIGURE 17.19B). Common symptoms include abdominal and chest pain, and coughing up blood.

CONCEPT AND REASONING CHECKS

17.17 Draw simple life cycles for *Taenia* and *Echinococcus* parasites.

Humans Are Hosts to at Least 50 Roundworm Species

KEY CONCEPT

- Roundworm diseases affect the digestive system and muscles.

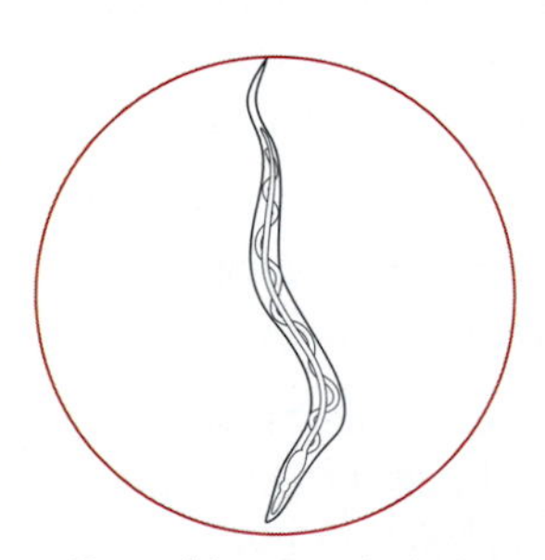
Enterobius vermicularis

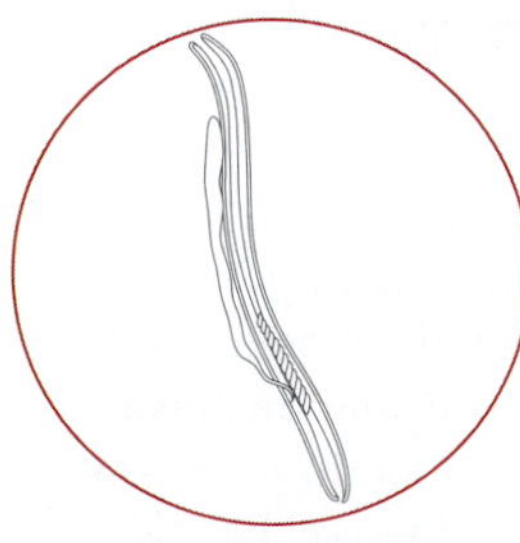
Trichinella spiralis

Here, we discuss a few significant parasitic infections caused by roundworms.

Pinworm Disease. The most prevalent helminthic infection in the United States is **pinworm disease**, where an estimated 30 percent of children and 16 percent of adults serve as hosts.

Pinworm disease is caused by *Enterobius vermicularis*. The male and female worms live in the distant part of the small intestine and in the large intestine, where the symptoms of infection include diarrhea and itching in the anal region. The female worm is about 10 mm long, and the male is about half that size.

The life cycle of the pinworm is relatively simple. Females migrate to the anal region at night and lay a considerable number of eggs. The area itches intensely and scratching contaminates the hands and bed linens with eggs. Reinfection can take place if the hands are brought to the mouth or if eggs are deposited in foods by the hands. The eggs are swallowed, whereupon they hatch in the duodenum and mature in the regions beyond.

Diagnosis of pinworm disease may be made accurately by applying the sticky side of cellophane tape to the area about the anus and examining the tape microscopically for pinworm eggs (FIGURE 17.20). Mebendazole is effective for controlling the disease, and all members of an infected person's family should be treated because transfer of the parasite probably has taken place. Even without medication, however, the worms will die in a few weeks, and the infection will disappear as long as reinfection is prevented.

Trichinellosis. Most of us are familiar with the term **trichinellosis** because packages of pork usually contain warnings to cook the meat thoroughly to avoid this disease. The disease is rare in the United States.

Trichinellosis is caused by the small roundworm *Trichinella spiralis*. The worm lives in the

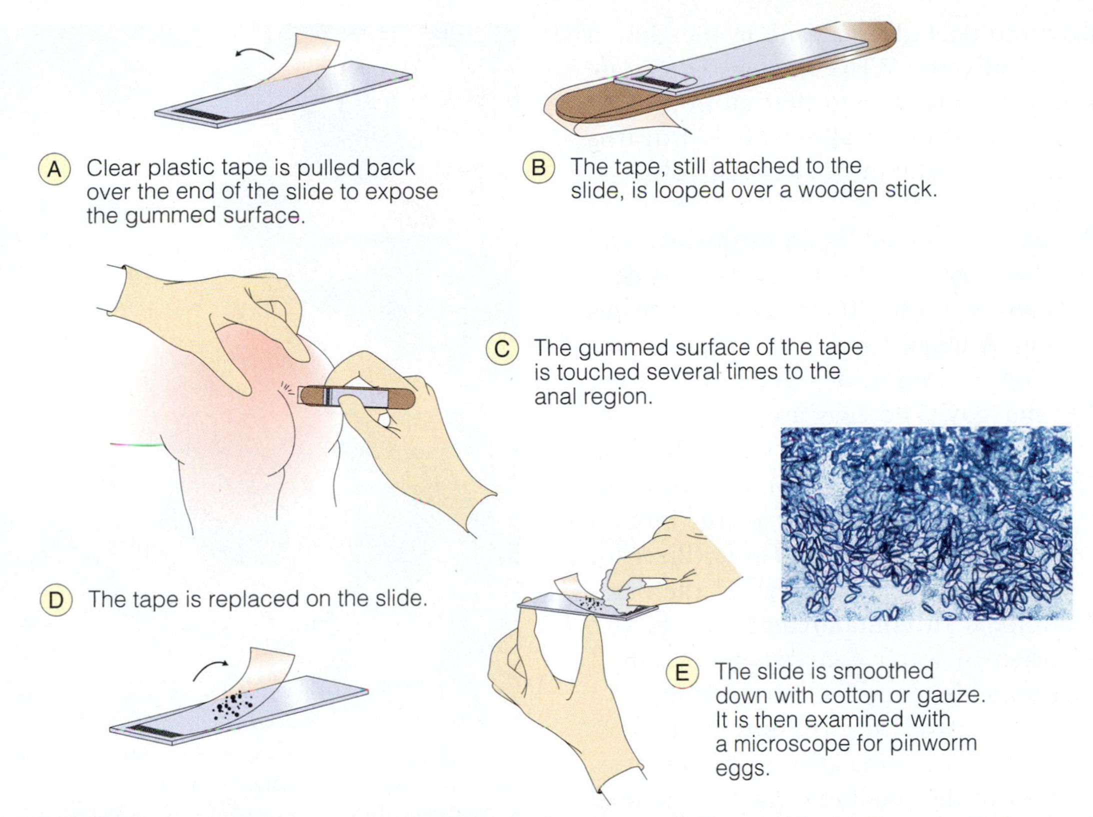

FIGURE 17.20 **Diagnosing Pinworm Disease.** The transparent tape technique used in the diagnosis of pinworm disease.

Q: What time of day might be best to use the tape technique?

intestines of pigs and several other mammals. Larvae of the worm migrate through the blood and penetrate the pig's skeletal muscles, where they remain in cysts. When raw or poorly cooked pork is consumed, the cysts pass into the human intestines and the worms emerge. Intestinal pain, vomiting, nausea, and constipation are common symptoms.

Complications of trichinellosis occur when *T. spiralis* adults mate and the female produces larvae in the intestinal wall. The tiny larvae migrate to the muscles primarily in the tongue, eyes, and ribs where they form cysts (FIGURE 17.21). The patient commonly experiences pain in the breathing muscles of the ribs and loss of eye movement.

The cycle of trichinellosis is completed as cysts are transmitted back to nature in the human feces. Consumption of human waste and garbage then brings the cysts to the pig. Drugs have little effect on cysts, although mebendazole can be used to kill larvae.

CONCEPT AND REASONING CHECKS

17.18 Why do you suppose pinworm disease is so common in the United States?

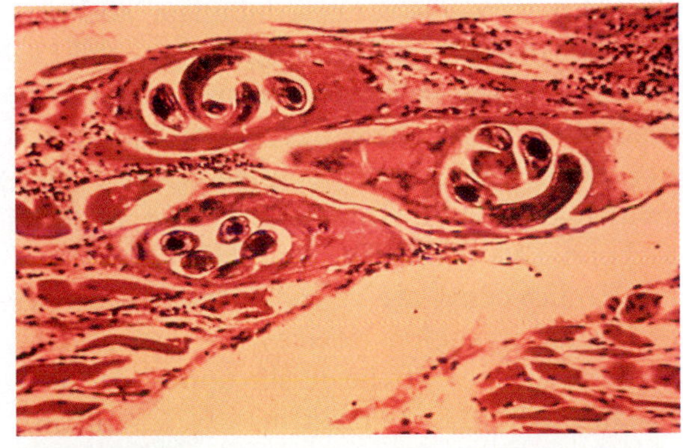

FIGURE 17.21 ***Trichinella spiralis.*** A false-color light micrograph of a *T. spiralis* larva (dark red) coiled inside a cyst in muscle tissue (red).

Q: What is the purpose of forming a cyst in muscle, which represents a good nutrient source for a parasite?

Soil-Transmitted Diseases. These diseases are among the most significant parasites in humans. They are associated with poverty, lack of adequate sanitation and hygiene, and overpopulation. One parasite, caused by whipworms, was mentioned in MicroFocus 17.5.

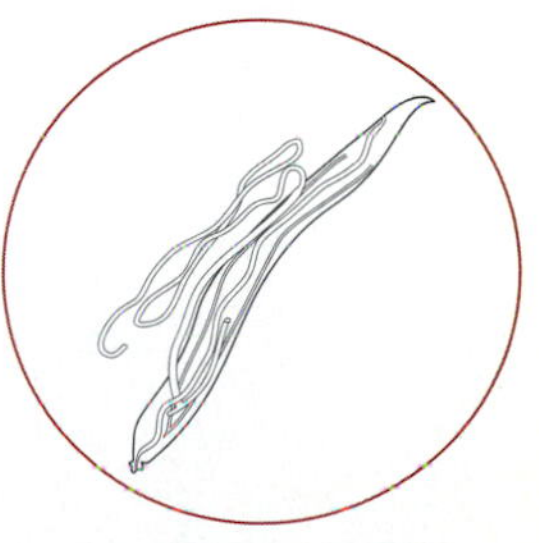

Ascaris lumbricoides

Ascariasis is an infection with *Ascaris lumbricoides*, a parasite that is the second most prevalent multicellular parasite in the United States. Globally, the WHO estimates there are 1.4 billion infections and 380 million cases worldwide, leading to about 60,000 deaths every year, especially in tropical and subtropical regions.

The parasite resembles an earthworm and one of the largest intestinal nematodes; females may be up to 30 cm long, and males 20 cm long. A female *Ascaris* is a prolific producer of eggs, sometimes generating over 200,000 per day. The eggs are fertilized and passed to the soil in the feces, where they can remain viable for several months. Unfortunately, in many parts of the world, human feces, called nightsoil, are used as fertilizer for crops, which adds to the spread of the parasite. Contact with contaminated fingers and consumption of water containing soil runoff are other possible modes of transmission.

When ingested in contaminated food, each egg releases a larva that grows but does not multiply in the small intestine. Abdominal symptoms develop as the worms reach maturity in about two months. Intestinal blockage may be a consequence when tightly compacted masses of worms accumulate, and perforation of the small intestine is possible. In addition, roundworm larvae may pass to the blood and infect the lungs, causing pneumonia. If the larvae are coughed up and then swallowed, intestinal reinfection occurs. Mebendazole is the drug for treatment.

Hookworms are roundworms with a set of hooks or sucker devices for firm attachment to tissues of the host's upper intestine (FIGURE 17.22). Approximately 1.3 billion people around the globe are be infected by hookworms. There are 150 million cases and 65,000 deaths each year from this disease.

FIGURE 17.22 **A Hookworm.** False-color scanning electron micrograph of the parasitic hookworm *A. duodenale*. Inset: the head of the hookworm.

Q: Why are these parasites called hookworms?

Two hookworms, both about 10 mm in length, may be involved in human disease. The first is the Old World hookworm, *Ancylostoma duodenale,* which is found in Europe, Asia, and the United States; the second is the New World hookworm, *Necator americanus,* which is prevalent in the Caribbean islands.

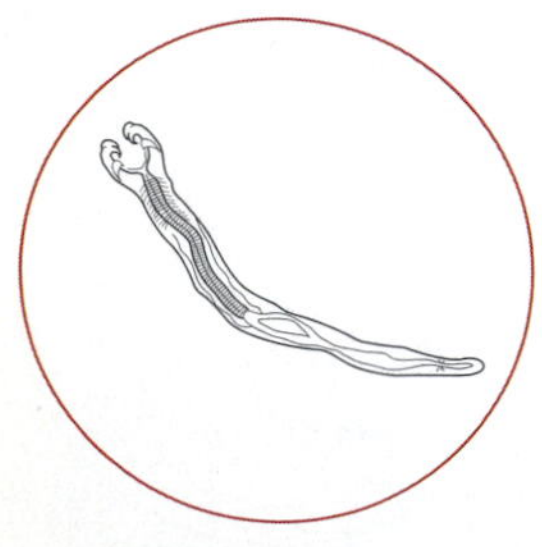

Necator americanus

These parasites live in the human intestine, where they suck blood from the ruptured capillaries. **Hookworm disease** therefore is accompanied by blood loss and is generally manifested by anemia. Cysts also may become lodged in the intestinal wall, and ulcer-like symptoms may develop.

The life cycle of a hookworm involves only a single host, the human (FIGURE 17.23). Female hookworms can release 5,000 to 20,000 eggs, which are excreted to the soil and remain viable for months. Eventually, larvae emerge as long, rod-like **rhabditiform** (*rhabdo* = "rod"; *form* = "shape") larvae. These later become thread-like **filariform** (*filum* = "thread") larvae that attach themselves to vegetation in the soil. When contact with bare feet is made, the filariform larvae penetrate the skin layers and enter the bloodstream. Soon, they localize in the lungs and are carried up to the pharynx in secretions, and then swallowed into the intestines.

Hookworms are common where the soil is warm, wet, and contaminated with human feces and the disease is prevalent where people go barefoot. Mebendazole may be used to reduce the worm burden and the diet may be supplemented with iron to replace that in the

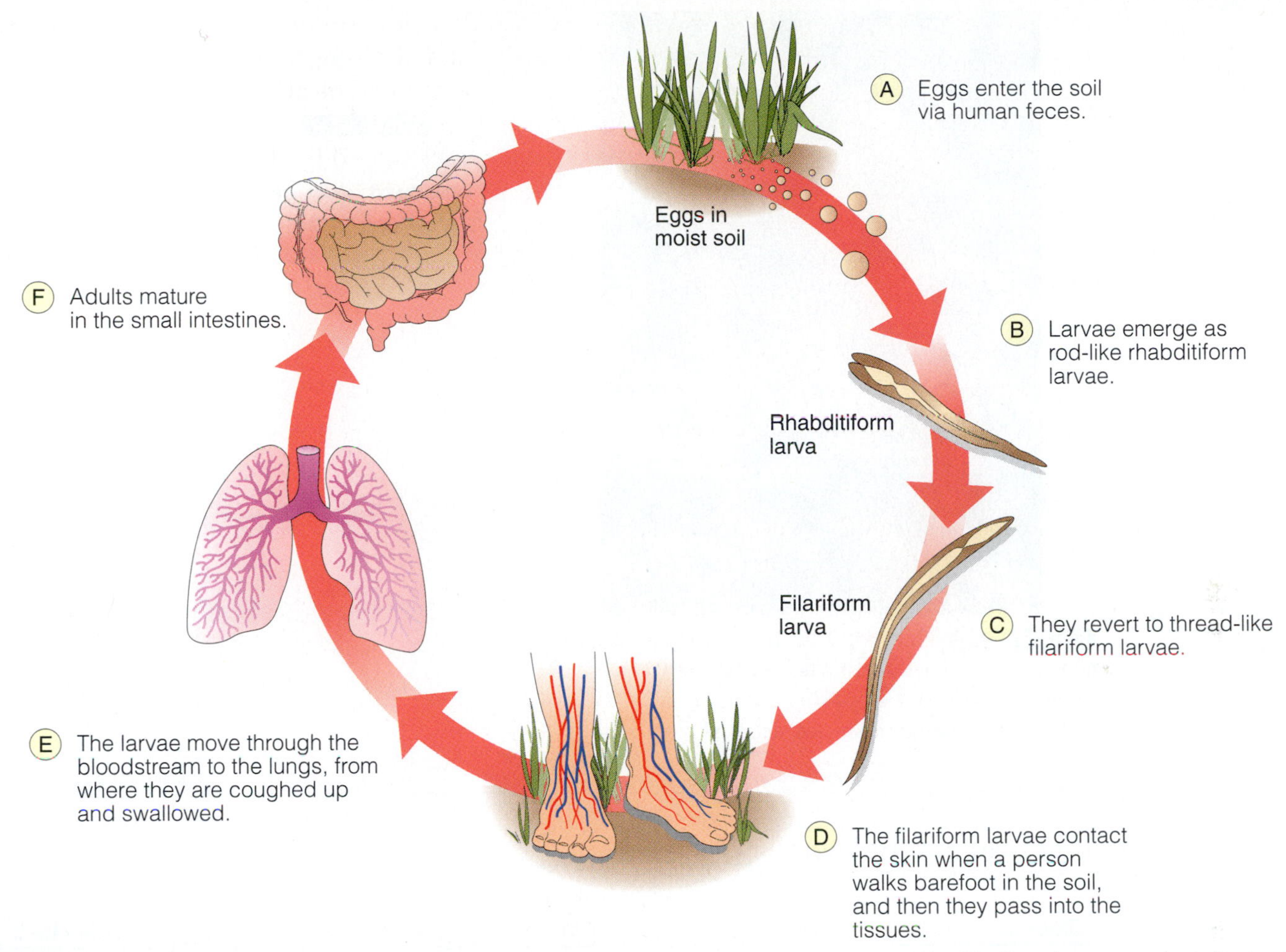

FIGURE 17.23 **The Life Cycle of the Hookworms *Ancylostoma duodenale* and *Necator americanus*.** The filariform larvae are the infective form.

Q: What other helminth can be coughed up and swallowed?

blood loss. It should be noted that dogs and cats also harbor hookworm eggs and pass them in the feces.

CONCEPT AND REASONING CHECKS

17.19 Why are ascariasis and hookworm disease so prevalent worldwide?

Roundworms Also Infect the Lymphatic System

KEY CONCEPT

- Filariasis is a swelling of the lymphatic tissues.

Filariasis is a parasitic disease affecting over 120 million people in 80 countries throughout the tropics and subtropics. The disease is caused by the thread-like roundworm *Wuchereria bancrofti*, which is transmitted by mosquitoes.

The female worm is about 100 mm long and carries larvae called **microfilariae**. Injected in a blood meal, the larvae grow to adults and infect the lymphatic system where they can survive for up to seven years, causing an extensive inflammation and damage to the lymphatic vessels and lymph glands. After years of infestation, the arms, legs, and scrotum swell enormously and become distorted with fluid (FIGURE 17.24). This condition is known as **elephantiasis** because of the gross swelling of lymphatic tissues, called **lymphedema**, and the resemblance of the skin to elephant hide.

The adult worms mate and release millions of microfilariae into the blood, which are ingested in a mosquito's blood meal. There

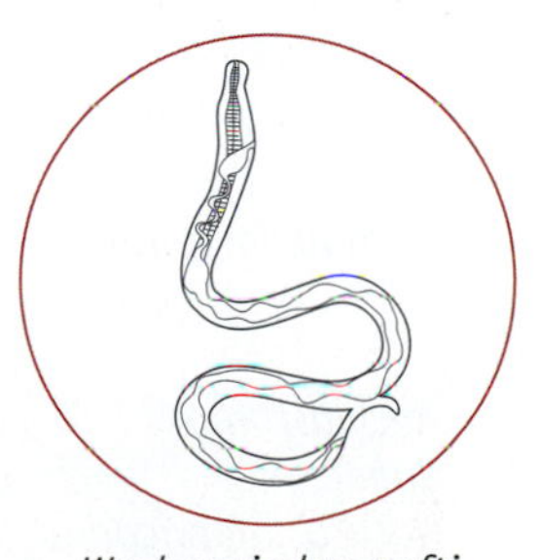

Wuchereria bancrofti

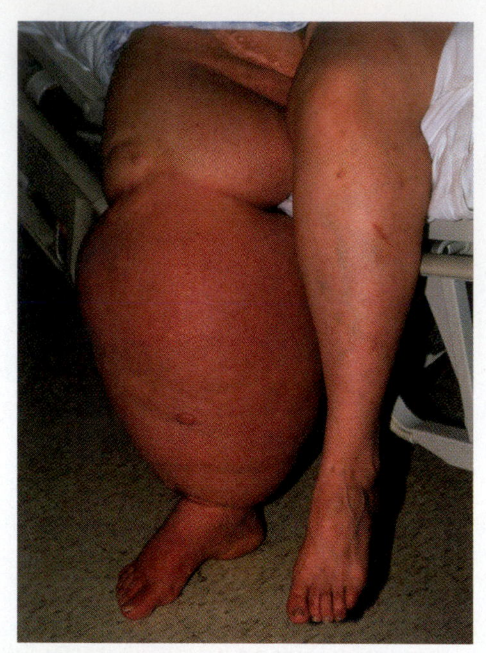

FIGURE 17.24 **Elephantiasis.** The severely swollen leg in a patient with lymphatic filariasis (elephantiasis) caused by *Wuchereria bancrofti*.

Q: Why is the disease called elephantiasis?

they develop into infective larvae, ready to be passed along to another human during the next blood meal.

TABLE 17.4 summarizes the human diseases caused by the helminths.

CONCEPT AND REASONING CHECKS

17.20 What is the relationship between the nematode infection and limb swelling?

TABLE 17.4 A Summary of Human Helminthic Diseases

Organism	Disease	Transmission	Organs Affected	Characteristic Sign	Animal Host
Schistosoma mansoni *S. japonicum* *S. haematobium*	Blood fluke disease	Water contact	Skin Liver Blood	Rash Liver damage Fever	Snail
Paragonimus westermani	Lung fluke disease	Consumption of crabs	Lungs	Cough Poor breathing	Snail Crab
Fasciola hepatica	Liver fluke disease	Consumption of water plants	Liver	Liver damage	Snail Cattle
Taenia saginata	Beef tapeworm disease	Beef consumption	Intestine	Diarrhea	Cattle
Taenia solium	Pork tapeworm disease	Pork consumption	Intestine	Diarrhea	Pig
Echinococcus granulosus	Dog tapeworm disease	Contact	Liver	Liver damage	Dog, other canines
Enterobius vermicularis	Pinworm disease	Contact Food, clothing	Intestine	Anal itching	None significant
Ascaris lumbricoides	Roundworm disease	Food, water Contact	Intestine Lungs	Emaciation Pneumonia	None significant
Trichinella spiralis	Trichinellosis	Pork consumption	Intestine Muscles Eyes	Diarrhea Muscle pain Loss of eye movement	Pig
Ancylostoma duodenale *Necator americanus*	Hookworm disease	Contact with moist vegetation	Intestine Lungs Lymph	Anemia Abdominal pain	None significant
Wuchereria bancrofti	Filariasis	Mosquito	Lymph vessels	Edema Elephantiasis	Mosquito

SUMMARY OF KEY CONCEPTS

17.1 Classification and Characteristics of Protists

- Protists are a diverse group. The majority are unicellular and free-living organisms inhabiting moist areas or water.
- Some protists, such as the green algae are photosynthetic and other protists, like the dinoflagellates, radiolarians, and foraminiferans, are part of the marine phytoplankton. Other protists are heterotrophic and have a fungus-like structure.
- The so-called protozoa include the heterotrophic, predatory or parasitic protists. Protozoa are of interest to microbiologists because they are unicellular, have a microscopic size, and cause infectious disease.
- The protozoa have many life styles but can be cataloged into one of six groups. The parabasalids and diplomonads are motile but have modified mitochondria; the kinetoplastids are motile, have an undulating membrane, and a mass of DNA called the kinetoplast; the amoebozoans move by means of pseudopods and reproduce by binary fission; the ciliates have their cell surfaces covered by cilia and contain two different types of nuclei—macronuclei, which control metabolic events and micronuclei, which play a critical role in genetic recombination through conjugation; and apicomplexans, which are nonmotile as adults and are obligate parasites.
- Protozoa typically exist in the trophozoite and cyst forms, the latter often a very resistant form. Some protozoa (and most helminths) require at least two hosts to complete their life cycle, the definitive host, where sexual reproduction occurs, and the intermediary host, where asexual reproduction takes place.

17.2 Protozoal Diseases of the Skin, and Digestive and Urinary Tracts

- Leishmaniasis is transmitted by sandflies and involves blood (visceral leishmaniasis; kala azar) and tissue (cutaneous leishmaniasis) invasion. About 90 percent of cases involving the visceral form are fatal if not treated.
- Amoebiasis is caused by *Entamoeba histolytica*. Intestinal ulcers and sharp, appendicitis-like pain accompanies the disease.
- Giardiasis is an intestinal disease, but the protozoa do not penetrate the tissue. The agent, a flagellate called *Giardia intestinalis,* is acquired from contaminated food and water.
- Cryptosporidiosis causes a cholera-like diarrhea. The protozoan, *Cryptosporidium parvum,* generates a more serious disease in people with suppressed immune systems.
- Cyclosporiasis is an intestinal disease contracted from eating fresh produce contaminated with *Cyclospora cayetanensis.* Symptoms include watery diarrhea, nausea, abdominal cramps, bloating, and vomiting.
- Trichomoniasis, a sexually transmitted disease caused by the flagellate *Trichomonas vaginalis,* is among the most common diseases in the United States. Females suffer a burning pain during urination; males may be asymptomatic.

17.3 Protozoal Diseases of the Blood and Nervous System

- The nonmotile protozoan, *Plasmodium,* is the cause of malaria, one of the most serious global health problems. Transmitted by mosquitoes, *Plasmodium* species destroy the red blood cells and often bring on death.
- Human African sleeping sickness (African trypanosomiasis) is transmitted by tsetse flies carrying *Trypanosoma brucei* variety *gambiense* or *Trypanosoma brucei* variety *rhodesiense.* Blood invasion and tissue involvement characterize the disease. Chagas disease (American trypanosomiasis) is characterized by fever and heart tissue damage. The protozoan, *Trypanosoma cruzi,* is transmitted by the triatomid ("kissing") bugs.
- A similar disease to malaria is called babesiosis. The protozoan, *Babesia microti,* is transmitted by ticks. The disease is rarely fatal and remains more localized in occurrence.
- An apicomplexan, *Toxoplasma gondii,* causes serious disease in AIDS patients. *Toxoplasma* affects various organs, especially those of the nervous system. Infections in pregnant women can lead to congenital infections of the fetus.
- The amoeba *Naegleria fowleri* enters the nose and follows the olfactory tracts to the brain. Most patients die within seven days on the infection.

17.4 The Multicellular Helminths and Helminthic Infections

- Among the helminths, there are two groups responsible for parasitic diseases in humans—flatworms and roundworms.
- Flatworms are multicellular animals lacking respiratory and circulatory structures, and having a gastrovascular cavity. Flatworms include the trematodes (flukes), which have a complex life cycle with sexual and asexual reproductive stages. The cestodes (tapeworms) have a ribbon-like body made of proglottids. The worms consist of segments called proglottids and a head region called the scolex, often with hooks or sucker devices for attachment to the host tissue. Tapeworms can infect all mammals and generally are transmitted to humans in foods.
- Roundworms (nematodes) are among the most prevalent animals worldwide. They are multicellular with separate sexes. Disease usually is the result of large worm burdens within the individual.
- One of the major fluke parasites of humans is the blood fluke *(Schistosoma)* that causes schistosomiasis. Symptoms include fever and chills. If eggs are carried by the blood to the liver, liver damage results. Bladder infections produce bloody urine.
- The lung fluke *(Paragonimus westermani)* is common in Asia and the South Pacific when people eat undercooked, infected crabmeat. The flukes then pass to the lungs and cause breathing difficulties. The human liver fluke *(Fasciola hepatica)* migrates to the liver, causing substantial damage.
- Foodborne tapeworms are the beef tapeworm *(Taenia saginata)* and the pork tapeworm (*T. solium*). Living in the intestines, the tapeworms expel many gravid proglottids daily. Humans acquire the parasite as cysts in poorly cooked beef or pork. Symptoms are mild.
- The dog tapeworm *(Echinococcus granulosus)* can be spread to humans, who are an intermediate host.
- The pinworm *Enterobius vermicularis* is a roundworm acquired as eggs from the soil. Symptoms include diarrhea and anal itching.
- The pork roundworm *Trichinella spiralis* is acquired in undercooked pork. Symptoms include intestinal pain, nausea, vomiting, and constipation. Complications occur when the parasite larvae move to the muscles and form cysts.
- Soil-transmitted helminths include the roundworm *Ascaris lumbricoides,* which develops larvae that grow in the small intestine. Intestinal blockage may occur and the small intestine may be perforated. Spread to the blood and lungs can cause pneumonia.
- Hookworm disease is caused by *Ancylostoma duodenale* and *Necator americanus*. On infection, they suck blood from the human intestinal wall, causing anemia. Cysts can form in the intestinal wall.
- The filarial worm *(Wuchereria bancrofti)* is mosquito-borne and causes elephantiasis. Blockage of the lymphatic system produces swelling as fluid accumulates in the legs.

LEARNING OBJECTIVES

After understanding the textbook reading, you should be capable of writing a paragraph that includes the appropriate terms and pertinent information to answer the objective.

1. List the characteristics of the protista, including those of the green algae and members of the phytoplankton.
2. Justify the study of protozoa by microbiologists.
3. Identify the six informal groups of protozoa.
4. Describe the unique characteristics of the parabasalids, diplomonads, and kinetoplastids.
5. Summarize the characteristics of the amoebozoans.
6. Compare the characteristics of the ciliates and apicomplexans.
7. Distinguish definitive and intermediate hosts.
8. Describe the mode of transmission and forms of leishmaniasis.
9. Identify the agent of amoebiasis and discuss the mode of transmission of the disease.
10. Summarize the characteristics and transmission of giardiasis.
11. Assess the significance of cryptosporidiosis and cyclosporiasis as human diseases.
12. Explain how trichomoniasis is a sexually-transmitted disease.
13. Construct a simple life cycle for *Plasmodium*.
14. Assess the role of antimalarial drugs and vaccines to combat malaria.
15. Compare and contrast human African sleeping sickness and Chagas disease.
16. Infer why babesiosis is a malaria-like disease.
17. Identify the role of trophozoites, cysts, and oocysts to the infection cycle of *Toxoplasma*.
18. Name the agent and summarize the characteristics of primary amoebic meningoencephalitis (PAM).
19. Identify the two groups of helminths of medical importance and summarize their characteristics.
20. Explain the life cycle and infection process of *Schistosoma*.
21. Describe the features of human lung and liver fluke diseases.
22. Compare and contrast beef and pork as well as dog tapeworm diseases.
23. Assess the significance of pinworm disease in the United States and describe the infectious cycle.
24. Summarize the disease cycle and agent responsible for trichinellosis.
25. Identify the soil-transmitted helminths, the diseases they cause, and their mode of transmission.
26. Discuss how *Wuchereria bancrofti* causes filariasis (elephantiasis).

SELF-TEST

Answer each of the following questions by selecting the ***one*** answer that best fits the question or statement. Answers to even-numbered questions can be found in **Appendix C**.

1. Which one of the following statements does *not* apply to parasites?
   A. They are free-living organisms.
   B. They must live in or on another species.
   C. Helminths are multicellular examples.
   D. Parasites provide nothing for the host.
   E. Protozoa are unicellular examples.
2. The _____ are members of the _____.
   A. green algae; protozoa
   B. radiolarians; apicomplexans
   C. dinoflagellates; phytoplankton
   D. radiolarians; fungus-like protists
   E. protozoa; photosynthetic protists
3. This group of protozoa has primitive mitochondria.
   A. Kinetoplastids
   B. Ciliates
   C. Apicomplexans
   D. Diplomonads
   E. None of the above (**A–D**) is correct.
4. The _____ group of protozoa have food vacuoles and pseudopods. If all are correct, please select **E**.
   A. Amoebozoans
   B. *Naegleria*
   C. *Entamoeba*
   D. *Acanthamoeba*
   E. All the above (**A–D**) are correct.
5. Which one of the following is *not* found in the ciliates?
   A. A contractile vacuole
   B. Macronuclei and micronuclei
   C. A pellicle
   D. A complex of organelles in the tip
   E. Mitochondria
6. An intermediate host is
   A. where parasite asexual cycle occurs.
   B. always a nonhuman host.
   C. always some form of insect vector.
   D. where parasite sexual cycle occurs.
   E. the human host between two other animal hosts.
7. The vector transmitting leishmaniasis is the
   A. mosquito.
   B. sandfly.
   C. tsetse fly.
   D. sand flea.
   E. tick.
8. _____ enters the human body as a cyst and develops into a trophozoite in the small intestine; a severe form of dysentery may occur.
   A. *Cryptosporidium parvum*
   B. *Entamoeba histolytica*
   C. *Giardia intestinalis*
   D. *Leishmania major*
   E. *Cyclospora cayetanensis*

9. Sucker-like devices allow this protozoan to adhere to the intestinal lining.
   A. *Cyclospora cayetanensis*
   B. *Entamoeba histolytica*
   C. *Giardia intestinalis*
   D. *Leishmania major*
   E. *Cryptosporidium parvum*

10. _____ causes an infection of the _____.
   A. *Giardia;* blood
   B. *Plasmodium;* nervous system
   C. *Trypanosoma;* urinary tract
   D. *Trichomonas;* urinary tract
   E. *Babesia;* digestive tract

11. This disease sickened more than 400,000 residents of Milwaukee in 1993.
   A. Giardiasis
   B. Cryptosporidiosis
   C. Trypanosomiasis
   D. Cyclosporiasis
   E. Schistosomiasis

12. This genus of parabasalid affects over 2.5 million Americans annually and, in females, the resulting disease causes a burning pain on urination.
   A. *Toxoplasma*
   B. *Cryptosporidium*
   C. *Cyclospora*
   D. *Trichomonas*
   E. *Trypanosoma*

13. The _____ form of the malarial parasite enters the human blood while the _____ enters the mosquito.
   A. sporozoites; merozoites
   B. merozoites; gametocytes
   C. gametocytes; sporozoites
   D. merozoites; sporozoites
   E. sporozoites; gametocytes

14. Chagas disease is caused by
   A. *Trypanosome cruzi.*
   B. *Toxoplasma gondii.*
   C. *Babesia microti.*
   D. *Plasmodium falciparum.*
   E. *Trypanosoma brucei*

15. The T in TORCH stands for the disease caused by
   A. *Trypanosome brucei.*
   B. *Toxoplasma gondii.*
   C. *Taenia spiralis.*
   D. *Trypanosoma cruzi.*
   E. *Trichomonas vaginalis.*

16. This opportunistic protozoan causes primary meningoencephalitis (PAM).
   A. *Plasmodium vivax*
   B. *Toxoplasma gondii*
   C. *Naegleria fowleri*
   D. *Paragonimus westermani*
   E. *Entamoeba histolytica*

QUESTIONS FOR THOUGHT AND DISCUSSION

Answers to even-numbered questions can be found in **Appendix C**.

1. A newspaper article written in the 1980s asserted that parasitology is a "subject of low priority in medical schools because the diseases are exotic infections only occurring in remote parts of the world." Would you agree with that statement today? Explain.

2. Many historians believe malaria contributed to the downfall of ancient Rome. Over the decades, malaria incidence increased with expansion of the Roman Empire, which stretched from the Sahara desert to the borders of Scotland, and from the Persian Gulf to the western shores of Portugal. How do you suspect the disease and the expansion are connected?

3. It has been said that until recent times, many victims of a particular protozoal disease were buried alive because their life processes had slowed to the point where they could not be detected with the primitive technology available. Which disease was probably present?

4. In 2005, the World Health Organization reported that, after malaria, schistosomiasis and filariasis (200 million and 90 million annual cases, respectively) are the most prevalent tropical diseases. How do you believe the incidence of these diseases can be reduced on a global scale?

5. Some restaurants offer a menu item called steak tartare, which is a dish served with raw ground beef. What hazard might this meal present to the restaurant patron?

6. Justify this statement: "Perhaps the most important reason for discussing parasitical diseases is that they highlight just how enmeshed we are in the web of life."

APPLICATIONS

Answers to even-numbered questions can be found in **Appendix C**.

1. You and a friend who is three months pregnant stop at a hamburger stand for lunch. Based on your knowledge of toxoplasmosis, what helpful advice can you give your friend? On returning to her home, you notice she has two cats. What additional information might you share with her?
2. Cardiologists at a local hospital hypothesized that a few patients with a certain protozoal disease easily could be lost among the far larger population of heart disease sufferers patronizing county clinics. They proved their theory by finding 25 patients with this protozoal disease among patients previously diagnosed as having coronary heart disease. Which protozoal disease was involved?
3. Federal law stipulates that food scraps fed to pigs must be cooked to kill any parasites present. It also is known that feedlots for swine are generally more sanitary than they have been in the past. As a result of these and other measures, the incidence of trichinellosis in the United States has declined, and the acceptance of "pink pork" has increased. Do you think this is a dangerous situation? Why?

REVIEW

Read the statement concerning the helminths, and then select the answer or answers that best apply to the statement. Place the letter(s) next to the statement. The answers to even-numbered statements are listed in **Appendix C**.

_____ **1.** Transmitted by an arthropod.
A. Filariasis
B. Trichinosis
C. Hookworm disease

_____ **2.** Beef tapeworm species.
A. *Echinococcus granulosus*
B. *Fasciola hepatica*
C. *Taenia saginata*

_____ **3.** Type of fluke.
A. *Schistosoma*
B. *Necator*
C. *Fasciola*

_____ **4.** Causes human lung fluke disease.
A. *Wuchereria*
B. *Paragonimus*
C. *Echinococcus*

_____ **5.** Infects the human intestines.
A. *Trichinella spiralis*
B. *Ascaris lumbricoides*
C. *Taenia saginata*

_____ **6.** Type of tapeworm.
A. *Paragonimus*
B. *Echinococcus*
C. *Fasciola*

_____ **7.** Snail is the intermediate host.
A. Blood fluke
B. Dog tapeworm
C. Intestinal fluke

_____ **8.** Attaches to host tissue by hooks.
A. *Necator*
B. *Trichinella*
C. *Fasciola*

_____ **9.** Affects pigs as well as humans.
A. *Wuchereria*
B. *Trichinella*
C. *Schistosoma*

_____ **10.** Life cycle includes miracidia and cercaria.
A. *Schistosoma*
B. *Fasciola*
C. *Paragonimus*

_____ **11.** Acquired by consuming contaminated pork.
A. *Taenia solium*
B. *Echinococcus*
C. *Necator americanus*

_____ **12.** Classified in the phylum Nematoda.
A. *Ascaris*
B. *Wuchereria*
C. *Schistosoma*

_____ **13.** Causes inflammation and damage to the lymphatic vessels.
A. *Paragonimus*
B. *Ascaris*
C. *Wuchereria*

_____ **14.** Male and female forms exist.
A. *Fasciola hepatica*
B. *Ascaris*
C. *Schistosoma*

_____ **15.** Forms hydatid cysts.
A. Blood fluke
B. Lung fluke
C. Dog tapeworm

HTTP://MICROBIOLOGY.JBPUB.COM/

The site features learning, an on-line review area that provides quizzes and other tools to help you study for your class. You can also follow useful links for in-depth information, read more MicroFocus stories, or just find out the latest microbiology news.

REVIEW—PARASITE LIFE CYCLES

Label the two parasite life cycles, (1) malaria and (2) schistosomiasis, using the term lists provided.

(1) *Plasmodium* Life Cycle Term List

Definitive host

Gametes

Gameyocyte

Intermediate host

Merozoites

Oocyst

Sporozoites

Zygote

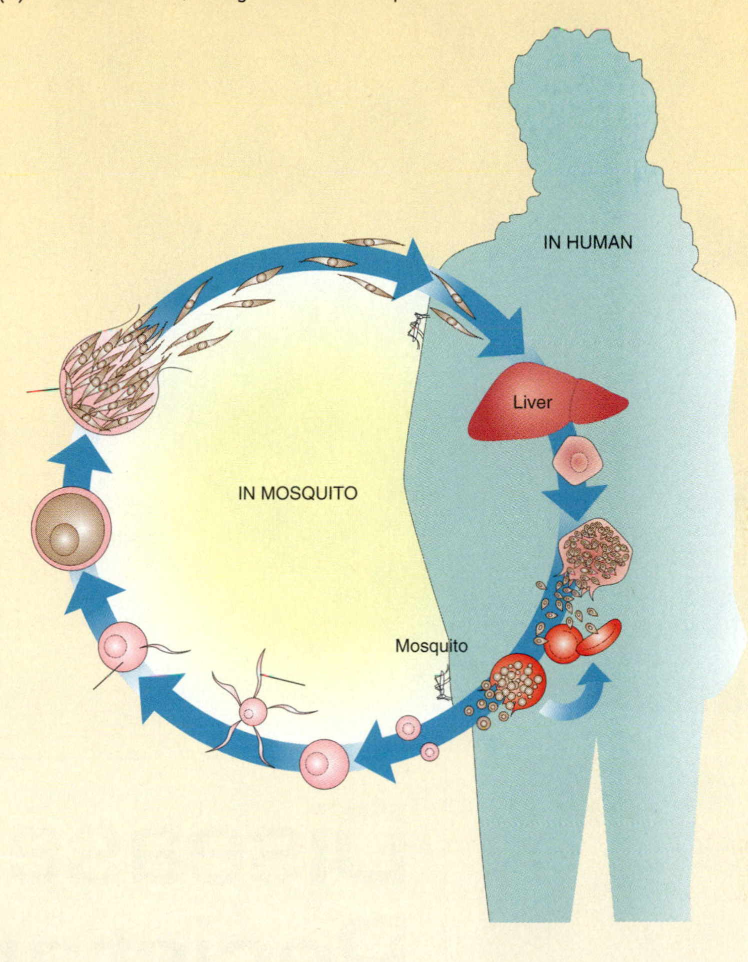

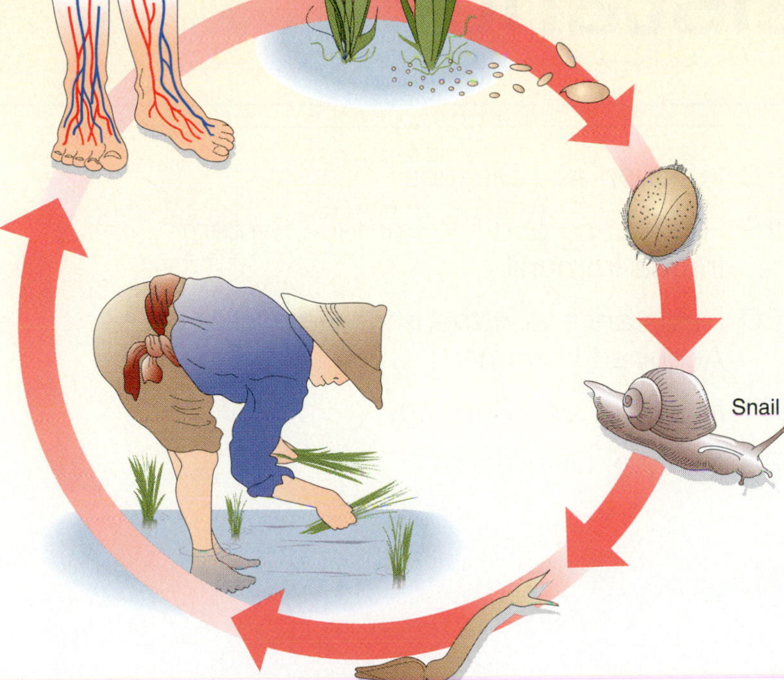

(2) *Schistosoma* Life Cycle Term List

Cercaria

Definitive host

Eggs

Intermediate host

Miracidium

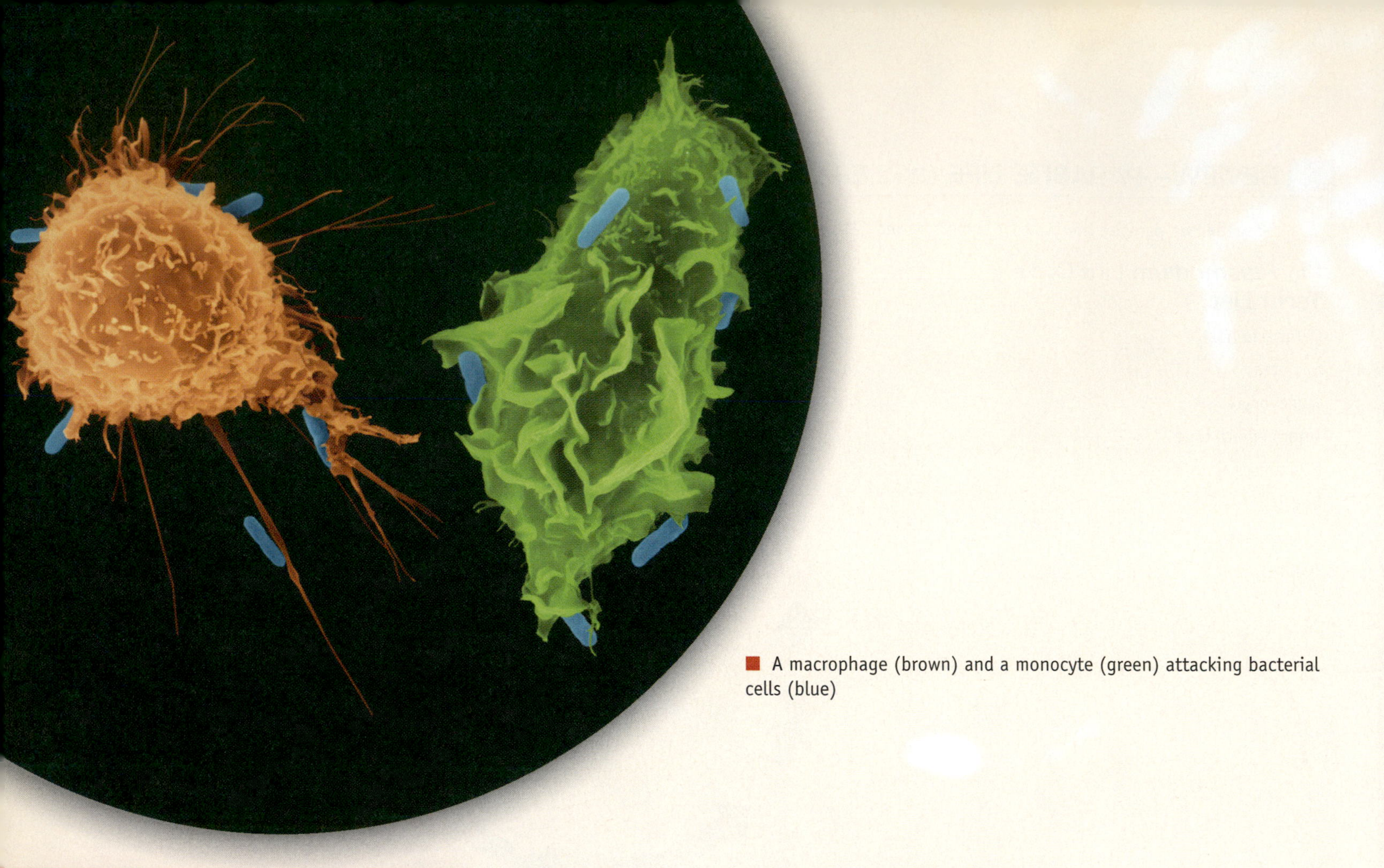

A macrophage (brown) and a monocyte (green) attacking bacterial cells (blue)

PART 5 Disease and Resistance

In past centuries, the spread of disease appeared to be willfully erratic. Illnesses would attack some members of a population while leaving others untouched. A disease that for many generations had taken small, steady tolls would suddenly flare up in epidemic proportions. And strange, horrifying plagues descended unexpectedly on whole nations.

Scientists now know humans live in a precarious equilibrium with the microorganisms surrounding them. Generally the relationship is harmonious, because humans can come in contact with most microorganisms and develop resistance to them. However, when the natural resistance is unable to overcome the aggressiveness of microorganisms, disease sets in. In other instances, the resistance is diminished by a pattern of human life that gives microorganisms the edge. For example, during the Industrial Revolution of the 1800s, many thousands of Europeans moved from rural areas to the cities. They sought new jobs, adventure, and prosperity. Instead, they found endless labor, unventilated factories, and wretched living conditions—and they found disease.

In Part 5 of this text, we explore the infectious disease process and the mechanisms by which the body responds to disease. Chapter 18 opens with an overview of the host–microbe relationship and the factors contributing to the establishment of disease. Epidemiology and diseases within populations also are explored. In Chapters 19 and 20, the discussion turns to nonspecific and specific methods by which body resistance develops, with emphasis on the immune system. Various types of immunity are explored in Chapter 21, together with a survey of laboratory methods using the immune reaction in the diagnosis of disease. Finally, in Chapter 22 the discussion centers on immune disorders leading to serious problems in humans. In these chapters, we uncover the roots of infectious disease and resistance, and come to understand them at the fundamental level.

MICROBIOLOGY PATHWAYS

Epidemiology

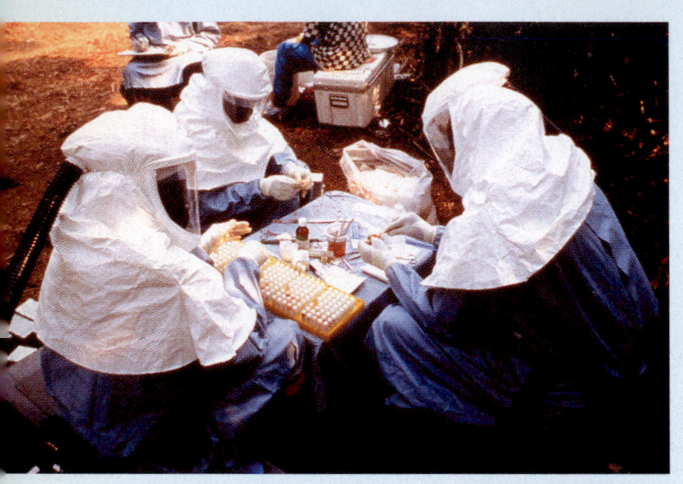

Flying to an impoverished African country on your second day of work to battle Ebola, one of the most deadly viruses, isn't most people's idea of a dream assignment. But it was for Marta Guerra. In fact, the trip to Uganda in 2000 was the assignment she had been coveting. "I wasn't that worried," Guerra says. "This particular strain has only a 65% death rate instead of the Congo strain which is 85%."

Guerra is a disease epidemiologist, popularly known as a "disease detective," with the Epidemic Intelligence Service (EIS) of the Centers for Disease Control and Prevention (CDC) in Atlanta.

Growing up in the multicultural environments of Havana, Cuba, and Washington D.C., Marta Guerra developed a keen interest in other cultures and teaching people about health risks. She was fascinated by stories her professors told about working overseas. After seeing the movie *Outbreak,* "I thought, that's what I want to do—help contain a deadly epidemic," recalls Guerra. So she obtained a Master's in public health and a Ph.D. in tropical medicine.

Like all EIS officers, Guerra's job is to isolate the cause of an outbreak, prevent its spread, and get out public health messages to people who could have been exposed. When Guerra flew to Uganda in November 2000, the Ebola outbreak had already been identified, so her task was to go from village to village, trying to locate family members and friends, and educate them about symptoms and treatment. "In every corner of Africa people know the word Ebola, and they are terrified of it," Guerra explains. "Sometimes they hide sick family members; sometimes they're frightened of survivors."

The job of a disease detective can be difficult—and dangerous. Although Guerra seldom considered she might acquire a disease she was investigating, she was concerned about political violence. In Uganda, Guerra's team needed military escorts on their travels through villages. In Ethiopia, while on a polio-eradication mission in the summer of 2001, she recalls, "There was rebel activity in all the areas we traveled through—plus land mines. It was pretty scary."

Perhaps you might be interested in a similar career. "You have to be highly motivated, with the ability to think fast on your feet and make quick decisions. You have to be able to walk into a chaotic situation and deal with whatever is thrown at you," Guerra says. "Sometimes I barely drop my bags at home before I'm called out again. Being adaptable is really essential."

To get started, you need a bachelor's degree in a biological science. In addition to required courses in chemistry and biology, undergraduates should study microbiology, mathematics, and computer science. A master of science in epidemiology or public health also is required; many have a Ph.D. or a medical or veterinary degree. Most American disease epidemiologists then apply to EIS's two-year, post-graduate program of service and on-the-job training, where they work with mentors like Marta. She says, "I like the fact that I am contributing to science in the sense that what I do will affect people far into the future."

(Essay modified from *Disease Detective* by Carol Sonenklar in MedHunters.com)

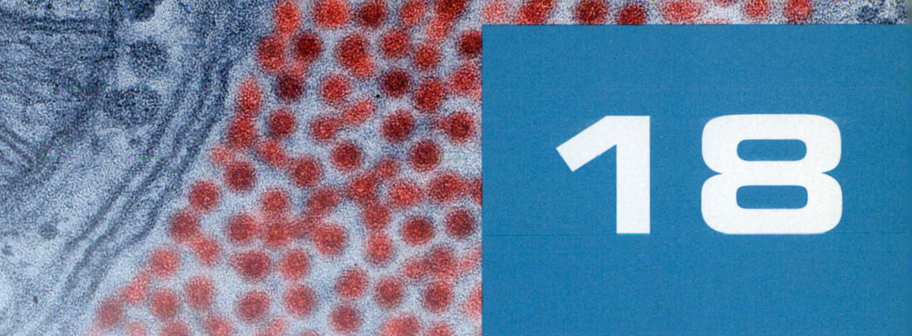

Infection and Disease

"Health care matters to all of us some of the time, public health matters to all of us all of the time"
—C. Everett Koop, former Surgeon General of the United States

Chapter Preview and Key Concepts

18.1 The Host–Microbe Relationship
- Infection and disease occur when a host–microbe relationship tilts in favor of the microbe.
- Microbes vary greatly in their pathogenicity.
- Contact with a potential pathogen can have several outcomes.

18.2 Establishment of Infection and Disease
- Disease progression involves incubation, prodromal, acme, decline, and convalescent stages.
- Pathogens gain access to the host through portals of entry.
- Invasiveness is critical for some pathogens.
- Virulence factors include enzymes and toxins.
- Pathogens leave the host through portals of exit.

18.3 Infectious Disease Epidemiology
- Diseases have certain behaviors in populations.
- Disease transmission involves contact, airborne, vector, or vehicle methods.
- Reservoirs are places in the environment where a pathogen can be found.
- Diseases are identified as being endemic, epidemic, or pandemic.
- Nosocomial infections are contracted as a result of being treated for another illness in a hospital or other health care setting.
- Diseases emerging or reemerging anywhere in the world can become a global health menace.

MicroInquiry 18: Epidemiological Investigations

By late July 1999, crows were literally dropping out of the sky in New York City and dead crows were found in surrounding areas as well. By early September, officials at the Bronx Zoo discovered other birds, including a cormorant, two red Chilean flamingoes, and an Asian pheasant at the zoo had died of the same brain and heart inflammations as found in the crows.

On August 23, 1999, an infectious disease physician at a hospital in northern Queens reported to the New York City Department of Health that two patients had been admitted with encephalitis. In fact, on further investigation, the health department identified a cluster of six patients with encephalitis. Testing by the Centers for Disease Control and Prevention (CDC) of these initial cases for antibodies to common North American arboviruses (see Chapter 15) was positive for St. Louis encephalitis (SLE)—a virus carried by mosquitoes. These findings prompted the health department to begin aerial and ground application of insecticides.

News of SLE and spraying caught the attention of the Bronx Zoo officials. If the birds were dying from the same encephalitis disease as in humans, it could not be caused by the SLE virus because birds do not contract SLE.

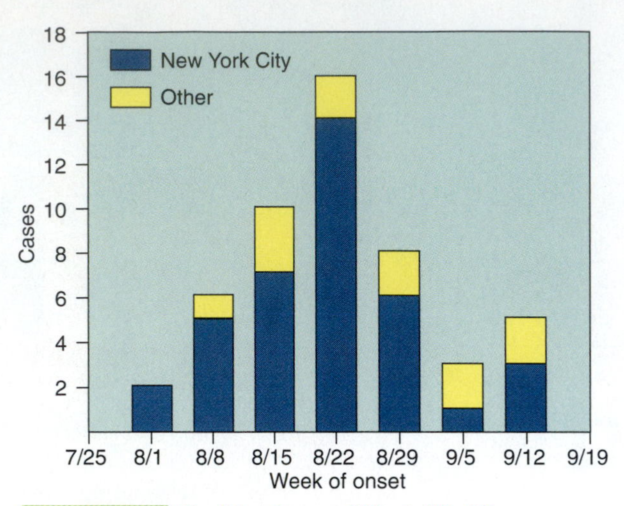

FIGURE 18.1 Positive Cases of West Nile Virus Infection, New York—1999.
Source: www.cdc.gov/mmwr/preview/mmwrthml/mm4839a5.htm.
Q: What does this graph tell you about the frequency with which WNV occurred?

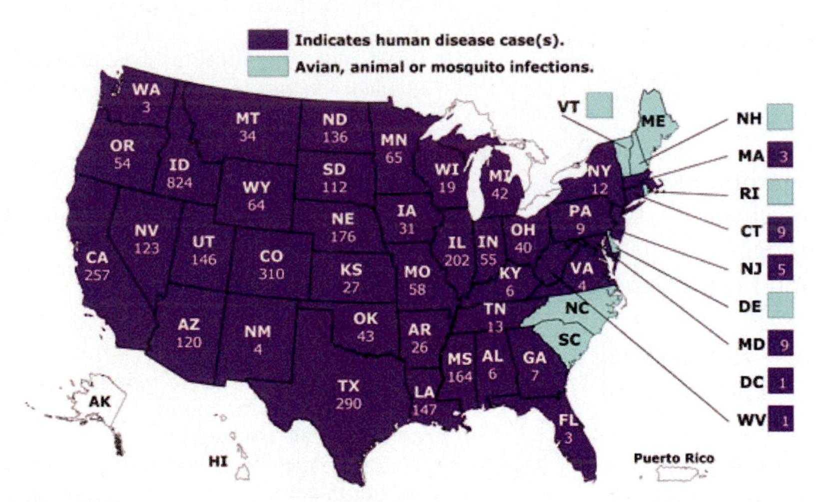

FIGURE 18.2 West Nile Virus Activity, United States—2006. The map indicates the distribution of avian, animal, or mosquito infections in 2006. It also identifies the number of reported human cases (purple) in each state (a state is shaded no matter in what area of the state the disease case(s) occurred).
Source: www.cdc.gov/ncidod/dvbid/westnile/mapsactivity/surv&contol06Maps.htm.
Q: Why is this epidemiological map referred to as a geographical distribution?

So, a reinvestigation by the CDC of virus samples taken from humans, birds, and mosquitoes was undertaken. Results indicated all viral isolates were closely related to West Nile virus (WNV), which had never been isolated in the western hemisphere. It soon became evident that indeed, the disease in birds and humans was caused by WNV.

By early fall, mosquito activity waned and the number of human cases declined (FIGURE 18.1). In all, 61 people would be infected and seven would die. Although New York City and the surroundings could breathe a sigh of relief, the WNV outbreak was only the beginning of a virus march across the United States. By 2006, human cases of West Nile encephalitis would be reported across America (FIGURE 18.2).

The outbreak of West Nile encephalitis is one example describing the epidemiology of infection and disease. In this chapter, we discuss the complex mechanisms underlying the spread and development of infectious disease. Because individual diseases are considered in many previous chapters of this text, our purpose here is to bring together many concepts of disease and synthesize an overview of the host–microbe relationship. We summarize much of the important terminology used in medical microbiology and outline the factors used by microorganisms to establish themselves in tissues. An understanding of the topics concerning the host–microbe relationship will be essential preparation for the detailed discussion of host resistance mechanisms in Chapters 19 and 20.

18.1 The Host–Microbe Relationship

■ **Epidemiology:** The scientific study of the causes and transmission of disease within a population.

By the early 1970s, the Surgeon General of the United States claimed we could "close the books on infectious diseases" (see Chapter 1). The development and use of antibiotics and vaccines had made the threat of infectious disease of little consequence. However, antibiotic resistance and new emerging diseases, including Legionnaires' disease, AIDS, Lyme disease, hantavirus pulmonary syndrome, and SARS, have thwarted such optimism. In 2005, of the approximately 57 million humans who died, more than 25 percent (15 million) died from infectious diseases, making them the second leading cause of death (behind cardiovascular disease) (FIGURE 18.3). In fact, infectious diseases are the leading cause of death in children under 5 years of age.

The Human Body Maintains a Symbiosis with Microbes

KEY CONCEPT

- Infection and disease occur when a host-microbe relationship tilts in favor of the microbe.

Infection refers to the relationship between two organisms—the host and the microbe—and the competition for supremacy taking place between them. (Note: in this chapter, for simplicity of discussion, "microbe" includes the viruses.) A host whose resistance is strong remains healthy, and the microbe is either driven from the host or assumes a benign relationship with the host. By contrast, if the host loses the competition, disease develops. The term **disease** refers to any change from the general state of good health. It is important to note that disease and infection are not synonymous; a person may be infected without suffering a disease.

Whether host or microbe gets the upper hand often is due in part to the 100 trillion microbes found on and in the human body. This remarkable number amounts to almost 3 pounds of human weight! These microbes represent the **microbiota** (*biota* = "life"), a population of microorganisms and viruses residing in the body without causing disease. Some, called the **indigenous microbiota** establish a permanent relationship with various parts of the body, while others, the **transient microbiota**, are more temporary and found only for limited periods of time. In the large intestine of humans, for example, *Escherichia coli* and *Candida albicans* are almost always found, but species of *Streptococcus* are transient.

The relationship between the body and its microbiota is an example of a **symbiosis**, or living together. If the symbiosis is beneficial to both the host and the microbe, the relationship is called **mutualism**. For example, species of *Lactobacillus* live in the female vagina and derive nutrients from the environment while producing acid to prevent the overgrowth of other organisms.

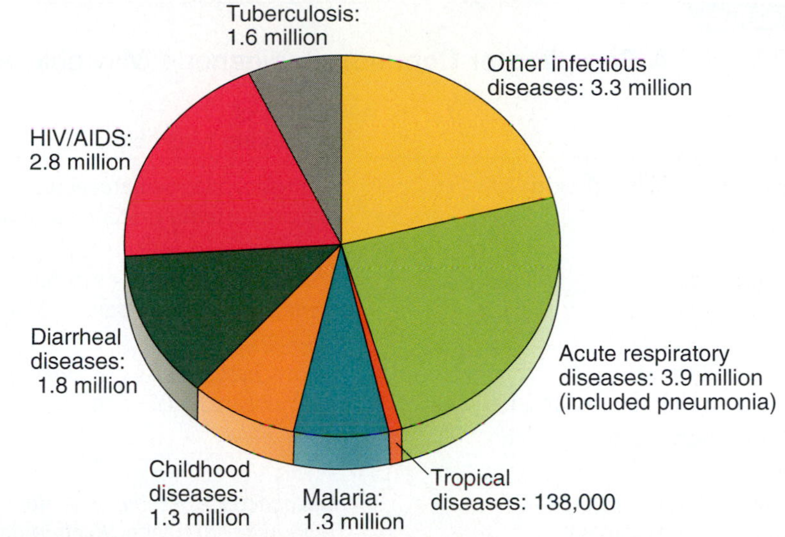

FIGURE 18.3 **Infectious Disease Deaths Worldwide.** This pie chart depicts the leading causes of infectious diseases and the number of worldwide deaths as reported by the World Health Organization. Tropical diseases: trypanosomiasis, Chagas disease, schistosomiasis, leishmaniasis, filariasis, and onchocerciasis. Childhood diseases: diphtheria, measles, pertussis, polio, and tetanus.

Q: Only one of the "pie slices" has grown explosively in mortality numbers since 1993. Identify the disease and explain why that is so.

A symbiosis also can be beneficial only to the microbe and the host is unaffected, in which case the symbiosis is called **commensalism**. *E. coli* is generally presumed to be a commensal in the human intestine, although some evidence exists for mutualism because the bacterial cells produce certain amounts of vitamins B and K. In addition, the microbiota usually will out-compete invading microbial pathogens, thereby protecting the body from dangerous infections.

Microbiota may be found in several body tissues (TABLE 18.1). These include the skin, the external ears and eyes, and upper respiratory tract. Most of the digestive tract, from oral cavity to rectum, is heavily populated with indigenous microbiota. In fact, many of these are thought to play an important functional role for humans (MicroFocus 18.1). Microbiota also make up a population at the urogenital orifices in both males and females.

Most other tissues of the body remain sterile; the blood and internal organs, such as the kidneys, liver, muscles, bone, and brain, are sterile unless disease is in progress.

Sterile: Devoid of living microorganisms, viruses, and spores.

TABLE
18.1 A Sampling of Common, Indigenous Microbiota of the Human Body

Site	Genera of Indigenous Microbiota	Remarks
Eyes (conjunctiva); Skin	*Staphylococcus, Streptococcus; Corynebacterium, Propionibacterium, Staphylococcus, Micrococcus, Candida*	Sparsely populated; primarily gram+ microbes on skin
Oral Cavity	*Streptococcus, Treponema, Neisseria, Haemophilus, Lactobacillus, Staphylococcus, Propionibacterium*	Dense and diverse groups of microbes
Upper Respiratory (nose, nasopharynx, oropharynx)	*Streptococcus, Neisseria, Haemophilus, Staphylococcus, Candida*	Diverse microbes that vary with site
Lower Respiratory (lungs, alveoli)	None	Normally sterile
Digestive (stomach, small and large intestines)	*Helicobacter, Lactobacillus, Haemophilus, Treponema, Neisseria, Bacteroides, Streptococcus, Escherichia, Clostridium, Enterococcus, Shigella, Candida, Entamoeba, Trichomonas*	Stomach sparsely populated; very dense and diverse population of microbes in intestines
Urinary (kidneys, ureters, bladder, urethra)	Female: *Lactobacillus, Corynebacterium, Streptococcus, Bacteroides* Male: *Corynebacterium, Streptococcus*	Mostly sterile, except urethra
Female reproductive (ovaries, uterus, vagina)	*Lactobacillus, Staphylococcus, Corynebacterium, Streptococcus, Enterococcus, Candida*	Vagina densely populated
Internal tissues (blood, lymph, nerves, etc.)	None	Normally sterile

The first nine months of human development within a mother's womb is the only time when the human body is truly a sterile organism. The indigenous microbiota are introduced when the newborn passes through the birth canal. Additional organisms enter when breathing begins and upon first feeding (FIGURE 18.4). Within a few days, many common microbiota organisms have appeared. During the next weeks, contact with the mother and other individuals will expose the child to additional microbes. By one year, the infant's indigenous microbiota is adult-like and remains throughout life, undergoing small changes in response to the internal and external environment of the individual.

CONCEPT AND REASONING CHECKS

18.1 Distinguish between a mutualistic relationship and a commensalistic one.

Pathogens Differ in Their Ability to Cause Disease

KEY CONCEPT

- Microbes vary greatly in their pathogenicity.

There also are symbiotic relationships, called **parasitism**, where the pathogen causes damage to the host and disease can result. Microbiologists once believed microbes were either pathogenic or nonpathogenic; they either caused disease or they did not. From the previous section, we know that is not true.

Pathogenicity refers to the ability of a microorganism to gain entry to the host's tissues and bring about a physiological or anatomical change, resulting in altered health and leading to disease. Certain pathogens, such as the cholera, plague, and typhoid bacilli, are well known for their ability to cause serious human disease. Others, such as common cold viruses, are considered less pathogenic because they induce milder illnesses. Still other microorganisms are opportunistic.

This degree of pathogenicity is called **virulence** (*virul* = "poisonous"). For example, an organism invariably causing disease, such as the typhoid bacillus, is said to be "highly virulent." By comparison, an organism sometimes causing disease, such as *Candida albicans*, is

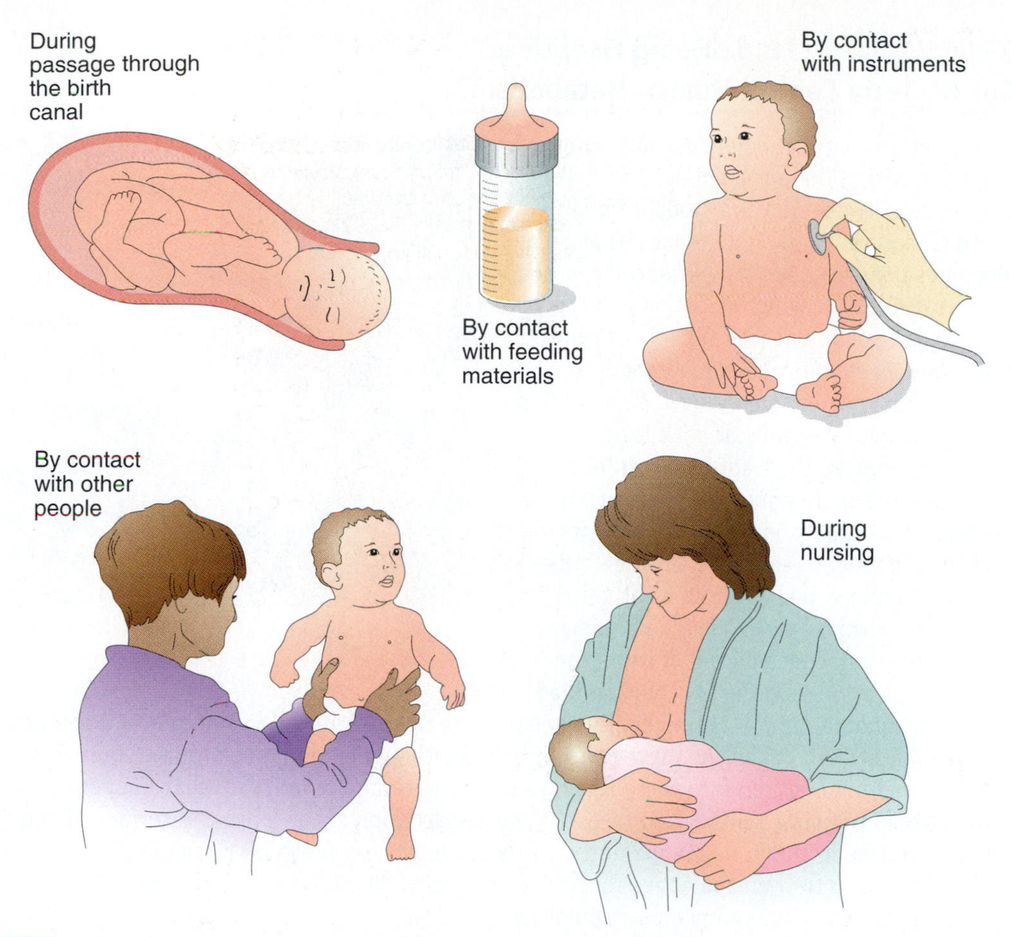

FIGURE 18.4 **Five Possible Origins of the Microbiota in a Newborn.** Barring an infection by the mother, for the nine-month gestation period, the human fetus is sterile; there are no microbes associated with its developing body. *Q: How would each of the origins introduce microbes onto or into the newborn?*

labeled "moderately virulent." Certain organisms, described as **avirulent**, are not regarded as disease agents. The lactobacilli and streptococci found in yogurt are examples. However, it should be noted that any microorganism has the ability to change genetically and become virulent.

In recent years, a new term, **pathogenicity islands**, has been used to refer to clusters of genes responsible for virulence (see Chapter 8). The genes, present on the bacterial cell's chromosome or plasmids, encode many of the virulence factors making a microbe more virulent. These unstable islands are fairly large segments of a pathogen's genome and are absent in nonpathogenic strains. Pathogenicity islands have many of the properties of intervening sequences, suggesting they move by horizontal gene transfer. A copy of these blocs of genetic information can move from a pathogenic strain into an avirulent (harmless) organism, converting it to a pathogen. Such horizontal transfer processes show how the evolution of pathogenicity can make quick jumps.

Before examining disease progression, realize that an infection and a resulting disease can be caused by a single microbe. The diseases identified by Pasteur, Koch, and their contemporaries are examples. However, it is now clear that some diseases are caused by two or more microbes acting together or in succession. Such **polymicrobial diseases** include tooth decay, gastroenteritis, urinary tract infections, otitis media, and many others.

■ Virulence factors: Pathogen-produced molecules or structures that allow the cell to invade or evade the immune system and possibly cause disease.

CONCEPT AND REASONING CHECKS

18.2 Distinguish between pathogenicity and virulence.

Several Events Must Occur for Disease to Develop in the Host

KEY CONCEPT

- Contact with a potential pathogen can have several outcomes.

MICROFOCUS 18.1: Being Skeptical
Can Gut Bacteria Control Human Metabolism?

There is a great diversity of indigenous microbiota present on and within the human body. Adults contain somewhere between 500 and 1,000 different species of prokaryotes in their gut alone. What do all these microbes do? We know *Escherichia coli* can help with water reabsorption and produce some of the vitamin K we use in our diet (see figure). Can these species actually be essential to our metabolism?

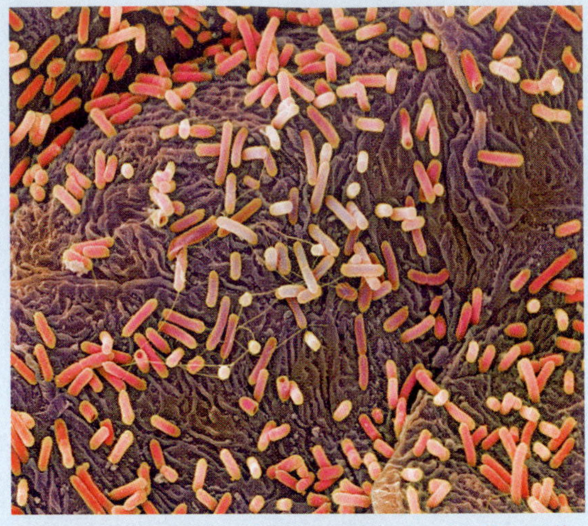

False-color scanning electron micrograph of *E. coli* cells on the rat intestinal lining.

Since the 1950s, investigators have manipulated and engineered special strains of mice that have germ-free guts; that is, their guts are sterile. So, one way to determine if various members of the normal microbiota help us in some way is to introduce each species separately into germ-free mice and see what happens. One assumes what happens in a mouse may mirror what happens in a human.

Jeffrey Gordon and colleagues at Washington University School of Medicine in St. Louis studied *Bacteroides thetaiotaomicron*, a gram-negative anaerobe that exists in the human gut at concentrations 1,000 times greater than *E. coli*. At these concentrations, it must be doing something. Gordon's team introduced *B. thetaiotaomicron* into the guts of germ-free mice and monitored what happened. The team quickly discovered the mice synthesized fucose, a monosaccharide sugar, onto the surface of the intestinal cells. Germ-free mice did not. Apparently, *B. thetaiotaomicron* provided a stimulus to the intestinal cells "telling" them to turn on the genes for fucose synthesis. Why fucose? This is the sugar that *B. thetaiotaomicron* uses for energy and metabolism.

Using DNA techniques that allow large numbers of genes to be monitored all at once, Gordon's group realized *B. thetaiotaomicron* actually triggered the intestinal cells to turn on or turn off some 100 of the 35,000 genes in the cell's genome. Some of these genes helped the mice absorb and metabolize sugars and fats. Other genes activated by the bacterial cells produced products helping protect the intestinal wall from being penetrated by other normal microbiota or pathogens. So, the indigenous microbiota may do more than simple commensals. Yet other genes stimulated the growth of new blood vessels, explaining why germ-free mice had to eat 30 percent more calories to maintain body weight than ordinary mice—germ-free mice have a less well developed blood vessel system and are inefficient at absorbing nutrients. *B. thetaiotaomicron* cells made the human digestive metabolic processes more efficient.

Conclusion: Just from this one bacterial species, the physical development of the normal gut in mice (and extrapolating to the human gut, too) appears to depend on the normal microbiota. Microbes might not only rule the world, they may control our gut physiology and development as well.

For disease to occur, a potential pathogen must first come in contact with exposed parts of the body (FIGURE 18.5). Several outcomes are possible: the pathogen may be lost to the environment or it may colonize the normal microbiota and remain as a transient member. Depending on the nature of the pathogen, it could also become a commensal.

An **exogenous infection** is established if the pathogen breaches the host's external defenses and enters sterile tissue. Likewise, if a microbial member of the normal microbiota should gain access to sterile tissue, an **endogenous infection** ensues. In both cases, the infection may trigger additional host defenses capable of eliminating the invader.

Should the pathogen cause injury or dysfunction to host tissues, then a disease is established. Again, additional host defenses may eliminate the pathogen, in which case the disease declines and the host recovers. In other cases, the pathogen and host reach a

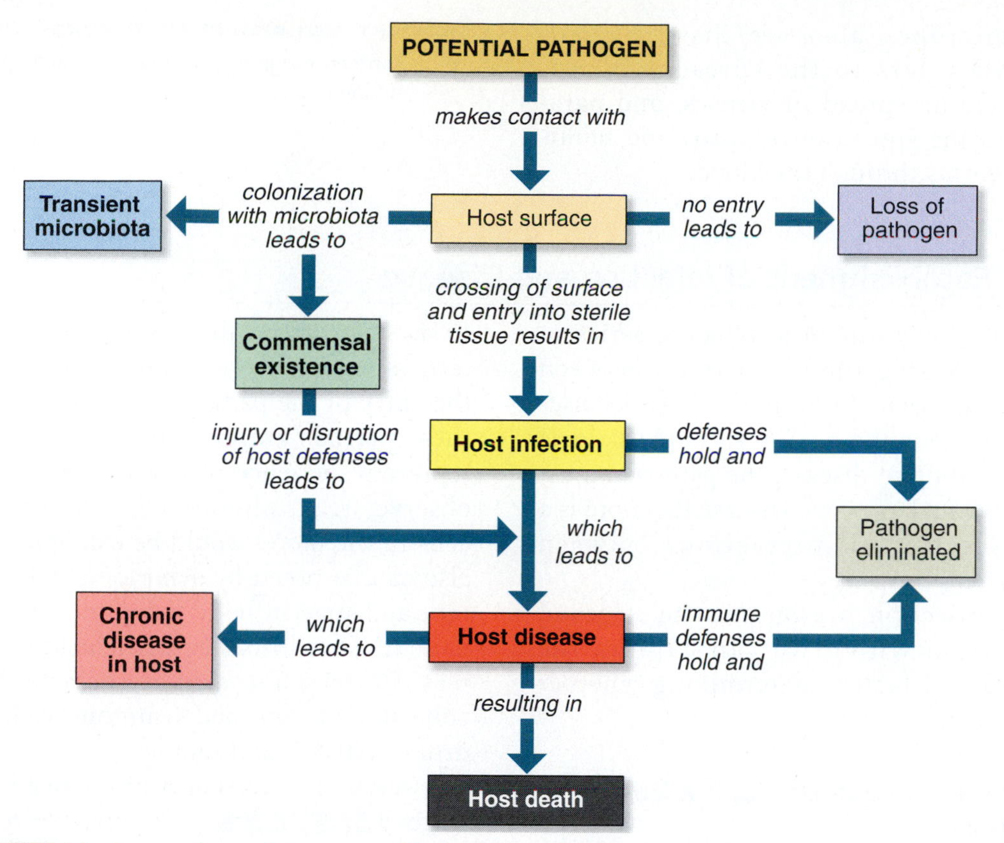

FIGURE 18.5 **The Progression and Outcomes of Infection and Disease.** A concept map illustrating possible outcomes resulting from the contact between host and pathogen.

Q: Propose some ways that a pathogen would gain entry into sterile tissue.

stalemate where neither has the advantage. Tuberculosis is an example of such a chronic state. Lastly, the inability to eliminate the pathogen may lead to death.

Opportunistic infections often are caused by commensals taking advantage of a shift in the body's delicate balance to one favoring the microbe. If the indigenous microbiota is reduced or the host's immune system is weakened, some commensals seize the "opportunity" to invade the tissues and cause disease. AIDS is an example where crippling of the immune system makes the patient highly susceptible to opportunistic organisms, such as *Pneumocystis jiroveci* (see Chapter 16) and *Toxoplasma gondii* (see Chapter 17). Thus, an upset in resistance mechanisms or microbiota control may enhance the ability of organisms to establish disease.

Infections may develop in one of two ways. A **primary infection** occurs in an otherwise healthy body, while a **secondary infection** develops in an individual weakened by the primary infection. In the influenza pandemic of 1918 and 1919, hundreds of millions of individuals contracted influenza as a primary infection and many developed pneumonia as a secondary infection. Numerous deaths in the pandemic were due to pneumonia's complications.

As the names imply, **local diseases** are restricted to a single area of the body, while **systemic diseases** are those disseminating to the deeper organs and systems. Thus, a staphylococcal skin boil beginning as a localized skin lesion may become more serious when staphylococci spread and cause systemic disease of the bones, meninges, or heart tissue.

The dissemination of living bacterial cells occurring through the bloodstream is referred to as **bacteremia**. **Septicemia** refers to the multiplication of bacterial cells in the blood.

Other microbes also are disseminated. **Fungemia** refers to the spread of fungi, **viremia** to the spread of viruses, and **parasitemia** to the spread of protozoa and multicellular worms through the blood.

CONCEPT AND REASONING CHECKS

18.3 Contrast exogenous, endogenous, and opportunistic infections.

18.2 Establishment of Infection and Disease

Disease is the result of a dynamic series of events expressing the competition between host and pathogen. To overcome host defenses and bring about the anatomic or physiologic changes leading to disease, the pathogen must possess unusual abilities. Disease therefore is a complex series of interactions between pathogen and host.

In this section, we outline the stages of disease from which we can explore the processes and factors determining whether disease can occur.

Diseases Progress through a Series of Stages

KEY CONCEPT

- Disease progression involves incubation, prodromal, acme, decline, and convalescent stages.

In most instances, there is a recognizable pattern in the progress of the disease following the entry of the pathogen into the host. Often these periods are identified by **signs**, which represent evidence of disease detected by an observer (e.g., physician). Fever or bacterial cells in the blood would be examples. Disease also can be noted by **symptoms**, which represent a change in body function sensed by the patient. Sore throat and headache are examples. Diseases also may be characterized by a collection of signs and symptoms called a **syndrome**. AIDS is an example.

Disease progression is distinguished by five stages (FIGURE 18.6). The episode of disease begins with an **incubation period**, reflecting the time elapsing between the entry of the microbe into the host and the appearance of the first

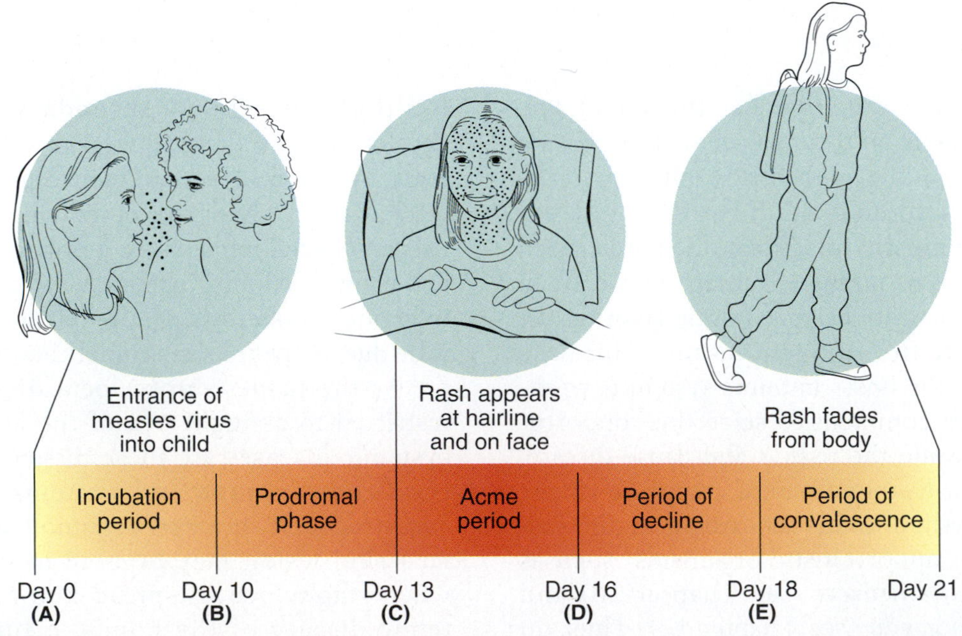

FIGURE 18.6 **The Course of Disease, as Typified by Measles.** (**A**) A child is exposed to measles viruses in respiratory droplets at which time the incubation period begins. (**B**) At the end of this period, the child experiences fever, respiratory distress, and general weakness as the prodromal phase ensues. (**C**) The acme period begins with the appearance of specific measles symptoms, such as the body rash. The period reaches a peak as the rash covers the body. (**D**) The rash fades first from the face and then the body trunk as the period of decline takes place. (**E**) With the period of convalescence, the body returns to normal.

Q: What are the signs and symptoms of measles?

symptoms. For example, an incubation period may be as short as one to three days for cholera; one to two weeks for measles; or three to six years for leprosy. Such factors as the number of organisms, their generation time and virulence, and the level of host resistance determine the incubation period's length. The location of entry also may be a determining factor. For instance, the incubation period for rabies may be as short as several days or as long as a year, depending on how close to the central nervous system the viruses enter the body.

The next phase in disease is a time of mild signs or symptoms, called the **prodromal phase**. For many diseases, this period is characterized by indistinct and general symptoms such as nausea, headache, and muscle aches, which indicate the competition between host and microbe has begun. During the onset and progression of a disease, it can be described as clinical or subclinical. A **clinical disease** is one in which the symptoms are apparent, while a **subclinical disease** is accompanied by few obvious symptoms. Many people, for example, have experienced subclinical cases of mumps or infectious mononucleosis and in the process developed immunity to future attacks. By contrast, certain diseases are invariably accompanied by clearly recognized clinical symptoms. Measles is one example.

The **acme period** or **climax** follows. This is the stage of the disease when signs and symptoms are of greatest intensity. Examples are the skin rash in scarlet fever, jaundice in hepatitis, swollen lymph nodes in infectious mononucleosis, and Koplik spots and rash in measles (see Chapter 14). Often, patients suffer high fever and chills, the latter reflecting differences in temperature between the superficial and deep areas of the body. Dry skin and a pale complexion may result from constriction of the skin's blood vessels to conserve heat. The length of this period can be quite variable, depending on the body's response to the pathogen and the nature of the pathogen. Although the patient feels miserable, there is some evidence some signs and symptoms can be beneficial (MicroFocus 18.2).

As the signs and symptoms begin to subside, the host enters a **period of decline**. Sweating may be common as the body releases excessive amounts of heat and the normal skin color soon returns as the blood vessels dilate. The sequence comes to a conclusion after the body passes through a **period of convalescence**. During this time, the body's systems return to normal.

When studying the course of a disease, it often can be defined by its severity or duration. An **disease** develops rapidly, is usually accompanied by severe symptoms, comes to a climax, and then fades rather quickly. Cholera, epidemic typhus, and SARS are examples of acute diseases. A **chronic disease**, by contrast, often lingers for long periods of time. The symptoms are slower to develop, a climax is rarely reached, and convalescence may continue for several months. Hepatitis A, trichomoniasis, and infectious mononucleosis are examples of chronic diseases. Sometimes an acute disease may become chronic when the body is unable to rid itself completely of the microbe. For example, one who has giardiasis or amoebiasis may experience sporadic symptoms for many years.

CONCEPT AND REASONING CHECKS

18.4 Assign the following signs and symptoms of the flu to the appropriate disease stage (fever, headache, chills, and muscle pain).

With this understanding of the disease stages, we now can examine several factors required for the establishment of disease, as outlined in FIGURE 18.7.

Pathogen Entry into the Host Depends on Cell Adhesion and the Infectious Dose

KEY CONCEPT

- Pathogens gain access to the host through portals of entry.

A **portal of entry** refers to the characteristic route by which an exogenous pathogen enters the host. It varies considerably for different organisms and is a key factor leading to the establishment of disease. Abrasions or mechanical injury to the skin may allow entry. For example, tetanus may occur if *Clostridium tetani* spores on a sharp object in the soil puncture the skin and enter the anaerobic tissue of a wound. Tetanus will not develop if

MICROFOCUS 18.2: Public Health

Illness May Be Good for You

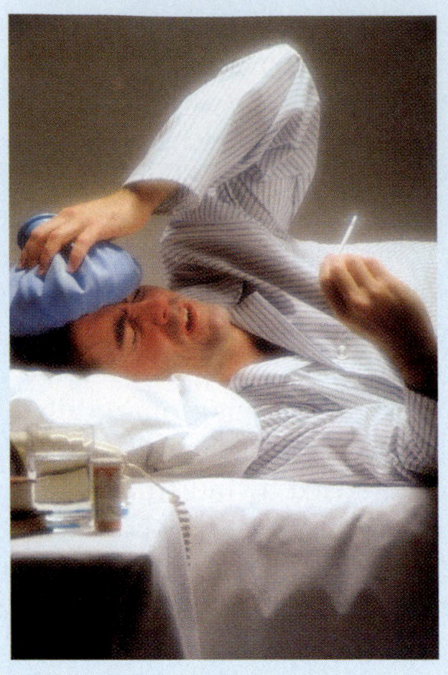

For most of the twentieth century, medicine's approach to infectious disease was relatively straightforward: Note the symptoms and eliminate them. However, that approach may change in the future, as Darwinian medicine gains a stronger foothold. Proponents of Darwinian medicine ask why the body has evolved its symptoms, and question whether relieving the symptoms may leave the body at greater risk.

Consider coughing, for example. In the rush to stop a cough, we may be neutralizing the body's mechanism for clearing pathogens from the respiratory tract. Nor may it be in our best interest to stifle a fever (at least a low grade fever), because fever enhances the immune response to disease. Many physicians view iron insufficiency in the blood as a symptom of disease, yet many bacterial species (e.g., tubercle bacilli) require this element, and as long as iron is sequestered out of the blood in the liver, the bacterial cells cannot grow well. Even diarrhea can be useful—it helps propel pathogens from the intestine and assists the elimination of the toxins responsible for the illness.

Darwinian biologists point out that disease symptoms have evolved over the vast expanse of time and probably have other benefits waiting to be understood. They are not suggesting a major change in how doctors treat their patients, but they are pushing for more studies on whether symptoms are part of the body's natural defenses. So, don't throw out the Nyquil, Tylenol, or Imodium quite yet.

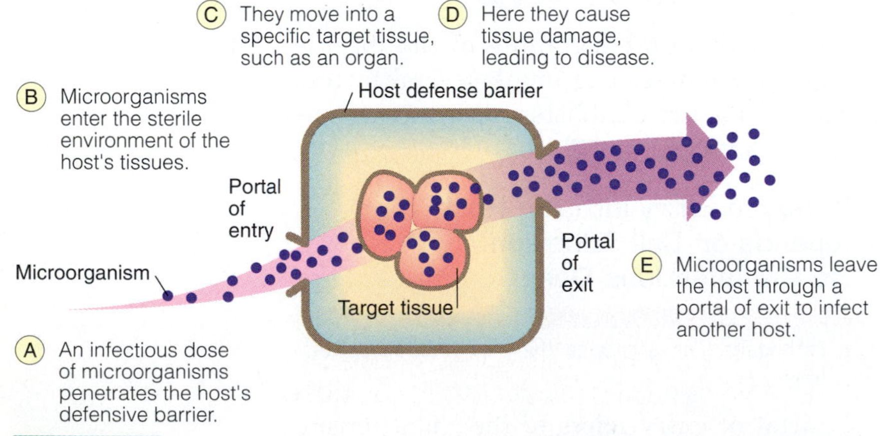

FIGURE 18.7 The Generalized Events on the Establishment of Disease. The infectious dose and adhesion to cells or tissues are required to initiate infection and disease.

Q: Identify which events would not occur if an infection but not a disease occurred.

spores are consumed with food because the spores do not germinate in the human intestinal tract.

The ability of a pathogen to establish an infection and possible disease usually depends on the **infectious dose**, the numbers of microbes taken into the body. The consumption of a few hundred thousand typhoid bacilli will lead to disease. By contrast, many millions of cholera bacilli must be ingested if cholera is to be established. One explanation for the difference is the high resistance of typhoid bacilli to the acidic conditions in the stomach, in contrast to the low resistance of cholera bacilli. Also, it may be safe to eat fish when the water contains hepatitis A viruses, but eating raw clams from the same water can be dangerous because clams are filter-feeders, concentrating hepatitis A viruses in these animals.

Often the host is exposed to low doses of a pathogen and, as a result, develops immunity. For instance, many people can tolerate low numbers of mumps viruses without exhibiting disease. They may be surprised to find they are immune to mumps when it breaks out in their family at some later date.

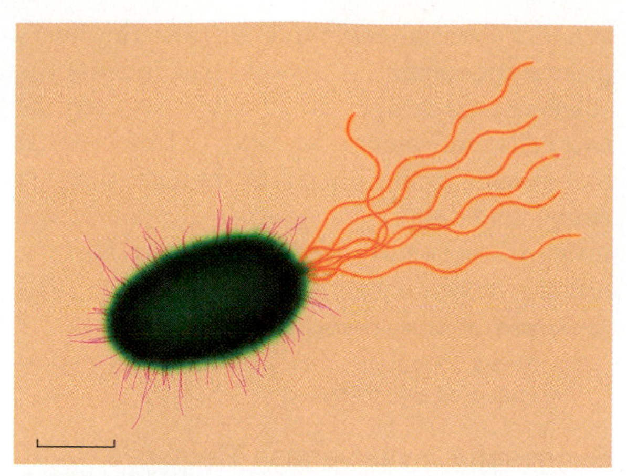

FIGURE 18.8 **Pili on *Escherichia coli*.** A false-color transmission electron micrograph of *E. coli* displaying the pili used for adhesion to the tissue. Adhesins in pili such as these increase the pathogenicity of the organism by allowing it to localize at its appropriate tissue site. (Bar = 1 μm.)

Q: What are the long structures projecting from the right side of the cell?

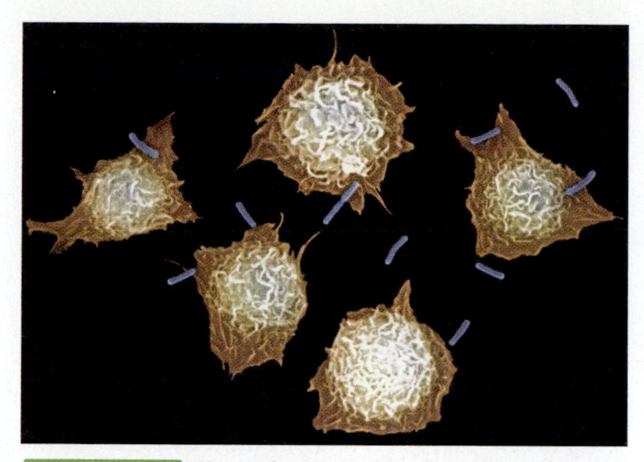

FIGURE 18.9 **Macrophages Undergoing Phagocytosis.** In this false-color scanning electron micrograph, macrophages, a type of immune cell, are capturing bacterial cells for phagocytosis.

Q: Why might phagocytosis be an important activity of the human immune system when infection occurs?

Many pathogens enter at specific, natural portals of entry because these microbes contain on their surface adhesive factors, called **adhesins**, that allow pathogens to adhere to appropriate tissue sites. A variety of adhesins often are associated with capsules, flagella, or pili (see Chapter 4) (FIGURE 18.8). For example, the gonococci and some other agents of sexually-transmitted diseases often attach by means of pili to specific receptor sites only found on tissues of the urogenital system. The host cell is often an active partner in the adhesion because the pathogen triggers it to express target receptor sites for adhesin binding.

Other species, such as *Streptococcus*, have independent adhesion proteins, such as the M protein, on its surface (see Chapter 9). Also, many viruses have spikes on the capsid or envelope, allowing for attachment (see Chapter 13).

Some pathogens have multiple portals of entry. The tubercle bacillus, for instance, may enter the body in respiratory droplets, contaminated food or milk, or skin wounds. The bacterial species causing Q fever can enter by any of these portals, as well as by an arthropod bite. The tularemia bacillus may enter the eye by contact, the skin by an abrasion, the respiratory tract by droplets, the intestines by contaminated meat, or the blood by an arthropod bite.

CONCEPT AND REASONING CHECKS

18.5 Assess the role of the infectious dose and adhesion in establishing an infection.

Breaching the Host Barriers Can Establish Infection and Disease

KEY CONCEPT

- Invasiveness is critical for some pathogens.

Some pathogens do not need to penetrate cells or tissues to cause disease. The pertussis bacillus, for example, adheres to the surface layers of the respiratory tract while producing the toxins causing disease. Likewise, the cholera bacillus attaches to the surface of the intestine, where it produces toxins.

However, the ability of a pathogen to penetrate tissues and cause structural damage is an important component for many pathogens. The ability to penetrate and spread is called **invasiveness** and the bacilli of typhoid fever and the protozoan causing amoebiasis are examples of pathogens that depend on their invasiveness. By penetrating the tissue of the gastrointestinal tract, these microorganisms cause ulcers and sharp, appendicitis-like pain characteristic of the respective diseases.

Invasiveness often is facilitated through the pathogen's internalization by immune cells (FIGURE 18.9). These cells, including

■ Macrophages: Large white blood cells that remove waste products, microorganisms, and foreign material from the bloodstream.

macrophages, undergo **phagocytosis** by engulfing pathogens, taking them into the cell cytoplasm in vacuoles, and then attempting to destroy them in lysosomes. In addition, some bacterial pathogens are internalized by inducing nonphagocytic cells to undergo phagocytosis. If the pathogen can evade destruction by lysosomes, the cell interior provides a protective niche or a vehicle to pass through otherwise impenetrable defenses, such as the blood-brain barrier.

Pathogens, such as *Listeria*, have cell membrane adhesive proteins that form a zipper-like binding of pathogen to host cell. As a result of this molecular adhesion and cross-talk, the pathogen triggers the host cell to undergo phagocytosis. Once in the cell, the pathogens escape from the vacuole and eliminate any chance of their destruction by host lysosomes (FIGURE 18.10). In the cytoplasm, the bacterial cells trigger the host cell to synthesize an actin tail on the bacterial cells, which propels the organism through the cell's cytoplasm. When a listerial cell thuds against the host's plasma membrane, it distorts and indents the adjacent cell, bridges the junction between the two cells, and enters the next cell (somewhat like moving from train car to train car through connecting doors). The system allows bacterial invasion to occur without the bacterial cells leaving the cellular environment.

■ Actin: A cytoskeletal protein essential for cell movement and the maintenance of cell shape in most eukaryotic cells.

CONCEPT AND REASONING CHECKS

18.6 Assess phagocytosis as an invasiveness mechanism used by pathogens.

Successful Invasiveness Requires Pathogens to Have Virulence Factors

KEY CONCEPT

- Virulence factors include enzymes and toxins.

It should be noted that upon entry into a host, a pathogen is confronted with a profoundly different environment. Adaptation to this environment requires the genetic machinery enabling the pathogen to adapt and withstand the resistance put forward by the host. Several factors may be present to overcome host defenses.

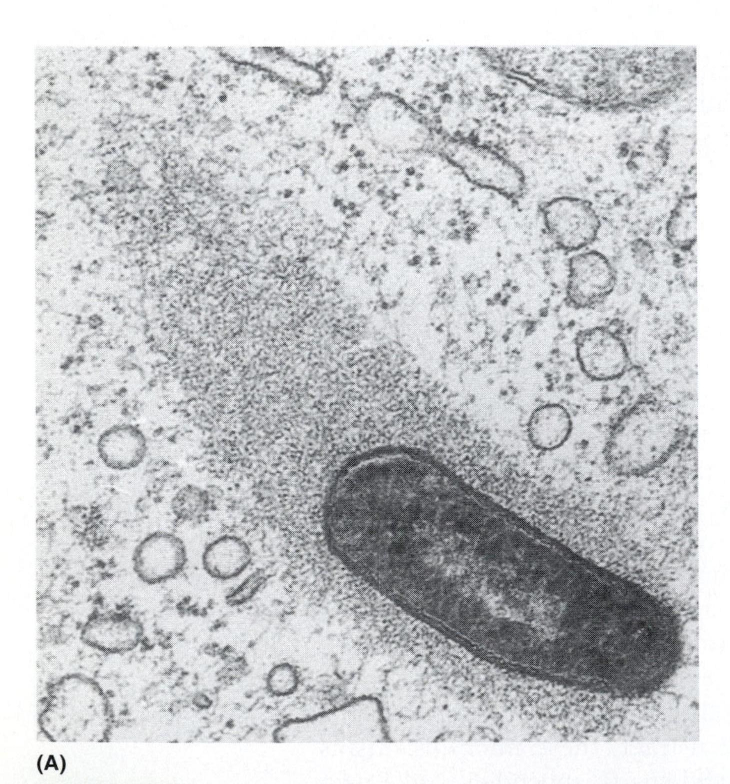

(A)

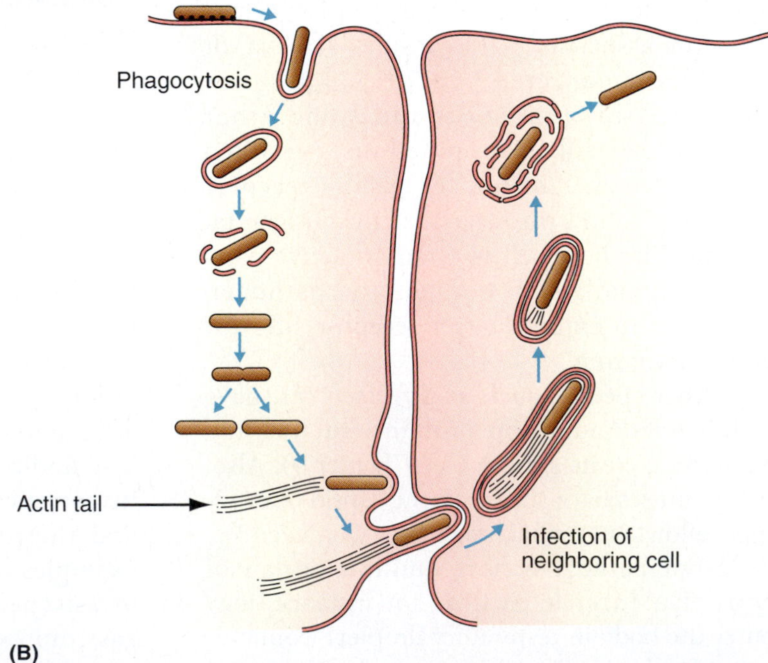

(B)

FIGURE 18.10 **Invasion by *Listeria monocytogenes*.** (**A**) A transmission electron micrograph of a *Listeria* cell with its actin tail. (**B**) After entering a host cell by phagocytosis, the bacterial cell moves about by forming a polar tail composed of filaments of actin protein. The cells divide to form a larger population. Invasion of neighboring cells uses the actin tail to "drive" the cells into the adjacent cell. In that cell, the cell loses the surrounding cell membranes and initiates another infection.

Q: What invasion process is negated by the Listeria *cell's ability to generate an actin tail?*

Enzymes. The virulence of a microbe depends to some degree on its ability to produce a series of extracellular enzymes to help the pathogen resist body defenses. A few examples illustrate how bacterial enzymes act on host cells and interfere with certain functions or barriers meant to retard invasion.

One bacterial enzyme is the **coagulase** produced by virulent staphylococci (FIGURE 18.11A). Coagulase catalyzes the formation of a blood clot from fibrinogen proteins in human blood. The clot sticks to staphylococci, protecting them from phagocytosis. Part of the walling-off process observed in a staphylococcal skin boil is due to the clot formation. Coagulase-positive staphylococci may be identified in the laboratory by combining staphylococci with human or rabbit plasma. The formation of a clot in the plasma indicates coagulase activity.

Many streptococci have the ability to produce the enzyme **streptokinase**. This substance dissolves fibrin clots used by the body to restrict and isolate an infected area. Streptokinase thus overcomes an important host defense and allows further tissue invasion by the bacterial cells.

Hyaluronidase is sometimes called the spreading factor because it enhances penetration of a pathogen through the tissues. The enzyme digests hyaluronic acid, a polysaccharide that binds cells together in a tissue (FIGURE 18.11B). The term tissue cement is occasionally applied to this polysaccharide. Hyaluronidase is an important virulence factor in pneumococci and certain species of streptococci and staphylococci. In addition, gas gangrene bacilli use it to facilitate spread through the muscle tissues.

Some pathogens also produce enzymes that destroy blood cells. **Leukocidins** are products of staphylococci, streptococci, and certain bacterial rods. The enzymes disintegrate circulating neutrophils and tissue macrophages, both of which are cells designed to phagocytize and destroy the pathogens. The enzymes attach to the immune cells' membrane and

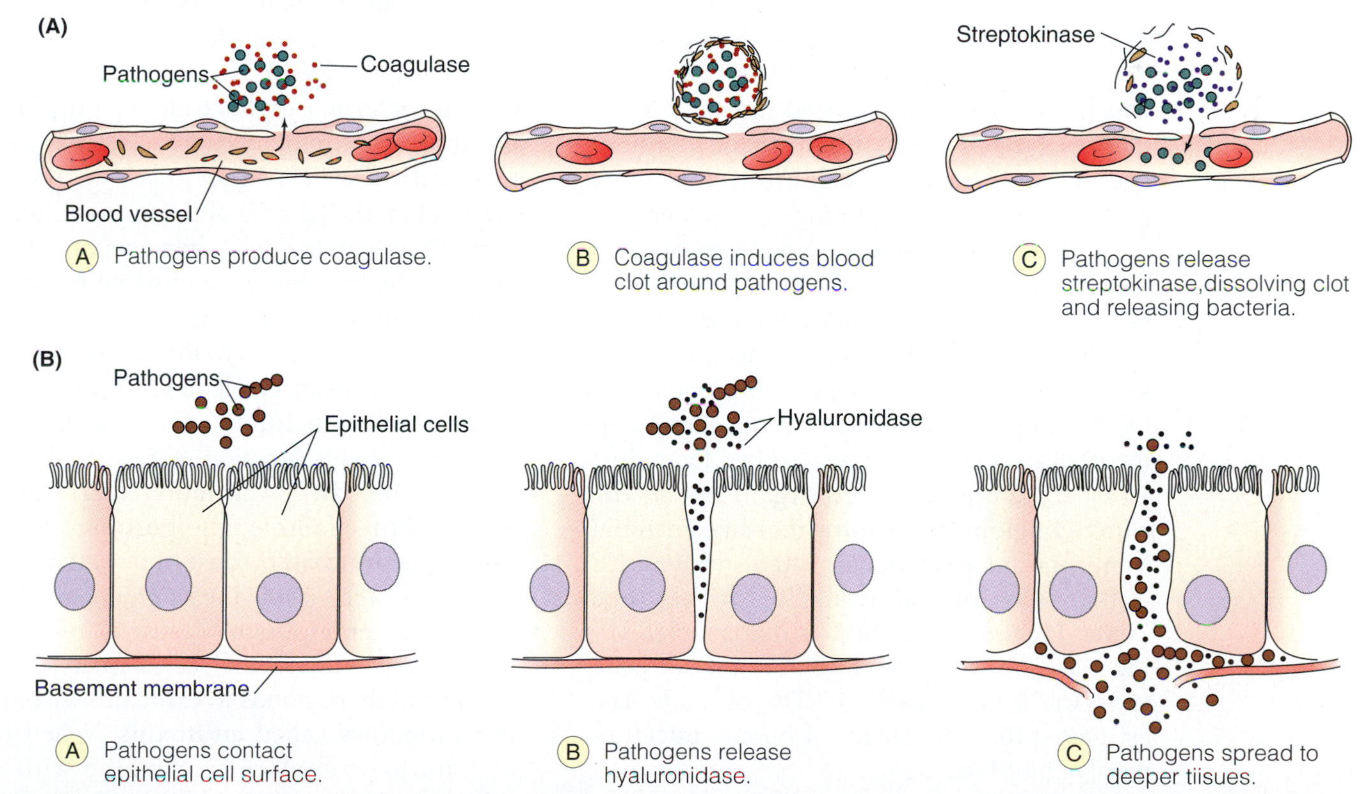

FIGURE 18.11 **Enzyme Virulence Factors.** (**A**) Some bacterial cells produce the enzyme coagulase, which triggers clotting of blood plasma. Bacterial cells within the clot can break free by producing streptokinase. (**B**) Some invasive bacterial species produce the enzyme hyaluronidase, which degrades the cementing polymer holding cells of the intestinal lining together.

Q: From these examples, how do virulence factors protect bacterial cells and increase their virulence?

TABLE 18.2 A Summary of Enzymes that Contribute to Virulence

Enzyme	Source	Action	Effect
Coagulase	Staphylococci	Forms a fibrin clot	Allows resistance to phagocytosis
Streptokinase	Streptococci	Dissolves a fibrin clot	Prevents isolation of infection
Hyaluronidase	Pneumococci Streptococci Staphylococci	Digests hyaluronic acid	Allows tissue penetration
Leukocidin	Staphylococci Streptococci Certain rods	Disintegrates phagocytes	Limits phagocytosis
Hemolysins	Clostridia Staphylococci	Dissolves red blood cells	Induces anemia and limits oxygen delivery

trigger changes leading to the release of lysosomal enzymes in the cytoplasm. The phagocytes quickly disintegrate.

Hemolysins dissolve red blood cells by combining with the membranes of erythrocytes and causing lysis to take place. Staphylococci and streptococci also are known to produce these virulence factors. Lysis by these enzymes gives the pathogens the iron in hemoglobin, which the bacterial cells need for metabolism. In the laboratory, hemolysin producers can be detected by **hemolysis**, a destruction of blood cells in a blood agar medium (see Chapter 10).

Furthermore, if a pathogen exists in a biofilm, its virulence can be enhanced because here it can resist body defenses and drugs. A biofilm is a sticky layer of extracellular polysaccharides and proteins enclosing a colony of bacterial cells at the tissue surface (see Chapter 4). Phagocytes and antibodies have difficulty reaching the microorganisms in this slimy conglomeration of armor-like material. Moreover, microorganisms often survive without dividing in a biofilm. This makes them impervious to the antibiotics that attack dividing cells. (Indeed, the antibiotics do not penetrate the biofilm easily.) CDC officials have estimated that 65 percent of human infections involve biofilms.

TABLE 18.2 summarizes the activities of these enzymes.

Toxins. Microbial poisons, called **toxins**, can profoundly affect the establishment and course of disease. The ability of pathogens to produce toxins is referred to as **toxigenicity**. Toxins present in the blood are referred to as **toxemia** and the person is considered **intoxicated** (see Chapter 10). Two types of toxins are recognized: exotoxins and endotoxins.

Exotoxins are protein molecules, manufactured during bacterial metabolism. They are produced by gram-positive and gram-negative bacterial cells and released into the surrounding environment. They dissolve in the blood fluid and circulate to their site of activity where symptoms of disease soon develop.

The exotoxin produced by the botulism bacillus *Clostridium botulinum* is among the most lethal toxins known. One pint of the pure toxin is believed sufficient to destroy the world's population. In humans, the toxin inhibits the release of acetylcholine at the synaptic junction, a process leading to the paralysis seen in botulism (see Chapter 9). Another exotoxin is produced by the tetanus bacillus, *C. tetani*. In this case, the exotoxin blocks the relaxation pathway that follows muscle contraction, thereby permitting volleys of spontaneous nerve impulses and uncontrolled muscular contractions.

Another exotoxin is produced by *Corynebacterium diphtheriae*, the diphtheria bacillus (see Chapter 9). The exotoxin interferes with the protein synthesis in the cytoplasm of epithelial cells of the upper respiratory tract. Disintegrated cells then accumulate with mucus, fibrous material, and white blood cells, resulting in life-threatening respiratory blockages. Other exotoxins are formed by the bacterial species causing scarlet fever, staphylococcal food poisoning, pertussis, and cholera.

When toxins function in a particular organ system, they are given more clearly defined names. For example, the botulism toxin is called a **neurotoxin** because of its activity in the nervous system, while the staphylococcal toxin is called an **enterotoxin** because it functions in the gastrointestinal tract (see Chapter 10).

The body responds to exotoxins by producing antibodies called **antitoxins**. When toxin and antitoxin molecules combine with each other, the toxin is neutralized (Chapter 20). This process represents an important defensive measure in the body. Therapy for people who have botulism, tetanus, or diphtheria often

TABLE
18.3 A Comparison of Exotoxins and Endotoxins

Characteristic	Exotoxins	Endotoxins
Source	Gram-positive and gram-negative bacteria	Gram-negative bacteria
Location in bacterium	Cytoplasm	Cell wall
Chemical composition	Protein	Lipid-polysaccharide
Antibodies elicited	Yes	No
Conversion to toxoid	Possible	Not possible
Liberation of toxin	On production by the cell	On disintegration of the cell
Representative effects	Interfere with synaptic activity Interrupt protein synthesis Increase capillary permeability Increase water elimination	Increase body temperature Increase hemorrhaging Increase swelling in tissues Induce vomiting, diarrhea

includes injections of antitoxins (immune globulin) to neutralize the toxins.

Because exotoxins are proteins, they are susceptible to the heat and chemicals that normally denature proteins. A chemical such as formaldehyde may be used to alter the toxin and destroy its toxicity without hindering its ability to elicit an immune response in the body. The result is a **toxoid**. When the toxoid is injected into the body, the immune system responds with antitoxins, which circulate and provide a measure of defense against disease. Toxoids are used for diphtheria and tetanus immunizations in the diphtheria-tetanus-acellular pertussis (DTaP) vaccine.

Endotoxins, are released only upon disintegration of gram-negative cells. They are present in the outer membrane in many gram-negative bacilli and are composed of lipid-polysaccharide complexes (see Chapter 4). The lipid portion of the lipopolysaccharide (LPS) of the outer membrane is the toxic portion. Endotoxins do not stimulate an immune response in the body, nor can they be altered to prepare toxoids. They function by activating a blood-clotting factor to initiate blood coagulation and by influencing the complement system (Chapter 19). The toxins of plague bacilli are especially powerful.

Endotoxins manifest their presence by certain signs and symptoms. Usually an individual experiences an increase in body temperature, substantial body weakness and aches, and general malaise. Damage to the circulatory system and shock may occur. In this case, the permeability of the blood vessels changes and blood leaks into the intercellular spaces. The tissues swell, the blood pressure drops, and the patient may lapse into a coma. This condition, commonly called **endotoxin shock**, may accompany antibiotic treatment of diseases due to gram-negative bacilli because endotoxins are released as the bacilli are killed by the antibiotic.

Endotoxins usually play a contributing rather than a primary role in the disease process. Certain endotoxins reduce platelet counts in the host and thereby hinder clot formation. Other endotoxins are known to increase hemorrhaging. Like exotoxins, endotoxins add to the virulence of a microbe and enhance its ability to establish disease.

TABLE 18.3 summarizes the characteristics of the bacterial toxins.

CONCEPT AND REASONING CHECKS

18.7 Evaluate the role of enzymes and toxins as important virulence factors in the establishment of disease.

Pathogens Must Be Able to Leave the Host to Spread Disease

KEY CONCEPT

- Pathogens leave the host through portals of exit.

At the conclusion of its pathogenicity cycle, pathogens or their toxins exit the host through some suitable **portal of exit** (FIGURE 18.12). This is of more than passing importance because easy transmission permits the pathogen to continue its pathogenic existence in the world.

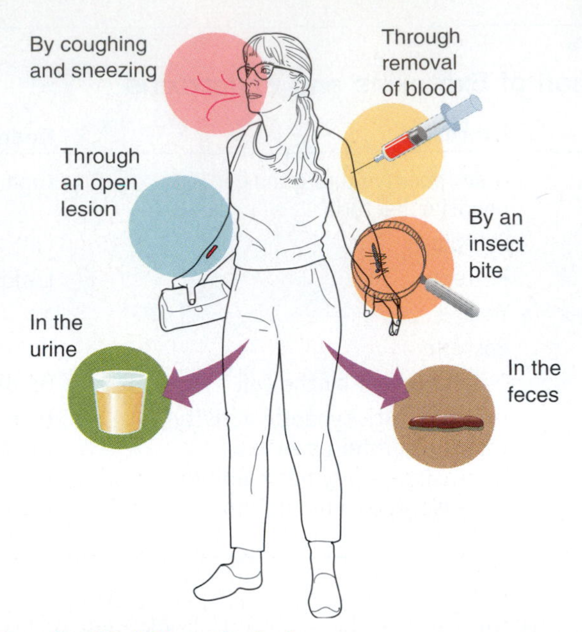

FIGURE 18.12 **Six Different Portals of Exit from the Body.**
Q: Why do pathogens need specific portals of exit?

18.3 Infectious Disease Epidemiology

Infectious disease epidemiology is concerned with how infectious diseases are distributed in a population and the factors influencing or determining that distribution. In this final section, we examine the factors putting population groups at risk of contracting infectious disease. Not only do epidemiologic investigations look at current outbreaks, they also consider future outbreaks, including bioterrorism (Microfocus 18.3).

We also look at special environments, such as health care settings, and the public agencies saddled with the job of disease identification, control, and prevention. First, let's examine the ways diseases are transmitted.

Epidemiologists Have Several Terms that Apply to the Infectious Disease Process

KEY CONCEPT

- Diseases have certain behaviors in populations.

Most diseases studied in this text are **communicable diseases**; that is, infectious diseases transmissible among hosts in a population. Certain communicable diseases are described as being **contagious** because they pass with particular ease among hosts and are highly infectious. Chickenpox and measles fall into this category.

Noncommunicable diseases are singular events in which the agent is acquired directly from the environment and are not easily transmitted to the next host. In tetanus, for example, penetration of soil containing *Clostridium tetani* spores to the anaerobic tissue of a wound must occur before this disease develops. It cannot be spread person-to-person.

Infectious Diseases Can Be Transmitted in Several Ways

KEY CONCEPT

- Disease transmission involves contact, airborne, vector, or vehicle methods.

Diseases can be transmitted by a broad variety of methods. One way is through direct or indirect contact (FIGURE 18.13).

Direct Contact Methods. Person-to-person transmission implies close or personal contact with someone who has the disease. Hand-shaking or kissing an infected person can spread bacterial cells, viruses, or other pathogens to an uninfected person. The exchange of body fluids, such as through sexual contact or blood transfusions, is another

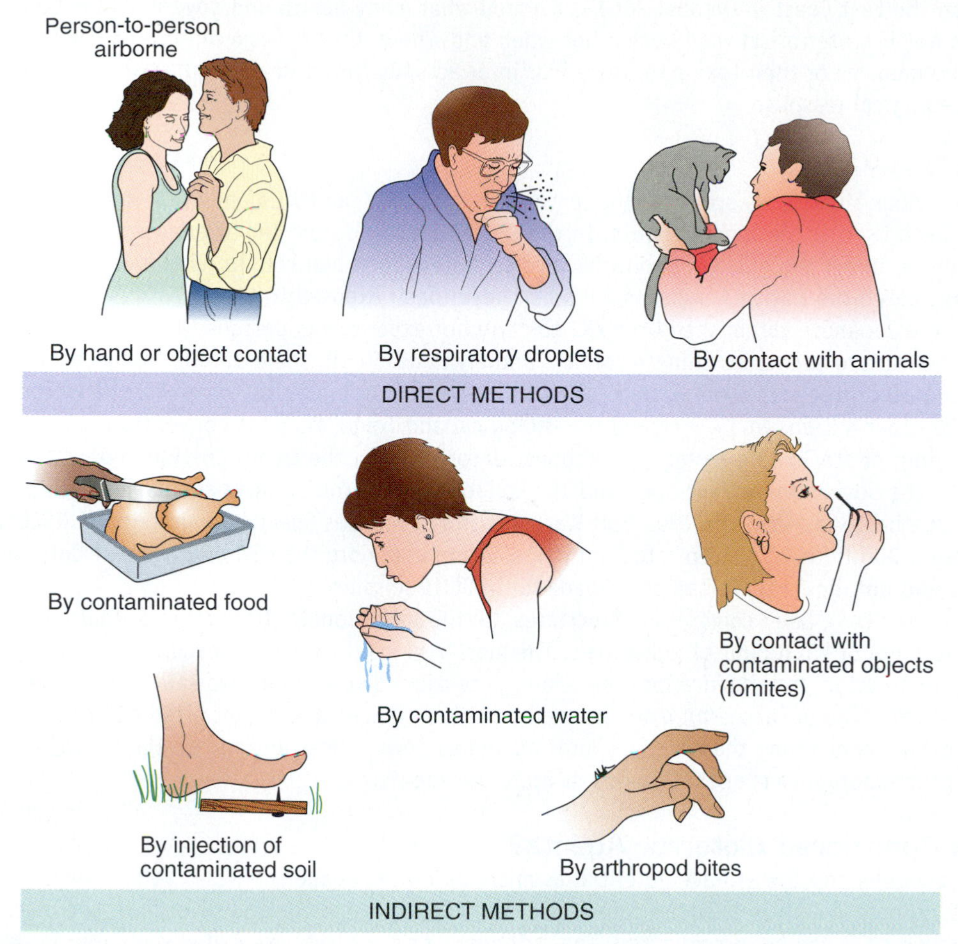

FIGURE 18.13 **Methods of Transmitting Disease through Contact.** Infectious diseases can be transmitted by direct methods and indirect methods.

Q: Without specifying a specific example, how would you determine if disease transmission was direct or indirect?

example of direct contact transmission for diseases like gonorrhea and hepatitis B.

Direct transmission also can involve the violent expulsion of airborne **respiratory droplets** through sneezing or coughing (FIGURE 18.14). Transmission through the air requires the "recipient" be in close proximity to an infected individual. In a sneeze, the droplets can travel 150 feet per second. However, the droplets are fairly large and fall out of the air within about 12 feet of their source. However, if an uninfected person is within that distance, the eyes, mouth, or nose may be portals of entry for the airborne pathogens.

Pathogens also are transmitted through the air on smaller particles called aerosols. These particles can remain suspended in the air for longer periods of time and can be moved some distance by air currents. The

FIGURE 18.14 **Droplet Transmission.** Sneezing or coughing represents a method for airborne transmission of pathogens.

Q: Identify some portals of entry for pathogen-containing airborne droplets?

MICROFOCUS 18.3: History/Public Health

Bioterrorism: The Weaponization and Purposeful Dissemination of Human Pathogens

The anthrax attacks that occurred on the East Coast in October 2001 confirmed what many health and governmental experts had been saying for over 10 years—it is not if bioterrorism would occur but when and where. Bioterrorism represents the intentional or threatened use of primarily microorganisms or their toxins to cause fear in or actually inflict death or disease upon a large population for political, religious, or ideological reasons.

Is Bioterrorism Something New?

Bioterrorism is not new, and two previous MicroFocus boxes in this text (MicroFocus 10.5 and 11.4) mention historical examples. Such bioterrorism agents also have been used as biowarfare agents. In the United States, during the aftermath of the French and Indian Wars (1754–1763), British forces, under the guise of goodwill, gave smallpox-laden blankets to rebellious tribes sympathetic to the French. The disease decimated the Native Americans, who had never been exposed to the disease before and had no immunity. Between 1937 and 1945, the Japanese established Unit 731 to carry out experiments designed to test the lethality of several microbiological weapons as biowarfare agents on Chinese soldiers and civilians. In all, some 10,000 "subjects" died of bubonic plague, cholera, anthrax, and other diseases. After years of their own research on biological weapons, the United States, the Soviet Union, and more than 100 other nations in 1973 signed the Biological and Toxin Weapons Convention, which prohibited nations from developing, deploying, or stockpiling biological weapons. Unfortunately, the treaty provided no way to monitor compliance. As a result, in the 1980s the Soviet Union developed and stockpiled many microbiological agents, including the smallpox virus, and anthrax and plague bacteria. After the 1991 Gulf War, the United Nations Special Commission (UNSCOM) analysts reported that Iraq had produced 8,000 liters of concentrated anthrax solution and more than 20,000 liters of botulinum toxin solution. In addition, anthrax and botulinum toxin had been loaded into SCUD missiles.

In the United States, several biocrimes have been committed. **Biocrimes** are the intentional introduction of biological agents into food or water, or by injection, to harm or kill groups of individuals. The most well known biocrime occurred in Oregon in 1984 when the Rajneeshee religious cult, in an effort to influence local elections, intentionally contaminated salad bars of several restaurants with the bacterium *Salmonella*. The unsuccessful plan sickened over 750 citizens and hospitalized 40. Whether biocrime or bioterrorism, the 2001 events concerning the anthrax spores mailed to news offices and to two US congressmen only increases our concern over the use of microorganisms or their toxins as bioterror agents.

What Microorganisms Are Considered Bioterror Agents?

A considerable number of human pathogens and toxins have potential as microbiological weapons. These "select agents" include bacterial organisms, bacterial toxins, fungi, and viruses. The seriousness of the agent depends on the severity of the disease it causes (virulence) and the ease with which it can be disseminated. The pathogens of most concern, called the Category A Select Agents, are those that can be spread by aerosol contact, such as anthrax and smallpox, and toxins that can be added to food or water supplies, such as the botulinum toxin (see the table below).

Why Use Microorganisms?

At least 15 nations are believed to have the capability of producing bioweapons from microorganisms. Such microbiological weapons offer clear advantages to these nations and terrorist organizations in general. Perhaps most important, biological weapons represent "The Poor Nation's Equalizer." Microbiological weapons are cheap to produce compared to chemical and nuclear

TABLE

Category A Select Agents and Perceived Risk of Use

Type of Microbe	Disease (Microbe Species or Virus Name)	Perceived Risk
Bacteria	Anthrax (*Bacillus anthracis*)	High
	Plague (*Yersinia pestis*)	Moderate
	Tularemia (*Francisella tularensis*)	Moderate
Viruses	Smallpox (Variola)	Moderate
	Hemorrhagic fevers (Ebola, Marburg, Lassa, Machupo)	Low
Toxins	Botulinum toxin (*Clostridium botulinum*)	Moderate

weapons and provide those nations with a deterrent every bit as dangerous and deadly as the nuclear weapons possessed by other nations. With biological weapons, you get high impact and the most "bang for the buck." In addition, microorganisms can be deadly in minute amounts to a defenseless (nonimmune) population. They are odorless, colorless, and tasteless, and unlike conventional and nuclear weapons, microbiological weapons do not damage infrastructure, yet they can contaminate such areas for extended periods. Without rapid medical treatment, most of the select agents can produce high numbers of casualties that would overwhelm medical facilities. Lastly, the threatened use of microbiological agents creates panic/anxiety, which often is at the heart of terrorism.

How Would Microbiological Weapons Be Used?

All known microbiological agents (except smallpox) represent organisms naturally found in the environment. For example, the bacterium causing anthrax is found in soils around the world (**Figure A**). Assuming one has the agent, the microorganisms can be grown (cultured) easily in large amounts. However, most of the select agents must be "weaponized"; that is, they must be modified into a form that is deliverable, stable, and has increased infectivity and/or lethality. Nearly all of the microbiological agents in category A are infective as an inhaled aerosol. Weaponization, therefore, requires the agents be small enough in size so inhalation would bring the organism deep into the respiratory system and prepared so that the particles do not stick together or form clumps. Several of the anthrax letters of October 2001 involved such weaponized spores.

Dissemination of biological agents by conventional means would be a difficult task. Aerosol transmission, the most likely form for dissemination, exposes microbiological weapons to environmental conditions to which they are usually very sensitive. Excessive heat, ultraviolet light, and oxidation would limit the potency and persistence of the agent in the environment. Although anthrax spores are relatively resistant to typical environmental conditions, the bacterial cells causing tularemia become ineffective after just a few minutes in sunlight. The possibility also exists that some nations have developed or are developing more lethal bioweapons through genetic engineering and biotechnology. The former Soviet Union may have done so. Commonly used techniques in biotechnology could create new, never before seen bioweapons, making the resulting "designer diseases" true doomsday weapons.

Conclusions

In May 2000, Ken Alibek, a scientist and defector from the Soviet bioweapons program, testified before the House Armed Services Committee. He strongly suggested the best biodefense is to concentrate on developing appropriate medical defenses that will minimize the impact of bioterrorism agents. If these agents are ineffective, they will cease to be a threat; therefore, the threat of using human pathogens or toxins for bioterrorism, like that for emerging diseases such as SARS and West Nile fever, is being addressed by careful monitoring of sudden and unusual disease outbreaks. Extensive research studies are being carried out to determine the effectiveness of various antibiotic treatments (**Figure B**) and how best to develop effective vaccines or administer antitoxins. To that end, vaccination perhaps offers the best defense. The United States has stated it has stockpiled sufficient smallpox vaccine to vaccinate the entire population if a smallpox outbreak occurred. Other vaccines for other agents are in development.

This primer is not intended to scare or frighten; rather, it is intended to provide an understanding of why microbiological agents have been developed as weapons for bioterrorism. We cannot control the events that occur in the world, but by understanding bioterrorism, we can control how we should react to those events—should they occur in the future.

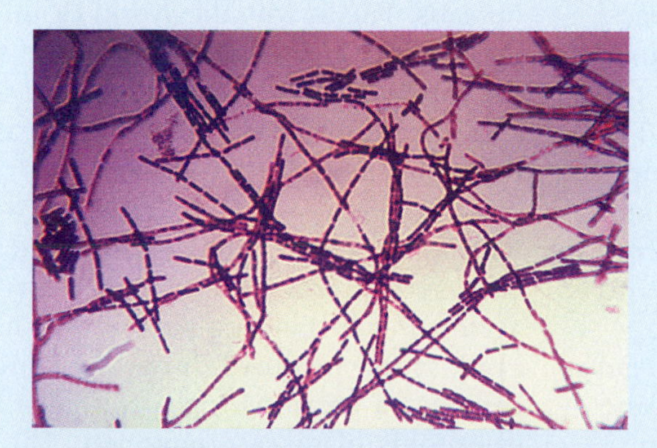

FIGURE A **Anthrax Bacteria.** Light micrograph of gram-stained *Bacillus anthracis*, the causative agent of anthrax. There is concern that terrorists could release large quantities of anthrax spores in a populated area, which potentially could cause many deaths.

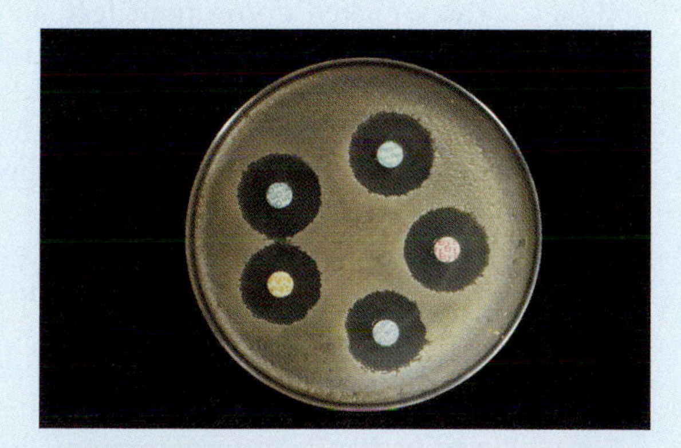

FIGURE B **Anthrax Antibiotics Research.** Antibiotic drugs in paper discs are used to test the sensitivity of anthrax bacteria (*Bacillus anthracis*) cultured on an agar growth medium. The clear zone surrounding each disc indicates the bacterial cells are sensitive to the antibiotic.

virus of SARS and the bacterial cells of tuberculosis are two pathogens that can be carried in the air by droplets or aerosols.

For some diseases, such as rabies, leptospirosis, and toxoplasmosis, direct contact with an animal is necessary. An animal bite or scratch by an infected animal can spread the disease to an uninfected person.

Direct transmission also includes the spread of pathogens, such as HIV or *Toxoplasma gondii*, from a pregnant mother to her unborn child. Transmission of gonorrhea from mother to newborn can occur during labor or delivery.

CONCEPT AND REASONING CHECKS

18.8 Identify four methods by which infectious disease can be transmitted directly.

Indirect Contact Methods. Indirect transmission can be the result of contact with **fomites**, inanimate objects on which or in which disease organisms linger for some period of time. For instance, bed linens may be contaminated with pinworm eggs, and contaminated syringes and needles may transport the viruses of hepatitis B and AIDS.

■ Niches: Environmental areas that ensure an organism's survival.

Arthropods represent another indirect method of transmission. Many pathogens hitch a ride on arthropods, such as mosquitoes, ticks, fleas, and lice, which act as **vectors**, living organisms carrying disease agents from one host to another. **Mechanical vectors** represent arthropods passively transporting microbes on their legs and other body parts. For example, house flies can carry diseases picked up on their feet. In other cases, arthropods represent **biological vectors**, where the pathogen must multiply in the insect before it can infect another host. The malarial protozoan and the yellow fever virus infect and reproduce in mosquitoes and accumulate in their salivary glands, from which the pathogens are infective and injected during the next bite of a host.

Vehicle transmission involves the spread of disease through contaminated food and water. Foods are contaminated during processing or handling, or they may be dangerous when made from diseased animals. Poultry products, for example, are often a source of salmonellosis because *Salmonella* species frequently infect chickens while pork may spread trichinellosis because *Trichinella* parasites may live in muscles of the pig. Other examples include the cholera bacterium, *Vibrio cholerae*, which can contaminate many water supplies in developing nations lacking proper water sanitation and the parasite causing giardiasis, *Giardia intestinalis*, which is found in some recreational waters.

MicroFocus 18.4 reports on two transmission modes: fomites and airborne transmission.

CONCEPT AND REASONING CHECKS

18.9 Identify the methods by which infectious disease can be transmitted indirectly.

Epidemiologists Often Have to Identify the Reservoir of an Infectious Disease

KEY CONCEPT

- Reservoirs are places in the environment where a pathogen can be found.

For many diseases to perpetuate themselves, the disease-causing microbes must exist somewhere in the environment. These ecological niches where a microbe lives and multiplies are called **reservoirs** of infection. Animals are one type of reservoir. A domestic house cat that is infected with *Toxoplasma gondii*, for example, usually shows no symptoms of toxoplasmosis but it can transmit the protozoan to humans, where the disease manifests itself. Water and soil also can be reservoirs because they often are contaminated with disease agents, such as the cholera bacterium or *Giardia* protozoan.

Not all diseases have a nonhuman reservoir though. The smallpox virus only exists in humans. This is why the World Health Organization (WHO) was able to limit the spread of the virus through vaccination and eradicate smallpox from the world by locating all human reservoirs (see Chapter 14).

A special type of reservoir is a **carrier**, which is a person who has recovered from the disease but continues to shed the disease agents. For instance, a person who has recovered from typhoid fever or amoebiasis becomes a carrier for many weeks after the symptoms of disease have left. The feces of this individual may spread the disease to others via contaminated food or water.

Recently, epidemiological investigations have identified rodents as possible reservoirs for a disease of tremendous significance during the colonial period in Mexico (MicroFocus 18.5).

MICROFOCUS 18.4: Public Health

Speeding Red Lights and Planes

Vehicles that can save our lives and others that can take us to great destinations at supersonic speeds can harbor and transmit infectious disease.

From November 2004 to April 2005, a decontamination firm in the United Kingdom examined the ambulances from 12 firms for microbial contamination. They swabbed several fomites, including stretcher rails, the stretcher tracks below the stretcher, the paramedic's utility bag, and five other sites within the vehicle. The swabs were streaked on nutrient agar plates to see what bacterial species would grow.

Examination of the plates indicated the ambulances were heavily contaminated with a diverse group of bacterial species. In fact, in many cases, there were so many bacterial colonies present, they could not be counted. (One assumes one colony arose from one bacterial cell that reproduced many times on the nutrient agar plate.) The bacterial species included antibiotic-resistant *Staphylococcus aureus* and a variety of species typically found in the human colon. More surprising, after the ambulances were cleaned by standard procedures, there was little reduction in the numbers of bacterial cells present. In fact, another study done the previous year showed that cleaning actually spread the bacterial cells onto previously "clean" surfaces. Such contaminated fomites could be dangerous to a person in the ambulance with open wounds.

In 2004, almost 2 billion people traveled by aircraft. With this number of people flying each year, a sick person could spread a disease across the globe quite quickly. One epidemiological study of commercial airlines examined the chances of spreading or catching an infectious disease on a given flight. Although the risk is very low of catching an airborne infection, it is not zero because most airlines do not have the top-grade filters able to trap all bacterial pathogens and aggregates of viruses.

Epidemiologists thought the most likely way to catch an infectious disease would be if passengers are within two rows of an ill person and on flights eight hours or longer in length. Then, in 2003, the SARS outbreak occurred. During the outbreak, 40 airline flights were monitored for passengers carrying SARS. Five flights showed evidence of transmission of SARS. On one 3-hour flight, from Hong Kong to Beijing, China, 22 people contracted the disease from a single infected passenger. Surprisingly, some passengers who became ill were a full seven rows away from the infected individual.

Still, this so-called superspreading of infectious disease by airborne droplets or particles is rare and airline flights are relatively safe from infectious disease spread. Most are still spread person-to-person, such as by hand contact, so hand washing remains the best prevention method. Just don't touch the handle on the door of the stall when leaving the airliner bathroom.

CONCEPT AND REASONING CHECKS

18.10 Identify the different reservoirs of disease transmission.

Diseases Also Are Described by How They Occur Within a Population

KEY CONCEPT

- Diseases are identified as being endemic, epidemic, or pandemic.

When epidemiologists investigate an infectious disease, they need to determine if it is localized or spread through a community or region. **Endemic** refers to a disease habitually present at a low level in a certain geographic area. Plague in the American Southwest is an example.

By comparison, an **epidemic** refers a disease that occurs in a community or region in excess of what is normally found within that population. Influenza often causes widespread

MICROFOCUS 18.5: History/Public Health/Epidemiology

When the Rains Came

Historians have proposed for some time that the infectious diseases brought to the New World by Spanish explorers and armies were responsible for the demise of the native peoples who would be susceptible to those diseases (see figure). For example, between 1519, when Hernando Cortés' armies arrived in Mexico, and 1599, the Aztec population in Mexico was reduced from 22 million to only 2 million. Historians attribute most of the deaths to imported infectious diseases. Smallpox was the major disease thought to have brought the downfall of the Aztec empire. However, recent epidemiological investigations are disputing that claim, and Rudolfo Acuña-Soto, a Mexican epidemiologist, is leading the reinvestigation.

Aztecs battling the Spaniards.

Acuña-Soto says the Aztecs had a word for smallpox *(zahuatl)* before the Spanish arrived and the epidemics in 1545 and 1576 were caused by what the Aztecs called *cocolitzli,* apparently a very different disease from smallpox. Acuña-Soto believes this disease is what killed millions of Aztecs and was not due to a disease brought by the Spanish invasion. Additionally, *cocolitzli* occurred decades after the Spanish arrived. If the Aztecs were susceptible to this "imported" disease, then they should have become ill much sooner and have been immune by the time the disease actually occurred.

When Acuña-Soto searched through old Spanish records, the disease symptoms of *cocolitzli* (high fever, headache, and bleeding from the nose, ears, and mouth) did not match those of smallpox, but rather sounded very similar to a viral hemorrhagic fever. Again, the Spanish could not have brought the disease from Europe—it must have been a local viral disease.

Looking at the circumstances surrounding each *cocolitzli* epidemic and comparing the events in each epidemic, Acuña-Soto discovered *cocolitzli* followed a period of drought and it was only in wet, rainy periods when people became ill. Because the old Spanish and Aztec records were somewhat scanty on rainy and drought periods, Acuña-Soto turned to another accurate record of weather conditions—tree rings. Botanists know they can predict climatic conditions by analyzing tree rings. Years of heavy moisture are reflected in expansive tree growth and thick tree rings, while drought produces little tree growth and therefore very thin tree rings. Using the tree-ring data, Acuña-Soto's team substantiated the Spanish and Aztec records—wet periods occurred at the time of each *cocolitzli* outbreak.

So, Acuña-Soto believes *cocolitzli* was caused by a hemorrhagic fever virus that had laid dormant in rodents during the drought periods. During these periods, the rodents passed the virus to other uninfected rodents. Then, when the rains came, the rodents bred quickly and spread the virus in their feces and urine to humans through inhalation—a scenario very similar to Hantavirus hemorrhagic fever in the American Southwest in 1993.

Finally, if Acuña-Soto's theory is correct, why didn't the Spanish become ill? They should have been just as susceptible to the virus. Acuña-Soto suspects here is where the Spanish conquest played a significant role. Being enslaved by the Spanish, the Aztecs survived in impoverished conditions, including near starvation and unsanitary conditions, which would be very attractive to the rodent population. Transmission of virus in rodent feces or urine would be a logical result. The Spanish, on the other hand, were the upper class, living in relatively sanitary conditions that would not be associated with rodent infestations.

Acuña-Soto believes *cocolitzli* brought the final decline to the Aztec population in Mexico. His concern is the virus could still be out there today!

epidemics. This should be contrasted with an **outbreak**, which is a more contained epidemic. An abnormally high number of measles cases in one American city would be classified as an outbreak.

A **pandemic** is a worldwide epidemic, affecting populations around the globe. The most obvious example here would be AIDS.

As in the opening quote, "*Health care matters to all of us some of the time, public health matters to all of us all of the time,*" maintaining vigilance against infectious disease is extremely important. For this reason, national and international public health organizations, such as the CDC and the WHO, learn a lot about diseases by analyzing disease data reported to them. The CDC, for example, has a list of infectious diseases that must be reported to state health departments, which then report them to the CDC (TABLE 18.4). These are published in the *Morbidity and Mortality Weekly Report.*

MicroInquiry 18 explores the use of epidemiological data as a tool for understanding disease occurrence.

CONCEPT AND REASONING CHECKS

18.11 Why do you think the term "outbreak" is typically used in news releases rather than epidemic?

Nosocomial Infections Are Serious Health Threats within the Health Care System

KEY CONCEPT

- Nosocomial infections are contracted as a result of being treated for another illness in a hospital or other health care setting.

Health care-associated infections (HAIs) occur when patients acquire an infection as a result of receiving treatment for other conditions or that health care workers acquire while performing their duties within a health care setting (for example, hospitals, nursing homes, or outpatient facilities).

Nosocomial infections are those HAIs associated with hospitals and account for an estimated 2 million infections and 90,000 deaths each year in the United States. FIGURE 18.15 shows one way diseases can spread in a hospital setting. Nosocomial infections are the result of several factors that often occur together in what are known as **chains of transmission**. These include:

- A high prevalence of pathogens in a hospital.
- A high number of patients who represent compromised hosts and are susceptible to pathogens.
- Efficient mechanisms of transmission from patient to patient in a hospital setting.

This shows the need for **universal precautions** when working with blood or other body fluids (Figure 18.15).

The most common type of nosocomial infection results from infection of surgical wounds, or infections to the respiratory, urogenital, or gastrointestinal tracts. These infections are often caused by breaches of infection control practices and procedures, unclean and non-sterile environmental surfaces, and/or ill employees.

According to the CDC, the most common pathogens responsible for nosocomial infections are *Staphylococcus aureus, Pseudomonas aeruginosa*, and *Escherichia coli*. Nosocomial infections are not just limited to bacterial species; certain fungi such as *Candida albicans* and *Aspergillus* as well as viruses such as respiratory syncytial virus and influenza have also been implicated.

Specific procedures should always be in place to prevent or limit a chain of transmission.

CONCEPT AND REASONING CHECKS

18.12 How do universal precautions limit the chains of transmission?

Infectious Diseases Continue to Challenge Public Health Organizations

KEY CONCEPT

- Diseases emerging or reemerging anywhere in the world can become a global health menace.

In an era of globalization characterized with supersonic jet travel and international commerce, it is not possible to adequately protect the health of the United States without focusing on diseases and epidemics elsewhere in the world. In 2006, we only have to look at AIDS and the potential bird flu to realize the seriousness of infectious diseases and their threats to global health.

TABLE 18.5 identifies several emergent and resurgent infectious diseases that have caught national attention. Globally, there have been 40 new diseases and 20 resurgent diseases identified

TABLE 18.4 CDC's Summary of Notifiable Diseases in the United States in 2004

- Acquired immunodeficiency syndrome (AIDS)
- Anthrax
- Botulism
- Brucellosis
- Chancroid
- *Chlamydia trachomatis*, genital infections
- Cholera
- Coccidioidomycosis
- Cryptosporidiosis
- Cyclosporiasis
- Diphtheria
- Ehrlichiosis
  - Human granulocytic
  - Human monocytic
  - Human, other or unspecified agent
- Encephalitis/meningitis, arboviral
  - California serogroup
  - Eastern equine
  - Powassan
  - St. Louis
  - Western equine
  - West Nile
- Enterohemorrhagic *Escherichia coli* (EHEC)
  - EHEC 0157:H7
  - EHEC, serogroup non-0157
  - EHEC not serogrouped
- Giardiasis
- Gonorrhea
- *Haemophilus influenzae*, invasive disease
- Hansen disease (leprosy)
- Hantavirus pulmonary syndrome
- Hemolytic uremic syndrome, postdirrheal
- Hepatitis A, acute
- Hepatitis B, acute
- Hepatitis B, chronic
- Hepatitis B, perinatal infection
- Hepatitis C, acute
- Hepatitis C, infection (past or present)
- Human immunodeficiency virus (HIV) infection
  - Adult (age $\geq$ 13 yrs)
  - Pediatric (age $<$ 13 yrs)
- Influenza-associated pediatric mortality
- Legionellosis
- Listeriosis
- Lyme disease
- Malaria
- Measles
- Menigococcal disease
- Mumps
- Pertussis
- Plague
- Poliomaelitis, paralytic
- Psittacosis
- Q fever
- Rabies
  - Animal
  - Human
- Rocky Mountain spotted fever
- Rubella
- Rubella, congenital syndrome
- Salmonellosis
- Sever acute respiratory syndrome-associated coronavirus (SARS-CoV) disease
- Shigellosis
- Smallpox
- Streptococcal disease, invasive, group A
- Streptococcal toxic-shock syndrome
- *Streprococcus pneumoniae*, invasive disease
  - Drug-resistant, all ages
  - Age $<$5 yrs
- Syphilis
- Syphilis, congenital
- Tetanus
- Toxic-shock syndrome (other than streptococcal)
- Trichinellosis
- Tuberculosis
- Tularemia
- Typhoid fever
- Vancomycin-intermediate *Staphylococcus aureus* infection (VISA)
- Vancomycin-resistant *Staphylococcus aureus* infection (VRSA)
- Varicella
- Varicella deaths
- Yellow fever

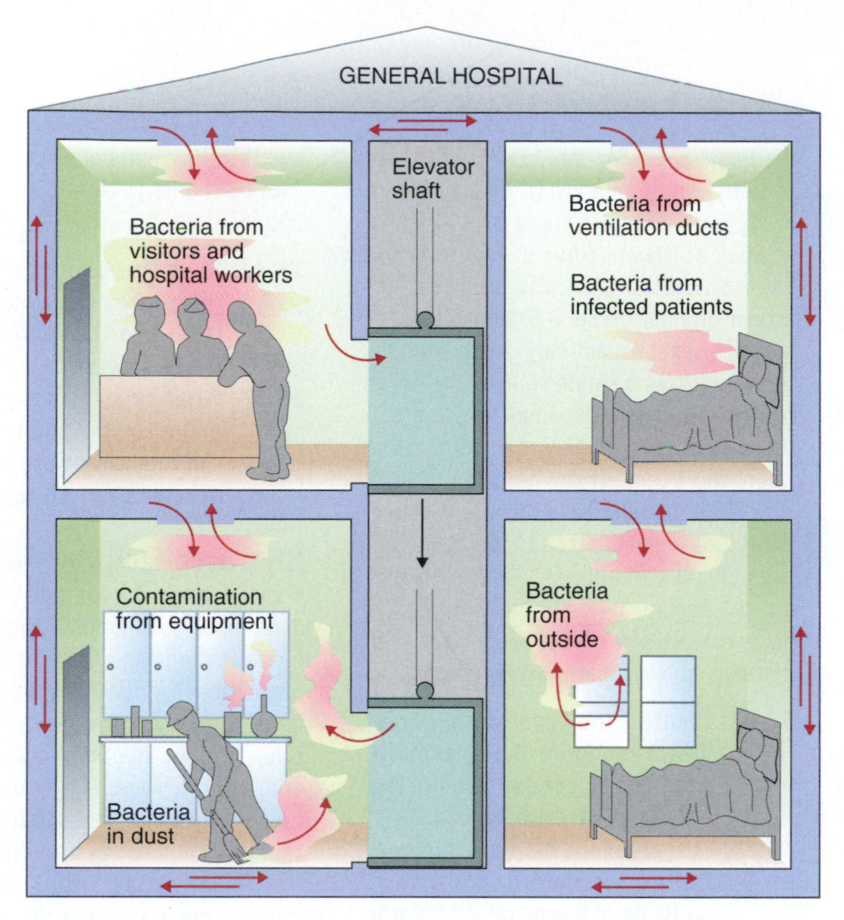

FIGURE 18.15 **Microbe Transmission in a Hospital and Universal Precautions.** Potential pathogens can be spread through several means within the hospital or health care environment.

Protecting patients and other health care workers also means using universal precautions if working with a patient who may be infected. The precautions when handling blood or other body fluids that may harbor pathogens (HIV, hepatitis B or C) include:

- Washing hands
- Wearing personal protective equipment (gloves, mask, and eye protection).
- Handling and disposing of sharps (hypodermic needles) properly.
- Disposing of all hazardous and contaminated materials in approved and labeled biohazard containers.
- Cleaning up all spills with disinfectant or diluted bleach solution to kill any pathogens present.

Q: Why is hand washing always at the top of the list for preventing disease transmission?

since 1973. In addition, outbreaks of foodborne diseases have increased fivefold since 1985.

In 2005, there were more than 1,400 human reported pathogens. Over half represent **zoonoses** (*zoo* = "animal"; *noso* = "disease"), diseases transmitted from vertebrate animals to humans. Some 177 (13 percent) of the 1,400 represent emerging or resurgent diseases. The largest single number (37 percent) are RNA viruses (see Chapter 13).

Numerous reasons help explain how disease emergence and resurgence is driven (in decreasing rank of pathogens involved).

1. **Changes in land use or agriculture practices.** Urbanization, deforestation, and water projects can bring new diseases (Dengue fever, schistosomiasis).
2. **Changes in human demographics.** The migration of many peoples to cities have brought new diseases to a susceptible population (malaria).
3. **Poor population health.** In many developing nations, large numbers of people suffer from malnutrition or poor public health infrastructure, making disease eruption much more likely (cholera).
4. **Pathogen evolution.** Pathogens have developed resistance to antibiotics and antimicrobial drugs, typical of many resurgent diseases (tuberculosis, yellow fever).

MICROINQUIRY 18

Epidemiological Investigations

Infectious disease epidemiology is a scientific study from which health problems are identified. In this inquiry, we are going to look at just a few of the applications and investigative strategies for analyzing the patterns of illness. Answers can be found in **Appendix D**.

One of the important measures is to assess disease occurrence. The **incidence** of a disease is the number of reported cases in a given time frame. **Figure A** is a line graph showing the number of new cases of AIDS per year in the United States.

The **prevalence** of a disease refers to the percentage of the population that is affected at a given time.

18.1a. What was the incidence of AIDS in 1993 and 2003?

18.1b. Assuming that 264,000 were living with AIDS in 1993 and 380,000 in 2003, how has the prevalence of AIDS changed between 1993 and 2003? (Assume the population of the United States has remained at 290 million).

Descriptive epidemiology describes activities (time, place, people) regarding the distribution of diseases within a population. Once some data have been collected on a disease, epidemiologists can analyze these data to characterize disease occurrence. Often a comprehensive description can be provided by showing the disease trend over time, its geographic extent (place), and the populations (people) affected by the disease.

Characterizing by Time

Traditionally, drawing a graph of the number of cases by the date of onset shows the time course of an epidemic. An epidemic curve, or "epi curve," is a histogram providing a visual display of the magnitude and time trend of a disease.

Look at the epi curve in **Figure B** for an Ebola outbreak in Africa. One important aspect of a bar graph is to consider its overall shape. An epi curve with a single peak indicates a single source (or "point source") epidemic in which people are exposed to the same source over a relatively short time. If the duration of exposure is prolonged, the epidemic is called a continuous common source epidemic, and the epi curve will have a plateau instead of a peak. Person-to-person transmission is likely and its spread may have a series of plateaus one incubation period apart.

18.2a. Identify the type of epi curve drawn in Figure B and explain what the onset says about the nature of disease spread.

18.2b. Is there more than one plateau? Explain the significance that multiple plateaus might have in interpreting the spread of the Ebola hemorrhagic fever outbreak.

Characterizing by Place

Analysis of a disease or outbreak by place provides information on the geographic extent of a problem and may show clusters or patterns that provide clues to the identity and origins of the problem. It is a simple and useful technique to look for geographic patterns where the affected people live, work, or may have been exposed. A geographic distribution for Lyme disease is shown in **Figure C**. It is similar to a spot map, where each case of a disease in a community or state may be shown to reflect clusters or patterns of disease. Figure C identifies the number of cases of Lyme disease by county in the United States in 2001.

18.3a. From this county map, what inferences can you draw with regard to the reported cases of Lyme disease?

18.3b. What suggestions could you make for the high incidence in the one county in Texas that had greater than 15 reported cases?

Characterizing by Person

Populations at risk for a disease can be determined by characterizing a disease or outbreak by person. Persons also refer to populations identified by personal characteristics (e.g., age, race, gender) or by exposures (e.g., occupation, leisure activities, drug intake). These factors are important because they may be related to disease susceptibility and to opportunities for exposure.

Age and gender often are the characteristics most strongly related to exposure and to the risk of disease. For example, **Figure D** is a histogram showing the incidence of pertussis (whooping cough) in the United States in 2003.

18.4a. Look at the histogram and describe what important information is conveyed in terms of the majority of cases and relative incidence in 2003.

18.4b. As a health care provider, what role do you see for vaccinations and booster shots with regard to this disease?

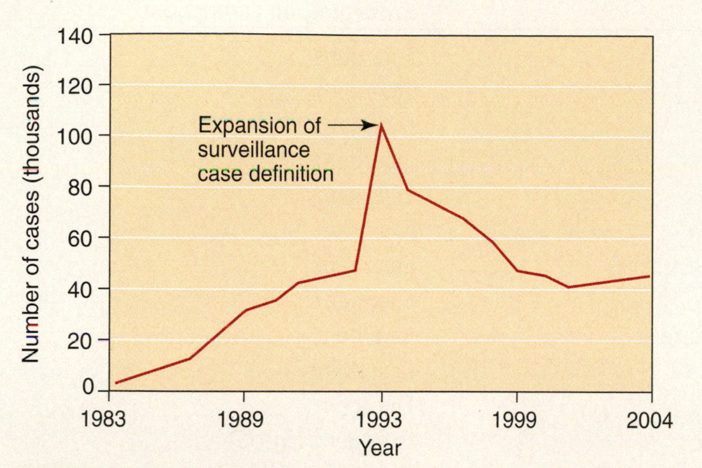

FIGURE A **Acquired Immunodeficiency Syndrome (AIDS).** Number of cases reported by year in the United States and US territories for the years 1983 to 2003. *Source:* CDC.

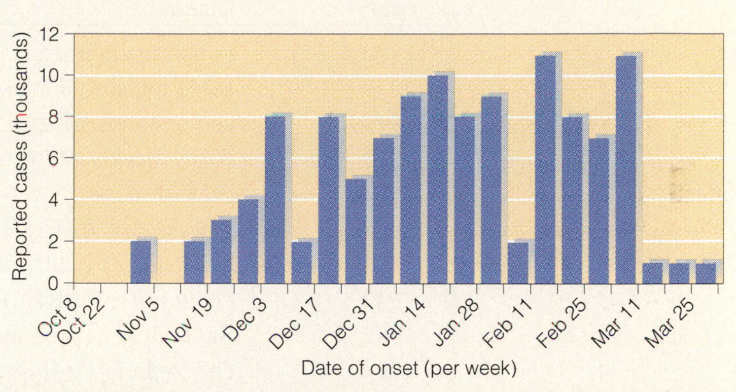

FIGURE B **Ebola Hemorrhagic Fever (Congo and Gabon).** Number of cases of Ebola hemorrhagic fever by week from October 2001 to March 2002. *Data from: Weekly Epidemiological Record,* No. 26, June 27, 2003.

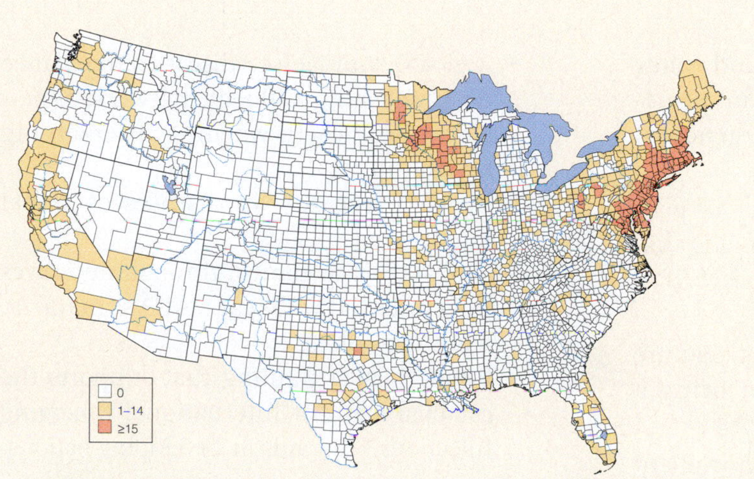

FIGURE C **Lyme Disease.** Counties in the mainland United States showing the incidence of Lyme disease in 2001. Note that the counties in orange represent 90 percent of the total number of cases reported in 2001. *Source:* CDC.

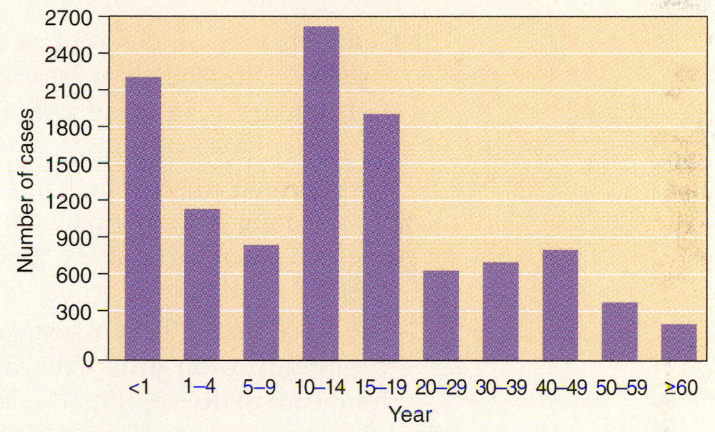

FIGURE D **Pertussis.** The reported number of cases of pertussis by age group in the United States in 2003. *Source:* CDC.

TABLE
18.5 Emerging and Resurging Diseases

Year	Disease	Emerging or Resurgent
1973	Rotavirus diarrhea	Emergent
1977	Ebola hemorrhagic fever	Emergent
1977	Legionnaires' disease	Emergent
1981	Toxic shock syndrome	Emergent
1982	Lyme disease	Emergent
1983	HIV disease/AIDS	Emergent
1991	Tuberculosis (multi-drug resistant)	Emergent
1993	Cholera (0139 strain)	Emergent
1993	Hantavirus pulmonary syndrome	Emergent
1994	*Cryptosporidium* infections	Resurgent
1998	Avian flu	Emergent
1999	West Nile fever/encephalitis	Emergent (in US)
2003	SARS (severe acute respiratory syndrome)	Emergent
2004	Marburg hemorrhagic fever	Resurgent

5. **Contamination of food sources and water supplies.** Substandard or lapses in sanitation practices can bring a resurgence of disease (cryptosporidiosis).
6. **International travel.** The number of people traveling internationally spreads diseases to other parts of the globe (SARS, West Nile fever).
7. **Failure of public health systems.** Failure of immunization programs can bring a resurgence of disease (diphtheria).
8. **International trade.** The global movement of produce can introduce new or resurgent diseases (hepatitis A, cyclosporiasis).
9. **Climate change.** Global warming and changes in weather patterns bring new diseases to new latitudes and elevations (hantavirus pulmonary syndrome).

By understanding these drivers, health organizations such as the CDC and WHO can develop plans to limit or stop emerging disease threats. The CDC has had a history of serving American public health (MicroFocus 18.6). Regarding emerging and resurgent diseases, the CDC's priority areas include:

- International outbreak assistance to host-countries to maintain control of new pathogens when an outbreak is over.
- A global approach to disease surveillance by establishing a global "network of networks" for early warning of emerging health threats.
- Applied research on diseases of global importance.
- Global disease control through initiatives to reduce HIV disease/AIDS, malaria, and tuberculosis.
- Public health training that supports the establishment of International Emerging Infections Programs in developing nations.

However, a new global health movement requires the involvement of more governmental and non-governmental organizations. Happily, today there is an array of new organizations committed to this cause, including: the Bill and Melinda Gates Foundation; The Global Fund to Fight AIDS, Tuberculosis, and Malaria; President's Emergency Plan for AIDS Relief; and the Global Alliance for Vaccines and Immunization.

CONCEPT AND REASONING CHECKS

18.13 Evaluate the usefulness of each of the five universal precautions in limiting disease transmission.

MICROFOCUS 18.6: History/Public Health

The Centers for Disease Control and Prevention

The Centers for Disease Control and Prevention (CDC) located in Atlanta, Georgia is one of six major agencies of the US Public Health Service. Originally established as the Communicable Disease Center in 1946, the CDC was the first governmental health organization ever set up to coordinate a national control program against infectious diseases. Eventually, all communicable diseases came under its aegis.

In April 1955, two weeks after release of the Salk vaccine for polio, the CDC received reports of six cases of polio in vaccinated children. It immediately established the Polio Surveillance Unit and identified a vaccine from a single vaccine manufacturer as the cause of more than 80 percent of the vaccine-associated polio cases. This incident established the role of the CDC in health emergencies, and soon it became a national resource for the development and dissemination of information on communicable disease. Reorganized with its current name in 1980 (the words "and Prevention" were added in 1992 but the CDC acronym was retained), the CDC is charged with protecting the public health of the US populace by providing leadership and direction in the prevention and control of infectious disease, and other preventable conditions, such as cancer.

It is concerned with urban rat control, quarantine measures, health education, and the upgrading and licensing of clinical laboratories. The CDC also provides international consultation on disease and participates with other nations in the control and eradication of communicable infections. It employs 3,500 physicians and scientists, the largest group in the world, and processes 170,000 samples of tissue annually. Its publication, *The Morbidity and Mortality Weekly Report,* is distributed each week to over 100,000 health professionals. You can find it on the Internet at: http://www.cdc.gov/mmwr/.

SUMMARY OF KEY CONCEPTS

18.1 The Host–Microbe Relationship

- Infection refers to competition between host and microbe for supremacy, while disease results when the microbe wins the competition. However, the human body contains large resident populations of indigenous microbiota, which usually out-compete invading pathogens. Transient microbiota are temporary residents of the body.
- The host maintains a symbiotic relationship with microbes. These relationships may reflect mutualism and commensalism, both of which do not harm the host.
- Parasitism is the symbiotic relationship occurring when the microbe does harm to the host. Pathogenicity, the ability of the pathogen to cause disease, and resistance go hand-in-hand. Virulence refers to the degree of pathogenicity a microbe displays.
- Infections may come from an exogenous or endogenous source. Both usually trigger a response from the host's immune system. If the immune system is compromised, microbes normally acting as commensals may cause opportunistic infections.
- A primary infection is an illness caused by a pathogen in an otherwise healthy host, while a secondary infection involves the development of other diseases as a result of the primary infection lowering host resistance. Local diseases are restricted to a specific part of the body while systemic diseases spread to several parts of the body and deeper tissues.

18.2 Establishment of Infection and Disease

- Most diseases have certain signs and symptoms, making it possible to follow the course of a disease through an incubation period, prodromal phase, and periods of acme, decline, and convalescence.
- To cause disease, most pathogens must enter the body through an appropriate portal of entry.
- The dose represents the number of pathogens taken into the body that can cause a disease.
- The possibility of disease is enhanced if a microbe can penetrate host tissues. The pathogen's ability to penetrate tissues and cause damage is referred to as invasiveness. Virulence and invasiveness are strongly dependent on the spectrum of virulence factors a pathogen possesses.
- Bacterial adhesins or viral spikes allow bacterial cells or viruses to adhere to specific cells. Adhesion or attachment usually leads to phagocytosis of the pathogens.
- Many bacterial species produce enzymes to overcome the body's defenses. These include coagulase, streptokinase, and hyaluronidase enzymes. In addition, some species produce lytic enzymes, such as leukocidins and hemolysins. The latter lyse red blood cells, providing bacterial cells with access to the iron-containing hemoglobin.
- Many bacterial species synthesize toxins that interfere with body processes. Exotoxins are proteins released by gram-positive and gram-negative cells. Their effects depend on the enzyme produced and host system affected. Endotoxins are the lipopolysaccharides released from dead gram-negative cells. Their effects on the host are more universal.
- To efficiently spread the disease to other hosts, the pathogen also must leave the body through an appropriate portal of exit.

18.3 Infectious Disease Epidemiology

- A disease may be communicable, such as measles, or noncommunicable, such as tetanus.
- Diseases may be transmitted by direct or indirect methods. Indirect methods include consumption of contaminated food or water, contaminated inanimate objects (fomites), and arthropods. Arthropods can be mechanical or biological vectors for the transmission of disease.
- Reservoirs include humans, who represent carriers of a disease, arthropods, and food and water in which some parasites survive.
- The occurrence of diseases falls into three categories: endemic, epidemic, and pandemic. An "outbreak" is essentially the same as an epidemic, although usually an outbreak is more confined in terms of disease spread.
- Nosocomial infections or health care-associated infections are getting an infection as a result of being treated for some other injury or medical problem.
- Universal precautions limit the chain of transmission.
- Public health organizations, such as the CDC, are charged with the duty of limiting or stopping disease threats, including emerging and resurgent diseases.

LEARNING OBJECTIVES

After understanding the textbook reading, you should be capable of writing a paragraph that includes the appropriate terms and pertinent information to answer the objective.

1. Distinguish between infection and disease, and between indigenous and transient microbiota.
2. Identify the three forms of symbiosis and the effects of the microbial partner upon the human host.
3. Contrast pathogenicity and virulence, explaining how each affects the establishment of disease.
4. Discuss the consequences of exogenous and endogenous (including opportunistic) infections on the progression and outcomes of infection and disease.
5. Distinguish between (a) primary and secondary infections, (b) local and systemic diseases, and (c) bacteremia and septicemia.
6. Explain the differences between signs, symptoms, and syndromes.
7. Identify the characteristics that compose the five stages in the course of disease development.
8. Assess the role of the infectious dose and pathogen penetration to establishing an infection and disease.
9. Name five enzymes and describe their roles as virulence factors.
10. Summarize the differences between exotoxins and endotoxins as virulence factors associated with disease.
11. Identify six portals of exit from the human body.

12. Distinguish between a communicable, contagious, and noncommunicable infectious disease and give an example of each.
13. Identify the direct contact methods of disease transmission.
14. Evaluate the indirect contact methods of disease transmission.
15. Summarize the characteristics of reservoirs as applied to infectious disease.
16. Discuss the three types of disease occurrence within populations.
17. Explain how nosocomial infections can be controlled or eliminated through using the universal precautions to break the chain of transmission.
18. Identify the drivers responsible for emerging and resurgent infectious diseases.

SELF-TEST

Answer each of the following questions by selecting the ***one*** answer that best fits the question or statement. Answers to even-numbered questions can be found in **Appendix C**.

1. Which one of the following statements (**A–D**) about indigenous microbiota is *not* true? If all are true, then select **E.**
   A. The microbiota includes bacteria and viruses.
   B. The microbiota are transient on the skin.
   C. They are part of the symbiotic relationship with the host.
   D. They can cause infections in a sterile part of the body.
   E. All the above (**A–D**) are correct.
2. A newborn
   A. contains indigenous microbiota before birth.
   B. remains sterile for many weeks after birth.
   C. becomes colonized soon after conception.
   D. is colonized with many common microbiota within a few days after birth.
   E. None of the above (**A–D**) is correct.
3. Factors affecting virulence may include
   A. the presence of pathogenicity islands.
   B. its ability to penetrate the host.
   C. the infectious dose.
   D. virulence factors.
   E. All the above (**A–D**) are correct.
4. A healthy person is diagnosed as having a _____ infection with _____, the multiplication of bacterial cells in the blood.
   A. primary; bacteremia
   B. secondary; parasitemia
   C. primary; viremia
   D. primary; septicemia
   E. secondary; parasitemia
5. Changes in body function sensed by the patient are called
   A. symptoms.
   B. syndromes.
   C. patterns.
   D. prodromes.
   E. signs.
6. The prodrome phase of disease is characterized by
   A. recovery.
   B. mild signs and symptoms.
   C. lessening symptoms.
   D. specific signs and symptoms.
   E. no symptoms.
7. Adhesins can be found on
   A. host cells.
   B. bacterial flagella.
   C. bacterial pili and capsules.
   D. cells at the portal of entry.
   E. Both **B** and **C** are correct.
8. Some virulent _____ form a blood clot by producing _____.
   A. bacterial cells; streptokinase
   B. viruses; hyaluronidase
   C. staphylococci; coagulase
   D. viruses; hemolysins
   E. staphylococci; leukocidins
9. Bacterial pathogens needing iron often produce
   A. streptokinase enzymes.
   B. leukocidins.
   C. hemolysins.
   D. hyaluronidase enzymes.
   E. coagulase enzymes.
10. Bacterial toxins present in the blood is referred to as
    A. toxemia.
    B. toxoid presence.
    C. intoxication.
    D. toxicidal.
    E. toxigenicity.
11. Which one of the following is *not* true of exotoxins?
    A. They are proteins.
    B. They are part of cell wall structure.
    C. They are released from live bacterial cells.
    D. They trigger antibody production.
    E. They can be used to produce toxoids.

12. Endotoxins are composed of _____ and are located in the _____ of a gram-negative cell.
    - A. carbohydrate; cell membrane
    - B. protein; cell wall
    - C. lipid; pili
    - D. protein; glycocalyx
    - E. lipid and polysaccharide; cell wall
13. All of the following are examples of communicable diseases *except:*
    - A. chickenpox.
    - B. measles.
    - C. the common cold.
    - D. tetanus.
    - E. AIDS.
14. Which one of the following is an example of an indirect method of disease transmission?
    - A. Coughing
    - B. Droplet transmission
    - C. Neonatal gonorrhea
    - D. A mosquito bite
    - E. An animal bite
15. Fifty cases of hepatitis A during one week in a community would most likely be described as a/an
    - A. outbreak.
    - B. pandemic.
    - C. noncontagious infection.
    - D. endemic disease.
    - E. epidemic.

QUESTIONS FOR THOUGHT AND DISCUSSION

Answers to even-numbered questions can be found in **Appendix C**.

1. In 1840, Great Britain introduced penny postage and issued the first adhesive stamps. However, politicians did not like the idea because it deprived them of the free postage they were used to. Soon, a rumor campaign was started, saying that these gummed labels could spread disease among the population. Can you see any wisdom in their contention? Would their concern "apply" today?
2. In 1892, a critic of the germ theory of disease named Max von Pettenkofer sought to discredit Robert Koch's work by drinking a culture of cholera bacilli diluted in water. Von Pettenkofer suffered nothing more than mild diarrhea. What factors may have contributed to the failure of the bacilli to cause cholera in von Pettenkofer's body?
3. A man takes a roll of dollar bills out of his pocket and "peels" off a few to pay the restaurant tab. Each time he peels, he wets his thumb with saliva. What is the hazard involved?
4. When Ebola fever broke out in Africa in 1995, disease epidemiologists noted how quickly the responsible virus killed its victims and suggested the epidemic would end shortly. Sure enough, within three weeks it was over. What was the basis for their prediction? What other conditions had to apply for them to be accurate in their guesswork?
5. A woman takes an antibiotic to relieve a urinary tract infection caused by *Escherichia coli*. The infection resolves, but in two weeks, she develops a *Candida albicans* ("yeast") infection of the vaginal tract. What conditions may have caused this to happen?

APPLICATIONS

Answers to even-numbered questions can be found in **Appendix C**.

1. The transparent covering over salad bars is commonly called a "sneeze bar" because it helps prevent nasal droplets from reaching the salad items. As a community health inspector, what other suggestions might you make to prevent disease transmission via the salad bar?
2. While slicing a piece of garden hose, your friend cut himself with a sharp knife. The wound was deep, but it closed quickly. Shortly thereafter, he reported to the emergency room of the community hospital, where he received a tetanus shot. What did the tetanus shot contain, and why was it necessary?
3. After reading this chapter, you decide to make a list of the ten worst "hot zones" in your home. The title of your top-ten list will be "Germs, Germs Everywhere." What places will make your list, and why?
4. As a state epidemiologist responsible for identifying any disease occurrances, would an epidemic disease or an endemic disease pose a greater threat to public health in the community? Explain.

REVIEW

Test your knowledge of this chapter's contents by determining whether the following statements are true or false. If the statement is true, write "True" in the space. If false, substitute a word for the underlined word to make the statement true. The answers to even-numbered statements are listed in **Appendix C**.

_____ **1.** An epidemic disease occurs at a low level in a certain geographic area.

_____ **2.** An invasive microbe can penetrate tissues and cause structural damage.

_____ **3.** Among the microbial enzymes able to destroy blood cells are hemolysins and leukocidins.

_____ **4.** The term disease refers to a symbiotic relationship between two organisms and the competition taking place between them for supremacy.

_____ **5.** Organs of the human body lacking a normal microbiota include the blood and the small intestine.

_____ **6.** Commensalism is a form of symbiosis where one organism benefits with no damage occurring to the other.

_____ **7.** A sign is a change in body function experienced by the patient.

_____ **8.** A biological vector is an arthropod that carries pathogenic microorganisms on its feet and body parts.

_____ **9.** Organisms causing disease when the immune system is depressed are known as opportunistic organisms.

_____ **10.** The human body responds to the presence of exotoxins by producing endotoxins.

_____ **11.** The term bacteremia refers to the spread of bacteria through the bloodstream.

_____ **12.** A pathogenic species of *Staphylococcus* can form a fibrin clot through its production of hyaluronidase.

_____ **13.** A toxoid is an immunizing agent prepared from an exotoxin.

_____ **14.** Few symptoms are exhibited by a person who has a subclinical disease.

_____ **15.** Indirect methods of disease transmission include kissing and handshaking.

_____ **16.** A chronic disease develops rapidly, is usually accompanied by severe symptoms, and comes to a climax.

_____ **17.** The acme period is the time between the entry of the pathogen into the host and the appearance of symptoms.

_____ **18.** An organism with high virulence generally is unable to cause disease.

HTTP://MICROBIOLOGY.JBPUB.COM/

The site features learning, an on-line review area that provides quizzes and other tools to help you study for your class. You can also follow useful links for in-depth information, read more MicroFocus stories, or just find out the latest microbiology news.

19

Chapter Preview and Key Concepts

19.1 An Overview to Host Immune Defenses
- Leukocytes perform a variety of defensive roles in the host.
- The lymphatic system maintains and distributes lymphocytes necessary for defense against pathogens.
- The interactions between innate and acquired host defenses make infection and disease establishment difficult.

19.2 The Innate Immune Response
- The skin, mucous membranes, chemical inhibitors, and microbes represent barriers to infection.
- Cellular defenses can phagocytize pathogens.
- The inflammatory response brings effector molecules and cells to the infection site.
- Fever, usually one sign of a possible infection, can help immune defenses.
- Natural killer cells are nonspecific defensive lymphocytes.
- Complement proteins assist innate immunity in identifying and eliminating pathogens.
- Pattern-recognition receptors trigger the innate immune response.

MicroInquiry 19: Immune System Outcomes

Resistance and the Immune System: Innate Immunity

". . . medical science today has set itself the task of attempting to prevent disease. In order to achieve this aim one must attempt, on the one hand to find the disease germ and destroy it, and on the other hand to give the body the strength to resist attack."

—K. A. H. Mörner, Rector of the Royal Caroline Institute on presenting the 1908 Nobel Prize in Physiology or Medicine to Elie Metchnikoff and Paul Ehrlich for their work on the theory of immunity

Edward Jenner's epoch-making moment in history came in 1798 when he demonstrated the protective action of vaccination against smallpox (see Chapter 1). As great as Jenner's discovery was, it did not advance the development or understanding of **immunity**, which is how the human body can generate resistance to a particular disease, whether by recovering from the disease or as a result of vaccination, such as devised by Jenner.

It would be another 90 years before Russian zoologist and immunologist Elie Metchnikoff devised the first experiments to study immunity by investigating how an organism could destroy a disease-causing microbe in the body. Through landmark studies with invertebrates, Metchnikoff proposed that certain types of cells could attack, engulf, and destroy foreign material, including infectious microbes. These were among many experiments in a chain of investigations that studied immunity and could be applied to mammals and humans. Metchnikoff's work culminated in the theory of phagocytosis (see Chapter 18), which suggested certain types of human white blood cells could capture and destroy disease-causing microbes that had penetrated the body.

Around the same time, Louis Pasteur developed vaccines for chicken cholera and rabies, further stimulating an interest in the mechanism for protective immunity. In Pasteur's lab, Émile Roux and Alexandre Yersin discovered that the bacterial species causing diphtheria actually produced a toxin causing the disease. Then, Koch's coworker, Emil von Behring modified the diphtheria toxin to produce a substance with "antitoxic activity" that successfully immunized animals for a short period against diphtheria. The substance gave the body the power to resist infection.

With these early discoveries, the science of immunology was off and running. Today, Metchnikoff's work forms part of the basis for innate immunity, which consists of several nonspecific defenses present in all humans from the time of birth. The discoveries of Roux, Yersin, and von Behring are the basis for part of what is referred to today as acquired immunity. This form of protective immunity or resistance is a response to a particular microbe and is directed only against that microbe. Together, innate and acquired immunity are equally important arms that form a "microbiological umbrella" protecting us against the torrent of potential microbial pathogens to which we are exposed daily (FIGURE 19.1).

This chapter examines the immune resistance to microbes (again, for brevity, we will include viruses as microbes), toxins, and other foreign substances by studying innate immunity. As we will see, the opening quote to this chapter made in 1908 is as true today as it was then.

19.1 An Overview to Host Immune Defenses

The work of Metchnikoff, Roux, Yersin, and von Behring as well as that of their contemporaries and followers, opened up the whole field of **immunology**, the scientific study of how the immune system functions in the body to prevent or destroy foreign material, such as pathogens.

Blood Cells Form an Important Defense for Innate and Acquired Immunity

KEY CONCEPT

- Leukocytes perform a variety of defensive roles in the host.

In the circulatory system, blood consists of three major components: the fluid, the clotting agents, and the formed elements or cells. The fluid portion, called **serum**, is an aqueous solution of minerals, salts, proteins, and other organic substances. When clotting agents, such as fibrinogen and prothrombin, are present, the fluid is referred to as **plasma**. Blood platelets are small, disk-shaped ("plate-let") cells that originate in the bone marrow. Platelets have no nucleus and function chiefly in the blood-clotting mechanism.

The other formed elements in the blood are the erythrocytes (red blood cells) and leukocytes, the latter being significant to immune resistance.

Leukocytes (White Blood Cells). As their name suggests, **leukocytes** (*leuko* = "white") have no pigment in their cytoplasm and therefore appear gray when unstained. These cells also are produced in the bone marrow. They number about 4,000 to 12,000 per microliter of blood and have different lifespans, depending on the type of cell. Most can be identified in a blood smear or tissue sample by the size and shape of the cell nucleus and the staining of visible cytoplasmic granules, if present, when observed with the light microscope. TABLE 19.1 summarizes the six major types of white blood cells.

Neutrophils have a multilobed nucleus and therefore are referred to as polymorphonuclear leukocytes (PMNs). Their cytoplasm contains many granules (lysosomes), which

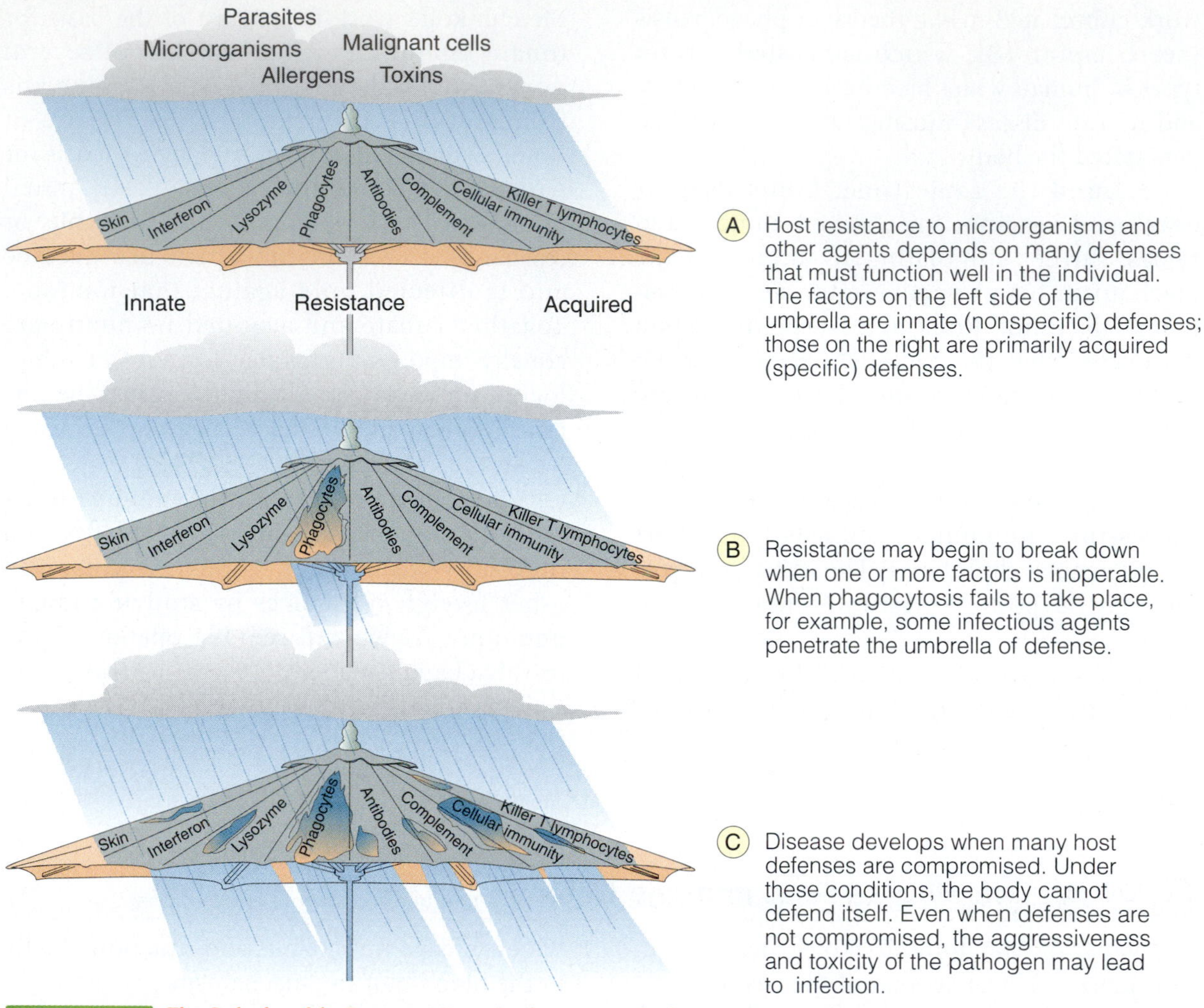

FIGURE 19.1 The Relationship Between Host Resistance and Disease. Host resistance can be likened to a "microbiological umbrella" that forms a barrier or defense against infectious disease. Should tears or holes develop in the umbrella (e.g., the host becomes compromised), the body may be exposed to infection and disease.

Q: Why are the defenses on the left side of the umbrella considered innate and those on the right acquired?

contain digestive enzymes and other substances capable of killing pathogens. Neutrophils function chiefly as **phagocytes**, cells that carry out phagocytosis. Although usually found in the blood, neutrophils can move out of the circulation to engulf foreign particles or pathogens causing infection. Their lifespan is short, only about one to two days, so they are continually replenished from the bone marrow.

Eosinophils exhibit red or pink cytoplasmic granules when a dye such as eosin is applied. Substances in the granules contain toxic compounds to defend against multicellular parasites, such as flukes and tapeworms. Eosinophils have some phagocytic activity.

■ Eosin:
A red anionic (acidic) dye.

Basophils are leukocytes whose cytoplasmic granules stain blue with methylene blue. Basophils are similar to **mast cells** found in tissues. Both function in allergic reactions, as will be discussed in detail in Chapter 22.

■ Methylene blue:
A blue cationic (basic) dye.

Eosinophils, basophils, and mast cells are concentrated in the skin, lungs, and the gastrointestinal tract, locations where breaks in the skin, inhalation, or ingestion might bring pathogens. Because neutrophils, eosinophils, and basophils contain cytoplasmic granules, they often are called **granulocytes**.

A fourth major group of leukocytes is the **monocytes**. These phagocytic cells have a single, bean-shaped nucleus and lack visible cytoplasmic granules. In the tissues, monocytes mature into **macrophages**. In contrast to the two-week lifespan of monocytes, macrophages may live for several months.

TABLE
19.1 Major Leukocytes of the Human Immune System

Types of Leukocytes		Approximate Percentage	Activated Function
Granulocytes	Neutrophil	50–70	Phagocytosis
	Eosinophil (Red-stained granules)	2–4	Phagocytosis of antibody-coated parasites
	Basophil & Mast cell (Blue-stained granules)	<1	Release of chemical mediators of inflammation and allergies
Agranulocytes	Monocyte & Macrophage	2–8	Phagocytosis and cytokine secretion
	Lymphocyte	20–30	Antibody production; cytotoxic properties; cytokine secretion
	Dendritic Cell	—	Activate lymphocytes; cytokine secretion

Macrophages (*macro* = "big"; *phage* = "eat") are more effective as phagocytes than neutrophils and are one of the key cells in acquired immunity.

The **lymphocytes** migrate from the bone marrow to the lymph nodes after maturation. They have a single, large nucleus and no granules. One type, the **natural killer** (NK) **cells** play a key role innate immunity by destroying virus-infected cells and abnormal cells, such as cancer cells. The two other types, **B lymphocytes** and **T lymphocytes**, are key cells of acquired immunity. They increase in number dramatically during the course of an infectious disease but have very different functions, as we will describe in the next chapter. Lacking visible granules, monocytes and lymphocytes often are called **agranulocytes**.

The last group of leukocytes is the **dendritic cells**. The name comes from their resemblance to the long, thin extensions (dendrites) seen on neurons. Dendritic cells are found in the skin, where they are called Langerhans cells, and in tissues where pathogens may enter. These cells also are crucial to the activation of acquired immunity.

CONCEPT AND REASONING CHECKS

19.1 Draw a simple concept map for leukocytes; besides the different leukocytes, include the terms granulocytes, agranulocytes, and phagocytes in your map.

The Lymphatic System Is Composed of Cells and Tissues Essential to Immune Function

KEY CONCEPT

- The lymphatic system maintains and distributes lymphocytes necessary for defense against pathogens.

In the human body, the clear fluid surrounding the tissue cells and filling the intercellular spaces is called tissue fluid, or **lymph** (*lymph* = "water"). Lymph bathes the body cells, supplying oxygen and nutrients while collecting wastes. To carry these materials, the lymph must be pumped through tiny vessels by the contractions of skeletal muscle cells. Eventually the lymph vessels unite to form larger lymphatic vessels. Along the way, lymph nodes filter the lymph before the lymph returns to the bloodstream. Should lymphatic vessels be blocked due to a parasitic infection (see Chapter 17), swelling of the extremities can result in a disease called elephantiasis (FIGURE 19.2).

It is the lymphatic system that maintains and distributes lymphocytes necessary for monitoring and defending against pathogens (FIGURE 19.3). The **primary lymphoid tissues** consist of the **thymus**, which lies behind the breast bone, and the **bone marrow**. Both are sites where immune cells form or mature.

The **secondary lymphoid tissues** are sites where mature immune cells interact with pathogens and carry out the acquired immune response. The **spleen**, a flattened organ at the upper left of the abdomen, contains immune cells to monitor and fight infectious microbes entering the body. Likewise, the **lymph nodes**, prevalent in the neck, armpits, and groin, are bean-shaped structures containing phagocytes, which engulf particles in the lymph, and lymphocytes, which respond specifically to substances in the circulation (FIGURE 19.4). Because resistance mechanisms are closely associated with the lymph nodes, it is not surprising they become enlarged (often called "swollen glands") during infections.

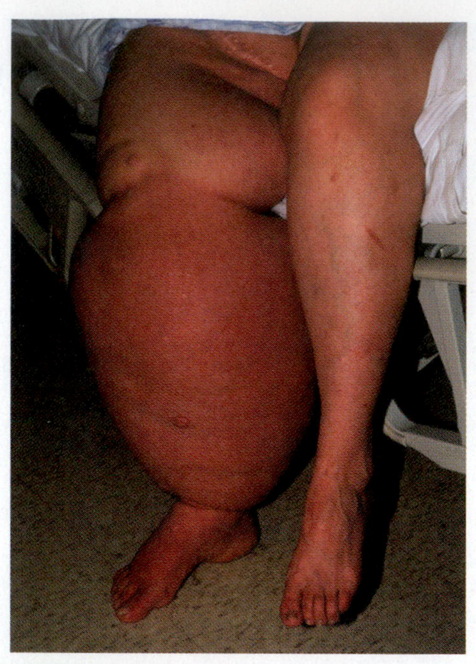

FIGURE 19.2 **Elephantiasis.** A woman with gross swelling of lymphatic tissues due to severe lymphatic filariasis (elephantiasis). The condition is caused by a parasitic worm infection *(Wuchereria bancrofti)* of the lymphatic system.

Q: Referring to Figure 19.3, which lymphatic vessels are blocked in this patient?

Other organ systems in the body contain additional secondary lymphoid tissues. In the intestine, the Peyer patches and appendix form part of the **mucosa-associated lymphoid tissue** (MALT), while associated with the respiratory tract are the **tonsils** and **adenoids**.

CONCEPT AND REASONING CHECKS

19.2 Assess the role of the lymphatic system to defense against pathogens.

Innate and Acquired Immunity Are Essential Components of a Fully Functional Human Immune System

KEY CONCEPT

- The interactions between innate and acquired host defenses make infection and disease establishment difficult.

When one thinks about the body fighting an infection, one usually envisions antibodies and white blood cells as the "knights in shining armor." Although these molecules and cells are critical, they represent only the second half of the total immune system defense. First, and foremost, immune defense depends on nonspecific resistance.

Nonspecific resistance is referred to as **innate immunity** because genetically-encoded molecules, present in the body from birth, recognize microbial features common to many pathogens and foreign substances. Metchnikoff's studies were the first to suggest innate immunity is evolution's defense mechanism against infectious disease and represents an ancient one common to most animals. Therefore, innate immunity, as exemplified by the lymphoid tissues and the cells they contain, represents the host's early-warning system against potentially-harmful pathogens, foreign cells, or even our own cells should they become damaged or cancerous. This defense system responds within minutes or a few hours after infection.

Upon recognition of microbes by innate immune defenses, these defenses try to eliminate or hold an infection in check while sending chemical signals to tissues involved with initiating acquired immunity. These chemical signals, called **cytokines**, are small proteins released by various defensive cells in response to an activating substance, such as an invading microbe. Cytokines are produced by many cells, including macrophages, lymphocytes, mast cells, and dendritic cells. We will see specific examples as we discuss innate and acquired immunity.

The knights in shining armor are part of the second line of defense—what is called specific resistance or **acquired immunity** because the response produces lymphocytes and antibodies specific only to the pathogen or foreign substance causing the infection. This form of immunity, which evolved more recently and only exists in vertebrates, is relatively slow compared to the innate response; in fact, acquired immunity develops over the course of the infection, taking several days to more than a week to mount an effective response and a protective defense.

In the rest of this chapter, we will examine innate immunity in more detail.

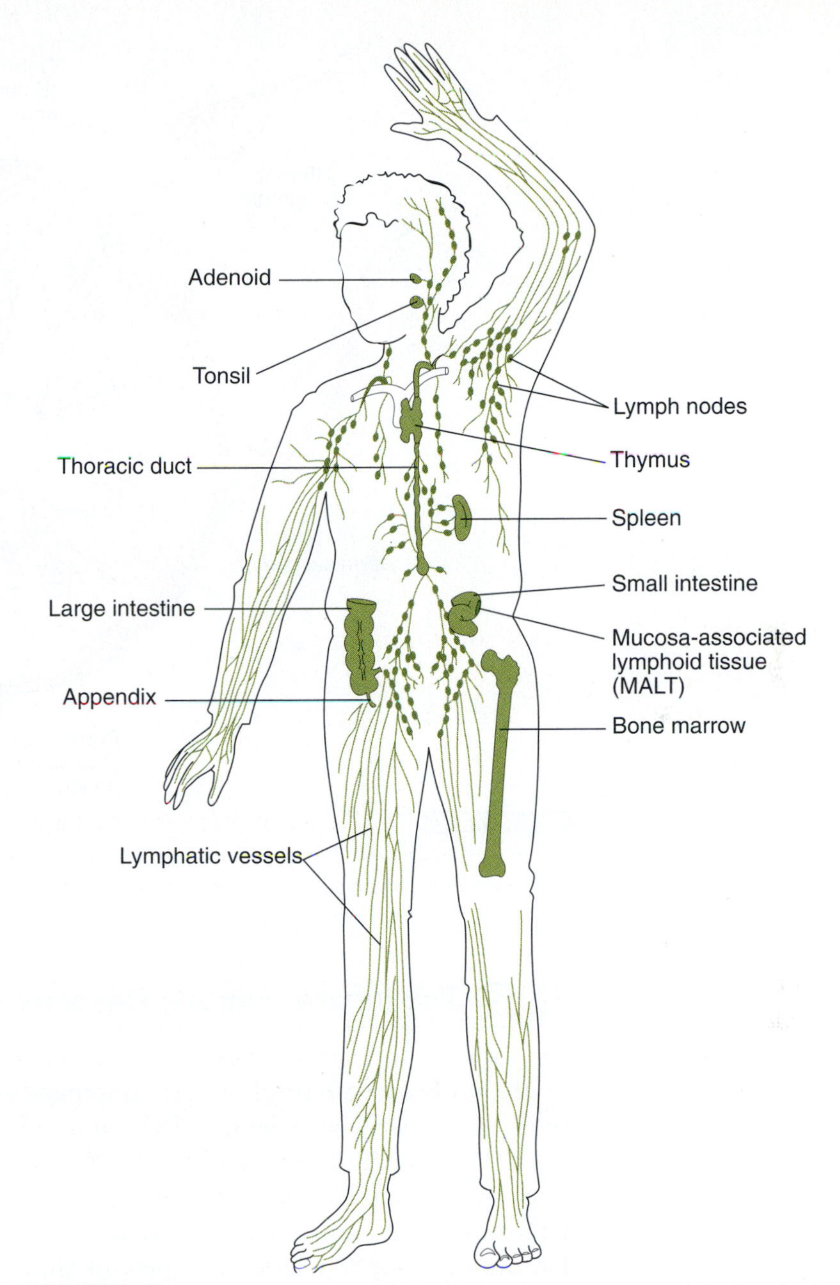

FIGURE 19.3 The Human Lymphatic System. The components of the lymphatic system consist of lymphocytes, lymphatic organs, lymph vessels, and lymph nodes located along the vessels. The lymphatic organs are illustrated, and the preponderance of lymph nodes in the neck, armpits, and groin is apparent.

Q: Why is there such a diversity of lymphatic organs in the human body?

CONCEPT AND REASONING CHECKS

19.3 Summarize the roles for innate and acquired immunity in the human host.

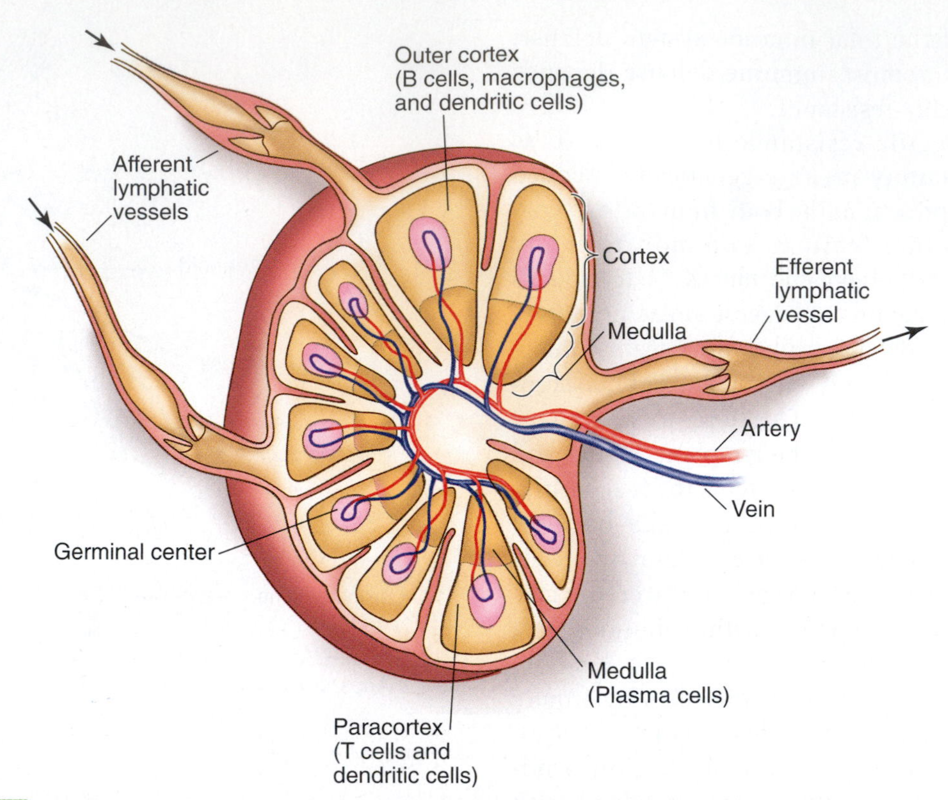

FIGURE 19.4 **The Structure of a Lymph Node.** Immune cells, including B cells, T cells, macrophages, dendritic cells, and plasma cells populate different areas in a lymph node. Lymph enters at the afferent vessels and exits at the efferent vessel.

Q: How would lymph nodes detect, intercept, and eliminate bacterial cells during an infection?

19.2 The Innate Immune Response

The innate immune response to disease involves a broad group of factors. It depends upon the general well-being of the individual and proper functioning of the body's systems. Accordingly, it takes into account such determinants as nutrition, fatigue, age, gender, and climate. Specific examples of these factors are highlighted in discussions of individual diseases, so we will not pause to delineate them here.

Mechanical, Chemical, and Microbiological Barriers Are Exposed First Lines of Defense

KEY CONCEPT

- The skin, mucous membranes, chemical inhibitors, and microbes represent barriers to infection.

There are several portals of entry by which pathogens can enter the body (see Chapter 18). The intact skin and the mucous membranes, which extend over the body surface or into the body cavities, and their indigenous microbiota, are defenses providing barriers to infection at these exposed sites.

Mechanical Barriers. The skin's epidermis is constantly being sloughed off, and with it goes microorganisms. In addition, the skin is a poor source of nutrition for invading microbes and, from the viewpoint of a pathogen, the low water content of the skin presents a veritable desert. Unless microbes can penetrate the skin barrier, disease is rare, although some dermatophytes can lead to diseases like athlete's foot (see Chapter 16).

A fact of everyday life is cuts or abrasions. These breaches to the skin may allow microbes to enter the blood; bites from infected arthropods, acting as hypodermic needles, may inject the pathogens responsible for such diseases as West Nile fever, malaria, or plague. Other means of penetrating this mechanical barrier include splinters, tooth extractions, burns, shaving nicks, war wounds, and injections.

Possible penetration of the skin barrier is patrolled by dendritic cells (called Langerhans cells in the skin tissues). If a pathogen is detected by these cells, they phagocytize the pathogen and migrate to a regional lymph node where they induce an acquired immune response. The dendritic cells also secrete cytokines, which influence both innate and acquired immune responses.

The **mucous membranes** (**mucosa**) are the moist epithelial lining of the digestive, urogenital, and respiratory tracts. Because these exposed sites represent potential portals of entry, the mucous membranes provide resistance to microbial invasion. For instance, cells of the mucous membranes that line the respiratory passageways secrete **mucus**, a sticky secretion of glycoproteins that traps heavy particles and microorganisms. The epithelial cilia then move the microbe-laden mucus up to the throat, where the material is swallowed.

Chemical Barriers. Several host chemicals also provide an antimicrobial defense to infection. Resistance in the vaginal tract is enhanced by the low pH. This develops when *Lactobacillus* species in the microbiota break down glycogen to various acids. Many researchers believe the disappearance of lactobacilli during antibiotic treatment encourages opportunistic diseases such as candidiasis and trichomoniasis to develop (see Chapters 16 and 17). In the urinary tract, the slightly acidic pH of the urine promotes resistance to pathogens, and the flow of urine flushes the microbes away.

A natural barrier to the gastrointestinal tract is provided by stomach acid, which with a pH of approximately 2.0, destroys most pathogens. However, there can be notable survivors, including polio and hepatitis A viruses, the typhoid and tubercle bacilli, and *Helicobacter pylori,* a major cause of peptic ulcers (see Chapter 10). Bile from the gallbladder enters the system at the duodenum and serves as an inhibitory substance. In addition, duodenal enzymes hydrolyze the proteins, carbohydrates, lipids, and other large molecules of microorganisms.

The chemical barriers also consist of numerous small, antimicrobial peptides, called **defensins**, which are found in various secretions throughout the body. They are produced by phagocytic cells and lymphocytes as well as epithelial cells of the respiratory, gastrointestinal, and urogenital tracts. The sebum on the skin surface, for example, produces at least 12 defensins. Some defensins are continually produced, while others are induced by microbial products. They are active against many gram-negative and gram-positive bacterial cells, fungi, and viruses such as HIV. Many of the defensins damage membranes, killing the pathogens through cell lysis.

■ Sebum: An oily substance secreted by the sebacceous glands to lubricate the hair and skin.

In a somewhat similar manner, **lysozyme** is a chemical inhibitor found in human tears, sweat, and saliva. The enzyme disrupts the cell walls of gram-positive bacterial cells by digesting peptidoglycan. Osmotic lysis kills the cells (see Chapter 4).

Interferons are the group of cytokines produced by T cells, NK cells, and dendritic cells in response to invasion by RNA viruses. Interferons trigger macrophage activation and the production of inhibitory substances to "interfere" with viral reproduction. A thorough account of the interferons is presented in Chapter 13.

Microbiological Barriers. Defensive barriers also include the normal microbiota of the body surfaces (see Chapter 18). These indigenous, nonpathogenic microbes form a cellular barrier by out competing pathogens for nutrients and attachment sites on the skin or mucosa.

TABLE 19.2 summarizes the nonspecific resistance mechanisms described above.

CONCEPT AND REASONING CHECKS

19.4 Explain how the skin and mucous membranes provide a mechanical defense against infection.

Phagocytosis Is a Nonspecific Defense Mechanism to Clear Microbes from Infected Tissues

KEY CONCEPT

- Cellular defenses can phagocytize pathogens.

One of Metchnikoff's key observations involved the larvae of starfish. When jabbed with a splinter of wood, motile cells in the starfish gathered around the splinter. From observations like these, Metchnikoff suggested cells could actively seek out and engulf (ingest) foreign particles in the body. Metchnikoff's theory of phagocytosis laid the foundation for one of the most important functions of innate immunity.

■ Bile: A yellowish-green fluid, which in the small intestine plays an essential role in emulsifying fats.

Phagocytosis (*phago* = "eat") is the capturing and digesting of foreign particles,

TABLE

19.2 Nonspecific Mechanical, Chemical, and Microbiological Barriers to Disease

Resistance Mechanism	Activity
Skin layers	Provide a protective covering to all external body tissues
Mucous membranes of body cavities	Trap airborne particles in mucus Sweep particles along by cilia
Acidity in the vagina and stomach	Acidic pH toxic to pathogens
Bile	Inhibitory to most pathogens
Duodenal enzymes	Digest structural and metabolic chemical components of microorganisms
Defensins	Disrupt microbial membranes
Lysozyme in tears, saliva, secretions	Digests cell walls of gram-positive bacterial cells
Interferons	Inhibit replication of viruses
Normal microbiota	Compete for nutrients and attachment sites Produce antimicrobial substances

including pathogens, by phagocytes. This may occur at the site of the infection as well as in the lymphoid tissues.

If a bacterial or viral pathogen breaches the exterior defenses, the intruder usually is recognized, ingested, and killed by macrophages residing in the tissue. Therefore, they will be the first defensive cell to interact with an invading pathogen. On binding the pathogen, macrophages release cytokines and lipid mediators that will trigger an inflammatory response (to be described next) as well as activating an acquired immune response. Some cytokines, called **chemokines**, stimulate the migration of neutrophils to the infection site. The events of phagocytosis involving a bacterial pathogen are outlined in FIGURE 19.5.

The process begins with the attachment of the microbe to the cell surface of the phagocyte. The plasma membrane then surrounds the microbe and an invagination, or folding in, of the membrane produces an internalized phagocytic vesicle, called a **phagosome**. The phagosome now becomes acidified and the lowered pH aids in killing or inactivating the pathogen. In addition, the phagosome fuses with several lysosomes. Within this **phagolysosome**, lysosome products contribute in two ways to the destruction of the ingested pathogen. Lysosome enzymes, such as lysozyme and acid hydrolases, digest the pathogen. In addition, other products, including hydrogen peroxide (H_2O_2), nitric oxide (NO), and superoxide anions (O_2^-) generated by the lysosome or in the cytoplasm also help in the killing.

Because neutrophils are short-lived cells whereas macrophages are longer lived, the latter are more likely to kill the majority of pathogens. In this case, the host clears the pathogen and the infection ends, as waste materials are eliminated from the phagocyte. However, the host-microbe relationship is always evolving to give either the host or the microbe the advantage. Sometimes the pathogen gets the upper hand by "using" phagocytosis to its advantage (MicroFocus 19.1).

Although phagocytes are effective at killing pathogens and clearing many infections, collateral damage often occurs. During the fierce phagocytic activity occurring at the infection site, lysosome enzymes may be released inadvertently into the surrounding tissue. In addition, neutrophils release lysosome enzymes when dying. The released lysosome products have damaging effects on surrounding tissue and often are the major reason for tissue damage from an infection. Indeed, **pus**, which is composed of dead and dying neutrophils and damaged tissue arising from lysosome damage, often is a sign of infection.

The internalization of microbe into phagocyte often is enhanced when the microbe is

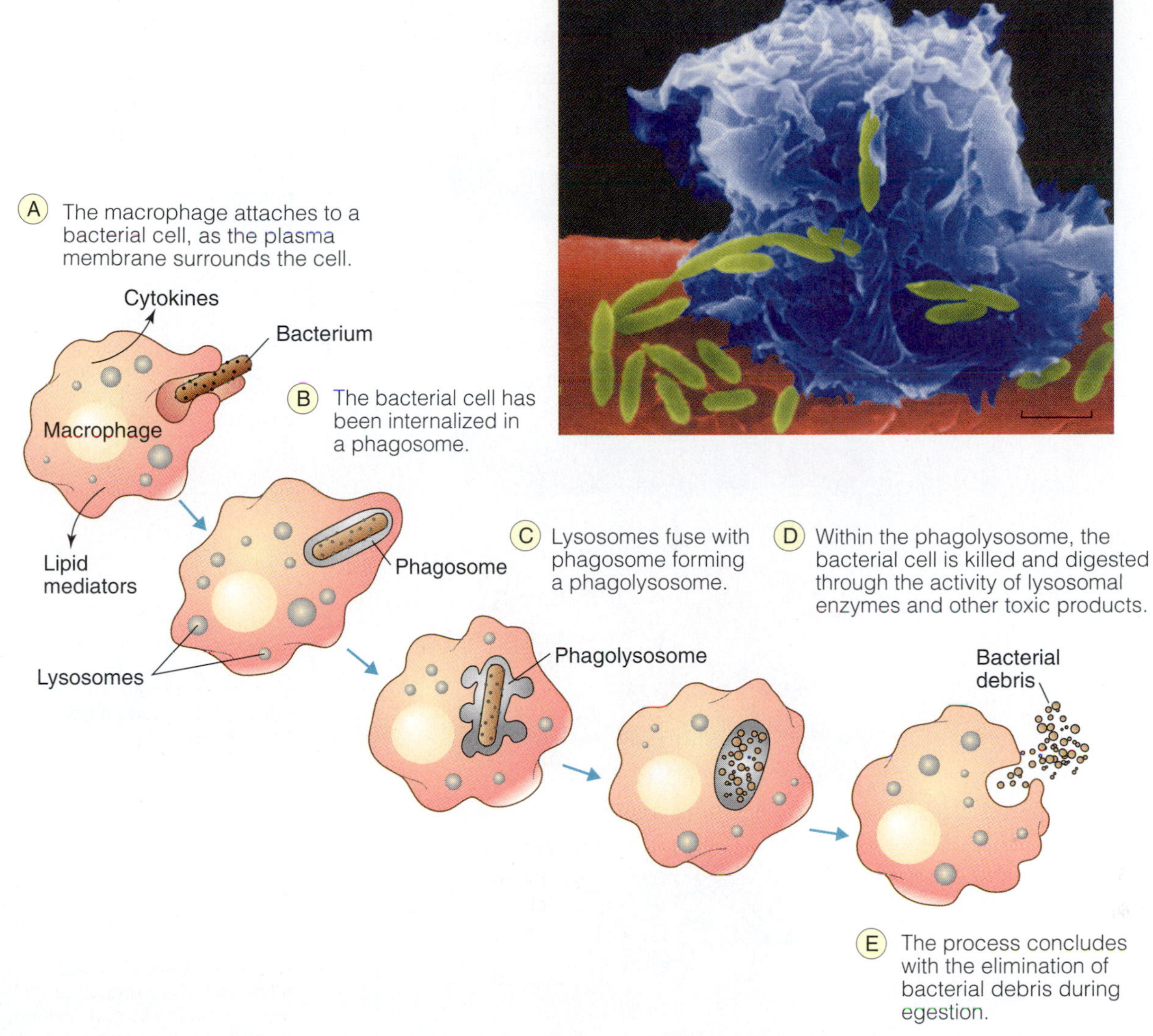

FIGURE 19.5 **The Mechanism of Phagocytosis.** The stages of phagocytosis are shown. Inset: a false-color scanning electron micrograph of a macrophage (blue) engulfing *E. coli* cells (green) on the surface of a blood vessel (red). (Bar = 5 µm.)

Q: What might the bacterial debris in (E) or waste products be?

coated with certain serum proteins, such as antibodies or complement (discussed shortly). These protein molecules, called **opsonins**, attach to microbes and increase the ability of phagocytes to adhere to the pathogen. The enhanced phagocytic process is called **opsonization**.

CONCEPT AND REASONING CHECKS

19.5 Discuss the importance of lysosomes to innate immunity.

Inflammation Plays an Important Role in Fighting Infection

KEY CONCEPT

- The inflammatory response brings effector molecules and cells to the infection site.

Inflammation is a nonspecific defensive response by the body to trauma. It develops after a mechanical injury, such as an injury or blow to the skin, or from exposure to a chemical agent, such as acid or bee venom. An inflammatory response also may be due to an infection by a pathogen.

In the case of an infection, the pathogen's presence and tissue injury sets into motion an innate process to limit the spread of the infection (FIGURE 19.6). At the site of tissue damage (infection), resident tissue macrophages secrete several cytokines triggering a local dilation of the blood vessels (**vasodilation**) and increasing capillary permeability. This allows the flow of

■ Antibodies: Proteins produced by plasma cells in response to the presence of a foreign material or cell, such as a bacterium or virus.

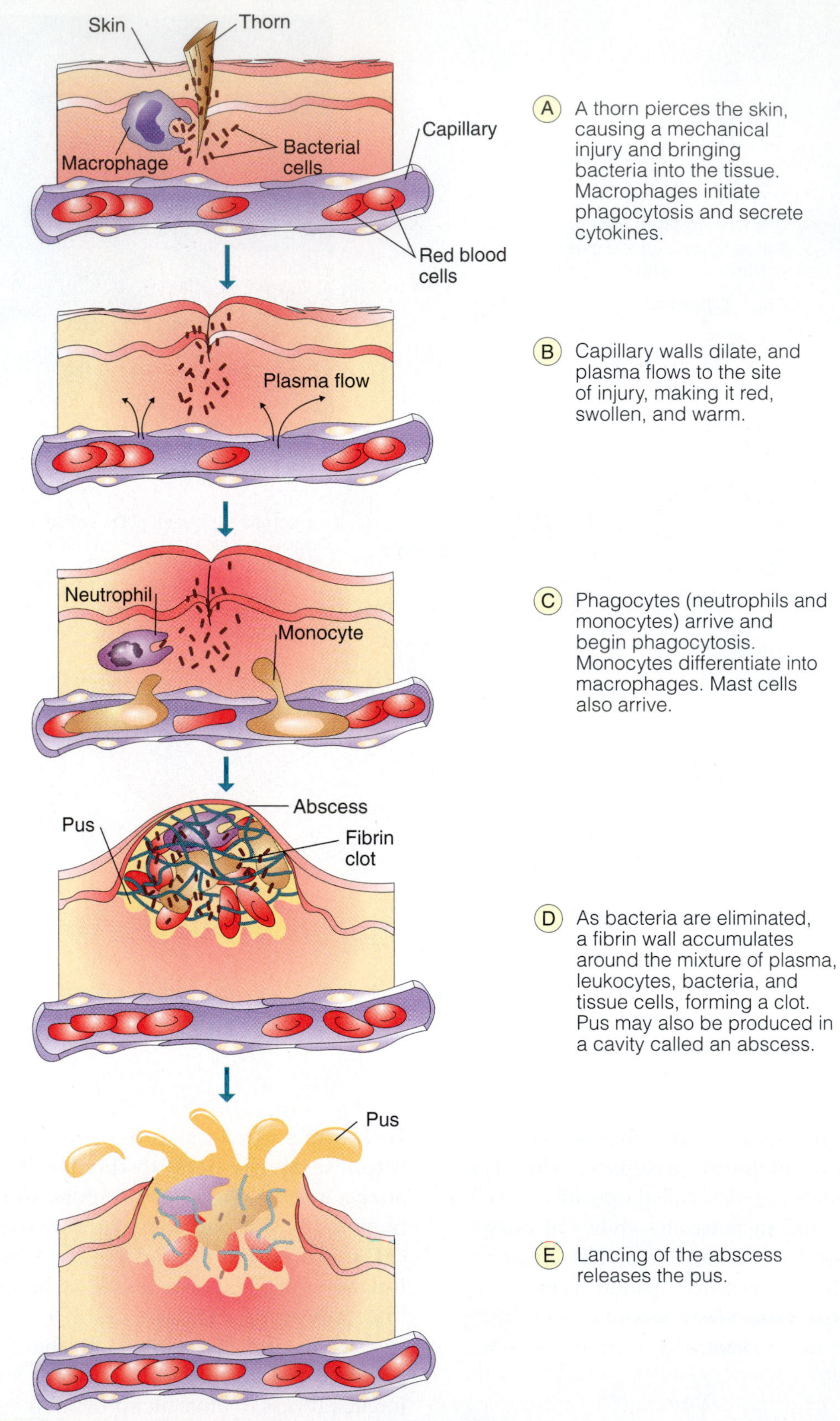

FIGURE 19.6 **The Process of Inflammation.** A series of nonspecific steps generate an inflammatory response to some form of trauma or infection, such as being pierced by a plant throne harboring bacterial cells.

Q: Why does the inflammatory response generate the same set of events no matter what the nature of the trauma?

MICROFOCUS 19.1: Evolution

Avoiding the "Black Hole" of the Phagocyte

Lysosomes contain quite a battery of enzymes, proteins, and other substances capable of efficiently digesting or being toxic to almost anything with which they come in contact. However, not all bacterial species are automatically engulfed by phagocytes.

Some species, such as *Streptococcus pneumoniae,* have a thick capsule (see Chapter 4), which is not easily internalized by phagocytes. Other pathogens simply invade in such large numbers they overwhelm the host defenses and the ability of phagocytes to "sweep" them all up. More interesting though are pathogens that have evolved strategies to evade the lysosome—the "black hole" of the phagocyte (see figure).

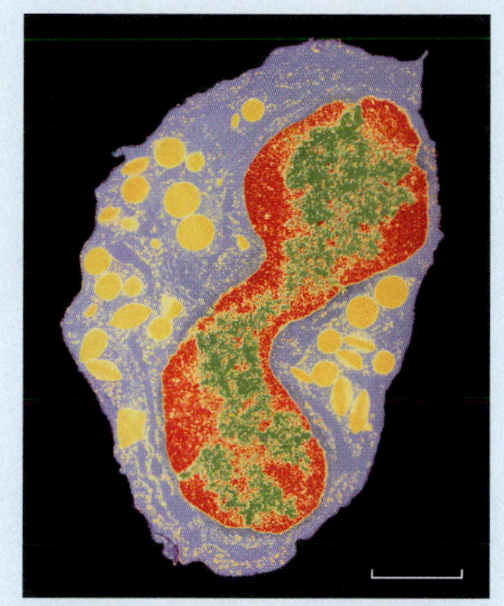

False-color transmission electron micrograph of lysosomes (yellow) in a macrophage. (Bar = 4 µm.)

Listeria monocytogenes is internalized by phagocytes. However, before lysosomes can fuse with the phagosome membrane, the bacterial cells releases a pore-forming toxin that lyses the phagosome membrane. This allows the *L. monocytogenes* cells to enter and reproduce in the phagocyte cytoplasm. The bacterial cells then spread to adjacent host cells by the formation of actin tails (see Chapter 18).

Other pathogens prevent lysosome fusion with the phagosome in which they are contained. The causative agent of tuberculosis, *Mycobacterium tuberculosis,* is internalized by macrophages. However, once in a phagosome, the bacterial cells prevent fusion of lysosomes with the phagosome. In fact, the tubercle bacteria actually reproduce within the phagosome. A similar strategy is used by *Legionella pneumophila.*

Toxoplasma gondii, the protozoan responsible for toxoplasmosis (see Chapter 17), evades the lysosomal "black hole" by enclosing itself in its own membrane vesicle that does not fuse with lysosomes. Avoiding digestion by lysosomes is key to survival for these pathogens.

plasma into the injured tissue and fluid accumulation (**edema**) at the site of infection. Chemokines attract additional phagocytes (neutrophils and monocytes) toward the infected tissue. These cells adhere to the blood vessels near the infection and then migrate between capillary cells (**diapedesis**) into the infected tissue. There the neutrophils augment phagocytosis of the macrophages and new monocytes differentiate into more macrophages. Chemokines also attract mast cells, which secrete histamine, bolstering the inflammatory response.

The inflamed area thus exhibits four characteristic signs: redness (from blood accumulation); heat (from the warmth of the blood); swelling (from the accumulation of fluid); and pain (from local tissue damage). Sometimes itching occurs, as MicroFocus 19.2 describes.

Eventually the formation of a fibrin clot occurs, which among other things prevents the spread of pathogens to the blood. Again, pus may be a product during inflammation. When pus becomes enclosed in a wall of fibrin through the clotting mechanism, an **abscess**, or boil, may form (see Chapter 12).

To limit tissue damage, the body tries to regulate the number of phagocytes and the level of cytokine production in the defense against infection. Therefore, the inflammatory response must be managed so it is only short-lived. Should the inflammatory response become long-lived, chronic inflammation may occur. The persisting cytokines can produce excessive damage, leading to a number of dissimilar diseases, including rheumatoid arthritis, coronary heart disease, atherosclerosis, Alzheimer disease, or some forms of cancer.

Another consequence of an overwhelming "cytokine storm" is **septic shock**, characterized by a collapse of the circulatory and respiratory systems. In some gram-negative bacterial infections, large amounts of cell wall

- **Histamine:** A chemical mediator causing contraction of smooth muscle and dilation of blood vessels.
- **Atherosclerosis:** An arterial disease in which degeneration and cholesterol deposits (plaques) form on the inner surfaces of the arteries.

MICROFOCUS 19.2: Evolution

But It Doesn't Itch!

How many of us at some time have had a small red spot or welt on our skin that looks like an insect bite—but it doesn't itch? When most of us are bitten by an insect, such as an ant, bee, or mosquito, the site of the bite starts itching almost immediately. Yet, there are cases where bites do not itch, such as a tick bite. So, what's the difference?

Most insect bites trigger a typical inflammatory reaction. Take a mosquito as an example. When it bites and takes a blood meal, saliva injected into the skin contains substances so the blood does not coagulate. The bite causes tissue damage and activation of chemical mediators that trigger the typical characteristics of an inflammatory response: redness, warmth, swelling, and pain at the injured site.

One of the major chemical mediators is histamine, released by the damaged tissue and immune cells in the tissue. An itch actually is a form of pain and arises when histamine affects nearby nerves. Itching is the result.

In a tick bite, the scenario is slightly different. The bite triggers the same set of inflammatory reactions and histamine is produced. However, in the tick saliva is another molecule called histamine-binding protein (HBP). HBP binds to histamine, preventing the mediator from affecting the nearby nerves—and no itching occurs.

So why is that an advantage for the tick? Probably because ticks take long blood meals that can last for three to six days. By the bite being painless and not itching, ticks may go unnoticed for a few days; there is less chance of them being noticed and removed.

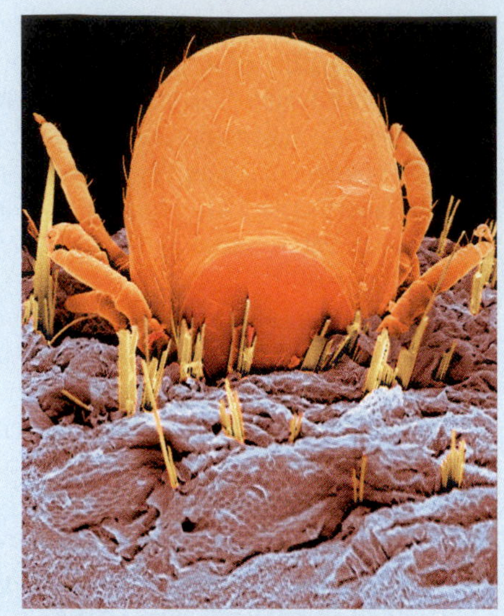

False-color scanning electron micrograph of a tick (orange) feeding head-down in human skin.

lipopolysaccharide (LPS) are released, which is referred to as an endotoxin (see Chapter 18). The systemic spread of LPS through the body triggers macrophages to secrete the "overdoses" of cytokines. Shock can lead to death.

Hypothalamus: The part of the brain controlling involuntary functions.

CONCEPT AND REASONING CHECKS

19.6 Explain how inflammation serves as an innate immune defense.

Moderate Fever Benefits Host Defenses

KEY CONCEPT

- Fever, usually one sign of a possible infection, can help immune defenses.

Fever is an abnormally high body temperature that remains elevated above the normal 37°C. Fever supports the immune system response in trying to gain an advantage over the pathogens by making the body a less favorable host.

Fever-producing substances, called **pyrogens**, represent endogenous cytokines produced by activated macrophages and other leukocytes, and exogenous microbial fragments from bacterial cells, viruses, and other microorganisms. Pyrogens move through the blood and affect the anterior hypothalamus such that the body temperature becomes elevated. As this takes place, cell metabolism increases and blood vessels constrict, thus denying blood to the skin and keeping its heat within the body. Patients thus experience cold skin and chills along with the fever.

A low to moderate fever in adults may be beneficial to immune defense because the elevated temperature inhibits the rapid growth of many microbes, encourages rapid tissue repair, and heightens phagocytosis. However, if the temperature rises above 40.6°C (105°F), convulsions and death may result from host metabolic inhibition. Also, infants with a fever above 38°C (100°F), or older children with a fever of 39°C (102°F), need medical attention. MicroFocus 19.3 examines the saying that it is good to "starve a fever and feed a cold."

MICROFOCUS 19.3: Being Skeptical

Feed a Cold, Starve a Fever?

One of the typical symptoms of the flu is a high fever (39–40°C) while a cold seldom produces a fever. So, is there any truth to the adage, "You should feed a cold and starve a fever?" The definite answer is—yes!

There are three reasons you should reduce food intake with a fever. First, food absorbed by the intestines could be misidentified by the body as an allergen, worsening the illness and fever. Also, with the body already under stress, heavy eating can contribute on rare occasions to body seizures, collapse, and delirium.

Perhaps most common though is that eating stimulates digestion of the foods. During times of increased physiologic stress, digestion overstimulates the parasympathetic nervous system, when the sympathetic nervous system (involved in constricting blood vessels and conserving heat = fever) already is active. In other words, eating tends to drop body temperature, possibly eliminating fever as a protective mechanism.

So, with the flu or other illness-induced fever, stick to bed rest and just drink plenty of liquids.

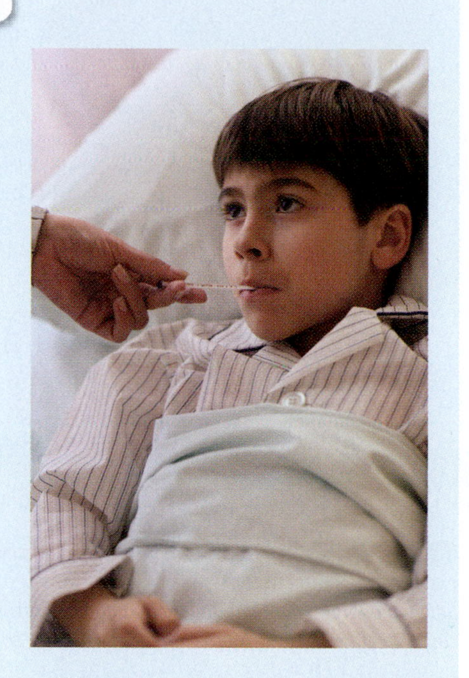

CONCEPT AND REASONING CHECKS

19.7 Assess the advantage of a low grade fever during an infection.

Natural Killer Cells Recognize and Kill Abnormal Cells

KEY CONCEPT

- Natural killer cells are nonspecific defensive lymphocytes.

Natural killer (NK) **cells** are another type of defensive lymphocyte. They are formed in the bone marrow and then migrate to the tonsils, lymph nodes, and spleen, where they await activation. On stimulation, they secrete several cytokines, triggering acquired immune responses by macrophages and other immune cells. While those events develop, the NK cells move into the blood and lymph where they act as potent killers of cancer cells and virus-infected cells.

NK cells are not phagocytic; rather, they contain on their surfaces a set of special receptor sites capable of forming cell-to-cell interactions with a target cell (FIGURE 19.7). If the receptor sites match up with a group of class I MHC proteins (to be discussed in Chapter 20) on the target cell, the NK cell recognizes the target cell as one of the body's own and leaves it alone. The NK cell does not release the cytolytic mediators.

However, when these MHC proteins are absent or in reduced amounts (as on a cancer cell or virus-infected cell), then the matchup is incomplete and the NK cell secretes the cytotoxic mediators that damage the tumor or virus-infected cell. The cytolytic mediators include **perforins**, which drill holes in the target cell. They also secrete **granzymes**, which are cytotoxic enzymes inducing the target cell to undergo apoptosis or cell suicide. The DNA condenses, the cell shrinks, and breaks into tiny pieces that can be mopped up by wandering phagocytes.

■ Apoptosis: A type of cell death activated by an internal death program.

CONCEPT AND REASONING CHECKS

19.8 How do natural killer cells assist the other innate defense mechanisms during a viral infection?

Complement Marks Pathogens for Destruction

KEY CONCEPT

- Complement proteins assist innate immunity in identifying and eliminating pathogens.

Complement (also called the complement system) is a series of nearly 30 inactive proteins circulating in the bloodstream that represents

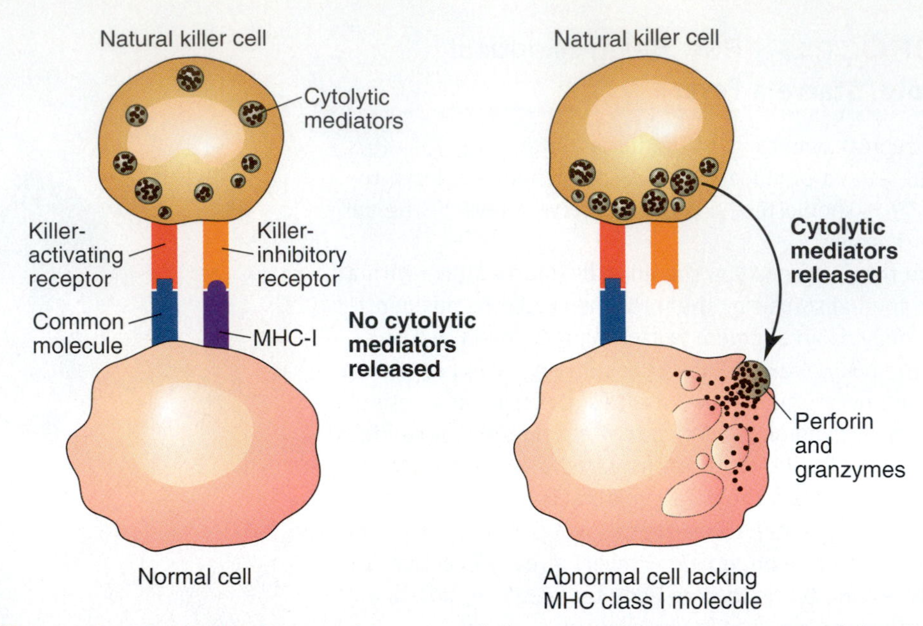

FIGURE 19.7 **Natural Killer (NK) Cell Recognition.** NK cells have the ability to destroy any cell they "see" as abnormal (for example, tumor cells or virus-infected cells).

Source: Delves PJ, Roitt IM. The Immune System. First of Two Parts. *New England Journal of Medicine*. 343: 37–49. Copyright © 2006 [2000] Massachusetts Medical Society. All rights reserved.

Q: How does the presence or absence of class I MHC receptors affect NK cell activity?

another innate defense to disease. If a microbe penetrates the body, complement proteins become sequentially activated through a cascade of steps that assist in the inflammatory response and phagocytosis. Additional complement proteins in the cascade bring about the destruction of the invading bacterial cells or viruses through cell or virion lysis. Two of the complementary pathways for complement activation are illustrated in FIGURE 19.8.

One pathway, called the **classical pathway**, involves antibody-microbe complexes activating a cascade of complement proteins that culminate in the activation of an enzyme called C3 convertase. The other pathway, called the **alternative pathway**, involves the binding of complement protein C3 directly to the pathogen cell surface. So, both pathways converge by the activation of the C3 convertase.

C3 convertase splits C3 into a C3a and C3b fragment. C3b activates another complement protein C5, which also is hydrolyzed into C5a and C5b fragments. Complement fragments C3a and C5a trigger an inflammatory response through vasodilation and increased capillary permeability. C3b also acts as an opsonin by binding to the pathogen surface (opsonization) and enhancing phagocytosis.

C5b plays one more important role. The complement fragment triggers another complement cascade that assembles into what are called **membrane attack complexes** (MACs). These complexes of complement proteins form large holes in the membranes of many microbes, especially gram-negative bacteria and enveloped viruses. The MACs are so prevalent that lysis occurs, killing the pathogen. This teamwork within the complement system and inflammation provides a formidable obstacle to pathogen invasion.

CONCEPT AND REASONING CHECKS

19.9 Explain the outcome to someone who has an infectious disease but has a genetic defect making the person unable to produce C5 complement protein.

Innate Immunity Depends on Receptor Recognition of Common Pathogen-Associated Molecules

KEY CONCEPT

- Pattern-recognition receptors trigger the innate immune response.

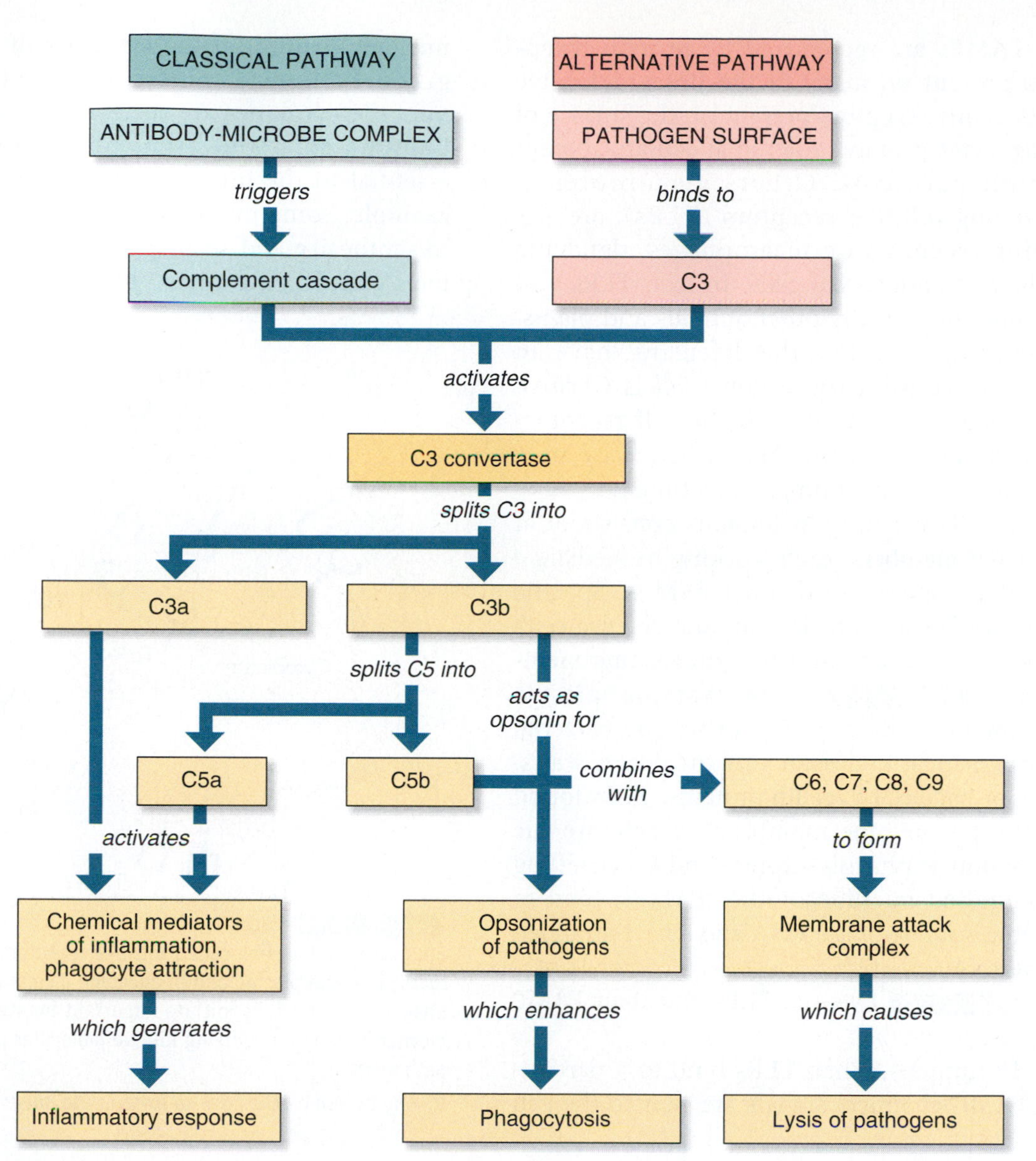

FIGURE 19.8 **A Concept Map of the Complement Pathways.** The early events of both the classical and alternative pathways of complement activation culminate in the activation of complement protein 3 (C3) and the effector functions of complement. Complement fragment C3b binds to the membrane and opsonizes bacterial cells, allowing phagocytes to internalize them (phagocytosis). The small fragments, C5a and C3a, are mediators of local inflammation. C5b also activates another complement cascade, whereby the terminal components of complement assemble into a membrane-attack complex (MAC) that can damage the membrane of gram-negative and virus-infected cells.

Q: What is the advantage of having two pathways leading to C3 activation?

The immune system responds to pathogen invasion by mounting a rapid innate response, followed several days later by the development of antibodies and lymphocytes as part of the acquired immune response. It is now clear that innate immunity is crucial for the generation of the acquired responses.

The innate immune system recognizes pathogens by identifying unique microbial molecular sequences not found on host cells. This system of recognition is called **pathogen-associated molecular patterns** (**PAMPs**) and includes small molecular sequences in such structures as the lipopolysaccharide (LPS) layer of gram-negative cell walls, peptidoglycans of gram-positive cell walls, flagella, fungal cell walls, and bacterial and viral nucleic acids. Currently, microbiologists believe innate immunity recognizes at least 1,000 different PAMPs.

PAMPs are recognized by protein receptors present on many of the host's defensive cells. Some receptors present on the surface of phagocytes promote microbial cell attachment for phagocytosis. Others glycoproteins, including **toll-like receptors (TLRs)**, are signaling receptors on macrophages, dendritic cells, and endothelial cells. In fact, TLRs also are present in many other animals and plants, suggesting they, like the defensins, have an ancient evolutionary origin. (*Toll* is German for "amazing," referring to the toll receptors first discovered in fruit flies where they are a fly's defense against fungal infections).

The TLR family in humans consists of at least ten members, each working to mediate a specific response to distinct PAMPs. Specific TLRs are located on the immune defense cell's plasma membrane and the phagosome membrane (FIGURE 19.9). For example, plasma membrane TLRs can bind to unique bacterial lipopeptides, zymosan in fungal cell walls, LPS, or bacterial flagellin proteins. TLRs found in the phagosome membrane accelerate the formation of phagolysosomes and faster killing of engulfed microbes. Other PAMPs, bind to double-stranded viral RNA and single-stranded viral RNA.

TABLE 19.3 lists the TLRs and their PAMP targets.

In humans, when TLRs bind to a particular PAMP, chemical signals are sent to the cell nucleus turning on the synthesis of cytokine genes. Following transcription and translation, the cytokines are secreted and sensed by immune cells and other nonimmune cells essential to the innate immune response. For example, some cytokines stimulate liver cells to synthesize and secrete defensive blood proteins called **acute phase proteins**, which ele-

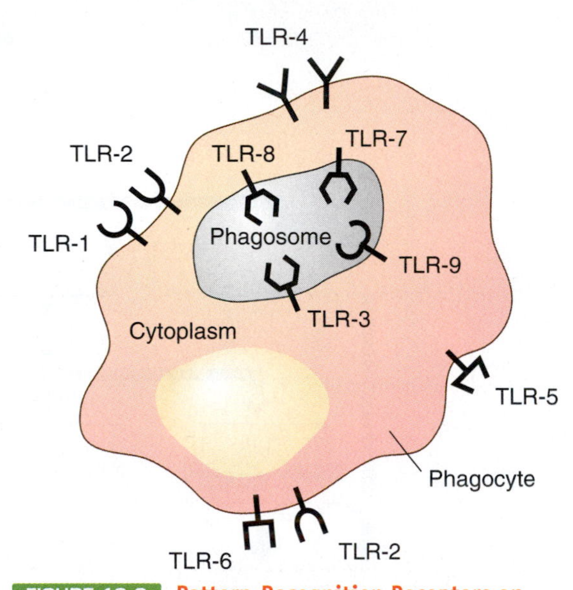

FIGURE 19.9 **Pattern-Recognition Receptors on Phagocytes.** Proteins called toll-like receptors (TLRs) are located on the plasma and phagosome membranes of phagocytes. The TLRs initiate important innate defense mechanisms by recognizing unique molecular patterns on pathogens.

Q: Why do some TLRs have to exist in the phagosome?

TABLE 19.3 The Recognized Human Toll-Like Receptors

Toll-Like Receptor	Location	PAMP (Molecular Sequences)
1+2	Plasma membrane	Bacterial cells (lipoproteins) Parasites (membrane proteins)
2+6	Plasma membrane	Gram-positive bacterial cell walls (peptidoglycan) Gram-positive bacterial cell walls (lipoteichoic acids) Fungal cell walls (polysaccharide)
3	Phagosome membrane	Viruses (double-stranded RNA)
4	Plasma membrane	Gram-negative bacterial cell walls (LPS)
5	Plasma membrane	Bacterial flagella (flagellin protein)
7, 8	Phagosome membrane	Viruses (single-stranded RNA)
9	Phagosome membrane	Bacterial and viral DNA (CpG sites: regions where a cytosine nucleotide occurs next to a guanine nucleotide in one polynucleotide strand)

vate inflammation and activate the complement cascade. TLR activation to PAMPs in macrophages and dendritic cells also induce these cells to differentiate into cells active in the acquired immune response.

Once again, the TLR response must be regulated. An ineffective response can leave the body wide open to infection and disease. An excessive response by TLRs and cytokines can bring about immune disorders, including heart disease and rheumatoid arthritis.

CONCEPT AND REASONING CHECKS

19.10 Justify the need for the spectrum of toll-like receptors found on and in human immune cells.

In summary, physical barriers, immune cells, antimicrobial chemicals, and signaling chemicals to needed are protect the human host, generate an effective innate immune response, and induce the acquired immune response.

If the physical barriers of skin or mucous membranes are breached, innate immunity involves several cells and chemical mediators:

- Macrophages, neutrophils, and dendritic cells (phagocytosis; cytokine production and secretion);
- Mast cells (histamine secretion);
- Natural killer cells (perforins and granzymes);
- Defensive proteins (defensins, complement, acute phase proteins); and
- Signaling receptors and proteins (toll-like receptors and cytokines).

With the innate response behind us, we now are ready in the next chapter to examine the events inducing acquired immunity and how this specific immune response works. MicroInquiry 19 investigates some alternative immune system outcomes.

MICROINQUIRY 19

Immune System Outcomes

As potent as the immue system can be, genetic deficiencies within the host can make the immune system less efficient, and in some cases can have overt consequences in attempting to fight infection and disease caused by pathogens.

Let's look at three cases. Based on your understanding of innate immunity described in this chapter, answer the questions posed for the three cases of immune system dysfunction within the host. Answers can be found in **Appendix D.**

Case 1

A 35-year-old female has a normally functioning immune system. She has had the typical flus and colds in her life and an occasional bacterial infection. In all cases, she has recovered without medical intervention. In one of the bacterial infections from which she recovered, she was infected with two species of gram-negative bacteria. One species contained a capsule and the other did not have a capsule.

19.1a. Describe how the complement system will respond to these two bacterial species in terms of membrane-attack complexes.

19.1b. Explain how the complement system will facilitate the disposal of these two species.

Case 2

An 8-year-old boy suffers from frequent and severe bacterial infections. The family physician thinks he suffers from agammaglobulinemia (absence of antibodies). The tests performed indicate he has normal immunoglobulin levels. However, in doing the blood work, a deficiency in complement protein C3 was discovered.

19.2a. What is the role of C3 in the normal complement cascade?

19.2b. Explain why a deficiency in C3 causes the frequent bacterial infections experienced by the patient.

Case 3

A boy was delivered by cesarean section without complications. However, by the 25th day, he had developed a cough, fever, and rapid breathing. Antimicrobial drug treatment was without effect. By four months, the child had developed pneumonia as identified by a computed tomography scan of the lungs. Respiratory syncytial virus was recovered.

19.3a. How could an absence of macrophages lead to this disease?

19.3b. What other types of pathogens could pose an infective threat to this boy?

SUMMARY OF KEY CONCEPTS

19.1 An Overview to Host Immune Defenses

- In terms of infection, the most important cells are the leukocytes, which consist of the neutrophils, eosinophils, basophils, monocytes, dendritic cells, and lymphocytes (NK cell, B lymphocytes, and T lymphocytes).
- The lymphatic system functions in disease to filter and trap pathogens in lymph nodes populated with B cells, T cells, macrophages, and dendritic cells.
- A fully functional immune defense system consists of the innate immune response that is genetically-encoded from birth and the acquired immune responses that one develops during one's lifetime and exposure to pathogens.

19.2 The Innate Immune Response

- The skin and mucous membranes are mechanical barriers, while chemical defenses include the acidic environment in the urinary tract and stomach. Defensins damage cell membranes and lysozyme and interferon destroy bacterial cells and block virus replication, respectively. The normal microbiota limits pathogen spread and infection.
- Phagocytosis by phagocytes (neutrophils, macrophages, dendritic cells) internalizes and kills pathogens.
- Inflammation is a defensive mechanism to tissue injury by pathogens. Chemical mediators, including cytokines, trigger vasodilation and increase capillary permeability at the site of infection. The inflamed area exhibits redness, warmth, swelling, and pain.
- Moderate fever inhibits the growth of pathogens while increasing metabolism for tissue repair and enhancing phagocytosis.
- Natural killer (NK) cells are defensive lymphocytes acting nonspecifically on virus-infected cells and abnormal host cells.
- Complement proteins enhance innate defense mechanisms and lyse cells with membrane attack complexes.
- Toll-like receptors are responsible for the ability of innate immunity be recognize nonspecific molecular patterns of pathogens.

LEARNING OBJECTIVES

After understanding the textbook reading, you should be capable of writing a paragraph that includes the appropriate terms and pertinent information to answer the objective.

1. Summarize the functions for the six groups of leukocytes.
2. Compare and contrast the primary and secondary lymphoid tissues in terms of function.
3. Distinguish between innate and acquired immunity.
4. Describe how mechanical barriers protect against pathogen invasion.
5. List the types of chemical barriers that provide a poor environment for pathogen colonization and infection.
6. Explain the role of defensins in innate defense.
7. Assess the importance of phagocytosis as a nonspecific defense mechanism.
8. Outline the steps in the uptake and killing of a bacterial pathogen.
9. Explain why the inflammatory response is a key factor in the identification of and defense against an invading microbe.
10. Describe the role of cytokines in an inflammatory response.
11. Assess the value of a low grade fever in the nonspecific responses to infection.
12. Discuss how natural killer (NK) cells would identify and kill virus-infected cells in a host.
13. Identify what is unique and what is common in the complement pathways activated in response to a bacterial infection.
14. Propose what microbes would most likely be susceptible membrane attack complexes.
15. Explain the role of toll-like receptors to innate immunity.

SELF-TEST

Answer each of the following questions by selecting the ***one*** answer that best fits the question or statement. Answers to even-numbered questions can be found in **Appendix C**.

1. All of the following are innate immune defenses *except*:
A. antibodies.
B. the skin.
C. phagocytosis.
D. interferon.
E. lysozyme.

2. Which one of the following is *not* able to carry out phagocytosis?
A. Neutrophils
B. Macrophages
C. Eosinophils
D. Basophils
E. Monocytes

3. Which pair of cells represents granulocytes?
A. Basophils and lymphocytes
B. Neutrophils and dendritic cells
C. Neutrophils and eosinophils
D. Eosinophils and monocytes
E. Lymphocytes and monocytes

4. The secondary lymphoid tissues include the _____ and _____.
A. thymus; bone marrow
B. bone marrow; tonsils
C. spleen; thymus
D. thymus; lymph nodes
E. spleen; lymph nodes

5. Which type of leukocyte would *not* be found in a lymph node?
A. B cell
B. Dendritic cell
C. Mast cell
D. Macrophage
E. T cell

6. These small, antimicrobial peptides are secreted throughout the body and kill pathogens by damaging cell membranes.
A. Bile
B. Interferons
C. Mucus
D. Defensins
E. Lysozymes

7. Which one of the following statements is *not* true of innate immunity?
A. It is an early-warning system against pathogens.
B. It is a form of immunity found only in vertebrates.
C. It is a nonspecific response.
D. It responds within minutes to many infections.
E. Metchnikoff was the first to study an aspect of the response.

8. The stomach is a chemical barrier to infection because the stomach
A. contains bile.
B. harbors *Helicobacter pylori,* a member of the host microbiota.
C. contains defensins.
D. possesses defensive cells.
E. has an acid pH.

9. Which one of the following is the correct sequence for the events of phagocytosis?
A. Cell attachment, acidification, phagosome formation, lysosome degradation
B. Cell attachment, phagosome formation, acidification, lysosome degradation
C. Phagosome formation, cell attachment, acidification, lysosome degradation
D. Acidification, phagosome formation, cell attachment, lysosome degradation
E. Cell attachment, phagosome formation, lysosome degradation, acidification

10. Which characteristic sign of inflammation (**A–D**) is *not* correctly associated with its cause? If all are correct, select **E**.
A. Edema—nerve damage
B. Heat—blood warmth
C. Swelling—fluid accumulation
D. Redness—blood accumulation
E. All the above (**A–D**) are correct.

11. Diapedesis refers to
A. blood vessel dilation.
B. fluid accumulation at the infection site.
C. blood vessel constriction.
D. neutrophil migration out of the capillaries.
E. the events that raise the host's body temperature.

12. Pyrogens are
A. proteins affecting the hypothalamus.
B. bacterial fragments.
C. fever-producing substances.
D. a type of cytokine.
E. All the above (**A–D**) are correct.

13. Natural killer (NK) cells kill by secreting
A. lysozyme.
B. MHC proteins.
C. granzymes.
D. defensins.
E. interferons.

14. Which one of the following is *not* a function of complement?
A. Stimulation of inflammation.
B. Lysis of pathogens.
C. Stimulation of antibody formation.
D. Formation of membrane attack complexes.
E. Heightened level of phagocytosis.

15. _____ bind to _____ on microbial invaders.
A. Toll-like receptors; PAMPs
B. Macrophages; cytokines
C. Mast cells; histamine
D. Toll-like receptors; complement
E. Macrophages; defensins

QUESTIONS FOR THOUGHT AND DISCUSSION

Answers to even-numbered questions can be found in **Appendix C**.

1. The opening of this chapter suggests that for many diseases, a penetration of the mechanical barriers surrounding the human body must take place. Can you think of any diseases where penetration is not a prerequisite to illness?
2. It has been said that no other system in the human body depends and relies on signals as greatly as the immune system. From the discussion of innate immunity, what evidence can you offer to support or reject this concept?
3. Phagocytes have been described as "bloodhounds searching for a scent" as they browse through the tissues of the body. The scent they usually seek is a chemotactic factor, a peptide released by bacterial cells. Does it strike you as unusual that bacterial cells would release a substance to attract the "bloodhound" that will eventually lead to the bacterial cell's demise? Explain.

APPLICATIONS

Answers to even-numbered questions can be found in **Appendix C**.

1. A roommate cuts his finger and develops an inflammation at the cut site. Having taken microbiology, he asks you to explain exactly what is causing the throbbing pain and the warmth at the cut site. How would you reply to your roommate's question?
2. On a windy day, some dust blows in your eye and your eye waters. Why does your eye water when something gets into your eye and how does this relate to immune defenses?
3. A friend is ill with an infection and asks you why she has "swollen glands" behind the jaw. As a microbiology student, what would you tell her?

REVIEW

To test your knowledge of the immune response and innate immunity, match the statement on the left to the term on the right by placing the correct letter in the available space. A term may be used once, more than once, or not at all. Answers to even-numbered statements are listed in **Appendix C**.

Statement

____ **1.** Peyer patches are part of this tissue.

____ **2.** Recognizes cancer and virus-infected cells.

____ **3.** A basophil is similar to this cell.

____ **4.** A flattened organ found in the upper left of the abdomen.

____ **5.** Small protein released by various defensive cells in response to an activating substance.

____ **6.** A cytotoxic enzyme that induces apoptosis.

____ **7.** A T cell or B cell is one.

____ **8.** A monocyte matures into one of these.

____ **9.** Found in human tears and sweat.

____ **10.** Also called a PMN.

____ **11.** Forms holes in a target cell membrane.

____ **12.** Septic shock can be caused by "overdoses" of this.

____ **13.** Is called a Langerhans cell if it is found in the skin.

____ **14.** Refers to a substance causing fever.

____ **15.** A short-lived phagocyte.

Term

A. Agranulocyte
B. Cytokine
C. Dendritic cell
D. Granulocyte
E. Granzyme
F. Interferon
G. Lymphocyte
H. Lysozyme
I. MALT
J. Macrophage
K. Mast cell
L. Monocyte
M. Natural killer (NK) cell
N. Neutrophil
O. Perforin
P. Pyrogen
Q. Spleen
R. Thymus

HTTP://MICROBIOLOGY.JBPUB.COM/

The site features learning, an on-line review area that provides quizzes and other tools to help you study for your class. You can also follow useful links for in-depth information, read more MicroFocus stories, or just find out the latest microbiology news.

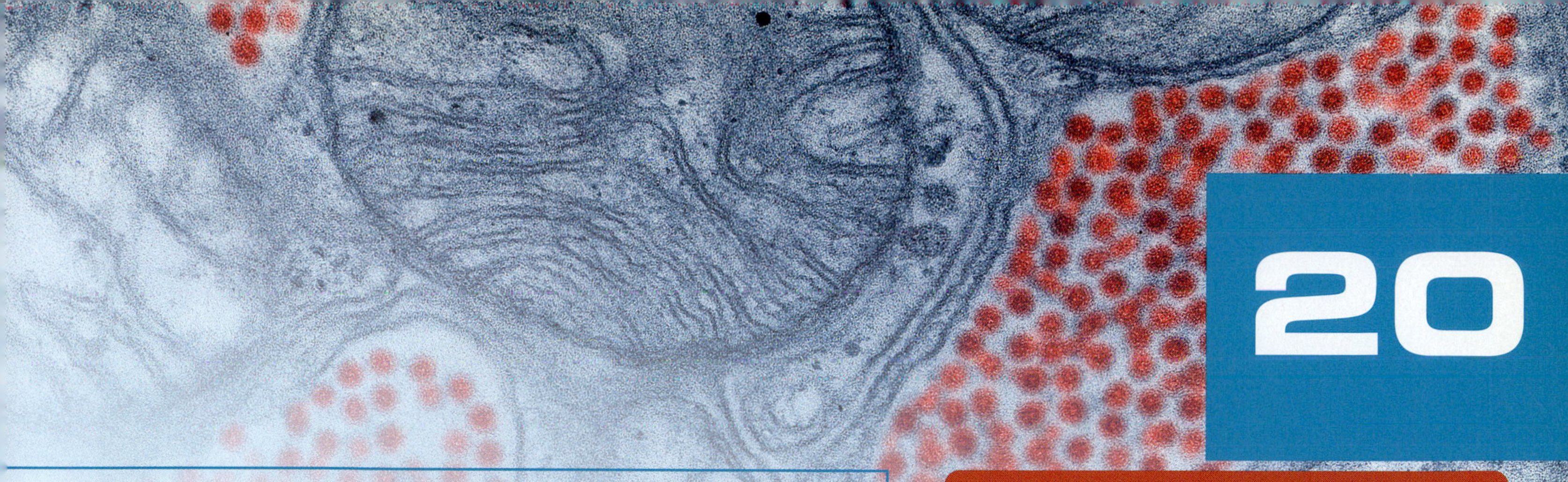

Resistance and the Immune System: Acquired Immunity

Chapter Preview and Key Concepts

20.1 An Overview of the Acquired Immune Response
- Acquired immunity responds to, distinguishes between, and remembers specific pathogens it has encountered.
- Acquired immunity involves humoral and cell mediated responses through receptor recognition of "nonself."
- Immune recognition only activates the appropriate B and T cell clones that recognize "nonself" epitopes.
- Hematopoietic stem cells give rise to all blood cells.

20.2 The Humoral Immune Response
- Antibodies are a group of protein molecules circulating in the body's fluids that recognize specific epitopes.
- Immunoglobulin classes differ in their heavy chains of the constant region.
- Memory cells produce a rapid antibody response.
- The mixing and matching of immunoglobulin gene segments produces unique antibodies.
- Antigen-antibody complexes are cleared by phagocytes.

20.3 The Cell Mediated Immune Response
- Protein receptors found on different populations of T cells define their role in cellular immunity.
  MicroInquiry 20: Can Thinking "Well" Keep You Healthy?
- Naive T-cell activation stimulates memory cell and helper T-cell production.
- Activated cytotoxic T cells seek out and kill virus-infected and abnormal cells.
- The active T_H2 cells co-stimulate B-cell activation.

Identifying HIV was the critical first step in defining the cause of AIDS, but, as Robert Koch so elegantly pointed out more than a century ago, showing that a particular infectious agent causes a specific disease can be an arduous process.
—Stanley Prusiner in *Historical Essay: Discovering the Cause of AIDS*

The great breakthroughs in any field of science are infrequent. Although most experiments advance our understanding of a phenomenon, more often than not they are the result of a "yes" or "no" answer. But once in awhile an experiment or observation not only gives a yes or no answer, it also is unexpected and makes a giant leap forward in scientific understanding. It is these rare "aha" moments, often coming after much debate, for which every research scientist hopes. The observations and work carried out by Pasteur and Koch on the germ theory is just one historical breakthrough that comes to mind.

In 1981, an immunodeficiency disease among gay men was reported. Once there was an alarming rise in the number of new cases of what became known as acquired immunodeficiency syndrome (AIDS), the search was on to discover its cause and how the disease propagated itself.

The appearance of AIDS in distinctly different populations including homosexual and heterosexual individuals, children, intravenous drug users, hemophiliacs, and blood transfusion recipients suggested it must be an infectious agent. And as the number of AIDS cases continued to rise, so did the hypotheses about its possible causation.

Luc Montagnier and his colleagues at the Pasteur Institute in Paris were the first to report the discovery of the virus now called the human immunodeficiency virus (HIV). Coupled with this breakthrough was the work of Robert Gallo and his research team at the National Institutes of Health in Bethesda, Maryland. They were able to show that the virus identified by Montagnier was the cause of AIDS. Yet despite all of the evidence confirming Montagnier and Gallo's findings, it would take more than ten years before HIV would be recognized by most scientists as the cause of AIDS.

So, what was the virus doing to cause the disease? As the clinical descriptions of AIDS solidified in the early 1980s, it became apparent the virus caused a dramatic decrease in the number of T lymphocytes in infected individuals. This supported an earlier discovery by Gallo's group showing that the cytokine interleukin-2 was necessary to support HIV replication. In fact, interleukin-2 stimulation made possible the growth and proliferation of HIV in cultured T lymphocytes (FIGURE 20.1).

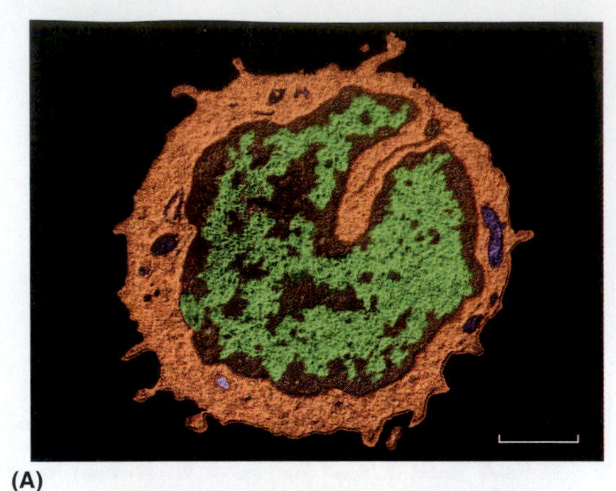

(A)

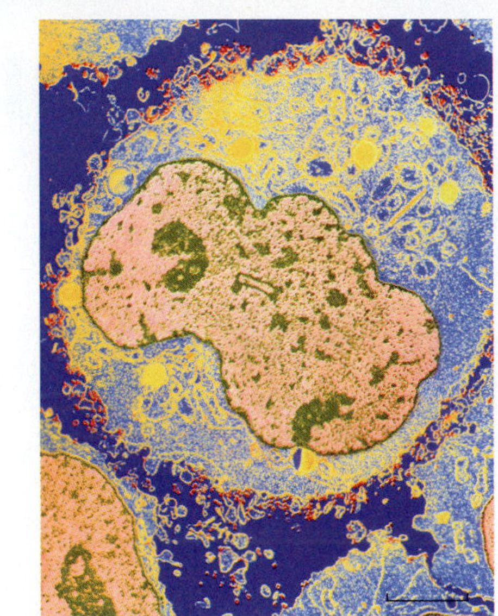

(B)

FIGURE 20.1 Uninfected and HIV-Infected T Lymphocytes. False-color transmission electron micrographs of T lymphocytes. (**A**) An uninfected T cell showing the typical large cell nucleus. (**B**) An infected T cell showing HIV virions (red) budding off the cell surface. (A & B: Bar = 2 μm.)

Q: Why is it significant that HIV primarily infects CD4 T cells?

Scientific breakthroughs not only advance a field but also spin off, in this case, many medical applications. Finding the cause of AIDS led to the development of a blood test to identify HIV, preventing millions of people from being infected through the transfusion of tainted blood. Knowing how the virus interacts and replicates in T cells stimulated the identification and design of many antiretroviral drugs to combat the disease, allowing infected individuals to live longer and more productive lives.

Most scientists today would agree that the discoveries made by Montagnier and Gallo rank as one of the major scientific breakthroughs of the 20th century and set the stage for influential discoveries made by many others.

If one set out to design a virus to cripple the immune system and the very cells designed to defend against the virus, one could not top the naturally evolved HIV. By destroying what are called CD4 T cells, the entire immune system is essentially nonfunctional and unable to respond to any infectious disease exposure.

How is this possible? How can a simple virus inactivate a complex and powerful immune defense?

This chapter provides the answer as we examine how the actions of innate immunity, discussed in the previous chapter, supply the acquired immune response with the needed boost to fight pathogens.

20.1 An Overview of the Acquired Immune Response

With regard to infectious disease, innate immunity, the topic of the previous chapter, is designed to eliminate pathogens—or at least limit their spread until the other half of the immune response, acquired (specific) immunity, can mobilize the cells and molecules to eliminate the pathogen. Before we consider the components and events of acquired immunity, it is best to consider the key features of acquired immunity.

The Ability to Eliminate Pathogens Requires a Multifaceted Approach

KEY CONCEPT

- Acquired immunity responds to, distinguishes between, and remembers specific pathogens it has encountered.

Acquired immunity is a defense against a tremendous variety of potential pathogens. This defense is based on four important attributes.

Specificity. As we saw in the previous chapter, innate immunity has the ability to recognize a diversity of foreign substances, including pathogens. Such microbes or their molecular parts that are capable of mobilizing the immune system and provoking an immune response are called **antigens** (FIGURE 20.2). Most antigens represent toxins, viruses, cells, or large, complex molecules not normally found in the body and are consequently referred to as "nonself."

Antigens do not stimulate the immune system directly. Rather, the immune system recognizes small parts of the antigen called **antigenic determinants**, or **epitopes** (FIGURE 20.3). An antigen may have numerous epitopes. For example, a structure such as a bacterial flagellum may have hundreds of epitopes, each having a characteristic three-dimensional shape. It is the ability of the immune system to identify these epitopes that generates the specificity possessed by acquired immunity. Epitopes are unique and specific microbial fingerprints to which the immune system responds.

In addition, the molecular make up and assembly of epitopes represents tremendous diversity. Proteins are the most potent antigens because their amino acids have the greatest array of building blocks. Polysaccharides are less potent antigens than proteins because they lack chemical diversity and rapidly break down in the body. Nucleic acids and lipids also can be antigenic. It is estimated that the human immune system can respond to at least 10 million diverse epitopes. If the body loses its ability to respond to antigens and antigenic determinants, an **immune deficiency** will occur (Chapter 22).

Tolerance of "Self." Under normal circumstances, one's own cells and molecules

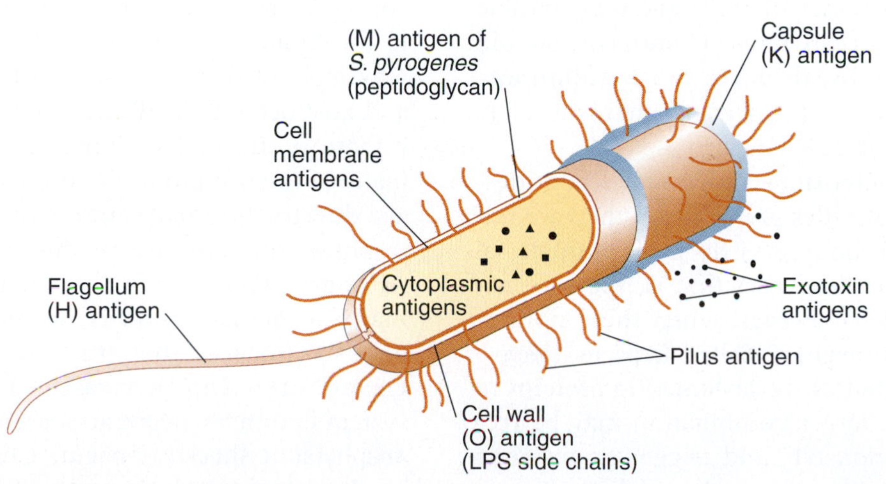

FIGURE 20.2 **The Various Antigens Possible on a Bacterial Cell.** Each different antigen in this idealized bacterial cell is capable of stimulating an immune response.

Q: Using this figure, what do you know about the Escherichia coli *strain called 0157:H7?*

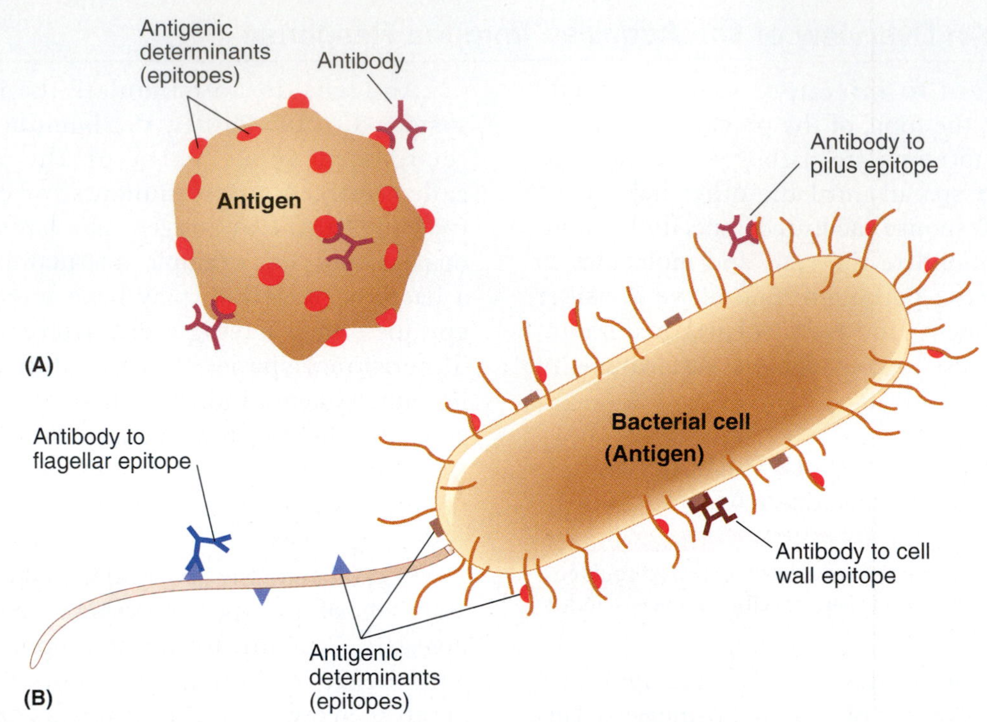

FIGURE 20.3 **Antigens and Antigenic Determinants.** (**A**) An antigen (toxin or virus) showing a number of antigenic determinants. These are the sites where products of the immune system, such as antibodies, will interact. (**B**) A bacterial cell with various sources of antigens and the antibodies that react with the antigenic determinants on the antigens. The antibody structures drawn in this chapter are based on the structure shown in Figure 20.7

Q: Why are there different antibodies seen on these antigens?

with their antigenic determinants do not stimulate an immune response. This failure occurs because the epitopes are interpreted as "self." A population of T cells called **regulatory T cells** prevents other T cells from attacking self. In fact, those self T cells are destroyed through apoptosis. The immune system thereby develops a tolerance of "self" and remains able to respond only to "nonself" antigens. Should self-tolerance break down, an **autoimmune disease**, such as lupus erythematosus, may occur (Chapter 22).

Many nucleotides, hormones, peptides, and other molecules are so small they are tolerated by the body and do not stimulate an immune system defense; that is, they are not **immunogenic**. However, when these nonimmunogenic molecules, called **haptens** (*hapt* = "fastened") are linked (fastened) to proteins in the body, the larger combination may be recognized as "nonself" and trigger an immune system response. Examples of haptens are penicillin molecules, molecules in poison ivy plants, and molecules in certain cosmetics and dyes. These often are the cause of allergies. Some vaccine antigens also are haptens but when chemically linked to a protein carrier, they become immunogenic (Chapter 21).

■ Anaphylactic shock: A sudden drop in blood pressure, itching, swelling, and difficulty in breathing typical of some allergic reactions.

Minimal "Self" Damage. In an immune response, it is important that the response be strong enough to eliminate the pathogen, yet controlled so as not to cause extensive damage to the tissues and organs of the body. That is why most of the encounters between microbe and host occur at local sites, such as the actual infection site and in lymph nodes. In fact, many of the symptoms of an infectious disease are due to the innate and acquired immune responses and not directly due to the invading pathogen. One reason the avian flu kills so many human victims is because the virus over-stimulates the immune system (see Chapter 14). In other cases, if the immune system extremely overreacts, a potentially fatal anaphylactic shock can occur (Chapter 22).

Immunological Memory. Most of us during our lives have become immune to certain diseases, such as chickenpox or measles, from which we have recovered or to which we have been immunized. This long-term ability to "remember" past pathogen exposures is called **immunological memory**. Thanks to the acquired memory, a second or ensuing exposure to the same pathogen produces such a rapid and potent response that the person does not even know they were infected. A loss of immunological memory can lead to recurring infectious diseases.

CONCEPT AND REASONING CHECKS

20.1 Justify the statement, "The four characteristics of acquired immunity are really variations of specificity."

Acquired Immunity Generates Two Complementary Responses to Most Pathogens

KEY CONCEPT

- Acquired immunity involves humoral and cell mediated responses through receptor recognition of "nonself."

The cornerstones of acquired immunity are the lymphocytes. Lymphocytes are small cells, about 10 to 20 μm in diameter, each with a large cell nucleus taking up almost the entire space of the cytoplasm. Viewed with the microscope, all lymphocytes look similar. However, two types of lymphocytes can be distinguished on the basis of developmental history, cellular function, and unique molecular markers. The two types are B lymphocytes and T lymphocytes. **B lymphocytes (B cells)** are involved in the process by which antibodies are produced against epitopes. **T lymphocytes (T cells)** provide resistance through direct cell-to-cell contact with and lysis of infected or otherwise abnormal cells (for example, cancer cells). MicroFocus 20.1 suggests how one might help boost their T cell population to stay healthier.

Pathogens within the host trigger the innate response, which in turn launches two complementary acquired immune responses: the humoral and cell mediated immune responses, both dependent on lymphocyte activity.

Humoral Immune Response. One of the immune behaviors to antigens is called the **humoral immune response** (*humor* = "a fluid"). The result of this response is the activation of B cells and the ensuing production of antibodies that recognize epitopes on the identified antigen in the blood or lymphatic fluids. The response is so specific that the body can generate antibodies to just about any antigen or epitope it encounters.

Cell Mediated Immune Response. Should microbes or antigens leave the body fluids and enter cells, antibodies are useless. Therefore, another immune behavior, called the **cell mediated immune response**, becomes activated to eliminate "nonself" cells, such as virus-infected cells or cancer cells. T cells are involved in these activities.

Both T cells and B cells have surface receptors spanning the width of the plasma membrane (FIGURE 20.4). After these receptors recognize and bind antigens or antigenic fragments, the lymphocyte is said to be "committed." In B cells, the receptor protein is an antibody molecule. In T cells, the receptor protein at the cell surface is composed of two antibody-like chains of glycoproteins. About 100,000 surface receptors are found on each B or T cell.

The presence of highly specific receptor proteins on the lymphocyte surface implies that even before an antigen enters the body, immunocompetent B and T cells are already waiting "in the wings." Moreover, the genetic code for synthesizing the surface receptor is present in an individual even though the individual has not been exposed to that antigen. We may never experience malaria, for example, yet we already have surface receptor proteins for recognizing and binding to the antigens of malaria parasites.

■ Immunocompetent: Referring to lymphocytes capable of reacting with a specific epitope.

CONCEPT AND REASONING CHECKS

20.2 How do the two arms of acquired immunity differ from one another?

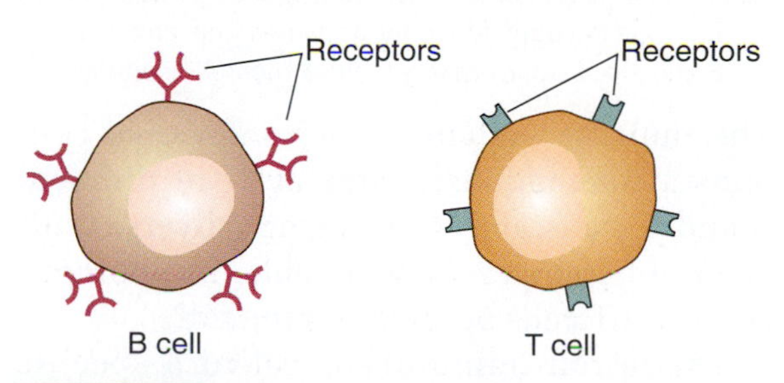

FIGURE 20.4 Receptors of B and T Lymphocytes. (**A**) B cells have antibodies on their surface that perform receptor functions by binding to antigen (epitope). (**B**) T cells have an antibody-like receptor to bind antigenic fragments.

Q: How does the characteristic of specificity apply to B and T cells?

MICROFOCUS 20.1: Being Skeptical

Starving the Apothecary

The healthy aspect of tea has been appreciated for centuries, but within the last 15 years scientific evidence has accumulated to back up the age-old claim. It is known that tea contains flavonoids and antioxidants that researchers believe can inhibit the formation of cancer cells as well as having positive protective effects on the cardiovascular system by preventing atherosclerosis.

In 2003, Jack Bukowski at Harvard Medical School's Brigham and Woman's Hospital in Boston published a paper suggesting tea also can help defend against infectious disease. It seems just three to five cups of ordinary tea made from tea leaves each day will do the trick. Really? What's the evidence?

Bukowski's team was studying the actions of a special subset of T cells called gamma delta T cells. When exposed to ethylamine, the T cells secrete large amounts of interferon, which is involved in inhibition of viral replication (see Chapter 13). Ethylamine is found in some bacterial cells, fungi, and parasites. Looking for other sources of the chemical, Bukowski discovered that tea also contains the same chemical.

Further experiments showed that exposing blood to these chemicals (simulated infection) in the test tube could increase the numbers of gamma delta T cells by up to five times. In contrast, human blood cells not exposed to the chemical showed a much less significant cell response.

Using human volunteers, the investigators had 21 non-tea-drinkers drink either five to six small cups of black tea or five to six small cups of instant black coffee daily for four weeks. Coffee doesn't contain the chemical. The results indicated the blood of the tea-drinking volunteers had a fivefold increase in interferon over the course of the study. Coffee drinking had no such effect.

So, it appears drinking tea can bring gamma delta T cells to a state of readiness. Should infection then occur, they are ready to defend the body.

Whether such stimulation would actually be useful in fighting a real infection remains to be discovered. But at least these early findings present intriguing evidence that tea can affect the immune system.

Perhaps the Chinese proverb is correct: "Drinking a daily cup of tea will surely starve the apothecary."

Clonal Selection Activates the Appropriate B and T Cells

KEY CONCEPT

- Immune recognition only activates the appropriate B and T cell clones that recognize "nonself" epitopes.

The immune system of each individual contains a tremendously large array of different B and T cells able to recognize diverse antigens and epitopes. How do only the appropriate B and T cells become committed?

Acquired immunity is called a specific response because only the specific B and T cells having receptors that recognize epitopes on a specific pathogen are committed. The vast majority of the B and T cells remain unreactive. The explanation for commitment was first proposed in the 1950s by Nobel laureates Frank MacFarlane Burnet and Peter Medawar. The theory, called **clonal selection**, suggested that exposure to an antigen only activates those T and B cells with receptors recognizing specific epitopes on the antigen. A detailed discussion of the contemporary theory follows for B cells.

Activation of the appropriate B cells begins when antigens enter a lymphoid organ, bringing the antigenic determinants close to the appropriate B cells (FIGURE 20.5). Antigenic determinants must match with surface receptor proteins on the B lymphocytes and then crosslink antibody molecules. Therefore, haptens may be poor stimulators of B cells because of their inability to perform cross-linking. However, when cross-linking does occur, it

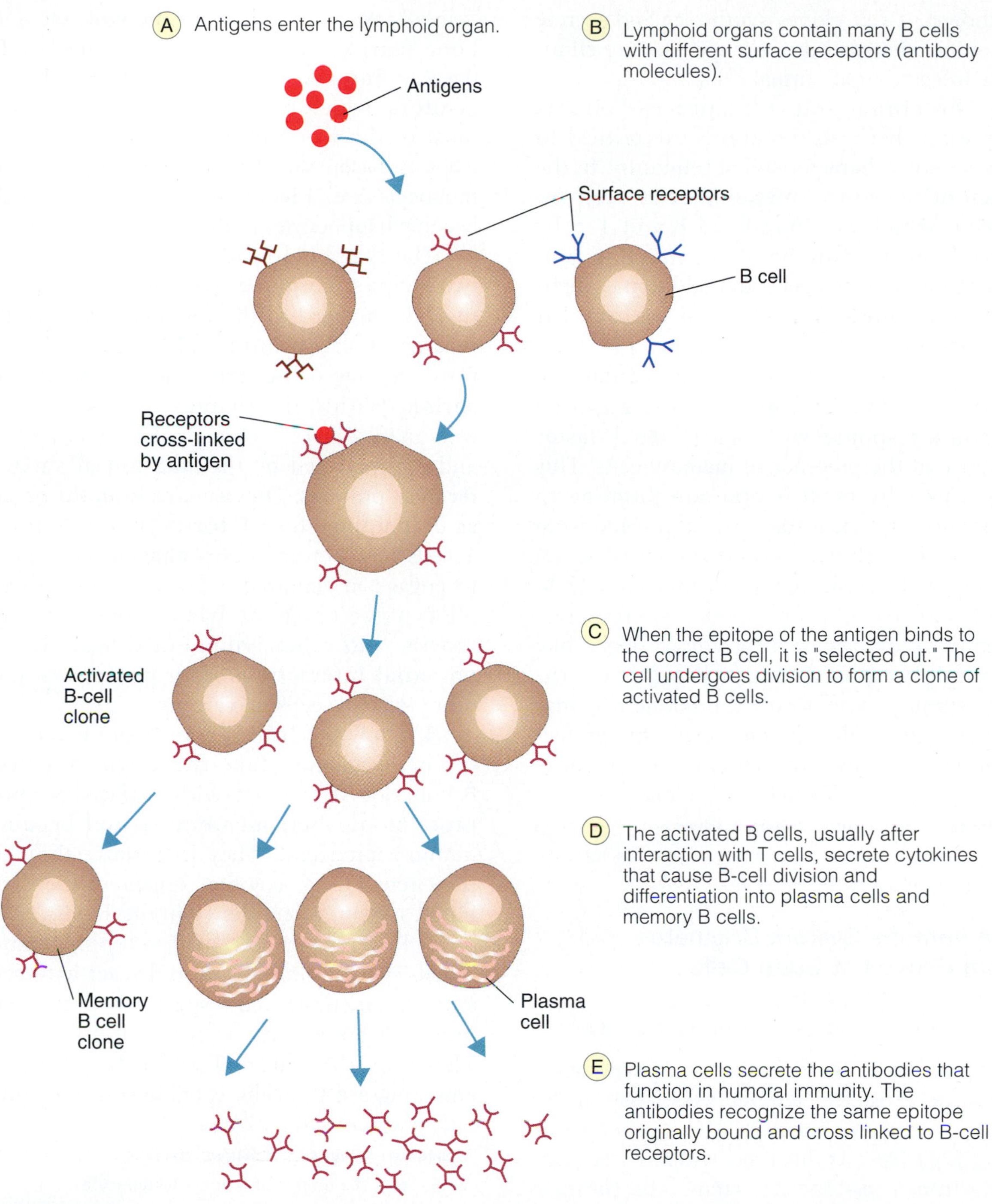

FIGURE 20.5 Clonal Selection of B Cells. In clonal selection, only those B cells that have a surface receptor that matches the shape of an antigenic determinant will be stimulated to divide, producing a clone of identical B cells.
Q: What is the type of effector cell produced from a B cell clone?

stimulates the B cell to divide to form a **clone**; that is, a population of genetically-identical B cells. A similar set of events occurs to activate T cells, only the T-cell receptor recognizes a specific antigenic fragment presented to it. This process will be described later.

Both B- and T-cell clones contain activated lymphocytes that will develop into effector cells and memory cells. **Effector cells** target the pathogen. For example, the B cell clones develop into **plasma cells**, which synthesize and secrete antibodies against the invading

■ **Undifferentiated:** Referring to cells that have yet to acquire specific characteristics.

pathogen. T-cell clones synthesize and secrete cytokines that will provoke reactions to eliminate infected or abnormal cells.

The clonal selection process also is important because the events are critical to the memory characteristic of immunity. In the immune response, **memory cells** are produced, which are long-lived B and T cells capable of dividing on short notice to produce more effector cells and additional memory cells. Memory cells can survive for decades in the body.

Should a second or ensuing contact be made with the same antigen, the acquired immune response will kick in much faster because of the presence of memory cells. This explains why most people are immune to many diseases once they have recovered from diseases like chickenpox or the measles. Yet having had measles does not prevent a child from contracting chickenpox. Likewise, immunization against a specific pathogen like the diphtheria bacterium or hepatitis B virus only stimulates an acquired immune response to the agent in the vaccine. Again, the protection acquired by experiencing one of these agents is specific for that agent alone.

CONCEPT AND REASONING CHECKS

20.3 Compare and contrast effector cells and memory cells in the clonal selection of B cells.

The Immune System Originates from Groups of Stem Cells

KEY CONCEPT

- Hematopoietic stem cells give rise to all blood cells.

All immune cells, including the phagocytes, dendritic cells, and lymphocytes, arise in the fetus about two months after conception (FIGURE 20.6). At this time, lymphocytes originate from hematopoietic **stem cells** (*hema* = "blood"; *poiet* = "make") in the yolk sac and bone marrow. These undifferentiated cells develop into two types of cells: **myeloid progenitors**, which become red blood cells and most of the white blood cells (that is, neutrophils, basophils, mast cells, eosinophils, and monocytes); and **lymphoid progenitors**, which become lymphocytes of the immune system.

■ **Hematopoietic:** Referring to cellular components formed in the blood.

The lymphoid progenitors take either of two courses. From the bone marrow, some of the cells proceed to the thymus. The thymus is large in size at birth and increases in size until the age of puberty, when it begins to shrink. Within the thymus, the progenitor cells mature over a two- or three-day period and are modified by the addition of surface receptor proteins. They emerge from the organ as T lymphocytes (T for thymus). Mature T cells now are immunocompetent and ready to engage in acquired immunity. The T lymphocytes colonize the lymph nodes, spleen, tonsils, and other lymphoid tissues where potential interactions with pathogens and other immune cells may occur.

The B lymphocytes (B for bone) mature in the bone marrow. Like the T lymphocytes, B lymphocytes mature with surface receptor proteins on their membranes and become immunocompetent. They then move through the circulation to colonize organs of the lymphoid system, where they join the T cells.

A large percentage of developing T and B cells, which normally would react with self antigens, are destroyed respectively in the thymus and bone marrow through apoptosis. Therefore, the mature T and B lymphocytes emerging are the cells capable of interacting only with nonself antigens.

CONCEPT AND REASONING CHECKS

20.4 Trace the origins of B and T lymphocytes.

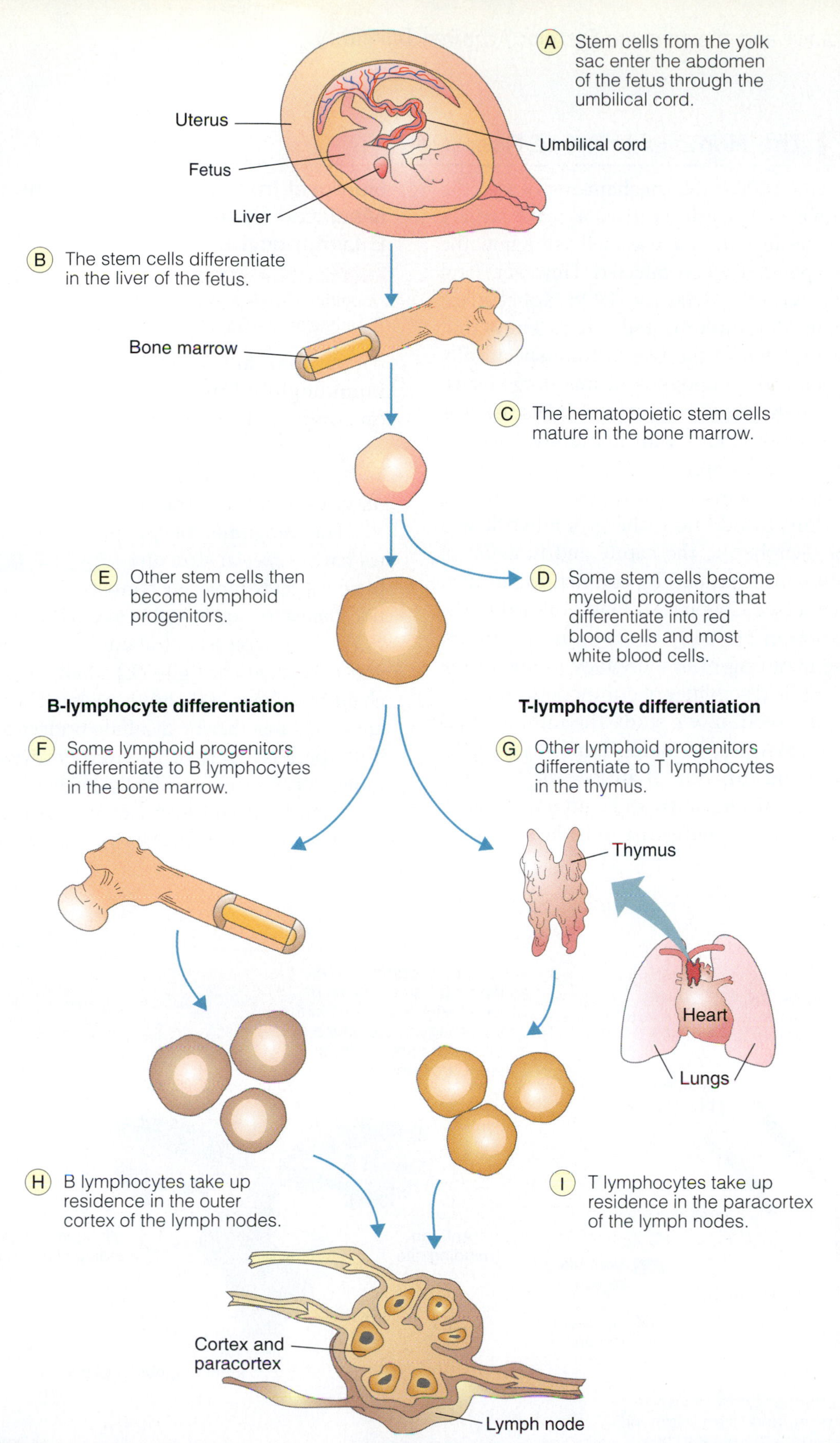

FIGURE 20.6 **The Origin of B and T Lymphocytes.** Lymophocytes arise in the bone marrow and mature in the bone marrow (B cells) or thymus (T cells).

Q: Why are the letters "B" and "T" used for different groups of lymphocytes?

20.2 The Humoral Immune Response

In the late 1800s, the mechanisms of specific resistance to infectious disease were largely obscure because no one was really sure how the body responded when infected. However, Paul Ehrlich, who received the 1908 Nobel Prize with Elie Metchnikoff, and others knew that certain proteins of the blood unite specifically with chemical compounds of microorganisms. By the 1950s, specific resistance to disease was virtually synonymous with immunity. Vaccines were available for numerous diseases and immunologists saw themselves as specialists in disease prevention. In addition, the groundwork was laid for deciphering the nature and function of the antitoxins first used by von Behring to treat diphtheria (see Chapter 1). This work led to the elucidation of antibody structure in the 1960s, and the maturing of immunology to one of the key scientific disciplines of our modern times.

In this section, we study the immune system as it relates to humoral immunity and the structure and function of antibodies. How the interaction of antibody and antigen leads to elimination of the antigen also is discussed.

Humoral Immunity Is a Response Mediated by Antigen-Specific B Lymphocytes

KEY CONCEPT

- Antibodies are a group of protein molecules circulating in the body's fluids that recognize specific epitopes.

Antibodies are a class of proteins called **immunoglobulins**. They react with epitopes on toxin molecules in the bloodstream, on antigen surfaces or microbial structures (such as flagella, pili, capsules), and with epitopes on viruses in the extracellular fluid.

The recognition of epitopes requires antibodies have a special structure (FIGURE 20.7). The basic antibody molecule consists of four polypeptide chains: two identical **heavy** (H) **chains** (each heavy chain consists of about 400 amino acids) and two identical **light** (L) **chains** (each light chain has about 200 amino acids). These chains are joined together by disulfide bridges to form a Y-shaped structure, which represents a **monomer**, having two identical halves.

Each light and heavy chain has a constant and variable region within the polypeptide

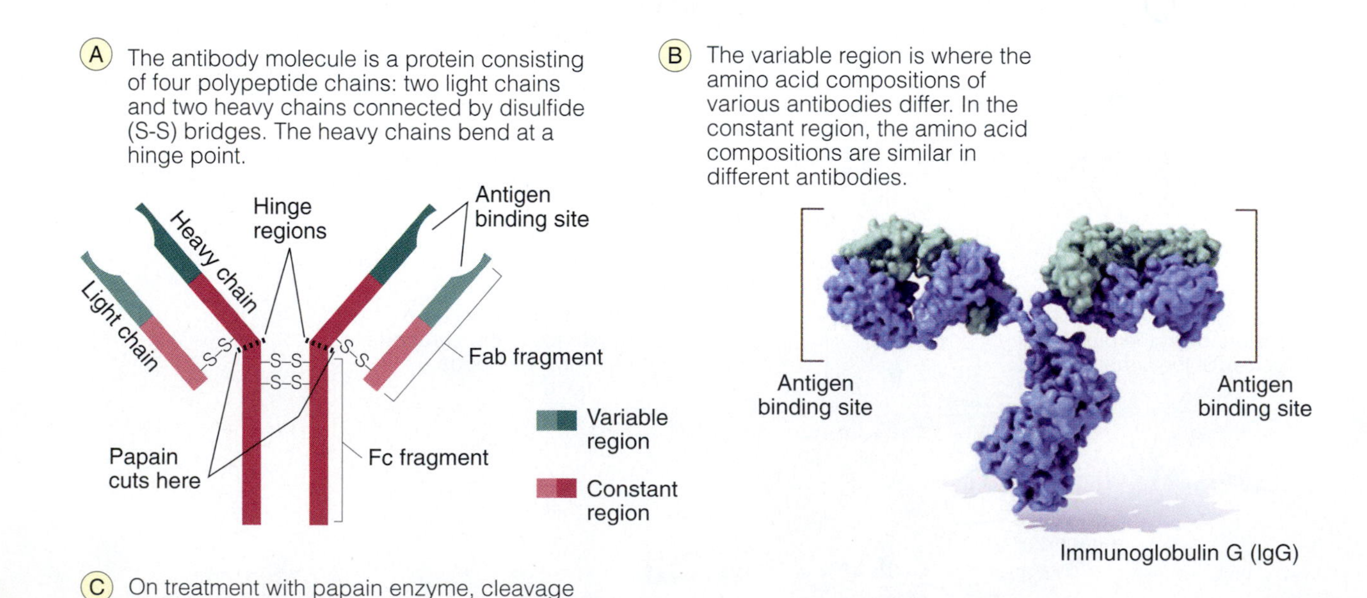

FIGURE 20.7 **Structure of an Antibody.** (**A**) Antibodies consist of two light and two heavy polypeptide chains. (**B**) Molecular model of the antibody immunoglobulin G (IgG).

Q: Using the illustration in (a), identify the light and heavy chains in (b).

chain. The **constant region**, which contains virtually identical amino acids in both light and heavy chains, determines the destination (body location) and functional class to which the antibody belongs. However, the amino acids of the **variable region** are different across the hundreds of thousands of antibodies produced in response to antigens (epitopes). This variability in a light and heavy chain is what forms a highly specific, three-dimensional structure, called the **antigen binding site**. It is to this region on the antibody that the antigen with its epitope binds.

The antigen-binding site is uniquely shaped to "fit" a specific epitope. Moreover, the two "arms" of the antibody are identical, so a single antibody molecule can combine with two identical epitopes on the same or separate antigens. As we will see, these combinations may lead to a complex of antibody and antigen molecules, which are targets for elimination by phagocytes.

When an antibody molecule is treated with papain, an enzyme from the papaya fruit, the enzyme splits the antibody at the hinge joint, and two functionally different segments are isolated (Figure 20.7). One segment (actually two identical segments), called the **Fab fragment**, for fragment-antigen-binding, is the portion that will combine with the epitope. The second segment, called the **Fc fragment**, for fragment able to be crystallized, performs various functions, depending on the antibody class: it can combine with phagocytes in opsonization; it may activate the complement system in resistance mechanisms; or it might attach to certain cells in allergic reactions.

CONCEPT AND REASONING CHECKS

20.5 How is an antibody's antigen binding site similar to the active site on an enzyme?

There Are Five Immunoglobulin Classes

KEY CONCEPT

- Immunoglobulin classes differ in their heavy chains of the constant region.

Classes of immunoglobulins are based on whether the antibody will be secreted into the bloodstream, attached onto a cell as a receptor, or deposited in body secretions. Using the abbreviation Ig (immunoglobulin), the five classes are designated IgM, IgG, IgA, IgE, and IgD. The five classes are outlined in TABLE 20.1.

IgM. One class of immunoglobulins, the IgM (M stands for macroglobulin), is the largest antibody molecule and consists of a pentamer (five monomers) whose tail segments are connected by a glycoprotein. It is the first, but short-lived, immunoglobulin to appear in the circulation after the stimulation

TABLE 20.1 The Five Immunoglobulin Classes

Class		Percentage of Antibody in Serum	Location in Body	Molecular Weight (Daltons)	Number of Monomers	Complement Binding	Crosses Placenta	Characteristic
IgM		5–10	Blood Lymph	900,000	5	Yes	No	First antibody formed in primary antibody response
IgG		80	Blood Lymph	150,000	1	Yes	Yes	Principal component of secondary antibody response
IgA		10	Secretions Body cavities	350,000	2	No	No	Protection in body cavities
IgE		<1	Blood Lymph	200,000	1	No	No	Role in allergic reactions
IgD		0.05	Blood Lymph	180,000	1	No	No	Receptor sites on B lymphocytes

of B cells. Thus, the presence of IgM in the serum of a patient indicates a very recent infection. Because of its size, most IgM remains in the circulation. IgM is formed during fetal infections with rubella or toxoplasmosis, indicating a certain immunological competence exists in the fetus.

IgG. The antibody commonly referred to as gamma globulin represents the IgG class. This antibody is the major circulating antibody, comprising about 80 percent of the total antibody content in normal serum. IgG appears about 24 to 48 hours after antigenic stimulation and continues the antigen-antibody interaction begun by IgM. Booster injections of a vaccine raise the level of this antibody considerably in the serum. IgG also is the maternal antibody that crosses the placenta and renders immunity to the fetus until the child can make antibodies at about six months of age.

IgA. Approximately 10 percent of the total antibody in normal serum is IgA and is similar to IgG. A second, dimer form accumulates in body secretions and is referred to as secretory IgA. This antibody provides resistance in the respiratory and gastrointestinal tracts, possibly by inhibiting the attachment of pathogens to the tissues. The secretory component comes from epithelial cells and helps move the antibody into the secretions of the mucosa (MicroFocus 20.2). IgA also is located in tears and saliva, and in the colostrum, the first milk secreted by a nursing mother. When consumed by an infant, the antibodies provide resistance to potential gastrointestinal disorders.

IgE. The IgE class is another monomer antibody. It plays a major role in allergic reactions by sensitizing cells to certain antigens. This process is discussed in Chapter 22.

IgD. The last class of immunoglobulins is the IgD antibody. The monomer exists as a cell surface receptor on the B cells. For this reason, it is called a membrane antibody and is important to the activation of B cells.

CONCEPT AND REASONING CHECKS

20.6 Where does each of the antibody classes function in the body?

Antibody Responses to Pathogens Are of Two Types

KEY CONCEPT

- Memory cells produce a rapid antibody response.

The first time the body encounters a pathogen or antigen, a **primary antibody response**

MICROFOCUS 20.2

The Other Army

Most people are familiar with the body's army, headquartered downtown—the antibodies and T lymphocytes that constitute the blood-centered system of immunity. Few, however, know much about the army of the suburbs—the specialized cells and antibodies that protect the body at its vulnerable portals of entry, in locations such as its respiratory, gastrointestinal, and urinary tracts. The cells and antibodies in the fragile mucous membranes lining these systems constitute a network now known as the **mucosal immune system.**

The extent of communication between the blood and mucosal systems is uncertain at this writing. Injected vaccines, for example, rarely evoke IgA production for mucosal immunity. However, stimulating one region of mucosal tissue to elicit IgA appears to produce an immune response at a distant mucosal surface. Thus, a vaccine in a nasal spray can elicit an antibody response along the gastrointestinal tract. This interrelatedness has led immunologists to think in terms of a single mucosal system.

Investigations of mucosal immunity have lagged behind those of bloodborne immunity in part because tissue samples from the mucosa are difficult to study. Studies are being encouraged, though, because stopping infection at its point of entry is highly desirable. Also, vaccine research would benefit considerably because oral vaccines are preferred to injectable types, and developing an oral vaccine requires a basic understanding of mucosal immunity. Indeed, the output of the mucosal system appears to outdistance that of the bloodborne system. It has been estimated, for instance, that over 3 grams of IgA are secreted as compared to only 1 gram of IgA released into the bloodstream. On this basis alone, the "suburban army" merits a greater share of attention.

occurs (FIGURE 20.8). B cells are activated and the effector cells, the plasma cells, start producing and secreting antibody. There is a lag of several days before any measurable antibodies appear. IgM are the first on the scene, but they are soon replaced by a stronger and longer-lasting IgG response. During the primary antibody response, memory cells also are produced, which provide the immunological memory needed for future infections.

A second or subsequent infection by the same pathogen or antigen produces a more powerful and sustained **secondary antibody response**. Due to the presence of memory cells, a rapid resonse to antigen leads to the production of IgG, the principal antibody. The secondary antibody response has a shorter time lag and greater antibody production than the primary response. It also provides a much longer period of resistance against disease.

CONCEPT AND REASONING CHECKS

20.7 Justify the need for the two antibody responses to pathogens.

Antibody Diversity Is a Result of Gene Rearrangements

KEY CONCEPT

- The mixing and matching of immunoglobulin gene segments produces unique antibodies.

For decades, immunologists were puzzled as to how an enormous variety of antibodies (perhaps a million or more different types) could be encoded by the limited number of genes associated with the immune system. Because, like all proteins, antibodies are specified by genes, it would be reasonable to assume an individual must have a million or more genes for antibodies. However, genetic analysis indicates human cells have only about 35,000 genes for all their functions (see Chapter 8).

The antibody diversity problem has been resolved by showing that embryonic cells contain about 300 genetic segments, which are shuffled (like transposons) and combined in each B cell as it matures. The process, known as **somatic recombination**, is a random mixing and matching of gene segments to form unique antibody genes. Information encoded by these genes then is expressed in the surface receptor proteins of B lymphocytes and in the antibodies later expressed by the stimulated clone of plasma cells.

The process of somatic recombination was first postulated in the late 1970s by Susumu Tonegawa, winner of the 1987 Nobel Prize in Physiology or Medicine. According to the process, the gene segments coding for the light and heavy chains of an antibody are located on different chromosomes. The light

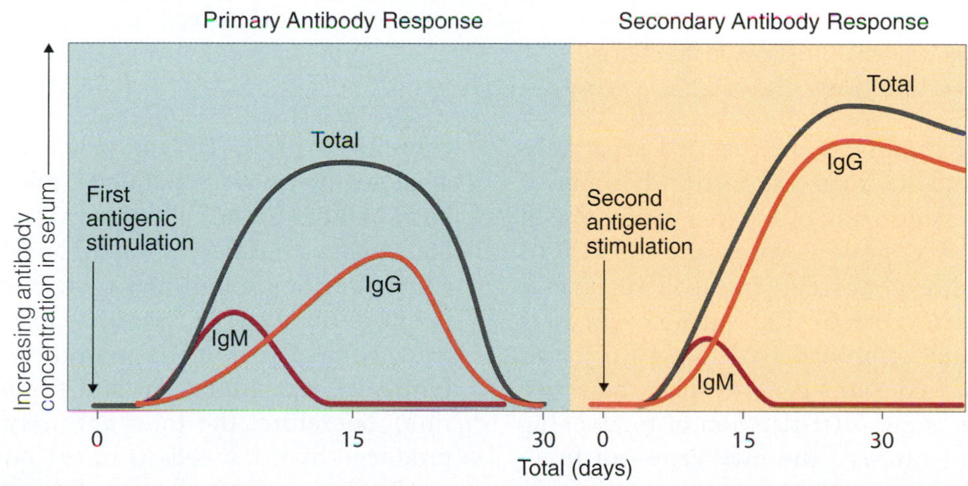

FIGURE 20.8 **The Primary and Secondary Antibody Responses.** After the initial antigenic stimulation, IgM is the first antibody to appear in the circulation. Later, IgM is supplemented by IgG. On a second or ensuing exposure to the same antigen, the production of IgG is more rapid, and the concentration in the serum reaches a higher level than previously. *Q: Why is the total antibody concentration greater than the sum of the IgM and IgG concentrations?*

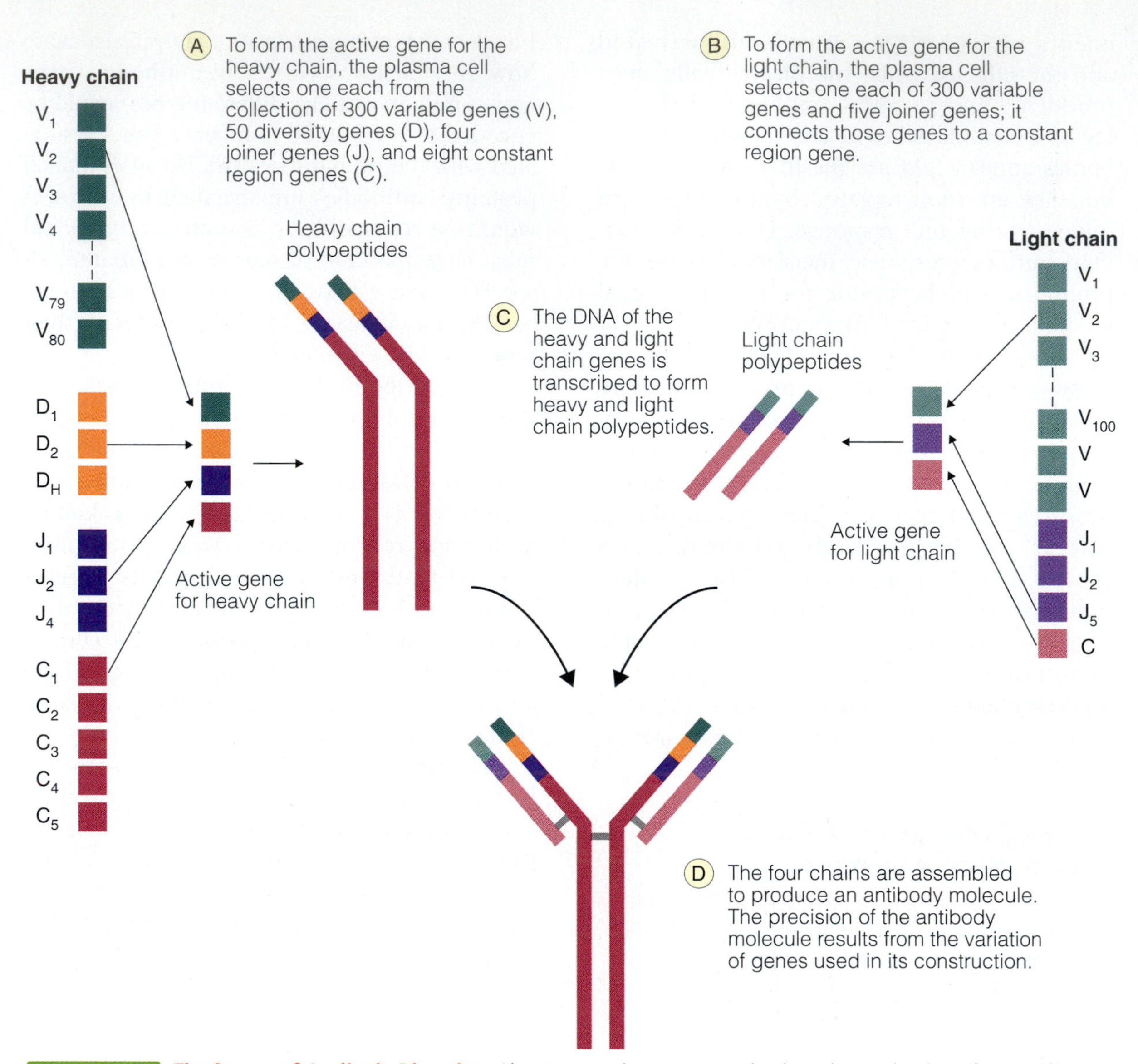

FIGURE 20.9 **The Source of Antibody Diversity.** Almost any substance can stimulate the production of a specific antibody. Immunoglobulin gene rearrangements provide the mechanism for this diversity. Note: Not all potential heavy and light chain segments are shown in the diagram.

Q: How can gene arrangements generate such antibody diversity?

and heavy chains are synthesized separately, then joined to form the antibody. One of 8 constant genes, one of 4 joiner genes, one of 50 diversity genes, and one of up to 300 variable genes can be used to form a heavy chain. One of 300 variable genes is selected and combined with one of 5 joiner genes and a constant gene to form the active light-chain gene. After deletion of intervening sequences (introns), the new gene can function in protein synthesis, as shown in FIGURE 20.9. Tonegawa's discovery was revolutionary because it questioned two dogmas of biology: (1) DNA for a protein needs to be one continuous piece (for antibody synthesis, the gene segments are separated); and (2) every body cell has exactly the same DNA (the antibody genes for different B lymphocytes differ). Current evidence indicates more than 600 different antibody gene segments exist per cell. Additional diversity is generated through imprecise recombination and somatic mutation. Therefore, the total antibody diversity produced by the B cells is in the range of 10^8 to 10^{11} immunoglobulin possibilities.

CONCEPT AND REASONING CHECKS

20.8 Why don't the constant genes for an antibody show the same diversity as the variable genes?

Antibody Interactions Mediate the Disposal of Antigens (Pathogens)

KEY CONCEPT

- Antigen-antibody complexes are cleared by phagocytes.

In order for specific resistance to develop, antibodies must interact with antigens so that the antigen is changed in some way. Thus, the formation of antigen-antibody complexes may result in death to the microorganism possessing the antigen, inactivation of the antigen, or increased susceptibility of the antigen to other body defenses. FIGURE 20.10 summarizes these mechanisms.

Certain antibodies can inhibit viral attachment to host cells through **viral inhibition**. By reacting with and covering capsid proteins or spikes, antibodies prevent viruses from entering their host cells. Influenza viruses are inhibited by neuraminidase antibodies in this way. **Neutralization** represents a vital mechanism of defense against toxins. Since the antibodies, called **antitoxins**, alter the toxin molecules near their active sites and mask their toxicity, the toxins are unable to bind to cells. Viral inhibition and neutralization also increase the size of the antigen-antibody complex, thus encouraging phagocytosis while lessening their ability to diffuse through the tissues.

As discussed earlier in the chapter, antibodies can coat bacterial cells (**opsonization**), preventing bacterial attachment to host cells. Because antibodies are bivalent, they can cross-link two separate antigens with the same epitope. This action causes clumping, or **agglutination**, of the bacterial cells. Movement is inhibited if antibodies react with antigens on the flagella of microorganisms, and the organisms may even be clumped together by their flagella. The reaction of antibodies with

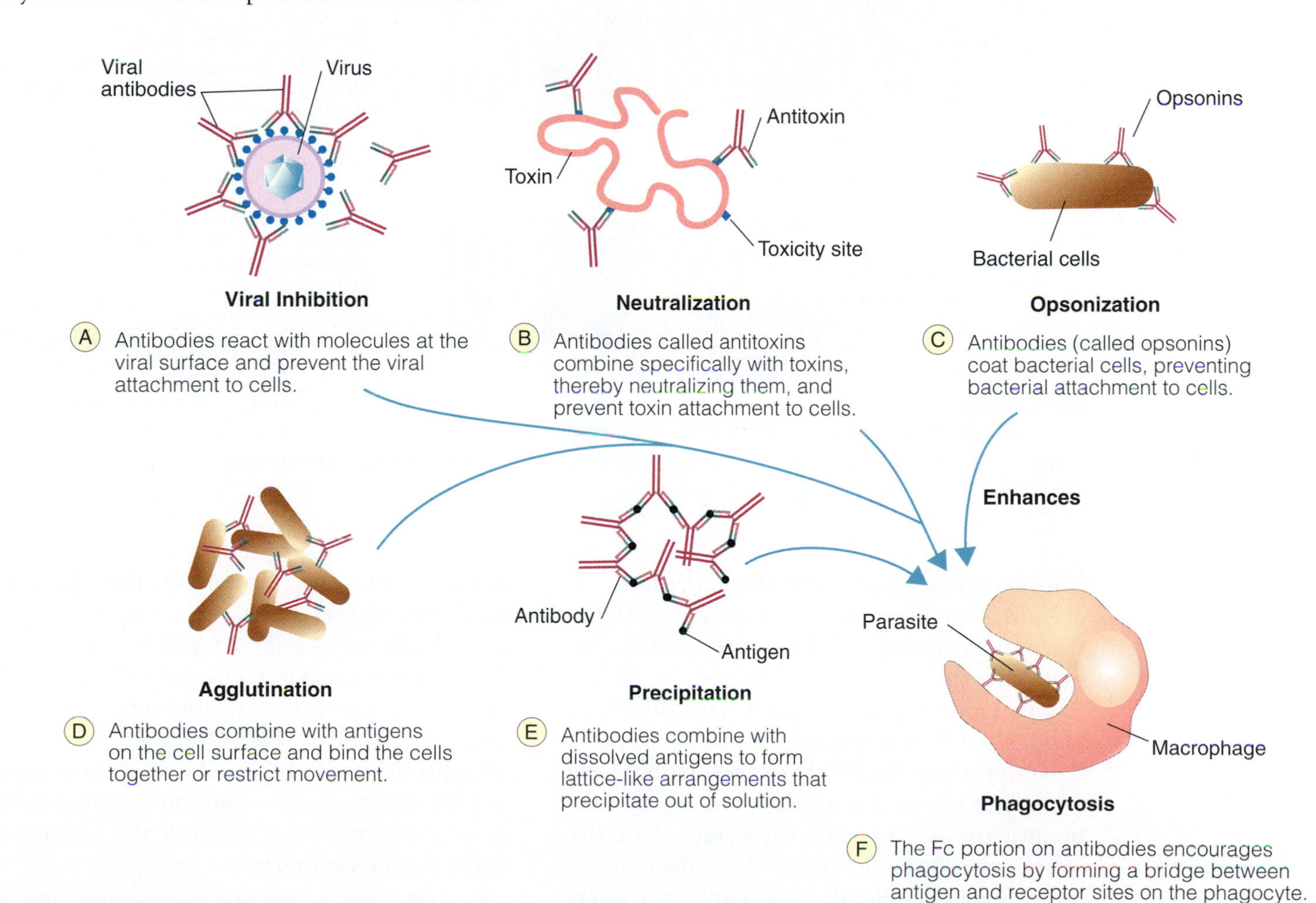

FIGURE 20.10 **Mechanisms by Which Antibodies Interact with Antigens.** In all cases, the interaction facilitates phagocytosis by macrophages.

Q: Reorganize these mechanisms for antigen clearance into two basic strategies.

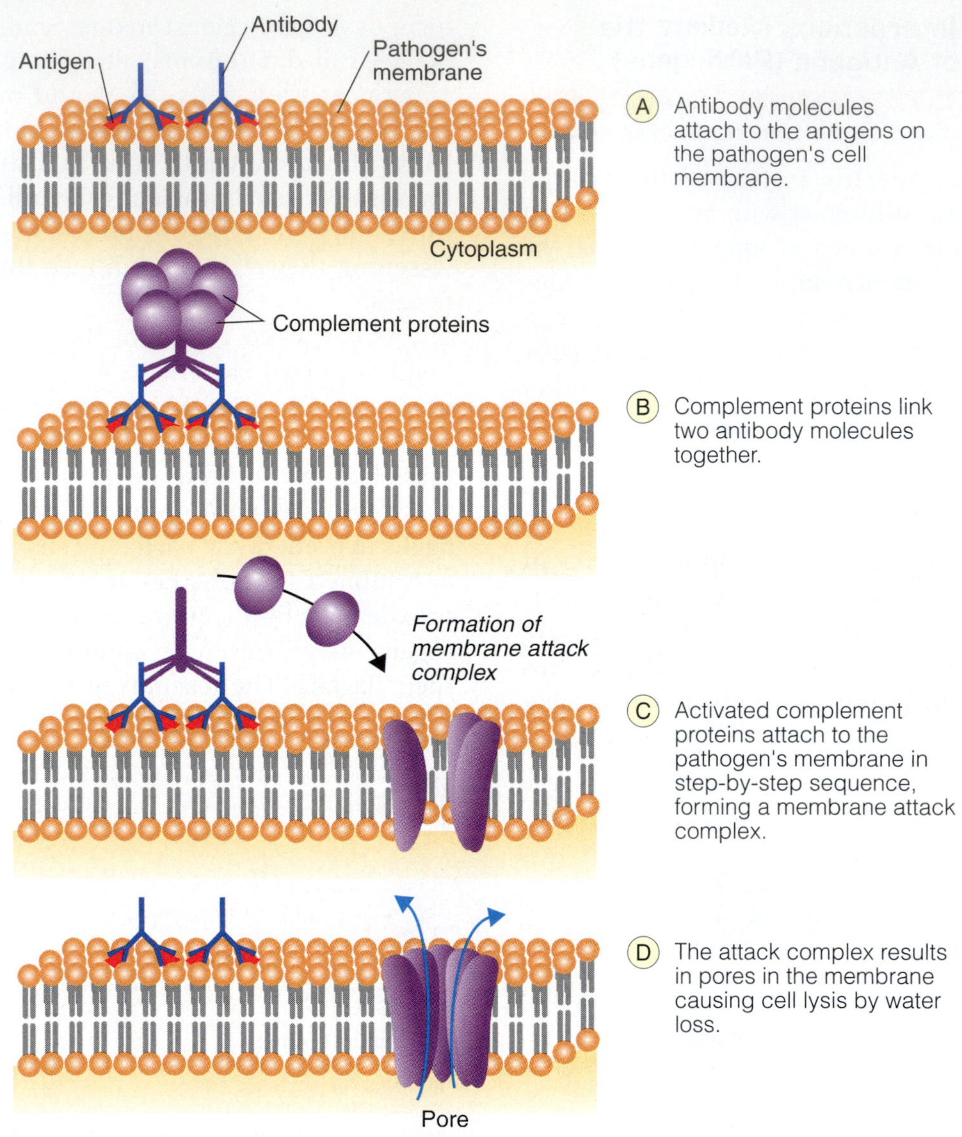

FIGURE 20.11 **The Formation of a Membrane Attack Complex (MAC).** Through the activation of a complement cascade, MACs are formed in the cell membrane of a pathogen.

Q: Which of the mechanisms in Figure 19.8 must occur before complement proteins can be assembled into MACs?

pilus antigens prohibits attachment of an organism to the tissues while agglutinating them and increasing their susceptibility to phagocytosis.

Antibodies also can react with dissolved antigens and convert them to solid precipitates (**precipitation**). In all these interactions, the antigen-antibody complexes formed enhance phagocytosis and thus the removal of the infecting pathogen or antigen from the body.

A final example of antigen-antibody interaction involves the complement system, introduced in Chapter 19. The complement system is a series of nearly 30 proteins that functions in a cascading set of reactions. One set of reactions leads to lysis of bacterial cells through a complement cascade at the cell surface, resulting in the formation of a **membrane attack complex** (MAC) (FIGURE 20.11). This complex forms pores in the membrane, increasing cell membrane permeability and inducing the cell to undergo lysis through the leakage of water from its cytoplasm.

CONCEPT AND REASONING CHECKS

20.9 To this point, summarize the roles that phagocytes have had in the acquired immune response.

20.3 The Cell Mediated Immune Response

The humoral immune response is one arm of acquired immunity. The production of antibodies and their association in antigen-antibody complexes can effectively clear an infection from the fluids of the body. However, many pathogens such as viruses, rickettsiae, certain bacterial species (including *Mycobacterium tuberculosis*) as well as fungi and protozoa have the ability to enter inside cells. Once in the host cell cytoplasm, they are hidden from the onslaught that otherwise would be leveled by antibodies, which are too large to enter cells.

The body's defense against microorganisms infecting cells and other "nonself" cells is centered in the cell mediated immune response. In this final section, we describe the roles for the T cells in recognizing and eliminating virus-infected cells and "nonself" cells as well as "priming the pump" for the humoral immune response.

Cellular Immunity Relies on T-Lymphocyte Receptors and Recognition

KEY CONCEPT

- Protein receptors found on different populations of T Cells define their role in cellular immunity.

As we have seen, B cell receptors directly recognize antigenic determinants on an antigen. However, the interaction between antigenic determinants and T cells depends on sets of surface receptors recognizing antigenic (peptide) fragments.

There are two subpopulations of effector T cells and their receptors determine how they will function in cellular immunity. **Cytotoxic T cells** have T-cell receptors (TCRs) and coreceptor proteins, called **CD8**, on their cell surface (FIGURE 20.12). This combination will allow them to recognize and eliminate "nonself," such as virus-infected and cancer cells.

Naive T cells have TCRs and coreceptors called **CD4**. As their name suggests, these "immature" T cells will help with both cellular and humoral immunity. Thus, as we will see, they play a central role for both cell mediated and humoral immune responses. It is to this CD4 receptor that HIV attaches and is able to infect the cell.

The combination of TCRs and coreceptors allows T cells to recognize and bind to another set of glycoprotein molecules called the **major histocompatibility complex (MHC)**. MHC proteins are embedded in the membranes of all cells of the body. At least 20 different genes encode MHC proteins, and at least 50 different forms of the genes exist. Thus, the variety of MHC proteins existing in the human population is enormous, and the chance of two individuals having the same MHC proteins is incredibly small. (The notable exception is identical twins.) The MHC proteins define the uniqueness of the individual and play a role in the immune response. Recent evidence also suggests MHC proteins may play more "amorous" roles (MicroFocus 20.3).

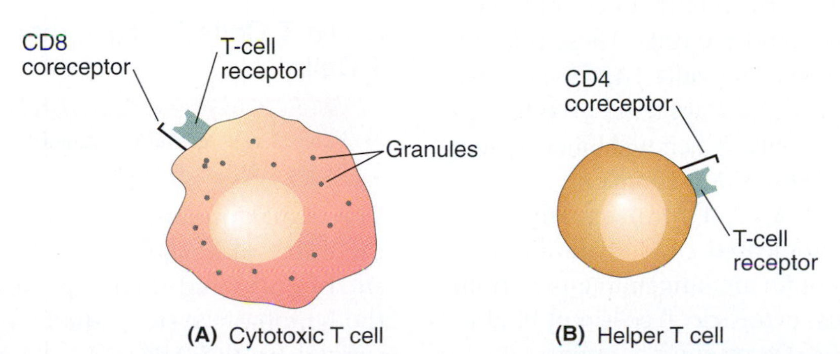

FIGURE 20.12 **T-Cell Receptors.** (**A**) Cytotoxic T cells have T-cell receptors and CD8 coreceptors that recognize antigen peptides associated with class I MHC molecules on virus-infected cells, tumor cells, and foreign tissue transplants. (**B**) Naive T cells have T-cell receptors and CD4 coreceptors that recognize antigen peptides associated with class II MHC molecules on antigen-presenting cells.

Q: What is common to the receptors of both types of T cells?

MICROFOCUS 20.3: Being Skeptical

Of Mice and Men—Smelling a Mate's MHC

Would you pick a mate by sight or smell? At least in the nonhuman world, it might be by smell.

The major histocompatibility complex (MHC) is a group of immune system molecules distinguishing essentially every individual. Recent research suggests mammals, which include humans, might consciously or unconsciously select a mate who has a different MHC than themselves.

In mice, there is a region of the nose that detects chemicals in the air and other molecules important to mouse reproduction. Frank Zufall and colleagues at the University of Maryland at College Park have discovered that mice also use the nasal region to recognize MHC molecules found in mouse urine. The researchers believe mice might use such "markers" to identify individuals through smell—and perhaps a mate as well. In fact, mice prefer to smell other mice that have a different MHC than themselves and prefer mating with mice of another MHC.

Now to the human "experience." In a human study, people were given sweaty T-shirts to smell. In recorded responses, the participants indicated they preferred smells that the researchers then linked back to different MHCs.

Zufall's theory is that in both mice and humans, individuals preferring a different MHC are preventing inbreeding. Therefore, mating between two individuals with different MHCs will produce offspring having a unique set of MHCs that might make that individual more resistant to factors like infectious disease.

However, most humans don't pick their mates by smelling sweaty T-shirts. So, Craig Roberts at the University of Liverpool wondered if there was a relationship between human faces and MHC. After determining the MHC for a large group, he selected 92 women and showed them photographs of six men, three with similar MHCs and three with different MHCs as compared to the women. The women were asked if they would prefer a long- or short-term relationship with each.

Roberts was surprised to find that the women preferred the faces of men with similar MHCs for a long-term relationship.

There certainly is much, much more research to be done. However, Roberts proposes two mechanisms are at work to select a mate. First, select a mate facially with a similar MHC. Then, smell will make sure the mate is not too similar so more "successful" offspring will be produced. How all this plays out requires much more investigation.

There are two important classes of MHC proteins and it is to these that the cytotoxic and naive T cells bind. **Class II MHC (MHC-II)** proteins are found primarily on the surface of B cells, macrophages, and dendritic cells. These cells are called **antigen-presenting cells (APCs)** because they "present" antigen fragments attached to MHC-II to naive T cells, which will bind to them with their TCR/CD4 receptor complex.

Class I MHC (MHC-I) proteins are found on virtually all nucleated cells of the body. If these proteins contain an antigen fragment from an infecting virus, cytotoxic T cells will bind to it with their TCR/CD8 receptor complex.

Confused? The next section explains how these responses occur and what they accomplish. As a break, read MicroInquiry 20, which talks about the possibility of "thinking healthy."

CONCEPT AND REASONING CHECKS

20.10 How can T cells be separated based on receptors and function?

Naive T Cells Mature into Effector T Cells

KEY CONCEPT

- Naive T-cell activation stimulates memory cell and helper T-cell production.

The actual immune response originates with the entry of antigens into the body. At their site of entry and their capture there or within the lymphatic system, the antigens are phagocytized by the APCs. As the cells migrate to the lymph nodes or other secondary lymphoid tissues (see Chapter 19), the antigens are broken down and peptide fragments are displayed on the surface as MHC-II/peptide complexes.

MICROINQUIRY 20

Can Thinking "Well" Keep You Healthy?

The idea that mental states can influence the body's susceptibility to and recovery from disease has a long history. The Greek physician Galen thought cancer struck more frequently in melancholy women than in cheerful women. During the twentieth century, the concept of mental state and disease was researched more thoroughly, and a firm foundation was established linking the nervous system and the immune system. A new field called psychoneuroimmunology has emerged. One such link exists between the hypothalamus and the T lymphocytes. The hypothalamus is a portion of the brain located beneath the cerebrum. It produces a chemical-releasing factor inducing the pituitary gland, positioned just below the hypothalamus, to secrete the hormone ACTH, which targets the adrenal glands. The adrenal glands, in turn, secrete steroid hormones (glucocorticoids) that influence the activity of T cells in the thymus gland.

Another link is established by the branches of the autonomic nervous system extending into the lymph nodes and spleen tissues. The autonomic nervous system automatically regulates the functioning of such organs as the heart, stomach, and lungs through myriad nerve fibers. The direct anatomical link between the immune and nervous systems allows a direct two-way communication.

A third link between the nervous and immune systems starts with thymosins, a family of substances originating in the thymus gland. When experimentally injected into brain tissue, thymosins stimulate the pituitary gland via the hypothalamus to release hormones, including the one stimulating the adrenal gland. Although the precise functions of thymosins are yet to be determined, they may serve as specific molecular signals between the thymus and the pituitary gland. A circuit apparently stimulates the brain to adjust immune responses and the immune system to alter nerve cell activity.

The outcome of these discoveries is the emergence of a strong correlation between a patient's mental attitude and the progress of disease. Rigorously controlled studies conducted in recent years have suggested that a person's aggressive determination to conquer a disease can increase one's lifespan. Therapies can consist of relaxation techniques, as well as mental imagery that disease organisms are being crushed by the body's stalwart defenses. Behavioral therapies of this nature can amplify the body's response to disease and accelerate the mobilization of its defenses.

Few reputable practitioners of behavioral therapies believe such therapies should replace drug therapy. However, the psychological devastation associated with many diseases such as AIDS cannot be denied, and it is this intense stress that the "thinking well" movement attempts to address. Very often, for instance, a person learning of a positive HIV test goes into severe depression, and because depression can adversely affect the immune system, a double dose of immune suppression ensues. Perhaps by relieving the psychological trauma, the remaining body defenses can adequately handle the virus.

As with any emerging treatment method, there are numerous opponents of behavioral therapies. Some opponents argue that naive patients might abandon conventional therapy; another argument suggests therapists might cause enormous guilt to develop in patients whose will to live cannot overcome failing health. Proponents counter with the growing body of evidence showing that patients with strong commitments and a willingness to face challenges—signs of psychological hardiness—have relatively greater numbers of T cells than passive, nonexpressive patients. To date, no study has proven that mood or personality has a life-prolonging effect on immunity. Still, doctors and patients are generally inspired by the possibility of using one's mind to help stave off the effects of infectious disease. Though unsure of what it is, they generally agree *something* is going on.

Discussion Point

Perhaps you agree or disagree with the studies reported here. Discuss the value of providing a patient with information on psychoneuroimmunology and its accompanying therapies.

When the APCs enter the lymphoid tissue, a hunt begins. With its exposed class II MHC/peptide complex, the APCs mingle among the myriad groups of naive T cells, searching for the cluster having the surface receptors that recognize the MHC-II/peptide. This process requires considerable time and energy because only one cluster of T lymphocytes may have matching T cell/CD4 receptors (FIGURE 20.13A).

It is important to remember that recognition is between the MHC-II proteins on the APC surface and receptors on the T-cell surface. It is as if the T lymphocyte must first ensure it and the APC are from the same body (same MHC proteins) before it will respond. The coreceptor enhances binding of the naive T cell to the macrophage because the CD4 protein recognizes the peptide associated with the class II MHC protein.

Bound to the T cell, the APC secretes specific cytokines, such as **interleukin-1** (**IL-1**), which binds with the naive T cell and stimulates T-cell activation. Interleukin-1 causes the naive T cell to secrete other cytokines, including **interleukin-2** (**IL-2**), which stimulates cell division of that T cell and others activated by the APCs. The result is the production of a clone of antigen-specific immature T cells (FIGURE 20.13B). Some of these mature into memory T cells (FIGURE 20.13C) and await a future encounter with the same epitope.

The majority of the immature T cells mature into one of two types of effector T cells, depending on the influence of specific cytokines. Some mature into **helper T2** (**T_H2**) **cells** (FIGURE 20.13D). T_H2 cells *help* in the activation of humoral immunity, which we will discuss just ahead.

Other immature T cells mature into **helper T1** (**T_H1**) **cells**. These cells recognize and bind to macrophages infected with bacterial cells, such as *Mycobacterium tuberculosis* (see Chapter 9). The tubercle bacillus resides and survives in the phagosomes of these macrophages by preventing lysosome fusion with the phagosomes. T_H1 cells stimulate *(help)* lysosome fusion in these cells, resulting in the destruction of the bacterial invaders (FIGURE 20.13E).

CONCEPT AND REASONING CHECKS

20.11 Justify the need for two populations of "helper T cells."

Cytotoxic T Cells Recognize MHC-I Peptide Complexes

KEY CONCEPT

- Activated cytotoxic T cells seek out and kill virus-infected and abnormal cells.

Because body cells also can become infected with viruses, the immune system has evolved a way to attack virus-infected cells. During the infection, the host cells manage to degrade some of the viral antigens into small peptides. These are attached onto MHC-I proteins and transported to the cell surface, where they are displayed like "red flags" to denote an infected cell.

Cytotoxic T cells are produced in the thymus and only need to be activated to become effector cells able to identify and destroy the virus-infected cells. Activation can occur through interaction with dendritic cells or association with an APC linked to CD8 T cell that have been exposed to and processed the same viral antigens as the virus-infected body cells. Either situation activates cytotoxic T cells to divide into a mature T-cell clone able to recognize the antigen being "red flagged" on infected body cells (FIGURE 20.13F).

The cytotoxic T cells leave the lymphoid tissue and enter the lymph and blood vessels. They circulate until they come upon their target cells, the infected cells displaying the telltale MHC-I/peptide on their surface (FIGURE 20.13G). The cytotoxic T cell joins its T cell/CD8 receptor complex with the MHC-I/peptide on the virus-infected cell surface. Then, the cytotoxic T cell releases a number of active substances, identical to those secreted by the NK cells described in Chapter 19.

The toxic proteins include **perforin**, which inserts into the membrane of the infected cell, forming cylindrical pores in the membrane. This "lethal hit" releases ions, fluids, and cell structures. In addition, the cytotoxic T cells release **granzymes** that enter the target cell and trigger apoptosis. Cell death not only deprives the viral pathogen of a place to survive and replicate, but it also exposes the pathogen to antibodies in the extracellular fluid.

Cytotoxic T cells also are active against tumor cells because these cells often display distinctive molecules on their surfaces. The molecules are not present in other body cells, so they are viewed as antigen peptides. Harbored within MHC-I proteins at the cell surface, the antigen peptides react with receptors on cytotoxic T cells and the tumor cells are subsequently killed through apoptosis (FIGURE 20.14). However, some cancers reduce the level of MHC-I proteins at the cell surface, which impedes the ability of cytotoxic cells to "find" the abnormal cells. Thus, cancer cells escape immunologic surveillance and survive.

Like naive T cells, cytotoxic T cells also produce a population of memory cells during clonal selection. The memory cells distribute themselves to all parts of the body and remain in the tissues to provide a type of long-term immunity. Should the antigens be detected once again in the tissues, the memory T cells will multiply rapidly, interact with the infected

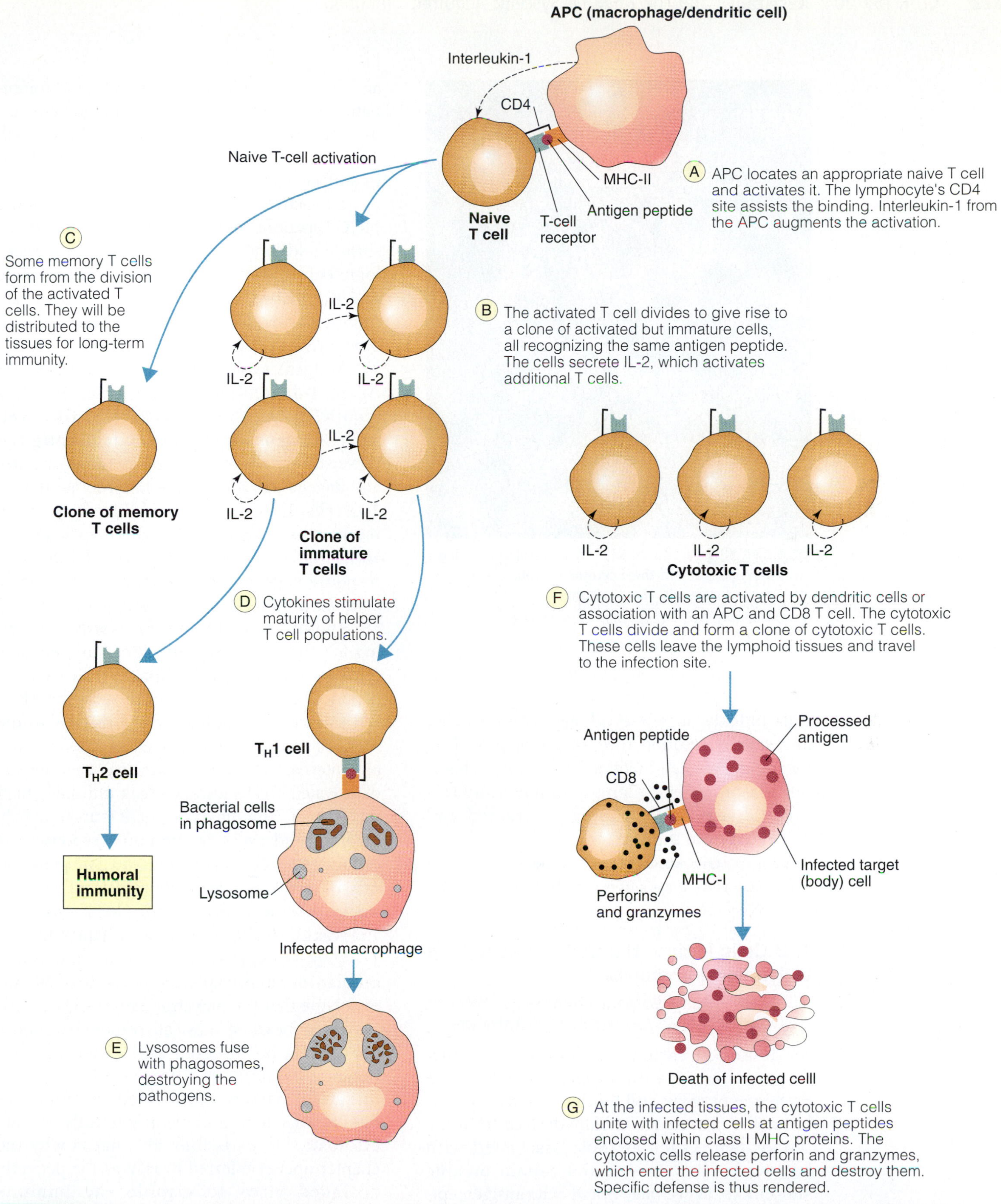

FIGURE 20.13 **The Process of Cell Mediated Immunity and Cytotoxic T-Cell Action.** Antigen-presenting cells (APCs), such as dendritic cells or macrophages, activate naive T cells. The activated immature T cells stimulate humoral immunity and destruction of intracellular pathogens, and help activate the cytotoxic T cells. These cells secrete proteins (perforins and granzymes) to kill virus-infected cells or cancer cells through apoptosis.

Q: What is the key recognition difference between helper T cells and cytotoxic T cells?

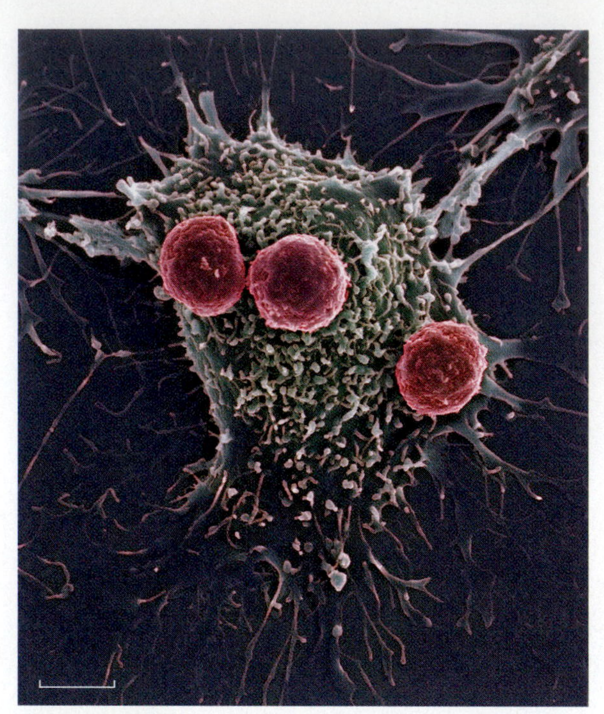

FIGURE 20.14 **A "Lethal Hit".** A false-color scanning electron micrograph of three cytotoxic T cells (pink) attacking a tumor cell. (Bar = 10 µm.)

Q: How do cytotoxic T cells "find" a tumor cell?

cells quickly, secrete cytokines without delay, and set into motion the process of providing instantaneous cell mediated immunity. This is one reason we enjoy long-term immunity to a given disease after having contracted and recovered from the disease.

CONCEPT AND REASONING CHECKS

20.12 Summarize the events leading to apoptosis of targeted cells.

T_H2 Cells Initiate the Cellular Response to Humoral Immunity

KEY CONCEPT

- The active T_H2 cells co-stimulate B-cell activation.

As described earlier, the humoral immune response begins with an encounter between pathogen or toxin and B cells. Antigens stimulating the process are usually derived from the bloodstream, such as those associated with bacterial cells, viruses, and certain organic substances, as we have noted. The antigen epitopes interact with B cells (FIGURE 20.15).

Once the antigenic determinant has bound with its corresponding surface receptor protein, and after cross-linking has occurred, the combination of receptor protein and antigenic determinant is taken into the cytoplasm of the B cell. Then, the B cell displays the antigen peptide on its surface within a MHC-II molecule.

The activated T_H2 cells recognize the same MHC-II/peptide molecule complex on the B-cell surface and bind to the B cells. This immunologic cooperation between the B lymphocyte and the helper T lymphocyte continues the immune response. Interleukin-2, along with other cytokines, assists B-cell activation.

Antigens evoking this sort of response are called **T-dependent antigens** because they require the services of T_H2 cells. However, some antigens are **T-independent antigens**; these substances (such as in bacterial capsules and flagella) do not require the intervention of T_H2 cells. Instead, they bind directly with the receptor proteins on the B-lymphocyte surface and stimulate the cells. However, the immune response is generally weaker, and no memory cells are produced.

Another group of antigens worth mentioning are the **superantigens**. "Regular" antigens must be broken down and processed to antigen peptides before they are presented on the APC's cell surface. Superantigens bind directly to the MHC proteins and to the T-cell receptor without any internal processing. Thus, massive numbers of activated T cells form, with an unusually high secretion of cytokines. The result is an extremely vigorous and excessive immune response that can lead to shock and death. Superantigens and massive cytokine release are associated with the staphylococcal toxins of toxic shock syndrome and scalded skin syndrome (see Chapter 12).

FIGURE 20.16 summarizes the humor and cellular immune responses of acquired immunity. Note that the naive/immature CD4 T cells are at the heart of most all responses. Should these cells be infected and killed by HIV, humoral immunity is adversely affected, stimulation of bacterially-infected macrophages does not occur, and the ability to fully activate cytotoxic T cells is limited. That is why the chapter opener referred to HIV as the perfectly designed virus to cripple the immune response. If you look back at Figure 15.4 in Chapter 15, notice how the CD4 + T-cell population declines with time of HIV infection. As

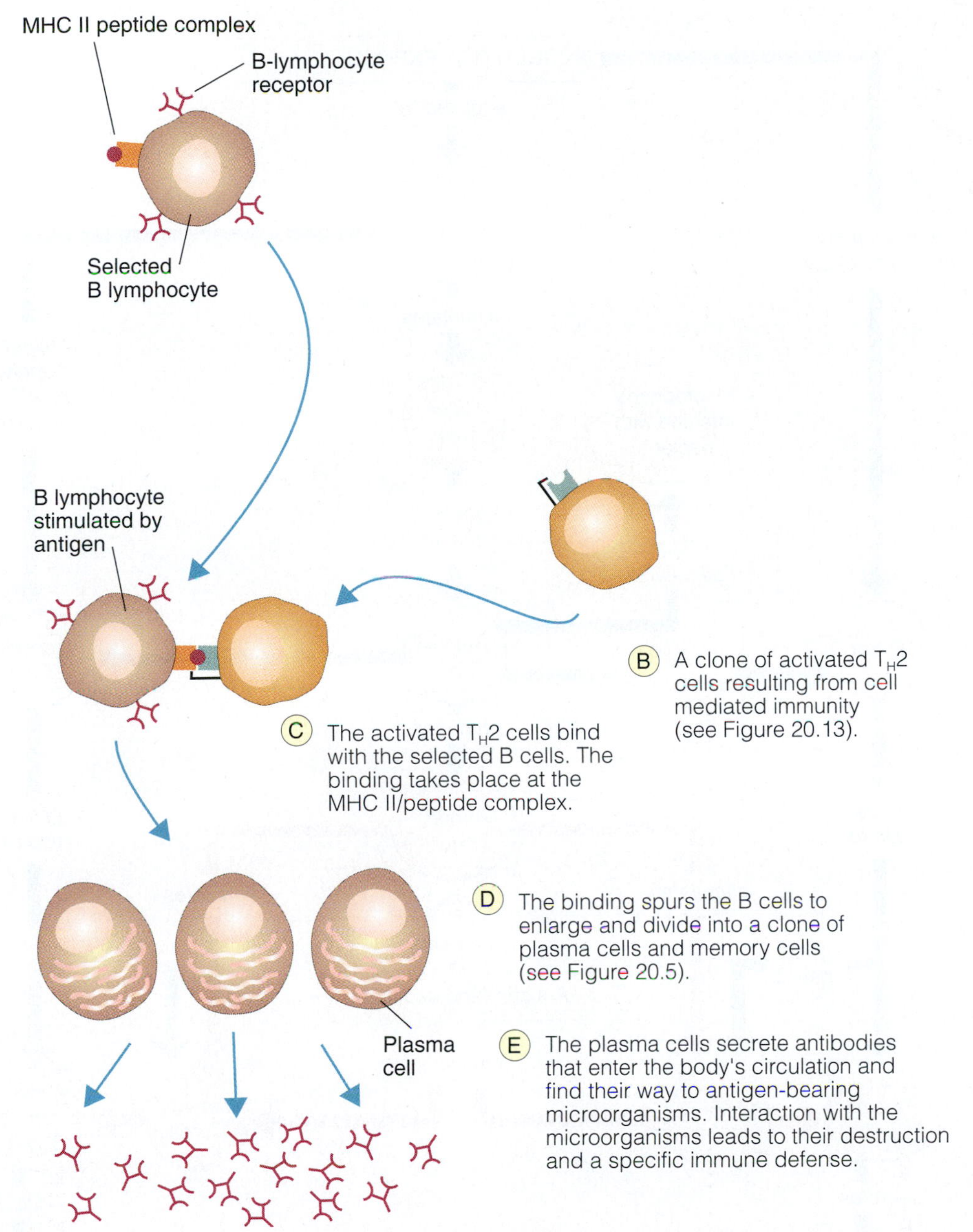

FIGURE 20.15 **T Cell Activation of Humoral Immunity.** The secretion of specific antibodies from plasma cells depends on B-cell receptors being cross-linked by antigen, presentation of MHC-II/peptides, and binding of T_H2 activated cells to the MHC-II/peptide complex.

Q: What is the similarity between the MHC-II/peptides complex on a B cell and the antigen presenting cell that activated the T_H2 cell?

the T-cell population drops, cell mediated immunity becomes less able to respond to pathogens. Eventually there are so few T cells that the entire immune response collapses.

CONCEPT AND REASONING CHECKS

20.13 Distinguish between the essential roles played by T_H2 cells during acquired immunity.

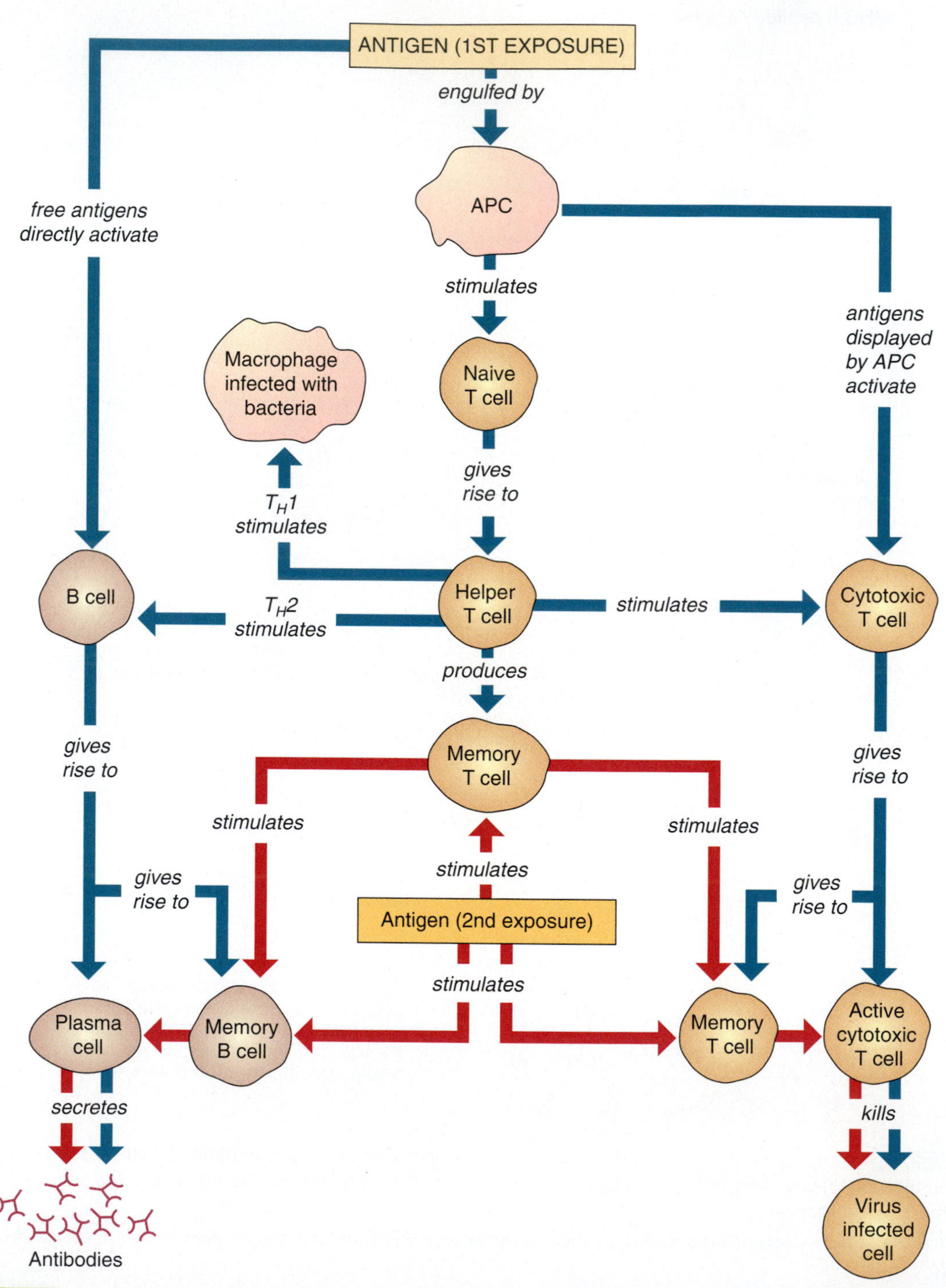

FIGURE 20.16 **A Concept Map Summarizing Acquired Immunity.** Humoral immunity produces antibodies that respond and bind to antigens. Cell mediated immunity stimulates helper T cells that activate B cells and cytotoxic T cells. Cytotoxic T cells bind to and kill infected cells or other abnormal cells. Memory B and T cells are important in a second or ensuing exposure to the same antigen.

Q: How would an HIV infection affect the immune response illustrated in this figure?

SUMMARY OF KEY CONCEPTS

20.1 An Overview of the Acquired Immune Response

- Acquired immunity requires specificity for epitopes, tolerance of and minimal damage to "self," and memory of past infections. It consists of humoral immunity maintained by B cells and antibodies, and cell mediated immunity controlled by T cells.
- Appropriate B and T cell populations are activated through clonal selection involving recognition of specific epitopes on antigens or antigen fragments.
- Immune system cells arise from stem cells in the bone marrow. Lymphoid progenitor cells give rise to T cells that mature in the thymus and B cells that mature in the bone marrow. Immunocompetent B and T cells colonize the lymphoid organs.

20.2 The Humoral Immune Response

- Antibodies consist of two identical light chains and two identical heavy chains. The variable regions of the light and heavy chains form two identical antigen-binding sites where the epitope (antigen) binds.
- Of the five classes of antibodies, IgM and IgG are primary disease fighters, while secretory IgA is found on body surfaces.
- A primary immune response produces primarily IgG antibodies and memory cells, the latter providing immunity in a secondary immune response.
- Antibodies can be produced that recognize almost any antigen. To do this, somatic recombination between a large number of genes (variable, diversity, joiner, constant) for the heavy and light chains occurs.
- The actual disposal or removal of antigen is facilitated by antibody-antigen interactions (inhibition, neutralization, opsonization, agglutination, precipitation) that result in phagocytosis by phagocytes. In addition, the formation of membrane attack complexes by complement directly lyses and kills pathogens.

20.3 The Cell Mediated Immune Response

- T-cell receptors (TCRs) on naive T cells recognize peptides contained within MHC-II proteins. The activated immature T cells then mature into T_H2 cells that stimulate humoral immunity or T_H1 cells that activate lysosomal killing in bacterially-infected macrophages.
- Cytotoxic T cells are activated by dendritic cells or APC/naive T-cell combination. Active cells recognize abnormal cells (virus-infected cells or tumors) presented as MHC-I proteins with bound antigen peptide. Binding of cytotoxic T cells triggers the release of perforins and granzymes, which lyse and kill the abnormal cells.
- T_H2 cells bind to MHC-II/peptides presented on the surface of B cells. Binding along with antigen cross-linking costimulate B-cell activation and the humoral response.

LEARNING OBJECTIVES

After understanding the textbook reading, you should be capable of writing a paragraph that includes the appropriate terms and pertinent information to answer the objective.

1. Identify and describe the four characteristics of an acquired immune response.
2. Distinguish between the cells responsible for a humoral immune response and a cell mediated immune response.
3. List the steps in clonal selection of B and T cells.
4. Trace the origins of lymphocytes versus other leukocytes.
5. Draw the structure of a monomeric antibody and label the parts.
6. Indentify the advantage for two antigen-binding sites on a monomeric antibody.
7. Summarize the characteristics for each of the five immunoglobulin classes.
8. Differentiate between a primary and secondary antibody response in terms of antibodies and responding cells.
9. Explain how antibody diversity is generated through gene arrangements.
10. Distinguish between the antibody mechanisms used to clear antigens (pathogens) from the body.
11. Summarize how complement can lyse a viral or bacterial pathogen.
12. Compare receptors, receptor binding, and function for cytotoxic T cells and naive T cells.
13. Assess the role of the major histocompatibility complex (MHC) to T-cell activation.
14. Summarize how naive T cells are activated and identify how these effects contribute to cellular immunity.
15. Discuss how cytotoxic T cells are activated and how they eliminate virus-infected cells.
16. Explain how T_H2 cells co-stimulate B-cell activation and the humoral response.

SELF-TEST

Answer each of the following questions by selecting the ***one*** answer that best fits the question or statement. Answers to even-numbered questions can be found in **Appendix C**.

1. The loss of self-tolerance could lead to
A. immune deficiencies.
B. autoimmune diseases.
C. damage to "self."
D. allergies.
E. Both **B** and **C** are correct.

2. All the following are immunogenic *except*:
A. bacterial flagella.
B. haptens.
C. "non-self" antigens.
D. bacterial pili.
E. viral spikes.

3. _____ cells are associated with _____ immunity while _____ cells part of _____ immunity.
A. B; cell mediated; T; innate
B. T; humoral; B; cellular
C. T; cell mediated; B; humoral
D. T; humoral; B; nonspecific
E. B; nonspecific; T; specific

4. Clonal selection includes
A. cross linking of antigen on B cells.
B. antigen-receptor binding on B cells.
C. antibody secretion recognizing same epitope as on B cell receptors.
D. differentiation of B cells into plasma cells and memory cells.
E. All the above (**A–D**) are correct.

5. Which one of the following cell types is *not* derived from myeloid stem cells?
A. Lymphocytes
B. Monocytes
C. Eosinophils
D. Basophils
E. Neutrophils

6. The elimination of lymphocytes recognizing "self-cells" is accomplished through
A. loss of self antigens.
B. apoptosis of the cells.
C. phagocytosis by macrophages.
D. control by regulatory T cells.
E. Both **B** and **D** are correct.

7. A monomeric antibody, such as IgG, consists of
A. one light and one heavy chain.
B. one light and two heavy chains.
C. two light and two heavy chains.
D. four light chains.
E. four heavy chains.

8. An antigen binding site on the IgG antibody is a combination of
A. one variable region from a light chain and one from a heavy chain.
B. two variable regions from two light chains.
C. two variable regions from two heavy chains.
D. one variable region from a constant region and one from a variable region.
E. four variable regions, two from a light chain and two from a heavy chain.

9. This antibody class crosses the placenta.
A. IgA
B. IgE
C. IgM
D. IgG
E. Both **C** and **D** are correct.

10. This dimeric antibody class often occurs in secretions of the respiratory and gastrointestinal tracts.
A. IgE
B. IgD
C. IgM
D. IgA
E. IgG

11. The presence of IgM antibodies in the blood indicates
A. an early stage of an infection.
B. a secondary infection.
C. cellular immunity is not activated.
D. humor immunity has yet to start.
E. a primary infection.

12. A/an _____ mechanism facilitates the elimination of toxins from the body.
A. opsonization
B. precipitation
C. agglutination
D. membrane attack
E. neutralization

13. T_H1 cells activate
A. B cells.
B. killing of pathogens in macrophages.
C. cytotoxic T cells.
D. humoral immunity.
E. memory cells.

14. Perforins and granzymes are found in
A. T_H2 cells.
B. antigen-presenting cells.
C. cytotoxic T cells.
D. B cells.
E. T_H1 cells.

15. T_H2 cells bind to
A. MHC-II/peptide complex on APC cells.
B. B cell receptors.
C. MHC-I/peptide complex on infected cells.
D. MHC-II/peptide complex on B cells.
E. MHC-II/peptide complex on infected cells.

QUESTIONS FOR THOUGHT AND DISCUSSION

Answers to even-numbered questions can be found in **Appendix C**.

1. The ancestors of modern humans lived in a sparsely settled world where communicable diseases were probably very rare. Suppose that by using some magical scientific invention, one of those individuals was thrust into the contemporary world. How do you suppose he or she would fare in relation to infectious disease? What is the immunological basis for your answer?
2. Your brother's high school biology text contains the following statement: "T cells do not produce circulating antibodies. Rather, they carry cellular antibodies on their surface." What is fundamentally incorrect about this statement?
3. Some time ago, an immunologist reported that cockroaches injected with small doses of honeybee venom develop resistance to future injections of venom that would ordinarily be lethal. Does this finding imply that cockroaches have an immune system? What might be the next steps for the research to take? What does this research tell you about the cockroach's ability to survive for three or four years, far longer than most other insects?

APPLICATIONS

Answers to even-numbered questions can be found in **Appendix C**.

1. Your microbiology professor suggests that an antigenic determinant arriving in the lymphoid tissue is like a parent searching for the face of a lost child in a crowd of a million children. Do you agree with this analogy? Why or why not?
2. In the book and classic movie, *Fantastic Voyage,* a group of scientists is miniaturized in a submarine (the Proteus) and sent into the human body to dissolve a blood clot. The odyssey begins when the miniature submarine carrying the scientists is injected into the bloodstream. Today, microscopic robots called nanorobots are being designed that would be injected into the body to fight diseases, including infectious ones. What do you think about this future microscopic robot technology?
3. As a consultant for the company Acme Nanobots, what immunological hurdles need to be considered before such microscopic robots could be fully developed?

REVIEW

Test your knowledge of this chapter's contents by determining whether the following statements are true or false. If the statement is true, write "True" in the space. If false, substitute a word for the underlined word to make the statement true. The answers to even-numbered statements are listed in **Appendix C**.

_____ 1. Cell mediated immunity involves T lymphocyte activity.

_____ 2. Small molecules called antigens are not immunogenic.

_____ 3. Cytokine is an alternate name for an antibody.

_____ 4. The end of an antibody molecule where an antigen binds is called the Fc fragment.

_____ 5. Antigenic materials are classified as "nonself."

_____ 6. Cells that secrete antibody molecules are B cells.

_____ 7. IgM consists of five monomers.

_____ 8. A secondary antibody response primarily involves IgG.

_____ 9. IgA are the largest antibody.

_____ 10. IgD has four heavy chains in the antibody molecule.

_____ 11. Epitope is an alternate name for an antigenic determinant.

_____ 12. Basophils phagocytize microorganisms and begin an immune response.

_____ 13. Dendritic cells are involved in an immune response.

_____ 14. Antibodies are transported in the blood.

_____ 15. Nucleic acids are part of an antibody molecule.

_____ 16. There are four polypeptide chains in a monomeric antibody molecule.

_____ 17. Secretory IgA has two antigen binding sites.

_____ 18. The humoral immune response depends on antibody activity.

_____ 19. The Fc region of an antibody consists of light chains.

_____ 20. Lysozyme is secreted by cytotoxic T cells.

_____ 21. IgA crosses the placenta.

_____ 22. The IgG antibody molecule has two identical halves.

_____ 23. Monocytes secrete antibody.

_____ 24. The IgE antibody is found on the surface of B lymphocytes.

_____ 25. T_H2 cells activate lysosome killing of intracellular bacterial cells.

HTTP://MICROBIOLOGY.JBPUB.COM/

The site features learning, an on-line review area that provides quizzes and other tools to help you study for your class. You can also follow useful links for in-depth information, read more MicroFocus stories, or just find out the latest microbiology news.

21

Immunity and Serology

Chapter Preview and Key Concepts

21.1 Immunity to Disease

- The form of active immunity depends on whether the host experiences a naturally occurring or artificially exposed antigen.
- All vaccines are designed to ultimately generate memory cells.
- The form of passive immunity depends on the patient receiving antibodies from an outside source, either naturally or artificially.
- When most of a population is vaccinated, disease spread is effectively stopped.
- Some individuals' immune system can react differently to a vaccine.

21.2 Serological Reactions

- Serological reactions generally consist of an antigen and a serum sample.
- Neutralization is a serological reaction in which antigens and antibodies neutralize each other.
- Precipitation is a serological reaction in which antigens and antibodies form a visible precipitate.
- Agglutination is a serological reaction in which antibodies interact with antigens on the surface of particular objects.
- Complement fixation depends on complement inactivation and hemolysis of sheep red blood cells.
- Fluorescent or radioactive antibodies can identify antigens (pathogens).

MicroInquiry 21: Applications of Immunology: Disease Diagnosis

21.3 Additional Laboratory Tests

- Monoclonal antibodies recognize specific cells or substances.
- Gene probes can detect pathogen DNA in a clinical sample.

"Vaccination is one of the most powerful means of protecting the public health."

—Marta A. Balinska, Institut National de Prévention et d'Education pour de Santé

Prior to the age of modern vaccines, the only way one could become immune to a disease was to contract the disease and hope for recovery. Unfortunately, the symptoms often were severe and perhaps disfiguring. There was the risk of complications, which if the disease did not kill, the complication might. In addition, ill individuals could be contagious and spread the disease to other unsuspecting and susceptible individuals. Epidemics and even pandemics could result.

As early as the eleventh century, Chinese doctors ground up smallpox scabs and blew the powder into the noses of healthy people to protect them against the ravages of smallpox (see Chapter 1). Jenner improved on this technique with his smallpox vaccination using a preparation from cowpox. Other diseases of past or recent history, including diphtheria and polio, were equally deadly to large numbers of people.

However, with the development of modern vaccines these and many other infectious diseases have been either eradicated (smallpox), almost eradicated (polio), or brought under control in much of the world (diphtheria). Equally important, vaccination is a technique to prevent disease from occurring—and preventing a disease from occurring is much safer and cheaper than trying to cure a disease after it occurs. For example, in one report for the United States, every dollar spent to vaccinate American children against rubella (German measles; see Chapter 14) saves more than $8 in treatment costs.

The value of vaccination can best be appreciated by examining how vaccines have changed the face of disease globally and in the United States. Cases of diseases like diphtheria and rubella have greatly declined

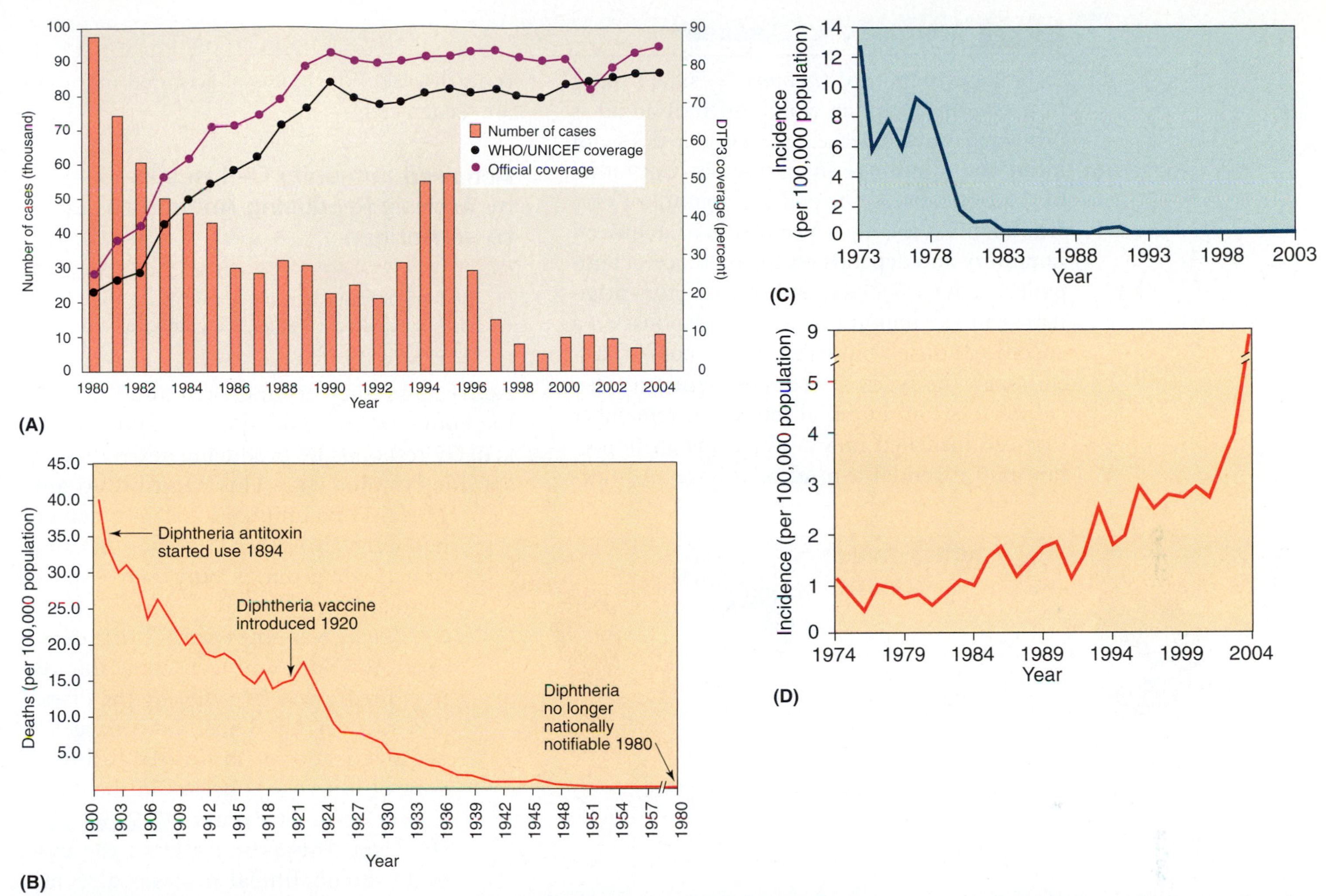

FIGURE 21.1 Vaccine Effects on Incidence and Mortality. (**A**) Global cases of diphtheria (bars) and vaccine coverage. With the establishment of vaccine coverage, the number of cases has dropped to all-time lows. *Source*: Data from the World Health Organization. (**B, C**) Deaths from diphtheria and rubella in the United States. After use of the antitoxin vaccine and then the diphtheria and rubella vaccines, mortality rates per 100,000 population have dropped to zero. *B Source*: Data from www.alternativehealth.co.nz. *C Source*: Data from the CDC, *Summary Notifiable Diseases*, US, 2003. p. 59. (**D**) However, even with a pertussis vaccine, pertussis incidence currently is at its highest level since 1964. *Source*: Data from the CDC, *Summary Notifiable Diseases*, US, 2003. p. 56.

Q: From these graphs, how do vaccines and vaccinations contribute to community health in preventing infectious diseases?

globally and in the United States with the introduction of vaccination (FIGURE 21.1A, B, C).

Unfortunately, many people have not been vaccinated against vaccine-preventable diseases. Many live in regions of the world where vaccines are not available while others have elected not to use the opportunity of vaccination to protect themselves and their family. Yet vaccines are a key to public health and its goal of preventing infectious disease.

However, vaccine-preventable diseases once under control in the United States also are resurfacing. Thanks to vaccination against pertussis, cases in the United States reached an all-time low in the mid-1970s (FIGURE 21.1D). Since 1981, for a set of reasons (see Chapter 9), there has been a slow but steady increase in the incidence of pertussis; which now is at its highest since 1964.

In this chapter, we study the role of vaccines in the immune response while examining the four major mechanisms by which immunity comes about. We also examine how antibodies may be detected in a disease patient by a variety of laboratory tests. These diagnostic procedures help the physician understand the disease and prescribe a course of treatment. Immune mechanisms generally protect against disease, but when they fail, the laboratory tests provide a clue about what is taking place in the patient.

21.1 Immunity to Disease

In biology, **immunity** (*immuno* = "safe") refers to a condition under which an individual is protected from disease. However, it does not mean one is immune to all diseases, but rather to a specific disease or perhaps a group of very similar diseases. It is the result of **acquired immunity** and depends on the presence of antibodies, T lymphocytes, and other factors originating in the immune system in response to a specific "nonself" antigen (see Chapter 20). Although the types of acquired immunity discussed focus on humoral immunity, remember that cell mediated immunity also is an important and essential arm of resistance to infectious disease. Four types of acquired immunity are recognized.

Acquired Immunity Can Result by Actively Producing Antibodies to an Antigen

KEY CONCEPT

- The form of active immunity depends on whether the host experiences a naturally occurring or artificially exposed antigen.

Active immunity occurs when antigens enter the body and the individual's immune system actively responds by producing antibodies and specific lymphocytes. This exposure to antigens may be unintentional, as when one becomes ill with a disease, or intentional, when one is purposely exposed to an antigen.

Before vaccines were fully developed, the only way to become immune to a disease was by suffering the disease and recovering. Thus, such **naturally acquired active immunity** follows a bout of illness and occurs in the "natural" scheme of events (FIGURE 21.2A). However, this is not always the case, because subclinical diseases also may bring on the immunity. For example, many people have acquired immunity from subclinical cases of mumps or from subclinical fungal diseases such as cryptococcosis. So, this active production of antibodies represents the primary antibody response described in Chapter 20.

Memory cells residing in the lymphoid tissues are responsible for the production of antibodies that yield naturally acquired active immunity. The cells remain active for many years and produce IgG almost immediately upon an ensuing exposure to the same antigen or pathogen as triggered the primary antibody response. This ensuing, or secondary, antibody response also was described in Chapter 20.

Artificially acquired active immunity is less risky and represents an easier way to become immune to an infectious dis-

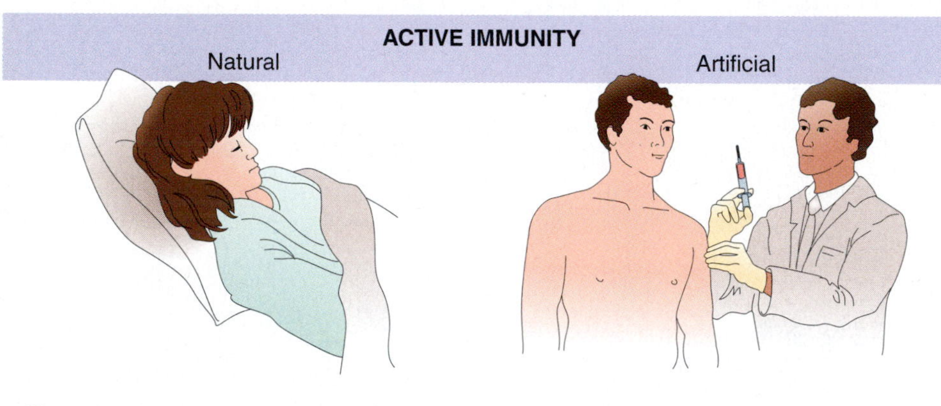

(A) Naturally acquired active immunity arises from an exposure to antigens and often follows a disease.

(B) Artificially acquired active immunity results from an inoculation of toxoid or vaccine.

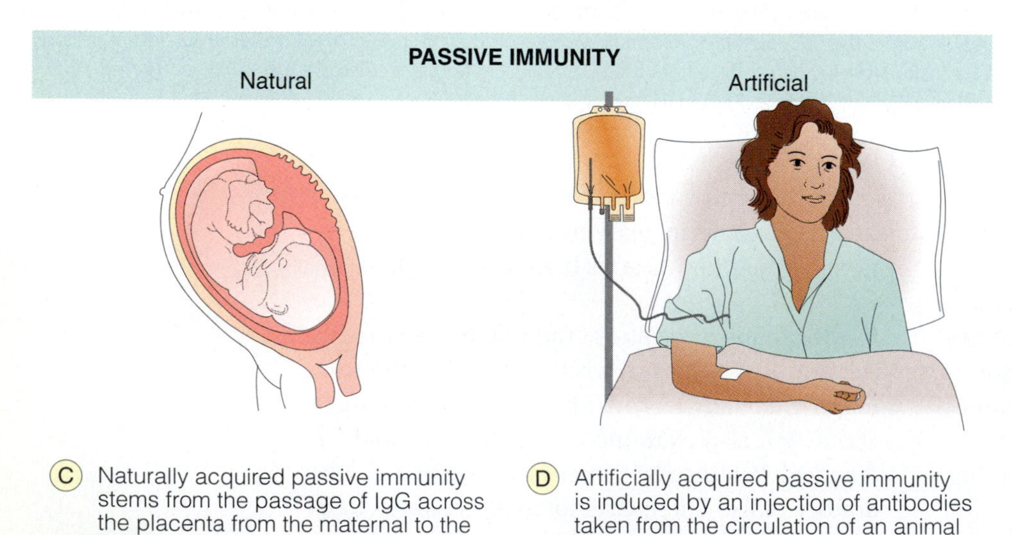

(C) Naturally acquired passive immunity stems from the passage of IgG across the placenta from the maternal to the fetal circulation.

(D) Artificially acquired passive immunity is induced by an injection of antibodies taken from the circulation of an animal or another person.

FIGURE 21.2 **The Four Types of Acquired Immunity.** Immunity can be natural or artificial and acquired in an active or passive form.

Q: Why do these forms of immunity represent an acquired response rather than an innate response to an antigen?

ease. This form of active immunity develops after the immune system produces antibodies following an intentional exposure to antigens; that is, through **vaccination** (FIGURE 21.2B). Because the antigens usually are contained in an immunizing agent, such as an inactivated or toxoid vaccine, the exposure is called "artificial."

Vaccines are composed of treated microorganisms or viruses, chemically altered toxins, or chemical parts of microorganisms. Such vaccines work by mimicking a "natural" infection. By exploiting the immune system's ability to recognize antigens and respond with antibodies and lymphocytes, a vaccine triggers a primary antibody response. However, because the vaccine has been altered in some way (see below), the pathogen or toxin usually does not trigger the disease and the person vaccinated does not become ill. Importantly, memory cells are formed, which now establish active immunity. If the vaccinated person is exposed to the same antigen at some later time, the immune system acts swiftly and produces a secondary antibody response, stopping the infection before it can make the individual sick. The person stays healthy.

Vaccines may be administered by injection, oral consumption, or nasal spray, as currently used for some respiratory viral diseases.

Let's now examine the different types of vaccines and find out how they work.

CONCEPT AND REASONING CHECKS

21.1 How do the naturally-acquired and artificially-acquired forms of active immunity differ?

There Are Several Types of Vaccine Strategies

KEY CONCEPT

- All vaccines are designed to ultimately generate memory cells.

There are seven strategies for producing vaccines. The viral and bacterial vaccines currently in use in the United States are summarized in TABLE 21.1.

Live, Attenuated Vaccines. Some microbes can be weakened in the lab such that they should not cause disease. They are still able to grow or replicate, but **attenuated** means they will multiply only at low rates in the body and fail to cause symptoms of disease. Because the attenuated microbes multiply or replicate for a period of time within the body, they increase the dose of antigen to which the immune system will respond. Such vaccines are the closest to the natural pathogens and, therefore, they generate the strongest immune response. Often, the person vaccinated will have lifelong immunity. Also, attenuated organisms can be spread to other people and reimmunize them, or immunize them for the first time.

The downside of attenuated vaccines results from their continued multiplication. Because the vaccines contain living organisms, there is a remote chance one of them could mutate (see Chapter 7) and revert back to a virulent form capable of causing disease. Usually a healthy person with a fully functioning immune system (immunocompetent) will be able to clear the infection without serious consequence. However, individuals with a compromised immune system, such as patients with AIDS, should not be given attenuated vaccines, if possible.

Today, there are many viral vaccines that consist of attenuated viruses. The Sabin oral polio vaccine as well as the measles, mumps, and chickenpox vaccines contain attenuated viruses. To avoid multiple injections of immunizing agents, it is sometimes advantageous to combine vaccines into a **single-dose vaccine**. The measles-mumps-rubella (MMR) vaccine is one example. In 2005, the US Food and Drug Administration (FDA) approved a combination vaccine (Proquad) for children 12 months to 12 years old. This single-dose vaccine protects against chickenpox, measles, mumps, and rubella.

Making a vaccine with attenuated bacterial cells is more difficult. In fact, there are no such vaccines routinely used in the United States. The BCG tuberculosis vaccine (see Chapter 9), which is used in some other countries, is composed of attenuated *Mycobacterium bovis* bacterial cells.

On a global scale, live, attenuated vaccines may not be the vaccine strategy of choice. These vaccines require refrigeration to retain their effectiveness, which could present a problem in many developing nations lacking widespread refrigeration facilities.

Inactivated Vaccines. Another strategy for preparing vaccines is to kill the pathogen.

TABLE 21.1 The Principal Bacterial and Viral Vaccines Currently in Use

Disease	Route of Administration	Recommended Usage/Comments
Contain Killed Whole Bacteria		
Cholera	Subcutaneous (SQ) injection	For travelers; short-term protection
Typhoid	SQ and intramuscular (IM)	For travelers only; variable protection
Plague	SQ	For exposed individuals and animal workers; variable protection
Contain Live, Attenuated Bacteria		
Tuberculosis (BCG)	Intradermal (ID) injection	For high-risk occupations only; protection variable
Subunit Bacterial Vaccines (Capsular Polysaccharides)		
Meningitis (meningococcal)	SQ	For protection in high-risk individuals, such as military recruits; short-term protection
Meningitis (*H. influenzae*)	IM	For infants and children; may be administered with DTaP
Pneumococcal pneumonia	IM or SQ	Important for people at high risk: the young, elderly, and immunocompromised; moderate protection
Pertussis	IM	For newborns and children
Subunit Bacterial Vaccine (Protective Antigen)		
Anthrax	SQ	For lab workers and military personnel
Toxoids (Formaldehyde-Inactivated Bacterial Exotoxins)		
Diphtheria	IM	A routine childhood vaccination; highly effective in systemic protection
Tetanus	IM	A routine childhood vaccination; highly effective
Botulism	IM	For high-risk individuals, such as laboratory workers
Contain Inactivated Whole Viruses		
Polio (Salk)	IM	Routine childhood vaccine; highly effective; safer than Sabin vaccine
Rabies	IM	For individuals sustaining animal bites or otherwise exposed; highly effective
Influenza	IM	Persons at high risk and those living with or caring for persons at high risk
Hepatitis A	IM	Protection for travelers and anyone at risk; effectiveness not established
Contain Attenuated Viruses		
Adenovirus infection	Oral	For immunizing military recruits
Measles (rubeola)	SQ	Routine childhood vaccine; highly effective
Mumps (parotitis)	SQ	Routine childhood vaccine; highly effective
Polio (Sabin)	Oral	Routine childhood vaccine; highly effective; possible vaccine-induced polio
Smallpox (vaccinia)	Pierce outer layers of skin	For lab workers, military personnel, health care workers
Rubella	SQ	Routine childhood vaccine; highly effective
Chickenpox (varicella)	SQ	Routine childhood vaccine; immunity can diminish over time; effectiveness not yet established
Yellow fever	SQ	For travelers, military personnel in endemic areas
Influenza	IM	Persons at high risk and those living with or caring for persons at high risk
Recombinant Viral Vaccine		
Hepatitis B	IM	Medical, dental, laboratory personnel; newborns, others at risk; highly effective

MICROFOCUS 21.1: Public Health/Tools

Preparing for Battle

Chicken eggs being "inoculated" with a flu virus.

Each flu season approximately 10 to 20 percent of Americans get the flu. Of this number, more than 114,000 are hospitalized and some 36,000 die from the complications of flu. Many of these hospitalizations and deaths could be prevented with a yearly flu shot, especially for those people at increased risk (people over 50, immunocompromised individuals, and health care workers in close contact with flu patients). Because influenza viruses change often (see Chapter 14), the influenza vaccine is updated each year to make sure it is as effective as possible. How is each year's vaccine designed?

Each flu season, information on circulating influenza strains and epidemiological trends is gathered by the World Health Organization (WHO) Global Influenza Surveillance Network. The network consists of 112 national influenza centers in 83 countries and four WHO Collaborating Centers for Reference and Research on Influenza located in Atlanta, United States; London, United Kingdom; Melbourne, Australia; and Tokyo, Japan. The national influenza centers sample patients with influenza-like illness and submit representative isolates to WHO Collaborating Centres for immediate strain identification.

Twice a year (February: northern hemisphere; September: southern hemisphere), WHO meets with the Directors of the Collaborating Centres and representatives of key national laboratories to review the results of their strain identifications and to recommend the composition of the influenza vaccine for the next flu season. Since 1972, WHO has recommended 39 changes in the influenza vaccine formulation.

In the United States, the Food and Drug Administration (FDA) or the Centers for Disease Control and Prevention (CDC) provide the viral strains to American vaccine manufacturers in February. Each virus strain is grown separately in chicken eggs (see figure). After it has replicated many times, the fluid containing the viruses is removed, and the viruses purified and attenuated or inactivated. Then, the appropriate strains are mixed together with a carrier fluid. For the 2006–2007 season, the American trivalent vaccine consisted of A/Wisconsin/67/2005 (H3N2)-like, A/New Caledonia/20/99 (H1N1)-like, and B/Shanghai/361/2002-like antigens. Production usually is completed in August and ready for shipment in October.

These vaccines are relatively easy to create because the pathogen is killed by simply using certain chemicals, heat, or radiation. However, the inactivation process alters the antigen so it produces a weaker immune response.

The Salk polio vaccine and the hepatitis A vaccine typify such preparations of inactivated whole viruses. For protection from diseases like hepatitis A, **booster shots** are required to maintain immunity (memory cells) for long periods of time. In the case of influenza, the virus genetically changes from year to year (see Chapter 14), so a different vaccine must be provided annually. MicroFocus 21.1 describes how the components of the flu vaccine are decided each year.

Some whole organism (bacterial) vaccines are used for short-term protection. For instance, bubonic plague and cholera vaccines are available to limit an epidemic. In these cases, the immunity lasts only for several months because the material in the vaccine is weakly antigenic. Weakly antigenic vaccines are available also for laboratory workers who deal with rickettsial diseases such as Rocky Mountain spotted fever, Q fever, and typhus (see Chapters 9 and 11).

Compared to attenuated vaccines, inactivated vaccines are safer as they cannot mutate

MICROFOCUS 21.2: Public Health

Quantity May Not Always Equate with Quality

In the United States today, most young children receive some 20 vaccinations against various childhood diseases. For this reason, vaccine immunologists have tried to combine vaccines as single-dose vaccines to reduce the number of trips to the doctor and make it easier for parents to "remember" a child's vaccination schedule.

Most combined vaccines are as effective as single disease vaccines in stimulating the immune system to produce antibodies for immunity. However, in 2005, British researchers randomly vaccinated 119 infants through a series of shots against meningococcal meningitis (accounts for about 40 percent of meningitis cases in the United Kingdom). They also vaccinated another 115 infants with a combined nine-valent vaccine (for meningococcal meningitis and eight other forms of meningitis and pneumonia).

Then, each month for three months all infants were vaccinated with vaccines for diphtheria, pertussis, tetanus, and *Haemophilus influenzae* type b (Hib) meningitis. They also received shots for polio.

At 5 months of age, the infants had blood drawn and antibody levels to meningococcal meningitis, diphtheria, and Hib meningitis tested. The infants receiving the nine-valent vaccine had only 25 percent of the anti-meningococcal meningitis antibodies as did those infants receiving the meningococcal meningitis-only vaccine. They also produced fewer antibodies against diphtheria and Hib antigens.

Scientists were left scratching their heads to figure out why this was so. They discovered that the source of meningococcal meningitis antigens in the nine-valent vaccine was different from that in the meningitis-only vaccine. Perhaps the quality was different in the two vaccines.

Other researchers believe it might be the quantity of antigens and carrier proteins in the two vaccines. The nine-valent vaccine had much higher levels of these carrier proteins, which are added to increase the immunogenicity of the vaccine and boost immunity. Could these proteins be interfering with the immune system's ability to recognize and respond to the actual antigens?

These types of technical problems need to be determined before further multi-valent, combined vaccines are developed. However, one drug company is working on a 13-valent vaccine to protect children against group C meningitis and 12 other forms of meningitis and pneumonia.

and therefore cannot cause the disease in a vaccinated individual. The vaccines can be stored in a freeze-dried form at room temperature, making them a vaccine of choice in developing nations. However, the need for booster shots can be a drawback if people do not keep up their booster schedule.

Toxoid Vaccines. For some bacterial diseases, such as diphtheria and tetanus, a bacterial toxin is the main cause of illness. So, a third strategy is to inactivate these toxins and use them as a vaccine. Such toxins can be inactivated with formalin, and the inactivated toxin is called a **toxoid**. Immunity of a toxoid vaccine allows the body to generate antibodies and memory cells to recognize the natural toxin, should the individual come in contact with it. Again, being an inactivated product, booster shots are necessary.

■ Formalin: A solution of formaldehyde in water.

Single-dose vaccines include diphtheria-pertussis-tetanus vaccine (DPT) and the newer diphtheria-tetanus-acellular pertussis (DTaP) vaccine. For other vaccines, however, a combination single-dose vaccine may not be useful because the antibody response may be lower for the combination than for each vaccine taken separately (MicroFocus 21.2).

Subunit Vaccines. Unlike the whole agent attenuated or inactivated vaccines, the strategy for a subunit vaccine is to have the vaccine contain only those parts or subunits of the antigen that stimulate a strong immune response. These subunits may be epitopes (see Chapter 20). For example, the subunit vaccine for pneumococcal pneumonia contains 23 different polysaccharides from the capsules of 23 strains of *Streptococcus pneumoniae*.

One way of producing a subunit vaccine is to use recombinant DNA technology (see Chapter 8) and the resulting vaccine is called a **recombinant subunit vaccine**. The hepatitis B vaccine (Recombivax HB or Engerix-B) is an example. Several hepatitis B virus genes are

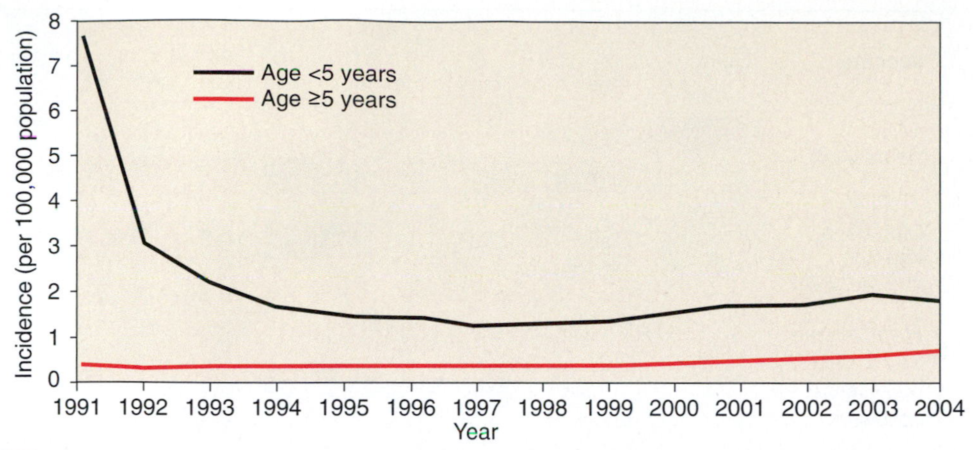

FIGURE 21.3 ***Haemophilus influenzae.*** Incidence, by age group—United States, 1991–2004. *Source*: Courtesy of the CDC. *Q: Although the Hib vaccine was introduced in 1987, why didn't the incidence rate start to dramatically drop until 1992?*

isolated and inserted into yeast cells, which then synthesize the antigens. These antigens are collected and purified to make the vaccine.

Adverse reactions to such subunit vaccines are very rare because only the important subunits of the antigen are included in the vaccine. These subunits cannot produce disease illness in the person vaccinated. Also, the vaccine is not made from blood fragments (as was a previous hepatitis B vaccine), so it relieves the fear of contracting human immunodeficiency virus (HIV) from contaminated blood.

Conjugate Vaccines. *Haemophilus influenzae* b (Hib), which is responsible for a form of childhood meningitis, produces an external glycocalyx coat called a capsule (see Chapter 4). Since the capsular polysaccharides by themselves are not strongly immunogenic, the strategy here is to conjugate (attach) capsular polysaccharides to tetanus or diphtheria toxoid, which will stimulate a strong immune response. The result is the Hib vaccine, which has been a critical factor in reducing the incidence of *Haemophilus* meningitis from 18,000 cases annually in 1986 to 19 cases in 2004 (FIGURE 21.3). In 2005, a conjugate vaccine (MCV-4) was licensed for a meningococcal vaccine against meningitis caused by *Neisseria meningitidis*.

FIGURE 21.4 outlines the recommendations for these vaccines and others in healthy infants, children, and adolescents.

DNA Vaccines. Part of the renaissance in vaccine development is a strategy to use DNA as a vaccine. It appears some cells in the body will take up injected foreign DNA, commence to make proteins encoded by the DNA, and display these antigens on its surface, much like an infected cell presents antigen fragments to T cells (see Chapter 20). Such a display should stimulate a strong antibody (humoral) and cell mediated immune response.

These experimental "**naked DNA vaccines**" consist of plasmids engineered to contain a protein-encoding gene from a viral or bacterial pathogen. They are simply injected in a saline solution. Unlike replicating viruses or live bacteria, plasmids are not infectious or replicative, nor do they encode any proteins other than those specified by the plasmid genes, so they have a measure of safety; someone vaccinated with a DNA vaccine could not contract the disease. They are relatively easy to construct and produce, and the vaccines are more stable than conventional vaccines at low and high temperatures, making shipping easier. At this writing, DNA vaccines against several diseases, including herpes, influenza, and malaria, are in human clinical trials.

Recombinant Vector Vaccines. The seventh and last vaccine strategy builds on the DNA vaccines. Rather than injecting the naked plasmid DNA, the DNA is first incorporated (recombined) into a vector (attenuated virus or bacterium), which is used to carry the DNA into the person being vaccinated.

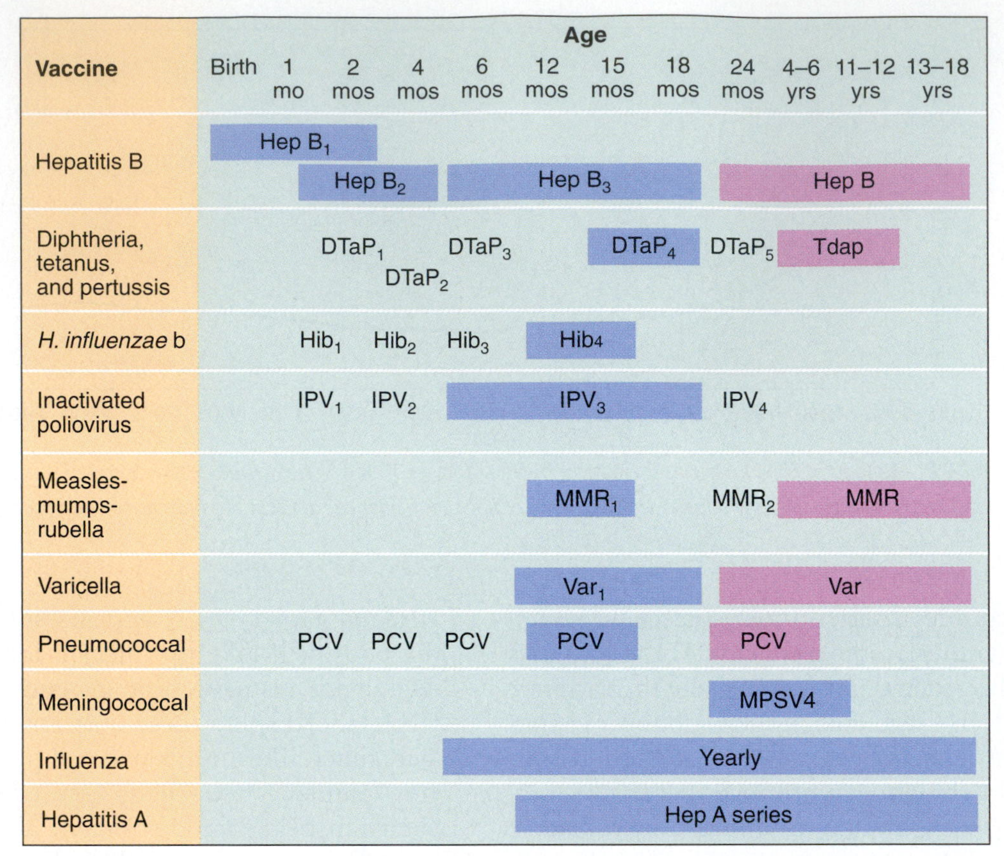

FIGURE 21.4 **Childhood and Adolescent Immunization Schedule—2006.** Each year the Advisory Committee on Immunization Practices of the CDC reviews the childhood and adolescent immunization schedule. This schedule indicates the recommended ages for routine administration of licensed vaccines. "Catch-up" vaccination may be administered if a visit to the doctor is missed. Additional recommendations concerning the schedule are available from the CDC at http://www.cdc.gov/nip/recs/child-schedule.htm.

An experimental recombinant vector vaccine using a virus takes the DNA into the body cells whereas a bacterial vector will incorporate and display the antigen on its surface. Both vectors will be seen as foreign and trigger a strong immune response.

MicroFocus 21.3 discusses substances often added to vaccines to increase their effectiveness.

TABLE 21.2 summarizes the vaccines universally recommended by the Centers for Disease Control and Prevention (CDC) as of 2006. A missing vaccine on this list is one for HIV disease/AIDS. It has been over 20 years since the virus was discovered, yet there still is no vaccine. MicroFocus 21.4 outlines some reasons for the lack of a vaccine.

CONCEPT AND REASONING CHECKS

21.2 How do attenuated, inactivated, subunit, and conjugate vaccines differ from one another?

Acquired Immunity Also Can Result by Passively Receiving Antibodies to an Antigen

KEY CONCEPT

- The form of passive immunity depends on the patient receiving antibodies from an outside source, either naturally or artificially.

Recall that in this section we have been discussing how one develops acquired immunity. The previous discussion looked at the active forms of acquired immunity. We finish this section by examining the ways one acquires antibodies in a passive and temporary manner.

Passive immunity develops when antibodies enter the body from an outside source (in contrast to active immunity, in which individuals synthesize their own antibodies). Again, the source of antibodies may be unintentional, such as a fetus receiving antibodies from the

TABLE 21.2 Universally Recommended Vaccinations for Children, Adolescents, and Adults

Population	Vaccination	Dosage
All young children	Measles, mumps, and rubella	2 doses
	Diphtheria-tetanus toxoid and pertussis vaccine	5 doses
	Polio	4 doses
	Haemophilus influenzae b	3–4 doses
	Hepatitis B	3 doses
	Varicella	1 dose
Previously unvaccinated or partially vaccinated adolescents	Hepatitis B	3 doses, total
	Varicella	If no previous history of varicella, 1 dose for children <12 years, 2 doses for children ≥12 years
	Measles, mumps, and rubella	2 doses, total
	Tetanus-diphtheria toxoid	If not vaccinated during previous 5 years, 1 combined booster during ages 11–16
All adults (19–49)	Varicella	2 doses for susceptible
	Tetanus-diphtheria (Td) toxoid	1 dose administered every 10 years
	Hepatitis A	2 doses total
	Hepatitis B	3 doses total
All adults >50 years	Influenza	1 dose administered annually
All adults >65 years	Pneumococcal	1 dose

mother or intentional, such as the transfer of antibodies from one individual to another (Figure 21.2).

Naturally acquired passive immunity, also called congenital immunity, develops when antibodies pass into the fetal circulation from the mother's bloodstream via the placenta and umbilical cord. The process occurs in the "natural" scheme of events.

The maternal IgG antibodies (see Chapter 20) remain with the child for approximately three to six months after birth and play an important role during these months of life by providing resistance to diseases such as pertussis, staphylococcal infections, and viral respiratory diseases. Certain antibodies, such as measles antibodies, remain for 12 to 15 months.

Maternal antibodies also pass to the newborn through the first milk, or **colostrum**, of a nursing mother as well as during future breast-feedings. In this instance, IgA is the predominant antibody, although IgG and IgM also have been found in the milk. The antibodies accumulate in the respiratory and gastrointestinal tracts of the child and lend increased disease resistance.

Artificially acquired passive immunity arises from the intentional injection of antibody-rich serum into the patient's circulation. The exposure to antibodies is thus "artificial." In the decades before the development of antibiotics, such an injection was an important therapeutic tool for the treatment of disease. The practice still is used for viral diseases such as Lassa fever and arthropod-borne encephalitis, and for bacterial diseases in which a toxin is involved. For example, established cases of botulism, diphtheria, and tetanus are treated with serum containing the respective antitoxins.

Various terms are used for the serum that renders artificially acquired passive immunity. **Antiserum** is one such term. Another is **hyperimmune serum**, which indicates the serum has a higher-than-normal level of a

MICROFOCUS 21.3: Tools

Enhancing Vaccine Effectiveness

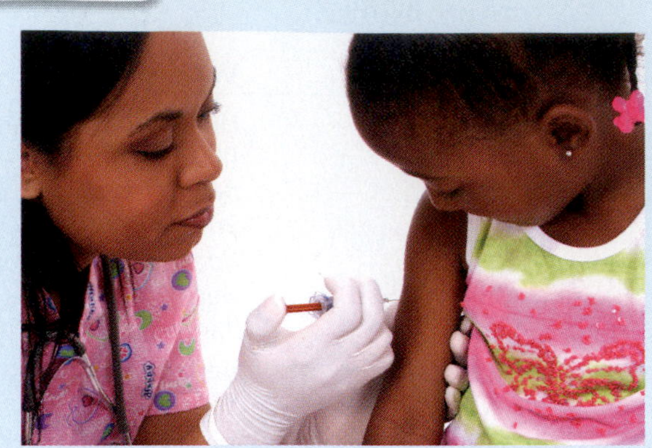

Often it may be necessary to add a substance to a vaccine to improve its ability to stimulate the immune system. One of those substances is **adjuvants**, which increase the efficiency of a vaccine by increasing the availability of the antigen in the lymphatic system.

The adjuvants licensed for human use in the United States are aluminum salts, such as aluminum sulfate ("alum") and aluminum hydroxide in toxoid preparations as well as mineral oil or peanut oil in viral vaccines.

The particles of adjuvant linked to antigen are taken up by macrophages and presented to lymphocytes more efficiently than dissolved antigens by themselves. Experiments also suggest that adjuvants may stimulate the macrophage to produce interleukin-1 (see Chapter 20), a lymphocyte-activating factor, and thereby reduce the necessity for helper T-cell activity. Adjuvants also provide slow release of the antigen from the site of entry and provoke a more sustained immune response than the antigen by itself.

Besides adjuvants, some vaccines may contain antibiotics. These may be added during the manufacturing process to prevent bacterial contamination. The flu vaccine shortage of 2004, for example, was due to contamination by *Serratia marcescens* during the manufacturing process in Liverpool, England. About half of the American flu vaccine was to come from that source, so the supply had to be thrown out.

Sometimes stabilizers are added to vaccines to maintain their effectiveness in less-than-optimal conditions. However, another substance has been of concern. This is **thimerosal**, a substance that had been put in some vaccines (not MMR) as a preservative. Thimerosal contains high concentrations of ethylmercury, which is a neurotoxin. Although considered a low-level exposure, the American Academy of Pediatrics, the US Public Health Service, and vaccine manufacturers stated in 1999 that as a precautionary measure thimerosal should be removed from or eliminated in all vaccines except some flu vaccines and the tetanus-diphtheria (Td) vaccine, which is given to children 7 years and older. Thimerosal is not found in any of the pediatric vaccines given to preschool children to protect them from 12 infectious diseases.

Although thimerosal was removed as a precautionary measure, it should be noted that almost all major studies on thimerosal, which have involved hundreds of thousands of children around the world, have not found any association between thimerosal exposure and harm to children. Only minor reactions, including redness and swelling at the injection site have been documented.

■ Prophylactic: Refers to a drug or agent preventing the development of a disease.

particular antibody. If the serum is used to protect against a disease such as hepatitis A, it is called prophylactic **serum**. When the serum is used in the therapy of an established disease, it is called **therapeutic serum** and when taken from the blood of a convalescing patient, physicians refer to it as **convalescent serum**. Another common term, **gamma globulin**, takes its name from the fraction of blood serum in which most antibodies are found. Gamma globulin usually consists of a pool of sera from different human donors, and thus it contains a mixture of antibodies (usually IgG), including those for the disease to be treated.

Passive immunity must be used with caution because in many individuals, the immune system recognizes foreign serum proteins as non-self antigens and synthesizes antibodies against them in an allergic reaction. When antibodies interact with the proteins, a series of chemical molecules called **immune complexes** may form and, with the activation of complement, the person develops a condition

MICROFOCUS 21.4: Public Health

An AIDS Vaccine—Why Isn't There One After All These Years?

Producing an AIDS vaccine might appear rather straightforward: Cultivate a huge batch of human immunodeficiency virus (HIV), inactivate it with chemicals, purify it, and prepare it for marketing. In fact, this was the mentality and approach in the mid-1980s. Unfortunately, things are not quite so simple when HIV is involved. Here are the major reasons why a vaccine remains elusive after all these years.

The effects of a bad batch of weakened or inactivated vaccine would be catastrophic, and people generally are reluctant to be immunized with whole HIV particles, no matter how reassuring the scientists. In addition, how the body actually protects itself from pathogens such as HIV is not yet completely understood. Without that understanding, an effective vaccine cannot be developed.

Vaccine development also has been slow because of the high mutation rate of the virus. Some HIV strains around the world vary by as much as 35 percent in the capsid and envelope proteins they possess. Thus, HIV is more mutable than the influenza viruses, and no vaccine has been developed yet to make one completely immune to influenza. It is hard to make a vaccine targeted at one strain when the virus keeps changing its coat.

Additional reasons why a vaccine has not been developed include factors such as a short-lived vaccine would protect an individual only for a short time, necessitating an endless series of scheduled booster shots to which few would adhere. A vaccine also might act like dengue fever, where vaccination could actually make someone more at risk if they actually were to contract the disease. In 2002, an HIV patient who was holding the virus in check became infected with another strain of HIV through unprotected sex. This patient became "superinfected," meaning his immune system could keep the original strain in check, but was powerless to control the new strain. So, could a vaccine produce the same result if the individual was infected with another strain of HIV? Almost every AIDS vaccine researcher around the world is concerned about the unknown factors concerning an AIDS vaccine.

As of 2006, no vaccine trials have been shown to stimulate cell mediated immunity to a level necessary to destroy HIV. The number of cytotoxic T cells and memory T cells produced is not up to the job of combating HIV. Since the inception of HIV vaccine research, there has been a lack of funds and leadership needed for rapid progress. In March 2002, Anthony Fauci, Director of the National Institute for Allergy and Infectious Diseases, reported to the Presidential Advisory Council on HIV and AIDS that a "broadly effective AIDS vaccine could be a decade or more away." Let's keep our fingers crossed and hope some breakthroughs are made soon.

called **serum sickness** (Chapter 22). This often is characterized by a hive-like rash at the injection site, accompanied by labored breathing and swollen joints.

Although artificially-acquired passive immunity provides substantial and immediate protection against disease, it is only a temporary mea-sure. The immunity developing from antibody-rich serum usually wears off within weeks or months. For example, a person traveling to a country where hepatitis A is prevalent can obtain a serum preparation of hepatitis A several weeks prior to departing. The antiserum usually comes from blood donors routinely screened for hepatitis A.

The four types of immunity are summarized in TABLE 21.3.

CONCEPT AND REASONING CHECKS

21.3 How do the naturally-acquired and artificially-acquired forms of passive immunity differ?

Herd Immunity Results from Effective Vaccination Programs

KEY CONCEPT

- When most of a population is vaccinated, disease spread is effectively stopped.

A population without a vaccination program is vulnerable to disease epidemics. Many people will suffer from the disease; some may die while others could be left with a permanent

TABLE
21.3 Characteristics of the Four Types of Immunity

Type of Immunity	Immunizing Agent	Exposure to Immunizing Agent	Effective Dose Required	Relative Time Until Immunity	Relative Duration of Immunity
Naturally acquired active	Antigens	Unintentional	Small	Long	Long (lifetime)
Artificially acquired active	Antigens	Intentional	Small	Long	Long (months to years)
Naturally acquired passive	Antibodies	Unintentional	Large	Short	Short (4–6 months)
Artificially acquired passive	Antibodies	Intentional	Large	Short	Short (up to 6 weeks)

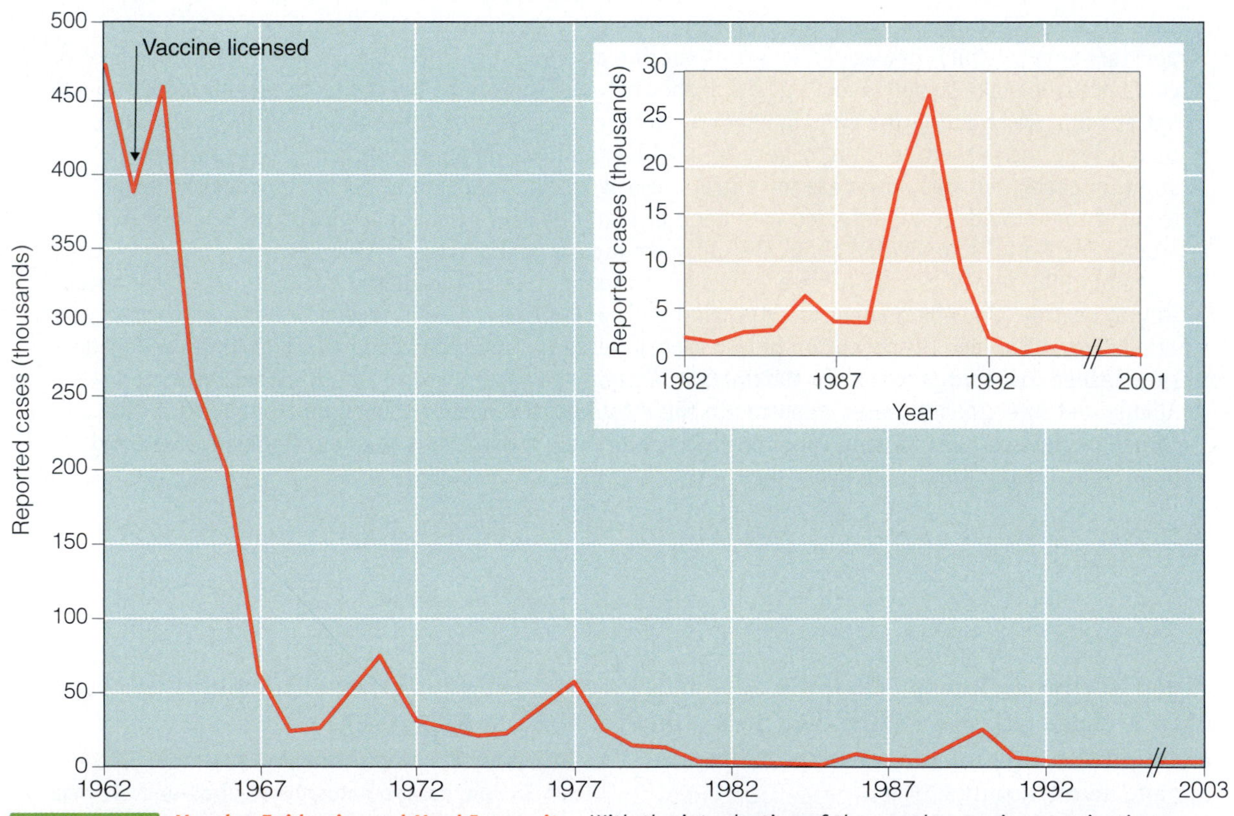

FIGURE 21.5 **Measles Epidemics and Herd Immunity.** With the introduction of the measles vaccine, vaccinations quickly built up a high herd immunity and few measles epidemics occurred. However, between 1989 and 1991 (inset), a measles epidemic occurred among college students and some groups of preschool children. *Source*: Courtesy of the CDC.
Q: Propose a hypothesis to explain the measles epidemic in 1989–1991.

disability. Even with a vaccination program, if insufficient numbers of citizens get the vaccination, the pathogen still can infect those who are not protected (FIGURE 21.5).

Vaccinations are never meant to reach 100 percent of the population. This is partly due to some people simply not being vaccinated and because some individuals simply respond poorly to a vaccine. They remain susceptible and unprotected. Still, the lack of 100 percent vaccination is not seen as a problem as long as most of the population is immune to an infectious disease. This makes it unlikely a susceptible person will come in con-

Usual Route of Introduction	Source of Antibodies	Function	Effectiveness in Newborn	Effectiveness in Adult	Origin
Various tissues	Self	Therapeutic Prophylactic	Low	High	Clinical or subclinical disease
Intramuscular or intradermal	Self	Prophylactic	Low	High	Toxoid or vaccine
Intravenous	Other than self	Prophylactic	High	Low	Transplacental passage of antibodies
Intravenous	Other than self	Prophylactic Therapeutic	High	Moderate	Serum that contains antibodies

tact with an infected individual. This phenomenon, called **herd immunity**, implies that if enough people in a population are immunized against certain diseases, then it is very difficult for those diseases to spread. Microbiologists and epidemiologists have found when approximately 90 percent of the population is vaccinated, the spread of the disease is effectively stopped. The other 10 percent of the "herd" or population remains susceptible, which is allowable because it is very hard for pathogens "to find someone" who isn't vaccinated. Susceptible individuals are protected from catching the disease and if the person should catch the disease, there are so many vaccinated people that it is unlikely the person would be able to spread it.

Herd immunity can be affected by several factors. One is the environment. People living in crowded cities are more likely to catch a disease if they are not vaccinated than nonvaccinated people living in rural areas because of the constant close contact with other people in the city.

Another factor is the strength of an individual's immune system. People whose immune systems are compromised, either because they have a disease or because of medical treatment (anticancer or antirejection drugs), may not be able to be immunized. They stand at a greater risk of catching the disease to which others are immunized.

MicroFocus 21.5 looks at the consequences of declining vaccinations and herd immunity.

CONCEPT AND REASONING CHECKS

21.4 What are the major factors required for a population to maintain high herd immunity?

Do Vaccines Have Dangerous Side Effects?

KEY CONCEPT

- Some individuals' immune system can react differently to a vaccine.

If you look back at Figure 21.4, notice the current immunization schedule in the United States for young children consists of up to 12 shots of five vaccines by six months of age and an additional four shots of six more vaccines by 18 months. Records show that vaccinations have been very successful at eliminating or greatly reducing the incidence of childhood diseases in the United States. Still, many parents wonder if this number of vaccines can have adverse effects on a developing young child, and if so, should they avoid taking their child to the doctor for routine vaccinations.

The FDA requires vaccine manufacturers to follow extensive safety procedures when producing vaccines in the United States to ensure they are safe and effective. After lab and animal tests have been completed, a promising vaccine must undergo thorough human clinical trials before being licensed by the FDA for use. Still, a reaction may occur that can cause mild fever, soreness at the injection site, or malaise from a vaccination.

It is estimated that one individual in 100,000 vaccinated may develop a serious

MICROFOCUS 21.5: Public Health

The Risk of Not Vaccinating

Data from epidemiological studies clearly show that vaccinations essentially eliminate outbreaks of infectious disease. Yet, about half the states in the United States allow parents to apply for vaccination waivers to enter their children in school. In mid-2003, the Centers for Disease Control and Prevention (CDC) reported that only about 75 percent of American children have been vaccinated on schedule against the nine diseases mandated by federal law. The CDC said that although coverage varied widely among states and major cities, there were areas in the country where too many children are not up-to-date on their shots. In addition, only a small percentage of adults receive booster shots for diseases like tetanus and diphtheria. Two examples point out the risk of not vaccinating.

In 1998, Colorado allowed vaccination exemptions for 2 percent of eligible children. This was double the exemption rate nationally. The analysis of Colorado and national health records showed that between 1987 and 1998, Colorado children three to eight years of age had a 22-times higher risk of contracting measles and nearly 6-times the risk of contracting pertussis (whooping cough) than did children nationally. One assumes the large number of unvaccinated children (lower herd immunity) was the main reason for the higher number of disease cases.

In 2003, a report indicated that measles outbreaks were becoming more prevalent in the United Kingdom. The reason was that more parents were not having their children vaccinated with the MMR vaccine because of its alleged side effects. In 1998, vaccinations dropped as there were reports that the MMR vaccine could cause autism. Even though such reports have been shown to be incorrect, vaccinations with the MMR vaccine have continued to drop. Herd immunity is losing its prescribed effects and, unless reversed, health officials fear that a continued drop could lead to the reestablishment of measles as an endemic disease in the United Kingdom.

reaction to a vaccination. To "catch" these rare occurrences, the FDA and the CDC have established the **Vaccine Adverse Events Reporting System** (**VAERS**) to which anyone, including doctors, patients, and parents, can report adverse vaccine reactions. The FDA monitors the system weekly.

One vaccine that was permanently removed from the US market was RotaShield®, which was administered to infants to provide immunity against rotavirus infections (see Chapter 15). In 1999, the Advisory Committee on Immunization Practices (ACIP) suggested that the vaccine licensed in the United States should no longer be recommended for infants based on a review of scientific studies indicating the virus might be associated with intussusception among some infants during the first one to two weeks following vaccination. Although such scientific studies may have been incorrect, new and better vaccines are now in place to replace the suspect rotavirus vaccine.

People who are allergic to eggs should not be vaccinated against the flu virus because the virus is replicated in chicken eggs. In 2002–2003, there was much discussion weighing the risks of the smallpox vaccine's potential side effects (mostly mild but potentially deadly for a few) versus the risk of a smallpox bioterrorist attack. The smallpox vaccine certainly is not recommended for young children.

Perhaps the vaccine of most concern regarding standard childhood vaccines is MMR. In 1998, a paper was published suggesting there might be a link between the MMR vaccine and autism in children. The thought was that perhaps the sheer number of vaccinations could overwhelm a child's immune system and cause neurological damage. To make a long story short, as of 2006, numerous studies carried out by other scientists and researchers around the world have found no evidence to support the autism claim.

Through vaccination and booster shots, vaccines have been essential to maintaining public health and controlling infectious diseases. Although a few of the millions of people vaccinated each year suffer a serious conse-

- **Autism:** A neurodevelopmental disturbance in which the use of communication, and normal behavior and social relationships, are not fully established and follow unusual patterns.
- **Intussusception:** A rare, potentially life-threatening condition where bowel sliding creates swelling and intestinal obstruction.

quence from vaccination, the risks of contracting a disease (especially in infants and children) from not being vaccinated are thousands of times greater than the risks associated with any vaccine. In addition, the licensed vaccines always are being examined for ways to improve their safety and effectiveness.

CONCEPT AND REASONING CHECKS

21.5 Assess the need for and effectiveness of maintaining vaccinations.

21.2 Serological Reactions

Antigen-antibody interactions studied under laboratory conditions are known as **serological reactions** because they commonly involve serum from a patient. In the late 1800s, serological reactions first were adapted to laboratory tests used in the diagnosis of disease. The principle was simple and straightforward: If the patient had an abnormal level of a specific antibody in the serum, a suspected disease agent probably was present. Today, **serology**, the study of blood serum and its constituents, especially its role in protecting the human body against disease, have diagnostic significance as well as more broad-ranging applications. For example, they are used to confirm identifications made by other procedures and detect organisms in body tissues. In addition, they help the physician to follow the course of disease and determine the immune status, and they aid in determining taxonomic groupings (serotypes) of microorganisms below the species level.

Serological Reactions Have Certain Characteristics

KEY CONCEPT

- Serological reactions generally consist of an antigen and a serum sample.

Diagnostic microbiology makes great use of the serological reactions between antigen and antibody. Antisera can be used to determine the nature of their antigens, or antigens used to detect antibodies present in a blood sample as evidence of a current or previous infection by that antigen.

In some cases, the unknown can be determined merely by placing the reactants on a slide and observing the presence or absence of a reaction. However, serological tests are not always quite so direct. For example, an antigen may have only one antigenic determinant, and the combination with an antibody molecule on a one-to-one basis may be invisible. To solve this dilemma, a second-stage reaction using an indicator system may be required, or a labeled molecule may be necessary. We shall see how this works in several tests to amplify detection.

A successful serological reaction may require the antigen or antibody solution to be greatly diluted to reach a concentration at which a reaction will be most favorable. The process, called **titration**, may be used to the physician's advantage because the dilution series is a valuable way of determining the titer of antibodies. The **titer** is the most dilute concentration of serum antibody yielding a detectable reaction with its specific antigen (FIGURE 21.6). This number is expressed as a ratio of antibody to total fluid (for example, 1:50) and is used to indicate the amount of antibodies in a patient's serum. For instance, the titer of influenza antibodies may rise from 1:20 to 1:320 as an episode of influenza progresses, and then continue upward, stabilizing at 1:1,280 as the disease reaches its peak. A rise in the titer also indicates an individual has a disease, an important factor in diagnosis.

Haptens may pose a problem in a serological reaction because of their small size (see Chapter 20). This problem has been solved by conjugating the haptens to carrier particles, such as polystyrene beads. When the hapten unites with an antibody, the entire bead is involved in the complex, and a visible reaction occurs.

Serology has become a highly sophisticated and often automated branch of immunology. As the following tests illustrate, the serological reactions have direct application to the clinical laboratory as well as in other fields.

CONCEPT AND REASONING CHECKS

21.6 What does titration accomplish?

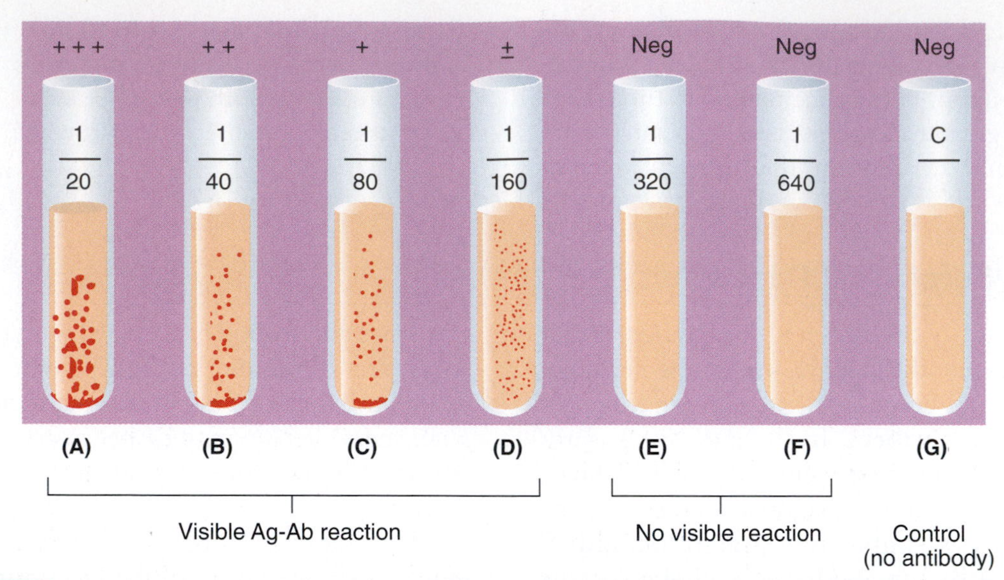

FIGURE 21.6 The Determination of Titer. A sample of antibody (Ab)-containing serum was diluted in saline solution to yield the dilutions shown. An equal amount of antigen (Ag) was then added to each tube, and the tubes were incubated. An antigen-antibody interaction may be seen in tubes (**A**) through (**D**), but not in tubes (**E**) or (**F**) or the control tube, (**G**). The titer of antibody is the highest dilution of serum antibody in which a reaction is visible.

Q: What is the titer of antibody in this example?

Neutralization Involves Antigen-Antibody Reactions

KEY CONCEPT

- Neutralization is a serological reaction in which antigens and antibodies neutralize each other.

Neutralization is a serological reaction used to identify toxins and antitoxins as well as viruses and viral antibodies (see Chapter 20). Normally, little or no visible evidence of a neutralization reaction is present, and the test mixture therefore must be injected into a laboratory animal to determine whether neutralization has taken place.

An example of a neutralization test is one used to detect botulism toxin in food. Normally the toxin is lethal to a laboratory animal, and if a sample of the food contains the toxin, the animal will succumb after an injection. However, if the food is first mixed with botulism antitoxins, the antitoxin neutralizes the toxin, and the mixture has no effect on the animal.

Conversely, if the toxin was produced by some other organism, no neutralization will occur, and the mixture will still be lethal to the animal. A similar test for diphtheria is the **Schick test**, in which a person's immunity to diphtheria can be determined by injecting diphtheria toxin intradermally. No skin reaction will occur if the person has neutralizing antibodies. A local edema occurs if no neutralizing antibodies are present, indicating the person is susceptible to diphtheria.

Precipitation Requires the Formation of a Lattice Between Soluble Antigen and Antibody

KEY CONCEPT

- Precipitation is a serological reaction in which antigens and antibodies form a visible precipitate.

Precipitation reactions are serological reactions involving thousands of antigen and antibody molecules cross-linked at multiple determinant sites forming a lattice. The lattices are so huge that particles of precipitate form and the reaction product is observed visually.

Precipitation tests are performed in either fluid media or gels. In fluids, the antibody and antigen solutions are layered over each other in a thin tube. The molecules then diffuse through the fluid until they reach a **zone of equivalence**, the ideal concentration for precipitation. A visible mass of particles now forms at the interface or at the bottom of the tube. Fluid precipitation is used frequently in forensic medicine to learn the origin of albumin proteins in bloodstains.

In **immunodiffusion**, the diffusion of antigens and antibodies takes place through a semisolid gel, such as agar. As the molecules diffuse through the gel, they eventually reach the zone of equivalence, where they interact and form a visible precipitate. Variations of this technique are called the Ouchterlony plate technique, named for Orjan Ouchterlony, who devised it in 1953, and the **double diffusion assay** because both reactants diffuse.

In immunodiffusion, antigen and antibody solutions are placed in wells cut into agar in Petri dishes. The plates are incubated and precipitation lines form at the zone of equivalence (FIGURE 21.7). The test has been used to detect fungal antigens of *Histoplasma*, *Blastomyces*, and *Coccidioides* (see Chapter 16).

In the procedure known as **immunoelectrophoresis**, the techniques of gel electrophoresis and diffusion are combined for the detection of antigens. A mixture of antigens is placed in a reservoir on an agarose slide, and an electrical field is applied to the ends of the slide. The different antigens then move through the agarose at different rates of speed, depending on their electrical charges. This process is gel electrophoresis (FIGURE 21.8). A trough is then cut into the agarose along the same axis, and a known antibody solution is added. During incubation, antigens and antibodies diffuse toward each other and precipitation lines form, as in the immunodiffusion technique.

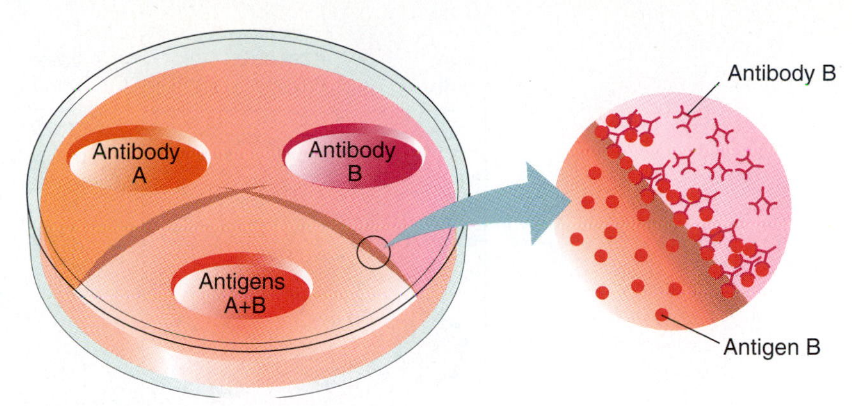

FIGURE 21.7 **A Precipitation Test.** Wells are cut into a plate of purified agar. Different known antibodies then are placed into the two upper wells, and a mixture of unknown antigens is placed into the lower well. During incubation, the reactants diffuse outward from the wells, and cloudy lines of precipitate form. The lines cross each other because each antigen has reacted only with its complementary antibody.

Q: Why does the precipitin line form where it does?

CONCEPT AND REASONING CHECKS

21.7 What do lines of precipitation indicate?

Agglutination Involves the Clumping of Antigens

KEY CONCEPT

- Agglutination is a serological reaction in which antibodies interact with antigens on the surface of particular objects.

The amount of antibody or antigen needed to form a visible reaction can be reduced if either

(A) Gel electrophoresis:
On a gel-coated slide, an antigen sample is placed in a central well. An electrical current is run through the gel to separate antigens by their electrical charge (electrophoresis). Unlike the diagram, the separate antigens cannot be detected visually at this point.

(B) Addition of antibodies:
A trough is made on the slide and a known antibody solution is added.

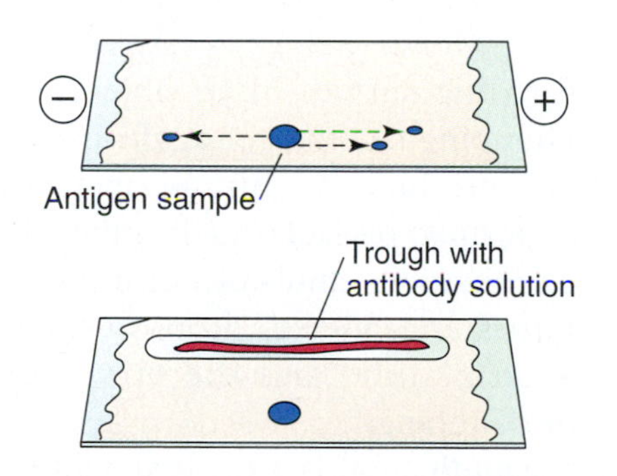

(C) Diffusion of antigens and antibodies:
As antigens and antibodies diffuse toward one another through the gel, precipitin lines are seen where optimal concentrations of antigen and antibodies meet.

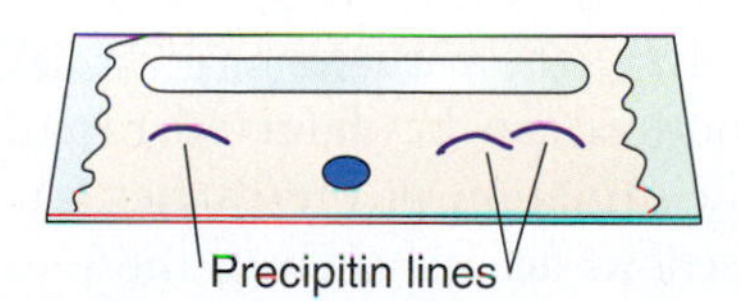

FIGURE 21.8 **Immunoelectrophoresis.** By applying an electrical field, immunoelectrophoresis separates antigens by their electrical charge before antibodies are added.

Q: In (A), what is the electrical charge (+ or –) on the three antigens shown?

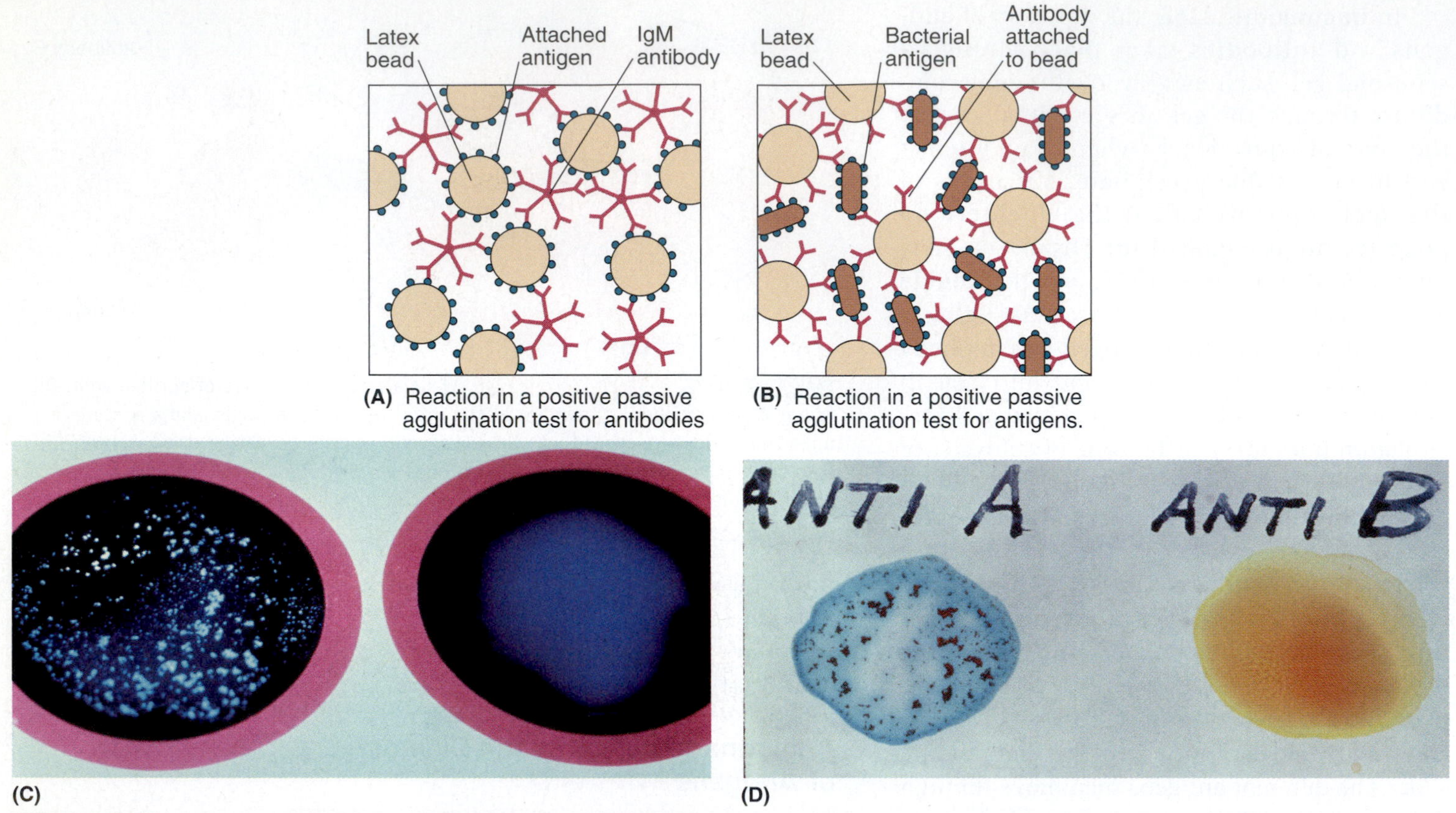

(A) Reaction in a positive passive agglutination test for antibodies

(B) Reaction in a positive passive agglutination test for antigens.

(C)

(D)

FIGURE 21.9 **Passive Agglutination Tests.** (**A**) When particles are bound with antigens, agglutination indicates the presence of antibodies, such as the IgM shown here. (**B**) When particles are bound with antibodies, agglutination indicates the presence of antigens. (**C**) Latex agglutination test for *Staphylococcus*. If group A antigen is present, it will bind to antibody-coated latex beads and agglutinate them for a positive test (left). A negative test is shown for comparison on the right. (**D**) Blood type can be determined using the agglutination test. Red blood cells are diluted with saline and mixed with anti-A agglutinin (at left), and anti-B agglutinin (at right). For some minutes they are allowed to react. The clumping (agglutination) is due to antibody-antigen reactions. These tests use antigens or antibodies adsorbed onto the surface of latex spheres.

Q: What is the blood type in the example shown in (D)?

is attached to the surface of an object. The result is a clumping together, or **agglutination**, of the linked product. Agglutination procedures are performed on slides or in tubes. For bacterial agglutination, emulsions of unknown bacterial species are added to drops of known antibodies on a slide, and the mixture is observed for clumping.

Passive agglutination is a modern approach to traditional agglutination methods. Most often, antigens are adsorbed onto the surface of latex spheres or polystyrene particles (FIGURE 21.9A). Serum antibodies can be detected rapidly by observing agglutination of the carrier particle. Bacterial infectious agents, such as those caused by *Streptococcus*, can be detected rapidly by mixing the bacterial sample with latex spheres containing streptococcus antibody (FIGURE 21.9B, C).

Hemagglutination refers to the agglutination of red blood cells. This process is particularly important in the determination of blood types prior to blood transfusion (FIGURE 21.9D). In addition, certain viruses, such as measles and mumps viruses, agglutinate red blood cells. Antibodies for these viruses may be detected by a procedure in which the serum is first combined with laboratory-cultivated viruses and then added to the red blood cells. If serum antibodies neutralize the viruses, agglutination fails to occur. This test, called the **hemagglutination inhibition (HAI) test**, is discussed in Chapter 13. A hemagglutination test called the **Coombs test** is used to detect Rh antibodies involved in hemolytic disease of the newborn (Chapter 22). A **slide agglutination test** called the **Venereal Disease Research Laboratory**

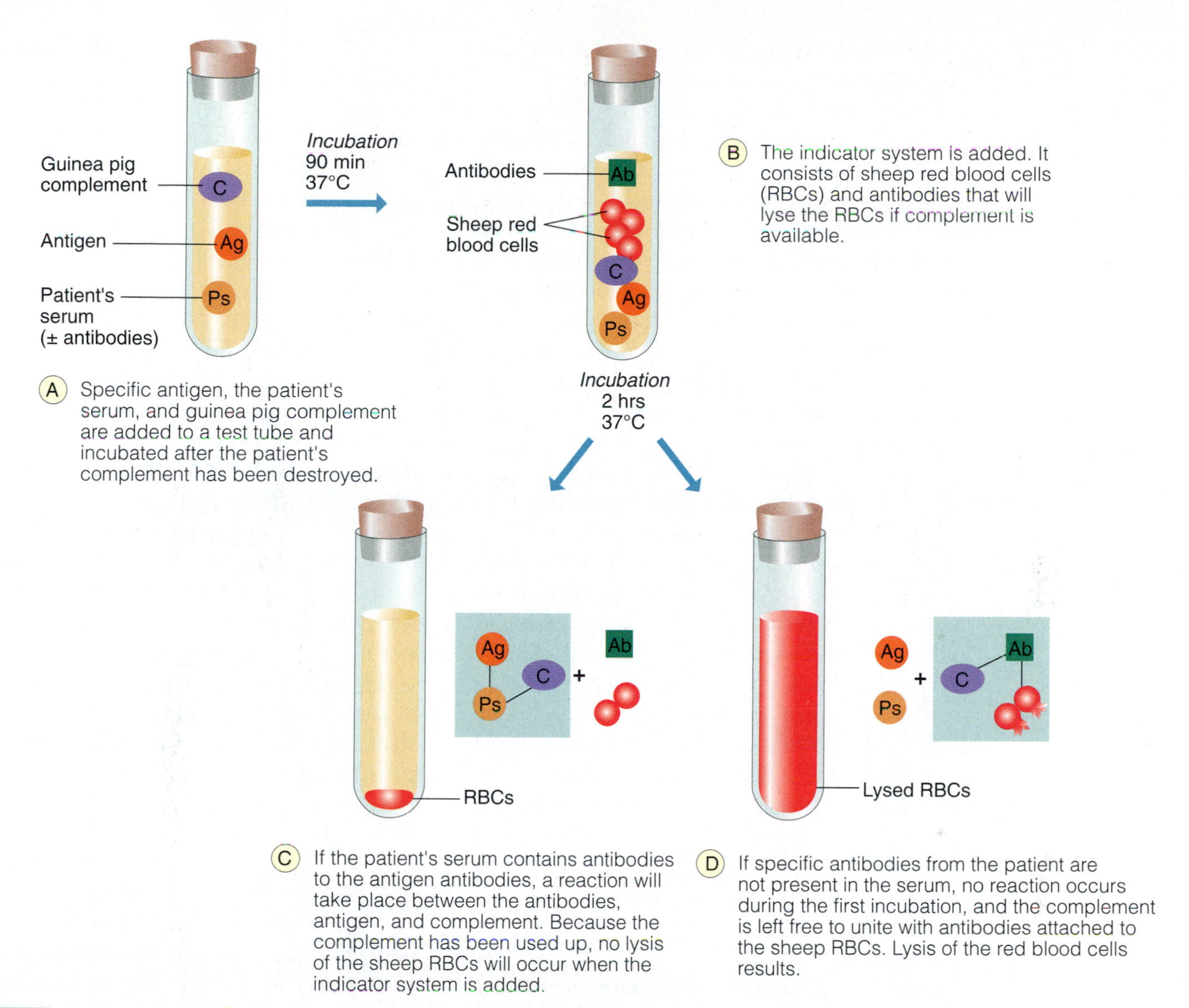

FIGURE 21.10 **The Complement Fixation Test.** Complement fixation tests are carried out in two stages.

Q: What would happen if the patient's complement proteins were not inactivated?

(VDRL) **test** is used for the rapid screening of patients to detect syphilis.

CONCEPT AND REASONING CHECKS

21.8 Summarize the uses for neutralization, precipitation, and agglutination tests.

Complement Fixation Can Detect Antibodies to a Variety of Pathogens

KEY CONCEPT

- Complement fixation depends on complement inactivation and hemolysis of sheep red blood cells.

The **complement fixation test** is performed in two parts. The first part—the test system—uses the patient's serum (± antibodies), a preparation of antigen from the suspected pathogen, and complement derived from guinea pigs. The second part—the indicator system—requires sheep red blood cells and a preparation of antibodies that recognize the sheep red blood cells. The first step in the test is to heat the patient's serum to destroy any complement present in the serum. Next, carefully measured amounts of antigen and guinea pig complement are added to the serum (FIGURE 21.10). This test system then is incubated at 37°C for 90 minutes. If antibodies specific for the antigen are present in the patient's serum, an antibody-antigen interaction takes place, and the complement is used up, or "fixed." However, there is no visible sign of whether a reaction has occurred.

Now the indicator system (sheep red blood cells and antisheep antibodies) is added to the tube, and the tube is reincubated. If the complement was previously fixed, lysis of the sheep red blood cells cannot take place. The blood cells therefore would remain intact, and when the tube is centrifuged, the technician observes clear fluid with a "button" of blood cells at the bottom. Conclusion: The patient's serum contained antibodies that reacted with the antigen and fixed the complement.

If the complement was not fixed in the test system, it will still be available to react with the antibodies bound to sheep red blood cells and, as a result, the sheep red blood cells will lyse. When the tube is centrifuged, the technician sees red fluid, colored by the hemoglobin of the broken blood cells, and no evidence of blood cells at the bottom of the tube. Conclusion: The serum lacked antibodies for the antigen tested.

The complement fixation test is valuable because it may be adapted by varying the antigen. In this way, tests may be conducted for such diverse diseases as encephalitis, Rocky Mountain spotted fever, meningococcal meningitis, and histoplasmosis. The versatility of the test, together with its sensitivity and relative accuracy, has secured its continuing role in diagnostic medicine.

CONCEPT AND REASONING CHECKS

21.9 Describe the uses for the complement fixation test.

Labeling Methods Are Used to Detect Antigen-Antibody Binding

KEY CONCEPT

- Fluorescent or radioactive antibodies can identify antigens (pathogens).

The detection of antigen-antibody binding can be enhanced (amplified) visually by attaching a label to the antigen or antibody. The label may be a fluorescent dye, a radioisotope, or an enzyme.

Fluorescent Antibody Technique. The detection of antigen-antibody binding can be done on a slide using a **fluorescent antibody technique**. Two commonly used dyes are fluorescein, which emits an apple-green glow, and rhodamine, which gives off orange-red light.

Fluorescent antibody techniques may be direct or indirect. In the direct method, the fluorescent dye is linked to a known antibody. After combining with particles having complementary antigens, the three components react, causing the complex to glow on illumination with ultraviolet (UV) light when viewed with a fluorescence microscope.

For example, suppose you want to know if spirochetes were present in a serum sample. The sample would be combined with antispirochete-labeled antibodies. If spirochetes are present, the tagged antibodies accumulate on the particle surface and the particle glows when viewed with fluorescence microscopy (FIGURE 21.11).

With the indirect method, the fluorescent dye is linked to an antibody recognizing human antibodies in a patient's serum. An example is the diagnostic procedure used for detecting syphilis antibodies in the blood of a patient (FIGURE 21.12). A sample of commercially available syphilis spirochetes is placed on a slide, and the slide is then flooded with the patient's serum. Next, a sample of fluorescein-labeled antiglobulin (antihuman) antibodies is added. These are the antibodies recognizing human antibodies. The slide then is observed using the fluorescence microscope.

The test is interpreted as follows. If the patient's serum contains antisyphilis antibodies, the antibodies bind to the surfaces of spirochetes and the labeled antiglobulin antibodies are attracted to them. The spirochetes then glow from the dye. However, if no antibodies are present in the serum, nothing accumulates on the spirochete's surface, and labeled antiglobulin antibodies also fail to gather on the surface. The labeled antibodies remain in the fluid and the spirochetes do not glow.

Fluorescent antibody techniques are adaptable to a broad variety of antigens and antibodies and are widely used in serology. Antigens may be detected in bacterial smears, cell smears, and viruses fixed to carrier particles. The value of the technique is enhanced because the materials are sold in kits and are readily available to small laboratories.

Radioimmunoassay. An extremely sensitive serological procedure used to measure the concentration of very small antigens, such as

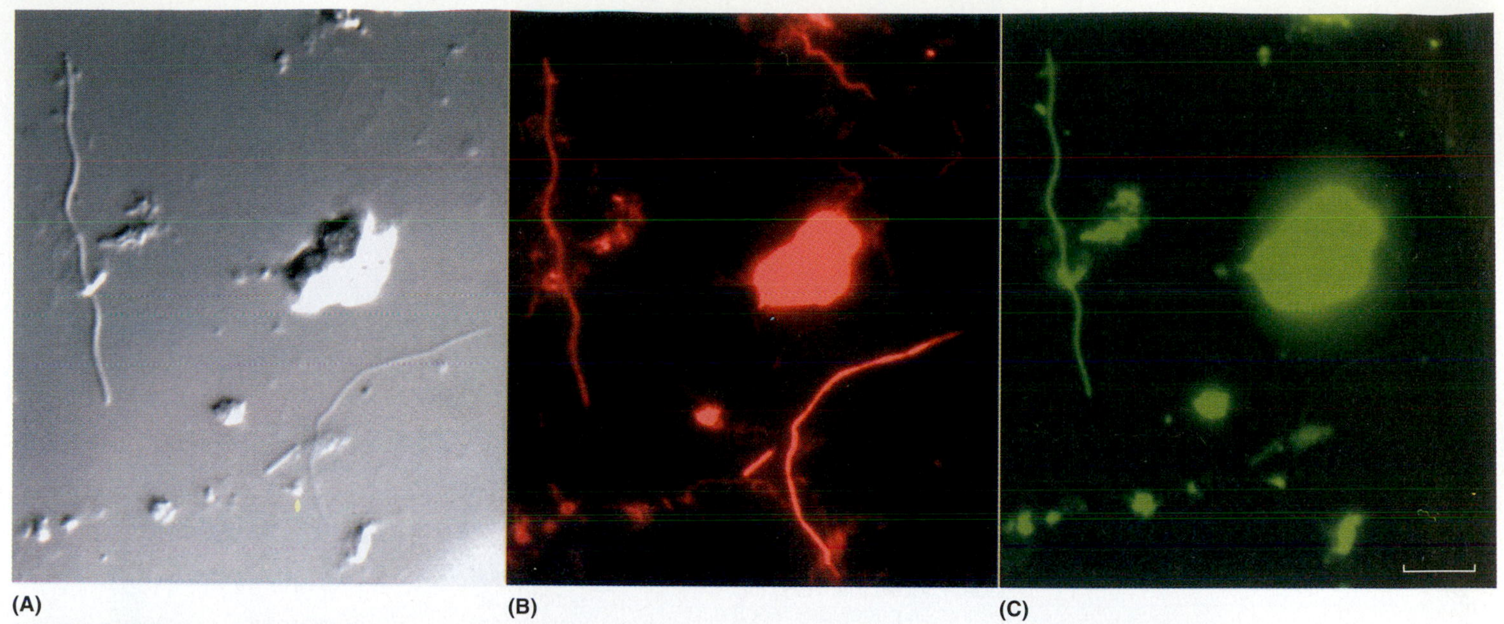

FIGURE 21.11 **Direct Fluorescent Antibody Staining.** Three views of spirochetes in the hindgut of a termite identified by a direct fluorescent antibody technique. (**A**) An unstained area displayed by differential interference contrast microscopy. (**B**) The same viewing area seen by fluorescence microscopy using rhodamine B as a stain. (**C**) The same area viewed after staining with fluorescein. (Bar = 10 μm.)

Q: Why is this method referred to as direct fluorescent antibody technique?

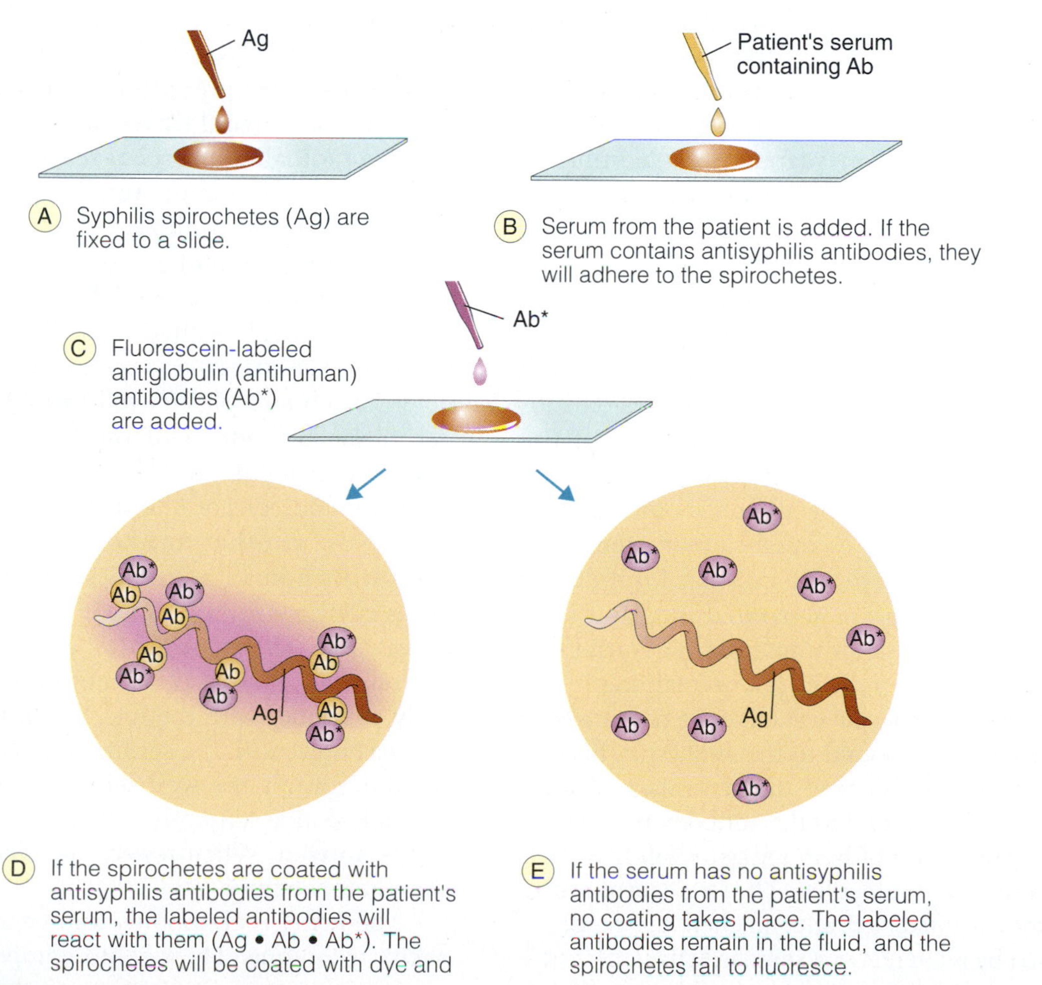

FIGURE 21.12 **The Indirect Fluorescent Antibody Technique for Diagnosing Syphilis.** With the indirect technique, a larger visible signal (fluorescence) is observed.

Q: Why is this method referred to as the indirect fluorescent antibody technique?

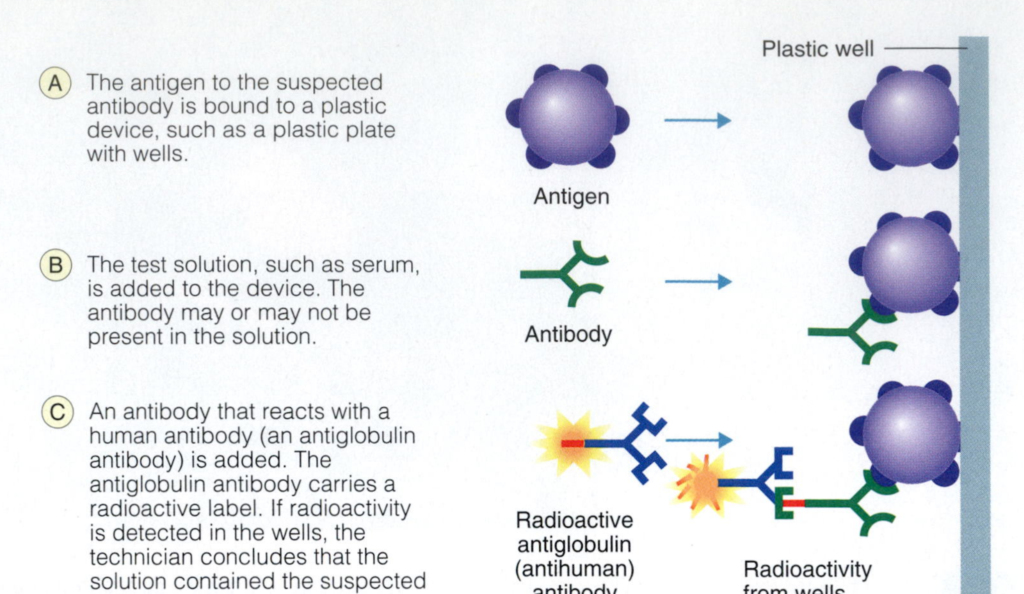

FIGURE 21.13 **The Radioallergosorbent Test (RAST).** This technique provides an effective way of detecting very tiny amounts of antibody in a preparation. A known antigen is used. The objective is to determine whether the complementary antibody is present.

Q: Would this test represent direct or indirect radioactive antibody labeling? Explain.

haptens, is the **radioimmunoassay** (**RIA**). Since its development in the 1960s, the technique has been adapted for quantitating hepatitis antigens as well as reproductive hormones, insulin, and certain drugs. One of its major advantages is that it can detect trillionths of a gram of a substance.

The RIA procedure is based on the competition between radioactive-labeled antigens and unlabeled antigens for the reactive sites on antibody molecules. A known amount of the radioactive (labeled) antigens is mixed with a known amount of specific antibodies and an unknown amount of unlabeled antigens. The antigen-antibody complexes that form during incubation then are separated out, and their radioactivity is determined. By measuring the radioactivity of free antigens remaining in the leftover fluid, one can calculate the percentage of labeled antigen bound to the antibody. This percentage is equivalent to the percentage of unlabeled antigen bound to the antibody because the same proportion of both antigens will find spots on antibody molecules. The concentration of unknown unlabeled antigen then can be determined by reference to a standard curve.

Radioimmunoassay procedures require substantial investment in sophisticated equipment and carry a certain amount of risk because radioactive isotopes are used. For these reasons, the procedure is not widely used in routine serological laboratories.

The **radioallergosorbent test** (**RAST**) is an extension of the radioimmunoassay. The test may be used, for example, to detect IgE antibodies in the serum of a person possibly allergic to compounds like penicillin.

To detect IgE against penicillin, penicillin antigens are attached to a suitable plastic device, such as a plastic well (FIGURE 21.13). Serum is then added. If the serum contains antipenicillin IgE, it will combine with the penicillin antigens on the surface of the plastic well. Now another antibody, one that will react with human antibodies, is added. This antiglobulin antibody carries a radioactive label. The entire complex therefore will become radioactive if the antiglobulin antibody combines with the IgE. By contrast, if no IgE was present in the serum, no reaction with the antigen on the well will take place, and the radioactive antibody will not be attracted to the antigen. When tested, the well will not show radioactivity.

The RAST is commonly known as a "sandwich" technique. There is no competition for an active site as in RIA, and the type of unknown antibody, as well as its amount, may be learned by determining the amount of radioactivity deposited.

Enzyme-Linked Immunosorbent Assay. Another test to detect antigens or antibodies is the **enzyme-linked immunosorbent assay** (ELISA). It has virtually the same sensitivity as radioimmunoassay and the RAST, but does not require expensive equipment or radioactivity. The procedure involves attaching antibodies or antigens to a solid surface and combining (immunosorbing) the coated surfaces with the test material. An enzyme system then is linked to the complex, the remaining enzyme is washed away, and the extent of enzyme activity is measured. This gives an indication that antigens or antibodies are present in the test material.

An application of the ELISA is found in the highly efficient laboratory test used to detect antibodies against the human immunodeficiency virus (HIV) (FIGURE 21.14). A serum sample is obtained from the patient and mixed with a solution of plastic or polystyrene beads coated with antigens from an HIV antigen. If antibodies to HIV are present in the serum, these "primary antibodies" will adhere to the antigens on the surface of the beads. The beads then are washed and incubated with antiglobulin (antihuman) antibodies chemically tagged with molecules of horseradish peroxidase or a similar enzyme. These antibodies are referred to as the "secondary antibody." The preparation is washed, and a solution of substrate molecules for the peroxidase enzyme is added. Initially the solution is clear, but if enzyme molecules react with the substrate, the solution will become yellow orange in color. The enzyme molecules will be present only if HIV antibodies are present in the serum. If no HIV antibodies are in the serum, no enzyme molecules could concentrate on the bead surface, no change in the substrate molecules could occur, and no color change would be observed.

ELISA procedures may be varied depending on whether one wishes to detect antigens or antibodies. The solid phase may consist of beads, paper disks, or other suitable supporting mechanisms, and alternate enzyme systems such as the alkaline phosphatase system may be used that produce a different color solution in a positive reaction. In addition, the results of the test may be quantified by noting

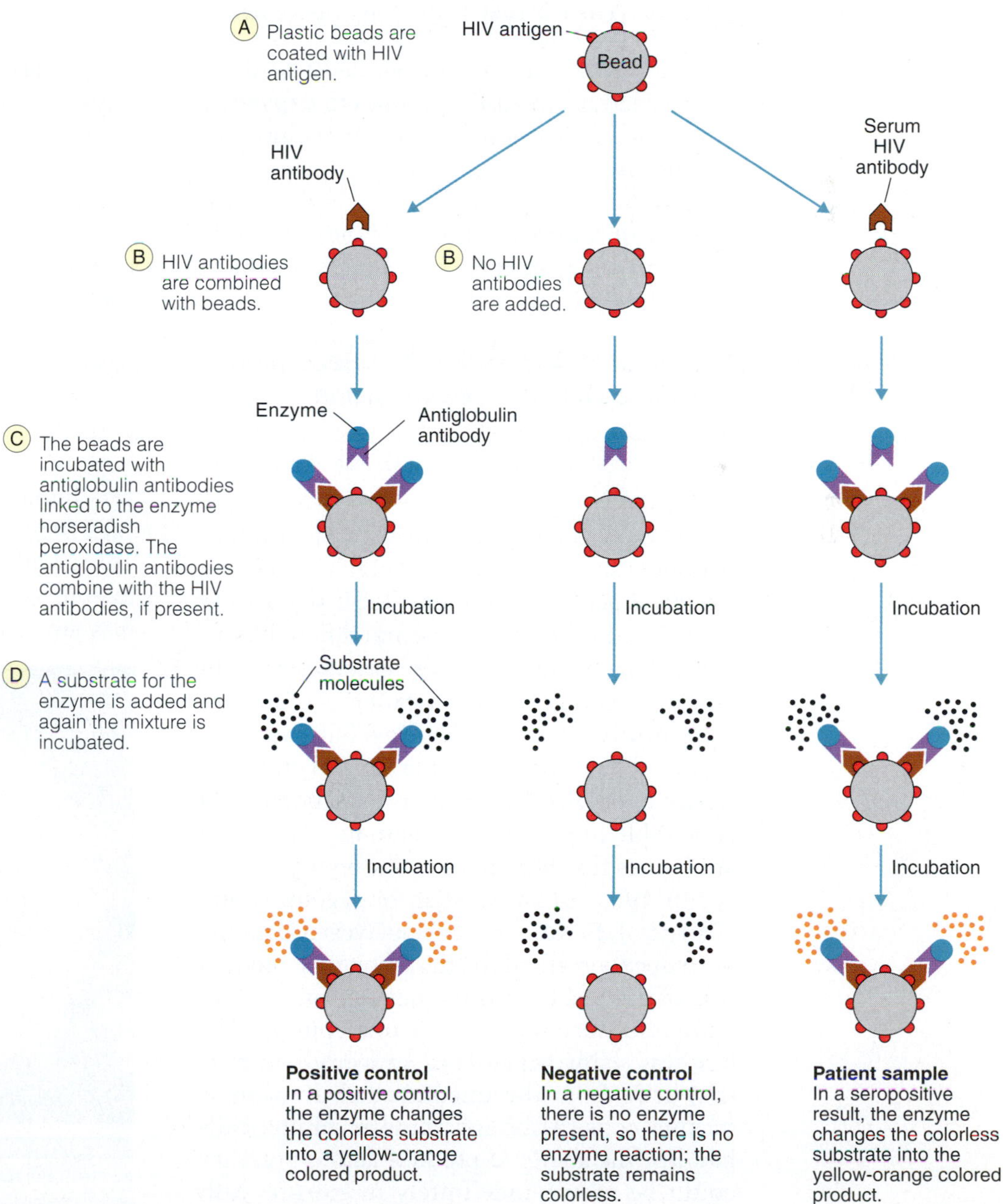

FIGURE 21.14 **The ELISA.** The enzyme-linked immunosorbent assay (ELISA) as it is used to detect antibodies (in this case HIV) in a patient's serum.

Q: What are positive and negative controls, and why are they needed in the test?

the degree of enzyme-substrate reactions as a measure of the amount of antigen or antibody in the test sample. The availability of inexpensive ELISA kits has brought the procedure into the doctor's office and routine serological laboratory. Importantly, a positive ELISA for HIV, or for any pathogen being tested, only indicates that antibodies to the pathogen are present. A positive ELISA does not necessarily mean the person has the disease. The presence of antibodies only says the person has been exposed to the pathogen. However, in the case of HIV, a positive ELISA probably does indicate an infection, as no one has ever been known to be cured of an HIV infection. MicroInquiry 21 looks at the ELISA test.

CONCEPT AND REASONING CHECKS

21.10 Assess the need to label antigen or antibody in the RAST and ELISA procedures.

21.3 Additional Laboratory Tests

The variety of laboratory tests used to diagnose infectious disease continues to expand as new biochemical and molecular techniques are adapted to special clinical situations. Some, such as monoclonal antibodies, still depend on antibody-antigen binding, while others involve the analysis of microorganism DNA. Both are touched upon here.

■ Screen: To evaluate someone or something for a particular characteristic or disease.

Monoclonal Antibodies Are Becoming a "Magic Bullet" in Biomedicine

KEY CONCEPT

- Monoclonal antibodies recognize specific cells or substances.

Usually pathogens contain several different epitopes (see Chapter 20). Therefore, when an infection occurs, a different B-cell population is activated for each epitope and different antibodies specific to each of the epitopes are produced. Serum from such a patient would contain **polyclonal antibodies** because the antibodies are derived from several different clones of B cells. However, in the laboratory it is possible to produce populations of identical antibodies that bind to only one epitope.

In 1975, César Milstein of Argentina and Georges J. F. Köhler of then–West Germany developed a method for the laboratory production of **monoclonal antibodies (mAbs)**; that is, antibodies recognizing only one epitope. The key was using **myelomas**, cancerous tumors that arise from the uncontrolled division of plasma cells. Although these myeloma cells had lost the ability to produce antibodies, they could be grown indefinitely in culture. Milstein and Köhler were able to fuse myeloma cells with B cells from which mAbs were produced. In 1984, Milstein and Köhler (along with Niels Jerne) shared the Nobel Prize in Physiology or Medicine for the development of the monoclonal antibody technique.

There are two basic steps to generating mAbs. First, a myeloma cell is fused to an activated (antibody secreting) B cell (plasma cell) to form a hybrid cell called a **hybridoma** (FIGURE 21.15). Then each hybridoma is screened for the desired mAb and cultured to produce hybridoma clones.

To produce hybridomas, a mouse is injected with the antigen of interest against which a mAb is needed (FIGURE 21.16). Injection triggers humoral immunity in the mouse. Plasma cells from the mouse's spleen are removed and mixed with myeloma cells from another mouse. By forcing many cells to fuse together, hybridomas are produced. Hybridomas are placed in a

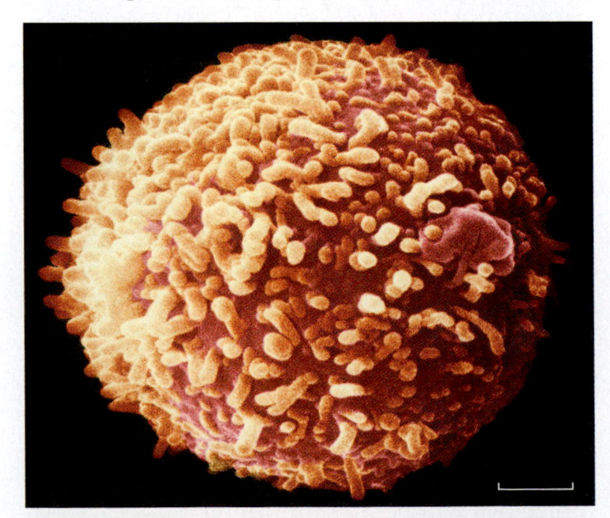

FIGURE 21.15 **A Hybridoma Cell.** False-color scanning electron micrograph of a hybridoma cell screened for its ability to produce an monoclonal antibody to a human cytoskeleton protein. (Bar = 3 µm.)

Q: What antibody must have been produced by the B cell used to form this hybridoma cell?

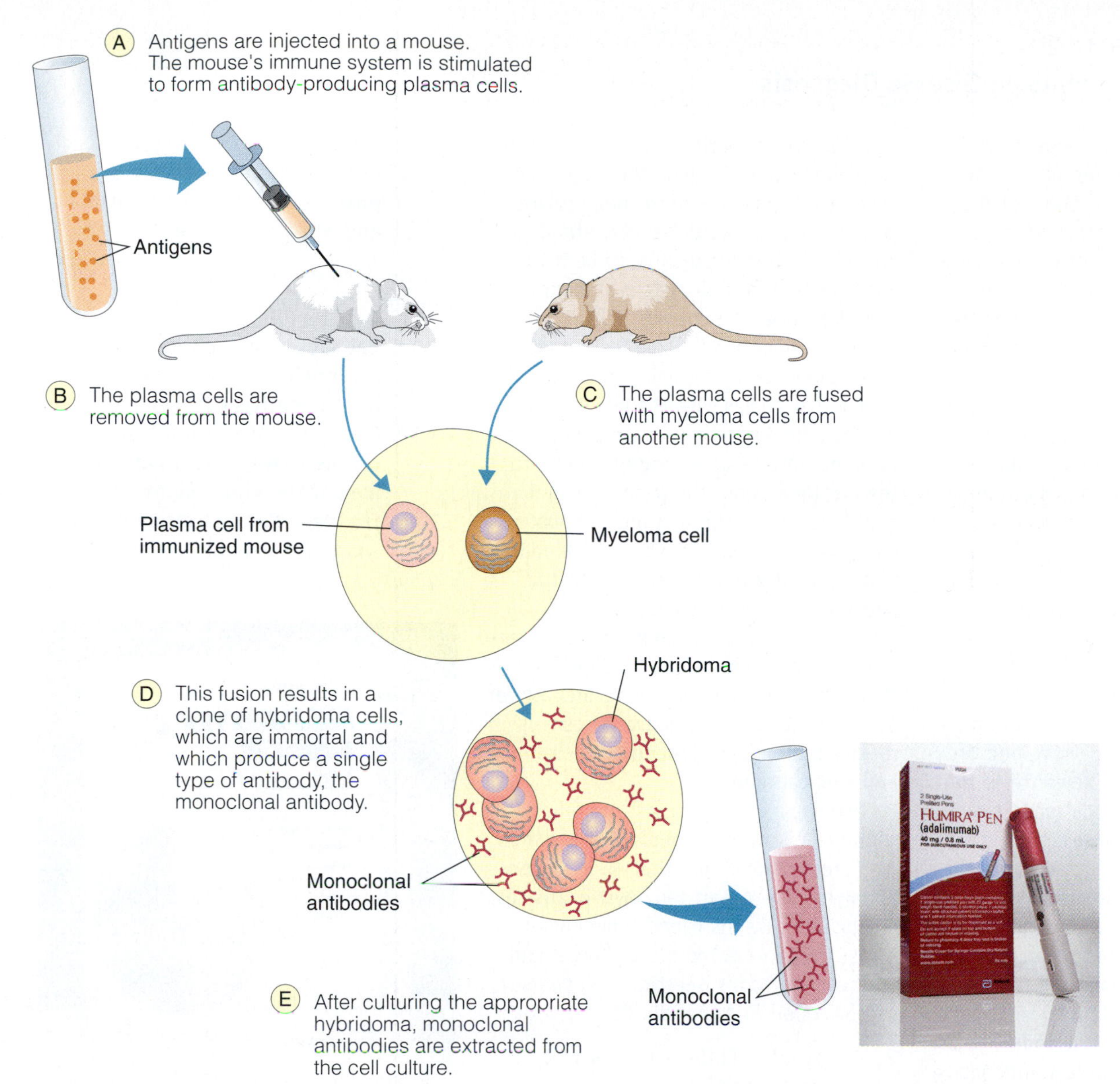

FIGURE 21.16 The Production of Monoclonal Antibodies. The production of a monoclonal antibody (mAb) requires fusing a myeloma cell with plasma cells. After screening for the mAb needed, those clones can be cultivated to produce more of the same mAb. (inset) Humira is the first fully human mAb approved for use in the United States.

Q: What are the unique features of the plasma cell from the mouse and the myeloma cell from another mouse?

special tissue culture medium and in seven to ten days, hybridomas form small clusters while any unfused cells (B cells or myelomas) cannot survive in the medium and die.

Because the antigen originally injected may have several epitopes, different hybridoma clusters may be producing different antibodies. Therefore, individual hybridoma cells are separated from one another and allowed to grow. Each hybridoma cell can multiply indefinitely (myeloma characteristic), producing antibodies recognizing but one epitope (B-cell characteristic).

Each clone of hybridomas is screened for the synthesis of the desired antibody. The hybridoma producing the desired mAb can be propagated (cloned) in tissue culture flasks.

Initially, mAbs were of mouse origin, meaning the mAbs produced would be seen as foreign when injected into humans. The antibodies, therefore, would be destroyed and in the process kidney damage might occur. More recently, humanized mAbs have been developed, whose structure is about 90 to 95 percent human immunoglobulin. Then, in 2002, the first totally human mAb was produced,

MICROINQUIRY 21

Applications of Immunology: Disease Diagnosis

Ancient Egyptian medical papyri from 1500 BC refer to many different disease symptoms and treatments. Some of these symptoms still can be used to identify diseases today. Thus, one of the most traditional "tools" of diagnosis over the centuries is a patient's signs and symptoms. However, there are problems when relying solely on signs and symptoms. Initially, many diseases display common symptoms. For example, the initial symptoms of a hantavirus infection are very similar to those of the flu. Some diseases do not display symptoms for perhaps weeks or months—or years in the case of AIDS. Yet, it is important to identify these diseases rapidly so appropriate treatment, if possible, can be started.

Serological (blood) tests have been used in the United States since about 1910 both to diagnose and control infectious disease. Today's understanding of immunology has brought newer tests that rely on identifying antibody-mediated (humoral) immune responses; that is, antibody reactions with antigens. Serology laboratories work with serum or blood from patients suspected of having an infectious disease. The lab tests look for the presence of antibodies to known microbial antigens. Serological tests that are seropositive indicate antibodies to the microbe were detected while seronegative means no antibodies were detected in a patient's serum.

Let's use a hypothetical person, named Pat, who "believes" that 12 days ago he may have been exposed to the hepatitis C virus through unprotected sex. Afraid to go to a neighborhood clinic to be tested, Pat goes to a local drugstore and purchases an over-the-counter, FDA-approved, hepatitis C home testing kit (see figure). Using a small spring-loaded device that comes with the kit, Pat pricks his little finger and puts a couple of drops of blood onto a paper strip included with the kit. He fills out the paperwork and mails the paper strip in a prepaid mailer (supplied in the kit) to a specific blood testing facility where the ELISA test is performed. In ten business days, Pat can call a toll-free number anonymously, identify himself by his unique 14-digit code that came with the testing kit, and ask for his test results. If necessary, he can receive professional post-test counseling and medical referrals.

Two days later, Pat runs into a good friend of his who is a nurse. Pat explains his "predicament" and that he is anxious about having to wait ten days for the test results. He asks his friend some questions. If you were the nurse, how would you respond to his questions? Answers can be found in **Appendix D**.

21.1a. What is ELISA and how is it performed?

21.1b. You mention positive and negative controls in your explanation of the ELISA process. Pat asks, "What are positive and negative controls and why are they necessary?"

After your explanations, Pat still has more questions for you to answer.

21.1c. Pat says, "So suppose I am seropositive. Does it mean I have hepatitis C?"

21.1d. Pat says, "Okay, so if I am seronegative, then I must be free of the virus. Right?"

21.1e. If indeed Pat was seronegative, what advice would you give him?

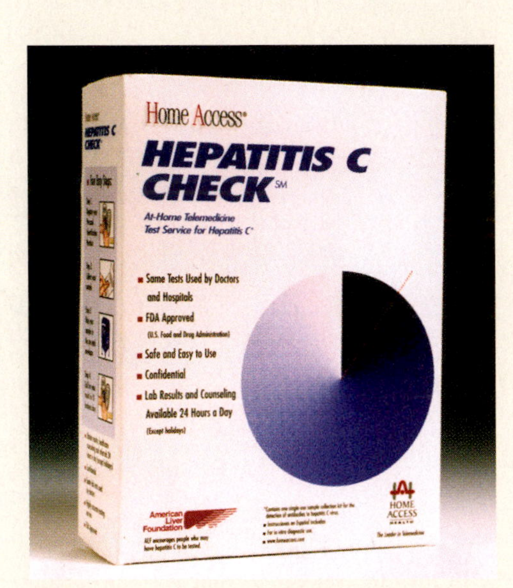

An FDA-approved home testing kit for hepatitis C.

which finally solved the problem of immune system recognition.

Currently, more than 100 mAbs are in drug clinical trials and some 18 have been approved for use in the United States. So far, the most prominent mAb targeted at infectious disease is the humanized palivizumab (Synagis), which is designed to prevent respiratory syncytial virus infections (see Chapter 14) in high risk groups. Monoclonal antibodies also are in the pipeline for treating hepatitis B and C infections.

Monoclonal antibodies also are being used for **immunomodulation**, which is controlling overactive inflammatory responses (see Chapter 19). In fact, the first truly human mAb product, called adalimumab (Humira), is directed against inflammatory responses resulting in rheumatoid arthritis.

Perhaps most prominent are those mAb drugs targeted at some forms of cancer. Monoclonal antibody products have been approved for treating malignant lymphoid tumors, leukemia, colorectal cancer, and pancreatic cancer. Many additional products are in clinical trials.

More recent developments have devised ways to add a "payload" to a mAb. For example, scientists have developed a technique in which tumor cells are removed from a patient and injected into a mouse, whereupon the mouse's spleen begins producing tumor antibodies. Spleen cells then are fused with myeloma cells to produce a hybridoma producing mAbs for that specific tumor. Toxins (the payload) are attached to the tail (Fc fragment) of the antibodies (see Chapter 20). When the antibodies are injected back into the patient, they act like "stealth missiles." They react specifically with the tumor cells and the toxin kills the cells without destroying other tissue cells.

The use of mAbs represents one of the most elegant expressions of modern biotechnology—and their potential uses in the clinic and medicine remain to be fully appreciated.

CONCEPT AND REASONING CHECKS

21.11 Explain how monoclonal antibodies are produced.

Gene Probes Are Single-Stranded DNA Segments

KEY CONCEPT

- Gene probes can detect pathogen DNA in a clinical sample.

Although antibody tests are a valuable resource in the clinical laboratory, a variety of new diagnostic tests permits the identification of an organism and its antigens (MicroFocus 21.6). One new test is based on the use of a DNA fragment called the gene probe and a procedure known as the polymerase chain reaction. A **gene probe** is a relatively small, single-stranded DNA segment that can hunt for a complementary fragment of DNA within a morass of cellular material, much like finding the proverbial needle in a haystack. When the probe locates its complementary fragment, it emits a signal such as a pulse of radioactivity. If the complementary fragment cannot be found, then no signal is sent. The procedure is remarkable for its accuracy. For example, if we were to unwind and line up the DNA from the 46 human chromosomes as a two-lane highway, the highway would stretch around planet Earth 300 times. A gene probe could locate and unite with a few-mile stretch of this highway. More information on gene probes is presented in Chapter 7.

To use a gene probe effectively, it is valuable to increase the amount of DNA to be searched. The **polymerase chain reaction** (**PCR**) accomplishes this task. The procedure takes a segment of DNA and reproduces it to a billion copies in just a few hours. The PCR process is a repeating three-step cycle (FIGURE 21.17). Target DNA is mixed with DNA polymerase (the enzyme that synthesizes DNA), short strands of primer DNA, and a mixture of nucleotides. The mixture is then alternately heated and cooled during which time the double-stranded DNA unravels, is duplicated, and then reforms the double helix. The process is repeated over and over again in a highly automated PCR machine, which is the biochemist's equivalent of an office copier. Each cycle takes about five minutes, and each new DNA segment serves as a model for producing many additional copies, which in turn serve as models for producing more copies. Instead of looking for a needle in a haystack, the gene probe now has a huge number of needles.

One place where gene probes and PCR have been useful is in the detection of human immunodeficiency virus (HIV). T lymphocytes are obtained from the patient and disrupted to obtain the cellular DNA. The DNA then is amplified by PCR and the gene probe is added. The probe is a segment of DNA that complements the DNA in the provirus synthesized from the genome of HIV (see Chapter 15). If the person is infected with HIV, the probe will locate the proviral DNA, bind to it, and emit radioactivity. An accumulation of radioactivity thus constitutes a positive test. Because the test identifies viral DNA rather than viral antibodies, the physician can be more confident of the patient's health status.

A gene probe test also is available for detecting human papilloma virus (HPV), the virus that causes genital warts (see Chapter 13). The test uses a gene probe to detect viral DNA in a sample of tissue obtained from a woman's cervix. Because certain forms of HPV have been linked to cervical tumors, the test has won acceptance as an important preventive technique, and it has been licensed by the FDA. It is commercially available as the ViraPap test.

A similar technique can be used to conduct water-quality tests based on the detection of coliform bacteria such as *E. coli* (Chapter 26). Traditionally, *E. coli* had to be cultivated in the laboratory and identified biochemically. With gene probe technology, a sample of water can be filtered, and the bacterial cells trapped on the

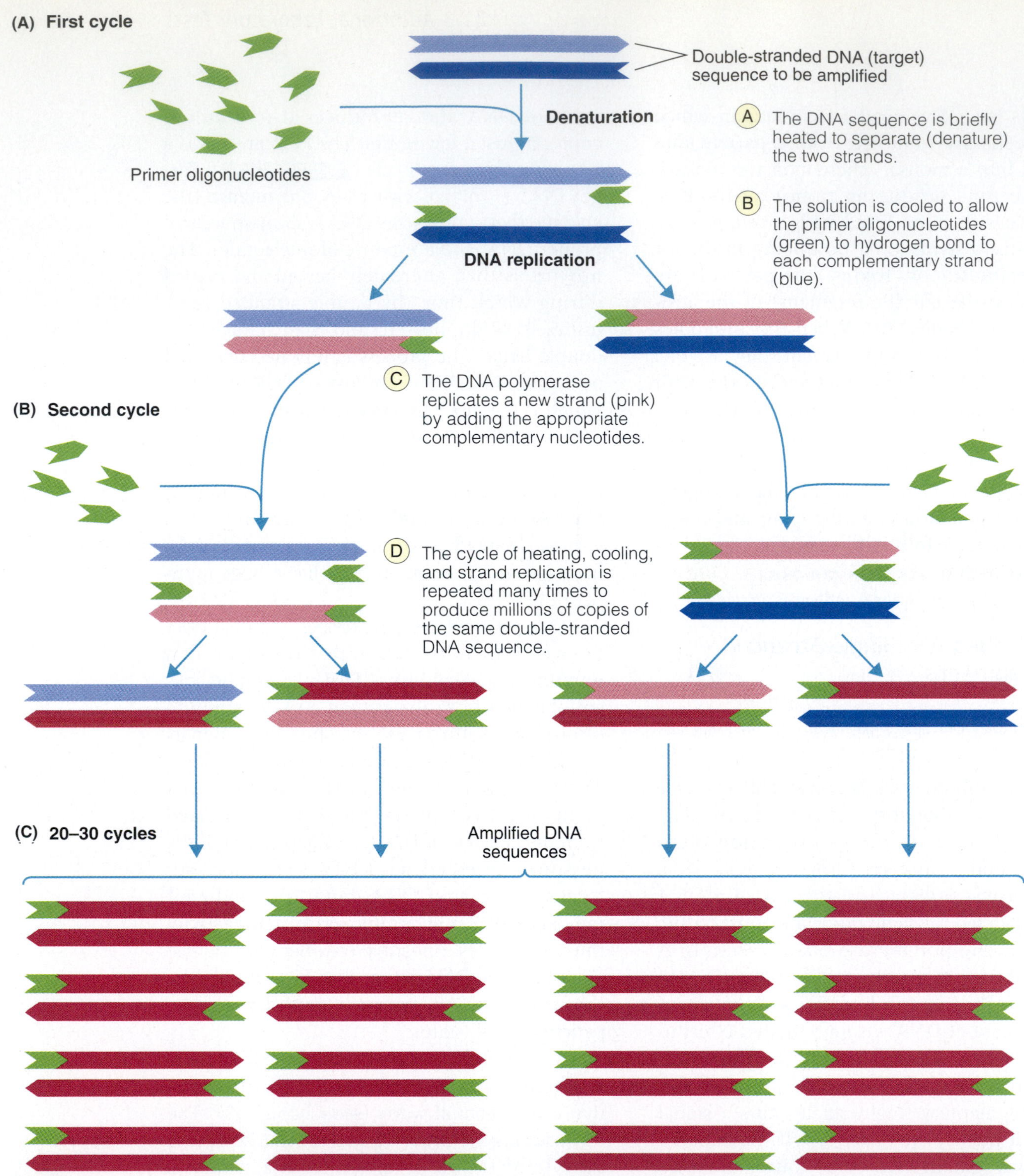

FIGURE 21.17 **The Polymerase Chain Reaction.** The polymerase chain reaction (PCR) is a method to copy (amplify) a specific DNA sequence. The sequence to be amplified is placed in a test tube with a solution of primer oligonucleotides, a supply of all four DNA nucleotides, and DNA polymerase enzymes that are heat resistant. The primer oligonucleotides are needed to start DNA synthesis. The number of copies of the target sequence doubles in each cycle of PCR, so a billion DNA sequences, identical to the target sequence, can be generated in 30 cycles of PCR.

Q: How is PCR similar to the generation time of a bacterial cell?

filter can be broken open to release their DNA for PCR and gene probe analysis. Not only is the process time saving (many days by the older method, but a few short hours by the newer method), it also is extremely sensitive: A single *E. coli* cell can be detected in a 100-mL sample of water. Moreover, the pathogens transmitted by water, rather than the "indicator" *E. coli*, can be detected by DNA analysis. Thus, the identification of *Salmonella*, *Shigella*, and *Vibrio* species

MICROFOCUS 21.6: Biotechnology

Caught in the Spotlight

Current diagnostic tests for tuberculosis can take several weeks to complete because the tubercle bacillus *Mycobacterium tuberculosis* multiplies very slowly, a binary fission taking place every 24 hours or so. While waiting for a definitive diagnosis, physicians must make treatment decisions based on very limited information. It therefore is possible that ineffective drugs may be prescribed during this interval, and the patient's illness may worsen; also, the patient may transmit the disease to others as the wait goes on.

With help from the firefly, researchers have developed an innovative and imaginative diagnostic test for tuberculosis that could shorten considerably the time interval for detection. Only a few days may be required, and the test could help determine whether that particular strain of *M. tuberculosis* is drug resistant.

A firefly whose abdomen is glowing due to luciferase activity.

The new approach relies on the firefly enzyme luciferase to produce a flash of light in living *M. tuberculosis* cells, just as happens in the firefly (see figure). The process works this way: A bacteriophage specific for *M. tuberculosis* is genetically engineered to carry the gene for luciferase. This phage is then mixed with a culture of the bacterial cells. If the culture contains *M. tuberculosis,* the phage penetrates the bacterial cell and inserts itself into the bacterial chromosome, carrying the luciferase gene along. The bacterial cell promptly begins producing luciferase.

Now luciferin, the substrate for luciferase, is added to the culture together with ATP (see Chapter 5). If luciferase is present, the enzyme breaks down luciferin, and the reaction results in a flash of light. A sensitive instrument detects the light flash, and the culture is confirmed to contain *M. tuberculosis*. The report is made to the physician, and the diagnosis is complete.

To determine drug susceptibility or resistance, the same procedure is used, except a drug is added to the culture. If the bacterial cells are sensitive to the drug, they die and, quite literally, their "lights go out." If they are resistant, they continue to live, and they produce luciferase—and they give off light.

When this test was first announced to the media, headline writers from numerous publications had a field day as word of the successful test filtered through various journals and newspapers. The well-worn cliché is particularly appropriate in this instance—the future of tuberculosis diagnosis "appears bright."

will become more feasible in the future as gene probe analysis becomes more widely accepted.

Gene probe assays are widely available in kit forms for a variety of bacterial species (for example, streptococci, *Haemophilus*, *Listeria*, *Mycobacterium*, and *Neisseria*) as well as for fungi (such as *Blastomyces*, *Coccidioides*, and *Histoplasma*). In many cases, the tests are described as "exquisitely accurate," with a high degree of discrimination and reliability as strong as older identification methods. Since first introduced in the 1970s, gene probe tests have been met with periods of unbridled enthusiasm counterbalanced by periods of disappointment. The future value of gene probes will depend in part on the development of ways to minimize false-positive reactions due to contamination, on methods of increasing the sensitivity of tests, and on mechanisms for enhancing the signals from probes bound to their target molecules.

CONCEPT AND REASONING CHECKS

21.12 Explain how PCR works.

SUMMARY OF KEY CONCEPTS

21.1 Immunity to Disease

- In naturally acquired active immunity antibodies are produced and lymphocytes activated in response to an infectious agent. Immunity arises from contracting the disease and recovering, or from a subclinical infection.
- Artificially acquired active immunity is when antibodies and lymphocytes are produced as a result of a vaccination and immunity is established for some length of time.
- There are seven strategies for producing vaccines. Often booster shots are required to maintain immunological memory.
- Naturally acquired passive immunity comes from the passage of antibodies from the mother to fetus, or mother to newborn through colostrum which confers immunity for a short period of time.
- Artificially acquired passive immunity is when one receives an antiserum (antibodies) produced in another human or animal. This form confers immunity for a short period of time. Serum sickness can develop if the recipient produces antibodies against the antiserum.
- Herd immunity results from a vaccination program that lowers the number of susceptible members within a population capable of contracting a disease. If there are few susceptible individuals, the probability of disease spread is minimal.
- Established vaccines are quite safe, although some people do experience side effects that may include mild fever, soreness at the injection site, or malaise after the vaccination. Few people suffer serious consequences.

21.2 Serological Reactions

- Serological reactions consist of antigens and antibodies (serum). Successful reactions require the correct dilution (titer) of antigens and antibodies.
- Neutralization is a serological reaction in which antigens and antibodies neutralize each other. Often there is no visible reaction, so injection into an animal is required to see if the reaction has occurred.
- In a precipitation reaction, antigens and antibodies react to form a matrix that is visible to the naked eye. Different microbial antigens can be detected using this technique.
- Agglutination involves the clumping of antigens. The cross-linking of antigens and antibodies causes the complex to clump. Some infectious agents can be detected readily by this method.
- Complement fixation can detect antibodies to a variety of pathogens. This serological method involves antigen-antibody complexes that are detected by the fixation (binding) of complement.

21.3 Additional Laboratory Tests

- The fluorescent antibody technique involves the addition of fluorescently tagged antibodies to a known pathogen to a slide containing unknown antigen (pathogen). If the antibody binds to the antigen, the pathogen will glow (fluoresce) when observed with a fluorescence microscope, and the pathogen is identified.
- The radioimmunoassay (RIA) uses radioactivity to quantitate radioactive antigens by competition with nonradioactive antigens for reactive sites on antibody molecules.
- The enzyme-linked immunosorbent assay (ELISA) can be used to detect if a patient's serum contains antibodies to a specific pathogen or to detect or measure antigens in serum. A positive ELISA is seen as a colored reaction product.
- Monoclonal antibodies result from the fusion of a specific B cell (plasma cell) with a myeloma cell to produce a hybridoma that secretes an antibody recognizing a single epitope.
- Gene probes are single-stranded DNA segments that recognize and bind to complementary sequences. Such binding is detected by radioactive means. Such probes can locate or identify specific disease-causing organisms.

LEARNING OBJECTIVES

After understanding the textbook reading, you should be capable of writing a paragraph that includes the appropriate terms and pertinent information to answer the objective.

1. Summarize how naturally acquired active immunity provides immunity to infectious diseases.
2. Identify how artificially acquired active immunity produces immunity.
3. Differentiate between vaccines consisting of live, attenuated antigens and those containing inactivated antigens. Provide examples for each.
4. Explain how toxoid vaccines are produced, giving several examples.
5. Describe why subunit vaccines cannot produce illness.
6. Discuss how conjugate vaccines are produced and give an example.
7. Compare and contrast experimental DNA vaccines and recombinant vector vaccines.
8. Distinguish between naturally-acquired passive immunity and artificially-acquired passive immunity.
9. Assess the importance of herd immunity to community and national health.
10. Judge the safety of vaccines and the reporting system to identify possible adverse affects.
11. Identify the role of titration in identifying disease.
12. Compare and contrast neutralization, precipitation, and agglutination as examples of serological reactions.
13. Explain how the complement fixation test works.
14. Differentiate between fluorescent antibody, RAST, and ELISA as labeling methods to detect antigen-antibody binding.
15. Summarize the characteristics of monoclonal antibodies and assess their role in identifying infectious disease.
16. Summarize the process of PCR for producing large numbers of identical DNA segments.

SELF-TEST

Answer each of the following questions by selecting the ***one*** answer that best fits the question or statement. Answers to even-numbered questions can be found in **Appendix C**.

1. Exposure to the flu virus, contracting the flu, and recovering from the disease would be an example of
A. artificially-acquired passive immunity.
B. naturally-acquired active immunity.
C. artificially-acquired active immunity.
D. naturally-acquired passive immunity.
E. Both **A** and **D** are correct.

2. Which one of the following (**A–D**) is *not* used to make a vaccine? If all are used, then select **E.**
A. Viruses
B. Toxins
C. Antibiotics
D. Bacteria
E. All the above (**A–D**) are correct.

3. An attenuated vaccine contains
A. inactive toxins.
B. living, but slow growing (replicating) antigens.
C. killed bacteria.
D. nonreplicating viruses.
E. noninfective antigen subunits.

4. All of the following are examples of single-dose vaccines *except*:
A. MMR.
B. Proquad.
C. DTaP.
D. hepatitis vaccine.
E. DPT.

5. An example of a conjugated vaccine is
A. Hib.
B. Sabin polio vaccine.
C. hepatitis B vaccine.
D. DPT.
E. any toxoid vaccine.

6. Immune complex formation and serum sickness are dangers of
A. artificially-acquired passive immunity.
B. naturally-acquired active immunity.
C. artificially-acquired active immunity.
D. naturally-acquired passive immunity.
E. Both **A** and **D** are correct.

7. Herd immunity is affected by
A. the percentage of a population that is vaccinated.
B. a population's living environment.
C. the strength of an individual's immune system.
D. the number of susceptible individuals.
E. All the above (**A–D**) are correct.

8. Approximately _____ of 100,000 vaccinated individuals are likely to suffer a serious reaction to the vaccination.
A. 1
B. 50
C. 100
D. 500
E. 1,000

9. Titer refers to
A. the most concentrated antigen-antibody concentration showing a reaction.
B. the first diluted antigen-antibody concentration showing a reaction.
C. the precipitation line formed between an antigen-antibody reaction.
D. the most dilute antigen-antibody concentration showing a reaction.
E. an undiluted antigen-antibody concentration showing a reaction.

10. The serological reaction where antigens and antibodies form an extensive lattice of large particles is called
A. fixation.
B. precipitation.
C. neutralization.
D. intoxication.
E. agglutination.

11. When antigens are attached to the surface of latex beads and then reacted with an appropriate antibody, a/an _____ reaction occurs.
A. inhibition
B. complement fixation
C. agglutination
D. neutralization
E. precipitation

12. What serological technique is used to measure the concentration of very small antigens, such as haptens?
A. ELISA
B. Gene probes
C. Radioimmunoassay
D. Immunodiffusion
E. Fluorescent antibody technique

13. In an ELISA, the primary antibody represents
A. the patient's serum.
B. the antibody recognizing the secondary antibody.
C. the enzyme-linked (labeled) antibody.
D. the antibodies washed away.
E. the antibodies binding to complement.

14. A hybridoma cell
A. secretes monoclonal antibodies.
B. presents antigens on its surface.
C. secretes polyclonal antibodies.
D. is formed from the fusion of myeloma and plasma cells.
E. Both **A** and **D** are correct.

15. A _____ produces billions of identical fragments of DNA.
A. monoclonal antibody procedure
B. serological reaction
C. polymerase chain reaction
D. gene probe technique
E. complement fixation reaction

QUESTIONS FOR THOUGHT AND DISCUSSION

Answers to even-numbered questions can be found in **Appendix C**.

1. It is estimated that when at least 90 percent of the individuals in a given population have been immunized against a disease, the chances of an epidemic occurring are very slight. The population is said to exhibit "herd immunity," because members of the population (or herd) unknowingly transfer the immunizing agent to other members and eventually immunize the entire population. What are some ways by which the immunizing agent can be transferred?
2. When children are born in Great Britain, they are assigned a doctor. Two weeks later, a social services worker visits the home, enrolls the child on a national computer registry for immunization, and explains immunization to the parents. When a child is due for an immunization, a notice is automatically sent to the home, and if the child is not brought to the doctor, the nurse goes to the home to learn why. Do you believe a method similar to this can work (or should be used) in the United States to achieve uniform national immunization?
3. Since 1981, the incidence of pertussis has been increasing in the United States with the greatest increase found in adolescents and adults. Why are adolescents and adults targets of the bacterial pathogen, considering these individuals were usually considered immune to the disease?
4. Children between the ages of 5 and 15 are said to pass through the "golden age of resistance" because their resistance to disease is much higher than that of infants and adults. What factors may contribute to this resistance?

APPLICATIONS

Answers to even-numbered questions can be found in **Appendix C**.

1. A friend is heading on a trip to a foreign country where hepatitis A outbreaks are quite common. For passive immunity, he gets a shot of an antiserum containing IgG antibodies. Why do you suppose IgM is not used, especially since the immunoglobulins are the important components of the primary antibody response?
2. A complement fixation test is performed with serum from a patient with an active case of syphilis. In the process, however, the technician neglects to add the syphilis antigen to the tube. Would lysis of the sheep red blood cells occur at the test's conclusion? Why?
3. Suppose the titer of mumps antibodies from your blood was higher than that for your fellow student. What are some of the possible reasons that could have contributed to this? Try to be imaginative on this one.
4. Given a choice, which of the four general types of acquired immunity would it be safest to obtain? Why?

REVIEW

On completing Section 21.1 (Immunity to Disease), test your comprehension of the section's contents by filling in the following blanks with two terms that answer the description best. **Appendix C** contains the answers to the even-numbered statements.

1. Two general forms of immunity: ____________ and ____________.
2. Two types of natural immunity: ____________ and ____________.
3. Two diseases that MMR is used against: ____________ and ____________.
4. Two diseases that DPT is used against: ____________ and ____________.
5. Two types of passive immunity: ____________ and ____________.
6. Two names for antibody-containing serum: ____________ and ____________.
7. Two antibody types formed on antigen stimulation: ____________ and ____________.
8. Two ways newborns have acquired maternal antibodies: ____________ and ____________.
9. Two types of viruses in viral vaccines: ____________ and ____________.
10. Two bacterial diseases for which toxoids are used: ____________ and ____________.
11. Two methods for administering vaccines: ____________ and ____________.
12. Two tracts in which IgA accumulates: ____________ and ____________.
13. Two functions of antibodies in antiserum: ____________ and ____________.
14. Two bacterial diseases where passive immunity is used: ____________ and ____________.

HTTP://MICROBIOLOGY.JBPUB.COM/

The site features learning, an on-line review area that provides quizzes and other tools to help you study for your class. You can also follow useful links for in-depth information, read more MicroFocus stories, or just find out the latest microbiology news.

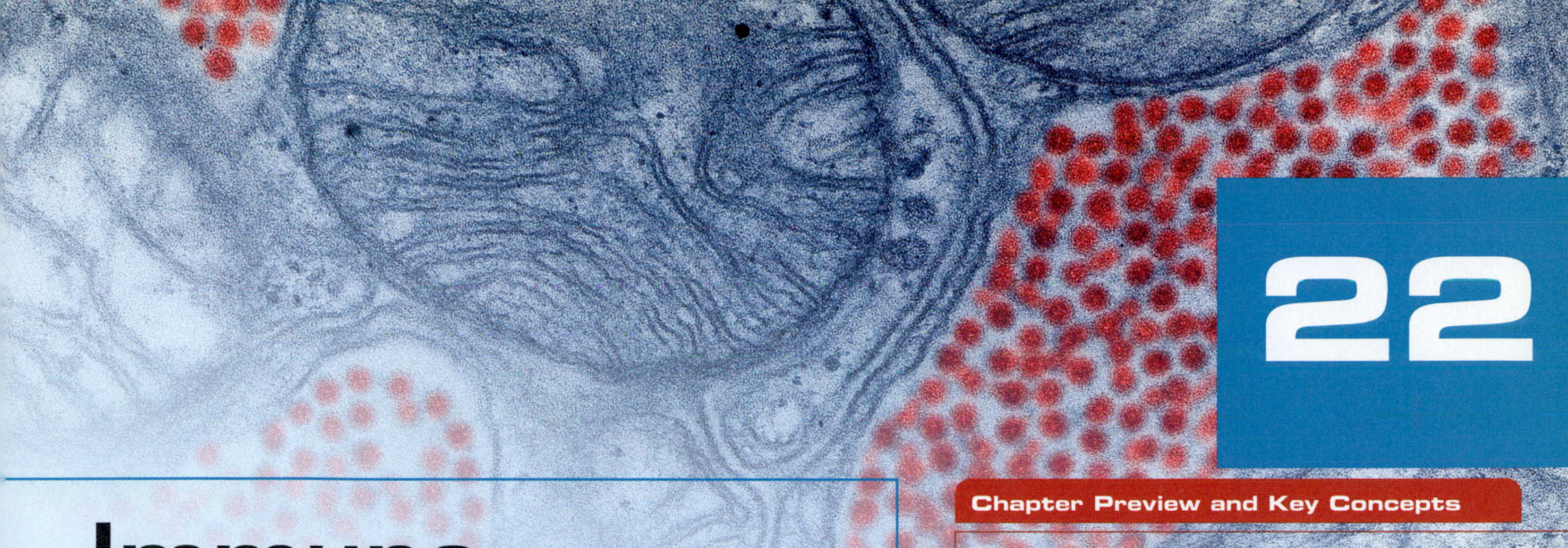

Immune Disorders

You may not think of asthma as a killer disease. Yet each year, nearly 500,000 Americans with asthma are hospitalized, and more than 4,000 die.
—MayoClinic.com

Chapter Preview and Key Concepts

22.1 Type I IgE-Mediated Hypersensitivity
- A type I hypersensitivity can vary from simply a runny nose and watery eyes to a life-threatening disorder.
- Anaphylactic reactions can be sudden, systemic allergic reactions that are life-threatening.
- Common allergies are examples of a localized anaphylaxis.
- Asthma is characterized by a constriction of the airways.
- Genetics and the environment are responsible for most allergies.

  MicroInquiry 22: Allergies, Dirt, and the Hygiene Hypothesis
- Physical avoidance and drug treatment can limit allergic attacks.

22.2 Other Types of Hypersensitivity
- Antibodies can be the cause of cytotoxic (type II) hypersensitivities.
- Immune complexes are responsible for type III hypersensitivities.
- Type IV hypersensitivity is an exaggeration of a cell mediated immune response.

22.3 Autoimmune Disorders and Transplantation
- Autoimmune disorders generate an immune attack on self cells and tissues.
- Adaptive immune responses can hinder successful tissue transplants or grafts.

22.4 Immunodeficiency Disorders
- Immunodeficiency disorders allow recurrent and often severe, longer lasting infections.

During the classical golden age of microbiology, investigations were carried out on certain diseases not as dangerous as microbial diseases, but nevertheless were a source of great discomfort and inconvenience. One such disease was hay fever. In the 1870s, British scientist Charles Harrison Blackley placed crude pollen preparations in the eyes of hay fever sufferers; the result was swelling of the membranes. Blackley also rubbed pollen grains into a skin scratch; he observed a local skin reaction at the scratch site. Some of Blackley's critics said the reactions were not due to the pollen but rather to mechanical injury inflicted by the introduced pollen grains. However, in 1903 another British investigator, William Philipps Dunbar, supported Blackley's work by using saline extracts of pollen grains; the same skin inflammation was observed.

Research on hay fever and other allergies have come a long way since the experiments of Blackley and Dunbar. The term and concept of "allergy" (*allo* = "other"; *erg* = "work") was coined by an Austrian pediatrician in 1906. He suggested the symptoms experienced by some of his patients might have been caused by normally harmless substances in the environment, such as pollen, dust, or even food.

Today we regarded allergies, which are more appropriately called hypersensitivities, as a disorder of the immune system. The common denominator among these hypersensitivities is a state of increased reactivity by the immune system involving both humoral and cell mediated aspects of immunity. In other words, the immune system is doing "work" other than what it should be doing.

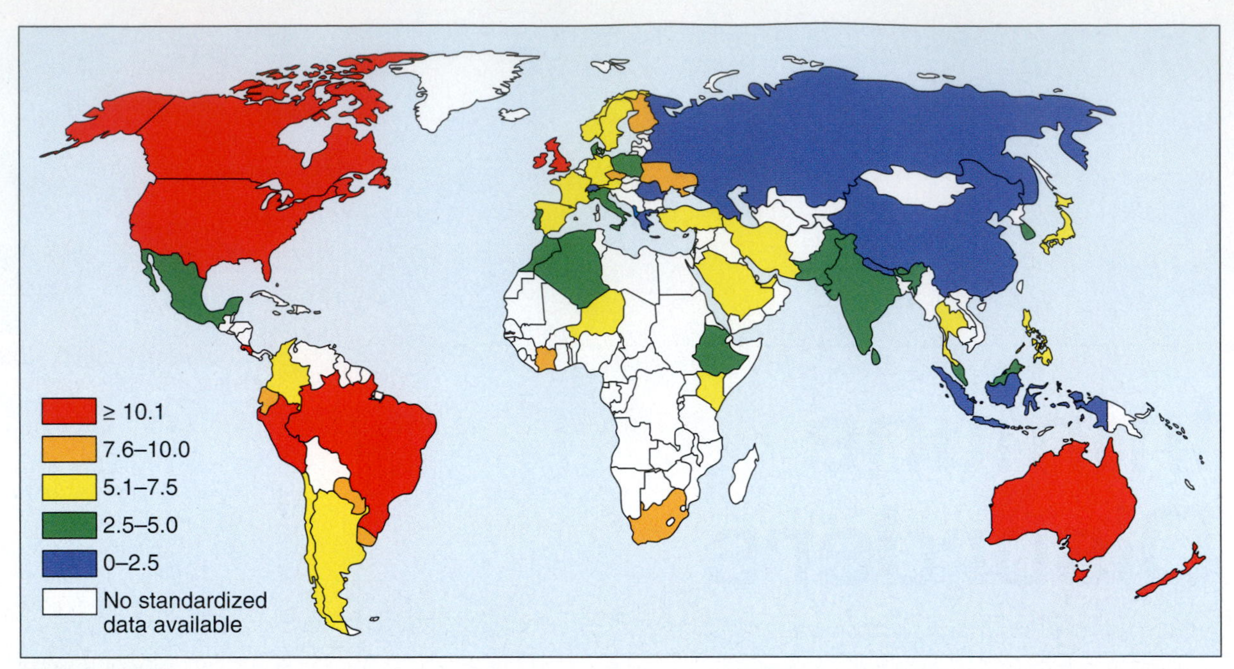

FIGURE 22.1 **Populations Suffering from Clinical Asthma in 2004.** Asthma can be difficult to diagnose, especially in children under 5 years of age. Clinical asthma refers to cases where physical exams have noted lung dysfunction associated with asthma and not due to an infectious disease. *Source*: Drawn with data from http://www.ginasthma.com.

Q: Do industrialized nations or developing nations suffer from higher rates of asthma? Explain.

About 35 million Americans currently suffer from hay fever, and thousands more are allergic to foods, cosmetics, leather, or metals. Many people cannot keep pets because of severe sensitivities to pet dander. One of the more serious forms of hypersensitivities is asthma. More than 300 million people around the world suffer from asthma and, according to the Global Initiative on Asthma, this number will increase to over 400 million by 2025 (FIGURE 22.1). In the United States, some 4,000 people die of asthma annually. All told, an estimated 40 to 50 million people in the United States have some type of allergy. We are in the midst of an allergy epidemic.

As serious as this may sound, you might wonder why we are going to talk about allergies in a microbiology textbook. Good question. The answer lies in what many believe is a consequence of children not fighting the types of infectious diseases experienced by past generations. This is not referring to deadly diseases like plague and smallpox, but rather to the relatively innocuous microorganisms our parents and grandparents picked up from water, dirt, and animals. If the developing immune system does not "experience" common microbes, there is a good chance it will fail to function properly—and allergies and hypersensitivities may be the result.

Hypersensitivities and allergies represent immune defenses that are in disorder. Also included in the broad category of immune disorders are the autoimmune diseases and various immune deficiency diseases that often have a microbial component associated with the disorder. So, immune disorders represent the major topics of this chapter and are currently a subject of intense research in immunology. We survey each of these in the following sections.

22.1 Type I IgE-Mediated Hypersensitivity

Hypersensitivity is a multistep phenomenon triggered by exposure to an antigen. It consists of a dormant (latent) stage, during which an individual becomes sensitized, and an active stage (hypersensitivity) following a subsequent exposure to the antigen. The process may involve elements of humoral or cell mediated immunity, or sometimes both.

In the early 1970s, P. G. H. Gell and R. A. Coombs proposed a method for classifying hypersensitivities into four types. **Immediate hypersensitivity** refers to an antibody response to the second or ensuing dose of the same antigens (FIGURE 22.2). Gell and Coombs classified immediate hypersensitivity into three types:

- Type I IgE-mediated hypersensitivity, which is a process involving IgE, mast cells, basophils, and cell mediators inducing smooth muscle contraction;
- Type II cytotoxic hypersensitivity, which involves IgG, IgM, complement, and the destruction of host cells; and
- Type III immune complex hypersensitivity, which involves IgG, IgM, complement, and the formation of antigen-antibody aggregates in the tissues.

Delayed hypersensitivity is a cell mediated immune response developing over two to three days. Gell and Coombs defined delayed hypersensitivity as type IV cellular hypersensitivity, involving cytokines and T lymphocytes.

Realize a hypersensitivity is an exaggerated or inappropriate immune defense that is causing the problems in the affected individual.

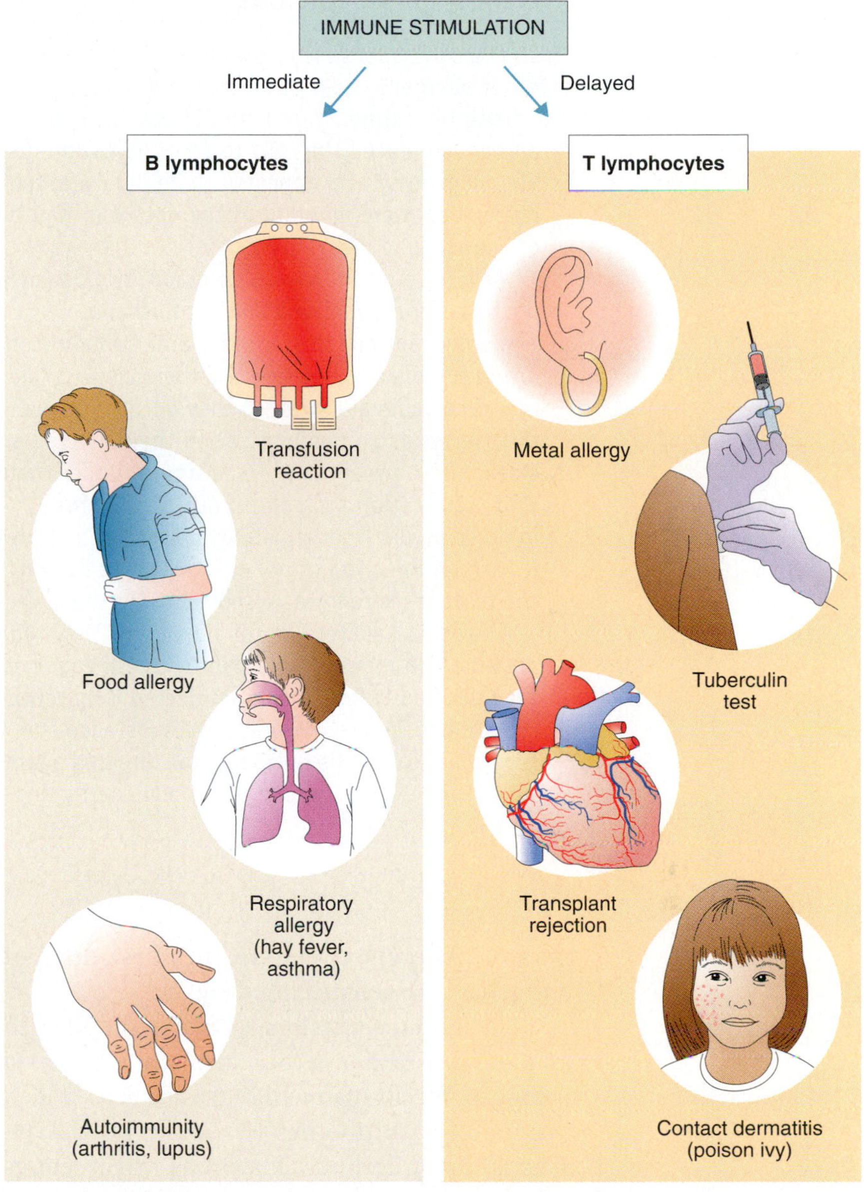

FIGURE 22.2 **The Various Forms of Hypersensitivities.** Immune system hypersensitivities may be related to B lymphocytes or T lymphocytes. Stimulation of the immune system is the starting point.

Q: Why are all these examples of immune stimulation referred to as hypersensitivities?

Type I Hypersensitivity Is Induced by Allergens

KEY CONCEPT

- A type I hypersensitivity can vary from simply a runny nose and watery eyes to a life-threatening disorder.

Type I hypersensitivity has all the characteristics of a humoral immune response. It begins with the entry of a substance, called an **allergen**, which triggers an allergic reaction in the body. The allergen may be any of a wide variety of materials such as plant pollen, certain foods, bee venom, serum proteins, or a drug, such as penicillin. In the case of penicillin, the drug molecule itself is the allergen, but the molecule does not stimulate the immune system until after it has combined with tissue proteins to form an allergenic complex (see Chapter 20). Doses of antigen as low as 0.001 mg have been known to sensitize a

MICROFOCUS 22.1 History

Itches and Antibodies

In the early 1960s, scientists knew that pollen and other allergens cause mast cells and basophils to release histamine. What they did not know was how allergens induced the cells to spill their contents. The immune system appeared to be involved, but researchers were ignorant of the nature or function of the immune mechanism.

At a hospital in Denver, Colorado, two Japanese doctors, Teruko Ishizaka and her husband, Kimishige, set out to search for an antibody stimulating the allergic reaction. Neither scientist had any observable allergies, so they decided to use themselves as guinea pigs. When they injected an extract of ragweed pollen (see figure) under their skin, no reaction took place. But, if they first injected serum from an allergy patient and followed it with an injection of pollen extract, a raised itchy welt appeared. It was apparent that something in the patient's serum was responsible for the allergy.

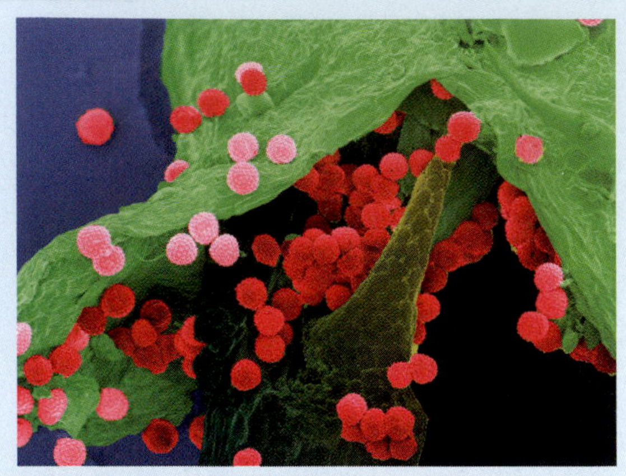

False-color scanning electron micrograph of ragweed pollen.

The Ishizakas went to the next step. They separated the serum into as many different components as possible and repeated the skin test with each component. When a particular component caused welts, they purified it further and reinjected themselves. After four years of experiments (and lots of itching), they finally isolated their elusive substance—an antibody. The Ishizakas named it immunoglobulin E (or IgE) because the antibody was directed against antigen E of ragweed pollen. Then, they breathed a sigh of relief—the investigation and the itching was over.

person. Allergists refer to this first dose of antigen as the **sensitizing dose**.

The immune system responds to the allergen as if it were a dangerous antigen, such as a pathogen. B cells mature into plasma cells, which produce IgE antibodies (MicroFocus 22.1). This antibody, formerly known as reagin, enters the circulation and attaches itself to the surface of mast cells and basophils (FIGURE 22.3). Mast cells are connective tissue cells found in the respiratory and gastrointestinal tracts, and near the blood vessels. They measure about 10 μm to 15 μm in diameter and are filled with 500 to 1,500 granules containing histamine and other physiologically active substances.

Basophils are circulating leukocytes in the blood that also are rich in granules. They represent about 1 percent of the total leukocyte count in the circulation and measure about 15 μm in diameter. Mast cells and basophils each have over 100,000 receptor sites where IgE antibodies can attach by the Fc tail portion (see Chapter 20). As IgE antibodies attach to mast cells and basophils, the individual becomes **sensitized**. Multiple stimuli by allergen molecules may be required to sensitize a person fully. This is why penicillin often must be taken several times before a penicillin allergy manifests itself.

Sensitization usually requires a minimum of one week, during which time millions of molecules of IgE attach to thousands of mast cells and basophils. Since attachment occurs at the Fc end of the antibody, the Fab ends point outward from the cell as shown in Figure 22.3.

On subsequent exposure to the same allergen, the allergen molecules bind to the Fab ends and cross-link IgE antibodies. This cross-linking triggers **degranulation**, a release of granule contents at the cell surface.

Degranulation first requires an inflow of calcium ions (Ca^{2+}) and a brief stimulation of the enzyme **adenylyl cyclase**, which catalyzes the conversion of ATP to cyclic AMP (cAMP).

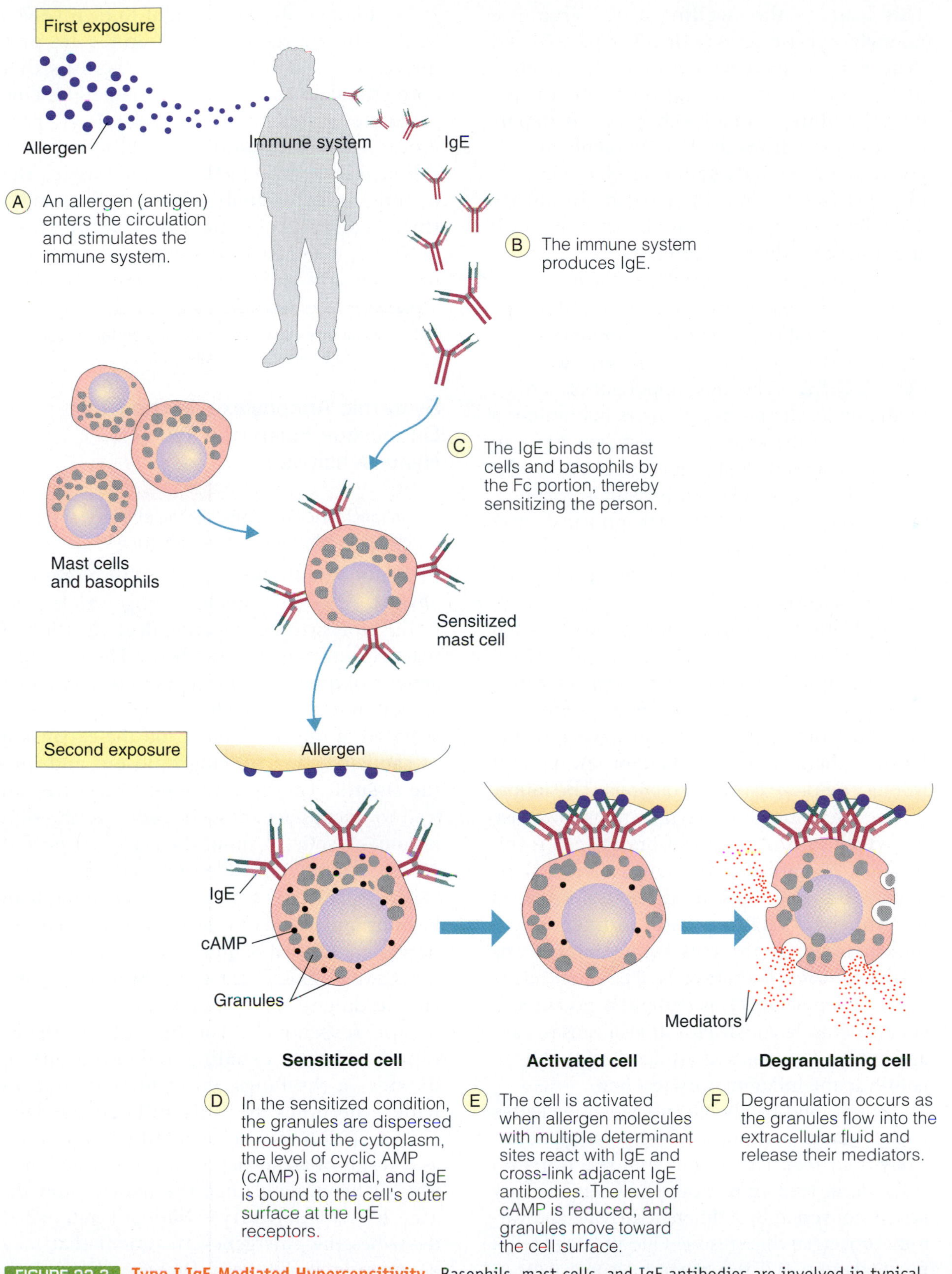

FIGURE 22.3 **Type I IgE-Mediated Hypersensitivity.** Basophils, mast cells, and IgE antibodies are involved in typical allergies like hay fever.

Q: Does a person experience the symptoms of an allergy on first or second exposure to an allergen? Explain.

This leads to the swelling of the granules. Adenylyl cyclase activity then is repressed and there is a sudden decrease in cAMP, which is necessary for degranulation to occur. An understanding of this biochemistry is important because drugs such as epinephrine are given to individuals suffering systemic anaphylaxis (see below). Epinephrine stimulates adenylyl cyclase, increasing the level of cAMP and inhibiting degranulation.

As granules fuse with the plasma membrane of the basophils and mast cells, they release a number of mediator substances having substantial pharmacologic activity (FIGURE 22.4). The most important preformed mediator of allergic reactions is **histamine**, a derivative of the amino acid histidine. Once in the bloodstream, histamine circulates to the body cells and attaches to histamine receptors present on most body cells. The principal effect will be to constrict smooth muscle cells. Serotonin and bradykinin are other preformed mediators released.

Still other mediators must be synthesized after the antigen-IgE reaction (Figure 22.4). One example is a series of substances called **leukotrienes** (so named because they are derived from leukocytes and have a triene [triple] chemical bond). Leukotrienes result from a complex set of interactions. The inflow of Ca^{2+} activates an enzyme (phospholipase A), which in turn releases from the cell membrane a 20-carbon fatty acid called arachidonic acid. The arachidonic acid is acted on by another enzyme and converted to leukotriene A, which is immediately converted to other leukotrienes. The latter, especially leukotriene D4, is extremely potent as a smooth muscle constrictor. It also causes leakage in blood vessels and attracts eosinophils to continue the inflammatory reaction.

The second family of synthesized mediators is the human hormones called **prostaglandins**. They also result from enzyme reactions on arachidonic acid. In this case, however, the fatty acid is converted by a different enzyme and various prostaglandins result. One prostaglandin, prostaglandin D2, is a powerful constrictor of the bronchial tubes.

Cytokines also are thought to be involved in the allergic response. Cytokines are produced by most cells and have actions that both stimulate and inhibit inflammation. One cytokine, called interleukin-4 (IL-4), promotes IgE production by B cells; another, called interleukin-5 (IL-5), encourages the maturation and activity of eosinophils; and a third, called tumor necrosis factor alpha (TNF-α), is released from mast cells and may be responsible for shock in systemic reactions.

CONCEPT AND REASONING CHECKS

22.1 Summarize the events occurring during a first and second exposure to an allergen.

Systemic Anaphylaxis Is the Most Dangerous Form of a Type I Hypersensitivity

KEY CONCEPT

- Anaphylactic reactions can be sudden, systemic allergic reactions that are life-threatening.

Systemic anaphylaxis (*ana* = "throughout"; *phylaxi* = "watch") involves antigen (allergen) in the bloodstream triggering degranulation of mast cells throughout the body. The principal activity of the released mediators is to contract smooth muscles in the body. One effect is constriction of the small veins and the expansion of capillary pores, forcing fluid out and into the tissues. The drop in blood pressure can lead to unconsciousness. In addition, the skin becomes swollen around the eyes, wrists, and ankles, a condition called **edema**. The edema is accompanied by a hive-like rash, along with burning and itching in the skin, as the sensory nerves are excited. Contractions also occur in the gastrointestinal tract and bronchial muscles, leading to sharp cramps and shortness of breath, respectively. The individual inhales rapidly without exhaling and traps carbon dioxide in the lungs, an ironic situation in which the lungs are fully inflated but lack oxygen. Death may occur in 10 to 15 minutes as a result of asphyxiation if prompt action is not forthcoming (hence the name "immediate" hypersensitivity). MicroFocus 22.2 describes the emergency treatment that may be given in such a case. Diverse allergies can trigger a systemic anaphylactic reaction,

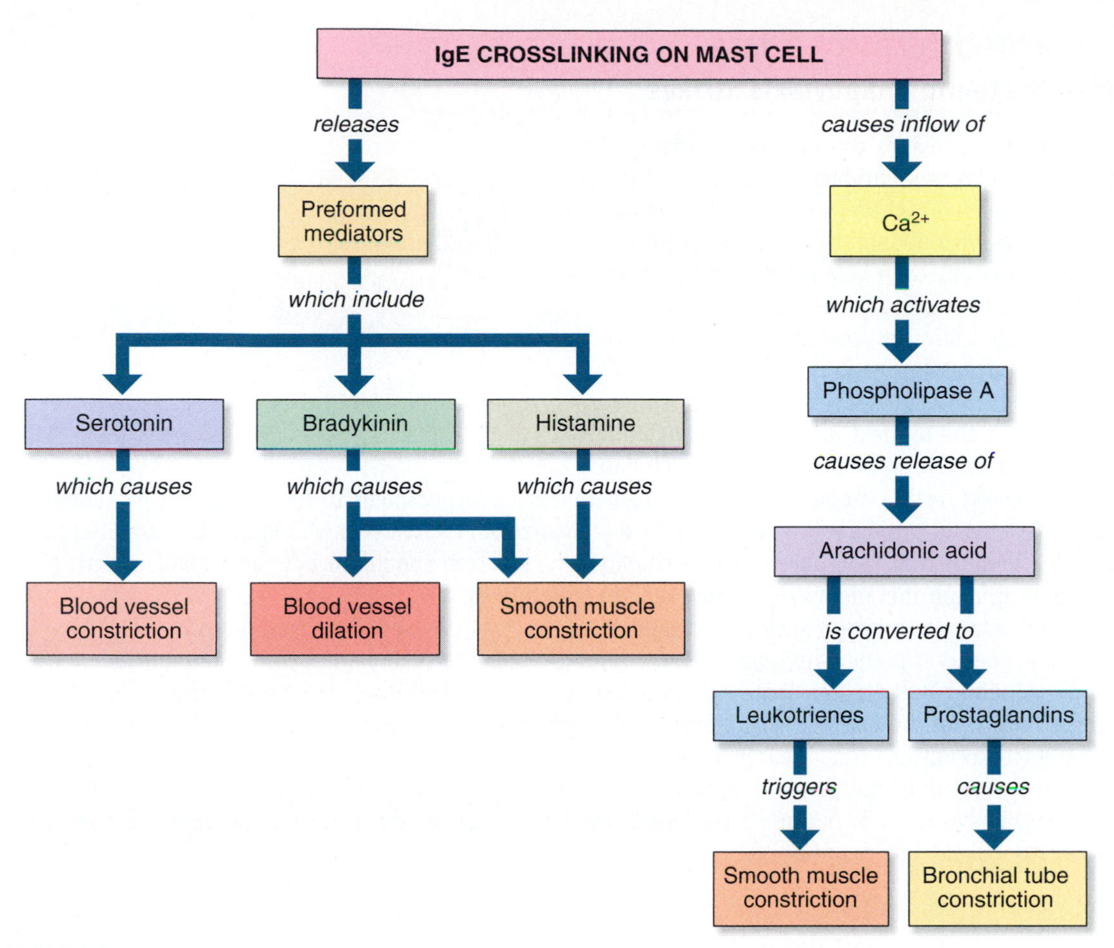

FIGURE 22.4 **Concept Map for Type I Hypersensitivity and Mediator Substance Production.** Several chemical mediators are stimulated upon IgE cross-linking on mast cells and basophils.

Q: Because IgE is an antibody normally targeting parasitic helminths, what might these mediator substances have done if it was a parasite infection rather than an allergen?

including bee stings, penicillin, antitoxins, and foods such as nuts and seafood.

Atopic Disorders Are the Most Common Form of a Type I Hypersensitivity

KEY CONCEPT

- Common allergies are examples of a localized anaphylaxis.

Type I hypersensitivity reactions need not result in the whole-body (systemic) involvement. In fact, the vast majority of hypersensitivity reactions are accompanied by limited production of IgE and the sensitization of mast cells only in localized areas of the body.

Common Allergies. According to public health estimates, up to 50 million Americans (almost 20 percent of the population) have some type of common (seasonal) allergy or **atopic disease** (*atopo* = "out of place"). An example of an atopic disease is hay fever, technically referred to as **allergic rhinitis**. Realize hay fever is not caused by hay itself nor does it cause a fever! The condition develops from seasonal inhalations of tree, grass, or weed pollens. Immune stimulation by pollen antigens leads to IgE production, and a sensitization of mast cells follows in the eyes, nose, and upper respiratory tract. Subsequent exposures bring on sneezing, tearing, swollen mucous membranes, and other well-known symptoms.

There also are year-round allergies. These perennial hypersensitivities usually result from

MICROFOCUS 22.2 Tools

When Systemic Anaphylaxis Strikes

Systemic anaphylaxis is a terrifying experience. The skin itches intensely and breaks into hives, the eyes and joints become red and puffy, and the person doubles over with abdominal pains. Breathing becomes difficult, then belabored, and finally is reduced to life-sucking gasps. The symptoms develop within minutes, and the individual usually faints and quickly lapses into a coma.

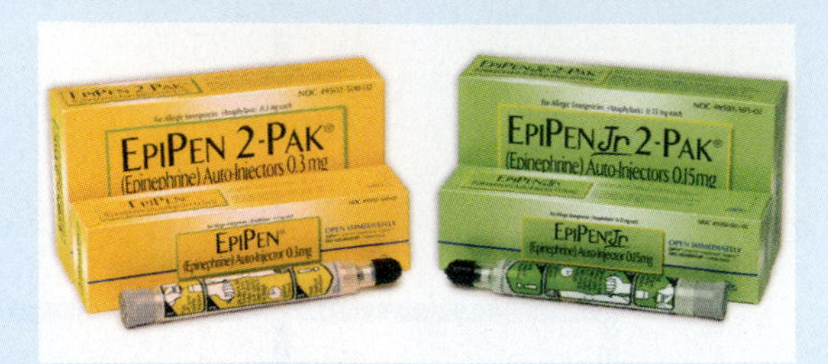

An anaphylaxis emergency kit.

The key to survival is swift action. Epinephrine (adrenalin) is the highest priority drug. Within minutes of injection, it stabilizes basophils and mast cells to prevent further mediator release. It also dilates the bronchioles to reopen the air passageways, and constricts the capillaries to keep fluid in the circulation. Individuals who know they are allergic to specific allergens that can cause a severe anaphylactic reaction should always carry a self-injecting syringe of epinephrine and antihistamine tablets (see figure).

A smooth muscle relaxant such as aminophylline also may be used. This drug helps dilate the bronchial tubes and pulmonary blood vessels. An antihistamine such as diphenhydramine (Benadryl) may be valuable. This drug competes with histamine for the active sites on smooth muscle receptors, thereby inhibiting the action of histamine. Hydrocortisone can be used to reduce swelling in the tissue, and an expectorant can help clear laryngeal edema.

If the patient does not respond rapidly to the drugs, it may be necessary to insert a tube into the respiratory passageway or perform a tracheostomy. Either must be done quickly, because life now is reduced to a scant few minutes.

chronic exposure to the substances such as house dust, mold spores, dust mites, detergent enzymes, and the particles of animal skin and hair called dander. (Dander itself is not the allergen; the actual allergens are proteins deposited in the dander from the animal's saliva when it grooms itself.) Included here are cockroaches and substances carried on their bodies.

Some people experience a **late-phase anaphylaxis**. In this case, it takes several hours for the tissue to become hot, tender, red, and swollen. The mast cells induce this reaction by releasing chemokines that attract other cells to the site to bring about the changes. For example, eosinophils exist in unusually high numbers in allergic individuals; they arrive at the site and release leukotrienes as well as toxic substances contributing to tissue damage. Neutrophils are normally the phagocytes of the bloodstream, but in allergic reactions, they too liberate a number of enzymes causing local tissue damage. T lymphocytes produce interleukin-4 (IL-4), which augments the allergic response.

■ Allergenic: A substance causing an allergic reaction.

Food Allergies. A variety of foods can cause allergies, which usually are accompanied by symptoms in the gastrointestinal tract, including swollen lips, abdominal cramps, nausea, and diarrhea. The skin may break out in a rash containing **hives**, each hive consisting of a central puffiness, called a **wheal**, surrounded by a zone of redness known as a **flare**. Such a rash is called **urticaria** (*urtica* = "stinging needle"). Allergenic foods include fish, shellfish, eggs, wheat, cow's milk, soy, tree nuts and even peanuts (MicroFocus 22.3). Some allergic reactions to foods can cause a potentially lethal anaphylactic reaction.

Physical and Exercise-Induced Allergies. Physical factors such as cold, heat, sunlight, or sweating also can cause allergies. Exactly what causes the reaction is unknown. Some immunologists believe a physical stimulus causes a change in a protein in the skin, which the immune system then "sees" as foreign and

MICROFOCUS 22.3

The Peanut Dilemma

Americans love peanuts—salted, unsalted, oil roasted, dry roasted, Spanish, honey crusted, in shells, out of shells, and on and on.

Yet, more than one million Americans fear the peanut because they risk a rather nasty allergic reaction: One can break out in hives, develop a serious headache, experience a racing heartbeat, or double over with intestinal cramps.

And, if that's not bad enough, someone eating peanuts on a plane can release small peanut particles into the air that can cause a reaction in a sensitive person seated nearby. Moreover, peanut-specific antibodies can be transferred from an organ donor to an organ recipient. (In the recipient, the skin reaction is not threatening, but it does complicate matters.) However, a research report in 2003 suggested that peanut allergies can be avoided if babies are not given peanuts until their immune systems are mature at age three.

The answer to the peanut allergy dilemma can best be summed up as "V plus V." The first V is for vigilance. Vigilance means avoiding peanuts or peanut butter; but it also means being cautious about egg roll wrappers, chili fillers, and protein extenders in cake mixes, all of which may contain peanuts in one form or another. Drinking liquid charcoal (available in pharmacies) will absorb peanut particles and prevent triggering of increased symptoms. A 2003 report in the *Journal of Allergy and Clinical Immunology* suggested that parents should keep liquid charcoal at home in case of a peanut allergy in young children. It means vigilance that manufacturers clearly label their products and insist that peanut-detection tests be performed routinely.

The second V is for vaccine. In 1999, investigators from Johns Hopkins University tested a peanut vaccine and showed that it protects sensitized mice against peanut proteins. The vaccine consists of DNA segments that encode the peanut proteins. Encased in protective molecules and delivered orally, the vaccine decreased the mice's capacity for producing peanut-related IgE. The developers postulated that the vaccine elicits the so-called "blocking antibodies" that bind the peanut antigens before they reach the animal's immune system.

So, does that mean we can expect health officials to distribute vaccine injections where we buy peanut butter, peanut brittle, or beer nuts? Not likely, say the researchers, at least not in the immediate future.

mounts an immune response. Exercise sometimes also causes allergies, usually in the form of an asthma attack.

CONCEPT AND REASONING CHECKS

22.2 Distinguish between the different types of typical allergies.

Allergic Reactions Also Are Responsible for Triggering Many Cases of Asthma

KEY CONCEPT

- Asthma is characterized by a constriction of the airways.

Some 300 million people live with asthma worldwide, and the disorder is responsible for one in 250 deaths. **Asthma**, which is characterized by wheezing and stressed breathing, can be caused by airborne allergens such as pollen, mold spores, or products from insects like dust mites. Attacks also can be induced by physical exercise or cold temperatures, without any known allergen being present (FIGURE 22.5).

Asthma primarily is a chronic inflammatory disorder that occurs in two parts. First, there is an early response to allergen exposure. Like atopic diseases, the degranulation of mast cells releases a variety of chemical mediators. However, the mediators are not released in the nose or eyes, but rather the lower respiratory tract, resulting in bronchoconstriction, vasodilation, and mucus buildup (FIGURE 22.6).

FIGURE 22.5 **Using an Asthma Inhaler.** For some asthma sufferers, simple exercise might bring on an asthma attack. Using an inhaler that delivers medication directly to the lungs, allows sufferers to live active lives without fear of an attack.

Q: How could physical exercise trigger an asthma attack?

The synthesis and release of other mediators, including IL-4, IL-5, and TNF-α, result in the recruitment of eosinophils and neutrophils into the lower respiratory tract. These events represent the late response because they occur hours after the initial exposure to allergen. The presence of the eosinophils and neutrophils can cause tissue injury and potentially cause blockage of the airways (bronchioles). Bronchodilators can be used to widen the bronchioles. More recently, the use of anti-inflammatory agents such as inhaled steroids or nonsteroidal cromolyn sodium have been prescribed. Cromolyn sodium blocks Ca^{2+} inflow into mast cells and helps prevent degranulation.

CONCEPT AND REASONING CHECKS

22.3 How can asthma be a life-threatening allergic reaction?

Why Do Only Some People Have IgE-Mediated Hypersensitivities?

KEY CONCEPT

- Genetics and the environment are responsible for most allergies.

Not everyone suffers from allergies. An interesting avenue of research was opened when it was discovered that the B cells responsible for IgE and IgA production lie close to one another in the lymphoid tissue, and that the IgA level and

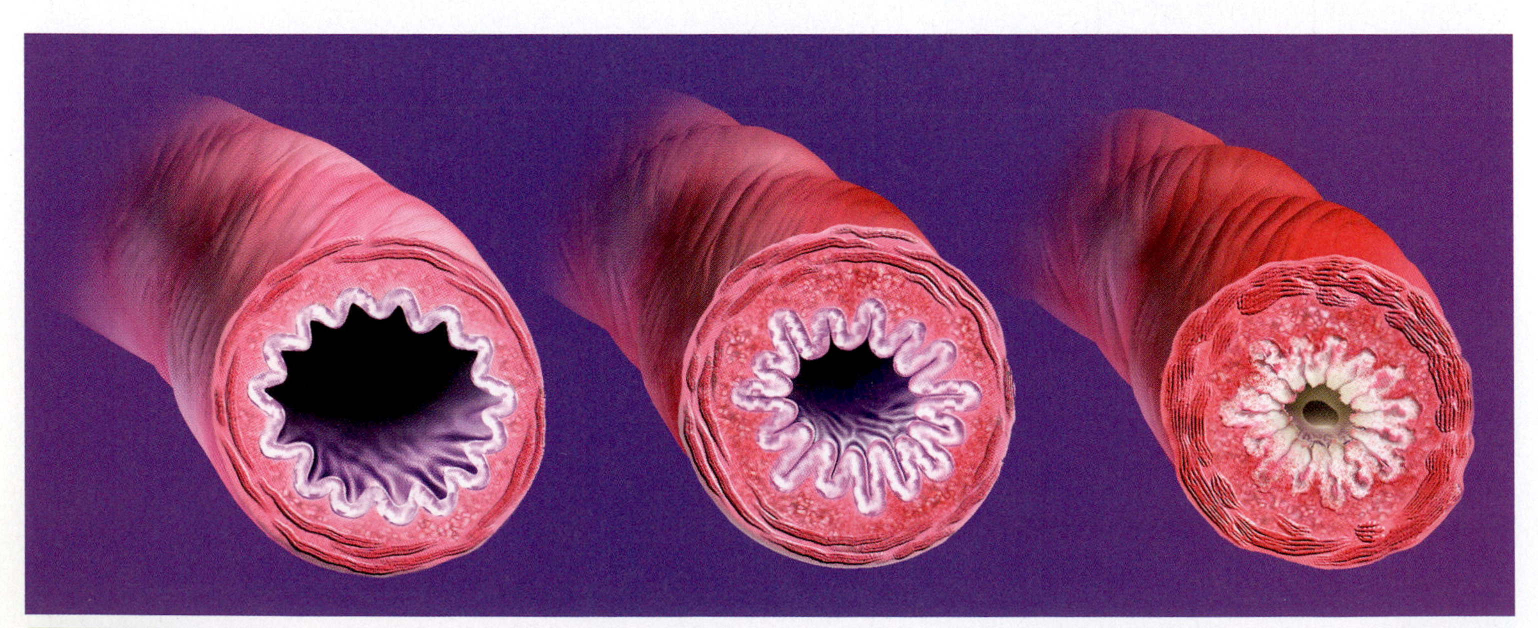

FIGURE 22.6 **Airway Narrowing During an Asthma Attack.** During an asthma attack, chemical mediators cause the smooth muscle layers of the bronchial tubes to spasm, constrict, and form a thick mucus that can block the airway.

Q: What types of cells are responsible for releasing the asthma-causing chemical mediators?

its corresponding lymphocytes are greatly reduced in atopic individuals. Immunologists have suggested that in nonallergic individuals, the B lymphocytes and plasma cells that produce IgA shield lymphocytes producing IgE from antigenic stimulation, but atopic people may lack sufficient IgA-secreting lymphocytes to block the antigens.

Another theory of atopic disease maintains an allergy results from a breakdown of feedback mechanisms in the immune system. Research findings indicate that B cells and the plasma cells that synthesize IgE may be controlled by suppressor T lymphocytes, and regulated in turn by IgE. Under normal conditions, IgE may limit its own production by stimulating **suppressor T-cell** activity. However, in atopic individuals the mechanism malfunctions, possibly because the T cells are defective, and IgE is produced in massive quantities. Allergic people are known to possess almost 100 times the IgE level of people who do not have allergies. Radioimmunoassay (RAI) techniques and radioallergosorbent tests (RAST) are used to detect the nature and quantity of IgE to specific substances known to cause allergies (see Chapter 21).

Some researchers postulate a positive role for the allergic response. They suggest, for example, that sneezing expels respiratory pathogens and contractions of the gastrointestinal tract force parasites out of the body. Others suggest allergy was once a survival mechanism, and atopic individuals are the modern generation of people who developed this ability to resist pathogens and passed the trait along. Hygiene also might affect allergic responses and demonstrates the connection between allergies and microorganisms (MicroInquiry 22).

CONCEPT AND REASONING CHECKS

22.4 Summarize the factors that can contribute to the development of allergies.

Therapies Sometimes Can Control Type I Hypersensitivities

KEY CONCEPT

- Physical avoidance and drug treatment can limit allergic attacks.

First, the best way to avoid hypersensitivities is to identify and avoid contact with the allergen. In other cases, immunotherapy may help control allergies.

A person sensitized to an allergen may undergo **desensitization therapy** to reduce the possibility of an allergic attack. For example, a person suffering from allergic rhinitis could undergo a procedure involving injections of tiny but increasing amounts of allergen over a period of hours or weeks, to effect a gradual reduction of granules in sensitized mast cells and basophils. Such treatment prevents a massive degranulation later.

Another approach to desensitization is to give a series of injections of allergens over a period of weeks. Allergists believe that these exposures cause the immune system to produce IgG antibodies, which circulate and neutralize allergens before they contact sensitized cells. These **blocking antibodies**, as they are called, appear to be an effective device for individuals sensitized to bee stings. They also may be used for people who have food allergies. A promising alternative is to inject Fc fragments of IgE to fill the receptor sites on mast cells and basophils, thereby making the sites unavailable to the person's IgE.

A novel approach to desensitization is to develop monoclonal antibodies (see Chapter 21) that recognize and react with IgE. These anti-IgE antibodies can be used to dislodge IgE from the surfaces of mast cells and basophils, thereby disarming the cells and preventing the allergic reaction from occurring. Also, in 2006, an allergy vaccine has been developed that reduces hay fever symptoms.

There are numerous over-the-counter (OTC) and prescription drugs used to relieve the symptoms of allergies or prevent the release of the chemical mediators that cause the allergies. **Antihistamines** are OTC and prescription drugs, such as Allegra and Claritin, that are used systemically to relieve or prevent the symptoms of hay fever and other types of common allergies. They block the effect of histamine. Cromolyn can help control allergic symptoms by inhibiting mast cells from releasing the chemical mediators that damage nearby tissues. Thus, products such as Nasalcrom, are only effective where they are applied (lung inhalation or eyes).

MICROINQUIRY 22

Allergies, Dirt, and the Hygiene Hypothesis

For reasons that are not completely understood, the incidence and severity of asthma—and allergies in general—are increasing in developed nations. In fact, between 1980 and 1994, the prevalence of asthma rose 71 percent in the United States. Today, about 15 million Americans suffer from asthma, including 5 million children and adolescents. Many scientists have suggested the increase is in large part due to our overly clean lifestyle. We use disinfectants for almost everything in the home, and antibacterial products have flooded the commercial markets (Chapter 23). In other words, maintaining overly good hygiene is making us sick. We need to eat dirt! Well, not literally.

The hygiene hypothesis, first proposed in 1989, suggests that a lack of early childhood exposure to dirt, microbes, and other infectious agents can lead to immune system weakness and an increased risk of developing asthma and allergies. In the early 1900s, infants and their immune systems had to battle all sorts of infectious diseases—from typhoid fever and polio to diphtheria and tuberculosis—as well as ones that were more mundane. Such interactions and recovery "pumped up" the immune system and prepared it to act in a controlled manner. Today, most children in developed nations are exposed to far fewer pathogens, and their immune systems remain "wimpy," often unable to respond properly to nonpathogenic substances like pollen and cat dander. Their immune systems have not had the proper "basic training."

The hygiene hypothesis has been the subject of debate since 1989. But new research studies are providing evidence that may make the hygiene hypothesis a theory. In 2003, researchers at the National Jewish Medical and Research Center in Denver reported that mice infected with the bacterium *Mycoplasma pneumoniae* had less severe immunological responses when challenged with an allergen. However, if mice were exposed to allergens first, they developed more severe allergic responses. Also, the allergy-producing mediators were at lower levels in mice first exposed to *M. pneumoniae*. So, early "basic training" of the immune system seems to temper allergic responses.

Two other studies also bolster the hygiene hypothesis. One study used data from the Third National Health and Nutrition Examination Survey conducted from 1988 through 1994 by the Centers for Disease Control and Prevention (CDC). The survey included 33,994 American residents ages 1 year to older than 90. The CDC analysis concluded that humans who were seropositive for hepatitis A virus, *Toxoplasma gondii*, and herpes simplex virus type 1—that is, markers for previous microbial exposures—were at a decreased risk of developing hay fever, asthma, and other atopic diseases.

A second study used data collected from 812 European children ages 6 to 13 who either lived on farms or did not live on farms. Using another marker—an endotoxin found in dust samples from bedding—the investigators reported that children who did not live on farms were more than twice as likely to have asthma or allergies as children growing up on farms. Presumably, on farms children are exposed to more "immune-strengthening" microbes.

So, all in all, microbial challenges to the immune system as it develops in young children can drive the system to a balanced response to allergens. If the hypothesis proves correct, eating dirt or moving to a farm is not a practical solution, nor is returning to pre-hygiene days. However, a number of environmental factors can help lower incidence of allergic disease early in life. These include the presence of a dog or other pet in the home before birth, attending day care during the first year of life, and simply allowing children to do what comes naturally—play together and get dirty.

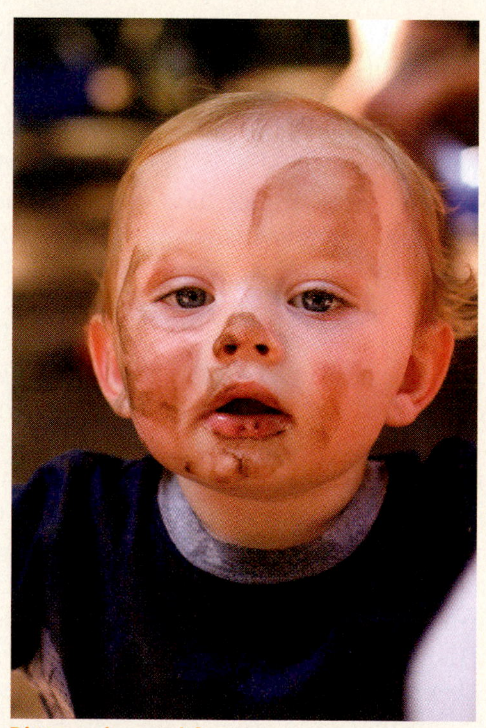

Dirt may be good for you.

Discussion Point

Do you suffer from common allergies? Discuss whether or not your allergies fit the description in this MicroInquiry

Corticosteroids are medicines sprayed or inhaled into the nose to help relieve the stuffy nose, irritation, and discomfort of hay fever, other allergies, or asthma. Drugs like Flonase and Nasonex also block the release of chemical mediators.

CONCEPT AND REASONING CHECKS

22.5 Evaluate the treatments and medications available for allergies.

22.2 Other Types of Hypersensitivity

Immunological responses involving IgG antibodies or T cells also can lead to adverse hypersensitivity reactions.

Type II Cytotoxic Hypersensitivity Involves Antibody-Mediated Cell Destruction

KEY CONCEPT

- Antibodies can be the cause of cytotoxic (type II) hypersensitivities.

A **cytotoxic hypersensitivity** is a cell-damaging reaction occurring when IgG reacts with antigens on the surfaces of cells (FIGURE 22.7). Complement often is activated and IgM may be involved, but IgE does not participate, nor is there any degranulation of mast cells.

A well-known example of cytotoxic hypersensitivity is the transfusion reaction arising from the mixing of incompatible blood types. Four major human blood types are

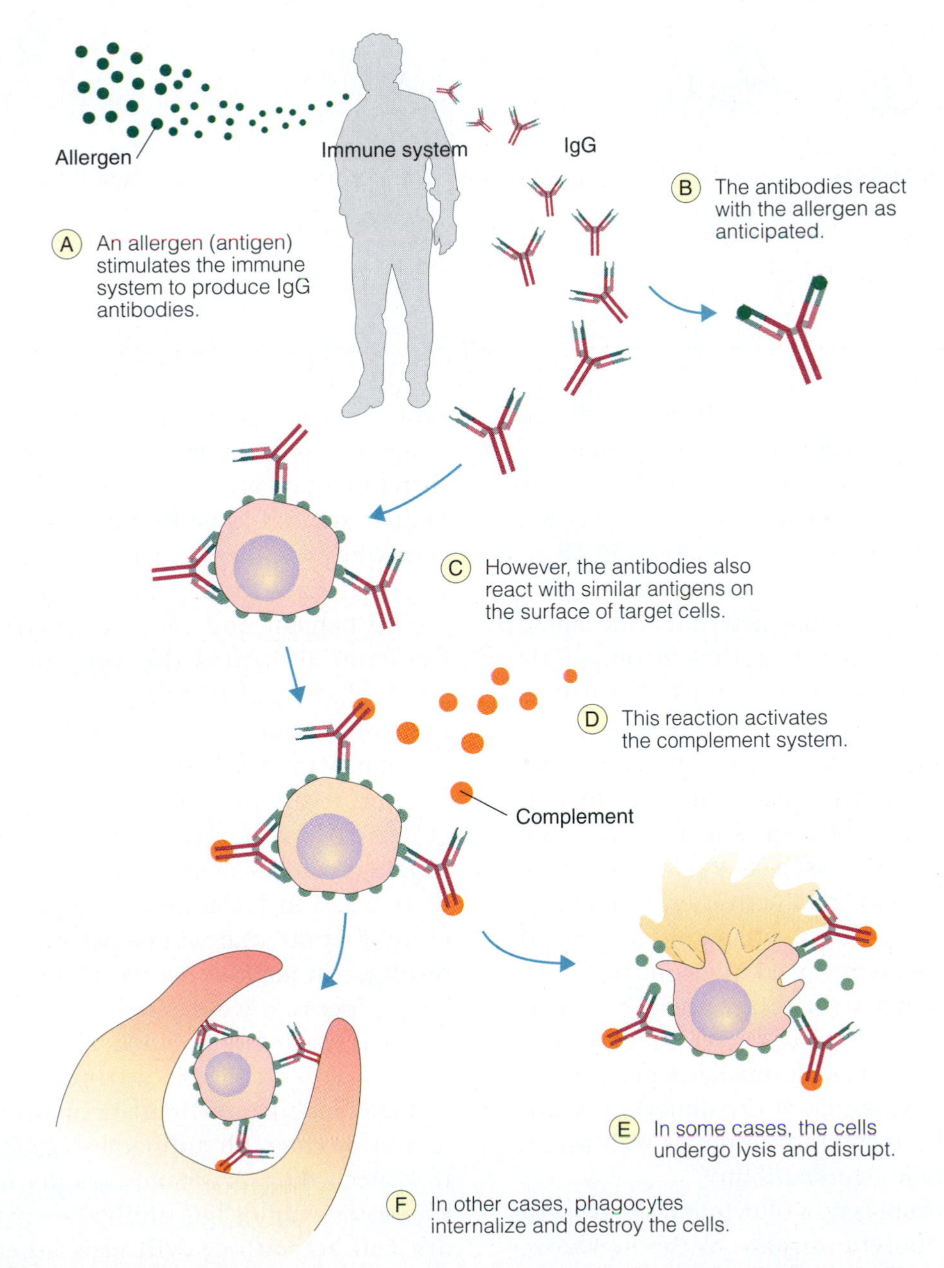

FIGURE 22.7 **Type II Cytotoxic Hypersensitivity.** Type II hypersensitivities involve the lysis of cells.

Q: How does complement cause cell lysis?

TABLE

22.1 Some Characteristics of the Major Blood Groups

	Type A	Type B	Type AB	Type O
Red blood cells	Antigen A	Antigen B	Antigens A and B	Neither A nor B antigens
Serum	B antibody	A antibody	Neither A nor B antibody	A and B antibodies
Approximate incidence, US Caucasian population	40%	10%	5%	45%

known: A, B, AB, and O. Each type is distinguished by unique antigens on the surface of erythrocytes and certain antibodies in the plasma that are directed against antigens not present in the individual's cells (TABLE 22.1).

Before a transfusion is attempted, the laboratory technician must determine the blood type of all participants so that incompatible types are not mixed. For example, if a person with type A blood donates to a recipient with type O blood, the A antigens on the donor's erythrocytes will react with anti-A antibodies in the recipient's plasma, and the cytotoxic effect will be expressed as agglutination of donor erythrocytes and activation of complement in the recipient's circulatory system. If the conditions are reversed, the donor's A antibodies will react with the recipient's A antigens, although to a lesser degree because dilution in the recipient's plasma takes place. Most blood banks cross-match the donor's erythrocytes with the recipient's serum as well as the reverse, to ensure compatibility.

Another expression of cytotoxic hypersensitivity is **hemolytic disease of the newborn**, or **Rh disease**. This problem arises from the fact that erythrocytes of approximately 85 percent of Caucasian Americans contain a surface antigen, first described in rhesus monkeys and therefore known as the Rh antigen. Such individuals are said to be Rh-positive. The 15 percent who lack the antigen are considered Rh-negative. For African Americans, the figures are 90 percent and 10 percent, respectively. Evidence indicates the antigen is really a group of antigens that vary among Rh-positive individuals, but we shall consider the group as a single factor for the purposes of discussion.

The ability to produce the Rh antigen is a genetically inherited trait. When an Rh-negative woman marries an Rh-positive man, there is a 3 to 1 chance (or 75 percent probability) that the trait will be passed to the child, resulting in an Rh-positive child. During the birth process, a woman's circulatory system is exposed to her child's blood, and if the child is Rh-positive, the Rh antigens enter the woman's blood and stimulate her immune system to produce Rh antibodies (FIGURE 22.8). If a succeeding pregnancy results in another Rh-positive child, IgG antibodies (from memory cell activation) will cross the placenta (along with other antibodies) and enter the fetal circulation. There they will react with

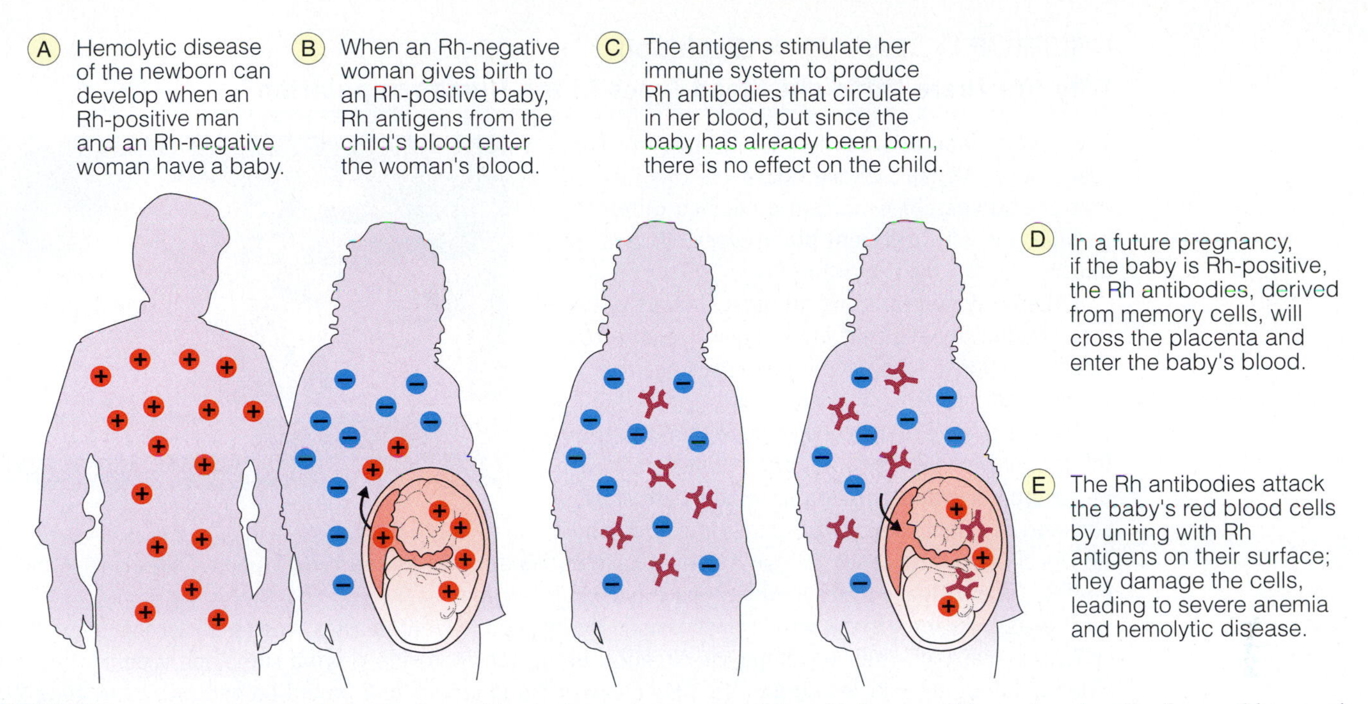

FIGURE 22.8 **Hemolytic Disease of the Newborn.** If a mother and a second child are Rh incompatible, maternal antibodies would cause the lysis and removal of the fetal red blood cells.

Q: Why is this type II hypersensitivity also called erythroblastosis fetalis?

Rh antigens on the fetal erythrocytes and cause complement-mediated lysis of the cells. The fetal circulatory system rapidly releases immature erythroblasts to replace the lysed blood cells, but these cells are also destroyed. The result may be stillbirth or, in a less extreme form, a baby with jaundice.

Modern treatment for hemolytic disease of the newborn consists of the mother receiving an injection of Rh antibodies (**RhoGAM**). The injection is given within 72 hours of delivery of an Rh-positive child (no injection is necessary if the child is Rh-negative). Antibodies in the preparation interact with Rh antigens on any fetal red blood cells in the mother and remove them from the circulation, thereby preventing the cells from stimulating the woman's immune system. This includes the prevention of memory cell production. The success of this procedure has virtually eliminated expectant parents' concerns about disease in their newborns. It should be noted, however, that an Rh-negative woman may produce Rh antibodies as a result of miscarriage or abortion of an Rh-positive fetus, or after a transfusion with Rh-positive blood.

Although cytotoxic hypersensitivity is generally cast in a negative role with deleterious effects on the body, the cytotoxic activity may contribute to the body's resistance to disease. For example, the antigen-antibody interaction occurring on the surface of a parasite leads to destruction of the parasite. The interaction also may encourage chemotaxis or histamine release through the activity of C3a and C5a components of the complement system (see Chapter 20). Increased phagocytosis and membrane damage from the complement attack complex are other by-products of complement activation. These activities probably account for resistance to many disorders. On this point, MicroFocus 22.4 examines a microbial reason why there are different human blood types.

CONCEPT AND REASONING CHECKS

22.6 Summarize the factors responsible for type II hypersensitivities.

Type III Immune Complex Hypersensitivity Is Caused by Antigen-Antibody Aggregates

KEY CONCEPT

- Immune complexes are responsible for type III hypersensitivities.

Immune complex hypersensitivity develops when antibodies combine with antigens and form aggregates that accumulate in blood

MICROFOCUS 22.4: Evolution

Why Are There Different Blood Types in the Human Population?

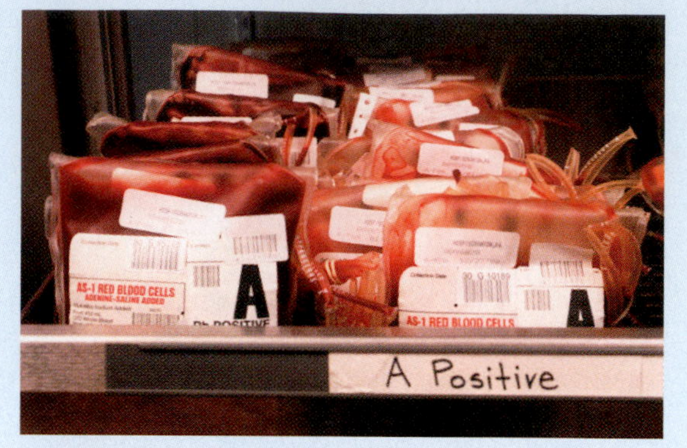

A photograph of blood types being stored.

The human population contains four different blood types: A, B, AB, and O (see figure). Is this just a matter of divergent evolution producing different populations with different blood types? Or is there another reason for the difference?

Robert Seymour and his group at University College London believe that blood types are an evolutionary response to balancing defenses against viruses and bacteria.

Here is their evidence. They have found that viral infections (not disease) dominate populations with blood type O, while bacterial infections are more common in populations with blood type A, B, and AB. As shown in Table 22.1, type A people have anti-B antibodies and type B people have anti-A antibodies; Type AB have neither while type O have both antibodies.

This association affects virus transmission. For example, when measles viruses break out of infected cells, they carry on their envelope the chemical group (antigen) identifying the blood type of that individual. Therefore, measles viruses from people with blood type A or B would be neutralized by type O blood because O blood has antibodies to antigens A and B. In reverse, measles viruses emerging from someone with blood type O carries neither the A nor B antigen, so the viruses would not be neutralized by the blood of people with blood types A, B, or AB. Therefore, people with blood type O are better prepared to defend against viruses coming from people with other blood types—and they are better at transmitting viruses to those blood types.

If this is all true, then type O people should have outcompeted all other blood types. So, here is the other side of the coin. A, B, and AB blood types are less likely to be infected by bacterial pathogens. Since these pathogens must evolve to attach to sugars on human cells, if blood type O is the majority, probability says bacterial pathogens will infect people with blood type O because that blood type is most common. According to Seymour, this restores the balance of blood types.

If all this is validated, here is yet another way that microorganisms and viruses have affected the human species. Once again, they do rule the world!

vessels or on tissue surfaces (FIGURE 22.9). As complement is activated, the C3a and C5a components increase vascular permeability and exert a chemotactic effect on phagocytic neutrophils, drawing them to the target site. Here the neutrophils release lysosomal enzymes, which cause tissue damage. Local inflammation is common, and fibrin clots may complicate the problem. The antibodies are predominantly IgG, with IgM also found in certain cases.

Serum sickness is a common manifestation of immune complex hypersensitivity (see Chapter 21). It develops when the immune system produces IgG against residual proteins in serum preparations. The IgG then reacts with the proteins, and immune complexes gather in the kidney over a period of days. The problem is compounded when IgE, also from the immune system, attaches to mast cells and basophils, thereby inducing a type I anaphylactic hypersensitivity. The sum total of these events is kidney damage, along with hives and swelling in the face, neck, and joints. Another form of immune complex hypersensitivity is the **Arthus phenomenon**, named for Nicolas Maurice Arthus, the French physiologist who described it in 1903. In this process, excessively large amounts of IgG form complexes with antigens, either in the blood vessels or near the site of antigen entry into the body. Antigens in dust from moldy

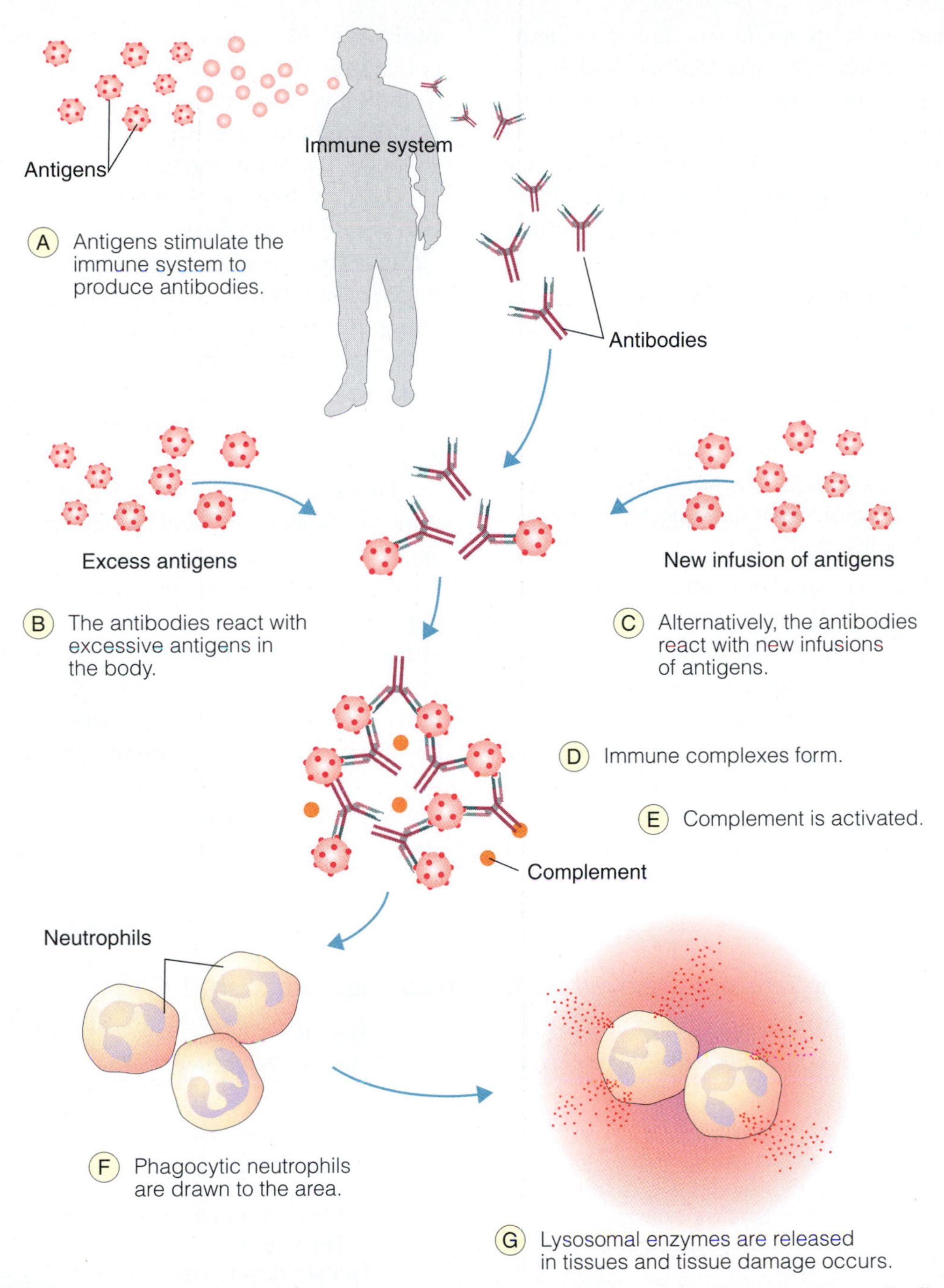

FIGURE 22.9 **Type III Immune Complex Hypersensitivity.** Type III hypersensitivities occur when antibodies combine with antigens to form aggregates (immune complexes) that accumulate in blood vessels or on tissue surfaces.

Q: What types of physical problems can develop from immune complex formation?

hay and in dried pigeon feces are known to cause this phenomenon. The names farmer's lung and pigeon fancier's disease are applied to the conditions, respectively. Thromboses in the blood vessels may lead to oxygen starvation and cell death.

Several microbial diseases also are complicated by immune complex formation. For example, the glomerulonephritis and rheumatic fevers that follow streptococcal diseases (see Chapter 9) appear to be consequences of immune complex formation in the kidneys and heart, respectively. In these cases, the deposit of complexes relates to common antigens in streptococci and the tissues. Other immune complex diseases include hemorrhagic shock,

Thromboses: The formation or presence of blood clots that partially or completely block one or more arteries.

which may accompany dengue fever; subacute sclerosing panencephalitis (SSPE), which follows cases of measles; and the slow-forming kidney deposits associated with lymphocytic choriomeningitis (LCM). Research evidence also suggests that Reye syndrome and Guillain-Barré syndrome may be related to immune complex formation.

CONCEPT AND REASONING CHECKS

22.7 Summarize the factors responsible for type III hypersensitivities.

Type IV Cellular Hypersensitivity Is Mediated by Antigen-Specific T Cells

KEY CONCEPT

- Type IV hypersensitivity is an exaggeration of a cell mediated immune response.

Cellular hypersensitivity is an exaggeration of the process of cell mediated immunity, discussed in Chapter 20. The adjective cell mediated was originally applied because contact between T cells and antigens was thought to be necessary for the reaction. However, the later identification of cytokines as mediators of the process challenged this assumption. The hypersensitivity cannot be transferred to a normal individual by serum cytokines because their concentration is too low in transfused serum. Transfer is accomplished only with T cells. TABLE 22.2 compares this delayed hypersensitivity with type I immediate hypersensitivity.

Type IV hypersensitivity is a delayed reaction whose maximal effect is not seen until 24 to 72 hours have elapsed (hence the name, delayed hypersensitivity). It is characterized by a thickening and drying of the skin tissue, a process called **induration**, and a surrounding zone of erythema (redness). Two major forms of type IV hypersensitivity are recognized: infection allergy and contact dermatitis.

Infection allergy develops when the immune system responds to certain microbial agents. Effector cells known as delayed hypersensitivity T lymphocytes migrate to the antigen site and release cytokines. The cytokines attract phagocytes and encourage phagocytosis (see Chapter 19). Sensitized lymphocytes then remain in the tissue and provide immunity to successive episodes of infection. Among the microbial agents stimulating this type of immunity are the bacterial agents of tuberculosis, leprosy, and brucellosis; the fungi involved

TABLE 22.2 Immediate and Delayed Hypersensitivities Compared

	Type I Immediate Hypersensitivity	Type IV Delayed Hypersensitivity
Clinical state:	Hay fever Asthma Urticaria Allergic skin conditions Serum sickness Anaphylactic shock	Drug allergies Infectious allergies Tuberculosis Rheumatic fever Histoplasmosis Trichinosis Contact dermatitis
Onset:	Immediate	Delayed
Duration:	Short: hours	Prolonged: days or longer
Allergens:	Pollen Molds House dust Danders Drugs Antibiotics Soluble proteins and carbohydrates Foods	Drugs Antibiotics Microorganisms: bacteria, viruses, fungi, animal parasites Poison ivy/oak and plant oils Plastics and other chemicals Fabrics, furs Cosmetics
Passive transfer of sensitivity:	With serum	With cells or cell fractions of lymphoid series

in blastomycosis, histoplasmosis, and candidiasis; the viruses of smallpox and mumps; and the chlamydiae of lymphogranuloma.

Infection allergy is demonstrated by injecting an extract of the microbial agent into the skin. As the immune response takes place, the area develops induration and erythema, and fibrin is deposited by activation of the clotting system. An important application of infection allergy is the **tuberculin test** for tuberculosis (FIGURE 22.10). A purified protein derivative (PPD) of *Mycobacterium tuberculosis* is applied to the skin by intradermal injection (the Mantoux test) or multiple punctures (tines). Individuals sensitized by a previous exposure to *Mycobacterium* species develop a vesicle, erythema, and induration.

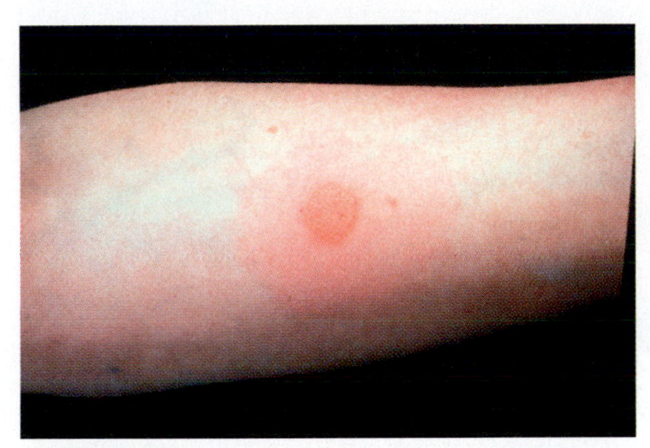

FIGURE 22.10 **A Positive Tuberculin Test.** The raised induration and zone of inflammation indicate that antigens have reacted with T cells, probably sensitized by a previous exposure to tubercle bacilli.

Q: Does this test mean the person has tuberculosis? Explain.

Skin tests based on infection allergy are available for many diseases. Immunologists caution, however, that a positive result does not constitute a final diagnosis. Rather, sensitivity may have developed from a subclinical exposure to the organisms, from clinical disease years before, or from a former screening test in which the test antigens elicited a T-cell response.

Contact dermatitis develops after exposure to a variety of antigens, such as the allergens in clothing, insecticides, coins, and cosmetics. The offending substances also may include formaldehyde, copper, dyes, bacterial enzymes, and protein fibers. In poison ivy, the allergen has been identified as urushiol, a low-molecular-weight chemical on the surface of the leaf. In the body, urushiol attaches to tissue proteins to form allergenic compounds. Within 48 hours, a rash appears and consists of very itchy pinhead-sized blisters usually occurring in a straight row (FIGURE 22.11).

The course of contact dermatitis is typical of the type IV reaction. Repeated exposures cause a drying of the skin, with erythema and scaling. Examples are on the scalp when allergenic shampoo is used, on the hands when

(A) When a sensitized person touches poison ivy, a substance called urushiol stimulates T lymphocytes in the skin; within 48 hours, a type IV reaction takes place.

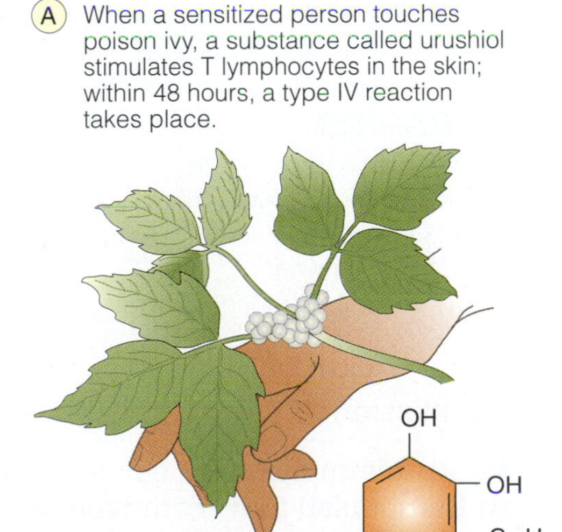

The chemical structure of urushiol

(B) The reaction is characterized by pinhead-sized blisters that usually occur in a straight row.

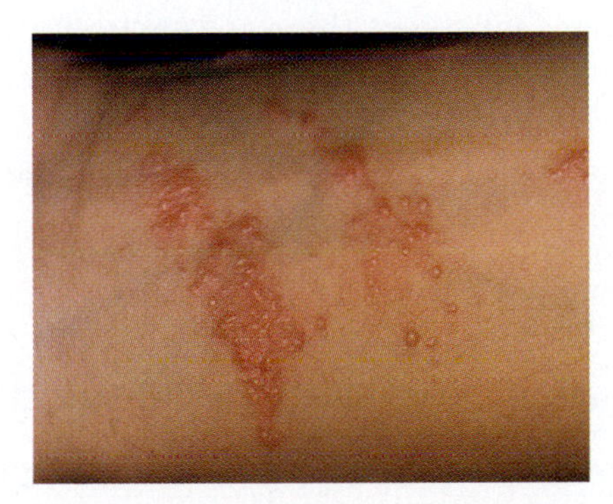

FIGURE 22.11 **The Poison Ivy Rash.** When sensitized skin makes contact with substances like urushiol in poison ivy, the chemical combines with tissue proteins to form allergenic compounds. The result is a rash on the skin surface.

Q: Why is this form of hypersensitivity called delayed hypersensitivity?

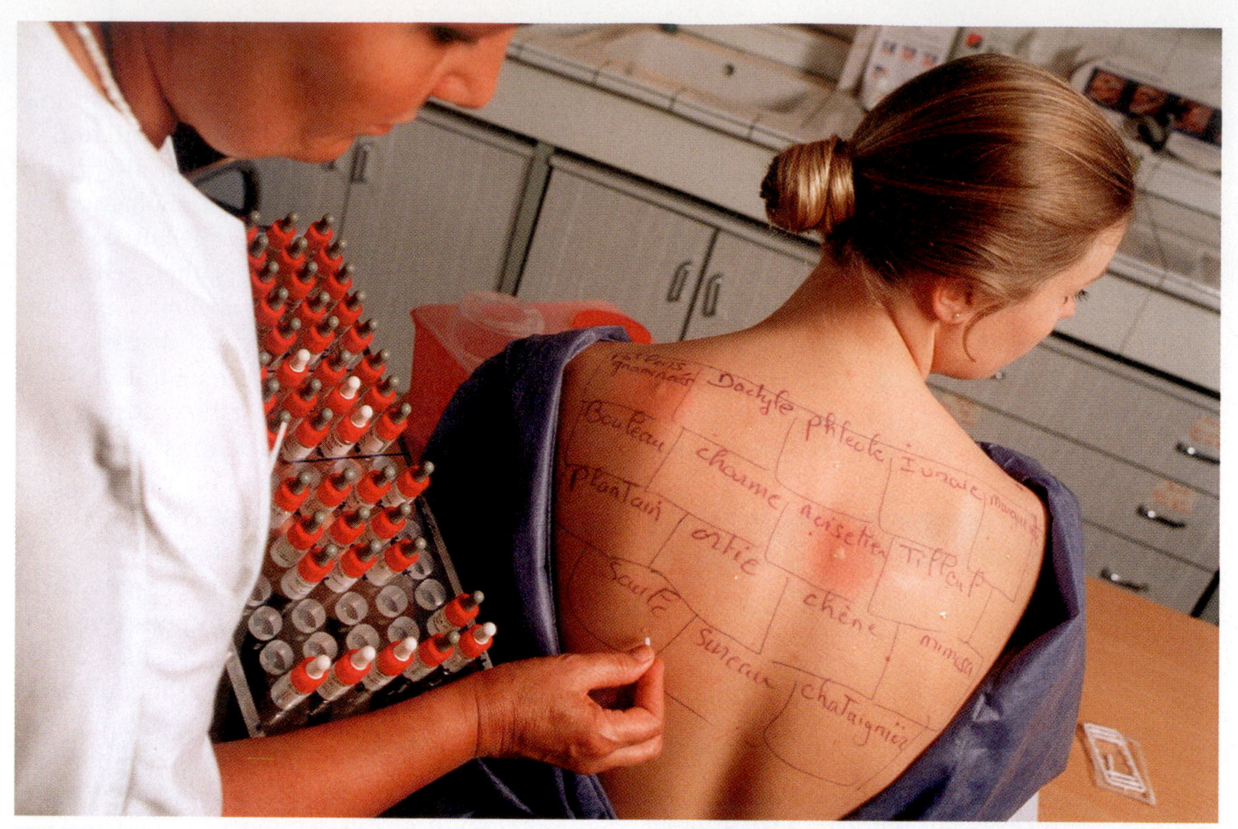

FIGURE 22.12 **A Test for Skin Allergies.** The back is marked and injected with various test allergens. In a few minutes, the allergist notes where a skin wheal has occurred. The reaction can be assessed as slight, mild, moderate, or severe. In other instances, the skin on the arm is used.

Q: Identify how many positive reactions are displayed by this patient.

contact is made with detergent enzymes, and on the wrists when an allergy to costume jewelry exists. Contact dermatitis also may occur on the face where contact is made with cosmetics, on areas of the skin where chemicals in permanent-press fabrics have accumulated, and on the feet when there is sensitivity to dyes in leather shoes. Factory workers exposed to photographic materials, hair dyes, or sewing materials may experience allergies. The list of possibilities is endless.

Applying a sample of the suspected substance to the skin (a patch test) and leaving it in place for 24 to 48 hours will help pinpoint the source of the allergy. Another method is illustrated in FIGURE 22.12. Relief generally consists of avoiding the inciting agents.

TABLE 22.3 summarizes the four types of hypersensitivities.

CONCEPT AND REASONING CHECKS

22.8 Summarize the factors responsible for type IV hypersensitivities.

22.3 Autoimmune Disorders and Transplantation

In Chapter 20, we mentioned that one of the properties of the immune system was tolerance of "self;" that is, one's own cells and molecules with their antigenic determinants do not stimulate an immune response. Should self-tolerance break down, an **autoimmune disorder** may occur.

An Autoimmune Disorder Is a Failure to Distinguish Self from Non-self

KEY CONCEPT

- Autoimmune disorders generate an immune attack on self cells and tissues.

The idea that an individual's immune system is incapable of recognizing "self" antigens is not

TABLE 22.3 Overview of the Gell and Coombs Classification of Hypersensitivity Reactions

Hypersensitivity Type	Origin of Hypersensitivity	Antibody Involved	Cells Involved	Mediators Involved	Transfer of Sensitivity	Evidence of Hypersensitivity	Skin Reaction	Examples
Type I IgE-mediated	B lymphocytes	IgE	Mast cells Basophils	Histamine Serotonin Leukotrienes Prostaglandins	By serum	30 minutes or less	Urticaria	Systemic anaphylaxis Hay fever Asthma
Type II Cytotoxic	B lymphocytes	IgG IgM	RBC WBC Platelets	Complement	By serum	5–8 hours	Usually none	Transfusion reactions Hemolytic disease of newborns
Type III Immune complex	B lymphocytes	IgG IgM	Host tissue cells	Complement	By serum	2–8 hours	Usually none	Serum sickness Arthus phenomenon SLE Rheumatic fever LCM
Type IV Cellular	T lymphocytes	None	Host tissue cells	Cytokines	By lymphoid cells	1–3 days	Induration Tissue death	Contact dermatitis Infection allergy

new. At the beginning of the 20th century, the German immunologist Paul Ehrlich suggested that the human body has an innate aversion to immunological self-destruction, what he called *horror autotoxicus* (terror of self-toxicity). Any autoimmune response therefore must be abnormal and due to human disease.

Today, we know autoimmune responses are critical to the normal development and functioning of the human immune system—and essential to the development of immunological tolerance. Each individual's immune system must have a mechanism to differentiate "self" from "non-self," so it will only react to foreign antigens, such as pathogens. Such tolerance is thought to develop in the following ways:

- **Clonal Deletion Theory**. This theory says that self-reactive lymphoid cells are destroyed during the development of the immune system in an individual.
- **Clonal Anergy Theory**. This theory proposes that self-reactive T or B cells become inactivated in the normal individual and cannot differentiate into effector cells when presented with antigen.
- **Regulatory T Cell Theory**. According to this theory, specific regulatory T cells function to suppress exaggerated immune responses.

In fact, all of these theories may be correct and several mechanisms actively contribute to the development of immunological tolerance.

About eight percent of the American population (22 million individuals) suffers from an autoimmune disorder, and far more women than men are affected. So, what causes the loss of tolerance? Part of the answer lies in human heredity. Various gene mutations have been identified that affect cell division and apoptosis—and therefore clonal deletion, anergy, and regulatory T cell activity. There also are several other ways autoimmune disorders can be triggered.

- **Access to privileged sites.** Certain parts of the body (e.g., brain, eye) are "hidden" from the immune system. If such sites become "accessible" to the immune system through injury, an immune response will be mounted.
- **Antigenic mimicry.** A foreign substance or microbe may enter the body that closely resembles (mimics) a similar body substance. The immune system then sees the "self" substance as foreign and attacks it. Rheumatoid arthritis and type I diabetes may be examples.

As a result of the loss of tolerance, the immune system attempts to mount an immune response against its own cells and tissues. Although many such disorders cause

TABLE 22.4 A Summary of Some Immune Disorders

Disorder	Target Tissue	Stimulating Antigen	Effect
Thrombocytopenia	Thrombocytes (blood platelets)	Aspirin Antibiotics Antihistamine	Impaired blood clotting Hemorrhages
Agranulocytosis	Neutrophils	Drugs	Reduced phagocytosis
Goodpasture syndrome	Kidneys	Own antigens (?)	Kidney failure
Myasthenia gravis	Membranes of muscle fibers	Not established	Loss of muscle activity
Graves disease	Thyroid gland	Not established	Abundant thyroxine High metabolic rate
Hashimoto disease	Thyroid gland	Not established	Thyroxine deficiency Low metabolic rate
Type I diabetes	Beta cells of pancreas	Not established (viral infection?)	Inability to produce insulin
Systemic lupus erythematosus	Skin, heart, kidney, blood vessels	Own nucleo-proteins	Butterfly rash, skin rash Heart, kidney failure
Rheumatoid arthritis	Joints	Bacterial infection (?)	Swollen joints

fever, specific symptoms of an autoimmune disorder depend on the disorder and the part of the body affected. The resulting inflammation and tissue damage can cause pain, deformed joints, jaundice, itching, breathing difficulties, and even death. To date, more than 80 clinically-distinct autoimmune disorders have been identified (TABLE 22.4).

Among the more notable organ-specific disorders is **myasthenia gravis**. In this disorder, antibodies react with acetylcholine receptors on membranes covering the muscle fibers. This interaction reduces nerve impulse transfer to the fibers and results in a loss of muscle activity, manifested as weakness and fatigue. In **Graves disease**, antibodies unite with receptors on the surfaces of thyroid gland cells, causing an overabundant secretion of thyroxine. The patient experiences goiter and a rise in the metabolic rate. In **type I diabetes**, the beta cells of the pancreas are destroyed, resulting in a lack of insulin production. Because type I diabetes typically appears after an infection, the immune response to insulin-producing cells may result from the immune system's response to the earlier viral infection. More than 5 million people worldwide depend on insulin injections to replace the insulin their bodies no longer produce (see Chapter 8).

■ Thyroxine: A hormone secreted by the thyroid gland to stimulate metabolism and control growth and development.

A systemic autoimmune disorder is **systemic lupus erythematosus** (SLE), also known as "lupus." With SLE, plasma cells produce IgG upon stimulation by nuclear components of disintegrating white blood cells. Immune complexes accumulate in the skin and body organs, and complement is activated. The patient experiences a butterfly rash, a facial skin condition across the nose and cheeks, and body rashes (FIGURE 22.13). Lesions also form in the heart, kidneys, and blood vessels. Another systemic disorder is **rheumatoid arthritis** (RA). Unlike osteoarthritis, which results from wear and tear on joints, RA is an inflammatory condition resulting in the accumulation of immune complexes in the joints. Some researchers suspect rheumatoid arthritis is triggered by an infection by a virus or bacterial pathogen.

Treatment of most autoimmune disorders requires suppressing the immune system, which means interrupting the system's ability to fight infectious disease. Immunosuppressants include corticosteroids, such as prednisone. Some disorders resolve as spontaneously as they appeared

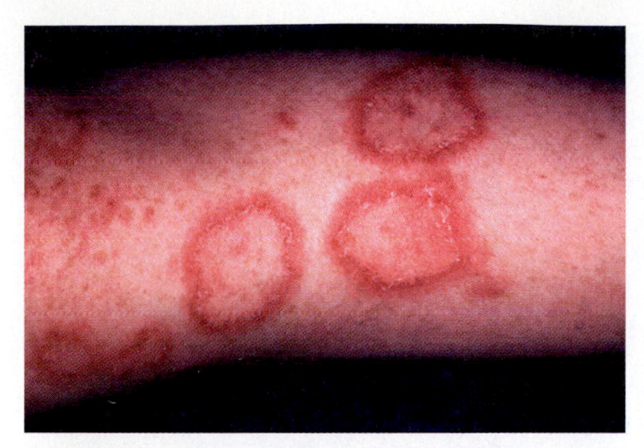

FIGURE 22.13 **Skin Lesions of Systemic Lupus Erythematosus.** This autoimmune disorder can affect the skin and other body organs.

Q: What causes the inflammation seen with lupus?

while others become chronic, life-long disorders. Thus, the prognosis depends on the particular disorder.

CONCEPT AND REASONING CHECKS

22.9 Identify the common attributes to all autoimmune disorders.

Transplantation of Tissues or Organs Is an Important Medical Therapy

KEY CONCEPT

- Adaptive immune responses can hinder successful tissue transplants or grafts.

Modern techniques for the transplantation of tissues and organs trace their origins to Jacques Reverdin, who in 1870 successfully grafted bits of skin to wounded tissues. Enthusiasm for the technique rose after his reports were published, but it waned when doctors found that most transplants were rapidly rejected by the body. Then, in 1954, a kidney was transplanted between identical twins, and again, interest grew. The graft survived for several years, until ultimately it was destroyed by a recurrence of the recipient's original kidney disease. During that time, attempts to transplant kidneys between unrelated individuals were less successful.

Transplantation technology improved considerably during the next few decades, and today, four types of transplantations, or grafts, are recognized, depending upon the genetic relationship between donor and recipient. A graft taken from one part of the body and transplanted to another part of the same body is called an **autograft**. This graft is never rejected because it is the person's own tissue. A tissue taken from an identical twin and grafted to the other twin is an **isograft**. This, too, is not rejected because the genetic constitutions of identical twins are the same.

Rejection mechanisms become more vigorous as the genetic constitutions of donor and recipient become more varied (MicroFocus 22.5). For instance, grafts between brothers and sisters, or between fraternal twins, may lead to only mild rejection because many of their genes are similar. Grafts between cousins may be rejected more rapidly, and as the relationship becomes more distant, the vigor of rejection increases proportionally. **Allografts**, or grafts between genetically different members of the same species, such as two humans, have variable degrees of success. Most transplants are allografts. **Xenografts**, or grafts between members of different species, such as a pig and a human, are rarely successful.

Transplanted tissue is rejected by the body if the immune system interprets the tissue as "nonself." Cytotoxic T cells will attack and destroy transplanted cells. T lymphocytes also release cytokines that stimulate phagocytes to enter the graft tissue. The phagocytes secrete lysosomal enzymes, which digest the tissues, leading to a dryness and thickening, as in type IV hypersensitivity. Cell death, or necrosis, follows.

A rejection mechanism of a completely different sort is sometimes observed in bone marrow transplants. In this case, the transplanted marrow may contain immune system cells that form immune products against the host after the host's immune system has been suppressed during transplant therapy. Essentially, the graft is rejecting the host. This phenomenon is called a **graft-versus-host reaction** (GVHR). It sometimes can lead to fatal consequences in the host body.

The rejection process is stimulated by the recognition of foreign **major histocompatibility complex** (MHC) proteins (see Chapter 20) on the surface of the graft cells. Thus, acceptance or rejection of a graft depends largely on a relatively small number of genes that encode a series of cell-surface glycoproteins also known as **human leukocyte antigens** (HLAs).

MICROFOCUS 22.5

Acceptance

Ordinarily a woman's body will reject a foreign organ, such as a kidney or heart, but it will accept the fetus growing within her womb. This acceptance exists, even though half of the fetus's genetic information has come from a "foreigner"—namely, the father. Has her immune system failed?

Apparently not. It seems that the sperm carries an antigenic signal that induces the woman's immune system to produce a series of so-called blocking antibodies. The blocking antibodies form a type of protective screen that protects the fetus and prevents its antigens from stimulating the production of rejection antibodies by the mother. So protected, the fetus grows to term and "escapes" before any immunological damage can be done to its tissues.

Sometimes, however, a rejection in the form of a miscarriage occurs. A number of physicians now believe at least certain miscarriages have an immunological basis. Their research indicates the level of blocking antibodies in some pregnant women is too low to protect the fetus, and that the miscarriage is due to the woman's antibodies. Ironically, scientists have discovered the blocking antibody level may be low because the father's tissue is very similar to the mother's. In such a case, the sperm's antigens elicit a weak antibody response, too low to protect the fetus.

With this knowledge in hand, physicians are now attempting to boost the level of blocking antibodies as a way of preventing miscarriage. They inject white blood cells from the father into the mother, thereby stimulating her immune system to produce antibodies to the cells. These antibodies exhibit the blocking effect.

In other experiments, injections of blocking antibodies are administered to augment the woman's normal supply. Both approaches have been successful in trial experiments, and continuing research has given cause for optimism that cases of fetal rejection can give way to acceptance and a full-term delivery.

Two individuals chosen at random (a husband and wife, for example) are not expected to have many MHC genes in common. Identical twins, by contrast, have identical MHC genes. MHC proteins are of two types: class I and class II. Class II MHC proteins are important in the recognition of non-self antigens when T lymphocytes combine with macrophages in cell-mediated immunity (see Chapter 20). Class I MHC proteins are present on every nucleated cell in the human body and help define the uniqueness of a person's tissue. For this reason, they are important subjects of transplantation immunology.

The nature of MHC proteins is a key element in transplant acceptance or rejection. The closer the match between donor and recipient MHC proteins, the greater the chance of a successful transplant. The matching of donor and recipient is performed by tissue typing (FIGURE 22.14). In this procedure, the laboratory uses standardized MHC antibodies for particular MHC proteins. Lymphocytes from the donor are incubated with a selected type of MHC antibodies. Complement and a dye, such as trypan blue, are then added. If the selected MHC antibodies react with the MHC proteins of the lymphocytes, the cell becomes permeable and dye enters the cells (living cells do not normally take up the dye). Similar tests then are performed with recipient lymphocytes to determine which

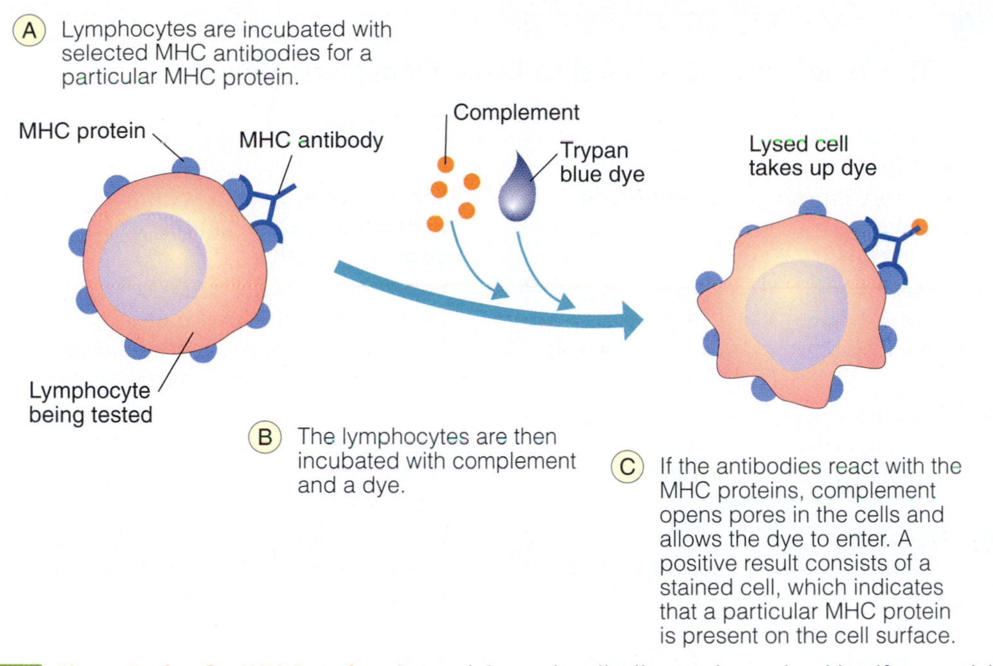

FIGURE 22.14 **Tissue Typing for MHC Proteins.** Dye staining and antibodies can be used to identify potentially successful tissue transplants.

Q: What color will lymphocytes appear if an antibody triggers complement activity?

MHC proteins are present and how closely the tissues match one another. The blood types of donor and recipient must also be identical.

Tissue typing and histocompatibility screening help reduce the rejection mechanism in allograft transfers, but they do not eliminate the mechanism completely. To inhibit rejection, it is necessary to suppress activity of the immune system (TABLE 22.5). One method uses prednisone, a steroid hormone suppressing the inflammatory response. Another means of diminishing the rejection mechanism is to introduce antilymphocyte antibodies. To obtain these antibodies, lymphocytes from the transplant recipient are injected into animals and later the animals are bled to obtain the antibody-rich serum. When introduced to the transplant recipient (serum sickness notwithstanding), the antibodies interact with the local lymphocytes, thereby slowing the rejection process and increasing the survival time of the transplant.

Antimitotic drugs are another therapy. These drugs prevent multiplication of lymphocytes in the lymph nodes. The drug azathioprine is a nucleic acid antagonist used for this purpose.

Another drug is cyclosporine. Originally isolated from strains of fungi as an antimicrobial substance, this drug now is produced synthetically. It appears to suppress cell mediated immunity without killing T cells or interfering with antibody formation. Immunologists believe that cyclosporine prevents the division of naive T cells by inhibiting transcription of the genes encoding IL-2 and IL-2 receptors.

Additional methods of reducing rejection include the use of monoclonal antibodies (see Chapter 21) and the bombardment of lymphocyte centers with radiations such as X rays.

Realize that virtually all drug treatments have possible side effects and leave the recipient in an immunosuppressed condition, thereby increasing the susceptibility to a variety of infectious diseases and cancer. Immunosuppression is, at best, a poor expedient and is viewed as only a stopgap measure until the rejection mechanism can be understood and exploited.

CONCEPT AND REASONING CHECKS

22.10 Assess the need for tissue typing before a tissue or organ transplant.

TABLE

22.5 The Types of Drugs Used to Limit Transplant Rejection

Type	Drug	Comments
Corticosteroids Anti-inflammatory drugs suppressing immune system responses	Prednisone	Given by IV at time of transplantation in high dose; then monthly at reduced dose indefinitely
Immunoglobulins Antibodies suppressing specific immune system responses	Antilymphocyte Antithymocyte	Given by IV and use with other immunosuppressants so these latter drugs can be started later or at reduced dosage
Mitotic inhibitors Drugs suppressing cell division of lymphocytes	Azathioprine Methotrexate	Given by IV or orally at time of transplantation in high dose; then monthly at reduced dose indefinitely
Fungal metabolites Drugs that inhibit T cell activity	Cyclosporine	Given first by IV and later orally; used with prednisone or azathioprine
Monoclonal antibodies Target and suppress specific immune cells	Basiliximab Infliximab Muromonab	Given by IV at time of transplantation or if a rejection event is initiated

22.4 Immunodeficiency Disorders

The spectrum of immunodeficiency diseases ranges from major abnormalities that are life-threatening to relatively minor deficiency states. The latter may be serious in populations where malnutrition and frequent contact with pathogenic organisms are common. Diagnostic techniques for determining immune deficiency diseases include measurements of antibody types, detection tests for B cell function, and enumeration of T cells. Assays of complement activity and phagocytosis also are useful in diagnosis.

Immunodeficiencies Can Involve Any Aspect of the Immune System

KEY CONCEPT

- Immunodeficiency disorders allow recurrent and often severe, longer lasting infections.

If there is an immune system malfunction or developmental abnormality in immune system function, an **immunodeficiency disorder** may result. Such disorders hamper the immune system's ability to provide a strong response to any viral or microbial infection. The immunodeficiency may be congenital (**primary immunodeficiency**) as a result of a genetic abnormality. Such disorders present from birth are rare, although more than 70 different congenital disorders have been documented.

An immunodeficiency may be acquired later in life (**secondary immunodeficiency**). The most common acquired immunodeficiency is AIDS, which is caused by infection with the human immunodeficiency virus (HIV). The virus and the diseases associated with AIDS are covered in Chapters 13 and 15, so AIDS will not be further covered here.

Immunodeficiency disorders are identified by the part of the innate or acquired immune system affected (see Chapters 19 and 20). Thus, the disorders may involve B cells and antibodies, T cells, both B and T cells, phagocytes, or complement proteins. The affected immune component may be absent, present in reduced numbers or amounts, or functioning abnormally.

People with immunodeficiencies have multiple infections, especially recurring respiratory infections. Bacterial infections are common, often severe, and lead to complications. Several congenital disorders are described below.

MICROFOCUS 22.6: History

Something from Nothing

In the 1940s and early 1950s, scientists frequently debated whether antibodies were essential to immunity. Amid this controversy, a pediatrician at Washington, D.C.'s Walter Reed Hospital made a momentous discovery.

The story began when a US Air Force officer's son was admitted to the hospital with acute streptococcal disease. The child's pediatrician, Ogden C. Bruton, used penicillin to control the infection, but two weeks later, the child was back, sick again. Wishing to determine the level of antibodies in the boy's system, Bruton sent a sample of blood for testing on a new machine recently acquired by the hospital.

The next day, the laboratory called Bruton to report that something must be wrong with the machine because it could not detect any antibodies in the blood. Bruton responded by sending over a second sample, but the results were the same: no antibodies. It occurred to Bruton that maybe the trouble was not in the machine but in the blood. Perhaps the blood had no antibodies. Conceivably, this could account for the recurring infections. So, Bruton began giving the boy monthly injections of antibodies, with outstanding success.

In 1952, Bruton reported his remarkable findings. His report was a medical bombshell because it established the concept of immunodeficiency disorders, while helping to solidify the role played by antibodies in resistance to infection.

The story was not finished, however, because certain patients with immunodeficiency disorders still produced the lymphocytes that immunize against skin grafts. This observation led to the notion that the immune system was actually a dual system—one branch centered in antibodies, a second branch centered in lymphocytes. The sources of antibodies and lymphocytes would be found almost simultaneously, 13 years later.

In 1965, at the meeting of the American Academy of Pediatrics, Robert A. Good's research group was presenting evidence on the importance of the bursa of Fabricius to antibody production. The audience was skeptical, but the evidence appeared substantial. When the speaker finished, a Philadelphia pediatrician named Angelo DiGeorge stepped to the microphone to add a footnote. DiGeorge described how four children at his hospital were struck repeatedly with severe infections. The children had antibodies in their blood, but curiously, each lacked a thymus. DiGeorge's observation was lost in the shuffle, but a question arose in his mind: Were the children susceptible to skin grafts, or could they possibly be immune?

DiGeorge hurried back to the hospital and devised a set of experiments to determine whether the children could successfully receive skin grafts. After weeks of study, he found they could. The thymus was apparently the key to the production of lymphocytes and hence to graft immunity. Bruton's patient lacked the bursa type of immunity, but DiGeorge's patients lacked the thymus type of immunity. The duality of the immune system was thus strengthened, and a second immune deficiency disease, DiGeorge syndrome, entered the dictionary of medical science.

An example of a humoral immunodeficiency is **X-linked (Bruton) agammaglobulinemia**, first described by Ogden C. Bruton in 1952. In this disease, B cells fail to develop from pre-B cells in the bone marrow. The patient's lymphoid tissues lack mature B cells and plasma cells, all five classes of antibodies are either low in level or absent, and antibody responses to infectious disease are undetectable. Infections from staphylococci, pneumococci, and streptococci are common between the ages of six months and two years. As the name suggests, the disorder is a sex-linked inherited trait, much more frequently observed in males than in females. Artificially acquired passive immunity is used to treat infectious disease in agammaglobulinemia patients.

DiGeorge syndrome is a cellular immunodeficiency in which T cells fail to develop. The deficiency is linked to failure of the thymus gland to mature in the embryo. Cellular immunity is defective in such individuals, and susceptibility is high to many fungal and protozoal diseases and certain viral diseases. Correction of the defect by grafts of thymus tissue corrects the disorder (MicroFocus 22.6).

Children suffering from **ataxia-telangiectasia** have malfunctioning B and T cells as well as

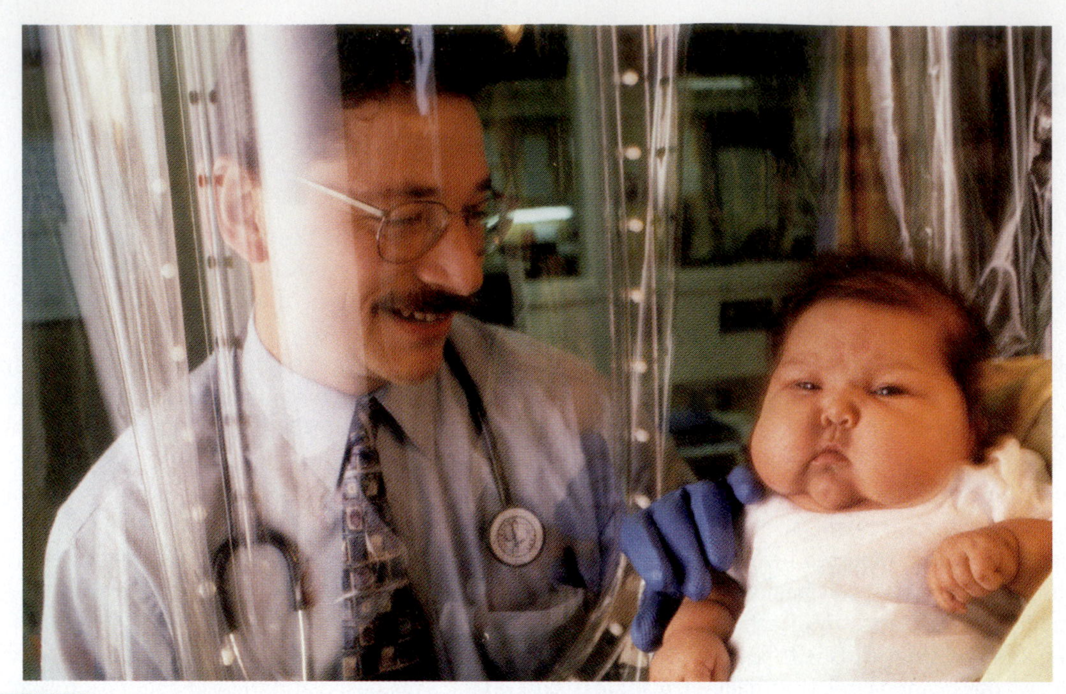

FIGURE 22.15 **SCID and Gene Therapy.** A five-month-old baby suffering from severe combined immunodeficiency disease (SCID) is protected in a sterile tent to prevent infection. He is receiving gene therapy to insert a gene for this enzyme deficiency into stem cells in the bone marrow. The stem cells are then transplanted into the baby. With this enzyme, stem cells may produce normal immune system blood cells.

Q: Why are stem cells in the bone marrow being used?

deficiencies in antibodies IgA and IgE. They have increased susceptibility to diseases, particularly pneumonia and bronchitis. Eventually paralysis and dementia lead to death by age 30.

Perhaps the most severe and life-threatening immunodeficiency is one combining both humoral and cell mediated immunity disorders. **Severe combined immunodeficiency disease (SCID)** involves lymph nodes depleted in both B and T cells. Without a population of B and T cells, all acquired immune functions are suppressed. One form of the disorder is caused by an enzyme deficiency, and individuals with the disorder have to be kept in strict isolation to prevent infections and diseases from occurring. If such measures are not taken, most children die before the age of two. Gene therapy also is being used to correct the gene defect (FIGURE 22.15).

■ Gene therapy: A genetic treatment to insert a normal or genetically altered gene into cells in order to replace or make up for the nonfunctional or missing genes.

Other immune deficiency diseases are linked to leukocyte malfunction. In the rare disease known as **Chédiak-Higashi syndrome**, the phagocytes fail to kill microorganisms because of the inability of lysosomes to release their contents into a phagosome (see Chapter 19). In **chronic granulomatous disease**, the phagocytes fail to produce hydrogen peroxide and other substances needed to kill ingested bacterial cells and fungi. Antibiotics and interferon are given to reduce the number and severity of infections.

Deficiencies in the complement system may be life-threatening. As noted in Chapter 19, complement is a series of proteins any of which the body may fail to produce. For reasons not currently understood, many patients with complement component deficiencies suffer systemic lupus erythematosus (SLE) or an SLE-like syndrome. Meningococcal and pneumococcal diseases are observed often in patients who lack C3, probably as a result of poor opsonization.

CONCEPT AND REASONING CHECKS

22.11 Compare and contrast the symptoms of the discussed immunodeficiency disorders.

SUMMARY OF KEY CONCEPTS

22.1 Type I IgE-Mediated Hypersensitivity

- Type I hypersensitivity is caused by IgE antibodies produced in response to certain antigens called allergens. The antibodies can attach to the surface of mast cells and basophils, and trigger the release of the mediators on subsequent exposure to the same allergen.
- Type I hypersensitivities can involve systemic anaphylaxis, a whole-body reaction in which a series of mediators induces vigorous and life-threatening contractions by the smooth muscles of the body.
- Atopic diseases involve localized anaphylaxis (common allergies), such as hay fever or a food allergy.
- Several ideas have been put forth concerning the nature of allergies. These include the shielding of IgE-secreting B cells (plasma cells) by IgA-secreting plasma cells and feedback mechanisms where IgE limits its own production in nonatopic individuals.
- Desensitization therapy attempts to limit the possibility of an anaphylactic reaction through the injection of tiny amounts of allergen over time. Therapy also can involve the presence of IgG antibodies as blocking antibodies.

22.2 Other Types of Hypersensitivity

- In type II cytotoxic hypersensitivity, the immune system produces IgG and IgM antibodies, both of which react with the body's cells and often destroy the latter.
- No cells are involved in type III hypersensitivity. Rather, the body's IgG and IgM antibodies interact with dissolved antigen molecules to form visible immune complexes. The accumulation of immune complexes in various organs leads to local tissue destruction in such illnesses as serum sickness, Arthus phenomenon, and systemic lupus erythematosus.
- Type IV cellular hypersensitivity involves no antibodies, but is an exaggeration of cell-mediated immunity based in T lymphocytes. Contact dermatitis is a manifestation of this hypersensitivity.

22.3 Autoimmune Disorders and Transplantation

- Autoimmune disorders can occur through defects in clonal deletion, clonal anergy, or regulatory T-cell activity.
- Human heredity as well as access to privileged sites and antigenic mimicry also can trigger an autoimmune response.
- Some of the more notable organ-specific autoimmune diseases are myasthenia gravis, Graves disease, and type I diabetes; systemic ones include lupus erythematosus and rheumatoid arthritis.
- Four types of grafts or transplants can be performed: autografts, isografts, allografts, and xenografts. Allografts are the most common. Rejection of grafts or transplants involves cytotoxic T cells and naive T cells. The graft also can be rejected by immune cells in the graft that reject the recipient (graft-versus-host reaction).
- The major histocompatibility complex (MHC) proteins are involved in transplant rejection, so tissue typing is important to match as closely as possible donor and recipient MHC proteins to prevent rejection.
- Preventing rejection is strengthened by using antirejection strategies, including anti-inflammatory drugs, antilymphocyte antibodies, antimitotic drugs, cyclosporine, monoclonal antibodies, or radiation.

22.4 Immunodeficiency Disorders

- Immunodeficiency disorders may be congenital (primary immunodeficiencies) or acquired later in life (secondary immunodeficiencies).
- Congenital disorders include X-linked agammaglobulinemia, DiGeorge syndrome, and immune deficiency disease. Deficiencies in phagocytic cells or complement components also may occur.

LEARNING OBJECTIVES

After understanding the textbook reading, you should be capable of writing a paragraph that includes the appropriate terms and pertinent information to answer the objective.

1. Distinguish between immediate and delayed hypersensitivity.
2. Summarize the events of and the role of humoral immunity in type I hypersensitivities.
3. Describe the roles for and pharmacological activities for mast cell chemical mediators.
4. Explain the initiation of and outcomes of systemic anaphylaxis.
5. Distinguish between the different forms of atopic disease.
6. Compare asthma to other forms of type I hypersensitivities.
7. Discuss the reasons why allergies develop, and the therapies and treatments available.
8. Summarize the characteristics of type II hypersensitivity and its relationship to blood transfusions and hemolytic disease of the newborn.
9. Summarize the characteristics of type III hypersensitivity and its relationship to serum sickness and the Arthus reaction.
10. Summarize the characteristics of type IV hypersensitivity and its relationship to infection allergies and contact dermatitis.
11. Identify the ways in which an autoimmune disease can arise.
12. Contrast organ-specific and systemic autoimmune disorders.
13. Describe the immunological reasons why organ transplants are rejected and list the four types of grafts (transplants).
14. Contrast primary and secondary immunodeficiencies and list several primary disorders and the immunological deficiency.

HTTP://MICROBIOLOGY.JBPUB.COM/

The site features learning, an on-line review area that provides quizzes and other tools to help you study for your class. You can also follow useful links for in-depth information, read more MicroFocus stories, or just find out the latest microbiology news.

SELF-TEST

Answer each of the following questions by selecting the ***one*** answer that best fits the question or statement. Answers to even-numbered questions can be found in **Appendix C**.

1. All the following are types of immediate hypersensitivities *except*:
A. asthma.
B. contact dermatitis.
C. blood transfusions.
D. food allergies.
E. hay fever.

2. In a type I hypersensitivity _____ antibodies bind to _____.
A. IgE; T cells
B. IgG; B cells
C. IgE; mast cells
D. IgA; basophils
E. IgD; mast cells

3. Which one of the following is *not* a cell mediator released from mast cells? If all are cell mediators, select **E.**
A. Serotonin
B. Histamine
C. IgE antibodies
D. Bradykinin
E. All the above (**A–D**) are correct.

4. Systemic anaphylaxis is characterized by
A. contraction of smooth muscles.
B. a red rash.
C. mast cell degranulation.
D. hives.
E. Both **A** and **C** are correct.

5. A wheal and flare reaction is typical in some cases of
A. food allergies.
B. Rh disease.
C. asthma.
D. exercise-induced allergies.
E. blood transfusions.

6. Desensitization therapy can involve
A. the use of blocking antibodies.
B. monoclonal antibodies.
C. injections of small amounts of allergen.
D. allergen injections of several weeks.
E. All the above (**A–D**) are correct.

7. A cytotoxic hypersensitivity would occur if blood type _____ is transfused into a person with blood type _____.
A. A; AB
B. O; AB
C. A; O
D. O; B
E. All the above (**A–D**) are correct.

8. Rh antigens are found on the surface of
A. phagocytes.
B. red blood cells.
C. pathogens.
D. tissue cells.
E. lymphocytes.

9. Serum sickness is a common symptom of
A. the Arthus reaction.
B. contact dermatitis.
C. hemolytic disease of the newborn.
D. immune complex hypersensitivity.
E. food allergies.

10. Immune complex formation is a complication in all the following diseases *except*:
A. myasthenia gravis.
B. dengue fever.
C. rheumatic fever.
D. glomerulonephritis.
E. lymphocytic choriomeningitis (LCM).

11. An induration is
A. a lung lesion caused by asthma.
B. the result of IgE cross-linking on mast cells.
C. a thickening and drying of the skin tissue.
D. typical of a sensitized individual undergoing a tuberculin skin test.
E. Both **C** and **D** are correct.

12. Which one of the following allergens is *not* associated with contact dermatitis?
A. Foods
B. Cosmetics
C. Coins
D. Poison ivy
E. Jewelry

13. Immunological tolerance to "self" is established by
A. regulatory T cells.
B. destruction of self-reactive lymphoid cells.
C. clonal anergy.
D. clonal deletion.
E. All the above (**A–D**) are correct.

14. All the following are examples of autoimmune disorders *except*:
A. Graves disease.
B. DiGeorge syndrome.
C. systemic lupus erythematosus.
D. type I diabetes.
E. myasthenia gravis.

15. A _____ is a graft between genetically different members of the same species.
A. xenograft
B. transgraft
C. autograft
D. allograft
E. isograft

16. What immunodeficiency disorder is associated with a lack of T and B cells and complete immune dysfunction?
A. Ataxia-telangiectasia
B. DiGeorge syndrome
C. Severe combined immunodeficiency disease
D. Chronic granulomatous disease
E. Chédiak-Higashi syndrome

QUESTIONS FOR THOUGHT AND DISCUSSION

Answers to even-numbered questions can be found in **Appendix C.**

1. As part of an experiment, one animal is fed a raw egg while a second animal is injected intravenously with a raw egg. Which animal is in greater danger? Why?
2. A woman is having the fifth injection in a weekly series of hay fever shots. Shortly after leaving the allergist's office, she develops a flush on her face, itching sensations of the skin, and shortness of breath. She becomes dizzy, then faints. What is taking place in her body, and why has it not happened after the first four injections?
3. You may have noted that brothers and sisters are allowed to be organ donors for one another, but that a person cannot always donate to his or her spouse. Many people feel bad about being unable to help a loved one in time of need. How might you explain to someone in such a situation the basis for becoming an organ donor and why it may be impossible to serve as one?
4. The immune system is commonly regarded as one that provides protection against disease. This chapter, however, seems to indicate that the immune system is responsible for numerous afflictions. Even the title is "Immune Disorders." Does this mean that the immune system should be given a new name? On the other hand, is it possible that all these afflictions are actually the result of the body's attempts to protect itself? Finally, why can the phrase "immune disorder" be considered an oxymoron?

APPLICATIONS

Answers to even-numbered questions can be found in **Appendix C.**

1. During war and under emergency conditions, a soldier whose blood type is O donates blood to save the life of a fellow soldier with type B blood. The soldier lives, and after the war becomes a police officer. One day he is called to donate blood to a brother officer who has been wounded and finds that it is his old friend from the war. He gladly rolls up his sleeve and prepares for the transfusion. Should it be allowed to proceed? Why?
2. Coming from the anatomy lab, you notice that your hands are red and raw and have begun peeling in several spots. This was your third period of dissection. What is happening to your hands, and what could be causing the condition? How will you solve the problem?
3. "He had a history of nasal congestion, swelling of his eyes, and difficulty breathing through his nose. He gave a history of blowing his nose frequently, and the congestion was so severe during the spring he had difficulty running." The person in this description is a certain former president of the United States, and the writer is an allergist from Little Rock, Arkansas. What condition (technically known as allergic rhinitis) is probably being described?

REVIEW

This chapter has summarized some of the disorders associated with the immune system. To gauge your understanding, rearrange the scrambled letters to form the correct word for each of the spaces in the statements. The answers to the even-numbered statements are in **Appendix C.**

1. The simple compound ______________ is one of the major mediators released during allergy reactions.
   I T M E H I A S N
2. An immune deficiency called ______________ syndrome is characterized by the failure of T lymphocytes to develop.
   I D O G G E R E
3. In type IV hypersensitivity, a drying and thickening of the skin known as ______________ is a symptom.
   A N U I R O N I D T
4. Cases of rheumatoid arthritis are accompanied by immune complex formation in the body's ______________.
   N I S T J O
5. In cases of ______________ disease, antibody molecules unite with receptors on the surface of thyroid gland cells.
   V S R E A G
6. In a ______________ hypersensitivity, antibodies unite with cells and trigger a reaction that results in cell destruction.
   X Y O T C C I T O
7. Hay fever is an example of an ______________ disease, one in which a local allergy takes place.
   O C T I A P
8. Immune complex hypersensitivities develop when antibody molecules interact with ______________ molecules and form aggregates in the tissues.
   E N I G N A T
9. In people suffering from X-linked agammaglobulinemia, the lymph nodes are noticeably deficient in ______________ cells.
   S L M A A P
10. The skin test for ______________ relies on a response by T lymphocytes to PPD placed in the skin tissues.
    U U I C T R L S B E O S
11. Mast cells and ______________ are the two principal cells that function in anaphylactic responses.
    S S B I O H P A L
12. Rh disease can develop in a fetus if the father's blood type was Rh-______________ and the mother's type was Rh-negative.
    O S I P I E T V
13. A key element in transplant acceptance or rejection is a set of molecules abbreviated as ______________ proteins.
    H C M
14. Urticaria is a form of skin ______________ occurring in a person having an allergic reaction.
    A H S R

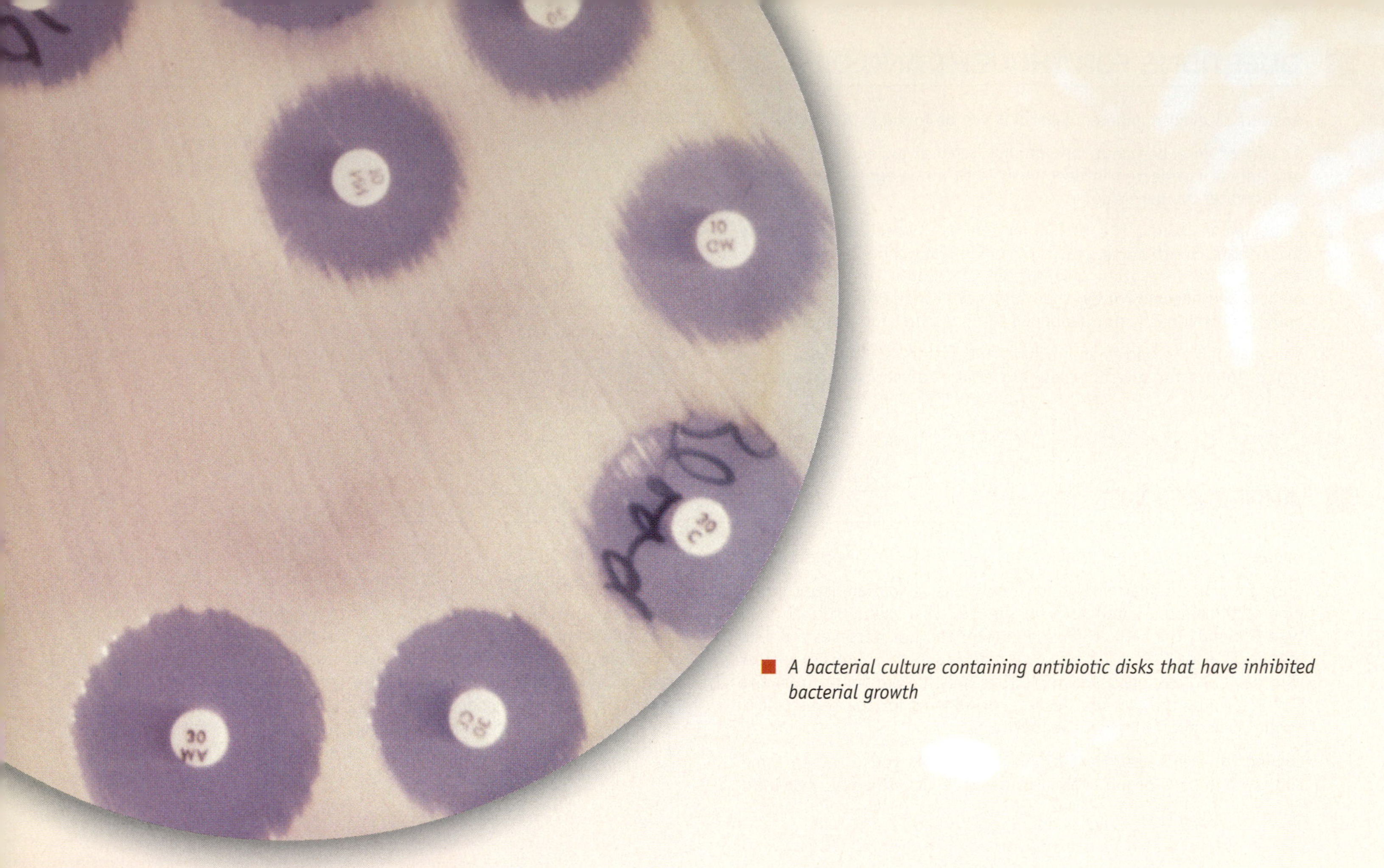

■ *A bacterial culture containing antibiotic disks that have inhibited bacterial growth*

PART 6 Control of Microorganisms

In the late 1800s and early 1900s, growing support for the germ theory of disease led to a dramatic reduction in the frequency of epidemics. Scientists reasoned that if microorganisms cause disease, then it was possible to control disease by controlling the microorganisms. The idea had been proposed by Pasteur decades before, but not until the turn of the century did it gather momentum and achieve a firm footing in the scientific community.

Two types of control gradually evolved: physical control and chemical control. To achieve physical control, scientists used heat, radiations, filters, and other physical agents to remove microorganisms from instruments, equipment, and fluids. Pasteurization of dairy products was in wide use by 1895, and food preservation methods were gradually updated. To achieve chemical control, doctors used antiseptics and disinfectants in medicine, surgery, and wound treatment. Moreover, municipalities began adding chlorine to their water supplies to protect citizens from waterborne diseases. As control methods became a way of life in public health, the chains of microbial transmission broke down, and disease outbreaks declined.

Little could be done for the patient who already was ill. Here, another type of control was necessary, one supplementing the body's own natural defenses. This form of control would emerge in the 1940s with the development of chemotherapeutic agents and antibiotics. Now, physicians could do something about diseases they could not hope to treat before. The new drugs and medicines ushered in a second dramatic reduction in the incidence of infectious disease.

The control of microorganisms is an essential factor in maintaining good health. In Part 6, we consider a variety of control methods and examine their uses and modes of action. Our survey begins in Chapter 23 with an exploration of physical methods of control used for objects outside the body. It continues with an examination of chemical methods for objects coming in contact with the body (e.g., instruments) and on the body surface itself. In Chapter 24, we move from treating the environment to treating the patient by discussing chemotherapeutic agents and antibiotics. Although Part 6 is only two chapters long, applied on a broad scale, the control methods described, along with vaccines for disease prevention, remain a major deterrent to infection and disease.

MICROBIOLOGY PATHWAYS

Sales and Research

Not all salespeople sell vacuum cleaners or brushes. And not all microbiologists wear white coats and work in ultra-clean laboratories. It is quite commonplace for a person trained in microbiology to find a comfortable and enjoyable future with a company that sells and distributes instruments, scientific chemicals, or pharmaceuticals. For example, it is valuable for a sterilizer salesperson to understand the microbiological basis for sterilizing such things as instruments, microbial media, and patient materials. Also, it makes sense for a disinfectant salesperson to realize why microorganisms must be eliminated from a particular surface. Finally, without question, an individual selling new antibiotics should have a strong familiarity with the microorganisms that the antibiotic is intended to eliminate. The bottom line is: Know your product, and know its uses.

Before a product is available for sale, however, it must be developed, and once again, the microbiologist is a member of the team. The microbiologist can appreciate why new methods must be developed for preserving dairy products and for storing foods. The microbiologist can set the direction for developing new sterilizing instruments and have significant input into the development of pharmaceuticals for treating infectious disease. The diagnostic lab depends heavily on instruments produced under the guidance of microbiologists because the lab's objective is to detect microorganisms.

As you might suspect, besides microbiology and other life science courses, a healthy dose of chemistry, physics, and mathematics is valuable, depending upon which road one chooses to follow. To be successful requires more than academic qualifications. A salesperson or service technician must have great motivation and drive. Also, in the business arena, enthusiasm should be at the top of one's personal skills list. Companies want people who can move a customer forward and enthusiasm is a key. Other positive characteristic include the ability to think on one's feet, strong problem-solving skills, and a sense of humor. The ability to tinker with instruments is important to research and development, while strong interpersonal skills and effective verbal and written skills are important for the sales phase.

The contemporary health care industry depends heavily on the talents of microbiologists for the development and sale of innovative and novel approaches to diagnosis and treatment. What may seem like service and sales in the health care industry often has microbiology at its core.

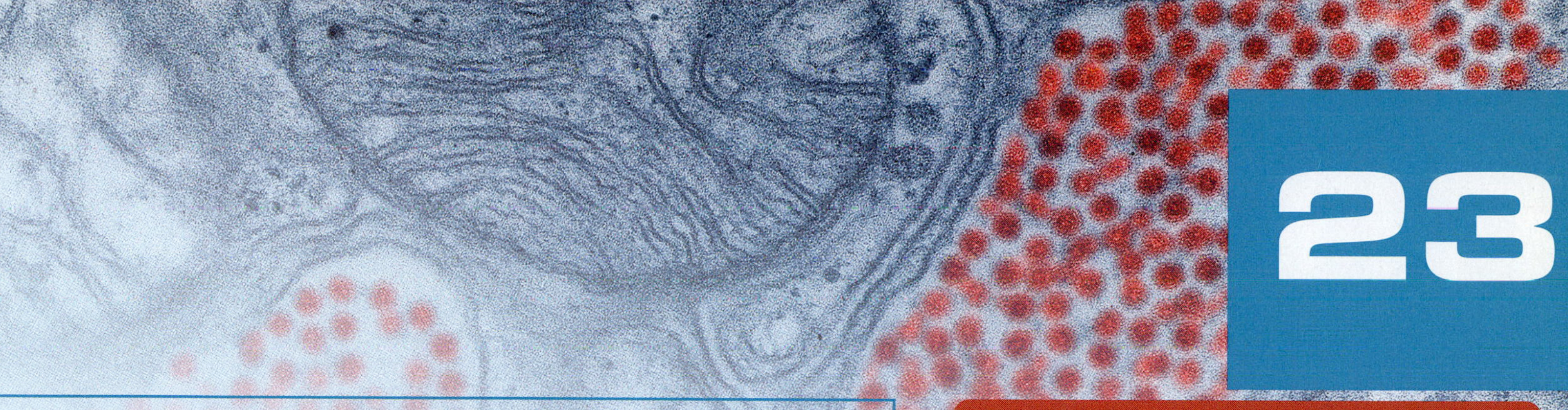

23

Physical and Chemical Control of Microorganisms

I do not believe that we can hope to prevent puerperal fever entirely . . . but I feel certain that by strict attention to antiseptics we shall be able to reduce its occurrence to a minimum and render its presence in hospital practice . . . a rarity.
—*On the Systematic Use of Antiseptic in Midwifery Practice*, which was a treatise by the Edinburgh Obstetrical Society 1880–1881

Chapter Preview and Key Concepts

23.1 General Principles of Microbial Control
- Microbes are kept under control either by eliminating them or reducing their numbers.

23.2 Physical Methods of Control
- Microbes and viruses are killed at temperatures above their temperature range for growth or replication.

  MicroInquiry 23: Exploring Heat as an Effective Control Method
- Dry heat does not penetrate material as well as moist heat.
- Filtration removes microbes from the air or water.
- Ultraviolet light can be bactericidal.
- X rays and gamma rays also are microbicidal.
- Dehydration and cold temperatures slow microbial growth.

23.3 General Principles of Chemical Control
- Disinfectants and antiseptics are key to improved sanitation and public health.
- Disinfectants and antiseptics are defined by their properties.
- Standards have been established to know the relative effectiveness of a chemical agent.

23.4 Chemical Methods of Control
- Chlorine and iodine are good disinfectant agents.
- Many phenolic derivatives are used as disinfectants or antiseptics.
- Mercury, copper, and silver compounds can be useful disinfectants.
- Alcohols are effective skin antiseptics.
- Cationic detergents are bacteriostatic.
- Hydrogen peroxide can be used as an antiseptic rinse.
- Aldehydes and gasses can be used for sterilization.

For personal hygiene, washing our hands, taking regular showers or baths, brushing our teeth with fluoride toothpaste, and using an underarm deodorant are common practices we use to control microorganisms on our bodies. In our homes, we try to keep microbes in check by cooking and refrigerating foods, cleaning our kitchen counters and bathrooms with disinfectant chemicals, and washing our clothes with detergents.

In our attempt to be hygiene-minded consumers, sometimes we have become excessively "germphobic." The news media regularly report about this scientific study or that survey identifying places in our homes (e.g., toilets, kitchen drains) or environment (e.g., public bathrooms, drinking-water fountains) that represent infectious dangers. Consumer groups distribute pamphlets on "microbial awareness" and numerous companies have responded by manufacturing dozens of household chemical products—some useful, many unnecessary (FIGURE 23.1A).

Our desire to protect ourselves from microbes also stems from events beyond our doorstep. The news media again report about dangerous disease outbreaks, many of which result from a lack of sanitary controls or a lack of vigilance to maintain those controls. In our communities, we expect our drinking water to be clean. That goes for our hospitals as well. Nowhere is this more important than in the operating rooms and surgical wards. Here, hospital personnel must maintain scrupulous levels of cleanliness and have surgical instruments that are sterile. Yet, nosocomial (hospital-acquired) infections do occur when hygiene barriers are breached.

Microbial control also is a global endeavor. Government and health agencies in many developing nations often lack the means (financial, medical, social) to maintain proper sanitary conditions. These circumstances can result in outbreaks of diseases such as diphtheria, malaria, measles, meningitis, and cholera. Cholera, as an example, tends to be associated with poverty-stricken areas where overcrowding and inadequate sanitation practices generate contaminated water supplies and food (FIGURE 23.1B).

We do need to be hygiene conscious. If the procedures and methods to control pathogens fail or are not monitored properly, serious threats to health and well-being may occur. Importantly, the successful control of microorganisms usually requires simple methods and procedures. These methods are not products of the modern era. As the opening quote demonstrates, they were used by Semmelweis, and other microbiologists over a century ago to prevent infection and contamination of their materials.

In this chapter, we examine a variety of physical and chemical methods used for controlling the growth and spread of microorganisms. Our study begins by outlining some general principles and terminology and then identifies physical methods commonly used today. We also explore chemical methods for microbial control and discuss of the spectrum of antiseptics and disinfectants. Whether the methods are physical or chemical, they are integral in public health practices to ensure continued good health and protection from infectious disease.

(A)

(B)

FIGURE 23.1 **Controlling Microorganisms.** (**A**) Household cleaning products are diverse and formulated for every cleaning need to maintain a sanitary condition. (**B**) This African shantytown has slum houses, open sewage, and littered walkways. It is not surprising that in these unsanitary conditions diseases such as cholera and typhoid are common.

Q: In these examples, does controlling microorganisms mean sterilization or simply reducing the number of microbes to a safe level? Explain.

23.1 General Principles of Microbial Control

As we have described in the previous chapters, your immune system usually does a great job keeping pathogens out of the body. But just as important are external control measures designed to eliminate or reduce the potential of such threats. To guard against these potential invaders, specific control and preventative measures have been devised and special materials developed to discourage the growth of, or to outright kill, microorganisms before they can become a health menace.

The effective control of disease-causing microbes requires an understanding of the procedures and agents available today, including the physical and chemical methods used to limit microbial growth and/or microbial transmission.

First, let's establish some basic vocabulary generally used by the public and by health officials when talking about microbial control.

Sterilization and Sanitization Are Key to Good Public Health

KEY CONCEPT

- Microbes are kept under control either by eliminating them or reducing their numbers.

Sterilization involves the destruction of all living microbes, including spores and viruses, on an object or in an area. Thus, microbiologically speaking, when the word "sterile" is used, we always mean there are no living microbes, viruses, or spores inhabiting the object or area. For example, in a surgical operation, the surgeon uses sterile instruments previously treated in some way to kill any microbes present on them (FIGURE 23.2).

Everyday experiences bring us in contact with sterile materials. An unopened can of corn or peas is sterile inside. During the canning process, companies use special sterilization procedures to kill all the microbes on the vegetables and in the tin can. Agents that kill microbes are **microbicidal** (*-cide* = "kill") or more simply called *germicides*. If the agent specifically kills bacteria, it is **bactericidal**; if it kills fungi, it is **fungicidal**. Many physical methods and chemical agents are capable of destroying microbes on nonliving materials or on the skin surface. However, once exposed to the air and surroundings, sterile objects will again be contaminated with microbes in the air or the surrounding area.

■ Contaminated: In microbiology, a sterile object that is again harboring microorganisms and/or viruses.

More often, in our daily experiences we are likely to encounter materials where microbial populations have been reduced or where their growth has been inhibited. **Sanitization** involves those procedures reducing the numbers of pathogenic microbes or discouraging (inhibiting) their growth. Given enough time, these pathogens will grow and some could possibly cause spoilage or a health problem. So, a tasty wedge of cheese in the refrigerator might look fine today, but in a few weeks it may have a mold growing on it. The toilet bowl *sanitized* with a disinfectant today contains few pathogens. Tomorrow it may again be an area with increased numbers of bacteria species. Many chemical agents are **microbiostatic** (*-static* = "remain in place");

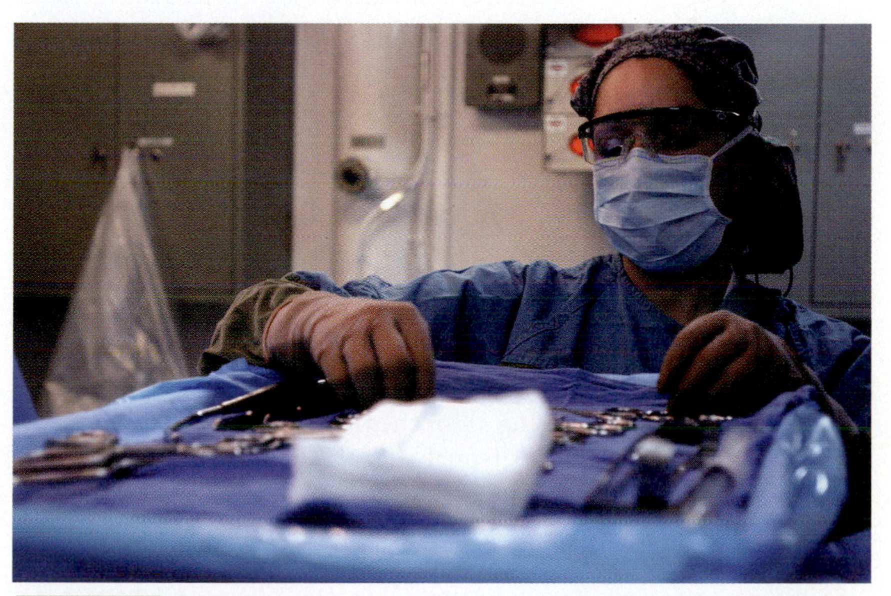

FIGURE 23.2 **Sterile Surgical Instruments.** Here a surgical nurse is unwrapping sterile surgical instruments.

Q: How would you sterilize these instruments?

they reduce microbial numbers or inhibit their growth. Again, agents can be **bacteriostatic** or **fungistatic**.

Sanitary measures to control pathogens are very important in areas frequented by the public. City and state sanitation agencies monitor drinking water quality and the preparation of food in public eating establishments to ensure pathogen control. Public health depends on good sanitary practices at home and in the workplace.

CONCEPT AND REASONING CHECKS

23.1 Assess the importance of maintaining sterility or sanitary conditions.

23.2 Physical Methods of Control

With this background, let's now examine some specific agents that kill microorganisms or inhibit their growth. Although there are several methods and agents used today that affect microbial survival, they generally fall into two categories. Physical methods and agents include temperature, radiation, osmotic pressure, and filtration. Chemical agents include antiseptics and disinfectants. We will consider the physical methods and agents first.

Heat Is One of the Most Common Physical Control Methods

KEY CONCEPT

- Microbes and viruses are killed at temperatures above their temperature range for growth or replication.

The Citadel is a novel by A. J. Cronin that follows the life of a young British physician, beginning in the 1920s. Early in the story, the physician, Andrew Manson, begins his practice in a small coal-mining town in Wales. Almost immediately, he encounters an epidemic of typhoid fever. When his first patient dies of the disease, Manson becomes terribly distraught. However, he realizes the epidemic can be halted, and in the next scene, he is tossing all of the patient's bed sheets, clothing, and personal effects into a huge bonfire.

The killing effect of heat on microorganisms has long been known. Heat is fast, reliable, and relatively inexpensive. Above the growth range temperature for a microbe (see Chapter 5), enzymes and other proteins as well as nucleic acids are denatured (see Chapter 2). Heat also drives off water, and because all organisms depend on water, this loss may be fatal.

■ Foot-and-mouth disease: A highly contageous viral disease affecting cattle, sheep, and pigs, in which the animal develops ulcers in the mouth and near the hooves.

The killing rate of heat may be expressed as a function of time and temperature. For example, bacilli of *Mycobacterium tuberculosis* are destroyed in 30 minutes at 58°C, but in only 2 minutes at 65°C, and in a few seconds at 72°C. Each microbial species has a **thermal death time**, the time necessary for killing the population at a given temperature. Each species also has a **thermal death point**, the minimal temperature at which it dies in a given time. These measurements are particularly important in the food industry, where heat is used for preservation (Chapter 25). MicroInquiry 23 examines heat and microbial killing.

Incineration using a direct flame can kill microbes very rapidly. For example, the flame of the Bunsen burner is employed for a few seconds to sterilize the bacteriological loop before removing a sample from a culture tube and after preparing a smear (FIGURE 23.3). Flaming the lip of the tube also destroys organisms that happen to contact the lip, while burning away lint and dust.

Disposable hospital gowns and certain plastic apparatus are examples of materials that may be incinerated. In past centuries, the bodies of disease victims were burned to prevent spread of the plague. It still is common practice to incinerate the carcasses of cattle that have died of anthrax and to put the contaminated field to the torch because anthrax spores cannot be destroyed adequately by other means. The 2001 outbreak of foot-and-mouth disease in British cattle required the mass incineration of thousands of cattle as a means to stop the spread of the disease (FIGURE 23.4).

CONCEPT AND REASONING CHECKS

23.2 Summarize the uses of incineration as a sterilization method.

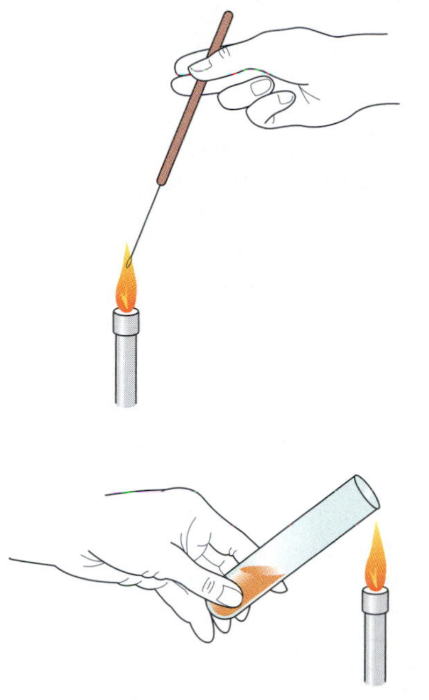

FIGURE 23.3 **Use of the Direct Flame as a Sterilizing Agent.** A few seconds in the flame of a laboratory Bunsen burner is usually sufficient to effect sterilization of an inoculation loop or culture tube lip.

Q: Why is it necessary to flame a culture tube lip?

FIGURE 23.4 **Incineration Is an Extreme Form of Heat.** Nearly four million hoofed animals infected with or exposed to foot-and-mouth disease in England in 2001 were incinerated and buried.

Q: How is incineration similar to flame sterilization?

Dry and Moist Heat Are Applied Differently

KEY CONCEPT

- Dry heat does not penetrate material as well as moist heat.

The application of heat and the form in which it is used depend on the nature of the substance to be treated.

Dry Heat. The hot-air oven uses radiating dry heat for sterilization. This type of energy does not penetrate materials easily, and therefore, long periods of exposure to high temperatures are necessary. For example, at a temperature of 160°C (320°F), a period of two hours is required for the destruction of bacterial spores. Higher temperatures are not recommended because the wrapping paper used for equipment tends to char at 180°C. The hot-air method is useful for sterilizing dry powders and water-free oily substances, as well as for many types of glassware, such as pipettes, flasks, and syringes. Dry heat does not corrode sharp instruments as steam often does, nor does it erode the ground glass surfaces of nondisposable syringes.

The effect of dry heat on microorganisms is equivalent to that of baking. The heat changes microbial proteins by oxidation reactions and creates an arid internal environment, thereby burning microorganisms slowly. It is essential that organic matter such as oil or grease films be removed from the materials, because such substances insulate against dry heat. Moreover, the time required for heat to reach sterilizing temperatures varies according to the material. This factor must be considered in determining the total exposure time.

Moist Heat. Boiling water is an example of moist heat. Moist heat penetrates materials

MICROINQUIRY 23

Exploring Heat as an Effective Control Method

Is that can of unopened peas in your pantry sterile? Yes, because companies like General Mills (manufacturer of Green Giant® products) and the food industry in general have established appropriate procedures for sterilizing commercial foods. Sterilization depends on several factors. Identifying the type(s) of microbes in a product can determine whether the heating process will sterilize or eliminate only potential disease-causing species. In many foods, the microbes usually are not in water but rather in the food material. Microbes in powders or dry materials will require a different length of time to sterilize the product than microbes in organic matter. Environmental conditions also influence the sterilization time. Microbes in acidic or alkaline materials decrease sterilization times while microbes in fats and oils, which slow heat penetration, increase sterilization times. It must be remembered that sterilization times are not precise values. However, by knowing these factors, heating the product to temperatures above the maximal range for microbial growth will kill microbes rapidly and effectively. Let's explore the factors of time and temperature. Answers can be found in **Appendix D**.

As we described for the microbial growth curve in Chapter 4, microbial death occurs in an exponential fashion. Look at the table to the right. The table records the death of a microbial population by heating. Notice that the cells die at a constant rate. In this generalized example, each minute 90 percent of the cells die (10 percent survive). Therefore, if you know the initial number of microorganisms, you can predict the **thermal death time** (**TDT**), which is defined as the time, at a specified temperature, required to kill a population of microorganisms.

The food industry depends on knowing a microorganism's heat sensitivity when planning the canning or packaging of many foods. One way the industry determines this sensitivity is by using standard curves that take into account the factors mentioned above. The graph drawn below represents three such curves, each representing a different bacterial species treated at the same temperature (60°C) in the same food material (curve B represents the plotted data from the Table).

22a. If you had to sterilize this food product that initially contained 10^6 bacteria, how long would it take for each bacterial species? (Hint: $10^0 = 1$)

Another value of importance is the **decimal reduction time** (**D value**), which is the time required at a specific temperature to kill 90 percent of the viable organisms. These are the values typically used in the canning industry. D values are usually identified by the temperature used for killing. In the graph to the right, the temperature was 60°C, so D is written as D60. Look at the graph again.

22b. Calculate the D60 values for the three bacterial populations depicted in curves A, B, and C.

On another day in the canning plant, you need to sterilize a food product. However, the only information you have is a D70 = 12 minutes for the microorganism in the food product. Assuming the D value is for the volume you have to sterilize and there are 10^8 bacteria in the food product:

22c. At what temperature will you treat the food product?

22d. How long will it take to sterilize the product?

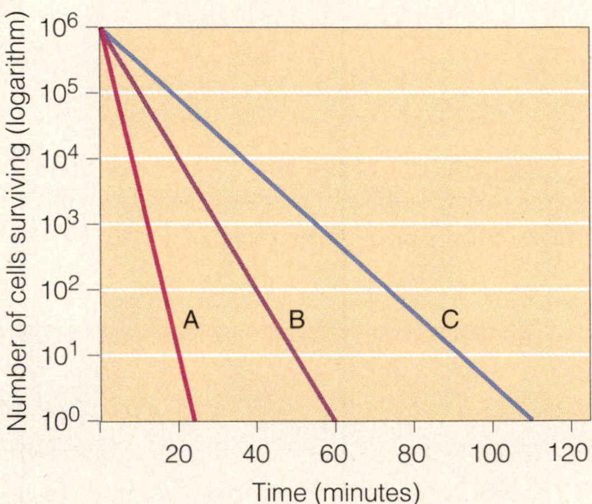

Standard curves for death of three microbial species (A, B, C).

TABLE

Microbial Death Rate

Time (minutes)	Number of Cells Surviving	% Killed
0	1,000,000	—
10	100,000	90
20	10,000	90
30	1,000	90
40	100	90
50	10	90
60	1	90

much more rapidly than dry heat because water molecules conduct heat better than air. Therefore, moist heat can be used at a lower temperature and shorter exposure time than for dry heat.

Moist heat kills microorganisms by denaturing their proteins. **Denaturation** is a change in the chemical or physical property of a protein. It includes structural alterations due to destruction of the chemical bonds holding proteins in a three-dimensional form. As proteins revert to a two-dimensional structure, they coagulate (denature) and become nonfunctional. Egg protein undergoes a similar transformation when it is boiled. (You might find reviewing the chemical structure of proteins in Chapter 2 helpful in understanding this process.) The coagulation of proteins requires less energy than oxidation, and, therefore, less heat need be applied.

Boiling water is not considered a sterilizing agent because the destruction of bacterial spores and the inactivation of viruses cannot always be assured. Under ordinary circumstances, with microorganisms at concentrations of less than 1 million per milliliter, most species of microorganisms can be killed within 10 minutes. Indeed, the process may require only a few seconds. However, fungal spores, protozoal cysts, and large concentrations of hepatitis A viruses require up to 30 minutes' exposure. Bacterial spores often require two hours or more. Because inadequate information exists on the heat tolerance of many species of microorganisms, boiling water is not reliable for sterilization purposes (FIGURE 23.5).

If boiling water must be used to destroy microorganisms, then materials must be thoroughly cleaned to remove traces of organic matter, such as blood or feces. The minimum exposure period should be 30 minutes, except at high altitudes, where it should be increased to compensate for the lower boiling point of water. All materials should be well covered. Baking soda may be added at a two percent concentration to increase the efficiency of the process.

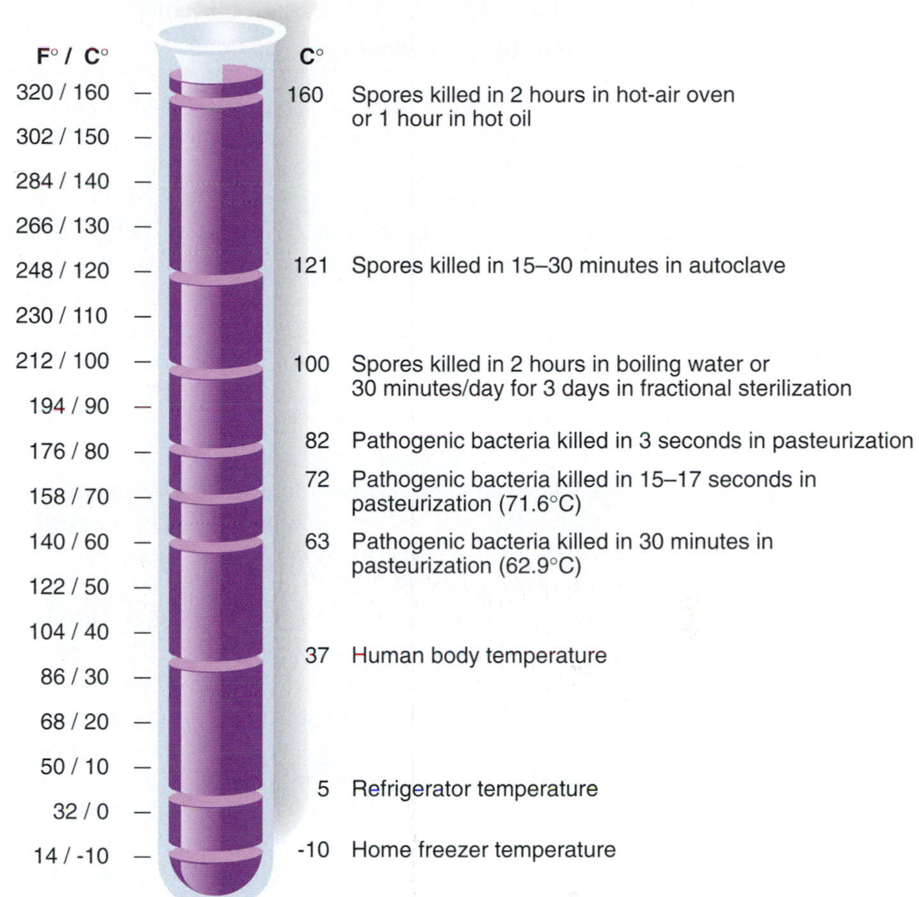

FIGURE 23.5 **Temperature and the Physical Control of Microorganisms.** Notice that materials containing bacterial endospores require longer exposure times and higher temperatures for killing.

Q: Pure water boils and freezes at what temperatures on the Celsius scale?

Moist heat in the form of pressurized steam is regarded as the most dependable method for sterilization, including the destruction of bacterial spores. This method is incorporated into a device called the **autoclave** (MicroFocus 23.1). When the pressure of a gas increases, the temperature of the gas also increases proportionally. Because steam is a gas, increasing its pressure in a closed system increases its temperature. As the water molecules in steam become more energized, their penetration increases substantially. This principle is used to reduce cooking time in the home pressure cooker and to reduce sterilizing time in the autoclave. During autoclaving, the sterilizing agent is the moist heat, not the pressure.

MICROFOCUS 23.1: History

A Heated Controversy

In an attempt to disprove spontaneous generation, Louis Pasteur had stated that boiled urine failed to support bacterial growth. However, among the last defenders of spontaneous generation was the British physician Harry Carleton Bastian who, in 1876, claimed urine was alkaline, allowing microorganisms occasionally to appear. So, Pasteur repeated Bastian's work and found it correct—certain microorganisms could resist death by boiling. The spores of *Bacillus subtilis,* discovered coincidentally in 1876 by Ferdinand Cohn, were an example.

Pasteur realized he would have to heat his broths at a temperature higher than 100°C to achieve sterilization. He put his pupil and collaborator Charles Chamberland in charge of developing a new sterilizer. Chamberland responded by constructing a pressure steam apparatus patterned after a steam "digester" invented in 1680 by the French physician Denys Papin. The sterilizer resembled a modern pressure cooker. It attained temperatures of 120°C and higher, and established the basis for the modern autoclave. Chamberland also would achieve fame in later years for his work with porcelain filters.

But Chamberland's invention was not universally accepted. A German group of investigators, led by Robert Koch, criticized the pressurized steam sterilizer because they believed its higher temperatures would destroy critical laboratory media. Instead they preferred an unpressurized steam sterilizer. In 1881, the German group developed a free-flowing steam sterilizer of the type used in tyndallization. In time, however, they came to appreciate the benefits of pressurized steam as a sterilizing agent, so much so that they modified Chamberland's device to an upright model. Ironically, the instrument became known as the Koch autoclave.

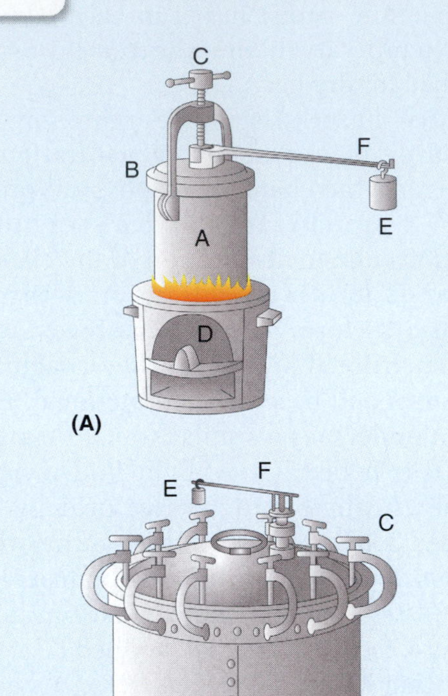

(**A**) Denis Papin's steam digester, designed in 1680. The digester consisted of a vessel (A) into which food was placed. The lid (B) and screw (C) sealed the vessel. A furnace (D) raised the temperature, and a weight (E) and safety lever (F) controlled the pressure of steam in the vessel. (**B**) Chamberland's autoclave, built in 1880 according to the principles of Papin's digester. The lid is held in place by screws (C), and a weight (E) and safety lever (F) are used to control pressure. This autoclave is basically similar to a home pressure cooker.

Autoclaves contain a sterilizing chamber into which articles are placed and a steam jacket where steam is maintained (FIGURE 23.6). As steam flows from the steam jacket into the sterilizing chamber, cool air is forced out and a special valve increases the pressure to 15 pounds/square inch (lb/in^2) above normal atmospheric pressure. The temperature rises to 121.5°C, and the superheated steam rapidly conducts heat into microorganisms. The time for destruction of the most resistant bacterial species is about 15 minutes. For denser objects, up to 30 minutes of exposure may be required. The conditions must be carefully controlled to assure sterilization has been accomplished (MicroFocus 23.2).

The autoclave is used to control microorganisms in both hospitals and laboratories. It is employed for blankets, bedding, utensils, instruments, intravenous solutions, and a broad variety of other objects. The laboratory technician uses it to sterilize bacteriological media and destroy pathogenic cultures. The autoclave is equally valuable for glassware and metalware, and is among the first instruments ordered when a microbiology laboratory is established.

The autoclave has certain limitations. For example, prions may not be destroyed; some

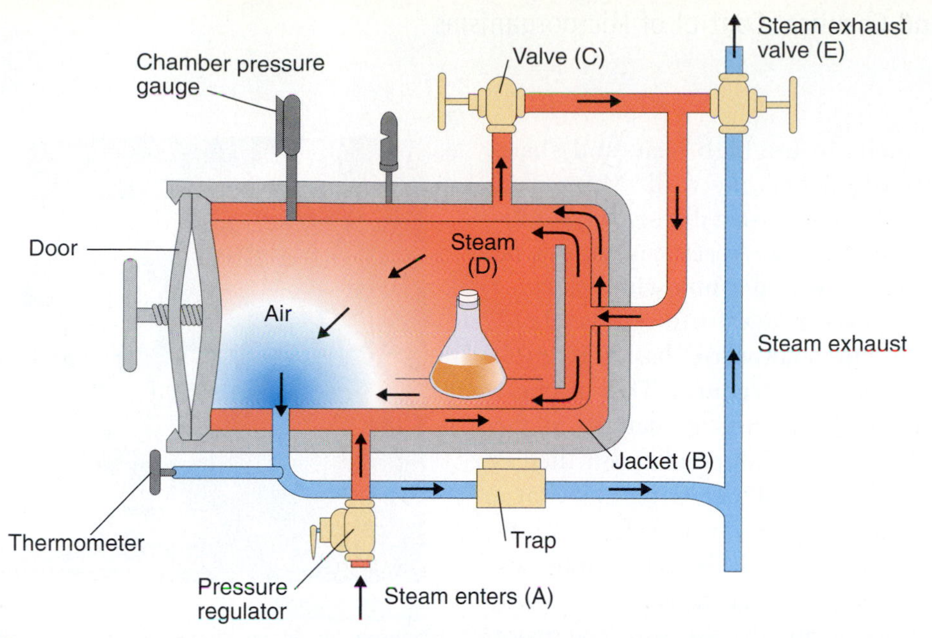

FIGURE 23.6 Operating an Autoclave. Steam enters through the port (A) and passes into the jacket (B). After the air has been exhausted through the vent, a valve (C) opens to admit pressurized steam (D), which circulates among and through the materials, thereby sterilizing them. At the conclusion of the cycle, steam is exhausted through the steam exhaust valve (E).

Q: Is it the steam or the pressure that kills microorganisms in an autoclave? Explain.

MICROFOCUS 23.2: Tools

Autoclave Quality Assurance

A nosocomial outbreak of *Pseudomonas* in a Thailand hospital in 1992 illustrates the need to carefully monitor the autoclave during use.

The problem began when hospital pharmacists prepared bottles of basal salts solution for use in the hospital operating rooms. To sterilize the solutions, the bottles were placed in the autoclave and left to run on its automatic cycle.

The bottles were then delivered to surgery to be used to irrigate the eyes of patients undergoing cataract surgery. Some bottles were left unused.

Within 30 hours, three cataract patients developed eye inflammations. The organism isolated from the patients was a pathogenic strain of *Pseudomonas* and the infected patients were treated with antibiotics.

An autoclave can be used to sterilize many dry and liquid materials.

Health investigators tested the unused bottles of salt solution as well as the tubes attached to the now-empty bottles. They found the identical strain of *Pseudomonas*.

Examining the pharmacy records, investigators noted that the autoclave pressure had reached only 10 to 12 lb/in^2, rather than the required 15 lb/in^2. The salts solution apparently was not sterilized.

There are several ways to assure that materials are properly sterilized. Autoclaves have temperature and pressure gauges visible from the outside and most models can produce a paper record of the temperature, time, and pressure.

To gauge the success of sterilization, materials usually are autoclaved with autoclave tape, which turns color if the object inside the material has been autoclaved correctly. Biological indicators also can be used. A strip containing spores of *Geobacillus stearothermophilus* can be included with the objects treated. At the conclusion of the cycle, the strip is placed in a nutrient broth medium and incubated. If the sterilization process has been unsuccessful, the spores will germinate and their metabolism will change the color of a pH indicator in the growth medium.

plasticware melts in the high heat; and sharp instruments often become dull. Moreover, many chemicals break down during the sterilization process and oily substances cannot be treated because they do not mix with water.

In recent years a new form of autoclave, called the **prevacuum autoclave**, has been developed for sterilization procedures. This machine draws air out of the sterilizing chamber at the beginning of the cycle. Saturated steam then is used at a temperature of 132°C to 134°C at a pressure of 28 to 30 lb/in^2. The time for sterilization is now reduced to as little as four minutes. A vacuum pump operates at the end of the cycle to remove the steam and dry the load. The major advantages of the prevacuum autoclave are the minimal exposure time for sterilization and the reduced time to complete the cycle.

In the years before the development of the autoclave, liquids and other objects were sterilized by exposure to free-flowing steam at 100°C for 30 minutes on each of three successive days, with incubation periods at room temperature between the steaming. The method was called **fractional sterilization** because a fraction of the sterilization was accomplished on each day. It was also called **tyndallization** after its developer, John Tyndall.

Sterilization by the fractional method is achieved by an interesting series of events. During the first day's exposure, steam kills virtually all organisms except bacterial spores, and it stimulates spores to germinate to vegetative cells. During overnight incubation, the cells multiply and are killed on the second day's exposure. Again, the material is cooled and the few remaining spores germinate, only to be killed on the third day. Although the method usually results in sterilization, occasions arise when several spores fail to germinate. The method also requires that spores be in a suitable medium, such as broth, for germination.

Fractional sterilization has assumed renewed importance in modern microbiology with the development of high-technology instrumentation and new chemical substances. Often, these materials cannot be sterilized at autoclave temperatures, or by long periods of boiling or baking, or with chemicals.

The final example of moist heat involves the process of pasteurization. **Pasteurization** reduces the bacterial population of a liquid such as milk and destroys organisms that may cause spoilage and human disease (FIGURE 23.7). Spores are not affected by pasteurization.

FIGURE 23.7 **The Pasteurization of Milk.** Milk is pasteurized by passing the liquid through a heat exchanger. The flow rate and temperature are monitored carefully. Following heating, the liquid is rapidly cooled.

Q: Why is the liquid rapidly cooled?

One method for milk pasteurization, called the **holding method**, involves heating at 62.9°C for 30 minutes. Although any thermophilic bacteria would thrive at this temperature, they are of little consequence because they cannot grow at body temperature. For decades, pasteurization has been aimed at destroying *Mycobacterium tuberculosis*, long considered the most heat-resistant bacterial species. More recently, however, attention has shifted to destruction of *Coxiella burnetii*, the agent of Q fever (see Chapter 9), because this rickettsial organism has a higher resistance to heat. Because both organisms are eliminated by pasteurization, dairy microbiologists assume other pathogenic bacteria also are destroyed. Pasteurization also is used to eliminate the *Salmonella* and *Escherichia coli* that can contaminate fruit juices. Two other methods of pasteurization are the **flash pasteurization method** at 71.6°C for 15 seconds and the **ultrapasteurization method** at 82°C for 3 seconds. These methods are discussed in Chapter 25.

Although heat is a valuable physical agent for controlling microorganisms, sometimes it is impractical to use. For example, no one would suggest removing the microbial population from a tabletop by using a Bunsen burner, nor can heat-sensitive solutions be subjected to an autoclave. In instances such as these and numerous others, a heat-free method must be used.

CONCEPT AND REASONING CHECKS

23.3 Summarize how the autoclave sterilizes materials.

Filtration Traps Microorganisms

KEY CONCEPT

- Filtration removes microbes from the air or water.

Filters came into prominent use in microbiology as interest in viruses grew during the 1890s. Previous to that time, filters were used to trap airborne organisms and sterilize bacteriological media, but now they became essential for separating viruses from other microorganisms. Among the early pioneers of filter technology was Charles Chamberland, an associate of Pasteur (MicroFocus 23.1). His porcelain filter was important to early virus research, as noted in Chapter 13. Another pioneer was Julius Petri (inventor of the Petri dish), who developed a sand filter to separate bacterial cells from the air.

Filtration is a mechanical method used to remove microorganisms from a solution. As fluid passes through the filter, organisms are trapped in the pores of the filtering material. The solution dripping through the filter into the receiving container is decontaminated or, in some cases, sterilized. Filters are used to purify such liquids as beverages, some bacteriological media, toxoids, many pharmaceuticals, and blood solutions (FIGURE 23.8A).

Several types of filters are used in the microbiology laboratory. Inorganic filters are typified by the Seitz filter, which consists of a pad of porcelain or ground glass mounted in a filter

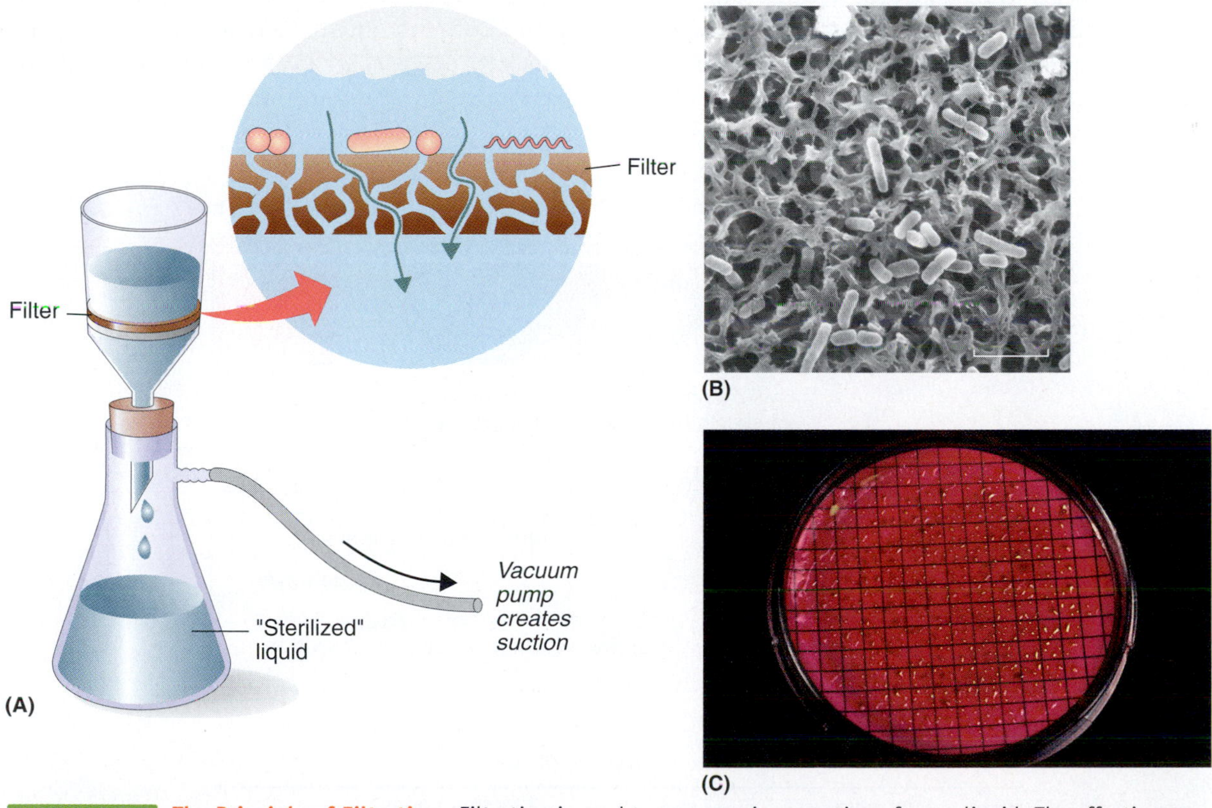

FIGURE 23.8 **The Principle of Filtration.** Filtration is used to remove microorganisms from a liquid. The effectiveness of the filter is proportional to the size of its pores. (**A**) Bacteria-laden liquid is poured into a filter, and a vacuum pump helps pull the liquid through and into the flask below. The bacterial cells are larger than the pores of the filter, and they become trapped. (**B**) A view of *Escherichia coli* cells trapped in the pores of a 0.45-μm nylon membrane filter. (Bar = 5 μm.) (**C**) *E. coli* colonies growing on a membrane filter.

Q: Why wouldn't most viruses be filtered out from a solution?

flask. Organic filters are advantageous because the organic molecules of the filter attract organic components in microorganisms. One example, the Berkefeld filter, uses a substance called diatomaceous earth (see Chapter 17). Their remains are gathered for use in swimming pool and aquarium filters as well as for microbiological filters used in laboratories.

The **membrane filter** consists of a pad of cellulose acetate or polycarbonate, mounted in a holding device. This filter is particularly valuable because bacterial cells multiply and form colonies on the filter pad when the pad is placed on a plate of culture medium. Microbiologists then can count the colonies to determine the number of bacteria originally present (FIGURE 23.8B, C).

Air also can be filtered to remove microorganisms. The filter generally used is a **high-efficiency particulate air (HEPA) filter**. This apparatus can remove over 99 percent of all particles, including microorganisms with a diameter larger than 0.3 μm. The air entering surgical units and specialized treatment facilities, such as burn units, is filtered to exclude microorganisms. In some hospital wards, such as for respiratory diseases and in certain pharmaceutical filling rooms, the air is recirculated through HEPA filters to ensure air purity.

CONCEPT AND REASONING CHECKS

23.4 Determine the uses for filtration in a health care setting.

Ultraviolet Light Can Be Used to Control Microbial Growth

KEY CONCEPT

- UV light can be bactericidal.

Visible light is a type of radiant energy detected by the light-sensitive cells of the eye. The wavelength of this energy is between 400 and 800 nanometers (nm). Other types of radiation have wavelengths longer or shorter than that of visible light, and therefore they cannot be detected by the human eye.

One type of radiant energy, **ultraviolet (UV) light**, is useful for controlling microorganisms. Ultraviolet light has a wavelength between 100 and 400 nm, with the energy at about 265 nm most destructive to bacterial cells (FIGURE 23.9). When microorganisms

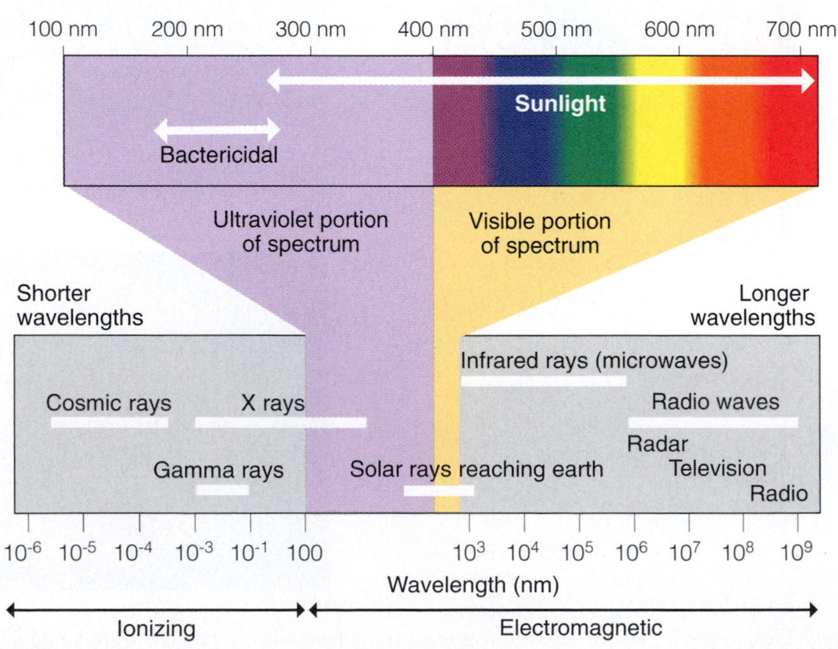

FIGURE 23.9 **The Ionizing and Electromagnetic Spectrum of Energies.** The complete spectrum is presented at the bottom of the chart, and the ultraviolet and visible sections are expanded at the top. Notice how the bactericidal energies overlap with the UV portion of sunlight. This may account for the destruction of microorganisms in the air and in upper layers of soil.

Q: How does UV light kill bacteria?

are subjected to UV light, cellular DNA absorbs the energy, and adjacent thymine molecules link together. Linked thymine molecules are unable to encode adenine on messenger RNA molecules during the process of protein synthesis (see Chapter 7). Moreover, replication of the chromosome prior to binary fission is impaired. The damaged organism can no longer produce critical proteins or reproduce, and it quickly dies.

Ultraviolet light effectively reduces the microbial population where direct exposure takes place. It is used to limit airborne or surface contamination in a hospital room, morgue, pharmacy, toilet facility, or food service operation. It is noteworthy that UV light from the sun may be an important factor in controlling microorganisms in the air and upper layers of the soil, but it may not be effective against all bacterial spores. Ultraviolet light does not penetrate liquids or solids, and it may cause damage in human skin cells.

CONCEPT AND REASONING CHECKS

23.5 Identify some uses for UV light as a physical control method.

Other Types of Radiation Also Can Sterilize Materials

KEY CONCEPT

- X rays and gamma rays also are microbicidal.

Looking back at Figure 23.9, there are two other forms of radiation useful for destroying microorganisms. These are **X rays** and **gamma rays**. Both have wavelengths shorter than the wavelength of UV light. As X rays and gamma rays pass through microbial molecules, they force electrons out of their shells, thereby creating ions. For this reason, the radiations are called **ionizing radiations**. The ions quickly combine mainly with cellular water and the free radicals generated affect cell metabolism and physiology. The radiation does not have a direct affect in nucleic acids such as DNA. Ionizing radiations currently are used to sterilize such heat-sensitive pharmaceuticals as vitamins, hormones, and antibiotics as well as certain plastics and suture materials.

Ionizing radiations also have been approved for controlling microorganisms, and for preserving foods, as noted in MicroFocus 23.3. The approval has generated much controversy, fueled by activists concerned with the safety of factory workers and consumers. First used in 1921 to inactivate *Trichinella spiralis,* the agent of trichinellosis (see Chapter 17), irradiation now is used as a preservative in more than 40 countries for over 100 food items, including potatoes, onions, cereals, flour, fresh fruit, and poultry (FIGURE 23.10A). The US Food and Drug Administration (FDA) approved cobalt-60 and cesium-137 irradiation to preserve poultry in the early 1990s, and in 1997, it extended the approval to preserve red meats such as beef, lamb, and pork. Irradiation is used to prepare many meals for the US military and the American astronauts (FIGURE 23.10B). What is called a **pasteurizing dose** is used on meats, poultry, and other foods. Such levels are not intended to eliminate all microbes in the food, but, like pasteurization of milk, to eliminate the pathogens. The foods are not necessarily sterile.

Another form of energy, the microwave, has a wavelength longer than that of ultraviolet light and visible light. In a microwave oven, microwaves are absorbed by water molecules, which are set into high-speed motion, and the heat of friction is transferred to foods. Other than the heat generated, there is no specific activity against microorganisms.

CONCEPT AND REASONING CHECKS

23.6 Identify some uses for X rays and gamma rays as a physical control method.

Preservation Methods Retard Spoilage by Microorganisms in Foods

KEY CONCEPT

- Dehydration and cold temperatures slow microbial growth.

Over the course of many centuries, various physical methods have evolved for controlling microorganisms in food. Though valuable for preventing the spread of infectious agents, these procedures are used mainly to retard spoilage and prolong the shelf life of foods, rather than for sterilization. Irradiation is an example of a preservation method.

Drying is useful in the preservation of various meats, fish, cereals, and other foods. Because water is necessary for life, it follows that where there is no water, there is virtually no life. Many of the foods in the kitchen pantry typify this principle.

MICROFOCUS 23.3: Public Health

"No, the Food Does Not Glow!"

In the United States, there are more than 76 million cases of foodborne disease accounting for more than 325,000 hospitalizations and 5,000 deaths each year. One major source of foodborne disease is agricultural produce contaminated with intestinal pathogens. Another source is from improperly cooked or handled meats or poultry harboring human intestinal pathogens, such as *Escherichia coli* O157:H7, *Campylobacter, Listeria,* and *Salmonella*.

Irradiated strawberries displaying the Radura symbol, meaning they have been treated by irradiation.

Irradiation has the potential to greatly limit such illnesses. The Centers for Disease Control and Prevention (CDC) have estimated that if just 50 percent of the meat and poultry sold in the United States was irradiated, there would be 900,000 fewer cases of foodborne illness and 350 fewer deaths each year.

Yet today manufacturers continue to wrestle with the concept of food irradiation as they constantly confront a leery public, some of who still have visions of Hiroshima and Nagasaki. In the United States, just 10 percent of the herbs and spices are irradiated and only 0.002 percent of fruits, vegetables, meats, and poultry are irradiated.

Food irradiation is entirely different from atomic radiation. The irradiation comes from gamma rays produced during the natural decay of cobalt-60 or cesium-137. Alternately, it may come from the same type of X rays used in X-ray machines, or it may come from an electron beam not unlike that used in an electron microscope. None of these types of radiations produce radioactivity—the irradiated food does not glow (see figure).

Low doses of irradiation are used for disinfestations and extending shelf life of packaged foods. As mentioned in the chapter narrative, a pasteurizing dose is used on meats, poultry, and other foods. Such levels do not eliminate all microbes in the food, but, similar to pasteurization, helps to reduce the dangers of pathogen-contaminated or cross-contaminated meats and poultry.

During the radiation, the gamma rays penetrate the food, and, just as in cooking, cause molecular changes in any contaminating microorganisms, which leads to their death.

Irradiation also has its limitations. The irradiation dose will not kill bacterial endospores, inactivate viruses, or neutralize toxins and prions. Therefore, irradiated food still must be treated in a sanitary fashion. Nutritional losses are similar to those occurring in cooking and/or freezing. Otherwise, there are virtually no changes in the food, and there is no residue.

The Department of Agriculture and the Food and Drug Administration currently are studying extending irradiation to other processed foods.

(A)

(B)

FIGURE 23.10 **Food Irradiation.** (**A**) The FDA has approved irradiation as a preservation method for numerous foods, including many fruits and vegetables, as well as poultry and red meats. (**B**) Many otherwise perishable foods eaten by NASA astronauts are prepared by irradiation.

Q: What is the irradiation killing in the food products?

Preservation by salting is based upon the principle of osmotic pressure. When food is salted (usually sodium chloride), water diffuses out of microorganisms toward the higher salt concentration and lower water concentration in the surrounding environment. This flow of water, called **osmosis**, leaves the microorganisms dehydrated and they die. The same phenomenon occurs in highly sugared foods (usually sucrose) such as syrups, jams, and jellies. However, fungal contamination (molds and yeasts) may remain at the surface because they can tolerate low water and high sugar concentrations.

Low temperatures found in the refrigerator and freezer retard spoilage by lowering the metabolic rate of microorganisms and thereby reducing their rate of growth (see Chapter 5). Spoilage is not totally eliminated in cold foods, however, and many microorganisms remain alive, even at freezer temperatures. These organisms multiply rapidly when food thaws, which is why prompt cooking is recommended.

Note in these examples that there are significant differences between killing microorganisms, holding them in check, and reducing their numbers. The preservation methods are described as bacteriostatic because they prevent the further multiplication of food-borne pathogens such as *Salmonella* and *Clostridium*. A more complete discussion of food preservation as it relates to public health is presented in Chapter 25.

TABLE 23.1 summarizes the physical agents used for controlling microorganisms.

Osmotic pressure: The pressure applied to a solution to stop the inward diffusion (osmosis) of a solvent through a semipermeable membrane.

CONCEPT AND REASONING CHECKS

23.7 Explain how salting foods acts as a preservation method.

TABLE 23.1 A Summary of Physical Agents Used to Control Microorganisms

Physical Method	Conditions	Instrument	Object of Treatment	Examples of Uses	Comment
Incineration	A few seconds	Flame	All microorganisms	Laboratory instruments	Object must be disposable or heat-resistant
Hot air	160°C for 2 hr	Oven	Bacterial spores	Glassware Powders Oily substances	Not useful for fluid materials
Boiling water	100°C for 10 min 100°C for 2 hr+	— —	Vegetative microorganisms Bacterial spores	Wide variety of objects	Total immersion and precleaning necessary
Pressurized steam	121°C for 15 min at 15 lb/in^2	Autoclave	Bacterial spores	Instruments Surgical materials Solutions and media	Broad application in microbiology
Fractional sterilization	30 min/day for 3 successive days	Arnold sterilizer	Bacterial spores	Materials not sterilized by other methods	Long process Sterilization not assured
Pasteurization	62.9°C for 30 min 71.6°C for 15 sec	Pasteurizer	Pathogenic microorganisms	Dairy products	Sterilization not achieved
Filtration	Entrapment in pores	Berkefeld filter Membrane filter	All microorganisms	Fluids	Many adaptations
Ultraviolet light	265 nm energy	Generator	All microorganisms	Surface and air sterilization	Not useful in fluids
X rays Gamma rays	Short-wave length energy	Generator	All microorganisms	Heat-sensitive materials	Extending food shelf life
Dehydration	Osmotic conditions	—	All microorganisms	Salted and sugared foods	Food preservation
Refrigeration/ Freezing	5°C / −10°C	Refrigerator/ Freezer	All microorganisms	Numerous foods	Spoilage/Food preservation

23.3 General Principles of Chemical Control

Tincture: A substance dissolved in alcohol.

The notions of sanitation and disinfection are not unique to the modern era. The Bible refers often to cleanliness and prescribes certain dietary laws to prevent consumption of what was believed to be contaminated food. Egyptians used resins and aromatics for embalming, and ancient peoples burned sulfur for deodorizing and sanitary purposes. Arabian physicians first suggested using mercury to treat syphilis. Over the centuries, spices were used as preservatives as well as masks for foul odors, making Marco Polo's trips to Asia for new spices a necessity as well as an adventure. And fans of Western movies probably have noted that American cowboys practiced disinfection by pouring whiskey onto gunshot wounds.

Medicinal Chemicals Came into Widespread Use in the 1800s

KEY CONCEPT

- Disinfectants and antiseptics are key to improved sanitation and public health.

As early as 1830, the United States Pharmacopoeia listed tincture of iodine as a valuable antiseptic, and soldiers in the Civil War used it in plentiful amounts. Joseph Lister established the principles of aseptic surgery using carbolic acid (phenol) for treating wounds (see Chapter 1).

As we have discussed, the physical agents for controlling microorganisms generally are intended to achieve sterilization. Chemical agents, by contrast, rarely achieve sterilization. Instead, they are expected only to destroy the pathogenic organisms on or in an object. The process of destroying pathogens is called **disinfection** and the object is said to be disinfected. If the object is lifeless, such as a tabletop, the chemical agent used is called a **disinfectant**. However, if the object is living, such as a tissue of the human body, the chemical agent used is an **antiseptic** (FIGURE 23.11). It is important to note that even though a particular chemical may be used as a disinfectant as well as an

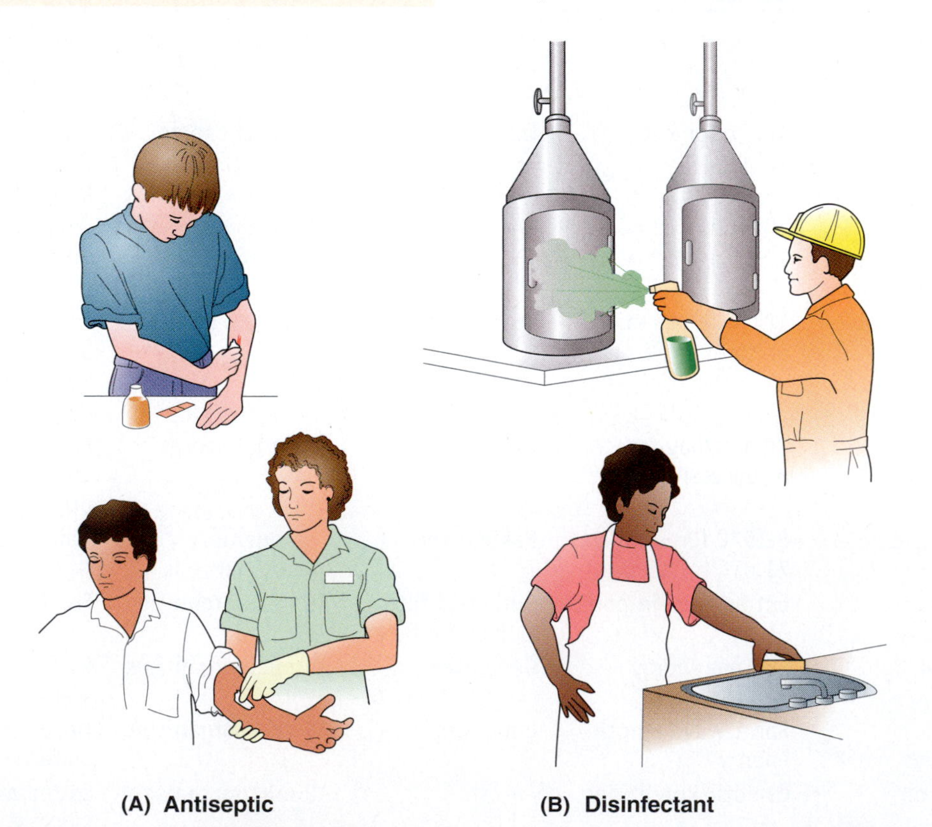

FIGURE 23.11 **Sample Uses of Antiseptics and Disinfectants.** (**A**) Antiseptics are used on body tissues, such as on a wound or before piercing the skin to take blood. (**B**) A disinfectant is used on inanimate objects, such as equipment used in an industrial process or tabletops and sinks.

Q: Why aren't disinfectants normally used as antiseptics?

antiseptic (e.g., iodine), the precise formulations are so different that its ability to kill or inactivate microorganisms differs substantially in the two products.

Antiseptics and disinfectants are usually bactericidal; they inactivate the major enzymes of an organism and interfere with its metabolism so that it dies. A chemical also may be bacteriostatic disrupting minor chemical reactions and slowing the metabolism, which results in a longer time between cell divisions. Although a delicate difference sometimes exists between the bactericidal and bacteriostatic agents, the terms indicate effectiveness in a particular situation.

The word **sepsis** (seps = "putrid") refers to the contamination of an object by microorganisms; thus, we have septicemia, meaning "microbial infection of the blood," and antiseptic, "against infection." It also is the origin of the term **asepsis**, meaning "free of contaminating microorganisms."

Other expressions are associated with chemical control. To **sanitize** an object is to reduce the microbial population to a safe level as determined by public health standards. For example, in dairy and food-processing plants, the equipment usually is sanitized through the process of sanitization. To **degerm** an object is merely to remove organisms from its surface. Washing with soap and water degerms the skin surface but has little effect on microorganisms deep in the skin pores.

CONCEPT AND REASONING CHECKS

23.8 Distinguish between (a) an antiseptic and a disinfectant and (b) disinfection and sanitization.

Antiseptics and Disinfectants Have Distinctive Properties

KEY CONCEPT

- Disinfectants and antiseptics are defined by their properties.

To be useful as an antiseptic or disinfectant, a chemical agent must have a number of properties. The agent should be:

- Able to kill or slow the growth of microorganisms.
- Nontoxic to animals or humans, especially if it is used as an antiseptic.
- Soluble in water and have a substantial shelf life during which its activity is retained.
- Useful in much diluted form and able to perform its job in a relatively short time. Both factors tend to minimize toxic side effects.

Other characteristics also will contribute to the value of a chemical agent: It should not separate on standing, it should penetrate well, and it should not corrode instruments. The chemical will have a distinct advantage if it does not combine with organic matter such as blood or feces, since organic matter can bind and "use up" the chemical. Of course, the chemical should be easy to obtain and relatively inexpensive.

Since disinfection is essentially a chemical process, several chemical parameters should be considered when selecting an antiseptic or disinfectant.

- **Temperature.** It important to know at what temperature the disinfection is to take place because a chemical reaction occurring at 37°C (body temperature) may not occur at 25°C (room temperature).
- **pH.** A particular chemical may be effective at a certain pH but not another.
- **Stability.** The chemical reaction may be very rapid with one agent and slower with another. Thus, if long-term disinfection is desired, the second agent may be preferable.

Two other considerations are the type of microorganism to be eliminated and the surface treated. For instance, the removal of bacterial spores requires more vigorous treatment than the removal of vegetative cells. Also, a chemical applied to a laboratory bench is considerably different from one used on a wound or for sterilizing an object.

It therefore is imperative to distinguish the antiseptic or disinfectant nature of a chemical before proceeding with its use. Indeed, chemical agents formulated as disinfectants are regulated and registered by the US Environmental Protection Agency (EPA), while chemicals formulated as antiseptics are regulated by the FDA.

CONCEPT AND REASONING CHECKS

23.9 Summarize the properties important in the selection of a disinfectant or antiseptic.

Antiseptics and Disinfectants Can Be Evaluated for Effectiveness

KEY CONCEPT

- Standards have been established to know the relative effectiveness of a chemical agent.

At the current time, there are more than 8,000 disinfectants and antiseptics for hospital use and thousands more for general use. Evaluating these chemical agents is a tedious process because of the broad diversity of conditions under which they are used.

A standard of effectiveness that has been used for the chemical agents is the **phenol coefficient** (**PC**). This is a number indicating the disinfecting ability of an antiseptic or disinfectant in comparison to phenol under identical conditions (TABLE 23.2). A PC higher than 1 indicates the chemical is more effective than phenol; a number less than 1 indicates poorer disinfecting ability than phenol. For example, antiseptic A may have a PC of 78.5, while antiseptic B has a PC of 0.28. These numbers are used relative to each other rather than to phenol because phenol is allergenic and irritating to tissues and thus is rarely used in a concentrated form.

The phenol coefficient is determined by a laboratory procedure in which dilutions of phenol and the test chemical are mixed with standardized bacterial species, such as *Staphylococcus aureus*, *Salmonella typhi*, or other species. The laboratory technician then determines which dilutions have killed the organisms after a 10-minute exposure but not after a 5-minute exposure. The test has many drawbacks, especially since it is performed in the laboratory rather than in a real-life situation. Nor does it take into account many of the factors cited above, such as tissue toxicity, activity in the presence of organic matter, or temperature variations.

A more practical way of determining the value of a chemical agent is by an **in-use test**. For example, swab samples from a floor are taken before and after the application of a disinfectant to determine the level of disinfection. Another method is to dry standardized cultures of a bacterial species on small stainless steel cylinders and then expose the cylinders to the test chemical. After an established period of time, the organism is tested for survival rates. These methods of standardization have value under certain circumstances. However, it is conceivable that a universal test never may be developed, in view of the huge variety of chemical agents available and the numerous conditions under which they are used.

CONCEPT AND REASONING CHECKS

23.10 Assess the need to know a chemical agent's effectiveness.

TABLE 23.2 Phenol Coefficients of Some Common Antiseptics and Disinfectants

Chemical Agent	*Staphylococcus aureus*	*Salmonella typhi*
Phenol	1.0	1.0
Chloramine	133.0	100.0
Tincture of iodine	6.3	5.8
Lysol	5.0	3.2
Mercury chloride	100.0	143.0
Ethyl alcohol	6.3	6.3
Formalin	0.3	0.7
Hydrogen peroxide	—	0.01

23.4 Chemical Methods of Control

The chemical agents currently in use for controlling microorganisms range from very simple substances, such as halogen ions, to very complex compounds, typified by detergents. Many of these agents in nature have been used for generations, while others represent the latest products of chemical companies. In this section, we shall survey several groups of chemical agents and indicate how they are best applied in the chemical control of microorganisms. MicroFocus 23.4 identifies some common but surprising antiseptics.

Halogens Oxidize Proteins

KEY CONCEPT

- Chlorine and iodine are good disinfectant agents.

The **halogens** are a group of highly reactive elements whose atoms have seven electrons in the outer shell (see Chapter 2). Two halogens, chlorine and iodine, are commonly used for disinfection. In microorganisms, halogens are believed to cause the release of atomic oxygen, which then combines with and inactivates certain cytoplasmic proteins, such as enzymes. Another theory suggests halogens change the structure of cell membranes, thus leading to cell leakage.

Chlorine is available in a gaseous form and as both organic and inorganic compounds (FIGURE 23.12). It is widely used in municipal water supplies, where it keeps bacterial populations at low levels. Chlorine combines readily with numerous ions in water; therefore, enough chlorine must be added to ensure a residue remains for antibacterial activity. In municipal water, the residue of chlorine is usually about 0.2 to 1.0 parts per million (ppm) of free chlorine. One ppm is equivalent to 0.0001 percent, an extremely small amount.

Chlorine also is available as sodium hypochlorite (NaOCl; Clorox®) or calcium hypochlorite ($Ca(OCl)_2$). The latter, also known as chlorinated lime, was used by Semmelweis in his studies (see Chapter 1). Hypochlorite compounds release free chlorine in solution and are used as a bleaching

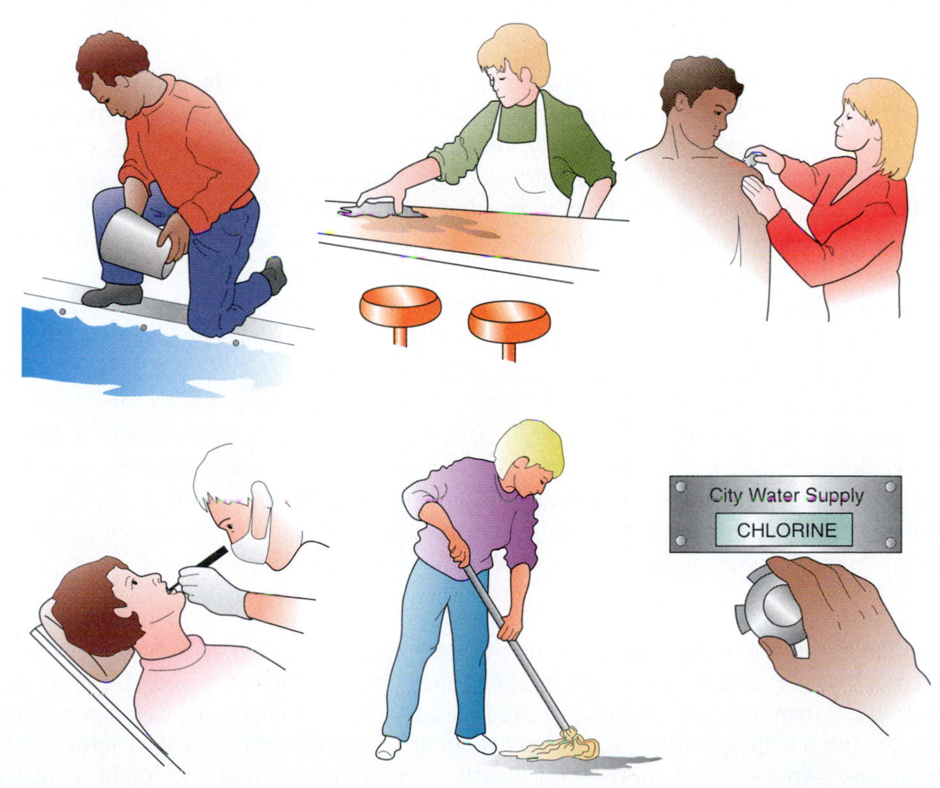

FIGURE 23.12 **Some Practical Applications of Disinfection with Chlorine Compounds.** Different chlorine compounds have been used as both disinfectants and antiseptics.

Q: How do the chlorine products differ as antiseptics and disinfectants?

MICROFOCUS 23.4: Being Skeptical

Antiseptics in Your Pantry?

Today, we live in an age when alternative and herbal medicine claims are always in the news, and these reports have generated a whole industry of health products that make often-unbelievable claims. With regard to "natural products," are there some that have genuine medicinal and antiseptic properties?

Cinnamon

Professor Daniel Y. C. Fung, Professor of Food Science and Food Microbiology at Kansas State University in Manhattan, Kansas, believes cinnamon might be an antiseptic that can control pathogens, at least in fruit beverages. Fung's group added cinnamon to commercially pasteurized apple juice. They then added typical foodborne pathogens (*Salmonella typhimurium, Yersinia enterocolitica,* and *Staphylococcus aureus*) and viruses. After one week of monitoring the juice at refrigerated and room temperatures, the investigators discovered the pathogens were killed more readily in the cinnamon blend than in the cinnamon-free juice. In addition, more bacterial organisms and viruses were killed in the juice at room temperature than when refrigerated.

Garlic

In 1858, Louis Pasteur examined the properties of garlic as an antiseptic. During World War II, when penicillin and sulfa drugs were in short supply, garlic was used as an antiseptic to disinfect open wounds and prevent gangrene. Since then, numerous scientific studies have tried to discover garlic's antiseptic powers.

Many research studies have identified a group of sulfur compounds as one key to garlic's antiseptic properties. When a raw garlic clove is crushed or chewed, the active antiseptic compound is produced. Studies using garlic, at least in the laboratory, suggest that this compound is responsible for combating the microbes causing the common cold, flu, sore throat, sinusitis, and bronchitis. The findings indicate that the compound blocks key enzymes that bacterial cells and viruses need to invade and damage host cells.

Honey

For the past two decades, Professor Peter Molan, associate professor of biochemistry and director of the Waikato Honey Research Unit at the University of Waikato, New Zealand, has been studying the medicinal properties of and uses for honey. Its acidity, between 3.2 and 4.5, is low enough to inhibit many pathogens. Its low water content (15 to 21 percent by weight) means that it osmotically ties up free water and "drains water" from wounds, helping to deprive pathogens of an ideal environment in which to grow. In addition, when honey encounters fluid from a wound, it slowly releases small quantities of hydrogen peroxide that are not damaging to skin tissues. It also speeds wound healing.

If that isn't enough, there also is evidence that honey protects against tooth decay. Professor Molan's group has shown that, in the lab, honey completely inhibits the growth of plaque-forming bacterial species, including *Streptococcus mitis, S. sobrinus,* and *Lactobacillus caseii.* Honey cut acid production to almost zero and stopped the bacteria from producing dextran, which is a component of dental plaque. Like its use for wound infections, hydrogen peroxide probably is, in part, responsible for the antimicrobial activity.

But beware! Not all honey is alike. The antibacterial properties of honey depend on the kind of nectar, or plant pollen, that bees use to make honey. At least manuka honey from New Zealand and honeydew from central Europe are thought to contain useful levels of antiseptic potency. Professor Molan is convinced that "honey belongs in the medicine cabinet as well as the pantry."

Wasabi

The green, pungent, Japanese horseradish called wasabi may be more than a spicy condiment for sushi. Professor Hedeki Masuda, director of the Material Research and Development Laboratories at Ogawa & Co. Ltd., in Tokyo, Japan, and his colleagues have found that natural chemicals in wasabi, called isothiocyanates, inhibit the growth of *Streptococcus mutans*—one of the bacterial species causing tooth decay. Researchers tested wasabi's tooth-decay fighting ability in test tubes and found the substance interferes with the way sugar affects teeth. At this point, these are only test-tube laboratory studies and the results will need to be proven in clinical trials.

So, are there products having genuine antimicrobial properties? It appears so—and there are more than can be described here.

agent in the textile industry. To disinfect clear water for drinking, the Centers for Disease Control and Prevention (CDC) recommends a half-teaspoon of household chlorine bleach in two gallons of water, with 30 minutes of contact time before consumption. Hypochlorites also are useful in very dilute solutions for disinfecting swimming pools and sanitizing factory equipment.

The chloramines, such as chloramine-T, are organic compounds containing chlorine and ammonia. These compounds release free chlorine more slowly than hypochlorite solutions and are more stable. They are valuable for general wound antisepsis and root canal therapy.

Chlorine is effective against a broad variety of organisms, including most gram-positive and gram-negative bacteria, and many viruses, fungi, and protozoa. However, it is not sporicidal.

The iodine atom is slightly larger than the chlorine atom and is more reactive and more germicidal. A tincture of iodine, a commonly used antiseptic for wounds, consists of 2 percent iodine. For the disinfection of clear water, the CDC recommends five drops of tincture of iodine in one quart of water, with 30 minutes of contact time before consumption. Iodine compounds in different forms are also valuable sanitizers for restaurant equipment and eating utensils.

Iodophors are iodine-detergent complexes that release iodine over a long period of time and have the added advantage of not staining tissues or fabrics. The detergent portion of the complex loosens the organisms from the surface and the halogen kills them. Some examples of iodophors are Wescodyne, used in preoperative skin preparations; Ioprep, for presurgical scrubbing; and Betadine, for local wounds. Iodophors also may be combined with nondetergent carrier molecules. The best known carrier is povidone, which stabilizes the iodine and releases it slowly. However, compounds like these are not self-sterilizing.

CONCEPT AND REASONING CHECKS

23.11 Compare the uses for chlorine and iodine chemical agents.

Phenol and Phenolic Compounds Denature Proteins

KEY CONCEPT

- Many phenolic derivatives are used as disinfectants or antiseptics.

Phenol (carbolic acid) and phenolic compounds have played a key role in disinfection practices since Joseph Lister used them in the 1860s. Phenol remains the standard against which other antiseptics and disinfectants are evaluated in the phenol coefficient test. It is active against gram-positive bacteria, but its activity is reduced in the presence of organic matter. Phenol and its derivatives act by denaturing proteins, especially in the cell membrane.

Phenol is expensive, has a pungent odor, and is caustic to the skin; therefore, the role of phenol as an antiseptic has diminished (FIGURE 23.13). However, phenol derivatives have greater germicidal activity and lower toxicity than the parent compound. Hexylresorcinol is used in some mouthwashes, topical antiseptics, and throat lozenges. It has the added advantage of reducing surface tension, thereby loosening bacterial cells from tissue and allowing greater penetration of the germicidal agent.

Combinations of two phenol molecules called **bisphenols** are prominent in modern disinfection and antisepsis. Orthophenylphenol, for example, is used in Lysol and Amphyl. Another bisphenol, hexachlorophene, was used extensively during the 1950s and 1960s in toothpastes, underarm deodorants, and bath soaps. One product, pHisoHex, combined hexachlorophene with a pH-balanced detergent cream. Pediatricians recommended it to retard staphylococcal infections of the scalp and umbilical stump, and for general cleansing of the newborn. However, a late 1960s study indicated that excessive amounts could be absorbed through the skin and cause neurological damage, so hexachlorophene was subsequently removed from over-the-counter products. The product pHisoHex is still available, but only by prescription.

An important bisphenol relative is chlorhexidine. This compound is used as a surgical scrub, hand wash, and superficial skin wound

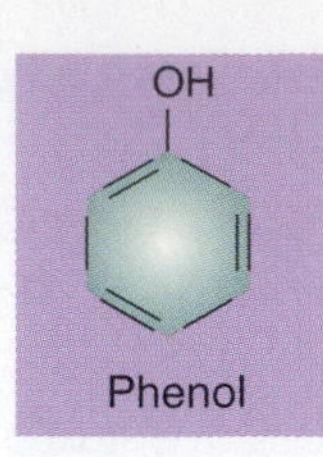

OH

Orthophenylphenol

Cl OH OH Cl
H — CH2 — H
Cl Cl Cl Cl
Hexachlorophene

OH Cl
Cl — O — Cl

Triclosan

FIGURE 23.13 Phenol and Some Derivatives. The chemical structure of phenol and some important derivatives.

Q: Why are most phenolic compounds only used as disinfectants?

cleanser. A four percent chlorhexidine solution in isopropyl alcohol is commercially available as Hibiclens. Another bisphenol in widespread use is **triclosan**, a broad-spectrum antimicrobial agent that destroys bacterial cells by disrupting cell membranes (and possibly, cell walls) by blocking the synthesis of lipids. Triclosan is fairly mild and nontoxic, and it is effective against pathogenic bacteria (but only partially against viruses and fungi). The chemical is included in antibacterial soaps, lotions, mouthwashes, toothpastes, toys, food trays, underwear, kitchen sponges, utensils, and cutting boards. A negative side to extensive triclosan use is the possibility of bacterial species developing resistance to the chemical, just as they have developed resistance to antibiotics.

CONCEPT AND REASONING CHECKS

23.12 Explain why bisphenols are preferred as disinfectants and antiseptics.

Heavy Metals Interfere with Microbial Metabolism

KEY CONCEPT

- Mercury, copper, and silver compounds can be useful disinfectants.

Mercury, silver, and copper are called **heavy metals** because of their large atomic weights and complex electron configurations. They are very reactive with proteins, particularly at the protein's sulfhydryl groups (–SH), and they are believed to bind protein molecules together by forming bridges between the groups. Because many of the proteins involved are enzymes, cellular metabolism is disrupted, and the microorganism dies. However, heavy metals are not sporicidal.

Mercury is very toxic to the host and the antimicrobial activity of mercury is reduced when other organic matter is present. In such products as merbromin (Mercurochrome) and thimersol (Merthiolate), mercury is combined with carrier compounds and is less toxic when applied to the skin, especially after surgical incisions. Thimerosol was previously used as a preservative in vaccines (see Chapter 21).

Copper is active against chlorophyll-containing organisms and is a potent inhibitor of algae. As copper sulfate ($CuSO_4$), it is incorporated into algicides and is used in swimming pools and municipal water supplies.

Silver in the form of silver nitrate ($AgNO_3$) is useful as an antiseptic and disinfectant. For example, one drop of a one percent silver nitrate solution used to be placed in the eyes of newborns to protect against infection by *Neisseria gonorrhoeae*. This gram-negative diplococcus can cause blindness if contracted by a newborn during passage through an infected mother's birth canal (see Chapter 12).

CONCEPT AND REASONING CHECKS

23.13 Evaluate the use of heavy metals as antiseptics and disinfectants.

Alcohols Denature Proteins and Disrupt Membranes

KEY CONCEPT

- Alcohols are effective skin antiseptics.

For practical use, the preferred alcohol is ethyl alcohol (ethanol), which is active against vegetative bacterial cells, including the tubercle bacillus, but it has no effect on spores. It denatures proteins and dissolves lipids, an action leading to cell membrane disintegration. Ethyl alcohol also is a strong dehydrating agent.

Because ethyl alcohol reacts readily with any organic matter, medical instruments and thermometers must be thoroughly cleaned before exposure. Usually, a 10-minute immersion in 50- to 80-percent alcohol solution is recommended to disinfect because water prevents rapid evaporation. Ethyl alcohol is used in many popular hand sanitizers.

Alcohol is used to treat skin before a venipuncture or injection. It mechanically removes bacterial cells from the skin and dissolves lipids. Isopropyl alcohol, or rubbing alcohol, has high bactericidal activity in concentrations as high as 99 percent.

■ Venipuncture: The piercing of a vein to take blood, to feed somebody intravenously, or to administer a drug.

CONCEPT AND REASONING CHECKS

23.14 Why is 70 percent ethanol preferable to 95 percent ethanol as an antiseptic?

Soaps and Detergents Act as Surface-Active Agents

KEY CONCEPT

- Cationic detergents are bacteriostatic.

Soaps are chemical compounds of fatty acids combined with potassium or sodium hydroxide. The pH of the compounds is usually about 8.0, and some microbial destruction is due to the alkaline conditions they establish on the skin. However, the major activity of soaps act as degerming agents for the mechanical removal of microorganisms from the skin surface.

Soaps therefore are surface-active agents called **surfactants**; that is, they emulsify and solubilize particles clinging to a surface and reduce the surface tension. Soaps also remove skin oils, further reducing the surface tension and increasing the cleaning action. MicroFocus 23.5 discusses the antibacterial soaps.

Detergents are synthetic chemicals acting as strong surfactants. Because they are actively attracted to the phosphate groups of cellular membranes, they also alter the membranes and encourage leakage from the cytoplasm. When used to clean cutting boards, they can reduce the possibility of transmitting contaminants.

The most useful detergents to control microorganisms are cationic (positively-charged) derivatives of ammonium chloride. In the detergents, four organic radicals replace the four hydrogens, and at least one radical is a long hydrocarbon chain (FIGURE 23.14). Such compounds often are called **quaternary ammonium compounds** or, simply, **quats**. They react with cell membranes and can destroy some bacterial species and enveloped viruses.

Quats have rather long, complex names, such as benzalkonium chloride in Zephiran and cetylpyridium chloride in Ceepryn. Quats are bacteriostatic, especially on gram-positive bacteria, and are relatively stable, with little odor. They are used as sanitizing agents for industrial equipment and food utensils; as skin antiseptics; as disinfectants in mouthwashes and in storage solutions for contact lenses; and as disinfectants for use on hospital walls and floors. Their use as disinfectants for food-preparation surfaces can help reduce contamination incidents. Mixing with soap, however, reduces their activity, and certain gram-negative bacteria, such as *Burkholderia (Pseudomonas) cepacia*, can grow in them.

CONCEPT AND REASONING CHECKS

23.15 How do soaps differ from quats as chemical agents of control?

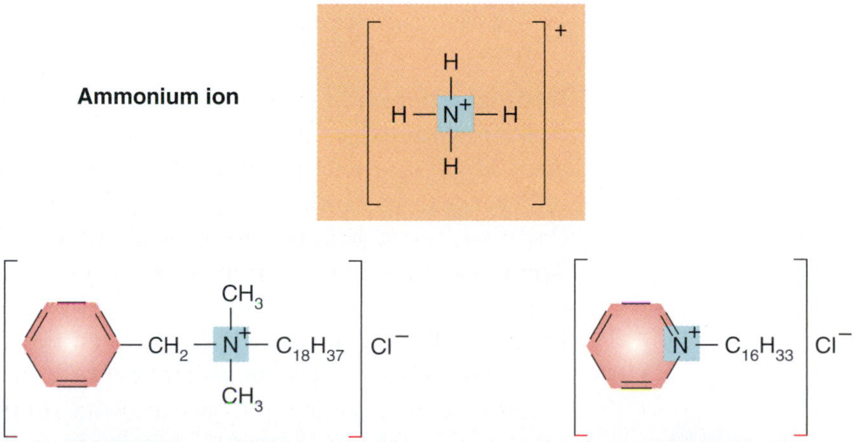

Ammonium ion

$[NH_4]^+$

Benzalkonium chloride (Zephiran)

$[C_6H_5–CH_2–N^+(CH_3)_2–C_{18}H_{37}]\ Cl^-$

Cetylpyridium chloride (Ceepryn)

$[C_5H_5N^+–C_{16}H_{33}]\ Cl^-$

FIGURE 23.14 **Cationic Detergents.** The chemical structures of some important quaternary ammonium compounds (quats) used in disinfection and antisepsis.

Q: How has the basic ammonium ion been modified to generate these two quats?

MICROFOCUS 23.5: Public Health/Being Skeptical

Are Hand Sanitizers Worth the Money?

All of us want to be as clean and bacterial safe as possible. In fact, hand washing is one of the best ways to protect oneself and prevent the spread of disease-causing microbes. To that end, numerous commercial companies have provided us with many different types of antimicrobial cleaning and hygiene products. Perhaps one of the most pervasive is the antibacterial soaps, which usually contain about 0.2 percent triclosan.

Washing hands with soap and water is a key to preventing disease transmission.

It is estimated that 75 percent of liquid and 30 percent of bar soaps on the market today are of the antibacterial type. The question though is: Are these products any better than regular soaps? The short answer is—no.

Numerous studies have shown these antibacterial soaps do little against foodborne pathogens such as *Salmonella* and *Escherichia coli*. In addition, they do nothing to reduced the chances of picking up and carrying infectious microbes.

A 2005 study gathered together over 200 families with children. Each family was given cleaning and hygiene supplies, such as soaps, detergents, and household cleaners, to use for one year. Half of the families (controls) received regular products without added antibacterial chemicals while the other half used products with the antibacterial chemicals.

When the families were surveyed after one year, those using the antibacterial products were just as likely to get sick as identified by symptoms such as coughs, fevers, sore throats, vomiting, and diarrhea.

You may say that this is not surprising, as many of these symptoms are the result of a viral infection—and the antibacterial products are not effective on viruses. However, further analysis of the families indicated there were just as many bacterial infections in the antibacterial group as there were in the control group.

Antibacterial soaps and hand sanitizers may be useful in a hospital environment, but they certainly are not worth the extra cost for home use.

Hydrogen Peroxide Damages Cellular Components

KEY CONCEPT

- Hydrogen peroxide can be used as an antiseptic rinse.

Hydrogen peroxide (H_2O_2) has been used as a rinse in wounds, scrapes, and abrasions. However, H_2O_2 applied to such areas foams and effervesces, as catalase in the tissue breaks down hydrogen peroxide to oxygen and water. Therefore, it is not recommended as an antiseptic for open wounds. However, the furious bubbling that produces oxygen gas is effective against anaerobic bacterial species because the sudden release of oxygen gas inhibits their growth. Hydrogen peroxide decomposition also results in a reactive form of oxygen—the superoxide radical—which is highly toxic to microorganisms.

New forms of H_2O_2 are more stable than traditional forms, do not decompose spontaneously, and therefore can be used topically. Such inanimate materials as soft contact lenses, utensils, heat-sensitive plastics, and food-processing equipment can be disinfected within 30 minutes. Sterilization also can be achieved after six hours of exposure to a 6 percent solution.

CONCEPT AND REASONING CHECKS

23.16 Judge the advantages and disadvantages of using hydrogen peroxide as an antiseptic.

Some Chemical Agents Can Be Used for Sterilization

KEY CONCEPT

- Aldehydes and gasses can be used for sterilization.

The chemical agents we discussed in previous sections are used as disinfectants and antiseptics. In addition, there are some chemicals that can be used for sterilization purposes, especially for modern high-technology equipment. Several such agents are considered.

Aldehydes. Aldehydes are agents that react with amino and hydroxyl groups of nucleic acids and proteins. The resulting cross linking inactivates the proteins and nucleic acids.

Formaldehyde is a gas at high temperatures and a solid at room temperatures. As a 37 percent solution it is called **formalin**. For over a century, formalin was used in embalming fluid for anatomical specimens (though rarely used anymore) and by morticians as well as for disinfecting purposes. In microbiology, formalin is used for inactivating viruses in certain vaccines and producing toxoids from toxins (see Chapter 21).

Instruments can be sterilized by placing them in a 20 percent solution of formaldehyde in 70 percent alcohol for 18 hours. Formaldehyde, however, leaves a residue, and instruments must be rinsed before use. Many allergic individuals develop a contact dermatitis to this compound (see Chapter 22).

Glutaraldehyde has become one of the most effective chemical liquids for sterilization purposes. This small molecule destroys vegetative cells within 10 to 30 minutes and spores in 10 hours. As a two percent solution, glutaraldehyde can be used for sterilization purposes. Materials have to be precleaned, then immersed for 10 hours, rinsed thoroughly with sterile water, dried in a special cabinet with sterile air, and stored in a sterile container to ensure that the material remains sterile. If any of these parameters is altered, the materials may be disinfected but may not be considered sterile.

Glutaraldehyde does not damage delicate objects, and therefore it can be used to sterilize optical equipment, such as the fiber-optic endoscopes used for arthroscopic surgery. It gives off irritating fumes, however, and instruments must be rinsed thoroughly in sterile water.

Ethylene Oxide and Chlorine Dioxide. The development of plastics for use in microbiology required a suitable method for sterilizing these heat-sensitive materials. In the 1950s, research scientists discovered the antimicrobial abilities of ethylene oxide, which essentially made the plastic Petri dish and plastic syringe possible.

Ethylene oxide is a small molecule with excellent penetration capacity and sporicidal ability. However, it is both carcinogenic and highly explosive. Its explosiveness is reduced by mixing it with Freon gas in Cryoxide or carbon dioxide gas in Carboxide, but its toxicity remains a problem for those who work with it. The gas is released into a tightly sealed chamber where it circulates for up to four hours with carefully controlled humidity. The chamber then must be flushed with inert gas for 8 to 12 hours to ensure that all traces of ethylene oxide are removed; otherwise the chemical will cause "cold burns" on contact with the skin.

Ethylene oxide is used to sterilize paper, leather, wood, metal, and rubber products as well as plastics. In hospitals, it is used to sterilize catheters, artificial heart valves, heart-lung machine components, and optical equipment. The National Aeronautics and Space Administration (NASA) uses the gas for sterilization of interplanetary space capsules. Ethylene oxide chambers, called *gas autoclaves*, have become chemical counterparts of autoclaves for sterilization procedures.

Chlorine dioxide has properties very similar to chloride gas and sodium hypochlorite but, unlike ethylene oxide, it produces nontoxic by-products and is not a carcinogen. Chlorine dioxide can be used as a gas or liquid. In a gaseous form, with proper containment and humidity, a 15-hour fumigation can be used to sanitize air ducts, food and meat processing plants, and hospital areas. It was the gas used to decontaminate the 2001 anthrax-contaminated mail and office buildings (MicroFocus 23.6).

TABLE 23.3 summarizes the chemical agents used in controlling microorganisms.

CONCEPT AND REASONING CHECKS

23.17 Summarize the uses for aldehydes, ethylene oxide, and chloride dioxide for sterilization.

MICROFOCUS 23.6: Tools/Public Health

Decontamination of Anthrax-Contaminated Mail and Office Buildings

This chapter has examined the chemical procedures and methods used to control the numbers of microorganisms on inanimate and living objects. These control measures usually involve a level of sanitation, although some procedures may sterilize. Some examples were given for their use in the home and workplace. However, what about real instances where large-scale and extensive decontamination has to be carried out?

(A)

(B)

In October 2001, the United States experienced a bioterrorist attack. The perpetrator(s) used anthrax spores as the bioterror agent. Four anthrax-contaminated letters were sent through the mail on the same day, addressed to NBC newscaster Tom Brokaw, the New York Post, and to two United States Senators, Senator Patrick Leahy and Senate Majority Leader Tom Daschle (see **Figure A**). The Centers for Disease Control and Prevention (CDC) confirmed that anthrax spores from at least the Daschle letter contaminated the Hart Senate Office Building and several post office sorting facilities in Trenton, New Jersey (from where the letters were mailed), and Washington, D.C., areas. This resulted in the closing of the Senate building and the postal sorting facilities. With the Senate building and postal sorting facilities closed, the CDC, the Environmental Protection Agency (EPA), other governmental agencies, and commercial companies had to devise and implement a strategy to decontaminate these buildings and the mail sorting machines (see **Figure B**).

As mentioned in this chapter, most sanitation procedures do not require a high technology solution. In fact, all of the methods actually used for this situation are described in this chapter.

The Hart Senate Office Building and the post office sorting facilities were contaminated with *Bacillus anthracis* endospores. These are large, multi-room facilities with many pieces of furniture and instruments, including computers, copy machines, and mail sorting machines. To decontaminate these buildings, chemical disinfectants such as bleach or phenol solutions could have been used. However, spores may have gotten into the office machinery and sorting machines. Liquids would not work here.

Therefore, a gas was needed that could permeate the air ducts as well as all the office machinery and sorting machines. The gas chosen was chlorine dioxide. Essentially, the buildings were sealed like they were going to be fumigated for termites. The gas was pumped in and after a time that was thought to have killed any anthrax spores, swabs were taken from the buildings and plated on nutrient media. If any spores were still alive, they would germinate on the plates and the vegetative cells would grow into visible colonies. Such results would require retreatment.

To protect the mail from future similar attacks, a system was devised using ultraviolet (UV) light to kill any spores that might be found in a piece of mail moving through the sorting machines.

It took months, and even years for some of the postal facilities, to be declared safe and free of anthrax spores. Still, simple physical and chemical methods worked to decontaminate the buildings and equipment.

TABLE

23.3 Summary of Chemical Agents Used to Control Microorganisms

Chemical Agent	Antiseptic or Disinfectant	Mechanism of Activity	Applications	Limitations	Antimicrobial Spectrum
Chlorine	Chlorine gas Sodium hypochlorite Chloramines	Protein oxidation Membrane leakage	Water treatment Skin antisepsis Equipment spraying Food processing	Inactivated by organic matter Objectionable taste, odor	Broad variety of bacteria, fungi, protozoa, viruses
Iodine	Tincture of iodine Iodophors	Halogenates tyrosine in proteins	Skin antisepsis Food processing Preoperative preparation	Inactivated by organic matter Objectionable taste, odor	Broad variety of bacteria, fungi, protozoa, viruses
Phenol and derivatives	Cresols Trichlosan Hexachlorophene Hexylresorcinol Chlorhexidine	Coagulates proteins Disrupts cell membranes	General preservatives Skin antisepsis with detergent	Toxic to tissues Disagreeable odor	Gram-positive bacteria Some fungi
Mercury	Mercuric chloride Merthiolate Merbromin	Combines with —SH groups in proteins	Skin antiseptics Disinfectants	Inactivated by organic matter Toxic to tissues Slow acting	Broad variety of bacteria, fungi, protozoa, viruses
Copper	Copper sulfate	Combines with proteins	Algicide in swimming pools Municipal water supplies	Inactivated by organic matter	Algae Some fungi
Silver	Silver nitrate	Binds proteins	Skin antiseptic Eyes of newborns	Skin irritation	Organisms in burned tissue Gonococci
Alcohol	70% ethyl alcohol	Denatures proteins Dissolves lipids Dehydrating agent	Instrument disinfectant Skin antiseptic	Precleaning necessary Skin irritation	Vegetative bacterial cells, fungi, protozoa, viruses
Cationic detergents	Commercial detergents	Dissolve lipids in cell membranes	Industrial sanitization Skin antiseptic Disinfectant	Neutralized by soap	Broad variety of microorganisms
Hydrogen peroxide	Hydrogen peroxide	Creates aerobic environment Oxidizes protein groups	Wound treatment	Limited use	Anaerobic bacteria
Formaldehyde	Formaldehyde gas Formalin	Reacts with functional groups in proteins and nucleic acids	Embalming Vaccine production Gaseous sterilant	Poor penetration Allergenic Toxic to tissues Neutralized by organic matter	Broad variety of bacteria, fungi, protozoa, viruses
Glutaraldehyde	Glutaraldehyde	Reacts with functional groups in proteins and nucleic acids	Sterilization of surgical supplies	Unstable Toxic to skin	All microorganisms, including spores
Ethylene oxide	Ethylene oxide gas	Reacts with functional groups in proteins and nucleic acids	Sterilization of instruments, equipment, heat-sensitive objects	Explosive Toxic to skin Requires constant humidity	All microorganisms, including spores
Chlorine dioxide	Chlorine dioxide gas	Reacts with functional groups in proteins and nucleic acids	Sanitizes equipment, rooms, buildings	Explosive	All microorganisms, including spores

SUMMARY OF KEY CONCEPTS

23.1 General Principles of Microbial Control

- The physical methods for controlling microorganisms are generally intended to achieve sterilization—the destruction or removal of all life-forms, including bacterial spores.
- Sanitization involves methods to reduce the numbers of or inhibit the growth of microbes.

23.2 Physical Methods of Control

- Heat is a common control method. Incineration using a direct flame achieves sterilization in a few seconds.
- Dry heat produces oxidation reactions in microbial proteins, while moist heat has better penetration properties.
- The exposure to boiling water at 100° C for two hours also can result in sterilization, but spore destruction cannot always be assured.
- The autoclave sterilizes materials in about 15 minutes, while a prevacuum sterilizer shortens this time using higher temperatures and pressures. Fractional sterilization is another method for sterilization through successive exposures to steam on three days.
- Pasteurization reduces the microbial population in a liquid and is not intended to be a sterilization method.
- Filtration uses various materials to trap microorganisms within or on a filter. Membrane filters are the most common. Air can be filtered using a high-efficiency particulate air (HEPA) filter.
- Ultraviolet (UV) light is an effective way of killing microorganisms on a dry surface and in the air. X rays and gamma rays are two forms of ionizing radiation used to sterilize heat-sensitive objects. Irradiation also is used in the food industry to control microorganisms on perishable foods.
- For food preservation, drying, salting, and low temperatures can be used to control microorganisms.

23.3 General Principles of Chemical Control

- Chemical agents are effectively used to control the growth of microorganisms, even though disinfection. A chemical agent used on a living object is an antiseptic; one used on a nonliving object is a disinfectant.
- Both antiseptics and disinfectants are selected according to certain criteria, including an ability to kill microorganisms or interfere with their metabolism.
- The phenol coefficient test can be used to evaluate antiseptics and disinfectants. Chemicals are contrasted based on their effectiveness compared to phenol. In-use tests are more practical for everyday applications of antiseptics and disinfectants.

23.4 Chemical Methods of Control

- Halogens (chlorine and iodine) are useful for water disinfection, wound antisepsis, and various forms of sanitation.
- Phenol derivatives, such as hexachlorophene, are valuable skin antiseptics and active ingredients in presurgical scrubs.
- Heavy metals (silver and copper) are useful as antiseptics and disinfectants, respectively.
- Alcohol (70 percent ethyl alcohol) is an effective skin antiseptic.
- Soaps and detergents are effective degerming agents. Quats are more effective as a disinfectant than as an antiseptic.
- Hydrogen peroxide acts by releasing oxygen to cause an effervescing cleansing action. It is better as a disinfectant than an antiseptic.
- Formaldehyde and glutaraldehyde are sterilants that cross-link amino and hydroxyl groups in proteins and nucleic acids to alter the biochemistry of microorganisms.
- Ethylene oxide gas is carcinogenic and highly explosive. Under controlled conditions, the gas is an effective sterilant for plasticware. Chlorine dioxide gas can be used to sanitize air ducts, food and meat processing plants, and hospital areas.

LEARNING OBJECTIVES

After understanding the textbook reading, you should be capable of writing a paragraph that includes the appropriate terms and pertinent information to answer the objective.

1. Distinguish between sterilization and sanitization.
2. Contrast thermal death time and thermal death point.
3. Distinguish between incineration and dry heat.
4. Assess the advantages and disadvantages of dry heat and moist heat for sterilization.
5. Explain how autoclaving sterilizes materials.
6. Describe the purpose for pasteurization.
7. Summarize the filtration methods used to sterilize a liquid and decontaminate air.
8. Summarize the uses for ultraviolet (UV) light, X rays, and gamma rays as physical control agents.
9. Explain how dehydration and cold temperatures preserve foods.
10. Compare and contrast an antiseptic and a disinfectant.
11. List the desirable and chemical properties of antiseptics and disinfectants.
12. Describe how the effectiveness of a chemical agent can be measured.
13. Evaluate the usefulness of (1) halogens and (2) phenolics as disinfectants.
14. Summarize the uses for heavy metals in the chemical control of microorganisms.
15. Justify why alcohol is not a method for skin sterilization.
16. Distinguish between a soap and a detergent.
17. Estimate the value of quats and hydrogen peroxide as bacteriostatic agents.
18. Identify the uses of aldehydes and other gasses as sterilants.

SELF-TEST

Answer each of the following questions by selecting the ***one*** answer that best fits the question or statement. Answers to even-numbered questions can be found in **Appendix C**.

1. All the following terms apply to microbial killing *except*:
A. sterilization.
B. microbicidal.
C. bactericidal.
D. fungistatic.
E. virucidal.

2. The thermal death time is
A. the temperature needed to kill a microorganism.
B. the time to kill a microbial population at a given temperature.
C. the time to kill a microbial population in boiling water.
D. the temperature to kill all pathogens.
E. the minimal temperature need to kill a microbial population.

3. At 160°C, it takes about _____ minutes to kill bacterial spores.
A. 10
B. 30
C. 60
D. 90
E. 120

4. An autoclave normally sterilizes material by heating the material to _____°C for _____ minutes at _____ psi.
A. 100; 10; 30
B. 121.5; 15; 15
C. 100; 15; 0
D. 110; 30; 5
E. 120; 30; 10

5. Heating milk to 62.9°C for 30 minutes represents pasteurization using the _____ method.
A. holding
B. short phase
C. flash
D. fractional
E. ultrapasteurization

6. Air filtration typically uses a _____ filter.
A. HEPA
B. membrane
C. Seitz
D. inorganic
E. diatomaceous earth

7. For bactericidal activity, _____ has/have the shortest wavelength.
A. sunlight
B. ultraviolet light
C. gamma rays
D. radio waves
E. microwaves

8. Preservation methods such as salting result in the _____ microbial cells.
A. loss of salt from
B. gain of water into
C. loss of water from
D. lysis of
E. Both **B** and **D** are correct.

9. Which one of the following (**A–D**) does *not* apply to antiseptics? If all apply, then select **E.**
A. They are used on living objects.
B. They usually are bactericidal.
C. They should be useful as dilute solutions.
D. They can sanitize objects.
E. All the above (**A–D**) are correct.

10. If a chemical has a phenol coefficient (PC) of 63, it means the chemical
A. is better than one with a PC of 22.
B. will kill 63 percent of bacteria.
C. kills microbes at 63°C.
D. must be used at a concentration of 63 percent.
E. will kill all bacteria in 63 minutes.

11. Which one of the following is *not* a halogen? If all are halogens, then select **E.**
A. Iodine
B. Mercury
C. Clorox bleach
D. Chlorine
E. All the above (**A–D**) are correct.

12. Phenolics include chemical agents
A. derived from carbolic acid.
B. that combine two phenol molecules.
C. used in skin cleansers.
D. such as triclosan.
E. All the above (**A–D**) are correct.

13. Heavy metals, such as _____ work by _____.
A. mercury; disrupting membranes
B. copper; producing toxins
C. iodine; denaturing proteins
D. silver; binding protein molecules together
E. chlorine; releasing oxygen gas

14. All the following statements apply to quats *except*:
A. they react with cell membranes.
B. they are positively-charged molecules.
C. they are types of soaps.
D. they are used as sanitizing agents.
E. they can be used as disinfectants.

15. Aldehydes are chemical agents that
A. are very explosive.
B. cannot be used for sterilization.
C. are used as skin rinses.
D. are nontoxic.
E. inactivate proteins and nucleic acids.

16. Ethylene oxide can be used to
A. kill bacterial spores.
B. sterilize plastics.
C. sterilize space capsules.
D. sterilize hospital machine components.
E. All the above (**A–D**) are correct.

QUESTIONS FOR THOUGHT AND DISCUSSION

Answers to even-numbered questions can be found in **Appendix C**.

1. Instead of saying that food has been irradiated, manufacturers indicate that it has been "cold pasteurized." Why do you believe they must use this deception?
2. The label on the container of a product in the dairy case proudly proclaims, "This dairy product is sterilized for your protection." However, a statement in small letters below reads: "Use within 30 days of purchase." Should this statement arouse your suspicion about the sterility of the product? Why?
3. In view of all the sterilization methods we have discussed in this chapter, why do you think none has been widely adapted to the sterilization of milk?
4. A liquid that has been sterilized may be considered pasteurized, but one that has been pasteurized may not be considered sterilized. Why not?
5. Before taking a blood sample from the finger, the blood bank technician commonly rubs the skin with a pad soaked in alcohol. Many people think that this procedure sterilizes the skin. Are they correct? Why?
6. Suppose you had just removed the thermometer from the mouth of your sick child and confirmed your suspicion of fever. Before checking the temperature of the next child, how would you treat the thermometer to disinfect it?
7. The water in your home aquarium always seems to resemble pea soup, but your friend's is crystal clear. Not wanting to appear stupid, you avoid asking him his secret. But one day, in a moment of desperation, you break down and ask, whereupon he knowledgeably points to a few pennies among the gravel. What is the secret of the pennies?

APPLICATIONS

Answers to even-numbered questions can be found in **Appendix C**.

1. When the local drinking water is believed to be contaminated, area residents are advised to boil their water before drinking. Often, however, they are not told how long to boil it. As a student of microbiology, what is your recommendation?
2. You need to sterilize a liquid. What methods could you devise using only the materials found in the average household?
3. Suppose you were in charge of a clinical laboratory where instruments are routinely disinfected and equipment is sanitized. A salesperson from a disinfectant company stops in to spur your interest in a new chemical agent. What questions might you ask the salesperson about the product?
4. A portable room humidifier can incubate and disseminate infectious microorganisms. If a friend asked for your recommendations on disinfecting the humidifier, what do you suggest?
5. A student has finished his work in the laboratory and is preparing to leave. He remembers the instructor's precautions to wash his hands and disinfect the lab bench before leaving. However, he cannot remember whether to wash first then disinfect, or to disinfect then wash. What advice would you give?

HTTP://MICROBIOLOGY.JBPUB.COM/

The site features learning, an on-line review area that provides quizzes and other tools to help you study for your class. You can also follow useful links for in-depth information, read more MicroFocus stories, or just find out the latest microbiology news.

REVIEW—PHYSICAL METHODS

Use the following syllables to form the term that answers the clue pertaining to physical methods of control. The number of letters in the term is indicated by the dashes, and the number of syllables in the term is shown by the number in parentheses. Each syllable is used only once. The answers to even-numbered terms are listed in **Appendix C.**

A AU BA BER BRANE BUN CIL CLAVE CU DE DER DRY GAM HOLD ING ING LET LO LUS MA MEM MO NA O OS PLAS POW PRES SEN SIS SIS SPORE SURE TIC TION TO TRA TU TUR UL VI

1. Instrument for sterilization. (3) _ _ _ _ _ _ _ _ _
2. Type of filter. (2) _ _ _ _ _ _ _ _
3. Sterilized in an oven. (2) _ _ _ _ _ _
4. Occurs with moist heat. (5) _ _ _ _ _ _ _ _ _ _ _ _
5. Preserves meat, fish. (2) _ _ _ _ _ _
6. Short-wavelength rays. (2) _ _ _ _ _
7. Most resistant life-form. (1) _ _ _ _ _
8. Raised in the autoclave. (2) _ _ _ _ _ _ _ _
9. Method of pasteurization. (2) _ _ _ _ _ _ _
10. Light for air sterilization. (5) _ _ _ _ _ _ _ _ _ _ _
11. Melts in the autoclave. (2) _ _ _ _ _ _ _
12. Water flow from salting. (3) _ _ _ _ _ _ _
13. Genus of spore formers. (3) _ _ _ _ _ _ _ _
14. Direct flame burner. (2) _ _ _ _ _ _
15. Disease prevented by pasteurization. (5) _ _ _ _ _ _ _ _ _ _ _ _

REVIEW—CHEMICAL METHODS

Chemical agents are a broad and diverse group, as this chapter has demonstrated. To test your knowledge over the chemical methods of control, match the chemical agent on the right to the statement on the left by placing the correct letter in the available space. A letter may be used once, more than once, or not at all. The answers to even-numbered statements are listed in **Appendix C.**

Statement

____ 1. The halogen in bleach.
____ 2. Sterilizes heat-sensitive materials.
____ 3. Used to prevent gonococcal eye disease.
____ 4. Part of chloramine molecule.
____ 5. Oxygen retards anaerobic bacteria.
____ 6. Seventy percent concentration recommended.
____ 7. Active ingredient in Betadine.
____ 8. Quaternary compounds, or quats.
____ 9. Can induce a contact dermatitis.
____ 10. Often used as a tincture.
____ 11. Rinse for wounds and scrapes.
____ 12. Example of a heavy metal.
____ 13. Two molecules in hexachlorophene.
____ 14. Aids mechanical removal of organisms.
____ 15. Triclosan is a derivative.
____ 16. Used by Joseph Lister.
____ 17. Used for plastic Petri dishes.
____ 18. Found in Zephiran.
____ 19. Derivatives of ammonium compounds.
____ 20. Broken down by catalase.

Chemical agent

A. Cationic detergent
B. Chlorine
C. Ethyl alcohol
D. Ethylene oxide
E. Formaldehyde
F. Glutaraldehyde
G. Hydrogen peroxide
H. Iodine
I. Phenol
J. Silver
K. Soap

24

Chapter Preview and Key Concepts

24.1 The History and Properties of Antimicrobial Agents
- Ehrlich believed in a selective toxicity for specific chemical agents against microbes.
- Penicillin was the first clinically effective antibiotic.
- Antimicrobial agents vary in their form and activity.

24.2 The Synthetic Antibacterial Agents
- Sulfonamides affect DNA synthesis.
- Cell walls or DNA are the targets for other synthetic drugs.

24.3 The Beta-Lactam Family of Antibiotics
- Penicillin is a bactericidal agent effective against gram-positive cells.
- Cephalosporins and carbapenems also contain a beta-lactam ring and inhibit peptidoglycan cross-linking.

24.4 Other Bacterially Produced Antibiotics
- Vancomycin is a glycopeptide antibiotic and cell wall assembly inhibitor.
- Antibiotics targeting the cell membrane either interfere with cell wall synthesis or disrupt membrane structure.
- Many antibiotics target the ribosome's ability to translate a mRNA.
- Rifampin inhibits transcription by inhibiting RNA synthesis.

24.5 Antifungal and Antiparasitic Agents
- Selective toxicity is important when treating fungal infections and diseases.
- Antiprotozoal agents also target unique aspects of these parasites.
- Antihelminthic drugs affect helminth neuromuscular coordination, carbohydrate metabolism, or microtubule structure.

24.6 Antibiotic Assays and Resistance
- Antibiotic susceptibility assays are used to study the inhibition of a test organism by one or more antimicrobial agents.
- Bacterial species have several ways to generate resistance to antibacterial agents.
- Antibiotic misuse and abuse encourage the emergence of resistant forms.
- The discovery of new targets and the development of new antibiotics may help fight antibiotic resistance.

MicroInquiry 24: Testing Drugs—Clinical Trials

Antimicrobial Drugs

We are facing a crisis because doctors are pressured to prescribe antibiotics for the common cold and inner ear infection, yet we know that it is not prudent to do so. We must collectively inform our patients about the reasons why overprescribing antibiotics will not help patients return to work sooner, and that in the long run, could make them more susceptible to drug-resistant diseases.

—Richard Besser, M.D., Centers for Disease Control and Prevention

For centuries, physicians believed heroic measures were necessary to save patients from the ravages of infectious disease. They prescribed frightening courses of purges (bowel emptying), enormous doses of strange chemical concoctions, blood-curdling ice water baths, deadly starvations, and bloodlettings. These treatments probably complicated an already bad situation by reducing the natural body defenses to the point of exhaustion. In fact, the death of George Washington in 1799 is believed to have been due to a streptococcal infection of the throat, perhaps exacerbated by the bloodletting treatment that removed almost 2 liters of his blood within a 24 hour period.

When the germ theory of disease emerged in the late 1800s, insights about microorganisms added considerably to the understanding of disease and increased the storehouse of knowledge available to the doctor. However, it did not change the fact that little, if anything, could be done for the infected patient. Then, in the 1940s, antimicrobial agents, including antibiotics, burst on the scene, and another revolution in medicine began.

Doctors were astonished to learn they could kill microorganisms in the body without doing substantial harm to the body itself. Medicine had a period of powerful, decisive growth, as doctors found they could successfully alter the course of infectious disease. Antibiotics effected a radical change in medicine and charted a new course for treating infectious disease. Since the 1940s, millions of lives have been saved. In 1969, then US Surgeon General William Stewart and many pharmaceutical companies believed it was time to "close the books on infectious diseases."

Unfortunately, this time of euphoria in treating infectious disease was short lived. It soon became clear bacterial species and other microorganisms could quite quickly develop resistance to first line antimicrobials. Today, antibiotic resistance is a major concern in the fight against infectious disease. In fact, in the United States and throughout the world there is an increasing number of antibiotic resistant organisms and a declining arsenal of antimicrobial drugs to fight infectious diseases. This is no more critical than in hospital intensive care units (FIGURE 24.1A).

There have been several reasons for this increase in antibiotic resistance. Widespread use of antibiotics promotes the spread of antibiotic resistance. The general public often misuses or abuses antibiotics. Individuals, for example, might stop taking an antibiotic as soon as they start feeling better; they do not finish the course of antibiotic treatment; they take an antibiotic for a viral infection like a cold or the flu; they take an antibiotic that was prescribed for some other illness; or they take an antibiotic that was prescribed for someone else. Antibiotic resistance allows bacterial species to survive and continue to multiply causing more harm.

In addition, many doctors still prescribe antibiotics for diseases that are untreatable with such drugs. This includes viral infections such as colds and the flu. Often this is at the insistence of the patient, as the opening quote for this chapter states.

Smart use of antibiotics is the key to controlling the spread of resistance. The Centers for Disease Control and Prevention (CDC) has started a campaign to prevent antimicrobial resistance. The campaign focuses on four main strategies: preventing infections, diagnosing and correctly treating infections, preventing transmission, and using antimicrobials wisely (FIGURE 24.1B).

In this chapter, we discuss the antimicrobial drugs as the mainstays of our health-care delivery system to treat bacterial, fungal, and parasitic infections and diseases. Some drugs have been known for generations, but most are of recent development. We explore their discovery and examine their uses (and abuses). Antiviral drugs were addressed in Chapter 13, so they will not be elaborated on here.

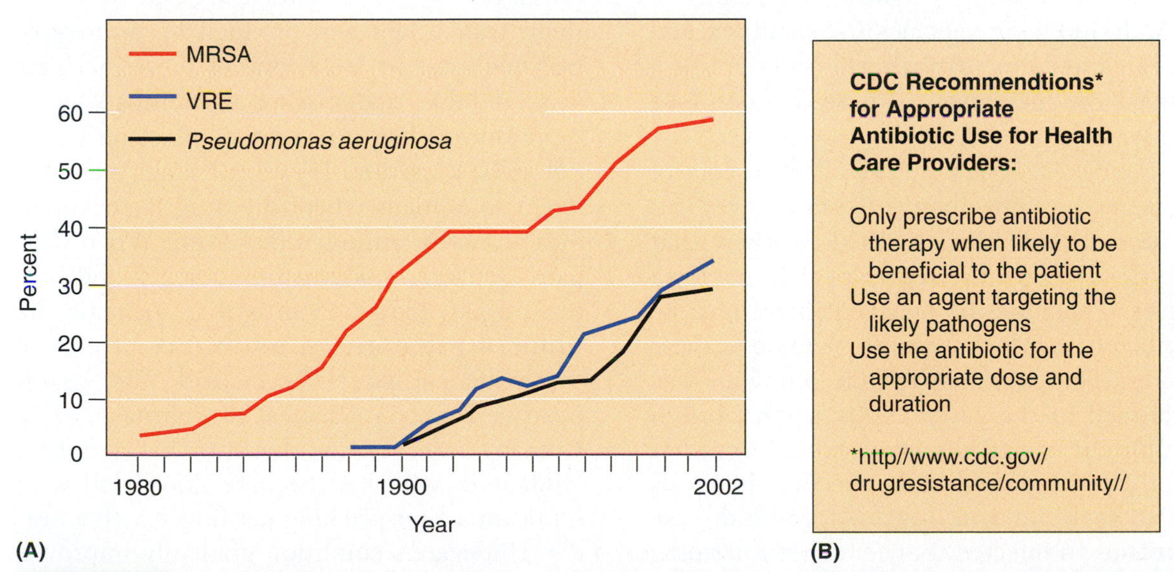

FIGURE 24.1 **Increase in Antibiotic Resistance, 1980–2002.** (**A**) This graph shows the increase in antibiotic strains of methicillin-resistant *Staphylococcus aureus* (MRSA), vancomycin-resistant *Enterococcus* (VRE), and antibiotic resistant *Pseudomonas aeruginosa* among ICU patients between 1980 and 2002. (**B**) The CDC recommendations to curb the increase in antibiotic resistant strains of microbes.
Source: CDC.

Q: Why is there an increase in the percent of resistance by these and other bacteria?

24.1 The History and Properties of Antimicrobial Agents

In the previous chapter, we discussed chemical methods to control microbes and viruses on surfaces or objects. Because these chemicals are too toxic to be taken internally to treat infections, another set of chemical agents is needed. Those chemical agents used to treat infections, diseases, and other disorders, such as cancer, are called **chemotherapeutic agents**. These drugs are used within the body for therapeutic purposes. In microbiology, we are concerned with the **antimicrobial agents**, those chemotherapeutic agents used to treat infectious disease.

Our discussion will begin with a brief review of the discovery and development of chemotherapy, as applied to infectious disease.

■ Chemotherapy: The use of chemical agents to treat diseases or disorders.

The History of Chemotherapy Originated with Paul Ehrlich

KEY CONCEPT

- Ehrlich believed in a selective toxicity for specific chemical agents against microbes.

In the drive to control and cure infectious disease, the efforts of microbiologists in the early 1900s were primarily directed toward enhancing the body's natural defenses. Sera containing antibodies lessened the impact of diphtheria, typhoid fever, and tetanus; and effective antibody-inducing vaccines for smallpox and rabies (and later, diphtheria and tetanus) reduced the incidence of these diseases (see Chapter 21).

Among the leaders in the effort to control disease was an imaginative German investigator named Paul Ehrlich. Ehrlich knew that specific dyes would stain specific bacteria species. Therefore, he believed there must be specific chemicals that would be toxic to these species. This selective toxicity concept was developed in the early 1900s when Ehrlich thought he could discover molecules that would be "magic bullets"—specific chemicals able to seek out and destroy specific disease organisms in infected tissues without harming those tissues.

Ehrlich and his staff had synthesized hundreds of arsenic-phenol compounds, any one of which might be such a magic bullet. One of Ehrlich's collaborators, the Japanese investigator Sahachiro Hata, set out to test the chemicals for their ability to destroy the syphilis spirochete *Treponema pallidum*. After months of painstaking study, Hata's attention focused on arsphenamine, compound #606 in the series. Hata and Ehrlich successfully tested arsphenamine against *T. pallidum* in animals and human subjects, and in 1910, they gave a derivative of the drug to doctors for use against syphilis. Arsphenamine, the first modern synthetic antimicrobial agent, was given the brand name **Salvarsan** because it offered salvation from syphilis and contained arsenic.

Salvarsan met with mixed success during the ensuing years. Its value against syphilis was without question, but local reactions at the injection site, and indiscriminate use by some physicians, brought adverse publicity. Ehrlich's death in 1915, together with the general ignorance of organic chemistry and the impending World War I, further eroded enthusiasm for chemotherapy. However, the team approach to drug discovery used by Ehrlich would become the model for modern pharmaceutical research.

Over the next 20 years, German chemists continued to synthesize and manufacture dyes for fabrics and other industries, and they routinely tested their new products for antimicrobial qualities. In 1932, one of these products was a red dye, trademarked as **Prontosil**.

Prontosil had no apparent effect on bacterial cells in culture. However, things were different in animals where the drug is converted into an active antimicrobial form. When Gerhard Domagk, a German pathologist and bacteriologist, tested Prontosil in animals, he found a pronounced inhibitory effect on staphylococci, streptococci, and other gram-positive bacterial species. In February 1935, Domagk injected the dye into his daughter Hildegard, who had become gravely ill with septicemia after pricking her finger with a needle. Hildegard's condition gradually improved and her arm did not have to be amputated. Many historians see her recovery as setting into motion the age of modern chemotherapy. For his discovery, Domagk was awarded the 1939 Nobel Prize in Physiology or Medicine.

CONCEPT AND REASONING CHECKS

24.1 Evaluate the early discoveries of and uses for chemotherapeutic agents for selective toxicity.

Alexander Fleming's Serendipitous Discovery of Penicillin Ushered in the Era of Antibiotics

KEY CONCEPT

- Penicillin was the first clinically effective antibiotic.

One of the first to postulate the existence and value of antibiotics was the British microbiologist Alexander Fleming (FIGURE 24.2A). During his early years, Fleming experienced the excitement of the classical Golden Age of microbiology and spoke up for the therapeutic value of Salvarsan. In a series of experiments in 1921, he described lysozyme, the nonspecific enzyme that breaks down cell walls in gram-positive bacteria (see Chapter 19).

The discovery of antibiotics is an elegant expression of Pasteur's dictum, "Chance favors the prepared mind." In 1928, Fleming was performing research on staphylococci at St. Mary's Hospital in London. Before going on vacation, he inoculated staphylococci onto plates of nutrient agar, and on his return, he noted that one plate was contaminated by a green mold (FIGURE 24.2B). His interest was piqued by the failure of staphylococci to grow near the mold. Fleming isolated the mold, identified it as a species of *Penicillium*, and found it produced a substance that kills gram-positive organisms. Though he failed to isolate the elusive substance, he named it **penicillin**.

Fleming was not the first to note the antibacterial qualities of *Penicillium* species. Joseph Lister had observed a similar phenomenon in 1871; John Tyndall did likewise in 1876; and a French medical student, Ernest Duchesne, wrote a research paper on the subject in 1896. However, it was Fleming who first proposed that penicillin could be used to eliminate gram-positive bacteria from mixed cultures. Further, he unsuccessfully tried to use a filtered broth from the fungus on infected wound tissue. At the time, vaccines and sera were viewed as essential to disease therapy, so Fleming's request for financial support for an antimicrobial went unheeded. Moreover, biochemistry was not sufficiently advanced to make complex separations possible and Fleming's discovery soon was forgotten.

(A)

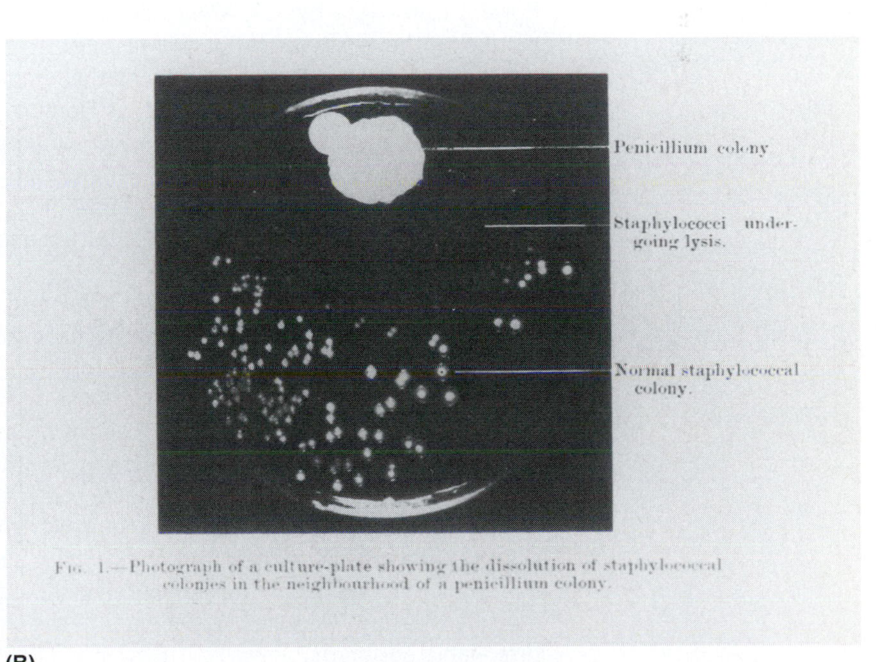

(B)

FIGURE 24.2 Alexander Fleming and His Culture of *Penicillium*. (**A**) Alexander Fleming reported the existence of penicillin in 1928 but was unable to purify it for use as an antimicrobial agent. (**B**) The actual photograph of Fleming's culture plate shows how staphylococci in the region of the *Penicillium* colony have been killed (they are "undergoing lysis") by some unknown substance produced by the mold. Fleming called the substance penicillin.

Q: How can you tell the bacteria have lysed near the Penicillium *colony?*

In 1935, Gerhard Domagk's dramatic announcement of the antimicrobial effects of Prontosil fueled speculation that chemicals could be used to fight disease in the body. Then, in 1939, a group at England's Oxford University, led by pathologist Howard Florey and biochemist Ernst Chain, reisolated Fleming's penicillin and conducted trials with highly purified samples. However, England was involved in World War II, so a group of American companies developed the techniques for the large-scale production of penicillin and made the drug available for commercial use (MicroFocus 24.1).

CONCEPT AND REASONING CHECKS

24.2 Identify how penicillin was discovered and later developed into a useful drug.

Antimicrobial Agents Have a Number of Important Properties

KEY CONCEPT

- Antimicrobial agents vary in their form and activity.

The preceding historical accounts illustrate the origins for the two groups of antimicrobial agents. Those drugs, like Salvarsan and Prontosil, which are made (synthesized) in the pharmaceutical laboratory, are called **synthetic agents**. By contrast, drugs like penicillin, which are products of or derived from the metabolism of living microorganisms are called **antibiotics** (*anti* = "against;" *biosis* = "life"); that is, the agents work to kill or inhibit living organisms.

Today, many antibiotics are produced by a process that is partly of laboratory origin and

MICROFOCUS 24.1: History

"The Mold in Dr. Florey's Coat"

Though Fleming initially reported the discovery of penicillin in 1928, it wasn't until 1940 that Oxford University scientists, led by Howard Florey (see figure), worked on reisolating, purifying, and producing the antibiotic for human testing.

However, the work was extremely hard and not much money was available. It was wartime in England, so Florey and his colleague Ernst Chain never had the funds necessary to properly carry out their work—they received a £25 ($100) grant for materials from the Medical Research Council. Funds from America and the Rockefeller Foundation were more generous, but ingenuity and improvisation remained essential.

Thus it was left to Norman Heatley, a chemist on the team, to design a mold juice extraction apparatus made from using glass tubing, assorted pumps, copper coils, colored warning lights, and even an old doorbell.

Howard Florey (right).

The meager amounts of penicillin the team was able to produce showed therapeutic potential, but larger quantities were needed to run the necessary clinical trials. Unable to interest British pharmaceutical companies, they turned to the United States.

Besides the problem of funding in England, Florey's team worried about the threat of a Nazi invasion and soldiers destroying their work. So, Heatley suggested they should preserve the mold spores by rubbing some into the fabric of their clothing should they be caught by the Germans. Fortunately, such extreme measures were not needed and, in the United States, penicillin was developed as the world's first viable antibiotic.

Although Fleming, Florey, and Chain shared a Nobel Prize in 1945 for their revolutionary work, accolades and media attention were disproportionately bestowed on Fleming, and in the popular imagination he was transformed into the sole creator of penicillin. Nobody in the group profited from the Prize.

If you want to read the whole story, pick up a copy of *The Mold in Dr. Florey's Coat* (2004) by Eric Lax.

partly of microbial origin. Such **semisynthetic drugs**, for simplicity, will be included with the discussions of the true antibiotics.

The synthetic and antibiotic drugs have certain important properties that need to be considered when prescribing a drug for an infection or disease.

Selective Toxicity. Ehrlich's idea of a magic bullet was based on **selective toxicity**, which says that a chemotherapeutic agent should harm the infectious agent but not the host. Today, two terms are used when considering the toxicity of a drug. The **toxic dose** refers to the concentration of the drug causing harm to the host. The **therapeutic dose** refers to the concentration of the drug that effectively destroys or eliminates the pathogen from the host. Together these can be used to formulate the **chemotherapeutic index**, which is the highest level (per kilogram of body weight) of the drug tolerated by the host divided by the lowest level (per kilogram body weight) of the drug that will eliminate the infection or disease agent.

The chemotherapeutic index of each antimicrobial agent must be considered, and the efficacy in eliminating disease and providing symptom relief must outweigh the associated toxicity and adverse events (FIGURE 24.3).

As we will see below, the best way to accomplish this is to develop drugs that target a specific component of microbial cells, such as the cell wall or a certain metabolic pathway, which is absent in human cells.

Antimicrobial Spectrum. Another important property in prescribing an appropriate drug is identifying the range of pathogens to which a particular drug will work. This range of antimicrobial action is the **antimicrobial spectrum**. Those drugs affecting many taxonomic groups are considered as having a **broad spectrum** of action, while those affecting few pathogens have a **narrow spectrum** of action (FIGURE 24.4). We will identify the drug spectrum with each antimicrobial drug discussed.

CONCEPT AND REASONING CHECKS

24.3 Assess the importance of the chemotherapeutic index and the antimicrobial drug spectrum in prescribing treatment for an infectious disease.

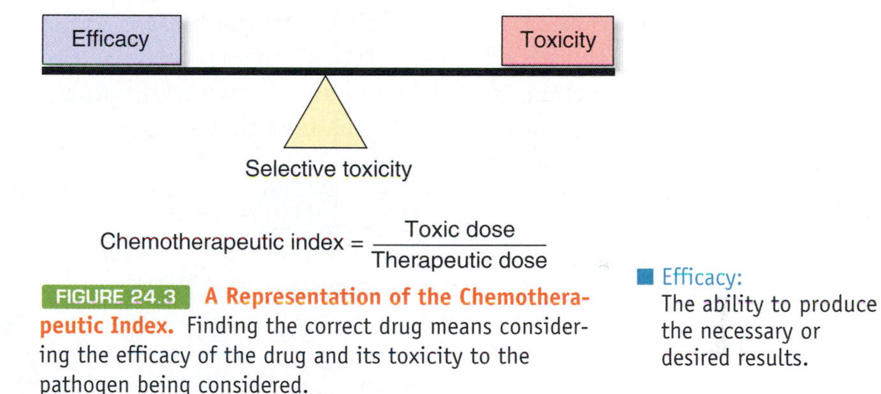

FIGURE 24.3 A Representation of the Chemotherapeutic Index. Finding the correct drug means considering the efficacy of the drug and its toxicity to the pathogen being considered.

Q: Determine whether a drug with a high or low chemotherapeutic index would be the more desirable.

■ Efficacy: The ability to produce the necessary or desired results.

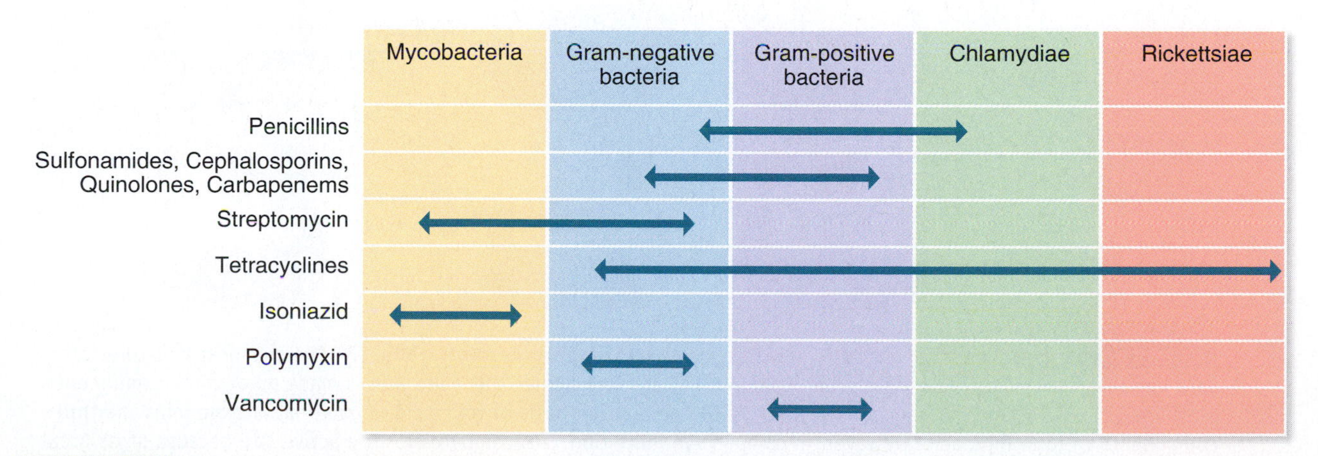

FIGURE 24.4 The Antimicrobial Spectrum of Activity. Antimicrobial drugs have a limited range of activity in the domain *Bacteria*.

Q: According to the figure, which drugs have a broad spectrum and which ones have a narrow spectrum?

24.2 The Synthetic Antibacterial Agents

The work of Gerhard Domagk in identifying the usefulness for Prontosil set the stage for the identification of many other synthetic antimicrobials. In 1935, a group at the Pasteur Institute, headed by Jacques and Therese Tréfouël, isolated the active form of Prontosil. They found it to be a chemical called sulfanilamide, which was highly active against gram-positive bacteria; it quickly became a mainstay for treating wound-related infections during World War II.

Sulfanilamide and Other Sulfonamides Target Specific Metabolic Reactions

KEY CONCEPT

- Sulfonamides affect DNA synthesis.

Sulfanilamide was the first of a group of synthetic agents known as **sulfonamides**. These so-called "sulfa drugs" interfere with the metabolism of bacterial cells without damaging body tissues. Here's how they work.

Bacterial cells synthesize an important vitamin called **folic acid** for use in nucleic acid synthesis. These organisms possess the necessary enzyme to manufacture folic acid; in fact, until recently they were incapable of absorbing folic acid from the surrounding environment. Humans cannot synthesize folic acid and must consume it in foods or vitamin capsules.

To produce folic acid, a bacterial enzyme joins together three important components, one of which is **para-aminobenzoic acid (PABA)**. This molecule is similar to sulfanilamide in chemical structure (FIGURE 24.5). However, if the cell contains large amounts of sulfanilamide, the sulfanilamide competes with PABA for the active site in the bacterial enzyme. Such **competitive inhibition** results in a reduction of folic acid, which stops nucleic acid synthesis and DNA replication (see Chapter 5). The bacterial cells eventually die.

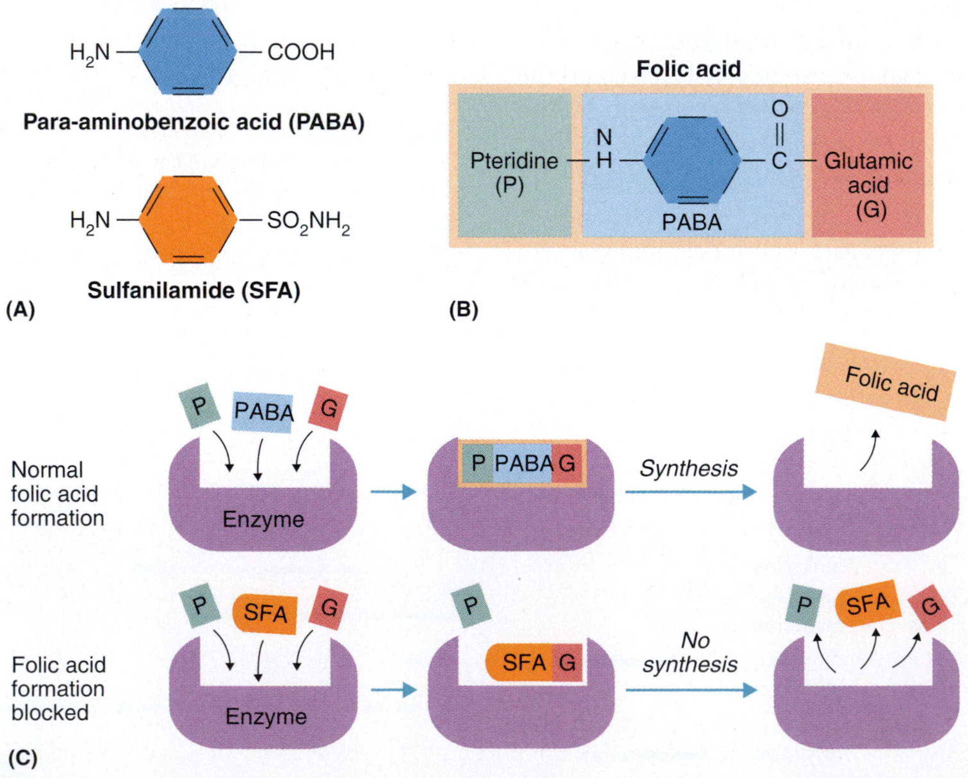

FIGURE 24.5 **The Disruption of Folic Acid Synthesis by Competitive Inhibition.** (**A**) The chemical structures of para-aminobenzoic acid (PABA) and sulfanilamide (SFA) are very similar. (**B**) Folic acid is made up of three components: pteridine (P), PABA, and glutamic acid (G). (**C**) In the normal synthesis of folic acid, a bacterial enzyme joins the three components to form folic acid. However, in competitive inhibition, SFA competes for the active site because of its great abundance. The SFA assumes the position normally reserved for PABA, and folic acid cannot form.

Q: Why does a bacterial cell die without folic acid?

Sulfa drugs have a broad spectrum and once were commonly used to treat streptococcal infections. Drug resistance now is quite common as mutations have occurred that allow the microbes to absorb folic acid from outside sources. Therefore, modern sulfonamides are typified by sulfamethoxazole, which is typically combined with trimethoprim, another synthetic agent that inhibits a different step in folic acid synthesis. The *synergistic* drug, called co-trimoxazole (Bactrim) is prescribed for urinary tract infections due to gram-negative rods.

In some patients, a drug allergy to sulfonamides develops, with a skin rash, gastrointestinal distress, or type II cytotoxic hypersensitivity (see Chapter 22).

Two other important synthetic drugs blocking PABA metabolism in *Mycobacterium* species are *p*-aminosalicylic acid (PAS), which is used for treating tuberculosis, and dapsone (diaminodiphenylsulfone), which is effective against leprosy (see Chapter 12).

CONCEPT AND REASONING CHECKS

24.4 Explain how sulfanilamide interferes with DNA replication.

Other Synthetic Antimicrobials Have Additional Bacterial Cell Targets

KEY CONCEPT

- Cell walls or DNA are the targets for other synthetic drugs.

Several synthetic agents that target other bacterial structures are currently in wide use.

Isoniazid (isonicotinic acid hydrazide, or INH) has a very narrow drug spectrum as the active form of the drug specifically interferes with cell wall synthesis in *Mycobacterium* species by inhibiting the production of mycolic acid, a component of the wall (MicroFocus 24.2). Isoniazid often is combined in therapy with such drugs as rifampin and

Sulfamethoxazole

Trimethoprim

Isoniazid

MICROFOCUS 24.2: Public Health

The Trojan Horse

In Greek mythology, the Trojan horse was a colossal statue of a horse in which Greek soldiers hid to gain access to the city of Troy (see figure). Now, scientists have uncovered a similar plot used by the drug isoniazid to kill *Mycobacterium tuberculosis,* the agent responsible for tuberculosis (TB).

Isoniazid, a small, synthetic molecule, enters the bacterial cytoplasm as a benign, nontoxic chemical substance no different than any nutrient passing through the cell wall and membrane. Once in the cytoplasm, the drug reveals its true identity. A common cytoplasmic enzyme called catalase activates the drug and converts it to a toxic form. The activated, toxic drug now attacks a protein used by the tubercle bacillus to synthesize mycolic acid, a key component of its cell wall. Without a strong cell wall, the organism is left vulnerable to osmotic lysis (see Chapter 4).

In the Greek tale, the city of Troy fell to invaders. However, bacterial cells don't read books, and in some cases, the mycobacteria fight back and resist the drug. What happens is the bacterial cell stops producing catalase by switching off the gene encoding the enzyme. Thus, the isoniazid remains inactive. However, the action also leaves the bacterial cell in a tenuous position because catalase breaks down hydrogen peroxide, a corrosive compound normally produced during bacterial metabolism.

To resolve this dilemma, a second bacterial gene encodes a second enzyme (alkyl hydroperoxidase), which takes over the job of catalase and destroys any hydrogen peroxide generated.

In mythology, the Greeks carried the day and won the city of Troy. And in the early halcyon days of antibiotic use, isoniazid was a prime weapon in the fight against TB. In 2004, almost 8 percent of TB cases in the United States were resistant to isoniazid and some 50 million people worldwide may be infected with drug-resistant strains of *M. Tuberculosis*. Tubercle bacilli are learning to resist the Trojan horse. Let's hope other antituberculoid drugs can reverse the trend.

ethambutol. **Ethambutol** is a synthetic, well-absorbed drug but side effects include visual disturbances, limiting its use in the treatment of tuberculosis.

Ciprofloxacin

Another group of synthetic drugs is the **quinolones**, which block DNA synthesis in gram-positive and gram-negative bacterial cells. Derivatives called **fluoroquinolones** are used to treat urinary tract infections, gonorrhea and chlamydia, and intestinal tract infections. An example of a fluoroquinolone is ciprofloxacin (Cipro), which became the drug of choice for treating people exposed to anthrax spores during the 2001 anthrax bioterror incidents. The fluoroquinolones are expected to be the number one selling class of antibiotics by 2011.

CONCEPT AND REASONING CHECKS

24.5 Describe the mode of action for (A) isoniazid and (B) the fluoroquinolones.

24.3 The Beta-Lactam Family of Antibiotics

One of the most common mechanisms of antibiotic action is to block the synthesis of the bacterial cell wall. Besides the synthetic drugs isoniazid and ethambutol, which target mycobacterial species, there are a large number of bacterially-synthesized antibiotics or semisynthetic ones targeting the assembly of the peptidoglycan component of bacterial cell walls. Most belong to the beta-lactam family of antibiotics because they share a common **beta-lactam nucleus** (FIGURE 24.6).

Penicillin Has Remained the Most Widely Used Antibiotic

KEY CONCEPT

- Penicillin is a bactericidal agent effective against gram-positive cells.

Thanks to the purification and production work of Florey, Chain, Heatley in the 1940s, penicillin has saved the lives of millions of individuals. Its high chemotherapeutic index has made it the drug of choice in eradicating many infections.

Penicillin G has been the most popular penicillin antibiotic and is usually the one intended when doctors prescribe "penicillin." It is sensitive to acid, so it is primarily given intravenously. Penicillin V is more acid resistant and can be given orally. Both forms of penicillin have the same basic structure of a beta-lactam nucleus and differ in the side groups attached (see Figure 24.6).

The penicillins are active against a variety of gram-positive bacteria, including staphylo-

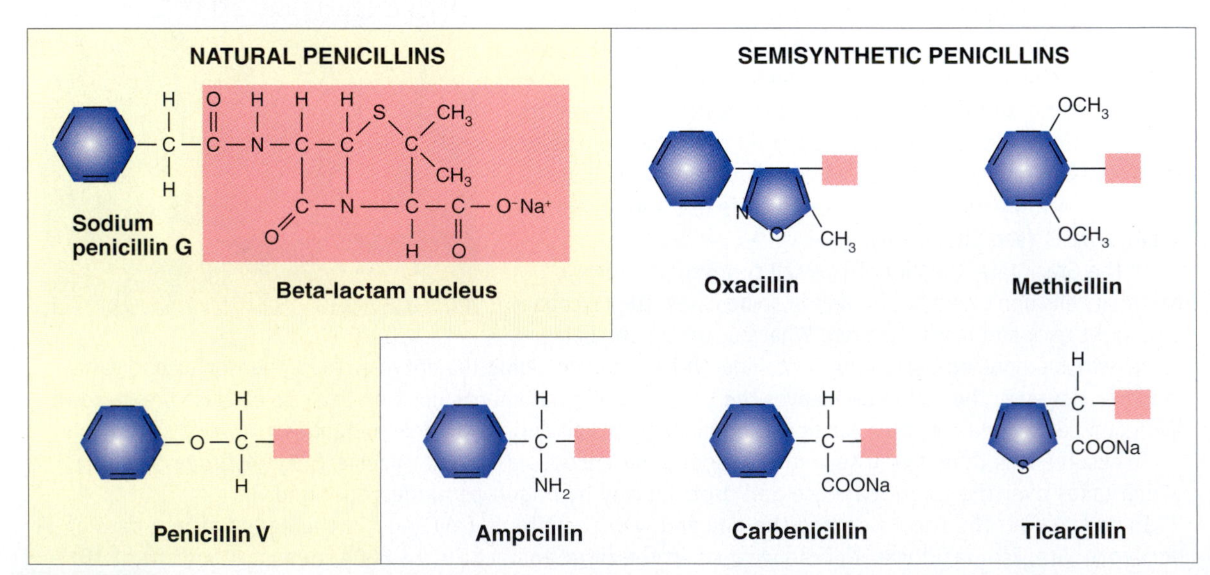

FIGURE 24.6 **Some Members of the Penicillin Group of Antibiotics.** The beta-lactam nucleus is common to all the penicillins. Different penicillins are formed by varying the side group on the molecule.

Q: Why are there so many semisynthetic penicillins?

cocci, streptococci, clostridia, and pneumococci. In higher concentrations, they also are inhibitory to the gram-negative diplococci causing gonorrhea and meningitis, and they are useful against syphilis spirochetes.

Penicillin functions during the synthesis of the bacterial cell wall. It blocks the cross-linking of carbohydrates in the peptidoglycan layer during wall formation. This results in such a weak wall that internal osmotic pressure causes the cell to swell and burst. Penicillin is therefore bactericidal in rapidly multiplying bacteria (as in an infection). Where bacterial cells are multiplying slowly or are dormant, the drug may have only a bacteriostatic effect or no effect at all.

Over the years, two major drawbacks to the use of penicillin have surfaced. The first is the anaphylactic reaction occurring in allergic individuals (see Chapter 22). This allergy applies to all compounds related to penicillin. Swelling around the eyes or wrists, flushed or itchy skin, shortness of breath, and a series of hives are signals of a hypersensitivity; penicillin therapy should cease immediately.

The second disadvantage is the evolution of penicillin-resistant bacterial species. Many of these organisms produce **beta-lactamases**, which inactivate beta-lactam antibiotics. For example, penicillinase opens the beta-lactam ring, converting penicillin G into harmless penicilloic acid (FIGURE 24.7). The ability to produce penicillinase probably has existed in certain bacterial mutants, but the ability manifests itself when the organisms are confronted with the drug. Thus, a process of natural selectivity takes place, and the rapid multiplication of penicillinase-producing bacterial cells yields organisms over which penicillin has no effect.

In the late 1950s, the beta-lactam nucleus of the penicillin molecule was identified and synthesized, and scientists found they could attach various groups to this nucleus and create new penicillins. In the following years, numerous "semisynthetic" penicillins emerged (Figure 24.6). Oxacillin and methicillin are penicillinase-resistant penicillins, while ampicillin is a broad spectrum drug of value against some gram-negative rods (*Escherichia, Proteus, Haemophilus*) as well as gonococci and meningococci. The drugs resist stomach acid and are absorbed from the intestine after oral consumption. Other broad spectrum drugs, such as carbenicillin and ticarcillin, are effective even against a broader range of gram-negative bacteria and can be used for infections of the urinary tract. Penicillins, such as amoxicillin, have been used in combination with chemicals like clavulanic acid (the combination is called Augmentin). The clavulanic acid inactivates penicillinase, allowing the antibiotic drug to affect cell wall synthesis.

CONCEPT AND REASONING CHECKS

24.6 Explain how penicillin works and how penicillinase affects penicillin.

Other Beta-Lactam Antibiotics Also Inhibit Cell Wall Synthesis

KEY CONCEPT

- Cephalosporins and carbapenems also contain a beta-lactam ring and inhibit peptidoglycan cross-linking.

Besides the penicillins, there are other beta-lactam antibiotics affecting cell wall synthesis.

Cephalosporins. While evaluating seawater samples along the coast of Sardinia in

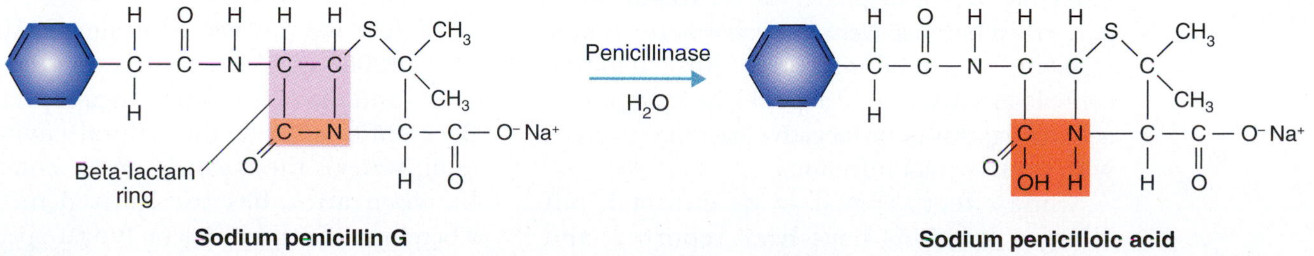

FIGURE 24.7 **The Action of Penicillinase on Sodium Penicillin G.** The enzyme penicillinase converts penicillin to harmless penicilloic acid by opening the beta-lactam ring and inserting a hydroxyl group to the carbon and a hydrogen to the nitrogen.

Q: What does the action of penicillinase tell you about the role of the beta-lactam ring in the structure of penicillin?

1945, an Italian microbiologist named Giuseppe Brotzu observed a striking difference in the amount of *E. coli* in two adjoining areas. Subsequently he discovered that a fungus, *Cephalosporium acremonium,* was producing an antibacterial substance in the water. The substance, named cephalosporin C, was later isolated and characterized by scientists as having a beta-lactam nucleus. Today, the cephalosporins and penicillins make up more than 50 percent of all antibiotics produced and used worldwide.

Cephalosporins resemble penicillins in chemical structure, except the beta-lactam nucleus has a slightly different composition. They are used as alternatives to penicillin where resistance is encountered, or in cases where penicillin allergy exists. Cephalosporins also have a broader antibacterial spectrum, are longer lasting in the body, and are resistant to many beta-lactamases (cephalosporinases).

First-generation cephalosporins are variably absorbed from the intestines and are useful against gram-positive cocci and certain gram-negative rods. They include cephalexin (Keflex) and cephalothin (Keflin). Modifications to the basic cephalosporin chemical structure have generated newer semisynthetic generation drugs that are more active and longer lasting.

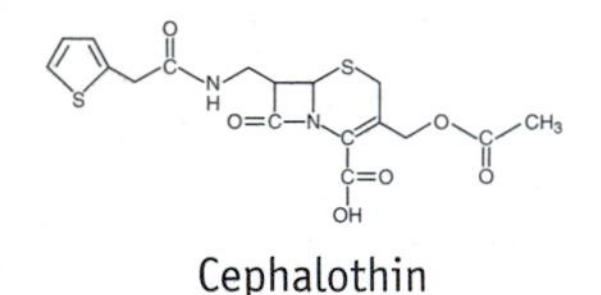

Cephalothin

Second-generation drugs have an expanded activity against gram-positive cocci as well as numerous gram-negative rods (e.g., *Haemophilus influenzae*). They include cefaclor and cefuroxime (Zinacef). The third-generation cephalosporins are used primarily against gram-negative rods (e.g., *Pseudomonas aeruginosa*) and for treating diseases of the central nervous system. Cefotaxime (Claforan) and ceftriaxone (Rocephin) are in the group. The latter has replaced penicillin for treatment of gonorrhea because *Neisseria gonorrhoeae* is now resistant to penicillin. The fourth generation cephalosporins (e.g., cefepime) have improved activity against gram-negative bacteria involved with urinary tract infections.

Imipenem

Side effects appear to be minimal, but allergic reactions have been reported, and thrombophlebitis can occur. Cephalosporins have a broader spectrum of activity than penicillins and are resistant to penicillinases; they are sensitive, however, to another group of beta-lactamases.

■ Thrombophlebitis: A venous inflammation with an associated blood clot.

CONCEPT AND REASONING CHECKS

24.7 List the advantages of the cephalosporins over the penicillins.

Monobactams and Carbapenems. A group of narrow-spectrum antibiotics first produced in the early 1980s is the **monobactams**. Isolated from the bacterium *Chromobacter violaceum*, a purple-pigmented bacterial species, monobactams, like moxalactam, are active against aerobic, gram-negative rods, especially those involved in nosocomial diseases and bacterial meningitis. Interference with platelet function and severe bleeding has limited their use.

Another set of beta-lactam drugs are the broad spectrum **carbapenems**, one of the most important groups of clinically-useful antibiotics today. The representative of this group, imipenem, is derived from a compound produced by the bacterium *Streptomyces cattley.* Imipenem is effective against a variety of aerobic gram-positive bacteria and gram-negative rods, as well as anaerobes (e.g., *Bacteroides fragilis*). It is not active against methicillin-resistant *Staphylococcus aureus*. Imipenem is prescribed often in cases where resistances occur, and it appears to have minimal side effects, although people allergic to penicillin and other beta-lactam antibiotics should not take imipenem. The drug normally is degraded by the kidneys before it can affect its action in the body. Therefore, imipenem usually is prescribed in combination with cilastatin. Cilastatin prevents the kidneys from degrading the drug. The imipenem /cilastatin combination is known as Primaxin.

CONCEPT AND REASONING CHECKS

24.8 List the advantages of the carbapenems.

Before we proceed to examine other specific antibiotics, it is worth noting the reasons why some bacterial and fungal species produce antibiotics. In the natural environment (soil, water), there can be fierce competition between microbes for limited nutrients. Therefore, if a bacterial or fungal species has the ability to secrete an antibiotic, it may gain a competitive advantage over those species unable to produce antibiotics or are sensitive to the chemicals. Antibiotic production is one way to enhance their fitness.

In response to antibiotic secretion, some species may develop mutations or gain plasmids capable of making the bacterial cell resistant to one or more antibiotics. So, in a natural habitat, it has been a continuous evolutionary process for bacterial species to compete against one another either through antibiotic production or by developing resistance.

24.4 Other Bacterially Produced Antibiotics

Besides antibiotics targeting the bacterial cell wall, there are many others that affect the cell membrane, protein synthesis, or nucleic acid synthesis. However, we will start with some non-beta-lactam drugs affecting cell walls.

Vancomycin Also Inhibits Cell Wall Synthesis

KEY CONCEPT

- Vancomycin is a glycopeptide antibiotic and cell wall assembly inhibitor.

Vancomycin, a cell wall inhibitor of gram-positive bacteria, is a product of *Amycolatopsis* (formerly *Streptomyces*) *orientalis*. It is administered by intravenous injection against diseases caused by gram-positive bacteria, especially severe staphylococcal diseases where penicillin allergy or bacterial resistance is found. It also is used against *Clostridium* and *Enterococcus* species (enterococci). It is inactive against gram-negative bacterial species. Because side effects include damage to the ears and kidneys, the drug is not routinely prescribed for trivial conditions.

As drug resistance has developed and spread among staphylococci, the choice of antibiotics has gradually diminished, and vancomycin has emerged as a key treatment in therapy; it often is referred to as the "drug of last resort." Unfortunately, resistance to vancomycin among bacterial groups such as the enterococci (*Enterococcus faecium* and *E. faecalis*) populations also has been observed, as indicated in the chapter introduction. The enterococci account for over 10 percent of all hospital acquired infections in the United States.

CONCEPT AND REASONING CHECKS

24.9 What problems are associated with vancomycin use?

Polypeptide Antibiotics Affect the Cell Membrane

KEY CONCEPT

- Antibiotics targeting the cell membrane either interfere with cell wall synthesis or disrupt membrane structure.

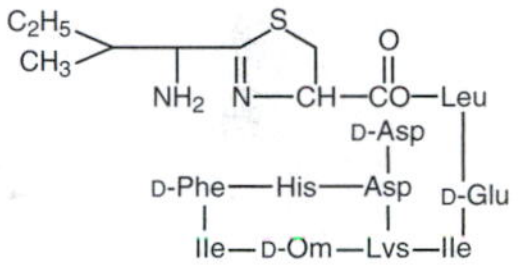

Both bacitracin and polymyxin B are polypeptide antibiotics produced by *Bacillus* species. These antibiotics are quite toxic internally and can cause kidney damage. Therefore, they generally are restricted to topical use, such as on the skin.

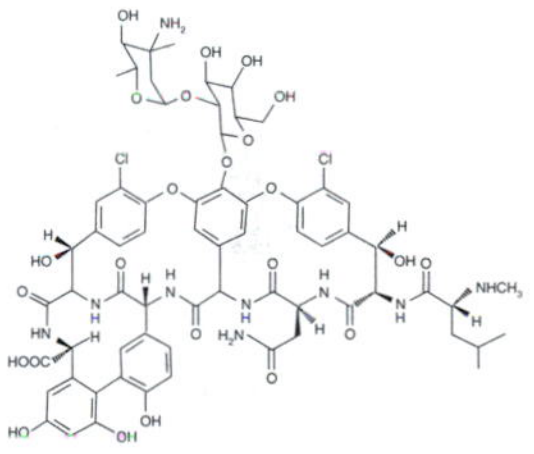

Bacitracin

Bacitracin is a cyclic polypeptide that interferes with the transport of cell wall precursors through the cell membrane. Being very toxic internally, bacitracin is available in ointments for topical treatment of skin infections caused by gram-positive bacteria as well as for the prevention of wound infections. When combined with neomycin (see below) and polymyxin B, it is sold under the brand Neosporin.

Vancomycin

Polymyxins are cyclic polypeptides that insert into the cell membrane. Acting like detergents, they increase permeability and lead to cell death. They are most active against gram-negative rods. Polymyxin B is valuable against *Pseudomonas aeruginosa* and other gram-negative bacilli, particularly those causing superficial infections in wounds, abrasions, and burns. The two antibiotics, bacitracin and polymyxin B, are combined with gramicidin (increases permeability of the cell membrane) in Polysporin.

CONCEPT AND REASONING CHECKS

24.10 Why are the polypeptide antibiotics only used topically?

Many Antibiotics Affect Protein Synthesis

KEY CONCEPT

- Many antibiotics target the ribosome's ability to translate a mRNA.

MICROFOCUS 24.3: History

Serendipity

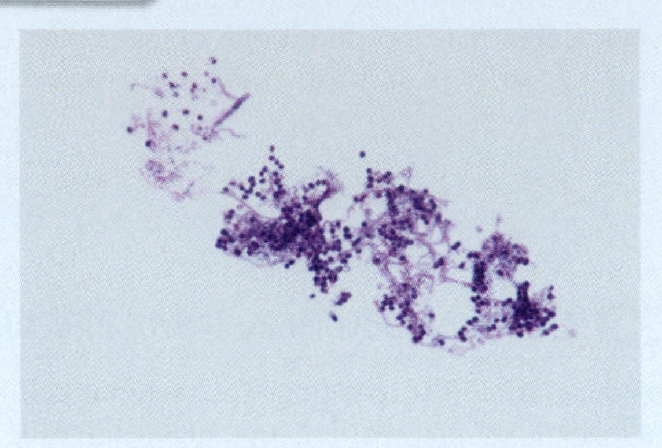

A gram-stained preparation of *Streptomyces griseus.*

They met by chance on a ship sailing from France to the United States: Rene Dubos, a 23-year-old French student interested in soil science, and Selman A. Waksman, a professor of soil microbiology at Rutgers University in New Jersey. The year was 1924. For the next two decades, their lives would intertwine as each carved out a niche in modern microbiology.

As they chatted aboard the ship, Waksman invited Dubos to come to Rutgers to earn a doctorate in microbiology. Dubos took the advice, and by 1927, he had his Ph.D. and a job at Rockefeller Institute in New York City. There he discovered a bacterial enzyme that destroys the capsules of pneumococci and hastens their death.

Knowing of Domagk's work on prontosil, Dubos began searching for ways to destroy whole organisms, not just capsules. He isolated a soil bacterium, *Bacillus brevis,* and in 1939 he extracted from it an antibiotic called tyrothricin. Further extractions yielded a second antibiotic, gramicidin. Both substances killed a variety of gram-positive bacteria, but both were too toxic for use in the body.

Selman Waksman followed his former pupil's research and in 1939 Waksman began testing soil bacteria for antimicrobial compounds. The work was long, systematic, and plodding. In 1940, Waksman's group isolated the toxic antibiotic actinomycin, and in 1942, they found another antibiotic, streptothricin. However, in August 1943, working with Albert Shatz and Elizabeth Bugie, Waksman isolated the mold-like bacillus called *Streptomyces griseus* from the throat of a chicken (see figure). The bacterium produced streptomycin, an antibiotic with extraordinary capabilities.

Preliminary studies showed streptomycin was effective against tubercle bacilli. Merck and Company soon began industrial production of the antibiotic, and within a decade, 26 companies throughout the world were manufacturing it.

Serendipity is a word derived from Horace Walpole's fairy tale *The Three Princes of Serendip.* In the tale, desirable things happen by accident or chance. Many antibiotics are the products of serendipity, but even more fundamental are the serendipitous meetings of two people such as once happened on a ship traveling from France to the United States.

There are several groups of bacterially produced antibiotics that target the protein synthesizing machinery in bacterial cells.

Aminoglycosides. The **aminoglycosides** are a group of bactericidal antibiotic compounds in which amino groups are bonded to carbohydrate molecules (glycosides) that are bonded to other carbohydrate molecules. All aminoglycosides attach irreversibly to bacterial ribosomes, thereby blocking the reading of the genetic code on messenger RNA (mRNA) molecules. Because oral absorption is negligible, the antibiotics must be administered by intramuscular injection. Their use has declined in recent years with the introduction of second- and third-generation cephalosporins, and with the introduction and development of quinolone drugs, such as the fluoroquinolones.

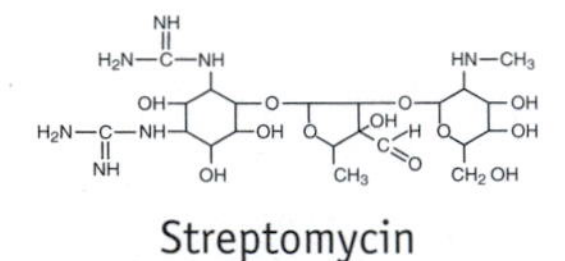

Streptomycin

In 1943, the first aminoglycoside was discovered by Selman Waksman's group at Rutgers University. In Waksman's group, a graduate student, Albert Schatz, isolated an antibacterial substance from *Streptomyces griseus* and named the substance **streptomycin** (MicroFocus 24.3). At the time, the discovery was sensational because streptomycin was useful against tuberculosis and numerous diseases caused by gram-negative bacteria. Since then it has been largely replaced by safer drugs, although streptomycin still is prescribed on occasion for tuberculosis. The major side effect is damage to the auditory branch of the nerve extending from the inner ear; deafness may result.

Gentamicin, a still-useful aminoglycoside, is administered for serious infections of the urinary tract caused by gram-negative bacteria. The antibiotic is produced by a species of *Micromonospora*, a bacterial species related to *Streptomyces*. Damage can occur to the kidney and the drug can cause deafness in genetically-susceptible individuals.

Neomycin, which also was discovered in Waksman's lab, sometimes is used to control intestinal infections because it is poorly absorbed and it is prescribed as an ointment for bacterial conjunctivitis. As already mentioned, it is available in combination with polymyxin B and bacitracin as Neosporin. Another aminoglycoside, kanamycin, is used primarily against gram-negative bacteria in wound tissue. A synthetic derivative of kanamycin called amikacin is used for controlling numerous nosocomial diseases and for intestinal and urinary tract diseases. Physicians use an aerosolized version of tobramycin (Tobi®) to treat *Pseudomonas*-caused respiratory infections in patients with cystic fibrosis.

Chloramphenicol. An antibiotic with a broad spectrum is **chloramphenicol**. Its discovery from *Streptomyces venezuelae* was hailed as a milestone in microbiology because the drug is capable of inhibiting a wide variety of gram-positive and gram-negative bacteria as well as several species of rickettsiae and fungi.

Chloramphenicol is a small molecule that passes into the tissues, where it interferes with protein synthesis. It diffuses into the nervous system and is thus useful in treating meningitis. It also remains the drug of choice in the treatment of typhoid fever and is an alternative to tetracycline for typhus fever and Rocky Mountain spotted fever (see Chapter 12).

The drugs usually are reserved for treating serious and life-threatening infections because of their side effects. In the bone marrow, it prevents hemoglobin incorporation into the red blood cells, causing a condition called aplastic anemia. Chloramphenicol also accumulates in the blood of newborns, causing a toxic reaction and sudden breakdown of the cardiovascular system known as the **gray syndrome**. Still, it is used to treat cholera in endemic parts of the world.

Tetracyclines. In 1948, scientists discovered chlortetracycline, the first of the **tetracycline** antibiotics. This finding completed the initial quartet of "wonder drugs": penicillin, streptomycin, chloramphenicol, and tetracycline.

Modern tetracyclines are a group of broad-spectrum bacteriostatic antibiotics with a range of activity similar to chloramphenicol. There are naturally occurring chlortetracyclines, isolated from species of *Streptomyces*, and semisynthetic tetracyclines, such as minocycline and doxycycline (FIGURE 24.8A). All have the four benzene ring chemical structure.

Tetracycline antibiotics may be taken orally, a factor that led to their indiscriminate use in the 1950s and 1960s. The antibiotics were consumed in huge quantities by tens of millions of people, and in some people, the normal microbiota of the intestine was destroyed. With these natural controls eliminated, fungi such as *Candida albicans* flourished. Tetracyclines also cause a yellow-gray-brown discoloration of teeth and stunted bones in children (FIGURE 24.8B). These problems are minimized by restricting use of the

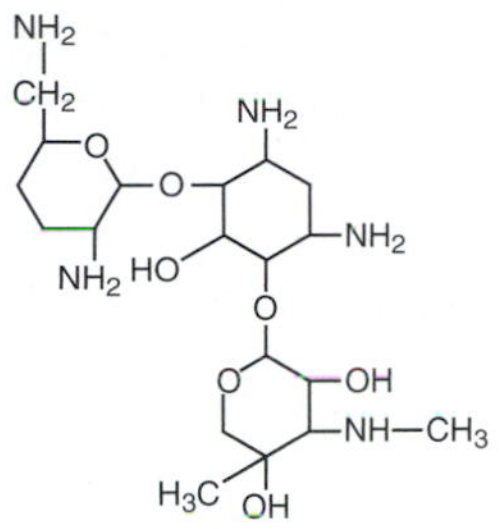

Gentamicin

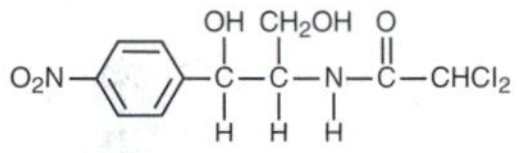

Chloramphenicol

Doxycycline

(A)

(B)

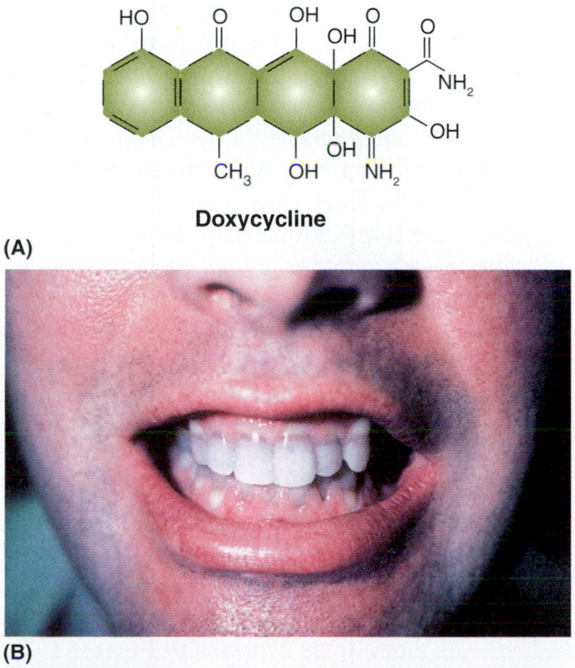

FIGURE 24.8 **The Tetracyclines.** (**A**) The chemical structure of doxycycline. (**B**) The staining of teeth associated with tetracycline use.

Q: What chemical feature characterizes all tetracycline antibiotics?

■ Cystic fibrosis: A hereditary disease starting in infancy and affecting various glands, resulting in secretion of thick mucus that blocks internal passages of the lungs, causing respiratory infections.

■ Aplastic anemia: Refers to the inability of the bone marrow to produce new blood cells.

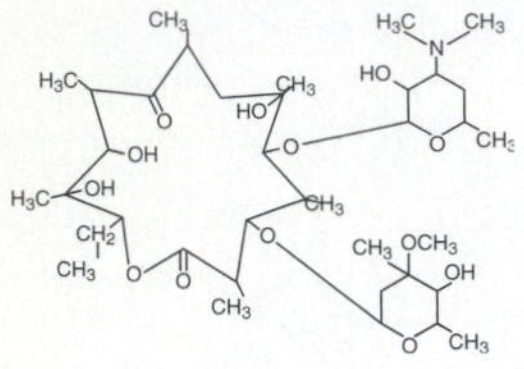

antibiotic in pregnant women and children through the teen years.

Erythromycin

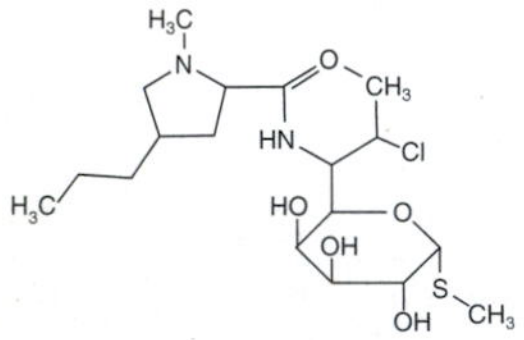

Despite these side effects, tetracyclines remain the drugs of choice for most rickettsial and chlamydial diseases (see Chapter 12). They are used against a wide range of gram-negative bacteria, and they are valuable for treating primary atypical pneumonia, syphilis, gonorrhea, and pneumococcal pneumonia. Although resistances have occurred, newer tetracyclines such as minocycline and doxycycline appear to circumvent these. Evidence indicates that tetracycline may have been present in the food of ancient people (MicroFocus 24.4).

Clindamycin

In 2005, the U.S. Food and Drug Administration (FDA) approved a new class of antibiotics related to the tetracyclines, called the **glycylcyclines**. The new drug, tigecycline (Tygacil) is effective against antibiotic resistant *S. aureus* infections.

Macrolides. The **macrolides** are broad spectrum, bacteriostatic antibiotics consisting of large carbon rings attached to unusual carbohydrate molecules. In 1949, Abelardo Aguilar, a Filipino scientist, sent some soil samples to his employer Eli Lilly. In the soil was the species *Streptomyces erythreus* (now called *Saccaropolyspora erythraea*), from which the Lilly scientists isolated a drug called erythromycin. In the 1970s, researchers discovered that erythromycin was effective for treating primary atypical pneumonia and Legionnaires' disease. It is recommended for use against gram-positive bacteria in patients with penicillin allergy and against both *Neisseria* and *Chlamydia* species affecting the eyes of newborns. Although it has few side effects, it sometimes interferes with gastrointestinal functions, presumably by combining with receptor sites used by body hormones to control nutrient transport in the gastrointestinal tract.

MICROFOCUS 24.4: History

Wonder Bread

In September 1980, a chance observation led to the discovery that antibiotics were protecting humans from disease long before anyone suspected.

The remarkable find was made by Debra L. Martin, a graduate student at Detroit's Henry Ford Hospital. After preparing thin bone sections for microscopic observation, Martin placed her slides under a fluorescence microscope because no other microscope was available for her use at the time. When illuminated with ultraviolet light, the sections glowed with a peculiar yellow-green color. Her colleagues identified the glow as due to the antibiotic tetracycline.

These were no ordinary bone sections. Rather, they were from the mummified remains of Nubian people excavated along the floodplain of the Nile River. Anthropologists from the University of Massachusetts, led by George Armelagos, had previously established that the Nubian population was remarkably free of infectious disease, and now Martin's discovery gave a possible reason why.

A modern-day Nubian mother and child.

Streptomyces species are very common in desert soil, and the anthropologists postulated that the bacterial cells may have contaminated the grain bins and deposited tetracycline. Bread made from the antibiotic-rich grain then conferred freedom from disease. The theory was strengthened when the amount of tetracycline in the ancient bone was shown to be equivalent to that in therapeutic doses used in medicine.

Another practice, reported in 1944, indicates that modern people were more deliberate in their use of contaminated bread. A doctor traveling in Europe noted that a loaf of moldy bread hung in the kitchens of many homes. He inquired about it and was told that when a wound or abrasion was sustained, a sliver of the bread was mixed with water to form a paste; the paste was then applied to the skin. A wound so treated was less likely to become infected. Presumably the modern bread, like the ancient bread, contained a chemical that would be recognized today as an antibiotic.

Other macrolides include clarithromycin, (Biaxin) and azithromycin (Zithromax), one of the world's best-selling antibiotics. Both macrolides are semisynthetic drugs associated with serious allergic and dermatologic reactions.

Clindamycin. *Streptomyces lincolnensis* produces an antibiotic called lincomycin from which the semisynthetic drug **clindamycin** is derived. This bacteriostatic drug is an alternative in cases where penicillin resistance is encountered. Clindamycin is active against aerobic, gram-positive cocci and anaerobic, gram-negative bacilli (e.g., *Bacteroides* species). Use of the antibiotic is limited to serious infections, however, because the drugs eliminate competing organisms from the intestine and permit *Clostridium difficile* to overgrow the area. The clostridial toxins then may induce a potentially lethal condition called **pseudomembranous colitis**, in which membranous lesions cover the intestinal wall (see Chapter 12).

Streptogramins. Another group of cyclic peptides are the **streptogramins**. Discovered in 1962 in another species of *Streptomyces*, these antibiotics are prescribed as a combination of two streptogramins, called quinupristin-dalfopristin (Synercid). Both components interfere with protein synthesis, but the interference is synergistic and bactericidal when used together. The drugs are effective against gram-positive bacteria, including *Staphylococcus aureus*, and respiratory pathogens.

Oxazolidinones. After the identification of the streptogramins in 1962, no new structural classes of antibiotics were discovered until 2000. These were the **Oxazolidinones**, such as linezolid, which were effective in treating gram-positive bacteria, including multidrug-resistant (MDR) *S. aureus*. Linezolid can produce allergic reactions and is toxic to mitochondria. Therefore, the oxazolidinones are drugs of last resort, being used only where every other antibiotic has failed.

CONCEPT AND REASONING CHECKS

24.11 Identify the bacterial source for the aminoglycosides, chloramphenicol, tetracyclines, macrolides, clindamycin, and the streptogramins.

Some Antibiotics Inhibit Nucleic Acid Synthesis

KEY CONCEPT

- Rifampin inhibits transcription by inhibiting RNA synthesis.

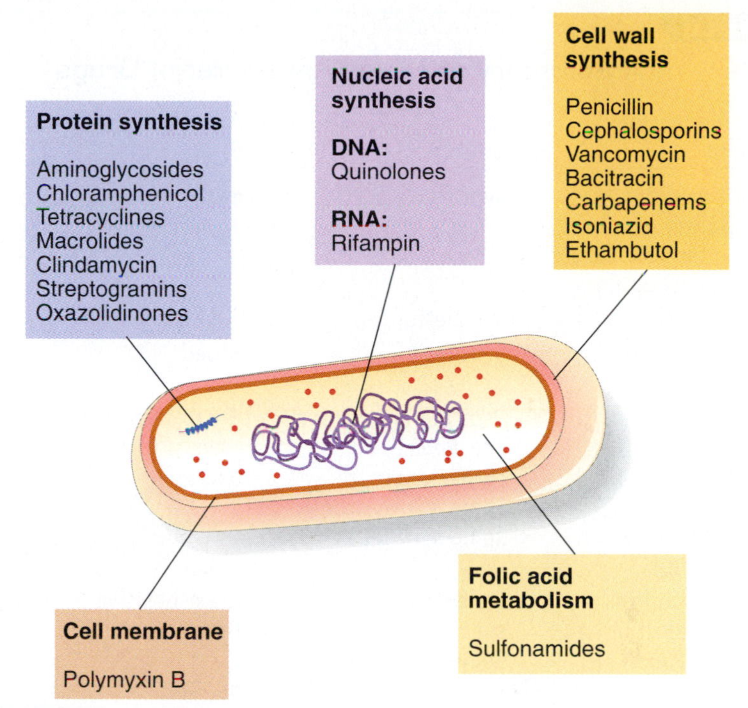

FIGURE 24.9 The Targets for Antibacterial Agents. There are six major targets for antibacterial agents: the cell wall, cell membrane, ribosomes (protein synthesis), nucleic acid synthesis (RNA and DNA synthesis), and metabolic reactions.

Q: Hypothesize why so many antibiotics have evolved that affect protein synthesis.

The quinolones have already been described. These synthetic drugs inhibit bacterial DNA synthesis.

Other drugs affect transcription. **Rifampin**, a semisynthetic bactericidal drug derived from *Streptomyces mediterranei*, interferes with RNA synthesis. Because mycobacterial resistance to rifampin can develop quickly, it is prescribed in combination with isoniazid and ethambutol for tuberculosis and leprosy patients. It also is administered to carriers of *Neisseria* and *Haemophilus* species that cause meningitis and as a prophylactic when exposure has occurred. Rifampin therapy may cause the urine, feces, tears, and other body secretions to assume an orange-red color and may cause liver damage. The drug is administered orally and is well absorbed.

FIGURE 24.9 summarizes the sites of activity for the antibacterial agents. TABLE 24.1 reviews those antibacterial drugs currently in use.

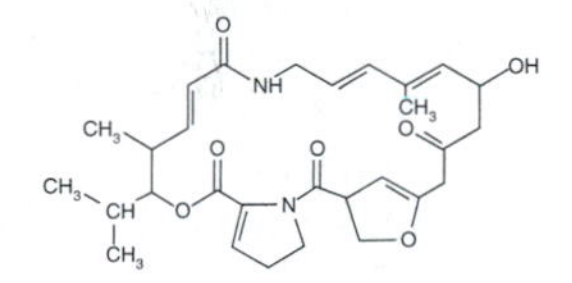

Streptogramin A

Linezolid

Rifampin

CONCEPT AND REASONING CHECKS

24.12 To what enzyme must rifampin bind to inhibit transcription?

TABLE 24.1 A Summary of Major Antibacterial Drugs

Antibacterial Agent	Source	Antibacterial Spectrum	Activity Impeded	Side Effects
Sulfonamides Sulfanilamide Sulfamethoxazole/Trimethoprim	Synthetic	Broad	Folic acid metabolism	Kidney and liver damage Allergic reactions
Isoniazid	Synthetic	Tubercle bacilli	Cell wall synthesis	Liver damage
Fluoroquinolones Ciprofloxacin	Synthetic	Broad	DNA synthesis	Few reported
Penicillins Penicillin G Ampicillin Amoxicillin Nafcillin Oxacillin	*Penicillium notatum* and *Penicillium chrysogenum* Some semisynthetic Some synthetic	Broad, especially gram-positive bacteria	Cell wall synthesis	Allergic reactions Selection of penicillinase-producing strains
Cephalosporins Cephalothin Cephalexin	*Cephalosporium* species Some semisynthetic	Gram-positive bacteria Broad	Cell wall synthesis	Occasional allergic reactions
Carbapenims Imipenum	*Chromobacter violaceum*	Broad	Cell wall synthesis	Few reported
Vancomycin	*Streptomyces orientalis*	Gram-positive bacteria, especially staphylococci	Cell wall synthesis	Ear and kidney damage
Bacitracin	*Bacillus subtilis*	Gram-positive bacteria, especially staphylococci	Cell wall synthesis	Kidney damage
Polymyxin	*Bacillus polymyxa*	Gram-negative rods, especially in wounds	Cell membrane function	Kidney damage
Aminoglycosides Gentamicin Neomycin Amikacin Streptomycin	*Micromonospora* species *Streptomyces* species	Broad, especially gram-negative bacteria	Protein synthesis	Hearing defects Kidney damage
Chloramphenicol	*Streptomyces venezuelae*	Broad, especially typhoid bacilli	Protein synthesis	Aplastic anemia Gray syndrome
Tetracyclines Chlortetracycline Oxytetracycline Doxycycline Minocycline	*Streptomyces* species Some semisynthetic	Broad, rickettsiae, chlamydiae, gram-negative bacteria	Protein synthesis	Destruction of natural flora Discoloration of teeth Stunted bones
Macrolides Erythromycin Clarithromycin Azithromycin	*Streptomyces erythreus* Some semisynthetic	Broad, gram-positive bacteria, *Mycoplasma*	Protein synthesis	Gastrointestinal distress
Clindamycin	*Streptomyces lincolnesis*	Gram-positive bacteria	Protein synthesis	Pseudomembranous colitis
Rifampin	*Streptomyces mediterranei* Semisynthetic	Tubercle bacilli Gram-positive cocci	RNA synthesis	Liver damage
Streptogramins Quinupristin/Dalfopristin	*Streptomyces* species	Resistant strains of *Staphylococcus aureus*	Proten synthesis	Muscle aches, rash, headache
Oxazolidinones Linezolid	Synthetic	Last resort antibiotics, gram-positive bacteria	Protein synthesis	Toxic to mitochondria

24.5 Antifungal and Antiparasitic Agents

The ability to use antimicrobial agents effectively against fungi presents some problems not encountered with bacterial pathogens. The major problem facing medical mycologists is finding drugs affecting the fungal pathogen but not the host tissue. Because fungi are members of the *Eukarya*, they possess much of the cellular machinery common to animals and humans; thus, many drugs targeting fungi will also target host tissues and potentially be quite toxic. Therefore, most antifungal agents are used only topically on the skin. The few drugs that are selectively toxic and systemic are targeted at unique fungal structures or metabolic processes.

Several Classes of Antifungal Drugs Cause Membrane Damage

KEY CONCEPT

- Selective toxicity is important when treating fungal infections and diseases.

Polyenes. The **polyenes** are large, ring-shaped organic compounds containing alternating double and single carbon-carbon bonds on one side of the ring and multiple hydroxyl (–OH) groups bonded to carbons on the other side of the ring. The polyenes bind to ergosterol, a sterol found in the fungal plasma membrane, causing the cell's contents to leak out and cell death. Human and animal cells lack this sterol.

For infections of the intestine, vagina, or oral cavity due to *Candida albicans*, physicians often prescribe **nystatin**. This product of *Streptomyces noursei* is commercially sold in ointment, cream, or suppository form. Often it is combined with antibacterial antibiotics to retard *Candida* overgrowth of the intestines during the treatment of bacterial diseases.

For serious systemic fungal infections, the drug of choice is amphotericin B. This broad spectrum antibiotic, extracted from *Streptomyces nodosus*, is used intravenously with immunocompromised patients, and for treating aspergillosis, cryptococcosis, and candidiasis. However, it causes a wide variety of side effects, including kidney damage, and therefore is used only in progressive and potentially fatal cases.

Imidazoles. Synthetic antifungal drugs, called **imidazoles**, inhibit an enzyme needed to form ergosterol in the fungal plasma membrane. Again, the lack of the sterol causes the cell's contents to leak out and, depending on the fungus and the drug, either brings about an inhibition of fungal growth (fungistatic) or cell death (fungicidal).

The imidazoles include clotrimazole, miconazole, and ketoconazole. Clotrimazole (Gyne-Lotrimin) is used topically for *Candida* skin infections, while the other drugs are used topically as well as internally for systemic diseases. Side effects are uncommon. Miconazole is commercially available as Micatin for athlete's foot and Monistat for yeast infections. Ketoconazole has been used to treat fungal infections in immunocompromised patients, including those with AIDS.

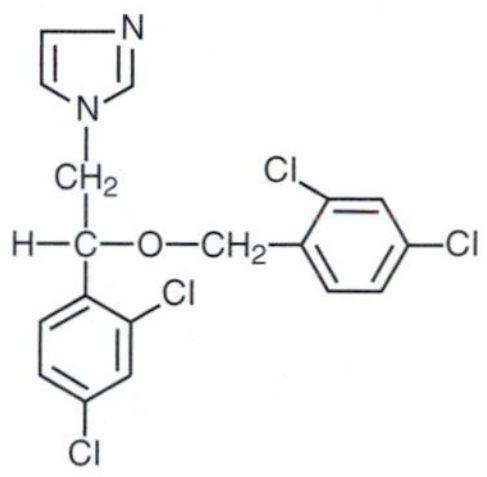

Miconazole

Echinocandins. Because fungi have cell walls, this structure represents another unique site for antifungal drug activity. The semisynthetic **echinocandins** inhibit the synthesis of the fungal cell wall. The drugs are fungistatic against *Aspergillus* and fungicidal against *Candida*. Currently, the only echinocandin approved by the FDA is caspofungin, which is used to treat invasive candidiasis or aspergillosis.

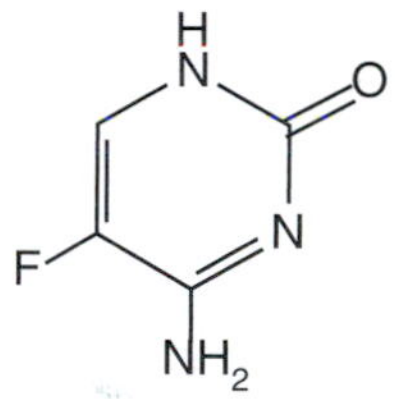

Flucytosine

Flucytosine. The antimetabolite drug **flucytosine** is converted in fungal cells to an inhibitor that interrupts nucleic acid synthesis. The drug is active against some strains of *Candida* and *Cryptococcus*. The drug usually is used in combination with amphotericin B.

Griseofulvin. The drug **griseofulvin** is a product of *Penicillium griseofulvum* and is taken orally. Griseofulvin interferes with mitosis by binding to microtubules (see Chapter 3). It is used for fungal infections of the skin, hair, and nails, such as ringworm and athlete's foot.

Griseofulvin

CONCEPT AND REASONING CHECK

24.13 Assess the emphasis for developing antifungal drugs targeting the fungal plasma membrane or cell wall.

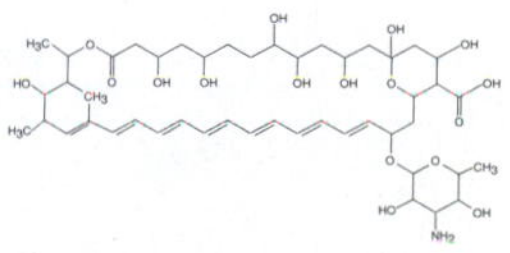

Amphotericin B

The Goal of Antiprotozoal Agents Is to Eradicate the Parasite

KEY CONCEPT

- Antiprotozoal agents also target unique aspects of these parasites.

MICROFOCUS 24.5: History

The Fever Tree

Rarely had a tree caused such a stir in Europe. In the 1500s, Spaniards returning from the New World told of its magical powers for malaria patients, and before long, the tree was dubbed "the fever tree." The tall evergreen grew only on the eastern slopes of the Andes Mountains (see figure). According to legend, the Countess of Chinchón, wife of the Spanish ambassador to Peru, developed malaria in 1638 and agreed to be treated with its bark. When she recovered, she spread news of the tree throughout Europe, and a century later, Linnaeus named it *Cinchona* after her.

Leaves of a cinchona tree.

For the next two centuries, cinchona bark remained a staple for malaria treatment. Peruvian Indians called the bark quina-quina (bark of bark), and the term quinine gradually evolved. In 1820, two French chemists, Pierre Pelletier and Joseph Caventou, extracted pure quinine from the bark and increased its availability still further. The ensuing rush to stockpile the chemical led to a rapid decline in the supply of cinchona trees from Peru, but Dutch farmers made new plantings in Indonesia, where the climate was similar. The island of Java eventually became the primary source of quinine for the world.

During World War II, Southeast Asia came under Japanese domination, and the supply of quinine to the West was drastically reduced. Scientists synthesized quinine shortly thereafter, but production costs were prohibitive. Finally, two useful substitutes were synthesized in chloroquine and primaquine. Today, as resistance to these drugs is increasingly observed in malarial parasites, scientists once again are looking to another "fever tree" to help control malaria.

$H_2C{=}CH$, HO, CH_3O, N

Quinine

N, N^+, O^-, O, N, OH

Metronidazole

Because protozoa also are members of the domain *Eukarya*, antiprotozoal agents attempt to target unique aspects of nucleic acid synthesis, protein, synthesis, or metabolic pathways.

Aminoquinolines. The **aminoquinolines** are antimalarial drugs that accumulate in parasitized red blood cells. The aminoquinolines appear to be toxic to the malaria parasite, interfering with the parasite's ability to break down and digest hemoglobin (see Chapter 17). The parasite thus starves or the drug causes the accumulation of toxic products resulting from the degradation of hemoglobin in the parasite. Quinine was one of the first natural antimicrobials and is derived from the bark of the South American cinchona tree (MicroFocus 24.5). It was the primary agent used to treat malaria until substantial resistance developed. More effective synthetic drugs, such as chloroquine, mefloquine, and primaquine, are now used. Chloroquine remains the drug of choice to treat all species of *Plasmodium* while mefloquine (Lariam) is primarily used for malaria caused by *P. falciparum*. Reports have been made that mefloquine can have rare but serious side-effects, including severe depression, anxiety, paranoia, nightmares, and insomnia. Besides its use in treating malaria, primaquine is used with clindamycin to treat pneumocystis pneumonia (see Chapter 16).

Sulfonamides. Similar to bacterial cells, protozoal cells require folic acid for the synthesis of nucleic acids but are unable to absorb it from the environment. Therefore, the same sulfonamides that are used against bacterial pathogens will produce a similar result with the protozoa. Another drug, diaminopyrimidine, can be used with trimethoprim to achieve a synergistic effect. These two drugs in combination with sulfamethoxazole are effective in treating toxoplasmosis (see Chapter 17).

Nitroimidazoles. Two common drugs in the **nitroimidazole** family, metronidazole (Flagyl) and tinidazole, interfere with DNA synthesis. These drugs are effective in the treatment of amebiasis, giardiasis, and trichomoniasis (see Chapter 17). Metronidazole can be mutagenic to the host.

Heavy Metals. Arsenic and antimony derivatives have been used since ancient times. The arsenic derivative, melarsoprol (Mel B) is used to treat African trypanosomiasis, while antimoniate is effective against leishmaniasis (see Chapter 17). Although the drugs are toxic in the nervous system and kidneys, they primarily affect the intense metabolism of protozoal cells.

Other Drugs. Several of the antibiotics that affect bacterial protein synthesis also inhibit protein synthesis in protozoa. Clindamycin and the tetracyclines, such as doxycycline, are used to treat malaria. Another protein synthesis inhibitor, **paromomycin**, can be used against cryptosporidiosis (see Chapter 17).

Pentamidine, a drug of unknown action, has been used against the parasites causing leishmaniasis. **Artemisinin** is used to treat multi-drug resistant strains of *P. falciparum*. The drug, isolated from the shrub *Artemisia annua* (sweet wormwood) has been used by Chinese herbalists (FIGURE 24.10). In red blood cells, the drug releases free radicals that destroy the malarial parasites.

FIGURE 24.10 **The Source of Artemisinin.** *Artemisia* shrubs are grown in Africa, as indicated in this harvest by a Kenyan farmer.

Q: What is the mode of action of artemisinin?

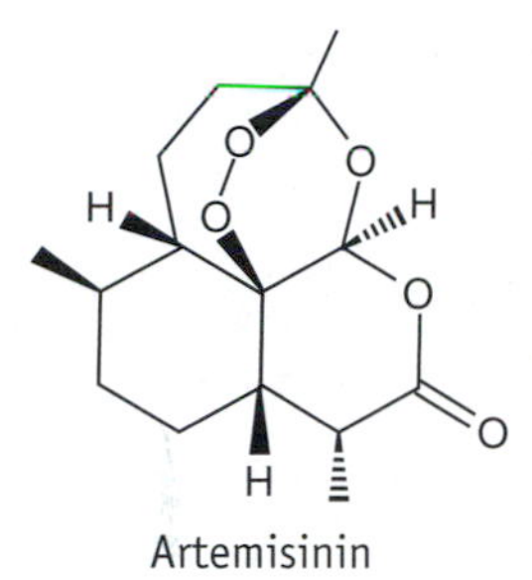

■ Free radicals: Highly reactive atoms or groups of atoms with unpaired electrons.

CONCEPT AND REASONING CHECKS

24.14 Identify the cellular structures targeted by the antiprotozoal agents.

Antihelminthic Agents Are Targeted at Nondividing Helminths

KEY CONCEPT

- Antihelminthic drugs affect helminth neuromuscular coordination, carbohydrate metabolism, or microtubule structure.

The helminths are eukaryotic, multicellular parasites. Although most antimicrobial drugs are targeted at actively dividing cells of the pathogen, the mechanism of action of most antihelminthic drugs targets the nondividing organisms.

Praziquantel is thought to change the permeability of the parasite plasma membranes of cestodes and trematodes. The permeability change causes the inflow of calcium ions, which leads to muscle contraction and paralysis in the parasite. Parasites then cannot feed and eventually die. Praziquantel has been the drug of choice for mass therapy campaigns, including the treatment of schistosomiasis (see Chapter 17; see also MicroFocus 15.4).

Mebendazole inhibits uptake of glucose and other nutrients by adult and larval worms from the host intestine where helminths are located. Without the ability to carry out ATP synthesis, the parasites will die. The drug has a wide spectrum, affecting many nematodes and cestodes. The drug also disrupts microtubules and cell division.

The **avermectins** are antihelminthic drugs derived from *Streptomyces avermitilis*. The drugs are effective in extremely low doses and against a wide variety of roundworms. The avermectins have an effect of the nematode nervous system such that muscle paralysis results. Ivermectin (Stromectol) is one drug currently used to treat onchocerciasis (river blindness) and is being investigated for treatment of a number of nematode infections.

TABLE 24.2 summarizes the characteristics of the antifungal and antiparasitic agents.

Praziquantel

CONCEPT AND REASONING CHECKS

24.15 Provide several reasons why there are fewer effective antifungal and antiparasitic drugs to treat infections.

TABLE
24.2 A Summary of Antifungal and Antiparasitic Drugs

Antifungal Agent	Source	Antifungal Spectrum	Activity Impeded	Side Effects
Polyenes Nystatin	*Streptomyces noursei*	*Candida albicans*	Cell membrane function	Few reported
Amphotericin B	*Streptomyces nodosus*	Systemic infections	Cell membrane function	Fever, Gastrointestinal distress
Imidazoles Clotrimazole Ketoconazole Miconazole	Synthetic	Superficial infections	Inhibit sterol synthesis	Few reported
Echinocandins Caspofungin	Semisynthetic	*Aspergillus, Candida*	Cell wall synthesis	Low incidence
Flucytosine	Synthetic	*Candida, Cryptococcus*	Nucleic acid synthesis	Adverse renal and liver effects
Griseofulvin	*Penicillium griseofulvum*	Superficial infection	Microtubules	Occasional allergic reactions

Antiparasite Agent	Source	Antiparasite Spectrum	Activity Impeded	Side Effects
Aminoquinolines Quinine Chloroquine Mefloquine Primaquine	Synthetic	*Plasmodium* species	Parasite digestion of hemoglobin	Severe neurological and behavioral side effect possible with mefloquine
Sulforamides Diaminopyramidine	Synthetic	*Toxoplasma, Plasmodium*	Folic acid metabolism	Kidney and liver damage; allergic reactions
Nitroimidazoles Metronidazole Tinidazole	Synthetic	*Trichomonas*	DNA synthesis	Tumors in mice
Artemisinin	*Artemisia annua*	*Plasmodium* species	Free radicals destroy parasites	Few reports of adverse effects
Praziquantel	Synthetic	*Schistosoma*	Membrane permeability	Result from killing of parasites
Mebendazole	Synthetic	Broad	Glucose uptake	Diarrhea, stomach pain
Ivermectin	*Streptomyces avermitilis*	*Ascaris, Enterobius*	Nervous system	Low incidence

24.6 Antibiotic Assays and Resistance

The substantial variety of antibiotics and chemotherapeutic agents developed since the 1930s necessitates that medical professionals know which one is best for the patient under the circumstances of the infection. Accordingly, antibiotic sensitivity assays can be performed. The assay also helps determine whether a microorganism is resistant to a particular antibiotic, a problem that has developed into a major concern of modern medicine, especially in health care settings as described in the introduction to this chapter.

There Are Several Antibiotic Susceptibility Assays

KEY CONCEPT

- Antibiotic susceptibility assays are used to study the inhibition of a test organism by one or more antimicrobial agents.

Two general methods are in common use to test the susceptibility of a bacterial pathogen to specific antibiotics. These are the tube dilution method and the agar disk diffusion method.

The **tube dilution method** determines the smallest amount of antibiotic needed to prevent

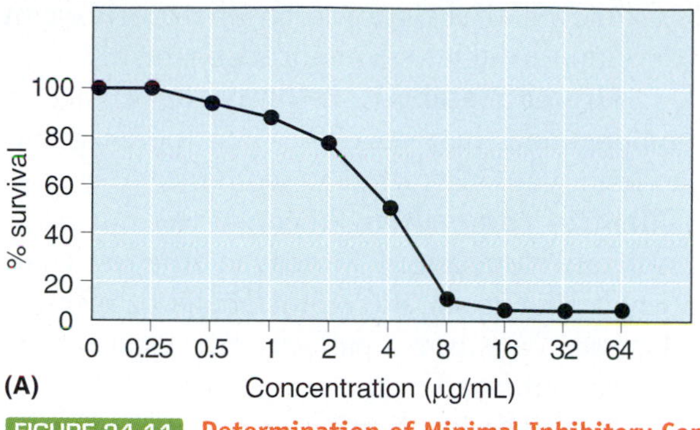

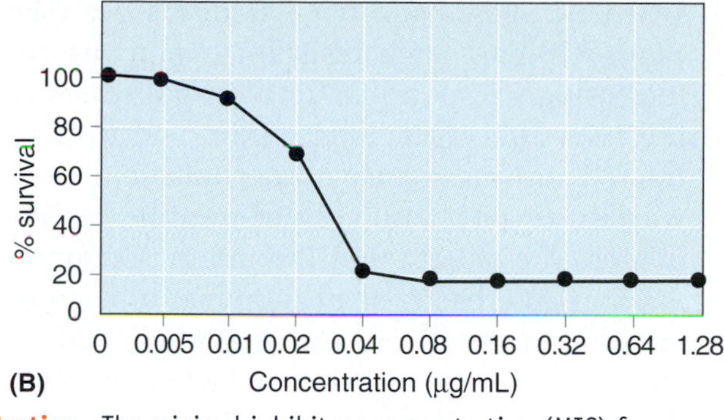

FIGURE 24.11 **Determination of Minimal Inhibitory Concentration.** The minimal inhibitory concentration (MIC) for ampicillin against (**A**) *Staphylococcus aureus* and (**B**) *Streptococcus pyogenes*.

Source: Data are modified from *the International Journal of Applied Research in Veterinary Medicine.*

Q: What are the MICs for Staphylococcus aureus *and* Streptococcus pyogenes*?*

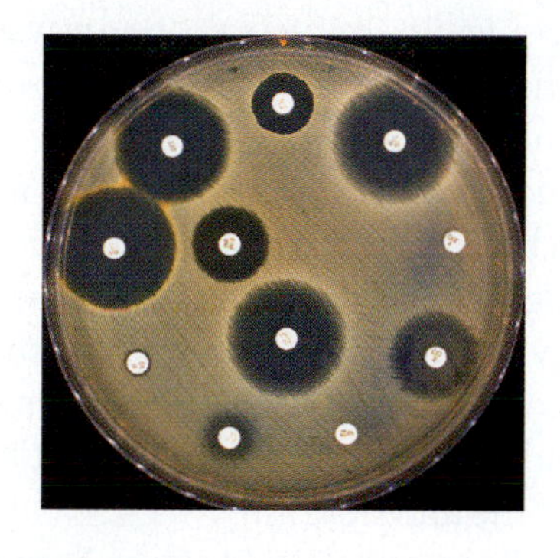
(A)

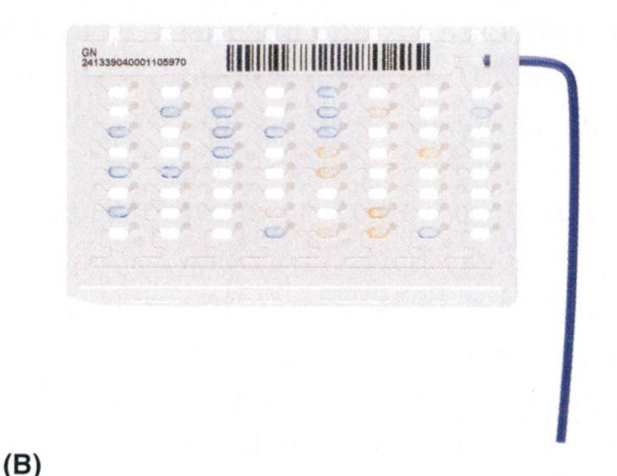
(B)

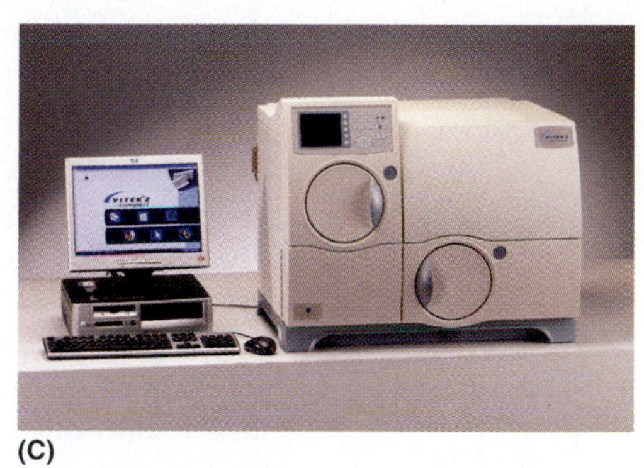
(C)

FIGURE 24.12 **Antibiotic Susceptibility Testing.** Two methods for determining an organism's antibiotic susceptibility are shown. (**A**) The more traditional method is the agar disk diffusion method. A bacterial species is spread on a plate and paper disks each containing a different antibiotic are added. The plate is incubated for 24 to 48 hours. (**B**) In a modern automated system, bacterial cells are automatically inoculated into wells on a card, each well containing a different antibiotic. The card is then incubated, and the presence or absence of growth is assayed by a computer (**C**). A printout relates the organism's susceptibility (no growth) to the antibiotic, or its resistance (growth).

Q: What do the clear zones around some paper disks in (A) signify?

growth of the pathogen. This amount is known as the **minimum inhibitory concentration (MIC)**. To determine the MIC, the microbiologist prepares a set of tubes with different concentrations (dilutions) of a particular antibiotic. Each tube is inoculated with an identical number of cells, incubated, and examined for the growth of bacterial cells. The extent of growth diminishes as the concentration of antibiotic increases, and eventually an antibiotic concentration is observed at which growth no longer occurs (FIGURE 24.11). This is the MIC.

The second method, the **agar disk diffusion method**, operates on the principle that antibiotics will diffuse from a paper disk or small cylinder into an agar medium containing a test organism (FIGURE 24.12A). Inhibition is observed as a failure of the organism to grow in the region of the antibiotic, forming a halo around the disk. A common application of the agar disk diffusion method is the **Kirby-Bauer test**, named after W. M. Kirby and A. W. Bauer, who developed it in the 1960s. This procedure determines the susceptibility of a microorganism to a series of antibiotics and is performed according to standards established by the FDA. A more sophisticated procedure is noted in FIGURE 24.12B, C.

CONCEPT AND REASONING CHECKS

24.16 Explain the significance of knowing a drug's minimal inhibitory concentration.

There Are Four Mechanisms of Antibiotic Resistance

KEY CONCEPT

- Bacterial species have several ways to generate resistance to antibacterial agents.

Since the 1980s, an alarming number of bacterial strains have evolved resistance to synthetic agents and natural antibiotics (see Figure 24.1a). Such resistant organisms are increasingly responsible for human diseases of the intestinal tract, lungs, skin, and urinary tract. Those in intensive care units and burn wards are particularly vulnerable, as are infants, the elderly, and the infirm. Common diseases like bacterial pneumonia, tuberculosis, streptococcal sore throat, and gonorrhea, which until recently succumbed to a single dose of antibiotics, are now among the most difficult to treat.

One of the major concerns of public health officials is the bacterial species *Staphylococcus aureus*. Capable of causing staphylococcal septicemia, pneumonia, endocarditis, and meningitis, *S. aureus* is involved in over 250,000 infections per year, primarily in hospitals and nursing homes—but now becoming more common as a community-acquired infection. Over the years, strains of *S. aureus* have developed resistance to penicillin and numerous other broad-spectrum drugs. Then, several decades ago *S. aureus* strains arose that were resistant to methicillin, one of the few drugs to which the bacterial species had been sensitive. Termed **MRSA**, for **methicillin-resistant *S. aureus***, the strain could only be treated with more potent but toxic drugs like vancomycin.

Then, in 1997, an MRSA strain evolved with intermediate (partial) vancomycin resistance; scientists named it **VISA**, for **vancomycin intermediately resistant *Staphylococcus aureus***. Although researchers have found useful alternatives in drug combinations, they are grappling with the possibility that nothing will be left in the antimicrobial arsenal to treat patients infected by this strain of staphylococci. Indeed, in 2001, reports of **vancomycin-resistant *S. aureus*** (**VRSA**) were reported in many countries, indicating global spread.

As already mentioned, microorganisms in the environment have been involved in chemical warfare with their neighbors for hundreds of millions of years. To counteract the production of antibiotics by some microbes, others developed self-protective (resistance) mechanisms to the natural antibiotics. Thus, it is not surprising that microbes quickly develop resistance to the natural antibiotics used to combat infectious diseases.

Bacterial resistance to antibiotics can develop through four major mechanisms.

Altered Metabolic Pathway. Resistance to sulfonamides may develop when the drug fails to bind with enzymes that synthesize folic acid. This can come about because the enzyme's structure has changed. Moreover, drug resistance may be due to an altered metabolic pathway in the microorganism, a pathway that bypasses the reaction inhibited by the drug.

Antibiotic Inactivation. Resistance can arise from the microorganism's ability to enzymatically split apart the antibiotic. The production of penicillinases (beta-lactamases) by penicillin-resistant gonococci is an example. By breaking the beta-lactam ring, penicillin- or cephalosporin-resistant bacterial cells have a mechanism to prevent blockage of cell wall synthesis.

The aminoglycosides normally block mRNA translation on bacterial ribosomes. Aminoglycoside-resistant bacterial cells have developed ways to enzymatically modify aminoglycosides so they cannot bind to ribosomes. The enzymatic modifications include phosphorylation, acetylation, or adenylation of the antibiotic.

■ Phosphorylation, acetylation, or adenylation: The addition, respectively, of a phosphate group, an acetyl group, or an adenyl group to the structure of an antibiotic.

Reduced Permeability/Active Export of Antibiotics. Another resistance mechanism involves the ability of microbes to prevent drug entry into the cytoplasm. For example, changes in membrane permeability in penicillin-resistant *Pseudomonas* prevent the antibiotic from entering the cytoplasm. Bacterial species such as *E. coli* and *S. aureus*, which are resistant to tetracyclines, actively export (pump out) the drug. Cytoplasmic and membrane proteins in these bacterial cells act as pumps to remove the antibiotic before it can affect the ribosomes in the cytoplasm.

Target Alteration. A fourth route leading to microbial resistance involves altering the drug target. Some streptomycin-resistant bacterial species can modify the structure of their ribosomes so the antibiotic cannot bind to the ribosome and protein synthesis is not inhibited. Other targets include RNA polymerase and enzymes involved in DNA replication.

TABLE 24.3 summarizes the resistance mechanisms in relation to the antibiotics mode of action and class of drugs.

CONCEPT AND REASONING CHECKS

24.17 How do the resistance mechanisms confer resistance to an antibiotic?

TABLE 24.3 A Summary of Bacterial Resistance Mechanisms

Antibiotic Target	Antibiotic Class	Resistance Mechanisms
Inhibition of cell wall synthesis	Penicillins, cephalosporins, carbapenems, vancomycin	Altered wall composition, drug destruction by beta-lactamases
Membrane permeability and transport	Polypeptide antibiotics	Altered membrane structure
Inhibition of DNA synthesis	Fluoroquinolones	Inactivation of drug, altered drug target
Inhibition of RNA synthesis	Rifampin	Altered drug target
Inhibition of protein synthesis	Aminoglycosides, chloramphenicol, tetracyclines, macrolides, clindamycin, streptogramins, oxazolidinones	Altered membrane permeability, drug pumping, antibiotic inactivation, altered drug target
Inhibition of folic acid synthesis	Sulfonamides	Alternate metabolic pathway, altered drug target

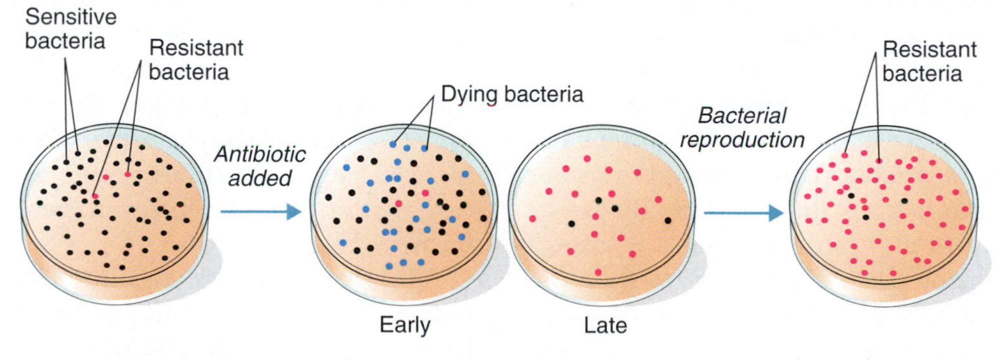

FIGURE 24.13 **The Expansion of Drug Resistance.** Drug resistance can occur widely in a population of bacterial cells as a result of natural selection during exposure to an antibiotic.

Q: Why is this example of drug resistance expansion often called a superinfection?

Antibiotic Resistance Is of Grave Concern in the Medical Community

KEY CONCEPT

- Antibiotic misuse and abuse encourage the emergence of resistant forms.

Antibiotic resistance is the result of improper use of antibiotics. For example, drug companies promote antibiotics heavily, patients pressure doctors for quick cures, and physicians sometimes write prescriptions without ordering costly tests to pinpoint the patient's illness. In addition, people may diagnose their own illness and take leftover antibiotics from their medicine chests for ailments where antibiotics are useless. Moreover, many people fail to complete their prescription, and some organisms remain alive to evolve to resistant forms by gathering "genetic debris" from the local area. And the survivors proliferate well because they face reduced competition from susceptible organisms FIGURE 24.13.

Hospitals are another forcing ground for the emergence of resistant bacterial species. In many cases, physicians use unnecessarily large doses of antibiotics to prevent infection during and following surgery. This increases the possibility that resistant strains will overgrow susceptible strains and subsequently spread to other patients, causing a **superinfection**. Antibiotic-resistant *E. coli, P. aeruginosa, Serratia marcescens,* and *Proteus* species now are widely encountered causes of illness in hospital settings.

Antibiotics also are abused in developing countries where they often are available without prescription, even though they have toxic side effects. Some countries permit the over-the-counter sale of potent antibiotics, and large doses encourage resistance to develop. Between 1968 and 1971, some 12,000 people died in Guatemala from shigellosis attributed to antibiotic-resistant *Shigella dysenteriae.*

Moreover, antibiotic use is widespread in livestock feeds. An astonishing 40 percent of all the antibiotics produced in the United States find their way into animal feeds to check disease and promote growth. By killing off less hardy bacterial species, chronic low doses of antibiotics create ideal growth environments for resistant strains. If these strains can be transferred to humans through meat, the resistant organisms may cause intractable illnesses. Allied to the problem of antibiotic resistance is the concern for transfer of the resistance. Researchers have demonstrated that plasmids and transposons account for the movement of antibiotic-resistant genes among bacterial species (see Chapter 7). Thus, the resistance in a relatively harmless bacterial species may be passed to a pathogenic one where the potential for disease is then increased by resistance to standardized treatment.

Antibiotics have been known as miracle drugs, but they are becoming overworked miracles. Some researchers suggest that antibiotics should be controlled as strictly as narcotics. Curbing the overuse of antibiotics through public and physician education and governmental regulation is a useful approach as well (although this approach will be particularly difficult because the antibiotics market is currently worth more than $25 billion per year in sales). The antibiotic roulette that is currently taking place should be a matter of discussion to all individuals concerned about infectious disease, be they scientist or student.

MicroFocus 24.6 recounts a recent example of drug resistance among American soldiers in Iraq.

CONCEPT AND REASONING CHECKS

24.18 Summarize the misuses and abuses of antibiotics.

New Approaches to Antibiotic Therapy Are Needed

KEY CONCEPT

- The discovery of new targets and the development of new antibiotics may help fight antibiotic resistance.

Dealing with the emerging antibiotic resistance is a major problem confronting contemporary researchers. As reports of antibiotic resistance and harmful side effects appear in the literature, scientists hasten their search for new approaches to antibiotic therapy as well as identifying new targets to which pathogens are susceptible. For example, it should be possible to prevent a bacterial cell from pumping an antibiotic like tetracycline out of its cytoplasm.

One targeted approach is to focus on the lipopolysaccharide (LPS) in the outer membrane of gram-negative bacteria (see Chapter 4). The LPS contains several unusual carbohydrates representing possible targets for new drugs.

Another approach is to interfere with an enzyme called DNA adenine methylase, which bacterial cells use to coat DNA with methyl groups and regulate DNA replication and repair. Researchers have significantly reduced the virulence of a *Salmonella* species by disabling the gene encoding the methylating enzyme. The researchers now hope to develop a chemical compound (synthetic antibiotic?) to neutralize the enzyme, thereby pinpointing a protein found only in the bacterial cell.

The current explosion in microbial genome sequencing (see Chapter 8) affords an opportunity to discover essential bacterial genes that may be targets for antimicrobial drugs and to identify other forms of antibiotic resistance.

Because many of the newer antibiotics are simply minor modifications of existing drugs, there is a need to find new and unique antibiotics. The marine microbial community has barely been taped as a source of new antibiotics. In addition, because only a very small portion of bacterial species can be cultured, 90 to 99 percent of the metagenome has not been sampled (see Chapter 8)—and there may be many beneficial antibiotics awaiting discovery.

Many microbiology labs around the world are sequencing the genomes of dozens of different pathogens thought to contain genes for undiscoved antibiotics. For example, the genus *Streptomyces* is the source of more than two thirds of the antibiotics in current use. Almost certainly there are more antibiotics produced by this genus, as the opener to Chapter 8 describes. Through genome sequencing, researchers have identified clusters of genes in *Streptomyces coelicolor* that may be sources for new antibiotic agents. What other, perhaps novel, natural products exist is an area of active research.

Although antimicrobial drugs probably are not the magic bullets once perceived by Ehrlich, new drug discovery may provide us with many new antimicrobial agents to fight pathogens and resistance.

MicroInquiry 24 looks at what is involved in bringing a drug to the pharmacy shelf and medicine cabinet.

CONCEPT AND REASONING CHECKS

24.19 Assess the need for new and novel antibiotics and targets for those antibiotics.

MICROFOCUS 24.6: Public Health

Coming Soon to a Hospital Near You?

In September, 2004, the Centers for Disease Control and Prevention (CDC) reported an outbreak of *Acinetobacter baumannii* bloodstream infections in soldiers at military hospitals in the Middle East. Most acinetobacters are found in soil and water, and *A. baumannii* accounts for 80 percent of infections in health care centers and military hospitals. Infections can occur in the skin and in wounds with complications leading to pneumonia and meningitis. Left untreated, infections can be fatal.

In 2003, a few cases of *A. baumannii* infections had been reported in military hospitals in Iraq and Afghanistan, but by 2005, the Department of Defense had reported more than 400 *A. baumannii* colonized soldiers. *Forbes* magazine stated there had been more than 280 infection cases. What threw everyone into panic was that these strains of *A. baumannii*, unlike the ones that infected soldiers during the Vietnam War, were resistant to many classes of antibiotics, including penicillin, chloramphenicol, and some aminoglycosides, the latter normally being used in combination for treatment of such infections. The antibiotic resistance events became even more of concern when *A. baumannii* strains started showing resistance to the carbapenems that had been the drug of last resort. Among one set of isolates from samples sent to the Walter Reed Army Medical Center, 35 percent were susceptible only to imipenem and 4 percent were resistant to all drugs tested.

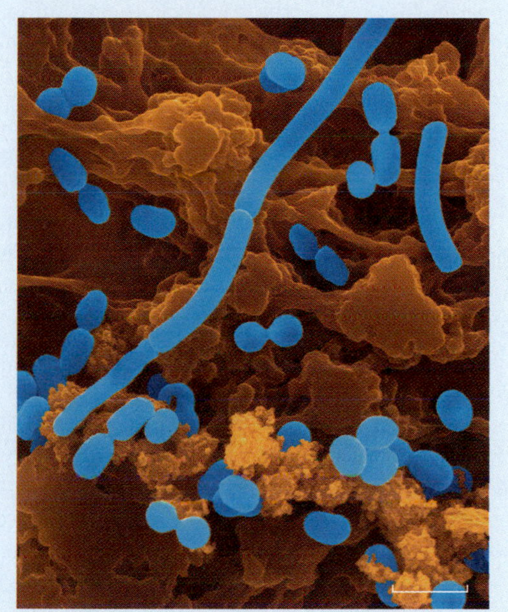

False-color scanning electron micrograph of *Acinetobacter* cells (Bar = 3 μm.)

The antibacterial drug resistance seen in *A. baumannii* leaves few treatment options for infections. In addition, *A. baumannii* can survive on dry surfaces for up to 20 days. This puts immunocompromised and other hospital patients at risk for drug resistant infections and makes it more likely soldiers wounded in battle will be vulnerable to infection. In fact, at least five soldiers have died from *A. baumannii* infections.

Also of concern is the reporting of more cases of *A. baumannii* infections and deaths within the United States. These cases involve pneumonia in hospitalized patients, especially those on ventilators in ICUs. In fact, as many as 25 percent of healthy adults harbor the bacterial cells and up to 27 percent of hospital sink traps and 20 percent of hospital floor swabs have yielded isolates of *Acinetobacter*. In early 2006, the Arizona Department of Health Services identified a potential outbreak in at least 236 cases of *A. baumannii* in patients ranging form 3 weeks to 90 years old. The antibiotic resistance remains to be reported at this writing. However, it seems it is only a matter of time before serious antibiotic-resistant strains come to a hospital near you.

MICROINQUIRY 24

Testing Drugs—Clinical Trials

Before any antibiotic or chemotherapeutic agent can be used in therapy, it must be tested to ensure its safety and efficacy. The process runs from basic biomedical laboratory studies to approval of the product to improve health care. On average it takes over eight years to study the drug in the laboratory (preclinical testing), test it in animals, and finally run human clinical trials.

In this MicroInquiry, we examine a condensed version of the steps through which a typical antibiotic would pass. Some questions asked are followed by the answer so you can progress further in your analysis. Please try to answer these questions before progressing through the scenario. Also, methods are used that you have studied in this and previous chapters. A real pharmaceutical company might have other, more sophisticated methods available. Answers can be found in **Appendix D**.

The Scenario

You are head of a drug research group with a pharmaceutical company that has isolated a chemical compound (let's call it FM04) that is believed to have antibacterial properties and thus potential chemotherapeutic benefit. As head, you must move this drug through the development pipeline from testing to clinical trials, evaluating at each step whether to proceed with further testing.

Preclinical Testing

Many experiments need to be done before the chemical compound can be tested in humans. Because FM04 is "believed" to have antibacterial properties,

24.1a. What would be the first experimental tests to be carried out?

You would experimentally test the drug on bacterial cells in culture to ascertain its relative strength and potency as an antibacterial agent. This testing might include the agar disk diffusion method described in this chapter. The figure to the right shows three diffusion disk plates, each plated with a different bacterial species (*Escherichia coli, Staphylococcus aureus,* and *Pseudomonas aeruginosa*) and disks with different concentrations of FM04 (0, 1, 10, 100 μg).

24.1b. What can you conclude from these disk diffusion studies? Should drug testing continue? Explain.

Let's assume the bacterial studies look promising and several months of additional studies confirm the antibacterial properties. FM04 now would be tested on human cells in culture to see if there are any toxic effects on metabolism and growth. **Table A** presents the results using the same drug concentrations from the bacterial studies. (The drug was prepared as a liquid stock solution to prepare the final drug concentration in culture.)

24.1c. What would you conclude from the studies of human cells in culture? Should drug testing continue? Explain.

24.1d. What was the point of including a solution with 0 μg of the drug?

Animal Testing

When drugs are injected or ingested into the whole body, concentrations required for a desired chemotherapeutic effect may be very different from the experimental results with cells. Higher concentrations often are required and these may have toxic side effects. In animal testing, efforts should be made to use as few animals as possible and they should be subjected to humane and proper care. Let's assume the research studies of FM04 on mice used oral drug concentrations of 1, 10, 100, and 1,000 mg. **Table B** presents the simplified results.

24.2a. What would you conclude from the mouse studies? Should drug testing continue? Explain.

Because the results look positive and higher doses can be administered without serious toxic side effects (except at 1,000 mg), additional studies would be carried out to determine how much of the drug is actually absorbed into the blood, how it is degraded and excreted in the animal, and if there are any toxic breakdown products produced. These studies could take a few years to complete. Again, for the sake of the exercise, let's say the studies with FM04 remain promising.

Although mice are quite similar to humans physiologically, they are not identical, so human clinical trials need to be done. This is the reason the US Food and Drug Administration (FDA) requires all new drugs to be tested through a *three-phase clinical trial period.*

First, as head of the research team, you must provide all the preclinical and animal testing results as part of an FDA *Investigational New Drug* (IND) *Application* before human tests can begin. The application must spell out how the clinical investigations will be conducted. As head of the investigational team, you must design the protocol for these human trials.

24.2b. How would you conduct the human tests to see if FM04 has positive therapeutic effects as an antibiotic in humans?

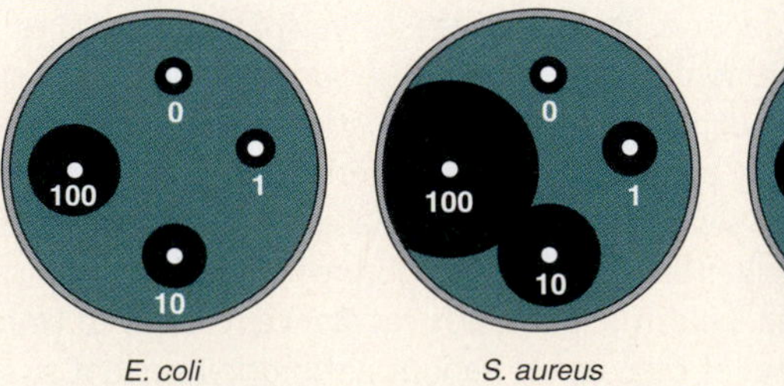

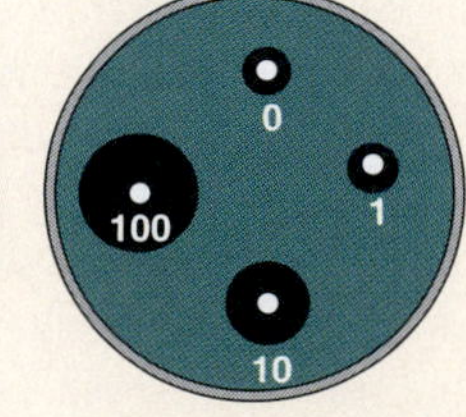

Briefly, there are many factors to take into consideration in designing the protocol.

(i) How will the drug be administered? Let's assume it will be oral in the form of pills.
(ii) You need to recruit groups of patients (volunteers) of similar age, weight, and health status (healthy as well as patients with the infectious condition to which the drug may be therapeutic). Note that the bacterial studies suggested FM04 had its most potent effect on *Staphylococcus aureus*, a gram-positive species. Perhaps patients with a staph infection would be recruited.
(iii) Eventually, volunteers (patients) need to be split into two groups, one that receives the drug and one that receives a placebo (an inactive compound that looks like the FM04 pill).
(iv) The clinical studies should be carried out as "double blind" studies. In these studies, neither the patients nor the investigators know which patients are getting the drug or placebo. This prevents patients and investigators from "wishful thinking" as to the outcome of the studies.

In actual clinical studies there are even more factors to consider, but we will assume that FM04 wins approval for clinical testing.

Clinical Trials

In *Phase 1 studies*, a small number (20 to 100) of healthy patients are treated with the test drug at different concentrations (doses) to see if there are any adverse side effects. These studies normally take at least several months. If unfavorable side effects are minimal, Phase 2 trials begin.

Phase 2 studies use a larger population of patients (several hundred) who have the disease the drug is designed to treat. Here, it is especially important to split the patients into the test and control groups because the major purpose of these trials is to study drug effectiveness. There needs to be a control group to accurately contrast effectiveness. These trials can take anywhere from several months to two years to complete. Provided there are no serious side effects or toxic reactions, or a lack of effectiveness, Phase 3 trials begin. There is an ethical question that can arise at this stage in the clinical trials, especially if the drug shows signs of reversing an illness that might be life-threatening. Even medical experts disagree as to the answer.

24.3 Is it ethical to give ill patients placebos when effective treatment appears available?

Phase 3 studies include several hundred to a few thousand patients. The trials can take several years to complete because the purpose is to evaluate safety, dosage, and effectiveness of the drug.

If the clinical trials are positive, as head you can recommend that the pharmaceutical manufacturer apply to the FDA for a New Drug Application (NDA). The application is reviewed by an internal FDA committee that examines all clinical data, proposed labeling, and manufacturing procedures. The most important questions that need to be addressed are:

(i) Were the clinical studies well controlled to provide "substantial evidence of effectiveness?"
(ii) Did the clinical study results demonstrate that the drug was safe; that is, do the benefits outweigh the risks?

Based on its findings, the committee then makes its recommendation to the FDA. If approved, an approval letter is given to the company to market the drug.

Note that very few drugs (perhaps 5 in 5,000 tested compounds) actually make it to human clinical trials. Then, only about 1 of those 5 are found to be safe and effective enough to reach the pharmacy.

TABLE A Adverse Cellular Effects Based on FM04 Drug Concentration

Final Drug Concentration (μg/culture)	Adverse Cellular Effects
0	None
1	None
10	None
100	Abnormal cells and cell death

TABLE B Toxicity Results on 50 Mice Given Different Concentrations of FM04

Drug Concentration (mg)	Number of Mice	Adverse Reactions
0	10	None
1	10	None
10	10	None
100	10	None
1,000	10	Tremors and seizures in 8 mice

SUMMARY OF KEY CONCEPTS

24.1 The History and Properties of Antimicrobial Agents

- Ehrlich saw antibiotics as a chemotherapeutic approach to alleviating disease. One of his students isolated the first chemotherapeutic agent, Salvarsan. Domagk used Prontosil to treat bacterial infections.
- Penicillin was discovered by Fleming, and purified and prepared for chemotherapy by Florey and Chain.
- All antimicrobial agents represent synthetic drugs or antibiotics, demonstrate selective toxicity, and have a broad or narrow drug spectrum.

24.2 The Synthetic Antibacterial Agents

- The sulfonamides interfere with the production of folic acid through competitive inhibition. Modern sulfonamides are trimethoprim and sulfamethoxazole.
- Isoniazid and ethambutol are antituberculosis drugs affecting wall synthesis. Ciprofloxacin is a fluoroquinolone that interacts with DNA to inhibit replication.

24.3 The Beta-Lactam Family of Antibiotics

- Penicillin interferes with cell wall synthesis in gram-positive bacterial cells. It can cause an anaphylactic reaction in sensitive individuals.
- Numerous synthetic and semisynthetic forms of penicillin are more resistant to beta-lactamase activity.
- Certain cephalosporin drugs are first-choice antibiotics for penicillin-resistant bacterial species and a wide variety of these drugs is currently in use.
- The carbapenems (imipenem) inhibit cell wall synthesis and are broad spectrum drugs.

24.4 Other Bacterially Produced Antibiotics

- Vancomycin inhibits cell wall synthesis, while bacitracin and polymyxin B affect the permeability of the cell membrane.
- Aminoglycosides (gentamicin, neomycin, and kanamycin) inhibit protein synthesis in gram-negative bacterial cells.
- Chloramphenicol is a broad-spectrum antibiotic used against gram-positive and gram-negative bacteria. Less severe side effects accompany tetracycline use, and this antibiotic is recommended against gram-negative bacteria as well as rickettsiae and chlamydiae.
- The macrolides, clindamycin, streptogramins inhibit protein synthesis, while rifampin blocks DNA transcription.

24.5 Antifungal and Antiparasitic Agents

- The polyenes (nystatin, amphotericin B) and imidazoles are valuable against fungal infections. The echinocandins inhibit fungal cell wall synthesis, flucytosine interrupts nucleic acid synthesis, and griseofulvin interferes with mitosis.
- Antiprotozoal agents include the aminoquinolines (quinine and chloroquine) that are used to treat malaria; sulfonamides interfere with folic acid synthesis; and nitroimidazoles inhibit nucleic acid synthesis.
- Antihelminthic drugs include praziquantel, mebendazole, and avermectins.

24.6 Antibiotic Assays and Resistance

- Antibiotic assays include the tube dilution assay, which measures the minimal inhibitory concentration, and the agar disk diffusion method, which determines the susceptibility of a bacterial species to a series of antibiotics.
- Bacterial species have developed resistance to antimicrobial agents by altering metabolic pathways, inactivating antibiotics, reducing membrane permeability, or modifying the drug target.
- Arising from any of several sources such as changes in microbial biochemistry, antibiotic resistance threatens to put an end to the cures of infectious disease that have come to be expected in contemporary medicine.
- New approaches include the discovery of new drug targets and developing new antibiotics to fight infectious disease and antibiotic resistance.

LEARNING OBJECTIVES

After understanding the textbook reading, you should be capable of writing a paragraph that includes the appropriate terms and pertinent information to answer the objective.

1. Summarize the accomplishments of Paul Ehrlich, Gerhard Domagk, and Alexander Fleming to early advances in antimicrobial chemotherapy.
2. Identify the important properties of antimicrobial agents.
3. Explain how sulfonamide drugs block the folic acid metabolic pathway in bacterial cells.
4. Compare and contrast the mechanism of action for isoniazid and the quinolones.
5. Summarize the mechanism of action of penicillin and its semisynthetic derivatives.
6. Discuss the other beta-lactam antibiotics that affect cell wall synthesis.
7. Describe why bacterial and fungal species naturally produce antibiotics.
8. Indicate how vancomycin differs from the beta-lactam antibiotics.
9. Distinguish between the mechanisms of action for the polypeptide antibiotics.
10. Hypothesize why there is such a large number of natural antibiotics that target the ribosome and protein synthesis.
11. Identify the synthetic drugs and antibiotics affecting DNA synthesis.
12. Explain why selective toxicity is a critical consideration when developing antifungal and antiparasitic drugs.
13. Indicate the targets for antifungal drugs.
14. Identify the drug targets in the protozoal and helminthic parasites.
15. Contrast between the two antibiotic susceptibility assays.
16. Describe the four mechanisms used by bacterial species to generate resistance to antibiotics.
17. Justify the statement that antibiotic resistance has resulted from their misuse and abuse.
18. Identify ways researchers are attempting to discover new drug targets and antibiotics.

SELF-TEST

Answer each of the following questions by selecting the ***one*** answer that best fits the question or statement. Answers to even-numbered questions can be found in **Appendix C**.

1. Ehrlich and Hata discovered _____ that was used to treat _____.
A. streptomycin; cholera
B. Salvarsan; syphilis
C. penicillin; surgical wounds
D. Salvarsan; malaria
E. Prontosil; malaria

2. The re-isolation and purification of penicillin was carried out by
A. Waksman.
B. Ehrlich.
C. Florey and Chain.
D. Domagk.
E. Fleming.

3. The concentration of an antibiotic causing harm to the host is called the
A. toxic dosage level.
B. therapeutic dosage level.
C. minimal inhibitory concentration.
D. effective dosage level.
E. chemotherapeutic index.

4. Trimethoprim and sulfamethoxazole are examples of _____ that block _____ synthesis.
A. sulfonamides; PABA
B. penicillins; cell wall
C. sulfonamides; folic acid
D. macrolides; protein
E. fluoroquinolones; cell wall

5. All the following are drugs or drug classes blocking cell wall synthesis *except*:
A. cephalosporins.
B. carbapenems.
C. penicillins.
D. vancomycin.
E. tetracyclines.

6. Which one of the following (**A–D**) would *not* be inactivated by a beta-lactamase? If all would be inactivated, select **E**.
A. Penicillin
B. Vancomycin
C. Cephalosporin
D. Carbapenems
E. All the above (**A–D**) would be inactivated.

7. All the following are aminoglycosides *except*:
A. gentamicin.
B. streptomycin.
C. kanamycin.
D. chloramphenicol.
E. neomycin.

8. The antibacterial drugs with four benzene rings are the
A. tetracyclines.
B. macrolides.
C. streptogramins.
D. aminoglycosides.
E. oxazolidinones.

9. Which one of the following is *not* an inhibitor of protein synthesis?
A. Clindamycin
B. Macrolides
C. Rifampin
D. Chloramphenicol
E. Streptogramins

10. Filamentous fungi, protozoa, and helminths are
A. eukaryotes.
B. multicellular.
C. organisms that produce antibiotics.
D. resistant to all known antibiotics.
E. Both **A** and **B** are correct.

11. Antifungal drugs, such as _____, inhibit proper formation of _____.
A. nystatin; a cell wall
B. miconazole; a plasma membrane
C. griseofulvin; DNA
D. ketoconazole; a cell wall
E. flucytosine; a microtubule

12. All of the following antiprotozoal drugs have been used to treat malaria *except*:
A. primaquine.
B. melarsoprol.
C. quinine.
D. mefloquine.
E. chloroquine.

13. This group of synthetic drugs is also used to inhibit folic acid synthesis in protozoa.
A. Penicillins
B. Heavy metals
C. Aminoquinolines
D. Imidazoles
E. Sulfonamides

14. This antihelminthic agent makes the membrane permeable to calcium ions.
A. Mebendazole
B. Bithionol
C. Pentamidine
D. Ivermectin
E. Praziquantel

15. The _____ is used to determine an antibiotic's minimal inhibitory concentration (MIC).
A. Ames test
B. expansion test
C. tube dilution method
D. agar disk diffusion method
E. Kirby-Bauer test

16. The phosphorylation of an antibiotic is an example of which mechanism of resistance?
A. Target modification
B. Reduced permeability
C. Active export
D. Antibiotic inactivation
E. Altered metabolic pathway

QUESTIONS FOR THOUGHT AND DISCUSSION

Answers to even-numbered questions can be found in **Appendix C**.

1. According to historians, 2,500 years ago the Chinese learned to treat superficial infections such as boils by applying moldy soybean curds to the skin. Can you suggest what this implies?
2. Most naturally occurring antibiotics appear to be products of the soil bacteria belonging to the genus *Streptomyces*. Can you draw any connection between the habitat of these organisms and their ability to produce antibiotics?
3. Of the thousands and thousands of types of organisms screened for antibiotics since 1940, only five bacterial genera appear capable of producing these chemicals. Does this strike you as unusual? What factors might eliminate potentially useful antibiotics?
4. Is an antibiotic that cannot be absorbed from the gastrointestinal tract necessarily useless? How about one that is rapidly expelled from the blood into the urine? Explain your answers?
5. The antibiotic resistance issue can be argued from two perspectives. Some people contend that because of side effects and microbial resistance, antibiotics will eventually be abandoned in medicine. Others see the future development of a superantibiotic, a type of "miracle drug." What arguments can you offer for either view? Which direction in medicine do you support?
6. History shows that over and over, creativity is a communal act, not an individual one. How does the 1945 Nobel Prize to Fleming, Florey, and Chain reflect this notion?

APPLICATIONS

Answers to even-numbered questions can be found in **Appendix C**.

1. In 1877, Pasteur and his assistant Joubert observed that anthrax bacilli grew vigorously in sterile urine but failed to grow when the urine was contaminated with other bacilli. What was happening?
2. In May 1953, Edmund Hillary and Tenzing Norgay were the first to reach the summit of Mount Everest, the world's highest mountain. Since that time over 150 other mountaineers have reached the summit, and groups have gone to Nepal from all over the world on expeditions. The arrival of "civilization" has brought a drastic change to the lifestyle of Nepal's Sherpa mountain people. For example, half of all Sherpas used to die before the age of 20, but partly due to antibiotics available to fight disease, the population has grown from 9 million to more than 23 million. Medical enthusiasts are proud of this increase in the life expectancy, but population ecologists see a bleaker side. What do you suspect they foresee, and what does this tell you about the impact antibiotics have on a culture?
3. Why would a synthetically produced antibiotic be more advantageous than a naturally occurring antibiotic? Why would it be less advantageous?

REVIEW—ANTIMICROBIAL DRUGS

On completing your study of these pages, test your understanding of their contents by deciding whether the following statements are true or false. If the statement is true, write "True" in the space. If false, substitute a word for the underlined word to make the statement true. The answers to even-numbered statements are listed in **Appendix C**.

1. _____ Penicillin is a synthetic drug.
2. _____ Tetracycline has four benzene rings in the molecule.
3. _____ Sulfonamides block folic acid formation in fungi.
4. _____ *Candida* is a fungal genus susceptible to nystatin therapy.
5. _____ Gentamicin is an antibiotic that interferes with transcription.
6. _____ Streptomycin is an example of an aminoglycoside.
7. _____ A side effect of chloramphenicol is a discoloration of the teeth.
8. _____ Isoniazid is a drug recommended for tuberculosis.
9. _____ Penicillinase also is known as an alpha-lactamase.
10. _____ Polymyxin B affects membrane permeability.
11. _____ Imidazoles are antifungal drugs that block protein synthesis.
12. _____ Folic acid synthesis is inhibited by rifampin.
13. _____ Ciprofloxacin is used to treat all species of *Plasmodium* that cause malaria.
14. _____ Vancomycin inhibits bacterial ribosomes.
15. _____ Penicillin has no effect on viruses
16. _____ Artemisinin is a drug that releases free radicals.
17. _____ Florey and Chain discovered prontosil.
18. _____ Vancomycin and penicillin are beta-lactam antibiotics.
19. _____ Neosporin contains bacitracin, polymyxin B, and tetracycline.
20. _____ Isoniazid interferes with RNA synthesis in bacterial cells.

REVIEW—IDENTIFICATION

Identify the bacterial structure inhibited or affected by each of the antibiotics in the description below. Answers to even-numbered descriptons can be found in **Appendix C**.

Descriptions

1. Sulfonamides act here.
2. Polymyxin B affects this structure.
3. Rifampin blocks this process.
4. Quinolones act here.
5. Tetracyclines inhibit the function of this structure.
6. Cephalosporins target this structure.
7. The marcrolides act here.
8. Vancomycin inhibits the assembly of this structure.
9. Aminoglycosides inhibit this structure.
10. Penicillin acts here.

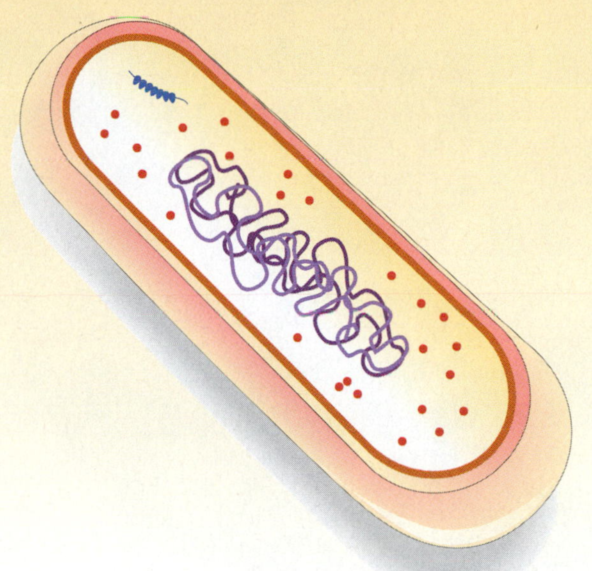

HTTP://MICROBIOLOGY.JBPUB.COM/

The site features learning, an on-line review area that provides quizzes and other tools to help you study for your class. You can also follow useful links for in-depth information, read more MicroFocus stories, or just find out the latest microbiology news.

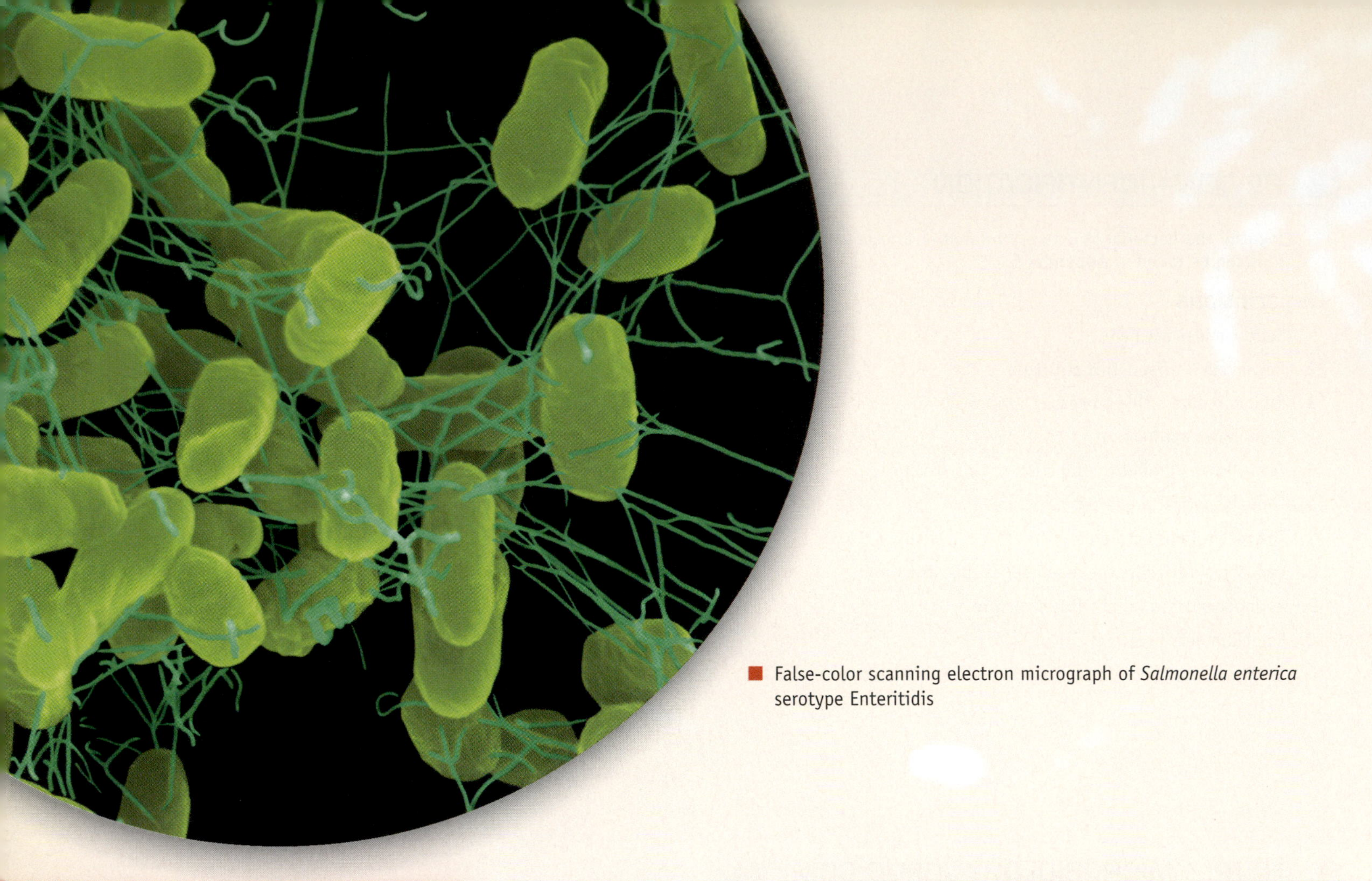

False-color scanning electron micrograph of *Salmonella enterica* serotype Enteritidis

PART 7

Microbiology and Public Health

Public health has many facets. Several—such as mental health, nutritional health, and environmental health—lie outside the realm of this text. Other facets, however, are concerned with communicable disease and therefore fall within the scope of microbiology. Immunization, for example, is a public health measure because it creates a barrier within the individual to lessen susceptibility to infection. Case findings and antibiotic treatments are also under the umbrella of public health because they are used to limit the spread of communicable disease.

In Part 7, we are concerned with sanitation, a facet of public health designed to prevent microorganisms from reaching the body. Sanitation came into full flower in the mid-1800s, and it is a relatively recent phenomenon. Before that time, living conditions in some Western European and American cities were almost indescribably grim: Garbage and dead animals littered the streets; human feces and sewage stagnated in open sewers; rivers were used for washing, drinking, and excreting; and filth was rampant.

The belief that filth was a catalyst to disease fueled the sanitary movement of the mid-1800s (see Chapter 1). As the Industrial Revolution sparked great population increases in the cities, health problems mounted, and sanitary reformers spoke up for effective sewage treatment, water purification, and food preservation. The germ theory of disease, developed in the 1870s, strengthened the movement and justified its actions because communicable disease could now be blamed on microorganisms in food and water. As sanitary reformers merged with bacteriologists, they issued loud calls for government support of public health and stimulated the development of new methods for dealing with food and water, which we study in Chapters 25 and 26.

As you know, only a very small minority of microorganisms are pathogens. In fact, some contribute mightily to public health. For example, certain microorganisms are used to produce a broad assortment of foods and dairy products making up a regular part of our diet. In addition, microorganisms play a dominant role in numerous cycles of elements in the environment and forge a link between what is useless and what is useful to other living organisms. Moreover, scientists employ microorganisms on industrial scales to synthesize a variety of products we could not obtain otherwise. We see examples of these contributions in Chapters 25, 26, and 27.

MICROBIOLOGY PATHWAYS

Public Health Inspector

How often have you seen on the television or in the newspaper a report on "dirty restaurants" where there is "slime in the ice machine" or "meat kept at inappropriate temperatures"? These reports are the result of visits by state, county, or city food or public health inspectors.

Nationally, the U.S. Department of Agriculture's Food Safety and Inspection Service (FSIS) employs more than 7,500 inspectors who comprise an essential component of the nation's public health mission. The food inspectors in private commercial slaughtering plants provide the expertise to ensure diseased and contaminated meat and poultry does not enter the food supply. Therefore, they are responsible for most of a slaughtering plant's daily inspections of animals—both pre- and post-slaughter.

One career path for a food inspector is as a consumer safety inspector in private meat, poultry, or egg processing plants. Such inspectors ensure the plant is operating within approved food-safety guidelines.

Another career path is as an import inspector. Import inspectors work at seaports and other points of entry to the United States. They make sure products imported from other countries are as safe as those produced domestically.

To qualify for an entry-level position, you must pass a written test and have either a Bachelors degree or one year of job-related experience in the food industry. This experience must demonstrate knowledge of sanitation practices and control measures used in the commercial handling and preparation of food products for human consumption.

Another career path is as a public health inspector, sometimes called an environmental health officer. These individuals help prevent infectious disease and promote health by educating people about public health issues and enforcing health legislation. They are the ones looking for "slime in the ice machine" and other dirty restaurant violations. Inspectors monitor conditions in restaurants and schools, food processing plants, hotels and motels, child care and long-term care facilities, animal facilities, swimming pools, and recreational camps. Public health inspectors also may investigate reportable diseases.

To become a public health inspector you must have a community college or university degree in an appropriate specialization such as public and environmental health, food sciences or water chemistry. You must have a strong interest in the applied sciences and an interest in community service; be curious, decisive, observant; and have good judgment and excellent interpersonal, investigative, and communication skills. And of course, having some microbiology background and course studies is a real bonus!

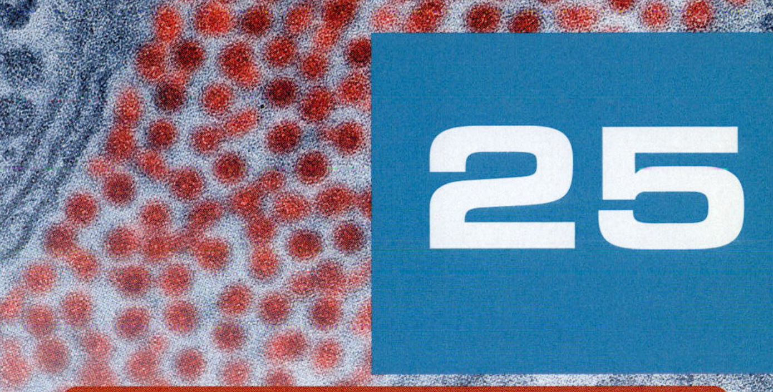

Microbiology of Foods

We are not losing the "war." Unfortunately, we probably can't win it either. Food safety folks will always have a job.
—Arthur P. Liang, MD, MPH, Centers for Disease Control and Prevention

Chapter Preview and Key Concepts

25.1 Food Spoilage
- Food spoilage occurs when foods are altered from their expected form.
- The chemical and physical properties of food have a significant bearing on the type of microorganisms growing on or in it.
- Food spoilage is due to the metabolism of contaminating microorganisms.
- Meats and fish become contaminated through handling and processing.
- Contamination in poultry and eggs may have a human or "natural" source.
- Flour, eggs, and sugar are generally the sources of spoilage organisms.
- *Aspergillus* and *Claviceps* species of fungi can spoil cereal grains.
- Microbial acids or enzymes can curdle milk products.

25.2 Food Preservation
- Heat kills microorganisms by changing the physical and chemical properties of their proteins.
- Microbial enzyme activity is greatly reduced at low temperatures.
- Drying and osmotic pressure remove water.
- Chemical preservatives can inhibit the growth of bacterial cells, yeast, and molds.
- Irradiated food provides the same benefits of preservation as does heat, refrigeration, or freezing.
- People can be affected by the organism or the toxin contaminating food.

  MicroInquiry 25: Keeping Microorganisms Under Control
- Food safety systems are in place to identify microbial contamination points.

25.3 Foods from Microorganisms
- Fermentation can retard spoilage and the growth of microbes in the food product.
- Microbial acids are important to milk products and cheese manufacture.

The idea was appealing and the price was right: a patty melt sandwich and a soft drink for lunch. The rye bread was toasted, the hamburgers were stacked and waiting to be cooked, the American cheese slices were lined up next to the grill, and the aroma from the sautéed onions was irresistible. It was October 1983, at the Skewer Inn, a restaurant at the Northwoods Mall in Peoria, Illinois. The stage was set for one of the worst outbreaks of botulism in the United States.

Between October 14 and 16, numerous people stopped by the restaurant and enjoyed patty melt sandwiches. Soon, however, 36 individuals began experiencing the paralyzing signs of botulism. They suffered blurred vision, difficulty swallowing and chewing, and labored breathing (see Chapter 10). One by one they called their doctors, and within a week, all were hospitalized. Twelve patients had to be placed on respirators. After many anxious hours, all but one recovered.

Investigators from the Centers for Disease Control and Prevention (CDC) arrived in Peoria shortly thereafter. They obtained detailed food histories from patients and from others who ate at the restaurant during the same three-day period. First, they identified patty melt sandwiches as the probable cause (24 of 28 patients interviewed specifically recalled eating the sandwiches); then they began a search to pinpoint the item that might be responsible. The data pointed to the onions. Investigators isolated *Clostridium botulinum* spores from the skins of fresh onions at the restaurant, and learned that after sautéing, the onions were left uncovered on the warm stove for hours. Furthermore, the onions were not reheated before serving. Spores had probably germinated within anaerobic mounds of warm onions and the vegetative cells deposited their deadly toxins.

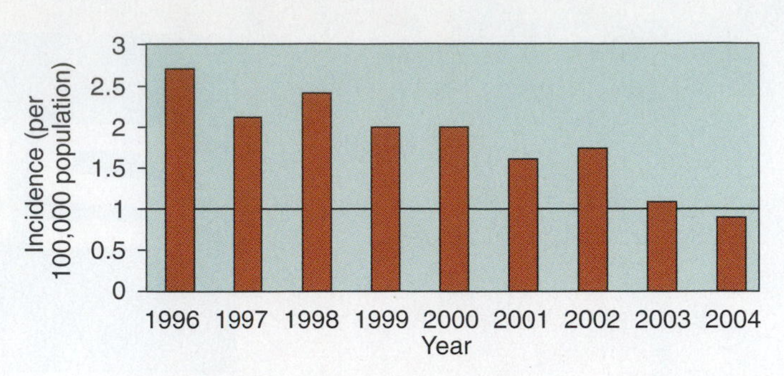

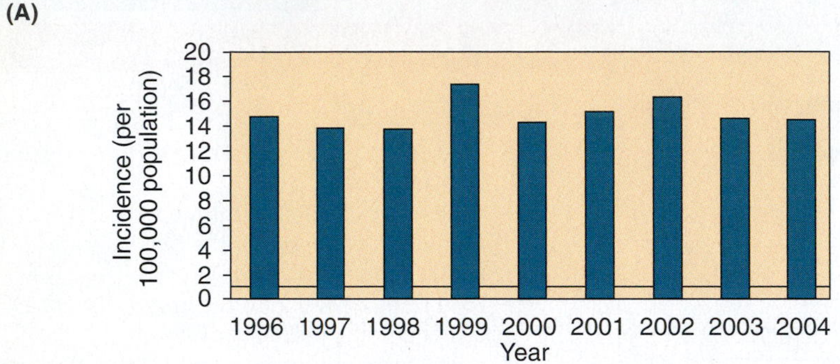

FIGURE 25.1 Incidence of Selected Foodborne Illnesses in the United States from 1996–2004. The incidence of foodborne illness caused by (**A**) *E. coli* 0157:H7 and (**B**) *Salmonella* in the United States from 1996 to 2004. The national health objective is to reach a level where the incidence is no more than 1 per 100,000 people. *Source:* FoodNet/CDC.

Q: Provide some ways in which the incidence of foodborne diseases caused by E. coli *and* Salmonella *can be reduced.*

Incidents like this one occur every so often in the United States. The CDC estimates more than 76 million Americans get sick (26 people in every 100), over 300,000 are hospitalized, and 5,000 people die each year from foodborne illnesses. However, efforts by the meat industry are attempting to reduce the number of foodborne illnesses (FIGURE 25.1). For example, *Escherichia coli* O157:H7 infections declined 36 percent between 2002 and 2003, the largest decline ever. Since 1996, *E. coli* O157:H7 infections have declined 42 percent, while others, such as those caused by *Salmonella* species, have shown less of a reduction. The national health objective is to reach a level where the incidence of these and other foodborne illnesses, such as *Campylobacter* and *Listeria,* is no more than 1 per 100,000 people. However, *E. Coli* O157:H7 infections transmitted by produce have been increasing. The 2006 outbreak of *E. coli* O157:H7 in fresh, bagged spinach is but one example (see Chapter 10).

The numbers of Americans infected every year illustrates how most foods, even cooked foods, provide excellent conditions for the growth of microorganisms and the depositing of their toxins. The organic matter in food is plentiful, the water content is usually sufficient, and the pH is either neutral or only slightly acidic. To the food manufacturer or restaurant owner, the growth of microorganisms may spell economic loss or a reputation for bad business. To the consumer, it may mean illness or, in some cases, death.

The Foodborne Diseases Active Surveillance Network (FoodNet) within the CDC is a collaborative project administered at ten U.S. sites along with the Department of Agriculture and the Food and Drug Administration (FDA). FoodNet actively surveys for foodborne diseases and carries out epidemiologic studies to help public health officials better understand the epidemiology of foodborne diseases in the United States.

The primary focus of this chapter is to examine the types of microorganisms that contaminate various foods and to point out the consequences of contamination. We examine food spoilage and the preservation methods used to prevent spoilage. Toward the end of the chapter, we describe some forms of microbial growth in food that are actually desirable because they lead to numerous fermented foods we consume regularly.

25.1 Food Spoilage

Food spoilage has been a continuing problem ever since humans first discovered they could produce more food than they could eat in a single meal. We all know of Marco Polo's travels to China in the 13th century to obtain spices and explore new trade routes. What often is not mentioned is that the spices were more than just a luxury of the time. Spices were essential for improving the smell and taste of spoiled food. Refrigeration was virtually unknown, and canning was yet to be invented.

Food Spoilage Comes from Many Microbial Sources

KEY CONCEPT

- Food spoilage occurs when foods are altered from their expected form.

Spoiled foods usually have an unpleasant appearance, aroma, and taste. Sometimes, however, these signs may be difficult to detect, such as when staphylococci deposit enterotoxins in food or when too few bacteria grow to cause a perceptible change. The consumption of toxins or microorganisms may cause a number of food poisonings (intoxications) or infections, including those noted in Chapter 10.

■ Enterotoxins: Bacterial exotoxins that can cause gastroenteritis.

Contaminating microorganisms enter foods from a variety of sources. Airborne organisms, for example, fall onto fruits and vegetables and then penetrate the product through an abrasion of the skin or rind; crops carry soilborne bacterial species to the processing plant; shellfish concentrate organisms by straining contaminated water and catching the organisms in their filtering apparatus; and rodents and arthropods transport microorganisms on their feet and body parts as they move about among foods.

Human handling of foods also provides a source of contamination. For example, bacterial species from an animal's intestine contaminate meat handled carelessly by a butcher. Raw vegetables, such as those typically found at salad bars, also are of great concern for spoilage.

CONCEPT AND REASONING CHECKS

25.1 Identify potential microbial sources leading to food spoilage.

Several Conditions Can Determine If Spoilage Will Occur

KEY CONCEPT

- The chemical and physical properties of food have a significant bearing on the type of microorganisms growing on or in it.

Because food is basically a culture medium for microorganisms, the chemical and physical properties of food have a significant bearing on the type of microorganisms growing on or in it (FIGURE 25.2).

Water. One of the prerequisites for all life is water (see Chapter 2). Therefore, food must be moist, with a minimum water content of 18 to 20 percent before contamination by microorganisms and spoilage can occur. Microorganisms do not grow in foods such as dried beans, rice, and flour because of their low water content.

pH. Another important factor is a food's acidity. Most foods fall into the slightly acidic range on the pH scale, and numerous bacterial species multiply under these conditions. In foods with a pH of 5.0 or below, acid-loving molds often replace the bacterial species. Citrus fruits, for example, generally escape bacterial spoilage but yield to mold contamination.

Oxygen and Temperature. Two other considerations are oxygen and temperature. Vacuum-sealed cans of food do not support the growth of aerobic bacteria, nor do vegetables or most bakery products support anaerobes. Similarly, the refrigerator is usually too cold for the growth of human pathogens, but the warm hold of a ship or a humid, hot warehouse storeroom is an environment conducive to the growth of these pathogens. It is common knowledge that contamination is more likely in cooked food at warm temperatures than in refrigerated cooked food.

Physical Structure. Another property of a food is its physical structure. A raw steak, for example, is not likely to spoil quickly because microorganisms cannot penetrate the meat easily. However, an uncooked hamburger can deteriorate rapidly because microorganisms exist within the loosely packed ground meat as well as on the surface.

Chemical Composition. A food's chemical composition may be another determining factor in the type of spoilage possible. Fruits, for instance, support organisms metabolizing carbohydrates, whereas meats attract protein decomposers. Starch-hydrolyzing bacterial cells and molds often are found on potatoes, corn, and rice products. The presence of certain vitamins encourages particular microorganisms to proliferate, while the absence of other vitamins provides natural resistance to decay.

The food industry recognizes three groups of foods loosely defined on the basis of their physical and chemical properties. Highly perishable foods are those that spoil rapidly. They include poultry, eggs, meats, most vegetables and fruits, and dairy products. Foods such as nutmeats, potatoes, and some apples are considered semiperishable, because they spoil less quickly. Orange juice is usually considered

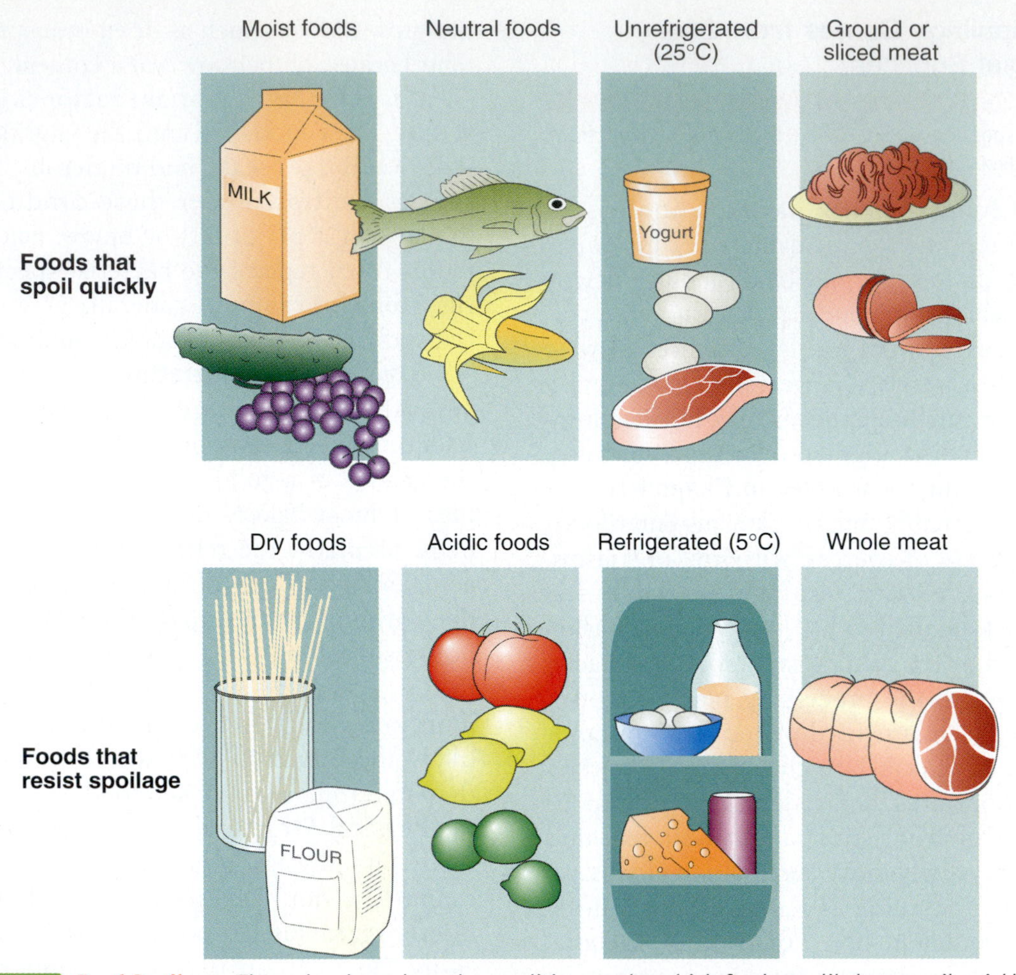

FIGURE 25.2 **Food Spoilage.** These drawings show the conditions under which foods are likely to spoil quickly or resist spoilage.

Q: Why are unrefrigerated foods more likely to spoil than refrigerated ones? Acidic ones versus neutral ones?

FIGURE 25.3 **The Perishability of Foods.** Examples of highly perishable, semiperishable, and nonperishable foods. The physical and chemical properties of these foods are reliable indicators of their rate of perishability.

Q: How do highly perishable foods differ from semiperishable foods?

semiperishable because of its high acidity. Nonperishable foods are stored often in the kitchen pantry. Included in this group are cereals, rice, dried beans, macaroni and pasta products, flour, and sugar. FIGURE 25.3 summarizes the three groups.

CONCEPT AND REASONING CHECKS

25.2 List the conditions that make foods susceptible to growth and spoilage by microorganisms.

MICROFOCUS 25.1: History

The Blood of History

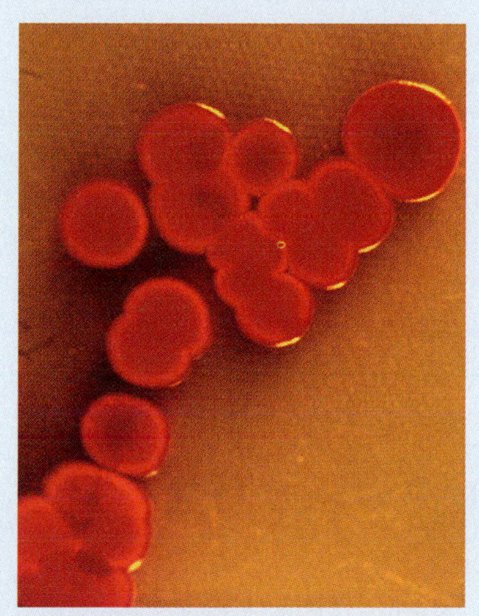

Serratia marcescens colonies growing on agar have the look of blood.

Because of its characteristic blood-red pigment and propensity for contaminating bread, *Serratia marcescens* has had a notable place in history. For example, the dark, damp environments of medieval churches provided optimal conditions for growth on sacramental wafers used in Holy Communion. At times, the appearance of "blood" was construed to be a miracle. One such event happened in 1264 when "blood dripped" on a priest's robe. The event was later commemorated by Raphael in his fresco The Mass of Bolsena. Unfortunately, religious fanatics used such episodes to institute persecutions because they believed that heretical acts caused the "blood" to flow.

It was not until 1819 that Bartholemeo Bizio, an Italian pharmacist, demonstrated the bloody miracles were caused by a living organism. Bizio thought the organism was a fungus. He named it *Serratia* after Serafino Serrati, a countryman whom he considered the inventor of the steamboat. The name *marcescens* came from the Latin word for "decaying," a reference to the decaying of bread.

In 332 BC, Alexander the Great and his army of Macedonians laid siege to the city of Tyre in what is now Lebanon. The siege was not going well. Then one morning, blood-red spots appeared on several pieces of bread. At first it was thought to be an evil omen, but a soothsayer named Aristander indicated the "blood" was coming from within the bread. This suggested that blood would be spilled within Tyre and the city would fall. Alexander's troops were buoyed by this interpretation and with renewed confidence they charged headlong into battle and captured the city. The victory opened the Middle East to the Macedonians. Their march did not stop until they reached India.

The Microorganisms Responsible for Spoilage Produce Specific Products

KEY CONCEPT

- Food spoilage is due to the metabolism of contaminating microorganisms.

Often the unpleasant appearance, aroma, and taste of spoiled foods is the result of natural products produced by the contaminating microorganisms. Yeasts, for example, live in apple juice and convert the carbohydrate into ethyl alcohol, a product giving spoiled juice an alcoholic taste. Certain bacterial species convert food proteins into amino acids, then break down the amino acids into foul-smelling end products. Cysteine digestion, for example, yields hydrogen sulfide, which imparts a rotten egg smell to food. Tryptophan digestion yields indole and skatole, which give food a fecal odor.

Two other possible products of the microbial metabolism of carbohydrates are acid, which causes food to become sour, and gas, which causes sealed cans to swell. Moreover, when fats break down to fatty acids as in spoiled butter, a rancid odor or taste may evolve. Capsule production by certain bacterial species causes food to become slimy, and pigment production imparts color. In numerous historical incidents, the red pigment from *Serratia marcescens* in bread has been interpreted as a sign of blood (MicroFocus 25.1).

Some foods resist spoilage naturally because they contain antimicrobial chemicals. The white of an egg has lysozyme, an enzyme that digests the cell wall of gram-positive bacteria (see Chapter 4). Garlic contains certain compounds that inhibit many bacterial species (see MicroFocus 23.4). Oil of cloves appears to have healing tendencies, while radish and onion extracts both retard bacterial growth.

CONCEPT AND REASONING CHECKS

25.3 What type of microbial products lead to spoilage?

Meats and Fish Can Become Contaminated in Several Ways

KEY CONCEPT

- Meats and fish become contaminated through handling and processing.

Sterile: Devoid of living microorganisms, viruses, and spores.

Meats and fish normally are free of contamination because the muscle tissues of living animals normally are sterile. Spoilage organisms enter during handling, processing, packaging, and storage. For example, if a piece of meat is ground for hamburger, microorganisms from the surface accumulate in the teeth of the grinder along with other dustborne organisms. Bacterial cells from preparers' hands or from a sneeze compound the problem. The grinder may be cleaned well and refrigerated, but rarely is it sterilized. The importance of foodborne infection in ground meat is pointed up by the recalls of ground beef that occur periodically in the United States.

There also is the problem of the so-called "choke points," or places where animals come together and epidemics spread. In the United States, for example, about 9,000 farms produce calves for beef; the animals then are sent to about 46,000 feedlots for developing, and then, to about 80 plants for slaughter. Microorganisms can spread at any of these points. Moreover, animal feeds often contain the entrails of poultry as added protein sources, and another opportunity for pathogen spread is presented.

Processed meats, such as luncheon meats, sausages, and frankfurters, represent special hazards because they are handled often. Also, natural sausage and salami casings made from animal intestines may contain residual bacterial cells, especially botulism spores. When preparers pack such casings tightly with meat, *Clostridium botulinum* may multiply and produce its powerful toxins. As early as the 1820s, people recognized the symptoms of "sausage poisoning" and coined the name "botulism" (*botulus* = "sausage").

Organ meats, such as livers, kidneys, and sweetbreads (thymus and pancreas), are less compact tissue than muscle and thus spoil more quickly. Moreover, because the organs contain many natural filtering tissues, bacterial cells tend to be trapped here. Foods like these therefore should be cooked as soon as possible after purchase.

The extent of contamination in meats often consists of a harmless "greening" seen on the surface of a steak. This discoloration is commonly due to the gram-positive rod *Lactobacillus*, or the gram-positive coccus *Leuconostoc*. The green color results from pigment alteration in the meat and represents no hazard to the consumer. Slime formation on the outer casings and souring in frankfurters, bologna, or processed meats also may result from these organisms as well as from *Streptococcus* species.

Microbiologists often can trace spoilage in fish to the water from which it was taken or held. Fish tissues deteriorate rapidly and the filleting of fish on a bloodstained block encourages contamination. Interestingly, the bacterial species in fish are naturally adapted to the cold environment in which most fish live and thus, cooling will not affect them; freezing is preferred. Shellfish are of particular concern because they commonly obtain their food by filtering particles from the water. Clams, oysters, and mussels concentrate such pathogens as hepatitis A viruses, typhoid bacilli, cholera vibrios, or amoebic cysts. Many cases of cholera have been related to raw oysters.

CONCEPT AND REASONING CHECKS

25.4 Summarize how meats and fish can become contaminated.

Poultry and Eggs Can Spoil Quickly

KEY CONCEPT

- Contamination in poultry and eggs may have a human or "natural" source.

Contamination in poultry and eggs may reflect human contamination, but it usually stems from microorganisms that have infected the bird. Members of the genus *Salmonella*, with over 2,400 pathogenic serotypes (see Chapter 10), may cause diseases in chickens and turkeys, then pass to consumers via poultry and egg products. Processed foods such as chicken pot pies, whole egg custard, mayonnaise, eggnog, and egg salad also may be sources of salmonellosis. Psittacosis (see Chapter 9) is another problem because *Chlamydia psittaci* infects poultry. In one instance, 27 employees in a turkey-processing plant contracted psittacosis while preparing consumer products from diseased turkeys.

Eggs are normally sterile when laid, but the outer waxy membrane as well as the shell and inner shell membrane may be penetrated by bacterial species. *Proteus* species cause black rot in eggs as hydrogen sulfide gas accumulates from the breakdown of cysteine. This gives eggs a rotten odor. Other spoilages in eggs include green rot from *Pseudomonas*

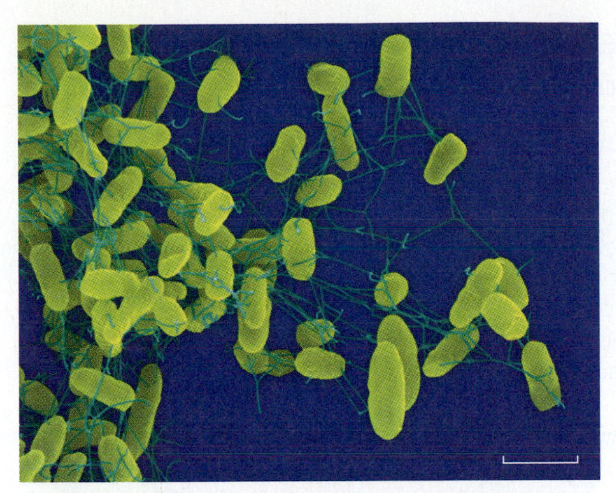

FIGURE 25.4 ***S. enteritidis.*** *Salmonella enterica* serotype Enteritidis is one of the most widely found contaminants in poultry and egg products. (Bar = 2 μm.)

Q: How might S. enterica *infect eggs or poultry?*

species and red rot from the growth of *Serratia marcescens*. Green rot results from a fluorescent pigment that causes the egg to glow when placed in front of an ultraviolet light. Red rot gives the yolk a blood-red appearance.

The primary focus of contamination in an egg is the yolk rather than the white. This is because of the nutritious quality of the yolk and because the white has a pH of approximately 9.0. Also, the lysozyme in egg white is inhibitory to gram-positive bacteria. However, *Salmonella enterica* is a common problem (FIGURE 25.4).

CONCEPT AND REASONING CHECKS

25.5 Why are eggs and poultry so susceptible to spoilage?

Breads and Bakery Products Can Support Bacterial and Fungal Growth

KEY CONCEPT

- Flour, eggs, and sugar are generally the sources of spoilage organisms.

In the production of bakery products, ingredients such as flour, eggs, and sugar can be a source of spoilage organisms, although most contaminants are killed during baking or competed out (MicroFocus 25.2). Some bacterial and mold spores may survive because bread is heated internally to 100°C for only about nine minutes. Members of the spore-forming genus *Bacillus* commonly survive, and as they proliferate, their capsular material accumulates, giving the bread a soft cheesy texture with long, stringy threads. The bread is said to be ropy.

Cream fillings and toppings in bakery products provide excellent chemical and physical conditions for bacterial growth. For example, custards made with whole eggs may be contaminated with *Salmonella* species, and whipped cream may contain dairy organisms such as species of *Lactobacillus* and *Streptococcus*. The acid produced by these bacteria results in a sour taste. High sugar environments of chocolate toppings and sweet icings support the growth of fungi. Most bakery products should be refrigerated during warm summer months.

Some Grains Are Susceptible to Spoilage

KEY CONCEPT

- *Aspergillus* and *Claviceps* species of fungi can spoil cereal grains.

Two types of grain spoilage are important in public health microbiology. The first type of spoilage is caused by the ascomycete *Aspergillus flavus*. This mold produces **aflatoxins**, a series of toxins that accumulate in stored grains such as wheat as well as peanuts, soybeans, and corn. Scientists have implicated aflatoxins in liver and colon cancers in humans (see Chapter 16). The toxins are consumed in grain products, as well as meat from animals that feed on contaminated grain.

The second type of grain spoilage is caused by *Claviceps purpurea*, the cause of ergot poisoning (**ergotism**). Rye plants are particularly susceptible to this type of spoilage, but wheat and barley grains may be affected also. The toxins deposited by *C. purpurea* may induce convulsions and hallucinations when consumed (see MicroFocus 16.6).

CONCEPT AND REASONING CHECKS

25.6 What type of microbial products lead to spoilage in bakery products and grains?

Milk and Dairy Products Sometimes Sour

KEY CONCEPT

- Microbial acids or enzymes can curdle milk products.

Milk is an extremely nutritious food. It is an aqueous solution of proteins, fats, and carbohydrates containing numerous vitamins and minerals. Milk has a pH of about 7.0 and is an excellent nutrient source for humans and animals, as well as microorganisms.

MICROFOCUS 25.2: Biotechnology/Environmental Microbiology

Sourdough Bread—Don't Leave San Francisco Without It

A sourdough baguette.

Most of us are familiar with the unique taste of sourdough bread. It has a crusty surface and a chewy, but sharply acidic taste. This most distinctive bread, which certainly tastes nothing like typical white sandwich bread, supposedly has its origins with San Francisco. Local stories say the recipe was brought to the city by Basque immigrants from the Pyrenees during the heyday of the California gold rush. So, what gives the bread it distinctive flavor and taste? It is, of course, the resident microbes!

To make sourdough you need a starter culture and when making sourdough it is necessary to save a piece of unused dough as a starter culture for another batch. In fact, the San Francisco bakeries say they have kept sourdough cultures going for over 100 years.

Besides the starter, making sourdough only requires unbleached flour, water, and salt. Most sandwich breads have at least 25 ingredients and additives—and no starter culture, meaning yeast is the only microbe present in the rising process for the dough. The result will be relatively bland tasting bread.

Therefore, the most important "ingredient"—and key to sourdough bread—is the microorganisms living in the starter. The sourdough starter is a mixed microbial population of wild yeasts and bacterial species. A teaspoon of sourdough starter contains some 50 million yeasts and 5 billion lactobacilli.

Sourdough is sour because of the acids produced by the lactobacilli. Initially, there may be other bacterial species in the dough, but they are quickly competed out because of the acids produced by the lactobacilli; the pH reaches about 3.8. The bacterial cells also add favor and produce the characteristic smell of sourdough bread.

So, a sourdough starter represents a mutualistic symbiosis between the principal bacterium *Lactobacillus sanfranciscensis* and the principal yeast, *Candida milleri,* which tolerates the acidic conditions. So, the next time you pass through the San Francisco airport, be sure to take some *Lactobacillus sanfranciscensis*-laced bread back home—or a starter culture and make your own bread at home.

About 87 percent of the substance of milk is water. Another 2.5 percent is a protein called **casein**, actually a mixture of three long chains of amino acids suspended in fluid. A second protein in milk is lactalbumin. This protein forms the surface skin when milk is heated to boiling. Lactalbumin is a whey protein, one that remains in the watery fluid after the casein curdles during milk spoilage or fermentation. Carbohydrates make up about 5 percent of the milk. The major carbohydrate is lactose, sometimes referred to as milk sugar (see Chapter 2). Rarely found elsewhere, lactose is a disaccharide digested by relatively few bacterial species, and these are usually harmless. The last major component of milk is butterfat, a mixture of fats usually churned into butter. Butterfat comprises about 4 percent of the milk and is removed in the preparation of nonfat (skim) milk or low-fat milk. When bacterial enzymes digest fats into fatty acids, the milk develops a sour taste and becomes unfit to drink.

■ Whey:
The watery liquid that separates from the solid part (casein) of milk when it turns sour or when enzymes are added in cheese making.

A common type of milk spoilage often takes place in the kitchen refrigerator or dairy case at the supermarket. Here, *Lactobacillus* or *Streptococcus* species multiply slowly and ferment the lactose in milk. Soon, large quantities of lactic and acetic acids accumulate. Enough acid may develop to change the structure of the protein and cause it to solidify as a curd. Dairy microbiologists refer to such an acidic curd as a **sour curd**. The lactobacilli and streptococci causing it usually have survived the pasteurization process.

Sweet curdling in milk may result when enzymes from species of *Bacillus, Proteus, Micrococcus,* or certain other bacterial genera attack casein. As weak hydrogen bonds break, casein loses its three-dimensional structure and curdles. The reaction is said to be sweet because little acid production occurs. It is an essential step in the production of cheese. The clear liquid is whey, an aqueous solution of lactose, minerals, vitamins, lactalbumin, and other milk components. Whey is used to make processed cheeses and "cheese foods."

Milk also may be contaminated by gram-negative rods of the coliform group, including *E. coli* and *Enterobacter aerogenes*. These bacteria produce acid and gas from lactose. The acid curdles the protein, and the gas forces the curds apart, sometimes so violently that they explode out of the container. The result is a stormy fermentation. *Clostridium* species also cause this reaction.

Ropiness in milk is similar to that in bread. It develops from capsule-producing organisms, such as *Alcaligenes*, *Klebsiella*, and *Enterobacter*. These gram-negative rods multiply in milk, even at low temperatures, and deposit gummy material that appears as stringy threads and slime.

Another form of spoilage is caused by species of *Pseudomonas* and *Achromobacter*, which produce the enzyme lipase. Lipase attacks butterfats in milk and digests them into glycerol and fatty acids, giving milk a sour taste and a putrid smell. A similar problem may develop in butter.

Additional types of milk spoilage result from the red pigment deposited in dairy products by *Serratia marcescens*, the blue or green pigment of *Pseudomonas* species, and the gray rot caused by certain *Clostridium* species. In gray rot, hydrogen sulfide (H_2S) from cysteine imparts a rotten-egg smell to milk, and the H_2S reacts with minerals to yield a gray or black sulfide compound. Spoilage due to wild yeasts is usually characterized by a pink, orange, or yellow coloration in the milk. Acid conditions stimulate mold decay in cheese products.

FIGURE 25.5 **A Dairy Milking Facility.** A man is cleaning a dairy cow milking facility to prevent contamination of the milking equipment.

Q: How would one sanitize a building such a milking facility?

Milk normally is sterile in the udder of the cow, but contamination can occur as it enters the ducts leading from the udder, as well as from other unlikely sources. The colostrum, or first milk, is laden with organisms. Species of soilborne *Lactobacillus* and *Streptococcus* are acquired during the passage, together with various coliform bacteria from dust, manure, and polluted water. The milk may derive other organisms from dairy plant equipment and unsanitary handling of dairy products by plant employees (FIGURE 25.5).

Colostrum: A fluid rich in antibodies and minerals that a mother animal produces after giving birth and before the production of true milk.

CONCEPT AND REASONING CHECKS

25.7 Summarize the ways milk can become spoiled.

25.2 Food Preservation

Centuries ago, humans battled the elements to keep a steady supply of food at hand. Sometimes there was a short growing season; at other times, locusts descended on their crops; at still other times, they underestimated their needs and had to cope with scarcity. However, experience taught humans they could overcome these difficult times by preserving foods.

MICROFOCUS 25.3: History

To Feed an Army

Part of Napoleon's genius was understanding the finer points of warfare, including how to feed an army. Recognizing that thousands of men-at-arms were a glut on the countryside, he broke up his army into smaller units that foraged on their own as they moved. When the time for battle neared, he reassembled his forces and engaged the enemy.

The shortcomings of this system became painfully clear when Napoleon crossed into northern Italy in 1800 and engaged the Austrians at the Battle of Marengo. Knowing the French army was scattered about, the Austrians charged before Napoleon could bring his forces together. Disaster was averted only when units arrived on the flanks to repel the Austrians. Napoleon had learned an important lesson: The next time he went to war, food would go with him.

Included in Napoleon's plan to resurrect France was a ministry that encouraged industry by offering prizes for imaginative inventions. A winemaker named Nicholas Appert attracted attention with his process of preserving food. Appert placed fruits, vegetables, soups, and stews in thick bottles, then boiled the bottles for several hours. He used wax and cork to seal the bottles and wine cages to prevent inadvertent opening of the bottle. By 1805, Appert had set up a bottling industry outside Paris and had a thriving business.

The Ministry of Industry encouraged Appert to publish his methods and submit samples of bottled foods for government testing. The French navy took numerous bottles on long voyages and reported excellent food preservation. In 1810, Appert was awarded 12,000 francs for his invention. Two years later, Napoleon assembled hundreds of cannons, thousands of men, and countless bottles of food, and marched off to war with Russia.

Among the earliest methods was drying vegetables and strips of meat and fish in the sun. Foods also could be preserved by salting, smoking, and fermenting. Individuals could now trek far from their native habitat, and taking preserved food with them, they took to the sea and moved overland to explore new lands.

The next great advance did not come until the mid-1700s. In 1767, Lazaro Spallanzani attempted to disprove spontaneous generation by showing that beef broth would remain unspoiled after being subjected to heat (see Chapter 1). Nicholas Appert applied this principle to a variety of foods (MicroFocus 25.3). Neither Appert nor any of his contemporaries was quite sure why food was being preserved, but it was clear that the spoilage could be retarded by prolonged heating. The significance of microorganisms as agents of spoilage awaited Pasteur's classic experiments with wine several generations later.

Through the centuries, preservation methods have had a common objective: to reduce the microbial population and maintain it at a low level until food can be consumed. Modern preservation methods are mere extensions of these principles. Though today's methods are sophisticated and technologically dynamic, advances in preservation processes are counterbalanced by the great volumes of food that must be preserved and the complexity of food products. Thus, the problems that early humans faced do not differ fundamentally from those confronting modern food technologists. In this section, we examine the methods of food preservation currently in use. Because the methods depend on physical and chemical control practices, a review of Chapter 23 might be useful.

Heat Denatures Proteins

KEY CONCEPT

- Heat kills microorganisms by changing the physical and chemical properties of their proteins.

In a moist heat environment, proteins are denatured and lose their three-dimensional structure (see Chapter 2). As structural proteins and enzymes undergo this change, the organisms die. The most useful application of heat is in the process of canning. Shortly after Appert established the use of heat in preservation, an English engineer named Bryan Donkin substituted iron cans coated with tin for Appert's bottles. Soon he was

FIGURE 25.6 **The Industrial Canning Process.** An inspector watches over the canning process to ensure sanitary conditions are maintained.

Q: How would tin cans be sterilized?

supplying canned meat to the British navy. In the United States, the tin can was virtually ignored until the Civil War period. In the years thereafter, mass production began and soon the tin can became the symbol of prepackaged convenience.

Modern canning processes are complex. For produce, machines wash, sort, and grade the food product, and then subject it to steam heat for three to five minutes. This last process, called **blanching**, destroys many enzymes in the food product and prevents any further cellular metabolism from taking place. Then the food is peeled and cored, and its diseased sections are removed. Canning comes next, after which the air is evacuated and placed in a pressured steam sterilizer similar to an autoclave at a temperature of 121°C or lower, depending on the pH, density, and heat penetration rate (FIGURE 25.6).

The sterilizing process is designed to eliminate the most resistant bacterial spores, especially those of the genera *Bacillus* and *Clostridium*. However, the process is considered **commercial sterilization**, which is not as rigorous as true sterilization, and some spores may survive. Moreover, should a machine error lead to improper heating temperatures, a small hole allow airborne bacterial cells to enter, or a proper seal not form, contamination may result.

Contamination of canned food is commonly due to facultative or anaerobic bacterial species that produce gas and cause the ends of the can to bulge. Food microbiologists call a can a flipper if the bulge can be flattened easily. It is a springer if pushing the bulge pushes out the opposite end of the can. A soft swell occurs when both ends bulge. If neither end can be pushed in because of the large amount of gas, a hard swell is present. The organisms often responsible for gas production are *Clostridium* species as well as coliform bacteria. Contamination is usually obvious since the spoiled product generally has a putrid odor.

■ Coliform bacteria: A group of gram-negative non–spore-forming rods that ferment lactose to acid and gas.

Growth of acid-producing bacterial species presents a different problem because spoilage cannot be discerned from the can's shape. Food has a flat-sour taste from the acid and has probably been contaminated by a *Bacillus* species, a coliform, or another acid-producing species that survived the heating.

The process of pasteurization was developed by Louis Pasteur in the 1850s to eliminate bacterial contamination of wines. His method was first applied to milk in Denmark about 1870, and by 1895 the process was widely employed. Although the primary object of **pasteurization** is to eliminate pathogenic bacterial species from milk, the process also lowers the total number of bacterial organisms and thereby reduces the chance of spoilage (see Chapter 22). The more traditional method involves heating the milk in a large bulk tank at 62.9°C (145°F) for 30 minutes. This is the **holding method**, also known as the LTLT method for "low temperature, long time." Machines stir the milk constantly during the pasteurization to ensure uniform heating, and cool it quickly when the heating is completed. More concentrated products, such as cream, often are heated at the higher temperature of 69.5°C (155°F) to ensure successful pasteurization. An innovative method of egg pasteurization recently has been introduced, as MicroFocus 25.4 explains.

The more modern method of pasteurization is called the **flash method**. In this process, raw milk is first warmed using the heat of previously pasteurized milk. Machines then pass the milk through a hot cylinder at 71.6°C (161°F) for a period of 15 seconds. Next, the milk is cooled rapidly, in part by transferring its heat to the incoming milk.

MICROFOCUS 25.4: Biotechnology

"Bring on the Caesar Salad!"

Some people have it Hawaiian-style with pineapple and ham; some have it Italian-style with pasta and tomatoes; and some enjoy it Southwestern-style with roasted chili peppers. Some toss it with grilled chicken or flank steak or grilled calamari or Cajun scallops. Regardless of the addition, however, the salad remains the same—Caesar salad.

For history buffs, the first Caesar salad is reported to have evolved on July 4, 1924, in the mind of Caesar Cardini, the proprietor of a restaurant (Caesar's Place) in Tijuana, Mexico. Cardini was desperate for a fill-in during a particularly busy day, so he threw together some Romaine lettuce, Parmesan cheese, lemon, garlic, oil, and raw eggs. His customers were enchanted.

Over the years, the reputations of the salad and its inventor grew. There was only one problem, however—the eggs. For true Caesar salad, raw eggs are used to add creaminess to the dressing. But that became a problem when increasing cases of *Salmonella* infections were traced to raw eggs. Eggless Caesar dressings (with heavy cream) were tried, but it just wasn't the same.

Caesar salad aficionados, take heart—the pasteurized egg is on the way. Purdue University microbiologists have found that *Salmonella* can be eliminated by heating eggs in hot water or a microwave oven, then maintaining them at 134°F in a hot-air oven for one hour. A New Hampshire company (Pasteurized Eggs, LP) is touting the benefits of its new machine for destroying egg-borne *Salmonella*. The machine heats eggs slowly and then directs them to successive baths in water ranging from 62°C to 72°C. Indeed, the process works so well that the US Department of Agriculture (USDA) has issued a new stamp certifying that eggs treated by the company meet standards for egg pasteurization established by the Food and Drug Administration (FDA).

Agricultural officials estimate that almost 50 billion eggs are produced for American consumption each year, and that over 2 million are infected with *Salmonella*. In the years ahead, that second number should dwindle considerably as egg pasteurization becomes standard practice (as milk pasteurization already has). Then, it will be safe to sample the cookie dough, or to have eggs over easy, or to enjoy Caesar salad the way it was meant to be enjoyed.

This is the HTST method, for "high temperature, short time." It is useful for high-quality raw milk, in which the bacterial count is consistently low.

A new method called **ultrapasteurization** is used in some dairy plants. In this process, milk and milk products are subjected to a temperature 82°C (180°F) for 3 seconds. Following pasteurization, a set of laboratory tests is performed to assay the quality of the milk and the success of pasteurization (TABLE 25.1). Sealed in sterile containers, the unopened milk will last about a month when refrigerated.

The bacterial cells surviving pasteurization may be involved in spoilage. *Streptococcus lactis*, for example, grows slowly in refrigerated milk, and when its numbers reach 20 million per milliliter, enough lactic acid has been produced to make the milk sour. Organisms surviving the heat of pasteurization are described as **thermoduric**. Pasteurization has no effect on spores.

Although milk in the United States is normally pasteurized and refrigerated, exposure to steam at 300°C for 4 to 15 seconds can sterilize it. This ultra-high temperature results in milk (e.g., Parmalat) with an indefinite shelf life as long as the container remains sealed. However after opening, it must be kept in the refrigerator. These milk products have been popular in

TABLE 25.1 Laboratory Tests Used to Assay the Quality of Milk

Test	Purpose	Importance
Phosphatase test	Determines whether phosphatase is present. Phosphatase is an enzyme normally destroyed during pasteurization.	To determine whether sufficient heat was used during pasteurization. If phosphatase is present, pathogens also might be present.
Reductase test	Estimates the number of bacterial cells in milk. The rate at which methylene blue is reduced to its colorless form is proportional to the number of cells present in a milk sample.	High-quality milk contains so few bacterial cells that a standard concentration of methylene blue will not be reduced in 6 hours. Low-quality milk has many bacterial cells, and methylene blue is reduced in 2 hours or less.
Standard plate count	Determines the total number of viable bacterial cells per ml of milk. Diluted milk is mixed with nutrient agar and incubated 48 hours; colonies are counted and the number of cells in the original sample is calculated.	The number per milliliter may not exceed 100,000 in raw milk before being pooled with other milk. It may not exceed 20,000 after pasteurization.
Test for coliforms	Determines the number of viable coliform bacteria per ml of milk. Similar to the standard plate count except that special media for coliform bacteria are used.	A positive coliform test indicates contamination with fecal material. The number per ml may not exceed established standards.
Test for pathogens	Detects the presence of pathogens. Methods depend on the pathogens suspected.	Helps locate the source of infectious agents that may be present in milk.

Europe, as they save on refrigeration. The product has a bit of a cooked taste though.

CONCEPT AND REASONING CHECKS

25.8 Distinguish between the various forms of heat in the process of pasteurization.

Low Temperatures Slow Microbial Growth

KEY CONCEPT

- Microbial enzyme activity is greatly reduced at low temperatures.

By lowering the environmental temperature, one can reduce the rate of enzyme activity in a microorganism and thus lower the rate of growth and reproduction. This principle underlies the process of refrigeration and freezing (see Chapter 23). Although the organisms are not killed, their numbers are kept low and spoilage is minimized. Ironically, the food is preserved by preserving the microorganisms.

Long before the invention of the refrigerator, the Greeks and Romans had partially solved the problem of keeping things cold. They simply dug a snow cellar in the basements of their homes, lined the cellar with logs, insulated it with heavy layers of straw, and packed it densely with snow delivered from far-off mountaintops. The compressed snow turned to a block of ice, and foods would remain unspoiled for long periods when left in this makeshift refrigerator.

The modern refrigerator at 5°C (41°F) provides a suitable environment for preserving food without destroying appearance, taste, or cellular integrity (FIGURE 25.7). However, psychrotrophic microorganisms survive and cause green meat surfaces, rotten eggs, moldy fruits, and sour milk. Pathogens such as *Listeria monocytogenes* and *Yersinia enterocolitica* also grow at low temperatures.

When food is placed in the freezer at –10°C (14°F), ice crystals form rapidly. These crystals tear and shred microorganisms and kill a significant number. However, many microorganisms survive, and the ice crystals are equally destructive to food cells. Therefore, when the food thaws, the surviving bacterial cells multiply quickly. Organisms such as staphylococci produce substantial amounts

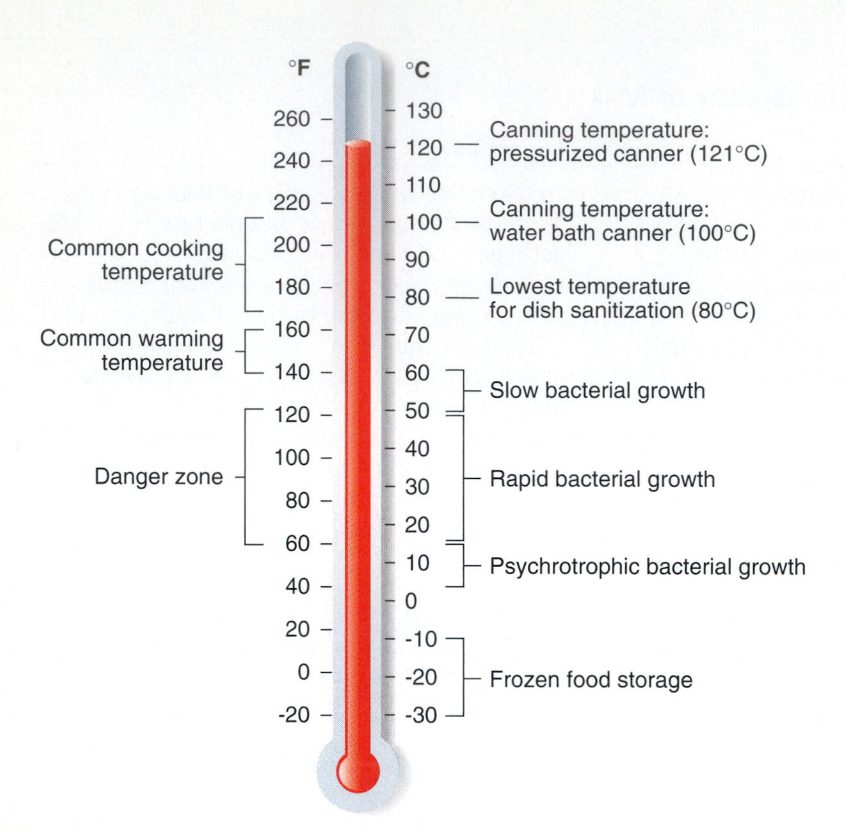

FIGURE 25.7 **Important Temperature Considerations in Food Microbiology.** Shown are several temperatures at which canning or sanitization procedures are carried out.

Q: Why is 15° to about 50°C called the "danger zone?"

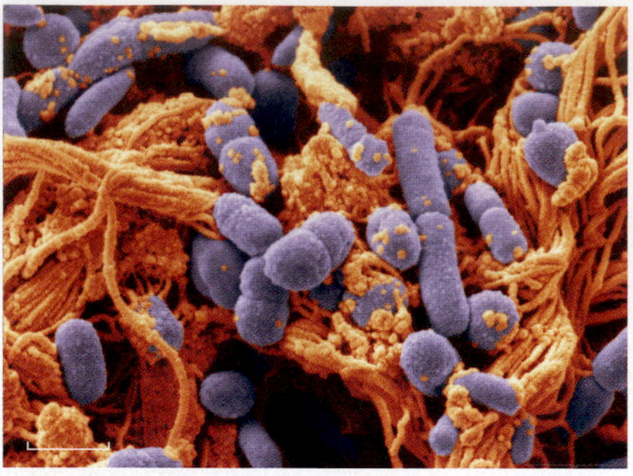

FIGURE 25.8 **Food Contamination.** False-color scanning electron micrograph of chicken skin from a retail chicken package. The skin is contaminated with spoilage bacteria (purple) after having been kept for 24 hours at 22°C. The filaments are degraded chicken tissue, an example of how fast food can spoil outside the refrigerator. (Bar = 2 µm.)

Q: How did these spoilage bacteria come to contaminate the chicken skin?

of enterotoxins, and *Salmonella* serotypes, streptococci, and other bacterial species grow to large numbers (FIGURE 25.8). Rapid thawing and cooking are recommended. Moreover, food should not be refrozen, because during thawing and refreezing, bacteria deposit sufficient enterotoxin to cause food poisoning the next time the food is thawed. Microwave cooking, which requires minimal thawing, may eliminate some of these problems.

Deep freezing at –60°C results in smaller ice crystals, and although the physical damage to microorganisms is less severe, their biochemical activity is reduced considerably. Small ice crystals do not damage food cells as severely as do the larger crystals formed at higher temperatures. Some food producers blanch their product before deep freezing, a process that further reduces the number of microorganisms.

Freezing has been a mainstay of preservation since Clarence Birdseye first offered frozen foods for retail purchase in the 1920s. Approximately 33 percent of all preserved food in the United States is frozen.

CONCEPT AND REASONING CHECKS

25.9 How does refrigeration temperature slow down the growth of microbes?

Drying and Osmotic Pressure Help Preserve Foods

KEY CONCEPT

- Drying and osmotic pressure remove water.

The advantage of drying foods is best expressed by the phrase, "Where there is no water, there is no natural life." Indeed, dry foods cannot support microbial growth. In past centuries, people used the sun for drying, but modern technology has developed specific machinery for drying foods.

The spray dryer expels a fine mist of liquid such as coffee into a barrel cylinder containing hot air. The water evaporates quickly, and the coffee powder falls to the bottom of the cylinder. Another method uses a heated drum over which liquids such as soup are poured. The water evaporates rapidly, leaving dried soup to be scraped off the drum. A third machine uses a belt heater that exposes liquids such as milk to a stream of hot air. The air evaporates any water and leaves dried milk

MICROFOCUS 25.5: History

It Started With "Stomped Potatoes"

Freeze-drying, or lyophilization, is the removal (sublimation) of water from a substance without going through a liquid state. Such freeze-dried foods last longer than other preserved foods and are very light, which makes them perfect for everyone from backpackers to astronauts. However, the origins of freeze-dried foods go back to the ancient Incas of Peru.

The basic process of freeze-drying food was known to the Peruvian Incas that lived in the high altitudes of the Andes Mountains of South America. The Incas stored some of their food crops, including potatoes, on the mountain heights. First, most of the moisture was crushed out of the potatoes when the Incas stomped on them with their feet. Then, the cold mountain temperatures froze the stomped potatoes and the remaining water inside slowly vaporized under the low air pressure of the high altitudes in the Andes.

Today, a similar procedure still is used in Peru for a dried potato product called chuno. It is a light powder that retains most of the nutritional properties of the original potatoes but, being freeze-dried, chuno can be stored for up to four years.

On the commercial market, the Nestlé Company was asked by Brazil in 1938 to help find a solution to save their coffee bean surpluses. Nestlé developed a process to convert the coffee surpluses into a freeze-dried powder that could be stored for long periods. The result was Nescafe® coffee, which was first introduced in Switzerland. However, during World War II the process of freeze-drying developed into an industrial process. Blood plasma and penicillin were needed for the armed forces and freeze-drying was found to be the best way to preserve and transport these materials.

In 1964, Nestlé developed an improved method of producing instant coffee by freeze-drying. Today, several other coffee producers use a similar process of preservation. In the late 1970s, freeze-drying was used for taxidermy, food preservation, museum conservation, and pharmaceutical production.

Today, more than 400 commercial food products are preserved by freeze-drying. In the medical field, freeze-drying is the method of choice for extending and preserving the shelf life of enzymes, antibodies, vaccines, pharmaceuticals, blood fractions, and diagnostics. The benefit of all freeze-dried products is that they rehydrate easily and quickly because of the porous structure created by the freeze-drying process.

But it all started with the Incas stomping on their potatoes.

solids. Unfortunately, spore-forming and capsule-producing bacterial species can be problems because they resist drying.

During the past 20 years, freeze-drying, or **lyophilization**, has emerged as a valuable preservation method, although it has a long history (MicroFocus 25.5). In this process, food is deep frozen, and then a vacuum pump draws off the water in a machine like the one pictured in FIGURE 25.9. (Water passes from its solid phase [ice] to its gaseous phase [water vapor] without passing through its liquid phase [water].) The dry product is sealed in foil and easily reconstituted with water. Hikers and campers find considerable value in freeze-dried food because of its light weight and durability. Lyophilization also is a useful method for storing, transporting, and preserving bacterial cultures.

Osmotic pressure also helps preserve foods. When living cells are immersed in large quantities of a compound such as salt or sugar, water diffuses out of cells through the

FIGURE 25.9 **An Industrial Model Lyophilizer.** This lyophilizer removes 500 pounds of product moisture in 24 hours of freeze-drying. Using vacuum and heat, water is drawn off from ice without passing through the liquid phase to produce the freeze-dried product.

Q: How are lyophilized products normally stored on the market shelf?

cell membranes and into the surrounding environment, where it dilutes the high concentration of the compound. This diffusion of water is called **osmosis**, and the tendency of water to move across a membrane is termed **osmotic pressure**.

Osmotic pressure can be used to preserve foods because water flows out of microorganisms as well as food cells (see Chapter 23). For example, in highly salted or sugared foods, microorganisms dehydrate, shrink, and die. Jams, jellies, fruits, maple syrups, honey, and similar products typify foods preserved by high sugar concentrations. Salted foods include ham, cod, bacon, and beef as well as certain vegetables such as sauerkraut, which has the added benefit of large quantities of acid. It should be noted, however, that staphylococci tolerate salt and may survive to cause staphylococcal food poisoning.

CONCEPT AND REASONING CHECKS

25.10 How do lyophilization and osmotic pressure preserve foods?

Chemical Preservatives Help Keep Foods Fresh

KEY CONCEPT

- Chemical preservatives can inhibit the growth of bacterial cells, yeast, and molds.

For a chemical preservative to be useful in foods, it must be inhibitory to microorganisms while easily broken down and eliminated by the human body without side effects. These requirements have limited the number of available chemicals to a select few.

A major group of chemical preservatives are organic acids, which microbiologists believe damage microbial membranes and interfere with the uptake of certain essential organic substances such as amino acids. The chemical preservatives primarily affect molds and yeasts, but their acidity also deters bacterial growth.

In 1908, **benzoic acid** was the first chemical to be approved by the FDA. Naturally found in cranberries, today benzoic acid is used to protect beverages, as well as foods like catsup and margarine. **Propionic acid** occurs naturally in apples, strawberries, and grains. The acid is incorporated in wrappings for butter and cheese, and is added to breads and bakery products, where it inhibits the ropiness commonly due to *Bacillus* species and prevents the growth of fungi. Other natural acids in food add flavor while serving as preservatives. Examples are lactic acid in sauerkraut and yogurt, and acetic acid in vinegar.

The process of smoking with hickory or other woods accomplishes the dual purposes of drying food and depositing chemical preservatives. By-products of smoke, such as aldehydes, acids, and certain phenol compounds, effectively inhibit microbial growth for long periods of time. Smoked fish and meats have been staples of the diet for many centuries.

Sulfur dioxide has gained popularity as a preservative for dried fruits, molasses, and juice concentrates. Used in either gas or liquid form, the chemical retards color changes on the fruit surface and adds to the aesthetic quality of the product. Another gas, **ethylene oxide**, is employed for the preservation of spices, nuts, and dried fruits, especially those packaged in cellophane bags. This same gas is used for the chemical sterilization of packaged Petri dishes and other plastic devices (see Chapter 23).

CONCEPT AND REASONING CHECKS

25.11 How do chemical preservatives work to prevent spoilage?

Radiation Can Sterilize Foods

KEY CONCEPT

- Irradiated food provides the same benefits of preservation as does heat, refrigeration, or freezing.

Though much of the public is apprehensive about foods exposed to radiation, various forms of radiation can be used to safely irradiate and sterilize foods (see Chapter 23). Taste and appearance are preserved by freezing food in liquid nitrogen and exhausting oxygen from the package before irradiation. Meat storage facilities use **ultraviolet (UV) light** to reduce surface contamination, and water can be treated with UV light when chlorine is not useful.

Gamma rays are used to extend the shelf life of fruits, vegetables, fish, and poultry from several days to several weeks. This form of radiation also increases the distances fresh food can be transported and significantly extends the storage time for food in the home.

Gamma rays are high-frequency forms of electromagnetic energy emitted by a radioactive isotope called cobalt-60. The radiations are not radioactive, and they cannot cause food to become radioactive. They kill microorganisms by reacting with and destroying microbial DNA (see Chapter 23).

Interest in gamma radiation for food preservation grew during the 1950s under President Eisenhower's Atoms for Peace program. For the next quarter-century, the FDA conducted extensive tests to determine whether the process was safe. In March 1981, the FDA approved radiated foods such as spices, condiments, fruits, and vegetables for sale to American consumers. Gamma radiation of pork to prevent trichinellosis won approval in 1985, and irradiated strawberries made it to market in 1992. Irradiation of red meat was approved in 1997, as further explored in Chapter 23.

CONCEPT AND REASONING CHECKS

25.12 What type of radiation is used to irradiate foods?

Foodborne Disease Can Result from an Infection or Intoxification

KEY CONCEPT

- People can be affected by the organism or the toxin contaminating food.

Food may be a mechanical vector for infectious microorganisms or a culture medium for growth. People then are affected by the organism or the toxin it has produced in the food. In the former case, a food infection is established; in the latter, a food poisoning or intoxication occurs (see Chapter 10).

Food infections are typified by typhoid fever, salmonellosis, cholera, and shigellosis, all of which are of bacterial origin. Amoebiasis and giardiasis represent protozoal foodborne diseases, while viral infections are exemplified by hepatitis A. **Food intoxications** include botulism, staphylococcal food poisoning, and clostridial food poisoning. Because full discussions of these diseases are presented elsewhere in this text (see Chapters 10, 15, and 17), we shall not examine them individually here.

In the United States, public health microbiologists estimate that between 2 and 10 million people are affected by foodborne illnesses annually. However, the number of outbreaks is increasing as globalization affects food production and transport (MicroFocus 25.6). Many illnesses require medical attention, but the vast majority of patients recover rapidly without serious complications. In many cases, the incident might have been avoided by taking some basic precautions. For example, unrefrigerated foods are a prime source of staphylococci and *Salmonella* serotypes, and perishable groceries such as meats and dairy products should not be allowed to warm up while other errands are performed. Also, a thermometer should be used to ensure that the refrigerator temperature is at or below 5°C (40°F) at all times.

Another way to avoid foodborne disease is to cover any skin boils while working with foods because boils are a common source of staphylococci (see Chapter 12). The hands always should be washed thoroughly before and after handling raw vegetables or salad fixings to avoid cross-contamination of other foods. It is wise to cook meat from a frozen or partly frozen state; if this is impossible, the meat should be thawed in the refrigerator. Cutting boards should be cleaned with hot, soapy water after use, and old cutting boards with cracks and pits should be discarded. MicroInquiry 25 quizzes you on your sanitary precautions.

Studies indicate that leftovers are implicated in most outbreaks of foodborne disease. It is therefore important to refrigerate leftovers promptly and keep them no more than a few days. Thorough reheating of leftovers, preferably to boiling, also reduces the possibility of illness.

Many instances of foodborne disease occur during the summer months, when foods are taken on picnics where they cannot be refrigerated. As a general principle, dairy foods, such as custards, cream pies, pastries, and deli salads should be excluded from the picnic menu. For outdoor barbecues, one dish should be used for carrying hamburgers to the grill and another dish for serving them. Many of these principles apply equally well to fall and winter tailgate parties.

Over 90 percent of botulism outbreaks reported to the CDC are traced to home-canned food. To prevent this sometimes fatal foodborne disease, health officials urge

MICROFOCUS 25.6: Public Health

Globalization and Foodborne Diseases

Today, you can go into just about any supermarket and find fresh fruit that seems out of season. You can find fruits such as peaches and grapes in the middle of winter. Thanks to advances in agriculture and transportation, there has been a globalization of the food market. Peaches from Chile or raspberries from Mexico can be heading to US markets within hours after being picked. It's great to have fresh fruits and vegetables, but it brings along the risk of a foodborne illness.

About 70 percent of many fresh fruits and vegetables we find in our local supermarket come from other countries. Sometimes these countries do not have the regulated and safe food practices found in the United States and other developed countries. Although outbreaks of foodborne illnesses are rare, they are occurring with more regularity today. Actual data can be hard to collect because many people never visit a doctor for a foodborne illness. Still, some data indicate that the number of outbreaks each year associated with fresh produce is increasing.

Health scientists say it is no longer necessary to travel to a foreign country to get a foodborne illness; the illness now can come to you—via the globalization of the food market.

Detecting foodborne outbreaks is not easy because fresh produce is distributed across the nation and not to just one local community. In fact, physicians and local health departments may not see a local outbreak as part of a national event. Most local health departments are accustomed to responding to a foodborne outbreak at a social function or commercial establishment. Globalization now means that local, state, and national food safety programs need to coordinate their surveillance methods and establish clear communication channels to properly recognize and quickly respond in a timely fashion to a foodborne outbreak.

Importantly, should you shy away from imported produce? Certainly not. In most all cases, proper washing of fresh fruits and vegetables, or cooking produce, will eliminate almost any chance of contracting a foodborne illness. Keep consuming those fruits and vegetables—5-times a day!

homemakers to use the pressure method to can foods. Reliable canning instructions should be obtained and followed stringently. Foods suspected of contamination should not be tasted to confirm the suspicion, but should be discarded immediately. When in doubt, boiling the food for a minimum of 10 minutes and thoroughly washing the utensil used to stir the food are recommended. Bulging or leaking cans must be discarded in a way that will not endanger other people or animals.

Milk is an unusually good vehicle for the transmission of pathogenic microorganisms because its fat content protects organisms from stomach acid, and, being a fluid, it remains in the stomach a relatively short period of time. The diseases of cows transmitted by milk include bovine tuberculosis, brucellosis, and Q fever (see Chapter 10). The biggest threat comes from drinking unpasteurized milk. In late 2002 and early 2003, the CDC received reports of a multistate outbreak of *Salmonella enterica* serotype Typhimurium that affected more than 94 individuals who had consumed raw, unpasteurized milk. Another smaller case was reported in 2006 from drinking raw milk contaminated with *E. coli* O157:H7.

Another milkborne organism of significance is *Campylobacter jejuni*, the cause of campylobacteriosis (see Chapter 10). In 2001, 75 persons in Wisconsin contracted campylobacteriosis from consuming raw,

MICROINQUIRY 25

Keeping Microorganisms Under Control

How knowledgeable are you concerning food safety in your home or apartment? If you are, do you actually practice what you know? Take the following quiz (honestly) and then we will see where microbes lurk and where you need to eliminate or keep them under control. Answers can be found in **Appendix D**.

25.1. Your refrigerator should be kept at what temperature?
a. 0°C (32°F).
b. 5°C (41°F).
c. 10°C (50°F).
d. Do not know.

25.2. Frozen fish, meat, and poultry products should be defrosted by
a. setting them on the counter for several hours.
b. microwaving.
c. placing them in the refrigerator.

25.3. After cutting raw fish, meat, or poultry on a cutting board, you can safely
a. reuse the board as is.
b. wipe the board with a damp cloth and reuse it.
c. wash the board with soapy hot water, sanitize it with a mild bleach, and reuse it.

25.4. After handling raw fish, meat, or poultry, how do you clean your hands?
a. Wipe them on a towel.
b. Rinse them under hot, cold, or warm tap water.
c. Wash them with soap and warm water.

25.5. When was the last time you sanitized your kitchen sink drain and garbage disposal?
a. Last night
b. Several weeks ago
c. Several months ago
d. Cannot remember

25.6. Leftover cooked food should be
a. cooled to room temperature before being put in the refrigerator.
b. put in the refrigerator immediately after the food is served.
c. left at room temperature overnight or longer.

25.7. How do you clean your kitchen counter surfaces that are exposed to raw foods?
a. Use hot water and soap, then a bleach solution.
b. Use hot water and soap.
c. Use warm water.

25.8. How was the last hamburger you ate cooked?
a. Rare
b. Medium
c. Well-done

25.9. How often do you clean or replace your kitchen sponge or dishcloth?
a. Daily
b. Every few weeks
c. Every few months

25.10 Normally dishes in your home are cleaned
a. by an automatic dishwasher and air-dried.
b. after several hours soaking and then washed with soap in the same water.
c. right away with hot water and soap in the sink and then air-dried.

So, how did you do?

unpasteurized milk. Health officials estimate *Campylobacter jejuni* is present in the intestinal tracts of about 40 percent of dairy cattle.

Despite the known association of raw milk with disease-causing organisms, some consumers believe that raw milk is of better quality than pasteurized milk. Public health microbiologists seek to limit milkborne disease by inspecting food and dairy plants regularly, and making recommendations on improved sanitary practices.

CONCEPT AND REASONING CHECKS

25.13 How do food infections differ from food intoxications?

HACCP Systems Attempt to Identify Potential Contamination Points

KEY CONCEPT

- Food safety systems are in place to identify microbial contamination points.

Fueled by consumer awareness, the entire food industry has been placed under a food-safety spotlight overseen by the FDA and the US Department of Agriculture (USDA).

Among the most important food safety systems is **Hazard Analysis and Critical Control Point** (HACCP), a set of scientifically-based safety regulations enforced in the seafood, meat, and poultry industries. In an HACCP system, manufacturers identify individual processing points, called **critical control points** (**CCPs**) where the safety of a product could be affected, such as by contaminating microorganisms. The CCPs are supervised to ensure that any hazards associated with the operation are contained or, preferably, eliminated (FIGURE 25.10). When all possible hazards are controlled at the CCPs, the safety of the product can be assumed without further testing or inspection.

The standard regulations require food processors to monitor and control eight key sanitation areas, including the condition and cleanliness of utensils, gloves, outer garments, and other food contact surfaces; the prevention of cross-contamination from raw products and unsanitary objects to foods; and the control of employee health conditions that could result in food contamination.

CONCEPT AND REASONING CHECKS

25.14 What are critical control points in the seafood, meat, and poultry industries?

FIGURE 25.10 **Meat Inspections.** A meat inspector checks beef prior to sale. The long-standing "sniff-and-poke" method of inspection is being replaced by a newly instituted HACCP system of ensuring meat safety.

Q: What CCP is being supervised in this photo?

25.3 Foods from Microorganisms

Over the centuries, social customs and traditions have brought acceptance of a variety of foods produced by microorganisms. Some individuals regard these foods as "spoiled," but to many people, the food is "fermented" (FIGURE 25.11).

Many Foods Are Fermented Products

KEY CONCEPT

- Fermentation can retard spoilage and the growth of microbes in the food product.

Fermentation precedes human history, as many fruits naturally ferment. The first known examples of humans controlling fermentation date back to at least 3000 BC, when the Babylonians and Egyptians fermented beverages like wine and beer. As described in Chapter 6, **fermentation** is the process for

FIGURE 25.11 **An Array of Foods and Beverages Produced by Microorganisms.** Many foods and beverages are the result of microbial fermentations.

Q: What is common to all these fermentation processes?

producing ATP using endogenous organic compounds as both electron donors and acceptors—exogenous electron acceptors are absent. The chemical process of fermentation makes a few ATP molecules in the absence of aerobic respiration.

Fermentation can produce a number of end (waste) products, depending on the organism doing the fermentation. End products include lactic and acetic acids, ethanol (ethyl alcohol), or other reduced end products.

Fermented foods have four things in common.

- Foods are less vulnerable to extensive spoilage than unfermented foods because of the acids or alcohol produced;
- Foods are less likely to be vectors of foodborne illness because of the acids and/or alcohol present;
- Foods may become enriched with protein, amino acids, and vitamins;
- Foods are enriched in their taste, texture, and flavor.

Here, we will mention a few typical fermented food products. Sauerkraut (German for "sour cabbage") is not only a well-preserved and tasty form of cabbage, but also nutritionally sound. For instance, the vitamin C content of sauerkraut is equivalent to that of citrus fruits, and sauerkraut often was taken on British sea voyages to prevent scurvy because citrus fruits were too expensive.

Sauerkraut is prepared commercially by adding salt to shredded cabbage and packing the cabbage tightly to encourage anaerobic conditions. The first organisms to multiply are species of *Leuconostoc,* a gram-positive coccus found naturally in the cabbage. The bacterial cells ferment carbohydrates in the plant cells and produce acetic and lactic acids. After some days, the acids lower the pH of the cabbage to about 3.5. Species of *Lactobacillus* then proliferate, and the additional lactic acid they produce reduces the pH to about 2.0. The salt helps retard mold contamination while drawing juices out of the plant cells. A compound called diacetyl (the flavoring agent in butter) is produced by *Leuconostoc,* adding aroma and flavor.

In the United States, "pickle" is practically synonymous with "pickled cucumber." Over 37 types of dill, sour, and sweet pickles have been categorized, but essentially the fermentations are similar. Cucumbers are placed in brine, a salt solution of eight percent or higher, at which point the cucumber changes color from bright green to olive green.

Three groups of microorganisms are important to the fermentation and curing of cucumbers. *Enterobacter aerogenes,* a gram-negative rod, produces large amounts of CO_2, which takes up all the air space and establishes anaerobic conditions. *Lactobacillus* and *Leuconostoc* species then dominate and form abundant amounts of acid, which softens the tissues and sours the cucumbers. Finally, certain yeasts grow and establish many flavors associated with ripe pickles. Dill, garlic, and other herbs and spices are added to finish the product. Pickled peppers, tomatoes, and other vegetables undergo a somewhat similar process. Most pickles are heat pasteurized or further acidified to increase their shelf life, but "kosher-style" pickles are given no further treatment.

Vinegar (derived from the French *vinaigre* for "sour wine") is a fermented liquid traditionally made by the spontaneous souring of wine. Pasteur's ideas concerning the germ theory of disease were built in part from his examinations of the souring of French wines (see Chapter 1).

A widely used method for industrial vinegar production uses yeasts to ferment the fruit juice to alcohol until the alcohol concentration is about 10 to 20 percent. Machines then spray the alcoholic juice into a tank containing the bacterial species *Acetobacter aceti* growing on the surface of wood shavings, gravel, or other substrate. As alcohol percolates through, bacterial enzymes convert it to acetaldehyde and then acetic acid. The vinegar recirculates several times before collection at the bottom of the tank. Residual alcohol evaporates in the heat, and the product usually has an acetic acid content of about 3 to 5 percent. The flavor of vinegar is determined by oils, sugars, and other compounds produced by the bacterial cells, plus the residue of organic compounds in the wine from which it was made.

Several other fermented foods also are worthy of note. One example, soy sauce, is made

■ Scurvy:
A vitamin C deficiency that causes spongy gums and bleeding into the skin.

FIGURE 25.12 **The Production of Soy Sauce.** Roasted soybeans are inoculated with *Aspergillus oryzae* and then mixed with a salt solution containing other microorganisms.

Q: Why would an opened bottle of soy sauce that has been left unrefrigerated become slightly bitter tasting?

■ Aldehydes: Organic compounds containing a carbon atom double bonded to an oxygen atom, and single bonded to a carbon atom and a hydrogen atom.

■ Ketones: Organic compounds containing a carbon atom double bonded to an oxygen atom, and single bonded to two carbon atoms.

from roasted soybeans and wheat inoculated with the fungus *Aspergillus oryzae* and allowed to stand for three days. The fungus-covered product, called koji, then is added to a solution of salt and microorganisms, and aged for about a year (FIGURE 25.12). During this time, lactobacilli produce acid, and yeasts produce small amounts of alcohol. Together with the fungal products, the acid and alcohol determine the flavor. The liquid pressed from the mixture is soy sauce.

Another example is fermented sausages. These are produced generally as dry or semidry products and include pepperoni from Italy, thuringer from Germany, and polsa from Sweden. Curing and seasoning agents first are added to ground meats, followed by stuffing into casings and incubation at warm temperatures. Mixed acids produced from carbohydrates in the meat give the sausage its unique flavor and aroma.

Cocoa and coffee also owe their flavor in part to microorganisms. Cocoa derives some of its taste from the microbial fermentation that helps remove cocoa beans from the pulp covering them in the pod. Likewise, coffee is believed to obtain some of its flavor from the fermentation of coffee berries when the beans are soaked in water to loosen the berry skins before roasting.

CONCEPT AND REASONING CHECKS

25.15 Identify the common end products to fermented foods, such as sauerkraut, vinegar, and soy sauce.

Many Milk Products Are the Result of Fermentation

KEY CONCEPT

- Microbial acids are important to milk products and cheese manufacture.

Over the centuries, fermented milk products have assumed a key place in our diet. The sour milks are typical examples of fermented milk products. Buttermilk is made by adding starter cultures of *Streptococcus cremoris* and *Leuconostoc citrovorum* to vats of skim milk. The *Streptococcus* ferments lactose to lactic and acetic acids, and the *Leuconostoc* continues the fermentation to yield various aldehydes and ketones, and the compound diacetyl. These substances, especially diacetyl, give buttermilk its flavor, aroma, and acidity. For sour cream, the same process is used, except that pasteurized light cream is the starting point.

Acidophilus milk is produced in much the same way as buttermilk, except that the skim milk is inoculated with *Lactobacillus acidophilus*. This bacterium is a normal member of the human intestinal flora. Many health care practitioners believe that the lactobacilli in acidophilus milk augment naturally occurring lactobacilli and help the digestion process, while keeping molds in check. Another type of acidophilus milk, sweet acidophilus milk, lacks the sour taste. It is prepared by adding *L. acidophilus* to pasteurized milk.

Yogurt is a form of sour milk, made by adding dry milk solids to boiled milk to achieve a custard-like consistency. The two starter cultures are *Streptococcus thermophilus* and *Lactobacillus bulgaricus*. Many microbiologists hold that these bacterial species in yogurt support good health similar to the bacterial species in acidophilus milk.

Cheese production begins when the casein curdles out of milk. Usually this accompanies a souring of the milk by streptococci, but the process may be accelerated by adding rennin, an enzyme obtained from the stomach lining of a calf (FIGURE 25.13). The milk curd is essentially an unripened cheese. It may be marketed as cottage cheese, or pot cheese. Cream cheese also is unripened cheese with a butterfat content of up to 20 percent.

To prepare ripened cheese, the milk curds are washed, pressed, sometimes cooked, and

FIGURE 25.13 Cheese Manufacture. Cheese manufacture begins by adding rennin to a vat of milk during cheese manufacture. This is the first process of cheese making.

Q: Why is rennin added to cheese manufacture?

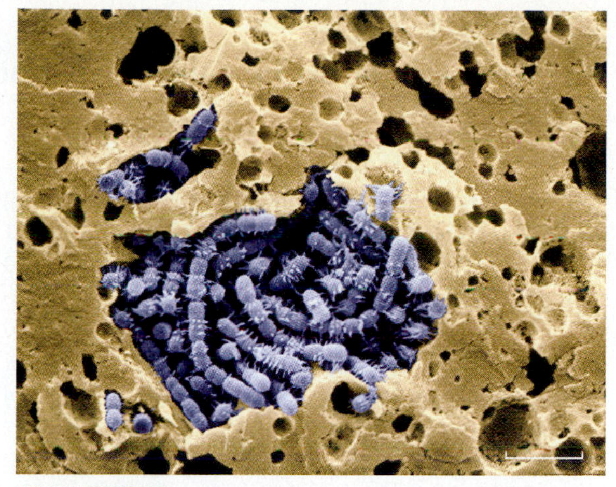

FIGURE 25.14 Goat Cheese. A false-color scanning electron micrograph of goat cheese showing the lactic acid streptococci (blue). (Bar = 10 µm.)

Q: Why does the starter culture contain rennet?

cut to the desired shape. Often the curds are salted to add flavor, control moisture, and prevent contamination by molds. If Swiss cheese is to be made, two bacterial species grow within the cheese: *Lactobacillus* species, which ferment the lactose to lactic acid; and *Propionibacterium* species, which produce organic compounds and carbon dioxide, which seeks out weak spots in the curd and accumulates as holes, or eyes. Cheddar cheese is scalded at a lower temperature than Swiss. Cheddar and Swiss are examples of cheese ripened internally by bacteria; Provolone, Edam, and Gouda are others.

Another group of cheeses is somewhat softer in texture, a characteristic deriving from the partial breakdown of the protein curds by microbial enzymes. Growth takes place primarily at the surface of these cheeses, and the products tend to be pungent. Within the group of soft cheeses are Muenster, Port du Salut, and Limburger. Yeasts and species of the gram-positive rod *Brevibacterium* are among the microbes on the surface of the cheese. The rind of the cheese is derived from microbial pigments.

Milk from other animals than cows can be used to make cheeses. Goat cheese (Chèvre; French for *goat*) uses the milk from domesticated goats. A bacterial starter culture of rennet and streptococci coagulate the milk (FIGURE 25.14). The water in the milk is strained away as whey and the remaining casein protein is ripened to obtain the structure and flavor of the cheese. The higher proportion of fatty acids in goat's milk imparts a tart flavor to the cheese.

The mold-ripened cheeses are represented by camembert. Camembert cheese is made by dipping salted curds into *Penicillium camemberti* spores to stimulate a surface growth. The fungus grows on the outside of the curd and digests the proteins, thus softening the curd. Stilton is another mold-ripened cheese where the mold penetrates cracks within the curd, creating its distinctive veins. Such cheeses have a unique community of microbes (MicroFocus 25.7).

CONCEPT AND REASONING CHECKS

25.16 What types of microbes can be found in cheeses?

MICROFOCUS 25.7: Biotechnology/Environmental Microbiology

Stilton Cheese—Slicing through a Microbial Community

Cheeses represent a unique community of microorganisms and, like human communities, temperaments and personalities may differ within a community. Take Stilton cheese, whose strong aroma and intense flavor some find delightful and others find repellent.

Stilton blue sheese.

Stilton blue cheese (see figure) is made by only six dairies in the English Midlands. It is made by seeding a milk culture with *Lactococcus lactis* and the blue mold *Penicillium*. So, how do these microbes give Stilton cheese its gourmet or peculiar characteristics? Christine Dodd and her colleagues at the University of Nottingham in England decided to find the answer.

Using the latest techniques of DNA sequencing analysis, the microbiologists identified a whole new community of bacterial species within the cheese after nine weeks of ripening. It must be these microbes and *Penicillium* that give Stilton its unique taste and color.

The research group sectioned up the slabs of cheese and discovered there was a distinct distribution of microbial species in different regions of the cheese. Two harmless *Staphylococcus* species were identified on the cheese surface. Mixed *Lactobacillus* species were found in the internal parts of the veins, while single species colonies were found in the core. Microbial populations were much less dense in the core. *Lactobacillus plantarum* was detected only underneath the surface (crust), while *Leuconostoc mesenteroides* was evenly spread through all parts of the cheese. The investigators aren't certain whether these bacterial species arrive on the scene by surviving the milk pasteurization process or by being introduced through equipment or other sources. It is likely each species establishes its regions of growth for ecological reasons; that is, each species finds optimal growth conditions associated with specific regions of the cheese and the other microbes present there. In the presence of other bacterial species and molds, microbial metabolites probably determine whether a specific species will survive.

The role of each microbial species in Stilton is not unknown, but their presence or absence may help explain why different batches of the cheese made at the same dairy can have highly different characteristics.

SUMMARY OF KEY CONCEPTS

25.1 Food Spoilage

- Food spoilage has been a continuing problem since ancient times. Certain conditions, such as water content, pH, physical structure, and chemical composition determine the extent of food spoilage because they influence the growth of microorganisms.
- Products of spoilage include alcohol, foul-smelling products such as indole and skatole, and acids and gases. Capsules and pigments also are involved with spoilage.
- In meats and fish, spoilage organisms enter during processing; in fish, water often is the source.
- In poultry, the spoilage may reflect human contamination, but often it is due to members of the genus *Salmonella*, that infected the bird.
- Bakery ingredients generally bring contaminants to bread and other bakery products.
- Grains, including peanuts and rye, may be spoiled by toxin-producing fungi.
- Dairy products are spoiled by microorganisms surviving pasteurization, such as curd, capsule, and pigment producers.

25.2 Food Preservation

- To preserve food from spoilage, a number of methods are used including heat. The most useful application of heat is in the process of canning, while pasteurization is used for milk and other liquid products.
- Low temperatures are achieved in the refrigerator and freezer. Low temperatures do not kill all microorganisms contaminating a refrigerated or frozen food product.
- Drying is useful because water is an absolute necessity for life, and dried foods therefore are unable to support microbial life. An ever-increasing list of foods preserved by freeze-drying is appearing on the market.
- Foods high in sugar or salt preserve foods by drawing water out of microorganisms contaminating the food product.
- Various chemicals, such as propionic and benzoic acids, are used to preserve foods.
- Ultraviolet light and gamma rays typify the radiations used in processing certain foods.
- Individuals can contract a foodborne disease either through an infection or by ingesting a microbial toxin.
- The Hazard Analysis and Critical Control Point (HACCP) are a set of scientifically based safety regulations that attempt to identify food processing points that are subject to contamination.

25.3 Foods from Microorganisms

- Certain foods "spoiled" by microorganisms have come to be accepted as the norm. Among these microbial products are sauerkraut, vinegar, and fermented sausages. The spoilage in these foods causes no harm to consumers, and the foods reflect the helpful activities that microorganisms perform to add to the quality of our lives. Other foods involving fermentation products include soy sauce, sausages, cocoa, and coffee.
- Buttermilk, sour cream, and yogurt are products where microbial action produces a fermented milk product.
- Numerous dairy products, such as cheeses, also are products of microorganisms.

LEARNING OBJECTIVES

After understanding the textbook reading, you should be capable of writing a paragraph that includes the appropriate terms and pertinent information to answer the objective.

1. Explain what is meant by food spoilage and identify the source for most food spoilage.
2. Describe the conditions that promote food spoilage by microorganisms.
3. Identify the microbial products responsible for food spoilage.
4. Assess how meats and fish become contaminated.
5. Describe the ways poultry and eggs become contaminated with microorganisms.
6. Determine how breads and bakery products can support microbial growth.
7. Identify the two types of fungi that can spoil grain by producing toxins.
8. Discuss how milk and diary products become soured or spoiled.
9. Explain how heat contributes to the preservation of foods.
10. Describe the methods used to preserve milk and other beverages.
11. Assess the effectiveness of low temperatures for preserving foods and beverages.
12. Summarize how conventional drying and freeze drying preserve foods.
13. Identify the chemical preservatives used in foods and beverages and describe how they contribute to preservation.
14. Discuss how radiation can preserve foods and contribute to their shelf life.
15. Contrast between a foodborne infection and an foodborne intoxication.
16. Assess the importance of the HACCP system to food safety.
17. Summarize how fermentation can act as a food preservation mechanism and name several common fermented foods.
18. Explain the purpose of fermented milk products.
19. Discuss the role of microorganisms to the cheese manufacture process.

SELF-TEST

Answer each of the following questions by selecting the ***one*** answer that best fits the question or statement. Answers to even-numbered questions can be found in **Appendix C**.

1. Foods can become spoiled by
A. human handling.
B. rodent or insect contamination.
C. airborne microorganisms.
D. contamination during packaging.
E. All the above (**A–D**) are true.

2. Which one of the following would *not* be considered a highly perishable food?
A. Milk
B. Eggs
C. Meat
D. Dried rice and beans
E. Fish

3. All of the following are spoilage products produced by microorganisms *except*:
A. lysozyme.
B. hydrogen sulfide.
C. acids.
D. ethyl alcohol.
E. skatole.

4. Meats and fish normally become contaminated by
A. handling and processing.
B. the muscle tissue from the living animal.
C. airborne particles.
D. preserving agents.
E. blood from the live animal.

5. The bacterial genus _____ is a common source of poultry and egg contamination.
A. *Escherichia*
B. *Listeria*
C. *Salmonella*
D. *Clostridium*
E. *Enterobacter*

6. Red rot in eggs is caused by
A. *Serratia*.
B. *Salmonella*.
C. *Pseudomonas*.
D. *Escherichia*.
E. *Proteus*.

7. The souring of bakery cream fillings and topping is due to microbial
A. protein degradation.
B. acids.
C. carbohydrate hydrolysis.
D. fat breakdown.
E. capsular accumulation.

8. The fungal aflatoxins are typically found in
A. rye plants.
B. stored grains.
C. rotten eggs.
D. sour milk.
E. spoiled meat.

9. Milk spoilage is usually due to species of
A. *Streptococcus*.
B. *Escherichia*.
C. *Salmonella*.
D. *Lactobacillus*.
E. Both **A** and **D** are correct.

10. Commercial sterilization
A. only kills fungal pathogens.
B. reduces the levels of dangerous pathogens.
C. eliminates the most resistant bacterial spores.
D. only eliminates vegetative bacteria.
E. is used for preserving milk.

11. The flash method of pasteurization subjects milk to a temperature of _____°C for _____ seconds.
A. 62.9; 30
B. 71.6; 15
C. 140; 3
D. 161; 5
E. 300;10

12. To keep foods fresh longer or frozen, a home refrigerator should keep foods at _____°C, while the freezer should be kept at _____°C.
A. 4; –10
B. 10; –4
C. 10; 0
D. 32; 4
E. 32; 0

13. Lyophilized foods have been
A. heat-dried.
B. irradiated.
C. chemically-treated.
D. freeze-dried.
E. Both **A** and **B** are correct.

14. Which one of the following (**A–D**) is *not* a chemical preservative? If all are preservatives, select **E**.
A. Propionic acid
B. Penicilloic acid
C. Sulfur dioxide
D. Benzoic acid
E. All the above (**A–D**) are correct.

15. All of the following foods or beverages involve a fermentation step *except*:
A. sauerkraut.
B. vinegar.
C. soft drinks.
D. pickles.
E. soy sauce.

16. Fermented dairy products include
A. eggs.
B. buttermilk.
C. yogurt.
D. ice cream.
E. Both **B** and **C** are correct.

QUESTIONS FOR THOUGHT AND DISCUSSION

Answers to even-numbered questions can be found in **Appendix C**.

1. Suppose you had the choice of purchasing "yogurt made with pasteurized milk" or "pasteurized yogurt." Which would you choose? Why? What are the "active cultures" in a cup of yogurt?
2. Why might blue (Stilton) cheese pose a possible threat to someone who has an acute allergy to penicillin?
3. Which principles of preservation ensure that each of the following remains uncontaminated on the pantry shelf: vinegar, olive oil, brown sugar, tea bags, spaghetti, hot cocoa mix, pancake syrup, soy sauce, rice?
4. On Saturday, a man buys a steak and a pound of liver and places them in the refrigerator. On Monday, he must decide which to cook for dinner. Microbiologically, which is the better choice? Why?
5. Foods from tropical nations such as Mexico tend to be very spicy, with lots of hot peppers, spices, garlic, and lemon juice. By contrast, foods from cooler countries such as Norway and Sweden tend to be much less spicy. Why do you think this pattern has evolved over the ages?
6. A standard set of recommendations and regulations exists for individuals who work in restaurants and cook food for customers (i.e., food handlers). However, very few regulations exist for individuals who pick fruits or vegetables in the fields. What regulations would you recommend for such workers?

APPLICATIONS

Answers to even-numbered questions can be found in **Appendix C**.

1. Having read this chapter, you decide to outfit your refrigerator with ultraviolet lights. What are you trying to accomplish?
2. Chicken and salad are two items on the dinner menu at home, and you are put in charge of preparing both. You have a cutting board and knife for slicing up the salad items and cutting the chicken into pieces. Which task should you perform first? Why? What other precautions might you take to ensure that the meal is not remembered for the wrong reason?
3. One day you discover an unopened container of sour cream that had been hidden in the back of the refrigerator for some three months. The sour cream appeared satisfactory, and there was no unusual smell. You proceeded to spoon it onto a baked potato and dig in. What factors might have contributed to the sour cream's preservation so long after the expiration date?
4. You and a friend go to an orchard to pick apples with the intention of pressing them in your new cider mill. However, you are aware of the recent outbreaks of infection with *E. coli* O157:H7 that were traced to fresh apple cider. What precautions can you take to ensure a "healthy" experience?

REVIEW

On completing this chapter on food microbiology, test your knowledge of its contents by using the following syllables to compose the term that answers the clue. Each term is a genus of microorganism important in food microbiology. The number of letters in the genus is indicated by the dashes, and the number of syllables in the genus is shown by the number in parentheses. Each syllable is used only once, and the answers to even-numbered terms are listed in **Appendix C**.

A A A AS AS BA BA BAC BAC CE CEPS CHLA CIL CIL CLAV CLO CLO CO COC CUS DI DO EN GIL I I I LA LAC LEU LUS LUS LUS MO MON MY NEL NOS O PER PRO PSEU RA SAL SER STREP STRID STRID TE TER TER TER TI TO TO TO TOC UM UM US

1. Common poultry contaminant (4) ___ ___ ___ ___ ___ ___ ___ ___ ___ ___
2. Discolors meat surface (5) ___ ___ ___ ___ ___ ___ ___ ___ ___ ___ ___ ___ ___
3. Destroyed in canning (4) ___ ___ ___ ___ ___ ___ ___ ___ ___ ___ ___
4. Red pigment in bread (4) ___ ___ ___ ___ ___ ___ ___ ___
5. Causes psittacosis (4) ___ ___ ___ ___ ___ ___ ___ ___ ___
6. Black rot in eggs (3) ___ ___ ___ ___ ___ ___ ___
7. Sours dairy products (4) ___ ___ ___ ___ ___ ___ ___ ___ ___ ___ ___ ___ ___
8. Spore-forming contaminant (3) ___ ___ ___ ___ ___ ___ ___ ___
9. Used to make vinegar (5) ___ ___ ___ ___ ___ ___ ___ ___ ___ ___ ___
10. Sauerkraut producer (4) ___ ___ ___ ___ ___ ___ ___ ___ ___ ___ ___
11. Causes ergot poisoning (3) ___ ___ ___ ___ ___ ___ ___ ___ ___
12. Produces potent exotoxins (4) ___ ___ ___ ___ ___ ___ ___ ___ ___ ___ ___
13. Green rot in foods (4) ___ ___ ___ ___ ___ ___ ___ ___ ___ ___ ___
14. Used to cure cucumbers (5) ___ ___ ___ ___ ___ ___ ___ ___ ___ ___ ___ ___
15. Aflatoxins in grains (4) ___ ___ ___ ___ ___ ___ ___ ___ ___ ___ ___

HTTP://MICROBIOLOGY.JBPUB.COM/

The site features learning, an on-line review area that provides quizzes and other tools to help you study for your class. You can also follow useful links for in-depth information, read more MicroFocus stories, or just find out the latest microbiology news.

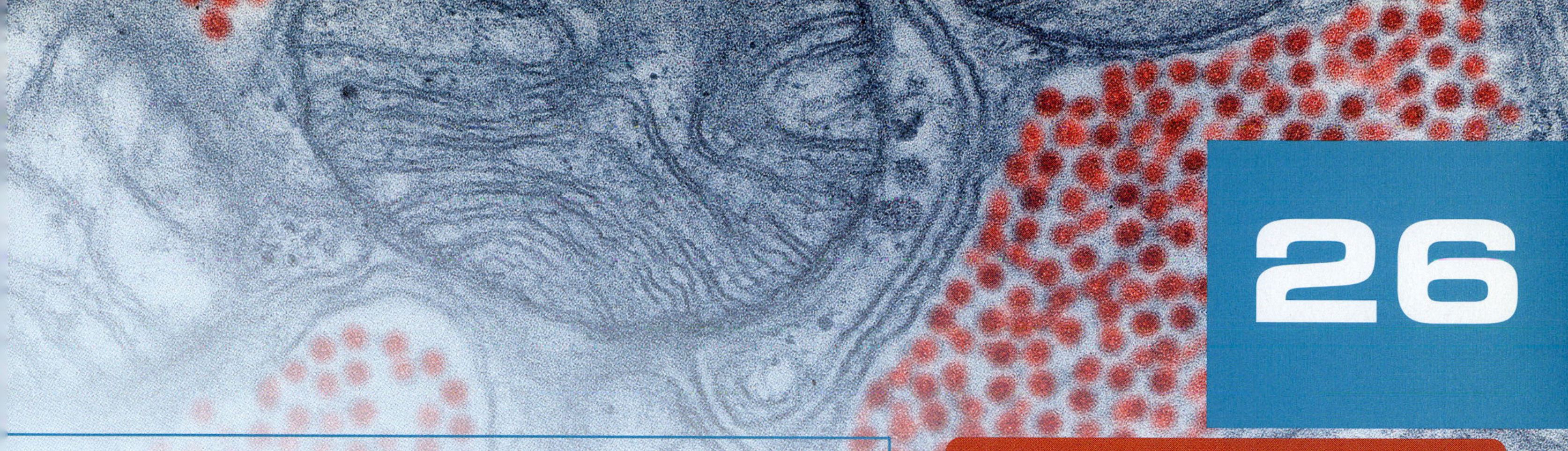

Environmental Microbiology

Access to safe water is a fundamental human need and, therefore, a basic human right. Contaminated water jeopardizes both the physical and social health of all people. It is an affront to human dignity.
—Kofi Annan, United Nations Secretary-General

In the early 1800s, the steam engine and its product, the Industrial Revolution, brought crowds of rural inhabitants to European cities. To accommodate the rising tide, row houses and apartment blocks were hastily erected, and owners of existing houses took in tenants. Not surprisingly, the bills of mortality from typhoid fever, cholera, tuberculosis, dysentery, and other diseases mounted in alarming proportions.

As the death rates rose, a few activists spoke up for reform. Among them was an English lawyer and journalist named Edwin Chadwick. Chadwick subscribed to the then-novel idea that humans could shape their environment and could eliminate diseases of filth by doing away with filth. In 1842, he published a landmark report suggesting that poverty-stricken laborers suffered a far higher incidence of disease than people from middle or upper classes. Chadwick attributed the difference to the abominable living conditions of workers, and he believed most of their diseases were preventable. His report established the basis for the Great Sanitary Movement, a wave of reform that began in Europe.

Chadwick was not a medical man, but his ideas captured the imagination of both scientists and social reformers. He proposed that sewers be constructed using smooth ceramic pipes and then enough water be flushed through the system to carry waste to some distant depository. To work efficiently, the system required the installation of new water and sewer pipes, the development of powerful pumps to bring water into homes, and the elimination of older sewage systems. The cost would be formidable.

Chadwick's vision eventually came to reality, but it might have taken decades longer without the intervention of cholera. In 1849, a cholera

Chapter Preview and Key Concepts

epidemic broke out in London and terrified so many people that public opinion began to form in favor of Chadwick's proposal. Another epidemic occurred in 1853, during which John Snow was able to trace the transmission of the disease back to a contaminating water pump (see Chapter 1). In both outbreaks, the disease reached the affluent as well as the poor, and the mortality rate exceeded 50 percent. Construction of the sewer system began shortly thereafter.

The proverbial "icing on the cake" came in 1892 when a devastating epidemic of cholera erupted in Hamburg, Germany. For the most part, Hamburg drew its water directly from the polluted Elbe River. Adjacent to Hamburg lay Altona, a city where the German government had previously installed a water filtration plant. Altona remained free of cholera. The contrast was sharpened further by a street that divided Hamburg and Altona. On the Hamburg side of the street, multiple cases of cholera broke out; across the street, none occurred. Chadwick and his fellow sanitarians could not have imagined a more clear-cut demonstration of the importance of water purification and sewage treatment.

Adequate safe water remains a serious problem in much of the world today (FIGURE 26.1). In fact, one billion people lack access to safe and sufficient drinking water; four billion people lack adequate sanitation; and four million deaths occur each year, mostly the poor and children, from insufficient and unsafe drinking water.

To address this and other health issues, in 2000 the leaders of 189 countries adopted the United Nations Millennium Declaration, which is a set of goals to solve many global environmental problems by 2015. One of these goals is to reduce by 50 percent the proportion of people without sustainable access to safe drinking water and sanitation.

Dealing with water pollution and treating sewage are but two of the myriad activities involving microbiology in public health. Other environmental issues biogeochemical cycles, whose microorganisms are critical to the recycling of nitrogen, sulfur, and carbon. Thus, this chapter focuses on environmental microbiology and the study of microorganisms as they affect the natural environment.

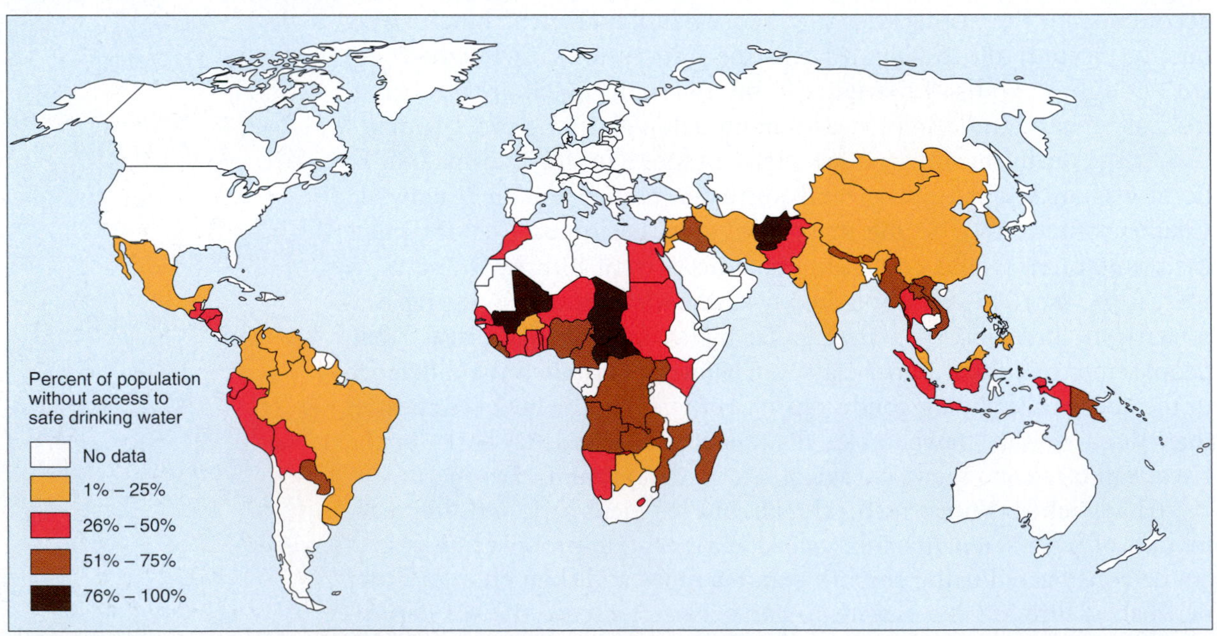

FIGURE 26.1 **Populations Lacking Access to Safe Drinking Water.** In 2005, more than one billion people lacked access to clean and sufficient drinking water.
Source: From *The World's Water*. The Biennial Report on Freshwater Resources (Gleick 1998).
Q: Provide several reasons why many developing nations lack clean water.

26.1 Water Pollution

For purposes of simplification, scientists classify water into two major types: groundwater and surface water. **Groundwater** originates from deep wells and subterranean springs, and, because of the filtering action of soil, deep sand, and rock, it is virtually free of microorganisms. As the water flows up along channels, contaminants may enter it and alter its quality. **Surface water** is found in lakes, streams, and shallow wells. Its microbial population may reflect the air through which rain has passed, the meat-packing plant near which a stream flows, or the sewage-treatment facility located along a riverbank.

Therefore, water is considered contaminated when it contains a chemical or biological poison, or an infectious agent. In polluted water, the same conditions apply, but the poison or agent is obvious, as it has an unpleasant taste, smell, or appearance. **Potability** (*pot* = "drink") by contrast, refers to water that is safe to drink; unpotable water is unfit.

Unpolluted and Polluted Water Contain Different Microbial Populations

KEY CONCEPT

- The microbial population in water reflects the organic matter present.

A body of unpolluted water, such as a mountain lake or stream, is usually low in organic nutrients, and thus, only a limited number of bacterial species are present, perhaps a few thousand per milliliter. Most are soil organisms that have run off into the water during a rainfall. An example is the actinomycetes, a group of mold-like bacterial species responsible for the musty odor to soil. Other inhabitants include yeasts, along with bacterial and mold spores. Cellulose digesters such as members of the genus *Cellulomonas* also are found. These bacterial species digest cellulose in plant cell walls. Autotrophic bacterial cells are common, and free-living protozoa such as *Paramecium*, *Amoeba*, and *Euglena* abound.

A polluted body of water, such as a polluted lake or river, presents a totally different picture (FIGURE 26.2A). The water contains large amounts of organic matter from sewage, feces, and industrial sources, and the population of microorganisms is usually heterotrophic. Such polluted water usually contains **coliform bacteria**, a group of gram-negative, non–spore-forming bacilli usually found in the human intestine (FIGURE 26.2B). Coliform bacteria, which ferment lactose to acid and gas, include *Escherichia coli* and species of *Enterobacter*. Noncoliform bacteria also common in polluted water include *Streptococcus*, *Proteus*, and *Pseudomonas* species.

In polluted water, microorganisms contribute to a chain of events that drastically alters the ecology of the environment (FIGURE 26.3). When phosphates accumulate in the water, algae reproduce rapidly. These **algal blooms** supply nutrients to other microorganisms, which multiply rapidly and use up the available oxygen. Soon, protozoa, small fish, crustaceans, and plants die and accumulate on the bottom. Anaerobic bacterial species such as *Desulfovibrio* and *Clostridium* then thrive in the mud and produce gases, giving the water a stench reminiscent of rotten eggs.

The marine environment of the oceans illustrates another view of microbial populations in water. In the high salt concentration of ocean water, salt-tolerant organisms, called **halophiles**, survive. In addition, the organisms must be **psychrophilic** because it is very cold below the surface. Those at the bottom must withstand great pressure and are therefore **barophilic**, or pressure tolerant. The organisms in these environments pose no threat to humans because they cannot grow in the body tissues.

Marine microorganisms are vital to ecological cycles because they form the foundations of many food chains. For example, marine algae such as **diatoms** (FIGURE 26.4A) and organisms such as the **dinoflagellates** capture the sun's energy and, using carbon dioxide, convert the energy to chemical

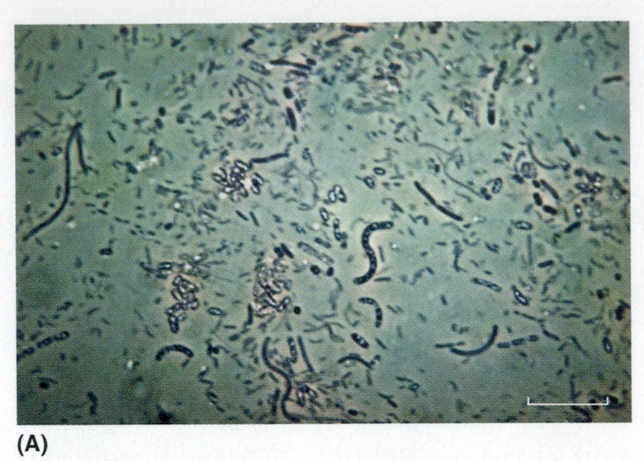

(A)

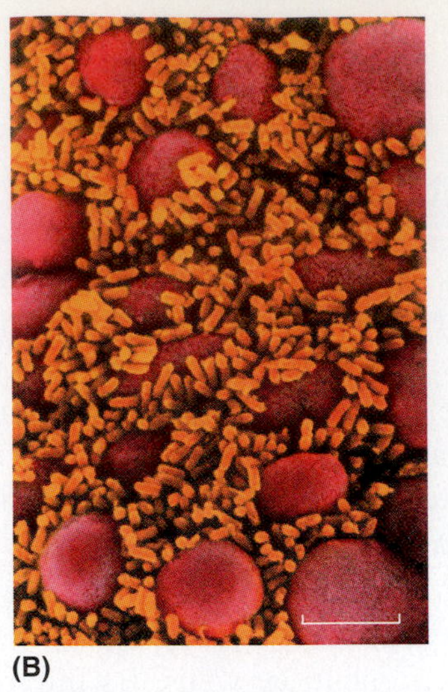

(B)

FIGURE 26.2 **Microorganisms in Water Environments.** (**A**) A light micrograph of bacterial species in polluted water. Many different bacterial shapes can be seen. (Bar = 10 μm.) (**B**) A false-color scanning electron micrograph of coliform bacteria. These *E. coli* bacteria on the intestinal lining can be defecated into the water. (Bar = 4 μm.)

Q: What bacterial shapes can you identify in the light micrograph?

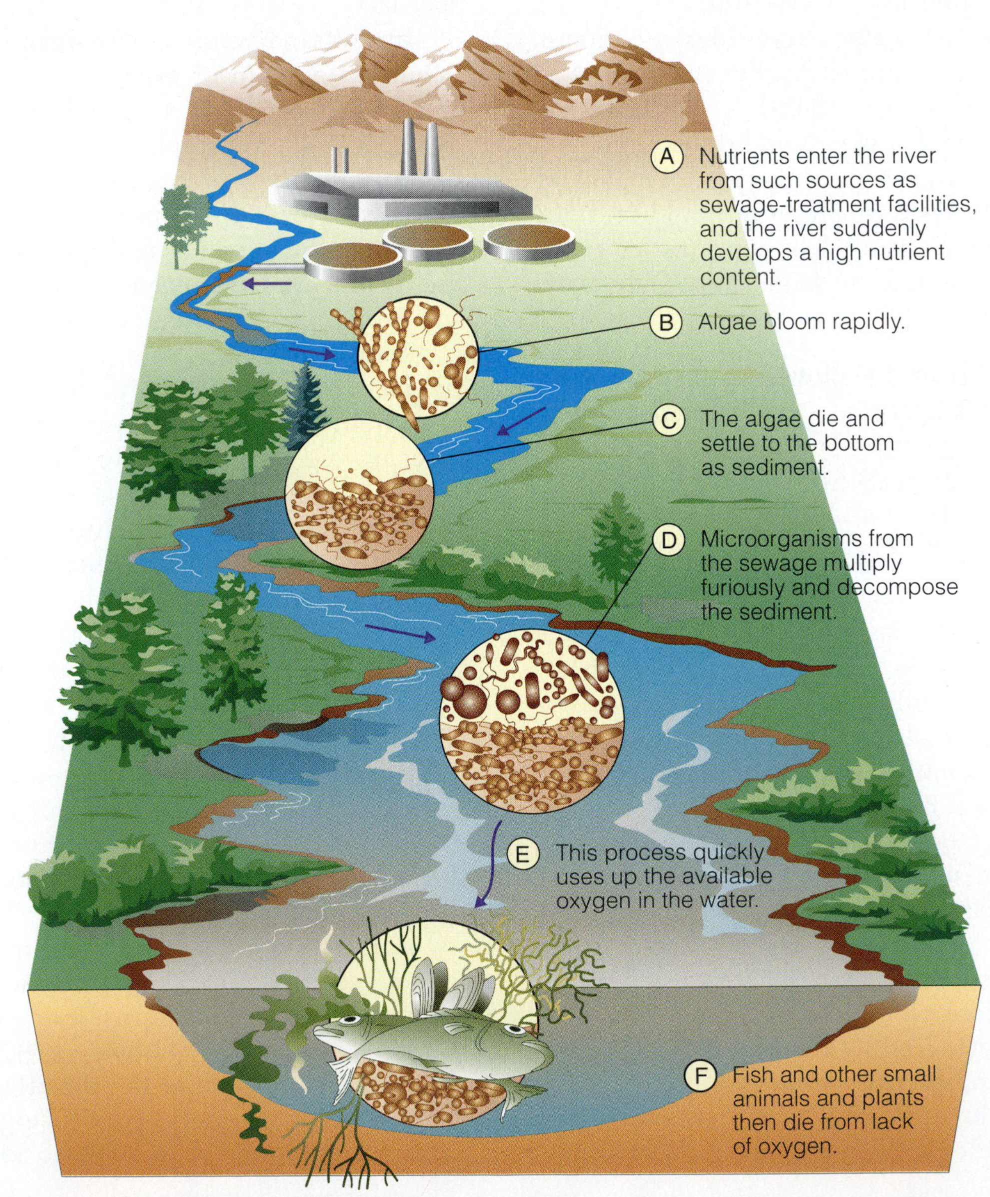

FIGURE 26.3 **The Death of a River.** The introduction of nutrients can lead to an algal bloom from which other microorganisms derive nutrients. This metabolism can quickly deplete the oxygen in the water, leading to the death of larger organisms, such as fish.

Q: Why would a lack of oxygen kill the fish?

energy in carbohydrates (see Chapter 6). The microorganisms then are consumed by other animals in the food chain. Dinoflagellates frequently make the news when certain species of *Gonyaulax* and *Gymnodinium* produce blooms called red tides (FIGURE 26.4B). The toxins produced by these organisms can kill many fish species.

Most marine microorganisms are found along the shoreline, or **littoral zone**, because this is where nutrients are plentiful. Certain unusual types of microorganisms have been found on the ocean floor in the **benthic zone** and even at the bottom of six-mile-deep trenches, the **abyssal zone**. Two types of protozoa, the foraminiferans and radiolarians, are of special interest to oil companies because these protozoa were dominant species during formation of the oil fields, and their fossils serve as markers for oil-bearing layers of rock (see Chapter 17).

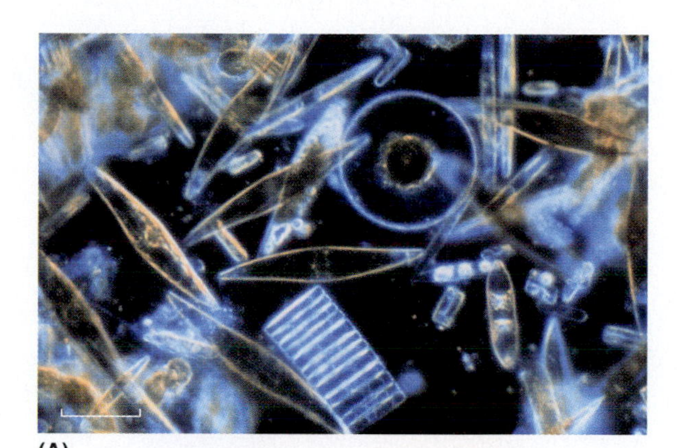

(A)

(B)

FIGURE 26.4 **Marine Microorganisms.** (**A**) A dark field micrograph of a collection of diatoms. Note the broad variety of shapes and sizes in this group. Diatoms trap the sun's energy in photosynthesis and use it to form carbohydrates that are passed on to other marine organisms as food. (Bar = 25 µm.) (**B**) Blooms of dinoflagellates, such as this one off the coast of Africa, can have such high rates of metabolism that they release substantial amounts of toxins that poison fish.

Q: What causes an algal bloom such as in (B)?

CONCEPT AND REASONING CHECKS

26.1 Explain why pathogens tend to be found in polluted water.

There Are Three Types of Water Pollution

KEY CONCEPT

- Biological pollution results from deposits of human waste.

Water is vital to such industries as food processing, meat packing, and paper manufacturing. Water also is used extensively in pharmaceutical plants and mines, and for cooling purposes in power-generating units. It irrigates agricultural lands and provides the focus for many recreational facilities. Uses such as these, however, commonly add to contamination and pollution of water.

Physical pollution of water occurs when particulate matter such as sand or soil makes the water cloudy, or when cyanobacteria bloom during midsummer and their remains give water the consistency of pea soup. **Chemical pollution** results from the introduction of inorganic and organic waste to the water. For example, water passing out of a mine contains large amounts of copper or iron. Other chemical pollutants in water include phosphates and nitrates from laundry detergents as well as acids such as sulfuric acid.

The third type of pollution, **biological pollution**, is the main concern of our discussion. This type of pollution develops from microorganisms entering water from human waste, food-processing and meat-packing plants, medical facilities, and similar sources. Normally, water can handle biological material because heterotrophic microorganisms digest organic matter to carbon dioxide, water, and useful ions (phosphates, nitrates, and sulfates). With the rapid movement of the water, aeration is constant, and waste is soon diluted and eliminated. However, when water stagnates or is overloaded with waste, it becomes polluted.

A critical measurement in polluted water is the **biochemical oxygen demand** (BOD). This refers to the amount of oxygen microorganisms require to decompose the organic matter in water. As the number of microorganisms increases, the demand for oxygen increases proportionally. In the laboratory, the BOD is determined by measuring the dissolved oxygen content of water immediately after collection and then after incubation at 20° C for five days. The difference in oxygen content represents the amount used up by microorganisms in the water sample. Results are generally expressed as parts per million (ppm), with a BOD of several hundred ppm usually considered high.

CONCEPT AND REASONING CHECKS

26.2 Identify the three forms of water pollution and describe how they differ.

Diseases Can Be Transmitted by Water

KEY CONCEPT

- Contaminated water can be a vehicle for intestinal disease and wound infections.

Water consumption may be the vehicle for transfer of a broad variety of intestinal diseases, including bacterial diseases such as typhoid fever, cholera, shigellosis, and others described in Chapter 10. Waterborne epidemics, however, are rare because of continual surveillance. Many illnesses transmitted by drinking water are due to less familiar bacterial genera such as *Campylobacter, Yersinia,* and *Legionella* (FIGURE 26.5) as well as by toxin-producing strains of *E. coli.* Septicemia, necrotizing fasciitis, and gangrene may develop from the initial infection. An emerging pathogen associated with contaminated water is *Vibrio vulnificus*, a gram-negative bacterial species that can cause serious intestinal illness and septicemia in individuals with preexisting liver disease or compromised immune systems (see Chapter 10). Raw oyster consumption has been implicated in many deaths due to *V. vulnificus*.

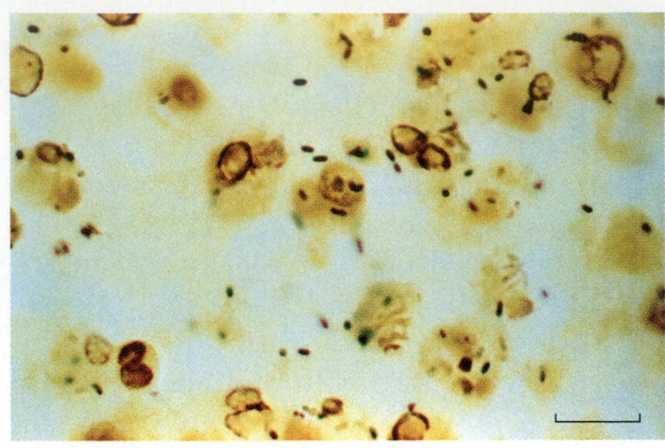

(A)

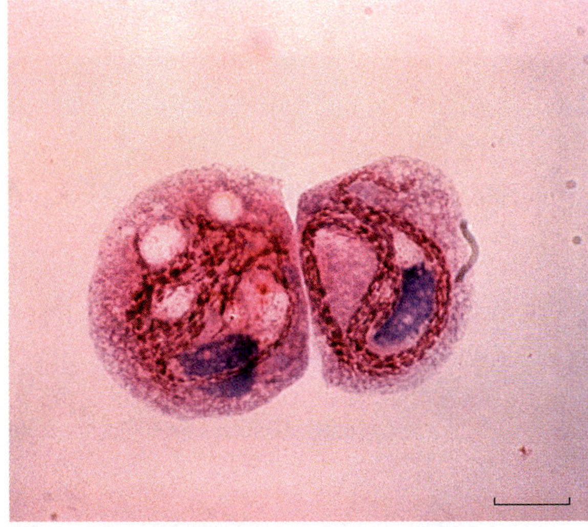

(B)

FIGURE 26.5 ***Legionella* in the Environment.** *Legionella pneumophila* is the gram-negative rod responsible for Legionnaires' disease. (**A**) It infects the respiratory passages and cells in this biopsied long tissue specimen (dark rods). (**B**) In nature, the bacterial cells live within the cytoplasm of the waterborne protozoa. Here, chains of cells (dark red) can be seen in cells of *Tetrahymena*. (Bar = 10 µm.)

Q: Because Legionella *can be transmitted in airborne water droplets from a contaminated air conditioner, how would the bacterial cells survive while in the air conditioning system?*

Several bacterial species in marine waters also contribute to wound infections when contaminated seawater has contacted the exposed tissue. Among the marine pathogens is *Erysipelothrix rhusiopathiae*, a gram-negative pleomorphic rod. Infections in humans, called **erysipeloid**, usually occur on the hands or feet and are accompanied by a bright-red, well-demarcated lesion that burns or itches. *Mycobacterium marinum*, an acid-fast rod related to the tubercle bacillus, causes a small papular lesion or **granuloma**, at the wound site, particularly near the knuckles of the

MICROFOCUS 26.1: Environmental Microbiology

Fish Handler's Disease

It was the start of the rockfish (striped bass) season in 2004 and recreational fishermen were flocking to the Chesapeake Bay. The previous year brought in a record catch of the popular game fish, so everyone was looking forward to another great year. What wasn't publicized was the appearance of fish handler's disease on some fishermen.

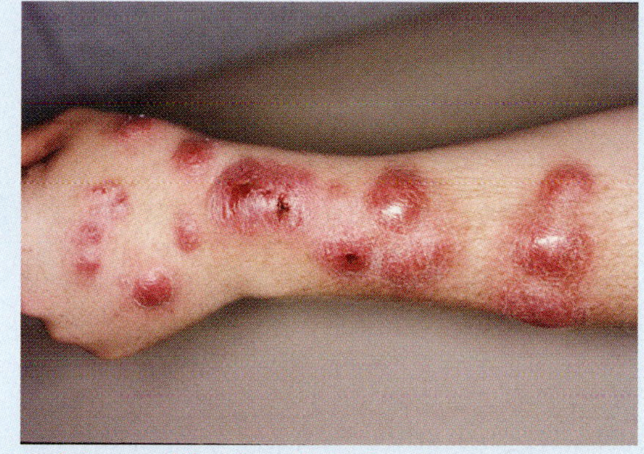

Nodules resulting from infection by *Mycobacterium marinum*.

Fish handler's disease usually develops on the skin of the hands or feet. Although there are several infections commonly referred to as fish handler's disease, abrasions on the skin most commonly become infected with *Mycobacterium marinum* present in polluted water or in infected fish. Human infections produce nodular lesions, which although not life-threatening, can spread to the wrist and elbow, producing arthritis-like joint problems. The disease also occurs in aquarium shop personnel who clean infected aquaria and do not wear gloves (see figure).

In the 1980s, overfishing in the Chesapeake Bay had so decimated the rockfish number that a moratorium on fishing was imposed. Then by the mid-1990s, the population had recovered and the ban was lifted. However, the polluted nature of the Chesapeake Bay lead to the identification of diseased rockfish. Now, in 2006, it is estimated that 75 percent of the fish in the bay are diseased with *M. marinum*.

Fisheries biologists and researchers are not sure how the fish become infected. They know most infections occur in the Chesapeake Bay, which is where the fish breed. Pollution of the bay certainly plays a major role. Fish may look healthy but *M. marinum* infects the spleen and can spread to other organs. Sores develop and the fish dies.

Because fish handler's disease is not a reportable disease, exactly how many humans become infected is unclear. In 2004, there were 46 reported cases. More cases appear in the summer and fall when food for rockfish becomes more scarce and heat degrades the bay's water quality. Such stresses produce more diseased fish.

Researchers have found 10 different strains of *M. marinum* in the rockfish. Two are newly identified strains, so their effects on humans remain to be determined.

hand. The infection may spread via the lymphatic system. Fish handlers, seafood workers, and aquatic sports enthusiasts may suffer infection from these organisms, and antibiotics are used to limit the growth of both (MicroFocus 26.1). *V. vulnificus* can cause wound infections involving gangrene and necrotizing fasciitis.

Viral diseases transmitted by water include hepatitis A, rotavirus disease, gastroenteritis due to Coxsackie viruses or noroviruses, and in rare instances, the polio virus. These diseases are generally related to fecal contamination of water.

Water may be a vehicle for the transfer of *Entamoeba histolytica*, *Giardia lamblia*, and *Cryptosporidium* (see Chapter 17). The dinoflagellate *Gonyaulax catanella* produces a toxin causing muscular paralysis and death from asphyxiation. The toxin is ingested from shellfish that feed on the dinoflagellate. Another dinoflagellate, *Gambierdiscus toxicus* is consumed by small fish that concentrate the toxin and pass it to larger fish, such as sea bass and red snapper. Human consumption of the fish leads to neurological and muscular intoxication and a condition called **ciguatera** fish poisoning (from *cigua* for "poisonous snail," originally thought to be the cause).

CONCEPT AND REASONING CHECKS

26.3 Summarize the types of infections caused by marine microorganisms.

26.2 The Treatment of Water and Sewage

Some years ago, health care workers in Africa asked villagers to name their single greatest need. The answer was almost unanimous: "Water." A startling survey by the World Health Organization (WHO) indicates that three of every four humans alive today do not have enough water to drink, or, if water is available, the supplies are contaminated (see chapter opener and quote).

Water is unfit to drink when it contains human sewage, animal waste, or disease-causing microorganisms. However, the situation can be reversed through the proper management of water resources. Water-purification procedures prevent pathogenic microorganisms from reaching the body, while sewage-treatment processes remove pathogens from body waste products. In this section, we will examine how these are accomplished.

Water Purification Is a Three-Step Process

KEY CONCEPT

- Sedimentation, filtration, and chlorination produce potable water.

Three basic steps are included in the preparation of water for drinking: sedimentation, filtration, and chlorination. FIGURE 26.6 illustrates the steps in the water-purification process.

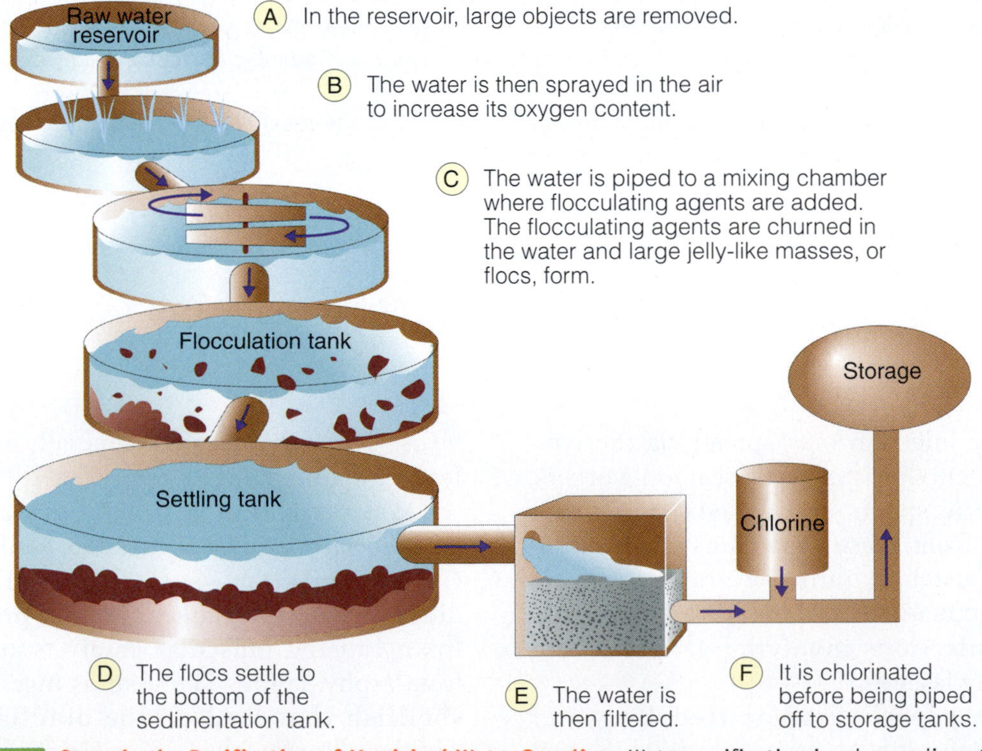

FIGURE 26.6 Steps in the Purification of Municipal Water Supplies. Water purification involves sedimentation, filtration, and chlorination.

Q: What would happen if the filtration step preceded the sedimentation step?

Sedimentation uses large reservoirs or settling tanks to remove leaves, particles of sand and gravel, and other materials from the soil. Chemicals such as aluminum sulfate (alum) or iron sulfate are dropped as a powder onto water and they form jelly-like masses of coagulated material called **flocs**. In the process of **flocculation**, these flocs fall through the water and cling to organic particles and microorganisms, dragging a major portion to the bottom sediment. MicroFocus 26.2 describes a new, environmentally safe flocculant.

The **filtration** step is next. Although different types of filtering materials are available, most filters use a layer of sand and gravel to trap microorganisms. A slow sand filter, containing fine particles of sand several feet deep, is efficient for smaller-scale operations. Within the sand, a biofilm of microorganisms (bacterial, fungal, and protozoal cells) acts as a supplementary filter. This layer is called a **schmutzdecke**, or dirty layer (the word is German for "grime or filth cover"). A slow sand filter may purify over 3 million gallons of water per acre per day. To clean the filter, the top layer is removed and replaced with fresh sand.

A rapid sand filter contains coarser particles of gravel. A schmutzdecke does not develop in this filter, but the rate of filtration is much higher, with over 200 million gallons purified per acre per day. This type of filter is commonly used in municipal water systems. It must be cleaned more often than the slow sand filter, a process accomplished by forcing water back through the filter by mechanical pressure. Both slow and rapid sand filters remove approximately 99 percent of the microorganisms from water.

The final step is **chlorination**, in which chlorine gas is added to the water. Chlorine is an active oxidizing agent that reacts with any organic matter in water (see Chapter 23). It is important, therefore, to continue adding chlorine until a residue is present. A residue of

MICROFOCUS 26.2: Environmental Microbiology

Purifying Water with the "Miracle Tree"

In developed nations, water purification usually uses chemical powders such as aluminum sulfate or iron sulfate for the flocculation process. Many developing nations do not have such chemicals available, nor can they readily afford to purchase them. Can something else be used for the flocculation step in water purification? Yes. It's the "miracle tree."

Scientifically known as *Moringa oleifera*, this tropical tree was given the name "miracle tree" because of the many uses the tree has—including an environmentally friendly way to purify water.

M. oleifera survives in arid areas and produces elongated seedpods. The local people use the leaves and roots for food, the wood is used in building, and some parts of the tree are used in traditional medicines. High-grade oil for lubrication and cosmetics is produced from the seeds. However, the seeds also have another use—purifying water. In those parts of Africa and India where the trees grow, people grind up the seeds and add the powder to cloudy water to precipitate all the solid particles. Clean drinking water results.

Ian Marison, research director at École Polytechnique Fédérale de Lausanne in Switzerland and his colleagues have examined the ground-up seed residue and have isolated a charged peptide. When they add this peptide to cloudy water, within two minutes the water goes from cloudy to clear.

More exciting is the discovery that the peptide also has bactericidal properties and can disinfect heavily contaminated water. According to Marison, it even works against some drug-resistant strains of *Staphylococcus, Streptococcus,* and *Legionella,* and is effective against some nonbacterial waterborne pathogens. In fact, Marison's group has discovered if they tweak the peptide's structure, they can increase the antimicrobial effect of the peptide.

This is certainly good news for developing nations where the trees mainly grow. However, if commercially produced, the plant flocculant also would be useful to developed nations because the traditional chemical flocculants often have safety and environmental problems.

As Marison says, "It's a biological, biodegradable, sustainable resource. *Moringa* grows where there is very little water; it grows very, very fast and it costs almost nothing." It truly is a "miracle tree."

0.2 to 1.0 parts of chlorine per million (ppm) of water often is the standard used. Under these conditions, most remaining microorganisms die within 30 minutes.

At this point, some communities soften water by removing magnesium, calcium, and other salts. Softened water mixes more easily with soap, and soap curds do not form. Water also may be fluoridated to help prevent tooth decay. Scientists believe fluoride strengthens tooth enamel and makes the enamel more resistant to the acid produced by anaerobic bacterial species in the mouth.

CONCEPT AND REASONING CHECKS

26.4 Assess the need for three steps in water purification.

Sewage Treatment Can Be a Multistep Process

KEY CONCEPT

- For sewage treatment, water must be separated from the waste.

Systems for the treatment of human waste range from the primitive outhouse, which is nothing more than a hole in the ground, to the sophisticated sewage-treatment facilities used by many large cities. All operate under the same basic principle: Water is separated from the waste, and the solid matter is broken down by microorganisms to simple compounds for return to the soil and water.

In many homes, human waste is emptied into underground **cesspools**, which are concrete cylindrical rings with pores in the wall. Water passes into the soil through the bottom and pores of the cesspool, while solid waste accumulates on the bottom. Microorganisms, especially anaerobic bacterial species, digest the solid matter into soluble products that enter the soil and enrich it. Some hardware stores sell enzymes and dried *Bacillus subtilis* spores to accelerate digestion.

Some homes have a **septic tank**, an enclosed concrete box that collects waste from the house. Organic matter accumulates on the bottom of the tank, while water rises to the outlet pipe and flows to a distribution box. The water is then separated into pipes that empty into the surrounding soil. Because digested organic matter is not absorbed into the ground, the septic tank must be pumped out regularly.

Sewers are at least as old as the Cloaca Maxima ("Great Sewer") of Roman times. Until the mid-1800s, however, sewers were simply elongated cesspools with overflow pipes at one end. They collected filth and had to be pumped out regularly. Finally, in 1842, Edwin Chadwick's report raised the possibility that sewage spreads disease, and soon thereafter a movement (fueled by a cholera outbreak) sprang up to sanitize European cities.

Small towns collect sewage into large ponds called **oxidation lagoons**, where the sewage is left undisturbed for up to three months to allow natural digestion of the organic mater to occur. During that time, aerobic bacterial species digest organic matter in the water, while anaerobic organisms break down sedimented material. Under controlled conditions, the waste may be totally converted to simple salts such as carbonates, nitrates, phosphates, and sulfates. At the cycle's conclusion, the bacterial cells die naturally, the water clarifies, and the pond may be emptied into a nearby river or stream.

Large municipalities rely on a mechanized sewage-treatment facility to handle domestic wastewater, which contains massive amounts of waste and garbage (FIGURE 26.7). The process involves primary, secondary, and tertiary treatment.

Primary treatment. In the first step, a screen is used to remove grit and large insoluble waste. Then, the raw sewage is piped into huge open tanks for organic waste removal. This waste, called **sludge**, is passed into sludge tanks for further treatment. Flocculating materials, such as aluminum and iron sulfate, then are added to the raw sewage to drag microorganisms and debris to the bottom.

Secondary treatment. The second stage of sewage treatment is designed to degrade the biological content of the sewage. It occurs in two phases. The liquid phase involves aeration of the water to encourage aerobic growth of microorganisms. As they grow, microorganisms digest proteins into simple amino acids, carbohydrates into simple sugars, and fats into fatty acids and glycerol. Acids and alcohols also are produced, and carbon dioxide gas is given off. The water then is passed through a clarifier and filter to remove the microorgan-

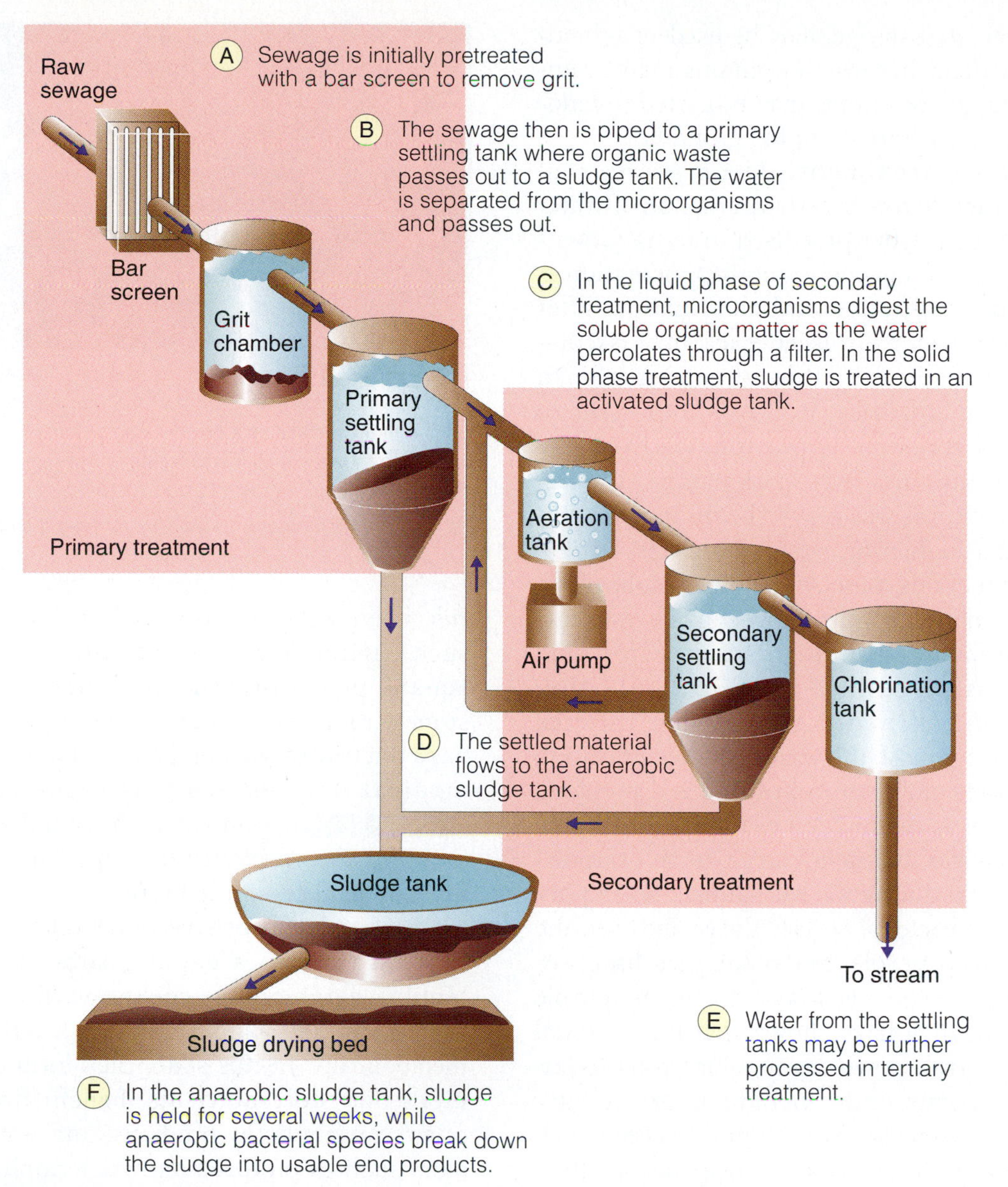

FIGURE 26.7 **A Sewage-Treatment Facility.** Sewage-treatment facilities use physical, chemical, and biological processes to treat and remove all water contaminants.

Q: Why must there be such an extensive sequence of steps to treat sewage?

isms and remaining organic matter, after which it flows into a stream or river.

The solid phase of secondary treatment is carried on in a **sludge tank**, within which microbial growth is encouraged by either an aerobic or an anaerobic process. In the aerobic process, compressed air is forced into the sludge, and the suspended particles form tiny gelatinous masses (biofilm) swarming with microorganisms, which thrive on the organic matter. *Zoogloea ramigera,* a gram-negative rod, produces the slime to which other microorganisms attach and congregate. This **activated sludge**, as it is termed, gathers to itself much of the microorganisms, organic material, color, and smell of the sewage. The activated sludge is drawn off and dried.

In the anaerobic method of sludge digestion, sewage is held in the tank for up to 30 days while the sludge ferments. Gases such as methane, carbon dioxide, and nitrogen are produced. The methane may be captured and used to run the machinery of the sewage facility. Other gases, such as ammonia and hydrogen sulfide, are of value to chemical industries. The digested sludge, together with the dried activated sludge, may be used as agricultural fertilizer because it contains many valuable salts, or the sludge may be carted to landfill sites or offshore dumping grounds.

Tertiary treatment. The separation of solid sludge leaves a certain amount of water that may be further processed in tertiary treatment by purifying the water. Sedimentation is followed by filtration and chlorination, after which the water is placed back into circulation and made available to consumers. In many municipalities, it also is important to remove salts such as phosphates from the water because they may spark blooms of algae.

Agricultural waste, which can include animal and slaughtering waste, pesticides, fertilizers, and fuel oils, pose still another problem for sanitary microbiologists. Most animal waste currently is handled through systems in which manure is piped into clay-lined oxidation lagoons and allowed to remain while bacterial species decompose the waste. The accumulation of ammonia nitrogen from urine is a problem, however, as is the buildup of phosphorus in the soil. Moreover, the stench tends to be overpowering, except during the late summer when photosynthetic bacterial species thrive and turn the water a deep purple. Microbiologists have isolated a species of *Rhodobacter* from the purple water, and they have shown that the bacterial cells can break down many odoriferous (odor-causing) compounds, including volatile fatty acids and phenols. In the future, seeding such organisms in the lagoons may help increase their efficiency and minimize their noxious odors.

CONCEPT AND REASONING CHECKS

26.5 Summarize the steps in the treatment of sewage.

Biofilms Are Prevalent in the Environment

KEY CONCEPT

- Biofilms are complex organizations of microorganisms.

For decades, microbiologists have studied free-floating bacterial species growing in laboratory cultures as examples of how bacterial organisms live and behave in nature. In recent years, however, they have come to realize the importance of bacterial species living in biofilms. A **biofilm** is an immobilized population of bacterial species (or other microorganisms) living in a web of tangled polysaccharide fibers adhering to a surface. Examples of such environments are the surfaces of teeth, aquatic plants or animals, water pipes, and stones. Biofilms also are responsible for urinary tract infections and for contamination of medical devices, such as catheter tubes (FIGURE 26.8). Contact with a fluid environment ensures a plentiful supply of nutrients for the members of the biofilm.

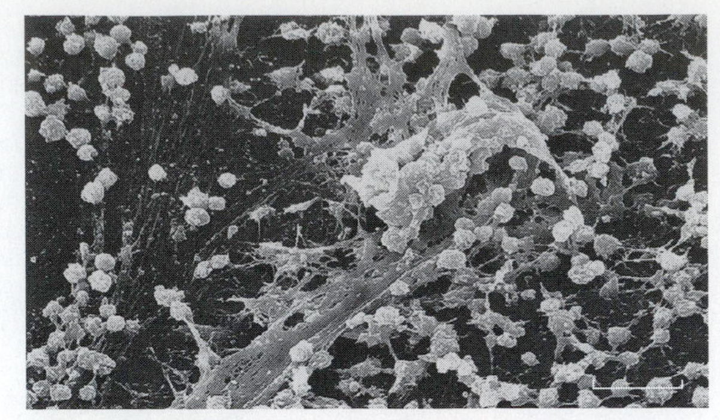

FIGURE 26.8 **Biofilm Contamination.** A scanning electron micrograph of a biofilm of *Staphylococcus aureus* on the inner surface of a needleless connector, which allows direct access to intravascular catheters. (Bar = 10 µm.)

Q: What do microorganisms gain by forming a biofilm?

Bacterial cells behave much differently in a biofilm than in a culture tube. Within a biofilm, aerobic ones coexist with anaerobic bacteria as they share passageways and interact metabolically. In this state, they form communities that store nutrients, benefit from each other's metabolic by-products, and resist predators such as protozoa and bacteriophages. By contrast, **planktonic bacteria** live as individuals and do not enjoy these benefits. They tend to be of smaller size and have a lower metabolic rate of respiration. Thus, the members of a biofilm are morphologically and physiologically distinct from their planktonic neighbors.

Microorganisms within a biofilm can chemically communicate with one another. Researchers have studied several bacterial species that form biofilms and have identified numerous extracellular signals. This so-called **quorum sensing** is discussed in Chapter 3.

Biofilms carry both positive and negative connotations. Among the benefits of biofilms are their uses in **bioremediation**, where biofilm populations are used to degrade toxic wastes and other synthetic products of industry. They also are employed in the oil industry to fill empty spaces after oil has been pumped out. In natural settings, they degrade organic compounds, thus retarding pollutant buildup.

On the negative side, concern continues to mount about the tendency of biofilms to form on contact lens surfaces, catheters, and medical implements (e.g., artificial hearts). Biofilms are important considerations in the development of dental caries and urinary tract infections (see Chapter 12). The concern is particularly acute because the bacterial species in biofilms resist phagocytosis by white blood cells, display enhanced resistance to antibiotics, and withstand the destructive action of disinfectants and antiseptics.

In the industrial setting, biofilms pose problems when their sulfate-reducing members anaerobically convert sulfur compounds to hydrogen sulfide. The latter corrodes water pipes, especially those that carry seawater (where sulfate abounds). When they contaminate computer chips, biofilms act as conductors and thereby interfere with electronic signals. Indeed, one researcher has called biofilms the "venereal disease of industry." Some scientists estimate that in nature, 99 percent of all microbial activities occur in biofilms.

CONCEPT AND REASONING CHECKS

26.6 Assess the positive and negative effects of biofilms.

FIGURE 26.9 Collection of Water for Analysis. A water sample is collected and potential indicator organisms identified by the membrane filtration technique.

Q: What indicator organism would indicate if this pond is contaminated?

The Bacteriological Analysis of Water Tests for Indicator Organisms

KEY CONCEPT

- Coliform bacteria indicate a fecal contamination of water.

Many methods are available for detecting bacterial contamination of water, and various ones are selected according to the resources of the testing laboratory. Because it is impossible to test for all pathogenic microorganisms, water-quality bacteriologists have adopted the practice of testing for certain **indicator organisms**. If these bacterial species, normally found in the human intestinal tract, are present in a water sample, fecal contamination has probably taken place in the original body of water.

Among the most frequently used indicator organisms are the coliform bacteria. **Coliform bacteria** normally are found in the intestinal tracts (colon) of humans and many warm-blooded animals. They can survive for extensive periods of time in the environment, and they are relatively easy to cultivate in the laboratory. *E. coli* is the most important indicator organism within the group.

The **membrane filter technique** is a popular laboratory test in water microbiology because it is straightforward and can be used in the field (see Chapter 23). A technician places a specially designed collecting bottle in the water and takes a 100-ml sample (FIGURE 26.9). The water then is passed through a membrane filter, and the filter pad is transferred to a plate of bacteriological medium. Bacterial cells trapped in the filter will form colonies, and by counting the colonies, the technician may determine the original number of bacterial cells in the sample.

Another method for testing water is the **standard plate count** (SPC) technique (see Chapter 5). Samples of water are diluted in sterile buffer solution, and carefully measured amounts are transferred to culture dishes. Agar medium is added, and the plates are set aside at incubation temperatures. A count of the colonies multiplied by the reciprocal of the dilution (the dilution factor) yields the total number of bacterial cells per ml of the original sample.

A third test is a statistical evaluation called the **most probable number** (MPN) test (see Chapter 5). In this procedure, a technician inoculates water in 10-mL, 1-mL, and 0.1-mL amounts into lactose broth tubes. The tubes are incubated and coliform organisms are identified by their production of carbon dioxide gas. Referring to an MPN table, a statistical range of the number of coliform bacteria is determined by observing how many broth tubes showed gas. MicroInquiry 26 (pages 804–805) gives you a chance to carry out a standard qualitative water analysis.

CONCEPT AND REASONING CHECKS

26.7 Describe the three lab tests for water quality.

26.3 The Cycles of Elements in the Environment

The thought of microorganisms often conjures a negative reaction because of their disease implication, and the contents of this chapter have undoubtedly supported that notion. It would be unwise, however, to neglect the positive role of microorganisms in the environment because it is a substantial one. We have alluded to their role in the decay of organic matter, and we shall expand the idea by briefly examining the place of microorganisms in three vital cycles of nature: the carbon, sulfur, and nitrogen cycles.

The Carbon Cycle Is Influenced by Microorganisms

KEY CONCEPT

- Photosynthetic organisms and microorganisms are keys to the carbon cycle.

Planet Earth is composed of numerous elements, among which is a defined amount of carbon, which is constantly recycled via the **carbon cycle** to allow the formation of organic compounds of which all living things are made. Photosynthetic organisms take carbon in the form of carbon dioxide and convert it into carbohydrates using the sun's energy and chlorophyll pigments (see Chapter 6). The vast jungles of the world, the grassy plains of the temperate zones, and the cyanobacteria of the oceans show the results of this process.

Photosynthetic organisms, in turn, are consumed by grazing animals, fish, and humans, who use some of the carbohydrates for energy and convert the remainder to cell parts. To be sure, some carbon dioxide is released back to the atmosphere in cellular respiration, but a major portion of the carbon is returned to the ground when the animal or plant dies (FIGURE 26.10). It is here that the microorganisms exert their influence, for they are the primary **decomposers** of dead organic matter. Working in their countless billions in the water and soil, many bacterial cells, fungi, and other microorganisms consume the organic substances and release atmospheric carbon dioxide for reuse by the plants. This activity results from the concerted action of a huge variety of microorganisms, each with its own nutritional pattern of protein, carbohydrate, or lipid digestion. Without microorganisms, the Earth would be a veritable garbage dump of animal waste, dead plants, and organic debris accumulating in implausible amounts.

Microorganisms accomplish a similar goal in compost, where they decompose manure and other natural materials and convert them to compost for crop fertilization. Although both conventional and organic agriculture use manure as part of regular farm soil fertilization programs, only certified organic farmers are required to have a plan detailing the methods for building soil fertility using raw or aged manure. Furthermore, certified organic farmers are prohibited from using raw manure for at least 60 days prior to harvesting crops for human consumption.

Microorganisms also break down the carbon-based chemicals produced by industrial

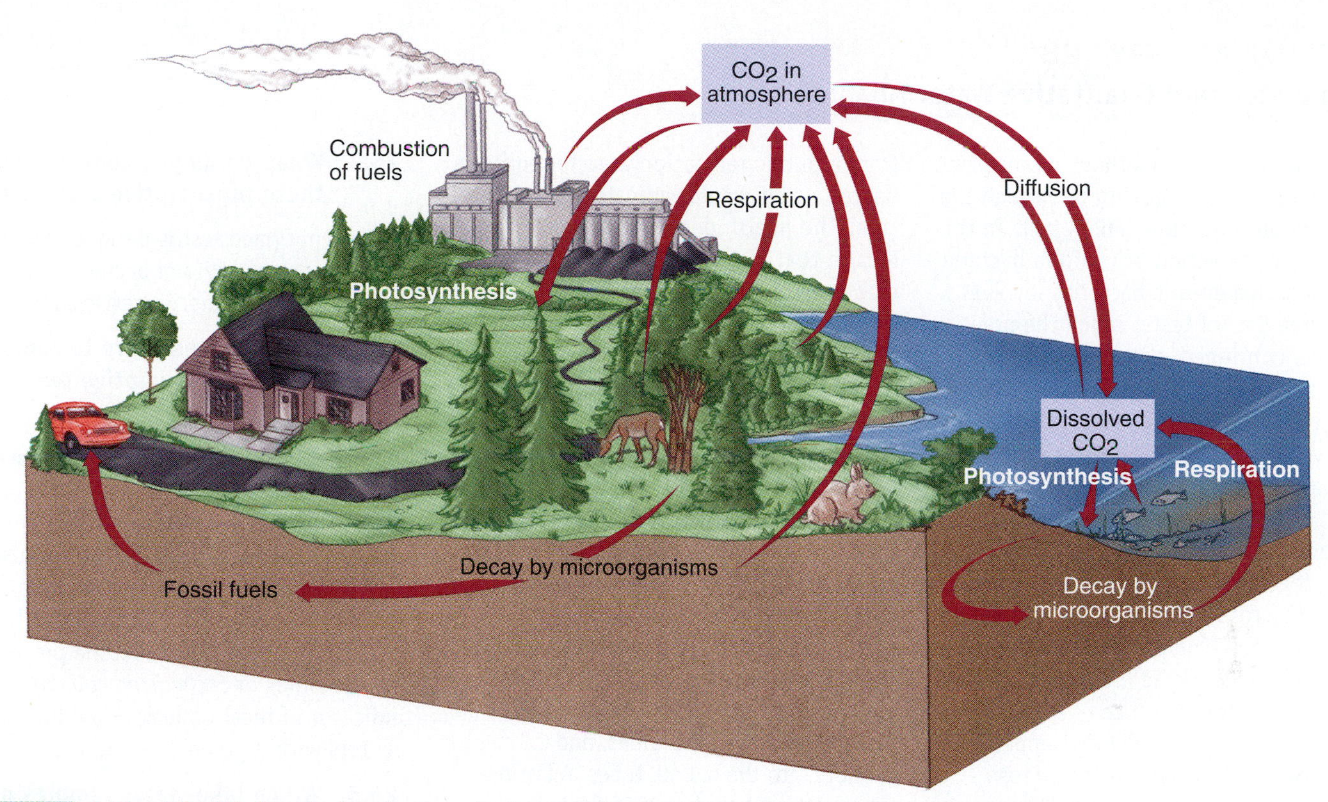

FIGURE 26.10 **A Simplified Carbon Cycle.** Photosynthesis represents the major method for incorporating carbon dioxide to organic matter, and cellular respiration accounts for its return to the atmosphere.

Q: Identify the roles played by microorganisms to decay in soil and ocean sediments.

processes, including herbicides, pesticides, and plastics. In addition, they produce methane, or natural gas, from organic matter and are probably responsible for the conversion of plants to petroleum and coal deep within the recesses of the Earth. Moreover, many microorganisms trap CO_2 from the atmosphere and form carbohydrates to supplement the results of photosynthesis. In these activities, the microorganisms represent a fundamental underpinning of organic life.

CONCEPT AND REASONING CHECKS

26.8 Identify the role of microorganisms in the carbon cycle.

The Sulfur Cycle Recycles Sulfate Molecules

KEY CONCEPT

- Some anaerobic microbes convert sulfate into hydrogen sulfide.

The **sulfur cycle** may be defined in more specific terms than the carbon cycle. Sulfur is a key constituent of such amino acids as cysteine and methionine, which are important components of proteins. Proteins are deposited in water and soil as living organisms die, and bacterial and fungal species decompose the proteins and break down the sulfur-containing amino acids to yield various compounds, including hydrogen sulfide. Sulfur also may be released in the form of sulfate molecules commonly found in organic matter. Anaerobic genera, such as *Desulfovibrio*, subsequently convert sulfate molecules to hydrogen sulfide.

The next set of conversions involves several bacterial genera, including members of the genera *Thiobacillus*, *Beggiatoa*, and *Thiothrix*. These bacterial genera release sulfur from hydrogen sulfide during their metabolism and convert it into sulfate. The sulfate now is available to plants, where it is incorporated into the sulfur-containing amino acids. Consumption by animals and humans completes the cycle.

CONCEPT AND REASONING CHECKS

26.9 Identify the role of microorganisms in the sulfur cycle.

MICROINQUIRY 26

Doing a Standard Qualitative Water Analysis

One of the traditional methods of analyzing bacteriological water quality is through the **most probable number** (**MPN**) test. In this analysis, the detection of coliform bacteria (gram-negative enteric bacteria) involves up to three sets of tests, called the presumptive, confirmed, and completed tests.

The presumptive test is designed to detect coliform bacteria in a water sample. Of the enteric bacteria, only the coliform bacteria use lactose as a carbon and energy source. Therefore, a lactose broth medium represents a selective medium for coliforms. If coliforms are present, they ferment the lactose and produce acid and carbon dioxide gas. The presence of acid is seen by a change in the broth color (red to yellow) due to the presence of a pH indicator and gas production will be trapped in a small, inverted tube within the lactose broth tube (see figure).

Three sets of five lactose broth tubes each are used. Set 1 tubes contain 10 mL of double-strength lactose broth, while set 2 tubes contain 1 mL and set 3 tubes contain 0.1 mL of single-strength lactose broth. Each tube then is inoculated with a known volume of the water sample. If any of the tubes produce gas, you can presume that coliform bacteria are present in the original water sample.

If a presumptive test indicates the presence of coliforms, you can determine the MPN by noting how many of the tubes in each group have produced acid and gas. Then, using a standard MPN table, you can obtain a statistical estimate for the number of coliforms present (see Table on page 805). For example, if three of the 10 mL tubes, two of the 1 mL tubes, and zero of the 0.1 mL tubes had positive results, the presumptive test would be read as 3-2-0, indicating approximately 14 coliforms per 100 mL of water (95 percent confidence that there were actually between 5.7 and 36 bacteria present).

Scenario

Working for the state health department, your job is to regularly analyze lake water from a public recreation area for coliform bacteria that would indicate fecal pollution. The health department needs to ensure that the water contains no appreciable number of coliform bacteria that otherwise could make people sick after swimming or playing in the water. Let's explore the presumptive test and then mention a few details about the confirmed and completed tests. Answers can be found in **Appendix D**.

You have obtained six lake-water samples from the public recreation area. You set up a presumptive test using 15 lactose broth tubes (5 each with 10 mL, 1 mL, and 0.1 mL of broth) for each of the six water samples. Using sterile pipettes, you transfer 10 mL aliquots into each of the 10 mL tubes, 1 mL aliquots into the 1 mL tubes, and 0.1 mL aliquots into the 0.1 mL tubes. All of the tubes are incubated at 37°C for 24 hours.

26.1. Why do the 10 mL tubes contain double-strength lactose broth?

26.2. Why were sterile pipettes used? Is this absolutely necessary for this test?

Here are the results for the six samples:

Sample Number	Positive Tubes 10 ml	1 ml	0.1 ml
1	0	0	1
2	0	0	0
3	0	1	0
4	5	4	5
5	0	2	0
6	2	0	1

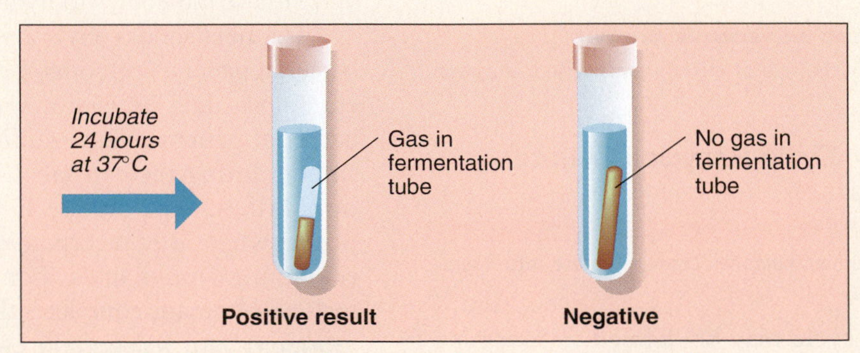

Detection of acid and gas in the presumptive test for water quality.

26.3. What would you conclude from these presumptive test results?

The confirmed test is designed to confirm the presence of coliform bacteria in positive or uncertain presumptive test results.

26.4. Why is it necessary to confirm a positive presumptive test?

In a confirmed test, a sample from each identified lactose broth tube is streaked onto an eosin-methylene blue (EMB) agar plate. This is a selective and differential growth medium. It is selective because it inhibits the growth of gram-positive (nonenteric) bacteria and is differential because of any enteric bacteria present, only colonies of *Escherichia coli*, the prime indicator of fecal pollution, produce dark centers with a green metallic sheen.

26.5. Which lake-water samples need to be confirmed? What was your reasoning for selecting these samples?

If needed, a completed test can be run for any samples that gave a doubtful result in the confirmed test.

26.6. Suppose that only sample 4 gives a positive confirmed test; as the water analysis expert what would be your recommendation to the state department of health regarding this recreational lake?

TABLE

The MPN index per 100 ml for combinations of positive and negative presumptive test results when five 10-ml, five 1-ml, and five 0.1-ml portions of sample are used.

Number of Tubes with Positive Results						Number of Tubes with Positive Results					
Five of 10 ml Each	Five of 1 ml Each	Five of 0.1 ml Each	MPN Index per 100 ml	Confidence Limit Low	Confidence Limit High	Five of 10 ml Each	Five of 1 ml Each	Five of 0.1 ml Each	MPN Index per 100 ml	Confidence Limit Low	Confidence Limit High
0	0	0	<1.8	—	6.8	4	2	1	26	9.8	70
0	0	1	1.8	0.09	6.8	4	3	0	27	9.9	70
0	1	0	1.8	0.09	6.9	4	3	1	33	10	70
0	2	0	3.7	0.7	10	4	4	0	34	14	100
1	0	0	2	0.1	10	5	0	0	23	6.8	70
1	0	1	4	0.7	10	5	0	1	31	10	70
1	1	0	4	0.7	12	5	0	2	43	14	100
1	1	1	6.1	1.8	15	5	1	0	33	10	100
1	2	0	6.1	1.8	15	5	1	1	46	14	120
2	0	0	4.5	0.79	15	5	1	2	63	22	150
2	0	1	6.8	1.8	15	5	2	0	49	15	150
2	1	0	6.8	1.8	17	5	2	1	70	22	170
2	1	1	9.2	3.4	22	5	2	2	94	34	230
2	2	0	9.3	3.4	22	5	3	0	79	22	220
2	3	0	12	4.1	26	5	3	1	110	34	250
3	0	0	7.8	2.1	22	5	3	2	140	52	400
3	0	1	11	3.5	23	5	3	3	180	70	400
3	1	0	11	3.5	26	5	4	0	130	36	400
3	1	1	14	5.6	36	5	4	1	170	58	400
3	2	0	14	5.7	36	5	4	2	220	70	440
3	2	1	17	6.8	40	5	4	3	280	100	710
3	3	0	17	6.8	40	5	4	5	430	150	1100
4	0	0	13	4.1	35	5	5	0	240	70	710
4	0	1	17	5.9	36	5	5	1	350	100	1100
4	1	0	17	6	40	5	5	2	540	150	1700
4	1	1	21	6.8	42	5	5	3	920	220	2600
4	1	2	26	9.8	70	5	5	4	1600	400	4600
4	2	0	22	6.8	50	5	5	5	1600	700	—

MICROFOCUS 26.3: History

Of Lumps and Bumps

The man from the Delft laboratory had an audacious proposal to the assembled farmers: "Don't plant your crops in the same field as last year," he said. "Leave the field alone for the next two years; let it lie fallow." The year was 1887. The man was Martinus Willem Beijerinck (Chapter 1). The country was the Netherlands. And, because agricultural land was at a premium, the proposal was revolutionary.

Beijerinck was a local bacteriologist. While his medical colleagues were investigating the germ theory of disease and its implications, Beijerinck was out in the fields. He was observing that fields are very productive when the land had just been cleared and freshly planted. In fact, these fields yielded bountiful crops when the farmer is away for a couple of years. Now he thought he had the answer: great populations of bacterial organisms.

Beijerinck was an expert on plants, but he also had something most other botanists lacked: a solid background in chemistry. He believed nitrogen was essential for plant growth, but he had no idea how nitrogen bridges the gap between atmosphere and plant. Then it dawned on him that bacterial organisms might be the bridge. And the little lumps and bumps on tree roots were the key. Time and again he observed great hordes of bacterial cells in the little lumps and bumps ("nodules") on plant roots. He didn't see the nodules as often on tended crops, but they always seemed to be on wild plants growing in untended fields.

Beijerinck performed the laboratory experiments to strengthen his views: He took bacterial cells from the nodules and inoculated them into seedlings of various plants. In many cases, the plants developed nodules, and when he planted the seedlings, the nitrogen content of the soil rose dramatically. So, his advice in 1887 was straightforward: Leave the field alone for a spell; plant elsewhere; let the wild plants thrive; and when the field is finally planted, the crop yield will be worth the wait.

Indeed, he was right. Modern farmers know it is important to let a field lie fallow and "refresh" itself. To the macrobiologist, the lumps and bumps on the roots of wild plants are the important parts. To the microbiologist, it's what is inside that count. Either way, we all benefit.

The Nitrogen Cycle Is Dependent on Microorganisms

KEY CONCEPT

- The microbial fixation of gaseous nitrogen into forms usable by living organisms depends on microorganisms.

The cyclic transformation of nitrogen is of paramount importance to life on Earth. Nitrogen is an essential element in nucleic acids and amino acids. Although it is the most common gas in the atmosphere (about 80 percent of air), animals cannot use nitrogen in its gaseous form, nor can any but a few species of plants. The animals and plants thus require the assistance of microorganisms to trap the nitrogen. Among the first to recognize this relationship was Martinus Beijerinck, who did his work at the end of the 1800s (MicroFocus 26.3).

The **nitrogen cycle** begins with the deposit of dead plants and animals in the soil. In addition, nitrogen reaches the soil in urea contained in urine. A process of digestion and putrefaction by soil bacteria and other microorganisms follows, thus yielding a mixture of amino acids (FIGURE 26.11). Amino acids are broken down further by microbial metabolism, and the ammonia that accumulates may be used directly by plants.

Next, **mineralization** takes place. In this process, complex organic compounds are converted to inorganic compounds and additional ammonia. Much of the ammonia is converted to nitrite ions by *Nitrosomonas* species, a group of nitrifying, aerobic gram-negative rods. In the process, the bacterial cells obtain energy for their metabolic needs. The nitrite ions then are converted to nitrate ions by *Nitrobacter,* another group of aerobic gram-negative rods. *Nitrobacter* obtains energy from the process. Nitrate is a crossroads compound: it can be used by plants for their nutritional needs, or it can be liberated as atmospheric nitrogen by certain microorganisms, including the denitrifying bacteria.

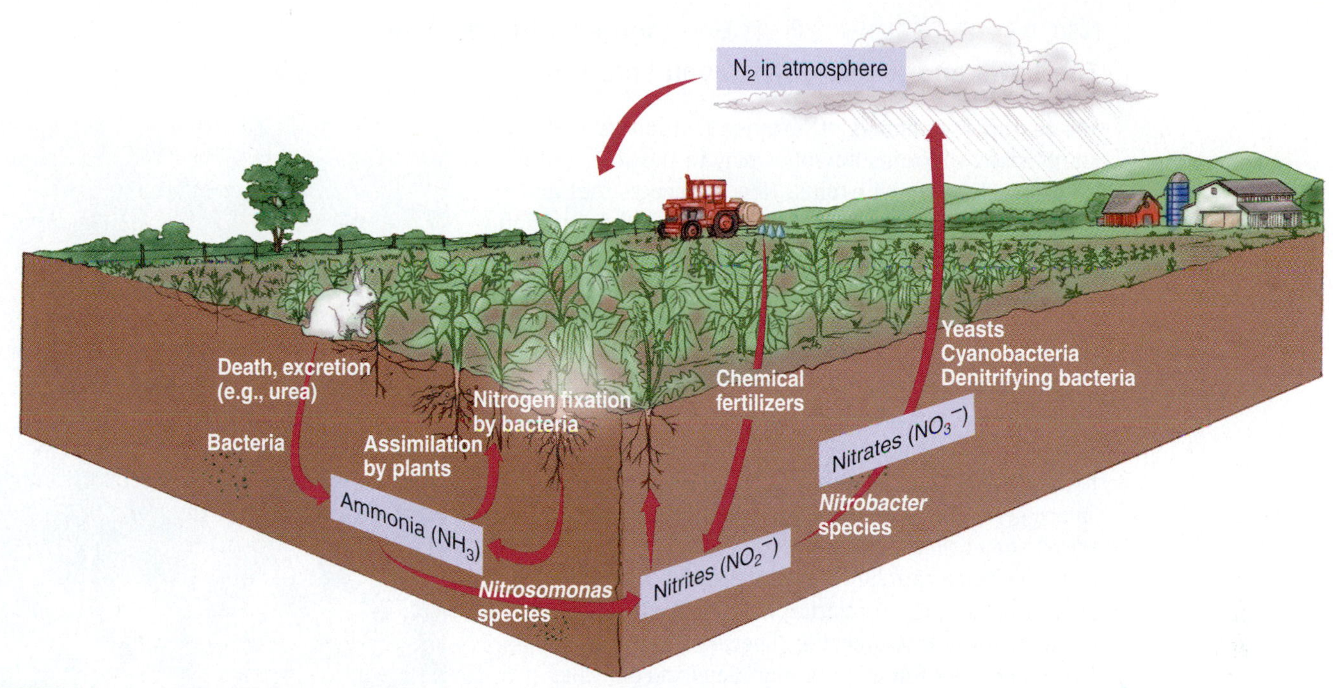

FIGURE 26.11 **A Simplified Nitrogen Cycle.** Plant and animal protein and metabolic wastes are decomposed by bacteria into ammonia. The ammonia may be used by plants, or it may be converted by *Nitrosomonas* and *Nitrobacter* species to nitrate, which also is used by plants. Some nitrate is broken down to atmospheric nitrogen. This nitrogen is returned to the leguminous plants by nitrogen-fixing microorganisms as nitrate, which is converted to ammonia. Animals consume the plants to nitrogen from the proteins plants contain.

Q: If plants could directly carry out nitrogen fixation, would it make microorganisms unnecessary for the nitrogen cycle? Explain.

For the nitrogen released to the atmosphere, a reverse trip back to living things is an absolute necessity for life to continue as we know it. The process is called **nitrogen fixation**. Once again microorganisms in water and soil play a key role because they possess the enzyme systems to trap atmospheric nitrogen and convert it to compounds useful to plants. In nitrogen fixation, gaseous nitrogen is incorporated to ammonia, which fertilizes plants.

Two general types of microorganisms are involved in nitrogen fixation: free-living species and symbiotic species. Free-living species include bacterial genera such as *Bacillus, Clostridium, Pseudomonas, Spirillum,* and *Azotobacter* as well as types of cyanobacteria and certain yeasts. Generally, the free-living species fix nitrogen during their growth cycles. The nitrogen-fixing ability of these species cannot be overemphasized.

Symbiotic species of nitrogen-fixing microorganisms live in association with plants that bear their seeds in pods. These plants, known as legumes, include peas, beans, soybeans, alfalfa, peanuts, and clover. Species of gram-negative rods known as *Rhizobium* infect the roots of the plants and live within swellings, or nodules, in the roots. Although complex factors are involved, the central theme of the mutualistic relationship is that *Rhizobium* fixes nitrogen and makes nitrogen compounds available to the plant while taking energy-rich carbon compounds in return. The bulk of the nitrogen compounds accumulates when *Rhizobium* cells die. Legumes then use the compounds to construct amino acids and, ultimately, protein. Animals consume the soybeans, alfalfa, and other legumes and convert plant protein to animal protein, thereby completing the cycle. In some case, however, such *Rhizobium* nodules may not exist in the soil (MicroFocus 26.4).

■ Mutualistic: Referring to a symbiotic relationship between two organisms of different species that benefits both.

Humans have long recognized that soil fertility can be maintained by rotating crops and including a legume. The explanation lies in the ability of rhizobia to fix nitrogen within the nodules of legumes (FIGURE 26.12). So

■ Symbiotic: Refers to an association between two species that often is beneficial to both.

MICROFOCUS 26.4: Environmental Microbiology

The Lows and Highs of Nitrogen Fixation

The textbook examples of nitrogen fixation have always emphasized the roles microbes play in the soils and in association with plant roots. Now it appears that at least in some tropical trees, the process can occur high up in the tree canopies.

The koa tree (see figure) is quite special to the Hawaiian people. It is from this tree that islanders made canoes for traveling between islands. As such, the tree became quite sacred and was protected by the god Kupulupulu.

A koa tree in Hawaii.

While conducting a 2003 statewide survey of nitrogen fixation bacteria in Hawaii's koa trees, graduate student James Leary of the University of Hawaii, Manoa, came across something startling. He discovered that in a few koa trees some nitrogen-fixing bacterial species reside not only in the soil in association with the tree's shallow roots but also in the tree branches. Species of the bacterium *Rhizobium* were found quite high up in the branches.

When these trees were examined more closely, Leary found rhizobia nodules in clumps of decomposing wood or soil-like debris where the tree branches fork and on tree limbs. In these clumps, Leary found rhizobia associated with an amazing community of other microbes, including molds, other fungi, and lichens.

Further analysis indicated these clumps of detritus were better places for rhizobia to exist because there was a higher concentration of carbon and lower concentrations of aluminum, which can interfere with nitrogen fixation. In fact, most soil-dwelling microbes were parasitic while the canopy-dwelling ones were specialized for a mutualistic form of symbiosis.

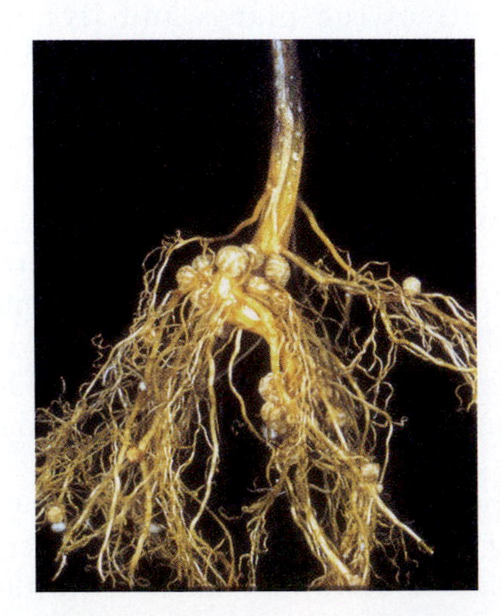

(A)

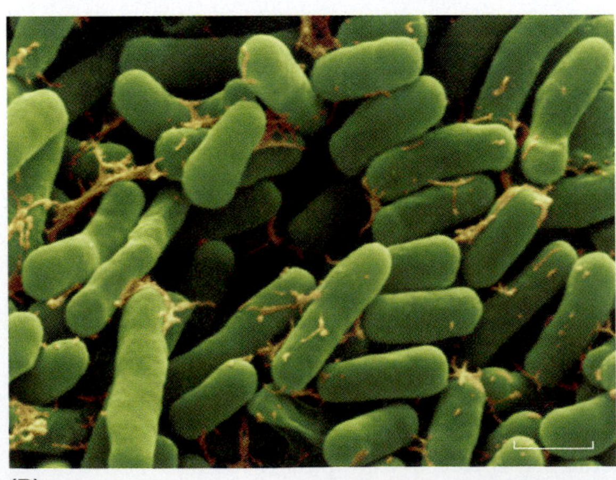

(B)

FIGURE 26.12 **Nitrogen Fixation.** (**A**) Nodules on the roots of a cowpea, a legume plant. Species of *Rhizobium* live within the nodules and fix nitrogen to nitrogen-containing compounds. When the bacterial cells die, the compounds are used by the legume to synthesize amino acids. (**B**) A false-color scanning electron micrograph of a *Rhizobium* species comprising a root nodule. (Bar = 10 µm.)

Q: Compared to a leguminous plant with root nodules, what might a leguminous plant look like if it lacked nodules containing Rhizobium*?*

much nitrogen is captured, in fact, that the net amount of nitrogen in the soil actually increases after a crop of legumes has been grown. When cultivating legumes, there is no need to add nitrogen fertilizer to the soil. In addition, when crops such as clover or alfalfa are plowed under, they markedly enrich the soil's nitrogen content. Thus, humans are indebted to microorganisms for such edible plants as peas and beans, as well as for the indirect products of nitrogen fixation—namely, steaks, hamburgers, and milk.

CONCEPT AND REASONING CHECKS

26.10 Identify the role of microorganisms in the nitrogen cycle.

SUMMARY OF KEY CONCEPTS

26.1 Water Pollution

- An unpolluted water environment is inhabited by limited numbers of soil bacteria, but a polluted environment contains an enormous variety of heterotrophic organisms from sewage, feces, and industrial sources. Coliform bacteria, the gram-negative rods of human and animal intestinal tracts, abound in polluted water. By contrast, a marine environment has halophilic microorganisms as well as psychrophilic and barophilic organisms.
- Of the three types of pollution, biological pollution is of primary interest to the water microbiologist. The biochemical oxygen demand (BOD) is a measure of the amount of biological pollution. A high BOD indicates the presence of large numbers of organisms needing oxygen for decomposition processes.
- Concern exists for a variety of bacterial, viral, and protozoal diseases transmitted by water. Most of the diseases were described in Chapter 9.

26.2 The Treatment of Water and Sewage

- To prepare water for drinking purposes, municipalities employ various levels of water purification, including sedimentation, filtration, and chlorination.
- Sewage also can be treated by different steps according to the needs of the municipalities. Cesspools and septic tanks are used for local treatment, and variations of oxidation lagoons and secondary and tertiary treatments are used on larger scales.
- As much as 99 percent of microbial activities in nature may occur in biofilms. The immobilized populations adhere to many different surfaces. They can be used in such processes as bioremediation and oil spill clean up. Biofilms also can cause disease and be quite resistant to chemicals and antibiotics.
- To test the effectiveness of purification procedures, several bacteriological tests are available, including the membrane filter technique, the standard plate count, and the most probable number test.

26.3 The Cycles of Elements in the Environment

- In the carbon cycle, bacterial and fungal cells are essential to the breakdown of organic matter and the release of carbon back to the atmosphere for recycling.
- Anaerobic bacteria are important to the sulfur cycle by producing and degrading sulfate.
- In the nitrogen cycle, many microorganisms release nitrogen from urea, amino acids, and nitrogenous organic matter. Many types of nitrogen-based conversions are performed by microorganisms, and specific bacterial species are essential in the steps of nitrogen fixation where nitrogen is brought back into the cycle. Indeed, the processes performed by nitrogen-fixing bacterial species are so essential that life as we know it probably would not exist without bacterial intervention.

LEARNING OBJECTIVES

After understanding the textbook reading, you should be capable of writing a paragraph that includes the appropriate terms and pertinent information to answer the objective.

1. Compare groundwater and surface water with regard to numbers of microorganisms.
2. Evaluate the microbiological differences between unpolluted and polluted water.
3. Provide examples of genera of coliform bacteria and noncoliform bacteria.
4. Explain how the terms halophilic, psychrophilic, and barophilic apply to marine microorganisms.
5. Identify three marine zones where microorganisms can be found.
6. Differentiate between physical, chemical, and biological pollution of water.
7. Assess the importance of the biological oxygen demand (BOD) when measuring polluted water.
8. Identify several intestinal diseases and wound infections transmitted by contaminated water.
9. Summarize the three sequential steps for water purification.
10. Distinguish between a cesspool and a septic tank.
11. Discuss what happens in each step of sewage treatment.
12. Describe biofilms and how these organisms differ from planktonic forms.
13. Identify three ways that water can be tested for indicator organism contamination.
14. Summarize the carbon cycle and identify the role of microorganisms in the cycle.
15. Describe how microorganisms affect the sulfur cycle.
16. Assess the critical role of microorganisms to the nitrogen cycle.
17. Distinguish between the two groups of nitrogen fixing bacteria.

SELF-TEST

Answer each of the following questions by selecting the ***one*** answer that best fits the question or statement. Answers to even-numbered questions can be found in **Appendix C**.

1. Potable water
A. is drinkable.
B. contains pollutants.
C. contains pesticides.
D. is contaminated with microbes.
E. All the above (**A–D**) are true.

2. Which one of the following would *not* be considered a coliform?
A. *Escherichia*
B. *Proteus*
C. *Enterobacter*
D. *Streptococcus*
E. Both **B** and **D** are correct.

3. A barophilic marine bacterial species would most likely be found in
A. the littoral zone.
B. extremely saline water.
C. warm water.
D. the abyssal zone.
E. surface waters.

4. The foundation of many marine food chains includes the
A. dinoflagellates.
B. indicator organisms.
C. diatoms.
D. coliform bacteria.
E. Both **A** and **C** are correct.

5. Chemical pollution in water would include all the following *except*:
A. pesticides.
B. particulate matter.
C. phosphates.
D. organic waste.
E. nitrates.

6. Clean water would have
A. a high BOD.
B. high numbers of indicator organisms.
C. a low BOD.
D. a low oxygen content.
E. a diverse microbial community.

7. Intestinal diseases transmitted in water include all the following *except*:
A. Legionnaires' disease.
B. cholera.
C. typhoid fever.
D. giardiasis.
E. tetanus.

8. During water purification, the jelly-like masses of coagulated material are called
A. sediments.
B. flocs.
C. MPNs.
D. schmutzdecke.
E. sludge.

9. A schmutzdecke is a/an
A. sediment in a cesspool.
B. powder used for sedimentation.
C. biofilm of microorganisms.
D. enclosed concrete box to collect waste.
E. mass of sewage left undisturbed.

10. Raw sewage piped into tanks for waste removal is called
A. activated sludge.
B. grit.
C. sludge.
D. slime.
E. detritus.

11. Planktonic bacteria are
A. individual and independent cells.
B. pathogens in open marine waters.
C. immobilized cells.
D. the cells in a biofilm.
E. more likely to communicate by quorum sensing.

12. The most probable number test applies to
A. detecting bacterial contamination of water.
B. estimating the number of nitrogen-fixing bacteria in soil.
C. predicting the amount of pollution in a stream.
D. the standard number of coliforms allowed in clean water.
E. determining the number of microbes in organic waste.

13. Which one of the following (**A–D**) is *not* critical to the recycling of carbon? If all are critical, select **E**.
A. Photosynthetic microorganisms
B. Decomposers
C. Green plants
D. Animals
E. All the above (**A–D**) are important.

14. Anaerobic microbes important to the sulfur cycle include all the following *except*:
A. *Desulfovibrio*.
B. *Rhizobium*.
C. *Thiobacillus*.
D. *Thiothrix*.
E. *Beggiatoa*.

15. _____ is a genus that carries out nitrogen fixation, which is the conversion of ______________.
A. *Azotobacter;* ammonia to nitrogen gas
B. *Rhizobium;* nitrite to nitrate
C. *Nitrosomonas;* nitrate to nitrite
D. *Nitrobacter;* nitrate to nitrogen gas
E. *Rhizobium;* nitrogen gas into ammonia

QUESTIONS FOR THOUGHT AND DISCUSSION

Answers to even-numbered questions can be found in **Appendix C**.

1. When sewers were constructed in New York City in the early 1900s, engineers decided to join storm sewers carrying water from the streets together with sanitary sewers bringing waste from the homes. The result was one gigantic sewer system. In retrospect, was this a good idea? Why?
2. The victims of disease, both animals and people, are buried underground, yet the soil is generally free of pathogenic organisms. Why?
3. What information might you offer to dispute the following four adages common among campers and hikers? (1) Water in streams is safe to drink if there are no humans or large animals upstream. (2) Melted ice and snow is safer than running water. (3) Water gurgling directly out of the ground or running out from behind rocks is safe to drink. (4) Rapidly moving water is germ free.
4. Some years ago, the syndicated columnist Erma Bombeck wrote a humorous book entitled "The Grass Is Always Greener Over the Septic Tank." (The title was an adaptation of the expression "The grass is always greener on the other side of the fence.") Indeed, the grass is often greener over the septic tank. Why is this so?
5. The author of a biology textbook writes: "Because the microorganisms are not observed as easily as the plants and animals, we tend to forget about them, or to think only of the harmful ones . . . and thus overlook the others, many of which are indispensable to our continued existence." How do the carbon, sulfur, and nitrogen cycles support this outlook?

APPLICATIONS

Answers to even-numbered questions can be found in **Appendix C**.

1. In August 1989, two men were tragically killed while working in an industrial septic tank during cleaning. The cause of death was listed as methane poisoning. What was the source of the methane, and how was it related to the septic tank?
2. In 1978, Legionnaires' disease broke out in the garment district of New York City and was given front-page treatment in the press. During that same period, the coliform count rose dramatically in water in the Murray Hill section of the city, but this was given minor coverage in the papers. Many public health officials believed that the Murray Hill problem posed the more substantial threat. Do you agree? Why?
3. You decide to build a new home in the country and you have the choice of installing a cesspool or a septic tank. Which might you be inclined to choose? Why?
4. Working for the city parks and recreation department, you have two swimming pools to inspect: an Olympic-sized pool for swimmers, and a small wading pool for toddlers. In which pool does the greater potential for disease transmission exist, and what precautions should you take to limit the transmission of microorganisms?

REVIEW

Using your knowledge of environmental microbiology, consider each characteristic and the three possible choices below. In the space, place the letter or letters of the most appropriate choice(s). The answers to even-numbered choices can be found in **Appendix C**.

____ **1.** Genus (genera) of coliform bacteria
A. *Escherichia*
B. *Staphylococcus*
C. *Enterobacter*

____ **2.** Where halophilic bacteria live
A. Lake
B. Ocean
C. Mountain stream

____ **3.** Used as markers for oil drilling
A. Bacteriophages
B. Radiolarians
C. Foraminifera

____ **4.** Waterborne microbial disease(s)
A. Hepatitis A
B. Amoebiasis
C. Hepatitis B

____ **5.** Step(s) in water purification
A. Filtration
B. Chlorination
C. Sedimentation

____ **6.** Found in polluted water
A. *Proteus* species
B. *E. coli*
C. HIV

____ **7.** Test(s) for oxygen consumption in water
A. SPC
B. BOD
C. MPN

____ **8.** Type(s) of pollution when microorganisms are present
A. Biological
B. Physical
C. Chemical

____ **9.** Toxin-producing dinoflagellate(s)
A. *Entamoeba*
B. *Gambierdiscus*
C. *Gonyaulax*

____ **10.** Produce(s) flocs in water
A. Iron sulfate
B. Copper sulfate
C. Aluminum sulfate

____ **11.** Needed to perform the standard plate count
A. Culture dishes
B. Pipettes
C. Agar medium

____ **12.** Possible cause(s) of red tide
A. *Gonyaulax*
B. *Streptococcus*
C. *Gymnodinium*

____ **13.** Found in anaerobic mud at lake bottom
A. *Giardia*
B. Diatoms
C. *Clostridium*

____ **14.** Release(s) carbon in carbon cycle
A. Viruses
B. Fungi
C. Bacteria

____ **15.** Function(s) in nitrogen cycle
A. *Thiobacillus*
B. *Nitrobacter*
C. *Thiothrix*

HTTP://MICROBIOLOGY.JBPUB.COM/

The site features learning, an on-line review area that provides quizzes and other tools to help you study for your class. You can also follow useful links for in-depth information, read more MicroFocus stories, or just find out the latest microbiology news.

27

Industrial Microbiology and Biotechnology

Chapter Preview and Key Concepts

27.1 Microorganisms in Industry
- Microbial metabolic products may represent primary or secondary metabolites.
- Microbial enzymes can be important as dietary supplements or to the textile industry.

27.2 Alcoholic Beverages
- Brewing is the production of alcoholic beverages from malted grains.
- Wine fermentation depends of the yeast *Saccharomyces ellipsoideus*.
- Distilled spirits are produced by concentrating the alcohol from fermentation.

27.3 Other Microbial Products
- Commercially useful antibiotics come from a few mold and bacterial species.
- Bacterial toxins can be produced that kill insect larvae.
- Yeast cells and mushrooms represent potential food or food supplement sources.
- Natural or engineered bacterial species can break down toxic waste.
  MicroInquiry 27: Working with Microorganisms
- Plasmids can be used to transform the properties of a bacterial cell.

Never underestimate the power of the microbe.
—Microbiologist Jackson W. Foster of the University of Texas

Humans probably discovered alcoholic beverages by accident. It is conceivable that sunlight warmed some sort of fallen grape or other fruit and accelerated the fermentation of its juices by yeasts. Humans must have sampled this "spoiled" fruit with curiosity, and if the taste of the aromatic concoction was not especially pleasing, the euphoric feeling that followed probably brought them back for more. By trial and error, humans discovered the important factors in fermentation and soon learned to control the process. In doing so, they became the first industrial microbiologists.

How fermentation occurs was worked out by Louis Pasteur when he studied the souring of French wines (see Chapter 1). With this understanding, the process of **industrial fermentation**, using the metabolic products of microorganisms for commercial purposes developed. By the 1930s, several types of fermentations were being carried out to produce ethyl alcohol and butyl alcohol. For the British war effort, Chaim Weizmann's development of a microbial fermentation process for acetone using *Clostridium acetobutylicum* was critical toward the manufacture of gunpowder (see Chapter 6).

However, industrial microbiology is not restricted to alcohol fermentations. It includes bread baking and cheese production as well as many of the foods resulting from fermentation (see Chapter 26). Industrial fermentation enhances the metabolic reactions already present in (or genetically engineered into) microbes. **Biocatalysis** is the term used to describe the metabolic reactions carried out at this industrial scale.

To produce a significantly useful amount of product from microorganisms, tremendously large numbers of microorganisms are needed. This means scaling up microbial metabolic processes to a level significant enough to produce the valuable commercial product (FIGURE 27.1A). On this large scale, the term fermentation has come to have a slightly different definition. **Industrial fermentation** procedures are not necessarily anaerobic processes; rather, they are large-scale microbial methods that may be aerobic or anaerobic.

Such biocatalytic processes are carried out in large containers or tanks called **bioreactors** or **fermentors** (FIGURE 27.1B). Technicians or scientists must continually monitor the growth conditions in these containers to ensure microbes continue to either grow or reach a state in which maximal product is produced. Temperature, pH, oxygen, and nutrients—all of the typical physical and chemical conditions that apply to growing microbes in a culture tube (see Chapter 5)—apply here, just on a much more grand scale.

The exploitation of microbes for use in industrial processes is closely linked to **biotechnology**. Thus, many of the products of industrial microbiology contribute to public health, interrupt the spread of disease, or improve the quality of life. Microbial cells or their products are used, for example, to produce enzymes, insecticides, food additives, protein hormones, antibiotics, monoclonal antibodies, interferon, and the myriad products of genetic engineering that are the heart of the biotechnology industry today.

As we see in this chapter, industrial microbiology is an extremely diversified field, in which inexpensive raw materials are converted to valuable commodities through the metabolism of microorganisms.

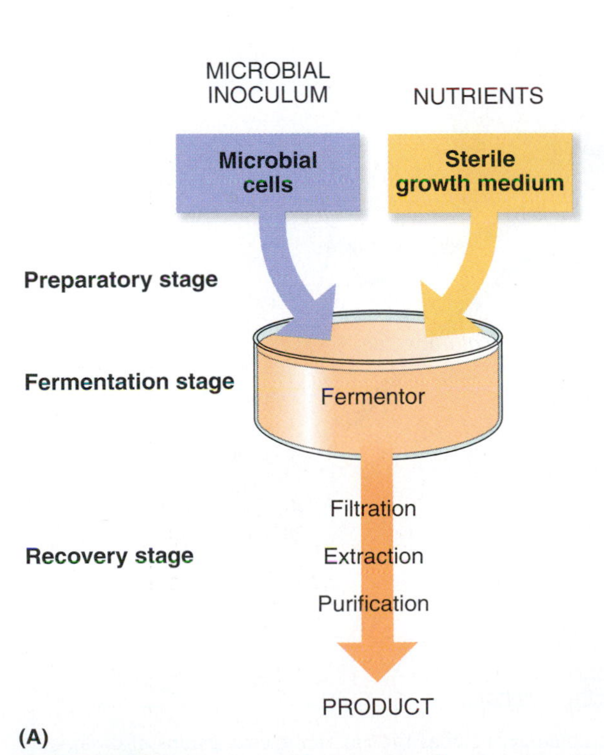

FIGURE 27.1 **The Industrial Fermentation Process.** (**A**) A general industrial process scheme for biocatalysis and product recovery. (**B**) A pharmaceutical technician monitors a series of fermentors to ensure a maximum yield of microbial product.

Q: What types of physical conditions must this technician monitor?

27.1 Microorganisms in Industry

Certain properties make microorganisms well suited for industrial processes. Microorganisms not only possess a broad variety of enzymes to make an array of chemical conversions possible, they also have a relatively high metabolic activity that allows conversions to take place rapidly. In addition, they have a large surface area for the quick absorption of nutrients and release of end products. Moreover, they usually multiply at a high rate, as evidenced by the 20-minute generation time for *Escherichia coli* under ideal conditions (see Chapter 5).

In the industrial process, microorganisms act like chemical factories. To be effective, they should liberate a large amount of a single product that can be extracted and purified efficiently. The organisms should be easy to maintain and cultivate, and should have genetic stability with infrequent mutations. Their value is enhanced if they can grow on an inexpensive, readily available medium that is a byproduct of other industrial processes. For example, a large amount of whey is produced in cheese manufacturing, and microorganisms that convert whey components to lactic acid add to the overall profit of the cheese industry. FIGURE 27.2 displays some possible conversions from a single metabolic substance.

■ Metabolites: Small molecules involved in or byproducts of metabolism.

Microorganisms Produce Many Useful Organic Compounds

KEY CONCEPT

- Microbial metabolic products may represent primary or secondary metabolites.

Microorganisms are used in industry to produce a variety of organic compounds, including acids, growth stimulants, and enzymes. In some cases, the production results from a microorganism manufacturing many thousands of times the amount of product necessary for its own metabolism.

Microorganisms may produce one of two types of metabolites that are important to industrial microbiology. **Primary metabolites** are directly involved in the normal growth,

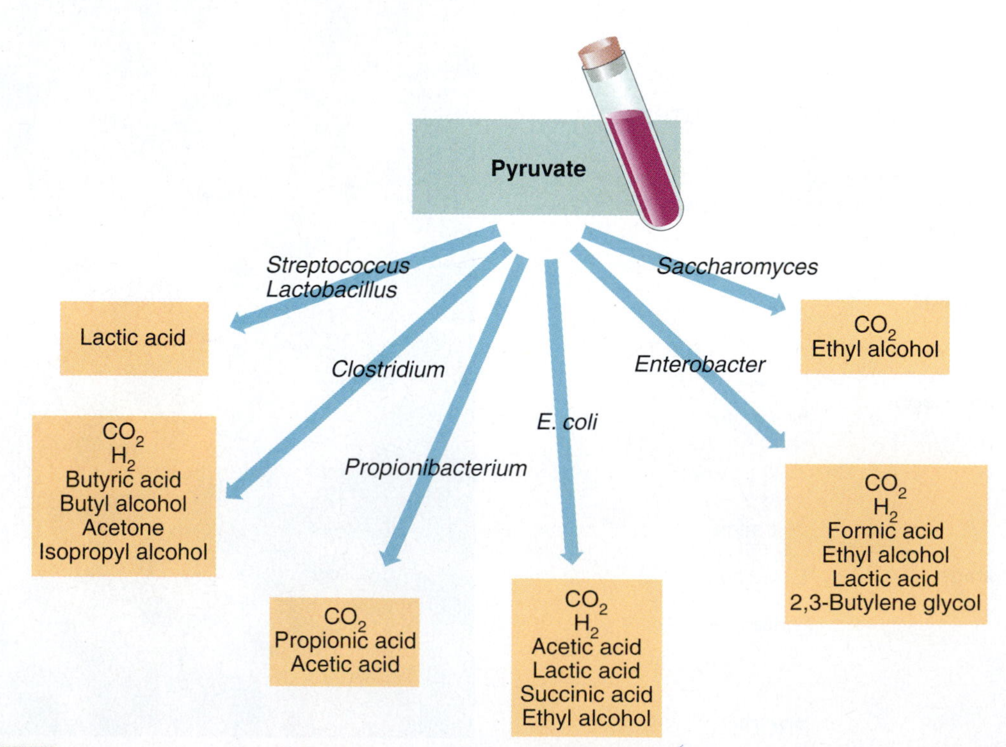

FIGURE 27.2 The Products of Fermentation. Industrial microbiology includes the metabolic conversions of organic compounds to other organic compounds by microorganisms. These conversions are called fermentations, which in this case are taking place in the absence of oxygen gas. Pyruvate is the starting point for the conversions shown here.

Q: What common metabolic pathway would contribute to the production of pyruvate?

development, and reproduction of the microbe. Pyruvate and the end products of the fermentation pathways shown in Figure 27.2 are examples of primary metabolites.

Secondary metabolites form the bulk of products of industrial interest. These metabolites are not directly involved in growth and reproduction, but usually have important ecological functions and often are species specific. Secondary metabolites usually are produced near or at the end of the microbe's growth phase; that is, in the stationary phase (see Chapter 5).

One of the first secondary metabolites to be made in bulk by industrial fermentation was citric acid. Manufacturers use this organic compound in soft drinks, candies, inks, engraving materials, and a variety of pharmaceuticals, such as anticoagulants and effervescent tablets (e.g., Alka-Seltzer). The organism most widely used for citric acid production is the fungus *Aspergillus niger.* Microbiologists inoculate the mold into a medium of cornmeal, molasses, salts, and inorganic nitrogen in huge aerobic fermentors. The absence of a Krebs cycle enzyme in the fungus prevents the metabolism of citric acid (citrate) into the next component of the cycle, and the citric acid accumulates in the medium. FIGURE 27.3 outlines this chemistry.

Another important microbial product is lactic acid, a compound employed to preserve foods, finish fabrics, prepare hides for leather, and dissolve lacquers. Lactic acid is commonly produced by bacterial activity on the whey portion of milk. *Lactobacillus bulgaricus* is widely used in the fermentation because it produces only lactic acid from lactose.

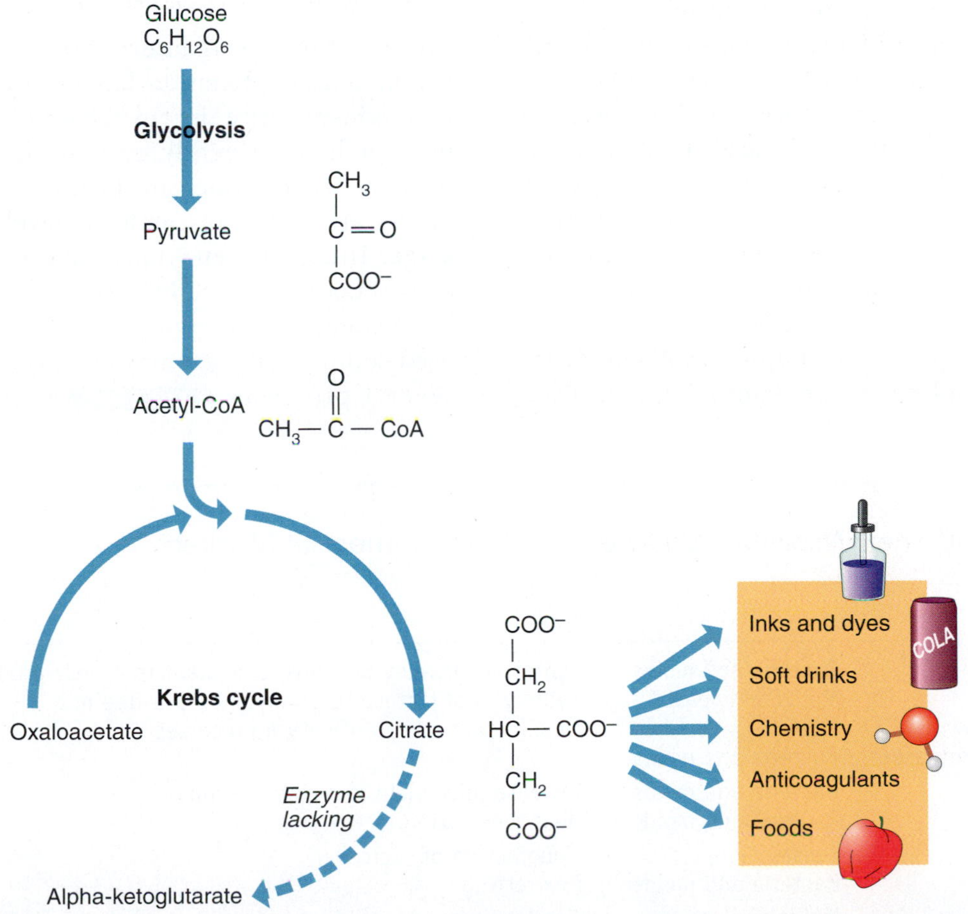

FIGURE 27.3 **The Chemistry of Citrate Production.** *Aspergillus niger* is grown in a mixture of nutrients, where it digests glucose into pyruvate. The pyruvate then is converted to acetyl-CoA, which condenses with oxaloacetate in the Krebs cycle to yield citric acid or citrate. However, the chemistry goes no further, because the next enzyme in the cycle is absent. Citric acid therefore accumulates and is isolated for use in various products, as shown.

Q: Because the rest of the Krebs cycle does not occur in this strain of Aspergillus, *how does the organism make ATP?*

Gluconic acid (gluconate), another valuable organic acid, is useful in medicine as a carrier for calcium, because gluconic acid is easily metabolized in the body, leaving a store of calcium for distribution. This acid is produced from carbohydrates by *A. niger* and by species of the bacterial genus *Gluconobacter* cultivated in fermentors. Calcium gluconate also is added to the feed of laying hens to provide calcium to strengthen the eggshells.

When the amount of amino acid produced by a microorganism exceeds what the microbe needs, the remainder is excreted into the environment. Such is the case with glutamic acid (glutamate) produced by certain species of *Micrococcus, Arthrobacter,* and *Brevibacterium*. Glutamic acid is a valuable food supplement for humans and animals, and its sodium salt, monosodium glutamate, is used in food preparations.

In the industrial production of lysine, another amino acid, two organisms are involved. *E. coli* is first cultivated in a medium of glycerol, corn steep liquor, and other ingredients, and the compound diaminopimelic acid (DAP) accumulates. Several days later, *Enterobacter aerogenes* is added to the fermentor. This organism produces an enzyme that removes the carboxyl group from DAP to produce the lysine used in breads, breakfast cereals, and other foods.

■ Corn steep liquor: A pure extract from corn that contains high concentrations of free amino acids.

Two important vitamins, **riboflavin** (vitamin B_2) and **cyanocobalamin** (vitamin B_{12}), also are products of microbial growth. Riboflavin is a product of *Ashbya gossypii*, a mold able to produce 20,000 times the amount of riboflavin it needs for metabolism. Cyanocobalamin is produced by selected species of *Pseudomonas, Propionibacterium,* and *Streptomyces* grown in a cobalt-supplemented medium. The vitamin prevents pernicious anemia in humans and is used in bread, flour, cereal products, and animal feeds.

CONCEPT AND REASONING CHECKS

27.1 Assess the importance of microbial secondary metabolites to industry and commercial markets.

Microorganisms Also Produce Important Enzymes and Other Products

KEY CONCEPT

- Microbial enzymes can be important as dietary supplements or to the textile industry.

The production of microbial enzymes for commercial exploitation has been an important industry since the emergence of industrial microbiology. Currently, over two dozen types of microbial enzymes are in use, and several others are in the research or developmental stage. Industrial enzymes have reached an annual market of $1.6 billion.

Among the important microbial enzymes used commercially are amylase, pectinase, and several proteases (TABLE 27.1). Amylase is

TABLE 27.1 Some Enzymes and Their Microbial Sources Used in Commercial Markets

Commercial Market	Enzyme	Source	Use
Dairy	Proteases	Bacteria and molds	Hydrolysis of whey proteins; coagulant in cheese production
	Lactase	Molds and yeasts	Hydrolysis of lactose to produce lactose-free milk
Brewing	Cellulases	Molds	Liquefaction and clarification processes
	Amylases and proteases	Bacteria and molds	
Wine and juices	Pectinases	Bacteria and molds	Increase juice yield and clarification
Meat	Proteases	Bacteria and molds	Meat tenderizing
Confectionary	Invertase	Yeasts	Liquefaction of sucrose
Textiles	Pectinase	Bacteria and molds	Flax retting
Pulp and paper	Xylanases, hemicellulases, lipases	Molds and yeasts	Cleaner and more efficient pulp and paper processing
Detergents	Proteases, amylases, lipases, cellulases	Bacteria and molds	Better cleansing of laundry and dishes

produced by the mold *Aspergillus oryzae.* It is used as a spot remover in laundry presoaks, as an adhesive, and in baking, where it digests starch to glucose. Pectinase, a product of a *Clostridium* species, is employed to ret flax for linen. In this process, manufacturers mix the flax plant with pectinase to decompose the pectin "cement" holding cellulose fibers together. The cellulose fibers then are spun into linen. MicroFocus 27.1 describes a more traditional process for retting. Pectinase also is used to clarify fruit juices.

Proteases are a group of protein-digesting enzymes produced by *Bacillus subtilis, A. oryzae,* and other microorganisms. Certain proteases are used for bating hides in leather manufacturing, a process in which organic tissue is removed from the skin to yield a finer texture and grain. Other proteases find value as liquid glues, laundry presoaks, meat tenderizers, drain openers, and spot removers.

One of the most appreciated but lesser known uses of a microbial enzyme is in making soft-centered chocolates. Invertase, an enzyme from yeast, is mixed with flavoring agents and solid sucrose, and then covered with chocolate. The enzyme converts some of the sucrose to liquid glucose and fructose, forming the soft center of the chocolate.

In medical microbiology, doctors use another microbial enzyme, **streptokinase**, to break down blood clots formed during a heart attack. Still another enzyme, **hyaluronidase**, is used to facilitate the absorption of fluids injected under the skin. The microbial roles for these enzymes as virulence factors are discussed in Chapter 18.

Although most natural food flavoring ingredients are produced by traditional processes from plant origins, new biotechnology methods have made it possible to produce novel flavoring ingredients by converting relatively cheap starting materials into higher-value flavor and aroma additives. The latter are used in foods, beverages, cosmetics, and other consumer items; examples include fruit, peach, and coconut flavoring agents called **lactones**. Although lactones can be generated from long-chain fatty acids from sweet potatoes, the fatty acids occur in limited quantity and are expensive to modify. To circumvent this problem, microbiologists use species of *Mucor* and other fungi to convert medium-chain fatty acids to compounds other microorganisms can easily transform to lactones. Another example is the methylketones, which confer strong cheese-associated flavors in dairy products. The flavoring agents are derived industrially from *Penicillium roqueforti* incubated in lipase-treated milk fats.

In addition to the major products we have surveyed, microorganisms provide a number of specialized materials. Typical of the miscellaneous microbial products is **alginate**, a sticky substance used as a thickener in ice cream, soups, and other foods. Another product of microbial origin is some perfumes. Musk oil, for example, is prepared from ustilagic acid, a product of the mold *Ustilago zeae,* which, ironically,

MICROFOCUS 27.1: History
It Smelled Bad, but It Worked

In past centuries, industrial pectinase was not available for retting flax, nor was protease available for bating hides. Nevertheless, the processes were carried on efficiently and successfully.

The retting process began by bundling flax plants and drying them in stacks. The stacks were then placed in a long trench several feet deep, covered with water, and weighted down with stones to exclude as much air as possible. After a few days, the water turned black, and an unmistakable stench signaled that retting was taking place. Two weeks later the flax was so soft and pliable that the fibers could be removed easily by pounding with wooden blocks. Today's microbiologists point out that *Clostridium* species were probably producing pectinase in the trenches.

The method for treating hides was equally messy. Skins were mixed with dog or fowl manure and set aside to cure. Fragments of tissue and hair gradually dissolved in the muck, and soon the hide became soft and pliable. Apparently, the proteases from fecal bacteria were responsible for the digestion. Bating hides was another smelly process, to be sure, but like the method for retting, it was usually reliable.

causes smut disease (see Chapter 16). Moreover, there are numerous pharmaceutical products derived from the ergot poisons of the mold *Claviceps purpurea.* These derivatives are prescribed to induce labor, treat menstrual disorders, and control migraine headaches.

CONCEPT AND REASONING CHECKS

27.2 Identify some of the enzymes for industry that are produced by microorganisms.

27.2 Alcoholic Beverages

The fermentation of beer, wine, and other alcoholic beverages is one of the most venerable and universal of human domestic activities. The origin of beer fermentation, for example, has been traced as far back as 4000 BC, when legend tells us that Osiris, the god of agriculture, taught Egyptians the art of brewing once they had learned how to farm the land (MicroFocus 27.2). Wine production apparently has an equally long history because archaeologists have discovered evidence of grape cultivation in the Nile Valley during the same period. Furthermore, scientists have found evidence of wine in jars excavated from an Iranian site 7,000 years old. Thus, it is conceivable that the Egyptians, Sumerians, Assyrians, and other Near East peoples were among the earliest consumers, if not connoisseurs, of alcoholic beverages.

Beer Is Produced by the Fermentation of Malted Barley

KEY CONCEPT

- Brewing is the production of alcoholic beverages from malted grains.

As early as 3400 BC, a tax was placed on beer in the ancient Egyptian city of Memphis on the Nile River. The Greeks later brought the art of

MICROFOCUS 27.2: History

This hqt's for You!

Beer, called hqt by the ancient Egyptians, was a very important drink. They often used beer in religious ceremonies and, because water could be a source of illness, they served beer at mealtimes to both adults and children. In fact, archaeologists have discovered that workmen at the great pyramids had five types of beer that they drank three times a day. It was the staple drink of the poor (wages often were paid in beer), it was a drink of the pharaohs, and a drink offered to the gods.

Because of the prevalence of beer in Egyptian life, many Egyptologists have studied beer residue from ancient Egyptian vessels. The traditional view held that their beer was made by crumbling lightly baked well-leavened bread into water. After straining through a vat, the water was allowed to ferment because of the yeast from the bread. It then was flavored with date juice or honey.

In 1996, these traditional views were challenged. Delwen Samuel, an archaeobotanist at the University of Cambridge, examined just what cereals and grains the Egyptians used to brew beer by looking at 2,000-year-old beer residues using a scanning electron microscope. Her findings suggested that the ancient Egyptians used barley to make malt and a type of wheat, called emmer, instead of hops. She says they heated the mixture and then added yeast and uncooked malt to the cooked malt. After adding the second batch of malt, the mixture was allowed to ferment. In her analysis, Samuel says she found no traces of flavorings.

Samuel and her colleagues tried brewing the beer using the recipe derived by the analysis. They brewed it at a modern brewery and found the beer to be fruity and sweet because it lacked the bitterness of hops. The beer was reported to have an alcoholic content of between 5 and 6 percent. Samuel gave her recipe to Scottish and Newcastle Breweries that then made a limited edition, 1,000-bottle batch of "King Tut Ale." It was sold at Harrods department store for $100 per bottle, the proceeds going toward further research into Egyptian beer making.

In 2002, a major Japanese brewery claimed to have recreated a 4,000-year-old Egyptian beer by following a recipe from ancient hieroglyphs in Egyptian tomb paintings. Kirin Brewery Company Ltd., Japan's second largest beer maker, said the eight gallons of brewed beer was dark brown, contained no froth, and had a strong sour taste and an alcohol content of about 10 percent. The company does not plan to sell the beer commercially; it was developed for research purposes only!

brewing to Western Europe, and the Romans refined it. Indeed, the main drink of Caesar's legions was beer. During the Middle Ages, monasteries were the centers of brewing, and by the 1200s, breweries and taverns were commonplace in Great Britain.

The word beer is derived from the Anglo-Saxon *baere,* meaning "barley." Thus, beer is traditionally a product of yeast fermentations of barley grains. However, yeasts cannot digest barley starch, and therefore it must be predigested for them (FIGURE 27.4). This is accomplished in the process of **malting**, where barley grains are steeped in water while naturally occurring enzymes digest the starch to simpler carbohydrates, principally maltose (malt sugar).

The malt is ground with water to achieve further digestion of the starch. This process, called **mashing**, often includes corn as a starch supplement. Brewers then remove the liquid portion, or **wort**, and boil it to inactivate the enzymes. Dried petals of the vine *Humulus lupulus,* called **hops**, are added to the wort, giving it flavor, color, and stability. Hops also prevent contamination of the wort, because the leaves contain at least two antimicrobial substances. At this point, the fluid is filtered, and yeast is added in large quantities.

The yeast usually employed in beer fermentation is one of two species of *Saccharomyces* developed for centuries by brewers. One species, *S. cerevisiae,* gives a uniform dark cloudiness to beer and is carried to the top of the fermentation vat by foaming carbon dioxide. This yeast therefore is called a **top yeast**. It is used primarily in English-type brews such as ale and stout. The second species, *S. carlsbergensis,* ferments the malt more slowly and produces a lighter, clearer beer, having less alcohol. This yeast sediments and is thus called a **bottom yeast**. Its product is pilsner or lager beer. Almost three-quarters of the world's beer is lager beer.

A normal fermentation requires approximately seven days in a fermentation tank. The young beer then is transferred to vats for secondary aging, or **lagering**, which may take an additional six months. If the beer is intended for canning or bottling, it is pasteurized at 60°C (140°F) for 55 minutes to kill the yeasts, or filtered through a membrane filter. Some yeast is used to seed new wort, and the remainder may be dried for animal feed or pressed to tablets for human consumption. The alcoholic content of beer is approximately 4 percent.

CONCEPT AND REASONING CHECKS

27.3 How does a lager differ from an ale microbiologically speaking?

Wine Is Produced by the Fermentation of Fruit or Plant Extracts

KEY CONCEPT

- Wine fermentation depends of the yeast *Saccharomyces ellipsoideus.*

During the Middle Ages and the centuries thereafter, wine was called *aqua vitae,* the "water of life." The title was appropriate because wine was one of the few safe things to drink. Indeed, until the late 1800s, safe drinking water was virtually nonexistent in the Western world (note that the Bible makes no references to water for drinking purposes). Wine, by contrast, was generally free of pathogens due to its acidity and alcohol content, and it provided a few minerals and vitamins to the diet, while serving as a pain reliever. Of course, it also bred a society somewhat inebriated much of the time.

Essentially, all wines are derived from the natural conversion of grape or other fruit sugars to ethyl alcohol by fermentation with *Saccharomyces.* Wild yeasts, which occur naturally on the grapes, can produce undesirable qualities in a wine, so they are killed by adding sulfur dioxide ("sulfites") during the production process.

Wine may be made from fruit, fruit juice, or plant extracts such as dandelions. Among the grapes, the species *Vitis vinifera* is recognized as the highest-quality fruit. The winemaking process begins with crushing to produce the juice, or **must** (FIGURE 27.5). For red wine, black grapes are used, including skins and sometimes, the stems. White wine, by contrast, is made from black or white grapes without their skins or stems. Cultured yeasts *(S. ellipsoideus)* are added to begin the alcoholic fermentation process. Anaerobic conditions soon are established as carbon dioxide evolves and takes up all the air space within the vats or steel tanks.

Alcohol production requires only a few days, but the aging process in wooden casks may go on for weeks or months. During this

■ Steeped: Immersed in liquid to soften or extract something.

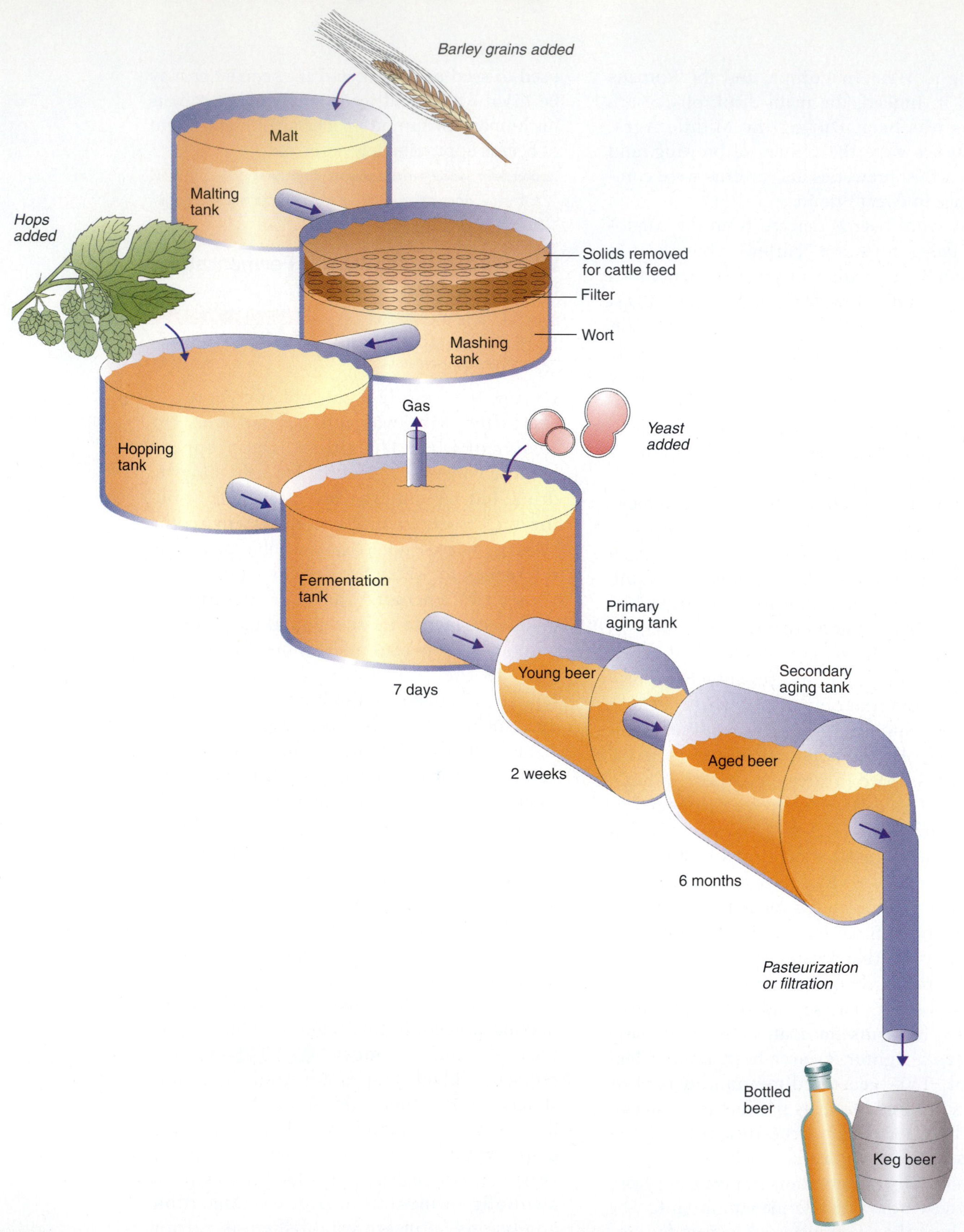

FIGURE 27.4 **A Generalized Process for Producing Beer.** Barley grains are held in malting tanks while the seeds germinate to yield fermentable sugars. The digested grain, or malt, is then mashed in a mashing tank and the fluid portion, the wort, is removed. Hops are added to the wort in the next step, followed by the yeast growth and alcohol production in fermentation. The young beer is aged in primary and secondary aging tanks. When it is ready for consumption, it is transferred to kegs, bottles, or cans.

Q: Why must beer be pasteurized or filtered before packaging?

(A)

(C)

(B)

FIGURE 27.5 **The Large-Scale Production of Wine.** (**A**) A view of industrial wine presses. The press is open for loading with crushed grapes. Once loaded, a rubber bag is inflated in the center of the press to gently squeeze the grapes against the inside walls of the press. This extracts as much juice as possible from crushed grapes without breaking the seeds. (**B**) Large, steel tanks in which the juice ferments. (**C**) Wooden casks of wine are allowed to age (for months or years) in a cool cellar.

Q: Why are wooden casks used for aging?

time wine develops its unique flavor, aroma, and bouquet. These result from the array of alcohols, acids, aldehydes, and other organic compounds produced by the yeast during aging. In fact, wine contains thousands of compounds, a few of which have been identified (MicroFocus 27.3). Soil and climate conditions (the *terroir,* in French) also contribute to the wine because they determine what organic compounds are present in the grape. The type of yeast and nature of wood derivatives from the fermentation casks are other determining factors. Thus, there are "vintage years" and poor years. For mass production, wine is pasteurized to increase its shelf life, filtered, and bottled.

The broad variety of available wines results from modifications of the basic fermentation process. In dry wines, for example, most or all of the sugar is metabolized, while in sweet wines, fermentation is stopped while there is residual sugar. Sparkling wines, including champagne, sparkle because of a second fermentation taking place inside the bottle. For a sweet sauterne, vintners enhance the sugar content of grapes by a controlled infection with the mold *Botrytis cinerea.* The mold literally sucks water out of the grapes, thereby increasing the sugar concentration.

The strongest natural wines measure about 15 percent alcohol because yeasts cannot tolerate alcohol much above this level.

MICROFOCUS 27.3: Being Skeptical
The Benefits of Red Wine

In 1992, scientists from the Bordeaux region of France did a population and epidemiological study and noted that while French and other Mediterranean peoples ate large amounts of fatty foods, they suffered a relatively low incidence of coronary artery disease. A 1996 report by the television news magazine program *60 Minutes* also pointed out that fatty meats, creams, butters, and sauces had little apparent effect on French hearts. This so-called "French Paradox" was due, in part, to the apparent medicinal properties of wine, especially drinking red wine. Is red wine good for your health?

In 1996, researchers identified phenol-based compounds in red wine that inhibited the oxidation of low-density lipoproteins (LDLs) in the blood, and by doing so, prevented the buildup of cholesterol and blood platelets in the arteries. The scientists pointed out that red wine contains more phenolics than white wine, and far more than beer. The phenolics also were present in other foods (e.g., raisins and onions), but not in the quantity found in the skins of grapes used for red wine, which concentrate even more after fermentation has taken place.

Scientists have known for decades that putting organisms on a calorie-restricted diet dramatically reduces the incidence of age-related illnesses such as cancer, osteoporosis, and heart disease. In 2003, David Sinclair, an assistant professor of pathology at Harvard Medical School, and his colleagues suggested an easier way to live longer—drink red wine. A compound known as resveratrol, which belongs to the family of phenols, naturally exists in grapes and red wine. This compound was shown to extend the lifespan of yeast cells by up to 80 percent. The molecule apparently mimics the life-extending effects of calorie restriction. Sinclair said he hoped resveratrol would prove to prolong life in other organisms, including humans. He believes his study might help explain why moderate consumption of red wine has been linked to lower incidence of heart disease and why resveratrol prevents cancer in mice. He also noted that red wines coming from harsher growing areas, such as Spain, Chile, Argentina and Australia, contain higher levels of resveratrol than those wines produced where grapes are not highly stressed or dehydrated (France and the United States).

However, red wine is not for everyone. Indeed, alcohol should be avoided by pregnant women, people taking medications, those under the legal drinking age, and anyone with a family history of alcoholism. For these and for anyone else wishing to stay away from alcohol, alcohol-free red wines are appearing in the marketplace. They offer the opportunity to take advantage of a natural health ingredient while enjoying a glass of nature's bounty. "Ah, a glass of wine, thou, a healthy heart—and more years to contemplate it."

Most table wines average about 10 to 12 percent alcohol, with fortified wines reaching 22 percent alcohol. In fortified wines, brandy or other spirits are added to produce such wines as Port, Sherry, and Madeira.

CONCEPT AND REASONING CHECKS

27.4 Explain why the alcohol produced in wine represents a primary metabolite.

Distilled Spirits Contain More Alcohol than Beer or Wine

KEY CONCEPT

- Distilled spirits are produced by concentrating the alcohol from fermentation.

Distilled spirits are alcoholic beverages containing ethyl alcohol resulting from distillation of a fermented substance, such as fruits or grains. Gin, vodka, tequila, and whiskey are a few types of spirits. Such spirits contain considerably more alcohol than beer or wine. Most of the world measures the alcohol content by its percentage of the total liquid; that is, by volume. However, in the United States, distilled beverages are designated with a **proof number**, which is twice the percentage of the alcohol content. For example, a 90 proof product contains 45 percent alcohol.

The production of distilled spirits begins like a wine fermentation. A raw product is fermented by *Saccharomyces* species, then aged, and finally matured in casks. At this point, the process diverges as manufacturers concentrate the alcohol by a distillation apparatus using heat and vacuum. Next, they mature the product in wooden casks to introduce unique flavors from various chemicals, such as aldehydes and volatile acids. Finally, the alcohol is standardized by diluting it with water before bottling.

Four basic types of distilled spirits are produced: brandy, whiskey, rum, and neutral spirits. Brandy is made from fruit or fruit juice, while rum is produced from molasses. Whiskey is a product of various malted cereal grains, such as scotch from barley, rye from rye grain, and bourbon from corn. The final type, neutral spirits, includes vodka, made from potato starch and left unflavored; gin, flavored with the oils of juniper berries; and tequila, containing the fermented juice of the blue agave.

CONCEPT AND REASONING CHECKS

27.5 Summarize how distilled spirit production produces a higher alcohol content than beer or wine.

27.3 Other Microbial Products

In addition to the products we have discussed, microorganisms are the sources of antibiotics and a number of valuable insecticides. Moreover, they are the producers of enzymes that break down natural and synthetic wastes in bioremediation, and they are the biological factories for the genetic engineering technology revolutionizing the biotechnology industry today. In the final section of this text, we study the methods for antibiotic and insecticide production and bioremediation, while discussing some details of the genetic engineering process.

Many Antibiotics Are the Result of Industrial Production

KEY CONCEPT

- Commercially useful antibiotics come from a few mold and bacterial species.

Penicillin was the first antibiotic to be produced on an industrial scale. In 1941, Robert H. Coghill of the Fermentation Division of the United States Department of Agriculture (USDA) suggested adapting the deep-tank method used to produce vitamins to produce penicillin. In the ensuing months, he offered several modifications to stimulate the growth of *Penicillium notatum* and increase the penicillin yield. For example, corn steep liquor in the culture medium increased the output 20 times, and the substitution of lactose for glucose made penicillin production still more efficient. Moreover, the search for a higher-yielding producer of the drug led researchers to *Penicillium chrysogenum*, a mold isolated from a rotten cantaloupe purchased in a Peoria, Illinois, supermarket. Treatment with ultraviolet light resulted in a mutant with still higher penicillin yields. By 1943, the United States was producing enough penicillin for the Allied forces, and by 1945, sufficient amounts were available for the civilian population.

To the present time, over 8,000 antibiotic substances have been described and approximately 100 such drugs are available to the medical practitioner. Although most antibiotics are produced by species of *Streptomyces* (FIGURE 27.6), several are also products of *Penicillium* or *Bacillus* species. The worldwide annual production of antibiotics exceeds 25,000 tons, two thirds of which is penicillins. Half of all antibiotics are prescribed for human use.

Antibiotic production may involve fermentation processes producing natural antibiotics or **semisynthetic drugs**, which represent a modification to a natural antibiotic. This is especially true for the penicillins (see Chapter 24). The fermentation process is carried out in large aerated, stainless steel fermentors. A typical fermentor may hold 30,000 gallons of medium. Older methods employed enormous mats of fungi or actinomycetes on the surface of the tank. Newer technology, however, employs small fragments of submerged fungal hyphae or bacterial cells, rotated and agitated in the medium with a constant stream of oxygen. After several weeks of growth, the stationary phase microorganisms produce the antibiotic. At the appropriate time, the antibiotic is filtered out, extracted, and purified from the growth medium. The remaining brown mash of

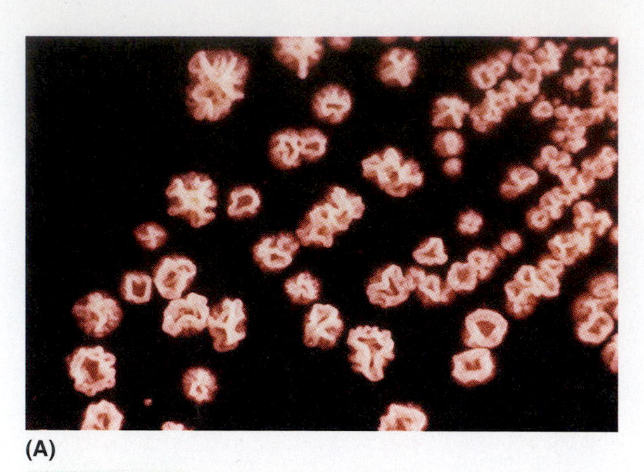

(A)

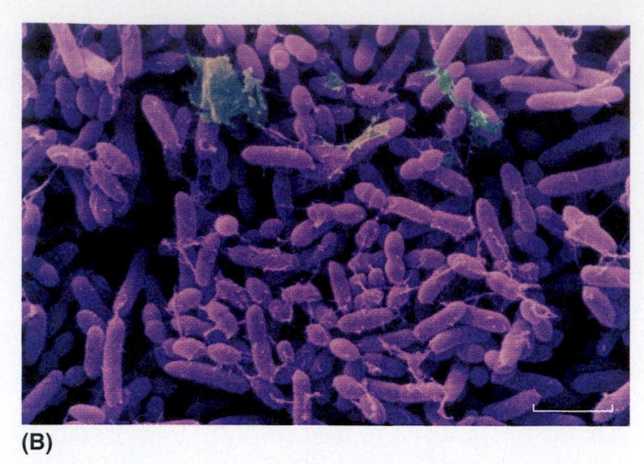

(B)

FIGURE 27.6 **An Antibiotic Producer.** Many antibiotics are produced by species of the soilborne rod *Streptomyces*. **(A)** Colonies of *S. griseus*, the organism from which Selman Waksman isolated streptomycin in the 1940s. **(B)** A false-color scanning electron micrograph of *S. griseus*, showing the chains of cells. (Bar = 10 μm.)

Q: Why is Streptomyces *such a prolific antibiotic producer?*

microorganisms may be dried and sold as an animal feed additive. Another alternative is to process it for use as human food.

CONCEPT AND REASONING CHECKS

27.6 Why has it been necessary for industry to develop semisynthetic antibiotics, such as many of the penicillins?

Some Microbial Products Can Be Used to Control Insects

KEY CONCEPT

- Bacterial toxins can be produced that kill insect larvae.

To be useful as an insecticide, a toxin from a microorganism should be relatively specific for an insect pest and should act rapidly. It should be stable in the environment and easily dispensed as well as inexpensive to produce. It helps if its odor is pleasant.

In the early part of this century, a scientist named G. S. Berliner discovered that sporulating cells of a *Bacillus* species were inhibitory to moth larvae. Berliner named the organism *Bacillus thuringiensis* after the German state Thuringia where he lived. The *Bacillus* remained in relative obscurity until recent years, when scientists learned that *B. thuringiensis* produces toxic crystals in older cells during the process of sporulation (FIGURE 27.7A). The toxic substance, an alkaline protein, is deposited on leaves and ingested by caterpillars (the larval forms of butterflies, moths, and related insects). In the caterpillar gut, the insecticide lyses the cells of the gut wall, possibly by inhibiting ATP phosphorylase or by forming pores in the cells lining the gut. As gut liquid diffuses between the cells, the larvae experience paralysis, and bacterial invasion soon follows (FIGURE 27.7B).

Bacillus thuringiensis toxin (Bt-toxin) appears to be harmless to plants and other animals. It is produced by harvesting bacterial cells at the onset of sporulation and drying them into a commercially available dusting powder. The product is useful on butterfly and moth caterpillars. Other strains of the bacterial species produce toxins to beetle and fly larvae, and to mosquitoes.

However, bacterial cells sprayed onto plants can soon wash off, so the protective effect of the insecticide can be limited. DNA technologists realized that long-term protection can be provided by inserting the genes for Bt-toxin directly into plants. To date, Bt-cotton and Bt-corn plants have been developed in the United States. In both cases, the isolated *Bt* gene has been spliced into plant genes using the Ti plasmid of *Agrobacterium tumefaciens* as the carrier (or vector), as noted in Chapter 8.

Genetic engineering also is being used to develop a single Bt-toxin to control many different plant pests. Genetic engineers can engineer a gene encoding the toxic factor along with the factor able to infect different plant pests. Such transformed bacterial cells can be grown in large fermentors to produce the Bt-toxin.

Another biotechnology approach to gene-related plant resistance has been used in corn infested with corn borers. Researchers isolated

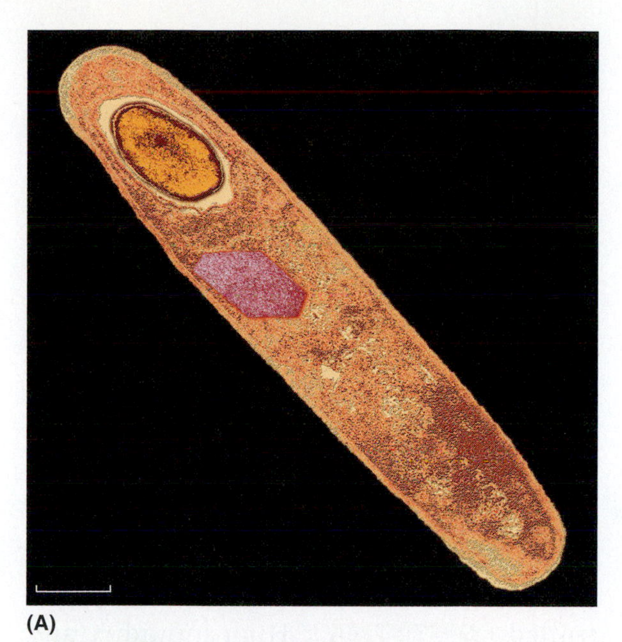
(A)

(B)

FIGURE 27.7 ***Bacillus thuringiensis* and Its Toxic Effects.** (**A**) A false-color transmission electron micrograph of *Bacillus thuringiensis,* showing the development of a spore (orange). (Bar = 0.5 μm.) (**B**) A photograph of a dead cabbage butterfly larva (caterpillar) on a leaf treated with Bt-toxin.

Q: How did the caterpillar become intoxicated?

the *B. thuringiensis* gene for toxin production and spliced it into a bacterial species living harmlessly with corn plants. The researchers then forced a colony of gene-altered bacterial cells into corn seeds. When the seeds were sown, the Bt-toxin-producing bacterial cells flourished along with the plant, and when an insect ate the plant, it consumed the toxin. This approach is advantageous because only the insect attacking the plant is subjected to the toxin.

Another useful species is *Bacillus sphaericus,* which kills at least two species of mosquitoes that ingest its poison. To increase the bacterial cell's efficiency, researchers inserted two of its genes into the bacterium *Asticcacaulis excentris* and achieved insecticidal activity against the mosquitoes transmitting malaria, filariasis, and St. Louis encephalitis. Using the gene carrier *A. excentris* is advantageous because it is easier to grow in large quantity; it tolerates sunlight better than *B. sphaericus;* and it floats in water, where mosquitoes feed (the heavier *Bacillus* species sinks because of its spores).

Many bacteriologists also are investigating the insecticidal abilities of the gram-negative rod *Photorhabdus luminescens.* Its toxin, known as Pht, attacks the gut lining of larvae (as Bt does), and it is contained in large cytoplasmic crystals; however, its spectrum of activity is wider than for Bt and includes numerous caterpillars as well as cockroaches. The Pht genes have been isolated, and efforts are underway to introduce them to plant cells. Normally, *P. luminescens* lives in the intestines of soilborne nematodes. The latter invade insect tissues in the soil, and the bacteria-derived toxin kills the insect.

Viruses also show promise as pest-control devices, partly because they are more selective in their activity than bacterial species. Once released in the field, the viruses spread naturally. It also is possible to harvest infected insects, grind them up, and use them to disseminate the virus to new locations. Among the insects successfully controlled with viruses are the cotton bollworm, cabbage looper, and alfalfa caterpillar.

Researchers also can develop insecticides by using a toxin from the venom of a scorpion. The toxin paralyzes the larvae of moths and other lepidopteran insects. It is attached to a baculovirus, a virus with a high affinity for lepidopteran tissues. Then, the virus is sprayed on lettuce and cotton plants infested with moth larvae. At the conclusion of the field trial, the plot is sprayed with 1 percent bleach to destroy any remaining viruses.

Viral genes also have been used to protect grapevines. French biotechnologists have incorporated genes from the grape fan-leaf virus (GFLV) into champagne grapevines. This virus is transmitted by a nematode and is endemic in the soils of many French regions. It causes malformation of the plant's leaf ("fan-leaf") and induces the plant to lose chlorophyll and become yellow. To protect the vines, researchers inserted the genes for viral capsids into *A. tumefaciens* and infected the plants with this bacterium. Soon the cells were producing viral capsid proteins, and they became resistant to the virus.

Even a fungus is being employed in the pesticide wars. California researchers have used *Lagenidium giganteum* to protect against crop-damaging mosquitoes in soybeans and rice, and in mosquito-infested nonagricultural settings such as wetlands. The fungus forces its spores into mosquito larvae, which die in a day or two. Marketed as Laginex, the fungal preparation has been approved for certain uses by the US Environmental Protection Agency (EPA).

CONCEPT AND REASONING CHECKS

27.7 How are bacterial toxins being produced industrially to kill insect larvae?

Fungal Organisms Also Are Being Commercially Developed

KEY CONCEPT

- Yeast cells and mushrooms represent potential food or food supplement sources.

From the previous sections, it should be obvious that yeasts are involved in many fermentation processes, including alcoholic beverages and baking. Today, the yeast cells or their products are of great commercial value.

Yeast cells can be grown in fermentors that are kept aerated. To grow at a maximal rate, plenty of ATP must be made, so aerobic respiration is essential. To keep the cells growing, the energy and carbon source is molasses, which is introduced into the fermentor in small amounts to prevent the yeast cells from switching to fermentation and producing alcohol.

After the growth period, the yeast cells are collected and processed for the commercial market as either dry yeast or as compressed yeast cakes for baking. The yeast cells also can be dried and marketed as a nutritional supplement. Yeast cells are rich in proteins and B vitamins. The product may also be added to commercial wheat or corn flour, or sold at health food stores as a nutritional supplement.

Besides the unicellular fungal yeasts, filamentous fungi also have nutritional value. The most commercial form for human consumption is the **mushrooms** (FIGURE 27.8A). In recent years, more and more species of mushrooms are being grown commercially and now include crimini, shitake, oyster, and portabella mushrooms. Mushrooms are not grown in fermentors, but rather in special buildings called "mushroom farms" where the temperature and humidity can be closely controlled (FIGURE 27.8B). High humidity and cool temperatures are necessary to trigger mushroom formation.

(A)

(B)

FIGURE 27.8 **Commercial Mushroom Farming.** (**A**) Many types of mushrooms are grown commercially today. (**B**) A mushroom farm often consists of a rich bed of soil in which the fungal mycelium grows and produces mushrooms in flushes, such as these *Agaricus bisporis*.

Q: Why must the beds be kept in moist, damp places?

The beds in which the mushrooms will be grown consist of soil mixed with rich organic matter. The soil is inoculated with a **spawn**, which is a pure culture of the mushroom mycelium. In the bed, the spawn spreads throughout the soil for several weeks. **Casing soil**, a non-nutritious soil layer providing the mycelium with more moisture for mushroom formation, is then added. The appearance of mushrooms on the bed is called a **flush**. Once the flush appears, the mushrooms must be collected while they remain fresh. Commercial farms then package and keep the mushrooms cool for delivery to the market.

CONCEPT AND REASONING CHECKS

27.8 Summarize the steps involved in mushroom farming.

Bioremediation Helps Clean Up Pollution Naturally

KEY CONCEPT

- Natural or engineered bacterial species can break down toxic waste.

Recruiting bacterial species and other microorganisms to break down synthetic waste is an immensely appealing idea. It signals a willingness to work with nature and adapt to its sophisticated sanitation systems, rather than trying to reinvent them. Putting microorganisms to work in this manner is the crux of **bioremediation**, which is the use of microorganisms to degrade or neutralize contaminants in the soil or water.

The concept of bioremediation is not new. In the 1800s, for example, night-soil men would, for a small fee, travel from house to house collecting sewage and excrement. After making their rounds, they would scatter their collections on fields, to be broken down by naturally occurring bacterial species in the soil. Although modern waste-disposal systems have replaced the night-soil men, a new concern is the plethora of environmental pollutants contaminating the land. Bioremediation seeks to exploit microorganisms to degrade these pollutants.

The advantages of bioremediation were displayed following a major oil spill from the tanker Exxon Valdez in 1987 along the Alaska coastline (FIGURE 27.9). Previous studies showed that where oil is spilled, bacterial species of *Pseudomonas* already present there could degrade the oil; technologists only needed to encourage bacterial growth. Thus, when the oil spill occurred, technologists "fertilized" the oil-soaked water with nitrogen sources (e.g., urea), phosphorus compounds, and other mineral nutrients to modify the environment and stimulate the growth of naturally occurring microorganisms. Areas treated this way were cleared of oil significantly faster than nonremediated shorelines. Indeed, the oil degraded five times faster when microorganisms were put to work.

Bioremediation also can be applied to help eliminate polychlorinated biphenyls (PCBs) from the environment. PCBs were used widely in industrial and electrical machinery before their threat to environmental quality was realized. These inert compounds contain numerous chlorine atoms and chlorine-containing

FIGURE 27.9 The Exxon Valdez Oil Spill. In 1987, the oil tanker Exxon Valdez ran aground on the shoreline of Alaska. Numerous naturally-occurring bacterial species demonstrated their value by digesting the oil during the ensuing cleanup efforts. This was one of the first large-scale attempts to use microorganisms in bioremediation.

Q: How would these naturally-occurring bacterial species be stimulated to rapidly grow to be a significant factor in the clean up effort?

Night soil: Human excrement collected at night from toilets or cesspools, especially for use as fertilizer.

groups, and researchers have identified anaerobic bacterial species able to remove the atoms and their daughter groups, and reduce the compounds to smaller molecules. Aerobic bacterial species now take over and reduce the molecular size still further. Field demonstrations in New York's Hudson River have shown the value of the combination anaerobic–aerobic degradation; where once there was an accumulation of PCBs, now carbon dioxide, water, and hydrogen chloride have evolved.

Many years ago, trichloroethylene (TCE) was a much-used cleaning agent and solvent. At the time, scientists did not realize TCE would diffuse through the soil and contaminate underground wells and water reservoirs (aquifers). To combat the problem, scientists have exploited bacterial species that grow on methane to degrade the TCE. During their metabolism, the bacterial cells produce a methane-digesting enzyme that breaks down TCE. Technologists pump methane and other nutrients into the TCE-contaminated water, and as the bacterial cells grow, they digest the TCE as well as the methane. The deliberate enhancement of microbial growth yields an environmental cleanup.

Deinococcus radiodurans is a bacterial species able to withstand 3,000 times more radiation than humans. The bacterial cells were found in a tin of irradiation-sterilized ground beef. Researchers are hoping to use this organism in the daunting task of cleaning up thousands of toxic-waste sites where radioactive materials such as plutonium and uranium are present. Genetic engineering methods have produced a *D. radiodurans* strain that degrades ionic mercury compounds common to these sites. The strain uses a gene cluster from *E. coli* to reduce mercury to a less toxic form.

During the 1940s through the 1960s, a major component of weaponry was the explosive compound 2,4,6-trinitrotoluene (TNT). Like other synthetic wastes, this compound has contaminated the soil from residues deposited around weapons plants. Scientists have discovered they can reduce the level of contamination by encouraging bacterial growth with molasses. In a pilot study, researchers mixed water with TNT-laced soil and added molasses at regular intervals. In a matter of weeks, the TNT concentration plummeted.

Plants also can conscript bacterial species for the task of environmental cleanup: Following the Gulf War of 1991, oily devastation remained in much of the Arabian Desert. Within four years, however, plant life returned, aided in large measure by *Arthrobacter* species. When researchers dug into oil-soaked desert, they found healthy plant roots surrounded by reservoirs of these oil-degrading gram-negative rods.

For many years, the "haul-and-bury" technique was the prevailing method for disposing of synthetic waste. As the public becomes increasingly intolerant of such an approach, the importance of bioremediation will become more apparent. Technologists are testing microorganisms for their ability to degrade flame-retardants, phenols, chemical warfare agents, and numerous other waste products of industry.

MicroInquiry 27 explores a few more examples with microorganisms.

CONCEPT AND REASONING CHECKS

27.9 Describe some of the ways bacterial species can be used in bioremediation.

Industrial Genetic Engineering Continues to Make Advances

KEY CONCEPT

- Plasmids can be used to transform the properties of a bacterial cell.

In 1973, Herbert Boyer, of the University of California at San Francisco, and Stanley Cohen, of Stanford University, performed the first practical experiment in genetic engineering. Working with *E. coli,* they removed the genes for kanamycin resistance and spliced them into a plasmid already carrying genes for tetracycline resistance (see Chapter 8). The plasmids then were mixed with *E. coli* cells not resistant to either antibiotic. Finally, the cells were streaked on plates of culture medium containing both kanamycin and tetracycline. The bacterial cells that grew had taken up the plasmids and now were resistant to both antibiotics. The cells had been transformed; they could now do something they could not do before. Feats like these launched the modern era of biotechnology.

Genetic engineering based in plasmid technology has been hailed as the beginning

MICROINQUIRY 27

Working with Microorganisms

As we approach the end of this text, you now should have a deep appreciation for the diverse roles microorganisms can have. Certainly, their roles go beyond causing human disease. In this last MicroInquiry, let's examine some additional ways microbes are being put to beneficial use. Answers can be found in **Appendix D.**

Oil Exploration through Microbial Chemistry

For decades, we have been hearing about the impending oil shortage. Indeed, it is becoming harder for oil companies to find deposits of light ("sweet") crude oil, which is the form usually refined. This means about 60 percent of the world's reserves are a very thick and sticky form called heavy ("sour") crude oil. Much of it remains below ground because it is very difficult to recover and costly to refine into the liquid gasoline you put in your automobile.

27.1a. Outline a microbial way to recover this sour crude oil more reasonably.

Another problem with heavy crude oil is that it contains large amounts of potential contaminants, including sulfur compounds and heavy metals, which could be spread into the atmosphere through refining. This would reverse the trend to cut sulfur emissions in the air. Add to this the high cost of refining such sulfur-rich crude and you have a dilemma.

27.1b. How can this heavy crude be more economically and safely extracted and refined?

Prospecting with Microbes

The technology of metal extraction has been around for a long time. More than 2,000 years ago, the Romans used iron, copper, lead, gold, silver, and alloys such as bronze and brass for tools, weapons, coins, and jewelry. Unknown to them, microbes had solubilized these metal ores, which the Roman metallurgists collected from mines in Spain and Wales.

The twentieth century saw an explosion in the production of such high-grade ores, which, like crude oil, are becoming depleted. With this expansion of metal production, besides disfiguring the land, comes the environmental pollution from the smelting process that produces sulfur dioxide in the atmosphere and toxic chemicals in the water and soil.

27.2 Outline a biotechnological scheme to provide for a continued supply of metals.

Superfund and Bioremediation

Historically and still in use today are chemical and physical methods, such as excavation and incineration, to degrade many of the toxic materials found at Superfund sites. However, the Environmental Protection Agency (EPA) has tested the feasibility of bioremediation at several of these sites.

Creosote Contamination

Between 1973 and 1985, a lumber company in southern Missouri operated a wood treating facility that preserved railroad ties with a creosote/diesel fuel mixture. These operations contaminated the soil at the site with polynuclear aromatic hydrocarbons (PAHs), which were major components of the creosote/diesel mixture. After the facility was shut down and designated as a Superfund site in 1987, the EPA oversaw construction and operation of a land treatment unit to remove the PAH-contaminated soils at the site. EPA studies indicated indigenous microorganisms could digest aromatic hydrocarbons existing in the soil.

27.3 Outline a process whereby excavated contaminated soil could be decontaminated.

Petrochemical Wastes

In Texas, an industrial waste disposal facility contained an estimated 70 million gallons of petrochemical wastes that were disposed of onsite between 1966 and 1971. Contaminants in the lagoon sediments included PAHs, chlorinated organics, and metals.

27.4 Outline a process where by lagoon sediments could be decontaminated.

■ Superfund: Common name for an environmental law giving the government broad authority to respond directly to releases or threatened releases of hazardous substances endangering public health or the environment.

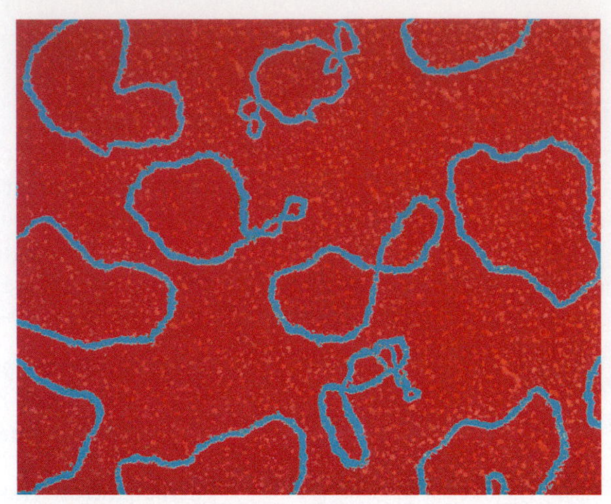

FIGURE 27.10 **Plasmids.** A false-color transmission electron micrograph of bacterial plasmids. Plasmid technology is key to many processes in industrial microbiology and biotechnology.

Q: What types of products may be carried by the information spliced into a plasmid?

FIGURE 27.11 **On the Horizon.** An artist's fanciful conception of a futuristic super plant produced by genetic engineering. Microorganisms or microbial products are key to many of its attributes.

Q: What concept is depicted by showing this plant growing out of a test tube?

of modern industrial microbiology. **Plasmids** are small, circular, double-stranded DNA molecules that exist apart from the bacterial chromosome (FIGURE 27.10). A single plasmid may contain between 2 and 250 genes. With **plasmid technology**, a gene of interest is spliced into the plasmid DNA and then the plasmid serves as the vector to carry the foreign gene into the targeted cell. The mechanics of this genetic engineering process are explored in Chapter 8. Some contemporary products already obtained by plasmid technology include interferon, insulin, and human growth hormone.

When microbiologists first developed plasmid technology, the likely choice for the prototype "bacterial factory" was *E. coli*. Nonpathogenic strains had long been used as test organisms in the laboratory, and the genetics of *E. coli* were well understood. In the 1980s, however, attention shifted to *Bacillus subtilis* and yeasts as host organisms. *B. subtilis* is a gram-positive rod that normally secretes the proteins it makes, while *E. coli* retains them. Also, *B. subtilis* is not regarded as a human pathogen, in contrast to some strains of *E. coli*, nor does it contain endotoxins in its cell wall (see Chapter 18). Yeast also is a key "factory" due to its traditional role in the fermentation processes and as an organism without disease potential. In fact, *Saccharomyces* species have been reengineered to produce a synthetic vaccine for hepatitis B and to obtain rennin, the enzyme used to make cheese (see Chapter 26).

The implications of genetic engineering are almost limitless. Biotechnologists are trying to introduce the antibiotic-producing genes of *Streptomyces* species directly into more rapidly growing organisms. They also are trying to amplify the number of plasmids in antibiotic producers to obtain a higher yield of product.

In the 1960s, a "Green Revolution" took place in which genetically-modified high-yielding supergrains were exported to poor countries throughout the world to encourage self-sufficiency. However, the program stalled when stocks of essential petroleum fertilizers declined due to high oil prices. Agriculturalists now hope for a second Green Revolution sparked by genetic engineering (FIGURE 27.11). They foresee the day when the genes for nitrogen fixation can be extracted from bacterial organisms such as *Rhizobium* and inserted into grain plants such as wheat, rye, and barley. The most optimistic planners look to the future and foresee fertilizers becoming obsolete, plants using microbial toxins to drive off insects, and crops living for weeks without water.

Another dream of biotechnologists is using replacement organisms to interrupt disease cycles in nature. For example, a transgenic snail that resists invasion of *Schistosoma* species conceivably could interrupt the life cycle of the parasite causing schistosomiasis (see Chapter 17). A slightly different strategy is being employed by DNA technologists who are attempting to produce transgenic cotton bollworms by inserting a gene activating a toxin-producing gene in offspring of the bollworm. Work continues to progress on the engineering of a transgenic mosquito unable to harbor and transmit malarial parasites. Released in large numbers, the mosquitoes could dilute or overwhelm native mosquito populations and break the chain of disease transmission.

Other research projects in genetic engineering have equally important goals. Research in genetic engineering holds promise for new strategies for cancer prevention, new diagnostic procedures for microbial diseases and genetic abnormalities, new methods to correct genetic disorders, new hormones, antibiotics, and vaccines, and a generally improved quality of life. The discoveries and insights made possible by genetic engineering have been described as breathtaking. In the future, we can expect startling developments in medicine, agriculture, and the pharmaceutical and chemical industries. Indeed, it is an exciting time to be a microbiologist.

CONCEPT AND REASONING CHECKS

27.10 Justify the need for industrial microbiology to the advances in biotechnology.

Transgenic: Referring to an organism containing a stable gene from another organism.

SUMMARY OF KEY CONCEPTS

27.1 Microorganisms in Industry

- Among the organic compounds synthesized on an industrial scale are organic acids (such as citric, lactic, and gluconic acids), along with various amino acids and vitamins.
- Microbes also produce a variety of enzymes such as amylase, pectinase, and protease. Microorganisms produce other key products and save much expense and tedium for the chemist.

27.2 Alcoholic Beverages

- Yeasts of the genus *Saccharomyces* ferment barley grains to produce beer under anaerobic conditions. The process requires an aging step to develop the full flavor of the product.
- Yeasts of the genus *Saccharomyces* ferment grape juice to wine under anaerobic conditions. This process also requires an aging step to develop the full flavor of the product.
- To produce spirits, alcohol produced from fermentation is distilled off to produce brandy, rum, whiskey, or neutral spirits.

27.3 Other Microbial Products

- The spores of *Bacillus thuringiensis* are used in the live form as insecticides for plants and are successfully employed against the caterpillars of various insect pests. In addition, scientists are investigating the use of bacterial insecticides, such as the toxin produced by *Photorhabdus*. Scorpion toxins attached to insect viruses are being developed as potent insecticides.
- Several fungi have uses as food or food additives. Besides their use in the alcoholic beverage industries, yeasts have commercial value in a dried or cake form in the baking industries. Nutritional yeasts can be a source of protein and B vitamins in the human diet. Mushroom farms have developed the technology to grow many commercial types of wild mushrooms.
- Bioremediation is still another innovative use for industrial microorganisms. In this process, naturally occurring microorganisms are encouraged to grow in a polluted environment and break down the pollutants. Bioremediation has been used successfully to degrade the oil in oil spills and to help eliminate polychlorinated biphenyls and trichloroethylene from the environment. Researchers are hoping to use the process in the daunting task of cleaning toxic waste sites that contain radioactive materials.
- The future of industrial microbiology will center on genetic engineering and plasmid technology. By inserting foreign genes into vector organisms such as bacterial species and yeast, biotechnologists can produce rare proteins on an industrial scale and provide treatments for such diseases as diabetes, hemophilia, and cancer. Genetic engineers predict plants will be developed with inborn resistance to disease, animals will be engineered to produce human proteins, and novel treatments will be designed to cure or prevent specific illnesses. Medicine, agriculture, and industry eagerly anticipate the fruits of modern biotechnology that are made possible with microorganisms.

LEARNING OBJECTIVES

After understanding the textbook reading, you should be capable of writing a paragraph that includes the appropriate terms and pertinent information to answer the objective.

1. Identify how natural fermentation differs from industrial fermentation.
2. Assess the need for fermentors (bioreactors) in industrial fermentation processes.
3. Distinguish between primary and secondary metabolites.
4. Describe the need for the industrial fermentation of organic acids.
5. Summarize the types of enzymes produced by the industrial fermentation of microorganisms for commercial and medical needs.
6. Construct a concept map for beer production.
7. Describe the steps in the production, fermentation, and aging of wine.
8. Compare and contrast the production of distilled spirits from wine making.
9. Explain why antibiotics produced through industrial fermentation represent secondary metabolites.
10. Evaluate the role of microorganisms to the industrial production of insecticides.
11. Identify the usefulness of fungi to food microbiology and as a food supplement.
12. Explain how bioremediation can be accelerated using microorganisms.
13. Describe how plasmid technology is advancing biotechnology.
14. Estimate the role of microorganisms to the expanding enterprise of biotechnology.

HTTP://MICROBIOLOGY.JBPUB.COM/

The site features learning, an on-line review area that provides quizzes and other tools to help you study for your class. You can also follow useful links for in-depth information, read more MicroFocus stories, or just find out the latest microbiology news.

SELF-TEST

Answer each of the following questions by selecting the ***one*** answer that best fits the question or statement. Answers to even-numbered questions can be found in **Appendix C**.

1. Industrial fermentation is a/an
A. process requiring alcohol.
B. aerobic process.
C. anaerobic process.
D. process always requiring yeast.
E. Both **B** and **C** are correct.

2. Which one of the following is *not* true of a primary metabolite?
A. It is directly involved with normal growth.
B. Pyruvate is an example.
C. It is normally produced during the stationary growth phase.
D. Alcohol is a primary metabolite.
E. It is a product essential to the survival of the microbe.

3. Citric acid is used in all the following products *except*:
A. inks.
B. effervescent tablets.
C. soft drinks.
D. candies.
E. wines.

4. Which one of the following (**A–D**) is *not* an organic acid produced through industrial fermentation? If all are organic acids, select **E**.
A. Gluconic acid
B. Lactic acid
C. Nitric acid
D. Glutamic acid
E. All the above (**A–D**) are correct.

5. Industrially-produced streptokinase is a/an _____ used to _____.
A. enzyme; dissolve blood clots
B. substrate; ret flax
C. vitamin; prevent pernicious anemia
D. enzyme; facilitate the absorption of fluids
E. enzyme; digest glucose

6. In the brewing of beer, wort is the
A. starch found in barley.
B. the carbohydrate yeasts cannot digest.
C. product of the fermentation process.
D. the liquid removed after mashing.
E. dried petals that add flavor to the beer.

7. A normal fermentation process for beer requires about _____ days.
A. 2
B. 7
C. 14
D. 30
E. 60

8. The brewing of a lager requires
A. a bottom yeast.
B. *Saccharomyces cerevisiae*.
C. a top yeast.
D. wild yeasts.
E. Both **B** and **C** are correct.

9. In wine making, wild yeasts often
A. are killed with sulfur dioxide.
B. produce undesirable wine qualities.
C. are found on the grapes.
D. are a species of *Saccharomyces*.
E. All the above (**A–D**) are correct.

10. In the wine making process, must is
A. the preservatives added to the wine.
B. the juice from the crushed grapes.
C. the sediment resulting from the fermentation process.
D. a degradative product of wild yeasts.
E. the additional alcohol added to fortified wines.

11. Most table wines have an alcohol percentage between
A. 1 and 3.
B. 5 and 8.
C. 10 and 12.
D. 15 and 20.
E. 25 and 30.

12. A distilled spirit that is 60 proof contains
A. 15 percent alcohol.
B. 15 percent water.
C. 30 percent water.
D. 30 percent alcohol.
E. 60 percent alcohol.

13. Most antibiotics are the products of _____ species.
A. *Streptomyces*
B. *Penicillium*
C. *Saccharomyces*
D. *Streptococcus*
E. *Bacillus*

14. The toxin produced by *Bacillus thuringiensis* is
A. a protein.
B. a lipid.
C. an acid.
D. a sugar.
E. a cell wall component.

15. A spawn is a
A. complex set of nutrients.
B. nonnutritious soil layer.
C. cluster of mushrooms.
D. beds in which mushrooms grow.
E. mushroom mycelium.

16. For plasmid technology, *Bacillus* _____ may be the "bacterial factory" because it produces no _____.
A. *cereus*; cell wall
B. *subtilis*; endotoxins
C. *megaterium*; endospores
D. *thuringiensis*; exotoxins
E. *subtilis*; cell wall

QUESTIONS FOR THOUGHT AND DISCUSSION

Answers to even-numbered questions can be found in **Appendix C**.

1. The poet John Donne once wrote: "No man is an island, entirely of itself." This maxim applies not only to humans, but to all living things in the natural world. What are some roles the microorganisms play in the interrelationships among living things?
2. When the Mayflower set sail for the New World, its intended destination was Virginia. Instead, it landed at Plymouth, Massachusetts, because, as one diarist put it, "We could not now take time for further search or consideration, our victuals being much spent, especially our beer." What do these last few words tell you about the Pilgrims?
3. In several places in this text, we have noted how apparently harmless organisms have been discovered later to be dangerous in humans. Yeasts have been consumed in breads and alcoholic beverages for centuries, and still no pathogenic signs have been observed. Can you postulate why?
4. Certain bacterial species produce many thousands of times their required amount of specific vitamins. Some biologists suggest that this makes little sense because the excess is wasted. Can you suggest a reason for this apparent overproduction in nature?
5. How many times in the last 24 hours have you had the opportunity to use or consume the industrial product of a microorganism?

APPLICATIONS

Answers to even-numbered questions can be found in **Appendix C**.

1. Many beer companies have developed strains of yeasts that break down more of the carbohydrate in barley malt than traditional yeasts. What do you think their product is called?
2. The discovery of the extremely high resistance of *Deinococcus radiodurans* to radioactivity has prompted hopes that this organism can be used in bioremediation. Suppose you were to go on a hunt for novel organisms that could be used for other environmental cleanups. Where might you look?
3. A product called Dipel dust contains *Bacillus thuringiensis*. It is used in the Northeast for destruction of the gypsy moth caterpillar. What dangers might result from extensive application of this product?
4. A friend who also has taken a class in microbiology wonders if it would be possible to engineer certain bacterial species to produce antibiotics and then give these species to diseased people to serve as antibiotic producers within the body. How would you respond to your friend's idea?

REVIEW

Many different microorganisms find value in industrial microbiology. To test your knowledge of these organisms, match the microorganism on the right to the characteristic on the left by placing the correct letter in the space. A letter may be used once, more than once, or not at all. The answers to even-numbered characteristics are listed in **Appendix C**.

Characteristic

____ **1.** Used to ferment grapes.
____ **2.** Produces penicillin.
____ **3.** Used in riboflavin production.
____ **4.** Produces many antibiotics.
____ **5.** Enhances grape sugar content.
____ **6.** Lactic acid producer.
____ **7.** Used in lysine production.
____ **8.** Biological insecticide.
____ **9.** Source of musk oil.
____ **10.** Used for citric acid production.
____ **11.** Fungal source of amylase.
____ **12.** Synthesizes pectinase.
____ **13.** Produces ergot poisons.
____ **14.** Used in top fermentation of beer.
____ **15.** Yeast for making pilsner beer.
____ **16.** Able to degrade oil.
____ **17.** Used for genetic engineering.
____ **18.** Use in the fermentation of lagers.
____ **19.** Degrades ionic mercury compounds.
____ **20.** Gluconic acid from carbohydrates.

Microorganism

A. *Acetobacter aceti*
B. *Agaricus phalloides*
C. *Ashbya gossypii*
D. *Aspergillus niger*
E. *Aspergillus oryzae*
F. *Bacillus thuringiensis*
G. *Bacillus subtilis*
H. *Botrytis cinerea*
I. *Claviceps purpurea*
J. *Clostridium* species
K. *Deinococcus radiodurans*
L. *Enterobacter aerogenes*
M. *Gibberella fujikuroi*
N. *Lactobacillus bulgaricus*
O. *Penicillium chrysogenum*
P. *Pseudomonas* species
Q. *Rhizopus nigricans*
R. *Saccharomyces carlsbergensis*
S. *Saccharomyces cerevisiae*
T. *Saccharomyces ellipsoideus*
U. *Staphylococcus aureus*
V. *Streptomyces* species
W. *Ustilago zeae*

Appendix A: Metric Measurement

TABLE

	Fundamental Unit	Quantity	Symbol	Numerical Unit	Scientific Notation
Length	Meter (m)				
		kilometer	km	1,000 m	10^{3}
		centimeter	cm	0.01 m	10^{-2}
		millimeter	mm	0.001 m	10^{-3}
		micrometer	µm	0.000001 m	10^{-6}
		nanometer	nm	0.000000001 m	10^{-9}
Volume (liquids)	Liter (l)				
		milliliter	ml	0.001 l	10^{-3}
		microliter	µl	0.000001 l	10^{-6}
Mass	Gram (g)				
		kilogram	kg	1,000 g	10^{3}
		milligram	mg	0.001 g	10^{-3}
		microgram	µg	0.000001 g	10^{-6}

Appendix B: Temperature Conversion Chart

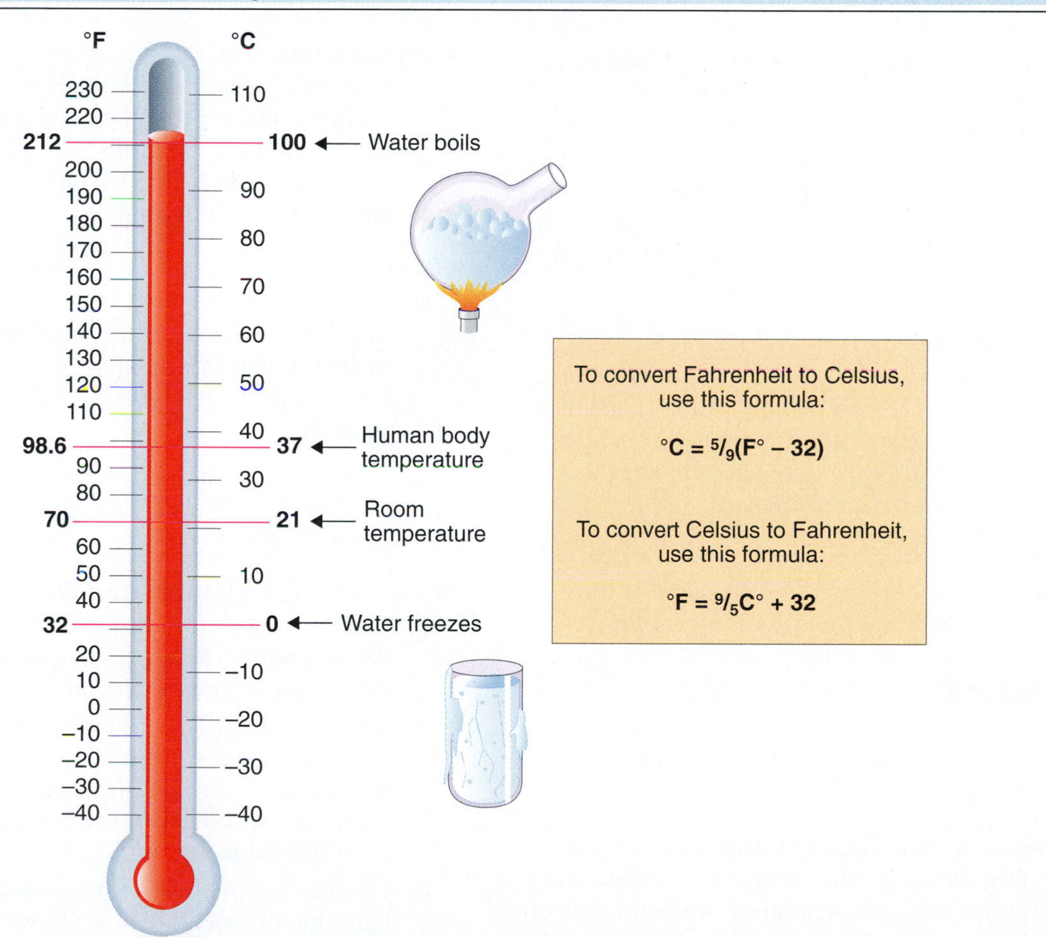

Appendix C: Answers to Even-Numbered End-of-Chapter Questions

Chapter 1

Self-Test

2. A; **4.** E; **6.** C; **8.** E; **10.** E; **12.** C; **14.** B.

Questions for Thought and Discussion

2. The control group does not receive the treatment to which the experimental group is exposed. Comparisons then can be made between the two groups as to whether the treatment had any affect. You can discover the controls for the Redi-Needham-Spallanzani experiments on spontaneous generation. Controls for Pasteur's experiments to refute the doctrine are identified in the MicroInquiry 1 exercise.
4. Bacterial species like *Escherichia coli* have many advantages over rats or guinea pigs as model systems. Bacterial cells are much easier to grow and much less costly to "feed." Substantially less space is needed to culture or maintain the organisms. The bacterial cells reproduce more rapidly, so several generations can be studied in much less time than a single generation of animals.
6. Chance certainly favored the prepared mind of Semmelweis in his recognition of the need for hand washing; Snow in recognizing the source for the London cholera outbreak; and Fleming in realizing that a mold might hold the cure for treating bacterial diseases.

Applications

2. Eukaryotes would have a cell nucleus that could be seen with a light microscope. If it were a prokaryote, rRNA base sequencing would place the organism in either the domain *Bacteria* or *Archaea*.

Review

2. F (bacterial cells); **4.** F (on a solid culture surface) **6.** T; **8.** F (German); **10.** F (virology).

Chapter 2

Self-Test

2. C; **4.** B; **6.** A; **8.** C; **10.** D; **12.** D; **14.** A; **16.** A; **18.** C.

Questions for Thought and Discussion

2. Organic molecules invariably contain carbon atoms, and carbon atoms form four covalent bonds with other atoms. Soon these evolve a Tinkertoy-like arrangement of atoms that is quite a large molecule, especially if multiple carbon atoms are present.
4. Different arguments can be presented for destroying different groups of chemical substances. For example, destroying proteins would eliminate enzyme and chemical structures; carbohydrate destruction would remove vital energy sources for life and the ability to form a cell wall; lipid destruction would cause leakage through the cell membrane and consequent death; eliminating nucleic acids would disrupt the chromosome and genes, thereby preventing protein synthesis.
6. Proteins are denatured by heat, thereby losing their tertiary structure and assuming a nonfunctional form that would be nontoxic.
8. In all organisms, including microorganisms, chemical reactions require the interaction of "unstable" molecules. This means there are atoms with unfilled electron shells. Filling these shells produces new combinations of atoms, which is what metabolism is all about.

Applications

2. First, determine the pH of the two buffers using the pH papers. Then add a drop of acid to the buffered broth. The pH will not drop. Then add a drop to the unbuffered broth. The pH will drop.

Review

2. (i). Hydroxyl; (ii). Hydroxyl; (iii). No functional groups; (iv). Amino, carboxyl, and hydroxyl; (v). Hydroxyl.

Chapter 3

Self-Test

2. E; **4.** E; **6.** A; **8.** E; **10.** E; **12.** C; **14.** D; **16.** D; **18.** A; **20.** D.

Questions for Thought and Discussion

2. They are: the "most small;" the most nutritionally hardy; the most ancient; the most able to cause disease; the most . . .
4. No doubt an order for *Bergey's Manual of Systematic Bacteriology* will shortly be placed, since the identification of unknown bacterial species depends on information from this book.
6. Oil does not increase the magnification of the light microscope. It merely allows the gathering of enough light by the oil-immersion lens to resolve the specimen.

Applications

2. Blue light because it has the shortest wavelength. When plugged into the resolving power equation, 500 nm gives the smallest resolvable object.

Review

2. R; 4. F; 6. G; 8. N; 10. Q; 12. L; 14. I; 16. W; 18. P; 20. S.

Review: Cell Structure

A: 2. cytoplasm; 4. cell wall. B: 6. flagellum; 8. mitochondrion; 10. DNA or chromosome; 12. smooth endoplasmic reticulum; 14. actin filaments; 16. Golgi apparatus.

Chapter 4

Self-Test

2. C; 4. A; 6. D; 8. E; 10. B; 12. B; 14. C.

Questions for Thought and Discussion

2. In the pre-electron microscope days, the chapter on bacterial cell structure would probably be very brief. Certainly there would be no discussion of pili, plasmids, magnetosomes, ribosomes, or the fluid mosaic model.

Applications

2. Being gram negative, it will have an outer membrane containing lipopolysaccharide (LPS). Providing an antibiotic may kill the pathogen but in the process free LPS material into the body, which can act as a toxin.

Review

2. plasmid; 4. pili; 6. capsule; 8. cell wall; 10. cell membrane; 12. cell membrane; 14. glycocalyx.

Chapter 5

Self-Test

2. B; 4. A; 6. B; 8. C; 10. E; 12. B; 14. E.

Questions for Thought and Discussion

2. Laundry detergents can use enzymes from extremophiles to break down blood, plant, and other organic molecule stains in hot water and at the high pH of soapy wash water. In addition, since the extremophiles thrive under conditions where pollutants are found, their enzymes might be used to break down these pollutants. The polymerase chain reaction (Chapter 21) uses these enzymes as well.
4. Thermophiles may produce toxins that interfere with metabolic reactions in the body. Thus, consuming food that contains toxins may be dangerous.

Applications

2. Boiling may kill the majority of bacterial cells, but bacterial spores survive 2 hours or more of boiling. It is wrong to believe the water is sterile after a few minutes of boiling. It has been disinfected, however, and is perfectly safe to drink under normal circumstances.

Review

2. T; 4. F (minority); 6. F (enriched medium); 8. T; 10. F (log phase).

Chapter 6

Self-Test

2. E; 4. E; 6. B; 8. A; 10. B; 12. C; 14. C.

Questions for Thought and Discussion

2. ATP could not be absorbed into the cytoplasm of a bacterial cell. More importantly, ATP cannot be stored as ATP; it will be quickly broken down.
4. Stopping glycolysis would mean that the pyruvate fuel for the Krebs cycle would cease being made from carbohydrate. However, pyruvate could be produced from amino acids by deamination of certain amino acids, and fatty acids might convert to acetyl-CoA for entry to the cycle.

Applications

2. The individual probably assumes that the green sulfur bacteria will use the hydrogen sulfide as a hydrogen source in photosynthesis, thus liberating elemental sulfur. Whether it will work is a point for conjecture, and students might like to discuss the pros and cons of such a scheme.

Review

2. protein/speed up/substrate; 4. fermentation/oxygen/glucose/an organic end product; 6. pyruvate/ carbon/carbon dioxide/NAD^+; 8. carbon dioxide/photosynthesis/glucose; 10. fatty acids/beta oxidation/two-carbon units/cellular respiration.

Chapter 7

Self-Test

2. E; 4. C; 6. A; 8. D; 10. B; 12. C; 14. B.

Questions for Thought and Discussion

2. One would have to believe that double-stranded DNA viruses would be better at DNA

repair because if a mutation occurs in one strand, the complementary strand is a template for mismatch repair. Single-stranded RNA viruses would lack this ability.

Applications

2. The lack of colonies near the center of the plate is probably due to the very high concentration of mutagen. Here the mutagen is so strong it caused so many mutations that they were lethal to any cells that were present. Farther out, where the mutagen is more diluted (less concentrated), fewer mutations would occur and some bacterial cells would survive.
4. You could take the plate with benzene-metabolizing colonies and replica plate it onto a similar synthetic medium plate containing the ^{32}P material. Any colonies sensitive to the radioactive material will not grow (negative selection). These can be identified by looking back at the master plate with benzene-metabolizing colonies.

Review

2. (a) DNA because it contains Ts; (b) met-cys-tyr-gln-asn-phe-asn-ala = silent because the fourth base remains gln; (c) met-cys-tyr-gln-asn- = nonsense because the insertion generated a stop codon.

Chapter 8

Self-Test

2. E; **4.** B; **6.** A; **8.** E; **10.** B; **12.** E; **14.** B; **16.** E.

Questions for Thought and Discussion

2. It might have been an adaptive advantage because some fragments might contain genes of use to the recipient cell. Such genes could be for antibiotic resistance or the ability to breakdown a nutrient source.
4. Reproduction leads to "more" cells, while recombination leads to "different" cells.

Applications

2. The plasmid only has a restriction cut site for *Eco*R1. Therefore, mixing plasmid with *Eco*R1 will produce a linear plasmid, while mixing the plasmid with *Pvu*1 will cause no change.

Review

2. PALINDROME; **4.** BACTERIOPHAGE; **6.** PNEUMOCOCCUS; **8.** LYSOGENY; **10.** INTERFERON; **12.** CONJUGATION; **14.** GENOME.

Chapter 9

Self-Test

2. A; **4.** D; **6.** E; **8.** B; **10.** B; **12.** B; **14.** C.

Questions for Thought and Discussion

2. This question assumes that a pathogen is in an advantageous condition, a situation with which you might take issue. Perhaps the virus places the bacillus in a difficult position because pathogenicity is not always desirable, especially after the host dies. On the other hand, the virus permits active growth in the tissues, a situation not possible without the toxin.
4. Rheumatic fever is a complication of streptococcal pharyngitis (strep throat).

Applications

2. The spread of diphtheria in the former Soviet Union was probably the result of the population movements and migrations associated with the breakup of the country and the reestablishment of new countries. At such a time, concern for one's life is primary, and the opportunity for receiving immunizations is minimal. A large population of children had probably emerged during the previous years with virtually no protection against diphtheria. The major effort to stop the epidemic would be to have mass vaccinations of children and adults.
4. It would be wise to target nursing homes and long-term care facilities. Impress upon them the numbers of older adults who die each year from a vaccine-preventable disease. Tell them it is a one-time immunization and they can receive it when they get their flu vaccination.

Review

2. tuberculosis; **4.** legionellosis; **6.** Miliary; **8.** pneumococcus; **10.** cell wall; **12.** tap; **14.** positive.

Chapter 10

Self-Test

2. D; **4.** C; **6.** D; **8.** C; **10.** B; **12.** A.

Questions for Thought and Discussion

2. Chicken barbecues are much more common in summertime. What other poultry-related products are related more to summer than to other seasons?
4. To make the cheese filling, one uses ricotta, mozzarella, and Parmesan cheeses with

appropriate spices (lots of basil and oregano), and a couple of fresh eggs to "hold things together." The association of *Salmonella* with eggs is well established, and it is quite likely that the eggs used by the manufacturer were contaminated.

Applications

2. Mushrooms are cultivated in dark, humid caves on trays of fresh manure and other organic matter. Clostridial spores probably enter the manure from the animal's intestine and cling to the mushrooms as they grow tall. When the mushrooms are bottled, anaerobic conditions can be established following a failed sterilization procedure with steam. Now the toxin is produced, and assuming the mushrooms are not boiled, the toxin will pass to consumers.
4. The disease was brucellosis. All recovered and returned to work. The health department also made several recommendations to prevent further outbreaks, including the use of rubber gloves, face shields, and negative air pressure on the floor.

Review

2. dysentery; 4. symptoms; 6. etiology; 8. serotypes; 10. intoxicated.

Chapter 11

Self-Test

2. D; 4. D; 6. E; 8. C; 10. B; 12. D; 14. D; 16. B; 18. A; 20. C.

Questions for Thought and Discussion

2. Endemic typhus is caused by rickettsiae transmitted by the fleas of rodents. When the rat poison was used, the rats died and the fleas quickly left the dead animal bodies, only to infest the nearby human bodies.
4. Read MicroFocus 18.3.
6. When leaves and debris pile at the curbside, soilborne arthropods flourish, including ticks that commonly occur on grass and leaves. A child's chance playing in the leaves may bring him/her in contact with the ticks, and any of the three diseases indicated may follow.

Applications

2. During civil war, there is little regard for sanitation, and lice flourish on the skin and in the hair of humans. The lice carry and transmit the bacterial agents of typhus.
4. It was Rocky Mountain spotted fever. The progression of the rash from extremities to body trunk was the key.

Review

2. D; 4. E; 6. L; 8. H; 10. F; 12. H; 14. I; 16. A; 18. D; 20. G.

Review—Diseases and Their Agents

2. Plague; 4. *typhi*; 6. *Clostridium*; 8. gas gangrene; 10. *Rickettsia*; 12. *Bartonella*; 14. *akari*.

Chapter 12

Self-Test

2. E; 4. B; 6. B; 8. E; 10. D; 12. C; 14. B; 16. C.

Questions for Thought and Discussion

2. Impetigo is a skin disease often caused by staphylococci and occurring most commonly among children. Children tend to have more contact with one another during the summer months than during any other time of the year, and the skin is often unclothed at this time.
4. Historic writings, public images and pictures, and stories about leprosy have all added to the social stigma accompanying the disease.
6. I can see pluses and minuses in this method. What do students think?

Applications

2. The disease was leprosy (Hansen disease).
4. At our college, we have had the hollow tubes removed because the practical benefit does not seem to outweigh the danger of contracting conjunctivitis (pinkeye), particularly when so many different students use the microscope. Disinfection is a useful alternative to removal, but this is not always possible.
6. The woman probably had syphilis. Congenital syphilis often leads to miscarriage after the fourth month. The symptoms in the newborn, including Hutchinson's triad, also point to syphilis.

Review

2. Pinkeye (conjunctivitis); 4. cell wall; 6. bacterial vaginosis (vaginitis); 8. Carbuncles; 10. *Escherichia coli*; 12. chain of transmission.

Chapter 13

Self-Test

2. D; 4. A; 6. B; 8. C; 10. E; 12. D; 14. E; 16. C.

Questions for Thought and Discussion

2. Oncogenes appear to function in the production of substances that are involved in the metabolism of body cells ("Jekyll"). When these substances are overproduced, however, they may transform the cell into a tumor cell ("Hyde").
4. Reproduction has a connotation in biology that implies the generation of new individuals by asexual or sexual processes. Although new individuals are generated in viral replication, the process is neither asexual nor sexual, but a completely separate process seen nowhere else in biology.
6. Viroids and prions are of special interest because they appear to lack protein and nucleic acid, respectively. Even at the level of viruses, both components seem to be necessary. Viruses do or do not conform to other living things, depending on how one defines living things. Discussions like these are fruitful because they challenge the perception of what is a living thing.

Applications

2. The brain tissue would contain Negri bodies, a cytopathic effect of rabies replication.

Review

2. BACTERIOPHAGE; 4. ACYCLOVIR; 6. TUMOR; 8. VIRULENT; 10. GENOME; 12. AMANTADINE; 14. PRIONS.

Chapter 14

Self-Test

2. D; 4. C; 6. A; 8. A; 10. D; 12. A; 14. D; 16. E.

Questions for Thought and Discussion

2. Herpes simplex viruses are associated with tumors of the cervix, so there is great reluctance to use these viruses in a vaccine. However, genetically engineered viral fragments (such as those used for hepatitis B) or genetically engineered viral antigens remain viable alternatives for a vaccine.
4. Children are easier to reach at 15 months for immunization, but after 20 years or so, the immunity has probably worn thin. By contrast, teenagers may be harder to reach and certify for immunization, but the level of immunity will be much higher.

Applications

2. The evidence points to chickenpox.

Review

2. contact/blisters/emotional stress; 4. Reye/testes/orchitis; 6. influenza/vaccine/bacteria; 8. coronavirus/enveloped/person-to-person; 10. icosahedral/HSV-2/thin/weeks/stress.

Chapter 15

Self-Test

2. A; 4. C; 6. E; 8. A; 10. A; 12. E; 14. D; 16. E.

Questions for Thought and Discussion

2. Disease reflects a competition between host and parasite. If the pathogen overcomes the body defenses, disease ensues; but if the body defenses overcome the pathogen, the latter is driven away. In each case, the pathogen and defenses compete until one wins out. AIDS is dramatically different. HIV eliminates body defenses by destroying portions of the body's immune system. Left defenseless, the body is subjected to "marauding pathogens" in the form of opportunistic microorganisms. Few other diseases take this approach.
4. The viral disease anticipated is hantavirus pulmonary syndrome. Residents are advised to avoid rodent nests, use disinfectant when and after handling a dead rodent, and to go to the local hospital quickly if symptoms develop.

Applications

2. The Sicilian barbers in the study used traditional razors that are nondisposable and unsterilized. They probably shaved themselves with the razors after shaving their patrons and transmitted hepatitis C viruses in bits of blood remaining on the razor. No doubt they also contributed to a high incidence of hepatitis C in their patrons.
4. This anomaly is probably the result of the cross-reactivity of the antibodies for yellow fever and dengue fever. The viruses are very similar, so the antibodies produced against one virus yield protection against the other virus.

Review

2. F(no symptoms); 4. F(rabies); 6. F(gastroenteritis); 8. T; 10. F(polio); 12. T; 14. T; 16. T.

Chapter 16

Self-Test

2. E; 4. C; 6. D; 8. C; 10. B; 12. C; 14. E; 16. B.

Questions for Thought and Discussion

2. Lime raised the pH of Mr. A. 's soil and prevented fungal disease in the turf. The mushrooms in Mr. B.'s lawn in June are a signal that fungi can grow in the acidic environment, and the brown spots are proof. Mr. B. should invest in lime next year.
4. According to public health epidemiologists, the outbreak of coccidioidomycosis was most likely due to dust made airborne by the earthquake. The disease is not transmitted from person to person but from the inhalation of airborne *Coccidioides immitis*.

Applications

2. The tests and information point to *Histoplasma capsulatum* and histoplasmosis.

Review

2. P; 4. O; 6. R; 8. O; 10. G; 12. K; 14. B; 16. N.

Chapter 17

Self-Test

2. C; 4. E; 6. A; 8. B; 10. D; 12. D; 14. A; 16. C.

Questions for Thought and Discussion

2. As the borders of Rome expanded, Romans ventured into far-off lands where mosquitoes thrived and where the malaria parasite was prevalent. Infected Romans probably brought the disease back to Rome, and the mosquito populations of the aqueducts and pools propagated the parasite and spread the disease.
4. On a global scale, diseases as these "impede national and individual development, make fertile land inhospitable, impair intellectual and physical growth, and exact a huge cost in treatment and control programs." Solutions are straightforward: Develop new drugs, vaccines, diagnostic tests, and control methods. Can you suggest any novel approaches?
6. The concept of studying parasitology to appreciate the web of life is intriguing and worthy of note. Hundreds of thousands of individuals are infected with multicellular parasites. The relationship is benign in huge numbers of cases but parasitical in many others, as this chapter shows. Perhaps a geographical summary of the parasites might help us appreciate the global perspective for parasites.

Applications

2. The disease is American trypanosomiasis (Chagas disease).

Review

2. C; 4. B; 6. B; 8. A; 10. A,B,C; 12. A,B; 14. B,C.

Review—Parasite Life Cycles

2. See figure 17.18

Chapter 18

Self-Test

2. D; 4. D; 6. B; 8. C; 10. A; 12. E; 14. D.

Questions for Thought and Discussion

2. Cholera bacilli are susceptible to stomach acid, so the bacilli may never have reached von Pettenkofer's intestine to cause disease. It is also conceivable that von Pettenkofer's anxiety caused his stomach to put out a higher-than-normal amount of acid, which would further reduce the bacterial population. Perhaps the laboratory isolate was less infectious than one directly from a diseased patient.
4. If a virus kills its victims quickly, there is no way it will be around long enough to spread to new hosts. Of course, this theory assumes that the virus does not exist anywhere in nature where it can be contracted easily, and that it relies on human-to-human transfer to remain in existence. In this particular case, the more virulent the virus, the less likely it is to be a slate-wiper (i.e., a disease that kills huge numbers of victims).

Applications

2. At the emergency room, your friend was given a "tetanus shot," a preparation of tetanus toxoid to induce his immune system to produce tetanus antitoxins. These antitoxins would protect him against tetanus toxins, since tetanus spores had probably entered the wound from the soil.
4. You could make a case either way. You could envisage an epidemic disease as the greater threat because of its explosiveness, but a case also could be made for an endemic disease since it often is hidden from detection and strikes without warning.

Review
2. T; 4. F(infection); 6. T; 8. F(mechanical); 10. F(antitoxins); 12. F(coagulase); 14. T; 16. F(acute); 18. F(low).

Chapter 19

Self-Test
2. D; 4. E; 6. D; 8. E; 10. A; 12. E; 14. C.

Questions for Thought and Discussion
2. There are chemical signals that stimulate phagocytosis, the defensins, interferons, inflammatory signals, signals triggering fever, signals involved with complement activation, and chemical interactions with the toll-like receptors.

Applications
2. The tears contain lysozyme, which is active on the cell walls of gram-positive bacteria. In addition, the flow of tears carries away any pathogens present.

Review
2. M; 4. Q; 6. E; 8. J; 10. N; 12. B; 14. P.

Chapter 20

Self-Test
2. B; 4. E; 6. E; 8. A; 10. D; 12. E; 14. C.

Questions for Thought and Discussion
2. In short, T cells carry receptor proteins that resemble antibodies. However, B cells do carry IgD and antibodies on their surface.

Applications
2. Once the techniques are perfected and the roles of nanorobots have been demonstrated, this "science fiction" technology may be very useful.

Review
2. F(haptens); 4. F(antigen binding site, or Fab fragment); 6. F(plasma cells); 8. T; 10. F(two); 12. F(macrophages); 14. T; 16. T; 18. T; 20. F(perforins, granzymes); 22. T; 24. F(IgD).

Chapter 21

Self-Test
2. C; 4. A; 6. A; 8. A; 10. B; 12. C; 14. E.

Questions for Thought and Discussion
2. It would seem to be a good idea, but it will take lots of money to implement. And the willingness for the American taxpayer to part with the funds will depend in part on how serious the threat of disease is perceived to be. Students should continue the debate from here.
4. Children between the ages of 5 and 15 generally eat well and are adequately clothed. Their cells are actively metabolizing, and their tissues are growing rapidly. Their immune systems are fully functional. They enjoy exercise for physical fitness and are relatively free of cares for mental fitness. At first signs of disease, they are given home or hospital care.

Applications
2. Lysis of the red blood cells would occur because without syphilis antigen, no activity would take place in the test system, and the complement would be left over for the indicator system. Without the mistake, the expected result would be no lysis.
4. With some insight and imagination, a case can be made for each type of immunity as being safest to obtain. Similarly, reasons may be offered for each type as being most helpful. Discussions such as these put immunity into perspective and help you make choices supported by what you have learned. Ultimately, I would think that the individual situation would dictate the choice.

Review
2. active/passive; 4. diphtheria/pertussis/tetanus; 6. hyperimmune/antiserum; 8. transplacental passage/breast feeding; 10. diphtheria/tetanus; 12. gastrointestinal/respiratory; 14. tetanus/diphtheria.

Chapter 22

Self-Test
2. C; 4. E; 6. E; 8. B; 10. A; 12. A; 14. B; 16. C.

Questions for Thought and Discussion
2. She is suffering an allergic reaction. Insufficient sensitization had taken place after the previous injections but now, after the fifth injection, she was fully sensitized and basophils and mast cells began to degranulate. Most allergists recommend that a patient spend several minutes in the office after injections to be available in case an allergic reaction takes place.
4. It is interesting to question whether the immune system is actually protecting the body during immune disorders. Allergies

can be interpreted as protective mechanisms for ridding the body of antigens, and the theory can be extended to other types of hypersensitivity, as well as to transplants and tumors. Autoimmune diseases stand in stark contrast because the body appears to be attacking itself. An oxymoron is two terms that do not fit together (the dictionary definition is "two mutually exclusive juxtaposed words"). "Immune disorder" appears to be an oxymoron because *immune* means "free of" and a disorder is a problem. Therefore, how can you have a problem that you don't have?

Applications

2. Formaldehyde used for preservation of the animals is probably causing a contact dermatitis. The sensitivity began to develop during the first two exposures, and now, during the third exposure, it is manifesting itself. Rubber gloves would be helpful for protecting the hands.

Review

2. DIGEORGE; 4. JOINTS; 6. CYTOTOXICITY; 8. ANTIGEN; 10. TUBERCULOSIS; 12. POSITIVE; 14. RASH.

Chapter 23

Self-Test

2. B; 4. B; 6. A; 8. C; 10. A; 12. E; 14. C; 16. E.

Questions for Thought and Discussion

2. A suspicious person might inquire what will happen within 30 days. Will spontaneous generation take place? Is it possible that the contents were sterilized at the manufacturing plant, but that the porous container is now permitting airborne microorganisms to enter? If so, then the product was once sterilized but is now contaminated, and evidence of contamination will appear by the expiration date.
4. Pasteurization merely implies the destruction of pathogenic microorganisms from milk or other liquid. The process does not affect bacterial endospores and it may be tolerated by protozoal or worm cysts. Therefore, a pasteurized product cannot be considered sterilized. By contrast, a sterilized product contains no life form of any type (including viruses) and is considered both pasteurized and sterilized.
6. One effective way of treating the thermometer would be to wash it well in hot soapy water, rinse it thoroughly, and immerse it in a tray of ethyl alcohol or rubbing alcohol for a minimum of 10 minutes. Admittedly, this would require an interval of time before taking the next child's temperature, but the wait might be worth it, especially if the second child was not already sick.

Applications

2. Some possibilities for sterilizing agents in an average household might be a pressure cooker, boiling water, wads of sterilized cotton used for wounds, an ultrasonic device for cleaning dental plates, a microwave oven, filtering material from an aquarium or swimming pool, or a steam sterilizer used for an infant's formula.
4. Some suggestions: a spoonful of bleach in a pail of water and several rinses with the diluted bleach; if the water reservoir is detachable, a scrubbing with a chlorine cleanser; a wipe-down with a disinfectant.

Review—Physical Methods

2. MEMBRANE; 4. DENATURATION; 6. GAMMA; 8. PRESSURE; 10. ULTRAVIOLET; 12. OSMOSIS; 14. BUNSEN.

Review—Chemical Methods

2. D; 4. B; 6. C; 8. A; 10. H; 12. J; 14. K; 16. I; 18. A; 20. G.

Chapter 24

Self-Test

2. C; 4. C; 6. B; 8. A; 10. E; 12. B; 14. E; 16. D.

Questions for Thought and Discussion

2. The soil is a highly volatile and complex environment where organisms compete for survival. Being able to produce an antibiotic gives an organism a selective advantage in this competition. *Streptomyces* species may owe their survival to their ability to destroy other organisms via antibiotic production.
4. An antibiotic that is not absorbed from the gastrointestinal tract is useful in treating intestinal tract infections. One that is rapidly expelled in the urine would be valuable for urinary tract infections. In these examples, you should be able to understand that the characteristics of a drug can be adapted to the use of the drug.

6. The three scientists worked together to bring about the development of penicillin. Fleming, the microbiologist, discovered the antibiotic; Florey, the physiologist, studied its effects in tissue; and Chain, the biochemist, performed the purification.

Applications

2. The arrival of antibiotics and modern medical practices in Nepal typifies how medical advances can disrupt the lives of a population. With more mouths to feed, forests had to be cleared, and great pressure was placed in preparing the land for crops. Living space soon became a premium, and natural resources, such as water supplies, were tapped to their limit. Greater populations also meant greater sanitation problems and, consequently, more opportunity for the spread of disease. Discussions such as these help us understand the negative aspects of medical advances.

Review

2. T; **4.** T; **6.** T; **8.** T; **10.** T; **12.** F(sulfonamides); **14.** F(aminoglycosides, chloramphenicol, tetracyclines, macrolides, etc.); **16.** T; **18.** F(cephalosporins); **20.** F(rifampin).

Identification

2. Membrane; **4.** DNA; **6.** Cell wall; **8.** Cell wall; **10.** Cell wall.

Chapter 25

Self-Test

2. D; **4.** A; **6.** A; **8.** B; **10.** C; **12.** A; **14.** B; **16.** E.

Questions for Thought and Discussion

2. Blue cheese is made with *Penicillium roqueforti*. On very rare occasions, eating food with penicillin in it can cause reactions (hives, itches) for penicillin-sensitive individuals.
4. The liver would probably be the better choice because it will spoil more rapidly than the steak. Liver is an organ meat, with a looser tissue consistency and richer blood supply than muscle tissue. Contamination is therefore more probable in the liver.
6. Individuals who pick fruits and vegetables in the fields should be vaccinated against any diseases they could transmit. Toilet facilities should be made available to them, and they should be discouraged from working if they are ill. The water used to wash the produce should be purified. Students should recommend other such regulations.

Applications

2. The correct sequence would be to prepare the salad before the chicken. *Salmonella* serotypes may be present in the chicken, and cross-contamination to the salad may take place. Cooking eliminates *Salmonella* in chicken, but salad is eaten raw and is potentially dangerous.
4. Precautions to take when gathering apples include not picking apples off the ground, just off the tree. Also, move far into the orchard away from any animals grazing nearby because their feces may have contaminated the soil. Moreover, wash and brush the apples before pressing them and briefly boil the cider before drinking it, or use a preservative to prepare the cider for storage.

Review

2. *LACTOBACILLUS*; **4.** *SERRATIA*; **6.** *PROTEUS*; **8.** *BACILLUS*; **10.** *LEUCONOSTOC*; **12.** *CLOSTRIDIUM*; **14.** *ENTEROBACTER*.

Chapter 26

Self-Test

2. E; **4.** E; **6.** C; **8.** B; **10.** C; **12.** A; **14.** B.

Questions for Thought and Discussion

2. The pathogens of disease are probably ill equipped to compete with vast populations of other organisms in the soil environment. Possibly they do not possess the enzyme systems for soil metabolites nor the structural components to contend with environmental fluctuations in the soil. Antibiotic-producing soil organisms are also present.
4. The grass is greener over the septic tank because the organic-rich water flowing from it provides a natural lawn fertilizer. In the wintertime, the hot water from a cesspool or septic tank warms the soil and melts the snow over it. To locate a cesspool or septic tank in climates that have snow, one need only watch to see where the snow melts first.

Applications

2. The presence of high coliform counts indicated that the water had been contaminated with sewage and that typhoid fever, bacterial dysentery, amoebiasis, hepatitis, or other serious diseases were imminent. Certainly,

the danger was equivalent to that posed by Legionnaires' disease.

4. The greater hazard probably exists in the smaller wading pool. This is where diaper-clad toddlers spread microorganisms to the water, where children are more likely to take a mouthful of water, where flatulence may eliminate microorganisms from the intestine, and where greater stagnation of the water may occur. Park officials may seek to limit disease transmission by providing constant water exchange and aeration, keeping vigilance on the chlorine content, watching for babies with soiled diapers, and posting signs advising parents not to allow children in the water if the children have diarrhea.

Review

2. B; **4.** A,B; **6.** A,B; **8.** A; **10.** A,C; **12.** A,C; **14.** B,C.

Chapter 27

Self-Test

2. C; **4.** C; **6.** D; **8.** A; **10.** B; **12.** D; **14.** A; **16.** B.

Questions for Thought and Discussion

2. They indicate that beer was an important drink among the Pilgrims. Since water would become contaminated, alcoholic beverages would more likely stay drinkable.

4. Perhaps a defect in the enzyme system prevents the metabolism of the vitamin to the next step; perhaps the organisms live in a symbiotic or synergistic relationship (biofilm) with other organisms that use the vitamin.

Applications

2. The place to look for organisms for bioremediation would be in extreme environments. For example, the deep subterranean regions would yield organisms that live in oil, and the ocean would contain organisms that survive in high-salt environments. Ultimately, solutions of the pollutants would have to be assayed for organisms. Other examples given in this chapter should give students leads on where to find the organisms.

4. One should realize by this point that anything is possible as long as one has vision and imagination. Antibiotic production in the body is conceivable, and you should outline the problems that require attention before this happens. Certainly, the fear of consuming microorganisms would have to be overcome.

Review

2. O; **4.** V; **6.** N; **8.** F; **10.** D; **12.** J; **14.** S; **16.** P; **18.** R; **20.** D.

Appendix D: Answers to MicroInquiry Questions

MICROINQUIRY 2

2a. Uracil is the base specific to DNA and thymine is the base specific to DNA.

2b. Radioactively label uracil with isotope of one element (say ^{3}H) and radioactively label thymine with the isotope of another element (^{14}C).

2c. You cannot use sulfur (^{35}S) to label nucleic acids because this element is only found in a few amino acids of proteins.

2d. You should have concluded that the chemical does affect growing bacterial cells. Specifically, it affects protein synthesis because as time passes, the radioactive label in proteins levels off as shown in the figure. This means that proteins are no longer being made and, therefore, the ^{35}S cannot be incorporated. DNA and RNA are not affected since there is continued incorporation of radioactive isotope into new DNA and RNA molecules.

MICROINQUIRY 5

5e. The results show in which tubes acid was produced. Therefore, tubes 4, 7, and 9 are *S. aureus*.

MICROINQUIRY 7

When tryptophan is absent from the growth medium, the repressor protein of the *trp* operon fails to bind to the operator (Figure A). The RNA polymerase is free to transcribe the five genes coding for tryptophan synthesis. When tryptophan is added to the growth medium, the cells preferentially use what is available rather than expend energy to make their own amino acids.

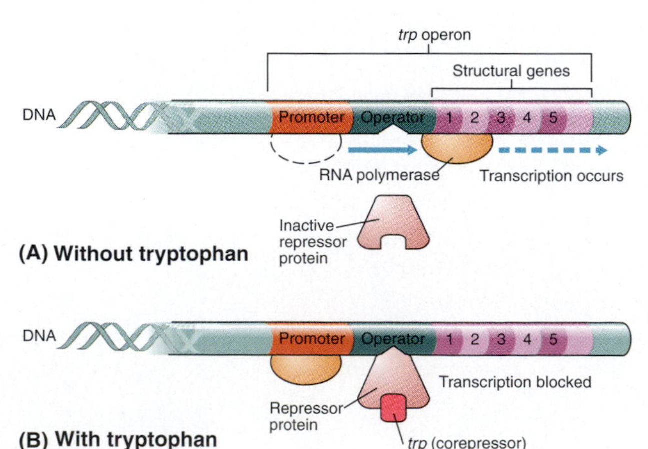

So, in the cell tryptophan binds to the repressor protein, which can now bind to the operator and block transcription by the RNA polymerase (Figure B). In this example, tryptophan is said to be a corepressor because it is needed to repress or "turn off" gene transcription.

MICROINQUIRY 8

Bacterial colonies without a plasmid and those containing the recombinant plasmid will not grow on (a) because they lack tetracycline resistance genes. Only the bacterial cells with intact plasmids (tetR) grow. So, we have identified those clones that lack the recombinant plasmid. Bacterial colonies lacking plasmids will not grow on medium (b) because they are sensitive to ampicillin. Colonies containing the recombinant plasmid and intact plasmid will grow because they contain the ampR gene. Colonies with recombinant (insulin-containing) plasmids are identified because they grow on ampicillin-containing but not on tetracycline-containing growth media.

MICROINQUIRY 9

Case 1

9.1a. Often a tubercle is visible in a chest X ray.

9.1b. Identification of *M. tuberculosis* by acid-fast staining from sputum usually is sufficient for a diagnosis of TB because these bacterial cells are not normally found in respiratory secretions.

9.1c. Skin becomes thick with a raised red welt developing within 48 to 72 hours.

9.1d. A tuberculin skin test does not necessarily mean the person has tuberculosis. It simply indicates the person has been exposed to *Mycobacterium tuberculosis*, been immunized recently, or had a previous tuberculin skin test.

9.1e. Tuberculosis; *Mycobacterium tuberculosis*.

9.1f. Most likely from his former roommate.

9.1g. Tuberculosis is more virulent in HIV-infected patients because their immune system is compromised from fighting the HIV infection.

9.1h. The patient was on INH for an extended period because of his HIV infection. With a lowered immune system, the patient's immune system will not mount a strong response to the TB infection. Drugs, such as INH, are the best hope for control and recovery.

Case 2

9.2a. *Corynebacterium diphtheriae.*

9.2b. Dead tissue accumulates with mucus, white blood cells, and fibrous material (pseudomembrane) in the throat. Respiratory blockage can result, which can lead to death.

9.2c. Having the proper vaccinations, in this case the DTaP vaccine.

9.2d. Treatment requires both antibiotics to kill the bacterial cells and antitoxins to neutralize the diphtheria toxins.

Case 3

9.3a. *Streptococcus pneumoniae.*

9.3b. The patient's underlying medical conditions (heavy smoker and alcoholic) place him at high risk for the disease.

9.3c. Vaccination with the pneumococcal 23-valent vaccine.

9.3d. Polysaccharide capsule and toxin production.

9.3e. Alpha-hemolytic.

MICROINQUIRY 10

Case 1

10.1a. *Campylobacter jejuni.*

10.1b. Drinking raw milk or contaminated water; consumption of contaminated food, especially chicken.

10.1c. Foods, especially chicken, contaminated during processing or preparation.

10.1d. Guillain-Barré syndrome where immune system attacks nerves.

10.1e. Erythromycin.

Case 2

10.2a. *Salmonella typhi, Vibrio cholerae,* and *Escherichia coli.*

10.2b. All gram-positive bacteria.

10.2c. Typhoid fever.

10.2d. Contaminated food or water; raw milk.

10.2e. Don't drink raw milk; drink purified water; eat cooked foods.

Case 3

10.3a. *Listeria monocytogenes.*

10.3b. One of the forms of listeriosis is meningoencephalitis, which is characterized by headache, stiff neck, and delirium.

10.3c. Penicillin or tetracycline.

10.3d. They should avoid foods (deli meats, soft cheeses) that potentially could be contaminated with the bacterial species. People with a strong immune system can fight off the bacterium.

10.3e. Psychrotrophic refers to the ability of the bacterial cells to grow at refrigerator temperatures. Therefore, refrigeration does not stop growth of *Listeria* in contaminated refrigerated foods, like feta cheese.

Case 4

10.4a. *Escherichia coli* O157:H7.

10.4b. Bloody diarrhea and hamburger meat that had been in the refrigerator for "some time."

10.4c. Other *E. coli* strains because they lack the bloody diarrhea symptom.

10.4d. It can lead to complications called hemorrhagic colitis and hemolytic uremic syndrome.

10.4e. On MacConkey agar, *Escherichia coli* O157:H7 produces white colonies while other strains produce red or pink colonies.

MICROINQUIRY 11

Case 1

11.1a. Headache, rash on thighs, fever, hiking near Seattle.

11.1b. Lyme disease; *Borrelia burgdorferi.*

11.1c. Tick bite.

11.1d. Without treatment, infection can lead to an early disseminated stage characterized by meningitis, facial palsy, and peripheral nerve disorders. Joint and muscle pain also can occur.

11.1e. Wear protective clothing and avoid tick-infested areas. If bitten by a tick, remove whole tick and wash the skin wound with soap and water, and apply antiseptic.

Case 2

11.2a. *Clostridium tetani;* tetanus.

11.2b. Sedatives and muscle relaxants; penicillin and tetanus antitoxin.

11.2c. No, because symptoms develop rapidly and the patient could die waiting for lab results.

11.2d. If the spores entered via a foot wound, they will germinate and growth will produce dead tissue.

11.2e. See 10.2d.

Case 3

11.3a. *Rickettsia rickettsii* and *Ehrlichia chaffeensis.*

11.3b. *R. rickettsii.* A pink rash on palms and feet; tick bite; geography (South Carolina).

11.3c. Lyme disease, relapsing fever, tularemia, and ehrlichiosis.

11.3d. Weil-Felix test.

11.3e. *E. chaffeensis* does not usually cause a rash and it is associated with a lowered WBC count, not a raised count.

Case 4

11.4a. Gas gangrene; *Clostridium perfringens.*

11.4b. Anaerobic conditions, as the bacterial species is an anaerobe.

11.4c. Debridement.

11.4d. Use of a hyperbaric chamber, which introduces oxygen gas into necrotic tissues. Oxygen gas will kill *C. perfringens* cells.

MICROINQUIRY 12

Case 1

12.1a. *E. coli, Proteus mirabilis, Enterobacter faecalis, Klebsiella pneumoniae, Pseudomonas aeruginosa, Staphylococcus aureus.*

12.1b. Due to anatomical differences. The female urethra is relatively short; proximity of urethra to anus.

12.1c. Infections of the urethra, bladder, and possibly the kidneys.

12.1d. Avoiding tight-fitting clothes and urinating as soon after intercourse as possible.

12.1e. Biofilms are a key factor. Bacterial cells are protected from immune attack by a slimy coating. The coating also makes antibiotic therapy difficult since the drug does not penetrate the biofilm.

Case 2

12.2a. *Chlamydia trachomatis* and *Neisseria gonorrhoeae.*

12.2b. *C. trachomatis* is an obligate pathogen and only grows in tissue culture. The reproductive cycle is described on pages 339–340.

12.2c. Laboratory tests include the use of fluorescent antibodies on a cervical swab or an immunoassay test.

12.2d. A *C. trachomatis* infection can be spread to others through sexual intercourse.

12.2e. Recovery from a gonorrhea infection does not generate lifelong immunity because of the weak immune response to the bacterium.

Case 3

12.3a. *Pasteurella multocida.*

12.3b. Cat bite, swollen and painful wrist, tenderness at the site, small puncture wound, and small abscess. Gram-negative rods.

12.3c. The pharynx.

12.3d. Limiting contact with the animal and washing the site thoroughly.

Case 4

12.4a. Gonorrhea; *Neisseria gonorrhoeae.*

12.4b. Gonorrhea can be spread by sexual intercourse. Complications in females include salpingitis and pelvic inflammatory disease.

12.4c. HIV, syphilis, and chlamydial infections, because they all are sexually transmitted diseases.

12.4d. Although these are additional signs of gonorrhea in males, the infection has not affected these tissues.

12.4e. Ceftriaxone or cefixime.

MICROINQUIRY 13

13.1a. 10 hours. During this time, the viral nucleic acid is being replicated and new capsid parts are being synthesized.

13.1b. 15 hours. During this time, the viral nucleic acid is being replicated, new capsid parts are being synthesized, and nucleic acid and capsids are being assembled into new virions (maturation stage).

13.1c. 10^8 virions are released from 10^5 cells in culture. Therefore, the burst size is 10^3 virions per infected cell.

13.1d. The decline reflects the loss of phages as they are slowly taken into host cells.

MICROINQUIRY 15

Case 1

15.1a. Hepatitis C; hepatitis C virus.

15.1b. History of intravenous drug abuse, nausea, vomiting, fever, headache, abdominal pain, and slight jaundice. Also, negative HBsAg and serum antibodies to HAV.

15.1c. This is indicative of liver damage, which would be the infection site for the virus.

15.1d. Being a drug user, contaminated needles are the probable source.

15.1e. Most likely hepatitis B and HIV, both of which can be transmitted by contaminated needles.

Case 2

15.2a. Infectious mononucleosis; Epstein-Barr virus.

15.2b. The elevated B lymphocyte count is because the Epstein-Barr virus infects B lymphocytes and the immune system is attempting to replace the damaged cells.

15.2c. Heart defects, paralysis of the face, and rupture of the spleen.

15.2d. Burkitt lymphoma. The malarial parasite may stimulate tumor development.

Case 3

15.3a. Hepatitis A and E.

15.3b. Hepatitis A because of the jaundice, short incubation period, initial symptoms, and travel history.

15.3c. A transmission route could be from contaminated water or food eaten at local restaurants.

15.3d. Hepatitis A antibodies in serum.

15.3e. Administering hepatitis A immune globulin.

Case 4

15.4a. Dengue fever.

15.4b. Clues included fever, backache, headache, bone and joint pain, and eye pain. Her trip to Bangladesh (where dengue is prevalent) and mosquito bites also were clues.

15.4c. She has a viral disease for which antibiotics are useless.

15.4d. Returning to Southeast Asia puts her at risk of being infected by another strain of dengue virus, which could lead to a severe or deadly illness called dengue hemorrhagic fever.

MICROINQUIRY 16

16.1 Fungi are more closely related to animals as many of the organismal, cellular, and biochemical characteristics listed are common only to animals and fungi.

16.2. See Figure 1.20.

MICROINQUIRY 17

Answers following each question.

MICROINQUIRY 18

18.1a. Approximately 110,000 cases in 1993 and 45,000 cases in 2003.

18.1b. Prevalence: 1993 = 0.09%; 2003 = 0.13%.

18.2a. Continuous common source epidemic. Plateau indicates person-to-person transmission.

18.2b. Possibly three plateaus, suggesting spread within different communities or groups of individuals.

18.3a. Highest reported cases (≥15) in Northeast states and upper Midwest. Fewer cases (1–14) in Eastern and Pacific states.

18.3b. Perhaps the infected host is newly introduced in this area. Another possibility is the area recently has become an area that is frequently visited by hikers or outdoors people. Depending on how the data are reported, the cases could have been the result of visiting another endemic area in the United States, but not reported until returning home.

18.4a. Highest number of cases in infants less than 1 year old, with higher incidence in 10–14 and 15–19 age groups.

18.4b. Vaccinations would eliminate the high numbers of infant cases, and booster shots would reduce the cases in all age groups.

MICROINQUIRY 19

Case 1

19.1a. The alternate complement pathway can be stimulated by pathogen surfaces while the classical pathway is stimulated by antigen-antibody complexes. Membrane attack complexes would be of significance only with nonencapsulated bacteria since these complement proteins assemble into membrane complexes on the cell membrane of the bacterium.

19.1b. The alternate pathway is activated spontaneously by the bacterial surface and can assist in an elevated phagocytic response by phagocytes. Membrane attack complexes can stimulate the destruction of nonencapsulated bacteria.

Case 2

19.3a. C3 is critical for all three of the complement's effector functions: inflammatory response, opsonization and phagocytosis, and membrane attack complex formation.

19.3b. A C3 deficiency cannot trigger any of the functions mentioned above, so the patient will experience more bacterial infections.

Case 3

19.3a. A lack of macrophages, which are a type of phagocyte, would make it hard for the innate immune system to eliminate the virus.

19.3b. Most infectious diseases with which he comes in contact could be a threat since the phagocytosis process carried out by macrophages would be absent.

MICROINQUIRY 21

21.1a. ELISA is a serological test for the presence of antibodies or antigens in serum. In your case, Pat, the hepatitis C kit will assay for the presence or absence of hepatitis C antibodies in the blood sample you sent.

21.1b. To make sure all the reagents are functioning properly, positive and negative controls are run along with your blood sample. A positive control contains antibodies specific to hepatitis C antigen so a positive color reaction will be produced. A negative control lacks hepatitis C antibodies, so no color reaction will develop.

21.1c. If your blood sample tests positive, it means you are seropositive. That is, your blood contains anti-HCV antibodies.

21.1d. A seronegative result would mean you are "probably" not infected with HCV. However, you still should be retested in six months because sometimes it can take that long for the immune system to produce enough antibodies to be detected by ELISA.

21.1e. Do another HCV test in six months. Even if that one is positive, lab errors or some other infection may mimic an HCV antigen. If that ELISA test proves positive, another type of test, called a Western blot, is run that detects the actual presence of HCV in your blood. If all tests come back positive, then I am afraid you are infected with the microbe because when ELISA is used with the Western blot, the results are more than 99.9 percent accurate.

MICROINQUIRY 23

23.1. A = 90 percent kill every 4 minutes, so in 24 minutes there theoretically would be one cell remaining alive. Another 4 minutes should kill the last cell, so 28 minutes would be required. Likewise, B = 70 minutes and C = 126 minutes, both one time interval beyond the point where one cell remains alive.

23.2. A = 4 min.; B = 10 min.; C = 18 min.

23.3. 70°C.

23.4. 108 minutes.

MICROINQUIRY 24

24.1a. Self-explanatory.

24.1b. The drug appears most effective against *S. aureus* as seen by the large rings where growth was inhibited. The drug looks promising against *S. aureus*.

24.1c. The drug looks good at 10 μg but not at 100 μg. Need to do intermediate concentration tests.

24.1d. The control was to make sure there wasn't something in the solution used to dissolve the drug that could cause an adverse cellular effect. Drug testing should continue.

24.2a. A concentration somewhere between 100 mg and 1,000 mg was toxic to the animals. Need to test intermediate concentrations to narrow down the toxicity range.

24.2b. Self-explanatory.

24.3. There are pros and cons to this decision.For example, without a control group, it might not be possible to accurately evaluate the efficacy of the drug. On the other hand, if early drug trials show obvious benefit, the control group could be given the drug. Discuss with classmates.

MICROINQUIRY 25

25.1. Your refrigerator should be kept at 5° C (41° F). Although this temperature will not kill most microbes, it will slow their growth, making it less likely you might spread or could get sick from food stored there.

25.2. Freezing also does not kill all microbes. So, foods like frozen chicken should not be set on the counter to defrost. You risk illness as bacteria grow rapidly at room temperature. **Microwaving the food following package directions or thawing it in the refrigerator will keep any microbes from growing.** Incidentally, that kitchen sponge you used to clean up spilled food juices should be rinsed and sanitized by microwaving it 20 seconds.

25.3. Raw foods can be contaminated with several disease-causing microbes that can cling to a cutting board. So, **clean that board with soap and hot water and then sanitize it with a mild bleach solution** (or put it in the dishwasher).

25.4. **Wash your hands with soap and warm water for about 20 seconds so** you don't deposit harmful bacteria on other surfaces. Also, keep tabby away from kitchen counters and where food is being prepared. Pets carry and spread diseases to foods.

25.5. Your sink drain and garbage disposal can harbor several species of viruses and bacteria. Some sinks can contain more bacteria than in a flushed toilet. So, **every week you should sanitize your drain: pour a solution of 1 teaspoon of chlorine bleach in 1 quart of water down the drain.**

25.6. In the oven, the high temperatures used in cooking kill microbes. However, **cooled leftovers should be refrigerated within two hours after cooking.** Once refrigerated, they are safe to eat for three to five days. Read discussion on pages 773–774.

25.7. Did you clean your kitchen counters after preparing raw food? **Hot, soapy water and a dilute chlorine bleach are recommended.** Hot water and soap alone do not get rid of all possible bacteria.

25.8. Meats should be cooked until there is no red seen and the juices run clear. **Well-done meats that reach 160°F kill food-borne microbes.** Meats and other improperly cooked foods can retain food-borne microbes.

25.9. Most kitchen sponges contain some bacteria that can make people sick. In one study, 20 percent of sponges and dishrags collected from many of the 1,000 kitchens tested in five American cities contained *Salmonella* bacteria, which can cause food poisoning, typhoid fever, and gastrointestinal diseases. **So, microwave that damp kitchen sponge for 20 seconds and wash dishrags.** Also, change to a new sponge often or use a germ-resistant sponge.

25.10. Now that dinner is over, what to do with the dishes? Don't soak them for several hours because the soaking water becomes "nutrient broth" for bacteria. **Either wash them in the dishwasher and air-dry them, or wash them within two hours in hot, soapy water and let them air-dry.**

MICROINQUIRY 26

26.1. Provides a richer growth medium, ensuring the growth of coliforms, if present.

26.2. Do not want to introduce any contaminating bacteria. It is unlikely that pipettes would contain coliforms.

26.3. Sample 4 appears to contain high numbers of coliforms. For this sample, the presumptive test read 5–4–5, indicating an MPN of 430 per 100 ml (confidence limits of 150–1,100 organisms).

26.4. Double-check on the presumptive test results.

26.5. Sample 4 and possibly sample 6 (MPN 14).

26.6. Close off the area of the lake where sample 4 was collected until further testing indicates the water again is safe. Also, investigate the cause for the high coliform numbers.

MICROINQUIRY 27

27.1a. Identify or genetically engineer bacterial species that can break down the thick crude into lighter forms.

27.1b. There are bacterial species that are known to produce enzymes that break down sulfur compounds into water-soluble products while other bacterial species can remove heavy metal contaminants. Since each oil field is slightly different in terms of the heavy crude, different "bacterial cocktails" may be needed. In some cases, this probably would require genetic modifications to give the microbes the precise genes to deal with crude-oil digestion. However, if the right mix of bacterial species can be assembled, the result could be a lighter, cleaner crude for refining.

27.2. Chemolithotrophic ("rock-eating") bacterial species represent groups that survive by using inorganic compounds (minerals) in rocks as their source of nutrients and energy. Bacterial species associated with iron pyrite break the minerals into acidic solutions of iron that then can be used to dissolve out usable forms of copper. Also, using natural bacterial species or genetically engineered ones, it may become possible to convert low-grade ores into a sustainable yield of metals for the future.

27.3. Contaminated soil was excavated and only mineral nutrients had to be added for growth since indigenous microorganisms already existed in the soil. Such biostimulation was performed from December 1989 through September 1991, and approximately 16,000 tons of soil were treated. Through bioremediation, PAH concentrations were reduced by 70 percent.

27.4. From January 1992 through November 1993, approximately 300,000 tons of sediment and subsoil from the lagoon were treated using the indigenous microorganisms that were biostimulated with added oxygen and growth nutrients. The bioremediation process achieved the specified soil cleanup goals for the contaminants within 11 months of treatment. For example, benzene levels were reduced from 608.0 mg/kg to 4.4 mg/kg.

Glossary

This glossary contains concise definitions for microbiological terms and concepts only. **Please refer to the index for specific infectious agents, infectious diseases, and immune disorders.**

A

abscess A circumscribed pus-filled lesion characteristic of staphylococcal skin disease; also called a boil.

abyssal zone The environment at the bottom of oceanic trenches.

acetyl CoA One of the starting compounds for the Krebs cycle.

acid A substance that releases hydrogen ions (H^+) in solution; *see also* base.

acid-fast technique A staining process in which mycobacteria resist decolorization with acid alcohol.

acidic dye A negatively charged colored substance in solution that is used to stain an area around cells.

acidocalcisome A membrane-enclosed cytoplasmic compartment found in some prokaryotic cells that contains calcium, polyphosphate, and other ions.

acidophile A microorganism that grows at acidic pHs below 4.

acme period The phase of a disease during which specific symptoms occur and the disease is at its height.

acquired immunity A response to a specific immune stimulus that involves immune defensive cells and frequently leads to the establishment of host immunity.

actinomycete A soil bacterium that exhibits fungus-like properties when cultivated in the laboratory.

activated sludge Aerated sewage containing microorganisms added to untreated sewage to purify it by accelerating its bacterial decomposition.

activation energy The energy required for a chemical reaction to occur.

active immunity The immune system responds to antigen by producing antibodies and specific lymphocytes.

active site The region of an enzyme where the substrate binds.

active transport An energy-requiring movement of substances from an area of lower concentration across a biological membrane to a region of higher concentration by means of a membrane-spanning carrier protein.

acute disease A disease that develops rapidly, exhibits substantial symptoms, and lasts only a short time.

acute phase protein A defensive blood protein secreted by liver cells that elevates the inflammatory and complement responses to infection.

acyclovir A drug that binds to guanine bases and is used as a topical ointment to treat herpes simplex and injected for herpes encephalitis.

adenoid A mass of secondary lymphoid tissue at the back of the throat that helps in immune system activities.

adenosine diphosphate (ADP) A molecule in cells that is the product of ATP hydrolysis.

adenosine triphosphate (ATP) A molecule in cells that provides most of the energy for metabolism.

adenylyl cyclase An enzyme that catalyzes the conversion of ATP to cyclic adenosine monophosphate (cAMP).

adhesin A protein in bacterial pili that assists in attachment to the surface molecules of cells.

adjuvant An agent added to a vaccine to increase the vaccine's effectiveness.

aerobe An organism that uses oxygen gas (O_2) for metabolism.

aerobic respiration The process for transforming energy to ATP in which the final electron acceptor in the electron transport chain is oxygen gas (O_2).

aerotolerant A bacterium not inhibited by oxygen gas (O_2).

aflatoxin A toxin produced by *Aspergillus flavus* that is cancer causing in vertebrates.

agar A polysaccharide derived from marine seaweed that is used as a solidifying agent in many microbiological culture media.

agar disk diffusion method A procedure for determining bacterial susceptibility to an antibiotic by determining if bacterial growth occurs around an antibiotic disk; also called the Kirby-Bauer test.

agglutination A type of antigen-antibody reaction that results in visible clumps of organisms or other material.

agranulocyte A white blood cell lacking visible granules; includes the lymphocytes and monocytes; *see also* granulocyte.

alcoholic fermentation A catabolic process that forms ethyl alcohol during the reoxidation of NADH to NAD^+ for reuse in glycolysis to generate ATP.

alga (pl. **algae**) An organism in the kingdom Protista that performs photosynthesis.

algaecide A chemical that kills algae.

algal bloom An excessive growth of algae on or near the surface of water, often the result of an oversupply of nutrients from organic pollution.

alginate A sticky substance used as a thickener in foods and beverages.

allergen An antigenic substance that stimulates an allergic reaction in the body.

allograft A tissue graft between two members of the same species, such as between two humans (not identical twins).

alpha (α) helix The spiral structure of a polypeptide consisting of amino acids stabilized by hydrogen bonds.

alpha (α) hemolytic Referring to those bacterial species that when plated on blood agar cause a partial destruction of red blood cells as seen by an olive green color in the agar around colonies.

alternative pathway A complement-activating sequence of steps involving a pathogen cell surface.

Ames test A diagnostic procedure used to detect potential cancer-causing agents in humans by the ability of the agent to cause mutations in bacterial cells.

amino acid An organic acid containing one or more amino groups; the monomers that build proteins in all living cells.

aminoglycoside An antibiotic that contains amino groups bonded to carbohydrate groups that inhibit protein synthesis; examples are gentamicin, streptomycin, and neomycin.

aminoquinoline An antiprotozoal drug that is toxic to the malarial parasite.

amoeboid motion A crawling type of movement caused by the flow of cytoplasm into plasma membrane projections; typical of the amoebozoans.

amoebozoan A protozoan that undergoes a crawling movement by forming cytoplasmic projections into the environment.

anabolism An energy-requiring process involving the synthesis of larger organic compounds from smaller ones; *see also* catabolism.

anaerobe An organism that does not require or cannot use oxygen gas (O_2) for metabolism.

anaerobic respiration The production of ATP where the final electron acceptor is an inorganic molecule other than oxygen gas (O_2); examples include nitrate and sulfate.

animalcule A microscopic organism observed by Leeuwenhoek.

anion An ion with a negative charge; *see also* cation.

antibiotic A substance naturally produced by a few bacterial or fungal species that inhibits or kills other microorganisms.

antibody A highly specific protein produced by the body in response to a foreign substance, such as a bacterium or virus, and capable of binding to the substance.

anticodon A three-base sequence on the tRNA molecule that binds to the codon on the mRNA molecule during translation.

antigen A chemical substance that stimulates the production of antibodies by the body's immune system.

antigen binding site The region on an antibody that binds to an antigen.

antigen presenting cell (APC) A macrophage or dendritic cell that exposes antigen peptide fragments on its surface to T cells.

antigenic determinant A section of an antigen molecule that stimulates antibody formation and to which the antibody binds; also called epitope.

antigenic drift A minor variation over time in the antigenic composition of influenza viruses.

antigenic shift A major change over time in the antigenic composition of influenza viruses.

antihistamine A drug that blocks cell receptors for histamine, preventing allergic effects such as sneezing and itching.

antimicrobial agent (drug) A chemical that inhibits or kills the growth of microorganisms.

antimicrobial spectrum The range of antimicrobial drug action.

antisense molecule RNA segments that complementarily bind to a messenger RNA and block translation.

antisepsis The use of chemical methods for eliminating or reducing microorganisms on the skin.

antiseptic A chemical used to reduce or kill pathogenic microorganisms on a living object, such as the surface of the human body.

antiserum (pl. **antisera**) A blood-derived fluid containing antibodies and used to provide temporary immunity.

antitoxin An antibody produced by the body that circulates in the bloodstream to provide protection against toxins by neutralizing them.

antiviral protein A protein made in response to interferon and that blocks viral replication.

antiviral state A cell capable of inhibiting viral protein synthesis due to interferon activation.

apicomplexan A protozoan containing a number of organelles at one end of the cell that are used for host penetration; no motion is observed in adult forms.

apoptosis A genetically programmed form of cell death.

aqueous solution One or more substances dissolved in water.

Archaea The domain of living organisms that excludes the *Bacteria* and *Eukarya*.

archaebacterium The former term for a unicellular organism in the domain *Archaea*.

artemisinin An anti-protozoal drug used to treat malaria.

Arthropoda A large phylum of animals having jointed appendages and a segmented body; includes lice, mosquitoes, fleas, ticks, and mites.

arthrospore An asexual fungal spore formed by fragmentation of a septate hypha.

Arthus phenomenon An immune complex hypersensitivity when large amounts of IgG antibody for complexes with antigens in blood vessels or near the site of antigen entry.

artificially acquired active immunity The production of antibodies by the body in response to antigens in a vaccination.

artificially acquired passive immunity The transfer of antibodies formed in one individual or animal to another susceptible person.

Ascomycota A phylum of fungi whose members have septate hyphae and form ascospores within saclike asci, among other notable characteristics.

ascospore A sexually produced fungal spore formed by members of the ascomycetes.

ascus (pl. **asci**) A saclike structure containing ascospores; formed by the ascomycetes.

asepsis The process or method of bringing about a condition in which no unwanted microbes are present.

asexual reproduction The form of reproduction that maintains genetic constancy while increasing cell numbers.

asymptomatic Without obvious indications of infection or disease.

atom The smallest portion into which an element can be divided and still enter into a chemical reaction.

atomic nucleus The positively charged core of an atom, consisting of protons and neutrons that make up most of the mass.

atomic number The number of protons in the nucleus of an atom.

atopic disease A condition resulting from the body's response to certain allergens and producing a localized reaction in the body; examples include hay fever and food allergies.

ATP synthase The enzyme involved in forming ATP by using the energy in a proton gradient.

attenuate Reduced ability of bacterial cells or viruses to do damage to the exposed individual.

attractant A substance that attracts cells through motility.

autoclave An instrument used to sterilize microbiological materials by means of high temperature using steam under pressure.

autograft Tissue taken from one part of the body and grafted to another part of the same body.

autoimmune disorder (**disease**) A reaction in which antibodies react with an individual's own chemical substances and cells.

autotroph An organism that uses carbon dioxide (CO_2) as a carbon source; *see also* chemoautotroph *and* photoautotroph.

auxotroph A mutant strain of an organism lacking the ability to synthesize a nutritional need; *see also* prototroph.

avermectin An antihelminthic drug that causes muscle paralysis in nematodes.

avirulent Referring to an organism that is not likely to cause disease.

B

bacillé Calmette Guérin (**BCG**) A strain of attenuated *Mycobacterium bovis* used for immunization against tuberculosis and, on occasion, leprosy.

bacillus (pl. **bacilli**) (1) Any rod-shaped prokaryotic cell. (2) When referring to the genus *Bacillus*, it refers to an aerobic or facultatively anaerobic, rod-shaped, endospore-producing, gram-positive bacterial cell.

bacitracin An antibiotic derived from a *Bacillus* species, effective against gram-positive bacteria when used topically.

bacteremia The presence of live bacterial cells in the blood.

Bacteria The domain of living things that includes all organisms not classified as *Archaea* or *Eukarya*.

bacterial growth curve The events occurring over time within a population of growing and dividing prokaryotic cells.

bactericidal Referring to any agent that kills bacterial cells.

bacteriochlorophyll A pigment located in the membrane systems of purple sulfur bacteria that upon excitement by light, loses electrons and initiates photosynthetic reactions.

bacteriocin One of a group of bacterial proteins toxic to other bacterial cells.

bacteriology The scientific study of prokaryotes; originally used to describe the study of bacteria.

bacteriophage (**phage**) A virus that infects and replicates within bacterial cells.

bacteriorhodopsin A photosynthetic pigment found in the extreme salt-loving archaeal cells.

bacteriostatic Referring to any substance that prevents the growth of bacteria.

bacterium (pl. **bacteria**) A single-celled microorganism lacking a cell nucleus and membrane-enclosed compartments, and often having peptidoglycan in the cell wall.

barophile A microorganism that lives under conditions of high atmospheric pressure.

basal body A structure at the base of a bacterial flagellum consisting of a central rod and set of enclosing rings.

base A chemical compound that accepts hydrogen ions (H^+) in solution; *see also* acid.

base analog A nitrogenous base with a similar structure to a natural base but differing slightly in composition.

basic dye A positively charged colored substance in solution that is used to stain cells.

Basidiomycota The phylum of fungi whose members have septate hyphae and form basidiospores on supportive basidia, among other notable characteristics.

basidiospore A sexually produced fungal spore formed by members of the basidiomycetes.

basidium (pl. **basidia**) A club-like structure containing basidiospores; formed by the basidiomycetes.

basophil A type of white blood cell with granules that functions in allergic reactions.

B cell *See* B lymphocyte.

benign Referring to a tumor that usually is not life threatening or likely to spread to another part of the body.

benthic zone The environment at the bottom of a deep river, lake, or sea.

benzoic acid A chemical preservative used to protect beverages, catsup, and margarine.

beta (β) **hemolytic** Referring to those bacterial species that when plated on blood agar completely destroy the red blood cells as seen by a clearing in the agar around the colonies.

beta-lactamase The enzyme that converts the beta-lactam antibiotics (penicillins, cephalosporins, and carbapenems) into inactive forms.

beta oxidation The breakdown of fatty acids during cellular metabolism through the successive removal from one end of two carbon units.

binary fission An asexual process in prokaryotic cells by which a cell divides to form two new cells while maintaining genetic constancy.

binomial system The method of nomenclature that uses two names (genus and specific epithet) to refer to organisms.

biocatalysis The metabolic reactions of microorganisms carried out at an industrial scale.

biochemical oxygen demand (**BOD**) A number referring to the amount of oxygen used by the microorganisms in a sample of water during a 5-day period of incubation.

biofilm A complex community of microorganisms that form a protective and adhesive matrix that attaches to a surface, such as a catheter or industrial pipeline.

bioinformatics The use of computers and statistical techniques to manage and analyze biological information, especially nucleotide sequences in genes.

biological pollution The presence of microorganisms from human waste in water.

biological vector An infected arthropod, such as a mosquito or tick, that transmits disease-causing organisms between hosts; *see also* mechanical vector.

bioreactor A large fermentation tank for growing microorganisms used in industrial production; also called a fermentor.

bioremediation The use of microorganisms to degrade toxic wastes and other synthetic products of industrial pollution.

biosphere The areas of the earth that are inhabited by living organisms.

biotechnology The commercial application of genetic engineering using living organisms.

bioterrorism The intentional or threatened use of biological agents to cause fear in or actually inflict death or disease upon a large population.

biotype A naturally occurring group of microorganisms with the same genetic makeup; also called biovar.

bipolar staining A characteristic of *Yersinia* and *Francisella* species in which stain gathers at the poles of the cells, yielding the appearance of safety pins.

bisphenol A combination of two phenol molecules used in disinfection.

blanching A process of putting food in boiling water for a few seconds to destroy enzymes.

blastospore A fungal spore formed by budding.

blocking antibody Usually IgG antibodies that bind to and neutralize allergens before they can contact sensitized mast cells.

B lymphocyte (B cell) A white blood cell that matures into memory cells and plasma cells that secrete antibody.

bone marrow A soft reddish substance inside some bones that is involved in the production of blood cells.

booster shot A repeat dose of a vaccine given some years after the initial course to maintain a high level of immunity.

bright-field microscope An instrument that magnifies an object by passing visible light directly through the lenses and object; *see also* light microscope.

broad spectrum Referring to an antimicrobial drug useful for treating many groups of microorganisms, including gram-positive and gram-negative bacteria; *see also* narrow-spectrum.

5-Bromouracil A chemical mutagen that competes with thymine for complementary base pairing with adenine.

bubo A swelling of the lymph nodes due to inflammation.

budding (1) An asexual process of reproduction in fungi, in which a new cell forms as a swelling at the border of the parent cell and then breaks free to live independently. (2) The controlled release of virus particles from an infected animal cell.

buffer (1) A compound that minimizes pH changes in a solution by neutralizing added acids and bases. (2) Refers to a solution containing such a substance.

boll's eye rash A circular lesion on the skin with a red border and central clearing; characteristic of Lyme disease.

burst size The number of virus particles released from an infected bacterial cell.

C

cancer A disease characterized by the radiating spread of malignant cells that reproduce at an uncontrolled rate.

capnophilic Referring to a prokaryotic cell requiring low oxygen gas (O_2) and a high concentration of carbon dioxide gas (CO_2) for metabolism.

capsid The protein coat that encloses the genome of a virus.

capsomere Any of the protein subunits of a capsid.

capsule A layer of polysaccharides and small proteins covalently bound some prokaryotic cells; *see also* slime layer *and* glycocalyx.

carbapenem An antibacterial drug derived from a species of *Streptomyces* that is effective against gram-positive and gram-negative bacterial cells by inhibiting cell wall synthesis.

carbohydrate An organic compound consisting of carbon, hydrogen, and oxygen that is an important source of carbon and energy for all organisms; examples include simple sugars, starch, and cellulose.

carbon cycle A series of interlinked processes involving carbon compound exchange between living organisms and the nonliving environment.

carbon-fixing reactions The stage of photosynthesis where electrons and ATP are used to reduce carbon dioxide gas (CO_2) to sugars; *see also* energy-fixing reactions.

carbuncle An enlarged abscess formed from the union of several smaller abscesses or boils.

carcinogen Any physical or chemical substance that causes tumor formation.

carrier An individual who has recovered from a disease but retains the infectious agents in the body and continues to shed them.

casein The major protein in milk.

casing soil A non-nutritious soil used to provide moisture for mushroom formation.

catabolism An energy-liberating process in which larger organic compounds are broken down into smaller ones; *see also* anabolism.

cation A positively charged ion; *see also* anion.

$CD4^+$ T cell A lymphocyte expressing the CD4 receptor on the cell surface.

$CD8^+$ T cell A lymphocyte expressing the CD8 receptor on the cell surface.

cell envelope The cell wall and cell membrane of a prokaryotic cell.

cell line A group of identical cells in culture and derived from a single cell.

cell-mediated immune response The body's ability to resist infection through the activity of T-lymphocyte recognition of antigen peptides presented on macrophages and dendritic cells and on infected cells.

cell membrane A thin bilayer of phospholipids and proteins that surrounds the prokaryotic cell cytoplasm. *See also* plasma membrane.

cell theory The tenet that all organisms are made of cells and arise from preexisting cells.

cellular hypersensitivity A type IV allergy characterized by an exaggerated cell mediated immune response.

cellular respiration The process of converting chemical energy into cellular energy in the form of ATP.

cell wall A carbohydrate-containing structure surrounding fungal, algal, and most prokaryotic cells.

central dogma The doctrine that DNA codes for RNA through transcription and RNA is converted to protein through translation.

cephalosporin An antibiotic derived from the mold *Cephalosporium* that inhibits cell wall synthesis in gram-positive bacteria and certain gram-negative bacteria.

cercaria (pl. **cercariae**) A tadpole-like larva form in the life cycle of a trematode.

cestode A flatworm, commonly known as a tapeworm, that lives as a parasite in the gut of vertebrates.

chain of transmission How infectious diseases can be spread from human to human (or animal to human).

chancre A painless, circular, purplish hard ulcer with a raised margin that occurs during primary syphilis.

chaperone A protein that ensures a polypeptide folds into the proper shape.

chemical bond A force between two or more atoms that tends to bind those atoms together.

chemical element Any substance that cannot be broken down into a simpler one by a chemical reaction.

chemical pollution The presence of inorganic and organic waste in water.

chemical reaction A process that changes the molecular composition of a substance by redistributing atoms or groups of atoms without altering the number of atoms.

chemiosmosis The use of a proton gradient across a membrane to generate cellular energy in the form of ATP.

chemoautotroph An organism that derives energy from inorganic chemicals and uses the energy to synthesize nutrients from carbon dioxide gas (CO_2).

chemoheterotroph An organism that derives energy from organic chemicals and uses the energy to synthesize nutrients from carbon compounds other than carbon dioxide gas (CO_2).

chemokine A protein that prompts specific white blood cells to migrate to an infection site and carry out their immune system functions.

chemotaxis A movement of a cell or organism toward a chemical or nutrient.

chemotherapeutic agent A chemical compound used to treat diseases and infections in the body.

chemotherapeutic index A number that represents the highest level of an antimicrobial drug tolerated by the host divided by the lowest level of the drug that eliminates the infectious agent.

chemotherapy The process of using chemical agents to treat diseases and infections, or other disorders, such as cancer.

chitin A polymer of acetylglucosamine units that provides rigidity in the cell walls of fungi.

chlamydia (pl. **chlamydiae**) A very small, round pathogenic bacterial genus visible only with the electron microscope and cultivated within living cells.

chloramphenicol A broad-spectrum antibiotic derived from a *Streptomyces* species that interferes with protein synthesis.

chlorination The process of treating water with chlorine to kill harmful organisms.

chlorine dioxide A gas used to sterilize objects or instruments.

chlorophyll A green or purple pigment in algae and some bacterial cells that functions in capturing light for photosynthesis.

chloroplast A double membrane-enclosed compartment in algae that contains chlorophyll and other pigments for photosynthesis.

cholera toxin An enterotoxin that triggers an unrelenting loss of fluid.

chromosome A structure in the nucleoid or cell nucleus that carries hereditary information in the form of genes.

chronic disease A disease that develops slowly, tends to linger for a long time, and requires a long convalescence.

chytrid A fungus in the phylum Chytridiomycota.

Chytridiomycota A phylum of predominantly aquatic fungi.

ciliate A protozoan that moves with the aid of cilia.

cilium (pl. **cilia**) A hair-like projection on some eukaryotic cells that along with many others assist in the motion of some protozoa and beat rhythmically to aid the movement of a fluid past the respiratory epithelial cells in humans.

cirrhosis Extensive injury of cells of the liver.

citric acid cycle *See* Krebs cycle.

classical pathway A complement-activating sequence of steps involving antibody–microbe complexes.

climax *See* acme period.

clindamycin An antibiotic that inhibits protein synthesis and is used as a penicillin substitute for certain anaerobic bacterial diseases.

clinical disease A disease in which the symptoms are apparent.

clonal selection The theory that certain lymphocytes are activated from the mixed population of B or T lymphocytes when stimulated by antigen or antigen peptide fragments.

clone A population of cells genetically identical to the parent cell.

cloning vector A plasmid used to introduce genes into a bacterial cell.

coagulase An enzyme produced by some staphylococci that catalyzes the formation of a fibrin clot.

coccus (pl. **cocci**) A spherical-shaped prokaryotic cell.

codon A three-base sequence on the mRNA molecule that specifies a particular amino acid insertion in a polypeptide.

coenocytic Referring to a fungus containing no septa (cross-walls) and multinucleate hyphae.

coenzyme A small, organic molecule that forms the nonprotein part of an enzyme molecule; together they form the active enzyme.

coenzyme A (CoA) A small, organic molecule of cellular respiration that functions in release of carbon dioxide gas (CO_2) and the transfer of electrons and protons to another coenzyme.

cofactor A inorganic substance that acts with and is essential to the activity of an enzyme; examples include metal ions.

cold agglutinin screening test (CAST) A laboratory procedure in which *Mycoplasma* antibodies agglutinate human red blood cells at cold temperatures.

coliform bacterium A gram-negative, nonsporeforming, rod-shaped cell that ferments lactose to acid and gas and usually is found in the human and animal intestine; high numbers in water is an indicator of contamination.

colony A visible mass of microorganisms of one type.

colony forming unit (CFU) A measure of the viable cells by counting the number of colonies on a plate; each colony presumably started from one viable cell.

colostrum The yellowish fluid rich in antibodies secreted from the mammary glands of animals or humans prior to the production of true milk.

commensalism A close and permanent association between two species of organisms in which one species benefits and the other remains unharmed and unaffected.

commercial sterilization A canning process to eliminate the most resistant bacterial spores.

communicable disease A disease that is readily transmissible between hosts.

comparative genomics The comparison of DNA sequences between organisms.

competence Referring to the ability of a cell to take up naked DNA from the environment.

competitive inhibition The prevention of a chemical reaction by a chemical that competes with the normal substrate for an enzyme's active site; *see also* noncompetitive inhibition.

complement A group of blood proteins that functions in a cascading series of reactions with antibodies to recognize and help eliminate certain antigens or infectious agents.

complement fixation test A serological procedure to detect antibodies to any of a variety of pathogens by identifying antibody-antigen-complement complexes.

complex medium A chemically undefined medium in which the nature and quantity of each component has not been identified; *see also* synthetic medium.

compound A substance made by the combination of two or more different chemical elements.

compound microscope *See* light microscope.

conidiophore The supportive structure on which conidia form.

conidium (pl. **conidia**) An asexually produced fungal spore formed on a supportive structure without an enclosing sac.

conjugate vaccine An antigen preparation consisting of the antigen bound to a carrier protein.

conjugation (1) In prokaryotes, a unidirectional transfer of genetic material from a live donor cell into a live recipient cell during a period of cell contact. (2) In the protozoan ciliates, a sexual process involving the reciprocal transfer of micronuclei between cells in contact.

conjugation pilus (pl. **pili**) A hollow projection for DNA transfer between the cytoplasms of donor and recipient bacterial cells.

conjunctivitis A general term for disease of the conjunctiva, the thin mucous membrane that covers the cornea and forms the inner eyelid; also called pinkeye.

constant region The invariable amino acids in the light and heavy chains of an antibody.

contact dermatitis A type IV hypersensitivity in which the immune system responds to allergens such as clothing materials, metals, and insecticides; the reaction is usually characterized by an hard, raised, red region on the skin.

contagious Referring to a disease whose agent passes with particular ease among hosts.

contractile vacuole A membrane-enclosed structure within a cell's cytoplasm that regulates the water content by absorbing water and then contracting to expel it.

control That part of an experiment not exposed to or treated with the factor being tested.

contrast In microscopy, to be able to see an object against the background.

convalescent serum Antibody-rich serum obtained from a convalescing patient.

Coombs test An antibody test used to detect Rh antibodies involved in hemolytic disease of the newborn.

corticosteroid A synthetic drug used to control allergic disorders by blocking the release of chemical mediators.

covalent bond A chemical linkage formed by the sharing of electrons between atoms or molecules.

Crenarchaeota A group within the domain *Archaea* that tend to grow in hot, acidic environments.

critical control point (CCP) In the food processing industry, a place where contamination of the food product could occur.

cyanobacterium (pl. **cyanobacteria**) An oxygen-producing, pigmented bacterial cell in a unicellular and filamentous form that carries out photosynthesis.

cyst A dormant and very resistant form of a protozoan and multicellular parasite.

cystitis An inflammation of the urinary bladder.

cytochrome A compound containing protein and iron that plays a role as an electron carrier in cellular respiration and photosynthesis; *see also* electron transport chain.

cytokine Small proteins released by immune defensive cells that affects other cells and the immune response to an infectious agent.

cytokinesis Division of the cytoplasm of a cell during binary fission or mitosis.

cytology The examination of cells obtained from body tissue or fluids.

cytopathic effect (CPE) Visible effect that can be seen in a virus-infected host cell.

cytoplasm The complex of chemicals and structures within a cell; in plant and animal cells excluding the nucleus.

cytoskeleton (1) The structural proteins in a prokaryotic cell that help control cell shape and cell division. (2) In a eukaryotic cell, the internal network of protein filaments and microtubules that control the cell's shape and movement.

cytosol The fluid, ions, and compounds of a cell's cytoplasm excluding organelles and other structures.

cytotoxic hypersensitivity A cell-damaging or cell-destroying hypersensitivity that develops when IgG reacts with antigens on the surfaces of cells.

cytotoxic T cell The type of T lymphocyte that searches out and destroys infected cells.

D

dalton A unit of weight equal to the mass of one hydrogen atom; used to measure molecular weights of biological molecules and compounds.

dapsone A chemotherapeutic agent used to treat leprosy patients.

dark-field microscopy An optical system on the light microscope that scatters light such that the specimen appears white on a black background.

deamination A biochemical process in which amino groups are enzymatically removed from amino acids or other organic compound.

decimal reduction time (**D valve**) The time required to kill 90 percent of the viable organisms at a specified temperature.

decline phase The final portion of a bacterial growth curve in which environmental factors overwhelm the population and induce death; also called death phase.

decomposer An organism, such as a bacterium or fungus, that recycles dead or decaying matter.

defensin An antimicrobial peptide present in white blood cells that plays a role in the prevention or elimination of infection.

definitive host An organism that harbors the adult, sexually mature form of a parasite.

degerm To mechanically remove organisms from a surface.

degranulation The release of cell mediators from mast cells and basophils.

dehydration synthesis reaction A process of bonding two molecules together by removing the products of water and joining the open bonds.

delayed hypersensitivity An allergy reaction that occurs over two to three days; *see also* immediate hypersensitivity.

denaturation A process caused by heat or pH in which proteins lose their function due to changes in their molecular structure.

dendritic cell A white blood cell having long finger-like extensions and found within all tissues; it engulfs and digests foreign material, such as bacterial cells and viruses, and presents antigen peptides on its surface.

deoxyribonucleic acid (**DNA**) The genetic material of all cells and many viruses.

dermatophyte A pathogenic fungus that affects the skin, hair, or nails.

desensitization therapy A process in which minute doses of antigens are used to remove antibodies from the body tissues to prevent a later allergic reaction.

detergent A synthetic cleansing substance that dissolves dirt and oil.

diapedesis A process by which phagocytes move out of the blood vessels by migrating between capillary cells.

diarrhea Excessive loss of fluid from the gastrointestinal tract.

diatomaceous earth Filtering material composed of the remains of diatoms.

diatom One of a group of microscopic marine algae that performs photosynthesis.

dichotomous key A method of deducing the correct species assignment of a living organism by offering two alternatives at each juncture, with the choice of one of those alternatives determining the next step.

differential medium A growth medium in which different species of microorganisms can be distinguished visually.

differential stain technique A procedure using two dyes to differentiate cells or cellular objects based on their staining; *see also* simple stain technique.

dikaryon A fungal cell in which two genetically different haploid nuclei closely pair.

dimorphic Referring to pathogenic fungi that take a yeast form in the human body and a filamentous form when cultivated in the laboratory.

dinoflagellate A microscopic photosynthetic marine alga that forms one of the foundations of the food chain in the ocean.

dipicolinic acid An organic substance that helps stabilize the proteins and DNA in a bacterial spore, thereby increasing spore resistance.

diplococcus (pl. **diplococci**) A pair of spherical-shaped prokaryotic cells.

diplomonad A protozoan that contains four pair of flagella, two haploid nuclei, and live in low oxygen or anaerobic environments; most members are symbiotic in animals.

direct contact The form of disease transmission involving close association between hosts; *see also* indirect contact.

direct microscopic count Estimation of the number of cells by observation with the light microscope.

disaccharide A sugar formed from two single sugar molecules; examples include sucrose and lactose.

disease Any change from the general state of good health.

disinfectant A chemical used to kill or inhibit pathogenic microorganisms on a lifeless object such as a tabletop.

disinfection The process of killing or inhibiting the growth of pathogens.

disulfide bridge A covalent bond between sulfur-containing R groups in amino acids.

DNA ligase An enzyme that binds together DNA fragments.

DNA polymerase III An enzyme that catalyzes DNA replication by combining complementary nucleotides to an existing strand.

DNA probe A short segment of single stranded DNA used to locate a complementary strand among many other DNA strands.

DNA replication The process of copying the genetic material in a cell.

DNA vaccine A preparation that consists of a DNA plasmid containing the gene for a pathogen protein.

domain (1) The most inclusive taxonomic level of classification; consists of the *Archaea*, *Bacteria*, and *Eukarya*.

double diffusion assay Another name for the diffusion of antibodies and antigens.

double helix The structure of DNA, in which the two complementary strands are connected by hydrogen bonds between complementary nitrogenous bases and wound in opposing spirals.

Downey cell A swollen lymphocyte with foamy cytoplasm and many vacuoles that develops as a result of infection with infectious mononucleosis viruses.

droplet An airborne particle of mucus and sputum from the respiratory tract that contains disease-causing microorganisms.

dysentery A condition marked by frequent, watery stools, often with blood and mucus.

E

echinocandin An antifungal drug that inhibits cell wall synthesis.

eclipse period The period of a viral infection when no viruses can be found inside the infected cell.

ecotype A subgroup of a species whose members have adaptations to certain environmental conditions for survival in its environment.

edema A swelling of the tissues brought about by an accumulation of fluid.

effector cell An activated immune cell targeting a pathogen.

electron A negatively charged particle with a small mass that moves around the nucleus of an atom.

electron microscope An instrument that uses electrons and a system of electromagnetic lenses to produce a greatly magnified image of an object; *see also* transmission electron microscope *and* scanning electron microscope.

electron shell An energy level surrounding the atomic nucleus that contains one or more electrons.

electron transport chain A series of proteins that transfer electrons in cellular respiration to generate ATP.

electrophoresis A laboratory technique involving the movement of charged organic molecules through an electrical field; used to separate DNA fragments in diagnostic procedures.

elementary body An infectious form of *Chlamydia* in the early stage of reproduction.

ELISA *See* enzyme-linked immunosorbent assay.

elongation (1) The addition of complementary nucleotides to a parental DNA strand. (2) The addition of additional amino acids onto the forming polypeptide during translation.

emerging infectious disease A new disease or changing disease that is seen for the first time; *see also* reemerging infectious disease.

encapsulated Referring to a prokaryotic cell surrounded by a capsule.

encephalitis Inflammation of the tissue of the brain or infection of the brain.

endemic Referring to a disease that is constantly present in a specific area or region.

endergonic reaction A chemical process that requires energy; *see also* exergonic.

endocarditis An inflammation of the membranous lining of the heart's cavities.

endoflagellum A microscopic fiber located along cell walls in certain species of spirochetes; contractions of the filaments yield undulating motion in the cell; also called axial filament.

endogenous infection A disorder that starts with a microbe or virus that already was in or on the body as part of the microbiota; *see also* exogenous infection.

endomembrane system A cytoplasmic set of membranes that function in the transport, modification, and sorting of proteins and lipids in eukaryotic cells.

endophyte A fungus that lives within plants and does not cause any known disease.

endoplasmic reticulum A network of membranous plates and tubes in the eukaryotic cell cytoplasm responsible for the synthesis and transport of materials from the cell.

endospore An extremely resistant dormant cell produced by some gram-positive bacterial species.

endosymbiotic theory The idea that mitochondria and chloroplasts originated from bacterial cells and cyanobacteria that took up residence in a primitive eukaryotic cell.

endotoxin A metabolic poison, produced chiefly by gram-negative bacteria, that are part of the bacterial cell wall and consequently are released on cell disintegration; composed of lipid-polysaccharide-peptide complexes.

endotoxin shock A drop in blood pressure due to an endotoxin.

energy-fixing reactions The stage of photosynthesis where light is trapped and used to generate ATP; *see also* carbon-fixing reactions.

enriched medium A growth medium in which special nutrients must be added to get an species to grow.

enterotoxin A toxin that is active in the gastrointestinal tract of the host.

envelope The flexible membrane of protein and lipid that surrounds many types of viruses.

environmental genomics The sequencing, identification, and study of genes from microbial communities.

enzyme A reusable protein molecule that brings about a chemical change while itself remaining unchanged.

enzyme-linked immunosorbent assay (ELISA) A serological test in which an enzyme system is used to detect an individual's exposure to a pathogen.

enzyme-substrate complex The association of an enzyme with its substrate at the active site.

eosinophil A type of white blood cell with granules that stains with the dye eosin and plays a role in allergic reactions and the body's response to parasitic infections.

epidemic Referring to a disease that spreads more quickly and more extensively within a population than normally expected.

epidemiology The scientific study of the source, cause, and transmission of disease within a population.

epitope *See* antigenic determinant.

erythema A zone of redness in the skin due to a widening of blood vessels near the skin surface.

erythema migrans (EM) An expanding circular red rash that occurs on the skin of patients with Lyme disease.

erythrogenic Referring to a streptococcal poison that leads to the rash in scarlet fever.

ethylene oxide A chemical gas that is used to sterilize many objects and instruments.

Eukarya The taxonomic domain encompassing all eukaryotic organisms.

eukaryote An organism whose cells contain a cell nucleus with multiple chromosomes, a nuclear envelope, and membrane-bound compartments; *see also* prokaryote.

eukaryotic Referring to a cell or organism containing a cell nucleus with multiple chromosomes, a nuclear envelope, and membrane-bound compartments.

Euryarchaeota A group within the domain *Archaea*.

exanthema A maculopapular rash occurring on the skin surface.

excision repair A mechanism to correct improperly bonded bases in a DNA sequence; *see also* mismatch repair.

exergonic reaction A chemical process releasing energy; *see also* endergonic.

exogenous infection A disorder that starts with a microbe or virus that entered the body from the environment; *see also* endogenous infection.

exotoxin A bacterial metabolic poison composed of protein that is released to the environment; in the human body, it can affect various organs and systems.

experiment A test or trial to verify or refute a hypothesis.

extreme halophile An archaeal organism that grows at very high salt concentrations.

extremophile A microorganism that lives in extreme environments, such as high temperature, high acidity, or high salt.

F

Fab fragment The branched portion of an antibody molecule that combines with an antigenic determinant of an antigen.

facilitated diffusion The movement of substances from an area of higher concentration across a biological membrane to a region of lower concentration by means of a membrane-spanning channel or carrier protein.

facultative Referring to an organism that grows in the presence or absence of oxygen gas (O_2).

Fc fragment The stem portion of an antibody molecule that combines with phagocytes, mast cells, or complement.

feedback inhibition The slowing down or prevention of a metabolic pathway when excess end product binds noncompetitively to an enzyme in the pathway.

fermentation A metabolic pathway in which carbohydrates serve as electron donors, the final electron acceptor is not oxygen gas (O_2), and NADH is reoxidized to NAD^+ for reuse in glycolysis for generation of ATP; *see also* industrial fermentation.

fermentor *See* bioreactor.

fever An abnormally high body temperature that is usually caused by a bacterial or viral infection.

F factor A plasmid containing genes for plasmid replication and conjugation pilus formation.

filariform Referring to the thread-like shape of the larvae of the hookworm.

filtration A mechanical method to remove microorganisms by passing a liquid or air through a filter.

firmicutes A group of bacterial species containing many of the gram-positive species.

five kingdom system The classification scheme placing all living organisms into one of five groups.

flaccid paralysis A loss of voluntary movement in which the limbs have little tone and become flabby.

flagellum (pl. **flagella**) A long, hair-like appendage composed of protein and responsible for motion in microorganisms; found in some bacterial, protozoal, algal, and fungal cells.

flare A spreading zone of redness around a wheal that occurs during an allergic reaction; *also see* wheal.

flash pasteurization method A treatment in which milk is heated at 71.6°C for 15 seconds and then cooled rapidly to eliminate harmful bacteria; also called HTST ("high temperature, short time") method.

flatworm A multicellular parasite with a flattened body; examples include the tapeworms.

floc A jelly-like mass that forms in a liquid and made up of coagulated particles.

flocculation (1) A serological reaction in which particulate antigens react with antibodies to form visible aggregates of material. (2) The formation of jelly-like masses of coagulated material in the water-purification process.

flucytosine An antifungal drug that interrupts nucleic acid synthesis.

fluid mosaic model The representation for the cell (plasma) membrane where proteins "float" within or on a bilayer of phospholipid.

fluke *See* trematode.

fluorescence The emission of one color of light after being exposed to light of another wavelength.

fluorescence microscopy An optical system on the light microscope that uses ultraviolet light to excite dye-containing objects to fluoresce.

fluorescent antibody technique A diagnostic tool that uses fluorescent antibodies with the fluorescence microscope to identify an unknown organism.

fluoroquinolone A drug used to treat urinary and intestinal tract infections.

flush A burst of mushroom growth.

folic acid The organic compound in bacteria whose synthesis is blocked by sulfonamide drugs.

folliculitis An inflammation of one or more hair follicles, producing small boils.

fomite An inanimate object, such as clothing or a utensil, that carries disease organisms.

food vacuole A membrane-enclosed compartment in some eukaryotic that results from the intake of large molecules, particles, or cells, for digestion.

foraminiferan A shell-containing amoeboid protozoan having a chalky skeleton with window-like openings between sections of the shell.

formalin A solution of formaldehyde used as embalming fluid, in the inactivation of viruses, and as a disinfectant.

F plasmid A DNA plasmid in the cytoplasm of an F^+ bacterial cell that may be transferred to a recipient bacterial cell during conjugation.

F' plasmid An F plasmid carrying a bacterial chromosome fragment.

fractional sterilization A sterilization method in which materials are heated in free-flowing steam for 30 minutes on each of three successive days; also called tyndallization.

fruiting body The general name for a reproductive structure of a fungus from which spores are produced.

functional genomics The identification of gene function from a gene sequence.

functional group A group of atoms on hydrocarbons that function in chemical reactions.

fungemia The dissemination of fungi through the circulatory system.

Fungi One of the five kingdoms in the Whittaker classification of living organisms; composed of the molds and yeasts.

fungicidal Referring to any agent that kills fungi.

fungistatic Referring to any substance that inhibits the growth of fungi.

furuncle An infection of a hair follicle.

G

gametocyte The stage in the life cycle of the malaria parasite during which it reproduces sexually in the blood of a mosquito.

gamma globulin A general term for antibody-rich serum.

gamma ray An ionizing radiation that can be used to sterilize objects.

gangrene A physiological process in which the enzymes from wounded tissue digest the surrounding layer of cells, inducing a spreading death to the tissue cells.

gastroenteritis Infection of the stomach and intestinal tract often due to a virus.

gas vesicle A cytoplasmic compartment in some prokaryotic cells used to regulate buoyancy.

gene A segment of a DNA molecule that provides the biochemical information for a function product.

gene probe A small, single-stranded DNA fragment labeled for identification of a specific DNA segment.

generalized transduction A process by which a bacteriophage carries a bacterial chromosome fragment from one cell to another; *see also* specialized transduction.

generation time The time interval for a cell population to double in number.

genetically modified organism (GMO) An organism produced by genetic engineering.

genetic code The specific order of nucleotide sequences in DNA or RNA that encode specific amino acids for protein synthesis.

genetic engineering The use of bacterial and microbial genetics to isolate, manipulate, recombine, and express genes.

genetic recombination The process of bring together different segments of DNA.

genome The complete set of genes in a virus or an organism.

genomic island A series of up to 25 genes absent in other strains of the same prokaryotic species.

genomics The study of an organism's gene structure and gene function in viruses and organisms.

genus (pl. **genera**) A rank in the classification system of organisms composed of one or more species; a collection of genera constitute a family.

germicide Any agent that kills microorganisms.

germ theory The principle formulated by Pasteur and proved by Koch that microorganisms are responsible for infectious diseases.

Glomeromycota A group of mycorrhizal fungi that exist within the roots of most land plants.

glucose A six-carbon sugar used as a major energy source for metabolism.

glutaraldehyde A liquid chemical used for sterilization.

glycocalyx A viscous polysaccharide material covering many prokaryotic cells to assist in attachment to a surface and impart resistance to desiccation; *see also* capsule *and* slime layer.

glycolysis A metabolic pathway in which glucose is broken down into two molecules of pyruvate with a net gain of two ATP molecules.

glycylcycline A drug that is effective against antibiotic resistant *Staphylococcus aureus*.

Golgi apparatus A stack of flattened, membrane-enclosed compartments in eukaryotic cells involved in the modification and sorting of lipids and proteins.

gonococcus A colloquial name for *Neisseria gonorrhoeae*.

graft-versus-host (GVH) reaction A phenomenon in which a tissue graft or transplant produces immune substances against the recipient.

gram-negative Referring to a bacterial cell that stains red after Gram staining.

gram-positive Referring to a bacterial cell that stains purple after Gram staining.

Gram stain A staining procedure used to identify bacterial cells as gram-positive or gram-negative.

granulocyte A white blood cell with visible granules in the cytoplasm; includes neutrophils, eosinophils, and basophils; *see also* agranulocyte.

granuloma A small lesion caused by an infection.

granzyme A cytotoxic enzyme that causes infected cells to undergo a programmed cell death.

gray syndrome A side effect of chloramphenicol therapy, characterized by a sudden breakdown of the cardiovascular system.

green alga A unicellular protist containing chloroplasts for photosynthesis.

griseofulvin An antifungal drug used against infections of the skin, hair, and nails.

groundwater Water originating from deep wells and subterranean springs; *see also* surface water.

group A streptococci (GAS) Contains those streptococcal species that are β-hemolytic organisms.

Guillain-Barré syndrome (GBS) A complication of influenza and chickenpox, characterized by nerve damage and polio-like paralysis.

gumma A soft, granular lesion that forms in the cardiovascular and/or nervous systems during tertiary syphilis.

H

HAART *See* highly active antiretroviral therapy.

HACCP *See* hazard analysis critical control point.

halogen A chemical element whose atoms have seven electrons in their outer shell; examples include iodine and chlorine.

halophile An organism that lives in environments with high concentrations of salt.

hapten A small molecule that combines with tissue proteins or polysaccharides to form an antigen.

Hazard Analysis Critical Control Point (HACCP) A set of federally enforced regulations to ensure the dietary safety of seafood, meat, and poultry.

health care-associated infection (HAI) An infection resulting from treatment for another condition.

heat fixation The use of warm temperatures to prepare microorganisms for staining and viewing with the light microscope.

heavy (H) chain The larger polypeptide in an antibody.

heavy metal A chemical element often toxic to microorganisms; examples include mercury, copper, and silver.

helminth A term referring to a multicellular parasite; includes roundworms and flatworms.

helper T1 (T_H1) cell A T lymphocyte that enhances the activity of B lymphocytes.

helper T2 (T_H2) cell A T lymphocyte that stimulates destruction of macrophages infected with bacterial cells.

hemagglutination The formation of clumps of red blood cells.

hemagglutination-inhibition (HAI) test A test using a patient's serum to detect the presence of antibodies against a specific infectious agent.

hemagglutinin (1) An enzyme composing one type of surface spike on influenza viruses that enables the viruses to bind to the host cell. (2) An agent such as a virus or an antibody that causes red blood cells to clump together.

hemolysin A bacterial toxin that destroys red blood cells.

hemolysis The destruction of red blood cells.

hemolytic disease of the newborn A type II hypersensitivity reaction disease in which Rh antibodies from a pregnant woman combine with Rh antigens on the surface of fetal erythrocytes and destroy them; also called Rh disease and erythroblastosis fetalis.

hepatitis B core antigen (HBcAG) An antigen located in the inner lipoprotein coat enclosing the DNA of a hepatitis B virus.

hepatitis B surface antigen (HBsAg) An antigen located in the outer surface coat of a hepatitis B virus.

hepatocellular carcinoma (HCC) Cancer of the liver tissue.

herd immunity The proportion of a population that is immune to a disease.

hermaphroditic Referring to an organism that possesses both male and female reproductive organs; examples include the parasitic flatworms.

heterokaryon A fungal cell that has two or more genetically different nuclei.

heterophile antigen An antigen that occurs in apparently unrelated species of organisms.

heterotroph An organism that requires preformed organic matter for its energy and carbon needs; *see also* photoheterotroph *and* chemoheterotroph.

high frequency of recombination (Hfr) Referring to a bacterial cell containing an F factor incorporated into the bacterial chromosome.

high efficiency particulate air (HEPA) filter A type of air filter that removed particles larger than 0.3 micrometers.

highly active antiretroviral therapy (HAART) The combination of several (typically three or four) antiretroviral drugs for the treatment of infections caused by retroviruses, especially the human immunodeficiency virus.

histamine A mediator in type I hypersensitivity reactions that is released from the granules in mast cells and basophils and causes the contraction of smooth muscles.

histone One of several proteins that organize and pack the DNA in a chromosome.

homeostasis The tendency of an organism to maintain a steady state or equilibrium with respect to specific functions and processes.

horizontal gene transfer (HGT) The movement of genes from one organism to another within the same generation; also called lateral gene transfer.

host An organism on or in which a microorganism lives and grows, or a virus replicates.

host range The variety of species that a disease-causing microorganism can infect.

human genome The complete set a genetic information in a human cell.

human leukocyte antigen (HLA) Cell surface antigens involved with tissue transplantation.

humoral immune response The immune reaction of producing antibodies directed against antigens in the body fluids.

hyaluronidase An enzyme that digests hyaluronic acid and thereby permits the penetration of pathogens through connective tissue.

hybridoma The cell resulting from the fusion of a B cell and a cancer cell.

hydatid cyst A thick-walled body formed in the human liver by *Echinococcus granulosus* and containing larvae.

hydrocarbon An organic molecule containing only hydrogen and carbon atoms that are connected by a sharing of electrons.

hydrogen bond A weak attraction between a positively charged hydrogen atom (covalently bonded to oxygen or nitrogen) and a covalently bonded, negatively charged oxygen or nitrogen atom in the same or separate molecules.

hydrogen peroxide An unstable liquid that readily decomposes in water and oxygen gas (O_2).

hydrolysis reaction A process in which a molecule is split into two parts through the interaction of H^+ and $(OH)^-$ of a water molecule.

hydrophilic Referring to a substance that dissolves in or mixes easily with water; *see also* hydrophobic.

hydrophobia An emotional condition ("fear of water") arising from the inability to swallow as a consequence of rabies.

hydrophobic Referring to a substance that does not dissolve in or mixing easily with water; *see also* hydrophilic.

hyperimmune serum The fluid portion of the blood containing a higher than normal amount of a particular antibody.

hypersensitivity An immunological response to exposure to an allergen or antigen.

hyperthermophile A prokaryote that has an optimal growth temperature above 80°C.

hypha (pl. **hyphae**) A microscopic filament of cells representing the vegetative portion of a fungus.

hypothesis An educated guess or answer to a properly framed question.

I

icosahedral Referring to a symmetrical figure composed of 20 triangular faces and 12 points; one of the major shapes of some viral capsids.

IgA The class of antibodies found in respiratory and gastrointestinal secretions that help neutralize pathogens.

IgD The class of antibodies found on the surface of B cells that act as receptors for binding antigen.

IgE The class of antibodies responsible for type I hypersensitivities.

IgG The class of antibodies abundant in serum that are major diseases fighters.

IgM The first class of antibodies to appear in serum in helping fight pathogens.

imidazole An antifungal drug that interferes with sterol synthesis in fungal cell membranes; examples are miconazole and ketoconazole.

immediate hypersensitivity An allergy reaction that occurs within minutes of the second or ensuing exposure; *see also* delayed hypersensitivity.

immune complex A combination of antibody and antigen capable of complement activation and characteristic of type III hypersensitivity reactions.

immune complex hypersensitivity Type III hypersensitivity, in which antigens combine with antibodies to form aggregates that are deposited in blood vessels or on tissue surfaces.

immune deficiency The lack of an adequate immune system response.

immunity The body's ability to resist infectious disease through innate and acquired mechanisms.

immunization The process of making an individual resistant to a particular disease by administering a vaccine; *see also* vaccination.

immunocompetent The ability of the body to develop an immune response in the presence of a disease-causing agent.

immunocompromised Referring to an inadequate immune response as a result of disease, exposure to radiation, or treatment with immunosuppressive drugs.

immunodeficiency disorder The inability, either inborn or acquired, of the body to produce an adequate immune response to fight disease.

immunodiffusion The movement of antigen and antibody toward one another through a gel or agar to produce a visible precipitate.

immunoelectrophoresis A laboratory diagnostic procedure in which antigen molecules move through an electric field and then diffuse to meet antibody molecules to form a precipitation line; *see also* electrophoresis.

immunogenic Capable of generating an immune response.

immunoglobulin (Ig) The class of immunological proteins that react with an antigen; an alternate term for antibody.

immunological memory The long-term ability of the immune system to remember past pathogen exposures.

immunology The scientific study of how the immune system works and responds to non-self agents.

immunomodulation The modification of some aspect of the immune system as part of a treatment, especially the suppression of an overactive inflammatory response.

incineration To burn to ashes, or cause something to burn to ashes.

inclusion body (1) A granule-like storage structure found in the prokaryotic cell cytoplasm. (2) A virus in the cytoplasm or nucleus of an infected cell.

incubation period The time that elapses between the entry of a pathogen into the host and the appearance of signs and symptoms.

index of refraction A measure of the light bending ability of a medium through which light passes.

indicator organism A microorganism whose presence signals fecal contamination of water.

indigenous microbiota The microbial agents that are associated with an animal for long periods of time without causing disease; *see also* transient microbiota.

indirect contact The mode of disease transmission involving nonliving objects; *see also* direct contact.

induced mutation A change in the sequence of nucleotide bases in a DNA molecule arising from a mutagenic agent used under controlled laboratory conditions.

induration A thickening and drying of the skin tissue that occurs in type IV hypersensitivity reactions.

industrial fermentation Any large scale industrial process, with or without oxygen gas (O_2), for growing microorganisms; *see also* fermentation.

infection The relationship between two organisms and the competition for supremacy that takes place between them.

infection allergy A type IV hypersensitivity reaction in which the immune system responds to the presence of certain microbial agents.

infectious disease A disorder arising from a pathogen invading a susceptible host and inducing medically significant symptoms.

infectious dose The number of microorganisms needed to bring about infection.

inflammation A nonspecific defensive response to injury; usually characterized by redness, warmth, swelling, and pain.

initiation (1) The unwinding and separating of DNA strands during replication. (2) The beginning of translation.

innate immunity An inborn set of the pre-existing defenses against infectious agents; includes the skin, mucous membranes, and secretions.

insertion sequence (IS) A segment of DNA that forms a copy of itself, after which the copy moves into areas of gene activity to interrupt the genetic coding sequence.

interferon An antiviral protein produced by body cells on exposure to viruses and which trigger the synthesis of antiviral proteins.

interleukin A chemical cytokine produced by white blood cells that causes other white blood cells to divide; *see also* cytokine.

intermediate host The host in which the larval or asexual stage of a parasite is found.

intoxication The presence of microbial toxins in the body.

in use test A procedure used to determine the value of a disinfectant or antiseptic.

invasiveness The ability of a pathogen to spread from one point to adjacent areas in the host and cause structural damage to those tissues.

iodophor A complex of iodine and detergents that is used as an antiseptic and disinfectant.

ion An electrically charged atom.

ionic bond The electrical attraction between oppositely charged ions.

ionizing radiation A type of radiation such as gamma rays and X rays that causes the separation of atoms or a molecule into ions.

isograft Tissue taken from one identical twin and grafted to the other twin.

isomer A molecule with the same molecular formula but different structural formula.

isoniazid (INH) An antimicrobial drug effective against the tubercle bacillus.

isotope An atom of the same element in which the number of neutrons differs; *see also* radioisotope.

J

jaundice A condition in which bile seeps into the circulatory system, causing the complexion to have a dull yellow color.

K

Kaposi sarcoma A type of cancer in immunocompromised individuals, such as AIDS patients, where cancer cells and an abnormal growth of blood vessels form solid lesions in connective tissue.

keratitis Infection of the cornea of the eye.

keratoconjunctivitis Eye inflammation accompanied by infection of the cornea and conjunctiva; the condition is characterized by tearing, swelling, and sensitivity to light.

kinetoplastid A protozoan with one flagellum; most are parasitic in aerobic or anaerobic environments.

Kirby-Bauer test *See* agar disk diffusion method.

Koch's postulates A set of procedures by which a specific organism can be related to a specific disease.

Koplik spots Red patches with white central lesions that form on the gums and walls of the pharynx during the early stages of measles.

Korarchaeota A group within the domain *Archaea* that tend to be hyperthermophiles.

Krebs cycle A cyclic metabolic pathway in which carbon from acetyl-CoA is released as carbon dioxide; the reactions yield ATP as well as protons and high-energy electrons that are transferred to coenzymes; also called citric acid cycle.

L

lactic acid fermentation A catabolic process that produces lactic acid during the reoxidation of NADH to NAD^+ for reuse in glycolysis to generate ATP.

lactone A chemical compound used as a flavoring ingredient in foods and beverages.

lactose A milk sugar composed of one molecule of glucose and one molecule of galactose.

lagging strand During DNA replication, the new strand that is synthesized discontinuously; *see also* leading strand.

lag phase A portion of a bacterial growth curve encompassing the first few hours of the population's history when no growth occurs.

larva (pl. **larvae**) A sexually immature stage in the life cycle of a multicellular parasite.

latency A condition in which a virus integrates into a host chromosome without immediately causing a disease.

latent phase The period during a viral infection when virus particles cannot be found outside the infected cells.

late-phase anaphylaxis After an acute IgE mediated reaction, a second response occurs many hours after the initial response, and appears to be based on the activity of eosinophils.

leading strand During DNA replication, the new strand that is synthesized continuously; *see also* lagging strand.

leproma A tumor-like growth on the skin associated with leprosy.

leukemia Cancer of the white blood cells.

leukocidin A bacterial enzyme that destroys phagocytes, thereby preventing phagocytosis.

leukocyte Any of a number of types of white blood cells.

leukopenia A condition characterized by an abnormal drop in the normal number of white blood cells.

leukotriene A substance that acts as a mediator during type I hypersensitivity reactions; formed from arachidonic acid after the antigen-antibody reaction has taken place.

lichen An association between a fungal mycelium and a cyanobacterium or alga.

light (L) chain A smaller polypeptide in an antibody.

light microscope An instrument that uses visible light and a system of glass lenses to produce a magnified image of an object; also called a compound microscope.

lipid A nonpolar organic compound composed of carbon, hydrogen, and oxygen; examples include triglycerides and phospholipids.

lipid A A component in the outer membrane of the gram-negative cell wall.

lipopolysaccharide (LPS) A molecule composed of lipid and polysaccharide that is found in the outer membrane of the gram-negative cell wall of bacterial cells.

littoral zone The environment along the shoreline of an ocean.

local disease A disease restricted to a single area of the body.

lockjaw *See* trismus.

logarithmic (log) phase The portion of a bacterial growth curve during which active growth leads to a rapid rise in cell numbers.

looped domain structure The term used to describe organization and packing of the prokaryotic chromosome.

lymph The tissue fluid that contains white blood cells and drains tissue spaces through the lymphatic system.

lymphedema An abnormal swelling due to the loss of normal lymph vessel drainage of the affected part; typical of elephantiasis.

lymph node A bean-shaped organ located along lymph vessels that is involved in the immune response and contains phagocytes and lymphocytes.

lymphocyte A type of white blood cell that functions in the immune system.

lymphoid progenitor A bone marrow cell that gives rise to lymphocytes; *see also* myeloid progenitor.

lyophilization A process in which food or other material is deep frozen, after which its liquid is drawn off by a vacuum; also called freeze-drying.

lysis The rupture of a cell and the loss of cell contents.

lysogenic cycle The events of a bacterial virus infection that result in the integration of its DNA into the bacterial chromosome.

lysosome A membrane-enclosed compartment in many eukaryotic cells that contains enzymes to degrade or digest substances.

lysozyme An enzyme found in tears and saliva that digests the peptidoglycan of gram-positive bacterial cell walls.

lytic cycle A process by which a bacterial virus replicates within a host cell and ultimately destroys the host cell.

M

macrolide An antibiotic that blocks protein synthesis.

macronucleus The larger of two nuclei in most ciliates that is involved in controlling metabolism; *see also* micronucleus.

macrophage A large cell derived from monocytes that is found within various tissues and actively engulfs foreign material, including infecting bacterial cells and viruses.

macule A pink-red skin spot associated with infectious disease.

maculopapular rash A rash consisting of pink-red spots that later become dark red before fading; occurs in rickettsial diseases.

magnetosome A cytoplasmic inclusion body in some prokaryotic cells that assists orientation to the environment by aligning with the magnetic field.

major histocompatibility complex (MHC) A set of genes that controls the expression of MHC proteins; involved in transplant rejection.

malignant Referring to a tumor that invades the tissue around it and may spread to other parts of the body.

Mantoux test The infection of a tuberculosis purified protein derivative into the forearm.

mashing A fermentable mixture of hot water and barley grain from which alcohol is distilled.

mass The amount of matter in a sample.

mass number The total number of protons and neutrons in an atom.

mast cell A type of cell in connective tissue which release histamine during allergic attacks.

matrix protein A protein shell found in some viruses between the genome and capsid.

matter Any substance that has mass and occupies space.

mebendazole An antihelminthic drug that inhibits the uptake of nutrients.

mechanical vector A living organism, or an object, that transmits disease agents on its surface; *see also* biological vector.

membrane attack complex A set of complement proteins that forms holes in a bacterial cell membrane and leads to cell destruction.

membrane filter A pad of cellulose acetate or polycarbonate.

memory cell A cell derived from B lymphocytes or T lymphocytes that reacts rapidly upon re-exposure to antigen.

meninges The covering layers of the brain and spinal cord.

meningitis A general term for inflammation of the covering layers of the brain and spinal cord due to any of several bacteria, fungi, viruses, or protozoa.

merozoite A stage in the life cycle of the malaria parasite that invades red blood cells in the human host.

mesophile An organism that grows in temperature ranges of 20°C to 40°C.

messenger RNA (mRNA) An RNA transcript containing the information for synthesizing a specific polypeptide.

metabolic pathway A sequence of linked enzyme-catalyzed reactions in a cell.

metabolism The sum of all biochemical processes taking place in a living cell; *see also* anabolism and catabolism.

metabolomics The measurement and analysis of metabolites in cells or organisms.

metacercaria (pl. **metacercariae**) An encysted intermediary stage in the life cycle of a trematode.

metachromatic granule A polyphosphate-storing granule commonly found in *Corynebacterium diphtheriae* that stains deeply with methylene blue; also called volutin.

metagenome The collective genomes from a population of organisms.

metastasize Referring to a tumor that spreads from the site of origin to other tissues in the body.

methanogen An archaeal organism that lives on simple compounds in anaerobic environments and produces methane during its metabolism.

miasma An ill-defined idea of the 1700s and 1800s that suggested diseases were caused by an altered chemical quality of the atmosphere.

microaerophile An organism that grows best in an oxygen-reduced environment.

microbe *See* microorganism.

microbicidal Referring to any agent that kills microbes.

microbiology The scientific study of microscopic organisms and viruses, and their roles in human disease as well as beneficial processes.

microbiostatic Referring to any agent that inhibits microbial growth.

microbiota The population of microorganisms that colonize various parts of the human body and do not cause disease in a healthy individual.

microcompartment Cellular compartments in prokaryotic cells delimited by a protein shell.

microenvironment A cell's or organism's physical and chemical surroundings.

microfilaria (pl. **microfilariae**) The larva stage of the parasitic nematode *Wuchereria bancrofti*.

micrometer (μm) A unit of measurement equivalent to one millionth of a meter; commonly used in measuring the size of microorganisms.

micronucleus The smaller of the two nuclei in most ciliates that contains genetic material and is involved in sexual reproduction; *see also* macronucleus.

microorganism (microbe) A microscopic form of life including bacterial, archaeal, fungal, and protozoal cells.

microtubule A hollow protein tube making up part of the cytoskeleton, flagella, and cilia in eukaryotic organisms.

mineralization The conversion of organic compounds to inorganic compounds and ammonia.

minimum inhibitory concentration (MIC) The lowest concentration of an antimicrobial agent that will inhibit its growth.

miracidium (pl. **miracidia**) A ciliated larva representing an intermediary stage in the life cycle of a trematode.

mismatch repair A mechanism to correct mismatched bases in the DNA; *see also* excision repair.

mitochondrion (pl. **mitochondria**) A double membrane-enclosed compartment in eukaryotic cells that carries out aerobic respiration.

mitosporic fungus A fungus without a known sexual stage of reproduction.

mold A type of fungus that consists of chains of cells and appears as a fuzzy mass of thin filaments in culture.

molecular epidemiology The study of the sources, causes, and mode of transmission of diseases by using molecular diagnostic techniques.

molecular formula The representation of the kinds and numbers of each atom in a molecule.

molecular taxonomy The systematized arrangement of related organisms based on molecular characteristics, such as ribosomal RNA nucleotide sequences.

molecular weight The sum of the atomic masses of all atoms in a molecule.

molecule Two or more atoms held together by a sharing of electron pairs.

monobactam A synthetic antibacterial drug used to inhibit cell wall synthesis in gram-negative bacteria.

monoclonal antibody A type of antibody produced by a clone of hybridoma cells, consisting of antigen-stimulated B cells fused to myeloma cells.

monocyte A circulating white blood cell with a large bean-shaped nucleus that is the precursor to a macrophage.

monolayer A single layer of cultured.

monomer A simple organic molecule that can join in long chains with other molecules to form a more complex molecule; *see also* polymer.

monosaccharide A simple sugar that cannot be broken down into simpler sugars; examples include glucose and fructose.

monospot test A method used to detect the presence of heterophile antibodies, which is indicative of infectious mononucleosis.

most probable number (MPN) A laboratory test in which a statistical evaluation is used to estimate the number of bacterial cells in a sample of fluid; often employed in determinations of coliform bacteria in water.

M protein A protein that enhances the pathogenicity of streptococci by allowing organisms to resist phagocytosis and adhere firmly to tissue.

mucosa-associated lymphoid tissue (MALT) An intestinal example of secondary lymphoid tissue.

mucous membrane (mucosa) A moist lining in the body passages of all mammals that contains mucus-secreting cells and is open directly or indirectly to the external environment.

mucus A sticky secretion of glycoproteins.

mutagen A chemical or physical agent that causes a mutation.

mutant An organism carrying a mutation.

mutation A change in a characteristic of an organism arising from a permanent alteration of a DNA sequence.

mutualism A close and permanent association between two populations of organisms in which both benefit from the association.

mycelium (pl. **mycelia**) A mass of fungal filaments from which most fungi are built.

mycology The scientific study of fungi.

mycoplasma One of a group of tiny submicroscopic bacteria that lacks cell walls and is visible only with an electron microscope.

mycorrhiza (pl. **mycorrhizae**) A close association between a fungus and the roots of many plants.

mycotoxin A poison produce by a fungus that adversely affects other organisms.

myeloid progenitor A bone marrow cell that gives rise to red blood cells and all white blood cells except lymphocytes; *see also* lymphoid progenitor.

myeloma A malignant tumor that develops in the blood-cell-producing cells of the bone marrow; the cells are used in the procedure to produce monoclonal antibodies.

myxobacteria A group of soil-dwelling bacterial species that exhibit multicellular behaviors.

N

naive T cell An immature T cell.

Nanoarchaeota A group within the domain *Archaea* that tend to be hyperthermophiles.

nanometer (nm) A unit of measurement equivalent to one billionth of a meter; the unit used in measuring viruses and the wavelength of energy forms.

narrow spectrum Referring to an antimicrobial drug that is useful for a restricted group of microorganisms; *see also* broad-spectrum antibiotic.

natural killer (NK) cell A type of defensive body cell that attacks and destroys cancer cells and infected cells without the involvement of the immune system.

naturally acquired active immunity A host response resulting in antibody production as a result of experiencing the disease agent.

naturally acquired passive immunity A process resulting from the passage of antibodies to the fetus via the placenta or the milk of a nursing mother.

negative control A form of gene regulation where a repressor protein binds to an operator and blocks transcription.

negative selection A method for identifying mutations by selecting cells or colonies that do not grow when replica plated; *see also* positive selection.

negative stain technique A staining process that results in colorless bacterial cells on a stained background when viewed with the light microscope.

negative strand Referring to those RNA viruses whose genome cannot be directly transcribed into protein.

Negri body A cytoplasmic inclusion that occurs in brain cells infected with rabies viruses.

neuraminidase An enzyme composing one type of surface spike of influenza viruses that facilitates viral release from the host cell.

neuraminidase inhibitor A compound that inhibits neuraminidase, preventing the release of flu viruses from an infected cell.

neurotoxin A toxin that is active in the nervous system of the host.

neutralization A type of antigen-antibody reaction in which the activity of a toxin is inactivated.

neutron An uncharged particle in the atomic nucleus.

neutrophil The most common type of white blood cell; functions chiefly to engulf and destroy foreign material, including bacterial cells and viruses that have entered the body.

night soil Human feces sometimes used as an agricultural fertilizer.

nitrogen cycle The processes that convert nitrogen gas (N_2) to nitrogen-containing substances in soil and living organisms, then reconverted to the gas.

nitrogen fixation The chemical process by which microorganisms convert nitrogen gas (N_2) into ammonia.

nitrogenous base Any of five nitrogen-containing compounds found in nucleic acids, including adenine, guanine, cytosine, thymine, and uracil.

nitroimidazole An antiprotozoal drug that interferes with DNA synthesis.

nitrous acid A chemical mutagen that converts adenine bases to hypoxanthine bases.

noncommunicable disease A disease whose causative agent is acquired from the environment and is not transmitted to another individual.

noncompetitive inhibition The prevention of a chemical reaction by a chemical that binds elsewhere than to active site of an enzyme; *see also* competitive inhibition.

nonhemolytic Referring to those bacterial species that when plated on blood agar cause no destruction of the red blood cells.

nonpolar molecule A covalently bonded substance in which there is no electrical charge; *see also* polar molecule.

nonseptate Referring to the lack of cross-walls in the filaments of many fungi.

nosocomial infection A disorder acquired during an individual's stay at a hospital or chronic care facility.

nucleic acid A high-molecular-weight molecule consisting of nucleotide chains that convey genetic information and are found in all living cells and viruses; *see* deoxyribonucleic acid *and* ribonucleic acid.

nucleocapsid The combination of genome and capsid of a virus.

nucleoid The chromosomal region of a prokaryotic cell.

nucleotide A component of a nucleic acid consisting of a carbohydrate molecule, a phosphate group, and a nitrogenous base.

nucleus (pl. **nuclei**) (1) The portion of an atom consisting of protons and neutrons. (2) A membrane-enclosed compartment in eukaryotic cells that contains the chromosomes.

nutrient agar A growth medium containing nutrients in a solidified medium.

nutrient broth A growth medium containing nutrients in a liquid medium.

nystatin A polyene drug used to treat infections caused by *Candida albicans*.

O

obligate aerobe An organism that requires oxygen gas (O_2) for metabolism.

obligate anaerobe An organism that cannot use oxygen gas (O_2) for metabolism.

observation The use of the senses or instruments to gather information on which science inquiry is based.

Okazaki fragment A segment of DNA resulting from discontinuous DNA replication.

oncogene A segment of DNA that can induce uncontrolled growth of a cell if permitted to function.

oncogenic Referring to any agent such as viruses that can cause tumors.

oocyst An oval body in the reproduction cycle of certain protozoa that develops by a complex series of asexual and sexual processes.

operator A sequences of bases in the DNA to which a repressor protein can bind.

operon The unit of bacterial DNA consisting of a promoter, operator, and a set of structural genes.

ophthalmia Severe inflammation of the eye.

opisthotonus An arching of the back that is characteristic of tetanus.

opportunist A microorganism that invades the tissues when body defenses are suppressed.

opportunistic infection A disorder caused by a microorganism that does not cause disease but that can become pathogenic or life-threatening if the host has a low level of immunity.

opsonin An antibody or complement component that encourages phagocytosis.

opsonization Enhanced phagocytosis due to the activity of antibodies or complement.

oral rehydration solution (**ORS**) A mixture of blood salts and glucose in water.

orchitis A condition caused by the mumps virus in which the virus damages the testes.

organelle A specialized compartment in cells that has a particular function.

organic compound A substance characterized by chains or rings of carbon atoms that are linked to atoms of hydrogen and sometimes oxygen, nitrogen, and other elements.

origin of replication The fixed point on a DNA molecule where copying of the molecule starts.

origin of transfer The fixed point on an F plasmid (factor) where one strand is nicked and transferred to a recipient cell.

osmosis The net movement of water molecules from where they are in a high concentration through a semipermeable membrane to a region where they are in a lower concentration.

osmotic pressure The force that must be applied to a solution to inhibit the inward movement of water across a membrane.

outbreak A small, localized epidemic.

outer membrane A bilayer membrane forming part of the cell wall of gram-negative bacteria.

oxazolidinone An antibiotic that blocks protein synthesis and is effective in treating gram-positive bacteria.

oxidation A chemical change in which electrons are lost by an atom; *see also* reduction.

oxidation lagoon A large pond in which sewage is allowed to remain undisturbed so that digestion of organic matter can occur.

oxidative phosphorylation A series of sequential steps in which energy is released from electrons as they pass from coenzymes to cytochromes, and ultimately, to oxygen gas (O_2); the energy is used to combine phosphate ions with ADP molecules to form ATP molecules.

P

pandemic A worldwide epidemic.

papule A pink pimple on the skin.

para-aminobenzoic acid (**PABA**) A precursor for folic acid synthesis.

parabasalid A protozoan that contains numerous of flagella and lives in low oxygen or anaerobic environments; most members are symbiotic in animals.

parasite A type of heterotrophic organism that feeds on live organic matter such as another organism.

parasitemia The spread of protozoa and multicellular worms through the circulatory system.

parasitism A close association between two organisms in which one (the parasite) feeds on the other (the host) and may cause injury to the host.

parasitology The scientific study of parasites.

paroxysm A sudden intensification of symptoms, such as a severe bout of coughing.

passive agglutination An immunological procedure in which antigen molecules are adsorbed to the surface of latex spheres or other carriers that agglutinate when combined with antibodies.

passive immunity The temporary immunity that comes from receiving antibodies from another source.

pasteurization A heating process that destroys pathogenic bacteria in a fluid such as milk and lowers the overall number of bacterial cells in the fluid.

pasteurizing dose The amount of irradiation used to eliminate pathogens.

pathogen A microorganism or virus that causes disease in a host organism.

pathogen-associated molecular pattern (**PAMP**) A unique microbial molecular sequence recognized by innate immune system receptors.

pathogenicity The ability of a disease-causing agent to gain entry to a host and bring about a physiological or anatomical change interpreted as disease.

pathogenicity island A set of adjacent genes that encode virulence factors.

pellicle A flexible covering layer typical of the protozoan ciliates.

pelvic inflammatory disease (**PID**) A disease of the pelvic organs; often a complication of a sexually transmitted disease.

penicillin Any of a group of antibiotics derived from *Penicillium* species or produced synthetically; effective against gram-positive bacteria and several gram-negative bacteria by interfering with cell wall synthesis.

penicillinase An enzyme produced by certain microorganisms that converts penicillin to penicilloic acid and thereby confers resistance against penicillin.

peptide bond A linkage between the amino group on one amino acid and the carboxyl group on another amino acid.

peptidoglycan A complex molecule of the bacterial cell wall composed of alternating units of N-acetylglucosamine and N-acetylmuramic acid cross linked by short peptides.

perforin A protein secreted by cytotoxic T lymphocytes and natural killer cells that forms holes in the plasma membrane of a targeted infected cell.

period of convalescence The phase of a disease during which the body's systems return to normal.

period of decline The phase of a disease during which symptoms subside.

periplasmic space A metabolic region between the cell membrane and outer membrane of gram-negative cells.

pH An abbreviation for the hydrogen ion concentration [H^+] of a solution.

phage *See* bacteriophage.

phage typing A procedure of using specific bacterial viruses to identify a particular strain of a bacterial species.

phagocyte A white blood cell capable of engulfing and destroying foreign materials, including bacterial cells and viruses.

phagocytosis A process by which foreign material or cells are taken into a white blood cell and destroyed.

phagolyososme A membrane-enclosed compartment resulting from the fusion of a phagosome with lysosomes and in which the foreign material is digested.

phagosome A membrane-enclosed compartment containing foreign material or infectious agents that the cell has engulfed.

pharyngitis An inflammation of the pharynx; commonly called a sore throat.

phase-contrast microscopy An optical system on the light microscope that uses a special condenser and objective lenses to examine cell structure.

phenol A chemical compound that has one or more hydroxyl groups attached to a benzene ring and derivatives are used as an antiseptic or disinfectant; also called carbolic acid.

phenol coefficient (**PC**) A number that indicates the effectiveness of an antiseptic or disinfectant compared to phenol.

phenotype The visible (physical) appearance of an organism resulting from the interaction between its genetic makeup and the environment.

phospholipid A water-insoluble compound containing glycerol, two fatty acids, and a phosphate head group; forms part of the membrane in all cells.

phosphorylation The addition of a phosphate group to a molecule.

photoautotroph An organism that uses light energy to synthesize nutrients from carbon dioxide gas (CO_2).

photoheterotroph An organism that uses light energy to synthesize nutrients from organic carbon compounds.

photophobia Sensitivity to bright light.

photophosphorylatyion The generation of ATP through the trapping of light.

photosynthesis A biochemical process in which light energy is converted to chemical energy, which is then used for carbohydrate synthesis.

photosystem A group of pigments that act as a light trapping system for photosynthesis.

pH scale A range of values that extends from 1 to 14 and indicates the degree of acidity or alkalinity of a solution.

phycology The scientific study of algae.

phylogeny The evolutionary history and relationships among a group of organisms.

physical pollution The presence of particulate matter in water.

phytoplankton Microscopic free-floating communities of cyanobacteria and unicellular algae.

pilus (pl. **pili**) A short hair-like structure used by bacterial cells for attachment.

pinkeye *See* conjunctivitis.

planktonic bacteria Referring to bacterial cells that live as individual cells.

plaque (1) A clear area on a lawn of bacterial cells where viruses have destroyed the bacterial cells. (2) The gummy layer of gelatinous material consisting of bacterial cells and organic matter on the teeth.

plasma The fluid portion of blood remaining after the cells have been removed; *see also* serum.

plasma cell An antibody-producing cell derived from B lymphocytes.

plasma membrane The phospholipid bilayer with proteins that surrounds the eukaryotic cell cytoplasm; *see also* cell membrane.

plasmid A small, closed-loop molecule of DNA apart from the chromosome that replicates independently and carries non-essential genetic information.

pleomorphic Referring to an organism that occurs in a variety of shapes.

pneumonia An inflammation of the bronchial tubes and one or both lungs.

point mutation The replacement of one base in a DNA strand with another base.

polar molecule A substance with electrically-charged poles; *see also* nonpolar molecule.

polyclonal antibodies Antibodies produced from different clones of B cell.

polyene An antifungal drug that destroys the plasma membrane; examples include nystatin.

polymer A substance formed by combining smaller molecules into larger ones; *see also* monomer.

polymerase chain reaction (PCR) A technique used to replicate a fragment of DNA many times.

polymicrobial disease A disorder caused by more than one infectious agent.

polymyxin An antibiotic derived from a *Bacillus* species that disrupts the cell membrane of gram-negative rods.

polynucleotide A chain of linked nucleotides.

polypeptide A chain of linked amino acids.

polysaccharide A complex carbohydrate made up of sugar molecules linked into a branched or chain structure; examples include starch and cellulose.

polysome A cluster of ribosomes linked by a strand of mRNA and all translating the mRNA.

polyvalent serum Blood fluid that contains a mixture of antibodies.

porin A protein in the outer membrane of gram-negative bacteria that acts as a channel for the passage of small molecules.

portal of entry The site at which a pathogen enters the host.

portal of exit The site at which a pathogen leaves the host.

positive selection A method for selecting mutant cells by their growth as colonies on agar.

positive strand Referring to the RNA viruses whose genome consists of a mRNA molecule.

post-exposure immunization The receiving of a vaccine after contracting the pathogen.

post-polio syndrome A condition that affects polio survivors years after recovery from an initial acute attack by the polio virus.

potability Referring to water that is safe to drink because it contains no harmful material or microbes.

pour plate method A process by which a mixed culture can be separated into pure colonies and the colonies isolated; *see also* streak plate method.

pox Pitted scars remaining on the skin of individuals who have recovered from smallpox.

praziquantel An antihelminthic drug that alters the permeability of the plasma membrane.

precipitation A type of antigen-antibody reaction in which thousands of molecules of antigen and antibody cross-link to form visible aggregates.

prevacuum autoclave An instrument that uses saturated steam at high temperatures and pressure for short time periods to sterilize materials.

primary antibody response The first contact between an antigen and the immune system, characterized by the synthesis of IgM and then IgG antibodies; *see also* secondary antibody response.

primary cell culture Animal cells separated from tissue and grown in cell culture.

primary immunodeficiency A congenital origin for the inability to of the body to produce an adequate immune response to fight disease.

primary infection A disease that develops in an otherwise healthy individual.

primary lymphoid tissue A site where immune cells form and mature; examples are the thymus and bone marrow; *see also* secondary lymphoid tissue.

primary metabolite A small molecule essential to the survival and growth of an organism; *see also* secondary metabolite.

primary structure The sequence of amino acids in a polypeptide.

prion An infectious, self-replicating protein involved in human and animal diseases of the brain.

proctitis Infection of the rectum.

prodromal phase The phase of a disease during which general symptoms occur in the body.

product A substance or substances resulting from a chemical reaction.

proglottid One of a series of segments that make up the body of a tapeworm.

prokaryote A microorganism in the domain *Bacteria* or *Archaea* composed of single cells having a single chromosome but no cell nucleus or other membrane-bound compartments; *see also* eukaryote.

prokaryotic Referring to cells or organisms having a single chromosome but no cell nucleus or other membrane-bound compartments.

promoter The region of a template DNA strand or operon to which RNA polymerase binds.

Prontosil A red dye found by Domagk to have significant antimicrobial activity when tested in live animals, and from which sulfanilamide was later isolated.

prophage The viral DNA of a bacterial virus that is inserted into the bacterial DNA and is passed on from one generation to the next during binary fission.

prophylactic serum Blood fluid rich in antibodies and used to protect against the development of a disease.

propionic acid A chemical preservative used in cheese, breads, and other bakery products.

prostaglandin A hormone-like substance that acts as a mediator in type I hypersensitivity reactions.

protease inhibitor A compound that breaks down the enzyme protease, inhibiting the replication of some viruses, such as HIV.

protein A chain or chains of linked amino acids used as a structural material or enzyme in living cells.

protein-only hypothesis The idea that prions are composed solely of protein and contain no nucleic acid.

protein synthesis The process of forming a polypeptide or protein through a series of chemical reactions involving amino acids.

proteobacteria A large group of gram-negative, chemoheterotrophic species in the domain *Bacteria* that are defined primarily in terms of their ribosomal RNA (rRNA) sequences; examples include *Escherichia coli*, *Salmonella*, and the rickettsia.

Protista One of the five kingdoms in the Whittaker classification of living things, composed of the protozoa and unicellular algae.

proton A positively charge particle in the atomic nucleus.

proto-oncogene A region of DNA in the chromosome of human cells; they are altered by carcinogens into oncogenes that transform cells.

prototroph An organism that contains all its nutritional needs; *see also* auxotroph.

protozoan (pl. **protozoa**) A single-celled eukaryotic organism that lacks a cell wall and usually exhibits chemoheterotrophic metabolism.

protozoology The scientific study of protozoa.

provirus The viral DNA that has integrated into a eukaryotic host chromosome and is then passed on from one generation to the next through cell division.

pruritis Itching sensation.

pseudomembrane An accumulation of mucus, leukocytes, bacteria, and dead tissue in the respiratory passages of diphtheria patients.

pseudopeptidoglycan A complex molecule of some archaeal cell walls composed of alternating units of N-acetylglucosamine and N-acetyltalosamine uronic acid.

pseudopod A projection of the plasma membrane that allows movement in members of the amoebozoans.

psychrophile An organism that lives at cold temperature ranges of 0°C to 20°C.

pure culture An accumulation or colony of microorganisms of one species.

pus A mixture of dead tissue cells, leukocytes, and bacteria that accumulates at the site of infection.

pustule A raised bump on the skin that contains pus.

pyelonephritis An inflammation of the kidney.

pyrogen A fever-producing substance.

pyruvate The end product of the glycolysis metabolic pathway.

Q

quat *See* quaternary ammonium compound.

quaternary ammonium compound (**quat**) A positively charge detergent with four organic groups attached to a central nitrogen atom; used as a disinfectant.

quaternary structure The association of two or more polypeptides in a protein.

question A statement written from an observation and used to formulate a hypothesis.

quinolone A synthetic antimicrobial drug that blocks DNA synthesis.

quorum sensing The ability of bacteria to chemically communicate and coordinate behavior via signaling molecules.

R

radioallergosorbent test (**RAST**) A type of radioimmunoassay in which antigens for the unknown antibody are attached to matrix particles.

radioimmunoassay (**RIA**) An immunological procedure that uses radioactive-tagged antigens to determine the identity and amount of antibodies in a sample.

radioisotope A unstable form of a chemical element that is radioactive; *see also* isotope.

radiolarian A single-celled marine organism with a round silica-containing shell that has radiating arms to catch prey.

reactant A substance that interacts with another in a chemical reaction.

recombinant DNA molecule A DNA molecule containing DNA from two different sources.

recombinant F^- A cell that received a few chromosomal genes and partial F factor genes from a donor cell during conjugation.

recombinant subunit vaccine The synthesis of antigens in a microorganism using recombined genes for the purpose of producing a vaccine.

red tide A brownish-red discoloration in seawater caused by increased numbers of dinoflagellates; *see also* algal bloom.

reduction The gain of electrons by a molecule.

regulatory gene A DNA segment that codes for a repressor protein.

regulatory T cell A population of lymphocytes that prevent other T lymphocytes from attacking self.

releasing factor A protein that triggers the release of the polypeptide from the ribosome.

remerging infectious disease A disease showing a resurgence in incidence or a spread in its geographical area; *see also* emerging infectious disease.

rennin An enzyme that accelerates the curdling of protein in milk.

replication fork The point where complementary strands of DNA separate and new complementary strands are synthesized.

repressor protein A protein that when bound to the operator blocks transcription.

reservoir of infection The location or organism where disease-causing agents exist and maintain their ability for infection.

resolving power The numerical value of a lens system indicating the size of the smallest object that can be seen clearly when using that system.

respiratory droplet Small liquid droplets expelled by sneezing or coughing.

restriction endonuclease A type of enzyme that splits open a DNA molecule at a specific restricted point; important in genetic engineering techniques.

reticulate body The replicating, intracellular, noninfectious stage of *Chlamydia trachomatis*.

reverse transcriptase An enzyme that synthesizes a DNA molecule from the code supplied by an RNA molecule.

reverse transcriptase inhibitor A compound that inhibits the action of reverse transcriptase, preventing the viral genome from being replicated.

revertant Referring to a mutant organism or cell that has reacquired its original phenotype or metabolic ability.

Reye syndrome A complication of influenza and chickenpox, characterized by vomiting and convulsions as well as liver and brain damage.

R group The side chain on an amino acid.

rhabditiform Referring to the elongated, rod-like shape of the larvae of the hookworm.

Rh disease *See* hemolytic disease of the newborn.

rheumatoid arthritis An autoimmune disorder characterized by immune complex formation in the joints.

RhoGAM Rh-positive antibodies.

ribonucleic acid (RNA) The nucleic acid involved in protein synthesis and gene control; also the genetic information in some viruses.

ribosomal RNA (rRNA) An RNA transcript that forms part of the ribosome's structure.

ribosome A cellular structure made of RNA and protein that participates in protein synthesis.

rice-water stool A colorless, watery diarrhea containing particles of intestinal tissue in cholera patients.

rickettsia (pl. **rickettsiae**) A very small bacterial cell generally transmitted by arthropods; most rickettsiae are cultivated only within living tissues.

rifampin An antibiotic prescribed for tuberculosis and leprosy patients and for carriers of *Neisseria* and *Haemophilus* species.

Rivers' postulates A set of procedures by which a specific virus can be associated with a specific disease.

RNA interference (RNAi) A process whereby translation is silenced by the binding of double-stranded RNA to specific messenger RNA molecules.

RNA polymerase The enzyme that synthesizes an RNA polynucleotide from a DNA template.

rolling circle mechanism A type of DNA replication in which a strand of DNA "rolls off" the loop and serves as a template for the synthesis of a complementary strand of DNA.

rose spots Bright red skin spots associated with diseases such as typhoid fever and relapsing fever.

roundworm A multicellular parasite with a round body; examples include the nematodes.

R plasmid A small, circular DNA molecule that occurs frequently in bacterial cells and carries genes for drug resistance.

S

salpingitis Blockage of the fallopian tubes; a possible complication of a sexually transmitted disease.

Salvarsan The first modern synthetic antimicrobial agent.

sanitization To remove microbes or reduce their populations to a safe level as determined by public health standards.

saprobe A type of heterotrophic organism that feeds on dead organic matter, such as rotting wood or compost.

sarcina (pl. **sarcinae**) (1) A packet of eight spherical-shaped prokaryotic cells. (2) A genus of gram-positive, anaerobic spheres.

saturated fat A water-insoluble compound that cannot incorporate any additional hydrogen atoms; *see also* unsaturated fat.

scanning electron microscope (SEM) The type of electron microscope that allows electrons to scan across an object, generating a three-dimensional image of the object.

Schick test A skin test used to determine the effectiveness of diphtheria immunization.

schmutzdecke In water purification, a slimy layer of microorganisms that develops in a slow sand filter.

scientific inquiry The way a science problem is investigated by formulating a question, collecting data about it through observation and experiment, and interpreting the results; also called scientific method.

sclerotium (pl. **sclerotia**) A hard purple body that forms in grains contaminated with *Claviceps purpurea*.

scolex The head region of a tapeworm where the attachment organ is located.

secondary antibody response A second or ensuing response triggered by memory cells to an antigen and characterized by substantial production of IgG antibodies; *see also* primary antibody response.

secondary immunodeficiency An acquired origin for the inability to of the body to produce an adequate immune response to fight disease.

secondary infection A disorder caused by an opportunistic microbe as a result of a primary infection weakening the host.

secondary lymphoid tissue A site where mature immune cells interact with pathogens; examples includes the spleen and lymph nodes; *see also* primary lymphoid tissue.

secondary metabolite A small molecule not essential to the survival and growth of an organism; *see also* primary metabolite.

secondary structure The region of a polypepetide folded into an alpha helix or pleated sheet.

sedimentation The removal of soil particulates from water.

selective medium A growth medium that contains ingredients to inhibit certain microorganisms while encouraging the growth of others.

selective toxicity A property of many antimicrobial drugs that harm the infectious agent but not the host.

semiconservative replication The DNA copying process where each parent (old) strand serves as a template for a new complementary strand.

semisynthetic drug A chemical substance synthesized from natural and lab components used to treat disease.

sensitized Referring to an individual sensitive to an allergen.

senescence Deteriorative changes in a cell or organism with aging.

sensitizing dose The first exposure to an allergy-causing antigen.

sepsis The growth and spreading of bacteria or their toxins in the blood and tissues.

septate Referring to the cross-walls formed in the filaments of many fungi.

septic shock A collapse of the circulatory and respiratory systems caused by an overwhelming immune response.

septicemia A growth and spreading of bacterial cells in the bloodstream.

septum (pl. **septa**) A cross-wall in the hypha of a fungus.

seroconversion The time when antibodies to a disease agent can be detected in the blood.

serogroup A cluster of microorganisms that only differing from one another their antigen.

serological reaction An antigen-antibody reaction studied under laboratory conditions and involving serum.

serology A branch of immunology that studies serological reactions.

serotype A rank of classification below the species level based on an organism's reaction with antibodies in serum; also called serovar.

serovar *See* serotype.

serum (pl. **sera**) The fluid portion of the blood consisting of water, minerals, salts, proteins, and other organic substances, including antibodies; contains no clotting agents; *see also* plasma.

serum sickness A type of hypersensitivity reaction in which the body responds to proteins contained in foreign serum.

sexually transmitted disease (STD) A disease such as gonorrhea or chlamydia that is normally passed from one person to another through sexual activity.

Shiga toxin An exotoxin that triggers gastroenteritis.

sign An indication of the presence of a disease, especially one observed by a doctor but not apparent to the patient; *see also* symptom.

simple stain technique The use of a single cationic dye to contrast cells; *see also* differential stain technique.

single-dose vaccine The combination of several vaccines into one measured quantity.

sinusitis An inflammation of the membrane lining a sinus.

S-layer The cell wall of most archaeal species consisting of protein or glycoprotein assembled in a crystalline lattice.

slide agglutination test *See* VDRL test.

slime layer A thin, loosely bound layer of polysaccharide covering some prokaryotic cells; *see also* capsule *and* glycocalyx.

sludge The solids in sewage that separate out during sewage treatment.

sludge tank The area in which secondary water treatment occurs.

soap A compound made by potassium or sodium hydroxide reacting with fatty acids.

solute A substance dissolved in another substance; the latter usually is water in biological systems.

solvent A substance able to dissolve other substances; the former usually is water in biological systems.

somatic recombination The reshuffling of antibody genetic segments in a B lymphocyte as it matures.

sour curd The acidification of milk, causing a change in the structure of milk proteins.

spawn A mushroom mycelium used to start a new culture of the fungus.

specialized transduction The transfer of a few bacterial genes by a bacterial virus that carries the genes to another bacterial cell; *see also* generalized transduction.

species The fundamental rank in the classification system of organisms.

specific epithet The second of the two scientific names for a species; *see also* genus.

spherule A stage in the life cycle of the fungus *Coccidioides immitis*.

spike A protein projecting from the viral envelope or capsid that aids in attachment and penetration of a host cell.

spiral A shape of many prokaryotic cells.

spirillum (pl. **spirilla**) (1) A bacterial cell shape characterized by twisted or curved rods. (2) A genus of aerobic, helical cells usually with many flagella.

spirochete A twisted bacterial rod with a flexible cell wall containing endoflagella for motility.

spleen An organ in the left upper abdomen of humans that helps to destroy pathogens.

spontaneous generation The doctrine that nonliving matter could spontaneously give rise to living things.

spontaneous mutation A mutation that arises from natural phenomena in the environment.

spore (1) A reproductive structure formed by a fungus. (2) A highly resistant dormant structure formed from vegetative cells in several genera of bacteria, including *Bacillus* and *Clostridium*; *see also* endospore.

sporicide An agent that kills bacterial spores.

sporozoite A stage in the life cycle of the malaria parasite that enters the human body.

sporulation The process of spore formation.

sputum Thick, expectorated matter from the lower respiratory tract.

standard plate count A procedure to estimate the number of cells in a sample dilution spread on an agar plate.

staphylococcus (pl. **staphylococci**) (1) An arrangement of bacterial cells characterized by spheres in a grapelike cluster. (2) A genus of facultatively anaerobic, nonmotile, nonsporeforming, gram-positive spheres in clusters.

start codon The starting nucleotide sequence (AUG) in translation.

stationary phase The portion of a bacterial growth curve in which the reproductive and death rates of cells are equal.

stem cell An undifferentiated cell from which specialized cells arise.

sterile Free from living microorganisms, spores, and viruses.

sterilization The removal of all life forms, including bacterial spores.

sterol A organic solid containing several carbon rings with side chains; examples include cholesterol.

stop codon The nucleotide sequence that terminates translation.

streak plate method A process by which a mixed culture can be streaked onto an agar plate and pure colonies isolated; *see also* pour plate method.

streptobacillus (pl. **streptobacilli**) (1) A chain of bacterial rods. (2) A genus of facultatively anaerobic, nonmotile, gram-negative rods.

streptococcus (pl. **streptococci**) (1) A chain of bacterial cocci. (2) A genus of facultatively anaerobic, nonmotile, nonspore-forming, gram-positive spheres in chains.

streptogramin An antibiotic that block protein synthesis.

streptokinase An enzyme that dissolves blood clots; produced by virulent streptococci.

streptomycin A drug still used to treat tuberculosis.

structural formula A chemical diagram representing the arrangement of atoms and bonds within a molecule.

structural gene A segment of a DNA molecule that provides the biochemical information for a polypeptide.

subclinical disease A disease in which there are few or inapparent symptoms.

substrate The substance or substances upon which an enzyme acts.

subunit vaccine A vaccine that contains parts of microorganisms, such as capsular polysaccharides or purified fimbriae.

sulfonamide A synthetic, sulfur-containing antibacterial agent; also called sulfa drug.

sulfur cycle The processes by which sulfur moves through and is recycled in the environment.

sulfur dioxide A chemical preservative used in dried fruits.

superantigen An antigen that stimulates an immune response without any prior processing.

supercoiled domain A loop of wound DNA consisting of 10,000 bases.

supercoiling The process by which a chromosome is twisted and packed.

superinfection The overgrowth of susceptible strains by antibiotic resistant ones.

suppressor T cell A group of lymphocytes that regulate IgE antibody production.

surface water The water in lakes, streams, ands hallow wells.

surfactant A synthetic chemical, such as a detergent, that emulsifies and solubilizes particles attached to surfaces by reducing the surface tension.

symbiosis An interrelationship between two populations of organisms where there is a close and permanent association.

symptom An indication of some disease or other disorder that is experienced by the patient; *see also* sign.

syncytium (pl. **syncytia**) A giant tissue cell formed by the fusion of cells infected with respiratory syncytial viruses.

syndrome A collection of signs or symptoms that together are characteristic of a disease.

synthetic biology A field of study that attempts to "build" new living organisms by combining parts of other species.

synthetic drug (**agent**) A substance made in the lab to prevent illness or treat disease.

synthetic medium A chemically defined medium in which the nature and quantity of each component is identified; *see also* complex medium.

synthetic vaccine A vaccine that contains chemically synthesized parts of microorganisms, such as proteins normally found in viral capsids.

systemic anaphylaxis The release of cell mediators throughout the body.

systemic disease A disorder that disseminates to the deeper organs and systems of the body.

T

tapeworm *See* cestode.

taxon Subdivisions used to classify organisms.

taxonomy The science dealing with the systematized arrangements of related living things in categories.

T cell *See* T lymphocyte.

T-dependent antigen An antigen that requires the assistance of T_H2 lymphocytes to stimulate antibody-mediated immunity.

teichoic acid A negatively charged polysaccharide in the cell wall of gram-positive bacteria.

temperate Referring to a bacterial virus that enters a bacterial cell and then the viral DNA integrates into the bacterial cell's chromosome.

termination (1) The completion of DNA synthesis during DNA replication. (2) The release of a polypeptide from a ribosome during translation.

tertiary structure The folding of a polypeptide back on itself.

tetanospasmin An exotoxin produced by *Clostridium tetani* that acts at synapses, thereby stimulating muscle contractions.

tetracycline An antibiotic characterized by four benzene rings with attached side groups that blocks protein synthesis in many gram-negative bacteria, rickettsiae, and chlamydiae.

tetrad An arrangement of four bacterial cells in a cube shape.

theory A scientific explanation supported by many experiments done by separate individuals.

therapeutic dose The concentration of an antimicrobial drug that effectively destroys an infectious agent.

therapeutic serum Antibody-rich serum used to treat a specified condition.

thermal death point (**TDP**) The temperature required to kill a bacterial population in a given length of time.

thermal death time (**TDT**) The length of time required to kill a bacterial population at a given temperature.

thermoacidophile An archaeal organism living under high temperature and very low pH conditions.

thermoduric Referring to an organism that tolerates the heat of the pasteurization process.

thermophile An organism that lives at high temperature ranges of 40°C to 90°C.

thioglycollate broth A microbiological medium containing a chemical that binds oxygen from the atmosphere and creates an environment suitable for anaerobic growth.

three domain system The classification scheme placing all living organisms into one of three groups based, in part, on ribosomal RNA sequences.

thymus A flat, bilobed organ where T lymphocytes mature.

tincture A substance dissolved in ethyl alcohol.

T-independent antigen An antigen that does not require the assistance of T_H2 lymphocytes to stimulate antibody production.

tinea Any of a group of fungal infections of the skin, feet, or scalp.

tissue tropism Refers to the specific tissues within a host that a virus infects.

titer A measurement of the amount of antibody in a sample of serum that is determined by the most dilute concentration of antibody that will yield a positive reaction with a specific antigen.

titration A method of calculating the concentration of a dissolved substance, such as an antibody, by adding quantities of a reagent of known concentration to a known volume of test solution until a reaction occurs.

T lymphocyte (**T cell**) A type of white blood cell that matures in the thymus gland and is associated with cell-mediated immunity.

toll-like receptor (**TLR**) A signaling molecule on immune cells that recognizes a unique molecular pattern on an infectious agent.

TORCH An acronym for four diseases that pass from the mother to the unborn child: toxoplasmosis, rubella, cytomegalovirus disease, and herpes simplex; the O stands for other diseases.

total magnification The enlargement of an object using the ocular lens and a specific objective lens of the microscope.

toxemia The presence of toxins in the blood.

toxic dose (1) The amount of toxin need to cause a disease. (2) The amount of an antimicrobial drug that causes harm to the host.

toxigenicity The ability of an organism to produce a toxin.

toxin A poisonous chemical substance produced by an organism.

toxoid A preparation of a microbial toxin that has been rendered harmless by chemical treatment but that is capable of stimulating antibodies; used as vaccines.

transcription The biochemical process in which RNA is synthesized according to a code supplied by the bases of a gene in the DNA molecule.

transduction The transfer of a few bacterial genes from a donor cell to a recipient cell via a bacterial virus.

transfer RNA (**tRNA**) A molecule of RNA that unites with amino acids and transports them to the ribosome in protein synthesis.

transformation (1) The transfer and integration of DNA fragments from a dead and lysed donor cells to a recipient cell's chromosome. (2) The conversion of a normal cell into a malignant cell due to the action of a carcinogen or virus.

transgenic Referring to organisms containing DNA from another source.

transient microbiota The microbial agents that are associated with an animal for short periods of time without causing disease; *see also* indigenous microbiota.

translation The biochemical process in which the code on the mRNA molecule is converted into a sequence of amino acids in a polypeptide.

transmission electron microscope (**TEM**) The type of electron microscope that allows electrons to pass through the object, resulting in a detailed view of the object's structure.

transposable genetic element Fragments of DNA called insertion sequences or transposons that can cause mutations.

transposon A segment of DNA that moves from one site on a DNA molecule to another site, carrying information for protein synthesis; also called jumping gene.

trematode A flatworm, commonly known as a fluke, that lives as a parasite in the liver, gut, lungs, or blood vessels of vertebrates.

triclosan A phenol derivative incorporated as an antimicrobial agent into a wide variety of household products.

triglyceride A lipid consisting of a glycerol and three fatty acids.

trismus A sustained spasm of the jaw muscles, characteristic of the early stages of tetanus; also called lockjaw.

trivalent vaccine A vaccine consisting of three components, each of which stimulates immunity.

trophozoite The feeding form of a microorganism, such as a protozoan.

tube dilution method A procedure for determining bacterial susceptibility to an antibiotic by determining the minimal amount of the drug needed to inhibit growth of the pathogen; *see* minimal inhibitory concentration.

tubercle A hard nodule that develops in tissue infected with *Mycobacterium tuberculosis*.

tuberculin test A procedure performed by applying purified protein derivative from *Mycobacterium tuberculosis* to the skin and noting if a thickening of the skin with a raised vesicle appears within a few days; used to establish if someone has been exposed to the bacterium.

tumor An abnormal uncontrolled growth of cells that has no physiological function.

turbidity The cloudiness of a broth culture due to bacterial growth.

tyndallization *See* fractional sterilization.

U

ultrapasteurization method A treatment in which milk is heated at 82°C for 3 seconds to destroy pathogens.

ultrastructure The detailed structure of an cell, virus, or other object when viewed with the electron microscope.

ultraviolet (UV) light A type of electromagnetic radiation of short wavelengths that damages DNA.

uncoating Referring to the loss of the viral capsid inside an infected eukaryotic cell.

universal precautions Using those measures to avoid contact with a patient's bodily fluids; examples include wearing gloves, goggles, and proper disposal of used hypodermic needles.

unsaturated fat A water-insoluble compound that can incorporate additional hydrogen atoms; *see also* saturated fat.

urethritis An inflammation of the urethra.

urinary tract infection (UTI) A common infection, particularly in young women, caused by a variety of bacterial, fungal, or protozoal species.

urticaria A hive-like rash of the skin.

V

vaccination Inoculation with weakened or dead microbes, or viruses, in order to generate immunity; *see also* immunization.

vaccine A preparation containing weakened or dead microorganisms or viruses, treated toxins, or parts of microorganisms or viruses to stimulate immune resistance.

vaccine adverse events reporting system (VAERS) A reporting system designed to identify any serious adverse reactions to a vaccination.

vaginitis A general term for infection of the vagina.

vancomycin An antibacterial drug that inhibits cell wall synthesis and is used in treating diseases caused by gram-positive bacteria, especially staphylococci.

variable That part of an experiment exposed to or treated with the factor being tested.

variable region The different amino acids in different antibody light and heavy chains.

variolation A 14th to 18th century method to inoculate a susceptible person with material from a smallpox vesicle to render that person resistant to infection.

vasodilation A widening of the blood vessels, especially the arteries, leading to increased blood flow.

VBNC Referring to prokaryotes that are viable but not culturable.

VDRL test A screening procedure used in the detection of syphilis antibodies.

vector (1) An arthropod that transmits the agents of disease from an infected host to a susceptible host. (2) A plasmid used in genetic engineering to carry a DNA segment into a bacterium or other cell.

vertical gene transfer The passing of genes from one cell generation to the next; *see also* horizontal gene transfer.

vesicle (1) A fluid-filled skin lesion, such as that occurring in chickenpox. (2) A small, membrane-enclosed compartment found in many eukaryotic cells.

viable count The living cells identified from a standard plate count.

vibrio (1) A prokaryotic cell shape occurring as a curved rod. (2) A genus of facultatively anaerobic, gram-negative curved rods with flagella.

viral inhibition The prevention of a virus infection by antibodies binding to molecules on the viral surface.

viral load test A method used to detect the RNA genome of HIV.

viremia The presence and spread of viruses through the blood.

virion A completely assembled virus outside its host cell.

viroid An infectious RNA segment associated with certain plant diseases.

virology The scientific study of viruses.

virucide A drug or chemical that inactivates viruses.

virulence The degree to which a pathogen is capable of causing a disease.

virulence factor A structure or molecule possessed by a pathogen that increases its ability to invade or cause disease to a host.

virulent Referring to a virus or microorganism that can be extremely damaging when in the host.

virus An infectious agent consisting of DNA or RNA and surrounded by a protein sheath; in some cases, a membranous envelope surrounds the coat.

W

wart A small, usually benign skin growth commonly due to a virus.

wheal An enlarged, hive-like zone of puffiness on the skin, often due to an allergic reaction; *see also* flare.

white blood cell *See* leukocyte.

wild type The form of an organism or gene isolated from nature.

wort A sugary liquid produced from crushed malted grain and water to which is added yeast and hops for the brewing of beer.

X

xenograft A tissue graft between members of different species, such as between a pig and a human.

X ray An ionizing radiation that can be used to sterilize objects.

Y

yeast (1) A type of unicellular, nonfilamentous fungus that resembles bacterial colonies when grown in culture. (2) A term sometimes used to denote the unicellular form of pathogenic fungi.

Z

zone of equivalence The region in a precipitation reaction where ideal concentrations of antigen and antibody occur.

zoonosis (pl. **zoonoses**) An animal disease that may be transmitted to humans.

Zygomycota A phylum of fungi whose members have coenocytic hyphae and form zygospores, among other notable characteristics.

zygospore A sexually produced spore formed by members of the Zygomycota.

Index

NOTE: page numbers followed by *b* indicate MicroInquiry or MicroFocus boxes; *f*, figures, and *t*, tables.

A

C

D

E

F

H

N

O

P

Q

R

T

U

V

W

X

Y

Z

Photograph Acknowledgments

Note: All chapter-opening images are courtesy of Fred Murphy and Sylvia Whitfield/CDC.

CDC = Centers for Disease Control and Prevention, Public Health Service, U.S. Department of Health and Human Services, Atlanta, Georgia.

PART 1 OPENER © Arthur Siegelman/Visuals Unlimited.

PART 1 MICROBIOLOGY PATHWAYS Courtesy of the CDC.

CHAPTER 1

1.1A © Vincent Yu/AP Photos; **1.1B** Courtesy of CDC; **1.2A and B** © Library of Congress [LC-USZ62-95187]; **1.3A** Courtesy of Pfizer, Inc.; **1.3C** Courtesy of Royal Society, London; **1.4A** © National Library of Medicine; **1.5A** © National Library of Medicine; **1.5B** Courtesy of Pfizer, Inc.; **1.6** Courtesy of Frerichs, R. R., John Snow website: http://www.ph.ucla.edu/epi/snow.html. 2006; **1.6 insert** © National Library of Medicine; **1.7** Courtesy of Pfizer, Inc.; **1.8A** © National Library of Medicine; **1.9** © Mary Evans Picture Library/Alamy Images; **1.9 insert** © National Library of Medicine; **1.10A** © National Library of Medicine; **1.11** © Institut Pasteur, Paris; **1.13** © Time Magazine/Time & Life Pictures/Getty Images; **1.14A** © Dr. Arthur Siegelman/Visuals Unlimited; **1.14B** Courtesy of Roger Burks (University of California at Riverside), Mark Schneegurt (Wichita State University), and Cyanosite (www.cyanosite.bio.purdue.edu); **1.14C** © Dr. Hans Gelderblom/Visuals Unlimited; **1.14D** © Jones and Bartlett Publishers. Photographed by Kimberly Potvin; **1.14E** © Wim van Egmond/Visuals Unlimited; **1.14F** © Phototake/Alamy Images; **1.15A** © Professor P. Motta and T. Naguro/Photo Researchers, Inc.; **1.15B** © Dr. Dennis Kunkel/Visuals Unlimited; **1.16A** Courtesy of Pfizer, Inc.; **1.16B** © St. Mary's Hospital Medical School/Photo Researchers, Inc.; **1.17** Courtesy of Pfizer, Inc.; **1.18A** Reproduced with permission of the New York State Department of Health; **1.18B** © Photodisc; **1.19A** Courtesy of Mike Cox, President of Anaerobe Systems; **1.19B** Courtesy of the Exxon Valdez Oil Spill Trustee Council/NOAA; **MicroFocus 1** © Bettman/Corbis; **MicroFocus 2A** © Dr. Dennis Kunkel/Visuals Unlimited; **MicroFocus 2B** Copyright (2004) National Academy of Sciences, U.S.A. Photo courtesy of Vincent Noireaux and Albert Libchaber; **MicroFocus 3** Courtesy of the Public Health Foundation/CDC; **MicroFocus 4** © National Library of Medicine.

CHAPTER 2

2.1A Courtesy of Jim Peaco/Yellowstone National Park; **2.1B** Courtesy of Burkhard Büedel; **2.1C** Courtesy of D. Hardesty/USGS; **MicroFocus 1** Courtesy of ESA–D. Ducros; **MicroFocus 4** © Dr. David Phillips/Visuals Unlimited; **Textbook Case** Courtesy of CDC.

CHAPTER 3

3.1 Courtesy of SeaWiFS Project and NASA GSFC Scientific Visualization Studio; **3.3** © MycoAlbum CD, George Barron; **3.4** © Dr. Dennis Kunkel/Visuals Unlimited; **3.8** Courtesy of Biolog; **3.10A** Courtesy of Carl Zeiss MicroImaging, Inc.; **3.12B** © Jack Bostrack/Visuals Unlimited; **3.13** Courtesy of Dr. George P. Kubica/CDC; **3.14A, B, and C** © David M. Phillips/Visuals Unlimited; **3.15** © Dr. Peter Lewis, The University of Newcastle, Australia; **3.16A** Courtesy of LEO Electron Microscopy; **3.17A** © CNRI/Photo Researchers, Inc.; **3.17B** © Dr. Dennis Kunkel/Visuals Unlimited; **Application question** © Andrew Syred/Photo Researchers, Inc.; **MicroFocus 1** © Michael Abbey/Visuals Unlimited; **MicroFocus 5** Courtesy of Heide Schulz, Max Planck Institute of Marine Microbiology, Germany; **Textbook Case** Courtesy of Dr. W. A. Clark/CDC.

PART 2 OPENER © Dr. David Phillips/Visuals Unlimited.

PART 2 MICROBIOLOGY PATHWAYS Courtesy of Bill Branson/National Cancer Institute.

CHAPTER 4

4.2A © Barrie Rokeach/Alamy Images; **4.2B** Courtesy of OAR/National Undersea Research Program (NURP)/NOAA; **4.4A** © Dr. David M. Phillips/Visuals Unlimited; **4.4B** © David B. Fankhauser. University of Cincinnati; http://biology.clc.uc.edu/Fankhauser/; **4.4C** © Dr. Fred Hossler/Visuals Unlimited; **4.5** © Michael Abbey/Visuals Unlimited; **4.6** © George Chapman/Visuals Unlimited; **4.8** © Dr. Dennis Kunkel/Visuals Unlimited; **4.9A** Photograph by Gary Gaard. Courtesy of Dr. A. Kelman (Department of Plant Pathology, University of Wisconsin–Madison); **4.11A** © Omikron/Photo Researchers, Inc.; **4.11B** Reprinted with permission from the American Society for Microbiology (Russell C. Johnson, Donna M. Ritzi, and Brian P. Livermore. *Infection and Immunity* (1973) 8(2): 291–295.) Photo courtesy of Russell C. Johnson, Professor at the Department of Microbiology at University of Minnesota; **4.12A** Courtesy of Elliot Juni, Department of Microbiology and Immunology, The University of Michigan; **4.12B** © George Musil/Visuals Unlimited; **4.13** © CNRI/Photo Researchers, Inc.; **4.17** © Dr. Dennis Kunkel/Visuals Unlimited; **4.18A** Courtesy of Rut Carballido-López. University of Oxford; **4.18B** Reprinted from *Cell*, vol. 115. Jacobs-Wagner, C., cover. Copyright 2003, with permission from Elsevier. Photo courtesy of Christine Jacobs-Wagner, Yale University; **4.19** Courtesy of Dr. Peter Lewis; School of Environmental and Life Sciences, University of Newcastle; **MicroFocus 1** Courtesy of Dr. Hans Paerl and Dr. Pia Moisander, University of North Carolina at Chapel Hill, Institute of Marine Sciences; **MicroFocus 4** © Dr. Dennis Kunkel/Visuals Unlimited; **MicroFocus 5** © Dr. David M. Phillips/Visuals Unlimited; **MicroInquiry** Reprinted with permission from the American Society for Microbiology (Seufferheld. M., et al. *J. Biol. Chem.* 278. 32: 29971–29978). Photo courtesy of Roberto Docampo, M.D., Ph.D., Professor of Cellular Biology at University of Georgia, Athens; **Textbook Case** Courtesy of Dr. Rodney M. Donlan and Janice Carr/CDC.

CHAPTER 5

5.1 © Photodisc/age fotostock; **5.2B** © Lee Simon/Photo Researchers, Inc.; **5.5A** Courtesy of Dr. J. J. Farmer/CDC; **5.5B** Courtesy of CDC; **5.6a** Courtesy of CDC; **5.6B** © Scott Camazine/Alamy Images; **5.6C** Courtesy of Janice Carr/CDC; **5.9A, B, and C**© Scott Coutts/Alamy Images; **5.11** Courtesy of Giles Scientific, Inc., CA. www.biomic.com; **5.12A–E** Courtesy of James Gathany/CDC; **5.14** © R.A. Longuehaye/Photo Researchers, Inc.; **MicroFocus 5** © CDC/Science Source/Photo Researchers, Inc.; **Textbook Case** © Dr. Terrence Beveridge/Visuals Unlimited.

CHAPTER 6

6.1 © Dr. Dennis Kunkel/Visuals Unlimited; **6.15** © Dr. Dennis Kunkel/Visuals Unlimited.

CHAPTER 7

7.1 Courtesy of Abigail Allwood, Geologist at Australian Centre for Astrobiology, Macquarie University; **7.3B** © H. Potter–D. Dressler/Visuals Unlimited; **7.9** Reprinted with permission from O. L. Miller, B. A. Hamkalo, and C. A. Thomas. *Science* 169: 392. Copyright 1977 American Association for the Advancement of Science; **7.11** Courtesy of Dr. Peter Lewis; School of Environmental and Life Sciences, University of Newcastle; **MicroFocus 1** © Vittorio Luzzati/Photo Researchers, Inc.; **MicroFocus 6** © National Library of Medicine/Photo Researchers, Inc.

CHAPTER 8

8.1 Courtesy of John Innes Centre; **8.5** © Dr. Dennis Kunkel/Visuals Unlimited; **8.13** Courtesy of Michael J. Daly; **MicroFocus 1** © AbleStock; **MicroFocus 2** Courtesy of Chris Gotschalk, University of California, Santa Barbara; **MicroFocus 5** Courtesy of The Wellcome Trust Sanger Institute—Photography by Richard Summers.

PART 3 OPENER © Dr. Dennis Kunkel/Visuals Unlimited.

PART 3 MICROBIOLOGY PATHWAYS © Viktor Pryymachuk/ShutterStock, Inc.

CHAPTER 9

9.1 © Dr. George J. Wilder/Visuals Unlimited; **9.3A** © Medical-on-Line/Alamy Images; **9.3B** © Gladden Willis, M.D./Visuals Unlimited; **9.3C** Courtesy of Vincent A. Fischetti. Ph.D., Head of the Laboratory of Bacterial Pathogenesis at Rockefeller University; **9.4** © NIBSC/Photo Researchers, Inc.; **9.5** © CAMR/A.B. DOWSETT/Photo Researchers, Inc.; **9.9** © Dr. Dennis Kunkel/Visuals Unlimited; **9.10A and B** © Michael Gabridge/Visuals Unlimited; **9.11A** © Phototake/Alamy Images; **9.11B** Courtesy of Don Howard/CDC; **9.12** © Scientifica/Visuals Unlimited; **MicroFocus 1** © Dr. David Phillips/Visuals Unlimited; **MicroFocus 3** © Scott Camazine/Alamy Images; **Textbook Case** Courtesy of Dr. Jim Feeley/CDC.

CHAPTER 10

10.1 © SPL/Photo Researchers, Inc.; **10.2** © Dr. Hans Ackermann/Visuals Unlimited; **10.4A** © Dr. Gary Gaugler/Visuals Unlimited; **10.4B** Courtesy of Julio Martin, University of Florida; **10.5** © CNRI/Photo Researchers, Inc.; **10.7A** © Dr. John D. Cunningham/Visuals Unlimited; **10.7B** © Scimat/Photo Researchers, Inc.; **10.8A** © SAS/Alamy Images; **10.9** From B. Kenny, S. Ellis, A.D. Leard, J. Warawa, H. Mellor, and M. A. Jepson. Co-ordinate regulation of distinct host cell signaling pathways by malfunctional enteropathogenic *E. coli* (EPEC) effector molecules. *Mol. Microbiol.* 44:4 (2002). © Blackwell Publishing; **10.10** © Jeff Chiu/AP Photos; **10.11** © P. Hawtin/Photo Researchers, Inc.; **10.12** © Dr. Terrence Beveridge/Visuals Unlimited; **10.13** © Science VU/Visuals Unlimited; **10.14 (left)** Reprinted with permission from the American Society for Microbiology (*ASM News*. January 2002. p. 20–24.); **10.14 (right)** Reprinted with permission from the American Society for Microbiology (Virginia L. Miller and Stanley Falkow. *Infection and Immunity* (1988) 56:1242–1248.) Photo courtesy of Doctor Virginia L. Miller, Professor of the Departments of Molecular Microbiology and Pediatrics at Washington University School of Medicine; **MicroFocus 4A** © Adrienne Marshall 2006; **MicroFocus 4B** © SPL/Photo Researchers, Inc.; **Textbook Case** Courtesy of CDC.

CHAPTER 11

11.1 © Mary Evans/Photo Researchers, Inc.; **11.2A** © Michael Abbey/Visuals Unlimited; **11.2B** © Phototake/Alamy Images; **11.3** Courtesy of James H. Steele/CDC; **11.4** Courtesy of Dr. Jack Poland/CDC; **11.5** © Eye of Science/Photo Researchers, Inc.; **11.6A** © Gary Gaugler/Visuals Unlimited; **11.6B** © Science Source/Photo Researchers, Inc.; **11.7** Courtesy of Dr. Brachman/CDC; **11.8A** © Phototake/Alamy Images; **11.8B** Courtesy of Scott Bauer/USDA; **11.10** © Science VU/Visuals Unlimited; **11.12A** Courtesy of Timothy J. Lysyk/Agriculture and Agri-Food Canada; **11.12B** Courtesy of WHO/CDC; **11.12C** Courtesy of World Health Organization/CDC; **11.13A** © Science VU/Visuals Unlimited; **11.13B** Courtesy of CDC; **11.14** Courtesy of Armed Forces Institute of Pathology; **MicroFocus 1** © Popperfoto/Alamy Images; **MicroFocus 2** © Oscar Knott/FogStock/Alamy Images; **MicroFocus 3** © NASA/Photo Researchers, Inc.; **MicroFocus 6** © National Library of Medicine; **Textbook Case** © Arthur Siegelman/Visuals Unlimited.

CHAPTER 12

12.2A Courtesy of M. Rein, VD/CDC; **12.2B** Courtesy of Dr. Gavin Hart/CDC; **12.2C** Courtesy of Susan Lindsley/CDC; **12.3** Courtesy of the CDC; **12.4** © Dr. David M. Phillips/Visuals Unlimited; **12.5** Courtesy of Joe Miller/CDC; **12.7** © Dr. R. Dourmashkin/Photo Researchers, Inc.; **12.8** © Don W. Fawcett/Photo Researchers, Inc.; **12.9A and B** Courtesy of American Leprosy Mission, 1 Broadway, Elmwood Park, NJ; **12.10B** Courtesy of the CDC; **12.11** Courtesy of the CDC; **12.13** © David M. Phillips/Visuals Unlimited; **12.14A** Courtesy of Armed Forces Institute of Pathology; **12.14B** Image courtesy of Division of Pediatric Surgery, Brown Medical School; **12.15B** © Dr. David Phillips/Visuals Unlimited; **12.16** Reprinted with permission from the American Society for Microbiology (S. L. Spruance; *J Clin Microbiol.* 1985 September; 22(3): 366–368). Photo courtesy of Doctor Ulf B. Göbel, Professor at the Institute of Microbiology and Hygiene CCM/CBF at Charité, University of Berlin; **12.18A** © Dr. Fred Hossler/Visuals Unlimited; **12.18B** Courtesy of CHROMagar; **MicroFocus 2** Photo courtesy of US Department of Health and Human Services, Health Resources Services Administration, Bureau of Primary Health Care, National Hansen's Disease Programs; **MicroFocus 3** © Jeff R. Clow/ShutterStock, Inc.; **MicroFocus 5** © Nina Shannon/ShutterStock, Inc.; **Textbook Case** Courtesy of Dr. E. Arum/Dr. N. Jacobs/CDC.

PART 4 OPENER © Dr. Gopal Murti/Visuals Unlimited.

PART 4 MICROBIOLOGY PATHWAYS Courtesy of James Gathany/CDC.

CHAPTER 13

13.1 © AP Photos; **13.3A** © Science Photo Library/Photo Researchers, Inc.; **13.3B** © Brad Mogen/Visuals Unlimited; **13.4** © Phototake/Alamy Images; **13.7B** © Eye of Science/Photo Researchers, Inc.; **13.13** © Phototake/Alamy Images; **13.14A** Courtesy of Greg Knobloch/CDC; **13.14B** © James King-Holmes/Photo Researchers, Inc.; **13.14C** Courtesy of Giles

Scientific, Inc., CA, www.biomic.com; **13.19A** Courtesy of APHIS photo by Dr. Al Jenny/CDC; **MicroFocus 2** © CNRS PhotothPque/RAOULT. Didier/ALDROVANDI. N; **MicroFocus 3** © Lee D. Simon/Photo Researchers, Inc.; **MicroFocus 4** © Marilyn Barbone/ShutterStock, Inc.; **MicroFocus 6** © Phototake/Alamy Images; **MicroFocus 7** © CDC/Photo Researchers, Inc.; **MicroFocus 8** © Jamie Wilson/ShutterStock, Inc.

CHAPTER 14

14.1 © Science Photo Library/Photo Researchers, Inc.; **14.3** Human rhinovirus 16: Picornaviridae; Rhinovirus; Human rhinovirus A; strain (NA). Hadfield, A.T., Lee, W. M., Zhao, R., Oliveira, M. A., Minor. I., Rueckert, R. R., and Rossmann, M. G. (1997). The refined structure of human rhinovirus 16 at 2.15 C resolution: implications for the viral life cycle. *Structure*, 5: 427–441. (PDB-ID: 1AYM) image by J. Y. Sgro, UW-Madison; **14.5** © Dr. Linda Stannard. UCT/Photo Researchers, Inc.; **14.7** © Phototake/Alamy Images; **14.8** © Phototake/Alamy Images; **14.10A** © Phototake/Alamy Images; **14.10B** Courtesy of Dr. Hermann/CDC; **14.11A–D** Reprinted with permission from the American Society for Microbiology (S. L. Spruance, *J. Clin. Microbiol.* 1985 22: 366–368.) Photo courtesy of Woody Spruance, M.D., Professor of Internal Medicine at University of Utah; **14.12A** © plainpicture GmbH & Co. KG/Alamy Images; **14.12B** © Science Photo Library/Photo Researchers, Inc.; **14.13** © Scott Camazine & Sue Trainor/Photo Researchers, Inc.; **14.14** Courtesy of National Cancer Institute; **14.15A** © Medical-on-Line/Alamy Images; **14.15B** Courtesy of CDC; **14.16** Courtesy of NIP/Barbara Rice/CDC; **14.17** © Dr. K. G. Murti/Visuals Unlimited; **14.18** © Dr. Ken Greer/Visuals Unlimited; **14.19B** © Science VU/Visuals Unlimited; **14.19C** © Dr. F. A. Murphy/Visuals Unlimited; **14.20A** Courtesy of World Health Organization; Diagnosis of Smallpox Slide Series/CDC; **14.20B** Courtesy of James Hicks/CDC; **MicroFocus 1** © Cristina Fumi/ShutterStock, Inc.; **MicroFocus 2** © Tihis/ShutterStock, Inc.; **MicroFocus 3** © Leah-Anne Thompson/ShutterStock, Inc.; **MicroFocus 4** © Martin Bowker/ShutterStock, Inc.; **MicroFocus 5** © Medical-on-Line/Alamy Images; **MicroInquiry B** Modified from Belshe, Robert. *New England Journal of Medicine*, 353: 2209–2211. Copyright © 2006 [2005] Massachusetts Medical Society. All rights reserved. **Textbook Case** Courtesy of Dr. Sherif Zaki/CDC.

CHAPTER 15

15.8A © Chris Bjornberg/Photo Researchers, Inc.; **15.8B** © Chris Bjornberg/Photo Researchers, Inc.; **15.9** Source: Pan American Health Organization (PAHO). Available at: http://www.paho.org. Accessed July 2003; **15.10** © Medical-on-Line/Alamy Images; **15.12** © James Cavallini/Photo Researchers, Inc.; **15.13** © Science Source/Photo Researchers, Inc.; **15.14** © Eye of Science/Photo Researchers, Inc.; **15.16** © Science VU/CDC/Visuals Unlimited; **15.17** © CDC/Photo Researchers, Inc.; **MicroFocus 1** © Dana Ward/ShutterStock, Inc.; **MicroFocus 2** Courtesy of OraSure Technologies, Inc.; **MicroFocus 3** © Richard Renaldi/Visuals Unlimited; **MicroFocus 6** Courtesy of Chris Zahniser/CDC; **MicroFocus 7** Courtesy of Robert S. Craig/CDC.

CHAPTER 16

16.1A © mes-Picayune/NNS /Landov; **16.1B** © REUTERS/ Charles W. Luzier/Landov; **16.2A** Courtesy of John Pitt of CSIRO Food Science, Australia; **16.2B** Courtesy of M.A. Lachance; **16.3A** © Dr. Dennis Kunkel/Visuals Unlimited; **16.4** © Jones and Bartlett Publishers. Photographed by Kimberly Potvin; **16.5A** © Dr. Gerald Van Dyke/Visuals Unlimited; **16.5B** © Science VU/R.Roncadori/Visuals Unlimited; **16.6A** © Andrew Syred/Photo Researchers, Inc.; **16.6B** © Phototake/Alamy Images; **16.7** © Medical-on-Line/Alamy Images; **16.8** © Science Photo Library/Photo Researchers, Inc.; **16.10** © Jones and Bartlett Publishers. Photographed by Kimberly Potvin; **16.11** © Phototake/Alamy Images; **16.12A** © Dr. John D. Cunningham/Visuals Unlimited; **16.12B** © Andrew Darrington/Alamy Images; **16.12C** © Jones and Bartlett Publishers. Photographed by Kimberly Potvin; **16.13A** © Biodisc/Visuals Unlimited; **16.13B** © Dr. John D. Cunningham/Visuals Unlimited; **16.14B** © V. Ahmadjian/ Visuals Unlimited; **16.14C** © David Kay/ShutterStock, Inc.; **16.15A** © Chris Hellyar/ShutterStock, Inc.; **16.15B** © Jim Zipp/Photo Researchers, Inc.; **16.15C** © Anthony Collins/Alamy Images; **16.16A** © Cephas Picture Library/Alamy Images; **16.16B** © Dr. Richard Kessel & Dr. Gene Shih/Visuals Unlimited; **16.16C** © Eye of Science/Photo Researchers, Inc.; **16.18A** © Dr. Arthur Siegelman/Visuals Unlimited; **16.18B** Courtesy of CDC; **16.19A** Courtesy of Dr. Godon Roberstad/CDC; **16.19B** Courtesy of CDC; **16.20A** © E. Gueho–CNRI/Photo Researchers, Inc.; **16.20B** © Everett Beneke/Visuals Unlimited; **16.21A** Courtesy of Dr. Leanor Haley/CDC; **16.21B** © Dr. G. W. Willis/Visuals Unlimited; **16.22A** © Dr. Arthur Siegelman/Visuals Unlimited; **16.22B** © Scott Camazine/Alamy Images; **MicroFocus 1** © Izatul Lail bin Mohd Yasar/ShutterStock, Inc.; **MicroFocus 2** © Jana R. Jirak/Visuals Unlimited; **MicroFocus 3** Courtesy of Dr. A. Elizabeth Arnold, Assistant Professor & Curator, Gilbertson Mycological Herbarium, Department of Plant Sciences, University of Arizona; **MicroFocus 4** © Bruce Coleman, Inc./Alamy Images; **MicroFocus 5** Courtesy Fleishmann's Yeast; **MicroFocus 6** © North Wind Picture Archives.

CHAPTER 17

17.2A © M. I. (Spike) Walker/Alamy Images; **17.2B** © Manfred Kage/Peter Arnold, Inc.; **17.2C** © Mark Bond/ShutterStock, Inc.; **17.3A** © Wim van Egmond/Visuals Unlimited; **17.3B** © Michael Abbey/Visuals Unlimited; **17.3C** Courtesy of the Laboratory of Parasitology, University of Pennsylvania School of Veterinary Medicine; **17.4A** © M. I. (Spike) Walker/Alamy Images; **17.4B** © Phototake/Alamy Images; **17.6A** © Phototake/Alamy Images; **17.6B** Courtesy of WHO/CDC; **17.6C** © Medical-on-Line/Alamy Images; **17.8** © Jerome Paulin/Visuals Unlimited; **17.9** © Science Source/Photo Researchers, Inc.; **17.10** © Dr. Dennis Kunkel/ Visuals Unlimited; **17.12A** © Dr. Arthur Siegelman/Visuals Unlimited; **17.12B** © Holt Studios International Ltd/Alamy Images; **17.13** © Phototake/Alamy Images; **17.15** Courtesy of Dr. Govinda S. Visvesvara/CDC; **17.16A** © Sinclair Stammers/ Photo Researchers, Inc.; **17.16B** © Medical-on-Line/Alamy Images; **17.16C** © Phototake/Alamy Images; **17.17** © R. F. Ashley/Visuals Unlimited; **17.19A** © Biodisc/Visuals Unlimited; **17.19B** © Alfred Pasieka/Photo Researchers, Inc.; **17.20insert** © Pr. Bouree/Photo Researchers, Inc.; **17.21** Courtesy of CDC; **17.22** © Phototake/Alamy Images; **17.24** © John Greim/Photo Researchers, Inc.; **MicroFocus 1** © Holt Studios International, Ltd/Alamy Images; **MicroFocus 2** © Leslie E. Kossoff/AP Photos; **MicroFocus 3** Courtesy of Christopher J. Rapuano, M.D., Cornea Service, Wills Eye. Professor, Jefferson Medical College of Thomas Jefferson University, Philadelphia, PA;

MicroFocus 4 © Mark Pink/Alamy Images; **MicroFocus 5** © CNRI/Photo Researchers, Inc.; **MicroInquiry** © Eye of Science/Photo Researchers, Inc.

PART 5 OPENER © Dr. Dennis Kunkel/Visuals Unlimited.

PART 5 MICROBIOLOGY PATHWAYS Courtesy of Ethleen Lloyd/CDC.

CHAPTER 18

18.8 © Phototake/Alamy Images; **18.9** © Dr. K. G. Murti/Visuals Unlimited; **18.10** Reprinted from *Trends in Microbiology*, vol. 1. Lewis G. Tilney and Mary S. Tilney. The wily ways of a parasite: induction of actin assembly by *Listeria*. 25–31. Copyright April 1993, with permission from Elsevier; **18.14** © Matt Meadows/Peter Arnold, Inc.; **MicroFocus 1** © SPL/Photo Researchers, Inc.; **MicroFocus 2** © Comstock Images/Jupiterimages; **MicroFocus 3A** Courtesy of the CDC; **MicroFocus 3B** © CNRI/Photo Researchers, Inc.; **MicroFocus 4** © Christophe Testi/ShutterStock, Inc.; **MicroFocus 5** © North Wind Picture Archives; **MicroFocus 6** Courtesy of CDC.

CHAPTER 19

19.2 © John Greim/Photo Researchers, Inc.; **19.5 top** © Phototake/Alamy Images; **19.7** Delves PJ, Roitt IM. The Immune System. First of Two Parts. *New England Journal of Medicine*, 343: 37–49. Copyright © 2006 [2000] Massachusetts Medical Society. All rights reserved; **MicroFocus 1** © Dr. Gopal Murti/Photo Researchers, Inc.; **MicroFocus 2** © Volker Steger/Photo Researchers, Inc.; **MicroFocus 3** © Blend Images/Jupiterimages.

CHAPTER 20

20.1A © Dr. Klaus Boller/Photo Researchers, Inc.; **20.1B** © NIBSC/Photo Researchers, Inc.; **20.7B** *Genetics Home Reference*. [Internet] Bethesda (MD): National Library of Medicine (US); 2003– [updated 2004 Aug 4; cited 2006 Aug 08] Immunoglobulin G (IgG). Available from http://ghr.nlm.nih.gov/handbook/illustrations/igg; **20.14** © Steve Gschmeissner/Photo Researchers, Inc.; **MicroFocus 1** © Jones and Bartlett Publishers. Photographed by Kimberly Potvin.

CHAPTER 21

21.5 Courtesy of the CDC; **21.9C** © Science VU/Wellcome/Visuals Unlimited; **21.9D** © Ed Reschke/Peter Arnold, Inc.; **21.11** Reprinted with permission from the American Society for Microbiology (B. J. Paster, F. E. Dewhirst, S. M. Cooke, V. Fussing, L. K. Poulsen, and J. A. Breznak; *Appli. Environ. Microbiol.* 1996 February; 62(2): 347–352.) Photo courtesy of Doctor Bruce J. Paster, Senior Member at Department of Molecular Genetics, The Forsyth Institute; **21.15** © Dr. Jeremy Burgess/Photo Researchers, Inc.; **21.16 insert** Image courtesy of Abbott Laboratories; **MicroFocus 1** © James King-Holmes/Photo Researchers, Inc.; **MicroFocus 3** © Jaimie Duplass/ShutterStock, Inc.; **MicroFocus 6** © Phil Degginger/Alamy Images; **MicroInquiry** Courtesy of Home Access Health Corporation.

CHAPTER 22

22.5 © Michael Donne/Photo Researchers, Inc.; **22.6** © Gary Carlson/Photo Researchers, Inc.; **22.10** © Phototake/Alamy Images; **22.11B** © Bill Beatty/Visuals Unlimited; **22.12** © BURGER/PHANIE/Photo Researchers, Inc.; **22.13** © Scott Camazine/Alamy Images; **22.15** © Peter Menzel/Photo Researchers, Inc.; **MicroFocus 1** © Gary Gaugler/The Medical File/Peter Arnold, Inc.; **MicroFocus 2** Courtesy of Dey, L.P.; **MicroFocus 3** Courtesy of USDA-ARS; **MicroFocus 4** © Stockbyte Silver/Getty Images; **MicroFocus 5** © Photodisc; **MicroInquiry** © D. Hurst/Alamy Images.

PART 6 OPENER © Fred Marsik/Visuals Unlimited.

PART 6 MICROBIOLOGY PATHWAYS © Jon Feingersh, Photogr/Blend Images/age fotostock.

CHAPTER 23

23.1A © Jones and Bartlett Publishers. Photographed by Kimberly Potvin; **23.1B** © Flat Earth/FotoSearch; **23.2** Courtesy of Journalist 2nd Class Shane Tuc/U.S Navy; **23.3** © Dieter Melhorn/Alamy Images; **23.4** © Simon Fraser/Photo Researchers, Inc.; **23.7** © Apply Pictures/Alamy Images; **23.8B** Courtesy of Pall Corporation; **23.8C** © Christine Case/Visuals Unlimited; **23.10A** © Jones and Bartlett Publishers. Photographed by Kimberly Potvin; **23.10B** Courtesy of US Army Natick Soldier Center; **MicroFocus 2** © Robert A. Levy Photography. LLC/ShutterStock. Inc.; **MicroFocus 3** © Jones and Bartlett Publishers. Photographed by Kimberly Potvin; **MicroFocus 4** © Photodisc; **MicroFocus 5** © Jones and Bartlett Publishers. Photographed by Kimberly Potvin; **MicroFocus 6A** © ReutersS/Department of Justice/Handout/Landov; **MicroFocus 6B** Courtesy of the US Census Bureau, Public Information Office.

CHAPTER 24

24.2A Courtesy of Pfizer, Inc.; **24.2B** © National Library of Medicine; **24.8B** © Kenneth E. Greer/Visuals Unlimited; **24.10** © Jack Barker/Alamy Images; **24.12A** Courtesy of Giles Scientific, Inc., CA. www.biomic.com; **24.12B and C** Courtesy of bioMerieux, Inc.; **MicroFocus 1** © Mary Evans Picture Library/Alamy Images; **MicroFocus 2** © imagebroker/Alamy Images; **MicroFocus 3** © Michael Abbey/Visuals Unlimited; **MicroFocus 4** © Gary Cook/Alamy Images; **MicroFocus 5** Courtesy of Forest and Kim Starr/Haleakala Field Station/USGS/BRD/PIERC; **MicroFocus 6** © Dr. Dennis Kunkel/Visuals Unlimited.

PART 7 OPENER © Dr. Dennis Kunkel/Visuals Unlimited.

PART 7 MICROBIOLOGY PATHWAYS © SHOUT/Alamy Images.

CHAPTER 25

25.4 © Phototake/Alamy Images; **25.5** © Inga Spence/Visuals Unlimited; **25.6** © The Laurinburg Exchange. Brian Klimek/AP Photos; **25.8** © Scimat/Photo Researchers, Inc.;**25.9** © Robert Longuehaye. NIBSC/SPL/Photo Researchers, Inc.; **25.10** Courtesy of USDA; **25.11** © Jones and Bartlett Publishers. Photographed by Kimberly Potvin; **25.12** © Pacific Press Service/Alamy Images; **25.13** © David R. Frazier Photolibrary, Inc./Alamy Images; **25.14** © Scimat/Photo Researchers, Inc.; **MicroFocus 1** © Christine Case/Visuals Unlimited; **MicroFocus 2** © Jones and Bartlett Publishers. Photographed by Kimberly Potvin; **MicroFocus 4** © Jones and Bartlett Publishers. Photographed by Kimberly Potvin; **MicroFocus 6** © Stephen Coburn/ShutterStock, Inc.; **MicroFocus 7** © Jones and Bartlett Publishers. Photographed by Kimberly Potvin.

CHAPTER 26

26.2A © Eric Grave/Photo Researchers, Inc.; **26.2B** © EM Unit, VLA/Photo Researchers, Inc.; **26.4A** Courtesy of Dr. Neil Sullivan, University of Sourthern California/NOAA Corps Collection; **26.4B** Courtesy of GSFC/LaRC/JPL, MISR Team/NASA; **26.5A** Courtesy of CDC; **26.5B** Courtesy of Don Howard/CDC; **26.8** Courtesy of Dr. Rodney M. Donlan and Janice Carr/CDC; **26.9** Courtesy of Scott Bauer/USDA; **26.12A** Courtesy of Harold Evans; **26.12B** © Medical-on-Line/Alamy Images; **MicroFocus 1** Courtesy of Chau Nguyen, M.D., Memorial University of Newfoundland; **MicroFocus 4** Courtesy of Forest and Kim Starr/USGS.

CHAPTER 27

27.1B © Phototake/Alamy Images; **27.5A** © Norman Price/Alamy Images; **27.5B** © Natalia Bratslavsky/ShutterStock, Inc.; **27.5C** © Darren Baker/ShutterStock, Inc.; **27.6A** © Christine Case/Visuals Unlimited; **27.6B** © SciMAT/Photo Researchers, Inc.; **27.7A** © George Chapman/Visuals Unlimited; **27.7B** © Holt Studios International Ltd/Alamy Images; **27.8A** © Jones and Bartlett Publishers. Photographed by Kimberly Potvin; **27.8B** © brt PHOTO/Alamy Images; **27.9** Courtesy of the Exxon Valdez Oil Spill Trust Council/NOAA; **27.10** © Dr. Gopal Murti/Visuals Unlimited.

Pronouncing Organism Names (*continued*)

Methanobacterium meth-a-nō-bak-tėr'ē-um
Methanococcus jannaschii meth-a-nō-kok'kus jan-nä'shē-ē
Micrococcus luteus mī-krō-kok'kus lū'tē-us
Micromonospora mī-krō-mō-nos'pōr-ä
Microsporum racemosum mī-krō-spô'rum ras-ē-mōs'um
Morchella esculentum môr-che'lä es-kyū-len'tum
Moriga oleifera mor-e'-ga ō-lif'ėr-ä
Moritella môr-i-tel'lä
Mucor mū'kôr
Mycobacterium avium mī-kō-bak-ti'rē-um ā'vē-um
M. bovis bō'vis
M. cheloni kē-lō'ē
M. haemophilum hē-mo'fil-um
M. kansasii kan-sä-sē'ī
M. leprae lep'rī
M. marinum măr'in-um
M. tuberculosis tü-bėr-kū-lō'sis
Mycoplasma genitalium mī-kō-plaz'mä jen'i-tä-lē-um
M. hominis ho'mi-nis
M. pneumoniae nu-mō'nē-ī
Myxococcus xanthus micks-ō-kok'kus zan'thus
Naegleria fowleri nī-gle'rē-ä fou'lėr-ē
Nannizzia nan'nė-zė-ä
Nanoarchaeum equitans na-nō-ärk-ē-um ē-kwi-tänz
Necator americanus ne-kā'tôr ä-me-ri-ka'nus
Neisseria gonorrhoeae nī-se'rē-ä go-nôr-rē'ī
N. meningitidis me-nin ji'ti-dis
Neurospora nū-ros'pōr-ä
Nitrobacter nī-trō-bak'tėr
Nitrosomonas nī-trō-sō-mō'näs
Nocardia asteroids nō-kär'dē-a as-tėr-oi'dēz
Nostoc nos'tok
Ornithodoros ôr-nith-ō'dô-rōs
Paragonimus westermani păr-ä-gōn'e-mus we-stėr-ma'nē
Paramecium păr-ä-mē'sē-um
Pasteurella multocida pas-tyėr-el'lä mul-tō'si-dä
Pediculus ped-ik'ū-lus
Pedomicrobium ped-ō-mī-krō'bē-um
Pelagibacter ubique pe-lag-i-bak'tėr ū'bi-kwē
Penicillium camemberti pen-i-sil'lē-um kam-am-bėr'tē
P. chrysogenum krī-so'gen-um
P. griseofulvin gris-ē-ō-fül'vin
P. notatum nō-tä'tum
P. roqueforti rō-kō-fôr'tē
Pfiesteria piscicida fes-ter'ē-ä pis-si-sē'dä
Phlebotomus fle-bot'ō-mus
Photorhabdus luminescens fō-tō-rab'dus lü-mi-nes'senz
Phytophthora infestans fī-tof'thô-rä in-fes'tans
P. ramorum rä-môr'-um
Picrophilus pik-rä'fil-us
Plasmodium falciparum plaz-mō'dē-um fal-sip'är-um
P. malariae mä-lā'rē-ī
P. ovale ō'va'lē
P. vivax vī'vaks
Plesiomonas shigelloides ple-sē-ō-mō'näs shi-gel-loi'dēs
Pneumocystis carinii nü-mō-sis'tis kär-i'nē-ī (or kar-i'nē-ē)
P. jiroveci jėr-ō-vek'ē
Prochlorococcus prō-klôr-ō-kok'kus
Porphyromonas gingivalis pôr'fī-rō-mō-näs jin-ji-val'is
Propionibacterium acnes prō-pē-on'ē-bak-ti-rē-um ak'nēz
Proteus mirabilis prō'tē-us mi-ra'bi-lis
Providencia stuartii prō-vi-den'sē-ä stū-är'tē-ē
Pseudomonas aeruginosa sū-dō-mō'näs ā-rü ji-nō'sä
P. cepacia se-pā'sē-ä
P. marginalis mär-gin-al'is
P. pseudomallei sū-dō-mal'lē-ī
P. putida pyū-tē'dä
Pyrolobus fumarii pī-rōl'ō-bus fū-mär'ē-ē
Rhizobium rī-zō'bē-um
Rhizopus stolonifer rī'zo-pus stō-lon-i-fėr
Rhodobacter capsulatus rō-dō-bac'tėr kap-sü-la'tus
Rhodoferax ferrireducens rō-dō-fer'aks fer-i-re-dü'sens
Rhodospirillum rubrum rō-dō-spī-ril'um rūb'rum
Rickettsia akari ri-ket'sē-ä ä-kăr'ī
R. prowazekii prou-wa-ze'kē-ē
R. rickettsii ri-ket'sē-ē
R. tsutsugamushi tsü-tsü-gäm-ü'shē
R. typhi tī'fē
Rochalimaea quintana rōk-ä-li-mē'ä kwin-tä'nä
Saccharomyces carlsbergensis sak-ä-rō-mī'sēs kä-rls-bėr-gen'sis
S. cerevisiae se-ri-vis'ē-ī
S. ellipsoideus ē-lip-soi'dē-us
Saccharopolyspora erythraea sak-kăr-ō-pol'ē-spo-rä ē-rith'rä-ē
Salmonella enterica säl-mōn-el'lä en-tėr-i'kä
S. enterica serotype Enteritidis en-tėr-i-tī'dis
S. enterica serotype Typhi tī'fē
S. enterica serotype Typhimurium tī-fi-mur'ē-um
Sarcocystis sär-kō-sis'tis